Y0-AFY-569

The chemistry of the

metal—carbon bond

Volume 4

THE CHEMISTRY OF FUNCTIONAL GROUPS

A series of advanced treatises under the general editorship of Professor Saul Patai

The chemistry of alkenes (2 volumes)
The chemistry of the carbonyl group (2 volumes)
The chemistry of the ether linkage
The chemistry of the amino group
The chemistry of the nitro and nitroso groups (2 parts)
The chemistry of carboxylic acids and esters
The chemistry of the carbon–nitrogen double bond
The chemistry of amides
The chemistry of the cyano group
The chemistry of the hydroxyl group (2 parts)
The chemistry of the azido group
The chemistry of acyl halides
The chemistry of the carbon–halogen bond (2 parts)
The chemistry of the quinonoid compounds (2 parts)
The chemistry of the thiol group (2 parts)
The chemistry of the hydrazo, azo and azoxy groups (2 parts)
The chemistry of amidines and imidates
The chemistry of cyanates and their thio derivatives (2 parts)
The chemistry of diazonium and diazo groups (2 parts)
The chemistry of the carbon–carbon triple bond (2 parts)
The chemistry of ketenes, allenes and related compounds (2 parts)
The chemistry of the sulphonium group (2 parts)
Supplement A: The chemistry of double-bonded functional groups (2 parts)
Supplement B: The chemistry of acid derivatives (2 parts)
Supplement C: The chemistry of triple-bonded functional groups (2 parts)
Supplement D: The chemistry of halides, pseudo-halides and azides (2 parts)
Supplement E: The chemistry of ethers, crown ethers, hydroxyl groups and their sulphur analogues (2 parts)
Supplement F: The chemistry of amino, nitroso and nitro compounds and their derivatives (2 parts)
The chemistry of the metal—carbon bond Volume 1
The chemistry of peroxides
The chemistry of the metal—carbon bond Volume 2
The chemistry of the metal—carbon bond Volume 3
The chemistry of organic selenium and tellurium compounds Volume 1
The chemistry of the metal—carbon bond Volume 4

The chemistry of the **metal—carbon bond**
Volume 4
The use of organometallic compounds in organic synthesis

Edited by

FRANK R. HARTLEY
The Royal Military College of Science, Shrivenham, England

1987
JOHN WILEY & SONS
CHICHESTER – NEW YORK – BRISBANE – TORONTO – SINGAPORE
An Interscience ® Publication

Library of Congress Cataloging-in-Publication Data:
Main entry under title:

The use of organometallic compounds in organic
synthesis.

(The Chemistry of the metal–carbon bond; v. 4)
(The Chemistry of functional groups)
'An Interscience publication.'
Includes indexes.
1. Chemistry, Organic—Synthesis.
2. Organometallic compounds. I. Hartley, F. R.
II. Patai, Saul. III. Series. IV. Series:
Chemistry of functional groups.
QD410.C43 1982 vol. 4 547′.05 s 85-19082
[QD262] [547′.050459]

ISBN 0 471 90888 6

British Library Cataloguing in Publication Data:

The Chemistry of the metal–carbon bond.—(The
Chemistry of functional groups)
Vol. 4: The use of organometallic compounds in
organic synthesis
1. Organometallic compounds
I. Hartley, Frank II. Patai, Saul
III. Series
547′.05 QD411

ISBN 0 471 90888 6

Printed and bound in Great Britain

Volume 4—Contributing authors

Didier Astruc	Laboratoire de Chimie Moléculaire des Métaux de Transition, U.A. 35, Université de Bordeaux I, 351 Cours de la Libération, 33405 Talence Cédex, France
Penny A. Chaloner	School of Chemistry and Molecular Sciences, University of Sussex, Falmer, Brighton BN1 9QJ, UK
John R. Chipperfield	Department of Chemistry, The University, Hull HU6 7RX, UK
Ernest W. Colvin	Chemistry Department, University of Glasgow, Glasgow G12 8QQ, Scotland, UK
Frank R. Hartley	Royal Military College of Science, Shrivenham, Swindon, Wiltshire SN6 8LA, UK
Fred H. Jardine	Department of Chemistry, North East London Polytechnic, Romford Road, London E15 4LZ, UK
Makoto Kumada	Department of Synthetic Chemistry, Kyoto University, Kyoto 606, Japan
Donald S. Matteson	Department of Chemistry, Washington State University, Pullman, Washington 99164–4630, USA
David Parker	Department of Chemistry, University Science Laboratories, South Road, Durham DH1 3LE, UK
Anthony J. Pearson	Department of Chemistry, Case Western Reserve University, Cleveland, Ohio 44106, USA
Colin J. L. Raston	Department of Physical and Inorganic Chemistry, University of Western Australia, Nedlands, Western Australia, 6009
Geoffrey Salem	Research School of Chemistry, Australian National University, Canberra, A.C.T. 2600, Australia
Kohei Tamao	Department of Synthetic Chemistry, Kyoto University, Kyoto 606, Japan

Sakae Uemura — Institute for Chemical Research, Kyoto University, Uji, Kyoto 611, Japan

Jim L. Wardell — Department of Chemistry, University of Aberdeen, Meston Walk, Old Aberdeen AB9 2UE, Scotland, UK

David E. Webster — Department of Chemistry, The University, Hull HU6 7RX, UK

Foreword

The Chemistry of the Metal—Carbon Bond is a multi-volume work within the well established series of books covering *The Chemistry of Functional Groups.* It aims to cover the chemistry of the metal—carbon bond as a whole, but lays emphasis on the carbon end. It should therefore be of particular interest to the organic chemist. The general plan of the material is the same as in previous books in the series with the exception that, because of the large amount of material involved, this is a multi-volume work.

The first volume was concerned with:

(a) Structure and thermochemistry of organometallic compounds.

(b) The preparation of organometallic compounds.

(c) The analysis and spectroscopic characterization of organometallic compounds.

The second volume was concerned with cleavage of the metal—carbon bond, insertions into metal—carbon bonds, nucleophilic and electrophilic attack of metal—carbon bonds, oxidative addition, and reductive elimination. It also included a chapter on the structure and bonding of Main Group organometallic compounds. The third volume was concerned with the use of organometallic compounds to create carbon—carbon bonds.

The present volume is concerned with the use of organometallic compounds in organic synthesis. It includes material not available when the third volume 'went to press' concerned with carbon—carbon bond formation, together with chapters concerned with the formation of carbon—hydrogen and other carbon—element bonds. The material is divided into two parts. The first part is concerned with the preparation of Main Group organometallic compounds and their use in organic synthesis. The second part includes the use of transition metal organometallics in organic synthesis and chapters on hydrogenation, saturated carbon—hydrogen bond activation, and the rapidly expanding field of supported metal complex catalysts.

In classifying organometallic compounds we have used Cotton's haptonomenclature (η-) to indicate the number of carbon atoms directly linked to a single metal atom.

In common with other volumes in *The Chemistry of the Functional Groups* series, the emphasis is laid on the functional group treated and on the effects which it exerts on the chemical and physical properties, primarily in the immediate vicinity of the group in question, and secondarily on the behaviour of the whole molecule. The coverage is restricted in that material included in easily and generally available secondary or tertiary sources, such as *Chemical Reviews* and various 'Advances' and 'Progress' series, as well as textbooks (i.e. in books which are usually found in the chemical libraries of universities and research institutes) is not, as a rule, repeated in detail, unless it is necessary for the balanced treatment of the subject. Therefore each of the authors has been asked *not* to give an encyclopaedic coverage of his or her subject, but to concentrate on the most important recent developments and mainly on material that has not been adequately covered by reviews or other secondary sources by the time of writing of the chapter, and to address himself or herself to a reader who is assumed to be at a fairly advanced postgraduate level. With these restrictions, it is realised that no plan can be devised for a volume that would give a *complete* coverage of the subject with *no* overlap between the chapters, while at the same time preserving the readability of the text. The Editors set themselves the goal of attaining *reasonable* coverage with *moderate* overlap, with a minimum of cross-references between the chapters of each volume. In this manner sufficient freedom is given to each author to produce readable quasi-monographic chapters. Such a plan necessarily means that the breadth, depth and thought-provoking nature of each chapter will differ with the views and inclinations of the author.

The publication of the Functional Group Series would never have started without the support of many people. Foremost among these is Dr Arnold Weissberger, whose reassurance and trust encouraged the start of the task. This volume would never have reached fruition without Mrs Baylis's help with typing and the efficient and patient cooperation of several staff members of the Publisher, whose code of ethics does not allow us to thank them by name. Many of our colleagues in England, Israel and elsewhere gave help in solving many problems, especially Professor Z. Rappoport. Finally, that the project ever reached completion is due to the essential support and partnership of our wives and families.

Shrivenham, England FRANK HARTLEY

Contents

List of Abbreviations Used

ac	acrylonitrile
Ac	acetyl
acac	acetylacetone
acacen	bis(acetylacetonate)ethylenediamine
aibn	azobisisobutyronitrile
all	allyl
An	actinide metal
ap	antiplanar
appe	$Ph_2AsCH_2CH_2PPh_2$
Ar	aryl
bae	bis(acetylacetonate)ethylenediamine
9-bbn	9-borabicyclo[3.3.1]nonane
bda	benzylideneacetone
bipy	2, 2′-bipyridyl
bnah	*N*-benzyl-1, 4-dihydronicotinamide
Btz	benzothiazole 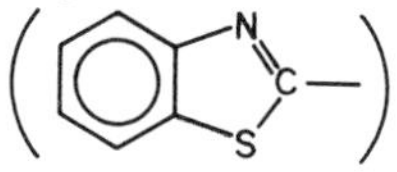
Bu	butyl
Bz	benzyl
cd	circular dichroism
cdt	(*E*, *E*, *E*) cyclododeca-1, 5, 9-triene
cht	cycloheptatriene
CI	chemical ionization
CIDNP	chemically induced dynamic nuclear polarization
CNDO	complete neglect of differential overlap
coct	cyclooctene
cod	cycloocta-1, 5-diene
cot	cyclooctatetraene
Cp	η^5-cyclopentadienyl
Cp*	η^5-pentamethylcyclopentadienyl
C.P.	cross-polarization
Cy	cyclohexyl
dabco	1, 4-diazobicyclo[2.2.2]octane

dba	dibenzylideneacetone
dbn	1, 5-diazabicyclo[5.4.0]non-5-ene
dbp	dibenzophosphole
dbu	1, 8-diazabicyclo[5.4.0]undec-7-ene
dccd	dicylohexylcarbodiimide
dcpe	1, 2-bis(dicyclohexylphosphino)ethane
ddq	2, 3-dichloro-5, 6-dicyano-1, 4-benzoquinone 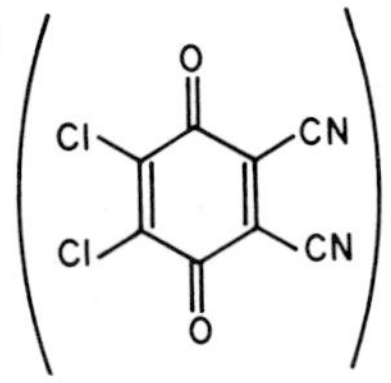
def	diethyl fumarate
DEPT	distortionless enhancement by polarisation transfer
diars	*o*-bis(dimethylarsino)benzene
dibah dibal	diisobutylaluminium hydride
dien	$H_2NCH_2CH_2NHCH_2CH_2NH_2$
diop	2, 3-*o*-isopropylidene-2, 3-dihydroxy-1, 4-bis(diphenylphosphino)butane
dma	*N, N*-dimethylacetamide
dme	1, 2-dimethoxyethane
dmfm	dimethyl fumarate
dmg	dimethyl glyoximate
dmm	dimethyl maleate
dmpe	bis(1, 2-dimethylphosphino)ethane
dmpf	1, 1′-bis(dimethylphosphino)ferrocene
dotnH	bis(diacetylmonoxime)propylene-1, 3-diamine
dpm	dipivaloylmethanato
dppb	bis(1, 4-diphenylphosphino)butane
dppe	bis(1, 2-diphenylphosphino)ethane
dppf	1, 1′-bis(diphenylphosphino)ferrocene
dppm	bis(1, 1-diphenylphosphino)methane
dppp	bis(1, 3-diphenylphosphino)propane
dmso	dimethyl sulphoxide
ee	enantiomeric excess
EI	electron impact
E_p	peak potential
ESCA	electron spectroscopy for chemical analysis
Et	ethyl
eV	electronvolt
Fc	ferrocene
FD	field desorption
FI	field ionization
fmn	fumaronitrile
FMO	frontier molecular orbital
fod	$F_3C(CF_2)_2COCH{=}C(O)C(CH_3)_3$
Fp	$Fe(\eta^5\text{-}C_5H_5)(CO)_2$
Fp*	$Fe(\eta^5\text{-}C_5H_5)(CO)(PPh_3)$
FT	Fourier transform

Hex	hexyl
c-hex	cyclohexyl
hfac	hexafluoroacetone
hfacac	hexafluoroacetylacetonato
hmdb	hexamethyl(Dewar)benzene
hmpa	hexamethylphosphoramide
hmpt	hexamethylphosphorotriamide
HOMO	highest occupied molecular orbital
INDOR	inter-nuclear double resonance
INEPT	inter-sensitive nuclei enhanced by polarisation transfer
LCAO	linear combination of atomic orbitals
lda	lithium diisopropylamide
LiCA	lithium *N*-isopropylcyclohexylamide
Ln	lanthanide metal
LUMO	lowest unoccupied molecular orbital
M	metal
M	parent molecule
ma	maleic anhydride
map	2-methyl-2-nitrosopropane
m-cpba	*m*-chloroperbenzoic acid
Me	methyl
Mes	methanesulphonyl
meSal	*N*-methylsalicylaldiminato
MNDO	modified neglect of diatomic overlap
ms	millisecond
Ms	mesityl
nadh	nicotinamide adenine dinucleotide
nbd	norbornadiene
nbs	*N*-bromosuccinimide
ncs	*N*-chlorosuccinimide
nmp	*N*-methylpyrrolidone
Non	nonyl
Np	naphthyl
oA	*o*-allylphenyldimethylarsine
Oct	octyl
Pc	Phthalocyanine
Pe	pentenyl
Ph	phenyl
phen	*o*-phenanthroline
phth	phthalimide
pmdeta	pentamethyldiethylenetriamine
ppm	parts per million
Pr	propyl
PRDDO	partial retention of diatomic differential overlap
psi	pounds per square inch

pvc	poly(vinyl chloride)
py	pyridyl
pz	pyrazolyl
R	any radical
RT	room temperature
salen	bis(salicylaldehyde)ethylenediamine
salophen	bis(salicylaldehyde)-*o*-phenylenediamine
SCE	saturated calomel electrode
{Si}	silica (used as a support)
sia	sianyl (3-methyl-2-butyl)
S_Ni	substitution nucleophilic internal
SOMO	singly occupied molecular orbital
sp	synplanar
SPT	selective population transfer
tba	tribenzylideneacetylacetone
tbdms	*tert*-butyldimethylsilyl
tcod	tricyclooctadiene
tcne	tetracyanoethylene
teta	5, 5, 7, 12, 12, 14-hexamethyl-1, 4, 8, 11-tetraazacyclotetradecane
tfa	trifluoroacetic acid
tfbb	tetrafluorobenzobarrelene
Tfo	triflate
thf	tetrahydrofuran
thp	tetrahydropyranyl
thpo	tetrahydropyranyloxy
Thx	thexyl (—CMe_2CHMe_2)
tmed	tetramethylethylenediamine
tmof	trimethyl orthoformate
tms	trimethylsilyl
tmtu	tetramethylthiourea
Tol	tolyl
tond	1, 3, 5, 7-tetramethyl-2, 6, 9-trioxobicyclo[3.3.1]nona-3, 7-diene
tos	tosyl
tpp	tetraphenylporphyrin
triphos	1, 1, 1-tris(diphenylphosphinomethyl)ethane
tta	thallium(III) acetate
ttfa	thallium(III) trifluoroacetate
ttn	thallium(III) nitrate
tu	thiourea
un	olefin or acetylene
X	halide

Part 1

Preparation and Use of Main Group Organometallics in Organic Synthesis

The Chemistry of the Metal—Carbon Bond, Vol. 4
Edited by F. R. Hartley

CHAPTER **1**

Preparation and use in organic synthesis of organolithium and Group IA organometallics

J. L. WARDELL

Department of Chemistry, University of Aberdeen, Meston Walk, Old Aberdeen AB9 2UE, Scotland, UK

I. INTRODUCTION

Organic derivatives of all the alkali metals, Li, Na, K, Rb and Cs, are known. By far the greater part of the reported work has been concerned with organolithiums, owing primarily to their having adequate reactivity for most synthetic purposes, and being the easiest to handle of all the organoalkali metal compounds.

A. Stabilities

Organoalkali metal compounds are air and moisture sensitive; their reactivity towards oxygen, water, and carbon dioxide increases with increasing electropositivity of the metal. The lower alkyls, even of lithium, can inflame in air. The use of inert atmospheres, nitrogen or preferably argon, and rigorously dried reagents and apparatus is essential for the handling of all organoalkali metal compounds.

Simple alkyl and aryl derivatives are thermally stable at ambient temperature with the thermal stability following the sequence RLi > RNa > RK... Thermal decomposition of alkyl—M compounds containing β-hydrogens (e.g. ethyllithium and -sodium) occurs via β-elimination to give MH, alkenes, and small amounts of alkenes, typically at temperatures between 80 and 100 °C. Greater thermal stability is experienced by alkyl—M having no β-hydrogens, e.g. methyllithium (decomposition to dilithiomethane at 240 °C) and neopentylsodium (decomposition at 144 °C). The thermal stabilities of functionally substituted organometallics cover a wide range. Particularly sensitive organolithium compounds are those containing $>C{=}O$, $-C{\equiv}N$, or other functional groups able to react with organolithiums. For these compounds the use of very low temperatures, even −110 °C, is necessary to prevent their self-destruction. Other thermally labile derivatives decompose via eliminations of MX (X = halide, alkoxide, etc.); both α-eliminations, e.g. with Cl_3CLi (providing Cl_2C:) and β-eliminations [e.g. with *o*-XC_6H_4Li to give benzyne and LiX, (*E*)-LiCH=CHOEt (loss of LiOEt at −80 °C to give HC≡CH), and $Me_2CHCHLiCH_2OLi$ (decomposition at −100 °C to Li_2O and $Me_2CHCH{=}CH_2$)] are known.

Chiral secondary alkyllithiums have configurational stability only at low temperatures (e.g. −40 °C for *sec*-butyllithium) in hydrocarbon solutions. Higher temperatures and the presence of ethers result in rapid racemization. Chiral *sec*- and *tert*-cyclopropyllithiums, -sodiums and -potassiums have been shown to be stable, considerably so in some cases. Again, greater configurational stability is found in hydrocarbon than in ethereal solutions. Another class of configurationally stable organolithiums is the α-alkoxyalkyl derivatives, RCH(OR′)Li, intra-aggregate coordination of the alkoxy groups and lithiums probably being an important factor here; these species are stable even in thf solution.

The geometric stability of vinyllithiums is very dependent on the substituents present, with alkyl derivatives, e.g. propenyllithiums, being particularly stable. (*Z*)-Arylvinyllithiums have a marked tendency to isomerize to the *E* isomers. Stabilities are greater in hydrocarbon solvents than in ethers or in the presence of donors.

B. Solvents

Solvent systems for the handling of organolithiums at very low temperatures include the Trapp solvent, a mixture of pentane, thf and Et_2O. Organolithiums are generally soluble in ethers. However, alkyl- and aryllithiums can react with etheral solvents (via proton abstraction) and even at ambient temperatures the lifetime of an organolithium in an ether solution can be limited, e.g. the lifetime of butyllithium in Et_2O is *ca.* 150 h at ambient temperature, with a considerably faster decomposition occurring in thf solution. Some simple organolithiums are soluble in hydrocarbons (both aliphatic and aromatic). Examples of such compounds are ethyllithium, butyllithium, and *t*-butyllithium; in contrast, methyllithium and phenyllithium are essentially insoluble. Alkyl and aryl derivatives of the heavier alkali metals (RM) have little solubility in hydrocarbons and they usually react with ethers. Suspensions in hydrocarbons are frequently employed; however, only hydrocarbons having a lower carbon acidity than RH should be used to prevent metallation of the solvent (see Section II.B).

Delocalised organometallic species, including allylic and benzylic compounds, are generally soluble and stable in ethers.

C. Structures

Considerable work has been performed on the structures of organolithiums; the structures of the other alkali metal compounds have attracted less attention.

Although RLi is frequently written to represent organolithium compounds, it should be remembered that alkyl- and aryllithiums are electron-deficient compounds and exist as aggregates in the gas, solution and solid phases. Aggregates found in the solid state include $(MeLi)_4$, $(EtLi)_4$ and $(cyclohexylLi)_6.2PhH$. In solution, the degree of aggregation depends on various factors; it decreases as the steric bulk about the α-carbon increases and as the coordinating ability of the solvent increases. Thus ethyllithium is hexameric in alkane and aromatic hydrocarbon solutions but is tetrameric in both Et_2O and thf; *tert*-butyllithium is tetrameric in hydrocarbons and in ethers whereas menthyllithium is dimeric in hydrocarbons. In the presence of donors (D) such as tmed and dabco, 1:1 RLi.D complexes are obtained in solution and which for R = Bu have been found to exist as monomers and dimers. The methyllithium tetramer is a particularly stable array and survives even in the presence of tmed.

As a general rule, the reactivity of an organolithium increases as the degree of aggregation decreases and so varies markedly with changes in solvent. Further, the reactivity increases when donors such as tmed and dabco are added to the medium. Lithium salts, e.g. halides and alkoxides (and also other salts), can also become incorporated into the organolithium aggregate. Although this will affect certain physical properties and possibly the reactivity, the avoidance or elimination of LiX in synthetic work has seldom been thought necessary. Lithium bromide and iodide are more soluble than lithium chloride in ethers and so a preparation using an organic chloride should provide a sample of an organolithium containing less lithium halide. Routes to organolithiums, free from LiX, are available, however.

The solid-state structure of methylsodium is similar to that of methyllithium; in contrast, methylpotassium and methylrubidium have ionic structures.

For delocalized systems, including benzylic and allylic compounds, ionic bonding may result in solution. In the solvents usually used, e.g. thf, dme, etc., ion pairs or aggregates will dominate. The different species present in solution, (such as contact ion pairs and solvent-separated ion pairs and free ions) may have different reactivities.

D. Availability

Several alkyllithiums are commercially available; these include methyl-, butyl-, *sec*-butyl-, *tert*-butyl-, and phenyl-lithiums and also amides, e.g. $LiNPr^i_2$ and $LiNEt_2$.

E. Bibliography

Reviews on organoalkali metal chemistry have appeared in articles in the series *Houben-Weyl Methoden der Organischen Chemie*[1], *Comprehensive Organometallic Chemistry*[2,3], and *Organometallic Compounds*[4]. Organolithium compounds have been dealt with separately in a book by Wakefield[5] and sodium and potassium compounds have featured in works by Schlosser[6,7].

II. SYNTHESIS OF ORGANOALKALI METAL COMPOUNDS

A. From Organic Halides

1. Using Alkali Metals[1,2,5–7]

The direct preparation of organoalkali metal compounds from alkali metals and organic halides:

$$RX + 2M \rightarrow RM + MX \tag{1}$$

TABLE 1. Formation of Organometallics from Organic Halides and Alkali Metals

$RX + 2M \rightarrow RM + MX$

Organometallic compound, RM	Organic halide, RX	Conditions	Yield (%)	Ref.
MeLi	MeCl	Li shavings, Et_2O	89	649
	MeBr	Li shavings, Et_2O	93–98	650
	MeI	Li shavings, Et_2O	80–82	651
BuLi	BuCl	Li wire, pentane, reflux	93–98	652
	BuCl	Li chips, Et_2O, reflux	75–80	653
	BuCl	Li wire, thf, −25 °C	74	654
Bu^sLi	Bu^sCl	Li wire, pentane, reflux	93–98	652
Bu^tLi	Bu^tCl	Li (1% Na) dispersion, pentane	70–80	655
$MeOCH_2Li$	$MeOCH_2Cl$	Li powder (0.8% Na), $(MeO)_2CH_2$	80–88	656
$Me_2NCH_2CH_2CH_2Li$	$Me_2NCH_2CH_2CH_2Cl$	Li powder, Et_2O	66	657
Me_3SiCH_2Li	Me_3SiCH_2Cl	Li powder, Et_2O	52	658
$Me_3SiCH_2CH_2CH_2Li$	$Me_3SiCH_2CH_2CH_2Br$	Li powder, Et_2O	93	658
Me_3CCH_2Li	Me_3CCH_2Cl	Li powder, Et_2O, −40 °C	85	659
$EtCClLiCO_2Pr^i$	$EtCCl_2CO_2Pr^i$	Li slices, thf, 0 °C		660
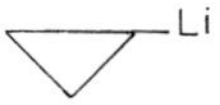 Li	 Br	Li wire, Et_2O, 0 °C	88	661
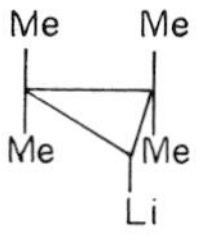 Me Me Me Me Li	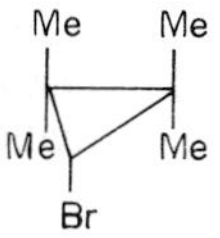 Me Me Me Me Br	Li, Et_2O		662
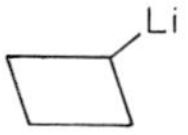 Li	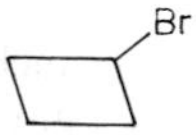 Br	Li (Cu), pentane, reflux	48	663

TABLE 1. (*Contd.*)

Organometallic compound, RM	Organic halide, RX	Conditions	Yield (%)	Ref.
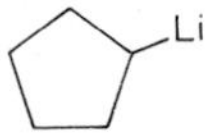 Li	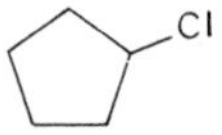 Cl	Li (Cu), pentane, reflux	51	663
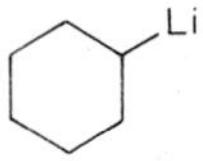 Li	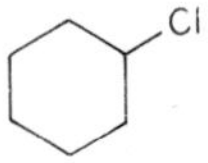 Cl	Li shot, light petroleum, reflux	70	664
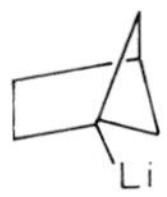 Li	 Cl	Li dispersion, cyclohexane, reflux	83	665
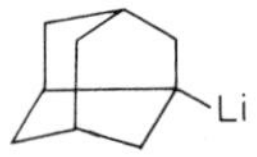 Li	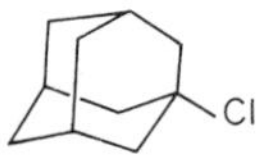 Cl	Li (2% Na), pentane, reflux, vigorous stirring	82	666
CH_2=CHLi	CH_2=CHCl	Li (20% Na) dispersion, thf	60	667
PhLi	PhCl	Li dispersion, Et_2O	90	668
	PhBr	Li wire, thf, −60 °C	90	669
	PhI	Li chips, Et_2O, reflux	80	650
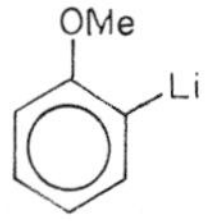 OMe Li	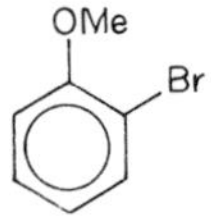 OMe Br	Li, Et_2O	85	670
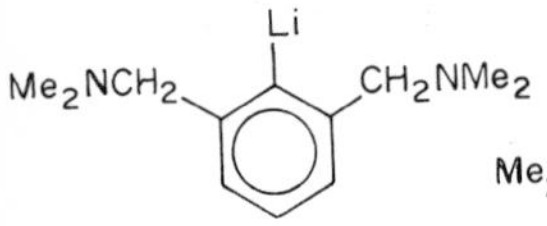 Li Me_2NCH_2 CH_2NMe_2	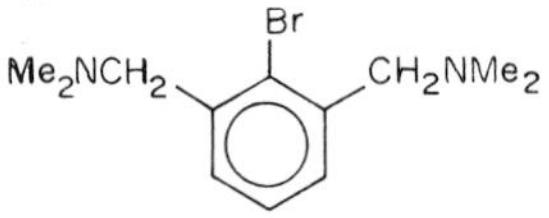 Br Me_2NCH_2 CH_2NMe_2	Li, Et_2O, reflux	80	671
$Li(CH_2)_4Li$	$Br(CH_2)_4Br$	Li powder, Et_2O, −10 °C	63	672
PrNa	PrCl	Na	26	673
$C_5H_{11}Na$	$C_5H_{11}Cl$	Na dispersion, heptane	89	674
Ph_3CNa	Ph_3CCl	Na amalgam, light petroleum	90	675
CH_2=CHNa	CH_2=CHBr	Na dispersion, heptane	65	676
	CH_2=CHCl		90	677
PhNa	PhCl	Na dispersion, PhH	75	678
p-PhC_6H_4Na	*p*-PhC_6H_4Cl	Na	78	679
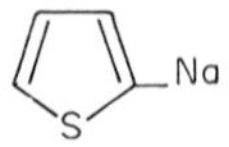 S Na	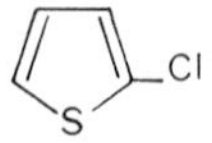 S Cl	Na amalgam	84	680
$C_5H_{11}K$	$C_5H_{11}Cl$	Finely divided K, pentane	35	681
CH_2=CHK	CH_2=CHCl	Na–K alloy, $Bu^t{}_2O$	90	682
PhK	PhCl	K dispersion, methylcyclohexane, 20 °C		683

is a valuable method, particularly for organolithiums. Various forms of solid lithium have been used, including chips, wires, powders, dispersions, and alloys (containing low amounts of sodium). In addition, lithium vapour has been employed to give polylithioalkanes[8,9], e.g. trilithiomethane was obtained in *ca.* 16% yield by co-condensing lithium vapour at 750 °C with chloroform on a cryogenic surface[8].

For sodium, potassium[6], and the other alkali metals, the use of finely divided metal, dispersions[10], amalgams, and alloys, with high-speed stirring and temperature control, have been recommended. Particularly reactive dispersions have been obtained by the use of ultrasonics[11].

Both ethereal and hydrocarbon media have been used; the latter are frequently required for secondary and tertiary alkyllithiums and for the other alkali metal compounds.

Important by-products are the coupled products, RR. This homo-coupling (Wurtz coupling) becomes easier as the electropositivity of the metal increases and in the halide sequence I > Br > Cl. So much coupled product arises from reactions of any alkali metal with allylic or benzylic halides that alternative routes to allylic or benzylic alkali metal compounds have to be followed. Exceptions appear to be Ph_2CHLi[12] and Ph_3CM[1] (M = Li, Na, K, Rb, or Cs); in both cases cleavage of the initial coupled intermediates, $Ph_2CHCHPh_2$ and Ph_3CCPh_3, by the metals apparently occurs.

Alkyl, alkenyl, and aryl compounds have all been obtained (Table 1). Both simple and functionally substituted organolithium species have been obtained from organic chlorides or bromides; alkyl iodides, except MeI, are not normally used. This direct route has been especially recommended for secondary and tertiary adamantyllithiums and related compounds, including 1-twistyl-, 1-triptycyl-, and 3-homoadamantyllithium[13], using either the organic chloride in pentane at 35 °C or the chloride or bromide in diethyl ether at − 45 °C. Alternative routes to these compounds have had only limited success.

a. Mechanism

Apart from RR, by-products are RH and possibly alkenes (R − H). These products suggest a free radical nature to the reaction. An electron transfer mechanism, illustrated in Scheme 1 for organolithiums, has proved popular. Electron transfer to the CX bond provides initially the radical anion, $RX^{-\bullet}$, which can either lead to the formation on RLi on

R–X / Li Surface ⟶ $RX^{-\bullet}Li^+$ / Li ⟶ $R^\bullet, X^-, Li^+$ / Li (diffusion of $R^\bullet$ into solution) ⟶ RLi + LiX / Li

SCHEME 1

the surface or to the release of free $R^\bullet$ into the bulk of the solution. The greater the stability of $R^\bullet$, the greater will be the possibility of diffusion of $R^\bullet$ away from the surface and greater will be the amounts of radical derived products, RR, RH and (R − H).

A number of stereochemical studies point to the involvement of radicals, e.g. partial recemization has been reported[14,15] in reactions of lithium with chiral cyclopropyl halides, such as 1-X-1-methyl-2, 2-diphenylcyclopropane (**1**). The extent of racemization in the products, 1-Li-1-Me-2, 2-Ph_2-cyclopropane, increased in the sequence X = I > Br > Cl. Other factors influencing racemization are the sodium content in the lithium sample and its particle size. Compound (**1**, X = F), also reacts with lithium to give largely racemized products; however, the reaction was considered to occur via electron transfer to the phenyl ring and not to the carbon—halogen bond[16].

Other examples indicating equilibration, via radicals, are (i) the two isomers of 2, 2, 6-d_3-cyclohexyl bromide providing the same isomeric mixture of 2, 2, 6-d_3-cyclohexyllithium products[17], (ii) *exo*- and *endo*-norbornyl chloride giving identical mixture of norbornyllithiums[18], (iii) menthyl and neomenthyl chloride with lithium sand in refluxing pentane producing the same mixture of epimeric lithium reagents[19], and (iv) (*E*)- or (*Z*)-4-chlorohept-3-ene with lithium containing 1% sodium in thf producing mixtures of (*E*)- and (*Z*)-vinyllithiums[20] (Scheme 2).

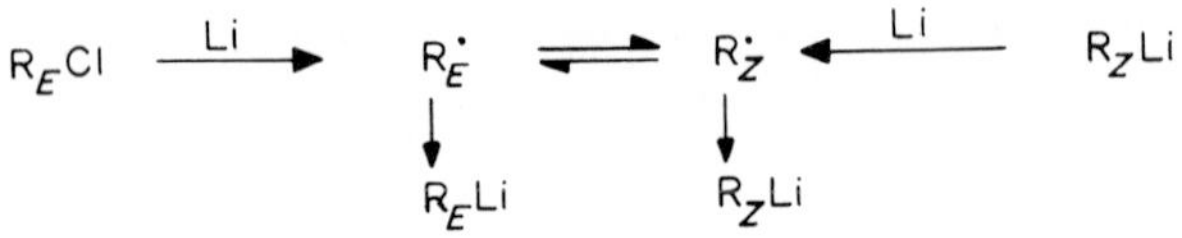

$$R_ECl \xrightarrow{Li} R_E^{\bullet} \rightleftharpoons R_Z^{\bullet} \xleftarrow{Li} R_ZLi$$

$$R_E^{\bullet} \downarrow R_ELi \qquad R_Z^{\bullet} \downarrow R_ZLi$$

R_ECl = (*E*)-EtCH=CClPr; R_ZCl = (*Z*)-EtCH=ClPr

SCHEME 2

In contrast to the last example, a number of other vinyl halides react with Li with retention[21,22]:

$$(Z)\text{-MeCH=CHCl} + \text{Li } (1\%\ \text{Na}) \xrightarrow[\text{(ref. 21)}]{\text{Et}_2\text{O}} (Z)\text{-MeCH=CHLi} \quad (2)$$

$$(E)\text{-MeCH=CHBr} + \text{Li } (1\%\ \text{Na}) \xrightarrow[\text{(ref. 21)}]{\text{Et}_2\text{O}} (E)\text{-MeCH=CHLi} \quad (3)$$

2. Using Alkali Metal Radical Anion Compounds, $ArH^{-\bullet}M^+$ and Dianion Compounds ArH^{2-} $2M^+$[23,24]

Use has been made of alkali metal arene radical anion compounds, $ArH^{-\bullet}M^+$, in place of the metal, M, in the formation[25–38] of organoalkali metal compounds, RM, from organic halides, RX. In certain cases, there appear to be advantages in the use of $ArH^{-\bullet}M^+$.

The $ArH^{-\bullet}M^+$ compounds are simply prepared by reaction of ArH with M, usually in an ethereal solvent, although preparations in hydrocarbons appear possible if ultrasonics are used to disperse the metal[39].

The reaction scheme involving RX and $ArH^{-\bullet}M^+$ is shown in Scheme 3, using $(\text{naphthalene})^{-\bullet}M^+$ as the reagent. Scheme 3 illustrates the formation of both RM and an

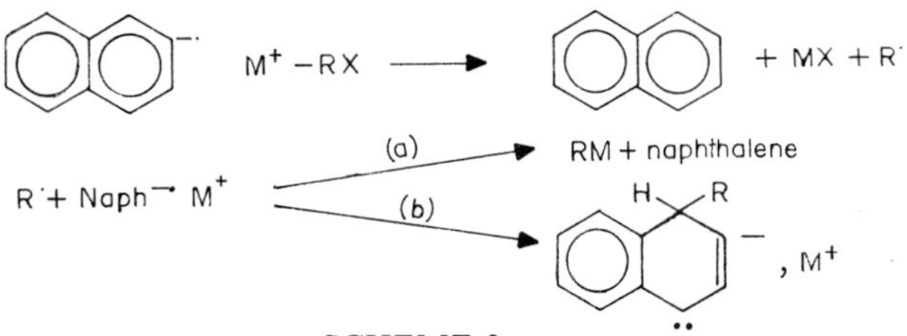

$$(\text{naphthalene})^{-\bullet}\ M^+ - RX \longrightarrow \text{naphthalene} + MX + R^{\bullet}$$

$$R^{\bullet} + \text{Naph}^{-\bullet}\ M^+ \xrightarrow{(a)} RM + \text{naphthalene}$$

$$R^{\bullet} + \text{Naph}^{-\bullet}\ M^+ \xrightarrow{(b)} \text{(1-H, 1-R-dihydronaphthalenide anion)}^{-},\ M^+$$

SCHEME 3

alkylated dihydroarene anion; for M = sodium, these latter products can become significant and so limit the yields of RNa. Additional problems with sodium (and the heavier alkali metal) systems are the reactions of RM with ether solvents (to give RH)[23,24,40,41] and also with RX (to give RR)[23,24,42,43]. For lithium, such problems are not so pronounced and the good yields of a variety of RLi compounds indicate the synthetic value of this method (see Table 2).

A step involving the reaction of an organic chloride with $Naph^{-\bullet}Li^-$ has been incorporated into a synthesis of β-alkoxyalkyllithiums[32] (Scheme 4).

$$RCOCH_2Cl \xrightarrow{(i)} RR'C(OLi)CH_2Cl \xrightarrow{(ii)}$$

$$RR'C(OH)CH_2Cl \xrightarrow{(i)} RR'C(OLi)CH_2Cl \xrightarrow{(ii)} RR'C(OLi)CH_2Li$$

(i) R′Li (e.g. BuLi), −78 °C; (ii) $Naph^{-\bullet}Li^+$, −78 °C.

R = H, R′ = Me, Pr^i, Bu^i, Ph or $PhCH_2$

R = Me, allyl, R′ = allyl

R = allyl, phenyl, R′ = phenyl

SCHEME 4

TABLE 2. Formation of RM from reactions of alkyl halides, RX, and $ArH^{-\cdot}M^+$

RX	$ArH^{-\cdot}M^{+a}$	Conditions	Yield of RLi (%)	Ref.
BuCl	$Phen^{-\cdot}Li^+$	thf, $-100\,°C$	45	25
$C_8H_{17}Cl$	$Naph^{-\cdot}Li^+$	thf, $-78\,°C$	45	26
	$Li^+dbn^{-\cdot}$	thf, $-78\,°C$	49	26
	$Li^+dbb^{-\cdot}$	thf, $-78\,°C$	94	26
$C_8H_{17}Br$	$Li^+dbb^{-\cdot}$	thf, $-78\,°C$	91	26
$C_6H_{13}CHMeCl$	$Li^+dbb^{-\cdot}$	thf, $-78\,°C$	87	26
BuMeCEtCl	$Li^+dbb^{-\cdot}$	thf, $-78\,°C$	88	26
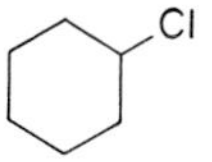	$Naph^{-\cdot}Li^+$	thf, $-50\,°C$	70	25
Ph_3Cl	$Phen^{-\cdot}Li^+$	thf, 25 °C	70	25
	$Naph^{-\cdot}Li^+$	thf, $-78\,°C$	100	37
$PhSCH_2CH_2CH_2Cl$	$Naph^{-\cdot}Li^+$	thf, 65 °C	59	25
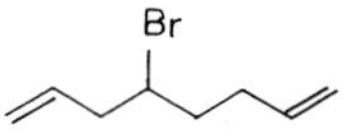	$Li^+dbb^{-\cdot}$	thf, $-78\,°C$	—	31
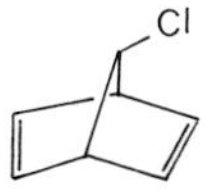	$Li^+dbb^{-\cdot}$	thf, $-78\,°C$	96	29
PhF	$Naph^{-\cdot}Li^+$	thf, $-50\,°C$	85	25
PhCl	$Naph^{-\cdot}Li^+$	thf, $-50\,°C$	85	25
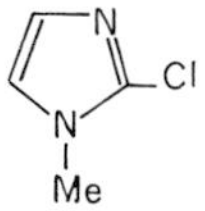	$Naph^{-\cdot}Li^+$	thf, 20 °C	88	33
	$Naph^{-\cdot}Na^+$	thf, 20 °C	73	33
2(4)-Chlorosemibullvalene	$Li^+dbb^{-\cdot}$	thf, $-78\,°C$	—	30

[a] dbn = Bu^t_2-naphthalene; dbb = p-$\mathrm{Bu}^tC_6H_4C_6H_4\mathrm{Bu}^t$-$p$.

The competition, illustrated in Scheme 3, between the electron transfer (step a) and the alkylation (step b) reactions can be diverted in favour of a by using hindered arenes[26]. The rationale for this is that electron transfer processes can proceed between species separated by much longer distances than demanded by the transition state for alkylation. This is well illustrated by the success of $(p\text{-}\mathrm{Bu}^tC_6H_4C_6H_4\mathrm{Bu}^t\text{-}p)^{-\cdot}Li^+$ in thf at $-78\,°C$ in providing good yields of RLi with only small amounts of alkylation products[26].

Another significant factor must be the reduction potential of ArH; the higher the reduction potential, the greater will be the prospect for electron transfer. From a number of sources, it appears that the effectiveness in forming RLi from $ArH^{-\cdot}Li^+$ is in the sequence ArH = p-$\mathrm{Bu}^tC_6H_4C_6H_4\mathrm{Bu}^t$-$p$ > $\mathrm{Bu}_2{}^t$-naphthalene (a mixture of 2, 6- and 2, 7-isomers) > PhH > naphthalene > anthracene[26].

Solvent-separated ion-paired (s.s.i.p.) $ArH^{-\cdot}M^+$ allow more electron transfer than do the contact ion-paired (c.i.p.) forms; hence, since the solvent, the cation, and the temperature affect the s.s.i.p.–c.i.p. ratios, these clearly are important factors[2].

The arene dianionic compounds $ArH^{2-}\,2M^+$ (which exist as very tight ion triples) are generally less effective than $ArH^{-\cdot}M^+$ in electron transfer reactions, as shown by the following two examples[26]: (i) $Naph^{2-}\,2Li^+$ and $C_6H_{13}CHMeBr$ in Et_2O provided only

TABLE 3. Stereochemistry of product RM from reactions of alkyl halides and $ArH^{-\bullet}M^{+}$

Alkyl halide (RX)	Reagent[a] conditions	Product (RM)	Ref.
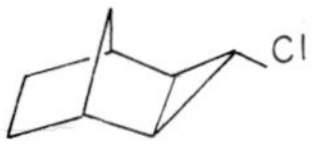 Cl; *anti*	$Li^{+}dbb^{-\bullet}$, thf $-78\,^{\circ}C$	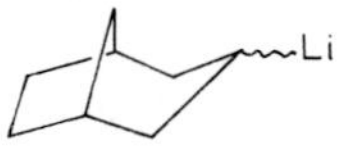 Li; *Syn:anti* > 200:1	28
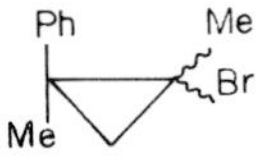 Ph, Me, Me, Br; either isomer	$Naph^{-\bullet}Li^{+}$, thf, 20 °C	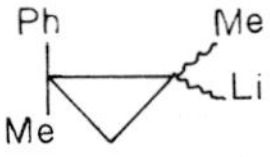 Ph, Me, Me, Li; *trans:cis* = 45:55	35
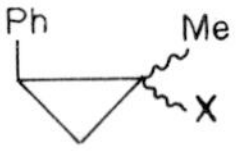 Ph, Me, X; X = Cl or Br	$Naph^{-\bullet}M^{+}$ (M = Li, Na or K)	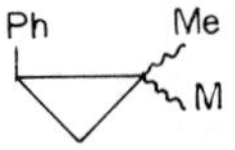 Ph, Me, M; *trans:cis* = 55–60:45–40	35
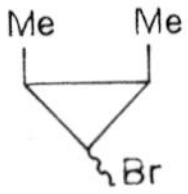 Me, Me, Br; either isomer	$Naph^{-\bullet}Li^{+}$, thf, room temp.	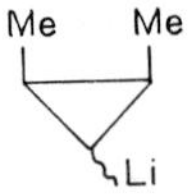 Me, Me, Li; *cis,cis:trans,trans* = 8:92	34
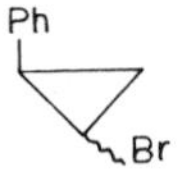 Ph, Br; either isomer	$Naph^{-\bullet}Li^{+}$, thf, room temp.	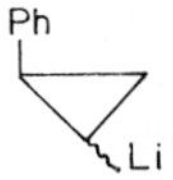 Ph, Li; *cis,trans* = 21:79	34
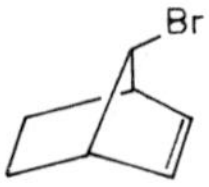 Br	$Li^{+}dbb^{-\bullet}$, thf, $-78\,^{\circ}C$	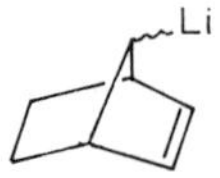 Li; *Syn:anti* = 1:3.8	28
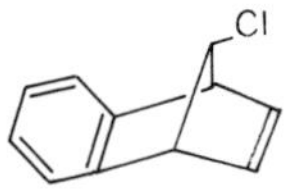 Cl; *anti*	$Li^{+}dbb^{-\bullet}$, thf, $-78\,^{\circ}C$	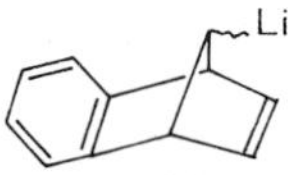 Li; *Syn:anti* = 1:1.33	28
 Cl	$Li^{+}dbb^{-\bullet}$, thf, $-78\,^{\circ}C$	 Li; *exo:endo* = 10:1	28
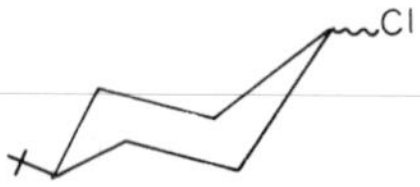 Cl	$Li^{+}dbb^{-\bullet}$, thf, $-78\,^{\circ}C$	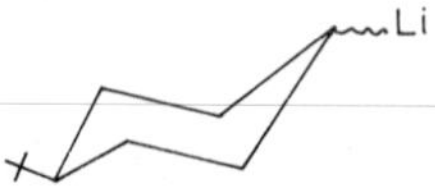 Li; *cis:trans* = 14:1	28

TABLE 3. (*Contd.*)

Alkyl Halide (RX)	Reagent[a] conditions	Product (RM)	Ref.
[structure: dichloro bicyclic compound]	$Li^+dbb^{-\bullet}$, thf, −78 °C	[structure: Cl, Li bicyclic product] major product 61%	30

[a] dbb = p-Bu^{t}C$_6$H$_4$C$_6$H$_4$Bu^t-p.

9% electron transfer compared with the 57% when $Naph^{-\bullet}Li^+$ in thf was used, and (ii) Bu^t_2-naphthalene^{2-} $2Li^+$ and $C_6H_{13}CHMeBr$ in Et_2O gave 34% of $C_6H_{13}CHMeLi$ compared with 96% for reaction of Bu^t_2-naphthalene$^{-\bullet}$ Li^+ in thf.

The benzynoid reagent[44] o-FC_6H_4Na has been prepared from o-BrC_6H_4F and $[Ph_2CCPh_2]^{2-}$ $2Na^+$ in thf at −70 °C.

Normally in reactions of RX and $Naph^{-\bullet}M^+$, the yields of RM are independent of the halogen, X (as indicated in Scheme 3). However, halogen effects were noticed[45] in reactions of a number of primary alkyl halides (RX) with the dianion $[Ph_2CCPh_2]^{2-}2Na^+$ in 2-methyltetrahydrofuran at ambient temperature: the average yields of RNa (and/or RNa derived products) were 34 ± 5, 52 ± 3, and 66 ± 3% for X = Cl, Br and I, respectively. The explanation for the $[Ph_2CCPh_2]^{2-}2Na^+$ results was based on the intermediacy of radical anions, $RX^{-\bullet}$, having finite if only very short lifetimes, and a competition between the decomposition of $RX^{-\bullet}$, within geminate radical pairs, $[(Ph_2CCPh_2)^{-\bullet}RX^{-\bullet}]$, and the diffusive separation of the pair. The longer the lifetime of $RX^{-\bullet}$, the greater will be the fraction of the geminate radical pairs undergoing separation.

a. Stereochemistry

Only if the free radical $R^\bullet$, formed as shown in Scheme 3, could be trapped before any possible isomerization occurred would the stereochemistry of the organic halide be retained. From many examples, it appears[28,29,34,35] that equilibration of radical intermediates proceeds at a faster rate than reactions with $ArH^{-\bullet}Li^+$ (see, for example Scheme 5 and Table 3). Thus in these reactions, the ratios of isomeric organolithium products reflect the equilibrium ratios of the intermediate free radicals.

SCHEME 5[28]

A different case emerges with phenyl-substituted cyclopropanes, **3**[38] and **4**[35]. For these compounds, a net retention of configuration results on reaction with $ArH^{-\bullet}M^+$. The

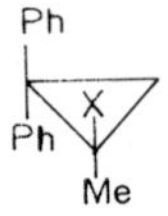

(**3**)

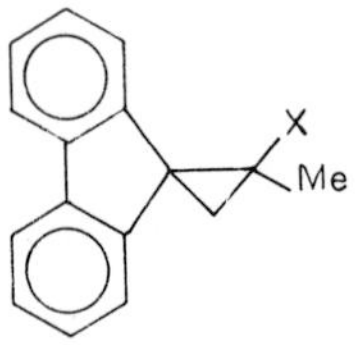

(**4**)

extent depends on the halide and M; e.g. from **3** (X = Cl, Br or I), net retention of configuration obtained on reaction with $Naph^{-\bullet}K^+$ in thf at 20 °C was 3, 53 and 41%, respectively, and from **3** (X = Br), the net retention of configuration obtained using $Naph^{-\bullet}M^+$ was 30, 49 and 53% for M = Li, Na, and K, respectively. For these reactions of **3** and **4**, it seems that the phenyl substituents play some role, as yet undefined[35].

Differences in stereochemistry of the products, RLi, have been found between reactions of RX with R′Li and with $ArH^{-\bullet}Li^+$, the R′Li reactions proceeding essentially with retention; e.g. in contrast to the results shown in Scheme 5, compound **2** reacts with BuLi in thf at − 15 °C with retention of configuration[28]. Another example is shown by *syn*-7-bromonorbornene[28,30], equation 4.

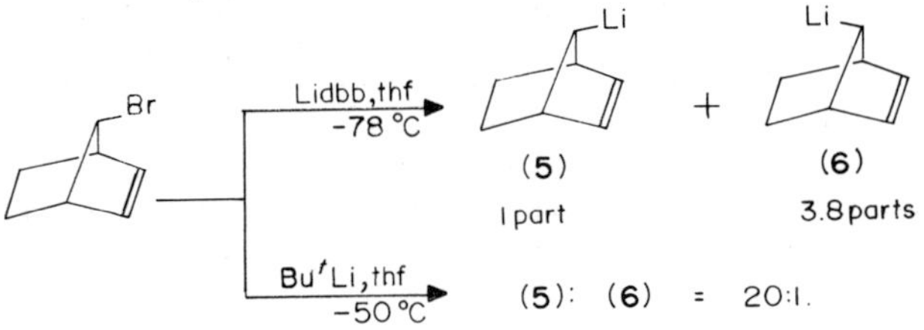

Lidbb, thf, −78 °C → (**5**) 1 part + (**6**) 3.8 parts (4)

Bu^tLi, thf, −50 °C → (**5**): (**6**) = 20:1.

However, retention of configuration was found in the reactions of both $(p\text{-}Bu^tC_6H_4C_6H_4Bu^t\text{-}p)^{-\bullet}Li^+$ and Li with *syn*-7-methoxy-2-*exo*-bromonorbornane[30].

Differences in products have also been recorded[46], e.g. equation 5 for reactions of (*E*)-BrCH=CHOEt.

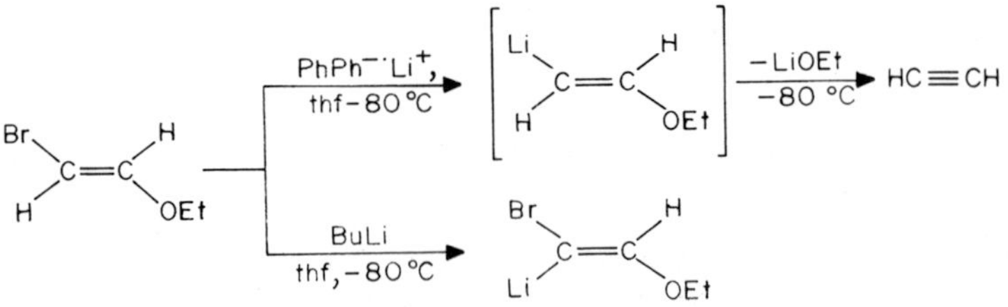

(*E*)-BrCH=CHOEt —($PhPh^{-\bullet}Li^+$, thf −80 °C)→ [(*E*)-LiCH=CHOEt] —(−LiOEt, −80 °C)→ HC≡CH (5)

(*E*)-BrCH=CHOEt —(BuLi, thf, −80 °C)→ BrLiC=CHOEt

3. Using Organometallic Compounds: Metal–Halogen Exchange[1,2,5,47]

The metal–halogen exchange reaction, equation 6, is a very valuable method of synthesis of organolithiums. Much less use has been made of this method for the other alkali metal compounds.

$$RX + R'M \rightleftharpoons RM + R'X \qquad (6)$$

Lithium–halogen exchanges are rapid reactions even at low temperatures. They are also reversible reactions and equilibrium constants have been determined for some systems[48,49], e.g. for interactions of RLi and PhI (equation 7):

$$RLi + PhI \rightleftharpoons PhLi + RI \qquad (7)$$
$$K_7 = [PhLi][RI]/[RLi][PhI]$$

at $-70\,^{\circ}C$ in Et_2O; log K (R) = 2.41 (vinyl), 0.98 (cyclopropyl), 3.5 (Et), 6.1 (Me_3CCH_2), and 6.9 (cyclopentyl), and for interactions of YC_6H_4Br and PhLi (equation 8):

$$YC_6H_4Br + PhLi \rightleftharpoons YC_6H_4Li + PhBr \qquad (8)$$
$$K_8 = [PhBr][YC_6H_4Li]/[PhLi][YC_6H_4Br]$$

at 25 °C in Et_2O; K_8(Y) = 0.6 (*p*-Me), 0.8 (*m*-Me), 5.3 (*p*-Cl), and 2.89 (*m*-CF_3).

Values of K_8 were found to be almost independent of the temperature and are also similar in Et_2O and thf. The aggregation of the organolithium was not considered in calculating the equilibrium constants.

As can be deduced from the quoted values of the equilibrium constants, lithium at equilibrium is preferentially attached to the organic residue better able to stabilize the negative charge. The larger the difference in the stabilizing abilities of the two organic fragments in a given exchange reaction, the further the equilibrium will lie to one side. Reactions of simple alkyllithiums and alkyl halides will not lead to complete exchange; the reaction of ethyllithium and methyl iodide in benzene solution gave only a 1:1 complex of methyl- and ethyllithium[50]. However, reactions of alkyllithiums with aryl halides, cyclopropyl halides, 1-alkenyl halides, alkynyl halides, and some α-substituted alkyl halides do lead to extensive exchanges and hence to useful preparations of new organolithiums. These will be referred to again later.

The general reactivity of the organic halide, RX, decreases in the order X = I > Br > Cl > F, e.g. the bromide is exchanged[51] in 2-F-4-ClC_6H_3Br, equation 9.

Br, F, Cl (benzene) —BuLi, Et_2O, hexane, −80 °C→ Li, F, Cl (benzene) (9)

Few exchanges have been realized with chlorides and rarely any with fluorides. For fluorides, especially, and chlorides (and occasionally also for bromides), an alternative reaction to metal–halogen exchange may occur, namely metallation of *o*-hydrogens (in the case of aromatic compounds) or α-hydrogens, made acidic by the adjacent halogen atom(s). There are, however, a number of instances of Li–Cl exchanges occurring with polychloroorganics[5].

Competitive metallation of organic bromides and chlorides can be reduced by employing electron-donating solvents at low temperatures; at low temperatures, the rates of Li–Br exchanges in ethereal solvents are generally greater than rates of metallations. The presence of tmed, however, has been found to promote metallations rather more than metal–halogen exchanges[52].

A further problem can be alkylations, i.e. the production of coupled products, RR′, from RLi and R′X (see Section III.C.2). Such coupling is more probable in thf than in Et_2O[53,54,55] or other poorly polar solvents, see equation 10[55]. The 1-butylnaphthalene product in equation 10 is apparently formed indirectly, that is, from the initial exchange products, 1-lithionaphthalene and BuBr, rather than directly from the reagents. Vinylic

Br; thf-hexane (5:1); + BuLi; Bu; 72%; Et_2O-hexane (5:1); Bu; Li; 3%

(10)

iodides give vinylic lithiums with RLi (R = Et or Bu) in Et_2O but in thf coupled products occur. Coupling occurs more readily with primary than with secondary or tertiary alkyllithiums. More alkylation results with MeLi or PhLi than with BuLi or Bu^tLi; see, for example, equations 11 and 12.

Cl; C≡C; BuLi, −100 °C, thf, Et_2O, hexane (ref. 56); Li; C≡C; PhLi, −8 °C, thf; Ph; C≡C

(11)

Br; BuLi, −78 °C, thf; Li; (ref. 57); MeLi, 0 °C, thf; Me

(12)

One method of overcoming alkylations is the use of two equivalents of RLi; the extra equivalent of RLi reacts with the RX formed in the Li–X exchange.

$$\mathrm{BuCH_2C{\equiv}CI + 2BuLi \xrightarrow[\text{(ref. 58)}]{Et_2O-hexane} BuCH_2C{\equiv}CLi + [BuBr] + LiI} \quad (13)$$

$$(E)\text{-}\mathrm{RCH{=}CHBr + 2Bu^tLi \xrightarrow[-110\,^{\circ}C\ \text{(ref. 59)}]{thf,\ Et_2O,\ pentane,} (\mathit{E})\text{-}RCH{=}CHLi + LiBr + Bu^tH + Me_2C{=}CH_2} \quad (14)$$

Lithium–halogen exchanges are faster in ethereal solvents than in hydrocarbons[48,49,60]. Evidence has been found for the presence of lithium halide retarding the rates of lithium–halogen exchanges[60].

The reactivity of different types of organic halides has also been investigated[61]; towards BuLi in thf–hexane at − 100 °C, the reactivity decreased in the order $ArCH_2Br > ArBr > Ar(CH_2)_nBr$ ($n > 2$) $> Ar(CH_2)_nCl$, as illustrated by equations 15–17. Even at − 100 °C, o-$BrC_6H_4CH_2Li$, formed in equation 17, undergoes coupling with the o-bromobenzyl bromide present; this illustrates the major problem of preparing benzyllithiums by this route.

$$\text{2-BrC}_6\text{H}_4\text{CRR}'\text{Cl} + \text{BuLi} \xrightarrow[\text{(refs.61,62)}]{\text{thf, hexane, }-100\,^\circ\text{C}} \text{2-LiC}_6\text{H}_4\text{CRR}'\text{Cl} \tag{15}$$

R,R′ = H, Me; R = R′ = H 88%

$$\text{2-BrC}_6\text{H}_4\text{CH}_2\text{Br} + \text{BuLi} \xrightarrow[\text{(ref.61)}]{\text{thf, hexane, }-100\,^\circ\text{C}} \text{2-BrC}_6\text{H}_4\text{CH}_2\text{Li} \tag{16}$$

$$\text{2-BrC}_6\text{H}_4\text{CH}_2\text{CH}_2\text{CHRBr} \xrightarrow[\text{(ref.62)}]{\text{thf, hexane, }-100\,^\circ\text{C}} \text{2-LiC}_6\text{H}_4\text{CH}_2\text{CH}_2\text{CHRBr} \tag{17}$$

R = H or Me; R = H (100%)

a. Mechanism[5]

Particularly well suited for kinetic study are the exchanges between aryl halides and aryllithiums; second-order reactions are generally found[5,49,63]. The ρ value for exchanges between PhLi and ArBr in Et_2O was calculated to be 4.0. Reactions of butyllithium and ArBr in hexane at 40 °C were shown to be first order in ArBr and in $(BuLi)_6$ (butyllithium exists as a hexamer in hexane)[64]. The ρ value was calculated to be *ca.* 2. Possible transition states include a four-centred transition state and one arising from an S_N2 type attack of Bu^- on ArBr.

A free radical component has been detected in some reactions, e.g. between alkyl halides and alkyllithiums[5,65]. However, the complete retention of configuration in some reactions of cyclopropyl[66] and 1-alkenyl halides[59], e.g. equations 18 and 19, and the partial retention in others (e.g. in the reaction of 2-octyl iodide and Bu^sLi in hexane–Et_2O at − 70 °C)[67], suggest that the extent of the free radical nature cannot be significant in all such reactions.

$$\text{(Ph)}_2\text{C-cyclopropyl-C(Me)(Br)} \xrightarrow{\text{BuLi}} \text{(Ph)}_2\text{C-cyclopropyl-C(Me)(Li)} \tag{18}$$

$$\text{Pr(Cl)C=CH(Et)} \xrightarrow{\text{BuLi}} \text{Pr(Li)C=CH(Et)} \tag{19}$$

Use of Li, in place of BuLi, with the reagents in equations 18 and 19 leads to loss of the stereochemistry.

b. Alkyl halides

As shown in equation 18, lithium–halogen exchanges provide cyclopropyllithiums with retention of configuration. A further example is shown in equation 20[68]; (*E*)-**7** also reacts with retention of configuration.

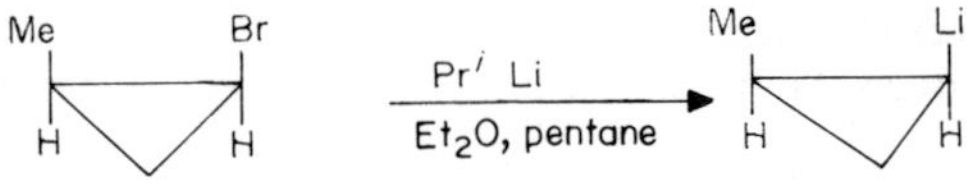

(20)

Compound **8** can be prepared[69] by a variety of routes, including Li–Br exchange, equation 21; however trimethylsilylcyclopropane could not be metallated to **8** using Bu^{s}Li and tmed in thf.

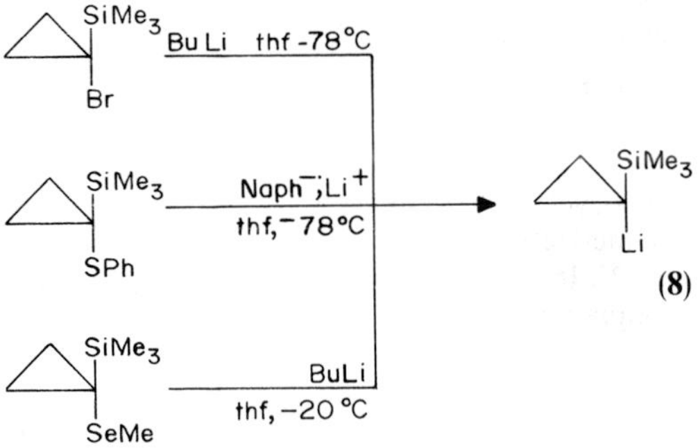

(21)

Bridgehead tertiary alkyllithiums, e.g. 1-Li-triptycene (equation 22)[70], 1-Li-norbornane[71], 1-Li-bicyclo[2.2.2]octane[70,71], and 1-Li-adamantane[71], have been prepared via lithium–halogen exchanges.

(22)

Suitably sited methoxy groups can enhance[72] rates of lithium–halogen exchange; e.g. **9** is more reactive towards BuLi than either **10** or **11**. This is due to the stabilization of the product organolithium arising from intramolecular coordination, a feature not available to the products from **10** and **11**.

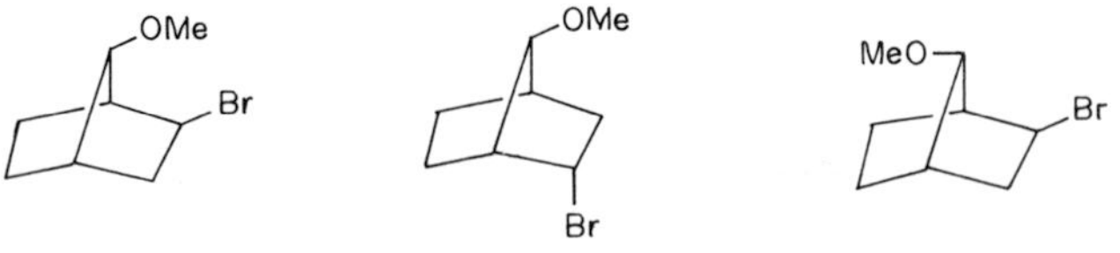

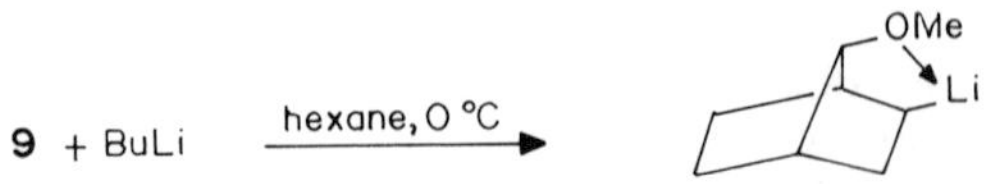

(9) (10) (11)

9 + BuLi $\xrightarrow{\text{hexane, 0 °C}}$ (23)

c. Polyhaloalkanes

Polyhaloalkanes, including polychloroderivatives, undergo Li–X exchanges to give carbenoid reagents. Very low temperatures must be used to prevent decomposition of the haloalkyllithium products. Some examples are listed in equations 24 and 25.

$$CCl_4 + BuLi \xrightarrow[\text{(ref. 73)}]{\text{thf, } -100\,^\circ C} CCl_3Li \quad (24)$$

$$CFCl_3 + BuLi \xrightarrow[\text{(ref. 74)}]{\text{thf, } -116\,^\circ C} CFCl_2Li \quad (25)$$

When both Cl and Br are present in the polyhaloalkane reagents, products arising from formal exchange of either halogen can be obtained, equation 26; however, this may be a consequence of scrambling of halogens. Halogen exchange can occur[76] between LiX and the polyhaloalkyllithium, as shown with LiBr and $[(EtO)_2PO]CCl_2Li$

$$CCl_3Br + MeLi \xrightarrow[-115\,^\circ C]{Et_2O} \underset{65\%:}{CCl_3Li} + \underset{11\%}{CCl_2BrLi} \quad (26)$$

(ref. 75)

Reactions of *gem*-dihalocyclopropanes with RLi occur stereoselectively if not stereospecifically. Subsequent isomerizations may occur[77]. Compounds **12** (X = Br)[78,79], **13**[78], **14** (X = Cl)[78], **15** (X = Br)[80,81], **16**[82], and **17**[83] all undergo Li–X exchange stereospecifically at the *endo* position, e.g. equations 27 and 28.

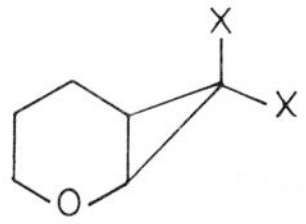

(12) 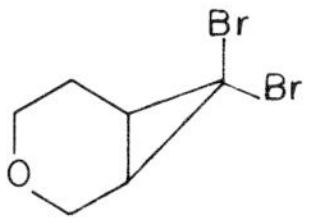 (13) 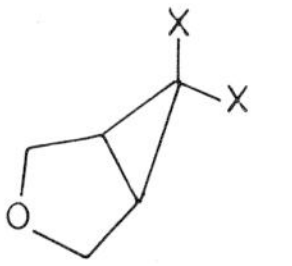 (14)

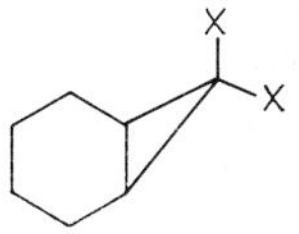

(15) 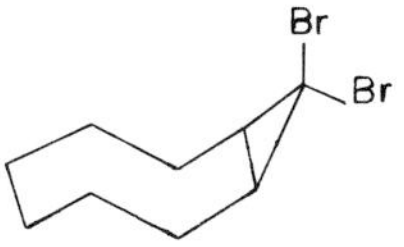 (16) 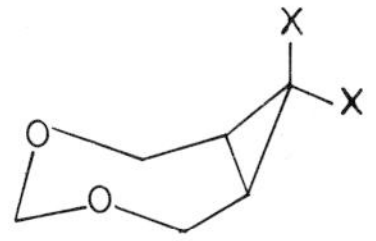 (17)

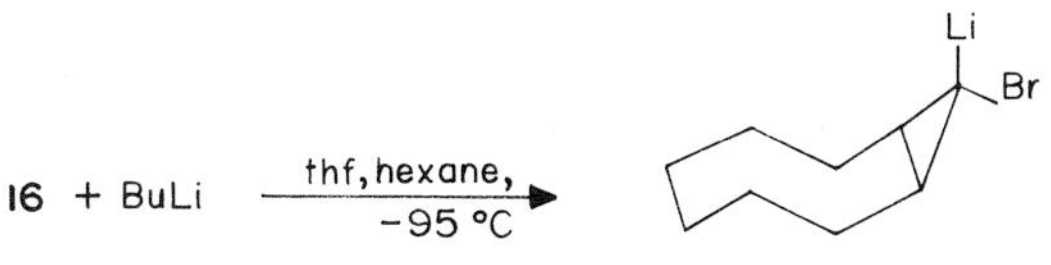

16 + BuLi $\xrightarrow[-95\,^\circ C]{\text{thf, hexane,}}$ (27)

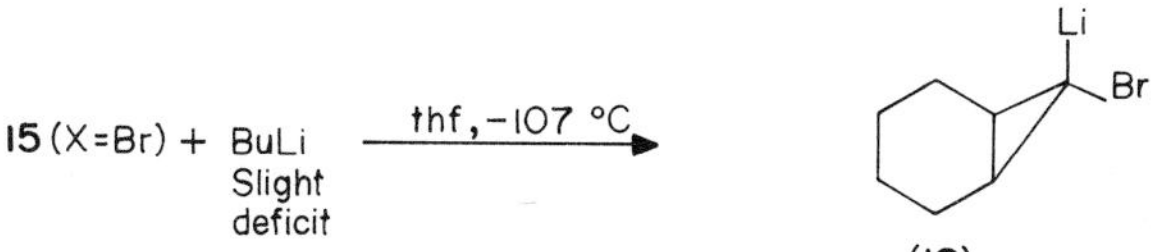

15 (X=Br) + BuLi (Slight deficit) $\xrightarrow{\text{thf, } -107\,^\circ C}$ (18) (28)

Of interest, the product **18**, *anti*-7-bromo-*syn*-7-lithionorcarane, isomerizes on standing or in the presence of excess of BuLi. Both isomers are obtained[84] from **15** (X = Cl) and BuLi in the Trapp solvent at −115 °C. *Endo* lithiation of **12** (X = Cl) occurs in Et_2O but both isomers result in thf. Both isomers are obtained from **14** (X = Br).

Further examples of substituted alkyllithiums, including α-RS[85], α-RSe[86], α-RO_2C[87], and α-R_3M (M = Si[88,89], Ge[88], Sn[90], or Pb[91]), are listed in Table 4.

TABLE 4. Formation of organolithiums via halogen–lithium exchange reactions

$$RX + R'Li \rightarrow RLi + R'X$$

Organic halide, RX	Organolithium reagent R′Li, and conditions	Product (%)	Ref.
$CFBr_3$	BuLi, thf, hexane, −116 °C	$CFBr_2Li$ (70)	684
CBr_4	BuLi, thf, pentane, −100 °C	CBr_3Li (80)	685
$Cl_3CP(O)(OEt)_2$	BuLi, thf, Et_2O, −105 °C	$LiCCl_2P(O)(OEt)_2$ (76)	686
$PhCCl_3$	BuLi, thf, hexane, −100 °C	$PhCCl_2Li$ (75)	687
$BuCBr_3$	BuLi, thf, Et_2O, −105 °C	$BuCBr_2Li$ (60)	688
$BuCHBr_2$	BuLi, thf, Et_2O, pentane	BuCHBrLi (64)	689
CH_2Br_2	Bu^sLi, LiBr, thf, Et_2O, pentane, −110 °C	$BrCH_2Li$	690
Ph_3SiCH_2Br	BuLi, Et_2O, hexane, −78 °C	Ph_3SiCH_2Li (78)	691
$(Me_3Si)_2CBr_2$	BuLi, thf, hexane, −115 °C	$(Me_3Si)_2CBrLi$ (70)	692
$(Me_3Si)_3CBr$	BuLi, Et_2O, hexane, −75 °C	$(Me_3Si)_3CLi$ (77)	692
Ph_3GeCH_2Br	BuLi, Et_2O, 20 °C	Ph_3GeCH_2Li (90)	690
Ph_3SnCHI_2	BuLi, Et_2O, −60 °C	$Ph_3SnCHILi$	693
Me_3SnCBr_3	BuLi, thf, Et_2O, hexane−105 °C	Me_3SnCBr_2Li	694
Ph_3PbCH_2I	BuLi, Et_2O, −50 °C	Ph_3PbCH_2Li (58)	695
$PhSCH_2Br$	BuLi, thf, −78 °C	$PhSCH_2Li$ (55)	696
$PhSeCH_2Br$	BuLi, thf, −78 °C	$PhSeCH_2Li$ (75)	697
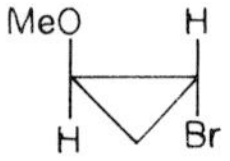 MeO H H Br	Bu^tLi, pentane, −78 °C	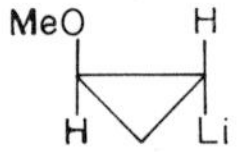 MeO H H Li (90)	698
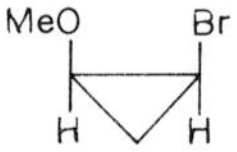 MeO Br H H	Bu^tLi, pentane, −78 °C	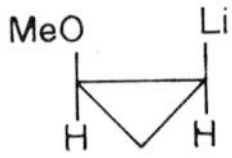 MeO Li H H (90)	698
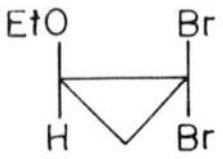 EtO Br H Br	BuLi, thf, hexane, −95 °C	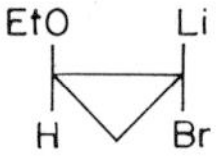 EtO Li H Br (77)	699
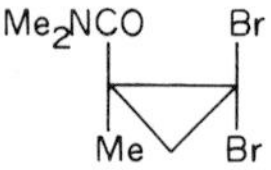 Me_2NCO Br Me Br	MeLi, Et_2O, −60 °C	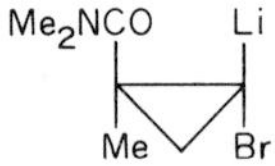 Me_2NCO Li Me Br (89)	700
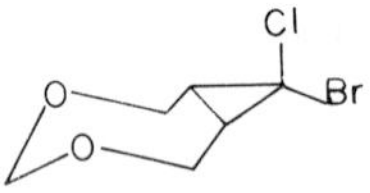 O O Cl Br	MeLi, Et_2O, thf, −78 °C	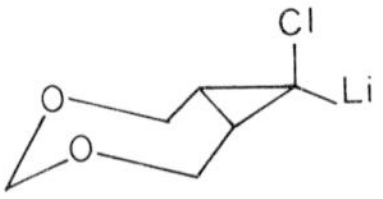 O O Cl Li	701
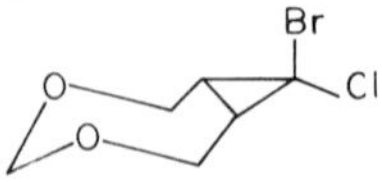 O O Br Cl	MeLi, Et_2O, thf, −78 °C	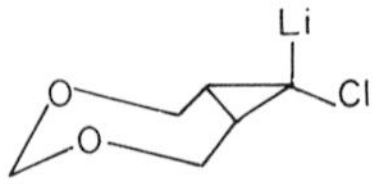 O O Li Cl	701
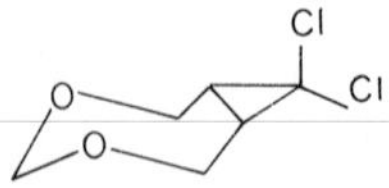 O O Cl Cl	BuLi (2 equiv.), Et_2O, −78 °C	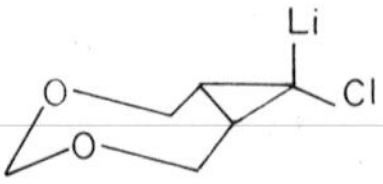 O O Li Cl	702
$HCBr_2CH{=}NBu^t$	BuLi, thf, −70 °C	$LiCHBrCH{=}NBu^t$ (80)	703

TABLE 4. (*Contd.*)

Organic halide, RX	Organolithium reagent R′Li, and conditions	Product (%)	Ref.
$CH_2{=}CHBr$	Bu^tLi (2 equiv.), thf, Et_2O, pentane, −110 °C	$CH_2{=}CHLi$ (85)	704
Ph, H / C=C / H, Br	Bu^tLi (2 equiv.), thf, Et_2O, pentane, −110 °C	Ph, H / C=C / H, Li (71)	704
Ph, Br / C=C / H, H	Bu^tLi (2 equiv.), thf, Et_2O, pentane, −110 °C	Ph, Li / C=C / H, H (81)	704
$CH_2{=}CPhBr$	Bu^tLi (2 equiv.), thf, Et_2O, pentane, −110 °C	$CH_2{=}CPhLi$ (71)	704
Bu, $SiMe_3$ / C=C / H, Br	BuLi, Et_2O, −70 °C, 20 min	Bu, $SiMe_3$ / C=C / H, Li (88)	705
Bu, SiMe / C=C / H, I	BuLi, thf, Et_2O, hexane, −95 °C, 30 min	$BuCH{=}C(SiMe_3)Li$ (*E*):(*Z*) = 93:7	705
Bu, Br / C=C / H, $SiMe_3$	BuLi, thf, −70 °C, 2 h	Bu, Li / C=C / H, $SiMe_3$ (96)	705
$CH_2{=}CBrCH_2OH$	Bu^tLi (2.5 equiv.), Et_2O, pentane, −78 to 0 °C	$CH_2{=}CLiCH_2OLi$ (73)	706
H, H / C=C / EtO, Br	BuLi, Et_2O, hexane, −80 °C	H, H / C=C / EtO, Li (84)	707
$Me_2C{=}CBrCH(OEt)_2$	BuLi, −90 °C	$Me_2C{=}CLiCH(OEt)_2$ (70)	708
MeO, Br / C=C / H, Br	BuLi, Et_2O, −78 °C	(*E*) and (*Z*) $MeOCH{=}CBrLi$	709
O (ring), Br	Bu^tLi, thf, Et_2O, pentane, −110 °C	O (ring), Li (90)	710
O, O (ring), Br, Me	BuLi (1.3 equiv.), thf, −78 °C	O, O (ring), Li, Me	711
$Me_2C{=}CBrCO_2H$	BuLi (2 equiv.), thf, hexane, −100 °C	$Me_2C{=}CLiCO_2Li$ (98)	712

TABLE 4. (*Contd.*)

Organic halide, RX	Organolithium reagent R′Li, and conditions	Product (%)	Ref.
BrHC=C(CO₂H)Me	BuLi (2 equiv.), Et_2O, hexane, − 78 °C	LiHC=C(CO₂Li)Me (71)	713
1-Br-2-CONMe₂-cyclohexene	Bu^tLi, thf, pentane, − 75 °C	1-Li-2-CONMe₂-cyclohexene (85)	714
H(Br)C=C(NMe₂)Ph	BuLi, thf, pentane, − 70 °C	H(Li)C=C(NMe₂)Ph (60)	715
$CH_2{=}CBrCF_3$	BuLi, Et_2O, − 78 to − 90 °C	$CH_2{=}CLiCF_3$	716
$CCl_2{=}CF_2$	BuLi, Et_2O, thf (1 equiv.), − 120 to 90 °C	$LiCCl{=}CF_2$ (86)	717
$CF_2{=}CFBr$	MeLi (2 equiv.), thf, hexane, Et_2O, − 110 °C	$CF_2{=}CFLi$ (96)	704
$CCl_2{=}CClBr$	BuLi, Et_2O, − 110 °C	$CCl_2{=}CClLi$ (92)	718
$Ph_2C{=}CBr_2$	BuLi, thf, pentane, − 100 °C	$Ph_2C{=}CBrLi$ (85)	719
1,2-dibromocyclopentene	BuLi, hexane, − 78 °C	1-Li-2-Br-cyclopentene	720
hexachloro-dibromo-norbornadiene (Cl, Cl, Cl / Cl Cl / Cl; Br, Br)	BuLi, Et_2O, − 75 °C	corresponding Li, Br compound (94)	721
$Me_2C{=}C{=}CHBr$	BuLi, Et_2O, − 70 °C	$Me_2C{=}C{=}CHLi$ (91)	722
PhBr	BuLi, PhMe, 50 °C	PhLi (95)	723
o-FC_6H_4Br	BuLi, Et_2O, − 70 °C	*o*-FC_6H_4Li (84)	724
o-BrC_6H_4Br	BuLi, Et_2O, thf, hexane,− 100 °C	*o*-BrC_6H_4Li (95)	725
m-FC_6H_4Br	BuLi, Et_2O, − 45 °C	*m*-FC_6H_4Li (65)	726
p-BrC_6H_4Cl	BuLi, Et_2O	*p*-ClC_6H_4Li (90)	727
p-BrC_6H_4Br	BuLi, Et_2O	*p*-BrC_6H_4Li (78)	727
	Excess BuLi	*p*-LiC_6H_4Li (89)	
1,3,5-$Br_3C_6H_3$	BuLi, Et_2O, − 78 °C	1-Li-3,5-$Br_2C_6H_3$ (97)	728
2,4,6-$Bu^t_3C_6H_2Br$	BuLi, thf, hexane, − 78 °C	2,4,6-$Bu^t_3C_6H_2Li$ (59)	729
o-$BrC_6H_4CH_2CH_2Cl$	BuLi, thf, − 100 °C	*o*-$LiC_6H_4CH_2CH_2CH_2Cl$ (81)	730

TABLE 4. (*Contd.*)

Organic halide, RX	Organolithium reagent R′Li, and conditions	Product (%)	Ref.
o-BrC_6H_4OH	BuLi (2 equiv.), Et_2O, 25 °C	*o*-LiC_6H_4OLi (62)	731
o-$BrC_6H_4CH_2OH$	BuLi (2 equiv.), Et_2O, − 20 °C	*o*-$LiC_6H_4CH_2OLi$ (50)	732
o-$BrC_6H_4CH_2SH$	BuLi (2 equiv.), thf, hexane, − 100 °C	*o*-$LiC_6H_4CH_2SLi$ (60)	732
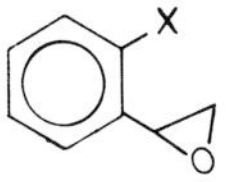 X; O; X = Br or I	BuLi, thf, − 78 °C	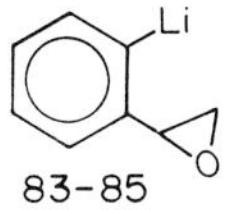 Li; O; 83-85	733
o-$BrC_6H_4CO_2H$	BuLi (2 equiv.), thf, − 78 °C	*o*-$LiC_6H_4CO_2Li$ (80)	732
o-$BrC_6H_4CO_2Me$	BuLi, thf, − 100 °C	*o*-$LiC_6H_4CO_2Me$ (88)	734
p-$BrC_6H_4CO_2Bu^t$	Bu^tLi, thf, hexane, − 100 °C	*p*-$LiC_6H_4CO_2Bu^t$ (75)	735
p-$BrC_6H_4CH_2CH_2CO_2H$	BuLi (2 equiv.), thf, − 90 °C, hexane	*p*-$LiC_6H_4CH_2CH_2CO_2Li$ (80)	736
o-$BrC_6H_4NHCOBu^t$	(i) MeLi, − 78 °C (ii) Bu^tLi, − 78 °C	*o*-$LiC_6H_4NLiCOBu^t$ (92)	737
BrC_6H_4CN	BuLi, thf, − 100 °C	LiC_6H_4CN *o*-(82), *m*-(88), *p*-(83)	738
m-$BrC_6H_4CF_3$	BuLi, Et_2O, 0 °C	*m*-$LiC_6H_4CF_3$ (62)	739
o-$BrC_6H_4NO_2$	PhLi, thf, − 100 °C	*o*-$LiC_6H_4NO_2$ (87)	740
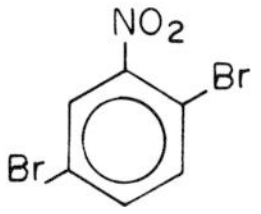 NO_2; Br; Br	BuLi, thf, hexane, − 100 °C	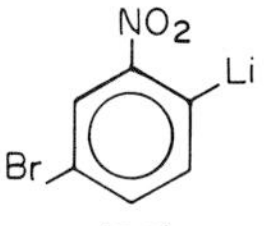 NO_2; Li; Br; (50)	741
p-$BrC_6H_4NH_2$	excess BuLi, Et_2O, − 60 °C	*p*-$LiC_6H_4NLi_2$ (68)	742
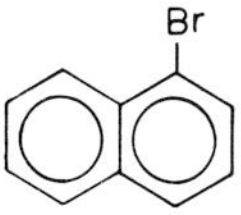 Br	PrLi, Et_2O, 25 °C	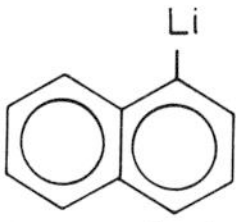 Li; (97)	743
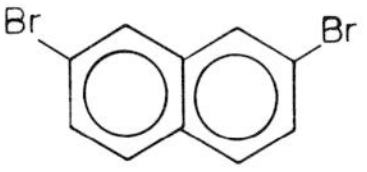 Br; Br	BuLi, Et_2O, 20 °C	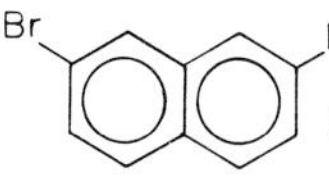 Br; Li; (91)	744
	BuLi, thf, − 35 °C	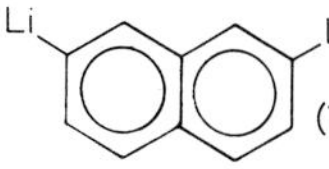 Li; Li; (72)	744
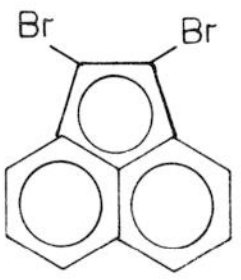 Br; Br	BuLi, Et_2O, hexane, tmed, − 10 °C	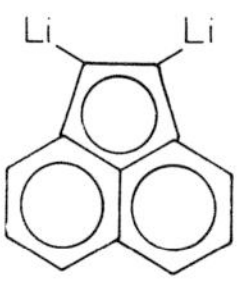 Li; Li; (99)	745

TABLE 4. (*Contd.*)

Organic halide, RX	Organolithium reagent R′Li, and conditions	Product (%)	Ref.
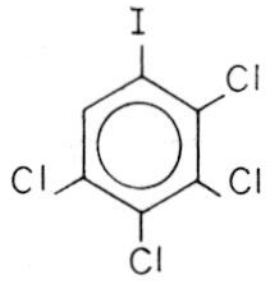	BuLi, Et_2O, −70 °C	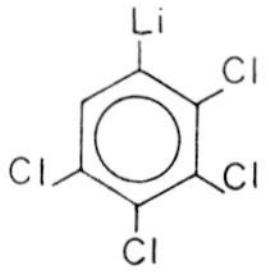 (59)	746
C_6Cl_6	BuLi, Et_2O, −78 to −10 °C	C_6Cl_5Li(79)	747
	Bu^tLi (3 equiv.), thf, −78 °C −78 °C	*p*-LiC_6Cl_4Li(72)	748
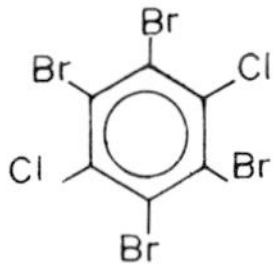	BuLi (2.2 equiv.), hexane, Et_2O, −78 °C	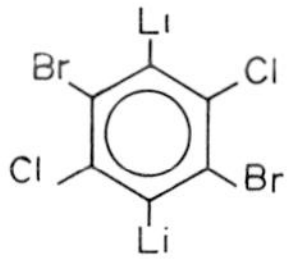	749
C_6Br_6	BuLi, Et_2O, −75 °C	C_6Br_5Li (17)	750
	BuLi (2 equiv.), hexane, Et_2O −78 °C	*p*-LiC_6Br_4Li	749
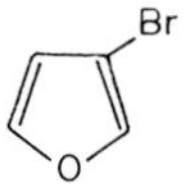	BuLi, heptane, Et_2O, −70 °C	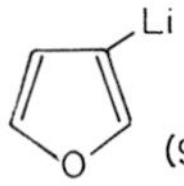 (97)	751
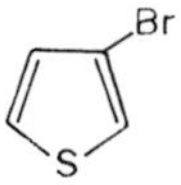	BuLi, Et_2O, −70 °C	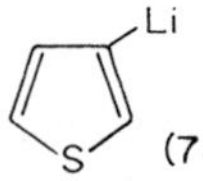 (78)	752
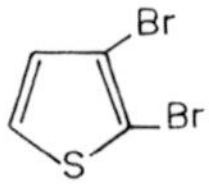	BuLi, Et_2O, −70 °C	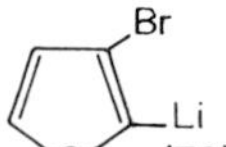 (70)	752
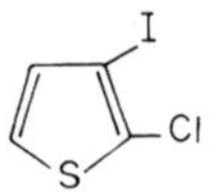	BuLi, Et_2O, −70 °C	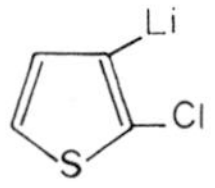 (51)	753
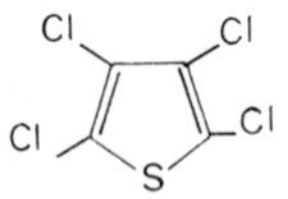	BuLi, Et_2O, hexane−25 °C	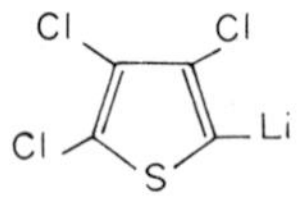	754
	BuLi (2 equiv.), Et_2O, 50 °C	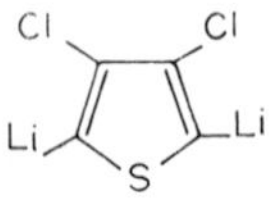 (68)	753
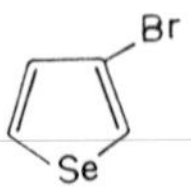	PhLi, Et_2O, reflux	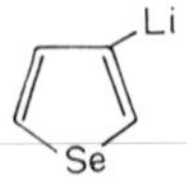 (65)	755
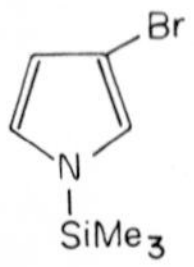	2 equiv. Bu^tLi, thf, −78 °C	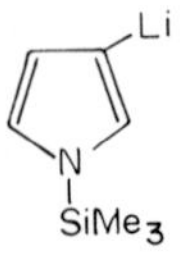	756

TABLE 4. (*Contd.*)

Organic halide, RX	Organolithium reagent R′Li, and conditions	Product (%)	Ref.
2-bromopyridine	BuLi, Et_2O, − 18 °C	2-lithiopyridine (69)	757
3-bromopyridine	BuLi, light petroleum, − 20 °C	3-lithiopyridine (34)	757
2,6-dibromopyridine	BuLi, Et_2O, − 30 °C	2-bromo-6-lithiopyridine (43)	757
	BuLi (2 equiv.), thf, − 90 °C	2,6-dilithiopyridine (52–58)	758

d. Alk-1-enyl halides

Lithium–halogen exchange reactions of alkyl-substituted vinyl halides with RLi proceed with retention of configuration[54,59,92,93], see equations 19 and 29.

$$\mathrm{R'(R)C{=}CH(I)} \xrightarrow[\substack{-50\ \text{to}\ -60\,^\circ\text{C}\\ (\text{ref.}92)}]{\text{BuLi or EtLi},\ Et_2O} \mathrm{R'(R)C{=}CH(Li)}$$

$$R = \text{Et, heptyl, not Me};\ R' = \text{H, Et or Bu} \tag{29}$$

Aryl-substituted vinyllithiums are also formed with retention, but as these are configurationally less stable than the alkyl-substituted compounds, controlled conditions have to be used e.g. equation 30. Either raising the temperature to above − 78 °C in thf

$$\mathrm{Ph(Br)C{=}C(Et)H} \xrightarrow[\text{or hexane, RT}]{\text{thf}, -78\,^\circ\text{C}} \mathrm{Ph(Li)C{=}C(Et)H} \tag{30}$$

solution or increasing the amount of thf in the hexane solution at room temperature results in the formation of both the PhLiC=CHEt isomers[93].

The more hindered *E*-isomers of $Me_3SiCX{=}CHR$ (X = Br or I; R = Bu, cyclohexyl, or Bu^t) react more slowly with BuLi than do the *Z*-isomers[94].

Vinyl chlorides react more sluggishly than do the corresponding bromides and iodides. As shown in equation 19, vinyl chlorides can react, but α-metallations of vinyl chlorides frequently occur[95]. Vinyl fluorides in the absence of other halogens do not undergo lithium–halogen exchange.

An illustration of the difference in reactivity of C—Br and C—Cl bonds is given[96] in equations 31 and 32.

$$\mathrm{H(Br)C{=}CH(OSiMe_3)} + \mathrm{Bu^tLi} \xrightarrow{\mathrm{Et_2O,\ -70\,^{\circ}C}} \mathrm{H(Li)C{=}CH(OSiMe_3)} \tag{31}$$

$$\mathrm{H(Cl)C{=}CH(OSiMe_3)} + \mathrm{Bu^tLi} \xrightarrow{\mathrm{Et_2O,\ -70\,^{\circ}C}} \mathrm{H(Cl)C{=}CH(OLi)} \tag{32}$$

For the vinylic bromide BrCH=CHOEt, the reaction followed with BuLi depends on the geometry, equations 33 and 34; stabilization due to the strong coordination between EtO and Li in the products appears to be the overriding factor[97]. The use of $PhPh^{-\bullet}Li^{+}$ on the *E*-isomer does provides the unstable (*E*)-LiCH=CHOEt.

$$\mathrm{H(Br)C{=}CH(OEt)} + \mathrm{BuLi} \xrightarrow{\mathrm{Et_2O,\ -80\,^{\circ}C}} \mathrm{H(Li)C{=}CH(OEt)} \tag{33}$$

$$\mathrm{H(Br)C{=}C(OEt)H} + \mathrm{BuLi} \xrightarrow{\mathrm{Et_2O,\ -80\,^{\circ}C}} \mathrm{Li(Br)C{=}C(OEt)H} \tag{34}$$

Halogen–lithium exchanges occur for a variety of substituted vinylic halides, including those with alkoxy[98,99], halo (including poly- and perhaloalkenes)[100–103], carboxy[104–106], amino[107], carboxyamido[108], and Me_3Sn[94,109] groups, see Table 4. Low temperatures are frequently required in order to preserve the organolithium product.

Ketal and thioketal derivatives survive[110], e.g. equation 35.

$$\text{(dithiolane, } (CH_2)_n\text{, vinyl Br)} + \mathrm{BuLi} \xrightarrow{\mathrm{thf,\ -78\,^{\circ}C}} \textbf{(19)} \text{ (Li)} \xrightarrow[\text{(ii) } HgCl_2,\ HgO,\ MeCN,\ H_2O]{\text{(i) } E^+} \text{(enone, } (CH_2)_n\text{, E)} \tag{35}$$

$n = 1$ or 2

The organolithiums **19** and the ketal analogues are in effect *β*-acylvinyl anion equivalents [as are the organolithium species obtained by Li–X exchange reactions of 1-Br-2-$CONMe_2$-cyclohexene[108], equation 36, and of (*Z*)-BrCR=CR′CO_2H, equation 37[106]], and find good use in synthesis.

$$\text{1-Br-2-CONMe}_2\text{-cyclohexene} \xrightarrow[\mathrm{thf,\ pentane,\ -75\,^{\circ}C}]{\mathrm{Bu^tLi\ (2.1\ equiv.)}} \text{1-Li-2-CONMe}_2\text{-cyclohexene}\ (\sim 90\%) \tag{36}$$

$$\mathrm{Br(R)C{=}C(R')CO_2H} \xrightarrow[\mathrm{or\ thf,\ -100\,^{\circ}C}]{\mathrm{BuLi\ (2\ equiv.),\ Et_2O,\ -78\,^{\circ}C,}} \mathrm{Li(R)C{=}C(R')CO_2Li} \tag{37}$$

R=R′=H; R=H, R′=Me; R=Me, R′=H

α-Acylvinyl anion equivalents may also be obtained[104,105], see equation 38.

$$\mathrm{Me_2C{=}C(Br)CO_2H} \xrightarrow[\text{thf (ref. 105)}]{\text{BuLi (2 equiv.)} - 100\,^\circ\text{C}} \underset{\text{major product}}{\mathrm{Me_2C{=}C(Li)CO_2Li}} + \underset{\substack{\text{minor product}\\ \textbf{(20)}}}{\mathrm{CH_2{=}C(Me){-}CH(Li)CO_2Li}} \qquad (38)$$

Small amounts of the dianion **20** are also formed. As expected from the general reactivity sequence for halides, when more than one vinylic halogen is present, the preference for exchange is in the sequence I > Br > Cl > F, e.g. see equation 39[100].

$$\mathrm{F_2C{=}CFCl} \xrightarrow[\text{thf , Et}_2\text{O}]{\text{BuLi, } -120\,^\circ\text{C,}} \mathrm{F_2C{=}CFLi} \qquad (39)$$

e. Other unsaturated organic halides

Allyllithiums are not normally prepared by Li–X exchange; however, $CF_2\text{┄}\bar{C}H\text{┄}CH_2, Li^+$ has been prepared and trapped *in situ* at −95 °C from the reaction of $CH_2{=}CHCF_2Br$ and BuLi in thf, Et_2O and pentane[111].

Allenyllithiums, e.g. equation 40[112], alkynyllithiums[113], e.g. equation 41, cyclooctatetraenyllithium[114], and α,ω-dilithiodienes, equations 42 and 43, are also obtained by Li–X exchanges.

$$\mathrm{C_8H_{17}(H)C{=}C{=}C(H)Br} + \mathrm{BuLi} \xrightarrow{\text{Et}_2\text{O}, -70\,^\circ\text{C}} \mathrm{C_8H_{17}(H)C{=}C{=}C(H)Li} \qquad (40)$$

$$\underset{\text{X = Cl or Br}}{\mathrm{PhC{\equiv}CX + BuLi}} \longrightarrow \mathrm{PhC{\equiv}CLi} \qquad (41)$$

$$\mathrm{PhC(Br){=}CH{-}CH{=}C(Br)Ph} + \mathrm{BuLi} \xrightarrow[\substack{-78\,^\circ\text{C}\\ \text{(ref. 115)}}]{\text{thf, hexane}} \mathrm{PhC(Li){=}CH{-}CH{=}C(Li)Ph} \qquad (42)$$

$$\mathrm{HC(I){=}CH{-}CH_2{-}CH{=}C(I)H} + \mathrm{BuLi} \xrightarrow[\substack{-50\,^\circ\text{C}\\ \text{(ref. 116)}}]{\text{pentane,}} \mathrm{HC(Li){=}CH{-}CH_2{-}CH{=}C(Li)H} \qquad (43)$$

f. Aryl halides

Aryl bromides and iodides react readily with alkyllithiums to give aryllithiums. Normally ethereal solutions are used. However, reactions in hydrocarbons although slower than in ethers, do have the advantage that the aryllithiums precipitate out and can be obtained in high yields and with good purity[117,118].

Both halogens in dibromo- or diiodobenzenes can be replaced by lithium in lithium–

halogen exchanges[5]. If the two halogens are in an *ortho* arrangement, then aryne formation can result from the mono exchange product. However, *o*-LiC_6H_4X can be trapped[119].

$$o\text{-}C_6H_4Br_2 + \text{BuLi} \xrightarrow[-110\,^\circ\text{C}]{\text{thf, Et}_2\text{O}} o\text{-}LiC_6H_4Br \xrightarrow[\text{temp.}]{\text{higher}} [C_6H_4] \quad (44)$$

A variety of functional groups are tolerated in the Li–X exchange, although side-reactions may reduce the yields of Li–X exchange products[120]. The use of low temperatures and electron-donating solvents can be used to minimize the competitive reactions, such as *ortho*-metallations or reaction with the functional group. *Ortho*-metallation to a chloro or fluoro group is favoured over exchange of these groups by lithium (except in polychloro compounds). Bromoarenes, containing powerful *ortho*-directing groups (Y) (see Section II.B.2.b.ii), may undergo metallations at a site *ortho* to Y, in addition to the Li–Br exchange. Good examples are the $MeOC_6H_4Br$ compounds[5]; higher temperatures and prolonged reaction times lead to complex reaction mixtures.

Lithium–bromine exchanges have been reported for aryl bromides substituted with the following groups: OH[47], SH[47,121], NO_2[122–124], NH_2[47], NHCOR (R = Bu^t or CF_3)[125], SO_2NH_2[47], SO_2NR_2[47], CN[126,127] CO_2H[121,128], CO_2R (R = Me[129] or Bu^t[130]), $(CH_2)_nCO_2H$[131], CR_2CN ($R \neq H$)[126], and epoxides[132,133].

Groups that have acidic hydrogens, e.g. OH, SH, NH_2, and CO_2H, are metallated during the course of the reaction with RLi (and hence additional equivalents of RLi must be added to allow for this). These groups are recovered on work-up, however.

Whereas *o*-$NO_2C_6H_4Br$ and other *o*-nitrobromoarenes smoothly undergo Li–Br exchange (e.g. using BuLi in thf at $-100\,^\circ$C)[123], redox reactions arise with the *m*- and *p*-analogues and so Li–X exchanges are not used to prepare the aryllithiums from these compounds.

The halogen–metal exchange reaction is suitable[129] for the preparation of *o*-$LiC_6H_4CO_2Me$ at $-100\,^\circ$C; however the *m*- and *p*-analogues undergo condensation reactions even at this temperature and so cannot be trapped. The more hindered *p*-$LiC_6H_4CO_2Bu^t$ has a greater lifetime at $-100\,^\circ$C and can be trapped. The product from the reaction of *o*-$HO_2CCH_2CH_2C_6H_4Br$ with BuLi, namely *o*-$LiO_2CCH_2CH_2C_6H_4Li$, cyclizes even at $-100\,^\circ$C; on the other hand, *p*-$LiO_2CCH_2CH_2C_6H_4Li$ and *o*-$LiC_6H_4CH_2CH_2CONHLi$ are stable at this temperature[131].

Complex reaction mixtures arise[131] with *o*- or *p*-$BrC_6H_4CH_2CO_2H$. Reactions with BuLi provide both Li–Br and Li–H exchanges and among the products are the trianions, *o*- and *p*-$LiC_6H_4CHLiCO_2Li$. The trianions react slowly with the solvent (thf) to provide $LiC_6H_4CH_2CO_2Li$, the product expected from the direct Li–Br exchanges of $BrC_6H_4CH_2CO_2H$.

Benzylic metallation results exclusively with *o*-, *m*-, or *p*-$BrC_6H_4CH_2CN$ using BuLi in thf–hexane at $-100\,^\circ$C to give[126] $BrC_6H_4CHLiCN$. When there are no benzylic protons, as in *o*-$BrC_6H_4CMe_2CN$, Li–Br exchange occurs with good yields at $-100\,^\circ$C.

Reaction of *o*-$BrC_6H_4CH_2PPh_2$ with BuLi in Et_2O at room temperature provides *o*-$LiC_6H_4CH_2PPh_2$; in contrast[134], the chloro analogue, *o*-$ClC_6H_4CH_2PPh_2$, undergoes benzylic metallation to give *o*-$ClC_6H_4CHLiPPh_2$.

The orientation of Li–Cl exchanges in C_6Cl_5Y and of Li–H exchanges in C_6H_5Y show interesting differences. Metallations (Li–H exchanges) of $C_6H_5CH_2NMe_2$ and $C_6H_5NMe_2$ occur *ortho* to the substituent groups; in contrast, the sites of Li–Cl exchanges of the perchloro analogues using BuLi in Et_2O at $-70\,^\circ$C are indicated[135,136] by the arrows:

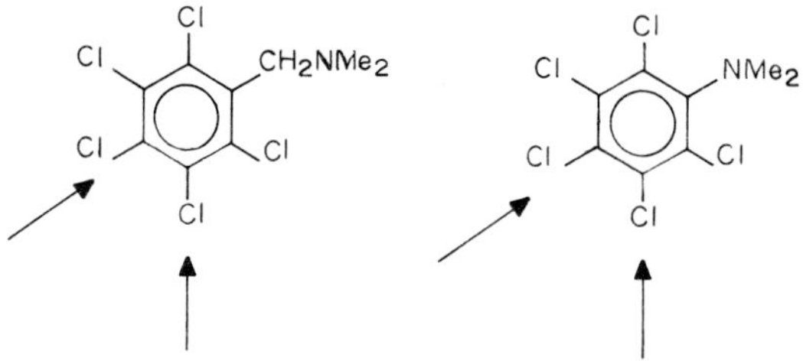

Cl CH2NMe2 Cl Cl Cl Cl

Cl NMe2 Cl Cl Cl Cl

g. Heteroaryl halides

Lithium–halogen exchange reactions are particularly useful methods of synthesis of five- and six-membered heterocyclic aryllithium, see Table 4.

i. *Thiophene derivatives.* Exchanges of α-Br or α-I groups in thiophenes occur more readily than the halogens in the β-position. *Ortho*-lithiations of β-bromo- or β-iodothiophenes have been recorded, see Section II.B.2.b.iii. For chlorothiophenes, Li–Cl exchange results only when no α-position is free, as with 2, 5-dichlorothiophene; 2-chlorothiophene reacts with BuLi to give 2-Cl-5-Li-thiophene. Lithium–halogen exchange reactions of 2-Cl-3-X-thiophene (X = Br or I), 2-Br-4-I-thiophene, and 2-Cl-4-Br-thiophene with BuLi at $-70\,^{\circ}$C all take place with the β-halogens[136]. Tetrachlorothiophene reacts with BuLi in Et_2O to provide 2, 5-Li_2-3, 4-Cl_2-thiophene[137]; the use of more controlled conditions lead to the mono exchange product[138].

ii. *Pyridine derivatives.* Lithium–halogen exchanges are valuable routes to lithiopyridine, since reaction of RLi with pyridine frequently results in additions to the C=N rather than metallations. Lithium–bromine exchanges have been used to prepare[122] 2- and 3-lithiopyridines using BuLi in thf–hexane at $-100\,^{\circ}$C. 2, 5-Dibromopyridine reacts[122] with BuLi (1 equiv.) in thf at $-100\,^{\circ}$C to give 2-Br-3-Li-pyridine. One[139] or both[140] of the halogens in 2, 6-X_2-pyridine (X = Br or I) can be exchanged, equation 45. 2, 6-Dichloropyridine is resistant to Li–Cl exchange.

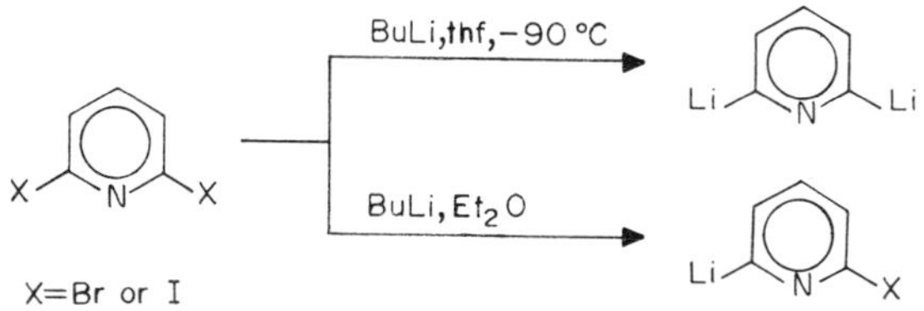

BuLi, thf, −90 °C

BuLi, Et_2O

X=Br or I

(45)

B. Replacement of Hydrogen in Organic Compounds by Metals: Metallation[1–7]

The formation of organoalkali metal compounds by the replacement of hydrogen in organic compounds has been achieved using a variety of reagents, including the metals themselves. An important factor is the acidity of the organic compound. Listings of the acidities of organic compounds have been variously made; one for the more common hydrocarbons and some substituted derivatives is given in Table 5.

TABLE 5. pK_a values (at sites shown in italics)

Compound	pK_a	Compound	pK_a
$Me_3\mathit{CH}$	47	$CH_2{=}CH_2$	36.5
Cyclohexane	45	$\mathit{CH}_3CH{=}CH_2$	35.5
$Me_2\mathit{CH}_2$	44	$Ph\mathit{CH}_3$	35
C_2H_6	42	$Ph_2\mathit{CH}_2$	33.5
CH_4	40	$Ph_3\mathit{CH}$	32
Cyclopropane	39	$HC{\equiv}CH$	25
$Ph\mathit{CH}Me_2$	37	Fluorene (9-position)	23
PhH	37	Indene (1-position)	18.5
		Cyclopentadiene	15
$PhSO_2\mathit{CH}_3$	29	$NC\mathit{CH}_2CH_3$	32.5
$O_2N\mathit{CH}_3$	17.2	$PhCO\mathit{CH}_2CH_3$	31
		$O_2N\mathit{CH}_2CH_3$	16.7
$(PhSO_2)_2\mathit{CH}_2$	12.2		
$(PhS)_2\mathit{CH}_2$	30.8	$PhSO_2\mathit{CH}_2OPh$	27.9
		$PhSO_2\mathit{CH}_2SPh$	20.3
$(PhSe)_2\mathit{CH}_2$	35.0	$PhSO_2\mathit{CH}_2PPh_2$	20.2
$(PhS)_3\mathit{CH}$	22.8		
1,3-dithiane (2-position, H)	31.2	$CH_3S(O)CH_3$	35.1

1. Using the Alkali Metals[1–3,5–7] and Their Arene Radical Anions or Dianion Compounds[141,142]

Formation of organoalkali metal compounds by metallation of organic compounds using the metals has only been profitably achieved with the more acidic hydrocarbons, such as Ph_nCH_{4-n} ($n = 1$–3), fluorene, indene, cyclopentadiene, and alk-1-ynes, in a variety of solvents including ethers, hydrocarbons, and liquid ammonia. In the last medium, the formation of amides, MNH_2, occurs and metallations proceed via these species[1].

The ease of metallation of phenylmethanes, Ph_nCH_{4-n}, at benzylic sites by metals follows the acidity sequence, namely $Ph_3CH > Ph_2CH_2 > PhCH_3$.

$$Ph_nCH_{4-n} + M \rightarrow Ph_nCH_{3-n}M \tag{46}$$

Metallation of the least acidic of these compounds, toluene, has been achieved using caesium[143] or potassium in the presence of Na_2O[144]. Caesium also reacts at the benzylic positions of ethylbenzene, cumene, xylenes, and mesitylene, with yields of 50–90% being obtained in thf–hexane solutions at 20 °C. Hydrogen is evolved in the caesium metallations[143]. The K–Na_2O combination successfully metallates other alkylarenes, including p-$MeC_6H_4Pr^i$. This compound is preferentially metallated at the methyl position rather than at the isopropyl group, in keeping with the relative acidities of the two sites[143].

$$Me{-}C_6H_4{-}Pr^i \xrightarrow{K{-}Na_2O} KCH_2{-}C_6H_4{-}Pr^i \tag{47}$$

Diphenylmethane has been metallated by potassium and by caesium in thf (e.g. K provides Ph_2CHK in 76% yield) and also by MNH_2 (M = Na or K) in liquid ammonia. Triphenylmethane reacts also with M (M = Na or K) in liquid ammonia[145] and in ethers[146]. However, complex series of products, including Ph_3CK and Ph_2CHK, have been obtained from the reactions in ethers. The formation of side-products may be suppressed on addition of butadiene. Radical anions are probable intermediates of those reactions in which no hydrogen is evolved.

Fluorene[147–150], indene[1,151], and cyclopentadiene[152–154] have all been directly metallated. Fluorene (FlH), the most acidic of these compounds, is particularly easily metallated[147]; Li[148], Na[149], or K[150] in ethereal solvents, K in benzene and metal amides have all been successful, e.g. a 71% yield of 9-lithiofluorene (9-LiFl) was obtained using Li in thf. The solvent has been shown to affect the rate of metallation, e.g. the ease of metallation of fluorene by potassium is in the solvent sequence $MeOCH_2CH_2OMe >$ thf $>$ dioxane[150]. The course of all the fluorene metallations by metals has been shown to occur[147] via radical anions, $FlH^{-\bullet}M^+$. These radical anions are stable at low temperatures but provide 9-FlM at higher temperatures. No hydrogen is evolved; instead, reduction of some fluorene to hexahydrofluorene occurs. This should be contrasted with the evolution of hydrogen in the cyclopentadiene[154] and indene reactions with K[147a] in benzene. Pentaphenylcyclopentadiene is metallated by Na, K, or Cs at 100 °C in toluene solution[147b]. Indene is metallated at the 1-position by K in dme, and cyclopentadiene has also been metallated by Li in thf[152], by Na in decalin, toluene, thf, or liquid ammonia[153], and by K in liquid ammonia[154]. In the Na–liquid NH_3–cyclopentadiene reaction, reduction of some cyclopentadiene to cyclopentene results; again, radical anions seem to be implicated as intermediates in the reaction.

1-Alkynes, RC≡CH, are sufficiently acidic to provide RC≡CM on reaction with alkali metals in liquid ammonia[1] and in ethereal solutions[1,155]. Acetylene reacts with various systems including sodium in liquid ammonia[1], xylene[156], or ethers. Allyl protons can be replaced directly on reaction with metals, e.g. allylpotassium is obtained from propene and potassium[157].

Acyclic and cyclic dienyl anions may be prepared from both conjugated and non-conjugated dienes on reaction with all the alkali metals (Li → Cs) in thf solution and in the presence of a tertiary amine, in particular Et_3N or tmed[158], e.g. equations 48 and 49.

$$\left\{\begin{array}{l} MeCH{=}CHCH{=}CH_2 \\ Z \text{ or } E \\ \text{or } CH_2{=}CH\,CH_2CH{=}CH_2 \end{array}\right. \xrightarrow{M,\text{thf},NEt_3} CH_2{\overset{\cdots}{=}}CH{\overset{\cdots}{=}}\bar{C}H{\overset{\cdots}{=}}CH{\overset{\cdots}{=}}CH_2,\ M^+ \tag{48}$$

$$\text{1,3-, 1,4-, or 1,5-cyclooctadiene} \xrightarrow{\text{K-thf-NEt}_3} \text{(cyclooctadienyl anion)}\ K^+ \tag{49}$$

Various functionally substituted hydrocarbons, bearing acidic hydrogens, also react directly with alkali metals; some examples are Me_2SO (with Na or K)[159], $RR'CHCO_2R^2$, $RCH(CO_2R')_2$, and RCH_2COR', equations 50–52. Direct metallation of thiophene in the 2-position has been reported using lithium[163].

$$H_2C(CO_2Et)_2 + Na \xrightarrow[\text{(ref. 160)}]{Et_2O} \left[NaCH(CO_2Et)_2 \right] \tag{50}$$

$$\text{2-(ethoxycarbonyl)cyclopentanone} + Na \xrightarrow[\text{(ref. 161)}]{PhH} \left[\text{2-Na-2-}CO_2Et\text{-cyclopentanone} \right] \tag{51}$$

$$\text{cyclohexanone} + K \xrightarrow[\text{(ref. 162)}]{Et_2O, HMPT} [\text{2-K-cyclohexanone}] \quad (52)$$

Radical anion species, $ArH^{-\bullet}M^+$, are also useful metallating agents. Fluorene has been metallated by a number of radical anions, including $Anth^{-\bullet}M^+$. The rates of metallation of fluorene by $Anth^{-\bullet}M^+$ have been found to increase as the solvent basicity decreases, which indicates that contact ion paired $Anth^{-\bullet}$ is more reactive than the solvent-separated ion paired form[164].

Reactions of $PhPh^{-\bullet}Li^+$ with a series of phenylmethanes, Ph_nCH_{4-n}, were studied[147b] in thf solution at *ca.* 30 °C; the yields of organolithium products were near quantitative for Ph_3CLi (from $PhCH_3$), *ca.* 50% for Ph_2CHLi (from Ph_2CH_2), but only about 1% for $PhCH_2Li$ (from $PhCH_3$). As in all metallations involving radical anions, hydrogen was not evolved with phenylcyclohexene derivatives being obtained instead. $PhPh^{-\bullet}Li$ also successfully metallates quinaldine and cyclohexanone (at the 2-position), in fact $PhPh^{-\bullet}Li^+$ appears to be a more effective lithiating agent than dispersed lithium.

The use of naphthalene radical anions has proved popular in these reactions. An extensive range of compounds has been metallated[1] by $Naph^{-\bullet}Na^+$ in thf, including indene (44% yield of 1-indenylsodium), fluorene (62% yield of 9-FlNa), Ph_3CH (30% yield of Ph_3CNa), Ph_2CH_2 (70% yield of Ph_2CHNa), 9, 10-dihydroanthracene (metallation at the 9-position), allylbenzene (which provides $PhCH{\cdots}\bar{C}H{\cdots}CH_2Na^+$), 1-alkynes (including acetylene), acetophenone (equation 53), 2-methylpyridine (equation 54), and $RR'CHCO_2H$. Compounds $RR'CHCO_2H$ are also metallated[1,165] by $Naph^{-\bullet}Li^+$ (equation 55) and $Naph^{-\bullet}K^+$.

$$PhCOMe + NaPh^{-\bullet}Na^+ \rightarrow [PhCOCH_2Na] \quad (53)$$

$$\text{2-Me-pyridine} + Naph^{-\bullet}Na^+ \longrightarrow \text{2-}CH_2Na\text{-pyridine} \quad (54)$$

$$RR'CHCO_2H \xrightarrow[-75\,^\circ C]{Naph^{-\bullet}Li^+,\ thf} RR'CLiCO_2Li \quad (55)$$

$$R, R = Me, Me;\ Ph, H;\ Me, H;\ —(CH_2)_5—$$

$Naph^{-\bullet}Li^+$ appears to be as equally reactive as $LiNPr^i_2$. However, the much reduced nucleophilicity of naphthalene compared with Pr^i_2NH (by-products formed in the metallation reaction) makes $Naph^{-\bullet}Li^+$ a better reagent to use for the formation of carbanions sensitive to nucleophiles. $Naph^{-\bullet}Cs^+$ also has been used, e.g. Ph_2CHCs was obtained in 75% yield from Ph_2CH_2 in thf–hexane.

Nitrogen heteroaromatic radical anion species, including sodium[166], potassium[167], and caesium[168] mono-, di- and triphenylquinolines and -pyridines, have proved effective metallating agents in thf solution. Such compounds metallate, Ph_3CH, Ph_2CH_2, $PhC{\equiv}CH$, 2-phenyl-1, 3-dithiane, e.g. equation 56, and $PhCH_2CO_2H$. The cations

$$\text{2-Ph-2-H-1,3-dithiane} \xrightarrow[\text{thf}]{\text{2,6-Ph}_2\text{-pyridine}^{-\bullet}Na^+} \text{2-Ph-2-Na-1,3-dithiane} \quad (56)$$

had an effect on the reactivity; the general sequence of decreasing reactivity was established as $Na^+ > Cs^+ \approx K^+ \gg Li^+$.

The effect of the arene on the reactivity of $ArH^{-\bullet}M^+$ has also been studied; the rates of metallation of thiophene (at the 2-position) by $ArH^{-\bullet}Li^+$ decrease in the order ArH = PhPh > Naph > PhCH=CHPh > phenanthrene > anthracene[163].

$$\text{Naph}^{-\bullet}\text{Li}^+ + \text{thiophene} \longrightarrow \text{2-lithiothiophene} \quad (50\%) \tag{57}$$

Reaction 57 is a two-electron process with yields of 2-lithiothiophene being *ca.* 50%. However, in the presence of $Ph_2C{=}CH_2$ or $PhCMe{=}CH_2$, a one-electron process results and yields of greater than 90% can be obtained.

Alkali metal arene dianions, $ArH^{2-}2M^+$, are considered more effective metallating reagents than the corresponding radical anions[169]. The metallating ability of $ArH^{2-}2M^+$ increases as the solvating power of the reaction media increases, e.g. as shown in reactions of $Naph^{2-}2M^+$, $PhPh^{2-}2M^+$, and $Anth^{2-}2M^+$ with Ph_3CH, Ph_2CH_2 and 2-methylnaphthalene. The rates of reaction of $ArH^{2-}2M^+$ also increase as the size of M^+ increases.

Functionally substituted alkanes, e.g. RCH_2CN (R = H or Ph), also are metallated[170] by $Naph^{2-}2Li^+$.

2. *Using Alkali Metal Compounds*[1,171,172]

Formation of organoalkali metal compounds via metal–hydrogen exchange has been achieved using a variety of alkali metal compounds, equations 58 and 59.

$$RH + R'M \rightleftharpoons RM + R'H \tag{58}$$

$$RH + MY \rightleftharpoons RM + YH \tag{59}$$

$$\text{e.g. } Y = OR',\ NR'_2,\ H$$

Compounds used as metallating agents include alkyl- and arylalkali metal compounds, either alone or in the presence of a donor, such as tmed, pmdt, dabco, or hmpt, metal amides, including the parent species, MNH_2, metal alkoxides or oxides, and metal hydrides. Also employed are metal alkoxide–organometal combinations, e.g. BuLi–Bu^{t}OK and BuLi–Me_2CEtOK. Some examples for lithium are RLi (R = Me, Bu, Bu^s, or Bu^t), Pr^i_2NLi (lda), 2,2,6,6-tetramethylpiperidine (ltmp), and $(Me_3Si)_2NLi$, for sodium RNa (R = Et, Bu, C_5H_{11}, Ph_3C or Ph) and $(Me_3Si)_2NNa$, for potassium RK (R = C_5H_{11} or Me_3SiCH_2), Pr^i_2NK (kda), $(Me_3Si)_2NK$, BuLi–Bu^{t}OK, and C_5H_{11}Na–Bu^{t}OK, for caesium Me_3SiCH_2Cs and for rubidium BuLi–RbOR.

Certain reagents may be favoured for particular metallations, for example it has been reported that either Me_3SiCH_2K or BuLi–Bu^{t}OK is especially useful for the preparation of allyl- and benzylpotassiums, whereas C_5H_{11}Na–Bu^{t}OK is favoured for the formation of vinyl and cyclopropyl derivatives[173]; of interest, BuK, prepared from Bu_2Hg and a K–Na alloy, has a different reactivity[174] to that of the BuLi–Bu^{t}OK combination. Metal dialkylamides are poor nucleophiles and can be used more widely and safely than can the strongly nucleophilic RM reagents. Particularly good use has been made of reactions of $LiNR_2$ with carbon acids bearing groups sensitive to nucleophiles.

Reactions 58 and 59 are in fact equilibria with the equilibrium constants being dependent on the relative acidities of the RH–R'H and RH–YH pairs. Equilibrium constants have been calculated in some cases[11,175,176]. Further, use has been made of these equilibria to obtain pK_a values for hydrocarbons[177]; for example, extensive use has been made of systems involving caesium cyclohexamide. Equilibria are not so easy to follow for potassium and sodium systems, owing to the poor solubility of RNa and RK compounds.

At equilibrium, the alkali metal is preferentially attached to the residue best able to support a negative charge, that is, the more acidic compound within the RH–R′H or RH–YH pairs provides the metal derivative in the greater amount. It follows that the greater the difference in the acidities of the proton sources, the further the equilibrium should lie to one side. For the purpose of synthesis, the equilibrium should lie far to the right, although removal of either R′H or YH or insolubility of the product, RM, would drive the exchange to completion.

Although thermodynamic measures of the acidity (pK_a values) will play dominant roles in deciding the positions of equilibrium as well as the sites of metallation, kinetic factors have also to be considered. Alkanes have pK_a values of greater than 40 and it would therefore be expected that alkylalkali metal compounds would react practically completely with organic compounds having pK_a values of less than 40. However, while butyllithium reacts readily with the more acidic hydrocarbons, butyllithium in hexane (in which it is hexameric) or in Et_2O (tetrameric) is unable to metallate benzene (pK_a 37) and provides only a poor yield of metallated product from toluene (pK_a 35)[178]. Changing the solvent to thf or addition of a donor molecule, such as tmed[179–181] or dabco[182], or addition of Bu^{t}OK[183] results in a more reactive metallating system and one able to metallate PhH and PhMe. Indeed, the BuLi–tmed combination, when used in excess, can polymetallate toluene[184].

The abilities of donor molecules to enhance the metallating ability of BuLi have been studied; towards benzene, the sequence was established as tmed > $Me_2NCHMeCH_2CH_2NMe_2$ > $Me_2NCH_2CH_2CH_2NMe_2$ > dabco > $Me_2NCH_2CH_2CH_2CH_2NMe_2$. Again towards PhH, Bu^{s}Li–tmed was found to be more reactive than BuLi–tmed[180].

The compound tmed can be metallated by BuLi (and other RLi derived from RH having a pK_a value greater than 35). This can lead to unwanted by-products although the metallated tmed, $LiCH_2NMeCH_2CH_2NMe_2$, can itself function as a metallating agent[185].

Generally, the reactivity of RLi is higher in ethers than in hydrocarbons. The reactivity increases with increasing Lewis basicity of the solvent, probably as a result of the depolymerization of the organolithium aggregates leading to increased carbanionic character and also probably to the increased stabilization of the transition state.

Alkylsodium and -potassium compounds are generally more reactive than the corresponding alkyllithiums (the reactivity sequence is RK > RNa > RLi); RNa and RK react, for example, with benzene and toluene[173]. Usually RNa and RK are used as suspensions in aliphatic hydrocarbon media; ethers are not normally used owing to their reactivity towards RNa and RK. As with RLi, the reactivity of RNa increases on addition of a donor such as tmed; one effect of the tmed has been described as a peptising or disposing effect[186].

Kinetic effects are also observed in the sites of metallation. For example, initial lithiation of alkylbenzenes occurs in the ring as well as at benzylic sites. With increasing reaction times, and in some cases in the presence of excesses of the alkylbenzenes, isomerizations to the thermodynamically more stable benzyllithiums occur; see Section II.B.2.a.iv. Other compounds that show kinetic and thermodynamic effects on metallation include PhSMe, PhOMe, Ph_2PMe, and PhSeCHRR′[187].

The mechanism of metallation of hydrocarbons by RLi is considered to be based on a simple acid–base interaction with substitution at the most acidic site; for heterosubstituted compounds, an alternative mechanism is based on prior coordination to the hetero atom, which leads to lithiation of neighbouring sites (*ortho*, alpha, or even beta)[188]. Lithium dialkylamides, $LiNR_2$, have reduced thermodynamic basicities relative to RLi, with pK_a values of HNR_2 in the region of 30. However, $LiNR_2$ have been found to be more effective lithiating agents, i.e. they show an increased kinetic basicity, probably as a consequence of the use of the nitrogen lone pair in the transition state.

$$\gt C{-}H + LiNR_2 \longrightarrow \left[\gt C \cdots H \cdots NR_2 \cdots Li \cdots \right]^{\ddagger} \longrightarrow \gt C{-}Li + HNR_2 \quad (60)$$

a. Metallation of hydrocarbons

i. *Metallations of alkenes.* Organolithiums and -sodiums do not metallate alkanes. Butyl- and arylpotassiums had been reported to metallate pentane, hexane, and cyclohexane; however, a contrary result has subsequently been published[189].

ii. *Metallations of cyclopropanes.* Cyclopropyl protons are more acidic than simple alkyl protons; cyclopropane, for example, is metallated by $C_5H_{11}Na$ or more readily by $C_5H_{11}Na$–Pr^iONa[190a]. Metallations of other cyclopropyl compounds by alkylsodiums are shown in equations 61[191] and 62[192]. Although BuLi does not react with **21**, there are a

(61)

(62)

few examples of lithium–hydrogen exchanges occurring at strained bridgehead cyclopropane sites, see e.g. equations 63–66.

(63)

(64)

(65)

(66)

Suitably placed hydroxy, alkoxy, or *tert*-amino groups[197] can activate cyclopropanes towards lithiation. Compounds **23**–**27** are examples of compounds metallated by alkyllithiums (e.g. Pr^iLi in Et_2O–pentane or BuLi in hexane at ambient temperature); the sites of lithiation are indicated by arrows. The hydroxy groups will themselves be initially metallated and assistance to the ring metallation will then proceed via the OLi group.

The assistance given to the cyclopropane metallations by the donor group is apparent; for example, quadicyclene (**22**), although metallated by BuLi–tmed and by BuLi–Bu^tOK in hexane[174b], is not lithiated by Pr^iLi in Et_2O–hexane under conditions successfully employed for the hydroxy derivative **23** (R = H)[197]. Further, **28**, in which the MeO substituent is directed away from the cyclopropyl ring, is not lithiated under the conditions used for **27**. Such assistance arises from the coordination of the lithiating agent by the donor group (RO or NMe_2) holding the lithiating agent close to the metallation site.

(67)

Another example is shown in equation 67, in which the major, if not exclusive, product is *cis*-1-RCH(OLi)-2-Li-cyclopropane (**29**); other products can be the *trans*-isomer and 1-RCH(OLi)-1-Li-cyclopropane[197a]. Alkylcyclopropanes do not react under comparable conditions.

iii. *Metallation of benzene and fused benzenoid aromatics.* As has already been pointed out, BuLi in the presence of a good donor, such as tmed[178–180,198] or dabco[181], but not simply in Et_2O solution, lithiates benzene to provide good yields (> 85%) of PhLi; Bu^sLi–tmed is even more effective[180] than BuLi–tmed. Various organosodiums, RNa (R = Et, Bu, C_5H_{11}, C_8H_{17}, or $CH_2{=}CH$), alkylpotassiums[185,199] and the combination[182] BuLi–Bu^tOK have also been found to metallate benzene successfully. Dimetallation of benzene occurs on prolonged treatment with $C_5H_{11}Na$; mixtures of *m*- and *p*-$Na_2C_6H_4$ are obtained in proportions dependent on the reaction conditions[200]; conditions for the production of *m*-$Na_2C_6H_4$ in an 85% yield have been published.

Metallation of polycyclic arenes, including naphthalene, biphenyl, and anthracene, occurs using BuLi–tmed in hexane; polylithiation can also result. Reaction of PhPh using BuLi (2·4 equiv.) in the presence of tmed in hexane solution provides *o*-$LiC_6H_4C_6H_4Li$-*o*[201]. Alkylsodiums, and no doubt alkylpotassiums, are also able to provide polymetallated products, e.g. naphthalene[202a] and $C_5H_{11}Na$ provides α- and β- mono-, di-, and trilithiated products. Acenaphthrene and $C_5H_{11}Na$ react to form mixtures of 1-sodio- and 1, 5- and 1, 6-disodioacenaphthrenes[202b].

iv. *Metallation of alkylbenzenes.* Toluene is lithiated to only a small extent by BuLi in Et_2O; a better yield (25% after 24 h) resulted from the use of BuLi (in excess) in an Et_2O–thf solution[203]. The presence of tmed[178–180,198] or dabco[181] led to ready metallation, e.g. BuLi–dabco in hexane at 80 °C provided[181] an 85% yield of $PhCH_2Li$ after 30 min. The combination BuLi–Bu^tOK is also particularly effective[182]. Amides, MNH_2, in liquid ammonia are too weakly basic to metallate PhMe; $PhCH_2M$ is in fact protonated by NH_3.

The reaction of PhMe with BuLi–tmed initially provides not only $PhCH_2Li$ but also small amounts of ring metallated products[178], e.g. equation 68; see also Table 6. The ring metallated products rearrange slowly[178,180] to the thermodynamically more stable

$$\mathrm{PhMe + BuLi \xrightarrow[30\,^\circ C,\ 30\ min]{tmed,\ hexane,} PhCH_2Li + MeC_6H_4Li} \qquad (68)$$

$$89\% \qquad o:m+p = 3:9$$

benzyllithium, e.g. on standing, on heating, on addition of more tmed, and in the presence of excess of toluene. Under the conditions used in reaction 68, such rearrangements were not significant. Clearly the initial metallation of $PhCH_3$, to provide the mixture of benzylic and tolyllithiums, occurs at much faster rates than any subsequent transmetallations.

Metallation of toluene by BuNa in benzene also initially leads to benzyl–metal and minor amounts of ring metallated products[204]; the conversion of *p*-MeC_6H_4Na to $PhCH_2Na$ occurred almost quantitatively at room temperature within 69 h; it has also been reported that $PhCH_2Na$ can be obtained from *p*-MeC_6H_4Cl and sodium sand in refluxing benzene or light petroleum solution[178]. Alkylpotassiums, the most reactive of RLi, RNa, and RK compounds, apparently provide only the benzylic product even with short reaction times[204]. Metallation of toluene by a series of Me_3SiCH_2M (M = Na, K, Rb, and Cs) has been reported[205].

Some dimetallation of toluene (to give *m*-$NaC_6H_4CH_2Na$) has been reported[206] using $C_5H_{11}Na$. Polylithiation of toluene has also been achieved using BuLi–tmed under vigorous conditions[183].

Metallations of ethyl-[178,207,208] and isopropylbenzenes[174b178,209,210] have been achieved using organolithium, -sodium, and -potassium reagents. (see Table 6). As

TABLE 6. Metallation of alkylbenzenes by organoalkalimetals

Substrate	Conditions	Reaction time (h)	Relative yields: PhCRR′M	Relative yields: $MC_6H_4CRR'H$ *o*	*m*-	*p*-	Overall yield	Ref.
$PhCH_3$	BuLi–tmed, hexane, 30 °C	0.25–2	89–92	2–3		6–9		759
	BuNa, PhH, 0 °C	0.5	92	—	4	4		760
	BuK, PhH 0 °C	0.5	> 99	—	—	—		760
$PhCH_2Me$	BuLi–tmed, hexane, 30 °C	0.5–6.5	37–38	9	36	17		759
	$C_5H_{11}Na$, $PhCH_2Me$, room temp.	3	18	—	52	30	32	761
		20	68	—	19	13	46	
	$C_5H_{11}Na$, octane, room temp.	24	26	—	41	33	55	762
	$C_5H_{11}Na$–tmed, octane, room temp.	0.25	12	2	55	28	—	762
		1	100	—	—	—	—	
	$C_5H_{11}K$, heptane, − 10 °C,	0.5	93	—	6	1	10	761
	room temp.	20	100	—	—	—	51	
$PhCHMe_2$	BuLi–tmed hexane, 30 °C	2–24	3	8–10	57–59	30		759
	$C_5H_{11}Na$, cumene	20	2	—	55	43	48	761
$PhCHMe_2$	$C_5H_{11}Na$, octane, room temp.	24	—	—	55	44	40	763
	$C_5H_{11}Na$–tmed, octane, room temp.	1	7	3	57	33		
		4	80	0	11	9		763
		24	97	0	2	1	95	
	$C_5H_{11}K$, cumene, room temp.	3	42	—	39	19		
		20	92	—	—	8	43	764
$PhCMe_3$	BuLi–tmed hexane, 30 °C	4	—	—	68	32		759

expected from the electron-releasing effects of the methyl group, the relative rates of metallation are 60:10:1 for PhMe, $PhCH_2Me$, and $PhCHMe_2$. Both benzylic and ring metallation initially occur, with relatively more ring metallation in the sequences $PhCHMe_2 > PhCH_2Me > PhCH_3$ and RLi > RNa > RK. Whereas the LiC_6H_4CHMeR (R = H, Me) products did not rearrange to PhCLiMeR under the reaction conditions employed, both the sodium and potassium analogues did, the isomerizations occurring via reaction, i.e. transmetallation with free alkylbenzenes. The rates of rearrangement decreased in the sequence RK > RNa (> RLi). Sodiocumene, $PhCMe_2Na$, is not thermally stable and decomposes[210] to $PhCMe{=}CH_2$ and NaH. However, it can be stabilized[207] on complexation with tmed and is obtained in near quantitative yields from $PhCHMe_2$ and $C_5H_{11}Na$–tmed in octane after 24 h at 0 °C. The considerable increase in reactivity of $C_5H_{11}Na$ on addition of tmed is also apparent from entries in Table 6.

Dimetallations of ethyl- and isopropylbenzenes have been reported[211] on refluxing for 24 h with 4 equiv. of BuLi–$EtCMe_2OK$, e.g. equation 69; $PhCHKCH_2K$ is unstable and decomposes to give $[PhCHKCH_2]_2$.

$$PhCH_2Me \longrightarrow PhCH(K)Me + [PhCH(K)CH_2K] \qquad (69)$$

tert-Butylbenzene, having no benzylic protons, is only lithiated in the ring; an approximate 2:1 ratio of *meta* to *para* products is obtained.

Cyclopropylbenzene, another alkylbenzene containing a benzylic hydrogen, is metallated by organopotassium reagents at the benzylic site. In contrast, $C_5H_{11}Na$ in pentane reacts only at *meta*- or *para*-ring positions[212], equation 70.

$C_5H_{11}Na$, pentane

Na Na

K

BuLi–Bu^tOK or Me_3SiCH_2K, thf

(70)

The reactivity of xylenes towards BuLi–tmed[213] or $C_5H_{11}Na$–tmed in hexane[214] decreases in the sequence *m*- > *o*- > *p*-xylenes. In contrast to the monometallation using BuLi or $C_5H_{11}Na$ alone at room temperature, metallation of both methyl groups of *o*- and *m*-xylenes, but not *p*-xylene, can result when tmed is also present, e.g. quantitative dimetallation of *m*-xylene occurred within 24 h at room temperature. Additional polymetallation products obtained from *m*-xylene and BuLi–tmed are the *gem*-dilithio

$$o\text{- or } m\text{-}Me_2C_6H_4 \xrightarrow{\text{RM–tmed}} o\text{- or } m\text{-}(LiCH_2)_2C_6H_4 \qquad (71)$$

dimetallation of *m*-xylene occurred within 24 h at room temperature. Additional polymetallation products obtained from *m*-xylene and BuLi–tmed are the *gem*-dilithio derivatives, *m*-$Li_2CHC_6H_4Me$ and *m*-$Li_2CHC_6H_4CH_2Li$; increasing amounts of tmed suppress *gem*-dimetallation. No *gem*-dimetallation of *m*-xylenes occurs when $C_5H_{11}Na$–tmed is used. Metallation of *p*-xylenes by BuLi–tmed or $C_5H_{11}Na$–tmed produces *p*-$MCH_2C_6H_4Me$ but not *p*-$(MCH_2)_2C_6H_4$ (**30**, M = Li or Na); however, **30** (M = K) can be produced from the more powerful metallating agent BuLi–Bu^tOK.

An explanation for the xylene results has been based on the relative charges delocalized into the ring of the benzyl anion: the magnitudes decrease in the sequence *p*- > *o*- > *m*-, which is the reverse of the reactivities towards metallation. In addition, the second metallation will occur preferentially to give the additional negative charges on the same set of carbons as did the first, i.e. *m*- > *o*- and *p*; the easier formation of *o*-$(MCH_2)_2C_6H_4$ compared with *p*-$(MCH_2)_2C_6H_4$ is considered to be due to the lone pair attractive interactions between adjacent benzylic sites. Another consideration is that lithium can bridge the benzylic and an *ortho* site during metallation and so the greater the negative charge on the *ortho* site, the stronger will be this interaction and the easier will be the substitution. The rates of abstraction of a proton by BuLi–Bu^tOK are greater with the transition state closer to the reactant than the product and hence *para*-dimetallation is not so difficult.

Tri- and tetramethylbenzenes provide[213,215] results in keeping with those of xylenes, e.g. 1, 3, 5-$Me_3C_6H_3$ and BuLi–tmed in hexane produce mono-, di-, and trilithiated mesitylenes; the major product of reaction of BuLi with mesitylene (molar ratio 6:1) after 24 h at room temperature is 1, 3, 5-$(LiCH_2)_3C_6H_3$ (60%); dilithiation occurs partially at the same methyl group but preferentially at different methyls. No *p*-$(LiCH_2)_2$ benzene products are obtained from *p*-dimethylbenzenes, such as 1, 2, 4-$Me_3C_6H_3$ or 1, 2, 4, 5-$Me_4C_6H_2$, using BuLi–tmed; however, BuLi–Bu^tOK can provide *p*-$(KCH_2)_2$-substituted benzenes. 1, 2, 3, 5-tetramethylbenzene can be tetrametallated using BuLi–Bu^tOK. No ring metallations occur.

TABLE 7. Formation of benzylic alkali metal compounds

Compound	Conditions	Product (yield, %)	Ref.
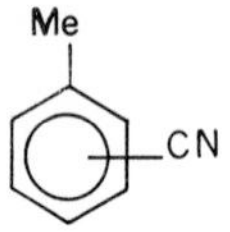 Me; CN	$LiNMe_2$, thf, hpmt, $-78\,°C$	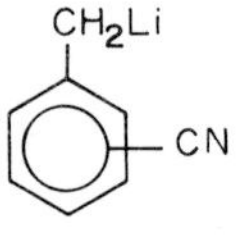 CH_2Li; CN (*o*,53, *m*,81, *p*=68)	765
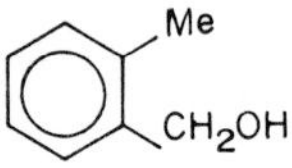 Me; CH_2OH	BuLi (2 equiv.)	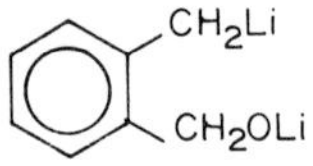 CH_2Li; CH_2OLi	766
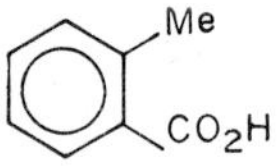 Me; CO_2H	$LiNPr^i_2$ (2 equiv.) hmpt, thf, hexane, $-78\,°C$	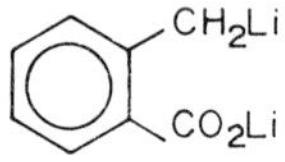 CH_2Li; CO_2Li (88)	767
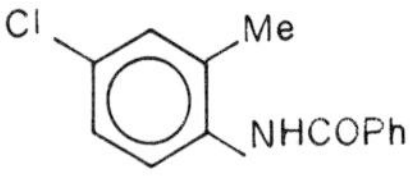 Cl; Me; NHCOPh	BuLi (2 equiv.), thf, hexane	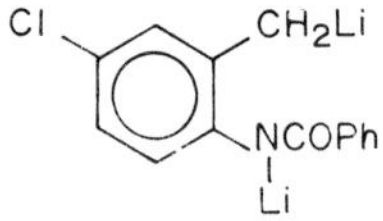 Cl; CH_2Li; N(Li)COPh	768
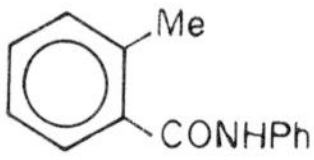 Me; CONHPh	BuLi (2 equiv.), thf, hexane	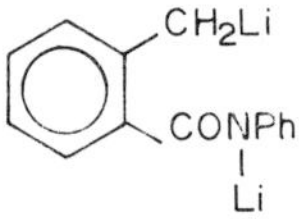 CH_2Li; CON(Li)Ph (50)	769
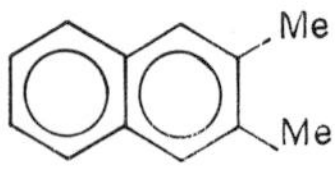 Me; Me	BuLi, tmed	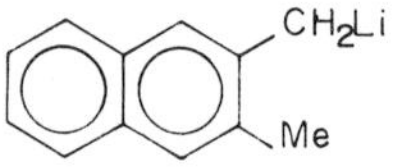 CH_2Li; Me	770
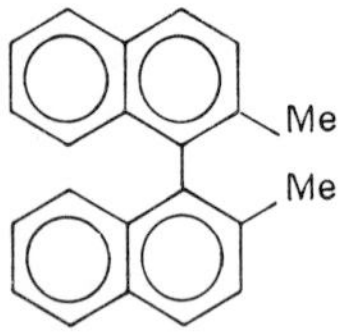 Me; Me	BuLi (3 equiv.), tmed, hexane, $-20\,°C$	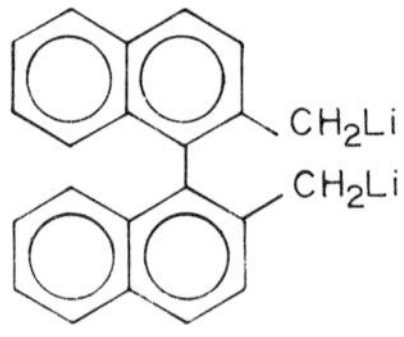 CH_2Li; CH_2Li	771
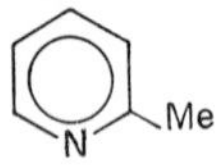 N; Me	BuLi, thf, hexane or $NaNH_2$, liq. NH_3 or KNH_2, liq. NH_3	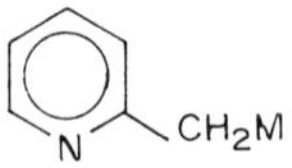 N; CH_2M	772 773 774
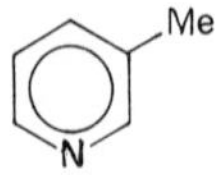 Me; N	$LiNPr^i_2$, hmpt, thf, hexane $0\,°C$	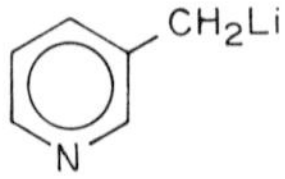 CH_2Li; N (90)	775
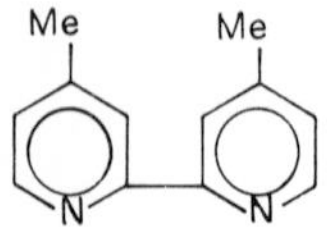 Me; Me; N; N	$LiNPr^i_2$ (2.5 equiv.), thf, hexane, $0\,°C$	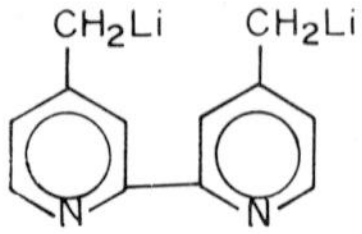 CH_2Li; CH_2Li; N; N (93)	776

TABLE 7. (*Contd.*)

Compound	Conditions	Product (yield, %)	Ref.
1,3-dimethylisoquinoline	BuLi, thf, hexane (75) $NaNH_2$, liq. NH_3 (42) $LiNPr^i_2$, thf, hexane (69)	1-(CH_2M)-3-methylisoquinoline	777
3-methylquinoline	$LiNPr^i_2$, hmpt, thf, hexane, −78 °C	3-(CH_2Li)quinoline (93)	778
3,5-dimethylisoxazole	BuLi, thf, −78 °C	3-methyl-5-(CH_2Li)isoxazole (>98)	779
2,4,5-trimethyloxazole	$LiNPr^i_2$, thf, −78 °C	4,5-dimethyl-2-(CH_2Li)oxazole (92)	780
2,4-dimethylthiazole	BuLi, hexane, thf, −78 °C	4-methyl-2-(CH_2Li)thiazole (>80)	781
1,2-dimethylimidazole	BuLi, Et_2O, hexane, 0 °C (inverse addition)	1-methyl-2-(CH_2Li)imidazole (82)	782
$PhCH_2SiMe_3$	MeLi, hmpt, 0 °C	$PhCHLiSiMe_3$ (85)	783
$PhCH_2NHCOPh$	$LiNPr^i_2$ (2 equiv.), diglyme, −78 °C	PhCHLiNLiCOPh (95)	784
isoindoline-N–P(O)$(NMe_2)_2$	BuLi, thf, 20 °C	1-Li-isoindoline-N–P(O)$(NMe_2)_2$ (95)	785
$PhCH_2PO(OEt)_2$	BuLi, thf, −78 °C	$PhCHLiPO(OEt)_2$ (81)	786
$PhCH(OSiMe_3)P(O)(OEt)_2$	$LiNPr^i_2$, thf, −78 °C	$PhC(Li)(OSiMe_3)P(O)(OEt)_2$ (88)	787
$PhCH_2CN$	BuLi, hexane, −70 °C	PhCHLiCN (95)	788
	BuLi (2 equiv.), thf, hexane	$PhCHLi_2CN$ (81)	789
$PhCH(CN)OSiMe_3$	$LiNPr^i_2$, thf, −78 °C	$PhC(Li)(CN)OSiMe_3$ (90)	790

TABLE 7. (*Contd.*)

Compound	Conditions	Product (yield, %)	Ref.
$PhCH_2OMe$	BuLi, tmed, hexane, $-10\,^\circ C$	PhCHLiOMe (85)	791
PhCH, CH_2, O	Bu^tLi, thf, -95 to $-50\,^\circ C$	Ph, Li, C, CH_2, O	792
$PhCH_2SH$	BuLi, thf, tmed, $0\,^\circ C$	PhCHLiSLi (72)	793
$PhCH_2SMe$	BuLi, tmed, hexane	PhCHLiSMe	794
$PhCH_2S(O)Me$	MeLi, thf, $-60\,^\circ C$	PhCHLiS(O)Me (85) (RS:SS = 1:15)	795
$PhCH_2S(O)_2NMe_2$	BuLi (2.5 equiv.), thf, hexane	$PhCLi_2S(O)_2Ph$ (83)	789
$PhCH_2SC(S)NMe_2$	$LiNPr^i{}_2$, thf, $-60\,^\circ C$	$PhCHLiSC(S)NMe_2$ (>99)	796
Ph, S, S, H	BuLi, thf	Ph, S, S, Li (*ca*·100)	797
$PhCH_2SePh$	$LiNPr^i{}_2$, thf, $-78\,^\circ C$	PhCHLiSePh (81)	798
$PhCH_2Se(O)Ph$	$LiNPr^i{}_2$, thf, Et_2O, $-78\,^\circ C$	PhCHLiSe(O)Ph (88)	798
X, $-CH_2PO(OEt)_2$	$NaNH_2$, liq. NH_3	X, $-CH(Na)PO(OEt)_2$ (X=O, 97; X=S, 95)	799
N, $-CH_2P(O)(OEt)_2$	$NaNH_2$, liq. NH_3	N, $-CH(Na)PO(OEt)_2$ (94)	799
CH_2SiMe_3, CH_2SiMe_3	BuLi, tmed, hexane	$CHLiSiMe_3$, $CHLiSiMe_3$	800
N, $CH(SiMe_3)_2$	BuLi, Et_2O, hexane, $20\,^\circ C$	N, $C(SiMe_3)_2$, Li	801

Substituted toluenes, XC_6H_4Me, and other methylaromatics can be metallated at benzylic sites; examples are given in Table 7. Metallations of methylnaphthalenes have been well studied; a 2-methyl group is more easily metallated by BuLi–tmed in hexane than a 1-methyl group as found in 1, 2, 3, 4-tetramethylnaphthalene[216]. 1, 8-Dimethylnaphthalene is monometallated by excess[216] of BuLi, whereas with $C_5H_{11}Na$–tmed, 1, 8-$(NaCH_2)_2$ naphthalene is quantitatively formed. The related compound acenaphthene (**31**) is dilithiated[217].

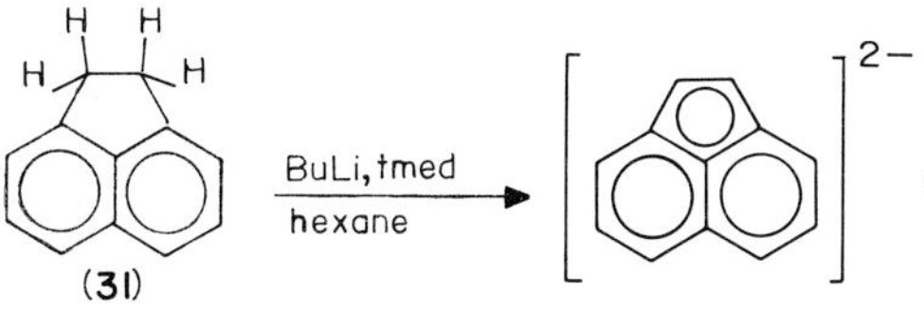

$$\text{(31)} \xrightarrow[\text{hexane}]{\text{BuLi, tmed}} [\,\ldots\,]^{2-}\ 2\,Li^+ \qquad (72)$$

The ease of dimetallation (α, α') of dimethylnaphthalenes using an excess of $C_5H_{11}Na$–tmed was found to be 1,3- > 1,2- > 1,6- ≈ 1,8- ≫ 1,4-dimethylnaphthalene, the last compound being only monometallated[186,214].

Alkylpyridines and related nitrogen heterocycles are considerably more acidic than the related benzenoid compounds and the methyl groups on the nitrogen heterocycles can be readily metallated by a number of reagents[218,219]. Although the acidity of a 4-methyl group is greater than that of a 2-methylsubstituent, the sites of metallation of 2,4-dimethylpyridine, 2,4,6-trimethylpyridine, and 2,4-dimethylquinoline depend on the conditions used and on whether kinetic or thermodynamic control operates. These heterocycles provided 2-MCH_2 derivatives, **32** or **34**, on treatment with BuLi in Et_2O–

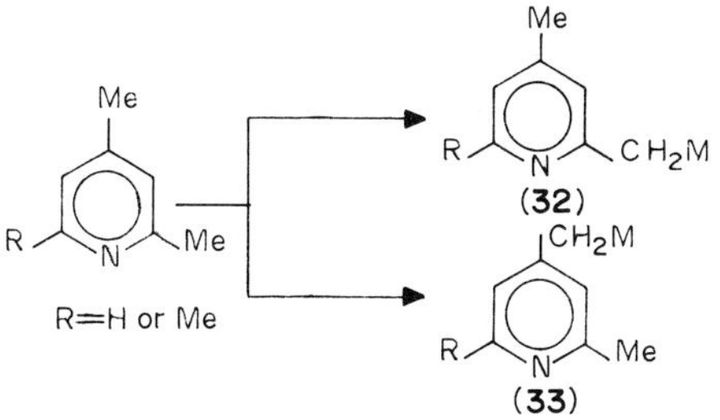

R=H or Me → **(32)** or **(33)** (73)

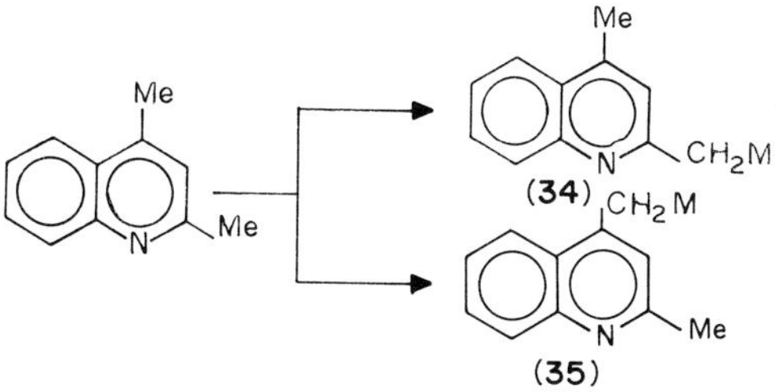

→ **(34)** or **(35)** (74)

hexane, regardless of the reaction time, or with PhM (M = Li or K) in Et_2O. The use of thf as the reaction medium for RLi metallations promoted the formation of 2-$LiCH_2$ derivatives, which isomerized to the 4-$LiCH_2$ derivatives, **33** or **35**, on standing. Such isomerizations [**32** → **33** and **34** → **35**] in thf were even faster in the presence of the free heterocycles or when a chelating agent, tmed or hmpt, was added.

The explanation for the RLi results is based on prior complexation of RLi with the ring nitrogen which holds the RLi close to the 2-methyl substituent (coordination-only limited mechanism). Such complexation lessens when alkali cations other than Li^+ or when basic solvents (NH_3 or other amines) are employed. In these cases, the metallating agent is able to attack the more acidic 4-methyl groups [acid–base limited mechanism].

The benzylic compounds, Ar_3CH, Ar_2CH_2, and 9,10-dihydroanthracenes, are particularly easily metallated. Diphenylmethane provides Ph_2CHM using, among other systems, (i) BuLi in thf (but not in hexane)[220], (ii) BuLi–cryptate in hexane[220], (iii) MNH_2

(M = Na or K) in liquid ammonia or KNH_2 in Et_2O, (iv) $C_5H_{11}Na$[185], (v) NaH–hmpt in thf[221], and (vi) KH in the presence of 18-crown ether[222].

Among the metallating systems to react with Ph_3CH are (i) BuLi in thf[220], (ii) BuLi-cryptate in hexane[220], (iii) $C_5H_{11}Na$[185], (iv) BuNa–tmed in PhH[223], (v) NaH–hmpt in thf[221], (vi) MNH_2 (M = Na or K) in liquid NH_3, (vii) KH in the presence of 18-crown ether[222], (vii) PhCMeK[185], and (vii) KOH in dme[224]. While Ph_2CH_2 or Ph_3CH is metallated in liquid ammonia by $NaNH_2$, evaporation of the solvent only leaves the starting materials; however, Ph_2CHK or Ph_3CK can be isolated from liquid ammonia. This probably arises from a slower attainment of the amminolyses equilibrium for the potassium system[6].

$(p\text{-}MeC_6H_4)_2CH_2$ is metallated at the central methylene group by either BuLi in thf solution or KH–18-crown ether[222]. The ease of metallation by BuLi in thf solution at benzylic sites decreases in the order $Ph_3CH > Ph_2CH_2 > (p\text{-}MeC_6H_4)_2CH_2 > p\text{-}PhC_6H_4Me$[220].

Both mono- and dimetallation of 9, 10-dihydroanthracenes have been reported. The

H Bu^t / H Me (**36**) —BuLi–tmed, 25 °C (ref. 225)→ H Bu^t / Me Li^+ (75)

Me H / Et H —2BuLi–tmed (ref. 226)→ Me / Et $2Li^+$ (76)

relative reactivities of (*Z*)- and (*E*)-**36** are 20:1. Metallation of 9, 10-dihydroanthracenes also occurs using NaH–hmpt in thf[221] or MNH_2[1].

v. *Metallations of cyclopentadienes, indenes, and fluorenes.* Cyclopentadiene, indene, and fluorene (pK_a values = 15, 18.5 and 23, respectively) and their derivatives are metallated by a number of systems. Cyclopentadiene is metallated, for example, by RLi in Et_2O[227], $NaNH_2$ in liquid ammonia[1], Bu^tONa, and KOH in dme[224]. Metallation of indene at the 1-position has been achieved with BuLi[228], RNa (R = C_5H_{11} or Ph_3C)[185], NaH–hmpt in thf[221], and KOH in dme[224], while fluorene reacts at the 9-position with PhLi in Et_2O[229], BuLi in thf[230], RNa (R = C_5H_{11} or Ph_3C)[185], NaH–hmpt in thf[221], $NaNH_2$ in decalin[1], KOH in dme[224], and Bu^tOK in dmso[1].

Polylithiations of indene and fluorene have been reported using BuLi–tmed at 70 °C; up to 6 and 9 lithiums may be incorporated for indene and fluorene, respectively[231]. Substituted derivatives and more complex derivatives have also been successfully metallated, see equations 77–80.

Other polycyclic fluorene derivatives that react include 7b*H*-indeno[1,2,3-*jk*]fluorene (with BuLi in PhH)[235], 9,9′-bifluorenyl (with BuLi–tmed in hexane at ambient temperature)[236], and cycloocta[*def*]fluorene (with BuLi in thf)[237].

$$\text{1,2,3,4,5-pentaphenylcyclopentadiene} + \text{BuLi} \xrightarrow[\text{(ref. 147b)}]{\text{PhMe, 100 °C}} \text{pentaphenylcyclopentadienide}\ Li^+ \tag{77}$$

$$\text{dihydropentalene} \xrightarrow[\text{hexane, −70 to 0 °C (ref. 232)}]{\text{2 BuLi, thf}} \text{pentalene dianion}\ 2\,Li^+ \tag{78}$$

$$\xrightarrow[\text{hexane, 25 °C (ref. 233)}]{\text{2 BuLi, hmpt}} 2\,Li^+ \tag{79}$$

$$\xrightarrow[\text{(ref. 234)}]{\text{2 BuLi−dmso}} 2\,Li^+ \tag{80}$$

vi. *Metallations of alkenes.* Metallation of alkenes could in principle provide vinyl- and/or allylmetal compounds. For an alkene without allylic protons, metallation could only give vinyl compounds, e.g. as with $CH_2{=}CH_2$ (using $C_5H_{11}Na{-}Pr^iONa$) and $Bu^tCH{=}CH_2$ (using $C_5H_{11}Na$)[238]. Organolithiums do not appear to metallate simple alkenes, which have no allylic protons; typical reactions with such compounds would be additions of RLi to the double bond or polymerizations. The acidity of allylic protons is

$$RCH{=}CH_2 + R'Na \longrightarrow RCH{=}CHNa \tag{81}$$
$$(R = H \text{ or } Bu^t)$$

greater than that of vinylic protons in alkenes (cf. the pK_a values of $CH_3CH{=}CH_2$ and $CH_2{=}CH_2$ are 35.5 and 36.5, respectively). When allylic protons are present, it would therefore be expected that metallations of alkenes would preferentially occur to give allylmetal compounds. This is usually borne out with strain-free alkenes and when there are no contrary kinetic effects.

Kinetic effects have been observed in metallations of alk-1-enes by alkylsodium and -potassium in which both allyl and vinyl products are obtained[239–241]. Metallation of $Me(CH_2)_9CH{=}CH_2$ by $C_5H_{11}Na$ in octane provided[239] $Me(CH_2)_8CH{\cdots}\bar{C}H{\cdots}CH_2\,Na^+$ and $Me(CH_2)_9CH{=}CHNa$. The latter appears as the major product in the early stages but as the reaction time increases so its relative yield decreases. Very little isomerization of $Me(CH_2)_9CH{=}CHNa$ to $Me(CH_2)_8CH\text{=}\bar{C}H\text{=}CH_2Na^+$ occurred under the reaction conditions; the relative increase in the allylic product is due to the very ready deprotonation of $Me(CH_2)_8CH{=}CHMe$ formed by the catalysed isomerization of the starting alkene, $Me(CH_2)_9CH{=}CH_2$ (see Scheme 6).

$$Me(CH_2)_9CH{=}CH_2 + C_5H_{11}Na \begin{cases} \rightarrow Me(CH_2)_9CH{=}CHNa + C_5H_{12} \\ \rightarrow Me(CH_2)_8CH\text{=}\bar{C}H\text{=}CH_2Na^+ + C_5H_{12} \end{cases}$$

$$Me(CH_2)_8CH{=}CH_2 + Me(CH_2)_8CH\text{=}\bar{C}H\text{=}CH_2Na^+ \longrightarrow$$
$$Me(CH_2)_8CH{=}CHMe + Me(CH_2)_8CH\text{=}\bar{C}H\text{=}CH_2Na^+$$
$$Me(CH_2)_8CH{=}CHMe + C_5H_{11}Na \longrightarrow Me(CH_2)_8CH\text{=}\bar{C}H\text{=}CH_2Na^+ + C_5H_{12}$$

SCHEME 6.

The relative ease of metallation was established as

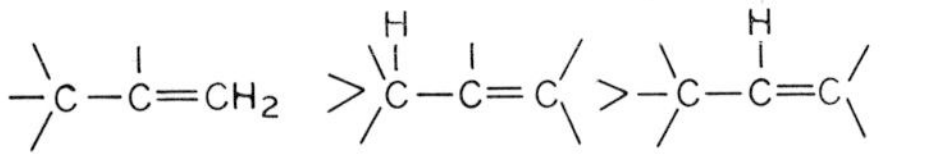

$$-\overset{|}{\underset{|}{C}}-\overset{|}{C}{=}CH_2 > \overset{H}{\underset{|}{>C}}-\overset{|}{C}{=}C< \; > -\overset{|}{\underset{|}{C}}-\overset{H}{\overset{|}{C}}{=}C<$$

The major product of BuK metallation of $Me(CH_2)_9CH{=}CH_2$ at all stages is the allyl product[241]. The catalysed isomerization of $Me(CH_2)_9CH{=}CH_2$ to $Me(CH_2)_8CH{=}CHMe$, a prominent feature of the RNa reaction, has less significance in the BuK reaction. A further difference between RK and RNa reactions is the much faster isomerization of $Me(CH_2)_9CH{=}CHM$ to $Me(CH_2)_8CH\text{=}\bar{C}H\text{=}CH_2M^+$ for M = K.

Deprotonations of bicyclic alkenes do not occur at bridge-head sites, even if these would lead to allylmetal; for example, compounds **37**–**40** react with alkylsodiums at the indicated sites to give vinylmetal products[191]. The metallation of **40** occurs at the vinyl site nearest to

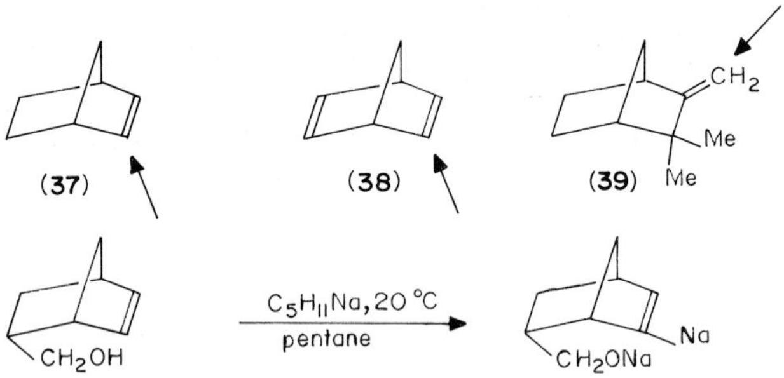

(37) (38) (39)

CH_2, Me, Me

(40) CH_2OH $\xrightarrow[\text{pentane}]{C_5H_{11}Na,\ 20\,^\circ C}$ CH_2ONa, Na (82)

the CH_2O^- group, which suggests assistance by this group. Allylic deprotonation of **41** and **42** occurs, but for **43** both vinylic and allylic metallation results.

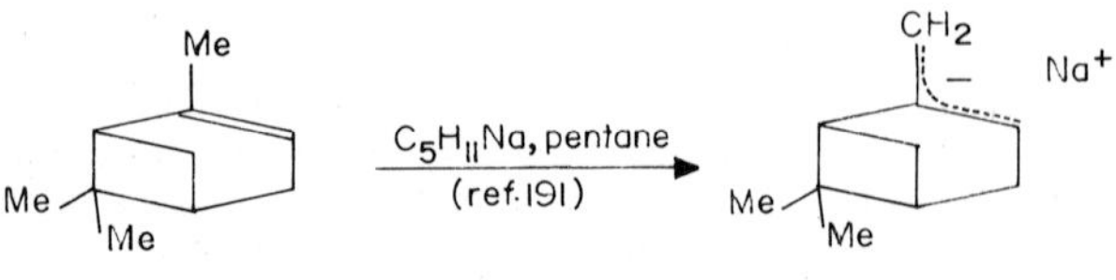

(41) Me, Me, Me $\xrightarrow[\text{(ref. 191)}]{C_5H_{11}Na,\ \text{pentane}}$ CH_2, Na^+, Me, Me (83)

(42) $\xrightarrow[\text{thf (ref. 242)}]{\text{BuLi, Bu}^t\text{OK}}$ K^+ (84)

(43) CH_2 $\xrightarrow[\text{pentane (ref. 191)}]{C_5H_{11}Na,}$ CHNa + CH_2 Na^+ (85)

There are also examples of vinyl lithiations of strained alkenes, e.g. equations 86 and 87.

Me Me Me $\xrightarrow[\text{or BuLi hexane, 20 °C (ref. 243)}]{\text{PhLi or MeLi } Et_2O}$ Me Me Me Li (86)

$\xrightarrow[\text{−70 °C (ref. 244)}]{\text{BuLi, thf}}$ Li (87)

For simple alkenes, bearing allylic protons, allylic deprotonation results; for example, metallation of propene occurs readily to give $CH_2{\overset{\cdots}{-}}\bar{C}H{\overset{\cdots}{-}}CH_2M^+$ using a variety of reagents, including BuLi in thf[245], RLi–tmed (R = Bu[181,245] or Bu^s[181,246]), $C_5H_{11}Na$[247], RK ($R = C_5H_{11}$, Me_3SiCH_2, or $PhCMe_2$)[174,248–250], BuLi–Bu^tOK[183,251], and Me_3SiCH_2Cs[246]. Crotylmetal, $MeCH{\overset{\cdots}{-}}\bar{C}H{\overset{\cdots}{-}}CH_2M^+$ (from but-1- or -2-ene, equation 88)[181,247,252], 2-methylallylmetal (from $Me_2C{=}CH_2$)[246,247,250,251], and phenylallyl-

$$\begin{matrix} MeCH_2CH{=}CH_2 \\ \text{or} \\ (E)\text{- or }(Z)\text{-}MeCH{=}CHMe \end{matrix} \xrightarrow{Bu^tLi\text{--}\text{tmed}} MeCH{\overset{\cdots}{-}}\bar{C}H{\overset{\cdots}{-}}CH_2Li^+ \qquad (88)$$

metal (from $PhCH_2CH{=}CH_2$)[253] are similarly prepared. Other examples are a series of compounds, $RCH{\overset{\cdots}{-}}\bar{C}H{\overset{\cdots}{-}}CH_2K^+$ [from RCH=CHMe (R = H, Me, Et, Pr^i, or Bu^t) in cyclohexane][248] and $CH_2{\overset{\cdots}{-}}\bar{C}R{\overset{\cdots}{-}}CH_2K^+$ [from $CH_2{=}CRMe$ (R = H, Pr^i, or Bu^t) in thf] prepared using Me_3SiCH_2K.

Metal amides have been used to obtain allylmetals, e.g. 1,3-diphenylallyllithium, Scheme 7[254]. Of interest, (*E*)-$PhCH{=}CHCH_2Ph$ is lithiated by $LiNPr^i$(cyclo-C_6H_{11}) in thf at −45 °C within a few minutes to give the less thermodynamically stable isomer (*E*, *Z*)-**44**, whereas BuLi in thf–hexane only reacts at −30 °C, at which temperature (*E*, *E*)-**44** is obtained.

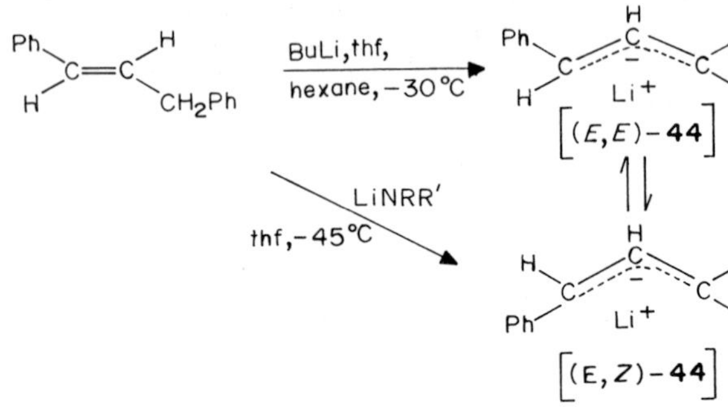

$$\text{PhCH=CHCH}_2\text{Ph} \xrightarrow[\text{hexane, } -30\,^\circ\text{C}]{\text{BuLi, thf,}} [(E,E)\text{-}\mathbf{44}] \rightleftharpoons [(E,Z)\text{-}\mathbf{44}]$$

$$\xrightarrow[\text{thf, } -45\,^\circ\text{C}]{\text{LiNRR}'} [(E,Z)\text{-}\mathbf{44}]$$

R=Pr^i, R$'$=cyclohexyl

SCHEME 7

(Z)-1-Alkylpropenes undergo allylic metallations at faster rates than the E isomers, equation 89[255]. A larger rate difference (15:1) was found for Z and E isomers of Bu^{t}CH=CHMe. The equilibration of 2-alkenylpotassium isomers, in particular, is very slow in thf

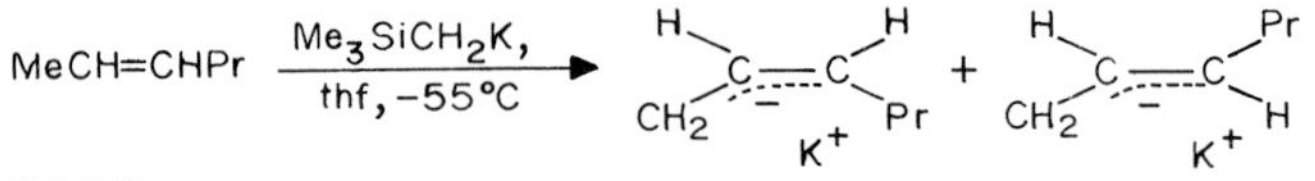

$$\text{MeCH=CHPr} \xrightarrow[\text{thf, } -55\,^\circ\text{C}]{\text{Me}_3\text{SiCH}_2\text{K,}} (\mathbf{45}) + (\mathbf{46}) \qquad (89)$$

(Z):(E) = 1:1 **45**:**46** = 4·6:1

at low temperatures; several days are required for equilibrium to be reached. Traces of oxygen, however, catalyse the isomerization[255].

In contrast to the monometallation of $Me_2C{=}CH_2$ and methylallylbenzenes using BuLi–thf, dimetallation can occur at both allylic sites using BuLi–tmed[256], $C_5H_{11}Na$[257], or BuLi–ROK[258] (equation 90). Other compounds to have been similarly dimetallated to give Y-shaped dicarbanions under relatively mild conditions include $(PhCH_2)_2C{=}CHR$ (R = H or Ph)[7], (Z)- or (E)-$PhCH_2CMe{=}CHPh$[259], and $Me_2C{=}CHMe$[260], equation 91.

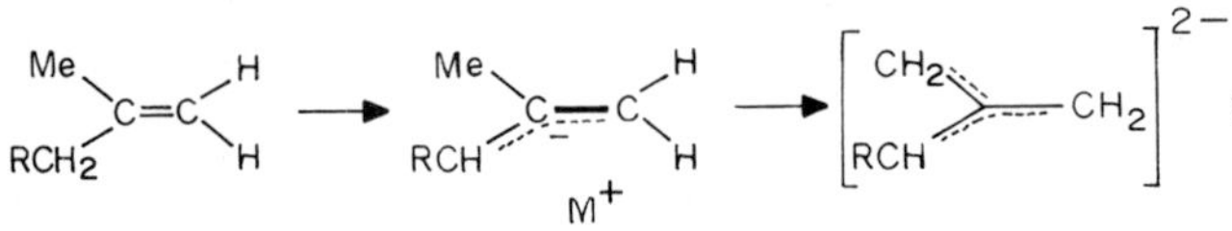

$$\text{Me(RCH}_2\text{)C=CH}_2 \longrightarrow \text{Me(RCH)C}{\cdots}\text{CH}_2\ \text{M}^+ \longrightarrow [\text{CH}_2(\text{RCH})\text{C}{\cdots}\text{CH}_2]^{2-}\ 2\text{M}^+ \qquad (90)$$

R=H or Ph

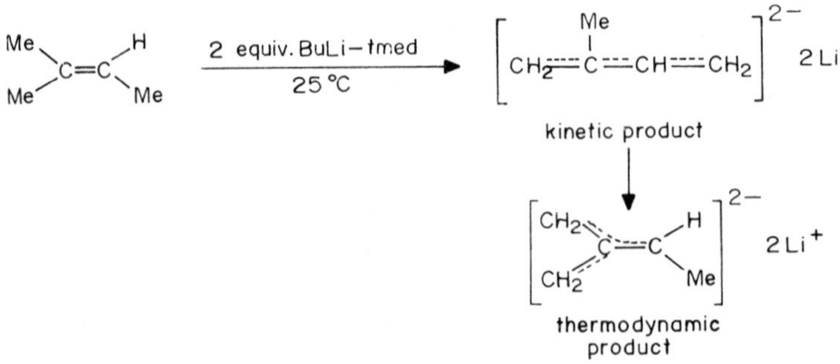

$$\text{Me}_2\text{C=CHMe} \xrightarrow[25\,^\circ\text{C}]{\text{2 equiv. BuLi–tmed}} [\text{CH}_2{\cdots}\text{C(Me)}{\cdots}\text{CH}{\cdots}\text{CH}_2]^{2-}\ 2\text{Li}^+ \text{ (kinetic product)} \longrightarrow [(\text{CH}_2)_2\text{C}{\cdots}\text{CHMe}]^{2-}\ 2\text{Li}^+ \text{ (thermodynamic product)} \qquad (91)$$

Alkenes such as $CH_2{=}CHMe$, $CH_2{=}CHCH_2Me$, $MeCH{=}CHMe$, and $PhCH_2CH{=}CH_2$ have also been dimetallated[261] by BuLi–tmed but much more forcing conditions are required than for $RCH_2CMe{=}CH_2$ (R = H or Ph), the second deprotonation occurring at a vinyl site:

$$PhCH_2CH{=}CH_2 \xrightarrow{+\text{BuLi–tmed}} PhCH{\cdots}\bar{C}H{\cdots}CH_2Li^+ \longrightarrow PhCH{\cdots}\bar{C}H{\cdots}CHLi\,Li^+ \quad (92)$$

Other substituted allyllithiums have been prepared from alkenes, although competing reactions may arise, e.g. equations 93 and 94.

(93): vinylcyclopropane + BuLi – tmed $\xrightarrow[\text{(ref. 262)}]{0\,°C}$ cyclopropylallyl anion Li⁺ (major product) + 1-lithio-1-vinylcyclopropane (minor product)

(94): methylenecyclopropane (47) + BuLi $\xrightarrow[\text{thf, hexane (ref. 263)}]{-10 \text{ to } +10\,°C}$ allyl anion Li⁺ + polymer

(47)

The presence of tmed during the metallation of **47** resulted in extensive polymerization. By comparing reactions 94, 95, and 96, it is clear that changes in the type of reaction undergone can be brought about by slight changes in structures. Whereas cyclohexene is

(95): (diphenylmethylene)cyclopropane + BuLi $\xrightarrow[\text{(ref. 264)}]{\text{thf, hexane}}$ 1-butyl-1-(diphenyllithiomethyl)cyclopropane (Bu, Li, Ph, Ph)

(96): alkylidenecyclobutane ($=CR_2$) + BuLi- tmed $\xrightarrow[\text{R=Ph; thf, hexane, }-100\,°C\text{ (ref. 264)}]{\text{R=H; hexane, 20 °C (ref. 265)}}$ allyl anion (R, R) Li⁺

metallated at an allylic site, for example by $C_5H_{11}Na$[241] or Me_3SiCH_2K[266], the smaller ring compound, cyclopentene, undergoes deprotonation at a vinyl site. The less reactive BuLi does not react with cyclohexene[267].

Conjugated carbanions are particularly readily obtained by metallation of appropriate mono-, di-, or polyenes. Pentadienylmetal compounds are available[268–270] by metallation of penta-1, 3- or 1, 4-dienes using reagents such as lithium amides, BuLi, Bu^sLi, BuLi–Bu^tOK, or Me_3SiCH_2K, e.g. equations 97 and 98.

$$\left.\begin{array}{l} CH_2{=}CHCH_2CH{=}CH_2 + \text{BuLi} \\ CH_2{=}CHCH{=}CHMe \quad + \text{BuLi-}Bu^tOK \end{array}\right\} \longrightarrow CH_2{\cdots}CH{\cdots}\bar{C}H{\cdots}CH{\cdots}CH_2\ M^+ \quad (97)$$

$$PhCH{=}CHCPh{=}CHCH_2Ph \xrightarrow[\text{20 °C (ref. 269)}]{\text{Litmp, hexane,}} PhCH{\cdots}CH{\cdots}\bar{C}Ph{\cdots}CH{\cdots}CHPh\ Li^+ \quad (98)$$

Allylic metallation of conjugated trienes and higher polyenes has also been reported[271], e.g. equation 99.

$$PhCH_2(CH{=}CH)_nCH_2Ph \xrightarrow{BuLi,\ thf} [PhCH_2(CH{\cdots}CH)_n{\cdots}CHPh]^- Li^+ n \geq 3 \qquad (99)$$

A variety of dianionic compounds have been obtained from suitable diene precursors, including 1,3-dienes, e.g. $CH_2{=}CMeCMe{=}CH_2$, equation 100, 1,4-dienes, including

$$CH_2{=}CMeCMe{=}CH_2 \xrightarrow[\text{thf (ref. 270)}]{2\,BuLi.Bu^tOK} \left[(CH_2)_2C{\cdots}C(CH_2)_2\right]^{2-}, 2K^+ \qquad (100)$$

$CH_2{=}CHCH_2CH{=}CHMe$, $CH_2{=}CMe_2CH_2CMe{=}CH_2$, $Me_2C{=}CHCH{=}CMe_2$, and $CH_2{=}CMeCH_2CH{=}CH_2$, and 1,5-dienes, e.g. $CH_2{=}CHCH_2CH_2CH{=}CH_2$ and $CH_2{=}CMeCH_2CH_2CH{=}CH_2$ (**48**)[261,272–274].

$$\begin{matrix} CH_2{=}CHCH_2CH{=}CHMe \\ \text{or} \\ CH_2{=}CHCH_2CH_2CH{=}CH_2 \end{matrix} \xrightarrow{2\,BuLi\text{-}tmed} \left[CH_2{\cdots}CH{\cdots}CH{\cdots}CH{\cdots}CH{\cdots}CH_2\right]^{2-}\ 2\,Li^+ \qquad (101)$$

Metallation of **48** by 2 equiv. of BuLi–tmed (or BuLi–Bu^{t}OK) provided mono- and dianions[273]; the dianions obtained are indicated in equation 102. The thermodynamic

$$(\mathbf{48}) \xrightarrow[25\,°C]{2\ \text{equiv. BuLi–tmed}} (\mathbf{49})^{2-}\ 2\,Li^+ + (\mathbf{50})^{2-}\ 2\,Li^+ + [\ ]^{2-}\ 2\,Li^+ + (\mathbf{51})\ 2\,Li^+ \qquad (102)$$

product is **50** and its yield increases with time at the expense of **49** and **51**; **51** is formed initially in the greater yield.

Trienes (**52**) are also easily dimetallated[275].

$$\underset{\substack{R = H \text{ or } Ph \\ (\mathbf{52})}}{RCH{=}C(CH_2CH{=}CH_2)_2} \xrightarrow[25\,°C]{2\,BuLi,\ thf} \left[RCH{\cdots}C(CH{\cdots}CH{\cdots}CH_2)_2\right]^{2-}\ 2\,Li^+ \qquad (103)$$

Trianionic compounds may also be obtained, equations 104 and 105.

$$CH_2{=}CHCH_2CH{=}CHCH_2CH_3 \xrightarrow[\text{hexane, 20 °C (ref. 276)}]{3\,BuLi\text{-}tmed} [\ ]^{3-}\ 3\,Li^+ \qquad (104)$$

$$\text{CH}_2\text{=C(Me)CH}_2\text{C(Me)=CH}_2 \xrightarrow[\text{hexane, 20 °C (ref. 272)}]{\text{3 BuLi-tmed}} [\ldots]^{3-} \quad 3\,\text{Li}^+ \qquad (105)$$

The examples so far quoted have involved acyclic compounds; some representative examples of cyclic delocalized mono-,· di-, and trianionic compounds are given in equations 106–109.

$$\text{C}_5\text{H}_6\text{SnBu}_2 \xrightarrow[\text{thf, }-40\text{ °C (ref. 277)}]{\text{LiNPr}^i\text{(cyclo-C}_6\text{H}_{11}\text{),}} [\text{C}_5\text{H}_5\text{SnBu}_2]^- \quad \text{Li}^+ \qquad (106)$$

$$\text{cycloheptadiene} \xrightarrow[\text{(ref. 270)}]{\text{BuLi-Bu}^t\text{OK}} [\text{C}_7\text{H}_9]^- \quad \text{K}^+ \qquad (107)$$

$$\text{cyclooctatriene} \xrightarrow[\text{(ref. 278)}]{\text{2 equiv. BuLi-tmed}} [\text{C}_8\text{H}_8]^{2-} \quad 2\,\text{Li}^+ \qquad (108)$$

$$\text{cycloheptadiene} \xrightarrow[\text{hexane, 20 °C (ref. 276)}]{\text{3 BuLi-tmed}} [\text{C}_7\text{H}_7]^{3-} \quad 3\,\text{Li}^+ \qquad (109)$$

vii. *Metallations of alkynes and allenes.* Acetylene can be dimetallated, for example by PhLi in Et_2O[279]. Monolithioacetylene has been obtained by the disproportionation of LiC≡CLi and HC≡CH in liquid ammonia[280], and from reaction[281] of HC≡CH with BuLi in thf at −78 °C. Metal amides have also been used[1].

Terminal alkynes, RC≡CH, are metallated[1] to give RC≡CM by a number of reagents, including BuLi in hydrocarbon[282] or ethereal solutions or in the presence of a donor[283], and metal amides in liquid ammonia or ethers. Replacement of an allylic proton and an alkynyl hydrogen occurs[284] in HC≡CCMe=CH_2, equation 110. Propargylic hydrogens

$$\text{HC}\equiv\text{CCMe=CH}_2 \xrightarrow[\text{thf, hexane, }-70\text{ °C}]{\text{2 equiv. BuLi, Bu}^t\text{OK,}} \text{KC}\equiv\text{CC}(\text{CH}_2)_2^- \quad \text{K}^+ \qquad (110)$$

are sufficiently acidic to be metallated by organometals[285,286], equation 111. Dimetallation can occur[287], at propargylic sites as in PhC≡CH_2R (R = H or Me) using BuLi in Et_2O. Replacement of propargyl and alkynyl hydrogens in the same compound can arise to give di- and polylithio products[184,288]. Thus MeC≡CH with BuLi (1 equiv.) provides MeC≡CLi and with excess of BuLi successive replacements of hydrogens occur eventually to provide C_3Li_4, equation 112.

$$\text{Me}_3\text{SiC}\equiv\text{CMe} \xrightarrow[\text{(ref. 285)}]{\text{BuLi, tmed}} \text{Me}_3\text{SiC}\equiv\text{CCH}_2\text{Li} \qquad (111)$$

$$\text{MeC}{\equiv}\text{CH} \rightarrow \text{MeC}{\equiv}\text{CLi} \rightarrow \text{LiCH}_2\text{C}{\equiv}\text{CLi} \rightarrow \text{Li}_2\text{CCH}{\equiv}\text{CLi} \rightarrow \text{C}_3\text{Li}_4 \quad (112)$$

The terminal alkyne, $HC{\equiv}CCH_2Me$, is lithiated by BuLi or Bu^{t}Li (3 equiv.) to give MeC_3Li_3, which can also be obtained from $MeC{\equiv}CMe$ using the more powerful metallating agent BuLi–tmed (3 equiv.)[289]. Dimetallations of $R_2CHC{\equiv}CH$ to $R_2CMC{\equiv}CM$ have been achieved using BuLi–tmed for R = Me[289] and $NaNH_2$ in liquid ammonia for R = Ph[290].

Metallations of **53** can occur at either propargylic site[291]. Monometallation results at the alkyl site for R = R′ = H, at the cyclopropyl ring for R = R′ = Me and at either site for R = H, R′ = Me. The diyne, $HC{\equiv}CCH_2C{\equiv}CH$, can be completely perlithiated to C_5Li_4 using BuLi–tmed; $MeC{\equiv}CC{\equiv}CMe$ provides MeC_5Li_3 under similar conditions[289].

In solution, equilibria between propargylic and allenic carbanions are indicated from the products obtained on reaction with electrophiles, for example as found in reactions of $PhCHRC{\equiv}CPh$ (R = H or Ph), equation 114.

$$\underset{(\mathbf{53})}{\text{Cyclopropyl-C}{\equiv}\text{CCHRR}'} \xrightarrow{\text{BuLi}} \text{cyclopropyl-C}{\equiv}\text{CCLiRR}' + \text{1-Li-cyclopropyl-C}{\equiv}\text{CCHRR}' \quad (113)$$

$$\text{PhCHRC}{\equiv}\text{CPh} \xrightarrow[\text{R = H, Et}_2\text{NLi (ref. 293)}]{\text{R = Ph, BuLi (ref. 292)}} \text{PhCRLiC}{\equiv}\text{CPh} \rightleftharpoons \text{PhCR}{=}\text{C}{=}\text{CLiPh} \quad (114)$$

Allenes are also metallated. Allene itself is metallated[112,294] by BuLi in thf, equation 115. Dilithiation apparently provides $LiCH_2C{\equiv}CLi$.

$$\text{CH}_2{=}\text{C}{=}\text{CH}_2 + \text{BuLi} \xrightarrow[-70\,^\circ\text{C}]{\text{thf,}} \text{CH}_2{=}\text{C}{=}\text{CHLi} \rightleftharpoons \text{HC}{\equiv}\text{CCH}_2\text{Li} \quad (115)$$

Other examples of metallation of substituted allenes are shown in equations 116 and 117.

$$\text{R}^2\text{R}^3\text{C}{=}\text{C}{=}\text{C(R}^1\text{)H} \xrightarrow[\text{Bu}^t\text{Li or BuLi, hmpt (ref. 295)}]{\text{BuLi, hexane, thf, }-78\,^\circ\text{C}} \text{R}^2\text{R}^3\text{C}{=}\text{C}{=}\text{C(R}^1\text{)Li} \quad (116)$$

$$\text{(1-cyclohexenyl)CH}{=}\text{C}{=}\text{CH}_2 \xrightarrow[\text{thf, Et}_2\text{O (ref. 296)}]{\text{BuLi, 70 °C}} \text{(1-cyclohexenyl)CH}{=}\text{C}{=}\text{CHLi} \quad (117)$$

Successive reactions[297] of **54** (R = R′ = H) with Bu^{t}Li in Et_2O at −78 °C and with electrophiles provide product ratios arising from replacement of R:R′ of up to 13.5:1. Metallation of (*S*)-**54**, (R = Me or Bu^t; R′ = H) using BuLi–Bu^{t}OK in thf and

C_8H_{17}, H, C, H, R, R^1

(**54**)

TABLE 8. Formation of α-nitrogen-substituted alkylmetal compounds

Compound	Conditions	Product (yield, %)	Ref.
N-methylpiperidine (N–Me)	Bu^sLi, Bu^tOK, isopentane, 0 °C	N–CH_2K piperidine (70)	802
pyrrolidine N–CH=NBu^t	Bu^tLi, thf, −78 to −20 °C	2-Li pyrrolidine N–CH=NBu^t (90)	803
Et_2NCH_2CN	$LiNPr^i{}_2$, hmpt, thf, −78 °C	$Et_2NCHLiCN$ (90)	804
2,4,6-$Pr^i{}_3C_6H_2$–C(O)NMe_2	Bu^sLi, tmed, thf, −78 °C	2,4,6-$Pr^i{}_3C_6H_2$–C(O)N(Me)CH_2Li (82)	805
$PhCH_2$, $PhCH_2$, O, NMe, O (bicyclic imide)	Bu^sLi, hmpt, thf, −100 °C	$PhCH_2$, $PhCH_2$, O, NCH_2Li, O (90)	806
2,4,6-$Pr^i{}_3C_6H_2$–OC(O)NMe_2	Bu^sLi, tmed, thf, 0 °C	2,4,6-$Pr^i{}_3C_6H_2$–OC(O)N(Me)CH_2Li	807
$Bu^tC(S)NMe$	Bu^sLi, tmed, thf, −78 °C	$Bu^tC(S)NMeCH_2Li$ (82)	808
1-CN-2,2-Ph$_2$-cyclopropane (CN, Ph, Ph)	$LiNPr^i{}_2$, thf, −78 °C	CN, Li, Ph, Ph cyclopropane (96)	809
N-nitrosoazetidine (N–N=O)	$LiNPr^i{}_2$, thf, −78 °C	2-Li N–N=O azetidine (65)	810
$(Me_3N)_3P{=}O$	Bu^sLi, dme, −78 °C	$(Me_2N)_2P(O)NMeCH_2Li$ (83)	811
$PhN{=}NMe$	BuLi, thf, hexane, −70 °C	$PhN{=}NCH_2Li$ (88)	812
$C_6H_{13}CHMeN{=}N(O)$ Me	$LiN(SiMe_3)_2$, thf, 0 °C	$C_6H_{13}CHMeN{=}N(O)CH_2Li$ (60)	813
$Ph_2C{=}NMe$	$LiNPr^i{}_2$, thf, Et_2O	$Ph_2C{=}NCH_2Li$ (85)	814
Me_3SiCHN_2	BuLi, thf, hexane, −78 °C	$Me_3SiCLiN_2$ (88)	815
$MeCOCHN_2$	$LiNPr^i{}_2$, thf, −78 °C	$MeCOCLiN_2$ (71)	816
$MeNO_2$	2 equiv. BuLi, hmpt, thf, −90 °C	$(CH_2NO_2)Li_2$ (65)	817

subsequent reaction of the carbanion with an electrophile occurs[298] primarily with retention of configuration, while (*R*)-**54** (R = H, R′ = Me or Bu^t) reacts mainly with inversion.

The compound $Me_2C{=}C{=}CH_2$ has been lithiated[296,299] by a number of reagents (see equation 118); compound **55**, in the presence of $Me_2C{=}C{=}CH_2$, is slowly converted to $Me_2CHC{\equiv}CLi$.

$$Me_2C{=}C{=}CH_2 \xrightarrow[\text{BuLi}]{(\text{MeLi} + \text{HNPr}^i{}_2),\ \text{Litmp or}} \underset{(\mathbf{55})}{Me_2C{=}C{=}CHLi} \qquad (118)$$

Metallation of $RC{\equiv}CMe$ by Bu^{s}Li in thf–cyclohexane at 0 °C proceeds to give a mixture of $RC{\equiv}CCH_2Li$ and $RCLi{=}C{=}CH_2$. On addition of at least 5 equiv. of hmpt, these products rearrange[300] to $RCH{=}C{=}CHLi$.

b. Metallation of functionally substituted hydrocarbons

The discussion so far has been concerned with hydrocarbons without functionally substituted groups. Attention is now turned to metallations of such substituted compounds. These have been particularly well studied[3,187,188,301,302], especially since many of the carbanion products have found considerable use in organic synthesis[188,303].

Various heterosubstituted functional groups enhance the acidity (thermodynamic and/or kinetic) of α-protons in alkyl and alkenyl compounds[301] or *ortho*-protons in

TABLE 9. Formation of α-phosphorus-substituted alkylmetal compounds

Compound	Conditions	Product (yield, %)	Ref.
Me_3P	Bu^{t}Li, pentane, 20 °C	Me_2PCH_2Li (93)	818
	BuLi, tmed		819
$(Ph_2P)_2CH_2$	BuLi, tmed, hexane, PhH, 20 °C	$(Ph_2P)_2CHLi$ (68)	820
$(Me_2P)_3CH$	Bu^{t}Li, thf, pentane	$(Me_2P)_3CLi$ (73)	821
$Ph_2PCH_2SiMe_3$	BuLi, tmed	$Ph_2PCHLiSiMe_3$	822
Ph_2PCH_2OMe	Bu^{s}Li, thf, −95 °C	$Ph_2PCHLiOMe$	823
$Ph_2P(O)Pr$	BuLi, tmed, thf, hexane	$Ph_2P(O)CHLiEt$ (73)	824
$Ph_2P(O)CH_2CMe_2NHCOPh$	2 equiv. BuLi, thf, −30 °C	$Ph_2P(O)CHLiCMe_2NLiCOPh$ (84)	825
$PhP(S)Me_2$	BuLi, tmed, thf, −78 °C	$PhS(O)(Me)CH_2Li$ (90)	826
$[Ph_2P(O)]_2CH_2$	BuLi, PhH, 20 °C	$[Ph_2P(O)]_2CHLi$ (73)	827
$(MeO)_2P(O)Me$	BuLi, thf, −78 °C	$(MeO)_2P(O)CH_2Li$	828
$(MeO)_2P(S)Me$	BuLi, thf, −78 °C	$(MeO)_2P(S)CH_2Li$ (81)	828
$(EtO)_2P(O)CH_2SMe$	BuLi, thf, −78 °C	$(EtO)_2P(O)CHLiSMe$	829
$(EtO)_2P(O)CH_2Cl$	BuLi, thf, −78 °C	$(EtO)_2P(O)CHLiCl$ (90)	830
$(EtO)_2P(O)CH$–CH_2 (epoxide, O bridging)	$LiNPr^i{}_2$, thf, hexane, Et_2O, −110 °C	$(EtO)_2P(O)C(Li)$–CH_2 (epoxide, O bridging) (84)	831
$(EtO)_2P(O)CH_2CN$	BuLi, thf	$(EtO)_2P(O)CHLiCN$	832
$(EtO)_2P(O)CH_2NC$	BuLi, thf, hexane, −70 °C	$(EtO)_2P(O)CHLiNC$	833
$(EtO)_2P(O)CH_2CO_2Me$	BuLi, thf	$(EtO)_2P(O)CHLiCO_2Me$	837
$(EtO)_2P(O)CF_2H$	$LiNPr^i{}_2$, thf, −78 °C	$(EtO)_2P(O)CF_2Li$ (87)	835
$(Me_2N)_2P(O)Me$	BuLi, thf, −78 °C	$(Me_2N)_2P(O)CH_2Li$	836
$[Me_2P(S)]_3CH$	Bu^{t}Li, thf, −20 °C	$[Me_2P(S)]_3CLi$ (76)	837
$MePh_2P{=}CHPPh_2$	MeLi, thf, Et_2O; $NaNH_2$, thf; KH, thf	$Ph_2PCH{=}PPh_2CH_2M$	838

TABLE 10. Formation of α-sulphur-substituted alkylmetal compounds

Compound	Conditions	Product (yield, %)	Ref.
Me_2S	BuLi, tmed, hexane, 20 °C	$MeSCH_2Li$ (84)	839
PhSMe	BuLi, dabco, thf, hexane, 0 °C	$PhSCH_2Li$ (93)	840
PhSCH (cyclopropyl, CH_2–CH_2)	BuLi, thf, 0 °C	PhSCLi (cyclopropyl, CH_2–CH_2) (95)	841
$PhSCH_2SiMe_3$	BuLi, tmed, hexane, 0 °C	$PhSCHLiSiMe_3$ (99)	842
$PhSCH_2OMe$	BuLi, thf, −30 °C	PhSCHLiOMe (88)	843
$PhSCH_2NC$	BuLi, thf, −60 °C	PhSCHLiNC (85)	844
$PhSCH_2CN$	$LiNPr^i_2$, thf, −78 °C	PhSCHLiCN (93)	845
$PhSCH_2CO_2H$	$LiNPr^i_2$, thf, 0 °C	$PhSCHLiCO_2Li$ (99)	846
$PhSCH_2CO_2Me$	$LiNPr^i_2$, thf, −60 °C	$PhSCHLiCO_2Me$ (88)	847
$HSCH_2CO_2Et$	2.2 equiv. $LiNPr^i_2$, tmed, thf, −78 °C	$LiSCHLiCO_2Et$ (90)	848
$(MeS)_2CHSnMe_3$	$LiNPr^i_2$, hmpt, thf, hexane, −78 °C	$(MeS)_2CLiSnMe_3$ (67)	849
$MeSCH_2S(O)_2Me$	BuLi, thf, hexane, −20 °C	$MeSCHLiS(O)_2Me$	850
1,3-dithiane	BuLi, thf, hexane, −40 °C	2-lithio-1,3-dithiane (92)	851
1,3-dithiane 1-oxide (S=O)	BuLi, thf, hexane, −10 °C	2-lithio-1,3-dithiane 1-oxide (79)	852
1,3-oxathiane	Bu^sLi, thf, −78 °C	2-lithio-1,3-oxathiane (86)	853
$(PrS)_3CH$	BuLi, thf, −78 °C	$(PrS)_3CLi$ (67)	854
$[Et_2NC(S)S]_2CH_2$	BuLi, thf, −78 °C	$[Et_2NC(S)S]_2CHLi$ (74)	854
MeS(O)Me	MNH_2 (M = Li, Na, K or CS)	$MeS(O)CH_2M$	855
	BuLi, thf,		856
	NaH		857
$PhS(O)CH_2Me$	MeLi, thf, −60 °C	PhS(O)CHLiMe (81) (13:1 mixture of diastereomers)	858
$PhS(O)CH_2I$	$LiNPr^i_2$, thf, hexane, −78 °C	PhS(O)CHLiI	859
2,3-diphenylcyclopropyl phenyl sulphoxide (Ph, Ph, Ph–S=O)	BuLi, thf, −30 °C	1-lithio derivative (Li, Ph, Ph, Ph–S=O) (95)	860
Bu^t-thiane S=O (chair)	MeLi, thf, −78 °C	Bu^t-thiane S=O, α-Li (100)	861

TABLE 10. (*Contd.*)

Compound	Conditions	Product (yield, %)	Ref.
$MeS(O)_2Me$	BuLi, PhH	$MeS(O)_2CH_2Li$	862
	2 equiv. MNH_2, liq. NH_3 (M = Li or Na)	$MCH_2S(O)_2CH_2M$ (780)	863
$PhS(O)_2CH_2OMe$	BuLi, thf, hexane, − 78 °C	$PhS(O)_2CHMOMe$ (100)	864
	NaH, thf, 25 °C, 1 h	(40)	
	Bu^tOK, thf, 25 °C, 1 h	(100)	
Ph; Ph; O=S=O; Ph	$LiNPr^i{}_2$, thf, − 20 °C	Ph; Ph; O=S=O; Ph (90)	860
$C_6H_{13}CH$; O; H; SO_2Ph	1.8 equiv. BuLi, thf, Et_2O hexane, − 110 °C	$C_6H_{13}CH$; O; C; Li; O=S=O; Ph (96)	865
$MeS(O)_2NHBu^t$	2 equiv. BuLi, thf, − 30 °C	$LiCH_2S(O)_2N(Li)Bu^t$ (87)	866
$PhS(O)(Me)$=NMe	BuLi, thf, 0 °C	$PhS(O)(CH_2Li)$=NMe (90)	867
$PhS(O)CHCl_2$	$LiNPr^i{}_2$, thf, − 78 °C	$PhS(O)C(Li)Cl_2$	868

TABLE 11. Formation of α-selenium-substituted alkylmetal compounds

Compound	Conditions	Product (yield, %)	Ref
PhSeMe	BuLi, tmed, thf, − 78 to − 50 °C	$PhSeCH_2Li$ (40)	869
$PhSeCH_2SiMe_3$	Bu^sLi, tmed, hexane, 25 °C	$PhSeCHLiSiMe_3$ (94)	870
	$LiNPr^i{}_2$, thf, hexane		871
m-$CF_3C_6H_4SeMe$	Litmp, thf, − 55 °C	*m*-$CF_3C_6H_4SeCH_2Li$	872
m-$CF_3C_6H_4SeCH_2OMe$	Litmp, thf, − 78 °C	*m*-$CF_3C_6H_4SeCHLiOMe$ (83)	872
$PhSeCH_2COPh$	$LiNPr^i{}_2$, thf, − 78 °C	PhSeCHLiCOPh	872
$PhSeCH_2CO_2Me$	$LiNPr^i{}_2$, thf, − 78 °C	$PhSeCHLiCO_2Me$ (93)	873
$(PhSe)_2CH_2$	$LiNPr^i{}_2$, thf, − 78 °C	$(PhSe)_2CHLi$ (77)	872
	(BuLi, thf, − 78 °C	$PhSeCH_2Li$	872)
$(PhSe)_3CH$	$LiNBu^t{}_2$, thf, − 78 °C	$(PhSe)_3CLi$ (93)	869
PhSe(O)Me	$LiNPr^i{}_2$, thf, − 78 °C	$PhSe(O)CH_2Li$	874
$PhSe(O)CHMe_2$	$LiNPr^i{}_2$, thf, − 78 °C	$PhSe(O)CLiMe_2$	874

arenes[188,304,305]. Coordination of the metallating agent to the functional group may also be a significant factor in directing attack to an adjacent site. Coordination of the lithiating agent to the heterosubstituted group was detected during lithiations of $RCH_2NR'CH$=NBu^t by Bu^tLi[306a] and of 2, 4, 6-$Pr^i{}_3C_6H_2CONMe_2$ by Bu^sLi in cyclohexane[306b]. These features, together or alone, result in the metallations being achieved under milder conditions than are required for the parent hydrocarbons, and may allow more sensitive functional groups to survive the metallations, although low temperatures and the employment of less nucleophilic reagents, e.g. MNR_2 in place of RM, are frequently

TABLE 12. Formation of α-halo-substituted alkylmetal compounds

Compound	Conditions	Product (yield, %)	Ref.
Me_3SiCH_2Cl	Bu^sLi, tmed, thf, −78 °C	$Me_3SiCHLiCl$ (> 90)	875
CH_2Cl_2	BuLi, tmed, thf, Et_2O, −90 °C	$LiCHCl_2$ (*ca.* 100)	876
$MeCHCl_2$	BuLi, tmed, thf, Et_2O, −90 °C	$MeCLiCl_2$ (*ca.* 100)	876
$CHCl_3$	$LiNPr^i{}_2$, Et_2O, −108 °C	$LiCCl_3$	877
CH_2Br_2	$LiNPr^i{}_2$, thf, Et_2O, −90 °C	$LiCHBr_2$ (63)	878
$Me_3SiCHBr_2$	$LiNPr^i{}_2$, thf, Et_2O, −80 °C	$Me_3SiCLiBr_2$ (93)	878
$CHBr_3$	$LiNPr^i{}_2$, thf, Et_2O, −110 °C	$LiCBr_3$ (90)	878

TABLE 13. Formation of α-metallo-substituted alkylmetal compounds

Compound	Conditions	Product (yield, %)	Ref.
BMe (bicyclic borane)	Litmp, PhH, 20 °C	BCH_2Li (bicyclic borane) (87)	879
$CH_2(B(O\text{–})_2)_2$ (cyclic boronate)	Litmp, thf, tmed, −75 °C	LiCHB(O–)₂ (cyclic boronate) (70)	880
Me_3SiCH_2B(OCMe₂CMe₂O)	Litmp, tmed, thf, 0 °C	$Me_3SiCH(Li)B$(OCMe₂CMe₂O) (87)	881
Me_4Si	BuLi, pmdt, PhH	Me_3SiCH_2Li	882
$(Me_3Si)_2CH_2$	Bu^tLi, hmpt, thf, −78 °C BuLi, pmdt, PhH	$(Me_3Si)_2CHLi$ (65)	883
$(Me_3Si)_3CH$	MeLi, Et_2O, thf	$(Me_3Si)_3CLi$ (95)	884
$(Me_3Si)_4C$	Bu^tLi, tmed, pentane	$(Me_3Si)_3CSiMe_2CH_2Li$ (57)	885
Me_3SiCH_2OMe	Bu^sLi, thf, hexane, −70 °C	$Me_3SiCHOMeLi$ (> 90)	886
$(Ph_3M)_2CH_2$	$LiNR_2$[a], hmpt, Et_2O, 20 °C	$(Ph_3M)_2CHLi$ M=Sn (92) Pb (67)	887
$Ph_3SnCH_2AsPh_2$	$LiNR_2$[a], hmpt, Et_2O, 20 °C	$Ph_3SnCHLiAsPh_2$ (67)	887
$(Ph_2M)_2CH_2$	$LiNR_2$[a], hmpt, Et_2O, 20 °C	$(Ph_2M)_2CHLi$ M=As (63) Sb (68)	887
$Ph_2As(O)Me$	$LiNPr^i{}_2$, thf, −40 °C	$Ph_2As(O)CH_2Li$ (81)	888
$(PhTe)_2CH_2$	$LiNPr^i{}_2$, thf, hexane, −78 °C (MeLi, thf, −78 °C	$(PhTe)_2CHLi$ (quant.) $PhTeCH_2Li$	889 889)

[a] R = cyclohexyl.

TABLE 14. Formation of α-carbonyl-substituted and related alkalimetal compounds

Compound	Conditions	Product (yield, %)	Ref.
$C_{12}H_{25}CH_2CO_2H$	2 equiv. $LiNPr^i_2$, hmpt, thf −78 °C	$C_{12}H_{25}CHLiCO_2Li$ (89)	890
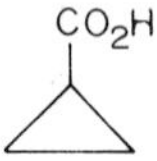 CO_2H (cyclopropane)	2.2 equiv. $LiNPr^i_2$, thf, 0 °C	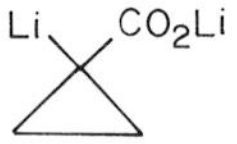 Li, CO_2Li (cyclopropane)	891
$PhCH_2OCH_2CO_2H$	2 equiv. $LiNPr^i_2$, thf, −78 °C	$PhCH_2OCHLiCO_2Li$ (80)	892
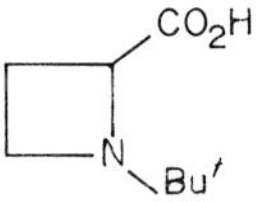 CO_2H, N, Bu^t (azetidine)	2 equiv. $LiNPr^i_2$, thf, 0 °C	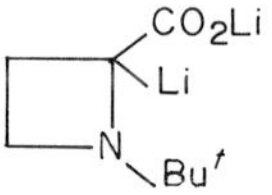 1-Bu^t-2-Li-2-CO_2Li- CO_2Li, Li, N, Bu^t (azetidine) (97)	893
Cl_2CHCO_2H	2 equiv. $LiNPr^i_2$, thf, −78 °C	Cl_2CLiCO_2Li (85)	894
MeCOSH	2 equiv. $LiNPr^i_2$, thf, −78 °C	$LiCH_2COSLi$ (90)	895
$MeCO_2Et$	NaH, hmpt	$NaCH_2CO_2Et$	896
$MeCO_2Bu^t$	$LiNPr^i_2$, hexane, −78 °C	$LiCH_2CO_2Bu^t$ (95)	897
Me_2CHCO_2Me	$LiNPr^i_2$, PhH, −10 °C	Me_2CLiCO_2Me	898
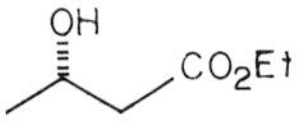 OH, CO_2Et	2equiv. $LiNPr^i_2$, thf, −50 °C	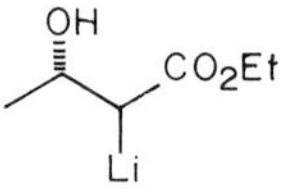 OH, CO_2Et, Li (82)	899
$(MeO)_2CHCO_2Me$	$LiNBu^t_2$, thf, −70 to −10 °C	$(MeO)_2CLiCO_2Me$ (70)	900
Me_3SiCH_2CSOEt	BuLi, Et_2O, −40 °C	$Me_3SiCHLiCSOEt$	901
$MeCONMe_2$	$LiNPr^i_2$, thf, pentane, −78 °C	MCH_2CONMe_2 (99)	902
	$NaNH_2$, PhMe	(60)	903
$MeCONHSiMe_3$	2 equiv. BuLi, thf, hexane, 0 °C	$LiCH_2CONLiSiMe_3$ (82)	904
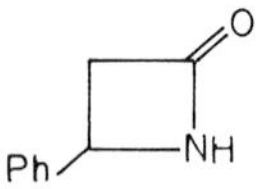 O, NH, Ph (azetidinone)	2 equiv. BuLi, thf, hexane, 0 °C	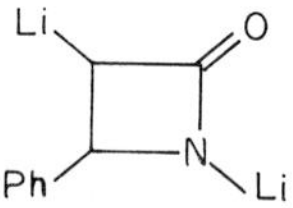 Li, O, N, Li, Ph (azetidinone) (66)	905
MeCOMe	$NaNH_2$, Et_2O	$MeCOCH_2Li$	906
MeCN	BuLi, hexane, thf, −70 °C	MCH_2CN (> 90)	907
	$NaNH_2$, liq. NH_3	(94)	908
	$KN(SiMe_3)_2$, thf		
Me_3SiCH_2CN	$LiNPr^i_2$, thf, hexane, −78 °C	$Me_3SiCHLiCN$ (95)	909
$(MeO)_2CHCN$	$LiNPr^i_2$, hmpt, thf, hexane, −78 °C	$(MeO)_2CLiCN$	910

necessary in order to avoid side reactions such as additions, couplings, and eliminations. By far the largest amount of work has been carried out with lithium reagents; by comparison, sodium, potassium, and the other alkali metals have received scant attention.

i. *Metallations of substituted alkenes.* The heteroatom in the functional group could be a metalloid (e.g. R_2B and R_3Si), from Group V (e.g. N, P, As, and Sb), from Group VI (e.g. O, S, Se, and Te) or from Group VII (e.g. F, Cl, Br, and I). As shown in Tables 8–13, compounds bearing a variety of functional groups have been metallated. Combinations of the same or different groups result in even easier metallations. Substituted benzyllithiums are readily obtained[307], see Table 7, as are α-carbonyl-substituted derivatives (Table 14).

The stability of the heterosubstituted alkyl carbanion can be due to several effects. The stability can arise from the electron-withdrawing effects of the heteroatom or group,

$$\mathrm{Y} \leftarrow \overset{\delta-}{\mathrm{C}} - \overset{\delta+}{\mathrm{Li}} \longleftrightarrow \overset{\delta-}{\mathrm{Y}} - \mathrm{C} - \overset{\delta+}{\mathrm{Li}} \tag{119}$$

e.g. Y = X or CX_3
X = halogen

equation 119. Charge–dipole interaction[308] is another method of stabilizing a carbanion, equations 120 and 121.

$$-\overset{\mathrm{Y}}{\overset{\|}{\mathrm{C}}}-\ddot{\mathrm{X}}-\overset{\mathrm{Li}}{\overset{|}{\mathrm{C}}}- \longleftrightarrow -\overset{\mathrm{Y^-}}{\overset{|}{\mathrm{C}}}=\overset{+}{\mathrm{X}}-\overset{\mathrm{Li}}{\mathrm{C}}\lt \tag{120}$$

X = O, S, or NR; Y = O or S

$$\mathrm{Z}-\mathrm{Y}-\overset{\mathrm{Li}}{\overset{|}{\mathrm{C}}}- \longleftrightarrow \bar{\mathrm{Z}}=\overset{+}{\mathrm{Y}}-\bar{\mathrm{C}}\lt \;\; \mathrm{Li^+} \longleftrightarrow \bar{\mathrm{Z}}=\mathrm{Y}=\mathrm{C}\lt \;\; \mathrm{Li^+} \tag{121}$$

Examples of stabilization arising via charge–dipole interactions are the carbanions derived from hindered amides[309], equations 122 and 123, thioamides (e.g. $Bu^tC(S)NMe_2$)[310], esters and thioesters[309,311], equation 124, formamidines[306,312] (e.g. $R_2NCH{=}NBu^t$), succinimides[313], thioimidates[314], dithiocarbonates[315], formamides[316], thioformamides[317] (equation 125), nitrosamines[318], e.g. equations 126 and 127, amine oxides[319], nitroalkanes[320], and isocyanides[321], equation 128.

$$\mathrm{Ar}-\overset{\mathrm{O}}{\overset{\|}{\mathrm{C}}}-\mathrm{N}(\text{piperidine}) \xrightarrow[\text{thf, }-78\,^{\circ}\mathrm{C}\text{ (ref. 309b)}]{\mathrm{Bu^sLi\text{-}tmed}} \mathrm{Ar}\overset{\mathrm{O}}{\overset{\|}{\mathrm{C}}}-\mathrm{N}(\text{2-Li-piperidine}) \longleftrightarrow \mathrm{Ar}-\overset{\mathrm{O^-}}{\overset{|}{\mathrm{C}}}=\overset{+}{\mathrm{N}}(\text{2-Li-piperidine}) \tag{122}$$

Ar = 2,4,6-$Pr^i_3C_6H_2$

$$\text{2-Ph-piperidine-}\mathrm{N{-}C({=}O)Et_3C} \xrightarrow[\text{thf, }-78\,^{\circ}\mathrm{C}\text{ (ref. 309b)}]{\mathrm{Bu^sLi-tmed}} \text{Li-piperidine(Ph)-}\mathrm{N{-}C({=}O)Et_3C} \longleftrightarrow \mathrm{Et_3C{-}C(O^-){=}\overset{+}{N}}\text{(Ph-piperidine)}\,\ddot{\ }\;\mathrm{Li^+} \tag{123}$$

$$\mathrm{ArC(=O)-X-CH_2R} \xrightarrow[\text{thf}, -95\,^\circ\text{C}]{\text{Bu}^s\text{Li-tmed}} \mathrm{ArC(O^-)\overset{+}{X}CH_2Li} \tag{124}$$

$\mathrm{Ar} = 2,4,6\text{-}\mathrm{Pr}^i_3\mathrm{C_6H_2}$; X = O, R = H[311]; X = S, R = Me[309a]

$$\mathrm{HC(=Y)NMe_2} \xrightarrow[-78\,^\circ\text{C}]{\text{LiNPr}^i_2, \text{Et}_2\text{O},} \mathrm{Li-C(=Y)-NMe_2} \longleftrightarrow \mathrm{Li-C(Y^-)=\overset{+}{N}Me_2} \tag{125}$$

Y = O[316] or S[317]

$$\mathrm{Me(ArCH_2CH_2)N-N=O} \xrightarrow[\substack{-80\,^\circ\text{C} \\ \text{(ref. 318b)}}]{\text{LiNPr}^i_2, \text{thf},} \mathrm{Me(ArCH_2CH(Li))N-N=O} \tag{126}$$

$$\mathrm{Me(ArCH_2CH_2)N-N=O} \xrightarrow[\substack{-80\,^\circ\text{C} \\ \text{(ref. 318b)}}]{\text{LiNPr}^i_2, \text{thf},} \mathrm{LiCH_2(ArCH_2CH_2)N-N=O} \tag{127}$$

$$\mathrm{MeN{\equiv}C} \xrightarrow[\text{(ref. 321)}]{\text{BuLi}, -70\,^\circ\text{C}} \mathrm{LiCH_2N{=}C} \longleftrightarrow \mathrm{Li^+\ CH_2{=}NC^-} \tag{128}$$

As shown in equations 123, 126, and 127, and also with metallations of aldimines[322] and $\mathrm{Me_2C{=}NOMe}$[323], lithiations occur at the *syn*-carbon. Groups that have partially or fully charge heteroatoms can also stabilize the carbanion, equation 129. Examples of such

$$-\overset{Z^-}{\overset{|}{Y^+}}-\overset{|}{\underset{|}{C}}-\mathrm{M} \longleftrightarrow -\overset{Z^-}{\overset{|}{Y}}=\mathrm{C}\langle\ \ \mathrm{M^+} \tag{129}$$

Y = S, S(O), Se, P or As; Z = O or S or NR

groups are $\mathrm{P(O)R_2}$[324,325], $\mathrm{P(O)(OR)_2}$[324], $\mathrm{P(S)(OEt)_2}$[326], $\mathrm{P(O)(NMe_2)_2}$[327], S(O)R[328], $\mathrm{S(O)NR_2}$[329], $\mathrm{S(O)_2R}$[330], S(=NR)R[331], Se(O)R[332], or $\mathrm{As(O)R_2}$[333].

A fourth method of stabilization is via delocalization as shown generally in equation 130 and exemplified by lithiated imines[334], equation 131.

$$\rangle\mathrm{C{=}X{-}C}\langle\ \ \mathrm{M^+} \longleftrightarrow \rangle\mathrm{(-)C{-}\overset{M^+}{X}{=}C}\langle \tag{130}$$

$$\mathrm{Ph_2C{=}NMe} \xrightarrow[\text{Et}_2\text{O}]{\text{LiNPr}^i_2} \mathrm{Ph_2C{=}NCH_2Li} \longleftrightarrow \mathrm{Ph_2CLi{-}N{=}CH_2} \tag{131}$$

The carbanionic stabilizing ability of sulphur and other second-row elements is considerable and explanations have been based on the involvement of d orbitals (i.e. stabilization is due to the delocalization of the negative charge into low-lying d orbitals)[301]. This theory is now losing favour to one based on polarization[335]. The polarizability of the heteroatom is considered to allow diffusion of the carbanion lone pair into the carbon framework. This diffusion is higher when the carbanion lone pair (or carbon—metal bond) is *syn*-periplanar to the carbon—heteroatom bond. This accounts for the greater reactivity of equatorial over axial hydrogens in acetals and thianes, e.g. as shown by **56** being metallated more readily than **57**[336] and **58** but not **59**[337] reacting with BuLi–tmed in hexane at −50 °C.

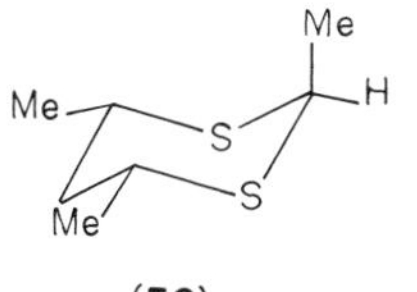

(**56**) 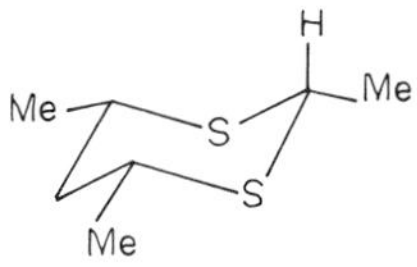 (**57**)

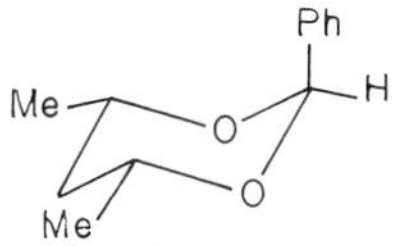

(**58**) 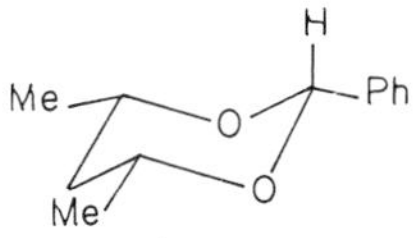 (**59**)

Alkyl substitution at the potential carbanion centre results in a reduced acidity as a consequence of its electron-releasing effect. Thus a MeX group is more easily metallated than a RCH_2X group; see, for example, equation 132. The less vigorous metallating conditions required for PhSMe than those used for $PhSCH_2Pr^i$, equation 133, and the different sites of lithiation of $3\text{-}CF_3C_6H_4SMe$ and $3\text{-}CF_3C_6H_4SCH_2Me$, equation 134, also arise from the reduced acidity on alkyl substitution.

$$Me(CH_2)_5PMe_2 \xrightarrow[\text{(ref. 338)}]{\text{BuLi, pentane}} Me(CH_2)_5PMeCH_2Li \qquad (132)$$

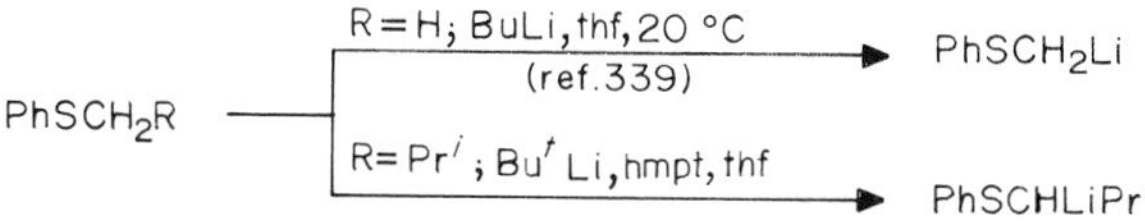

$$PhSCH_2R \begin{cases} \xrightarrow[\text{(ref. 339)}]{R=H;\ \text{BuLi, thf, 20 °C}} PhSCH_2Li \\ \xrightarrow{R=Pr^i;\ Bu^tLi,\ \text{hmpt, thf}} PhSCHLiPr^i \end{cases} \qquad (133)$$

$$3\text{-}CF_3C_6H_4SCH_2R + LiNPr^i_2 \xrightarrow[\text{PhMe}]{\text{thf}} \begin{cases} \xrightarrow[\text{(ref. 340)}]{R=H} 3\text{-}CF_3C_6H_4SCH_2Li \\ \xrightarrow{R=Me} 2\text{-}Li\text{-}3\text{-}(EtS)C_6H_3\text{-}5\text{-}CF_3 \end{cases} \qquad (134)$$

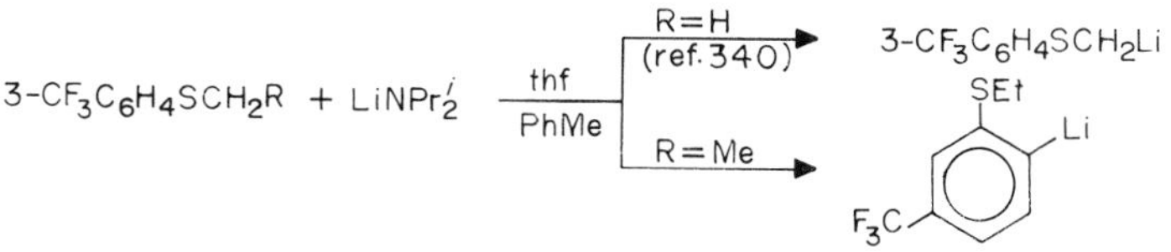

The *ortho*-metallation of $3\text{-}CF_3C_6H_4CH_2SCH_2Me$ is one of a number of reported *ortho*-metallations of ArXCHRR′. Greater electron-attracting groups aid deprotonation, e.g. $3\text{-}CF_3C_6H_4SCH_2OMe$ is more easily lithiated than $PhSCH_2OMe$ by Litmp.

ii. *Metallations of substituted benzenes.* various functional groups in benzenoid compounds have been found to direct the metal into their *ortho* sites[304] (see Table 15). Among the *ortho*-directing groups are the following: NR_2[341], CH_2NR_2[342], $CH(NR_2)_2$[343],

TABLE 15. Metallation of substituted benzenes

Compound	Metallating conditions	Product (yield, %)	Ref.
$PhNMe_2$	BuLi, hexane, reflux, 16 h, or BuLi, hexane, tmed, 25 °C, 4 h	o-$LiC_6H_4NMe_2$	911
	$C_5H_{11}Na$	o-$NaC_6H_4NMe_2$ (18)	912
PhNC	(i) Bu^tLi, −78 °C; (ii) Bu^tLi, tmed	o-$LiC_6H_4N{=}CLiBu^t$	913
$PhNHCOBu^t$	2 equiv. BuLi, thf, hexane, 0 °C	o-$LiC_6H_4NLiCOBu^t$	914
$PhNHCO_2Bu^t$	2 equiv. Bu^tLi, pentane, thf, −78 °C	o-$LiC_6H_4NLiCO_2Bu^t$	915
$PhCH_2NMe_2$	BuLi, Et_2O	o-$LiC_6H_4CH_2NMe_2$	916
Ph–(1,3-dimethylimidazolidin-2-yl) (Me, N, N, Me)	BuLi, tmed, Et_2O, hexane, 25 °C	o-LiC_6H_4–(1,3-dimethylimidazolidin-2-yl) (Me, N, N, Me)	917
$PhCHOHCH_2NMe_2$	2 equiv. BuLi, Et_2O, 20 °C	o-$LiC_6H_4CHOLiCH_2NMe_2$	918
$PhCONEt_2$	Bu^sLi, tmed, thf, −78 °C	o-$LiC_6H_4CONEt_2$	919
PhOMe	BuLi, Et_2O	o-LiC_6H_4OMe	920
	$C_5H_{11}Na$	o-NaC_6H_4OMe (80)	921
$PhOBu^t$	Bu^tLi, cyclohexane, reflux	o-$LiC_6H_4OBu^t$	922
$PhSPr^i$	BuLi, tmed, hexane, 10–25 °C	o-$LiC_6H_4SPr^i$ (81)	923
PhSMe	PhNa	o-NaC_6H_4SMe (45)	924
$PhCH_2OH$	BuLi, tmed, hexane, reflux	o-$LiC_6H_4CH_2OLi$	925
$PhCH(OMe)_2$	Bu^tLi, Et_2O, −78 °C	o-$LiC_6H_4CH(OMe)_2$	926
$PhSO_3H$	2 equiv. BuLi, thf, 0 °C	o-$LiC_6H_4SO_3Li$	927
PhF	BuLi, thf, −50 °C	o-LiC_6H_4F	928
$PhCF_3$	BuLi	o- + m-$LiC_6H_4CF_3$	920
2-MeO-C_6H_4-$CONMe_2$ (MeO, $CONMe_2$)	1.2 equiv. Bu^sLi, tmed, thf, −90 °C	2-OMe-6-Li-C_6H_3-$CONMe_2$ (OMe, $CONMe_2$, Li)	929
2-Cl-C_6H_4-CH_2NMe_2 (Cl, CH_2NMe_2)	BuLi	2-Cl-6-Li-C_6H_3-CH_2NMe_2 (Cl, CH_2NMe_2, Li)	930
1,3-$C_6H_4(CN)_2$ (CN, CN)	$LiNPr^i_2$, thf, −96 °C	2-Li-1,3-$C_6H_3(CN)_2$ (CN, Li, CN)	931
3-Br-C_6H_4-SO_2Ph (Br, S(=O)(=O)Ph)	BuLi	3-Br-2-Li-C_6H_3-SO_2Ph (Br, Li, S(=O)(=O)Ph)	932

TABLE 15. (*Contd.*)

Compound	Metallating conditions	Product (yield, %)	Ref.
OMe, F	BuLi, thf, hexane, – 65 °C	OMe, Li, F	933
MeO, $CONEt_2$	Bu^sLi, tmed, thf, – 78 °C	MeO, Li, $CONEt_2$	919
N, O, Cl	BuLi, thf, – 78 °C	O, N, Li, Cl	934
OCH_2OMe, Me	Bu^tLi	OCH_2OMe, Li, Me	935
F, Me	BuLi, thf	F, Li, Me	928
Cl, OPh	BuLi	Cl, Li, OPh	936
OMe, O, N	BuLi, Et_2O, hexane, – 78 °C	OMe, Li, O, N	937

$CR_2CH_2NR'_2$[344], $CONR_2$[345], pyrazolyl[346], 2-oxazolinyls[347], imines[348], CN[349], OR[350] (including OCH_2OR), $OCONEt_2$[351], CR_2OR[352], $CH_2CH(OR)_2$[353], $CH(OR)_2$[353,354], SR[355], SO_2NR_2[356], SO_2Ar[357], SO_2OR[356a], F[358], Cl[345c], CF_3[359], $PPh_2(=NPh)$[360] (R = alkyl or aryl), as well as those groups which have replaceable hydrogens and are themselves metallated, e.g. $NHCOBu^t$[361], $NHCO_2Bu^t$[362], CONHR[363], CSNHR[364], CR_2OH[365], $CHOHCH_2NR_2$[366], $CHOHNR_2$[367], SO_2NHR[368], SO_3H[356b,369], and those which can combine with RLi prior to substitution[370], e.g. N=C.

$$\mathrm{PhN{=}C} \xrightarrow[\mathrm{Et_2O}]{\mathrm{Bu^tLi,\ tmed,}} \mathrm{PhN{=}C(Li)Bu^t} \xrightarrow[\mathrm{Et_2O}]{\mathrm{Bu^tLi,\ tmed,}} \mathrm{2\text{-}LiC_6H_4N{=}C(Li)Bu^t} \quad (135)$$

For those *ortho*-directing groups which react with RLi, additional equivalents of the metallating agents must be employed. Many of the listed functional groups, e.g. those containing C=O and C=S units, can also react with RLi; to avoid this, appropriate conditions must be used, e.g. low temperatures, poorly nucleophilic reagents or hindered systems. However, an addition of RLi to a C=O unit has been incorporated[371,372] into syntheses, equation 136.

$$\mathrm{PhCONMe_2} \xrightarrow[-70\,^\circ\mathrm{C}]{\mathrm{MeLi,thf}} \mathrm{PhC(Me)(OLi)NMe_2} \xrightarrow[\text{hexane (ref. 371)}]{\mathrm{BuLi,}-70\,^\circ\mathrm{C}}$$

$$\mathrm{2\text{-}Li\text{-}C_6H_4C(OLi)(Me)NMe_2} \xrightarrow[\mathrm{(ii)\ H_3O^+}]{\mathrm{(i)\ E^+}} \mathrm{2\text{-}E\text{-}C_6H_4COMe} \quad (136)$$

Most of these reactions have involved lithium reagents; only very limited work has been performed with the other alkali metals. Studies with organosodiums and -potassiums have involved aryl ethers, thioethers and amines[1,6,173].

Metallations of benzene derivatives containing *ortho*-directing groups are more readily achieved than those of benzene itself, irrespective of the nature of the electronic effects of the directing group. *Ortho*-directing groups having electron-withdrawing effects will clearly enhance the acidity of *ortho*-hydrogens and will thereby facilitate metallations at the *ortho* sites via an acid-base mechanism. Many directing groups have donor properties and are able to coordinate to the RM metallating agent and hold it in a suitable position for reaction at an *ortho* site (such coordination has been detected by ^{1}H n.m.r. spectroscopy for $2,4,6\text{-}Pr^i_3\text{-}C_6H_2CONMe_2\text{–}Bu^sLi$ interactions)[307]. Metallations then proceed via a coordination mechanism. The acid–base and coordination mechanisms are two extremes and their relative importance in a given situation depends on the group. Groups acting as *ortho* directors via electron-withdrawing effects include trihalomethyl and halo groups, while $CR_2NR'_2$, $CR_2CH_2NR'_2$, CR_2OR', and $CR(OR')_2$ are examples of good coordinating groups. For other *ortho*-directing groups both effects operate. Clearly the most successful *ortho*-directing groups possess both electron-releasing groups and are good donors. A consequence of coordination of the metallating agent to the directing group will be an enhancement of the electron-withdrawing effect, e.g. coordination of RM to OMe (a − I and + M group) weakens at least, the + M effect and so leaves the inductive withdrawing effect to aid substitution at the *ortho* site. Substitution occurs at only one of the *ortho* sites of a donor directing group, even in the presence of excess metallating agent, e.g. as found with $PhCH_2NMe_2$[342b]. The donor group should be complexed within the product organolithium aggregate and will be unable to assist in further metallations. Such a restriction would not apply to a strongly electron-withdrawing group and, for example, O,O'-dimetallations of $PhSO_3R$ have been reported[356b].

Sequences of *ortho*-directing abilities of group, Y, have been established from both intramolecular competitions (e.g. using *p*-YC_6H_4OMe[373] and *p*-$YC_6H_4CONEt_2$)[345c] and intermolecular competition (e.g. using PhY and 2-oxazolonyl-Ph)[374] for alkyllithiums under suitable conditions. While there are differences between the sequences found (see, for example, ref. 375), the following listing of *ortho*-directing abilities is considered to be acceptable: $SO_2NR_2 > SO_2Ar >$ 2-oxazolinyl $>$ C(O)NHR, C(S)NHR, CH_2NMe_2 $> CR(OLi)CH_2NMe_2 > OMe > OAr > NHAr > SAr > NRAr > NMe_2 > CRR'OLi$. The OCH_2OMe group is a considerably better directing group than the OMe group (owing to the chelating effect of the two oxygens)[350a].

It must be stressed that the listing above applies to kinetically controlled metallations. When thermodynamic control applies, differences can arise, as shown in equation 137[373].

p-$MeOC_6H_4CH_2NMe_2$ —(BuLi, hexane, 27 °C)→ 2-Li-4-MeO-$C_6H_3CH_2NMe_2$ (Li *ortho* to CH_2NMe_2)
p-$MeOC_6H_4CH_2NMe_2$ —(BuLi, tmed, 27 °C)→ 3-Li-4-MeO-$C_6H_3CH_2NMe_2$ (Li *ortho* to OMe) (137)

Lithiation of *p*-$MeOC_6H_4CH_2NMe_2$ by $(BuLi)_6$ aggregate (a good Lewis acid) occurs *ortho* to CH_2NMe_2 (kinetically controlled conditions—a consequence of butyllithium coordination to the nitrogen; in contrast the BuLi–tmed complex, a poor Lewis acid with a poor residual coordinating ability, attacks at the most acidic ring position, that is, *ortho* to the MeO group (thermodynamically controlled conditions). Thioanisole is kinetically lithiated at an *ortho* ring position; this product then rearranges to the substituted side chain product, $PhSCH_2Li$[376]. In contrast, $PhSeCH_2Pr^i$ is metallated kinetically in the side chain (by Bu^tLi–hmpt in thf at -78 °C), but at 0 °C $PhSeCHLiPr^i$ rearranges[377] to the ring-lithiated product, *o*-$LiC_6H_4SeCH_2Pr^i$. The latter can be obtained directly from $PhSeCH_2Pr^i$ using BuLi–dabco or BuLi–tmed.

A study has been made of the thermodynamic acidities of monosubstituted benzenes, PhY; there was no correlation between the activating effects[305] of the Y groups and the pK_a values of PhY; some pK_a values of PhY are > 40.3 ($Y = CH_2NMe_2$), 39.0 (OMe), 38.2 (SO_2NMe_2), 38.1 (2-oxazolinyl), 38.1 (CN), 37.8 ($CONPr^i_2$), and 37.2 ($OCONEt_2$).

Butyllithium in hexane[378] metallates $PhCH_2NMe_2$ at the *ortho* site, but the more reactive BuLi–Bu^tOK provides $PhCHKNMe_2$. Butyllithium in hexane also reacts at the *ortho* sites of $PhNMe_2$ and $PhCH_2CH_2NMe_2$; in contrast, it reacts with $PhCH_2CH_2CH_2NMe_2$ at the benzylic site[379].

Lithiations of *o*-MeC_6H_4Y, in which Y is an *ortho*-directing group, such as OR, CR_2CONR_2, SO_2NR_2, 2-oxazolinyl, or CH_2NMe_2, occurs partially (for $Y = OR$ or NR_2) or completely (other cited Y groups) at the methyl group to give the benzylic carbanion species, *o*-Y-$C_6H_4CH_2Li$. The *ortho*-directing groups clearly enhance the ease of metallation of the methyl groups. For *p*-$MeC_6H_4CONPr^i_2$, reaction can occur *ortho* to the amide group or at the methyl group[380], equation 138.

$CONNPr^i_2$-C_6H_4-Me (para) → (Bu^sLi, tmed, thf, −78 °C) → 2-Li-4-Me-$C_6H_3CONPr^i_2$

→ ($LiNPr^i_2$, thf, 0 °C) → 4-$LiCH_2$-$C_6H_4CONPr^i_2$ (138)

Halobenzenes can take part in metallation or halogen–metal exchanges reactions. Fluoro- and chlorobenzenes are metallated *ortho* to the halogen, with metallation becoming easier with increasing halogen substitution; such products decompose to give benzynes via elimination of MX, but can be obtained in good yields at low temperatures, e.g. below $-50\,^{\circ}C$ for o-F-C_6H_4Li but much lower for the chloro analogue[381]. However, for bromo- and iodobenzenes, the most probable reactions are metal–halogen exchanges. Metal–halogen exchanges are also more probable using Bu^tLi rather than MeLi or PhLi, equation 139[382].

1,2,3,4-$Cl_4C_6H_2$ → (RLi, thf, −70 °C) → (60) + (61) (139)

R	[60]:[61]
Bu^t	0:100
Me	93:7
Ph	89:11

Halophenyl ethers, XC_6H_4OR, provide mixtures of products (e.g. arising from metal–halogen exchanges and metallations) and which depend on various factors, including X, orientation of the substituents and the lithiating conditions[5].

Another type of reaction undergone by halogenobenzenes is alkylation; this can occur even for fluoro compounds, e.g. 1, 3, 5-$F_3C_6H_3$ using Bu^tLi but not BuLi[383].

Metallations of 1, 2-, 1, 3- and 1, 4-disubstituted and more highly substituted compounds have been well studied[188]; some have already been mentioned. The site of lithiation, under kinetically controlled conditions, is generally governed by the relative directing abilities of the groups present, i.e. *ortho* to the more effective directing group in 1, 2- and 1, 4-disubstituted compounds. For 1, 3-disubstituted compounds in which both groups have *ortho*-directing abilities, even weak ones, metallations occur to the greatest extents at the 2-sites; see equation 140. Some exceptions to this include 1, 3-$(CF_3)_2$-C_6H_4[384], equation 141, and 1, 3-$(BuNHCO)_2C_6H_4$[188].

$$3\text{-}YC_6H_4CN \xrightarrow[\text{y = CN or Cl}]{\text{Litmp}} 2\text{-Li-}3\text{-}YC_6H_3CN \qquad (140)$$

(ref. 188)

The site of metallation of 3-Me_2N-$C_6H_4OCH_2OMe$ depends on the metallating agent[385], equation 142.

$$\text{1,3-}(CF_3)_2C_6H_4 \xrightarrow[\text{hexane}]{\text{BuLi, }Et_2O} \textbf{(62)} + \textbf{(63)} \qquad (141)$$

[62]:[63] = 3:2

$$\text{3-}Me_2NC_6H_4OCH_2OMe \xrightarrow{Bu^tLi, Et_2O, 0\,°C} \textbf{(64)} + \textbf{(65)} \qquad (142)$$

[64] : [65] = 99:1

BuLi, hexane, 20 °C → 64 + 65

[64]:[65] = 2:98

Another compound to undergo[386] lithiation at different sites is 3-MeO-C_6H_4Me, equation 143. The ratios of the products **66**:**67** depend on the size of the butyllithium aggregate, e.g. [**66**]:[**67**] decreases in the sequence $(BuLi)_6$ in hydrocarbons {[**66**]:[**67**] = 9:1} > $(BuLi)_4$ in hydrocarbon > $(BuLi)_2$ in Et_2O > monomeric BuLi–tmed {[**66**]:[**67**] = 13:12}.

Of interest, the sites of ring lithiations of arenechromium tricarbonyl complexes are different from those of the uncomplexed arenes, e.g. metallation[387] of 4-F-$C_6H_4NMe_2$ provides 2-Li-4-F$C_6H_3NMe_2$, whereas [4-F-C_6H_4NMeSiPr^i_3)Cr$(CO)_3$] with Bu^tLi in thf at −78 °C forms [(1-Me_2N-3-Li-4-F-C_6H_3)Cr$(O)_3$], equation 144. Both the NMeSiPr^i_3 and OSiPr^i_3 groups *meta*-direct lithium in [(arene)Cr$(CO)_3$] complexes[387].

$$\text{3-MeOC}_6H_4Me \xrightarrow{BuLi} \textbf{(66)} + \textbf{(67)} \qquad (143)$$

$$[(OC)_3Cr(4\text{-F-}C_6H_4NMeSiPr^i_3)] \longrightarrow [(\text{Li, F-}C_6H_3NMeSiPr^i_3)Cr(CO)_3] \qquad (144)$$

Lithiations of [(1-$HOCH_2$-3-MeOC_6H_4)Cr$(CO)_3$] occur primarily at position 4; the selectivity increases with increasing bulk of the organolithium agent[388]. Many examples of *ortho*-lithiation of arene derivatives are given in ref. 188.

iii. *Metallations of heteroaromatics*[188]. Metallation of five-membered heteroaromatics (heteroatom O, N, S, or Se) proceed very readily and far milder conditions can be employed than are required for benzenoid derivatives. One consequence of the milder metallating conditions is that more sensitive substituents, e.g. iodo, can survive the metallations of five-membered heteroaromatics. Metallations occur *ortho* to an hetero atom if free.

a. Thiophenes. Thiophene can be mono-[173,188,389] or dimetallated[390], depending on the conditions. Lithiation of thiophene derivatives have had considerable attention and

2,5-dimetallothiophene ← (BuLi, tmed, hexane, reflux) — thiophene — ($BuLi, Et_2O$ −35 °C or $LiNPr^i_2$, Et_2O, 0 °C or PhNa or BuM, thf, dme −60 °C (M′ = K or Cs)) → 2-metallothiophene (145)

some general conclusions regarding the of lithiation can be made, as follows. (i) 2-Substituted thiophenes are completely lithiated in the 5-position, with the exceptions of pyridyl[391] and 2-oxazolinyl groups[392]. (ii) Thiophenes bearing *ortho*-directing groups,

2-(2-thienyl)pyridine —(BuLi, 0 °C (ref. 391))→ **(68)** (3-lithio) + **(69)** (5-lithio) (146)

in thf: [68]:[69] = 4:93
in Et_2O: [68]:[69] = 62:13

including cyano[393], bromo[394], and iodo[395] groups, in the 3-position are lithiated predominantly at least in the 2-position, e.g. equation 147, while other 3-substituted

3-iodothiophene + 1-methyl-2-lithiopyrrole —(Et_2O, tmed, −10 °C)→ **(70)** (3-iodo-2-lithiothiophene) + **(71)** (3-iodo-5-lithiothiophene) (147)

[70] : [71] = 4:1

thiophenes (including 3-alkyl or 3-aryl derivatives) are mainly lithiated in the 5-position[396]. (iii) 3,4- and 2,5-disubstituted thiophenes are lithiated *ortho* to the more powerful directing group[397].

Benzothiophenes and dibenzothiophenes react at the α and the *ortho* sites, respectively[398]. Much more limited studies with sellenophenes and tellurophenes suggest similar patterns[188].

b. Furans. Although furan is not as reactive as thiophene, it can be 2,5-dimetallated[390a] using BuLi–tmed; monometallation has been achieved with BuLi in refluxing Et_2O, and with $PhCH_2Na$, BuK, or BuCs in thf, dme at −60 °C[173,399]. As for thiophene, *ortho*-directing groups, including Br[400], in the 3-position of furan direct lithium into the 2-position, while 2-substituted furans react at the 5-position. An exception to the latter is the

3- and 5-lithiation of 2-[2-(4,4-dimethyl)oxazolinyl] furan[401], equation 148, and 2-(2-pyridyl)furan (**72**)[402]. The 5-lithio product of **72** is the thermodynamic product.

(148)

(**73**) (**74**)

in Et_2O at 0 °C : [**73**] : [**74**] = 1:3
in DME at −78 °C: [**73**] : [**74**] = 9.5:1

Lithiations of benzofurans and dibenzofurans occur at α and *ortho* sites, respectively[398b,403].

c. Pyrroles. These are less readily lithiated than furan, but 1-substituted pyrroles are mono- and dilithiated by BuLi–tmed in the α-positions[404]. However, bulky 1-substituents can result also in some 2,4-dilithiation[405]. 1-Phenylpyrroles can undergo metallations in the phenyl ring (*ortho* site) as well as at the α-pyrrole site[406]. Metallations of 1-Me_2N-[407] and 1-Bu^tOCO-pyrroles[408] occur more readily than 1-alkylpyrroles and since these 1-substituents can be replaced by H they are particularly useful in pyrrole syntheses. 2-(2-Oxazolinyl)pyrroles are metallated at the 5- and 3-sites in proportions dependent on the 1-substituent and on the reaction conditions[401,409].

In the indole series, 1-$MeOCH_2$[410] and 1-$PhSO_2$[411] derivatives enhance the ease of α-metallations compared with 1-Me-indole. The reduced *ortho*-directing influence of the ring nitrogen in pyrroles can be seen in the lithiation of 5-MeO-1-Me-indole occurring *ortho* to the MeO group whereas that of 5-MeO-benzofuran happens at the α site[412].

Lithiations of five-membered heteroaromatics containing two or more heterocarbons have also been reported[188]. Various examples of lithiation of heteroatomic derivatives are given in ref. 188.

d. Pyridines. A frequently met reaction of RLi with pyridine derivatives is addition to provide 1,2- or 1,4-adducts[5]. However, lithiations of substituted pyridines have been reported recently; these include *ortho*-lithiations to halogens[413], $CONPr^i_2$[414], $CONHCH_2R$[415], $NHCOBu^t$[416], 2-oxazolinyl[417], 1-SO_2-piperidyl[418], OR[419], and OCH_2OMe[420].

Lithiations of 2-substituted pyridines (2-Y-C_5H_4N; e.g. Y = F, Cl, Br, $CONPr^i_2$, $CONHCH_2R$, $NHCOBu^t$) and 4-substituted pyridines (4-Y-C_5H_4N; e.g. Y = halogen, $CONPr^i_2$, 2-oxazolinyl, $NHCOBu^t$) both occur in the 3-position. For 3-substituted pyridines (3-Y-C_5H_4N) lithiation has been reported to occur in the 2-position [for Y = Br, $CONPr^i_2$, 2-oxazolinyl, $SO_2\overline{N(CH_2)_4CH_2}$, OR (R = Et, Me or $PhCH_2$)], in the 4-position (for Y = $NHCOBu^t$) and in both sites for (Y = OCH_2OMe, F, Cl).

The complex between pyridine and $(CF_3)_2CO$ is regiospecifically lithiated at the 2-position[421a] by Litmp in thf, Et_2O at −107 °C. Pyridine has also been reported to be metallated directly using BuLi–Bu^tOK[421b]. Polychloropyridines are sufficiently acidic to

BuLi, Et_2O, −75 °C (ref. 422) (149)

be easily metallated, e.g. equation 149. It should be borne in mind that halopyridines can also undergo lithium–halogen exchanges.

TABLE 16. Vinyl metallation of functionally substituted alkenes

Alkene	Conditions	Product (yield, %)	Ref.
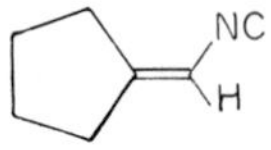 NC, H	BuLi, thf, Et_2O, pentane, $-110\,^\circ C$	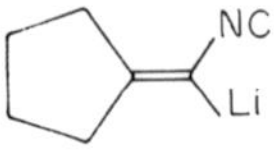 NC, Li (63)	938
$CH_2=CHOMe$	Bu^tLi, thf, pentane, $-65\,^\circ C$	$CH_2=CLiOMe$ (quant)	939
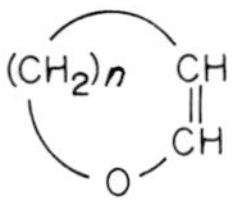 $(CH_2)_n$ CH=CH, O	Bu^tLi, thf, $-78\,^\circ C$	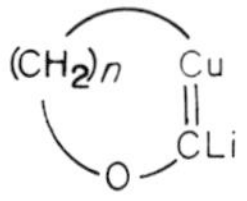 $(CH_2)_n$ Cu=CLi, O ($n = 2$, 67) ($n = 3$, 68)	940
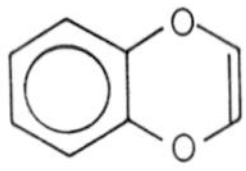 O, O	BuLi, thf, hexane, $-78\,^\circ C$	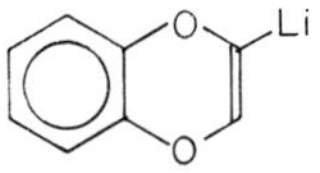 O, Li, O (80)	941
$CH_2=CHSEt$	Bu^sLi, hmpt, thf, $-78\,^\circ C$	$CH_2=CLiSEt$ (90)	942
(Z)-$EtSCH=CHSEt$	$LiNPr^i_2$, thf, $-80\,^\circ C$	(Z)-$EtSCH=CLiSEt$ (100)	943
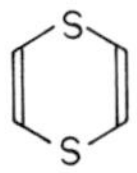 S, S	BuLi, thf, hexane, $-110\,^\circ C$	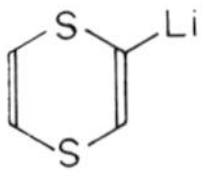 S, Li, S (98)	944
(E)-$PhCH=CHSO_2Ph$	MeLi, thf, $-95\,^\circ C$	(E)-$PhCH=CLiSO_2Ph$ (79)	945
$CH_2=CHCl$	BuLi, thf, Et_2O, pentane, $-110\,^\circ C$	$CH_2=CClLi$ (99)	946
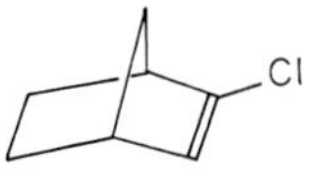 Cl	BuLi, thf, pentane, $-45\,^\circ C$	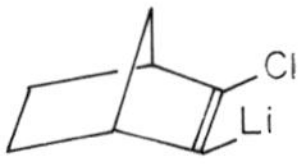 Cl, Li (70)	947
(E)-$ClCH=CHCl$	BuLi, thf, Et_2O, hexane, $-110\,^\circ C$	(E)-$ClCH=CClLi$ (99)	946
$CF_2=CH_2$	Bu^sLi, thf, $-110\,^\circ C$	$CF_2=CHLi$ (100)	948
$CF_2=CFH$	BuLi, Et_2O, $-100\,^\circ C$	$CF_2=CFLi$ (79)	949
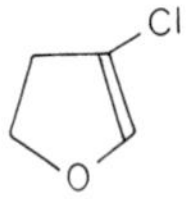 Cl, O	BuLi, thf, hexane, $-78\,^\circ C$	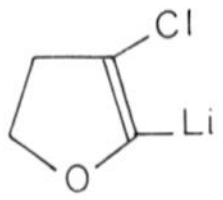 Cl, Li, O (78)	950
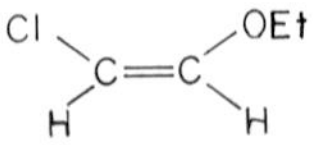 Cl(H)C=C(OEt)(H)	BuLi, thf, hexane, $-100\,^\circ C$	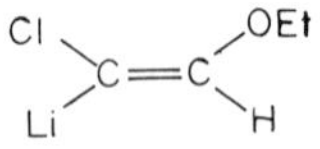 Cl(Li)C=C(OEt)(H) (40)	951
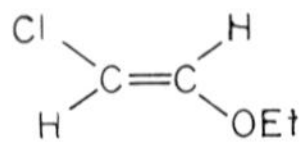 Cl(H)C=C(H)(OEt)	BuLi, thf, hexane, $-100\,^\circ C$	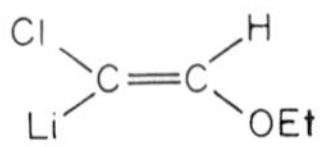 Cl(Li)C=C(H)(OEt) (100)	951

TABLE 16. (*Contd.*)

Alkene	Conditions	Product (yield, %)	Ref.
PhS(H)C=C(Me)CO_2Me	$LiNPr^i_2$, −80 °C	PhS(Li)C=C(Me)CO_2Me (100)	952
MeO(H)C=C(H)CO_2Me	$LiNPr^i_2$, thf, −90 °C	MeO(Li)C=C(H)CO_2Me	953

iv. *Metallation of substituted alkenes.* Vinyl lithiation of an alkene does not occur unless the alkene is strained, see for example equations 86 and 87, or is made acidic (or reactive) by appropriate functional groups, such as RO[423], RS[424], RSe[425a,b], RTe[425c], RSO[426], RSO_2[427], CN[428,429], CO_2R[429,430], $CONR_2$[429], NC[431], NR_2[432], and halogens[433] (see Table 16). Additions to alkenes (even polymerization) and allylic deprotonation can occur instead of vinyl deprotonations (allylic deprotonations will be referred to later). The use of $LiNR_2$ rather than LiR promotes deprotonation over addition, e.g. as shown with RSCH=CH_2 and RSeCH=CH_2.

Many metallations of heterosubstituted alkenes arise with retention of configuration; however, isomerization to thermodynamically more stable products have been noted, for example equations 150 and 151.

Ph(H)C=C(CN)H —[$LiNPr^i_2$, −80 °C, Et_2O, pentane (ref. 434)]→ Ph(H)C=C(CN)Li —[thf or crown ether]→ Ph(H)C=C(Li)CN (**75**) (150)

H(piperidino)C=C(CN)H (**76**, RCN) —[$LiNPr^i_2$, −113 °C, thf, pentane (ref. 432)]→ Li(piperidino)C=C(CN)Li —[−78 °C]→ H(piperidino)C=C(CN)Li (151)

The product **75** can be achieved directly from (*E*)-PhCH=CHCN and $LiNPr^i_2$ at −113 °C.

Many of the vinyllithium isomerizations, exemplified by equation 151, occur when there are two functional groups present at the alkene centre—one a good donor group (which complexes the lithiating agent and so directs attack to its α-position to give the kinetic product) and the other group, a strongly electron-withdrawing group, which increases the acidity of its α-proton, replacement of which gives the thermodynamic product[435]. In contrast to equation 151, the ketone, the amide, and the ester analogues of **76**, i.e. RCOR′, $RCONR'_2$, RCO_2R', provide configurationally stable vinyllithiums, equation 152.

H(N)C=C(C(=O)Y)H —[Bu^tLi or $LiNPr^i_2$, thf, −113 to −65 °C]→ Li(N)C=C(C(=O)Y)H (152)

Y = Ph, NEt_2 or OEt

Intramolecular coordination (even dipole stabilization) can be important factors in stabilizing these vinylic carbanions[432]. As shown in equations 153 and 154, intramolecular

$$\text{Ph(H)C=CH–C(=O)OEt} \xrightarrow[-80\,^{\circ}\text{C}]{\text{LiNPr}^i_2\text{, thf}} \text{Ph(H)C=C(Li)–C(OEt)=O} \quad (153)$$

$$\text{H(Ph)C=CH–C(=O)OEt} \xrightarrow[-100\,^{\circ}\text{C}]{\text{LiNPr}^i_2\text{ thf}} \text{Li(Ph)C=CH–C(=O)OEt} \quad (154)$$

coordination can control the sites of lithiation; even addition of thf to solutions of the carbanions produced in equations 153 and 154 do not lead to isomerizations[430].

The directing influence of RS is superior to EtO but inferior to CN, as shown by the metallations of (*Z*)-RSCH=CHOEt[424a] and (*Z*)-EtSCH=CHCN[432b] by Bu^{t}Li in thf, initially providing (*Z*)-RSCLi=CHOEt and (*Z*)-EtSCH=CLiCN (**77**). Subsequent isomerization of these products arise; **77** even isomerizes at −113 °C.

The same vinylic products, (*E*)-RCH$_2$CH=CLiSOAr, are obtained[426b] on metallation of either (*Z*)- or (*E*)-RCH$_2$CH=CHSOAr, equation 155. Equation 155 points to the low

$$\underset{[(Z)\text{or}(E)-\mathbf{78}]}{\text{RCH}_2\text{CH=CHSOAr}} \xrightarrow[-78\,^{\circ}\text{C}]{\text{LiNPr}^i_2\text{, thf}} \text{RCH}_2\text{(H)C=C(Li)–S}^{+}\text{(O}^{-}\text{)Ar} \quad (155)$$

R = C_5H_{11} or C_8H_7 Ar = Ph or *p*-MeC_6H_4

barriers to isomerization of α-SO-substituted carbanions as well as to the significant α-directing influence of this group, so much so that it directs metallations to a vinylic rather than an allylic site in **78**.

As shown by this example, the conflict between allylic and vinyl deprotonation is not always won by the former. The compounds RCH(OMe)CH=CHSBu^t (**79**) can be vinylic or allylic deprotonated, depending on the geometry[436]. As shown by equation 156,

$$\underset{\substack{R = \text{Me, Pr or } C_5H_{11} \\ [(Z)-\mathbf{79}]}}{\text{RCH(OMe)(H)C=C(H)SBu}^t} \begin{cases} \xrightarrow[-78\,^{\circ}\text{C}]{\text{Bu}^s\text{Li, thf}} \text{RCH(OMe)(Li)C=C(H)SBu}^t \\ \xrightarrow[\text{thf, PhH, }-78\,^{\circ}\text{C}]{\text{BuLi, Bu}^t\text{OK}} \text{RCH(OMe)(H)C=C(K)SBu}^t \end{cases} \quad (156)$$

$$\underset{\substack{(E)-\mathbf{79}\text{ or} \\ (Z)-\mathbf{79};\ R=H}}{\text{RCH(OMe)(H)C=C(H)SBu}^t} \xrightarrow{\text{Bu}^s\text{Li, thf or LiNPr}^i_2} [\text{MeO(R)C}\cdots\text{CH}\cdots\text{CHSBu}^t]^{-}\ \text{Li}^+ \quad (157)$$

metallation at either vinyl site can be achieved; with the more powerful system, BuLi–Bu^{t}OK, metallation occurs at the most acidic protons that is, α to the Bu^{t}S group.

A further example is shown by metallations of PhSCH=CMeCO$_2$R (**80**), equation 158[437]. Clearly, several factors are important in deciding the site and type of metallation.

$$\begin{array}{l} \text{PhSCH=CMeCO}_2\text{R } (\mathbf{80}) \xrightarrow[\text{LiNPr}^i_2\text{, thf, }-80\,^\circ\text{C}]{(E)\text{-isomer}} \text{(PhS)(Li)C=C(Me)(CO}_2\text{R)} \\ \text{PhSCH=CMeCO}_2\text{R } (\mathbf{80}) \xrightarrow[\text{2.2 equiv. Litmp, thf, }-97\,^\circ\text{C}]{(Z)\text{-isomer, R = H}} \text{PhS}\text{---}\text{Li}^+ \text{ allyl anion, CO}_2\text{Li} \end{array} \quad (158)$$

(*E*)-**80**

Slight changes in structures can change matters, e.g. compare (*E*)-**80** with (*E*)-PhSCH=CMeCONHBu^t, which gives the allylic dianion PhSCH$\overline{\cdots}\bar{\text{C}}$(CONLiBu^t)$\overline{\cdots}CH_2^-Li^+$ with 2.2 equiv. of LiNPr$^i_2$ in thf at $-80\,^\circ$C.

Both vinylic and allylic deprotonations of ArSeCH=CHMe (*Z* or *E*; Ar = Ph or *m*-CF$_3$-C$_6$H$_4$) occur[425b] using LiNPr$^i_2$ in thf at $-78\,^\circ$C; the vinylic products, ArSeCLi=CHMe, isomerize to the allylic compounds, ArSeCH$\overline{\cdots}\bar{\text{C}}H\overline{\cdots}CH_2$, Li$^+$, on standing. The related compounds, ArSeCH=CHR (R = Et or Pr (**81**) and ArSeCH=CMe$_2$ (**82**) illustrate the effect of changes in structure. Using Litmp in thf at $-50\,^\circ$C **81** is solely vinyl deprotonated (to ArSeCLi=CHR isomers) while **82** provides the allyl anion, ArSeCH$\overline{\cdots}\bar{\text{C}}Me\overline{\cdots}CH_2Li^+$.

Other examples of alkenes undergoing preferential vinyl deprotonation (at the site indicated in italics) include the following: PhS(O)C*H*=CHCH$_2$OMe (by LiNPr$^i_2$)[438], BuOC*H*=CHCH$_2$OBu (by BuLi–hmpt)[439], C*H*$_2$=CHCH$_2$NHBu^t (**83**) by BuLi–tmed[440], C*H*$_2$=CHCH$_2$NHSiMe$_3$[441], and PhC*H*=CHCH$_2$NMe$_2$ (**84**) by BuLi[442]. In contrast, (*E*)-RSCH=CHCH$_2$XR$'$ (**85**; X = S or O)[442,443] are metallated at the allylic site. The differences in the results with **83** and **85** is rationalized in terms of N (i) being less electronegative than O or S (and hence being a less powerful α-directing group), (ii) being able to depolymerize the BuLi aggregate, and (iii) being the most powerful internal donor atom.

In contrast to **83** and **84**, PhCH=CHCH$_2$NHBu^t[440] and CH$_2$=CHCH$_2$OH undergo RLi additions rather than deprotonation[444]. Table 17 lists some functionally substituted allylic alkali metal derivatives prepared by metallation.

v. *Metallations of substituted alkynes, allenes and conjugated alkenes.* The parent hydrocarbons are themselves readily metallated; compounds substituted by the groups mentioned in the previous sections are even easier to deprotonate. Some examples are shown in Table 18.

C. Transmetallation Reactions

Transmetallations provide useful routes to organoalkali metal compounds. These may involve the reaction of an alkali metal with an organic derivative of another element, especially of mercury, equation 159 (M$'$ = Hg) or an exchange between an organoalkali metal and an organic derivative of another metal, such as mercury, silicon, germanium, tin, lead, arsenic, antimony, or bismuth, equation 160.

$$n\text{M} + \text{R}_n\text{M}' \rightleftharpoons n\text{RM} + \text{M}' \quad (159)$$

$$n\text{R}'\text{M} + \text{R}_n\text{M}' \rightleftharpoons n\text{RM} + \text{R}'_n\text{M}' \quad (160)$$

TABLE 17. Formation of functionally substituted allylalkalimetal compounds by Metallation

Compound	Metallating conditions	Product (yield, %)	Ref.
(1-piperidyl)-$CH_2CH{=}CH_2$	2 equiv. Bu^sLi, thf, − 78 to − 10 °C	(1-piperidyl)-$CH{\cdots}C\bar{C}H{\cdots}CH_2Li^+$ (7)	954
$Me_2NC(CN){=}CHCH_3$	$LiNPr^i{}_2$, thf, hexane, − 78 °C	$Me_2NC(CN){\cdots}C\bar{H}{\cdots}CH_2Li^+$ (46)	955
(1-piperidyl)-$CH(CN)CH{=}CHPh$	$LiNPr^i{}_2$, thf, − 78 °C	(1-piperidyl)-$C(CN){\cdots}\bar{C}H{\cdots}CHPh$ (87)	956
$PhCH_2CH{=}C(CN)NEt_2$	$LiNPr^i{}_2$, thf, − 60 °C	$PhCH{\cdots}\bar{C}H{\cdots}C(CN)NEt_2Li^+$	957
$(RRN)(Me)C{=}C(H)(CO_2Me)$	$LiNPr^i{}_2$, thf, tmed, − 78 °C	$CH_2{\cdots}\bar{C}(NRR'){\cdots}CHCO_2MeLi^+$	958
$(Me)(H)C{=}C(NMePh)(CO_2Me)$	$LiNPr^i{}_2$, thf, − 78 °C	$CH_2{\cdots}\bar{C}H{\cdots}C(CO_2Me)NPhMeLi^+$	959
$MeCH{=}C(NR_2)P(O)(OEt)_2$	Bu^tLi, thf, − 78 °C	$CH_2{\cdots}C\bar{H}{\cdots}C(NR)_2P(O)(OEt)_2Li^+$	960
$CH_2{=}CHCH_2NHCOBu^t$	2 equiv. $LiNPr^i{}_2$, diglyme, − 78 °C	$CH_2{\cdots}\bar{C}H{\cdots}CHNLiCOBu^tLi^+$ (85)	961
$CH_2{=}CHCH_2NC$	BuLi, thf, − 70 °C	$CH_2{\cdots}\bar{C}H{\cdots}CHNCLi^+$	962
$CH_2{=}CHCH_2NO_2$	2 equiv. BuLi, hmpt, thf, − 80 °C	$CH_2{\cdots}\bar{C}H{\cdots}CHNO_2{}^{2-}2Li^+$ (82)	963
$CH_2{=}CHCH_2N(NO)Bu^t$	$LiNPr^i{}_2$, thf, − 78 °C	$CH_2{\cdots}\bar{C}H{\cdots}CHN(NO)MeLi^+$ (95)	964
$(Me)(H)C{=}C(H)CH_2P(O)Ph_2$	BuLi, thf, − 70 °C	$MeCH{\cdots}\bar{C}H{\cdots}CHP(O)Ph_2Li^+$ (72)	965
$CH_2{=}CHCH_2NMeP(O)(NMe_2)_2$	BuLi, thf, − 50 °C	$CH_2{\cdots}\bar{C}H{\cdots}CHNMeP(O)(NMe_2)_2Li^+$	966
$CH_2{=}CHCH_2P(O)(NMe)_2$	PhNa, PhMe, thf, 0 °C BuLi, thf, − 78 °C	$CH_2{\cdots}\bar{C}H{\cdots}CHP(O)(NMe_2)_2M^+$	967

Substrate	Conditions	Product	No.
$MeCH{=}CHCH_2P(O)(OEt)_2$	BuLi, thf, $-60\,°C$	$MeCH{\cdots}\bar{C}H{\cdots}CHP(O)(OEt)_2Li^+$ (9)	968
$CH_2{=}CHCH_2OMe$	Bu^sLi, thf, $-65\,°C$	$CH_2{\cdots}\bar{C}H{\cdots}CHOMeLi^+$ (93)	969
$Me(H)C{=}C(OSiBu^tMe_2)Ph$	Bu^tLi, tmed, hexane, $-10\,°C$	$Me_2Bu^tSi{-}O{-}C(Ph){=}CHLi$	970
$CH_2{=}CHCH(OR)_2$	Bu^sLi, thf, Et_2O, pentane, $-95\,°C$	$CH_2{\cdots}\bar{C}H{\cdots}C(OR)_2Li^+$	971
$CH_2{=}CH{-}CH(OMe)P(O)(OEt)_2$	$LiNPr^i{}_2$, thf, $-70\,°C$	$CH_2{\cdots}\bar{C}H{\cdots}C(OMe)P(O)(OEt)_2Li^+$	972
$MeCH{=}CHCH(CN)OSiMe_3$	$LiNPr^i{}_2$, thf, $-78\,°C$	$MeCH{\cdots}\bar{C}H{\cdots}C(CN)OSiMe_3Li^+$ (87)	973
$CH_2{=}CHCH_2SH$	2 equiv. BuLi, tmed, thf, hexane, $0\,°C$	$CH_2{\cdots}\bar{C}H{\cdots}CHSLiLi^+$	974
(2-pyridyl)-$SCH_2CH{=}CH_2$	BuLi, thf, $-30\,°C$	(2-pyridyl)—$SCH{\cdots}\bar{C}H{\cdots}CH_2Li^+$	975
$CH_2{=}CHCH_2SCH_2CO_2Me$	(i) $LiNPr^i{}_2$; (ii) Bu^sLi, thf, $-78\,°C$	$CH_2{\cdots}\bar{C}H{\cdots}CHSCHLiCO_2MeLi^+$	976
$PhCOCMe{=}C(SMe)_2$	$LiNPr_2{}^i$	$CH_2{\cdots}C(\bar{C}OPh){\cdots}C(SMe)_2Li^+$	977
2-vinyl-1,3-dithiane (2-($CH{=}CH_2$)-1,3-dithiane)	BuLi, thf, -78 to $0\,°C$	1,3-dithian-2-ylidene${\cdots}\bar{C}H{\cdots}CH_2Li^+$	978
$MeCH{=}CHCH_2SCONMe_2$	$LiNPr^i{}_2$, thf, $-78\,°C$	$MeCH{\cdots}\bar{C}H{\cdots}CHSCONMe_2Li^+$	979
$MeCH{=}CHCH_2S{-}C(S)NMe_2$	$LiNPr^i{}_2$, thf, $-78\,°C$	$MeCH{\cdots}\bar{C}H{\cdots}CHSC(S)NMe_2Li^+$	980
$CH_2{=}CHCH_2SOPh$	BuLi, thf, $-50\,°C$	$CH_2{\cdots}\bar{C}H{\cdots}CHSOPhLi^+$	981
3-(SOPh)cyclohexene	$LiNPr^i$, thf, $-60\,°C$	cyclohexenyl anion–SOPh Li^+	982
$CH_2{=}CHCH_2SO_2Ph$	BuLi, hmpt, thf, $-78\,°C$	$CH_2{\cdots}\bar{C}H{\cdots}CHSO_2PhLi^+$	983
$Me_3SiCH_2CH{=}CHCH_2SO_2Ph$	BuLi, thf, hexane, $-78\,°C$	$Me_2SiCH_2CH{\cdots}\bar{C}H{\cdots}CHSO_2PhLi^+$	984

TABLE 17. (*Contd.*)

Compound	Metallating conditions	Product (yield, %)	Ref.
$CH_2{=}CHCH_2SePh$	$LiNPr^i{}_2$, thf, $-78\,°C$	$CH_2{\doteq}\bar{C}H{\doteq}CHSePhLi^+$	985
$PhSeCH_2CH{=}CHSePh$	$LiNPr^i{}_2$	$PhSeCH{\doteq}\bar{C}H{\doteq}CHSePhLi^+$	986
$CH_2{=}CHCH_2Cl$	$LiNPr^i{}_2$, thf, $-78\,°C$	$CH_2{\doteq}\bar{C}H{\doteq}CHClLi^+$	987
$CH_2{=}CHCHCl_2$	$LiNPr^i{}_2$, thf, $-78\,°C$	$CH_2{\doteq}\bar{C}H{\doteq}CCl_2Li^+$ (68)	988
$CH_2{=}CHCH_2CN$	MeLi, Et_2O, thf, $-100\,°C$	$CH_2{\doteq}\bar{C}H{\doteq}CHCNLi^+$	989
$CONR_2$	Bu^sLi	$CONR_2$ Li^+	990
$Me_3SiCH_2CH{=}CH_2$	BuLi, Bu^tOK, hexane	$Me_3SiCH{\doteq}\bar{C}H{\doteq}CH_2$ K^+	991
MeO, OMe	Bu^tLi, thf, $-78\,°C$	MeO, OMe Li^+ (99)	992
$MeCH{=}C(OSiEt_3)CH{=}CH_2$	Bu^sLi, cyclohexane, thf, $-78\,°C$	$OSiEt_3$ Li^+ (81)	993
$MeC(O)CH_2CH_2CH{=}CH_2$	(i) KH, thf, $20\,°C$ (ii) Bu^sLi, tmed, isopentane, $-78\,°C$	Me, O^- K^+, Li^+ (72)	993
$MeS(S)CCH_2CH_2CH{=}CH_2$	(i) KH, thf, $20\,°C$ (ii) Bu^sLi, tmed, isopentane, $-78\,°C$	MeS, S^- K^+, Li^+ (71)	994

$Me_2C{=}CHCH{=}NBu^t$	$LiNPr^i{}_2$, Et_2O	$CH_2{\cdots}CMe{\cdots}\bar{C}H{\cdots}CH{\cdots}NBu^tLi^+$	995
$EtCH_2CH{=}N,NMe_2$	$LiNEt_2$, thf, hmpt	$EtCH{\cdots}\bar{C}H{\cdots}NNMe_2$, Li^+	996
$EtMeC{=}N{-}C_6H_{11}$	$LiNEt_2$, hmpt, thf, $-30\,°C$	$CH_2{\cdots}\bar{C}(Et){\cdots}N{-}C_6H_{11}$ Li^+ (98)	997
$Et(CH_2Me)C{=}N{-}N$ (2-CH_2OMe-pyrrolidine)	$LiNPr^i{}_2$, Et_2O, $0\,°C$	$Et(CHLiMe)C{=}N{-}N$ (2-CH_2OMe-pyrrolidine) (82)	998
$Me_2C{=}NPr^i$	Li, 5-crown-15, graphite, thf	$H_2C{\cdots}\bar{C}Me{\cdots}NPr^i$, Li^+	999

TABLE 18. Formation of substituted alkynyl- propargyl- and allenylalkylmetal compounds

Compound	Lithiating Conditions	Product (yield, %)	Ref.
$HC\equiv CCO_2H$	2 equiv. $LiNPr^i{}_2$, hmpt, hexane, thf, −45 °C	$LiC\equiv CCO_2Li$ (55)	1000
$HC\equiv CCH_2Othp$	BuLi, thf, hexane, −40 °C	$LiCH{=}C{=}CHOthp$ (76)	1001
$MeC\equiv CSiPr^i{}_3$	BuLi, tmed, Et_2O, −15 °C BuLi, Et_2O, pentane, −10 °C, or BuLi, thf, −20 °C	$LiCH_2C\equiv CSiCPr^i{}_3$ (quant.)	1002
$Me_3SiCH_2C\equiv CMe$	BuLi, thf, hexane, −60 to 20 °C, 3 h, 20 °C, 4 h	$Me_3SiC\equiv CCHLiMe$ ⇌ $Me_3SiCLi{=}C{=}CHMe$	1003
$MeC\equiv CCO_2H$	2 equiv. BuLi, tmed	$LiCH_2C\equiv CCO_2Li$ (90)	1004
$C\equiv CSiMe_3$ [on C-2 of a 1,3-dioxane ring]	BuLi, thf, hexane, −65 °C	$HC\equiv CSiMe_3$ [on a 1,3-dioxolane ring] ⇌ [1,3-dioxane ring]$C{=}C{=}CLi\,SiMe_3$ (95)	1005
$PhC\equiv CCH_2OMe$	BuLi, Et_2O, −75 °C	$PhLiC{=}C{=}CLiOMe$ (87)	1006
$C_5H_{11}C\equiv CCH_2OMe$	BuLi, tmed, Et_2O, −78 °C	$C_5H_{11}CLi{=}C{=}CHOMe$ ⇌ $C_5H_{11}C\equiv CCHLiOMe$	1007
$RC\equiv CCH_2OR'$	BuLi, Bu^tOK, hmpt, thf	$RCH{=}C{=}CKOR'$ (75)	1008
$MeSC\equiv CCH_2OMe$	$LiNPr^i{}_2$, thf, −60 °C	$MeSCLi{=}C{=}CHOMe$ (94)	1009
$C_5H_{11}C\equiv CCH_2NMe_2$	BuLi, thf, −70 °C	$C_5H_{11}C\equiv CCHLiNMe_2$ (40)	1010
$CH_2{=}C{=}CHOMe$	BuLi, Et_2O, −25 °C	$CH_2{=}C{=}CLiOMe$ (88)	1011
$CH_2{=}C{=}C(OMe)SiMe_3$	Bu^tLi, thf, −78 °C	$CHLi{=}C{=}C(OMe)SiMe_3$ (95)	1012
$CH_2{=}C{=}CHNMeP(O)(OEt)_2$	BuLi, thf, −78 °C	$CH_2{=}C{=}CLiNMeP(O)(OEt)_2$ (80)	1013
$HC\equiv CCH_2SPh$	2 equiv. BuLi, thf, tmed, −60 °C	$LiC\equiv CCHLiSPh$ (89)	1014
$HC{=}CCH_2SePh$	$LiNBu^t{}_2$, thf, −78 °C	$LiC\equiv CCH_2SePh$ (92)	1015
	2 equiv. $LiNPr^i{}_2$, dme, −78 °C	$LiC\equiv CCHLiSePh$ (93)	1015

The alkali metals and their derivatives also react with certain ethers, sulphides, selenides, and tellurides; included in these reactions are those involving radical anionic species, $ArH^{-\bullet}Li^+$, and sulphides. In general, these reactions have value both as routes to organoalkali metal compounds difficult to prepare by other means (e.g. allyl- and benzyl- as well as α-alkoxy - and α-aminoalkyl derivatives) and as routes to derivatives required free of metal halides, alkoxides, or donor molecules, such as tmed.

1. Use of Alkali Metals

Alkali metals react under mild conditions with organic derivatives of heavy metals, e.g. mercury. The reactions may be slow and reversible, however. An excess of the alkali metal is often considered prudent to ensure that the reaction (equation 159) goes far to the right and to limit contamination from residual organomercurials.

All types of organic groups can be transferred from mercury to the alkali metal[1,2,5,6,173]; reported examples include simple alkyl[445] (e.g. equation 161), functionally substituted alkyl (e.g. equation 162)[446], 1-alkenyl (e.g. equation 163)[447], benzyl[448], and aryl groups (e.g. equation 164)[449].

$$Bu_2Hg + 2M \xrightarrow[0\ ^\circ C]{\text{ether}} 2BuM + Hg \qquad (161)$$

$$M = Li, Na, K, \text{ or } Cs$$

$$(MeOCH_2CH_2CHMe_2)_2Hg + 2Li \xrightarrow[25\ ^\circ C,\ 2\ d]{\text{pentane}} 2MeOCH_2CH_2CHMeLi + Hg \qquad (162)$$

$$(RCH{=}CHCH_2)_2Hg + 2M \xrightarrow[-20\ ^\circ C]{\text{thf,}} 2RCH{\cdots}\bar{C}H{\cdots}CH_2Li^+ + Hg \qquad (163)$$

$$M = Li, Na, K, Cs, \text{ or } Rb;\ R = H, Bu^tCH_2, \text{ etc.}$$

$$(o\text{-}C_6H_4Hg)_3 + Li \xrightarrow[RT]{Et_2O,\ Li_4d} o\text{-}C_6H_4Li_2 \qquad (164)$$

Although mercury compounds are most frequently employed, compounds of other metals have found use, including organosilicon, -tin, and -lead compounds, for example to prepare $CH_2{=}CHLi$ [from $(CH_2{=}CH)_4M$ (M = Sn or Pb)][4], $CH_2{=}CHCH_2Li$ [from $(CH_2{=}CHCH_2)_4Sn$][450], and $PhCH_2Li$ [from $(PhCH_2)_3SnCl$[451] or $PhCH_2SiPh_3$][452].

Polylithioalkanes and -alkenes have been generated from appropriate mercurated compounds and lithiums[453] (as well as with organolithiums), equations 165–167.

$$MeCH(HgCl)_2 + 2Li \xrightarrow{Et_2O,\ 20\ ^\circ C} MeCHLi_2 \qquad (165)$$

$$HC(HgCl)_3 + 3Li \xrightarrow{\text{thf}} HCLi_3 \qquad (166)$$

$$C_6H_{10}{=}C(HgBu^t)_2 \xrightarrow[2\ Bu^tLi,\ Et_2O]{Li\text{(excess) or}} C_6H_{10}{=}CLi_2 \qquad (167)$$

Use has been made of transmetallation reactions to prepare β-substituted alkylalkali metal compounds[454], equations 168 and 169.

$$HYCHRCH_2HgBr + PhM \xrightarrow{thf,\ -78\ °C} MYCHRCH_2HgBr \xrightarrow[-78\ °C]{M'} MYCHRCH_2M' \quad (>75\%)$$

Y = PhN, R = H or Ph; M, M′ = Li, Na or K

Y = O, R = Ph, M = M′ = Li

$$Hg(CH_2CH_2CH{=}CH_2)_2 \xrightarrow[(ii)\ KBr]{(i)\ PhNH_2,\ Hg(OAC)_2} Hg(CH_2CH_2CH(NHPh)CH_2HgBr)_2 \xrightarrow[(ii)\ M']{(i)\ PhM} M'CH_2CH_2CH(NPhM)CH_2M' \xrightarrow[(ii)\ H_3O^+]{(i)\ D_2O} DCH_2CH_2CH(NHPh)CH_2D \quad (169)$$

M, M′ = Li, Na, or K.

Transmetallations have in general greater utility for the heavier alkali metals than for lithiums, although they are valuable routes to benzyl- and alkyllithiums.

Benzylic and allylic alkali metal compounds are also available from the cleavage of appropriate ethers, equation 170. In general, ether cleavage results only if at least one of the organic groups can provide a stable carbanion. However, potassium has been reported[1] to provide PhK from PhOMe in heptane. Some specific examples are given in equations 171–173.

$$ROR' + 2M \longrightarrow RM + MOR' \quad (170)$$

R = allylic or benzylic group; M = Li → Cs

$$CH_2{=}CHCH_2OMe + 2Li \xrightarrow[(ref.\ 455)]{thf,\ -15\ °C} CH_2{\overset{---}{=}}\bar{C}H{\overset{---}{=}}CH_2Li^+ + LiOMe \quad (171)$$

$$PhCMeROMe + Na\text{–}K\ alloy \xrightarrow[(ref.\ 456)]{ethers} PhCMeR^-K^+ \quad (172)$$

$$\text{7-MeO-7-Ph-norbornane} \xrightarrow[or\ Cs/Na/K,\ thf,\ 0\ °C\ (ref.\ 457)]{Na/K,\ thf,\ 0\ °C} \text{7-Ph-norbornyl}^-\ K^+ \quad (173)$$

As shown by these examples, the direction of cleavage of ethers is such to provide the organometallic, RM, having the most stable carbanion grouping, R^-. Organic sulphides, RSR′, are also cleaved by alkali metals but here the most stable organometallic is not necessary obtained[25,458]. For example, the reactions of PhSR with dispersed lithium

$$PhSR + 2Li \longrightarrow PhSLi + RLi \quad (174)$$

provide a range of RLi, including R = primary alkyl [e.g. $Me(CH_2)_n$ ($n = 6$ or 7), $Bu^tCH_2CH_2$, $Ph(CH_2)_n$ ($n = 2$–4), $PhCH_2O(CH_2)_n$ ($n = 3$–4), and $PhS(CH_2)_n$ ($n = 3$–6)], secondary and tertiary alkyl [e.g. $Me(CH_2)_nCHMe$ ($n = 4$ or 5), Bu^s, cyclohexyl, Bu^t and $Bu^tCH_2CMe_2$], aryl, e.g. Ph, benzyl [e.g. PhCHR (R = Me or Ph), $Ph_{3-n}Me_n$ ($n = 0$, 1, or 2), and PhCR(OLi)CHPh (R = Me or Ph)], and α-substituted alkyl, e.g. Bu(PhS)CH. Also

TABLE 19. Formation of organolithiums from reactions of organic sulphides with lithium aromatic-radical anion species

$$RSPh + ArH^{-\bullet}Li^{+} \xrightarrow{thf,\ -78\ ^{\circ}C} RLi + LiSPh$$

Sulphide		ArH[a]	Product (yield, %)	Ref.
$CH_3(CH_2)_6SPh$		A	$CH_3(CH_2)_6Li$ (87)	1016
$Ph(CH_2)_3SPh$		A	$Ph(CH_2)_3Li$ (87)	1016
PhCHMeSPh		A	PhCHMeLi (54)	1017
Ph_3CSPh		A	Ph_3CLi (95)	1017
$[PhS(CH_2)_3]_2O$		A	$[Li(CH_2)_3]_2O$ (88)	1016
SPh, O (2-SPh tetrahydropyran ring)		A or B	Li, O (2-Li tetrahydropyran ring) (65)	1018
R, R', C, MeO, SPh			R, R', C, MeO, Li	
R	R'			
Et	H	B	(85)	1018
Me	Me		(69)	1018
R, R', C, PhS, SPh			R, R', C, PhS, Li	
R	R'			
Et	H	B	(86)	1019
Bu	H	A	(90)	1017
Me	Me	B	(92)	1019
R, R', C, Me_3Si, SPh			R, R', C, Me_3Si, Li	
R	R'			
Et	H	B	(65)	1019
Me	Me	B	(79)	1019
H	PhS	A	(73)	1020
Me_3Si	Me	A	(69)	1020
Me_3Si	Me_3Si	A	(68)	1020
X, SPh (cyclopropane)			X, Li (cyclopropane)	
X				
MeO		B	(90)	1018
PhS		B	(95)	1019
Me_3Si		A	(94)	1021, 1022

TABLE 19. (*Contd.*)

Sulphide	ArH[a]	Product (yield, %)	Ref.
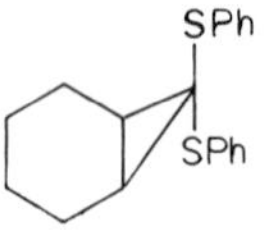 SPh, SPh (bicyclic)	B	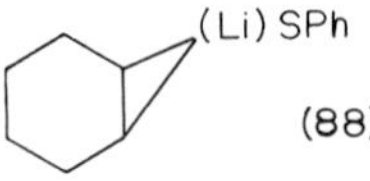 (Li) SPh (88)	1019
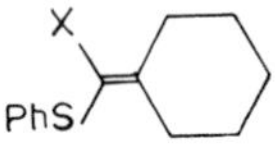 X, PhS (cyclohexylidene)		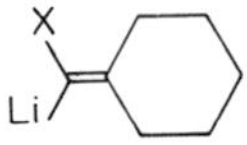 X, Li (cyclohexylidene)	
X			
Me_3Si	B	(96)	1019
PhS	B	(96)	1019
	A	(91)	1023
$Ph_2C{=}C(SPh)_2$	B	$Ph_2C{=}CLiPh$ (87)	1019
$Me_2C{=}C(SPh)_2$	A	$Me_2C{=}CLiSPh$ (77)	1023
$PhSCH_2CH{=}CMe(OSiMe_3)$	B	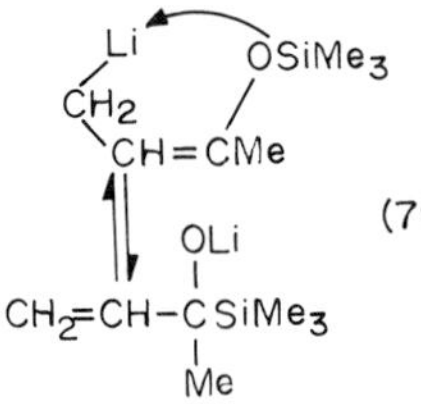 Li, CH_2, $CH{=}CMe$, $OSiMe_3$ ⇌ $CH_2{=}CH{-}C(OLi)(Me)SiMe_3$ (79)	1018

[a] A = $Naph^{-\cdot}Li^+$; B = 1-$Me_2NNaph^{-\cdot}Li^+$.

formed this way are the dimetallated species $(LiCHRCH_2CH_2)_2O$ (R = H or Ph), $LiCHPh(CH_2)_nCHPhLi$ (n = 3–6, 10), and p-$(LiCHRCH_2)_2C_6H_4$ (R = H or Ph). As well as lithium dispersion, the radical anionic species $Naph^{-\cdot}Li^+$ (or simply lithium in the presence of catalytic quantities of naphthalene) also work well to give good yields (40–90%) of the organolithium. Such sulphide cleavage reactions are not restricted to lithium, for example sodium and potassium also were shown to cleave $Ph_2CMeSPh$ to Ph_2CMeM (M = Na or K).

Allyllithiums have also been generated[459] by reaction of allylic mesitoates with lithium in thf.

2. *Use of Lithium Arene Radical Anions, $ArH^{-\cdot}Li^+$*

As was referred to in the previous section[458], lithium arene radical anion species react with phenyl sulphides to give organolithiums. A variety of interesting organolithiums, RLi, have been obtained[69,460–464]; for example R can be an alkyl, 1-alkenyl or cyclopropyl group substituted at an α-position by an RO, PhS or Me_3Si unit. Some examples are given in Table 19.

$$2ArH^{-\cdot}Li^+ + PhSR \longrightarrow RLi + LiSPh \qquad (175)$$

A disadvantage of this method of preparation of organolithiums for subsequent elaboration could be the presence of the arene as a by-product. This could lead to separation problems, as have been reported with naphthalene. The use of 1-dimethylaminonaphthalene rather than naphthalene has been recommended[46], since a simple acid wash removes the arene and enables the desired product to be collected more easily.

The $ArH^{-\bullet}Li^{+}$–RSPh reactions probably involve an initial electron transfer from $ArH^{-\bullet}Li^{+}$ to the substrate, followed by the homolytic cleavage of the R–S bond to give $R^{\bullet}$ and PhS^{-}. Further reduction of the radical, $R^{\bullet}$, then provides the carbanion, R^{-}.

3. Use of Organoalkali Metal Compounds

Transmetallations equation 160, have particular value[465,466] for lithium systems, i.e. M = Li. Very limited work has been reported for the other alkali metals. While many metals (M′) may be used, most synthetic utility has been found for mercury, lead, and tin. Organotin compounds, in particular, have become valuable precursors of substituted allyl-, α-alkoxyalkyl-, and α-aminoalkyllithiums.

Transmetallations are in principle reversible reactions; in specific cases, equilibria have been established, for example as with arylmercury–aryllithium[467] and vinyltin–phenyllithium[468] systems, equations 176 and 177.

$$Ph_2Hg + 2\,p\text{-}MeC_6H_4Li \overset{Et_2O}{\rightleftharpoons} (p\text{-}MeC_6H_4)_2Hg + 2PhLi \qquad (176)$$

$$[(Z)\text{-}MeCH{=}CH]_4Sn + 4PhLi \overset{Et_2O}{\rightleftharpoons} 4(Z)\text{-}MeCH{=}CHLi + Ph_4Sn \qquad (177)$$

At equilibrium, it is generally found that the more stabilized of the carbanions, R^{-} or R'^{-}, equation 160, forms the organolithium to the greater extent. Thus the use of alkyllithiums, e.g. BuLi, will result in extensive transfers from metals, such as tin and mercury, of such groups as vinyl, allyl, aryl, alkynyl, cyclopropyl, and some α-substituted alkyl groups, e.g. R_2NCH_2, $RSCH_2$, and $ROCH_2$, but not simple alkyllithiums. Hence in general simple alkyllithiums are not prepared by this transmetallation route, although the insolubility of MeLi in hydrocarbons does mean that it can be obtained in good yield from Me_2Hg and EtLi in such media[469].

In addition to thermodynamic considerations, kinetic effects are also important, as shown for some tin–lithium exchanges. No reaction occurred[301] between $Bu_3SnCH_2SiMe_3$ and BuLi in hexane at 20 °C even after 24 h; however, the more reactive BuLi–thf system did produce[470] a good yield of $LiCH_2SiMe$ at 0 °C within 30 min. The reactivity of BuLi in ethereal solutions was found to decrease[471] in the sequence dme > thf > Et_2O (≫ hydrocarbon). However, as shown for the preparation of MeLi, and for both $CH_2{=}CHLi$ and $CH_2{\overset{\cdots}{=}}\bar{C}H{\overset{\cdots}{=}}CH_2Li^{+}$, the poor solubility of some organolithiums in hydrocarbons does allow such media to be used to give reasonable yields of isolated (and solvent-free) products. The passive groups attached to tin also effect the reactivity of $R_3Sn{-}R'$ compounds; the rates of cleavage of the $Sn{-}R'$ bond increases in the sequence $(cyclo\text{-}C_6H_{11})_3Sn \ll Bu_3Sn \ll Me_3Sn$.

Transmetallations proceed with retention of configuration as shown by transfer of secondary alkyl[472], α-alkoxyalkyl[473], cyclopropyl[474] and alk-1-enyl[468] groups, equations 177–181.

$$(EtCHMe)_2Hg + Me(CH_2)_5CHMeLi \xrightarrow[-40\,^{\circ}C]{pentane} EtCHMeLi \qquad (178)$$

$$Et(H)C(SnBu_3)OCH_2OCH_2Ph + BuLi \xrightarrow[-78\,^{\circ}C]{BuLi,\ thf} Et(H)C(Li)OCH_2OCH_2Ph \qquad (179)$$

$$\text{Bu}_3\text{Sn(H)C}\triangle\text{C(H)Et} \xrightarrow[0\,^\circ\text{C}]{2\text{ equiv. BuLi, thf}} \text{Li(H)C}\triangle\text{C(H)Et} \quad (76\%) \qquad (180)$$

$$\text{Bu}_3\text{SnCH=CH-(C}_4\text{H}_6\text{O)(Me)(OSiPr}^i_3) \xrightarrow[\text{(ref. 475)}]{\text{BuLi, thf, } 78\,^\circ\text{C}} \text{LiCH=CH-(C}_4\text{H}_6\text{O)(Me)(OSiPr}^i_3) \quad (89\%) \qquad (181)$$

Good use has been made of the transmetallation reaction to obtain α-functionally-substituted alkyllithiums, especially those with R_2N[301,476] (e.g. equation 182). RO[1,301,471,473,477–481] (e.g. equations 179 and 183), RS[301,482,483] (e.g. equation 184) (all via Sn–Li exchange), and RSe[484–489] (via Se–Li exchange in di- and polyselenoalkanes, e.g. equation 185); this route to α-alkoxy- and α-aminoalkyllithiums is particularly valuable as alternatives are inferior.

$$\text{Bu}_3\text{SnCH}_2\text{NMe}_2 + \text{BuLi} \xrightarrow[\text{(ref. 476b)}]{\text{BuLi, thf, } -78\,^\circ\text{C}} \text{LiCH}_2\text{NMe}_2 \quad 95\% \qquad (182)$$

$$\text{PhCH}_2(\text{Me})\text{CH-CH(OCH}_2\text{OMe)SnBu}_3 \xrightarrow[\text{(ref. 473)}]{\text{BuLi, thf, } -78\,^\circ\text{C}} \text{PhCH}_2(\text{Me})\text{CH-CH(OCH}_2\text{OMe)Li} \qquad (183)$$

$$\text{(CH}_2)_3\text{S}_2\text{C(SnBu}_3)\text{CH}_2\text{CH}_2\text{CH}_2\overset{\text{O}}{\text{CH-CH}_2} \xrightarrow[\text{(ref. 482)}]{\text{BuLi}} \text{(CH}_2)_3\text{S}_2\text{C(Li)CH}_2\text{CH}_2\text{CH}_2\overset{\text{O}}{\text{CH-CH}_2} \qquad (184)$$

$$\text{(PhSe)}_2\text{CH}_2 \xrightarrow[\text{(ref. 484)}]{\text{BuLi, thf, } -78\,^\circ\text{C}} \text{PhSeCH}_2\text{Li} \qquad (185)$$

An interesting and simple preparation of the α-alkoxyalkyltin precursors has been reported[480], equation 186.

$$\text{ROCH}_2\text{Cl} + \text{LiBr.SnCl}_2 \xrightarrow[20\,^\circ\text{C}]{\text{thf, 30 min,}} \text{ROCH}_2\text{SnX}_3 \xrightarrow{\text{BuLi, 1 h, } -78\,^\circ\text{C}} \text{ROCH}_2\text{Li} \qquad (186)$$

$R = Bu^t$, $PhCH_2$, $MeOCH_2CH_2$, Ph or Me; X = halogen

Tertiary groups, such as $R_2C(OR)$, apparently are not transferred from tin to lithium. Substituents further from the metal, i.e. in β- or γ-positions, also may enable transmetallations to occur more extensively than occur for simple alkyl groups, e.g. equations 187–189.

$$Ph_3SnCH_2CH_2CONHPh \xrightarrow[\substack{thf, -78\,°C \\ (ref. 490)}]{2\,BuLi, dabco} \text{[cyclic: PhN(Li)–C(=O)–CH}_2\text{–CH}_2\text{–Li]} \quad 90\% \qquad (187)$$

$$(Pr^iO)_2P(O)CH_2COCH_2CH_2SnBu_3 \xrightarrow[\substack{(ii)\ BuLi, hexane, -78\,°C \\ (ref. 471)}]{(i)\ NaH, thf} (Pr^iO)_2P(O)CH(Na)C(O)CH_2CH_2Li \quad 75\% \qquad (188)$$

$$Bu_3Sn(CH_2)_3OH \xrightarrow[(ref. 478)]{2\,BuLi} LiCH_2CH_2CH_2OLi \qquad (189)$$

Tetrasulphidomethanes[492,493] can also react with alkyllithiums:

$$\text{2,2-bis(methylthio)-1,3-dithiane} \xrightarrow[\substack{thf-hexane \\ (ref. 492)}]{BuLi, -78\,°C} \text{2-lithio-2-(methylthio)-1,3-dithiane} \qquad (190)$$

$$(PhS)_4C \xrightarrow[\substack{-30\,°C \\ (ref. 493)}]{BuLi, thf} (PhS)_3CLi \qquad (191)$$

Cleavage of a single carbon—metal bond[91,494] (or a carbon–metalloid, e.g. carbon—boron bond[495,496]) in di- and polymetallated methanes by organolithiums provide a variety of metallo-substituted methyllithiums. For these reactions the use of BuLi in thf at a low temperature has proved successful (see Table 20). Of interest, $(Ph_3Pb)_3CH$, $(Ph_3Pb)_2CHAsPh_2$[494], and $(PhSe)_3CH$[493] are reported to undergo transmetallations whereas $(Ph_3Sn)_3CH$ and $(Ph_2M)_3CH$ (M = As or Sb)[494] do not; the explanation for the non-reactivity of the latter compounds was that their strong complexation with BuLi effectively reduces the reactivity of BuLi.

Reactions of $Ph_3PbCCl_2MPh_3$ (M = Si or Ge) with BuLi occurred at the Pb—C(Cl) bond and the C—Cl bond[497]; however, for the Sn and Pb analogues only metal—carbon bond cleavage occurred.

Di- and polylithioalkanes have been generated on treatment of appropriate mercuriated precursors with alkyllithiums[453,498]:

$$BrHgCH_2CR_2CH_2HgBr + Bu^tLi \xrightarrow[(ref. 498)]{pentane} LiCH_2CR_2CH_2Li \qquad (192)$$
$$R = H \text{ or } Me$$

$$CH_2(HgI)_2 + Bu^tLi \xrightarrow{hexane} CH_2Li_2 \qquad (193)$$

Cyclopropyllithiums[474], including functionally substituted derivatives[499,500], equations 194 and 195, have also been obtained by transmetallation. Cyclopropyllithium itself is available from (cyclo-$C_3H_5)_4Sn$ and BuLi in pentane or Et_2O; good yields are obtained using a 1:2 molar ratio of tin to lithium reagent[501].

TABLE 20. Formation of organoalkalimetals via transmetallations

Reagent	Conditions	Product (yield, %)	Ref.
$Bu_3SnCH_2NMe_2$	BuLi, thf, −98 °C	$LiCH_2NMe_2$ (95)	1024
$Bu_3SnCH_2N(Me)CH_2CH_2NMe$	BuLi, thf, −65 °C	$LiCH_2NMeCH_2CH_2NMe_2$ (72)	1025
Bu_3SnCH_2OMe	BuLi, hexane	$LiCH_2OMe$ (86)	1026
$X_3SnCH_2OBu^t$ X = halide	BuLi, thf, −78 °C	$LiCH_2OBu^t$ (95)	1027
$Bu_3SnCH(OMe)C_6H_{13}$	BuLi, thf, −70 °C	$LiCH(OMe)C_6H_{13}$ (98)	1028
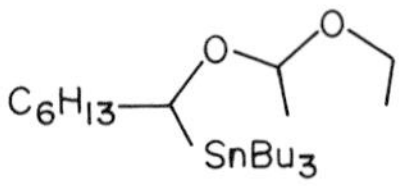 C_6H_{13}, O, O, $SnBu_3$	BuLi, thf, −78 °C	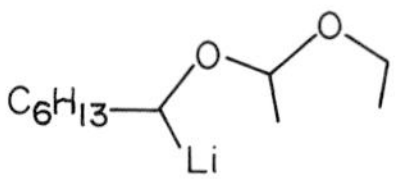 C_6H_{13}, O, O, Li (81)	1028
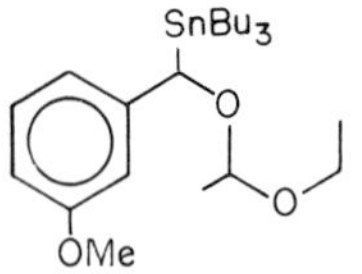 $SnBu_3$, O, O, OMe	BuLi, thf, −78 °C	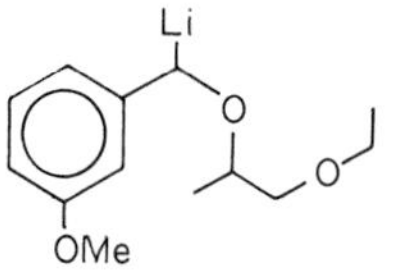 Li, O, O, OMe (98)	1029
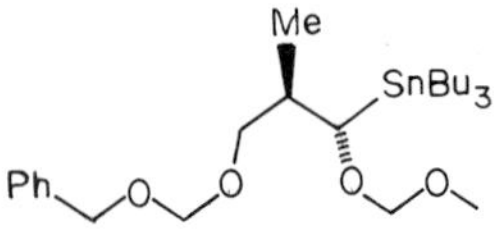 Me, $SnBu_3$, Ph, O, O, O, O	BuLi, thf, −78 °C	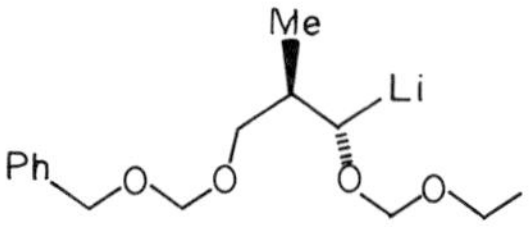 Me, Li, Ph, O, O, O, O	1030
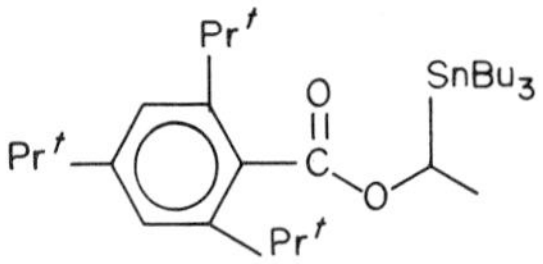 Pr^i, Pr^i, Pr^i, O, C, O, $SnBu_3$	MeLi, thf, −78 °C	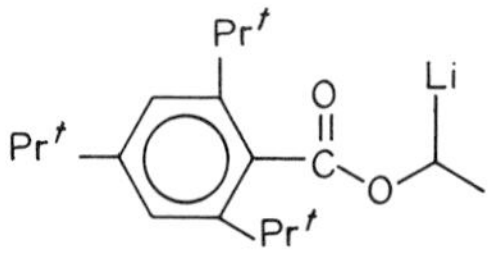 Pr^i, Pr^i, Pr^i, O, C, O, Li (76)	1031
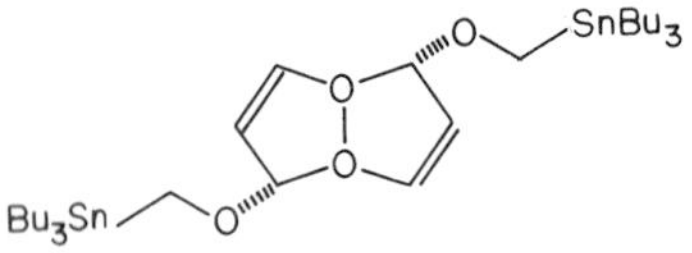 O, $SnBu_3$, O, O, Bu_3Sn, O	2BuLi, thf, −78 °C	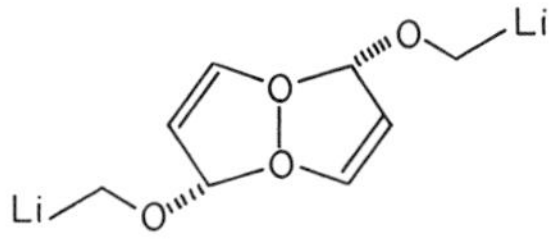 O, Li, O, O, Li, O	1032
 Bu^t, $SnBu_3$, O, O	BuLi	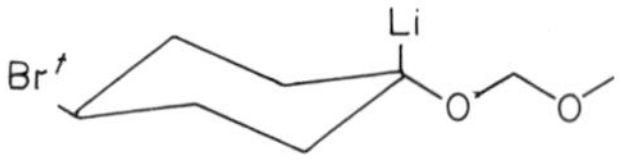 Bu^t, Li, O, O	1033
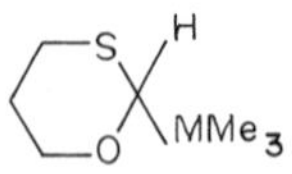 S, H, O, MMe_3 M = Sn or Ph	Bu^sLi, thf, −78 °C	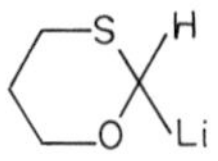 S, H, O, Li (95)	1034
Bu_3SnCH_2SMe	BuLi, hexane	$LiCH_2SMe$ (85)	1026
$Ph_3SnCH_2SC_6H_4Me$-*p*	BuLi, hexane	$LiCH_2SC_6H_4Me$-*p* (80)	1035
PhSeCHMeSPh	BuLi, thf, −78 °C	LiCHMeSPh (92)	1036
$(PhSe)_2CMe_2$	BuLi, thf, −78 °C	$(PhSe)CLiMe_2$ (80)	1037
$(PhSe)_2CMeSiMe_3$	BuLi, thf, −78 °C	$PhSeCMe(SiMe_3)Li$ (95)	1038
$(MeSe)_3CMe$	BuLi, thf, −78 °C	$(MeSe)_2CLiMe$ (70)	1039

TABLE 20. (*Contd.*)

Reagent	Conditions	Product (yield, %)	Ref.
$(PhSe)_4C$	BuLi, thf, −78 °C	$(PhSe)_3CLi$	1040
$(PhTe)_2CH_2$	MeLi, thf, −78 °C	$PhTeCH_2Li$	1041
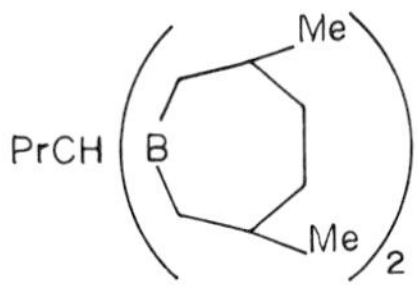 PrCH(B-cyclooctyl, 2,6-Me)₂ [structure]	MeLi, Et_2O, 0 °C	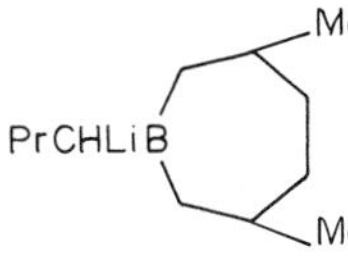 PrCHLiB(cyclooctyl, 2,6-Me) [structure] (84)	1042
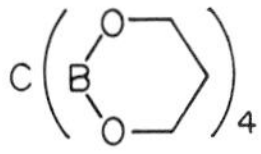 C(1,3,2-dioxaborinan-2-yl)₄ [structure]	BuLi, thf, −70 °C	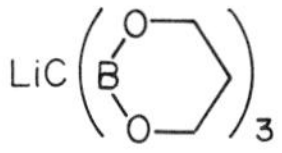 LiC(1,3,2-dioxaborinan-2-yl)₃ [structure] (75)	1043
$Me_3Si)_3CH$	NaOMe, hmpt, 60 °C	$(Me_3Si)_2CHNa$ (83)	1044
$(Me_3Si)_4C$	NaOMe, hmpt, 60 °C	$(Me_3Si)_3CNa$ (18)	1044
$(Ph_3Sn)_2CH_2$	PhLi, thf, −70 °C	$LiCH_2SnPh_3$ (36)	1045
$(Me_3Sn)_2CClBr$	BuLi, thf, Et_2O, methylal, pentane, −97 °C	$Me_3SnCClBrLi$ (49)	1046
$Ph_3SnCH_2AsPh_2$	PhLi, Et_2O, −40 °C	$LiCH_2AsPh_2$ (36)	1045
$(Ph_3Pb)_2CH_2$	PhLi, thf, −70 °C	$LiCH_2PbPh_3$ (100)	1047
$Ph_3PbCH_2GePh_3$	PhLi, thf, −70 °C	$LiCH_2GePh_3$ (87)	1047
$(Ph_3Pb)_3CH$	PhLi	$LiCH(PbPh_3)_2$ (98)	1048
$(Ph_3Pb)_2CHSiMe_3$	PhLi	$LiCH(SiMe_3)(PbPh_3)$ (70)	1048
$(Ph_3Pb)_2CHGePh_3$	PhLi	$LiCH(GePh_3)(PbPh_3)$ (87)	1048
$(Ph_2As)_2CH_2$	BuLi, thf, −40 to +20 °C	$LiCH_2AsPh_2$ (72)	1047
$(Ph_2Sb)_2CH_2$	PhLi, thf, −78 °C	$LiCH_2SbPh_2$ (82)	1049
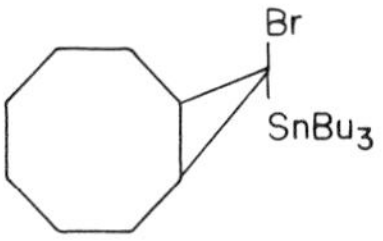 bicyclo[6.1.0]nonane, 9-Br, 9-$SnBu_3$ [structure]	BuLi, −102 °C	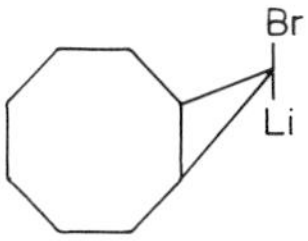 bicyclo[6.1.0]nonane, 9-Br, 9-Li [structure] (88)	1050
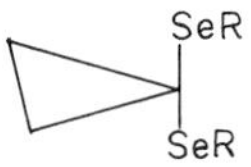 cyclobutane, 1,1-(SeR)₂ [structure]	BuLi, thf, −78 °C	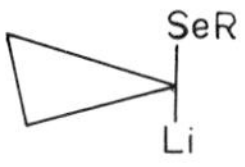 cyclobutane, 1-SeR, 1-Li [structure] R = Ph (72) R = Me (75)	1051
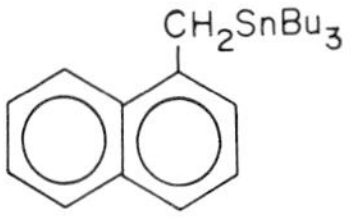 1-(CH_2SnBu_3)naphthalene [structure]	BuLi, hexane, Et_2O	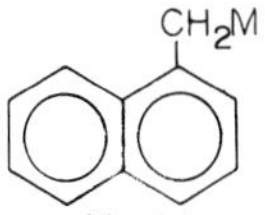 1-(CH_2M)naphthalene [structure] M = Li	1052
	NaOBu^t, BuLi, hexane	M = Na	
	KOBu^t, BuLi, hexane	M = K	
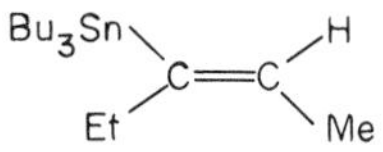 $Bu_3Sn(Et)C{=}C(H)Me$ [structure]	BuLi, thf, −20 °C	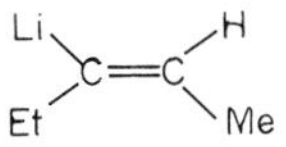 $Li(Et)C{=}C(H)Me$ [structure] (42)	1053

TABLE 20. (*Contd.*)

Reagent	Conditions	Product (yield, %)	Ref.
$Bu_3Sn(H)C{=}C(NMePh)H$	BuLi, thf, hexane, −70 °C	$Li(H)C{=}C(NMePh)H$ (80)	1054
$Bu_3Sn(H)C{=}C(OEt)H$	BuLi, thf, −78 °C	$Li(H)C{=}C(OEt)H$ (97)	1055
$Me_3Sn(MeO)C{=}CH_2$	BuLi, hexane	$Li(MeO)C{=}CH_2$	1056
$Bu_3Sn(H)C{=}C(H)CH_2O\text{thp}$	BuLi, thf, −78 °C	$Li(H)C{=}C(H)CH_2O\text{thp}$ (85)	1057
$Bu_3Sn(H)C{=}C(H)CH_2OCH_2SMe$	BuLi, thf, −78 °C	$Li(H)C{=}C(H)CH_2OCH_2SMe$ (82)	1058
$Bu_3Sn(H)C{=}C(H)CH(OSiMe_3)C_5H_{11}$	BuLi, thf, −50 °C	$Li(H)C{=}C(H)CH(OSiMe_3)C_6H_{11}$ (87)	1059
1-$SnBu_3$-5-Bu^tMe_2SiO-3-thpO-cyclopentene	BuLi, thf, −45 °C	1-Li-5-Bu^tMe_2SiO-3-thpO-cyclopentene (100)	1060
$Me_3Sn(PhS)C{=}CH_2$	BuLi, thf, −78 °C	$Li(PhS)C{=}CH_2$ (78)	1061
$Me_3Sn(PhS)C{=}CPh_2$	BuLi, thf, −78 °C	$Li(PhS)C{=}CH_2$	1061
$Ph(H)C{=}C(SePh)_2$	MeLi, thf, −70 °C	$Ph(H)C{=}C(Li)SePh$ (61)	1061
$Ph(H)C{=}C(SnMe_3)_2$	MeLi, thf, −78 °C	$Ph(H)C{=}C(Li)SnMe_3$ (45)	1062

TABLE 20. (*Contd.*)

Reagent	Conditions	Product (yield, %)	Ref.
$CH_2{=}C(SnMe_3)CH_2CH_2Cl$	MeLi, thf, −78 °C	$CH_2{=}C(Li)CH_2CH_2Cl$	1063
$CH_2{=}CHCH_2SePh$	BuLi, thf, −78 °C	$CH_2{\cdots}\bar{C}H{\cdots}CH_2\,Li^+$ (88)	1064
$Me_3SnCH_2CH{=}CHMe$ (*E*) or (*Z*)-	MeLi, Et_2O	$MeCH{\cdots}\bar{C}H{\cdots}CH_2\,Li^+$ (>90)	1065
$(MeCH{=}CHCH_2)_4Sn$	EtLi, PhH	$MeCH{\cdots}\bar{C}H{\cdots}CH_2\,Li^+$	1065
$Me_3SnCH_2CMe{=}CHEt$ (*E*):(*Z*) = 1:1	MeLi, thf, 0 °C	$CH_2{\cdots}\bar{C}Me{\cdots}CHEt\,Li^+$ (91)	1066
$Me_3SnCH_2CH{=}$(cyclohexylidene)	MeLi, thf, 0 °C	$CH_2{\cdots}\bar{C}H{\cdots}$(cyclohexylidene) Li^+ (92)	1066
$Ph_3PbCH_2CH{=}CHCl$	BuLi, thf, −90 °C	$CH_2{\cdots}\bar{C}H{\cdots}CHCl\,Li^+$	1067
$Ph_3PbCH_2CH{=}CHSIMe_3$	BuLi, thf, −90 °C	$CH_2{\cdots}CH{\cdots}\bar{C}ClSiMe\,Li^+$ (98)	1068
$Ph_3PbCH_2CH{=}CClMe$	BuLi, thf, −90 °C	$CH_2{\cdots}\bar{C}H{\cdots}CClMe\,Li^+$ (87)	1069
$Ph_3PbCH_2CH{=}CCl_2$	BuLi, thf, −95 °C	$CH_2{\cdots}\bar{C}H{\cdots}CCl_2\,Li^+$	1070
$Me_3SnCH_2CH{=}CF_2$	BuLi, thf, −95 °C	$CH_2{\cdots}\bar{C}H{\cdots}CF_2\,Li^+$	1071
$Me_3SiC{\equiv}CC{\equiv}CSiMe_3$	MeLi, thf, 20 °C	$LiC{\equiv}CC{\equiv}CSiMe_3$ (65)	1072
1-Bu^tMe_2SiO-3-$SnMe_3$-cyclohexene	MeLi, thf, −78 °C	1-Bu^tMe_2SiO-3-Li-cyclohexene (84)	1073

$$\text{1-}SnMe_3\text{-1-Br-2-Me-3-Me-cyclopropane} \xrightarrow{BuLi,\,-102\,°C} \text{1-Li-1-Br-2-Me-3-Me-cyclopropane} \tag{194}$$

$$\text{1-SeR-1-Y-cyclopropane} \xrightarrow{BuLi,\,thf} \text{1-Li-1-Y-cyclopropane} \tag{195}$$

Y = SeR, $SiMe_3$ $CR{=}CR'R^2$

Vinyllithiums have been obtained in a number of transmetallation reactions, including (i) $(CH_2{=}CH)_4Sn$ with PhLi in thf[502] or Et_2O[468] or with BuLi in pentane or Et_2O[468], (ii) $Ph_3MCH{=}CH_2$ (**86**, M = Sn or Pb) with PhLi, and (iii) $(CH_2{=}CH)_4Pb$ with PhLi in Et_2O[503]. In contrast to the transmetallation reactions of **86** (M = Sn or Pb), PhLi adds to **86** (M = Sn[504] or Ge[468]) to provide $Ph_3MCHLiCH_2Ph$. However, the styryl derivative, $Ph_3SiCH{=}CHPh$, is cleaved[505] by PhLi to give $LiCH{=}CHPh$.

As shown in equation 177, transmetallations involving alk-1-enyl groups generally occur with retention of configuration; see also equation 196.

$$R'_3M(H)C{=}C(H)SnR_3 \ (\mathbf{87}) \xrightarrow{\text{RLi, ether}} R'_3M(H)C{=}C(H)Li \ (\mathbf{88}) \tag{196}$$

M = Si[506] or Sn[507]; R = Ph or Bu

Use of 2- equivalents of RLi with **87** (M = Sn) does not lead to the formation of (*E*)-$LiCH{=}CHLi$; only the mono exchange product (**88**, M = Sn) is obtained; similarly, both Me_3Sn groups are not cleaved[508] by MeLi from $RCH{=}C(SnMe_3)_2$ in thf at − 78 °C. In contrast, gem-C—Hg bonds can be cleaved, see equation 167.

Trifluorovinyllithium has been prepared from $BuSn(CF{=}CF_2)_3$ and BuLi in Et_2O from $(CF_2{=}CF)_4Sn$ and PhLi at − 40 °C or from $PhSi(CF{=}CF_2)_3$ in Et_2O–pentane[509]. Other functionalized vinyllithiums are listed in Table 20.

Preparations of 1-lithiobuta-1, 3-dienes and penta-1, 4-dienes also include transmetallations, e.g. equations 197 and 198.

$$Bu_3SnCH{=}CHCHCHOEt \xrightarrow[\text{(ref. 510)}]{\text{BuLi, thf}, -78\,^\circ\text{C}, 1\text{ h}} LiCH{=}CHCH{=}CHOEt \ (90\%) \tag{197}$$

$$\text{(stannacyclohexa-2,5-diene, } SnR_2) + MeLi \xrightarrow[\text{(ref. 511)}]{Et_2O\text{–hexane}, -78\,^\circ\text{C}} LiCH{=}CHCH_2CH{=}CHSnR_2Me \tag{198}$$

R = Me or Bu

Transmetallations of allyltin and -lead compounds provide useful routes to allyllithiums; allyllithium has been produced[512] from $(CH_2{=}CHCH_2)_4Sn$ and BuLi or PhLi in pentane or Et_2O, and also from $Ph_3SnCH_2CH{=}CH_2$ and PhLi in Et_2O; see Table 20 for examples of functionalized allyllithiums prepared by transmetallations.

In contrast to the ready transmetallations of vinyl or allyl groups, but-3-enyl and pent-4-enyl groups are not transferred from $Bu_3Sn(CH_2)_nCH{=}CH_2$ ($n = 3$ or 4) using BuLi in Et_2O[468].

Other groups to be exchanged are alkynyl[513,514], allenyl[515], and benzyl[451,516,517] equations 199–201.

$$Me_3SiC{\equiv}CC(OEt)_3 + BuLi \xrightarrow[\text{(ref. 513)}]{\text{thf, 0 °C}} LiC{\equiv}CC(OEt)_3 \ (94\%) \tag{199}$$

$$Me_2C{=}C{=}C(SPh)SiMe_3 \xrightarrow[\text{25 °C (ref. 515)}]{\text{MeLi, thf}} Me_2C{=}C{=}C(SPh)Li \tag{200}$$

$$Me_2C{=}C{=}C(SPh)SiMe_3 \xrightarrow[-25\,^\circ\text{C}]{Bu^tLi} Me_2C{=}C{=}C(Li)SiMe_3$$

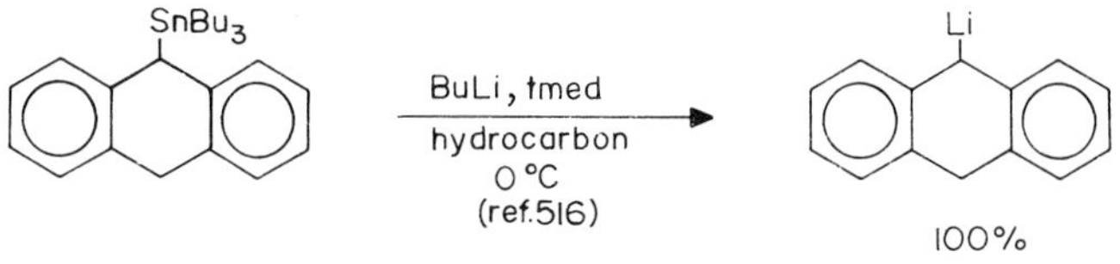

(201)

4. *Exchanges of Alkali Metals*

Treatment of an organolithium with an alkali metal alkoxide has been used to produce organic derivatives of the heavier alkali metals (Na → Rb). This approach has been used, for example, for simple alkyl derivatives, e.g. MeM (M = Na, Rb or Cs)[1,518] from MeLi and $NaOBu^t$, $ROBu^t$, and $CsOMe_2Pr$, respectively, for substituted alkyls [e.g. Me_2CMCO_2R from MOR′ (M = Na, K, or Cs) and Me_2CLiCO_2R, and also MCH_2CMe_2COMe[519]], and for allyl compounds, e.g. $Bu^tCH_2CH{=}CHCH_2Na$[520].

Exchange in the reverse sense, i.e. from a heavy alkali metal to lithium, has been realized using LiBr, as with benzyl[183] and allyl derivatives[447].

D. Other Methods of Preparation

There are various other methods of synthesis; a few of the more important of these are described in this section.

1. *Additions to Alkenes and Alkynes*

Additions of organolithiums to alkenes and alkynes have been used to obtain new organolithiums; see Volume 2, Chapter 4.

2. *Alk-1-enyllithiums from Arenesulphonyl Hydrazones*

The Shapiro synthesis, e.g. Scheme 8, provides vinyllithiums, $RCLi{=}CR^1R^2$, from arenesulphonyl hydrazones, $R(R^1R^2CH){=}NNHSO_2Ar$ (Ar = p-MeC_6H_4 or 2,4,6-$Pr^i_3C_6H_2$)[521]. The compound $CH_2{=}CLiCH{=}CH_2$ was similarly prepared[522] from 2,4,6-$Pr^i_3C_6H_2SO_2NHN{=}CMeCH{=}CH_2$.

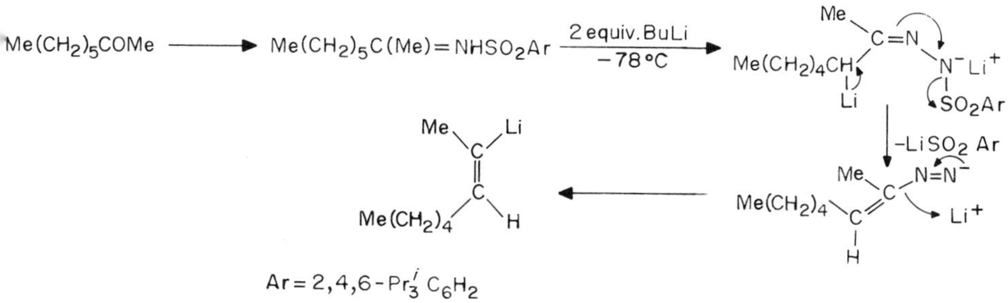

SCHEME 8

3. *Ring-opening of Cyclopropylalkali Metal Compounds: Preparation of Substituted Allyl Metal Derivatives*

Allylalkali metal compounds have been prepared by ring opening of cyclopropyl anions **89** (R = Ph; X = CN, PhSO, $PhSO_2$, CO_2H, or CO_2Me, but not NC, PhS, or H) (Scheme (*trans*-**89**) (*exo,exo*-**90**) + *endo*,endo (**90**)

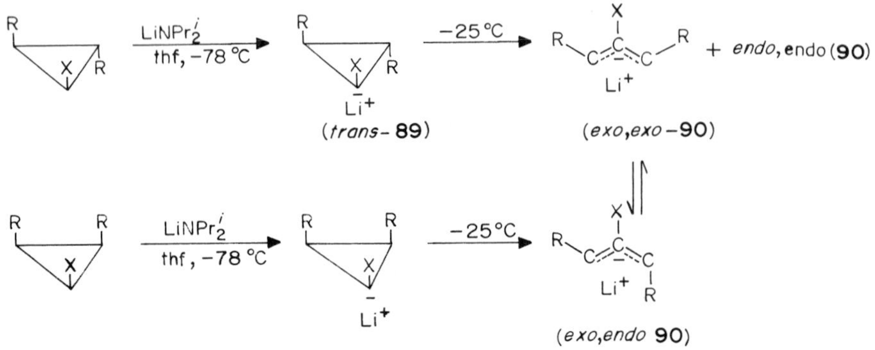

(*exo,endo* **90**)

SCHEME 9

9). Thermal ring openings are conrotatory processes, e.g. *trans*-**89** opens directly to *exo, exo*-**90** (and/or its *endo, endo* isomer). However, rapid isomerization of *exo, exo*-**90** occurs to give thermodynamically more stable *exo, endo*-**90**, the direct product of ring opening of *cis*-**89**[523]. For rigid cyclopropyl systems, in which the geometry prevents conrotatory openings, only slow reactions result, as with **91**; carbanion **91** ring opening

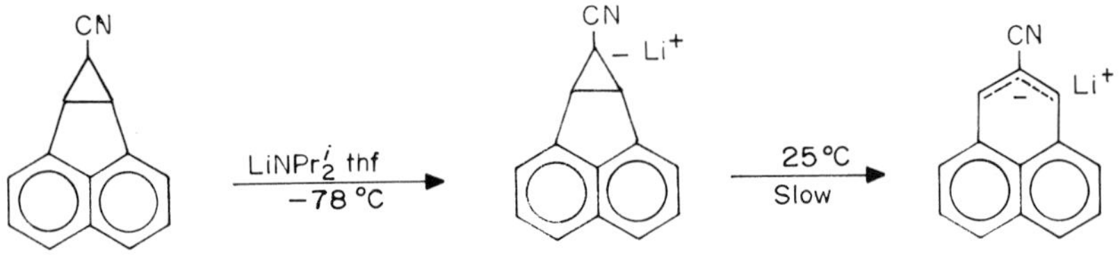

(**91**)

take places 740 times more slowly than for *cis*-**89** (R = Ph, X = CN). The ready opening of the rigid anion from 3-X-2, 4-Ph_2-*endo*-tricyclo[3.2.$10^{2,4}$]octane has been found to occur in a disrotatory manner but not, however, in a synchronous process[524a]. Ring opening of *trans*-**89** (R = Ph, X = CN or CO_2H) can also result on irradition[524b].

III. REACTIONS OF ORGANOALKALI METAL COMPOUNDS

A. General Considerations

Organoalkali metal derivatives, RM, have found extensive use in synthesis as sources of carbanions, as bases (for example, in the formation of alkoxides, ylides, and metal amides), and as sources of such reactive intermediates as arynes and carbenes.

Reactions of organoalkali metal compounds with electrophiles have been used to generate various carbon—element bonds, including carbon—hydrogen (—deuterium), carbon—carbon, carbon—nitrogen, carbon—oxygen and carbon—sulphur bonds, as well as various carbon—metal and carbon—metalloid bonds. These reactions are general reactions in that they are successful for a great variety of R groups. However, certain organic groupings would not survive the sequence of formation and elaboration of the organoalkali metal species, or alternatively would not provide a targeted product unless modified, protected, or masked. The use of such groupings or synthetic equivalents

TABLE 21. Synthetic equivalents: acyl anions RCO^-

Synthetic equivalent	Ref.	Synthetic equivalent	Ref.
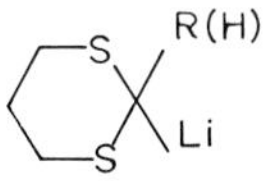 S, S, R(H), Li	1074	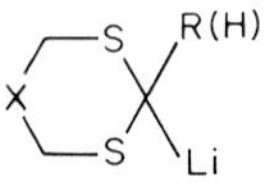 S, X, S, R(H), Li X=S or MeN	1075
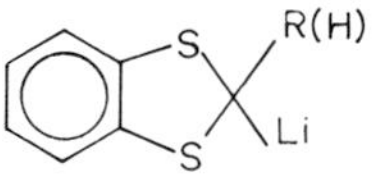 S, S, R(H), Li	1076	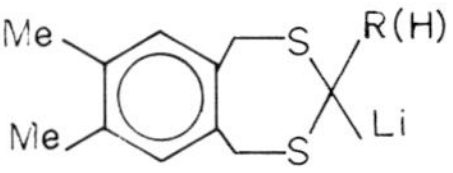 Me, Me, S, S, R(H), Li	1077
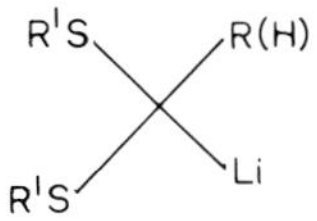 R^1S, R^1S, R(H), Li	1078	$(PhS)_3CLi$	1079
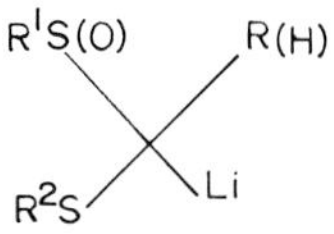 $R^1S(O)$, R^2S, R(H), Li	1080	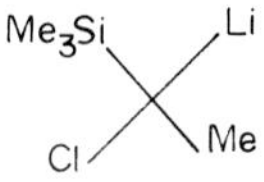 Me_3Si, Cl, Li, Me	1081
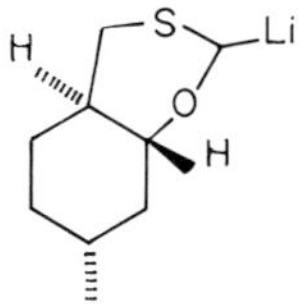 H, S, Li, O, H	1082	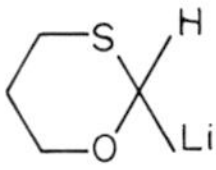 S, O, H, Li	1083
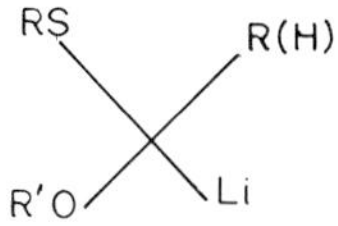 RS, R'O, R(H), Li	1084	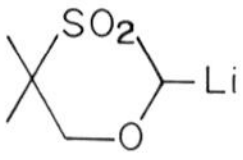 SO_2, O, Li	1085
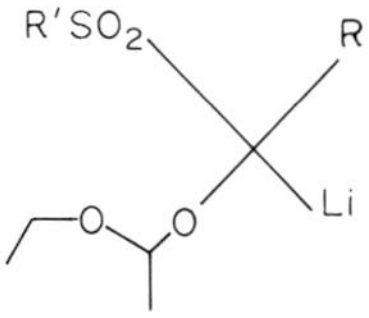 $R'SO_2$, O, O, R, Li	1086	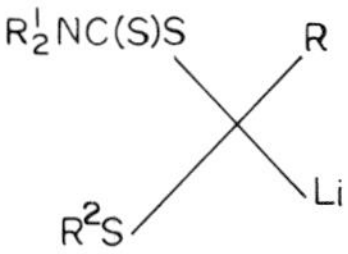 $R^1_2NC(S)S$, R^2S, R, Li	1087
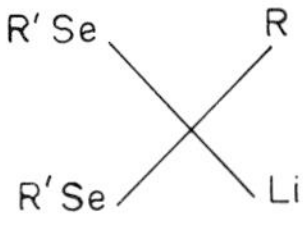 R'Se, R'Se, R, Li	1088	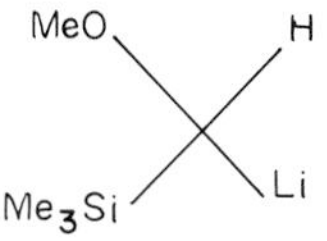 MeO, Me_3Si, H, Li	1089
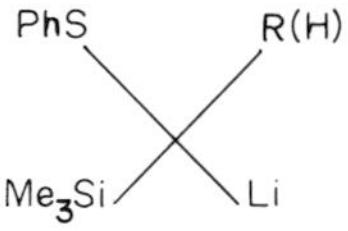 PhS, Me_3Si, R(H), Li	1090	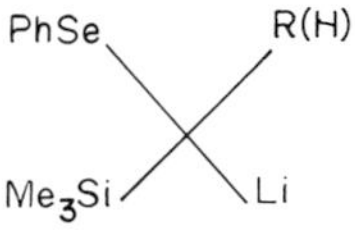 PhSe, Me_3Si, R(H), Li	1091

TABLE 21. (*Contd.*)

Synthetic equivalent	Ref.	Synthetic equivalent	Ref.
$RO(Ph_2P(O))C(H)Li$	1092	$R'S(Ph_2P(O))C[R(H)]Li$	1093
benzothiazole (N, S) with Li at C-2	1094	$R'S(HO_2C)C(R)Li$	1095
$Me_2NC(S)S(NC)C(H)Li$	1096	$RS(NC)C(H)Li$	1097
$EtOCH(CH_3)O(NC)C[R(H)]Li$	1098	$R^1_2N(NC)C(R)M$	1099
$(EtO)_2P(O)(R_2N)C(H)Li$	1100	$LiC(R)=NR^1$	1101
$PhCH=N(RO_2C)C(H)Li$	1102	$PhCH=N((EtO)_2P(O))C(H)Li$	1103
$RN(N=O)CH_2Li$	1104	$ArSO_2(C=N)C(H)Na$	1105
$(EtO)_2P(O)(NC)C(H)M$	1106	$(EtO)_2P(O)(Me_3SiO)C(R)Li$	1107
$Me_3SiO(NC)C(R)Li$	1108	$CH_2=C(OR^1)Li$	1109

TABLE 21. (*Contd.*)

Synthetic equivalent	Ref.	Synthetic equivalent	Ref.
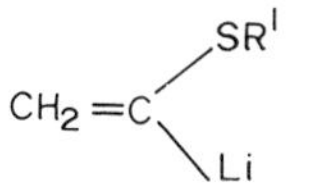 $CH_2{=}C(SR^1)Li$	1110	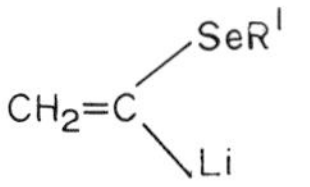 $CH_2{=}C(SeR^1)Li$	1111,1112
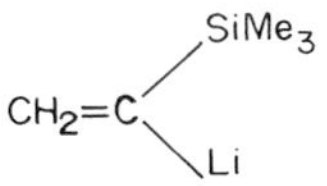 $CH_2{=}C(SiMe_3)Li$	1112	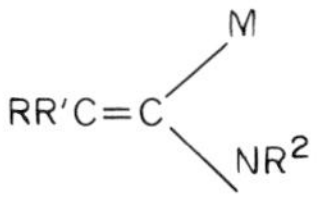 $RR'C{=}C(M)NR^2$	1113

TABLE 22. List of synthons

Synthon	Ref.	Synthon	Ref.
$HOCHR^-$		(oxazolidinone: $=C$, O, $CH(Ph)$, $CH(OMe)$, NLi ring)	1125
Bu_4Sn–C(–O–CH_2) $^{2-}$ $2Li^+$	1114		
$R^1_2BCHRLi$	1115	$LiCH(CO_2Li)CO_2Et$	1126
2,4,6-$Pr^i_3C_6H_2CO_2CHRLi$	1116		
$EtOCHMeOCHRLi$	1117	$PhSCO^-$	
$HSCH_2^-$		$PhS(O)C(H)(Cl)Li$	1127
(cyclohexane-1,1-diyl)$(S(O)CH_2)(SMe)$	1118		
$R^1R^2NCHR^-$		$EtSCH_2CO^-$	
$R_2C{\cdots}\bar{N}{\cdots}CHR\ Li^+$	1119	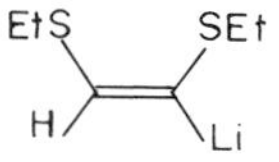 $(EtS)HC{=}C(SEt)Li$	1128
$CNCHRLi$	1120		
$ONNR'CHRLi$	1121	RO_2CCO^-	
 (2,2,6,6-tetramethylpiperidino)$NCONMeCH_2Li$	1122	$RO_2CC(OMe)Li$	1129
2,4,6-$Bu^t_3C_6H_2O_2CNMeCH_2Li$	1123	$HC(O)CH_2^-$	
$RO_2CCH_2^-$		 $ROCH{=}CHLi$	1130
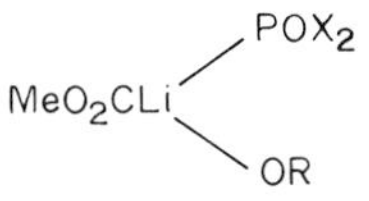 $MeO_2CLi(POX_2)(OR)$	1124	$HCOCHR^-$	
		$RCH{\cdots}\bar{C}H{\cdots}NR'\ Li^+$	1131

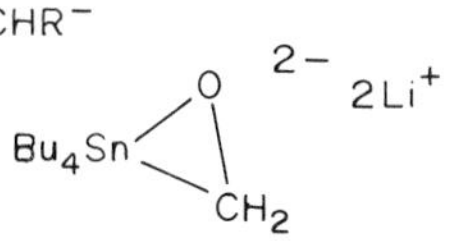

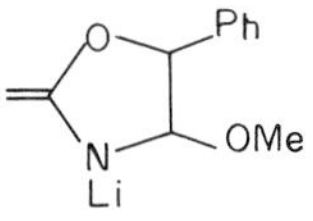

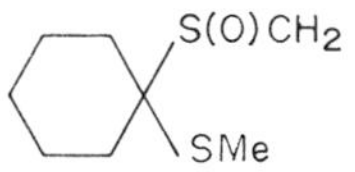

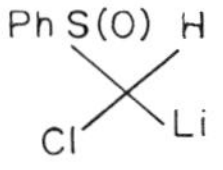

TABLE 22. (*Contd.*)

Synthon	Ref.	Synthon	Ref.
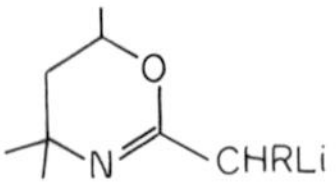		$HCO\bar{C}{=}CH_2$	
O, N, CHRLi	1132	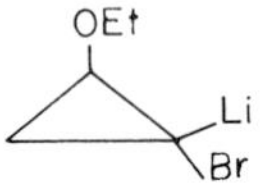 OEt, Li, Br	1145
$R'COCHR^{2-}$		$CH_2{=}CHCO^-$	
$RN{\cdots}\bar{C}R^1{\cdots}CHR^2Li^+$	1133		
$R_2NN{\cdots}\bar{C}R^1{\cdots}CHR^2Li^+$	1134	$(EtO)_2P(O)NRCLi{=}C{=}CH_2$	1146
$HCOCH_2CH_2^-$		$HCOCH{=}CHCH_2^-$	
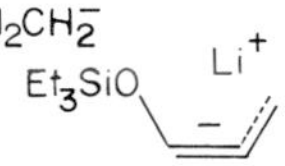 Et_3SiO, Li^+	1135	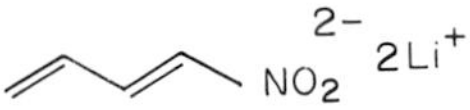 NO_2 2− $2Li^+$	1147
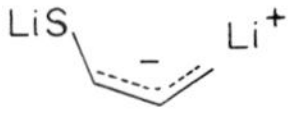 LiS, Li^+	1136	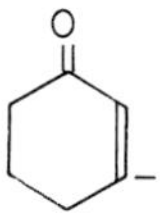 O	
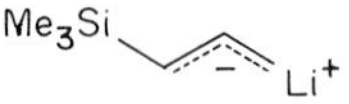 Me_3Si, Li^+	1137	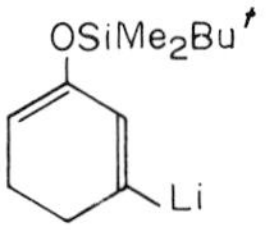 $OSiMe_2Bu^t$, Li	1148
$HCOCH_2CHR^-$			
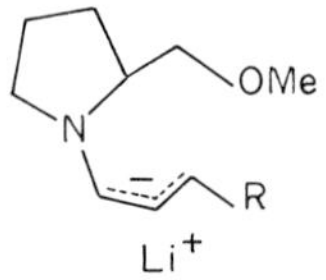 OMe, N, R, Li^+	1138	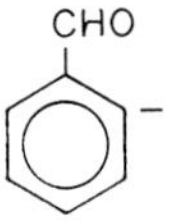 CHO	
$RCOCH_2CH_2^-$		$HC(NR_2)_2$, Li	
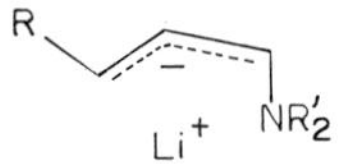 R, Li^+, NR'_2	1139	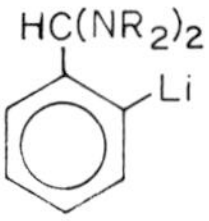	1149
$R^3COCHR^2CHR^{1-}$		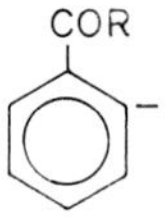 COR	
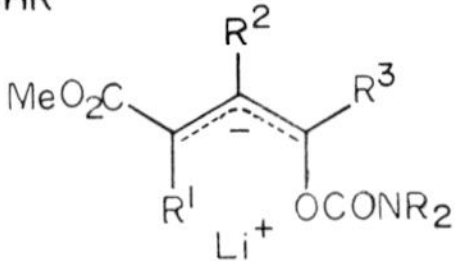 R^2, MeO_2C, R^3, R^1, $OCONR_2$, Li^+	1140	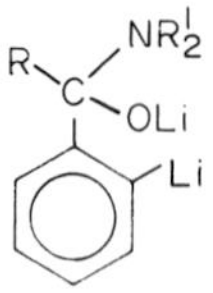 R, NR^1_2, C, OLi, Li	
 $Ph_2P(O)C(Li)R^1CHR^2CR^3$ (O, O)	1141		1150
$RCOCH{=}CH^-$		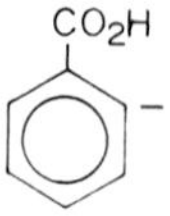 CO_2H	
$EtO_2CCH_2\bar{C}HNO_2\ K^+$	1142		
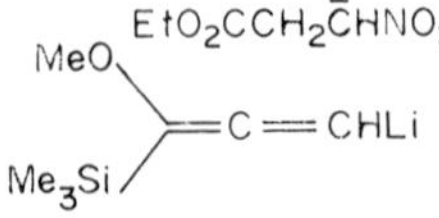 MeO, Me_3Si, $C{=}C{=}CHLi$	1143	O, N, Li	1151
 MeS, SMe, Li^+	1144	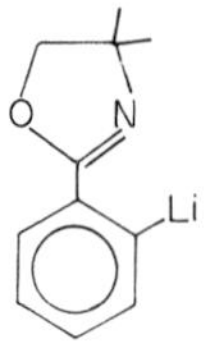	

(synthons), which have of necessity to be readily unveiled at the end of the reaction scheme, have greatly expanded the synthetic utility of organoalkali metal reactions[187]. Some synthons are listed in Tables 21 and 22.

Many of the reactions of the organoalkali metals with electrophiles involve an initial, formal insertion of the electrophile into the carbon—alkali metal bond. These reactions were discussed in Volume 2, Chapter 4, and no further detailed consideration will be given here. Mention of such reactions will however be made at appropriate places in the following sections.

B. Formation of Carbon–Hydrogen Bonds. Reactions with Proton Sources. Deuteriation and Tritiation[3,5]

Organoalkali metal compounds, RM, provide RH on reaction with a variety of proton sources, including water, alcohols, carboxylic acids, inorganic acids, amines, and carbon acids, see equations 58 and 59. Reactions with deuteriated analogues[3,5], or the tritium source, HTO[5], similarly provide RD or RT.

As discussed in Section II.B.2, reactions of RM with proton sources are, in fact, equilibria with the position at equilibrium dependent to a major extent on the relative acidities of the proton acid pairings, RH–R'H and RH–HY, in equilibria 58 and 59. With sufficiently different acidities, the equilibria lie essentially completely to one side. Hydrolysis of organoalkali metal compounds using aqueous media proceeds for all practical purposes to completion and is used for most synthetic purposes. Reactions of RM with carbon acids, R'H, alcohols, R'OH, or amines, R'_2NH, have little if any synthetic utility as routes to RH but are used instead as valuable sources of R'M, R'OM, and R'_2NM, respectively[3,5].

Reaction with hydrogen also can lead to RH, e.g. equation 203; however, hydrogenolysis has had little synthetic application[525].

$$CH_3(CH_2)_7Li + H_2 \longrightarrow CH_3(CH_2)_7CH_3 + LiH \tag{203}$$

Any synthetic value of hydrolysis is clearly restricted to those RM species prepared by routes other than metallation of RH, for example by transmetallation, by halogen–metal or sulphur–metal exchanges, equations 204[61] and 205[460], and by addition or rearrangement reactions. Reaction 204 illustrates an overall reduction of an organic halide.

$$\text{2-BrC}_6\text{H}_4\text{CH}_2\text{CH}_2\text{Br} \xrightarrow[\text{hexane, }-100\,^\circ\text{C (ref. 61)}]{\text{BuLi, thf}} \text{2-LiC}_6\text{H}_4\text{CH}_2\text{CH}_2\text{Br} \xrightarrow[-100\,^\circ\text{C}]{\text{H}_2\text{O}} \text{PhCH}_2\text{CH}_2\text{Br} \tag{204}$$

$$\text{BuCH}_2\text{C(O)SPh} \xrightarrow[\text{(ii) Me}_3\text{SiCl (ref. 460)}]{\text{(i) LiNPr}^i_2\ -78\,^\circ\text{C, thf}} \text{BuCH}{=}\text{C(SPh)OSiMe}_3 \xrightarrow[-78\,^\circ\text{C}]{\text{1-Me}_2\text{NNaph}^-\text{Li}^+}$$

$$\text{BuCH}_2\text{C(O)SiMe}_3 \xleftarrow{\text{H}_2\text{O}} \text{BuCH}{=}\text{C(OLi)SiMe}_3 \rightleftharpoons \text{BuCH}{=}\text{CLi(OSiMe}_3) \tag{205}$$

Considerable use has been made of deuteriolysis, not only as a route to isotopically labelled compounds but also as a means (as also with hydrolyses) of identifying sites and extents of metallation[8], equations 169, 205, and 206.

$$\text{Bu}^t\text{CH}_2\text{CMe}_2\text{N}{=}\text{C} \xrightarrow[\text{(ref. 526)}]{\text{RLi}} \text{Bu}^t\text{CH}_2\text{CMe}_2\text{N}{=}\text{CLiR} \xrightarrow{\text{D}_2\text{O}}$$

$$\text{Bu}^t\text{CH}_2\text{CMe}_2\text{N}{=}\text{CDR} \xrightarrow{\text{H}_3\text{O}^+} \text{RCDO} \tag{206}$$

$$HC{\equiv}CLi \xrightarrow[\text{(ref. 527)}]{\text{AcOD}} HC{\equiv}CD \qquad (207)$$

While the deuterium source most frequently used is D_2O, others, including AcOD[527] and MeOD[345b,375], have been employed to good effect. One advantage of using the latter two reagents is the resulting homogeneous media. Deuteriolysis, using D_2O, AcOD, or ROD, can be assumed to proceed quantitatively; however, the presence of adventitious moisture can reduce the extent of deuterium incorporation. It has been suggested[5] that the use of a D_2O-saturated medium would largely prevent this. Other sources of protons, which lead to reduced extents of deuteriation, have been indicated to be the solvents and alkyl halides, either those used to prepare the organoalkali metals or those obtained via halogen–lithium exchanges. For example, sequential reactions of **92** with MeLi and D_2O provided **93** with only 70% deuterium incorporation; the MeBr formed in the Li–Br exchange was considered to be the source of protons, which provided 30% of the protium incorporated product.

MeLi, Et_2O, −60 °C [−MeBr] (ref. 528); D_2O [MeBr]

(**92**) (**93**) X = D 70%, X = H 30%

(208)

Deuteriolysis (as well as hydrolysis) generally proceeds regiospecifically for alkyl, functionally-substituted alkyl[32], aryl[305,345c,529], and benzyl[218] derivatives; however, for some extensively delocalized carbanions, e.g. **94**[36], protonations ($E^+ = H_3O^+$ in equation 209) and other reactions with electrophiles can occur at any site having appreciable electron density.

[RCl] → (**94**, R^-M^+) —E–Y or E^+→ ArCHE + ...

(209)

Deuteriolysis of cyclopropyl-[66,79,530], cyclohexyl-[531,532], and alkenyllithiums[107,429] proceed with extensive, if not complete, retention of configuration.

LiNPr$^i_2$, thf, −78 °C (ref. 530); MeOD (210)

LiNPr$^i_2$ (ref. 107); D_2O, −70 °C (211)

(i) 2 equiv. Bu^tLi, thf, −90 °C
(ii) MeOD
(ref. 533)

(S)-(+) → 90% (100% optical yield) (212)

The products of deuteriolysis really reflect the composition of the organoalkali metal species present in solution, and indeed deuteriation has been used to establish the composition of metallation reactions. When isomerization, partial or complete, of the initially formed organoalkali metal compound results before it is attacked, the deuteriated product mixture mirrors this and so an overall loss of stereochemistry arises. For example, (*Z*)-$RCH_2CH{=}CLiSOAr$, formed initially from (*Z*)-$RCH_2CH{=}CHSOAr$, immediately isomerizes to the more stable *E* isomer, which is then trapped[426b] by D_2O as (*E*)-$RCH_2CH{=}CDSOAr$.

The situation for ion-paired species can be different, as shown, for example, with $[RS(O)\bar{C}PhH\ Li^+]$. The stereochemistry of deuteriolysis was found to depend on the deuteron source and on its ability to complex with the Li^+ within the ion pair. The more strongly donating deuteron sources, such as D_2O, MeOD, and even AcOD, lead to predominant retention of configuration (*ca.* 90%), whereas deuteron sponge, unable to complex with Li^+, provided a 1:1 mixture of the diastereomers **95** and **96**. The presence of lithium salts also leads to less stereospecificity[534].

$PhCH_2SOMe$ —(MeLi, thf, −60 °C)→ Ph—C(H)(⁻ Li⁺)—S(O)(:)—Me + E—Y

(a) retention → Ph—C(E)(H)—S(O)(:)—Me (95)

(b) inversion → Ph—C(H)(E)—S(O)(:)—Me (96)

(213)

Lithiated dithianes, no matter what the stereochemistry of the parent dithianes is, having fixed conformations are attacked by deuterons (protons) almost completely from the equatorial direction[336].

Far greater study has been made of organolithium reagents; however, deuteriations (protonations) of derivatives of the other alkali metals have also been studied[1,535].

C. Formation of Carbon—Carbon Bonds

1. *Via Insertion Reactions*

Reactions of organoalkali metals, RM, with a number of carbon electrophiles produce new carbon—carbon bonds. Many of these reactions involve a formal initial insertion of the electrophile into the carbon—alkali bond of RM (see Table 23), and as such were discussed in Volume 2, Chapter 4. No further consideration of these reactions will be given here.

TABLE 23. Insertion reactions of RM leading to new carbon—carbon bonds

Reagent	Insertion product(s)	General products after hydrolyses
$R^1R^2C{=}CR^3R^4$	$RR^1R^2CCR^3R^4M$	$RR^1R^2CCR^3R^4H$
	$R[R^1R^2CCR^3R^4]_nM$	Polymer
$R^1C{\equiv}CR^2$	$RR^1C{=}CR^2M$	$RR^1C{=}CHR^2$
$R^1C{\equiv}N$	$RR^1C{=}NM$	$RR^1C{=}NH$
		RR^1CO
$R^1R^2C{=}NR^3$	$RR^1R^2CNR^3M$	$RR^1R^2CNHR^3$
R^1NCO	$R^1N{=}CR(OM)$	R^1NHCOR
$R^1N{=}C$	$R^1N{=}CRM$	$R^1N{=}CHR$
		$O{=}CHR$
CO	[RC(O)M]	
CO_2	RCO_2M	RCO_2H
	$R_2C(OM)_2$	R_2CO
	R_3COM	R_3COH
$O{=}C{=}C{=}C{=}O$	$R(LiO)C{=}C{=}CR(OLi)$	$RCOCH_2COR$
R^1R^2CO	RR^1R^2COM	RR^1R^2COH
$R^1{}_2C{=}C{=}O$	$R^1{}_2C{=}CR(OLi)$	$R^1{}_2CHCOR$
R^1COX	$RR^1C(OM)X$	RR^1CO
(X = OH, OM, Cl, OR^2, $NR^2{}_2$, or $OCOR^2$)	R_2R^1COM	R_2R^1COH
$R^1R^2C{=}CR^3COR^4$	$RR^1R^2CCR^3{=}C(OM)R^4$	$RR^1R^2CCHR^3COR^4$
	$R^1R^2C{=}CR^3CRR^4OM$	$R^1R^2C{=}CR^3CRR^4OH$
R^1R^2C—CR^3R^4 epoxide (O bridging the two carbons)	$RR^1R^2CCR^3R^4OM$	$RR^1R^2CR^3R^4OH$
oxetane (O—CH_2, CH_2—CH_2 four-membered ring)	$R(CH_2)_3OM$	$R(CH_2)_3OH$
$R^1R^2C{=}S$	$RR^1R^2CSM + [R^1R^2CMSR]$	RR^1R^2CSH
CS_2	RCS_2M	RCS_2H

2. *Cross-coupling Reactions with Organic Halides*[1,3,5,536]

An important carbon—carbon bond-forming reaction is the coupling reaction, equation 214.

$$RM + R'X \longrightarrow RR' \tag{214}$$

Halogen–metal exchange reactions between RM and R′X were discussed in Section II.A.3 as a route to new organolithium compounds. However, as referred to in that Section, alternative reactions between RM and R′X are the coupling reactions, equation 214. Some of the conditions which favour couplings over halogen metal exchanges have been found and include, for example, the use of thf and other polar solvents rather than Et_2O or other less polar solvents[53–55]. The presence of hmpt[424a,537] and increased temperatures have also been found to promote couplings.

The reactivity of halides is generally I > Br ≫ Cl with relatively few couplings known for organic chlorides (a number are listed in the Tables of ref. 188). For chlorides, reactions other than coupling reactions tend to dominate; these include α-metallations and α- and β-dehydrohalogenations. More dehydrohalogenations (and metal–halide exchanges) occur with sodium and potassium derivatives than with those of lithium.

Various transition metal compounds catalyse cross-coupling reactions[536]. Such catalysed reactions are especially useful for couplings which would otherwise be difficult. Particularly useful are the copper(I)-catalysed coupling reactions involving aryl, alkenyl, and alkynyl halides and organoalkali metal compounds. Nickel and palladium species also catalyse, for example, couplings involving sp^2 and/or sp hybridized organic groups[536].

a. *Alkylations*

Alkyl halides readily alkylate among others cyclopropyl-[77,531], vinyl-[107,425b,539], allyl-[302,540,541], benzyl-[218,302], aryl-[530,372], propargyl-, alkynyl-[542], and certain functionally substituted alkyllithiums (see Table 24), but generally not simple alkyllithiums. However, alkyl halides do react with the alkyl (and other organic) derivatives of the heavier alkali metals. Indeed, such is the ease of coupling (homo-coupling) of organic halides, RX, with the organic derivatives of the heavier alkali metals that reactions of RX with alkali metals (especially sodium) are important sources of RR (Wurtz coupling), via the intermediacy of RM.

In general, the ease of production of coupled products from alkyl halides is primary > secondary > tertiary. For hindered alkyl halides, β-eliminations tend seriously to reduce the amounts of cross-coupling products. Alkylations have also been achieved with alkyl arenesulphonates[5,307b,543,544], $ROSO_2Ar$, and dialkyl sulphates[5,188,307b,396,473,545], $ROSO_2OR$.

Intramolecular couplings occur particularly readily, e.g. equations 215 and 216.

(i) BuLi; (ii) $Br(CH_2)_nCl$ (ref. 546); $(CH_2)_nCl$; BuLi; $(CH_2)_nCl$, Li; Δ; $(CH_2)_n$ (215)

$(CH_2)_2Br$, Br; $BuLi, Et_2O$, $-100\,^{\circ}C$ (ref. 547); $(CH_2)_2Br$, Li; 25 °C (216)

TABLE 24. Alkylations of organolithium reagents

Initial reagent	Conditions	Alkylated product (yield, %)	Ref.
$PhNHCH_2CH_2HgBr$	(i) PhLi (ii) Li (iii) EtBr (iv) H_2O	$PhNHCH_2CH_2Et$ (62)	1152
Bu_3SnCH_2OH	(i) 2 equiv. BuLi, hexane, $-20\,°C$ (ii) $PhCH_2Br$	$PhCH_2CH_2OH$ (45)	1153
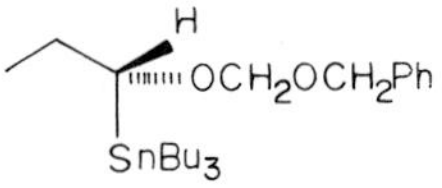 H, OCH_2OCH_2Ph, $SnBu_3$	(i) BuLi, thf, $-78\,°C$ (ii) Me_2SO_4	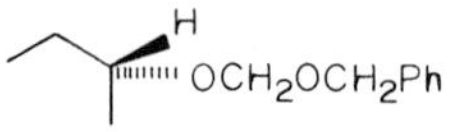 H, OCH_2OCH_2Ph	1154
$PhSCH_2CHMe_2$	(i) Bu^tLi, thf, hmpt, $-78\,°C$ (ii) EtBr, -78 to $25\,°C$	$PhSCH(Et)CHMe_2$ (74)	1155
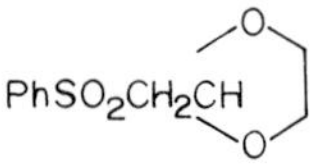 $PhSO_2CH_2CH$ (O, O)	(i) BuLi, thf, $-75\,°C$ (ii) $C_8H_{17}Br$	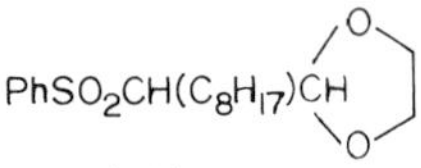 $PhSO_2CH(C_8H_{17})CH$ (O, O) (92)	1156
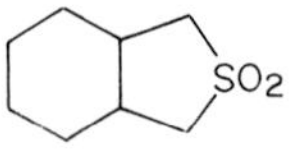 SO_2	(i) 2.5 equiv. BuLi, hexane (ii) MeI (iii) H_2O	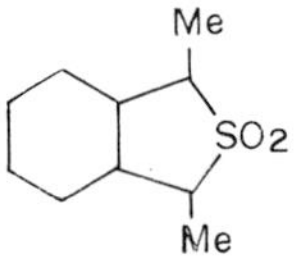 Me, SO_2, Me (87–89.5)	1157
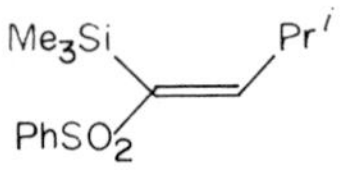 Me_3Si, Pr^i, $PhSO_2$	(i) MeLi, thf, $-78\,°C$ (ii) $CH_2{=}CHCH_2Br$	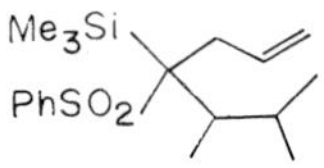 Me_3Si, $PhSO_2$ (55)	1158
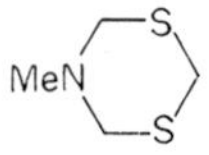 MeN, S, S	(i) BuLi, thf, $-78\,°C$ (ii) $C_{10}H_{21}I$ (iii) H_2O	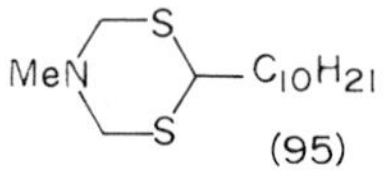 MeN, S, S, $C_{10}H_{21}$ (95)	1159
$EtS(O)CH_2SEt$	(i) $LiNPr^i{}_2$, thf, $0\,°C$ (ii) BuBr, $25\,°C$	$EtS(O)CHBuSEt$ (95)	1160
$PhSe(O)(CH_2)_3Ph$	(i) $LiNPr^i{}_2$, thf, $-78\,°C$ (ii) $Me_2C{=}CHCH_2Br$	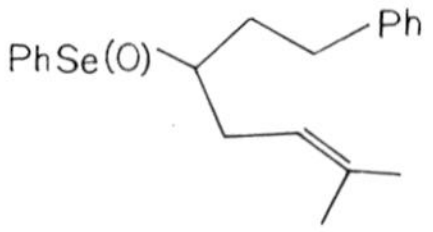 PhSe(O), Ph (88)	1161
$MeCHCl_2$	(i) BuLi, tmed, thf, $-95\,°C$ (ii) $C_7H_{15}Br$, hmpt, $-100\,°C$	$MeCCl_2C_7H_{15}$ (88)	1162
$CH_3CH_2CH_2CN$	(i) $LiNPr^i{}_2$, thf, $-78\,°C$ (ii) $MeCHBrCMe(OMe)_2$	$CH_3CH_2CH(CN)CH(Me)CMe(OMe)_2$ (61)	1163

TABLE 24. (*Contd.*)

Initial reagent	Conditions	Alkylated product (yield, %)	Ref.
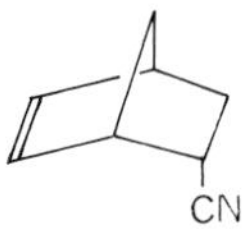 CN	(i) LiBu, thf, $-78\,^{\circ}C$ (ii) BuI	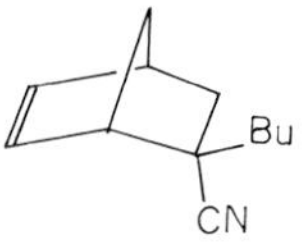 Bu CN (90)	1164
$MeCO_2Bu^t$	(i) $LiNPr^i$ (cyclo-C_6H_{11}), thf, $-78\,^{\circ}C$ (ii) o-$(BrCH_2)_2C_6H_4$	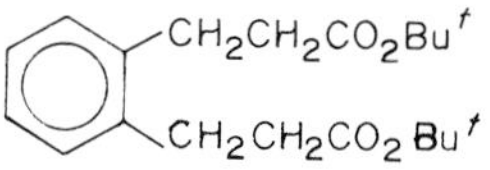 $CH_2CH_2CO_2Bu^t$ $CH_2CH_2CO_2Bu^t$ (93)	1165
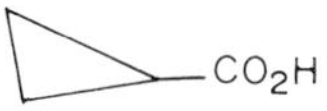 CO_2H	(i) 2.2 equiv. $LiNPr^i_2$, thf, $0\,^{\circ}C$ (ii) MeI (iii) H_3O^+	 CO_2H Me (89)	1166
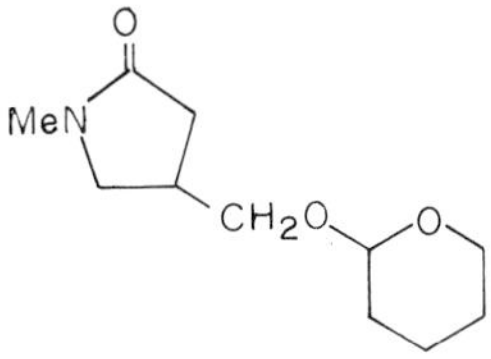 O MeN CH_2O O	(i) $LiNPr^i_2$, thf, $-78\,^{\circ}C$ (ii) $CH_2{=}CH(CH_2)_6I$	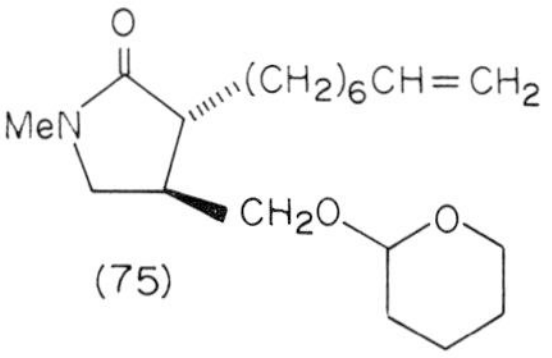 O MeN $(CH_2)_6CH{=}CH_2$ CH_2O O (75)	1167
$HC(S)NMe_2$	(i) $LiNPr^i_2$, thf, $-100\,^{\circ}C$, 3 min (ii) MeI	$MeC(S)NMe_2$ (50)	1168
$HC(O)NPr^i_2$	(i) $LiNPr^i_2$, thf, $-78\,^{\circ}C$ (ii) MeI 5 mm	$MeC(O)NPr^i_2$ (20)	1169
$(MeO)_2P(O)CH_2COMe$	(i) NaH, thf, $20\,^{\circ}C$ (ii) BuLi, $0\,^{\circ}C$ (iii) $BrCHMeC_6H_4Me$-p	$(MeO)_2P(O)CH_2COCH_2CH(Me)C_6H_4Me$-$p$ (50)	1170
$CH_3CH_2CH_2NO_2$	(i) 2 equiv. BuLi, thf, hmpt, $-90\,^{\circ}C$ (ii) $CH_3(CH_2)_5I$, -90 to $-15\,^{\circ}C$ (iii) AcOH, $-90\,^{\circ}C$	$CH_3CH_2CHNO_2(CH_2)_5CH_3$ (51)	1171
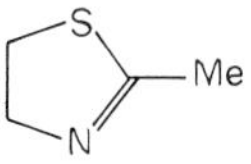 S N Me	(i) BuLi, thf, $-78\,^{\circ}C$ (ii) $PhCH_2Cl$, -78 to $25\,^{\circ}C$	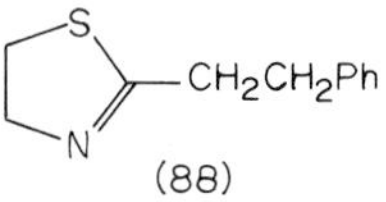 S N CH_2CH_2Ph (88)	1172
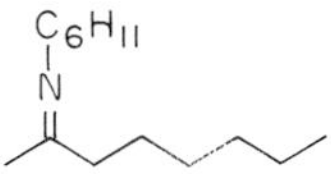 C_6H_{11} N	(i) $LiNPr^i_2$, thf, -40 to $-10\,^{\circ}C$ (ii) $BrCH_2CBr{=}CH_2$, thf, $-20\,^{\circ}C$ (iii) aq. HCl, $-40\,^{\circ}C$	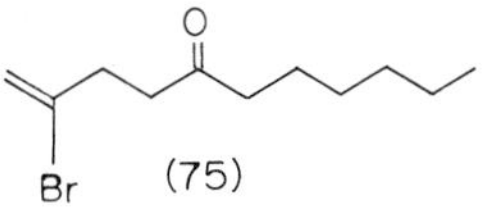 O Br (75)	1173

TABLE 24. (*Contd.*)

Initial reagent	Conditions	Alkylated product (yield, %)	Ref.
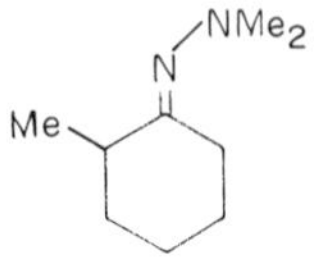 NMe$_2$, N, Me	(i) LiNPr$^i_2$, thf, 0 °C (ii) MeI, − 78 to 0 °C (iii) MeOH, NaIO$_4$, pH 25 °C	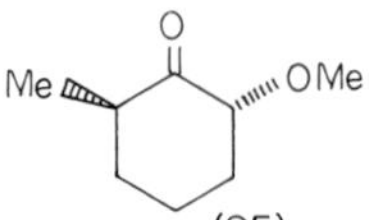 O, Me, OMe (95) *trans*:*cis* = 97:3	1174
$Ph_2As(O)Me$	(i) LiNPr$^i_2$, thf, − 40 °C (ii) EtBr, thf	$Ph_2As(O)CH_2CH_2CH_3$ (72)	1175
$(Ph_2Sb)_2CH_2$	(i) PhLi, thf, − 78 °C (ii) PrBr, thf, − 40 °C (iii) H_2O	$Ph_2SbCH_2CH_2CH_2CH_3$ (12)	1176
$CH_3(CH_2)_3C{\equiv}CH$	(i) BuLi, pentane, − 35 °C (ii) EtBr, 0 °C (iii) 4 N HCl	$CH_3(CH_2)_2CH(Et)C{\equiv}CH$ (64)	1177
$PhSCH_2C{\equiv}CH$	(i) equiv. BuLi, tmed, thf (ii) $Me_2C{=}CHCH_2Br$ (iii) H_3O^+	$HC{\equiv}CCHSPhCH_2CH{=}CMe_2$ (83)	1178
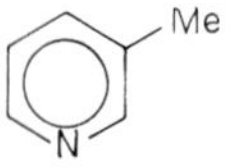 Me, N	(i) LiNPr$^i_2$, hmpt, thf (ii) Pr^{n}Br	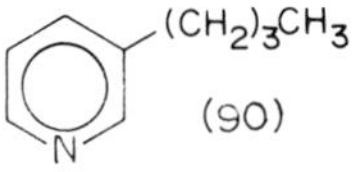 $(CH_2)_3CH_3$, N (90)	1179
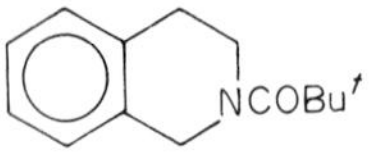 NCOBu^t	(i) Bu^{t}Li, tmed, thf, − 78 °C (ii) $CH_3(CH_2)_7Br$, 2 h or $CH_3(CH_2)_7Cl$, 24 h	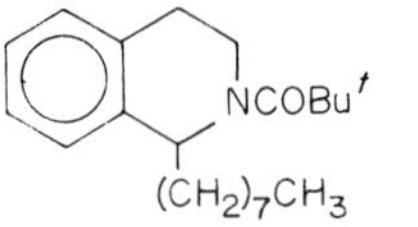 NCOBu^t, $(CH_2)_7CH_3$ (85–6)	1180
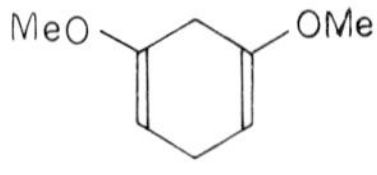 MeO, OMe	(i) Bu^{t}Li, thf, − 78 °C (ii) hmpt (iii) $CH_2{=}CH(CH_2)_2Br$, − 78 to 25 °C	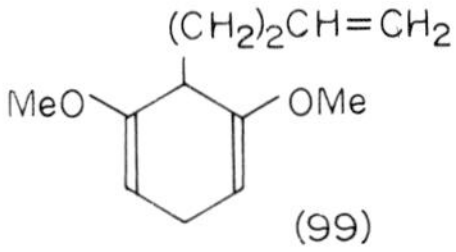 $(CH_2)_2CH{=}CH_2$, MeO, OMe (99)	1181
Bu^t $CH_2CMe_2N{=}C$	(i) Bu^{t}Li (ii) $CH_2{=}CHCH_2Br$, thf, − 70 to 25 °C	Bu^t $CH_2CMe_2N{=}C$(Bu^t)$CH{=}CHMe$ (46)	1182
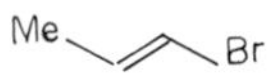 Me, Br	(i) 2 equiv. Bu^{t}Li, thf, Et_2O, pentane, − 120 °C (ii) $CH_3(CH_2)_7I$	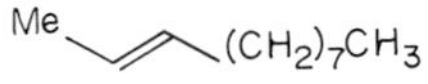 Me, $(CH_2)_7CH_3$ (62)	1183
$H_2C{=}CHOMe$	(i) Bu^{t}Li, thf, − 65 to − 50 °C (ii) $CH_3(CH_2)_7I$	$H_2C{=}C(OMe)(CH_2)_7CH_3$ (80)	1184

TABLE 24. (*Contd.*)

Initial reagent	Conditions	Alkylated product (yield, %)	Ref.
	(i) Bu^tLi, − 78 to 50 °C (ii) $CH_3(CH_2)_5I$	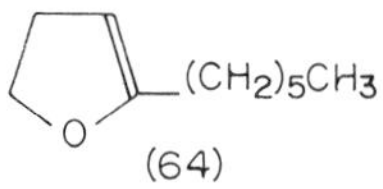 $(CH_2)_5CH_3$ (64)	1185
$CH_2{=}CHSCH_2CH_3$	(i) Bu^sLi, thf, hmpt, (ii) $\frac{1}{2}$ $Br(CH_2)_4Br$ (iii) $HgCl_2$	$CH_3CO(CH_2)_4COCH_3$ (60)	1186
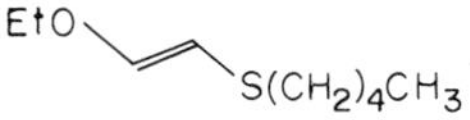 EtO, $S(CH_2)_4CH_3$	(i) Bu^tLi, thf, − 70 °C (ii) BuI	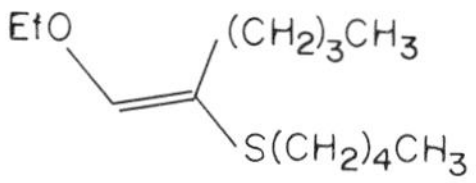 EtO, $(CH_2)_3CH_3$, $S(CH_2)_4CH_3$ (60)	1187
$H_2C{=}CHSeC_6H_4CF_3$-*m*	(i) $LiNPr^i{}_2$, thf, − 78 °C (ii) MeI	$H_2C{=}CMeSeC_6H_4CF_3$-*m* (90)	1188
$CH_3(CH_2)_9CH{=}C(SeMe)_2$	(i) BuLi, thf, − 78 °C (ii) MeI	$CH_3(CH_2)_9CH{=}CMeSeMe$ (80)	1189
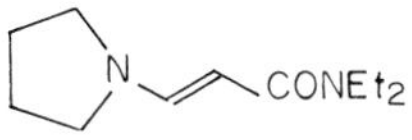 N, $CONEt_2$	(i) Bu^tLi, thf, − 115 °C (ii) MeI	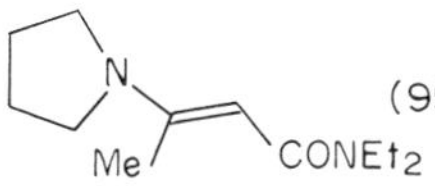 N, Me, $CONEt_2$ (95)	1190
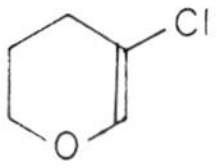 Cl	(i) BuLi, thf, 25 °C, 2 h (ii) BuI	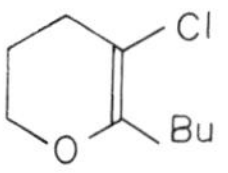 Cl, Bu (65)	1191
$PhCH(OH)CH_2NMe_2$	(i) BuLi, Et_2O, 25 °C, 24 h (ii) MeI (iii) H_2O	*o*-$MeC_6H_4CHOHCH_2NMe_2$ (47)	1192
$PhCONHBu^t$	(i) BuLi, thf, 0 °C, 1 h (ii) MeI (iii) H_2O	*o*-$MeC_6H_4CONHBu^t$ (50)	1193
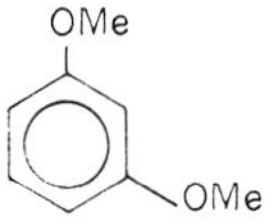 OMe, OMe	(i) PhLi, Et_2O, 25 °C (ii) 1-CH_2Cl-pyrrolidine	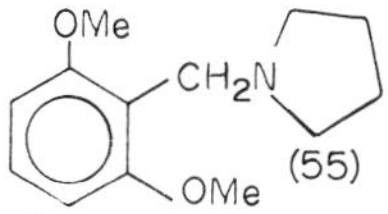 OMe, CH_2N, OMe (55)	1194
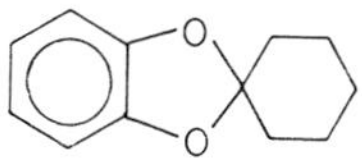	(i) BuLi, thf, Et_2O, 0 to 25 °C (ii) $Br(CH_2)_{11}Othp$	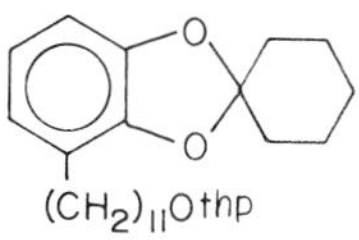 $(CH_2)_{11}Othp$ (84)	1195

TABLE 24. (*Contd.*)

Initial reagent	Conditions	Alkylated product (yield, %)	Ref.
$C_6H_5CH_2OH$	(i) BuLi, petrol, tmed, 11 h (ii) MeI	2-Me-$C_6H_4CH_2OH$ (30)	1196
3-chlorofuran	(i) $LiNPr^i_2$, thf, −80 °C 2.5 h (ii) $Me_2C{=}CHCH_2Br$	3-Cl-2-($CH_2CH{=}CMe_2$)furan (41)	1197
3-OBu^t-thiophene	(i) BuLi, Et_2O, 2 h (ii) Me_2SO_4	3-OBu^t-2-Me-thiophene (87)	1198
2-(4-pyridyl)-4,4-dimethyloxazoline	(i) MeLi, thf, −78 to 0 °C (ii) MeI	2-(3-Me-4-pyridyl)-4,4-dimethyloxazoline (63)	1199

There appears to be conflicting evidence regarding the mechanism and stereochemistry of alkylations of organoalkali metals[1,3,5,536]. Second-order kinetics—first order in both the organolithium and the alkyl halide—are often met.[5] Net inversions of configuration (the extents of which are not always known) have been reported, for example in the reactions between (i) 2-lithiothiane[542] and $MeCH_2CHBrMe$ (> 85% inversion), (ii) PhLi and $MeCH_2CH_2CHDCl$ in PhH[548], (iii) Ph_2CHLi and (*R*)-$PhCHClCMe_3$, equation 217[549], and (iv) for allyl– and benzyl–metal reactions with secondary alkyl halides and tosylates[536]. Inversion of configuration (> 93%) was noted[514] in the reaction between $CH_2{=}CHCH_2Li$ and (−)-2-octyl tosylate.

$$Ph_2CHLi + Ph\text{–}C(Bu^t)(H)\text{–}Cl \xrightarrow{\text{thf, 25 °C}} Ph_2CH\text{–}C(Bu^t)(H)\text{–}Ph + PhCHBu^tCHBu^tPh + Ph_2CHCHPh_2 \quad (217)$$

50 %; *meso* 10 %, α,α' 7 %; 18 %

Such findings on the order of the reaction and the stereochemistry suggest an S_N2 type reaction. However, in other cases racemization, particularly for primary and secondary alkyl metals with secondary alkyl halides and sulphonates, and the intermediacy of free radicals, e.g. for primary alkyllithiums with primary alkyl halides, have been noted. A radical nature is also suggested for the reaction of PhC≡C(CH$_2$)$_4$Br with BuLi by the formation of PhCH=$\overline{\text{C(CH}_2)_3\text{C}}H_2$.

Alkylation of ion-paired [MeS(O)CHPh$^-$ Li$^+$] by MeI in solution occurs predominantly[534,550] with inversion of configuration to give **96** (E = Me), whereas (MeO)$_3$PO reacts with retention. This suggests for the acyclic anion that strong coordination to Li$^+$ [as with (MeO)$_3$PO] leads to retention whereas weak or no coordination (as with MeI) provides inversion; however, this did not seem to occur with alicyclic sulphoxides[551].

Alkylation of 2-lithiodithianes occur preferentially to give equatorial alkylated products[336], equation 218.

(i) BuLi; (ii) MeI (218)

The overall stereochemistry of successive lithiations and alkylation of carboxylates have also been reported[552], equation 219.

endo- or *exo*- —CO$_2$H; (i) LiNPr$^i_2$, thf, −75 °C; (ii) RX, −75 °C; (iii) H$_3$O$^+$ → **(97)** + **(98)**

[97] : [98]	66:34	88:12
RX	Me I	BuBr

(219)

MeS—CH=CH—(CH$_2$)$_7$Me; (i) Bu^{s}Li, thf, hmpt; (ii) Me(CH$_2$)$_7$Br, −78 to 25 °C → Me(CH$_2$)$_7$(MeS)C=CH(CH$_2$)$_7$Me; HgCl, MeCN; H$_2$O, Δ → Me(CH$_2$)$_7$CO(CH$_2$)$_8$Me (220)

Cyclopropyl-[538] and vinyllithiums[107,425b,539], e.g. equation 220[424d], react with alkyl halides with retention of configuration. However, isomerization of the initial organolithium can result. The stereochemistry of the products of successive lithiations (by BuLi in thf at −95 °C) and methylations (by MeI) of 1-R-2, 2-Br$_2$-cyclopropanes depends on the ageing of the carbenoid intermediates, 1-R-2-Br-2-Li-cyclopropanes. Only the most rapid trapping of the carbenoid provided some *trans*-methylated product[77].

Alkylations of 9-R-10-Li-9, 10-dihydroanthracenes, **99**, occur preferentially by axial attack, especially by primary alkyl halides; e.g. EtBr and **99** (R = Et) in thf provide 92% (*cis*)-9, 10-Et$_2$-9, 10-dihydroanthracene. With greater steric hindrance, the amount of *trans*-product increases[553]. Alkylation of the delocalized carbanion **94** occurs at C$_9$ and C$_{10}$ (equation 209, E = alkyl)[36]. Other studies on delocalized carbanions have revealed different reactivities for different ion-pair forms[554].

As indicated in Section B.2.a.vii, equilibria exist in solution between propargyl- and allenyllithiums[112,299]. Both forms can be alkylated; indeed alkylations have been used to provide measures of propargyl–allenyl equilibria. Of interest, CIDNP was observed[299] in the reaction between $PhCH_2Br$ and $Me_2C{=}C{=}CHLi$.

$$CH_2{=}C{=}CH_2 \xrightarrow[\text{(ii) } C_8H_{17}I]{\text{(i) BuLi, thf, } -70\,^\circ C} \underset{(\mathbf{100})}{C_8H_{17}CH{=}C{=}CH_2} + \underset{(\mathbf{101})}{C_8H_{17}CH_2C{\equiv}CH} \qquad (221)$$

$$[\mathbf{100}] : [\mathbf{101}] = 87{:}13$$

Substituted allyllithiums can in principle be alkylated at either allylic position[302]: steric hindrance and internal coordination are among the more important factors controlling the sites of alkylation; compare equations 222 with 223[450a] and 224 with 225[540b].

$$PhSCH_2CH_2 \xrightarrow{LiNPr^i_2} PhSCH\text{-----}\bar{C}H\text{-----}CH_2Li^+ \xrightarrow[\text{thf}, -65\,^\circ C]{C_6H_{11}I,} \underset{(\mathbf{102})}{PhSCH(C_6H_{11})CH{=}CH_2} + \underset{(\mathbf{103})}{PhSCH{=}CHCH_2C_6H_{11}}$$

$$[\mathbf{102}] : [\mathbf{103}] = 3{:}1 \qquad (222)$$

$$\text{1-Me-imidazol-2-yl}{-}SCH_2CH{=}CH_2 \longrightarrow \text{1-Me-imidazol-2-yl}{-}S\text{-}CH\text{---}CH\text{---}CH_2\ (N \rightarrow Li^+) \xrightarrow[\text{thf}, -65\,^\circ C]{C_6H_{11}I} \text{1-Me-imidazol-2-yl}{-}S{-}CH(C_6H_{11})CH{=}CH_2 \quad >99\% \qquad (223)$$

$$MeCH{=}CHMe \xrightarrow[\text{(ii) } Ph(CH_2)_2Br]{\text{(i) BuLi, } Et_2O} \underset{(\mathbf{104})}{Ph(CH_2)_2CH(Me)CH{=}CH_2} + \underset{(\mathbf{105})}{MeCH{=}CH(CH_2)_3Ph} \qquad (224)$$

$$[\mathbf{104}]{:}[\mathbf{105}] = 68{:}32$$

$$Me_2CH{=}CHMe \xrightarrow[\text{(ii) } Ph(CH_2)_2Br]{\text{(i) BuLi, } Et_2O} Me_2C{=}CH(CH_2)_3Ph \quad \textit{ca. } 100\% \qquad (225)$$

b. Reactions with other organic halides

Benzyl and allyl halides are especially reactive in coupling reactions and for this reason are not normally used as precursors of benzyl- and allyllithiums. Reactions of chloro and fluoro derivatives can be complicated by some metallation[5] occurring α to the halo group (**106** in equation 226 is probably derived from the α-lithiated species). As indicated in equation 226, reaction of substituted allyl halides can take place at either allylic site. In addition, it can be seen that the geometry about the double bond is retained in **108**[555,556].

$$MeCH{=}CHCH_2Cl \xrightarrow[\text{(ii) } H_2O]{\text{(i) PhLi}} \underset{(\mathbf{107})}{PhCH(Me)CH{=}CH_2} + \underset{(\mathbf{108})}{MeCH{=}CHCH_2Ph} + \underset{(\mathbf{106})}{Ph\text{-cyclopropyl-}Me} \qquad (226)$$

$$[\mathbf{106}] : [\mathbf{107}] : [\mathbf{108}] = 11{:}23{:}66$$

Other complications arise from metal–halide exchanges, which would provide, for example, the symmetric coupled products in equation 217. Synthetic use[557] can however be made of such metal–halide exchanges, equation 227.

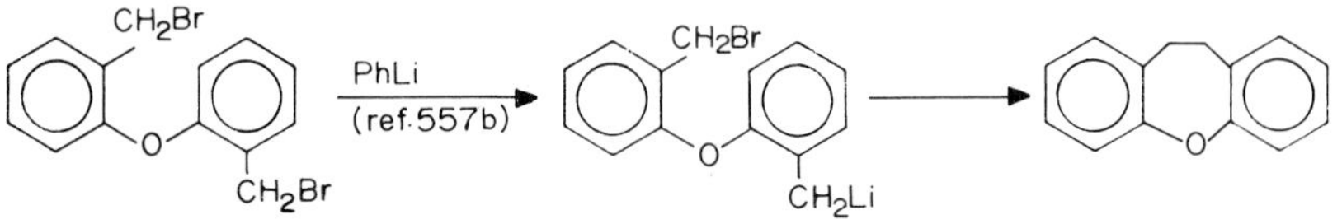

(227)

Vinyl, alkynyl, and aryl halides, including per- and polyhalo derivatives, although much less reactive than alkyl halides, can also take part in coupling reactions[5]. Direct coupling of aryl or vinyl halides with aryl- or vinyllithium species is difficult at the very least. Direct coupling, e.g. equation 214, between aryl halides and RLi has only been reported for aryl halides[5] containing strongly electron-withdrawing groups, e.g. oxazolinyl[558] and nitro[559]. Other mechanisms have been detected for vinyl and aryl halides. These mechanisms involve (i) addition–elimination steps, e.g. as in alkylation of arenes, including pyridines[5], and haloalkenes, equations 228[560] and 229[561].

$$PhLi + CF_2{=}CF_2 \longrightarrow LiCF_2CF_2Ph \xrightarrow{-LiF} CF_2{=}CFPh \quad (228)$$

$$RLi + PhCCl{=}CH_2 \longrightarrow PhCClLiCH_2R \xrightarrow{-LiCl} [Ph\ddot{C}CH_2R] \rightarrow PhCH{=}CHR \quad (229)$$

(ii) metallation–elimination–addition steps[562–564], e.g. equations 230 and 231:

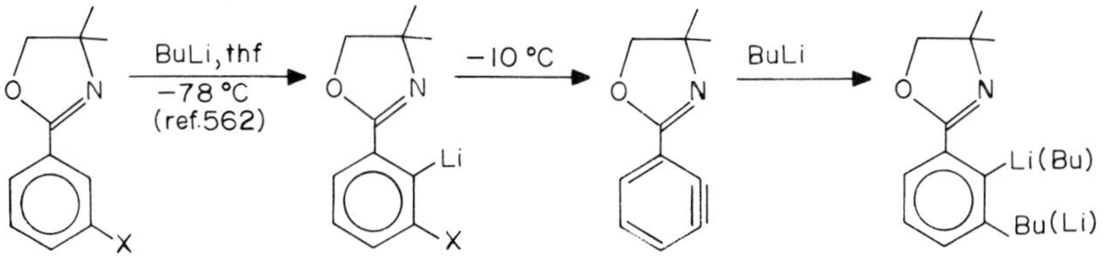

(230)

X = F or Cl

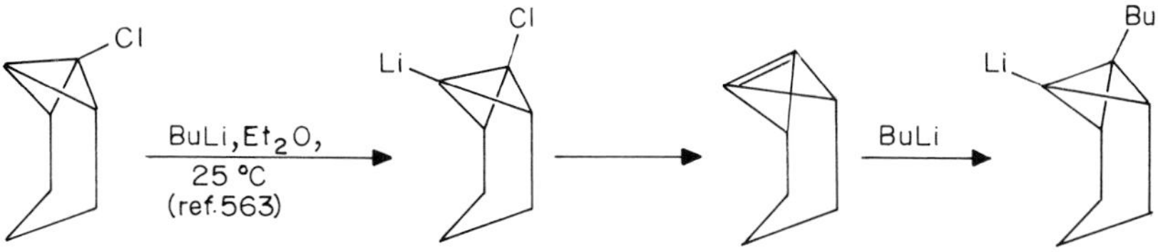

(231)

and (iii) metal–halide exchanges[55,565], equation 10. Whereas the reaction of BuLi and ArBr does eventually lead to coupled products, those of Bu^{s}Li or Bu^{t}Li merely produce ArLi[536]. Cyclopropyllithium and halobenzenes in refluxing Et_2O provide cyclopropylbenzene as well as other products, including biphenyl, via mechanisms including metal–halide exchanges and the formation of benzyne.

c. Asymmetric synthesis

Asymmetric synthesis involving alkylation of metallated chiral molecules has been extensively developed in recent times[566]. Considerable success has been realized using compounds in which intramolecular coordination creates a chiral environment at the reaction centre. Some examples follow.

Introduction of a chiral group α to a carbonyl group has been made via the intermediacy of chiral hydrazones[567] and imines[568], Schemes 10 and 11.

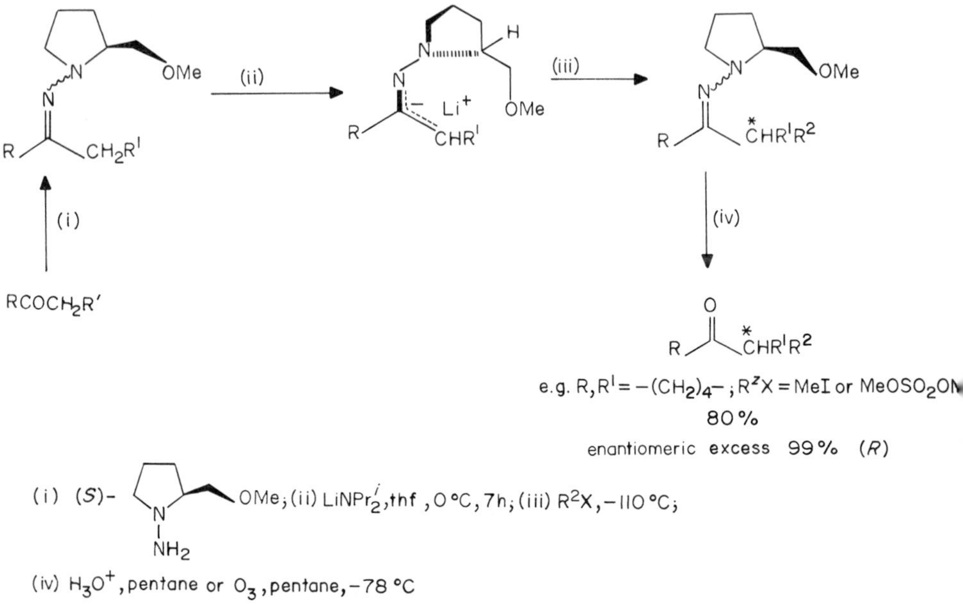

$RCOCH_2R'$

e.g. $R,R^1 = -(CH_2)_4-$; $R^2X = MeI$ or $MeOSO_2OMe$

80%

enantiomeric excess 99% (*R*)

(i) (*S*)- [N-amino-2-(methoxymethyl)pyrrolidine]; (ii) $LiNPr^i_2$, thf, 0 °C, 7h; (iii) R^2X, −110 °C;

(iv) H_3O^+, pentane or O_3, pentane, −78 °C

SCHEME 10

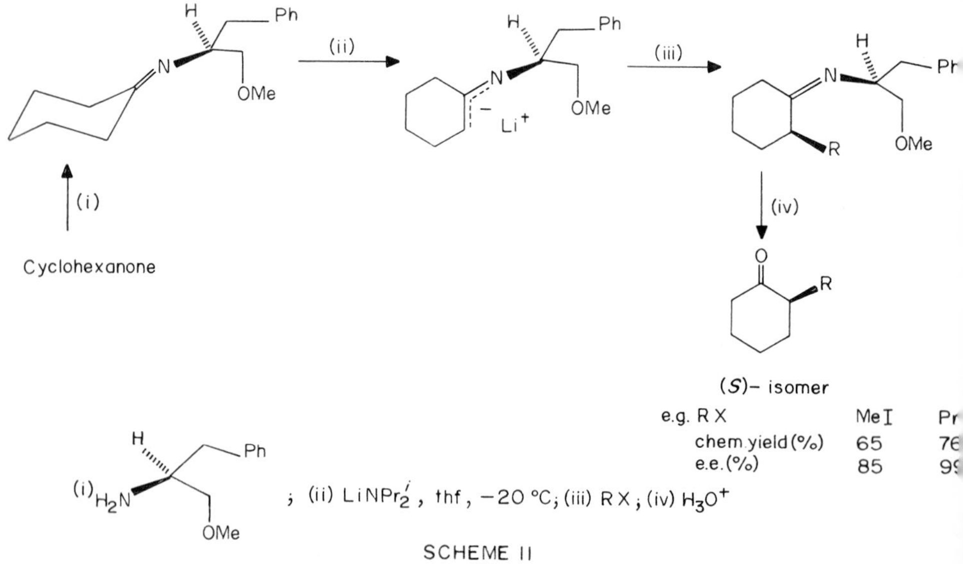

Cyclohexanone

(*S*)- isomer

e.g. RX	MeI	Pr
chem.yield (%)	65	76
e.e. (%)	85	99

(i) [H_2N–CH(CH$_2$Ph)(CH$_2$OMe)]; (ii) $LiNPr^i_2$, thf, −20 °C; (iii) RX; (iv) H_3O^+

SCHEME 11

Introduction of a chiral group in the 1-position of a tetrahydroisoquinoline was achieved as shown in Scheme 12 with greater than 90% enantiomeric excess[307c,569].

(i) HCO_2Et; $Et_3O^+BF_4^-$; R^*NH_2; (ii) $LiNPr^i_2$, thf, –78 °C; (iii) RX, –78 or –100°C (iv) H_2NNH_2, HOAC.

R^*	RX	Overall chem yield %	e.e.(%)	Configuration of product
(R)-(–)	MeI	85	10	(R)
	Bu^iBr	84	27	(R)
	$PhCH_2CH_2Br$	89	52	(S)
(S),(S)-(+)	MeI	79	99	(S)
	Bu^iBr	85	91	(S)
	$PhCH_2CH_2Br$	85	99	(S)

SCHEME 12

A number of schemes[570,571] have involved asymmetric synthesis of α-alkylated carboxylic acids, as shown for example in Scheme 13, which provides an overall substitution of the α-proton in α-hydroxy- and α-mercaptocarboxylic acids, $RCH(XH)CO_2H$ (**109**; X = O or S) by an alkyl group with retention of configuration, via the *cis*-isomers of 2-Bu^t-5-R-1, 3-dioxolanones or -1, 3-oxothiolanones (**110**), the products of condensation of **109** with Bu^tCHO. The *trans*-isomer of **110** could be used to obtain $RR'C(XH)CO_2H$ with inversion of configuration.

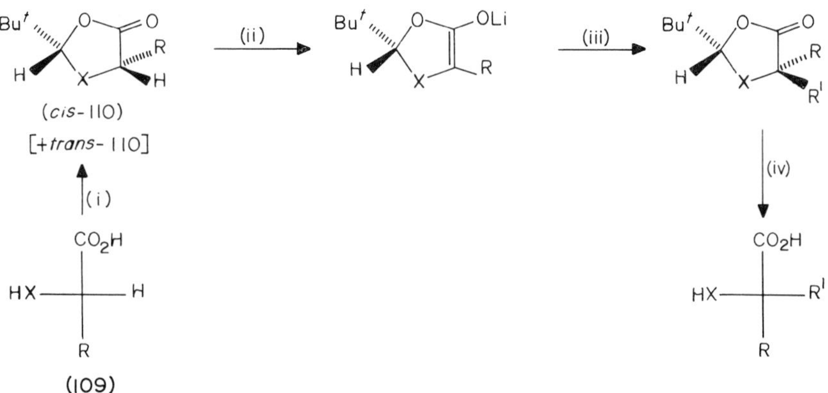

(i) Bu^tCHO; (ii) $LiNPr^i_2$, thf, hexane, –78 °C; (iii) R′X, –78 °C; (iv) hydrolysis

	Chem yield (%)	Diastereoselectivity %
X = S; R = Me; R′X = CH_2=$CHCH_2Br$	92	>98
X = O; R = Me; R′X = EtBr	82	>97

SCHEME 13

Syntheses of chiral α-alkyl-β-hydroxy esters[572] and α-alkyl-α-amino esters[573] have also been developed e.g. equations 232 and 233. Dianion **111** has a rigid structure, which arises from chelation of the lithium cation by the two oxygen anions. Asymmetric alkylations have also been reported for metalloenamines[574] and metallated amides[575], in addition to metallated **112**[576], **113**[577], and **114**[578]—the arrows indicate sites of metallation and alkylation.

(S)-isomer —2 equiv. $LiNPr^i_2$ thf, −50 to 0 °C→ (111) —Li^+ (i) RX (ii) H_3O^+→ 2-(S)-3R(S), RX = MeI, chem. yield 67%, diastereoselectivity >95% (232)

(i) BuLi, thf, −78 °C (ii) RX; H_3O^+; R = Pr^i e.e. >92% (233)

(112) (113) (114)

3. Formation of Alkenes[187,301,324,536,579]

Alkenes have been generated from reactions of certain functionally substituted alkyllithiums with carbonyl compounds. Such reactions compliment the Wittig reaction.

$$\mathrm{Y(X)CHLi + RR^1CO \rightarrow Y(X)CHC(R)(R^1)OLi \rightarrow YCH{=}CRR^1} \quad (234)$$

Typical substituents, X, in equation 234 include triorganosilyl groups[579] (Peterson reaction), sulphur-containing groups, e.g. RS, RS(O), and $RS(O)_2$, selenium-containing groups[187], amino groups[301], and phosphorus-containing groups[324], e.g. R_2P, $(R_2N)_2P(O)$, $(RO)_2P(O)$, and $(RO)_2P(S)$ (Wadsworth–Emmons reaction). Various substituted alkenes can be generated this way; Table 25 lists some examples. The formation of the alkene from the β-alkoxy adduct can occur spontaneously, or on simple hydrolysis or after further reaction and workup.

Use of α-trialkylsilylalkyllithiums in these alkene formations has been especially popular since Peterson's first report[579] and allows the formation of unsubstituted alkenes (equation 235) as well as vinyl cyanides, sulphides, sulphoxides, halides, silanes,

TABLE 25. Formation of alkenes

Reagent	Conditions	Product (yield, %)	Ref.
Me_3SiCH_2Ph	(i) BuLi, tmed (ii) Ph_2CO (iii) H_2O	$Ph_2C{=}CHPh$ (77)	1200
Ph_3SnCH_2I	(i) BuLi, Et_2O_3 $-50\,°C$ (ii) RR′CO, Et_2O (iii) H or H_3O^+	$CH_2{=}CRR'$ (95) $R = RR$; $R' = H$	1201
Ph_2PMe	(i) BuLi, tmed, hexane, thf (ii) Ph_2CO, thf (iii) MeI (iv) Bu^{t}OK, dme	$CH_2{=}CPh_2$ (37)	1202
$(MeO)_2P(S)CHMe_2$	(i) BuLi, thf, $-50\,°C$ (ii) 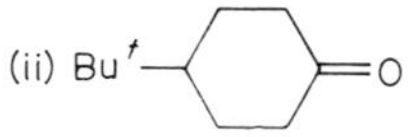 4-Bu^t-cyclohexanone (Bu^t–C$_6$H$_{9}$=O) (iii) $50\,°C$, 4 h	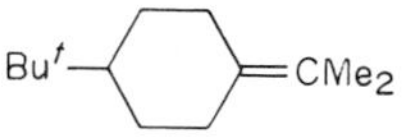 Bu^t–C$_6$H$_{9}$=CMe$_2$ (71)	1203
PhSCHMeLi	(i) $Me(CH_2)_{10}CHO$, thf, $-78\,°C$ (ii) PI_3, $20\,°C$, 4 h	$Me(CH_2)_{10}CH{=}CHMe$ (50)	1204
Bu^{t}S(O)CH_2Me	(i) MeLi, thf, $-60\,°C$ (ii) PhCOMe, thf, $-78\,°C$ (iii) H_2O (iv) NCS	$MeCH{=}CMePh$ (73)	1205
PhS(=O)(=NMe)Me	(i) BuLi, thf, $0\,°C$ (ii) RR^1CO, thf (iii) Al(Hg), AcOH, H_2O	$RR^1C{=}CH_2$ $R = C_{15}H_{31}$; $R^1 = Me$ (90)	1206
PhP(=S)(NMe$_2$)Me	(i) BuLi, thf, $-78\,°C$ (ii) RR^1CO (iii) MeI, py	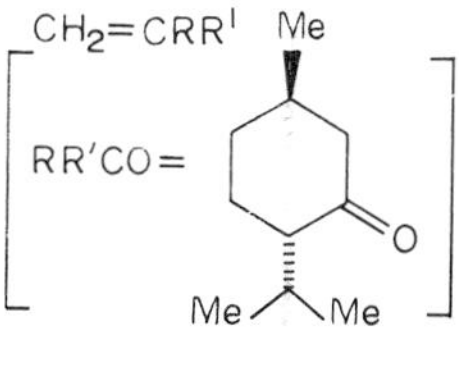 $CH_2{=}CRR^1$ [RR′CO = (menthone: Me-substituted, isopropyl-substituted cyclohexanone)] (99)	1207
Me_3SiCH_2CN	(i) BuLi, thf, $-78\,°C$ (ii) RR^1CO (iii) H_2O	$RR^1C{=}CHCN$ $R = Ph$; $R^1 = H$, (79); (*Z*):(*E*) = 1:1	1208
$(EtO)_2P(O)CH_2CN$	(i) NaH, dme (ii) R^1R^2CO (iii) H_2O	$RRC{=}CHCN$ $R = Ph$; $R^1 = Me$ (**E**):(**Z**) = 10:1	1209
$(EtO)_2P(O)CH_2CHCNSiMe_3$	(i) LiNPr$^i_2$, thf, $-78\,°C$ (ii) RR^1CO, $-78\,°C$ (iii) H_2O	$(EtO)_2P(O)CH_2CCN{=}CRR'$ $R = Pr^i$; $R^1 = H$ (48) (*E*)-isomer	1210
$(EtO)_2P(O)CHRCO_2Et$	(i) NaH, PhH (ii) R^1R^2CO, $20\,°C$ (iii) 60–5 °C	$R^1R^2C{=}CRCO_2Et$ $R = H$; $R^1,R^2 =$ $-(CH_2)_5-$ (67–77)	1211

TABLE 25. (*Contd.*)

Reagent	Conditions	Product (yield, %)	Ref.
$Ph_2P(O)CH_2N$(morpholino)	(i) BuLi, thf, 0 °C (ii) RR^1CO (iii) H_3O^+ (iv) KH	$RR^1C{=}CH$(morpholino); R = Ph; R^1 = H (90)	1212
$(EtO)_2P(O)CH_2N$(pyrrolidino)	(i) BuLi, thf, −78 °C (ii) RR^1CO	$RR^1C{=}CH$(pyrrolidino); R,R^1 = $-(CH_2)_5-$	1213
$(EtO)_2P(O)CHRN{=}CHPh$	(i) BuLi (ii) R^1R^2CO	$R^1R^2C{=}C(R)N{=}CHPh$; R = H; R^1,R^2 = $-(CH_2)_5-$	1214
$(EtO)_2P(O)CH_2NC$	(i) BuLi, thf, pentane, −70 °C (ii) RR^1CO, thf, −60 to 20 °C (iii) H_2O	$RR^1C{=}CH(NC)$; R = Ph; R^1 = H (75)	1215
Me_3SiCH_2OMe	(i) Bu^{s}Li, thf, −78 to −30 °C (ii) RR^1CO (iii) KH, thf, 60 °C	$RR^1C{=}CH(OMe)$; R,R^1 = $-(CH_2)_5$ (80)	1216
$Me_3SiCH_2S(O)Ph$	(i) Bu^{t}Li, thf, pentane, −70 °C (ii) RR^1CO, thf, −70 to 20 °C (iii) H_2O	$RR^1C{=}CHSOPh$; R = H; R^1 = Ph (87)	1217
$(EtO)_2P(O)CH_2S(O_2)Me$	(i) BuLi, thf, pentane, −78 °C (ii) RR^1CO, thf, −78 to 25 °C (iii) H_2O	$RR^1C{=}CHSO_2Me$; R = Ph; R^1 = H (87)	1218
$(PhSe)_2CH_2$	(i) $LiNPr^i_2$ (ii) RCHO (iii) H_2O	PhSeCH=CHR; R = Ph; (74); (*E*)-isomer; R = Me; (84); (*E*):(*Z*) = 1:1	1219
$MeSCH(SiMe_3)SnMe_3$	(i) $LiNPr^i_2$, thf, hmpt (ii) RR^1CO (iii) H_2O	$Me_3Sn(MeS)C{=}CRR^1$; R = H; R^1 = Ph (82)	1220

TABLE 25. (*Contd.*)

Reagent	Conditions	Product (yield, %)	Ref.
$(EtO)_2P(O)CHClSPh$	(i) $LiCCl_3$ (ii) RR^1CO (iii) H_2O	$RR^1C{=}C(Cl)SPh$; $R = p\text{-}FC_6H_4, R^1 = H$ (60)	1221
2-Me_3Si-1,3-dithiane	(i) BuLi, thf, hexane, 0 °C (ii) RR^1CO, thf, 0 to 25 °C (iii) H_3O^+	$RR^1C{=}$(1,3-dithian-2-ylidene); $R = R^1 = Ph$ (78); $R = Pr^i, R^1 = H$ (44)	1222
$Me_3SiCH(NC)\cdot SO_2C_6H_4Me\text{-}p$	(i) BuLi, thf, hexane, − 60 °C (ii) RR^1CO, thf, − 30 °C (iii) H_2O, MeOH	$RR^1C{=}C(N{=}C)SO_2C_6H_4Me\text{-}p$; $R = Ph$; $R^1 = H$ (> 80)	1223
$(EtO)_2PCH_2Cl$	(i) BuLi, Et_2O, thf, − 75 °C (ii) CCl_4 (iii) $LiCCl_3$, thf, − 70 °C (iv) RR′CO (v) H_2O	$RR'C{=}CCl_2$; $R = p\text{-}FC_6H_4, R^1 = H$ (80)	1224
$(Me_3Si)_2CBr_2$	(i) BuLi, thf, hexane, − 115 °C (ii) RCHO	$RCH{=}CBrSiMe_3$; R = Me (*E*):(*Z*) = 1:1	1225
$H_2C{=}CHCH_2SiMe_3$	(i) Bu^tLi, hmpt, − 78 °C (ii) $MgBr_2$ cat., $RR^1C{=}O$ (iii) MeCOCl, Δ	$CH_2{=}CHCH{=}CRR^1$; R = Ph; R^1 = Me	1226

carboxylates, phosphonates, etc. However, these are not usually highly stereoselective reactions. In contrast, reactions of $(EtO)_2P(O)CHYLi$, especially with aldehydes, RCHO, are frequently highly stereospecific, the products being (*E*)-RCH=CHX (X = SMe, SOMe, SO_2Ar, SePh, CN, CO_2Et, etc.)[324,325]. From *β*-hydroxy adducts, containing both trialkylsilyl and phosphorus (V) groups, it appears that the R_3Si group is the one to preferentially depart, e.g. equation 236[579].

$$Me_3SiCHLiPh + Ph_2CO \longrightarrow [Me_3SiCHPhCPh_2OLi] \longrightarrow PhCH{=}CPh_2 \quad (235)$$

$$Me_3SiCHLiP(S)Ph_2 + Ph_2CO \longrightarrow [(Me_3Si)Ph_2P(S)CHC(OLi)Ph_2] \longrightarrow Ph_2P(S)CH{=}CPh_2 \quad (236)$$

The adducts of reaction of $(Me_3Si)(MeSe)CR'Li$ and R^2R^3CO are particularly versatile [580] since conditions have been realized to provide $Me_3SiCR^1{=}CR^2R^3$, $MeSeCR^1{=}CR^2R^3$, and even $BrCR^1{=}CR^2R^3$ from these adducts (equation 237).

$$\text{MeSe(Me}_3\text{Si)CLiR}^1 \xrightarrow[\text{(ii) H}_2\text{O}]{\text{(i) R}^2\text{R}^3\text{CO}} \text{Me}_3\text{Si(R}^1\text{)(SeMe)C—C(OH)R}^2\text{R}^3$$

$$\xrightarrow[55\,^\circ\text{C}]{\text{Bu}^t\text{OK, thf,}} \text{MeSe(R}^1\text{)C=CR}^2\text{R}^3$$

$R^1 = R^2 = H$; $R^3 = C_{10}H_{21}$ 74%
$R^1 = R^3 = H$; $R^2 = C_{10}H_{21}$ 78%

$$\xrightarrow[\text{CH}_2\text{Cl}_2]{\text{POCl}_3\text{, Et}_3\text{N,}} \text{R}^1\text{(Me}_3\text{Si)C=CR}^2\text{R}^3 \qquad (237)$$

$R^1 = R^2 = H$; $R^3 = C_{10}H_{21}$ 83%
$R^1 = R^3 = H$; $R^2 = C_{10}H_{21}$ 75%

$$\xrightarrow{\text{Br}_2\text{, CCl}_4\text{, 20}\,^\circ\text{C}} \text{BrCR}^1\text{=CR}^2\text{R}^3$$

$R^1 = Me$; $R^2 = H$; $R^3 = C_{10}H_{21}$ 80%

In addition to aldehydes or ketones, amides have also been successfully used[581], e.g. equation 238.

$$\text{Me}_3\text{SiCHLiSR} + \text{R}^1\text{CONR}_2{}^2 \xrightarrow[\text{(ii) H}_2\text{O}]{\text{(i) thf, 0}\,^\circ\text{C}} \text{RSCH=CR}^1\text{NR}_2{}^2 \qquad (238)$$

$R = Ph$, $R^1 = H$, $R^2 = Me$
(*E*)-isomer, 64%

4. Formation of Cyclopropanes[536]

Reactions of alkenes with α-haloalkyl alkali metal derivatives are well established routes to cyclopropanes (equation 239)[433]. Some examples are given in equations 240–243.

$$\text{XYZCLi} + \text{>C=C<} \longrightarrow \text{cyclopropane (CZY)} \qquad (239)$$

X = helo

$$\text{Me—C}_6\text{H}_4\text{—CH}_2\text{Cl} \xrightarrow[\text{Et}_2\text{O}]{\text{Litmp}} \text{Me—C}_6\text{H}_4\text{—CHLiCl} \xrightarrow[\text{(ref.582)}]{\text{CH}_2\text{=CHOEt}}$$

$$\text{Me—C}_6\text{H}_4\text{—c-C}_3\text{H}_4\text{—OEt} \qquad (240)$$

75%

$$ClCH_2CH_2OCH_2Cl \xrightarrow[Et_2O, 0\,°C]{Li\,tmp} ClCH_2CH_2OCHLiCl \xrightarrow[(ref\ 583)]{cyclohexene} \text{bicyclo[4.1.0]heptane}{-}OCH_2CH_2Cl \tag{241}$$

55%; *Syn* : *anti* = 1·75:1

$$H_2C{=}CHCHCl_2 \xrightarrow[Et_2O, -5\,°C]{Li\,tmp} H_2C{\cdots}\bar{C}H{\cdots}CLiCl_2\ Li^+ \xrightarrow[(ref. 584)]{R^1R^2C{=}CR^3R^4} \text{cyclopropane } (R^1, R^2, R^3, R^4;\ Cl,\ CH{=}CH_2) \tag{242}$$

$R^1 = R^2 = R^3 = R^4 = Me$ 18%

$R^1 = R^4 = Me$; $R^2 = R^3 = H$ 30%

$$CFBr_3 \xrightarrow{BuLi} CFLiBr_2 \xrightarrow[(ref. 74)]{MeCH{=}CHMe} \text{cyclopropane (Me, Me, H, H; F, Br)} + \text{cyclopropane (Me, Me, H, H; Br, F)} \tag{243}$$

$$LiC(SPh)_3 + H_2C{=}C(SR)_2 \longrightarrow \text{cyclopropane (SPh, SPh; SR, SR)} \tag{244}$$

$$\left[Bu^tCH_2OCH_2Cl \xrightarrow{Li\,tmp}\right] Bu^tCH_2OCHLiCl + Me(CH_2)_2C{\equiv}C(CH_2)_2Me \xrightarrow{(ref. 585)} \text{cyclopropene } (OCH_2Bu^t;\ Me(CH_2)_2,\ (CH_2)_2Me) \tag{245}$$

As shown in equations 242 and 243, addition of the carbene to the alkenes maintains the stereochemistry of the alkene. Tris(phenylthio)alkyllithium, $LiS(CPh)_3$, also acts as a carbenoid agent (equation 244)[493]. Cyclopropenes may similarly be obtained from carbenoid reagents on reaction with alkynes, equation 245. Another route to cyclopropanes is the reaction of carbonyl compounds with lithiated sulphoximes, e.g. $RS(O)(NMe)CH_2Li$ and $RS(O)(NSO_2C_6H_4Me\text{-}p)CH_2Li$. The initial β-hydroxy adducts require successive treatments with alkylating agents and bases to generate the cy-

Ph–S(O)(=NMe)–Me —BuLi→ Ph–S(O)(=NMe)–CH_2Li (115) —(i) PhCOCH=CHPh, (ii) H_2O→ Ph–S(O)(=NMe)–$CH_2CH(Ph)CH_2COPh$ —(i) Separatoin of diastereomers, (ii) $Me_3O^+BF_4^-$→ $PhCH(CH_2COPh)CH_2S^+(O)(Ph)NMe_2$ BF_4^- → trans-1-benzoyl-2-phenylcyclopropane (Optically pure) (246)

clopropane[331]. Other syntheses of cyclopropanes involve reactions of α-metalloallylsulphones and -nitroalkalkanes (and various ylides) with electron-deficient alkenes[187].

5. Formation of Epoxides[187,536]

Epoxides have been produced from α-haloalkylalkali metals and carbonyl compounds[586–588] equation 247. Specific examples are given in equations 248–250.

$$R^1R^2CO + MCXYR^3 \longrightarrow R^1R^2C(OM)\text{—}CR^3Y(X) \longrightarrow R^1R^2C\overset{O}{\text{—}}CR^3Y \quad (247)$$

X = halo; Y = H, alkyl, aryl, CN, SR, SOR, SO_2R, $SiMe_3$, etc.

[Me_3SiCH_2Cl —Bu^sLi tmed, thf, −78 °C→ $Me_3SiCHLiCl$] + ketone → Me_3Si-epoxide, 85% (248)

$PhS(O)CH_2Cl$ —BuLi, thf, −78 °C→ $PhS(O)CHLiCl$ —(i) R^1R^2CO, (ii) KOH, MeOH, 25 °C (ref 588)→ PhS(O)-epoxide-R^1R^2; $R^1,R^2 = (CH_2)_5$ one isomer, 79% (249)

$$\mathrm{EtO_2CCHRX} \xrightarrow[\text{or LiN(SiMe}_3)_2;\ \text{M=Li}]{\text{Bu}^t\text{OK; M=K}} \mathrm{EtO_2CCRMX} \xrightarrow[\text{(ref.589)}]{\text{R}^1\text{R}^2\text{CO}} \mathrm{R^1R^2}\text{(epoxide)}\mathrm{RCO_2Et} \qquad (250)$$

Other lithiated derivatives that produce epoxides on reaction with carbonyl compounds are α-thioalkyllithiums[590,591] α-lithiated sulphoximes (equations 253–255), α-lithiosulphoxides (equation 256), and α-selenoalkyllithiums[187,592,593] (equation 256). Further treatment of the β-hydroxy adducts with alkylating agents and bases is necessary.

PhSCH$_2$Li + (2,2,6,6-tetramethylcyclohexanone) → (ref. 590) → HO CH$_2$SPh → Me$_3$O$^+$ BF$_4^-$ / MeNO$_2$ → HO, $\overset{+}{S}$(Ph)(Me), BF$_4^-$ → K$_2$CO$_3$ → (spiro-epoxide) (91%) (251)

RS—C(R^1)(R^2)—SePh → BuLi, thf, −78 °C (ref. 591) → RS—C(R^1)(R^2)—Li → (i) R^{3}R^{4}CO (ii) H$_2$O → RS—C(R^1)(R^2)—C(R^3)(R^4)—OH → (i) MeI, AgBF$_4$, 25 °C (ii) Bu^{t}OK, dmso → R^{1}R^{2}C—CR^{3}R^4 (epoxide, O) (252)

R^1 = R^2 = R^3 = H; R^4 = Ph 50%
R^1 = R^3 = H, R^2 = Me; R^4 = Ph 60%

(115) → (i) PhCHO (ii) H$_2$O (iii) Separation of diastereomers (ref 331) → Ph—S(O)(=NMe)—CH$_2$CHPhOH → (i) Me$_3$O$^+$; BF$_4^-$ (ii) Base → Ph (epoxide) 22% optical purity (253)

MeS(O)(=NSO$_2$C$_6$H$_4$Me-p)CH$_2$Na + Bu^t (cyclohexanone) → dmso → Bu^t (spiro-epoxide) 84% (254)

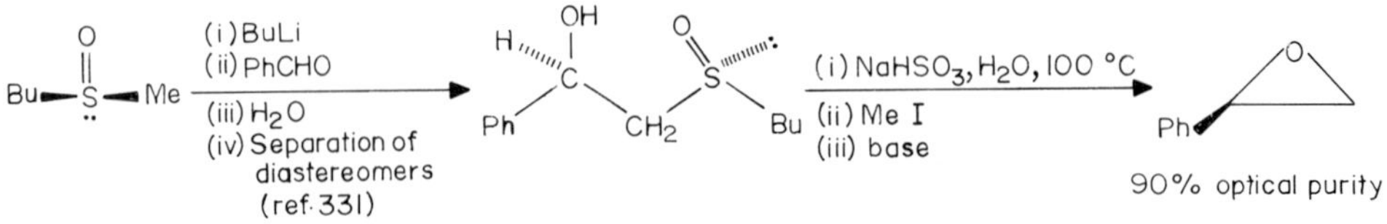

(255)

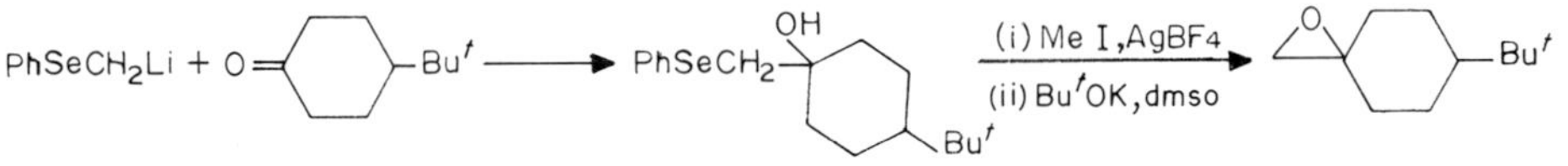

(256)

Mention should also be made of the Darzen's glycidic esters condensation, which often leads to the stereoselective formation, albeit in low yields, of (*E*)-epoxides from R^1R^2CO and $MCHClCO_2R$[589a]. Aziridines[594] and episulphides[595] have been obtained by similar routes.

6. Formation of Aldehydes and Ketones[3,5,187,536]

Ketones can be obtained by reaction of organoalkali metal compounds with various carbonyl species, such as RCOX (e.g. Y = OH, OR′, OCOR, SR′, NR_2 or Cl), orthoformates, dialkyl formamides, lactones, chloroformates, chlorocarbamates and ketenes and also with nitriles and isocyanides. These reactions, insertion reactions, were dealt with in Volume 2, Chapter 4. Further discussion of these reactions can be found in ref. 3, 5, 187, and 536.

The use of acyl anion equivalents (and other carbonyl synthons) has become a most valuable source of aldehydes and ketones[187,536,596,308]. These equivalents are included in the lists of synthons in Tables 21 and 22. The use of nucleophilic acylating agents allows the normal reactivity of acyl carbon atoms to be reversed (umpolung) (Scheme 14).

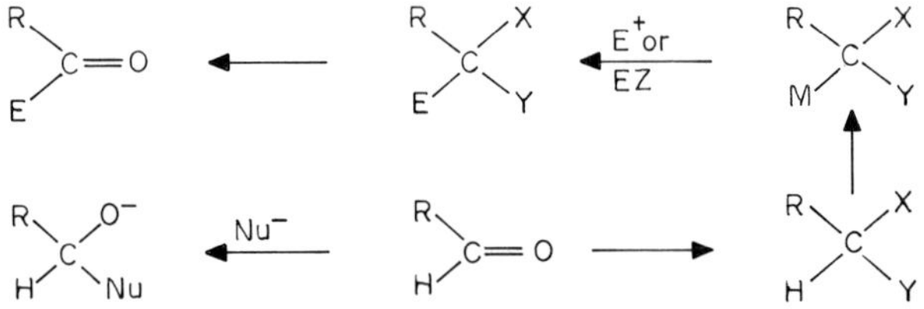

SCHEME 14

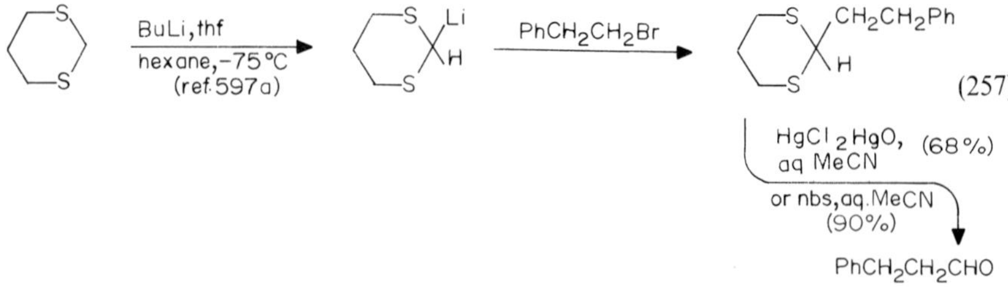

(257)

TABLE 26. Formation of aldehydes and ketones via the use of acyl anion equivalents

Reagent	Conditions	Product (yield, %)	Ref.
$PhSCH_2SiMe$	(i) BuLi, tmed, hexane (ii) PrBr (iii) m-ClC_6H_4COOOH, CH_2Cl_2 (iv) H_3O^+	PrCHO (70)	1227
$PhSeCH_2SiMe_3$	(i) $LiNPr^i_2$, thf, − 78 °C (ii) BuBr (iii) H_2O_2; 0–25 °C	BuCHO (80)	1228
O O S O	(i) BuLi, thf, − 80 °C (ii) $PhCH_2Br$	$PhCH_2CHO$ (100)	1229
S S S	(i) BuLi, thf, − 70 °C (ii) $Me(CH_2)_{13}Br$ (iii) $HgCl_2$, HgO, MeOH (iv) H_3O^+	$Me(CH_2)_{13}CHO$ (47–55)	1230
PhSe Me C PhSe H	(i) $KNPr^i_2$, thf, − 78 °C (ii) $Me(CH_2)_9Br$, thf, − 78 °C (iii) $CuCl_2$, CuO, Me_2CO, 0 °C	$MeCO(CH_2)_9Me$	1231
S S	(i) BuLi, thf, pentane, − 30 °C (ii) $B[(CH_2)_5Me]_3$, thf, − 30 °C (iii) H_2O_2, OH^-	$CO(CH_2)_5Me$ (76)	1232
Me_3SiO Ph C $(EtO)_2P$ H O	(i) $LiNPr^i_2$, thf, − 78 °C (ii) MeBr (iii) OH^-	PhCOMe (78)	1233
Me_3SiO R C NC H	(i) $LiNPr^i_2$, thf, − 78 °C (ii) R′X (iii) 0.6 N HCl	RCOR′ R = Ph, R′ = Me (98)	1234
EtO O H C NC $CH=CH_2$	*(i) $LiNPr^i_2$, thf, − 78 °C (ii) $Me(CH_2)_5Br$, thf, hmpt (iii) 5% H_2SO_4, MeOH	$CH_2{=}CHCO(CH_2)_5Me$ (80–85)	1235
SePh	(i) $LiNPr^i_2$, thf, − 78 °C (ii) $Me(CH_2)_9Br$, thf, hmpt (iii) dil. H_2SO_4	$MeCO(CH_2)_9Me$ (70)	1236
Me_3SiCH_2Cl	(i) Bu^sLi, tmed, thf, − 78 °C (ii) RR′CO (iii) dil. H_2SO_4, MeOH	RR′CHCHO [R, R′ = $(CH_2)_5$] (60)	1237

TABLE 26. (*Contd.*)

Reagent	Conditions	Product (yield, %)	Ref.
$MeOCH_2PPh_2$	(i) Bu^sLi, thf, −95 °C (ii) RR′CO, −78 °C (iii) MeOH (iv) MeI (v) CCl_3CO_2H	CHO (RR′CO =) (91)	1238
Br, $OSiMe_3$	(i) Bu^tLi, pentane, Et_2O, −70 °C (ii) PrCHO (iii) HCl, thf	Pr, O (46)	1239
$MeSiCH_2OMe$	(i) Bu^sLi, thf, −78 to −30 °C (ii) RR′C=O (iii) 90% HCO_2H, 25 °C	RR′CHCHO [R,R′ = $(CH_2)_5$] (80)	1240
O, C, Me, N, NMe_2	(i) MeLi, PhH, 0 °C (ii) BuLi, 25 °C (iii) MeI (iv) H_3O^+	COMe (77)	1241

As shown in Table 21, a wide range of acyl anion equivalents have been utilized (see also equations 35 and 136). Direct comparisons of the abilities of these synthons have seldom been made; perhaps the most important reason for choosing a particular synthon would be its availability. Particularly well studied are lithiated dithianes[597], e.g. equations 215, 257, and 258. A further example in the use of acyl anion equivalents in the carbohydrate field is given in equation 259[598]; see also Table 26. Reactions of acyl anion equivalents with aldehydes or ketones are also widely used (equation 260). Some specific examples are given in equations 261[599] and 262[600]; see also Table 26.

S, S, Me —(i) BuLi, thf, hexane, −75 °C; (ii) PhCN; (iii) H_3O^+ (ref. 597 b)→ S, S, Me, COPh —nbs→ PhCOCOPh (258)

CH_2I, O, OMe, O, O + Me, N, S, S, Li —(i) thf, hmpt, −78 °C; (ii) HgO, $HgCl_2$, aq. MeCN→ CHO, CH_2, O, OMe, O, O (259)

$$X(Y)C(R)M + R^1R^2C{=}O \longrightarrow X{-}C(R)(Y){-}C(R^2)(OLi){-}R^1 \longrightarrow RCOCHR^1R^2 \quad (260)$$

$$(EtO)_2P(O)C(Me)(H)SMe \xrightarrow[\text{pentane}, -75\,°C]{\text{BuLi, thf,}} (EtO)_2P(O)C(Me)(Li)SMe \xrightarrow[\text{thf}]{\text{PhCHO}} PhCH{=}CMeSMe \xrightarrow[\text{aq. MeCN}]{HgCl_2} PhCH_2COMe \quad (261)$$

(*E*):(*Z*) = 5:1, 80%; $PhCH_2COMe$ (84%)

$$Ph_2P(O)C(H)(Li)OMe + \text{cyclohexanone} \xrightarrow[\text{(ii) NaH, (iii) } H_3O^+]{\text{(i) thf}} \text{cyclohexyl-CHO} \quad (262)$$

(57%)

Metal enolates, or their synthons (see Table 22) clearly have great value in the synthesis of aldehydes and ketones. An excellent discussion of these reactions is given in ref. 536. The use of enolate synthons is illustrated in equations 263–265. Scheme 15 indicates a range of products obtained from the metal enolate synthon **116**[604]. See also Section III.C.2.c for a discussion of related synthons in asymmetric synthesis. The use of homoenolate synthons, $RCOCH_2CH_2^-$, is illustrated in equations 266[605] and 267[606].

$$\text{Li-CH=CH-OEt} \xrightarrow[\text{(ii) } H_2O]{\text{(i) } Me(CH_2)_9I, \text{ thf, hmpt}} Me(CH_2)_{10}CHO \quad (263)$$

(ref. 601)

$$EtCH{\cdots}\bar{C}H{\cdots}N(\text{piperidino})\ Li^+ \xrightarrow[\text{(ii) tartaric acid, 0 °C}]{\text{(i) 2-BrCH}_2\text{-dioxolane, thf, -60 °C}} \text{(1,3-dioxolan-2-yl)}CHCH_2CHEtCHO \quad (264)$$

(ref. 600) (83%)

$$\text{2-benzyl-4,4,6-trimethyl-5,6-dihydro-1,3-oxazine} \xrightarrow[\text{(iii) BuLi, -78 °C; (iv) NaBH}_4, H_3O^+; \text{(v) } H_3O^+]{\text{(i) BuLi, thf, -78 °C; (ii) Br(CH}_2)_4\text{Br, -78 °C}} \text{1-phenylcyclopentane-1-carbaldehyde} \quad (265)$$

(ref. 603) (50–55%)

(116)

$$R^1COCHRR^2$$ ($R^1 = C_5H_{11}$; $R^2 = H$; RY = MeI) (93%)

$$CH_2{=}CHCH_2OSiEt_3 \xrightarrow{Bu^sLi,\ thf} [CH_2{\cdots}CH{\cdots}CHOSiEt_3]^- Li^+ \xrightarrow[(ii)\ H_3O^+]{(i)\ MeI,\ thf,\ hmpt} PrCHO\ (90\%) \quad (266)$$

$$Me_2NP(O)NMe\text{-allyl-}Ph^-\ Li^+ \xrightarrow[(ii)\ 2\ N\ HCl,\ Et_2O]{(i)\ Me_2SO_4,\ thf} PhCH(Me)CH_2CHO\ (66\%) \quad (267)$$

7. Formation of Carboxylic Acids and Derivatives[3,5]

Carboxylic acids are generally obtained from organoalkali metals on treatment with CO_2 (equation 268); see Volume 2, Chapter 4. A number of synthons for RO_2C-containing

$$RM \xrightarrow[(ii)\ H_3O^+]{(i)\ CO_2} RCO_2H \quad (268)$$

units have been developed; see Table 22. Especially useful are oxazoline derivatives[607], e.g. equations 269 and 270. Also important are the malonic ester syntheses[536]; see also

$$\text{2-methyl-4,4-dimethyloxazoline} \xrightarrow[(ii)\ RX]{(i)\ BuLi,\ -78\ ^\circ C} \text{2-}CH_2R\text{-oxazoline} \xrightarrow{H_3O^+} HO_2CCH_2R \quad (269)$$

$$\xrightarrow{BuLi} \xrightarrow[(ii)\ H_3O^+]{(i)\ MeSSMe} CO_2H,\ SMe,\ OMe,\ OMe\ (93\%) \quad (270)$$

$$HO_2CCH_2CO_2Et \xrightarrow[\text{(ii) RX, hmpt, thf; (iii) } H_3O^+ \text{ (ref. 608)}]{\text{(i) LiNPr}^i\text{(cyclohexyl), thf, }-78\,^\circ\text{C}} HO_2CCH(R)CO_2Et \xrightarrow{\Delta} RCH_2CO_2Et \quad RX = BuBr, 80\% \tag{271}$$

equation 271. For asymmetric syntheses of carboxylic acids, see Section II.C.2.c. Amides are available from reaction of organoalkali metal compounds and isocyanates (see Volume 2, Chapter 4).

$$R'NCO \xrightarrow{RM} R'N{=}C(R)OM \xrightarrow{H_3O^+} R'NHCOR \tag{272}$$

8. Formation of Cyanides

Organoalkali metals can be converted into cyanides via reaction with cyanates[609], equation 273, isocyanides[610], equation 274, or cyanogen chloride, CNCl[611].

$$Me(CH_2)_5CH{=}CHLi + PhOCN \xrightarrow{Et_2O, -70\,^\circ C} Me(CH_2)_5CH{=}CHCN \; (80\%) \tag{273}$$

$$\text{2-lithio-1,3-dithiane} + Ph_3CNC \xrightarrow{thf, -30\,^\circ C} \text{2-cyano-1,3-dithiane} \; (83\%) \tag{274}$$

D. Formation of Carbon—Oxygen Bonded Compounds: Alcohols and Hydroperoxides[3,5]

As discussed in Volume 2, Chapter 4, alcohols and hydroperoxides can be formed from reaction of organoalkali metals and oxygen (equation 275). However, products other than alcohols of hydroperoxides can result. No other details of the reactions with oxygen will be

$$RM \longrightarrow ROOM \xrightarrow{RM} 2ROM \tag{275}$$

given here. As indicated in the second half of equation 275, hydroperoxides also react with organoalkali metals to give, after hydrolysis, alcohols, phenols[612], or, with vinyllithiums, ketones (equations 276 and 277)[59]. Reactions of vinyllithiums with peroxides proceed with

$$PhLi + Bu^tOOH \longrightarrow PhOH \; (93\%) \tag{276}$$

$$\text{1-lithiocyclooctene} \xrightarrow[\text{(ii) } H_3O^+]{\text{(i) } Bu^tOOH} \text{cyclooctanone} \tag{277}$$

$$R^1CH{=}C(R^2)Li \xrightarrow[\text{(ii) } Ac_2O \text{ (ref. 613)}]{\text{(i) } Bu^tOOLi} R^1CH{=}C(R^2)OAc \tag{278}$$

$$\text{R}^1\text{CH}{=}\text{C(R}^2\text{)Li} \xrightarrow[\text{(ref. 612)}]{\text{Me}_3\text{SiOOSiMe}_3} \text{R}^1\text{CH}{=}\text{C(R}^1\text{)OSiMe}_3 \qquad \text{R}^1,\text{R}^2 = -(\text{CH}_2)_6- \quad 52\% \tag{279}$$

retention of configuration (equations 278, 279). An indirect route to alcohols from organoalkali metals involves their conversion into organoboranes and subsequent treatment[614a] with alkaline hydrogen peroxide. Oxidation of aryllithiums to phenols has also been achieved using a peroxyborate, 2-*tert*-butylperoxy-1, 3, 2-dioxaborolane[614b]. Oxophilic reactions of organoalkali metal compounds with carbonyl compounds also provide carbon—oxygen bonded compounds; see Volume 2, Chapter 4.

TABLE 27. Formation of sulphides from reaction of organoalkali metal compounds with disulphides

Substrate	Reaction conditions	Product (yield, %)	Ref.
$Me(CH_2)_{14}CO_2H$	(i) $LiNPr^i_2$, thf, 0 °C, hmpt (ii) MeSSMe, 0 °C	$Me(CH_2)_{13}CH(SMe)CO_2H$ (90)	1242
Ph_2CHCO_2H	(i) $LiNPr^i_2$, thf, −78 to −25 °C (ii) MeSSMe	$Ph_2C(SMe)CO_2H$ (100)	1242
$R^2(H)C{=}C(R^1)Br$	(i) Bu^tLi, −80 °C (ii) PhSSPh	$R^2(H)C{=}C(R^1)SPh$ $R^1 = R^2 = H$ (74) $R^1 = H, R^2 = Bu$ (93)	1243
$(EtO)_2P(O)CH_2Me$	(i) BuLi, thf, −78 °C (ii) PhSSPh, thf	$(EtO)_2P(O)CH(SPh)Me$ (84)	1244
2-Me-3-CO_2Et-4-OMe-benzene (ethyl 2-methyl-6-methoxybenzoate)	(i) $LiNPr^i_2$, thf, −78 °C (ii) PhSSPh	2-CH_2SPh-3-CO_2Et-4-OMe-benzene	1245
4-MeO-C_6H_4-C(S)NHMe	(i) 2 equiv. BuLi, thf (ii) MeSSMe	2-SMe-4-MeO-C_6H_3-C(S)NHMe	1246
$PhCHOHCH_2Cl$	(i) BuLi, −78 °C (ii) $Naph^-$, Li^+ (iii) MeSSMe (iv) H_2O	$PhCHOHCH_2SMe$ (85)	1247

E. Formation of Carbon—Sulphur Bonded Compounds[3,5]

1. *Thiols and Sulphides*

Insertions of sulphur into C—M bonds provides thiolates, RSM, which can be converted into thiols or sulphides (Volume 2, Chapter 4). Other routes to sulphides involve reactions with sulphenyl halides, RSX, and more conveniently with disulphides (equation 280). Some examples of the disulphide reactions are given in Table 27.

$$RM + R'SSR' \longrightarrow RSR' + MSR' \qquad (280)$$

Thiophilic reactions of RM with thiocarbonyl compounds also provide sulphides (see Volume 2, Chapter 4).

2. *Other Carbon—Sulphur Bond-forming Reactions*

Formal insertion reactions of organoalkali metal compounds with SO_2 and CS_2 also provide carbon—sulphur bonded compounds; see Volume 2, Chapter 4.

F. Formation of Amines

The transformation of organoalkali metal compounds into amines has been achieved using azides and hydroxylamines (Table 28). The most useful azide is the readily available tosyl azide[615–618] which works well with alkyl-, benzyl-, phenyl-, and heteroaryllithiums (equation 281). Other azides to be used include $(PhO)_2P(O)N_3$[619] and vinyl azides[620] (*E*)-

$$RLi \xrightarrow[\text{(ii) } H_2S^{617} LiAlH_4{}^{617}, Ni-Al/aq.KOH^{618}, \text{ or } NaBh_4{}^{615}]{\text{(i) } p\text{-}MeC_6H_4SO_3N_3} RNH_2 \qquad (281)$$

$R'CH{=}CRN_3$ (e.g. R = Ph, R′ = H; R = H, R′ = Bu^t). The latter have been employed for aryl-, heteroaryl-, or stabilized alkyllithiums. However, simple alkyllithiums (R^2Li) react differently and provide ketones rather than amines on workup, equation 282. Another

$$R^2Li + RC(N_3){=}CHR^1 \longrightarrow RC(LiN{-}N{=}NR^2){=}CHR^1 \xrightarrow{H_3O^+} RC(O)CH(R^1)R^2 \qquad (282)$$

R^2 = Me, Bu, or Bu^t

azide reagent used with phenyl derivatives is $PhSCH_2N_3$. However, better yields are obtained if the organolithium is converted into the organomagnesium reagent, equation 283[621].

$$PhOMe \xrightarrow[\text{(iii) } PhSCH_2N_3 \text{ (iv) aq.} NH_4Cl]{\text{(i) BuLi, thf, } -78\,^{\circ}C \text{ (ii) } MgBr_2, Et_2O} 2\text{-}MeOC_6H_4NHN{=}NCH_2SPh \xrightarrow[\text{thf}, H_2O]{KOH, MeOH} 2\text{-}MeOC_6H_4NH_2 \qquad (283)$$

A number of hydroxylamine derivatives have also been used to generate primary, secondary, and tertiary amines. Primary amines have been obtained via H_2NOMe[622,623], $H_2NOP(O)Ph_2$[624], or $H_2NOSO_2C_6H_4Me_3$-2, 4, 6[625]. The compound $H_2NOP(O)Ph_2$ is of particular value for benzyl or other stabilized carbanions.

$$RLi \xrightarrow[\text{(iii) } OH^-]{\text{(i) } H_2NOMe, MeLi \text{ (ii) } H_3O^+} RNH_2 \qquad (284)$$

TABLE 28. Formation of amines from organolithiums

Organolithium	Reagents[a]	Product (yield, %)	Ref.
MeLi	(i) A (ii) H_3O^+ (iii) PhCOCl	MeNHCOPh (80)	616
Bu^{s}Li	(i) A (ii) H_3O^+ (iii) PhCOCl	Bu^{s}NHCOPh (67)	616
Bu^{t}Li	(i) A (ii) H_3O^+ (iii) PhCOCl	Bu^{t}NHCOPh (80)	616
PhLi	(i) B (ii) 10% HCl (iii) OH^-	$PhNH_2$ (68)	620
PhLi	(i) A (ii) H_3O^+ (iii) PhCOCl	PhNHCOPh (90)	616
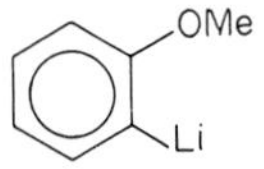 OMe, Li	(i) C (ii) aq. KOH, 0 °C, Ni–Al	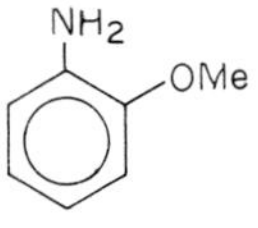 NH_2, OMe (80)	
	(i) A (ii) H_3O^+ (iii) PhCOCl	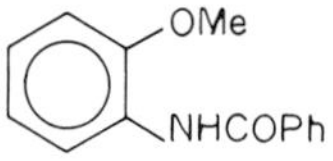 OMe, NHCOPh (96)	616
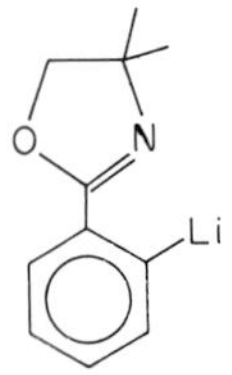 O, N, Li	(i) C (ii) $Bu_4N^+HSO_4^-$, $NaBH_4$	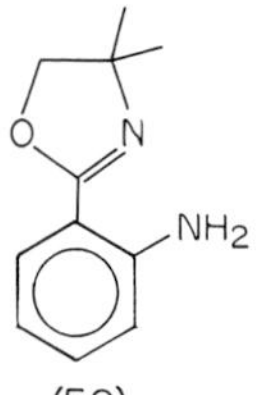 O, N, NH_2 (50)	615
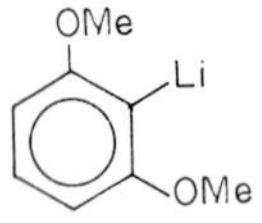 OMe, Li, OMe	(i) B (ii) 10% HCl (iii) OH^-	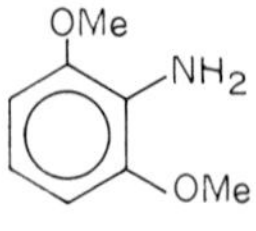 OMe, NH_2, OMe	620
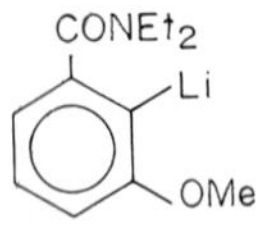 $CONEt_2$, Li, OMe	(i) C (ii) $Bu_4N^+HSO_4^-$, $NaBH_4$	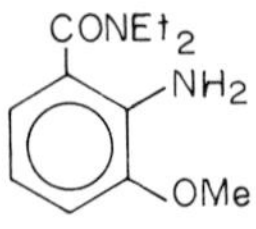 $CONEt_2$, NH_2, OMe (55)	615

TABLE 28. (*Contd.*)

Organolithium	Reagents[a]	Product (yield, %)	Ref.
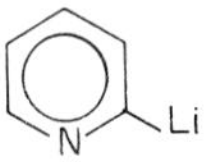 N Li	(i) B (ii) 10% HCl (iii) OH^-	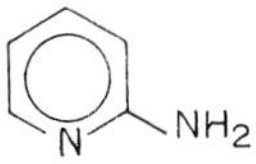 N NH_2 (45)	620
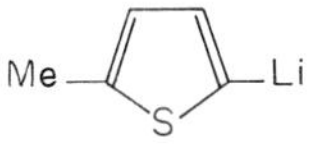 Me S Li	(i) B (ii) 10% HCl (iii) OH^-	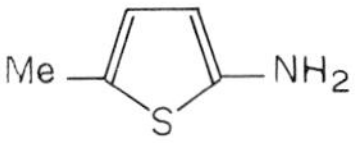 Me S NH_2 (58)	620
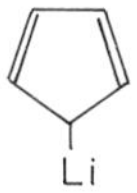 Li	(i) D	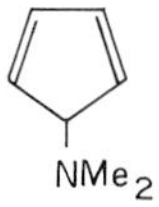 NMe_2 (47)	629
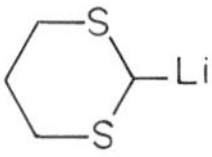 S S Li	(i) B (ii) 10% HCl (iii) OH^-	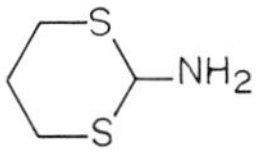 S S NH_2 (64)	620
$PhCHLiCO_2Et$	(i) E (ii) H_3O^+	$PhCH(NH_2)CO_2Et$ (45)	624
$PhCH_2Li$	(i) E (ii) H_3O^+	$PhCH_2NH_2$ (30)	624
	(i) B (ii) 10% HCl (iii) OH^-	$PhCH_2NH_2$ (60)	620
	(i) A (ii) H_3O^+ (iii) PhOCl	$PhCH_2NHCOPh$ (97)	616
Ph_2CHLi	(i) E (ii) H_3O^+	Ph_2CHNH_2 (41)	624
Ph_3CLi	(i) E (ii) H_3O^+	Ph_3CNH_2 (30)	624
PhCH=CHCHPhLi	(i) E (ii) H_3O^+	$PhCH_2CH_2COPh$ (30)	624
Bu^sLi	(i) F (ii) PhCOCl	$Bu^sMeCOPh$ (62)	626
Bu^tLi	(i) F (ii) PhCOCl	$Bu^tMeNCOPh$ (30)	626
PhLi	(i) F (ii) PhCOCl	PhMeNCOPh (67)	626
MeLi	(i) G	Me_3N (45)	628

TABLE 28. (*Contd.*)

Organolithium	Reagents[a]	Product (yield, %)	Ref.
1-NpLi	(i) H	1-NpNEt$_2$ (9)	628
Me–CH=CH–Li (cis)	(i) G	Me–CH=CH–NMe$_2$ (cis) (38)	628
9-Li-fluorene	(i) G	9-NMe$_2$-fluorene (61)	628
PhC(CN)(Li)CO$_2$Et	(i) G	PhC(CN)(NMe$_2$)CO$_2$Et (95)	628

[a] A = H_2NOMe–MeLi, hexane, Et_2O, −70 °C.
B = $H_2C{=}CPhN_3$, thf, −78 °C.
C = *p*-$MeC_6H_4SO_2ON_3$, thf, −78 °C.
D = Me_2NOSO_2Me, −20 °C, thf.
E = $Ph_2P(O)ONH_2$, thf, −20 to 25 °C.
F = MeNHOMe–MeLi, Et_2O, hexane, −78 °C.
G = Me_2NOSO_2R, R = 2,4,6-trimethylphenyl.
H = Et_2NOSO_2R, R = 2,4,6-trimethylphenyl.

Secondary amines (and amides) have similarly been produced from alkyl- and phenyllithiums (but not from 2-lithiothiophene or -*N,N*-diisopropylbenzamide) using R'NHOMe–MeLi (1:1), e.g. equation 285[626].

$$\mathrm{Bu^sLi} \xrightarrow[\text{(ii) PhCOCl}]{\text{(i) MeNHOMe–MeLi, Et}_2\text{O, hexane}} \underset{62\%}{\mathrm{Bu^sMeNCOPh}} \qquad (285)$$

Reagents for producing tertiary amines include Me_2NOSO_2R' (R' = 2,4,6-$Me_3C_6H_2$[627,628] or Me[629,630]), $Et_2NOSO_2C_6H_2Me_3$-2,4,6[628], and $Me_2NOP(O)Ph_2$[630]; however, Me_2NOMe did not provide $PhNMe_2$ from PhLi. Conversion of alkynyllithiums to the corresponding cuprates prior to reaction with $Me_2NOP(O)Ph_2$ or Me_2NOSO_2Me has been recommended[630]. Another use of copper reagents in the formation of amines is illustrated in equation 286[631]. Chiral aminating

$$\text{2,6-(EtO)}_2\text{C}_6\text{H}_3\text{Li} + \text{CuN(piperidino)} \xrightarrow[\text{(ii) O}_2]{\text{(i) thf, hexane}} \underset{51\%}{\text{2,6-(EtO)}_2\text{C}_6\text{H}_3\text{-N(piperidino)}} \qquad (286)$$

$$\text{PhCHLiCO}_2\text{Et} + \underset{[(-)\text{-}117]}{\text{(oxazaphospholidine, Ph, Me, N-Me, P(O)ONMe}_2)} \longrightarrow \underset{\substack{50\% \\ \text{e.e. } 23\%}}{\text{PhCH(NMe}_2)\text{CO}_2\text{Et}} \qquad (287)$$

agents, such as **117** obtained from (−)-ephedine, have been used but with only limited success[632].

Amines have also been obtained by addition of RLi to imines[3,5], e.g. equation 288. However, side reactions can result, e.g. by α-deprotonation and subsequent loss of LiH.

2-Me-1,3-dithiane $\xrightarrow[\text{(ii) PhCH=NPh; (iii) } H_2O]{\text{(i) BuLi}}$ 2-Me-2-(CHPhNHPh)-1,3-dithiane $\longrightarrow$ O=C(Me)CHPhNHPh (288)

G. Formation of Carbon—Halogen Bonds

Replacement of the alkali metal in an organoalkali metal by halogen (equation 289) can be achieved using either the halogens (Cl_2, Br_2, or I_2) or halogen-containing compounds in reverse halide–alkali metal exchanges. The latter reagents have particular advantages when the use of halogens (especially Cl_2 or Br_2) could lead to further reaction.

$$RM \longrightarrow RX \qquad X = \text{halogen} \tag{289}$$

$$Ph_2C{=}CHCl \xrightarrow[\text{(ii) } X_2 \text{ (ref. 633)}]{\text{(i) BuLi, thf, light petroleum, } -100\,^\circ C} Ph_2C{=}CClX \quad X = Br\ 94\%,\ X = I\ 97\% \tag{290}$$

1,3-(MeO)$_2$C$_6$H$_4$ $\xrightarrow[\text{(ii) XY (ref. 634)}]{\text{(i) PhLi, } Et_2O}$ 2-X-1,3-(MeO)$_2$C$_6$H$_3$ (291)

XY = Cl_2 39%
Br_2 18%
I_2 80%
BrCN 46%
ICN 46%
$Br(CH_2)_2Br$ 71%

Direct fluorination by F_2 is not feasible. Perchloryl fluoride and dinitrogen difluoride[636] have been used, however, although explosions with the former have been reported[637].

Alternative reagents to chlorine include CCl_4 (equation 292)[638], Cl_3CCCl_3 (equation 293)[639,640], *N*-chlorosuccinimide[641], and *p*-toluenesulphonyl chloride[642].

$$(EtO)_2P(O)CH_2Ar \xrightarrow[\text{(ii) } CCl_4\text{; (iii) } H_2O \text{ (ref. 638)}]{\text{(i) BuLi, thf, } -70\,^\circ C} (EtO)_2P(O)CHClAr \quad >80\% \tag{292}$$

3,6-dihydro-2H-thiopyran 1,1-dioxide $\xrightarrow[\text{(ii) } Cl_3CCCl_3 \text{ (ref 639)}]{\text{(i) BuLi, PhH, hexane, RT}}$ 2-Cl derivative (32%) + 2,2-Cl$_2$ derivative (12%) + 4-Cl-3,4-dihydro-2H-thiopyran 1,1-dioxide (5%) (293)

Alternative reagents to bromine are BrCN[634,643], *p*-toluenesulphonyl bromide[642], and 1,2-dibromoalkanes [e.g. RCHBrCHBrR (R = H or Me)]. However, the use of $BrCH_2CH_2Br$ under controlled conditions can lead to the $BrCH_2CH_2$ alkylated product[644], e.g. equation 294.

$$PhCH(OSiMe_3)(CN) \xrightarrow[\text{(ii) } BrCH_2CH_2Br \text{ (1/2 equiv.) (ref. 644)}]{\text{(i) } LiNPr^i_2, \text{ thf}, -78\,^\circ C} PhC(OSiMe_3)(CN)CH_2CH_2Br \quad (30\%) \tag{294}$$

Alternative sources to iodine are CH_2I_2[640], $ClCH_2CH_2I$[420], PhC≡CI[645], and ICN[634].

Halogenation of (+)-(*S*)-1-Li-1-Me-2,2-diphenylcyclopropane by Br_2 or I_2 in diethyl ether proceeds with retention of configuration[66]. Although the brominolysis of (*Z*)- and (*E*)-1-lithio-2-methylcyclopropane is highly stereospecific in pentane solution, in pentane–diethyl ether solution a much reduced stereospecificity results[68]. Considerable inversion of configuration occurs in the reaction of 2-norbornyllithium with bromine[18]. In reactions of halogens with methyllithium or 4-*tert*-butylcyclohexyllithium, inversion can predominate; in contrast, reactions with $BrCH_2CH_2Br$, PhBr, or the pyridine–bromine complex provide bromoproducts with predominant retention of configuration[532a].

H. Formation of Carbenes, Arynes, and Ylides[1,3,5]

1. *Carbenoid reagents*

The preparation of α-haloalkylalkali metals is given in Table 12. These and related compounds have had considerable use as carbene precursors (equation 295). For reviews on the use of such reagents, see references 433a, 646, and 647.

$$R^1R^2C(X)M \longrightarrow [R^1R^2C:] \tag{295}$$

2. *Arynoid reagents*[3,5]

o-Haloarylalkali metals have had an extensive use as aryne sources, e.g. equations 44 and 280. Their generation, chemistry, and use have been reviewed[646].

3. *Ylides*

Organoalkali metals have been widely used in the generation of ylides, for example for use in the Wittig reactions[648].

IV. REFERENCES

1. *Houben-Weyl. Methoden der Organischen Chemie, Band 13–I, Metallorganische Verbindungen: Li, Na, K, Rb, Cs, Cu, Ag, Au*, Georg Thieme Verlag, Stuttgart, 1970.
2. J. L. Wardell, *Alkali Metals*, in *Comprehensive Organometallic Chemistry* (Eds. G. Wilkinson, F. G. A. Stone, and E. W. Abel), Vol. 1, Pergamon Press, Oxford, 1982, Chapter 2.
3. B. J. Wakefield, *Compounds of the Alkali and Alkaline Earth Metals in Organic Synthesis*, in *Comprehensive Organometallic Chemistry* (Eds. G. Wilkinson, F. G. A. Stone, and E. W. Abel), Vol. 7, Pergamon Press, Oxford, 1982, Chapter 44.
4. G. E. Coates, M. L. H. Green, and K. Wade, *Organometallic Compounds*, 3rd ed., Vol. 1, Methuen, London, 1967, Chapter 1.
5. B. J. Wakefield, *The Chemistry of Organolithium Compounds*, Pergamon Press, Oxford, 1974.

6. M. Schlosser, *Angew, Chem., Int. Ed. Engl.*, **2**, 287, 362 (1964).
7. M. Schlosser, *Polare Organometalle*, Springer Verlag, Berlin, 1973.
8. F. J. Landro, J. A. Gurak, J. W. Chinn, Jr., R. M. Newman, and R. J. Lagow, *J. Am. Chem. Soc.*, **104**, 7345 (1982).
9. C. Chung and R. J. Lagow, *J. Chem. Soc., Chem. Commun.*, 1972, 1078; L. A. Shimp and R. J. Lagow, *J. Org. Chem.*, **44**, 2311 (1979).
10. I. Fatt and M. Tashima, *Alkali-Metal Dispersions*, Van Nostrand, Princeton, 1961.
11. G. Gau and S. Marques, *J. Am. Chem. Soc.*, **98**, 1538 (1976).
12. C. Tamborski, G. J. Moore, and E. J. Soleski, *Chem. Ind.* (*London*), 696 (1962).
13. G. Molle, P. Bauer, and J. E. Dubois, *J. Org. Chem.*, **48**, 2975 (1983).
14. M. J. S. Dewar and J. M. Harris, *J. Am. Chem. Soc.*, **91**, 3652 (1969).
15. H. M. Walborsky and M. S. Aronoff, *J. Organomet. Chem.*, **51**, 55 (1973).
16. H. M. Walborsky and E. J. Powers, *Isr. J. Chem.*, **21**, 210 (1981).
17. J. B. Lambert, M. W. Majchrzak, B. I. Rosen, K. P. Steele, and S. A. Oliver, *Isr. J. Chem.*, **20**, 177 (1980).
18. D. E. Applequist and G. N. Chmurny, *J. Am. Chem. Soc.*, **89**, 875 (1967).
19. W. H. Glaze and C. M. Selman, *J. Org. Chem.*, **33**, 1987 (1968).
20. C. Georgoulis, J. Meyet, and W. Smadja, *J. Organomet. Chem.*, **121**, 271 (1976).
21. D. Seyferth and L. G. Vaughan, *J. Organomet. Chem.*, **1**, 201 (1963).
22. N. L. Allinger and R. B. Hermann, *J. Org. Chem.*, **26**, 1040 (1961); D. Y. Curtin and J. W. Crump, *J. Am. Chem. Soc.*, **80**, 1922 (1958); F. D. Bordwell and P. S. Landis, *J. Am. Chem. Soc.*, **79**, 1593 (1957); E. A. Braude and J. A. Coles, *J. Chem. Soc.*, 2078 (1951); A. S. Dreiding and R. J. Prajj, *J. Am. Chem. Soc.*, **76**, 1902 (1954).
23. N. L. Holy, *Chem. Rev.*, **74**, 243 (1974).
24. J. F. Garst, *Acc. Chem. Res.*, **4**, 400 (1971).
25. C. G. Screttas, *J. Chem. Soc., Chem. Commun.*, 752 (1972); C. G. Screttas and M. Micha-Screttas, *J. Org. Chem.*, **43**, 1064 (1978).
26. P. K. Freeman and L. L. Hutchinson, *J. Org. Chem.*, **45**, 1924 (1980).
27. P. K. Freeman and L. L. Hutchinson, *J. Org. Chem.*, **45**, 3191 (1980).
28. P. K. Freeman and L. L. Hutchinson, *J. Org. Chem.*, **48**, 4705 (1983).
29. J. Stapesma and G. W. Klumpp, *Tetrahedron*, **37**, 187 (1981).
30. J. Stapesma, P. Kuipers and G. W. Klumpp, *Recl. Trav. Chim. Pays-Bas*, **101**, 213 (1982).
31. S. E. Wilson, *Tetrahedron Lett.*, 4651 (1975).
32. J. Barluenga, J. Florez, and M. Yus, *J. Chem. Soc., Perkin Trans. 1*, 3019 (1983); see also C. Najera, M. Yus, and D. Seebach, *Helv. Chim. Acta*, **67**, 289 (1984).
33. B. A. Tertov and A. S. Morkovnik, *Khim. Geterotsikl. Soedin.*, 392 (1975).
34. G. Boche and D. R. Schneider, *Tetrahedron Lett.*, 2327 (1978).
35. G. Boche, D. R. Schneider, and H. Wintermayr, *J. Am. Chem. Soc.*, **102**, 5697 (1980).
36. M. Takagi, M. Nojima, and S. Kusabayashi, *J. Am. Chem. Soc.*, **105**, 4676 (1983).
37. H. E. Zieger, I. Angres, and L. Maresca, *J. Am. Chem. Soc.*, **95**, 8201 (1973).
38. J. Jacobus and D. Pensak, *J. Chem. Soc. Chem. Commun.*, 400 (1969).
39. T. Fujita, S. Watanabe, K. Suga, K. Sugahara, and K. Tsuchimoto, *Chem. Ind.* (*London*), 167 (1983).
40. J. F. Garst and F. E. Barton, Jr., *J. Am. Chem. Soc.*, **96**, 523 (1974).
41. S. J. Cristol and R. V. Barbour, *J. Am. Chem. Soc.*, **88**, 4262 (1966).
42. J. F. Garst and J. T. Barbas, *J. Am. Chem. Soc.*, **96**, 3239 (1974).
43. Y. -J. Lee and W. D. Closson, *Tetrahedron Lett.*, 1395 (1974).
44. E. Müller and G. Röscheisen, *Chem. Ber.*, **90**, 543 (1957).
45. J. F. Garst, R. D. Roberts, and J. A. Pacifici, *J. Am. Chem. Soc.*, **99**, 3528 (1977).
46. K. S. Y. Lau and M. Schlosser, *J. Org. Chem.*, **43**, 1595 (1978).
47. R. G. Jones and H. Gilman, *J. Org. Chem.*, **6**, 339 (1951).
48. D. E. Applequist and D. F. O'Brien, *J. Am. Chem. Soc.*, **85**, 743 (1963).
49. H. J. S. Winkler and H. Winkler, *J. Am. Chem. Soc.*, **88**, 964, 969 (1966).
50. E. Weiss, *Chem. Ber.*, **97**, 3241 (1964).
51. C. W. Perry, G. J. Bader, A. A. Liebman, R. Barner, and J. Wuersch, *J. Org. Chem.*, **43**, 4391 (1978).
52. N. Gjös and S. Gronowitz, *Acta Chem. Scand.*, **25**, 2596 (1971); R. E. Merrill and E. I. Negishi, *J. Org. Chem.*, **39**, 3452 (1974).

53. H. Heaney, *Organomet. Chem. Rev.*, **1**, 27 (1966); V. S. Petrosyan and O. A. Reutov, *J. Organomet. Chem.*, **52**, 307 (1973).
54. G. Cahiez, D. Bernard and J. F. Normant, *Synthesis*, 245 (1976).
55. G. Porzi and C. Concilio, *J. Organomet. Chem.*, **128**, 95 (1977).
56. D. Merkel and G. Köbrich, *Chem. Ber.*, **106**, 2040 (1973).
57. N. A. Donskaya, T. V. Akhachinskaya, T. V. Leonova, E. V. Shulishov, and Yu. S. Shabarov, *J. Org. Chem. USSR* (*Engl. Transl.*), **16**, 487 (1980).
58. S. Bhanu, E. A. Khan, and F. Scheinmann, *J. Chem. Soc., Perkin Trans. 1*, 1609 (1976).
59. H. Neumann and D. Seebach, *Chem. Ber.*, **111**, 2785 (1978).
60. A. P. Batalov and G. A. Rostokin, *J. Gen. Chem. USSR* (*Engl. Transl.*), **43**, 959 (1973); A. P. Batalov, *Tr. Khim. Khim. Teknol.*, 51 (1975); A. P. Batalov, *Kinet. Katal.*, **17**, 310 (1976).
61. W. E. Parham, L. D. Jones, and Y. A. Sayed, *J. Org. Chem.*, **41**, 1184 (1976).
62. W. E. Parham, C. K. Bradsher, and D. C. Reames, *J. Org. Chem.*, **46**, 4804 (1981).
63. A. P. Batalov, G. Ya Mal'Kova, and G. A. Rostokin, *Tr. Khim. Khim., Tekhnol.*, 93 (1974); A. P. Batalov and G. A. Rostokin, *Tr. Khim. Khim. Tekhnol.*, 99 (1974); A. P. Batalov and G. A. Rostokin, *Tr. Khim. Khim. Teknol.*, 1168 (1971); A. P. Batalov and G. A. Rostokin, *J. Gen. Chem. USSR* (*Engl. Transl.*), **41**, 1740, 1743 (1971; A. P. Batalov, A. A. Kvasov, G. A. Rostokin, and I. A. Korshunov, *J. Gen. Chem. USSR* (*Engl. Transl.*), **41**, 2545 (1971); A. P. Batalov, *J. Gen. Chem. USSR* (*Engl. Transl.*), **44**, 2492 (1974).
64. H. R. Rogers and J. Houk, *J. Am. Chem. Soc.*, **104**, 522 (1982).
65. A. R. Lepley and R. L. Landau, *J. Am. Chem. Soc.*, **91**, 748 (1969); H. R. Ward, R. G. Lawler, and R. A. Cooper, *J. Am. Chem. Soc.*, **91**, 746 (1969); G. A. Russell and D. W. Lamson, *J. Am. Chem. Soc.*, **91**, 3967 (1969).
66. H. M. Walborsky, F. J. Impastato, and A. E. Young, *J. Am. Chem. Soc.*, **86**, 3283 (1964).
67. R. Letsinger, *J. Am. Chem. Soc.*, **72**, 4842 (1950).
68. D. E. Applequist and A. H. Peterson, *J. Am. Chem. Soc.*, **83**, 862 (1961).
69. L. A. Paquette, G. J. Wells, K. A. Horn, and T. -H. Yan, *Tetrahedron*, **39**, 913 (1983).
70. G. Wittig and W. Tochtermann, *Justus Liebigs Ann. Chem.*, **660**, 23 (1962).
71. T. T. Lansbury and J. D. Sidler, *Tetrahedron Lett.*, 691 (1965).
72. E. Boerhorst, R. F. Schmitz, and G. W. Klumpp, *Tetrahedron Lett.*, 3347 (1975).
73. D. F. Hoeg, D. I. Lusk, and A. L. Crumbliss, *J. Am. Chem. Soc.*, **87**, 4147 (1967).
74. D. J. Burton and J. L. Hahnfeld, *J. Org. Chem.*, **42**, 828 (1977).
75. R. G. Micetich and C. G. Chin, *Can. J. Chem.*, **48**, 1371 (1970).
76. P. Savignac, M. Dreux, and P. Coutrot, *Tetrahedron Lett.*, 609 (1975); P. Savignac, J. Petrova, M. Dreux, and P. Coutrot, *J. Organomet. Chem.*, **91**, C45 (1975).
77. K. Kitatani, T. Hiyama, and H. Nozaki, *Bull. Chem. Soc. Jpn.*, **50**, 3288 (1977).
78. K. G. Taylor, W. E. Hobbs, M. S. Clark, and J. Chaney, *J. Org. Chem.*, **37**, 2436 (1972).
79. K. G. Taylor, W. E. Hobbs, and M. Saquet, *J. Org. Chem.*, **36**, 369 (1971).
80. D. Seyferth and R. L. Lambert, Jr., *J. Organomet. Chem.*, **55**, C53 (1973).
81. T. Ando, W. Wakabayash, H. Yamanaka, and W. Funasaka, *Bull. Chem. Soc. Jpn.*, **45**, 1576 (1972).
82. H. J. J. Loozen, W. A. Castenmiller, E. J. M. Buter, and H. M. Buck, *J. Org. Chem.*, **41**, 2965 (1976).
83. K. G. Taylor and J. Chaney, *J. Am. Chem. Soc.*, **98**, 4158 (1976).
84. G. Köbrich and W. Goyert, *Tetrahedron*, **24**, 4327 (1968).
85. W. Dumont, A. Anciaux, A. Eman, J. N. Denis, and A. Krief, quoted in A. Krief, *Tetrahedron Lett.*, **36**, 2531 (1980).
86. W. Dumont, M. Sevrin, and A. Krief, *Angew. Chem., Int. Ed. Engl.*, **16**, 541 (1978).
87. W. M. Jones and R. S. Pyron, *Tetrahedron Lett.*, 479 (1965).
88. A. G. Brook, J. M. Duff, and D. G. Anderson, *Can. J. Chem.*, **48**, 561 (1970).
89. D. Seyferth, J. L. Lefferts, and R. L. Lambert, *J. Organomet. Chem.*, **142**, 39 (1977).
90. T. Kauffmann, R. Kriegsmann, and A. Woltermann, *Angew. Chem., Int. Ed. Engl.*, **16**, 862 (1977).
91. T. Kauffmann, K. J. Echsler, A. Hamsen, R. Kriegsmann, F. Steinseifer, and A. Vahrenhorst, *Tetrahedron Lett.*, 4391 (1978).
92. C. Georgoulis, J. Meyet, and W. Smadja, *J. Organomet. Chem.*, **121**, 271 (1976).
93. E. J. Panek, B. L. Neff, H. Chu, and M. G. Panek, *J. Am. Chem. Soc.*, **97**, 3996 (1975).
94. G. Zweifel and W. Lewis, *J. Org. Chem.*, **43**, 2739 (1978).

95. G. Köbrich and H. Trapp. *Chem. Ber.*, **99**, 680 (1966).
96. L. Duhamel and F. Tombret, *J. Org. Chem.*, **46**, 3741 (1981).
97. K. S. Y. Lau and M. Schlosser, *J. Org. Chem.*, **43**, 1595 (1978).
98. R. Knorr and E. Lattke, *Tetrahedron Lett.*, 3969 (1977).
99. J. Ficini, P. Kahn, S. Falou, and A. M. Touzin, *Tetrahedron Lett.*, 67 (1979).
100. D. Masuré, R. Sauvetre, J. F. Normant, and J. Villieras, *Synthesis*, 761 (1976).
101. J. F. Normant, J. P. Foulon, D. Masure, R. Sauvetre, and J. Villeras, *Synthesis*, 122 (1975).
102. D. Masuré, C. Chuit, R. Sauvetre, and J. F. Normant, *Synthesis*, 458 (1978).
103. K. Okuhara, *J. Org. Chem.*, **41**, 1487 (1976).
104. H. L. Elbe and G. Köbrich, *Tetrahedron Lett.*, 2557 (1974).
105. W. E. Parham and D. W. Boykin, *J. Org. Chem.*, **42**, 260 (1977).
106. D. Caine and A. S. Frobese, *Tetrahedron Lett.*, 5167 (1978).
107. L. Duhamel and J. M. Poirier, *J. Am. Chem. Soc.*, **99**, 8356 (1977).
108. W. R. Baker and R. M. Coates, *J. Org. Chem.*, **44**, 1022 (1979).
109. R. Amouroux and T. H. Chan, *Tetrahedron Lett.*, 4453 (1978).
110. C. Shih and J. S. Swenton, *J. Org. Chem.*, **47**, 2835 (1982).
111. D. Seyferth, R. M. Simon, D. J. Sepelak, and H. A. Klein, *J. Org. Chem.*, **45**, 2273 (1981).
112. G. Linstrumelle and D. Michelot, *J. Chem. Soc., Chem. Commun.*, 561 (1975).
113. H. Gilman and A. H. Hauben, *J. Am. Chem. Soc.*, **67**, 1420 (1945).
114. J. D. Edwards, J. A. K. Howard, S. A. R. Knox, V. Riera, and F. G. A. Stone, *J. Chem. Soc., Dalton Trans.*, 75 (1976); see also V. Bakthavachalam, M. d'Alarcao, and N. J. Leonard, *J. Org. Chem.*, **49**, 289 (1984).
115. H. J. Reich and I. L. Reich, *J. Org. Chem.*, **40**, 2248 (1975).
116. P. Jutzi, J. Baumgärtner, and W. Schraut, *J. Organomet. Chem.*, **132**, 333 (1977).
117. M. Schlosser, *J. Organomet. Chem.*, **8**, 9 (1967).
118. W. J. Trepka and R. J. Sonnenfeld, *J. Organomet. Chem.*, **16**, 317 (1969).
119. L. S. Chen, C. J. Chen, and C. Tamborski, *J. Organomet. Chem.*, **193**, 283 (1980).
120. W. E. Parham and C. K. Bradsher, *Acc. Chem. Res.*, **15**, 300 (1982).
121. W. E. Parham, E. C. Egberg, Y. A. Sayed, R. W. Thraikill, G. E. Keyser, M. Neu, W. E. Montgomery, and L. D. Jones, *J. Org. Chem.*, **41**, 2628 (1976).
122. W. E. Parham and R. M. Piccirilli, *J. Org. Chem.*, **42**, 257 (1977).
123. G. Köbrich and P. Buck, *Chem. Ber.*, **103**, 1412 (1970).
124. P. Buck, *Angew. Chem., Int. Ed. Engl.*, **8**, 120 (1969).
125. P. A. Wender and A. W. White, *Tetrahedron Lett.*, **22**, 1475 (1981).
126. W. E. Parham and L. D. Jones, *J. Org. Chem.*, **41**, 1187 (1976).
127. W. E. Parham and R. M. Piccirilli, *J. Org. Chem.*, **41**, 1268 (1976).
128. W. E. Parham and Y. A. Sayed, *J. Org. Chem.*, **39**, 2051 (1974).
129. W. E. Parham and Y. A. Sayed, *J. Org. Chem.*, **39**, 2053 (1974).
130. W. E. Parham and L. D. Jones, *J. Org. Chem.*, **41**, 2704 (1976).
131. W. E. Parham, L. D. Jones, and Y. A. Sayed, *J. Org. Chem.*, **40**, 2394 (1975).
132. C. K. Bradsher and D. C. Reames, *J. Org. Chem.*, **43**, 3800 (1978).
133. E. Akgun, M. B. Glinski, K. L. Dhawan, and T. Durst, *J. Org. Chem.*, **46**, 2730 (1981).
134. H. P. Abicht, U. Baumeister, H. Hartung, K. Issleb, R. A. Jacobson, J. Richardson, S. M. Socol, and J. G. Verkade, *Z. Anorg. Allg. Chem.*, **494**, 55 (1982).
135. N. J. Foulger and B. J. Wakefield, *J. Organomet. Chem.*, **69**, 321 (1974); D. J. Berry, I. Collins, S. M. Roberts, H. Suschitzky, and B. J. Wakefield, *J. Chem. Soc. C*, 1285 (1969).
136. S. Gronowitz and B. Holm, *Acta Chem. Scand., Part B*, **30**, 505 (1976).
137. M. R. Smith and H. Gilman, *J. Organomet. Chem.*, **42**, 1 (1972).
138. M. D. Rausch, T. R. Criswell, and A. K. Ignatowicz, *J. Organomet. Chem.*, **13**, 419 (1968).
139. H. Gilman and S. M. Spatz, *J. Org. Chem.*, **16**, 1485 (1956).
140. G. R. Newkome and J. M. Roper, *J. Organomet. Chem.*, **186**, 147 (1980).
141. E. S. Petrov, M. I. Terekhova, and A. I. Shatenshtein, *Russ. Chem. Rev.*, **42**, 713 (1973).
142. V. Kalyanaraman and M. V. George, *J. Organomet. Chem.*, **47**, 225 (1973).
143. N. Collignon, *J. Organomet. Chem.*, **96**, 139 (1975).
144. C. E. Claff and A. A. Morton, *J. Org. Chem.*, **20**, 440, 981 (1955).
145. C. A. Kraus and R. Rosen, *J. Am. Chem. Soc.*, **47**, 2739 (1925).
146. H. O. House, and V. Kramar, *J. Org. Chem.*, **27**, 4146 (1962); I. K. Brandes, R. Suhrmann, and R. J. Gerdes, *J. Org. Chem.*, **23**, 741 (1967).

147. (a) E. G. Janzen and J. L. Gerlock, *J. Organomet. Chem.*, **8**, 354 (1967); J. J. Eisch and W. C. Kaska, *J. Org. Chem.*, **27**, 3745 (1962); R. L. Kugel, W. G. Hodgson, and H. R. Allcock, *Chem. Ind. (London)*, 1649 (1962); (b) R. Zhang, M. Tsutsui, and D. E. Bergbreiter, *J. Organomet. Chem.*, **229**, 109 (1982).
148. H. Gilman and R. Gorsich, *J. Org. Chem.*, **23**, 550 (1958).
149. G. W. H. Scherf and R. K. Brown, *Can. J. Chem.*, **38**, 697, 2450 (1960).
150. R. Meier, *Chem. Ber.*, **86**, 1483 (1953).
151. A. Bosch and R. K. Brown, *Can. J. Chem.*, **42**, 1728 (1964).
152. K. Hafner and H. Kaiser, *Justus Liebigs Ann. Chem.*, **618**, 140 (1958).
153. K. Ziegler, H. Froitzheim-Kühlhorn, and K. Hafner, *Chem. Ber.*, **89**, 434 (1956); S. McLean and P. Haynes, *Tetrahedron*, **21**, 2313 (1965).
154. J. Thiele, *Chem. Ber.*, **34**, 68 (1901); E. O. Fischer, W. Hafner, and H. O. Stahl, *Z. Anorg. Allg. Chem.*, **282**, 45 (1955).
155. H. Normant, T. Cuvigny, J. F. Normant, and B. Angelo, *Bull. Soc. Chim. Fr.*, 3446 (1965).
156. T. F. Rutledge, *J. Org. Chem.*, **22**, 649 (1957).
157. J. B. Wilkes, *J. Org. Chem.*, **32**, 3231 (1967).
158. H. Yasuda, Y. Ohnuma, M. Yamauchi, H. Tani, and A. Nakamura, *Bull. Chem. Soc. Jpn.*, **52**, 2036 (1979).
159. A. Ledwith and N. McFarlane, *Proc. Chem. Soc.*, 108 (1964); D. E. O'Connor and W. I. Lyness, *J. Org. Chem.*, **30**, 1620 (1965).
160. H. Normant, *Bull. Soc. Chem. Fr.*, 3446 (1965).
161. J. P. Ferris, C. E. Sullivan, and B. G. Wright, *J. Org. Chem.*, **29**, 87 (1964).
162. H. Gross and J. Freiberg, *Chem. Ber.*, **99**, 3260 (1966).
163. C. G. Screttas, *J. Chem. Soc., Perkin Trans. 2*, 745 (1974).
164. R. N. Nasirov and S. P. Solodovnikov, *Bull. Acad. Sci. USSR (Engl. Trans.)*, **22**, 226 (1973).
165. A. P. Krapcho and W. P. Stephens, *J. Org. Chem.*, **45**, 1106 (1980).
166. B. Angelo, *Bull. Soc. Chim. Fr.*, 1710 (1969).
167. R. Weil and N. Collignon, *Bull. Soc. Chim. Fr.*, 253 (1974).
168. N. Collignon, *Bull. Soc. Chim. Fr.*, 1821 (1975).
169. M. I. Terekhova, E. S. Petrov, and A. I. Shatenshtein, *J. Org. Chem. USSR (Engl. Transl.)*, **9**, 883 (1973).
170. I. N. Juchnovski and I. G. Binev, *J. Organomet. Chem.*, **99**, 1, (1975).
171. H. Gilman and J. W. Morton, Jr., *Org. React.*, **8**, 253 (1954).
172. R. A. Benkeser, D. J. Foster, D. M. Sauve and J. F. Nobis, *Chem. Rev.*, **57**, 867 (1957).
173. J. Hartmann and M. Schlosser, *Helv. Chim. Acta*, **59**, 453 (1976).
174. M. Schlosser and S. Strunk, *Tetrahedron Lett.*, **25**, 741 (1984).
175. G. Gau, *J. Organomet. Chem.*, **121**, 1 (1976).
176. T. I. Lebedeva, E. S. Petrov, and A. I. Shatenshtein *J. Org. Chem. USSR (Engl. Transl.)*, **13**, 829 (1977).
177. A. Steitwieser, Jr., D. A. Bors, and M. J. Kaufman, *J. Chem. Soc., Chem. Commun.*, 1394 (1983); R. R. Fraser, M. Bresse, and T. S. Mansour, *J. Chem. Soc., Chem. Commun.*, 620 (1983).
178. H. Gilman, H. A. Pacewitz, and O. Baine, *J. Am. Chem. Soc.*, **62**, 1514 (1940); H. Gilman and H. A. Pacewitz, *J. Am. Chem. Soc.*, **62**, 672 (1940).
179. C. D. Broaddus, *J. Org. Chem.*, **35**, 10 (1970).
180. A. J. Chalk and T. J. Hoogeboom, *J. Organomet. Chem.*, **11**, 615 (1968); A. W. Langer, Jr., *Trans. N. Y. Acad. Sci.*, **27**, 741 (1965).
181. W. N. Smith, *Adv. Chem. Ser.*, No. 130, 23 (1974).
182. G. G. Eberhardt and W. A. Butte, *J. Org. Chem.*, **29**, 2928 (1964).
183. M. Schlosser and J. Hartmann, *Angew. Chem., Int. Ed. Engl.*, **12**, 508 (1973).
184. R. West, *Adv. Chem. Ser.*, No. 130, 211 (1974).
185. A. W. Langer, Jr., *Adv. Chem. Ser.*, No. 130, 1 (1974).
186. G. B. Trimitsii, A. Tuncay, R. D. Beyer, and K. J. Ketterman, *J. Org. Chem.*, **38**, 1491 (1973).
187. A. Krief, *Tetrahedron*, **36**, 2531 (1980).
188. H. W. Gschwend and H. R. Rodriquez, *Org. React.*, **26**, 1 (1979).
189. (a) R. A. Finnegan, *Tetrahedron Lett.*, 429 (1963); (b) R. Lehmann and M. Schlosser, *Tetrahedron Lett.*, **25**, 745 (1984).
190. E. J. Lanpher, L. M. Redman, and A. A. Morton, *J. Org. Chem.*, **23**, 1370 (1958).
191. R. A. Finnegan and R. S. McNees, *J. Org. Chem.*, **29**, 3234 (1964).

192. P. G. Gassman and K. T. Mansfield, *J. Org. Chem.*, **32**, 915 (1967).
193. R.P. Zerger and G. D. Stucky, *J. Chem. Soc., Chem. Commun.*, 44 (1973).
194. (a) J. Harnisch, O. Baumgärtel, G. Szeimies, M. van Meerssche, G. Germain and J. -P. Declercq, *J. Am. Chem. Soc.*, **101**, 3370 (1979); see also O. Baumgärtel and G. Szeimies, *Chem. Ber.*, **116**, 2180 (1983).
195. G. L. Closs and R. B. Larrabee, *Tetrahedron Lett.*, 287 (1965).
196. J. Stapersma, I. D. C. Rodd, and G. W. Klumpp, *Tetrahedron*, **38**, 191 (1982).
197. (a) G. W. Klumpp, M. Kool, M. Schakel, R. F. Schmitz, and C. Boutkan, *J. Am. Chem. Soc.*, **101**, 7065 (1979); (b) G. W. Klumpp, M. Kool, A. H. Veefkind, M. Schakel, and R. F. Schmitz, *Recl. Trav. Chim Pays-Bas*, **102**, 542 (1983); (c) S. von Norman, T. Butkowskyj-Walkiw, and G. Szeimies, *Tetrahedron Lett.*, **24**, 4813 (1983).
198. M. D. Rausch and A. J. Sarnelli, *Adv. Chem. Ser.*, No. 130 248 (1974).
199. H. Gilman and J. C. Baillie, *J. Org. Chem.*, **2**, 84 (1937).
200. D. Bryce-Smith, *J. Chem. Soc.*, 5983 (1963); A. A. Morton, E. L. Little, and W. O. Strong, *J. Am. Chem. Soc.*, **65**, 1335 (1943).
201. W. Neugebauer, A. J. Kos, and P. von R. Schleyer, *J. Organomet. Chem.*, **228**, 107 (1982).
202. (a) A. A. Morton, J. B. Davidson. T. R. P. Gibb, E. L. Little, E. F. Clarke, and A. G. Green, *J. Am. Chem. Soc.*, **64**, 2250 (1942); (b) A. A. Morton, J. T. Massengale, and G. M. Richards, *J. Am. Chem. Soc.*, **65**, 1346 (1943).
203. H. Gilman and B. J. Gaj, *J. Org. Chem.*, **28**, 1725 (1963).
204. C. D. Broaddus, *J. Am. Chem. Soc.*, **88**, 4174 (1966).
205. A. J. Hart, D. H. O'Brien, and C. R. Russell, *J. Organomet. Chem.*, **72**, C19 (1974).
206. A. A. Morton, E. L. Little, Jr., and W. O. Strong, Jr., *J. Am. Chem. Soc.*, **65**, 1339 (1943).
207. R. A. Benkeser, A. E. Trevillyan, and J. Hooz, *J. Am. Chem. Soc.*, **84**, 4971 (1962).
208. T. F. Crimmins and E. M. Rather, *J. Org. Chem.*, **43**, 2170 (1978).
209. T. F. Crimmins and C. M. Chan, *J. Org. Chem.*, **41**, 1870 (1976).
210. R. A. Benkeser, J. Hooz, T. V. Liston, and A. E. Trevillyan, *J. Am. Chem. Soc.*, **85**, 3984 (1963).
211. D. Wilheim, T. Clark, and P. von R. Schleyer, *J. Chem. Soc., Chem. Commun.*, 211 (1983).
212. M. Schlosser and P. Schneider, *Helv. Chim. Acta*, **63**, 2404 (1980).
213. J. Klein, A. Medlik, and A. Y. Meyer, *Tetrahedron*, **32**, 51 (1976).
214. G. B. Trimitsis, A. Tuncay, and R. D. Beyer, *J. Am. Chem. Soc.*, **94**, 2152 (1972).
215. J. Klein and A. Medlik, *J. Am. Chem. Soc.*, **99**, 1473 (1977).
216. E. Dunkelblum and H. Hart, *J. Org. Chem.*, **44**, 3482 (1979).
217. W. E. Rhine, J. H. Davis, and G. Stucky, *J. Organomet. Chem.*, **134**, 139 (1977).
218. E. M. Kaiser, *Tetrahedron*, **39**, 2055 (1983), and references cited therein.
219. P. L. Compagnon and T. Kimny, *J. Organomet. Chem.*, **184**, 297 (1980).
220. B. Menon and E. Buncel, *J. Organomet. Chem.*, **159**, 357 (1978).
221. T. Cuvigny and H. Normant, *Bull. Soc. Chim. Fr.*, 1872 (1965).
222. E. Buncel and B. Menon, *J. Am. Chem. Soc.*, **99**, 4457 (1977).
223. H. Koster and E. Weiss, *J. Organomet. Chem.*, **168**, 273 (1979).
224. W. L. Jolly, *Inorg. Chem.*, **6**, 1435 (1967).
225. E. J. Panek and T. J. Rodgers, *J. Am. Chem. Soc.*, **96**, 6921 (1974); see also M. Daney, H. Bouas-Laurent, B. Calas, L. Giral, and N. Platzer, *J. Organomet. Chem.*, **188**, 277 (1980).
226. R. G. Harvey, L. Nazareno and H. Cho, *J. Am. Chem. Soc.*, **95**, 2376 (1973); see also W. E. Rhine, J. Davis, and G. Stucky, *J. Am. Chem. Soc.*, **97**, 2079 (1975).
227. W. V. E. Doering and C. C. H. Depuy, *J. Am. Chem. Soc.*, **75**, 7955 (1953).
228. O. Meth-Cohn and S. Gronowitz, *J. Chem. Soc., Chem. Commun.*, 81 (1966).
229. G. Witting, D. Davis, and G. Koenig, *Chem. Ber.*, **84**, 627 (1951).
230. T. J. Barton and C. R. Tully, *J. Organomet. Chem.*, **172**, 11 (1979).
231. A. F. Halasa, *J. Organomet. Chem.*, **31**, 369 (1971).
232. T. J. Kartz, M. Rosenberger, and R. K. O'Hara, *J. Am. Chem. Soc.*, **86**, 249 (1964).
233. M. W. Haenel, *Tetrahedron Lett.*, 1273 (1977).
234. D. Hellwinkel and G. Haas, *Justus Liebigs Ann. Chem.*, 145 (1979).
235. D. Bladauski, H. Dietrich, H. J. Hecht, and D. Rewicki, *Angew. Chem., Int. Ed. Engl.*, **16**, 474 (1977).
236. M. Walczak and G. D. Stucky, *J. Organomet. Chem.*, **97**, 313 (1975).
237. I. Willner and M. Rabinovitz, *Tetrahedron Lett.*, 1223 (1976).
238. A. A. Morton, F. D. Marsh, R. D. Coombs, A. L. Lyons, S. E. Penner, H. E. Ramsden, V. B. Baker, E. L. Little, and R. L. Letsinger, *J. Am. Chem. Soc.*, **72**, 3785 (1950).

239. C. J. Broaddus, T. J. Logan, and T. J. Flautt, *J. Org. Chem.*, **28**, 1174 (1962).
240. A. A. Morton and M. E. T. Holden, *J. Am. Chem. Soc.*, **69**, 1675 (1974).
241. C. D. Broaddus and D. L. Muck, *J. Am. Chem. Soc.*, **89**, 6533 (1967).
242. F. H. Köhler and N. Hertkorn, *Chem. Ber.*, **116**, 3274 (1983).
243. R. Ya. Levina, I. B. Avezov, L. S. Surmina, and I. J. Bolesov, *J. Org. Chem. USSR* (*Engl. Transl.*), **8**, 1118 (1972).
244. C. F. Huebner, R. T. Puckett, M. Brzechffa, and S. L. Schwartz, *Tetrahedron Lett.*, 359 (1970).
245. S. Akiyama and J. Hooz, *Tetrahedron Lett.*, 4115 (1973).
246. T. B. Thompson and W. T. Ford, *J. Am. Chem. Soc.*, **101**, 5459 (1979).
247. E. J. Lanpher, *J. Am. Chem. Soc.*, **97**, 5579 (1975).
248. D. H. O'Brien, C. R. Russell, and A. J. Hart, *Tetrahedron Lett.*, 37 (1976).
249. M. Schlosser and M. Stähle, *Angew. Chem., Int. Ed. Engl.*, **19**, 487 (1980).
250. K. Ziegler and H. Dislich, *Chem. Ber.*, **90**, 1107 (1957).
251. M. Schlosser, J. Hartmann and V. David, *Helv. Chim. Acta*, **57**, 1567 (1974).
252. R. B. Bates and W. A. Beavers, *J. Am. Chem. Soc.*, **96**, 5001 (1974).
253. R. Y. Mixer and W. G. Young, *J. Am. Chem. Soc.*, **78**, 3379 (1956).
254. G. Boche and D. R. Schneider, *Tetrahedron Lett.*, 3657 (1976).
255. M. Stähle, J. Hartmann, and M. Schlosser, *Helv. Chim. Acta*, **60**, 1730 (1977).
256. J. Klein, A. Medlik, A. Y. Meyer, and M. Chorev, *Tetrahedron*, **32**, 1839 (1976); R. B. Bates, W. A. Beavers, B. Gordon, and N. S. Mills, *J. Org. Chem.*, **44**, 3800 (1979).
257. A. A. Morton and M. L. Brown, *J. Am. Chem. Soc.*, **69**, 160 (1947).
258. W. L. Trepka, J. A. Favre, and R. J. Sonnenfeld, *J. Organomet. Chem.*, **55**, 221 (1973).
259. D. Wilhelm, T. Clark, P. von R. Schleyer, K. Buckl, and G. Boche, *Chem. Ber.*, **116**, 1669 (1983); D. Wilhelm, T. Clark, and P. von R. Schleyer, *Tetrahedron Lett.*, **24**, 3985 (1983); D. Wilhelm, T. Clark, P. von R. Schleyer, J. L. Courtneidge, and A. G. Davies, *J. Chem. Soc., Perkin Trans. 2*, 915 (1984).
260. N. S. Mills, J. Shapiro, and M. Hollingsworth, *J. Am. Chem. Soc.*, **103**, 1263 (1981).
261. J. Klein and A. Medlik-Balan, *J. Chem. Soc., Chem. Commun.*, 877 (1975); R. B. Bates, W. A. Beavers, M. G. Greene, and J. H. Klein, *J. Am. Chem. Soc.*, **96**, 5640 (1974).
262. W. A. Beavers, S. E. Wilson, B. Gordon, R. B. Bates, and A. B. Romano, *Tetrahedron Lett.*, 1675 (1979).
263. E. W. Thomas, *Tetrahedron Lett.*, **24**, 1467 (1983).
264. E. Dunkelblum and S. Brenner, *Tetrahedron Lett.*, 669 (1973).
265. S. R. Wilson, L. R. Phillips, and K. J. Natalie, *J. Am. Chem. Soc.*, **101**, 3340 (1979).
266. J. Hartmann and M. Schlosser, *Synthesis*, 328 (1975).
267. H. Gilman and R. L. Bebb, *J. Am. Chem. Soc.*, **61**, 109 (1939).
268. R. D. Bates, D. W. Gosselink, and J. A. Koczynski, *Tetrahedron Lett.*, 1991 (1967); W. E. Paget, K. Smith, M. G. Hutchings, and G. E. Martin, *J. Chem. Res.* (*S*), 30 (1983); M. Schlosser and G. Rauchschwalbe, *J. Am. Chem. Soc.*, **100**, 3258 (1978); H. Bosshart and M. Schlosser, *Helv. Chim. Acta*, **63**, 2393 (1980).
269. D. H. Hunter, R. E. Klinck, R. P. Steiner, and J. B. Stothers, *Can. J. Chem.*, **84**, 1464 (1976).
270. J. J. Bahl, R. B. Bates, and B. Gordon, *J. Org. Chem.*, **44**, 2290 (1979).
271. H. M. Parkes and R. N. Young, *J. Chem. Soc., Perkin Trans. 2*, 1137 (1980).
272. J. J. Bahl, R. B. Bates, W. A. Beavers, and N. S. Mills, *J. Org. Chem.*, **41**, 1620 (1976); S. K. Arora, R. B. Bates, W. A. Beavers, and R. S. Cutler, *J. Am. Chem. Soc.*, **97**, 6271 (1975).
273. N. S. Mills, *J. Am. Chem. Soc.*, **104**, 5689 (1982).
274. A. Rusinko, III, N. C. Mills, and P. Morse, *J. Org. Chem.*, **47**, 5198 (1982).
275. J. Klein and A. Medlik-Balan, *Tetrahedron Lett.*, 279 (1978).
276. J. J. Bahl, R. B. Bates, W. A. Beavers, and C. R. Launer, *J. Am. Chem. Soc.*, **99**, 6126 (1977).
277. P. Jutzi and J. Bäumgartner, *J. Organomet. Chem.*, **148**, 247 (1978).
278. W. Gausing and G. Wilke, *Angew. Chem., Int. Ed. Engl.*, **17**, 371. (1978).
279. H. H. Inhoffen, H. Pommer, and E. Gmeth, *Justus Liebigs Ann. Chem.*, **565**, 45 (1949).
280. K. P. Martin, C. W. Kamienski, M. H. Dellinger, and R. O. Bach, *J. Org. Chem.*, **33**, 778 (1968).
281. M. M. Midland, *J. Org. Chem.*, **40**, 2250 (1975).
282. G. Fraenkel and P. Pramanik, *J. Chem. Soc., Chem. Commun.*, 1527 (1983).
283. B. Schubert and E. Weiss, *Chem. Ber.*, **116**, 3212 (1983).
284. W. Kulik, H. D. Verkruijsse, R. L. P. de Jong, H. Hommes, and L. Brandsma, *Tetrahedron Lett.*, **24**, 2203 (1983).

285. J. Klein, S. Brenner, and A. Medlik, *Isr. J. Chem.*, **9**, 177 (1971); E. J. Corey and D. E. Cane, *J. Org. Chem.*, **35**, 3405 (1970); G. I. Pis'mennaya, L. M. Zubritskii, and Kh. V. Bal'yan, *Zh. Org. Khim.*, **7**, 251 (1971).
286. E. J. Corey, H. A. Kirst, and J. A. Katzenellenbogen, *J. Am. Chem. Soc.*, **92**, 6314 (1970).
287. J. Y. Becker and J. Klein, *J. Organomet. Chem.*, **157**, 1 (1978); J. Y. Becker, *J. Organomet. Chem.*, **127**, 1 (1977).
288. S. Bhanu and F. Scheinmann, *J. Chem. Soc., Chem. Commun.*, 817 (1975).
289. T. L. Chwang and R. West, *J. Am. Chem. Soc.*, **95**, 3324 (1973); W. Priester, R. West, and T. L. Chwang, *J. Am. Chem. Soc.*, **98**, 8413 (1976); W. Priester and R. West, *J. Am. Chem. Soc.*, **98**, 8426 (1976).
290. N. M. Libman, V. I. Zlobina, and S. G. Kuznetsov, *J. Org. Chem. USSR* (*Engl. Transl.*), **10**, 2074 (1974).
291. D. Bauer and G. Köbrich, *Chem. Ber.*, **109**, 2185 (1976).
292. P. I. Dem'yanov, I. B. Fedot'eva, E. V. Babaev, V. S. Petrosyan, and O. A. Reutov, *Dokl. Akad. Nauk SSSR*, **268**, 1403 (1983).
293. Y. Leroux and R. Mantione, *C. R. Acad. Sci., Ser. C*, **281**, 313 (1975).
294. F. Jaffe, *J. Organomet. Chem.*, **23**, 53 (1970).
295. J.-C. Clinet and G. Linstrumelle, *Synthesis*, 875 (1981).
296. R. Baudouy, F. Delbecq, and J. Gore, *J. Organomet. Chem.*, **177**, 39 (1979).
297. P. Condran, Jr., M. L. Hammond, A. Mourino, and W. H. Okamura, *J. Am. Chem. Soc.*, **102**, 6259 (1980).
298. E. M. G. A. van Kruchten, A. Haces, and W. H. Okamura, *Tetrahedron Lett.*, **24**, 3939 (1983).
299. X. Creary, *J. Am. Chem. Soc.*, **99**, 7632 (1977); G. Balme, A. Doutheau, J. Gore, and M. Malacria, *Synthesis*, 508 (1979).
300. C. Huynh and G. Linstrumelle, *J. Chem. Soc., Chem. Commun.*, 1133 (1983).
301. D. J. Peterson, *Organomet. Chem. Rev. A*, **7**, 295 (1972).
302. J. F. Biellmann and J. B. Ducep, *Org. React.*, **27**, 1 (1982).
303. e.g. S. Arseniyadis, K. S. Kyler, and D. S. Watt, *Org. React.*, **31**, 1 (1984).
304. H. P. Abricht and K. Issleib, *Fresenivs Z. Chem.*, **17**, 1 (1977); D. W. Slocum and D. I. Sugerman, *Adv. Chem. Ser.*, **130**, 222 (1974); P. Beak and V. Snieckus, *Acc. Chem. Res.*, **15**, 306 (1982).
305. R. R. Fraser, M. Bresse, and T. S. Mansour, *J. Am. Chem. Soc.*, **105**, 7790 (1983).
306. (a) A. I. Meyers, W. F. Rieker, and L. M. Fuentes, *J. Am. Chem. Soc.*, **105**, 2082 (1983); (b) M. Al-Aseer, P. Beak, D. Hay, D. J. Kempf, S. Mills, and S. G. Smith, *J. Am. Chem. Soc.*, **105**, 2080 (1983).
307. (a) S. Cabiddu, C. Floris, S. Melis, and F. Sotgiu, *Phosphorus Sulfur*, **19**, 61 (1984); (b) D. Seebach, J. -J. Lohmann, M. A. Syfrig, and M. Yoshifuji, *Tetrahedron*, **39**, 1963; (1983); (c) A. I. Meyers, L. M. Fuentes, and Y. Kubota *Tetrahedron*, **40**, 1361 (1984).
308. P. Beak and D. B. Reitz, *Chem. Rev.*, **78**, 275 (1978).
309. (a) D. B. Reitz, P. Beak, R. F. Farney, and L. S. Helmick, *J. Am. Chem. Soc.*, **100**, 5428 (1978); (b) P. Beak and W. J. Zajdel, *J. Am. Chem. Soc.*, **106**, 1010 (1984), and references cited therein.
310. D. Seebach and W. Lubosch, *Angew. Chem., Int. Ed. Engl.*, **15**, 313 (1976).
311. P. Beak and B. G. McKinnie, *J. Am. Chem. Soc.*, **99**, 5213 (1977).
312. A. I. Meyers, P. D. Edwards, W. F. Rieker, and T. R. Bailey, *J. Am. Chem. Soc.*, **106**, 3270 (1984); A. I. Meyers and G. E. Jagdmann, *J. Am. Chem. Soc.*, **104**, 877 (1982).
313. R. Schlecker and D. Seebach, *Helv. Chim. Acta*, **60**, 1459 (1977).
314. A. I. Meyers and M. Ford, *J. Org. Chem.*, **41**, 1735 (1976).
315. C. R. Johnson and K. Tanaka, *Synthesis*, 413 (1976).
316. O. Schöllkopf and H. Beckhaus, *Angew. Chem., Int. Ed. Engl.*, **15**, 293 (1976).
317. D. Seebach, W. Lubosch, and D. Enders, *Chem. Ber.*, **109**, 1309 (1976).
318. (a) D. Seebach and D. Enders, *Angew. Chem., Int. Ed. Engl.*, **14**, 15 (1975); (b) D. H. R. Barton, R. D. Bracho, A. A. L. Guratilaka, and P. A. Widdowson, *J. Chem. Soc., Perkin Trans. 1*, 579 (1975).
319. R. A. Abramovitch, E. M. Smith, E. E. Knaus, and M. Saha, *J. Org. Chem.*, **37**, 1690 (1972).
320. D. Seebach, R. Henning, F. Lehr, and J. Gonnermann, *Tetrahedron Lett.*, 1161 (1977).
321. U. Schöllkopf, *Angew. Chem., Int. Ed. Engl.*, **16**, 339 (1977).
322. R. R. Fraser and J. Banville, *J. Chem. Soc., Chem. Commun.*, 47 (1979).
323. S. Shatzmiller and R. Lidor, *Synthesis*, 590 (1983).

324. W. S. Wadswroth, *Org. React.*, **25**, 73 (1977).
325. J. Boutagy and R. Thomas, *Chem. Rev.*, **74**, 87 (1974).
326. E. J. Corey and G. T. Kwiatkowski, *J. Am. Chem. Soc.*, **88**, 5654 (1966).
327. E. J. Corey, *Pure Appl. Chem.*, **14**, 19 (1967).
328. S. Lavielle, S. Bory, B. Moreau, J. Luche, and A. Marquet, *J. Am. Chem. Soc.*, **100**, 1558 (1978); E. J. Corey and T. Durst, *J. Am. Chem. Soc.*, **88**, 5656 (1966).
329. E. J. Corey and T. Durst, *J. Am. Chem. Soc.*, **90**, 5553 (1968).
330. P. D. Magnus, *Tetrahedron*, **33**, 2019 (1977).
331. C. R. Johnson, *Acc. Chem. Res.*, **6**, 341 (1973).
332. H. J. Reich and S. K. Shah, *J. Am. Chem. Soc.*, **97**, 3250 (1975).
333. T. Kauffmann, R. Joussen, and A. Woltermann, *Angew. Chem., Int. Ed. Engl.*, **16**, 709 (1977).
334. T. Kauffmann, *Angew. Chem., Int. Ed. Engl.*, **13**, 627 (1974).
335. A. Streitwieser, Jr., and J. E. Williams, *J. Am. Chem. Soc.*, **97**, 191 (1975); F. Bernardi, I. G. Csizmadia, A. Mangini, H. B. Schlegel, M. -H. Whangbo, and S. Wolfe, *J. Am. Chem. Soc.*, **97**, 2209 (1975); N. D. Epiotis, R. L. Yates, F. Bernandi, and S. Wolfe, *J. Am. Chem. Soc.*, **98**, 5435 (1976); J. M. Lehn and G. Wipff, *J. Am. Chem. Soc.*, **98**, 7498 (1976); J. M. Lehn, G. Wipff, and J. Demuynck, *Helv. Chim. Acta*, **60**, 1239 (1977).
336. E. L. Eliel, *Angew. Chem., Int. Ed. Engl.*, **11**, 739 (1972); *Tetrahedron*, **30**, 1503 (1974); E. L. Eliel, A. A. Hartmann and A. G. Abatjoglou, *J. Am. Chem. Soc.*, **96**, 1807 (1974).
337. A. I. Meyers, A. L. Campbell, A. G. Abatjoglou and E. L. Eliel, *Tetrahedron Lett.*, 4159 (1979).
338. D. J. Peterson and H. R. Hays, *J. Org. Chem.*, **30**, 1939 (1965).
339. E. J. Corey and D. Seebach, *J. Org. Chem.*, **31**, 4097 (1966); T. M. Dolak and T. A. Bryson, *Tetrahedron Lett.*, 1961 (1977).
340. H. J. Reich, F. Chow, and S. K. Shah, *J. Am. Chem. Soc.*, **101**, 6638 (1979).
341. A. G. Giumanini and G. Lercker, *J. Org. Chem.*, **35**, 3756 (1970); G. W. H. Cheeseman and S.G. Greenberg, *J. Organomet. Chem.*, **166**, 139 (1979); A. P. Lepley, W. A. Khan, A. B. Giumanini, and A. G. Guimanini, *J. Org. Chem.*, **31**, 2047 (1966); G. Friedmann, P. Linder, M. Brini, and A. Cheminat, *J. Org. Chem.*, **44**, 237 (1979).
342. (a) C. T. Viswanathan and C. A. Wilkie, *J. Organomet. Chem.*, **54**, 1 (1973); (b) F. N. Jones, M. F. Zinn, and C. R. Hauser, *J. Org. Chem.*, **28**, 663 (1963).
343. T. D. Harris and G. P. Roth, *J. Org. Chem.*, **44**, 2004 (1979).
344. D. W. Slocum, T. R. Engelmann, and C. A. Jennings, *Aust. J. Chem.*, **21**, 2319 (1968); N. S. Narasimhan and A. C. Ranade, *Tetrahedron Lett.*, 603 (1966); F. N. Jones, R. L. Vaulx, and C. R. Hauser, *J. Org. Chem.*, **28**, 3461 (1963).
345. (a) P. Beak, G. R. Brubaker, and R. F. Farney, *J. Am. Chem. Soc.*, **98**, 3621 (1976); (b) P. Beak and R. A. Brown, *J. Org. Chem.*, **42**, 1823 (1977); (c) **44**, 4463 (1979); (d) S. O. de Silva, I. Ahmad, and V. Snieckus, *Tetrahedron Lett.*, 5107 (1978); M. Iwao, K. K. Mahalanabis, M. Watanabe, S. O. de Silva, and V. Snieckus, *Tetrahedron*, **39**, 1955 (1983).
346. A. Marxer and M. Siegrist, *Helv. Chim. Acta.* **57**, 1988 (1974).
347. H. W. Gschwend and A. Hamdan, *J. Org. Chem.*, **40**, 2008 (1978); M. S. Newman and S. Kumar, *J. Org. Chem.*, **43**, 370 (1978); T. D. Harris, B. Neuschwander, and V. Boekelheide, *J. Org. Chem.*, **43**, 727 (1978); L. della Vecchia and I. Vlattas, *J. Org. Chem.*, **42**, 2649 (1977); A. I. Meyers and K. Lutomski, *J. Org. Chem.*, **44**, 4464 (1979).
348. F. E. Ziegler and K. W. Fowler, *J. Org. Chem.*, **41**, 1564 (1976).
349. T. D. Krizan and J. C. Martin, *J. Org. Chem.*, **47**, 2681 (1982); *J. Am. Chem. Soc.*, **105**, 6155 (1983).
350. (a) R. A. Ellison and F. N. Kotsonis, *J. Org. Chem.*, **38**, 4192 (1973); (b) R. C. Ronald, *Tetrahedron Lett.*, 3973 (1975); D. A. Shirley and J. P. Hendrix, *J. Organomet. Chem.*, **11**, 217 (1968); R. Muthukrishnan and M. Schlosser, *Helv. Chim. Acta*, **59**, 13 (1976).
351. M. P. Sibi and V. Sniekus, *J. Org. Chem.*, **48**, 1935 (1983).
352. H. Gilman, W. J. Meikle, and J. W. Morton, *J. Am. Chem. Soc.*, **74**, 6282 (1952).
353. E. Napolitano, E. Giannone, R. Fiaschi, and A. Marsili, *J. Org. Chem.*, **48**, 3653 (1983).
354. H. P. Plaumann, B. A. Keay, and R. Rodrigo, *Tetrahedron Lett.*, 4921 (1979).
355. L. Horner, A. J. Lawson, and G. Simons, *Phosphorus Sulfur*, **12**, 353 (1982).
356. (a) H. Watanabe, R. A. Schwarz, C. R. Hauser, J. Lewis, and D. W. Slocum, *Can. J. Chem.*, **47**, 1543 (1969); (b) F. M. Stoyanovich, Ya. L. Gol'dfarb, M. A. Marakatkina, and R. G. Karpenko, *Bull. Acad. Sci. USSR (Engl. Transl.)*, **29**, 129 (1980); (c) D. Hellwinkel and M. Supp, *Chem. Ber.*, **109**, 3749 (1979).

357. Ya. L. Gol'dfarb and F. M. Stoyanovich, *Bull. Acad. Sci. USSR* (*Engl. Transl.*), **24**, 1473 (1975).
358. H. Gilman and T. S. Soddy *J. Org. Chem.*, **22**, 1715 (1957).
359. A. M. Roe, R. A. Burton and D. R. Reavill, *J. Chem. Soc., Chem. Commun.*, 582 (1965); D. A. Shirley, J. R. Johnson, Jr., and J. P. Hendrix, *J. Organomet. Chem.*, **11**, 209 (1968).
360. C. G. Stuckwisch, *J. Org. Chem.*, **41**, 1173 (1976).
361. W. Fuhrer and H. W. Gschwend, *J. Org. Chem.*, **44**, 1133 (1979).
362. J. M. Muchowski and M. C. Venuti, *J. Org. Chem.*, **45**, 4798 (1980).
363. C. L. Mao, I. T. Barnish, and C. R. Hauser, *J. Heterocycl. Chem.*, **6**, 475 (1969); A. Marxer, H. R. Rodriquez, J. M. McKenna, and H. M. Tsai, *J. Org. Chem.*, **40**, 1427 (1975); J. E. Baldwin and K. W. Blair, *Tetrahedron Lett.*, 2559 (1978); A. S. Kende and J. P. Rizzi, *Tetrahedron Lett.*, **22**, 1779 (1981).
364. J. J. Fitt and H. W. Gschwend, *J. Org. Chem.*, **41**, 4029 (1976).
365. E. F. Perozzi and J. C. Martin, *J. Am. Chem. Soc.*, **101**, 1591 (1979); J. C. Martin and T. M. Balthazor, *J. Am. Chem. Soc.*, **99**, 152 (1977); N. Meyer and D. Seebach, *Chem. Ber.*, **113**, 1304 (1980).
366. D. W. Slocum and W. Achermann, *J. Chem. Soc. Chem. Commun.*, 968 (1974).
367. D. L. Collins, J. D. Brown, and N. B. Mantlo, *Tetrahedron Lett.*, **23**, 3979 (1982).
368. H. Watanabe, R. L. Gay, and C. R. Hauser, *J. Org. Chem.*, **33**, 900 (1968); J. G. Lombardino, *J. Org. Chem.*, **36**, 1843 (1971).
369. H. Watanabe, C. -L. Mao, and C. R. Hauser, *J. Org. Chem.*, **34**, 1786 (1969); G. D. Figuly and J. C. Martin, *J. Org. Chem.*, **45**, 3728 (1980).
370. H. M. Walborsky and P. Ronman, *J. Org. Chem.*, **43**, 731 (1978).
371. L. Barsky, H. W. Gschwend, J. McKenna, and H. R. Rodriquez, *J. Org. Chem.*, **41**, 3651 (1976).
372. D. L. Comins and J. D. Brown, *Tetrahedron Lett.*, **24**, 5465 (1983); *J. Org. Chem.*, **49**, 1078 (1984).
373. D. W. Slocum and C. A. Jennings, *J. Org. Chem.*, **41**, 3653 (1976).
374. A. I. Meyers and R. A. Gabel, *Tetrahedron Lett.*, 227 (1978).
375. P. Beak, A. Tse, J. Hawkins, C. -W. Chen, and S. Mills, *Tetrahedron*, **39**, 1983 (1983).
376. D. A. Shirley and B. J. Reeves, *J. Organomet. Chem.*, **16**, 1 (1969).
377. T. M. Dolak and T. A. Bryson, *Tetrahedron Lett.*, 1961 (1977).
378. F. T. Oakes and J. F. Sebastian, *J. Organomet. Chem.*, **159**, 363 (1978).
379. R. L. Vaulz, F. N. Jones, and C. R. Hauser, *J. Org. Chem.*, **30**, 58 (1965).
380. P. Beak and R. A. Brown, *J. Org. Chem.*, **47**, 34 (1982).
381. H. Gilman and R. D. Gorsich, *J. Am. Chem. Soc.*, **78**, 2217 (1956).
382. I. Haiduc and H. Gilman, *Chem. Ind.* (*London*), 1278 (1968).
383. S. S. Dua and H. Gilman, *Indian J. Chem.*, **17B**, 562 (1979).
384. P. Aeberli and W. J. Houlihan, *J. Organomet. Chem.*, **67**, 321 (1974).
385. M. R. Winkle and R. C. Ronald, *J. Org. Chem.*, **47**, 2101 (1982).
386. D. A. Shirley, T. E. Harmon, and C. F. Cheng, *J. Organomet. Chem.*, **69**, 327 (1974).
387. N. F. Masters and D. A. Widdowson, *J. Chem. Soc., Chem. Commun.*, 955 (1983); see also P. J. Beswick, S. J. Leach, N.F. Masters, and D. A. Widdowson, *J. Chem. Soc., Chem. Commun.*, 46 (1984).
388. M. Uemura, N. Nishikawa, K. Take, M. Ohniski, K. Hirotsu, T. Higuchi, and Y. Hayashi, *J. Org. Chem.*, **48**, 2349 (1983).
389. Y. L. Gol'dfarb, B. P. Fabrichnyl, and V. I. Rogovik, *Izv. Akad. Nauk SSSR, Ser. Khim.*, 515 (1965).
390. D. J. Chadwick and C. Willbe, *J. Chem. Soc., Perkin Trans. 1*, 887 (1977).
391. T. Kauffmann, A. Mitschker, and A. Woltermann, *Chem. Ber.*, **116**, 992 (1983).
392. G. M. Davies and P. S. Davies, *Tetrahedron Lett.*, 3507 (1972).
393. P. Dubus, B. Decroix, J. Morel, and P. Pastour, *Bull. Soc. Chim. Fr.*, 628 (1976).
394. S. Gronowitz, *Ark. Kemi*, **7**, 361 (1954).
395. N. J. Gjøs and S. Gronowitz, *Acta Chem. Scand.*, **25**, 2596 (1971).
396. S. Gronowitz, B. Cederlund, and A. B. Hörnfeldt, *Chem. Sci.*, **5**, 217 (1974).
397. J. Z. Mortensen, B. Hedegaard, and S. -O. Lawesson, *Tetrahedron*, **27**, 3839 (1971).
398. (a) D. A. Shirley and M. D. Cameron, *J. Am. Chem. Soc.*, **74**, 667 (1952); (b) E. B. McCall, A. J. Neale, and T. J. Rawlings, *J. Chem. Soc.*, 4900 (1962).
399. V. Ramanathan and R. Levine, *J. Org. Chem.*, **27**, 1216 (1962).
400. N. D. Ly and M. Schlosser, *Helv. Chim. Acta*, **60**, 2085 (1977).

401. D. J. Chadwick, M. V. McKnight, and R. Ngochindo, *J. Chem. Soc., Perkin Trans. 1*, 1343 (1982).
402. P. Ribereau and G. Queguiner, *Tetrahedron*, **39**, 3593 (1983).
403. P. Cagniant and G. Kirsch, *C. R. Acad. Sci., Ser. C*, **281**, 111 (1975).
404. T. Kauffmann and H. Lexy, *Chem. Ber.*, **114**, 2596 (1981).
405. D. J. Chadwick and I. A. Cliffe, *J. Chem. Soc., Perkin Trans. 1*, 2845 (1979).
406. D. A. Shirley, B. H. Gross, and P. A. Roussel, *J. Org. Chem.*, **20**, 225 (1955).
407. G. R. Martinez, P. A. Gieco, and C. V. Srinivisan, *J. Org. Chem.*, **46**, 3760 (1981).
408. I. Hasan, E. R. Marinelli, L. -C. Chang Lin, F. W. Fowler, and A. B. Levy, *J. Org. Chem.*, **46**, 157 (1981).
409. M. E. K. Cartoon and G. W. H. Cheeseman, *J. Organomet. Chem.*, **234**, 123 (1982).
410. R. J. Sundberg and H. R. Russell, *J. Org. Chem.*, **38**, 3324 (1973).
411. R. J. Sundberg and R. L. Parton, *J. Org. Chem.*, **41**, 163 (1976).
412. D. A. Shirley and P. A. Roussel, *J. Am. Chem. Soc.*, **75**, 375 (1953).
413. F. Marsais and G. Queguiner, *Tetrahedron*, **39**, 2009 (1983); G. W. Gribble and M. G. Saulnier, *Tetrahedron Lett.*, **21**, 4137 (1980); T. Güngör, F. Marsais, and G. Queguiner, *J. Organomet. Chem.*, **215**, 139 (1981); F. Marsais, P. Breant, A. Ginguene, and G. Quegunier, *J. Organomet. Chem.*, **216**, 139 (1981); F. Marsais, B. Laperdrix, T. Güngör, M. Mallet, and G. Queguiner *J. Chem. Res. (S)*, 278 (1982).
414. J. Epsztajn, Z. Berski, J. K. Brzezinski, and A. Jozwiak, *Tetrahedron Lett.*, **21**, 4739 (1980).
415. A. R. Katirtzky, S. Rahimi-Rastgoo and N. K. Ponkshe, *Synthesis*, 127 (1981).
416. J. A. Turner, *J. Org. Chem.*, **48**, 3401 (1983); T. Güngör, F. Marsais, and G. Queguiner, *Synthesis*, 499 (1982).
417. A. I. Meyers and R. A. Gabel, *Heterocycles*, **11**, 138 (1978); *J. Org. Chem.*, **47**, 2633 (1982).
418. P. Breant, F. Marsais, and G. Queguiner, *Synthesis*, 822 (1983).
419. F. Marsais, G. Le Nard and G. Queguiner, *Synthesis*, 235 (1982).
420. R. C. Ronald and M. R. Winkle, *Tetrahedron*, **39**, 2031 (1983).
421. (a) S. L. Taylor, D. Y. Lee, and J. C. Martin, *J. Org. Chem.*, **48**, 4156 (1983); (b) J. Verbeek, A. V. E. George, R. L. P. de Jong, and L. Brandsma, *J. Chem. Soc., Chem. Commun.*, 257 (1984).
422. J. D. Cook and B. J. Wakefield, *J. Chem. Soc. C*, 1973 (1969).
423. U. Schöllköpf and P. Hänssle, *Justus Liebigs Ann. Chem.*, **763**, 208 (1972); J. E. Baldwin, G. A. Höfle, and O. W. Lever, Jr., *J. Am. Chem. Soc.*, **96**, 7125 (1974); W. Adam, E. Schmidt, and K. Takayama, *Synthesis*, 322 (1982).
424. (a) I. Vlattas, L. della Vecchia, and A. O. Lee, *J. Am. Chem. Soc.*, **98**, 2008 (1976); (b) R. C. Cookson and P. J. Parsons, *J. Chem. Soc., Commun.*, 990 (1976); (c) B. Harirchian and P. Magnus, *J. Chem. Soc., Chem. Commun.*, 522 (1977); (d) K. Oshima, K. Shimoji, H. Takahashi, H. Yamomoto, and H. Nozaki, *J. Am. Chem. Soc.*, **95**, 2694 (1973); (e) R. R. Schmidt and B. Schmid, *Tetrahedron Lett.*, 3583 (1977).
425. (a) M. Sevrin, J. N. Denis, and A. Krief, *Angew. Chem., Int. Ed. Engl.*, **17**, 526 (1978); (b) H. J. Reich, W. W. Willis, Jr., and P. D. Clark, *J. Org. Chem.*, **46**, 2775 (1981); (c) T. Kauffmann and H. Ahlers, *Chem. Ber.*, **116**, 1001 (1983).
426. (a) H. Takei, H. Sugimura, M. Miura, and H. Okamura, *Chem. Lett.*, 1209 (1980); (b) G. H. Posner, P. W. Tang, and J. P. Mallamo, *Tetrahedron Lett.*, 3995 (1978).
427. J. J. Eisch and J. E. Galle, *J. Org. Chem.*, **44**, 3279 (1979).
428. R. R. Schmidt and H. Speer, *Synthesis*, 797 (1979); J. Melamed and B. A. Feit, *J. Chem. Soc., Perkin Trans. 1*, 1228 (1978).
429. R. R. Schmidt and H. Speer, *Tetrahedron Lett.*, **22**, 4259 (1981).
430. B. A. Feit, U. Melamed, R. R. Schmidt, and H. Speer, *J. Chem. Soc., Perkin Trans. 1*, 1329 (1981).
431. U. Schöllkoph, D. Stafforst and R. Jentsch, *Justus Liebigs Ann. Chem.*, 1167 (1977).
432. (a) R. R. Schmidt, J. Talbiersky, and P. Russegeer, *Tetrahedron Lett.*, 4273 (1979); (b) R. R. Schmidt, H. Speer, and B. Schmid, *Tetrahedron Lett.*, 4277 (1979).
433. (a) G. Köbrich, *Angew. Chem., Int. Ed. Engl.*, **6**, 41 (1967); **11**, 473 (1972); (b) G. Köbrich and K. Flory, *Chem. Ber.*, **99**, 1773 (1966); D. J. Nelson, *J. Org. Chem.*, **49**, 2059 (1984).
434. B. A. Feit, U. Melamed, R. R. Schmidt, and H. Speer, *Tetrahedron*, **37**, 2143 (1981).
435. see also for example O. Miyata, and R. R. Schmidt, *Tetrahedron Lett.*, **23**, 1793 (1982).
436. C. B. Bi Ekogha, O. Ruel, and S. A. Julia, *Tetrahedron Lett.*, **24**, 4825 (1983); O. Ruel, C. B. Bi Ekogha, and S. A. Julia, *Tetrahedron Lett.*, **24**, 4829 (1983).

437. M. Kitaoka, Y. Takahashi, H. Kosugi, and U. Uda, *Chem. Lett.*, 1065 (1983); Y. Takahashi, H. Hagiwaza, H. Uda, and H. Kosugi, *Heterocycles*, **15**, 225 (1981).
438. H. Okamura, Y. Mitsuhira, M. Miura, and H. Takei, *Chem. Lett.*, 517 (1978).
439. S. J. Gould and B. D. Remillard, *Tetrahedron Lett.*, 4353 (1978).
440. D. Hanssgen and E. Odenhausen, *Chem. Ber.*, **112**, 2389 (1979).
441. J. Schulze, R. Boese, and G. Schmid, *Chem. Ber.*, **114**, 1297 (1981).
442. J. J. Fitt and H. W. Gschwend, *J. Org. Chem.*, **44**, 303 (1979).
443. H. Wada, H. Nakumura, T. Taguchi, and H. Takei, *Chem. Lett.*, 345 (1977).
444. J. K. Crandall and A. C. Clark, *J. Org. Chem.*, **37**, 4236 (1972).
445. P. Benoit and N. Collignon, *Bull. Soc. Chim. Fr.*, 1302 (1975).
446. G. W. Klumpp. P. J. A. Geurink, A. L. Spek, and A. J. M. Duisenberg, *J. Chem. Soc., Chem. Commun.*, 814 (1983).
447. S. Bywater and D. J. Worsfold, *J. Organomet. Chem.*, **159**, 229 (1978); S. Brownstein, S. Bywater and D. J. Worsfold, *J. Organomet. Chem.*, **199**, 1 (1980).
448. S. Bywater and D. J. Worsfold, *J. Organomet. Chem.*, **33**, 273 (1971).
449. G. Wittig and F. Bickelhaupt, *Chem. Ber.*, **91**, 883 (1958).
450. W. Neugebauer and P. von R. Schleyer, *J. Organomet. Chem.*, **198**, C1 (1980).
451. D. Seyferth. R. Suzuki, C. T. Murphy, and C. R. Sabet, *J. Organomet. Chem.*, **2**, 431 (1964).
452. H. Gilman and H. A. McNinch, *J. Org. Chem.*, **26**, 3723 (1961).
453. (a) A. Maercker, M. Theis, A. J. Kos, and P. von R. Schleyer, *Angew. Chem., Int. Ed. Engl.*, **22**, 733 (1983); (b) V. S. Bleshinskii and S. V. Bleshinskii, *Izv. Akad. Nauk. Kirg. SSR*, 47 (1982); *Chem. Abstr.*, **97**, 216264 (1982); (c) A. Maercker and R. Dujardin, *Angew Chem., Int. Ed. Engl.*, **23**, 224 (1984).
454. J. Barluenga, F. J. Fananas, M. Yus, and G. Asensio, *Tetrahedron Lett.*, 2015 (1978); J. Barluenga, F. J. Fananas, and M. Yus, *J. Org. Chem.*, **46**, 1281 (1981); J. Barluenga, J. Villamana, F. J. Fananas, and M. Yus, *J. Chem. Soc., Chem. Commun.*, 355 (1982).
455. J. J. Eisch and A. M. Jacobs, *J. Org. Chem.*, **28**, 2145 (1963).
456. G. Pastor, R. Calas, B. Brun, and L. Giral, *C. R. Acad. Sci., Ser. C*, **277**, 1159 (1973).
457. P. R. Peoples and J. B. Grutzner, *J. Am. Chem. Soc.*, **102**, 4709 (1980).
458. C. S. Screttas and M. Micha-Screttas, *J. Org. Chem.*, **44**, 713 (1979).
459. J. A. Katzenellenbogen and R. S. Lenox, *J. Org. Chem.*, **38**, 326 (1972).
460. T. Cohen and J. R. Matz, *J. Am. Chem. Soc.*, **102**, 6900 (1980).
461. T. Cohen, J. P. Sherbine, J. R. Matz, R. R. Hutchins, B. M. McHenry, and P. R. Willey, *J. Am. Chem. Soc.*, **106**, 3245 (1984).
462. D. J. Ager, *J. Org. Chem.*, **49**, 168 (1984).
463. T. Cohen. W. M. Daniewski, and R. B. Weisenfeld, *Tetrahedron Lett.*, 4665 (1978).
464. T. Cohen and R. B. Wiesenfeld, *J. Org. Chem.*, **44**, 3601 (1979).
465. D. Seyferth, M. A. Weiner, L. G. Vaughan, G. Raab, D. E. Welch, H. M. Cohen, and D. L. Alleston, *Bull. Chim. Soc. Fr.*, 1364 (1963).
466. R. G. Jones and H. Gilman, *Chem. Rev.*, **54**, 863 (1954).
467. H. Gilman and R. G. Jones, *J. Am. Chem. Soc.*, **63**, 1439 (1941).
468. D. Seyferth and L. G. Vaughan, *J. Am. Chem. Soc.*, **86**, 883 (1964); D. Seyferth and M. A. Weiner, *J. Am. Chem. Soc.*, **84**, 361 (1962).
469. W. Schlenk and J. Holtz, *Chem. Ber.*, **50**, 262 (1917).
470. D. E. Seitz and A. Zapata, *Tetrahedron Lett.*, **21**, 3451 (1980).
471. J. S. Sawyer, T. L. MacDonald, and G. J. McGarvey, *J. Am. Chem. Soc.*, **106**, 3376 (1984).
472. D. Y. Curtin and W. J. Koehl, *J. Am. Chem. Soc.*, **84**, 1967 (1962).
473. W. C. Still and C. Sreekumar, *J. Am. Chem. Soc.*, **102**, 1201 (1980).
474. E. J. Corey and B. De, *J. Am. Chem. Soc.*, **106**, 2735 (1984).
475. E. J. Corey, B. -C. Pan, D. H. Hua, and D. R. Deardorff, *J. Am. Chem. Soc.*, **104**, 6816 (1982).
476. (a) D. J. Peterson and J. E. Ward, *J. Organomet. Chem.*, **66**, 209 (1974); (b) J. P. Quintard, B. Elissondo, and B. Jousseaume, *Synthesis*, 495 (1984).
477. W. C. Still, *J. Am. Chem. Soc.*, **100**, 1481 (1978).
478. S. D. Burke, S. A. Shearouse, D. J. Burch, and R. W. Sutton, *Tetrahedron Lett.*, **21**, 1285 (1980).
479. N. Meyer and D. Seebach, *Chem. Ber.*, **113**, 1290 (1980).
480. E. J. Corey and T. M. Eckrich. *Tetrahedron Lett.*, **24**, 3163 (1983).
481. D. G. Farnum and T. Monego, *Tetrahedron Lett.*, **24**, 1361 (1983).
482. D. Seebach and K. H. Geiss, *J. Organomet. Chem. Libr.*, **1**, 1 (1976).

483. R. D. Taylor and J. L. Wardell, *J. Organomet. Chem.*, **77**, 311 (1974).
484. D. Seebach and N. Peleties, *Chem. Ber.*, **105**, 511 (1972); *Angew. Chem., Int. Ed. Engl.*, **8**, 450 (1969).
485. W. Dumont and A. Krief, *Angew. Chem., Int. Ed. Engl.*, **15**, 161 (1976).
486. J. Lucchetti and A. Krief, *Tetrahedron Lett.*, **22**, 1623 (1981).
487. J. N. Denis, W. Dumont, and A. Krief, *Tetrahedron Lett.*, 453 (1976).
488. D. van Ende, A. Cravador, and A. Krief, *J. Organomet. Chem.*, **177**, 1 (1979).
489. D. Seebach and A. K. Beck, *Chem. Ber.*, **108**, 314 (1974).
490. R. Goswani and D. E. Corcoran, *Tetrahedron Lett.*, **23**, 1463 (1982).
491. R. Goswani, *J. Am. Chem. Soc.*, **102**, 5973 (1980).
492. D. Seebach, K. - H. Geiss, A. K. Beck, B. Graf, and H. Daum, *Chem. Ber.*, **105**, 3280 (1972).
493. D. Seebach, *Chem. Ber.*, **105**, 487 (1972).
494. A. Rensing, E. -J. Echsler, and T. Kauffmann, *Tetrahedron Lett.*, **21**, 2807 (1980).
495. D. S. Matteson, L. A. Hagelee, and R. J. Wilcsek, *J. Am. Chem. Soc.*, **95**, 5096 (1973).
496. G. Zweifel, R. P. Fisher, and A. Hornig, *Synthesis*, 37 (1973); D. S. Matteson, *Synthesis*, 147 (1975); D. S. Matteson, R. J. Moody, and P. K. Jesthi, *J. Am. Chem. Soc.*, **97**, 5608 (1975).
497. C. M. Warner and J. G. Noltes, *J. Chem. Soc., Chem. Commun.*, 694 (1970).
498. J. W. F. L. Seetz, G. Schat, O. S. Akkerman, and F. Bickelhaupt, *J. Am. Chem. Soc.*, **104**, 6848 (1982).
499. D. Seyferth, R. L. Lambert, and M. Massol, *J. Organomet. Chem.*, **88**, 255 (1975).
500. (a) S. Halazy and A. Krief, *Tetrahedron Lett.*, **22**, 4341 (1981); (b) S. Halazy, W. Dumont, and A. Krief, *Tetrahedron Lett.*, **22**, 4737 (1981); (c) S. Halazy, J. Lucchetti, and A. Krief, *Tetrahedron Lett.*, 3971 (1978).
501. D. Seyferth and H. M. Cohen, *Inorg. Chem.*, **2**, 625 (1963).
502. D. Seyfertg and M. A. Weiner, *J. Am. Chem. Soc.*, **83**, 3583 (1961).
503. E. C. Juenge and D. Seyferth, *J. Org. Chem.*, **26**, 563 (1961).
504. L. F. Cason and A. G. Brooks, *J. Org. Chem.*, **19**, 1278 (1954).
505. T. C. Wu, D. Wittenberg, and H. Gilman, *J. Org. Chem.*, **25**, 596 (1960).
506. R. F. Cunico and F. J. Clayton, *J. Org. Chem.*, **41**, 1480 (1976).
507. D. Seyferth and S. C. Vick, *J. Organomet. Chem.*, **144**, 1 (1978); E. J. Corey and R. H. Wollenberg, *J. Am. Chem. Soc.*, **96**, 5581 (1974).
508. T. N. Mitchell and A. Amamria, *J. Organomet. Chem.*, **252**, 47 (1983).
509. D. Seyferth, D. E. Welch, and G. Raab, *J. Am. Chem. Soc.*, **84**, 4266 (1962).
510. R. H. Wollenberg, *Tetrahedron Lett.*, 717 (1978).
511. P. Jutzi and J. Bäumgartner, *J. Organomet. Chem.*, **148**, 257 (1978).
512. D. Seyferth and M. A. Weiner, *J. Org. Chem.*, **26**, 4797 (1961); *Org. Synth.*, **41**, 30 (1961).
513. G. Boche and J. Bigalke, *Tetrahedron Lett.*, **25**, 955 (1984).
514. A. B. Holmes and G. E. Jones, *Tetrahedron Lett.*, **21**, 3111 (1980); J. A. Miller, and G. Zweifel, *Synthesis*, 128 (1983).
515. A. J. Bridges, V. Fredij, and E. C. Turowski, *J. Chem. Soc., Chem. Commun*, 1093 (1983).
516. W. Neugebauer, T. Clark, and P. von R. Schleyer, *Chem. Ber.*, **116**, 3283 (1983).
517. F. J. Kronzer and V. R. Sandel, *J. Am. Chem. Soc.*, **94**, 5750 (1972).
518. E. Weiss, G. Sauermann and G. Thirase, *Chem. Ber.*, **116**, 74 (1981); E. Weiss and H. Köster, *Chem. Ber.*, **110**, 717 (1977).
519. L. Lochmann and J. Trekoval, *J. Organomet. Chem.*, **99**, 329 (1975); L. Lochmann, R. L. De, and J. Trekoval, *J. Organomet. Chem.*, **156**, 307 (1978).
520. W. H. Glaze and D. P. Duncan, *J. Organomet. Chem.*, **99**, 11 (1975).
521. R. H. Shapiro, M. F. Lipton, K. J. Kolonko, R. L. Buswell, and L. A. Capuano, *Tetrahedron Lett.*, 1811 (1975); R. H. Shapiro, *Org. React.*, **23**, 405 (1976); R. H. Shapiro and J. H. Duncan, *Org. Synth.*, **51**, 66 (1971); J. E. Stemke, A. R. Chamberlin and F. T. Bond, *Org. Synth.*, 2947 (1976); A. R. Chamberlin, J. E. Stemke, and F. T. Bond, *J. Org. Chem.*, **43**, 147 (1978); A. R. Chamberlin and F. T. Bond, *Synthesis*, 44 (1979).
522. P. A. Brown and P. R. Jenkins, *Tetrahedron Lett.*, **23**, 3733 (1982).
523. W. T. Ford and M. Newcomb, *J. Am. Chem. Soc.*, **95**, 6277 (1973); **95**, 7186 (1973), and references cited therein; G. Boche, K. Buckl, D. Martens, D. R. Schneider, and H. U. Wagner, *Chem. Ber.*, **112**, 2961 (1979), and references cited therein.
524. (a) G. Boche and M. Marsch, *Tetrahedron Lett.*, **24**, 3225 (1983); (b) M. A. Fox, C. -C. Chen, and K. A. Campbell, *J. Org. Chem.*, **48**, 321 (1983).

525. A. A. Vitale and J. San Filippo, *J. Am. Chem. Soc.*, **104**, 7341 (1982), and references cited therein.
526. G. E. Niznik, W. H. Morrison, III, and H. M. Walborsky, *Org. Synth.*, **51**, 31 (1971).
527. L. Lompa-Krzymien and L. C. Leitch, *Synthesis*, 124 (1976).
528. M. S. Baird and A. G. W. Baxter, *J. Chem. Soc., Perkin Trans. 1*, 2317 (1979).
529. D. W. Slocum, B. P. Koonsvitsky, and C. R. Ernst, *J. Organomet. Chem.*, **38**, 125 (1972).
530. M. P. Periasamy and H. M. Walborsky, *J. Am. Chem. Soc.*, **99**, 2631 (1977).
531. W. H. Glaze, C. M. Selman, A. L. Ball, and L. E. Bray, *J. Org. Chem.*, **34**, 641 (1969).
532. J. D. Taylor and J. F. Wolf, *J. Chem. Soc., Chem. Commun.*, 876 (1972).
533. M. Duraisamy and H. M. Walborsky, *J. Am. Chem. Soc.*, **106**, 5035 (1984).
534. J. F. Biellmann and J. J. Vicens, *Tetrahedron Lett.*, 467 (1978); G. Chassaing, R. Lett, and A. Marquet, *Tetrahedron Lett.*, 471 (1978); T. Durst, R. Viau, and M. R. McClory, *J. Am. Chem. Soc.*, **93**, 3077 (1971); Y. Yamamoto and K. Maruyuma, *J. Chem. Soc., Chem. Commun.*, 239 (1980).
535. M. Schlosser, H. Bosshardt, A. Walde, and M. Stähle, *Angew. Chem., Int. Ed. Engl.*, **19**, 303 (1980).
536. E. I. Negishi, *Organometallics in Organic Synthesis*, Vol. 1, Wiley, New York, 1980.
537. J. Villieras, P. Perriot, and J. F. Normant, *Bull. Soc. Chim. Fr.*, 765 (1977); R. J. Cregge, J. L. Herrmann, C. S. Lee, J. E. Richman, and R. H. Schlessinger, *Tetrahedron Lett.*, 2425 (1973); W. Bos and H. J. J. Pabon, *Recl. Trav. Chim. Pays-Bas*, **99**, 141 (1980).
538. R. R. Schmidt and R. Hirsenkorn, *Tetrahedron*, **39**, 2043 (1983).
539. H. Neumann and D. Seebach, *Tetrahedron Lett.*, 4839 (1976).
540. W. D. Korte, K. Cripe, and R. Cooke, *J. Org. Chem.*, **39**, 1168 (1974).
541. D. A. Evans and G. C. Andrews, *Acc. Chem. Res.*, **7**, 147 (1974).
542. D. Seebach, D. Steinmüller, and F. Demuth, *Angew. Chem., Int. Ed. Engl.*, **7**, 620 (1968).
543. D. Seebach and E. M. Wilka, *Synthesis*, 476 (1976).
544. W. D. Korte, L. Kinner and W. C. Kaska, *Tetrahedron Lett.*, 603 (1970).
545. K. Schank, H. G. Schmitt, F. Schroeder, and A. Weber, *Justus Liebigs Ann. Chem.* 1116 (1977); N. J. Foulger and B. J. Wakefield, *J. Organomet. Chem.*, **69**, 161 (1974).
546. D. Seebach and A. K. Beck, *Org. Synth.*, **51**, 76 (1971).
547. P. D. Brewer, J. Tagat, C. A. Hergrueter, and P. Helquist, *Tetrahedron Lett.*, 4573 (1977).
548. A. Strietwieser, Jr., J. R. Wolfe, and W. D. Schaeffer, *Tetrahedron*, **6**, 338 (1959).
549. D. A. Bright, D. E. Mathisen, and H. E. Zieger, *J. Org. Chem.*, **47**, 3521 (1982).
550. J. F. Biellmann, J. F. Blanzat, and J. J. Vicens, *J. Am. Chem. Soc.*, **102**, 2460 (1980).
551. V. Cere, S. Pollicino, E. Sandri, and A. Fava, *Tetrahedron Lett.*, 5239 (1978).
552. A. P. Krapcho and E. A. Dundulis, *J. Org. Chem.*, **45**, 3236 (1980).
553. M. Daney, R. Lapouyade, M. Mary, and H. Bouas-Laurent, *J. Organomet. Chem.*, **92**, 267 (1975); S. Bank, J. Bank, M. Daney, B. Labrande, and H. Bouas-Laurent, *J. Org. Chem.*, **42**, 4058 (1977); P. P. Fu, R. G. Harvey, J. W. Paschal, and P. W. Rabideau, *J. Am. Chem. Soc.*, **97**, 1145 (1975); E. J. Panek, *J. Am. Chem. Soc.*, **96**, 7959 (1974); M. Daney and R. Lapouyade, *J. Organomet. Chem.*, **172**, 385 (1979).
554. A. A. Solovyanov, P. I. Demyanov, I. P. Beletskaya, and O. A. Reutov, *J. Org. Chem. USSR* (*Engl. Transl.*), **12**, 2411 (1976); *Bull. Acad. Sci. USSR* (*Engl. Transl.*), **24**, 2251 (1975).
555. R. N. Magid, *Tetrahedron*, **36**, 1901 (1980).
556. R. M. Magid and R. D. Gandour, *J. Org. Chem.*, **35**, 269 (1970).
557. H. Gilman and A. H. Haubein, *J. Am. Chem. Soc.*, **66**, 1515 (1944); E. D. Bergmann, I. Shabak, and Z. Aizenshtat, *Tetrahedron Lett.*, 2007 (1969).
558. A. I. Meyers and B. E. Williams, *Tetrahedron Lett.*, 223 (1978).
559. K. E. Chippendale, B. Iddon, and H. Suschitzky, *J. Chem. Soc., Perkin Trans. 1*, 2023 (1972).
560. S. Dixon, *J. Org. Chem.*, **21**, 400 (1956); see also W. T. Middleton, D. Metzger, and J. A. Snyder, *J. Med. Chem.*, **14**, 1193 (1971); O. P. Petrii, G. V. Timofeyuk, G. U. Zenina, T. V. Talaleeva, and K. A. Kocheshkov, *Zh. Obshch. Khim.*, **39**, 522 (1969).
561. R. F. Cunico, *Tetrahedron Lett.*, 2935 (1975).
562. A. I. Meyers and P. D. Pansegrau, *Tetrahedron Lett.*, **24**, 4935 (1983); A. I. Meyers, and W. Rieker, *Tetrahedron Lett.*, **23**, 2091 (1982).
563. G. Szeimies, J. Harnish, and O. Bäumgärtel, *J. Am. Chem. Soc.*, **99**, 5183 (1977).
564. L. K. Montgomery and L. E. Applegate, *J. Am. Chem. Soc.*, **89**, 2952 (1967).
565. W. Kurtz and E. Effenberger, *Chem. Ber.*, **106**, 560 (1973).
566. A. I. Meyers, *Pure Appl. Chem.*, **51**, 1255 (1979).

567. D. Enders and H. Eichenauer, *Chem. Ber.*, **112**, 2933 (1979); D. Enders, H. Eichenauer, U. Baus, H. Schubert, and K. A. M. Kremer, *Tetrahedron*, **40**, 1345 (1984); see also K. G. Davenport, H. Eichenauer, D. Enders, M. Newcomb, and D. E. Bergbreiter, *J. Am. Chem. Soc.*, **101**, 5654 (1979); R. R. Fraser, F. Akiyama, and I. Banville, *Tetrahedron Lett.*, 3929 (1979).
568. A. I. Meyers, D. R. Williams, G. W. Erickson, S. White, and M. Druelinger, *J. Am. Chem. Soc.*, **103**, 3081 (1981); A. I. Meyers, D. R. Williams, S. White, and G. W. Erickson, *J. Am. Chem. Soc.*, p. 3088.
569. A. I. Meyers and L. M. Fuentes, *J. Am. Chem. Soc.*, **105**, 117 (1983).
570. D. Seebach, R. Naef, and G. Calderari, *Tetrahedron*, **40**, 1313 (1984).
571. D. Seebach and R. Naef, *Helv. Chim Acta*, **64**, 2704 (1981); R. Naef and D. Seebach, *Angew, Chem., Int. Ed. Engl.*, **20**, 1030 (1981).
572. G. Frater, U. Müller, and W. Günther, *Tetrahedron*, **40**, 1269 (1984).
573. U. Schöllkopf, *Tetrahedron*, **39**, 2085 (1983).
574. H. Ahlbrecht, G. Bonnet, D. Enders, and G. Zimmermann, *Tetrahedron Lett.*, **21**, 3175 (1980).
575. M. Larchveque, E. Ignatova, and T. Cuvigny, *J. Organomet. Chem.*, **177**, 5(1979); P. E. Sonnet and R. R. Heath, *J. Org. Chem.*, **45**, 3137 (1980).
576. A. I. Meyers, Y. Yamamoto, E. D. Mihelich, and R. A. Bell, *J. Org. Chem.*, **45**, 2792 (1980).
577. A. I. Meyers, N. R. Natale, D. G. Wettlaufer, S. Rafii, and J. Clardy, *Tetrahedron Lett.*, **22**, 5123 (1981).
578. D. A. Evans and J. Bartroli, *Tetrahedron Lett.*, **23**, 807 (1982).
579. D. J. Peterson, *J. Org. Chem.*, **33**, 780 (1968).
580. W. Dumont, D. van Ende, and A. Krief, *Tetrahedron Lett.*, 485 (1979).
581. T. Agawa, M. Ishikawa, M, Komatsu, and Y. Oshiro, *Chem. Lett.*, 335 (1980); *Bull. Chem. Soc. Jpn.*, **55**, 1205 (1982).
582. C. M. Dougherty and R. A. Olofson, *Org. Synth.*, **58**, 37 (1978).
583. G. N. Barber and R. A. Olofson, *Tetrahedron Lett.*, 3783 (1976).
584. R. A. Moss and R. C. Munjal, *Synthesis*, 425 (1979).
585. R. A. Olofson, K. D. Lotts, and G. N. Barber, *Tetrahedron Lett.*, 3779 (1976).
586. C. Burford, F. Cooke, E. Ehlinger, and P. Magnus, *J. Am. Chem. Soc.*, **99**, 4536 (1977).
587. P. Magnus and G. Roy, *J. Chem. Soc., Chem. Commun.*, 297 (1978).
588. T. Durst and K. C. Tin, *Tetrahedron Lett.*, 2369 (1970).
589. (a) M. Ballester, *Chem. Rev.*, **55**, 283 (1955); (b) R. F. Borch, *Tetrahedron Lett.*, 3761 (1972).
590. J. R. Shanklin, C. R. Johnson, J. Ollinger, and R. M. Coates, *J. Am. Chem. Soc.*, **95**, 3429 (1973).
591. A. Anciauz, A. Eman, W. Dumont, and A. Krief, *Tetrahedron Lett.*, 1617 (1975).
592. W. Dumont and A. Krief, *Angew Chem., Int. Ed. Engl.*, **14**, 350 (1975); D. van Ende, W. Dumont, and A. Krief, *Angew. Chem., Int. Ed. Engl.*, **14**, 700 (1975).
593. D. van Ende and A. Krief, *Tetrahedron Lett.*, 457 (1976).
594. C. R. Johnson and G. F. Katekar, *J. Am. Chem. Soc.*, **92**, 5733 (1970).
595. A. I. Meyers and M. E. Ford, *J. Org. Chem.*, **41**, 1735 (1976).
596. D. Seebach, *Angew. Chem., Int. Ed. Engl.*, **18**, 239 (1979); O. W. Lever, Jr., *Tetrahedron*, **32**, 1943 (1976); S. F. Martin, *Synthesis*, 633 (1979). D. Seebach, M. Pohmakotr, C. Schregenberger, B. Weidmann, R. S. Mali, and S. Pohmakotr, *Helv. Chim. Acta*, **65**, 419 (1982).
597. (a) D. Seebach and E. J. Corey, *J. Org. Chem.*, **40**, 231 (1975); (b) E. J. Corey and B. W. Erickson, *J. Org. Chem.*, **36**, 3553 (1971).
598. H. Paulsen, M. Stubbe, and F. R. Heiker, *Justus Liebigs Ann. Chem.*, 825 (1980).
599. E. J. Corey and J. I. Shulman, *J. Org. Chem.*, **35**, 777 (1970).
600. C. Earnshaw, C. J. Wallis, and S. Warren, *J. Chem. Soc., Chem. Commun.*, 314 (1977).
601. R. H. Wollenberg, K. F. Albizati, and R. Peries, *J. Am. Chem. Soc.*, **99**, 7365 (1977).
602. J. F. Le Borgne, T. Cuvigny, M. Larchevêque, and H. Normant, *Tetrahedron Lett.*, 1379 (1976).
603. I. P. Politzer and A. I. Meyers, *Org. Synth.*, **51**, 24 (1971).
604. E. J. Corey and D. Enders, *Chem. Ber.*, **111**, 1337, 1362 (1978).
605. W. C. Still and T. L. MacDonald, *J. Am. Chem. Soc.*, **96**, 5561 (1974).
606. P. Coutrot and P. Savignac, *J. Chem. Res. (S)*, 308 (1977).
607. A. I. Meyers and E. D. Mihelich, *Angew. Chem., Int. Ed. Engl.*, **15**, 270 (1976).
608. J. E. McMurry and J. H. Musser, *J. Org. Chem.*, **40**, 2556 (1975).
609. R. E. Murray and G. Zweifel, *Synthesis*, 150 (1980); see also G. Köbrich, H. Trapp, and A. Akhtar, *Chem. Ber.*, **101**, 2644 (1968); G. Szeimies, F. Philipp, O. Baumgärtel, and J. Harnisch, *Tetrahedron Lett.*, 2135 (1977).
610. H. N. Khatri and H. M. Walborsky, *J. Org. Chem.*, **43**, 734 (1978).

611. R. A. van der Welle and L. Brandsma, *Recl. Trav. Chim. Pays-Bas*, **92**, 667 (1973).
612. H. S. Chang and J. T. Edward, *Can. J. Chem.*, **41**, 1233 (1963).
613. E. J. Panek, L. R. Kaiser, and G. M. Whitesides, *J. Am. Chem. Soc.*, **99**, 3708 (1977).
614. (a) G. Rauchschwalbe and M. Schlosser, *Helv. Chim. Acta*, **58**, 1094 (1975); F. G. Thorpe, G. M. Pickles and J. C. Podesta, *J. Organomet. Chem.*, **128**, 305 (1977); (b) R. W. Hoffmann and K. Ditrich, *Synthesis*, 107 (1983).
615. J. N. Reed and V. Snieckus, *Tetrahedron Lett.*, **24**, 3795 (1983).
616. P. Beak and B. J. Kokko, *J. Org. Chem.*, **47**, 2822 (1982).
617. P. Spagnolo, P. Zanirato, and Gronowitz, *J. Org. Chem.*, **47**, 3177 (1982).
618. N. S. Narasimhan and R. Ammanamanchi, *Tetrahedron Lett.*, **24**, 4733 (1983).
619. S. Mori, T. Aoyama, and T. Shioiri, *Tetrahedron Lett.*, **25**, 429 (1984).
620. A. Hassner, P. Munger, and B. A. Belinka, Jr., *Tetrahedron Lett.*, **23**, 699 (1982).
621. B. M. Trost and W. H. Pearson, *J. Am. Chem. Soc.*, **103**, 2483 (1981).
622. H. Gilman, M. H. van Ess, H. B. Willis, and C. G. Stuckwisch, *J. Am. Chem. Soc.*, **62**, 2602 (1940).
623. D. Seyferth, W. B. Hughes, and J. K. Heeren, *J. Am. Chem. Soc.*, **87**, 2847 (1965); H. Gilman and R. K. Ingram, *J. Am. Chem. Soc.*, **75**, 4843 (1953); H. Gilman, and C. G. Stuckwisch, *J. Am. Chem. Soc.*, **65**, 1461 (1943); H. Gilman and D. R. Swayampati, *J. Am. Chem. Soc.*, **79**, 208 (1957); E. M. Acton and R. M. Silverstein, *J. Org. Chem.*, **24**, 1487 (1959); T. Oguri, T. Shiori, and S. T. Yamada, *Chem. Pharm. Bull.*, **23**, 167 (1975).
624. G. Boche, M. Bernheim, and W. Schrott, *Tetrahedron Lett.*, **23**, 5399 (1982).
625. E. C. Taylor and J. -H. Sun, *Synthesis*, 801 (1980).
626. B. J. Kokko and P. Beak, *Tetrahedron Lett.*, **24**, 561 (1983).
627. T. Abraham and D. Curran, *Tetrahedron*, **38**, 1019 (1982).
628. G. Boche, N. Mayer, M. Bernheim, and K. Wagner, *Angew. Chem., Int. Ed. Engl.*, **17**, 687 (1978).
629. M. Bernheim and G. Boche, *Angew Chem., Int. Ed. Engl.*, **19**, 1010 (1980).
630. G. Boche, M. Bernheim, and M. Neissner, *Angew. Chem., Int. Ed. Engl.*, **22**, 53 (1983).
631. H. Yamamoto and K. Maruoka, *J. Org. Chem.*, **45**, 2739 (1980).
632. G. Boche and W. Schrott, *Tetrahedron Lett.*, **23**, 5403 (1982).
633. G. Köbrich and H. Trapp, *Chem. Ber.*, **99**, 670 (1966).
634. K. H. Boltze and H. -D. Dell, *Justus Liebigs Ann. Chem.*, **709**, 63 (1967).
635. R. D. Schuetz, D. D. Taft, J. P. O'Brien, J. L. Shea, and H. M. Mork, *J. Org. Chem.*, **28**, 1420 (1963); S. Gronowitz and U. Rosen, *Chem. Scr.*, **1**, 33 (1971); M. Schlosser and G. Heinz, *Chem. Ber.*, **102**, 1944 (1969).
636. J. Bensoam and F. Mathey, *Tetrahedron Lett.*, 2797 (1977).
637. W. Adcock and T. C. Khor, *J. Organomet. Chem.*, **91**, C20 (1975); J. H. J. Peet and B. W. Rockett, *J. Organomet. Chem.*, **82**, C57 (1974).
638. J. Petrova, P. Coutrot, M. Dreux, and P. Savignac, *Synthesis*, 658 (1975).
639. J. Kattenberg, E. R. de Waard, and H. O. Huisman, *Tetrahedron*, **29**, 4149 (1973).
640. R. L. Gay, T. F. Crimmins, and C. R. Hauser, *Chem. Ind.* (*London*), 1635 (1966).
641. A. I. Meyers and E. D. Mihelich, *J. Org. Chem.*, **40**, 3158 (1975); V. Verboom, H. Westmijze, L. J. de Noten, P. Vermeer, and H. J. T. Bos, *Synthesis*, 296 (1979).
642. D. W. Slocum and P. L. Gierer, *J. Org. Chem.*, **41**, 3668 (1976); G. Szeimies, F. Philipp, O. Baumgärtel, and J. Harnisch, *Tetrahedron Lett.*, 2135 (1977).
643. C. A. Townsend and L. M. Bloom, *Tetrahedron Lett.*, **22**, 3923 (1981).
644. L. Horner, H. Hoffmann, and V. G. Toscano, *Chem. Ber.*, **95**, 536 (1962); S. Hunig and G. Wehner, *Synthesis*, 180 (1975); U. Hertenstein, S. Hünig, and M. Oller, *Synthesis.*, 416 (1976).
645. B. A. Tertov, V. V. Barykin, and I. D. Sadekov, *Chem. Heterocycl. Compd.* (*Engl. Transl.*), **5**, 418 (1969).
646. O. M. Nefedov, A. I. D'yachenko, and A. K. Prokov'ev, *Russ. Chem. Rev.* (*Engl. Transl.*), **46**, 941 (1977).
647. W. Kirmse, *Carbene Chemistry*, 2nd ed., Academic Press, New York, 1971; E. Chinoporos, *Chem. Rev.*, **63**, 235 (1963); M. Jones and R. A. Moss, *Carbenes*, Wiley, New York, Vol. 1, 1973; Vol. 2, 1975.
648. A. W. Johnson, *Ylid Chemistry*, Academic Press, New York, 1966; A. Maercker, *Org. React.*, **14**, 270 (1965).
649. H. Gilman, E. A. Zoellner, and W. M. Selby, *J. Am. Chem. Soc.*, **55**, 1252. (1933).
650. E. Müller and D. Ludsteck, *Chem. Ber.*, **87**, 1887 (1954).

651. K. Ziegler, K. Nagel, and M. Pathelger, *Z. Anorg. Allg. Chem.*, **282**, 345 (1955).
652. D. Bryce-Smith and E. E. Turner, *J. Chem. Soc.*, 861 (1953).
653. G. Wittig, *Newer Methods of Preparative Organic Chemistry*, Vol. 1, Interscience, New York, 1948, p. 571.
654. H. M. Cohen, *J. Organomet. Chem.*, **9**, 375 (1967).
655. W. N. Smith, *J. Organomet. Chem.*, **82**, 1 (1974); see also C. Giancaspro and G. Sleiter, *J. Prakt. Chem.*, **321**, 871 (1979).
656. U. Schöllkopf, H. Küppers, H.-J. Traenckner and W. Pitteroff, *Justus Liebigs Ann. Chem.*, **704**, 120 (1967).
657. K.-H. Thiele, E. Langguth and G. E. Müller, *Z. Anorg. Allg. Chem.*, **462**, 152 (1980).
658. A. Maercker, M. Eckers, and M. Passlack, *J. Organomet. Chem.*, **186**, 193 (1980).
659. H. D. Zook, J. March, and D. F. Smith, *J. Am. Chem. Soc.*, **81**, 1617 (1959).
660. J. Villieras, P. Perriot, M. Bourgain and J. -F. Normant, *Synthesis*, 533 (1975).
661. D. Seyferth and H. M. Cohen, *J. Organomet. Chem.*, **1**, 15 (1963).
662. D. Seyferth and D. D. Dagani, *Synth. React. Inorg. Met. Org. Chem.*, **10**, 137 (1980).
663. D. E. Applequist and D. F. O'Brien, *J. Am. Chem. Soc.*, **85**, 743 (1963).
664. O. H. Johnson and W. H. Nebergall, *J. Am. Chem. Soc.*, **71**, 1720 (1949).
665. K. B. Wiberg and B. R. Lowry, *J. Am. Chem. Soc.*, **85**, 3188 (1963).
666. G. Molle, P. Bauer, and J. E. Dubois, *J. Org. Chem.*, **48**, 2975 (1983).
667. R. West and W. H. Glaze, *J. Org. Chem.*, **26**, 2096 (1961).
668. D. L. Esmay, *Adv. Chem. Ser.*, No. 23, 47 (1959).
669. H. Gilman and B. J. Caj, *J. Org. Chem.*, **22**, 1165 (1957).
670. H. Gilman, E. A. Zoellner, and W. M. Selby, *J. Am. Chem. Soc.*, **54**, 1957 (1932).
671. J. T. B. H. Jastrzebski, G. van Kotten, M. Konijn, and C. H. Stam, *J. Am. Chem. Soc.*, **104**, 5490 (1982).
672. R. West and E. G. Rouchow, *J. Org. Chem.*, **18**, 1739 (1953).
673. A. A. Morton, G. M. Richardson, and A. T. Hallowell, *J. Am. Chem. Soc.*, **63**, 327 (1941).
674. A. A. Morton and E. F. Cluff, *J. Am. Chem. Soc.*, **75**, 134 (1953).
675. B. Renfrow and C. R. Hauser, *Org. Synth.*, Coll. Vol. **2**, 607 (1943).
676. B. Bartocha, C. M. Boucher, and M. Y. Gray, *Z. Naturforsch., Teil B*, **14**, 809 (1959).
677. D. J. Forster to Union Carbide Corp., *Br. Pat.*, 886980, 1959.
678. I. Fatt and M. Tashima, *Alkali Metal Dispersion*, Van Nostrand, New York, 1961.
679. A. A. Morton, G. H. Patterson, J. J. Donovan, and E. L. Little, *J. Am. Chem. Soc.*, **68**, 93 (1946).
680. J. W. Schick and H. D. Hartough, *J. Am. Chem. Soc.*, **70**, 286 (1948).
681. A. A. Morton, M. L. Brown, M. E. T. Holden, R. L. Letsinger, and E. E. Magat, *J. Am. Chem. Soc.*, **67**, 2224 (1945).
682. R. G. Anderson, M. B. Silverman, and D. M. Ritter, *J. Org. Chem.*, **23**, 750 (1958).
683. G. Gau, *J. Organomet. Chem.*, **121**, 1. (1976).
684. D. J. Burton and J. L. Hahnfeld, *J. Org. Chem.*, **42**, 828 (1977).
685. R. H. Fischer and G. Köbrich, *Chem. Ber.*, **101**, 3230 (1968).
686. J. Villieras, P. Perriot, and J. F Normant, *Synthesis*, 29 (1978).
687. D. F. Hoeg, D. I. Lusk, and A. L. Crumbliss, *J. Am. Chem. Soc.*, **87**, 4147 (1965).
688. J. Villieras, P. Perriot, and J. F. Normant, *Synthesis*, 968 (1979).
689. J. Villieras, M. Rambaud, R. Tarhouni, and B. Kirschleger, *Synthesis*, 68 (1981).
690. R. Tarhouni, B. Kirschleger, M. Rambaud, and J. Villieras, *Tetrahedron Lett.*, **25**, 835 (1984).
691. A. G. Brook, J. M. Duff, and D. G. Anderson, *Can. J. Chem.*, **48**, 561 (1970).
692. D. Seyferth, J. L. Lefferts, and R. L. Lambert, *J. Organomet. Chem.*, **142**, 39 (1977).
693. T. Kauffmann, R. Konig, R. Kriegesmann, and M. Wensing, *Tetrahedron Lett.*, **25**, 641 (1984).
694. D' Seyferth, F. M. Armbrecht, Jr., R. L. Lambert, Jr., and W. Tronich, *J. Organomet. Chem.*, **44**, 299 (1972).
695. T. Kauffmann, K. -J. Echsler, A. Hamsen, R. Kriegesmann, F. Steinseifer, and A. Vahrenhorst, *Tetrahedron Lett.*, 4391 (1978).
696. W. Dumont, A. Anciaux, A. Eman, J. N. Denis, and A. Krief, cited in A. Krief, *Tetrahedron*, **36**, 2531 (1980).
697. W. Dumont, M. Sevrin, and A. Krief, *Angew. Chem., Int. Ed. Engl.*, **16**, 541 (1977).
698. E. J. Corey and P. Ulrich, *Tetrahedron Lett.*, 3685 (1975).
699. T. Hiyama, A. Kanakura, H. Yamamoto, and H. Nozaki, *Tetrahedron Lett.*, 3047 (1978).
700. M. S. Baird and G. W. Baxter, *J. Chem. Soc., Perkin Trans. 1*, 2317 (1979).

701. K. G. Taylor, J. Chaney, and J. C. Deck, *J. Am. Chem. Soc.*, **98**, 4163 (1976).
702. K. G. Taylor and J. Chaney, *J. Am. Chem. Soc.*, **98**, 4158 (1976).
703. L. Duhamel and J. -Y. Valnot, *Tetrahedron Lett.*, 3319 (1979).
704. H. Neumann and D. Seebach, *Chem. Ber.*, **111**, 2785 (1978).
705. G. Zweifel and W. Lewis, *J. Org. Chem.*, **43**, 2739 (1978).
706. E. J. Corey and G. N. Widiger, *J. Org. Chem.*, **40**, 2975 (1975).
707. K. S. Y. Lau and M. Schlosser, *J. Org. Chem.*, **43**, 1595 (1978).
708. J. Ficini and J.-C. Depezay, *Tetrahedron Lett.*, 4797 (1969).
709. R. H. Smithers, *J. Org. Chem.*, **48**, 2095 (1983).
710. J. Ficini, P. Kahn, S. Falou, and A. M. Touzin, *Tetrahedron Lett.*, 67 (1979).
711. C. G. M. Janssen, L. H. J. G. Simons, and E. F. Godefroi, *Synthesis*, 389 (1982).
712. W. E. Parham and D. W. Boykin, *J. Org. Chem.*, **42**, 260 (1977).
713. D. Caine and A. S. Frobese, *Tetrahedron Lett.*, 5167 (1978).
714. W. R. Baker and R. M. Coates, *J. Org. Chem.*, **44**, 1022 (1979).
715. L. Duhamel and J. M. Poirier, *J. Am. Chem. Soc.*, **99**, 8356 (1977).
716. F. G. Drakesmith, O. J. Stewart, and P. Tarrant, *J. Org. Chem.*, **33**, 280 (1968).
717. D. Masure, C. Chuit, R. Sauvetre, and J. F. Normant, *Synthesis*, 458 (1978).
718. G. Köbrich and K. Flory, *Chem. Ber.*, **99**, 1773 (1966).
719. G. Köbrich, *Angew. Chem., Int. Ed. Engl.*, **6**, 41 (1967).
720. L. Fitjer, U. Kliebisch, D. Wehle, and S. Modaressi, *Tetrahedron Lett.*, **23**, 1661 (1982).
721. D. Seyferth, A. B. Evnin, and D. R. Blank, *J. Organomet. Chem.*, **13**, 25 (1968).
722. G. Linstrumelle and D. Michelot, *J. Chem. Soc., Chem. Commun.*, 561 (1975).
723. W. J. Trepka and R. J. Sonnenfeld, *J. Organomet. Chem.*, **16**, 317 (1969).
724. H. Gilman and R. D. Gorsich, *J. Am. Chem. Soc.*, **77**, 3919 (1955).
725. L. S. Chen, G. J. Chen., and C. Tamborski, *J. Organomet. Chem.*, **193**, 283 (1980).
726. H. Gilman, R. D. Gorsich, and B. J. Gaj, *J. Org. Chem.*, **27**, 1023 (1962).
727. H. Gilman, W. Langham, and F. W. Moore, *J. Am. Chem. Soc.*, **62**, 2327 (1940).
728. L. S. Chen, G. J. Chen, and C. Tamborski, *J. Organomet. Chem.*, **215**, 281 (1981).
729. M. Yoshifuji, I. Shima, and N. Inamoto, *Tetrahedron Lett.*, 3963 (1979).
730. W. E. Parham, L. D. Jones and Y. A. Sayed, *J. Org. Chem.*, **41**, 1184 (1976).
731. H. Gilman, C. E. Arnitzen, and F. J. Webb, *J. Org. Chem.*, **10**, 374 (1945).
732. W. E. Parham, D. C. Egberg, Y. A. Sayed, R. W. Thraikill, G. E. Keyser, M. Neu, W. E. Montgomery, and L. D. Jones, *J. Org. Chem.*, **41**, 2628 (1976).
733. E. Akgün, M. D. Glinski, K. L. Dhawan, and T. Durst, *J. Org. Chem.*, **46**, 2730 (1981).
734. W. E. Parham and Y. A. Sayed, *J. Org. Chem.*, **39**, 2053 (1974).
735. W. E. Parham and L. D. Jones, *J. Org. Chem.*, **41**, 2704 (1976).
736. W. E. Parham, L. D. Jones, and Y. Sayed, *J. Org. Chem.*, **40**, 2394 (1975).
737. P. A. Wender and A. W. White, *Tetrahedron Lett.*, **22**, 1475 (1981).
738. W. E. Parham and L. D. Jones, *J. Org. Chem.*, **41**, 1187 (1976).
739. H. Gilman and L. A. Woods, *J. Am. Chem. Soc.*, **66**, 1981 (1944).
740. G. Köbrich and P. Buck, *Chem. Ber.*, **103**, 1412 (1970).
741. W. E. Parham and R. M. Piccirilli, *J. Org. Chem.*, **42**, 257 (1977).
742. H. Gilman and C. G. Stuckwisch, *J. Am. Chem. Soc.*, **63**, 2844 (1941).
743. H. Gilman and F. W. Moore, *J. Am. Chem. Soc.*, **62**, 1843 (1940).
744. G. Porzi and C. Concilio, *J. Organomet. Chem.*, **128**, 95 (1977).
745. N. Tanaka and T. Kasai, *Bull. Chem. Soc. Jpn.*, **54**, 3020 (1981).
746. N. J. Foulger and B. J. Wakefield, *J. Organomet. Chem.*, **69**, 321 (1974).
747. M. D. Rausch, G. A. Moser, and C. F. Meade, *J. Organomet. Chem.*, **51**, 1 (1973).
748. I. Haiduc and H. Gilman, *Rev. Roum. Chim.*, **16**, 907 (1971).
749. H. Hart and G. C. Nwokogu, *Tetrahedron Lett.*, **24**, 5721 (1983).
750. D. J. Berry and B. J. Wakefield, *J. Chem. Soc. C*, 2342 (1969).
751. Y. Fukuyama, Y. Kawashima, T. Miwa, and T. Tokoroyama, *Synthesis*, 443 (1974).
752. P. Moses and S. Gronowitz, *Ark. Kemi*, **18**, 119 (1962).
753. S. Gronowitz and B. Holm, *Acta Chem. Scand., Part B*, **30**, 505 (1976).
754. M. D. Rausch, T. R. Criswell, and A. K. Ignatiowicz, *J. Organomet. Chem.*, **13**, 419 (1968).
755. Yu. K. Yur'ev, N. K. Sadovaya, and E. A. Grekova, *J. Gen. Chem. USSR (Engl. Trans.)*, **34**, 841 (1964).
756. J. M. Muchowski and R. Naef, *Helv. Chim. Acta*, **67**, 1168 (1984).

757. H. Gilman and S. M. Spatz, *J. Org. Chem.*, **16**, 1485 (1951).
758. G. R. Newkome and J. M. Roper, *J. Organomet. Chem.*, **186**, 147 (1980).
759. C. D. Broaddus, *J. Org. Chem.*, **35**, 10 (1970).
760. C. D. Broaddus, *J. Am. Chem. Soc.*, **88**, 4174 (1966).
761. R. A. Benkeser, A. E. Trevillyan, and J. Hooz, *J. Am. Chem. Soc.*, **84**, 4971 (1962).
762. T. F. Crimmins and E. M. Rather, *J. Org. Chem.*, **43**, 2170 (1978).
763. T. F. Crimmins and C. M. Chan, *J. Org. Chem.*, **41**, 1870 (1976).
764. R. A. Benkeser, J. Hooz, T. V. Liston, and A. E. Trevillyan, *J. Am. Chem. Soc.*, **85**, 3984 (1963).
765. E. M. Kaiser and J. D. Petty, *J. Organomet. Chem.*, **107**, 219 (1976).
766. M. Braun and E. Ringer, *Tetrahedron Lett.*, **24**, 1233 (1983).
767. D. J. Coughlin and R. G. Salomon, *J. Org. Chem.*, **44**, 3784 (1979).
768. W. J. Houlihan, V. A. Parrino, and Y. Uike, *J. Org. Chem.*, **46**, 4511 (1981).
769. T. Abraham, *Monatsh. Chem.*, **113**, 371 (1982).
770. R. H. Mitchell, T. W. Dingle, and R. V. Williams, *J. Org. Chem.*, **48**, 903 (1983).
771. R. Noyori, N. Sano, S. Murata, Y. Okamoto, and H. Yuki, *Tetrahedron Lett.*, **23**, 2969 (1982).
772. K. Takahashi, K. Konishi, M. Ushio, M. Takaki, and R. Asami, *J. Organomet. Chem.*, **50**, 1 (1973).
773. H. C. Brown and W. A. Murray, *J. Am. Chem. Soc.*, **73**, 3308 (1951).
774. A. D. Miller and R. Levine, *J. Org. Chem.*, **24**, 1364 (1959).
775. E. M. Kaiser and J. D. Petty, *Synthesis*, 705 (1975).
776. C. G. Griggs and D. J. H. Smith, *J. Chem. Soc., Perkin Trans. 1*, 3041 (1982).
777. E. M. Kaiser and J.R. McClure, *J. Organomet. Chem.*, **175**, 11 (1979).
778. E. M. Kaiser and J. D. Petty, *J. Org. Chem.*, **41**, 716 (1976).
779. D. J. Brunelle, *Tetrahedron Lett.*, **22**, 3699 (1981).
780. B. H. Lipshutz and R. W. Hungate, *J. Org. Chem.*, **46**, 1410 (1981).
781. G. Knaus and A. I. Meyers, *J. Org. Chem.*, **39**, 1189 (1974).
782. D. S. Noyce, G. T. Stowe, and W. Wong, *J. Org. Chem.*, **39**, 2301 (1974).
783. T. H. Chan and E. Chang, *J. Org. Chem.*, **39**, 3264 (1974).
784. A. N. Tischler and M. H. Tischler, *Tetrahedron Lett.*, 3 (1978).
785. D. Seebach and M. Yoshifuji, *Helv. Chim. Acta*, **64**, 643 (1981).
786. M. Mikolajczyk, S. Grzejsczak, A. Chefczynska, and A. Zatorski, *J. Org. Chem.*, **44**, 2967 (1979).
787. M. Sekine, M. Nakajima, and T. Hata, *Bull. Soc. Chem. Jpn.*, **55**, 218 (1982).
788. R. Sauvetre and J. Seyden-Penne, *Tetrahedron Lett.*, 3949 (1976).
789. E. M. Kaiser, L. E. Solter, R. A. Schwarz, R. D. Beard, and C. R. Hauser, *J. Am. Chem. Soc.*, **93**, 4237 (1971).
790. K. Deuchert, U. Hertenstein, S. Hünig, and G. Wehner, *Chem. Ber.*, **112**, 2045 (1979).
791. M. K. Yeh, *J. Chem. Soc., Perkin Trans. 1*, 1652 (1981).
792. J. J. Eisch and J. E. Galle, *J. Organomet. Chem.*, **121**, C10 (1976).
793. K. -H. Geiss, D. Seebach, and B. Seuring, *Chem. Ber.*, **110**, 1833 (1977).
794. S. Cabiddu, C. Floris, S. Melis, and F. Sotgiu, *Phosphorus Sulfur*, **19**, 61 (1984).
795. T. Durst and M. Molin, *Tetrahedron Lett.*, 63 (1975).
796. T. Hayashi and H. Baba, *J. Am. Chem. Soc.*, **97**, 1608 (1975).
797. T. L. Emmick, *US Pat.*, 3 766 209; *Chem. Abstr.*, **79**, 146136 (1973).
798. H. J. Reich and S. K. Shah, *J. Am. Chem. Soc.*, **97**, 3250 (1975).
799. V. Lachkova and M. Kirilov, *Justus Liebigs Ann. Chem.*, 496 (1974).
800. M. F. Lappett, C. L. Raston, B. W. Skelton, and A. H. White, *J. Chem. Soc., Chem. Commun.*, 14 (1982).
801. R. I. Papasergio, C. L. Raston, and A. H. White, *J. Chem. Soc., Chem. Commun.*, 1419 (1983).
802. H. Ahlbrecht and H. Dollinger, *Tetrahedron Lett.*, **25**, 1353 (1984).
803. A. I. Meyers, P. D. Edwards, W. F. Rieker, and T. R. Bailey, *J. Am. Chem. Soc.*, **106**, 3270 (1984).
804. G. Stork, A. A. Ozorlo, and A. Y. W. Leong, *Tetrahedron Lett.*, 5175 (1978).
805. D. B. Reitz, P. Beak, R. F. Farney, and L. S. Helmick, *J. Am. Chem. Soc.*, **100**, 5428 (1978).
806. R. Schlecker and D. Seebach, *Helv. Chim. Acta*, **60**, 1459 (1977).
807. D. Seebach and T. Hassel, *Angew. Chem., Int. Ed. Engl.*, **17**, 274 (1978).
808. D. Seebach and W. Lubosch, *Angew. Chem., Int. Ed. Engl.*, **15**, 313 (1976).
809. M. P. Periasamy and H. M. Walborsky, *J. Am. Chem. Soc.*, **99**, 2631 (1977).
810. D. Seebach, D. Enders, and B. Renger, *Chem. Ber.*, **110**, 1852 (1977).
811. P. Magnus and G. Roy, *Synthesis*, 575 (1980).

812. T. Kauffmann, D. Berger, B. Scheerer, and A. Woltermann, *Chem. Ber.*, **110**, 3034 (1977).
813. R. A. Moss and G. M. Love, *Tetrahedron Lett.*, 4701 (1973).
814. T. Kauffmann, H. Berg, E. Köppelmann, and D. Kuhlmann, *Chem. Ber.*, **110**, 2659 (1977).
815. T. Aoyama, S. Inoue, and T. Shioiri, *Tetrahedron Lett.*, **25**, 433 (1984).
816. R. Pellicciari, E. Castagnino, and S. Corsano, *J. Chem. Res.* (*S*), 76 (1979).
817. D. Seebach and F. Lehr, *Angew. Chem., Int. Ed. Engl.*, **15**, 505 (1976).
818. H. H. Karsch and H. Schmidbauer, *Z. Naturforsch., Teil B*, **32**, 762 (1977).
819. L. M. Engelhardt, G. E. Jacobsen, C. L. Raston, and A. H. White, *J. Chem. Soc., Chem. Commun.*, 220 (1984).
820. R. Appel, M. Wander, and F. Knoll, *Chem. Ber.*, **112**, 1093 (1979).
821. H. H. Karsch, *Z. Naturforsch., Teil B*, **34**, 1178 (1979).
822. R. Appel, G. Haubrich, and F. Knoch, *Chem. Ber.*, **117**, 2063 (1984).
823. E. J. Corey and M. A. Tius, *Tetrahedron Lett.*, **21**, 3535 (1980).
824. J. I. Grayson and S. Warren, *J. Chem. Soc., Perkin Trans. 1*, 2263 (1977).
825. D. Cavalla and S. Warren, *Tetrahedron Lett.*, **24**, 295 (1983).
826. F. Mathey and F. Mercier, *J. Organomet. Chem.*, **177**, 255 (1979).
827. K. Issleib and P. Abricht, *J. Prakt. Chem.*, **312**, 456 (1970).
828. E. J. Corey and G. T. Kwiatkowski, *J. Am. Chem. Soc.*, **88**, 5654 (1966).
829. E. J. Corey and J. I. Shulman, *J. Org. Chem.*, **35**, 777 (1970).
830. Ph. Savignac, M. Dreux, and Ph. Coutrot, *Tetrahedron Lett.*, 609 (1975).
831. J. J. Eisch and J. E. Galle, *J. Organomet. Chem.*, **121**, C10 (1976).
832. B. Deschamps, *Tetrahedron*, **34**, 2009 (1978).
833. U. Schöllkopf and R. Schröder, *Tetrahedron Lett.*, 633 (1973).
834. T. Bottin-Strzalko, J. Corset, F. Froment, M. -J. Pouet, J. Seyden-Penne, and M. P. Simonnin, *J. Org. Chem.*, **45**, 1270 (1980).
835. M. Obayashi, E. Ito, K. Matsui, and K. Kondo, *Tetrahedron Lett.*, **23**, 2323 (1982).
836. E. J. Corey and G. T. Kwiatkowski, *J. Am. Chem. Soc.*, **88**, 5652 (1966).
837. H. H. Karsch, *Chem. Ber.*, **115**, 818 (1982).
838. H. Schmidbaur and U. Deschler, *Chem. Ber.*, **116**, 1386 (1983).
839. D. J. Peterson, *J. Org. Chem.*, **32**, 1717 (1967).
840. E. J. Corey and D. Seebach, *J. Org. Chem. Soc.*, **31**, 4097 (1966).
841. B. M. Trost, D. Keeley, and M. J. Bogdanowicz, *J. Am. Chem.Soc.*, **95**, 3068 (1973).
842. D. J. Ager and R. C. Cookson, *Tetrahedron Lett.*, **21**, 1677 (1980).
843. B.M. Trost and C. H. Miller, *J. Am. Chem. Soc.*, **97**, 7182 (1975).
844. U. Schöllkopf and E. Brume, *Tetrahedron Lett.*, 629 (1973).
845. N. Y. Wang, S. S. Su, and L. Y. Tsai, *Tetrahedron Lett.*, 1121 (1979).
846. P. A. Grieco and C. L. -J. Wang, *J. Chem. Soc., Chem. Commun.*, 714 (1975).
847. S. Yamagiwa, N. Hoshi, H. Sato, H. Kosugi and H. Uda, *J. Chem. Soc., Perkin Trans. 1*, 214 (1978).
848. S. Matsui, *Bull. Chem. Soc. Jpn.*, **57**, 426 (1984).
849. R. Bustinghaus and D. Seebach, *Chem. Ber.*, **110**, 841 (1977).
850. S. Tsuboi, T. Masuda, and A. Takeda, *Bull. Chem. Soc. Jpn.*, **56**, 3521 (1983).
851. D. Seebach and E. J. Corey, *J. Org. Chem.*, **40**, 231 (1975).
852. R. M. Carlson and P. M. Helquist, *J. Org. Chem.*, **33**, 2596 (1968).
853. K. Fuji, M. Ueda, K. Sumi, and E. Fujita, *Tetrahedron Lett.*, **22**, 2005 (1981).
854. R. A. J. Smith and A. R. Lal, *Aust. J. Chem.*, **32**, 353 (1979).
855. A. Ratajczak, F. A. L. Anet, and D. J. Cram, *J. Am. Chem. Soc.*, **89**, 2072 (1967); C. D. Ritchie and R. E. Uschold, *J. Am. Chem. Soc.*, **89**, 1721 (1967).
856. K. R. Martin, *J. Organomet. Chem.*, **24**, 7 (1970).
857. E. J. Corey and M. Chaykovsky, *J. Am. Chem. Soc.*, **87**, 1345 (1965).
858. T. Durst, R. Viau and M. R. McClory, *J. Am. Chem. Soc.*, **93**, 3077 (1971).
859. V. Reutrakul, C. Panyachotipun, V. Hahnvajanawong, and S. Sotheeswaran, *Tetrahedron Lett.*, **25**, 1825 (1984).
860. G. Boche, K. Buckl, D. Martens, D. R. Schneider, and H. U. Wagner, *Chem. Ber.*, **112**, 2961 (1979).
861. R. Lett and G. Chassaing, *Tetrahedron*, **34**, 2705 (1978).
862. W. E. Truce and K. R. Buser, *J. Am. Chem. Soc.*, **76**, 3577 (1954).
863. E. M. Kaiser, R. D. Beard, and C. R. Hauser, *J. Organomet. Chem.*, **59**, 53 (1973).

864. K. Schank, H. G. Schmitt, F. Schroeder, and A. Weber, *Justus Liebigs Ann. Chem.*, 1116 (1977).
865. J. J. Eisch and J. E. Galle, *J. Organomet. Chem.*, **121**, C10 (1976).
866. M. E. Thompson, *J. Org. Chem.*, **49**, 1700 (1984).
867. C. R. Johnson and J. R. Zeller, *J. Am. Chem. Soc.*, **104**, 4021 (1982).
868. V. Reutrakul and K. Herunsalee, *Tetrahedron Lett.*, **24**, 527 (1983).
869. D. Seebach and N. Peleties, *Chem. Ber.*, **105**, 511 (1972).
870. H. J. Reich and S. K. Shah, *J. Am. Chem. Soc.*, **97**, 3250 (1975).
871. K. Sachdev and H. S. Sachdev, *Tetrahedron Lett.*, 4223 (1976).
872. H. J. Reich, F. Chow, and S. K. Shah, *J. Am. Chem. Soc.*, **101**, 6638 (1979).
873. J. Lucchetti and A. Krief, *Terahedron Lett.*, 2693 (1978).
874. H. J. Reich, S. K. Shah, and E. Chow, *J. Am. Chem. Soc.*, **101**, 6648 (1979).
875. C. Burford, F. Cooke, E. Ehlinger, and P. Magnus, *J. Am. Chem. Soc.*, **99**, 4536 (1977); C. Burford, F. Cooke, G. Roy, and P. Magnus, *Tetrahedron*, **39**, 867 (1983).
876. J. Villieras, P. Perriot, and J. F. Normant, *Bull. Soc. Chim. Fr.*, 765 (1977).
877. G. Köbrich, K. Flory, and R. H. Fischer, *Chem. Ber.*, **99**, 1793 (1966).
878. J. Villieras, C. Bacquet, and J. F. Normant, *Bull. Soc. Chim. Fr.*, 1797 (1975).
879. M. W. Rathke and R. Kow, *J. Am. Chem. Soc.*, **94**, 6854 (1972).
880. D. S. Matteson and R. J. Moody, *J. Am. Chem. Soc.*, **99**, 3196 (1977).
881. D. S. Matteson and D. Majumdar, *J. Chem. Soc., Chem. Commun.*, 39 (1980).
882. M. Lappert, L. M. Engelhardt, C. L. Raston, and A. H. White, *J. Chem. Soc., Chem. Commun.*, 1323 (1982).
883. B. T. Gröbel and D. Seebach, *Chem. Ber.*, **110**, 852 (1977).
884. Z. H. Aiube and C. Eaborn, *J. Organomet. Chem.*, **269**, 217 (1984).
885. O. W. Steward and J. J. Johnson, *J. Organomet. Chem.*, **55**, 209 (1973).
886. P. Magnus and G. Roy, *J. Chem. Soc., Chem. Commun.*, 822 (1929).
887. T. Kauffmann, B. Altepeter, K. J. Echsler, J. Ennen, A. Hamsen, and R. Joussen, *Tetrahedron Lett.*, 501 (1979).
888. T. Kauffmann, H. Fischer, and A. Woltermann, *Angew. Chem., Int. Ed. Engl.*, **16**, 53 (1977).
889. D. Seebach and A. K. Beck, *Chem. Ber.*, **108**, 314 (1975).
890. P. E. Pfeffer, E. Kinsel, and L. S. Silbert, *J. Org. Chem.*, **37**, 1256 (1972).
891. P. M. Warner and D. Le, *J. Org. Chem.*, **47**, 893 (1982).
892. W. Adam, L. A. Encarnacion, and H. H. Fick, *Synthesis*, 828 (1978).
893. H. H. Wasserman, B. H. Lipshutz, A. W. Tremper, and J. S. Wu, *J. Org. Chem.*, **46**, 2991 (1981).
894. P. Coutrot and A. El. Gadi, *Synthesis*, 115 (1984).
895. A. M. Sarpeshkar, G. J. Gossick, and J. Wemple, *Tetrahedron Lett.*, 703 (1979).
896. H. Normant and T. Cuvigny, *Bull. Chim. Soc. Fr.*, 1881 (1965).
897. M. W. Rathke and D. F. Sullivan, *J. Am. Chem. Soc.*, **95**, 3050 (1973).
898. L. Lochmann and D. Lim, *J. Organomet. Chem.*, **50**, 9 (1973).
899. G. Frater, U. Müller, and W. Gunther, *Tetrahedron*, **40**, 1269 (1984).
900. F. Huet, M. Pellet, and J. M. Conia, *Synthesis.*, 33 (1979).
901. S. J. Harris and D. R. M. Walton, *J. Chem. Soc., Chem. Commun.*, 1008 (1976).
902. R. P. Woodbury and M. W. Rathke, *J. Org. Chem.*, **42**, 1688 (1977).
903. H. L. Needles and R. E. Whitfield, *J. Org. Chem.*, **31**, 989 (1966).
904. P. C. Kuzma, L. E. Brown, and T. M. Harris, *J. Org. Chem.*, **49**, 2015 (1984).
905. T. Durst, R. van der Elzen, and R. Legault, *Can. J. Chem.*, **52**, 3206 (1974).
906. R. Levine, *J. Am. Chem. Soc.*, **67**, 1510 (1945).
907. R. Sauvetre, M. C. Roux-Schmitt, and J. Seyden-Penne, *Tetrahedron*, **34**, 2135 (1978).
908. C. J. Eby and C. R. Hauser, *J. Am. Chem. Soc.*, **79**, 723 (1957).
909. I. Matsuda, S. Murata, and Y. Ishii, *J. Chem. Soc., Perkin Trans. 1*, 26 (1979).
910. K. Utimoto, Y. Wakabayashi, Y. Shishiyama, M. Inoue, and H. Nozaki, *Tetrahedron Lett.*, **22**, 4279 (1981).
911. A. R. Lepley, W. A. Khan, A. B. Giumanini, and A. G. Giumanini, *J. Org. Chem.*, **31**, 2047 (1966).
912. H. Gilman and R. L. Bebb, *J. Am. Chem. Soc.*, **61**, 109 (1939).
913. H. M. Walborsky and P. Ronman, *J. Org. Chem.*, **43**, 731 (1978).
914. W. Fuhrer and H. W. Gschwend, *J. Org. Chem.*, **44**, 1133 (1979).
915. J. M. Muchowski and M. C. Venuti, *J. Org. Chem.*, **45**, 4798 (1980).
916. F. N. Jones, M. F. Zinn, and C. R. Hauser, *J. Org. Chem.*, **28**, 663 (1963).

917. T. D. Harris and G. P. Roth, *J. Org. Chem.*, **44**, 2004 (1979).
918. D. W. Slocum and W. Achermann, *J. Chem. Soc., Chem. Commun.*, 968 (1974).
919. P. Beak and R. A. Brown, *J. Org. Chem.*, **42**, 1823 (1977).
920. D. A. Shirley, J.R. Johnson, Jr., and J. P. Hendrix, *J. Organomet. Chem.*, **11**, 209 (1968).
921. A. A. Morton and I. Hechenbleikner, *J. Am. Chem. Soc.*, **58**, 2599 (1936).
922. D. A. Shirley and J. P. Hendrix, *J. Organomet. Chem.*, **11**, 217 (1968).
923. L. Horner, A. J. Lawson, and G. Simons, *Phosphorus Sulfur*, **12**, 353 (1982).
924. H. Gilman and F. J. Webb, *J. Am. Chem. Soc.*, **71**, 4062 (1949).
925. N. Meyer and D. Seebach, *Angew. Chem., Int. Ed. Engl.*, **17**, 521 (1978).
926. H. P. Plaumann, B. A. Keay, and R. Rodrigo, *Tetrahedron Lett.*, 4921 (1979).
927. G. D. Figuly and J. C. Martin, *J. Org. Chem.*, **45**, 3728 (1980).
928. H. Gilman and T. S. Soddy, *J. Org. Chem.*, **22**, 1715 (1957).
929. M. Watanabe, M. Sahara, S. Furukawa, R. Billedeau, and V. Snieckus, *Tetrahedron Lett.*, **23**, 1647 (1982).
930. H. W. Gschwend, unpublished work, cited in H. W. Gschwend and H. R. Rodriquez, *Org. React.*, **26**, 1 (1979).
931. T. D. Krizan and J. C. Martin, *J. Org. Chem.*, **47**, 2681 (1982).
932. W. E. Truce and M. F. Amos, *J. Am. Chem. Soc.*, **73**, 3013 (1951).
933. D. L. Ladd and J. Weinstock, *J. Org. Chem.*, **46**, 203 (1981).
934. A. I. Meyers and W. Rieker, *Tetrahedron Lett.*, **23**, 2091 (1982).
935. R. C. Ronald, *Tetrahedron Lett.*, 3973 (1975).
936. K. Oita and H. Gilman, *J. Org. Chem.*, **21**, 1009 (1956).
937. H. W. Gschwend and A. Hamdan, *J. Org. Chem.*, **40**, 2008 (1975).
938. U. Schöllkopf, D. Stafforst, and R. Jentsch, *Justus Liebigs Ann. Chem.*, 1167 (1977).
939. J. E. Baldwin, G. A. Höfle, and O. W. Lever, Jr., *J. Am. Chem. Soc.*, **96**, 7125 (1974).
940. R. K. Boeckman, Jr., and K. J. Bruza, *Tetrahedron Lett.*, 4187 (1977).
941. W. Adam, E. Schmidt, and K. Takayama, *Synthesis*, 322 (1982).
942. K. Oshima, K. Shimoji, H. Takahashi, H. Yamamoto, and H. Nozaki, *J. Am. Chem. Soc.*, **95**, 2694 (1973).
943. R. R. Schmidt and B. Schmid, *Tetrahedron Lett.*, 3583 (1977).
944. M. Schoufs, J. Meyer, P. Vermeer, and L. Brandsma, *Recl. Trav. Chim. Pays-Bas*, **96**, 259 (1977).
945. J. J. Eisch and J. E. Galle, *J. Org. Chem.*, **44**, 3279 (1979).
946. G. Kobrich and K. Flory, *Chem. Ber.*, **99**, 1773 (1966).
947. P. G. Gassmann and I. Gennick, *J. Am. Chem. Soc.*, **102**, 6863 (1980).
948. R. Sauvetre and J. F. Normant, *Tetrahedron Lett.*, 957 (1981).
949. F. G. Drakesmith, R. D. Richardson, O. J. Stewart, and P. Tarrant, *J. Org. Chem.*, **33**, 286 (1968).
950. M. Schlosser, B. Schaub, B. Spahic, and G. Sleiter, *Helv. Chim. Acta*, **56**, 2166 (1973).
951. J. Ficini and J.C. Depezay, *Tetrahedron Lett.*, 937 (1968).
952. K. Isobe, M. Fuse, H. Kosugi, H. Hagiwara, and H. Uda, *Chem. Lett.*, 785 (1979).
953. O. Miyata and R. R. Schmidt, *Tetrahedron Lett.*, **23**, 1793 (1982).
954. J. E. Biellman and J. B. Ducep, *Tetrahedron Lett.*, 33 (1971).
955. B. Costisella, H. Gross, and H. Schick, *Tetrahedron*, **40**, 733 (1984).
956. H. Ahlbrecht and C. Vonderheid, *Synthesis*, 512 (1975).
957. S. De Lombaert, B. Lesur, and L. Ghosez, *Tetrahedron Lett.*, **23**, 4251 (1982).
958. T. H. Chan and G. J. Kang, *Tetrahedron Lett.*, **23**, 3011 (1982).
959. H. Ahlbrecht and H. Simon, *Synthesis*, 58 (1983).
960. H. Ahlbrecht and W. Farnung, *Chem. Ber.*, **117**, 1 (1984).
961. A. N. Tischler and M. H. Tischler, *Tetrahedron Lett.*, 3407 (1978).
962. F. Kienzle, *Helv. Chim. Acta*, **56**, 1671 (1973).
963. D. Seebach, R. Henning, and F. Lehr, *Angew. Chem., Int. Ed. Engl.*, **17**, 458 (1978).
964. B. Renger and D. Seebach, *Chem. Ber.*, **110**, 2334 (1977).
965. B. Lythgoe, T. A. Moran, M. E. N. Nambudiry, S. Ruston, J. Tideswell, and P. W. Wright, *Tetrahedron Lett.*, 3863 (1975).
966. P. Savignac, P. Coutrot, and Y. Leroux, *C. R. Acad. Sci., Ser. C*, **279**, 609 (1974).
967. E. J. Corey and D. E. Cane, *J. Org. Chem.*, **34**, 3053 (1969).
968. K. Kondo, A. Negishi, and D. Tunemoto, *Angew. Chem., Int. Ed. Engl.*, **13**, 407 (1974).

969. D. A. Evans, G. C. Andrews, and B. Buckwalter, *J. Am. Chem. Soc.*, **96**, 5560 (1973).
970. G. Trimitsis, S. Beers, J. Ridella, M. Carlon, D. Cullin, J. High, and D. Brutts, *J. Chem. Soc., Chem. Commun.*, 1088 (1984).
971. D. Seyferth, R. E. Mammarella, and H. A. Klein, *J. Organomet. Chem.*, **194**, 1 (1980).
972. M. Maleki, J. A. Miller, and O. W. Lever, Jr., *Tetrahedron Lett.*, **22**, 3789 (1981).
973. U. Hertenstein, S. Hünig, and M. Oller, *Synthesis*, 416 (1976).
974. K. H. Geiss, D. Seebach, and B. Seuring, *Chem. Ber.*, **110**, 1833 (1977).
975. D. A. Evans and G. C. Andrews, *Acc. Chem. Res.*, **7**, 147 (1974).
976. K. Tanaka, M. Terauchi, and A. Kaji, *Chem. Lett.*, 315 (1981).
977. J. P. Marino and J. L. Kostusyk, *Tetrahedron Lett.*, 2489 (1979).
978. D. Seebach, *Synthesis*, 17 (1969).
979. T. Nakai, T. Mimura, and T. Kurokawa, *Tetrahedron Lett.*, 2895 (1978).
980. T. Hayashi, N. Fujitaka, T. Oishi, and T. Takeshima, *Tetrahedron Lett.*, **21**, 303 (1980).
981. D. A. Evans, G. C. Andrews, and C. L. Sims, *J. Am. Chem. Soc.*, **93**, 4956 (1971).
982. D. A. Evans, G. C. Andrews, T. T. Fujimoto, and D. Wells, *Tetrahedron Lett.*, 1385 (1973).
983. M. Hirama, *Tetrahedron Lett.*, **22**, 1905 (1981).
984. C. N. Hsiao and H. Shechter, *Tetrahedron Lett.*, **25**, 1219 (1984).
985. H. J. Reich, W. W. Willis, Jr., and P. D. Clark, *J. Org. Chem.*, **46**, 2775 (1981).
986. H. J. Reich, M. C. Clark and W. W. Willis, Jr., *J. Org. Chem.*, **47**, 1618 (1982).
987. T. L. MacDonald, B. A. Narayanan, D. E. Odell, *J. Org. Chem.*, **46**, 1504 (1981).
988. T. Hiyama, M. Shinodo, and H. Nozaki, *Tetrahedron Lett.*, 771 (1978).
989. S. Brenner and M. Bovete, *Tetrahedron Lett.*, 1377 (1974).
990. P. Beak, J. E. Hunter, and Y. M. Jun, *J. Am. Chem. Soc.*, **105**, 6350 (1983).
991. K. Komlaglo and T. H. Chan, *Tetrahedron Lett.*, **25**, 717 (1984).
992. E. Piers and J. R. Grierson, *J. Org. Chem.*, **42**, 3755 (1977).
993. W. Oppolzer, R. L. Snowden, and P. H. Briner, *Helv. Chim. Acta*, **64**, 2022 (1981).
994. D. Seebach, M. Pohmakotr, C. Schregenberger, B. Weidmann, R. S. Mali, and S. Pohmakotr, *Helv. Chim. Acta*, **65**, 419 (1982).
995. K. Takabe, H. Fujiwara, T. Katagiri and J. Tanaka, *Tetrahedron Lett.*, 1237 (1975).
996. T. Cuvigny, J. F. le Borgne, M. Larcheveque and H. Normant, *Synthesis*, 237 (1976).
997. M. Larcheveque, G. Valette and T. Cuvigny, *Tetrahedron*, **35**, 1745 (1979).
998. D. Enders, H. Eichenauer, U. Baus, H. Schubert, and K. A. M. Kremer, *Tetrahedron*, **40**, 1345 (1984).
999. E. A. Mistryukov and I. K. Korshevets, *Bull. Acad. Sci. USSR* (*Engl. Transl.*), **34**, 448 (1985).
1000. H. J. Reich, M. J. Kelly, R. E. Olsen, and R. C. Holtan, *Tetrahedron*, **39**, 949 (1983).
1001. R. M. Carlson and A. R. Oyler, *Tetrahedron Lett.*, 2615 (1974).
1002. E. J. Corey and C. Rücker, *Tetrahedron Lett.*, **23**, 719 (1982).
1003. J. Pornet, D. Mesnard, and L. Miginiac, *Tetrahedron Lett.*, **23**, 4083 (1982).
1004. C. C. Shen and C. Ainsworth, *Tetrahedron Lett.*, 83 (1979).
1005. K. J. H. Kruithof, R. F. Schmitz, and G. W. Klumpp, *Tetrahedron*, **39**, 3073 (1983).
1006. Y. Leroux and R. Mantione, *Tetrahedron Lett.*, 591 (1971).
1007. Y. Leroux and C. Roman, *Tetrahedron Lett.*, 2585 (1973).
1008. H. D. Verkruijsse, W. Verboom, P. E. van Rijn, and L. Brandsma, *J. Organomet. Chem.*, **232**, C1 (1982).
1009. R. M. Carlson, R. W. Jones, and A. S. Hatcher, *Tetrahedron Lett.*, 1741 (1975).
1010. F. Mercier, R. Epsztein, and S. Holand, *Bull. Soc. Chim. Fr.*, 690 (1972).
1011. S. Hoff, L. Brandsma, and J. F. Arens, *Recl. Trav. Chem. Pays-Bas*, **87**, 1179 (1968); see also I. Kuwajiwa, S. Sugahara, and J. Enda, *Tetrahedron Lett.*, **24**, 1061 (1983).
1012. P. Pappalardo, E. Ehlinger, and P. Magnus, *Tetrahedron Lett.*, **23**, 309 (1982).
1013. B. Corbel, J. P. Paugam, M. Dreux, and P. Savignac, *Tetrahedron Lett.*, 835 (1976).
1014. E. I. Negishi, C. L. Rand, and K. P. Jadhau, *J. Org. Chem.*, **46**, 5041 (1981).
1015. H. J. Reich, S. K. Shah, P. M. Gold, and R. E. Olson, *J. Am. Chem. Soc.*, **103**, 3112 (1981).
1016. C. G. Screttas and M. Micha-Screttas, *J. Org. Chem.*, **43**, 1064 (1978).
1017. C. G. Screttas and M. Micha-Screttas, *J. Org. Chem.*, **44**, 713 (1979).
1018. T. Cohen and J. R. Matz, *J. Am. Chem. Soc.*, **102**, 6902 (1980).
1019. T. Cohen, J. P. Sherbine, J. R. Matz, R. R. Hutchins, B. M. McHenry, and P. R. Willey, *J. Am. Chem. Soc.*, **106**, 3245 (1984).
1020. D. J. Ager, *J. Org. Chem.*, **49**, 168 (1984).

1021. T. Cohen, W. M. Daniewski, and R. B. Weisenfeld, *Tetrahedron Lett.*, 4665 (1978).
1022. L. A. Paquette, G. J. Wells, K. A. Horn, and T.-H. Yan, *Tetrahedron*, **39**, 913 (1983).
1023. T. Cohen and R. B. Weisenfeld, *J. Org. Chem.*, **44**, 3601 (1979).
1024. J. P. Quintard, B. Elissondo, and B. Jousseaume, *Synthesis*, 494 (1984).
1025. D. J. Peterson and J. E. Ward, *J. Organomet. Chem.*, **66**, 209 (1974).
1026. D. J. Peterson, *Organomet. Chem. Rev. A*, **7**, 295 (1972).
1027. E. J. Corey and T. M. Eckrich, *Tetrahedron Lett.*, **24**, 3163 (1983).
1028. W. C. Still, *J. Am. Chem. Soc.*, **100**, 1481 (1978).
1029. S. D. Burke, S. A. Shearouse, D. J. Burch, and R. W. Sutton, *Tetrahedron Lett.*, **21**, 1285 (1980).
1030. W. C. Still and C. Sreekumar, *J. Am. Chem. Soc.*, **102**, 1201 (1980).
1031. P. Beak and L. G. Carter, *J. Org. Chem.*, **46**, 2363 (1981).
1032. D. G. Farnum and T. Monego, *Tetrahedron Lett.*, **24**, 1361 (1983).
1033. J. S. Sawyer, T. L. Macdonald, and G. J. McGarvey, *J. Am. Chem. Soc.*, **106**, 3376 (1984).
1034. K. Fuji, M. Ueda, K. Sumi, and E. Fujita, *Tetrahedron Lett.*, **22**, 2005 (1981).
1035. R. D. Taylor and J. L. Wardell, *J. Organomet. Chem.*, **77**, 311 (1974).
1036. A. Anciaux, A. Eman, W. Dumont, and A. Krief, *Tetrahedron Lett.*, 1617 (1975).
1037. J. N. Denis, W. Dumont, and A. Krief, *Tetrahedron Lett.*, 453 (1976).
1038. D. van Ende, W. Dumont, and A. Krief, *J. Organomet. Chem.*, **149**, C10 (1978).
1039. J. Lucchetti and A. Krief, *Tetrahedron Lett.*, **22**, 1623 (1981).
1040. D. Seebach and N. Peleties, *Chem. Ber.*, **105**, 511 (1972).
1041. D. Seebach and A. K. Beck, *Chem. Ber.*, **108**, 314 (1974).
1042. G. Zweifel, R. P. Fisher, and A. Hornig, *Synthesis*, 37 (1973).
1043. D. S. Matteson, L. A. Hagelee, and R. J. Wilcsek, *J. Am. Chem. Soc.*, **95**, 5096 (1973).
1044. H. Sakurai, K. Nishiwaki, and M. Kira, *Tetrahedron Lett.*, 4193 (1973).
1045. T. Kauffmann and R. Kriegesmann, *Chem. Ber.*, **115**, 1810 (1982).
1046. D. Seyferth, F. M. Armbrecht, R. L. Lambert, and W. Tronich, *J. Organomet. Chem.*, **44**, 299 (1972).
1047. T. Kauffmann, K.-J. Echsler, A. Hamsen, R. Kriegesmann, F. Steinseifer, and A. Vahrenhorst, *Tetrahedron Lett.*, 4391 (1978).
1048. A. Rensing, K.-J. Echsler, and T. Kauffmann, *Tetrahedron Lett.*, **21**, 2807 (1980).
1049. T. Kauffmann, R. Joussen, N. Klas, and A. Vahrenhorst, *Chem. Ber.*, **116**, 473 (1983).
1050. D. Seyferth, R. L. Lambert, and M. Massol, *J. Organomet. Chem.*, **88**, 255 (1975).
1051. S. Halazy, J. Lucchetti, and A. Krief, *Tetrahedron Lett.*, 3971 (1978).
1052. F. J. Kronzer and V R. Sandel, *J. Am. Chem. Soc.*, **94**, 5750 (1972).
1053. E. Piers, J. M. Chong, K. Gustafson, and R. J. Andersen, *Can. J. Chem.*, **62**, 1 (1984).
1054. J. Ficini, S. Falou, A. M. Touzin, and J. D'Angelo, *Tetrahedron Lett.*, 3589 (1977).
1055. R. H. Wollenberg, K. F. Albizati, and R. Peries, *J. Am. Chem. Soc.*, **99**, 7365 (1977).
1056. J. A. Sonderquist and G. J.-H. Hsu, *Organometallics*, **1**, 830 (1982).
1057. E. J. Corey and R. H. Wollenberg, *J. Org. Chem.*, **40**, 2265 (1975).
1058. E. J. Corey and R. H. Wollenberg, *Tetrahedron Lett.*, 4705 (1976).
1059. S. L. Chen, R. E. Schaub, and C. V. Grudzinskas, *J. Org. Chem.*, **43**, 3450 (1978).
1060. M. Gill, H. P. Bainton, and R. W. Rickards, *Tetrahedron Lett.*, **22**, 1437 (1981).
1061. B.-T. Gröbel and D. Seebach, *Chem. Ber.*, **110**, 867 (1977).
1062. T. N. Mitchell and A. Amamria, *J. Organomet. Chem.*, **252**, 47 (1983).
1063. E. Piers and V. Karunaratne, *J. Org. Chem.*, **48**, 1774 (1983).
1064. M. Clarembeau and A. Krief, *Tetrahedron Lett.*, **25**, 3629 (1984).
1065. D. Seyferth and T. F. Jula, *J. Organomet. Chem.*, **66**, 195 (1974).
1066. D. Seyferth and R. E. Mammarella, *J. Organomet. Chem.*, **177**, 53 (1979).
1067. A. Doucoure, B. Mauzé, and L. Miginiac, *J. Organomet. Chem.*, **236**, 139 (1982).
1068. D. Seyferth and R. E. Mammarella, *J. Organomet. Chem.*, **156**, 279 (1978).
1069. B. Mauzé, *J. Organomet. Chem.*, **170**, 265 (1979).
1070. D. Seyferth, G. J. Murphy, and R. A. Woodruff, *J. Organomet. Chem.*, **141**, 71 (1977).
1071. D. Seyferth and K. R. Wursthorn, *J. Organomet. Chem.*, **182**, 455 (1977).
1072. A. B. Holmes and G. E. Jones, *Tetrahedron Lett.*, **21**, 3111 (1980).
1073. E. Piers and H. E. Morton, *J. Org. Chem.*, **44**, 3437 (1979).
1074. D. Seebach, *Synthesis*, 17 (1969); D. Seebach and A. K. Beck, *Org. Synth.*, **51**, 76 (1971); D. Seebach and E. J. Corey, *J. Org. Chem.*, **40**, 231 (1975).
1075. X = S: D. Seebach, E. J. Corey, and A. K. Beck, *Chem. Ber.*, **107**, 367 (1974); D. Seebach and

A. K. Beck, *Org. Synth.*, **51**, 39 (1971). X = NMe: R. D. Balanson, V. M. Kobal, and R. R. Schumaker, *J. Org. Chem.*, **42**, 393 (1977); H. Paulsen, M. Stubbe, and F. R. Heiker, *Justus Liebigs Ann. Chem.*, 825 (1980).
1076. S. Ncube, A. Pelter, and K. Smith, *Tetrahedron Lett.*, 1893, 1895 (1979); S. Ncube, A. Pelter, K. Smith, P. Blatcher, and S. Warren, *Tetrahedron Lett.*, 2345 (1978).
1077. K. Mori, H. Hashimoto, Y. Takenaka, and T. Takigawa, *Synthesis*, 720 (1975).
1078. J. F. Arens, M. Froling, and A. Froling, *Recl. Trav. Chim. Pays-Bas*, **78**, 663 (1959); W. E. Truce and F. E. Roberts, *J. Org. Chem.*, **28**, 961 (1963); G. Schill and C. Merkel, *Synthesis*, 387 (1975); P. Blatcher, J. I. Grayson, and S. Warren, *J. Chem. Soc., Chem. Commun.*, 547 (1976).
1079. T. Cohen and S. M. Nolan, *Tetrahedron Lett.*, 3533 (1978); D. Seebach, *Chem. Ber.*, **105**, 487 (1972).
1080. G. E. Veenstra and B. Zwanenburg, *Tetrahedron*, **34**, 1585 (1978); L. Colombo, C. Gennari, C. Scolastico, G. Guanti, and E. Narisano, *J. Chem. Soc., Perkin Trans. 1*, 1278 (1981).
1081. F. Cooke and P. Magnus, *J. Chem, Soc., Chem. Commun.*, 513 (1977).
1082. E. L. Eliel and J. E. Lynch, *Tetrahedron Lett.*, **22**, 2855 (1981).
1083. K. Fuji, M. Ueda, and E. Fujita, *J. Chem. Soc., Chem. Commun.*, 814 (1977).
1084. A. de Groot and B. J. M. Jansen, *Tetrahedron Lett.*, **22**, 887 (1981); K. Ogura, M. Yamashita, and G.-I. Tsuchihashi, *Tetrahedron Lett.*, 1303 (1978).
1085. G. W. Gokel, H. M. Gerdes, D. E. Miles, J. M. Hufnal, and G. A. Zerby, *Tetrahedron Lett.*, 3375 (1979).
1086. K. Tanaka, S. Matsui, and A. Kaji, *Bull. Chem. Soc. Jpn.*, **53**, 3619 (1980).
1087. I. Hori, T. Hayashi, and H. Midorikawa, *Synthesis*, 705 (1974).
1088. D. van Ende, A. Cravador, and A. Krief, *J. Organomet. Chem.*, **177**, 1 (1979); A. Burton, L. Hevesi, W. Dumont, A. Cravador, and A. Krief, *Synthesis*, 877 (1979); S. Raucher and G. A. Koolpe, *J. Org. Chem.*, **43**, 3794 (1978).
1089. P. Magnus and G. Roy, *Organometallics.*, **1**, 553 (1982).
1090. P. J. Kocienski, *Tetrahedron Lett.*, **21**, 1559 (1980); D. J. Ager, *J. Chem. Soc., Perkin Trans. 1*, 1131 (1983).
1091. K. Sachdev and H. S. Sachdev, *Tetrahedron Lett.*, 4223 (1976); W. Dumont, D. van Ende and A. Krief, *Tetrahedron Lett.*, 485 (1979).
1092. C. Earnshaw, C. J. Wallis, and S. Warren, *J. Chem. Soc. Perkin Trans. 1*, 3099 (1979).
1093. J. I. Grayson and S. Warren, *J. Chem. Soc., Perkin Trans. 1*, 2263 (1977).
1094. E. J. Corey and D. Boger, *Tetrahedron Lett.*, 5,9,13 (1978).
1095. S. Yamagiwa, N. Hoshi, H. Sato, H. Kosugi, and H. Uda, *J. Chem. Soc., Perkin Trans. 1*, 214 (1978); B. M. Trost and Y. Tamaru, *Tetrahedron Lett.*, 3797 (1975).
1096. Y. Masuyama, Y. Ueno and M. Okawara, *Tetrahedron Lett.*, 2967 (1976).
1097. N.-Y. Wang, S.-S. Su and L.-Y. Tsai, *Tetrahedron Lett.*, 1121 (1979).
1098. E. Aufdefhaar, J. E. Baldwin, D. H. R. Barton, D. J. Faulkner, and M. Slaytor, *J. Chem. Soc. C*, 2175 (1971); G. Stork and L. Maldonado, *J. Am. Chem. Soc.*, **93**, 5286 (1971); see also J. D. Albright, *Tetrahedron*, **39**, 3207 (1983).
1099. F. J. McEvoy and J. D. Albright, *J. Org. Chem.*, **44**, 4597 (1979); H. Ahlbrecht, W. Raab, and C. Vonderheid, *Synthesis*, 127 (1979); G. Stork, A. A. Ozorio, and A. Y. W. Leong, *Tetrahedron Lett.*, 5175 (1978).
1100. S. F. Martin and R. Gompper, *J. Org. Chem.*, **39**, 2814 (1974).
1101. G. E. Niznik, W. H. Morrison, III, and H. M. Walborsky, *J. Org. Chem.*, **39**, 600 (1974).
1102. G. Stork, A. Y. W. Leong, and A. M. Touzin, *J. Org. Chem.*, **41**, 3491 (1976).
1103. S. F. Martin and G. W. Phillips, *J. Org. Chem.*, **43**, 3792 (1978).
1104. D. Seebach and D. Enders, *Angew. Chem., Int. Ed. Engl.*, **14**, 15 (1975).
1105. O. Possel and A. M van Leusen, *Tetrahedron Lett.*, 4229 (1977); D. van Leusen and A. M. van Leusen, *Tetrahedron Lett.*, 4233 (1977).
1106. B. Deschamps, *Tetrahedron*, **34**, 2009 (1978).
1107. T. Hata, A. Hashizume, M. Nakajima, and M. Sekine, *Chem. Lett.*, 519 (1979); M. Sekine, M. Nakajima, A. Kume, and T. Hata, *Tetrahedron Lett.*, 4475 (1979).
1108. S. Hünig and G. Wehner, *Chem. Ber.*, **113**, 302 (1980); K. Deuchert, U. Hertenstein, S. Hünig, and G. Wehner, *Chem. Ber.*, **112**, 2045 (1979); U. Hertenstein, S. Hünig, and M. Öller, *Chem. Ber.*, **113**, 3783 (1980).

1109. J. E. Baldwin, G. A. Höfle, and O. W. Lever, Jr., *J. Am. Chem. Soc.*, **96**, 7125 (1974); A. B. Levy and S. J. Schwartz, *Tetrahedron Lett.*, 2201 (1976).
1110. K. Oshima, K. Shimoji, H. Takahashi, H. Yamamoto, and Nozaki, *J. Am. Chem. Soc.*, **95**, 2694 (1973); I. Vlattas, L. Della Vecchia, and A. O. Lee, *J. Am. Chem. Soc.*, **98**, 2008 (1976); T. Cohen and R. B. Weisenfeld, *J. Org. Chem.*, **44**, 3601 (1979).
1111. M. Sevrin, J. N. Denis, and A. Krief, *Angew. Chem., Int. Ed. Engl.*, **17**, 526 (1978).
1112. B.-T. Gröbel and D. Seebach, *Chem. Ber.*, **110**, 867 (1977).
1113. J. E. Baldwin and J. C. Bottaro, *J. Chem. Soc., Chem. Commun.*, 1121 (1981); C. Wiaux-Zamar, J.-P. Dejonghe, L. Ghosez, J.-F. Normant, and J. Villieras, *Angew. Chem., Int. Ed. Engl.*, **15**, 371 (1976).
1114. N. Meyer and D. Seebach, *Chem. Ber.*, **113**, 1290 (1980).
1115. G. Zweifel, R. P. Fischer, and A. Hornig, *Synthesis*, 37 (1973).
1116. P. Beak, M. Baillargeon, and L. G. Carter, *J. Org. Chem.*, **43**, 4255 (1982).
1117. W. C. Still, *J. Am. Chem. Soc.*, **100**, 1481 (1978).
1118. K. Ogura, S. Furukawa, and G. Tsuchihashi, *Synthesis*, 202 (1976).
1119. T. Kauffmann, H. Berg, E. Köppelmann, and D. Kuhlmann, *Chem. Ber.*, **110**, 2659 (1977).
1120. U. Schollköpf, *Pure Appl. Chem.*, **51**, 1347 (1979).
1121. D. Enders, R. Pieter, B. Renger, and D. Seebach, *Org. Synth.*, **58**, 113 (1978).
1122. T. Hassel and D. Seebach, *Helv. Chim. Acta*, **61**, 2237 (1978).
1123. D. Seebach and T. Hassel, *Angew. Chem., Int. Ed. Engl.*, **17**, 274 (1978).
1124. E. Nakamura, *Tetrahedron Lett.*, **22**, 663 (1981).
1125. A. I. Meyers and E. D. Mihelich, *Angew. Chem., Int. Ed. Engl.*, **15**, 270 (1976).
1126. J. E. McMurry and J. H. Musser, *J. Org. Chem.*, **40**, 2556 (1975).
1127. K. M. More and J. Wemple, *J. Org. Chem.*, **43**, 2713 (1978).
1128. R. R. Schmidt and B. Schmid, *Tetrahedron Lett.*, 3583 (1977).
1129. G. Neef and U. Eder, *Tetrahedron Lett.*, 2825 (1977); R. E. Damon and R. H. Schlessinger, *Tetrahedron Lett.*, 4551 (1975).
1130. R. H. Wollenberg, *Tetrahedron Lett.*, 717 (1978); L. Duhamel and F. Tombret, *J. Org. Chem.*, **46**, 3741 (1981).
1131. J. F. LeBorgne, T. Cuvigny, M. Larchveque, and H. Normant, *Tetrahedron Lett.*, 1379 (1976).
1132. I. R. Politzer and A. I. Meyers, *Org. Synth.*, **51**, 24 (1971).
1133. A. I. Meyers, D. R. Williams, S. White, and G. W. Erickson, *J. Am. Chem. Soc.*, **103**, 3088 (1981).
1134. E. C. Corey and D. Enders, *Chem. Ber.*, **111**, 1337, 1362 (1978).
1135. W. C. Still and T. L. MacDonald, *J. Am. Chem. Soc.*, **96**, 5561 (1974).
1136. K. -H. Geiss, D. Seebach, and B. Seuring, *Chem. Ber.*, **110**, 1833 (1977).
1137. E. Ehlinger and P. Magnus, *J. Am. Chem. Soc.*, **102**, 5004 (1980).
1138. H. Ahlbrecht, G. Bonnet, D. Enders, and G. Zimmermann, *Tetrahedron Lett.*, **21**, 3175 (1980).
1139. H. Ahlbrecht, *Chimia*, **31**, 391 (1977).
1140. D. Hoppe, R. Hanko, and A. Brönneke, *Angew. Chem., Int. Ed. Engl.*, **19**, 625 (1980).
1141. A. Bell, A. H. Davidson, C. Earnshaw, H. K. Norrish, R. S. Torr, and S. Warren, *J. Chem. Soc., Chem. Commun.*, 988 (1978).
1142. P. Bakuzis, M. L. F. Bakuzis, and T. F. Weingartner, *Tetrahedron Lett.*, 2371 (1978).
1143. J. C. Clinet and G. Linstrumelle, *Tetrahedron Lett.*, **21**, 3987 (1980).
1144. E. C. Corey, B. W. Erickson, and R. Noyori, *J. Am. Chem. Soc.*, **93**, 1724 (1971); B. W. Erickson, *Org. Synth.*, **54**, 19 (1974).
1145. T. Hiyama, A. Kanakura, H. Yamamoto, and H. Nozaki, *Tetrahedron Lett.*, 3047, 3051 (1978).
1146. B. Corbel, J.-P. Paugam, M. Dreux, and P. Savignac, *Tetrahedron Lett.*, 835 (1976).
1147. D. Seebach, R. Henning, and F. Lehr, *Angew. Chem., Int. Ed. Engl.*, **17**, 458 (1978).
1148. E. Piers and H. E. Morton, *J. Org. Chem.*, **44**, 3437 (1979).
1149. T. D. Harris and G. P. Roth, *J. Org. Chem.*, **44**, 2004 (1979).
1150. L. Barsky, H. W. Gschwend, J. McKenna, and H. R. Rodriquez, *J. Org. Chem.*, **41**, 3651 (1976).
1151. H. W. Gschwend and A. Hamdan, *J. Org. Chem.*, **40**, 2008 (1975); A. I. Meyers and E. D. Mihelich, *J. Org. Chem.*, **40**, 3158 (1975).
1152. J. Barluenga, F. J. Fananas, J. Villamana, and M. Yus, *J. Org. Chem.*, **47**, 1560 (1982).
1153. D. Seebach and N. Meyer, *Angew. Chem., Int. Ed. Engl.*, **15**, 438 (1976).
1154. W. C. Still and C. Sreekumar, *J. Am. Chem. Soc.*, **102**, 1201 (1980).

1155. T. M. Dolak and T. A. Bryson, *Tetrahedron Lett.*, 1961 (1977).
1156. K. Kondo and D. Tunemoto, *Tetrahedron Lett.*, 1007 (1975).
1157. J. M. Photis and L. A. Paquette, *Org. Synth.*, **57**, 53 (1977).
1158. M. Isobe, M. Kitamura, and T. Goto, *Chem. Lett.*, 331 (1980).
1159. R. D. Balanson, V. M. Kobal, and R. R. Schumaker, *J. Org. Chem.*, **42**, 393 (1977).
1160. J. E. Richman, J. L. Herrmann, and R. H. Schlessinger, *Tetrahedron Lett.*, 3267 (1973).
1161. H. J. Reich and S. K. Shah, *J. Am. Chem. Soc.*, **97**, 3250 (1975).
1162. J. Villieras, P. Perriot, and J. F. Normant, *Bull. Soc. Chim. Fr.*, 765 (1977).
1163. R. M. Jacobson, A. Abbaspour, and G. P. Lahm, *J. Org. Chem.*, **43**, 4650 (1978).
1164. I. S. Ponticello, *J. Polym. Sci., Polym. Chem. Ed.*, **17**, 3499 (1979).
1165. G. D. Ewing and L. A. Paquette, *J. Org. Chem.*, **40**, 2965 (1975).
1166. P. M. Warner and D. Le, *J. Org. Chem.*, **47**, 893 (1982).
1167. P. A. Zoretic and F. Barcelos, *Tetrahedron Lett.*, 529 (1977).
1168. D. Seebach, W. Lubosch, and D. Enders, *Chem. Ber.*, **109**, 1309 (1976).
1169. R. R. Fraser and P. R. Hubert, *Can. J. Chem.*, **52**, 185 (1974).
1170. P. A. Grieco and R. S. Finkelhor, *J. Org. Chem.*, **38**, 2909 (1973).
1171. D. Seebach, R. Henning, F. Lehr, and J. Gonnermann, *Tetrahedron Lett.*, 1161 (1977).
1172. A. I. Meyers, J. L. Durandetta, and R. Munavu, *J. Org. Chem.*, **40**, 2025 (1975).
1173. T. Cuvigny, M. Larcheveque, and H. Normant, *Justus Liebig's Ann. Chem.*, 719 (1975).
1174. E. J. Corey and D. Enders, *Tetrahedron Lett.*, 3, 11 (1976).
1175. T. Kauffmann, H. Fischer, and A. Woltermann, *Chem. Ber.*, **115**, 645 (1982).
1176. T. Kauffmann, R. Joussen, N. Klas, and A. Vahrenhorst, *Chem. Ber.*, **116**, 473 (1983).
1177. A. J. Quillonan and F. Scheinmann, *Org. Synth.*, **58**, 1 (1978).
1178. E. Negishi, C. L. Rand, and K. P. Jadhav, *J. Org. Chem.*, **46**, 5041 (1981).
1179. E. M. Kaiser and J. P. Petty, *Synthesis*, 705 (1975).
1180. D. Seebach, J.-J. Lohmann, M. A. Syfrig, and M. Yoshifuji, *Tetrahedron*, **39**, 1963 (1983).
1181. E. Piers and J. R. Gierson, *J. Org. Chem.*, **42**, 3755 (1977).
1182. M. J. Marks and H. M. Walborsky, *J. Org. Chem.*, **46**, 5405 (1981).
1183. H. Neumann and D. Seebach, *Tetrahedron Lett.*, 4839 (1976).
1184. J. E. Baldwin, G. A. Höfle, and O. W. Lever, Jr., *J. Am. Chem. Soc.*, **96**, 7125 (1974).
1185. R. K. Boeckman, Jr. and K. J. Bruza, *Tetrahedron Lett.*, 4187 (1977).
1186. K. Oshima, K. Shimoji, H. Takahashi, H. Yamamoto, and H. Nozaki, *J. Am. Chem. Soc.*, **95**, 2694 (1973).
1187. I. Vlattas, L. Della Vecchia, and A. O. Lee, *J. Am. Chem. Soc.*, **98**, 2008 (1976).
1188. H. J. Reich, W. W. Willis, Jr., and P. D. Clark, *J. Org. Chem.*, **46**, 2775 (1981).
1189. R. R. Schmidt and J. Talbiersky, *Angew. Chem., Int. Ed. Engl.*, **15**, 171 (1976).
1190. J. N. Denis and A. Krief, *Tetrahedron Lett.*, **23**, 3411 (1982).
1191. O. Riobe, A. Leboux, J. Delaunay, and H. Normant, *C. R. Acad. Sci., Ser. C*, **284**, 281 (1977).
1192. G. E. Hall, R. Piccolini, and J. D. Roberts, *J. Am. Chem. Soc.*, **77**, 4540 (1955).
1193. H. R. Rodriquez, cited in H. W. Gschwend and H. R. Rodriquez, *Org. React.*, **26**, 1 (1979).
1194. H. Böhme and U. Bomke, *Arch. Pharm.*, **303**, 779 (1970).
1195. J. Boeckmann and G. Schill, *Chem. Ber.*, **110**, 703 (1977).
1196. N. Meyer and D. Seebach, *Angew. Chem., Int. Ed. Engl.*, **17**, 521 (1978).
1197. N. D. Ly and M. Schlosser, *Helv. Chim. Acta*, **60**, 2085 (1977).
1198. H. J. Jakobsen and S. O. Lawesson, *Tetrahedron*, **21**, 3331 (1965).
1199. A. I. Meyers and R. A. Gabel, *Tetrahedron Lett.*, 227 (1978).
1200. D. J. Peterson, *J. Org. Chem.*, **33**, 780 (1968).
1201. T. Kauffmann, R. Kriegesmann, and A. Woltermann, *Angew. Chem., Int. Ed. Engl.*, **16**, 862 (1977); T. Kauffmann, A. Ahlers, R. Joussen, R. Kriegesmann, A. Vahrenhorst, and A. Woltermann, *Tetrahedron Lett.*, 4399 (1978).
1202. D. J. Peterson, *J. Organomet. Chem.*, **8**, 199 (1967).
1203. E. J. Corey and G. T. Kwiatkowski, *J. Am. Chem. Soc.*, **88**, 5654 (1966).
1204. J. N. Denis, W. Dumont, and A. Krief, *Tetrahedron Lett.*, 4111 (1979).
1205. F. Jung, N. K. Sharma, and T. Durst, *J. Am. Chem. Soc.*, **95**, 3420 (1973).
1206. C. R. Johnson, J. R. Shanklin, and R. A. Kirchhoff, *J. Am. Chem. Soc.*, **95**, 6462 (1973).
1207. C. R. Johnson and R. C. Elliott, *J. Am. Chem. Soc.*, **104**, 7041 (1982).
1208. I. Matsuda, S. Murata, and Y. Ishii, *J. Chem. Soc., Perkin Trans. 1*, 26 (1979).

1209. G. Jones and R. F. Maisey, *J. Chem. Soc., Chem. Commun.*, 543 (1968).
1210. M. Nakano and Y. Okamoto, *Synthesis*, 917 (1983).
1211. W. S. Wadsworth, Jr., and W. D. Emmons, *Org. Synth.*, **45**, 44 (1965).
1212. N. L. J. M. Broekhof, F. L. Jonkers, and A. van der Gen, *Tetrahedron Lett.*, 2433 (1979).
1213. S. S. Martin and R. Gompper, *J. Org. Chem.*, **39**, 2814 (1974).
1214. S. S. Martin and G. W. Phillips, *J. Org. Chem.*, **43**, 3792 (1978).
1215. U. Schöllkopf, R. Schröder and D. Stafforst, *Justus Liebigs Ann. Chem.*, 44 (1974).
1216. P. Magnus and G. Roy, *Organometallics*, **1**, 553 (1982).
1217. F. A. Carey and D. Hernandez, *J. Org. Chem.*, **38**, 2670 (1973).
1218. G. H. Posner and D. J. Brunelle, *J. Org. Chem.*, **37**, 3547 (1972).
1219. H. J. Reich, W. W. Willis, Jr., and P. D. Clark, *J. Org. Chem.*, **46**, 2775 (1981).
1220. B. T. Gröbel and D. Seebach, *Chem. Ber.*, **110**, 852 (1977).
1221. P. Coutrot, C. Laurenco, J. Petrova, and P. Savignac, *Synthesis*, 107 (1976).
1222. F. A. Carey and A. S. Court, *J. Org. Chem.*, **37**, 1926 (1972).
1223. A. M. van Leusen and J. Wildeman, *Recl. Trav. Chim. Pays-Bas*, **101**, 202 (1982).
1224. P. Savignac, J. Petrova, M. Dreux, and P. Coutrot, *Synthesis*, 535 (1975).
1225. D. Seyferth, J. L. Lefferts, and R. L. Lambert, Jr., *J. Organomet. Chem.*, **142**, 39 (1977).
1226. P. W. K. Lau and T. H. Chan, *Tetrahedron Lett.*, 2383 (1978).
1227. D. J. Ager, *J. Chem. Soc., Perkin Trans. 1*, 1131 (1983).
1228. K. Sachdev and H. S. Sachdev, *Tetrahedron Lett.*, 4223 (1976).
1229. G. W. Gokel, H. M. Gerdes, D. E. Mills, J. M. Hufnal, and G. A. Zerby, *Tetrahedron Lett.*, 3375 (1979).
1230. D. Seebach and A. K. Beck, *Org. Synth.*, **51**, 39 (1971).
1231. S. Raucher and G. A. Koolpe, *J. Org. Chem.*, **43**, 3794 (1978).
1232. S. Ncube, A. Pelter, and K. Smith, *Tetrahedron Lett.*, 1893 (1979).
1233. T. Hata, A. Hashizume, M. Nakajima, and M. Sekine, *Tetrahedron Lett.*, 363 (1978).
1234. K. Deuchert, U. Hertenstein, and S. Hünig, *Synthesis*, 777 (1973).
1235. G. Stork and L. Maldonado, *J. Am. Chem. Soc.*, **93**, 5286 (1971).
1236. M. Sevrin, J. N. Denis, and A. Krief, *Angew. Chem., Int. Ed. Engl.*, **17**, 526 (1978).
1237. C. Burford, F. Cooke, E. Ehlinger, and P. Magnus, *J. Am. Chem. Soc.*, **99**, 4536 (1977).
1238. E. J. Corey and M. A. Tius, *Tetrahedron Lett.*, **21**, 3535 (1980).
1239. L. Duhamel and F. Tombret, *J. Org. Chem.*, **46**, 3741 (1981).
1240. P. Magnus and G. Roy, *Organometallics*, **1**, 553 (1982).
1241. D. L. Comins and J. D. Brown, *Tetrahedron Lett.*, **24**, 5465 (1983).
1242. B. M. Trost and Y. Tamaru, *J. Am. Chem. Soc.*, **99**, 3101 (1977).
1243. H. Neumann and D. Seebach, *Chem. Ber.*, **111**, 2785 (1978).
1244. M. Mikolajczyk, P. Balczewski, and G. Grzejszcak, *Synthesis*, 127 (1980).
1245. F. M. Hauser, R. P. Rhee, S. Prasanna, S. M. Weinreb, and J. H. Dodd, *Synthesis*, 72 (1980).
1246. J. J. Fitt and H. W. Gschwend, *J. Org. Chem.*, **41**, 4029 (1976).
1247. J. Barluenga, J. Florez, and M. Yus, *J. Chem. Soc., Perkin Trans.1*, 3019 (1983).

The Chemistry of the Metal—Carbon Bond, Vol. 4
Edited by F. R. Hartley

CHAPTER **2**

Preparation and use of Grignard and Group II organometallics in organic synthesis

C. L. RASTON

Department of Physical and Inorganic Chemistry, University of Western Australia, Nedlands, Western Australia, 6009

G. SALEM

Research School of Chemistry, Australian National University, Canberra, A.C.T. 2600, Australia

---

I. INTRODUCTION

The application of Group II organometallic reagents in organic synthesis is vast and varied, particularly that of Grignard reagents. There have been several reviews[1–5] on it, the most recent being three chapters in *Comprehensive Organometallic Chemistry* dealing with organoberyllium chemistry[6], organomagnesium through to organobarium chemistry[7], and compounds of the alkaline earth metals in organic synthesis[8], all covering the literature up to 1980. In this chapter we further discuss the synthesis and utility in organic chemistry of Group II organometallic species. Our criterion for inclusion is that an M—C interaction is present, either in the species or implied as an intermediate in a reaction of a Group II complex with an organic substrate. Group II metal amides, alkoxides, and aryloxides also feature in organic synthesis but they are not included in this chapter.

We focus on the literature covering the period 1980–83 inclusive, with emphasis on new developments in techniques and synthesis. Over 1600 papers dealing with the application of Group II chemistry in organic synthesis have appeared during this period. For the literature prior to 1980 only examples of well established reactions will be cited so that the chapter will be comprehensive for all reaction types of organo-Group II species thus far reported.

Group II organometallic reagents play a pivotal role in both organic and organometallic chemistry. A discussion of the latter is, however, beyond the scope of this review. Commercially available Group II reagents are restricted to those of magnesium, and include Grignard reagents of allyl, vinyl, ethyl, *n*- and *i*-Pr, *n*-, *s*-, *i*- and *t*-Bu, C_5H_5, Ph, $PhCH_2$, and mesityl chlorides and/or bromides as solutions in either Et_2O or thf.

The next section of this review deals with the synthesis, stability, and mechanism of formation of organomagnesium reagents. This is followed by a survey of their reactions, classified according to new bonds formed, the type of bond fission process, or the reaction of a particular functional group. A separate section deals with reactions of organomagnesium reagents in the presence of metal complexes, including catalysed reactions. The final section is devoted to the synthesis and utility of organo-beryllium, -calcium, -strontium, and -barium reagents; this constitutes only a small part of the review, a consequence of the slight attention that the elements Be, Ca, Sr, and Ba have received, which is related to the toxicity of Be and Ba, the reduced stability of their complexes compared with those of magnesium, and the fact that there is usually little or no advantage over using readily accessible and stable magnesium and/or lithium reagents. In short, organomagnesium reagents have the greatest application, ranging from bond formation in simple compounds, including the synthesis of ^{13}C and ^{2}H labelled compounds, to natural products where the key step is a regio- and stereo-selective reaction involving such a reagent.

Where possible, a comparison of reactions of organo-Group II reagents with alternative organometallic reagents will be given to assess their relative merits for a particular molecular transformation.

Beryllium almost exclusively forms covalent compounds, whereas magnesium compounds can range from covalent to ionic, and compounds of the heavier congeners of the Group II elements are ionic, for example $(CaCl)(Ph_3C)$ is fully dissociated in thf[7]. The nature of Mg—C single bonds is very important in considering the application of organomagnesium reagents in organic synthesis. They are thermodynamically stable, a fact which is demonstrated by the ability to sublime some organomagnesium complexes at temperatures in excess of 150 °C. Any instability of organomagnesium compounds is associated with either rearrangement and/or a facile elimination reaction, usually due to the presence of a functional group. For example, $R_2NCH_2CH_2MgX$ undergoes 1,2- (or β-) elimination, yielding $MgX(NR_2)$ and ethylene above *ca.* −90 °C[9]. Unlike transition metal—alkyl complexes, the β-hydrogen elimination decomposition pathway is a high-energy process, usually requiring temperatures above 100 °C. In consequence, magnesium—alkyl complexes with optical activity at the β-position are accessible (equation 1)[10].

$$\underset{(S)}{Et—\overset{*}{C}H(Me)—CH_2Cl} \longrightarrow Et—\overset{*}{C}H(Me)—CH_2MgCl \longrightarrow \underset{(S)}{Et—CH(Me)—CH_2\text{-(2-thienyl)}} \quad (1)$$

Carbon atoms attached to magnesium, however, are stereochemically unstable, rapidly inverting (primary > secondary), the mechanism of which has not been established[7]. Consider the general cross-coupling reaction shown in equation 2, catalysed by optically active nickel and palladium complexes and yielding a product rich in one enantiomer.

$$R^1R^2R^3C—MgX \underset{\text{racemization}}{\overset{\text{fast}}{\rightleftharpoons}} XMg—CR^1R^2R^3 \xrightarrow{R^4X} R^1R^2R^3\overset{*}{C}—R^4 \quad (2)$$

Inversion at the chiral centre in the magnesium complex must be faster than the coupling reaction for the optical purity of the product to be kept constant throughout the reaction. For magnesium attached to a carbon of an alicyclic ring the rate of inversion of configuration is, however, relatively slow. Cyclopropyl Grignard reagents, for example, are configurationally stable[11].

II. ORGANOMAGNESIUM REAGENTS

A. Grignard Reagents

These are reagents of the type RMX[†] and they are usually prepared by the direct oxidative addition of an organic halide to elemental magnesium (method A, Table 1). Victor Grignard reported this reaction in 1900[12], although the reactivity of magnesium with organic halides, in the presence of another organic substrate, was reported one year earlier[13]. The latter is the Grignard *in situ* trapping or Barbier reaction. Its application to synthesis has recently been reviewed[14], and is further discussed in Section II.A.4.

1. Synthesis

Grignard reagents are extremely sensitive to air and moisture and the use of an inert atmosphere (nitrogen or argon) for their preparation and manipulation is essential. A typical Grignard synthesis is the slow addition of a solution of an organic halide in an appropriate solvent to a suspension of activated magnesium (powder or turnings) in the same solvent, at such a rate as to control the temperature of the reaction. They are exothermic reactions and cooling may be necessary, particularly if there is an induction period followed by a vigorous reaction. After the addition of the halide is complete, stirring

[†] Grignard reagents prepared in coordinating solvents have the general formula RMgX(solvent)$_n$, and similarly for R_2M species, with solvent molecules as part of the coordination environment of magnesium. For simplicity, reference to RMgX and related species implies that solvent molecules are present unless stated otherwise.

is usually continued for about 1 h and/or the mixture is subjected to prolonged heating, particularly if the organic halide is difficult to react.

Factors controlling Grignard formation are extensively surveyed in ref. 7. The nature of the magnesium used, magnesium turnings or powder, condensed magnesium, or finely divided Rieke's[15] magnesium, can be important. The last two are the most reactive forms of magnesium and are particularly effective for slow-reacting or otherwise unreactive organic halides. There are several methods for activating magnesium turnings and powder; the most common is to treat it with $Br(CH_2)_2Br$ in a suitable solvent, the evidence for activation being the evolution of ethylene. The co-product is solvated $MgBr_2$, and it is noteworthy that this is a synthetically useful method for the synthesis of stock solutions of moisture-free $MgBr_2$. Moreover, the associated 1,2- or β-elimination reaction has featured in synthesis (equation 3)[16].

$$BrCH_2CHBr(CH_2)_4Br \xrightarrow[-MgBr_2]{Mg} CH_2{=}CH(CH_2)_4MgBr \qquad (3)$$

Other metals as impurities affect the yield of the Grignard synthesis. The di-Grignard reagents of o-$C_6H_4(CH_2Cl)_2$[17] and $[(o\text{-}C_6H_4CH_2Cl)_2]$[18] are accessible under fairly critical conditions and surprisingly using only one brand of magnesium powder. It may be that an impurity in the magnesium is responsible for the success of their syntheses.

The Grignard reagent of 2-bromothiophene is unique in that its formation is activated photolytically[19].

The choice of solvent and concentration can be decisive for a high-yield synthesis. Typical solvents are the ethers Et_2O and thf. The latter, which is the more polar of the two and is a more recent solvent in Grignard synthesis, is essential for some difficult reactions. A requirement of preparing the aforementioned di-Grignard reagents, which for a long time were thought to be inaccessible, and the reagent derived from pentamethylbenzyl chloride[20] is the use of thf as the solvent, together with high dilution, *ca.* 0.1 M. In all cases the yield was found to diminish by *ca.* 40% for a two-fold increase in the projected concentration of the Grignard reagent, which suggests the competing reaction is intermolecular coupling ($RX + RMgX \rightarrow RR + MgX_2$; Wurtz-type coupling) with elimination of $MgCl_2$.

Other solvents have been used to good effect in Grignard syntheses viz. diglyme, acetals, formals, and tertiary amines. The use of coordinating solvents other than Et_2O or thf may be necessary to solubilize a reagent. Mixtures of coordinating and non-coordinating solvents have also been found useful; in some cases, usually with alkyl chlorides, little or no polar solvent need be present, although for such a case the active species is predominantly R_2Mg (unsolvated) rather than RMgX.

Dioxane added to a solution of a Grignard reagent in Et_2O or thf yields a sparingly soluble dioxane complex of MgX_2, a reaction that is standard methodology for the redistribution of RMgX, e.g. equation 4[21]. The dimeric complex in equation 4 is sublimable *in vacuo*, but usually under such conditions the dioxane is removed to yield

$$4RMgCl + 5\,\text{O}\langle\text{dioxane}\rangle\text{O} \xrightarrow[R=\bar{C}H(SiMe_3)_2]{} R_2Mg{-}\text{O}\langle\text{dioxane}\rangle\text{O}{-}MgR_2 + 2\,MgCl_2(\text{dioxane})_2\downarrow \qquad (4)$$

dioxane-free R_2Mg species, although using a 1:1 mixture of Grignard to dioxane also yields dioxane-free MgR_2, but with other coordinating solvents usually present it would still be solvated. Diglyme, pyridine, and tmeda are also effective in generating R_2Mg from Grignard reagents. In effect, these strongly coordinating solvents alter the position of the Schlenk equilibrium (equation 5).

$$2RMgX \rightleftharpoons MgX_2 + R_2Mg \qquad (5)$$

TABLE 1. Common methods of preparing organo derivatives of magnesium

Reaction type	General equation	Comments
A Grignard	$RX + Mg \rightarrow RMgX$	R = alkyl, aryl, alkenyl. Critical conditions are required for R = arylmethyl and 2-alkenyl
B Grignard redistribution	$2RMgX \rightarrow R_2Mg + MgX_2$	Can be effective by change in temperature of the Grignard solution or change in solvent, especially adding dioxane
C 1,2-Grignard addition to alkenes and alkynes	$RMgX + R^1CH{=}CHR^2 \rightarrow RR'CHCHR^2(MgX)$ and/or $R^1CH(MgX)CHRR^2$ $RMgX + R^1C{\equiv}CR^2 \rightarrow$ $R(R^1)C{=}C(R^2)MgX$ *Z* and/or *E* isomer	Both reactions usually require temperatures > 100 °C, an activated multiple bond, or a transition metal catalyst
D Magnesium-ene reaction	RCH=CH$_2$ + XMg–C^{1}H(R^3)–C^{2}H=C^3(R^1)R^2 $\rightarrow$ RCH(MgX)CH$_2$C^3(R^1)(R^2)–C^{2}H=C^{1}HR^3	Temperature > 60 °C. Both intra- and intermolecular additions are known

E	Hydromagnesiation	$HMgX + R^1C{\equiv}CR^2 \longrightarrow$ (Z)-$R^1(XMg)C{=}C(H)R^2$ and/or E isomer	Usually transition metal catalysed. Similar reactions for terminal alkenes occur
F	Salt elimination	$RM + MgX_2 \xrightarrow{-MX} RMgX$ or R_2Mg	Suitable for difficult Grignard syntheses, e.g. R = arylmethyl, and for readily available organolithium species
G	Metal–halogen exchange	$RMgX + R^1X \rightarrow R^1MgX + RX$	For activated R^1, e.g. halogenated aryls, fluorinated alkyls
H	Metallation	$RH + R^1MgX \rightarrow RMgX + R^1H$	Usual route to $RC{\equiv}CMgX$ and magnesium cyclopentadienyl complexes, etc.
I	Oxidative–reductive transmetallation	$RMX + Mg \rightarrow RMgX + M$ or $R_2M + Mg \rightarrow R_2Mg + M$	For the synthesis of otherwise difficult di-Grignards and MgX_2-free species
J	Magnesium electron transfer reactions	nAr + Mg → $(Ar)_nMg$ ($n = 1$ or 2)	Restricted to aromatic molecules > C_{10}, e.g. naphthalene, and conjugated polyenes, e.g. butadiene.
K	Magnesium (anthracene) reaction	$RX + AMg \rightarrow RMgX + A$	For the synthesis of benzylic-type Grignard reagents

It is possible to prepare compounds of the same composition as that prepared from a Grignard reaction by mixing an equimolar mixture of R_2Mg and MX_2 (excluding MgF_2) in Et_2O and/or thf. However, some solutions of Grignard reagents in thf have the position of the Schlenk equilibrium predominantly to the right and magnesium halide-free species have been isolated without the addition of dioxane. For example, cooling a 0.1 M solution of the di-Grignard of *o*-$C_6H_4(CH_2Cl)_2$ in thf yields a magnesium macrometallacycle (equation 6) which has been structurally authenticated[17]. Other magnesium halide-free species, also derived from di-Grignard reagents, are discussed in Section II.C.

MgCl(thf)$_n$ MgCl(thf)$_n$ —(−30 °C)→ (thf)$_2$Mg ... (thf)$_2$Mg ... Mg(thf)$_2$ (6)

The corresponding *para*-isomeric di-Grignard reagent, prepared under identical conditions, is unstable with respect to the formation of a thf-insoluble species, of composition $Mg_{1.5}Cl(thf)_2(C_8H_8)$[22]. Its low solubility and the loss of half a $MgCl_2$ unit relative to 'RMgX' suggest that it is oligomeric with the *p*-xylenediyl entity bridging successive magnesium centres. In contrast, a magnesium halide-rich species, $[Mg_2Cl_3Et(thf)_3]$, has been crystallized from a thf solution of EtMgCl[23].

The choice of halide, RX, is another consideration in optimizing the yield and selectivity of Grignard formation. Where there are two different halogen residues in the same molecule, the Grignard formation proceeds according to the inequalities I > Br > Cl > F. This is demonstrated by the selective mono-Grignard synthesis derived from oxidative addition of Mg to C-8 in 1-bromo-8-iodonaphthalene[24]. Bromides and chlorides feature the most, since iodides other than MeI and ArI tend to give side-reaction products, alkyl fluorides are difficult to react, and aryl fluorides require highly activated magnesium prepared by Rieke's method[15]. This is finely divided magnesium prepared *in situ* by treating a solution of MgX_2 in thf with excess of potassium metal. Other methods of generating it include the reduction of MgX_2 with potassium–graphite or $Na[C_{10}H_8]$[25]. (More recently highly activated magnesium has been prepared by equilibration with its anthracene adduct in thf[26].) Grignards of aryl fluorides have been prepared in almost quantitative yield by condensing magnesium and PhF vapours at −196 °C[27]. It is noteworthy that this method is a general procedure for preparing unsolvated Grignard reagents.

The use of highly activated magnesium, either condensed or Rieke's magnesium, has been one of the recent major developments in Grignard chemistry. Their use has been effective in the synthesis of a variety of previously inaccessible Grignard reagents. The use of Rieke's magnesium has featured the most, presumably because of its simple preparative procedure compared with the generation of condensed magnesium. Both forms of highly reactive magnesium allow the preparation of Grignard reagents at low temperature which would otherwise decompose at ambient temperature. Steinborn[9] synthesized a variety of

Grignards of the type $R_2NCH_2CH_2MgBr$ at $-100\,^{\circ}C$. They are unstable above *ca.* $-90\,^{\circ}C$, decomposing to ethylene and $MX(NR_2)$[9].

Rieke's magnesium reacts with thf at *ca.* 65 °C[28] (equation 7), but is less reactive towards

$$\text{thf} + Mg^* \longrightarrow \text{(ring-opened)} Mg(thf)_n \tag{7}$$

other ethers. This is usually not a problem, however, as the metal is sufficiently active for Grignard formation at very low temperatures.

Another major development has been the synthesis of $ArCH_2MgCl$ reagents, in particular di-Grignard reagents of this type, e.g. *o*-$C_6H_4(CH_2Cl)_2$[17] and [(*o*-$C_6H_4CH_2Cl)_2$][18]. From earlier work on these systems, it was concluded that di-Grignard formation was unlikely owing to the facile elimination of MgX_2, forming *o*-quinodimethide and 9, 10-dihydrophenanthrene (equation 8), respectively.

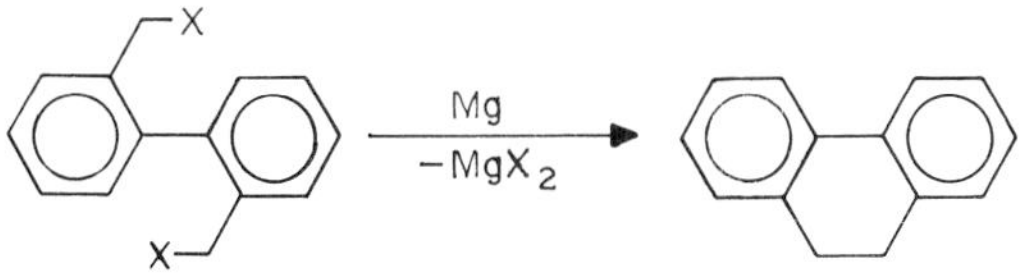

$$\xrightarrow[-MgX_2]{Mg} \tag{8}$$

The differences in requirements for a viable synthesis of these reagents with simple (alkyl)MgX reagents are (i) it is necessary to use organic chlorides rather than bromides; (ii) the use of thf as the solvent is essential for high yields (typically 90–96%); and (iii) high dilution, *ca.* 0.1 M. The last requirement precludes the possibility that the coupling decomposition pathway involves an intramolecular elimination of $MgCl_2$. However, in the case of dibromide, [(*o*-$C_6H_4CH_2Br)_2$], it is intramolecular coupling (equation 8)[18]. It may be that the differences between $ArCH_2X$ and alkyl-X systems is related to the enhanced stability of benzylic-type radicals such as *o*-$\dot{C}H_2C_6H_4(CH_2MgCl)$ or *o*-$\dot{C}H_2C_6H_4CH_2Cl$, allowing a greater probability of an intermolecular encounter with $ArCH_2Cl$ to yield a coupled species. The formation of Grignard reagents is now widely held to implicate free-radical intermediates (Section II.A.2).] The use of chlorides rather than bromides is well known, even for simple Grignard reagents, $C_6H_5CH_2MgX$, to reduce the extent of the coupling reaction which is favoured both kinetically (Br^- or $Br^\bullet$ are better leaving groups than Cl^- or $Cl^\bullet$) and thermodynamically (based on C—Cl, C—Br, Mg—Cl, and Mg—Br bond strengths)[17].

The use of the same conditions in an attempt to prepare a di-Grignard reagent of 1, 8-bis(halomethyl)naphthalene, however, failed; the product was that derived from intermolecular coupling for the dichloride and intramolecular for the dibromide[29a], a difference that is also consistent with both kinetic effects (the leaving group capabilities of $X^\bullet$ or X^-) and variation in bond energies. In contrast, the dichloride can be converted into the di-Grignard reagent in high yield by reaction with [Mg(anthracene)$(thf)_3$][29b].

The use of [Mg(anthracene)$(thf)_3$] in the synthesis of Grignard reagents is a very recent development. It is an orange 1:1 adduct of magnesium with anthracene (see Section II.B.6 for the proposed structure), which acts as a source of magnesium in the synthesis of benzylic-type Grignard reagents[29b], and is readily prepared by the reaction of activated magnesium (using $BrCH_2CH_2Br$ or CH_3CH_2Br)[28,29b] with a two-fold excess of anthracene in thf at room temperature for 48 h. The Grignard reagents are prepared by the slow addition of a solution of a benzylic halide in thf to a suspension of a stoicheiometric amount of magnesium anthracene in thf at *ca.* 20 °C. A deep green colour persists

($g_{av.}$ 2.0024, no hyperfine coupling) until the addition of the halide is complete. A range of benzylic-type di-Grignard reagents have been synthesized in this way in up to 96% yield (see Section II.C and Table 2). Also, the previously inaccessible Grignard reagent of 9-chloromethylanthracene has been prepared in 92% yield. In this case the resulting reaction mixture must be stirred for *ca.* 36 h after addition of the chloride. The ability of magnesium anthracene in attenuating coupling reactions that are common for benzylic halides may arise from it being a 'soluble' form of magnesium, since this would favour intermolecular encounter of magnesium anthracene with RX rather than preformed RMgX reacting with RX.

A well established method for the synthesis of Grignard reagents of relatively unreactive organic halides is the addition of a reactive halide, ideally $(CH_2)_2Br_2$, since it forms easily removed ethylene and $MgBr_2$ on reaction. Its role is presumably to produce an active magnesium surface and/or promote the radical reaction (Section II.A.2) between the inert halogen and magnesium[7]. It is usually called the entrainment method.

Competitiveness for the competing reaction to Grignard formation, Wurtz-type coupling, for which an intermediate Grignard species is implied, is more favoured for RX, X = Br or I and for R = allyl[30] or benzyl. Such coupling, either intra- (e.g. equation 9) or inter-molecular (e.g. equation 10)[31], has featured in organic synthesis.

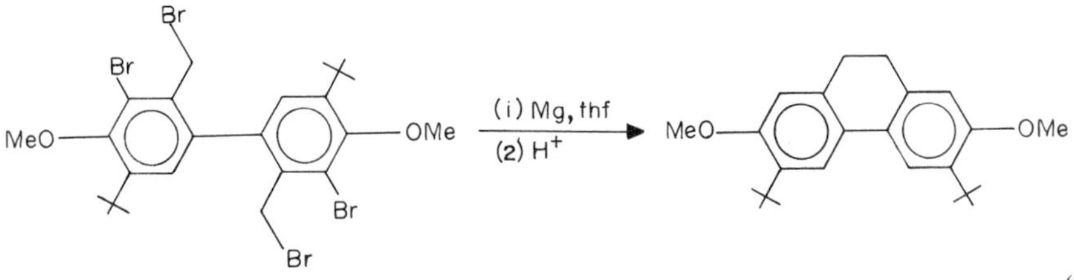

(i) Mg, thf; (2) H^+ (9)

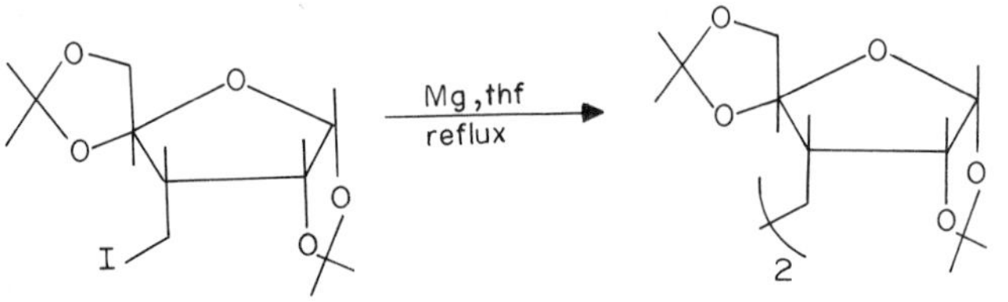

Mg, thf; reflux (10)

The synthesis of benzocyclobutane by the method shown in equation 11 has intramolecular coupling associated with the elimination of $MgBr_2$ as a key step[32].

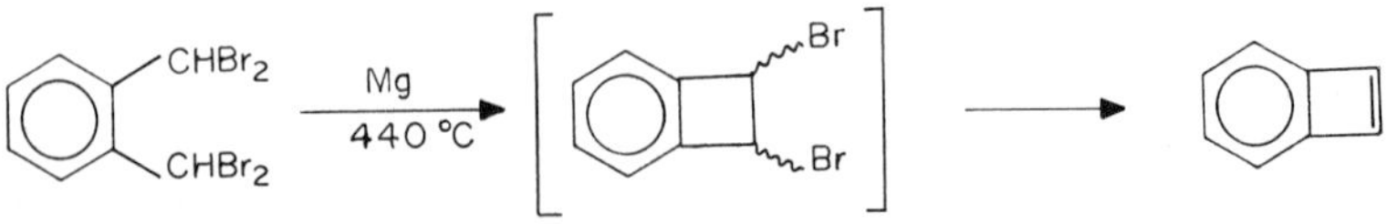

Mg; 440 °C (11)

1-Alkenyl Grignard reagents, $R^1R^2C{=}CR^3(MgX)$, are readily prepared from vinylic halides in thf using standard procedures, but unfortunately there is invariably some *cis*- or *trans*-isomerization during their formation. This is presumably a consequence of a radical reaction pathway (Section I.A.3).

There appears to be little difficulty in the synthesis of Grignard reagents of sterically hindered aryl or alkyl halides, which is consistent with the radical mechanism of Grignard formation. One example, a mesityl halide Grignard, is commercially available. Other

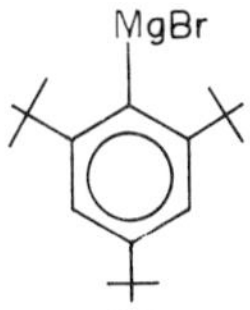

(1) 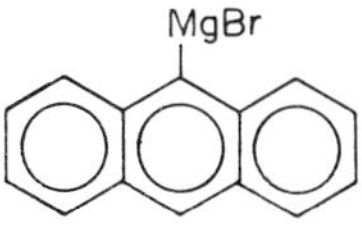 (2) 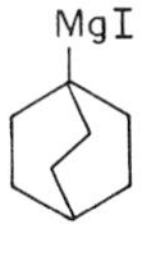 (3) 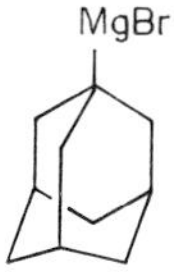 (4)

examples are **1**–**4**, which are described in refs. 33–36, respectively.

2. Mechanism

Mechanisms involving paramagnetic species $R^{\bullet}$, $RX^{-\bullet}$ and $Mg^{+\bullet}$, have been proposed for the Grignard reaction[7]. Whitesides and coworkers have demonstrated that alkyl halides react at the surface of the magnesium in ether solvents at transport limited rates[37]. More recent mechanistic studies are on reactions of RX with condensed magnesium at low temperatures. For aryl halides, the mechanism in Scheme 1 was proposed[38], the same as

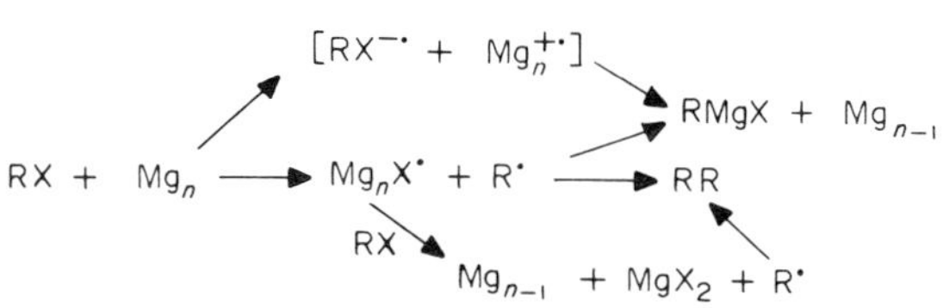

$$RX + Mg_n \longrightarrow [RX^{-\bullet} + Mg_n^{+\bullet}] \longrightarrow RMgX + Mg_{n-1}$$
$$RX + Mg_n \longrightarrow Mg_nX^{\bullet} + R^{\bullet} \longrightarrow RMgX + Mg_{n-1}; \quad \longrightarrow RR$$
$$Mg_nX^{\bullet} \xrightarrow{RX} Mg_{n-1} + MgX_2 + R^{\bullet}; \quad R^{\bullet} \longrightarrow RR$$

SCHEME 1

that initially suggested for the reaction of alkyl halides. Others have suggested that clusters of magnesium atoms with Mg–Mg interactions[39] may be involved. A theoretical paper suggests that strong Mg—Mg bonding would stabilize RMg_2X species and that larger clusters such as RMg_4X may be intermediates in these reactions[40].

3. Stability

If β-hydrogen atoms are present in Grignard reagents (or R_2Mg), then 1,2- or β-hydrogen elimination can occur, but at temperatures in excess of 100 °C. Grignard reagents are usually unreactive towards ethereal solvents at temperatures below 100 °C, unlike organolithium reagents, which readily cleave C—O bonds even at ambient temperature. They can be unstable with respect to redistribution to a mixture of $R_2Mg(solvent)_n$ and solvated MgX_2 or some intermediate species, although the overall composition of the solution would be 'RMgX'. There are several factors that determine the nature of the species present, including the solvent, concentration, and temperature. Attempts to prepare a di-Grignard reagent of **5** yielded a magnesium bromide-free species (equation 12), possibly owing to complexation of the functional group. Interestingly on warming a solution of the derived reagent to 50 °C, the product decomposed to 1-oxa-2-magnesiocyclopentane[41].

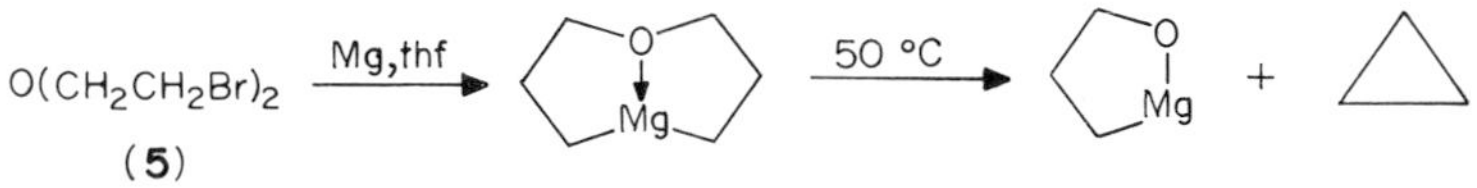

$$\underset{(\mathbf{5})}{O(CH_2CH_2Br)_2} \xrightarrow{Mg,\,thf} [\text{O}\rightarrow\text{Mg bicyclic}] \xrightarrow{50\ ^\circ C} [\text{1-oxa-2-magnesiocyclopentane}] + \triangle \qquad (12)$$

Facile decomposition pathways may prevail for Grignard reagents, RMgX, if functional groups are present on R. In certain cases, however, they can be suppressed by preparing and reacting the Grignard reagent at low temperature. Examples of α-functionalized Grignard reagents are those derived from chloromethyl ethers, which are prepared at $-30\,^{\circ}$C. At temperatures above $-15\,^{\circ}$C they rapidly decompose, yielding MgCl(OR) and ethylene. In contrast, the corresponding thioethers are more stable, being prepared in thf at 10–20 °C. There has been no evidence for the formation of carbenes in these reactions[42].

β-Elimination decomposition of Grignards (e.g. equation 13)[43] is more facile unless the generated olefin is an allene[44,45] or if the olefinic bond is at a bridgehead[46].

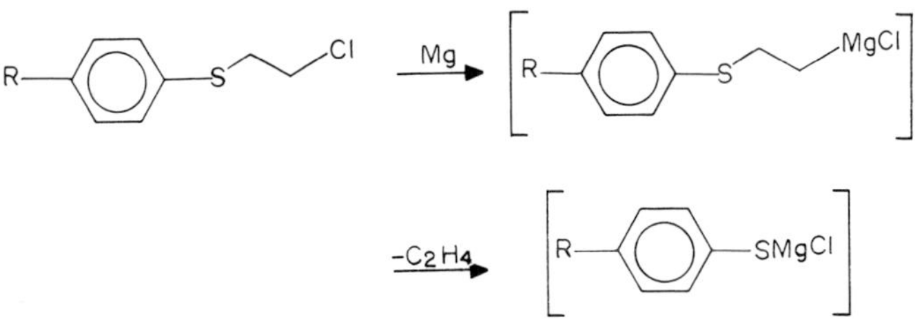

(13)

1, 2-Dihaloarenes react with magnesium to form Grignard reagents which only β-eliminate under forcing conditions and/or if the halogens are bromine or iodine. The decomposition species are highly reactive benzynes which have application in synthesis

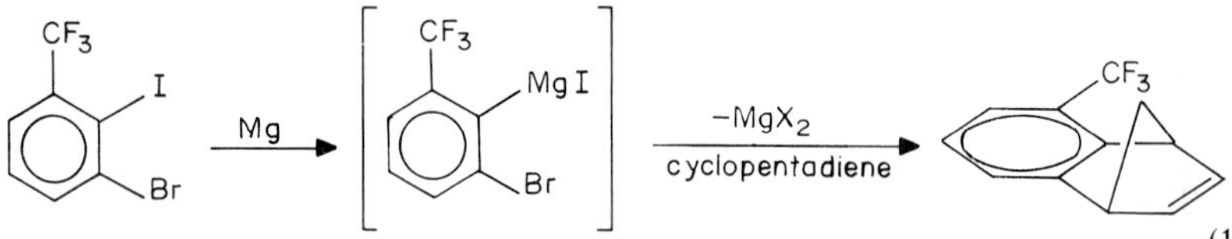

(14)

(equation 14)[47,48], although the use of elemental lithium (or RLi) rather than magnesium is more common as the elimination reaction is more facile.

On treating pentachlorobenzene with magnesium in thf at 10–15 °C, the Grignard **6**[49] is generated in modest yield, and similarly for pentachloropyridine (**7**)[50]. In both the site of Grignard formation is at a position with two halogen centres *ortho* to it.

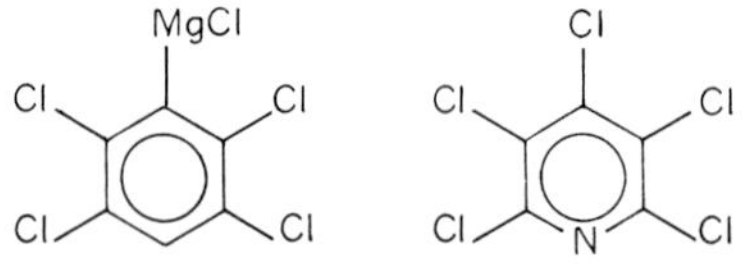

(**6**) (**7**)

Fluorinated aryl Grignard reagents have been known for a long time, but they are less stable than analogous chloro or bromo derivatives.

Perfluoroalkyl halides, $C_nF_{2n+1}X$, yield Grignard reagents that decompose via the β- rather than the α-elimination route at temperatures close to 0 °C. The thermal decomposition of Grignard reagents of the type $R^1CF_2CF_2MgBr$ (**8**) in the presence of R^2MgX ($R^1 = C_6F_{13}, C_4F_9$; $R^2 =$ aryl) is a novel method of preparing fluorinated alkynes.

Compound **8** decomposes above $-45\,^{\circ}C$ by intramolecular exchange followed by β-elimination (equation 15), then metathetical exchange with R^2MgX, β-elimination, and finally arylation (equation 16)[51].

$$R^1CF_2CF_2MgBr \ (\mathbf{8}) \longrightarrow [R^1CF_2CFBrMgF] \xrightarrow{-MgF_2} R^1(F)C{=}C(F)Br \ (\mathbf{9}) \qquad (15)$$

$$(\mathbf{9}) \xrightarrow{R^2MgBr} R^1(F)C{=}C(F)MgBr \xrightarrow{-MgBrF} R^1C{\equiv}CF \xrightarrow{R^2MgBr} R^1C{\equiv}CR^2 \qquad (16)$$

Compound (**10**) has remarkable thermal stability, being unchanged after several hours at 35 °C in Et_2O[52]. Its stability reflects that β-elimination involving a double bond at a bridgehead is unfavourable.

(10)

Greater stability for RMgX, where R is cyclopropyl with a β-leaving group relative to open-chain analogues, is conceivable since the β-elimination product is a strained cyclopropene. In this context, (*Z*)-and (*E*)-dibromocyclopropane both yield the (*Z*)-di-Grignard reagent (Table 2), which compares with ethylene and $MgBr_2$ formation for 1, 2-dibromoethane[53].

1, 3-dihaloalkyl compounds are susceptible to β-elimination of MgX_2 under the usual conditions of Grignard formation. The product of γ-elimination is cyclopropanes, which are usually formed in good yield, and in consequence this has been exploited as a general route to carbocyclic (e.g. equation 17)[54] and heterocyclic three-membered rings.

$$Cl(CH_2)_3C(Me)(CH_2Cl)_2 \xrightarrow[-MgCl_2]{Mg} ClMg(CH_2)_3C(Me)(cyclopropane) \qquad (17)$$

The di-Grignard reagent of 1, 3-dibromopropane is, however, accessible by the careful addition of an Et_2O solution of the dibromide to magnesium in Et_2O in 30% yield (Table 2)[55]. For 1, *n*-dihaloalkanes, $X(CH_2)_nX$ ($n \geqslant 4$), intramolecular elimination is not prevalent and their di-Grignard reagents are readily prepared in yields $> 60\%$[56]. Di-Grignards of benzylic-type halides are also accessible, albeit with difficulty[17,18], for which the competing reaction is intra- and inter-molecular coupling with loss of MgX_2. 3-Haloethers are susceptible to elimination, yielding cyclopropanes, unless the Grignard reaction is carried out at low temperature ($-78\,^{\circ}C$) and Rieke's magnesium is used[15].

Grignard reagents of *o*- and *p*-$MeOC_6H_4CH_2X$ are unstable with respect to the formation of quinodimethide (e.g. equation 18 for the *o*-isomer)[57].

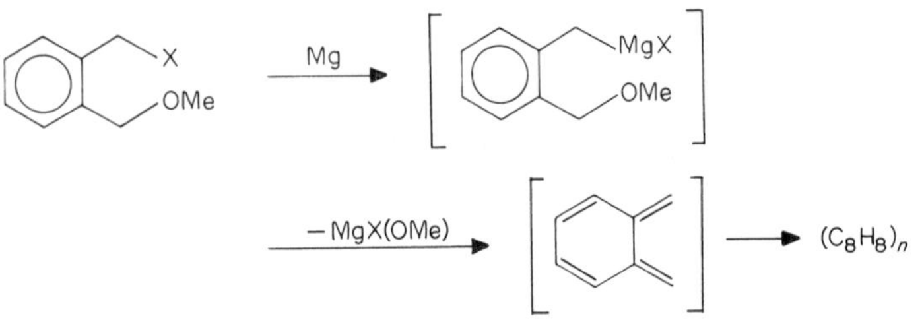

(18)

The Grignard reagent in equation 19 readily undergoes an O—C bond-cleavage reaction[58]. This type of reaction is common for a 1, 2-disposition of magnesium halide and an electropositive element and is a valuable route to various substituted alkynes.

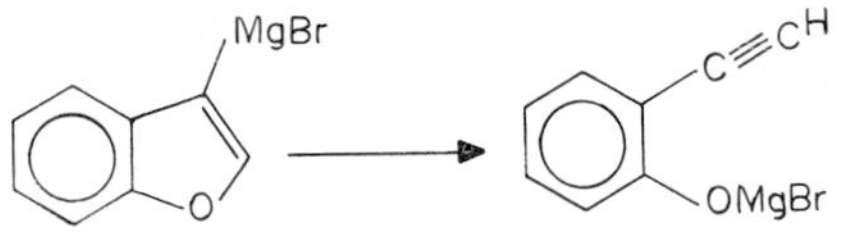

(19)

2-Bromomagnesiosulpholenes undergo S—C bond breakage to yield butadiene sulphonates (equation 20)[59].

(20)

Functional groups reactive towards Grignard reagents require either protection or neutralization prior to the Grignard reaction. For example, the way to prepare Grignards of $Cl(CH_2)_nOH$ ($n = 3, 4$, and 6) is to treat the alcohol first with RMgCl, which yields RH and a chloro-functionalized alkoxidomagnesium complex, then with Mg in thf, affording $ClMg(CH_2)_nOMgCl$ or $\overline{Mg(CH_2)_nO}$[60].

Grignards and organomagnesium reagents in general can be unstable with respect to rearrangements arising from C—C bond rupture (β-cleavage) or intramolecular addition of C—MgX to a carbon—carbon multiple bond (equation 21). This subject has been recently extensively reviewed[7,61] and only new developments are discussed here.

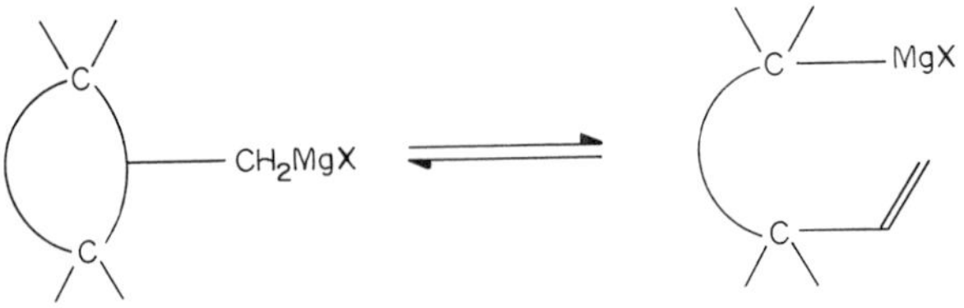

(21)

Magnesium/thf slurries, formed by the condensation of magnesium vapour, readily react at $-75\,^\circ C$ with cyclopropylmethyl bromide to yield the Grignard reagent, which is stable at this temperature[62], unlike the product prepared by the conventional method in Et_2O where ring cleavage, formally involving a 1, 2-vinyl shift, prevails (equation 22)[63].

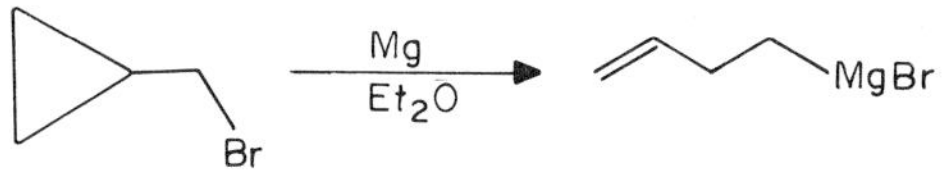

(22)

The condensed magnesium strategy is, however, less effective for a cyclobutenyl derivative (equation 23). The ratio expected to rearranged Grignard reagent is 1:1.2 at $-75\,^{\circ}C$, inferred from the analysis of the carboxylation products. Using Et_2O rather than thf as the solvent, and at a temperature of $-50\,^{\circ}C$, however, the rearrangement is suppressed, and (**11**) is the exclusive product[62]. Grignard reagents of cyclobutylmethyl halides ring cleave

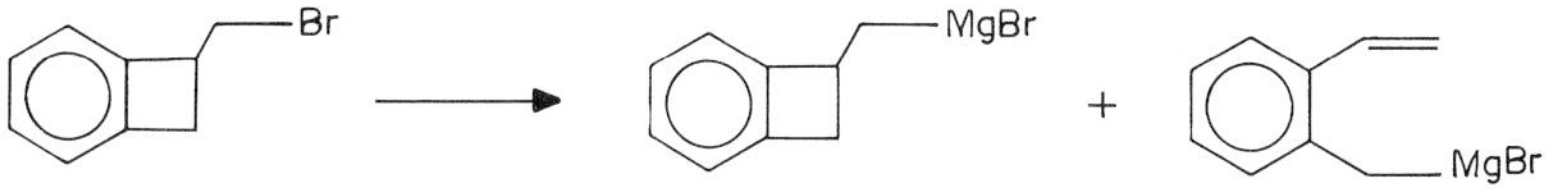

(23)

under more forcing conditions ($> 50\,^{\circ}C$)[64]. When ring strain is minimal, Grignards of cyclic alkyl halides are in equilibrium in solution with those of alkenyl halides and there may even be a minor active species present. For example, for the reaction in equation 24, 38–55% of the Grignard reagents present were the cyclic isomers under the conditions studied. It is noteworthy that hydrolysis of the reaction mixture yielded predominantly the product derived from protonation of the straight-chain Grignard[65].

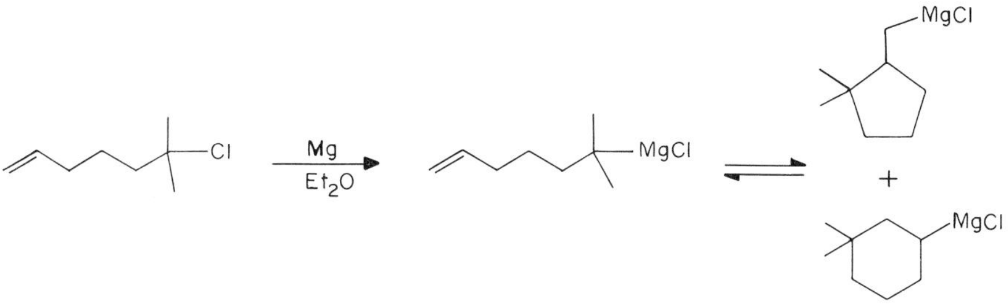

(24)

Intramolecular rearrangement of Grignard reagents has recently been applied to the synthesis of complex molecules where regio- and stereo-control is important. It is often called the intramolecular 'magnesium-ene' reaction, and its application in synthesis is the work of Oppolzer and colleagues. The formation of seven-, six-, and five-membered methylene-substituted carbocycles with high regio- and stereo-specificity has been possible (equation 25).

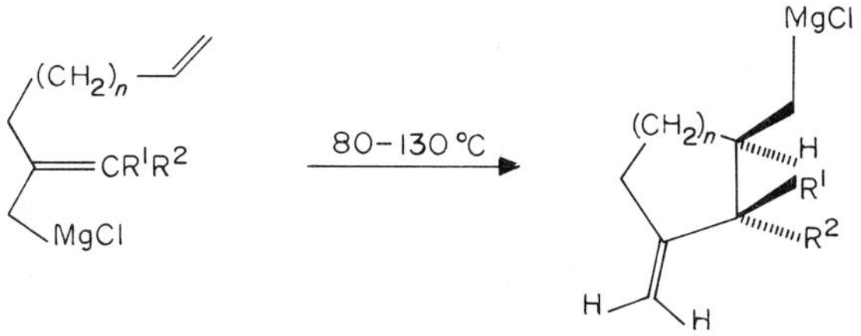

(25)

$R^1 = Me, n\text{-}C_6H_{13}$
$R^2 = H$
$n = 1, 2$ or 3

After preparing the allyl Grignard at $-65\,^{\circ}C$ in thf using precondensed magnesium, the subsequent 'magnesium-ene' reaction was found to be effective by gentle thermolysis[66]. It is the key step in the direct regio- and stereo-controlled total synthesis of ($\pm$)-khusimone (equation 26). Other applications to the synthesis of natural products or derivatives of natural products for which the key step is using this methodology are shown in equations 27[67], 28[68], 29[69], and 30[70].

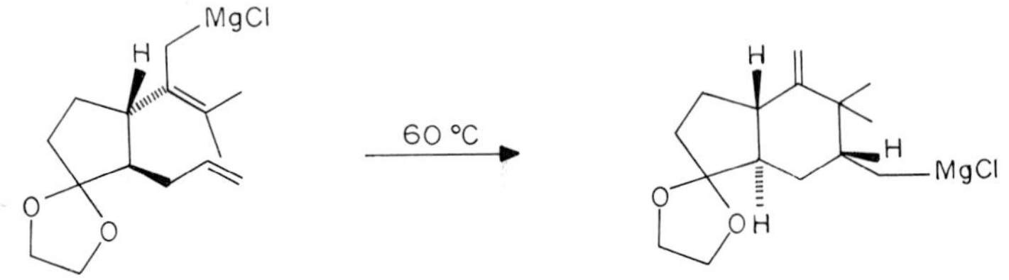

MgCl H 60 °C H H MgCl O O O OH (26)

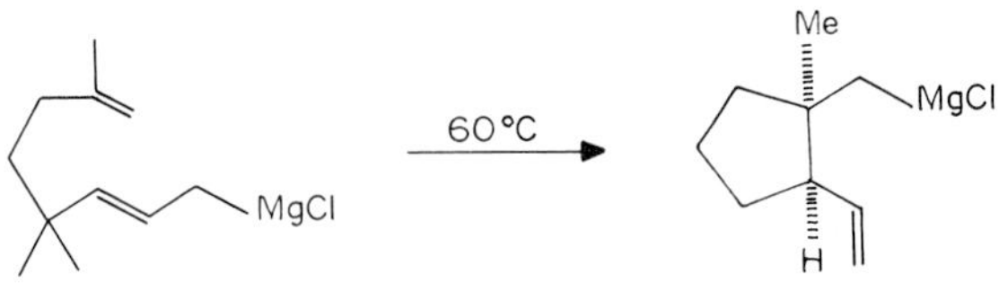

Me MgCl 60 °C MgCl H (27)

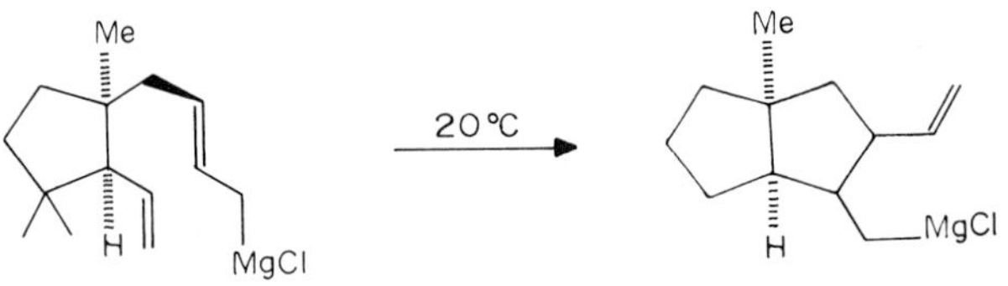

Me Me 20 °C H MgCl H MgCl (28)

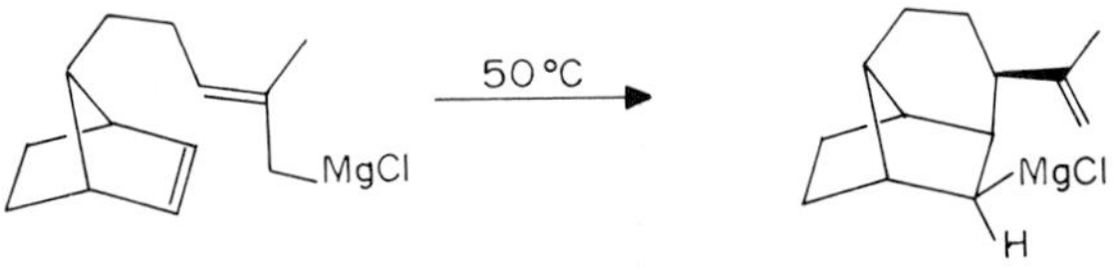

50 °C MgCl MgCl H (29)

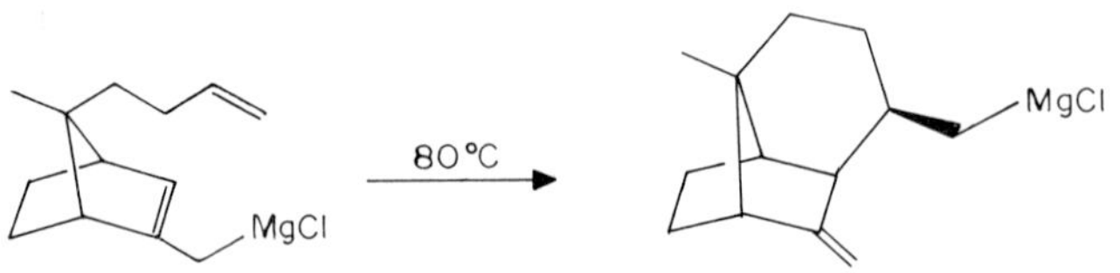

80 °C MgCl MgCl (30)

Clearly, the 'magnesium-ene' reaction has enormous scope in organic synthesis. The rearrangement products, δ,ε-unsaturated Grignard reagents, offer numerous possibilities for molecular modification. Prior to the work of Oppolyer and colleagues, intramolecular rearrangements were found to be of little use in synthesis because those studies involved equilibria and mixtures of products were obtained.

Unlike Grignard reagents of strained cycloalkylmethyl halides, which rapidly rearrange via C—C bond rupture in the carbocycle (equation 22), those of strained cycloalkyl halides are relatively stable[7]. However, Grignards of *n*-alkenyl halides do not rearrange to what appears to be the thermodynamically favoured product on heating (e.g. equation 31), but do so on photolysis[71].

R–CH=CH–CH$_2$–CH$_2$MgBr $\xrightarrow{h\nu}$ (R-substituted cyclopropyl)CH$_2$MgBr (31)

Exocyclic alk-1-enyl Grignard reagents possess similar stabilities to their saturated analogues, discussed above. Cyclopropenylmagnesium bromide decomposes according to equation 32, whereas Grignards of cyclobutenyl halides and those of larger rings are stable[72].

(cyclopropylidene)C(Ph)Br $\xrightarrow{Mg}$ (cyclopropylidene)C(Ph)MgBr $\longrightarrow$ PhC≡C–CH$_2$CH$_2$MgBr (32)

In discussing the stability of Grignard reagents with respect to rearrangements, it is important to remember that the same behaviour applies equally well to corresponding diorganomagnesium compounds.

4. *Grignard in situ trapping reaction*

This reaction, often called the Barbier reaction, is the reaction of RX, magnesium, and a substrate, usually under similar conditions to those for syntheses of Grignard reagents. Such an approach is very attractive if the Grignard is difficult to prepare or is inaccessible. The fact that it is a one-step reaction also is noteworthy. The disadvantages are, however, the difficulty of starting the reaction and that more byproducts are invariably formed[14]. For example, crotyl halides react with propanal in Et_2O to yield a mixture of the regular monoaddition product, **12**, as well as that derived from diaddition (**13**) (equation 33)[14]. Another common reaction is the coupling of ketyl radicals, $R_2\dot{C}O^-$, which are formed by a

MeCH=CHCH$_2$X + EtCHO $\xrightarrow[(2)\ H^+]{(1)\ Mg}$ CH$_2$=CH–CH(Me)–CH(OH)–Et (**12**) + CH$_2$=CH–CH(Me)–C(Et)(OH)–CH(Me)–CH=CH$_2$ (**13**) (33)

single electron transfer from magnesium to the ketone. Low temperatures can minimize ketyl radical formation, but for organic chlorides higher temperatures are required to effect the Barbier reaction, unless Rieke's or condensed magnesium is used.

Like the synthesis of Grignard reagents, the *in situ* method involves various radical species[17]. The question of whether intermediate organomagnesium compounds play a role in this reaction is not fully resolved. For the reaction of *N*-(2-haloethyl)-*N*-methylaniline with magnesium and an aldehyde, ethylene and products not consistent with an intermediate Grignard species are formed[14]. On the other hand, the products of reaction 34 clearly indicate that a Grignard reagent is generated.

More recent work[73] has shown that the Barbier reaction can occur without the *in situ* formation of the organometallic compound and that the reaction mechanism is probably the formation of $R_2\dot{C}O^-$, which reacts with $RX^{-\bullet}$ or R···MgX on the metal surface. Molle and Bauer[73] also suggested that from a knowledge of the stabilities of the ketyl radicals, the reaction yield can be optimized by using either the *in situ* method or the two-step Grignard approach.

The *in situ* method is useful for allyl and benzyl halides since their Grignard syntheses are difficult, requiring high dilution amongst other things. A recent application has been their one-step reactions in replacing the halogen with a silyl group, exemplified by equations 35[74], 36[29], 37[75], and 38[76]. It is significant that in these reactions the yields were in excess of 70%.

+ $PhCO_2H$ + Mg $\xrightarrow[-24\,°C]{Me_2O}$ [MgI ⇌ MgI] $\xrightarrow{H^+}$ 48% + 53% (34)

Br, X; $\xrightarrow[\text{RMeSiCl}_2,\ R=H, Me]{\text{Mg, thf}}$ SiMeR (35)

X = Cl, Br

X, X; $\xrightarrow[\text{Me}_2\text{SiCl}_2]{\text{Mg, thf}}$ Me_2Si (36)

(**14**) X = Cl

Cl, Br; $\xrightarrow[\text{Me}_2\text{SiCl}_2]{\text{Mg, thf}}$ $SiMe_2$ (37)

R^1, R^2, N, O, Me; $\xrightarrow[\text{Me}_3\text{SiCl}]{\text{Mg, thf,}}$ $SiMe_3$, N, O, Me; $SiMe_3$, N, O, Me (38)

Compound **14**, X = Cl, also reacts with Me_3SiCl and magnesium to yield 1,8-bis(trimethylsilyl)naphthalene in 90% yield. In contrast, **14** X = Br gives under the same conditions exclusively acenaphthylene, the intramolecular cyclization product. Thus the choice of halide even for the *in situ* reaction is important[29]. To date the di-Grignard reagent for X = Cl or Br has not been prepared from elemental magnesium, so that from a mechanistic point of view the formation of an intermediate di-Grignard reagent in these *in situ* reactions is unlikely. Grignard reactions involving magnesium and **14** gave inter- (X = Cl) and intra- (X = Br) Wurtz-type coupling. The difference in behaviour for a change from chloride has been attributed to both kinetic effects (leaving group capabilities of X$^{\bullet}$ or X$^-$, Br > Cl, noting that the mechanism of Grignard formation involves free radicals (Section II.A.2)), and variation in bond energies[17].

Silyl groups incorporated in organic molecules offer enormous potential for molecule modification.[77] In this context, 1,8-bis(trimethylsilyl)naphthalene can be dimetallated at the benzylic carbon atoms using Bu^{n}Li(tmeda)[29] yielding a lithium reagent based on 1,8-$(CHR)_2C_{10}H_6^{2-}$, R = H. Only recently has a di-Grignard reagent containing the dianion, R = H, been prepared, from Mg(anthracene)-$(thf)_3$[29b].

The *in situ* method is a novel and synthetically useful route to vinylallenes (> 60% yield) from 5-chloro-3-en-1-ynes and Me_3SiCl (equation 39)[78].

$Me_3SiCH{=}C{=}C(Me)CH{=}CH_2$

(39)

A related reaction to those in equations 35 and 36 is that in equation 40, and again the yield is high[74]. Other applications include its use in syntheses of tricyclic drugs (equation 41)[79] and the reaction shown in equation 42, which was effected using Rieke's magnesium[80].

(40)

(41)

X = CH_2, O

$$\text{MePhPr}^i\text{SnX} \xrightarrow[p\text{-BrC}_6\text{H}_4\text{CH(OCH}_2)_2]{\text{Mg}^*} p\text{-MePhPr}^i\text{Sn-}p\text{-C}_6\text{H}_4\text{CH(OCH}_2)_2 \qquad (42)$$

Related *in situ* trapping reactions of magnesium with conjugated polyenes and alkylhalosilane substrates are discussed in Section II.B.6.

The one-step Barbier reaction, using calcium rather than magnesium, has received some attention; thus ketones, after hydrolysis of the reaction mixture, yield tertiary aliphatic alcohols in 80% yield[14].

B. Other Methods

1. Reactions of Grignards with alkenes and alkynes

There are two established ways in which Grignard reagents add to multiple bonds, both of which have featured as key steps in many total syntheses. For alkenes the most common is the 1, 2-addition (method C, Table 1), yielding two isomeric 'Grignard reagents' unless steric and/or electronic effects dictate the course of the reaction to be stereospecific. The formation of these reagents has been recently reviewed[7,81] and only recent and novel examples are discussed here.

Alkenes are less reactive than alkynes and unless the double bond bears an activating substituent or the addition is catalysed by a transition metal complex (Section IV), a temperature greater than 60 °C is usually required. Complications can arise if in the product of addition there is scope for rearrangement or there is an equilibrium between several organomagnesium species (Section II.A.3). If the starting Grignard reagent is of the 2-alkenylhalogenomagnesium type, departure from 1, 2-addition prevails; the primary addition process may be coupling with the alkene via its C-3 position rather than to C-1. The general equation is given in Table 1, method D; it is often called the 'magnesium-ene' reaction. For the Grignard and substrate in equation 43 this process accounts for 60% of the product, determined by treating the mixture with formaldehyde[82].

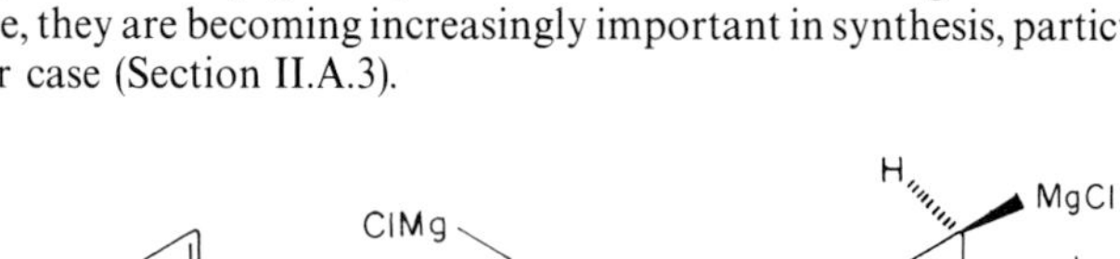

(43)

The examples in equations 44 and 45[82] highlight an important fundamental aspect of such reactions, that the conjugated *syn*-addition of the alkene is regio- and stereo-specific. In consequence, they are becoming increasingly important in synthesis, particularly for the intramolecular case (Section II.A.3).

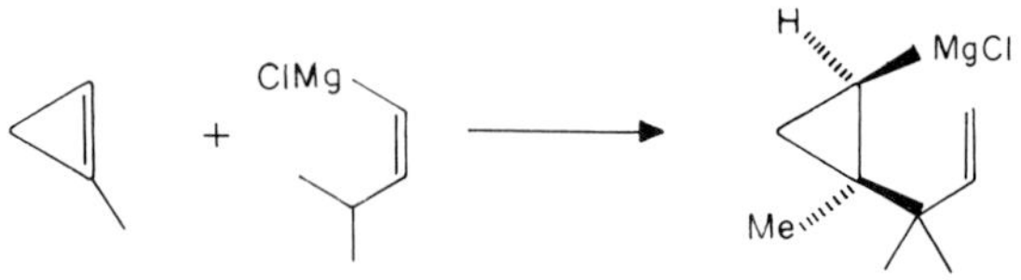

(44)

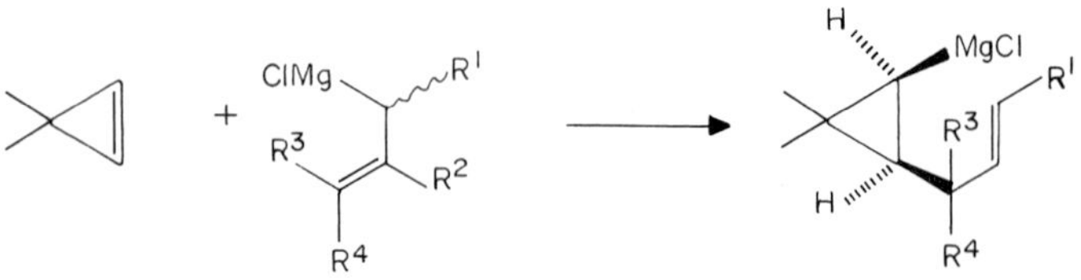

(45)

Acetylene and monosubstituted acetylenes are metallated by Grignard reagents (Section II.B.5). Disubstituted acetylenes react in a variety of ways. These include 1,2-additions which can be stereospecific, yielding either the *E*- or *Z*-isomer if there is (i) anchimeric assistance from a functional group[83], or (ii) if the reaction is transition metal catalysed[84]. 2-Alkenyl Grignard reagents yield cycloaddition products[7]. There is an example of a novel Diels–Alder type of addition for the reaction of $(\eta^1\text{-}C_5H_5)MgX$ with benzyne[85].

Allyl alcohols, alkynols, and allenols react with Grignard reagents, first to form alkoxomagnesium species, then *Z*- or *E*-1,2-addition of the Grignard across the multiple bond and ultimately the elimination of MgX_2, yielding cyclic oxomagnesium-alkenyls (equation 46) (ref. 7, p. 211).

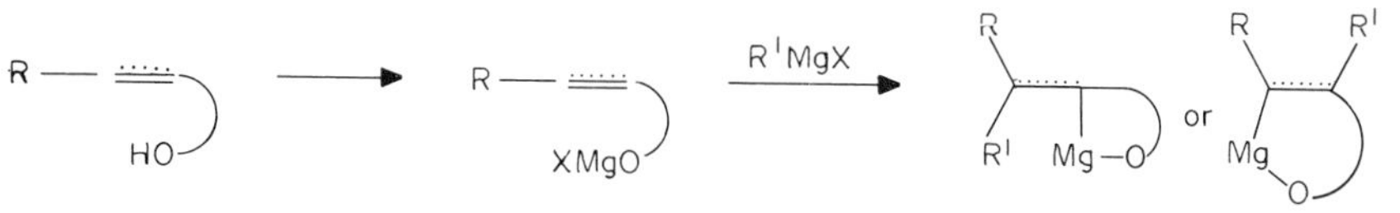

(46)

2. Hydromagnesiation

There have been only a few reports on the synthesis of organomagnesium reagents by the addition of Mg—H to a multiple bond (method E, Table 1). This 'hydromagnesiation' reaction, however, appears to have considerable potential in organic synthesis. HMgX (X = Cl or Br), prepared from MgX_2 and an 'active' form of MgH_2, in turn derived from $LiAlH_4$ reduction of Ph_2Mg in thf[86], is a reagent for the synthesis of a 'Grignard solution' using this method. Terminal and internal alkenes react with the reagent H_2Mg in thf at 60 °C, catalysed by $[TiCl_2Cp_2]$, affording solutions of solvated R_2Mg species[87], although none have been isolated to substantiate this claim.

The bulk of hydromagnesiation reactions reported are highly stereo- and regio-specific and are effective under mild conditions. They are reactions catalysed by $[TiCl_2Cp_2]$ and a typical experiment is to add it to a Grignard reagent and the unsaturated substrate, the active metal hydride being $[TiHCp_2]$, formed by sequential reduction of $[TiCl_2Cp_2]$ by the Grignard, alkylation, and β-hydrogen elimination (Scheme 2)[88].

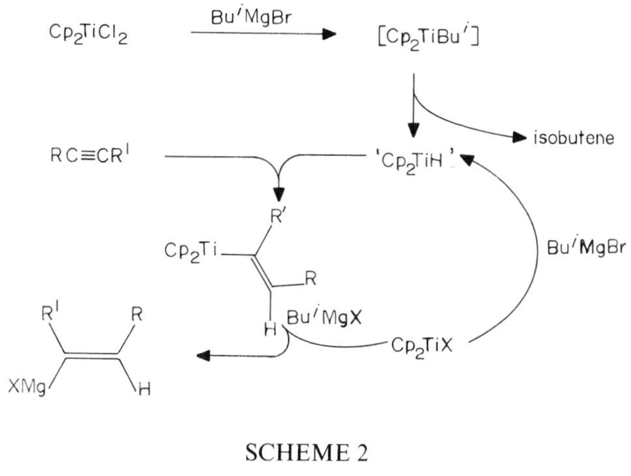

SCHEME 2

Some reactions yield the E-product rather than the expected 1,2- or Z-addition product. Hydromagnesiation of $Me_3SiCCCH_2OH$ with $[TiCl_2Cp_2]$ and two equivalents of Bu^iMgCl (one equivalent to neutralize the hydroxy group) in Et_2O afforded initially the Z-product, but this slowly isomerized over 6 h to the E-product[89]. Presumably the latter is favoured by chelation of the oxygen centre to the 'Grignard' magnesium, a process which is likely to favour elimination of MgX_2 (cf. equation 46)[90]. Compound **15**, however, under the same conditions gave exclusively the Z-addition products (equation 47)[91].

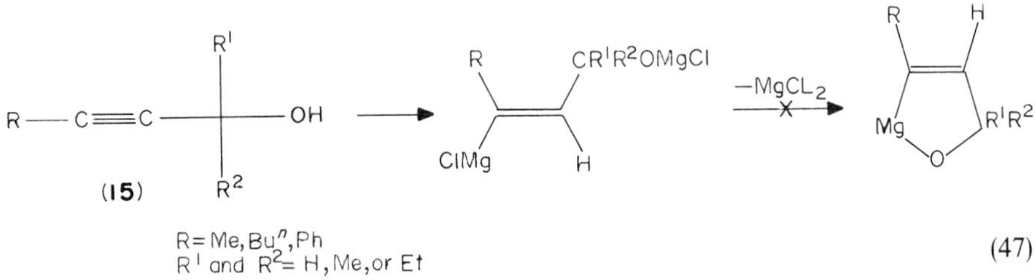

R—C≡C—C(R¹)(R²)—OH (**15**) ⟶ (R)(ClMg)C=C(H)(CR¹R²OMgCl) —(−MgCL₂, ✕)⟶ [Mg–O ring, R¹R², R, H]

R = Me, Buⁿ, Ph
R¹ and R² = H, Me, or Et (47)

Interestingly, the catalysed addition of Grignards to prop-2-ynylic alcohols is *anti* and is thought to proceed via a concerted chelation–elimination of MgX_2 and addition of RMg (equation 48)[92].

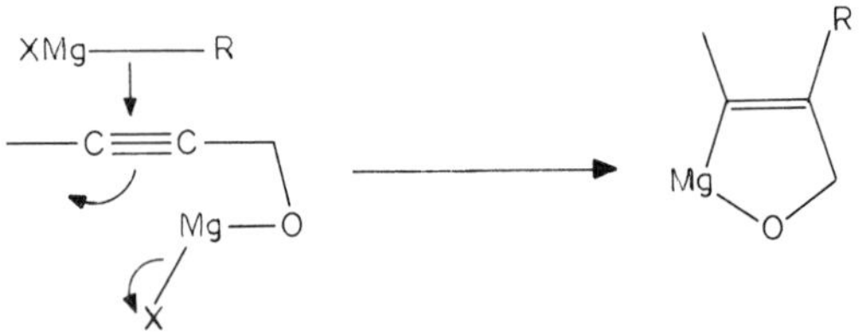

XMg—R → —C≡C—CH₂—O—Mg—X ⟶ [Mg–O ring, R] (48)

Hydromagnesiation of oct-1-ene using RMgX, R_2Mg, RMgH, H_2Mg, and HMgX and catalysed by either $[TiCl_2Cp_2]$ or $[TiCl_3Cp]$ yields up to 65% of the corresponding n-octylmagnesium species (determined by the percentage of octane formed on hydrolysis)[93]. For 1,3-dienes and styrenes, the product of the hydromagnesiation reaction are allylic (equation 49) and α-arylethyl Grignard reagents, respectively, and they are generated in almost quantitative yields. Like the reaction of alkynes, the active hydride is '$TiHCp_2$'[94].

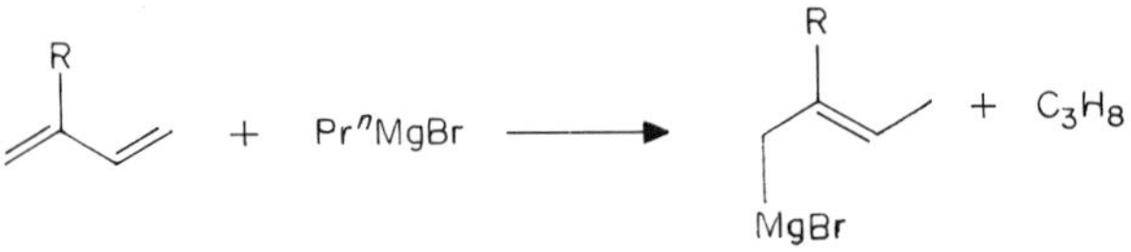

$$\text{CH}_2{=}\text{C(R)CH}{=}\text{CH}_2 + \text{Pr}^n\text{MgBr} \longrightarrow \text{BrMgCH}_2\text{C(R)}{=}\text{CHCH}_3 + \text{C}_3\text{H}_8 \quad (49)$$

Cyclopentadienylmagnesium hydride, readily prepared from MgH_2 and cyclopentadiene in thf, may have application in Grignard syntheses from alkynes. Interestingly, it reacts with Ph_3CX ($X = Cl$ or Br) to afford a radical intermediate $Ph_3C^{\bullet}$, then Ph_3CH and $CpMgX$[95].

3. Salt elimination

Organomagnesium reagents react with magnesium dihalides to form RMgX or R_2Mg compounds (method F, Table 1). It is a particularly appropriate synthon where Wurtz coupling is the major reaction using standard Grignard conditions, for example the Grignards of benzylic- and allylic-type halides. It is also useful in preparing solvent-free R_2Mg species, such as hydrocarbon-soluble $Bu^s{}_2Mg$. Furthermore, lithium–hydrocarbyl species, the most common reagent using this approach, are readily prepared by direct metallation of an organic substrate (e.g. equation 50) or by exchange reactions of an aryl halide with alkyllithium reagents. Organomagnesium reaction analogues of the latter are unusual (Section II.B.4).

The use of organomagnesium reagents rather than the precursor Group I reagents can greatly improve the yield and/or may result in a completely different reaction pathway. For example, conversion of a lithium reagent to a Grignard may be necessary for a subsequent palladium phosphine complex-catalysed coupling reaction (e.g. equation 50)[96]. The magnesium complex[†] in equation 51 was found to be more effective in reactions with lactones than the lithium reagent[97]; the lithium reagent in equation 52 did not give the cyclopropene, whereas the Grignard did[98].

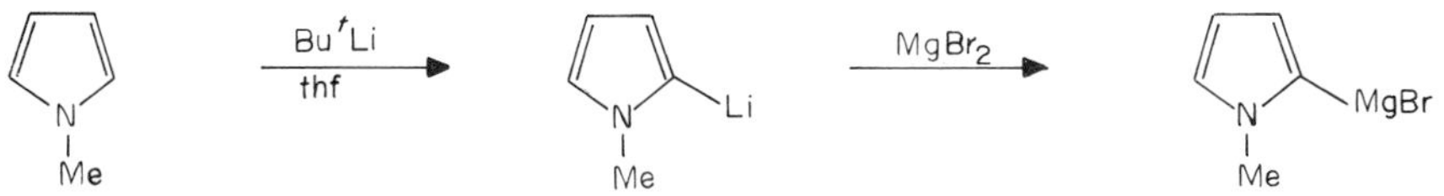

(50)

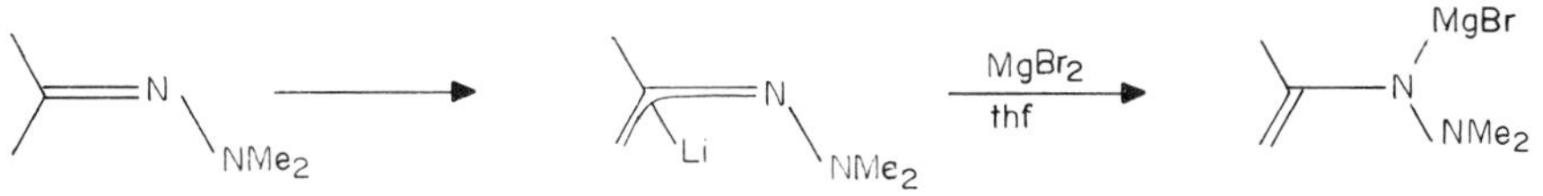

(51)

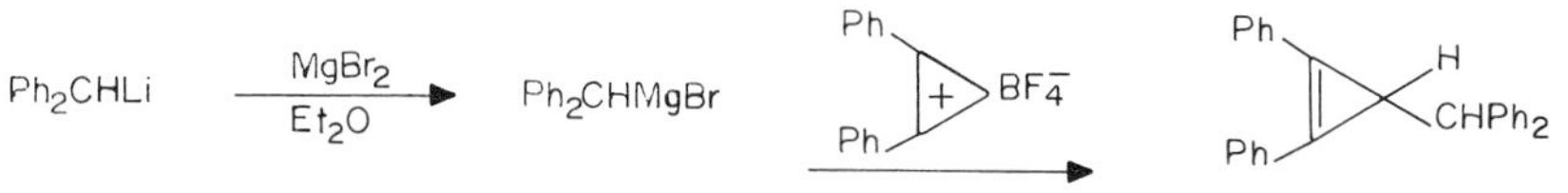

(52)

The following cyclization reaction for a Grignard but not an organolithium reagent (equation 53) is also noteworthy. It proceeds via ring opening of the epoxide or rearrangement of the epoxide function to a ketone or aldehyde, followed by ring closure[99].

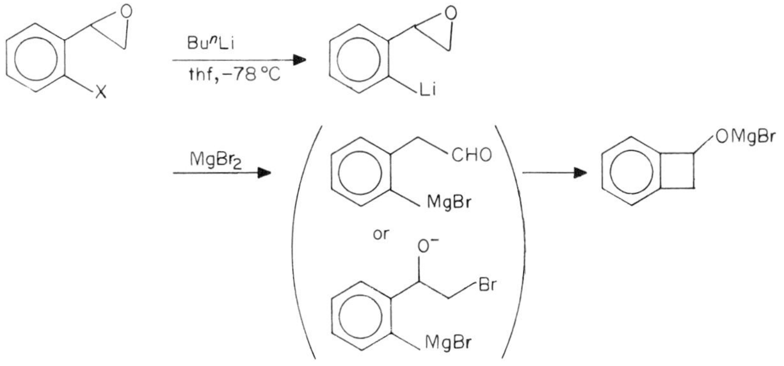

(53)

[†]For equation 51 the organic moiety is probably an η^3-aza-allyl species in the lithium reagent, whereas in the magnesium complex it is probably N-bound to the metal. Nevertheless, the reactions of the magnesium complex is as if the charge was C-centred.

The above examples clearly demonstrate that major changes in reactivity and selectivity occur in converting lithiumhydrocarbyl reagents to those of magnesium.

Lochman's reagent (Bu^nLi–Bu^tOK) is a very powerful metallating agent, yielding novel anionic hydrocarbyl species that are inaccessible using other more established metallating reagents. Treating these species with MgX_2 will prove valuable in preparing new classes of organomagnesium reagents. The synthesis of a novel 'di-Grignard' reagent (equation 54) illustrates this point[100].

$$H{-}C{\equiv}C{-}Ph \xrightarrow[Bu^tOK]{Bu^nLi} Li{-}C{\equiv}C{-}C_6H_4K \xrightarrow{MgBr_2} BrMg{-}C{\equiv}C{-}C_6H_4MgBr \quad (54)$$

Some bisdienylmagnesium (tmeda) complexes have been isolated from salt elimination reactions of the corresponding potassium dienides[101]. Magnesium salts containing the cyclooctadiene anion[102] and cyclooctatetraene dianion[103] have also been prepared using this approach.

4. Metal–halogen exchange

The metal–halogen exchange reaction (method G, Table 1) is a relatively new method for generating Grignard reagents. A prerequisite is that the organic substrate, RX, possesses electronegative substituents, otherwise complete conversion may be difficult to achieve and there may be an equilibrium between two Grignard species[7]. With this proviso, it works well for a variety of organic halides. For example, (i) the reagents $(CH_{3-n}X_n)MgCl$ ($X = Cl$, Br, I; $n = 1, 2, 3$), which have great application in synthesis, are accessible by this method using Pr^iMgCl and the appropriate methyl halide at *ca.* $-80\,°C$; the products precipitate from solution (Et_2O–thf mixture), possibly enhancing a high equilibrium yield[103]; (ii) selective Grignard formation is possible for polyhalodiaryl

$$\text{4,5-dibromo-1-methylimidazole} \xrightarrow{EtMgBr} \text{4-bromo-5-(BrMg)-1-methylimidazole} \quad (55)$$

compounds (equation 55)[104]; and (iii) selective halogen exchange for the least electronegative halide (equation 56).[105]

$$C_6F_{13}CF{=}CFBr \xrightarrow[-PhBr]{PhMgBr} C_6F_{13}CF{=}CFMgBr \quad (56)$$

The efficiency of the reaction also depends on the organic halide generated, R′X, with $X = I > Br > Cl$, which is reflected in C—X bond energies. Other than the problem of effecting complete conversion, which depends on the presence of electronegative groups, there can be complications due to competing Wurtz-type coupling (equation 57)[7].

$$RX + R'MgX \longrightarrow R'X + RMgX \xrightarrow[-MgX_2]{RX} RR \quad (57)$$

5. Metallation

The metallation of an organic substrate, R_3CH, by magnesium reagents, usually Grignards, is a powerful synthetic route to a wide variety of organomagnesium species (method H, Table 1). A requirement of course is that C—H is of greater kinetic acidity than the conjugate acid of the metallating agent. Strongly coordinating solvents such as thf and hmpt are the most effective in promoting these reactions.

The greatest application to synthesis is the metallation of acetylenes[106]; for the unique case of acetylene selective mono- or di-metallation is possible[107]. Hydroxy groups on the alkyne substituent lead to the formation of alkoxide complexes (e.g. equation 58)[108], provided that two equivalents of metallating agent are used and the alkoxide does not affect the next step in the organic sequence.[109] A common strategy, however, is to protect hydroxy groups by converting them to the corresponding tetrahydroxopyranyl ether derivative[110].

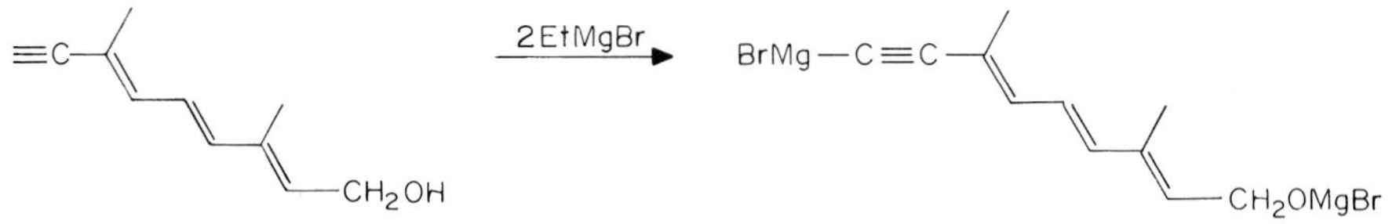

$$\equiv C\text{-}(\ldots)\text{-}CH_2OH \xrightarrow{2EtMgBr} BrMg\text{—}C\equiv C\text{-}(\ldots)\text{-}CH_2OMgBr \quad (58)$$

The reaction of EtMgBr with $CH{\equiv}C(CH_2)_8CH_2Cl$ gave the metallated product[111]; its subsequent reaction with magnesium, and that of similar compounds, may yield 'di-Grignard' reagents, possibly of great utility in synthesis. It is noteworthy that the alk-2-yne Grignard reagent, $CH{\equiv}CCH_2MgBr$, is stable with respect to rearrangement to $BrMgC{\equiv}CMe$[109].

A second type of metallation is that yielding anionic aromatic complexes of magnesium, e.g. $(\eta^1\text{-}C_5H_5)MgX$ and (indenyl)MgX and the product in equation 59[112]. They are well established reactions and the products have been used to good effect in synthesis (see ref. 113 for recent examples).

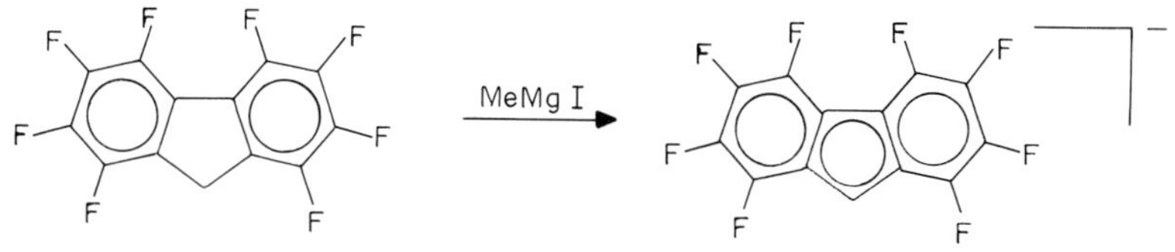

$$\text{octafluorofluorene} \xrightarrow{MeMgI} [\text{octafluorofluorenyl}]^- \quad (59)$$

The compound $[(\eta\text{-}C_5H_5)_2Mg]$ can be prepared by this route. It is, however, accessible by direct reaction between cyclopentadiene and magnesium at 0 °C, catalysed by $CpTiCl_3$ or $TiCl_4$[114].

A related form of metallation is the magnesiation of arylmethanes yielding (arylmethyl)MgX, 'Grignard'-type species, e.g. metallation of α-picoline, and toluene, albeit under forcing conditions, using Bu^tMgCl[115]. The scope for this approach is high since the products are usually difficult to prepare using the classical Grignard strategy, or are unknown.

There is an example of metallation as a competing reaction in the condensation of a ketone with a Grignard reagent (equation 60), ascertained by the dimeric compound obtained after acid work-up[116]. The expected product is that derived by nucleophilic attack of the ketone.

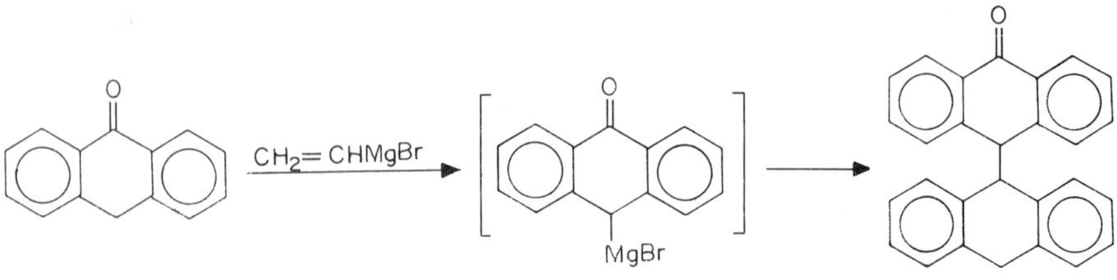

$\xrightarrow{CH_2=CHMgBr}$ [MgBr] $\longrightarrow$ (60)

Aromatic ring metallation is possible and is directionally controlled by strongly complexing oxygen and/or nitrogen centres in the molecule, the position of metallation

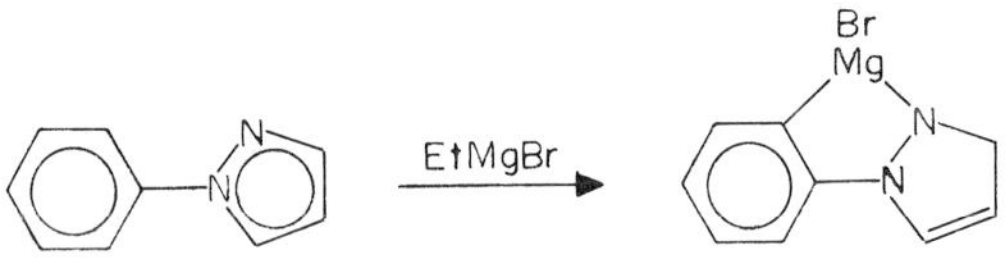

$\xrightarrow{EtMgBr}$ (61)

being *ortho* to substituents with such hetero atoms, e.g. equation 61[117]. The directional control of metallation is akin to the syntheses of aryllithium reagents.

Polyhalogenated compounds are susceptible to metallation, although competing reactions of metal–halogen exchange (Section II.B.4) and their decomposition by α- or β-elimination pathways may occur. The metallation of pentafluorobenzene[118] and of the halo forms CHX_3 (X = Cl or Br)[103] are examples of this approach.

Grignard reagents and other strong bases of magnesium (e.g. amides, dialkyls, and alkyl and aryl oxides) abstract α-hydrogens of imines[119,120] (some ketones, depending on the conditions since they tend to yield additional compounds[121]) carboxylic acids and esters[122], thiol esters[123], sulphonic acids, sulphones and sulphoxides[124], and phosphinates and phosphonates[125]. The derived species, however, are not organometallic compounds since the interaction to the metal is through the oxygen and/or nitrogen centres, a consequence of the relatively hard nature of magnesium(II). Some authors choose to depict them as C-bound, perhaps to emphasize that the site of their reactions is at that carbon. Consider the reaction in equation 62[120]. The enamine nitrogen is probably sp^2 and associated with lone pair–π-system overlap and possibly some π-bonding to magnesium.

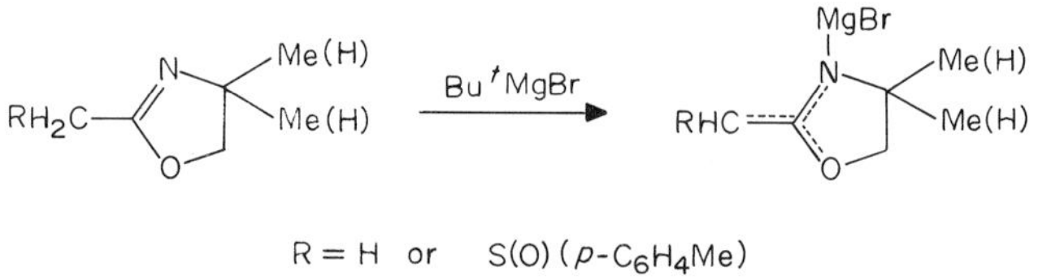

$\xrightarrow{Bu^tMgBr}$ (62)

R = H or $S(O)(p\text{-}C_6H_4Me)$

α-Metallated ketones, enolates, would have similar structures. Their formation can be promoted by hmpt[126]. Unlike metallation or addition for ketones, thioketones are prone to several competing reactions, including addition at the thioketone, and/or thiophilic addition at the sulphur, yielding α-functionalized Grignard reagents (equation 63).

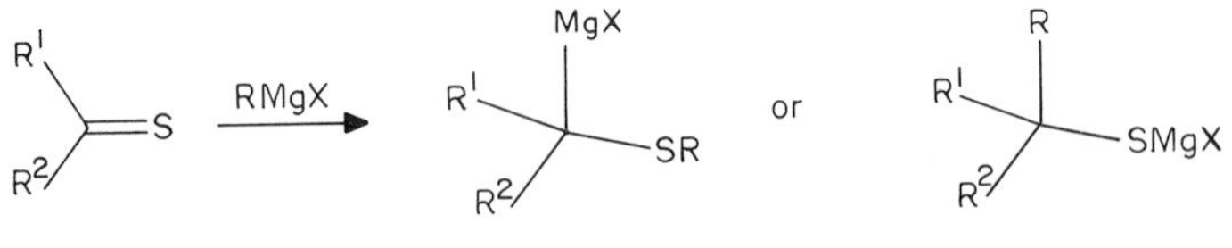

$\xrightarrow{RMgX}$ or (63)

Interestingly, an intermediate 'α-metallated' sulphonyl compound has been prepared by C—C bond rupture rather than by hydrogen abstraction (equation 64)[127].

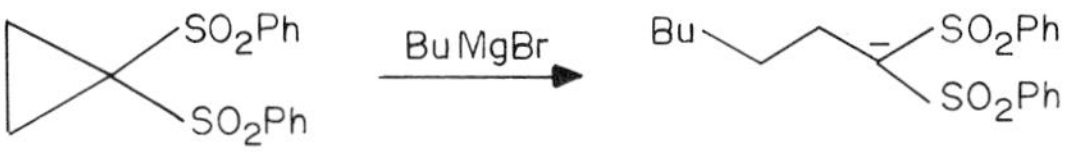

(64)

6. *Magnesium electron transfer reactions*

These are reactions involving electron transfer from magnesium to an unsaturated moiety (method J, Table 1), either aromatic molecules, or conjugated polyenes, including cyclooctatetraene, in a strongly coordinating polar solvent from which they can be isolated as solvates. The derived anions are either radicals or dianions. Such reactions have been extensively reviewed; the most recent review covers the literature upto 1980[7]. A new development is the synthesis of magnesium anthracene in thf; the magnesium–hydrocarbyl interaction is at the 9, 10-positions (equation 65)[26,128].

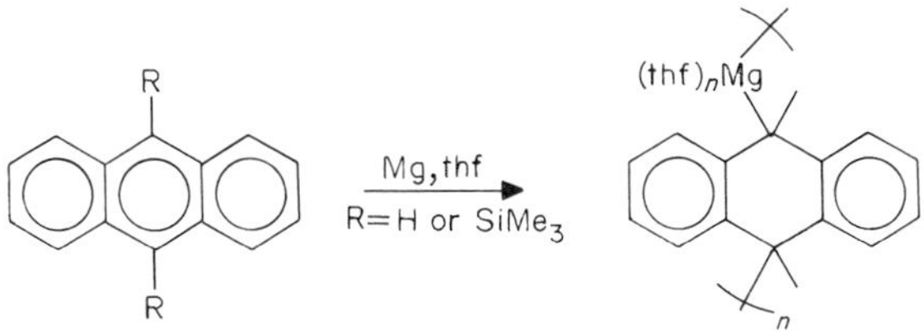

(65)

On addition of Et_2O, the compound decomposes into its constituents, anthracene and elemental magnesium in a form activated for Grignard syntheses. This highlights the importance of using a strongly coordinating solvent (thf > Et_2O). The analogous reagent derived from naphthalene, $[Mg(C_{10}H_8)_2]$, possesses radical anions. Its hmpt solutions are useful for preparing radical anions with more easily reduced aromatic compounds[129].

The reaction of conjugated polyenes invariably requires a catalyst, usually a compound of Fe(III), Ti(IV), Ni(II), Cu(II), or Zn(II). 1, 3-Allylic rearrangements, of the type discussed in Sections II.A.3 and II.B.1 (addition of alkenes to alk-2-enylmagnesium complexes), are likely and because of this there has been few applications to synthesis. One study has been on the head-to-tail and head-to-head combinations of dienyls derived from isoprene, myrcene, ocimene, and piperylene (e.g. equation 66)[130].

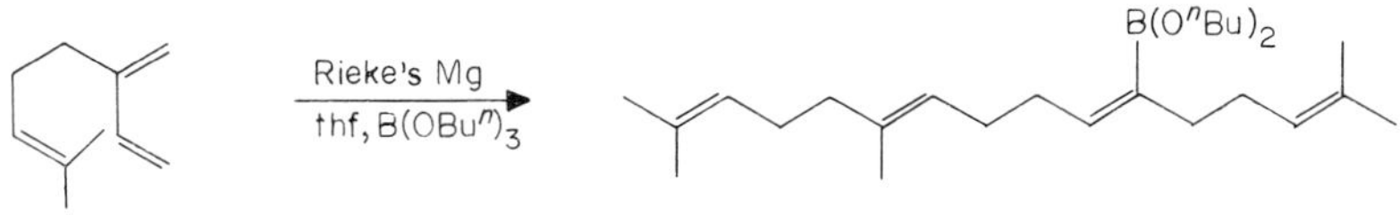

(66)

In situ trapping of organic anions formed by this method has some application (cf. *in situ* trapping of Grignards, the Barbier reaction, Section II.A.4). The following examples are illustrative: (i) the reductive silylation of butadienes (e.g. equation 67)[131] and of hexa-1, 3, 5-triene [reaction in equation 68 (70% yield)][132]; and (ii) a novel silylation reaction of acenaphthene (equation 69); interestingly, acenaphthalene, Li, and Me_3SiCl in thf yield a tetrakis silylated compound[133].

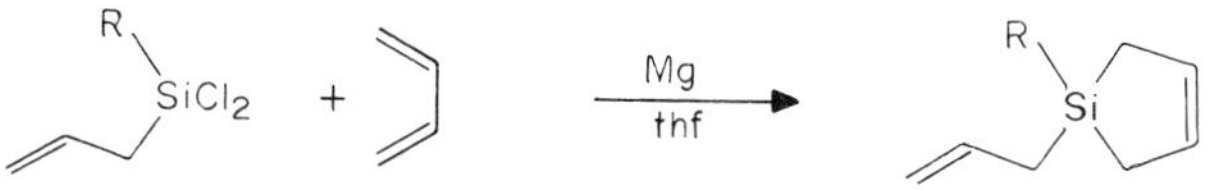

(67)

$$Me_3SiCl + \text{(isoprene)} \xrightarrow[\text{thf}]{\text{Mg}} Me_3Si\text{—}CH_2CH{=}C(Me)CH{=}CH\text{—}CH_2SiMe_3 \quad (68)$$

$$Me_3SiCl + \text{(acenaphthene)} \xrightarrow[\text{hmpa}]{\text{Mg}} \text{(bis-}SiMe_3\text{ dihydroacenaphthene)} \quad (69)$$

7. *Oxidative–reductive transmetallation*

This method is the exchange between organic derivatives of less electropositive metals than magnesium, predominantly those of mercury and to a lesser extent zinc, and elemental magnesium, yielding either RMgX or R_2Mg (method I, Table 1). It is a common route to unsolvated R_2Mg compounds, made possible by the highly exothermic reactions of R_2Hg being effective in the absence of a solvent. Another feature of this class of reaction has been the ability to prepare unusual di-Grignard reagents. Costa and Whitesides[134] prepared $BrMg(CH_2)_3MgBr$ by a five-step synthesis from allene; the final step is shown in equation 70. [There is, however, a recent more direct synthesis of the same reagent, from magnesium and $Br(CH_2)_3Br$[55].] A geminal di-Grignard has also been prepared, but from an organozinc reagent (equation 71)[135].

$$BrHg(CH_2)_3HgBr \xrightarrow[\text{thf}]{\text{Mg, MgBr}_2} BrMg(CH_2)_3MgBr \quad (70)$$

$$Me_3SiCHBr_2 \xrightarrow{\text{Zn/Cu}} Me_3SiCH(ZnBr)_2 \xrightarrow{\text{Mg}} Me_3SiCH(MgBr)_2 \quad (71)$$

8. *Miscellaneous methods*

Exchange reactions yielding RMgX or R_2Mg species from compounds of metalloids and non-metals are uncommon (ref. 7, p. 166). Heterobimetallic organometallic complexes of magnesium and alkali metals, boranes, aluminium, and zinc are known (ref. 7, pp. 209, 221), but they appear to be of limited utility in organic synthesis with no clear advantages over conventional organomagnesium reagents. However, those of transition metals have a rich chemistry, and are usually prepared by the addition of a transition metal complex to a solution of RMgX or R_2Mg. The nature of the active alkylating species formed is not always known. Complexes of the type $Mg[R_nM]_n$ or $XMg[R_nM]$ (transition metal metallates), magnesium halide-free reagents, R_nM, or intermediates between it and the simple metallates are possible. Their application in synthesis is discussed in Section IV.A.

Compounds of the type RMg (OR^1) (R^1 = alkyl or aryl) are prepared by the partial alcoholysis of R_2Mg (equation 72) (ref. 7, p. 210). (Such compounds are formed as intermediates in the nucleophilic attack of ketones using R_2Mg.) Similarly, secondary amines yield $RMg(NR^1R^2)$ and alkanethiols, $RMg(SR^1)$.

$$R_2Mg + R^1OH \longrightarrow RMg(OR^1) + RH \quad (72)$$

Allyl aryl ethers react with magnesium in coordinating solvents to yield aryloxoalkylmagnesium species, a pseudo-Grignard reaction (equation 73) with only a small amount of Wurtz coupling[136]. The corresponding thioethers undergo the same reaction[137].

$$ArOCH_2C(R){=}CH_2 + Mg \xrightarrow{\text{thf}} ArOMgCH_2C(R){=}CH_2 \quad (73)$$

C. Di-Grignard Reagents

Di-Grignards are valuable reagents in organic synthesis, including their stoichiometric and catalysed reactions with transition metal complexes. However, their preparations are often difficult because of strongly competing elimination/coupling reactions. A separate section on them is included here because of their utility and also because the syntheses of some are novel while for others the critical conditions for formation are usually more than adequate for related mono-Grignards. Moreover, some of the reagents possess unusual solubility properties that facilitate their purification. Details concerning di-Grignard syntheses and any special features are given in Table 2.

The reaction of the Grignard of CH_2Cl_2 and an organic substrate is usually carried out by the *in situ* one-step (or Barbier) method (Section II.A.4), but there is doubt as to whether the geminal dimagnesium species $CH_2(MgCl)_2$ is involved[138]. Bertini *et al.*[139] have, however, developed a reliable synthesis of solutions of such a di-Grignard, from CH_2X_2 ($X = Br$ or I) and elemental magnesium; they are remarkably stable, being unchanged after storage for several weeks at 0 °C. The reagent $Me_3SiCH(MgBr)_2$ has been prepared from the corresponding zinc reagent (equation 71) rather than the Grignard route used for the aforementioned example[135]. *In situ* trapping reactions involving $Me_3SiCHBr_2$ have been investigated (e.g. equation 74), but it is also unlikely that di-Grignard intermediates are formed. Interestingly, mono-Grignards of geminal dihalides are accessible by low-temperature exchange reactions (e.g. equation 75; see also Section II.A.3)[140].

(74)

(75)

The only 1,2-dimagnesiumalkane reagent in the literature is that derived from a Grignard reaction of (*Z*)- or (*E*)-1,2-dibromopropane, and although the yields are low, the pure di-Grignard precipitates from the reaction mixture. It can be solubilized in Et_2O by

TABLE 2. Synthesis of di-Grignard reagents

Reagent	Method of preparation[a]	Solvent	Yield (%)	Special features	Ref.
1. $CH_2(MgBr)_2$	A	Et_2O–benzene	50–60	Mg amalgam gives the most reliable results	139
2. $Me_3SiCH(MgBr)_2$	J	thf	—	Derived from a geminal di-zinc species	135
3. H H BrMg MgBr	A	Et_2O	15–17	The *Z*-product is obtained from either the (*Z*)- or (*E*)-dibromide	53
4. R R BrMg MgBr	A(R = H or Me), J(R = H)	Et_2O, thf	30, 98	Method J is the action of Mg on $RHg(CH_2)_3HgR$ (R = Me or Cl)	55, 134, 146
5. $XMg(CH_2)_nMgX$ ($n = 4$–12) and related species, e.g. CH_2MgCl CH_2MgCl MgCl MgCl	A	thf or Et_2O	> 60	The highest yields are obtained from alkyl chlorides with Mg in refluxing thf	56
6. $BrMg(CF_2)_xMgBr$ ($x = 6$ or 8)	Exchange method, $Me_{3-n}H_nSi(CF_2)_xSiH_nMe_{3-n}$ + MgBrEt ($n = 0, 1$)	thf	Good	Decomposition to dialkenes occurs > − 10 °C	143

7. F2 MgBr, F2, F2, F2, F2, BrMg (bicyclic structure)	A	Et_2O	> 63	Stable in refluxing ether	52
8. Me_2Si—$SiMe_2$, $BrMgCH_2$, CH_2MgBr	A	Et_2O	60	For the diiodide only an 8% yield was obtained, the major product being $\overline{Me_2SiCH_2SiMe_2CH_2}$	147
9. $B_{10}H_{10}$, C, C, MgBr, MgBr	A	Et_2O	—	—	148
10. $BrMg(CH_2)_nO(CH_2)_nMgBr$ ($n = 3$ or 4)	A	thf	96	Completely redistributes to $\overline{Mg(CH_2)_nO(CH_2)_{n-1}CH_2}$	41, 143
11. MgCl, MgCl, *o*,*m* and *p*, ; MgCl, MgCl	A, K	thf	> 90	Optimum yields are for di-Grignard concentrations of *ca.* 0.1 M	17, 22, 29 149
12. MgCl, MgCl; MgCl, MgCl	A, K	thf	40, 90(A), > 90(K)	Optimum yields are for di-Grignard concentrations of *ca.* 0.1 M. Method K works for the dibromides	18, 29, 150
13. BrMg, O, MgBr	A	thf	96	Redistributes to $MgBr_2$ and a cyclic organomagnesium compound with an intramolecular Mg—O linkage	143

TABLE 2. (*Contd.*)

Reagent	Method of preparation[a]	Solvent	Yield (%)	Special features	Ref.
14. 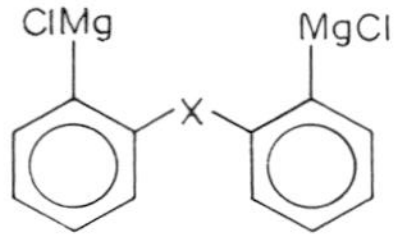 ClMg, MgCl (structure); $X = CH_2$, SiMe (H or Me), or CHPh(Cl → Br)	A	thf	60	For X = SiMe (H or Me) successful Grignard formations was possible only with Rieke's magnesium. For X = CHPh there was no reaction using conventional Mg	151
15. 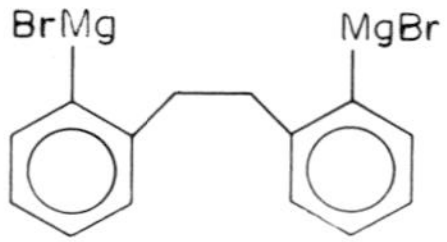 BrMg, MgBr (structure)	A	thf	90	—	152
16. 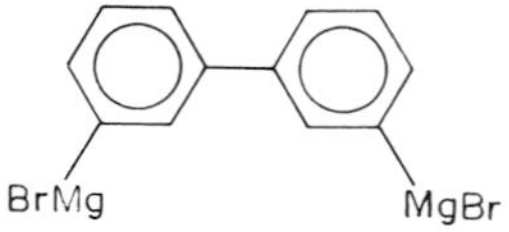 BrMg, MgBr (structure)	A	thf	—	Rieke's magnesium	153
17. 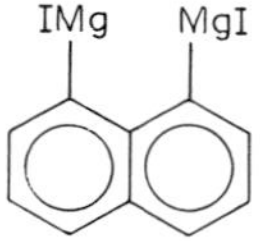 IMg, MgI (structure)	A	thf	—	Rieke's magnesium	154
18.  XMg, MgX′ (structure); X = X′ = Br; X = Cl, X′ = Br	A	thf	100	Rieke's magnesium. For X = X′ = Cl di-Grignard formation ≈ 30%	155

Compound	Method	Solvent	Yield (%)	Comments	Ref.
19. BrMg–C₆H₄–O–C₆H₄–MgBr	A	thf	90	Composition of solution is concentration dependent	156
20. BrMg, Z, Z, MgBr (structure); Z = O or NMe	A	thf	*ca.* 40	—	157
21. Br, Br, Br, Br, BrMg, S, S, MgBr (structure)	A	Et_2O	75	Entrainer method (C_2H_5Br added)	158
22. Cl, Cl, BrMg, S, MgBr (structure)	G	thf	41	Prepared from Cl, Cl, Cl, S, MgBr (structure) using EtMgBr (40% unreacted using 2 equiv.)	159
23. Br, Br, BrMg, MgBr, Br, Br (structure)	G	thf	> 50	25% mono-Grignard present in solution	160
24. (a) BrMgC≡CMgBr (b) XMgC≡CSi(Me_2)Si(Me_2)C≡CMgX	H	Et_2O	> 95	Compound (a) is only sparingly soluble in Et_2O or benzene	107(a), 161(b)

TABLE 2. (*Contd.*)

Reagent	Method of preparation[a]	Solvent	Yield (%)	Special features	Ref.
25. 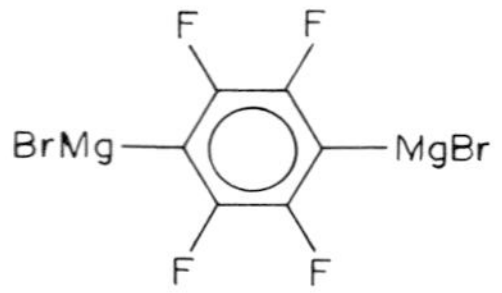 F F BrMg MgBr F F	H	thf	24	29% Mono-metallated 1, 2, 4, 5-tetrafluorobenzene	118
26. 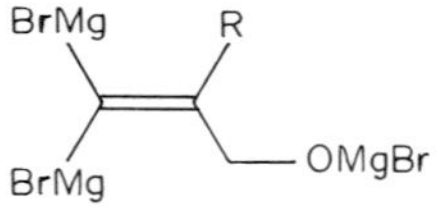 BrMg R BrMg OMgBr	C	Et_2O	> 50	Derived from BrMgC≡CCH_2OMgBr; CuI-catalysed	162
27. 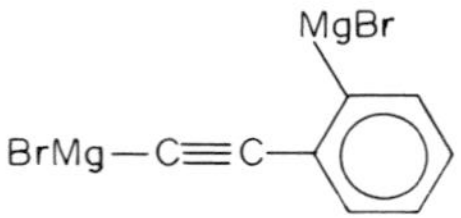 MgBr BrMg—C≡C—	F	thf	> 71	Prepared from a dimetallated complex and $MgBr_2$ (equation 54)	100
28. 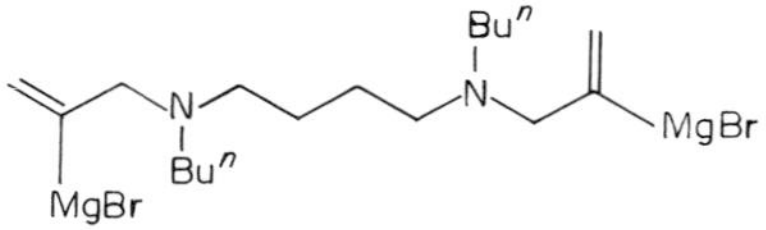 Bu^n N N Bu^n MgBr MgBr	A	thf	~ 60	Similar yields for related dibromides	163
29. 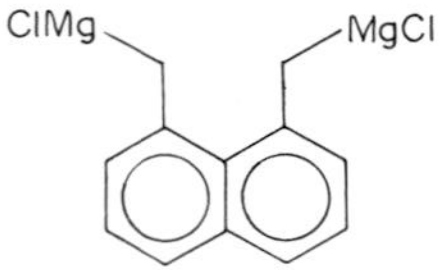 ClMg MgCl	K	thf	94	Optimum yield is for a di-Grignard concentration of *ca.* 0.1 M	29

[a] See Table 1 for definition of synthetic methods.

the addition of $MgBr_2$ and rapidly dissolves in thf, but after several minutes an insoluble oligomeric, $MgBr_2$-free species, $[(C_3H_4)Mg]_n$, forms[53].

The di-Grignard of 1, 3-dibromopropane is accessible using method A in Table 1[55]. Again, the yield is low but its purification is possible. Addition of thf to an etheral solution of the reagents yields an insoluble oligomeric material that is soluble in Et_2O on addition of one equivalent of $MgBr_2$ (equation 76). Alternatively, the di-Grignard can be prepared

$$\text{BrMg}\frown\text{MgBr} \underset{Et_2O}{\overset{thf}{\rightleftharpoons}} [\text{Mg}\frown]_n + MgBr_2 \qquad (76)$$

by a five-step synthesis starting with allene, the last step, with a 98% yield, involving an organomercury reagent (equation 70)[134]. In contrast, Grignards of the type $XMg(CH_2)_nMgX$ ($n \geqslant 4$) are readily prepared in high yield using the normal Grignard procedure. Like the aforementioned propylene analogue, they have limited solubility in Et_2O. In thf the predominant species are solvated metallacycles $\overline{Mg(CH_2)_{n-1}CH_2}$ or $\overline{Mg(CH_2)_nMg(CH_2)_{n-1}CH_2}$, depending on the value of n[141,142]. The thf, in effect, shifts the Schlenk equilibria to the right (e.g. equation 76). For di-Grignards derived from dihalo ethers, $Br(CH_2)_nO(CH_3)_nBr$ ($n = 3$ or 4), the equilibria are also to the right and the species are metallacycles with oxygen complexation to magnesium. The preference for the formation of chelate rings has been ascribed to the intramolecular coordination of oxygen[41,143]. Some perfluoroalkylene di-Grignard reagents have been prepared by exchange methods (Table 2)[144].

Di-Grignard reagents derived from benzylic-type dihalides (11 and 12, Table 2) are accessible under critical conditions of (i) concentration, typically 0.1 M, (ii) the use of magnesium powder, (iii) thf as the solvent, (iv) temperature kept below 35 °C, and (v) choice of chloride rather than dibromide or iodide. In contrast, di-Grignards of aryl dihalides are available without strict conditions, although aryl chlorides may be difficult to react unless highly reactive Rieke's magnesium is used.

Benzylic-type di-Grignard reagents (11, 12 and 29, Table 2) are also accessible by reacting $[Mg(anthracene)(thf)_3]$ with the appropriate dihalide[29b].

Metallation of fluorinated aromatics as a route to Grignard reagents appears to be effective if two fluorines, *ortho* to the site of metallation, are present. The treatment of 1, 2, 3, 4-tetrafluorobenzene under conditions effective for the dimetallation of the 1, 2, 4, 5-isomeric compound failed to yield mono- or di-metallated species[118].

The di-Grignard of 2, 6-dibromopyridine is formed in 1.2% yield in Et_2O using the entraining method, determined by trapping it with CO_2[145]. However, by using Rieke's magnesium, and at low temperature to minimize Wurtz-type coupling, or the rigid conditions found necessary for benzylic-type Grignard reagents, a viable synthesis of it and other novel di-Grignard reagents is possible.

III. REACTIONS OF ORGANOMAGNESIUM REAGENTS WITH ORGANIC COMPOUNDS

This section is devoted to reactions of magnesium reagents possessing a carbon—metal linkage with organic substrates. Such reactions are of great value in synthesis as molecular building blocks where new carbon—carbon, and to a lesser extent carbon—heteroatom, bond formation is desired. Moreover, they have utility in reactions where the key step is bond fission, and in the synthesis of labelled compounds. Organomagnesium reagents also find application in synthesis as strong bases, readily deprotonating amines, alcohols, etc., and the derived metal complexes have an extensive organic chemistry, being strong nucleophiles. A discussion of their use in synthesis is, however, beyond the scope of this review.

The causality of reactions of organomagnesium reagents with organic substrates is the ionic or at least covalent polarized nature of the metal–carbon interaction(s). Their reaction can be considered as nucleophilic attack of the carbon centre possessing the greatest ionic character. Strongly coordinating solvents lead to an increase in ionic character and thus an increase in reactivity. There is, however, a concomitant increase in electrophilicity of the metal centre, enhancing its attack on nucleophilic functional groups, but this can be attenuated by using exceptionally strong coordinating solvents. In short, the reactivity and selectivity of organomagnesium reagents can be greatly modified by changing the solvent of the reaction medium.

In principle, the order of reactivity of $R_{2-n}MgX_n$ ($n = 0, 1$) should reflect the variation in basicity of R, viz. $Bu^t > Pr^i > Et > Me > Ph$, and this should be exacerbated in strongly coordinating solvents. For simple metallation reactions where the carbanionic character is the rate-determining factor, this is observed. However, for reactions proceeding via a radical pathway involving $R^{\bullet}$, the R group forming the most stable carbanion will be the most reactive and the reverse sequence applies, $Me \approx Et > Pr^i > Bu^t$.

Organic chemists usually generate organomagnesium reagents, RMgX and R_2Mg, in solution for use in synthesis, with no attempt to isolate them. For Grignard reagents the exact nature of the species present in solution is controlled by the position of the Schlenk equilibrium, which depends on concentration, and the choice of solvent and halide (Section II.A.1). When preparing stock solutions of organomagnesium reagents, they should be standardized prior to use (ref. 7, p. 194). Base analysis after acid hydrolysis of aliquots of the reagent is only satisfactory for standardizing fresh solutions.

Relevant general references to this section are refs. 1–4, 7 (pp. 192–194), 8 and 81. The most recent covers the literature upto 1981. We focus attention mainly on new developments, but general reactions of all reaction types are included, and in this context the review is comprehensive.

A. Addition to Multiple Bonds

1. Carbon—oxygen multiple bonds

The reactions of Grignard reagents with carbon monoxide (carbonylations) usually require high partial pressures of CO. Several products are obtained and in consequence they have limited utility[8]. Other addition reactions of organomagnesium reagents are discussed below.

a. Carboxylation

The reaction of carbon dioxide with organomagnesium reagents is a well established route for preparing carboxylic acids, and is used to characterize new magnesium reagents. Usually the reagent is added to dry-ice or a slurry of it in an inert solvent. Yields are typically high, with only a few competing reactions likely. One is the formation of cyclic ketones for the carboxylation of di-Grignard reagents. Seetz *et al.*[164] have devised a method of preparing cyclobutanone, a potentially very valuable basic building block, by careful control of the carboxylation step, viz. passing a CO_2–N_2 gaseous mixture over a solution of the di-Grignard reagent at ambient temperature (Scheme 3).

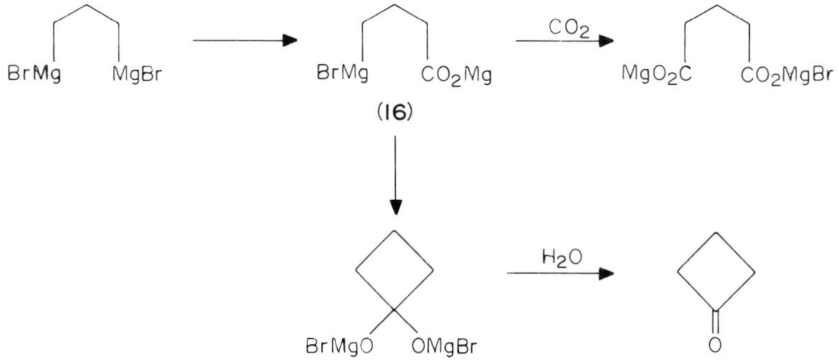

SCHEME 3

Uncontrolled carboxylation yields glutaric acid. Slow addition of CO_2 facilitates attack at the second Grignard centre, despite the inherently reduced reactivity of the carboxylate in **16**. It may be favoured by the close proximity of the other reactive centre[164]. Note that mono-Grignard reagents usually stop at the carboxylate formation stage.

A related competing reaction is the formation of ketones directly by the liberation of XMgOMgX, rather than during hydrolysis. The generated ketones can then react in either of several ways with residual Grignard reagent (Table 3).

b. Reactions with ketenes

Grignard reagents with disubstituted ketenes yield metal enolates, which on hydrolysis afford ketones (Table 3)[8]. However, there have been few applications of this reaction in synthesis. The 'pseudo ketene' carbon suboxide, C_3O_2, yields XMgOC(R)=C=C(R)OMgX and 1,3-diketones in good yield on hydrolysis[165].

c. Reactions with isocyanates

The synthesis of amides from isocyanates is well established in organic synthesis (Table 3). Treatment of **17** with one equivalent of PhMgBr afforded a heterocyclic compound, derived from attack initially at the isocyanate group; using excess of Grignard reagent metal–halogen exchange and nucleophilic attack at the benzoyl carbon—oxygen double bond, yielding (*Z*)-(2-benzamidophenyl)benzohydrol, prevailed[166].

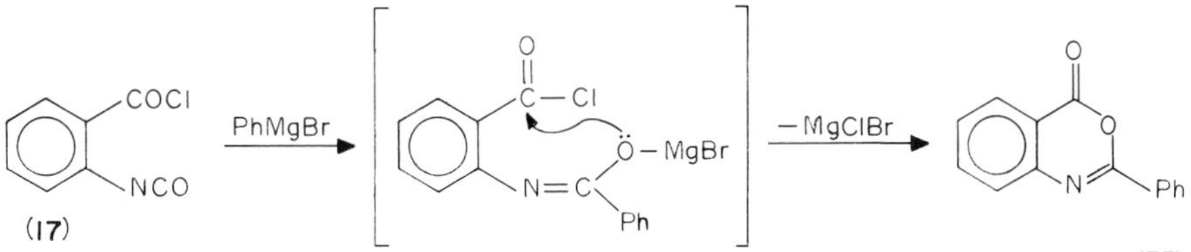

(77)

d. Reactions with aldehydes and ketones

These are the most utilized reactions of organic carbonyl compounds with magnesium reagents. The initial step in these and the aforementioned reactions is coordination of the oxygen of the functional group to magnesium. Three subsequent reaction pathways have

TABLE 3. Common reactions of organomagnesium reagents, $R_{2-n}MgCl_n$ ($n = 1$ or 0), with compounds possessing CX multiple bonds[a]

Description	Substrate	Product[b]	Hydrolysis Product	Comments
1. *Carbon—oxygen multiple bonds*				
(i) Carbon dioxide Carboxylation	CO_2	$R_2C(OMgL_n)_2$ or RCO_2MgL_n	Ketone or carboxylic acid	Useful for incorporating ^{13}C from $^{13}CO_2$
(ii) Ketenes 1,2-Addition	$R^1R^2C{=}C{=}O$	$R^1R^2C{=}C(R)OMgL_n$	Ketone	Similarly for C_3O_2
(iii) Isocyanates 1,2-Addition	$R^1N{=}C{=}O$	$R^1N{=}C(R)OMgL_n$	Carboxylic amide	
(iv) Ketones A. 1,2-Addition	R^1R^2CO	$RR^1R^2COMgL_n$	Alcohol	This is usually the main product in non-polar solvents and with MgX_2 present
B. Reduction–hydrogen transfer	R^1R^2CO	$R^1R^2C(H)OMgL_n$	Alcohol	β-Hydrogen transfer or ketyl radical hydrogen abstraction of R
C. α-Deprotonation (enolization)	$R^1COCHR^2_2$	$R_2C{=}CROMgL_n$	Aldehyde or ketone	More common for ketones and for bulky R^1, R^2, and/or R groups
(v) α,β-Unsaturated ketones A. 1,2-Addition	$R^3R^4C{=}C(R^2)C(R^1){=}O$	$R^3R^4C{=}C(R^2)C(R^1)(R)OMgL_n$	Alcohol	Favoured for aldehydes
B. 1,4-Addition	$R^3R^4C{=}C(R^2)C(R^1){=}O$	$R^3R^4C(R){-}C(R^2){=}C(R^1)OMgL_n$	Aldehydes or ketones	Favoured by bulky substituents on the carbonyl group and for solvents promoting electron transfer.

(vi) Acyl chlorides				
A. Exchange	R^1COCl	RR^1CO	Ketone	Formation of $[R^1(O)C]_2$ via a radical pathway can also occur
B. exchange then 1,2-addition	R^1COCl	$R_2R^1COMgL_n$	Alcohol	
(vii) Carboxylic anhydrides				
A. Ketone formation	R^1COCOR^1	$R^1COR + R^1CO_2MgL_n$	Ketone and a carboxylic acid	Favoured by a low reaction temperature, strongly coordinating solvents, and steric hindrance for R^1
B. Lactone formation	R^1COCOR^1	$R^1CO_2CR_2R^1$	Lactone	
(viii) Carboxylic esters (and lactones)				
A. Exchange	$R^1CO_2R^2$	R^1COR	Ketone	Mild conditions are required for ketone formation. α-Metallation (enolization) can also occur
B. Exchange then 1,2-addition	$R^1CO_2R^2$	$R_2C(R^1)OMgL_n + R^2OMgL_n$	Alcohol	
(ix) Carboxylic amides (and lactones and carbamates)	$R^1CONR^2_2$	$RR^1C(OMgL_n)NR^2_2$	Ketone	Hindered formamides are susceptible to formyl proton abstraction
(x) α-Keto esters 1,2-Addition	$R^1COCO_2R^2$	$RR^1C(OMgL_n)CO_2R^2$	α-Hydroxy ester	1,2-Addition of R^1CO- requires forcing conditions
(xi) Carboxylate salts	$R^1CO_2ML_n$	$RR^1C(OML_n)OMgL_n$	Ketone	
2. *Carbon—sulphur multiple bonds*				
(i) Carbon disulphide Carbophilic addition	CS_2	$RC(=S)SMgL_n$		The salt is usually used *in situ*
(ii) Isothiocyanates 1,2-Addition	R^1NCS	$R^1N=C(R^2)SMgL_n$	Thioamide	
(iii) Thiocyanates 1,2-Addition	R^1SCN	R^1SR	Thioether	
(iv) Sulphines 1,2-Addition	R^1R^2CSO	$R^1R^2C(MgL_n)S(R)=O$	Sulphoxide	Carbophilic addition can also occur
(v) Thioketone 1,2-Addition	R^1R^2CS	$R^1R^2C(SR)MgL_n$	Thioether	
(vi) Thioesters (trithiocarbonates)	$R^1C(=S)SR^2$	$R^1C(SR)(SR^2)MgL_n$	Dithioacetal	Carbophilic addition can occur, as for esters, yielding tertiary thiols

TABLE 3. (*Contd.*)

Description	Substrate	Product[b]	Hydrolysis Product	Comments
3. *Carbon—nitrogen multiple bonds*				
(i) Imines (and nitrogen aromatic heterocycles)				
1,2-Addition	$R^1R^2C{=}NR^3$	$RR^1R^2CN(R^3)MgL_n$	Amine	α-Hydrogen abstraction at R^1 or R^2 can occur. For nitrogen aromatic heterocycles (α, β-unsaturated imines) 1,4-addition is possible
(ii) Iminium salts				
1,2-Addition	$R^1R^2C{=}N^+R^3R^4\ X^-$	$RR^1R^2CNR^3R^4$	Tertiary amine	
(iii) Imidoyl chlorides				
Exchange	$R^1ClC{=}NR^2$	$R^1RC{=}NR^2$	Imine	Further reaction is possible (i)
(iv) Carbodiimines				
1,2-Addition	$R^1N{=}C{=}NR^1$	$R^1N(MgL_n)C(R){=}NR^1$	Amidines	
(v) Nitrones				
1,3-Addition	$R^1R^2C{=}NR^3$ (N→O)	$RR^1R^2CN(R^3)OMgL_n$	Hydroxylamines	
(vi) Nitriles				
A. 1,2-Addition	$R^1C{\equiv}N$	$RR^1C{=}NMgL_n$	Imine or ketone	The imine is the first product on hydrolysis
B. α-Deprotonation	$R^1R^2CHC{\equiv}N$	$R^1R^2C(MgL_n)C{\equiv}CN$	Nitrile	
(vii) Nitrile oxide				
1,3-Addition	$R^1C{\equiv}N\rightarrow O$	$RR^1C{=}NOMgL_n$	Ketoximes	
(viii) Isocyanates	$R^1N{\equiv}C{:}$	$R^1N{=}C(R)MgL_n$	Aldehyde (RCHO)	
(ix) Cyanates	$R^1OC{\equiv}N$	$R^1OC(R){=}NMgL_n \downarrow RCN + R^1OMgL_n$	Nitrile	
(x) Isocyanates	$R^1N{=}C{=}O$	$R^1N{=}C(R)OMgL_n$	Amide	
(xi) Isothiocyanates	$R^1N{=}C{=}S$	$R^1N(MgL_n)C(R){=}S$	Thioamide	

[a] For addition to carbon—carbon multiple bonds, see Table 1.
[b] L_n refers to the other ligands on magnesium, R or X and solvent.

been established: (a) a concerted mechanism involving a four-centre pericyclic species[167], which yields exclusively a 1,2-addition compound; (b) the formation of a cyclic transition state, leading to α-hydrogen abstraction (enolization)[121]; and (c) a radical pathway which is usually rapid (Scheme 4).

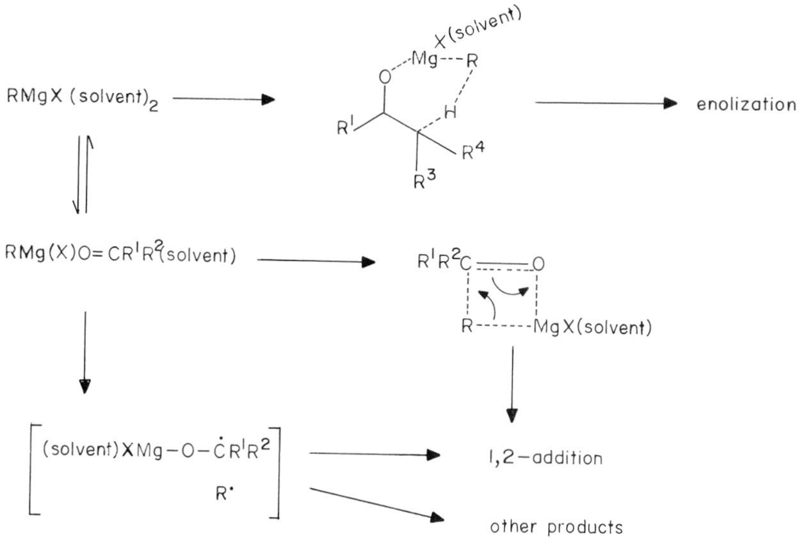

SCHEME 4

In Scheme 4, the electron transfer process is the rate-determining step; reactions other than 1,2-addition include hydrogen transfer to the ketyl radical and coupling of ketyl radicals (pinacol formation), which is favoured for the radical pair dissociating to form free radicals[8,168]. Ashby and coworkers suggested that all Grignard reagents with aromatic ketones proceed via an electron transfer pathway[169]. This is based on the formation of cyclized products in the reaction of **18**, which is consistent with the generation of a neopentyl type radical rearranging before combining with $Ph_2\dot{C}O^{-}$[65].

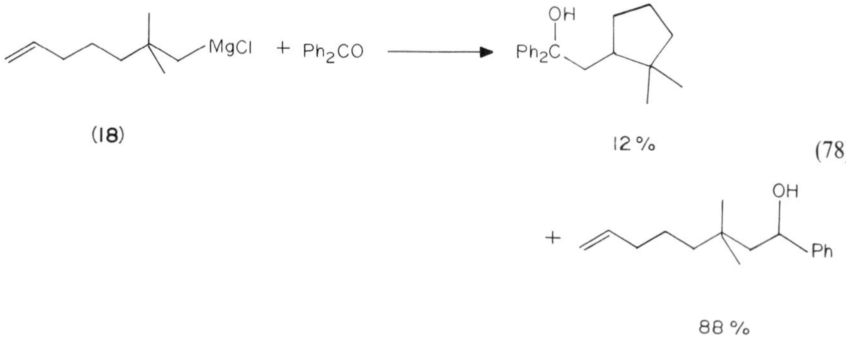

(78)

The Grignard of 1-chloromethylnaphthalene with ketones yields *ortho*-alcohols, the normal 1,2-addition product, and enolization (Scheme 5)[170]. The former originates from the ambient nature of the arylmethyl group, as has been noted for the reaction of a related lithium reagent with various electrophiles[171]. For unhindered aldehydes and ketones (Scheme 5), the initial reaction is the addition to the aromatic ring, but it is reversible[170].

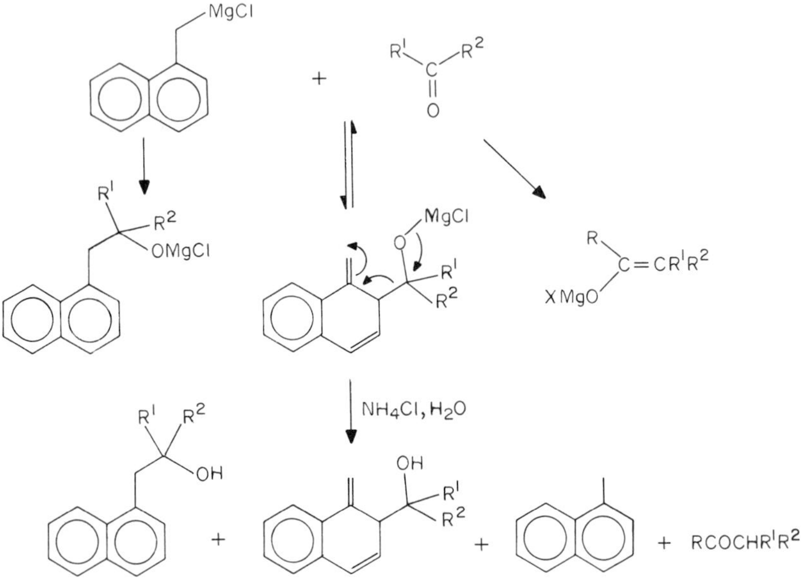

SCHEME 5

For ketones possessing functional groups, other reactions are possible, for example conjugated 1,4-addition to α,β-unsaturated ketones (see below), proceeding either by a radical pathway or a concerted process involving a cyclic transition state.

The reaction pathway appears to depend on the nature of R (the ability to form a stable radical, R$^{\bullet}$, will tend to favour the radical pathway) and steric compression of R and/or R^1 and R^2 (defined in Table 3). The latter will favour enolization. Choice of solvent and temperature can also be important. Low temperatures and non-coordinating solvents[172] favour 1,2-addition, which is usually the desired route. Nevertheless, for ketyl radicals, $R_2\dot{C}O^-$, hydrogen abstraction is usually a minor route and reduction by β-hydrogen transfer is only important for hindered, non-enolizable ketones, and by careful choice of the experimental conditions 1,2-addition is normally accessible.

An alternative approach to the reaction of ketones with Grignards is the Barbier reaction (Section II.A.4), the *in situ* Grignard trapping reaction. With a knowledge of the factors that affect the reactions of ketones with organomagnesium reagents and those which favour the *in situ* method, it may be possible to predict which method is suited for a particular substrate and/or Grignard reagent.

Various reactions of organomagnesium reagents with carbonyl compounds are tabulated in ref. 173. Reactions from *Organic Syntheses* are listed in ref. 8, p. 26. Some representative new examples are given in Table 4, together with some novel reactions. An alcohol is the expected product on hydrolysis of a 1,2-addition compound, but often spontaneous dehydration prevails. Attack of a ketone group is favoured over a carboxylic acid or ester group (e.g. entries 1 and 2, Table 4), unless the reaction is carried out in an aromatic solvent[174]. Epoxy ketones react selectively with Grignard reagents at low temperature, *ca.* $-10\,^{\circ}$C, to afford the 1,2-addition product of the ketone residue[175]. Halocarbonyl compounds undergo nucleophilic attack at the carbonyl group, and for more forcing conditions alkyl halogen exchange is possible[176]. α-Chlorocarbonyl

TABLE 4. Reactions of Grignard reagents with ketones and related compounds[a]

Substrate	Grignard	Product on hydrolysis	Ref.
1. 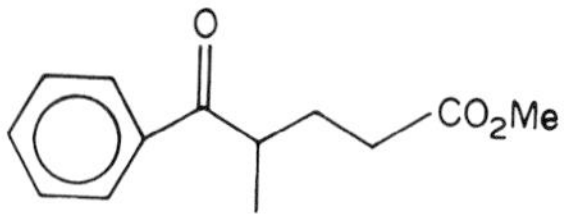	MeMgCl	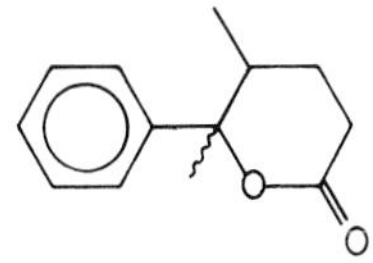 (*Z*)- + (*E*)-	193
2. 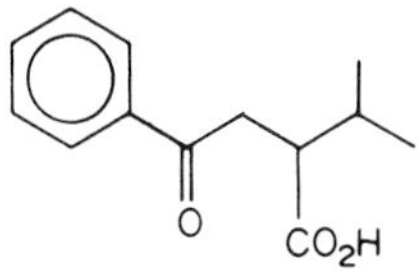	MeMgI	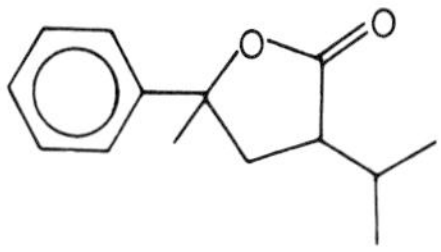	194
3. 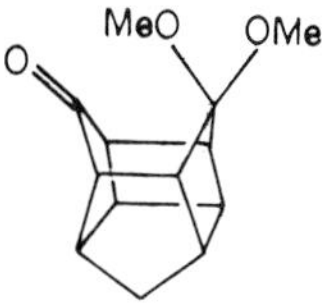	PhMgI	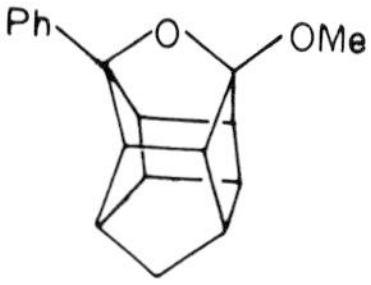	195
4. 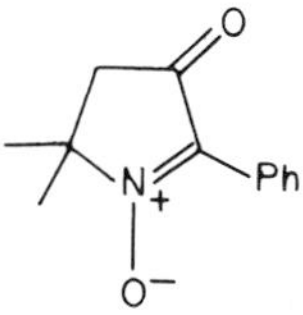	PhMgX	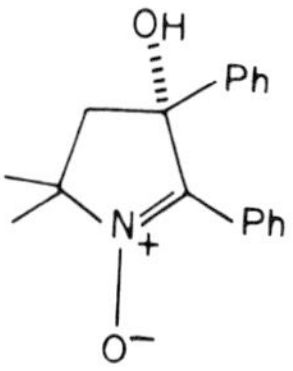	196
5. 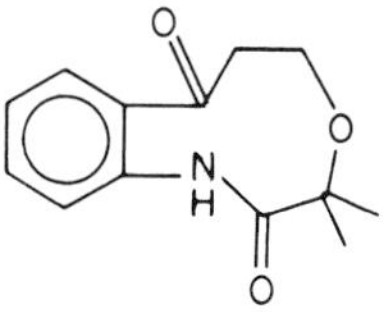	PhMgBr	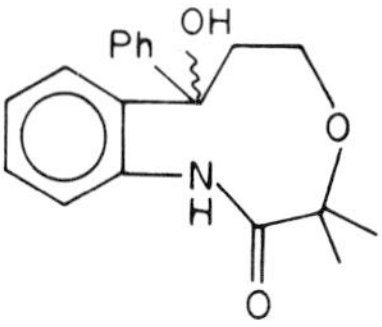	197
6. 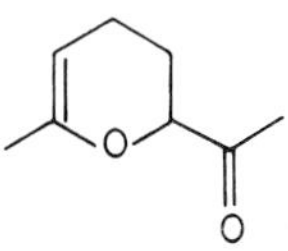	MeMgX	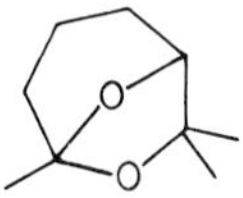	198
7. 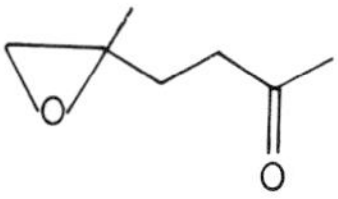	MeMgBr	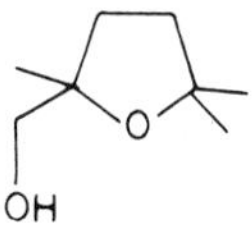	175

TABLE 4. (*Contd.*)

Substrate	Grignard	Product on hydrolysis	Ref.
8.	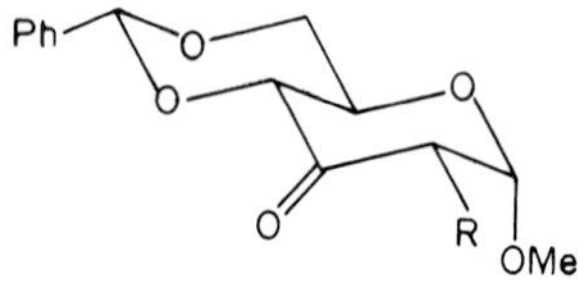 Me_3SiCH_2MgCl	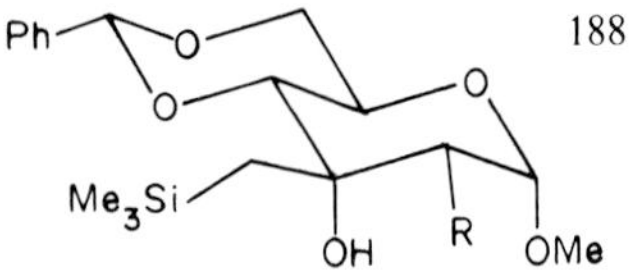	188
9.	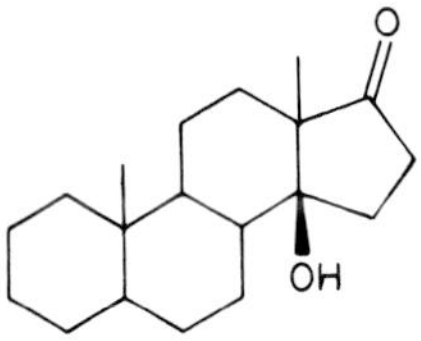 BrMgC≡CMgBr	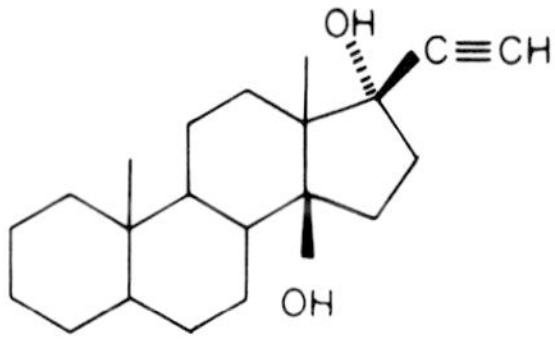	189
10.	H(Me)–CH(CH=CH₂)MgBr		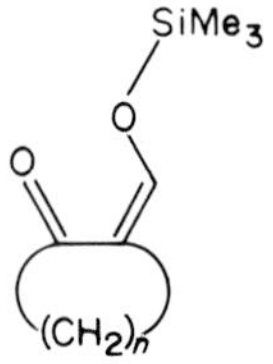 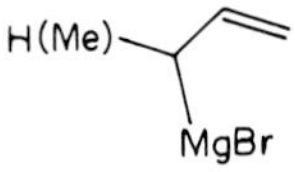 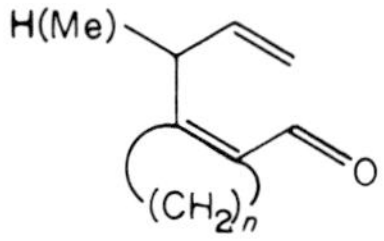 208
11.	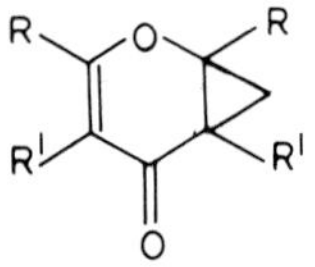 MeMgI	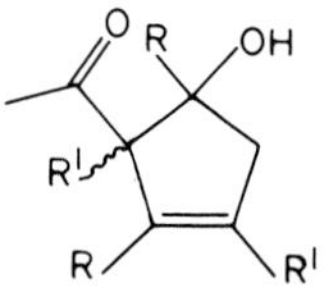	209
12.	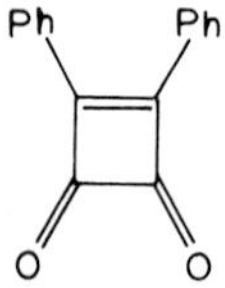 MeMgBr	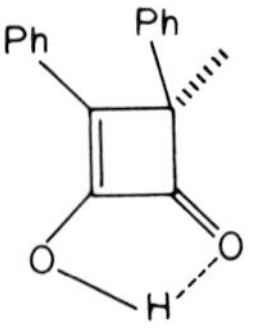	212
13.	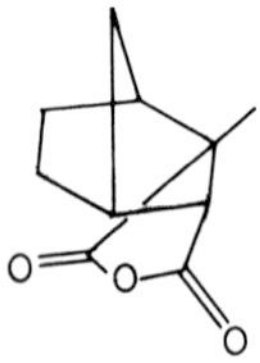 $BrMg(CH_2)_nMgBr$ ($n = 4$ or 5)	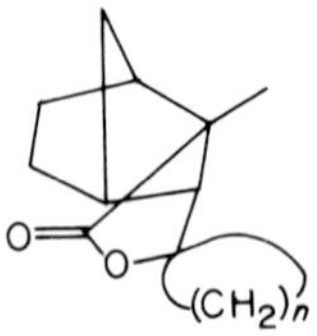	219
14.	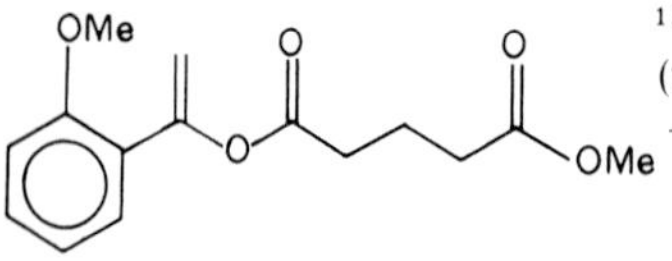 $^{13}CH_3^*MgI$, (thf–OEt_2), −78 °C	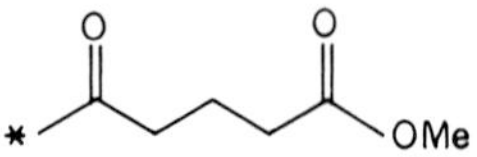	218

TABLE 4. (*Contd.*)

Substrate	Grignard	Product on hydrolysis	Ref.
15. $Me_3SiCH_2CO_2Et$	$PhCH_2MgCl$	$CH_2{=}C(CH_2Ph)_2$	222
16. 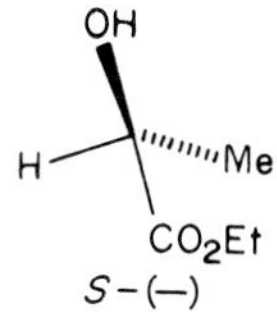 OH, H, Me, CO_2Et; S-(−)	ArMgBr	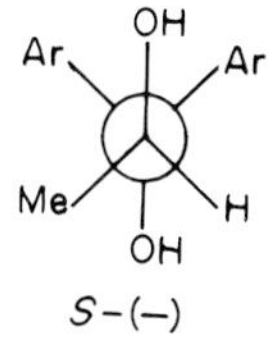 OH, Ar, Ar, Me, H, OH; S-(−)	223
17. 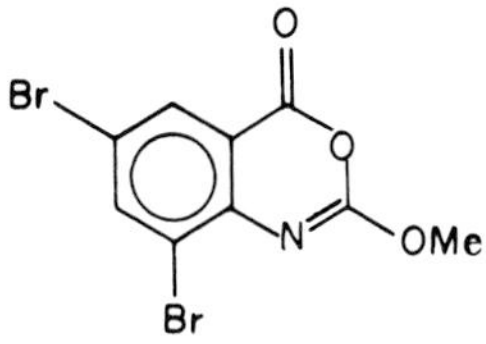 Br, O, O, N, OMe, Br	RMgX	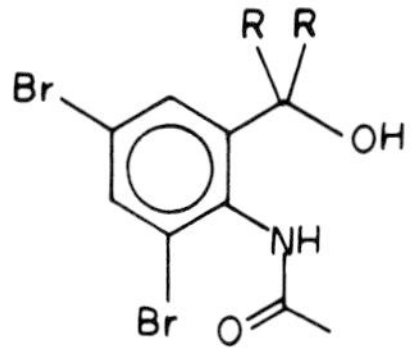 R R, Br, OH, NH, Br, O	226
18. 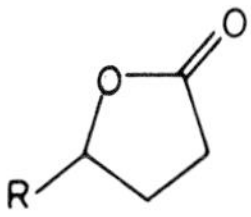 O, O, R	$2Me_3SiCH_2MgBr$	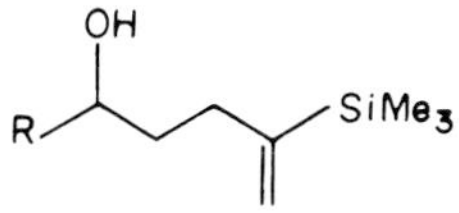 OH, R, $SiMe_3$	227
19. 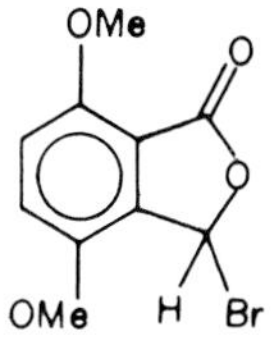 OMe, O, O, OMe, H, Br	PhMgBr	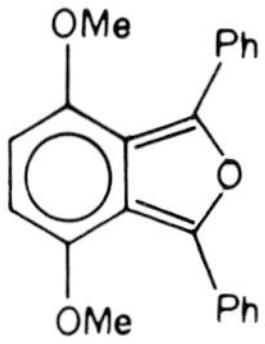 OMe, Ph, O, OMe, Ph	236
20. 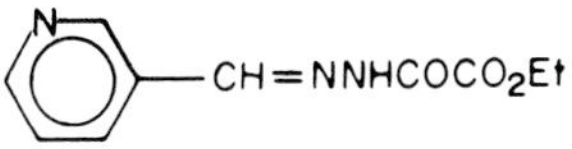 N, $CH{=}NNHCOCO_2Et$	PhMgBr	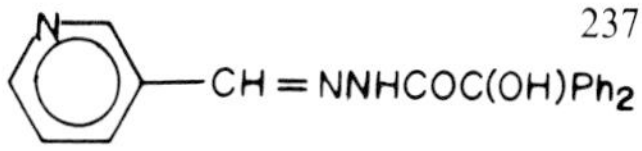 N, $CH{=}NNHCOC(OH)Ph_2$	237
21. 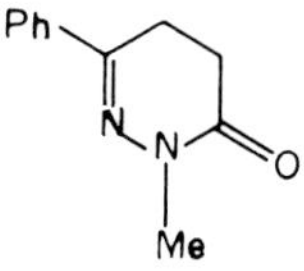 Ph, N, N, O, Me	PhMgBr	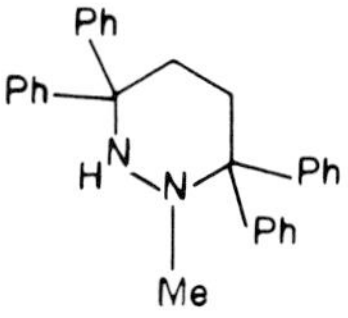 Ph, Ph, H, N, N, Ph, Ph, Me	238
22. 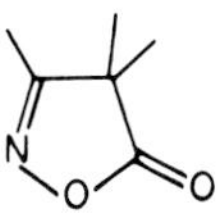 N, O, O	$BrMg(CH_2)_4MgBr$	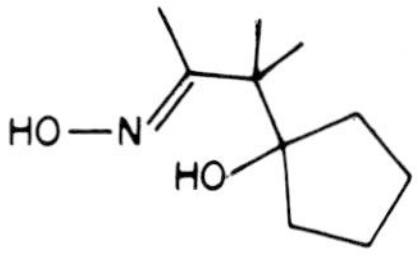 HO—N, HO	239

TABLE 4. (*Contd.*)

Substrate	Grignard	Product on hydrolysis	Ref.
23. 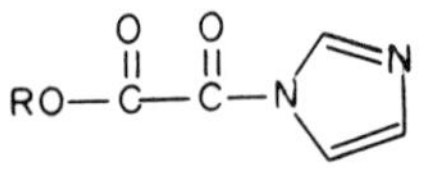 RO—C(=O)—C(=O)—N(imidazolyl)	R^1MgX R^1 = aryl	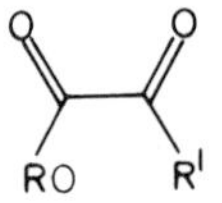 RO—C(=O)—C(=O)—R^1	240
24. 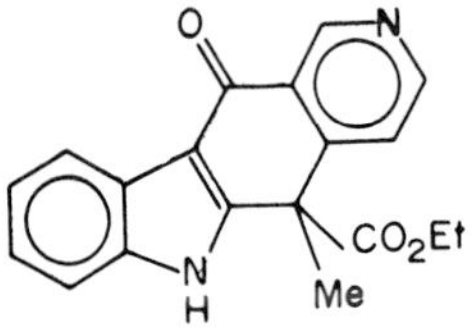 (indole-fused keto ester; CO$_2$Et, Me, N–H)	RMgX (5 equiv.)	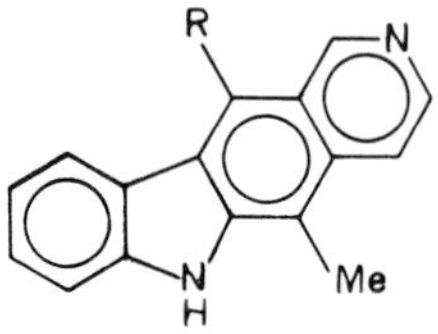 (R-, Me-substituted pyrido-carbazole; N–H)	241

[a]The hydrolysis products are all in useful yields. Typically > 60%.

compounds with Grignard reagents, followed by lithium-halogen exchange, offers a powerful route to classes of olefins (e.g. equation 79)[177].

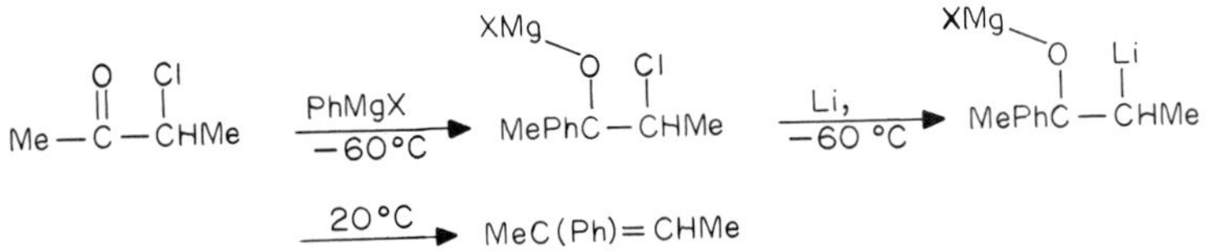

$$\mathrm{Me{-}C(O){-}CHClMe} \xrightarrow[-60\,^\circ\mathrm{C}]{\mathrm{PhMgX}} \mathrm{MePhC(OMgX){-}CHClMe} \xrightarrow[-60\,^\circ\mathrm{C}]{\mathrm{Li,}} \mathrm{MePhC(OMgX){-}CHLiMe} \quad (79)$$

$$\xrightarrow{20\,^\circ\mathrm{C}} \mathrm{MeC(Ph){=}CHMe}$$

Tabushi *et al.*[178] obtained good yields of unsymmetric ketones on hydrolysis of the mixture obtained by treating MeMgI with acyl metal complexes of *N*,*N*′-ethylenebis(salicylideneiminato)cobalt(III). The intermediate magnesium complex, $[L_nCoC(Me)(R)OMgX]$, resembles that of the normal 1,2-addition product of ketones with Grignard reagents.

1,2-Additions to prochiral carbonyl compounds or chiral carbonyl compounds possessing one or more chiral centres can give rise to enantiomeric mixtures and diastereoisomers, respectively, because of the formation of a new asymmetric carbon centre. Facial attack from either side of the R^1R^2CO plane (concerted pathway) or $R^1R^2\dot{C}O^-$ (radical pathway), is usually controlled by steric effects. Consider the examples in equations 80 and 81. In equation 80 the addition is exclusively in the *endo* direction owing to the steric hindrance of the benzyloxy group on the *exo* face of the ketone[179]. In equation 81 there is stereoselective control of addition by the steric hindrance of the π-bound $Cr(CO)_3$ moiety[180].

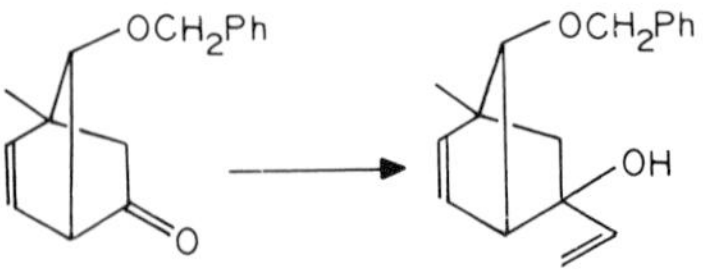

OCH$_2$Ph (bicyclic ketone) → OCH$_2$Ph, OH (vinyl adduct) (80)

Cr(CO)$_3$ indanone with CH$_2$Ph, Me → Cr(CO)$_3$ indanol with CH$_2$Ph, Me, Me, OH (81)

There has been considerable research into developing 1, 2-addition reactions that yield products of high optical purity.(Two main strategies have been considered, viz. (a) the reaction of a prochiral ketone or aldehyde in a chiral medium or using a chiral magnesium complexing agent and (b) relying on asymmetric induction from chiral centres in either the carbonyl compound or the hydrocarbyl anion of the magnesium reagent. A detailed discussion of the factors affecting the degree of asymmetric induction is given elsewhere[169,181].

The greatest success for chiral complexing agents is for polydentate ligands based on carbohydrates and is understood in terms of the stereochemical control around the magnesium environment on complexation[182]. For the strategy of having an asymmetric centre within the carbonyl compound, almost complete asymmetric synthesis is possible, even for that centre remote from the carbonyl group (e.g. equation 82)[183]. Complexation of an N- or O-centred functional group in the organic substrate may anchimerically assist the formation of one enantiomer[184].

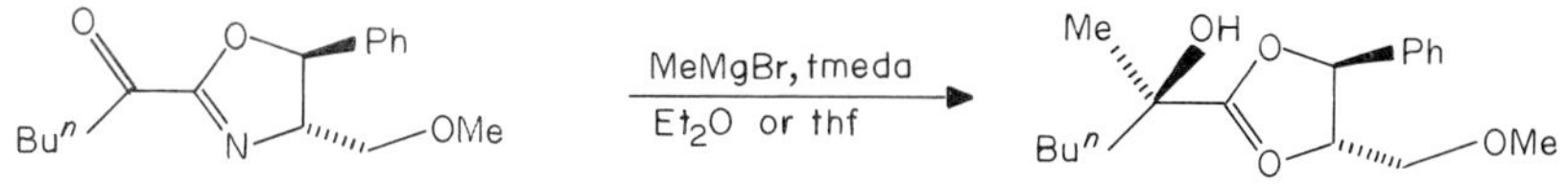

MeMgBr, tmeda / Et_2O or thf (82)

Compound **19** has yielded products of high enantiomeric excess and by changing the order of addition of two different Grignard reagents both enantiomers have been obtained (Scheme 6).

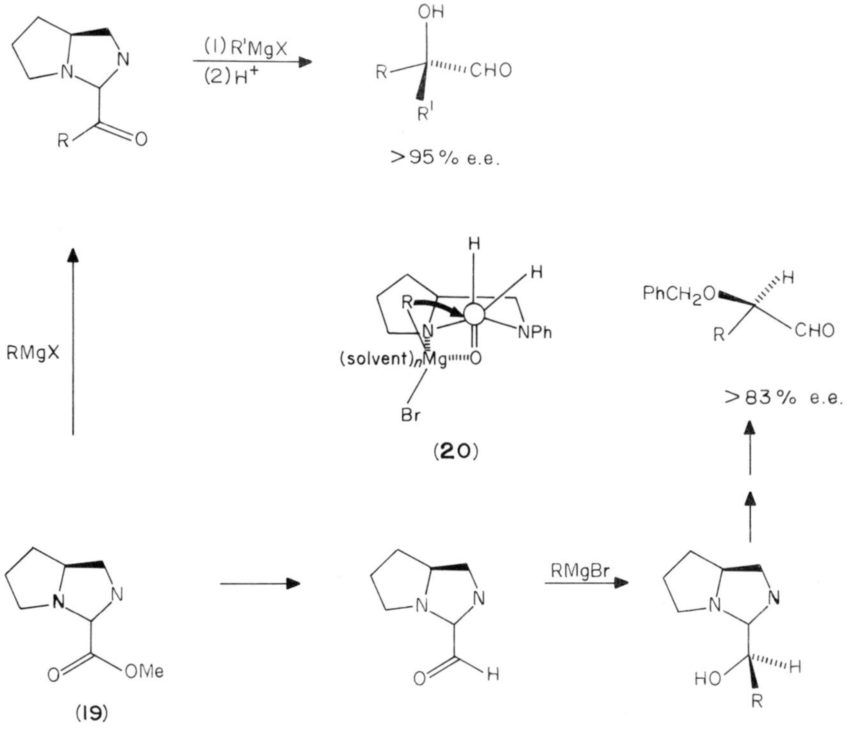

(1) R'MgX (2) H^+ >95 % e.e.

RMgX (solvent)$_n$Mg (20) >83 % e.e. (19) RMgBr

SCHEME 6

The high stereoselectivity of the reaction of the aminal in Scheme 6 has been interpreted mechanistically by considering the formation of a *cis*-fused bicyclic ring structure (**20**) controlling the addition of the Grignard reagent[185].

Stereoselective syntheses have featured as key steps in the total syntheses of natural products[186] and the syntheses of chiral tertiary alcohols of various sugars (entry 8, Table 4)[187]. Interestingly, the stereospecificity incorporating a side-chain in a 14β-hydroxy steroid (entry 9, Table 4) is reversed using a lithium reagent[188].

The generation of two new chiral centres is possible (e.g. equation 83)[189] and has potential in synthesis.

Ph–CH=CH–CH$_2$MgCl + PhHCO ⟶ (70%) + (30%) (83)

⇅ PhCH(MgCl)CH=CH$_2$

Allylic Grignard reagents usually exist as an equilibrium mixture in solution (e.g. equation 83) and two modes of addition are possible. Unhindered carbonyl compounds with a substituted allylic magnesium reagent give the product in which the allylic group is attached at the highly substituted position (equation 83), whereas with hindered ketones it is the least substituted position. In the presence of $AlCl_3$, however, even unhindered ketones give predominantly products in which the allylic group is attached at the least substituted position[190].

The stereoselectivity of a reaction can usually be improved at low temperatures, particularly if the environments of *endo* and *exo* attack of the carbonyl group are similar. It is noteworthy that low temperatures tend to favour 1, 2-addition, the thermodynamically controlled route, reducing the likelihood of competing reaction(s).

If a mixture of epimers is undesirable, their dehydration to the same olefin is often useful, having been used to good effect in the synthesis of steroids[191].

In addition to steric control of 1, 2-addition, there is evidence that the nucleophilicity of the organometallic reagent plays some role. The reaction of 3-phenylbutan-1-one with arylmetal species gives the highest stereoselectivity for the reagent of greatest nucleophilicity according to the following inequality[192]: $PhLi > Ph_2Mg > PhMgBr > Ph_3Al > PhMgBr–Cu^I$.

e. *Reactions of α, β-unsaturated carbonyl compounds*

The main types of reaction are 1, 2-addition as in (iii) and 1, 4-addition (Table 3); some cases of 1, 3-addition have been noted but the amount of product is usually small compared with that from 1, 2- and/or 1, 4-addition[199]. 1, 2-Addition may be reversible, so that if it is the kinetically favoured product, that of 1, 4-addition may be thermodynamically favoured[8]. Factors that favour 1, 4-addition are steric compression at the reactive nucleophilic centre of the magnesium reagent or around the carbonyl group, solvents favouring an electron transfer mechanism (e.g. hmpt) and delocalized hydrocarbyl groups of the organomagnesium reagent. A similar behaviour occurs for higher conjugated carbonyl compounds (e.g. 1,8-addition to tropone, yielding 2-substituted dihydrotropone)[200], but these have received little attention[201].

Aldehydes usually undergo 1,2-addition in accordance with low steric hindrance at the carbonyl group. As for ketones, consider the reaction of alkyl styryl ketones, PhCH=CHCOR. For various Grignard reagents the amount of 1,4-addition product increased with a concomitant decrease in that derived from 1,2-addition for an increase in the size of the alkyl group[202]. It is noteworthy that dehydration, on acid work-up of 1,2-addition products, yields conjugated dienes (equation 84)[203] or more complex reaction products (equation 85)[204].

CO_2Et —MeMgI→ IMgO Me, CO_2Et —H^+→ CO_2Et (84)

OEt —RMgBr→ BrMgO R, OEt —H^+→ R, O (85)

Copper halides, either as catalysts or in stoichiometric amounts with organomagnesium reagents, afford exclusively the 1,4-addition product for ketones and either 1,2- or 1,4-addition products for aldehydes.

Like 1,2-additions to simple ketones, 1,4-additions also have the feature of yielding diastereoisomers. One chiral centre is generated on the addition step and another on hydrolysis of the magnesium enolate (Table 3). For chiral centres already present, the stereocontrol can be high (e.g. equation 86)[205], whereas the approach of using a chiral complexing agent has met with limited success[206].

O —$CH_2{=}CHMgBr$, CuI, thf, −78 °C→ O (86)

Virtually complete asymmetric induction for the 1,4-addition to optically pure 2-arylsulphinylcyclopent-2-enones has been reported. Interestingly, the other diastereoisomer possible is obtained by first complexing the sulphinyl compound with divalent zinc[207].

Some unusual reactions of α,β-unsaturated ketones are shown in Table 4 (No. 10, 1,2-addition to vinylogous alkyl ethers then acid work-up[208]; No. 11, 1,2-addition then isomerization[209]). Quinones and related species react by a radical pathway and often yield a complex mixture of 1,4-addition products and compounds derived from electron transfer processes. Addition can occur at either or both carbonyl carbon centres and olefin centres and only by judicious choice of reaction conditions does one product predominate.

The reactions in Scheme 7 are representative of the reactions encountered[210].

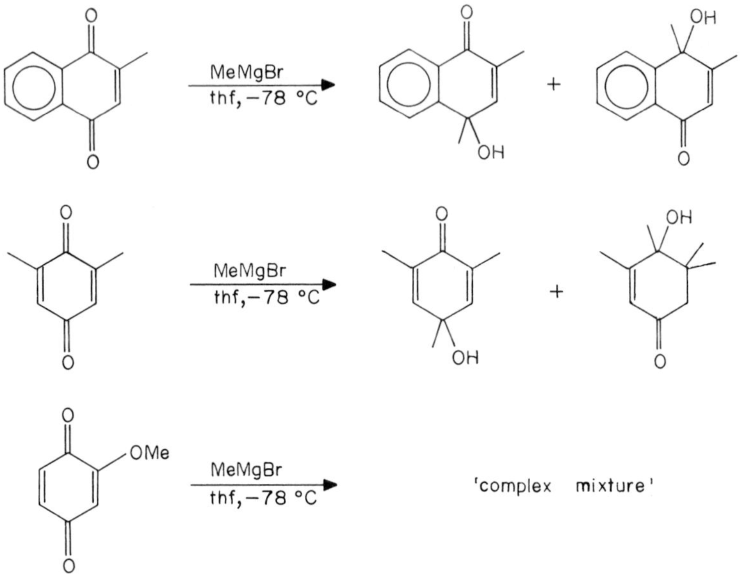

SCHEME 7

Quinone monoketals react with magnesium reagents with a greater regiochemical outcome relative to quinones and they are less susceptible to electron transfer reduction processes. For the ketal in Scheme 8, MeMgBr gave the 1, 2-addition product, (allyl)MgBr the ring alkylated product, and Bu^tMgBr the reduced ketal[211].

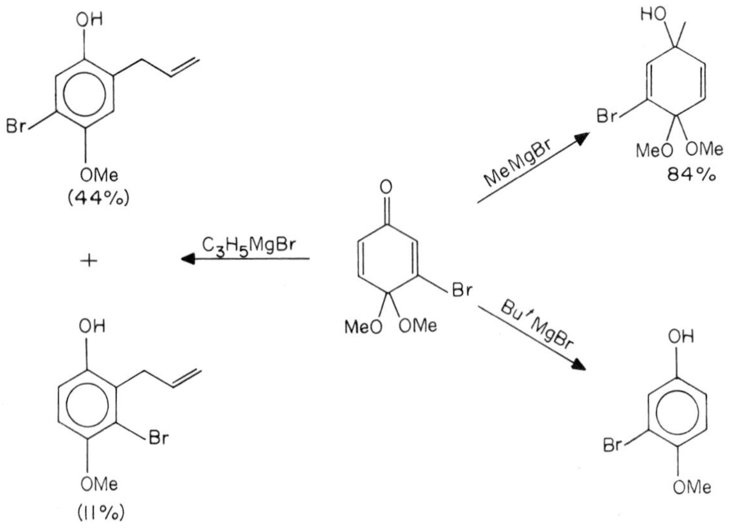

SCHEME 8

Cyclobutenediones react with alkylmagnesium bromide to yield 1,4-addition products (No. 12, Table 4)[212], whereas squaric acid, $\overline{OCC(O)C(OR)C}(OR)$, yields both 1,2- and 1,4-addition products[213].

f. Reactions with acyl halides

These and related compounds such as chloroformates, chlorocarbamates, and phosgene react vigorously with Grignard reagents. At low temperature (*ca.* $-70\,^{\circ}C$) with thf as a solvent, with or without a catalyst[214], or with a sterically hindered Grignard[215], it is possible to obtain excellent yields of ketones from acyl halides. Two equivalents of the alkylating agent usually afford alcohols in high yield. Barluenga *et al.*[216] prepared a series of γ-substituted organoalkali metal compounds by treating the intermediate complex with an alkali metal (equation 87).

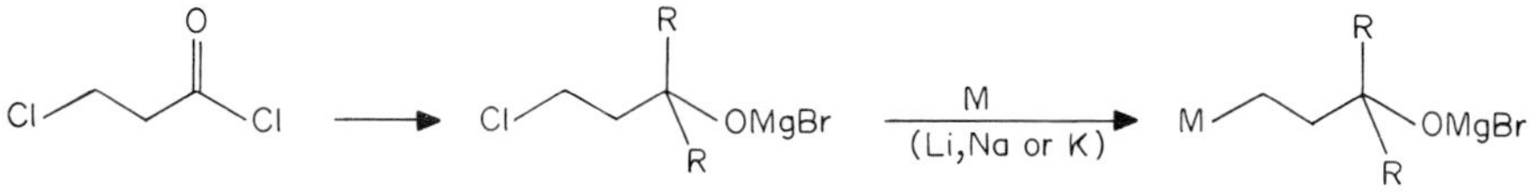

(87)

A competing reaction is the formation of radicals leading to α-diketones. Studies have shown that the main cause of radical formation is the electron transfer ability of RMgX. For R as an isopropyl group, electron transfer is favoured relative to it being a methyl group, consistent with isopropyl being more electron donating than methyl. Moreover, for X as iodide, electron transfer is favoured more than for a bromide (Scheme 9)[217].

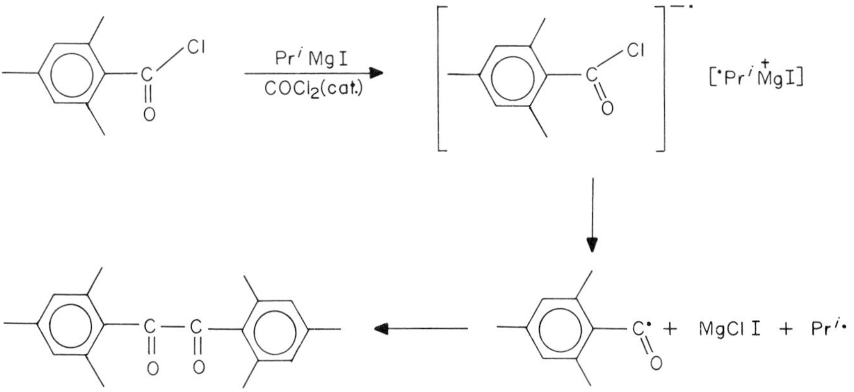

SCHEME 9

g. Reactions of carboxylic anhydrides

Several reaction products are possible, although under critical conditions and with one equivalent of Grignard reagent ketones and carboxylic acids are accessible, the formation of which is a consequence of the good leaving group capability of $-CO_2^-$ (Table 3). Preferential attack of one carbonyl group for unsymmetrical anhydrides, $RC(O)OC(O)R^1$, is electronically controlled (leaving group capability of RCO_2^- *vs.* $R^1CO_2^-$, e.g. No. 14, Table 4)[218] and/or is regiosterically controlled.

Two equivalents of Grignard reagent usually lead to lactone formation (Table 3). There has been much research into the reactions of di-Grignard reagents as a route to spirolactones (No. 13, Table 4)[219]. For unsymmetrical cyclic anhydrides the reactions (spiroannelation) are highly regioselective with nucleophilic addition at the least hindered carbonyl group. Canonne *et al.*[220] proposed a mechanism for this reaction which involves the formation of a stable intermediate carboxylate (Scheme 10).

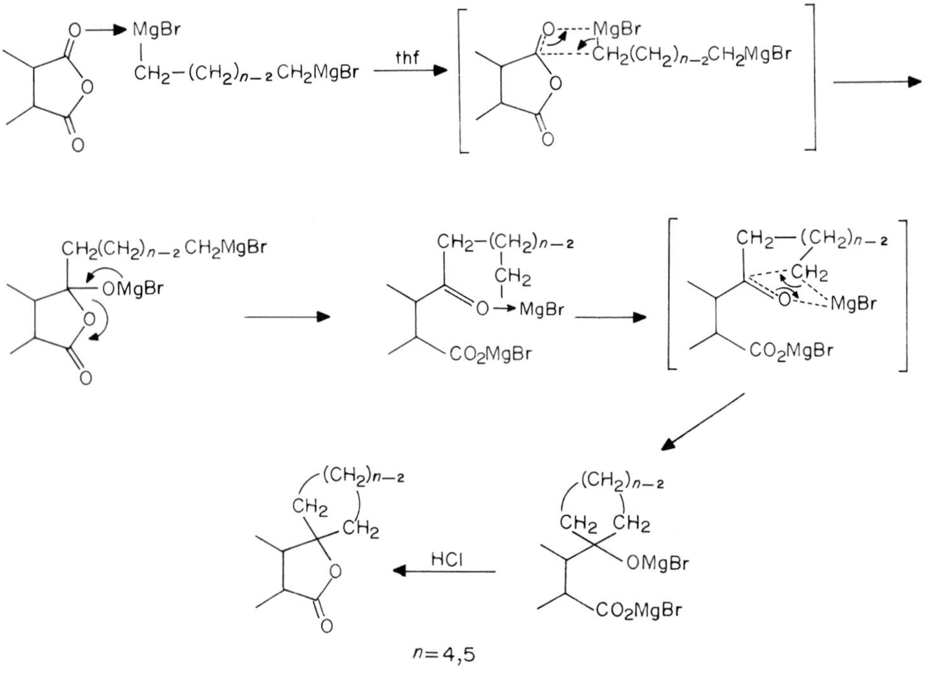

$n = 4,5$

SCHEME 10

h. Carboxylic esters, lactones, thiol esters, and thiol lactones

As for acyl halides and anhydrides, under controlled conditions ketones are accessible. Forcing conditions can yield tertiary alcohols or, in the case of formates, secondary alcohols (Table 3). Scheme 11 highlights reactions that can occur with di-Grignard reagents[221]. Varying the value of n in $BrMg(CH_2)_nMgBr$ dramatically affects the product distribution. For $n = 4$ annelation is $> 90\%$ whereas for $n = 5$ annelation, intramolecular reduction, and some enolization, depending on R, occur[221]. In contrast, for anhydrides (discussed above) annelation is the major reaction.

Reactions of α-silylated esters with Grignard reagents have recently been investigated. They yield olefins in which the double bond is introduced stereospecifically (No. 15, Table 4) and for highly polar solvents such as thf and hmpt some coupled Grignard

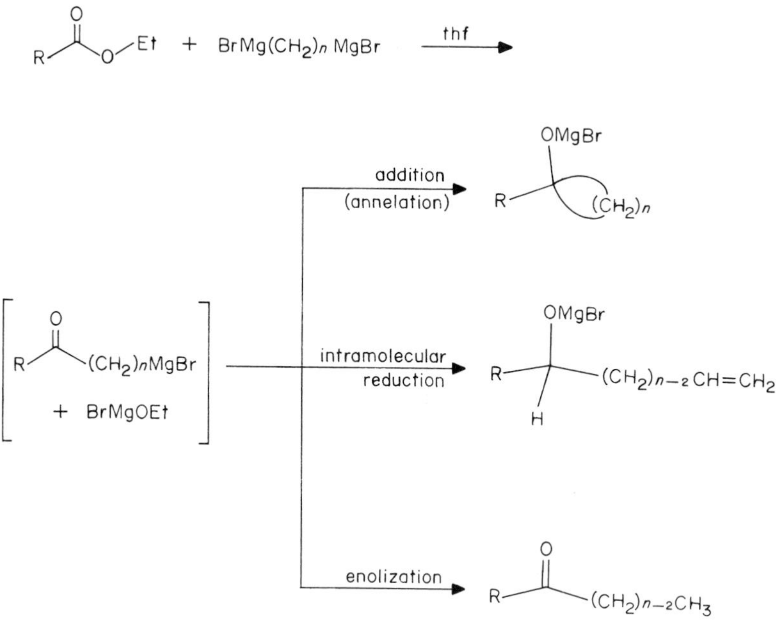

$RC(O)OEt + BrMg(CH_2)_nMgBr \xrightarrow{thf}$ [$RC(O)(CH_2)_nMgBr$ + BrMgOEt]

addition (annelation) → $R\text{–}C(OMgBr)(CH_2)_n$ (cyclic)

intramolecular reduction → $R\text{–}CH(OMgBr)\text{–}(CH_2)_{n-2}CH{=}CH_2$

enolization → $R\text{–}C(O)\text{–}(CH_2)_{n-2}CH_3$

SCHEME 11

product. The latter is attributed to a single electron transfer process involving $Me_3SiCH_2\dot{C}(\bar{O})OEt$ which may be stabilized by the β-silyl and $R^{\cdot}$ groups[222].

Dicarboxylic esters react with two equivalents of Grignard reagent to afford a product derived from addition to one ester group, yielding on hydrolysis a hydroxycarboxylic acid[223]. This is related to the behaviour of carboxylic acid anhydrides. The Barbier or *in situ* method (Section II.A.4) is also applicable for the reaction of esters and related carbonyl compounds. The condensation of esters with Grignard reagents has featured as the key step in the syntheses of natural products (e.g. No. 18, Table 4)[224].

α,β-Unsaturated esters, like α,β-unsaturated ketones, yield 1,2-addition products which can further react to yield an alcohol on hydrolysis, and/or 1,4-addition products. It is noteworthy that organolithium reagents tend to give 1,2-addition products, whereas copper-catalysed Grignard reactions favour 1,4-addition[8]. Dialkyl carbonates with Grignard reagents, RMgX, usually yield a tertiary alcohol, R_3COH.

Lactones react in a similar manner to esters. The simplest lactone, propiolactone, has been the subject of numerous studies, it being a useful synthon for the corresponding homologous acid with three more carbon atoms. Where the reactions are copper halide-catalysed, the primary process is C—O bond rupture. These reactions are discussed in Section II.B.2.

For uncatalysed reactions the initial step is nucleophilic addition to CO, which can result in three product types, depending on the reaction conditions and work-up procedure. At low temperature ketols (**21**) are generated, whereas more vigorous conditions yield diols and on hydrolysis cyclic ether formation is possible. This is exemplified by the reactions in Scheme 12[225,239].

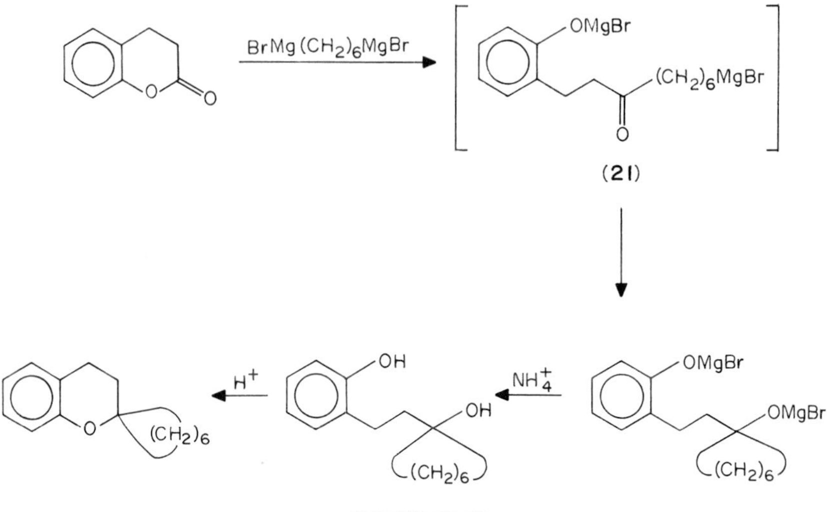

SCHEME 12

Other reactions of lactones, which include some novel types, are those listed in Table 4 (17 and 18), and 1,4-additions for α,β-unsaturated compounds. Copper halide-catalysed 1,4-additions to lactones bearing an asymmetric centre within the lactone ring can be highly stereospecific[228]. In a study of the reaction of di-Grignard reagents of the type $XMg(CH_2)_nMgX$ with α,β-unsaturated lactones, the value of n affected the route and products obtained, viz. annelation with $n = 4$ and conjugate addition with $n = 5$[229].

Trihaloacetates, CX_3CO_2R ($X = Cl$ or Br; $R = Me$ or Pr^i), undergo metal–halogen exchange, yielding magnesium enolates, rather than nucleophilic attack of the carbonyl group (equation 88)[230]. α-Metallation of esters, which is related to this, is often encountered.

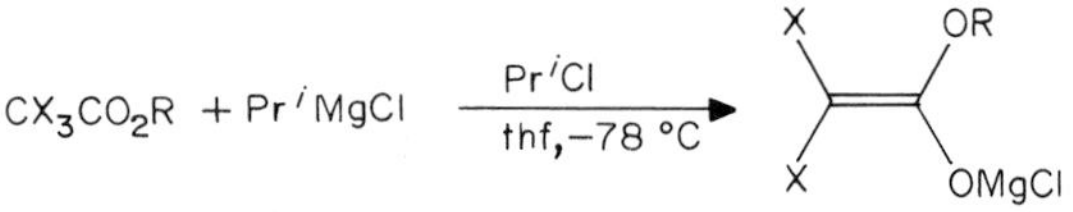

$$CX_3CO_2R + Pr^iMgCl \xrightarrow[\text{thf}, -78\,^\circ C]{Pr^iCl} X_2C{=}C(OR)(OMgCl) \qquad (88)$$

Reactions of thiol esters and thiol lactones with Grignard reagents have been little studied, but they appear to behave in an analogous way to the corresponding oxygen compounds[8]. Compounds of the type $CR(SR')_2CO_2R''$ undergo C—S bond cleavage (cf. equation 88), yielding magnesium enolates.

i. Carboxylic amides and related compounds

N,N-Disubstituted carboxylic amides give 1,2-addition compounds, yielding ketones on hydrolysis, or they may undergo elimination yielding enamines, $R_2C{=}CHNR_2$ (Table 2). Under more forcing conditions the intermediate addition product can further react to yield unsymmetrical alcohols in one pot (equation 89)[231].

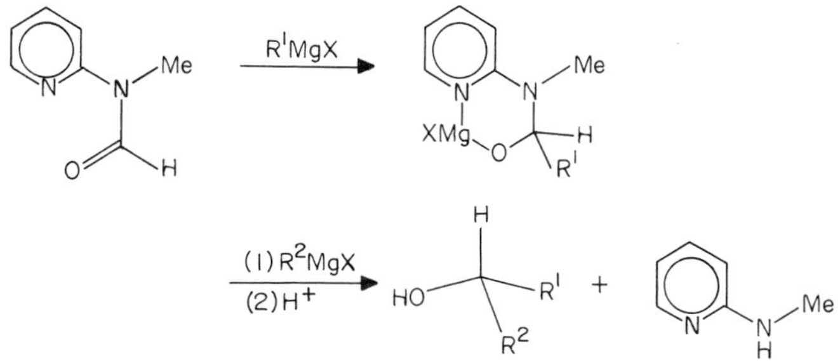

(89)

For α, β-unsaturated amides, 1,4-addition prevails[232]. Small-ring lactams show similar behaviour to lactones, yielding products derived from C—O bond rupture. *N*-Alkylsuccinimides afford simple addition products. For the strained ring compound **22**, however, ring cleavage occurs[233].

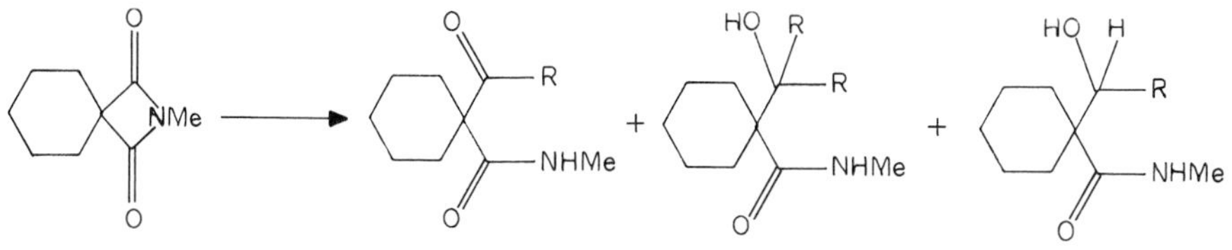

(**22**)

(90)

j. α-Keto esters

Keto esters generally react with one equivalent of an organomagnesium reagent regioselectively at the keto group (Table 3). A notable exception is when the reaction is carried out in non-coordinating solvents[174]. Presumably the greater complexing ability of the ester moiety will tend to direct the nucleophile to the closest carbonyl group, that of the ester. Whitesell *et al.*[234] studied the addition of Grignard reagents to α-keto esters for the case where the ester group is chiral, and found practical levels of asymmetric induction for 1,2-addition (> 90%).

k. Carboxylate salts

Addition products of carboxylate salts (Table 3) on hydrolysis yield geminal diols, which spontaneously dehydrate to ketones. Carboxylate salts, prepared by the carboxylation of Grignard reagents, can further react to form the ketone in this way. Carboxylic acids require two equivalents of the alkylating agent, one initially to form a magnesium carboxylato complex. Sato *et al.*[235] reinvestigated the reaction of formic acid with Grignard reagents and found that the use of thf rather than Et_2O as the solvent is a convenient method for preparing a variety of aldehydes such as alkyl, aryl, allyl, benzyl, and vinyl aldehydes. Vinyl Grignard reagents gave retention of configuration at the double bond.

l. Miscellaneous reactions

Unusual reactions of Grignard reagents, where the primary process is addition to a carbonyl group, are listed in Table 4 (Nos. 22–24). For No. 22[238], with a sterically hindered N-substituent on the substrate, the exclusive product is that derived from 1,2-addition to the CN multiple bond (Section III.A.3). The reaction associated with No. 23 is related to the annelation of anhydrides and lactones[239]. *N*-Nitroso-*N*-benzylformamide reacts with two equivalents of PhMgBr, yielding benzhydrol and desoxybenzoin, but from mechanistic considerations the reaction is poorly understood[242]. α,β-Unsaturated alkyl cyanoacetates undergo 1,4-addition[243].

In systematically considering a multitude of addition reactions of the various classes of carbonyl compounds, it appears that the following inequalities govern the reactivity of the functional groups: —NCO > —C(O)X > HC(O)R > RC(O)R′ > RC(O)OC(O)R > —C(O)ONR$_2$ > —C(O)OR. Although this is a general order of reactivity, it should be realized that factors such as changes in solvent, temperature (kinetic control *vs.* thermodynamic control), and leaving group capabilities may affect it.

2. *Carbon—sulphur multiple bonds*

The most common reactions are cited in Table 3. There are two modes of addition of organomagnesium reagents to thiocarbonyl compounds, the hydrocarbonyl group of the reagent becoming attached to either the carbon or the sulphur, referred to as carbophilic and thiophilic addition, respectively. For carbonyl compounds the addition is always carbophilic, a consequence of the hard nature of magnesium dictating complexation of the oxygen throughout the course of the reaction.

a. Reaction with carbon disulphide.

Grignard reagents treated with CS$_2$ yield dithiocarboxylates, which are usually reacted *in situ* and have been used to prepare a wide variety of compounds[244] including dithio esters (equation 91)[245], trithio peresters (equation 92)[246], and ketene dithioacetals (equation 93)[247].

$$\mathrm{PhCS_2MgBr} \xrightarrow{\mathrm{RR^1C{=}CHCOR^2}} \mathrm{PhC({=}S){-}S{-}CR_2{-}CH_2COR^2} \qquad (91)$$

$$\mathrm{RCS_2MgBr} \xrightarrow{\mathrm{MeSSO_2Me}} \mathrm{RC({=}S){-}S{-}S{-}Me} \qquad (92)$$

$$\mathrm{R^1R^2CHCS_2MgBr} \xrightarrow[\text{(2) } \mathrm{R^3X}]{\text{(1) } \mathrm{LiN(Pr^i)_2}} \mathrm{R^1R^2C{=}C(SR^2)(SR^3)} \qquad (93)$$

b. Reactions with monothio substrates

Isothiocyanates yield an addition product which is of utility in synthesis (ref. 8, p. 42) (e.g. equation 94)[248]. On hydrolysis thioamides are generated. Thiocyanates are prone to thiophilic attack, leading to S—C bond rupture (equation 95)[249].

$$\mathrm{Ph_3\overset{+}{P}N{=}C{=}S\ \ NCS^-} \xrightarrow{\mathrm{PhCH_2MgBr}} \left\{ \mathrm{PhCH_2C({=}S)N{=}PPh_3} \longleftrightarrow \mathrm{PhCH_2C(\overset{-}{S}){=}\overset{+}{N}PPh_3} \right\} \xrightarrow{-\mathrm{Ph_3PS}} \mathrm{PhCH_2CN} \qquad (94)$$

$$\mathrm{PhCHClCH_2SCN} \xrightarrow{\mathrm{C_4F_9MgBr}} \mathrm{PhCHClCH_2SC_4F_9} \qquad (95)$$

Sulphines ($R_2C{=}S{=}O$) undergo both thiophilic and carbophilic attack and appear to be of limited use in synthesis.

The favoured products for thioketones are those derived from thiophilic and/or carbophilic 1,2-addition (Table 3). Possible side reactions are those found in the reaction of ketones. The magnesium complexes formed by both types of 1,2-addition to the thioketone are useful intermediates for a variety of compounds, e.g. alkylation of the carbophilic complex yields thioethers.

There have been few studies on the reactions of thioamides. α-Metallation, *N*-alkyl metallation, and 1,2-addition (e.g. Scheme 13)[250] have been noted.

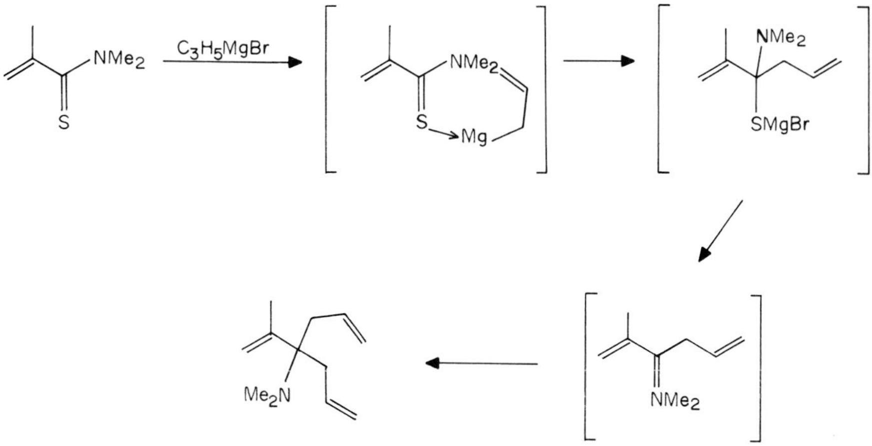

NMe2 S C3H5MgBr NMe2 S Mg NMe2 SMgBr NMe2 Me2N

SCHEME 13

c. Dithio substrates

Dithio esters are prone to both carbophilic and thiophilic 1,2-addition and, like the aforementioned substrates, it is the reaction of the magnesium complex with electrophiles, other than H^+, that has great synthetic utility. Alkylation of the carbophilic product, for example, yields a dithioacetal. Some recent and novel reactions of dithio esters are illustrated in equation 96 (thiophilic addition and elimination)[251] and equation 97[252]. It is noteworthy that by changing the conditions of the reaction, selective generation of the carbophilic or thiophilic product for the same substrate is possible[253].

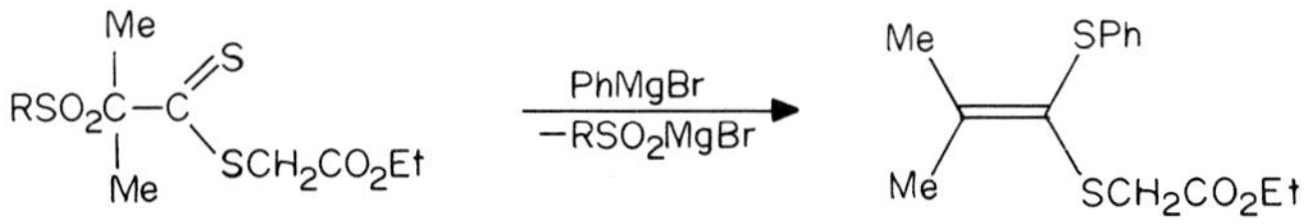

$$RSO_2C(Me)_2-C(=S)SCH_2CO_2Et \xrightarrow[-RSO_2MgBr]{PhMgBr} Me_2C{=}C(SPh)SCH_2CO_2Et \qquad (96)$$

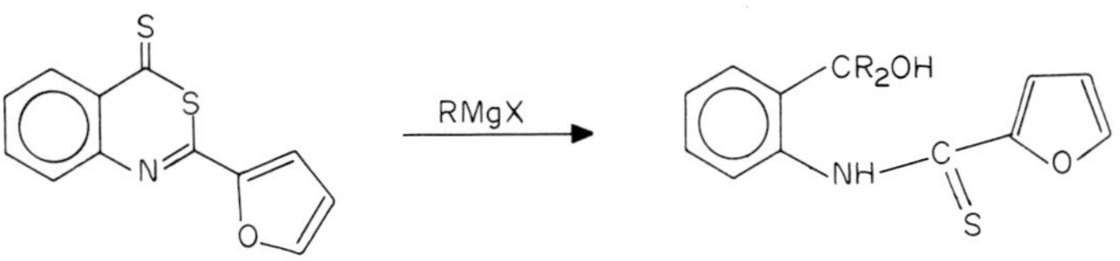

S S N O RMgX CR2OH NH C S O (97)

The reaction of **23** (equation 98) is a novel carbophilic addition to a carbon—sulphur aromatic bond. In general, cationic aromatic species are very reactive towards organomagnesium reagents[254].

(23) RMgX (98)

3. Carbon—carbon multiple bonds

Some discussion of the addition of organomagnesium reagents to carbon—carbon multiple bonds is given in Section II.A.1. 1,4-Additions to α,β-unsaturated functional groups, discussed above, are essentially additions to activated carbon—carbon double bonds.

Usually more forcing conditions than for additions to carbon—oxygen (and carbon—nitrogen) bonds are required, unless the reaction is transition metal catalysed (Section IV), the organic substrate is an aromatic cation, for example the propylium ion[255], or the multiple bond is activated. Factors affecting the reactivity of carbon—carbon multiple bonds and the stereochemistry of their reactions are discussed in Section II and in refs. 7, 8 and 81. An important new reaction which appears to have enormous application in synthesis is the intra-molecular 'magnesium-ene' reaction (Section II.A.3).

Anchimeric assistance by functional groups within the substrate can effectively 'activate' multiple bonds. This may be so for the metallated phenolic hydroxy group in **24**[256] and also for metallated amines in close proximity to the multiple bond[257].

(24) RMgX (99)

Fluorinated olefins readily give an addition product, but the reaction is followed by elimination (e.g. **25**)[258].

(25) RMgX −FMgX (100)

Addition to aromatic rings is possible with highly electron-withdrawing substituents present. Bartoli and coworkers[259] investigated the reactions of nitroarenes. The reactions (e.g. equation 101) proceed via a single electron transfer pathway yielding a nitronate complex (**26**) which is always reacted *in situ*.

(26)

(101)

Organo-magnesium (and -calcium and -barium) reagents can initiate the polymerization of olefins, despite the presence of a functional group which can react with such reagents (ref. 8, p. 8). A recent development is the enantiomeric selective polymerization of racemic methacrylates (equation 102). The (*S*)-monomer polymerizes in preference to the

racemic (S)

(102)

(*R*)-monomer, induced by a chiral bidentate ligand, (−)-sparteine (**27**) or its derivatives, complexed to the magnesium of a primary or secondary Grignard reagent[260].

(27)

4. Carbon—nitrogen multiple bonds

a. Imines and related compounds

Imines yield magnesium amide complexes with Grignard reagents, which are usually converted to amines by hydrolysis. Iminium salts, however, undergo salt elimination to yield an amine (tertiary) directly (Table 3). Simple imines react slowly whereas for conjugated imines the reactivity is enhanced. Silylamines are a useful synthon for primary amines since on hydrolysis Si—N bond cleavage prevails (equation 103)[261]. Imines of the

$$R^1CHO \xrightarrow{LiN(SiMe_3)_2} R^1CH{=}NSiMe_3 \xrightarrow[(2)\ H_2O]{(1)\ RMgX} RR^1CHNH_2 \qquad (103)$$

type $R^1R^2C{=}NX$ (X = OH, NHR, NR_2, NR_3^+, NHCONHR, SAr, and $N{=}CR^3R^4$) yield 1, 2-addition products, derived from nucleophilic attack on the imine carbon atom[8]. The reaction of these and of simple imines are of interest in that for unsymmetrical compounds a new asymmetric centre is created on 1, 2-addition, and there have been numerous studies aimed at devising experiments to achieve asymmetric induction[262]. α, β-Unsaturated imines usually afford the thermodynamically controlled product by 1,4-

addition, and this is also of interest in asymmetric synthesis (equation 104)[263].

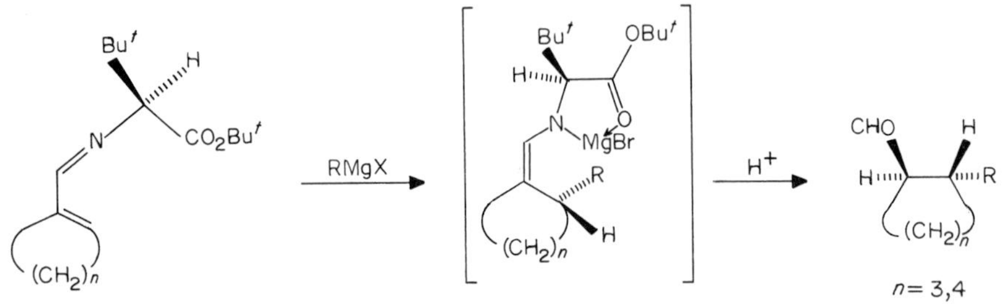

(104)

Imidoyl halides undergo salt elimination reactions[264], as for acyl chlorides, and subsequent addition is possible (e.g. equation 105)[265]. α-Dichloroiminium salts undergo

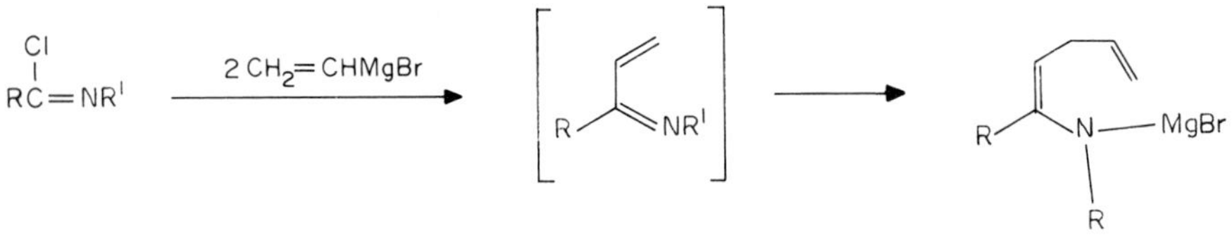

(105)

1, 2-addition, or reduction if the Grignard hydrocarbyl group is hindered, and if one chloro is replaced by a bromo group a β-haloenamine is produced[266]. Oximes afford 1, 2-addition products or cyclic compounds, formed by a deprotonation step (equation 106)[267].

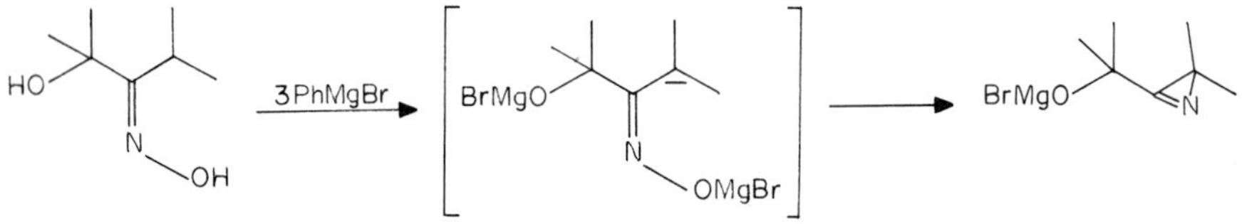

(106)

Imines with an α-carbonyl group undergo 1, 2-addition at the imine moiety or, under more forcing conditions, 1, 2-addition to both double bonds. Imines with an α-carboxylic ester give the imine addition product under mild conditions. Yields of primary amines of greater than 90% from phosphinyl formimidate compounds (**28**) are possible (equation 107)[268].

$$\underset{(\mathbf{28})}{Ph_2P(O)N{=}CHOEt} \xrightarrow[\text{(2) aq. } NH_4Cl]{\text{(1) RMgBr}} [Ph_2P(O)N{=}CHR] \xrightarrow[\text{(2) } H^+]{\text{(1) RMgBr}} H_3\overset{+}{N}CHR_2 \qquad (107)$$

b. Nitrones

These and aromatic nitrones, e.g. pyridine *N*-oxide, usually undergo 1, 3-addition, yielding hydroxylamines on hydrolysis (Table 3). Side reactions are elimination of Mg(OH)X (equation 108)[269] and HMgX[270] if an α-hydrogen is present, or MgOMeX and

MgCNX for α-OMe and α-CN substituents, respectively[270], yielding nitroxides. δ-CN groups are also susceptible to elimination[271]. The 1,3-addition complexes are readily oxidized to nitrones if another hydrogen is present, otherwise to a nitroxide radical[270].

PhMgBr; −Mg(OH)Br; H; Ph; OMgBr; N; O

(108)

Pyridazine *N*-oxide yields substituted olefins, presumably arising from addition followed by C—N bond cleavage[272].

c. Nitrogen heterocyclic aromatic compounds

Six-membered nitrogen aromatic compounds yield 1,2-addition products and to a lesser extent 1,4-addition products and metallation. Unlike aromatic compounds, the conditions for addition are mild, especially for their quaternized derivatives, which also favour addition as the primary process[8]. Only for hindered substrates is a catalyst required[273]. Barbier reaction conditions are also applicable for quaternary salts in forming 1,2-addition products[274].

Acridines and acridinium halides yield 9-substituted acridans (pseudo 1,4-additions) on hydrolysis of the intermediate complex[275]. Addition to pyridines and pyridinium salts can be highly regioselective. *N-tert*-Butyldimethylsilylpyridinium triflate undergoes 1,4-addition exclusively[276], whereas from *N*-methoxycarbonylpyridinium chloride the major product is that derived from 1,2-addition[277]. 1,4-Addition in general prevails in the presence of copper(I) iodide catalyst[278], as is the case for α,β-unsaturated ketones. Steric hindrance at the 2,6-positions of pyridinium salts seems to favour 1,4-addition[279].

Alkylations of simple nitrogen heterocycles with organomagnesium and lithium reagents are tabulated in ref. 8, p. 17.

d. Nitriles

A variety of reaction pathways are possible, some of which are undesirable. 1,2-Addition yields *N*-metallamines (Table 3), and on hydrolysis imines, then ketones. Activated nitriles such as methoxyacetonitrile permit a double addition of organometallic reagents, yielding primary amines on hydrolysis; organolithium reagents affect the second

$$(EtO)_2C(R^1)C{\equiv}N \xrightarrow{RMgX} (EtO)_2C(R^1)C(=NMgX){-}R \xrightarrow[(2)\ H^+]{(1)\ LiR^2} (EtO)_2C(R^1){-}C(R)(NH_2){-}R^2 \qquad (109)$$

addition (equation 109)[280]. Side reactions are significantly reduced using benzene plus one equivalent of Et_2O as the solvent rather than Et_2O alone.

The intermediate complex derived from 1,2-addition reacts with various electrophiles; some examples are illustrated in ref. 8, p. 20. It can further react with organomagnesium reagents by α-metallation to yield a new organomagnesium reagent, which can add to the unreacted nitrile[281].

Other reactions of nitriles are the displacement of the cyano group, as for α-cyano-substituted nitrones, metal cyanide exchange[282], and some very complex reactions yielding novel molecules. Whether the reagent adds to a nitrile group, favours elimination of Mg(CN)R, ring addition to an aromatic C—N bond, or some other reaction for cyano-substituted nitrogen heterocyclic compounds is difficult to predict.

Cyanogen gives a variety of products, including $R_2C(CN)NH_2$ from two successive additions, followed by hydrolysis[283]. Nitrile oxides afford 1,3-addition products, ketoximes, in useful yields[284].

e. Miscellaneous addition reactions

General reactions are given in Table 3 for 1,2-addition to isocyanides[285], isocyanates[286], isothiocyanates[287], and cyanates[288]. However, such reactions have received little attention.

5. Other C—X multiple bonds

Pyrilium species undergo addition with Grignard reagents, the mode of addition being sensitive to the nature of the solvent (equation 110)[289].

1,2-addition ← PhMgBr / C_6H_6 — [4-ethoxy-1-benzopyrylium, OEt] — PhMgBr / Et_2O → [4-Ph-4-OEt-4H-chromene] → (Ph)(OEt)C-aryl with OH and CH=CHPh

(110)

Addition to phosphabenzenes occurs with the alkyl of the Grignard reagent becoming attached to the phosphorus and magnesium to the carbon[290] (cf. thiophilic additions to C=S double bonds).

6. Other multiple bonds

The addition of Grignard reagents to multiple bonds between two non-carbon centres has had limited success in synthesis. Recent developments, however, suggest that such reactions have some potential.

a. Nitrogen—nitrogen multiple bonds

Many of the reactions of compounds with nitrogen—nitrogen double bonds are subject to several side reactions. This is not so for diphenylphosphorazide (**29**), a compound which is an excellent synthon for RN_2 species (equation 111)[291].

$$\underset{\textbf{(29)}}{(PhO)_2P(O)N_3} \xrightarrow{Me_3SiCH_2MgCl} Me_3SiCHN_2 \qquad (111)$$

Trialkyltriazines have been prepared in high yield (> 80%) from alkyl azides, shown in Scheme 14. Reactions of the tautomeric mixture **30** gives the isomers resulting from attack at either N-1 or N-3[292].

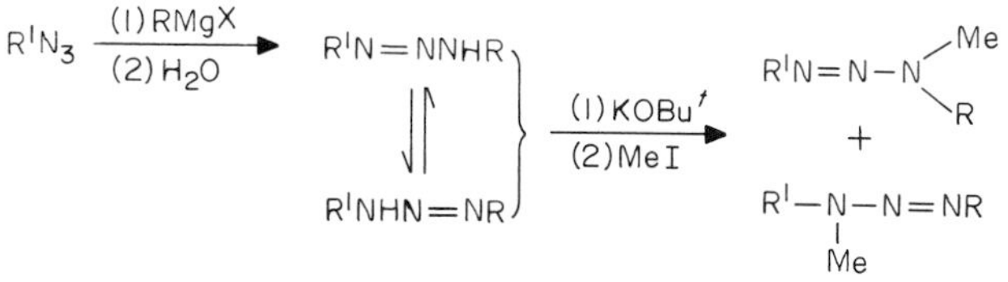

$$R^1N_3 \xrightarrow[(2)\ H_2O]{(1)\ RMgX} \left\{ \begin{array}{c} R^1N{=}NNHR \\ \rightleftharpoons \\ R^1NHN{=}NR \end{array} \right\} \xrightarrow[(2)\ MeI]{(1)\ KOBu^t} R^1N{=}N{-}N(Me)R \; + \; R^1{-}N(Me){-}N{=}NR$$

(30)

SCHEME 14

In a systematic study of the reaction of $RXCH_2N_2$ archetypes, X = O or S, with Grignard reagents it has been established that the activity effect for the reaction for X = S is greater than for X = O. The nature of the addition complexes (X = S) at two temperatures was probed by quenching the reaction mixtures with acetic anhydride (and acyl chlorate) (Scheme 15). Hydrolysis of the acetic anhydride-quenched reaction mixture was found to be a convenient route to the amination of aliphatic Grignard reagents[293].

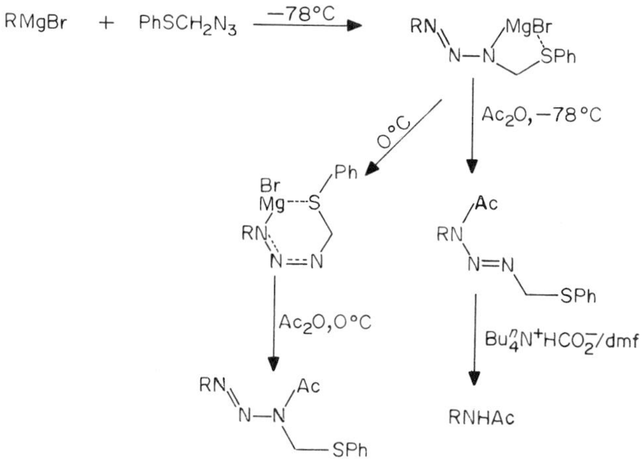

$$RMgBr + PhSCH_2N_3 \xrightarrow{-78\,°C} RN{=}N{-}N(MgBr)CH_2SPh$$

$$\xrightarrow{Ac_2O,\ -78\,°C} RN(Ac){-}N{=}N{-}CH_2SPh \xrightarrow{Bu^n_4N^+HCO_2^-/dmf} RNHAc$$

$$\xrightarrow{0\,°C} \text{cyclic } RN{\cdots}Mg(Br){\cdots}S(Ph)CH_2N{=}N \xrightarrow{Ac_2O,\ 0\,°C} RN{=}N{-}N(Ac)CH_2SPh$$

R = cyclohexyl

SCHEME 15

Other reactions of compounds with nitrogen—nitrogen double bond linkages are as follows: azo compounds ($R^1N{=}NR^2$) yield hydrazines (RR^1NNHR^2) after reaction work-up; diazoalkanes ($R^1R^2CN_2$) yield hydrazones ($R^1R^2C{=}NNHR$), which is unexpected, as addition to carbon—nitrogen double bonds is usually carbophilic owing to the preference of magnesium for hard ligands[294]; and aryldiazonium salts ($ArN_2^+X^-$) yield azo compounds ($ArN{=}NR$)[295].

b. Nitrogen—oxygen multiple bonds

There are only a few examples of the addition of organomagnesium reagents across a nitrogen—oxygen linkage of bond order greater than one. The criteria for addition is not fully understood, with the exception that the presence of electron-donating groups on the organic substrate is advantageous. Nevertheless, such reactions are likely to have a place in synthesis.

In one study, α-chloronitroso compounds yielded the addition/elimination product, labile ketonitrones, in modest yield[296] (equation 112). Note that ketonitrones can further react via a 1, 3-addition (Section III.A.4). For the hindered α-chloronitroso compound **31**,

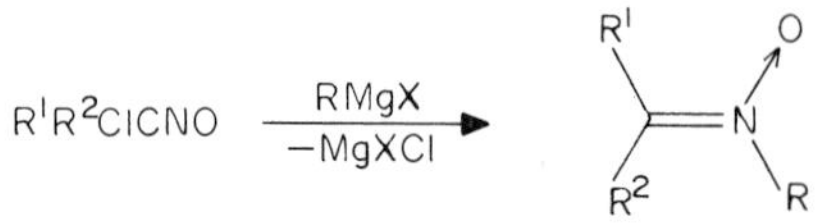

$$R^1R^2ClCNO \xrightarrow[-MgXCl]{RMgX} R^1R^2C{=}N(\to O)R \tag{112}$$

however, the ketonitrone yield is low and the major product is an oxime, thought to arise by an electron transfer pathway (Scheme 16)[297].

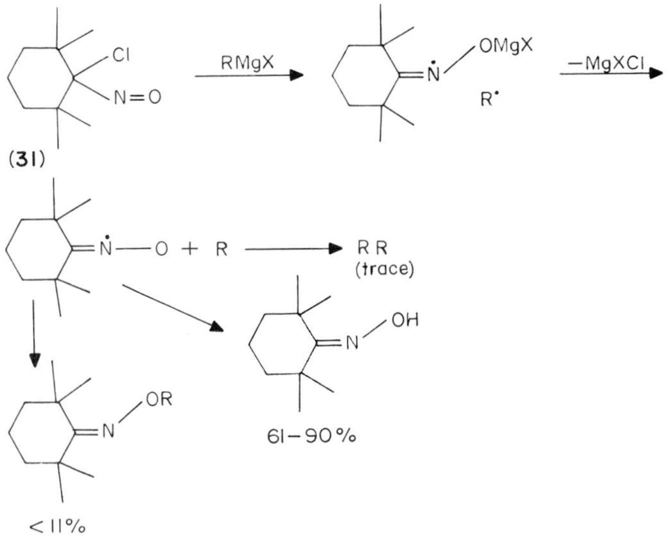

RMgX; −MgXCl; OMgX; R•; (31); RR (trace); OH; OR; 61−90%; <11%

SCHEME 16

Solutions of the di-*tert*-butyliminoxy radical with Grignard reagents yield the corresponding oxime and oxime ether[298]. This is consistent with the mechanism proposed for the reaction of **31** (Scheme 16).

The preparation of nitroso compounds by treating nitrosyl chloride with organomagnesium reagents has some potential, but it has been little studied. Catalysed reactions of alkylnitrates, R^1NO_2, yield addition/elimination products, hydroxylamines, R^1RNOH or R^1NHOH (> 90%), on reaction work-up[299].

c. Reactions with oxygen

Under controlled conditions oxygen yields either alkoxide or hydroperoxide complexes (equation 113). The latter is favoured by a high concentration of oxygen and a low temperature. Organolithium compounds react similarly. However, the yields are low but they can be improved by converting the reagent *in situ* to an organomagnesium compound by the addition of MgX_2 (Section II.B.3)[8].

$$RMgX \xrightarrow{O_2} ROOMgX \text{ or } ROMgX \qquad (113)$$

d. Sulphur (and selenium)—oxygen multiple bonds

Classifying reactions of substrates with sulphur—oxygen multiple bonds is often arbitrary, since invariably there is bond cleavage which may be concomitant with the thiophilic addition. The example of a reaction of a thiazine (**32**) illustrates this nicely[300].

PhMgX, −78°C; RLi, −78°C → PhS(O)R

(**32**) (114)

Reactions of sulphur dioxide, however, are readily classified as addition reactions. Sulphinate salts, RSO_2MgX, are generated, being useful reagents for preparing a variety of sulphonyl derivatives; for example, R′X yield sulphinate esters, RSO(OR′). The best yields of the sulphinate salts are obtained by adding the organometallic reagent to excess SO_2 (cf. carboxylate salts from CO_2, Section IV.A.1)[8]. Interestingly, SeO_2 affords dialkyl selenides in modest yields[301]. Selenium oxychloride undergoes metathetical exchange and 1,2-addition with PhMgX (equation 115)[302].

$$SeOCl_2 + PhMgX \longrightarrow Ph_3SeOMgX \qquad (115)$$

Addition across the N=S or O=S multiple bond is evident for compound **33**[303].

(1) excess Pr^iMgBr; (2) H_2O

(**33**) (116)

B. Displacement of Substituents

In Section III.A, attention was focused on addition reactions of multiple bonds, together with possible competing reactions. This section is concerned with reactions where the primary process is C—X bond rupture, i.e. nucleophilic displacement of a substituent at a carbon centre. Some aspects will be related to those in the previous section as some additions are followed by elimination, or elimination as the initial step may be a side reaction. General reactions are given in Table 5.

1. Carbon—halogen cleavage

The displacement of a halide by an organomagnesium reagent (coupling reactions) has a pivotal role in synthesis but unfortunately side reactions are possible. These include (a)

TABLE 5. Common displacement reactions resulting from C—X cleavage by organomagnesium reagents, $R_{2-n}MgCl_n$ ($n = 1$ or 0)

Description	Substrate	Product	Comment
1. X = Halide	R^1X	RR^1	More reliable if transition metal catalysed or by *in situ* conversion of the reagent to an organo-transition metal compound
2. X = *OY*			
(i) *Ethers*	R^1OR^2 R^1 or R^2 = aryl	RR^1 and R^2OMgX	One or both products may be synthetically useful
(ii) Acetals and orthoesters	$R^1_{n-1}C(OR^2)_n$ ($n = 2$ or 3)	$R^1_{n-1}RC(OR^2)_n + R^2OMgX$	Orthoesters are more reactive
(iii) *gem*-Amino ethers	$R^1_{n-1}C(OR^2)_nNR^3_2$ ($n = 1$ or 2)	$R^1_{n-1}R_nCNR^3_2$	C—O cleavage prevails
(iv) Oxiranes	$R^1R^2CCOR^3R^4$	$RR^1R^2CC(OMgX)R^3R^4$	Other isomer is possible for $R^1, R^2 = R^3, R^4$
(v) Sulphates and sulphonates	$R^1OSO_2R^2$ $R^2 = OR^3, R^3$	$RR^1 + L_nMgOSO_2R^2$	Competing reactions possible
(vi) Alkyl phosphates	$R^1OPO(OR^2)_2$	$RR^1 + L_nMgOP(OR^2)_2$	Also catalysed and with cuprate salts
(v) Propiolactone	[β-propiolactone structure]	$R(CH_2)_2CO_2MgX$	Transition metal catalysed
3. X = CN			
α-Aminonitriles	$R^1_2NCR^2R^3(CN)$	$R^1NCR^2R^3R$	Also for α-alkoxynitriles
4. X = S			
Thioethers	R^1SR^2	R^1R and R^2SMgL_n	R^1SR and R^2MgX formation is also possible
Sulphoxides	R^1SOR^2	$R^1SOR + R^2MgL_n$	Reduction is also likely (see text)
Sulphones	$R^1SO_2R^2$	$R^1R + R^2SO_2MgL_n$ or $R^1SOR + R^2OMgL_n$	Transition metal catalysed for S—C cleavage

deprotonation of the organic halide followed by elimination of MgX_2, yielding an alkene or carbene; (b) metal–halogen exchange (see Section II.B.4), resulting in a mixture of coupled products; and (c) unexpected products formed by an electron transfer reaction pathway. It is noteworthy that in general strong coordinating solvents such as hmpt favour coupling.

Alkyl and benzyl halides, particularly the bromides, give high yields of coupled products with various alkyl, perfluoroalkyl, phenyl, and perfluorophenyl Grignard reagents. Rearranged products are possible with allylic halides, a consequence of attack at the 3-position of the allyl group[304]. Displacement is promoted by electron-donating substituents in close proximity to the halogen group. Some examples are given in equations 117[305], 118[306] and 119[307].

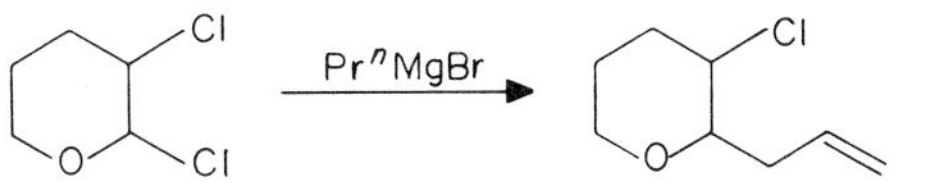

Pr^nMgBr (117)

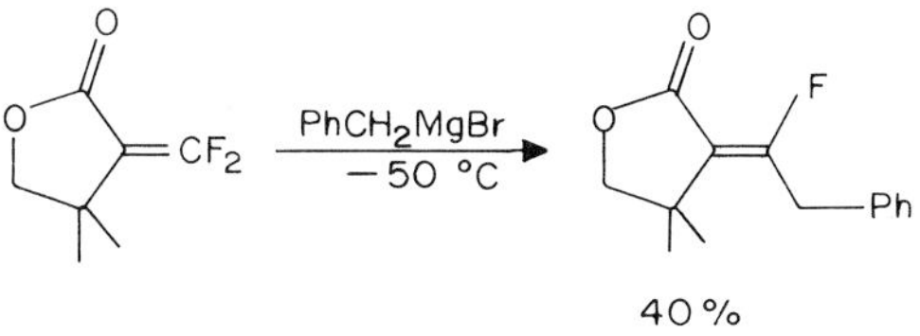

$PhCH_2MgBr$, $-50\ °C$ (118)

40%

$$F_2C{=}CHCO_2Li \xrightarrow[(2)\ H^+]{(1)\ EtMgBr} EtCF{=}CHCO_2H \tag{119}$$

Alkyl and aryl halides usually require forcing conditions, viz. temperatures in excess of 150 °C, without the use of a solvent[8]. Magnesium complexes of aromatic anions and conjugated polyene anions readily react with organic halides. 2-Methylbut-2-enedienylmagnesium (**34**) reacts with high regioselectivity (equation 120). Its transition metal-catalysed reactions are also regioselective, although some yield different isomers[308].

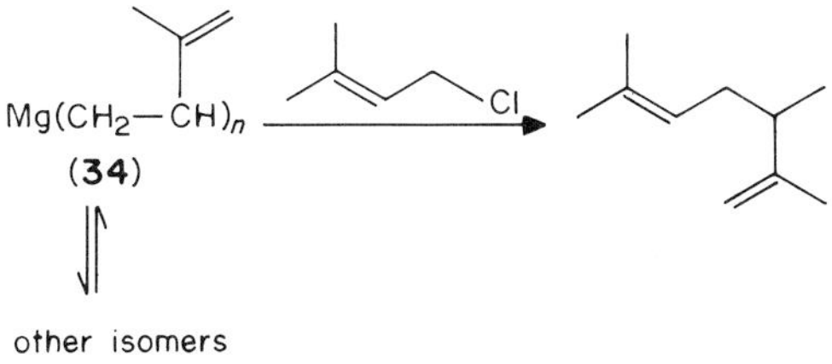

$Mg(CH_2{-}CH)_n$ (**34**) (120)

other isomers

Aspects of the stereochemistry of alkylation of organic halides are presented in ref. 8, p. 46. They include asymmetric synthesis by incorporating one or more chiral centres within the organic halide or the organomagnesium reagent.

Any difficulties associated with alkylation of organic halides can be overcome by converting the magnesium reagent into that of a transition metal, notably copper, or by using a transition metal catalyst. These strategies are discussed in detail in Section IV.

2. Carbon—oxygen cleavage

The displacement of an alkoxide group is favoured under special circumstances such as ring strain in the substrate, low electron density at the carbon bearing the alkoxide group, and/or forcing conditions. Simple ethers, for example thf, require high temperatures, usually in excess of 100 °C. Most of the useful reactions described below require similar treatment. It is noteworthy that solid Grignard reagents, free of ether, are more reactive towards cleaving ethers than 'solvated' Grignard reagents[309].

a. Reactions of aryl (and allyl) ethers

Aryl alkyl ethers are cleaved with either aryl oxide (equation 121)[310] or alkyl oxide (equation 122)[311] displacement.

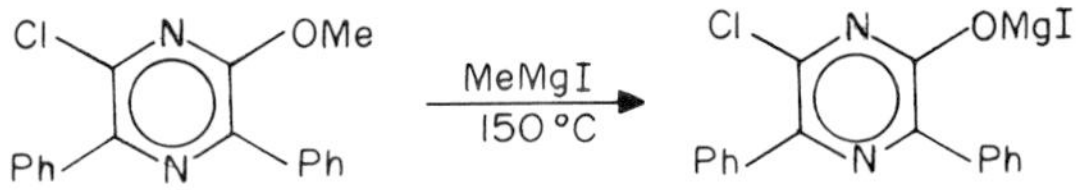

MeMgI, 150 °C (121)

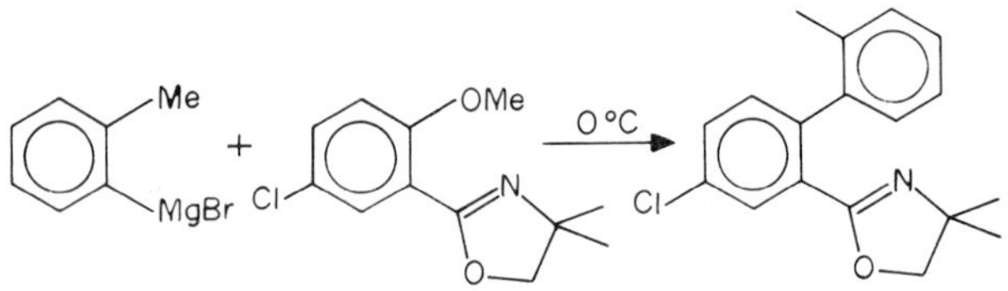

0 °C (122)

Selective methoxide displacement has featured as key steps in the synthesis of some complex molecules[312].

Aluminium trialkyls in some cases are more useful than Grignard reagents, whereas organolithium species promote *ortho*-metallation.

An interesting combination of organo-aluminium and -magnesium reagents is shown in equation 123[313].

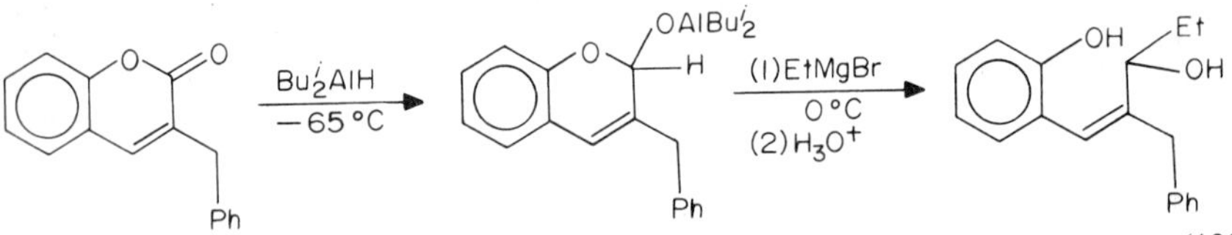

Bu^i_2AlH, −65 °C; (1) EtMgBr, 0 °C, (2) H_3O^+ (123)

b. Reactions of acetals and orthoesters

These reactions are facile and there is the prospect of displacing one or two ROMgX units (Table 5), particularly when R is an aryl group[314]. Hemiacetals react with excess of Grignard reagent according to equation 124[315].

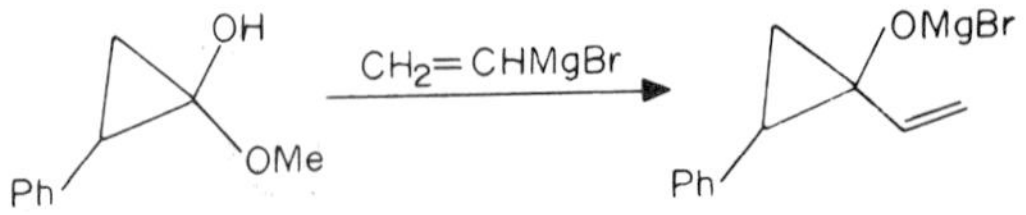

$CH_2{=}CHMgBr$ (124)

Monothioacetals favour displacement of a magnesium alkyl or aryl oxide complex, (OR)MgX, rather than (SR)MgX[8], presumably as a consequence of the preference of magnesium for hard ligands.

The reaction of paraformaldehyde, $(CH_2O)_n$, with Grignard reagents is a powerful method for a one-carbon homologation[316].

c. Reactions of gem-amino ethers

Compounds of the type $R^1CH(OR^2)NR^3_2$ (**35**) and $CH(OR^1)_2NR^2_2$ (**36**) displace alkoxy rather than amino groups. They are excellent synthons for preparing tertiary amines (equation 125)[317].

$$R^1{-}N(\text{ring with two O}) \xrightarrow{RMgX} R^1N(CH_2R)_2 \quad (125)$$

Thio analogues of **35** and **36** afford products derived from C—S cleavage[318].

d. Reactions of oxiranes and related compounds

These have previously been dealt with in depth[8] and only reaction types (Table 5) and recent developments are presented here. Oxiranes are useful for a two-carbon homologation, although some of their reactions give undesirable side reactions, in particular the addition of MgX_2. Ring opening of unsymmetrical oxiranes is by attack at the most electron-deficient carbon centre of the oxirane or is controlled by steric constraints. Nevertheless, it is highly regioselective with formation of a *trans* stereochemistry[319]. An oxirane moiety is invariably more reactive than a carbonyl group within the same molecule.

Copper halide-catalysed reactions of oxiranes are less susceptible to side reactions. The same applies to reactions involving magnesium cuprates as the alkylating agent[320]. Oxiranes, effective three-carbon homologation species, are usually reacted in the presence of a catalyst, CuI[321]. The reactions of larger cyclic ethers have limited application, usually requiring severe reaction conditions. An exception is the generation of open-chain compounds from various nucleocides with MeMgI[322].

The major products from oxiranes are alcohols formed by nucleophilic ring opening. Alcohols can also arise by attack of an intermediate carbonyl compound, formed by a stereospecific hydride shift (equation 126)[323].

thpo ... $\xrightarrow{EtOC\equiv CMgBr}$ [thpo ... CHO] $\xrightarrow{R^1MgX}$ thpo ... OH, R^1 (126)

Allylic Grignards can add at the 3-position of the allyl group. The reaction in equation 127 is such a case and, interestingly, if the same reaction is catalysed by CuI the 'normal' ring-cleaved product results[324].

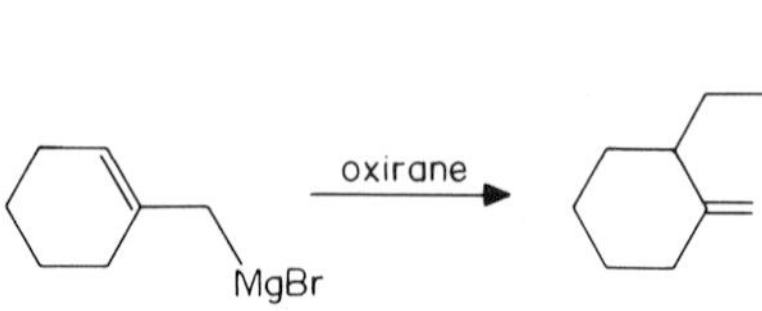

(127)

Sulphonyl oxiranes of the type **37** yield several species with MeMgI (Scheme 17). The *cis* stereochemistry in the cyclized product **38** is thought to arise by chelation of the magnesium ion at some intermediate stage in the reaction. Lithium reagents react with **37** to yield cyclopropylmethanol derivatives[324].

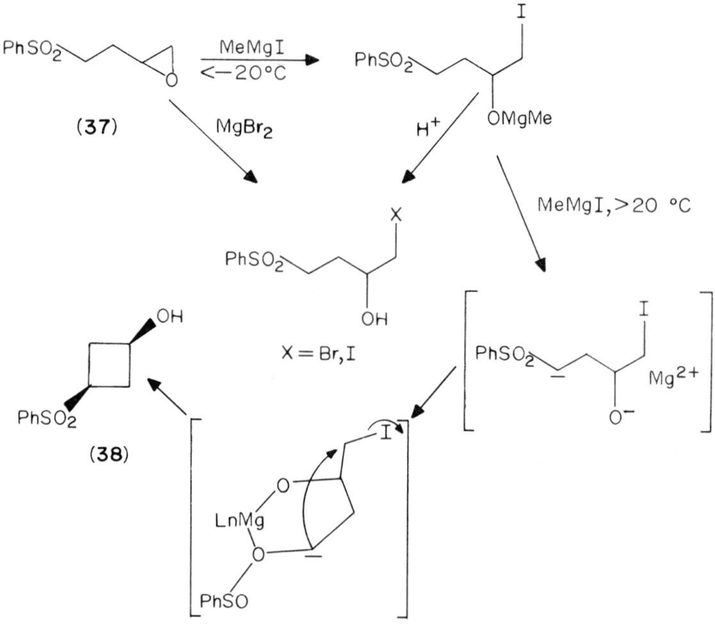

SCHEME 17

e. Reactions of sulphates and sulphonates

General reactions for which the primary process is carbon—oxygen rupture in a sulphate, sulphonate, and sulphinate are shown in Table 5. Metallation (Scheme 17), the formation of alkyl halides and sulphur—oxygen cleavage are possible competing reactions. Examples of the latter are the reaction of triflic anhydride, affording trifluoromethyl sulphone in modest yield (equation 128)[325] and a new synthesis of 2-hydroxyalkylsulphones (equation 129)[326]. Sulphones are readily accessible from the treatment of aryl tosylates with Grignard reagents.

$$\text{CH}_2{=}\text{CHCH}_2\text{MgCl} + (\text{CF}_3\text{SO}_2)_2\text{O} \longrightarrow \text{CH}_2{=}\text{CHCH}_2\text{SO}_2\text{CF}_3 \qquad (128)$$

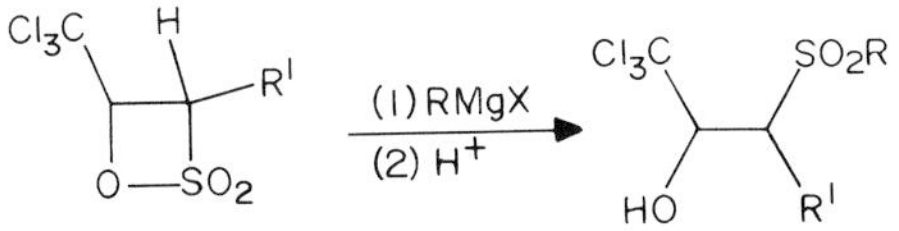

$$\text{Cl}_3\text{C(H)}\text{-oxetane(R}^1\text{)(O—SO}_2\text{)} \xrightarrow[(2)\ \text{H}^+]{(1)\ \text{RMgX}} \text{Cl}_3\text{C(HO)CH—CH(R}^1\text{)SO}_2\text{R} \quad (129)$$

f. Reactions of trialkyl phosphates

These reactions have been investigated in detail[327]. Recent work includes high regio- and stereo-specific 'coupling' reactions between allyl phosphates and Grignard reagents[328]. Vinylic phosphate coupling reactions are either transition metal-catalysed[329] or the Grignard reagent is converted into an organocuprate prior to the reaction[330].

g. Miscellaneous reactions

Strained lactones and substituted propiolactones are susceptible to carbon—oxygen bond cleavage, but they usually require a transition metal catalyst or a magnesium cuprate as the alkylating agent.

Substitution of acetate from a propargyl acetate has been noted[331]. Reactions of cyanates, although little studied, appear to have potential for the synthesis of nitriles (equation 130)[332].

$$R^1OCN \xrightarrow{\text{RMgX}} RCN + R^1OMgX \quad (130)$$

Alkoxy groups are readily displaced from allyl (and propargyl) ethers with attack most likely in the 3-position of the allyl group (cf. reactions of allyl halides)[8].

The utility of Grignard reagents in carbohydrate synthesis has recently been established. The reactions involve a carbon—oxygen bond cleavage step. Those in Scheme 18 proceed via elimination of Otos and 1,2-hydride shifts on the furanocide rings with some ring breakage in the case of **39**[333].

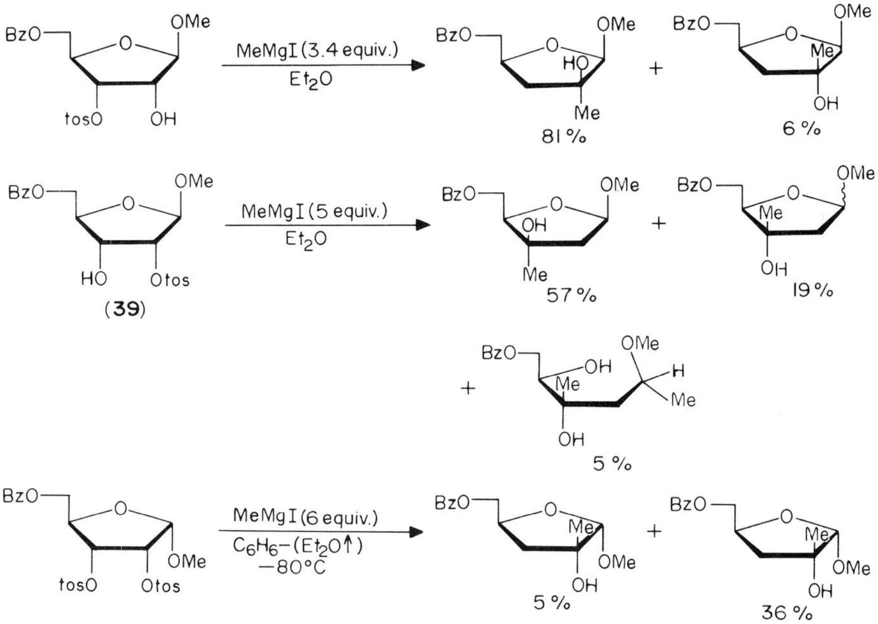

BzO, OMe, tosO, OH → MeMgI (3.4 equiv.), Et_2O → BzO, OMe, HO, Me (81 %) + BzO, OMe, Me, OH (6 %)

BzO, OMe, HO, Otos (**39**) → MeMgI (5 equiv.), Et_2O → BzO, OMe, OH, Me (57 %) + BzO, OMe, Me, OH (19 %) + BzO, OMe, OH, Me, H, Me, OH (5 %)

BzO, OMe, tosO, Otos → MeMgI (6 equiv.), C_6H_6–(Et_2O↑), −80°C → BzO, Me, OMe, OH (5 %) + BzO, Me, OMe, OH (36 %)

SCHEME 18

Anomerization is also possible, e.g. equation 131, for which the proposed mechanism is ring opening–reclosure by the coordination of the sugar oxygens to the magnesium of the Grignard reagent[322,334,335]. It is an equilibrium reaction and with careful reaction control the process can be reversed. References 334 and 335 cite other reactions of carbohydrates.

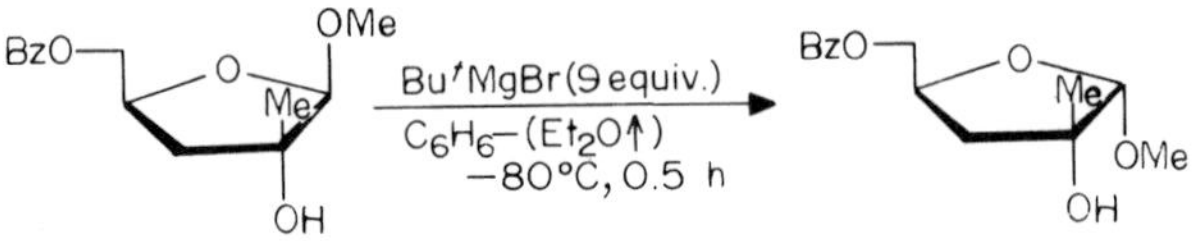

$$\xrightarrow[\substack{C_6H_6-(Et_2O\uparrow)\\ -80^\circ C,\ 0.5\ h}]{Bu^tMgBr\ (9\ equiv.)} \tag{131}$$

3. Carbon—carbon cleavage

The most common type of carbon—carbon cleavage reaction is that involving the elimination of CN^-. The organic substrates are usually α-functionalized nitrites, for example α-amino nitrites **40**[336] and **41**[337].

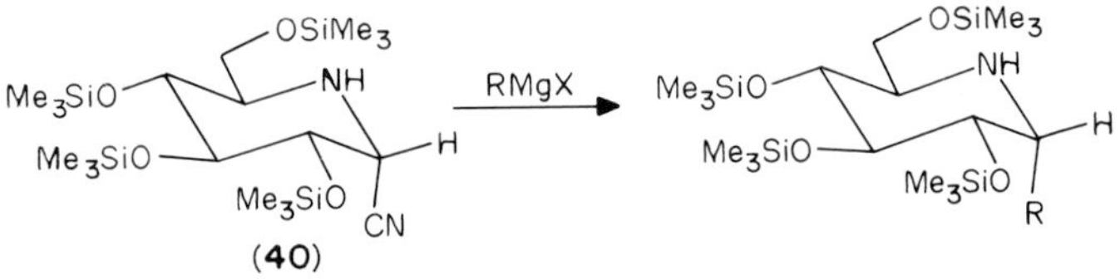

$$\mathbf{(40)} \xrightarrow{RMgX} \tag{132}$$

$$\underset{\mathbf{(41)}}{(RO)_3CC{\equiv}N} \xrightarrow[-MgXCN]{arylMgX} (RO)_3C(aryl) \tag{133}$$

Another form of carbon—carbon bond breakage is found in strained ring systems, e.g. **42**[338].

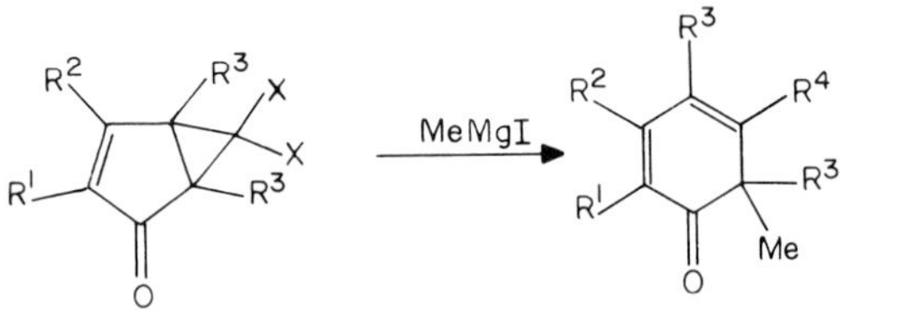

$$\mathbf{(42)} \xrightarrow{MeMgI} \tag{134}$$

R^4 = Me or halogen

4. Carbon—sulphur (selenium and tellurium) cleavage

a. Thioethers

Thioethers are susceptible to cleavage, more so than ordinary ethers but excluding strained ones such as oxiranes. Their reactions are usually nickel catalysed (equation 135) or the alkylating agent is a magnesium cuprate[339,340]. Sulphur—alkenyl rupture takes

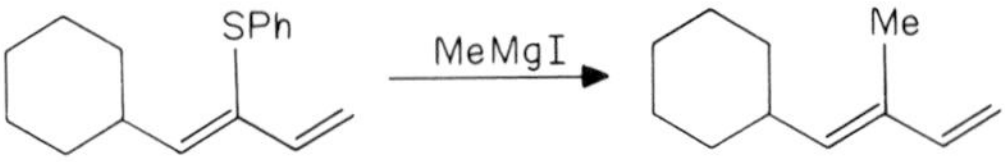

$$\xrightarrow{MeMgI} \tag{135}$$

precedence over sulphur—aryl rupture (equation 135). Reactions of alkenyl aryl tellurates (metal catalysed) however, give both product types under the conditions studied[341].

In addition to displacement of ^-SR (or ^-SeR), thiophilic attack can occur, yielding a new organometallic reagent (Table 5) and a compound with a new carbon—heteroatom bond[8].

b. Sulphoxides

Nucleophilic attack at the sulphur atom, bringing about a replacement reaction (Table 5), e.g. the stereospecific reaction in equation 136[342], and/or reduction coupled with alkylation (equation 137)[343] are possible.

$$\text{Ph(H)C=C(SMe)(SOMe)} \xrightarrow{\text{EtMgCl}} \text{Ph(H)C=C(SMe)(MgCl)} \tag{136}$$

$$\text{MeSOMe} \xrightarrow{\text{RMgX}} \text{MeSCH}_2\text{R} \tag{137}$$

c. Sulphones

Vinylic and aryl sulphones react under mild conditions, catalysed by transition metal complexes, affording good yields of substitution products derived from O—C rupture (e.g. $R = $ trisubstituted olefin, $R^2 = Bu^t$, $R = $ Me or Ph for the general reaction in Table 5)[344]. Uncatalysed cleavage of S—O bonds is also possible, the product being a sulphoxide[345].

Alkyl thiocyanates yield thioethers via thiophilic addition across the S—C bond/elimination of CN^-, as MgXCN (equation 138[345]; see also Section III.A.3).

$$\text{Bu}^n\text{SCN} + \text{PhMgBr} \longrightarrow \text{PhSBu}^n \tag{138}$$

5. Other displacement reactions

a. Oxygen—oxygen cleavage

The reaction of oxygen with Grignard reagents, yielding either addition products (hydroperoxides) and/or elimination products (alkoxides), is discussed in Section III.A.5. The cleavage of organic peroxides and related species (diacyl peroxides and peroxy esters) is well known (equation 139) and nothing further of significance has appeared since the last review on organomagnesium reactions[8].

$$\text{R}^1\text{OOR}^2 + \text{RMgX} \longrightarrow \text{R}^1\text{OR} + \text{XMgOR}^2 \tag{139}$$

b. Nitrogen—other heteroatom cleavage

Sulphur nitride, N_4S_4, affords disulphides, RS_2R, in almost quantitative yield[346]. Chloroamines with R_2Mg and RMgCl yield amines, whereas alkylmagnesium halides other than the chloride give predominantly the corresponding haloamine[8].

c. Phosphorus—other heteroatom cleavage

Various oxophosphorus compounds readily undergo P—O bond cleavage with either retention (equation 140)[347] or inversion (equation 141)[348] of configuration at the phosphorus centre. Interestingly, compound **43** yields some P—N cleaved product with inversion of configuration. Also compound **43**, X = S, but with opposite chirality at the carbon centre adjacent to X, gives exclusively the P—S cleaved product, with retention of configuration[349]. Compound **43**, X = O, on treatment with Bu^{t}MgBr, rearranges to the product with X and S interchanged and with change in configuration[350].

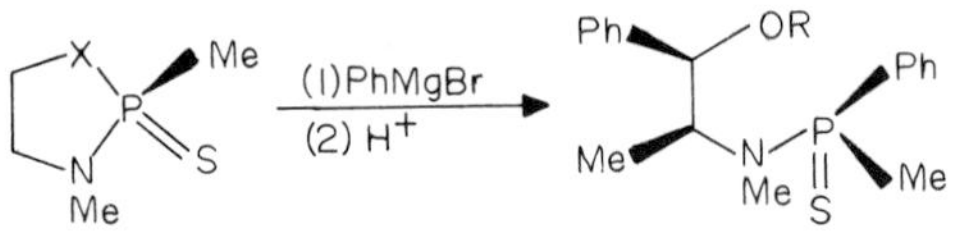

Me
S
N
Me
(1)PhMgBr
(2) H+
Ph
OR
Ph
Me
N
Me
S
Me

(X = O)
(**43**)

(140)

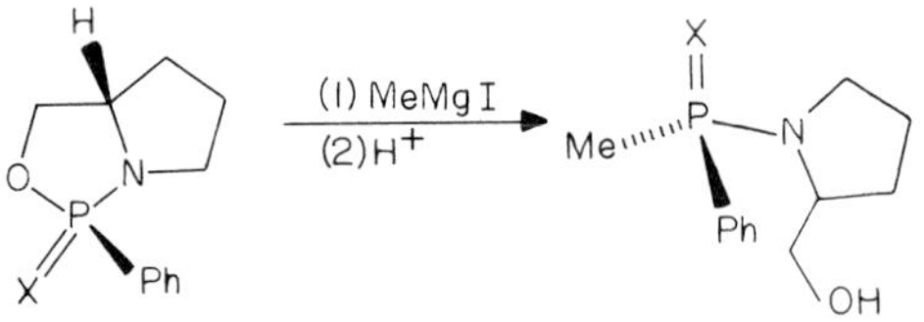

H
O
N
P
X
Ph
(1) MeMgI
(2)H+
X
Me
P
N
Ph
OH

(141)

The first convenient preparation of a chiral trialkylphosphine in reasonable yield, which involves sequential P—S and P—O bond cleavage by Grignard reagents, has recently been reported (Scheme 19)[351].

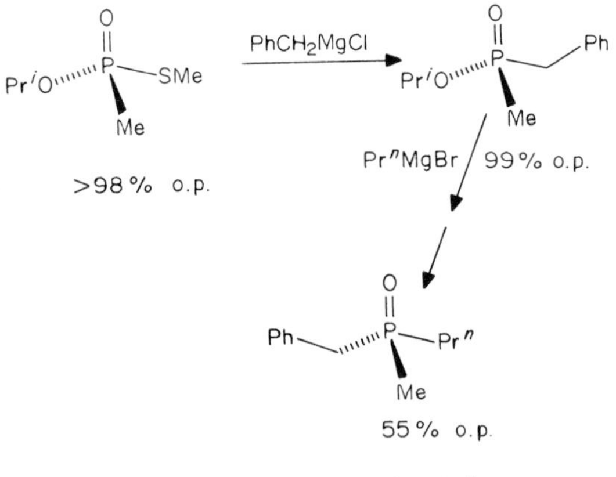

O
Pr^{i}O
P
SMe
Me
>98 % o.p.
PhCH2MgCl
O
Pr^{i}O
P
Ph
Me
Pr^{n}MgBr
99 % o.p.
O
Ph
P
Pr^n
Me
55 % o.p.
(o.p. = optical purity)

SCHEME 19

There are numerous reactions of substitution of halides by Grignard reagents for phosphorus(III) and -(V) in the literature. Aspects of this are discussed in detail in ref. 8.

d. Sulphur (and selenium and tellurium)—heteroatom cleavage

Elemental sulphur, selenium, and tellurium yield metal thiolates, selenolates and tellurolates, respectively, in good yield. These are hydrolysed[352] to thiols, etc., or are reacted with other electrophiles[8,353] or oxidized to disulphides, etc. (equation 142)[354].

$$\text{BrMg—C}_6\text{H}_4\text{—C(CH}_3\text{)(OCH}_2\text{CH}_2\text{O)} \xrightarrow[(2)\,[\text{O}]]{(1)\,\text{Te}} \text{ArTeTeAr} \quad (142)$$

(BrMgAr)

Disulphides and diselenides are cleaved by Grignard reagents, yielding thioethers (equation 143) and selenoethers. These reactions and those of halides, oxyhalides, and related species of the elements sulphur, selenium, and tellurium which give halide substitution products are discussed in ref. 8.

$$R^1SSR^2 + RMgX \longrightarrow R^1SR + XMgSR^2 \quad (143)$$

Alkyl sulphinamides, $R^1SONR^2_2$, undergo S—N cleavage with Grignard reagents, affording sulphoxides, R^1SOR[355].

e. Halogen—heteroatom cleavage

The reactions of metalloid and metal halides, yielding substitution of the halogen of the hydrocarbyl group of the organomagnesium reagent, are not discussed here. They are extensively covered in the appropriate chapters of *Comprehensive Organometallic Chemistry*.

Elemental halogens, Cl_2, Br_2, and I_2, react vigorously with organomagnesium reagents, yielding the corresponding organic halide. Several other reagents also convert organomagnesium reagents to organic halides, RX, X = F, Cl, Br, or I. These include $ArSO_2X$ (for chlorides and bromides), perchlorofluoride, dinitrogen difluoride (explosive reactions), amine halides, and triflic anhydrides[8].

f. Metal or metalloid—heteroatom cleavage

Reactions of metal and metalloid alkyl, aryl oxides, thiolates, etc., with organomagnesium reagents, resulting in the substitution of OR, SR, etc., for the hydrocarbyl of the reagent are also described in *Comprehensive Organometallic Chemistry*.

IV. REACTIONS OF GRIGNARD REAGENTS IN THE PRESENCE OF A TRANSITION METAL COMPOUND

This section is concerned with the reactivity of Grignard reagents towards various organic substrates in the presence of a transition metal compound. Both the use of transition metal complexes as stoichiometric reagents or as catalysts is described in detail.

A. Stoichiometric Reactions

It has long been known that many metal halides interact with a stoichiometric amount of an aryl Grignard reagent to form a symmetrical biaryl in high yield[1]. In the past decade research interests have focused on the interaction between organocopper(I) compounds and Grignard reagents and their uses in organic synthesis. With the advent of the Normant reagents a number of synthetically viable reactions have been realized. Not surprisingly, the use of copper(I) salts as stoichiometric reagents far outweighs studies involving other metal compounds. This topic has been the subject of two reviews[356,357].

The Normant reagents are typically represented as $RCu{\cdot}MgX_2$ (or $RCuX{\cdot}MgX$ according to the solvent), R_2CuMgX, or $RR'CuMgX$ and are prepared by reacting anhydrous copper(I) halides with the chosen Grignard reagent (equations 144–146)[358]. Both the purity and solubility of these organocuprates can be increased by using

$Me_2S{\cdot}CuX$[359], or indeed by using dimethyl sulphide as a co-solvent[360]. A similar effect is noted using other neutral ligands, such as $P(OR)_3$, in a 2:1 or 1:1 ratio to RCu. Enhanced reactivity can also be effected by the stoichiometric addition of LiX or MgX_2[358]. It is noteworthy that the exact nature of the Normant reagents is not known and indeed may be incorrectly formulated. In a recent study into the composition of these reagents, only the presence of various magnesium methylcuprates, $Cu_nMg_mMe_{2m+n}$ (where $m = 1, n = 1$–4 or 6 and where $m = 2$ and $n = 3$) having no halide interactions was observed[361]. These studies, however, were carried out in the absence of an organic substrate and therefore gave no indication as to the interaction between reagent and substrate.

$$RMgX + CuX \xrightarrow{Et_2O} [RCu{\cdot}MgX_2] \tag{144}$$

$$2RMgX + CuX \xrightarrow{Et_2O} [R_2CuMgX] \tag{145}$$

$$[RCu{\cdot}MgX_2] + R'MgX \xrightarrow{Et_2O} [RR'CuMgX] \tag{146}$$

A recent development has been the utilization of diorganoargentates of the type R_2AgMgX in organic synthesis. These complexes are prepared in the presence of two equivalents of LiBr, which considerably enhances their stability (equation 147)[362]. Again, the exact nature of these diorganoargentates has not been ascertained.

$$AgBr{\cdot}2LiBr + 2RMgCl \xrightarrow[-60\,^{\circ}C]{thf,} [R_2AgMgBr] \tag{147}$$

Stoichiometric reagents such as ZnX_2 or CdX_2 have been omitted from this survey as they are believed to displace totally the magnesium in the Grignard reagent and are therefore more correctly categorized as organozinc or organocadmium reagents rather than modified Grignard reagents.

1. *Addition to alkynes*

Although Grignard reagents do not add readily to the triple bond of simple alkynes, in the presence of a transition metal this process is quite facile. The most efficient reagents for carbometallation of alkynes are the Normant reagents (and R_2CuLi), and this has been the subject of a recent review[358]. For instance, $RCu{\cdot}MgX_2$ adds readily to acetylene, propyne, and phenylacetylene but does not react with higher terminal alkynes. In the same solvent, however, both R_2CuMgX and $RR'CuMgX$ add to higher alk-1-ynes. Only one of the R groups, however, is normally transferred when using these diorganocuprates[358].

The 1,2-addition of a Normant reagent to an alkyne can in principle occur in a (*Z*) (**44**) or (*E*) (**45**) fashion, provided carbometallation occurs at only one of the carbon centres of

$R^1C{\equiv}CR^2$ + $RCu{\cdot}MgX_2$ ⟶ [R^1, R^2 / R, $Cu{\cdot}MgX_2$] + [R^1, $Cu{\cdot}MgX_2$ / R, R^2]

(**44**) (**45**) (148)

the triple bond. For alk-1-ynes stereospecific addition can be assured by using diethyl ether as the solvent and carrying out the reaction at -35 to $-10\,^{\circ}C$. The R group in the Normant reagent, however, must be primary but β-branching or γ-saturation in R is allowed. Further, the terminal acetylene must be unsubstituted in the propargylic position. With thf as solvent both secondary and tertiary alkyl groups can be similarly added, but, at the expense of a diminished regio- and stereoselectivity[358]. A multitude of alk-1-enyl cuprates have been generated in the above manner and all of these react with simple inorganic electrophiles to form the expected (*Z*)-substituted alkenes (equations 149–151).

$$R^1C{\equiv}CH + RCu{\cdot}MgX_2 \longrightarrow \left[(R^1)(R)C{=}C(H)(Cu{\cdot}MgX_2) \right] \xrightarrow{H^+} (R^1)(R)C{=}C(H)(H) \qquad (149)$$

$$\xrightarrow{I_2} (R^1)(R)C{=}C(H)(I) \qquad (150)$$

$$\xrightarrow[H^+]{CO_2} (R^1)(R)C{=}C(H)(CO_2H) \qquad (151)$$

Specific examples can be found in refs. 356 and 357. The reaction of alk-1-enyl cuprates with carbon dioxide (equation 151) can be facilitated by using hmpt as a secondary solvent in the presence of $P(OEt)_3$[363]. Electrophilic substitution can also be aided by firstly converting the (alk-1-enyl) cuprate to a diorgano cuprate by reaction with $LiC{\equiv}CBu^n$. Subsequent reaction with an electrophile gives the expected (*Z*)-substituted alkene (equation 152)[364].

$$ClMgO(CH_2)_4Cu{\cdot}MgBr_2 \xrightarrow[thf,\ 15\,°C]{HC{\equiv}CH} \left[(H)(ClMgO(CH_2)_4)C{=}C(H)(Cu{\cdot}MgBr) \right] \xrightarrow[thf]{LiC{\equiv}CBu^n} \left[(H)(ClMgO(CH_2)_4)C{=}C(H)(CuC{\equiv}CBu^n\,MgBr) \right] \xrightarrow{H^+} \underset{55\%}{(H)(HO(CH_2)_4)C{=}C(H)(H)} \qquad (152)$$

These alk-1-enyl cuprates can react with a variety of organic substrates, such reactions being described in detail under the appropriate headings (see below). Again, the exact nature of these alk-1-enyl cuprates has not been determined and consequently they are prepared *in situ*.

Higher alk-1-enyl cuprates can be prepared by reacting the chosen terminal acetylene with $R_2Cu{\cdot}MgX$ or $RR'Cu{\cdot}MgX$ (equation 153). However, the necessity to use large

$$5Me_2CuMgBr{\cdot}LiBr + Hex^n{\equiv}CH \xrightarrow[30\,°C,\ 5\ h]{thf} \underset{95\%}{\left[(Hex^n)(Me)C{=}C(H)(CuMgBr{\cdot}LiBr\ Me) \right]} \qquad (153)$$

$$\text{MeMgBr} + \text{Hex}^n\text{C}{\equiv}\text{CH} \xrightarrow[\text{Et}_2\text{O, 5 d}]{\text{CuBr·Me}_2\text{S}} [(\text{Hex}^n)(\text{Me})\text{C}{=}\text{CH}(\text{Cu·MgBr}_2)] \quad 100\% \tag{154}$$

excesses of the Grignard reagent limits further applications[358]. One prime exception is illustrated in equation 154, the only drawback being the long reaction time[365].

It is noteworthy that the reaction between $Bu^nCu \cdot MgBr_2$ and $HC{\equiv}C(CH_2)_nZ$ (**46**) (where $Z = NEt_2$, SEt, or OR) gives almost exclusively (*Z*) products, but for $1 < n < 3$ the site of carbometallation is dependent on the nature of Z (equation 155). For instance,

$$\underset{(\mathbf{46})}{\text{HC}{\equiv}\text{C(CH}_2)_n\text{Z}} + \text{Bu}^n\text{Cu·MgBr}_2 \longrightarrow \underset{(\mathbf{47})}{[(\text{Br}_2\text{Mg·Cu})(\text{H})\text{C}{=}\text{C}(\text{Bu}^n)((\text{CH}_2)_n\text{Z})]} + [(\text{Bu}^n)(\text{H})\text{C}{=}\text{C}(\text{Cu·MgBr}_2)((\text{CH}_2)_n\text{Z})] \tag{155}$$

exclusive formation of **47** is observed for $n = 2$ and $Z = OSiMe_3$ (72% yield on hydrolysis)[363,366].

For $HC{\equiv}CSiMe_3$ (**48**), the site of carbometallation is exclusively the β-carbon atom of the alk-1-yne (equation 156)[367]. Using $Hex^nCu \cdot MgBr_2$ a 76% yield of the hydrolysed product is obtained.

$$\underset{(\mathbf{48})}{\text{HC}{\equiv}\text{CSiMe}_3} + \text{RCu·MgBr}_2 \longrightarrow [(\text{H})(\text{R})\text{C}{=}\text{C}(\text{SiMe}_3)(\text{Cu·MgBr}_2)] \tag{156}$$

Normant reagents add stereospecifically to the triple bond of conjugated enynes containing a terminal alkyne. Maximum yields are obtained using R_2CuMgX in preference to $RCu \cdot MgX_2$, but only one R group is transferred when using the former

$$\text{HC}{\equiv}\text{CCH}{=}\text{CR}^1_2 + \text{R}_2\text{CuMgBr} \xrightarrow[\text{H}^+]{\text{thf}} \text{R}^1_2\text{C}{=}\text{CH}{-}\text{C(R)}{=}\text{CH}_2 \quad 80{-}95\% \tag{157}$$

$$\underset{(\mathbf{49})}{\text{(cyclohex-1-enyl)}{-}\text{C}{\equiv}\text{CH}} + \text{Bu}^t_2\text{AgMgCl} \longrightarrow \underset{(\mathbf{50})}{\text{(cyclohex-1-enyl)}(\text{H})\text{C}{=}\text{C}(\text{H})(\text{Bu}^t)} \quad 55\% \tag{158}$$

reagent (equation 157)[358]. Interestingly, the addition of $Bu^t_2AgMgCl \cdot 2LiBr$ to enyne **49** gave diene **50** on hydrolysis, instead of the expected allene (Section IV.A.2)[362].

A number of disubstituted alkynes are known to react stereospecifically with Normant reagents. Compounds of the type R′C≡CSR″ (**51**) react to give the expected (*Z*)-substituted alkenes[368], whereas 1-phenylpropyne (**52**) is converted selectively to (*Z*)-1-phenylpropene (**53**) on hydrolysis[369]. Diorgano cuprates add in a (*Z*) fashion to R′C≡CCN (**54**)[370], whereas the analogous diorganoargentates add in an (*E*) mode (equations

$$R^1C{\equiv}CSR^2 + RCu{\cdot}MgBr_2 \longrightarrow [\text{(R)(R}^1\text{)C=C(CuMgBr}_2\text{)(SR}^2\text{)}] \quad (159)$$

(**51**)

$$5\,Bu^n_2CuMgBr + PhC{\equiv}CMe \xrightarrow{H^+} \text{(Ph)(H)C=C(Me)(H)} \quad (160)$$

(**52**) 100% (**53**)

$$R^1C{\equiv}CCN + R_2CuMgX \xrightarrow{thf} [\text{(R}^1\text{)(R)C=C(CN)(CuMgX R)}] \quad (161)$$

(**54**)

$$R^1C{\equiv}CCN + R_2AgMgCl \xrightarrow{thf} [\text{(R}^1\text{)(R)C=C(R AgMgCl)(CN)}] \quad (162)$$

(**54**)

161 and 162)[371]. In both cases only one of the R groups is transferred to the organic substrate.

Enynes of the type $R^2C{\equiv}CCN$ (**55**) (where R^2 = cyclohex-1-enyl or isopropenyl) undergo (*Z*) addition across the triple bond when reacted with a Normant reagent (equation 163)[372]. Similarly, diorgano argentates containing a primary R group add across the triple bond, but in an (*E*) fashion (equation 164). In this instance both R groups

$$R^2C{\equiv}CCN + RCu{\cdot}MgBrCl{\cdot}LiBr \xrightarrow{thf} [\text{(R}^2\text{)(R)C=C(CN)(Cu}{\cdot}\text{MgBrCl}{\cdot}\text{LiBr)}] \quad (163)$$

(**55**)

$$2\,R^2C{\equiv}CCN + R_2AgMgCl{\cdot}2\,LiBr \xrightarrow{thf} [\text{(R)(R}^2\text{)C=C(CN)(Ag}{\cdot}\text{MgClBr)}] \quad (164)$$

(**55**)

are transferred to the enyne. Using branched R groups allenes are generated (Section IV.A.2), one exception being shown in equation 165[372].

TABLE 6. Conversion of alkynes to allenes using Normant reagents

Alkyne	Allene	R	Yield (%)	Ref.
1. 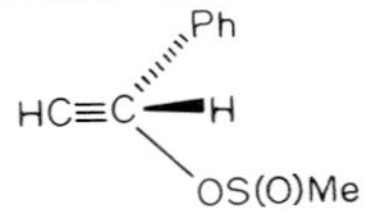 HC≡C, Ph, H, OS(O)Me	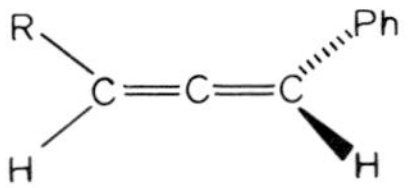 R, H, C=C=C, Ph, H	Me, Ph[a]	> 90	373
2. 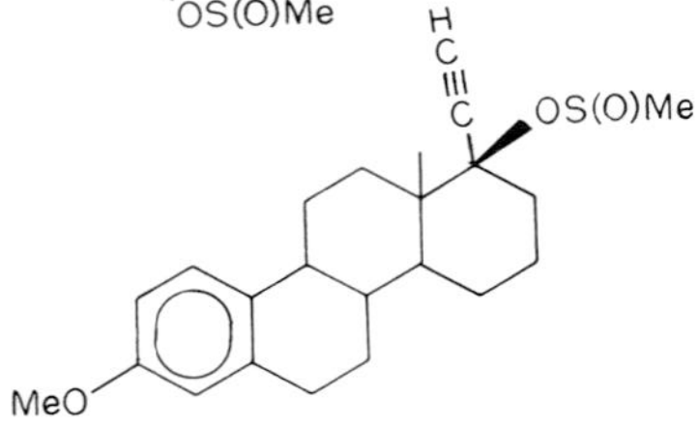 HC≡C, OS(O)Me, MeO	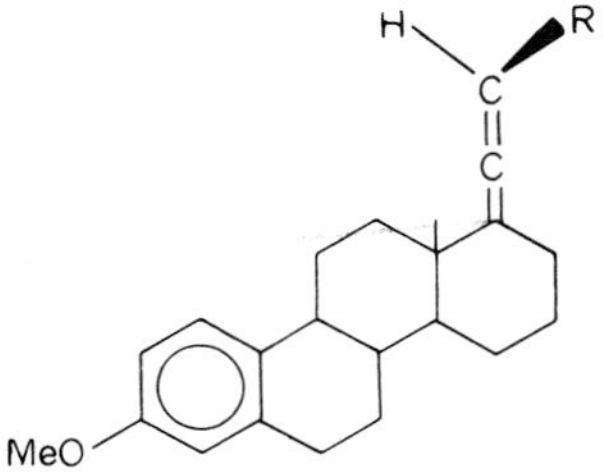 H, R, C=C, MeO	Me, Bu[t], Ph[b]	> 97	374
3. 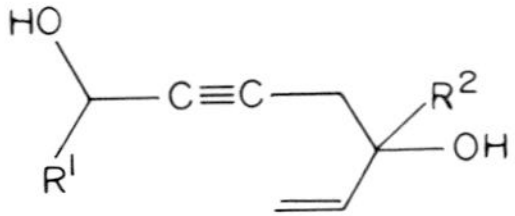 HO, C≡C, R^2, OH, R^1 R^1 = H, Me R^2 = H, Me	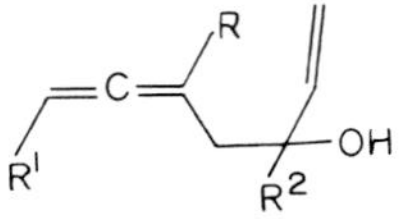 =C=, R, OH, R^1, R^2	Et, Pe[n c]	58–85	375
4. $R^1CH_2C{\equiv}CCH_2R^1$ R^1 = OS(O)Me, Otos	$R^1CH_2C(R){=}C{=}CH_2$	Bu[n], Pr[i], Bu[t], Ph[b]	—	376

5. 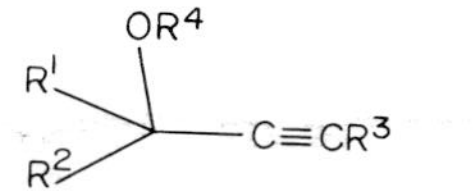 $R^1R^2C(OR^4)C{\equiv}CR^3$ $R^1 = H, Me, Pr^n, Ph; R^4 = Ac, tos$ $R^2 = H, Me; R^3 = H, Pe^n, SiMe_3$	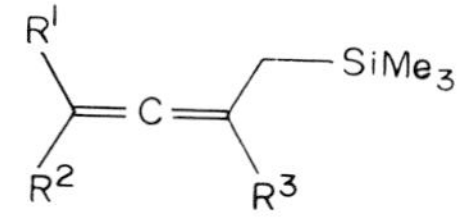 $R^1R^2C{=}C{=}C(R^3)CH_2SiMe_3$	$Me_3Si\bar{C}H_2{}^b$	50–85	377
6. $R^1CH[OS(O)Me]C{\equiv}C(CH_2)_nOCH(Me)OEt$ $R^1 = H, Pr^n; n = 1, 2$	$R^1CH{=}C{=}C(R)(CH_2)_nOH$	Me, Et, $Bu^{t\,b}$	60–77	378
7. 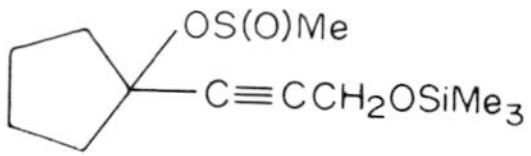 1-OS(O)Me-1-($C{\equiv}CCH_2OSiMe_3$)cyclopentane	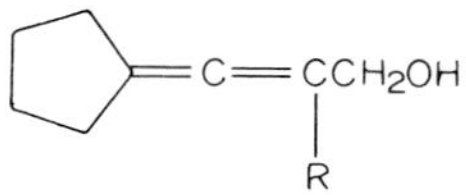 cyclopentylidene$=C{=}C(R)CH_2OH$	Me, Ph^b	70, 90	378
8. 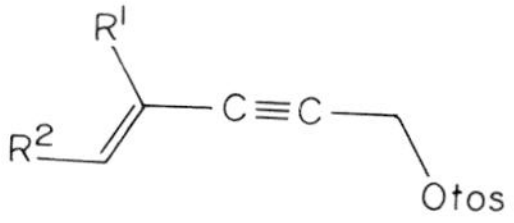 $R^2CH{=}C(R^1)C{\equiv}CCH_2Otos$ $R^1 = H, Me; R^2 = H, Me$	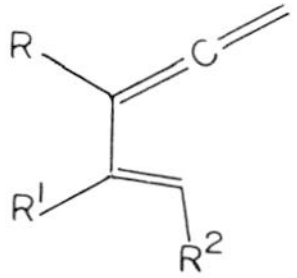 $R^2CH{=}C(R^1)C(R){=}C{=}CH_2$	$Hex^{N\,b}$	76, 78	379
9. 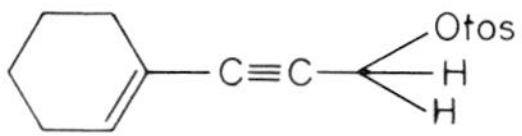 1-cyclohexenyl$-C{\equiv}C-CH_2Otos$	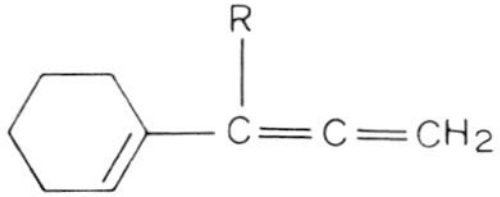 1-cyclohexenyl$-C(R){=}C{=}CH_2$	Me^b	80	379
10. 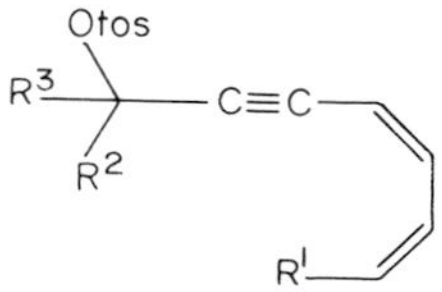 $R^3R^2C(Otos)C{\equiv}C-CH{=}CH-CH{=}CH-R^1$ $R^1 = H, Me; R^2 = H, Me; R^3 = H, Me$	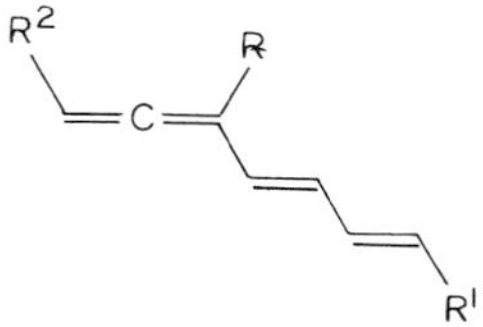 $R^2CH{=}C{=}C(R)-CH{=}CH-CH{=}CH-R^1$	Bu^i, Et, Hex^n, $Bu^{n\,b}$	50–77	380

TABLE 6. (*Contd.*)

Alkyne	Allene	R	Yield (%)	Ref.
11. 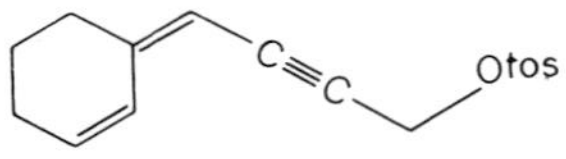 [structure: cyclohexenylidene–CH–C≡C–CH2–Otos]	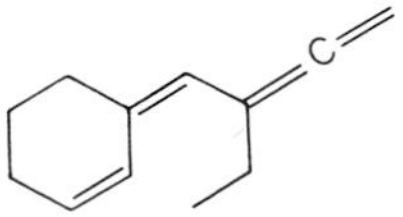 [structure]	Et^b	85	380
12. 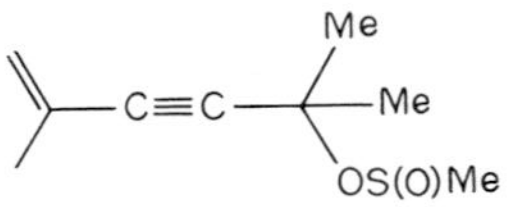 [structure: –C≡C–C(Me)(Me)OS(O)Me]	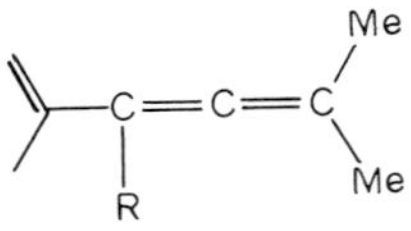 [structure: –C(R)=C=C(Me)Me]	$Bu^{t\,d}$	90	362
13. 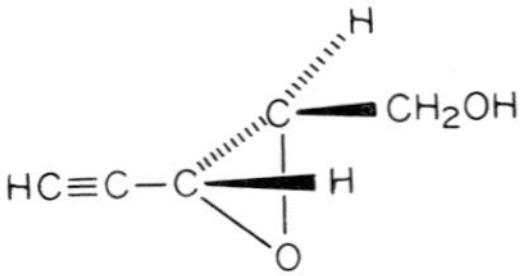 HC≡C–C(epoxide)–C(H)CH2OH	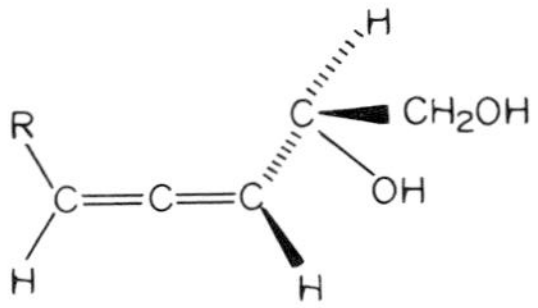 [structure: R(H)C=C=C(H)–C(H)(CH2OH)OH]	Bu^n, $Oct^{n\,e}$	57, 68	381

[a] thf, − 60 °C, 2 h; [b] thf; c Et_2O; [d] thf–hmpt; [e] Et_2O, − 60 °C.

$$\text{1-cyclohexenyl}-C\equiv CCN + Pr^i_2AgMgCl \xrightarrow[H^+]{thf} \text{1-cyclohexenyl}(Pr^i)C{=}CH(CN)\quad 80\% \tag{165}$$

The addition of a Normant reagent to an alkyne possessing a suitable leaving group in the propargylic position [e.g. Otos, OS(O)Me, or OAc] results in an S_N2' type of reaction, which yields an allene (equation 166).

$$R^1R^2C(OR^3)-C\equiv C-R^4 + R^5Cu{\cdot}MgX_2 \longrightarrow R^1R^2C{=}C{=}CR^4R^5 \tag{166}$$

Some specific examples are given in Table 6. It is noteworthy that a number of chiral allenes have been generated using this procedure. For instance, the allene **56** was isolated with an enantiomeric excess of 88% (equation 167)[373].

$$HC\equiv C-C(Ph)(H)OS(O)Me + PhCu{\cdot}MgBr_2 \longrightarrow Ph(H)C{=}C{=}C(Ph)H \quad [(R)\text{-}(-)\text{-}\mathbf{56}] \tag{167}$$

2. Addition to alkenes

For over 40 years it has been known that the 1,4-addition of Grignard reagents to α,β-unsaturated carbonyl compounds can be achieved by using copper salts as either catalysts or stoichiometric reagents[282,283]. In more recent times Normant reagents have been used as stoichiometric reagents in these reactions. For instance, the organocopper reagent derived from 2-(buta-1,3-dienyl)magnesium chloride reacts regiospecifically with enone **57** to give butadienone **58** on hydrolysis[384]. Similarly, lactone **59** undergoes a 1,4-

$$\mathbf{(57)} + CH_2{=}CH-C(=CH_2)Cu{\cdot}MgClBr \xrightarrow[H^+]{thf} \mathbf{(58)}\ 81\% \tag{168}$$

$$\mathbf{(59)} + RCu{\cdot}MgBr_2 \xrightarrow[H^+]{thf} \mathbf{(60)} \tag{169}$$

addition reaction with Normant reagents to generate lactone **60**[385]. Steroid **63** can be produced in 85% yield and > 99% stereochemical purity by reacting enone **61** with the

TABLE 7. Reaction of enones with Normant reagents based on R

Enone	Product	R	Reaction conditions[a]	Yield (%)[b]	Ref.
1. 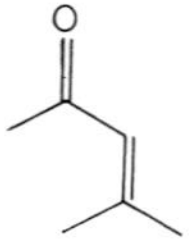	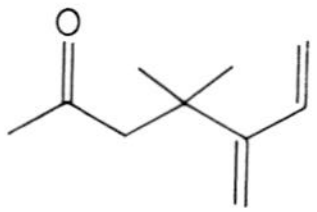	Buta-1, 3-dien-2-yl	A	81	384
2. 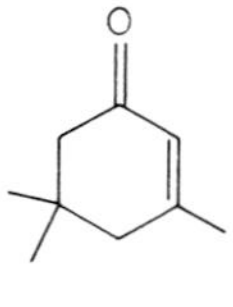	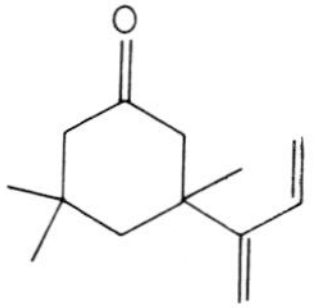	Buta-1, 3-dien-2-yl	B	52	384
3. 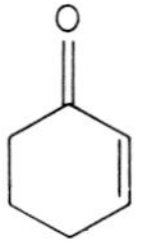	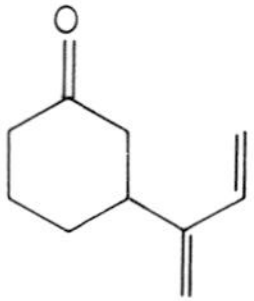	Buta-1, 3-dien-2-yl	A	73	384
4. 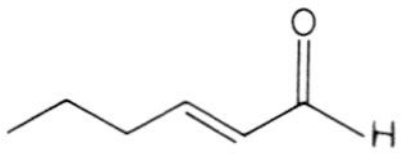	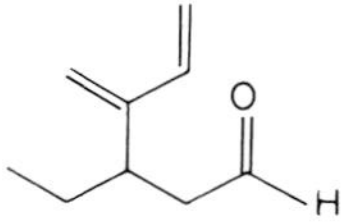	Buta-1, 3-dien-2-yl	A	84	384
5. 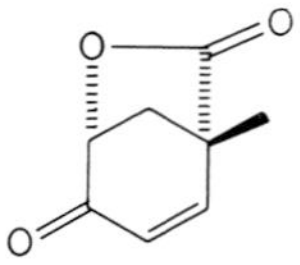	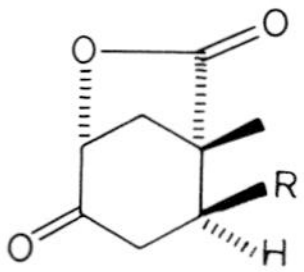	Vinyl But-3-enyl	C D	47 74	385 385

6.		[c]	E	53	386
				31	386
7.		[c]	F	85[g]	386
				84[g,h]	386
8. $CH_2{=}CHCO_2Et$		*p*-Cumyl	G	—	387
9.		Vinyl[d]	G	95	388
		Ph[d]		97	388
10.		Vinyl	F	40–55[i]	389
11. $CH_2{=}CHCO_2Bu^n$	$PhCH_2CH_2CO_2Bu^n$	Ph	H	—	390

TABLE 7. (*Contd.*)

Enone	Product	R	Reaction conditions[a]	Yield (%)[b]	Ref.
12. 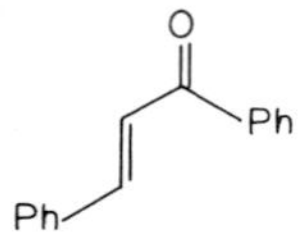	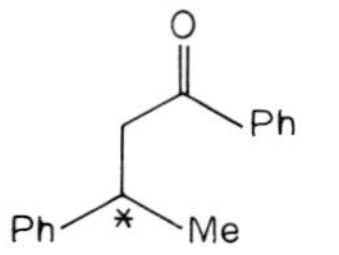	Me[e]	F	—	391
13. 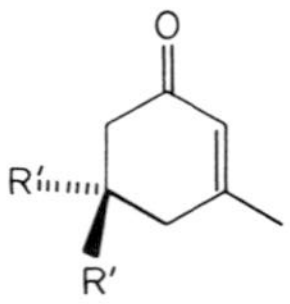	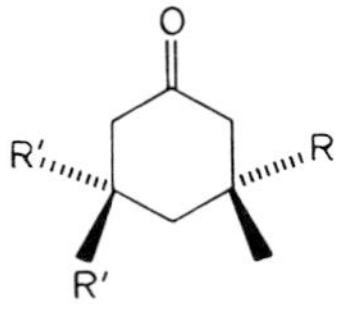	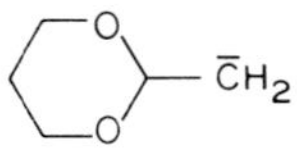 $Pr^i\bar{C}H_2$, But-3-enyl, isopropenyl	I	58–86	392
14. 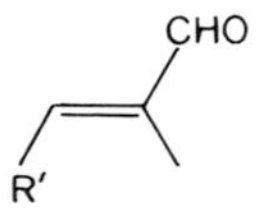 R^1 = Et, Pr^n	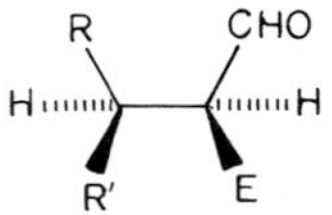	Me, Bu^n, allyl	J	43–71[j,k]	393
15. 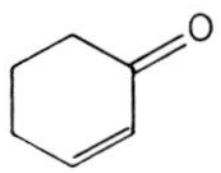	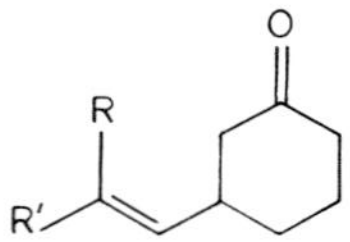 R^1 = Me, Et, Hex^n	Me, Et, Bu^n, Hex^n[f]	K	63–73	394

	Substrate	Product	R	Conditions	Yield (%)	Ref.
16.	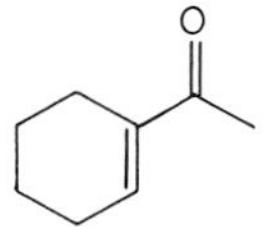 O	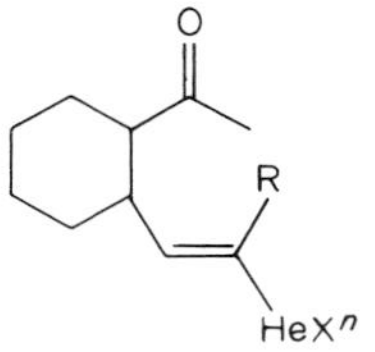 O, R, HeX^n	Et[c,f]	K	69	394
17.	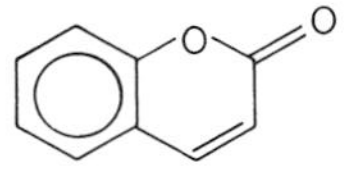 O, O	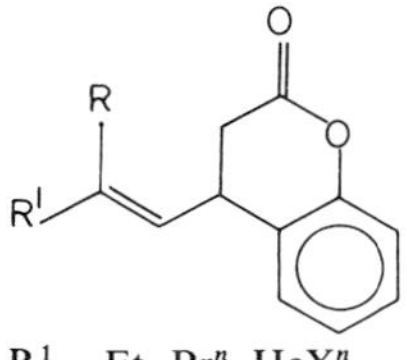 O, R, R^1, O R^1 = Et, Pr^n, HeX^n	Et, Pr^n [c,f]	K	50–62	394
18.	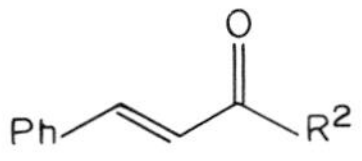 O, Ph, R^2 R^2 = Me, Ph	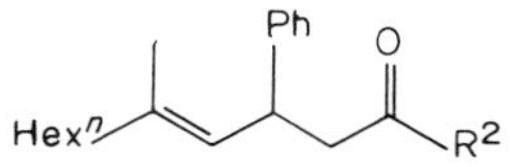 Ph, O, Hex^n, R^2	Et[c,f]	K	52, 53	394

[a] A = thf, − 25 to − 20 °C, 1.5–3 h; B = thf, − 60 to 0 °C, 18 h; C = thf, − 78 to 0 °C; D = thf, − 10 °C; E = thf, 25 °C, 4 h; F = thf; G = thf, − 40 °C; H = MeCN; I = Et_2O, − 30 °C, 1 h; = thf, − 50 °C; and K = Et_2O.
[b] H^+.
[c] $R(Pr^nC{\equiv}C)CuMgBr$.
[d] $R(MeSO_2CH_2)CuMgBr$.
[e] 1-R^2-2-$(CH_2)_n$OCu(Me)MgBr-pyrrolidine.
[f] $RR^1C{=}CHCu{\cdot}MgBr_2$.
[g] ICH_2CO_2Et.
[h] $BrCH_2CH{=}CH_2$.
[i] $BrCH_2C(OMe){=}CHCO_2Me$.
[j] Br_2, pyridine.
[k] Me_3SiCl.

diorgano cuprate **62**[386]. The reactions of a number of 1, 3-enones with Normant reagents are summarized in Table 7. By incorporating an optically active ligand in the cuprate it is

(61) + (62) $\xrightarrow[\text{ICH}_2\text{CO}_2\text{Et, H}^+]{\text{thf}}$ (63) (170)

possible to generate a chiral ketone from 1, 3-enones. For instance, enone **64** reacts with the chiral cuprate (*N*-methylprolinol)MeCuMgBr (**65**) to give ketone **66** with up to 88% enantiomeric excess[391].

(64) + (65) $\xrightarrow[\text{H}^+]{\text{thf}}$ (66) (171)

Other α, β-unsaturated carbonyl compounds also undergo 1, 4-addition by Normant reagents. For instance, both enal **67** and the chiral ethylenic ester **68** undergo the expected 1, 4-addition with these reagents[393,395]. In the latter example both (*E*)- and (*Z*)-**68** give rise to products of the same absolute configuration. This is consistent with a mono-electron

$$\text{EtCH=C(Me)CHO} + \text{R}_2\text{CuMgCl} \xrightarrow[\text{Br}_2]{\text{thf,}} \text{EtCH(R)C(Me)(Br)CHO} \quad (172)$$

(67)

MeCH=CHCO$_2$-menthyl (68) (*E* or *Z*) + $Ph_2CuMgBr$ $\xrightarrow[\text{H}^+]{\text{thf}}$ MeCH(Ph)CH$_2$CO$_2$-menthyl (173)

(69) + CH_2=CHCO$_2$Et (70) $\xrightarrow[\text{thf}]{\text{CuI}}$ (71) (174)

transfer mechanism[395]. 6-Bromopenicillanoylmagnesium bromide (**69**) reacts both stereo- and regio-selectivity with ethyl acrylate **70**, in the presence of a stoichiometric amount of copper(I) iodide, to give the 6-spirocyclopropylpenicillanate (**71**) in 73% yield[396]. Further examples of this type of reaction are included in Table 7.

Enones can also undergo 1,4-addition by alk-1-enyl cuprates. Marfat *et al.*[394] showed that both enone **72** and lactone **73** react with a number of alk-1-enyl cuprates to give the expected 1,4-addition products. Specific details are given in Table 7.

(**72**) + $R^2(R^1)C=CH(Cu\cdot MgBr_2)$ ⟶ 3-($R^2R^1C=CH$)cyclohexanone, 63–73% (175)

(**73**) + $R^2(R^1)C=CH(Cu\cdot MgBr_2)$ ⟶ product, 50–62% (176)

The addition of Normant reagents to buta-1,3-dienes results in the formation of alkenes, on further reaction with a suitable electrophile (see equation 177). A variety of alkenes have been prepared in this manner and isolated in good yield (75–90%)[397].

$$CH_2{=}CHCH{=}CH_2 + RCu\cdot MgBr_2 \longrightarrow (RCH_2CH \cdots CH \cdots CH_2)Cu\cdot MgBr_2$$

$$\xrightarrow{CO_2,\ H^+} CH_2{=}CHCH(CO_2H)CH_2R \qquad \xrightarrow{R^1X} RCH_2CH{=}CHCH_2R^1 \tag{177}$$

Spectinomycin derivative (OBz, MeN, OH, HO, NMe–OBz, vinyl enone) + $RCu\cdot MgBr_2$ $\xrightarrow{H^+,\ thf}$ 1,4-adduct $\xrightarrow{2\ steps}$ (**74**) (178)

White *et al.*[398] utilized this reaction in the synthesis of alkyl analogues of spectinomycin (**74**).

Diorgano argentates of the type $R_2AgMgCl$ can in some instances add to the double bond of enynes in preference to the triple bond (see Section IV.A.1). For instance, diorgano argentates react with the double bond in terminal enyne **75**[8], but for enyne **76** they only add in this manner when the R group in $R_2AgMgCl$ is branched[372]. In both cases the

$$\text{(75)}\ CH_2{=}C(R^1)C{\equiv}CH + R_2AgMgCl \xrightarrow{H^+} H_2C{=}C{=}C(CH_2R)R^1 \quad (179)$$

$$\text{(76)}\ CH_2{=}C(Me)C{\equiv}CCN + Bu^t_2AgMgCl \xrightarrow{H^+} (Bu^tCH_2)(Me)C{=}C{=}C(CN)H \quad 90\% \quad (180)$$

product on hydrolysis is an allene. Replacing the CN moiety in **76** with SMe or PPh_2, however, results in the formation of allenes regardless of whether the R group in $R_2AgMgCl$ is primary or branched[399].

Allenes containing a suitably positioned leaving group react with Normant reagents in an S_N2' mode (equations 181 and 182). The product obtained, however, depends on the type of allene employed. For instance, allene **77** gave a 1,3-diene[376], whereas allene **78**

$$\text{(77)}\ MeS(O)O{-}CH_2{-}C(R^1){=}C{=}CR^2R^3 + RCu{\cdot}MgBr_2 \xrightarrow{thf} H_2C{=}C(R^1){-}C(R){=}CR^2R^3 \quad (181)$$

$$\text{(78)}\ (MeO)I{-}C{=}C{=}CH_2 + RCu{\cdot}MgBr_2 \xrightarrow{thf} RCH_2C{\equiv}COMe \quad (182)$$

(**79**) $\xrightarrow[thf]{RMgX/CuBr}$ [XMgCu(R) coordinated to N of benzothiazole, S_N2' transition state] $\longrightarrow$ $R^1CH{=}CH{-}C(R)(R^2)R^3$ (183)

yielded an alkyne under these conditions[400]. In contrast, alkenes of the type **79** react with organo cuprates in an S_N2' fashion to yield (*E*)-olefins with $> 98\%$ stereoselectivity[401–403].

3. Ring-opening reactions

It is well established that the ring opening of lactones can be achieved by reaction with Grignard or alkyl lithium reagents and certain hetero atoms[404]. For β-propiolactones two types of ring opening are prevalent using these reagents, giving rise to a mixture of products (equation 184). Regiospecific ring opening, however, can be attained by use of

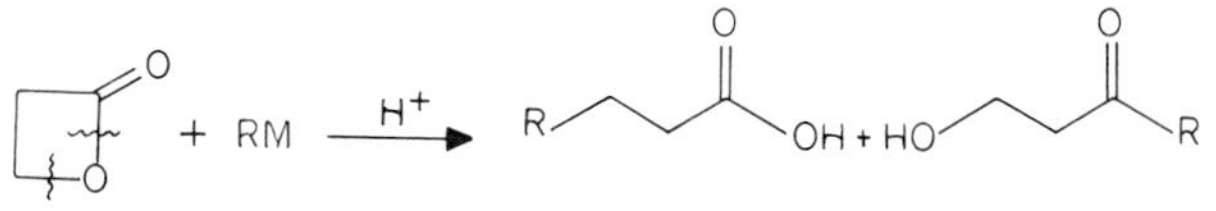

(184)

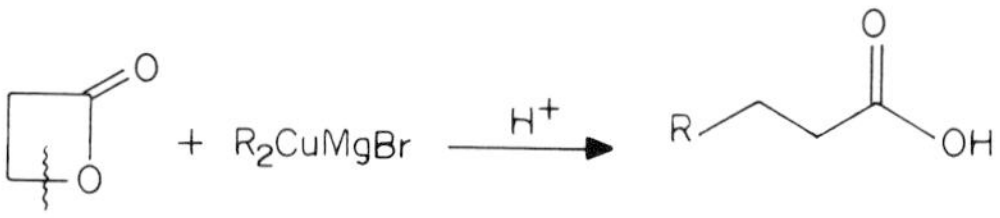

(185)

Normant reagents (and R_2CuLi), with cleavage only occurring at the methylene—oxygen junction (equation 185). Only one R group is transferred from the diorgano cuprate to the substrate during the reaction. Substituted β-propiolactones react in a similar manner. For instance, β-lactones **80** and (+)-**81** both react with $R_2CuMgBr$ to give the expected carboxylic acid[404,405]. In the latter case the chiral product is obtained with an

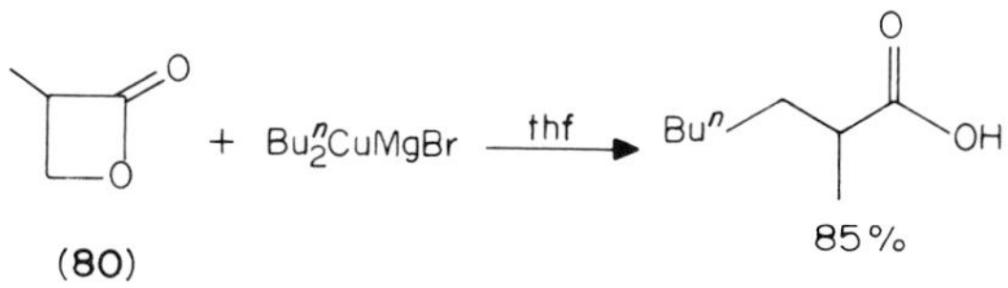

(186)

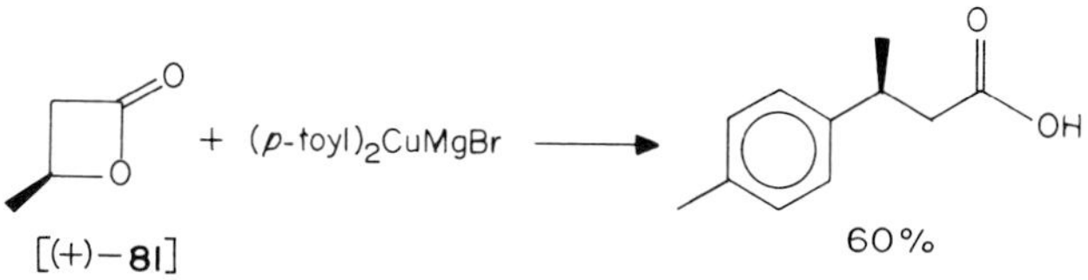

(187)

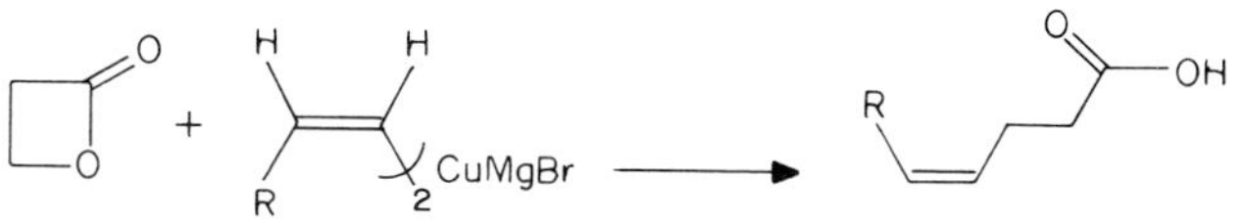

(188)

enantiomeric excess of 84%. Dialk-1-enyl cuprates also react with β-propiolactones regiospecifically (equation 188)[406].

Normant reagents react stereoselectively with β-vinyl-β-propiolactone (**82**) to afford (*E*)-alk-3-enoic acids (**83**) as the major product (*E*:*Z* ratio $\approx$ 8:1). Only trace amounts, if any, of the expected alk-4-enoic acid **84** were obtained. Evidently the diorgano cuprate adds to the terminal alkene of **82** via an S_N2 pathway, which results in the ring opening reaction (equation 189)[407]. A similar reaction has been observed for γ-vinyl-γ-butyrolactone and δ-vinyl-δ-valerolactone with Normant reagents to yield (*E*)-alk-4-

$R_2CuMgBr$ + (82) $\xrightarrow{H^+}$ (83) + ... + (84) (189)

and -5-enoic acids, respectively[408]. Although a diminished regioselectivity was evident (in general the E:Z ratio $\approx$ 4:1) no other product was obtained. This method has been successfully employed to prepare (E)-7-oxoalk-3-enoic acids (**86**) by reacting **82** with

(82) + (85) $\xrightarrow{H^+}$ (86) (190)

a bis (dimethylhydrazone) cuprate (**85**)[409]. A number of examples of lactones reacting with Normant reagents are listed in Table 8.

Oxiranes can be successfully converted to (Z)-homoallylic alcohols by reaction with alk-1-enyl-Cu(C≡CPr^n)MgBr (**87**), obtained by reacting alk-1-enyl cuprates with 1-lithiopent-1-yne[394]. Both oxirane and 2-substituted oxiranes react with **87** to give the

(87) $\xrightarrow[H^+]{thf}$ 74–95% (191)

+ 87 $\xrightarrow[H^+]{thf}$ 82–94% (192)

desired alcohol in high yield (equations 191 and 192)[394,410]. Specific examples are included in Table 8.

TABLE 8. Reaction of Normant reagents (based on R) with lactones and oxiranes

Substrate	Product	R	Reaction Conditions[a]	Yield (%)	Ref.
1. 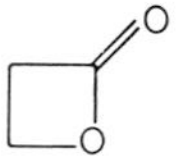	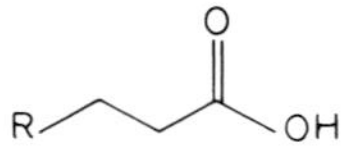	Me, Bu^n, Pr^i, Bu^t, Ph, allyl, vinyl, $Me_2C{=}CHCH_2{-}$	A	65–95	404, 411
2. 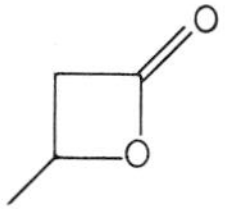	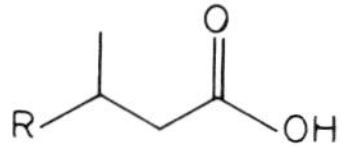	$Me_2C{=}CHCH_2CH_2{-}$, Bu^n, Me, *p*-tolyl	A	60–89	404, 405, 412
3. 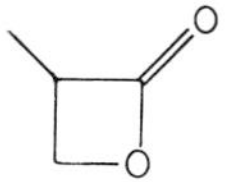	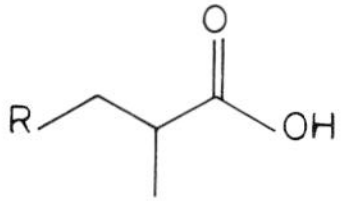	Bu^n, Me, vinyl, Ph	A	52–85	404
4. 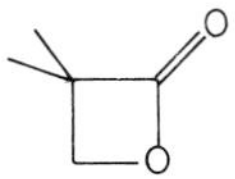	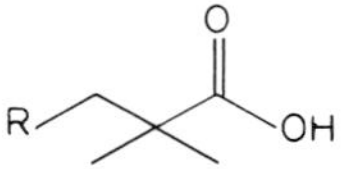	Bu^n, Me, vinyl	A	82, 91, 48	404
5. 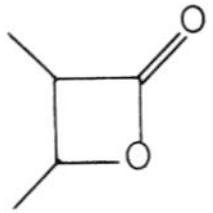	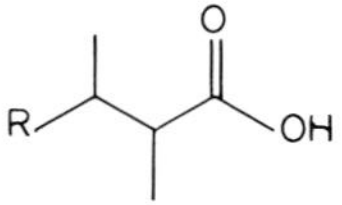	Bu^n, Me	A	82, 81	404

TABLE 8. (*Contd.*)

Substrate	Product	R	Reaction Conditions[a]	Yield (%)	Ref.
6.		Et, Bu^n, Pe^n, Hex^n, Hept^n, Oct^n, Non^n	—	—	406, 214b
7.	[c]	Bu^n, allyl	B	92, 88	407
8.	[c]	Ph, vinyl, allyl	C	91, 70, 41	408
9.	[b,c]	Pe^n	D	87	408
10.	[d]	Allyl	B	51	408
11.	[e]	R = Me, Et, Pr^n R^1 = Me, Pr^n, Bu^n, Hex^n	E	75–95	394, 410

12. 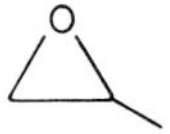		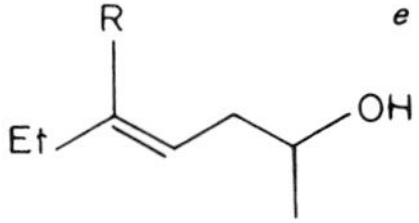 [e]	Bu^n	E	82	394
13. 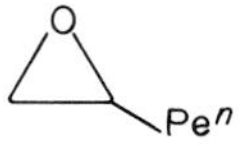		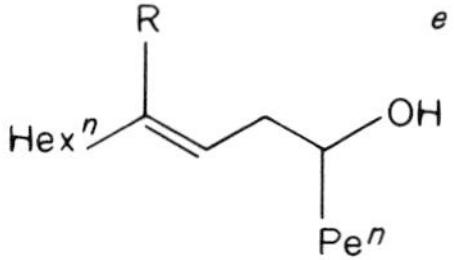 [e]	Et	E	94	394
14. 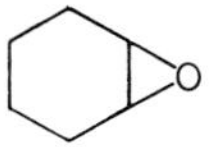		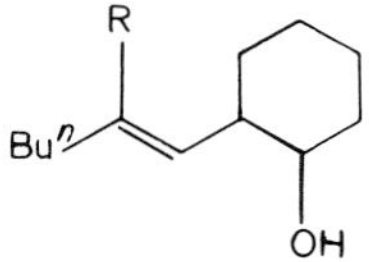 [e]	Pr^n	E	26	394

[a] A = thf, Me_2S, − 30 to 0 °C, 2 h; B = thf, Me_2S, − 78 to − 50 °C; C = thf, Me_2S, − 30 to − 25 °C, 1 h; D = thf, Me_2S, − 30 °C, 13 h; E = thf, Me_2S, − 23 °C.
[b] Using RC(Me)═C(Me)—.
[c] $E:Z \approx 8:1$.
[d] $E:Z \approx 11:1$.
[e] RR^1C═C(Me)—.

4. *Addition to acyl derivatives*

The direct 1, 2-addition of Grignard reagents to aldehydes or ketones can, in general, be achieved without the assistance of a transition metal[1,8]. In view of this, few synthetically useful reactions have been developed in which transition metals act as stoichiometric reagents in the 1, 2-addition of Grignard reagents to this type of substrate. Two such examples are the addition of the organo cuprate **88** to acetone and the conversion of enone **89** to the terphenyl derivative **90**[397,413]. $[ZrCp_2Cl_2]$ has been utilized as a stoichiometric

$$(Bu^tCH_2CH{=}CH{=}CH_2)Cu{\cdot}MgBr_2 + MeCOMe \longrightarrow Bu^tCH_2CH(CMe_2OH)CH{=}CH_2 \quad 62\% \qquad (193)$$

(**88**)

(**89**) $\xrightarrow[CuCl_2]{R^4MgCl}$ (**90**) (194)

$$[Cp_2ZrCl_2] + MeCH{=}CHCH_2MgCl \xrightarrow{thf} [Cp_2Zr(C_4H_7)_2] + [Cp_2ZrCl(C_4H_7)] + [Cp_2Zr(C_4H_7)_3]MgCl{\cdot}thf \xrightarrow{RCHO/H^+} \text{OH / R product} \qquad (195)$$

reagent in the 1, 2-addition of but-2-enylmagnesium chloride to a number of aldehydes to give selectively *threo* homoallylic alcohols[414]. Presumably the Grignard reagent initially reacts with $[ZrCp_2Cl_2]$ to give an alkylzirconocene species, which has the coordinated butene in an *E* configuration, and this then reacts with the aldehyde to give the observed product (equation 195).

One of the most intensively studied reactions between a Grignard reagent and an acyl derivative, with a transition metal salt as a stoichiometric reagent, has been that involving acyl chlorides. These substrates are capable of reacting with Normant reagents to yield very hindered ketones which are otherwise difficult or impossible to synthesise (equation 196)[415]. For example, hindered ketones **91** and **92** can be obtained in high yield via this method[215,415]. It is noteworthy that further alkylation of these hindered ketones can lead to even more sterically crowded compounds (equation 199)[215]. Phenyl ketone **93** can be prepared from the corresponding acyl chloride and a diphenyl cuprate; the product is an important intermediate in the synthesis of a thienamycin analogue (**94**)[416]. Similarly, the acyl chloride **95** [prepared by treating the analogous chiral alcohol (see equation 187) with thionyl chloride] can be converted to (*S*)-(+)-*iso*-*ar*-tumerone (**96**), which can in turn be isomerized to (*S*)-(+)-*ar*-tumerone (**97**), the latter being isolated with 94% enantiomeric excess[405].

$$R^1COCl + RCu{\cdot}MgX_2 \longrightarrow RCOR^1 \quad (196)$$

$$Pr^i{}_2CHCOCl + Et_3CCu{\cdot}MgX_2 \longrightarrow \underset{(\mathbf{91})\ 77\%}{Pr^i{}_2CHCOCEt_3} \quad (197)$$

$$Pr^i{}_3CCOCl + Pr^iEtCHCu{\cdot}MgX_2 \longrightarrow \underset{(\mathbf{92})}{Pr^i{}_3CCOCHEtPr^i} \quad (198)$$

$$Pr^i{}_3CCOCl + Bu^tCH_2Cu{\cdot}MgX_2 \longrightarrow Pr^i{}_3CCOCH_2Bu^t \xrightarrow{2EtX-NaNH_2} Pr^i{}_3CCOCEt_2Bu^t \quad (199)$$

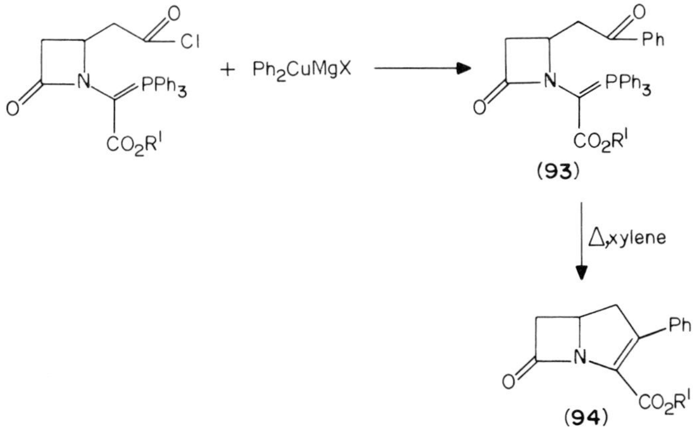

+ Ph_2CuMgX → (**93**) →[Δ, xylene] (**94**) (200)

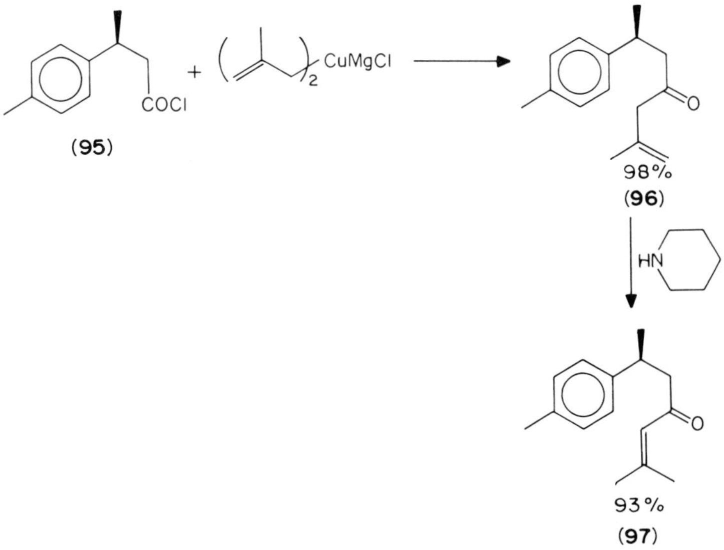

(**95**) + $(CH_2{=}C(CH_3)CH_2)_2CuMgCl$ → 98% (**96**) →[piperidine (HN)] 93% (**97**) (201)

In addition to organo and diorgano cuprates, a number of other organometallic reagents have been used to convert acyl chlorides into ketones. For instance, $[Rh(CO)ClL_2]$ can be used as a stoichiometric reagent in the addition of primary alkyl, aryl, or allyl Grignard reagents to acyl chlorides (equation 202)[417]. The initial step in the reaction is believed to be the generation of an alkylrhodium species, to which oxidative addition of the acid chloride occurs. Subsequent reductive elimination yields the unsymmetrical ketone. A cobalt(III) species has similarly been used to generate ketones (equation 203)[418].

$$\begin{array}{c} Rh(CO)ClL_2 + RMgX \longrightarrow [RhR(CO)L_2] \\ \downarrow R^1COCl \\ \underset{58\text{–}85\%}{RCOR^1} + [RhR(CO)L_2] \longleftarrow [RhR(Cl)(R^1CO)(CO)L_2] \end{array} \qquad (202)$$

$$\text{[Co(salen)(COR}^1\text{)]} \xrightarrow{MeMgI} \text{[Co(salen)(COR}^1\text{)(MeMgI)]} \xrightarrow{H^+} R'COMe \qquad (203)$$

5. Displacement of a halide

The reaction of an organic halide with a Grignard reagent, in the presence of a stoichiometric transition metal reagent, to yield the unsymmetrical coupled product has long been known[1]. A more recent example is the reaction of $I(CH_2)_{10}CO_2R'$ (**98**) with a variety of diorgano cuprates to give the expected coupled products in good yield[419]. Alk-1-enyl cuprates similarly react, not only with simple alkyl halides, but with more elaborate substrates to give a variety of di- and trisubstituted alkenes. For example, both $IC{\equiv}CCH_2Othp$ (**99**)[420] and $I(CH_2)_2C{\equiv}CSiMe_3$ (**100**)[421] react with alk-1-enyl cuprates to form enynes. Similarly, alkenyl halides react with the same organo cuprates to give dienes (equations 207 and 208)[422–424]. In the latter case $[Pd(PPh_3)_4]$ was used as a necessary catalyst[424].

$$\underset{(\mathbf{98})}{I(CH_2)_{10}CO_2R^1} + R(Me)CuMgX \longrightarrow \underset{54\text{–}85\%}{R(CH_2)_{10}CO_2R^1} \qquad (204)$$

$$\left[(Z)\text{-EtCH=CH–Cu}\cdot\text{MgBr}_2\right] + \underset{(\mathbf{99})}{IC{\equiv}CCH_2Othp} \longrightarrow (Z)\text{-EtCH=CH–C}{\equiv}\text{CCH}_2\text{OThp} \qquad (205)$$

$$\left[\text{Pr}^n\text{HC=C(SiMe}_3\text{)CuMgBr(Me)}\right] + \underset{(\mathbf{100})}{ICH_2C{\equiv}CSiMe_3} \longrightarrow \underset{76\%}{\text{Pr}^n\text{HC=C(SiMe}_3\text{)(CH}_2)_2\text{C}{\equiv}\text{CSiMe}_3} \qquad (206)$$

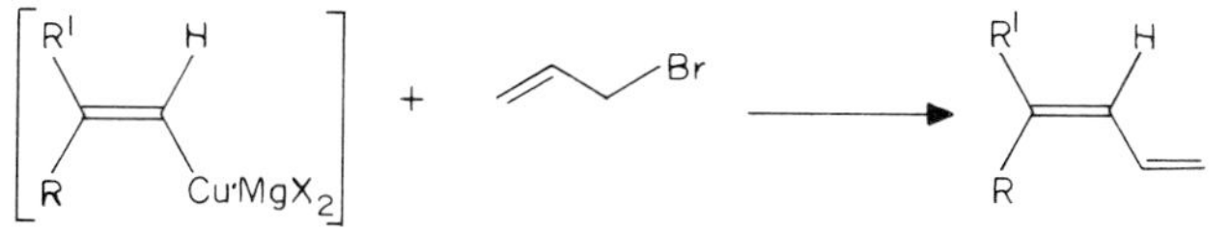

(207)

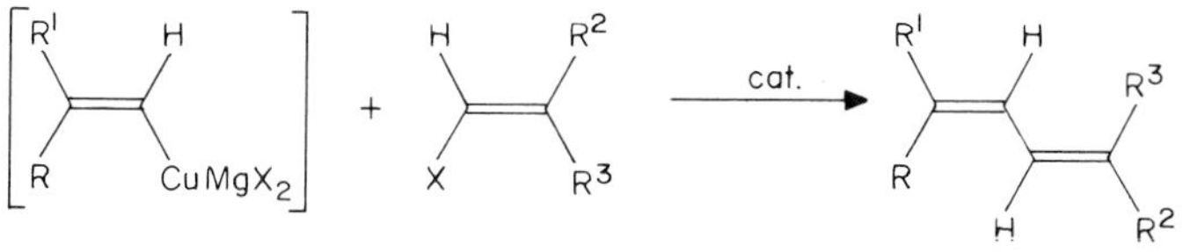

(208)

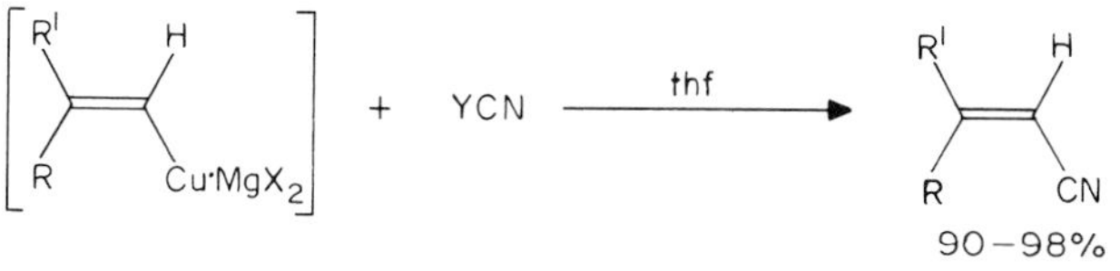

(209)

Cyanates of the type YCN (where Y = Cl, $PhSO_2$, or 4-$MeC_6H_4SO_2$) react with alk-1-enyl cuprates to give alk-1-enyl nitriles in good yields (equation 209)[425]. Interestingly, the analogous bromo- or iodocyanates, however, add the halide group to alk-1-enyl cuprates in preference to the cyano moiety[426].

6. *Displacement of a non-halide*

In addition to the halide group, a number of other moieties are capable of fulfilling the role of a suitable leaving group in these displacement reactions. As previously mentioned, alkynes possessing an alkoxy moiety in the propargylic position undergo an S_N2' type of reaction with Normant reagents, which results in the elimination of the alkoxide to generate allenes (equation 166, Section IV.A.1). Similarly, allenes containing an —S(O)OMe group or alkenes having an —Obtz group (reaction 210), in the appropriate position, react with organo cuprates in either an S_N2 or S_N2' fashion to give 1, 3-dienes or olefins, respectively, with high stereoselectivity (equations 181 and 183, Section IV.A.2). Direct substitution of an —Obtz group from an alkene by an organo cuprate, instead of initial attack on the double bond, has also been observed (equation 210)[402]. Alk-1-enyl

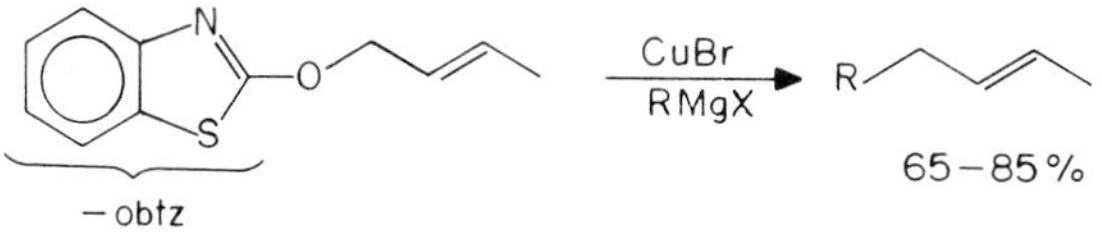

(210)

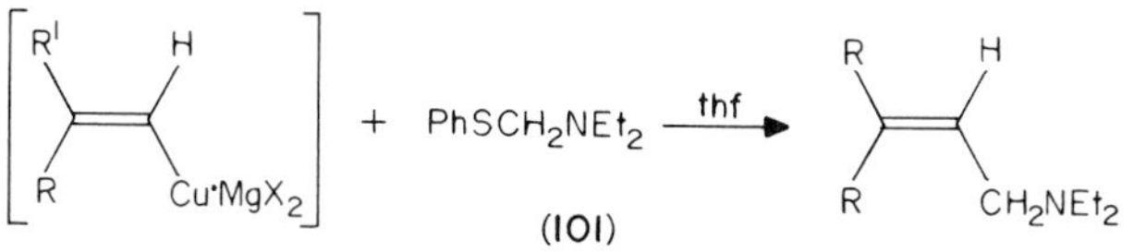

(211)

cuprates similarly react with organic sulphides of the type **101** resulting in rupture of the carbon—sulphur linkage to give enamines[427].

The displacement of a dimethylamino group by organo cuprates has been exemplified in the preparation of alkyl derivatives of spectinomycin (**74**) (see equation 178 for structure)[398].

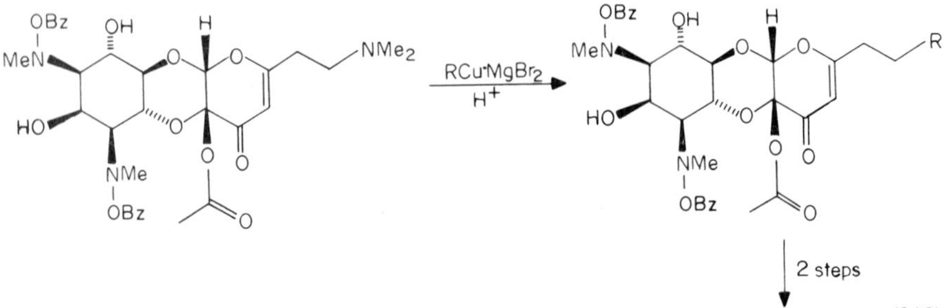

(212)

7. Homocoupling reactions

The interaction of transition metal halides with aryl Grignard reagents to give symmetrical biaryls in good yield is well established[1]. A recent example is the coupling of Grignard **102**, in the presence of a stoichiometric amount of copper(II) chloride, to give 4, 4′, 7, 7′-tetramethylbis(1-indenyl) (**103**). The resulting pair of diastereoisomers were separated and subsequently isomerized to 4, 4′, 7, 7′-tetramethylbis(3-indenyl) (**104**) by using triethylamine[113a]. Coupling of alk-1-enyl cuprates can be achieved by heating in the presence of oxygen (equation 214)[423].

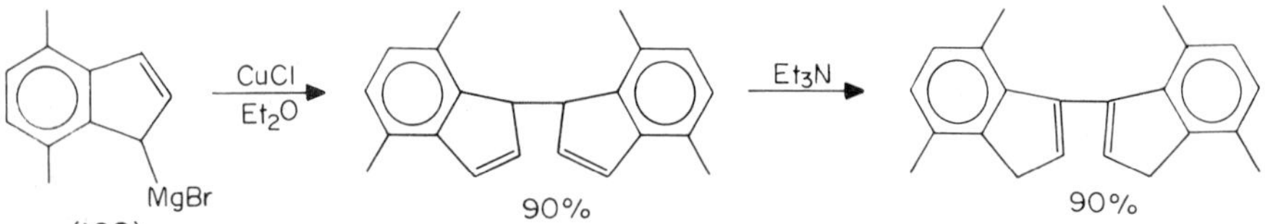

(213)

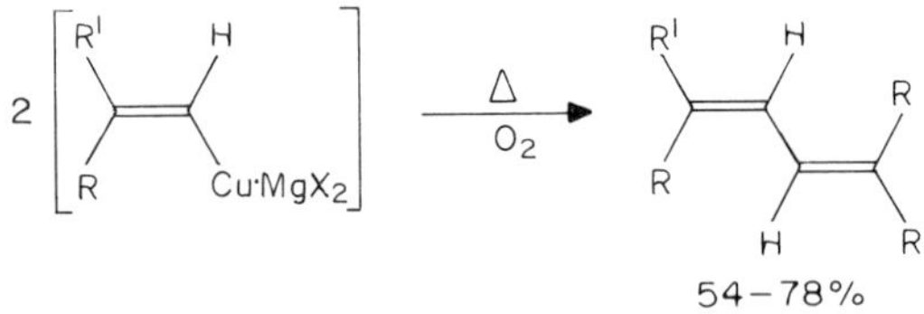

(214)

8. Carboxylation and carbonylation reactions

As previously described, alk-1-enyl cuprates react with carbon dioxide in the presence of hmpt–P(OEt)$_3$ to give carboxylic acids on hydrolysis (Section IV.A.1). In addition to copper(I), magnanese(II) compounds have also been used as stoichiometric reagents in the carboxylation of Grignard reagents. For instance, organomanganates, which can be prepared by reacting manganese(II) chloride with the appropriate Grignard reagent, react with carbon dioxide to give the expected carboxylic acid derivative (equation 215)[428]. Interestingly, a similar reaction with alk-1-enyl manganate (**105**) gave the tertiary carboxylic acid **106** in good yield; no mechanism was suggested[428]. Pentacarbonyliron(0) reacts with Grignard reagents, via a carbonylation reaction, to give aldehydes in high yields on hydrolysis (equation 217). This method has also been successfully used to prepare RCDO[429].

$$3RMgX + MnCl_2 \longrightarrow [R_3MnMgX] \xrightarrow[H]{CO_2} 3RCO_2H \quad 50\text{–}86\% \tag{215}$$

$$\left[\left((Pr^n)_2C{=}CH{-}CH_2\right)_3MnMgCl\right] \textbf{(105)} \xrightarrow[H^+]{CO_2} Pr^n{-}C(Pr^n)(CO_2H){-}CH_2CH{=}CH_2 \quad 71\% \ \textbf{(106)} \tag{216}$$

$$RMgBr + [Fe(CO)_5] \longrightarrow [RCOFe(CO)_4] \xrightarrow{H^+} RCHO \tag{217}$$

B. Catalytic Reactions

For over 40 years it has been known that copper(I) salts catalyse the 1,4-addition of Grignard reagents to α,β-unsaturated carbonyl compounds and that cobalt(II) salts catalyse the formation of biaryls from the appropriate organic halide and arylmagnesium derivative[382,383,430]. A multitude of synthetically useful transition metal-catalysed reactions of Grignard reagents with a variety of organic substrates have been discovered. A number of catalysts have been employed in these organic syntheses, ranging from simple metal halides to the more elaborate coordination and organometallic compounds. The main transition metals used are Cu(I), Ni(II), Pd(II), Co(II), Fe(III), and Ti(IV). A review of copper(I)-catalysed reactions has recently appeared[431].

1. *Addition to alkynes*

Although the addition of Grignard reagents to alkynes in the presence of a stoichiometric amount of a copper(I) halide is well established (see Section IV.A.1), a number of synthetic reactions of this type have been devised using catalytic amounts of a transition metal salt (see ref. 358 for a recent review). For example, Grignard reagents add to prop-2-ynylic alcohols in an (*E*) mode in the presence of a copper(I) catalyst[432], as indeed was found using Normant reagents (see Section IV.A.1). Presumably the reactive intermediate in these reactions is a homocuprate akin to that observed for the stoichiometric copper(I) reagents. Subsequent hydrolysis of these reaction mixtures gives the expected alkenols in excellent yields (equation 218)[432]. In contrast, if the copper(I) salt

$$R'C{\equiv}CCH_2OH + RMgBr \xrightarrow[\text{Cat},\ H^+]{\text{CuI},} (R^1)(H)C{=}C(R)(CH_2OH) \quad \text{up to } 97\% \tag{218}$$

$$R'C{\equiv}C{-}C(R^2)(R^3){-}OH + 2Bu^iMgCl \xrightarrow[E^+/H^+]{3\text{ mol-\% }[TiCp_2Cl_2]} (R^1)(E)C{=}C(H)(CR^2R^3OH) \quad 78\text{–}100\% \tag{219}$$

is replaced by a catalytic amount of dichlorobis(cyclopentadienyl)titanium(IV) then only Z addition is observed[91]. The alkyl group of the Grignard reagent, however, is not transferred to the substrate. This procedure has been successfully utilized in the preparation of the monoterpene nerol (**107**) and (E,Z)-farnesol (**108**)[91].

$$n\text{-}C_5H_{11}C{\equiv}CCH_2OH \xrightarrow[\text{3 mol-\% }[TiCp_2Cl_2]\text{, MeI}]{2\,Bu^iMgCl} \textbf{(107)}\ 95\% \qquad (220)$$

$$\xrightarrow[\text{3 mol-\% }[TiCp_2Cl_2]\text{, MeI}]{2\,Bu^iMgCl,} \textbf{(108)}\ 98\% \qquad (221)$$

$[TiCp_2Cl_2]$ can also be used to promote the Z addition of Grignard reagents to other disubstituted alkynes. Again, no transfer of the alkyl moiety of the Grignard reagent is observed (equation 222)[88].

$$R'C{\equiv}CR'' + Bu^iMgCl \xrightarrow[H^+]{\text{3 mol-\% }[TiCp_2Cl_2]} R'(H)C{=}C(H)R'' \quad 92\text{–}100\%\ \textbf{(108)} \qquad (222)$$

Interestingly, although this type of reaction proceeds with high stereoselectivity (96–100%), a low regioselectivity is observed on deuteriolysis or reaction with iodine[88].

These 'hydromagnesiation' reactions (Method E, Table 1, for Grignard syntheses) involving $[TiCp_2Cl_2]$ are believed to proceed via a multi-step mechanism, the key feature being the generation of a highly reactive titanocene hydride (Scheme 1, Section II.B.2). It is this species which is considered to add to the alkyne to form a vinyltitanium(III) species, which subsequently transmetallates to Mg by further reaction with the Grignard reagent[88,93].

(1-Trimethylsilyl)alkynes also react with the isobutylmagnesium bromide in a Z fashion in the presence of $[TiCp_2Cl_2]$. Both high stereoselectivity (above 94%) and high regiospecificity (95%) was observed, although the nature of the latter was dependent on R (equations 223 and 224)[88]. Treatment of alkyne **109** with ethylmagnesium bromide in the

$$RC{\equiv}CSiMe_3 + Bu^iMgCl \xrightarrow[E^+\ (R=Bu^n)]{\text{4 mol-\% }[TiCp_2Cl_2]} R(H)C{=}C(SiMe_3)E \qquad (223)$$

$$\xrightarrow[\text{(R=Ph)}]{\text{4 mol-\% [TiCp}_2\text{Cl}_2]} \text{(R)(E)C=C(SiMe}_3\text{)(H)} \qquad (224)$$

$$\underset{(109)}{\text{Hex}^n\text{C}\equiv\text{CSiMe}_3} + \text{EtMgBr} \xrightarrow[\substack{\text{10 mol-\% (1:1)} \\ \text{E}^+}]{\text{[Ni(acac)}_2\text{]-Bu}^i_2\text{AlH,}} \text{(Hex}^n\text{)(H)C=C(SiMe}_3\text{)(E)} \qquad (225)$$

presence of bis(acetylacetonato)nickel(II) and diisobutylhydridoaluminium(III) (1:1, 10 mol-%) gave preferentially the *Z* isomer in 58–65% yield (95% stereoselectivity). The reaction is believed to proceed via an ethylnickel species, which rapidly undergoes *β*-hydride elimination to form a nickel hydride. It is this species which adds to the substrate followed by transmetallation to magnesium (cf. [$TiCp_2Cl_2$])[433]. Although the addition occurs only in the presence of the aluminium species, its role is not clear.

Methylmagnesium bromide similarly adds to (1-trimethylsilyl)alkynes when using [Ni(acac)$_2$] and trimethylaluminium(III) (1:1, 10 mol-%) as the catalyst, but the methyl group is now transferred to the substrate. Primarily *Z* addition occurs, but the stereoselectivity was generally lower (85–95%) (equation 226)[433]. Enol **110** can be isolated

$$\text{Hex}^n\text{C}\equiv\text{CSiMe}_3 + \text{MeMgBr} \xrightarrow[\substack{\text{10 mol-\% (1:1)} \\ \text{E}^+}]{\text{[Ni(acac)}_2\text{]-Me}_3\text{Al,}} \underset{(Z)}{\text{(Hex}^n\text{)(Me)C=C(SiMe}_3\text{)(E)}} + \underset{(E)}{\text{(Hex}^n\text{)(Me)C=C(E)(SiMe}_3\text{)}} \qquad (226)$$

$$\text{RCH}_2\text{C}\equiv\text{CSiMe}_3 + \text{MeMgBr} \xrightarrow[\substack{\text{10 mol-\% (1:1)} \\ \text{oxirane}}]{\text{[Ni(acac)]-Me}_3\text{Al}} \underset{\substack{(52-3\%) \\ (110)}}{\text{(RCH}_2\text{)(Me)C=C(SiMe}_3\text{)CH}_2\text{CH}_2\text{OH}} + \underset{7\%}{\text{(RCH}_2\text{)(Me)C=C(SiMe}_3\text{)CH}_2\text{CH}_2\text{OH}} \qquad (227)$$

in 52–63% yield using this procedure, the products being separated by chromatography[434]. It is noteworthy that the *Z* product slowly isomerizes to the *E* form in solution and is believed to be catalysed by the nickel(II) species[433].

The synthesis of allenes or dienes from alkynes containing a suitably positioned leaving group, via an S_N2' type reaction with a Grignard reagent, can be achieved by using stoichiometric copper(I) salts (Section IV.A.1) or indeed in the presence of a transition metal catalyst. For instance, copper(I) halides have been used to catalyse the addition of Grignard reagents to alkynes **111**, **112**, and **114**[435–437]. Both **111** and **112** give rise to dienes, the latter being a useful precursor to 3,4-di-*tert*-butylthiophene (**113**)[436]. Alkyne **114** is stereoselectively converted into an allene[437]. *β*-Ethynyl-*β*-propiolactone (**115**)

$$\underset{(111)}{\text{MeOCH}_2\text{C}\equiv\text{CCH}_2\text{OMe}} + 2\,\text{RMgBr} \xrightarrow[\text{CuBr}]{\text{10 mol-\%}} \text{H}_2\text{C=C(R)-C(R)=CH}_2 \qquad (228)$$

$$\underset{(112)}{ClCH_2C{\equiv}CCH_2Cl} \xrightarrow[\text{CuBr}]{Bu^tMgCl,\ 2\text{–}5\ \text{mol-\%}} H_2C{=}C(Bu^t){-}C(Bu^t){=}CH_2 \xrightarrow[SCl_2]{CH_2Cl_2} \underset{(113)}{\text{3,4-di-}Bu^t\text{-thiophene}} \tag{229}$$

$$\underset{(114)}{MeOC(R)(R^1)C{\equiv}CR^2} + R^3MgBr \xrightarrow[\text{CuBr}]{10\ \text{mol-\%}} RR^1C{=}C{=}CR^2R^3 \quad 60\text{–}94\% \tag{230}$$

$$\underset{(115)}{HC{\equiv}C\text{-oxetan-2-one}} + RMgX \xrightarrow[\text{CuI}]{2\ \text{mol-\%}} R(H)C{=}C{=}C(H)CH_2CO_2H \tag{231}$$

$$(115) + Pe^nMgBr \xrightarrow[\text{CuI}]{2\ \text{mol-\%}} Pe^n(H)C{=}C{=}C(H)CH_2CO_2H \ (90\%) \xrightarrow{3\ \text{Steps}} \underset{(116)\ 93\%}{\text{pellitorine}} \tag{232}$$

similarly reacts with Grignards in the presence of copper(I) iodide to afford alka-3,4-dienoic acids in high yield and with high regioselectivity. This procedure has been utilized in the synthesis of the insecticidal compound pellitorine (**116**)[438]. The mechanism of these copper(I)-catalysed reactions is believed to proceed in a comparable fashion to that observed for the additions utilizing stoichiometric copper(I) reagents (see Section IV.A.1).

In addition to copper(I) catalysts, both iron(III) and palladium(II) compounds have been effectively used in the synthesis of allenes from alkynes (equations 233 and 234)[439,440]. Stereospecific addition (99%) occurs in both cases, the active catalytic species being an iron(I) or palladium(0) compound, respectively. The former catalyst is believed to arise from the facile reduction of $FeCl_3$ by the Grignard reagent[439].

$$R^1C(Cl)(R^2){-}C{\equiv}CR^3 + RMgX \xrightarrow[\text{cat.}]{FeCl_3} R^1R^2C{=}C{=}CR^3R \quad 70\text{–}90\% \tag{233}$$

$$R^1C(Cl)(R^2){-}C{\equiv}CH + RMgX \xrightarrow[Bu^i_2AlH(1:1),\ 10\ \text{mol-\%}]{[Pd(PPh_3)_2Cl_2],} R^1R^2C{=}C{=}C(H)R \quad 62\text{–}77\% \tag{234}$$

2. *Addition to alkenes*

The 1,4-addition of Grignard reagents to α,β-unsaturated carbonyl compounds in the presence of a copper(I) catalyst has been the subject of two thorough reviews by Posner[382,383], where numerous examples of this type of reaction were tabulated. Two current examples which exemplify the 1,4-addition reaction are shown in equations 235 and 236[441,442]. In the latter example, the acetal Grignard reagent **117** adds to the cyclic lactone **118** with copper(I) bromide as the catalyst, to give the expected 1,4-addition product, but on acid hydrolysis annulation occurs to yield the bicyclic compound **119**[442]. This procedure has been utilized in the synthesis of the naturally occurring tricyclic sesquiterpenes (±)-isocomene (**120**) and (±)-siliphene (**121**) (equations 237 and 238)[443,444].

The reaction of cyclohex-2-enone with the Grignard reagent of 2-bromo-3-(trimethylsilyl)propene, in the presence of copper(I) iodide, also gave the expected 1,4-

+ MeMgI —(10 mol-%, CuI, H^+)→ 78% (235)

(118) (n = 2–4) + (117) (m = 2 or 3) $(CH_2)_mMgX$ —(25 mol-%, CuBr, NH_4^+)→ 70–87% —H^+→ (119) 45–89% (236)

+ (117) (m = 2) —(CuBr cat.)→ 68% —(7 steps)→ [(±)-120] (237)

+ (117) (m = 2) —(CuBr cat.)→ 68% —(6 steps)→ [(±)-121] (238)

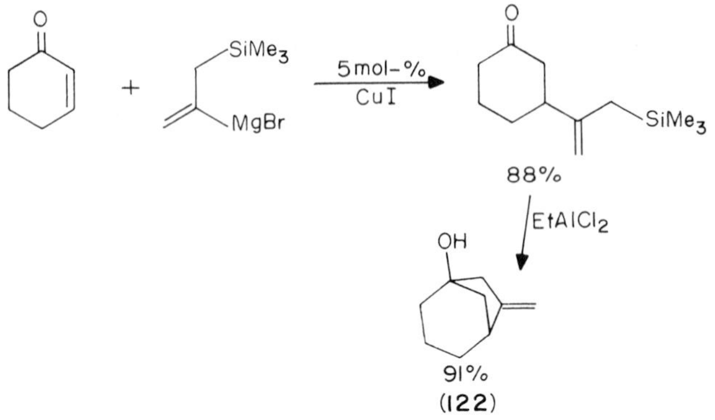

(239)

88%

91%

(122)

addition product, this being subsequently converted into the bicyclic compound **122**[445]. Stereoselective 1,4-addition has been observed in the reaction of aryl and *tert*-butyl Grignard reagents with enonate **123** in the presence of a catalytic amount of copper(I) iodide. Only products with the D-manno configuration resulted from this asymmetric

(240)

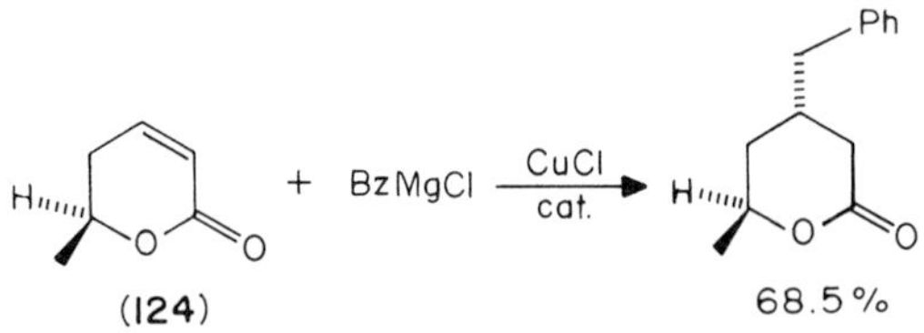

(123)

89%

(241)

(124)

68.5%

synthesis[446]. The chiral lactone **124** undergoes a similar reaction with benzylmagnesium chloride to give selectively the (*E*)-1,4-addition product[228].

In addition to α,β-unsaturated carbonyl compounds, a number of other olefinic substrates have been found to undergo addition reactions with Grignard reagents in the presence of a transition metal catalyst. For instance, alk-1-enes react with *n*-propylmagnesium bromide, using $TiCl_4$ as catalyst, to give predominantly the primary Grignard reagent **125**. The same substrates undergo an analogous reversible exchange reaction with a variety of Grignard reagents (having a relatively labile hydrogen atom on the β-carbon) in the presence of nickel(II) chloride, again to give preferentially primary organomagnesium compounds[447]. As previously mentioned in Section II.B.2, an irreversible reaction occurs on reacting Pr^nMgBr, in the presence of dichlorobis(cyclopentadienyl)titanium(IV), with 1,3-dienes or styrene to yield allylic or α-phenylethyl Grignard reagents, respectively (equations 49 and 243)[94]. The exchange reactions involving catalyst $[TiCp_2Cl_2]$ presumably proceed in a similar fashion to that outlined in Scheme 1, Section II.B.2, while the mechanisms of those involving $TiCl_4$ or $NiCl_2$ have been adequately covered in an earlier review[447].

$$\text{Pr}^n\text{MgBr} + \text{RCH}{=}\text{CH}_2 \xrightleftharpoons{\text{TiCl}_4} \underset{(\mathbf{125})}{\text{RCH}_2\text{CH}_2\text{MgBr}} \tag{242}$$

$$\text{Pr}^n\text{MgBr} + \text{PhCH}{=}\text{CH}_2 \xrightarrow[\text{3 mol-\%}]{[\text{TiCp}_2\text{Cl}_2]} \text{PhCH(Me)MgBr} \tag{243}$$

For over a decade it has been known that some allylic alcohols react with certain Grignard reagents, in the presence of a catalytic amount of dichloro-bis(triphenylphosphine)nickel(II), in an S_N2 mode to give substituted alkenes[447]. For instance, α-phenylallyl alcohol (**126**) reacts with methylmagnesium bromide, using $[Ni(PPh_3)_2Cl_2]$ as catalyst, to yield only the *E*-conjugated olefin (reaction 244). In general,

$$\underset{(\mathbf{126})}{\text{PhCH(OH)}{-}\text{CH}{=}\text{CH}_2} + \text{MeMgBr} \xrightarrow[\text{(2) H}^+]{\text{(1) [Ni(PPh}_3)_2\text{Cl}_2]} \underset{75-87\%}{(E)\text{-PhCH}{=}\text{CHCH}_2\text{Me}} \tag{244}$$

however, these reactions are not completely stereospecific[447]. Moreover, most allylic alcohols under the above conditions undergo direct substitution of the hydroxy group by the Grignard reagent with no apparent participation of the double bond (see Section IV.B.4). Some allylic alkoxides also react with Grignard reagents via addition to the double bond, followed by elimination of the alkoxy moiety, in preference to direct substitution of the said group. For example, the allyl ether **127** was converted selectively (100%) to alkene **128** using *n*-heptylmagnesium chloride in the presence of copper(I)

$$\underset{(\mathbf{127})}{\text{CH}_2{=}\text{CHC(Me)}_2\text{OMe}} + \text{Hept}^n\text{MgCl} \xrightarrow[\text{CuBr}]{\text{5 mol-\%}} \underset{\substack{70\% \\ (\mathbf{128})}}{\text{Oct}^n\text{CH}{=}\text{C(Me)}_2} \tag{245}$$

$$\underset{(\mathbf{129})}{\text{CH}_2{=}\text{CHC(OEt)}_3} + \text{RMgX} \xrightarrow[\text{CuBr}]{\text{5 mol-\%}} \text{RCH}_2(\text{XMg})\text{C}{=}\text{C(OEt)}_2 \xrightarrow{\text{H}^+} \underset{\substack{66-76\% \\ (\mathbf{130})}}{\text{RCH}_2\text{CH}_2\text{CO}_2\text{Et}} \tag{246}$$

bromide[448]. The same mode of addition was presumed in the reaction of Grignard reagents with the triether **129** as only the ester **130** was isolated on hydrolysis[449].

3. Ring-opening reactions

As previously stated in Section IV.A.3, the regioselective ring opening of β-propiolactones can be achieved by using Normant reagents or R_2CuLi. The use of Grignard reagents in the presence of a copper(I) catalyst, however, have also proved to be a success in this role. For instance, β-propiolactone reacts with a variety of Grignard reagents, with a copper(I) halide as catalyst, to give the expected alkanoic acids in good yields (equation 247)[411]. Racemic β-methyl-β-propiolactone (**81**) was converted into

+ RMgX —(10 mol-% CuBr)→ $RCH_2CH_2CO_2H$ 82–86% (247)

(81) + MgBr —(2 mol-% CuI)→ OH 90% (131) (248)

(S)-(–) 81 + $ClMg(CH_2)_4$ $(CH_2)_6$ H (132) —(3 mol-% CuI)→ HO_2C H $(CH_2)_4$ $(CH_2)_6$ H 80% → H $(CH_2)_4$ $(CH_2)_6CHO$ (133) (249)

citronellic acid (**131**) by using this method[412]. Optically active (*R*)-(+)-(**131**) was similarly synthesized from (*R*)-(+)-(**81**)[405], the enantiomer of which was reacted with Grignard **132** in the presence of CuI to give the precursor to the sex pheromone trogdermal (**133**)[450].

Grignard reagents also add stereoselectively to the double bond of β-vinyl-β-propiolactone (**82**), which initiates the ring opening of the lactone via an S_N2' pathway to yield primarily (*E*)-alk-3-enoic acids (*E*:*Z* ratio $\approx 9:1$) (equation 250)[407]. Copper(I) halides

(82) + RMgX —(2 mol-% CuI)→ R OH 70–96% (250)

(134) + MeMgBr $\xrightarrow[\text{CuI}]{\text{2 mol-\%}}$ Et–CH=CH–CH₂CH₂–COOH 87% (251)

(135) + (136) $ClMg(CH_2)_5$–(1,3-dioxan-2-yl) ⟶ $HO_2C(CH_2)_3CH{=}CH(CH_2)_6$–(1,3-dioxan-2-yl) 95% (137) (252)

are used as catalysts in these reactions, or indeed Normant reagents can be used as stoichiometric reagents (see Section IV.A.3). Both γ-vinyl-γ-butyrolactone (**134**) and δ-vinyl-δ-valerolactone (**135**) react with Grignards in a similar fashion, in the presence of a copper(I) halide. Methylmagnesium bromide adds stereoselectively to **134** to generate (*E*)-hept-4-enoic acid (*E*:*Z* ratio = 92:8)[408], while **135** reacts with Grignard **136** to yield solely the (*E*)-carboxylic acid **137**[450]. In general, however, the regioselectivity of these reactions involving **134** or **135** is lower than that observed for the related β-vinyl-β-propiolactones. Interestingly, the previously mentioned β-ethynyl-β-propiolactone (**115**) reacts both stereo- and regioselectively with Grignard **138** in an S_N2', mode to generate the allene **139**, which is the precursor to the antibiotic A26771B (**140**)[451].

HC≡C–(115) + (138) $CH_3CH(Othp)(CH_2)_9MgBr$ $\xrightarrow[\text{(2) H}^+]{\text{(1) CuI cat.}}$ $CH_3CH(OH)(CH_2)_9CH{=}C{=}CH–CH_2COOH$ (139) $\xrightarrow{\text{5 steps}}$ (140) (253)

Oxiranes react with a variety of Grignard reagents in the presence of a copper(I) catalyst to generate the ring-opened product. For instance, monosubstituted oxirane reacts with

R′-oxirane + RMgX $\xrightarrow[\text{CuBr, H}^+]{\text{10 mol-\%}}$ $RCH_2CH(OH)R'$ 81–95% (254)

R′-oxirane + $Me_3SiCH_2C(MgBr){=}CH_2$ $\xrightarrow[\text{CuI, H}^+]{\text{10 mol-\%}}$ $R'CH(OH)CH_2C({=}CH_2)CH_2SiMe_3$ 82–85% (255)

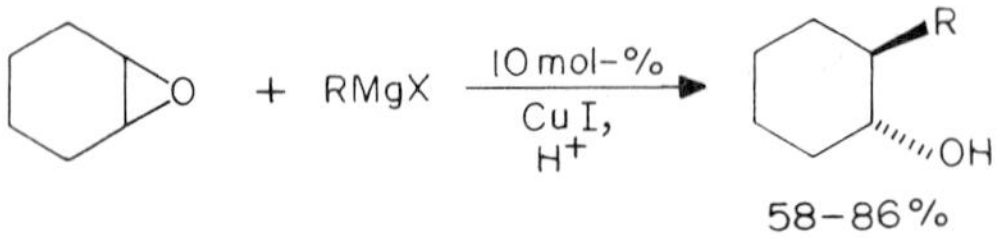

(256)

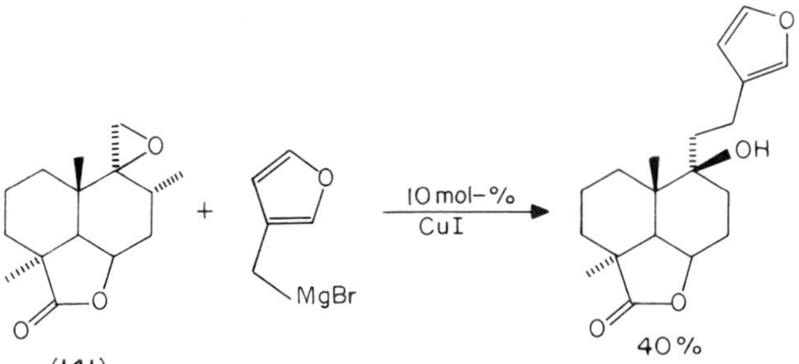

(257)

(141) [(+)-142]

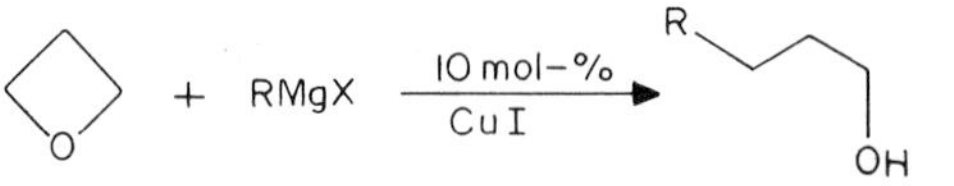

(258)

(143)

Grignard reagents to yield alcohols[321], while reaction with the Grignard derived from 2-bromoallyltrimethylsilane gives enols (equations 254 and 255)[320a]. Cyclohexane oxide undergoes a similar reaction with Grignard reagents to generate Z-products stereospecifically (equation 256)[321]. This overall ring-opening procedure has been adopted in the conversion of the oxirane **141** into isomarrubin (**142**)[320c]. Interestingly, the cyclic ether **143** similarly reacts with certain organomagnesium derivatives to yield the expected alcohols[321].

In addition to lactones and oxiranes, other small-membered ring systems can undergo ring-opening reactions by reaction with Grignard reagents in the presence of a copper(I) halide catalyst. 1, 1-Bis(benzenesulphonyl)cyclopropane (**144**), for instance, reacted with n-butylmagnesium bromide to give the ring-opened product[127], while 1-phenylsulphonylbicyclobutane (**145**) was converted stereospecifically into the cyclobutane **146** on reaction with methylmagnesium bromide in the presence of copper(I) bromide[452].

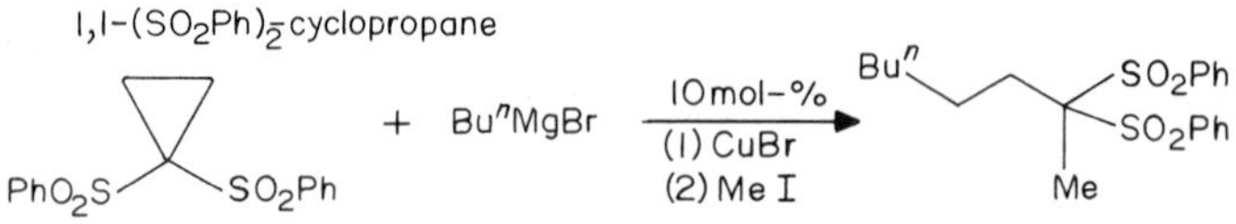

(259)

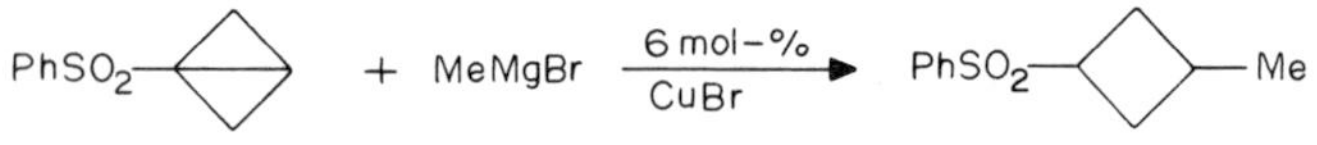

(144)

(260)

(145) (146)

4. *Addition to acyl derivatives*

The reaction between Grignard reagents and acyl chlorides to give ketones, in the presence of a copper(I) catalyst, has been known for some time[453]. In the past decade, however, few synthetically useful reactions involving an acyl derivative and an organomagnesium halide being coupled in the presence of a transition metal complex catalyst have been reported. Ketones have been shown to undergo 1,2-addition by Grignard reagents in the presence of a catalytic amount of dichlorobis(cyclopentadienyl)titanium(IV) (equation 261)[454]. A similar reaction was observed for esters, but here the addition reaction was followed by the elimination of an alkoxy group to generate an alcohol again (equation 262)[455]. The distribution of products in these reactions was found to be dependent on the amount of $[TiCp_2Cl_2]$ used, but it is noteworthy that most acyl substrates which undergo 1,2-addition by Grignard reagents (or lithium reagents) do so fairly effectively in the absence of a catalyst.

$$R^1COR^2 + RMgX \xrightarrow[(2)\ H^+]{(1)\ [TiCp_2Cl_2]} \underset{>83\%}{R^1RR^2COH} \qquad (261)$$

$$R^1CO_2R^2 + 2RMgX \xrightarrow[(2)\ H^+]{(1)\ [TiCp_2Cl_2]} \underset{74\text{–}94\%}{R^1RCHOH} \qquad (262)$$

5. *Displacement of a halide*

The cross-coupling reaction between organic halides and organomagnesium halides in the presence of a nickel salt (equation 263) has been known for over half a century and has recently been the subject of an extensive review[456]. In these reactions the catalyst is

$$R^1X + RMgX \xrightarrow[\text{cat.}]{\text{Ni(II) or Ni(O)}} R^1R + MgX_2 \qquad (263)$$

normally a $[NiCl_2L_2]$ complex, $[Ni(acac)_2]$ or another Ni(II) or Ni(0) salt. For the former L is invariably a phosphine ligand with bidentates generally being more active than monodentate ligands. The most frequently used diphosphine is bis(diphenylphosphino)propane, while triphenylphosphine shows the highest activity of the monodentate tertiary phosphines[456].

An example of the nickel-catalysed cross-coupling reaction is the addition of both alkyl and aryl Grignard reagents to alkenyl halides. For instance, trimethylsilylmethylmagnesium chloride adds stereoselectively to alkenyl bromides[457], *n*-butylmagnesium chloride to 1,2-dichloroalkenes[458], and phenylmagnesium bromide to the bicycloalkenyl

$$\text{MeCH=CHBr} + Me_3SiCH_2MgCl \xrightarrow[5\ \text{mol-\%}]{[NiCl_2(dppp)],} \text{MeCH=CHCH}_2SiMe_3 \qquad (264)$$

$$\text{ClCH=CHCl} + Bu^nMgCl \xrightarrow[10\ \text{mol-\%}]{[Ni(PPh_3)_4],} Bu^n\text{CH=CHCl} \qquad (265)$$

$$\underset{(147)}{\text{bicycloalkenyl–Cl}} + PhMgBr \xrightarrow[Et_2O]{[Ni(acac)_2],} \text{bicycloalkenyl–Ph} \qquad (266)$$

chloride **147**[459] (equations 264–266). Interestingly, a different nickel catalyst was employed in each case. Alkyl and aryl Grignard reagents also react with a variety of aryl halides in the presence of a nickel catalyst to give the cross-coupled product. For instance, methylmagnesium bromide adds selectively to 2-chloropyridine[460], phenylmagnesium bromide to chlorophenyl alkyl sulphides[461], and 4-methylpent-3-en-1-ylmagnesium bromide to 3-methyl-6-chlorobenzofuran (equations 267–269)[462]. The last reaction yields furoventalene (**148**), which is a marine natural benzofuran. Allyl and some vinyl Grignard reagents are also known to couple with organic halides[456].

2-chloropyridine + MeMgBr —[NiCl(2-Py)(PPh$_3$)]$_2$ cat.→ 2-methylpyridine (267)

ClC$_6$H$_4$SR + PhMgBr —[NiCl$_2$(PPh$_3$)$_2$] cat.→ PhC$_6$H$_4$SR (268)

4-methylpent-3-en-1-ylMgBr + 3-methyl-6-chlorobenzofuran —[NiCl$_2$(dppp)] cat.→ (148) (269)

Nickel catalysts containing an optically active phosphine ligand have been used extensively in the asymmetric cross-coupling of a racemic Grignard reagent with an organic halide[456,463]. To date, the best optical yields have been obtained using vinyl bromide and a chiral Grignard reagent, the highest attained being 94% e.e. in the synthesis of 3-phenylbut-1-ene (equation 270)[464]. Here a nickel(II) complex of (*R*)-(−)-1-dimethylamino-1-*tert*-butyl-2-diphenylphosphinoethane (**149**) was used as catalyst. Optical yields of 66–75% were obtained for *p*-substituted 3-phenylbut-1-enes in the presence of similar catalysts (equation 271)[465,466]. Thus far, however, the optical purities are generally ⩽ 50%[456].

$$\text{Ph(Me)HCMgCl} + \text{CH}_2{=}\text{CHBr} \xrightarrow{[\text{Ni}(149)\text{Cl}_2]} \text{PhCH(Me)CH}{=}\text{CH}_2 \quad (270)$$

$$p\text{-RC}_6\text{H}_4\text{(Me)HCMgCl} + \text{CH}_2{=}\text{CHBr} \xrightarrow{[\text{Ni}(150)\text{Cl}_2]} p\text{-RC}_6\text{H}_4\text{CH(Me)CH}{=}\text{CH}_2 \quad (271)$$

R = Me, Bu^i, or Ph

Me$_2$NCH(Bu^t)CH$_2$PPh$_2$ (149)   Me$_2$NCH(R^1)CH$_2$PPh$_2$ (150)

R^1 = Bz, Pr^i, Bu^s, Ph, Cy, or Bu^t

The mechanism of the coupling reaction is not known, but a discussion of the mechanistic considerations can be found in a recent review[456].

Both nickel(II) and palladium(II) catalysts have been successfully used in the cross-coupling of aryl- and *n*-butylmagnesium bromide with (*E*)- or (*Z*)-1-bromo-2-phenylthioethene (equation 272). The stereoselectivity was higher than 99% for the *E*-isomers and in the range 95–98% for the *Z*-isomers, with an overall yield of 70–100%[467]. The displacement of thioalkyl groups is discussed in Section IV.B.6. Similarly, nickel(0)

$$\mathrm{BrCH{=}CHSPh} + \mathrm{RMgX} \xrightarrow[\text{or } [\mathrm{PdCl_2(PPh_3)_2}]]{[\mathrm{NiCl_2(dppe)}]} \mathrm{RCH{=}CH\,SPh} \xrightarrow{\mathrm{R^1MgX,\ Ni\ cat.}} \mathrm{RCH{=}CHR^1}\ \ 70\text{–}100\% \tag{272}$$

$\mathrm{R = Bu}^n$, 1-naphthyl, Ph or Ph(Me)CH

$$\mathrm{R^1R^2C{=}CHI} + \mathrm{Me_3SiCH_2MgCl} \xrightarrow[\text{or } [\mathrm{Pd(PPh_3)_4}]]{[\mathrm{Ni(PPh_3)_4}]} \mathrm{R^1R^2C{=}CHCH_2SiMe_3} \tag{273}$$

and palladium(0) catalysts have been used in the cross-coupling of trimethylsilyl-methylmagnesium chloride with alkenyl halides (equation 273). Again, good stereo-selectivity (> 98%) was attained[468].

As for nickel, both palladium(0) and palladium(II) catalysts having phosphine ligands have been utilized in the coupling of Grignard reagents with organic halides, but to a much lesser extent. For instance, bis(triphenylphosphine)iodo(phenyl)palladium(II) catalyses the reaction of 2-phenylethyn-1-ylmagnesium bromide with iodobenzene[469], while tetrakis(triphenylphosphine)palladium(0) catalyses the addition of 2-methylethyn-1-ylmagnesium bromide to alkenyl iodides with 97% or more stereoselectivity (equations 274 and 275)[470].

$$\mathrm{PhC{\equiv}CMgBr} + \mathrm{PhI} \xrightarrow{[\mathrm{PdI(Ph)(PPh_3)_2}]} \mathrm{PhC{\equiv}CPh}\ \ 84\% \tag{274}$$

$$\mathrm{R^1R^2C{=}CHI} + \mathrm{MeC{\equiv}CMgBr} \xrightarrow{[\mathrm{Pd(PPh_3)_4}]} \mathrm{R^1R^2C{=}CHC{\equiv}CMe}\ \ 80\text{–}83\% \tag{275}$$

$$\mathrm{CH_2{=}CHC(MgCl){=}CH_2} + \mathrm{ArI} \xrightarrow{[\mathrm{Pd(PPh_3)_4}]} \mathrm{CH_2{=}CHC(Ar){=}CH_2} \tag{276}$$

$$\underset{(151)}{\mathrm{Me_2C{=}C{=}CHBr}} + \mathrm{RMgX} \xrightarrow[\mathrm{Bu^i_2AlH,\ 7.8\ mol\text{-}\%\ (1:1)}]{[\mathrm{PdCl_2(PPh_3)_2}]} \mathrm{Me_2C{=}C{=}CHR} \tag{277}$$

$\mathrm{R = Oct}^n$ 98%

R = Tol 67%

Palladium catalysts also provide a convenient route for the synthesis of 2-substituted buta-1,3-dienes as they effectively catalyse the cross-coupling reaction of 2-(buta-1,3-dienyl)magnesium chloride with aryl iodides (equation 276)[471]. Stereoselective (99%) coupling has also been observed in the reaction of the allenyl bromide **151** with Grignard reagents[440].

As previously mentioned (Section IV.A.5), palladium catalysts have been used to couple organic halides with organo cuprates, an example being the cross-coupling of alk-1-enyl cuprates with alkenyl halides in the presence of tetrakis(triphenylphosphine)palladium(0) to yield dienes (equation 208).

In addition to nickel and palladium, the other major catalyst which has been used in Grignard cross-coupling reactions is copper. Both copper(I) halides and $Li_2[CuCl_4]$[472] have been used extensively in this role. A number of these reactions have appeared in a review[431]. A variety of both alkyl and aryl Grignard reagents have been shown to couple with a range of organic halides in the presence of a copper catalyst. For instance, the Grignard reagent **152** reacts with 1,4-dibromobutane and the aryl Grignard **153** with a range of alkyl iodides to give the expected coupled products[450,473]. Stereospecific addition reactions of this type have been utilized in the synthesis of **154**, which is a precursor to (±)-lactaral (**155**)[474], and in the preparation of (−)-α-*cis*-bergamotene (**156**)[475]. 2-(Buta-1,3-dienyl)magnesium chloride has been successfully coupled with *n*-octyl halides, both

(152) + Br(CH₂)₄Br —[$Li_2[CuCl_4]$, 1 mol-%]→ 2-[(CH₂)₆Br]-1,3-dioxane, 86% (278)

(153) + RI —[$Li_2[CuCl_4]$, 1 mol-%]→ 3,5-(MeO)₂C₆H₃R, 66–74% (279)

Furyl Grignard (MgCl, Othp) + allylic chloride —[$Li_2[CuCl_4]$, 1 mol-%]→ (154) 75% —[2 steps]→ [(±) 155] (280)

$$\text{+ ClMg-prenyl} \xrightarrow{\text{CuI}} [(-)156] \qquad (281)$$

$$\text{MgCl-diene} + RX \xrightarrow[\text{cat.}]{Li_2[CuCl_4]} \text{2-R-diene } 60-95\% \qquad (282)$$

alkyl and aryl dihalides, and haloesters in the presence of a copper catalyst to yield 2-substituted 1, 3-dienes (equation 282)[476].

Although copper, nickel, and palladium catalysts are by far the most widely used in coupling reactions of Grignard reagents with organic halides, other transition metals have had limited success in this role. For instance, tris(dibenzoylmethido)iron(III), $[Fe(dbm)_3]$, catalyses the coupling of aryl Grignard reagents with alkenyl halides[477], while iron(III) chloride has been successfully used in the coupling of ethenylmagnesium bromide with **157** (equations 283 and 284)[478]. The reactive catalytic entity in these reactions is believed to be an iron(I) species formed by the facile reduction of the iron(III) precursor by the organomagnesium compound[439].

$$\text{PhCH=CHBr} + RMgX \xrightarrow[{[Fe(dbm)_3]}]{0.5-1.0\text{ mol-\%}} \text{PhCH=CHR } 75-100\% \qquad (283)$$

$$(157) + CH_2{=}CHMgBr \xrightarrow[\text{cat.}]{FeCl_3} 62\% \qquad (284)$$

$$\text{PhCH=CHBr} + MeMgBr \xrightarrow[\text{cat.}]{[RuCl_2(PPh_3)_3]} \text{PhCH=CHMe} \qquad (285)$$

Similarly, dichlorotris(triphenylphosphine)ruthenium(II) catalyses the stereospecific addition of methylmagnesium bromide to α-styryl bromide (equation 285)[479].

Further examples of halide displacement reactions using transition metal catalysed Grignard reagents are given in Table 9.

6. Displacement of a non-halide

Grignard cross-coupling reactions are not confined to organic halides but have been reported for a variety of organic alcohols, ethers, sulphides, selenides, and silyl ethers.

TABLE 9. Transition metal-catalysed displacement of a halide by Grignard reagents

$$RMgX + R'X \xrightarrow{cat.} RR' + MgX_2$$

R′X	R	Catalyst[a]	Yield (%)	Ref.
1. R^2–CH=CH–Br (*E*)-or (*z*)-; R^2 = Me, Ph, Hex^n	Me_3SiCH_2—, $(Pr^iO)_2MeSiCH_2$—	A	70–100	457, 484
2. Hex^n, H, R^2, *I*; R^2 = Me, H	Me_3SiCH_2—	B, I	74–85	468
3. H, H, Pr^n, *I*	Me_3SiCH_2—	B, I	89, 84	468
4. *Cl*, Cl (*E*) or (*Z*)	Oct^n, $Ph(CH_2)_3$—	B	60–72	458
5. Cl, N, N, N, N, O, R^2O, R^2O, OR^2; R^2 = $Bu^t(Me)_2Si$—	Et, Cy, Ph, $Ph(CH_2)_2$—, $Ph(CH_2)_3$—, $Me_2C{=}CH(CH_2)_2$—	A	40–50	480

6. 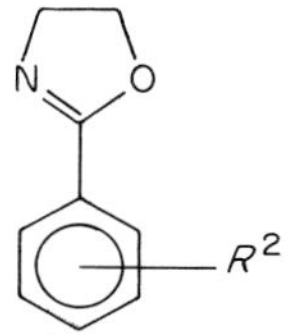 $R^2 = 2\text{-Cl}, 3\text{-Cl}, 4\text{-Cl}$	Ph, p-$MeOC_6H_4$—, Me, Et, Pr^n, Bz	A, C	56–94	481
7. 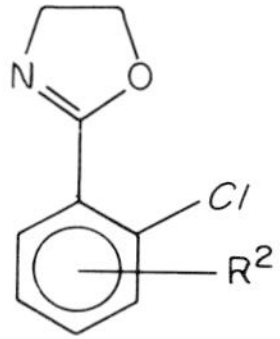 $R^2 = 4\text{-Cl}, 5\text{-Cl}$	Me	D	83	481
8. 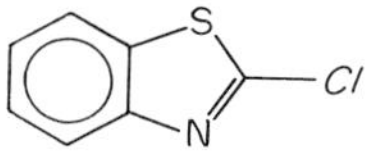	Me, Et, Pr^i, Bu^n, Ph	D, E	50–95	482
9. 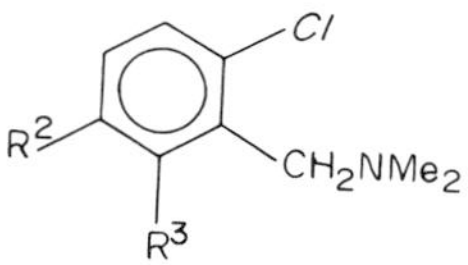 $R^2 \neq R^3 = H, OMe$	Me_3SiCH_2—	D, F	80–90	483
10. 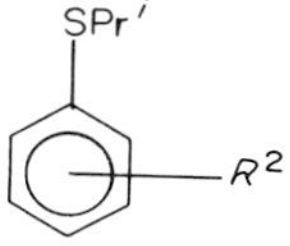 $R^2 = 2\text{-Cl}, 3\text{-Cl}, 4\text{-Cl}$	Ph, Bu^n, Me, isopropenyl, allyl,	D	50–84	461

TABLE 9. (*Contd.*)

R′X	R	Catalyst[a]	Yield (%)	Ref.
11. 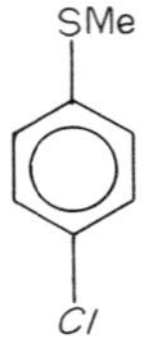 SMe, Cl	Ph, Me, isopropenyl, allyl	D	35–50	461
12. 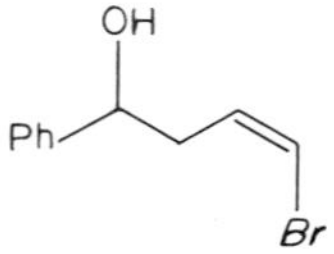 OH, Ph, Br	Et	A	86	305c
13. 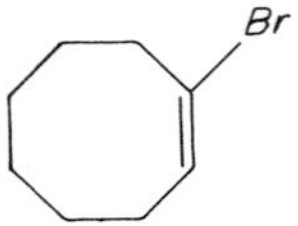 Br	$(Pr^iO)_2MeSiCH_2$—	A	88	484
14. Ph*I*	Buta-1, 3-dien-2-yl, Cy, mesityl, *p*-FC_6H_4—, *m*-FC_6H_4—	G, L	51–80	469, 476
15.  Cl, Br	Ph	G	73	469
16. 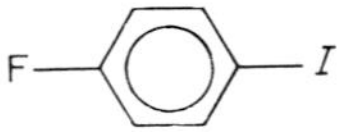 F, *I*	Ph	H	82	469

17. 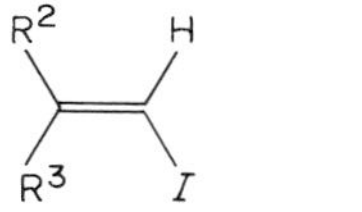 $R^2R^3C{=}CH$*I* $R^2 \neq R^3 = H$, Hex^n	Et, Ph, MeC≡C—	I	80–87	470
18. 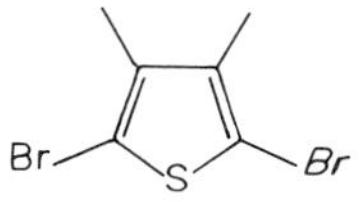 Br, S, *Br* (3,4-dimethylthiophene ring)	Ph	I	63	485
19. Ph*X* X = I, Br	1-Me-pyrrol-2-yl, 1-Me-indol-2-yl, Bu^s	J, K	79–93	96, 486
20. 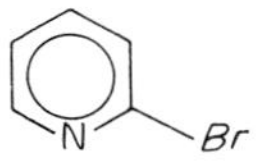 N, *Br* (pyridine ring)	1-Me-pyrrol-2-yl	J	71	96
21. PhCH=CH*Br* (*E*)- or (*Z*)-	Bu^s, Me, *p*-Tol, vinyl	I, K	78–99	479, 486
22. $H_2C{=}CR^2$*Br* R^2 = H, Me (*E*)-Bu^{n}C(Me)=CH*I*	Bu^s, Othp$(CH_2)_8$C≡C—	I, K	80, 66	110a, 486
23. 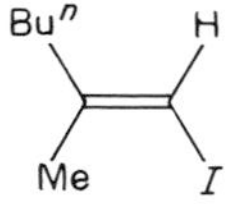 Bu^n(Me)C=CH*I*	Me_3SiCH_2—	I	85	468
24. 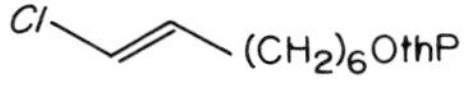 *Cl*CH=CH$(CH_2)_6$OthP	But-1-enyl	I	96	458
25. *Br*$(CH_2)_3R_2$ R^2 = Cl, $BrCH_2$—	Dioxan-2-yl-CH_2CH_2—, Othp$(CH_2)_n$— (*n* = 4, 6)	L	53–86	450, 487

TABLE 9. (*Contd.*)

R′X	R	Catalyst[a]	Yield (%)	Ref.
26. R^2I R^2 = Me, Et, Pr^i, Bu^n, Pe^n	3, 5-$(OMe)_2C_6H_3$—	L	66–74	473
27. 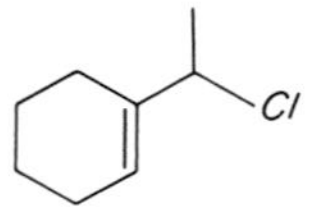 Cl	Fur-3-yl-CH_2—	L	47	474
28. Oct^nX X = Br, I	Buta-1, 3-dien-2-yl, Pr^i,Ph	L	76–95	476
29. p-$XC_6H_4R^2$ X = Cl, Br, I $R^2 = CH_2Br$, $(CH_2)_2I$, $(CH_2)_3I$	Buta-1, 3-dien-2-yl	L	60–87	476
30. $MeOCO(CH_2)_nI$ (n = 3, 4, 5)	Buta-1, 3-dien-2-yl	M	65–80	476
31. $R^2(CH_2)_2X$ X = Br, I $R^2 = PhCO_2$, HO, PhO	Buta-1, 3-dien-2-yl	L	75–86	476
32.  Cl	Othp$(CH_2)_8C{\equiv}C$—	N	54	488
33. 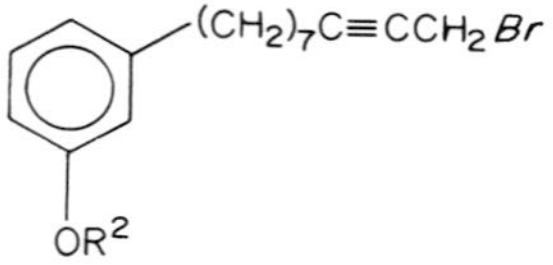 $(CH_2)_7C{\equiv}CCH_2Br$ OR^2 R^2 = H, Me	$Pr^nC{\equiv}C$—	O	83, 88	489

34. 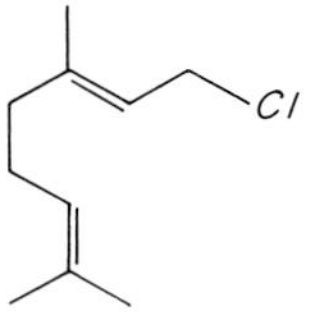 Cl	$(Pr^iO)_2MeSiCH_2$—	M	91	484
35. 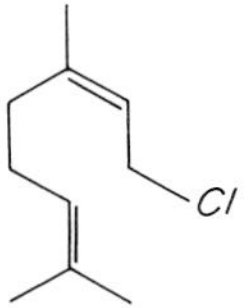 Cl	$(Pr^iO)_2MeSiCH_2$—	M	96	484
36. EtC≡CCH$_2$*Br*	CH_2=CH(OH)CHCH$_2$C≡C—	P	78	109c
37. *Br*$(CH_2)_nR^2$ (n = 3, 4, 5, 7) R^2 = Me, Cl, CN, CO_2Et	Dioxan-2-yl-CH_2CH_2—	L	74–89	490
38. 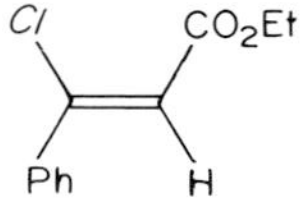 Cl, CO_2Et, Ph, H	Me, Et, Pr^n, Pr^i, Bu^i, Bu^s, Bu^b, 2-MeBu, C-pentyl, Cy	M	70–91	491

[a] A = [$NiCl_2$(dppp)]; B = [Ni($PPh_3)_4$]; C = [$NiCl_2$(dppe)]; D = [$NiCl_2(PPh_3)_2$]; E = [$NiBr_2$(dmso)PMe_3]; F = [Ni(acac)$_2$] G = [PdI(Ph)($PPh_3)_2$]; H = [PdI(p-FC_6H_4)($PPh_3)_2$]; I = [Pd($PPh_3)_4$]; J = [$PdCl_2$(dppb)]; K = [$PdCl_2$(dppf)]; L = $Li_2[CuCl_4]$; M = CuI; N = CuBr; O = CuCl; and P = CuCN.

Examples of nickel-catalysed reactions of this type have appeared in a review[456], as have copper-catalysed coupling reactions[366,431]. Allylic alcohols, for instance, couple with both alkyl and aryl Grignard reagents in the presence of nickel(II)[456] or palladium(II)[492] catalysts (equations 286 and 287). Interestingly, alcohol **158** reacts with *n*-

$$\text{CH}_2{=}\text{CHCH}_2\text{OH} + \text{PhCH}_2\text{MgBr} \xrightarrow{[\text{NiCl}_2(\text{PPh}_3)_2]} \text{CH}_2{=}\text{CHCH}_2\text{CH}_2\text{Ph} \quad 90\% \tag{286}$$

$$\text{CH}_2{=}\text{CHCH}_2\text{OH} + \text{RCH(Me)MgBr} \xrightarrow{[\text{PdCl}_2(\text{dppf})]} \text{RCH(Me)CH}_2\text{CH}{=}\text{CH}_2 \quad \text{R = Ph } 91\%;\ \text{R = Hex}^n\ 95\% \tag{287}$$

$$\text{Me(CH}_2)_{12}\text{CH}{=}\text{CHCH}_2\text{CH}_2\text{CH}_2\text{OH}\ (\mathbf{158}) + \text{Pe}^n\text{MgBr} \xrightarrow[17\text{ mol-}\%]{\text{Li}_2[\text{CuCl}_4]} \text{Me(CH}_2)_{12}\text{CH}{=}\text{CH(CH}_2)_7\text{Me}\ (\mathbf{159})\ 95\% \tag{288}$$

pentylmagnesium bromide in the presence of a copper catalyst to yield muscalure (**159**), the sex pheromone of the housefly, in high yield[493].

A variety of allylic alkoxides are known to couple primarily with alkyl Grignard reagents in the presence of a copper catalyst[366,431]. For instance, methyl hept-2-enyl ether reacts stereospecifically (99%) with *n*-heptylmagnesium chloride in the presence of copper(I) bromide to give the expected coupled product (equation 289)[448]. Similarly, the allylic acetate (**160**) coupled with phenylmagnesium bromide, using $Li_2[CuCl_4]$ as catalyst, to give the enyne (**161**) in 70% yield[494], while (*E*)-but-2-enyl acetate added to *n*-butylmagnesium bromide under similar conditions (equations 290 and 291)[495]. Certain

$$\text{Bu}^n\text{CH}{=}\text{CHCH}_2\text{OMe} + \text{Hept}^n\text{MgCl} \xrightarrow[\text{CuBr}]{5\text{ mol-}\%} \text{Bu}^n\text{CH}{=}\text{CHCH}_2\text{Hept}^n \quad 65\% \tag{289}$$

$$\text{HC}{\equiv}\text{C(Me)C}{=}\text{CHCH}_2\text{OAc}\ (\mathbf{160}) + \text{PhMgBr} \xrightarrow[\text{Li}_2[\text{CuCl}_4]]{4\text{ mol-}\%} \text{HC}{\equiv}\text{C(Me)C}{=}\text{CHCH}_2\text{Ph}\ (\mathbf{161}) \tag{290}$$

$$\text{MeCH}{=}\text{CHCH}_2\text{OAc} + \text{Bu}^n\text{MgBr} \xrightarrow{\text{Li}_2[\text{CuCl}_4]} \text{MeCH}{=}\text{CH(CH}_2)_4\text{Me} \quad 88\% \tag{291}$$

$$\text{CH}_2{=}\text{CHOPh} + \text{PhMgBr} \xrightarrow{[\text{NiCl}_2(\text{PPh}_3)_2]} \text{CH}_2{=}\text{CHPh} \tag{292}$$

$$\text{2-MeOC}_{10}\text{H}_7 + \text{PhMgBr} \xrightarrow{[\text{NiCl}_2(\text{PPh}_3)_2]} \text{2-PhC}_{10}\text{H}_7 \tag{293}$$

alkoxides also couple with Grignard reagents in the presence of a nickel(II) catalyst. For example, ethenyl phenyl ether and 2-methoxynaphthalene both react with phenylmagnesium bromide, using dichlorobis(triphenylphosphine)nickel(II) as catalyst, to give the expected coupled products[496].

Silyl ethers can couple with organomagnesium halides in the presence of a nickel(II) catalyst. For instance, but-1-enyl trimethylsilyl ether adds to phenylmagnesium bromide[497], while the silyl ether **162** couples with ethylmagnesium bromide in the presence of a catalytic amount of $[NiCl_2(PPh_3)_2]$ (equations 294 and 295)[498].

OSiMe3 + PhMgBr —[NiCl2(PPh3)2]→ Ph (294)

(162) OSiMe3 ⇌ OSiMe3 —2 EtMgBr, [NiCl2(PPh3)2]→ 42% (295)

Allyl, aryl, and alkenyl sulphides can be coupled with both alkyl or aryl Grignard reagents by using nickel(II) catalysts. For example, *iso*propyl allyl sulphide couples with phenylmagnesium bromide and diene **163** with methylmagnesium bromide, using $[NiCl_2(PPh_3)_2]$ as catalyst (equations 296 and 297)[499,339b]. The alkenyl disulphide **164** reacts with ethylmagnesium bromide stereoselectively to yield an alkenyl sulphide, which can further couple to methylmagnesium bromide, generating a trisubstituted alkene[500]. In both steps dichloro[bis(diphenylphosphino)propane]nickel(II) was utilized

S + PhMgBr —[NiCl2(PPh3)2]→ Ph 99% (296)

(163) SPh + MeMgBr —[NiCl2(PPh3)2]→ Me 78% (297)

(164) EtS, EtS, Ph, H —EtMgBr, [NiCl2(dppp)]→ EtS, Et, Ph, H 85% —MeMgBr, [NiCl2(dppp)]→ Me, Et, Ph, H 66% (298)

N SMe + PhMgBr —[NiCl2(dppp)]→ N Ph (299)

as a catalyst. The same catalyst was similarly used to promote the coupling of phenylmagnesium bromide to 2-thiomethylpyridine (equation 299)[501].

Alkyl and allyl sulphones are known to couple primarily with alkyl Grignard reagents in the presence of a copper catalyst. For example, allyl phenyl sulphone couples with *n*-hexylmagnesium bromide while the cyclopropyl disulphone **165** reacts with two equivalents of ethylmagnesium bromide, using bis(acetylacetonato)copper(II) and $Li_2[CuCl_4]$, respectively (equations 300 and 301)[495,502]. The Grignard reagent **166** similarly couples to sulphone **167** in the presence of $Li_2[CuCl_4]$[503]. Both nickel(II) and iron(III) catalysts have been effectively used in the cross-coupling of phenylmagnesium

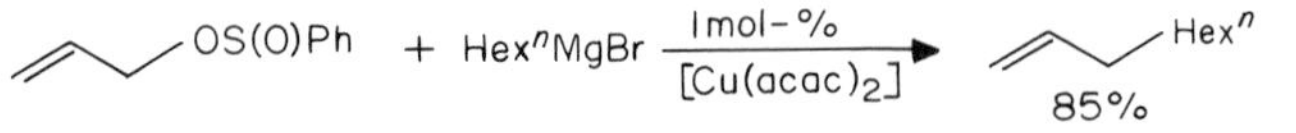

(300)

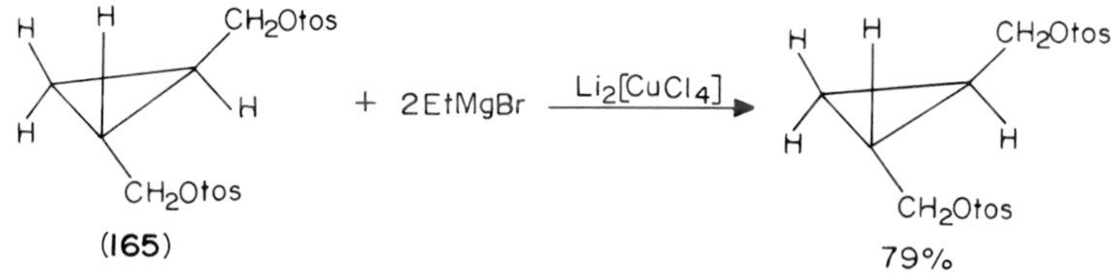

(301)

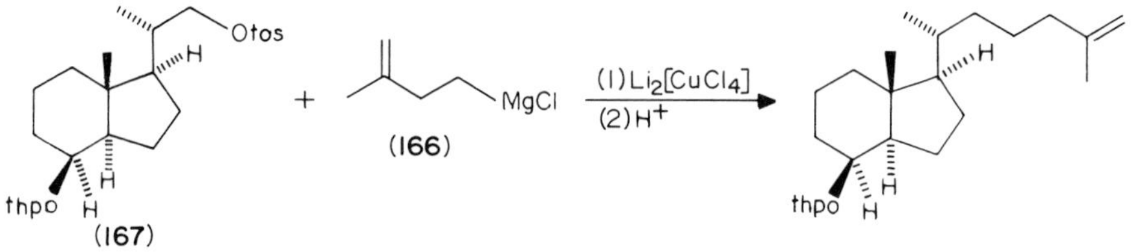

(302)

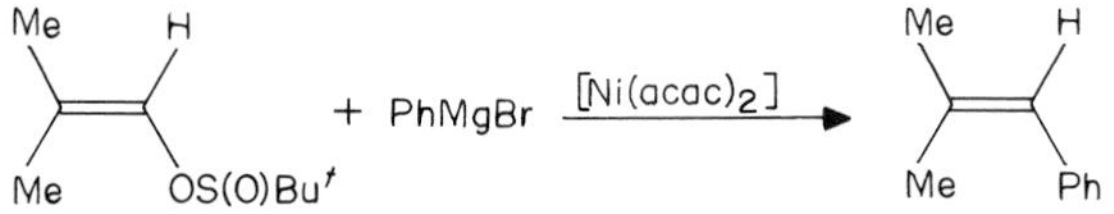

(303)

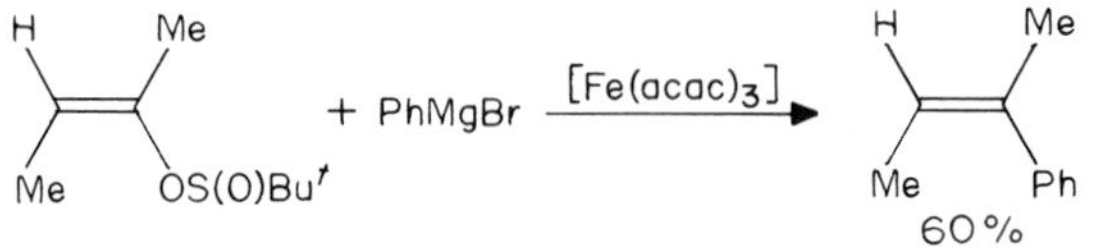

(304)

bromide with alkenyl sulphones (equations 303 and 304)[344b], the latter example proceeding with complete stereospecificity.

Allyl and alkenyl selenides undergo a coupling reaction with alkyl and aryl Grignard reagents in the presence of $[NiCl_2(PPh_3)_2]$ or $[NiCl_2(dppp)]$. Allyl phenyl selenide, for instance, and hex-1-enyl phenyl selenide both react with phenylmagnesium bromide to yield the expected coupled product (equations 305 and 306)[504].

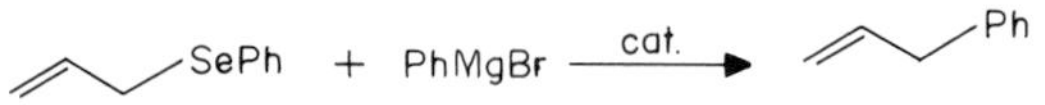

(305)

$$\text{Bu}^n\text{CH=CHSePh} + \text{PhMgBr} \xrightarrow{\text{cat.}} \text{Bu}^n\text{CH=CHPh} \qquad (306)$$

A variety of Grignard reagents are known to couple with allyl phosphates using copper(I) bromide as catalyst. For instance, ethynylmagnesium bromide couples with the alkenyl phosphate **168** to give the expected product in 85% yield[505]. Trimethylsilylmethylmagnesium chloride similarly couples to cyclohex-1-enyl phosphate (**169**) in the presence of bis(acetylacetonato)nickel(II), or with alkenyl phosphate **170** with tetrakis(triphenylphosphine)palladium(0) as catalyst[329].

$$\underset{(168)}{(\text{Bu}^n)_2\text{C=CHCH}_2\text{OPO(OEt)}_2} + \text{HC}\equiv\text{CMgBr} \xrightarrow[\text{CuBr}]{5\text{ mol-\%}} (\text{Bu}^n)_2\text{C=CHCH}_2\text{C}\equiv\text{CH} \qquad (307)$$

$$\underset{(169)}{\text{cyclohex-1-enyl–OPO(OEt)}_2} + \text{Me}_3\text{SiCH}_2\text{MgCl} \xrightarrow{[\text{Ni(acac)}_2]} \underset{81\%}{\text{cyclohex-1-enyl–CH}_2\text{SiMe}_3} \qquad (308)$$

$$\underset{(170)}{\text{MeCH=C(Ph)OPO(OEt)}_2} + \text{Me}_3\text{SiCH}_2\text{MgCl} \xrightarrow{[\text{Pd(PPh}_3)_4]} \underset{78\%}{\text{MeCH=C(Ph)CH}_2\text{SiMe}_3} \qquad (309)$$

Further examples of transition metal-catalysed Grignard cross-coupling reactions with organic substrates other than halides are given in Table 10.

7. Addition to nitrogen heterocyclic aromatic compounds

Although Grignard reagents are known to react with nitrogen heterocyclic aromatic compounds, few synthetically useful reactions of this type have been unearthed. Organolithium compounds are generally found to be more reactive and more regiospecific than their magnesium counterparts in this role (ref. 8, pp. 15–18). Recently, however, the regiospecific addition of transition metal-catalysed Grignard reagents to nitrogen heterocyclic aromatic compounds has appeared in the literature. For instance, 4-substituted pyridines can be prepared from 1-acylpyridinium salts by treatment with the chosen Grignard in the presence of copper(I) iodide, followed by aromatization (equation 310)[278]. Regiospecificity of the addition reaction was 99.1% using *n*-butylmagnesium bromide and 100% for cyclohexyl- or phenylmagnesium bromide. The intermediate 1-acyl-4-substituted-1,4-dihydropyridines can also be converted to 2,4-substituted pyridines (equation 311)[507]. Interestingly, 2,4,5-trichloropyrimidine undergoes selective addition to the unsubstituted position by phenylmagnesium bromide in the

$$\text{1-(COR')pyridinium Cl}^- \xrightarrow[5\text{ mol-\% CuI}]{\text{RMgX,}} \text{1-COR-4-H-4-R-1,4-dihydropyridine} \xrightarrow[\Delta]{\text{S}_8} \underset{62-77\%}{\text{4-R-pyridine}} \qquad (310)$$

TABLE 10. Transition-Metal catalysed displacement of a non-halide using Grignard reagents

$$RMgX + R'X \xrightarrow{\text{cat.}} RR' + MgX_2$$

R′X	R	Catalyst[a]	Yield (%)	Ref.
1. 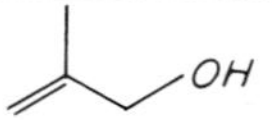 OH	Ph	A	80	492
2. 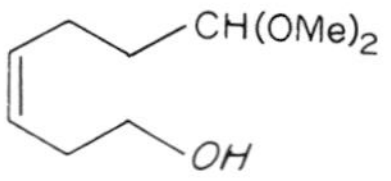 CH(OMe)$_2$ OH	$C_{10}H_{21}$	B	75	493
3. 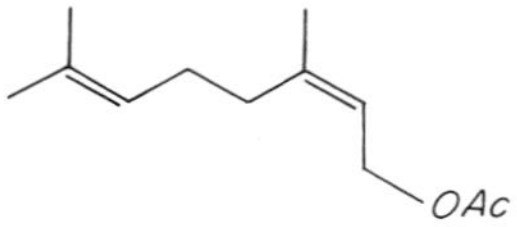 OAc	BzO$(CH_2)_2$CH(Me)CH_2—, EtOCH(Me)O$(CH_2)_2$CH(Me)CH_2—	B	79, 66	494
4. 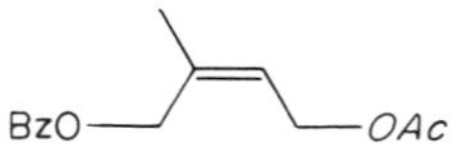 BzO OAc	*p*-BrC_6H_4—	B	49	494
5. 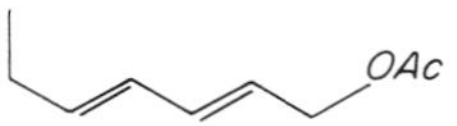 OAc	Othp$(CH_2)_5$—	B	50	506
6.  SR^2 $R^2 = Pr^i$, Ph	Ph$(CH_2)_3$—, *p*-$Pr^iC_6H_4$—, Ph	C	82–90	499
7. 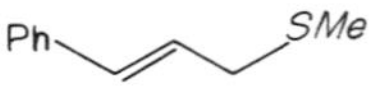 Ph SMe	Ph, Ph$(CH_2)_3$—	C	96, 83	499

8. 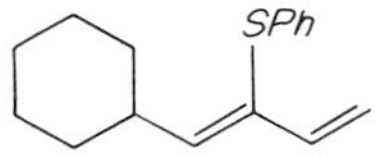 (structure: cyclohexyl–CH=C(SPh)–CH=CH$_2$)	Me	D	70	339b
9. p-Bu^{t}C$_6$H$_4$*SMe*	Cy	C	70	500
10. Me$_2$C=C(SEt)*SEt* 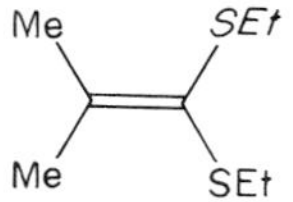 (structure: Me, Me, *SEt*, SEt)	Et	C	64	500
11. 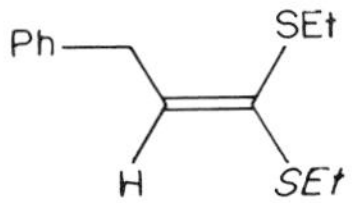 (structure: Ph, SEt, H, *SEt*)	Me, 2Me	E, C	60, 50	500
12. 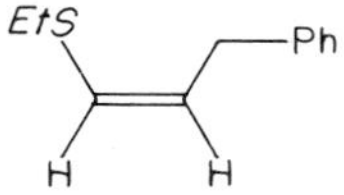 (structure: *EtS*, Ph, H, H)	Et, Me	C	85, 64	500
13. 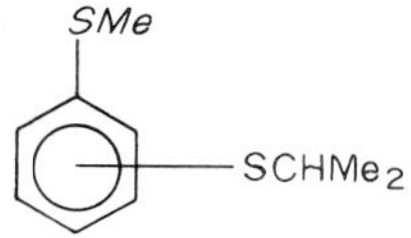 (structure: *SMe*, SCHMe$_2$)	4-H$_2$C=CHCH$_2$—, 3-H$_2$C=C(Me)—, 4-H$_2$C=C(Me)—, 3-Ph, 3-Me, 4-Ph, 4-Me	D	47–87	344a
14. p-Me$_3$CSC$_6$H$_4$*SMe*	Ph	D	76	344a
15. 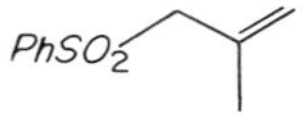 (structure: *PhSO$_2$*CH$_2$C(Me)=CH$_2$)	HeX^n	F	92	502

TABLE 10. (*Contd.*)

R′X	R	Catalyst[a]	Yield (%)	Ref.
16.  Otos	Hex^n	F	91	502
17. 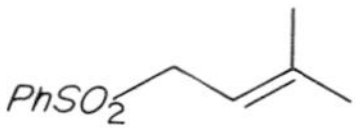 $PhSO_2$	Hex^n	F	65	502
18. 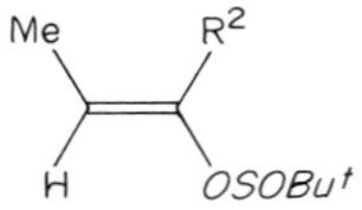 Me, R^2, H, $OSOBu^t$; $R^2 = Me, Me_2C{=}CH(CH_2)_2{-}$	Ph	G	68, 71	344b
19. 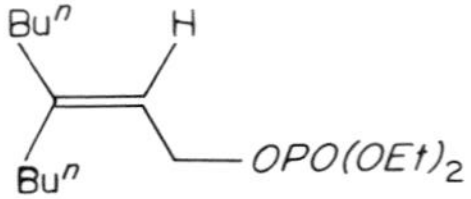 Bu^n, H, Bu^n, $OPO(OEt)_2$	Pr^i, Bu^t, $MeCH{=}CHCH_2{-}$, $H_2C{=}C{=}CH{-}$	H	80–92	505
20. 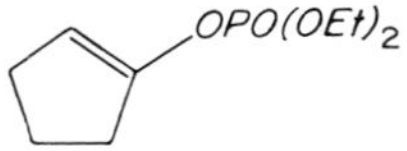 $OPO(OEt)_2$	$Me_3SiCH_2{-}$	G	75	329
21. 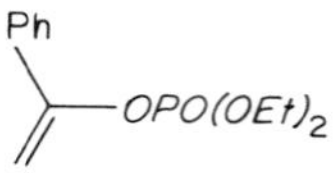 Ph, $OPO(OEt)_2$	$Me_3SiCH_2{-}$, $Me_3SiCH(Ph){-}$	I, G	82, 47	329
22. 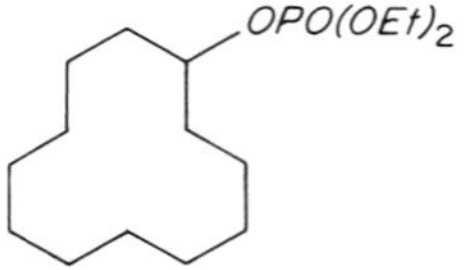 $OPO(OEt)_2$	$Me_3SiCH_2{-}$	G	87	329

23. Hex^n ... OPO(OEt)$_2$	Me_3SiCH_2—	I	92	329
24. ... OPO(OEt)$_2$	Me_3SiCH_2—	I	81	329
25. Ph ... OPO(OEt)$_2$	Me_3SiCH_2—	I	70	329
26. ... OPO(OEt)$_2$	Me_3SiCH_2—	I	48	329

[a] A = [$PdCl_2$(dppf)]; B = $Li_2[CuCl_4]$; C = [$NiCl_2$(dppp)]; D = [$NiCl_2(PPh_3)_2$]; E = $NiCl_2$–PPh_3(1:1); F = [$Cu(acac)_2$]; G = [$Ni(acac)_2$]; H = CuBr; and I = $NiBr_2$.

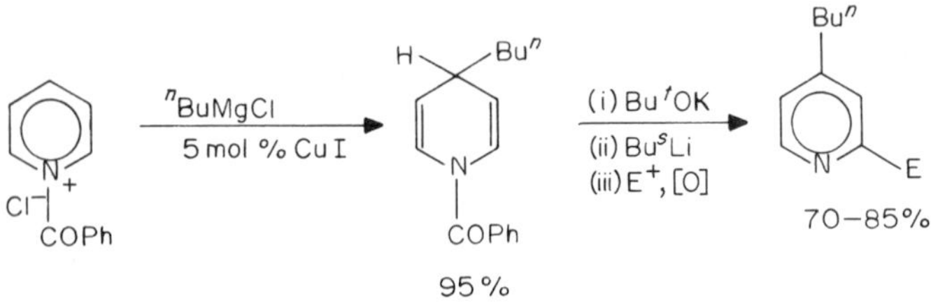

(311)

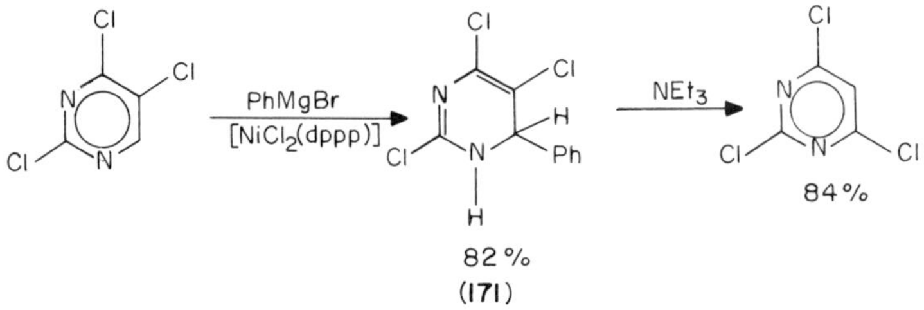

(312)

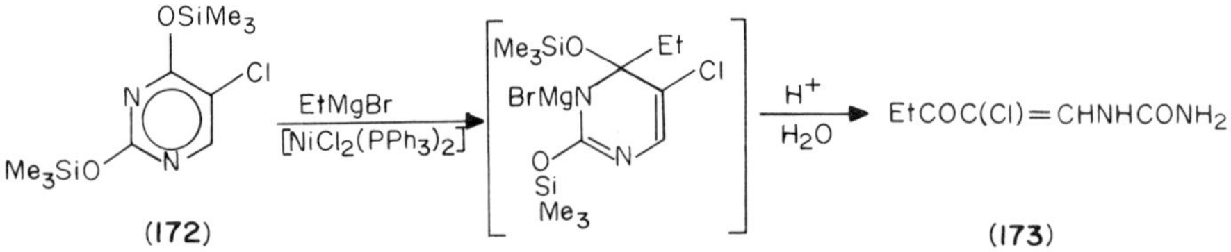

(313)

presence of a nickel catalyst to yield dihydropyrimidine (**171**)[273]. Subsequent treatment of (**171**) with triethylamine gave 2,4-dichloro-6-phenylpyrimidine in good yield. Similar treatment of halopyrimidine (**172**) with ethylmagnesium bromide in the presence of a nickel catalyst, however, resulted in attack of the 4-position. Acid hydrolysis of this intermediate gave rise to the ring-opened product **173**[508].

V. ORGANO-BERYLLIUM, -CALCIUM, -STRONTIUM, AND -BARIUM COMPOUNDS IN ORGANIC SYNTHESIS

This section is primarily concerned with the role of organo-beryllium, -calcium, -strontium, and -barium compounds in organic synthesis. To date, however, little attention has been focused on this area for reasons outlined in Section I. For ease of discussion, this section has been divided into two parts, the first on organo-calcium, -strontium, and -barium compounds and the second on organoberyllium reagents. In each sub-section a description of synthetic methods used to prepare the organometallic compounds precedes the discussion on reactivity.

A. Organo-calcium, -strontium, and -barium Reagents

The organometallic chemistry of calcium, strontium, and barium has been extensively covered in a recent review (ref. 7, pp. 223–240). Surveys of the field have also appeared in two textbooks[2,3] and in another review over the past decade[509]. Of the three elements, the organometallic chemistry of calcium has been investigated in the greatest detail. Both organo and diorgano derivates have been prepared for all three elements, but only the organo halides have made any contribution to organic synthesis. In view of this only the synthesis of organo-calcium, -strontium, and -barium halides will be covered in this chapter.

1. Synthesis

The halogenoorganometallic derivatives are usually formulated as M(R)X (M = Ca, Sr, Ba; X = Cl, Br, I) and, as is the case for the analogous Grignard reagents, the actual constitution is generally not clear. The usual method of preparation is the direct interaction of metal with organic halide in a suitable solvent in a manner similar to that employed for their magnesium counterpart (equation 314)[7].

$$M + RX \longrightarrow M(R)X \tag{314}$$

In a typical experiment the organic halide, dissolved in tetrahydrofuran, is added dropwise to a stirred solution of finely divided metal suspended in the same solvent. For calcium, the purity of the metal has been found to greatly influence the yield of the resulting organometallic species. The best yields have been obtained using calcium containing 0.0019% Na and 0.49% Mg[7,510]. Activation of the metal is generally not necessary when using *n*-alkyl iodides, but for other substrates the addition of a small amount of iodine or 1 mol-% of RI, or amalgamation of the metal with mercury or mercury(II) chloride is recommended to initiate the reaction. Low temperatures are also a key factor in the preparation, − 78 °C being used for strontium and barium[511] and between − 70 and 0 °C for calcium[510]. Other solvents such as diethyl ether and toluene have also been used with varied success. *n*-Alkyl- and phenylbarium iodides, *n*-alkyl- and arylstrontium iodides, and both alkyl- and arylcalcium bromides and iodides are accessible via this method (up to 97% yield)[7,510,511]. Further, triphenylmethyl chlorides of calcium, strontium, and barium are also accessible by this method[512,513], as are triarylmethyl and allylhalides of calcium and strontium[513–515].

Alternatively for calcium, the desolvation of $[Ca(NH_3)_6]$ or rapid cooling with argon gas of the vaporised metal can be employed instead of finely divided metal in these reactions[7]. The latter method has been used in the preparation of arylcalcium fluorides and chlorides in up to 50% yield[516].

Two indirect methods for the formation of halogeno-organometal compounds have also been employed (equations 315 and 316). Firstly, metallation of more acidic

$$RC{\equiv}CH + Ca(Ph)I \xrightarrow{Et_2O} \underset{46\text{–}91\%}{Ca(C{\equiv}CR)I} \tag{315}$$

$$\text{1-bromonaphthalene (Br)} + Ca(Ph)I \longrightarrow \text{1-naphthyl-CaI} \tag{316}$$

hydrocarbons has been observed for phenylstrontium iodide and a number of organocalcium halides, the phenyl derivatives being activated by complexation with tmeda or dabco[7]. For instance, alk-1-ynes are readily metallated by phenylcalcium iodide in diethyl ether (equation 315)[517]. Fluorene, indene, and thiophene are similarly metallated. Secondly, halogen–metal exchange, although less well documented, has been demonstrated for 1-bromonaphthalene using phenylcalcium iodide (equation 316)[518].

The organometallic halides of barium, calcium, and strontium are all sensitive to air and moisture. Most are thermally stable, the main exceptions being alkyl-barium and -strontium iodides[7,511].

2. Reactivity

The role of organo-calcium, -strontium, and -barium halides in organic synthesis is currently very limited, but they have been successfully utilized as initiators of polymerization reactions. This is particularly true of organobarium iodides, which have been extensively used in this role[7].

Both organo-barium and -calcium iodides have been shown to react with conjugated enynes to give either the 1,2- or 1,4-addition product, or both (equation 317)[509,519]. Stereoselective 1,4-addition of these organometallic derivatives to a variety of enynes has been observed to give solely the allene product on hydrolysis (equation 318)[519,520]. A similar reaction with the enyne **174** again resulted in the formation of an allene, but the alkoxy moiety was replaced in the process (equation 319)[521].

$$R'C{\equiv}C{-}CH{=}CH_2 + M(R)I \longrightarrow \begin{cases} \xrightarrow{1,4-} R'C(MI){=}C{=}CHCH_2R \\ \xrightarrow{1,2-} R'C{\equiv}CCH(MI)CH_2R \end{cases} \qquad (317)$$

$$R^1C{\equiv}CCH{=}CH_2 + M(R)I \xrightarrow{H^+} R^1HC{=}C{=}CHCH_2R \qquad (318)$$

R^1 = alkyl, $Me_2C(OH)$, RS, vinyl, allyl, isopropenyl, or $Me_2NCH_2^-$

$$\underset{\mathbf{(174)}}{R^1OCH_2C{\equiv}CCH{=}CH_2} + Ca(Et)I \xrightarrow{H^+} EtCH_2CH{=}C{=}CHCH_2Et \qquad (319)$$

A number of acyl substrates have been shown to react favourably with primarily organocalcium iodides. For instance, benzoyl chloride reacts with phenylcalcium iodide to give triphenylmethanol in high yield (equation 320)[518], while the same substrate reacts with methylcalcium iodide to give **175** in 77% yield on hydrolysis (equation 321)[522]. In addition to acyl chlorides, ketones and aldehydes have also been shown to undergo a 1,2-addition reaction with these organometallic reagents. For example, methylcalcium iodide adds to both acetone and benzophenone to give *tert*-butanol or 1,1-diphenylethanol, respectively, in good yield (equations 322 and 323)[522]. Methylstrontium iodide similarly reacts with benzophenone, although the product is obtained in lower yield (69%)[511]. Benzaldehyde similarly undergoes a 1,2-addition reaction with pentafluoroethylcalcium iodide, *in situ*, to give **176** in 96% yield (equation 324)[523]. In general, however, the reaction of ketones and aldehydes with organocalcium reagents is not stereospecific with mixtures of both addition and reduction products being formed[509,524].

$$PhCOCl + 2Ca(Ph)I \xrightarrow{H^+} \underset{94\%}{Ph_3COH} \qquad (320)$$

$$PhCOCl + 2Ca(Me)I \xrightarrow{H^+} \underset{77\%\ \mathbf{(175)}}{PhMe_2COH} \qquad (321)$$

$$Me_2CO + Ca(Me)I \xrightarrow{H^+} \underset{56\%}{Me_3COH} \qquad (322)$$

$$Ph_2CO + Ca(Me)I \xrightarrow{H^+} Ph_2MeCOH \quad 84\% \tag{323}$$

$$PhCHO + Ca(C_2F_5)I \xrightarrow{H^+} C_2F_5CH(OH)Ph \quad 96\% \; \textbf{(176)} \tag{324}$$

Virtually quantitative yields of carboxylic acids can be obtained on carboxylation of either aryl-calcium or -strontium halides[518,525]. For instance, (1-naphthyl)calcium iodide reacts with carbon dioxide to give 1-naphthoic acid in 98% yield (equation 325)[518].

CaI (1-naphthyl) + CO_2 $\xrightarrow{H^+}$ 1-naphthoic acid (CO_2H) 98% (325)

Carboxylation of the analogous alkyl derivatives, however, gives rise to a mixture of ketones and carboxylic acids[518].

Interestingly, phenylcalcium iodide has been observed to couple with pyridine to give primarily 2-phenylpyridine (equation 326). The side products 2, 5- and 2, 6-diphenylpyridine were also obtained in variable low yields depending on the reaction conditions

pyridine + Ca(Ph)I ⟶ 2-phenylpyridine (326)

2-phenylpyridine + Ca(Ph)I ⟶ 2,6-diphenylpyridine (327)

employed. Subsequent reaction of 2-phenylpyridine with phenylcalcium iodide, however, gave 2, 6-diphenylpyridine as the exclusive product (equation 327)[518].

It is noteworthy that acenaphthalene is known to react with calcium amalgam to form an organometallic derivative, which on rapid carboxylation, forms (*E*)-acenaphthene-1, 2-dicarboxylic acid in high yield (equation 328)[526].

acenaphthylene + Ca/Hg $\xrightarrow[H^+]{CO_2}$ (HO_2C, CO_2H) acenaphthene-1,2-dicarboxylic acid (328)

B. Organoberyllium Reagents

Much of the synthetic chemistry related to organoberyllium compounds was published prior to 1974, and very little has appeared in the literature in the interim period. The chemistry of organoberyllium compounds is adjudged to be intermediate between that of magnesium and the Group IIB elements, and mainly centres around diorganoberyllium species. In fact, in contrast to the other alkaline earth elements, very few organoberyllium halides have been isolated. The organometallic chemistry of beryllium has been the subject of a recent review[6].

1. Synthesis

As for organomagnesium compounds, the preparation of organoberyllium reagents must be carried out in an inert atmosphere. The most commonly used solvent is diethyl ether but varying levels of success have been achieved using hydrocarbon solvents. Diorganoberyllium compounds of the type R_2Be can be prepared in one of four ways (equations 329–332)[6]. Both dialkyl- and diarylberyllium reagents can be readily synthesized by employing either the Grignard or organolithium method (equations 329

$$2RMgX + BeX_2 \longrightarrow R_2Be + 2MgX_2 \tag{329}$$

$$2RLi + BeX_2 \longrightarrow R_2Be + 2LiX \tag{330}$$

$$R_2Hg + Be \longrightarrow R_2Be + Hg \tag{331}$$

$$R_3B + Et_2Be \longrightarrow R_2Be + REt_2B \tag{332}$$

and 330), respectively, cf. method F for Grignard synthesis, Table 1). In general, the appropriate organo-lithium or -magnesium reagent is added to a solution of beryllium halide in diethyl ether. The resulting precipitate (MgX_2 or LiX) is filtered off, the solvent removed by evaporation, and the product isolated by distillation or crystallization. Invariably the product is obtained as an etherate, a consequence of R_2Be being a good Lewis acid[6]. For thermally stable diorganoberyllium compounds the ether can normally be removed by prolonged heating of the ether complexes at low pressure before distillation[527]. The organolithium method is particularly useful in the synthesis of dialkynylorganoberyllium compounds[528,529].

Thermally robust dialkyl- and diarylberyllium reagents can be prepared via the organomercury method (equation 331), although this procedure is more applicable to small-scale syntheses. The general procedure involves heating beryllium powder with R_2Hg in the presence of a trace amount of I_2, $HgCl_2$, or Et_2Be[6].

The fourth alternative (equation 332), which involves an exchange reaction between a triorganoboron species and diethylberyllium, has been successfully used to prepare dialkyl-, diaryl-, and diallylberyllium compounds. The reaction is carried out at room temperature over a period of several days[530,531].

Mixed organoberyllium compounds of the type RR'Be are generally prepared via an exchange reaction between two different diorganoberyllium reagents (equation 333). Interestingly, no heteroaryl compounds have been isolated using this procedure[529].

Organoberyllium halides of the type RBeX are believed to be formed on heating haloalkanes with powdered beryllium (equation 334). The structure of these compounds is unknown, although, a polymeric constitution is presumed[6].

$$R_2Be + R_2'Be \longrightarrow 2RR'Be \tag{333}$$

$$RX + Be \xrightarrow{\Delta} \text{‘RBeX’} \tag{334}$$

$$RCOX + Be \longrightarrow \text{‘RCOBeX’} \tag{335}$$

Acylberyllium halides are similarly believed to be formed on reaction of an acyl halide with powdered beryllium (equation 335). The products, however, have not been isolated in a pure state, their formation arising by virtue of the nature of their chemical reactions[532].

Diorganoberyllium reagents are generally thermally robust, the main exception being branched-chain alkylberyllium compounds, which are thermally unstable above *ca.* 40–50 °C. Little is known about the thermal stability of the organoberyllium halides, although acylberyllium halides are known to react vigorously with both water and alcohols[6,532].

2. Reactivity

Diorganoberyllium compounds such as Et_2Be and Ph_2Be are effective as catalysts for the dimerization and polymerization of alkenes[6], but very little contribution has been made to the field of organic synthesis by these and other organoberyllium reagents.

Carboxylation of the optically active diorganoberyllium compound **177**, which was prepared via the organolithium method (equation 330), provides a moderate yield of the corresponding optically active carboxylic acid **178** (equation 336)[533]. Reagent **177** has also been utilized in the asymmetric synthesis of chiral secondary alcohols (97–99% purity) via reduction of alkyl phenyl ketones (equation 337). The highest optical yield obtained was 46% for the reduction of isopropyl phenyl ketone, the prevalent enantiomer produced having the *S* absolute configuration[534]. Further, **177** has been utilized in the nickel-catalysed displacement reaction between itself and α-olefins. The optically active ligand on beryllium, however, did not exert any significant asymmetric induction during the reaction[535].

$$\underset{[(R)\text{-}(\pm)\text{-}\mathbf{177}]}{(\mathrm{EtMeCHCH_2})_2\mathrm{Be} + 2\mathrm{CO_2}} \xrightarrow{\mathrm{H^+}} \underset{[(S)\text{-}(+)\text{-}\mathbf{178}]}{2\mathrm{EtMeCHCH_2CO_2H}} \qquad (336)$$

$$\mathrm{RCOPh} + \mathbf{177} \xrightarrow{\mathrm{H^+}} \underset{88\text{–}97\%}{\mathrm{PhRHCOH}} \qquad (337)$$

Apart from **177**, only the acylberyllium bromides have made any contribution to organic synthesis. These organoberyllium reagents undergo some interesting chemical transformations with a number of organic substrates. For instance, aliphatic ketones undergo a 1,2-addition reaction with acylberyllium bromides to yield semicarbazones on hydrolysis, or bromoketones on quenching with an acid bromide (equations 338 and 339). Acetone, ethyl methyl ketone, *n*-butyl methyl ketone and di-*n*-propyl ketone all react with acylberyllium bromides in this fashion[536]. Aromatic ketones on the other hand, react differently with acylberyllium bromides, generating pinacols on hydrolysis (equation 340). If an excess of the acyl bromide is present in solution, however, then only pinacolones are isolated (equation 341)[537].

$$\mathrm{RCOBeBr} + \mathrm{R^1COR^2} \xrightarrow{\mathrm{H^+}} \mathrm{RCOCR^1(R^2)OH} \qquad (338)$$

$$\mathrm{RCOBeBr} + \mathrm{R^1COR^2} \xrightarrow{\mathrm{R^3COBr}} \mathrm{RCOCR^1(R^2)Br} \qquad (339)$$

$\mathrm{R = Et, Me, Pr}^n$

$$\mathrm{Ar_2CO} + \mathrm{RCOBeBr} \xrightarrow{\mathrm{H^+}} 1/2\mathrm{Ar_2C(OH)CAr_2OH} \qquad (340)$$

$$\mathrm{Ar_2CO} + \mathrm{RCOBeBr} \xrightarrow[\mathrm{H^+}]{\mathrm{RCOBr}} \tfrac{1}{2}\mathrm{Ar_3COAr} \qquad (341)$$

Aromatic aldehydes react with acylberyllium bromides in ethyl acetate to generate alkenes in good yield (60–90%). For instance, the reaction of AcOBeBr with benzaldehyde gave (*E*)-stilbene (equation 342), while the same reaction with cinnamaldehyde yielded a diphenylhexatriene (equation 343)[538].

$$\text{RCOBeBr} + \text{PhCHO} \xrightarrow[\text{1 h, }\Delta]{\text{EtOAc}} \tfrac{1}{2}\ (E)\text{-PhCH=CHPh} \tag{342}$$

$$\text{PhCH=CHCHO} + \text{RCOBeBr} \xrightarrow[\text{1 h, }\Delta]{\text{EtOAc}} \tfrac{1}{2}\text{Ph(CH=CH)}_3\text{Ph} \tag{343}$$

Acid amides are readily prepared by reaction of acylberyllium bromides with aromatic nitro compounds. For instance, both nitrobenzene and *o*-nitrotoluene react with a range of acylberyllium bromides to give the corresponding acid amides in high yield (equations 344 and 345)[6,539].

$$\text{RCOBeBr} + \text{PhNO}_2 \xrightarrow{\text{EtOAc}} \text{PhNHCOR} \tag{344}$$

$$\text{RCOBeBr} + o\text{-MeC}_6\text{H}_4\text{NO}_2 \xrightarrow{\text{EtOAc}} o\text{-MeC}_6\text{H}_4\text{NHCOR} \tag{345}$$

R = Me, Et, Pr^n, Bu^n, or Cy

VI. REFERENCES

1. M. S. Kharash and O. Reinmuth, *Grignard Reactions of Nonmetallic Substances*, Prentice Hall, New York, 1954.
2. S. T. Ioffe and A. Nesmeyanov, *The Organic Chemistry of Magnesium, Beryllium, Calcium, Strontium and Barium*, North-Holland, Amsterdam, 1967.
3. K. Nützel, *Houben-Weyl, Methoden der Organischer Chemie, Vol. 13/2a, Metallorganische Verbindungen Be, Mg, Ca, Sr, Ba, Zn, Cd*, Georg Thieme Verlag, Stuttgart, 1973, p. 47.
4. T. Chaudron, *C. R. Acad. Sci.*, **264**, 1093 (1967); T. Chaudron and A. Sekera, *C. R. Acad. Sci.*, **265**, 277 (1967).
5. J. Sadet and R. Rumpf, *Bull. Soc. Chim. Fr.*, 2016 (1973).
6. N. A. Bell, *Comprehensive Organometallic Chemistry*, Vol. 1, Pergamon Press, Oxford, 1982, Chapter 3, pp. 121–154.
7. W. E. Lindsell, *Comprehensive Organometallic Chemistry*, Vol. 1, Pergamon Press, Oxford, 1982, Chapter 4, pp. 155–252.
8. B. J. Wakefield, *Comprehensive Organometallic Chemistry*, Vol. 7, Pergamon Press, Oxford, 1982, Chapter 44, pp. 1–110.
9. D. Steinborn, *J. Organomet. Chem.*, **182**, 313 (1979).
10. R. Menicagli, C. Botteghi, and M. Marchetti, *J. Heterocycl. Chem.*, **17**, 57 (1980).
11. E. Pechhold, D. G. Adams, and G. Fraenkel, *J. Org. Chem.*, **36**, 1368 (1971); H. G. Richey and L. M. Moses, *J. Org. Chem.*, **48**, 4013 (1983).
12. V. Grignard, *C. R. Acad. Sci.*, **130**, 1322 (1900).
13. P. Barbier, *C. R. Acad. Sci.*, **128**, 110 (1899).
14. C. Blomberg and F. A. Hartog, *Synthesis*, 18 (1977).
15. R. D. Rieke, S. E. Bales, P. M. Hudnall, and G. S. Poindexter, *Org. Synth.*, **59**, 85 (1980); T. P. Burns and R. D. Rieke, *J. Org. Chem.*, **48**, 4141 (1983).
16. J. M. Hornback and R. D. Barrows, *J. Org. Chem.*, **48**, 90 (1983).
17. M. F. Lappert, T. R. Martin, C. L. Raston, B. W. Skelton, and A. H. White, *J. Chem. Soc., Dalton Trans.*, 1959 (1982).

18. L. M. Engelhardt, W.-P. Leung, C. L. Raston, P. Twiss, and A. H. White, *J. Chem. Soc., Dalton Trans.*, 331 (1984).
19. B. Gandha and J. K. Sagden, *Synth. Commun.*, **9**, 851 (1979).
20. A. Duff and M. F. Lappert, personal communication.
21. M. F. Lappert and A. J. Thorne, personal communication.
22. W.-P. Leung, C. L. Raston, B. W. Skelton, and A. H. White *J. Chem. Soc., Dalton Trans.*, 1801 (1984).
23. D. Leibfritz, B. O. Wagner, and J. D. Roberts, *Justus Liebigs Ann. Chem.*, **763**, 173 (1972).
24. P. Boudjouk, J. S. Kiely, and R. Sooriyakumaran, *J. Organomet. Chem.*, **221**, 33 (1981).
25. C. Ungurenasu and M. Palie, *Synth. React. Inorg. Metal-Org. Chem.*, **7**, 581 (1977); R. T. Arnold and S. T. Kulenovic, *Synth. Commun.*, **7**, 223 (1977).
26. W. Oppolzer and P. Schneider, *Tetrahedron·Lett.*, **25**, 3305 (1984).
27. G. B. Sergeev, V. V. Smirnov, and F. Z. Badaev, *J. Organomet. Chem.*, **224**, C29 (1982).
28. F. Freijee, G. Schat, R. Mierop, C. Blomberg, and F. Bickelhaupt, *Heterocycles*, **7**, 237 (1977).
29. (a) L. M. Engelhardt, R. I. Papasergio, C. L. Raston, and A. H. White, *J. Chem. Soc., Dalton Trans.*, 311 (1984); (b) C. L. Raston and G. Salem, *J. Chem. Soc., Chem. Commun.*, 1702 (1984).
30. L. Fitjer, *Chem. Ber.*, **115**, 1035 (1982).
31. F. R. Hewgill and F. Legge, personal communication; M. Ariatti and J. Zemlicka, *J. Org. Chem.*, **46**, 5204 (1981).
32. Y. Sawaki and C. S. Foote, *J. Am. Chem. Soc.*, **101**, 6292 (1979).
33. W. Winter, T. Batters, A. Rieker, and Y. Butsugan, *Z. Naturforsch., Teil B*, **37**, 855 (1982).
34. H. Kikuchi, S. Seki, G. Yamamoto, T. Mitsuhashi, N. Nakamura, and M. Oki, *Bull. Chem. Soc. Jpn.*, **55**, 1514 (1982).
35. J. Battersby, R. Stephens, J. C. Tatlow, and L. F. Thomas, *J. Fluorine Chem.*, **15**, 139 (1980).
36. M. Bochmann, G. Wilkinson and G. B. Young, *J. Chem. Soc., Dalton Trans.*, 1879 (1980); G. Molle, P. Bauer, and J. E. Dubois, *J. Org. Chem.*, **47**, 4120 (1982).
37. H. R. Rogers, C. L. Hill, Y. Fujiwara, R. J. Rogers, H. L. Mitchell, and G. M. Whitesides, *J. Am. Chem. Soc.*, **102**, 217 (1980), and accompanying papers.
38. G. B. Sergeev, V. V. Zagorsky, and F. Z. Badaev, *J. Organomet. Chem.*, **243**, 123 (1983).
39. P. G. Jasien and C. E. Dykstra, *J. Am. Chem. Soc.*, **105**, 2089 (1983).
40. Y. Tanaka, S. C. Davis, and K. J. Klabunde, *J. Am. Chem. Soc.*, **104**, 1013 (1982).
41. C. Blomberg, G. Schat, H. H. Grootveld, A. D. Vreugdenhil, and F. Bickelhaupt, *Justus Liebigs Ann. Chem.*, **763**, 148 (1972); K. Ogura, M. Fujita, K. Takahashi, and H. Iida, *Chem. Lett.*, 1697 (1982).
42. M. de Bottom, *Bull. Soc. Chim. Fr.*, 1773 (1977).
43. G. Salbeck, *Phosphorus Sulfur*, **13**, 165 (1982).
44. D. Bar, P. Marcincal, and A. Marcincal-Lefebvre, *Bull. Soc. Chim. Fr.*, 2484 (1972).
45. J. Ficini and J.-C. Depezay, *Tetrahedron Lett.*, 4795 (1969).
46. H. H. Grootveld, C. Blomberg, and F. Bickelhaupt, *Tetrahedron Lett.*, 1999 (1971).
47. G. L. Grunewald, V. M. Paradkar, B. Pazhenchevsky, M. A. Pleiss, D. J. Sall, W. L. Seibel, and T. J. Reitz, *J. Org. Chem.*, **48**, 2321 (1983).
48. M. T. Rahman, *J. Organomet. Chem.*, **225**, 25 (1982).
49. M. T. Rahman, *Monatsh. Chem.*, **114**, 445 (1983).
50. A. G. Mack, H. Suschitzky, and B. J. Wakefield, *J. Chem. Soc., Perkin Trans. 1*, 1472 (1979).
51. R. Albadri, P. Moreau, and A. Commeyras, *Nouv. J. Chim.*, **6**, 581 (1982).
52. S. F. Campbell, R. Stephens, J. C. Tatlow, and W. T. Westwood, *J. Fluorine Chem.*, **1**, 439 (1972).
53. J. W. F. L. Seetz, O. S. Akkerman, and F. Bickelhaupt, *Tetrahedron Lett.*, **22**, 4857 (1981).
54. E. L. McCaffery and S. W. Shalaby, *J. Organomet. Chem.*, **44**, 227 (1972).
55. J. W. F. L. Seetz, F. A. Hartog, H. P. Böhm, C. Blomberg, O. S. Akkerman, and F. Bickelhaupt, *Tetrahedron Lett.*, **23**, 1497 (1982).
56. G. M. Whitesides and F. D. Gutowski, *J. Org. Chem.*, **41**, 2882 (1976).
57. F. G. Mann and F. H. C. Stewart, *J. Chem. Soc.*, 2826 (1954).
58. T. Reichstein and J. Baud, *Helv. Chim. Acta*, **20**, 892 (1937).
59. E. V. Polinin, I. M. Zaks, A. M.Moiseenkov, and A. V. Semenovski, *Izv. Akad. Nauk SSSR, Ser. Khim.*, **18**, 239 (1979).
60. G. Cahiez, A. Alexakis, and J. F. Normant, *Tetrahedron Lett.*, 3013 (1978).
61. E. A. Hill, *Adv. Organomet. Chem.*, **16**, 131 (1977).

62. E. P. Kundig and C. Perret, *Helv. Chim. Acta*, **64**, 2606 (1981).
63. L. I. Smith and S. McKenzie, Jr., *J. Org. Chem.*, **15**, 74 (1950).
64. D. J. Patel, C. L. Hamilton, and J. D. Roberts, *J. Am. Chem. Soc.*, **87**, 5144 (1965); E. A. Hill and J. Davidson, *J. Am. Chem. Soc.*, **86**, 4663 (1964).
65. E. C. Ashby and J. R. Bowers, Jr., *J. Am. Chem. Soc.*, **103**, 2242 (1981).
66. W. Oppolzer, R. Pitteloud, and H. F. Strauss, *J. Am. Chem. Soc.*, **104**, 6476 (1982).
67. W. Oppolzer and R. Pitteloud, *J. Am. Chem. Soc.*, **104**, 6478 (1982).
68. W. Oppolzer and K. Battig, *Tetrahedron Lett.*, **23**, 4669 (1982).
69. W. Oppolzer, H. F. Strauss, and D. P. Simmons, *Tetrahedron Lett.*, **23**, 4673 (1982).
70. W. Oppolzer, T. Begley, and A. Ashcroft, *Tetrahedron Lett.*, **25**, 825 (1984).
71. S. Cohen and A. Yogev, *J. Am. Chem. Soc.*, **98**, 2013 (1976).
72. J. L. Derocque and F.-B. Sundermann, *J. Org. Chem.*, **39**, 1411 (1974).
73. G. Molle and P. Bauer, *J. Am. Chem. Soc.*, **104**, 3481 (1982).
74. J. S. Kiely and P. Boudjouk, *J. Organomet. Chem.*, **182**, 173 (1979).
75. T. J. Barton and G. T. Burns, *Organometallics*, **1**, 1455 (1982).
76. R. Nesi, A. Ricci, M. Taddei, P. Tedeschi, and G. Seconi, *J. Organomet. Chem.*, **195**, 275 (1980).
77. W. P. Weber, *Silicon Reagents for Organic Synthesis*, Springer-Verlag, New York, 1983.
78. J. P. Dulcere, J. Grimaldi, and M. Santelli, *Tetrahedron Lett.*, **22**, 3179 (1981).
79. A. Miodownik, J. Kreisberger, M. Nussim, and D. Avnir, *Synth. Commun.* **11**, 241 (1981).
80. M. Lequan and S. Queroix, *J. Organomet. Chem.*, **194**, 33 (1980).
81. J. V. N. Vara Prasad and C. N. Pillai, *J. Organomet. Chem.*, **259**, 1 (1983).
82. H. Lehmkuhl, C. Naydowski, R. Benn, A. Rufinska, G. Schroth, R. Mynott, and C. Krüger, *Chem. Ber.*, **116**, 2447 (1983); A. M. Moiseenkov, B. A. Czeskis, and A. V. Semenovsky, *J. Chem. Soc., Chem. Commun.*, 109 (1982); A. M. Moiseenkov, B. A. Cheskis, T. Yu. Rudashevskii, O. A. Nesmeyanova, and A. V. Semenovskii, *Izv. Akad. Nauk. SSSR, Ser. Khim.*, 1088 (1982).
83. R. Mornet and L. Gouin, *Bull. Soc. Chim. Fr.*, 737 (1977).
84. B. B. Snider, R. S. E. Conn, and M. Karras, *Tetrahedron Lett.*, **20**, 1679 (1979).
85. G. R. Buske and W. T. Ford, *J. Org. Chem.*, **41**, 1995 (1976).
86. E. C. Ashby and A. B. Goel, *J. Am. Chem. Soc.*, **99**, 310 (1977).
87. E. C. Ashby and T. Smith, *J. Chem. Soc., Chem. Commun.*, 30 (1978).
88. F. Sato, H. Ishikawa, and M. Sato, *Tetrahedron Lett.*, **22**, 85 (1981).
89. F. Sato, H. Watanabe, Y. Tanaka, and M. Sato, *J. Chem. Soc., Chem. Commun.*, 1126 (1982).
90. J. J. Eisch and J. E. Galle, *J. Organomet. Chem.*, **160**, C8 (1978).
91. F. Sato, H. Ishikawa, H. Watanabe, T. Miyake, and M. Sato, *J. Chem. Soc., Chem. Commun.*, 1126 (1982).
92. J. G. Duboudin and B. Jousseaume, *J. Organomet. Chem.*, **168**, 1 (1979).
93. E. C. Ashby and R. D. Ainslie, *J. Organomet. Chem.*, **250**, 1 (1983).
94. F. Sato, H. Ishikawa and M. Sato, *Tetrahedron Lett.*, **21**, 365 (1980).
95. A. B. Goel and E. C. Ashby, *J. Organomet. Chem.*, **214**, C1 (1981).
96. A. Minato, K. Tamao, T. Hayashi, K. Suzuki, and M. Kumada, *Tetrahedron Lett.*, **22**, 5319 (1981).
97. T. Fujisawa, M. Takeuchi, and T. Sato, *Chem. Lett.*, 1795 (1982).
98. K. Komatsu, T. Moriyama, T. Nishiyama, and K. Okamoto, *Tetrahedron*, **37**, 721 (1981).
99. E. Akgun, M. B. Glinski, K. L. Dhawan, and T. Durst, *J. Org. Chem.*, **46**, 2730 (1981).
100. H. Hommes, H. D. Verkruijsse, and L. Brandsma, *Tetrahedron Lett.*, **22**, 2495 (1981).
101. M. A. Fox and N. J. Singletary, *J. Org. Chem.*, **47**, 3412 (1982).
102. H. Lehmkuhl, S. Kintopf, and K. Mehler, *J. Organomet. Chem.*, **46**, C1 (1972).
103. J. Villieras, *Organomet. Chem. Rev. A*, **7**, 81 (1971).
104. M. Elborai and M. Hassanein, *Org. Prep. Proced. Int.*, **14**, 409 (1982); M. Elborai, A. H. Moustata, M. Anwar, and F. I. A. Hay, *Pol. J. Chem.*, **55**, 1659 (1981).
105. N. Redwane, P. Moreau, and A. Commeyras, *J. Fluorine Chem.*, **20**, 6999 (1982).
106. R. Rossi, A. Carpita, and M. G. Quirici, *Gazz. Chim. Ital.*, **111**, 173 (1981); P. Babin, J. Dunogues, and F. Duboudin, *J. Heterocycl. Chem.*, **18**, 519 (1981); H. J. Bestmann, T. Brosche, K. H. Koschatzky, K. Michaelis, and H. Platz, *Tetrahedron Lett.*, **23**, 4007 (1982).
107. E. R. H. Jones, L. Skattebol, and M. C. Whiting, *J. Chem. Soc.*, 4765 (1956).
108. W. T. Colwell, C. Soohoo, and J. I. Degraw, *J. Labelled Compd. Radiopharm.*, **16**, 551 (1979).
109. S. F. Karaev, A. K. Aliev, M. E. Askerov, and D. S. Dzhatarov, *Zh. Org. Khim.*, **18**, 807 (1982);

J. G. Gleason, D. B. Bryan, and C. M. Kinzig, *Tetrahedron Lett.*, **21**, 1129 (1980); F.-J. Marner, W. Boland, and L. Jaenicke, *Justus Liebigs Ann. Chem.*, 579 (1982).
110. R. Rossi, A. Carpita, and M. L. Gaudenzi, *Synthesis*, 359 (1981); M. Bertrand, M. Gil, and A. Kumar, *Nouv. J. Chim.*, **4**, 69 (1980).
111. C. E. Bishop and G. W. Morrow, *J. Org. Chem.*, **48**, 657 (1983).
112. R. Filler, A. E. Fiebig, and M. Y. Pelister, *J. Org. Chem.*, **45**, 1290 (1980).
113. A. Benaboura, J.-Y. Sanchez, P. Nicolet, and M. J. M. Abadie, *Bull. Soc. Chim. Fr., Pt. 2*, 167 (1981); P. H. Mazzocchi and B. C. Stahly, *J. Med. Chem.*, **24**, 457 (1981).
114. E. A. Chernyshev, M. D. Reshetova, and I. A. Rodnikov, *Zh. Obshch. Khim.*, **50**, 1037 (1980).
115. P. Canonne, G. B. Foscolos, and G. Lemay, *Tetrahedron Lett.*, **21**, 155 (1980).
116. J. Coudane, *Synthesis*, 319 (1981).
117. G. Rihs, H. Fuhrer, and A. Marxer, *Helv. Chim. Acta*, **64**, 769 (1981).
118. R. J. Harper, Jr., E. J. Soloski, and C. Tamborski, *J. Org. Chem.*, **29**, 2385 (1964).
119. J. K. Whitesell and M. A. Whitesell, *Synthesis*, 517 (1983).
120. R. Annunziata, M. Cinquini, and A. Gilardi, *Synthesis*, 1016 (1983).
121. H. Asefi and J. G. Tillett, *J. Chem. Soc., Perkin Trans. 2*, 1579 (1979).
122. H. Kawa, F. Yamaguchi, and N. Ishikawa, *Chem. Lett.*, 153 (1982); C. Mioskowski and G. Solladie, *Tetrahedron*, **36**, 227 (1980).
123. G. E. Wilson, Jr., and A. Hess, *J. Org. Chem.*, **45**, 2766 (1980).
124. J. M. Decesare, B. Corbel, T. Durst, and F. J. Blount, *Can. J. Chem.*, **59**, 1415 (1981).
125. B. Costisella, I. Keitel, and H. Gross, *Tetrahedron*, **37**, 1227 (1981).
126. F. Huet, G. Emptoz, and A. Jubier, *Tetrahedron*, **29**, 479 (1973).
127. B. M. Trost, J. Cossy, and J. Burks, *J. Am. Chem. Soc.*, **105**, 1052 (1983).
128. H. Bönnemann, B. Bogdanovic, R. Brinkmann, D.-W. He, and B. Spliethoff, *Angew. Chem., Int. Ed. Engl.*, **22**, 728 (1983); H. Lehmkuhl, K. Mehler, R. Benn, A. Rufińska, G. Schroth, and C. Krüger, *Chem. Ber.*, **117**, 389 (1984).
129. J.-P. Pascault and J. Golé, *C. R. Acad. Sci.*, **264**, 115 (1967).
130. K. K. Balasubramanian and S. Selvaraj, *Tetrahedron Lett.*, **21**, 851 (1980).
131. G. Manuel, G. Bertrand, P. Mazerolles, and J. Ancelle, *J. Organomet. Chem.*, **212**, 311 (1981); G. Wickham and W. Kitching, *Organometallics*, **2**, 541 (1983).
132. L. Birkofer, M. Bockhorst, A. Steigel, and D. Eichstadt, *J. Organomet. Chem.*, **233**, 291 (1982).
133. G. Félix, M. Laguerre, J. Dunoguès, and R. Calas, *J. Org. Chem.*, **47**, 1423 (1982).
134. L. C. Costa and G. M. Whitesides, *J. Am. Chem. Soc.*, **99**, 2390 (1977).
135. B. Martel and M. Varache, *J. Organomet. Chem.*, **40**, C53 (1972).
136. A. Maercker, *J. Organomet. Chem.*, **18**, 249 (1969); A. Maercker and H.-J. Jaroschek, *J. Organomet. Chem.*, **108**, 145 (1976).
137. A. Maercker and H.-J. Jaroschek, *J. Organomet. Chem.*, **116**, 21 (1976).
138. W. T. Ford, *J. Organomet. Chem.*, **32**, 27 (1971); W. T. Ford and J. B. Grutzner, *J. Org. Chem.*, **37**, 2561 (1972).
139. F. Bertini, P. Grasselli, G. Zubiani, and G. Cainelli, *Tetrahedron*, **26**, 1281 (1970).
140. H. Koichi, Y. Iwano, and K. Fujimoto, *Heterocycles*, **17**, 201 (1982).
141. H. C. Holtkamp, G. Schat, C. Blomberg, and F. Bickelhaupt, *J. Organomet. Chem.*, **240**, 1 (1982); F. Bickelhaupt, *J. Organomet. Chem.*, **240**, 217 (1982).
142. F. J. M. Freijee, G. Schat, O. S. Akkerman, and F. Bickelhaupt, *J. Organomet. Chem.*, **240**, 217 (1982).
143. F. J. M. Freijee, G. van der Wal, G. Schat, O. S. Akkerman, and F. Bickelhaupt, *J. Organomet. Chem.*, **240**, 229 (1982).
144. H. Gilman, *J. Organomet. Chem.*, **100**, 83 (1975).
145. G. R. Newkome and J. M. Roper, *J. Organomet. Chem.*, **186**, 147 (1980).
146. J. W. F. L. Seetz, G. Schat, O. S. Akkerman, and F. Bickelhaupt, *J. Am. Chem. Soc.*, **104**, 6848 (1982); *Angew. Chem.*, **95**, 242 (1983).
147. K. Tamao, J. Yoshida, S. Okazaki, and M. Kamada, *Isr. J. Chem.*, **15**, 265 (1977).
148. L. I. Zakharkin, A. I. Kovredov, A. V. Kazantev, and M. G. Meiramov, *Zh. Obshch. Khim.*, **51**, 357 (1981).
149. B. Jousseaume, J. G. Duboudin, and M. Petraud, *J. Organomet. Chem.*, **238**, 171 (1982).
150. C. L. Raston and G. Salem, unpublished results.
151. F. Bickelhaupt, G. Jongsma, P. de Koe, R. Lourens, N. R. Mast, G. L. van Mourik, H. Vermeer,

and R. J. M. Weustink, *Tetrahedron*, **32**, 1921 (1976); W. Z. McCarthy, J. Y. Corey, and E. R. Corey, *Organometallics*, **3**, 255 (1984); P. Jutzi, *Chem. Ber.*, **104**, 1455 (1971).
152. M. A. Bennett and H. Neumann, *Aust. J. Chem.*, **33**, 1251 (1980).
153. K. Mukai and M. Inagaki, *Bull. Chem. Soc. Jpn.*, **53**, 2695 (1980).
154. L. S. Yang, T. A. Engler, and H. Shechter, *J. Chem. Soc., Chem. Commun.*, 866 (1983).
155. R. D. Rieke and S. E. Bales, *J. Chem. Soc., Chem. Commun.*, 879 (1973).
156. C. F. Blakely and G. J. Keech, *J. Chem. Soc. B*, 1435 (1970).
157. G. van Koten and J. G. Noltes, *J. Organomet. Chem.*, **104**, 127 (1976).
158. Y. V. Shklaev, Y. P. Dormidontov, and I. I. Lakpin, *Khim Geterotsikl. Soedin.*, 468 (1981).
159. M. T. Rahman, *J. Indian Chem. Soc.*, **58**, 21 (1981).
160. C. F. Smith, G. J. Moore, and C. Tamborski, *J. Organomet. Chem.*, **33**, C21 (1971).
161. H. Sakurai, Y. Nakadaira, A. Hosomi, Y. Eriyama, and C. Kabuto, *J. Am. Chem. Soc.*, **105**, 3359 (1983).
162. J.-G. Duboudin and B. Jousseaume, *Synth. Commun.*, **9**, 53 (1979).
163. T. Cuvigny and H. Normant, *J. Organomet. Chem.*, **1**, 120 (1963).
164. J. W. F. L. Seetz, R. Tol, O. S. Akkerman, and F. Bickelhaupt, *Synthesis*, 721 (1983).
165. L. B. Doshkevich and L. N. Kuzmenkov, *J. Gen. Chem. USSR* (*Engl. Transl.*) **29**, 2330 (1959).
166. B. K. Misra, Y. R. Rao, and S. N. Mahaputra, *Indian J. Chem.*, **19B**, 908, 1980.
167. M. Lasperas, A. Perez-Rubalcaba, and M. L. Quiroga-Feijoo, *Tetrahedron*, **36**, 3403 (1980).
168. M. Okubo, M. Yoshida, K. Horinouchi, H. Nishida, and Y. Fukuyama, *Bull. Chem. Soc. Jpn.*, **56**, 1196 (1983).
169. E. C. Ashby, J. Bowers, and R. Depriest, *Tetrahedron Lett.*, **21**, 3541 (1980).
170. C. Bernardon and A. Deberly, *J. Org. Chem.*, **47**, 463 (1982).
171. R. H. Mitchell, T. W. Dingle, and R. V. Williams, *J. Org. Chem.*, **48**, 903 (1983).
172. P. Canonne, G. B. Foscolos, and G. Lemay, *Tetrahedron Lett.*, 4383 (1979).
173. M. S. Kharasch and O. Reinmuth, *Grignard Reactions of Nonmetallic Substances*, Constable, London, 1954.
174. C. Cianetti, G. Di Maio, V. Pignatelli, P. Tagliatesta, E. Vecchi, and E. Zeuli, *Tetrahedron*, **39**, 657 (1983).
175. M. Chastrette and G. P. Axiotis, *J. Organomet. Chem.*, **206**, 139 (1981).
176. D. A. Holt, *Tetrahedron Lett.*, **22**, 2243 (1981).
177. J. Barluenga, M. Yus, J. M. Concellon, and P. Bernad, *J. Chem. Res.* (*M*), 677 (1980).
178. I. Tabushi, K. Seto, and Y. Kobuke, *Tetrahedron*, **37**, 863 (1981).
179. M. E. Jung and G. L. Hatfield, *Tetrahedron Lett.*, **24**, 2931 (1983).
180. J. M. Lamarche and B. Laude, *Bull. Soc. Chim. Fr., Pt. II*, 97 (1982).
181. E. C. Ashby and J. T. Laemmle, *Chem. Rev.*, **75**, 521 (1975).
182. T. D. Inch, G. L. Lewis, G. L. Sainsbury, and D. J. Sellers, *Tetrahedron Lett.*, 3657 (1969); N. Baggett and R. J. Simmonds, *J. Chem. Soc., Perkin Trans. 16*, 197 (1982).
183. A. Meyers and J. Slade, *J. Org. Chem.*, **45**, 2785 (1980).
184. L. Angiolini, P. Costa Bizzarri, G. Scapini, and M. Tramontini, *Tetrahedron*, **37**, 2137 (1981).
185. M. Asami and T. Mukaiyama, *Chem. Lett.*, 93 (1983); Y. Sakito and G. Suzukamo, *Tetrahedron Lett.*, **23**, 4953 (1982).
186. C. Fuganti, P. Grasselli, and Servi, *J. Chem. Soc., Perkin Trans. 1*, 241 (1983).
187. A. Rosenthal and S. N. Mikhailov, *Carbohydr. Res.*, **79**, 235 (1980); F. A. Carey and W. C. Frank, *J. Org. Chem.*, **47**, 3548 (1982).
188. J. C. Beloeil, M. Bertranne, M. Fetizon, and T. Prange, *J. Chem. Soc., Chem. Commun.*, 363 (1981).
189. J. M. Coxon, G. W. Simpson, P. J. Steel, and V. C. Trenerry, *Tetrahedron Lett.*, **24**, 1427 (1983).
190. Y. Yamamoto and K. Maruyama, *J. Org. Chem.*, **48**, 1564 (1983).
191. H. Berbalk, K. Eichinger, and R. Schuster, *Synthesis*, 613 (1981).
192. C. Alvarez-Ibarra, O. Arjona, R. Pérez-Ossorio, A. Pérez-Rubalcaba, M. L. Quiroga, and M. J. Santesmases, *J. Chem. Soc., Perkin Trans. 2*, 1645 (1983).
193. S. Lociuro, E. Vecchi, E. Zeuli, and M. Delfini, *Tetrahedron*, **38**, 3339 (1982).
194. F. Bohlmann and W. Mailahn, *Chem. Ber.*, **114**, 1091 (1981).
195. R. R. Sauers, M. Zampino, M. Stockl, J. Ferentz, and H. Shams, *J. Org. Chem.*, **48**, 1862 (1983).
196. D. S. C. Black, N. A. Blackman, and L. M. Johnstone, *Aust. J. Chem.*, **32**, 2025 (1979).
197. C. A. Demerson and L. G. Humber, *Can. J. Chem.*, **57**, 3296 (1979).

198. B. P. Mundy, G. W. Dirks, R. M. Larter, and A. C. Craig, *J. Org. Chem.*, **46**, 4005 (1981).
199. L. Jalander, *Acta Chem. Scand., Ser. B*, **35**, 419 (1981).
200. J. H. Rigby, *Tetrahedron Lett.*, **23**, 1863 (1982).
201. M. C. Lasne, J. L. Ripoll, and J. M. Denis, *Tetrahedron*, **37**, 503 (1981).
202. R. Sjoholm and V. Stahlstrom, *Finn. Chem. Lett.*, 46 (1982).
203. J. P. Gesson, J. C. Jacquesy, and B. Renoux, *Tetrahedron Lett.*, **24**, 2757 (1983).
204. M. C. Pirrung, *J. Am. Chem. Soc.*, **101**, 7130 (1979); S. A. Godleski and R. S. Valpey, *J. Org. Chem.*, **47**, 381 (1982).
205. K. E. Stevens and L. A. Paquette, *Tetrahedron Lett.*, **22**, 4393 (1981).
206. T. Imai and S. Nishida, *Chem. Lett.*, 41 (1980).
207. G. H. Posner, J. P. Mallamo, M. Hulce, and L. L. Frye, *J. Am. Chem. Soc.*, **104**, 4180 (1982).
208. M. A. Tius, A. Thurkauf, and J. W. Truesdell, *Tetrahedron Lett.*, **23**, 2819 (1982); M. A. Tius and S. Ali, *J. Org. Chem.*, **47**, 3163 (1982).
209. H. Yamaoka, I. Mishima, and T. Hanafusa, *Chem. Lett.*, 411 (1980).
210. D. Liotta, M. Saindane, and C. Barnum, *J. Org. Chem.*, **46**, 3369 (1981).
211. J. S. Swenton, *Acc. Chem. Res.*, **16**, 74 (1983).
212. W. Ried and M. Vogl, *Chem. Ber.*, **115**, 791 (1982).
213. E. V. Dehmlow and H. G. Schell, *Chem. Ber.*, **113**, 1 (1980).
214. M. K. Eberle and G. G. Kahle, *Tetrahedron Lett.*, **21**, 2303 (1980); T. Sato, T. Kawara, K. Sakata, and T. Fujisawa, *Bull. Chem. Soc. Jpn.*, **54**, 505 (1981).
215. J.-E. Dubois, I. Saumtally, and C. Lion, *Bull. Soc. Chim. Fr., Pt. 2*, 318 (1982).
216. J. Barluenga, J. Florez, and M. Yus, *Synthesis*, 378 (1983).
217. M. Okubo, Y. Morigami, and R. Suenaga, *Bull. Chem. Soc. Jpn.*, **53**, 3029 (1980).
218. S. D. Boyce, A. C. Barefoot, and J. F. Hornig *J. Labelled Compd. Radiopharm.*, **20**, 243 (1983).
219. P. Canonne, G. Lemay, and D. Bélanger, *Tetrahedron Lett.*, **21**, 4167 (1980) and *J. Org. Chem.*, **47**, 3953 (1982).
220. P. Canonne, D. Bélanger, G. Lemay, and G. B. Foscolos, *J. Org. Chem.*, **46**, 3091 (1981); P. Canonne, H. Caron, and M. Akssira, *Synth. Commun.*, **12**, 439 (1982); P. Canonne and D. Bélanger, *J. Chem. Soc., Chem. Commun.*, 125 (1980).
221. P. Canonne, D. Bélanger and G. Lemay, *Tetrahedron Lett.*, **22**, 4995 (1981).
222. G. L. Larson and D. Hernández, *Tetrahedron Lett.*, **23**, 1035 (1982).
223. L. P. Shadrina, Y. P. Dormidontov, and A. N. Nedugov, *Zh. Org. Khim.*, **17**, 1609 (1981).
224. I. G. Vasi and R. H. Acharya, *Indian J. Chem.*, **226**, 67 (1983); K. Tanaka, F. Uchiyama, T. Ikeda, and Y. Inubushi, *Chem. Pharm. Bull.*, **31**, 1958 (1983).
225. P. Canonne, D. Bélanger, and G. Lemay, *Heterocycles*, **15**, 455 (1981); L. M. Tolbert and D. P. Martone, *J. Org. Chem.*, **48**, 1185 (1983).
226. M. F. Ismail, N. A. Shams, M. R. Salem, and S. A. Emara, *J. Prakt. Chem.*, **325**, 417 (1983).
227. M. Ochiai, F. Fujita, M. Arimoto, and H. Yamaguchi, *J. Chem. Soc., Chem. Commun.*, 1108 (1982).
228. W. H. Pirkle and P. E. Adams, *J. Org. Chem.*, **45**, 4117 (1980).
229. P. Canonne, H. Caron and G. Foscolos, *Tetrahedron*, **40**, 865 (1984).
230. B. Rague, Y. Chapleur, and B. Castro, *J. Chem. Soc., Perkin Trans. 1*, 2063 (1982).
231. D. L. Comins and W. Dernell, *Tetrahedron Lett.*, **22**, 1085 (1981).
232. F. Barbot, A. Kadib-Elban, and P. Miginiac, *Tetrahedron Lett.*, **24**, 5089 (1983).
233. K. Maruyama and Y. Arita, *Bull. Chem. Soc. Jpn.*, **56**, 1977 (1983).
234. J. K. Whitesell, D. Deyo, and A. Bhattacharya, *J. Chem. Soc., Chem. Commun.*, 802 (1983).
235. F. Sato, K. Oguro, H. Watanabe, and M. Sato, *Tetrahedron Lett.*, **21**, 2869 (1980).
236. L. Lepage and Y. Lepage, *Synthesis*, 1018 (1983).
237. V. B. Avdeer, I. S. Berdinskii, Z. D. Belykh, and G. F. Rozenblat, *Zh. Org. Khim.*, **19**, 1184 (1983).
238. A. K. Fateen, M. F. Ismail, A. M. Kaddah, and N. A. Shams, *J. Prakt. Chem.*, **322**, 617 (1980).
239. P. Canonne, G. Fytas, and D. Thibeault, *Tetrahedron Lett.*, **24**, 2991 (1983).
240. J. S. Nimitz and H. S. Mosher, *J. Org. Chem.*, **46**, 211 (1981).
241. M. J. Wanner, G. J. Koomen, and U. K. Pandit, *Tetrahedron*, **39**, 3673 (1983).
242. M. Nakajima and J.-P. Anselme, *J. Org. Chem.*, **48**, 2492 (1983).
243. P. Stanetty, *J. Chem. Res. (M)*, 1043 (1981); G. Hund and W. Schneider, *Chem. Ber.*, **113**, 401 (1980).
244. D. Paquer, *Bull. Soc. Chim. Fr.*, 1439 (1975); P. Metzner, in *Organic Compounds of Sulphur,*

Selenium and Tellurium, Chemical Society Specialist Periodical Report, **4**, 131 (1977); **5**, 125 (1978).
245. J. Ramachandran, S. R. Ramadas, and C. N. Pillaic, *Org. Prep. Proced. Int.*, **13**, 71 (1981).
246. R. S. Sukhai and L. Brandsma, *Synthesis*, 888 (1977).
247. R. Kaya and N. R. Beller, *Synthesis*, 814 (1981).
248. Y. Tamura, M. Adachi, T. Kawasaki, H. Yasuda, and Y. Kita, *J. Chem. Soc., Perkin Trans. 1*, 1132 (1980).
249. N. Thoai, M. Rubinstein, and C. Wakselman, *J. Fluorine Chem.*, **21**, 437 (1982).
250. Y. Tamaru, T. Harada, S. Nishi, and Z. Yoshida, *Tetrahedron Lett.*, **23**, 2383 (1982).
251. J. M. van der Leij, H. J. M. Strijtveen, and B. Swanenburg, *Recl. Trav. Chim. Pays-Bus.*, **99**, 45 (1980).
252. A. Essawy, *Rev. Roum. Chim.*, **27**, 415 (1982).
253. S. Masson and A. Thuillier, *Tetrahedron Lett.*, **23**, 4087 (1982).
254. R. Neidlein and H. Zeiner, *Angew. Chem.*, **92**, 199 (1980).
255. K. Komatsu, K. Masumoto, Y. Waki, and K. Okamoto, *Bull. Chem. Soc. Jpn.*, **55**, 2470 (1982).
256. H. G. Richey, Jr., and M. S. Domalski, *J. Org. Chem.*, **46**, 3780 (1981).
257. H. G. Richey, Jr., L. M. Moses, M. S. Domalski, W. F. Erickson, and A. S. Heyn, *J. Org. Chem.*, **46**, 3773 (1981).
258. R. W. Saalfrank, W. Paul, and H. Liebenow, *Angew. Chem.*, **92**, 740 (1980).
259. G. Bartoli, M. Bosco, and A. C. Boicelli, *Synthesis*, 570 (1981); N. Armillotta, G. Bartoli, M. Bosco, and R. Dal Pozza, *Synthesis*, 836 (1982); G. Bartoli, M. Bosco, R. Dal Pozzo, and F. Ciminale, *J. Org. Chem.*, **47**, 5227 (1982).
260. Y. Okamoto, K. Suzuki, T. Kityama, H. Yuki, H. Kageyama, K. Miki, N. Tanaka, and N. Kasai, *J. Am. Chem. Soc.*, **104**, 4618 (1982).
261. D. J. Hart, K. Kanai, D. G. Thomas, and T. K. Yang, *J. Org. Chem.*, **48**, 289 (1983); D. J. Hart and K. Kanai, *J. Org. Chem.*, **47**, 1555 (1982).
262. H. Takahashi, K. Tomita, and H. Noguchi, *Chem. Pharm. Bull.*, **29**, 3387 (1981); H. Takashi and H. Inagaki, *Chem. Pharm. Bull.*, **30**, 922 (1982).
263. H. Kogen, K. Tomioka, S. Hashimoto, and K. Koga, *Tetrahedron*, **37**, 3951 (1981).
264. Y. Ishida, S. Sasatani, K. Maruoka, and H. Yamamoto, *Tetrahedron Lett.*, **24**, 3255 (1983); K. Maruoka, T. Miyazaki, M. Ando, Y. Matsumura, S. Sakane, K. Hattori, and H. Yamamoto, *J. Am. Chem. Soc.*, **105**, 2831 (1983).
265. K. S. Ng and H. Alper, *J. Org. Chem.*, **46**, 1039 (1981); *J. Organomet. Chem.*, **202**, 1 (1980).
266. J. M. Poirier, *Bull. Soc. Chim. Fr., Pt. II*, 17 (1982).
267. A. Laurent, A. Marsura, and J. L. Pierre, *J. Heterocycl. Chem.*, **17**, 1009 (1980).
268. A. Zwierzak and E. Ślusarska, *Synthesis*, 691 (1979).
269. D. Döpp, L. Greci, and A. M. Nour-el-Din, *Chem. Ber.*, **116**, 2049 (1983).
270. M. Colonna, L. Greci, and M. Poloni, *J. Heterocycl. Chem.*, **17**, 293 (1980).
271. M. Colonna, L. Greci, and M. Poloni, *J. Heterocycl. Chem.*, **17**, 1473 (1980).
272. E. Giraudi and P. Teisseire, *Tetrahedron Lett.*, **24**, 489 (1983).
273. M. R. H. Elmoghayar, P. Groth, and K. Undheim, *Acta Chem. Scand., Ser. B*, **37**, 109 (1983).
274. T. Iwakumo, K. Yamada, H. Itoh, and S. Sugasamo, *Heterocycles*, **15**, 1115 (1981).
275. D. Braun and R. Studenroth, *Justus Liebigs Ann. Chem.*, 999 (1981).
276. K. Akiba, Y. Iseki and M. Wada, *Tetrahedron Lett.*, **23**, 3935 (1982).
277. R. Yamaguchi, Y. Nakazono, and M. Kawanisi, *Tetrahedron Lett.*, **24**, 1801 (1983).
278. D. L. Comins and A. H. Abdullah, *J. Org. Chem.*, **47**, 4315 (1982).
279. A. R. Katritzky, H. Beltrami, and M. P. Sammes, *J. Chem. Res* (*M*), 1684 (1981).
280. M. Chastrette and G. P. Axiotis, *Synthesis*, 889 (1980).
281. J. Rabesiaka, *Bull. Soc. Chim. Fr.*, 1440 (1970); A. A. Scala, N. M. Bikales, and E. I. Becker, *J. Org. Chem.*, **30**, 303 (1965).
282. D. J. Berry and B. J. Wakefield, *J. Organomet. Chem.*, **23**, 1 (1970); N. J. Foulger and B. J. Wakefield, *Tetrahedron Lett.*, 4169 (1972).
283. R. Gauthier and M. Chastrette, *J. Organomet. Chem.*, **165**, 139 (1979).
284. T. Yamamori and I. Adachi, *Tetrahedron Lett.*, **21**, 1747 (1980). T. Yamamori, Y. Hiramatsu, and I. Adachi, *J. Heterocycl. Chem.*, **18**, 347 (1981).
285. G. E. Niznik, W. H. Morrison, III, and H. M. Walborsky, *Org. Synth.*, **51**, 31 (1971).
286. D. J. Hart and T. K. Yang, *Tetrahedron Lett.*, **23**, 2761 (1982).

287. A. Gossauer, F. Roessler, H. Zilch, and L. Ernst, *Justus Liebigs Ann. Chem.*, 1309 (1979).
288. N. Thoai, M. Rubinstein, and C. Wakselman, *J. Fluorine Chem.*, **20**, 271 (1982).
289. A. V. Koblik, K. F. Suzdalev, G. N. Dorefeenko, and A. A. Loktionov, *Zh. Org. Khim.*, **17**, 880 (1981).
290. A. J. Ashe, III, *Acc. Chem. Res.*, **11**, 153 (1978); G. Märkl, C. Martin, and W. Weber, *Tetrahedron Lett.*, **22**, 1207 (1981).
291. S. Mori, I. Sakai, T. Aoyama, and T. Shioiri, *Chem. Pharm. Bull.*, **30**, 3380 (1982).
292. D. H. Sieh, D. J. Wilbur, and C. J. Michejda, *J. Am. Chem. Soc.*, **102**, 3883 (1980).
293. B. M. Trost and W. H. Pearson, *J. Am. Chem. Soc.*, **105**, 1054 (1983).
294. T. Severin and H. Pehr, *Chem. Ber.*, **112**, 3559 (1979).
295. M. E. Garst and D. Lukton, *Synth. Commun.*, **10**, 155 (1980).
296. C. Schenk, M. L. Beekes, J. A. M. van der Drift, and T. J. de Boer, *Recl. Trav. Chim. Pays-Bas*, **99**, 278 (1980).
297. C. Schenk and T. J. de Boer, *Tetrahedron*, **36**, 1843 (1980); J. Lub, M. L. Beekes, and T. J. de Boer, *J. Chem. Soc., Perkin Trans. 2*, 721 (1983).
298. C. Schenk and T. J. de Boer, *Tetrahedron*, **35**, 2119, (1979).
299. H. Shosenji, A. Nagayoshi, T. Takemoto, and K. Yamada, *J. Chem. Soc., Chem. Commun.*, 770 (1979).
300. K. Hiroi, S. Sato and R. Kitayama, *Chem. Lett.*, 1595 (1980).
301. A. Arase and Y. Masuda, *Chem. Lett.*, 1331 (1975).
302. Y. Ishii, Y. Iwama, and M. Ogawa, *Synth. Commun.*, **8**, 93 (1978).
303. H. Suginome, N. Yonekura, and T. Masamune, *Bull. Chem. Soc. Jpn.*, **53**, 210 (1980).
304. A. Haas and J. Koehler, *J. Fluorine Chem.*, **17**, 531 (1981).
305. H. J. Bestmann, K.-H. Koschatzky, A. Plenchette, J. Suss, and O. Vostrowsky, *Justus Liebigs Ann. Chem.*, 536 (1982); J. P. Quintard, B. Elissondo, and M. Pereyre, *J. Org. Chem.*, **48**, 1559 (1983); R. W. Hoffmann and B. Landmann, *Tetrahedron Lett.*, **24**, 3209 (1983).
306. N. Ishikawa, S. Butler, and M. Maruta, *Bull. Chem. Soc. Jpn.*, **54**, 3084 (1981).
307. J. P. Gillet, R. Sauvetre, and J. F. Normant, *Synthesis*, 297 (1982).
308. Y. Kajihara, K. Ishikawa, H. Yasuda, and A. Nakamura, *Bull. Chem. Soc. Jpn.*, **53**, 3035 (1980).
309. M. V. Bhatt and S. U. Kulkarni, *Synthesis*, 249 (1983).
310. A. Ohta, A. Imazeki, Y. Itoigawa, H. Yamada, and C. Suga, *J. Heterocycl. Chem.*, **20**, 311 (1983).
311. H. W. Gschwend and A. Hamdan, *J. Org.*, **47**, 3652 (1982).
312. J. Novak and C. A. Salemink, *Tetrahedron Lett.*, **24**, 101 (1983).
313. A. Alberola, A. G. Ortega, R. Pedrosa, and J. L. P. Bragado, *J. Heterocycl. Chem.*, **20**, 715 (1983).
314. L. Ponini, *Bull. Soc. Chim. Belg.*, **92**, 215 (1983); F. Barbot and P. Miginiac, *J. Organomet. Chem.*, **222**, 1 (1981).
315. G. Rousseau and N. Slougui, *Tetrahedron Lett.*, **24**, 1215 (1983).
316. M. L. Mancini and J. F. Honek, *Tetrahedron Lett.*, **23**, 3249 (1982).
317. H. Takahashi, Y. Suzuki, and T. Kametani, *Heterocycles*, **20**, 607 (1983); G. Courtois and P. Miginiac, *Bull. Soc. Chim. Fr., Pt. II*, 148 (1983); H. Kapnang and G. Charles, *Tetrahedron Lett.*, **24**, 1597 (1983).
318. H. Walther, L. Haase, H. Gross, B. Costisella, and I. Keitel, *J. Prakt. Chem.*, **322**, 902 (1980).
319. T. Asano, S. Yokota, and O. Mitsunobu, *Chem. Lett.*, 343 (1983).
320. H. Nishiyama, H. Yokoyama, S. Narimatsu, and K. Itoh, *Tetrahedron Lett.*, **23**, 1267 (1982); S. G. Davies and S. Wollowitz, *Tetrahedron Lett.*, **21**, 4175 (1980); S. C. Welch, A. S. C. P. Rao, J. T. Lyon, and J. M. Assercq, *J. Am. Chem. Soc.*, **105**, 252 (1983).
321. C. Huynh, F. Derguini-Boumechal, and G. Linstrumelle, *Tetrahedron Lett.*, 1503 (1979).
322. M. Kawana, *Chem. Lett.*, 1541 (1981).
323. J. R. Schauder and A. Krief, *Tetrahedron Lett.*, **23**, 4389 (1982); P. F. Hudrlik, A. M. Hudrlik, R. N. Misra, D. Peterson, G. P. Withers, and A. K. Kulkarni, *J. Org. Chem.*, **45**, 4444 (1980).
324. G. Linstrumelle, R. Lorne, and H. P. Dang, *Tetrahedron Lett.*, 4069 (1978).
325. X. Creary, *J. Org. Chem.*, **45**, 2727 (1980).
326. W. Hanefeld and D. Kluck, *Synthesis*, 229 (1981).
327. K. D. Berlin and M. E. Peterson, *J. Org. Chem.*, **32**, 125 (1967).
328. S. Araki, T. Sato, and Y. Butsugan, *J. Chem. Soc., Chem. Commun.*, 285 (1982).
329. T. Hayashi, T. Fujiwa, Y. Okamoto, Y. Katsuro, and M. Kumada, *Synthesis*, 1001 (1981).
330. A. Claesson, A. Quader, and C. Sahlberg, *Tetrahedron Lett.*, **24**, 1297 (1983).

331. F. Coulomb-Delbecq, J. Gore, and M.-L. Roumestant, *Bull. Soc. Chim. Fr.*, 533 (1976).
332. A. Holm and E. Huge-Jensen, *Acta Chem. Scand., Ser. B*, **28**, 705 (1974).
333. M. Kawana, H. Kuzuhara, and S. Emoto, *Bull. Chem. Soc. Jpn.*, **54**, 1492 (1981); M. Kawana, T. Koresawa, and H. Kuzuhara, *Bull. Chem. Soc. Jpn.*, **56**, 1095 (1983).
334. L. Birkofer and D. Wundram, *Chem. Ber.*, **115**, 1132 (1982).
335. J. G. Buchanan, K. A. Maclean, H. Paulsen, and R. H. Wightman, *J. Chem. Soc., Chem. Commun.*, 486 (1983).
336. H. Boshagen, W. Geiger, and B. Junge, *Angew. Chem.*, **93**, 800 (1981).
337. W. Kantlehner, T. Maier, and J. J. Kapassakalidis, *Synthesis*, 380 (1981).
338. K. Steinbeck, T. Schenke, and J. Runsink, *Chem. Ber.*, **114**, 1836 (1981).
339. H. Okamura, M. Miura, and H. Taki, *Tetrahedron Lett.*, **43**, (1979); Y. Ikeda, K. Furuta, N. Meguriya, N. Ikeda, and H. Yamamoto, *J. Am. Chem. Soc.*, **104**, 7663 (1982); H. Takei, H. Sugimura, M. Miura, and H. Okamura, *Chem. Lett.*, 1209 (1980).
340. J. Drabowicz, B. Bujnicki, and M. Mikolajczyk, *J. Org. Chem.*, **47**, 3325 (1982); J.-G. Duboudin, B. Jousseaume, and E. Thoumazeau, *Bull. Soc. Chim. Fr., Pt. II*, 105 (1983).
341. S. Uemura and S. Fukuzawa, *Tetrahedron Lett.*, **23**, 1181 (1982).
342. K. Ogura, K. Arai, and G. Tsuchihashi, *Bull. Chem. Soc. Jpn.*, **55**, 3669 (1982); D. O. Spry, *Tetrahedron Lett.*, **21**, 1293 (1980).
343. R. Oda and K. Yamamoto, *J. Org. Chem.*, **26**, 4679 (1961).
344. M. Tiecco, L. Testaferri, M. Tingoli, and E. Wenkert, *Tetrahedron*, **39**, 2289 (1983); J. L. Fabre, M. Julia, and J. N. Verpeaux, *Tetrahedron Lett.*, **23**, 2469 (1982).
345. S. Groowitz and R. Hakansson, *Ark. Kemi*, **17**, 33 (1961).
346. S. Mataka, K. Takahashi, H. Yamamoto, and M. Tashiro, *J. Chem. Soc., Perkin Trans. 1*, 2417 (1980).
347. M. Segi, Y. Nakamura, T. Nakajima, and S. Suga, *Chem. Lett.*, 913 (1983); C. R. Hall and T. D. Inch, *Pol. J. Chem.*, **54**, 1489 (1980).
348. T. Koizumi, R. Yanada, H. Takagi, H. Hirai, and E. Yoshii, *Tetrahedron Lett.*, **22**, 571 (1981).
349. C. R. Hall and N. E. Williams, *J. Chem. Soc., Perkin Trans. 1*, 2746 (1981).
350. C. R. Hall and N. E. Williams, *Tetrahedron Lett.*, **21**, 4959 (1980).
351. M. Moriyama and W. G. Bentrude, *Tetrahedron Lett.*, **23**, 4547 (1982).
352. R. Pratap, R. Castle, and M. L. Lee, *J. Heterocycl. Chem.*, **19**, 439 (1982).
353. F. Wudl and D. Nalewajek, *J. Organomet. Chem.*, **217**, 329 (1981); M. Evers and L. Christiaens, *Chem. Scr.*, **18**, 143 (1981).
354. M. R. Detty, B. J. Murray, D. L. Smith, and N. Zumbulyadis, *J. Am. Chem. Soc.*, **105**, 875 (1983).
355. R. Bussas, H. Munsterer, and G. Kresze, *J. Org. Chem.*, **48**, 2828 (1983).
356. J. F. Normant, *J. Organomet. Chem. Libr.*, **1**, 219 (1976).
357. J. F. Normant, *Pure Appl. Chem.*, **50**, 709 (1978).
358. J. F. Normant and A. Alexakis, *Synthesis*, 841 (1981), and references cited therein.
359. H. O. House, C. Y. Chu, J. M. Wilkins, and M. J. Umen, *J. Org. Chem.*, **40**, 1460 (1975).
360. A. Marfat, P. R. McGuirk, R. Kramer, and P. Helquist, *J. Am. Chem. Soc.*, **99**, 253 (1977).
361. E. C. Ashby and A. B. Goel, *J. Org. Chem.*, **48**, 2125 (1983).
362. H. Westmijze, H. Kleijn, and P. Vermeer, *J. Organomet. Chem.*, **172**, 377 (1979).
363. A. Alexakis, J. F. Normant, and J. Villieras, *J. Organomet. Chem.*, **96**, 471 (1975).
364. M. Gardette, A. Alexakis, and J. F. Normant, *Tetrahedron Lett.*, **23**, 5155 (1982).
365. A. Marfat, P. R. McGuirk, and P. Helquist, *Tetrahedron Lett.*, 1363 (1978).
366. W. Carruthers, *Comprehensive Organometallic Chemistry*, Vol. 7, Pergamon Press, Oxford, 1982, Chapter 49, pp. 716–722.
367. M. Obayashi, K. Utimoto, and H. Nozaki, *Tetrahedron Lett.*, 1805 (1977).
368. A. Alexakis, G. Cahiez, J. F. Normant, and J. Villieras, *Bull. Soc. Chim. Fr.*, 693 (1977).
369. J. K. Crandall and F. Collonges, *J. Org. Chem.*, **41**, 4089 (1976).
370. H. Westmijze, H. Kleijn, and P. Vermeer, *Synthesis*, 454 (1978).
371. H. Westmijze, H. Kleijn, and P. Vermeer, *Tetrahedron Lett.*, 3327 (1979).
372. H. Kleijn, H. Westmijze, J. Meijer, and P. Vermeer, *J. Organomet. Chem.*, **206**, 257 (1981).
373. G. Tadema, R. H. Everhardus, H. Westmijze, and P. Vermeer, *Tetrahedron Lett.*, 3935 (1978).
374. H. Westmijze and P. Vermeer, *Tetrahedron Lett.*, 4101 (1979).
375. A. Doutheau, G. Balme, M. Malacria, and J. Gore, *Tetrahedron*, **36**, 1953 (1980).
376. H. Kleijn, H. Westmijze, J. Meijer, and P. Vermeer, *Recl. Trav. Chim. Pays-Bas*, **99**, 340 (1980).

377. M. Montury, B. Psaume, and J. Gore, *Tetrahedron Lett.*, **21**, 163 (1980).
378. H. Kleijn, C. J. Elsevier, H. Westmijze, J. Meijer, and P. Vermeer, *Tetrahedron Lett.*, 3101 (1979).
379. R. Baudouy, F. Delbecq, and J. Gore, *Tetrahedron*, **36**, 189 (1980).
380. G. Balme, M. Malacria, and J. Gore, *J. Chem. Res. (M)*, 2869 (1981).
381. A. C. Oehlschlager and E. Czyzewska, *Tetrahedron Lett.*, **24**, 5587 (1983).
382. G. H. Posner, *Org. React.*, **19**, 1 (1972).
383. G. H. Posner, *Org. React.*, **22**, 253 (1975).
384. K. J. Shea and P. Q. Pham, *Tetrahedron Lett.*, **24**, 1003 (1983).
385. G. Stork and E. W. Logusch, *Tetrahedron Lett.*, 3361 (1979).
386. N. E. Burlinson, M. E. Sitzman, L. A. Kaplan, and E. Kayser, *J. Org. Chem.*, **44**, 3695 (1979).
387. P. O. Vigo, S. S. Bari, K. Singh, and D. M. Dua, *Indian J. Chem., Sect. B*, **20**, 619 (1981).
388. C. R. Johnson and D. Dhanoa, *J. Chem. Soc., Chem. Commun.*, 358 (1982).
389. S. Danishefsky, K. Vaughan, R. C. Gadwood, and K. Tsuzuki, *J. Am. Chem. Soc.*, **102**, 4262 (1980).
390. N. T. Luong-Thi and H. Riviere, *Tetrahedron Lett.*, 4657 (1979).
391. F. Leyendecker, F. Jesser, and D. Laucher, *Tetrahedron Lett.*, **24**, 3513 (1983).
392. F. Leyendecker and F. Jesser, *Tetrahedron Lett.*, **21**, 1311 (1980).
393. M. Bourgain-Commercon, J.-P. Foulon, and J. F. Normant, *J. Organomet. Chem.*, **228**, 321 (1982).
394. A. Marfat, P. R. McGuirk, and P. Helquist, *J. Org. Chem.*, **44**, 3888 (1979).
395. D. Cabaret and Z. Welvart, *J. Organomet. Chem.*, **177**, 75 (1979).
396. J. E. Arrowsmith, C. W. Greengrass, and M. J. Newman, *Tetrahedron*, **39**, 2469 (1983).
397. J. F. Normant, G. Cahiez, and J. Villieras, *J. Organomet. Chem.*, **92**, C28 (1975).
398. D. R. White, C. J. Maring, and G. A. Cain, *J. Antibiot.*, **36**, 339 (1983).
399. H. Westmijze, H. Kleijn, and P. Vermeer, *J. Organomet. Chem.*, **234**, 117 (1982).
400. J. M. Oostveen, H. Westmijze, and P. Vermeer, *J. Org. Chem.*, **45**, 1158 (1980).
401. V. Calo, L. Lopez, G. Marchese, and G. Pesce, *Synthesis*, 885 (1979).
402. V. Calo, L. Lopez, and G. Pesce, *J. Organomet. Chem.*, **231**, 179 (1982).
403. V. Calo, L. Lopez, G. Pesce, and A. Calianno, *J. Org. Chem.*, **47**, 4482 (1982).
404. T. Fujisawa, T. Sato, T. Kawara, M. Kawashima, H. Shimizu, and Y. Ito, *Tetrahedron Lett.*, **21**, 2181 (1980), and references cited therein.
405. T. Sato, T. Kawara, A. Nishizawa, and T. Fujisawa, *Tetrahedon Lett.*, **21**, 3377 (1980).
406. T. Fujisawa, T. Sato, T. Kawara, and K. Naruse, *Chem. Lett.*, 1123 (1980).
407. T. Sato, M. Takeuchi, T. Itoh, M. Kawashima, and T. Fujisawa, *Tetrahedron Lett.*, **22**, 1817 (1981).
408. T. Fujisawa, T. Sato, M. Kawashima, K. Naruse, and K. Tamai, *Tetrahedron Lett.*, **23**, 3583 (1982).
409. T. Fujisawa, M. Takeuchi, and T. Sato, *Chem. Lett.*, 1521 (1982).
410. R. Baker, D. C. Billington, and N. Ekanayake, *J. Chem. Soc., Perkin Trans. 1*, 1387 (1983).
411. J. F. Normant, A. Alexakis, and G. Cahiez, *Tetrahedron Lett.*, **21**, 935 (1980).
412. T. Fujisawa, T. Sato, T. Kawara, A. Noda, and T. Obinata, *Tetrahedron Lett.*, **21**, 2553 (1980).
413. N. Arumugam and A. Kumaraswamy, *Synthesis*, 367 (1981).
414. K. Mashima, H. Yasuda, K. Asami, and A. Nakamura, *Chem. Lett.*, 219 (1983).
415. J. E. Dubois, M. Boussu, and C. Lion, *Tetrahedron Lett.*, 829 (1971).
416. L. Cama and B. G. Christensen, *Tetrahedron Lett.*, **21**, 2013 (1980).
417. L. S. Hegedus, P. M. Kendall, S. M. Lo, and J. R. Sheats, *J. Am. Chem. Soc.*, **97**, 5448 (1975).
418. I. Tabushi, K. Seto, and Y. Kobuke, *Tetrahedron Lett.*, **37**, 863 (1981).
419. D. E. Bergbreiter and G. M. Whitesides, *J. Org. Chem.*, **40**, 779 (1975).
420. D. W. Knight and A. P. Knott, *J. Chem. Soc., Perkin Trans. 1*, 623 (1982).
421. M. Obayashi, K. Utimoto, and H. Nozaki, *J. Organomet. Chem.*, **177**, 145 (1979).
422. J. F. Normant, G. Cahiez, M. Bourgain, C. Chuit, and J. Villieras, *Bull. Soc. Chim. Fr.*, 1656 (1974).
423. J. F. Normant, G. Cahiez, C. Chuit, and J. Villieras, *J. Organomet. Chem.*, **77**, 269 (1974).
424. N. Jabri, A. Alexakis, and J. F. Normant, *Tetrahedron Lett.*, **23**, 1589 (1982).
425. H. Westmijze and P. Vermeer, *Synthesis*, 784 (1977).
426. H. Westmijze, J. Meijer, and P. Vermeer, *Recl. Trav. Chim. Pays-Bas*, **96**, 168 (1977).
427. C. Germon, A. Alexakis, and J. F. Normant, *Tetrahedron Lett.*, **21**, 3763 (1980).

428. G. Friour, G. Cahiez, A. Alexakis, and J. F. Normant, *Bull. Soc. Chim. Fr., Pt. 2*, 515 (1979).
429. M. Yamashita, K. Miyoshi, Y. Nakazono, and R. Suemitsu, *Bull. Chem. Soc. Jpn.*, **55**, 1663 (1982).
430. L. F. Elsom, J. D. Hunt, and A. Mckillop, *Organomet. Chem. Rev. (A)*, **8**, 135 (1972).
431. E. Erdik, *Tetrahedron*, **40**, 641 (1984).
432. B. Jousseaume and J.-G. Duboudin, *J. Organomet. Chem.*, **91**, C1 (1975).
433. B. B. Snider, M. Karass, and R. S. E. Conn, *J. Am. Chem. Soc.*, **100**, 4624 (1978).
434. B. B. Snider and M. Karras, *J. Org. Chem.*, **47**, 4588 (1982).
435. Y. Ishino, I. Nishiguchi, F. Takihira, and T. Hirashima, *Tetrahedron Lett.*, **21**, 1527 (1980).
436. L. Brandsma, J. Meijer, H. D. Verkruijsse, G. Bokkers, A. J. M. Duisenberg, and J. Kroon, *J. Chem. Soc., Chem. Commun.*, 922 (1980).
437. J. L. Moreau and M. Gaudemar, *J. Organomet. Chem.*, **108**, 159 (1976).
438. T. Sato, M. Kawashima, and T. Fujisawa, *Tetrahedron Lett.*, **22**, 2375 (1981).
439. D. J. Pasto, G. F. Hennion, R. H. Shults, A. Waterhouse, and S.-K. Chou, *J. Org. Chem.*, **41**, 3496 (1976).
440. T. Jeffery-Luong and G. Linstrumelle, *Tetrahedron Lett.*, **21**, 5019 (1980).
441. E. L. Eliel and S. Chandrasekaran, *J. Org. Chem.*, **47**, 4783 (1982).
442. S. A. Bal, A. Marfat, and P. Helquist, *J. Org. Chem.*, **47**, 5045 (1982).
443. L. A. Paquette and Y.-K. Han, *J. Am. Chem. Soc.*, **103**, 1835 (1981).
444. A. Leone-Bay and L. A. Paquette, *J. Org. Chem.*, **47**, 4173 (1982).
445. B. M. Trost and B. P. Coppola, *J. Am. Chem. Soc.*, **104**, 6879 (1982).
446. I. W. Lawston and T. D. Inch, *J. Chem. Soc., Perkin Trans. 1*, 2629 (1983).
447. H. Felkin and G. Swierczewski, *Tetrahedron*, **31**, 2735 (1975), and references cited therein.
448. A. Commercon, M. Bourgain, M. Delaumeny, J. F. Normant, and J. Villieras, *Tetrahedron Lett.*, 3837 (1975).
449. Y. Gendreau and J. F. Normant, *Bull. Soc. Chim. Fr., Pt. 2*, 305 (1979).
450. T. Sato, K. Naruse, and T. Fujisawa, *Tetrahedron Lett.*, **23**, 3587 (1982).
451. T. Fujisawa, N. Okada, M. Takeuchi, and T. Sato, *Chem. Lett.*, 1271 (1983).
452. Y. Gaoni, *Tetrahedron Lett.*, **23**, 5215 (1982).
453. J. F. Normant, *Synthesis*, 63 (1972).
454. F. Sato, T. Jinbo, and M. Sato, *Tetrahedron Lett.*, **21**, 2171 (1980).
455. F. Sato, T. Jinbo, and M. Sato, *Tetrahedron Lett.*, **21**, 2175 (1980).
456. P. W. Jolly, *Comprehensive Organometallic Chemistry*, Vol. 8, Pergamon Press, Oxford, 1982, Chapter 56.5, pp. 738–772.
457. T. Hayashi, K. Kabeta, I. Hamachi, and M. Kumada, *Tetrahedron Lett.*, **24**, 2865 (1983).
458. V. Ratovelomanana and G. Linstrumelle, *Tetrahedron Lett.*, **22**, 315 (1981).
459. A. Harris, D. Middlemiss, K. Mills, A. J. Gower, and M. B. Tyers, *J. Med. Chem.*, **26**, 55 (1983).
460. K. Isobe, Y. Nakamura, and S. Kawaguchi, *Bull. Chem. Soc. Jpn.*, **53**, 139 (1980).
461. M. Tiecco, L. Testaferri, M. Tingoli, D. Chianelli, and E. Wenkert, *Tetrahedron Lett.*, **23**, 4629 (1982).
462. D. E. Bergstrom and P. A. Reddy, *J. Heterocycl. Chem.*, **20**, 469 (1983).
463. H. B. Kagan, *Comprehensive Organometallic Chemistry*, Vol. 8, Pergamon Press, Oxford, 1982, Chapter 53.10, pp. 492–493.
464. T. Hayashi, M. Fukushima, M. Konishi, and M. Kumada, *Tetrahedron Lett.*, **21**, 79 (1980).
465. K. Tamao, T. Hayashi, H. Matsumoto, H. Yamamoto, and M. Kumada, *Tetrahedron Lett.*, 2155 (1979).
466. T. Hayashi, M. Konishi, M. Fukushima, K. Kanehira, T. Hioki, and M. Kumada, *J. Org. Chem.*, **48**, 2195 (1983).
467. V. Fiandanese, G. Marchese, F. Naso, and L. Ronzini, *J. Chem. Soc., Chem. Commun.*, 647 (1982).
468. E. Negishi, F.-T. Luo, and C. L. Rand, *Tetrahedron Lett.*, **23**, 27 (1982).
469. A. Sekiya and N. Ishikawa, *J. Organomet. Chem.*, **118**, 349 (1976).
470. H. P. Dang and G. Linstrumelle, *Tetrahedron Lett.*, 191 (1978).
471. S. Nunomoto, Y. Kawakami, and Y. Yamashita, *Bull. Chem. Soc. Jpn.*, **54**, 2831 (1981).
472. M. Tamura and J. K. Kochi, *Synthesis*, 303 (1971).
473. J. Novak and C. A. Salemink, *Synthesis*, 597 (1983).
474. S. P. Tanis and D. B. Head, *Tetrahedron Lett.*, **23**, 5509 (1982).
475. F. Derguini, Y. Bessiere, and G. Linstrumelle, *Synth. Commun.*, **11**, 859 (1981).

476. S. Nunomoto, Y. Kawakami, and Y. Yamashita, *J. Org. Chem.*, **48**, 1912 (1983).
477. G. A. Molander, B. J. Rahn, D. C. Shubert, and S. E. Bonde, *Tetrahedron Lett.*, **24**, 5449 (1983).
478. U. H. Brinker and L. König, *Chem. Ber.*, **116**, 882 (1983).
479. S.-I. Murahashi, M. Yamamura, K.-I. Yanagisawa, N. Mita, and K. Kondo, *J. Org. Chem.*, **44**, 2408 (1979).
480. D. E. Bergstrom and P. A. Reday, *Tetrahedron Lett.*, **23**, 4191 (1982).
481. L. N. Pridgen, *J. Org. Chem.*, **47**, 4319 (1982).
482. F. Babudri, S. Florio, L. Ronzini, and M. Aresta, *Tetrahedron*, **39**, 1515 (1983).
483. Y. Ito, M. Nakatsuka, and T. Sagegusa, *J. Am. Chem. Soc.*, **104**, 7609 (1982).
484. K. Tamao, N. Ishida, and M. Kumada, *J. Org. Chem.*, **48**, 2120 (1983).
485. A. Minato, K. Tamao, K. Suzuki, and M. Kumada, *Tetrahedron Lett.*, **21**, 4017 (1980).
486. T. Hayashi, M. Konishi, and M. Kumada, *Tetrahedron Lett.*, 1871 (1979).
487. D. Michelot, *Synthesis*, 130 (1983).
488. R. Rossi, A. Carpita, L. Gaudenzi, and M. G. Quirici, *Gazz. Chim. Ital.*, **110**, 237 (1980).
489. J. Caplin and J. H. P. Tyman, *J. Chem. Res. (M)*, 321 (1982).
490. R. A. Volkmann, J. T. Davis, and C. N. Meltz, *J. Org. Chem.*, **48**, 1767 (1983).
491. L. Jalander and M. Broms, *Acta Chem. Scand., Ser. B*, **37**, 173 (1983).
492. T. Hayashi, M. Konishi, and M. Kumada, *J. Organomet. Chem.*, **186**, C1 (1980).
493. V. N. Odinokov, G. A. Tolstikov, R. I. Galeyeva, and T. A. Kargapol'tseva, *Tetrahedron Lett.*, **23**, 1371 (1982).
494. S. Suzuki, M. Shiono, and Y. Fujita, *Synthesis*, 804 (1983).
495. G. Fouquet and M. Schlosser, *Angew. Chem., Int. Ed. Engl.*, **13**, 82 (1974).
496. E. Wenkert, E. L. Michelotti, and C. S. Swindell, *J. Am. Chem. Soc.*, **101**, 2246 (1979).
497. T. Hayashi, Y. Katsuro, and M. Kumada, *Tetrahedron Lett.*, **21**, 3915 (1980).
498. G. D. Crouse and L. A. Paquette, *J. Org. Chem.*, **46**, 4272 (1981).
499. H. Okamura and H. Takei, *Tetrahedron Lett.*, 3425 (1979).
500. E. Wenkert and T. W. Ferreira, *J. Chem. Soc., Chem. Commun.*, 840 (1982).
501. H. Takei, M. Miura, H. Sugimura, and H. Okamura, *Chem. Lett.*, 1447 (1979).
502. M. Julia, A. Righini, and J. N. Verpeaux, *Tetrahedron Lett.*, 2393 (1979).
503. G. A. Leyes and W. H. Okamura, *J. Am. Chem. Soc.*, **104**, 6099 (1982).
504. H. Okamura, M. Miura, K. Kosugi, and H. Takei, *Tetrahedron Lett.*, **21**, 87 (1980).
505. M. Bourgain–Commercon, J. F. Normant, and J. Villieras, *J. Chem. Res. (M)*, 2101 (1977).
506. G. Cassani, P. Massardo, and P. Piccardi, *Tetrahedron Lett.*, **21**, 3497 (1980).
507. D. L. Comins, *Tetrahedron Lett.*, **24**, 2807 (1983).
508. M. R. H. Elmoghayar and K. Undheim, *Acta Chem. Scand., Ser. B*, **37**, 160 (1983).
509. B. G. Gowenlock and W. E. Lindswell, *J. Organomet. Chem. Libr., Organomet. Chem. Rev.*, **3**, 1 (1977).
510. N. Kawabata, A. Matsumara, and S. Yamashita, *Tetrahedron*, **29**, 1069 (1973).
511. B. G. Gowenlock, W. E. Lindswell, and B. Singh, *J. Chem. Soc., Dalton Trans.*, 657 (1978).
512. R. Masthoff, H. Schuler, and G. Krieg, *J. Organomet. Chem.*, **13**, 37 (1968).
513. W. E. Lindswell, F. C. Robertson, I. Soutar, and D. H. Richards, *Eur. Polym. J.*, **17**, 107 (1981).
514. R. Masthoff and H. Wirth, GDR Pat., 76 975 (1970); *Chem. Abstr.*, **76**, 46709 (1972).
515. C. Glacet, *Bull. Soc. Chim. Fr.*, **5**, 895 (1938).
516. K. Mochida and H. Ogawa, *J. Organomet. Chem.*, **243**, 131 (1983).
517. L. L. Ivanov, V. P. Napochatykh, and N. A. Smyslova, *Zh. Org. Khim.*, **7**, 2623 (1971).
518. D. Bryce-Smith and A. C. Skinner, *J. Chem. Soc.*, 577 (1963).
519. L. N. Cherkasov, S. I. Radchenko, and K. V. Bal'yan, *Zh. Org. Khim.*, **10**, 2618 (1974).
520. L. N. Cherkasov and R. S. Razina, *Zh. Org. Khim.*, **9**, 17 (1973).
521. L. N. Cherkasov, *Zh. Vses. Khim. Ova.*, **22**, 468 (1977); *Chem. Abstr.*, **87**, 200 702 (1977)).
522. N. Kawabata, H. Nakamura, and S. Yamashita, *J. Org. Chem.*, **38**, 3403 (1973).
523. G. Santini, M. Le Blanc, and J. G. Riess, *J. Organomet. Chem.*, **140**, 1 (1977).
524. M. Chastrette and R. Gauthier, *J. Organomet. Chem.*, **66**, 219 (1974).
525. L. F. Kozhemyakina, I. E. Paleeva, V. A. Chernoplekova, and K. A. Kocheskov, *Izv. Akad. Nauk SSR, Ser. Khim.*, 431 (1976).
526. D. Bryce-Smith and A. C. Skinner, *J. Chem. Soc. C*, 154 (1966).
527. G. E. Coates and P. D. Roberts, *J. Chem. Soc. A*, 2651 (1968).
528. G. E. Coates and B. R. Francies, *J. Chem. Soc. A*, 474 (1971).

529. G. E. Coates and B. R. Francis, *J. Chem. Soc. A*, 160 (1971).
530. G. E. Coates and B. R. Francis, *J. Chem. Soc. A*, 1308 (1971).
531. G. Wiegand and K.-H. Thiele, *Z. Anorg. Allg. Chem.*, **405**, 101 (1974).
532. I. I. Lapkin, G. Ya Anvarova, and T. N. Povarnitsyna, *J. Gen. Chem. USSR* (*Engl. Transl.*) **36**, 1945 (1966).
533. L. Lardicci, L. Lucarini, P. Palagi, and P. Pino, *J. Organomet. Chem.*, **4**, 341 (1965).
534. G. P. Giacomelli, R. Menicagli, and L. Lardicci, *Tetrahedron Lett.*, 4135 (1971).
535. L. Lardicci, G. P. Giacomelli, P. Salvadori, and P. Pino, *J. Am. Chem. Soc.*, **93**, 5794 (1971).
536. I. I. Lapkin and T. N. Povarnitsyna, *Zh. Obshch. Khim.*, **38**, 99 (1968); *Chem. Abstr.*, **69**, 19 233 (1970).
537. I. I. Lapkin and G. Ya. Zinnatullina, *Zh. Obshch. Khim.*, **39**, 2708 (1969); *Chem. Abstr.*, **72**, 89 966 (1970).
538. I. I. Lapkin and G. Ya. Zinnatullina, *Zh. Obshch. Khim.*, **39**, 1132 (1969); *Chem. Abstr.*, **71**, 60 884 (1969).
539. I. I. Lapkin, N. F. Tenenboim and N. E. Eustafeeva, *Zh. Obshch. Khim.*, **41**, 1554 (1971); *Chem. Abstr.*, **75**, 129 495 (1971).

The Chemistry of the Metal—Carbon Bond, Vol. 4
Edited by F. R. Hartley

CHAPTER **3**

Preparation and use of organoboranes in organic synthesis

D. S. Matteson

Department of Chemistry, Washington State University, Pullman, Washington 99164–4630, USA

I. PREPARATION OF ORGANOBORANES

A. Introduction

1. Scope

This chapter describes the chemistry of the carbon—boron bond as it relates to organic synthesis. Reductions of organic compounds by boron—hydrogen bonds are excluded, although some processes which involve hydride donation from carbon with simultaneous cleavage of a carbon—boron bond are included.

The major significance of organoborane chemistry is the high degree of stereoselectivity and regioselectivity that can be achieved. Boron may serve as a template on which organic groups are assembled and joined, and once the desired carbon—carbon connection has been achieved, the boron connection is easily severed in stereospecific ways. Auxiliary chiral groups can be attached to the boron atom, and several organoborane reactions rate among the highest in chiral selectivity of any known reactions. Asymmetric hydroboration was the first highly selective and truly practical chiral synthesis directed by a chiral auxiliary group[1].

By far the largest single contributor to this field is Professor Herbert C. Brown, Nobel Laureate, 1979, of Purdue University. Brown's own books have thoroughly covered his earlier contributions[1,2], and this review will necessarily omit many of the details of the history of hydroboration chemistry.

Hydroboration is not the only easy way to make organoboron compounds, and syntheses from other organometallics are given full attention in this review.

Those aspects of organoborane chemistry that seem most relevant to an understanding of the fundamentals or that are likely to have future applications are included here, and developments during the past decade are emphasized. Reactions of boranes that are useful in stereocontrolled synthesis are covered in considerable detail, and reactions that are fundamentally incapable of preserving stereochemical information are included but de-emphasized.

This chapter is divided into three broad sections. Section I covers methods of synthesizing organoboranes, Section II describes the various classes of organoborane reactions, and Section III describes the use of organoboranes in chiral synthesis. Other applications of borane chemistry to synthetic problems, for example stereoselective alkane syntheses, achiral or racemic insect pheromones, and isotopically labelled compounds, are included together with the relevant reactions of boranes in Section II.

2. General routes to carbon—boron bonds

Carbon—boron bonds are formed in processes which take advantage of the electrophilic character of tricoordinate boron. The reaction of trivalent boron compounds with organometallic reagents has more than a century of tradition[3–11], beginning with the preparation of triethylborane, **1**, ('boric ethide') from diethylzinc and triethoxyborane (equation 1)[3]. The use of Grignard reagents with trialkoxyboranes[6] remains, with modern improvements[8–11], among the best of general routes to organoboron compounds. Reactions of RMgX or RLi with $B(OR')_3$ can be carried out stepwise to yield $RB(OR')_2$, R_2BOR' (or $RR''BOR'$), or R_3B. This approach to carbon—boron bond synthesis is discussed in Section I.B.

$$B(OEt)_3 + ZnEt_2 \longrightarrow \underset{(1)}{Et_3B} + Zn(OEt)_2 \qquad (1)$$

The more recently discovered hydroboration of alkenes[1,2,12] also provides a wide variety of organoboron compounds, often highly regioselectively and generally stereospecifically. In an early typical example, oct-1-ene was converted into tri-*n*-octylborane, **2** (equation 2)[12]. The boron atom adds preferentially to the least sterically hindered site, and the B—H addition is always *syn* via a four-centre cyclic transition state. Hydroboration is discussed in Section I.C.

$$AlCl_3 + NaBH_4 \longrightarrow B_2H_6 \xrightarrow[\text{diglyme}]{\text{oct-1-ene}} B(C_8H_{17})_3 \; (2)$$

R = H, alkyl, aryl; Y = H, alkyl, halogen, OR, SR (2)

Dominating the chemistry of organoboranes is the strong tendency of boron to oxidize and thus to become attached preferentially to more electronegative elements. Thermodynamic measurements indicate that B—O bonds are much stronger than B—C bonds, while B—C, C—C, and C—O, bonds all have the same general strength[13,14]. The atomic arrangement B—O—C—H tends to be more stable than the alternative H—O—C—B by approximately 125–167 kJ mol^{-1}. One consequence is that 'hydroboration' of a carbonyl group results strictly in reduction, with carbon—hydrogen bond formation and not carbon—boron bond formation[2a]. Another is that no tricoordinate boron compound having the bonding arrangement O=C—C—B is known, because rearrangement to B—O—C=C is generally exothermic by *ca.* 105–125 kJ mol^{-1}. A more general consequence is that boron—carbon bonds are broken in a wide variety of oxidative processes, which provide the basis for the synthetic utility of organoboranes, discussed in Section II.

The other dominant factor in the chemistry of organoboranes is the Lewis acidity of boron, the general ability of any tricoordinate X_3B (X = alkyl, alkoxy, halogen, H, etc.) to add a Lewis base, Y^-, to form a tetracovalent borate complex, X_3BY^-. Such borate(1−) complexes are intermediates in a very wide range of reactions, from that of an organometallic with a trialkoxyborane to form an organoborane in the first place, to reactions which assemble organic moieties on boron and connect them in a stereocontrolled manner, to the final oxidation with hydrogen peroxide or other reagents used to remove the boron from the completed structure.

Other hypothetical possibilities for forming boron—carbon bonds would include free radical processes and reaction of metalloboranes with electrophilic carbons. Radical reactions generally result in oxidative dealkylation at boron, not reductive alkylation, as dictated by the thermodynamic relationships mentioned above. Carborane anions[15,16] have been alkylated with methyl iodide. However, the alleged dialkylboron anion, $KBBu_2$[17–19], has been shown to yield $PhCH_2OBBu_2$, not $PhCOBBu_2$, with PhCOCl, and to form C—D bonds on quenching with D_2O, suggesting an oligomeric alkylborohydride structure[20]. The reported preparation of CF_3BBu_2 from CF_3I and $KBBu_2$[18] offers reasonable although not exhaustive evidence for the structure of the product, but no proof of the nature of the intermediate.

3. Safety considerations

Spontaneous ignition in air and the beautiful green flame were the first properties of triethylborane, **1**, to be noticed[3]. A sweet taste was mentioned as a property of ethylboronic acid, $C_2H_5B(OH)_2$[3,6]. These early observations parallel modern knowledge of the relative hazards of reduced *versus* oxidized organoboranes.

The toxicities of water-soluble boronic acids or their esters are often fairly low, whereas fat-soluble boronic acids tend to have significant although not highly hazardous toxicities[21,22]. Diborane is much more toxic, with the maximum allowable concentration for workers set at 0.1 ppm[23]. Tributylborane has been tested for toxicity as a possible ingredient in acrylic dental cement. The intravenous LD_{50} in male rats was 104 $\mu l\ Kg^{-1}$ (i.e. *ca.* 80 $mg\ kg^{-1}$) and the oral LD_{50} 1.0–1.2 $ml\ kg^{-1}$[24]. These are moderate values, and in view of the air sensitivity of trialkylboranes it is hard to imagine any accidental ingestion sufficient to produce acute toxic effects.

Fire hazards with boronic esters, $RB(OR')_2$, are no greater than with ordinary organic compounds of similar volatility. Alkylboronic acids, $RB(OH)_2$, are stable in air if pure, and butylboronic acid and several arylboronic acids are commercially available[25,26]. However, butylboronic acid has been reported to be stable in air if moist but to autoxidize if dry[6], and benzylboronic acid, $PhCH_2B(OH)_2$, autoxidizes even when moist[7]. In the author's experience, freshly prepared and dried samples of low molecular weight boronic acids may, after an induction period, autoxidize exothermically and darken[27]. A possible cause is the presence of R_2BOH as an impurity, since it was observed by Johnson's group that Bu_2BOH is stable under moist conditions but autoxidizes if dry, and also that the easily formed anhydride, Bu_2BOBBu_2, chars on cotton[28]. Butylboronic acid is sold with[25] or without[26] added water as a stabilizer.

The spontaneous flammability of trialkylboranes[3,4] has been investigated with modern equipment, and trimethylborane–oxygen mixtures were found to ignite spontaneously at low pressures, for example, 5 Torr of Me_3B and 25 Torr of O_2[29]. As the molecular weight increases, the hazard decreases, and tributylborane does not normally ignite spontaneously[8,11]. For synthetic purposes, trialkylboranes are handled in solution under an inert atmosphere and not normally isolated[1,2]. Provided that normal precautions are taken, trialkylboranes present no more hazard to the chemist than typical Grignard reagents.

Borane–thf can decompose with pressure build-up on storage, and instances of bursting of bottles (without ignition) have been reported[30,31]. The alternative dimethyl sulphide complex appears to be stable (see Section I.C.1.b).

Environmental and disposal problems with organoboron compounds appear to be minor. Boranes and boronic acids readily oxidize to alcohols and boric acid, and are unlikely to persist very long in the presence of oxygen. Sodium borate and sodium perborate are common ingredients in laundry detergents and bleaches, and typical environmental studies do not suggest any serious hazard[32]. Boric acid is moderately toxic. An old estimate of the acute lethal dose to an adult human is 15–20 g, with 0.5 $g\ day^{-1}$ for 6 months likely to cause toxic effects[23]. These estimates are in line with recent long-term feeding studies with rats and dogs in which 350 ppm of boron contained in boric acid in the diet caused no apparent harm, although serious toxic effects became evident at three times this level[33]. The well known effectiveness of boric acid as an insecticide for cockroaches appears to depend on the fact that these insects can ingest large amounts of boric acid by preening themselves[34].

4. *Nomenclature*

Official nomenclature rules[35] designed by inorganic chemists call for the naming of $RB(OH)_2$ as 'alkyldihydroxyborane' and $RB(OR')_2$ as 'alkyldialkoxyborane'. The older organic 'alkaneboronic acid' system paralleled the naming of sulphonic acids. More recently, *Chemical Abstracts* has converted to a system in which 'boronic acid' is the hypothetical $HB(OH)_2$ and thus $RB(OH)_2$ is an 'alkylboronic acid'[36]. It is then convenient to name $RB(OR')_2$ as a 'dialkyl alkylboronate'. Since this approach tends to yield the simplest names, it will generally be used in this chapter. 'Dihydroxyboryl' and 'dialkoxyboryl' are systematic names for $(HO)_2B$ and $(R'O)_2B$ as substituents, and will be used where appropriate. Systematic names for cyclic boronic ester groups often become particularly unwieldy, for example 2, 4, 4, 5, 5-pentamethyl-1, 3, 2, -dioxaborolane, and descriptive common names will generally be used, as for the example just cited, pinacol methylboronate. In parallel with the 'boronic acid' nomenclature is the name 'borinic acid' for the hypothetical H_2BOH[36].

Trialkylboranes are named as such in a straightforward manner, and borinic esters can also be named as alkoxydialkylboranes. For complex structures, the boron connections can be indicated by the prefix *B*, for example *B*-butoxy-*B*-(1-bromo-3, 3, 3-trichloro)propyl-*B*-2, 5-dimethylphenylborane. Cyclic boranes are generally named as hydrocarbons, with the prefix 'bora' to signify a BH group. Tetraalkylborates, R_4B^-, are named formally as 'tetraalkylborate(1 –)'. Tetrasubstituted borates are sometimes called 'ate complexes'. This German terminology commits a perfect spoken English pun with '8-complexes', which sounds like a plausible interpretation of the technical jargon to the uninitiated. In written English, 'boron ate complexes' is a simple phrase, absurd but distracting. This review will use 'borate complexes' or 'tetraalkylborates'.

B. Boranes from Organometallic Reagents

1. *Boronic acids and esters*

As noted in Section I.A.1, the reaction of trialkoxyboranes or boron halides with reactive organometallic reagents has a long history[3–11]. It is generally easy to make any simple boronic acid by using the procedure described by Snyder *et al.*[8] for butylboronic acid. This has been further refined and described in *Organic Syntheses* for phenylboronic ('benzeneboronic') acid, **3** (equation 3)[11]. If a boronic ester is desired, exchange of hydroxy and alkoxy ligands is generally rapid and the equilibrium easily shifted by such techniques as azeotropic distillation, as for example in the author's preparation of dibutyl vinylboronate without isolation of the air-sensitive acid[27].

$$PhMgBr + B(OMe)_3 \xrightarrow[Et_2O]{-78\,^{\circ}C,} \underset{\text{(precipitates)}}{PhB(OMe)_3^- MgBr^+} \xrightarrow{H^+, H_2O} \underset{\mathbf{(3)}}{PhB(OH)_2} \qquad (3)$$

In the *Organic Syntheses* procedure[11], the Grignard reagent and the trimethyl borate are added simultaneously from separate dropping funnels to diethyl ether in a flask cooled in a $-78\,^{\circ}C$ (dry-ice–acetone) bath with efficient stirring. If a special high-speed stirrer is used, the yield of **3** is 90%[11]. With more ordinary Teflon paddle-type stirrers, yields of boronic acids or esters in the 70–80% range are commonly obtained with a variety of Grignard reagents[8,27,37,38]. The use of simultaneous addition of the reagents is not of measurable benefit in most cases, and we have often used the more convenient procedure of simply adding the Grignard reagent to the trimethyl borate solution. There may be no reason except precedent[8] for mixing the reagents in this order. Trimethyl borate has to be added to the thermally unstable (dichloromethyl) lithium[39], and yields are good. Addition of trimethyl borate to a Grignard reagent has been reported[40], and trimethyl borate has been added successfully to the lithium reagent $PhSCH_2Li$ with excellent results[41]. The initial reaction product is a borate complex, $RB(OR')_3^-$, which is inert until it exchanges a ligand to unreacted $B(OR')_3$ to form $RB(OR')_2$ and $B(OR')_4$, and the possibility that excess Grignard or lithium reagent might lead to less trialkylborane side product than does excess of borate ester has not been investigated. Finally, it might be noted that for the isolation of boronic acids, autoxidation can be a problem but is inhibited by water[8], and the author has had good success recrystallizing boronic acids from diethyl ether–hexane or dichloromethane–hexane to which a drop of water, just enough to maintain a visible separate phase, has been added.

One frustrating exception to the easy synthesis of simple boronic acids by the *Organic Syntheses* procedure has been methylboronic acid, $CH_3B(OH)_2$ (equation 4). This boronic acid is water soluble and volatile, and the corresponding boronic anhydride (trimethylboroxine), **4**, and methyl ester (dimethoxymethylborane) are also volatile. These problems were circumvented and high yields obtained by a long work-up procedure which led to the boronic anhydride–pyridine complex[38]. The 1:1 complexes of trialkylboroxines with pyridine are easily formed[42] and can serve as a stable form for storing reactive boronic acids[43].

$$\mathrm{MeMgI} + (\mathrm{MeO})_3\mathrm{B} \longrightarrow \mathrm{MeB(OMe)_3}^- + \mathrm{Me_3B}$$

$$\longrightarrow \mathrm{MeB(OH)_2} + \mathrm{MeB(OMe)_2} + (\mathrm{MeBO})_3 \;\; \textbf{(4)} \qquad (4)$$

Another problem, discovered when Matteson and Moody repeated the preparation years later and agilely avoided disaster[44], is that a spontaneously flammable by-product, presumably trimethylborane, is generated in the reaction of methylmagnesium iodide with trimethyl borate, and may ignite during work-up when ethereal solutions are exposed to air. Less volatile trialkylboranes formed as minor by-products in preparations of higher molecular weight boronic acids have not caused any such problem, although odours consistent with descriptions of those of trialkylboranes[3,7] commonly arise in these preparations.

An elegant solution to the methylboronic ester problem has recently been found by Brown and Cole[45]. Addition of alkyllithiums to triisopropyl borate generally yields monoalkyl triisopropoxyborate salts in nearly quantitative yields, except that *tert*-butyllithium gives lower yields with some disproportionation. Work-up consists of addition of anhydrous hydrogen chloride and distillation of the propan-2-ol followed by the diisopropyl alkylboronate, **5** (equation 5).

$$RLi + (Pr^iO)_3B \longrightarrow RB(OPr^i)_3Li \xrightarrow{HCl} \underset{(\mathbf{5})}{RB(OPr^i)_2} + Pr^iOH + LiCl \qquad (5)$$

Triisopropyl borate is commercially available, and the Brown–Cole route is particularly useful for diisopropyl methylboronate and diisopropyl dichloromethylboronate. It is also convenient for boronic esters derivable from any commercially available lithium reagent. In making diisopropyl methylboronate, we have observed that it is necessary to use a fractionating column in order to separate the propan-2-ol and achieve the reported boiling point[46], and in scaling up the procedure the approximate dilutions of reactants reported[45] should be maintained in order to avoid stirring difficulties and formation of spontaneously flammable trimethylborane, which in this case burned harmlessly at the argon exit during the distillation under argon.

Organometallic reagents other than Grignard or lithium reagents can also be used to prepare boronic esters, but there has not been much exploration of such chemistry for synthetic organic purposes. A preparation of $ICH_2B(OBu)_2$ utilizes the reaction of ICH_2HgI with BBr_3[47]. However, for practical purposes the preparation of $ICH_2B(OBu)_2$ from $PhSCH_2B(OBu)_2$[38,48], which is derived from $PhSCH_2Li$, or the preparation of $ClCH_2B(OR)_2$ by tributyltin hydride reduction of $Cl_2CHB(OR)_2$[49], is more convenient.

Preparations of alkyldihaloboranes, RBX_2, and also dialkylhaloboranes, R_2BX, from boron trihalides and organotin compounds, R_4Sn, appear to be especially facile and efficient[50,51], and in view of the availability of a variety of organotin compounds these have potential synthetic utility.

2. *Alkoxydialkylboranes and trialkylboranes*

Reaction of BX_3 (X = halogen, alkoxy, etc.) with sufficient RM (M = Li, MgX, AlR_2, etc.) often yields BR_3 directly. This type of route to symmetric trialkylboranes has industrial potential, but is not normally as convenient as hydroboration for laboratory use, and has been reviewed elsewhere[52].

Reaction of a boronic ester $RB(OR')_2$ with a Grignard reagent R″MgX readily yields the mixed dialkylalkoxyborane RR″BOR′. For example, where R was vinyl, OR′ was butoxy, and R″ was aryl, 70–75% yields were obtained[53]. Where R and R″ were both vinyl, even though the product polymerized so readily that it had to be transesterified with $Me_2NCH_2CH_2OH$ in order to obtain a stable chelated product before isolation, the yield was 65%[54].

This classical approach can be extended to the reaction of R_2BOR' with R″MgX or R″Li to form R_2BR''. Its potential utility for making boranes that are inaccessible by hydroboration has been noted by Brown and coworkers, who have prepared 9-substituted 9-borabicyclo[3.3.1]nonanes, **7**, (equation 6) and other mixed trialkylboranes[55–59]. In hydrocarbon solvents the lithium alkoxide separates from the tetracoordinate borate intermediate **6** and precipitates[55]. With alkynyllithiums, removal of the alkoxide by treatment with boron trifluoride etherate has been used[57,58].

$R = CH_3$, Bu^t, Ph, $CH_2{=}CH{-}CH_2$, etc.; various $R{-}C{\equiv}C{-}$ (6)

Negishi and Boardman's synthesis[60] of alkenylboranes from alkenylaluminium compounds and *B*-alkoxy-*B*-dialkylboranes is noteworthy because of its potential utility in the stereocontrolled synthesis of trisubstituted alkenes. The preparation of *B*-[(*E*)-2-methyloct-1-enyl]-9-bbn, **8**, is illustrative (equation 7)[60].

Al_2Me_6, Cp_2ZrCl_2; MeOB(9-bbn) (7)

(8)

3. *Bis-, tris-, and tetrakis(dialkoxyboryl)methanes*

The Wurtz coupling of di-, tri-, or tetra-chloromethane with dimethoxychloroborane in the presence of finely divided lithium metal in thf appears to involve initial reaction of the lithium with the chloromethane, and is thus a special case of an organolithium compound reacting with a trivalent boron species. Bis-, tris-, and tetrakis(dimethoxyboryl)methane (**9**, **10**, and **11**) are readily produced in this manner (equations 8–10)[61–65].

$$CH_2Cl_2 + 2(MeO)_2BCl + 4Li \longrightarrow \underset{\mathbf{(9)}}{CH_2\{B(OMe)_2\}_2} + 4LiCl \qquad (8)$$

$$CHCl_3 \longrightarrow \underset{\mathbf{(10)}}{HC\{B(OMe)_2\}_3} \qquad (9)$$

$$CCl_4 \longrightarrow \underset{\mathbf{(11)}}{C\{B(OMe)_2\}_4} \qquad (10)$$

Surprisingly, the yields are highest with the tetraborylmethane **11**. The diborylmethane **9** is most tolerant of variations in the reaction conditions, and can even be made from trimethyl borate, in contrast to the more base-sensitive tri- and tetra-borylmethanes **10** and **11**. It is absolutely essential that there be an excess of chloroborane and halomethane over lithium metal in the synthesis of **10** and **11** in order to avoid decomposition during work-up. An unfortunate error in reactant quantities in one review[64] has led to several frustrated chemists. Correct proportions are given in the original papers[61,62] and a later review[65]. Inexplicably, this process has failed completely in several attempts to use the cyclic 2-chloro-1, 3, 2-dioxaborinane in place of chlorodimethoxyborane.

C. Boranes by Hydroboration

1. *Trialkylboranes*

a. General principles

Hydroboration of alkenes to produce trialkylboranes, **12**, is a well known and widely useful reaction (equation 11)[1,2,66]. Herbert C. Brown was awarded the Nobel Prize in 1979 for his discovery of hydroboration and the development of the chemistry of the resulting organoboranes. The simple example illustrated may serve as a reminder of the basic

$$RCH{=}CH_2 + B_2H_6 \xrightarrow{Et_2O} (RCH_2CH_2)_3B \xrightarrow{H_2O_2,\,OH^-} RCH_2CH_2OH \quad (11)$$

(12)

principles of hydroboration[1,2b]. (1) The boron generally goes preferentially to the less sterically hindered site. Electronic effects of substituents are less influential than steric effects, with very few exceptions. (2) Ethereal solvents or other Lewis bases of moderate strength greatly accelerate hydroboration compared with the rates seen with diborane, B_2H_6, because the ether aids the dissociation of borane dimers. The mechanism is discussed in Section I.C.2. For certain substituted boranes such as catecholborane that are monomeric, ethereal solvents are irrelevant. (3) Primary alkenes yield trialkylboranes, and it is not possible to stop the reaction selectively at the mono- or di-alkylborane stage. However, more sterically hindered alkenes and cycloalkenes can yield mono- or di-alkylboranes, and these will hydroborate less hindered alkenes in subsequent steps. (4) Because organoboranes are oxygen sensitive, it is customary to carry out further transformations of them *in situ*, for example, the oxidation of **12** to the corresponding alcohol as illustrated in equation 11. (5) Although several types of functionality are tolerated, nucleofugic groups such as halogen or alkoxy if β to the boron in the hydroboration product are likely to be co-eliminated with the boron[67].

Trialkylboranes are usually stable at room temperature, but on heating, usually in refluxing diglyme [bis(2-methoxyethyl) ether] at *ca.* 165 °C, rearrangement which amounts to dehydroboration and rehydroboration occurs, with efficient migration of the boron from sterically hindered secondary sites to less hindered, preferably primary alkyl sites[1,2c]. A more detailed description of these rearrangements is provided in Section I.C.1.g.

b. Regioselectivity: steric

Regioselectivity in hydroboration is usually dominated by steric effects. Hydroboration of alk-1-enes with diborane in diethyl ether or tetrahydrofuran–borane yields trialkylboranes having three equivalent alkyl groups, and is perfectly satisfactory for such operations as conversion of an alk-1-ene to a primary alcohol, provided a few percent of the isomeric secondary alcohol can be tolerated. A typical primary to secondary ratio (for pent-1-ene in diethyl ether) is 94:6[68]. 1, 1-Disubstituted ethylenes give higher selectivities, typically 99:1. These figures are, of course, the average net result of three successive hydroborations by BH_3, RBH_2 and R_2BH as the reaction progresses. Of these, BH_3 is the least and R_2BH the most selective. Where a high degree of regiocontrol is sought, hydroboration with diborane itself is obsolete and a more selective RBH_2 or R_2BH reagent would be used.

For those reactions where BH_3 is a suitable hydroborating agent, a highly useful storable complex is dimethyl sulphide–borane, $(CH_3)_2S{-}BH_3$, first reported by Adams's group[69,70] and subsequently developed by Lane[71] and Brown's group[72]. The alkylborane reagents used to obtain improved regioselectivities are themselves generally most easily prepared from dimethyl sulphide–borane[72]. Thf–borane is also commercially available, but suitable precautions must be taken in storing the material, as bursting of bottles from pressure build-up has been reported[30,31]. For those who prefer a milder stench to dimethyl sulphide, 1, 4-oxathiane–borane is recommended[73]. It is of interest that the thioether function can be oxidized selectively with sodium hypochlorite to form a water-soluble sulphoxide without affecting the borane, and thus interference of either dimethyl sulphide or 1, 4-oxathiane with subsequent radical reactions can be circumvented and the problem of disposing of noxious sulphur compounds minimized[74]. Conversely, selective oxidation of the borane function by hydrogen peroxide without affecting the sulphide was achieved by making the medium more basic than normally necessary[74].

The construction of stable RBH_2 and R_2BH reagents is straightforward[2b]. 1, 2-Disubstituted alkenes can generally be converted to bis(*sec*-alkyl)boranes, and tetrasubstituted alkenes yield monoalkylboranes that are too hindered to react further, except with less hindered alkenes. Thus, disiamylborane, **13** (siamyl = 3-methyl-2-butyl)[66,75–78], dicyclohexylborane, **14**[66,78], thexylborane, **15**[66,75,79,80], and 9-borabicyclo[3.3.1]nonane, **16**, generally referred to as 9-bbn[66,81], are readily prepared and serve as useful reagents for the hydroboration of less hindered alkenes (equations 12–15).

$$Me_2C{=}CHMe + B_2H_6 \longrightarrow [Me_2CHCH(Me)]_2BH \text{ or } Sia_2BH \quad (12)$$

(13)

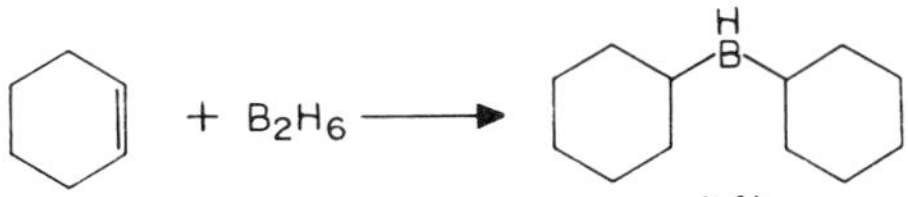

(13)

(14)

$$Me_2C{=}CMe_2 + B_2H_6 \longrightarrow Me_2CHC(Me_2)BH_2 \quad (14)$$

(15)

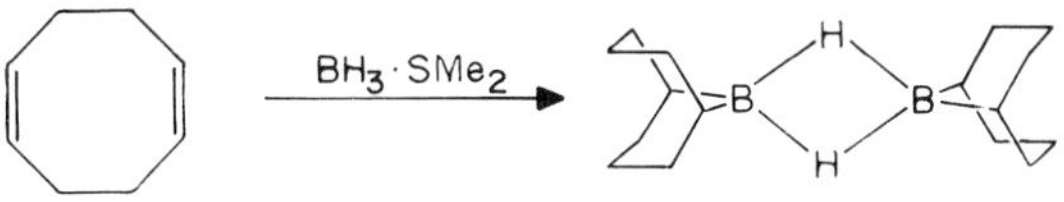

$BH_3 \cdot SMe_2$ (15)

(16)

Other representations of 9-bbn:

(16) or, abbreviated: (16)

The kinetically controlled hydroboration of cycloocta-1, 5-diene yields about a 3:1 mixture of 9-borabicyclo[3.3.1]nonane, **16**, and its [4.2.1] isomer, but in refluxing 1, 2-dimethoxyethane (85 °C) this all rearranges to the more stable [3.3.1] isomer, which crystallizes on cooling[82]. Samples of 9-bbn prepared in this manner are said to be stable in air, although it is prudent to keep the material under an inert atmosphere at all times. The compound is commercially available.

Reactions of 9-bbn, **16**[81], dicyclohexylborane, **14**[83], thexylborane, **15**[84], and other stable alkylboranes with alkenes that are not too hindered readily yield mixed trialkylboranes, **17**–**79** (equations 16–18). The high reactivity and high regioselectivity of 9-bbn are particularly useful, and the resulting 9-alkyl-9-bbn derivatives often react almost exclusively at the alkyl C—B bond and leave the 9-bbn unit intact during subsequent transformations.

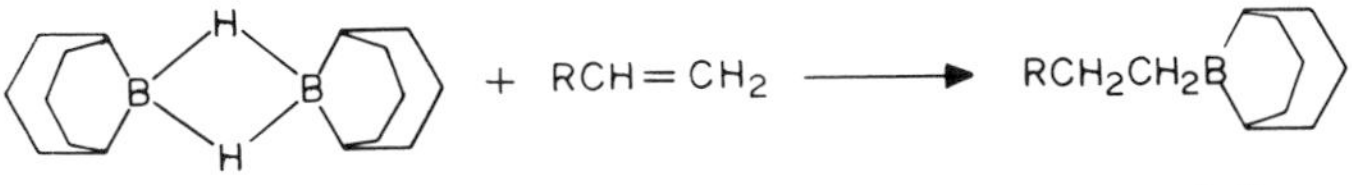

$$+ RCH{=}CH_2 \longrightarrow RCH_2CH_2B \quad (16)$$

(16) (17)

$$\text{(14)} + Me(CH_2)_3C{\equiv}CH \longrightarrow \underset{\text{(18)}}{Me(CH_2)_3(H)C{=}C(H)B(C_6H_{11})_2} \quad (17)$$

$$\underset{\text{(15)}}{Me_2CHC(Me_2)BH_2} + H_2C{=}C(Me)CH_2C(Me){=}CH_2 \longrightarrow \underset{\text{(19)}}{Me_2CHC(Me_2)\text{—}B} \quad (18)$$

The examples in equations 16–18 involves the use of bulky alkyl groups as blocking groups, both in the hydroboration process itself and in the further intended transformations. The synthesis of mixed trialkylboranes with sufficient general control of alkyl group structure for the purpose of incorporating two or all three of the alkyl groups into a desired structure requires special approaches and is discussed in Section I.C.1.e. A chiral directing blocking group is introduced in Section I.C.1.d. The use of groups other than alkyl as blocking groups is described in Section I.C.3.

c. *Regioselectivity: electronic*

Electronic influences on regioselectivity are in accord with expectation based on the boron being the positive end of the dipole in the B—H bond[2b]. For example, hydroboration of allyl chloride, **20**, with diborane does not follow the usual 94% preference for placement of the boron on the terminal carbon but yields an estimated 30% of 2-boryl product (equation 19)[85,86]. This relatively minor electronic influence is readily overcome by the use of 9-bbn, which results in total domination by the usual steric effects (equation 20)[87].

$$\underset{\text{(20)}}{ClCH_2CH{=}CH_2} + B_2H_6 \longrightarrow \underset{\sim 70\%}{ClCH_2CH_2CH_2BR_2} + \underset{\sim 30\%\ \text{(unstable)}}{ClCH_2CH(BR_2)CH_3} \quad (19)$$

$$\text{(16)} + 2ClCH_2CH{=}CH_2 \longrightarrow \underset{\text{(20)}}{2ClCH_2CH_2CH_2B} \quad (20)$$

Compounds of the general formula $CH_2{=}CHCH_2CH_2X$, where X = OMe, OAc, Cl, NH_2, etc., or $CH_2{=}CHCH_2CO_2Et$, undergo hydroboration to form the 1-boryl derivatives less selectively than simple hydrocarbons do, but again the use of a hindered borane, in this case disiamylborane, increases the regioselectivity to 98–99%[88]. In a series of but-2-enyl compounds, $CH_3CH{=}CHCH_2X$, where X = OEt, OAc, Cl, OH (which goes to OBR_2), or related groups, 84–100% of the boron of diborane attacked the 2-carbon[89]. The β-substituted boranes were unstable toward elimination except when X was alkoxy or boryloxy. Hydroboration of isobutenyl chloride, $Me_2C{=}CHCl$, has yielded mostly the 1-boryl product, $Me_2CHCHClBR_2$[90]. Where R = H, this product rearranges in thf to Me_2CHCH_2BHCl[91]. Isobutenyl ethyl ether, **21**, is converted into the 2-boryl derivative in spite of the steric hindrance at the tertiary carbon (equation 21)[90]. The major product from hydroboration of vinyl chloride, $CH_2{=}CHCl$, is evidently $B(CH_2CH_2Cl)_3$, which undergoes boron–chloride elimination to form ethylene on warming above $-78\,^{\circ}C$[85].

$$\underset{\mathbf{21}}{Me_2CH{=}CHOEt} + BH_3\text{–thf} \longrightarrow Me_2\underset{\displaystyle BR_2}{\underset{|}{C}}CH_2OEt \tag{21}$$

The directive effect of a dialkoxyboryl group places the incoming boron atom preferentially on the same carbon as the boron already present, as in the hydroboration of dibutyl vinylboronate, **22** (equation 22)[92–94]. Smaller amounts of the sterically preferred 1,2-diboryl derivative are formed.

$$\underset{(\mathbf{22})}{CH_2{=}CHB(OBu)_2} + BH_3\text{–thf} \longrightarrow \underset{\displaystyle H_2BCH_2CH_2B(OBu)_2}{\overset{\displaystyle MeCH(BH_2)B(OBu)_2}{+}} \xrightarrow{BuOH} \underset{\displaystyle (BuO)_2BCH_2CH_2B(OBu)_2}{\overset{\displaystyle MeCH[B(OBu)_2]_2}{+}} \tag{22}$$

Alkenylboranes formed as the initial products of hydroboration of acetylenic compounds also preferentially yield *gem*-diboryl compounds as products[95,96]. However, phenylacetylene yields substantial amounts of *vic*-diboryl derivative[96], which becomes the major product if the phenylacetylene is added to excess of BH_3–thf[97].

Vinyltrimethylsilane, $CH_2{=}CHSiMe_3$, yields a mixture of 1- and 2-boryl derivatives with diborane[88], but with 9-bbn is cleanly hydroborated to the 2-boryl product, $C_8H_{14}BCH_2CH_2SiMe_3$[99]. Triphenylvinylsilane with diborane has yielded 80% of 1-boryl and 20% of 2-boryl derivatives, and the diphenylsilacyclohexene **23** yielded exclusively the α-boryl product (equation 23)[100].

(23) $SiPh_2$ + B_2H_6 ⟶ $SiPh_2$ / BR_2 (23)

d. Stereoselectivity

The absolute *syn* stereospecificity of hydroboration together with the high susceptibility of the reaction to steric influences results in some highly useful diastereoselective additions to double bonds, and also geometrically specific additions to triple bonds. The hydroboration of an alk-1-yne to a *trans*-alk-1-enylborane, **18**[83], has been illustrated in equation 17 in the preceding Section I.C.1.b.

A diastereospecific, regioselective hydroboration of an open-chain double bond is illustrated by the reaction of diborane with (*Z*)-2-*p*-methoxyphenylbut-2-ene, **24** (equation

(24) Ar(Me)C=C(Me)H + B_2H_6 ⟶ H–C(Ar)(Me)–C(Me)(H)–BR_2

Ar = MeO-C$_6$H$_4$– (24)

24)[101]. The *syn* geometry of hydroboration applied to 1-methylcyclohexene results in exclusive formation of *trans*-l-boryl-2-methylcyclohexanes **25** (equation 25)[102]. Steric

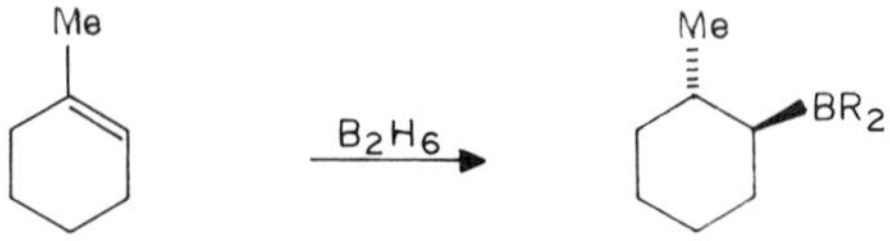

(25)

R = 2-methylcyclohexyl or H

(**25**)

factors can result in highly preferential hydroboration on one side of the double bond in cyclic systems[1,2b]. For example, hydroboration of norbornene, **26**, leads to > 99% of *exo*-borylnorbornane (equation 26)[103]. A methyl group *syn* to the double bond provides

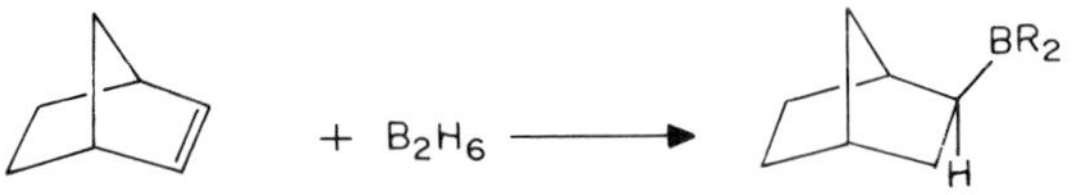

(26)

(**26**)

R = *exo*-2-norbornyl or H

sufficient steric interference to reverse the stereochemical preference, resulting in 78% *endo* hydroboration of 7,7-dimethylnorbornene, **27**, with BH_3–thf (equation 27)[104], and 97% *endo* with 9-bbn[105].

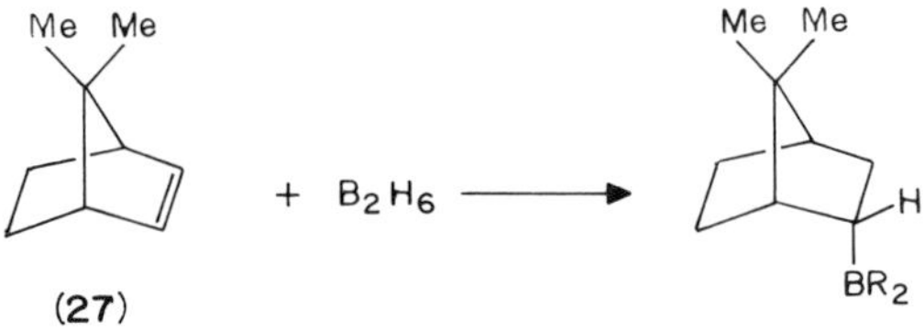

(27)

(**27**)

Hydroboration of (+)-α-pinene, **28**, illustrates both the high stereoselectivity and the stereospecific *syn* geometry of the addition. The resulting monoisopinocampheylborane, **29**, and diisopinocampheylborane, **30**, either of which can be made in high yield by adjusting the stoichiometry (equation 28), are highly useful asymmetric hydroborating agents (see Section III)[106,107].

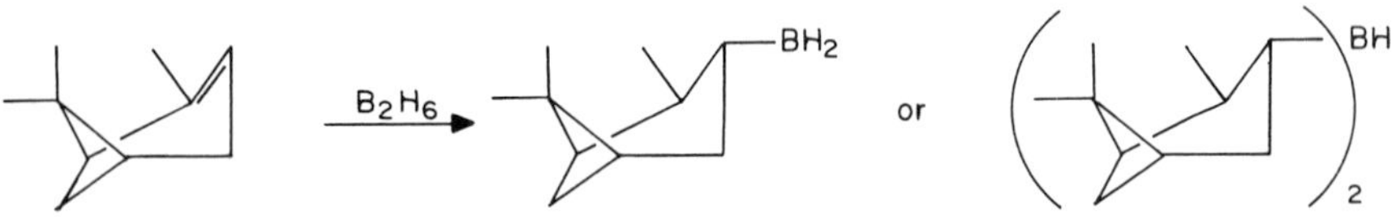

(**28**) (**29**) (**30**)

(28)

Both enantiomers of α-pinene are readily available from natural sources, but neither isomer is accessible with 100% enantiomeric excess at a reasonable price. Accordingly, Brown and coworkers have sought ways to upgrade the isomeric purities of the mono- and di-pinanylboranes **29** and **30**. For **30**, reaction of BH_3–thf with a *ca.* 15% excess of (+)-α-pinene of 97% ee in thf at 0 °C with equilibration for 3 days yielded 99% of **30** having 99.8% ee in the pinanyl groups[108]. The use of α-pinene of 92% ee with BH_3–SMe_2 requires removal of the dimethyl sulphide before the equilibration will proceed, but is then successful[109]. It is more practical to omit the equilibration and simply filter the precipitated **30**, which is of high purity, at the cost of lower yields[110]. Treatment of **30** with benzaldehyde liberates α-pinene of high enantiomeric purity[111].

Monoisopinocampheylborane, **29**, has been prepared via the reaction of diisopinocampheylborane, **30**, with tetramethylethylenediamine, which liberates α-pinene and yields the crystalline tmed complex of **29**, RBH_2–$Me_2NCH_2CH_2NMe_2$–BH_2R[112]. The crystallization process efficiently removes the small amount of complex having one R group of the 'wrong' chirality, and the **29** obtained on treatment of the amine complex with BF_3 etherate is optically pure[112]. An alternative preparation has used thexylborane and triethylamine, and the net reaction replaces 2, 3-dimethylbutene with pinene[113]. The problem with just hydroborating α-pinene with an equivalent amount of borane is that the product consists of a mixture of two types of borane dimers, RBH_2—BH_2R (**29** dimer) and R_2BH—BH_3 (**30**-BH_2), in which the latter predominates. The hydroboration mechanism involves dissociation of the dimers (see Section I.C.2), and the use of this mixture results in formation of large amounts of achiral products via the BH_2. This problem can be overcome by allowing the initial mixture to equilibrate for 96 h at 25 °C, which yields 90% of **29** dimer, or by heating it, which yields 86% of **29** dimer. The **29** can then be purified via the tmed complex[114,115].

e. Unsymmetrical trialkylboranes

Many unsymmetrical trialkylboranes can be made by hydroborating a relatively hindered olefin first to make a mono- or di-alkylborane, which is then used to hydroborate a less hindered olefin to provide a trialkylborane containing two different alkyl groups. Several of the most useful examples have been described in Sections I.C.1.b–d. However, the mechanism of hydroboration (Section I.C.2) makes it impossible to synthesize RBH_2 or R_2BH except in those fortuitous cases where the balance of steric factors stops the reaction at the right stage. There are a number of uses for boranes which bear two or three different alkyl groups, and accordingly several approaches to their general and controlled synthesis have been pursued. The controlled synthesis of RBH_2 and RR′BH is usually antecedent to the preparation of RR′R″B and is treated here as part of the same topic. The special utility of 9-bbn, **16**, as a hydroborating agent is sufficient that it is treated separately in Section I.C.1.f.

Unsymmetrical trialkylboranes are capable of disproportionation, and dimethylethylborane has been reported to be stable below −20 °C but to decompose rapidly to an equilibrium mixture of Me_3B, Me_2BEt, $MeBEt_2$, and BEt_3 above that temperature[116]. Dimethylvinylborane was stable at room temperature, which seemed a possible consequence of carbon—boron π-bonding at that time, but it is now known that electrophilic reagents (which would include boranes) attack alkenylboranes much faster than alkylboranes[2c], and it seems more plausible that the vinylborane was stabilized by scavenging adventitious boron hydride or that the organozinc used in preparing the dimethylethylborane (but not the dimethylvinylborane) may have contributed a disproportionation catalyst. Many of the unsymmetrical trialkylboranes to be discussed in the following paragraphs are presumably thermodynamically stable because of steric factors or cyclic structure, but a variety of simple RR′BH and RR′R″B that ought to be capable of disproportionation are now known to be stable enough for synthetic operations at room temperature. Perhaps it would be possible to prepare some kinds of RBR'_2 by reaction of R_3B with BR'_3, as has been done where R′ = halogen or alkoxy (see Section I.C.3) or

$R' = H$ (see the following paragraphs), but it would be unusual to encounter structures where such a thermodynamically controlled process would yield a single and thus synthetically useful product.

Thexylborane, **15**, hydroborates unsubstituted alk-1-enes to form thexyldialkylboranes even when excess of **15** remains unreacted, but yields monoalkylthexylboranes with any kind of disubstituted alkene, including isobutene, but-2-ene, or cycloalkenes. Trisubstituted alkenes such as 2-methylbut-2-ene or 1-methylcyclopentene also yield monoalkylthexylboranes. The monoalkylthexylboranes can be cleaved by treatment with triethylamine to form monoalkylboranes, for example *trans*-2-methylcyclopentylborane, **31**, as their triethylamine complexes (equation 29)[117], or can be used to hydroborate a second alkene to form a mixed trialkylborane which has thexyl as one of its alkyl groups, for example *B*-thexyl-*B*-cyclohexyl-*B*-(2-methylpentyl)borane, **32** (equation 30)[118]. The cleavage by triethylamine can also be accomplished by starting with triethylamine–thexylborane[119]. The use of thexylborane to make mixed trialkylboranes and cyclic

BH_2 (15) + 1-methylcyclopentene ⟶ thexyl(H)B–(2-methylcyclopentyl) $\xrightarrow{Et_3N}$ Et_3N-H_2B–(2-methylcyclopentyl) (31) + 2,3-dimethylbut-2-ene (29)

thexyl(H)B–cyclohexyl + $CH_2{=}C(Me)(CH_2)_2Me$ ⟶ thexyl–B(cyclohexyl)–$CH_2CH(Me)(CH_2)_2Me$ (32) (30)

boranes has been reviewed[120], as also has the preparation of boraheterocycles by this and other routes[121,122].

Hydroboration of non-conjugated dienes with thexylborane, **15**, has been found by Still and Darst[123] to provide 80–95% control of the diastereomeric relationship between non-adjacent chiral centres. The steric relationships are governed by cyclic borane intermediates. The contrasting results obtained from (4*Z*)-2, 6-dimethylhepta-1, 4-diene, **33**, and its (4*E*)-isomer, **34**, illustrate typical examples of the series of dienes investigated (equations 31 and 32). The synthetic utility of this hydroboration/oxidation process was further demonstrated with syntheses of dihydromyoporone and of the vitamin E side-chain.

thexyl-BH_2 (15) + (33) ⟶ $\xrightarrow{H_2O_2}$ diol (OH, OH) (31)

$$\text{(15)} + \text{(34)} \longrightarrow \xrightarrow{H_2O_2} \text{diol} \qquad (32)$$

The thexylborane route to acyclic unsymmetrical trialkylboranes has obvious limitations, being based on a delicate balance of steric factors that cannot necessarily be arranged for the synthesis someone really wants to achieve. An entirely different approach based on dibromoalkylboranes shows promise of wider generality. Treatment of an alkyldibromoborane–dimethyl sulphide complex, **35**, with an equivalent amount of lithium aluminium hydride in the presence of excess of dimethyl sulphide yields an alkylbromoborane, which can be used to hydroborate a second alkene (equation 33). The resulting bromodialkylborane cannot be reduced *in situ* without disproportionation, but if it is first converted into the borinic ester (alkyloxydialkylborane) with sodium methoxide, the second replacement by hydride and hydroboration of a third alkene proceeds smoothly[124].

$$\underset{(35)}{R^1BBr_2{-}SMe_2} + \tfrac{1}{4}\,LiAlH_4 \xrightarrow{Me_2S} R^1BHBr-SMe_2 \xrightarrow{R^2CH=CH_2} R^1B(Br)CH_2CH_2R^2 \xrightarrow{NaOMe}$$

$$R^1B(OMe)CH_2CH_2R^2 \xrightarrow{LiAlH_4} R^1BHCH_2CH_2R^2 \xrightarrow{R^3CH=CH_2} R^1B(CH_2CH_2R^3)CH_2CH_2R^2$$

R^1, R^2, and R^3 may be primary alkyl (33)

Pelter and coworkers[125] have prepared symmetrical R_2BH by sodium hydride reduction of R_2BBr and used them to hydroborate alkenes. Brown and Kulkarni[126] have found that potassium triisopropoxyborohydride and lithium aluminium hydride are particularly useful for reducing R_2BBr to R_2BH. Reduction of catecholboronic esters such as catechol butylboronate, **36**, with lithium aluminium hydride provides a simple route to monoalkylboranes such as 1-butylborane (equation 34)[127]. The stable complexes of

$$\underset{(36)}{C_6H_4O_2B{-}CH_2CH_2CH_2CH_3} + LiAlH_4 \longrightarrow (CH_3CH_2CH_2CH_2BH_2)_2 \qquad (34)$$

tetramethylethylenediamine with monoalkylboranes and dialkylboranes provide a particularly useful route for purifying and storing these reagents, which can be regenerated by treatment of the complexes with boron trifluoride etherate[128,129].

The synthesis of borinane (boracyclohexane, **40**) proved particularly elusive and has an interesting solution. Simple hydroboration of penta-1,4-diene yields predominantly the methylborolane derivative, **37**, as the kinetic product (equation 35). This can be rearranged to pentamethylenebis(borinane), **38**, at 170 °C, which can then be cleaved with BH_3–thf to form borinane, **40** (equation 36). However, a simpler procedure is achieved by reacting the penta-1,4-diene with 9-bbn, **16**, to form pentamethylenebis(9-bbn), **39** (equation 37), which on treatment with BH_3–thf or BH_3–SMe_2 yields borinane, **40**, in a mixture with 9-bbn (equation 38), from which the **40** can be separated either by distillation or its preferential complexation with Et_3N. Once borinane is available, it can be used to hydroborate penta-1,4-diene to form **37**, with the ultimate net result being the conversion of 2 mol of borinate to 3[130,131]. Borinane is a useful alternative to 9-bbn as a hydroborating agent, especially when radical cleavage of the third alkyl group is the ultimate goal, since 9-bbn fails as a blocking group for such purposes (see Section II.C.1.a).

+ BH3—thf → (37) → (38) (35)

38 $\xrightarrow{BH_3}$ 3 BH (36)

(40)(exists as dimer)

B–H + → (37)

(16) (39)

39 → BH + 2 BH (38)

(40) (16)

The use of 9-bbn with the appropriate diene can also lead to five- or seven-membered ring boracycloalkanes, but it has not proved possible to separate these from 9-bbn[132]. Borolane, **41**, rearranges to the relatively inert 1,6-diboracyclodecane, **42**, with a half-life of *ca.* 15 min at 25 °C, but has been prepared in solution at − 25 °C by acidification of the corresponding borohydride (equation 39)[133]. The redistribution reaction leads to polymeric boranes rather than eight-membered rings, and thermal depolymerization occurs with partial rearrangement to smaller rings[132].

Cl B thf $\xrightarrow{LiH}$ BH_2Li $\xrightarrow{MeSO_3H}$ B H B H $\xrightarrow{25\ °C}$ B H H B

(41) (42)

(39)

A different sort of hydroboration occurs when lithium triethylborohydride reacts with styrene or substituted styrenes[134]. The mechanism is probably anionic, and the boron atom adds at the benzylic carbon. The initial product is a borate complex, **43**, which can be cleaved with D_2O to form the α-deuterioalkylbenzene, or with methanesulphonic acid to yield ethane and the diethyl(α-alkylbenzyl)borane (equation 40).

$$PhCH{=}CH_2 + LiHBEt_3 \longrightarrow \underset{\mathbf{(43)}}{PhCH(Me)\bar{B}Et_3} \xrightarrow{MeSO_3H} PhCH(Me)BEt_2 \qquad (40)$$

Dimesitylborane, **44**, is an especially selective reagent for the hydroboration of alkynes[135]. The reagent is easily prepared from dimesitylfluoroborane and lithium aluminium hydride and is a solid that can be handled in air. It hydroborates linear alk-1-enes efficiently in a few hours at 25 °C, *cis*-pent-2-ene in 24 h at 65 °C, and cyclohexene only 15% in 24 h at 65 °C, but monohydroborates either alk-1-ynes or internal alkynes within a few minutes at 25 °C. Further, it is capable of distinguishing between the methyl and propyl groups of hex-2-yne with 90% selectively (equation 41). With 1-phenylpropyne, 9-bbn yields 65% electronically controlled attack at the 1-carbon, but dimesitylborane yields 98% sterically controlled attack at the 2-carbon. 1-Phenylpropyne was the slowest alkyne tested, requiring about 1 h at 25 °C.

(44)

(41)

The special utility of 9-bbn and the unsymmetrical trialkylboranes derived from it is such that it is covered separately in the following Section I.C.1.f. This section will be closed with a reminder that not all structural types can be made by hydroboration, and the synthesis of such structures from lithium or Grignard reagents is discussed in Section I.B.2.

f. Hydroborations with 9-bbn

9-Borabicyclononane, 9-bbn **16**, is the most regioselective of the readily available hydroborating agents[136]. The relative reactivities of 37 selected olefins toward 9-bbn have been determined by competition experiments and, where data were available, compared with reactivities toward disiamylborane[137]. Some relative rates measured for 9-bbn in thf at 25 °C include vinylcyclopropane, 2.3; hex-1-ene, 1.00; cycloheptene, 0.076; cyclopentene, 0.072; styrene, 0.025; *trans*-hex-3-ene, 0.12; *trans*-4-methylpent-2-ene, 0.076; *cis*-4-methylpent-2-ene, 0.061; *cis*-hex-3-ene, 0.056; *cis*-4, 4-dimethylpent-2-ene, 0.038; cyclohexene, 0.0067; 1-methylcyclohexene, 0.0011; and 2, 3-dimethylbut-2-ene, 0.0006. The behaviour of disiamylborane contrasts in that it shows a 10-fold preference for *cis*- over *trans*-hex-3-ene, and a 20-fold preference for cycloheptene over cyclopentane. Electronic effects in substituted styrenes are considerably larger for 9-bbn, a *p*-methoxy group enhancing the reactivity toward 9-bbn by a factor of 14 with 9-bbn, but only 1.4 with disiamylborane[137]. With substituted cycloalkenes, 9-bbn generally gives higher selectivity than BH_3 or even disiamylborane[105]. 1, 3-Dimethylcycloalkenes are hydroborated with high regio- and stereo-selectivity by 9-bbn. For example, 1, 3-dimethylcyclopentene yields 9-(*trans, trans*-1, 3-dimethyl-2-cyclopentyl)-9-bbn, **45** as the only detectable product (equation 42)[138]. Cyclooctylboranes are particularly prone to rearrangement (see Section I.C.1.g), even at room temperature, but 9-(2-methylcyclooctyl)-9-bbn does not rearrange[139].

(42)

(16) (45)

Conjugated dienes are generally attacked more slowly than simple alkenes by 9-bbn[140]. Thus, isoprene with 1 mol of 9-bbn, **16**, yields 0.5 mol of dihydroboration product **46** and 0.5 mol of unconverted isoprene (equation 43), although sufficient difference in hindrance at the two double bonds, as in 2,4-dimethylpenta-1,3-diene, allows the preparation of the monohydroboration product **47** (equation 44). The unusual inertness of cyclohexenes toward 9-bbn also provides a special case, and the 3-borylcyclohexene, **48**, is obtained in good yield (equation 45).

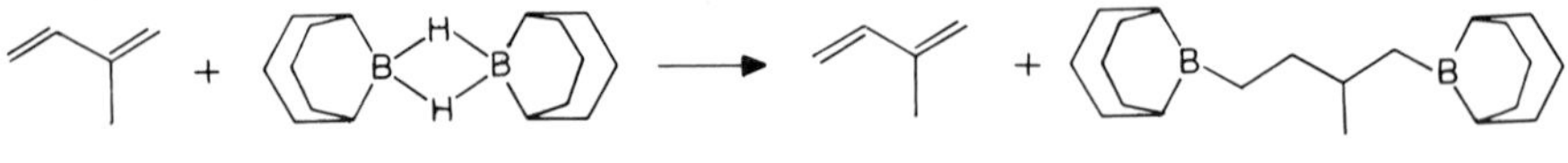

(**46**)

(43)

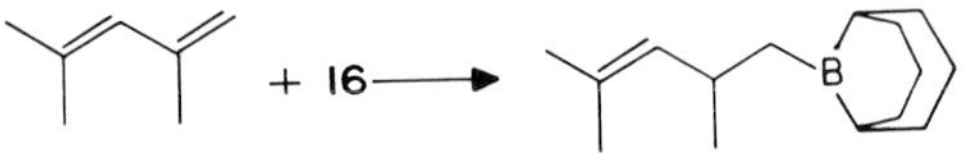

+ 16 → (**47**) (44)

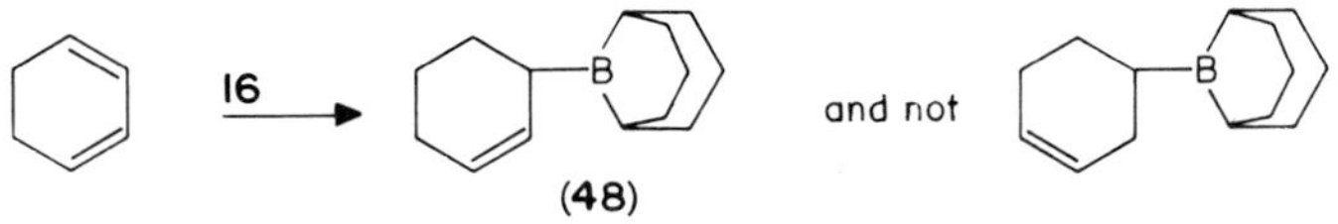

16 → (**48**) and not (45)

The reaction of allene with 9-bbn, **16**, cannot be controlled to give monohydroboration, but either mono-, 1,1-di-, or 1,3-di-substituted allenes can be hydroborated to form allylic derivatives as the major or exclusive products[141]. Even tetramethylallene yields the allylic product **49** and not the vinylic product with 9-bbn (equation 46). Cyclonona-1,2-diene yields 83% allylic and 17% vinylic product with 9-bbn. Disiamylborane yields gross mixtures with 1,3-disubstituted allenes, and mainly vinylic product with cyclonona-1,2-diene.

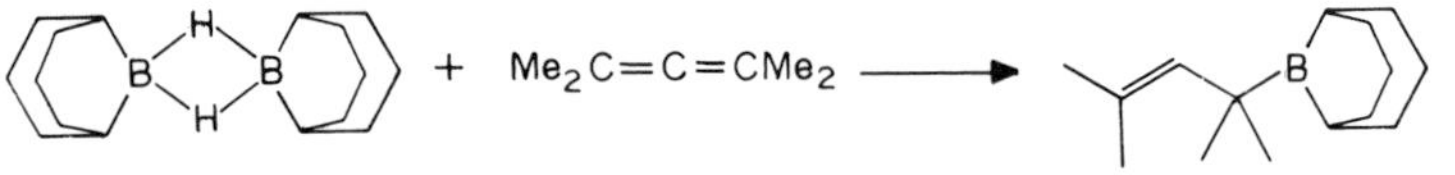

(**16**) + $Me_2C{=}C{=}CMe_2$ → (**49**) (46)

Non-conjugated dienes can generally be monohydroborated selectively with 9-bbn, with the position of hydroboration easily predictable on the basis of the relative reactivities of the separate olefinic functions[142]. If the diene is symmetrical, the product tends toward a statistical mixture of non-hydroborated and mono- and di-hydroborated products, although the second hydroboration tends to be slower than the first. With cyclohexa-1,4-diene or cycloocta-1,6-diene the second hydroboration is markedly hindered and good yields of monohydroboration product such as the 5-borylcyclooctane **50** are obtained (equation 47).

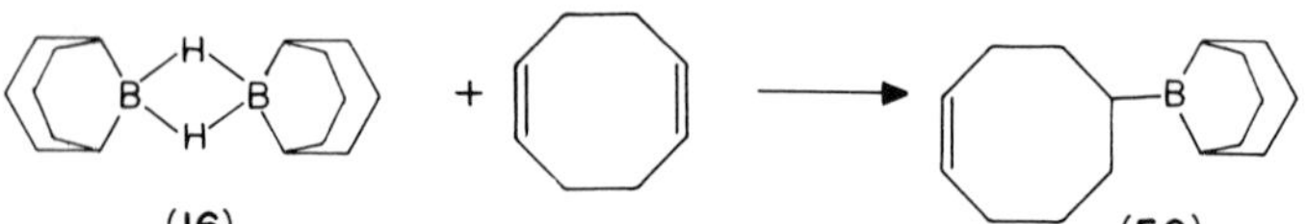

(**16**) + → (**50**) (47)

The hydroboration of a number of functionally substituted compounds with 9-bbn has been reported[143]. In general, the electronic effects of halogen or oxygen substituents in allylic systems $CH_2{=}CHCH_2X$ or $CH_2{=}CHCHX_2$ are overcome by the steric demands of the reagent, and the boron attacks the terminal carbon exclusively. In crotyl systems, MeCH=CHCH$_2$X (X = Cl or OR) the electronic influence predominates and the boron atom attacks the 2-position. With ethyl crotonate, a 50% yield of ethyl 2-ethyl-3-oxohexanoate was obtained, evidently the result of Claisen condensation of the initially formed boron enolate **51** (equation 48). As noted in Section I.A.1, structures of the type B—C—C=O are highly unstable with respect to rearrangement to C=C—O—B both thermodynamically and in having no Woodward–Hoffmann symmetry barrier to intramolecular rearrangement, and thus it is not expected that the α-boryl carboxylic ester from electronically directed hydroboration would survive if it were formed at all.

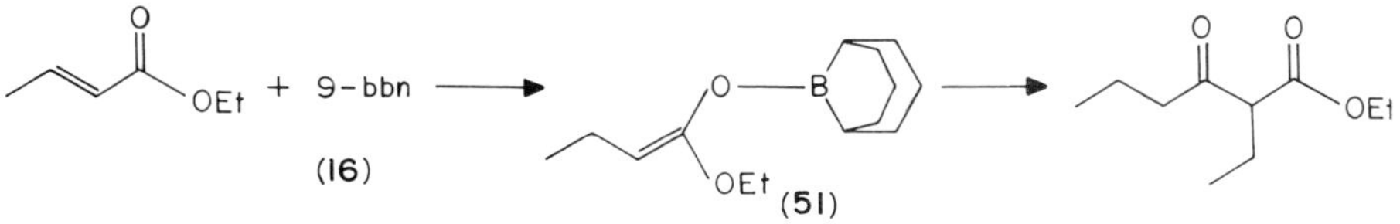

OEt + 9-bbn (16) → O—B ... OEt (51) → ... OEt

(48)

Acetylenes are hydroborated by 9-bbn to yield either vinylborane or *gem*-diboryl products, depending on the ratio of reactants, and 9-bbn shows greater regioselectivity than other reagents[144]. With alk-1-ynes, the boron attacks the 1-position exclusively, as illustrated with hex-1-yne, **52**, (equation 49) but if clean (*ca.* 95%) monohydroboration is desired a 2-fold excess of the alkyne is required. With internal alkynes, monohydroboration is readily achieved, and dihydroboration in the two cases studied (dec-5-yne and hex-3-yne) yielded exclusively *gem*-diboryl products. Unsymmetrical alkynes can be hydroborated with synthetically useful selectivity only if one of the alkyl groups in branched, as illustrated with 4-methylpent-2-yne, **53**, which yields 96% selectivity (equation 50). Selectivity is a marginal 78:22 with hex-2-yne.

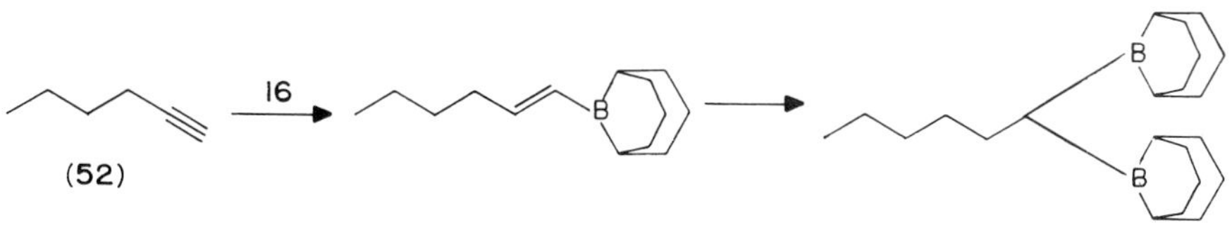

(52) —16→ B → B B

(49)

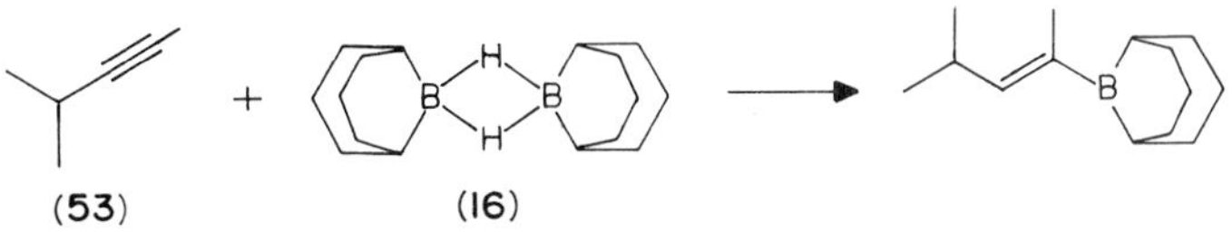

(53) + B H H B (16) → B (50)

A systematic survey of the use of several dialkylboranes for the hydroboration of alkynes has shown that monohydroboration of alk-1-ynes is favoured by lower temperatures, and that even as unhindered a hydroborating agent as di-*n*-hexylborane yields 82% mono- and 9% di-hydroboration product with an equimolar amount of oct-1-yne if the reaction temperature is $-50\,^{\circ}C$[145].

g. Borane rearrangements and displacements

Rapid rearrangement of tri-*tert*-butylborane to triisobutylborane at room temperature or slightly above was implied by the attempted preparation of the *tert*-butyl compound from the Grignard reagent and BCl_3, which yielded the isobutyl compound instead[146]. In this work by Hennion *et al.*[146], which predates hydroboration, it was conclusively demonstrated that tri-2-butylborane rearranges to tri-*n*-butylborane at or below its boiling point, *ca.* 210 °C at 1 atm. The much less reactive tri-2-butylboroxine, $\{MeCH_2CH(Me)BO\}_3$, rearranges in 24 h at reflux (242 °C) to tri-*n*-butylboroxine, $\{Me(CH_2)_3BO\}_3$[147].

The temperature required for alkyl group rearrangements in boranes is usually in the range 100–150 °C in ethereal solvents such as diglyme [bis(methoxyethyl)ether][148,149]. The reaction is catalysed by dialkylboranes and suppressed by excess of alkene[1,150]. This catalysis implies a mechanism more complex than mere separation of $RCH(CH_3)BR'_2$ into $RCH{=}CH_2$ and R'_2BH, followed by recombination to form the less sterically hindered $RCH_2CH_2BR'_2$, and appears inconsistent with the hypothesis that all hydroborations may follow the same detailed mechanism as those by 9-bbn (see Section I.C.2). It appears that the borane and olefin can remain bound in a π-complex in some cases, as indicated by the results of Rickborn and Wood[151] in which the product of hydroboration of *cis*-1,2-dimethylcyclohexene rearranged initially to *cis*-1-borylmethyl-2-methylcyclohexane, the product of migration of the boron along one side of the double bond. The kinetic hydroboration product would normally include a substantial amount of *trans*-isomer, and the thermodynamic product is 73% *trans*, which results after prolonged heating and obviously requires a dissociative mechanism[151]. Another system in which it appears that the borane–olefin π-complex remains bound is the isomerization of triisopropylborane to tri-*n*-propylborane, for which the entropy of activation was found to be -9 e.u.[152]. The entropy of activation for isomerization of *tert*-butyldiisobutylborane to tri-*tert*-butylborane in the same study was $+14$ e.u., which implies a dissociative mechanism.

One alkene can displace another from a trialkylborane under conditions similar to those used for alkyl group rearrangement[1,2c]. It has long been known that thermal cracking of trialkylboranes yields alkenes, which can be distilled out, and dialkylborane dimers[2b,153,154]. The displacement process is faster than borane rearrangement if an alk-1-ene is present, as shown for example by the liberation of α-pinene, **28**, from its hydroboration product, **54**, by displacement with dec-1-ene (equation 51)[155]. If **54** is heated in the presence of R_2BH before treatment with dec-1-ene, isomerization to the primary alkylborane **55** then leads to β-pinene, **56**. The borane rearrangement does not disturb even highly strained carbon skeletons, as exemplified by the rearrangement of **54** to **55**[150]. The process is generally good for rearranging the boron atom from a ring position to the end of a side-chain.

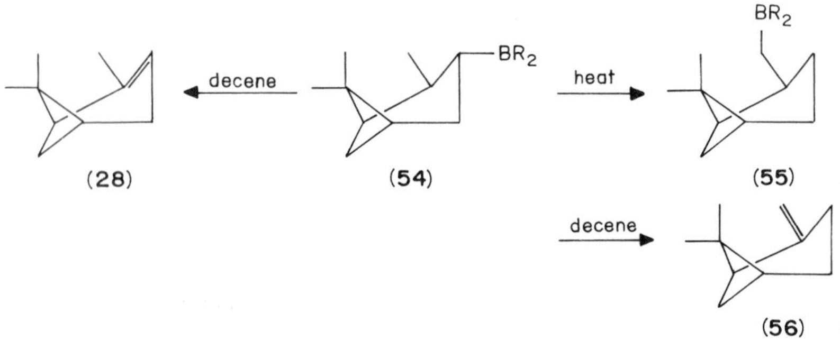

(28) ←decene— (54) —heat→ (55) —decene→ (56) (51)

The kinetics of replacement of isobutene from triisobutylborane by ethylene to form triethylborane in the gas phase require that the mechanism involves dissociation followed by recombination[156]. Kinetic data indicate a similar mechanism for exchange of alkenes with alkyl groups in 9-alkyl-9-bbns in refluxing thf[157]. In accord with the hypothesis of dehydroboration followed by rehydroboration, isomerization will not proceed past a carbon atom that does not have at least one hydrogen attached[149,158], as illustrated with the isomerization of 2, 2, 4-trimethyl-3-pentylborane, **57**, exclusively to 2, 4, 4-trimethyl-1-pentylborane and not to the 2, 2, 4-trimethyl isomer (equation 52).

$$(\mathrm{Me_3CCH(CHMe_2)})_3\mathrm{B} \xrightarrow{\text{heat}} (\mathrm{Me_3CCH_2CH(Me)CH_2})_3\mathrm{B} \qquad (52)$$

At higher temperatures, generally 200–300 °C, trialkylboranes eliminate hydrogen and alkenes with the formation of five- or six-membered rings[150,159]. Phenylethyl- or phenylpropyl-boranes close to boraindanes or boratetralins[160]. These cyclizations are seldom specific enough to be of much synthetic interest, but fortunately they are much slower than the borane isomerizations under discussion and do not interefere.

It was noted in the preceding Section I.C.1.f that 9-bbn derivatives rearrange relatively slowly, and that 9-bbn is therefore a good blocking group if prevention of rearrangement is sought[139]. This perhaps reflects the fact that 9-alkyl-9-bbn compounds suffer a relatively low degree of steric interaction, in contrast to the transition states for their formation from or decomposition to 9-bbn and alkene. The relative steric strain in the boranes themselves is evidently increased in bulky bis(cycloalkyl)alkylboranes, which undergo unusually rapid rearrangement. Rates have been studied with bis(cycloalkyl)-3-hexylboranes at 150 °C in diglyme[161–163]. Relative to tris(3-hexyl)borane as 1, relative rates for equilibration are 9-bbn 0.12, cyclohexyl 5, 2, 5-dimethylcyclohexyl 500[161], cyclopentyl 1.7, cycloheptyl 25, cyclooctyl 167[162], *exo*-norbornyl 60, and 2-bicyclo[2.2.2]octyl, **58**, 1500[163]. Thus, bis(2-bicyclo[2.2.2]octyl)borane is the hydroborating agent of choice if the ultimate goal is borane rearrangement (equation 53).

(**58**) (53)

Although 9-alkyl-9-bbns rearrange relatively slowly, it is unusually easy to recover the alkyl groups from them as alkenes by displacement with aldehydes. Midland and coworkers have shown that the displacement is first order in aldehyde and first order in 9-alkyl-9-bbn and that the reaction is much faster than dissociation of the 9-alkyl-9-bbn[164,165]. The very negative entropies of activation are in accord with a concerted cyclic transition state, and the 9-bbn group is sterically prevented from ring opening by such a mechanism. Secondary alkenes are displaced much faster than primary, and the reaction of *B*-3-pinanyl-9-bbn with carbonyl groups provides a highly useful method of enantioselective reduction, discussed in Section III.

2. Mechanism of hydroboration

The stereochemistry of hydroboration corresponds to direct addition of the B—H unit to the C=C group in a cyclic four-centre transition state, and catalysis of the reaction by ethers was apparent when hydroboration was discovered[1]. The first kinetic studies showed a first-order dependence on disiamylborane dimer[166], but this and other systems studied earlier were highly complex, and interpretations involving borane dimers or solvates[166,167] now appear erroneous.

A definitive series of studies of reactions of 9-bbn with alkenes and other substrates carried out by Brown and coworkers has recently shown that 9-bbn dimer, **16**, dissociates to 9-bbn monomer, **59**, and it is free **59** that attacks the substrate via a four-centre transition state, **60** (equation 54)[168–178]. The kinetic evidence may be summarized briefly

$$(16) \underset{k_{-1}}{\overset{k_1}{\rightleftharpoons}} 2\ \text{H—B} \ (59) \xrightarrow[k_2]{\text{RCH=CH}_2} (60) \longrightarrow \text{RCH}_2\text{CH}_2\text{B} \qquad (54)$$

as follows. If the alkene is highly reactive toward the borane monomer, **59**, then $k_2 \gg k_{-1}$ and the reaction is first order in 9-bbn dimer, **16**, with rate constant k_1, and independent of the nature of the reactive substrate or its concentration. If the alkene is sterically hindered or halogen substituted and of low reactivity, so that $k_{-1} \gg k_2$, then the reaction becomes of 0.5 order in 9-bbn dimer, **16**, and first order in the slowly reacting substrate. These relationships hold in a variety of solvents, including some such as carbon tetrachloride which cannot complex with the borane monomer, **59**[168–172,177,178]. Borinane (boracyclohexane) is a much faster hydroborating agent than 9-bbn but shows similar rate laws[179].

Lewis base complexes of 9-bbn, including those with thf, dimethyl sulphide, or amines, dissociate to yield 9-bbn monomer, **59**, which is the active species that hydroborates alkenes[173]. Reduction of aldehydes and ketones with 9-bbn in thf shows similar kinetic patterns to hydroboration of alkenes[174]. Substrates that are reactive enough can attack 9-bbn dimer directly and yield kinetics first order in substrate and first order in (9-bbn)$_2$, but this type of behaviour appears to be confined to protonolysis by unhindered alcohols[175] and complex formation with unhindered amines[176].

In general, borane–thf complexes are faster hydroborating agents than borane dimers. How can the formation of the borane–thf complex be exothermic, yet dissociation of the borane–thf complex requires less energy than dissociation of the original borane dimer? The straightforward answer is that the significant quantity of energy is that needed to liberate 1 mol of borane monomer. Dissociation of the dimer liberates 2 mol of monomer, and requires less energy than dissociation of 2 mol but more than dissociation of 1 mol of borane–thf complex without violating thermodynamics. Thus, complexing with the ether solvent lowers the activation energy for hydroboration[173].

Although it is not possible to study the B_2H_6–BH_3 system in the comprehensive detail done with 9-bbn, hydroborations with BH_3–NR_3 and BH_3–SMe_2 are inhibited by added

NR_3 or SMe_2, which makes sense only if free BH_3 is the active hydroborating agent[180]. There is an anomalous observation that an impurity in a substituted styrene can inhibit initiation of hydroboration by a large excess of BH_3–thf for up to six half-lives[181], which Brown has cited as further evidence that free BH_3 is the active hydroborating agent[180]. However, there is no way a minor impurity could remove the whole molar equivalent or more of BH_3 that would be generated by a first-order dissociation in such a long time, and these results imply either experimental error or a chain mechanism[167]. Conceivably RBH_2 + B_2H_6 could yield RBH_2–BH_3 + BH_3 in a chain reaction, and the anomalously low heat of activation[182] might be an artifact of an unrecognized chain process, but the data are insufficient to support any conclusion. Gas-phase kinetics have shown that free BH_3 hydroborates ethylene with an activation energy of *ca.* 8.3 kJ mol^{-1}, and direct attack by B_2H_6 does not occur[183].

Observations that alkyl group isomerization in trialkylboranes is catalysed by R_2BH[150] or ethereal solvents[1] (see Section I.C.1.g) seem inconsistent with the dissociation mechanism for hydroboration. If dehydroboration proceeds directly to a borane dimer or etherate, then the principle of microscopic reversibility requires hydroboration to proceed directly from the borane dimer or etherate. However, isomerization might also be aided by disproportionation of RBR'_2 to R_3B and BR'_3, and the R_2BH catalysis might work by accelerating ligand exchange.

The four-centre mechanism for the addition of the B—H unit to C=C is consistent with the H/D isotope effect[184]. Significantly for the synthetic chemist, this mechanism requires that the addition be strictly *syn*, with no rotation about a transient C—C single bond. Although the older evidence strongly supports this conclusion[1,101], a definitive proof with an acyclic substrate has been provided by Kabalka *et al.*[185], based on n.m.r. characterization of the trialkylboranes (not oxidation products) generated from (*E*)- and (*Z*)-1, 2-dideuteriohex-1-ene.

A practical application of this mechanistic information is the finding that hydroboration by $HBBr_2$–SMe_2 is greatly accelerated and helped to completion by the addition of a few percent of BBr_3[186]. It appears that the product $RBBr_2$ complexes less strongly than $HBBr_2$ with dimethyl sulphide, and the addition of a small amount of BBr_3 to the mixture shifts the equilibrium concentration of free dimethyl sulphide to a very low value and increases the concentration of the active hydroborating agent, $HBBr_2$.

3. Boronic esters and haloalkylboranes

a. Introduction

The presence of a halogen atom connected to boron deactivates, and the presence of an oxygen greatly deactivates, the borane toward B—H or B—C bond reactions. Accordingly haloboranes and alkoxyboranes are much less active than diborane or alkylboranes as hydroborating agents. However, there is considerable synthetic utility in having alkylboranes in which only one alkyl group is available for reactions. Such devices as the use of dialkylthexylboranes or 9-alkyl-9-bbns in order to block one or two positions of the borane Section I.C.1.e. and f) are often successful, but there are also a variety of circumstances in which reactions fail to distinguish the blocking alkyl group from the desired reaction site.

Boronic esters, $RB(OR')_2$, are generally readily available in molar and larger quantities from Grignard reagents and trialkyl borates (Section I.B.1). However, hydroboration provides a useful alternative route for certain types of structures and is compatible with some functional substituents where Grignard reagents are not. Alkylhaloboranes are more inconvenient to make by classical routes, and hydroboration makes them available as practical synthetic intermediates for the first time.

b. Boronic esters from catecholborane

Simple dialkoxyboranes such as $(MeO)_2BH$[187] or their cyclic analogues such as 1,3,2-dioxaborolane, $(CH_2)_2O_2BH$[188], are unstable toward disproportionation into B_2H_6 and $(RO)_3B$. The sterically hindered 4,4,6-trimethyl-1,3,2-dioxaborinane, **61**, has proved to be stable and hydroborates alkenes and alkynes above 100 °C, with mediocre yields[189,190]. The reagent hydroborates allenes efficiently at 130 °C, with boron attack preferentially on an unsubstituted terminal carbon (equation 55), or if the allene is 1,3-disubstituted, on the central carbon[191].

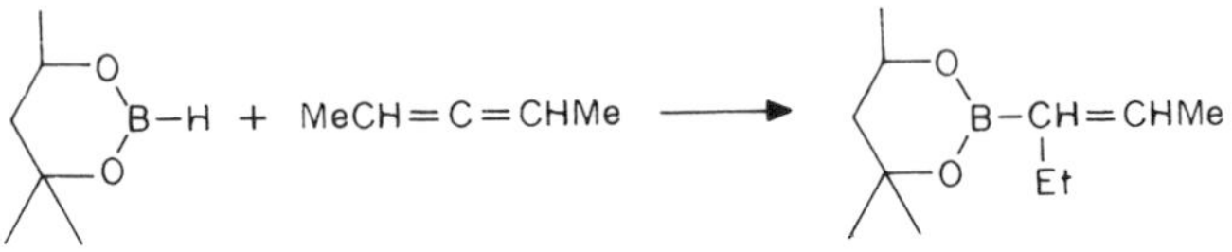

(61) B–H + MeCH=C=CHMe ⟶ B–CH(Et)–CH=CHMe (55)

In contrast to the sluggish reactivity of **61**, catecholborane, **62**, hydroborates alkenes and alkynes efficiently at 65–100 °C[192–195]. This is an especially valuable route to (*E*)-alk-1-eneboronic esters such as catechol (*E*)-hex-1-enylboronate, **63** (equation 56), the reaction being stereospecific and reasonably regioselective, *ca* 93% for unbranched alk-1-ynes[194]. The regioselectivity falls to 60:40 with hex-2-yne, $Pr^nC{\equiv}CMe$, but is a useful 92% with 1-cyclohexylpropyne, **64** (equation 57)[194]. The major isomer is usually easily purified by hydrolysis of the catechol ester to the boronic acid and recrystallization. Catecholborane hydroborates alk-1-enes slightly faster than alk-1-ynes, and an extensive series of competition experiments has indicated that catecholborane tends not to be as selective as 9-bbn or disiamylborane in distinguishing between different olefins[195].

In addition to being valuable for making (*E*)-alk-1-eneboronic esters, catecholborane provides a useful route to certain functionally substituted boronic esters that are not accessible from Grignard or lithium reagents. Thus, catecholborane, **62**, with 3-chloropropyne yields catechol (*E*)-3-chloroprop-1-enylboronate, **65** (equation 58), and although the selectivity of the boron for the terminal carbon is only 82%[194], any alternative route to this compound would be long. Similarly, catecholborane hydroborates allyl chloride or bromide to form $XCH_2CH_2CH_2BO_2C_6H_4$, and in this case it appears that the byproduct $XCH_2CH(CH_3)BO_2C_6H_4$ eliminates $XBO_2C_6H_4$ and propene, which is captured by catecholborane to yield *ca* 10% of catechol *n*-propylboronate[196]. Pinanediolborane was tested as an alternative hydroborating agent but gave low yields slowly[196]. Catecholborane hydroborates methyl pent-4-enoate, **66**, efficiently, but was found to reduce the ketone group first with hex-5-en-2-one (equation 59)[197]. An alternative to catecholborane which reacts with alk-1-ynes at 50 °C is

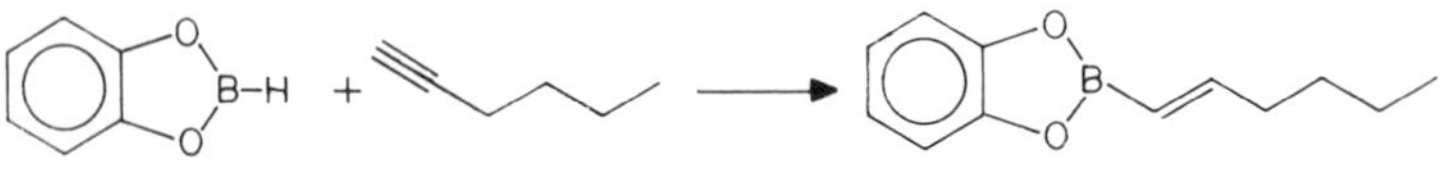

(62) B–H + hex-1-yne ⟶ (63) (56)

62 + (64) ⟶ (57)

62 + [propargyl chloride] → (65) (58)

62 + (66) CO_2Me → CO_2Me (59)

1, 3, 2-dithiaborolane, **67** (equation 60)[198], which is easily prepared as its trimethylamine complex by reaction of BH_3–thf with ethane-1, 2-dithiol in the presence of trimethylamine[199]. The free dithiaborolane is liberated from the amine complex by treatment with boron trifluoride etherate[198]. The hydroboration products are easily hydrolysed to boronic acids.

(67) B–H + → 91% (60)

c. Mono- and di-haloboranes

Monochlorodiborane was first prepared by Burg over 50 years ago[200], and proved difficult to isolate because of its tendency to disproportionate to boron trichloride and diborane. However, $ClBH_2$ and Cl_2BH are readily prepared in ethereal solvents by treating B_2H_6 with HCl or BCl_3, and these were introduced as hydroborating agents independently by Zweifel[201] and Pasto[202]. In thf, $ClBH_2$ and Cl_2BH do not disproportionate appreciably[203]. The similar reagents $PhSBH_2$ and $(PhS)_2BH$ had been found to be active hydroborating agents earlier[203]. The use of 1, 3, 2-dithiaborolane as an alternative to catecholborane has been described in the preceding Section I.C.3.b.

Chloroborane was subsequently investigated by Brown and Ravindran, who found that the reagent is most efficiently generated by treatment of lithium borohydride with boron trichloride in diethyl ether[205,206]. In contrast to $ClBH_2$–thf, which is a relatively slow hydroborating agent, $ClBH_2$–OEt_2 reacts rapidly and completely with olefins to form $ClBR_2$[206]. If 1.2–2.0 mol of thf are added to the reagent, the reactions can be stopped after one of the two hydrides has reacted to form ClBHR. Alkynes, RC≡CH, are readily converted into dialkenylchloroboranes, $(RCH{=}CH)_2BCl$, with the usual *E*-geometry. The reagent is useful in preparing cyclic boranes, for example 1-chloro-1-boracyclooctane, **68** (equation 61)[207]. Dichloroborane etherate is a very slow hydroborating agent, but on

$\xrightarrow{ClBH_2-OEt_2}$ $[(CH_2)_7BCl]_n$ $\xrightarrow{250\ °C}$ B–Cl + smaller ring by-products

68% (68)

(61)

addition of 1 mol of BCl_3 it becomes an efficient reagent for the hydroboration of alkenes and alkynes[208]. Methanolysis efficiently converts the dichloroboranes into boronic esters, for example dimethyl cyclopentylboronate, **69** (equation 62). Thus, dichloroborane provides an alternative to catecholborane for synthetic purposes. It differs from catecholborane in its capacity to dihydroborate alk-1-ynes, RC≡CH, to form 1,1-bis(chloroboryl)alkanes, $RCH_2CH(BCl_2)_2$[208].

$$\text{cyclopentene} \xrightarrow{HBCl_2} \text{cyclopentyl-}BCl_2 \xrightarrow{MeOH} \text{cyclopentyl-}B(OMe)_2 \ \ \mathbf{(69)} \qquad (62)$$

More recently, Brown's group has found that haloborane–dimethylsulphide complexes are superior hydroborating agents. These complexes are easily prepared by redistribution of BH_3–SMe_2 with BCl_3–SMe_2 or BBr_3–SMe_2[209]. H_2BCl–SMe_2 hydroborates olefins at 25 °C with high efficiency and regioselectivity. $HBCl_2$–SMe_2 requires 1 mol of BCl_3 to complex with the dimethyl sulphide in order to be an effective hydroborating agent[210]. H_2BBr–SMe_2 provides a simple route to R_2BBr[211]. Comparison of HBX_2–SMe_2 where X = Cl, Br, or I indicates that all are effective hydroborating agents, that the iodo compound is slower than the others, and that separation of pure R_2BX from dimethyl sulphide is easier when X = Cl than when X = Br[212]. A direct route to 9-halo-9-bbns is provided by reaction of cycloocta-1,5-diene with H_2BX–SMe_2[213].

Dibromoborane–dimethyl sulphide, $HBBr_2$–SMe_2, has proved to be a faster hydroborating agent than the chloro analogue, and reacts with many olefins at a satisfactory rate in refluxing dichloromethane[214]. In this case, the products are often isolable as distillable dimethyl sulphide complexes, for example *trans*-2-methylcyclopentyldibromoborane–dimethyl sulphide, **70** (equation 63)[215]. Alkynes yield alkenyldibromoboranes in the expected manner[217]. Efficient procedures for converting the alkyldibromoboranes in to various types of boronic esters have been reported[216]. An extensive study of relative reactivities of various alkenes and alkynes toward $HBBr_2$–SMe_2 has revealed that internal alkynes react fastest, closely followed by $CH_2{=}C(Me)R$, which react an order of magnitude faster than RC≡CH and $RCH{=}CH_2$, which react faster than internal alkenes or cyclopentene by factors of 3–8 and faster than cyclohexene by two orders of magnitude[218]. It was pointed out that $HBBr_2$–SMe_2 will hydroborate $CH_2{=}C(Me)R$ in the presence of $CH_2{=}CHR$, while disiamylborane shows a strong preference in the reverse direction, permitting the selective hydroboration of either double bond in a bifunctional structure. Similarly, $HBBr_2$–SMe_2 will selectively hydroborate an internal alkyne in the presence of a terminal alkene, and 9-bbn can be used if the reverse preference is desired[218].

$$\text{1-methylcyclopentene} + HBBr_2{-}S(CH_3)_2 \longrightarrow \textit{trans}\text{-2-methylcyclopentyl-}BBr_2{-}S(CH_3)_2 \ \ \mathbf{(70)} \qquad (63)$$

Thermal isomerization of $RBCl_2$–SMe_2 at 150 °C has been found to be unusually slow, even slower than rearrangement of the 9-bbn analogues[219]. Relative rates for R = 3-hexyl are $RBCl_2$–SMe_2 1,R-9-bbn 22, $RBBr_2$–SMe_2 45, and RBI_2–SMe_2 4380. The dimethyl sulphide does not have a significant effect on the rates. Hence the dichloroboranes are especially suitable reagents if highly labile systems such as 1-methylcyclooctene are hydroborated.

An interesting and useful synthesis of (*Z*)-alk-1-enylboronic esters, **71**, has been reported by Brown and Imai (equation 64)[220]. An earlier analogous (*Z*)-alkenyldialkylborane synthesis by Negishi and coworkers is described in Section II.B.1.b. Thexylchloroborane–

$$\mathrm{RC{\equiv}CBr} \xrightarrow{\mathrm{HBBr_2{-}SMe_2}} \mathrm{R(H)C{=}C(Br)BBr_2} \xrightarrow{\mathrm{R'OH}} \mathrm{R(H)C{=}C(Br)B(OR')_2} \xrightarrow{\mathrm{KHB(OPr^{i})_3}} \mathrm{R(H)C{=}C(H)B(OR')_2}\ \mathbf{(71)}$$

$R = Bu^n, Pr^i, Cl(CH_2)_3$; $R' = Me, Pr^i, HOCH_2CH_2, HO(CH_2)_3, HOCMe_2CMe_2$

(esters from diols are cyclic).

(64)

dimethyl sulphide, **72**, has been prepared from thexylborane–dimethyl sulphide and HCl in diethyl ether by Zweifel and Pearson[221], and by the direct hydroboration of tetramethylethylene with $H_2BCl–SMe_2$ by Brown *et al.*[222,223]. The regioselectivity of **72** is very high because of the thexyl group steric influence, typical figures for the position of boron attack being >99% terminal with oct-1-ene[221,223], *ca.* 99% terminal with styrene[221–223] 97% at the 2-position of (*Z*)-4-methylpent-2-ene, $MeCH{=}CHCHMe_2$[222], and 76% at the 2-position with the virtually undifferentiated (*Z*)-pent-2-ene[223]. A study of ^{11}B n.m.r. spectra has shown that thexylchloroborane etherate disproportionates to substantial proportions of thexylborane and thexyldichloroborane, but the thf and dimethyl sulphide complexes do not disproportionate, and consequently are more selective as hydroborating agents[224]. A study of relative rates of hydroboration of alkenes with **72** has yielded the approximate values alk-1-enes 1000, *cis*-alkenes *ca.* 100, styrene 11, norbornene 6, *trans*-alkenes 1.2, and cyclohexene 0.7[225]. The *cis*/*trans* ratio of *ca.* 100 is unusually high.

A useful synthetic application of thexylchloroborane methyl sulphide, **72**, is the synthesis of unsymmetrical trialkylboranes, one group of which is a thexyl blocking group[221]. The first of the other two alkyl groups is attached by hydroboration, the second with a Grignard or lithium reagent. Extension of this chemistry to the synthesis of thexylalkenylalkynylboranes, and to other alkylalkenylalkynylboranes, has been reported (equation 65)[226,227]. Surprisingly, reaction of the intermediate thexylalkenylchloroborane, **73**, with alkynyllithium is slow, and the process was speeded up by converting **73** to a methoxy derivative first (equation 66).

(72) + 1-octyne → **(73)**; (73) + (Z)-1-lithio-1-octene → thexyl(dialkenyl)borane (65)

73 —MeOH→ thexyl(alkenyl)B–OMe; + Li–C≡C–Bu → thexyl(alkenyl)(alkynyl)borane (66) *d. Redistributions*

Redistribution reactions of various types of alkylboranes have been mentioned in preceding sections as synthetic nuisances. However, redistributions between trialkylboranes and BX_3, where X is halogen or alkoxy, provide potentially useful routes to RBX_2 and R_2BX. For example, it has been known since 1957 that reaction of R_3B with BCl_3 at 100 °C yields $RBCl_2$ or R_2BCl, depending on the stoichiometry (R = 1-butyl, 2-butyl, isobutyl), and that the alkyl groups themselves are not isomerized in the exchange[228]. Tetraalkyldiboranes have been shown to catalyse the reaction[229,230]. Borane–dimethyl sulphide can be used to catalyse alkyl–halogen exchange, for example the reaction of tris(3-hexyl)borane and boron tribromide to form 3-hexylboron dibromide, **74** (equation 67)[231].

$$(\text{3-hexyl})_3B + BBr_3 \xrightarrow[70\ ^\circ C,\ 20\ h]{BH_3-SMe_2,} \text{3-hexyl}-BBr_2 \quad (79\%) \qquad (67)$$

(74)

Aryloxyboranes, $(ArO)_3B$, undergo redistribution with R_3B to form R_2B–OAr at 100 °C with the aid of BH_2–thf as catalyst[232]. The triaryl borates react much faster than trialkyl borates. This type of redistribution is useful for preparing catechol boronic esters from trialkylboranes of low molecular weight, where hydroboration of the volatile alkene with catecholborane would require high pressures[192].

e. Other borylations

Borane carbonyl, BH_3CO, which is derived from diborane and carbon monoxide, reacts with water or alcohols at low temperatures with hydride migration from boron to carbon to form cyclic dimeric derivatives of hydroxymethylboronic acid, $HOCH_2B(OH)_2$[233].

Although the addition of B—H across the carbon—carbon double bond is the well known and widely used reaction, analogous additions of B—B and B—Br are also possible. The oldest of these is the reaction of acetylene with boron trichloride in a gas-phase flow system at atmospheric pressure over a mercury(I) chloride on carbon catalyst, which yields $ClCH{=}CHBCl_2$ and $(ClCH{=}CH)_2BCl$, the geometry of which was not investigated[234]. The reaction easily yields large amounts of material, but the chlorovinyl groups undergo base-catalysed B—Cl elimination too easily for the material to have any obvious utility.

The facile *cis* addition of 9-bromo-9-bbn to alkynes has been reported by Hara *et al.*[235]. The resulting β-bromoalkenylboranes form borate complexes with alkynes that are stable enough to undergo typical iodine-initiated rearrangement with carbon—carbon bond formation (see Section II.B.3.c). Boron tribromide adds to cyclohexene to form cyclohexenylboron dibromide, **75**, the logical product from loss of HBr, and cyclohexylboron dibromide, for which the hydrogen source is unclear (equation 68)[236,237]. Ethylene yields ethylboron dibromide and polymer[236], and several other olefins react in an analogous manner to cyclohexene[237].

$$\text{cyclohexene} + BBr_3 \longrightarrow \text{cyclohexenyl}-BBr_2 + \text{cyclohexyl}-BBr_2 + \text{cyclohexyl}-Br \qquad (68)$$

(75)

The addition of $Cl_2B–BCl_2$ to C=C or C≡C is an efficient reaction related mechanistically to hydroboration, and shows *syn* geometry; for example, addition to acetylene produces (*Z*)-$Cl_2BCH{=}CHBCl_2$[238–240]. The reagent is difficult to prepare, and the reaction has not found synthetic use.

Finally, it might be noted that arylboron dichlorides such as $PhBCl_2$ can be made efficiently from boron trichloride, aromatic hydrocarbons such as benzene, toluene, or xylene, and aluminium metal[241]. Although not a very convenient laboratory process, this type of reaction could be useful for relatively inexpensive industrial production.

II. REACTIONS OF BORANES

A. Oxidative Replacement of Boron

1. Reagents which replace boron by oxygen

a. Oxygen

If spontaneous combustion of triethylborane, **1**, in air was the first organoborane reaction to be observed, the controlled air oxidation of **1** to diethyl ethylboronate was the second (equation 69)[3]. Hydrolysis yielded ethylboronic acid and, to balance, ethanol. In this century, air oxidation of tributylborane has been found to be controllable to cleave only one of the three alkyl groups and stop at Bu_2BOBu merely by adding water, which inhibits further oxidation[28]. The initial product formed from R_3B and O_2 is a peroxide, $R_2B{-}O{-}O{-}R$[242]. Treatment of thf solutions of R_3B with 1.5 O_2 at 0 °C followed by aqueous alkali results in quantitative conversion of all of the alkyl groups to alcohol, ROH[243]. However, this inexpensive oxidation of expensive organoboranes involves radical intermediates[242] and consequently fails to preserve the stereochemistry[243], which makes it of limited synthetic utility. The reaction does have synthetic potential as a (non-stereospecific) source of organic hydroperoxides. Trialkylboranes, R_3B, yield $(ROO)_2BR$, which on treatment with hydrogen peroxide yield 2 ROOH + ROH[244]. Alkyldichloroboranes, $RBCl_2$, are oxidized to $ROOBCl_2$, which hydrolyse to ROOH[245].

$$\underset{(\mathbf{1})}{Et_3B} + O_2 \longrightarrow EtB(OEt)_2 \xrightarrow{H_2O} EtB(OH)_2 + EtOH \tag{69}$$

Isotopically labelled oxygen is most easily obtained in elemental form, and the reaction of oxygen with organoboranes consequently is of special utility in preparing alcohols containing labelled oxygen. Kabalka *et al.*[246] have synthesized several ^{17}O-labelled primary alcohols in order to demonstrate the feasibility of the process and its compatibility with the presence of functional substituents. This chemistry becomes of unique value for the preparation of such simple compounds as ^{15}O-labelled butan-1-and-2-ol, which are of interest as agents for cerebral blood flow measurements[247]. The 2.04-min half-life of ^{15}O is ideal for medical diagnostic purposes, but provides the chemist with a considerable challenge to devise an ultra-fast synthetic technique.

b. Hydrogen peroxide

The reagent of choice for most conversions of boranes to alcohols is alkaline hydrogen peroxide[8,28], which reacts quantitatively at 0 °C and gives stereospecific replacement with retention of the configuration of the alkyl group[1,2c,248]. The kinetics of oxidation of phenylboronic acid have been studied in detail, and are consistent with the stereochemical results in requiring a hydroperoxyborate complex, **76**, as an intermediate, with migration of the phenyl group from carbon to electron deficient oxygen in the crucial step[249]. The initial product, $(HO)_2BOPh$, hydrolyses very rapidly to PhOH and $B(OH)_4^-$ under the reaction conditions (equation 70). In the presence of buffers, peroxy anions derived from

$$PhB(OH)_2 \underset{}{\overset{HOO^-}{\rightleftharpoons}} Ph-\overset{OH}{\underset{OH}{\overset{|}{\underset{|}{B^-}}}}-O-OH \longrightarrow (HO)_2\overset{-}{B}\cdots\overset{OH}{O}\cdots Ph \longrightarrow (HO)_2BOPh + OH^- \qquad (70)$$

(76)

the buffer also enter into the rate law, including peroxyborates. In acidic solution, a pH-independent term in the rate law evidently involves a protonated version of **76**, $PhB^-(OH)_2OOH_2^+$, with no net charge, and in very strong acid evidence for protonated species of undetermined hydration was obtained[249]. For a series of boronic acids, $RB(OH)_2$, at pH 5.23, which is on the basic side of the rate minimum near pH 3, second-order rate constants $\{[RB(OH)_2][H_2O_2]\}$ at 25 °C are R = $PhCH_2$ 0.088, Bu^t 0.072, 2-Bu 0.023, Ph 0.017, $CH_2{=}CH$ 0.007, 1-Bu 0.005, and Me 0.00013 $l\ mol^{-1}\ s^{-1}$ [250]. This is a narrow range of rates and shows the insensitivity of the reaction to steric or electronic factors in the substrate. Although the kinetic results with boronic acids show good behavior over a wide range of pH, it is customary to use strongly basic conditions with trialkylboranes[1,2c], perhaps to avoid any accidental radical-initiated processes that would not give full selectivity.

The net result of the *syn* geometry of hydroboration and the peroxidic oxidation with retention of configuration is the *syn* addition of water across the double bond, with a strong regioselective bias toward 'anti-Markownikoff' addition, the OH occupying the least sterically hindered position[1,2c]. Results dating back to shortly after discovery of hydroboration support this conclusion[101,248]. Some finishing touches on the stereochemical proof have been supplied by Kabalka and Bowman[251], who showed that deuterioboration of *cis*-but-2-ene yields at least 99% *erythro*-butan-2-ol-3-*d*, **77**, and deuterioboration of *trans*-but-2-ene yields *threo*-butan-2-ol-3-*d*, **78** (equations 71 and 72)[251]. These results require that the observed stereochemistry be inherent in the process, not an artifact of diastereoselection.

$$cis\text{-}MeCH{=}CHMe \xrightarrow[(2)\ H_2O_2,\ OH^-]{(1)\ BD_3-thf} D-\overset{Me}{\underset{H}{C}}-\overset{Me}{\underset{H}{C}}-OH \quad \text{(racemic)} \qquad (71)$$

(77)

$$trans\text{-}MeCH{=}CHMe \xrightarrow[(2)\ H_2O_2,\ OH^-]{(1)\ BD_3-thf} D-\overset{Me}{\underset{H}{C}}-\overset{H}{\underset{Me}{C}}-OH \quad \text{(racemic)} \qquad (72)$$

(78)

Peroxide oxidation of an alkenylborane leads to an enol and hence an aldehyde or ketone[252]. Although strongly basic conditions are customary for the oxidation of boranes to alcohols, slightly basic buffered media are necessary if labile aldehydes are to be isolated[252,253].

Carbon—carbon bond cleavage is a side reaction in the hydrogen peroxide oxidation of alkenylboronic acids[253]. Pure oct-1-enylboronic acid yielded 2% of heptanal in the derived octanal, and 2-phenylpropenylboronic acid, **79**, yielded up to 40% of acetophenone under buffered conditions (equation 73). This proportion was reduced to 4% by making the reaction mixture very strongly basic. Extensive carbon—carbon bond

$$\underset{\textbf{(79)}\text{ (isomer mixture)}}{\text{Ph(Me)C=CH}\text{-B(OH)}_2 + \text{Ph(Me)C=CH-B(OH)}_2} \xrightarrow{H_2O_2} \text{Ph(Me)CH-CHO} + \text{Ph(Me)C=O} \quad (73)$$

fragmentation has also been observed in peroxide oxidation of the diboryl compounds obtained from hydroboration of phenylacetylene[254]. Such problems are not common, and oxidation of organoboranes with hydrogen peroxide is among the most general and efficient of synthetic processes.

c. Other peroxidic reagents

Sodium perborate is an inexpensive peroxide with an excellent shelf-life and functions like hydrogen peroxide for most purposes. The cleavage of the alkenylboronic acid **79** to acetophenone was reduced to *ca* 1% by the use of sodium perborate together with a strongly basic reaction medium[253]. The only drawback to sodium perborate appears to be its low solubility, even in media of high water content, and slow reactions may be encountered with sterically hindered boronic esters.

Perbenzoic acid cleaves tributylborane quantitatively to butanol at 0 °C[28]. This reagent or *m*-chloroperbenzoic acid can be used for oxidation of certain cyclic β-chloroboranes which undergo boron–chloride elimination under the basic conditions of hydrogen peroxide oxidation but give good yields of β-chloro alcohols with peracids[255,256].

The molybdenum peroxide reagent MoO_5–py–hmpa oxidizes organoboranes to alcohols under mild conditions[257] and has been shown to do so with stereospecific retention of configuration[258].

d. Trimethylamine oxide

Anhydrous trimethylamine oxide is less reactive and more selective than hydrogen peroxide and can be used to convert R_3B in to R_2BOR or $RB(OR)_2$[259,260]. Trimethylamine oxide cleaves alkyl groups faster than aryl or alkenyl groups from boron, and has found special use in cleavage of one disiamyl group from a *B*-disiamyl-*B*-alkenylborane (1, 25 °C, thf–toluene) prior to stereospecific bromodeboronation[261].

Hydrated trimethylamine oxide is more convenient to use and gives high yields of alcohols when stirred as a suspension and heated with the trialkylborane in any of a variety of solvents[262,263]. With tricyclohexylborane, a relatively reactive substrate, one cyclohexyl group is converted in to alcohol in a few hours at 25 °C, all three are oxidized at 66–110 °C, and there appears to be no clean break for oxidation of two cyclohexyl groups. With tri-*n*-hexylborane, the first hexyl group is oxidized rapidly at 25 °C, the second at 66 °C, and the third requires 138 °C[263].

An interesting application of trimethylamine oxide is the selective oxidation of the hydroboration products from alkynylsilanes to acylsilanes. Oxidation of $RCH{=}C(BR_2)SiMe_3$ with hydrogen peroxide yields the carboxylic acid, thus providing the last link in an efficient route from $RC{\equiv}CH$ to RCH_2CO_2H[264]. The more interesting challenge is to prepare the acylsilane, which can be done with carbonate-buffered hydrogen peroxide with some loss[265]. Hydrated trimethylamine oxide is effective with the hydroboration products from alkynylsilanes with dichloroborane, but the yields of hydroboration products are low[266]. Borane–methyl sulphide hydroborates alkynylsilanes efficiently, but hydrated trimethylamine oxide oxidizes the resulting trialkenylboranes very slowly, and anhydrous trimethylamine oxide was found to be the unique reagent for obtaining high yields of acylsilanes[265]. The selectivity of the amine oxide is demonstrated dramatically with the synthesis of [(trimethylsilyl)acetyl]trimethylsilane, **81**[265], an interesting synthetic building block (equation 74)[267]. The geometry of the intermediate boron enolate, **80**, has not been determined, but retention of configuration is expected on mechanistic grounds.

$$Me_3SiC{\equiv}CSiMe_3 \xrightarrow{BH_3SMe_2} \left(\begin{matrix} Me_3Si \\ Me_3Si \end{matrix}\!\!>C{=}CH-\right)_3B \xrightarrow{Me_3NO} \underset{(80)}{\left(\begin{matrix} Me_3Si \\ Me_3Si \end{matrix}\!\!>C{=}CH-O-\right)_3B} \longrightarrow \underset{(81)}{Me_3Si-C(=O)-CH_2SiMe_3} \quad (74)$$

e. Chromium(VI)

Chromic acid oxidizes *sec*-alkylboranes to ketones in strongly acidic solution[268]. At pH 3–7, where alcohols are not oxidized by chromium(VI), the species $HCrO_4^-$ oxidizes boronic acids to alcohols with much more selectivity than that shown by hydrogen peroxide, relative rates for $RB(OH)_2$ being for R = Me 1, Et 28, and Bu^t 313000[269]. The reaction was said to have the same stereochemistry as that with hydrogen peroxide, but no data were given. Pyridinium chlorochromate readily oxidizes primary trialkylboranes, $(RCH_2)_3B$ or $RCH_2B(Sia)_2$, to aldehydes, RCHO[270,271]. Pyridinium chlorochromate also oxidizes borate esters, $(RCH_2O)_3B$, to aldehydes under strictly anhydrous conditions[272].

2. Replacement of boron by halogen

Bromine reacts only slowly with neat tributylborane, Bu_3B, and cleaves only one of the butyl groups to form Bu_2BBr and BuBr[10]. The reaction is much faster in dichloromethane and has been found to involve an unexpected mechanism[273]. The first step is radical bromination to an α-haloalkyl borane, for example diisobutyl(α-bromoisobutyl)borane, **82**, which is cleaved by the HBr formed in the reaction (equation 75). The intermediate **82** undergoes rearrangement with carbon—carbon bond formation in the presence of water (see Section II.B.1.b). This bromination cannot preserve the stereochemistry and is therefore of limited utility in synthesis. However, the use of bromine with methanolic sodium methoxide at 0 °C cleaves all three alkyl groups to alkyl bromide in high yield[274]. Iodine with methanolic sodium hydroxide cleaves two alkyl groups from the trialkylborane to yield the alkyl iodide[275]. Both bromination[276] and iodination[277] under these conditions proceed with predominant inversion of configuration. Bromination of tris(*exo*-norbornyl)borane, **83**, yields norbornyl bromide that is *ca.* 75% *endo* (equation 76)[276]. Iodination of **83** yields *endo*-norbornyl iodide that isomerizes in part to *exo* during work-up. Bis[(−)-pinanyl]-2*S*)-2-butylborane, **84**, of 86% ee [as shown by peroxide oxidation to (*S*)-butan-2-ol] yielded (*R*)-2-iodobutane, 84% ee, which corresponds to 98% inversion (equation 77)[277].

$$[Me_2CHCH_2]_3B + Br_2 \longrightarrow \underset{(\mathbf{82})}{[Me_2CHCH_2]_2BCHBrCHMe_2} + HBr \longrightarrow [Me_2CHCH_2]_2BBr + BrCH_2CHMe_2 \quad (75)$$

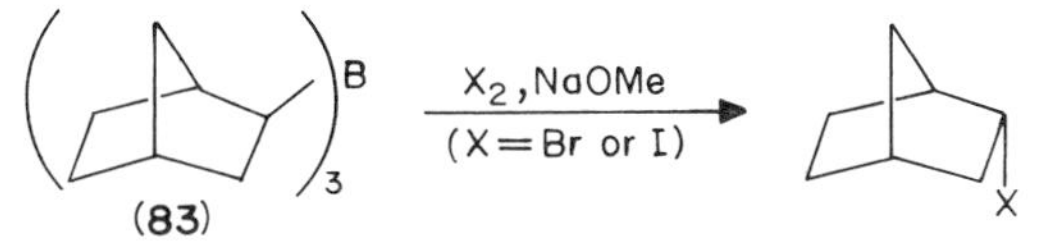

(83) (76)

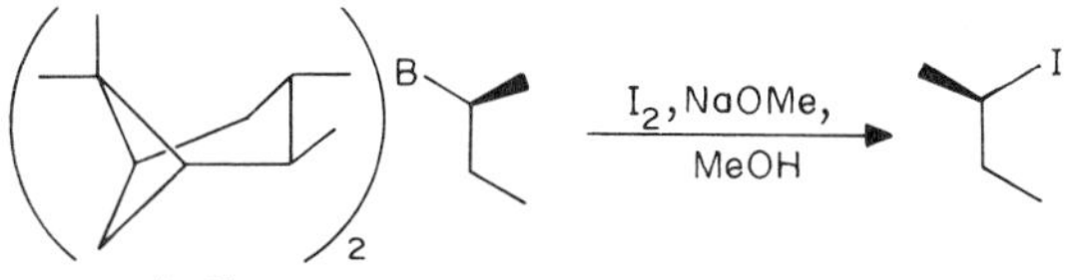

(84) (77)

Alkenylboron compounds undergo stereospecific replacement of boron by halogen. Bromination proceeds by way of a halogenation–deborohalogenation mechanism which results in inversion of the geometry, illustrated by the reactions of (*Z*)-**85** and (*E*)-but-2-enylboronic esters, **86** (equations 78 and 79)[278]. In the context of hydroboration

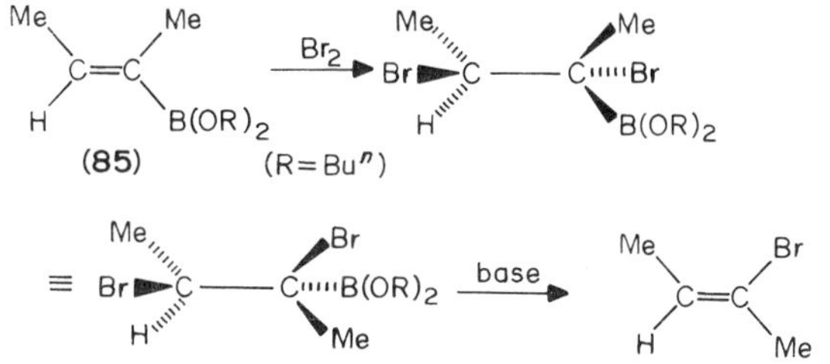

(85) (R=Bu^n)

(78)

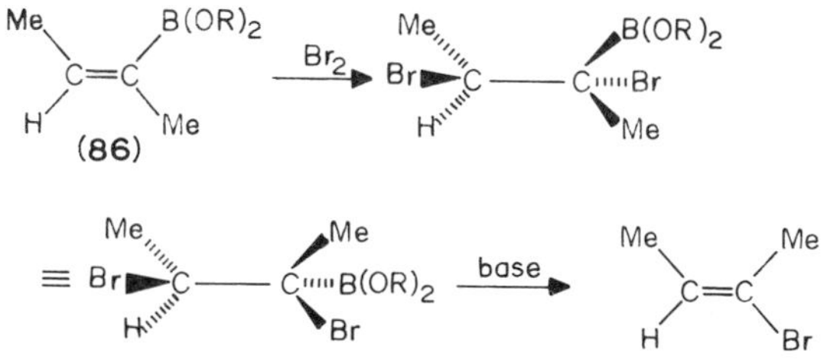

(86)

(79)

chemistry, this bromination can be used as the final step of a stereospecific conversion of alk-1-ynes to (*Z*)-1-bromoalkenes, **87** (equation 80)[279]. Iodine and sodium hydroxide together yield the (*E*)-1-iodoalkene, **88** (equation 81)[280]. Addition of iodine to the alkenylboronic ester followed by subsequent base treatment yields the (*Z*)-1-iodoalkene by the same mechanism as the bromo analogue, except that the procedure gives clean results only if the alkyl group of the alkenylboronic ester is unbranched[281]. Bromine and sodium methoxide together at −78 °C yield 2-bromoacetals, **89** (equation 82)[282].

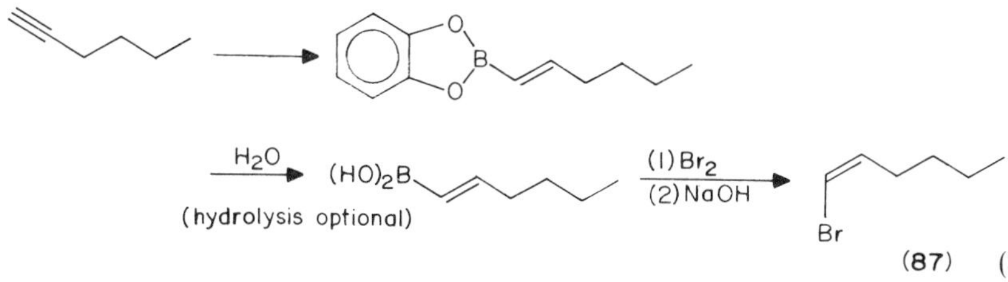

(87) (80)

$$(HO)_2B\text{-CH=CH-}(CH_2)_3CH_3 \xrightarrow{I_2 + NaOH} I\text{-CH=CH-}(CH_2)_3CH_3 \quad \textbf{(88)} \tag{81}$$

$$(HO)_2B\text{-CH=CH-}(CH_2)_3CH_3 + Br_2 + NaOMe + MeOH \longrightarrow (MeO)_2CH\text{-CHBr-}(CH_2)_3CH_3 \quad \textbf{(89)} \tag{82}$$

The conventional halogenation procedures are not suitable for introducing radio-labelled bromine or iodine because the radioactive species are obtained and safely handled as halide salts, and because it would be unacceptable to use only half of labelled I_2 while discarding the other half as I^-. Kabalka and coworkers have devised suitable procedures based on *in situ* oxidation of bromide or iodide with chloramine-T in the presence of the borane[283,284]. This approach works well for conversion of catechol alk-1-enylboronates to 1-iodoalkenes with retention of configuration[285]. The chemical stability of the vinylic iodide is particularly helpful for radioimaging in medicine, and fatty acid analogues of the general formula[125] $ICH{=}CH(CH_2)_xTe(CH_2)_yCO_2H$ show considerable promise as myocardial imaging agents[286]. Kabalka has reviewed his work in this field recently[287].

Reaction of α-(phenylthio)alkylboronic esters, $RCH(SPh)B(OR')_2$, with *N*-chlorosuccinimide yields α-chloroalkyl phenyl thioethers, RCHClSPh[288]. When the reaction is carried out in methanol buffered with triethylamine, the products are monothioacetals, RCH(SPh)OMe, or if two equivalents of *N*-chlorosuccinimide are used, acetals, $RCH(OMe)_2$. The reaction in methanolic solution evidently involves attack of the oxidizing agent at sulphur rather than boron, and no reaction occurs if the α-phenylthio substituent is not present.

3. Replacement of boron by nitrogen

a. Hydroxylaminesulphonic acid and chloramine

Direct replacement of boron by nitrogen can be accomplished efficiently for one group of a trialkylborane with hydroxylaminesulphonic acid[289]. The replacement occurs with retention of configuration, as illustrated by the conversion of tris(*exo*-norbornyl)borane, **90**, into *exo*-norbornylamine (equation 83). Chloramine reacts similarly, but the reagent is

$$(\text{exo-norbornyl})_3B \; \textbf{(90)} + H_3\overset{+}{N}\text{-O-}SO_3^- \longrightarrow \text{exo-norbornyl-}NH_2 + (\text{exo-norbornyl})_2BOH \tag{83}$$

difficult to prepare and the yields are not satisfactory[290]. However, Kabalka *et al.*[291] have found that chloramine works very well if generated *in situ* from ammonia and sodium hypochlorite, and two of the alkyl groups of the borane can often by used. This technique is especially useful for the efficient and rapid preparation of *N*-labelled primary amines[292]. *N*-Chloro primary amines usually give good yields of secondary amines with trialkylboranes[293].

N-Alkylsulphonamides can be obtained from trialkylboranes and *N*-

chlorosulphonamides[294]. *N*-Chlorodimethylamine, $ClNMe_2$, yields a mixture of $BuNMe_2$ and BuCl from Bu_3B, but the chlorination is a free radical reaction and can be suppressed by the use of galvinoxyl as inhibitor[295]. Nitrogen trichloride gives only radical chlorination[296].

b. Azides

Reaction of alkyl azides with trialkylboranes provides a controlled synthesis of secondary amines, for example the synthesis of butylethylamine, **91**, (equation 84)[297]. The reaction of trialkylboranes with azides uses only one of the alkyl groups, but alkyldichloroboranes can be used efficiently, as in the synthesis of (*trans*-2-methylcyclopentyl)cyclohexylamine, **92**, (equation 85)[298]. With β-iodoazides, the products are aziridines, for example 1-phenyl-*trans*-2, 3-dimethylaziridine, **93** (equation 86)[299].

$$Et_3B + BuN_3 \longrightarrow Et_3\bar{B}{-}N(N_2^+){-}Bu \longrightarrow Et_2B{-}N(Bu)(Et) \xrightarrow{H_2O_2} HN(Bu)(Et) \; (\mathbf{91}) + EtOH + B(OH)_3 \qquad (84)$$

HBCl₂ hydroboration of 1-methylcyclopentene → trans-2-methylcyclopentyl-BCl₂; (1) cyclohexyl azide, (2) H₂O → (**92**) (85)

$$PhBCl_2 + N_3{-}CH(Me){-}CH(Me)I \longrightarrow \mathbf{(93)}\ \text{(racemic)} \qquad (86)$$

4. Protodeboronation

Trialkylboranes are resistant to protodeboronation, and refluxing tributylborane with 48% hydrobromic acid for 1 h removes only one butyl group (as butane) to yield dibutylborinic acid, Bu_2BOH[10]. The hydrolysis of triethylborane by water is very slow at room temperature and is unaffected by hydrochloric acid but inhibited by sodium hydroxide[300]. Carboxylic acids cleave Et_3B to $EtH + Et_2BO_2CR$ at 25 °C, and the reaction stops at that point in the presence of hydroxylic solvents (water, ethylene glycol) or dimethylformamide. In diglyme or acetic anhydride the cleavage of a second alkyl group [yielding $2EtH + EtB(O_2CR)_2$] is essentially complete in 10–20 h[300]. Refluxing with propionic acid in diglyme cleaves all three alkyl groups from R_3B to form 3RH and $B(O_2CC_2H_5)_3$[301]. The mechanism of the reaction appears to involve attack of the carboxyl oxygen on the boron atom to form a cyclic transition state, as shown by the fact that weaker carboxylic acids react faster than stronger acids, Taft $\rho^* = -0.94$ for R of RCO_2H[302]. As a consequence, the reaction proceeds with retention of configuration, as illustrated for the deuteriolysis of trinorbornylborane, **94** (equation 87)[303].

$$(\text{norbornyl})_3\text{B}\ (\mathbf{94}) \xrightarrow{\text{EtCO}_2\text{D}} \text{norbornyl-D} \tag{87}$$

Trialkylboranes, R_3B, are cleaved by methanesulphonic acid to yield dialkylboron methanesulphonates, R_2BOSO_2Me, which are reactive and useful borylating agents[304]. Alkenyl-, aryl-, benzyl-, and allyl-boranes are much more easily cleaved by proton sources. The most synthetically useful of these cleavages is that of alkenyldialkylboranes, which proceeds rapidly with acetic acid at 0 °C and provides a useful route for the hydrogenation of alkynes to *cis*-alkenes, **95** (equation 88)[305]. Arylboronic acids undergo protodeboron-

$$\text{RC}\equiv\text{CR}' \xrightarrow{\text{SiaBH}_2} \text{R(H)C}{=}\text{C(R}')\text{BSia}_2 + \text{Sia}_2\text{B(R)C}{=}\text{C(R}')\text{H} \xrightarrow[\text{0 °C}]{\text{HOAc},} \text{R(H)C}{=}\text{C(R}')\text{H}\ (\mathbf{95}) \tag{88}$$

ation by several mechanisms, one of which is molecular attack of a carboxylic acid to form the same sort of cyclic transition state, **96**, believed to account for the stereochemistry of the reactions just discussed (equation 89)[306]. The base-catalysed deuteriodeboronation of

$$\text{PhB(OH)}_2 + \text{HCO}_2\text{H} \longrightarrow [\mathbf{96}] \longrightarrow \text{C}_6\text{H}_6 + (\text{HO})_2\text{BO}_2\text{CH} \tag{89}$$

1-phenylethylboronic acid has been shown to proceed with 54% net inversion of configuration[307]. This is not a synthetically significant process, except that it is a possible decomposition reaction of benzylic and presumably allylic boranes that does not occur with other types of boranes.

5. Replacement by mercury

One of the first reactions of benzylboronic acid to be discovered was mercuration to benzylmercury(II) chloride[6]. The reaction is first order in benzylboronic acid or ester, first order in mercury(II) chloride, and first order in hydroxide ion, and much faster if mercury(II) chloroacetate is used as the electrophile[37]. Addition of sodium hydroxide to a mixture of mercury(II) chloride and a primary trialkylborane at 70–80 °C yields the corresponding dialkylmercury[308], and mercury(II) acetate converts primary trialkylboranes into alkylmercury(II) acetates at 0–25 °C[309,310]. An efficient hydroboration, mercuration, and iodination sequence which converts alk-1-enes into esters, **97**, in the anti-Markovnikov sense has been reported (equation 90)[311]. Mercurideboronation at primary alkyl sites occurs with predominant (*ca.* 90%) inversion, detectable only in deuterium-labelled compounds, of course. Substrates were *erythro*- and *threo*-$(Bu^tCHDCHD)_3B$, prepared by logical hydroboration sequences from $Bu^tC{\equiv}CH$ or $Bu^tC{\equiv}CD$, and the diastereomers of the boranes and the $Bu^tCHDCHDHgX$ were identified by n.m.r.[312,313].

$$RCH{=}CH_2 \longrightarrow (RCH_2CH_2)_3B \xrightarrow{Hg(OAc)_2} RCH_2CH_2HgOAc \xrightarrow{I_2} RCH_2CH_2OAc \qquad (90)$$

(97)

Secondary alkyl groups on boron are inert to the usual mercuration conditions[309,310], but one secondary alkyl group of a trialkylborane can be replaced in a free radical reaction with mercury(II) methoxide[314], and two such groups can similarly be cleaved with mercury(I) *tert*-butoxide[315]. The radical mechanism with consequent loss of stereochemical integrity limits the possible utility of these reactions. Benzylic activation makes it possible to displace the boron from a 1-phenylethylboronic ester, and the reaction has been shown to proceed with some net retention of configuration[316], but the 1-phenylethylmercury(II) chloride produced is easily racemized, and in view of the somewhat erratic kinetics[317] the possibility of competing radical reactions has not been ruled out.

Addition of sodium hydroxide to a mixture of 1, 1-bis(dibutoxyboryl)ethane, $[(BuO)_2B]_2CHMe$, with mercury(II) chloride at 0 °C in thf–water results in rapid displacement of both boron atoms to form 1, 1-bis(chloromercuri)ethane, $(ClHg)_2CHMe$[92]. Replacement of the first boron atom is faster than that of the second, and the reaction can accordingly be controlled to produce α-(chloromercuri)boronic esters, for example $ClHgCH_2B(OMe)_2$[318] or $PhCH_2CH(HgCl)B(O_2C_2H_4)$[97]. Reaction of $C[B(OMe)_2]_4$ with mercury(II) acetate yielded $C(HgOAc)_4$[319], which is a useful staining agent for electron microscopy, but ironically, once it had been identified, the compound proved to be trivially simple to make by heating mercury(II) acetate in ethanol[320].

Alkenylboron compounds undergo replacement of boron by mercury with retention of configuration[321–324]. An example is the preparation of methyl (*E*-11-acetoxymercuri-10-undecenoate, **98** (equation 91)[323]. With 1, 1-bis(dialkoxyboryl)alkenes, both boronic ester groups are readily replaced, as for example in the preparation of cyclohexylidenebis(chloromercuri)methane, **99** (equation 92)[325,326].

$$MeO_2C(CH_2)_8(H)C{=}C(H)B(OH)_2 \xrightarrow{Hg(OAc)_2} MeO_2C(CH_2)_8(H)C{=}C(H)HgOAc \qquad (91)$$

(98)

$$C_6H_{10}{=}C\left(B\langle{}^{O}_{O}\rangle\right)_2 \xrightarrow[NaOAc]{HgCl_2} C_6H_{10}{=}C(HgCl)_2 \qquad (92)$$

(99)

6. Other boron replacements

Dimethyl disulphide cleaves R_3B to RSMe and $B(SMe)_3$ by a radical chain mechanism[327]. Iron(III) thiocyanate cleaves unsymmetrical trialkylboranes preferentially at the more highly branched alkyl group to form RSCN, and iron(III) selenocyanate behaves similarly[328]. Treatment of R_3B with $BrMg(CH_2)_5MgBr$ yields RMgBr, with formation of the stable spiroborate anion $(CH_2)_5B(CH_2)_5^-$ to balance[329]. This conversion of boranes to Grignard reagents was also successful with *B*-alkyl-9-bbns. Oxidation of boranes, R_3B, with silver nitrate results in alkyl coupling to form $R{-}R^1$.. With optically active

1-phenylethylboronic acid as substrate, the reaction is not stereoselective[307]. Organosilver intermediates seem likely. For alkenyl coupling by copper and palladium reagents, see Section II.B.3.f.

B. Carbon—Carbon Linkage Guided by Tetracoordinate Boron

1. α-Haloalkylborate rearrangements

a. α-Haloboronic esters

α-Haloalkylboronic esters are uniquely useful in directed chiral synthesis (see Section III). The chemistry reviewed in this section provides a background for understanding the chiral applications, and its major significance becomes apparent only in the chiral context.

The first α-haloalkylborane reported, FCH_2BF_2, was made from diazomethane and boron trifluoride and was unstable at 20 °C[330]. Shortly afterwards it was found that radical addition of bromotrichloromethane to dibutyl vinylboronate readily yields dibutyl 1-bromo-3, 3, 3-trichloropropylboronate, **100** (equation 93)[27]. Exploration of the chemistry of this compound by Matteson and Mah[331] revealed that nucleophilic displacement of the α-bromine is very facile and generally involves coordination of the boron atom with the attacking nucleophile. Most remarkable was the observation that Grignard reagents would add to the boron atom to form a borate complex **101** at −78 °C, which on immediate protonation yielded the corresponding dialkylborinic ester **102**, but which on standing at 25 °C would rearrange with displacement of bromide by the migrating aryl or alkyl group to form a secondary alkyl boronic ester, **103** (equation 94)[331]. The structure of **102** was confirmed by alternate synthesis from $CH_2{=}CHB(R)OBu$ and CCl_3Br[53], and it was observed that shaking an ethereal solution of **102** with aqueous sodium hydrogen carbonate sufficed to cause rearrangement to **103**[331].

$$CH_2{=}CHB(OBu)_2 + Cl_3CBr \xrightarrow{Cl_3C^{\cdot}} [Cl_3CCH_2\dot{C}HB(OBu)_2]$$

$$\xrightarrow{Cl_3CBr} Cl_3C^{\cdot} + \underset{(100)}{Cl_3CCH_2CH(Br)B(OBu)_2} \qquad (93)$$

$$100 \xrightarrow{RMgBr} \underset{(101)}{Cl_3CCH_2CH(Br)\overset{-}{B}(R)(OBu)_2} \begin{cases} \xrightarrow{H^+} \underset{(102)}{Cl_3CCH_2CH(Br)B(R)OBu} \\ \xrightarrow{25\ ^\circ C} \underset{(103)}{Cl_3CCH_2CH(R)B(OBu)_2} \end{cases} \qquad (94)$$

A remarkable feature of α-haloalkylboronic esters is their tendency not to undergo dehydrohalogenation in favour of nucleophilic displacement. The trichloromethyl group of **100** should encourage dehydrohalogenation, and it was found possible to accomplish this objective with triethylamine or, better, *tert*-butylamine to make $Cl_3CCH{=}CHB(OBu)_2$[332].

In its original form, α-haloboronic ester chemistry was a mere mechanistic curiosity because of the lack of any general synthesis of these compounds. Addition of hydrogen bromide to alkenylboronic esters was found to yield the α-bromo compound only if there was an α-alkyl group, as in the synthesis of dibutyl 2-bromo-2-propylboronate, **104** (equation 95)[333]. Dibutyl prop-1-enylboronate yielded the useless β-bromo compound $MeCHBrCH_2B(OBu)_2$[333], and radical addition of hydrogen bromide to dibutyl vinylboronate yielded the expected $BrCH_2CH_2B(OBu)_2$[334]. The only nucleophile found capable of displacing bromide from this β-bromoboronic ester was iodide, all others resulting in elimination of boron and bromine to form ethylene. The reaction of liquid hydrogen iodide with dibutyl vinylboronate was found to yield a gross mixture of the α- and β-iodoboronic esters, $CH_3CHIB(OBu)_2$ and $ICH_2CH_2B(OBu)_2$, and advantage was taken of the rapid destruction of the β-isomer by water in order to obtain the pure α-isomer[335]. It was also found that radical-catalysed addition of bromomalononitrile to dibutyl vinylboronate to form dibutyl 1-bromo-3, 3-dicyanopropylboronate, **105**, is a highly efficient process (equation 96). The only reaction of **105** with bases was deprotonation of the dicyanomethyl function and closure to dibutyl 2, 2-dicyanocyclopropylboronate, **106**.

$$CH_2{=}C(Me)B(OBu)_2 + HBr(liq.) \longrightarrow \underset{(104)}{(CH_3)_2C(Br)B(OBu)_2} \qquad (95)$$

$$CH_2{=}CHB(OBu)_2 + BrCH(CN)_2 \longrightarrow \underset{(105)}{(NC)_2CHCH_2CH(Br)B(OBu)_2} \longrightarrow \underset{(106)}{\text{cyclo-}[C(CN)_2CH_2CH]B(OBu)_2} \qquad (96)$$

An attempt to prepare a chloromethylboronic ester by chlorination of di-*tert*-butyl methylboronate with *tert*-butyl hypochlorite gave uselessly low yields, because radical attack on the *tert*-butyl hydrogens is almost as fast as that on the *B*-methyl hydrogens and carbon—boron bond cleavage also occurs[38]. (For practical syntheses of halomethylboronic esters, see Section I.B.1.) The situation proved much more favourable for light-initiated bromination of *sec*-alkylboronic esters[336–338]. The propane-1, 3-diol esters proved particularly useful for this purpose, as for example in the synthesis of propane-1,3-diol 1-bromocyclohexylboronate, **107** (equation 97)[338]. Reactions of these α-bromoboronic esters with Grignard reagents proved highly efficient even with fairly sterically hindered systems, as for example the reaction of propane-1, 3-diol isopropylboronate, **108**, with isopropylmagnesium bromide (88%) (equation 98)[339]. The corresponding reaction of *tert*-butylmagnesium chloride produced only an 11% yield, although even that is surprising considering the steric hindrance.

$$C_6H_{11}\text{-}B(OCH_2CH_2CH_2O) \xrightarrow[\text{light}]{Br_2} \underset{(107)}{1\text{-Br-}C_6H_{10}\text{-}B(OCH_2CH_2CH_2O)} \qquad (97)$$

$$\underset{(108)}{Me_2C(Br)\text{—}B(OCH_2CH_2CH_2O)} + Me_2CHMgBr \longrightarrow (CH_3)_2CHC(Me_2)\text{—}B(OCH_2CH_2CH_2O) \qquad (98)$$

By far the most useful route to α-halo boronic esters is homologation of boronic esters with $LiCHCl_2$ or its complement, the reaction of lithium or Grignard reagents with (dichloromethyl)boronic esters, $Cl_2CHB(OR')_2$. This route was discovered implicitly in its latter form by Rathke *et al.*[39], who reported the synthesis of diisopropyl dichloromethylboronate, **109**, and its reactions with alkyllithiums followed by oxidation to aldehydes (equation 99). This reaction clearly involved α-chloroboronic esters as intermediates, but the yields of aldehydes were variable and the potential synthetic value of the α-chloroboronic esters, which are more versatile synthetic intermediates than aldehydes even before the chiral synthesis applications are considered, was apparently not recognized.

$$LiCHCl_2 \xrightarrow{(MeO)_3B} \xrightarrow{Pr^iOH} \underset{(\mathbf{109})}{Cl_2CHB(OPr^i)_2} \xrightarrow{RLi} \xrightarrow{H_2O_2} RCHO \qquad (99)$$

It was but a small step to react boronic esters with (dichloromethyl)lithium to synthesize α-chloroboronic esters, and the high efficiency and generality of this process was demonstrated by Matteson and Majumdar[197,340]. An example of the possibilities inherent in this type of chemistry is the synthesis of pinacol 1-chloroallylboronate, **111**, and its conversion to pinacol 1-chloro-2-carbo-*tert*-butoxymethyl-3-butenylboronate, **112** (equation 100). The tetracoordinate borate complex **110** illustrated for the first step is the same type of intermediate that must arise in the reaction of diisopropyl dichloromethylboronate with lithium reagents.

$CH_2{=}CH{-}B(pinacolato)$ + $LiCHCl_2$ $\xrightarrow{-100\ °C}$ (110) $\xrightarrow{0-25\ °C}$ (111) $\xrightarrow{LiCH_2CO_2Bu^t}$ $\xrightarrow{LiCHCl_2}$ (112) (100)

The generality of this conversion of boronic esters, $RB(OR')_2$, into α-chloroboronic esters, $RCHClB(OR')_2$, was tested in a number of ways. It was found that R could be primary, secondary, or tertiary alkyl, or phenyl, and that the reaction tolerates the presence of an α-benzyloxy or a remote ketal function in R. The construction of highly complex structures in a few steps, as illustrated with the synthesis of **112**, suffers from the production of mixtures of diastereomers. This has been neatly solved by the use of chiral boronic esters which provide nearly pure absolute configuration as each chloromethylene group is introduced, discussed in Section III.

b. α-Halotrialkylboranes

It has been noted in Section II.A.2 that bromodeboronation of trialkylboranes is initiated by light and proceeds via α-bromoalkylboranes, which are cleaved by hydrogen bromide[273]. If the reaction is carried out in the presence of water, the intermediate (α-bromoalkyl) dialkylborane undergoes rearrangement, as illustrated by the conversion of tricyclohexylborane to 1-hydroxybicyclohexyl, **113** (equation 101[341]. A similar procedure starting from dicyclohexylhydroxyborane, $(C_6H_{13})_2BOH$, also yields **113** without the cyclohexanol by-product[342]. Removal of the hydrogen bromide during radical bromination of triethylborane, **1**, yielded (α-bromoethyl)diethylborane, which proved stable in the absence of oxygen or water but rearranged very rapidly on contact with water to form *B*-ethyl-*B*-2-butyl-*B*-hydroxyborane, **114** (equation 102)[343], or with aluminium bromide to form the corresponding bromoborane, EtB(Br)CH(Me)Et[344]. As might be expected with radical brominations, *N*-bromosuccinimide is a more efficient brominating agent than bromine[345].

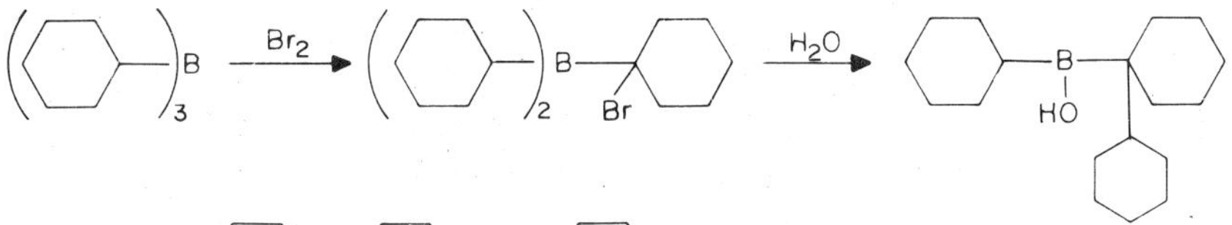

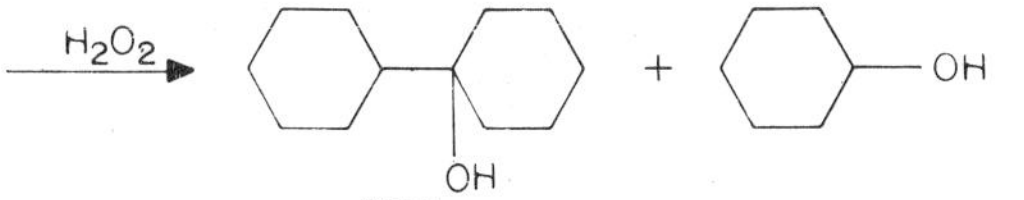

(113) (101)

$$Et_3B \ (1) \xrightarrow{Br_2} Et_2BCH(Br)Me \xrightarrow{H_2O} EtB(OH)CH(Me)Et \ (114) \qquad (102)$$

Removal of hydrogen by free radicals from 9-alkyl-9-borabicyclo[3.3.1]nonanes occurs on the alkyl substituent only if it is secondary. If the alkyl group is primary, then the α-position of the 9-bbn ring is attacked. Bromination is no exception, and rearrangement of the bromination products in the presence of water followed by oxidation leads to *cis*-1-bicyclo[3.3.0]octanol, **115** (equation 103)[346,347]. 9-Methoxy-9-bbn undergoes similar bromination and rearrangement, and a number of related ring contraction reactions with other boraheterocycles have been reported[347,348].

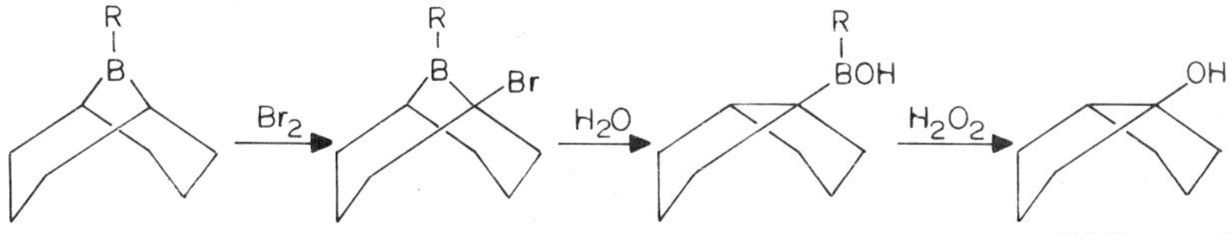

R = primary alkyl or alkoxy (115) (103)

The base-induced rearrangement of α-haloalkylboranes has been shown by Midland *et al.*[349] to invert the carbon from which the halide is displaced. For example, hydroboration of (*Z*)-1-iodo-2-methylbut-1-ene, **116**, with diethylborane followed by treatment with base and oxidation yielded > 99% pure (*RS*, *SR*)-4-methylhexane-3-ol, **117** (equation 104). The bromo analogue produced similar results, and the isomeric (*E*)-1-iodo-2-methylbut-1-ene yielded (*RR,SS*)-4-methylhexan-3-ol.

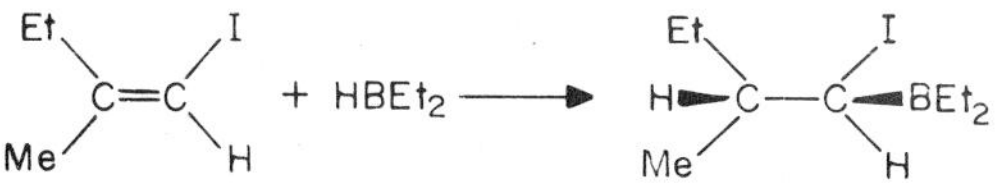

(116) 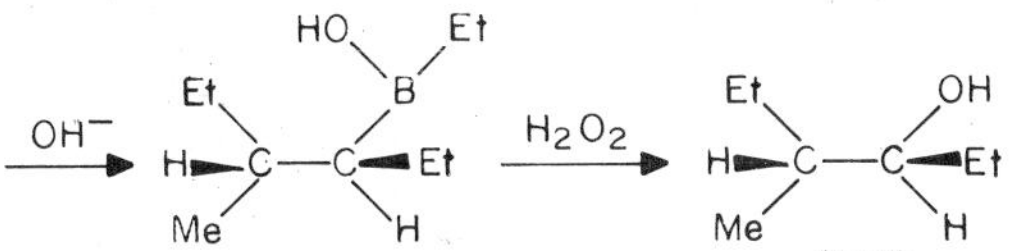 (117) (104)

α-Haloalkenylboranes are sufficiently activated that nucleophilic displacement of the halide occurs readily. It has been known for some time that the α-carbon is inverted in such displacements. An example of synthetic interest is the preparation of *cis*-alkenylboranes, **118**, by Negishi *et al.*[350], in which iodide is displaced by hydride from a trialkylborohydride (equation 105). A recent adaptation of this chemistry to prepare *cis*-alk-1-enylboronic esters[220] has been noted in another context in Section I.C.3.c. The use of this type of chemistry to prepare alkenes with a high degree of stereoselectivity is discussed in Section II.B.3.a.

$$\xrightarrow{HBR_2} \quad \xrightarrow{LiBHEt_3} \quad (118) \qquad (105)$$

c. Anions from α-halo esters, ketones, or nitriles

Trialkylboranes react with ethylbromoacetate and other α-halocarbonyl reagents in the presence of sterically hindered bases such as potassium 2, 6-di-*tert*-butylphenoxide to form borate complexes which rearrange with displacement of the halide[2c,351–357]. The probable mechanism is illustrated in equation 106 with the synthesis of ethyl cyclohexylacetate, **122**. The anion from ethyl bromoacetate and tricyclohexylborane presumably form the borate complex **119**, which rearranges to the α-boryl ester **120**, which is probably unstable and rearranges to the boron enolate **121** (see Section I.A.1), which is rapidly converted to **122** by the proton source, *tert*-butyl alcohol. These reactions generally work best with 9-bbn derivatives, and only a few of the many known examples are illustrated. The preparation of ethyl (*trans*-2-methyl-1-cyclopentyl)acetate, **123**, shows retention of stereochemistry in the migrating group (equation 107)[352]. Cyclopentylchloroacetonitrile, **124**, illustrates an α-chloronitrile which can be alkylated again with a different borane if desired (equation 108). α-Halo ketones work, as in the synthesis of cyclopentylacetone, **125**

$$(C_6H_{11})_3B + BrCH_2CO_2Et \xrightarrow[{[Br\bar{C}HCO_2Et]}]{KOCMe_3} (C_6H_{11})_3\bar{B}—CH(Br)CO_2Et \;(119)$$

$$\longrightarrow (C_6H_{11})_2B—CH(CO_2Et)—C_6H_{11} \;(120) \longrightarrow (C_6H_{11})_2B—O(EtO)C=CH(C_6H_{11}) \;(121)$$

$$\xrightarrow{Me_3COH} C_6H_{11}CH_2CO_2Et \;(122) \qquad (106)$$

$$\text{(2-methylcyclopentyl)-9-BBN} + BrCH_2CO_2Et \longrightarrow \text{2-methylcyclopentyl-}CH_2CO_2Et \quad \textbf{(123)} \tag{107}$$

$$\text{cyclopentyl-9-BBN} + Cl_2CHCN \longrightarrow \text{cyclopentyl-}CH(Cl)CN \quad \textbf{(124)} \tag{108}$$

$$\text{cyclopentyl-9-BBN} + BrCH_2C(O)CH_3 \longrightarrow \text{cyclopentyl-}CH_2C(O)CH_3 \quad \textbf{(125)} \tag{109}$$

$$Et_3B + BrCH_2CH{=}CHCO_2Et \longrightarrow EtCH{=}CHCH_2CO_2Et \quad \textbf{(126)} \tag{110}$$

(equation 109). Ethyl 4-bromocrotonate with triethylborane yields ethyl hex-3-enoate (79% *trans*), **126** (equation 110), the product consistent with the reviewer's postulate of an enolate intermediate analogous to **121**.

d. Haloform and dihalomethane anions

Dichloromethylborates from $LiCHCl_2$ and boronic esters rearrange to homologous α-chloroalkylboronic esters and have been discussed in Section II.B.1.a. This section covers several analogous reactions of other halocarbanions with various boranes, which do not include boronic esters because these fail to react with halocarbanions other than (dihalomethyl)lithiums as far as is known. Anions generated from haloforms with sterically hindered bases react with trialkylboranes, R_3B, to yield trialkylcarbinylboronic esters, $R_3CB(OR')_2$, which can be oxidized to trialkylcarbinols, R_3COH[358,359]. Yields of Bu_3COH from chloroform were 85%, from chlorodifluoromethane 98%, and from chloromethyl methyl ether 80%. The reaction with dichloromethyl methyl ether can be used to make highly hindered trialkylcarbinols, such as tricyclopentylcarbinol[359] and (cyclopentyl)(cyclohexyl)(thexyl) carbinol[360].

The reaction of dichloromethyl methyl ether with hindered dialkylalkoxyboranes has been reported to yield α-chloroboronic esters[361–364], although the n.m.r. evidence on which the structure assignment was based was not described explicitly, nor was the most likely type of alternative structure mentioned. The first intermediate has to have the structure RC(OMe)ClB(OR')R, and if migration of the second alkyl group is base catalysed, chloride and not methoxide would be displaced. Indeed, with 2 mol of $LiOCEt_3$, Bu_2BOMe yielded $Bu_2C(OMe)B(OMe)OCEt_3$[365]. With no excess of base, the second alkyl migration would displace chloride to form a boron halide, which would be highly acidic and probably catalyse the interchange of chlorine and methoxy between boron and carbon. The position of the equilibrium is unknown. The only compound reported which could (but might not) show an unequivocal distinction in the proton n.m.r. spectrum between $R_2CClB(OR')OMe$ and the isomer $R_2C(OMe)B(OR')Cl$ was that having R = cyclohexyl and R′ = Me, which in principle should have one methoxy peak if the former, two if the latter. Methoxide converted this compound into $R_2CH(OMe)B(OMe)_2$, which did show separate peaks at δ 3.28 ($COCH_3$) and 3.63 ($BOCH_3$)[365]. The only other reactions reported, oxidation to ketones[361,364] and solvolysis or pyrolysis to olefins such as cyclohexyl(cyclohexylidene)methane **27** (equation 111)[363,365], yield no information about this structure question, since either alternative might yield the same results. If the $R_2CHClB(OMe)OCEt_3$ structure is correct, it is unusual that it does not disproportionate, especially under conditions of Cl–OMe interchange, although steric hindrance could be invoked as an explanation. The alternative structure has the precedent of reported examples of stable RBFOR′ and RBClOR′[40,366]. Whatever the intermediates are, the procedure is well documented as a route from R_2BOR' to $R_2C{=}O$. A detailed procedure for the conversion of a 9-alkoxy-9-bbn into the corresponding ketone, bicyclo[3.3.1]nonan-9-one, **128**, has been published (equation 112)[367].

$$(C_6H_{11})_2BOMe \xrightarrow[\text{LiOCEt}_3]{Cl_2CHOMe,} (C_6H_{11})_2C(X)B(OMe)X \xrightarrow{250\ ^\circ C} C_6H_{11}{-}CH{=}C_6H_{10} \quad (127) \qquad (111)$$

X = Cl, OMe

$$\text{9-OR-9-borabicyclo[3.3.1]nonane} \xrightarrow[\text{LiOCEt}_3]{Cl_2CHOMe} \text{9-X-9-BXOR-bicyclo[3.3.1]nonane} \xrightarrow{H_2O_2} \text{bicyclo[3.3.1]nonan-9-one}\ (128) \qquad (112)$$

R = 2,6-dimethylphenoxy; X = OMe, Cl

Brown and Imai[368] have found that 2-alkyl-1, 3, 2-dithiaborolanes (ethanedithiol boronates) will react with (trichloromethyl)lithium, in contrast to boronic esters, which are inert towards this reagent. The alkyl group and both thiol groups migrate from boron to carbon, the order of migration being unknown, and the rearrangement is associated with cleavage of the thf solvent to produce an intermediate boronic ester, for example, bis(chlorobutyl) (*trans*-2-methylcyclopentyl)dichloromethylboronate, **129**, which may be oxidized with hydrogen peroxide to *trans*-2-methylcyclohexylpentanecarboxylic acid, **130**, or hydrolysed with aqueous sodium hydroxide to the thiocetal of *trans*-2-methylcyclohexanecarboxaldehyde, **131** (equation 113). The examples illustrated proved stereospecific, but in the oxidation to prepare 2-*exo*-norboranecarboxylic acid in an analogous manner, 14% epimerization to *endo* acid occurred.

e. Anions from (halomethyl) silanes

The reaction of trialkylboranes, R_3B, with (silylhalomethyl)lithiums, $R'_3SiCHXLi$, to form α-silylalkylboranes, $RCH(SiR')_3BR_2$, was first explored by Larson and coworkers[369].

$$\xrightarrow[\text{thf}]{\text{LiCCl}_3} \quad (129) \quad \longrightarrow \quad (130) \text{ or } (131) \qquad (113)$$

The efficient reaction of [(trimethylsilyl) (chloro)methyl] lithium[370] with boronic esters to form α-trimethylsilyl boronic esters, **132**, has been described by Matteson and Majumdar (equation 114)[371,372]. Application of this reaction to (+)-pinanediol phenylboronate yielded only a 73:27 diastereomeric ratio[373].

$$\text{R–B(pinacolate)} + \text{Me}_3\text{Si}\bar{\text{C}}\text{HCl Li}^+ \longrightarrow \text{Li}^+\ [\text{R}(\text{Me}_3\text{SiCHCl})\text{B(pinacolate)}]^- \longrightarrow \text{R–CH(SiMe}_3\text{)–B(pinacolate)}\ (132) + \text{LiCl} \qquad (114)$$

f. Diazo esters and ketones

Diazo esters, ketones, and aldehydes, N_2CHCOY (Y = OR, R, H), were found by Hooz and coworkers to function as the synthetic equivalents of α-halo ester, ketone, or related anions[374–377]. The initial product of reaction of R_3B with N_2CHCOR' is an enol borinate, $RCH{=}C(R')OBR_2$, which is a useful intermediate for aldol or related reactions[378]. The major product has been shown to be the *E*-isomer by Masamune *et al.*[379], who demonstrated the utility of these intermediates in stereocontrolled aldol condensations.

2. Carbonylation and related reactions

a. Carbon monoxide

The fundamental chemistry of the reactions of carbon monoxide with trialkylboranes was first reported by Hillman[380], who showed that at temperatures in the range 50–75 °C two alkyl groups would migrate from boron to carbon to produce 1,4-dioxa-2,5-diborinanes, **135**, which would rearrange further at *ca.* 150 °C to form boroxines (boronic anhydrides), **136** (equations 115)[380]. Hillman[380] also found that adding an aldehyde leads to a 4-bora-1,3-dioxolane, **137**, perhaps by capture of the presumed unstable intermediate

133 before it can dimerize to **134** (equation 116). Addition of ethylene glycol facilitates the third alkyl migration and yields a stable boronic ester, **138** (equation 117).

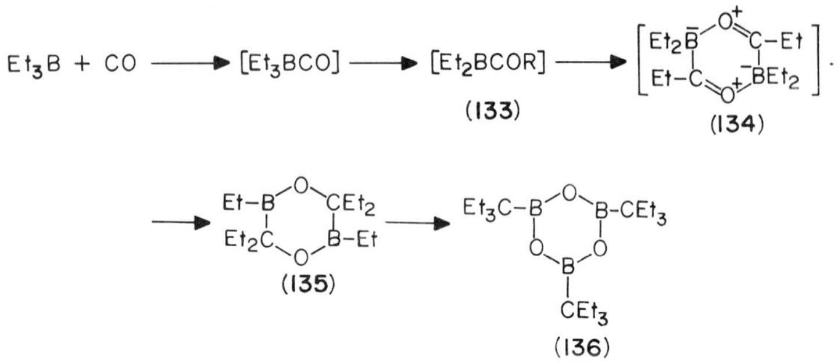

$$Et_3B + CO \longrightarrow [Et_3BCO] \longrightarrow \underset{(133)}{[Et_2BCOR]} \longrightarrow (134) \longrightarrow (135) \longrightarrow (136) \qquad (115)$$

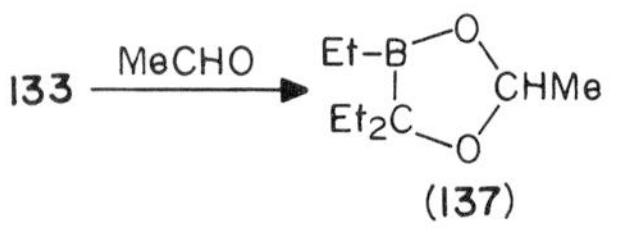

$$133 \xrightarrow{MeCHO} (137) \qquad (116)$$

$$136 \xrightarrow{HOCH_2CH_2OH} (138) \qquad (117)$$

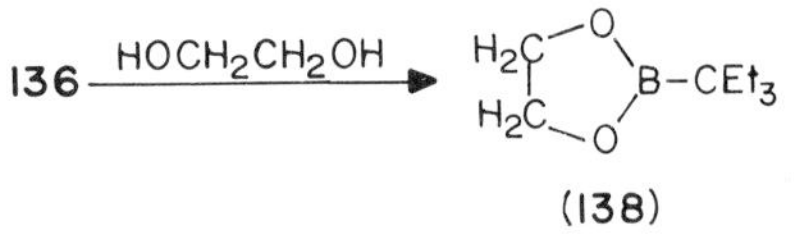

Isonitriles react with trialkylboranes in an analogous fashion, but require higher temperatures for each stage of rearrangement to occur, and the analogues of all structures **133**–**136** having NR in place of O are isolable if the substituents are bulky enough[381].

Hillman's reactions were carried out at high pressures of carbon monoxide, which was conveniently available in his industrial laboratory but which is unpopular with academic chemists. Brown and Rathke[382] subsequently found that the reactions occur under 1 atm of carbon monoxide in diglyme solvent if the temperature is increased to 100–125 °C, and followed up this discovery with an extensive investigation of the applications of the process. This chemistry has been reviewed by Brown[2c,66,383], and only a few of the highlights are covered here.

Hydride sources such as lithium borohydride or trialkoxyaluminium hydrides greatly increase the rate of absorption of carbon monoxide, allowing the reactions to be carried out at 25–45 °C but stopping the reaction after migration of one alkyl group[384]. The result of this reduction is a borane intermediate which can be hydrolysed to an alcohol[384] or oxidized to an aldehyde[385]. It is generally advantageous to use 9-alkyl-9-bbns for these reactions[386]. Functional groups such as esters and nitriles (separated from the boron by three or more carbons) are unaffected, especially if lithium tri-*tert*-butoxyaluminium hydride is used as the reducing agent[387]. Retention of configuration of the migrating alkyl group has been confirmed[352].

In more recent work, it has been found that potassium triisopropoxyborohydrides[388] are particularly active reagents for promoting the reaction of trialkylboranes with carbon monoxide. The process may be illustrated most succinctly with *trans*-2-methylcyclopentyl-9-bbn, **139**, which shows the usual stereospecific retention in the migrating group when the α-hydroxyalkylborane intermediate is formed as the potassium salt, **140** (equation 118). Buffered hydrogen peroxide converts **140** to the aldehyde, or alkaline hydrolysis yields the alcohol (equations 119 and 120).

(139) $\xrightarrow[CO]{KHB(OR)_3}$ (140) (probably a cyclic dimer) (118)

140 $\xrightarrow[pH\ 7]{H_2O_2,}$ CHO (119)

140 $\xrightarrow{NaOH}$ CH_2OH (120)

A useful alternative procedure for the carbonylation process utilizes the potassium trialkylborohydride[389], which is easily prepared from the trialkylborane and potassium hydride[390]. A catalytic amount of the free borane must be present in order for this reaction to proceed.

Hubbard and Smith[391] obtained evidence that the active hydride which captures the originally formed borane carbonyl and thus catalyses the uptake of carbon monoxide in these reactions is generally trialkylborohydride, even when the added hydride reagent is an aluminohydride.

Carbonylation of a 9-alkyl-9-bbn in the presence of $LiHAl(OMe)_3$ followed by reduction with lithium aluminium hydride yields the homologous 9-alkyl-CH_2-9-bbn, for example, 9-(*exo*-2-norbornylmethyl)-9-bbn, **141** (equation 121)[392]. It may be noted that **141** has the configuration opposite that of the hydroboration product from 2-methylenenorbornane. It is possible to generate the olefin from boranes such as **141** by treatment with benzaldehyde[393].

$\xrightarrow[CO]{LiHAl(OMe)_3,}$ Al(OMe)$_3$ O B

$\xrightarrow{LiAlH_4}$ B (121)

(141)

Mild acid treatment of the α-hydroxyborinic acid intermediates from reductive carbonylation of trialkylboranes will result in migration of a second alkyl group from boron to carbon, with displacement of the hydroxyl group as water. Subsequent oxidation yields dialkyl carbinols, **142** (equation 122)[394].

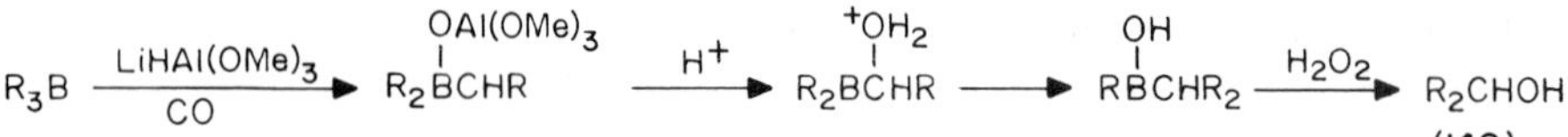

$$R_3B \xrightarrow[CO]{LiHAl(OMe)_3} R_2\overset{OAl(OMe)_3}{\overset{|}{B}}CHR \xrightarrow{H^+} R_2\overset{^+OH_2}{\overset{|}{B}}CHR \longrightarrow R\overset{OH}{\overset{|}{B}}CHR_2 \xrightarrow{H_2O_2} \underset{(142)}{R_2CHOH} \qquad (122)$$

R = ethyl, 1-octyl, 2-butyl, isobutyl, cyclohexyl, norbornyl

Oxidation of intermediates of structure **135** (see Hillman's process above) yields ketones, and Brown and coworkers have developed this chemistry extensively. One alkyl group is sacrificed, but if one of the groups is tertiary, most conveniently thexyl, then the primary or secondary alkyl groups migrate faster and the tertiary group serves as a blocking group. The original process used thexylborane as a hydroborating agent, which proved successful if the two other alkyl groups were alike, if a cyclic intermediate was involved, or if one relatively hindered and one primary alkyl group constituted the desired combination[2b]. Examples of ketones synthesized in this way include indanone, **143**[395] and juvabione, **144** (equation 123 and 124)[396]. More recently, the use of thexylchloroborane has made possible the stepwise synthesis of unsymmetrical boranes having two different primary alkyl groups (see Section I.C.3.c.), and the method consequently has wider generality[397], as for example in the synthesis of an intermediate, **145** (equation 125), which can be converted to dihydrojasmone[398]. For this last synthesis, carbonylation at 70 atm allowed the use of a lower temperature, 50 °C.

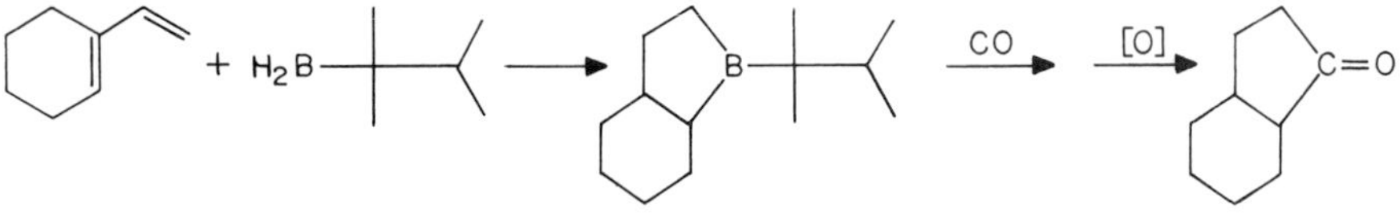

+ H_2B— → B— $\xrightarrow{CO}$ $\xrightarrow{[O]}$ C=O

(143)

(123)

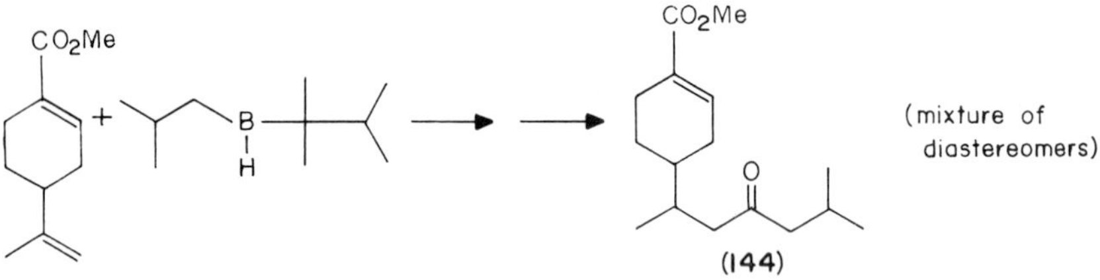

CO_2Me + B–H → → CO_2Me O (mixture of diastereomers)

(144)

(124)

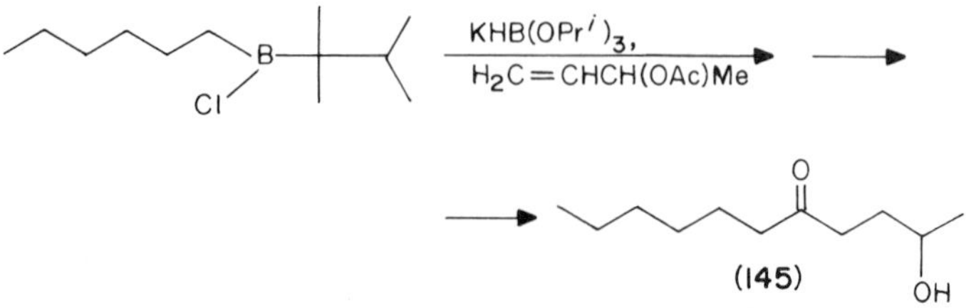

B— Cl $\xrightarrow[H_2C=CHCH(OAc)Me]{KHB(OPr^i)_3,}$ →

→ O (145) OH (125)

Carbonylation of boranes with carbon monoxide has been found to be an efficient route to ^{13}C-and ^{14}C-labelled aldehydes, carboxylic acids, alcohols, and ketones[287,399–403].

b. Cyanidation

The use of sodium cyanide and trifluoroacetic anhydride in order to accomplish the same types of synthetic transformations as carbon monoxide with boranes was discovered by Pelter and coworkers[404]. The advantage for laboratory operations is that the reactions proceed at room temperature and require no special apparatus. The first stable intermediate is apparently the 2-trifluoromethyl-4, 4, 5-trialkyl-1, 3, 5-oxazaboroline, **146**, formed on migration of two alkyl groups from boron to carbon, and **146** can be rearranged further by warming with trifluoroacetic anhydride, ultimately leading to the trialkylcarbinol, or oxidized with hydrogen peroxide to yield the ketone (equation 126).

$$R_3B + CN^- \longrightarrow R_3\bar{B}CN \xrightarrow{(CF_3CO)_2O} \mathbf{(146)} \xrightarrow{H_2O_2} R_2CO$$

$$\mathbf{(146)} \xrightarrow{(CF_3CO)_2O,\ \text{warm}} CF_3CO_2{-}B(CR_3){-}O{-}C(CF_3){=}N{-}COCF_3 \xrightarrow{H_2O_2} R_2COH \qquad (126)$$

Cyanidation of dialkylalkenylboranes, $R_2BCH{=}CHR'$, leads to an oxazaboroline, **146**, which can be oxidized to an α, β-unsaturated ketone, $RCOCH{=}CHR'$[405]. Ordinarily the products from cyanidation and those from carbonylation are the same, but cyanidation of *cis, cis*-perhydro[9*b*]boraphenalene, **147**, gives mainly the carbinol derived from attack on the opposite side of the boron atom from that observed with carbon monoxide (equation 127)[406]. Since the alkyl groups are known to migrate with retention of configuration, the stereochemistry must be determined by which side of the boron the cyanide or carbon monoxide is on at the time of the first irreversible step. It appears that the cyanide equilibrates to the thermodynamically favoured face before the first migration occurs, but carbon monoxide does not. The *cis, trans*-isomer of **147** yielded the *cis, trans, trans* carbonylation product with either cyanide or carbon monoxide. Hydroboration with thexylborane followed by cyanidation has provided a key transformation in a synthesis of (±)-estrone methyl ether, **148** (equation 128).[407]

(147) → (CN⁻ product) or (CO product) (127)

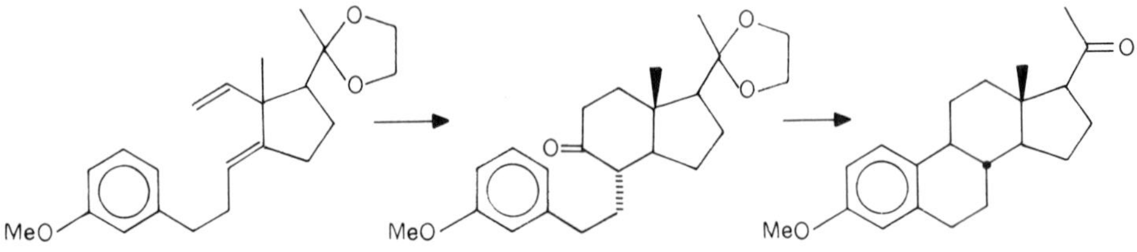

(148)

(128)

c. Thiol anions

The first sulphur-substituted carbanion to be reacted with trialkylboranes was the ylide $Me_2S^+—CH_2^-$, which Tufariello and Wojtkowski[408] used to convert R_3B into RCH_2BR_2. The nitrogen ylide $Me_3N^+—CH_2^-$ similarly yielded RCH_2BR_2–NMe_3[409]. It was also found that $Me_2S^+—CH^-—CO_2Et$ would convert R_3B into RCH_2CO_2Et and R_2BOH[410], a reaction equivalent in net result to that of trialkylboranes with ethyl bromoacetate and a hindered base discovered several years later (see Section II.B.1c). Negishi *et al.*[411] improved on the original Tufariello and Wojtkowski process by treating R_3B with $LiCH_2SMe$ followed by MeI to yield the equivalent intermediate. The reaction is of potential value for converting alkenyldisiamylboranes, $RCH{=}C(R')B(Sia)_2$, stereospecifically into the homologous allyldisiamylboranes, $RCH{=}C(R')CH_2B(Sia)_2$.

More useful results have been obtained with the anions derived from bis(phenylthio)methane and related compounds[412–414]. For example, 1-lithio-1, 1-bis(phenylthio)butane, **149**, reacts with trialkylboranes to produce α-(phenylthio)alkylborane intermediates, which can be oxidized to ketones or rearranged further with mercury(II) chloride to form borane precursors to propyldialkylcarbinols (equation 129)[413]. The reaction has been shown convert cyclopentyl-9-bbn, **150**, reasonably efficiently into the corresponding 9-[(cyclopentyl](phenylthio)methyl]-9-bbn, which was oxidized with *N*-chlorosuccinimide (ncs) is buffered methanol to yield a monothioacetal of cyclopentanecarboxaldehyde (equation 130)[288].

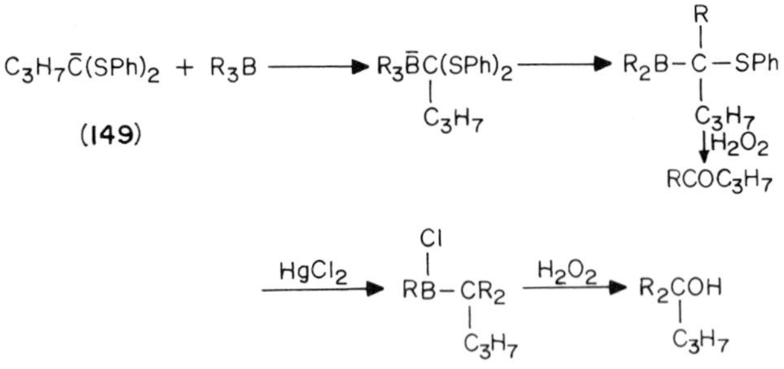

$$C_3H_7\bar{C}(SPh)_2 + R_3B \longrightarrow R_3\bar{B}C(SPh)_2(C_3H_7) \longrightarrow R_2B{-}C(R)(C_3H_7){-}SPh \xrightarrow{H_2O_2} RCOC_3H_7$$

(149)

$$\xrightarrow{HgCl_2} RB(Cl){-}CR_2(C_3H_7) \xrightarrow{H_2O_2} R_2C(C_3H_7)OH \qquad (129)$$

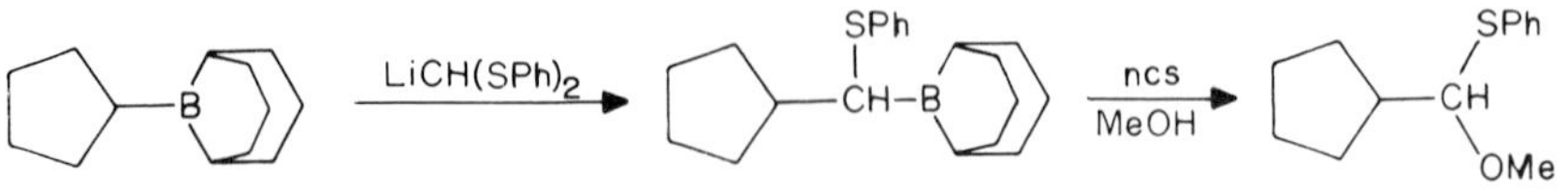

(150)

$$\text{LiCH(SPh)}_2;\ \text{ncs, MeOH} \qquad (130)$$

An attempt to adapt similar chemistry to boronic esters and either react $RB(OR')_2$ with $LiCH(SPh)_2$ or RLi with $(PhS)_2CHB(OR')_2$ was unsuccessful, although treatment of $(PhS)_2CHBO_2C_3H_6$ with the hazardously toxic reagent methyl fluorosulphonate followed by butyllithium did yield 30% of $BuCH(SPh)BO_2C_3H_6$[414]. The problem of converting boronic esters into aldehydes was effectively solved with the development of the reaction with $LiCHCl_2$ (see Section II.B.1.a), but a useful alternative with LiCH(OMe)SPh as the homologating agent has been developed by Brown and Imai[415]. As in all other intramolecular rearrangements of borate complexes, the configuration of the migrating group is retained, as shown by the conversion of propane-1,3-diol *trans*-2-methylcyclopentylboronate, **151**, into the corresponding aldehyde (equation 131).

LiCH(SPh)OMe / $HgCl_2$; H_2O_2 ; OMe ; CH–B ; CHO

(151)

(131)

3. *Alkenylborane chemistry*

a. *The Zweifel alkene synthesis*

The stereospecific synthesis of alkenes from alkenylboranes is among the most useful of the synthetic applications of organoborane chemistry. An example of Zweifel's alkene synthesis in its original form is the synthesis of *cis*-hex-1-enylcyclohexane, **154**, from cyclohexene and hex-1-yne via dicyclohexylborane and *trans*-hex-1-enyl(dicyclohexyl)borane (equation 132)[416]. The probable mechanistic steps involve formation of a cyclic iodonium ion, **152**, which undergoes base-induced migration of a cyclohexyl group to open the iodonium ion with inversion of the α-carbon and form a β-iodoalkylborinate, **153**, which then with base undergoes *anti*-elimination of iodide and cyclohexylboronic acid. (The evidence that such eliminations are *anti* is discussed in Section II.A.2.)

BH ; C≡CH ; C=C ; I_2 / NaOH

(152) (153) (154)

(132)

A complementary synthesis of *trans*-hex-1-enylcyclohexane, **157**, began with 1-iodohex-1-yne, which was hydroborated with dicyclohexylborane to form [(*Z*)-1-iodohex-1-enyl]dicyclohexylborane, **155**, which on treatment with sodium methoxide yielded the (*E*)-1-cyclohexylhex-1-enylborinic ester, **156**, which was cleaved with acetic acid to **157** (equation 133)[417]. It was also found that **155** could be cleaved with acetic acid to yield (*Z*)-1-iodohex-1-ene.

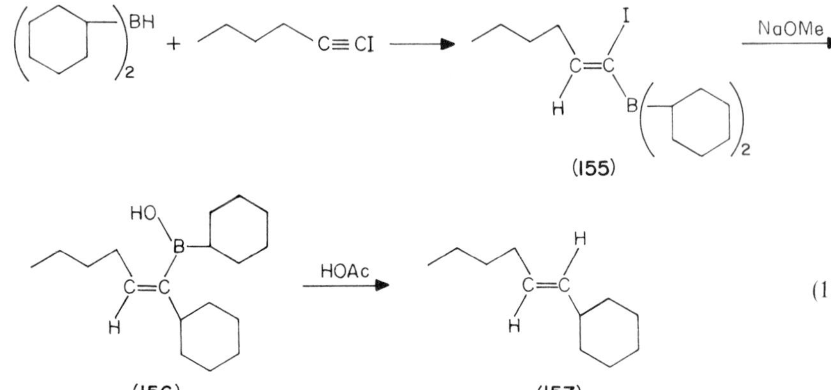

(133)

An alternative synthesis of the *trans*-alkene **157** uses cyanogen bromide, which leads to a β-bromoalkylborane intermediate, **158**, which undergoes *syn*-elimination (equation 134)[418]. This alternative is not limited to alk-1-enylboranes, although hydroboration is not a practical route to internal alkenylboranes unless symmetry or considerable steric hindrance makes regiospecific synthesis possible.

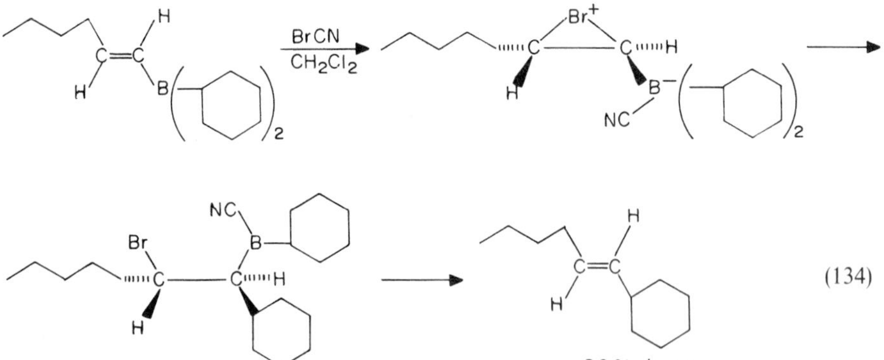

(134)

The original iodination procedure applied to the hydroboration product from hex-3-yne yielded (*Z, E*)-4, 5-diethylocta-3, 5-diene, **159** (equation 135)[419]. Trialk-1-enylboranes are not available because of dihydroboration, and with thexyldialkenylboranes some thexyl group migration occurred, which was remedied by cleaving the thexyl group with trimethylamine *N*-oxide (see Section II.A.1.d), followed by treatment of the resulting dialk-1-enylhydroxyborane with iodine and sodium hydroxide to make the *Z, E*-diene.

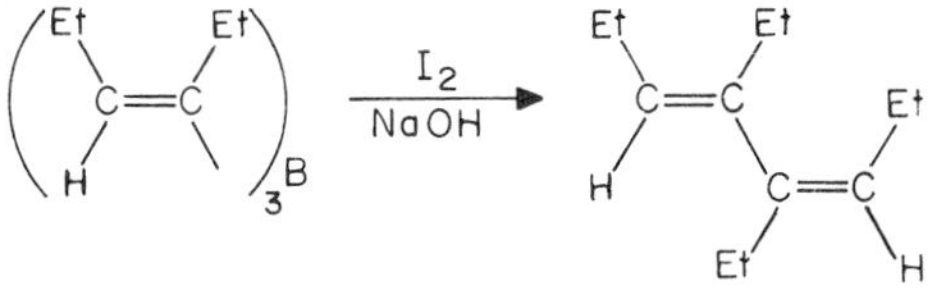

(135)

(159)

The *trans*-alkene synthesis was applied to the preparation of dienes by Negishi and Yoshida[420]. A 1-haloalkyne with thexylborane forms a (haloalkenyl)thexylborane, which can hydroborate a second alkyne to produce a *B*-haloalkenyl-*B*-alkenyl-*B*-thexylborane, which on treatment with sodium methoxide rearranges to the *E, E*-diene, **160** (equation 136)[420]. The use of 2 mol of 1-iodoalkyne with 1 mol of thexylborane results in a *B, B*-bis(1-iodoalkenyl)-*B*-thexylborane, which undergoes rearrangement and elimination to form the very sensitive butatriene **161** stereoselectively (equation 137)[421].

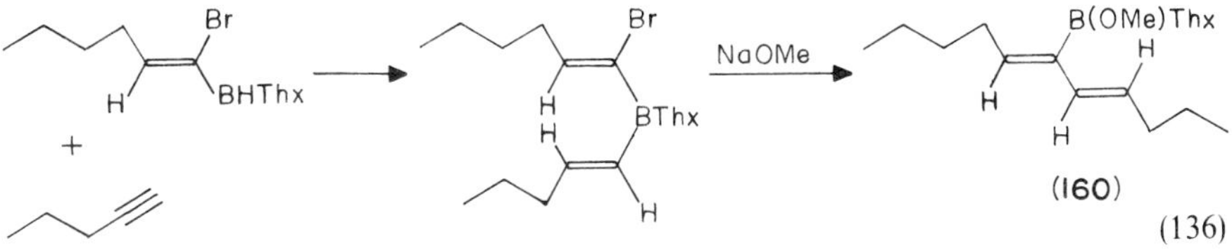

(160)

(136)

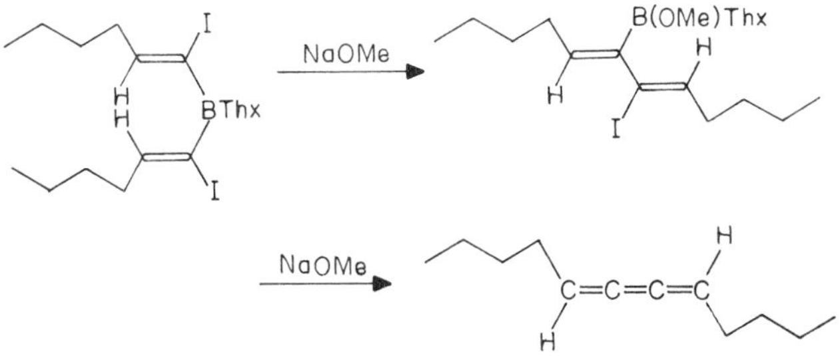

(137)

(161)

The Zweifel olefin syntheses have become much more general and more efficient with more recent developments in organoborane chemistry, and have been applied to the synthesis of a variety of natural products. Among the first of these was a synthesis of a prostaglandin, **162**, by Corey and Ravindranathan (equation 138)[422].

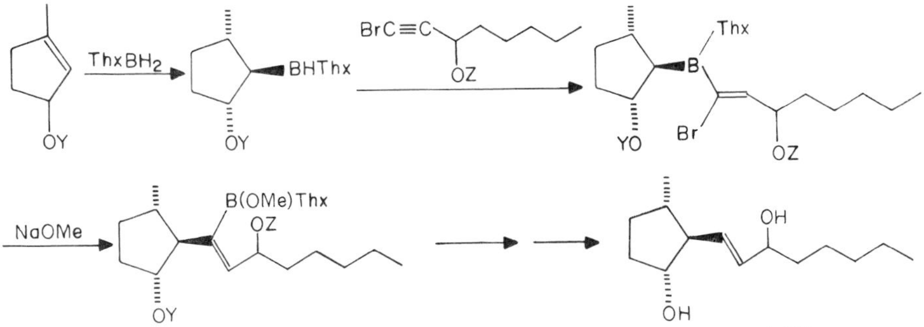

Y= $SiMe_2CMe_3$; Z= tetrahydropyranyl

(162)

(138)

In a model study for prostaglandin synthesis, Evans *et al.*[423] made significant improvements in the original Zweifel method. One problem is that if a thexyl blocking group is used, there is considerable thexyl migration in competition with that of the desired group. Another problem is that iodine causes some direct carbon–boron bond cleavage to form the 1-iodoalkene (see Section II.A.2). The problem was solved by removing the thexyl group by treatment with triethylamine to make the amine borane, which was converted into a boronic ester, for which there are now several alternative hydroboration approaches. The second side reaction was reduced to *ca.* 15% merely by changing the solvent to methanol. The significant steps of Evans *et al.*'s approach are illustrated in equation 139, starting from dimethyl *trans*-2-ethylcyclopentaneboronate, **163**. In addition to the *cis*-1-lithioalkene and *trans* rearrangement product illustrated, the sequence was also carried out with *trans*-1-lithioalkene and shown to yield pure *cis* product.

B(OMe)$_2$ (163) Li OSiMe$_2$CMe$_3$ → B OMe OSiMe$_2$CMe$_3$

$\xrightarrow[\text{NaOMe}]{I_2}$ H OSiMe$_2$CMe$_3$ (139)

The Zweifel synthesis is readily applicable to cyclic boranes such as borinane, **40**[424]. For example, borinane and 1-bromohexyne can be converted into pentylidenecyclohexane, **164** (equation 140). Recent advances in methods of preparation of unsymmetrical boranes

BH + Br → Br H B (40)

$\xrightarrow{\text{NaOH}}$ HO B H $\xrightarrow{I_2}$ (164) (140)

by Brown's group have greatly extended the generality of the Zweifel synthesis[425]. Basavaiah and Brown[426] have used this approach to synthesize (*Z*)-alk-7-en-1-ols, **165** (equation 141)[426]. Some of these alcohols or their acetates are moth sex pheromones. Another combination of unsymmetrical borane synthesis with the Zweifel olefin synthesis

BCl $\xrightarrow[\text{(2) RC}\equiv\text{CH}]{\text{(1) LiAlH}_4}$ B R $\xrightarrow[-78\ ^\circ\text{C}]{\text{NaOMe, I}_2}$ BOMe H R I →

(MeO)$_2$B R → HO R (141)

R = *n*-butyl or *n*-hexyl (165)

results in a simple preparation of (*Z*)-2-methyloctadec-7-ene, **166** (equation 142), which can be epoxidized to (±)-disparlure, the (+)-enantiomer of which is the sex pheromone of the gypsy moth[427]. A similar synthesis leads to (*Z*)-$Me(CH_2)_7CH{=}CH(CH_2)_{12}Me$, which is muscalure, a sex pheromone of the housefly[428].

BBr_2SMe_2 $LiAlH_4$ $BHBrSMe_2$ NaOMe I_2

(166)

(142)

b. Other alkenylborate rearrangements

The use of iodine as the electrophile to induce alkenylborate rearrangements, discussed in the preceding Section II.B.3.a, has parallels with other electrophiles. The results tend to be simpler when subsequent elimination of boron is not a possibility. For example, (*E*)-hexenyldicyclohexylborane from hydroboration of hex-1-yne with dicyclohexylborane was treated with methyllithium and then hydrogen chloride, resulting in hexyl group migration from boron to carbon. The final product after oxidation was 1-cyclohexyl-hexan-1-ol, **167** (equation 143)[429]. It should be noted that the methyl group shows a particularly low migratory aptitude, and only about 5% methyl migration was observed. The use of the usual *trans*-2-methylcyclopentyl migrating group established that the configuration of the migrating group is retained, as is usual for this type of mechanism. Hydrolysis of alkenyltrialkylborates with aqueous acid has been reported to yield alkenes from borane elimination after the rearrangement step[430].

CH_3Li Li^+ HCl OH

(167)

(143)

Reaction of (α-methoxyvinyl)lithium with trialkylboranes at $-78\,°C$ leads to *B*-(α-methoxyvinyl)-*B*, *B*, *B*-trialkylborates, **168**, which rearrange at 25 °C to *B*-(α-alkylvinyl)-*B*-methoxy-*B*, *B*-dialkylborates, **169** (equation 144). Acid hydrolysis of **169** results in migration of a second alkyl group to form a borinic acid, which can be oxidized to a tertiary alcohol, and iodination of either intermediate results in alkyl migration and deboronation analogous to the Zweifel alkene synthesis, **168** yielding $H_2C{=}C(OMe)R$ and **169** yielding $R_2C{=}CH_2$[431,432].

$$CH_2{=}C(OMe)Li \xrightarrow[-78\,°C]{BR_3} CH_2{=}C(OMe){-}\bar{B}R_3\ (168) \xrightarrow{25\,°C} CH_2{=}C(R){-}\bar{B}R(OMe)\ (169)$$

$$\xrightarrow{H_3O^+} MeCR_2B(OH)R \xrightarrow{H_2O_2} MeCR_2OH \qquad (144)$$

Aldehydes as electrophiles convert vinyltrialkylborates into oxaborolanes, **170**, which can be oxidized to 1, 3-diols[433] or treated with phosphorus pentachloride followed by base to yield cyclopropanes (equation 145)[434]. Vinyltrialkylborates also react with epoxides to generate $R_2BCHR(CH_2)_3OH$ and, of course, with iodine to form $RCH{=}CH_2$[435]. *B*-(1-Alkenyl)-9-bbns do not need to be converted into borate complexes in order to react with aldehydes, but in this case the reaction is direct electrophilic displacement of the boron by the aldehyde carbon to produce allylic alcohols, $RCH{=}CHCHOHR'$, with retained *trans* geometry[436].

$$CH_2{=}CH{-}\bar{B}R_3 \xrightarrow{R'CHO} (170) \begin{cases} \xrightarrow{H_2O_2} R'CH(OH)CH_2CH(OH)R \\ \xrightarrow{PCl_5;\,NaOH} R{-}\triangle{-}R' \end{cases} \qquad (145)$$

E.g. R = ethyl, R′ = phenyl

Two alkyl groups migrate in an alkenylborate type rearrangement of 2-furylborates reported by Suzuki *et al.*[437], who obtained *cis*-but-2-ene-1, 4-diols, **171**, on oxidation of the postulated borane intermediate (equation 146). 2-Lithio-6-bromopyridine forms

$$\text{2-furyl}{-}Li + BR_3 \longrightarrow \text{2-furyl}{-}\bar{B}R_3 \xrightarrow{H^+} \text{(cyclic B–R borane, }CR_2\text{)} \xrightarrow{H_2O_2} HOCH_2CH{=}CHCR_2OH\ (171) \qquad (146)$$

borate complexes which undergo a fragmentation type of ring opening with migration of a single alkyl group. The product after protolysis with acetic acid is a (2*Z*, 4*E*)-2, 4-dienonitrile, **172** (equation 147)[438].

$$Br{-}C_5H_3N{-}\bar{B}Pr_3 \longrightarrow N{\equiv}C{-}CH{=}CH{-}CH{=}C(Pr)BPr_2 \xrightarrow{HOAc} N{\equiv}C{-}CH{=}CH{-}CH{=}CH{-}Pr\ (172) \qquad (147)$$

N-Methylindole provides an example of an aromatic heterocycle that is stable under the borate rearrangement conditions, and Levy[439] has synthesized 2-alkyl-*N*-methylindoles by reaction of the lithiated indole with a trialkylborane followed by iodination. The use of carbon electrophiles in place of iodine leads to 2,3-disubstituted *N*-methylindoles, as in the synthesis of 1-methyl-2-ethylindole-3-acetamide, **173**, illustrated in equation 148[440].

$$\text{(N-Me-indol-2-yl)}\bar{B}Et_3 + ICH_2CONH_2 \longrightarrow \xrightarrow{H_2O_2} \text{1-Me-2-Et-indole-3-}CH_2CONH_2 \; \textbf{(173)} \tag{148}$$

Hydroboration of propargyl chloride leads to 3-chloroprop-1-enylboranes, which have a built-in electrophilic centre analogous to that of an α-haloborane. Treatment with methyllithium results in migration of one of the alkyl groups (other than methyl) with displacement of the allylic chloride. The initial rearrangement products are allylic boranes, which are easily cleaved with acetic acid to yield the allyl-substituted alkane, as in the synthesis of allylpinane, **174**, in equation 149[441]. An earlier example of this type of rearrangement process was provided by the reaction of $(BuO)_2BCH{=}CHCCl_3$ with RMgX to yield $(BuO)_2BCH(R)CH{=}CCl_2$[332].

$$(\text{pinanyl})_2BCH{=}CHCH_2Cl \xrightarrow[(2)\ HOAc]{(1)\ MeLi} \text{pinanyl-}CH_2CH{=}CH_2 \; \textbf{(174)} \tag{149}$$

Photochemical rearrangement of (1*E*)-dienylboranes, **175** (X = H), which can be prepared by hydroboration of alkenylacetylenes, provides another example of alkyl migration[442,443]. However, for synthetic purposes the same boracyclopentenes can be obtained more efficiently by hydroboration of 1-iodoalkenylacetylenes to form (1*Z*)-1-iododienylboranes, **175** (X = I), which react with potassium tri-2-butylborohydride to displace the iodide and form (1*Z*)-dienylboranes (176), which cyclize spontaneously with consequent alkyl migration (equation 150)[443]. Thus, it appears that the photochemical transformation consists only of isomerization of the (1*E*)-dienylborane, **175** (X = H), to the (1*Z*)-isomer **176**. The boracyclopentene products **177** can be oxidized to *cis*-but-2-ene-1,-3-diols, or can be cleaved sequentially with acetic acid and hydrogen peroxide to yield homoallylic alcohols, $H_2C{=}CMeCH_2CH(R)OH$[442,443].

$$\textbf{(175)} \xrightarrow{(i)} \textbf{(176)} \longrightarrow \text{(zwitterion, } \bar{B}R_2\text{)} \longrightarrow \textbf{(177)} \tag{150}$$

(i) If X = H, UV light; if X = I, $KHB(2\text{-}Bu)_3$

c. Alkynylborate rearrangements

Alkynyltrialkylborate rearrangements were first studied by Binger[444], who found that sodium triethylpropynylborate, **178**, and acetyl chloride yielded a rearranged heterocyclic product resulting from acetylation at the β-carbon followed by migration of two ethyl groups to the α-carbon (equation 151). Protonation of $RC{\equiv}C\bar{B}R'_3$ was found to cause migration of one alkyl group to form the alkenylborane, $RCH{=}C(R')BR'_2$[445]. The addition of electrophiles in this manner is fairly general, with a tendency toward predominant *trans*-addition of the electrophile and migrating alkyl group across the double bond, although the stereoselection is not necessarily high enough for synthetic purposes. An exception is the protodeboronation of *B*-phenylethynyl-*B*-thexyl-*B*, *B*-dibutylborate, **179**, with propionic acid, which yielded hex-1-enylbenzene that was 98% *cis* (equation 152)[446]. Hydrogen chloride causes migration of one alkyl group of $R_3\bar{B}C{\equiv}CR'$ at $-78\,°C$ and a second group at higher temperatures[447]. Ethynyltrialkylborates, **180**, rearrange on protonation to yield the opposite regioisomer of the borane from what hydroboration would produce, and vinyltrialkylboranes behave similarly (equation 153)[448].

MeCOCl + $Et_3B{-}C{\equiv}CMe$ ⟶ [Et–B ring product with O, Me, Me, Et, Et]
(178) (151)

$Bu_2\bar{B}(Sia){-}C{\equiv}CPh$ (179) $\xrightarrow{EtCO_2H}$ Bu(Bu–B(Sia))C=C(Ph)H ⟶ Bu(H)C=C(Ph)H (152)

$R_3\bar{B}C{\equiv}CH$ + H^+ ⟶ $R(R_2B)C{=}CH_2$
(180) (153)

Other electrophiles which have been added to alkynyltrialkylborates include dimethyl sulphate[449], epoxides[450], $Me_2N{=}CH_2{}^+$[451], acetylpyridinium ion[452], protonated anisoleiron tricarbonyl[453], dibromomethane[454], and chloromethyl methyl ether[455]. All of these reactions may be summarized as $RC{\equiv}C\bar{B}R'_3 + E^+ \rightarrow REC{=}C(R')BR_2$. The connection to the acetylpyridinium ion is regioselective at the 4-position of the pyridine ring, but the alkenyl group is a typical *cis, trans* mixture[452]. Triethyloxonium fluoborate with $MeOCH_2C{\equiv}C\bar{B}Et_3$ yields 97% *E*-isomer of $MeOCH_2C(Et){=}C(Et)BEt_2$[455]. Good *trans* selectivity has been observed in reactions of trimethylsilyl chloride[456] and tributyltin chloride[457]. Phenylselenyl chloride reacts with $R_3\bar{B}C{\equiv}CBu$ to form $R_2BC(R){=}C(SePh)Bu$, which can be oxidized with trimethylamine oxide to yield RCOCH(SePh)Bu, a precursor to an α, β-unsaturated ketone[458].

Treatment of $RC{\equiv}C\bar{B}R'_3$ with HCl to form $RCH{=}C(R')BR'_2$ followed by sodium hydroxide and iodine leads to trisubstituted olefins, $RCH{=}CR'_2$[459]. Retention of configuration of the migrating R′ groups was proved with R′ = *trans*-2-methylcyclopentyl by oxidative degradation of the olefin to known *trans*-2-methylcyclopentanol and *trans*-2-methylcyclopentylcarbinol.

B-Alkynyl-*B*-alkenyl-*B*, *B*-dialkylborates undergo 97–99% stereoselective alkenyl group migration on treatment with either boron trifluoride or tributyltin chloride[460]. An example is the synthesis of methyl (10*E*, 12*Z*)-hexadecadienoate, **181** (equation 154). Methyl undecynoate was hydroborated with dicyclohexylborane, the resulting alkenylborane was treated with 1-lithiopent-1-yne, and the borate complex was treated with boron trifluoride etherate, then with acetic acid at 50 °C, which yielded **181** (66%). It was already known that **181** can be converted into the sex pheromone of the silkworm moth, **182**, by reduction with lithium aluminium hydride.

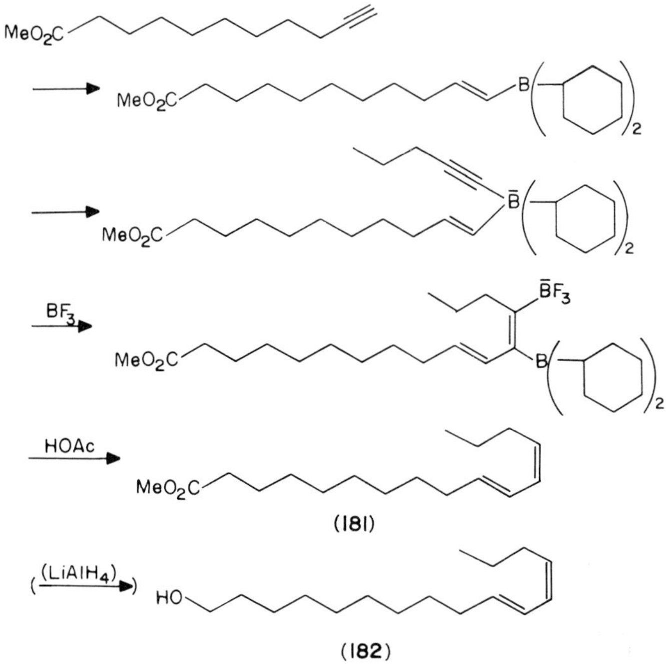

$$\text{(154)}$$

The use of thexylchloroborane makes it a simple matter to construct a *B*-(α-haloalkenyl)-*B*-alkynyl-*B*-thexyl-*B*-methoxyborate, which rearranges in the usual manner to provide an enynylborinic ester, **183**, that can be oxidized to an alkynyl alkyl ketone (equation 155)[461], a type of intermediate that is useful in chiral synthesis (see Section III.C).

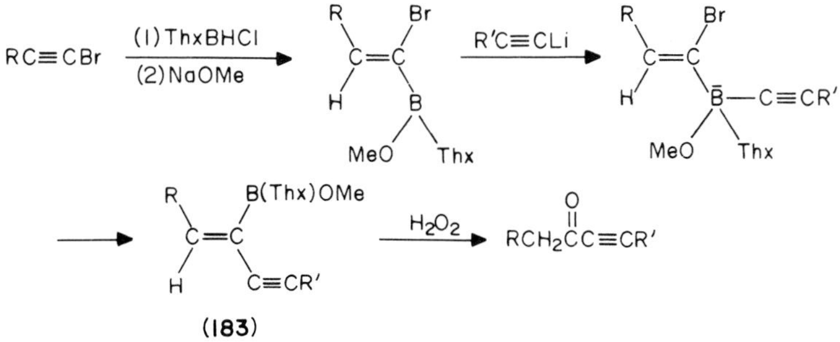

$$\text{(155)}$$

A less exotic but probably more widely useful acetylenic ketone synthesis is provided by the reaction of lithium acetylides with boron trifluoride etherate to form the borate complexes, $RC{\equiv}C\bar{B}F_3$, which react directly with acid anhydrides, $(R'CO)_2O$, to yield the acetylenic ketones, $RC{\equiv}CCOR'$[462]. Alkynyl-9-bbns react with 4-methoxybut-3-en-2-one, $MeOCH{=}CHCOMe$, to form enynones, $RC{\equiv}CCH{=}CHCOCH_2$, generally with a *trans* double bond[463].

Alkynyltrialkylborates, $Li[R_3BC{\equiv}CR']$, generally rearrange to alkynes, $RC{\equiv}CR'$, on treatment with iodine[464,465]. The migrating group can be alkenyl, which results in a synthesis of enynes[466]. The enyne synthesis can be extended to the use of (*Z*)-2-haloalk-1-enyl-9-bbns, which yield halogen-substituted enynes, **184**, stereospecifically (equation 156)[467]. The preparation of the haloalkenyl-9-bbns from 9-halo-9-bbn and alkynes was noted in Section I.C.3.e[235], and it is noteworthy that β-elimination of the boron and halogen does not interfere with the formation and rearrangement of the alkynylborate complex.

Br–B(9-BBN) —RC≡CH→ (R)(Br)C=C(H)(B-9-BBN) —RC≡CLi→ (R)(Br)C=C(H)(B⁻(9-BBN)(C≡CR)) —I_2→ (R)(Br)C=C(H)(C≡CR')

(184)

(156)

The reaction of disiamyldialkynylborates, $[Sia_2B(C{\equiv}CR)_2]^-$, with iodine results in an efficient preparation of diacetylenes, $RC{\equiv}CC{\equiv}R$[56,468]. The R groups may be different when the route to the borate complexes involves successive treatment of $(Sia)_2BOMe$ with $LiC{\equiv}CR$ followed by boron trifluoride etherate to make $(Sia)_2BC{\equiv}CR$, which may then be converted to the dialkynylborate with a different $LiC{\equiv}CR'$[56].

d. Allylborane chemistry

Electrophiles generally attack the γ-carbon of allylboranes to displace the boron, and the major significance of this chemistry lies in the possibility of chiral control (see Section III.D). The special reactivity of allylboranes has been known for some time, and tribute should be paid to B. M. Mikhailov for his extensive pioneering efforts with allylboranes and their complex chemistry. Much of his work was exploratory in nature and not of direct relevance to problems of stereospecifically and regiospecifically controlled synthesis of complex structures, which the reviewer considers to be the major current thrust of synthetic organic chemistry. However, the elaboration of triallylborane to 1-boraadamantane derivatives, **185**, is worthy of specific mention (equation 157)[469–471]. Carbonylation converts 1-boraadamantanes into 1-boryladamantanes, which can be oxidized to 1-adamantanols[472]. Allylborane derivatives have also been converted into 2-boraadamantanes[473]. Mikhailov has recently reviewed his work in this field in English[474].

$B(CH_2CH{=}CH_2)_3 + HC{\equiv}CCH_2Cl$ → (allyl-bicyclic borane with CH₂Cl) + (allyl-bicyclic borane with =CH₂ and Cl)

—$(Et_2BH)_2$→ (1-boraadamantane·thf, X) —CO→ —H_2O_2→ (1-adamantanol, X)

X=H or Cl

(185)

(157)

Mikhailov's work also provided an early example of the allylborane reaction of major current interest, the displacement of boron by aldehydes with allylic rearrangement[475]. At first tricrotylborane, $(MeCH{=}CHCH_2)_3B$, was reported to yield only 15% of the allylic rearrangement product, $\{CH_2{=}CHCH(Me)CH_2O\}_2BCH_2CH{=}CHMe$, with 85% of the product being $(MeCH{=}CHCH_2CH_2O)_2BCH_2CH{=}CHMe$. More recent work by Mikhailov's group has indicated that allylic rearrangement is the principal mode of reaction of various $R_2BCH_2CH{=}CR'R''$[476–479]. Cadiot's group was the first to study the reaction of allylic boronic esters with aldehydes[480].

Kramer and Brown[481] have found that *B*-allyl-9-bbn, **186**, readily allylates all of the common varieties of carbonyl compounds in a manner similar to the allyl Grignard reagent (equation 158). Complete allylic rearrangement was observed in the reaction of formaldehyde with *B*-crotyl-9-bbn, and the only alcohol produced on hydrolysis was 2-methylbut-3-en-1-ol. Rapid intramolecular isomerization of the allyl group of *B*-allyl-9-bbn has been verified by n.m.r. measurements, which show multiple allyl proton resonances at low temperatures but coalescence to a single type of CH_2 at *ca.* 10 °C[482].

$$Me_2CO + MeCH{=}CHCH_2B\text{(9-bbn)} \longrightarrow \xrightarrow{H_2O} Me_2C(OH)CH(Me)CH{=}CH_2 \qquad (158)$$

(186)

The important new development in allylborane chemistry has been the stereoselective displacement of boron by aldehydes, developed primarily by Hoffmann and coworkers. For example, the reaction of pinacol (*Z*)-but-2-enyl-1-boronate, **187**, with simple aldehydes leads to 94–97% '*erythro*' or '*syn*' homoallylic alcohol **189** (equation 159)[483,484]. The predominant stereochemistry of the reaction may be understood by reference to the postulated transition state (**188**), in which the *cis* geometry of the original double bond requires that the methyl group of **187** occupy an axial position in **188**, but the R group from the aldehyde can take an equatorial position and thus strongly favours **188** over alternative isomers. The initial product is a borate ester which can be cleaved with triethanolamine to yield the homoallylic alcohol **189**[484].

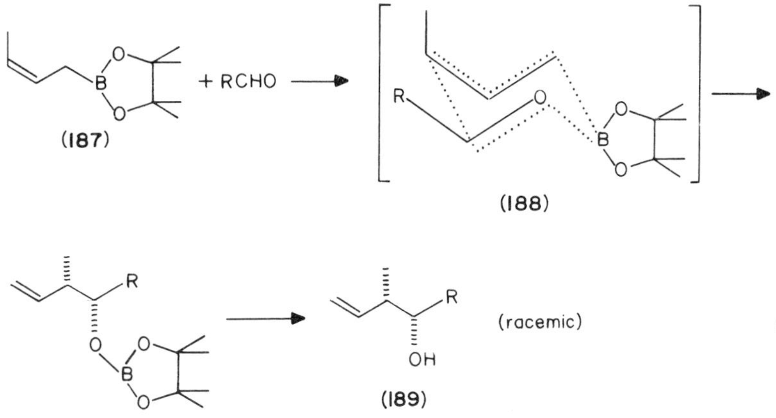

(187) + RCHO ⟶ [(188)] ⟶

⟶ (189) (racemic) (159)

R= Me, Et, Pr^i, Ph

The Hoffmann synthesis works essentially the same way with pinacol (*Z*)-3-methylthioallylboronate, **190**, to produce the *syn*-methylthio alcohols (equation 160)[485], and with (*Z*)-alkoxyallylboronates, **191**, to yield *syn*-alkoxy alcohols, which can be deprotected to diols (equation 161)[486]. Typical diastereoselectivities were in the range 80–95%, most often *ca.* 90%. The stereochemistry was proved by synthesis of the (racemic) insect pheromone brevicomin, **256**, which contains an internally ketalized diol (see Section III.B.1.a for structure)[486]. Similar chemistry of alkoxyallylboronic esters has been developed independently by Wuts and Bigelow[487], who have also synthesized brevicomin[488].

MeS ... (190) —RCHO→ MeS ... R, OH (160)

R′O ... (191) —RCHO→ R′O ... R, OH (161)

Attempts to apply Hoffmann's method to $H_2C{=}CHCH(Me)BO_2C_2Me_4$ resulted in low stereoselectivities[489]. The pinacol group forces the α-methyl group preferentially into the axial position, but only by a *ca.* 3:1 margin, and the result is a mixture of (*Z*)- and (*E*)-alkene products. Although racemic materials were used, the mechanism would require the geometric isomers to be formed with opposite chirality transfer, and the reaction is therefore not immediately useful without improvement of its selectivity.

A potentially useful synthesis of allylic boronic esters from the corresponding allyltin compounds has been reported[490]. Allylic boronic esters have also been synthesized from pinacol chloromethylboronate and alkenyllithiums[491]. A '*threo*'-selective analogue of the foregoing chemistry has been achieved by Yamamoto *et al.*[492] by reacting crotyllithium with triethylborane to form *trans*-$MeCH{=}CHCH_2BEt_3^-$, which with aldehydes yields the '*threo*' or '*anti*' diastereomers of **189** with *ca.* 6:1 diastereoselectivity. Higher stereoselectivity has been obtained with *B*-[α-trimethylsilyl- or α-trimethylstannyl-(*E*)-crotyl]-9-bbn, **192**, and aldehydes in the presence of pyridine (equation 162)[493]. Boronic ester analogues of **192**, studied briefly by Tsai and Matteson[373], showed similar patterns of diastereoselection, and pinanediol (*Z*)-1-trimethylsilylcrotyl-1-boronate yielded homoallylic alcohol that was all *erythro*- and 96% *E*-isomer.

Me_3M, M = Si or Sn (192) —RCHO, pyridine→ R ... MMe_3, OH (162)

Tsai and Matteson[494] have adapted Hoffmann's chemistry to provide a highly stereoselective synthesis of *anti*-(α-trimethylsilylallyl)carbinols **193**, which are easily converted into terminal (*Z*)-dienes, **194**, or (*E*)-dienes, **195** of ⩾98% isomeric purity (equation 163). The pheromone of the red bollworm moth consists of a 20:80 mixture of **194** and **195** having R = $AcO(CH_2)_8$, both of which were synthesized. If a methyl group is present α to the boron, the reaction produces a 75:25 ratio of Z- and E-isomers[495], as noted previously in other systems[489].

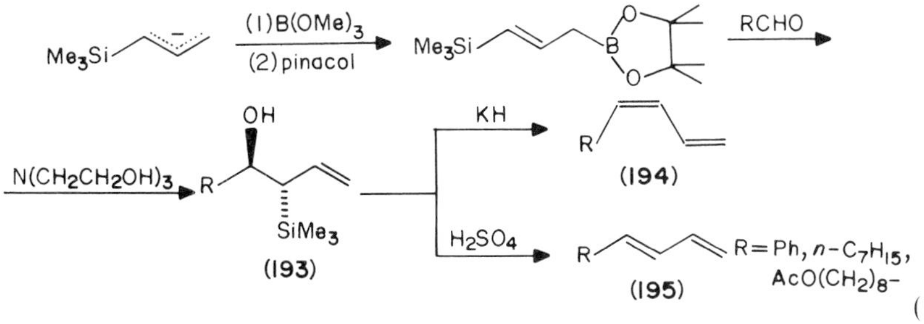

(163)

e. Allenic and propargylic boranes

Allylic rearrangement of allenylboranes yields propargylboranes, and either the allenic or propargylic compounds undergo allylic rearrangement on reaction with electrophiles. This chemistry is thus closely related to that in the preceding Section (II.B.3.d), but it has a degree of complexity all its own, and provides clean routes to some highly labile unsaturated systems that are hard to prepare in other ways.

Treatment of a trialkylborane such as tricyclopentylborane with lithiopropargyl chloride below $-60\,^{\circ}C$ results in the formation of a borate complex which loses chloride ion and rearranges to an allenic borane, for example, B-1-(cyclopentylallenyl)-B,B-dicyclopentylborane, **196**[496]. On warming to $20\,^{\circ}C$, the allenic borane isomerizes to a propargylic borane, **197**[497]. Treatment of the allenic borane **196** with an aldehyde such as acrolein yields a homopropargylic alcohol, **198**, and similar treatment of the rearranged propargylborane yields an allenic alcohol, **199** (equation 164)[497]. Protonolysis with acetic acid converts the propargylic borane **197** into the corresponding allene **200**. At first it was assumed that the allene arose from the allenic borane[496], but it was subsequently realized

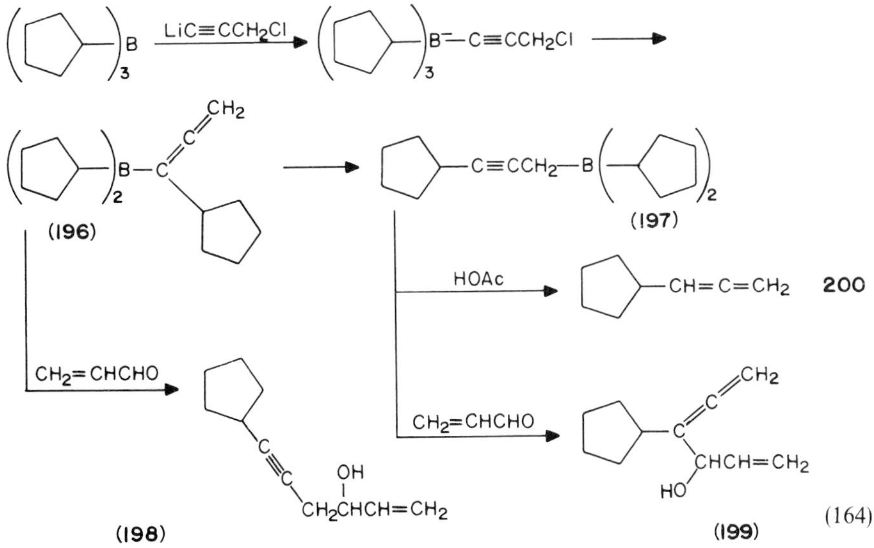

(164)

that **196** rearranges to **197** before protonation under the conditions used[497]. Extension of the foregoing chemistry to reactions of alkenylthexylchloroboranes with lithium chloropropargylide results in simple routes to 1,3-enynols, **201**, and 1,2,4-trienols, **202** (equation 165); R = Bu^n with R′ = Et, Pr^i, Bu^t, vinyl; R′ = Et with R = Cy, Bu^t, Ph, or with EtC≡CEt in place of RC≡CH)[498].

$$\text{ThxBHCl} + \text{RC}{\equiv}\text{CH} \longrightarrow (R)(H)\text{C}{=}\text{C}(H)(\text{BClThx}) \xrightarrow[-78\ ^\circ\text{C}]{\text{LiC}\equiv\text{CCH}_2\text{Cl}} (R)(H)\text{C}{=}\text{C}(H)\text{B}(\text{C}{\equiv}\text{CCH}_2\text{Cl})_2$$

$$\longrightarrow (R)(H)\text{C}{=}\text{C}(H)\text{C}(\text{ThxBC}{\equiv}\text{CCH}_2\text{Cl}){=}\text{C}{=}\text{CH}_2 \xrightarrow{25\ ^\circ\text{C}} (R)(H)\text{C}{=}\text{C}(H)\text{C}{\equiv}\text{CCH}_2\text{ThxBC}{\equiv}\text{CCH}_2\text{Cl}$$

$$\xrightarrow{\text{R}'\text{CHO}} (R)(H)\text{C}{=}\text{C}(H)\text{C}{\equiv}\text{CCH}_2\text{R}'\text{CHOH}\ \textbf{(201)} \qquad \xrightarrow{\text{R}'\text{CHO}} (R)(H)\text{C}{=}\text{C}(H)\text{C}(\text{R}'\text{CHOH}){=}\text{C}{=}\text{CH}_2\ \textbf{(202)} \qquad (165)$$

Carbon electrophiles other than aldehydes failed to react with the boranes in the foregoing scheme, but propargylborate complexes, $[(2\text{-Bu})_3\text{BCH}_2\text{C}{\equiv}\text{CR}]^-$, do react with $Me_2C{=}CCH_2Br$ to yield the allene $H_2C{=}C{=}CRCH_2CH{=}CMe_2$, or with CO_2 to yield $H_2C{=}C{=}CCRCO_2H$[499]. However, it appears that in general the analogous triisobutylaluminate complexes work better for this purpose.

Midland[500] prepared allenic boranes from ethynylalkanol acetates, **203**, and trialkylboranes (equation 166). The substitution pattern evidently makes this series of allenylboranes less prone to rearrange than those first studied by Zweifel. Protonolysis with acetic acid yielded the corresponding allene, and with water the product was the acetylene. When the reaction was carried out with (*R*)-(+)-oct-1-yn-3-ol acetate and tributylborane, the product was *ca.* 23–40% ee (*S*)-(+)-dodeca-5,6-diene, **204**, which corresponds to acetate loss *anti* to the migrating group (equation 167)[501]. The allenylborane is configurationally unstable. It also rearranges to the acetylene, but this does not account for the configurational instability because migration of the boron along one side of the allene yields a chiral acetylene, and configurational loss requires some kind of inversion process.

$$\underset{\textbf{(203)}}{\text{R}'\text{CH(OAc)C}{\equiv}\text{CLi}} + \text{R}_3\text{B} \longrightarrow \text{R}'\text{CH}{=}\text{C}{=}\text{C}(\text{BR}_2)\text{R} \begin{cases} \xrightarrow{\text{HOAc}} \text{R}'\text{CH}{=}\text{C}{=}\text{CHR} \\ \xrightarrow{\text{H}_2\text{O}} \text{R}'\text{C}{\equiv}\text{CCH}_2\text{R} \end{cases} \qquad (166)$$

$$\text{Bu}_3\bar{\text{B}}\text{C}{\equiv}\text{CC}(\text{H})(\text{C}_5\text{H}_{11})(\text{OAc}) \longrightarrow (\text{Bu})(\text{Bu}_2\text{B})\text{C}{=}\text{C}{=}\text{C}(\text{H})(\text{C}_5\text{H}_{11}) \xrightarrow{\text{HOAc}} \underset{\textbf{(204)}}{(\text{Bu})(\text{H})\text{C}{=}\text{C}{=}\text{C}(\text{H})(\text{C}_5\text{H}_{11})} \qquad (167)$$

f. Catalysed cross-coupling

Alkenylboranes can be cross-coupled with aryl, alkenyl, or allyl halides with the aid of methylcopper or tetrakis(triphenylphosphine)palladium(0) and base. These reactions appears to involve electrophilic replacement of the boron by the transition metal and do not necessarily involve boron after that point. Other organometallics, especially alkenylzincs, are generally more reactive than alkenylboranes and give better yields in these cross-couplings, although the ease of synthesis of alkenylboranes stereospecifically makes them useful substrates for this stereospecific process.

The palladium-catalysed cross-coupling process has been reviewed by its principal inventor, Negishi[502], who found alkenylboranes to be unreactive, but observed that an alkynylborate, $[Me(CH_2)_4C{\equiv}CBBu_3]^-$, coupled with *o*-tolyl iodide in the presence of $Pd(PPh_3)_4$ to form $Me(CH_2)_4C{\equiv}CC_6H_4Me$ on heating[502]. The discovery that alkenylboranes would react in the presence of bases such as ethoxide was first reported by Miyaura and Suzuki[503]. A typical example is the stereospecific reaction of *B*-[(*E*)-hex-1-enyl]-*B*, *B*-disiamylborane with bromobenzene to produce (*E*)-hex-1-enylbenzene, **205** (equation 168)[504]. The geometry of *cis*-alkenylboranes is also retained in this coupling[505]. Use of triethylamine as the base results in a rearranged coupling product, $RCH{=}CHBY_2$ ($Y = $ siamyl or $Y_2 = $ catechol), with R′Br yielding $RR'C{=}CH_2$[506].

$$\text{(E)-BuCH=CH-BSia}_2 + \text{PhBr} \xrightarrow[\text{NaOEt}]{\text{Pd(PPh}_3)_4} \text{(E)-BuCH=CH-Ph} \quad \textbf{(205)} \qquad (168)$$

Palladium acetate catalyses cross-coupling of *B*-alkenyl-9-bbns, **206**, with allyl chloride (equation 169)[507].

$$\textbf{(206)} + \text{CH}_2\text{=CHCH}_2\text{Cl} \xrightarrow{\text{Pd(OAc)}_2} \qquad (169)$$

The use of methylcopper as a cross-coupling reagent was developed slightly earlier[508]. An example is its use for the stereospecific alkylation of ethyl (*Z*)-*β*-bromoacrylate, **207** (equation 170). Lithium trialkylmethylborates, R_3BMeLi, are allylated by allyl chloride in the presence of copper(I) bromide to produce $RCH_2CH{=}CH_2$, or react with propargyl

$$R_3B + MeCu + \underset{\textbf{(207)}}{(Br)(H)C{=}C(CO_2Et)(H)} \longrightarrow (R)(H)C{=}C(CO_2Et)(H) \qquad (170)$$

chloride to produce the alkylallene, $RCH{=}C{=}CH_2$[509]. Methylcopper also couples dialkenylboranes such as methylbis[(*E*)-hex-1-enyl]borane, **208** (equation 171)[510].

$$\underset{\textbf{(208)}}{(\text{(E)-BuCH=CH})_2\text{BMe}} + \text{MeCu} \longrightarrow \text{BuCH=CH-CH=CHBu} \qquad (171)$$

Treatment of borate complexes from sodium methoxide and alkenyl-9-bbns or alkenyldicyclohexylboranes with copper(I) bromide–dimethyl sulphide results in coupling of pairs of alkenyl groups with complete retentions of configuration[511]. In the presence of allyl bromide, cross-coupling of the alkenyl group with the allyl group occurs, again with retention of the geometry of the alkenyl group[512]. 1-Haloalkynes can also be cross-coupled with alkenylboranes[513]. In all of these cross-coupling reactions involving copper, it appears possible that the boranes are converted into organocopper intermediates.

9. *Boron elimination and ring formation*

β-Elimination of boron and chlorine from β-chloroboranes and cyclopropane formation by base-initiated γ-elimination from tri(γ-chloropropyl)borane were reported by Hawthorne and Dupont[85]. The β-elimination process has been discussed in conjunction with the replacement of boron from alkenyl groups by halogen in Section II.A.2[278–282] and as a part of the Zweifel alkene synthesis in Section II.B.3.a[416–419], and receives very brief further extension here. The synthetic chemist must always keep in mind that such β-eliminations of boron and halogen or even oxygen substituents[67] may be very facile, and synthetic strategy must avoid opportunities for such decomposition of borane intermediates if alkene formation is not the objective.

A useful application of β-elimination is in the reductive transposition of a ketone to an alcohol at the adjacent less hindered site via hydroboration of the silyl enol ether. Larson and Fuentes[514] have demonstrated this process with the conversion of 1-cyclohexylpropan-1-one to 1-cyclohexylpropan-2-ol, **209** (equation 172).

$$\text{C}_6\text{H}_{11}\text{COEt} \longrightarrow \text{C}_6\text{H}_{11}\text{C(OSiMe}_3\text{)=CHMe} \xrightarrow{\text{9-bbn}} \text{C}_6\text{H}_{11}\text{CH(OSiMe}_3\text{)CH(Me)—B} \longrightarrow \text{C}_6\text{H}_{11}\text{CH=CHMe} \xrightarrow{\text{9-bbn}} \text{C}_6\text{H}_{11}\text{CH}_2\text{CH(Me)—B} \xrightarrow{\text{H}_2\text{O}_2} \text{C}_6\text{H}_{11}\text{CH}_2\text{CH(OH)Me} \quad \textbf{(209)} \tag{172}$$

The facility of β-elimination of boron makes possible an interesting fragmentation/ring expansion process reported by Marshall and Bundy[515,516]. The process leads stereospecifically to (*E, E*)-cyclododeca-1, 6-dienes such as **210**, which are of interest in themselves and which can also be used as hydroazulene precursors (equation 173)[517]. The stereochemistry at the hydroboration site is not fixed, except that the addition must be *syn* to either side of the ring. Either borane meets the geometric requirements for the fragmentation process only when aligned in a manner that will lead to the *trans* double bond of **210** at the site of boron elimination[517].

OMes; (1) BH_3–thf, (2) MeOH; $(MeO)_2BO$, $B(OMe)_2$, OMes; MeONa; HO, **(210)** (173)

The Hawthorne cyclopropane synthesis has been greatly improved by Brown and Rhodes[518], who generated *B*-(3-chloropropyl)-9-bbn as the cyclopropane precursor. Hydroboration of homopropargyl tosylate with 2 mol of 9-bbn followed by base treatment yielded *B*-cyclobutyl-9-bbn, **211** (equation 174)[519]. The cyclobutane ring could not be closed without the aid of the second boron atom in stabilizing carbanionic character (see Section II.C.2). *B*-Cyclopropyl-9-bbn, **212**, was prepared in an analogous manner (equation 175).

HB(9-bbn) + $HC{\equiv}CCH_2CH_2Otos$ ⟶ $tosOCH_2CH_2CH_2CH$(B-9-bbn)$_2$ $\xrightarrow{NaOH}$ cyclobutyl–B(9-bbn)

(211) (174)

HB(9-bbn) + $HC{\equiv}CCH_2Cl$ ⟶ $ClCH_2CH_2CH$(B-9-bbn)$_2$ $\xrightarrow{NaOH}$ cyclopropyl–B(9-bbn)

(212) (175)

Goering and Trenbeath[520] have shown that cyclopropane ring closure is stereospecific, with inversion of the carbon from which the boron is displaced. The synthesis of *trans*-1, 2-dimethylcyclopropane, **213**, is illustrated in equation 176, and similar results were obtained with the *cis*-isomer.

Me(ClCH$_2$)C=C(H)Me $\xrightarrow{9\text{-bbn}}$ Me(H)(ClCH$_2$)C–C(H)(Me)B(9-bbn) $\xrightarrow{NaOH}$ *trans*-1,2-dimethylcyclopropane

(213) (176)

C. Other Carbon—Carbon Bond-Forming Reactions

1. Free radical reactions of boranes

a. Alkylation by boranes

Radical reactions inherently lead to loss of steric integrity at the reacting centre and are therefore not as widely useful as the borate rearrangements discussed in Section II.B, but they do accomplish some transformations that are otherwise difficult. Accordingly, this topic is reviewed very briefly and incompletely here. Most of this chemistry has been reviewed in detail by Brown[2c].

Trialkylboranes react readily with a wide variety of α,β-unsaturated carbonyl compounds in the presence of a small amount of oxygen as initiator to yield the conjugate addition products[2c]. Acrolein reacts especially readily, and the postulated radical process is illustrated, which is followed by hydrolysis to yield the 3-alkylpropionaldehyde, **214** (Scheme 1)[521,522]. Another typical example is the reaction of methyl vinyl ketone with tricyclohexylborane to form 4-cyclohexylbutan-2-one, **215** (equation 177)[523]. Substitu-

Initiation:	$R'O\cdot + R_3B \longrightarrow R'OBR_2 + R\cdot$
Propagation:	$R\cdot + CH_2{=}CHCHO \longrightarrow RCH_2CH{=}CHO\cdot$
	$RCH_2CH{=}CHO\cdot + R_3B \longrightarrow RCH_2CH{=}CHOBR_2 + R\cdot$
Hydrolysis:	$RCH_2CH{=}CHOBR_2 + H_2O \longrightarrow RCH_2CH_2CHO + HOBR_2$

SCHEME 1

ents α to the carbonyl group do not cause any problem, and reagents such as 2-bromoacrolein[524] or ketones derived from Mannich bases[525] provide useful reactions. Crotonaldehyde reacts more sluggishly, but can be induced to react efficiently in the presence of controlled quantities of oxygen or other initiators[526]. The reaction with quinones to produce alkylhydroquinones is probably a similar radical reaction[527,528].

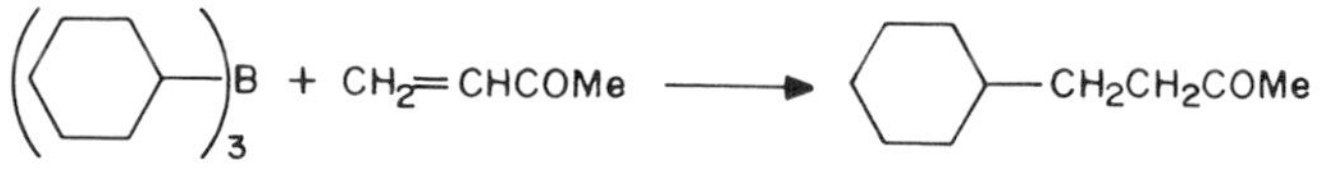

$$(C_6H_{11})_3B + CH_2{=}CHCOMe \longrightarrow C_6H_{11}{-}CH_2CH_2COMe \quad \textbf{(215)} \tag{177}$$

Butadiene monoxide, **216**, also undergoes radical alkylation by trialkylboranes[529]. The analogous reaction of 3,4-epoxy-1-butyne leads to allenic alcohols, $RCH{=}C{=}CHCH_2OH$ (equation 178)[530].

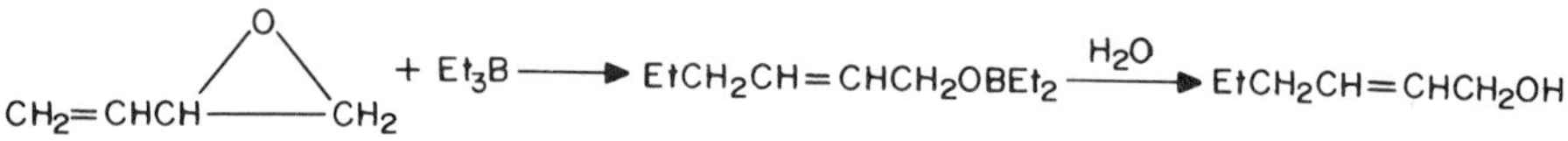

$$CH_2{=}CHCH\overset{O}{\frown}CH_2 \;\textbf{(216)} + Et_3B \longrightarrow EtCH_2CH{=}CHCH_2OBEt_2 \xrightarrow{H_2O} EtCH_2CH{=}CHCH_2OH \tag{178}$$

Because the relative stabilities of radicals are tertiary > secondary > primary, *B*-alkyl-9-bbns will undergo ring cleavage in competition with alkyl group cleavage during these reactions, except when the alkyl group is tertiary. The problem can be partially overcome by the use of *B*-alkylborinanes, which give efficient preferential cleavage of secondary alkyl groups[531]. Radical alkylation with symmetrical primary trialkylboranes selectively utilizes the secondary alkyl group impurities, and since only one of the three alkyl groups is consumed, the effect on product purity is decidedly deleterious[522,528].

Although configuration is lost at the radical centre, cases of strong diastereoselection are known, such as the 2-methylcyclopentyl radical, which yields mainly the *trans* product in its reaction with acrolein, to the point where the small amount of *cis*-isomer probably present was not noticed before the radical nature of the reaction was recognized[352].

b. Radical additions to alkenylboronic esters

Although the development of newer synthetic methods has made a wide variety of functionalized boronic esters available, radical additions to vinylboronic esters can still provide a few structures that are not easily accessible otherwise. For example, β-alkylthioboronic esters, **217**, are easily produced from dibutyl vinylboronate and a variety of mercaptans in the presence of ultraviolet light or azobisisobutyronitrile (equation 179)[22,27]. Similar addition of potassium bisulphite led to potassium 2-(dihydroxyboryl)ethanesulphonate, $KO_3SCH_2CH_2B(OH)_2$[22]. Vinylboronic esters appear to be particularly efficient substrates for a variety of radical addition reactions. The use of these radical additions to prepare the first α-halo boronic esters has been noted in Section II.B.1.a, and the efficient addition of bromomalononitrile to dibutyl vinylboronate to form $(BuO)_2BCHBrCH_2CH(CN)_2$[335] might find future use.

$$CH_2{=}CHB(OBu)_2 \xrightarrow{RS\cdot} RSCH_2\dot{C}HB(OBu)_2 \xrightarrow{RSH} RS\cdot + \underset{(\mathbf{217})}{RSCH_2CH_2B(OBu)_2} \qquad (179)$$

$R = n\text{-}C_6H_{13}$, MeCO, H, $H_3\overset{+}{N}CH_2CH_2$, $^{-}O_2CCH(\overset{+}{N}H_3)CH_2$, HO_2CCH_2, and others

2. Boron substituted carbanions

a. By deboronation

The first abstraction of boron from a *gem*-diboryl compound with an alkyllithium to produce an α-lithio borane, **218**, was reported by Cainelli *et al.*[532], who observed carboxylation. Wittig-type condensation with aldehydes and ketones was reported by Zweifel and Arzoumanian[533] and by Cainelli *et al.*[534] (equation 180). The preparation of methanetetraboronic esters and related compounds by Castle and Matteson[61,62] (Section I.A.3) opened the way for the development of a similar chemistry of boronic esters, much of which led to the synthesis of exotic organometallic compounds for their own sake and which has been reviewed elsewhere[64,65]. Lithium tris(trimethylenedioxyboryl)methide, **219**, is an isolable ionic compound[535], and the alkene-1, 1-diboronic esters formed by its reactions with carbonyl compounds may eventually prove useful as synthetic intermediates (equation 181). Lithium bis(ethylenedioxyboryl)methide, **220**[63,536], has been developed as a reagent for the efficient homologation of aldehydes[253]. The initial products are alk-1-enylboronic esters, **221**, which are 90–95% *E*-isomer (equation 182)[63,536]. It is usually possible to recrystallize and purify the boronic acid from hydrolysis of **221**, and this is therefore a potentially useful alternative route to the use of hydroboration of alkynes for preparation of this class of boronic acids.

$$BuC{\equiv}CH \xrightarrow{HBR_2} BuCH_2CH(BR_2)_2 \xrightarrow{BuLi} \underset{(\mathbf{218})}{BuCH_2CH(Li)BR_2} \xrightarrow{R'CHO} BuCH_2CH{=}CHR' \qquad (180)$$

$$C\left(-B\langle O(CH_2)_3O\rangle\right)_4 \xrightarrow{BuLi} BuB\langle O(CH_2)_3O\rangle + Li^+\ \underset{(\mathbf{219})}{^{-}C\left(-B\langle O(CH_2)_3O\rangle\right)_3} \xrightarrow{RCHO} RCH{=}C\left(-B\langle O(CH_2)_3O\rangle\right)_2 \qquad (181)$$

$$HC\left(-B\langle OCH_2CH_2O\rangle\right)_3 \xrightarrow{MeLi} \underset{(\mathbf{220})}{H\bar{C}\left(-B\langle OCH_2CH_2O\rangle\right)_2} \xrightarrow{RCHO} \underset{(\mathbf{221})}{(E)\text{-}RCH{=}CH{-}B\langle OCH_2CH_2O\rangle} \xrightarrow{NaBO_3} RCH_2CHO \qquad (182)$$

b. By deprotonation

Bases show a strong tendency to coordinate to boron rather than abstract a proton, and deprotonation therefore requires a considerable degree of steric blocking of the access of the base to the boron atom. The first success was reported by Rathke and Kow[537], who deprotonated *B*-methyl-9-bbn, **222**, with lithium 2, 2, 6, 6-tetramethylpiperidide[537], and also found that this base would deprotonate alkenylboranes, $RCH_2CH{=}CHBR_2$, to allylic anions (equation 183)[538]. Allylic anions, **223**, generated from (*E*)-alkenyl-9-bbns have recently been found by Yatagai *et al.*[539] to yield (*Z*)-allylic tin derivatives, **224**, on treatment with tributyltin chloride followed by deboronation with water and ethanolamine (equation 184). These tin compounds are of interest as '*erthro*'-selective reagents for the synthesis of homoallylic alcohols by reaction with aldehydes.

B–Me + LiN → B–$\bar{C}H_2$ —BuBr→ —H_2O_2→ $BuCH_2OH$ 42% (183)

(222)

(223) + Bu_3SnCl → Bu_3Sn —H_2O→ Bu_3Sn (224) (184)

Wilson[540] has deprotonated dimesitylmethylborane, **225**, with lithium dicyclohexylamide and alkylated the resulting anion with methyl iodide (equation 185). The resulting ethylborane can be deprotonated and methylated, and the process was repeated to replace all three protons of the methyl group. Further studies on **225**, which may be abbreviated to

$(Ms)_2BMe$ → $(Ms)_2B\bar{C}H_2$ —Me I→ $(Ms)_2BEt$ (185)

(225)

Ms_2BMe, have been reported by Pelter and coworkers. The anion $Ms_2BCH_2^-$ can be generated by treatment of **225** with either lithium dicyclohexylamide or mesityllithium, but unhindered bases such as butyllithium or sodium hydride add to the boron atom[541]. Alkylation of $Ms_2BCH_2^-$ with a number of primary alkyl bromides has been reported[542]. Dimesitylallylborane, $Ms_2BCH_2CH{=}CH_2$, can be deprotonated and undergoes alkylation at the terminal carbon by RI to yield $Ms_2BCH{=}CHCH_2R$ having a *trans* double bond[543]. Perhaps the most likely candidate for synthetic utility is the Wittig-type reaction of $Ms_2BCH_2^-$ with aldehydes and ketones, $RR'C{=}O$, to yield $RR'C{=}CH_2$[544]. Reaction of $Ms_2BCH_2^-$ with Me_3SiCl has yielded $Ms_2BCH_2SiMe_3$, and analogous reactions have been used to prepare $Ms_2BCH_2SnMe_3$, Ms_2BCH_2SPh, $(Ms_2B)_2CH_2$, and related compounds[545]. Several of these can themselves be deprotonated, and the anion from $Ms_2BCH_2SiMe_3$ reacts with benzaldehyde to produce a mixture of *ca.* 45% $PhCH{=}CHBMs_2$ and 55% $PhCH{=}CHSiMe_3$. The most likely obstacle to general synthetic utility of these mesitylboranes is the extreme degree of steric hindrance around the boron atom, which results in failure to react with moderately hindered substrates.

It has not proved possible to deprotonate methylboronic or alkylboronic esters[44], but deprotonation of a methylenediboronic ester, **226**, with lithium 2,2,6,6-tetramethylpiperidide (Litmp) has been reported by Matteson and Moody[44,546]. The resulting diborylmethide anion, **227**, can be alkylated with primary alkyl halides, and the 1,1-bis(trimethylenedioxyboryl)alkanes, **228**, can also be deprotonated and alkylated (equation 186). The base strength of the diborylmethide ion, **227**, appears to be less than that of triphenylmethide ion, which can generate **227** from **226**.

$$CH_2(-BO_2C_3H_6)_2 \xrightarrow{\text{Litmp}} H\bar{C}(-BO_2C_3H_6)_2 \xrightarrow{\text{RBr}} RCH(-BO_2C_3H_6)_2 \xrightarrow[\text{R'Br}]{\text{Litmp}} RR'C(-BO_2C_3H_6)_2$$

(226) (227) (228) (186)

Carbanions such as **229** from deprotonation of 1,1-bis(trimethylenedioxyboryl)alkanes, **228**, undergo Wittig condensation with aldehydes or ketones to yield alkenylboronic esters, **230**, regiospecifically but without stereochemical control (equation 187)[44]. Peroxidic oxidation leads to ketones at the site vacated by the boron. With carboxylic esters, the anions **299** replace the alkoxy group and eliminate boron to yield ketones at the site of the ester group, **231**, adjacent to the site vacated by the boron.

$$n\text{-}C_5H_{11}\bar{C}(-BO_2C_3H_6)_2 \; (229) \xrightarrow{\text{PhCHO}} PhCH{=}C(n\text{-}C_5H_{11})-BO_2C_3H_6 \; (230) \xrightarrow{H_2O_2} PhCH_2CC_5H_{11}\text{-}n \; (\text{C=O})$$

$$n\text{-}C_5H_{11}\bar{C}(-BO_2C_3H_6)_2 \; (229) \xrightarrow{\text{PhCO}_2\text{Me}} Ph(n\text{-}C_5H_{11}CH{=})C-O-BO_2C_3H_6 \longrightarrow PhCCH_2C_5H_{11}\text{-}n \; (\text{C=O}) \; (231)$$

(187)

Although the 1,1-bis(trimethylenedioxyboryl)methanes, **228**, available from alkylation of the anion **227**, are restricted to those having a primary alkyl group, similar compounds branched at the second carbon have been made by hydrogenation of 1,1-bis(trimethylenedioxyboryl)alk-1-enes, $RR'C{=}C(BO_2C_3H_6)_2$, which are available by condensation of ketones with the tris(trimethylenedioxyboryl)methide anion, **219** (see Section II.C.2.a)[44]. It might also be noted that 1,1-bis(dichloroalkylboryl)alkanes are readily available from dihydroboration of acetylenes with $BHCl_2$ (see Section I.C.3.c)[208], although the hydroboration process is not regiospecific and the amount of 1,2-isomer formed, which was not determined, might be enough to cause purification problems.

Although the *gem*-diboronic esters are versatile and interesting reagents, sufficient effort has to be expended in their preparation[62,65] that the synthetic chemist can usually find alternative routes that require less labour. It is easier to make heterosubstituted boronic esters such as pinacol (phenylthio)methylboronate, **232**[41,547], which can be deprotonated to the corresponding carbanion **233** (equation 188). Reactions of **233** include alkylation by alkyl halides, Wittig-type condensation with aldehydes or ketones, and reaction with esters to produce α-phenylthioketones (equation 189)[41] or with formate esters to form α-phenylthioaldehydes[548]. The alkylation products, α-phenylthioboronic esters, **234**, are easily oxidized directly to monothioacetals or acetals with *N*-chlorosuccinimide[288]. In general, it was found possible to deprotonate **234** and subject the resulting anions to the same reactions as **233**, except that the alkylation products, $RR'C(SPh)BO_2C_2Me_4$, were not successfully deboronated by hydrogen peroxide, evidently because of steric hindrance (equations 190 and 191)[41].

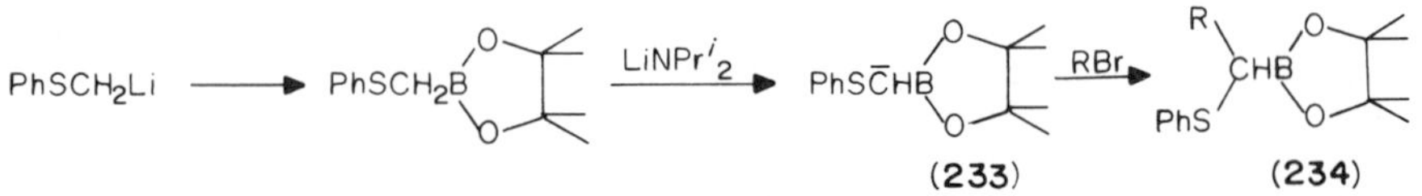

(188)

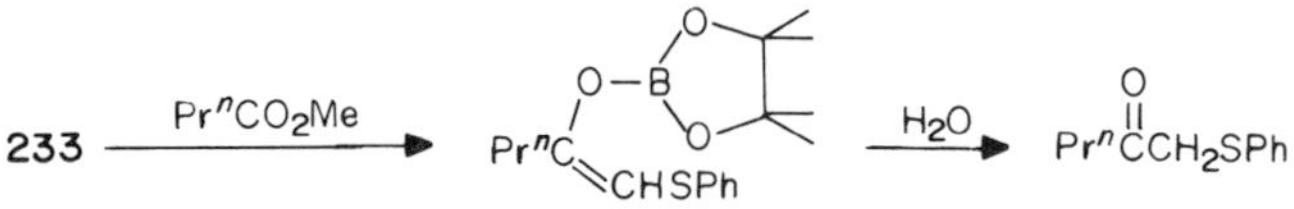

(189)

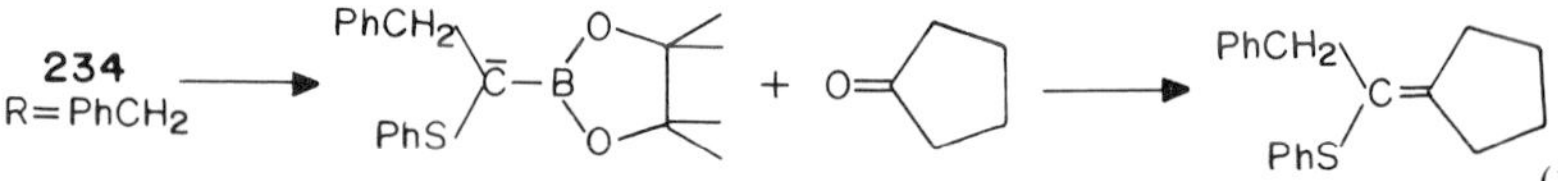

(190)

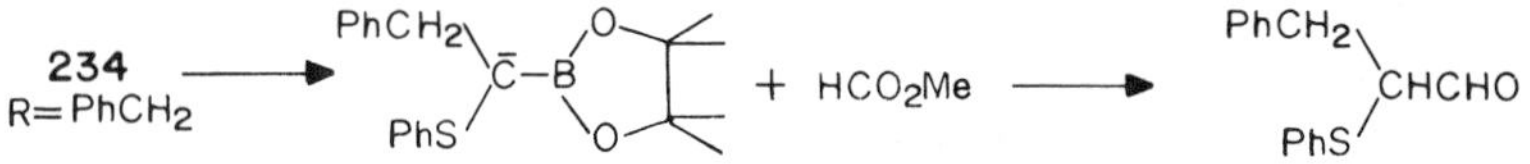

(191)

The preparation of propane-1, 3-diol bis(phenylthio)methylboronate, **235**, is straightforward, and on deprotonation and reaction with ketones, ketene thioacetals are obtained (equation 192)[414]. With enolizable ketones such as acetone and cyclohexanone, **235** was found to give good yields, in contrast to $(PhS)_2CHSiMe_3$, which gave little or no product with these ketones.

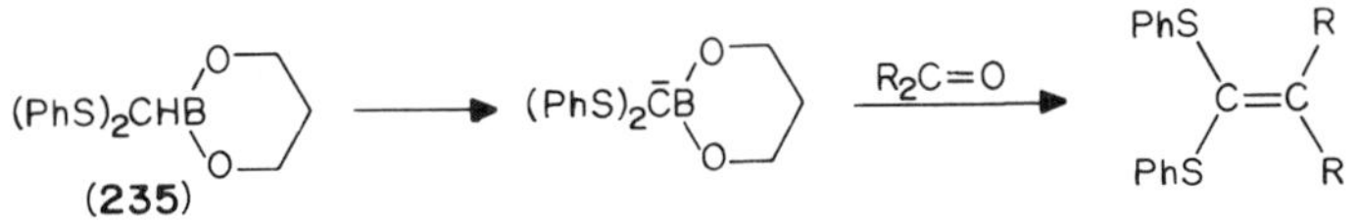

(192)

Pinacol (trimethylsilyl)methylboronate, **236**, has been prepared from the Grignard reagent, Me_3SiCH_2MgCl, and trimethyl borate, with subsequent esterification with pinacol[371,372]. Deprotonation and alkylation proved straightforward, but the higher homologues failed to undergo deprotonation with any base tried[372]. The most interesting reaction of the anion from **236** is its reaction with aldehydes or ketones, which exclusively eliminates silicon to yield the alkenylboronic ester (equation 193)[372,549]. It is amusing that the alkenylboronic esters produced are predominantly the *cis*-isomers, but the reaction is not selective enough to be truly useful in this regard, and an efficient synthesis of pure *cis*-alk-1-eneboronic esters has been described in Section I.C.3.c.

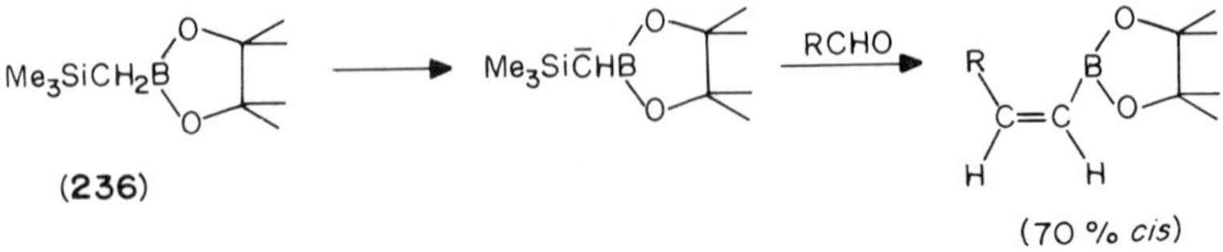

(193)

3. Cycloaddition reactions of vinylboronic esters

Vinylboronic esters have moderate dienophilic character, and dibutyl vinylboronate adds to cyclopentadiene under reflux to yield dibutyl norborneneboronate as a 60:40 mixture of *exo*- and *endo*-isomers, **237** and **238** (equation 194)[550]. The isomers have been separated by fractional crystallization of their *o*-phenylenediamine derivatives. Heating dibutyl vinylboronate with cyclohexadiene in a bomb at 200 °C gives a good yield of a 20:80 mixture of *exo*- and *endo*-dibutyl bicyclo[2.2.2]oct-2-ene-2-boronates, **239** and **240** (equation 195)[551].

$CH_2{=}CHB(OBu)_2$ + cyclopentadiene ⟶ (237) $B(OBu)_2$ + (238) $B(OBu)_2$ (194)

$CH_2{=}CHB(OBu)_2$ + cyclohexadiene ⟶ (239) $B(OBu)_2$ + (240) $B(OBu)_2$ (195)

Cycloaddition of ethyl diazoacetate to dibutyl vinylboronate proceeds readily, but the presumed initial adduct, **241**, is unstable and apparently undergoes rearrangement of the boron from carbon to nitrogen. On contact with water, the isolable product is 5-carbethoxy-2-pyrazoline, **242** (equation 196)[552]. Diphenyldiazomethane behaves in an analogous manner, but diazomethane yields only a small amount of pyrazoline and a large amount of polymethylene and nitrogen, the usual products of contact of diazomethane with borate esters[553]. Dibutyl ethynylboronate, $HC{\equiv}CB(OBu)_2$, also readily undergoes cycloaddition with ethyl diazoacetate, and the product aromatizes by proton tautomerization to yield 3-carbethoxypyrazolyl-5-boronic acid after hydrolysis[552].

$CH_2{=}CHB(OBu)_2$ + N_2CHCO_2Et ⟶ $(BuO)_2BCH$–N=N–$CHCO_2Et$ (ring closed by CH_2) (241) ⟶ HC=N–NH–$CHCO_2Et$ (ring closed by CH_2) (242) (196)

III. ASYMMETRIC SYNTHESIS WITH BORANES

A. Introduction

Organoboranes have uniquely favourable properties for use in directed chiral synthesis and have been employed in several unrelated processes. The boron atom is small, only slightly larger than carbon, and can serve as a template for assembling carbon—carbon or carbon—heteroatom bonds with high sensitivity to the steric demands of substituents. Once the desired stereochemistry has been established, it is generally possible to replace the boron with other elements in a stereospecific manner, as discussed in Sections II.A. and II.B.

The first truly successful asymmetric synthesis was the hydroboration/oxidation of *cis*-

alkenes with diisopinocampheylborane by Brown and Zweifel[106], which was developed into a practical synthesis of several optically active secondary alcohols[554]. Very high asymmetric inductions were observed at a time when other reactions employing what we would now call chiral auxiliary groups generally yielded small enantiomeric excesses of only theoretical interest. Recent developments in hydroboration chemistry have resulted in some very simple, highly enantioselective syntheses of several types of asymmetric structures, discussed in detail in Section III.C.

Other organoborane reactions now show great promise for the controlled construction of chiral centres, with a wider range of possible structures than hydroboration permits. Sequential construction of adjacent chiral centres has been carried out via the insertion of (dichloromethyl)lithium into boronic esters to form α-chloro boronic esters (Section II.B.1.a), which with the aid of certain chiral diol groups provides 95–99.5% control of the absolute configuration of the new chiral centre, to be discussed in the next Section, III.B. In choosing to discuss his own work out of historical order, the reviewer should first acknowledge his debt not only to H. C. Brown and G. Zweifel, who showed that pinene derivatives can be excellent chiral directors[1,106], but also to A. I. Meyers, who demonstrated the general value of chelated intermediates in achieving high enantioselectivities in his work on carbanion alkylations[555].

The reaction of allylic boranes with aldehydes can create two adjacent chiral centres at once with high control of relative configuration (Section II.B.3.d), and chiral auxiliaries can provide good control of absolute configuration, described in Section III.D. These reactions are mechanistically related to enantioselective aldol condensations of dialkylboron enolates, which fall outside the scope of this review but are referred to briefly at the end of Section III.D. Enantioselective reductions with *B*-isopinocampheyl-9-bbn are at the margin of topics reviewed here, but in view of their considerable synthetic utility are summarized briefly in Section III.E.

B. Chiral α-Chloroboronic Esters

1. Pinanediol as directing group

a. Synthetic applications

A practical preparation of (+)-pinanediol, **243**, or (−)-pinanediol, **244**, by osmium tetraoxide catalysed oxidation of the appropriate enantiomer of α-pinene with trimethylamine oxide was discovered by Ray and Matteson (equation 197)[556], and homologation of (+)-pinanediol boronic esters, **245**, with (dichloromethyl)lithium was then found to yield (α*S*)-α-chloroboronic esters, **247**, with generally good diastereoselectivity, *ca.* 97% in the most favourable case (equation 198)[557]. The reaction involves formation of a borate complex, **246**, at −100 °C, which rearranges to the α-chloroboronic ester **247** at 0–25 °C. Zinc chloride catalysis in the rearrangement step was found by Matteson and Sadhu[558] to increase the diastereoselectivities to ⩾99% generally and to 95% in the least favourable case, R = Me. Absolute configurations were determined by reacting **247** with Grignard reagents, which form borate complexes, **248** (see Section II.B.1.a[331]) that rearrange with inversion (Section II.B.1.b[349]) to expel chloride and form *sec*-alkylboronic esters, **249**. Oxidation of **249** with hydrogen peroxide yielded secondary alcohols **250** of known configuration[557]. Chiral purities were originally estimated from the rotations of esters of **249**[557], but the small amounts of diastereomers of the α-chloroboronic esters, **247**, encountered in the improved synthesis could only be detected by high-field n.m.r.[558]. A final factor which contributes to the efficiency of the synthesis is that the pinanediol is recoverable as its borate salt afterwards.

Me$_3$NO, OsO$_4$ → HO, HO **(243)** abbreviated: HO, (s), HO

or enantiomer HO, HO **(244)** abbreviated: HO, (r), HO (197)

R–B(O,O)(s) **(245)** —LiCHCl$_2$, −100 °C→ Cl$_2$CH, R–B(O,O)(s) **(246)** —ZnCl$_2$, 0–25 °C→ Cl, R–C(H)–B(O,O)(s) **(247)**

—R′MgBr, −78 °C→ Cl, R–C(H)–B(R′)(O,O)(s) **(248)** —25 °C→ H, R–C(R′)–B(O,O)(s) **(249)** —H$_2$O$_2$→ H, R–C(R′)–OH **(250)** (198)

A noteworthy feature of this synthesis is that the boronic ester product **249** is the same as the starting material **245**, except that the group R of **245** has been changed. Thus, it is a simple matter to repeat the cycle to introduce an additional chiral centre, as in the synthesis of (2*S*, 3*S*)-3-phenylbutan-2-ol, **251** (equation 199)[557]. If the opposite chirality is desired at the next chiral centre, it is possible to remove the original pinanediol group and then replace it with its enantiomer, as in the synthesis of (2*R*, 3*S*)-3-phenylbutan-2-ol, **253**. The intermediate (*S*)-1-phenylethylboronic acid, **252**, was purified via its crystalline

PhB(O,O)(s) —(1) LiCHCl$_2$, (2) MeMgBr→ Ph–C(H)(Me)–B(O,O)(s) —(1) LiCHCl$_2$, (2) MeMgBr→ Ph–C(H)(Me)–C(H)(Me)–B(O,O)(s) —H$_2$O$_2$→ Ph–C(H)(Me)–C(H)(Me)–OH **(251)**

Ph–C(H)(Me)–B(O,O)(s) —BCl$_3$→ Ph–C(H)(Me)–B(OH)$_2$ **(252)** —HO, (r), HO→ Ph–C(H)(Me)–B(O,O)(r) —(1) LiCHCl$_2$, (2) MeMgBr, (3) H$_2$O$_2$→ Ph–C(H)(Me)–C(Me)(H)–OH **(253)** (199)

chelated diethanolamine ester. The only serious problem with this route is that pinanediol esters are unusually stable thermodynamically and cannot be hydrolysed but must be cleaved with boron trichloride under conditions which destroy the pinanyl group and would be incompatible with sensitive substituents.

The improved procedure with zinc chloride catalysis makes it possible to start a synthesis from pinanediol methylboronate as well as any other alkylboronate, and the route outlined to **253** is already obsolete. Any simple secondary alcohol with an additional adjacent chiral centre can be made merely by choosing the proper enantiomer of pinanediol and the correct order of introducing the alkyl groups. An example is the synthesis of (3*S*, 4*S*)-4-methylheptan-3-ol, **254**, a component of the aggregation pheromone of the European elm bark beetle, *Scolytus multistriatus* (equation 200)[558,560].

(i) (ii) (iii) OH (254) (200)

(i) $LiCHCl_2, ZnCl_2$, then CH_3MgBr; (ii) $LiCHCl_2, ZnCl_2$, then C_2H_5MgBr; (iii) $H_2O_2/NaOH$

The new procedure has also been tested successfully with ketal and ether substituents present in a synthesis of brevicomin, **256** (equation 201), which is a component of the aggregation pheromone of the western pine beetle, *Dendroctonus brevicomis*[558]. The benzyl-protected precursor **255** had been reported previously as an oil[559], but was a low-melting crystalline solid when obtained by the borane route.

MgCl (i) (ii) Cl (iii) OBz (iv) OBz Cl (v) OBz (vi) OBz OH (vii) (255) (256) (201)

(i) $B(OMe)_3$, then (−)-pinanediol, **244**; (ii) $LiCHCl_2, ZnCl_2$; (iii) $LiOCH_2Ph$; (iv) $LiCHCl_2, ZnCl_2$; (v) EtMgBr; (vi) H_2O_2–NaOH, then H^+–SiO_2; (vii) H_2–Pd

Carboxylic ester substituents are known to be tolerated in reactions of achiral boronic esters with (dichloromethyl) lithium (see Section II.B.1.a)[197,340]. These and other functional groups are also tolerated in the pinanediol ester series[560].

This functional group tolerance has permitted an efficient synthesis of eldanolide, the wing gland pheromone of the African sugarcane borer[560].

b. The epimerization problem

Chloride can displace chloride from α-chloroboronic esters, with consequent inversion of the chiral centre. Not surprisingly, benzylic or allylic compounds are the most susceptible, and the phenomenon was first noticed when reaction of (+)-pinanediol phenylboronate with dichloromethyllithium yielded a mixture of (+)-pinanediol (α*S*)-α-chlorobenzylboronate, **257**, and the (α*R*)-isomer, **258**, in which the latter predominated slightly[557]. The source of the problem was prolonged exposure of the **257** to the lithium chloride generated in the reaction, and the problem was solved by lowering the temperature to 0 °C and shortening the time to 1 h.

(257) ⇌ (258) (202)

A quantitative study of this reaction has shown that it is first order in **257** and *ca.* 0.7 order (i.e. half order plus a salt effect) in lithium chloride in thf[561]. Free chloride ion appears to be the active epimerization agent. At 0.45 M lithium chloride (nearly saturated), the pseudo-first-order rate constant determined polarimetrically is $5.7 \times 10^{-5}\,s^{-1}$ at 24.9 °C, which corresponds to a half-life of 3.4 h or, more to the point, 1% randomization of the chiral centre every 3 min. Non-activated α-chloroboronic esters epimerized more slowly, with rates relative to **257** as 1.00 being 0.11 for pinanediol 1-chloro-2-phenylethylboronate and 0.47 for pinanediol 1-chloropentylboronate. Thus, the original conditions for preparing α-chloroboronic esters[557] would typically result in *ca.* 1% epimerization per hour, and several hours were generally required in order to complete the preparation process.

Zinc chloride not only accelerates the rearrangement of the intermediate borate complex **246** (Section III.B.1.a), as evidenced by higher yields in shorter reaction times[557], but also suppresses the epimerization rate by about an order of magnitude at the optimum composition, which corresponds to $LiZnCl_3$[561]. A small deficiency of zinc chloride is not critical, as the suppression of epimerization is still three-fold at the composition $Li_2[ZnCl_4]$. However, any excess can rapidly become deleterious, because there is a term in $[ZnCl_2][ZnCl_3^-]$ in the rate law, which becomes large at high concentrations and could result in major epimerization of the α-chloroboronic ester as the solution is concentrated during workup. Fortunately, as structures become more complex and sterically hindered, epimerization rates become slower. Even in the presence of water, which greatly accelerates the process[561], it can take several days to epimerize a sufficient fraction of a β-branched α-chloroboronic ester to detect the epimer unequivocally in the n.m.r. spectrum for analytical purposes[560].

It is also possible to suppress epimerization with mercury(II) chloride[561], but mercury(II) chloride did not appear to have any beneficial effect on yields in the preparation of α-chloroboronic esters[196].

Epimerization is not a problem in the reaction of Grignard or alkyllithium reagents with α-chloroboronic esters (**247**, Section III.B.1.a) because a stable borate complex, **248**, is formed as an intermediate. The boron atom in **248** lacks the requisite vacant p orbital in order to assist halide exchange[331], the steric hindrance to external displacements on **248** must resemble that of neopentyl halides, and the negative charge on the borate complex would be a further deterrent to chloride attack. Another strong base, $LiN(SiMe_3)_2$, evidently also forms stable borate complexes and yields displacement products without detectable epimerization[558,562]. However, the use of weaker bases or bases which can interchange between the borate **248** and the boronic ester product **249** can result in significant epimerization[560], and even with a base as strong as lithium benzyloxide evidence for epimerization at the 1–2% level has been observed[558].

c. Pinanediol (dichloromethyl)boronate

It would be desirable to be able to start a chiral synthesis from RLi or RMgX by reaction with $Cl_2CHB(OR')_2$ rather than first having to make $RB(OR')_2$ for reaction with $LiCHCl_2$. However, the borate salt **246** formed from attack of dichloromethide ion on the less hindered side of a pinanediol alkylboronate (**245**, Section III.B.1.a) differs diastereomerically from the borate salt **260** derived from pinanediol dichloromethylboronate, **259**, and a lithium or Grignard reagent (equation 203), and its rearrangement produces a grossly different diastereomeric mixture of (α*S*)- and (α*R*)-α-chloroboronic esters[563]. Unfortunately for synthetic purposes, the diastereomeric ratios produced from the uncatalysed rearrangement of **260** were usually between 1:2 and 2:1.

(203)

Application of zinc chloride catalysis to the rearrangement of **260** resulted in significant alterations in the diastereomer ratios of the α-chloroboronic ester products. For R = isobutyl, the (α*S*):(α*R*) ratio was 34:66 without zinc chloride, but 92:8 with zinc chloride. For R = Me, the (α*S*):(α*R*) ratio was 22:78 without but 51:49 with zinc chloride[563]. Previous results with **246** had not unequivocally ruled out the possibility that the improvement in diastereomeric ratio by zinc chloride was entirely due to suppression of epimerization[557]. These data support the idea that the metal cation influences the diastereoselection by some sort of chelation analogous to that proposed by Meyers[555].

2. (R, R)-Butane-2, 3-diol esters

A chiral directing group having C_2 symmetry would not have different diastereomeric borate complexes corresponding to **246** and **260** (Section III.B.1.c), and it would then make no difference whether the dichloromethyl group or the alkyl group was attached to the boron first. The most commonly used chiral diols having C_2 symmetry, tartrate esters, were tested unsuccessfully, with no yields of α-chloroboronic esters without zinc chloride[557] and a low yield resulting from the use of zinc chloride and diisopropyl tartrate[46]. Noting the commercial availability of (*R, R*)-butane-2, 3-diol and its reported successful use as a chiral directing group in cationic acetal–olefin cyclizations[564], Sadhu *et al.*[565] tested (*R, R*)-butane-2, 3-diol butylboronate, **261**, with lithium dichloromethide–zinc chloride and obtained 1-chloropentylboronic ester, **264**, which was 95% (1*S*)-isomer (equation 204). The intermediate borate complex **263** was also prepared from (*R, R*)-butane-2, 3-diol dichloromethylboronate, **262**, and butyllithium, and was shown to yield a similar (96%) diastereomeric preference, as required by the mechanism. This reaction was also tested with other alkyl groups in place of *n*-butyl, including methyl, isopropyl, phenyl, and benzyl, and uniformly produced 95–96% (αS)-α-chloroboronic esters (with one exception attributed to experimental error, isopropyl by one of two routes, 91%). The 95% diastereoselectivity achieved with the methyl group is particularly noteworthy. The diastereoselectivities were determined by converting the butanediol esters to pinanediol esters for n.m.r. analysis.

(261) $\xrightarrow{LiCHCl_2}$ (263); (262) $\xrightarrow{n\text{-}BuLi}$ (263) $\xrightarrow{ZnCl_2}$ (264) (204)

Treatment of (*R, R*)-butane-2, 3-diol (αS)-α-chlorobenzylboronate with water and diethyl ether resulted in hydrolysis of the boronic ester, with the butanediol going into the aqueous phase and the (αS)-α-chlorobenzylboronic acid into the ether. Recrystallization from diethyl ether–hexane yielded boronic acid of $\geqslant 99\%$ ee, as shown by the n.m.r. spectrum of its (+)-pinanediol ester[565]. This hydrolysis does not work efficiently with the water-soluble α-chloroethylboronic acid, but appears to be generally applicable to higher members of the series. The ease of hydrolysis of these butanediol esters contrasts with the extreme stability of the pinanediol esters, and basically solves the problem of being able to remove the chiral directing group so that it can be replaced by one that will direct construction of the next chiral carbon in the opposite chiral sense, which was discussed in Section III.B.1.a.

C. Asymmetric Hydroboration

1. *Asymmetric hydroborating agents*

a. Diisopinocampheylborane

The synthesis of diisopinocampheylborane, **30**, and its successful use in asymmetric synthesis by Brown and Zweifel[106] have been mentioned in Sections I.C.1.c and III.A, and Brown and coworkers have provided two recent reviews of this and related chemistry[566,567]. Recent advances have considerably expanded the applications of asymmetric hydroboration and have provided several organoborane intermediates of synthetic interest in very high chiral purities.

The major limitation to hydroboration as an approach to chiral synthesis is that it is useful with only a small range of structural types. For example, hydroboration of *cis*-but-2-ene with enantiomerically pure (−)-diisopinocampheylborane, **30**, and (+)-α-pinene[108–110] followed by oxidation of the intermediate borane **265** with hydrogen peroxide provides a highly efficient route to (*R*)-(−)-butan-2-ol, **266**, in $> 98\%$ enantiomeric excess, and the opposite enantiomer of **266** is made equally easily (equation 205).

Other symmetrical *cis*-alkenes behave similarly, although *cis*-hex-3-ene has yielded an ee of 93% in the derived hexan-3-ol, and other ees were lower[109]. Problems arise immediately if the *cis*-alkene is not symmetrical. For example, *cis*-pent-2-ene has yielded a mixture of pentan-2-ol, **267**, and pentan-3-ol, **268**, in proportion ranging from 76:24 to 59:41, and the 92% ee of the **267** is thus subverted by the regioselectivity problem (equation 206)[109,554]. Chiral 2-pentylboronic esters are available in high purity from the homologation process discussed in Section III.B[558], and hydroboration is obsolete as an approach to such structures. With *cis*-4-methylpent-2-ene, the regioselectivity problem is largely overcome and 96% 4-methylpentan-2-ol, **269**, was obtained, but the ee was only 76% (equation 207)[554]. This particular alcohol has not been made by the (dichloromethyl)lithium–boronic ester process (Section III.B), but the critical precursor, (+)-pinanediol (1*S*)-1-chloro-3-methylbutylboronate, is available in 99.5% chiral purity[558], and the enantiomer could be made similarly. The utility of **30** in chiral synthesis is further limited by the low ees of alcohols obtained from the hydroboration of *trans*-alkenes or 2-substituted alk-1-enes, although these limitations are in part overcome by the use of monoisopinocampheylborane (see the following Section, III.C.1.b).

BH (30) → (265) $\xrightarrow{H_2O_2}$ HO (266) (205)

30 + → → HO (267) + OH (268) (206)

30 + → → HO (269) (207)

In spite of its restricted range of structural applicability, asymmetric hydroboration can provide some highly useful starting materials. An example is the hydroboration of methyl cyclopentadiene-5-acetate, **271**, with (+)-diisopinocampheylborane, **270**, followed by oxidation to yield methyl (3*R*,4*R*)-4-hydroxycyclopentene-3-acetate, **272**, in 92% ee, a useful intermediate for prostaglandin syntheses (equation 208)[568]. Analogous hydroboration of 5-methylcyclopentadiene provides a useful intermediate for the synthesis of the natural product loganin[569]. For such cyclic systems, the (dichloromethyl)lithium–boronic ester reaction (Section III.B) is not applicable, and although ring closures of products from such reactions remain a future possibility, it would be difficult to improve upon the directions of the hydroboration route.

BH (270) + CH_2CO_2Me (271) → → OH CH_2CO_2Me (272) (208)

Surprisingly high selectivity was found by Masamune *et al.*[570] in the hydroboration–oxidation of a 2-methylalk-1-ene, **273**, with (−)-diisopinocampheylborane, **30**, to produce a chiral 2-methyl primary alcohol, **274**, used in the synthesis of tylonolide (equation 209). The ratio of **274** to its epimer **275** was at least 50:1. The chirality already present in the alkene, **273**, had little influence on the outcome, and hydroboration of **273** with (+)-diisopinocampheylborane, **270**, produced pure (> 50:1) epimer **275** (equation 210). The minimal influence of the chirality of **273** was further demonstrated by its hydroboration with 9-bbn to produce a 2:1 mixture of **274** and **275**. These results are especially noteworthy in view of the maximum 30% ees observed by Brown and Jadhav[567] in the hydroboration–oxidation of simple 2-methylalk-1-enes with diisopinocampheylborane. Incidentally, Masamune *et al.*[570] used *m*-chloroperbenzoic acid for the borane oxidation to prepare **274**.

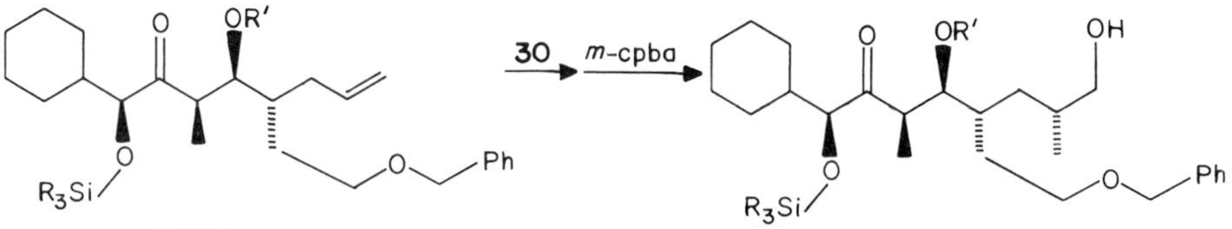

(209)

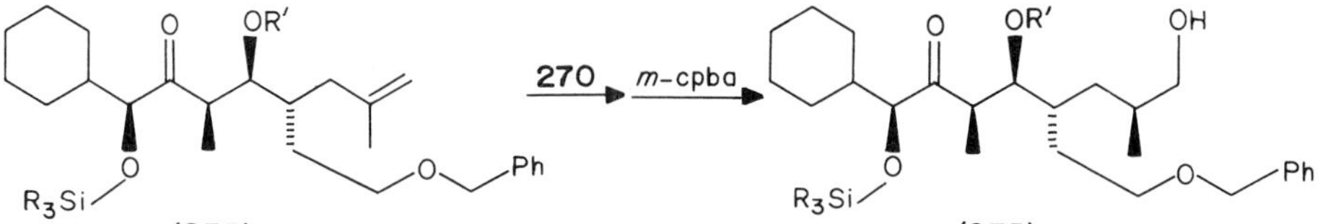

(210)

b. Monoisopinocampheylborane

(+)-Monoisopinocampheylborane, **29**, derived from (+)-α-pinene, hydroborates 2-methylbut-2-ene to provide (*S*)-3-methylbutan-2-ol in 53% ee[571], or *trans*-but-2-ene to yield (*S*)-(+)-butan-2-ol in 73% ee[572]. These results represented a great improvement over the 13–14% ees found in the same systems with diisopinocampheylborane, **30**. Phenyl substituents result in still higher ees, and the hydroboration of 1-phenylcyclopentene with **29** followed by oxidation of the intermediate borane **276** yielded (1*S*,2*R*)-2-phenylcyclopentanol, **277**, in 100% ee insofar as could be detected with the aid of a chiral shift reagent in the 90-MHz n.m.r. spectrum (equation 211)[573]. With 2-phenylcyclohexanol the ee is a still useful 88%. Full details of this work have been published[574].

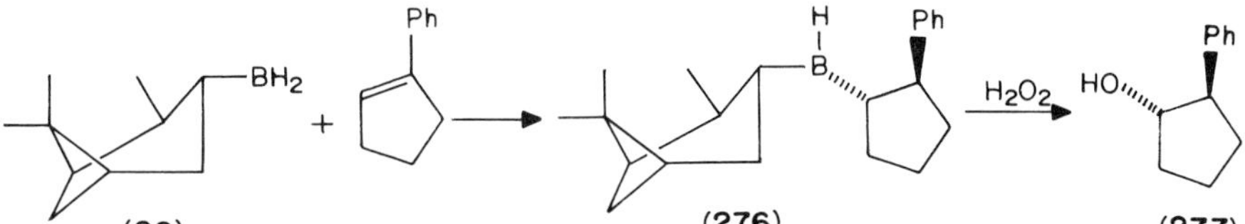

(211)

Treatment of isopinocampheylalkylboranes with acetaldehyde results in preferential cleavage of α-pinene (recyclable) and production of the diethyl ester of the corresponding boronic acid[575]. Thus, **276** is readily converted into diethyl (1*S*, 2*S*)-2-phenylcyclopentylboronate, **278**, in 66% yield (equation 212). (The 2-positions of **277** and **278** have the same absolute configurations but opposite notations because of relative group priorities.) Similarly, cleavage of bis(isopinocampheyl)-2-butylborane, **265**, with acetaldehyde leads to (*S*)-2-butylboronic acid, **279**, which was isolated as its dimethyl ester (71%) (equation 212). One possible application of these boronic acids, especially cyclic examples such as **278**, would be as starting materials for some of the chemistry described in Section III.B.

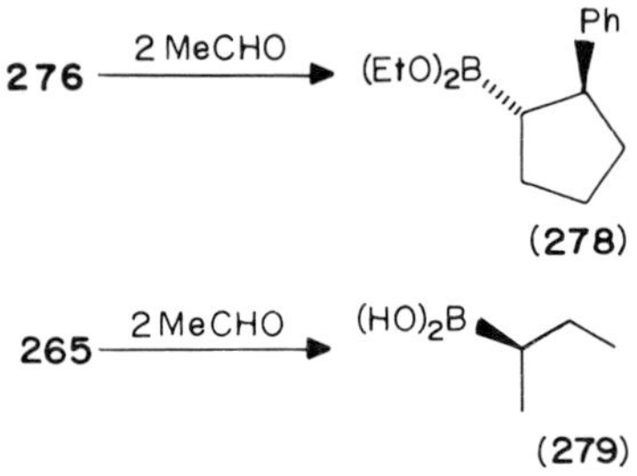

276 —(2 MeCHO)→ $(EtO)_2B$ (278)

265 —(2 MeCHO)→ $(HO)_2B$ (279) (212)

Another use of the cleavage by acetaldehyde is the preparation of chiral borinic esters, which can be converted into chiral ketones by reaction with $LiCCl_2OMe$ (see Section II.B.1.d)[576,577]. An example is the hydroboration of pent-1-ene with **280**, which is the enantiomer of **276**, followed by conversion of the resulting borane **281** to (1*R*, 2*S*)-2-phenylcyclopentyl pentyl ketone, **282**, 90% ee (equation 213). Another is the hydroboration of *trans*-but-2-ene with (−)-monoisopinocampheylborane followed by reaction of the resulting borane **283** with ethylene and conversion to (*R*)-4-methylhexan-3-one, **284**, 60% ee, which is an alarm pheromone of the ant *Manica mutica* (equation 214).

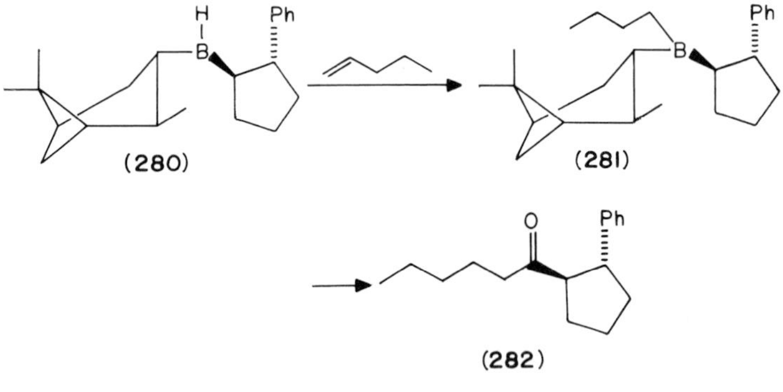

(280) → (281) → (282) (213)

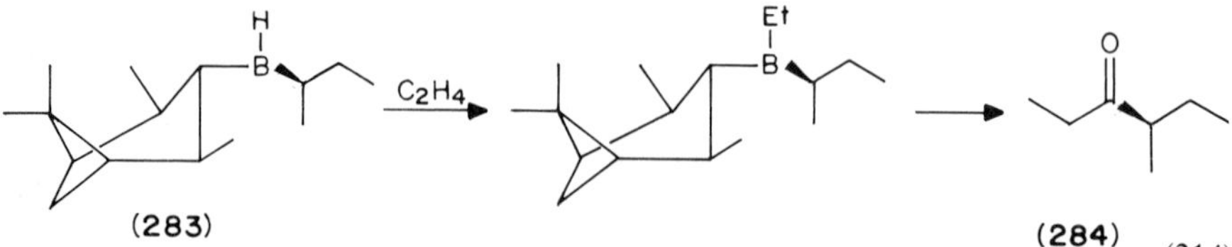

(283) —(C_2H_4)→ → (284) (214)

A final development which promises to make this hydroboration chemistry more useful than the observed enantiomeric excesses might suggest is the observation that several monoisopinocampheylboranes crystallize from the solutions in which they are made and can be purified to very high diastereomeric purity, so that the alcohols derived by oxidation have ees approaching 100%[578]. In the simplest cases, the crystals were diastereomerically pure or could be aged to high purity, and in other cases more elaborate approaches had to be used.

c. *Other asymmetric boranes*

Dilongifolylborane hydroborates *cis*-but-2-ene, 2-methylbut-2-ene, 1-methylcyclopentene, and other olefins of similar steric requirements to yield alcohols after oxidation which have ees in the 60–75% range[579]. Limonylborane is a cyclic borane derived from limonene and available in both enantiomers, and yields alcohol derivatives with ees in the 45–67% range with *cis*- or trans-but-2-ene, 2-methyl-but-2-ene, or 1-methylcyclopentene[580].

2. *Asymmetric substrates*

Once a chiral centre is in place, hydroboration can be used for the diastereoselective introduction of additional chiral centres. This is not asymmetric synthesis in the same sense as hydroboration with asymmetric reagents. Examples of diastereoselective hydroboration have been noted in Sections I.C.1.e and f[123,138], and the selective creation of new chiral centres during the hydroboration of terpenes and steroids has been noted in Brown's books[1,2b]. These generally depend on steric factors in a straightforward manner or, in the case of dienes, on cyclization to establish steric relationships. The examples which follow are not a comprehensive survey.

Hydroboration–oxidation of the substituted allylic furan derivative **285** with borane–thf yielded the alcohol **286** and its diastereomer resulting from attack at the opposite side of the double bond in an 8:1 ratio[581], thus providing an intermediate used in the monensin synthesis by Kishi and coworkers (equation 215).

$$\textbf{(285)} \xrightarrow{BH_3} \xrightarrow{H_2O_2} \textbf{(286)} \qquad (215)$$

Midland and Kwon have studied the stereocontrolled synthesis of side chains at the 17-position of steroids[582–584]. Approach of 9-bbn to the β-face of a 17-ethylidene group is sterically blocked by the angular methyl group, and as a result the Z-isomer **287** produces the 'natural' steroid configuration at C-20 (equation 216) and the E-isomer **288** the 'unnatural' configuration (equation 217)[582,583]. Further elaboration of the side chain was carried out via reaction of the borane with the carbanion from chloroacetonitrile (see Section II.B.1.c). Hydroboration of 20-alkylidene steroids, **289**, results in attack from the top face of the double bond as illustrated (probably not the reactive conformation) and produces the 'natural' isomer at C-20 (equation 218)[584]. The reaction works best if R′ is bulky, and if R is 3-methyl-1-butyl the side chain illustrated is that of the insect moulting hormone ecdysone.

$$\textbf{(287)} \xrightarrow{\text{9-bbn}} \xrightarrow{Cl\bar{C}HCN} \text{(}CH_2CN\text{ product)} \qquad (216)$$

(288) 9-bbn ClC̄HCN CH2CN (217)

R BR'2 R OAc R

(289) R=H or CH2CH2CHMe2 (218)

D. Chiral Sigmatropic Rearrangements

1. *Allylboranes*

a. *Boronic esters*

Allylic boronic esters have been developed as reagents for directed chiral synthesis by Hoffmann, who has reviewed this and related chemistry recently[585]. The reaction of an aldehyde with a γ-substituted allylic boronic ester creates two chiral centres with very good control of their relative configurations by the steric constraints of the cyclic transition state, described in Section II.B.3.d. Useful but not outstanding control of the absolute configuration has been achieved by the use of the ester of a phenylbornanediol with (*Z*)-but-2-enylboronic acid, **290**[586]. For example, reaction of **290** with acetaldehyde followed by cleavage of the boron yielded (3*S*, 4*S*)-3-methylpent-1-en-4-ol, **291**, in 95% diastereomeric purity but only 70% enantiomeric excess (equation 219). This compound was then converted into the pheromone **292** of the drugstore beetle, *Stegobium panaceum*, in several steps, and proved the absolute configuration.

O-B O Ph MeCHO OH O O

(+)-isomer of diol (290)

70% ee 95% this diastereomer (291)

(292) (219)

Similar chemistry has been used with the ester of the (−)-enantiomer of the phenylbornanediol with (*E*)-but-2-enylboronic acid, **293**, in a synthesis of multistriatin, **295**, a component of the aggregation pheromone of the elm bark beetle, *Scolytus multistriatus* (equation 220)[587]. In this case, the diastereoselectivity in forming the intermediate homoallylic alcohol **294** was still 95% but the ee was only 52%.

(220)

Although the ees obtainable to date in the Hoffmann synthesis have not been very encouraging, significantly better results are obtained if a single enantiomer of a chiral aldehyde is used with one of the chiral boronic esters **290** or **293** (equation 221 and 222)[588]. With the correct pairing of chiral aldehyde and chiral boronic ester, double stereodifferentiation results in good diastereoselectivity, and with the opposite pairing the selectivity is very low. This is not chiral synthesis directed by a chiral auxiliary, because it incorporates the chiral centre of the aldehyde into the product. The (*S*)-(+)-2-methylbutyraldehyde, **296**, used to illustrate this method was not made by organoborane chemistry, but it might be noted that it could be, since hydroboration of *cis*-but-2-ene with diisopinocampheylborane generates a borane that can be converted into **296** (see Section III.C) or the asymmetric boronic ester homologation (see Section III.B) could easily be adapted for this purpose.

(221)

(222)

Incorporation of a preformed chiral centre into the product has also been used by Wuts and Bigelow[589] in a synthesis of the sugar oleandrose. The reaction of *meso*-butane-2, 3-diol (*Z*)-3-methoxyallylboronate with (*S*)-2-benzyloxypropanal gave diastereomeric homoallylic alcohol products in the ratio 8.7:1.2:1, with the major isomer being the useful intermediate.

Efficient chirality transfer has been found by Hoffmann and Landmann[590] in the reaction of a chiral boronic ester with aldehydes. (*R, R*)-Butane-2, 3-diol (*S*)-

chloroallylboronate, **297**, was prepared by the method outlined in Section III.B.2 and converted into the pinacol ester **298** in order to achieve high stereoselectivity in the reaction with aldehydes. The (*Z*)-chloroallyl alcohols **300** were the predominant products (93–96%) for R = Me, Et, Pr^i, or Ph, which is in accord with the expectation that the chlorine atom will preferentially assume the axial orientation in the cyclic chair-form transition state **299** (equation 223). The ees of the predominant products **300** were 90–93%[590], in accord with the ees measured for other α-chloroboronic esters prepared in the same manner as **297**[565]. However, it should be noted that the (*E*)-chloroallyl by product **301** has the opposite chirality to **300** and must be separated in order to utilize the high ees available in **300**[590].

(297) (298) + RCHO (299)

(300) ~95% (301) ~5% (223)

The preference of the α-substituent for an axial or equatorial position in transition states analogous to **299** depends on the nature of the substituent itself and on the other substituents present. Rearrangement of an optically active (*Z*)-α-(trimethylsilyl)crotylboronic ester has yielded 96% (*E*)-*erythro*-homoallylic alcohol, described in Section II.B.1.d[373]. The ees were low and not determined, but the very high diastereoselectivity implies nearly total chirality transfer.

Optically active propargylic alcohols are readily available by reduction of acetylenic ketones with Midland's reagent, *B*-isopinocampheyl-9-bbn (see Section III.E). Hydroboration of their acetate esters with dialkylboranes yields 3-acetoxyalk-1-enylboranes, which on treatment with base rearrange with high stereoselectivity to chiral allylic borinic esters, **302**[591]. The borinic esters react with aldehydes according to the usual stereoselection rules to yield 96–99% *threo* homoallylic alcohols, **303**, having ees in the 50–85% range (equation 224)[592]. The R groups used for R_2BH included cyclohexyl and isopinocampheyl, R′ was primary alkyl, and R″ was methyl, phenyl, isopropyl, or *n*-pentyl. Although the R of R_2BH must be bulky, that is not a serious limitation for synthetic purposes, since useful chiral intermediates can be obtained by cleaving the double bond of **303** so that only R′ and R″ are retained.

R_2B–CH=CH–CH(OAc)–R′ $\xrightarrow{OH^-}$ R–CH(OBR)... (302) $\xrightarrow{R''CHO}$ (303)

(302) (303) (224)

b. Trialkylboranes

Brown and Jadhav[593,594] have shown that *B*-(allylic)-*B*, *B*-diisopinocampheylboranes react with aldehydes to give homoallylic alcohols in high enantiomeric purities (83–96%). The process is illustrated by the synthesis of (−)-artemisia alcohol, **304**, in 96% ee (equation 225)[594].

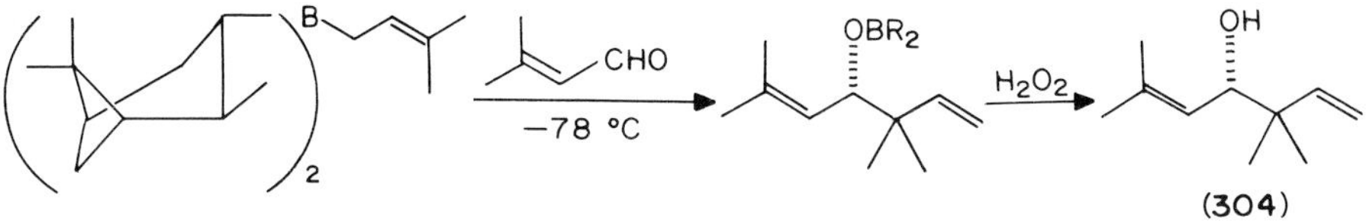

(304)

(225)

Yamamoto *et al*[595] have observed very high chirality transfer in reactions of *B*-allyl-9-bbn with imines derived from 2-phenylpropionaldehyde. The reaction of *B*-allyl-9-bbn with *N*-(1-phenylethyl) isobutyraldimine, **305**, was 92% diastereoselective (equation 226). Although the reaction was not carried out on optically active material, optically active 1-phenylethylamine is readily available, and the distinction between the 'asymmetric' syntheses included in this section and the 'diastereoselective' syntheses covered in Section II.B.1.d becomes arbitrary.

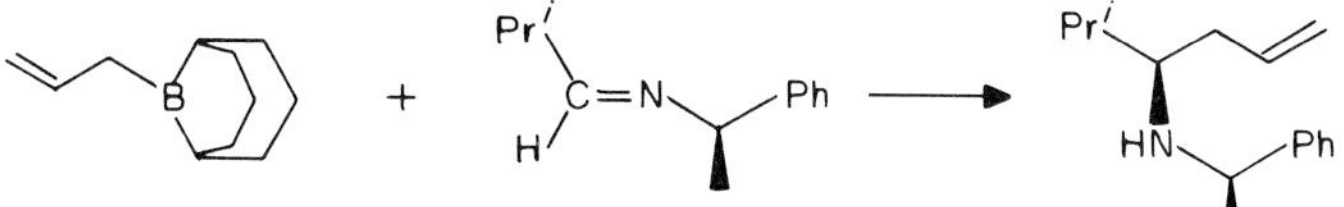

(305)

(226)

2. Allenylboronic esters

Allenylboronic acid has been prepared from propargylmagnesium bromide and trimethyl borate, and its esters with diethyl or diisopropyl tartrate react with aldehydes to form homopropargylic alcohols, **306**, having ees generally in the 60–95% range (equation 227)[596].

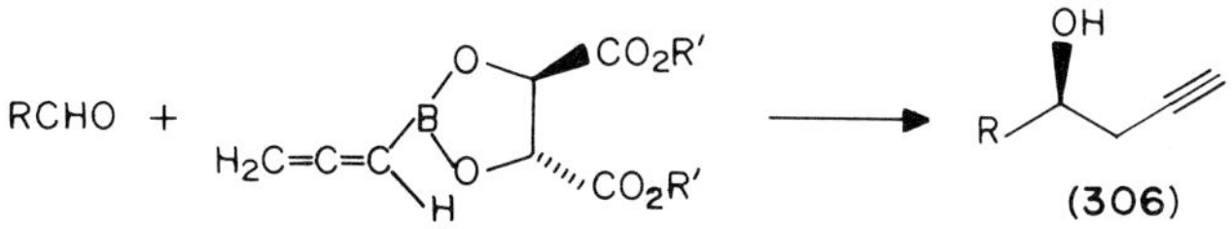

(306)

(227)

3. Enol borinates

Enol borinates have been used with great success in enantioselective aldol condensations. The carbon—boron bond is an incidental feature of these reactions and does not participate except insofar as the alkyl groups on boron influence the stereoselectivity. These reactions are therefore outside the scope of this review, but in view of the close mechanistic analogies to allylborane–aldehyde reactions, the work of Evans and coworkers[597,598] and Masamune and coworkers[570,599] is mentioned here. By incorporating a chiral substituent, later removed, in the boron enolate at the enol carbon, enantioselectivities of these reactions can be made very high.

E. Asymmetric Reductions

Reductions with Midland's reagent, *B*-isopinocampheyl-9-bbn, **307**, involve a carbon—boron bond cleavage and therefore fall marginally within the scope of this review.

Reductions with borohydrides that involve only the boron—hydrogen bond are not covered. Midland has recently reviewed reductions with chiral boron reagents[600]. The reagent is now commercially available as 'Alpine Borane' from the Aldrich Chemical Company.

The first successful application of Midland's reagent, **307**, was in the reduction of deuteriated aldehydes to produce stereoselectively labelled α-deuterio alcohols[601]. The **307** prepared from (+)-α-pinene reduced benzaldehyde to (*S*)-α-deuteriobenzyl alcohol having *ca.* 98% of the enantiomeric purity of the reducing agent. The alcohol is initially formed as its 9-bbn ester, which can be conveniently cleaved with ethanolamine. The hydrogen for the reduction comes from the isopinocampheyl group, which is eliminated as α-pinene. The steric interactions as judged by the experimental results correspond to the pinanyl bridge methylene group behaving as smaller than the 2-methyl group. If α-pinene is hydroborated with *B*-deuterio-9-bb, the resulting deuteriated **307** will reduce benzaldehyde to yield the (*R*)-enantiomer of α-deuteriobenzyl alcohol (equation 228). Reduction of a series of deuteriated aromatic and aliphatic aldehydes with **307** produced α-deuterio alcohols having 71–101% of the estimated ee of the reducing agent[602].

$$(307) + PhCHO \longrightarrow \longrightarrow Ph-C(OH)(H)-D \qquad (228)$$

Acetylenic functions interact with **307** as small groups, so that the same direction of chiral selectivity is found in reduction of acetylenic ketones as with aldehydes[603]. Reduction of a variety of acetylenic ketones with **307** produced propargylic alcohols, **308**, having ees in the 73–100% range (equation 228). The configurations of **308** are *R* if the priority of the other substituent *R* is lower than that of the acetylenic group, often but not always true. The opposite enantiomer of **307** is readily available, and the enantiomers of **308** are therefore also easily prepared.

$$307 + RC(=O)C{\equiv}CR' \longrightarrow R-C(OH)(H)-C{\equiv}CR' \ (308) \qquad (229)$$

It is necessary to carry out reductions with **307** at or below room temperature because the reagent dissociates easily at higher temperatures to generate free 9-bbn, a much faster and achiral reducing agent[604]. Reductions by **307** are slow, often taking several days, and Brown and Pai reported that the use of neat **307** rather than the dilute solutions originally used by Midland results in improved enantioselectivities and also permits reduction of a number of non-acetylenic ketones with fair to good ees[605], including aryl halomethyl ketones with high ees[606]. Midland and McLoughlin[607] found that high pressure (6000 atm) suppresses the dissociation of **307** and permits the reduction of acetophenone to 1-phenylethanol in essentially 100% ee and various other ketones highly selectively.

The propargyl ketone reduction remains the most facile and reliable application of Midland's reagent, **307**, and this chemistry has been explored in considerable detail[608]. The borane chemistry ends with production of the propargylic alcohols, **308**, except for the hydroboration chemistry of these alcohols noted in Section III.D.1.a. As would be expected, reduction of ketone functions by **307** can be carried out in the presence of other functional groups such as carboxylic esters, which allows syntheses of various lactones[608,609], including the Japanese beetle pheromone[610]. The acetylenic group of **308** can be oxidized to provide chiral β-hydroxy acids[611]. Conversion of **308**, R = isopropyl and R′ = methyl, to (*R*)-(*Z*)-5-methylhex-2-en-4-ol and then to the *O*-allyl or benzyl ether followed by [2.3]sigmatropic Wittig rearrangement has yielded *erythro* homoallylic alcohols with good diastereoselection[612]. An alternative reagent to **307** for chiral reductions is the 9-bbn adduct of nopol benzyl ether, which gives similar results to the enantiomer of **307** because nopol, a derivative of β-pinene, has the same absolute configuration as (−)-α-pinene[613]. Midland's reagent, **307**, has proved useful in a considerable variety of organic syntheses carried out by other research groups. Examples of these applications include syntheses of steroids[614,615], toad poisons[616,617], substrates for various biosynthetic pathaway studies[618–621], arachidonic acid metabolites[622], leukotrienes[623], α-hydroxy esters[624], and compounds for mechanistic studies[625].

IV. REFERENCES

1. H. C. Brown, *Hydroboration*, W. A. Benjamin, New York, 1962; second printing, with Nobel Lecture, Benjamin/Cummings Publishing Com., Reading, MA, 1980.
2. H. C. Brown, *Boranes in Organic Chemistry*, Cornell University Press, Ithaca, NY, 1972, (a) pp. 227–251; (b) pp. 255–297, (c) pp. 301–446.
3. E. Frankland and B. F. Duppa, *Justus Liebigs Ann. Chem.*, **115**, 319 (1860).
4. (a) E. Frankland, *J. Chem. Soc.*, **15**, 363 (1862); (b) E. Frankland, *Justus Liebigs Ann. Chem.*, 124, 129 (1862); (c) E. Frankland, *Proc. R. Soc. London.*, **25**, 165 (1876).
5. (a) A. Michaelis and P. Becker, *Chem. Ber.*, **13**, 58 (1880); (b) A. Michaelis and P. Becker, *Chem. Ber.*, **15**, 180 (1882); (c) A. Michaelis, *Chem. Ber.*, **27**, 244 (1894); (d) A. Michaelis, *Justus Liebigs Am. Chem.*, **315**, 19 (1901).
6. E. Khotinsky and M. Melamed, *Chem. Ber.*, **42**, 3090 (1909).
7. E. Krause and R. Nitsche, *Chem. Ber.*, **54**, 2784 (1921).
8. H. R. Snyder, J. A. Kuck, and J. R. Johnson, *J. Am. Chem. Soc.*, **60**, 105 (1938).
9. J. R. Johnson, M. G. Van Campen, Jr., and O. Grummitt, *J. Am. Chem. Soc.*, **60**, 111 (1938).
10. J. R. Johnson, H. R. Snyder, and M. G. Van Campen, Jr., *J. Am. Chem. Soc.*, **60**, 115 (1938).
11. R. M. Washburn, E. Levens, C. F. Albright, and F. A. Billing, *Organic Syntheses*, Coll. Vol. 4., Wiley, New York, 1963, p. 68.
12. H. C. Brown and B. C. Subba Rao, *J. Am. Chem. Soc.*, **78**, 5694 (1956).
13. A. Finch and P. J. Gardner, in *Progress in Boron Chemistry* (Eds. R. J. Brotherton and H. Steinberg), Vol. 3., Pergamon Press, Oxford, 1970, p. 177.
14. J. Hine, *Physical Organic Chemistry*, 2nd Ed., McGraw-Hill, New York, 1962, pp. 30–32.
15. (a) V. A. Brattsev, S. P. Knyazev, and V. I. Stanko, *Zh. Obshch. Khim.*, **45**, 1192 (1975); (b) S. P. Knyazev, V. A. Brattsev, and V. I. Stanko, *Zh. Obshch. Khim.*, **47**, 2627 (1977).
16. V. I. Stanko, Yu. V. Gol'tyapin, T. P. Klimova, N. I. Kirillova, and Yu. T. Struchkov, *Zh. Obshch. Khim.*, **47**, 2248 (1977).
17. R. W. Auten and C. A. Kraus, *J. Am. Chem. Soc.*, **74**, 3398 (1952).
18. T. D. Parsons, J. M. Self, and L. H. Schaad, *J. Am. Chem. Soc.*, **89**, 3446 (1967).
19. G. Schmid and H. Nöth, *Chem. Ber.*, 2502 (1968).
20. K. Smith and K. Swaminathan, *J. Chem. Soc., Dalton Trans.*, 2297 (1976).
21. (a) A. H. Soloway, *Science*, **128**, 1572 (1958); (b) A. H. Soloway, B. Whitman, and J. R. Messer, *J. Med. Pharm. Chem.*, **5**, 191 (1962).
22. D. S. Matteson, A. H. Soloway, D. W. Tomlinson, J. D. Campbell, and G. A. Nixon, *J. Med. Chem.*, **7**, 640 (1964).
23. R. L. Hughes, I. C. Smith, and E. W. Lawless, in *Production of the Boranes and Related Research* (Ed. R. T. Holzmann), Academic Press, New York, 1967, pp. 291–294.
24. M. Tagaki, K. Hisata, M. Baba, and E. Masuhara, *Nippon Kōkūka Gakkai Zasshi*, **22**, 533 (1973). [The units of the LD_{50} are misquoted by a frightening factor of 1000 in *Chem. Abstr.*, **83**, 72861 t (1975).]

25. Morton Thiokol, Inc., Alfa Products, Danvers, MA, Catalog, 1983–84.
26. Aldrich Chemical Company, P.O. Box 355, Milwaukee, WI, Catalog, 1984–85.
27. D. S. Matteson, *J. Am. Chem. Soc.*, **82**, 4228 (1960).
28. J. R. Johnson and M. G. Van Campen, Jr., *J. Am. Chem. Soc.*, **60**, 121 (1938).
29. C. H. Bamford and D. M. Newitt, *J. Chem. Soc.*, 695 (1946).
30. J. Kollonitsch, *Chem. Eng. News*, **57**, No. 47, 3 (1974).
31. M. I. Bruce, *Chem. Br.*, **11**, 237 (1975).
32. (a) H. Mann, *Arch. Fischereiwiss.*, **24**, 171 (1973); *Chem. Abstr.*, **81**, 403 f (1974); (b) N. J. Antia and J. Y. Cheng, *J. Fish. Res. Board Can.*, **32**, 2487 (1975).
33. R. J. Weir, Jr., and R. S. Fisher, *Toxicol. Appl. Pharmacol.*, **23**, 351 (1972).
34. W. Ebeling, D. A. Reierson, R. J. Pence, and M. S. Viray, *Pestic. Biochem. Physiol.*, **5**, 81 (1975).
35. (a) Advisory Subcommittee on the Nomenclature of Boron Compounds (J. Carter, Chairman), *Inorg. Chem.*, **7**, 1945 (1968); (b) IUPAC Commission on Nomenclature of Inorganic Chemistry (IUPAC, Oxford, UK), *Int. Union Pure Appl. Chem. Inf. Bull., Appendix, Tentative Nomenclature, Symbols, Units, Standards*, 1970, 25 pp.
36. *Chem. Abstr.*, 1972–76 Chemical Substance Index, p. 7971CS; p. 8035CS.
37. D. S. Matteson and E. Krämer, *J. Am. Chem. Soc.*, **90**, 7261 (1968).
38. D. S. Matteson, *J. Org. Chem.*, **29**, 3399 (1964).
39. M. W. Rathke, E. Chao, and G. Wu, *J. Organomet. Chem.*, **122**, 145 (1976).
40. B. M. Mikhailov and F. B. Tutorskaya, *Izv. Akad. Nauk SSSR, Otd. Khim. Nauk*, 1865 (1959).
41. D. S. Matteson and K. H. Arne, *Organometallics*, **1**, 280 (1982).
42. H. R. Snyder and M. S. Konecky, *J. Am. Chem. Soc.*, **80**, 3611 (1958).
43. D. S. Matteson, *J. Org. Chem.*, **27**, 3712 (1962).
44. D. S. Matteson and R. J. Moody, *Organometallics*, **1**, 20 (1982).
45. H. C. Brown and T. E. Cole, *Organometallics*, **2**, 1316 (1983).
46. K. M. Sadhu and D. S. Matteson, unpublished work.
47. (a) D. S. Matteson and T. C. Cheng, *J. Org. Chem.*, **33**, 3055 (1968); (b) D. S. Matteson and T. C. Cheng, *J. Organomet. Chem.*, **6**, 100 (1966).
48. D. S. Matteson and D. Majumdar, *J. Organomet. Chem.*, **170**, 259 (1979).
49. P. G. M. Wuts and P. A. Thompson, *J. Organomet. Chem.*, **234**, 137 (1982).
50. K. Niedenzu, J. W. Dawson, and P. Fritz, *Inorg. Synth.*, **10**, 126 (1967).
51. H. Nöth and H. Vahrenkamp, *J. Organomet. Chem.*, **11**, 399 (1968).
52. T. Onak, *Organoborane Chemistry*, Academic Press, New York, 1975.
53. D. S. Matteson and R. W. H. Mah, *J. Org. Chem.*, **28**, 2171 (1963).
54. D. S. Matteson, *J. Org. Chem.*, **27**, 275 (1962).
55. G. W. Kramer and H. C. Brown, *J. Organomet. Chem.*, **73**, 1 (1974).
56. J. A. Sinclair and H. C. Brown, *J. Org. Chem.*, **41**, 1078 (1976).
57. H. C. Brown and J. A. Sinclair, *J. Organomet. Chem.*, **131**, 163 (1977).
58. J. A. Sinclair, G. A. Molander, and H. C. Brown, *J. Am. Chem. Soc.*, **99**, 954 (1977).
59. G. A. Molander and H. C. Brown, *J. Org. Chem.*, **42**, 3106 (1977).
60. E. Negishi and L. D. Boardman, *Tetrahedron Lett.*, **23**, 3327 (1982).
61. R. B. Castle and D. S. Matteson, *J. Am. Chem. Soc.*, **90**, 2194 (1968).
62. R. B. Castle and D. S. Matteson, *J. Organomet. Chem.*, **20**, 19 (1969).
63. D. S. Matteson and P. K. Jesthi, *J. Organomet. Chem.*, **110**, 25 (1976).
64. D. S. Matteson, *Synthesis*, 147 (1975).
65. D. S. Matteson, *Gmelins Handbuch der Anorganischen Chemie*, 8th ed., New Supplement Series (Eds K. Niedenzu and K.-C. Buschbeck), Vol. 48, Part 16, Springer-Verlag, Berlin, 1977, p. 37.
66. H. C. Brown, G. W. Kramer, A. B. Levy, and M. M. Midland, *Organic Synthesis via Boranes*, Wiley–Interscience, New York, 1975.
67. D. J. Pasto and C. C. Cumbo, *J. Am. Chem. Soc.*, **86**, 4343 (1964).
68. H. C. Brown and G. Zweifel, *J. Am. Chem. Soc.*, **82**, 4708 (1960).
69. R. A. Braun, D. C. Braun, and R. M. Adams, *J. Am. Chem. Soc.*, **93**, 2823 (1971).
70. R. A. Braun, D. C. Braun, H. R. Crissman, M. Opperman, and R. M. Adams, *J. Org. Chem.*, **36**, 2388 (1971).
71. C. F. Lane, *J. Org. Chem.*, **39**, 1438 (1974).
72. H. C. Brown, A. K. Mandal, and S. U. Kulkarni, *J. Org. Chem.*, **42**, 1392 (1977).
73. H. C. Brown and A. K. Mandal, *Synthesis*, 153 (1980).
74. H. C. Brown and A. K. Mandal, *J. Org. Chem.*, **45**, 916 (1980).

75. H. C. Brown and B. C. Subba Rao, *J. Am. Chem. Soc.*, **81**, 6428 (1959).
76. H. C. Brown and G. Zweifel, *J. Am. Chem. Soc.*, **82**, 3222 (1960).
77. H. C. Brown and A. W. Moerikofer, *J. Am. Chem. Soc.*, **84**, 1478 (1962).
78. H. C. Brown and G. Zweifel, *J. Am. Chem. Soc.*, **83**, 3834 (1961).
79. G. Zweifel and H. C. Brown, *J. Am. Chem. Soc.*, **85**, 2066 (1963).
80. H. C. Brown and C. D. Pfaffenberger, *J. Am. Chem. Soc.*, **89**, 5475 (1967).
81. E. F. Knights and H. C. Brown, *J. Am. Chem. Soc.*, **90**, 5280, 5281, 5283 (1968).
82. J. A. Soderquist and H. C. Brown, *J. Org. Chem.*, **46**, 4599 (1981).
83. G. Zweifel, H. Arzoumanian, and C. C. Whitney, *J. Am. Chem. Soc.*, **89**, 3652 (1967).
84. H. C. Brown and E. Negishi, *J. Am. Chem. Soc.*, **89**, 5477, 5478 (1967).
85. M. F. Hawthorne and J. A. Dupont, *J. Am. Chem. Soc.*, **80**, 5830 (1958).
86. P. Binger and R. Köster, *Tetrahedron Lett.*, 156 (1961).
87. H. C. Brown and S. P. Rhodes, *J. Am. Chem. Soc.*, **91**, 2149 (1969).
88. H. C. Brown and R. M. Gallivan, Jr., *J. Am. Chem. Soc.*, **90**, 2906 (1968).
89. H. C. Brown and M. K. Unni, *J. Am. Chem. Soc.*, **90**, 2902 (1968).
90. H. C. Brown and R. L. Sharp, *J. Am. Chem. Soc.*, **90**, 2915 (1968).
91. D. J. Pasto, J. Hickman, and T.-C. Cheng, *J. Am. Chem. Soc.*, **90**, 6259 (1968).
92. (a) D. S. Matteson and J. G. Shdo, *J. Am. Chem. Soc.*, **85**, 2684 (1963); (b) D. S. Matteson and J. G. Shdo, *J. Org. Chem.*, **29**, 2742 (1964).
93. B. M. Mikhailov and P. M. Aronovich, *Izv. Akad. Nauk SSSR, Otd. Khim. Nauk*, 1233 (1963).
94. B. M. Mikhailov, P. M. Aronovich, and V. G. Kiselev, *Izv. Akad. Nauk SSSR, Ser. Khim.*, 146 (1968).
95. G. Zweifel and H. Arzoumanian, *J. Am. Chem. Soc.*, **89**, 291 (1967).
96. D. J. Pasto, *J. Am. Chem. Soc.*, **86**, 3039 (1964).
97. D. S. Matteson and P. K. Jesthi, *J. Organomet. Chem.*, **114**, 1 (1976).
98. D. Seyferth, *J. Inorg. Nucl. Chem.*, **7**, 152 (1958).
99. J. A. Soderquist and A. Hassner, *J. Organomet. Chem.*, **156**, C12 (1978).
100. A. G. Brook and J. B. Pierce, *J. Org. Chem.*, **30**, 2566 (1965).
101. E. L. Allred, J. Sonnenberg, and S. Winstein, *J. Org. Chem.*, **25**, 26 (1960).
102. H. C. Brown and G. Zweifel, *J. Am. Chem. Soc.*, **81**, 247 (1959).
103. H. C. Brown and G. Zweifel, *J. Am. Chem. Soc.*, **83**, 2544 (1961).
104. H. C. Brown and J. H. Kawakami, *J. Am. Chem. Soc.*, **92**, 1990 (1970).
105. H. C. Brown, R. Liotta, and L. Brener, *J. Am. Chem. Soc.*, **99**, 3427 (1977).
106. H. C. Brown and G. Zweifel, *J. Am. Chem. Soc.*, **83**, 486 (1961).
107. G. Zweifel and H. C. Brown, *J. Am. Chem. Soc.*, **86**, 393 (1964).
108. H. C. Brown and N. M. Yoon, *Isr. J. Chem.*, **15**, 12 (1977).
109. H. C. Brown, M. C. Desai and P. K. Jadhav, *J. Org. Chem.*, **47**, 5065 (1982).
110. H. C. Brown and B. Singaram, *J. Org. Chem.*, **49**, 945 (1984).
111. H. C. Brown, P. K. Jadhav, and M. C. Desai, *J. Org. Chem.*, **47**, 4583 (1982).
112. H. C. Brown, J. R. Schwier, and B. Singaram, *J. Org. Chem.*, **43**, 4395 (1978).
113. H. C. Brown and A. K. Mandal, *Synthesis*, 146 (1978).
114. A. Pelter, D. J. Ryder, J. H. Sheppard, C. Subrahmanyam, H. C. Brown, and A. K. Mandal, *Tetrahedron Lett.*, 4777 (1979).
115. H. C. Brown, A. K. Mandal, N. M. Yoon, B. Singaram, J. R. Schwier, and P. K. Jadhav, *J. Org. Chem.*, **47**, 5069 (1982).
116. T. D. Parsons, M. B. Silverman, and D. M. Ritter, *J. Am. Chem. Soc.*, **79**, 5091 (1957).
117. H. C. Brown, E. Negishi, and J.-J. Katz, *J. Am. Chem. Soc.*, **97**, 2791 (1975).
118. H. C. Brown, J.-J. Katz, C. F. Lane, and E. Negishi, *J. Am. Chem. Soc.*, **97**, 2799 (1975).
119. H. C. Brown, N. M. Yoon, and A. K. Mandal, *J. Organomet. Chem.*, **135**, C10 (1977).
120. E. Negishi and H. C. Brown, *Synthesis*, 77 (1974).
121. H. C. Brown and E. Negishi, *Tetrahedron*, **33**, 2331 (1977).
122. E. Negishi and H. C. Brown, *Heterocycles*, **5**, 883 (1976).
123. W. C. Still and K. P. Darst, *J. Am. Chem. Soc.*, **102**, 7385 (1980).
124. S. U. Kulkarni, D. Basavaiah, M. Zaidlewicz, and H. C. Brown, *Organometallics*, **1**, 212 (1982).
125. P. J. Maddocks, A. Pelter, K. Rowe, K. Smith, and C. Subramahniyam, *J. Chem. Soc., Perkin Trans. 1*, 653 (1981).
126. H. C. Brown and S. U. Kulkarni, *J. Organomet. Chem.*, **218**, 299 (1981).
127. H. C. Brown and S. K. Gupta, *J. Am. Chem. Soc.*, **93**, 4062 (1971).

128. H. C. Brown, J. R. Schwier, and B. Singaram, *J. Org. Chem.*, **44**, 465 (1979).
129. (a) H. C. Brown, B. Singaram, and J. R. Schwier, *Inorg. Chem.*, **18**, 51 (1979); (b) H. C. Brown and B. Singaram, *Inorg. Chem.*, **18**, 53 (1979).
130. H. C. Brown and G. G. Pai, *Heterocycles*, **17**, 77 (1982).
131. H. C. Brown and G. G. Pai, *J. Organomet. Chem.*, **250**, 13 (1983).
132. H. C. Brown, G. G. Pai, and R. G. Naik, *J. Org. Chem.*, **49**, 1072 (1984).
133. H. C. Brown and E. Negishi, *J. Am. Chem. Soc.*, **93**, 6682 (1981).
134. H. C. Brown and S.-C. Kim, *J. Org. Chem.*, **49**, 1064 (1984).
135. A. Pelter, S. Singaram, and H. C. Brown, *Tetrahedron Lett.*, **24**, 1433 (1983).
136. H. C. Brown, E. F. Knights, and C. G. Scouten, *J. Am. Chem. Soc.*, **96**, 7765 (1974).
137. H. C. Brown, R. Liotta, and C. G. Scouten, *J. Am. Chem. Soc.*, **98**, 5297 (1976).
138. L. Brener and H. C. Brown, *J. Org. Chem.*, **42**, 2702 (1977).
139. H. Taniguchi, L. Brener, and H. C. Brown, *J. Am. Chem. Soc.*, **98**, 7107 (1976).
140. H. C. Brown, R. Liotta, and G. W. Kramer, *J. Org. Chem.*, **43**, 1058 (1978).
141. H. C. Brown, R. Liotta, and G. W. Kramer, *J. Am. Chem. Soc.*, **101**, 2966 (1979).
142. R. Liotta and H. C. Brown, *J. Org. Chem.*, **42**, 2836 (1977).
143. H. C. Brown and J. C. Chen, *J. Org. Chem.*, **46**, 3978 (1981).
144. H. C. Brown, C. G. Scouten, and R. Liotta, *J. Am. Chem. Soc.*, **101**, 96 (1971).
145. H. C. Brown, D. Basavaiah, and S. U. Kulkarni, *J. Organomet. Chem.*, **225**, 63 (1982).
146. G. F. Hennion, P. A. McCusker, E. C. Ashby, and A. J. Rutkowski, *J. Am. Chem. Soc.*, **79**, 5190 (1957).
147. G. F. Hennion, P. A. McCusker, E. C. Ashby, and A. J. Rutkowski, *J. Am. Chem. Soc.*, **79**, 5194 (1957).
148. H. C. Brown and B. C. Subba Rao, *J. Am. Chem. Soc.*, **81**, 6434 (1959).
149. H. C. Brown and G. Zweifel, *J. Am. Chem. Soc.*, **82**, 1504 (1960).
150. H. C. Brown and G. Zweifel, *J. Am. Chem. Soc.*, **89**, 561 (1967).
151. B. Rickborn and S. E. Wood, *J. Am. Chem. Soc.*, **93**, 3940 (1971).
152. F. M. Rossi, P. A. McCusker, and G. F. Hennion, *J. Org. Chem.*, **32**, 450 (1967).
153. R. Köster, *Justus Liebigs Ann. Chem.*, **618**, 31 (1958).
154. E. C. Ashby, *J. Am. Chem. Soc.*, **81**, 4791 (1959).
155. H. C. Brown, M. V. Bhatt, T. Munekata, and G. Zweifel, *J. Am. Chem. Soc.*, **89**, 567 (1967).
156. A. T. Cocks and K. W. Egger, *J. Chem. Soc. A*, 3606 (1971).
157. M. M. Midland, J. E. Petre, and S. A. Zedric *J. Organomet. Chem.*, **182**, C53 (1979).
158. H. C. Brown and G. Zweifel, *J. Am. Chem. Soc.*, **88**, 1433 (1966).
159. R. Köster and G. Rotermund, *Angew. Chem.*, **72**, 138 (1960).
160. R. Köster and K. Reinert, *Angew. Chem*, **71**, 521 (1959).
161. H. C. Brown, U. S. Racherla, and H. Taniguchi, *J. Org. Chem.*, **46**, 4313 (1981).
162. H. C. Brown and U. S. Racherla, *Organometallics*, **1**, 765 (1982).
163. H. C. Brown and U. S. Racherla, *J. Organomet. Chem.*, **241**, C37 (1983).
164. M. M. Midland, A. Tramontano, and S. A. Zderic, *J. Organomet. Chem.*, **156**, 203 (1978).
165. M. M. Midland and S. A. Zderic, *J. Am. Chem. Soc.*, **104**, 525 (1982).
166. H. C. Brown and A. W. Moerikofer, *J. Am. Chem. Soc.*, **83**, 3417 (1961).
167. D. S. Matteson, *Organometallic Reaction Mechanisms*, Academic Press, New York, 1974, pp. 227–232.
168. H. C. Brown, C. G. Scouten, and K. K. Wang, *J. Org. Chem.*, **44**, 2589 (1979).
169. H. C. Brown, K. K. Wang, and C. G. Scouten, *Proc. Natl. Acad. Sci. USA*, **77**, 698 (1980).
170. K. K. Wang and H. C. Brown, *J. Org. Chem.*, **45**, 5303 (1980).
171. D. J. Nelson and H. C. Brown, *J. Am. Chem. Soc.*, **104**, 4907 (1982).
172. D. J. Nelson, C. D. Blue, and H. C. Brown, *J. Am. Chem. Soc.*, **104**, 4913 (1982).
173. K. K. Wang and H. C. Brown, *J. Am. Chem. Soc.*, **104**, 7148 (1982).
174. H. C. Brown, K. K. Wang, and J. Chandrasekharan, *J. Am. Chem. Soc.*, **105**, 2340 (1983).
175. H. C. Brown, J. Chandrasekharan, and K. K. Wang, *J. Org. Chem.*, **48**, 2901 (1983).
176. H. C. Brown, J. Chandrasekharan, and K. K. Wang, *J. Org. Chem.*, **48**, 3689 (1983).
177. H. C. Brown, D. J. Nelson, and C. G. Scouten, *J. Org. Chem.*, **48**, 641 (1983).
178. H. C. Brown, J. Chandrasekharan, and K. K. Wang, *Pure Appl. Chem.*, **55**, 1387 (1983).
179. H. C. Brown, J. Chandrasekharan, and D. J. Nelson, *J. Chem. Soc.*, **106**, 3768 (1984).
180. H. C. Brown and J. Chandrasekharan, *J. Am. Chem. Soc.*, **106**, 1863 (1983).
181. J. Klein, E. Dunkelblum, and M. A. Wolff, *J. Organomet. Chem.*, **7**, 377 (1967).

182. D. J. Pasto, B. Lepeska, and T.-C. Cheng, *J. Am. Chem. Soc.*, **94**, 6083 (1972).
183. T. P. Fehlner, *J. Am. Chem. Soc.*, **93**, 6366 (1971).
184. D. J. Pasto and S. Z. Kang, *J. Am. Chem. Soc.*, **90**, 3797 (1968).
185. G. W. Kabalka, R. J. Newton, Jr., and J. Jacobus, *J. Org. Chem.*, **43**, 1567 (1978).
186. H. C. Brown and J. Chandrasekharan, *Organometallics*, **2**, 1261 (1983).
187. A. B. Burg and H. I. Schlesinger, *J. Am. Chem. Soc.*, **55**, 4020 (1933).
188. S. H. Rose and S. G. Shore, *Inorg. Chem.*, **1**, 744 (1962).
189. W. G. Woods and P. L. Strong, *J. Am. Chem. Soc.*, **88**, 4667 (1966).
190. R. H. Fish, *J. Org. Chem.*, **38**, 158 (1973).
191. R. H. Fish, *J. Am. Chem. Soc.*, **90**, 4435 (1968).
192. H. C. Brown and S. K. Gupta, *J. Am. Chem. Soc.*, **93**, 1816 (1971).
193. H. C. Brown and S. K. Gupta, *J. Am. Chem. Soc.*, **94**, 4370 (1972).
194. H. C. Brown and S. K. Gupta, *J. Am. Chem. Soc.*, **97**, 5249 (1975).
195. H. C. Brown and J. Chandrasekharan, *J. Org. Chem.*, **48**, 5080 (1983).
196. D. S. Matteson, P. K. Jesthi, and K. M. Sadhu, *Organometallics*, **3**, 1284 (1984).
197. D. S. Matteson and D. Majumdar, *Organometallics*, **2**, 1529 (1983).
198. S. Thaisrivongs and J. D. Wuest, *J. Org. Chem.*, **42**, 3243 (1977).
199. B. Z. Egan, S. G. Shore, and J. E. Bonnell, *Inorg. Chem.*, **3**, 1024 (1964).
200. A. B. Burg, *J. Am. Chem. Soc.*, **56**, 499 (1934).
201. G. Zweifel, *J. Organomet. Chem.*, **9**, 215 (1967).
202. D. J. Pasto and P. Balasubramaniyan, *J. Am. Chem. Soc.*, **89**, 295 (1967).
203. D. J. Pasto, V. Balasubramanian, and P. W. Wojtkowski, *Inorg. Chem.*, **8**, 594 (1969).
204. D. J. Pasto, C. C. Cumbo, and P. Balasubramaniyan, *J. Am. Chem. Soc.*, **88**, 2187 (1966)
205. H. C. Brown and N. Ravindran, *J. Am. Chem. Soc.*, **94**, 2112 (1972).
206. H. C. Brown and N. Ravindran, *J. Am. Chem. Soc.*, **98**, 1785 (1976).
207. H. C. Brown and M. Zaidlewicz, *J. Am. Chem. Soc.*, **98**, 4917 (1976).
208. H. C. Brown and N. Ravindran, *J. Am. Chem. Soc.*, **98**, 1798 (1976).
209. H. C. Brown and N. Ravindran, *Inorg. Chem.*, **16**, 2938 (1977).
210. H. C. Brown and N. Ravindran, *J. Org. Chem.*, **42**, 2533 (1977).
211. H. C. Brown and N. Ravindran, *Synthesis*, 695 (1977).
212. H. C. Brown, N. Ravindran, and S. U. Kulkarni, *J. Org. Chem.*, **44**, 2417 (1979).
213. H. C. Brown and S. U. Kulkarni, *J. Org. Chem.*, **44**, 2422 (1979).
214. H. C. Brown and N. Ravindran, *J. Am. Chem. Soc.*, **99**, 7097 (1977).
215. H. C. Brown, N. Ravindran, and S. U. Kulkarni, *J. Org. Chem.*, **45**, 384 (1980).
216. H. C. Brown, N. G. Bhat, and V. Somayaji, *Organometallics*, **2**, 1311 (1983).
217. H. C. Brown and J. B. Campbell, Jr., *J. Org. Chem.*, **45**, 389 (1980).
218. H. C. Brown and J. Chandrasekharan, *J. Org. Chem.*, **48**, 644 (1983).
219. H. C. Brown and U. S. Racherla, *J. Org. Chem.*, **48**, 1389 (1983).
220. H. C. Brown and T. Imai, *Organometallics*, **3**, 1392 (1984).
221. G. Zweifel and N. R. Pearson, *J. Am. Chem. Soc.*, **102**, 5919 (1980).
222. H. C. Brown, J. A. Sikorski, S. U. Kulkarni, and H. D. Lee, *J. Org. Chem.*, **45**, 4540 (1980).
223. H. C. Brown, J. A. Sikorski, S. U. Kulkarni, and H. D. Lee, *J. Org. Chem.*, **47**, 863 (1982).
224. H. C. Brown and J. A. Sikorski, *Organometallics*, **1**, 28 (1982).
225. J. A. Sikorski and H. C. Brown, *J. Org. Chem.*, **47**, 872 (1982).
226. H. C. Brown, D. Basavaiah, and N. G. Bhat, *Organometallics*, **2**, 1468 (1983).
227. H. C. Brown, N. G. Bhat, and D. Basavaiah, *Isr. J. Chem.*, **24**, 72 (1984).
228. P. A. McCusker, G. F. Hennion, and E. C. Ashby, *J. Am. Chem. Soc.*, **79**, 5192 (1957).
229. R. Köster and M. A. Grassberger, *Justus Liebigs Ann. Chem.*, **719**, 169 (1968).
230. R. Köster and P. Binger, *Inorg. Synth.*, **15**, 149 (1974).
231. H. C. Brown, D. Basavaiah, and N. G. Bhat, *Organometallics*, **2**, 1309 (1983).
232. H. C. Brown and S. K. Gupta, *J. Am. Chem. Soc.*, **93**, 2802 (1971).
233. (a) L. J. Malone and M. R. Manley, *Inorg. Chem.*, **6**, 2260 (1967); (b) L. J. Malone, *Inorg. Chem.*, **7**, 1039 (1968).
234. H. R. Arnold, *US Pats.*, 2 402 589 and 2 402 590; W. A. Lazier and P. L. Salzberg, *US Pat.*, 2 402 591; *Chem. Abstr.*, **40**, 5769 (1946).
235. S. Hara, H. Dojo, S. Takinami, and A. Suzuki, *Tetrahedron Lett.*, 731 (1983).
236. G. W. Willcockson, *US Pat.*, 3060218, (1962).

237. F. Joy, M. F. Lappert, and B. Prokai, *J. Organomet. Chem.*, **5**, 506 (1966).
238. R. W. Rudolph, *J. Am. Chem. Soc.*, **89**, 4216 (1967).
239. M. Zeldin, A. R. Gatti, and T. Wartik, *J. Am. Chem. Soc.*, **89**, 4217 (1967).
240. T. D. Coyle and J. J. Ritter, *J. Am. Chem. Soc.*, **89**, 5739 (1967).
241. (a) E. L. Muetterties, *J. Am. Chem. Soc.*, **81**, 2597 (1959); (b) E. L. Muetterties, *J. Am. Chem. Soc.*, **82**, 4163 (1960); (c) E. L. Muetterties and F. N. Tebbe, *Inorg. Chem.*, **7**, 2663 (1968).
242. A. G. Davies and B. P. Roberts, *J. Chem. Soc. B*, 311 (1969).
243. H. C. Brown, M. M. Midland, and G. W. Kabalka, *J. Am. Chem. Soc.*, **93**, 1024 (1971).
244. M. M. Midland and H. C. Brown, *J. Am. Chem. Soc.*, **93**, 4078 (1971).
245. M. M. Midland and H. C. Brown, *J. Am. Chem. Soc.*, **95**, 4069 (1973).
246. G. W. Kabalka, T. J. Reed, and S. A. Kunda, *Synth. Commun.*, **13**, 737 (1983).
247. G. W. Kabalka, Ṙ. M. Lambrecht, M. Sajjad, J. S. Fowler, A. P. Wolf, S. A. Kunda, G. W. McCollum, and R. MacGregor, *Int. J. Appl. Radiat. Isot.*, in press; G. W. Kabalka, personal communication(1985).
248. H. C. Brown and G. Zweifel, *J. Am. Chem. Soc.*, **83**, 2544 (1961).
249. (a) H. G. Kuivila, *J. Am. Chem. Soc.*, **76**, 870 (1954; (b) H. G. Kuivila, *J. Am. Chem. Soc.*, **77**, 4014 (1955); (c) H. G. Kuivila and R. A. Wiles, *J. Am. Chem. Soc.*, **77**, 4830 (1955); (d) H. G. Kuivila and A. G. Armour, *J. Am. Chem. Soc.*, **79**, 5659 (1957).
250. H. Minato, J. C. Ware, and T. G. Traylor, *J. Am. Chem. Soc.*, **85**, 3024 (1963).
251. G. W. Kabalka and N. S. Bowman, *J. Org. Chem.*, **38**, 1607 (1973).
252. H. C. Brown and G. Zweifel, *J. Am. Chem. Soc.*, **81**, 1512 (1959).
253. D. S. Matteson and R. J. Moody, *J. Org. Chem.*, **45**, 1091 (1980).
254. D. J. Pasto, S. K. Arora, and J. Chow, *Tetrahedron*, **25**, 1571 (1969).
255. S. J. Cristol, F. P. Darungo, and D. E. Plorde, *J. Am. Chem. Soc.*, **87**, 2870 (1965).
256. D. J. Pasto and J. Hickman, *J. Am. Chem. Soc.*, **90**, 4445 (1968).
257. D. A. Evans, E. Vogel, and J. V. Nelson, *J. Am. Chem. Soc.*, **101**, 6120 (1979).
258. M. M. Midland and S. B. Preston, *J. Org. Chem.*, **45**, 4514 (1980).
259. R. Köster and Y. Morita, *Angew. Chem.*, **78**, 589 (1966); *Justus Liebigs Ann. Chem.*, **704**, 70 (1967).
260. A. G. Davies and B. P. Roberts, *J. Chem. Soc. C*, 1474 (1968).
261. R. P. Fisher, H. P. On, J. T. Snow, and G. Zweifel, *Synthesis*, 127 (1982).
262. G. W. Kabalka and H. C. Hedgecock, Jr., *J. Org. Chem.*, **40**, 1776 (1975).
263. G. W. Kabalka and S. W. Slayden, *J. Organomet. Chem.*, **125**, 273 (1977).
264. G. Zweifel and S. J. Backlund, *J. Am. Chem. Soc.*, **99**, 3184 (1977).
265. J. A. Miller and G. Zweifel, *Synthesis*, 288 (1981).
266. A. Hassner and J. Soderquist, *J. Organomet. Chem.*, **131**, CI (1977).
267. J. A. Miller and G. Zweifel, *J. Am. Chem. Soc.*, **103**, 6217 (1981).
268. H. C. Brown and C. P. Garg, *J. Am. Chem. Soc.*, **83**, 2951 (1961); 2952 (1961).
269. J. C. Ware and T. G. Traylor, *J. Am. Chem. Soc.*, **85**, 3026 (1963).
270. C. G. Rao, S. U. Kulkarni, and H. C. Brown, *J. Organomet. Chem.*, **172**, C20 (1979).
271. H. C. Brown, S. U. Kulkarni, and C. G. Rao, *Synthesis*, 151 (1980).
272. H. C. Brown, S. U. Kulkarni, and C. G. Rao, *Synthesis*, 702 (1979).
273. C. F. Lane and H. C. Brown, *J. Am. Chem. Soc.*, **92**, 7212 (1970).
274. H. C. Brown and C. F. Lane, *J. Am. Chem. Soc.*, **92**, 6660 (1970).
275. H. C. Brown, M. W. Rathke, and M. M. Rogit, *J. Am. Chem. Soc.*, **90**, 5038 (1968).
276. H. C. Brown and C. F. Lane, *J. Chem. Soc. D*, 521 (1970).
277. H. C. Brown, N. R. De Lue, G. W. Kabalka, and H. H. Hedgecock, Jr., *J. Am. Chem. Soc.*, **98**, 1290 (1976).
278. D. S. Matteson and J. D. Liedtke, *J. Am. Chem. Soc.*, **87**, 1526 (1965).
279. H. C. Brown, T. Hamaoka, and N. Ravindran, *J. Am. Chem. Soc.*, **95**, 6456 (1973).
280. H. C. Brown, T. Hamaoka, and N. Ravindran, *J. Am. Chem. Soc.*, **95**, 5786 (1973).
281. H. C. Brown, in *Comprehensive Organometallic Chemistry* (Ed. G. Wilkinson), Pergamon Press, Oxford, 1980, p. 111.
282. T. Hamaoka and H. C. Brown, *J. Org. Chem.*, **40**, 1189 (1975).
283. G. W. Kabalka, E. E. Gooch, and K. A. R. Sastry, *J. Nucl. Med.*, **22**, 908 (1981).
284. G. W. Kabalka, K. A. R. Sastry, and P. G. Pagni, *J. Radioanal. Chem.*, **74**, 315 (1982).
285. G. W. Kabalka, K. A. R. Sastry, and V. Somayaji, *Heterocycles*, **18**, 157 (1982).
286. (a) F. F. Knapp, Jr., M. M. Goodman, A. P. Callahan, L. A. Ferren, G. W. Kabalka, and

K. A. R. Sastry, *J. Med. Chem.*, **26**, 1293 (1983); (b) F. F. Knapp, Jr., P. C. Srivastava, A. P. Callahan, E. B. Cunningham, G. W. Kabalka, and K. A. R. Sastry, *J. Med. Chem.*, **27**, 57 (1984).
287. G. W. Kabalka, *Acc. Chem. Res.*, **17**, 215 (1984).
288. (a) A. Mendoza and D. S. Matteson, *J. Chem. Soc., Chem. Commun.*, 357 (1978); (b) A. Mendoza and D. S. Matteson, *J. Organomet. Chem.*, **156**, 149 (1978).
289. M. W. Rathke, N. Inoue, K. R. Varma, and H. C. Brown, *J. Am. Chem. Soc.*, **88**, 2870 (1966).
290. H. C. Brown, W. R. Heydkamp, E. Breuer, and W. S. Murphy, *J. Am. Chem. Soc.*, **86**, 3565 (1964).
291. G. W. Kabalka, K. A. R. Sastry, G. W. McCollum, and H. Yoshioka, *J. Org. Chem.*, **46**, 4296 (1981).
292. G. W. Kabalka, K. A. R. Sastry, G. W. McCollum, and C. A. Lane, *J. Chem. Soc., Chem. Commun.*, 62 (1982).
293. G. W. Kabalka, G. W. McCollum, and S. R. Kunda, *J. Org. Chem.*, **49**, 1656 (1984).
294. V. B. Jigajinni, A. Pelter, and K. Smith, *Tetrahedron Lett.*, 181 (1978).
295. A. G. Davies, S. C. W. Hook, and B. P. Roberts, *J. Organomet. Chem.*, **23**, C11 (1970).
296. H. C. Brown and N. R. De Lue, *J. Organomet. Chem.*, **135**, C57 (1977).
297. A. Suzuki, S. Sono, M. Itoh, H. C. Brown, and M. M. Midland, *J. Am. Chem. Soc.*, **93**, 4329 (1971).
298. H. C. Brown, M. M. Midland, and A. B. Levy, *J. Am. Chem. Soc.*, **95**, 2394 (1973).
299. A. B. Levy and H. C. Brown, *J. Am. Chem. Soc.*, **95**, 4067 (1973).
300. H. C. Brown and N. C. Hébert, *J. Organomet. Chem.*, **255**, 135 (1983).
301. H. C. Brown and K. Murray, *J. Am. Chem. Soc.*, **81**, 4108 (1959).
302. L. H. Toporcer, R. E. Dessy, and S. I. E. Green, *J. Am. Chem. Soc.*, **87**, 1236 (1965); *Inorg. Chem.*, **4**, 1649 (1965).
303. H. C. Brown and K. Murray, *J. Org. Chem.*, **26**, 631 (1961).
304. S. Trofimenko, *J. Am. Chem. Soc.*, **91**, 2139 (1969).
305. H. C. Brown and G. Zweifel, *J. Am. Chem. Soc.*, **81**, 1512 (1959).
306. H. G. Kuivila and K. V. Nahabedian, *J. Am. Chem. Soc.*, **83**, 2159, 2164, 2167 (1961).
307. A. G. Davies and B. P. Roberts, *J. Chem. Soc. C*, 1474 (1968).
308. J. B. Honeycutt, Jr., and J. M. Riddle, *J. Am. Chem. Soc.*, **82**, 2593 (1959); 3051 (1960).
309. R. C. Larock and H. C. Brown, *J. Am. Chem. Soc.*, **92**, 2467 (1970).
310. J. D. Buhler and H. C. Brown, *J. Organomet. Chem.*, **40**, 265 (1972).
311. R. C. Larock, *J. Org. Chem.*, **39**, 834 (1974).
312. M. Gielen and R. Fosty, *Bull. Soc. Chim. Belg.*, **83**, 333 (1974).
313. D. E. Bergbreiter and D. P. Rainville, *J. Org. Chem.*, **41**, 3031 (1976); *J. Organomet. Chem.*, **121**, 19 (1976).
314. R. C. Larock, *J. Organomet. Chem.*, **67**, 353 (1974).
315. R. C. Larock, *J. Organomet. Chem.*, **72**, 35 (1974).
316. D. S. Matteson and R. A. Bowie, *J. Am. Chem. Soc.*, **87**, 2587 (1965).
317. D. S. Matteson, R. A. Bowie, and G. Srivastava, *J. Organomet. Chem.*, **16**, 33 (1969).
318. D. S. Matteson and P. G. Allies, *J. Organomet. Chem.*, **54**, 35 (1973).
319. D. S. Matteson, R. B. Castle, and G. L. Larson, *J. Am. Chem. Soc.*, **92**, 231 (1970).
320. (a) K. G. Strothkamp, J. Lehmann, and S. J. Lippard, *Proc. Natl. Acad. Sci. U. S. A.*, **75**, 1181 (1978); (b) D. Grdenic, B. Kamenar, B. Korpar-Colig, M. Sikirika, and G. Jovanovski, *J. Chem. Soc., Chem. Commun.*, 646 (1974).
321. R. C. Larock, S. K. Gupta, and H. C. Brown, *J. Am. Chem. Soc.*, **94**, 4371 (1972).
322. R. C. Larock and H. C. Brown, *J. Organomet. Chem.*, **36**, 1 (1972).
323. S. A. Kunda, R. S. Varma, and G. W. Kabalka, *Synth. Commun.*, **14**, 755 (1984).
324. R. S. Varma, S. A. Kunda, and G. W. Kabalka, *J. Organomet. Chem.*, **276**, 311 (1984).
325. D. S. Matteson and P. B. Tripathy, *J. Organomet. Chem.*, **69**, 53 (1974).
326. A. Mendoza and D. S. Matteson, *J. Organomet. Chem.*, **152**, 1 (1978).
327. H. C. Brown and M. M. Midland, *J. Am. Chem. Soc.*, **93**, 3291 (1973).
328. A. Arase and Y. Masuda, *Chem. Lett.*, 1115 (1976).
329. K. Kondo and S. Murahashi, *Tetrahedron Lett.*, 1237 (1979).
330. J. Goubeau and K. H. Rohwedder, *Justus Liebigs Ann. Chem.*, **604**, 168 (1957).
331. D. S. Matteson and R. W. H. Mah, *J. Am. Chem. Soc.*, **85**, 2599 (1963).
332. D. S. Matteson and R. W. H. Mah, *J. Org. Chem.*, **28**, 2174 (1963).
333. D. S. Matteson and J. D. Liedtke, *Chem. Ind.* (*London*), 1241 (1963).

334. D. S. Matteson and J. D. Liedtke, *J. Org. Chem.*, **28**, 1924 (1963).
335. D. S. Matteson and G. D. Schaumberg, *J. Org. Chem.*, **31**, 726 (1966).
336. D. J. Pasto, J. Chow, and S. K. Arora, *Tetrahedron*, **25**, 1557 (1969).
337. D. J. Pasto and K. McReynolds, *Tetrahedron Lett.*, 801 (1971).
338. H. C. Brown, N. R. De Lue, Yamamoto, and K. Maruyama, *J. Org. Chem.*, **42**, 3252 (1977).
339. H. C. Brown, N. R. De Lue, Y. Yamamoto, K. Maruyama, T. Kasahara, S. Murahashi, and A. Sonoda, *J. Org. Chem.*, **42**, 4088 (1977).
340. D. S. Matteson and D. Majumdar, *J. Am. Chem. Soc.*, **102**, 7588 (1980).
341. C. F. Lane and H. C. Brown, *J. Am. Chem. Soc.*, **93**, 1025 (1971).
342. H. C. Brown and C. F. Lane, *Synthesis*, 303 (1972).
343. H. C. Brown and Y. Yamamoto, *J. Am. Chem. Soc.*, **93**, 2796 (1971).
344. H. C. Brown and Y. Yamamoto, *J. Chem. Soc., Chem. Commun.*, 71 (1972).
345. H. C. Brown and Y. Yamamoto, *Synthesis*, 699 (1972).
346. C. F. Lane and H. C. Brown, *J. Organomet. Chem.*, **26**, C51 (1971).
347. Y. Yamamoto and H. C. Brown, *J. Org. Chem.*, **39**, 861 (1974).
348. Y. Yamamoto and H. C. Brown, *J. Chem. Soc., Chem. Commun.*, 801 (1973).
349. M. M. Midland, A. R. Zolopa, and R. L. Halterman, *J. Am. Chem. Soc.*, **101**, 248 (1979).
350. E. Negishi, R. M. Williams, G. Lew, and T. Yoshida, *J. Organomet. Chem.*, **92**, C4 (1975).
351. H. C. Brown, M. M. Rogić, M. W. Rathke, and G. W. Kabalka, *J. Am. Chem. Soc.*, **90**, 818 (1968).
352. H. C. Brown, M. M. Rogić, M. W. Rathke, and G. W. Kabalka, *J. Am. Chem. Soc.*, **91**, 2150 (1969).
353. H. C. Brown, H. Nambu, and M. M. Rogić, *J. Am. Chem. Soc.*, **91**, 6852 (1969).
354. H. C. Brown, H. Nambu, and M. M. Rogić, *J. Am. Chem. Soc.*, **91**, 6854 (1969).
355. H. C., H. Nambu, and M. M. Rogić *J. Am. Chem. Soc.*, **91**, 6855 (1969).
356. H. C. Brown and H. Nambu, *J. Am. Chem. Soc.*, **92**, 1761 (1970).
357. H. Nambu and H. C. Brown, *J. Am. Chem. Soc.*, **92**, 5790 (1970).
358. H. C. Brown, B. A. Carlson, and R. H. Prager, *J. Am. Chem. Soc.*, **93**, 2070 (1971).
359. H. C. Brown and B. A. Carlson, *J. Org. Chem.*, **38**, 2422 (1973).
360. H. C. Brown, J.-J. Katz, and B. A. Carlson, *J. Org. Chem.*, **38**, 3968 (1973).
361. B. A. Carlson and H. C. Brown, *J. Am. Chem. Soc.*, **95**, 6876 (1973).
362. B. A. Carlson, J. J. Katz, and H. C. Brown, *J. Organomet. Chem.*, **67**, C39 (1974).
363. H. C. Brown, J.-J. Katz, and B. A. Carlson, *J. Org. Chem.*, **40**, 813 (1975).
364. B. A. Carlson and H. C. Brown, *Synthesis*, 776 (1973).
365. J.-J. Katz, B. A. Carlson, and H. C. Brown, **39**, 2817 (1974).
366. V. W. Buls, O. L. Davis, and R. I. Thomas, *J. Am. Chem. Soc.*, **79**, 337 (1957).
367. B. A. Carlson and H. C. Brown, *Org. Syntheses*, **58**, 25 (1978).
368. H. C. Brown and T. Imai, *J. Org. Chem.*, **49**, 892 (1984).
369. (a) O. Rosario, A. Oliva, and G. L. Larson, *J. Organomet. Chem.*, **146**, C8 (1978); (b) G. L. Larson, R. Arguelles, O. Rosario, and S. Sandoval, *J. Organomet. Chem.*, **198**, 15 (1980).
370. C. Burford, F. Cooke, E. Ehlinger, and P. Magnus, *J. Am. Chem. Soc.*, **99**, 4536 (1977).
371. D. S. Matteson and D. Majumdar, *J. Organomet. Chem.*, **184**, C41 (1980).
372. D. S. Matteson and D. Majumdar, *Organometallics*, **2**, 230 (1983).
373. D. J. S. Tsai and D. S. Matteson, *Organometallics*, **2**, 236 (1983).
374. J. Hooz and S. Linke, *J. Am. Chem. Soc.*, **90**, 5936 (1968).
375. J. Hooz and S. Linke, *J. Am. Chem. Soc.*, **90**, 6892 (1968).
376. J. Hooz and G. F. Morrison, *Can. J. Chem.*, **48**, 868 (1970).
377. J. Hooz and D. M. Gunn, *Tetrahedron Lett.*, 3455 (1969).
378. J. Hooz and J. N. Bridson, *J. Am. Chem. Soc.*, **95**, 602 (1973).
379. S. Masamune, S. Mori, D. Van Horn, and D. W. Brooks, *Tetrahedron Lett.*, 1665 (1979).
380. M. E. D. Hillman, *J. Am. Chem. Soc.*, **84**, 4715 (1962); *J. Am. Chem. Soc.*, **85**, 982 (1963); *J. Am. Chem. Soc.*, **85**, 1626 (1963).
381. (a) J. Casanova and R. E. Schuster, *Tetrahedron Lett.*, 405 (1964); (b) J. Casanova, H. Kiefer, D. Kuwada, and A. Boulton, *Tetrahedron Lett.*, 703 (1965).
382. H. C. Brown and M. W. Rathke, *J. Am. Chem. Soc.*, **89**, 2737 (1967).
383. H. C. Brown, *Science*, **210**, 485 (1980).
384. M. W. Rathke and H. C. Brown, *J. Am. Chem. Soc.*, **89**, 2740 (1967).
385. H. C. Brown, R. A. Coleman, and M. W. Rathke, *J. Am. Chem. Soc.*, **90**, 499 (1968).
386. H. C. Brown, E. F. Knights, and R. A. Coleman, *J. Am. Chem. Soc.*, **91**, 2144 (1969).

387. H. C. Brown and R. A. Coleman, *J. Am. Chem. Soc.*, **91**, 4606 (1969).
388. H. C. Brown, J. L. Hubbard, and K. Smith, *Synthesis*, 701 (1979).
389. H. C. Brown and J. L. Hubbard, *J. Org. Chem.*, **44**, 467 (1979).
390. C. A. Brown, *J. Am. Chem. Soc.*, **95**, 4100 (1973); *J. Org. Chem.*, **39**, 3913 (1974).
391. J. L. Hubbard and K. Smith, *J. Organomet. Chem.*, **276**, C41 (1984).
392. H. C. Brown, T. M. Ford, and J. L. Hubbard, *J. Org. Chem.*, **45**, 4067 (1980).
393. H. C. Brown and T. M. Ford, *J. Org. Chem.*, **46**, 647 (1981).
394. J. L. Hubbard and H. C. Brown, *Synthesis*, 676 (1978).
395. H. C. Brown and E. Negishi, *J. Chem. Soc., Chem. Commun.*, 594 (1968).
396. E. Negishi, M. Sabanski, J. J. Katz, and H. C. Brown, *Tetrahedron*, **32**, 925 (1976).
397. S. U. Kulkarni, H. D. Lee, and H. C. Brown, *J. Org. Chem.*, **45**, 4542 (1980).
398. H. C. Brown, D. Basavaiah, and U. S. Racherla, *Synthesis*, 886 (1983).
399. G. W. Kabalka, E. Gooch, C. J. Collins, and V. F. Raaen, *J. Chem. Soc., Chem. Commun.*, 607 (1979).
400. G. W. Kabalka, *Synth. Commun.*, **10**, 93 (1980).
401. G. W. Kabalka, M. C. Delgado, U. Sastry, and K. A. R. Sastry, *J. Chem. Soc., Chem. Commun.*, 1273 (1982).
402. G. W. Kabalka, M. C. Delgado, U. S. Kunda, and S. A. Kunda, *J. Org. Chem.*, **49**, 174 (1984).
403. G. W. Kabalka, M. Mohammadi, S. A. Kunda, and R. D. Finn, *Org. Prep. Proc. Int.*, **17**, 17 (1985).
404. (a) A. Pelter, M. G. Hutchings, and K. Smith, *J. Chem. Soc. D*, 1048 (1971); (b) A. Pelter, M. G. Hutchings, and K. Smith, *J. Chem. Soc., Perkin Trans. 1*, 142 (1975).
405. A. Pelter, A. Arase, and M. G. Hutchings, *J. Chem. Soc., Chem. Commun.*, 346 (1974).
406. A. Pelter, P. J. Maddocks, and K. Smith, *J. Chem. Soc., Chem. Commun.*, 805 (1978).
407. T. A. Bryson and C. J. Reichel, *Tetrahedron Lett.*, **21**, 2381 (1980).
408. J. J. Tufariello, P. Wojtkowski, and L. T. C. Lee, *J. Chem. Soc., Chem. Commun.*, 505 (1967).
409. W. K. Musker and R. R. Stevens, *Tetrahedron Lett.*, 995 (1967).
410. J. J. Tufariello, L. T. C. Lee, and P. Wojktowski, *J. Am. Chem. Soc.*, **89**, 6804 (1967).
411. E. Negishi, T. Yoshida, A. Silveira, Jr., and B. L. Chiou, *J. Org. Chem.*, **40**, 814 (1975).
412. S. Yamamoto, M. Shiono, and T. Mukaiyama, *Chem. Lett.*, 961 (1973).
413. R. J. Hughes, A. Pelter, and K. Smith, *J. Chem. Soc., Chem. Commun.*, 863 (1974).
414. A. Mendoza and D. S. Matteson, *J. Org. Chem.*, **44**, 1352 (1979).
415. H. C. Brown and T. Imai, *J. Am. Chem. Soc.*, **105**, 6285 (1983).
416. G. Zweifel, H. Arzoumanian, and C. C. Whitney, *J. Am. Chem. Soc.*, **89**, 3652 (1967).
417. G. Zweifel and H. Arzoumanian, *J. Am. Chem. Soc.*, **89**, 5086 (1967).
418. G. Zweifel, R. P. Fisher, J. T. Snow, and C. C. Whitney, *J. Am. Chem. Soc.*, **94**, 6560 (1972).
419. G. Zweifel, N. L. Polston, and C. C. Whitney, *J. Am. Chem. Soc.*, **90**, 6243 (1968).
420. E. Negishi and T. Yoshida, *J. Chem. Soc., Chem. Commun.*, 606 (1973).
421. T. Yoshida, R. M. Williams, and E. Negishi, *J. Am. Chem. Soc.*, **96**, 3688 (1974).
422. E. J. Corey and T. Ravindranathan, *J. Am. Chem. Soc.*, **94**, 4013 (1972).
423. D. A. Evans, T. C. Crawford, R. C. Thomas, and J. A. Walker, *J. Org. Chem.*, **41**, 3947 (1976).
424. G. Zweifel and R. P. Fisher, *Synthesis*, 557 (1972).
425. S. U. Kulkarni, D. Basavaiah, and H. C. Brown, *J. Organomet. Chem.*, **225**, C1 (1982).
426. D. Basavaiah and H. C. Brown, *J. Org. Chem.*, **47**, 1792 (1982).
427. H. C. Brown and D. Basavaiah, *Synthesis*, 283 (1983).
428. H. C. Brown and D. Basavaiah, *J. Org. Chem.*, **47**, 3806 (1982).
429. G. Zweifel and R. P. Fisher, *Synthesis*, 339 (1974).
430. K. W. Chiu, E. Negishi, M. S. Plante, and A. Silveira, Jr., *J. Organomet. Chem.*, **112**, C3 (1976).
431. A. B. Levy and S. J. Schwartz, *Tetrahedron Lett.*, 2201 (1976).
432. A. B. Levy, S. J. Schwartz, N. Wilson, and B. Christie, *J. Organomet. Chem.*, **156**, 123 (1978).
433. K. Utimoto, K. Uchida, and H. Nozaki, *Chem. Lett.*, 1493 (1974).
434. K. Utimoto, K. Uchida, and H. Nozaki, *Tetrahedron*, **33**, 1949 (1977).
435. K. Utimoto, K. Uchida, M. Yamaya, and H. Nozaki, *Tetrahedron*, **33**, 1945 (1977).
436. P. Jacob, III, and H. C. Brown, *J. Org. Chem.*, **42**, 579 (1977).
437. A. Suzuki, N. Miyaura, and M. Itoh, *Tetrahedron*, **27**, 2775 (1971).
438. K. Utimoto, N. Sakai, and H. Nozaki, *J. Am. Chem. Soc.*, **96**, 5601 (1974).
439. A. B. Levy, *J. Org. Chem.*, **43**, 4684 (1978).
440. A. B. Levy, *Tetrahedron Lett.*, 4021 (1979).

441. G. Zweifel and A. Horng, *Synthesis*, 672 (1973).
442. G. M. Clark, K. G. Hancock, and G. Zweifel, *J. Am. Chem. Soc.*, **93**, 1308 (1971).
443. G. Zweifel, S. J. Backlund, and T. Leung, *J. Am. Chem. Soc.*, **99**, 5192 (1977).
444. P. Binger, *Angew. Chem., Int. Ed. Engl.*, **6**, 84 (1967).
445. P. Binger, G. Benedikt, G. W. Rotermund, and R. Köster, *Justus Liebigs Ann. Chem.*, **717**, 21 (1968).
446. N. Miyaura, T. Yoshinari, M. Itoh, and A. Suzuki, *Tetrahedron Lett.*, 2961 (1975).
447. M. M. Midland and H. C. Brown, *J. Org. Chem.*, **40**, 2845 (1975).
448. H. C. Brown, A. B. Levy, and M. M. Midland, *J. Am. Chem. Soc.*, **97**, 5017 (1975).
449. A. Pelter, C. Subrahmanyam, R. J. Laub, K. J. Gould, and C. R. Harrison, *Tetrahedron Lett.*, 1633 (1975).
450. M. Naruse, K. Utimoto, and H. Nozaki, *Tetrahedron*, **30**, 3037 (1975).
451. P. Binger and R. Köster, *Chem. Ber.*, **108**, 395 (1975).
452. A. Pelter and K. J. Gould, *J. Chem. Soc., Chem. Commun.*, 347 (1974).
453. A. Pelter, K. J. Gould, and L. A. P. Kane-Maguire, *J. Chem. Soc., Chem. Commun.*, 1029 (1974).
454. A. Pelter and C. R. Harrison, *J. Chem. Soc., Chem. Commun.*, 828 (1974).
455. P. Binger and R. Köster, *Synthesis*, 350 (1974).
456. P. Binger and R. Köster, *Synthesis*, 309 (1973).
457. J. Hooz and R. Mortimer, *Tetrahedron Lett.*, 805 (1976).
458. J. Hooz and R. Mortimer, *Can. J. Chem.*, **56**, 2786 (1978).
459. G. Zweifel and R. P. Fisher, *Synthesis*, 376 (1975).
460. G. Zweifel and S. J. Backlund, *J. Organomet. Chem.*, **156**, 159 (1978).
461. H. C. Brown, N. G. Bhat, and D. Basavaiah, *Synthesis*, 885 (1983).
462. H. C. Brown, U. S. Racherla, and S. M. Singh, *Tetrahedron Lett.*, **25**, 2411 (1984).
463. G. A. Molander and H. C. Brown, *J. Org. Chem.*, **42**, 3106 (1977).
464. A. Suzuki, N. Miyaura, S. Abiko, M. Itoh, H. C. Brown, J. A. Sinclair, and M. M. Midland, *J. Am. Chem. Soc.*, **95**, 3080 (1973).
465. M. M. Midland, J. A. Sinclair, and H. C. Brown, *J. Org. Chem.*, **39**, 731 (1974).
466. E. Negishi, G. Lew, and T. Toshida, *J. Chem. Soc., Chem. Commun.*, 874 (1973).
467. S. Hara, Y. Satoh, H. Ishiguro, and A. Suzuki, *Tetrahedron Lett.*, **24**, 735 (1983).
468. A. Pelter, K. Smith, and M. Tabata, *J. Chem. Soc., Chem. Commun.*, 857 (1975).
469. B. M. Mikhailov and V. N. Smirnov, *Izv. Akad. Nauk SSSR, Ser. Khim.*, 2165 (1973).
470. B. M. Mikhailov and K. L. Cherkasova, *Izv. Akad. Nauk SSSR, Ser. Khim.*, 2056 (1976).
471. B. M. Mikhailov and T. K. Baryshnikova, *Dokl. Akad. Nauk SSSR*, **243**, 929 (1978).
472. B. M. Mikhailov, V. N. Smirnov, and V. A. Kasparov, *Izv. Akad. Nauk SSSR, Ser. Khim.*, 2302 (1976).
473. B. M. Mikhailov, T. A. Shchegoleva, E. M. Shashkova, and V. G. Kiselev, *Izv. Akad. Nauk SSSR, Ser. Khim.*, 894 (1977).
474. B. M. Mikhailov, *Pure Appl. Chem.*, **55**, 1439 (1983).
475. B. M. Mikhailov and V. F. Pozdnev, *Izv. Akad. Nauk SSSR, Ser. Khim.*, 1477 (1967).
476. B. M. Mikhailov, Yu. N. Bubnov, A. V. Tsyban, and M. Sh. Grigoryan, *J. Organomet. Chem.*, **154**, 131 (1978).
477. B. M. Mikhailov, Yu. N. Bubnov, and A. V. Tsyban, *J. Organomet. Chem.*, **154**, 113 (1978).
478. B. M. Mikhailov, Yu. N. Bubnov, A. V. Tsyban, and K. Base, *Izv. Akad. Nauk. SSSR, Ser. Khim.*, 1586 (1978).
479. B. M. Mikhailov, Yu. N. Bubnov, and A. V. Tsyban, *Izv. Akad. Nauk. SSSR, Ser. Khim.*, 1892 (1978).
480. J. Blais, A. L'Honoré, J. Soulié, and P. Cadiot, *J. Organomet. Chem.*, **78**, 323 (1974).
481. G. W. Kramer and H. C. Brown, *J. Org. Chem.*, **42**, 2292 (1977).
482. G. W. Kramer and H. C. Brown, *J. Organomet. Chem.*, **132**, 9 (1977).
483 R. W. Hoffmann and H. J. Zeiss, *Angew. Chem., Int. Engl.*, **18**, 306 (1979).
484. R. W. Hoffmann and H. J. Zeiss, *J. Org. Chem.*, **46**, 1309 (1981).
485. R. W. Hoffmann and B. Kemper, *Tetrahedron Lett.*, **21**, 4883 (1980).
486. R. W. Hoffmann and B. Kemper, *Tetrahedron Lett.*, **23**, 845 (1982).
487. P. G. M. Wuts and S. S. Bigelow, *J. Org. Chem.*, **47**, 2498 (1982).
488. P. G. M. Wuts and S. S. Bigelow, *Synth. Commun.*, **12**, 779 (1982).
489. R. S. Hoffmann and U. Weidmann, *J. Organomet. Chem.*, **195**, 137 (1980).
490. R. W. Hoffmann, G. Feussner, H.-J. Zeiss, and S. Schulz, *J. Organomet. Chem.*, **187**, 321 (1980).

491. P. G. M. Wuts, P. A. Thompson, and G. R. Callen, *J. Org. Chem.*, **48**, 5398 (1983).
492. Y. Yamamoto, H. Yatagai, and K. Maruyama, *J. Chem. Soc., Chem. Commun.*, 1072 (1980).
493. Y. Yamamoto, H. Yatagai, and K. Maruyama, *J. Am. Chem. Soc.*, **103**, 3229 (1981).
494. D. J. S. Tsai and D. S. Matteson, *Tetrahedron Lett.*, **22**, 2751 (1981).
495. D. J. S. Tsai, *PhD Thesis*, Washington State University, (1982).
496. T. Leung and G. Zweifel, *J. Am. Chem. Soc.*, **96**, 5620 (1974).
497. G. Zweifel, S. J. Backlund, and T. Leung, *J. Am. Chem. Soc.*, **100**, 5561 (1978).
498. G. Zweifel and N. R. Pearson, *J. Org. Chem.*, **46**, 829 (1981).
499. N. H. Pearson, G. Hahn, and G. Zweifel, *J. Org. Chem.*, **47**, 3334 (1982).
500. M. M. Midland, *J. Org. Chem.*, **42**, 2650 (1977).
501. M. M. Midland and D. C. McDowell, *J. Organomet. Chem.*, **156**, C5 (1959).
502. E. Negishi, *Acc. Chem. Res.*, **15**, 340 (1982).
503. N. Miyaura and A. Suzuki, *J. Chem. Soc., Chem. Commun.*, 866 (1979).
504. N. Miyaura, K. Yamada, and A. Suzuki, *Tetrahedron Lett.*, 3437 (1979).
505. N. Miyaura, H. Suginome, and A. Suzuki, *Tetrahedron Lett.*, **22**, 127 (1981).
506. N. Miyaura and A. Suzuki, *J. Organomet. Chem.*, **213**, C53 (1981).
507. H. Yatagai, *Bull. Chem. Soc. Jpn.*, **53**, 1670 (1980).
508. N. Miyaura, N. Sasaki, M. Itoh, and A. Suzuki, *Tetrahedron Lett.*, 3369 (1977).
509. N. Miyaura, M. Itoh, and A. Suzuki, *Bull. Chem. Soc. Jpn.*, **50**, 2199 (1977).
510. Y. Yamamoto, H. Yatagai, K. Maruyama, A. Sonoda, and S. Murahashi, *Bull. Chem. Soc. Jpn.*, **50**, 3427 (1977).
511. J. B. Campbell, Jr., and H. C. Brown, *J. Org. Chem.*, **45**, 549 (1980).
512. H. C. Brown and J. B. Campbell, Jr., *J. Org. Chem.*, **45**, 551 (1980).
513. H. C. Brown and G. A. Molander, *J. Org. Chem.*, **46**, 645 (1981).
514. G. L. Larson and L. M. Fuentes, *Synth. Commun.*, **9**, 841 (1979).
515. J. A. Marshall and G. L. Bundy, *J. Am. Chem. Soc.*, **88**, 4291 (1966).
516. J. A. Marshall and G. L. Bundy, *J. Chem. Soc., Chem. Commun.*, 854 (1967).
517. J. A. Marshall and W. F. Huffman, *J. Am. Chem. Soc.*, **92**, 6359 (1970).
518. H. C. Brown and S. P. Rhodes, *J. Am. Chem. Soc.*, **91**, 2149 (1969).
519. H. C. Brown and S. P. Rhodes, *J. Am. Chem. Soc.*, **91**, 4306 (1969).
520. H. L. Goering and S. L. Trenbeath, *J. Am. Chem. Soc.*, **98**, 5016 (1976).
521. H. C. Brown, M. M. Rogić, M. W. Rathke, and G. W. Kabalka, *J. Am. Chem. Soc.*, **89**, 5709 (1967).
522. G. W. Kabalka, H. C. Brown, A. Suzuki, S. Honma, A. Arase, and M. Itoh, *J. Am. Chem. Soc.*, **92**, 710 (1970).
523. A. Suzuki, A. Arase, H. Matsumoto, M. Itoh, H. C. Brown, M. M. Rogić, and M. W. Rathke, *J. Am. Chem. Soc.*, **89**, 5708 (1967).
524. H. C. Brown, G. W. Kabalka, M. W. Rathke, and M. M. Rogić, *J. Am. Chem. Soc.*, **90**, 4165 (1968).
525. H. C. Brown, M. W. Rathke, G. W. Kabalka, and M. M. Rogić, *J. Am. Chem. Soc.*, **90**, 4166 (1968).
526. H. C. Brown and G. W. Kabalka, *J. Am. Chem. Soc.*, **92**, 712 (1970); 714 (1970).
527. M. F. Hawthorne and M. Reintjes, *J. Am. Chem. Soc.*, **87**, 4585 (1965).
528. G. W. Kabalka, *Tetrahedron*, **29**, 1159 (1973).
529. A. Suzuki, N. Miyaura, M. Itoh, H. C. Brown, G. W. Holland, and E. Negishi, *J. Am. Chem. Soc.*, **93**, 2792 (1971).
530. A. Suzuki, N. Miyaura, M. Itoh, H. C. Brown, and P. Jacob, III, *Synthesis*, 305 (1973).
531. H. C. Brown and E. Negishi, *J. Am. Chem. Soc.*, **93**, 3777 (1971).
532. G. Cainelli, G. Dal Bello, and G. Zubiani, *Tetrahedron Lett.*, 4329 (1965).
533. G. Zweifel and H. Arzoumanian, *Tetrahedron Lett.*, 2535 (1966).
534. G. Cainelli, D. Dal Bello, and G. Zubiani, *Tetrahedron Lett.*, 4315 (1966).
535. (a) D. S. Matteson, L. A. Hagelee, and R. J. Wilcsek, *J. Am. Chem. Soc.*, **95**, 5096 (1973); (b) D. S. Matteson and L. A. Hagelee, *J. Organomet. Chem.*, **93**, 21 (1975).
536. D. S. Matteson, R. J. Moody, and P. K. Jesthi, *J. Am. Chem. Soc.*, **97**, 5608 (1975).
537. M. W. Rathke and R. Kow, *J. Am. Chem. Soc.*, **94**, 6854 (1972).
538. R. Kow and M. W. Rathke, *J. Am. Chem. Soc.*, **95**, 2715 (1973).
539. H. Yatagai, Y. Yamamoto, and K. Maruyama, *J. Am. Chem. Soc.*, **102**, 4548 (1980).
540. J. W. Wilson, *J. Organomet. Chem.*, **186**, 297 (1980).

541. A. Pelter, B. Singaram, L. Williams, and J. W. Wilson, *Tetrahedron Lett.*, **24**, 623 (1983).
542. A. Pelter, L. Williams, and J. W. Wilson, *Tetrahedron Lett.*, **24**, 627 (1983).
543. A. Pelter, B. Singaram, and J. W. Wilson, *Tetrahedron Lett.*, **24**, 631 (1983).
544. A. Pelter, B. Singaram, and J. W. Wilson, *Tetrahedron Lett.*, **24**, 635 (1983).
545. M. K. Garad, A. Pelter, B. Singaram, and J. W. Wilson, *Tetrahedron Lett.*, **24**, 637 (1983).
546. D. S. Matteson and R. J. Moody, *J. Am. Chem. Soc.*, **99**, 3196 (1977).
547. D. S. Matteson and K. Arne, *J. Am. Chem. Soc.*, **100**, 1325 (1978).
548. R. Ray and D. S. Matteson, *J. Org. Chem.*, **47**, 2479 (1982).
549. D. S. Matteson and D. Majumdar, *J. Chem. Soc., Chem. Commun.*, 39 (1980).
550. D. S. Matteson and J. O. Waldbilling, *J. Org. Chem.*, **28**, 366 (1963).
551. D. S. Matteson and M. L. Talbot, *J. Am. Chem. Soc.*, **89**, 1123 (1967).
552. D. S. Matteson, *J. Org. Chem.*, **27**, 4293 (1962).
553. H. Meerwein and W. Burneleit, *Chem. Ber.*, **61**, 1840 (1928).
554. H. C. Brown, N. R. Ayyanger, and G. Zweifel, *J. Am. Chem. Soc.*, **86**, 397 (1964).
555. A. I. Meyers, *Acc. Chem. Res.*, **11**, 375 (1978).
556. (a) R. Ray and D. S. Matteson, *Tetrahedron Lett.*, **21**, 449 (1980); (b) R. Ray and D. S. Matteson, *J. Indian Chem. Soc.*, **59**, 119 (1982).
557. (a) D. S. Matteson and R. Ray, *J. Am. Chem. Soc.*, **102**, 7590 (1980); (b) D. S. Matteson, R. Ray, R. R. Rocks, and D. J. Tsai, *Organometallics*, **2**, 1536 (1983).
558. D. S. Matteson and K. M. Sadhu, *J. Am. Chem. Soc.*, **105**, 2077 (1983); correction (identification of tabulated compounds), *J. Am. Chem. Soc.*, **105**, 6195 (1983).
559. A. E. Sherk and B. Fraser-Reid, *J. Org. Chem.*, **47**, 932 (1982).
560. D. S. Matteson, K. M. Sadhu, and M. L. Peterson, *J. Am. Chem. Soc.*, **108**, 810 (1986).
561. D. S. Matteson and E. Erdik, *Organometallics*, **2**, 1083 (1983).
562. (a) D. S. Matteson, K. M. Sadhu, and G. E. Lienhard, *J. Am. Chem. Soc.*, **103**, 5241 (1981). (b) D. S. Matteson and K. M. Sadhu, *Organometallics*, **3**, 614 (1984).
563. D. J. S. Tsai, P. K. Jesthi, and D. S. Matteson, *Organometallics*, **2**, 1543 (1983).
564. W. S. Johnson, C. A. Harbert, and R. D. Stipanovic, *J. Am. Chem. Soc.*, **90**, 5279 (1968).
565. K. M. Sadhu, D. S. Matteson, G. D. Hurst, and J. M. Kurosky, *Organometallics*, **3**, 804 (1984).
566. H. C. Brown, P. K. Jadhav, and A. K. Mandal, *Tetrahedron*, **37**, 3547 (1981).
567. H. C. Brown and P. K. Jadhav, *Asymmetric Synthesis*, Vol. 2, Academic Press, New York, 1983, p. 1.
568. J. J. Partridge, N. K. Chadha, and M. R. Uskoković, *J. Am. Chem. Soc.*, **95**, 7171 (1973).
569. J. J. Partridge, N. K. Chadha, and M. R. Uskoković *J. Am. Chem. Soc.*, **95**, 532 (1973).
570. S. Masamunẹ, L. D.-L. Lu, W. P. Jackson, T. Kaiho, and T. Toyoda, *J. Am. Chem. Soc.*, **104**, 5523 (1982).
571. H. C. Brown and N. M. Yoon, *J. Am. Chem. Soc.*, **99**, 5514 (1977).
572. H. C. Brown and P. K. Jadhav, *J. Org. Chem.*, **46**, 5047 (1981).
573. A. K. Mandal, P. K. Jadhav, and H. C. Brown, *J. Org. Chem.*, **45**, 3543 (1980).
574. H. C. Brown, P. K. Jadhav, and A. K. Mandal, *J. Org. Chem.*, **47**, 5074 (1982).
575. H. C. Brown, P. K. Jadhav, and M. C. Desai, *J. Am. Chem. Soc.*, **104**, 4303 (1982).
576. H. C. Brown, P. K. Jadhav, and M. C. Desai, *J. Am. Chem. Soc.*, **104**, 6844 (1982).
577. H. C. Brown, P. K. Jadhav, and M. C. Desai, *Tetrahedron*, **40**, 1325 (1984).
578. H. C. Brown and B. Singaram, *J. Am. Chem. Soc.*, **106**, 1797 (1984).
579. P. K. Jadhav and H. C. Brown, *J. Org. Chem.*, **46**, 2988 (1981).
580. P. K. Jadhav and M. K. Desai, *Heterocycles*, **18**, 233 (1982).
581. G. Schmid, T. Fukuyama, K. Akasaka, and Y. Kishi, *J. Am. Chem. Soc.*, **101**, 259 (1979).
582. M. M. Midland and Y. C. Kwon, *J. Org. Chem.*, **46**, 229 (1981).
583. M. M. Midland and Y. C. Kwon, *Tetrahedron Lett.*, **23**, 2077 (1982).
584. M. M. Midland and Y. C. Kwon, *J. Am. Chem. Soc.*, **105**, 3725 (1983).
585. R. W. Hoffmann, *Angew. Chem., Int. Ed. Engl.*, **21**, 555 (1982).
586. (a) R. W. Hoffmann, W. Ladner, K. Steinbach, W. Massa, R. Schmidt, and G. Snatzke, *Chem. Ber.*, **114**, 2786 (1981); (b) R. W. Hoffmann and W. Ladner, *Tetrahedron Lett.*, 4653 (1979).
587. R. W. Hoffmann and W. Helbig, *Chem. Ber.*, **114**, 2802 (1981).
588. R. W. Hoffmann, H. J. Zeiss, W. Ladner, and S. Tabche, *Chem. Ber.*, **115**, 2357 (1982).
589. P. G. M. Wuts and S. S. Bigelow, *J. Org. Chem.*, **48**, 3489 (1983).
590. R. W. Hoffmann and B. Landmann, *Angew. Chem., Int. Ed. Engl.*, **23**, 437 (1984).
591. M. M. Midland and S. B. Preston, *J. Org. Chem.*, **45**, 747 (1980).

592. M. M. Midland and S. B. Preston, *J. Am. Chem. Soc.*, **104**, 2330 (1982).
593. H. C. Brown and P. K. Jadhav, *J. Am. Chem. Soc.*, **105**, 2092 (1983).
594. H. C. Brown and P. K. Jadhav, *Tetrahedron Lett.*, **25**, 1215 (1984).
595. Y. Yamamoto, T. Komatsu, and K. Maruyama, *J. Am. Chem. Soc.*, **106**, 5031 (1984).
596. R. Haruta, M. Ishiguro, N. Ikeda, and H. Yamamoto, *J. Am. Chem. Soc.*, **104**, 7667 (1982).
597. D. A. Evans, J. V. Nelson, and T. R. Taber, in *Topics in Stereochemistry* (Eds. N. L. Allinger, E. L. Eliel, and S. H. Wilen), Vol. 13, Wiley, New York, 1982, p. 1.
598. D. A. Evans, J. Bartroli, and T. L. Shih, *J. Am. Chem. Soc.*, **103**, 2127 (1981).
599. S. Masamune, T. Kaiho, and D. S. Garvey, *J. Am. Chem. Soc.*, **104**, 5521 (1982).
600. M. M. Midland, in *Asymmetric Synthesis* (Eds. J. D. Morrison), Vol. 2, Academic Press, New York, 1983, p. 45.
601. M. M. Midland, A. Tramontano, and S. A. Zderic, *J. Am. Chem. Soc.*, **99**, 5211 (1977).
602. M. M. Midland, S. Greer, A. Tramontano, and S. A. Zderic, *J. Am. Chem. Soc.*, **101**, 2352 (1979).
603. M. M. Midland, D. C. McDowell, R. L. Hatch, and A. Tramontano, *J. Am. Chem. Soc.*, **102**, 867 (1980).
604. M. M. Midland, J. E. Petre, S. A. Zderic, and A. Kazubski, *J. Am. Chem. Soc.*, **104**, 528 (1982).
605. H. C. Brown and G. G. Pai, *J. Org. Chem.*, **47**, 1606 (1982).
606. H. C. Brown and G. G. Pai, *J. Org. Chem.*, **48**, 1784 (1983).
607. M. M. Midland and J. I. McLoughlin, *J. Org. Chem.*, **49**, 1316 (1984).
608. M. M. Midland, A. Tramontano, A. Kazubski, R. S. Graham, D. J. S. Tsai, and D. B. Cardin, *Tetrahedron*, **40**, 1371 (1984).
609. M. M. Midland and A. Tramontano, *Tetrahedron Lett.*, **21**, 3549 (1980).
610. M. M. Midland and N. H. Nguyen, *J. Org. Chem.*, **46**, 4107 (1981).
611. M. M. Midland and P. E. Lee, *J. Org. Chem.*, **46**, 3933 (1981).
612. D. J. S. Tsai and M. M. Midland, *J. Org. Chem.*, **49**, 1842 (1984).
613. M. M. Midland and A. Kazubski, *J. Org. Chem.*, **47**, 2814 (1982).
614. W. S. Johnson, B. Frei, and A. S. Gopalan, *J. Org. Chem.*, **46**, 1512 (1981).
615. E. Caspi, E. Santaniello, K. Patel, and T. Arunachalam, *J. Chem. Soc., Chem. Commun.*, 379 (1984).
616. L. E. Overman and K. L. Bell, *J. Am. Chem. Soc.*, **103**, 1851 (1981).
617. L. E. Overman, K. L. Bell, and F. Ito, *J. Am. Chem. Soc.*, **106**, 4192 (1984).
618. R. J. Parry, H. S. P. Rao, and J. Mueller, *J. Am. Chem. Soc.*, **104**, 339 (1982).
619. R. J. Parry and A. Minta, *J. Am. Chem. Soc.*, **104**, 871 (1982).
620. K. Kobayashi, P. K. Jadhav, T. M. Zydowsky, and H. G. Floss, *J. Org. Chem.*, **48**, 3510 (1983).
621. D. J. Aberhart, S. J. Gould, H.-J. Lin, T. K. Thiruvengadam, and B. H. Weiller, *J. Am. Chem. Soc.*, **105**, 5461 (1983).
622. K. C. Nicolau and S. E. Webber, *J. Am. Chem. Soc.*, **106**, 5734 (1984).
623. K. C. Nicolau, R. E. Zipkin, R. E. Dolle, and B. D. Harris, *J. Am. Chem. Soc.*, **106**, 3548 (1984).
624. H. C. Brown, G. G. Pai, and P. K. Jadhav, *J. Am. Chem. Soc.*, **106**, 1531 (1984).
625. J. W. Labadie and J. K. Stille, *J. Am. Chem. Soc.*, **105**, 6129 (1983).

The Chemistry of the Metal—Carbon Bond, Vol. 4
Edited by F. R. Hartley

CHAPTER **4**

Preparation and use of organoaluminium compounds in organic synthesis

P. A. CHALONER

School of Chemistry and Molecular Sciences, University of Sussex, Falmer, Brighton BNI 9QJ, UK

I. INTRODUCTION

Hard to handle, and in many cases difficult to prepare, organoaluminium compounds are not at first sight the organic chemist's first choice for synthetic procedures. However, in recent years there has been much wider use of these reagents and new areas of applicability are regularly being reported. The lower alkylalanes, especially Me_3Al and to a lesser extent Et_3Al and $Pr^n{}_3Al$, are pyrophoric and must be handled in an inert atmosphere, in a glove-box or on a vacuum line. Alkylaluminium halides and alkoxides may be handled under CO_2. As CCl_4 and water exacerbate fires in which alkylaluminiums are involved, only foam or sand should be used. Contact with skin may cause serious injuries and prompt medical attention is essential.

The most important organoaluminium compounds in organic synthesis are, or are derived from, commercially available compounds. These include Me_3Al, Et_3Al, $Bu^i{}_3Al$, $Bu^i{}_2AlH$, Me_2AlCl, $MeAlCl_2$, and $EtAlCl_2$. Trialkylalanes and alkylaluminium halides are available industrially from alkyl halides and aluminium metal in appropriate proportions. The synthesis of more complex alanes from these precursors will be discussed in the appropriate sections.

The use of organoaluminium compounds in synthesis has been regularly reviewed[1–5]; this account will concentrate on recent examples. The use of organoaluminium compounds in the catalysis of alkene oligomerization, polymerization, and metathesis is excluded and the many reactions involving $AlCl_3$ as a Lewis acid or $LiAlH_4$ as a reductant are treated only for comparison with organoaluminium compounds.

II. REACTIONS WITH CARBON—CARBON DOUBLE BONDS

A. Hydroalumination

The addition of H—Al to alk-1-enes (equation 1) is relatively complete, but internal alkenes give less satisfactory results. Selectivity to primary alkyls is high[6]. The kinetics and

$$Al{-}H + \rangle C{=}C\langle \longrightarrow Al{-}\overset{|}{\underset{|}{C}}{-}\overset{|}{\underset{|}{C}}{-}H \qquad (1)$$

equilibria of the reaction have been extensively studied[7–9]. Theoretical studies on the interaction of C_2H_4 with H_3Al suggest that the process is best visualized as occurring by association to form a π-complex, which then reacts in a four-centred transition state[10].

The presence of an aryl substituent in styrene gives a mixture of primary and secondary products (reaction 2)[11], but with 1, 1-diphenylethene only the primary product is found[12]. Allyl and vinyl ethers yield alkenes by routes to be discussed later. 1,2-Dienes react with Bu^i_2AlH (dibah) to give, after hydrolysis, alk-1-enes in excellent yield (equation 3)[13]. Reactions of 1,3-dienes with dibah give, after hydrolysis, monoenes and alkanes but the reaction is complicated by oligomerization[14]. Unconjugated dienes may give good yields of trialkylaluminiums provided that one of the double bonds is terminal (equation 4)[15]. Few reactions with internal alkenes have found synthetic application.

$$PhCH{=}CH_2 + Et_2AlH \longrightarrow PhCH_2CH_2AlEt_2 + PhCH(Me)AlEt_2 \qquad (2)$$

$$Hex^nCH{=}C{=}CH_2 \xrightarrow[(2)\ D_2O]{(1)\ dibah} Hex^nCHDCH{=}CH_2 \qquad (3)$$

$$CH_2{=}CHCH{=}CHCH(Me)CH{=}CH_2 + Bu^i_3Al \longrightarrow$$
$$[CH_2{=}CHCH{=}CH_2CH(Me)CH_2CH_2]_3Al \qquad (4)$$

The importance of hydroalumination in synthesis has been enhanced by the availability of catalysed reactions. Nickel catalysis was discovered by Ziegler, but cobalt and titanium chlorides also accelerate the transformation of equation 5[16]. The process is slower for di- and tri-substituted alkenes than for alk-1-enes. A variety of other nickel complexes are active in alkene/trialkylaluminium exchange reactions including $[(C_2H_4)_3Ni]$, $[(cod)_2Ni]$ and $[(PhCH{=}CHPh)_3Ni]$[17]. Mechanistic studies suggest that both the alkene and the organoaluminium compound are simultaneously coordinated to nickel. An unusual asymmetric addition of Bu^i_3Al to an alkene, with elimination of isobutene has been reported in the presence of *bis*(*N*-methylsalicylideneamine)nickel, $[Ni(mesal)_2]$, and (−)-*N*, *N*-dimethylmenthylamine (L) (equation 6). Hydrolysis of the intermediate alane gives the hydrocarbon in 67% *R* optical yield and the chiral base is recoverable.

$$R^1R^2C{=}CR^3R^4 + LiAlH_4 \longrightarrow Li[H_3AlCR^1R^2CHR^3R^4] \qquad (5)$$

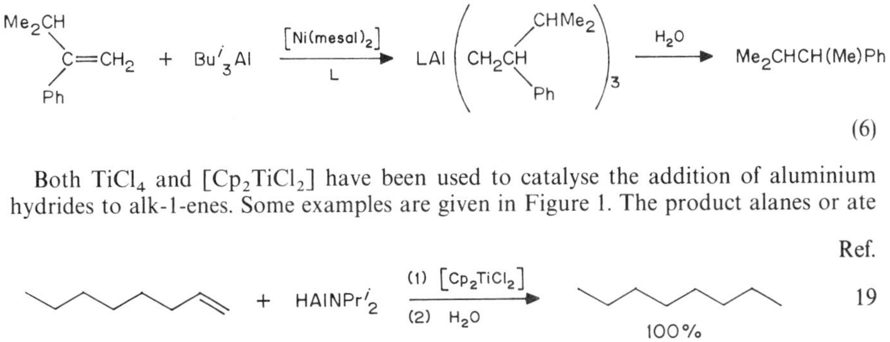

$$(Me_2CH)(Ph)C{=}CH_2 + Bu^i_3Al \xrightarrow[L]{[Ni(mesal)_2]} LAl\left(CH_2CH(CHMe_2)(Ph)\right)_3 \xrightarrow{H_2O} Me_2CHCH(Me)Ph \qquad (6)$$

Both $TiCl_4$ and $[Cp_2TiCl_2]$ have been used to catalyse the addition of aluminium hydrides to alk-1-enes. Some examples are given in Figure 1. The product alanes or ate

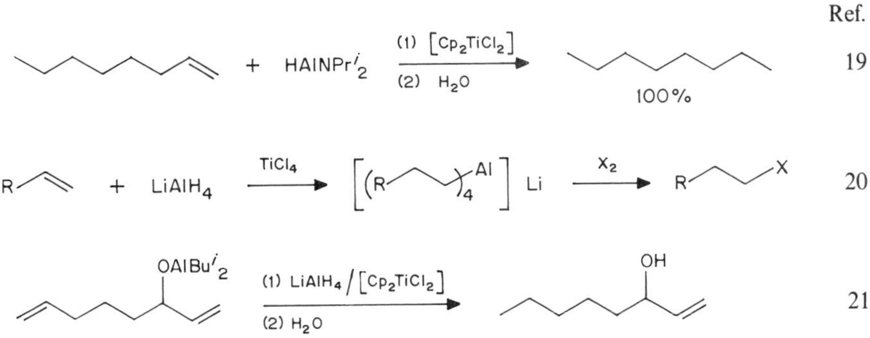

	Ref.
oct-1-ene + $HAlNPr^i_2$ $\xrightarrow[(2)\ H_2O]{(1)\ [Cp_2TiCl_2]}$ octane (100%)	19
$RCH{=}CH_2$ + $LiAlH_4$ $\xrightarrow{TiCl_4}$ $[(RCH_2CH_2)_4Al]Li$ $\xrightarrow{X_2}$ RCH_2CH_2X	20
$CH_2{=}CH(CH_2)_4CH(OAlBu^i_2)CH{=}CH_2$ $\xrightarrow[(2)\ H_2O]{(1)\ LiAlH_4/[Cp_2TiCl_2]}$ $CH_3(CH_2)_5CH(OH)CH{=}CH_2$	21

FIGURE 1. Titanium-catalysed hydroalumination of alkenes

complexes have been used in numerous syntheses. Coupling occurs directly with halogens to give primary halides (equation 7)[22], whilst copper salts catalyse the range of reactions

(1) $LiAlH_4/TiCl_4$ (2) I_2 (7)

shown in Figure 2[23–28]. Treatment with oxygen or *m*-chloroperbenzoic acid gives alcohols[29] whilst $Pb(OAc)_4$ gives acetates[30].

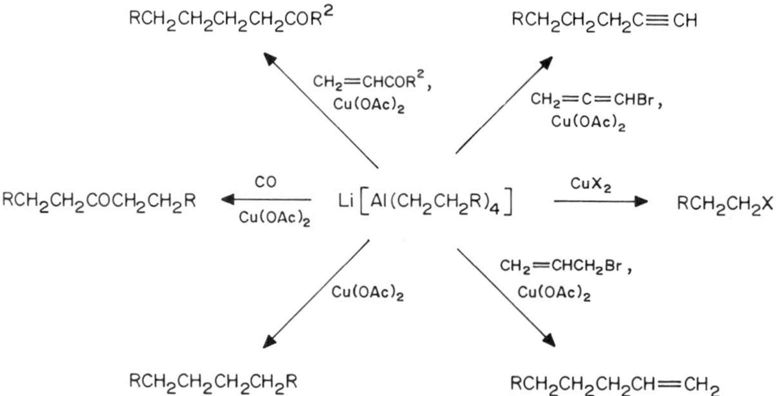

FIGURE 2. Reactions of ate complexes in the presence of copper salts

The relative reactivities of alkenes depend mainly on steric factors, with internal alkenes reacting more slowly than terminal alkenes[19]. With non-conjugated dienes the less substituted double bond reacts selectively (reactions 8[31] and 9[27]). The reaction of α,ω-

(1) $LiAlH_4/TiCl_4$ (2) H_2O (8)

97%

(1) $LiAlH_4/TiCl_4$ (2) $CuCl_2$ (9)

Cl

67%

dienes such as **1** depends on the catalyst. With $TiCl_4$ the *bis*-aluminium species **2** is obtained[19], but by a change of solvent (to C_6H_6) or catalyst to $ZrCl_4$ cyclization of **3** yields **4**, giving methylcyclopentane on hydrolysis (reaction 10). Dibah gives a similar reaction, but the main product is methylenecyclopentane by elimination from the cyclic intermediate **4**[32].

(10)

The reaction mechanism is thought to involve the formation of a titanium hydride, which is subsequently added to the double bond (equations 11–13).

$$LiAlH_4 + [Cp_2TiCl_2] \longrightarrow [Cp_2TiHCl] + LiAlH_3Cl \quad (11)$$
$$RCH{=}CH_2 + [Cp_2TiHCl] \longrightarrow [RCH_2CH_2TiClCp_2] \quad (12)$$
$$[RCH_2CH_2TiClCp_2] + LiAlH_4 \longrightarrow RCH_2CH_2AlH_3Li + [Cp_2TiHCl] \quad (13)$$

Reactions of Bu^i_3Al and dibah with internal alkenes to eliminate isobutene are best catalysed by $Ti(OBu)_4$, although many titanium (IV) species are active. Primary organoalanes are obtained by slow migration of aluminium to carbon down the chain to the end, and primary alcohols are obtained on oxidation.

Hydrometallation catalysed by zirconium complexes is equally successful, if less thoroughly explored. $Bu^i{}_3Al$ reacts with alk-1-enes in the presence of catalytic $[Cp_2ZrCl_2]$ and tolerates several functional groups[33], but the analogous reaction with dibah requires stoichiometric $[Cp_2ZrCl_2]$. The reaction of cyclohexene with dibah proceeds in good yield in the presence of $ZrCl_4/4ROH$[34]. Alk-1-enes undergo alkylation with $Et_2AlCl/Zr(OBu)_4$, often in excellent yield (equation 14)[35]. Addition of $LiAlH_4$ to alk-1-enes is also catalysed by $ZrCl_4$, to give trialkylaluminiums which react with water and halogens as before[36]. With allyl alcohols and allyl ethers both hydroalumination and/or hydrogenolysis occur in varying proportions[37].

$$Et_2AlCl + Bu^nCH{=}CH_2 \xrightarrow{Zr(OBu^n)_4} \underset{95\%}{Bu^nC(Et){=}CH_2} + Hex^n{}_2AlCl \quad (14)$$

B. Carboalumination

Carboalumination (equation 15) is not as important for alkenes as for alkynes, but interesting examples are known. The reactivity pattern for alkenes is similar to that for

$$Al{-}\overset{|}{\underset{|}{C}}{-} + \;>C{=}C<\; \longrightarrow Al{-}\overset{|}{\underset{|}{C}}{-}\overset{|}{\underset{|}{C}}{-}\overset{|}{\underset{|}{C}}{-} \quad (15)$$

hydrometallation but the reaction is more difficult to reverse. Some uncatalysed reactions are known, mainly involving especially reactive alkenes. A strained cyclic compound such as bicyclo[2.2.1]hept-2-ene is hydroaluminated and carboaluminated (*ca.*5%) with Et_2AlH and exclusive carboalumination occurs with Et_3Al[38]. Ethylene may then be inserted into the secondary carbon—aluminium bond. Norbornadiene and dicyclopentadiene are similarly reactive[39]. With **5** and **7** carboalumination proceeds by a radical reaction at low temperature (equation 16 and 17). The reaction rates decrease in the order

$$(NC)_2C{=}C(CN)_2\ (\mathbf{5}) + R_3Al \longrightarrow (NC)_2C(AlR_2){-}C(R)(CN)_2\ (\mathbf{6}) \qquad (16)$$

$$(NC)_2C{=}C_6H_4{=}C(CN)_2\ (\mathbf{7}) + R_3Al \longrightarrow R(NC)_2C{-}C_6H_4{-}C(CN)_2AlR_2\ (\mathbf{8}) \qquad (17)$$

R = Et > Bu^i > Me, paralleling the ease of ionization of the Al—R bond in the initial step (equation 18)[40].

$$\text{tcne} + R_3Al \longrightarrow \text{tcne}^{\cdot -} + R^{\cdot} + R_2Al^+ \qquad (18)$$

Catalysed carbometallations are more common, the major catalysts again being titanium and zirconium complexes. Carbometallation of alk-1-enes by R_3Al occurs at room temperature in the presence of stoichiometric $[Cp_2TiCl_2]$. The product is the 2-alkyl-1-ene **10**, formed by dehydroalumination of **9** (equation 19)[41]. The reaction is tolerant of halide, hydroxy and ester functions in the alkene. A similar reaction occurs in the presence of $Zr(OBu)_4$ with Et_2AlCl and $EtAlCl_2$ to give 2-ethyl-alk-1-enes. Transmetallation to yield alkylzirconium complexes seems to be the critical step[36].

$$R^1CH{=}CH_2 + R^2{}_3Al \longrightarrow \underset{(\mathbf{9})}{R^1R^2CHCH_2AlR^2{}_2}$$

$$\mathbf{9} \longrightarrow \underset{(\mathbf{10})}{R^1R^2C{=}CH_2} + R^2{}_2AlH \qquad (19)$$

III. REACTIONS WITH ALKYNES

A. Hydroalumination

Hydroalumination of alkynes occurs fairly readily and usually with high stereoselectivity, providing a route to vinylalanes in cases where the corresponding magnesium and lithium derivatives are difficult to prepare. The main side reactions are oligomerization (equation 20) and metallation of alk-1-ynes when R = aryl (equation 21). Alkynes react faster than alkenes and **12** is the sole product of the reaction of dibah with **11** (equation 22)[42]. By contrast vinylacetylene gives poor results[43]. With monosubstituted enynes such as **13** cyclization may occur; **15** is obtained in diethyl ether/hexane since the intermediate **14** is stabilized. Without ether stabilization **14** undergoes dehydroalumination between C-1 and C-5 to give an intermediate which quenches to **16** (reaction 23)[44].

$$2R^1C{\equiv}CR^1 + R^2{}_2AlH \longrightarrow RCH{=}CH(R)CH(R)CH(R)AlR^2{}_2 \qquad (20)$$

$$R^1C{\equiv}CH + R^2{}_2AlH \longrightarrow R^1C{\equiv}CAlR^2{}_2 + H_2 \qquad (21)$$

(11) + $Bu^i{}_2AlH$ ⟶ (12) (22)

(13) $\xrightarrow{Et_2AlH}$ (14); (14) $\xrightarrow{Et_2O,\ hexane}$ (15); (14) $\xrightarrow{hexane}$ (16) (23)

Other additions to alk-1-ynes have been used for reduction; dibah is the reagent of choice since its large bulk enhances stereoselectivity and it is commercially available and easy to handle (reaction 24)[45]. The vinylalane **17** has been cyclopropanated to **18** and subsequently protonated or brominated (reaction 25)[46]. A corresponding ate complex, **19**,

$$Hex^nC{\equiv}CH \xrightarrow[(2)\ H^+]{(1)\ 2\ Bu^i{}_2AlH,\ MeCu,\ MeMgBr,\ LiCl} Hex^nCH{=}CH_2 \qquad (24)$$

$Bu^nC{\equiv}CH \xrightarrow{dibah}$ (17) $\xrightarrow[Cu/Zn]{CH_2Br_2}$ (18); (18) $\xrightarrow{Br_2}$ bromocyclopropane; (18) $\xrightarrow{H_3O^+}$ Bu^n-cyclopropane (25) reacts with cyanogen to give a vinyl cyanide (reaction 26)[47] or the allyl derivative **20** to give a thioether with good stereochemical control in both the vinyl and allyl fragments

$$\mathbf{17} \xrightarrow{\text{RLi}} \left[\text{Bu}^n\text{CH=CH–Al(R)Bu}^i_2 \right]^- \text{Li}^+ \; \mathbf{(19)} \xrightarrow{(\text{CN})_2} \text{Bu}^n\text{CH=CHCN} \;\; 87\% \tag{26}$$

$$\mathbf{19} + \text{R}^2\text{CH=CHCH}_2\text{SSO}_2\text{Ph} \; \mathbf{(20)} \longrightarrow \text{Bu}^n\text{CH=CH–S–CH}_2\text{CH=CHR}^2 \; \mathbf{(21)} \tag{27}$$

(reaction 27)[48]. The better reactivity of alkynes is exploited in reaction 28 of **22** to yield, after treatment with a chiral epoxide, a precursor of the lactone **23**, isolated from the fungus *Cephalosporium recifei*[49].

(22) —(1) dibah, (2) Bu^nLi→ vinyl-$AlBu^i_2Bu^n$ ate complex —(chiral methyloxirane)→ secondary alcohol (H, OH) —→—→—→ (23) (28)

Disubstituted alkynes react similarly providing clean, stereospecific reductions in reactions 29[50] and 30[51]. Deuteriolysis gives deuteriated alkenes. Excess of alkyne may

$$\text{Hex}^n\text{C}\equiv\text{CCH}_2\text{C}\equiv\text{C(CH}_2)_5\text{CH}_2\text{Cl} \xrightarrow[(2)\,\text{H}_2\text{O}]{(1)\,\text{dibah}} \text{Hex}^n\text{CH=CHCH}_2\text{CH=CH(CH}_2)_5\text{CH}_2\text{Cl} \tag{29}$$

cyclododecyne —(1) dibah, (2) MeOH→ cyclododecene (30)

result in dimerization to dienes and cyclotrimerization to arenes. Regiospecificity with non-symmetric alkynes is variable[52] but certain substituents provide a directing effect. For example, 1-silylated alkynes, **24**, give regiospecific addition to yield **25**. The vinylalanes are converted into halides, **26**, using *N*-chloro- or *N*-bromo-succinimide, Br_2, or I_2. Compound **26** may be isomerized to **28** photochemically or obtained from **27**, which is the hydroalumination product in the absence of a solvent (reaction 31)[53,54]. The reaction has

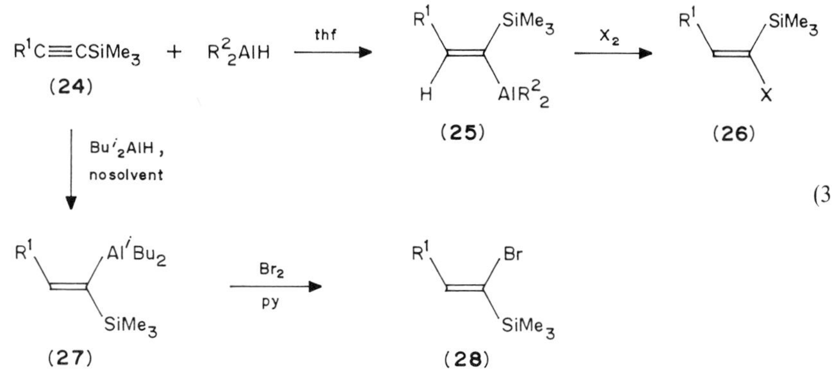

$R^1C{\equiv}CSiMe_3$ (24) + R^2_2AlH —thf→ (25) —X_2→ (26)

(24) —Bu^i_2AlH, no solvent→ (27) —Br_2 / py→ (28)

(31)

been used in various natural product syntheses (Figure 3); **29** is a precursor of the indoloquinolizine alkaloids[55] and **30** is a synthon for the *β*-formyl vinyl anion and cation[56].

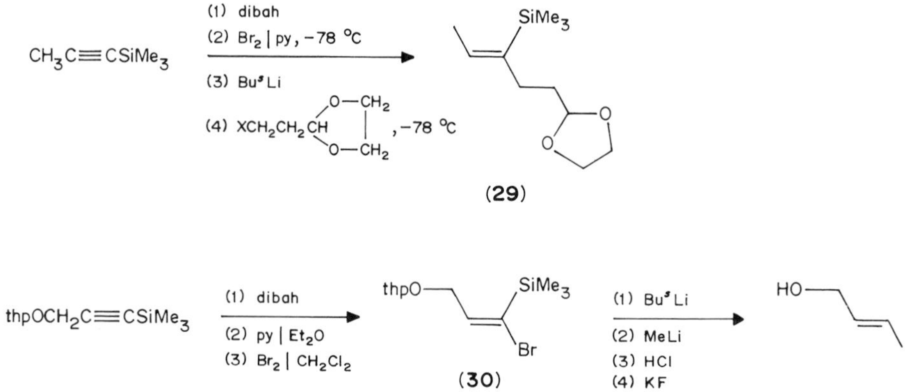

$CH_3C{\equiv}CSiMe_3$ —(1) dibah; (2) Br_2 | py, −78 °C; (3) Bu^sLi; (4) $XCH_2CH_2CH(OCH_2)_2$, −78 °C→ (29)

$thpOCH_2C{\equiv}CSiMe_3$ —(1) dibah; (2) py | Et_2O; (3) Br_2 | CH_2Cl_2→ (30) —(1) Bu^sLi; (2) MeLi; (3) HCl; (4) KF→

FIGURE 3. Hydroalumination of silylated alkynes in natural product synthesis

Addition of R_2AlH is usually stereospecific and *cis*, but $RAlH_2$, AlH_3, $LiAlH_4$, and $Li[Bu^i_2AlMeH]$ all give *trans*-addition (Figure 4). Alkynes bearing oxygen or nitrogen functions give very regiospecific results. For example, **31** adds dibah in an unusual *trans*-reaction to give only **32** (reaction 32)[60]. With **33**, an intramolecular *trans*-

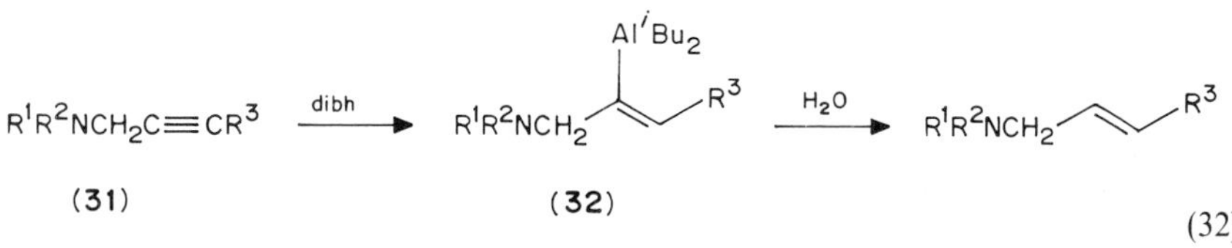

$R^1R^2NCH_2C{\equiv}CR^3$ (31) —dibh→ (32) —H_2O→ $R^1R^2NCH_2CH{=}CHR^3$

(32)

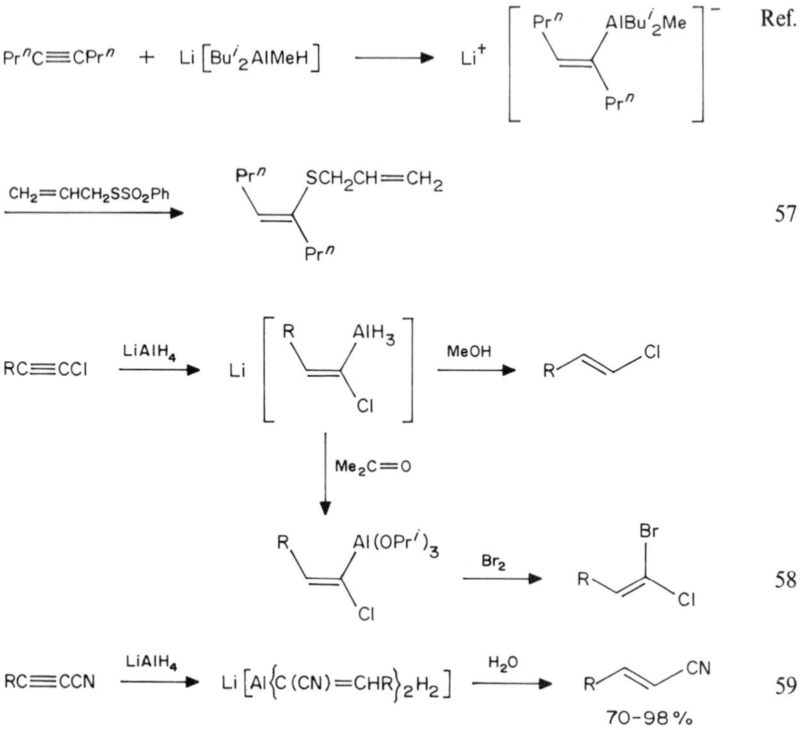

Ref.

$Pr^nC{\equiv}CPr^n$ + $Li[Bu^i_2AlMeH]$ ⟶ $Li^+[(Pr^n)CH{=}C(AlBu^i_2Me)(Pr^n)]^-$

$\xrightarrow{CH_2{=}CHCH_2SSO_2Ph}$ $(Pr^n)CH{=}C(SCH_2CH{=}CH_2)(Pr^n)$ 57

$RC{\equiv}CCl$ $\xrightarrow{LiAlH_4}$ $Li[RCH{=}C(AlH_3)Cl]$ $\xrightarrow{MeOH}$ $RCH{=}CHCl$

$Li[RCH{=}C(AlH_3)Cl]$ $\xrightarrow{Me_2C{=}O}$ $RCH{=}C(Al(OPr^i)_3)Cl$ $\xrightarrow{Br_2}$ $RCH{=}CBrCl$ 58

$RC{\equiv}CCN$ $\xrightarrow{LiAlH_4}$ $Li[Al\{C(CN){=}CHR\}_2H_2]$ $\xrightarrow{H_2O}$ $RCH{=}CHCN$ 59

70–98%

FIGURE 4. *Trans*-hydroalumination of alkynes

hydroalumination of the alkyne occurs, giving ultimately the chiral allenic alcohol **34** (reaction 33)[61].

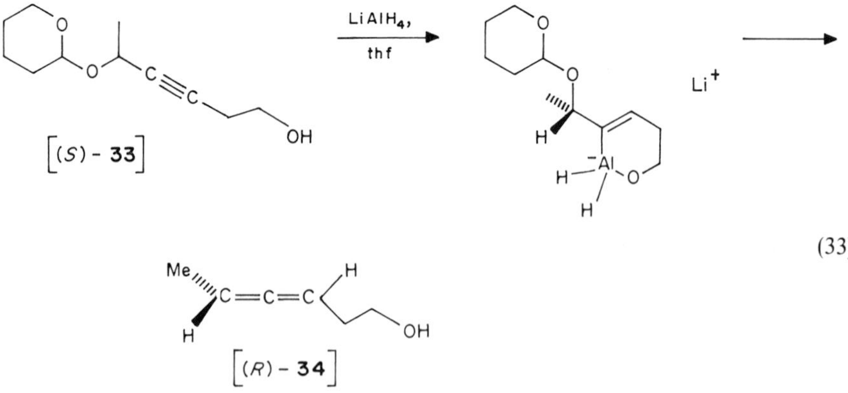

[(*S*) - **33**] $\xrightarrow{LiAlH_4,\ thf}$ Li^+ [cyclic alanate intermediate: H, $H{-}Al^-$, H, O] ⟶ [(*R*) - **34**] (Me)(H)C=C=C(H)CH_2CH_2OH (33)

Catalysed hydrometallations have been comparatively little studied since the reaction proceeds well in the absence of catalysts. $[Ni(acac)_2]$ accelerates the reaction of dibah with internal alkynes, allowing it to proceed at a temperature at which no *cis*$\rightleftharpoons$*trans* isomerization occurs; *cis*-alkenes are obtained on hydrolysis[62]. $(Pr^i_2N)_2AlH$ adds to internal alkynes in the presence of $[Cp_2TiCl_2]$ and the product vinylalanes may be deuteriated or iodinated. The reaction is not regiospecific except for arylalkynes, which complex the titanium hydride intermediate[63].

B. Carboalumination

Although carbometallation has the same orbital character as hydrometallation, uncatalysed reactions in this area are few and unpromising. Photochemical addition of Et_6Al_2 to diphenylacetylene gives both *cis*- and *trans*-dialkylaluminium compounds (**35**), the *cis* yielding **36** by dimerization (reaction 34)[64]. Attempted addition of Bu^i_3Al to alk-1-

$$PhC{\equiv}CPh + R_3Al \xrightarrow{h\nu} \text{(35)} \quad \text{(36)} \tag{34}$$

ynes gives mainly metallation, some hydroalumination and a small amount of stereospecifically *syn* carbometallation[65]. With [Ni(mesal)$_2$] as catalyst the main products are dimers and trimers. Enynes give a complex mixture of products[66].

Acetylene reacts with R_3Al to give the *cis*-addition product, $RCH{=}CHAlR_2$, but metallation is a major side reaction. Little use has been made of this or of the complex reactions of alk-1-ynes[67]. Disubstituted alkynes react more slowly. Hex-3-yne reacts only above 90 °C and gives largely a dimer, **37** (reaction 35). With diphenylacetylene clean

$$EtC{\equiv}CEt + Et_3Al \longrightarrow Et_2C{=}C(Et)AlEt_2 \xrightarrow{EtC{\equiv}CEt} \text{(37)} \tag{35}$$

carbometallation occurs with Et_3Al to give (*Z*)-PhCH=CHPh(Et) after hydrolysis[68]. The regiochemistry of addition in non-symmetric cases is not easy to predict[69].

The use of titanium compounds as carboalumination catalysts has received some attention. Alkynols react with Et_2AlCl in the presence of $[Cp_2TiCl_2]$ or bis(methylcyclopentadienyl)$TiCl_2$ (reaction 36). Yields are good and the stereochemistry

$$HC{\equiv}CCH_2CH_2OH + Et_2AlCl \xrightarrow[(2)\ H_2O]{(1)\ [(MeCp)_2TiCl_2]} Et\text{–}CH{=}CH\text{–}CH_2CH_2OH + CH_2{=}C(Et)CH_2CH_2OH \tag{36}$$

is cleanly *cis*, but both regioisomers are obtained in comparable amounts[70]. Carbometallation of alkynes with Me_3Al and stoichiometric $[Cp_2TiCl_2]$ gives the *cis* addition product

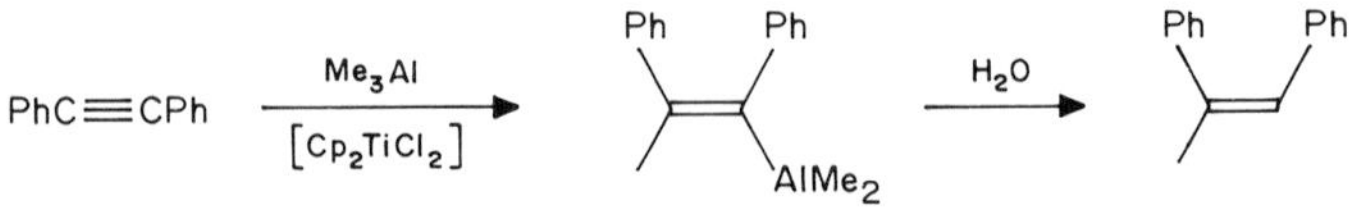

$$PhC\equiv CPh \xrightarrow[{[Cp_2TiCl_2]}]{Me_3Al} \text{(Ph)(Me)C=C(Ph)(AlMe}_2) \xrightarrow{H_2O} \text{(Ph)(Me)C=CH(Ph)} \quad (37)$$

in good yield (reaction 37)[71]. Silylated alkynes give regio- and stereo-selective addition with $Me_2AlCl_2[Cp_2TiCl]$ (reaction 38)[72,73]. The use of Me_3Al is less successful in this

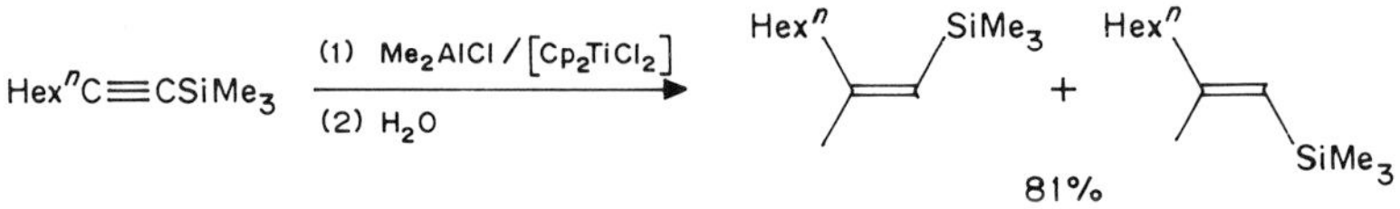

$$Hex^nC\equiv CSiMe_3 \xrightarrow[(2)\ H_2O]{(1)\ Me_2AlCl/[Cp_2TiCl_2]} \text{(Hex}^n\text{)(Me)C=CH(SiMe}_3) + \text{(Hex}^n\text{)(Me)C=CH... (SiMe}_3) \quad 81\%,\ 95:5 \quad (38)$$

case, giving the regioisomers in the ratio 70:30. It seems probable that alkyltitanium compounds are the attacking species.

Nickel complexes have also been studied as carbometallation catalysts, but the results are usually unsatisfactory. Silylated alkynes give reasonable rates but the metallated product isomerizes under the reaction conditions to give mixtures of *cis*- and *trans*-isomers[74]. With alk-1-ynes the main product in the presence of $[Ni(mesal)_2]$ is the head-to-tail dimer **38** (reaction 39)[66,75,76].

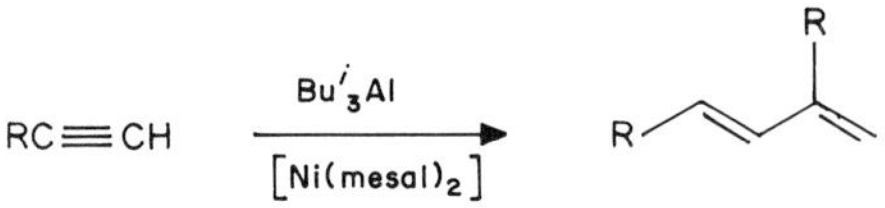

$$RC\equiv CH \xrightarrow[{[Ni(mesal)_2]}]{Bu^i_3Al} \text{RCH=CH–C(R)=CH}_2 \ \mathbf{(38)} \quad (39)$$

$$PhC\equiv CD \xrightarrow[{[Cp_2ZrCl_2]}]{Me_3Al,} \text{(Ph)(Me)C=C(D)(AlMe}_2) \xrightarrow{H_2O} \text{(Ph)(Me)C=CHD} \quad 98\%\ Z$$

$PhC\equiv CH$ → (1) Bu^nLi (2) D_2O → $PhC\equiv CD$

$$PhC\equiv CH \xrightarrow[{[Cp_2ZrCl_2]}]{Me_3Al,} \text{(Ph)(Me)C=CH(AlMe}_2) \xrightarrow{D_2O} \text{(Ph)(Me)C=CHD} \quad 96\%\ E$$

$$\text{(Ph)(Me)C=CH(AlMe}_2) \xrightarrow{I_2} \text{(Ph)(Me)C=CHI}$$

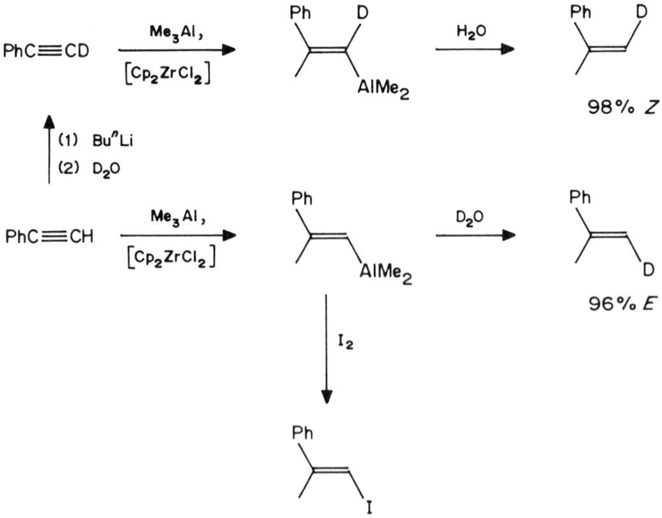

FIGURE 5. Zirconium-catalysed hydroalumination of alk-1-ynes

The problems of competing metallation are largely overcome when zirconium complexes are used as catalysts[71,77]. For alk-1-ynes and Me_3Al the reaction is both regio- and stereo-selective (Figure 5)[77]. Oct-1-yne reacts to give, after hydrolysis, a 95:5 2-methyloct-1-ene–non-2-ene mixture[78]. Fortunately, the terminal vinylalanes are much more reactive than the internal isomers and after subsequent treatment with electrophiles the minor isomer may effectively be neglected. Vinylalanes are potent nucleophiles towards a variety of centres (Figure 6); a discussion of the catalysed coupling reactions with halides will be deferred until a later section[78]. Such transformations have been widely used in natural product synthesis. For example, dolicholide, **70**, is prepared by the route in Figure 7[79] and udoteatrial, **71**, by that in Figure 8[80].

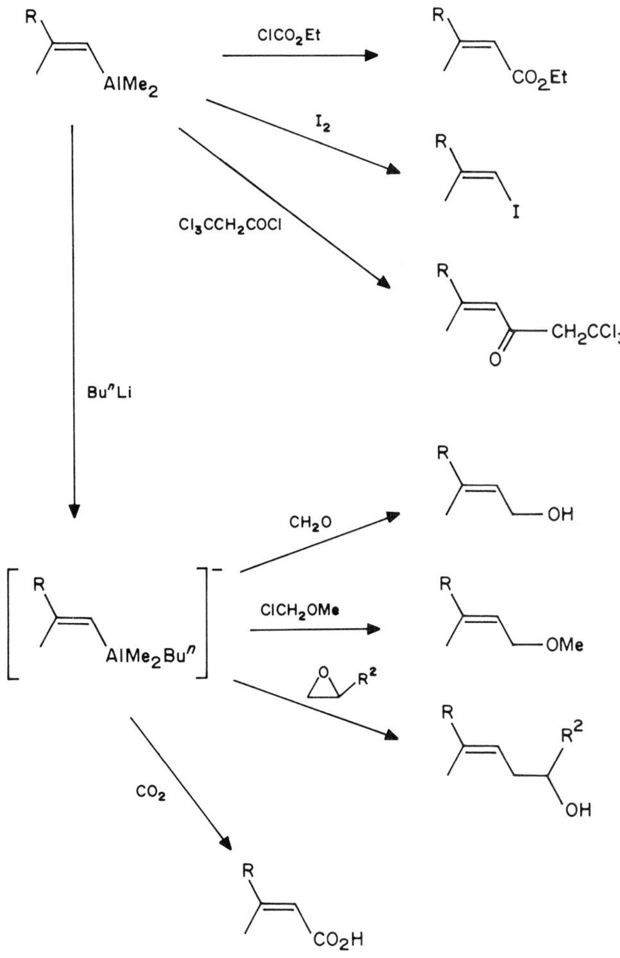

FIGURE 6. Reactions of vinylalanes with electrophiles

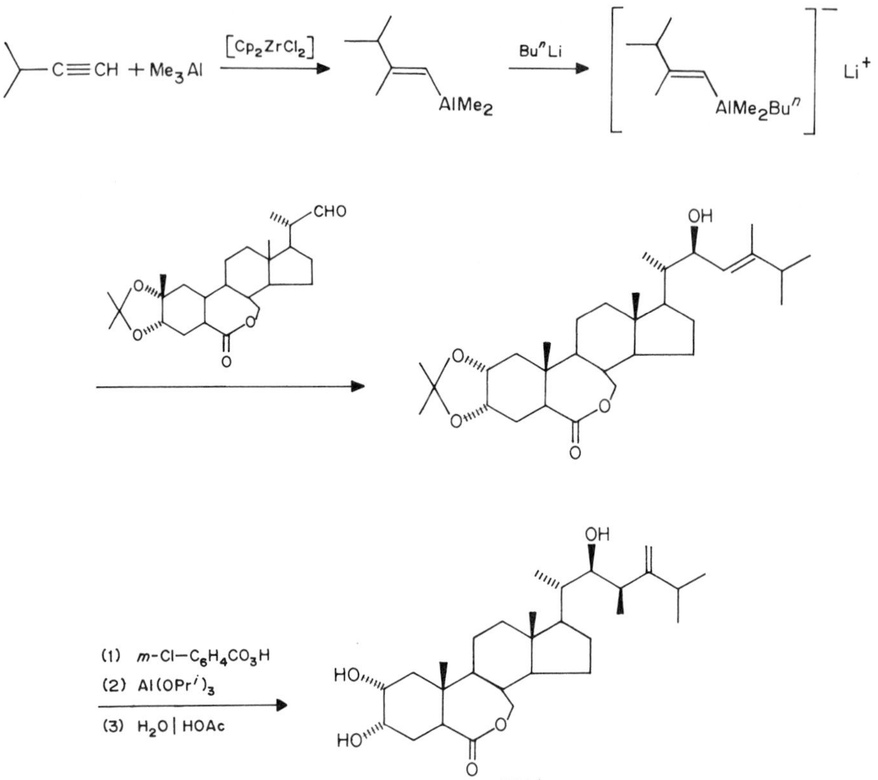

FIGURE 7. Synthesis of dolicholide

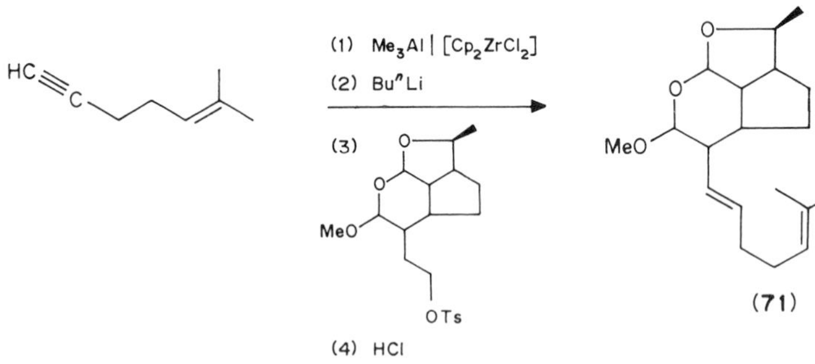

FIGURE 8. Synthesis of udoteatrial

Trialkylaluminiums other than the trimethyl compound have generally proved less satisfactory. For example, oct-1-yne and Pr^n_3Al give a mixture of products, 2-propyloct-1-ene (29%), (*E*)-undec-4-ene (8%) and oct-1-ene (47%), after hydrolysis, whereas Bu^i_3Al gives only hydrometallation. Hydrometallation may be suppressed by using Pr_2AlCl but the problem of regiochemistry is not solved[71].

Internal alkynes may be cleanly and stereospecifically carbometallated by Me_3Al (reaction 40)[77]. With non-symmetric alkynes the regioselectivity is variable[81]. Reactions

$$Bu^nC{\equiv}CBu^n \xrightarrow[(2)\ H_2O]{(1)\ Me_3Al/[Cp_2ZrCl_2]} (Z)\text{-}Bu^nC(Me){=}CHBu^n \quad 89\%,\ 98\%\ Z \tag{40}$$

with silylated alkynes have not always given the desired results. When **39** is treated with $Me_3Al/[Cp_2ZrCl_2]$ the product formed on work-up was not the vinylsilane **40**, but the silylated cyclobutene **41** (reaction 41)[82]. The generality of the reaction is confirmed by the conversion of **42** to **43** in reaction 42.

$$XCH_2CH_2C{\equiv}CSiMe_3\ (39) \xrightarrow[{[Cp_2ZrCl_2]}]{Me_3Al,} XCH_2CH_2C(Me){=}C(SiMe_3)AlMe_2 \not\rightarrow XCH_2CH_2C(Me){=}CHSiMe_3\ (40)$$

$$XCH_2CH_2C(Me){=}C(SiMe_3)AlMe_2 \rightarrow \text{cyclopropyl(Me)C}^+\text{–CH(SiMe}_3\text{)}\ \overset{-}{A}lMe_2X \rightarrow \text{cyclobutyl cation (Me, SiMe}_3\text{, }\overset{-}{A}lMe_2X) \rightarrow \text{1-SiMe}_3\text{-2-Me-cyclobutene}\ (41) \tag{41}$$

$$Me_3SiC{\equiv}CCH_2Br\ (42) \xrightarrow[(2)\ H^+]{(1)\ Me_3Al/[Cp_2ZrCl_2]} \text{1-SiMe}_3\text{-2,2-dimethylcyclopropane}\ (43) \tag{42}$$

The vinylalanes produced by this route have also been used as precursors to vinylmercury compounds by a general and simple method (reaction 43)[83].

$$RC{\equiv}CH \xrightarrow[\text{[}Cp_2ZrCl_2\text{]}]{Me_3Al} \text{R(H)C=C(H)AlMe}_2 \xrightarrow[\text{thf}]{HgCl_2} \text{R(H)C=C(H)HgCl} \quad (43)$$

R = Hex_n, 83%
> 98% *cis*-addition

There has been considerable debate as to whether the reaction is an aluminium-assisted carbozirconation or a zirconium-assisted carboalumination. N.m.r., deuterium labelling and product studies give unequivocal evidence for the latter[84]. Bimetallic intermediates are proposed, and by analogy **44** is very effective for carboalumination of silylacetylenes

$$Et_2Al(\mu\text{-}Bu)(\mu\text{-}Et)Mg(\mu\text{-}Et)(\mu\text{-}Bu)AlEt_2$$

(44)

(reaction 44)[85]. The use of $Hex^n{}_2Mg/Et_3Al$ improves the yield of the ethylated product (> 99% for $PhC{\equiv}CSiMe_3$). Stereoselection is strongly *cis*, but with enynes *trans*-addition occurs.

$$C_{10}H_{21}C{\equiv}CSiMe_3 \xrightarrow[(2)\ H_3O^+]{(1)\ \mathbf{44}} (C_{10}H_{21})(Et)C{=}C(SiMe_3)H + (C_{10}H_{21})(Bu^n)C{=}C(SiMe_3)H \quad (44)$$

5 : 1

IV. REACTION WITH CARBONYL GROUPS

A. Aldehydes and Ketones

1. *Alkylation*

Organoalanes are less nucleophilic and more Lewis acidic than Grignard reagents or organolithiums, so reactions involving their attack at carbonyl centres are more prone to elimination and rearrangement. Hence the uses of organoalanes in simple alkylations are relatively few. However, higher stereoselectivities are possible with aluminium reagents than in those with lithium and magnesium and the greater accessibility of vinylalanes has made them popular.

Simple aliphatic aldehydes give complex mixtures with trialkylaluminiums. Both addition and reduction occur, followed by Meerwein–Pondorff–Verley reduction, Oppenauer oxidation and the Tischenko reaction[86]. However, **45** reacts with $(Me_3Si)_3Al$ to give **46** in fair yield (reaction 45)[87] and **47** has been successfully alkenylated to **48** (reaction 46)[88]. The use of ate complexes gives better yields at the expense of increased complexity.

(45) + $(Me_3Si)_3Al \cdot Et_2O$ ⟶ (46) (45)

$MeCO_2CH_2CH(NR_2)CHO + Bu^i{}_2Al\text{–CH=CH–}(CH_2)_{12}Me$ (47) ⟶ $MeCO_2CH_2CH(NR_2)CH(OH)CH=CH(CH_2)_{12}Me$ (48) (46)

The reaction of Me_3Al with ketones has been well studied, the primary reaction in either benzene or ether being addition. Better yields are obtained in benzene[89]. With acetone **49** may be isolated as an intermediate (reaction 47)[90]. The mechanism of reaction with Ph_2CO has been suggested to involve a six-membered transition state, **50**[91]. Reactions

$Me_2Al(\mu\text{-}Me)_2AlMe_2$ + Me_2CO ⟶ (49) (47)

(50)

with cyclohexanones show that axial addition occurs when Me_3Al is in deficiency and equatorial addition when it is in excess in hydrocarbon solvents[92]. A ketone·Me_3Al complex is supposed to be the reacting species. This may undergo unimolecular reorganization *via* an electron-transfer path to give the axial alcohol. The equatorial alcohol (axial addition) results from bimolecular kinetics[93]. Similar results are obtained with other sterically hindered ketones.

Excess of Me_3Al under forcing conditions will yield a *gem*-dimethyl compound[94], but the same reaction occurs at or below room temperature using $[Me_2TiCl_2]$[95]. Very hindered ketones tend to give aluminium enolates, which are useful in aldol condensations. Few synthetic applications have been reported, but Heathcock *et al.*[96] converted the ketone **51** into **52** by elimination of an aluminoxane from the first-formed alcoholate complex (reaction 48).

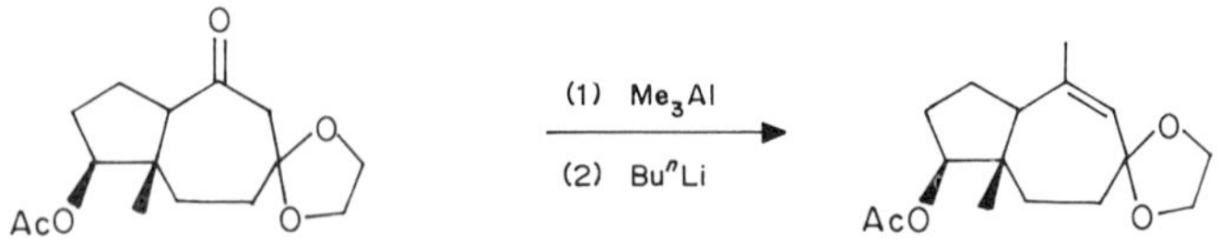

(51) $\xrightarrow[\text{(2) Bu}^n\text{Li}]{\text{(1) Me}_3\text{Al}}$ (52) (48)

Reaction of Et_3Al with ketones is complex, since both reduction and addition as well as aldol condensation are seen. A considerable body of evidence suggests that radicals may be involved in both these and the reactions of Ph_3Al[93,97].

Modification of alane structure and careful control of conditions have brought success in specific synthetic projects. Propargylaluminium compounds undergo addition in good yields to give branched alkynols (Figure 9), the degree of transposition being determined by the substitution pattern[98]. Acetylenic alanes undergo direct addition and the acetylene is transferred from $Me_2AlC{\equiv}CR$ twenty times more readily than the methyl group[99].

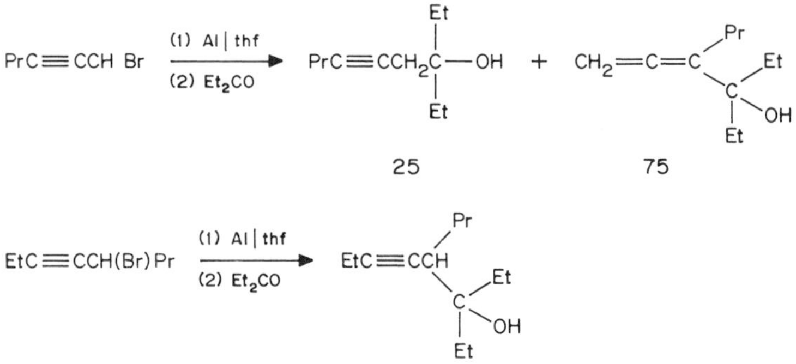

$$\text{PrC}{\equiv}\text{CCH Br} \xrightarrow[\text{(2) Et}_2\text{CO}]{\text{(1) Al | thf}} \text{PrC}{\equiv}\text{CCH}_2\text{C(Et)}_2\text{—OH} \ (25) + \text{CH}_2{=}\text{C}{=}\text{C(Pr)C(Et)}_2\text{OH} \ (75)$$

$$\text{EtC}{\equiv}\text{CCH(Br)Pr} \xrightarrow[\text{(2) Et}_2\text{CO}]{\text{(1) Al | thf}} \text{EtC}{\equiv}\text{CCH(Pr)C(Et)}_2\text{OH}$$

FIGURE 9. Reaction of propargylalanes with ketones

Allylalanes add to ketones with essentially complete allylic transposition (reaction 49)[100], and this has been used in a stereoselective synthesis of vicinal diols (Figure 10)[101], the products being employed in a route to *exo*-brevicomin.

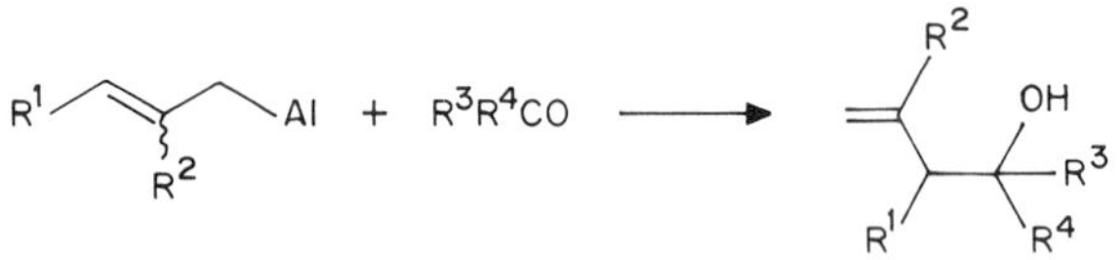

$$R^1CH{=}C(R^2)CH_2Al + R^3R^4CO \longrightarrow CH_2{=}C(R^2)CH(R^1)C(OH)R^3R^4 \qquad (49)$$

$$R^1O\text{CH}{=}\text{CHCH}_2\text{AlEt}_2 \xrightarrow[\text{(2) H}_2\text{O}]{\text{(1) R}^2\text{COR}^3} R^2R^3C(OH)CH(OR^1)CH{=}CH_2 \ (10) + R^2R^3C(OH)CH(OR^1)CH{=}CH_2 \ (1)$$

50-95%

10 : 1

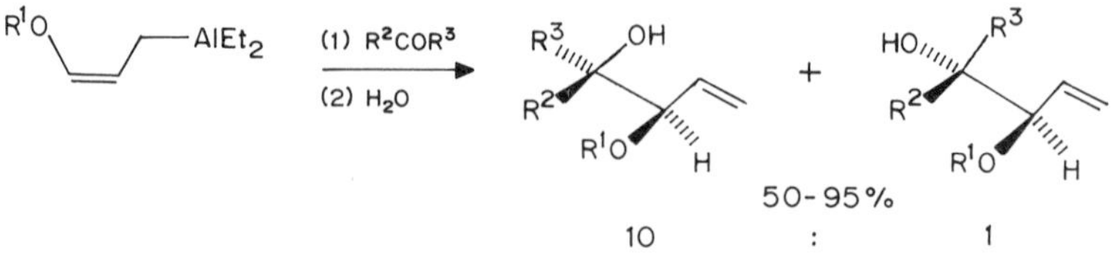

FIGURE 10. Synthesis of vicinal diols using allylalanes

Finally, using Et_2AlCN, cyanohydrins may be prepared; **53** does not react with HCN/KCN but may be converted by this route to **54** in good yield (reaction 50)[102].

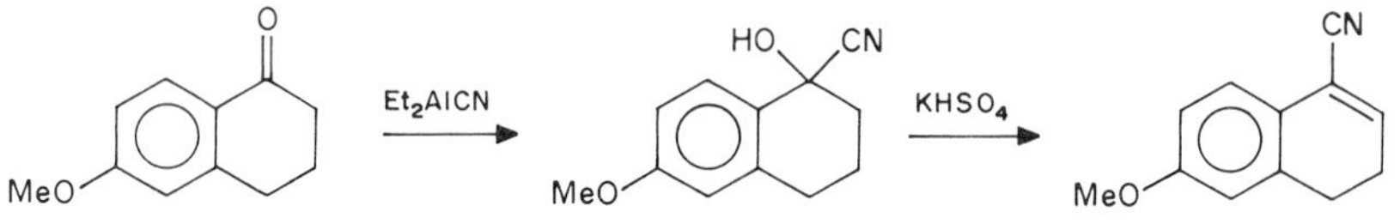

(50)

2. Reduction

Both aldehydes (reaction 51)[103] and ketones are readily reduced by $Bu^i{}_2AlH$ and we

(51)

noted earlier that reduction is a major side reaction in alkylation by Et_3Al. The steric bulk of dibah gives high stereoselectivity in the reduction of both cyclic and acyclic ketones (Figure 11). In the last example only diastereomer **55** is formed; $LiAlH_4$ gives both

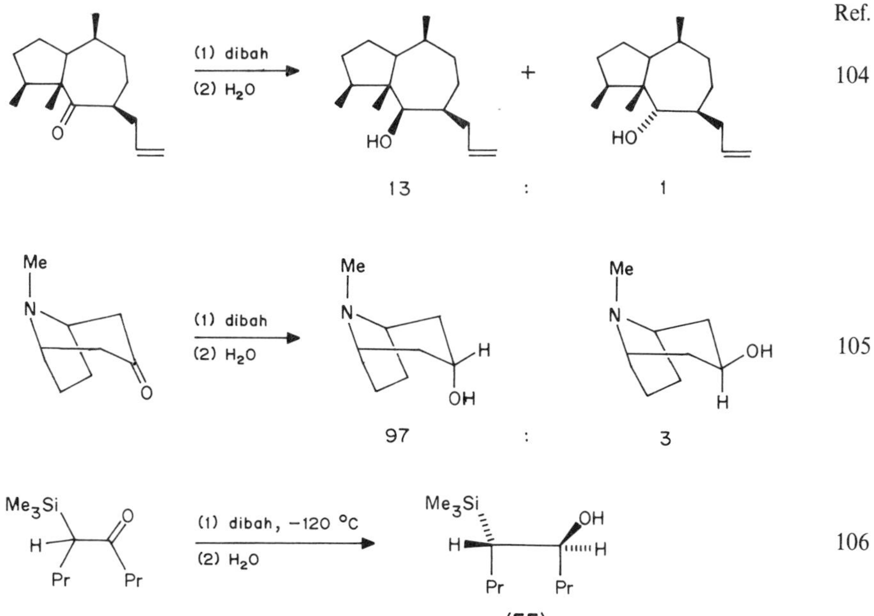

FIGURE 11. Stereoselective reduction of ketones by dibah

diastereomers whilst $NaBH_4$ causes desilylation. *Syn*-elimination (KH/thf) gives 95% *trans*-oct-4-ene whilst acid-catalysed *anti*-elimination gives the *cis*-isomer with 92% stereoselectivity.

Improved selectivity for specific reductions is obtained by modification of dibah with alcohols such as **56**. In the presence of 2 mol of **56**, reduction of **57** proceeds with 92% α-selectivity[107].

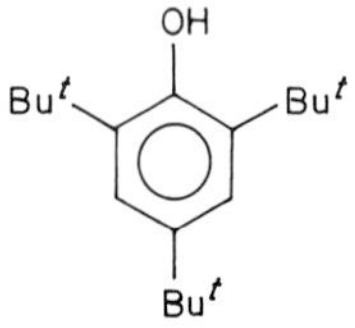

(56)

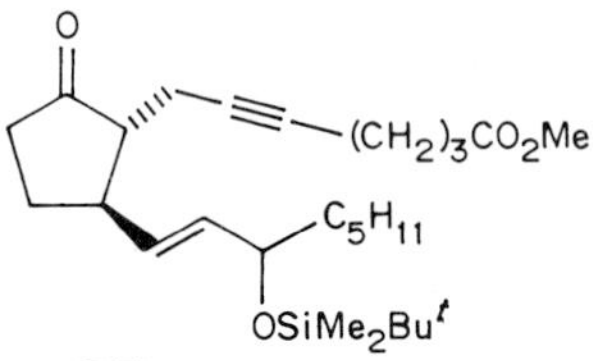

(57)

Reduction of cyclohexanones with lithium diisobutyl-*tert*-butylaluminium hydride (from dibah/Bu^{t}Li) is very stereoselective[108]. Yields are > 95% and the thermodynamically less stable isomer is formed. High selectivity for the *exo*-alcohol is found in the reduction of norcamphor. For α-hydroxyketones, $Bu^t{}_3Al$ was found to give the highest proportion of *erythro*-diol[109].

The potential for using chiral alanes for reduction has not been explored in detail. Hydrogen transfer usually occurs from the chiral centre, but asymmetric induction is modest. The reagents are trialkylaluminium compounds and the yields and mechanisms are similar to those with $Bu^i{}_3Al$. [(*S*)-2-Methylbutyl]$_3$Al reduces alkyl aryl ketones in good yield but low enantiomer excess[110–113]. If the mixture is worked up after 3 h, alcohols of *S*-configuration are obtained, but on standing for 1 day alkanes of *R*-configuration are recovered by reaction with the solvent[114].

3. *Aldol condensations*

Aluminium enolates may be generated from α-haloketones and react with aldehydes and ketones to give good yields of aldol products (reaction 52)[115,116]. In the other cases the

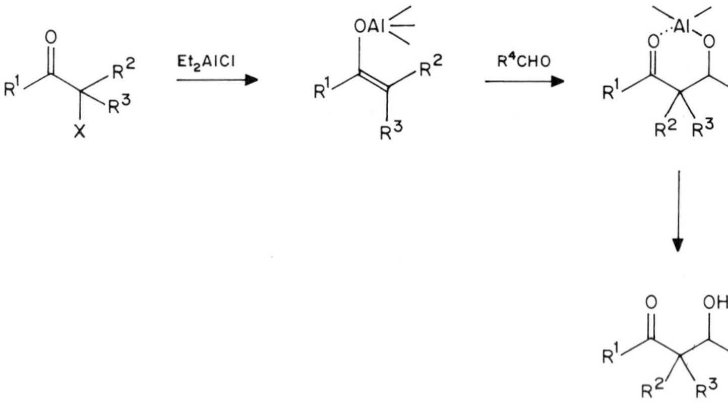

(52)

aldol condensation is often an undesired side reaction, but dibah/PhOH has been used to effect condensation of **58** to a mixture of aldol product isomers and **59** is cyclized with the same reagent to a muscone precursor[116].

B. Enones

Both alkylation and reduction of enones may occur in a 1, 2- or a 1, 4- manner and both types of reaction are known for organoaluminium reagents. $\alpha\beta$-Unsaturated aldehydes have not been much studied but propargylalanes add 1, 2- with good selectivity (reaction 53)[117]. An unusual reaction (54) is observed with allylalanes resulting from substitution with and without allylic transposition on **60**[100].

(53)

(54)

The reaction between Me_3Al and **61** has been studied in detail and gives both 1, 2- and 1, 4-addition[118]. Selective 1, 4-addition of Me_3Al to enones occurs in the presence of [Ni(acac)$_2$], some examples being given in Figure 12. By-products are high molecular weight aldol condensation products.

The reaction of alkenylaluminium compounds with enones proceeds in moderate yields; only the alkenyl group is transferred and double bond stereochemistry is maintained. Both 1, 2- and 1, 4-additions are known. For example, **62** gives **63** in 35% yield (reaction 55)[122]

(55)

and **64** reacts with **65** to give only 1, 2-addition, but the ate complex **66** gives clean 1, 4-addition to 11, 15-dideoxyprostaglandin E_1 (reaction 56)[123]. Trialkynyl aluminiums

Ref.

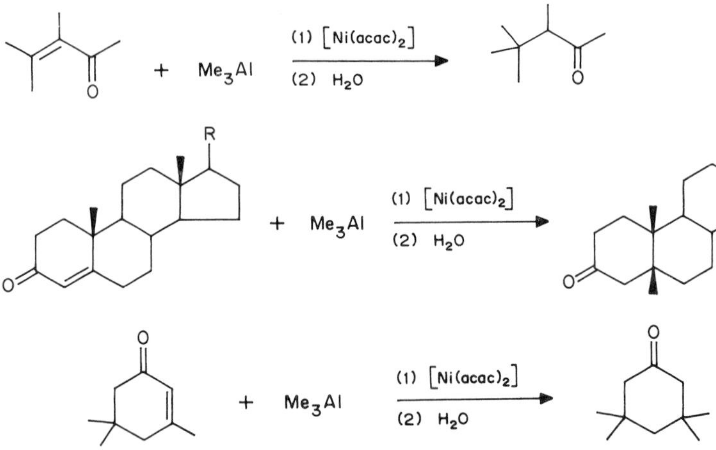

$+ \; Me_3Al \xrightarrow[(2)\ H_2O]{(1)\ [Ni(acac)_2]}$ 119

$+ \; Me_3Al \xrightarrow[(2)\ H_2O]{(1)\ [Ni(acac)_2]}$ 120

$+ \; Me_3Al \xrightarrow[(2)\ H_2O]{(1)\ [Ni(acac)_2]}$ 121

FIGURE 12. 1, 4-Addition of Me_3Al to enones

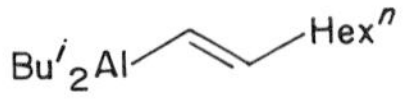

Bu^i_2Al–CH=CH–Hex^n

(64)

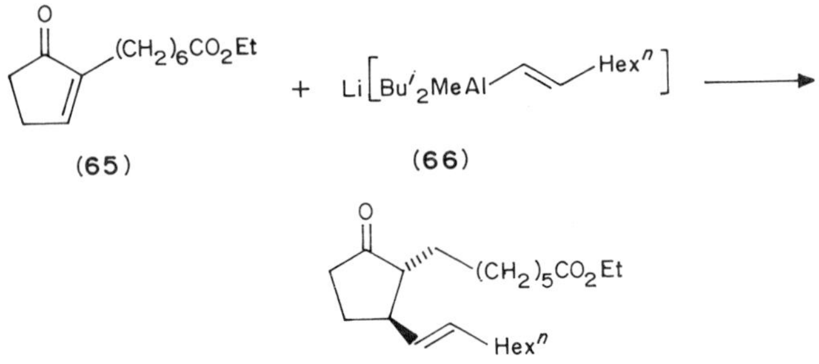

(65) $(CH_2)_6CO_2Et$ + $Li[Bu^i_2MeAl$–CH=CH–$Hex^n]$ (66) ⟶ $(CH_2)_5CO_2Et$, Hex^n (56)

have also been important in the prostaglandin field[124]. Compound **67** reacts with tri-octynylaluminium to give the 1, 4-addition products (reaction 57). Blocking the OH group prevents 1, 4-addition, suggesting that precoordination occurs.

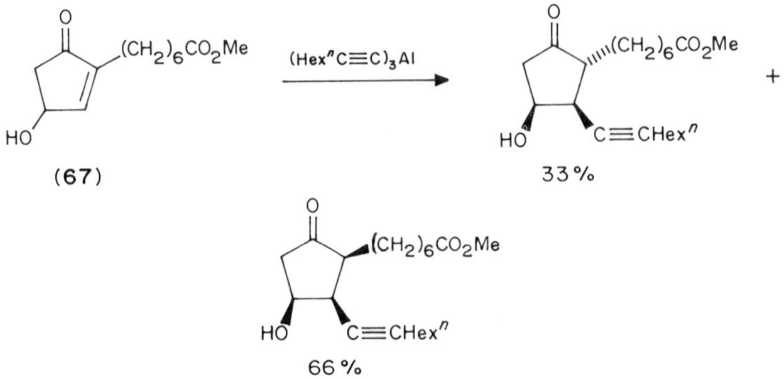

(67) HO, $(CH_2)_6CO_2Me$ $\xrightarrow{(Hex^nC{\equiv}C)_3Al}$ HO, $(CH_2)_6CO_2Me$, $C{\equiv}CHex^n$ 33% + HO, $(CH_2)_6CO_2Me$, $C{\equiv}CHex^n$ 66% (57)

Dialkylalkynylaluminiums (prepared by reaction 58)[125] transfer only the alkynyl group

$$\mathrm{RC{\equiv}CH \xrightarrow{BuLi} RC{\equiv}CLi \xrightarrow[-LiCl]{Et_2AlCl} RC{\equiv}CAlEt_2} \quad (58)$$

to enones and 1,4-addition occurs selectively. A typical example is provided by reaction 59; in order to obtain good yields it is essential that excess of alane is present to

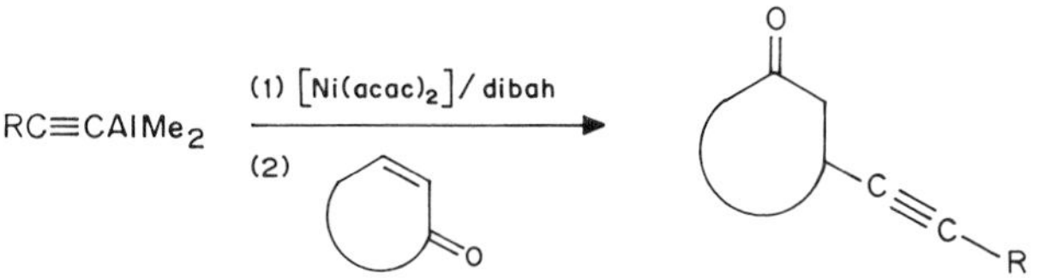

(59)

prevent aldol condensation. Without catalysis acetylenic alanes add 1,4- only to (*S*)-*cis*-enones *via* the cyclic transition state **68**, 1,2-addition occurring with the (*S*)-*trans*

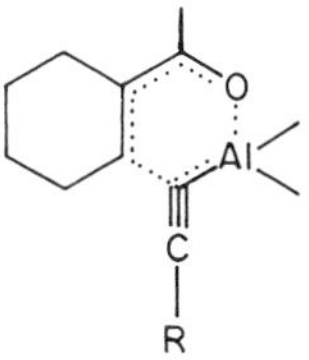

(68)

compounds[127]. A counter example is provided by reaction 60[128], and the uncatalysed reaction of **69** gives both 1,2- and 1,4-addition[129].

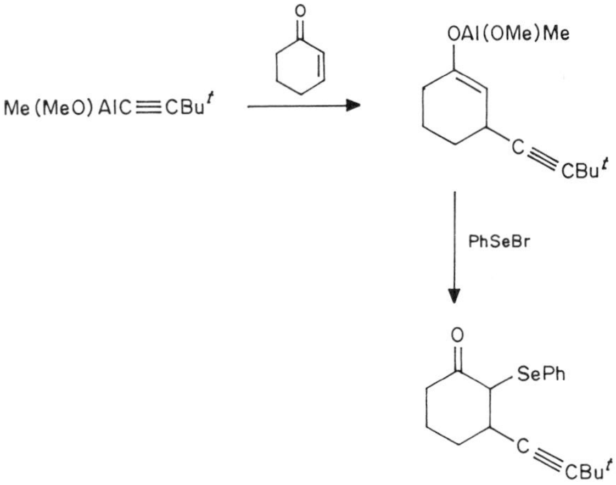

(60)

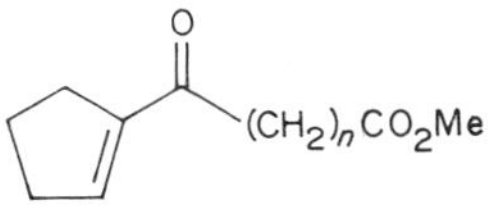

(69)

Cyclopropyl ketones also undergo homoconjugate addition (reactions 61[130] and 62[129]).

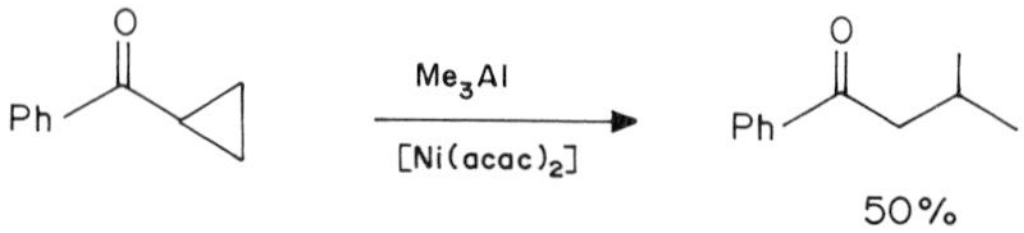

(61)

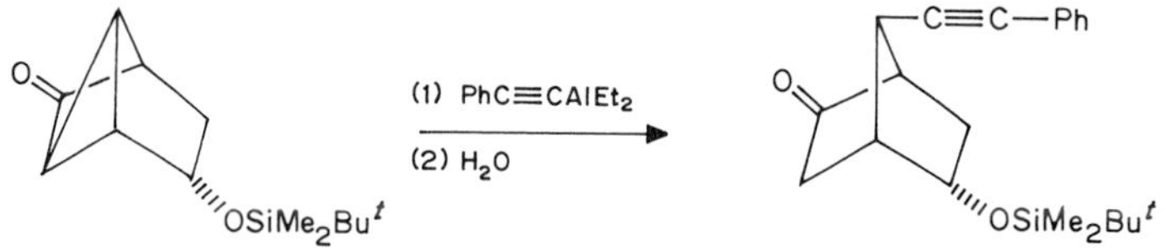

(62)

The theory of hard and soft acids and bases suggests that R_2AlX should add X to enones particularly easily when X is a soft atom such as sulphur or selenium. This is indeed the case, as Figure 13 shows. The initially formed enolate from 1,4-addition gives an aldol with R^4CHO and sulphur is removed oxidatively[131].

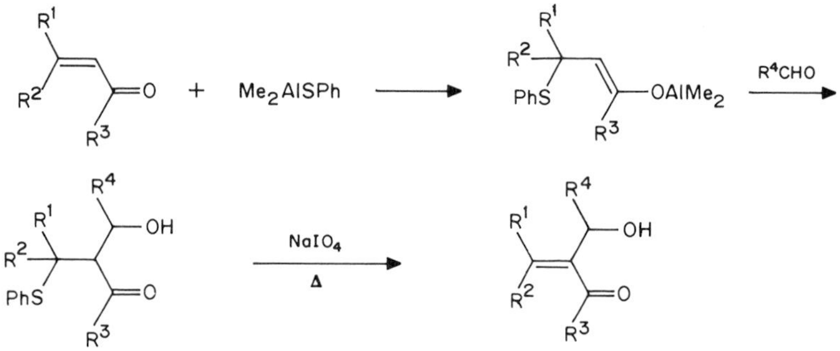

FIGURE 13. Reaction of enones with sulphur-containing alanes

There are many instances of transfer of cyanide to enones from Et_2AlCN, accomplishing hydrocyanation overall, 1,2-Addition to give the cyanohydrin is rapid and reversible and almost quantitative 1,4-addition product may eventually be obtained. Unusual addition specificities are often noted and it is believed that 1,4-addition may also be reversible. For example, cyanide is added *trans* to **70** whereas R_2CuLi gives only *cis*-products (reaction 63)[133]. Also, **71** gives only the *trans*-addition product **72** in 81% yield; KCN gives mainly the *cis*-product (reaction 64)[133]. The hydrocyanation of **73** is the first step in an efficient synthesis of the antitumour antibiotic sarkomycin (reaction 65)[134].

(63)

(70)

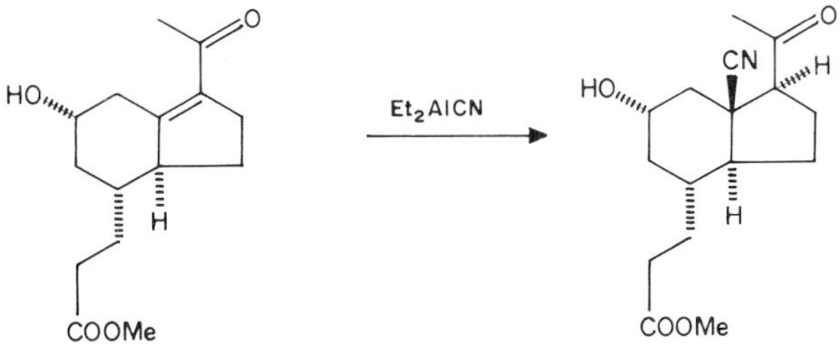

(71) $\xrightarrow{Et_2AlCN}$ (72) (64)

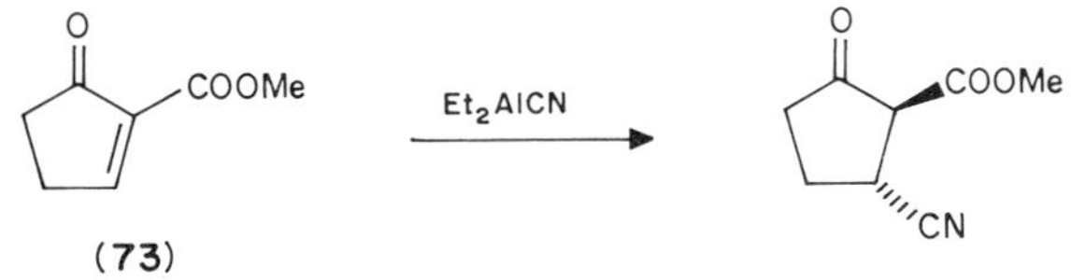

(73) $\xrightarrow{Et_2AlCN}$ (65)

Reduction of enones with $Bu^i{}_3Al$ or dibah gives mainly allylic alcohols[135]. Some asymmetric reductions have been noted using $Al(CH_2CHEtMe)_3$[136]. dibah/Bu^nLi gives 1,2-reduction of cyclic and acyclic enones and dibah/**57** gives stereospecific reduction in reaction 66. Coordination of the aluminium at the C-11 hydroxy group seems essential since the stereoselectivity is lowered when this position is blocked[137]. In the presence of $[Ni(mesal)_2]$, however, $Bu^i{}_3Al$ gives mainly conjugate reduction[138].

$\xrightarrow{\text{dibah / } \mathbf{57}}$ (66)

C. Other Carbonyl Compounds

1. Carboxylic acids and their salts

The initial product from the reaction of R_3Al with carboxylic acids is formation of R_2Al(carboxylate). With excess of organoaluminium compound alkylation occurs and carboxylic acids may be converted into *tert*-butyl groups by Me_3Al[193]. Treatment of **74** with an organoalane formed *in situ* gave **75** in 84% yield (reaction 67)[139]. With

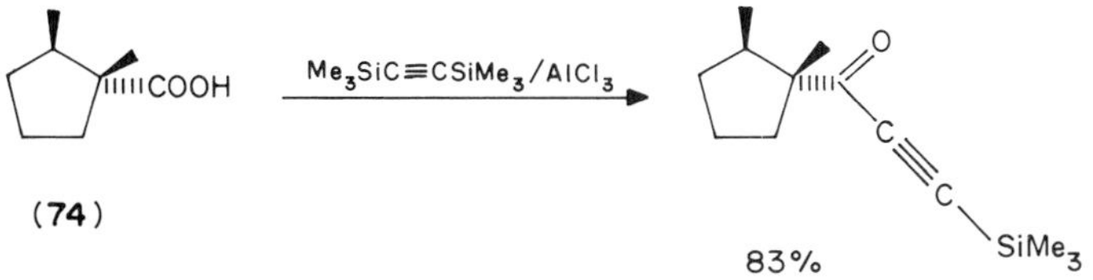

(74) $\xrightarrow{Me_3SiC\equiv CSiMe_3/AlCl_3}$ 83% (67)

allylaluminium compounds *bis* allylic transposition occurs[100]. Reduction of acids to aldehydes using 2 mol of dibah gives reasonable yields[140] but *bis*(*N*-methylpiperazinyl)AlH is better[141]. An alternative procedure *via* **75** is also satisfactory (reaction 68)[142].

RCOOH ⟶ R–C(=O)–S–(2-thiazolinyl) ⟶(dibah) RCHO (68)

(**75**)

2. *Acyl halides*

The reaction of acid chlorides with alkylaluminium dihalides is a standard method for ketone preparation[143]. Acyl halides also react with R_3Al and R_2AlCl_2, but with these species the product ketone is also reactive and overalkylation to alcohols occurs. The mechanism was proposed to involve an acylium ion, but later workers rejected this hypothesis[144].

3. *Esters and lactones*

Esters complex strongly to aluminium but are alkylated or reduced slowly by Et_3Al. Initial addition is followed by reduction of the ketone to a secondary alcohol[115]. Some specific reactions have been used in synthesis. Oxalate esters react with R_3Al to give alkyl 2-oxocarboxylates in 80% yield[146]. Lactones behave much as acyclic esters. As with enones, alkylaluminiums bearing soft heteroatoms are particularly prone to addition reactions. Thio and seleno esters are readily synthesized (reaction 69)[147]. Lactones and

Bu^t–C$_6$H$_{10}$–CO$_2$Me ⟶(MeSeAlMe$_2$) Bu^t–C$_6$H$_{10}$–COSeMe ⟶(R$_2$CuLi) Bu^t–C$_6$H$_{10}$–COR (69)

γ-butyrolactone ⟶($Me_2AlSCH_2CH_2SAlMe_2$) spiro dithiolan (70)

esters are converted to dithiolans using $Me_2AlSCH_2CH_2SAlMe_2$ (reaction 70)[148]. Amides are formed by reaction of esters with Me_2AlNMe_2 (reaction 71)[149] and in refluxing xylene cyanides are obtained[150]. An interesting amidoalane generated *in situ* from **76** cyclizes to give an entry into the clovine alkaloids (reaction 72)[151].

$$RCO_2Me \xrightarrow{Me_2AlNMe_2} \underset{83-93\%}{RCONMe_2} \qquad (71)$$

(1) tfa

(2) Me_3Al

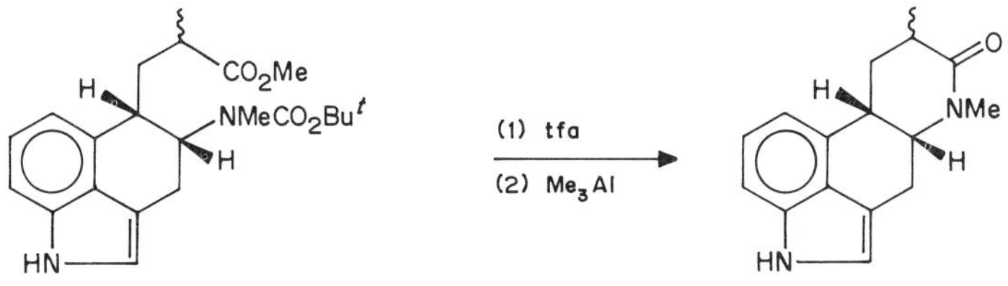

(76)

(72)

The reduction of esters and lactones to alcohols and hemiacetals has been widely exploited[152–154], examples being provided by reactions 73[155] and 74[156]. The reaction is

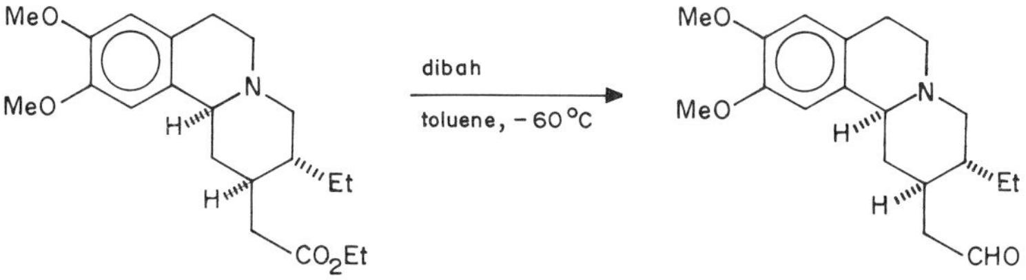

dibah

toluene, −60 °C

(73)

dibah

−70 °C

94%

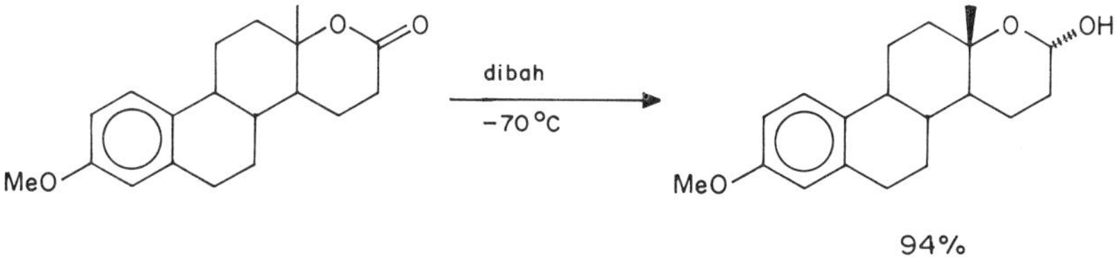

(74)

very selective and the best known examples come from the prostaglandin field (reaction 75)[152]. Ate complexes have been used for the reduction of unsaturated esters of allyl alcohols[137].

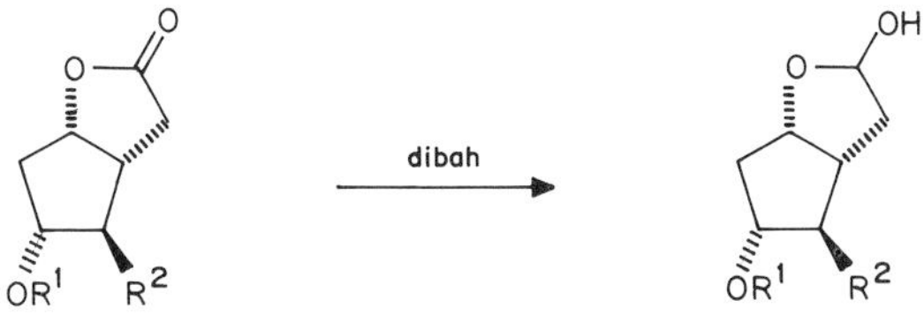

dibah

(75)

D. Bimetallic Reagents

The Tebbe reagent, $[Cp_2Ti\text{-}\mu\text{-}Cl\text{-}\mu\text{-}CH_2AlMe_2]$, **77**, prepared from $[Cp_2TiCl_2]$ and Me_3Al[157], has been widely used for methylenation of ketones and esters[158]. Some

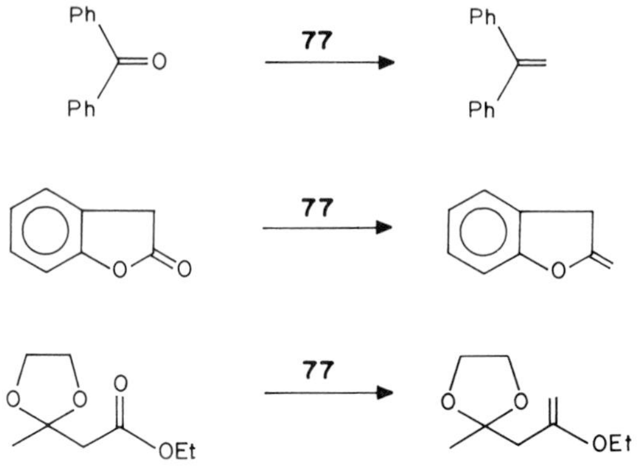

Ref.

$Ph_2C{=}O \xrightarrow{\mathbf{77}} Ph_2C{=}CH_2$ 157

159

159

FIGURE 14. Methylenation of carbonyl groups using the Tebbe reagent

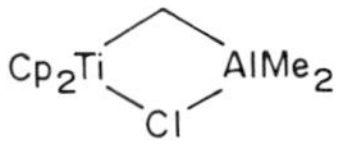

$Cp_2Ti(\mu\text{-}CH_2)(\mu\text{-}Cl)AlMe_2$

(**77**)

examples are shown in Figure 14. With acyl halides the expected Wittig product is not obtained but ketones are formed in moderate yields (reaction 76)[160]. It is suggested

$$RCOX + [Cp_2TiCl_2] + 2Me_3Al \longrightarrow RCOCH_3 \qquad (76)$$

that the reaction proceeds *via* the titanium enolate **78**, which loses halide and gives the ketone on protonation (reaction 77). A convenient synthesis of 1,5-enones from allyl

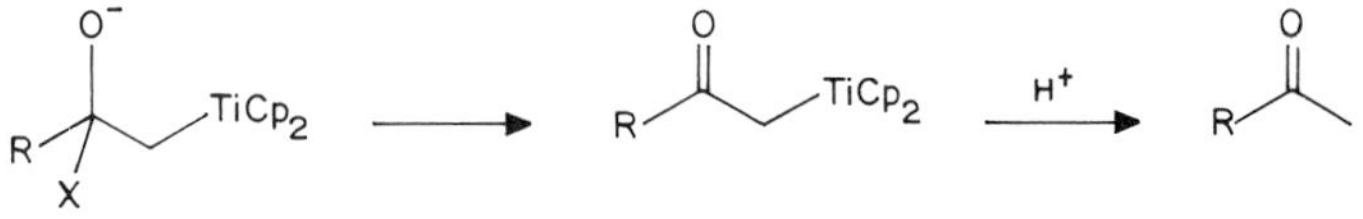

(77)

acetates uses the Tebbe reagent; methylenation is followed by a [3,3]-sigmatropic shift to give the 1,5-unsaturated carbonyl compounds (reaction 78)[161]. The Tebbe reagent is

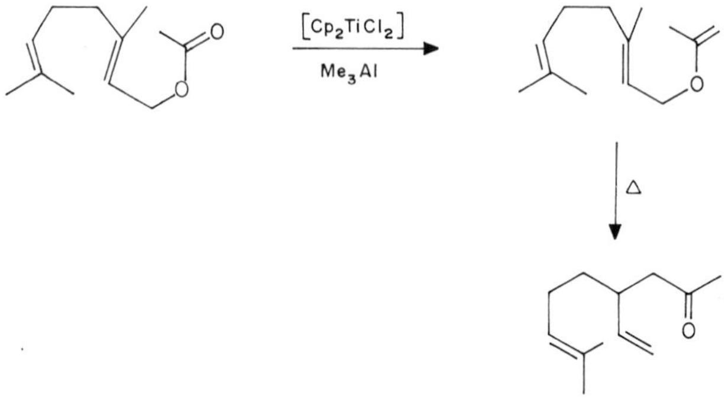

$[Cp_2TiCl_2]$

Me_3Al

△

(78)

tedious to prepare and, since it requires the use of Me_3Al with its associated hazards, there is considerable interest in alternatives. The zinc complex $[Cp_2Ti{=}CH_2 \cdot ZnI_2]$ is easy to prepare from zinc dust, CH_2I_2 and $[Cp_2TiCl_2]$. This effects methylenation of ketones but is generally less versatile than **77**[162].

V. REACTIONS AT CARBON—OXYGEN SIGMA-BONDS

A. Epoxides

Organoaluminium compounds open epoxides cleanly and easily and with clearly defined *trans*-stereochemistry. Reaction of vinylalanes with oxirane yields homoallylic

$$RCH{=}CHAlR_2 \xrightarrow[(2)\ H_2O]{(1)\ \text{oxirane}} RCH{=}CHCH_2CH_2OH \qquad (79)$$

alcohols according to equation 79[163], but in weakly basic solvents for R = Et, **79** is also formed. Ate complexes give better yields.

(79)

The reaction of propylene oxide with Me_3Al and Et_3Al depends on stoichiometry. With excess of R_3Al the secondary carbon is alkylated by reaction of R_3Al with the aluminium epoxide complex **80** (reaction 80). When there is less than 2 molar equivalents of R_3Al the initial alkylation product, **81**, removes R_3Al from **80** to give free propene oxide and **82**. The reaction of **82** with propene oxide occurs at the primary carbon yielding **83**. Uncomplexed oxiranes alkylate mainly at the less substituted carbon atom. Benzyloxirane reacts similarly (reaction 81)[164]. The clean inversion of stereochemistry obtained in these reactions has recently become more significant because of the accessibility of chiral epoxides *via* the Sharpless asymmetric epoxidation procedure[165].

Et_3Al + **(80)** (propylene oxide·$AlEt_3$) ⟶ Et–CH(Me)–CH₂–$OAlEt_2$ **(81)** + Et_3Al

propylene oxide + Et–CH(Me)–CH₂–$OAlEt_2$·$AlEt_3$ **(82)** ⟶ Et–CH₂CH(Me)... **(83)** ($OAlEt_2$) + **81** (80)

$$PhCH_2CH\overset{O}{—}CH_2 + 2Me_3Al \longrightarrow PhCH_2CH(Me)CH_2OH \qquad (81)$$

Recent focus has been on the generation of chemically complex structures by such alkylations and numerous examples exist. Me_3Al opens **84** (reaction 82)[166] and **85** is attacked only at the 3-position to give an aldehyde after decomposition of the cyanohydrin (reaction 83)[167]. Treatment with organoaluminiums was found to give a facile reaction

$$\xrightarrow{Me_3Al} \qquad (82)$$

(84)

$$\xrightarrow[(2)\ H_3O^+]{(1)\ Me_3Al} Bu^tCHO \qquad (83)$$

(85)

with epoxy alcohols generated from (*E*)-allylic alcohols **86**, but (*Z*)-compounds, **87**, react sluggishly. Interaction with the alcohol is thought to be important. The ate complex from $Et_3Al/EtLi$ gives alkylation of steroid epoxides but competing reduction is serious[168].

(86) **(87)**

Vinylalanes have received little attention (reaction 84)[78] and allyl- and propargylalanes give only reduction[100]. However, the reaction of alkynylalanes with epoxides is of

$$R^1C{\equiv}CH \xrightarrow[{[Cp_2ZrCl_2]}]{Me_3Al} R^1(Me)C{=}CH(AlMe_2) \xrightarrow[(2)\ R^2CH\overset{O}{-}CH_2]{(1)\ Bu^nLi} R^1(Me)C{=}CHCH_2CH(OH)R^2 \qquad (84)$$

considerable significance because epoxides are highly resistant to attack by alkynyl Grignard and lithium reagents. The process is stereospecific and regiochemical control is high. Most of the available examples are from the prostaglandin field. For example, **88** is allowed to react with diethylaluminium 8-tetrahydropyranyloxyhept-1-yn-1-ide **89** and the product converted into **90** in several steps (reaction 85). Compound **90** reacts with another alkynylalane to give an intermediate convertible in two steps into **91**, prostaglandin $F_{1\alpha}$-alcohol (reaction 86)[169]. Other examples are given in Figure 15, on page 442.

$$\mathbf{(88)} + Et_2AlC{\equiv}C(CH_2)_5Othp \longrightarrow\longrightarrow \mathbf{(90)} \qquad (85)$$

(88) **(89)** **(90)**

90 + $Me_2AlC{\equiv}CCH(OBu^t)(CH_2)_4Me$ → → →

(91) (92) (86)

Epoxides are resistant to attack by the classical cyanation reagent composed of KCN/NH_4Cl, and Et_2AlCN has been widely investigated as a reagent for opening the three-membered ring[174]. Cleavage to the *trans*-hydroxynitrile takes place under mild conditions, usually giving the more substituted nitrile[175]. The route has recently been used in the synthesis of 3-cyano-3-deoxy-D-galactopyranose (reaction 87)[176]. The reagent

(1) Et_2AlCN / benzene; (2) H_2O (87)

formed from Me_3SiCN and Et_2AlCl has been used to attack several epoxides and oxetanes[177]. Yields are variable but the selectivity for the less hindered site is high in all cases, as exemplified by reactions 88 and 89.

Me_3SiCN / Et_2AlCl (88)

Me_3SiCN / Et_2AlCl (89)

Dialkylaluminium amides add to epoxides to give amino alcohols[178] and cause isomerization to allyl alcohols (reaction 90)[179]. The reactivity pattern depends on initial

Et_2AlNR_2 (90)

aluminium coordination to oxygen. Diethylaluminium 2, 2, 6, 6-tetramethylpiperidide (datmp) is the reagent of choice for such isomerizations[180], usually operating with high regiospecificity; compare reactions 91 and 92[181]. Oxetanes are opened to homoallylic

datmp

(93) (91)

Ref.

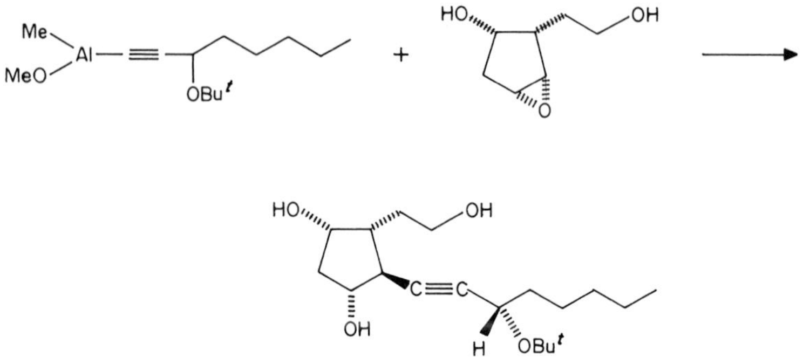

170

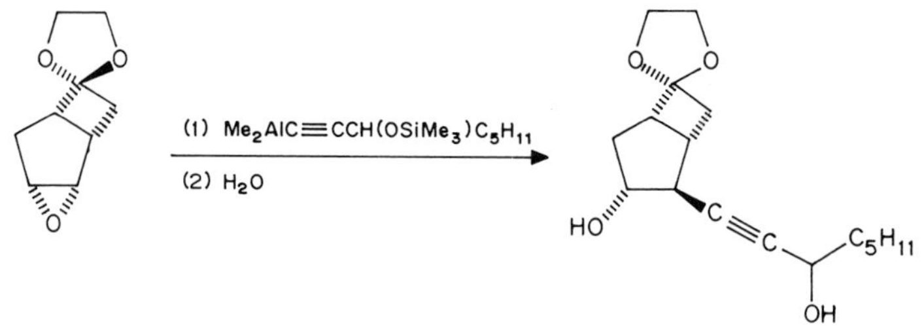

171

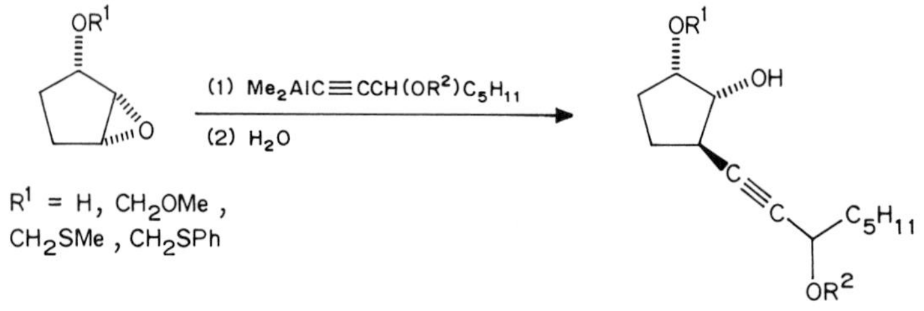

172

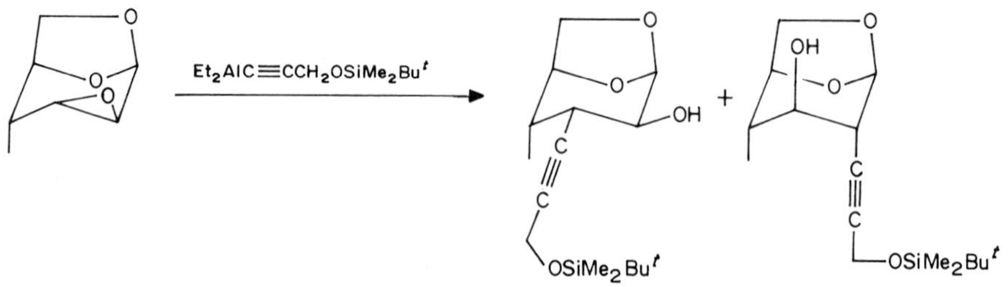

173

FIGURE 15. Reaction of alkynylalanes with epoxides

datmp

+ 12% **93**

(92)

datmp

(93)

alcohols (reaction 93)[182]. Both dibah and $Bu^i{}_3Al$ induce similar isomerizations (reactions 94)[152] and $(Pr^iO)_3Al$ has also recently been used (reaction 95)[79].

$Bu^i{}_3Al$

(94)

$(Pr^iO)_3Al$

(95)

Reduction of oxiranes to alcohols occurs with both dibah and $Bu^i{}_3Al$, although the reaction is slower than that for ketones. For example, **94** is reduced with good stereospecificity to give the alcohol **95** (reaction 96) in 99% yield[183]. The products of

$Bu^i{}_3Al$

(96)

99%

(94) **(95)**

epoxide reduction are not always easy to predict and $Bu^i{}_3Al$ may also give addition[184]. Again, the availability of chiral epoxides has provided a route to stereocontrolled synthesis. The polyol **96** is made by a series of four steps which may be repeated (Figure 16)[186].

When a polar group is available to interact with the aluminium compound, the regiochemistry of reduction may be well defined. For example, **97** gives a mixture of products with $LiAlH_4$, **98** only with Red-al and mainly **99** with dibah, the regiochemistry being solvent dependent (reaction 97)[187].

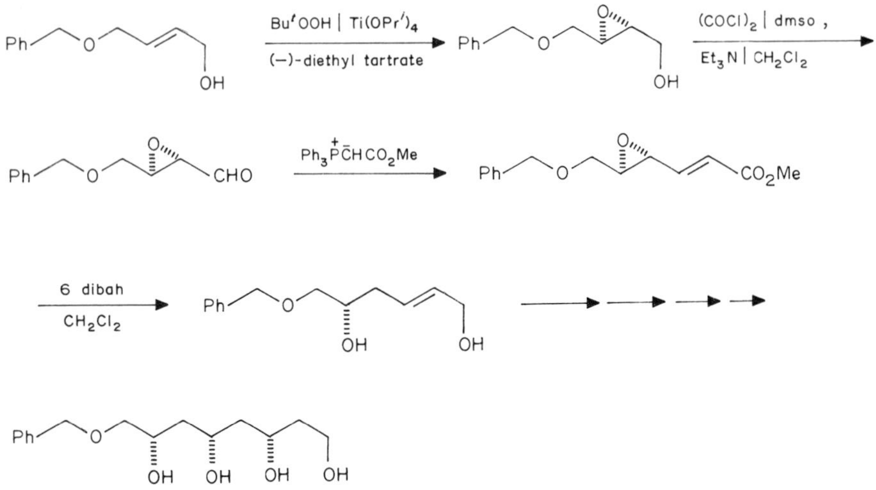

FIGURE 16. Asymmetric synthesis of polyols

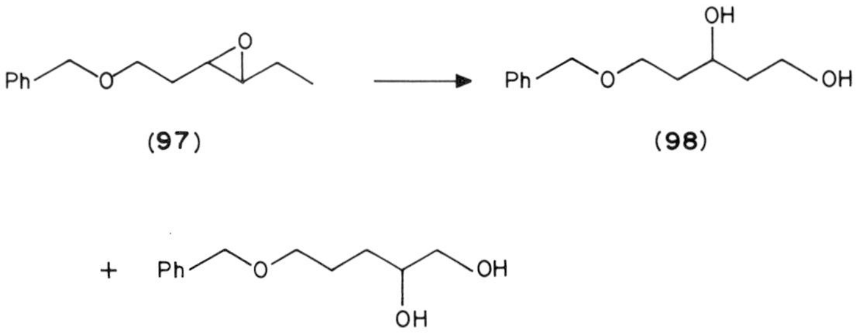

(97)

Vinyl epoxides provide a special case. Conjugate reduction often occurs with dibah (for example, reaction 98)[188]. In hexane this is thought to proceed *via* an oxirane–dibah

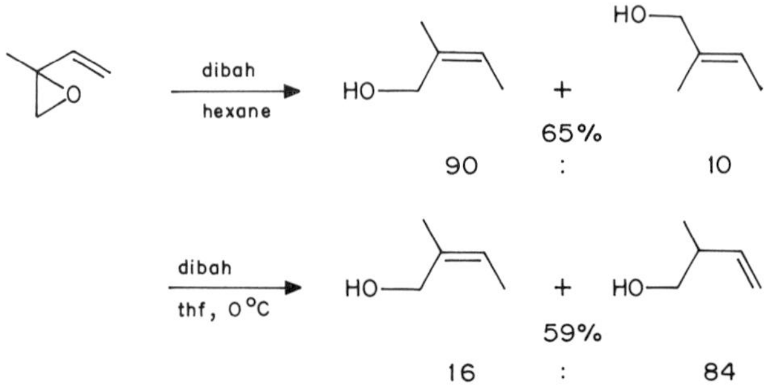

(98)

complex which transfers hydride to the double bond in a cyclic transition state. Conjugate addition is also known. For example, **100** reacts with $Me_3Al/MeCu$ to give the *anti*-

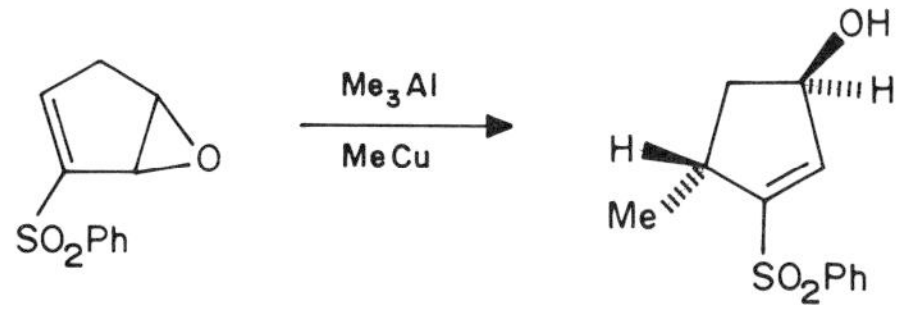

(99)

(100) (101)

product **101**, essentially uncontaminated by the *syn*-isomer (reaction 99). Pure *syn*-isomer is obtained with $MeLi/LiClO_4$[189]. Addition of Et_2AlSPh also proceeds in a conjugate manner (reaction 100)[190].

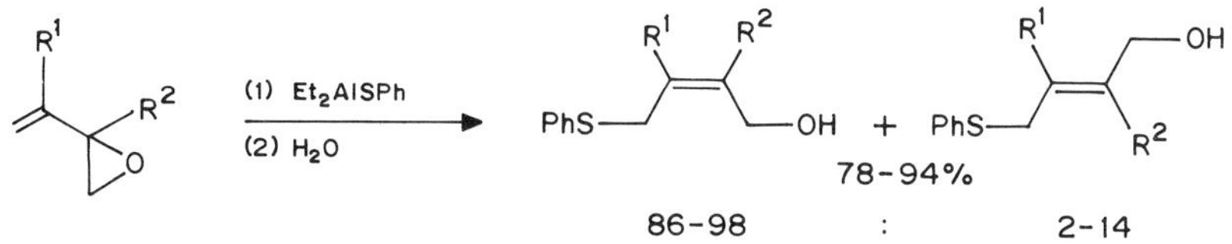

78–94%

86–98 : 2–14

(100)

B. Other Compounds Containing Carbon—Oxygen Sigma-Bonds

Cleavage of simple ethers occurs in the presence of dibah but has been little used except for the demethylation of aromatic steroidal ethers to phenols[191]. Vinyl ethers react according to equation 101 with retention of configuration in chiral R.

$$CH_2{=}CHOR + Bu^i{}_2AlH \longrightarrow Bu^i{}_2AlCH_2CH_2OR \longrightarrow C_2H_4 + ROAlBu^i{}_2 \quad (101)$$

Allyl ethers, **102**, may, after deprotonation with Bu^nLi, give reaction with electrophiles at either the α- or the γ-position; benzaldehyde gives an $\alpha{:}\gamma$ product ratio of 28:72. However, in the presence of Et_3Al the intermediate, **103**, is produced and α-selectivity to **104** is greater than 99% (reaction 102)[193]. Detritylation of **105** occurs in the presence of

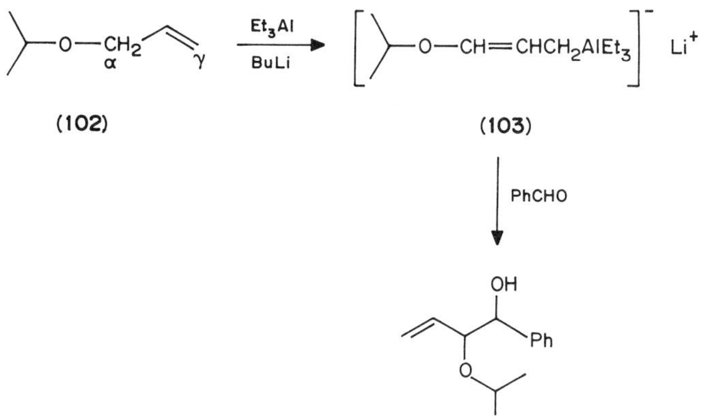

(102) (103)

(102)

(104)

Et_2AlCl or $Bu^i{}_2AlCl$ and does not suffer from the depurination which often accompanies simple protolysis[194].

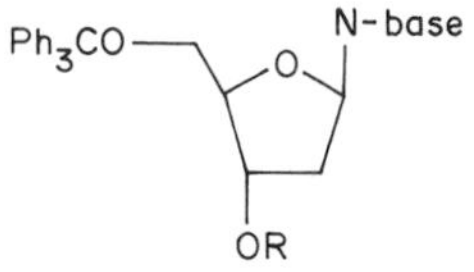

Ph3CO— N-base OR

(105)

The reaction of alkylaluminiums at the carbon—oxygen bond of esters has been studied with a wide range of compounds. Simple esters, such as **106**, are alkylated with more or less inversion *via* carbocation-like intermediates[195], and **107** is detosylated with dibah[196]. Allyl

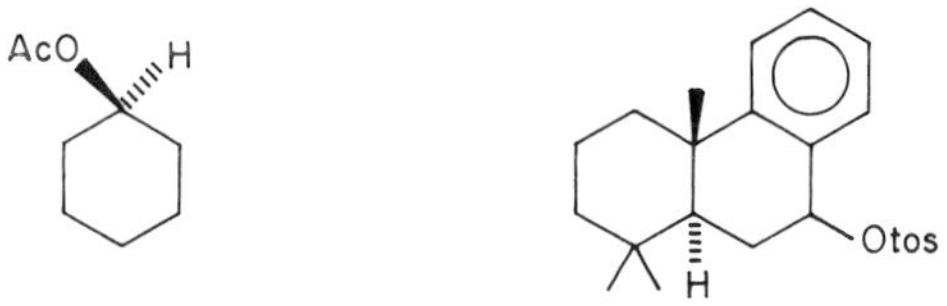

AcO H Otos H

(106) (107)

esters have, however, been more popular substrates. Compound **108** reacts with R_3Al to give **109** and **110** (reaction 103); **109** is the major product at $-78\,°C$ but **110** is also formed

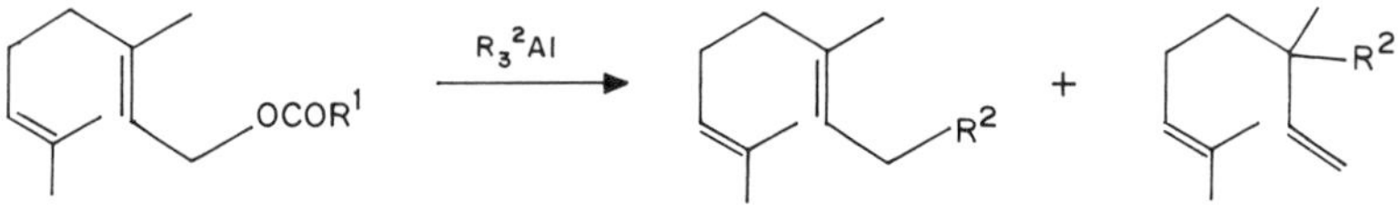

$$\text{(108)} \xrightarrow{R^2_3Al} \text{(109)} + \text{(110)} \quad (103)$$

OCOR1 R2 R2

in 20–30% yield at 0 °C. Palladium(0) complexes are particularly effective catalysts for this reaction, the intermediates being η^3-allyl complexes[197]. For example, **111** is converted into

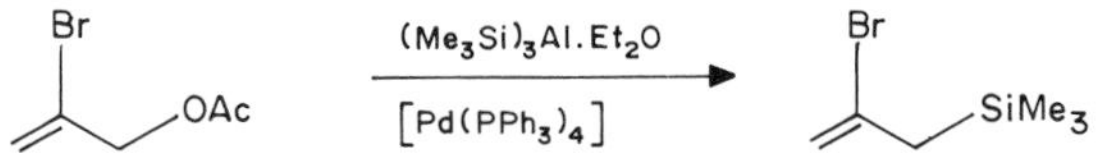

$$\text{(111)} \xrightarrow[\text{[Pd(PPh}_3)_4]]{(Me_3Si)_3Al.Et_2O} \text{(112)} \quad (104)$$

Br OAc Br SiMe3

112 in 87% yield (reaction 104)[198]. The reaction proceeds with inversion of configuration at the reacting centre as shown by reaction 105[199].

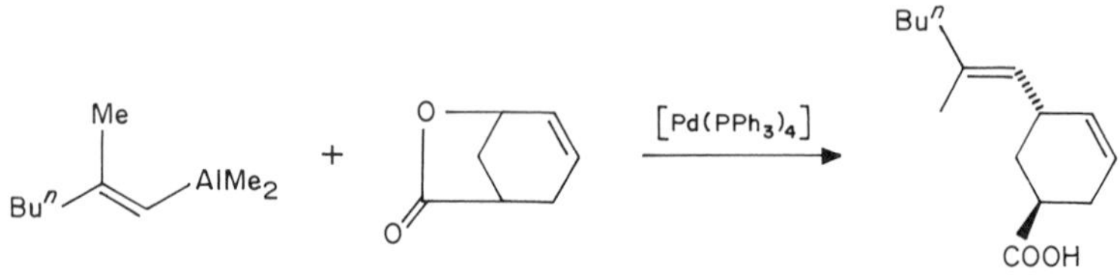

$$\text{Bu}^n\text{C(Me)=CHAlMe}_2 + \text{lactone} \xrightarrow{[Pd(PPh_3)_4]} \text{product} \quad (105)$$

Me Bun AlMe2 O O Bun COOH

The propargyl compound, **113**, is converted into **114** and the free 1,4-diyne obtained by treatment with ammonium cerium(IV) nitrate (reaction 106)[200]. This method is said to be superior to the copper-catalysed routes to 1,4-diynes, since here no conjugation occurs.

$$\left[HC{\equiv}C{-}C(R^1)(R^2){-}OAc \;\; (Co_2(CO)_6) \right] \xrightarrow{(R^3C{\equiv}C)_3Al} \left[HC{\equiv}C{-}C(R^1)(R^2){-}C{\equiv}C{-}R^3 \;\; (Co_2(CO)_6) \right] \tag{106}$$

(**113**) (**114**)

Whilst a few cyclopropyl methyl esters may be alkylated without ring opening (reaction 107)[195,201], conjugate addition (reaction 108) is more common. Compound **116** is obtained (reaction 109) from both stereoisomers of **115**, suggesting a carbocation

$$\text{(bicyclo[4.1.0]heptyl)}{-}CH(OCOR^1){-}Hex^n \xrightarrow{R^2_3Al} \text{(bicyclo[4.1.0]heptyl)}{-}CH(R^2){-}Hex^n \tag{107}$$

R^2 = Me, Et

$$\text{(1-phenylbicyclo[4.1.0]heptyl)}{-}CH_2OAc \xrightarrow{Me_3Al} \text{2-methyl-2-phenyl-1-allylcyclohexane} \tag{108}$$

$$Ph{\sim}\text{(cyclopropyl)}{-}CH(OAc){-}CH_2R^1 \xrightarrow{R^2_3Al} PhCH(R^2){-}CH_2CH_2CH{=}CH{-}R^1 \tag{109}$$

(**115**) (**116**)

R^1 = H, Me

R^2 = Me, Et, Bu^i

intermediate. Using the cyclopropane **117**, alkylation occurs predominantly at the cyclopropyl carbon atom (reaction 110).

$$\text{(117)} \xrightarrow{Me_3Al} \text{branched diene–Ph (83)} + \text{linear diene–Ph (17)} \tag{110}$$

(**117**) 83 17

Allyl phosphates such as **118** may be substituted by Me_2AlX (X = OPh, SPh, NHPh) in hexane, with clean inversion of configuration. In more polar solvents the stereochemical purity of the product is lower and γ-attack is increased[202]. With the geranyl derivative **119**,

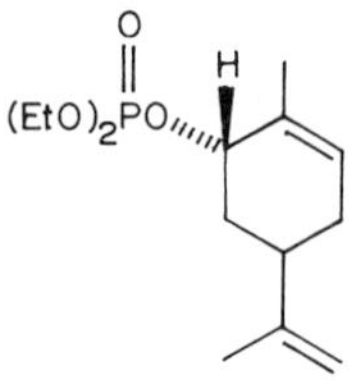

(118)

both α- and γ-attack (reaction 111) occur but with the neryl compound **120**, the only reaction is cyclization (reaction 112)[203].

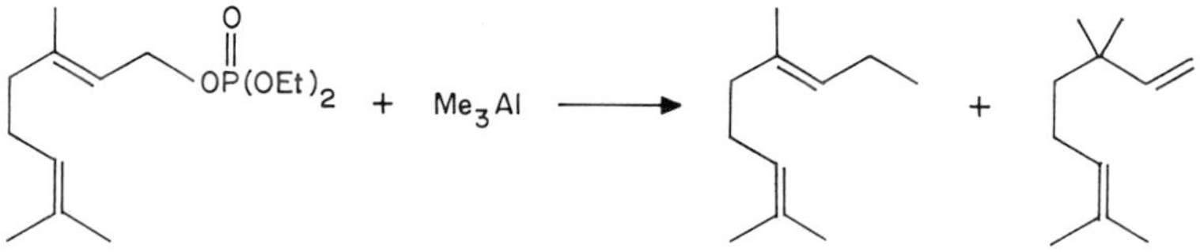

(119)

(111)

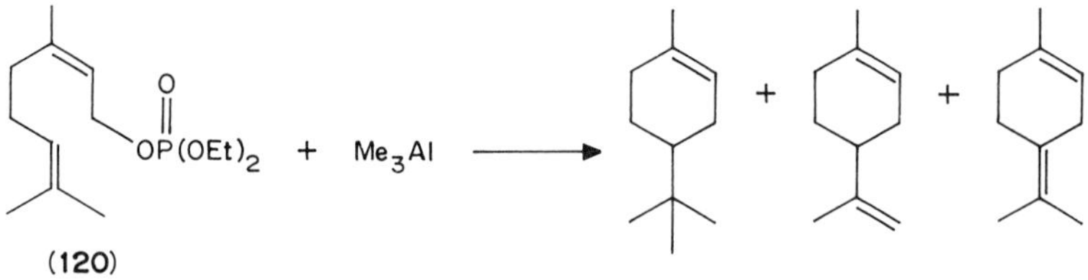

(120)

(112)

Substitution of vinyl phosphates occurs only in the presence of palladium(0) catalysts and there are numerous examples of the synthetic utility of this procedure, which occurs with essentially complete stereoselectivity (reactions 113 and 114)[204]. Vinyl- and

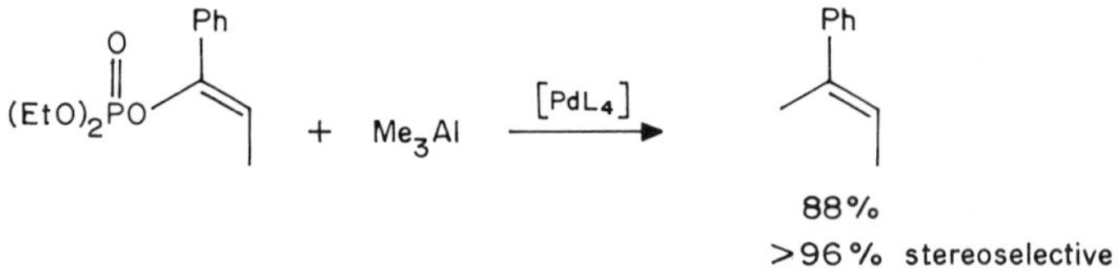

(113)

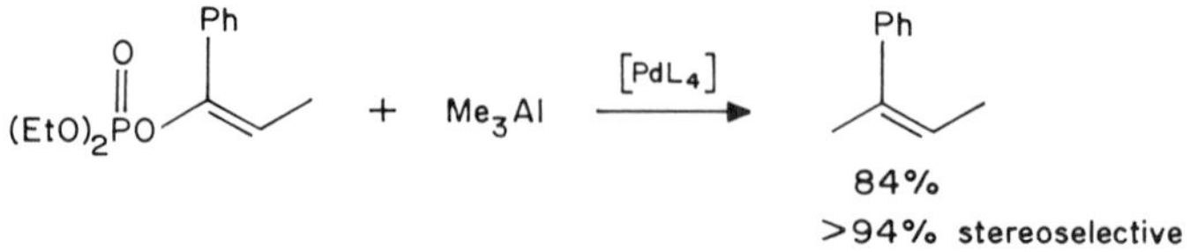

88%
>96% stereoselective

(114)

84%
>94% stereoselective

propargyl-alanes also react satisfactorily and $PhMe_2SiAlEt_2$ transfers only the silyl substituent to give vinylsilanes[205]. Sulphur substituents on the double bond do not interfere with the reaction[206], and this has been used in a route for alkylation/

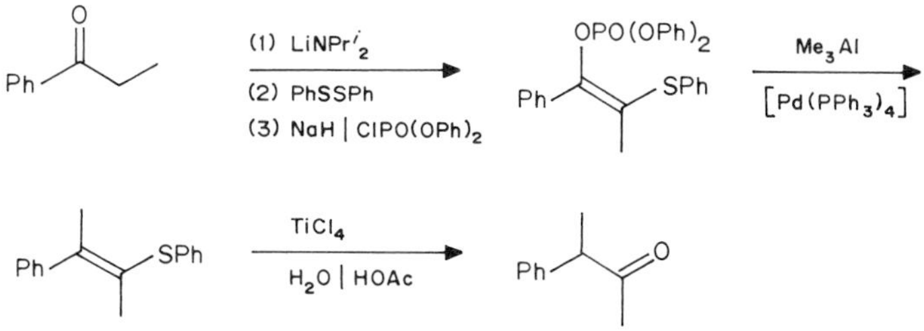

FIGURE 17. Alkylation and 1, 2-transposition of a ketone

1, 2-transposition of ketones (Figure 17)[207]. Aryl phosphates have been substituted in quantitative yields using nickel complexes as catalysts[208].

Acetals react with allylalanes with allylic transposition and displacement of one OR group (reaction 115) and ortho-esters give *bis*-allylated ethers, also in fair yields (equation 116)[100]. Propargylalanes give an analogous reaction, although isomerization to allenyl products is known (equation 117) and allenylalanes are also reactive.

$$\mathrm{RCH{=}CHR^1CH_2Al} \xrightarrow[\text{(2) } \mathrm{H_2O}]{\text{(1) } \mathrm{R^2R^3C(OR^4)_2}} \underset{80\text{–}90\%}{\mathrm{R^2R^3C(OR^4)CHRCHR^1{=}CH_2}} \quad (115)$$

$$\mathrm{RCH{=}CHR^1CH_2Al + HC(OEt)(OPh)_2} \xrightarrow{-50^\circ\mathrm{C}} \mathrm{EtOCH(CHRCHR^1{=}CH_2)_2} \quad (116)$$

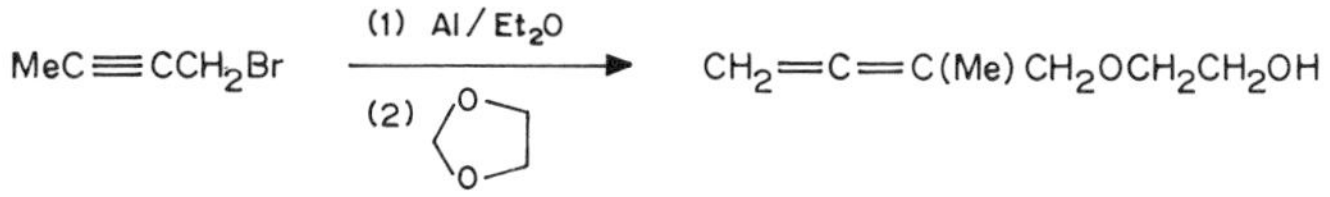

$$\mathrm{MeC{\equiv}CCH_2Br} \xrightarrow[\text{(2) 1,3-dioxolane}]{\text{(1) } \mathrm{Al/Et_2O}} \mathrm{CH_2{=}C{=}C(Me)\,CH_2OCH_2CH_2OH} \quad (117)$$

Reaction with the chiral cyclic acetal **121** gives mainly 1, 4-addition in dichloroethane and 1, 2-addition in chloroform. In both cases the optical yields after hydrolysis are excellent (reaction 118)[209].

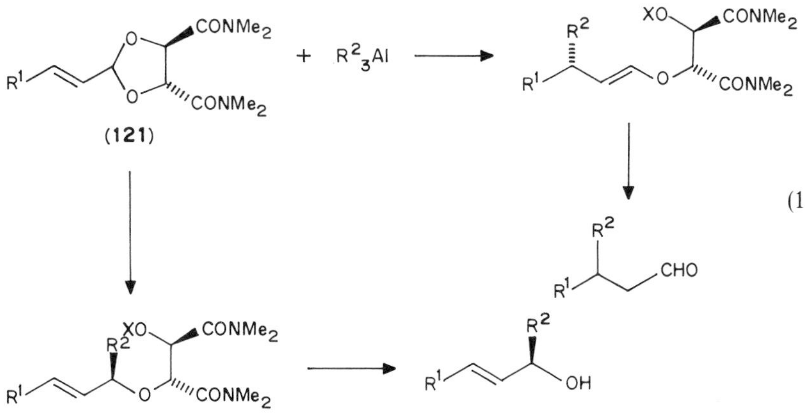

(121) (118)

Finally, alcohols react with R_3Al under forcing conditions (reaction 119)[210], and this has been used in an amide synthesis (reaction 120)[211].

$$R^1R^2R^3COH + Me_3Al \longrightarrow R^1R^2R^3CMe \tag{119}$$

$$RCONH_2 + RCHO \longrightarrow RCONHCH_2OH \xrightarrow{Me_3Al} RCONHCH_2Me \tag{120}$$

VI. REACTIONS WITH HALIDES

A. Alkyl Halides

Few of the reactions of triorganoaluminiums with primary alkyl halides are preparatively useful, but their consideration is important in synthetic planning. Primary alkyl chlorides react rapidly to give a mixture of products and halide exchange is common[212]. Tertiary halides may be converted into quaternary compounds in fair yields according to equation 121[213]. The reactions with other trialkylaluminium compounds, where elimination is possible, are less satisfactory. Reduction of primary alkyl bromides occurs in the presence of the ate complex from dibah/Bu^nLi[137].

$$Me_3CCl + Me_3Al \longrightarrow Me_3C^+ClAlMe_3^- \longrightarrow C(Me)_4 + Me_2AlCl \tag{121}$$

B. Benzyl Halides

The reaction of benzyl chloride with Me_3Al is unsatisfactory, giving mainly polybenzyl, but $Et_3Al \cdot Et_2O$ gives 1-phenylpropane in good yield[214]. A carbocation mechanism is proposed for these uncatalysed reactions. Vinylalanes react with benzyl halides successfully and stereospecifically in the presence of $[Pd(PPh_3)_4]$ (reaction 122)[215].

Hex^n–C(Me)=CH–AlMe$_2$ $\xrightarrow[\text{[Pd(PPh}_3)_4]]{\text{PhCH}_2\text{Br}}$ Hex^n–C(Me)=CH–CH$_2$Ph (93%) (122)

C. Aryl Halides

Uncatalysed couplings of aryl halides are unknown, but in the presence of $[Ni(PPh_3)_4]$ vinylalanes react with **122** to give **123** with excellent stereospecificity (reaction 123)[216].

1-bromonaphthalene **(122)** + Bu^{n}CH=CH–AlBu$^i_2$ $\xrightarrow{\text{[Ni(PPh}_3)_4]}$ 1-(CH=CHBu^n)naphthalene **(123)** (123)

D. Vinyl Halides

Uncatalysed reactions with this type of halide are known but are not preparatively useful[217]. Numerous catalysed reactions have, however, been successfully exploited.

Alkenyldiisobutylaluminiums have been the most popular reagents and both palladium and nickel complexes have been used as catalysts. Homocoupled products are usually limited to less than 15%. Retention of configuration in the halide component is usually better with palladium than with nickel; **124** is 95% *E, E* with $[NiL_n]$ but better than 99% *E, E* with $[PdL_n]$ (reaction 124)[218]. The procedure is not completely successful with more

C_5H_{11}–CH=CH–$AlBu^i_2$ + Bu^n–CH=CH–I $\xrightarrow{ML_n}$ C_5H_{11}–CH=CH–CH=CH–Bu^n (**124**) (124)

$Me_3Al/[Cp_2ZrCl_2]$ → (**125**) (AlMe₂ alane); then CH₂=CHBr, $ZnCl_2/[(Ph_3P)_2PdCl_2]$ → triene, 98% E (125)

substituted alanes such as **125**, derived from carbometallation; $ZnCl_2$ is needed as a cocatalyst (reaction 125)[53]. Allenyl halides such as $CH_2{=}C{=}CHBr$ have been coupled with alkylaluminiums in the presence of Cu_2Cl_2[28].

E. Allyl Halides

Reaction of 3-chloroprop-1-ene with Me_3Al is slow without a catalyst, giving but-1-ene[23,214]. The ate complex $Li[AlHex^n_4]$ reacts more rapidly (reaction 126) but preparative uses have been few[23]. Again, catalysed reactions are more important. Cu_2Cl_2 catalyses

$$E\text{—}MeCH{=}CHCH_2Cl \xrightarrow{Li[AlHex^n_4]} CH_2{=}CHCH(Me)Hex^n + Et_2MeCH{=}CHC_7H_{15} \quad (126)$$

reaction 127[219] but with substituted allyl halides some allylic transposition occurs (reaction 128)[220].

R^1–CH=C(R^2)–AlR^3_2 $\xrightarrow[Cu_2Cl_2]{CH_2=CHCH_2Br}$ R^1–CH=C(R^2)–CH₂CH=CH₂ (127)

Ph–CH=CH–CH₂Cl + R_3Al $\xrightarrow[PPh_3]{[Cu(acac)_2]}$ Ph–CH=CH–CH₂R + Ph–CH(R)–CH=CH₂ (128)

R = Me	59 :	41
R = Et	24 :	76

Allyl halides in the presence of $[Pd(PPh_3)_4]$ give excellent results *via* palladium–allyl complexes, as the synthesis of α-farnesene shows (reaction 129)[221].

$$HC{\equiv}CCH{=}CH_2 \xrightarrow[{[Cp_2ZrCl_2]}]{Me_3Al} \text{(dienyl)}AlMe_2 \xrightarrow[{[Pd(PPh_3)_4]}]{\text{Cl-(geranyl)}} \text{product} \quad 86\% \qquad (129)$$

VII. REACTIONS WITH NITROGEN-CONTAINING COMPOUNDS

A. Nitriles

Nitriles form strong complexes with organoaluminium compounds and may, like carbonyls, be either reduced or alkylated. For nitriles with α-hydrogens metallation is the principal reaction, whereas with organoaluminiums with a β-hydrogen reduction to an aldimine predominates.

Alkylation of benzonitrile to give, after hydrolysis, an ethyl ketone, requires 2 mol of Et_3Al and a cyclic transition state, **126**, is proposed. Isocyanates, RN=C=O, are

Ph—C≡N ··· AlEt₂ ··· CH₂(Me) ··· Al(Et₂) ··· CH₂(Me)

(**126**)

similarly converted into amides, RNHCOEt, in good yield[222]. $[Ni(acac)_2]$ has been used to catalyse the transformation shown in equation 130[223], but otherwise vigorous conditions are necessary (equation 131)[224].

$$PhCH_2CN + Me_3Al \xrightarrow{[Ni(acac)_2]} PhCH_2C(Me){=}NAlMe_2 \xrightarrow{H_3O^+} \underset{70\%}{PhCH_2COMe} \qquad (130)$$

$$Me_3CCN + Me_3Al \xrightarrow{150\,^{\circ}C} \underset{95\%}{Me_3CC(Me){=}NAlMe_2} \qquad (131)$$

B. Ammonia and Amines

Ammonia and primary and secondary amines react with trialkylaluminiums to protonate one of the alkyl groups and yield $R_2AlNR^1{}_2$. A number of such species have already been encountered as synthetic intermediates. Their structures have been determined and are characteristically oligomeric[239]. They are used for the conversions of esters to amides (reaction 132)[228] and have been implicated in an interesting reductive alkylation of **127** (reaction 133)[229].

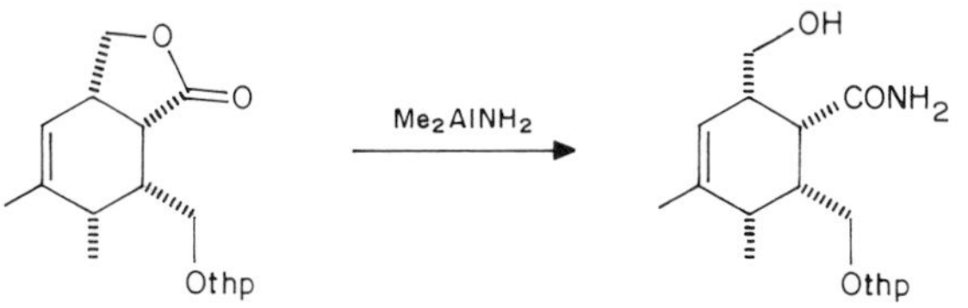

(132)

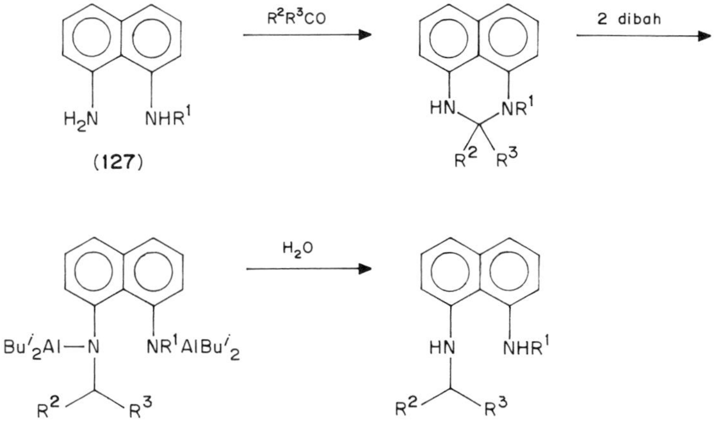

(133)

C. Imines

Imines form complexes with organoaluminium compounds and are slowly reduced to secondary amines (reaction 134)[230]. A special example is provided by reaction 135[231].

$$\text{PhCH{=}NR} \xrightarrow[\text{Bu}^i{}_3\text{Al}]{\text{Me}_3\text{Al or}} \text{PhCH}_2\text{NHR} \qquad (134)$$

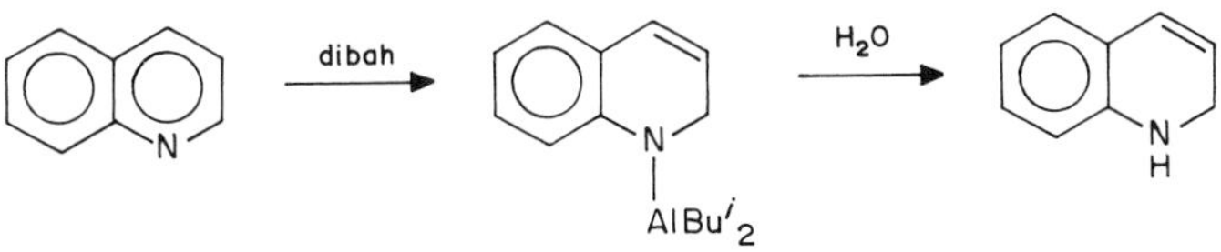

(135)

Alkylations are also known[232,233]. Metallation of the imine complexes can give, after appropriate work-up, reduction, alkylation, and condensation (reactions 136–138). It is clear from the results of alkylation that there is some charge delocalization in the intermediates[234].

$$\text{Et}_2\text{AlN{=}CHPh} \xrightarrow[\text{C}_6\text{H}_6]{\text{Na,}} \text{Et}_2\text{AlN(Na)CHPhCHPhN(Na)AlEt}_2$$

$$\xrightarrow{\text{H}_2\text{O}} \text{PhCH(NH}_2\text{)CH(NH}_2\text{)Ph} \qquad (136)$$

$$\text{Et}_2\text{AlN{=}CPh}_2 \xrightarrow{2\text{K}} \text{Et}_2\text{AlN(K)C(K)Ph}_2 \xrightarrow{\text{H}_2\text{O}} \text{Ph}_2\text{CHNH}_2 \qquad (137)$$

$$Et_2AlN(K)C(K)Ph_2 \xrightarrow{Me_2SO_4} Ph_2C(Me)NH_2 + PhC(Me)NHMe + Ph_2CHNHMe + MePhCH_2NH_2 \quad (138)$$

VIII. ALUMINIUM-PROMOTED REARRANGEMENT REACTIONS

Organoaluminium compounds offer a variation of Lewis acidity ranging from low in R_3Al to high in $RAlCl_2$. In causing rearrangements, organoaluminiums have the advantage over $AlCl_3$ that they scavenge protons which frequently cause deleterious side reactions, including polymerizations and double bond migrations.

A. Cycloadditions

The promotion of Diels–Alder reactions by Lewis acids has long been recognized[235], reactions proceeding under much milder conditions than are otherwise possible. The action of the Lewis acid is readily explained in frontier orbital terms.

$EtAlCl_2$ and $AlCl_3$ were compared for the reaction of butadiene with acrolein, methacrolein, and crotonaldehyde. $EtAlCl_2$ gives better results since it removes protic impurities to which aldehydes are sensitive[236]. **128** is cyclized with stoichiometric $EtAlCl_2$ or Et_2AlCl to give *trans*-perhydroindenes[237].

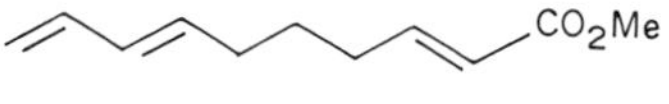

CO_2Me

(128)

In a reaction usually employed to effect annelation, stereochemical control is critical. The presence of heteroatoms able to interact with aluminium can profoundly affect the reaction stereochemistry. For example, **129** gives **130** as the sole product with methyl vinyl ketone in the presence of $EtAlCl_2$ (reaction 139)[238] and reaction 140 is analogous[239].

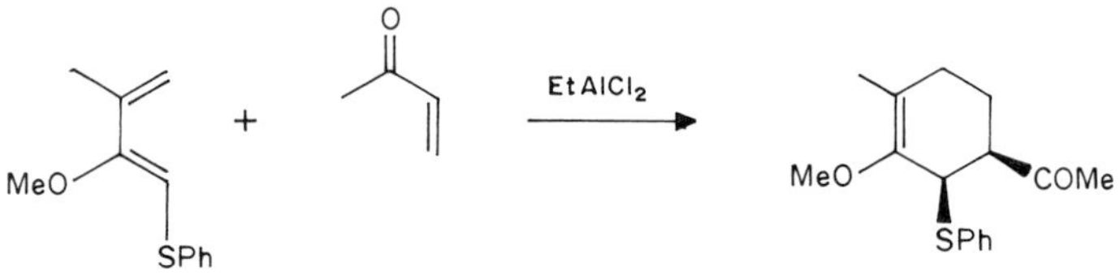

MeO, SPh + O → ($EtAlCl_2$) MeO, SPh, COMe (139)

(129) **(130)**

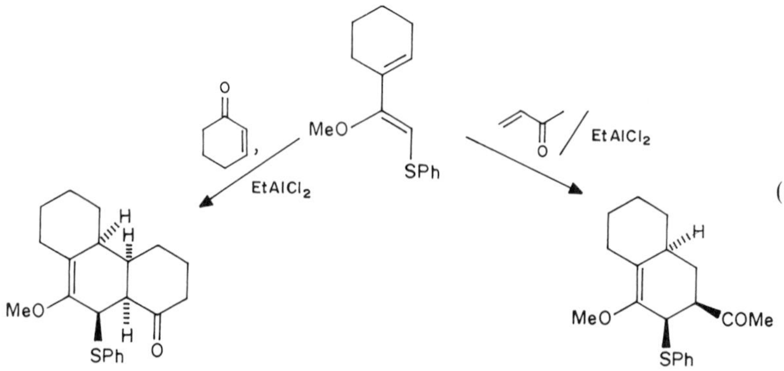

MeO, SPh; O, $EtAlCl_2$; $EtAlCl_2$ (140)

MeO, H, H, H, SPh, O; MeO, H, SPh, COMe

Experiments on similar dienes such as **131** with methyl acrylate, *p*-benzoquinone, methyl vinyl ketone, and maleic anhydride also give only *endo*-products, but **131** (R = Me) with methacrolein gives **132** by an aluminium-catalysed aldol condensation of the first formed product (reaction 141).

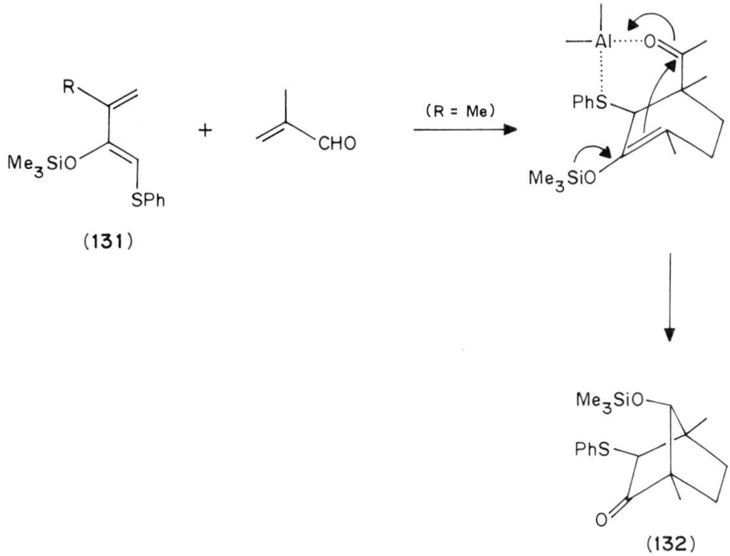

(131) (R = Me) (132) (141)

An organoaluminium-promoted Diels–Alder reaction was employed in the synthesis of the mycotoxin trichodiene. The key step was reaction 142, yielding the correct skeleton[240].

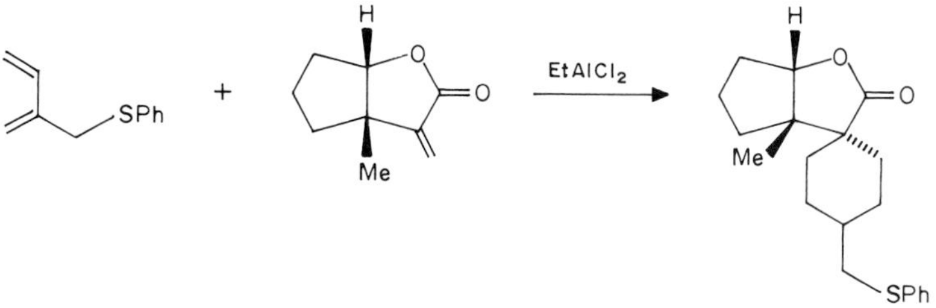

$EtAlCl_2$ (142)

Intramolecular reactions of undeca-2, 8, 10-trienoic esters have been studied. The thermal cyclization is stereorandom but the reaction in the presence of $EtAlCl_2$ shows excellent *endo*-selectivity (reactions 143 and 144)[241]. Bridgehead alkenes have also been synthesized

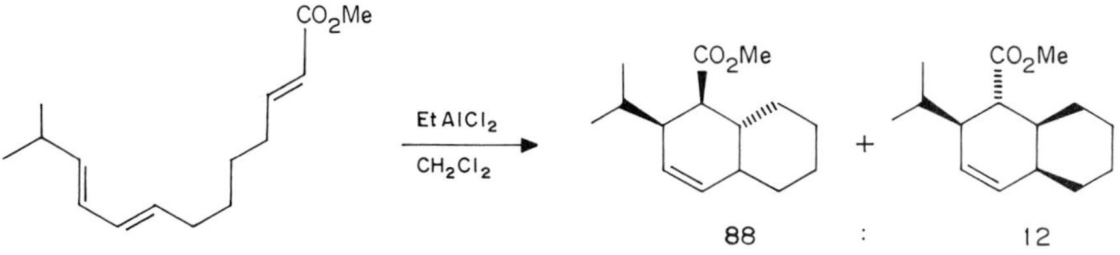

$EtAlCl_2$ / CH_2Cl_2 88 : 12 (143)

CO2Me EtAlCl2 CH2Cl2 CO2Me 92 : 8

(144)

by Et_2AlCl-catalysed intramolecular cyclization, the products, such as **133**, in this case being very similar to those obtained by the thermal reaction (reaction 145)[242].

Et_2AlCl (145)

(133)

A few 2 + 2 cycloadditions seem to be catalysed by $EtAlCl_2$, in particular those of allenes (reaction 146)[243] and alk-1-ynes (reaction 147)[244]. The reaction rates increase with substitution in the alkene moiety.

Et_2AlCl (146)

HC≡CCO2Me + Et_2AlCl CO2Me (147)

B. The Ene and Related Reactions

The ene reaction (148) has until recently found few uses in synthesis because of the

H Δ or Lewis acid H (148)

extreme conditions required to effect the reaction thermally and because of the many proton-catalysed side reactions (polymerization and double bond migration) encountered when conventional Lewis acids are used as promoters. Organoaluminium compounds, being both Lewis acids and Brønstead bases, are ideal promoters for the reaction[5].

Possibly the most widely used enophile is formaldehyde, which reacts with a range of alkenes. For example, the aryltetralin lignan skeleton, **134**, was prepared by reaction 149[245]. A similar reaction has been employed in the synthesis of pseudomonic acids A and C from diene **135** by sequential ene reactions (reaction 150)[246].

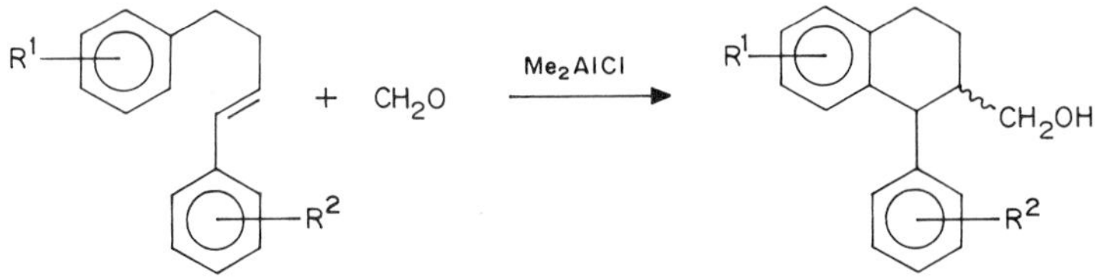

(149)

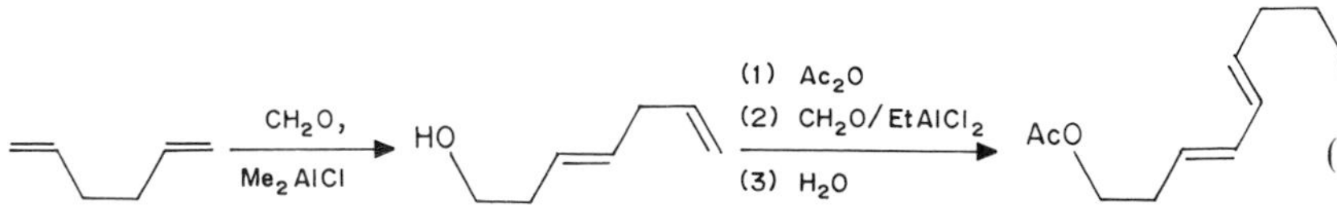

(134)

(150)

(135)

It has been found, however, that Me_3Al-promoted reaction of formaldehyde with some alkenes gives unexpected products (reaction 151)[247]. It was suggested that steric crowding

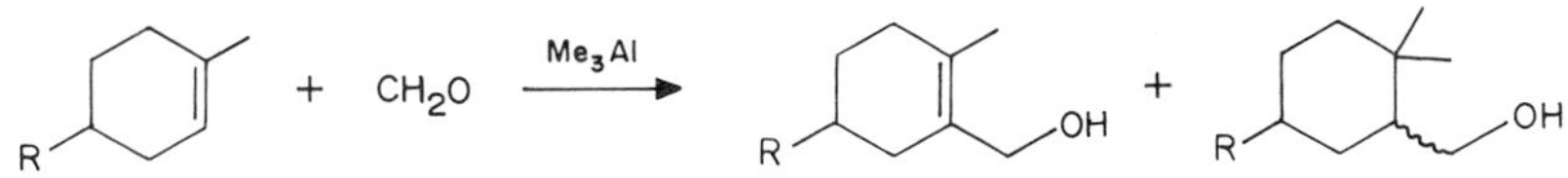

(151)

prevents the 1,5-hydrogen shift. Similar observations were made in the reactions of Me_3Al/CH_2O with **136**, **137**, and **138**. Reaction 152 with enol ethers provides a general, stereoselective route to 1,3-diols[248].

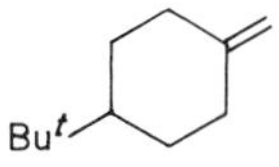

(136) 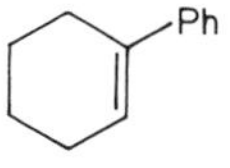 (137) 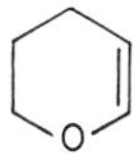 (138)

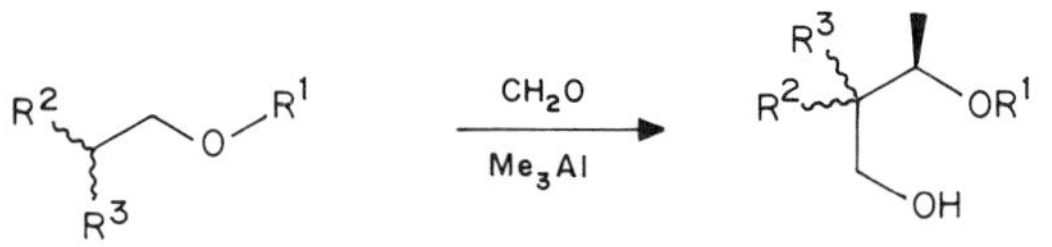

(152)

$EtAlCl_2$ is a better promoter of the reaction of acetate-functionalized alkenes such as **139** (reaction 153); Me_2AlCl complexes **139** and the double bond becomes less nucleophilic so that the methyl groups of Me_2AlCl attack formaldehyde[249]. The reaction is general

(153)

(139)

and has been used in pheromone synthesis. The reaction of formaldehyde with alkynes (reaction 154) is also catalysed by Me_2AlCl[250].

$$RCH_2C{\equiv}CH + CH_2O \xrightarrow{Me_2AlCl} RCH{=}C{=}CHCH_2OH + (E)\text{-}RCH_2C(Cl){=}CHCH_2OH \qquad (154)$$

The reactions of higher aldehydes with alkenes are often complicated by alkylation of the aldehyde, forming alcohols. In these cases 1,1-disubstituted alkenes are good substrates but with tri- and tetra-substituted alkenes alkylation is significant[251]. For example, **140** is converted into **141** (reaction 155) without significant products deriving from 1-methylcyclohexene, which is formed from **140** in the presence of traces of acid. The synthetic utility of the route was shown in the one step synthesis of ipsenol (**142**), a bark beetle pheromone (reaction 156)[252]. Later it was found that the less nucleophilic $EtAlCl_2$

(140) + RCHO —(Me_2AlCl)→ (141) (155)

+ CHO —(Me_2AlCl)→ OH (142) + O (156)

was more suitable for reactions with non-nucleophilic alkenes. This was utilized in a key step in the synthesis of ricinelaidic acid (reaction 157)[249]. Stereochemical considerations in the reaction are complex and *threo*/*erythro* selectivity is usually not high[253].

$$CH_2{=}CH(CH_2)_8CHO + Hex^nCHO \xrightarrow{EtAlCl_2} Hex^nCH(OH)CH_2CH_2CH{=}CH(CH_2)_7CO_2H \quad (157)$$

(143) —(Me_2AlCl, $-80°C$)→ (144)

(143) —(2 Me_2AlCl)→ (145) → (146) (158)

(147) (148) (149) (150)

Intramolecular reactions of aldehydes have also been studied but the results are complex[254]. For example, **143** undergoes a concerted ene reaction with 1 mol of Me_2AlCl yielding **144**, but with 2 molar equivalents **146** is the product *via* the zwitterion, **145** (reaction 158)[255]. Me_2AlCl at 0 °C gives **147** and **148**, whereas $MeAlCl_2$ at − 80 °C gives **149** and $EtAlCl_2$ gives **149** and **150**. The authors proposed reasons for the differences and other cyclizations including those of **151** and **152** were studied. This cyclization methodology has been further developed using sequential ene reactions in the preparation of bicyclic alcohols from alkylidene cycloalkanes (reaction 159)[256].

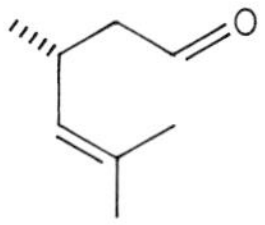

(**151**) 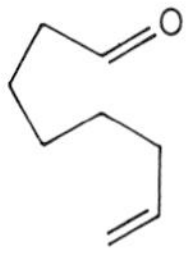 (**152**)

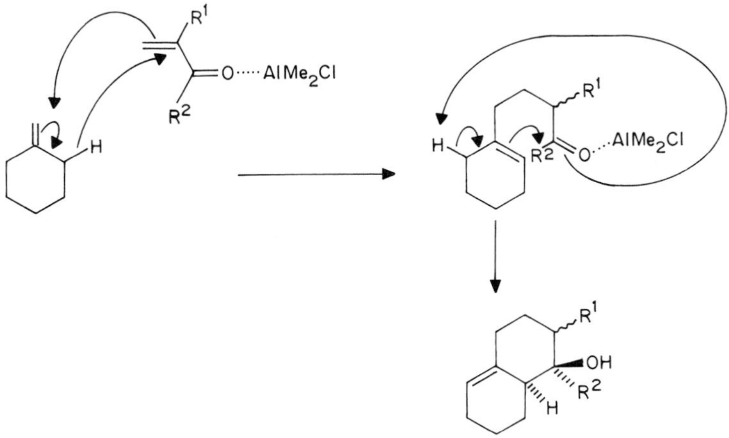

R¹ O····AlMe₂Cl R² H

H R² O····AlMe₂Cl R¹

R¹ OH R² H (159)

Classic ene reactions involved carbon—carbon double and triple bonds, and these too are promoted by organoaluminium compounds. For example, methyl α-chloroacrylate reacts with 2-methylbut-2-ene to give **153** in the presence of $EtAlCl_2$ with only a few percent of the other diastereomer present[257]. The hydrogen is transferred from the alkyl

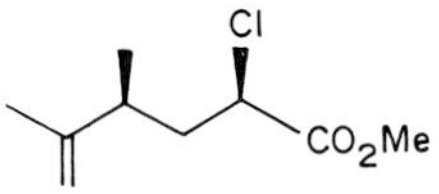

Cl CO_2Me

(**153**)

group *syn* to the alkenyl hydrogen. This has been applied to the synthesis of 24-oxycholesteryl acetate (reaction 160)[256]. Alkynes are similarly reactive, although in the

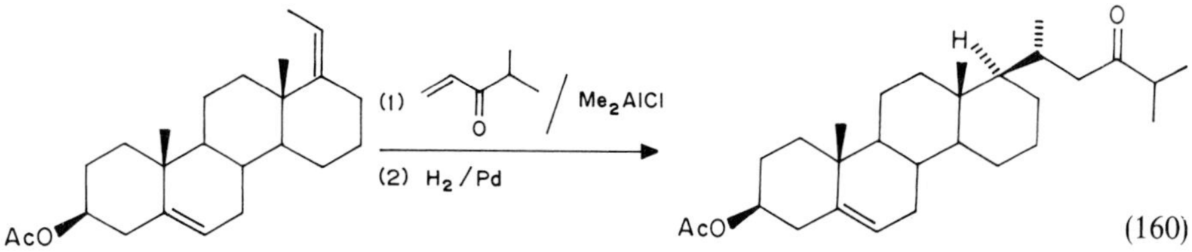

AcO (1) O / Me_2AlCl (2) H_2 / Pd AcO H O (160)

case of propynoates (reaction 161) the hydrogen is transferred from the alkyl group *anti* to the alkenyl hydrogen[258]. Finally, with a chiral acrylate ester asymmetric induction occurs to give a 3:1 mixture of diastereomers (reaction 162)[259].

+ HC≡CCO$_2$Me —RAlCl$_2$→ CO$_2$Me 92% (161)

+ Br CO$_2$ menthyl —EtAlCl$_2$→ Br CO$_2$ menthyl (162)

C. Friedel–Crafts Alkylations and Acylations

The $AlCl_3$-catalysed alkylation and acylation of benzene is too well known to require further comment. The range of Lewis acidities and Brønsted basicities of alkylaluminiums is, however, also very useful. Benzene is alkylated by alkyl chlorides in the presence of alkylaluminium halides *via* an initial complex such as $EtAlCl_2$—ClR[260]. $EtAlCl_2$ has also been used to alkylate alkenes[261] and acyl chlorides are also reactive under mild conditions[262].

Friedel–Crafts acylation of alkenes is also catalysed. For example, the reactions of cyclohexene, 1-methylcyclohexene, 2-methyl-but-2-ene, hex-1-ene, and isoprene with acetyl chloride, acetic anhydride, and maleic anhydride, promoted by Et_2AlCl, were studied as a route to β,γ-unsaturated ketones. The yields were variable[263].

D. Claisen Rearrangements

The rate of the Claisen rearrangement of allyl phenyl ethers (reaction 163) increases in

O + Et_2AlCl → OAlEtCl + C_2H_4 (163)

the presence of alkylaluminium halides, but the aluminium compound must be present in stoichiometric or greater amounts as it is complexed by the product. The best catalysts are Et_2AlCl and $Bu^i{}_2AlCl$; $RAlCl_2$ does catalyse the reaction but cyclization to the coumarin occurs[264].

The importance of the correct choice of Lewis acid was shown by Yamamura *et al.*[265] in a biomimetic synthesis of lavandulol, **155** (reaction 164). Weak Lewis acids do not induce

O —(Bu^t, Bu^t, OAl, OSO$_2$CF$_3$) **(154)**→ OH **(155)** (164)

the rearrangement whereas the more powerful ones affect the double bonds. Compound **154** was chosen for the degree of its Lewis acidity, its ability to function as a Brønsted base and its bulk, which renders the species monomeric. It has been used for a number of analogous transformations (reaction 165). For the reaction of **156**, Bu^i_3Al seems to be the

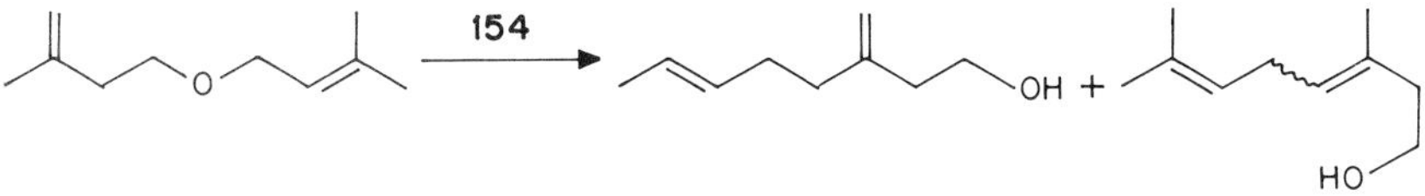

(165)

Lewis acid of choice; Me_3Al and Et_3Al give alkylated products (reaction 166)[266]. The use

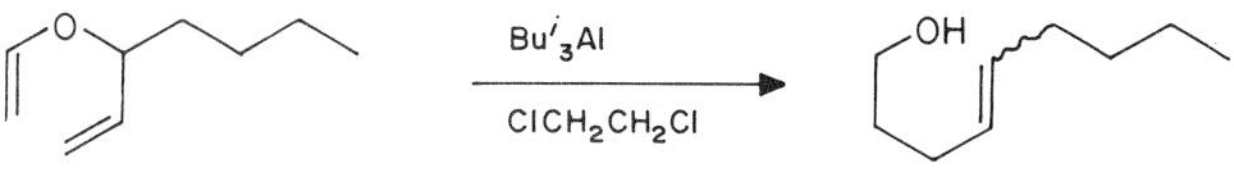

(166)

(156) E : Z *ca.* 1:1

of Et_2AlSPh or Et_2AlCl/PPh_3 suppresses the final reduction and **157** is converted into **158** in good yield (reaction 167). Compound **159** reacts with Et_3Al to give **160**, the

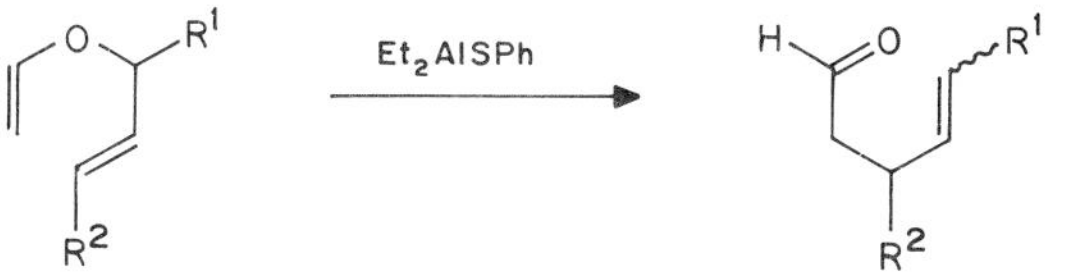

(167)

(157) (158)

configuration of the product being explained in terms of the Lewis acidity of the solvent (reaction 168)[267].

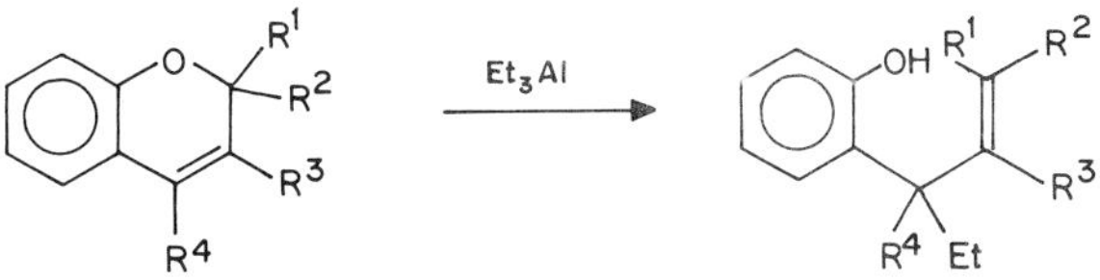

(168)

(159) (160)

E. Beckmann and Related Rearrangements

The reduction of oximes to secondary amines by $LiAlH_4$ has long been established. However, treatment of **161** with dibah gives the rearranged product **162** (reaction 169)[268].

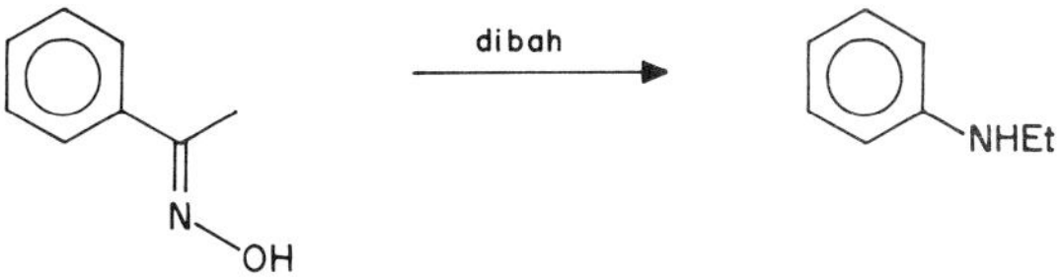

(169)

(161) (162)

The hydroxylamine **163**, is an intermediate. Functionalized hydroxylamines such as **164**

NHOH

(163)

are alkylated (reaction 170); the oxygenophilic aluminium reagent cleaves the N—O bond heterolytically to the anilinium ion, which is susceptible to nucleophilic attack by R_3Al. Regioselectivity is variable but no double alkylation occurs. The intermediate **165**

R_3Al (170)

(164)

synthesized in this way has been employed in an indole synthesis (reaction 171).

(1) CuI / dmf (2) Na/NH_3 (171)

(165)

The amphoteric nature of aluminium compounds has been employed to induce the Beckmann rearrangement and capture the iminocarbocation (reaction 172). Reduction of

R_2AlX (172)

X = alkyl, alkynyl, SR, CN

the imine then yields an amine. In this way the naturally occurring alkaloid **166** was produced with high stereoselectivity (reaction 173)[269]. Capture of the iminocarbocation

$Pr^n{}_3Al$ (173)

(166)

may also employ a silyl enol ether, giving a reaction regiospecific in both components (reaction 174).

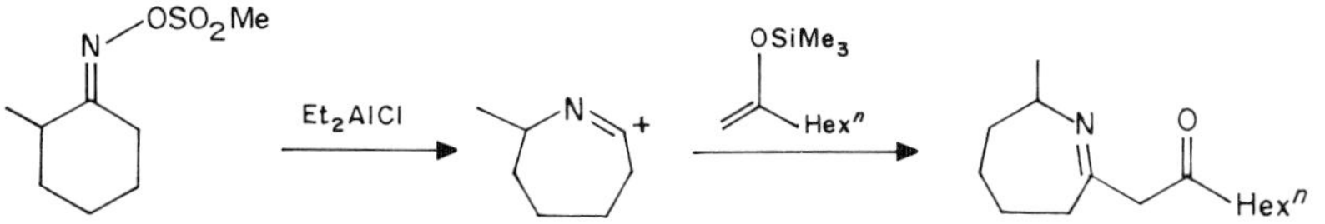

(174)

F. Other Rearrangements and Related Reactions

Numerous other types of rearrangements and cyclizations have been promoted by organoaluminium compounds. Unsaturated hydrocarbons are isomerized by reversible hydroaluminations and cyclizations are also known. Reaction 175 is thought to proceed *via* two additions to the alkyne and then cyclization[270]. Alanes may be synthesized directly

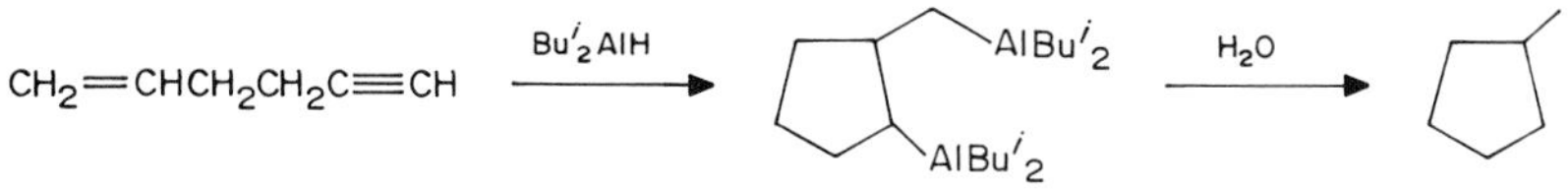

$CH_2{=}CHCH_2CH_2C{\equiv}CH$ (175)

from alkenes using aluminium metal and hydrogen (reaction 176), but in the case of **167** both rearrangement and oligomerization compete (reaction 177)[271]. Other cyclizations, such as that of reaction 178, have received relatively little attention in synthesis[272].

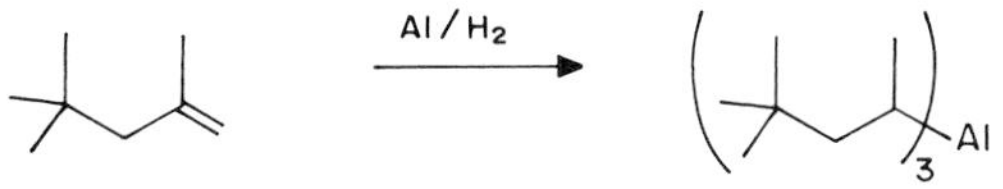

(176)

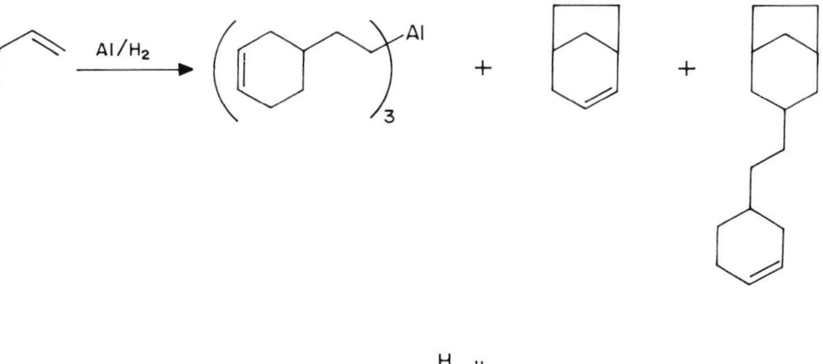

(167)

(177)

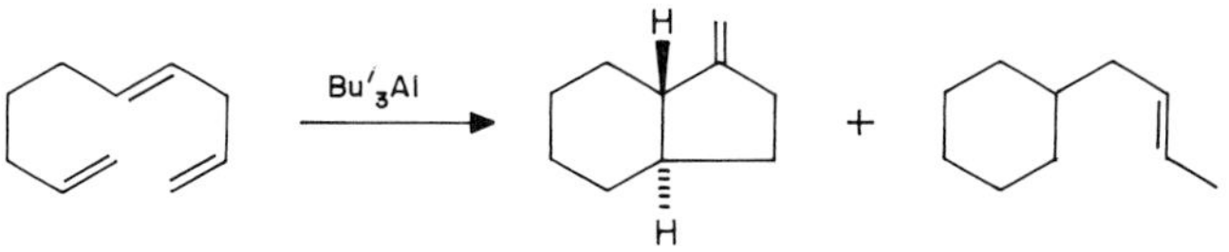

(178)

The reaction of vinyl acetals with Bu^i_3Al and Et_3Al has been studied, and it is found that whilst Bu^i_3Al stereospecifically effects ring contraction to the *trans*-alcohols **168**, the reaction with Et_3Al gives a complex mixture of products, including those from alkylation, reduction, and ring opening (reaction 179)[273]. Although the reaction mechanism is not known, the high stereospecificity may derive from the coordination of the aluminium to the ring oxygen with hydrogen or alkyl transfer occuring *via* a four-centred transition state. Compounds such as **169** are unreactive[274], indicating the importance of the exocyclic oxygen; the active complex may be **170**.

(168)

168 + ... + RCHCH$_2$CH$_2$CHEt(CHO)(OEt) + RCHCH$_2$CH$_2$CHEt(CH$_2$OH)(OEt) (179)

(169) (170)

Another reaction involving C—O bond cleavage as a prelude to rearrangement is the organoaluminium promoted rearrangement of vicinal diol monoacetates (reaction 180).

Et$_2$AlC≡CPh, 25 °C, 1 h → PhC≡C– ... OH 62% (180)

The product ketones are not isolated but alkylated *in situ* to give alcohols[275].

The use of a good leaving group in conjunction with an organoalane to cause rearrangement is a well known procedure and has been employed by Suzuki *et al.*[276] in another pinacol-type rearrangement of chiral α-hydroxymethylsulphonates to give optically pure α-aryl and α-vinyl ketones (reaction 181). The reaction is thought to proceed

Et$_3$Al, −78 °C (181)

R^1 = Ar, vinyl

via a ligand exchange at aluminium with concerted migration of R^1 whilst the OMs group is lost. By combining Et$_3$Al with dibah the reaction may be extended; the ketone[171] is reduced *in situ* and rearranged, and then the product aldehyde is reduced to an optically active alcohol (reaction 182)[277].

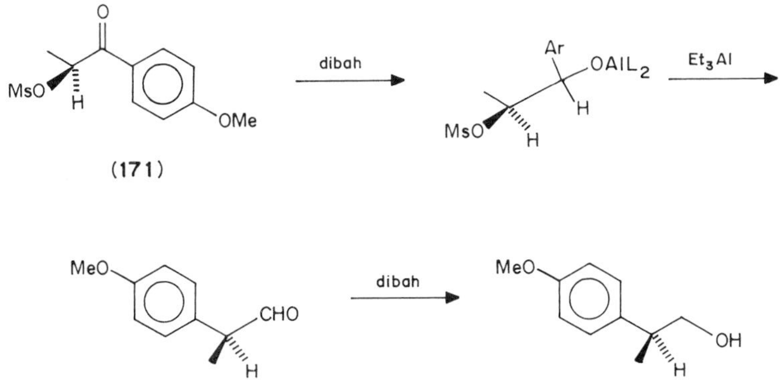

(171)

(182)

By using Et_2AlCl, which is more Lewis acidic than Et_3Al, alkyl groups may also be migrated with good stereospecificity. The mechanism of the reaction is thought to involve a cyclic transition state (**172**); the energy of the HOMO of the migrating group is increased by donation from AlO and then has an effective interaction with the LUMO of the carbocation developed by the Lewis acidity of the aluminium. The reaction has been used in the synthesis of an ant alarm pheromone (reaction 183)[278].

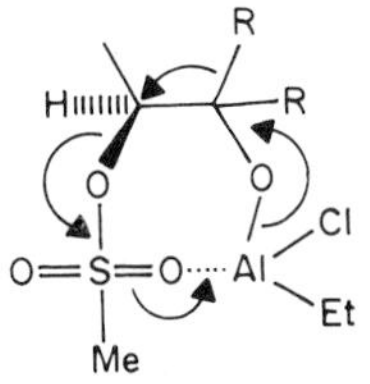

(172)

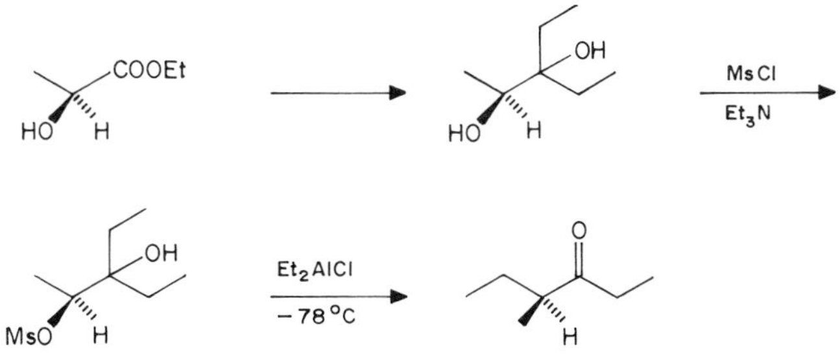

(183)

A number of addition reactions of enones involve the formation of an aluminium enolate. For example, **173** reacts with **174** in the presence of 2 mol of $EtAlCl_2$ to give initially **175**. At low temperature this collapses reversibly to **176** and **177** but the thermodynamic products are **178** and **179**, formed by alkyl and hydride shifts (reaction 184). Cyclohexenone reacts with 2-methyl-but-2-ene to give **180** whilst 2-methylpropene gives a double addition to **181**[279]. Intramolecular reactions, (185 and 186), give cyclizations with high regio- and stereo-specificity[255].

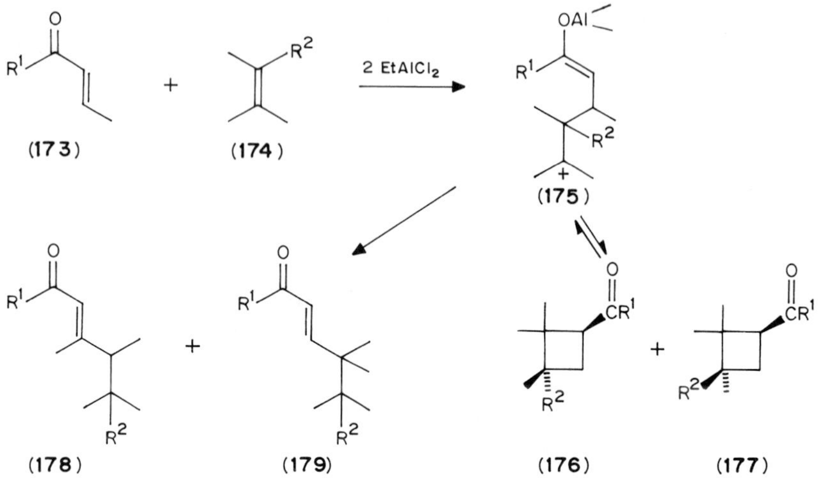

(173) (174) (175) (178) (179) (176) (177) (184)

(180) (181)

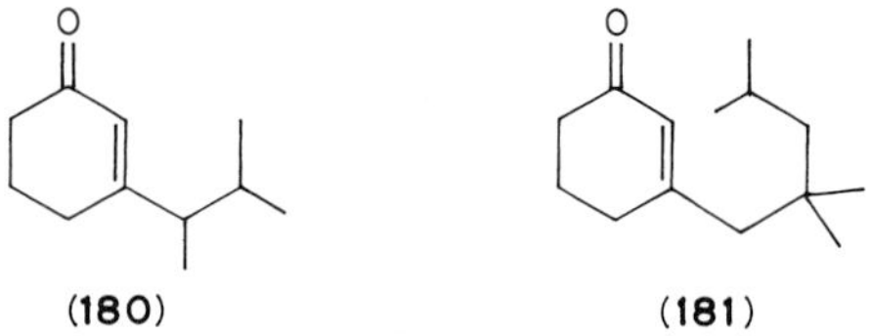

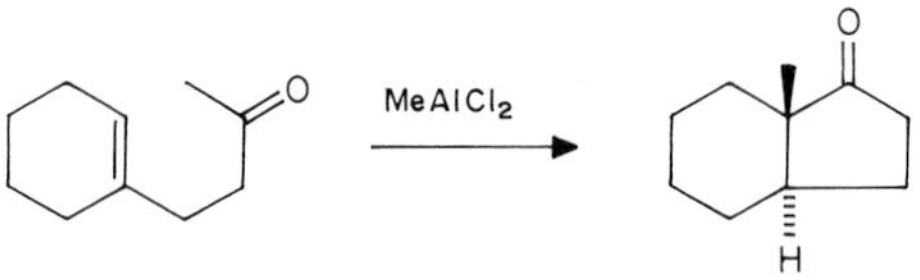

(185)

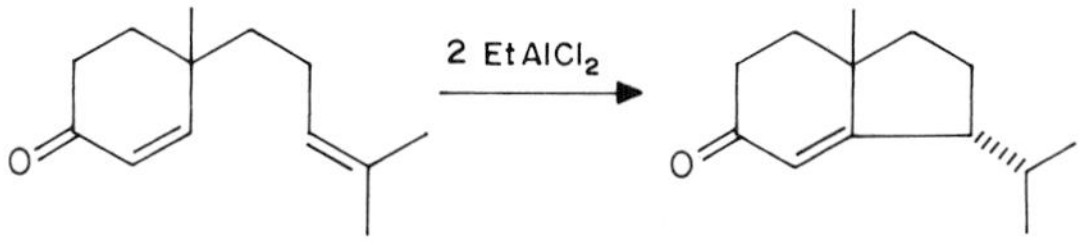

(186)

Trost and co-workers used $EtAlCl_2$ as a promoter in several complex cyclizations. For example, reaction 187 of enones with **182** gives **183**, which may be cyclized in excellent

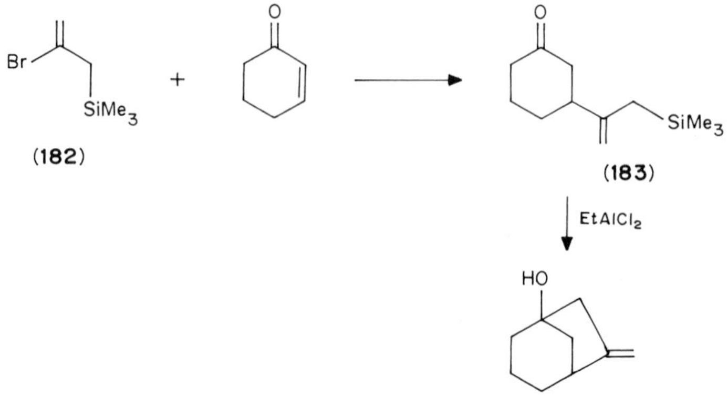

(182) (183) (187)

yield[280]. The driving force for the reaction is the strength of the silicon—halogen bond, but the fluoride ion-catalysed reaction suffers from competing desilylation. In the similar reaction 188 to construct the taxane skeleton, fluoride catalysis is unsatisfactory but $EtAlCl_2$ is an excellent promoter[281].

SiMe3 H MeS O (1) m-ClC6H4CO3H (2) EtAlCl2 OH SO2Me (188)

IX. ACKNOWLEDGEMENTS

I thank Duncan Carmichael for some preliminary work on this subject and Dr J. D. Smith for his interest and advice.

X. REFERENCES

1. J. J. Eisch, *Comprehensive Organometallic Chemistry*, Vol. 1, Pergamon Press, Oxford, 1982, p. 555.
2. J. R. Zietz, G. C. Robinson, and K. L. Lindsay, *Comprehensive Organometallic Chemistry*, Vol. 7, Pergamon Press, Oxford, 1982, p. 365.
3. E.-I. Negishi, *Organometallics in Organic Synthesis*, Volume 1, Wiley, New York, 1980, p. 350.
4. J. D. Smith, *J. Organomet. Chem.*, **257**, 17 (1983), and previous annual reviews in this series.
5. B. B. Snider, *Acc. Chem. Res.*, **13**, 426 (1980).
6. K. Ziegler, *Angew. Chem.*, **68**, 721 (1956).
7. J. N. Hay, P. G. Hooper, and J. C. Robb, *Trans. Faraday Soc.*, **66**, 2800 (1970).
8. P. E. M. Allen and R. M. Lough, *J. Chem. Soc., Faraday Trans. I*, **69**, 2087 (1973).
9. K. W. Egger, *Helv. Chim. Acta*, **55**, 1502 (1972).
10. O. Gropen and A. Haaland, *Acta Chem. Scand.*, **A36**, 435 (1982).
11. Esso Research, *Br. Pat.*, 876536 (1956); *Chem. Abstr.*, **56**, 6000e (1962).
12. K. Ziegler, W.-R. Kroll, W. Larbig, and O.-W. Steudel, *Justus Liebigs Ann. Chem.*, **629**, 53 (1960).
13. M. Montury and J. Goré, *Tetrahedron Lett.*, **21**, 51 (1980).
14. L. I. Zakharkin and L. A. Savina, *Bull. Acad. Sci. USSR. Div. Chem. Sci.*, 72 (1967).
15. V. P. Yur'ev, I. M. Salimgareeva, A. V. Kuchin, G. A. Tolstikov, and S. R. Rafikov, *Dokl. Akad. Nauk SSSR*, **203**, 356 (1972).
16. E. C. Ashby and J. J. Lin, *J. Org. Chem.*, **43**, 2567 (1978).
17. K. Fischer, K. Jonas, A. Mollbach, and G. Wilke, *Z. Naturforsch.*, **39B**, 1011 (1984).
18. G. Giacomelli, L. Bertero, and L. Lardicci, *Tetrahedron Lett.*, **22**, 883 (1981).
19. E. C. Ashby and S. A. Noding, *Tetrahedron Lett.*, 4579 (1977); *J. Org. Chem.*, **44**, 4364 (1979); **45**, 1035 (1980).
20. F. Sato, S. Sato, H. Kodama, and M. Sato, *J. Organomet. Chem.*, **142**, 71 (1977).
21. J. Tsuji and T. Mandai, *Chem. Lett.*, 975 (1977).
22. J. Tsuji, T. Yamakawa, and T. Mandai, *Tetrahedron Lett.*, 565 (1978).
23. F. Sato, H. Kodama, and M. Sato, *J. Organomet. Chem.*, **157**, C30 (1978).
24. F. Sato, H. Kodama, Y. Tomuro, and M. Sato, *Chem. Lett.*, 623 (1979).
25. F. Sato, T. Oikawa, and M. Sato, *Chem. Lett.*, 167 (1979).
26. K. Isagawa, M. Ohige, K. Tatsumi, and Y. Otsuji, *Chem. Lett.*, 1155 (1978).
27. F. Sato, Y. Mori, and M. Sato, *Chem. Lett.*, 833 (1978).
28. F. Sato, H. Kodama, and M. Sato, *Chem. Lett.*, 789 (1978).
29. K. Isagawa, K. Tatsumi, and Y. Otsuji, *Chem. Lett.*, 1117 (1977).
30. F. Sato, Y. Mori, and M. Sato, *Tetrahedron Lett.*, 1405 (1979).

31. F. Sato, S. Sato, and M. Sato, *J. Organomet. Chem.*, **131**, C26 (1977).
32. P. W. Chum and S. E. Wilson, *Tetrahedron Lett.*, 1257 (1976).
33. E.-I. Negishi and T. Yoshida, *Tetrahedron Lett.*, **21**, 1501 (1980).
34. G. A. Tolstikov, U. M. Dzhemilev, O. S. Vostrikova, and A. G. Tolstikov, *Bull. Acad. Sci. USSR, Div. Chem. Sci.*, **31**, 596 (1982).
35. U. M. Dzhemilev, A. G. Ibragimov, O. M. Vostrikova, G. A. Tolstikov, and L. M. Zelenova, *Bull. Acad. Sci. USSR, Div. Chem. Sci.*, **30**, 281 (1981).
36. M. Sato and F. Sato, *Jpn. Kokai Tokkyo Koho*, 78, 103426 (1978); *Chem. Abstr.*, **90**, 5881f (1979).
37. F. Sato, Y. Tomuro, H. Ishikawa, and M. Sato, *Chem. Lett.*, 99 (1980).
38. W. Pfohl, *Justus Liebigs Ann. Chem.*, **629**, 207 (1960).
39. R. Schimpf and P. Heimbach, *Chem. Ber.*, **103**, 2122 (1970).
40. Z. Florjańczyk and U. Iwaniak, *J. Organomet. Chem.*, **252**, 275 (1983).
41. J. J. Barber, C. Willis, and G. M. Whitesides, *J. Org. Chem.*, **44**, 3603 (1979).
42. A. P. Kozikowski and Y. Kitigawa, *Tetrahedron Lett.*, **23**, 2087 (1982).
43. V. V. Markova, V. A. Kormer, and A. A. Petrov, *J. Gen. Chem. USSR*, **37**, 208 (1967).
44. M. J. Smith and S. E. Wilson, *Tetrahedron Lett.*, **23**, 5013 (1982).
45. D. Masure, P. Coutrot, and J. F. Normant, *J. Organomet. Chem.*, **226**, C55 (1982).
46. G. Zweifel, G. M. Clark, and C. C. Whitney, *J. Am. Chem. Soc.*, **93**, 1305 (1971).
47. G. Zweifel, J. T. Snow, and C. C. Whitney, *J. Am. Chem. Soc.*, **90**, 7139 (1968).
48. A. P. Kozikowski, A. Ames, and H. Wetter, *J. Organomet. Chem.*, **164**, C33 (1979).
49. K. Utimoto, K. Uchida, M. Yamaya, and H. Nozaki, *Tetrahedron Lett.*, 3641 (1977).
50. W. J. Gensler and J. J. Bruno, *J. Org. Chem.*, **28**, 1254 (1963).
51. W. Ziegenbein and W. M. Schneider, *Chem. Ber.*, **98**, 824 (1965).
52. J. J. Eisch, H. Gopal, and S.-G. Rhee, *J. Org. Chem.*, **40**, 2064 (1975).
53. G. Zweifel and W. Lewis, *J. Org. Chem.*, **43**, 2739 (1978).
54. H. P. On, W. Lewis, and G. Zweifel, *Synthesis*, 999 (1981).
55. L. E. Overman and T. C. Malone, *J. Org. Chem.*, **47**, 5297 (1982).
56. R. B. Miller and M. I. Al-Hassan, *Tetrahedron Lett.*, **24**, 2055 (1983).
57. G. Zweifel, R. A. Lynd, and R. E. Murray, *Synthesis*, 52 (1977).
58. G. Zweifel, W. Lewis, and H. P. On, *J. Am. Chem. Soc.*, **101**, 5101 (1979).
59. H. Westmijze, H. Kleijn, and P. Vermeer, *Synthesis*, 430 (1979).
60. W. Granitzer and A. Stütz, *Tetrahedron Lett.*, 3145 (1979).
61. L.-I. Olsson and A. Claesson, *Acta Chem. Scand.*, **B31**, 614 (1977).
62. J. J. Eisch and M. W. Foxton, *J. Organomet. Chem.*, **12**, P33 (1968).
63. E. C. Ashby and S. R. Noding, *J. Organomet. Chem.*, **177**, 117 (1979).
64. H. Hoberg and F. Aznar, *Angew. Chem., Int. Ed. Engl.*, **19**, 146 (1980).
65. G. Giacomelli, A. M. Caporusso, and L. Lardicci, *J. Org. Chem.*, **44**, 231 (1979).
66. A. M. Caporusso, G. Giacomelli, and L. Lardicci, *J. Chem. Soc., Perkin Trans. 1*, 1900 (1981).
67. R. Reinäcker and D. Schwengers, *Justus Liebigs Ann. Chem.*, **737**, 182 (1970).
68. J. J. Eisch and J. M. Biedermann, *J. Organomet. Chem.*, **30**, 167 (1971).
69. J. J. Eisch and R. Amtmann, *J. Org. Chem.*, **37**, 3410 (1972).
70. D. C. Brown, S. A. Nichols, A. B. Gilpin, and D. W. Thompson, *J. Org. Chem.*, **44**, 3457 (1979).
71. E.-I. Negishi, *Pure Appl. Chem.*, **53**, 2333 (1981).
72. B. B. Snider and M. Karras, *J. Organomet. Chem.*, **179**, C37 (1979).
73. J. J. Eisch, R. J. Manfre, and D. A. Komer, *J. Organomet. Chem.*, **159**, C13 (1978).
74. B. B. Snider, R. S. E. Conn, and M. Karras, *Tetrahedron Lett.*, 1679 (1979).
75. A. M. Caporusso, G. Giacomelli, and L. Lardicci, *J. Org. Chem.*, **42**, 914 (1977).
76. A. M. Caporusso, G. Giacomelli, and L. Lardicci, *J. Org. Chem.*, **44**, 1496 (1979).
77. D. E. Van Horn and E.-I. Negishi, *J. Am. Chem. Soc.*, **100**, 2252 (1978).
78. N. Okukado and E.-I. Negishi, *Tetrahedron Lett.*, 2357 (1978); M. Kobayashi, L. F. Valente, E.-I. Negishi, W. Patterson, and A. Silveira, *Synthesis*, 1034 (1980).
79. K. Mori, M. Sakakibara, and K. Okada, *Tetrahedron*, **40**, 1767 (1984).
80. J. K. Whitesell, M. Fisher, and P. D. S. Jardine, *J. Org. Chem.*, **48**, 1556 (1983).
81. T. Yoshida, *Chem. Lett.*, 293 (1982).
82. E.-I. Negishi, L D. Boardman, J. M. Tour, H. Sawada, and C. L. Rand, *J. Am. Chem. Soc.*, **105**, 6344 (1983).
83. E.-I. Negishi and L. D. Boardman, *Tetrahedron Lett.*, **23**, 3327 (1982).

84. T. Yoshida and E.-I. Negishi, *J. Am. Chem. Soc.*, **103**, 4985 (1981).
85. H. Hayami, K. Oshima, and H. Nozaki, *Tetrahedron Lett.*, **25**, 4433 (1984).
86. T. Araki, K. Hayakawa, T. Aoyagi, Y. Nakano, and H. Tani, *J. Org. Chem.*, **38**, 1130 (1973).
87. L. Rösch, G. Altnau, and W. H. Otto, *Angew. Chem., Int. Ed. Engl.*, **20**, 581 (1981).
88. H. Newman, *J. Am. Chem. Soc.*, **95**, 4098 (1973).
89. E. C. Ashby and J. T. Laemmle, *Chem. Rev.*, **75**, 521 (1975).
90. E. C. Ashby, J. Laemmle, and G. E. Parris, *J. Organomet. Chem.*, **19**, P24 (1969).
91. E. C. Ashby and R. S. Smith, *J. Org. Chem.*, **42**, 425 (1977).
92. J.-L. Namy, E. Henry-Basch and P. Freon, *C.R. Acad. Sci., Ser. C*, **268**, 287 (1969).
93. E. C. Ashby and S. Yu, *J. Chem. Soc., Chem. Commun.*, 351 (1971); J. Laemmle, E. C. Ashby, and P. V. Roling, *J. Org. Chem.*, **38**, 2526 (1973).
94. A. Meisters and T. Mole, *J. Chem. Soc., Chem. Commun.*, 595 (1972).
95. M. T. Reetz, J. Westermann, and R. Steinbach, *J. Chem. Soc., Chem. Commun.*, 237 (1981).
96. C. H. Heathcock, E. G. Del Mar, and S. L. Graham, *J. Am. Chem. Soc.*, **104**, 1907 (1982).
97. E. C. Ashby and A. B. Goel, *J. Organomet. Chem.*, **221**, C15 (1981).
98. J. Flahaut and P. Migniac, *Helv. Chim. Acta*, **61**, 2275 (1978).
99. K. B. Starowieyski and A. Bęcalska, *J. Organomet. Chem.*, **269**, 223 (1984).
100. F. Barbot, *Bull. Soc. Chim. Fr.*, 83 (1984).
101. M. Koreeda and Y. Tanaka, *J. Chem. Soc., Chem. Commun.*, 845 (1982).
102. W. Nagata, M. Yoshioka, and M. Murakami, *Org. Synth.*, **52**, 96 (1972).
103. J. A. Oakleaf, M. T. Thomas, A. Wu, and V. Snieckus, *Tetrahedron Lett.*, 1645 (1978).
104. C. H. Heathcock, C. M. Tice, and T. C. Germroth, *J. Am. Chem. Soc.*, **104**, 6081 (1982).
105. Y. Hayakawa and R. Noyori, *Bull. Chem. Soc. Jpn.*, **47**, 2617 (1974).
106. P. F. Hudrlik and D. Peterson, *J. Am. Chem. Soc.*, **97**, 1464 (1975).
107. M. Suzuki, T. Kawagashi, and R. Noyori, *Tetrahedron Lett.*, **23**, 5563 (1982).
108. S. Kim, K. H. Ahn, and Y. W. Chung, *J. Org. Chem.*, **47**, 4581 (1982).
109. J. A. Katzenellenbogen and S. W. Bowlus, *J. Org. Chem.*, **38**, 627 (1973).
110. R. A. Kretchmer, *J. Org. Chem.*, **37**, 801 (1972).
111. G. P. Giacomelli, R. Menicagli, and L. Lardicci, *J. Am. Chem. Soc.*, **97**, 4009 (1975).
112. H. Haubenstock, *Top. Stereochem.*, **14**, 213 (1983).
113. G. Giacomelli, L. Lardicci, F. Palla, and A. M. Caporusso, *J. Org. Chem.*, **49**, 1725 (1984).
114. G. Giacomelli and L. Lardicci, *J. Org. Chem.*, **47**, 4335 (1982).
115. K. Maruoka, S. Hashimoto, Y. Kitagawa, H. Yamamoto, and H. Nozaki, *J. Am. Chem. Soc.*, **99**, 7705 (1977); *Bull. Chem. Soc. Jpn.*, **53**, 3301 (1980).
116. J. Tsuji, T. Yamada, M. Kaito, and T. Mandai, *Tetrahedron Lett.*, 2257 (1979); *Bull. Chem. Soc. Jpn.*, **53**, 1417 (1980).
117. C. Sato, S. Ikeda, H. Shirahama, and T. Matsumoto, *Tetrahedron Lett.*, **23**, 2099 (1982).
118. E. C. Ashby and T. L. Wiesemann, *J. Am. Chem. Soc.*, **100**, 3101 (1978).
119. L. Bagnell, E. A. Jeffrey, A. Meisters, and T. Mole, *Aust. J. Chem.*, **28**, 801 (1975).
120. L. Bagnell, A. Meisters, and T. Mole, *Aust. J. Chem.*, **28**, 817 (1975).
121. E. C. Ashby and G. Heinsohn, *J. Org. Chem.*, **39**, 3297 (1974).
122. J. Hooz and R. B. Layton, *Can. J. Chem.*, **51**, 2098 (1973).
123. K. F. Bernady, M. B. Floyd, J. F. Poletto, and M. J. Weiss, *J. Org. Chem.*, **44**, 1438 (1979).
124. R. Pappo and P. W. Collins, *Tetrahedron Lett.*, 2627 (1972).
125. J. Fried, C.-H. Lin, and S. H. Ford, *Tetrahedron Lett.*, 1379 (1969).
126. R. T. Hansen, D. B. Carr, and J. Schwartz, *J. Am. Chem. Soc.*, **100**, 2244 (1978).
127. J. Schwartz, D. B. Carr, R. T. Hansen, and F. M. Dayrit, *J. Org. Chem.*, **45**, 3053 (1980).
128. J. Schwartz and Y. Hayasi, *Tetrahedron Lett.*, **21**, 1497 (1980).
129. R. F. Newton, D. P. Reynolds, J. Greenwood, and F. Scheinmann, *J. Chem. Soc., Perkin Trans. 1*, 2346 (1980).
130. L. Bagnell, A. Meisters, and T. Mole, *Aust. J. Chem.*, **28**, 821 (1975).
131. A. Itoh, S. Ozawa, K. Oshima, and H. Nozaki, *Tetrahedron Lett.*, 361 (1980).
132. T. Tokeroyama, K. Fujimori, T. Shimizu, Y. Yamagiwa, M. Monden, and H. Iio, *J. Chem. Soc. Chem. Commun.*, 1516 (1983).
133. G. Stork and D. H. Sherman, *J. Am. Chem. Soc.*, **104**, 3758 (1982).
134. J. N. Marx and G. Minaskanian, *Tetrahedron Lett.*, 4175 (1979).
135. G. Giacomelli, A. M. Caporusso, and L. Lardicci, *Tetrahedron Lett.*, **22**, 3663 (1981).

136. S. Iguchi, H. Nakai, M. Hayashi, H. Yamamoto, and K. Maruoka, *Bull. Chem. Soc. Jpn.*, **54**, 3033 (1981).
137. S. Kim and K. H. Ahn, *J. Org. Chem.*, **49**, 1717 (1984).
138. A. M. Caporusso, G. Giacomelli, and L. Lardicci, *J. Org. Chem.*, **47**, 4640 (1982).
139. J. Huguet, M. Karpf, and A. S. Dreiding, *Tetrahedron Lett.*, **24**, 4177 (1983).
140. L. I. Zakharkin and I. M. Khorlina, *J. Gen. Chem. USSR*, **34**, 1021 (1964).
141. T. D. Hubert, D. P. Eyman, and D. F. Wiemer, *J. Org. Chem.*, **49**, 2279 (1984).
142. Y. Nagao, K. Kawabata, and E. Fujita, *J. Chem. Soc., Chem. Commun.*, 330 (1978).
143. H. Reinheckel, K. Haage, and D. Jahnke, *Organomet. Chem. Rev.*, **4**, 47 (1969).
144. G. Sonnek, K. G. Baumgarten, and H. Reinheckel, *J. Organomet. Chem.*, **142**, 23 (1977).
145. Y. Baba, *Bull. Chem. Soc. Jpn.*, **41**, 1022 (1968).
146. D. Jahnke and H. Reinheckel, *Ger. Offen.*, 2151867 (1973); *Chem. Abstr.*, **79**, 18133m (1973).
147. A. F. Sviridov, M. S. Ermolenko, D. V. Yashunsky, and N. K. Kochetkov, *Tetrahedron Lett.*, **24**, 4355 (1983).
148. E. J. Corey and D. J. Beames, *J. Am. Chem. Soc.*, **95**, 5829 (1973).
149. M. F. Lipton, A. Basha, and S. M. Weinreb, *Org. Synthe.*, **59**, 49 (1980).
150. J.-L. Wood, N. A. Khatri, and S. M. Wienreb, *Tetrahedron Lett.*, 4907 (1979).
151. W. Oppolzer, J. I. Grayson, H. Wegmann, and M. Urrea, *Tetrahedron*, **39**, 3695 (1983).
152. E. Winterfeldt, *Synthesis*, 617 (1975).
153. A. Barco, S. Benetti, G. P. Pollini, P. G. Baraldi, M. Guarneri, D. Simoni, and C. Gandolfi, *J. Org. Chem.*, **46**, 4518 (1981).
154. J.-P. Deprés, A. E. Greene, and P. Crabbé, *Tetrahedron*, **37**, 621 (1981).
155. C. Szántay, L. Töke, and P. Kolonits, *J. Org. Chem.*, **31**, 1447 (1966).
156. J. S. Baran, *J. Org. Chem.*, **30**, 3564 (1965).
157. F. N. Tebbe, G. W. Parshall, and G. S. Reddy, *J. Am. Chem. Soc.*, **100**, 3611 (1978).
158. F. N. Tebbe and R. L. Harlow, *J. Am. Chem. Soc.*, **102**, 6149 (1980).
159. S. H. Pine, R. Zahler, D. A. Evans, and R. H. Grubbs, *J. Am. Chem. Soc.*, **102**, 3270 (1980).
160. T.-S. Chou and S.-B. Huang, *Tetrahedron Lett.*, **24**, 2169 (1983).
161. J. W. S. Stevensen and T. A. Bryson, *Tetrahedron Lett.*, **23**, 3143 (1982).
162. J. J. Eisch and A. Piotrowski, *Tetrahedron Lett.*, **24**, 2043 (1983).
163. D. B. Malpass, S. C. Watson, and G. S. Yeargin, *J. Org. Chem.*, **42**, 2712 (1977).
164. J.-L. Namy, G. Boireau, and D. Abenhaim, *Bull. Soc. Chim. Fr.*, 3191 (1971).
165. P. Ma, V. S. Martin, S. Masamune, K. B. Sharpless, and S. M. Viti, *J. Org. Chem.*, **47**, 1378 (1982).
166. R. S. Bly, R. K. Bly, and T. S. Shibata, *J. Org. Chem.*, **48**, 101 (1983).
167. J. Cantacuzène and J. M. Normant, *Tetrahedron Lett.*, 2947 (1970).
168. G. Teutsch, *Tetrahedron Lett.*, **23**, 4697 (1982).
169. J. Fried, C. Lin, M. Mehra, W. Kao, and P. Dalven, *Ann. N. Y. Acad. Sci.*, **180**, 38 (1972).
170. J. Fried, C. H. Lin, J. C. Sih, P. Dalven, and G. F. Cooper, *J. Am. Chem. Soc.*, **94**, 4342 (1972).
171. R. F. Newton and S. M. Roberts, *Tetrahedron*, **36**, 2163 (1980); R. J. Cave, C. C. Howard, G. Klinkert, R. F. Newton, D. P. Reynolds, A. H. Wadsworth, and S. M. Roberts, *J. Chem. Soc., Perkin Trans. 1*, 2954 (1979).
172. R. S. Matthews, E. D. Mihelich, L. S. McGowan, and K. Daniels, *J. Org. Chem.*, **48**, 409 (1983).
173. L. Magdzinski, B. Cweiber, and B. Frazer-Reid, *Tetrahedron Lett.*, **24**, 5823 (1983).
174. W. Nagata, M. Yoshioka, and T. Okumura, *J. Chem. Soc. C*, 2365 (1970).
175. W. Lidy and W. Sundermeyer, *Tetrahedron Lett.*, 1449 (1973).
176. A. Mubarak and B. Frazer-Reid, *J. Org. Chem.*, **47**, 4265 (1982).
177. J. C. Mullis and W. P. Weber, *J. Org. Chem.*, **47**, 2873 (1982).
178. L. E. Overman and L. A. Flippin, *Tetrahedron Lett.*, **22**, 195 (1981).
179. A. Yasuda, H. Yamamoto, and H. Nozaki, *Bull. Chem. Soc. Jpn.*, **52**, 1705 (1979).
180. H. Yamamoto and H. Nozaki, *Angew. Chem., Int. Ed. Engl.*, **17**, 169 (1978).
181. A. Yasuda, S. Tanaka, K. Oshima, H. Yamamoto, and H. Nozaki, *J. Am. Chem. Soc.*, **96**, 6513 (1974).
182. Y. Kitagawa, A. Itoh, S. Hashimoto, H. Yamamoto, and H. Nozaki, *J. Am. Chem. Soc.*, **99**, 3864 (1977).
183. M. R. Roberts, W. H. Parsons, and R. H. Schlessinger, *J. Org. Chem.*, **43**, 3970 (1978).
184. L. I. Zakharkin and I. M. Khorlina, *Bull. Acad. Sci. USSR, Div. Chem. Sci.*, 834 (1965).
185. V. P. Yur'ev, A. V. Kuchin, T. Yu. Romanova, and G. A. Tolstikov, *J. Gen. Chem. USSR*, **48**, 2256 (1978).

186. K. C. Nicolaou and J. Uenishi, *J. Chem. Soc., Chem. Commun.*, 1292 (1982).
187. J. M. Finan and Y. Kishi, *Tetrahedron Lett.*, **23**, 2719 (1982).
188. R. S. Lenox and J. A. Katzenellenbogen, *J. Am. Chem. Soc.*, **95**, 957 (1973).
189. J. C. Saddler and P. L. Fuchs, *J. Am. Chem. Soc.*, **103**, 2112 (1981).
190. A. Yasuda, M. Takahashi, and H. Takaya, *Tetrahedron Lett.*, **22**, 2413 (1981).
191. A. Tröber, G. Gladigau, B. Richter, W. Dreschel, and D. Topf, *J. Prakt. Chem.*, **311**, 942 (1969).
192. E. Chiellini, *Gazz. Chim. Ital.*, **102**, 830 (1972).
193. Y. Yamamoto, H. Yatagi, and K. Maruyama, *J. Org. Chem.*, **45**, 195 (1980).
194. H. Köster and N. D. Sinha, *Tetrahedron Lett.*, **23**, 2641 (1982).
195. A. Itoh, K. Oshima, S. Sasaki, H. Yamamoto, T. Hiyama, and H. Nozaki, *Tetrahedron Lett.*, 4751 (1979).
196. C. G. M. Janssen and E. F. Godefroi, *J. Org. Chem.*, **47**, 3274 (1982).
197. H. Matsushita and E.-I. Negishi, *J. Org. Chem.*, **47**, 4161 (1982).
198. B. M. Trost, J.-I. Yoshida, and M. Lautens, *J. Am. Chem. Soc.*, **105**, 4494 (1983).
199. H. Matsushita and E.-I. Negishi, *J. Chem. Soc., Chem. Commun.*, 160 (1982).
200. S. Padmanabhan and K. M. Nicholas, *Tetrahedron Lett.*, **24**, 2239 (1983).
201. T. Hiyama, Y. Morizawa, H. Yamamoto, and H. Nozaki, *Bull. Chem. Soc. Jpn.*, **54**, 2151 (1981).
202. S. Ozawa, A. Itoh, K. Oshima, and H. Nozaki, *Tetrahedron Lett.*, 2909 (1979).
203. Y. Kitagawa, S. Hashimoto, S. Iemura, H. Yamamoto, and H. Nozaki, *J. Am. Chem. Soc.*, **98**, 5030 (1976).
204. K. Takai, K. Oshima, and H. Nozaki, *Tetrahedron Lett.*, **21**, 2531 (1980).
205. Y. Okuda, M. Sato, K. Oshima, and H. Nozaki, *Tetrahedron Lett.*, **24**, 2015 (1983).
206. M. Sato, K. Takai, K. Oshima, and H. Nozaki, *Tetrahedron Lett.*, **22**, 1609 (1981).
207. K. Takai, M. Sato, K. Oshima, and H. Nozaki, *Bull. Chem. Soc. Jpn.*, **57**, 108 (1984).
208. T. Hayashi, Y. Katsuro, Y. Okamoto, and M. Kumada, *Tetrahedron Lett.*, **22**, 4449 (1981).
209. J. Fujiwara, Y. Fukutani, M. Hasegawa, K. Maruoka, and H. Yamamoto, *J. Am. Chem. Soc.*, **106**, 5004 (1984).
210. D. W. Harney, A. Meisters, and T. Mole, *Aust. J. Chem.*, **27**, 1639 (1974).
211. A. Basha and S. M. Weinreb, *Tetrahedron Lett.*, 1465 (1977).
212. S. Pasynkiewicz, W. Dahlig, and L. Meszorer, *Rocz. Chem.*, **35**, 1301, (1961); *Chem. Abstr.*, **57**, 12519d (1962).
213. J. P. Kennedy, N. V. Desai, and S. Sivaram, *J. Am. Chem. Soc.*, **95**, 6386 (1973).
214. J. P. Kennedy, *J. Org. Chem.*, **35**, 532 (1970).
215. E.-I. Negishi, H. Matsushita, and N. Okukado, *Tetrahedron Lett.*, **22**, 2715 (1981).
216. E.-I. Negishi, *Acc. Chem. Res.*, **15**, 340 (1982).
217. S. Pasynkiewicz and W. Kuran, *J. Organomet. Chem.*, **15**, 307 (1968).
218. S. Baba and E.-I. Negishi, *J. Am. Chem. Soc.*, **98**, 6729 (1976); N. Okukado, D. E. Van Horn, W. L. Klima and E.-I. Negishi, *Tetrahedron Lett.*, 1027 (1978).
219. R. A. Lynd and G. Zweifel, *Synthesis*, 658 (1974).
220. G. A. Tolstikov, F. Kh. Valitov, and A. V. Kuchin, *J. Gen. Chem. USSR*, **50**, 2195 (1980); K. Takai, K. Oshima, and H. Nozaki, *Bull. Chem. Soc. Jpn.*, **54**, 1281 (1981).
221. H. Matsushita and E.-I. Negishi, *J. Am. Chem. Soc.*, **103**, 2882 (1981).
222. H. Reinheckel and D. Jahnke, *Chem. Ber.*, **97**, 2661 (1964).
223. L. Bagnell, E. A. Jeffrey, A. Meisters, and T. Mole, *Aust. J. Chem.*, **27**, 2577 (1974).
224. F. Barbot and P. Miginiac, *J. Organomet. Chem.*, **170**, 1 (1979).
225. R. S. Stevens, L. E. DuPree, and P. L. Loewenstein, *J. Org. Chem.*, **37**, 977 (1972).
226. S. Trofimenko, *J. Org. Chem.*, **29**, 3046 (1964).
227. S. Cucinella, T. Salvatori, and A. Mazzei, *Ger. Offen.*, 2700454 (1977); *Chem. Abstr.*, **88**, 7046v (1978).
228. A. Basha, M. Lipton, and S. M. Weinreb, *Tetrahedron Lett.*, 4171 (1977).
229. H. Yamamoto and K. Maruoka, *J. Am. Chem. Soc.*, **103**, 4186 (1981).
230. A. Alberola, F. A. Cermeno, and A. Anton, *Ann. Quím.*, **73**, 886 (1977); *Chem. Abstr.*, **87**, 184589z (1977).
231. W. P. Neumann, *Justus Liebigs Ann. Chem.*, **618**, 90 (1958).
232. R. Giezyński, S. Pasynkiewics, and A. Serwatowska, *J. Organomet. Chem.*, **69**, 345 (1974).
233. J. M. Klerks, D. J. Stufkens, G. Van Koten, and K. Vrieze, *J. Organomet. Chem.*, **181**, 271 (1979).
234. H. Hoberg and U. Griebsch, *Justus Liebigs Ann. Chem.*, 1516 (1977).
235. H. Wollweber, *Diels–Alder Reactions*, Georg Theime Verlag, Stuttgart, 1972.

236. S. Miyajima and T. Inukai, *Bull. Chem. Soc. Jpn.*, **45**, 1553 (1972).
237. W. R. Roush and H. R. Gillis, *J. Org. Chem.*, **45**, 4267 (1980).
238. T. Cohen and Z. Kosarych, *J. Org. Chem.*, **47**, 4005 (1982).
239. A. P. Kozikowski and E. M. Huie, *J. Am. Chem. Soc.*, **104**, 2923 (1982).
240. R. H. Schlessinger and J. A. Schultz, *J. Org. Chem.*, **48**, 407 (1983).
241. W. R. Rousch and H. R. Gillis, *J. Org. Chem.*, **47**, 4825 (1982).
242. K. J. Shea and J. W. Gilman, *Tetrahedron Lett.*, **24**, 657 (1983).
243. J. H. Lukas, A. P. Kouwenhoven, and F. Baardman, *Angew. Chem., Int. Ed. Engl.*, **14**, 709 (1975).
244. J. H. Lukas, F. Baardman, and A. P. Kouwenhoven, *Angew. Chem., Int. Ed. Engl.*, **15**, 369 (1976).
245. B. B. Snider and A. C. Jackson, *J. Org. Chem.*, **48**, 1471 (1983).
246. B. B. Snider, G. B. Phillips, and R. Cordova, *J. Org. Chem.*, **48**, 3003 (1983).
247. B. B. Snider, R. Cordova, and R. T. Price, *J. Org. Chem.*, **47**, 3643 (1982).
248. B. B. Snider and G. B. Phillips, *J. Org. Chem.*, **46**, 2789 (1983).
249. B. B. Snider and G. B. Phillips, *J. Org. Chem.*, **48**, 464 (1983).
250. B. B. Snider and D. J. Rodini, *Tetrahedron Lett.*, **21**, 1815, 3857 (1980).
251. B. B. Snider and D. K. Spindell, *J. Org. Chem.*, **45**, 5017 (1980).
252. B. B. Snider, D. J. Rodini, T. C. Kirk, and R. Cordova, *J. Am. Chem. Soc.*, **104**, 555 (1982).
253. C. P. Cartaya-Marin, A. C. Jackson, and B. B. Snider, *J. Org. Chem.*, **49**, 2443 (1984).
254. B. B. Snider, M. Karras, R. T. Price, and D. J. Rodini, *J. Org. Chem.*, **47**, 4538 (1982).
255. M. Karras and B. B. Snider, *J. Am. Chem. Soc.*, **102**, 7951 (1980).
256. B. B. Snider and E. A. Deutsch, *J. Org. Chem.*, **47**, 745 (1982).
257. B. B. Snider and J. V. Duncia, *J. Am. Chem. Soc.*, **102**, 5926 (1980).
258. B. B. Snider, D. M. Roush, D. J. Rodini, D. Gonzalez, and D. Spindell, *J. Org. Chem.*, **45**, 2773 (1980).
259. J. V. Duncia, P. T. Lansbury, T. Miller, and B. B. Snider, *J. Am. Chem. Soc.*, **104**, 1930 (1982).
260. S. Pasynkiewicz, *Vysokomol. Soedin.*, **5**, 1585 (1963); *Chem. Abstr.*, **61**, 675g (1964).
261. R. Wolovsky, N. Maoz, and Z. Nir, *Synthesis*, 659 (1970).
262. K. E. Wiegand, J. T. Kao, and M. J. Dagini, *Ger. Offen.*, 2524299 (1975); *Chem. Abstr.*, **84**, 164599n (1976).
263. B. B. Snider and A. C. Jackson, *J. Org. Chem.*, **47**, 5393 (1982).
264. F. M. Sonnenberg, *J. Org. Chem.*, **35**, 3166 (1970).
265. Y. Yamamura, K. Umeyama, K. Maruoka, and H. Yamamoto, *Tetrahedron Lett.*, **23**, 1933 (1982).
266. K. Takai, I. Mori, K. Oshima, and H. Nozaki, *Bull. Chem. Soc. Jpn.*, **57**, 446 (1984).
267. A. Alberola, A. G. Ortega, R. Pedrosa, J. L. P. Bragado, and M. Vicente, *J. Chem. Soc., Perkin Trans. 1*, 1259 (1984).
268. H. Yamamoto and K. Maruoka, *Pure Appl. Chem.*, **55**, 1953 (1981).
269. K. Hattori, Y. Matsumura, T. Miyazaki, K. Maruoka, and H. Yamamoto, *J. Am. Chem. Soc.*, **103**, 7368 (1981).
270. G. Zweifel, G. M. Clark, and R. Lynd, *J. Chem. Soc., Chem. Commun.*, 1593 (1971).
271. L. V. Gaponik, V. P. Mardykin, and P. N. Gaponik, *USSR Pat.*, 706417 (1979); *Chem. Abstr.*, **92**, 164077f (1980).
272. G. A. Tolstikov, U. M. Dzhemilev, and S. S. Shavanov, *Bull. Acad. Sci. USSR, Div. Chem. Sci.*, **25**, 2415 (1976).
273. R. Menicagli, C. Malanga, and L. Lardicci, *J. Org. Chem.*, **47**, 2288 (1982).
274. R. Menicagli, C. Malanga, L. Lardicci, and L. Tinucci, *Tetrahedron Lett.*, **21**, 4525 (1980).
275. S. Kanemoto, Y. Morizawa, K. Oshima, and H. Nozaki, *Bull. Chem. Soc. Jpn.*, **55**, 3941 (1982).
276. K. Suzuki, E. Katayama and G.-I. Tsuchihashi, *Tetrahedron Lett.*, **24**, 4997 (1983).
277. K. Suzuki, E. Katayama, T. Matsumoto and G.I. Tsuchihashi, *Tetrahedron Lett.*, **25**, 3715 (1984).
278. G.-I. Tsuchihashi, K. Tomooka, and K. Suzuki, *Tetrahedron Lett.*, **25**, 4253 (1984).
279. B. B. Snider, D. J. Rodini, and J. van Straten, *J. Am. Chem. Soc.*, **102**, 5872 (1980).
280. B. M. Trost and B. P. Coppola, *J. Am. Chem. Soc.*, **104**, 6879 (1982).
281. B. M. Trost and H. Hiemstra, *J. Am. Chem. Soc.*, **104**, 886 (1982).

The Chemistry of the Metal—Carbon Bond, Vol. 4
Edited by F. R. Hartley

CHAPTER **5**

Preparation and use of organothallium(III) compounds in organic synthesis

S. UEMURA

Institute for Chemical Research, Kyoto University, Uji, Kyoto 611, Japan

I. INTRODUCTION

Various preparative methods for organothallium compounds have been developed and a variety of tri-, di- and mono-organothallium(III) and organothallium(I) compounds have been synthesized and characterized[1–6]. Among these compounds only mono-organothallium(III) compounds prepared by direct aromatic thallation and by oxythall-

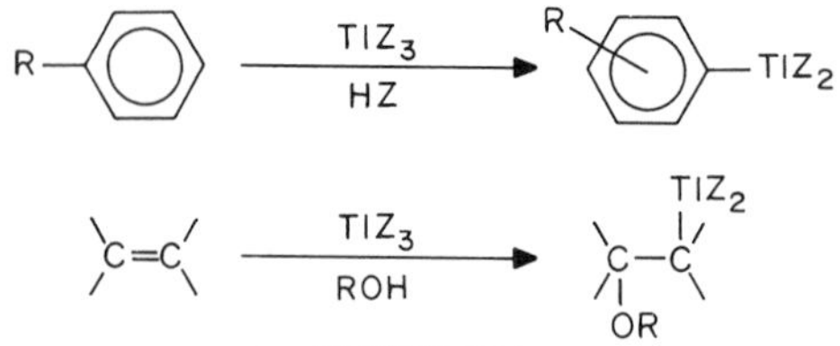

$$R-C_6H_5 \xrightarrow[HZ]{TlZ_3} R-C_6H_4-TlZ_2$$

$$>C=C< \xrightarrow[ROH]{TlZ_3} >C(OR)-C(TlZ_2)<$$

SCHEME 1

ation of olefins and related compounds with inorganic thallium(III) salts (Scheme I) seem to be generally useful in organic synthesis, considered from the viewpoint of simplicity of the method, high yield of the product, high regio- and stereo-selectivity of the reaction, and facile substitution of the thallium moiety of the resultant compounds by various other functional groups. The rate of the reactions and the stability of the mono-organothallium(III) compounds produced depend greatly on the nature of the thallium(III) salts employed. Thallium(III) trifluoroacetate [$Tl(OCOCF_3)_3$, abbreviated to ttfa] and acetate [$Tl(OAc)_3$, abbreviated to tta] are most commonly used for aromatic thallation and oxythallation, respectively, where the stable organothallium(III) compounds are to be isolated. The dissociation energy for the first C—Tl bond of trimethylthallium has been estimated as 115[7] or 152[8] kJ mol^{-1}, which is lower than the corresponding values for the indium and gallium analogues, but the C—Tl bond dissociation energy of mono- and di-organothallium(III) compounds is not yet known accurately. In this review the preparation of stable organothallium(III) compounds and their use in organic synthesis are surveyed, focusing especially on mono-organothallium(III) compounds prepared by direct aromatic thallation and oxythallation of olefins and related compounds.

For the purpose of organic synthesis it is not necessary to isolate the intermediate organothallium(III) compounds and a one-pot reaction will suffice to obtain the intended products. Thus, in oxythallation of olefins and related compounds there are many cases where the reaction proceeds through mono-organothallium(III) compounds (oxythallation adducts), but such intermediates cannot be isolated because of a facile C—Tl bond fission due to highly ionizable anions on thallium such as nitrate, perchlorate, sulphate, and trifluoroacetate. Since these reactions are usually encountered in the oxidation of olefins, cyclopropanes, acetylenes, allenes, and ketones and are synthetically very important, they are included in this review. Thallium(III) nitrate [$Tl(NO_3)_3 \cdot 3H_2O$. abbreviated to ttn], tta, and ttfa are most commonly used for this purpose, and ttn seems to be a most useful reagent in view of its very high reactivity and selectivity. The chemistry of diorganothallium(III) compounds is referred to occasionally, but that of triorganothallium(III) compounds and organothallium(I) compounds, mainly thallium(I) cyclopentadienyl derivatives, is not considered here because of their very limited significance in organic synthesis. The literature coverage is nearly complete up to the end of 1984. The reader should also consult a book on thallium chemistry[9] and many reviews on thallium in organic synthesis that have appeared previously[10–26].

II. AROMATIC THALLATION FOR ARYLTHALLIUM(III) COMPOUNDS AND ITS USE IN ORGANIC SYNTHESIS

A. Preparation of Arylthallium(III) Compounds

The preparation of arylthallium(III) halides has long been known from the transmetallation reaction between thallium(III) halides and various arylmetal compounds such as Mg, B, Hg, Pb, and Bi (Scheme 2)[1–6,27–31]. The halide anions of the compounds formed

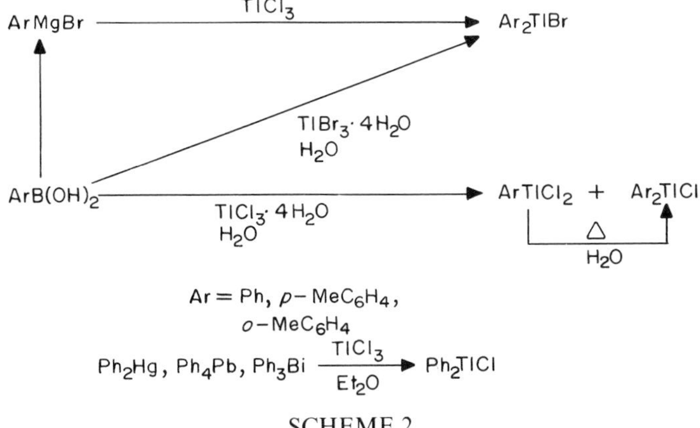

SCHEME 2

can be easily replaced by many other anions to produce various arylthallium(III) compounds, typical examples being shown in Scheme 3[27–32]. Aryllithium can also be used for this purpose[33], as well as some lithium and sodium arenesulphinates (Scheme 4)[34]. Thallium(III) carboxylates such as the acetate[34–36], isobutyrate[37,38], and trifluoroacetate[33] were also effective instead of halides.

Direct aromatic thallation by treatment of aromatic hydrocarbons with thallium(III) salts to give arylthallium(III) compounds has been developed and widely used in organic synthesis. The first example of such a thallation is the reaction of dibenzofuran with thallium(III) chloride (Scheme 5)[39]. A similar reaction occurs between benzene or anisole and thallium(III) isobutyrate under severe reaction conditions (Scheme 5)[40]. Aromatic thallation using tta is also possible (Scheme 6)[41,42], although the reaction is very slow. It is catalysed by strong acids such as perchloric, methanesulphonic, trifluoroacetic, and sulphuric acids[43–45], and this acid-catalysed reaction is found to be electrophilic from the relative reactivity of benzene and toluene and the isomer distributions in the products formed from toluene[45]. The acid–base interaction with tta produces $Tl(OAc)_2X$, $Tl(OAc)X_2$, and TlX_3 ($X = ClO_4$, HSO_4, etc.), which on ionization yield more reactive electrophiles than tta itself[45]. In fact, arylthallium(III) compounds containing a perchlorate anion were isolated from benzene, toluene, xylenes, and anisole in moderate to good yields as stable white solids (Scheme 6)[44,46]. Aromatic thallation was found to be

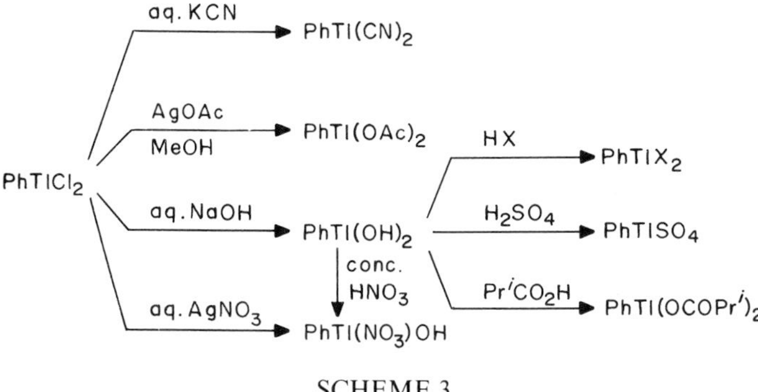

SCHEME 3

$$\text{ArLi} \xrightarrow[\text{Et}_2\text{O}]{\text{TlCl}_3} \text{Ar}_2\text{TlCl}$$

$$\text{Ar} = \text{Ph}, p\text{-MeC}_6\text{H}_4$$

$$\text{ArSO}_2\text{Li} \xrightarrow{\text{Tl(OAc)}_3} \text{Ar}_2\text{TlOAc}$$

$$\text{Ar} = \text{C}_6\text{F}_5, p\text{-HC}_6\text{F}_4, m\text{-HC}_6\text{F}_4$$

$$3\,\text{ArSO}_2\text{Na} \xrightarrow{\text{TlX}_3} \text{Ar}_2\text{TlO}_2\text{SAr}$$

$$\text{Ar} = 2,4,6\text{-Me}_3\text{C}_6\text{H}_2;\ \text{X} = \text{OAc, Cl}$$

SCHEME 4

similar to aromatic mercuration in nature but 200–400 times slower in aqueous $HClO_4$. The main differences between the two are that monothallation is nearly the sole reaction, whereas polymercuration is very facile, and disproportionation to afford diarylthallium(III) compounds occurs under many reaction conditions[43]. Direct aromatic thallation with ttn in carbon tetrachloride gives arylthallium(III) nitrate hydroxide, $ArTl(NO_3)OH$ [Ar = XC_6H_4 (X = H, Me, Et, *i*-Pr)], in 50–80% yield with a very high *para*-selectivity[47].

Aromatic thallation with ttfa in trifluoroacetic acid (abbreviated to tfa) is generally very fast, affording good to excellent yields of a wide range of arylthallium(III)

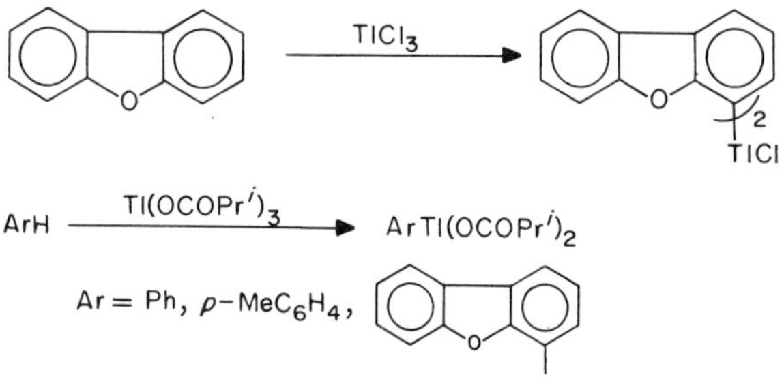

SCHEME 5

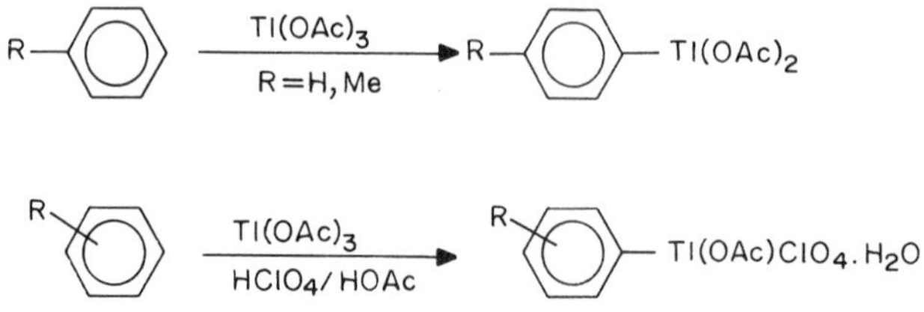

SCHEME 6

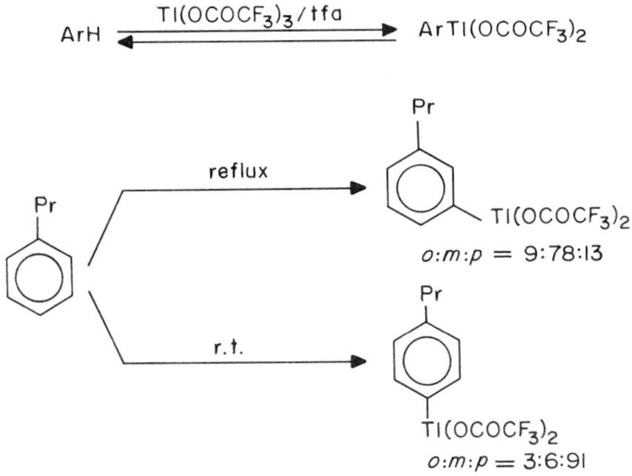

$$ArH \underset{}{\overset{Tl(OCOCF_3)_3/tfa}{\rightleftharpoons}} ArTl(OCOCF_3)_2$$

SCHEME 7

bis(trifluoroacetate)s [$ArTl(OCOCF_3)_2$] and is a reversible, electrophilic substitution (Scheme 7)[48,49]. It can be carried out similarly by using thallium(III) oxide in place of ttfa. Reaction with aromatic hydrocarbons having electron-releasing groups is generally complete within a few minutes at room temperature, whereas thallation of deactivated aromatics such as benzoic acid and trifluoromethylbenzene requires fairly vigorous conditions (refluxing tfa; 1–4 days). In addition to the simplicity and rapidity of these aromatic thallation reactions, a characteristic is that under conditions of thermodynamic and kinetic control *meta*- and *para*-substitution, respectively, are generally achieved (Scheme 7)[50,51]. Further, when chelation of thallium by ttfa with the basic centre in the side chain [CO_2R, CH_2CO_2R, CH_2OR, CH_2CH_2OR (R = H, Me)] permits intramolecular delivery of the electrophile, *ortho*-substitution occurs under conditions of kinetic control (Scheme 8)[50,51]. In the thallation of anisole and phenetole, on the other hand, *para* ⟶ *ortho* rearrangement of the resulting arylthallium(III) compounds was observed under thermodynamically controlled conditions and this was attributed to stabilization of the *ortho*-isomer by formation of a four-membered chelate ring (Scheme 9)[52,53].

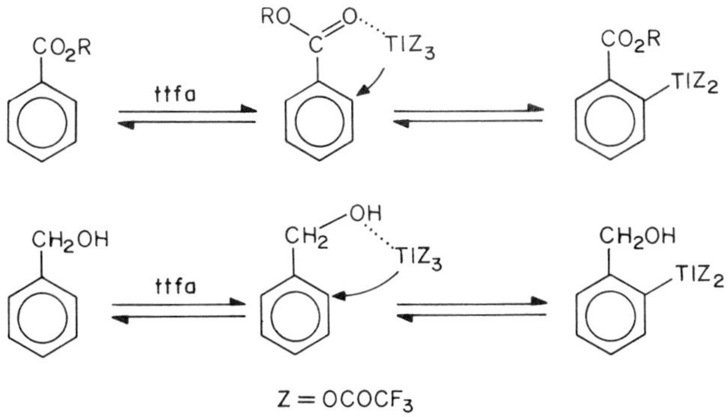

$Z = OCOCF_3$

SCHEME 8

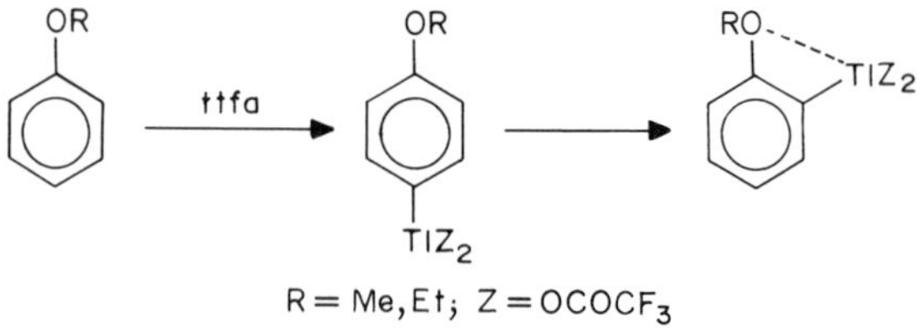

R = Me, Et; Z = OCOCF$_3$

SCHEME 9

The relative rates of thallation of toluene (k_T) and benzene (k_B) of $k_T/k_B = 33$[54] or 43.5[55] and partial rate factors for the thallation of toluene ($o_f/m_f/p_f = 9.6/5.7/168$[54] or 12.7/4.5/226[55]) at 25 °C in this aromatic thallation suggest that the reaction proceeds via a conventional mechanism, presumably through Whealand-type intermediate. There is also a report that the reaction is an electrophilic aromatic substitution (10–10^2 times slower than mercuration) with a ρ^+ value of -8.3 for a rate correlation with σ^+ indicating greater charge developement during the reaction than the corresponding mercuration process[56,57]. Recent detailed mechanistic studies of thallation of polymethylbenzenes by ttfa revealed that electrophilic (two-electron) and electron-transfer (one-electron) pathways occur simultaneously and the cationic $[Tl(OCOCF_3)_2]^+$ serves as the active electrophile as well as the active electron acceptor (Scheme 10)[58]. The extent of nuclear thallation decreases monotonically in the order mesitylene (*ca.* 100%) > durene (*ca.* 50%) > pen-

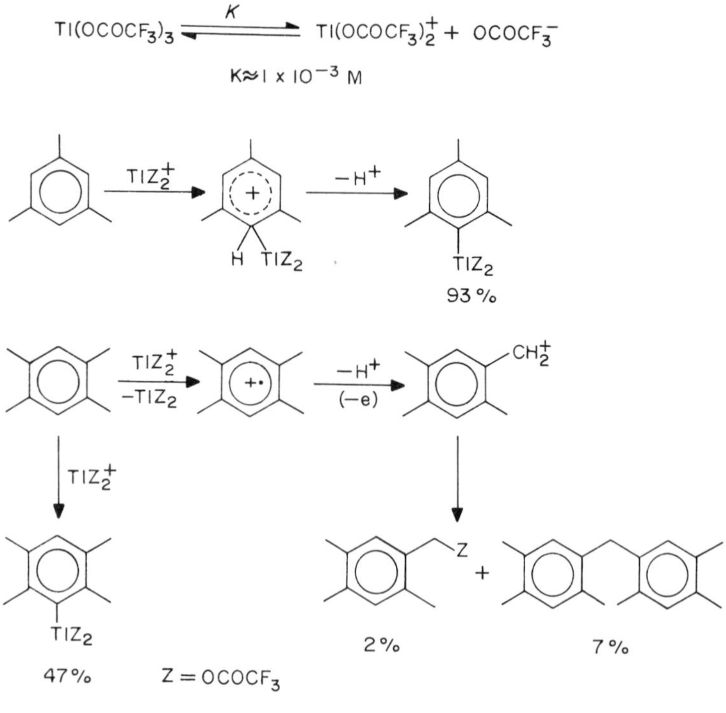

$$Tl(OCOCF_3)_3 \underset{}{\overset{K}{\rightleftharpoons}} Tl(OCOCF_3)_2^+ + OCOCF_3^-$$

$K \approx 1 \times 10^{-3}$ M

Z = OCOCF$_3$

SCHEME 10

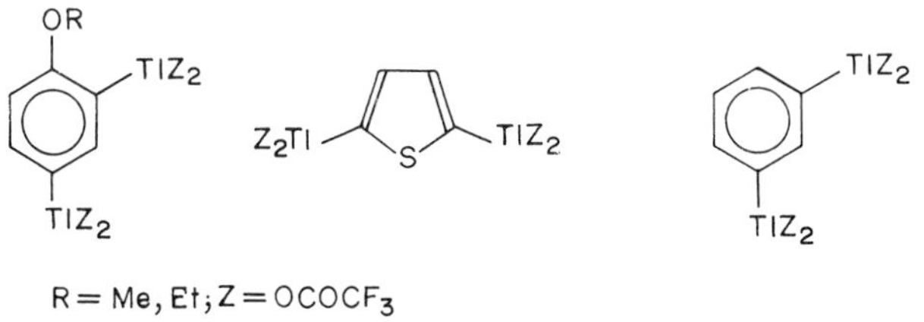

OR, TlZ$_2$, TlZ$_2$; Z_2Tl, S, TlZ$_2$; TlZ$_2$, TlZ$_2$

R = Me, Et; Z = OCOCF$_3$

SCHEME 11

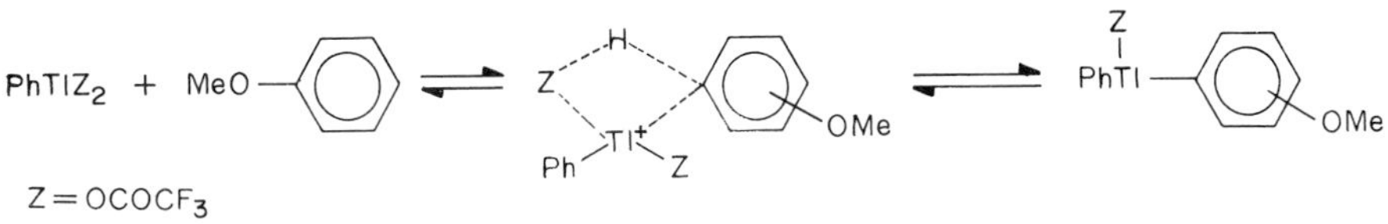

$PhTlZ_2$ + MeO–C$_6$H$_5$ ⇌ [Z···H···C$_6$H$_5$OMe, Ph–Tl$^+$–Z] ⇌ PhTl(Z)–C$_6$H$_4$–OMe

Z = OCOCF$_3$

SCHEME 12

tamethylbenzene (*ca.* 25%) with a concomitant increase in the proportion of products (dimeric aromatic hydrocarbon and benzylic and nuclear trifluoroacetates) derived from the aromatic cation radical[58].

In contrast to facile polymercuration, polythallation does not usually occur, probably because of the strong deactivating nature of the introduced $Tl(OCOR)_2$ group as described above[43]. However, when the reaction was carried out with activated aromatics by using an excess of ttfa and a longer reaction time, dithallated compounds such as those shown in Scheme 11 were produced[52,53]. Phenylthallium(III) bis(trifluoroacetate) reacts slowly with an excess of anisole to afford an isomeric mixture of methoxyphenyl (phenyl)thallium(III) trifluoroacetates (the *ortho*-isomer always predominating) (Scheme 12)[59,60].

Various $ArTl(OCOCF_3)_2$ compounds have also been prepared by *ipso*-substitution of the trimethylsilyl group of arylsilicon compounds by the $Tl(OCOCF_3)_2$ moiety in 40–95% yields (Scheme 13)[61]. The compounds undergo disproportionation to the corresponding diarylthallium(III) trifluoroacetate regiospecifically when heated with acetone and/or water or treated with trimethyl phosphite (Scheme 14)[62,63]. Although diarylthallium(III) compounds are widely believed to be chemically inert, they are useful, versatile intermediates for the synthesis of various substituted aromatic compounds[63]. Replacement of the $OCOCF_3^-$ group of $ArTl(OCOCF_3)_2$ by F^- or Cl^- results in the formation of the corresponding arylthallium(III) dihalides ($ArTlX_2$) (Scheme 14), the stability of which varies considerably with the nature of X; they are stable when X = F and Cl, unstable when

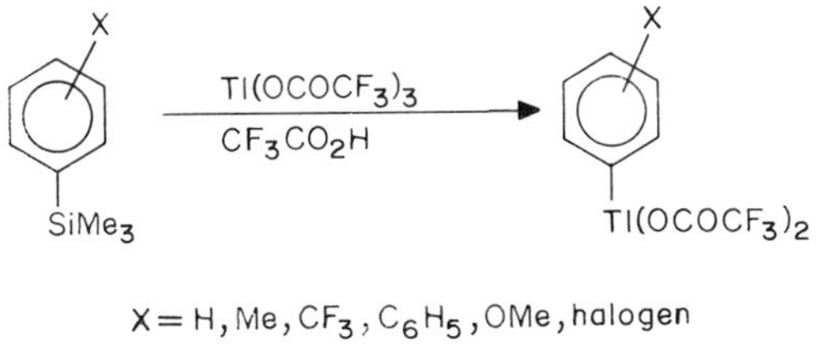

X–C$_6$H$_4$–SiMe$_3$ $\xrightarrow[CF_3CO_2H]{Tl(OCOCF_3)_3}$ X–C$_6$H$_4$–Tl(OCOCF$_3$)$_2$

X = H, Me, CF$_3$, C$_6$H$_5$, OMe, halogen

SCHEME 13

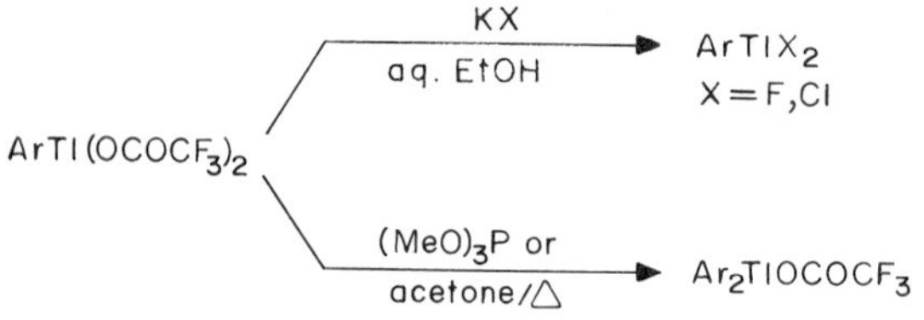

$$ArTl(OCOCF_3)_2 \xrightarrow[\text{aq. EtOH}]{KX} ArTlX_2 \quad X = F, Cl$$

$$ArTl(OCOCF_3)_2 \xrightarrow[\text{acetone}/\Delta]{(MeO)_3P \text{ or}} Ar_2TlOCOCF_3$$

SCHEME 14

X = Br, and not isolatable when X = I[30,62]. In the last case they decompose rapidly to aryl iodides and this reaction is a useful method for the regioselective introduction of iodine into aromatic rings, as will be described in the next section. Recently, a monophenylthallium(III) complex with a Tl—I bond was prepared in the presence of dibenzo-18-crown-6. This crown ether complex is stable in acetonitrile at room temperature for a long period, but at 100 °C it decomposes to give iodobenzene quantitatively (Scheme 15)[64]. Its photolysis also affords a good yield of iodobenzene, partly at least through a phenyl radical intermediate[65].

Thallium(III) trifluoromethanesulphonate, prepared from Tl_2O_3 and CF_3SO_3H, acts in tfa as a stronger aromatic thallation reagent than ttfa and it can even thallate strongly deactivated polyfluoroaromatic compounds to afford the corresponding polyfluoroaryl-thallium(III) bis(trifluoromethanesulphonate)s (Scheme 16)[66,67]. The reaction is facilitated by Lewis acids such as SbF_5 and BF_3Et_2O[68]. These polyfluoroarylthallium(III) compounds can also be prepared by normal Mg–Tl transmetallation reactions between the corresponding Grignard reagents and thallium(III) halides[69].

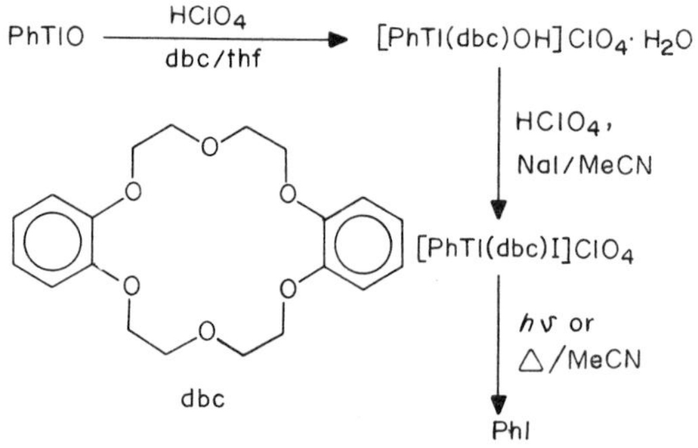

$$PhTlO \xrightarrow[\text{dbc/thf}]{HClO_4} [PhTl(dbc)OH]ClO_4 \cdot H_2O \xrightarrow[\text{NaI/MeCN}]{HClO_4,} [PhTl(dbc)I]ClO_4 \xrightarrow[\Delta/\text{MeCN}]{h\nu \text{ or}} PhI$$

dbc

SCHEME 15

$$Tl_2O_3 + 6\ CF_3SO_3H \longrightarrow 2\ Tl(O_3SCF_3)_3 + 3\ H_2O$$

$$C_6F_4H \xrightarrow[CF_3CO_2H]{Tl(O_3SCF_3)_3} C_6F_4\text{—}Tl(O_3SCF_3)_2$$

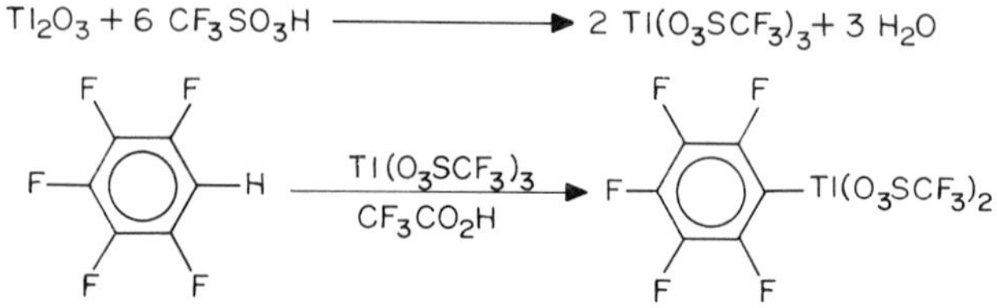

SCHEME 16

B. Arylthallium(III) Compounds in Organic Synthesis

The thallium moiety of arylthallium(III) compounds thus prepared can be easily substituted by various functional groups at the position where thallium was previously attached (*ipso*-substitution). As a result of much effort by many researchers to find useful dethallation methods, it is now known that the thallium moiety, mostly $Tl(OCOR)_2$ or TlX_2 (X = halogen), can be replaced with F, Cl, Br, I, CN, SCN, SeCN, NO, NO_2, NH_2, OH, SH, SO_2Ph, D, alkyl, aryl, vinyl and CO. These dethallations usually occur much more easily than do the reactions with the corresponding arylmetal compounds, including arylmercury(II) compounds, and are sometimes characteristic reactions of arylthallium(III) compounds themselves. Therefore, the developed dethallation method combined with a very facile aromatic thallation as described in the previous section makes arylthallium(III) compounds useful for organic synthesis.

The formation of iodobenzene observed by Challenger and coworkers[30,31] over 50 years ago in the treatment of several phenylthallium(III) compounds with potassium iodide seems to be the first example of such a reaction where neither reaction conditions nor yields were specified. McKillop and coworkers elegantly developed the synthetic utility of this spontaneous iododethallation reaction by combining a facile aromatic thallation with ttfa in tfa to prepare various kinds of aromatic iodides in good to excellent yields[48–51,70,71]. The reaction is completed by the addition of aqueous potassium iodide to tfa solution containing $ArTl(OCOCF_3)_2$ at room temperature. Since the path of aromatic thallation can be controlled by the reaction temperature and time (kinetic *vs.* thermodynamic control) and the kind of substituent (chelate etc.) as described in the previous section, this method permits the ready introduction of iodine into aromatic nuclei with all the potential for path control inherent in the initial thallation process. It is not necessary to isolate arylthallium(III) compounds prepared in tfa and the reaction is usually carried out in one flask. Typical examples are shown in Scheme 17[50,51]. This thallation–iododethallation method has been applied to the preparation of many aromatic iodides such as *o*-iodotoluic acids (75–80%)[72], l-iododibenzosuberone (71%)[73],

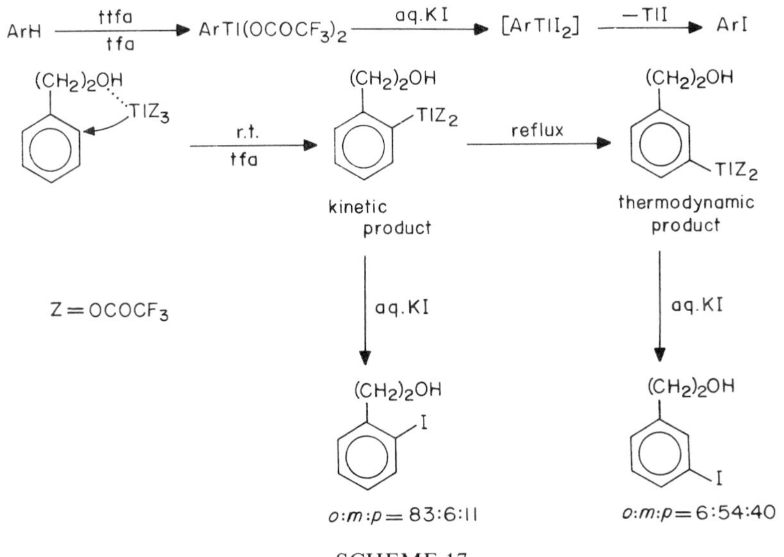

$$ArH \xrightarrow[\text{tfa}]{\text{ttfa}} ArTl(OCOCF_3)_2 \xrightarrow{\text{aq.KI}} [ArTlI_2] \xrightarrow{-\text{TlI}} ArI$$

SCHEME 17

SCHEME 18

iodopentafluorobenzene (90%)[68], and l-iodo-4-methoxytetrafluorobenzene (99%)[67]. In some instances, diiodo compounds such as 2, 5-diiodothiophene and 1-ethoxy-2, 4-diiodobenzene can be obtained selectively by using an excess of ttfa (cf. Scheme 11)[53].

Treatment of arylthallium(III) compounds with iodine in tfa[74] or $CHCl_3$[67] also results in iododethallation (74–93%) probably via electrophilic attack of iodine on the C—Tl bond. This method was applied to a high-yield synthesis of some nitroaryl iodides after nitration of $ArTl(OCOCF_3)_2$ with acetyl nitrate, which occurs *meta* to the thallium moiety[75].

Halogeno- or pseudohalogeno-dethallation of arylthallium(III) compounds occurs on treating them with the corresponding copper(II) or copper(I) salt in various organic solvents, giving good yields of the expected aromatic compounds. For example, aryl chlorides and bromides are obtained by treatment with copper(II) chloride and bromide, respectively, 1,4-dioxane being the best solvent[76,77]. Aryl cyanides are prepared by reaction with copper(II) or copper(I) cyanide in acetonitrile or pyridine[78]. Aryl thiocyanates[79] and selenocyanates[80] can be produced similarly by treatment with copper(II) thiocyanate and selenocyanate, respectively. Unsymmetrical diarylsulphones are synthesized by the reaction with copper(II) benzenesulphinate[18]. The precise reaction mechanism of these dethallations is not yet clear, and an attempt to trap phenyl or aryl radicals by nitrosodurene as the spin adduct $ArN(O^{\bullet})C_6HMe_4$ (e.s.r. technique) failed[82]. A tentatively proposed mechanism involves nucleophilic displacement assisted by coordination of copper on the ligand of thallium (Scheme 18)[77,81,83].

Although the reaction is not as fast as iododethallation, treatment of $ArTl(OCOCF_3)_2$ with potassium bromide also gives aryl bromides. It gives arylthallium(III) dibromides first, which on gentle heating decompose rapidly to aryl bromides and thallium(I) bromide[62]. Treatment of arylthallium(III) compounds with bromine in CCl_4 affords aryl bromides in excellent yields[84]. Similar electrophilic C—Tl bond fission occurs with bis(pentafluorophenyl) thallium(III) bromide[69], which gives pentafluorophenyl bromide and chloride slowly but almost quantitatively on treatment with excess of bromine and chlorine, respectively[85]. Although aryl fluorides cannot be obtained by treatment of arylthallium(III) compounds with copper(II) fluoride or tetrafluoroborate[77], transformation of the C—Tl bond to the C—F bond occurs when arylthallium(III) difluorides, prepared from the corresponding bis(trifluoroacetate)s and potassium fluoride in aqueous ethanol[62], are treated with gaseous BF_3 in non-polar solvents; the overall isolated yields are 50–70%[86]. Aryl cyanides[87] and thiocyanates[88,89] can also be prepared photochemically. Thus, irradiation of an aqueous KCN solution of $ArTl(OCOCF_3)_2$ with 300 nm light affords aryl cyanides in 27–80% yields by replacing the thallium moiety with the CN group mainly at the *ipso*-position. Aryl thiocyanates are similarly produced in 36–58% yield by using KSCN, here again the proportion of *ipso*-substitution being over 85%, irrespective of the great possibility of a homolytic process. Scheme 19 summarizes these halogeno- and pseudohalogeno-dethallations.

Electrophilic dethallations other than those by I_2, Br_2, and Cl_2 are also known. Arylthallium(III) dichlorides react with nitrosyl chloride in $CHCl_3$ to afford nitrosoarenes via a four-centred transition state (Scheme 18)[90], while arylthallium(III) compounds react

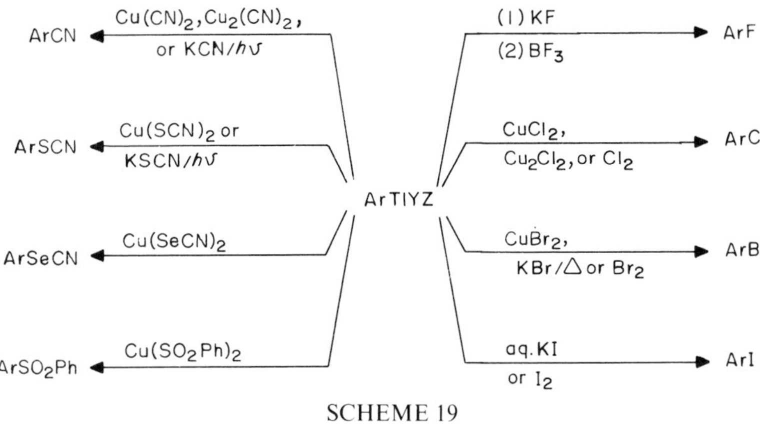

SCHEME 19

with nitrogen dioxide in tetrahydrofuran to give nitroarenes[91]. Treatment of arylthallium(III) compounds with metal nitrites such as $NaNO_2$, KNO_2, and $AgNO_2$ in tfa first produces nitrosoarenes, which are subsequently oxidized to nitroarenes in high yields, electrophilic attack of NO^+ or its carrier N_2O_3 on the carbon of C—Tl bond having been proposed[92,93].

Arylthallium(III) compounds react with palladium(II) salts in acetic acid to give aromatic coupling products. The reaction might involve an electrophilic transmetallation to give reactive arylpalladium(II) species[94]. This aromatic coupling can be carried out catalytically with respect to the palladium(II) salt, and thus oxidation of arenes by ttfa in the presence of catalytic amounts of palladium(II) acetate affords biaryls in good yields (Scheme 20)[95,96]. The thallation of arenes and substitution of the thallium moiety for palladium(II) in $ArTl(OCOCF_3)_2$ are characterized by Hammett plots with slopes of -5.6 (ρ^+) and -3.0 (ρ), respectively[96]. Similar aromatic coupling with a catalytic amount of lithium tetrachloropalladate has also been developed to produce 4,4′-biaryls highly selectively and in good yields[97]. These methods are not applicable to arenes with bulky substituents or arylthallium(III) compounds having a substituent at the *ortho*-position[95–97]. When suitable olefins are present in the reaction system, the arylpalladium(II) species derived from arylthallium(III) compounds add to olefins, followed by dehydropalladation to give arylated olefins, the overall reaction being the replacement of an olefinic hydrogen by the aryl group[98]. Recently Larock *et al.* developed a novel and general synthetic method for isocoumarins and 3,4-dihydroisocoumarins (a biologically important ring system) by the reaction of *ortho*-thallated benzoic acid with

$$\mathrm{ArH} + \mathrm{TlZ_3} \longrightarrow \mathrm{ArTlZ_2} + \mathrm{HZ}$$

$$\mathrm{ArTlZ_2} + \mathrm{Pd(OAc)_2} \longrightarrow \mathrm{ArPdOAc} + \mathrm{TlZ_2(OAc)}$$

$$2\mathrm{ArPdOAc} \xrightarrow{\text{fast}} \mathrm{ArAr} + \mathrm{Pd(OAc)_2} + \mathrm{Pd^0}$$

$$\mathrm{Pd^0} + \mathrm{Tl^{III}} \xrightarrow{\text{fast}} \mathrm{Pd^{II}} + \mathrm{Tl^{I}}$$

$$\mathrm{Z} = \mathrm{OCOCF_3}$$

SCHEME 20

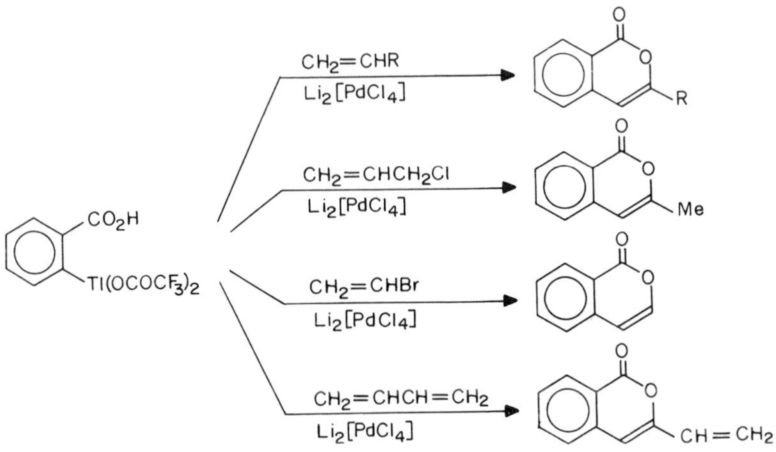

SCHEME 21

palladium(II) chloride in the presence of simple olefins, dienes, allylic halides, vinylic halides, or vinylic esters (Scheme 21)[99]. The yields are good and the reactions using organic halides or 1,2- and 1,3-dienes proceed catalytically with respect to palladium(II). Application of this thallation–olefination reaction to *p*-tolylacetic acid, *N*-methylbenzamide, benzamide, and acetanilide provides a novel route to a variety of important oxygen and nitrogen heterocycles (Scheme 22)[100].

The Tl–Pd transmetallation reaction has also been applied to a carbonyl insertion into the C—Tl bond of arylthallium(III) compounds. Phenylthallium(III) compound has been known to react with carbon monoxide to give benzoic acid or its methyl ester, but the reaction requires high temperatures (*ca.* 100 °C), high pressures (*ca.* 200 atm), and long reaction times (*ca.* 18 h)[101]. In the presence of a palladium(II) salt, however, the reaction proceeds very smoothly under lower CO pressures to give the carboxylic acid derivates[102,103]. The CO insertion occurs in the C—Pd bond of the arylpalladium(II) species

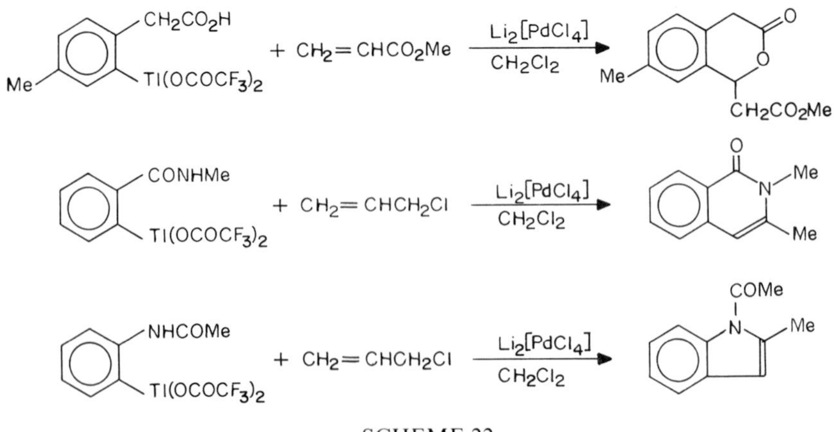

SCHEME 22

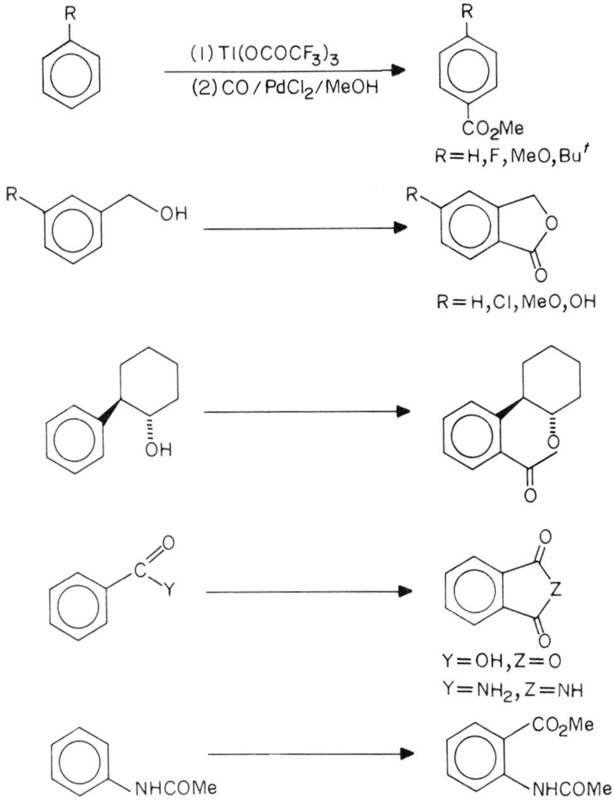

(1) $Tl(OCOCF_3)_3$
(2) $CO/PdCl_2/MeOH$

CO_2Me
R=H,F,MeO,Bu^t

R=H,Cl,MeO,OH

Y=OH,Z=O
Y=NH_2,Z=NH

NHCOMe

CO_2Me
NHCOMe

SCHEME 23

before coupling to form biaryls takes place. The aromatic thallation and subsequent palladium-catalysed carbonylation of various arenes (at room temperature under 1 atm of CO) provides a convenient new route to a wide variety of aromatic esters, lactones, anhydrides, and phthalimides (40–90% yields), the reaction being highly stereo- and regio-specific (Scheme 23)[103,104].

Another example of metal–metal exchange is the reaction of diborane with arylthallium(III) compounds in tetrahydrofuran to give arylboron intermediates, which on oxidation by alkaline H_2O_2 or on hydrolysis give good yields of phenols or arylboronic acids, respectively[105]. Treatment of arylthallium(III) compounds with lead (IV) acetate–triphenylphosphine followed by alkaline hydrolysis also gives phenols[87]. Thiophenols can be prepared via photolysis of $ArTl(S_2CNMe_2)_2$[10]. Reduction of arylthallium(III) compounds with $NaBH_4$ in EtOD is a convenient method for the specific introduction of a single deuterium atom (D content 73–85%) at the position where thallium was attached previously (protonodethallation)[106]. When $NaBD_4$ is used here, the percentage of D-incorporation increases.

Asymmetric biphenyls can be prepared photochemically in high yields when arylthallium(III) compounds are irradiated in benzene at room temperature[107]. Homolysis of the aryl C—Tl bond followed by capture of the resulting aryl radical by benzene

$$ArTl(OAc)_2 + R_2\underline{C}^{-}(NO_2)\;Li^{+} \xrightarrow[h\nu]{N_2} R_2C(NO_2)Ar$$

$$\begin{cases} ArTl(OAc)_2 + R_2\bar{C}NO_2 \longrightarrow ArTlOAc + OAc^{-} + R_2\dot{C}NO_2 \\ ArTlOAc \longrightarrow Ar\cdot + TlOAc \\ Ar\cdot + R_2\dot{C}NO_2 \longrightarrow R_2C(Ar)NO_2 \end{cases}$$

SCHEME 24

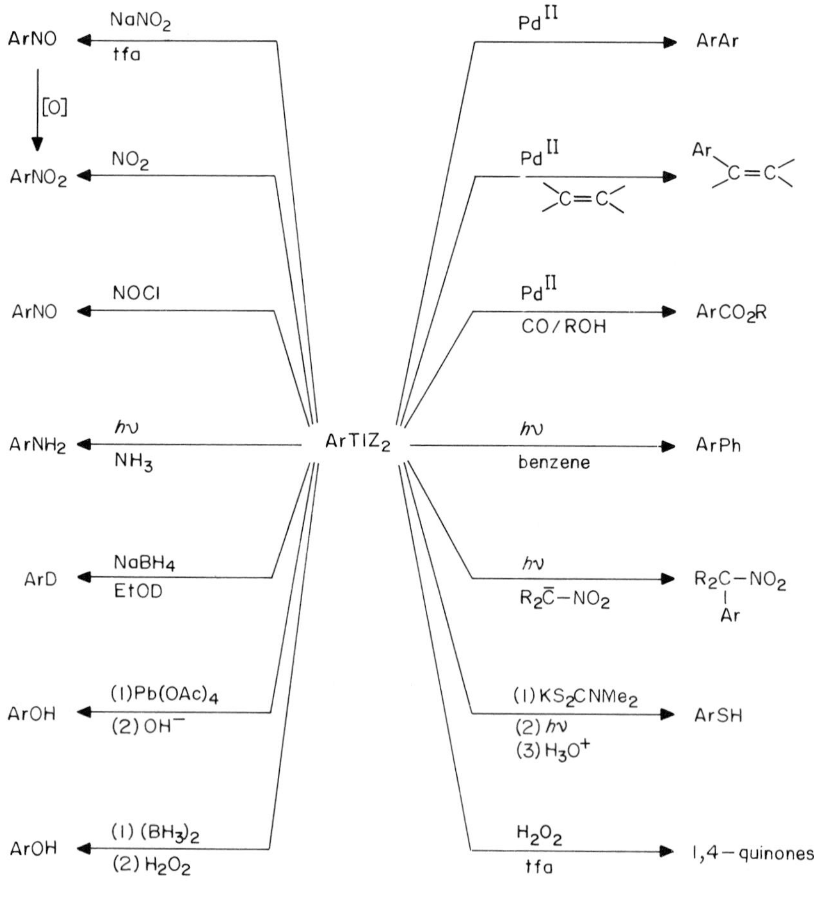

SCHEME 25

leads to the products. Photolysis of phenylthallium(III) bis(trifluoroacetate) in the presence of ammonia is said to give aniline[10]. The reaction of arylthallium(III) diacetates with the anions of nitroalkanes (nitronate ions) under photolysis gives the C—C bonded products in 60–70% yields. It proceeds through radical intermediates which are generated by electron-transfer activation of the C—Tl bond (Scheme 24)[108]. The reaction is also applicable to alkyl- and vinyl-thallium(III) compounds, as will be described in Section III.B.

Oxidation of $ArTl(OCOCF_3)_2$ with 90% H_2O_2 gives 40–70% yields of 1,4-quinones with either elimination or migration of the substituent group on the starting arenes, depending on the nature of the group and on the electron density of the ring. The first step of the reaction is thought to be hydroxydethallation and the phenols formed are readily oxidized *in situ* to the ensuing products. The oxidation of phenols to quinones by ttfa is known.[110]

All these substitution reactions other than halogeno- and pseudohalogeno-dethallations are summarized in Scheme 25. These dethallations can be applied to diarylthallium(III) trifluoroacetates[63] and $ArTl(OCOCF_3)_2$[111].

III. OXYTHALLATION OF ALKENES, ALKYNES, AND ALLENES FOR ORGANOTHALLIUM(III) COMPOUNDS AND ITS USE IN ORGANIC SYNTHESIS

A. Preparation of Alkyl- and Vinyl-thallium(III) Compounds

The transmetallation method employed for arylthallium(III) compounds is also applicable to the preparation of a variety of alkyl- and vinyl-thallium(III) compounds. Various combinations of organometallic compounds of Mg, B, Hg, Zn, Al, Si, Sn, Pb, Bi, Cr, etc., with thallium(III) halides and carboxylates have been developed to produce di- and mono-organo (alkyl or vinyl) thallium(III) compounds (Scheme 26)[1–6]. Some monoalkylthallium(III) compounds can be prepared by the reaction shown in Scheme 27[112–114]. The compounds thus prepared are often useful for

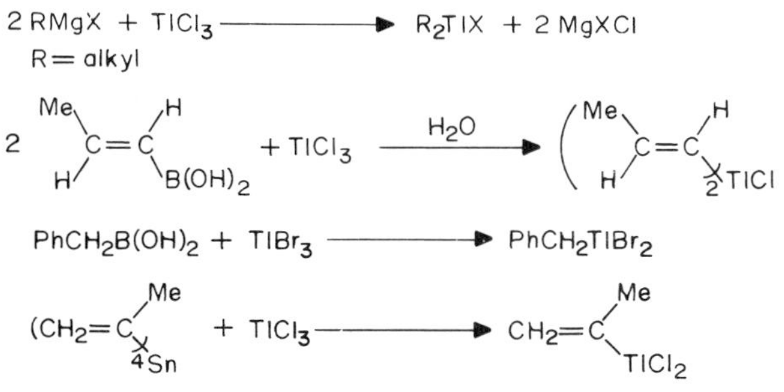

$$2\,RMgX + TlCl_3 \longrightarrow R_2TlX + 2\,MgXCl \quad R = \text{alkyl}$$

$$2\; \text{Me(H)C=C(H)B(OH)}_2 + TlCl_3 \xrightarrow{H_2O} (\text{Me(H)C=C(H)})_2TlCl$$

$$PhCH_2B(OH)_2 + TlBr_3 \longrightarrow PhCH_2TlBr_2$$

$$(CH_2{=}C(Me))_4Sn + TlCl_3 \longrightarrow CH_2{=}C(Me)TlCl_2$$

SCHEME 26

$$Me_2TlOAc + Hg(OAc)_2 \longrightarrow MeTl(OAc)_2$$

$$(Me_3YCH_2)_2TlCl + Br_2 \longrightarrow Me_3YCH_2Tl(Cl)Br + Me_3YCH_2Br \quad Y = C, Si$$

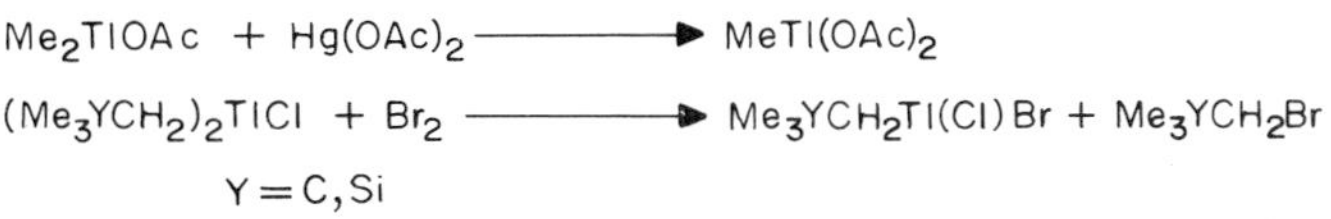

SCHEME 27

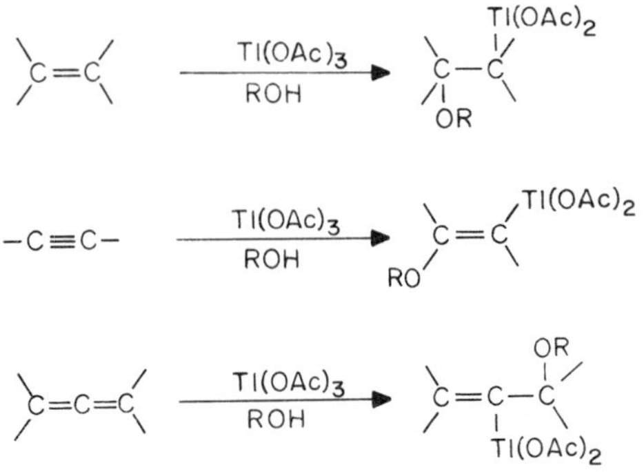

$$\mathrm{>C{=}C<} \xrightarrow[\mathrm{ROH}]{\mathrm{Tl(OAc)_3}} \mathrm{>C(OR){-}C(Tl(OAc)_2)<}$$

$$\mathrm{{-}C{\equiv}C{-}} \xrightarrow[\mathrm{ROH}]{\mathrm{Tl(OAc)_3}} \mathrm{(RO)C{=}C(Tl(OAc)_2)}$$

$$\mathrm{>C{=}C{=}C<} \xrightarrow[\mathrm{ROH}]{\mathrm{Tl(OAc)_3}} \mathrm{>C{=}C(Tl(OAc)_2){-}C(OR)<}$$

SCHEME 28

basic studies to clarify the nature of the C—Tl bond chemically and spectroscopically, but have not been frequently used in organic synthesis.

Oxythallation of alkenes, alkynes, and allenes with tta is a simple and unique method for the direct preparation of mono-organothallium(III) compounds (oxythallation adducts), which are very useful for organic synthesis (Scheme 28). The reaction closely resembles to the popular oxymercuration[18,115,116]. When thallium(III) salts such as nitrate (ttn), perchlorate, and trifluoroacetate (ttfa) are used, the corresponding oxythallation adducts cannot be isolated because of a facile C—Tl bond fission resulting in the formation of various oxidation products. These reactions will be described in the next section. Even with tta the oxythallation adducts produced [mono-organothallium(III) diacetates] are thermodynamically unstable compared with oxymercuration adducts and the number of isolated adducts is still limited. The first example of isolation is the methoxythallation adduct of styrene, **1**, which was prepared almost quantitatively in the reaction of tta with styrene in methanol at room temperature[36,117]. Other compounds so far isolated from olefins and characterized are **2**[118,119], **3**[118], **4**[120,121], **5**[121], **6**[36], **7**[122,123], **8**[124], **9**[125], **10**[121], **11**[121], **12**[126], **13**[127], **14**[127], **15**[128,129], **16**[129], **17**–**19**[130], **20**[131], and **21**[129] (Scheme 29). Similar compounds, **22**, have also been isolated from styrene, methanol and ethanol, and thallium(III) isobutyrate in place of tta[118]. The reactions have generally been carried out at or below room temperature by using acetic acid, alcohols, aqueous tetrahydrofuran, chloroform, and dichloromethane as the solvent to isolate these compounds. All of the compounds are white crystalline solids except **8** and **9**, which are viscous oils. The stereochemistry of oxythallation of norbornene and norbornadiene **13** and **14** and that of *trans-β*-deuteriostyrene (**4**) was shown to be *cis-exo*[127,132] and *trans*[120], respectively, by ^{1}Hn.m.r. conformational analysis. Oxythallation generally involves electrophilic attack of the $Tl(OCOR)_2{}^+$ species followed by nucleophilic attack by solvents such as acetic acid and alcohols. The addition obeys the Markownikoff rule and $Tl(OAc)_2{}^+$ is the only important reactive species in the oxidation of olefins with tta[133]. An intramolecular nucleophilic attack of oxygen atom occurs occasionally, as evidenced by the formation of **6**, **20**, and **21**. In the case of norbornene derivatives only acetoxythallation occurs even in methanol as the solvent, in sharp contrast to the oxymercuration of those olefins where the main reaction is methoxymercuration[129]. A concerted or near-concerted addition of tta to such strained olefins through a cyclic intermediate is proposed for this reaction[129].

Azidothallation adducts of norbornene and benzonorbornene, **23**–**25**, have been isolated by the treatment of the olefin with a mixture of tta and trimethylsilylazide in dichloromethane. A series of intermediate thallium(III) species, $Tl(OAc)_{3-n}(N_3)_n$ $(n = 0-3)$, are involved (Scheme 30)[128].

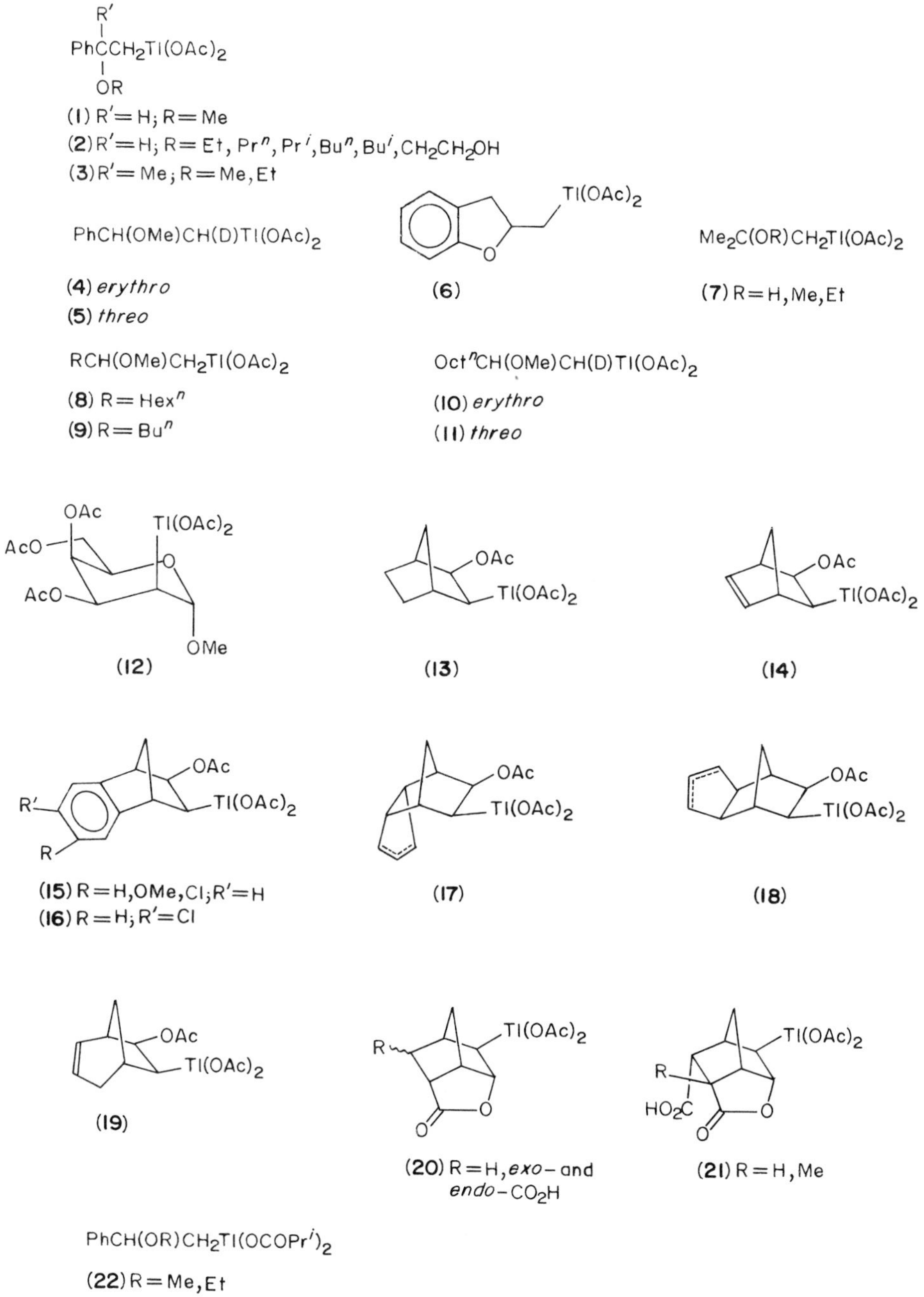
PhC(R')(OR)CH2Tl(OAc)2
(1) R'=H; R=Me
(2) R'=H; R=Et, Pr^n, Pr^i, Bu^n, Bu^i, CH2CH2OH
(3) R'=Me; R=Me, Et
PhCH(OMe)CH(D)Tl(OAc)2
(4) erythro
(5) threo
Tl(OAc)2
O
(6)
Me2C(OR)CH2Tl(OAc)2
(7) R=H, Me, Et
RCH(OMe)CH2Tl(OAc)2
(8) R=Hex^n
(9) R=Bu^n
Oct^nCH(OMe)CH(D)Tl(OAc)2
(10) erythro
(11) threo
OAc
AcO
AcO
Tl(OAc)2
O
OMe
(12)
OAc
Tl(OAc)2
(13)
OAc
Tl(OAc)2
(14)
R'
R
OAc
Tl(OAc)2
(15) R=H, OMe, Cl; R'=H
(16) R=H; R'=Cl
OAc
Tl(OAc)2
(17)
OAc
Tl(OAc)2
(18)
OAc
Tl(OAc)2
(19)
R
Tl(OAc)2
O
O
(20) R=H, exo- and endo-CO2H
Tl(OAc)2
R
HO2C
O
O
(21) R=H, Me
PhCH(OR)CH2Tl(OCOPr^i)2
(22) R=Me, Et

SCHEME 29

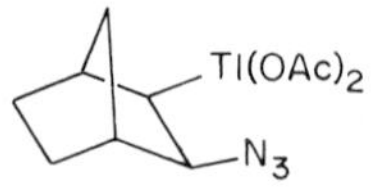

(23) 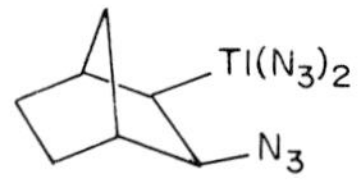 (24) 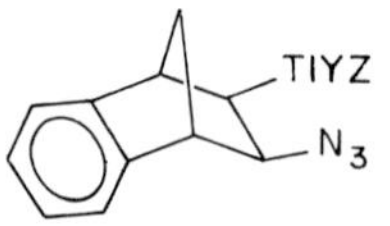 (25) Y = Z = OAc; Y = Z = N_3; Y = OAc, Z = NCO

SCHEME 30

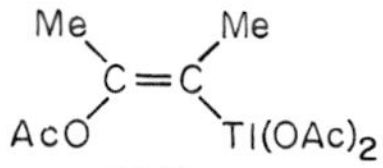

(26) 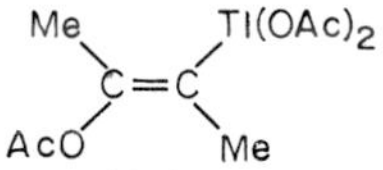 (27) 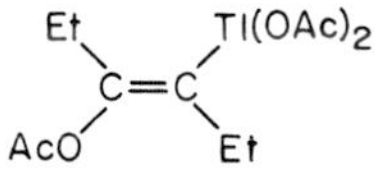 (28)

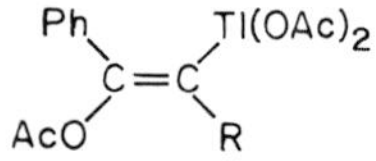

(29) (30) (31)

R = Me, Et, Pr^n, Bu^n R = Me, Et, Pr^n, Bu^n R = Ph, Me, Bu^n, Hex^n

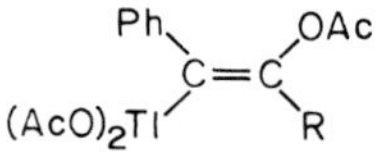

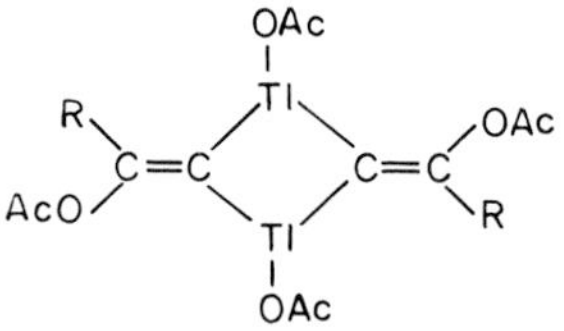

SCHEME 31

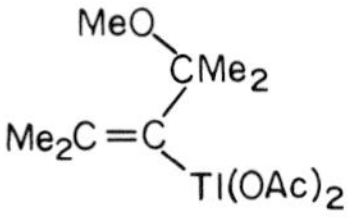

(32) 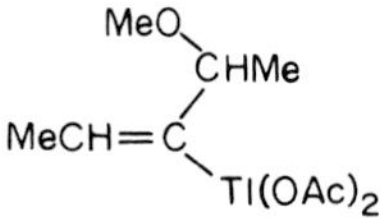 (33) 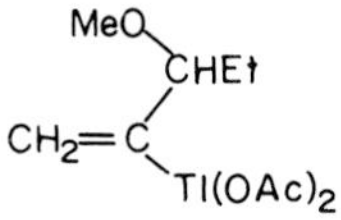 (34)

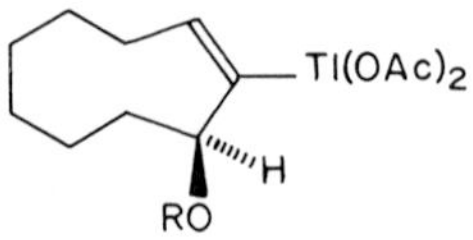

(35) R = Me, MeCO

SCHEME 32

Acteoxythallation of internal acetylenes with tta in acetic acid affords some stable monovinylthallium(III) compounds (acetoxythallation adducts) in high yields. The compounds so far isolated are **26** and **27** from dimethylacetylene[134], **28** from diethylacetylene[134], and **29** and **30** from alkylphenylacetylenes (Scheme 31)[135,136]. The stereochemistry of addition is *trans* except in the case of dimethylacetylene, from which a mixture of *cis*- and *trans*-adducts **26** and **27** is isolated[134–136]. From alkylphenylacetylenes the regioisomeric mixtures **29** and **30** are formed, the ratio depending slightly on R although **29** always predominates[136]. Similar treatment of terminal acetylenes in acetic acid, chloroform, or dichloromethane results in the direct formation of novel and stable divinylthallium(III) compounds, **31**, which are thought to be the intermediates of thallium(III)-catalysed conversion of terminal acetylenes to carbonyl compounds[137,138].

Methoxy- and acetoxy-thallation of acyclic and cyclic allenes is another direct method for monovinylthallium(III) compounds, although examples are very limited. The isolated stable compounds so far known are **32**–**34**[139] and **35**[140,141] (Scheme 32). Electrophilic thallium(III) species bind regiospecifically to the central sp carbon of the allenic moiety with nucleophilic attack of the solvent occurring at the terminal carbon. Methoxythallation proceeded by greater than 70% antarafacial addition for **35** (R = Me).

B. Alkyl- and Vinyl-thallium(III) Compounds (Oxythallation Adducts) in Organic Synthesis

Various reactivities of isolated oxythallation adducts [$RTl(OAc)_2$] have been studied and many useful methods for replacement of the thallium moiety by other functional groups have been explored in which the C—Tl bond fission occurs either heterolytically or homolytically. The thallium moiety $Tl(OAc)_2$ can be substituted by acetoxy or alkoxy groups to give primarily the corresponding alkyl acetates or ethers when heated in suitable solvents or treated with acid, as exemplified in the reactivities of **1**[36,117,118], **6**[36,42], **7** (R = H)[122,123], **13** and **14**[127], and **20** (R = H)[131] (Scheme 33). All these reactions involve dethallation, giving inorganic thallium(I) salts, some being accompanied by migration of a substituent such as phenyl, hydroxy, or alkoxy. These are closely related to the oxidation of olefins by thallium(III) salts via non-isolable oxythallation adducts, the chemistry of which will be described in the next section. The decomposition of **1** in aqueous methanol to phenylacetaldehyde and its dimethyl acetal follows a first-order rate law and $RTl(OAc)_2$ [$R = C_6H_5CH(OMe)CH_2$] is shown to be dissociated at low concentrations yielding two reactive species, $RTlOH^+$ and RTl^{2+}, the latter of which is much more reactive than the former[142]. The mode of decomposition of **1** depends on the ligand bound to thallium. At low Cl^- concentrations the rate of the oxidative decomposition decreases, whereas at high Cl^- concentrations $RTlCl_2$ or $RTlCl_3^-$ is formed and decomposed to the starting materials, styrene and thallium(III) salt, the chemical behaviour being analogous to that of the corresponding organomercury(II) compound (oxymercuration adduct)[143].

Treatment of **1** and **2** with copper(I) iodide, bromide, and chloride in acetonitrile results in the replacement of the $Tl(OAc)_2$ moiety by halogen to afford the corresponding halides in good yields (Scheme 34)[118]. Addition of potassium halide increases the product yield and iododethallation even proceeds automatically with only potassium iodide present, as found with arylthallium(III) compounds. Similarly, the thallium moiety can be replaced by CN with copper(I) cyanide[118], SCN with potassium and/or copper(I) thiocyanate[118,144], and SeCN with potassium selenocyanate[80]. This halogeno- and pseudohalogenodethallation occur at the position where thallium was attached previously to the alkyl carbon, no phenyl group migration being observed. For alkyl thiocyanates and selenocyanates it is not necessary to isolate the intermediate organothallium(III) compounds and in fact their high-yield syntheses are conducted by the *in situ* oxythallation of olefins followed by reaction with solid KSCN and KSeCN (Scheme 35)[80,144]. The

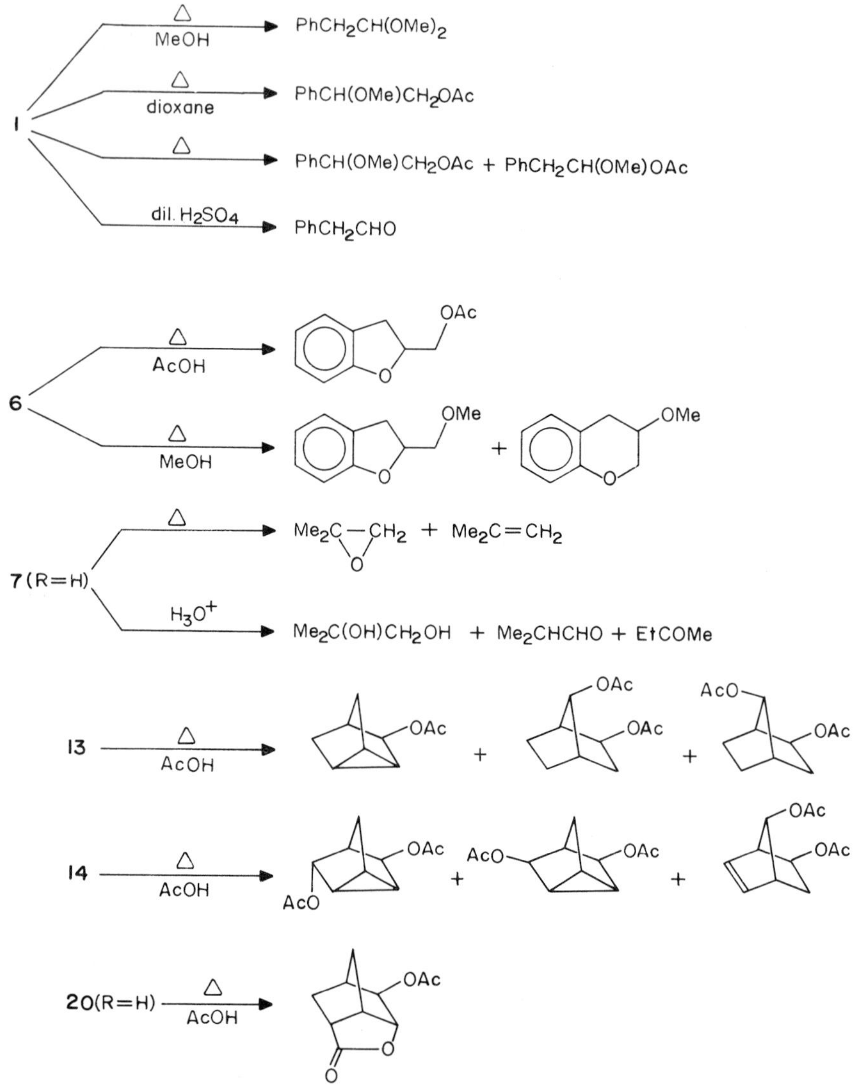

1 —Δ, MeOH→ $PhCH_2CH(OMe)_2$

1 —Δ, dioxane→ $PhCH(OMe)CH_2OAc$

1 —Δ→ $PhCH(OMe)CH_2OAc + PhCH_2CH(OMe)OAc$

1 —dil. H_2SO_4→ $PhCH_2CHO$

6 —Δ, AcOH→ 2-(acetoxymethyl)-2,3-dihydrobenzofuran (OAc)

6 —Δ, MeOH→ 2-(methoxymethyl)-2,3-dihydrobenzofuran (OMe) + 3-methoxychroman (OMe)

7 (R=H) —Δ→ $Me_2C{-}CH_2$ (epoxide, O) + $Me_2C{=}CH_2$

7 (R=H) —H_3O^+→ $Me_2C(OH)CH_2OH + Me_2CHCHO + EtCOMe$

13 —Δ, AcOH→ OAc + OAc, OAc + AcO, OAc

14 —Δ, AcOH→ OAc, AcO + AcO, OAc + OAc, OAc

20(R=H) —Δ, AcOH→ OAc, O, O

SCHEME 33

1,2 —CuX/KX, MeCN, reflux→ $PhCH(OR)CH_2X$

X=I,Br,Cl,CN,SCN,SeCN

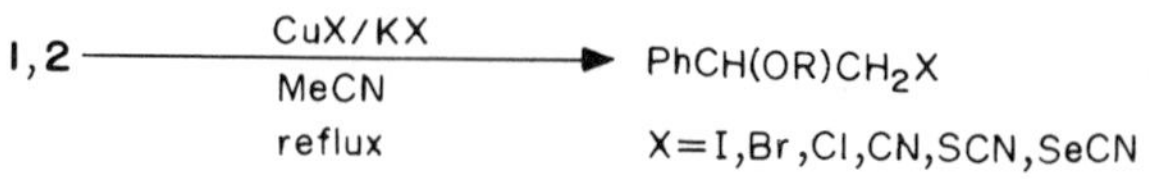

SCHEME 34

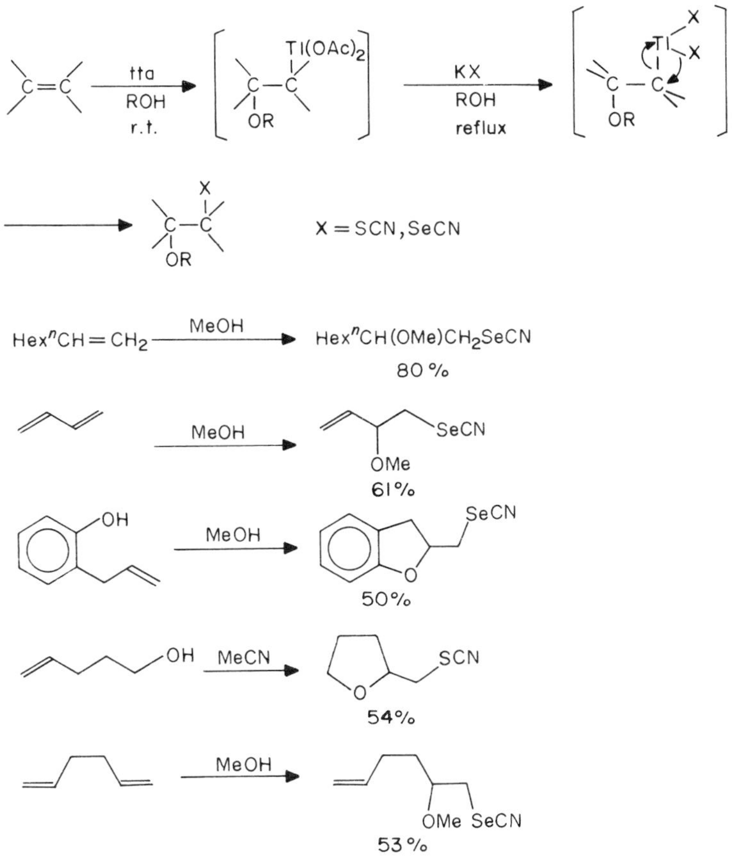

$X = SCN, SeCN$

$Hex^nCH{=}CH_2 \xrightarrow{MeOH} Hex^nCH(OMe)CH_2SeCN$ 80%

SCHEME 35

reaction is likely to proceed through the formation of organothallium(III) dithiocyanates and diselenocyanates followed by S_Ni-type decomposition. Bromodethallation of **9** and its non-isolable trifluoroacetate analogue occurs very smoothly to afford a quantitative yield of the corresponding alkyl bromide on treatment with solid KBr and a catalytic amount of 2, 6-dimethyl-18-crown-6, the most likely pathway of the reaction being S_N2 displacement after and/or before anion exchange on the carbon atom bearing the $Tl(OAc)_2$ moiety (Scheme 36)[125]. The mechanism of copper(I) halide-mediated halogenodethallation is more complicated and governed by the reaction temperature. In the reaction in acetonitrile at 80 °C a radical path involving $RCH(OMe)CH_2^{\bullet}$ [$R = C_6H_5$, *n*-Oct] free radicals accounts for approximately two thirds of the product, whereas at 60 °C an ionic process predominates[121,145]. The role of copper(I) is probably to act as a reducing agent by transferring one electron to thallium, producing a labile organothallium(II) compound.

The replacement of the thallium moiety of oxythallation adducts with hydrogen (protonodethallation) occurs by reduction with sodium amalgam in water to give good yields of alcohols[127]. However, it is difficult to effect this using $NaBH_4$ reduction, and in neutral solution the reduction gives the parent olefin solely and almost quantitatively

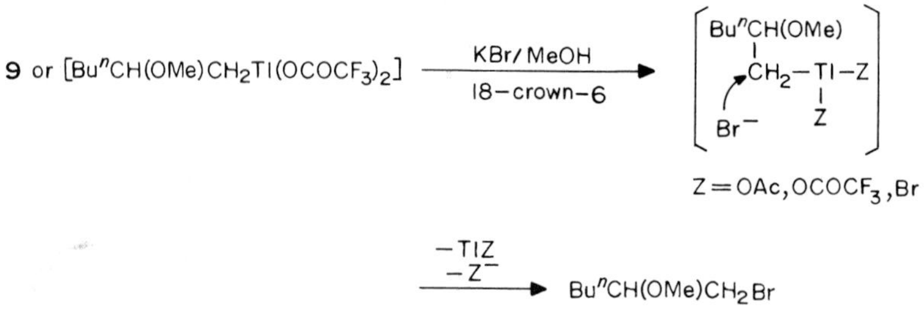

9 or [$Bu^nCH(OMe)CH_2Tl(OCOCF_3)_2$] $\xrightarrow[\text{18-crown-6}]{\text{KBr/MeOH}}$ [$Bu^nCH(OMe)CH_2$–Tl(Z)–Z ← Br^-]

Z = OAc, $OCOCF_3$, Br

$\xrightarrow[-Z^-]{-TlZ}$ $Bu^nCH(OMe)CH_2Br$

SCHEME 36

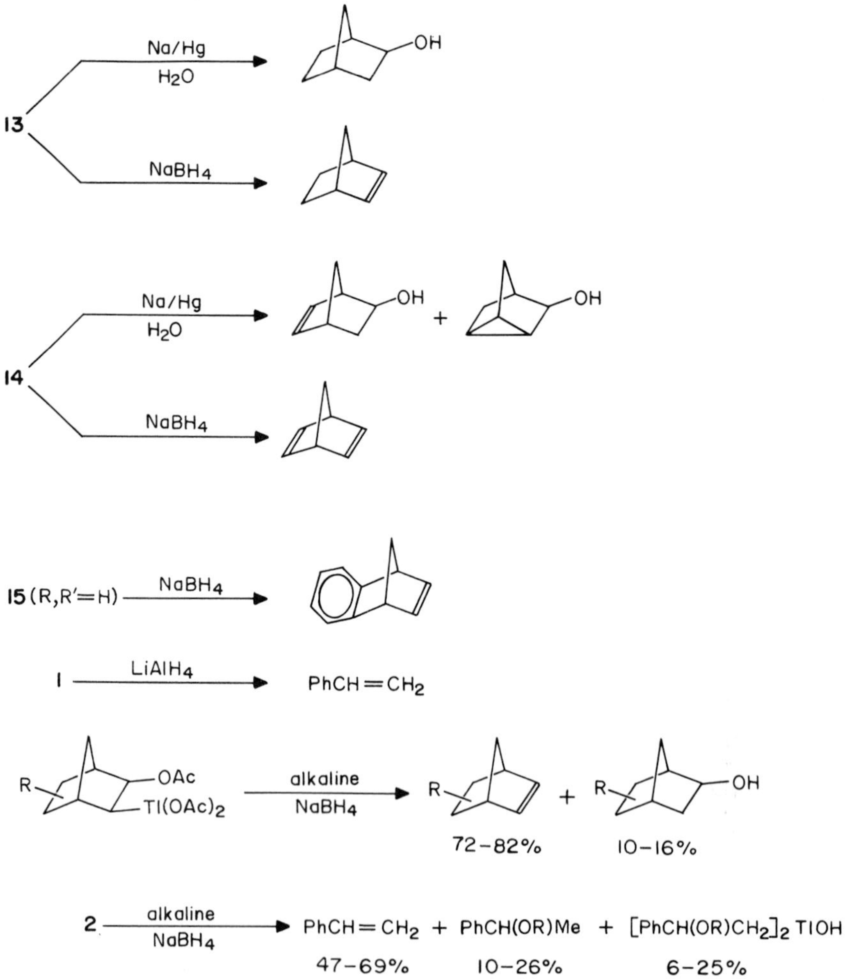

13 $\xrightarrow[H_2O]{Na/Hg}$ [norbornan-2-ol]

13 $\xrightarrow{NaBH_4}$ [norbornene]

14 $\xrightarrow[H_2O]{Na/Hg}$ [norbornen-5-ol] + [nortricyclanol]

14 $\xrightarrow{NaBH_4}$ [norbornadiene]

15 (R,R′=H) $\xrightarrow{NaBH_4}$ [benzonorbornadiene]

1 $\xrightarrow{LiAlH_4}$ $PhCH{=}CH_2$

[R-substituted 2-OAc-3-Tl(OAc)$_2$-norbornane] $\xrightarrow[NaBH_4]{\text{alkaline}}$ [R-norbornene] + [R-norbornan-2-ol]

72–82% 10–16%

2 $\xrightarrow[NaBH_4]{\text{alkaline}}$ $PhCH{=}CH_2$ + $PhCH(OR)Me$ + $[PhCH(OR)CH_2]_2TlOH$

47–69% 10–26% 6–25%

SCHEME 37

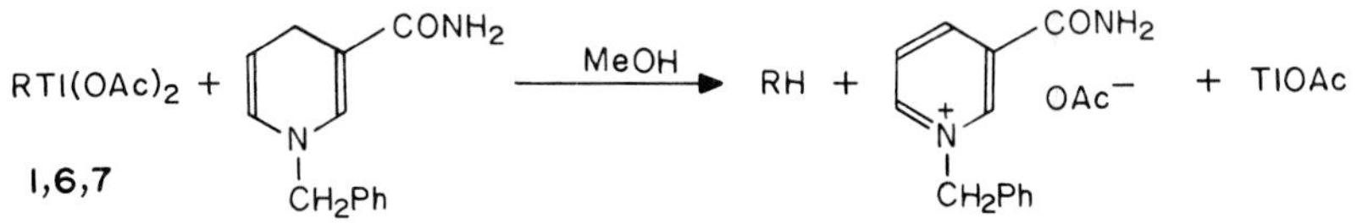

$$RTl(OAc)_2 + \text{[N-benzyl-1,4-dihydronicotinamide]} \xrightarrow{MeOH} RH + \text{[N-benzyl-3-carbamoylpyridinium]}\ OAc^- + TlOAc$$

1,6,7

SCHEME 38

(Scheme 37)[127,129]. The $LiAlH_4$ reduction of **1** is known to afford styrene and thallium metal[36]. Under alkaline conditions the $NaBH_4$ reduction gives a mixture of the parent olefins, alkyl ethers, or alcohols, and in some cases dialkylthallium(III) compounds[129,146]. The yield of the protonodethallation product is, however, very poor compared with the reduction of the corresponding oxymercuration adducts[147–149]. Hydrogen for replacement of the thallium moiety arises from the solvent and not from $NaBH_4$[146], in sharp contrast to the mercury case[147,150]. The reduction of oxythallation adducts such as **1**, **6**, and **7** with an nadh model *N*-benzyl-1,4-dihydronicotinamide (bnah) gives an 88–95% yield of the corresponding protonodethallation product under an N_2 atmosphere (Scheme 38)[151]. One-electron transfer from bnah to the organothallium(III) compound and homolysis of the C—Tl bond of the intermediate organothallium(II) compound are proposed. In the presence of oxygen, alkyl radicals are trapped to form alcohols in good yields[151]. Similarly, alcohol formation (32–67% yield) was also observed when ascorbic acid was added as the reducing agent and stereochemical studies using **4** showed that the product was a 1:1 mixture of **36-e** and **36-t** and that the reduction proceeded via homolysis of the C—Tl bond (Scheme 39)[152].

Treatment of ethylene, propylene, or styrene with active methylene compounds such as acetylacetone and ethyl acetoacetate in the presence of tta gives dihydrofuran derivatives in moderate yields. *In situ* formation of oxythallation adducts followed by their ionic reaction with active methylene carbons has been assumed for this novel C—C bond formation[153,154]. Another carbon alkylation reaction has been observed in the treatment of **4** and **6** with the anions of nitroalkanes (nitronate ions) under irradiation and a nitrogen atmosphere (Schemes 24 and 40)[108]. The reaction proceeds through radical intermediates which are generated by electron-transfer activation of the C—Tl bond. This C—C bond formation reaction is also applicable to vinylthallium(III) compounds, the reaction proceeding retentively and being unaffected by irradiation and/or oxygen (Scheme 40)[108].

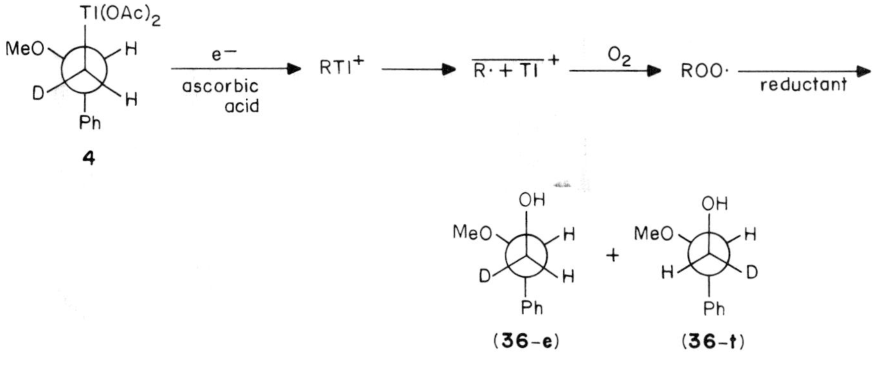

$$\mathbf{4} \xrightarrow[\text{ascorbic acid}]{e^-} RTl^+ \longrightarrow \overline{R\cdot + Tl}^{\,+} \xrightarrow{O_2} ROO\cdot \xrightarrow{\text{reductant}} (\mathbf{36\text{-}e}) + (\mathbf{36\text{-}t})$$

SCHEME 39

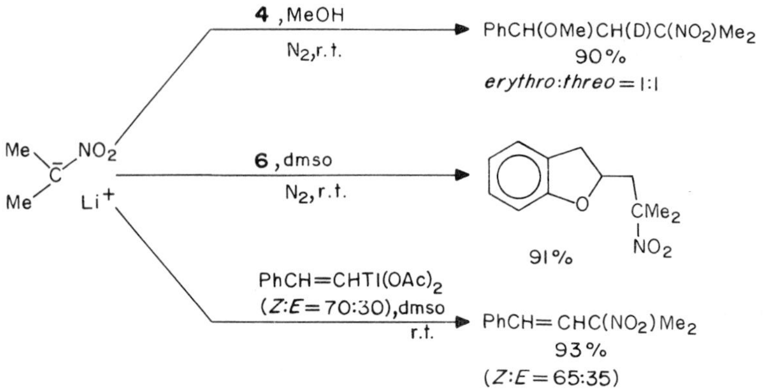

SCHEME 40

Reactions of **1** and **2** with palladium(II) species give the ketones and/or their acetals via Tl-Pd transmetallation. For example, **2** (R = CH_2CH_2OH) behaves completely differently in the absence and presence of a palladium(II) salt (Scheme 41)[119]. Hydride migration at the organopalladium(II) intermediate is postulated from the fact that no deuterium incorporation occurs in MeOD as the solvent in the initial stage of the reaction (Scheme 41)[119,155].

Several studies on the reactivities of monovinylthallium(III) compounds such as oxythallation adducts of acetylenes and allenes have been carried out and informations on protono-, halogeno-, and pseudohalogeno-dethallation are available. Thus, protonodethallation proceeds smoothly and retentively by heating in acetic acid or by reduction with $NaBH_4$ in a protic solvent under neutral conditions[136,138]. In the reduction hydrogen comes mainly from the solvent[136,138], as in reductive protonodethallation of oxythallation adducts of olefins[146] and arylthallium(III) compounds[106] as described above in this section. Halogeno- and pseudohalogeno-dethallation is conducted by treatment with the corresponding copper(II) and/or potassium salts in acetonitrile, although the product yields are sometimes low (11–74%)[136]. Bromodethallation also occurs in over 70% yield using bromine, but the stereochemistry of the product depends on the solvent employed:

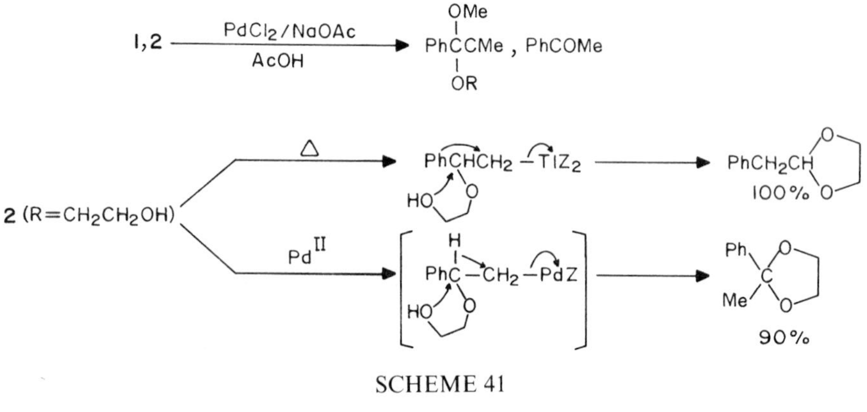

SCHEME 41

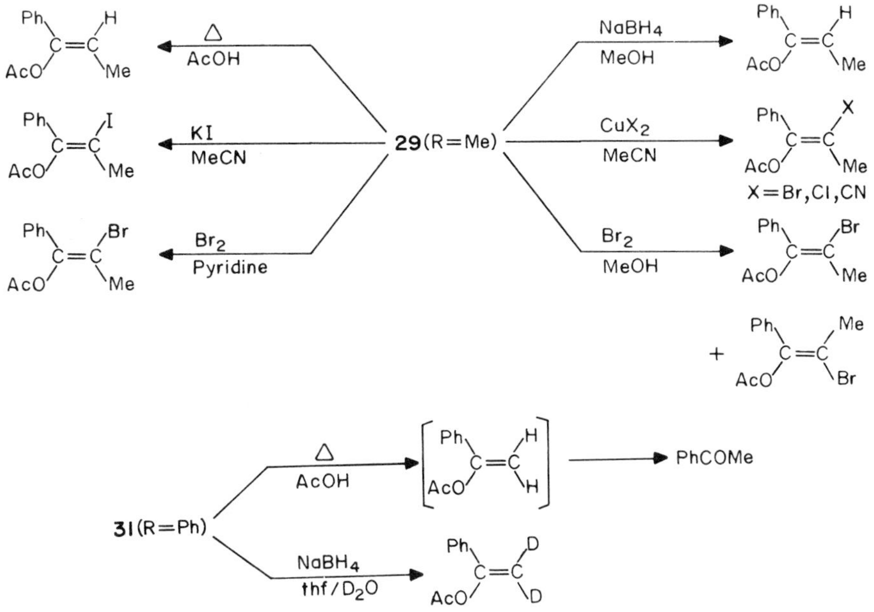

SCHEME 42

retention in pyridine and scrambling in methanol. Typical results are shown in Scheme 42. Similarly, protono- and bromo-dethallation are also known to occur when **35** is treated with alkaline $NaBH_4$–MeOH and Br_2–CCl_4, respectively (61–82% yields)[140,141].

C. Oxidation of Alkenes via Oxythallation and Related Reactions

In the previous section it was indicated that the C—Tl bond of oxythallation adducts of olefins is thermally cleaved when heated in some solvents, resulting in a replacement of the $Tl(OAc)_2$ moiety by oxygen functional groups. The overall reaction can be considered as the oxidation of olefins with tta; actually it involves the oxythallation of olefins followed by dethallation of the resultant monoalkylthallium(III) diacetates to give the oxidation products and thallium(I). So far many oxidation reactions of olefins by thallium(III) salts have been explored without intermediate oxythallation adducts being isolated; the products are usually glycols, their mono- and di-esters, aldehydes, ketones, and epoxy compounds. The product distribution depends greatly on the reaction conditions such as solvent, temperature and time, olefin structure, and the kind of thallium(III) salt used. The oxidation is often accompanied by migration of a substituent such as phenyl, vinyl, alkyl, hydroxy, alkoxy, or hydrogen. Alkyl group migration resulting in a facile ring contraction or ring enlargement has also been observed. In this section, synthetically useful and important oxidation reactions of a variety of alkenes and related reactions such as allylic oxidation and the oxidation of cyclopropanes are described, together with some mechanistic considerations of these reactions. Most reactions proceed via *in situ* oxythallation, but some are assumed to involve organothallium(III) compounds produced *in situ* in different ways.

1. *Simple olefins*

The first example of such a reaction is the oxidation of ethylene by thallium(III) hydroxide or oxide in aqueous HNO_3 and H_2SO_4 to afford ethylene glycol and some carbonyl compounds, an oxythallation–dethallation sequence giving a carbonium ion being proposed for the first time.[156] Similarly, treatment of hex-2-ene with tta in aqueous acetic acid gave a mixture of hexane-2, 3-diol monoacetates and hexan-2-one, a high water content favouring the latter[156]. Kinetic studies on these oxidations revealed the following: (1) the reaction is first order both in ethylene and in thallium(III) ion: (2) it is strongly accelerated on increasing the salt concentration, the order of magnitude of this effect, being perchlorate > sulphate > nitrate; (3) hydroxythallation is the rate-determining step; (4) the effect of olefin structure on rate decreases in the order isobutene ≫ propylene ~ but-1-ene > *cis*-but-2-ene > *trans*-but-2-ene > ethylene[133,157,158]. Deuterium isotope effects in thallium(III) perchlorate oxidation of ethylene also support the view that hydroxythallation is the rate-limiting step[159]. An example of the effect of olefin structure on product distribution in tta oxidation in aqueous acetic acid at 25 °C is shown in Scheme 43 together with the relative reaction rates[133]. Similar studies on the relationship between olefin structure and product distribution were also carried out for thallium(III) sulphate oxidation in aqueous H_2SO_4 or methanol[160,161] and for ttn oxidation in methanol[162,163]. Thus, in the decomposition of oxythallation adducts, the following results were found: (1) the hydrogen migration (as hydride) with the participation of the neighbouring hydroxy group is more favoured than the methyl migration; (2) with increasing temperature the quantity of carbonyl compounds decreases compared with diols or diethers in the oxidation of internal alkenes, whereas the reverse occurs with disubstituted terminal alkenes; (3) steric effects influence the ratio of ketone and aldehyde formation in the oxidation of disubstituted terminal alkenes in an aqueous medium, whereas in methanolic medium polar effects are decisive[161].

The tta oxidation in acetic acid or alcohols has been applied to many simple olefins such as cyclohexene[36,164–166], styrene and its derivatives[36,167], oct-1-ene[124,168], *trans*-oct-4-ene[169], vinylferrocene[170], and vinylic acetates and ethers[171]. Thus, with cyclohexene nearly equal amounts of *trans* and *cis*-1, 2-diacetoxycyclohexanes and ring-contracted cyclopentyl compounds are the main products (35–55% yield) in acetic acid as the solvent, whereas the ring-contracted compound is the sole product (62%) in methanol. The ratio of the *trans* and *cis* products depends on the water content of the solvent. The results were explained by the reaction pathway involving dethallation of the *trans*-acetoxythallation

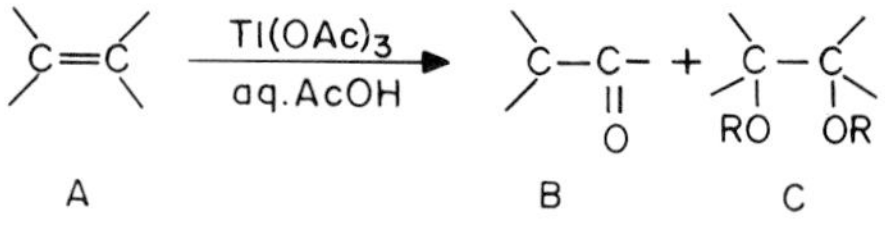

$$\underset{\mathbf{A}}{>C{=}C<} \xrightarrow[\text{aq.AcOH}]{Tl(OAc)_3} \underset{\mathbf{B}}{>C{-}\underset{O}{\overset{\|}{C}}{-}} + \underset{\mathbf{C}}{>\underset{RO}{C}{-}\underset{OR}{C}<}$$

A	Rel. rate	Yield (%) B		C
Ethylene	1	45	(MeCHO)	55
Propylene	152	81	(Me_2CO)	17
But-2-ene	157	75	(MeCOEt)	16
cis-But-2-ene	60	85–90	(MeCOEt)	< 0.5
trans-But-2-ene	35	85–90	(MeCOEt)	< 0.5
Isobutene	2.3×10^5	37	(Me_2CHCHO)	52

SCHEME 43

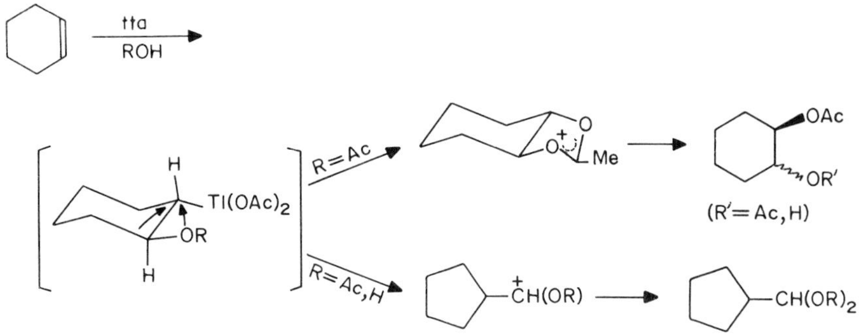

SCHEME 44

adduct to give an acetoxonium ion or ring-contracted intermediate (Scheme 44)[164–166]. From an accurate analysis of the products of the tta oxidation of oct-1-ene and *trans*-oct-4-ene in methanol or acetic acid, the following conclusions were drawn: (1) an acetate group in the $Tl(OAc)_2$ moiety can be transferred by an S_N1 process; (2) a carbonium ion is generated on heterolysis of the C—Tl bond and this undergoes competitive hydride shift and nucleophilic attack; (3) anchimeric assistance to heterolysis is provided by the neighbouring methoxy substituent[124,169]. Interestingly, in the ttfa oxidation of oct-1-ene, it is suggested that the neighbouring thallium substituent aids the hydrolysis of the introduced trifluoroacetoxy group to give several oxidation products when the initial oxythallation adduct is treated with water (Scheme 45)[172]. In a detailed study of the oxidation of six ring-substituted styrenes by tta in acetic acid, which involves aryl migration and gives a high yield of the products, a Hammett-type correlation with Brown σ^+ values has been established with $\rho = -2.2$, showing a carbonium ion character of the activated complex for oxythallation[167]. The decomposition of an oxythallation adduct has been revealed to proceed via a carbonium ion intermediate by loss of thallium(I) on the basis of the effect of the styrene structure on the product distribution (Scheme 46)[167]. Similar work on the ttn oxidation of 1, 1-diphenylethylenes in methanol, which gives

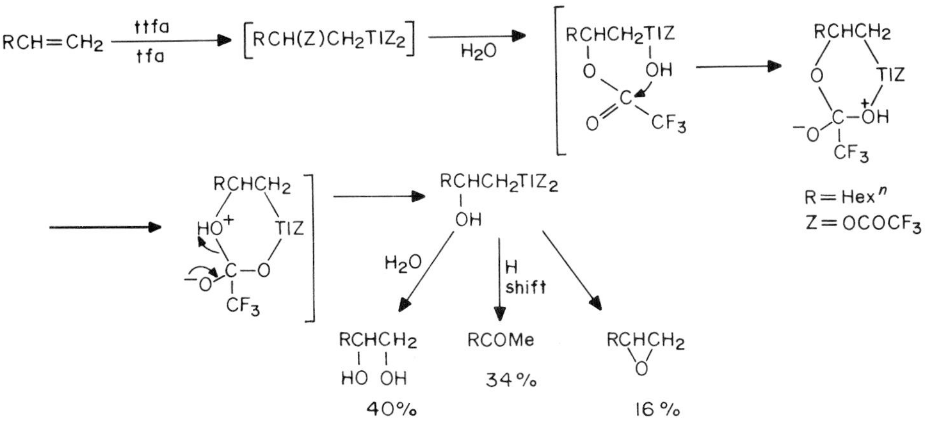

SCHEME 45

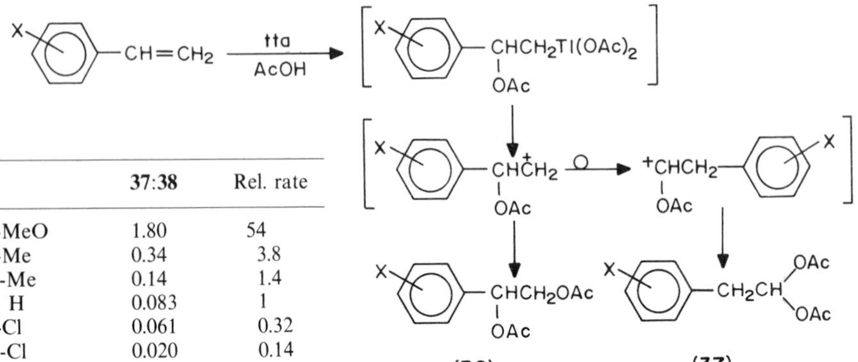

X	**37**:**38**	Rel. rate
p-MeO	1.80	54
p-Me	0.34	3.8
m-Me	0.14	1.4
H	0.083	1
p-Cl	0.061	0.32
m-Cl	0.020	0.14

SCHEME 46

rapidly and selectively the corresponding ketones in high yields, resulting from the migration of aryl groups, clearly revealed that electron-releasing substituents strongly favour the migration of the corresponding aryl group (Scheme 47)[173].

In aqueous tetrahydrofuran, tta promotes the epoxidation of propylene and isobutene to give good yields of the corresponding oxide (72 and 82% selectivity, respectively) via hydroxythallation, giving **7** (R = H), followed by its dethallation accompanied by neighbouring hydroxy group participations (Scheme 33)[122]. Similar epoxidation is also reported in the patent literature[174]. These epoxidations are stoichiometric reactions with respect to thallium(III) salts (Scheme 48). Recently, a different approach to olefin

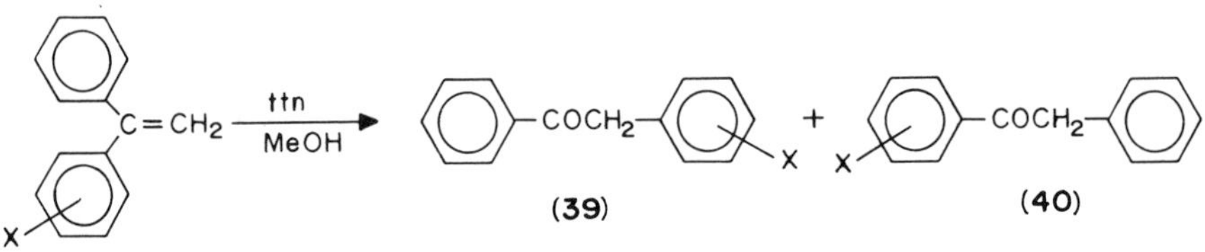

X	**39**:**40**
p-MeO	⩾ 50
p-Me	5.2
H	1
p-Cl	0.52
m-Cl	0.13

SCHEME 47

$$RCH{=}CH_2 + H_2O + TlZ_3 \longrightarrow RCH{-}CH_2\ (\text{epoxide, }-O-) + 2\,HZ + TlZ$$

SCHEME 48

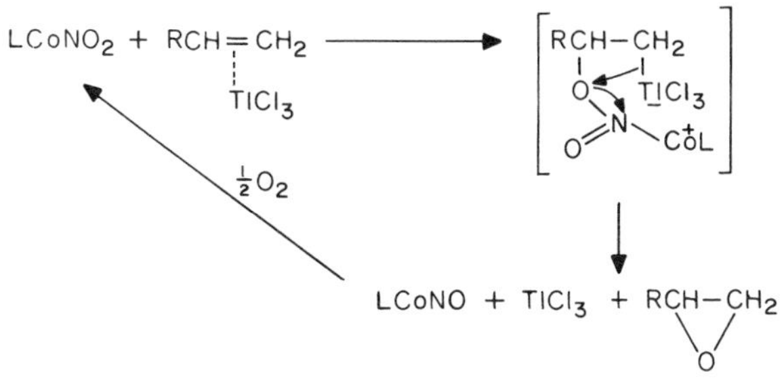

$$LCoNO_2 + RCH{=}CH_2 \cdots TlCl_3 \longrightarrow \left[RCH{-}CH_2 \text{ (O-N(=O)-}\overset{+}{Co}L\text{, } TlCl_3) \right] \longrightarrow LCoNO + TlCl_3 + RCH{-}CH_2 \text{ (epoxide)}; \quad LCoNO + \tfrac{1}{2}O_2 \rightarrow LCoNO_2$$

SCHEME 49

epoxidation by molecular oxygen has appeared where the epoxidation of oct-1-ene and propylene was carried out using cobalt nitro complexes as oxygen transfer agents in the presence of thallium(III) chloride[175]. The reaction is assumed to proceed via activation of olefin by thallium(III), as shown in Scheme 49. Importantly, during the epoxidation in theory thallium remains as thallium(III). Actually, however, reduction of thallium(III) to thallium(I) occurs through a side reaction and it is necessary to use an excess of $TlCl_3$.

The application of the tta oxidation to slightly complicated olefins at room temperature resulted in the formation of several interesting products; typical examples are shown in Scheme 50. Dibenzobicyclo[2.2.2]octatriene, **41**, is converted into **42** via *cis*-acetoxythallation[176]; 5α-cholest-2-en-5-ol, **43**, 5α-cholest-2-ene, 5α-cholest-2-en-6-one, and 5-hydroxy-5α-cholest-2-en-6-one give mainly the corresponding *cis*-hydroxylation products[177]; epicholesterol, **44**, affords **45** as the major product via a Westphalen-type methyl rearrangement of the acetoxythallation adduct of **44**[178]; thujopsene, **46**, gives a stereospecifically ring-contracted ketone **47** via acetoxythallation and an allylic acetate **48**, probably via an allylic thallium(III) compound, higher temperatures favouring allylic oxidation[179].

As exemplified in the reactivity of **6** and **20** (R = H) in Scheme 33, an intramolecular oxythallation followed by dethallation can afford interesting oxidation products. Thus, in the oxidation of styrenes in diols, $HO(CH_2)_nOH$, the main products were 1,3-dioxolane, 1,3-dioxane, and, 1,3-dioxepane derivatives for $n = 2, 3$, and 4, respectively (17–81%), the reaction involving both phenyl migration and interamolecular attack of the hydroxy group of the introduced 2-hydroxyethoxy group[168]. More clear-cut examples are shown in Scheme 51: norbornene-5-*endo*- and -*exo*-carboxylic acid and their analogues, **49**–**51**, afford various lactones[131,180,181], and a novel functionalization of the prostaglandin skeleton occurs with $PFG_{2\alpha}$ methyl ester, **52**, and related compounds, **53**, producing dioxatricyclic and oxabicyclic compounds, respectively[182,183].

Although tta and ttfa are very effective oxidizing agents of olefins as described so far, the reagent ttn has proved to be even more effective in several ways, including product selectivity and reaction rate, owing to the highly ionic nature of the nitrate anion leading to an extraordinarily facile C—Tl bond fission. Thus, the ttn oxidation of simple olefins results in a facile and selective oxidative rearrangement via oxythallation to give high yields of carbonyl compounds or their acetals[184,185]. Typical examples of ring contraction and aryl migration are shown in Scheme 52, together with some results of the unrearrangement reaction. Comparison of the oxidation of cyclohexene by these three salts revealed the approximate order of reactivity to be ttn (instantaneously) > ttfa (several minutes) > tta (several hours) for completion at room temperature[185]. The influence of the ligand on thallium on the product distribution can be clearly seen from the oxidation of dec-1-ene[185], hex-1-ene[125], and oct-1-ene[124] (Scheme 53). The ring-

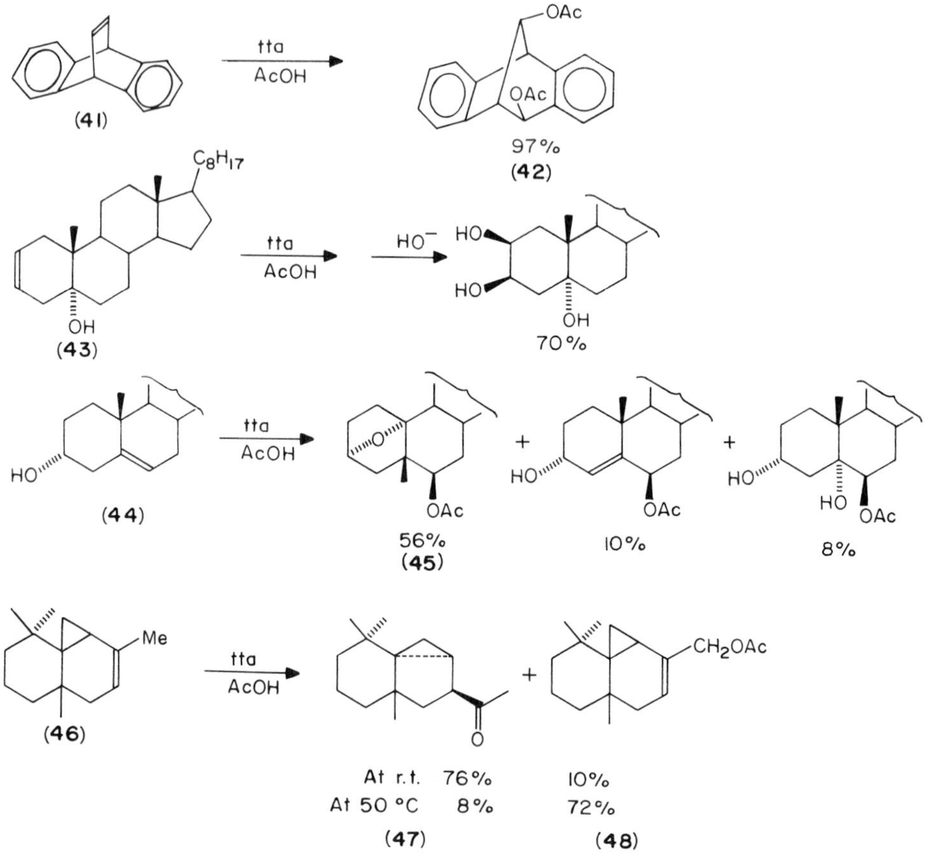

tta AcOH
(41)
97% (42)
tta AcOH; HO⁻
(43)
70%
tta AcOH
(44)
56% (45)
10%
8%
tta AcOH
(46)
At r.t. 76% 10%
At 50 °C 8% 72%
(47) (48)

SCHEME 50

contraction reaction has been applied to the preparation of a lactone aldehyde, **54**, which is a key intermediate for the synthesis of 11-desoxyprostaglandins (Scheme 54)[186]. Application of ttn-mediated oxidation to some *exo*-methylene compounds such as methylenecyclobutane, **55**[187], *ent*-16-kaurene, **56**[188], and 2-methylenenorbornane, **57**[189], resulted in the formation of a ring-expanded ketone which generally reacted further to give a ring-contracted cyclic carboxylic acid derivative (Scheme 55). The oxidation of ketones will be described in a later section. Extrapolation of aryl migration in the ttn oxidation of styrene derivatives provided a general and highly effective method for the ring expansion of cyclic aralkyl ketones (Scheme 56)[190]. The reaction occurs instantaneously at room temperature, a methylene carbon introduced by an appropriate Wittig reagent is inserted regiospecifically between the aromatic ring and the carbonyl group, and the product is obtained as a dimethyl ketal when trimethylorthoformate [$HC(OMe)_3$; tmof] is used as the solvent.

The ttn-mediated formation of *cis*-diols from olefins[191,192] has been reported as in the case of the tta-mediated reactions. Typical examples are *cis*-hydroxylation of 2-methyl-2*H*-chromen, **58**[191], and *cis*-dimethoxylation of 2,2-dimethyl-2*H*-chromen, **59**[191], and

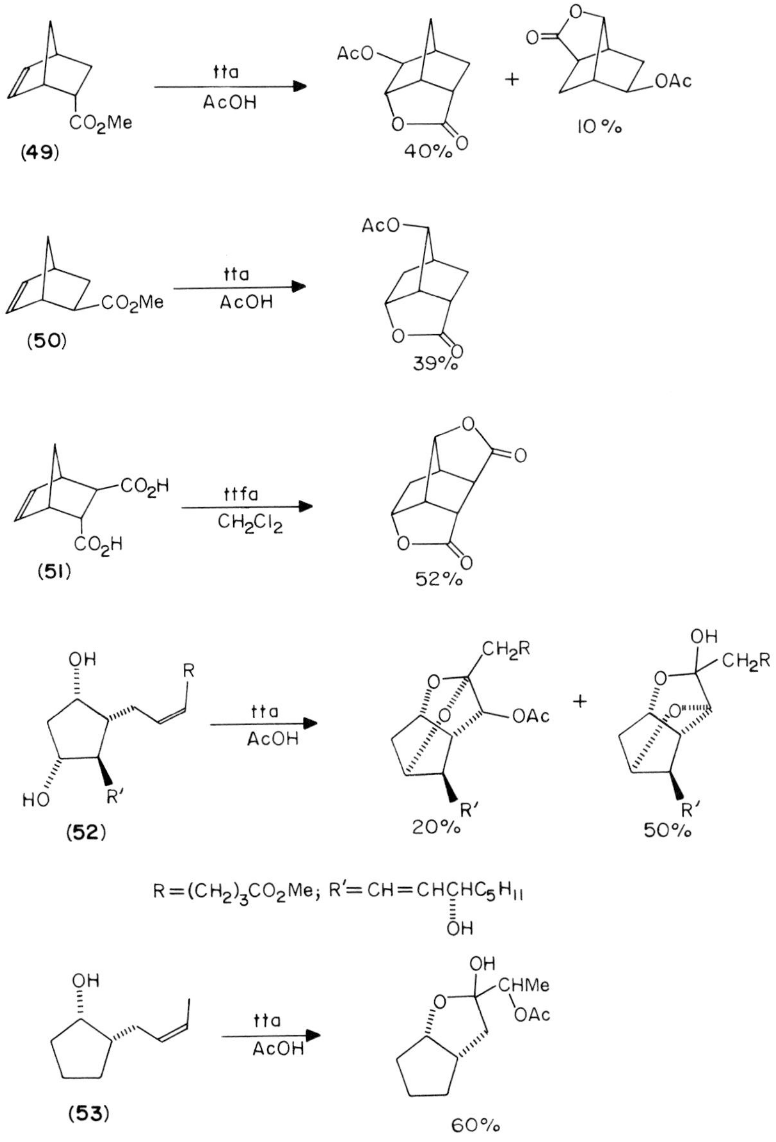

tta, AcOH: (49) → 40% + 10%

tta, AcOH: (50) → 39%

ttfa, CH_2Cl_2: (51) → 52%

tta, AcOH: (52) → 20% + 50%

$R=(CH_2)_3CO_2Me$; $R'=CH=CHCHC_5H_{11}$ (OH)

tta, AcOH: (53) → 60%

SCHEME 51

3-methoxyflavylinium salt, **60**[192] (Scheme 57). These reactions seem to be substituent dependent, since a similar chromen, **62**, and non-substituted flavylinium salt, **63**, afford a ring-contraction product[193] and a flavone, **64**[194], respectively. Acid hydrolysis of *cis*-diol derivatives from **60** gives flavone **61**[192]. A toxic diterpenoid, grayanotoxin-II, **65**, reacts with ttn instantaneously to give the derivative of grayanol B(**66**) by a remote hydroxy

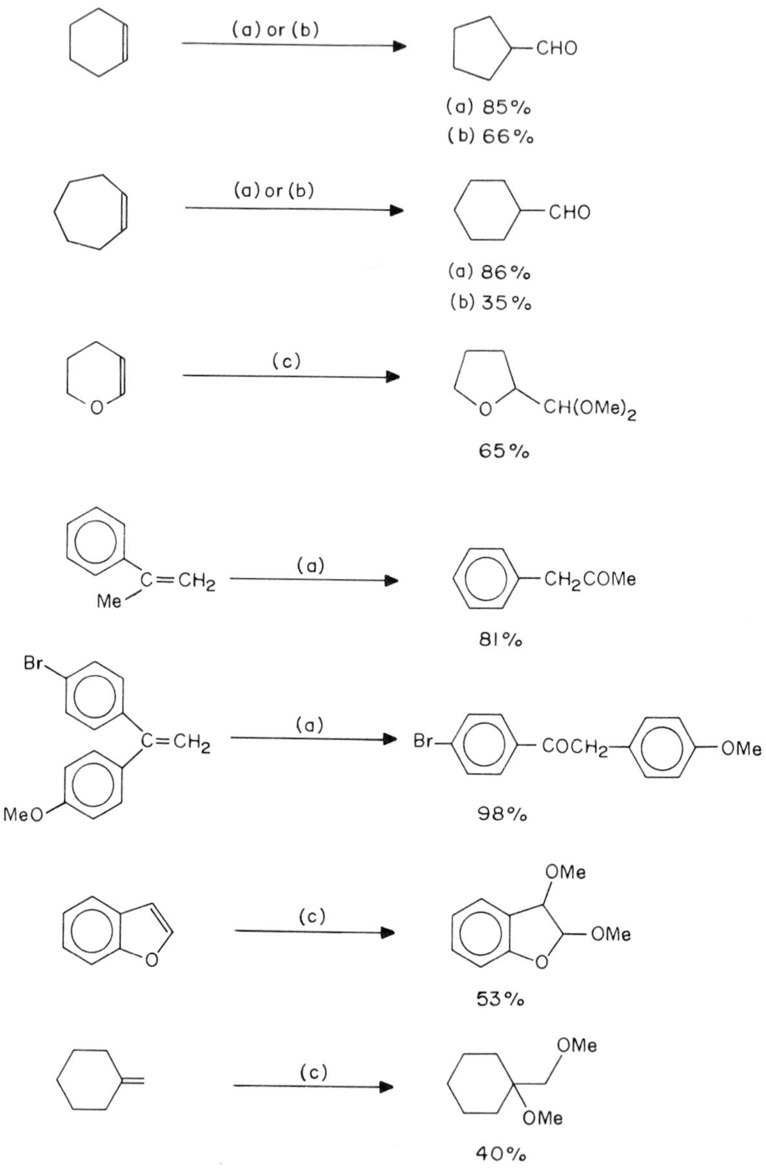

(a) or (b) → CHO
(a) 85%
(b) 66%

(a) or (b) → CHO
(a) 86%
(b) 35%

O (c) → O $CH(OMe)_2$
65%

$C{=}CH_2$ Me (a) → CH_2COMe
81%

Br $C{=}CH_2$ MeO (a) → Br $COCH_2$ OMe
98%

O (c) → OMe OMe O
53%

(c) → OMe OMe
40%

Conditions: (a) ttn/MeOH, H^+
(b) ttfa/Et_2O, H^+
(c) ttn/MeOH

SCHEME 52

$$\text{Oct}^n\text{CH}{=}\text{CH}_2 \xrightarrow[\text{MeOH, r.t.}]{\text{ttn}} \underset{52\%}{\text{Oct}^n\text{CH(OMe)CH}_2\text{OMe}} + \underset{28\%}{\text{Oct}^n\text{COMe}}$$

$$\text{Bu}^n\text{CH}{=}\text{CH}_2 \xrightarrow[\text{MeOH, r.t.}]{\text{ttfa}} \underset{38\%}{\text{Bu}^n\text{CH(OMe)CH}_2\text{OMe}} + \underset{51\%}{\text{Bu}^n\text{CH(OMe)CH}_2\text{OH}} + \underset{3\%}{\text{Bu}^n\text{COMe}}$$

$$\text{Hex}^n\text{CH}{=}\text{CH}_2 \xrightarrow[\text{MeOH, r.t.}]{\text{tta}} \underset{10\%}{\text{Hex}^n\text{CH(OMe)CH}_2\text{OMe}} + \underset{55\%}{\text{Hex}^n\text{CH(OMe)CH}_2\text{OAc}} + \underset{23\%}{\text{Hex}^n\text{CH(OAc)CH}_2\text{OAc}} + \underset{6\%}{\text{Hex}^n\text{COMe}}$$

SCHEME 53

ttn, aq. $HClO_4$; CHO; 70%; **(54)**; C_5H_{11}-*n*

SCHEME 54

NC; **(55)**; ttn, aq. glyme; CH_2; O; 81%

(56); CH_2; ttn, MeOH; 33%; ttn, AcOH; CO_2H; 86%

(57); CH; ttn, MeOH; 68%; ttn, MeOH; CO_2Me; 85%

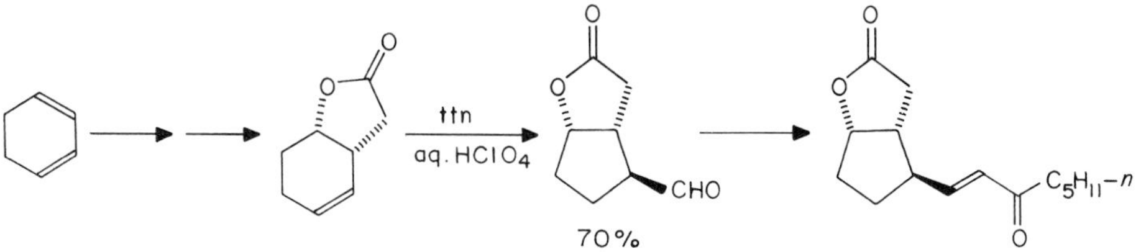

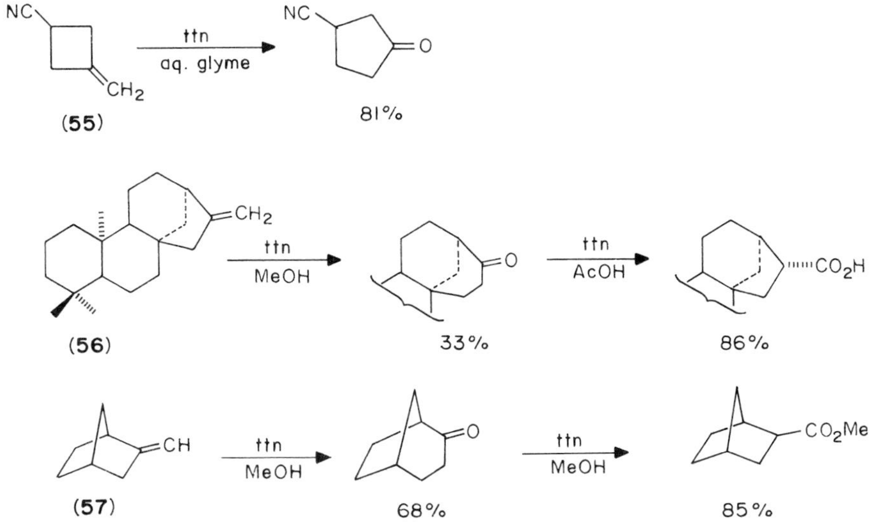

SCHEME 55

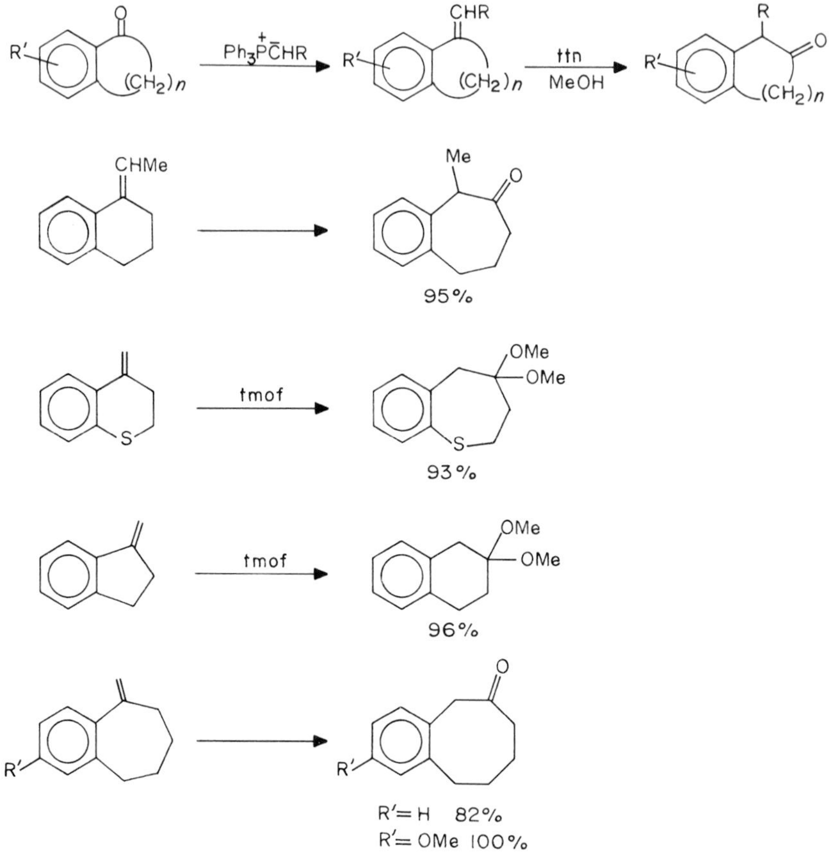

SCHEME 56

group participation (Scheme 58)[195]. The ttn oxidation of olefins in methanol sometimes gives a minor amount of nitrate ester[185,196,197], whereas the esters become the major or the sole products when *n*-pentane is used as the solvent (Scheme 59)[198].

Thallium(III) perchlorate is also an effective oxidizing agent for olefins. Kinetic and product studies on the oxidation of alkenols[199], cycloalkenes[200,201], and methylenecycloalkanes[201] have been carried out, and all the reactions have been explained by an oxythallation–dethallation mechanism. Some examples are shown in Schemes 60 and 61, the products being carbonyl compounds, diols, and/or cyclic ethers, as has been observed for tta oxidation in aqueous acetic acid[133]. The carbonyl products of the oxidation of cycloalkenes and methylenecycloalkanes are generally ring-contracted aldehydes and ring-expanded cyclic ketones, respectively. This ring enlargement reaction is useful for four- and five-membered ring *exo*-olefins, but not so useful for the six-membered analogues[202], and has been applied to 2-methylenenorbornene[187], tetracyclic adamantane derivatives[187], and 2-methylenenoradamantane, **67**[203], to afford the expected ketones in 50–60% yields (Scheme 62). Electrochemical oxidation of the thallium(I) salt produced at an anode has been attempted in the thallium(III) perchlorate oxidation of

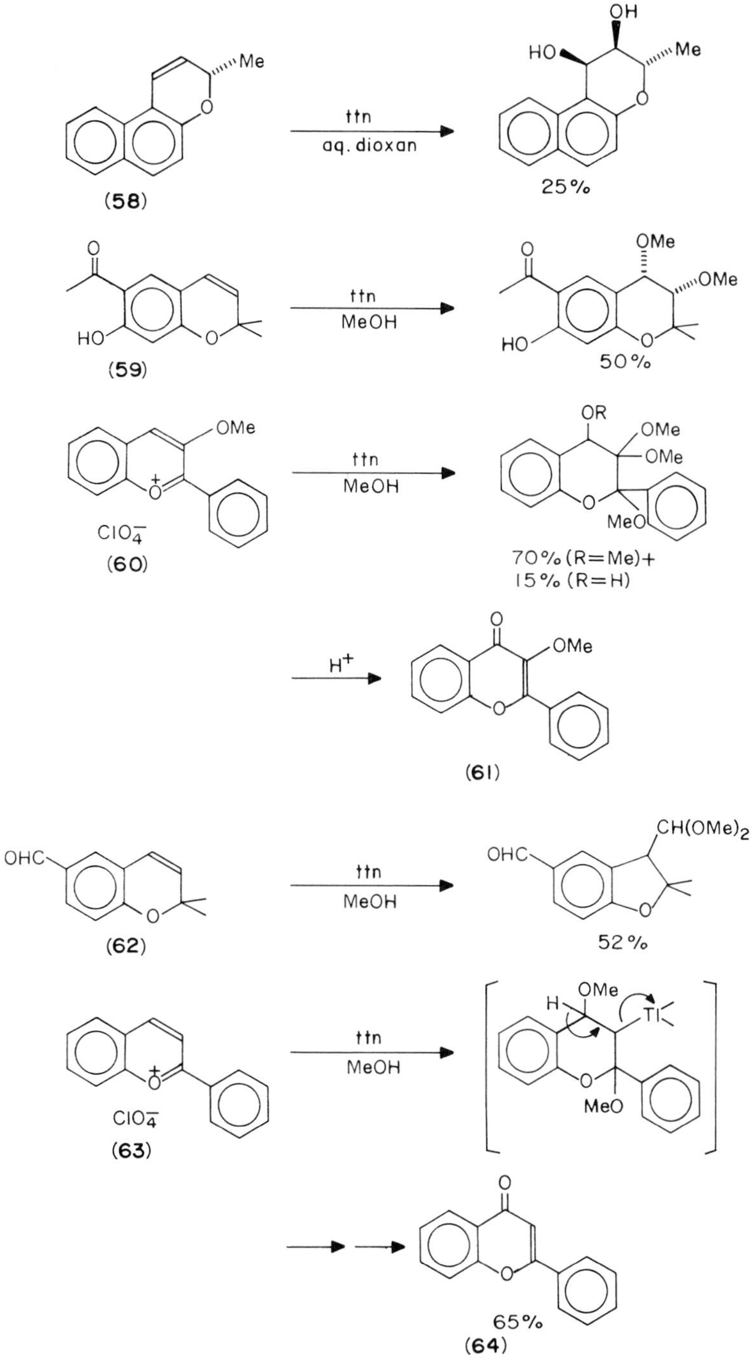

ttn
aq. dioxan
(58)
25%

ttn
MeOH
(59)
50%

ttn
MeOH
(60)
70% (R=Me)+
15% (R=H)

H^+
(61)

ttn
MeOH
(62)
52%

ttn
MeOH
(63)
65%
(64)

SCHEME 57

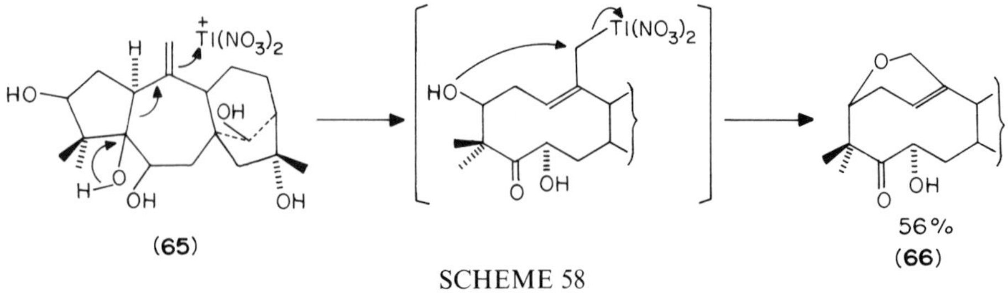

HO
H
$\overset{+}{Tl}(NO_3)_2$
OH
H
O
OH
OH
(65)
$Tl(NO_3)_2$
HO
O
OH
O
O
OH
56%
(66)

SCHEME 58

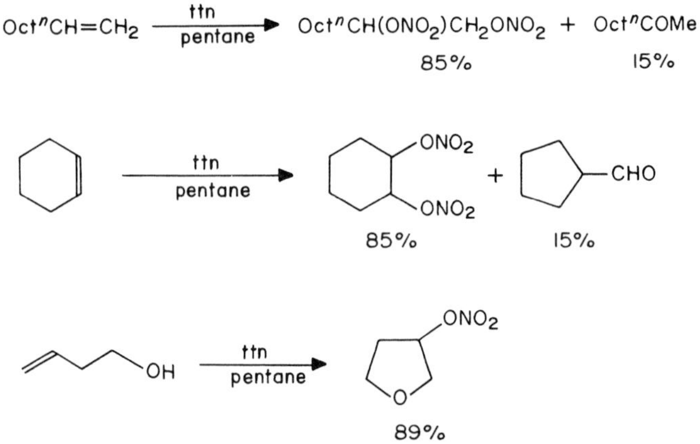

$Oct^nCH=CH_2$
ttn
pentane
$Oct^nCH(ONO_2)CH_2ONO_2$
+
Oct^nCOMe
85%
15%
ttn
pentane
ONO_2
ONO_2
+
CHO
85%
15%
OH
ttn
pentane
ONO_2
O
89%

SCHEME 59

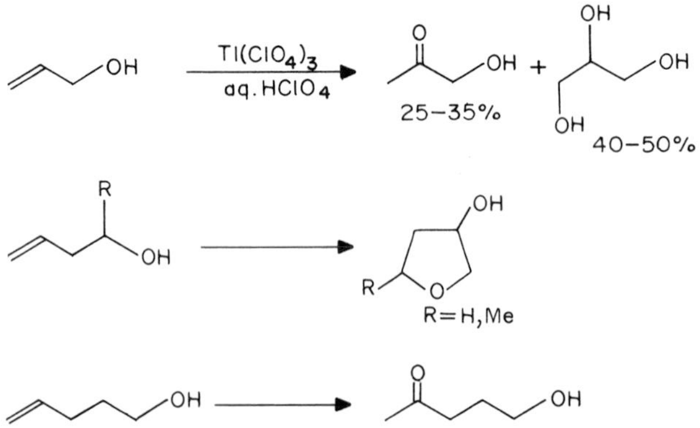

OH
$Tl(ClO_4)_3$
aq. $HClO_4$
O
OH
+
OH
OH
OH
25–35%
40–50%
R
OH
OH
R
O
R=H,Me
OH
O
OH

SCHEME 60

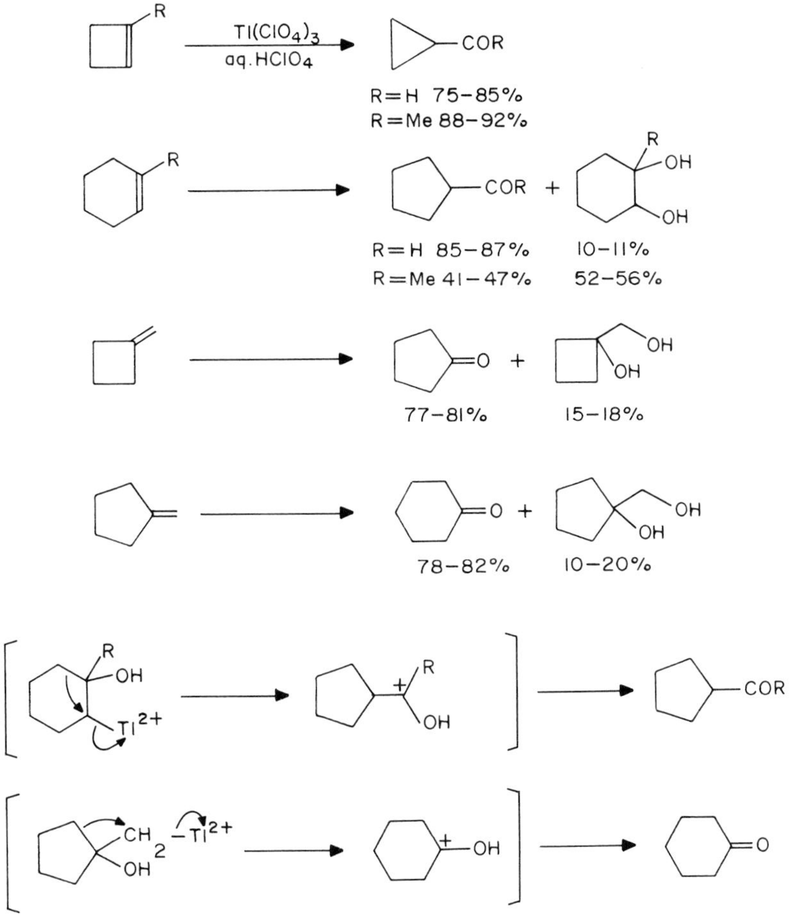

$Tl(ClO_4)_3$
aq. $HClO_4$
COR
R=H 75–85%
R=Me 88–92%

COR + OH OH
R=H 85–87% 10–11%
R=Me 41–47% 52–56%

=O + OH OH
77–81% 15–18%

=O + OH OH
78–82% 10–20%

OH Tl^{2+} → + R OH → COR

CH_2–Tl^{2+} OH → + OH → =O

SCHEME 61

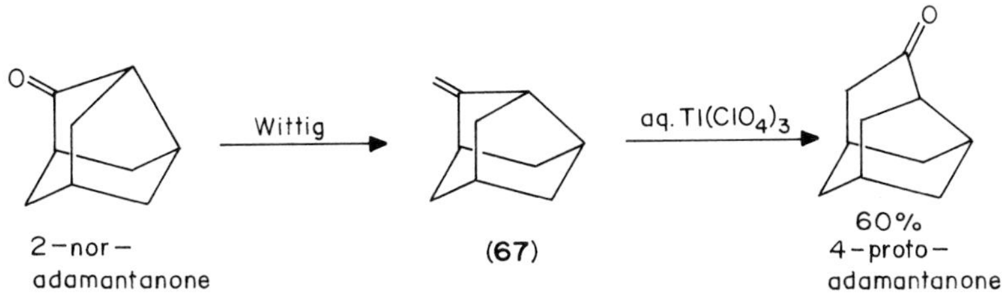

Wittig
aq. $Tl(ClO_4)_3$
2-nor-adamantanone
(**67**)
60%
4-proto-adamantanone

SCHEME 62

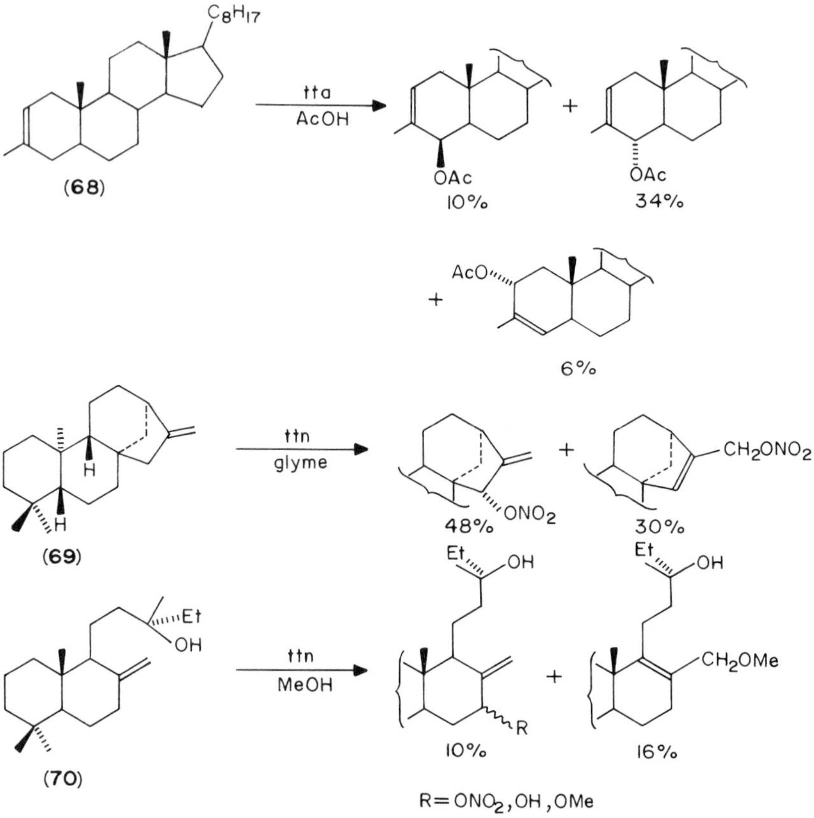

SCHEME 63

cyclohexene and 3-*tert*-butylcyclohexene[204]. Unfortunately, however, the product selectivity was lower than in the stoichiometric reaction[204,205].

In the tta oxidation of isobutene and cyclohexene, only small amounts of allylic oxidation products are formed[36,165,166]. In contrast, the tta oxidation of thujopsene, **46**, at higher temperatures gives an allylic acetate, **48**, in a high yield, as shown in Scheme 50. There are several other examples where allylic oxidation products are mainly formed. Thus, the ttfa oxidation of cyclooctene in dichloromethane affords 3-trifluoroacetoxycyclooctene in 75% yield[206]. Treatment of steroidal exo- and endo-cyclic olefins such as 3-methyl-2-cholestene, **68**[207], 5-cholestene[178], and 17-methylene-5α-androstan-3β-yl acetate[208] with tta in acetic acid or methanol results in the formation of a mixture of allylic acetates or ethers. Similarly, the ttn oxidation of some diterpenes such as ent-16-kaurene, **69**[209,210], ent-15-kaurene[209], 13β-kaur-16-ene[210], and labd-8(17)-en-13-ol, **70**[211], in glyme (1,2-dimethoxyethane) or methanol affords allylic nitrate esters and/or allylic ethers (Scheme 63). An allylic thallium(III) compound is proposed as the reactive intermediate, which may give the allylic cation by dethallation or suffer an S_N2' reaction[165,208–211].

2. Conjugated and non-conjugated dienes

The reaction of conjugated dienes such as buta-1, 3-diene, isoprene, 2, 3-dimethylbuta-1, 3-diene, 2, 5-dimethylbuta-2, 4-diene, cyclopenta-1, 3-diene, and cyclohexa-1, 3-diene with tta in acetic acid affords an isomeric mixture of the corresponding diacetoxyalkenes (1, 2- and 1, 4-addition products) in 10–92% yields. The 1, 2-addition products are predominantly formed in all cases examined except with cyclopenta-1, 3-diene[212,213]. Similar oxidation occurs with ttfa in dichloromethane, the stereochemistry of the bis(trifluoroacetoxyalkene) products being *cis* only[206]. Cyclic dienes give nearly equal amounts of both the 1, 2- and 1, 4-isomers, whereas linear dienes afford only the 1, 2-isomer[206]. These reactions are assumed to proceed through the oxythallation and dethallation steps and the formation of the *cis*-product has been explained by *trans*-oxythallation followed by the S_N2 attack of $OCOCF_3$ group of another ttfa molecule (Scheme 64). Under similar conditions *cis*- and *trans*-but-2-enes are transformed into *meso*- and racemic *threo*-2, 3-bis(trifluoroacetoxy)butanes, respectively, in 20–40% yield by *cis*-addition of two $OCOCF_3$ groups. Cholesta-3, 5-diene is oxidized by tta to a mixture of the corresponding 1, 2- and 1, 4-diol esters (total yield 55%), where the 1, 4-isomers predominate substantially[178]. The ttn oxidation of 1, 3-dienes normally gives a complex mixture of products (ketones, aldehydes, dimethoxyalkenes, cyclopropanes, etc.) with low selectivity because of the secondary reactions induced by HNO_3 formed *in situ* such as olefin isomerization and Michael-type addition of methanol[214]. A vinyl group migration has been shown to occur in some cases via a cyclopropane intermediate[214]. Application of the oxythallation–halogenodethallation method to conjugated linear dienes results in the formation of only 1, 2-addition products and in the case of hexa-1, 5-diene and diallyl ether only one double bond reacts to give the corresponding substituted alkanes, as shown in Scheme 35[80,144].

The tta and ttfa oxidation of some non-conjugated 1, 5-dienes (Scheme 65) results in a novel carbon—carbon bond formation such as a transannular cyclization of cycloocta-

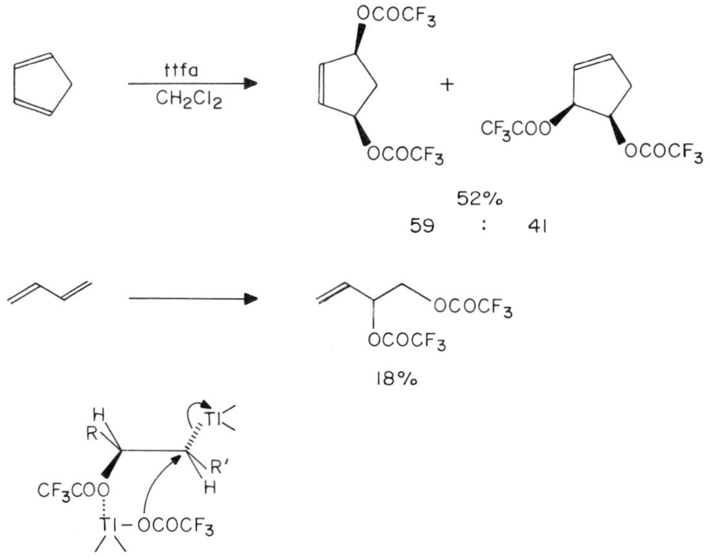

SCHEME 64

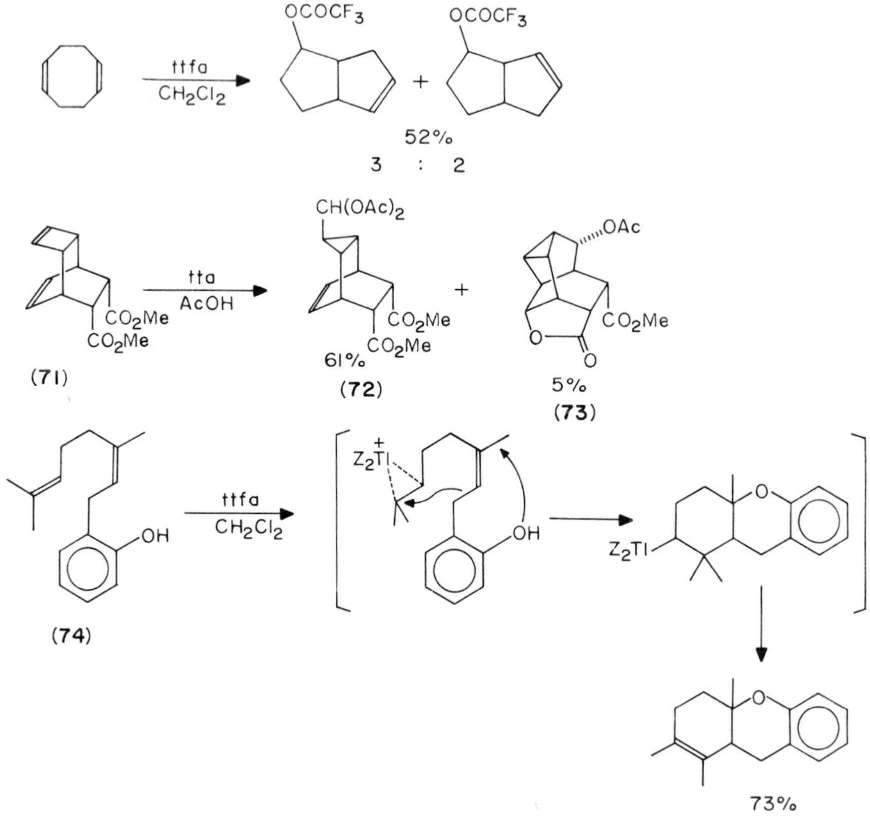

$OCOCF_3$ $OCOCF_3$

ttfa / CH_2Cl_2

52%

3 : 2

$CH(OAc)_2$ OAc

tta / AcOH

CO_2Me CO_2Me CO_2Me CO_2Me CO_2Me

(71) 61% (72) 5% (73)

ttfa / CH_2Cl_2

OH $Z_2\overset{+}{Tl}$ OH Z_2Tl

(74)

73%

SCHEME 65

HO OH OH

aq. $Tl(ClO_4)_3$ / CH_2Cl_2

(75) 49% 11% 31%

CHO Me CHO

(76) 80%

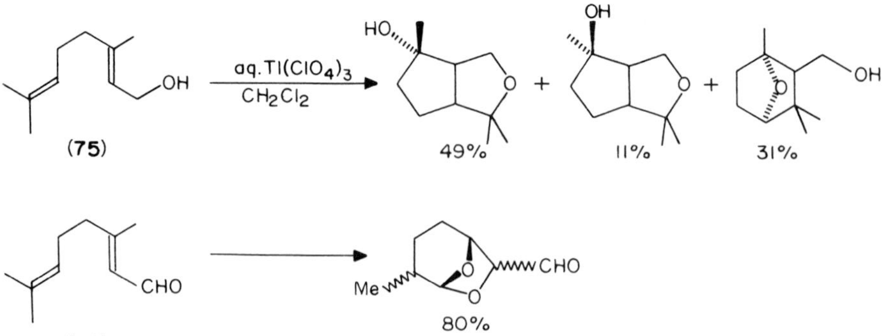

SCHEME 66

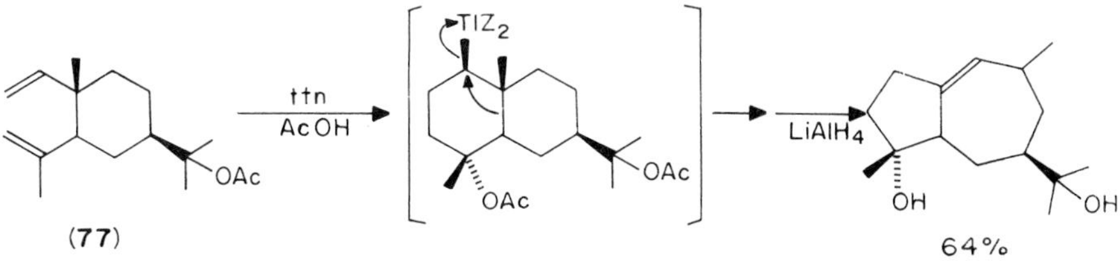

SCHEME 67

1,5-diene[215], a skeletal rearrangement of tricyclodiene, **71**, to **72** and **73**[216], and olefinic cyclization of *o*-nerylphenol (**74**) and *o*-geranylphenol[217]. The last reaction is assumed to be a thallium(III)-induced carbon—carbon bond formation as shown in Scheme 65, and does not proceed via oxythallation. A similar reaction has also been observed in the treatment of geraniol, **75**[218], nerol[219], and citral, **76**[220], with thallium(III) perchlorate, but the products are complex mixtures of diastereoisomeric isomers (Scheme 66). Another example of carbon—carbon bond formation is the ttn-mediated conversion of (−)-elemol acetate, **77**, to a bicyclic compound, cryptomeridol, **78** (Scheme 67)[221,222].

3. α, β-Unsaturated carbonyl compounds and chalcones

The tta oxidation of α,β-unsaturated carbonyl compounds such as acrolein, acrylic acid, and methyl acrylate is very slow in acetic acid as the solvent and does not afford any

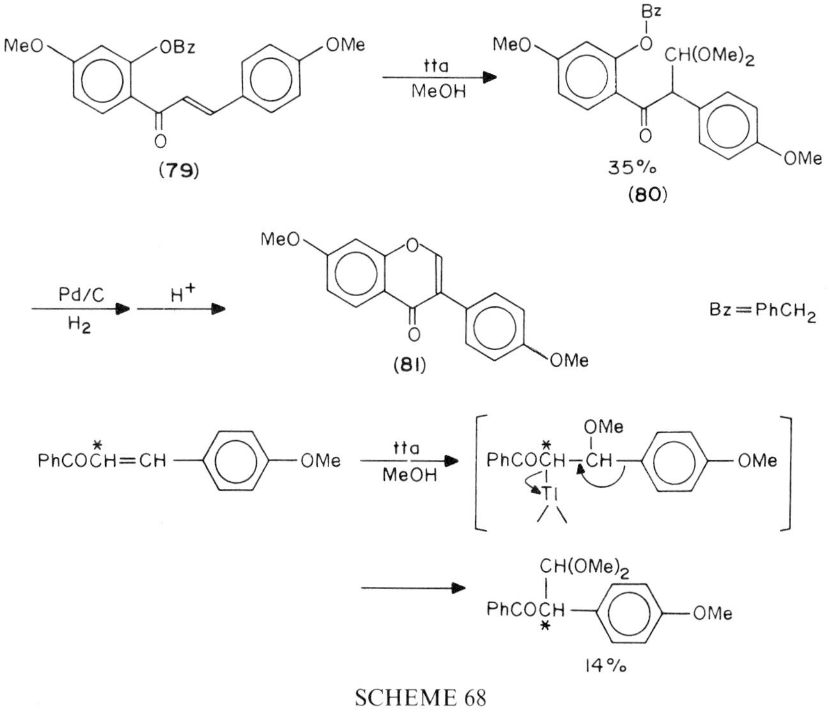

SCHEME 68

expected diacetoxylated compounds[171]. On the contrary, a substituted chalcone such as **79** reacts with tta in methanol to give the oxidatively rearranged product **80** in 35% yield from which 7, 4′-dimethoxyisoflavone, **81**, can be synthesized (Scheme 68)[223,224]. The method has been applied to the synthesis of other naturally occurring isoflavones such as milldurone and lettadurone[224]. A tracer (^{14}C) experiment on the oxidation of 4-methoxy [α-^{14}C]chalcone revealed that the reaction involves a 1, 2-aryl migration, in contrast to the 1, 2-aroyl migration, which has previously been observed in the BF_3-catalysed rearrangement of chalcone epoxides (Scheme 68)[223,225].

In contrast to the tta oxidation, which is slow and leads to low yields of products, the ttn oxidation of chalcones is very rapid, giving high yields of oxidation products, and in some cases constitutes a simple and convenient procedure for the preparation of benzils. Thus, treatment of chalcones with 1 equiv. of ttn in acidic methanol resulted in the formation of the rearranged acetal **92**[226,227], but when 3 equiv. of ttn were used in aqueous acidic glyme both symmetrical and unsymmetrical benzils were obtained directly by means of a discrete series of three independent oxidations involving the intermediacy of deoxybenzoins, **83**, and then benzoins, **84** (Scheme 69), provided that no deactivating substituents were present in the migrating aryl group[226,228]. The oxidation of deoxybenzoins and benzoins seems to proceed via a thallium(III) enolate instead of an organothallium(III) intermediate.[228] When this ttn oxidation is carried out in tmof with Ar′ groups of greater migratory aptitude than that of the Ar group, the ester **86** became the main or sole product via acetalization followed by oxythallation and Ar′ group migration (Scheme 70)[227,229]. A side reaction is the formation of the acetal **82** via the pathway shown in Scheme 69. This occurs competitively before the ttn induced acetalization of the starting chalcones in tmof. When the dimethylacetals **85** are used as starting materials, **86** is the sole product. The effectiveness of tmof as the solvent is also known in the oxidative rearrangement of

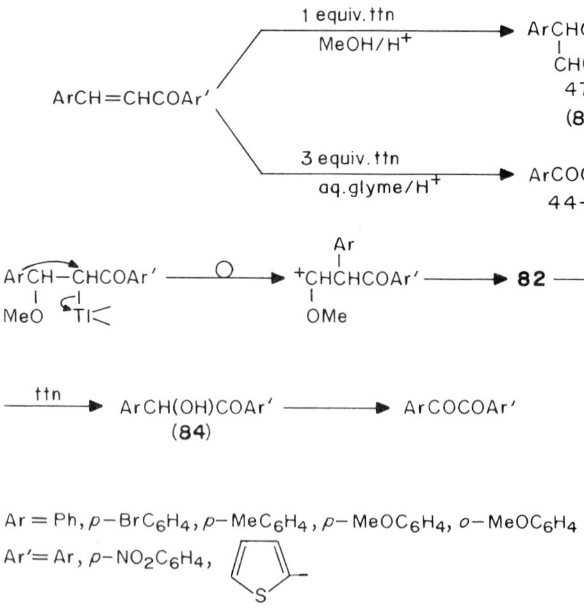

ArCH=CHCOAr′ —(1 equiv. ttn, MeOH/H$^+$)→ ArCH(COAr′)CH(OMe)$_2$ 47–100% (82)

ArCH=CHCOAr′ —(3 equiv. ttn, aq. glyme/H$^+$)→ ArCOCOAr′ 44–70%

ArCH(OMe)–CH(Tl<)COAr′ —→ $^+$CH(OMe)CH(Ar)COAr′ —→ **82** —→ ArCH$_2$COAr′ (83)

—(ttn)→ ArCH(OH)COAr′ (84) —→ ArCOCOAr′

Ar = Ph, p-BrC$_6$H$_4$, p-MeC$_6$H$_4$, p-MeOC$_6$H$_4$, o-MeOC$_6$H$_4$

Ar′ = Ar, p-NO$_2$C$_6$H$_4$, 2-thienyl

SCHEME 69

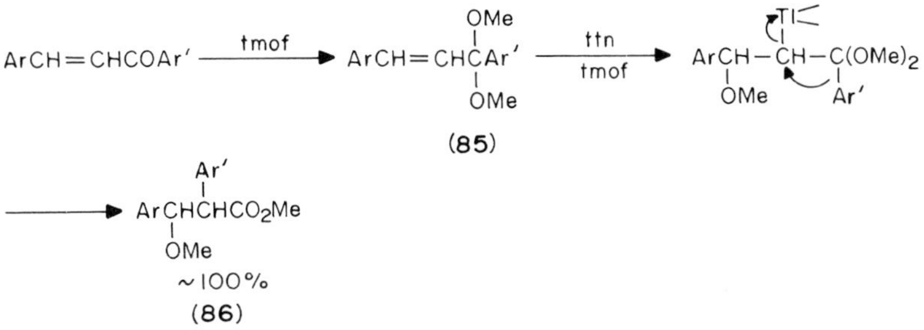

$$\mathrm{ArCH{=}CHCOAr'} \xrightarrow{\text{tmof}} \mathrm{ArCH{=}CHC(OMe)_2Ar'}\ \textbf{(85)} \xrightarrow[\text{tmof}]{\text{ttn}} \mathrm{ArCH(OMe){-}CH(Tl){-}C(OMe)_2Ar'} \longrightarrow \mathrm{ArCH(OMe)CH(Ar')CO_2Me}$$

~100%

(86)

Ar = Ph, p-ClC_6H_4, o-ClC_6H_4, p-$NO_2C_6H_4$, o-$NO_2C_6H_4$

Ar′ = Ph, p-MeC_6H_4, p-$MeOC_6H_4$

SCHEME 70

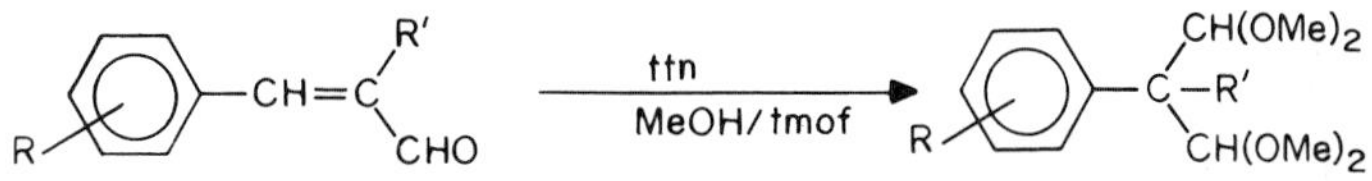

$$R{-}C_6H_4{-}CH{=}C(R')CHO \xrightarrow[\text{MeOH/tmof}]{\text{ttn}} R{-}C_6H_4{-}C(R')(CH(OMe)_2)_2$$

R = H, p-MeO, p-NO_2, m-NO_2

R′ = H, Me

50–84%

$$R{-}C_6H_4{-}CH{=}CHCO_2Me \xrightarrow[\text{tmof}]{\text{ttn}} R{-}C_6H_4{-}CH(CO_2Me)CH(OMe)_2$$

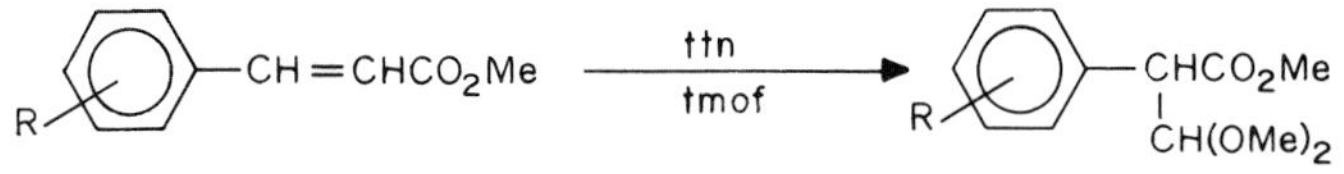

R = H, p-Me, p-MeO, o-F, p-Cl

83–96%

SCHEME 71

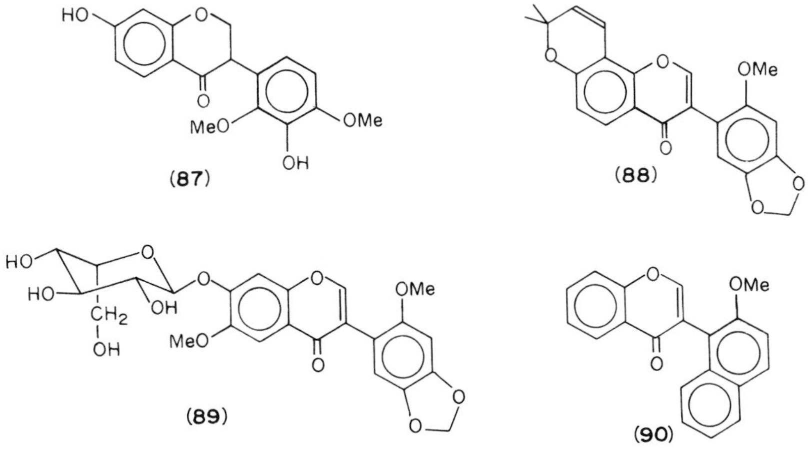

(87) **(88)**

(89) **(90)**

SCHEME 72

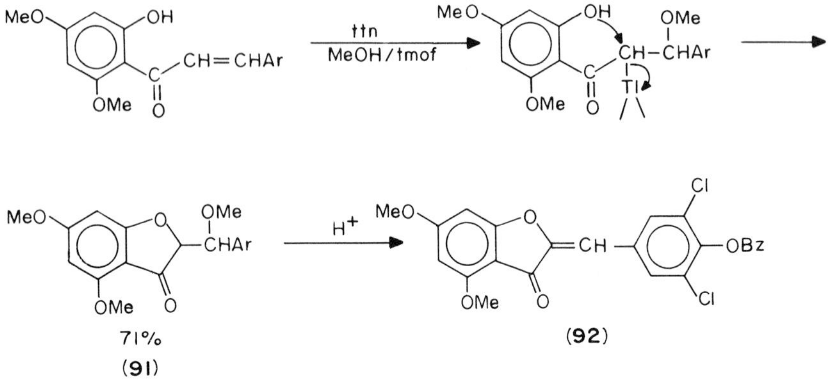

SCHEME 73

cinnamaldehydes and cinnamic esters, as shown in Scheme 71, the role of tmof being (1) the initial conversion of carbonyls to acetals or ketals and (2) the lowering of the dielectric constant of the reaction medium to favour S_N2 as opposed to S_N1 reactions of the methoxythallated intermediates[230].

The ttn-mediated facile oxidative rearrangement has been applied for the synthesis of many isoflavones and their derivatives directly or indirectly from 2′-hydroxy- or 2′-acetoxy-chalcones (Scheme 68). Isoflavones prepared in over 50% yield include (±)-mucronulatol, **87**[231], (±)-sophorol[231], pentamethoxyisoflavones[232], violanone[233], vestitol[233], classequinone[233], lonchocarpan[233], jamaicin, **88**[234], leiocarpin[234], isoflavone glycoside[235], dalptin, **89**[236], fujikinin[236], glycitein[236], naphthalene analogues of isoflavone **90**[237], and many other compounds (Scheme 72). In some cases the ttn oxidation of 2′-hydroxychalcones, depending on the substitution pattern, leads to the formation of coumaranone derivatives **91**, which can be converted into aurones **92** by acid treatment (Scheme 73)[238–240]. Other applications are in the oxidative rearrangement of α-benzylideneketones[241] and 1, 3-diarylpropane-1, 3-diones[242].

The reagent ttfa in the presence of BF_3 causes oxidative dimerization of 4-alkoxy-cinnamic acids to moderate yields of fused bislactones, e.g. 2, 6-diaryl-3, 7-dioxabicyclo[3.3.0]octane-4, 8-dione, **93**, which undergo acid-catalysed rearrangement to naturally occurring lignans, **94** (Scheme 74)[243,244], although a much lower yield (only 9%) has been claimed for some of these reactions[245,246]. The reaction is more reasonably explained by the oxythallation mechanism giving an intermediate **95**, followed by

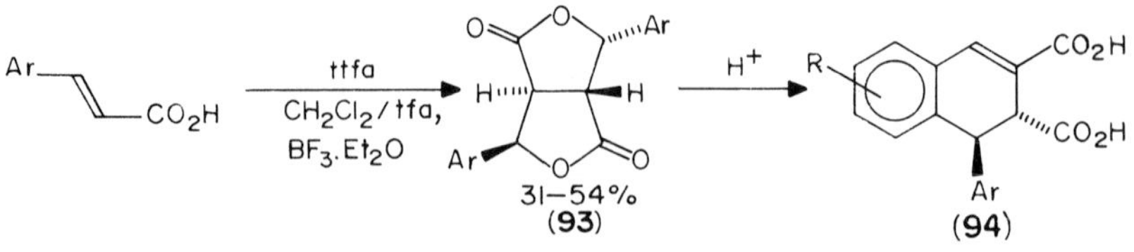

Ar = 3,4-$(MeO)_2C_6H_3$, 2,3,4-$(MeO)_3C_6H_2$, 3,4,5-$(MeO)_3C_6H_2$, 3,4-$(OCH_2O)C_6H_3$, etc.

SCHEME 74

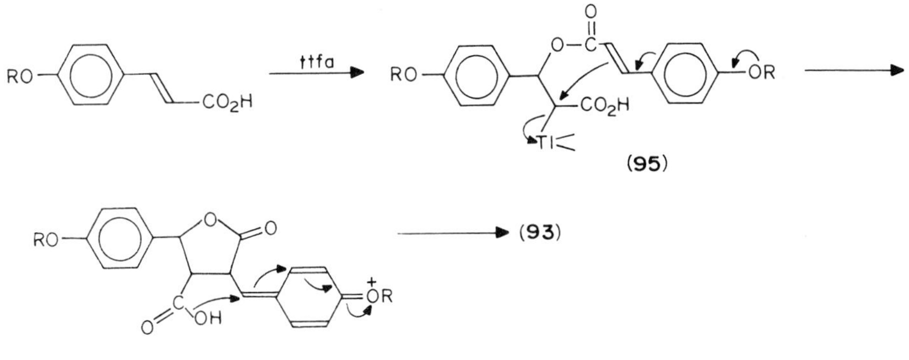

SCHEME 75

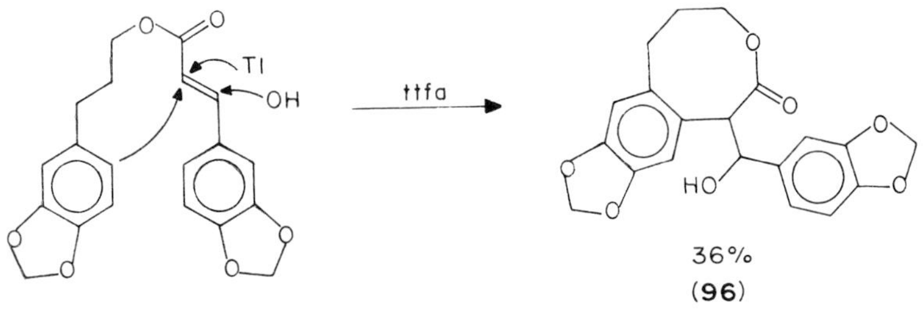

SCHEME 76

dethallation and intramolecular Michael addition (Scheme 75)[244], rather than the originally proposed electron-transfer mechanism involving a radical cation[243]. A similar reagent system gives the eight-membered ring lactone **96** from the corresponding α,β-unsaturated ester through oxythallation and dethallation followed by intramolecular nucleophilic attack of the benzene ring (Scheme 76)[247]. Such phenyl participation is also found in the tta oxidation of allyl-*m*-methoxybenzene which gives a complex mixture of products in lower yields[42].

4. *Cyclopropanes*

Various cyclopropanes react with tta in acetic acid to give mainly 1, 3-diacetoxyalkanes via organothallium(III) compounds such as **97**, which are formed by 1, 2-bond fission, although such compounds have never been isolated (Scheme 77)[248]. In the case of

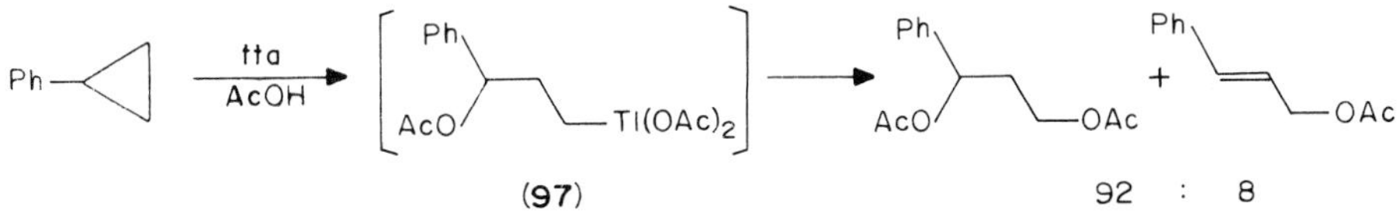

SCHEME 77

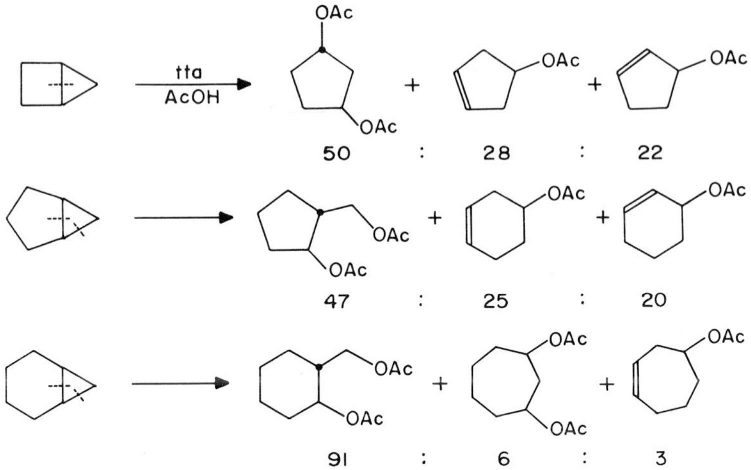

tta / AcOH

50 : 28 : 22

47 : 25 : 20

91 : 6 : 3

SCHEME 78

bicyclo[*n*.1.0]alkanes both internal and external bond cleavages occur to give *trans*-2-acetoxymethylcycloalkyl acetates and unsaturated mono- and 1,3-di-acetates, respectively, and the products resulting from internal bond cleavage increase with decreasing ring size (Scheme 78)[249]. A detailed kinetic study of the oxidative cleavage of arylcyclopropanes shows that the reaction is overall second order, first order in each reactant, that electron-releasing groups facilitate the reaction ($\rho = -4.3$, correlation with σ^+) and that the order of reactivity for cleavage is tta > $Hg(OAc)_2$ > $Pb(OAc)_4$[133,250]. Spiro compounds such as spiro[4.2]heptane and spiro[5.2]octane are oxidized more rapidly than any bicyclo[*n*.1.0]alkanes[251]. These cyclopropane cleavage reactions have been applied to methyl *ent*-trachyloban-19-oate[252,253], trioxane[254], and 3,5-dehydronoriceane, **98**[255], although the product selectivity is generally not high in either case (Scheme 79). A cyclopropene ring is cleaved oxidatively by tta to afford *gem*-diacetoxypropene derivatives such as **99** and **100**, formal carbene–tta complexes being proposed as reaction intermediates (Scheme 80)[256,257]. The oxidation of methylenecyclopropane by ttn in methanol is known to afford both products from oxythallation of the cyclopropane ring and the double bond, the former being predominant (Scheme 81)[258]. In the cases of methylenecyclobutane and methylenecyclopentane,

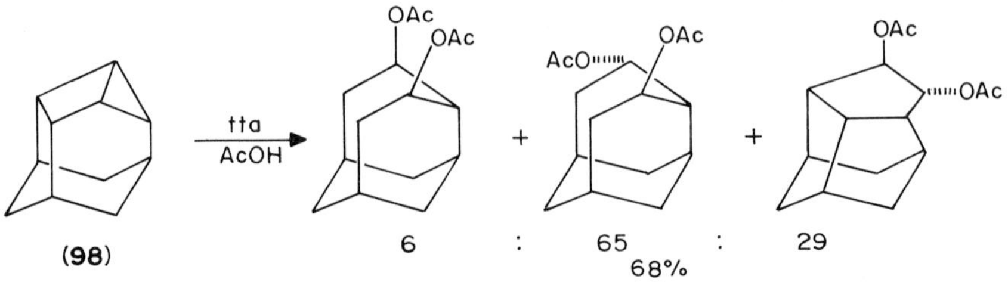

(98) tta / AcOH

6 : 65 : 29

68%

SCHEME 79

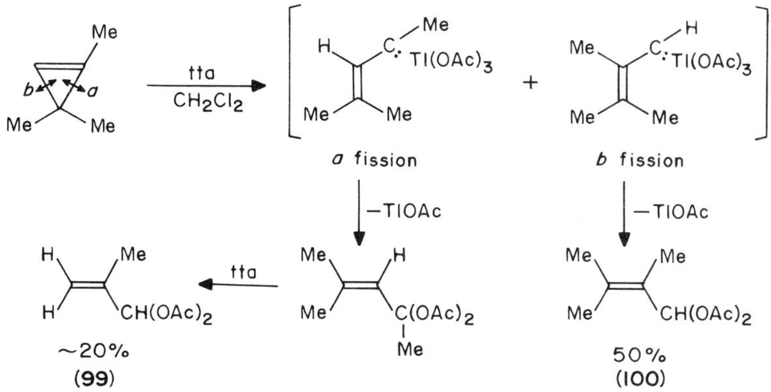

SCHEME 80

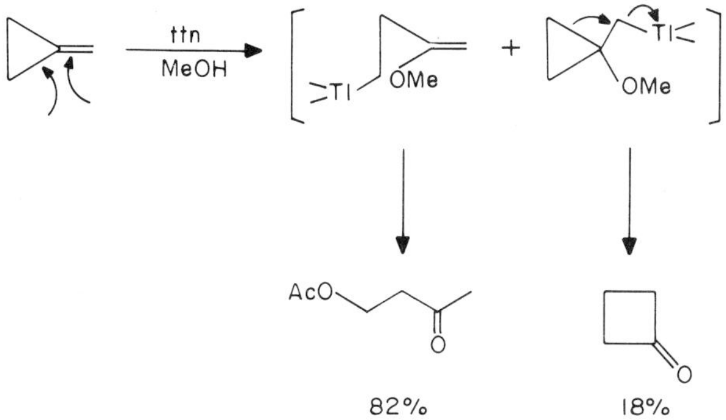

SCHEME 81

oxythallation occurs only on the double bond, resulting in ring enlargement as described in Section III.C.1 (cf. Schemes 55 and 61).

D. Oxidation of Alkynes and Allenes via Oxythallation and Related Reactions

Compared with the oxidation of olefins, examples of thallium(III) salt oxidation of acetylenes and allenes are still very limited. The most interesting and synthetically useful reactions are the selective oxidation of acetylenes to carboxylic acids, acyloins, benzils, and arylacetic acids with ttn[259,260]. The nature of the products depends greatly on the solvent employed and the structure of the acetylene used. Typical examples are shown in Scheme 82. All reactions are explained via oxythallation, although the intermediate organothallium(III)compounds have never been isolated in these cases (Scheme 83). Kinetic studies on the oxidation of phenylacetylene with thallium(III) oxide in aqueous $HClO_4$, in comparison with that of styrene, revealed that the decomposition of the intermediate oxythallation adducts of the acetylene is very slow compared with that of the adduct from the olefin; the relative rates (k_f/k_d) of adduct formation (k_f) and its decomposition (k_d) for styrene and phenylacetylene are 28 and 8200, respectively.[261]

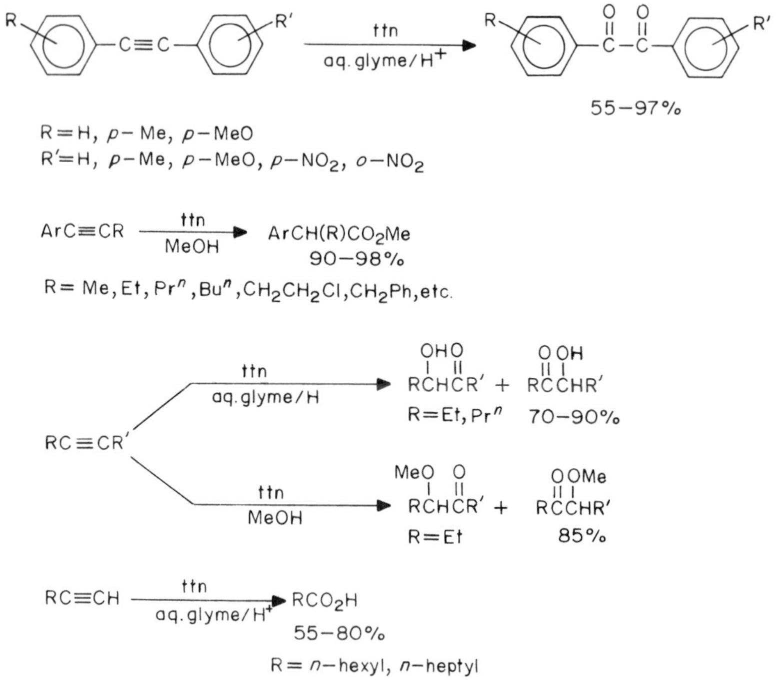

$$\text{R-C}_6\text{H}_4\text{-C}\equiv\text{C-C}_6\text{H}_4\text{-R}' \xrightarrow[\text{aq. glyme/H}^+]{\text{ttn}} \text{R-C}_6\text{H}_4\text{-CO-CO-C}_6\text{H}_4\text{-R}'$$

55–97%

R = H, *p*-Me, *p*-MeO

R′ = H, *p*-Me, *p*-MeO, *p*-NO_2, *o*-NO_2

$$\text{ArC}\equiv\text{CR} \xrightarrow[\text{MeOH}]{\text{ttn}} \text{ArCH(R)CO}_2\text{Me} \quad 90\text{–}98\%$$

R = Me, Et, Pr^n, Bu^n, CH_2CH_2Cl, CH_2Ph, etc.

$$\text{RC}\equiv\text{CR}' \xrightarrow[\text{aq. glyme/H}]{\text{ttn}} \text{RCH(OH)C(O)R}' + \text{RC(O)CH(OH)R}'$$

R = Et, Pr^n 70–90%

$$\text{RC}\equiv\text{CR}' \xrightarrow[\text{MeOH}]{\text{ttn}} \text{RCH(OMe)C(O)R}' + \text{RC(O)CH(OMe)R}'$$

R = Et 85%

$$\text{RC}\equiv\text{CH} \xrightarrow[\text{aq. glyme/H}^+]{\text{ttn}} \text{RCO}_2\text{H}$$

55–80%

R = *n*-hexyl, *n*-heptyl

SCHEME 82

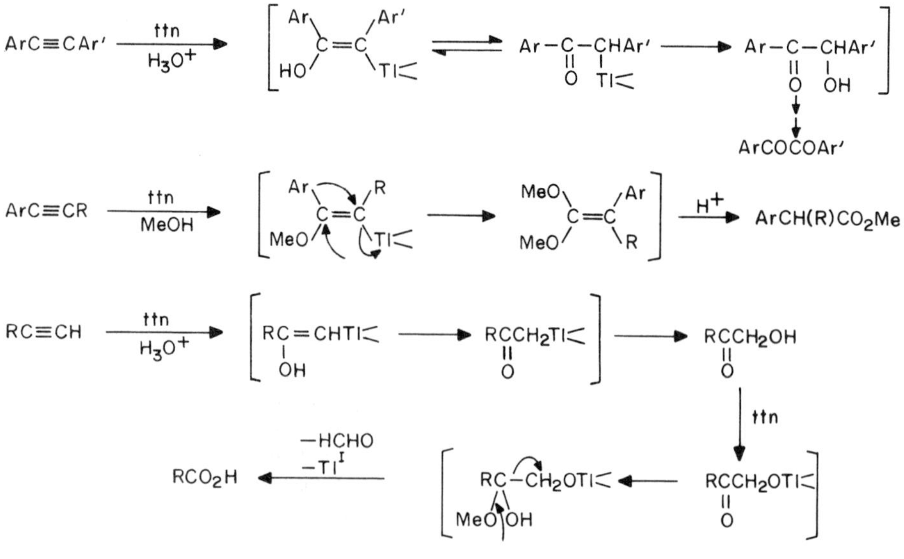

$$\text{ArC}\equiv\text{CAr}' \xrightarrow[\text{H}_3\text{O}^+]{\text{ttn}} \left[\text{Ar(HO)C=C(Ar}')\text{Tl}< \rightleftharpoons \text{Ar-C(=O)-CHAr}'(\text{Tl}<) \longrightarrow \text{Ar-C(=O)-CHAr}'(\text{OH}) \right] \longrightarrow \text{ArCOCOAr}'$$

$$\text{ArC}\equiv\text{CR} \xrightarrow[\text{MeOH}]{\text{ttn}} \left[\text{Ar(MeO)C=C(R)Tl}< \longrightarrow (\text{MeO})_2\text{C=C(Ar)R} \right] \xrightarrow{\text{H}^+} \text{ArCH(R)CO}_2\text{Me}$$

$$\text{RC}\equiv\text{CH} \xrightarrow[\text{H}_3\text{O}^+]{\text{ttn}} \left[\text{RC(OH)=CHTl}< \longrightarrow \text{RC(=O)CH}_2\text{Tl}< \right] \longrightarrow \text{RC(=O)CH}_2\text{OH}$$

$$\text{RC(=O)CH}_2\text{OH} \xrightarrow{\text{ttn}} \left[\text{RC(=O)CH}_2\text{OTl}< \longrightarrow \text{RC(OMe)(OH)-CH}_2\text{OTl}< \right] \xrightarrow[-\text{Tl}^{\text{I}}]{-\text{HCHO}} \text{RCO}_2\text{H}$$

SCHEME 83

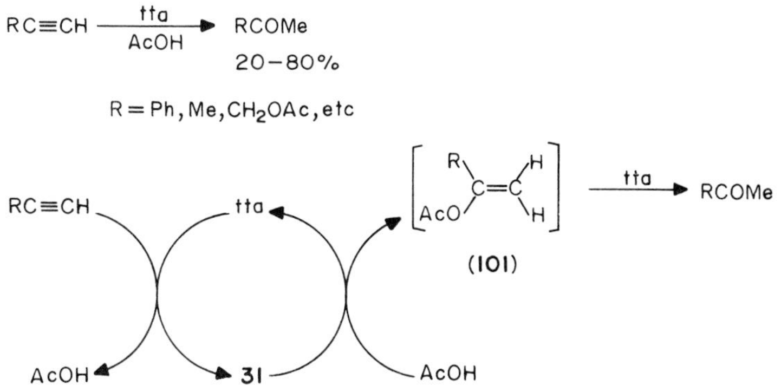

$$RC{\equiv}CH \xrightarrow[\text{AcOH}]{\text{tta}} RCOMe \quad 20\text{–}80\%$$

R = Ph, Me, CH_2OAc, etc

RC≡CH, tta, AcOH, 31, AcOH, (101), tta, RCOMe

SCHEME 84

The reagent tta works as a catalyst for the conversion of some terminal acetylenes to carbonyl compounds in acetic acid[262]. The reaction seems to proceed through acetoxythallation and, in fact, the separately prepared oxythallation adducts **31** afford the corresponding methyl ketones by heating in acetic acid[183,184]. Protonodethallation of **31** gives tta and vinyl acetate derivatives, **101**, the latter of which are easily converted into methyl ketones in the presence of a catalytic amount of tta. A proposed catalytic cycle is shown in Scheme 84. A similar catalytic reaction is observed on addition of aromatic amines to phenylacetylene to give Schiff bases, **102**[263,264], or enamines, **103**[264], depending on the starting amines, the yield varying from 6 to 45 mol per mol of tta used. Aminothallation of the acetylene followed by protonodethallation of the resultant vinylthallium(III) compounds is proposed for this catalytic reaction (Scheme 85). When this reaction system was applied to olefins, *vic*-diaminoalkanes were produced in high

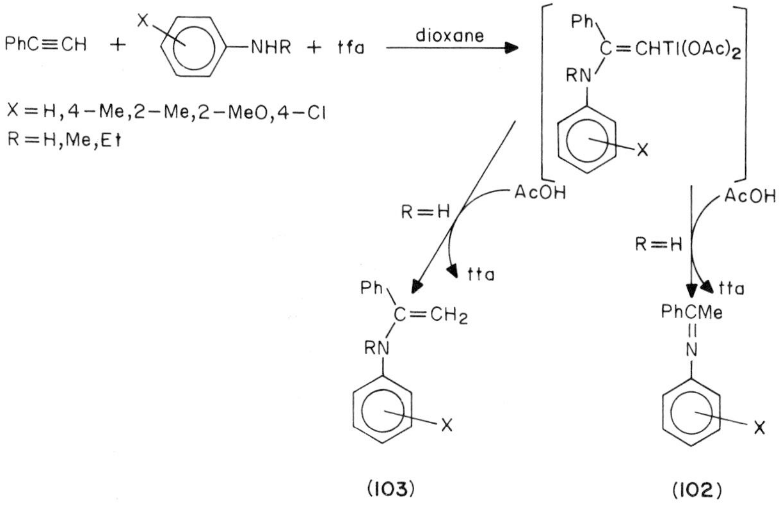

PhC≡CH + X-C_6H_4-NHR + tfa →(dioxane) [Ph(RN-C_6H_4X)C=CHTl(OAc)$_2$]

X = H, 4-Me, 2-Me, 2-MeO, 4-Cl
R = H, Me, Et

AcOH, R=H, tta → Ph(RN-C_6H_4X)C=CH_2 (103)

AcOH, R=H, tta → PhC(Me)=N-C_6H_4X (102)

SCHEME 85

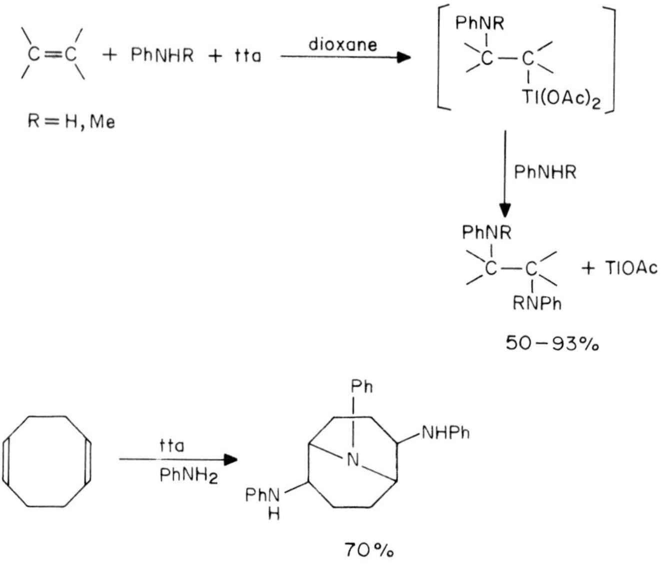

SCHEME 86

yields via aminothallation of olefins[265]. The reaction is stoichiometric with respect to tta, in contrast to the above enamine or Schiff base formation (Scheme 86).

The tta oxidation of substituted vinylallenes in acetic acid affords cyclopentenone derivatives in 25–68% yield via an organothallium(III) intermediate, as shown in Scheme 87[266]. The reaction also proceeds by use of mercury(II) acetate to afford the same products in slightly better yields (50–80%)[266,267].

IV. OXIDATION OF KETONES VIA OXYTHALLATION OF THEIR ENOLS AND RELATED REACTIONS

It is known that the reaction of acetophenone with thallium(III) isobutyrate gives $C_6H_5COCH[Tl(OCOPr^i)_2]_2$ as a stable solid[40]. The treatment of various methyl ketones with tta in methanol affords mono-α-thallation products **104** and **105** (observed by ^1H n.m.r. spectroscopy), which decompose to the corresponding α-acetoxylated ketones (Scheme 88)[268]. Analogous thallation products have also been observed in the reaction of acetone and ethyl methyl ketone with thallium(III) isobutyrate[268]. Similarly, treatment of various aliphatic and aromatic ketones with tta in acetic acid results in α-acetoxylation and α, α-diacetoxylation, probably via enolization followed by oxythallation and acetoxy-dethallation[269]. Kinetic studies of the thallium(III) perchlorate oxidation showed that the reaction is zero order in thallium(III) concentration, first order in ketone, and acid-dependent, indicating that the enolization step is rate-determining[270]. Separately, the product in this reaction was shown to be ring-contracted cyclopentanecarboxylic acid[271]. Closer investigation of the reaction revealed that the nature of the product formed was dependent on the temperature and work-up procedure, the ttn oxidation followed by treatment with aqueous $NaHCO_3$ being a good method for the synthesis of several adipoins (Scheme 89).

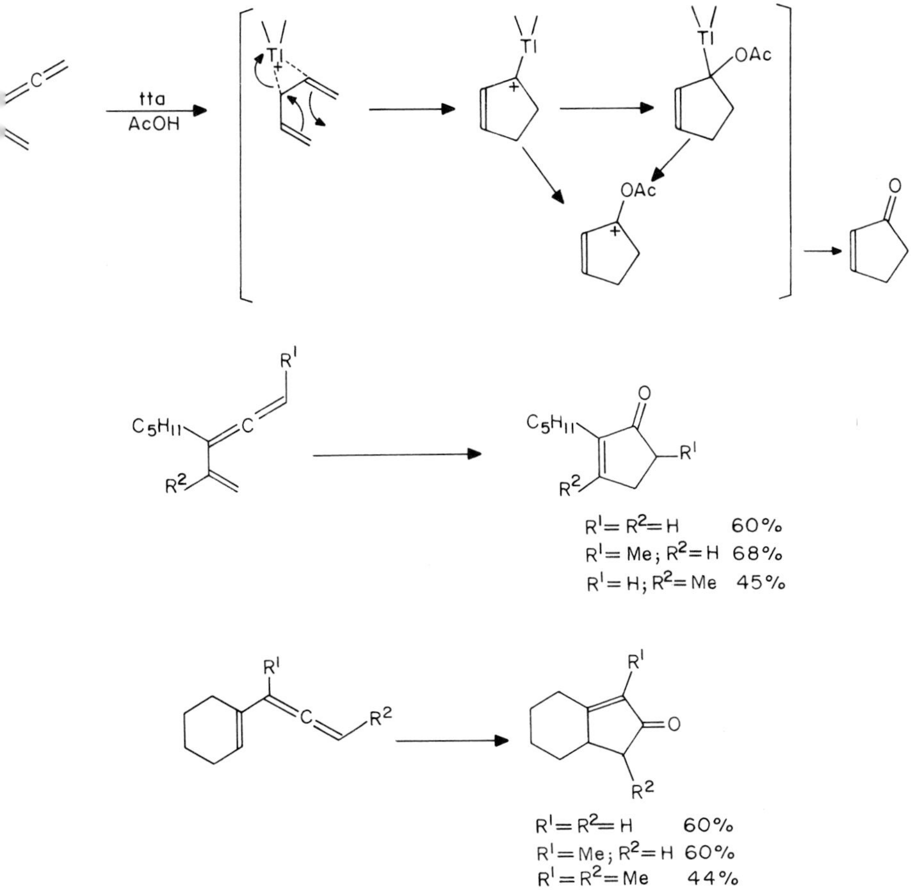

tta
AcOH

OAc

OAc

C_5H_{11}

R^2

R^1

C_5H_{11}

R^2

R^1

$R^1 = R^2 = H$ 60%
$R^1 = Me; R^2 = H$ 68%
$R^1 = H; R^2 = Me$ 45%

R^1

R^2

R^1

R^2

$R^1 = R^2 = H$ 60%
$R^1 = Me; R^2 = H$ 60%
$R^1 = R^2 = Me$ 44%

SCHEME 87

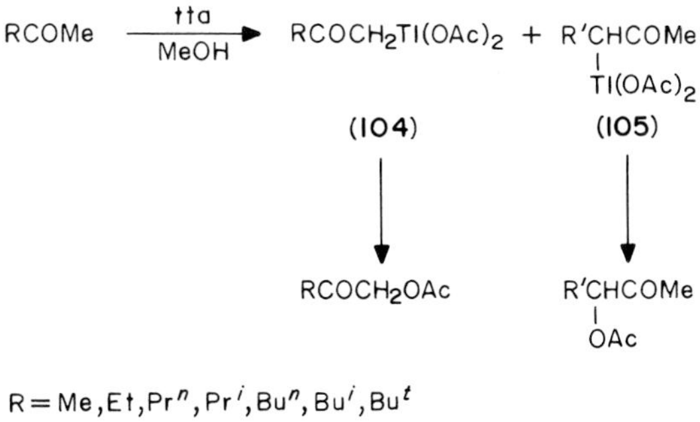

RCOMe $\xrightarrow[\text{MeOH}]{\text{tta}}$ $RCOCH_2Tl(OAc)_2$ + $R'CH(Tl(OAc)_2)COMe$

(104) (105)

$RCOCH_2OAc$ $R'CH(OAc)COMe$

$R = Me, Et, Pr^n, Pr^i, Bu^n, Bu^i, Bu^t$

SCHEME 88

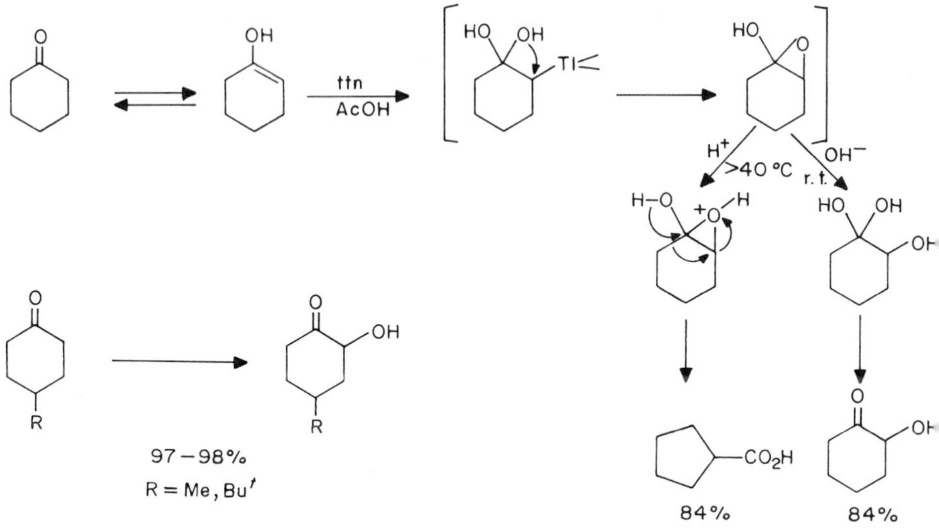

SCHEME 89

Treatment of acetophenones with ttn in acidic methanol resulted in smooth oxidative rearrangement to give methyl arylacetates in moderate to excellent yields. Acid-catalysed enolization of the ketone, oxythallation of the C—C double bond thus formed, and oxidative rearrangement by 1,2-aryl migration with simultaneous reduction of thallium(III) to thallium(I) are proposed for the reaction scheme (Scheme 90)[273,274]. α-Methoxylation occurs competitively as a side reaction in the oxidation of alkyl aryl ketones, but the rearrangement becomes the sole reaction when tmof is used as the solvent (Scheme 91)[230]. The oxidation in tmof proceeds via the methyl enol ether of the starting ketone[275]. Detailed investigations of the reaction mechanism for this oxidative rearrangement clarified that in the absence of additives an organothallium(III) intermediate, probably $C_6H_5COCH_2Tl(NO_3)_2$, persists and the key to an efficient rearrangement is the ready conversion of this compound into its acetal[276]. This oxidation can be applied to a convenient synthesis of α-arylsuccinic acids (70–80% yield) from β-aroylpropionic acid[277]. Ttn adsorbed on K-10 acidic montmorillonite clay is a remarkably effective reagent for these oxidative rearrangement of alkyl aryl ketones and olefins, and the superiority of ttn/K-10 over the usual ttn/MeOH system is found in the rapid, highly selective, high yield and r.t. oxidation with the former reagent[278]. Non-polar solvents such as n-heptane,

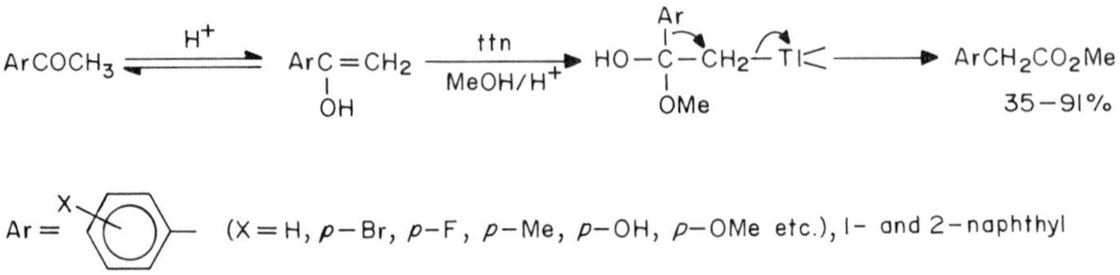

SCHEME 90

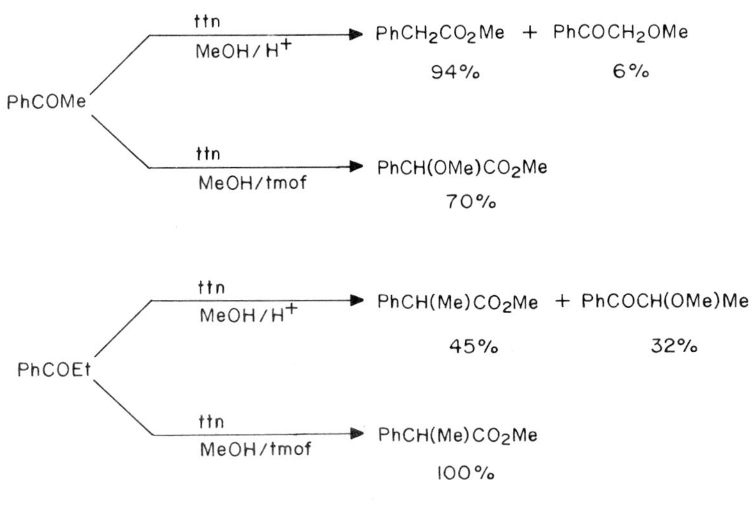

SCHEME 91

dichloromethane, carbon tetrachloride, toluene, and dioxane are generally used in this case. The ttn oxidation of enolizable ketones in non-nucleophilic solvents such as acetonitrile and diethyl carbonate affords high yields of α-nitrato ketones instead of the rearranged oxidation products obtained in methanol as described above (Scheme 92)[279]. The formation of an α-thallated ketone intermediate followed by intramolecular oxygen attack on the C—Tl bond is proposed for this reaction.

When aromatic ketones bearing a hydroxy or methoxy group in the *ortho*-position are oxidized with tta in acetic acid, oxidative cyclization occurs before aryl group rearrange-

$X\text{-}C_6H_4\text{-}COMe \xrightarrow[MeCN]{ttn} X\text{-}C_6H_4\text{-}COCH_2ONO_2$ (89–97%)

$RCH_2COCH_2R \longrightarrow RCH(ONO_2)COCH_2R$ (78–90%)

$RCOCH_2R' \rightleftharpoons RC(OH){=}CHR' \xrightarrow[MeCN]{ttn} R\text{-}\overset{+}{C}(OH)\text{-}CHR'(Tl\lt)$

$\xrightarrow{-H^+} R\text{-}C(=O)\text{-}CHR'\text{-}TlONO_2 \longrightarrow RCOCHR'(ONO_2)$

SCHEME 92

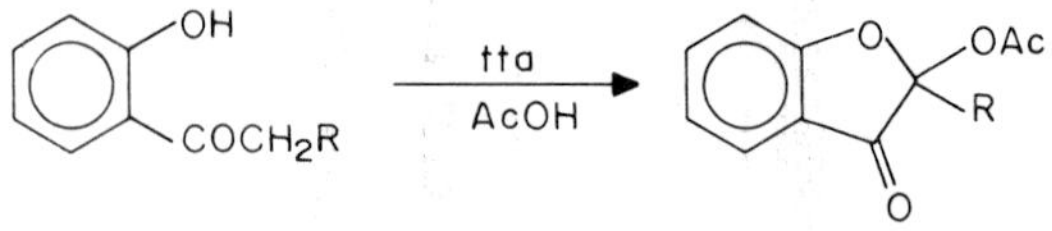

(106)
R=H 31%
R=Me 71%
R=CH2Ph 65%

SCHEME 93

ment to give 3(2*H*)-benzofuranone derivatives, **106** (Scheme 93)[225,280]. Acetoxythallation of the enol followed by intramolecular S_N2 attack of the aromatic OH or OMe group on the C—Tl bond is a proposed reaction pathway[280].

The tta oxidation of steroidal ketones gives more complicated product mixtures derived mainly from an acetoxylation, a dehydrogenation, and a rearrangement of the carbon skeleton[281,282]. The ttn-mediated oxidative rearrangement has been applied to kauranones[283,284] and morphinan derivatives[285]. Typical examples are shown in Scheme 94: 2α-carbomethoxy-A-nor-5α-cholestane, **108**, from 5α-cholestan-3-one, **107**[281], 9,

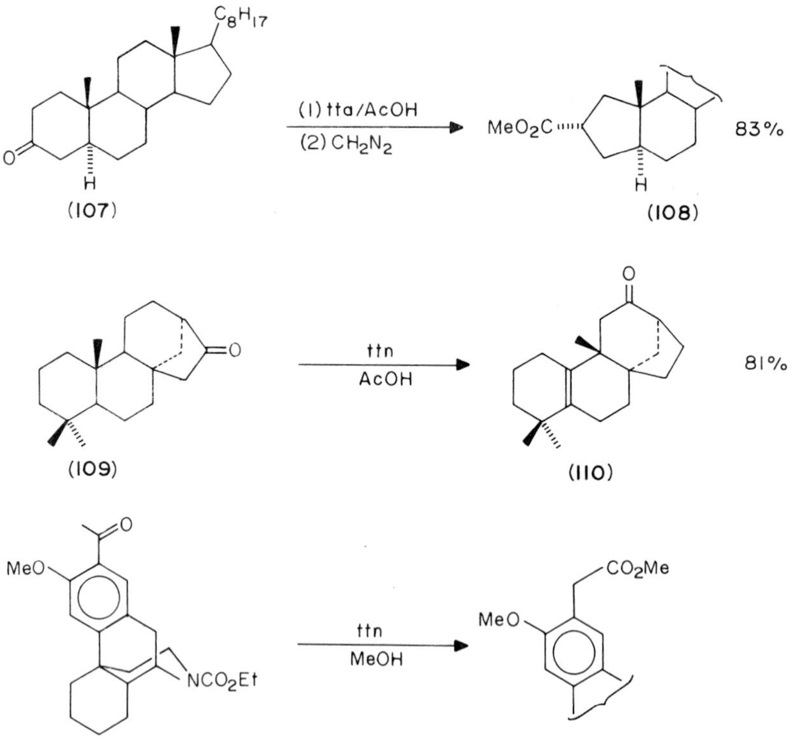

(107) (108)

(109) (110)

(111) (112)

SCHEME 94

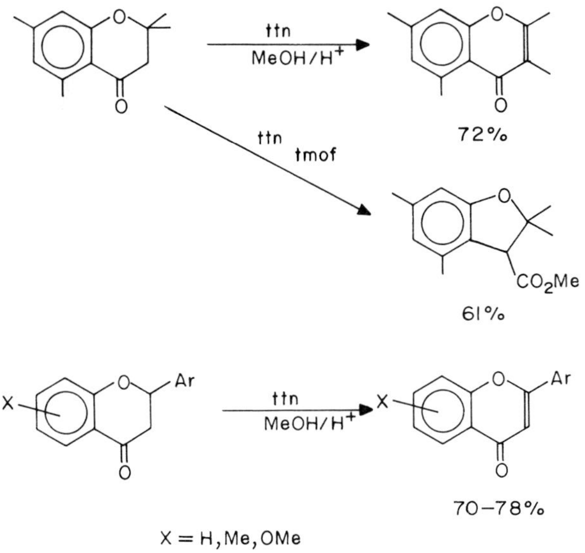

SCHEME 95

10-friedo-17-norkaur-5(10)-en-12-one, **110**, from 17-nor-13β-kauran-16-one, **109**[284], and the acidic morphinan derivative **112** from a ketone, **111**[285].

Treatment of chroman-4-ones with ttn in acid methanol results in dehydrogenation[286] or, more generally, the oxidation of flavanones to flavones[287], whereas α-methoxylation and/or oxidative rearrangement predominate in tmof (Scheme 95)[286].

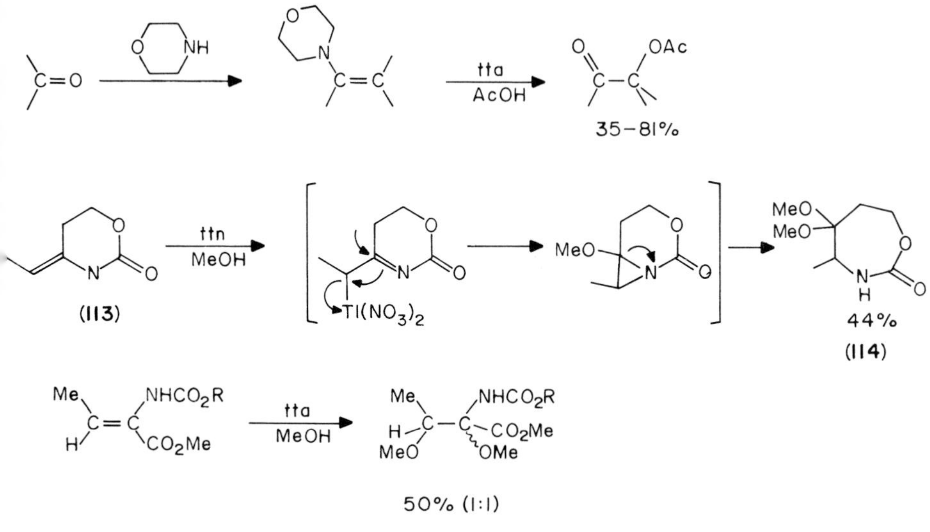

SCHEME 96

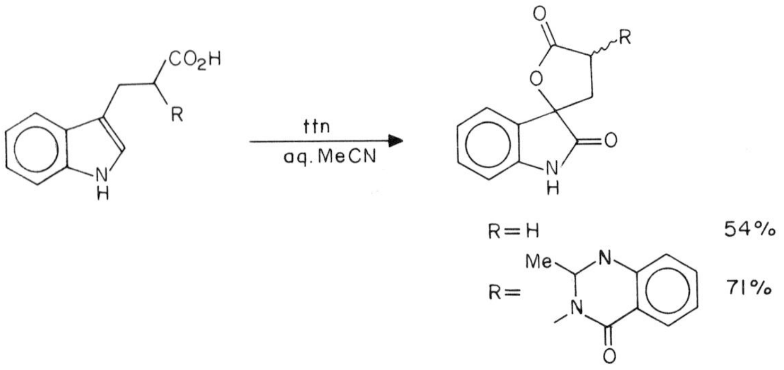

SCHEME 97

Several examples of the oxidation of enamines, imines, and related compounds are known, although the plausible organothallium(III) intermediates have never been isolated. The morphorin enamine derivatives of ketones are oxidized with tta to give the α-acetoxylation compounds in much higher yield than that of the direct oxidation of ketones[288], whilst the ttn oxidation of the enamide 1, 3-oxazin-2-one derivative **113** affords the pharmacologically interesting 1, 3-oxazepin-2-one derivative **114** as a result of ring enlargement[289]. On the other hand, the tta oxidation of methyl α-acylaminocrotonates in methanol, which bear both enamide and α,β-unsaturated carbonyl groups, affords *ca.* 1:1 diastereoisomeric mixtures of *vic*-dimethoxyalkanes (Scheme 96)[290]. Other types of enamine oxidation are found in one of the key steps for the preparation of antitumour alkaloids of the vinblastin group using tta[291], and in the formation of oxindole spirolactones by the ttn oxidation of indole-3-propionic acid and its derivatives (Scheme 97)[292]. Tautomeric imines are oxidized by ttn in methanol to give α-methoxy ketones and α-diketones after hydrolysis, rearrangements to carboxylic acid derivatives

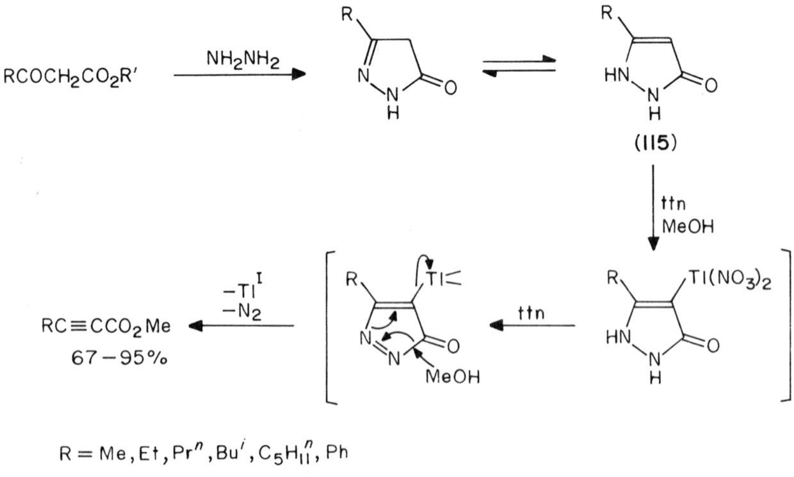

SCHEME 98

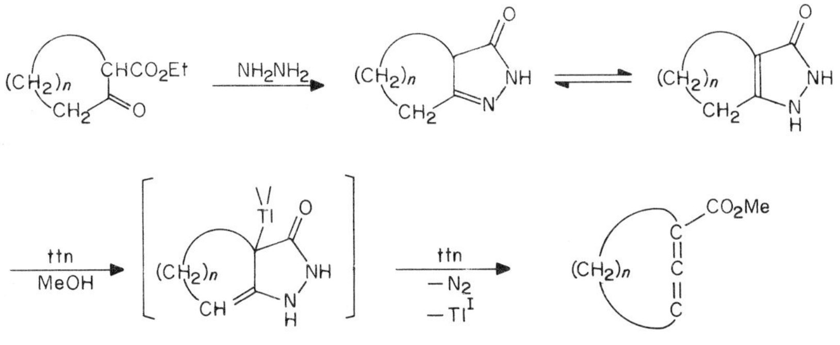

SCHEME 99

hardly being found, in contrast to the oxidation of corresponding ketones[293]. The enamine tautomers, **115**, of 5-pyrazolines prepared from β-keto esters and hydrazine react with ttn to give alk-2-ynoic acid esters in good to high yields, the intermediacy of vinylic thallium(III) compounds being proposed (Scheme 98)[294]. This conversion represents in a formal sense the dehydration of β-keto esters to alk-2-ynoic acid esters. Similar treatment of α-alkyl-β-keto esters affords allenic esters in 48–70% yield[245], and the application of this reaction to cyclic keto esters results in a formation of the difficult-to-obtain cyclic allenic esters **116** (Scheme 99)[296].

V. MISCELLANEOUS REACTIONS

The oxidation of tricarbonyl(cyclohexa-1, 3-diene)iron complexes with ttfa, ttn, and tta in alcohols results in the introduction of alkoxy group[297], and if the complex has a hydroxy group in a suitable position an intramolecular cyclization occurs to give cyclic ethers, probably via oxythallation (scheme 100)[298,299].

Treatment of isocyanides (RNC) with ttn or tta affords carbamates ($RNHCO_2R'$) in alcohols[300,301] and isocyanates (RNCO) in acetic acid[301] in good to excellent yields, oxythallation on the α-carbon of isocyanide (α-addition) being proposed.

The thallium(III)-catalysed hydrolysis of isopropenyl acetate to acetone is known to occur via an oxythallation–deoxythallation mechanism, diorganothallium(III) species acting as a catalyst[302].

Carbonylation of piperidine occurs in the presence of tta in methanol to give a small amount of methyl 1-piperidinecarboxylate, **117**, under a CO pressure of 80 atm[303]. Methoxycarbonylthallium(III) diacetate, **118**, which may be formed by insertion of CO in methoxythallium diacetate, is proposed as a reactive intermediate (Scheme 101).

Diethylthallium *tert*-butoxide reacts with cyclohexene in dichloromethane or chloroform to afford dichloronorcarane and diethylthallium chloride[304]. It also reacts with dihalogenated amide **119** to afford the α-halogeno-β-lactam **120**, probably via thermal decomposition of the intermediate diethylthallium compound **121** (Scheme 102)[305]. In relation to these reactions, triethylthallium reacts with the dichlorocarbene to produce the unstable intermediate $Et_2TlC(Cl)_2Et$, which decomposes to Et_2TlCl and 1-chloroprop-1-ene[306].

Allylation of aromatic compounds occurs by treating allylsilane, allylgermane, or allylstannane with aromatic hydrocarbons in the presence of ttfa in dichloromethane

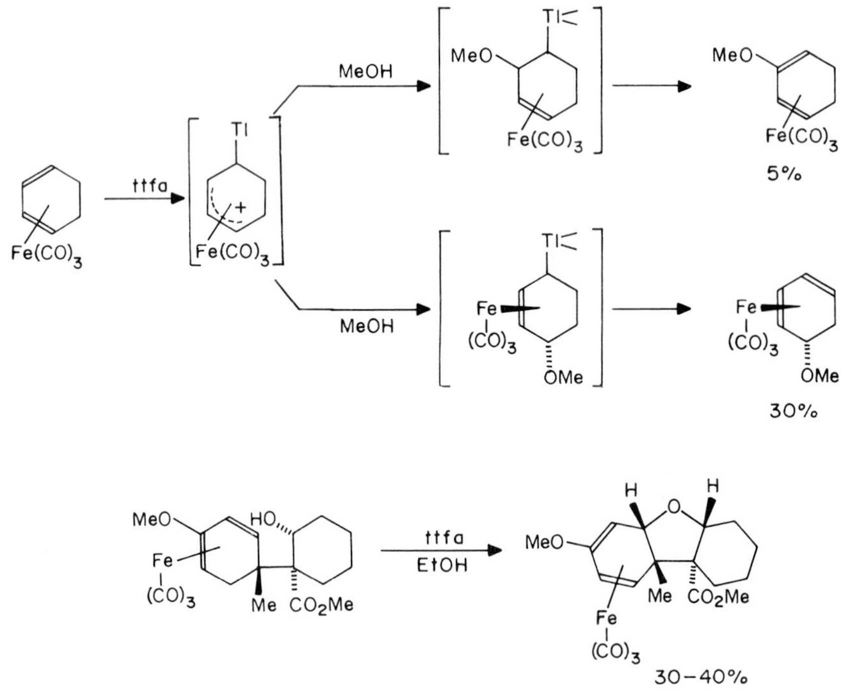

SCHEME 100

(Scheme 103)[307,308]. The reaction is assumed to proceed via allylthallium(III) compounds or allylcationic species, umpolung of the reactivity of these allylmetal compounds as allyl cation equivalents being established. The same reaction also proceeds by using $ArTl(OCOCF_3)_2$ instead of an inorganic thallium(III) salt, and the actual transmetallation reagent for allylmetal compounds in this reaction has been found to be ttfa formed *in situ* by disproportionation of the arylthallium(III) compound[309]. Reactions of allylsilanes or allylstannanes with thallium(III) salts in nitriles as the solvents afford the corresponding

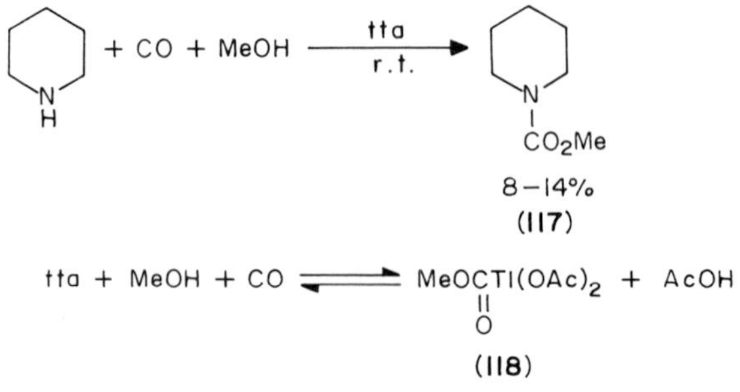

SCHEME 101

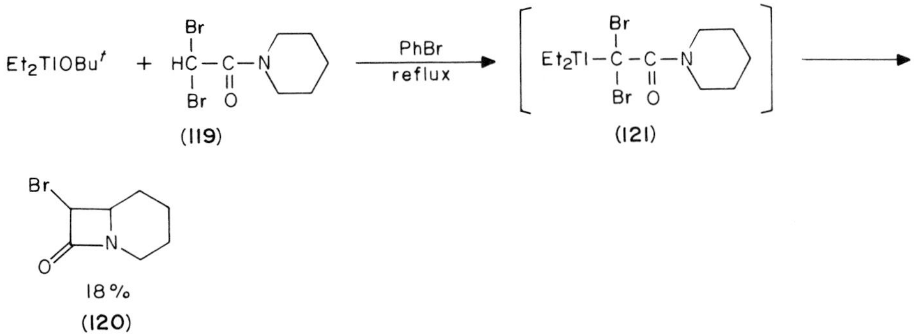

Et_2TlOBu^t + $HCBr_2C(O)N(CH_2)_5$ (119) $\xrightarrow[\text{reflux}]{\text{PhBr}}$ [$Et_2Tl-CBr_2-C(O)-N(CH_2)_5$ (121)] ⟶

18% (120)

SCHEME 102

N-allylamides via a similar reaction pathway to that above (Scheme 104), the order of effectiveness of thallium(III) salt being ttfa > ttn > tta[310].

For the structural determination of organothallium(III) compounds, ^1H and ^{13}C n.m.r. spectra are most useful. Since the pioneering work of Maher and Evans on ^1H n.m.r. spectra[311–314], many reports have appeared on this subject, some of them being included in the references in this chapter[44,62,118,120,131,132,134,136,315]. Many studies on ^{13}C n.m.r.

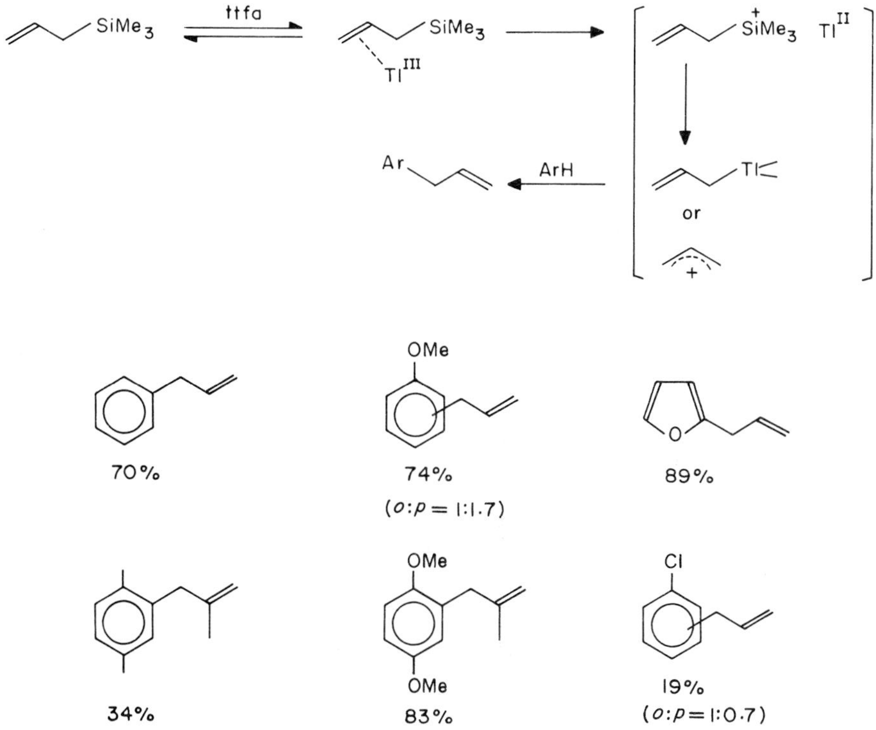

SiMe3 ttfa SiMe3 Tl^III ⟶ [SiMe3 Tl^II ; Tl< or +] ArH ⟶ Ar

70%

OMe 74% (*o*:*p* = 1:1.7)

89%

34%

OMe OMe 83%

Cl 19% (*o*:*p* = 1:0.7)

SCHEME 103

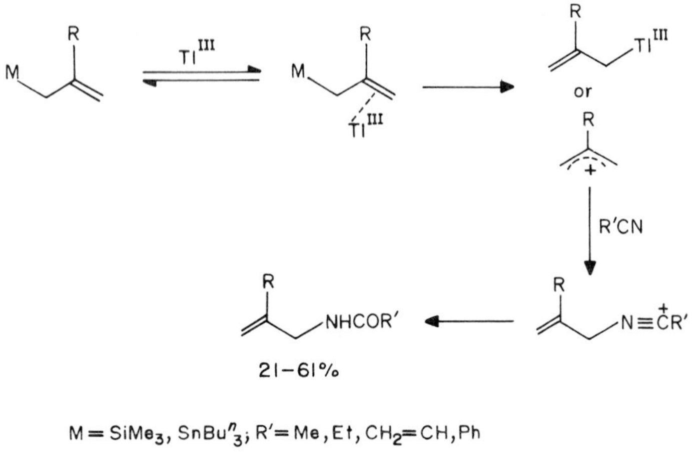

M = $SiMe_3$, $SnBu^n_3$; R′ = Me, Et, CH_2=CH, Ph

SCHEME 104

spectra have also been published[126,129,130,316–324]. It is worth consulting a recent report on a systematic study of ^{13}C and ^{1}H coupling constants (J_{Tl-C} and J_{Tl-H}) and chemical shifts in mono- and di-organothallium(III) compounds and the references cited in that report[325].

VI. REFERENCES

1. K. Yasuda and R. Okawara, *Organomet. Chem. Rev.*, **2**, 255 (1967).
2. H. Kurosawa and R. Okawara, *Organomet. Chem. Rev. A*, **6**, 65 (1970).
3. A. N. Nesmeyanov and K. A. Kocheskov, *Methods of Elemento-Organic Chemistry, Vol. 1, The Organic Compounds of Boron, Aluminum, Gallium, Indium and Thallium*, North-Holland, Amsterdam, 1967, p. 539.
4. A. G. Lee, Q. Rev. *Chem. Soc.*, **24**, 310 (1970).
5. A. G. Lee, in *Organometallic Reactions* (Eds. E. I. Becker and M. Tsutsui), Vol. 5, Wiley, New York, 1975, p. 1.
6. H. Kurosawa, in *Comprehensive Organometallic Chemistry* (Eds. G. Wilkinson, F. G. A. Stone, and E. W. Abel), Pergamon Press, Oxford, 1982, Ch. 8.
7. M. G. Jacko and S. J. W. Price, *Can. J. Chem.*, **43**, 1961 (1965).
8. S. J. Price, J. P. Richard, R. C. Rumfeldt, and M. G. Jacko, *Can. J. Chem.*, **51**, 1397 (1973).
9. A. G. Lee, *The Chemistry of Thallium*, Elsevier, Amsterdam, 1971.
10. E. C. Taylor and A. McKillop, *Acc. Chem. Res.*, **3**, 338 (1970).
11. E. C. Taylor and A. McKillop, *Aldrichim. Acta*, **3**, 4 (1970).
12. A. McKillop and E. C. Taylor, *Chem. Br.*, **9**, 4 (1973).
13. A. McKillop and E. C. Taylor, in *Advances in Organometallic Chemistry* (Eds. by F. G. A. Stone and R. West), Vol. 11, Academic Press, New York, 1973, p. 147.
14. A. McKillop, *Pure Appl. Chem.*, **43**, 463 (1975).
15. A. McKillop and E. C. Taylor, *Endeavour*, **35**, 88 (1976).
16. A. McKillop and E. C. Taylor, in *Comprehensive Organometallic Chemistry* (Eds. G. Wilkinson, F. G. A. Stone, and E. W. Abel), Pergamon Press, Oxford, 1982, Ch. 47.
17. K. Ichikawa and S. Uemura, *Yuki Gosei Kagaku Kyokai Shi*, **29**, 400 (1971).
18. S. Uemura and K. Ichikawa, *Kagaku Zokan*, Vol. 46, Kagaku Dojin, Kyoto, 1971, p. 159.
19. S. Uemura, *Kagaku (Kyoto)*, **34**, 786 (1979).

20. S. Uemura, *Yuki Gosei Kagaku Kyokai Shi*, **37**, 1002 (1979).
21. S. Uemura, in *Synthetic Reagents* (Ed. J. S. Pizey), Vol. 5, Ellis Horwood, Chichester, 1983, p. 165.
22. R. J. Ouellette, in *Oxidation in Organic Chemistry, Part B* (Ed. W. S. Trahanovsky), Academic Press, New York, 1973, p. 135.
23. A. Banerji, J. Banerji, and R. Das, *J. Sci. Ind. Res.*, **33**, 510 (1974).
24. K. Iguchi and Y. Yamada, *Yuki Gosei Kagaku Kyokai Shi*, **36**, 259 (1978).
25. M. Strasak, *Chem. Listy*, **72**, 673 (1978).
26. S. Toma, *Chem. Listy*, **73**, 281 (1979).
27. A. E. Goddard, *J. Chem. Soc.*, **121**, 36 (1922).
28. D. Goddard and A. E. Goddard, *J. Chem. Soc.*, **121**, 256 (1922).
29. A. E. Goddard and D. Goddard, *J. Chem. Soc.*, **121**, 482 (1922).
30. F. Challenger and B. Parker, *J. Chem. Soc.*, 1462 (1931).
31. F. Challenger and O. V. Richards, *J. Chem. Soc.*, 405 (1934).
32. A. G. Lee, *J. Organomet. Chem.*, **22**, 537 (1970).
33. N. N. Mel'nikov and G. P. Gracheva, *Zh. Obshch. Khim.*, **6**, 634 (1936).
34. G. B. Deacon and I. K. Johnson, *J. Fluorine Chem.*, **6**, 437 (1975).
35. G. B. Deacon and I. K. Johnson, *J. Organomet. Chem.*, **112**, 123 (1976).
36. H. J. Kabbe, *Justus Liebigs Ann. Chem.*, **656**, 204 (1962).
37. V. P. Glushkova and K. A. Kocheshkov, *Dokl. Akad. Nauk SSSR*, **103**, 615 (1955).
38. V. P. Glushkova and K. A. Kocheshkov, *Dokl. Akad. Nauk SSSR*, **116**, 233 (1957).
39. H. Gilman and R. K. Abbott, Jr., *J. Am. Chem. Soc.*, **65**, 122 (1943).
40. V. P. Glushkova and K. A. Kocheshkov, *Izv. Akad. Nauk SSSR, Otd. Khim. Nauk*, 1186 (1957).
41. J. M. Davidson and C. Triggs, *Chem. Ind.* (*London*), 457 (1966); *J. Chem. Soc. A*, 1324 (1968).
42. A. Lethbridge, R. O. C. Norman, and C. B. Thomas, *J. Chem. Soc., Perkin Trans. 1*, 2465 (1975).
43. P. M. Henry, *J. Org. Chem.*, **35**, 3083 (1970).
44. K. Ichikawa, S. Uemura, T. Nakano, and E. Uegaki, *Bull. Chem. Soc. Jpn.*, **44**, 545 (1971).
45. J. M. Briody and R. A. Moore, *J. Chem. Soc., Perkin Trans. 2*, 179 (1972).
46. S. Uemura, Y. Ikeda, O. Itoh, and K. Ichikawa, *Bull. Chem. Soc. Jpn.*, **44**, 2571 (1971).
47. S. Uemura, A. Toshimitsu, and M. Okano, unpublished results.
48. A. McKillop, J. S. Fowler, M. J. Zelesko, J. D. Hunt, E. C. Taylor, and G. McGillivray, *Tetrahedron Lett.*, 2423 (1969).
49. A. McKillop, J. D. Hunt, M. J. Zelesko, J. S. Fowler, E. C. Taylor, G. McGillivray, and F. Kienzle, *J. Am. Chem. Soc.*, **93**, 4841 (1971).
50. E. C. Taylor, F. Kienzle, R. L. Robey, and A. McKillop, *J. Am. Chem. Soc.*, **92**, 2175 (1970).
51. E. C. Taylor, F. Kienzle, R. L. Robey, A. McKillop, and J. D. Hunt, *J. Am. Chem. Soc.*, **93**, 4845 (1971).
52. G. B. Deacon, R. N. M. Smith, and D. Tunaley, *J. Organomet. Chem.*, **114**, C1 (1976).
53. G. B. Deacon, D. Tunaley, and R. N. M. Smith, *J. Organomet. Chem.*, **144**, 111 (1978).
54. G. A. Olah, I. Hashimoto, and H. C. Lin, *Proc. Natl. Acad. Sci. USA*, **74**, 4121 (1977).
55. P. Y. Kwok, L. M. Stock, and T. L. Wright, *J. Org. Chem.*, **44**, 2309 (1979).
56. R. M. G. Roberts, *Tetrahedron*, **36**, 3281 (1980).
57. S. F. Al-Azzawi and R. M. G. Roberts, *J. Chem. Soc., Perkin Trans. 2*, 677 (1982).
58. W. Lau and J. K. Kochi, *J. Am. Chem. Soc.*, **106**, 7100 (1984).
59. A. V. Huggens, J. Wolters, and E. C. Kooyman, *Tetrahedron Lett.*, 3341 (1970).
60. E. C. Kooyman, A. V. Huygens, J. P. J. de Lepper, D. de Vos, and J. Wolters, *Recl. Trav. Chim. Pays-Bas*, **100**, 24 (1981).
61. H. C. Bell, J. R. Kalman, J. T. Pinhey, and S. Sternhell, *Tetrahedron Lett.*, 3391 (1974).
62. A. McKillop, J. D. Hunt, and E. C. Taylor, *J. Organomet. Chem.*, **24**, 77 (1970).
63. E. C. Taylor, H. W. Altland, and A. McKillop, *J. Org. Chem.*, **40**, 2351 (1975).
64. Y. Kawasaki, W. Yokota, and N. Enomoto, *Chem. Lett.*, 941 (1982).
65. H. Kurosawa, N. Okuda, and Y. Kawasaki, *J. Organomet. Chem.*, **255**, 153 (1983).
66. G. B. Deacon and D. Tunaley, *J. Fluorine Chem.*, **10**, 177 (1977).
67. G. B. Deacon and D. Tunaley, *Aust. J. Chem.*, **32**, 737 (1979).
68. G. B. Deacon and R. N. M. Smith, *J. Fluorine Chem.*, **15**, 85 (1980).
69. G. B. Deacon, J. H. S. Green, and R. S. Nyholm, *J. Chem. Soc.*, 3411 (1965).
70. A. McKillop, J. S. Fowler, M. J. Zelesko, J. D. Hunt, E. C. Taylor, and G. McGillivray, *Tetrahedron Lett.*, 2427 (1969).

71. E. C. Taylor, F. Kienzle, and A. McKillop, *Org. Synth.*, **55**, 70 (1976).
72. W. Carruthers and R. Pooranamoorthy, *J. Chem. Soc., Perkin Trans. 1*, 2405 (1974).
73. R. A. Hollins and V. M. Salim, *Tetrahedron Lett.*, 591 (1979).
74. N. Ishikawa and A. Sekiya, *Bull. Chem. Soc. Jpn.*, **47**, 1680 (1974).
75. E. C. Taylor, H. W. Altland, and A. McKillop, *J. Org. Chem.*, **40**, 3441 (1975).
76. K. Ichikawa, Y. Ikeda, and S. Uemura, *J. Chem. Soc., Chem. Commun.*, 169 (1971).
77. S. Uemura, Y. Ikeda, and K. Ichikawa, *Tetrahedron*, **28**, 5499 (1972).
78. S. Uemura, Y. Ikeda, and K. Ichikawa, *Tetrahedron*, **28**, 3025 (1972).
79. S. Uemura, S. Uchida, M. Okano, and K. Ichikawa, *Bull. Chem. Soc. Jpn.*, **46**, 3254 (1973).
80. S. Uemura, A. Toshimitsu, M. Okano, and K. Ichikawa, *Bull. Chem. Soc. Jpn.*, **48**, 1925 (1975).
81. R. A. Hancock and S. T. Orszulik, *Tetrahedron Lett.*, 3789 (1979).
82. S. Uemura and T. Kawamura, unpublished data.
83. K. Ichikawa and S. Uemura, *Bull. Inst. Chem. Res., Kyoto Univ.*, **50**, 225 (1972).
84. S. Uemura, K. Sohma, M. Okano, and K. Ichikawa, *Bull. Chem. Soc. Jpn.*, **44**, 2490 (1971).
85. G. B. Deacon and J. C. Parrott, *J. Organomet. Chem.*, **22**, 287 (1970).
86. E. C. Taylor, E. C. Bigham, D. K. Johnson, and A. McKillop, *J. Org. Chem.*, **42**, 362 (1977).
87. E. C. Taylor, H. W. Altland, R. H. Danforth, G. McGillivray, and A. McKillop, *J. Am. Chem. Soc.*, **92**, 3520 (1970).
88. E. C. Taylor, F. Kienzle, and A. McKillop, *Synthesis*, 38 (1972).
89. H. Mitani, T. Ando, and Y. Yukawa, *Mem. Inst. Sci. Ind. Res., Osaka Univ.*, **30**, 81 (1973).
90. E. C. Taylor, R. H. Danforth, and A. McKillop, *J. Org. Chem.*, **38**, 2088 (1973).
91. B. Davies and C. B. Thomas, *J. Chem. Soc., Perkin Trans. 1*, 65 (1975).
92. S. Uemura, A. Toshimitsu, and M. Okano, *Bull. Chem. Soc. Jpn.*, **49**, 2582 (1976).
93. S. Uemura, A. Toshimitsu, and M. Okano, *J. Chem. Soc., Perkin Trans. 1*, 1076 (1978).
94. S. Uemura, Y. Ikeda, and K. Ichikawa, *J. Chem. Soc., Chem. Commun.*, 390 (1971).
95. A. D. Ryabov, S. A. Deiko, A. K. Yatsimirsky, and I. V. Berezin, *Tetrahedron Lett.*, **22**, 3793 (1981).
96. A. K. Yatsimirsky, S. A. Deiko, and A. D. Ryabov, *Tetrahedron*, **39**, 2381 (1983).
97. R. A. Kjonaas and D. C. Shubert, *J. Org. Chem.*, **48**, 1924 (1983).
98. T. Spencer and F. G. Thorpe, *J. Organomet. Chem.*, **99**, C8 (1975).
99. R. C. Larock, S. Varaprath, H. H. Lau, and C. A. Fellows, *J. Am. Chem. Soc.*, **106**, 5274 (1984).
100. R. C. Larock, C.-L. Liu, H. H. Lau, and S. Varaprath, *Tetrahedron Lett.*, **25**, 4459 (1984).
101. J. M. Davidson and G. Dyer, *J. Chem. Soc. A*, 1616 (1968).
102. J. J. Van Venroy, *US Pat.*, 4093647 (1978); *Chem. Abstr.*, **89**, 163274u (1978).
103. R. C. Larock and C. A. Fellows. *J. Org. Chem.*, **45**, 363 (1980).
104. R. C. Larock and C. A. Fellows, *J. Am. Chem. Soc.*, **104**, 1900 (1982).
105. S. W. Brener, G. M. Pickles, J. C. Podesta, and F. G. Thorpe, *J. Chem. Soc., Chem. Commun.*, 36 (1975).
106. R. G. Herbert, *Tetrahedron Lett.*, 1375 (1973).
107. E. C. Taylor, F. Kienzle, and A. McKillop, *J. Am. Chem. Soc.*, **92**, 6088 (1970).
108. H. Kurosawa, M. Sato, and H. Okada, *Tetrahedron Lett.*, **23**, 2965 (1982).
109. G. K. Chip and J. S. Grossert, *J. Chem. Soc., Perkin Trans. 1*, 1629 (1972).
110. A. McKillop, B. P. Swann, and E. C. Taylor, *Tetrahedron*, **26**, 4031 (1970).
111. S. Uemura, H. Miyoshi, M. Wakasugi, M. Okano, O. Itoh, T. Izumi, and K. Ichikawa, *Bull. Chem. Soc. Jpn.*, **53**, 553 (1980).
112. H. Kurosawa and R. Okawara, *J. Organomet. Chem.*, **10**, 211 (1967).
113. M. D. Johnson, *J. Chem. Soc., Chem. Commun.*, 1037 (1970).
114. H. Kurosawa, S. Numata, T. Konishi, and R. Okawara, *Bull. Chem. Soc. Jpn.*, **51**, 1397 (1978).
115. W. Kitching, *Organomet. Chem. Rev.*, **3**, 61 (1968).
116. R. N. Butler, in *Synthetic Reagents* (Ed. by J. S. Pizey), Vol. 4, Ellis Horwood, Chichester, 1977, p. 15.
117. R. Criegee, *Angew. Chem.*, **70**, 173 (1958).
118. S. Uemura, K. Zushi, A. Tabata, A. Toshimitsu, and M. Okano, *Bull. Chem. Soc. Jpn.*, **47**, 920 (1974).
119. S. Uemura, H. Miyoshi, A. Toshimitsu, and M. Okano, *Bull. Chem. Soc. Jpn.*, **49**, 3285 (1976).
120. H. Kurosawa, R. Kitano, and T. Sasaki, *J. Chem. Soc., Dalton Trans.*, 234 (1978).
121. J. E. Bäckvall, M. U. Ahmad, S. Uemura, A. Toshimitsu, and T. Kawamura, *Tetrahedron Lett.*, 2283 (1980), and unpublished work.

122. W. Kruse and T. M. Bednarski, *J. Org. Chem.*, **36**, 1154 (1971).
123. C. Beaudry and M. Zador, *J. Organomet. Chem.*, **102**, 265 (1975).
124. A. Lethbridge, R. O. C. Norman, and C. B. Thomas, *J. Chem. Soc., Perkins Trans. 1*, 1929 (1974).
125. A. J. Bloodworth and D. J. Lapham, *J. Chem. Soc., Perkin Trans. 1*, 3265 (1981).
126. V. G. Gibb and L. D. Hall, *Carbohydr. Res.*, **63**, C1 (1978).
127. K. C. Pande and S. Winstein, *Tetrahedron Lett.*, 3393 (1964).
128. E. Maxa, G. Schultz, and E. Zbiral, *Justus Liebigs Ann. Chem.*, 933 (1974).
129. S. Uemura, H. Miyoshi, M. Okano, I. Morishima, and T. Inubushi, *J. Organomet. Chem.*, **165**, 9 (1979).
130. S. Uemura, H. Miyoshi, M. Okano, I. Morishima, and T. Inubushi, *J. Organomet. Chem.*, **171**, 131 (1979).
131. A. McKillop, M. E. Ford, and E. C. Taylor, *J. Org. Chem.*, **39**, 2434 (1974).
132. F. A. L. Anet, *Tetrahedron Lett.*, 3399 (1964).
133. P. M. Henry, *J. Am. Chem. Soc.*, **88**, 1597 (1966).
134. R. K. Sharma and N. H. Fellers, *J. Organomet. Chem.*, **49**, C69 (1973).
135. S. Uemura, K. Sohma, H. Tara, and M. Okano, *Chem. Lett.*, 545 (1973).
136. S. Uemura, H. Tara, M. Okano, and K. Ichikawa, *Bull. Chem. Soc. Jpn.*, **47**, 2663 (1974).
137. S. Uemura, H. Miyoshi, H. Tara, M. Okano, and K. Ichikawa, *J. Chem. Soc., Chem. Commun.*, 218 (1976).
138. S. Uemura, H. Miyoshi, M. Okano, and K. Ichikawa, *J. Chem. Soc., Perkin Trans. 1*, 991 (1981).
139. R. K. Sharma and E. D. Martinez, *J. Chem. Soc., Chem. Commun.*, 1129 (1972).
140. R. D. Bach and J. W. Holubka, *J. Am. Chem. Soc.*, **96**, 7814 (1974).
141. R. D. Bach, J. W. Holubka, and C. L. Willis, *J. Am. Chem. Soc.*, **104**, 3980 (1982).
142. L. Nadon and M. Zador, *Can. J. Chem.*, **52**, 2667 (1974).
143. L. Nadon and M. Zador, *Can. J. Chem.*, **55**, 3590 (1977).
144. H. Mitani, T. Ando, and Y. Yukawa, *Chem. Lett.*, 455 (1972).
145. S. Uemura, A. Toshimitsu, M. Okano, T. Kawamura, T. Yonezawa, and K. Ichikawa, *J. Chem. Soc., Chem. Commun.*, 65 (1978).
146. S. Uemura, A. Tabata, and M. Okano, *J. Chem. Soc., Chem. Commun.*, 512 (1972).
147. F. G. Bordwell and M. L. Douglass, *J. Am. Chem. Soc.*, **88**, 993 (1966).
148. H. C. Brown and P. J. Geoghegan, Jr., *J. Am. Chem. Soc.*, **89**, 1522 (1967).
149. H. C. Brown and P. J. Geoghegan, Jr., *J. Org. Chem.*, **35**, 1844 (1970).
150. D. J. Pasto and J. A. Gontarz, *J. Am. Chem. Soc.*, **91**, 719 (1969).
151. H. Kurosawa, H. Okada, and M. Yasuda, *Tetrahedron Lett.*, 959 (1980).
152. H. Kurosawa and M. Yasuda, *J. Chem. Soc., Chem. Commun.*, 716 (1978); H. Kurosawa, H. Okada, M. Sato, and T. Hattori, *J. Organomet. Chem.*, **250**, 83 (1983).
153. K. Ichikawa, S. Uemura, and T. Sugita, *Tetrahedron*, **22**, 407 (1966).
154. S. Uemura, T. Nakano, and K. Ichikawa, *Nippon Kagaku Zasshi*, **89**, 203 (1968).
155. S. Uemura, K. Zushi, M. Okano, and K. Ichikawa, *J. Chem. Soc., Chem. Commun.*, 234 (1972).
156. R. R. Grinstead, *J. Org. Chem.*, **26**, 238 (1961).
157. P. M. Henry, *J. Am. Chem. Soc.*, **87**, 990 (1965).
158. P. M. Henry, *J. Am. Chem. Soc.*, **87**, 4423 (1965).
159. P. M. Henry, *J. Org. Chem.*, **38**, 2415 (1973).
160. J. Urbanec, M. Strašák, M. Hrušovský, and J. Vojtko, *Collect. Czech. Chem. Commun.*, **41**, 604 (1976).
161. M. Strašák and J. Majer, *J. Org. Chem.*, **45**, 3580 (1980).
162. C. Lion and J. E. Dubois, *C. R. Acad. Sci., Ser. C*, **274**, 1073 (1972).
163. J. Grignon and S. Fliszár, *Can. J. Chem.*, **52**, 3209 (1974).
164. J. B. Lee and M. J. Price, *Tetrahedron Lett.*, 156 (1962).
165. J. B. Price and M. J. Price, *Tetrahedron*, **20**, 1017 (1964).
166. C. B. Anderson and S. Winstein, *J. Org. Chem.*, **28**, 605 (1963).
167. R. J. Ouellette, G. Kordosky, C. Levin, and S. Williams, *J. Org. Chem.*, **34**, 4104 (1969).
168. S. Uemura, H. Miyoshi, A. Toshimitsu, and M. Okano, *Bull. Chem. Soc. Jpn.*, **49**, 3285 (1976).
169. A. Lethbridge, R. O. C. Norman, C. B. Thomas, and W. J. E. Parr, *J. Chem. Soc., Perkin Trans. 1*, 231 (1975).
170. B. Floris, *Gazz. Chim. Ital.*, **112**, 489 (1982).
171. S. Uemura, R. Kitoh, and K. Ichikawa, *Nippon Kagaku Zasshi*, **87**, 986 (1966).
172. A. Lethbridge, R. O. C. Norman, and C. B. Thomas, *J. Chem. Soc., Perkin Trans. 1*, 2763 (1973).
173. L. Nadon, N. Tardat, M. Zador, and S. Fliszár, *Can. J. Chem.*, **51**, 2366 (1973).

174. W. F. Brill, *US Pat.*, 4021453; *Chem. Abstr.*, **87**, p.22342m (1977).
175. S. E. Diamond, F. Mares, A. Szalkiewicz, D. A. Muccigrosso, and J. P. Solar, *J. Am. Chem. Soc.*, **104**, 4266 (1982).
176. D. D. Tanner, P. B. von Bostelen, and M. Lai, *Can. J. Chem.*, **54**, 2004 (1976).
177. E. Glotter and A. Schwartz, *J. Chem. Soc., Perkin Trans. 1*, 1660 (1976).
178. A. Schwartz and E. Glotter, *J. Chem. Soc., Perkin Trans. 1*, 2470 (1977).
179. H. Sekizaki, M. Ito, and S. Inoue, *Bull. Chem. Soc. Jpn.*, **51**, 2439 (1978).
180. R. M. Moriarty and H. Gopal, *Tetrahedron Lett.*, 347 (1972).
181. E. C. Taylor, G. E. Jagdmann, Jr., and A. McKillop, *J. Org. Chem.*, **45**, 3375 (1980).
182. V. Simonidesz, Z. Gombos-Visky, G. Kovacs, E. Baitz-Gacs, and L. Radics, *J. Am. Chem. Soc.*, **100**, 6756 (1978).
183. V. Simonidesz, A. Behr-Papp, J. Ivanics, G. Kovacs, E. Baitz-Gacs, and L. Radics, *J. Chem. Soc., Perkin Trans. 1*, 2572 (1980).
184. A. McKillop, J. D. Hunt, E. C. Taylor, and F. Kienzle, *Tetrahedron Lett.*, 5275 (1970).
185. A. McKillop, J. D. Hunt, F. Kienzle, E. Bigham, and E. C. Taylor, *J. Am. Chem. Soc.*, **95**, 3635 (1973).
186. E. J. Corey and T. Ravindranthan, *Tetrahedron Lett.*, 4753 (1971).
187. D. Fărcaşiu, P. von R. Schleyer, and D. B. Ledlie, *J. Org. Chem.*, **38**, 3455 (1973).
188. E. Fujita and M. Ochiai, *Chem. Pharm. Bull.*, **25**, 3013 (1977).
189. A. J. Irvin and J. B. Jones, *J. Org. Chem.*, **42**, 2176 (1977).
190. E. C. Taylor, C.-S. Chiang, and A. McKillop, *Tetrahedron Lett.*, 1827 (1977).
191. M. J. Begley, S. E. Mohamed, D. A. Whiting, F. D'Souza, and N. A. R. Hatam, *J. Chem. Soc., Perkin Trans. 1*, 883 (1983).
192. H. Meyer, B. Bodo, C. Deschamps, and D. Molho, *Tetrahedron Lett.*, **25**, 4519 (1984).
193. S. Antus, Á. Gottsegen, M. Nógrádi, and A. Gergely, *Chem. Ber.*, **112**, 3879 (1979).
194. M. Meyer-Dayan, B. Bodo, C. Deschamps-Vallet, and D. Molho, *Tetrahedron Lett.*, 3359 (1978).
195. T. Kaiya, N. Shirai, J. Sakakibara, and Y. Iitaka, *Tetrahedron Lett.*, 4297 (1979).
196. R. J. Bertsch and R. J. Ouellette, *J. Org. Chem.*, **39**, 2755 (1974).
197. M. Niwa, H. Noda, H. Kobayashi, and S. Yamamura, *Chem. Lett.*, 85 (1980).
198. R. J. Ouellette and R. J. Bertsch, *J. Org. Chem.*, **41**, 2782 (1976).
199. J. E. Byrd and J. Helpern, *J. Am. Chem. Soc.*, **95**, 2586 (1973).
200. J. E. Byrd, L. Cassar, P. E. Eaton, and J. Halpern, *J. Chem. Soc., Chem. Commun.*, 40 (1971).
201. P. Abley, J. E. Byrd, and J. Halpern, *J. Am. Chem. Soc.*, **95**, 2591 (1973).
202. For similar reaction with ttn, see refs. 187–190 and Schemes 55 and 56.
203. M. Fărcaşiu, D. Fărcaşiu, T. Slutsky, and P. von R. Schleyer, *Tetrahedron Lett.*, 4059 (1974).
204. R. Favier, C. Freppel, J.-C. Richer, and M. Zador, *Can. J. Chem.*, **49**, 2590 (1971).
205. C. Freppel, R. Favier, J.-C. Richer, and M. Zador, *Can. J. Chem.*, **49**, 2586 (1971).
206. G. Emmer and E. Zbiral, *Tetrahedron*, **33**, 1415 (1977).
207. B. Cocton and A. Crastes de Paulet, *Bull. Soc. Chim. Fr.*, 2947 (1966).
208. G. Ortar and I. Torrini, *Tetrahedron*, **38**, 859 (1977).
209. M. Ochiai and E. Fujita, *J. Chem. Soc., Chem. Commun.*, 967 (1975).
210. E. Fujita and M. Ochiai, *J. Chem. Soc., Perkin Trans. 1*, 1948 (1977).
211. P. K. Grant, H. T. L. Liau, and K. S. Low, *Aust. J. Chem.*, **28**, 903 (1975).
212. S. Uemura, A. Tabata, M. Okano, and K. Ichikawa, *J. Chem. Soc., Chem. Commun.*, 1630 (1970).
213. S. Uemura, H. Miyoshi, A. Tabata, and M. Okano, *Tetrahedron*, **37**, 291 (1981).
214. M. Murakami and S. Nishida, *Chem. Lett.*, 997 (1981).
215. Y. Yamada, A. Shibata, K. Iguchi, and H. Sanjoh, *Tetrahedron Lett.*, 2407 (1977).
216. T. Sasaki, K. Kanematsu, A. Kondo, and K. Okada, *J. Org. Chem.*, **41**, 2231 (1976).
217. Y. Yamada, S. Nakamura, K. Iguchi, and K. Hosaka, *Tetrahedron Lett.*, **22**, 1355 (1981).
218. Y. Yamada, H. Sanjoh, and K. Iguchi, *J. Chem. Soc., Chem. Commun.*, 997 (1976).
219. Y. Yamada, H. Sanjoh, and K. Iguchi, *Tetrahedron Lett.*, 1323 (1979).
220. Y. Yamada, H. Sanjoh, and K. Iguchi, *Tetrahedron Lett.*, 423 (1979).
221. W. Renold, G. Ohloff, and T. Norin, *Helv. Chim. Acta*, **62**, 985 (1979).
222. F. J. McQuillin and D. G. Parker, *J. Chem. Soc., Perkin Trans. 1*, 2092 (1975).
223. W. D. Ollis, K. L. Ormand, and I. O. Sutherland, *J. Chem. Soc., Chem. Commun.*, 1237 (1968).
224. W. D. Ollis, K. L. Ormand, B. T. Redman, R. J. Roberts, and I. O. Sutherland, *J. Chem. Soc. C*, 125 (1970).
225. W. D. Ollis, K. L. Ormand, and I. O. Sutherland, *J. Chem. Soc. C*, 119 (1970).

226. A. McKillop, B. P. Swann, and E. C. Taylor, *Tetrahedron Lett.*, 5281 (1970).
227. E. C. Taylor, R. A. Conley, D. K. Johnson, A. McKillop, and M. E. Ford, *J. Org. Chem.*, **45**, 3433 (1980).
228. A. McKillop, B. P. Swann, M. E. Ford, and E. C. Taylor, *J. Am. Chem. Soc.*, **95**, 3641 (1973).
229. E. C. Taylor, R. A. Conley, D. K. Johnson, and A. McKillop, *J. Org. Chem.*, **42**, 4167 (1977).
230. E. C. Taylor, R. L. Robey, K.-T. Liu, B. Favre, H. T. Bozimo, R. A. Conley, C.-S. Chiang, A. McKillop, and M. E. Ford, *J. Am. Chem. Soc.*, **98**, 3037 (1976).
231. L. Farkas, A. Gottsegen, M. Nógrádi, and S. Antus, *J. Chem. Soc., Chem. Commun.*, 825 (1972).
232. L. Farkas and A. Wolfner, *Chem. Ber.*, **107**, 3966 (1974).
233. L. Farkas, A. Gottsegen, M. Nógrádi, and S. Antus, *J. Chem. Soc., Perkin Trans. 1*, 305 (1974).
234. S. Antus, L. Farkas, A. Gottsegen, Z. Kardos-Balogh, and M. Nógrádi, *Chem. Ber.*, **109**, 3811 (1976).
235. S. Antus and M. Nógrádi, *Chem. Ber.*, **112**, 480 (1979).
236. S. Antus, L. Farkas, Z. Kardos-Balogh, and M. Nógrádi, *Chem. Ber.*, **108**, 3883 (1975).
237. R. S. Varma and M. Varma, *J. Chem. Res. (S)*, 255 (1982).
238. A. Levai and A. L. Tökés, *Synth. Commun.*, **12**, 701 (1982).
239. R. S. Varma and M. Varma, *Tetrahedron Lett.*, **23**, 3007 (1982).
240. R. S. Varma, *Chem. Ind. (London)*, 56 (1982).
241. S. Antus, F. Boross, M. Kajtár-Peredy, L. Radics, and M. Nógrádi, *Justus Liebigs Ann. Chem.*, 1283 (1980).
242. S. Antus, E. Baitz-Gacs, F. Boross, M. Nógrádi, and A. Solyom, *Justus Liebigs Ann. Chem.*, 1271 (1980).
243. E. C. Taylor, J. G. Andrade, G. J. H. Rall, and A. McKillop, *Tetrahedron Lett.*, 3623 (1978).
244. E. C. Taylor, J. G. Andrade, G. J. H. Rall, K. Steliou, G. E. Jagdmann, Jr., and A. McKillop, *J. Org. Chem.*, **46**, 3078 (1981).
245. A. Pelter, R. S. Ward, D. J. Watson, P. Collins, and I. T. Kay, *Tetrahedron Lett.*, 2275 (1979).
246. A. Pelter, R. S. Ward, D. J. Watson, P. Collins, and I. T. Kay, *J. Chem. Soc., Perkin Trans. 1*, 175 (1982).
247. S. Nishiyama and S. Yamamura, *Chem. Lett.*, 1511 (1981).
248. R. J. Ouellette, D. L. Shaw, and A. South, Jr., *J. Am. Chem. Soc.*, **86**, 2744 (1964).
249. R. J. Ouellette, A. South, Jr., and D. L. Shaw, *J. Am. Chem. Soc.*, **87**, 2602 (1965).
250. A. South, Jr., and R. J. Ouellette, *J. Am. Chem. Soc.*, **90**, 7064 (1968).
251. R. J. Ouellette and S. Williams, *J. Org. Chem.*, **35**, 3210 (1970).
252. H. M. Cambell, P. A. Gunn, A. J. McAlees, and R. McCrindle, *Can. J. Chem.*, **51**, 4167 (1973).
253. H. M. Cambell, P. A. Gunn, A. J. McAlees, and R. McCrindle, *Can. J. Chem.*, **53**, 20 (1975).
254. D. F. Covey and A. Nickon, *J. Org. Chem.*, **42**, 794 (1977).
255. T. Katsushima, R. Yamaguchi, S. Iemura, and M. Kawanisi, *Bull. Chem. Soc. Jpn.*, **53**, 3318 (1980).
256. T. Shirafuji, Y. Yamamoto, and H. Nozaki, *Tetrahedron Lett.*, 4713 (1971).
257. T. Shirafuji and H. Nozaki, *Tetrahedron*, **29**, 77 (1973).
258. J. Salaun, B. Garnier, and J. M. Conia, *Tetrahedron*, **30**, 1423 (1974).
259. A. McKillop, O. H. Oldenziel, B. P. Swann, E. C. Taylor, and R. L. Robey, *J. Am. Chem. Soc.*, **93**, 7331 (1971).
260. A. McKillop, O. H. Oldenziel, B. P. Swann, E. C. Taylor, and R. L. Robey, *J. Am. Chem. Soc.*, **95**, 1296 (1973).
261. C. Deschamps and M. Zador, *J. Organomet. Chem.*, **217**, 303 (1981).
262. S. Uemura, R. Kitoh, K. Fujita, and K. Ichikawa, *Bull. Chem. Soc. Jpn.*, **40**, 1499 (1967).
263. J. Barluenga and F. Aznar, *Synthesis*, 704 (1975).
264. J. Barluenga and F. Aznar, *Synthesis*, 195 (1977).
265. V. Gómez Aranda, J. Barluenga, and F. Aznar, *Synthesis*, 504 (1974).
266. R. Baudony, F. Delbecq, and J. Gore, *Tetrahedron*, **36**, 189 (1980).
267. F. Delbecq and J. Gore, *Tetrahedron Lett.*, 3459 (1976).
268. M. Vollmerhaus and F. Huber, *Z. Naturforsch., Teil B*, **36**, 141 (1981).
269. S. Uemura, T. Nakano, and K. Ichikawa, *Nippon Kagaku Zasshi*, **88**, 1111 (1967).
270. J. S. Littler, *J. Chem. Soc.*, 827 (1962).
271. K. B. Wiberg and W. Koch, *Tetrahedron Lett.*, 1779 (1966).
272. A. McKillop, J. D. Hunt, and E. C. Taylor, *J. Org. Chem.*, **37**, 3281 (1972).
273. A. McKillop, B. P. Swann, and E. C. Taylor, *J. Am. Chem. Soc.*, **93**, 4919 (1971).

274. A. McKillop, B. P. Swann, and E. C. Taylor, *J. Am. Chem. Soc.*, **95**, 3340 (1973).
275. J. A. Walker and M. D. Pillai, *Tetrahedron Lett.*, 3707 (1977).
276. S. D. Higgins and C. B. Thomas, *J. Chem. Soc., Perkin Trans. 1*, 235 (1982).
277. E. C. Taylor, R. A. Conley, A. H. Katz, and A. McKillop, *J. Org. Chem.*, **49**, 3840 (1984).
278. E. C. Taylor, C.-S. Chiang, A. McKillop, and J. F. White, *J. Am. Chem. Soc.*, **98**, 6750 (1976).
279. A. McKillop, D. W. Young, M. Edwards, R. P. Hug, and E. C. Taylor, *J. Org. Chem.*, **43**, 3773 (1978).
280. N. Malaitong and C. Thebtaranonth, *Chem. Lett.*, 305 (1980).
281. A. Romeo and G. Ortar, *Tetrahedron*, **28**, 5337 (1972).
282. G. Ortar and A. Romeo, *J. Chem. Soc., Perkin Trans. 1*, 111 (1976).
283. E. Fujita and M. Ochiai, *J. Chem. Soc., Perkin Trans. 1*, 1182 (1977).
284. E. Fujita and M. Ochiai, *Can. J. Chem.*, **56**, 246 (1978).
285. P. C. Belanger, C. S. Rooney, F. M. Robinson, and L. H. Sarett, *J. Org. Chem.*, **43**, 906 (1978).
286. R. G. Ciattini, E. Morera, and G. Ortar, *J. Heterocycl. Chem.*, **19**, 395 (1982).
287. R. S. Varma and M. Varma, *Synth. Commun.*, **12**, 927 (1982).
288. M. E. Kuehne and T. J. Giacobbe, *J. Org. Chem.* **33**, 3359 (1968).
289. T. G. Back, O. E. Edwards, and G. A. MacAlpine, *Tetrahedron Lett.*, 2651 (1977).
290. M. P. Paradisi and G. P. Zecchini, *Tetrahedron*, **33**, 1729 (1977).
291. P. Mangeney, R. Zo Andriamialisoa, N. Langlois, Y. Langlois, and P. Potier, *J. Am. Chem. Soc.*, **101**, 2243 (1979).
292. T. Ohnuma, Y. Kimura, and Y. Ban, *Tetrahedron Lett.*, **22**, 4969 (1981).
293. H. Ahlbrecht and D. Hagena, *Chem. Ber.*, **109**, 2345 (1976).
294. E. C. Taylor, R. L. Robey, and A. McKillop, *Angew. Chem.*, **84**, 60 (1972); *Angew. Chem., Int. Ed. Engl.*, **11**, 48 (1972).
295. E. C. Taylor, R. L. Robey, and A. McKillop, *J. Org. Chem.*, **37**, 2797 (1972).
296. A. Silveira, Jr., M. Angelastro, R. Israel, F. Totino, and P. Williamsen, *J. Org. Chem.*, **45**, 3522 (1980).
297. B. F. G. Johnson, J. Lewis, and D. G. Parker, *J. Organomet. Chem.*, **127**, C37 (1977).
298. A. J. Pearson, *J. Chem. Soc., Chem. Commun.*, 488 (1980).
299. A. J. Pearson and C. W. Ong, *J. Chem. Soc., Perkin Trans. 1*, 1614 (1981).
300. F. Kienzle, *Tetrahedron Lett.*, 1771 (1972).
301. S. Tanaka, H. Kido, S. Uemura, and M. Okano, *Bull. Chem. Soc. Jpn.*, **48**, 3415 (1975).
302. P. Abley, J. E. Byrd, and J. Halpern, *J. Am. Chem. Soc.*, **94**, 1985 (1972).
303. T. Saegusa, T. Tsuda, and K. Isayama, *Tetrahedron Lett.*, 4123 (1967).
304. V. I. Shcherbakov and S. F. Zhil'tsov, *Zh. Obshch. Khim.*, **40**, 2046 (1970).
305. B. Åkermark, J. Chodowska-Palicka, and I. Lagerlund, *J. Organomet. Chem.*, **113**, C4 (1976).
306. V. I. Shchervakov, *Zh. Obshch. Khim.*, **41**, 1095 (1971).
307. M. Ochiai, M. Arimoto, and E. Fujita, *Tetrahedron Lett.*, **22**, 4491 (1981).
308. M. Ochiai, E. Fujita, M. Arimoto, and H. Yamaguchi, *Chem. Pharm. Bull.*, **31**, 86 (1983).
309. M. Ochiai, E. Fujita, M. Arimoto, and H. Yamaguchi, *Chem. Pharm. Bull.*, **30**, 3994 (1982).
310. M. Ochiai, S. Tada, M. Arimoto, and E. Fujita, *Chem. Pharm. Bull.*, **30**, 2836 (1982).
311. J. P. Maher and D. F. Evans, *Proc. Chem. Soc. London*, 208 (1961).
312. J. P. Maher and D. F. Evans, *Proc. Chem. Soc. London*, 176 (1963).
313. J. P. Maher and D. F. Evans, *J. Chem. Soc.*, 5534 (1963).
314. J. P. Maher and D. F. Evans, *J. Chem. Soc.*, 637 (1965).
315. B. V. Cheesman and R. F. M. White, *Can. J. Chem.*, **62**, 521 (1984).
316. W. Kitching, D. Praeger, C. J. Moore, D. Doddrell, and W. Adcock, *J. Organomet. Chem.*, **70**, 339 (1974).
317. L. Ernst, *J. Organomet. Chem.*, **82**, 319 (1974).
318. L. Ernst, *Org. Magn. Reson.*, **6**, 540 (1974).
319. W. Kitching, C. J. Moore, D. Doddrell, and W. Adcock, *J. Organomet. Chem.*, **94**, 469 (1975).
320. R. J. Abraham, G. E. Hawkes, and K. M. Smith, *J. Chem. Soc., Perkin Trans. 2*, 627 (1974).
321. A. T. T. Hsieh, C. A. Rogers, and B. O. West, *Aust. J. Chem.*, **29**, 49 (1976).
322. T. Inubushi, I. Morishima, H. Miyoshi, and S. Uemura, *J. Am. Chem. Soc.*, **100**, 354 (1978).
323. P. F. Barron, D. Doddrell, and W. Kitching, *J. Organomet. Chem.*, **132**, 351 (1977).
324. C. S. Hoad, F. W. Matthews, M. M. Thakur, and D. G. Gillies, *J. Organomet. Chem.*, **124**, C31 (1977).
325. F. Brady, R. W. Matthews, M. M. Thakur, and D. G. Gillies, *J. Organomet. Chem.*, **252**, 1 (1983).

The Chemistry of the Metal—Carbon Bond, Vol. 4
Edited by F. R. Hartley

CHAPTER **6**

Preparation and use of organosilicon compounds in organic synthesis

E. W. COLVIN

Chemistry Department, University of Glasgow, Glasgow G12 8QQ, UK

I. INTRODUCTION

The basic aim of this chapter is to describe routes to, and the useful synthetic behaviour of, functionalized organosilanes, with constant emphasis on the organic moiety. In other words, silicon will be considered as a 'ferryman', mediating the transformation of one organic molecule into another, normally by temporarily replacing a hydrogen atom in the substrate or reagent. A consequence of this formal treatment is that only reactions of demonstrated or high potential utility will be discussed. Molecules of more theoretical interest[1], such as silaethenes, silanones, silylenes, and silenes, are outside its scope. Additionally, although some introductory material is provided in each subject area, the emphasis is on reactions reported in the last 4 years. Fleming's Tilden Lecture[2], two monographs[3,4], several general reviews[5], and annual surveys[6] of the silicon—carbon bond have all appeared in this period, and it is to these sources that the reader is directed for coverage of earlier work.

More details[7] on the FMO treatment of the stereochemistry of nucleophilic displacement at silicon have been published, as has a multi-author book[8] which contains, *inter alia*, a good chapter on n.m.r. studies of organosilicon compounds. Polarization transfer n.m.r. spectroscopy using the INEPT and DEPT techniques has been applied[9] to ^{29}Si: such sophisticated methods are required since ^{29}Si is distinctly more difficult to observe than ^{13}C in n.m.r., in spite of its greater relative abundance. Ager's review[10] of silicon-containing carbonyl equivalents covers vinylsilanes, α,β-epoxysilanes, α-silyl sulphides and selenides, and provides an excellent overview of acylsilanes (α-ketosilanes). Reactions of vinyl-, allyl-, ethynyl- and propargyl-silanes have been reviewed[11]. A computer-assisted mechanistic evaluation[12] of organosilicon chemistry may be considered as an indication of the attainment of the pinnacle of respectability.

II. PHYSICAL PROPERTIES OF ORGANOSILICON COMPOUNDS

Silicon's utility in organic synthesis can be ascribed to three main factors: its relative bond strengths with other elements, its relative electronegativity, and the possible involvement of its valence p- and empty d-orbitals.

A. Bond Dissociation Energies and Bond Lengths

A selection of bond dissociation energies[13] and bond lengths is given in Table 1. From even these limited thermochemical data, it can be seen that, relative to carbon, silicon makes strong bonds to oxygen, fluorine, and chlorine.

B. Relative Electronegativity

Relative electronegativity can be established on several scales. Regardless of the scale used (Table 2), silicon always appears markedly more electropositive than carbon,

TABLE 1. Bond dissociation energies and bond lengths

Bond	Energy (kJ mol^{-1})	Length (Å)	Bond	Energy (kJ mol^{-1})	Length (Å)
Si—C	318	1.89	C—C	334	1.54
Si—O	531	1.63	C—O	340	1.41
Si—Cl	471	2.05	C—Cl	335	1.78
Si—F	807	1.60	C—F	452	1.39

TABLE 2. Relative electronegativities

H	C	N	O	F
2.8	2.35	3.1	3.5	4.0
	Si	P	S	Cl
	1.64	2.1	2.5	2.8

resulting in polarization of Si—C bonds and a tendency for nucleophilic attack to occur at silicon, i.e. $Si^{\delta+}$—$C^{\delta-}$.

C. Cleavage of C—Si and O—Si Bonds

The Si—C bond is fairly stable towards homolytic fission, but is readily cleaved by ionic reagents, either by initial nucleophilic attack at Si or by electrophilic attack at C. Since C—H bonds break in the same direction, C^-H^+, as do C—Si bonds, C^-Si^+, then a good indication of the likely behaviour of a C—Si bond can be predicted by consideration of an analogous C—H bond. Just as Ar—H bonds are cleaved by electrophiles such as Br_2, so are Ar—Si bonds. Similarly, the 1, 2-elimination reactions displayed by H—C—C—X systems occur even more readily in the fragmentation reactions of β-functionalized silanes. As a broad generalization, it is usually the case that when a C—H bond can be cleaved by a particular ionic reagent, then the corresponding C—$SiMe_3$ bond will be cleaved even more readily by the same reagent. (In a competitive situation, a C—Si bond is the more reactive towards oxygen and halogen nucleophiles/bases, whereas a C—H bond is the more reactive towards carbon and nitrogen nucleophiles/bases). Similar parallels can also be drawn for O—H and O—Si bonds, although with the opposite emphasis, i.e., O—H bonds can be cleaved more readily than O—Si bonds. Indeed, Fleming[2] has suggested that Si bonded to C can be considered as a 'super-proton', whereas when bonded to O it should be considered as an 'enfeebled proton'.

Methods of formation of C—Si and O—Si bonds are discussed in the appropriate sections. Cleavages of C—Si and O—Si bonds, whether to create a new reactive species or to liberate the protio product, are of fundamental importance. 'Anhydrous' tetrabutylammonium fluoride has been used to great advantage on numerous occasions, but not without certain inconsistencies; the vagaries[14] associated with its dehydration have been quantitatively studied. No general route to anhydrous, organic-soluble and highly silicophilic fluoride ion which generates a reactive species has been revealed as yet. Full details[15] of the use of the complex fluoride $(R_2N)_3S^+Me_3SiF_2^-$ to generate 'naked' enolate ions from silyl enol ethers have been published, as has its extension[16] to silyl ketene

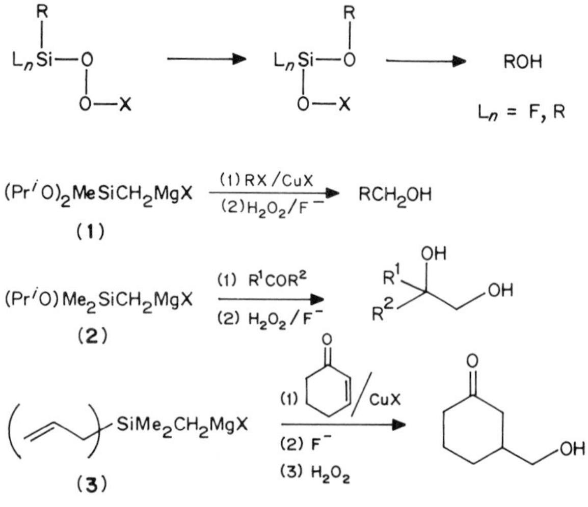

$L_nSi(R)$—O—O—X ⟶ $L_nSi(R)$—O—O—X ⟶ ROH L_n = F, R

$(Pr^iO)_2MeSiCH_2MgX$ (1) —(1) RX/CuX; (2) H_2O_2/F^-⟶ RCH_2OH

$(Pr^iO)Me_2SiCH_2MgX$ (2) —(1) R^1COR^2; (2) H_2O_2/F^-⟶ $R^1R^2C(OH)CH_2OH$

(allyl)$SiMe_2CH_2MgX$ (3) —(1) cyclohexenone/CuX; (2) F^-; (3) H_2O_2⟶ 3-(hydroxymethyl)cyclohexanone

SCHEME 1

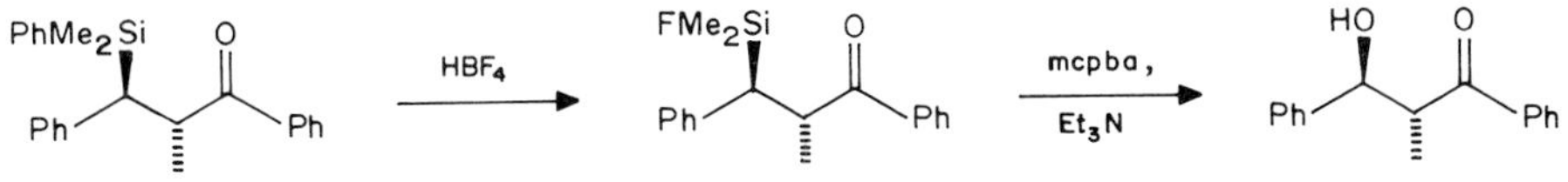

SCHEME 2

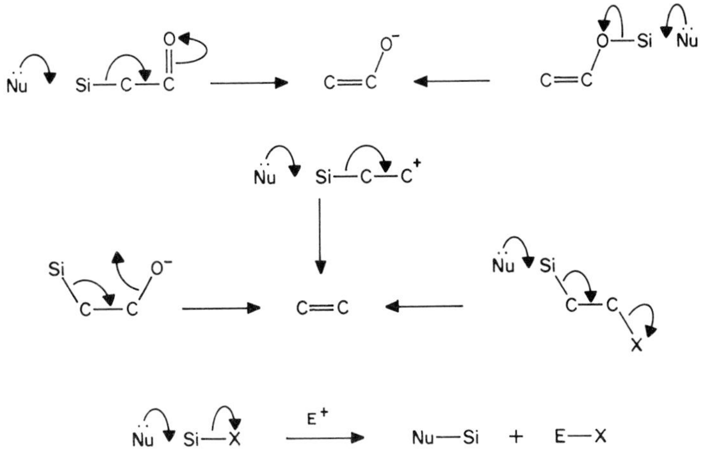

SCHEME 3

acetals. If anhydrous conditions are unnecessary, then an excellent oxidative method of cleavage can be applied, as can various HF systems[17,18]. The oxidative method[19] requires that the silane carry an electronegative substituent, such as fluoro, alkoxy, or amino. Either H_2O_2 or *m*-chloroperbenzoic acid may be used as the oxidant, and the silane is converted into an alcohol with retention[20] of configuration. Fluoride ion is normally an essential additive in what is probably a fluoride ion-assisted rearrangement[21] of a silyl peroxide, via a hypervalent silicon species (Scheme 1). This methodology can be clearly seen in the construction of the d^1-methanol synthons **1**[22], **2**[23], and **3**[24].

The otherwise possibly inconvenient substituent requirement can be created readily[20b] by protiodesilylation of phenyldimethylsilyl moieties by HF equivalents (Scheme 2).

Such C—Si cleavages must be seen in perspective. Silicon's bond to carbon, although certainly polarized, is only weakly so in comparison with those of other organometallic compounds. In general, organosilicon (i.e. C—Si) compounds can be handled readily, often without the necessity for inert atmospheres or exclusion of moisture. The C—Si bond can withstand varied reaction conditions, yet it has a latent lability which can be revealed at an appropriate moment. Some characteristic reactions are summarized in Scheme 3. As mentioned earlier, all of these reactions normally occur more readily than do the corresponding hydrogen (or, in the last case, carbon) analogues.

D. The β-Effect and α-Anionoids

Two other important properties of organosilicon compounds are that carbonium ions β and carbanions (or carbanionoids) α to silicon are often favoured by stabilization over alternatives. The first of these phenomena is known as the β-effect, and is due to an Si—C bond being better at stabilizing a neighbouring carbonium ion than a C—C or a C—H

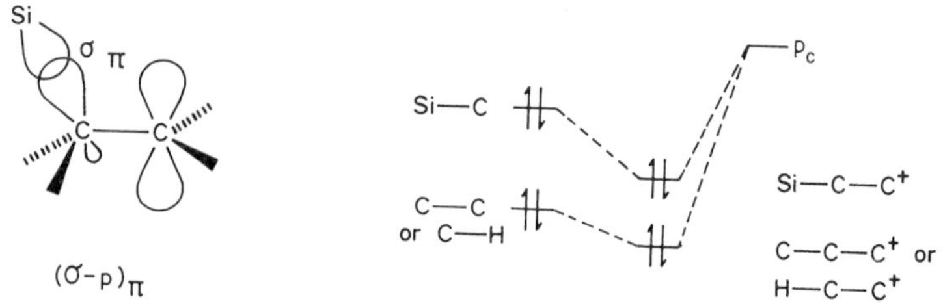

SCHEME 4

bond, because of its higher ground state energy and therefore better energy match with the p-orbital (Scheme 4). For good overlap, the Si—C bond must be able to adopt a co-parallel alignment with the empty p-orbital. Accordingly, this stabilization[25] is most effective in acyclic, conformationally mobile systems: it can, of course, be overwhelmed in complex, heteroatom-substituted cases. When operational, it will weaken the C—Si bond, making the Si atom more susceptible to nucleophilic attack.

The other phenomenon, that of relative stabilization of carbanionoids α to silicon, can also be explained by a molecular orbital π-overlap representation, by analogy with sulphur-containing molecules. An important mechanism of stabilization of carbanionoids or polarized metalloids by adjacent sulphur is polarization of the electron distribution, dispersing the charge over the whole molecule. Perturbational MO calculations have indicated that (n_c–σ^*_{SR}) interactions of the carbon lone pair with the antibonding σ^* orbital of the adjacent antiperiplanar SR bond can contribute strongly to carbanion stabilization. The ground-state polarization of the C—Si bond will ensure a relatively high coefficient on Si in the σ^* level, further enhancing the stabilizing effect of such an overlap (**4**).

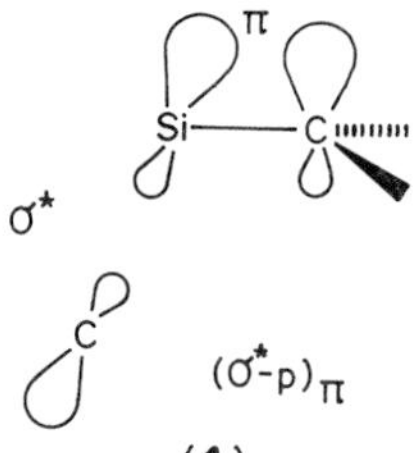

(**4**)

One thing that silicon cannot do with any success is form multiple bonds, so that silenes and silaethenes are rare and normally extremely unstable. In other words, silicon, in its ground state, greatly prefers to make four single covalent bonds.

III. REARRANGEMENT REACTIONS

Discussion of rearrangement reactions involving the migration of silyl groups is limited here to those which have demonstrable or potential synthetic utility. This area has been the subject of an excellent review[26], and only some of the more recent highlights are discussed in this section.

A. 1,2-Rearrangement Reactions

The best documented example of anionic C→O rearrangements is the well known Brook rearrangement[26], which occurs with the anions of silyl methanols (Scheme 5).

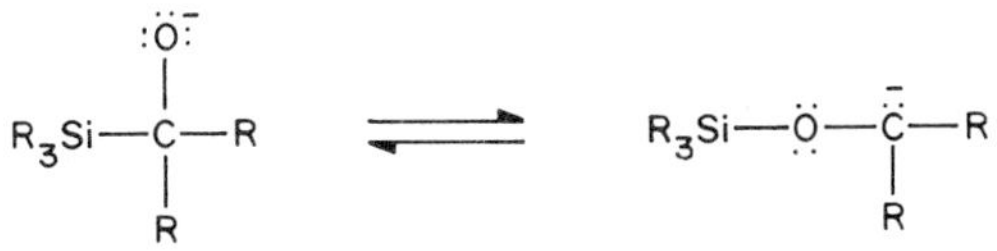

SCHEME 5

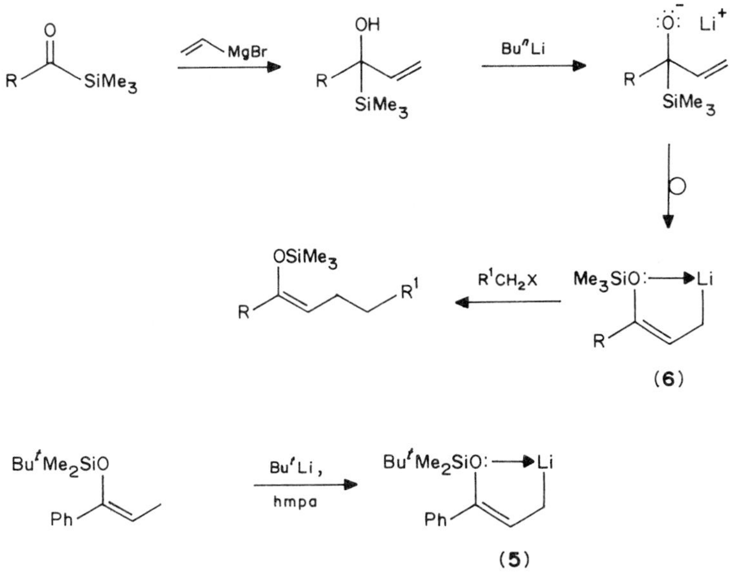

SCHEME 6

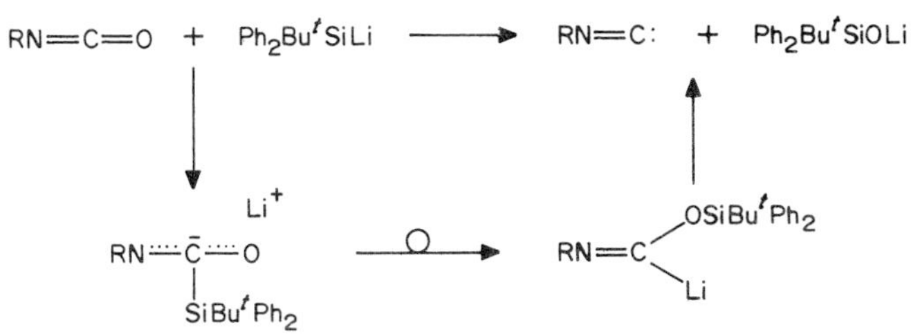

SCHEME 7

One of the many elegant applications of this rearrangement can be seen in the work of Kuwajima on the regio- and stereo-specific generation of silyl enol ethers (Scheme 6); full details[27] of this process have been reported. Interestingly, an intermediate (**5**) analogous to **6** can be generated[28] by direct deprotonation of the *tert*-butyldimethylsilyl enol ether of phenyl ethyl ketone.

A 1,2-anionic C$\rightarrow$O silyl migration has been implicated in a mechanistic investigation[29] of the deoxygenation[30] of isocyanates to isonitriles (Scheme 7).

B. 1,3-Rearrangement Reactions

β-Ketosilanes undergo a facile thermal rearrangement to silyl enol ethers (Section XIII.A). This does not normally provide an attractive route to such valuable species, since the required substrates have, until recently, been difficult to obtain. A new method[31] which employs the more accessible hydroxyalkylvinylsilanes as precursors to the (intermediate) β-ketosilanes has been described (Scheme 8) (see also Section XIII.A). However, one of the most useful applications of 1,3-C→0 migrations is seen in the sila-Pummerer rearrangement[32] of α-silyl-sulphoxides and -selenides, with phenylthio- or phenylseleno-methyltrimethylsilane[33] acting as a formaldehyde d^1-synthon (Scheme 9). More recent variants include the non-oxidative generation[34] of the intermediate α-silylsulphoxides by [2, 3]-sigmatropic rearrangement of allyl sulphenate esters (Scheme 10).

Two reaction pathways[35,36] involving 1,3-migration from sp^2-hybridized carbon to oxygen have been described; one application of this is seen in the protiodesilylation of suitable vinyl silanes (see Section IV.B).

On the other hand, several instances of 1,3-O→C silyl rearrangements have been discovered and these have significant synthetic promise as routes to β-ketosilanes of various structural types. α-Selenocyclohexanones, when converted into sterically hindered silyl enol ethers, undergo reductive cleavage of the seleno moiety when treated with lithium dimethylaminonaphthalenide. The resulting α-lithio silyl enol ethers rearrange[37] rapidly to silyl enolates, and thence to β-ketosilanes (Scheme 11; ldman = lithium-dimethylaminonaphthalenide).

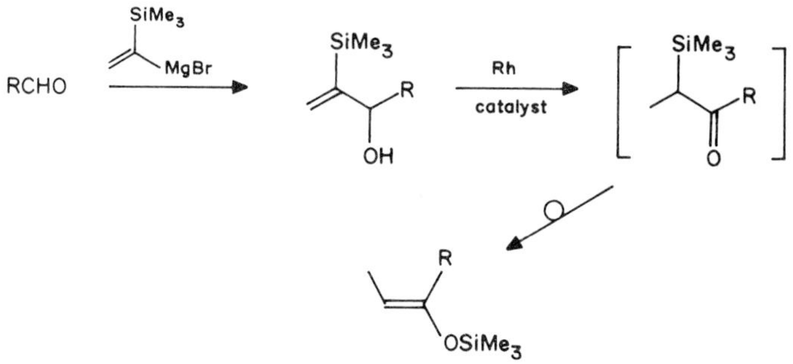

RCHO —(SiMe₃-vinyl-MgBr)→ SiMe₃ / R / OH —(Rh catalyst)→ [SiMe₃ / R / O] —(Δ)→ R / OSiMe₃

SCHEME 8

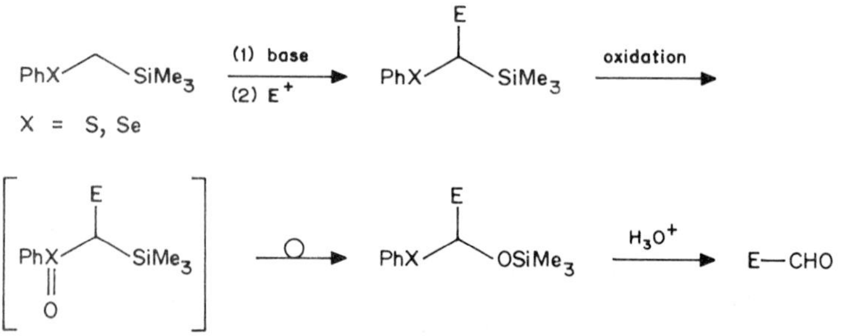

PhX–CH₂–SiMe₃ (X = S, Se) —((1) base, (2) E⁺)→ PhX–CH(E)–SiMe₃ —(oxidation)→ [PhX(=O)–CH(E)–SiMe₃] —(Δ)→ PhX–CH(E)–OSiMe₃ —(H₃O⁺)→ E—CHO

SCHEME 9

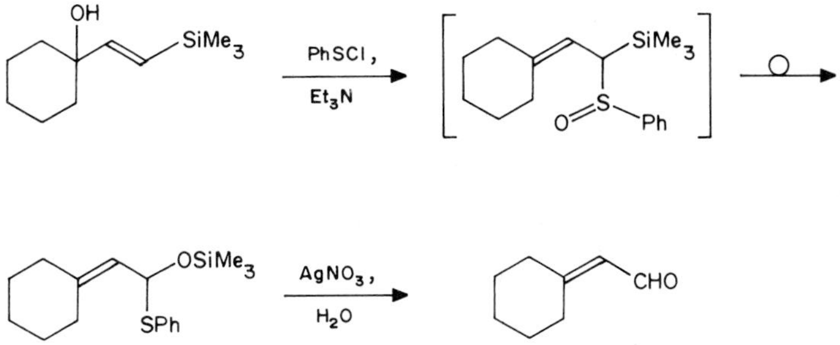

SCHEME 10

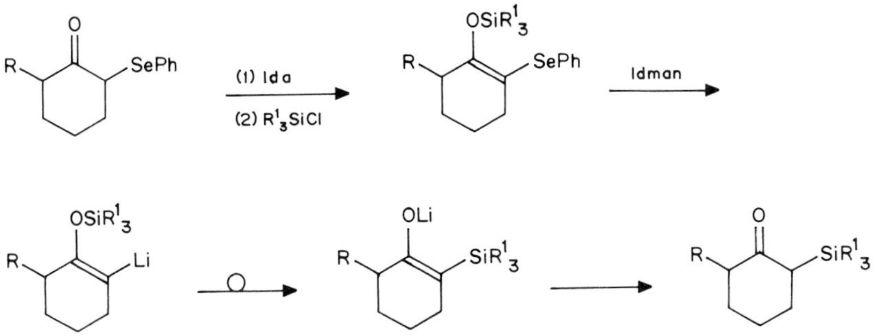

SCHEME 11

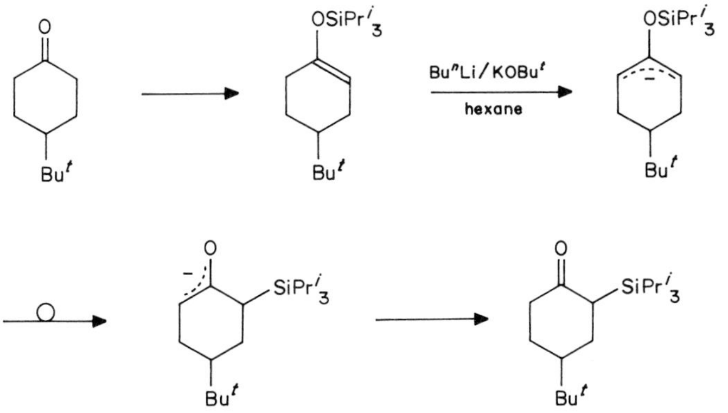

SCHEME 12

Even more simply, treatment[38] of the triisopropylsilyl enol ethers of a variety of cyclic and acyclic ketones with the strong base combination of $Bu^nLi/KOBu^t$ leads directly, after aqueous work-up, to β-ketosilanes in good yield (Scheme 12). In contrast to the previous method, this rearrangement seems to proceed by allylic, rather than vinylic, metallation, since enol ethers lacking an allylic α-proton are unreactive.

C. 1,4-Rearrangement Reactions

The tris(trimethylsilyl)alcohol **7** rearranges[39] to the silyl ether **8** on treatment with NaH in thf. This rearrangement did not occur in any other solvent tested, suggesting that thf was acting as a proton source, quenching the strongly basic anion **9** (Scheme 13). A similar 1,4-C → O silyl migration has been implicated[40] in a route to substituted cyclobutanols.

Allylsilanes possessing a suitably positioned hydroxy group undergo a base-induced 1,4-C → O silyl migration, thus providing a relatively mild method[41] for the generation of an allyl anion equivalent (Scheme 14). A related 1,4-C → O migration from sp^2 carbon has been observed[42].

Treatment of phenylthiomethyl-substituted silyl ethers such as **10** with lithium di-*tert*-butylbiphenyl[43] results in rapid reductive cleavage of the C—S bond, and equally rapid O → C silyl migration of the resulting non-stabilized carbanion (Scheme 15).

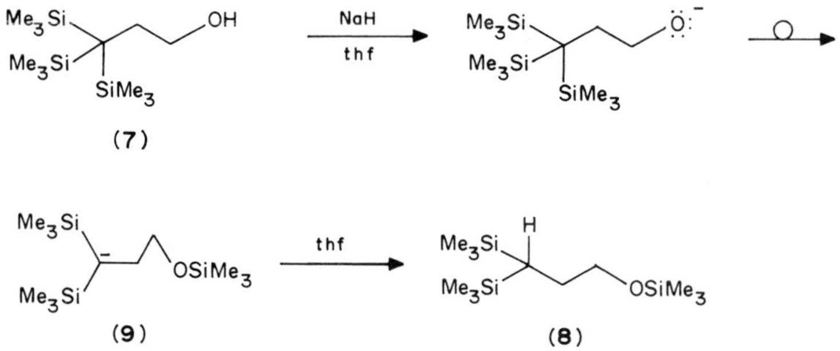

Me_3Si, Me_3Si, $SiMe_3$, OH (7) —NaH, thf→ Me_3Si, Me_3Si, $SiMe_3$, $\ddot{O}^-$ →

Me_3Si, Me_3Si, $OSiMe_3$ (9) —thf→ H, Me_3Si, Me_3Si, $OSiMe_3$ (8)

SCHEME 13

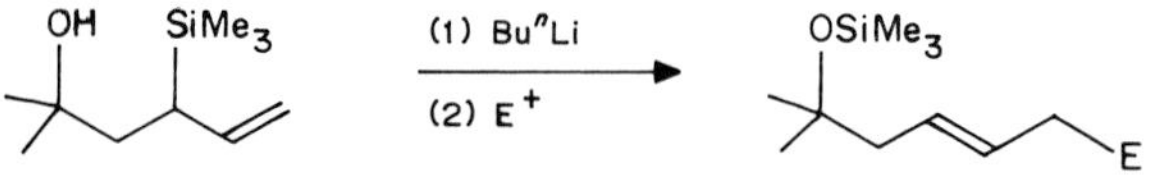

OH, $SiMe_3$ —(1) Bu^nLi, (2) E^+→ $OSiMe_3$, E

SCHEME 14

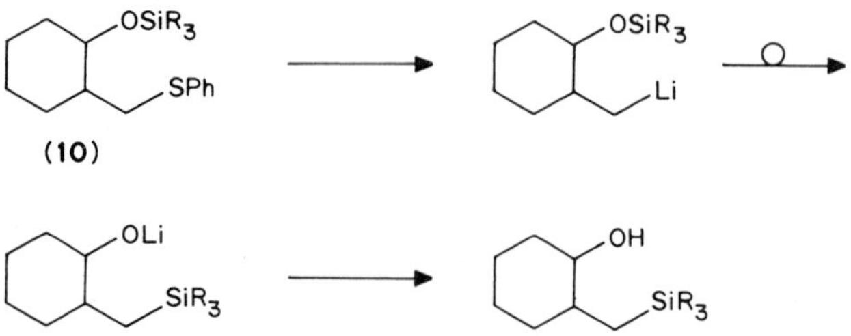

$OSiR_3$, SPh (10) → $OSiR_3$, Li →

OLi, SiR_3 → OH, SiR_3

SCHEME 15

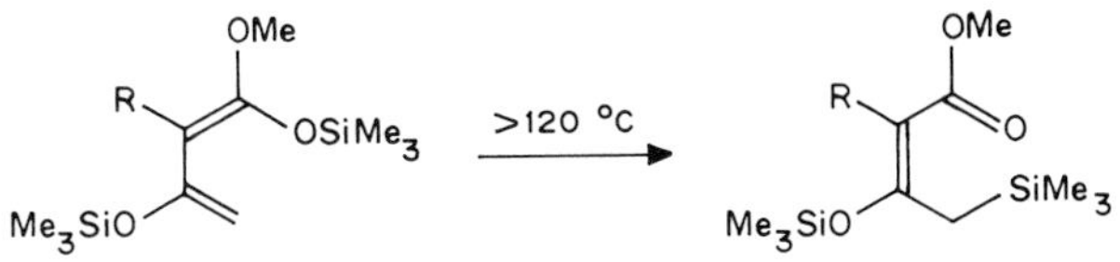

SCHEME 16

D. 1, 5-Rearrangement Reactions

Once again, the synthetically most relevant examples of this class of rearrangements involve migration of a silyl group between carbon and oxygen. Bis-tms enol ethers of alkyl acetoacetates undergo[44,45] a facile thermal 1,5-O$\rightarrow$C rearrangement (Scheme 16), suggesting necessary care in their use.

IV. VINYLSILANES

A. Preparation

Most routes to vinylsilanes utilize either alkynes, carbonyl compounds, or vinyl halides as starting materials, as detailed[3,4,10,11] elsewhere. Tris(trimethylsilyl)aluminium reacts with vinyl iodides to produce[46] the corresponding vinylsilanes, with retention of configuration. A general route of increasing application involves metallation/carbo- or protio-demetallation of alkynes and tms-alkynes. Representative examples[47–54] illustrated in Scheme 17 include cases of controlled regiospecificity and of intramolecular carbodemetallation. The last example[54] defines a new route to the allylic alcohol **11** which, as its halide, undergoes direct nucleophilic substitution, providing an alkylative equivalent to but-3-en-2-one.

Several other routes[55–58] to β-tms-allylic alcohols have been revealed, including one[59] which involves the lithiation of *tert*-allylic alcohols (Scheme 18). 2-tms-Allyl alcohol esters undergo enolate Claisen rearrangement to functionalized vinyl silanes.[60]

New routes[61] to the tms-enone **12**, an improved Michael acceptor, have been described, as have full details[62] of its utility in annelation processes.

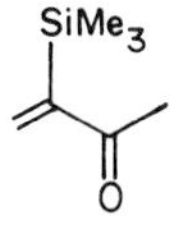

(12)

B. Reactions

Vinylsilanes react with a range of electrophiles to give products of substitution. The regiochemistry of reaction is normally controlled by the β-effect, encouraging carbonium ion formation at the β-carbon, and hence regiospecific cleavage of the C—Si bond (Scheme 19). The overall stereochemistry will depend on a number of factors, including the stereochemistry of addition and subsequent elimination. The few exceptions to this regiochemical generalization arise when the α-carbon atom carries a substituent which can well stabilize carbonium ion development, such as oxygen or sulphur. For example, the vinylsilane **13** reacts with α,β-unsaturated acid chlorides in a Nazarov cyclization[63] to

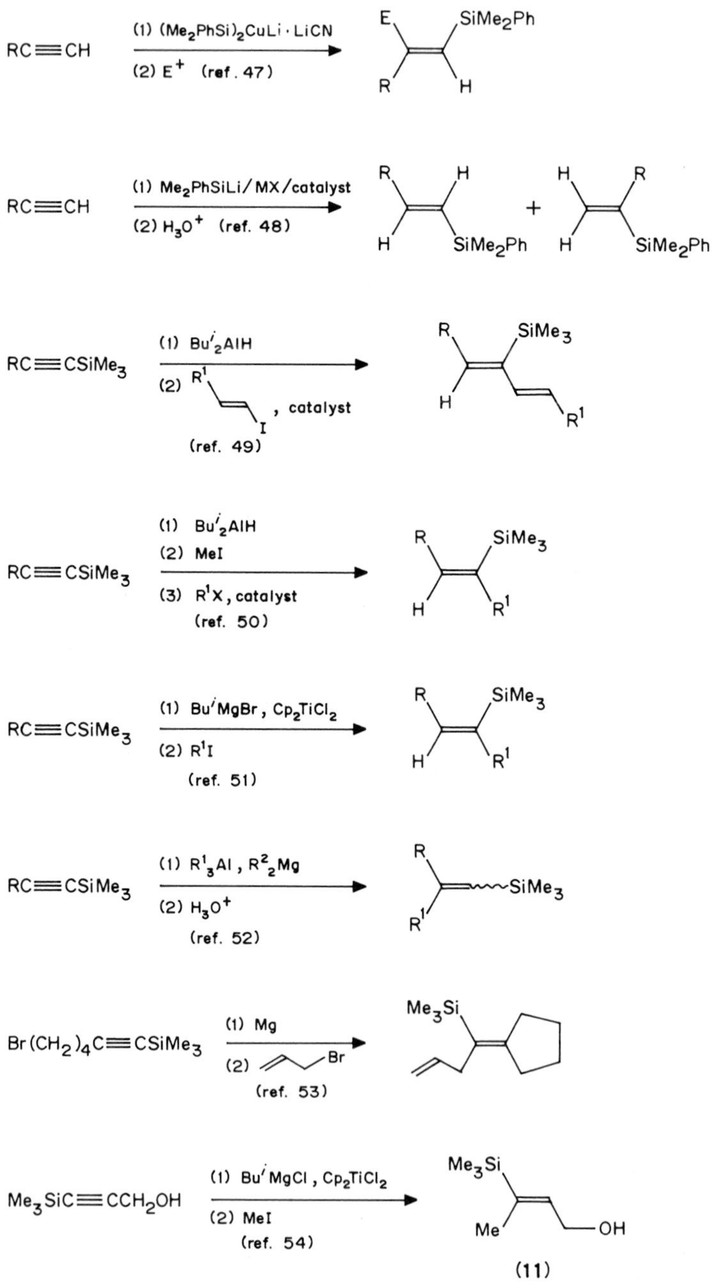
RC≡CH
(1) (Me2PhSi)2CuLi·LiCN
(2) E+ (ref. 47)
E
SiMe2Ph
R
H
RC≡CH
(1) Me2PhSiLi/MX/catalyst
(2) H3O+ (ref. 48)
R
H
H
SiMe2Ph
+
H
R
H
SiMe2Ph
RC≡CSiMe3
(1) Bui2AlH
(2) R1
I
, catalyst
(ref. 49)
R
SiMe3
H
R1
RC≡CSiMe3
(1) Bui2AlH
(2) MeI
(3) R1X, catalyst
(ref. 50)
R
SiMe3
H
R1
RC≡CSiMe3
(1) BuiMgBr, Cp2TiCl2
(2) R1I
(ref. 51)
R
SiMe3
H
R1
RC≡CSiMe3
(1) R13Al, R22Mg
(2) H3O+
(ref. 52)
R
SiMe3
R1
Br(CH2)4C≡CSiMe3
(1) Mg
(2) Br
(ref. 53)
Me3Si
Me3SiC≡CCH2OH
(1) BuiMgCl, Cp2TiCl2
(2) MeI
(ref. 54)
Me3Si
Me
OH
(11)

SCHEME 17

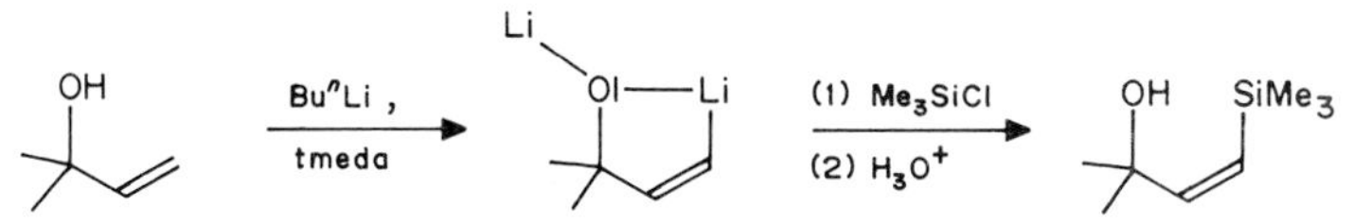

SCHEME 18

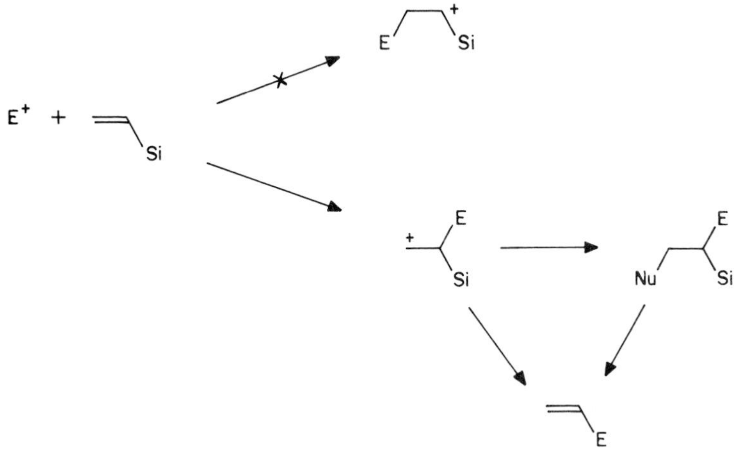

SCHEME 19

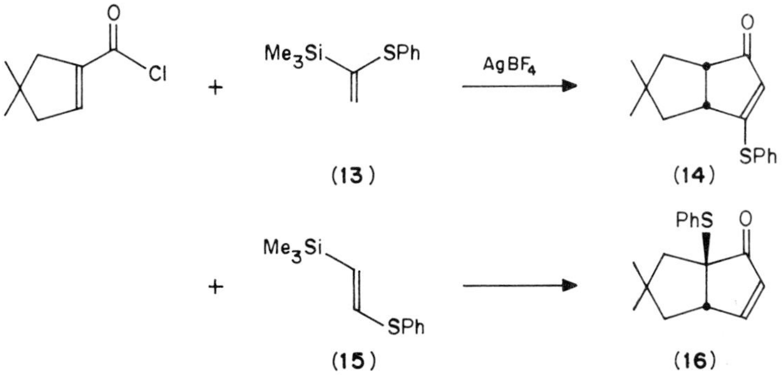

SCHEME 20

give cyclopentenones such as **14** (Scheme 20). The isomeric vinylsilane **15**, where both directing effects are in concert, gives the cyclopentenone **16**; Diels–Alder cycloaddition reactions of the sulphone derived from **15** have been revealed[64]. An intriguing case[65] of transannular halogenodesilylation has been described (Scheme 21).

Tms groups have been found to play a profound role in the regiochemistry of the ene reaction[66] of certain vinylsilanes. Vinylsilanes undergo oxidative transposition[67] on exposure to $^1O_2$, as exemplified in Scheme 22.

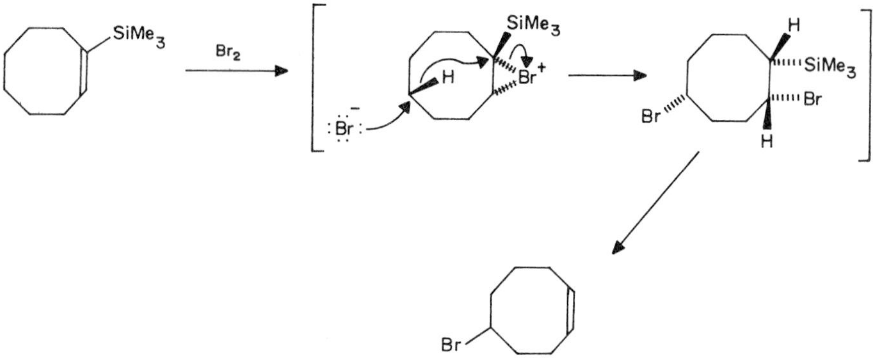

SCHEME 21

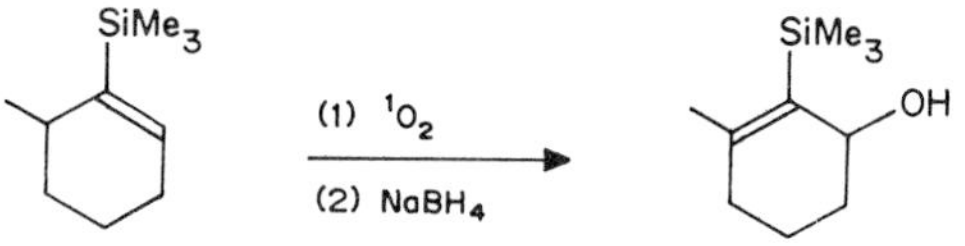

SCHEME 22

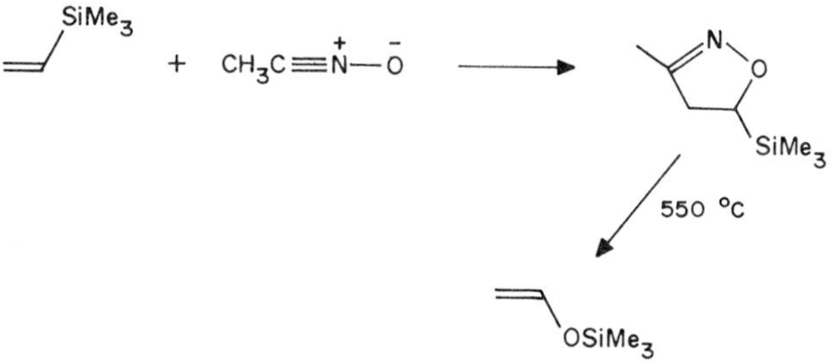

SCHEME 23

1,3-Dipolar cycloaddition reactions of vinylsilanes with nitrile oxides[68] and nitrones[69] have been described: the adducts of simple vinylsilanes with acetonitrile oxide undergo[70] a stereospecific silatropic cycloreversion, to produce silyl enol ethers (Scheme 23). In addition to the Nazarov studies just mentioned, several other examples of the reaction of vinylsilanes with carbon electrophiles have been described. Interesting intramolecular cases of these include cyclizations involving acetals[71], a simple example[72] of which is shown in Scheme 24, dithioacetals[73], acid chlorides, an elegant example[74] of which is also included in Scheme 24, and iminium ions[75].

Many methods have been recommended for the protiodesilylation of vinylsilanes. For those without adjacent participating functionality, aqueous HBF_4 in hot CH_3CN seems to be the system of choice[18]. With vinylsilanes possessing α- or β-hydroxy groups, 1,3- or 1,4-silyl migrations[36,42] from C to O can be induced to occur under basic conditions (Scheme 25); fluoride ion also effects this desilylation[76], possibly by a similar mechanism.

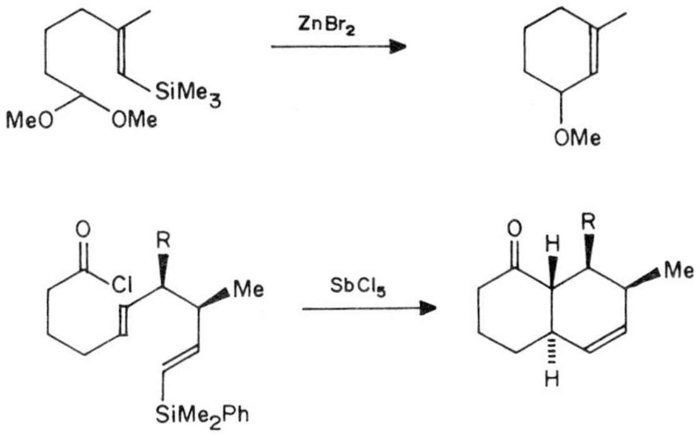

SCHEME 24

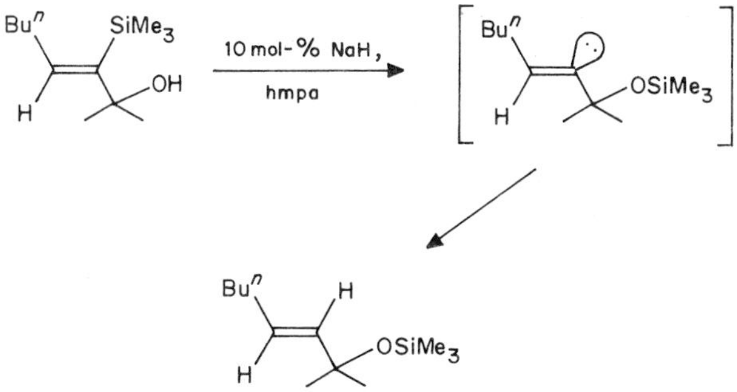

SCHEME 25

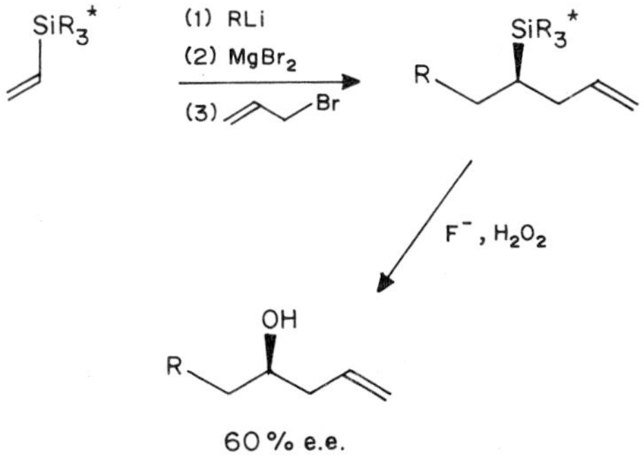

SCHEME 26

Interestingly, α-hydroxyvinylsilanes, and here the intermediate α-silyloxyvinyl anions, do not suffer elimination to allenes.

The possibility of chirality transfer from vinylsilanes bearing a chiral silicon function has been explored. Only low levels of asymmetric induction[77] were observed, implying that

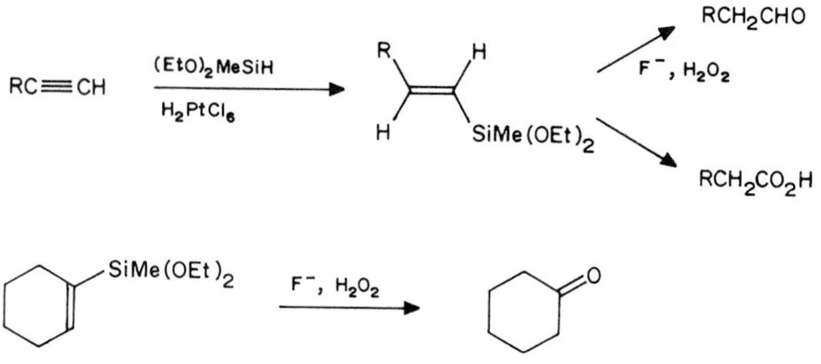

$$RC\equiv CH \xrightarrow[H_2PtCl_6]{(EtO)_2MeSiH} \text{(E)-RCH=CH-SiMe(OEt)}_2 \xrightarrow{F^-,\ H_2O_2} RCH_2CHO \text{ or } RCH_2CO_2H$$

$$\text{1-[SiMe(OEt)}_2\text{]cyclohexene} \xrightarrow{F^-,\ H_2O_2} \text{cyclohexanone}$$

SCHEME 27

chirality transfer from silicon to carbon may be inefficient generally in the absence of other factors. On the other hand, the reaction of silicon-chiral vinylsilanes[78] with organometallic reagents and, subsequently, suitable electrophiles proceeds with substantial chiral induction. Oxidative cleavage, with retention[20], produces chiral alcohols (Scheme 26).

Vinylsilanes in which the silicon atom carries electronegative substituents undergo similar fluoride-induced oxidative cleavage of the C—Si bond to produce carbonyl compounds in which the carbonyl carbon was the Si-bearing carbon of the vinylsilane; variation of the conditions[79] allows both selectivity and control of the oxidation level of the product. Unlike similar methodology proceeding via α,β-epoxysilanes (Section V), this direct method is equally applicable to acyclic and cyclic vinylsilanes (Scheme 27). In acyclic cases, the requisite vinylsilanes are readily available by hydrosilylation of alkynes.

V. α, β-EPOXYSILANES

The main synthetic utility of epoxysilanes, obtained by epoxidation of vinylsilanes or from aldehydes and ketones using reagents **17**[80] and **18**[81], lies in their ring-opening reactions when treated with nucleophiles. These reactions are normally stereo- and regio-specific, resulting in α-opening and producing diastereoisomerically pure β-hydroxysilanes (Scheme 28). In acyclic cases, subsequent Peterson elimination affords a variety of heteroatom-substituted alkenes of controlled geometry, as exemplified by a recent stereospecific route[82] to (*Z*)- and (*E*)-enamines. However, the most generally applied transformation[10] of this type produces enols, and thence carbonyl compounds, the reagents **17** and **18** having acted as nucleophilic acylating agents. This methodology can be seen in several recent total syntheses, including those of gymnomitrol[83] and qinghaosu[84]. The possibility of generating reagents related to **18**, but in chiral form, has been explored[85] without great success.

This regiospecific ring opening by cleavage of the C—O bond proximate to silicon has been subjected to considerable scrutiny, since the opposite regiospecificity would be expected if the β-effect were playing a dominant role. Hückel calculations have indicated[67] that the enhanced ground state electrophilicity of the α-carbon may be due to enhanced antibonding C–O interactions. An extensive study of 1,2-epoxy-1-tms-cyclohexane has confirmed[86] this propensity for ring opening to occur in a *trans* manner by attack of the nucleophile at the silicon-substituted carbon atom; similar results[87] have been obtained with 1,2-epoxy-tms-cyclooctane. In conjunction with the regioselectivity observed in related reactions of the parent vinylsilane, 1-tms-cyclohexene, it has been proposed that the major factor determining the regioselectivity in either case is the stability of the intermediate 'onium ion, and that assistance by an empty d-orbital on Si to the attacking

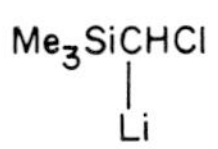

$Me_3SiCH(Cl)Li$ (17) $Me_3SiC(Me)(Cl)Li$ (18)

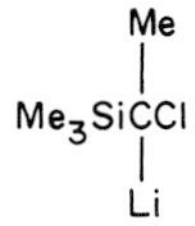

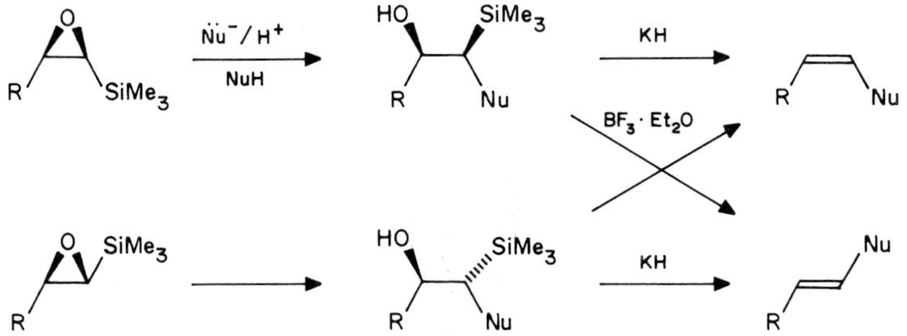

SCHEME 28

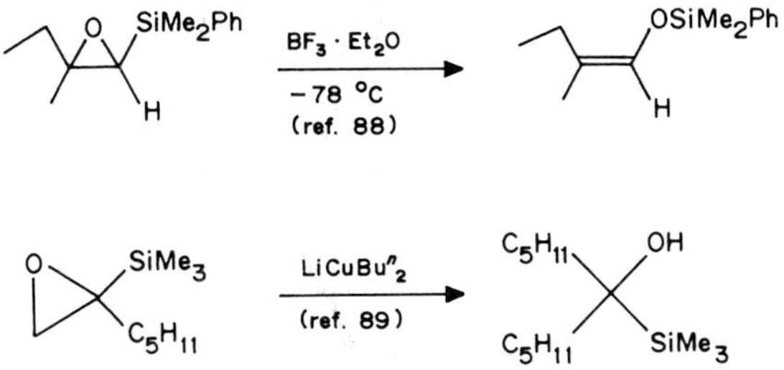

SCHEME 29

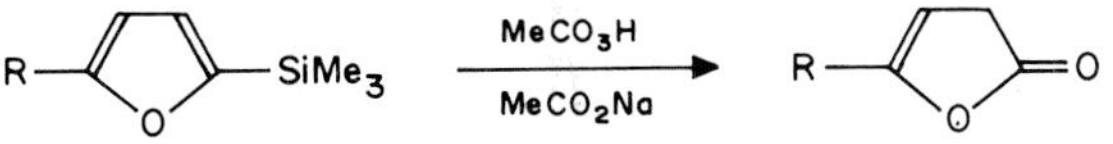

SCHEME 30

nucleophile will play a role only in those reactions involving less stable three-membered ring intermediates, such as the protonated epoxide.

However, and whatever the origin of this directing effect, it can be overcome; several reports[88,89] of regioselective β-opening of α,β-epoxysilanes have appeared (Scheme 29). 2-tms-Furans undergo regiospecific oxidation[90] to butenolides (Scheme 30); an intermediate epoxysilane has been implicated.

β'-Hydroxy-α,β-epoxysilanes, on treatment with $BF_3{\cdot}Et_2O$, undergo sequential rearrangement and Peterson olefination to provide α,β-unsaturated carbonyl compounds; a

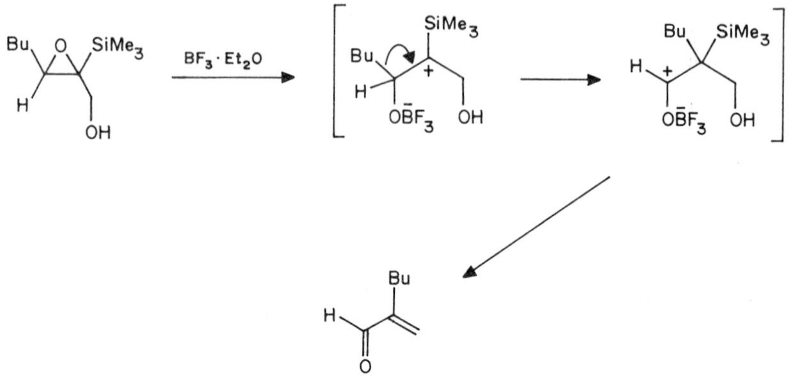

SCHEME 31

stepwise pathway involving preliminary ring opening is shown (Scheme 31), but a concerted process may be involved[91].

VI. ALLYLSILANES

A. Preparation

Alkynes of various types have been transformed[92–94] into functionalized allylsilanes, as illustrated in Scheme 32. An extensive study of S_N displacement reactions of allyl halides with silyl anions and their complexes has provided[95] the regioselective alternatives shown in Scheme 33. This scheme also illustrates the application[96] of an established route for the alkylative deoxygenation of allyl alcohols.

Allyl acetates[97] and allyl silyl ethers[98] can be transformed into allylsilanes with varying regio- and stereo-selectivity. In particular, the silyl cuprate **19** reacts[99] with tertiary allyl acetates in a stereospecifically *anti* manner (Scheme 34), providing a convergent synthesis of stereo-defined allylsilanes. The same cuprate has been employed[100] in a three-step

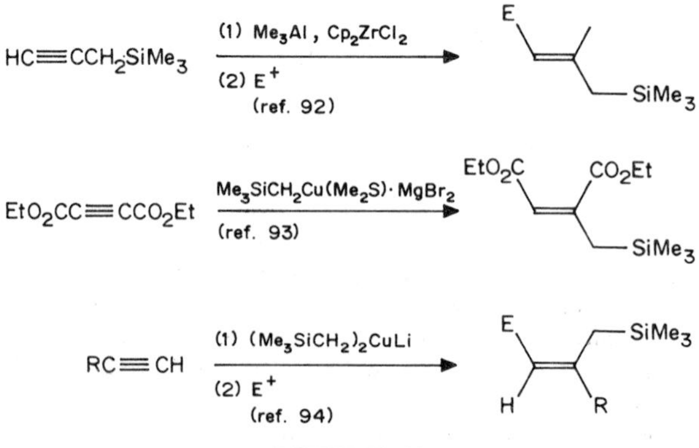

SCHEME 32

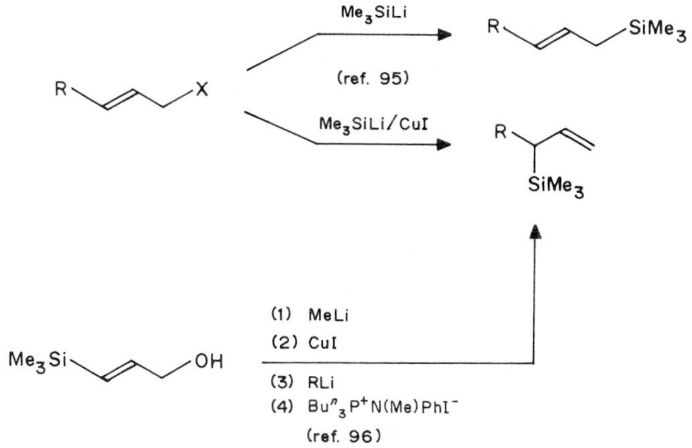
Me3SiLi
(ref. 95)
Me3SiLi/CuI
R
X
SiMe3
Me3Si
OH
(1) MeLi
(2) CuI
(3) RLi
(4) Bun3P+N(Me)PhI−
(ref. 96)

SCHEME 33

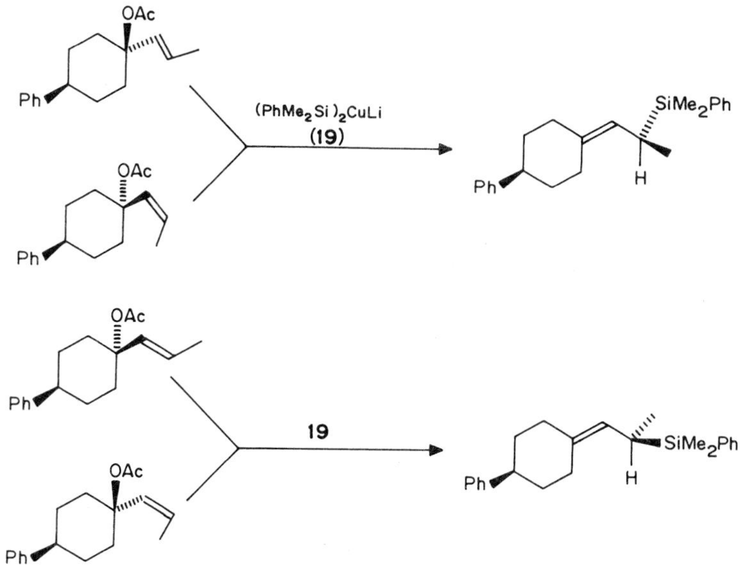
OAc
Ph
(PhMe2Si)2CuLi
(19)
SiMe2Ph
H
19

SCHEME 34

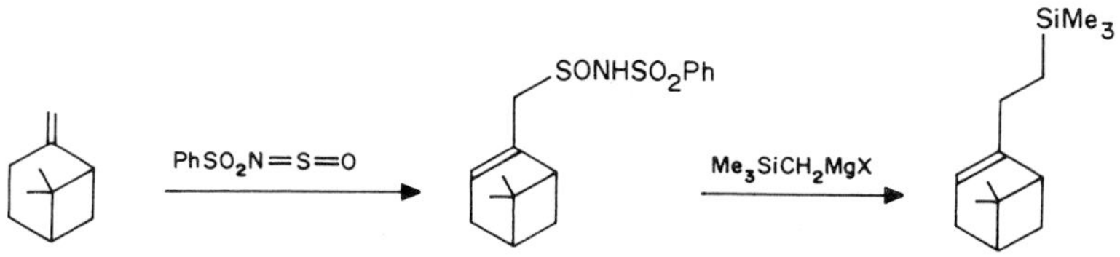
PhSO2N=S=O
SONHSO2Ph
Me3SiCH2MgX
SiMe3

SCHEME 35

transformation of α,β-unsaturated esters in which the silyl group appears at the more substituted end of the allyl group.

Various [3, 3]-sigmatropic rearrangement routes[101] to allylsilanes have been reported. An intriguing process[102] involving an ene reaction which leads to homoallyl silanes is exemplified in Scheme 35.

B. Reactions

Allylsilanes, being homologues of vinylsilanes, undergo a similar regio-controlled electrophilic attack, loss of the silyl group resulting in formation of a product of substitution with net shift of the double bond (Scheme 36).

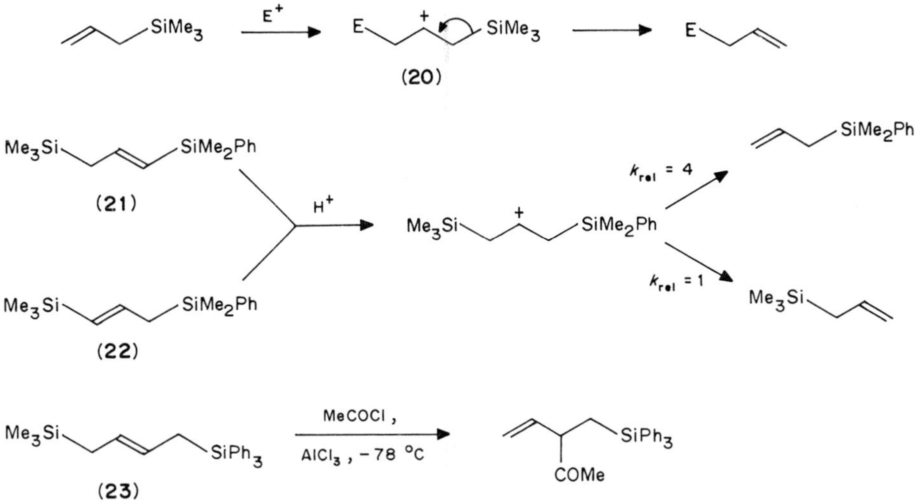

SCHEME 36

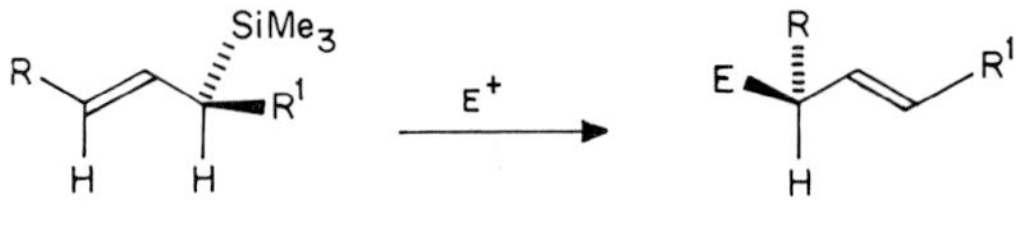

SCHEME 37

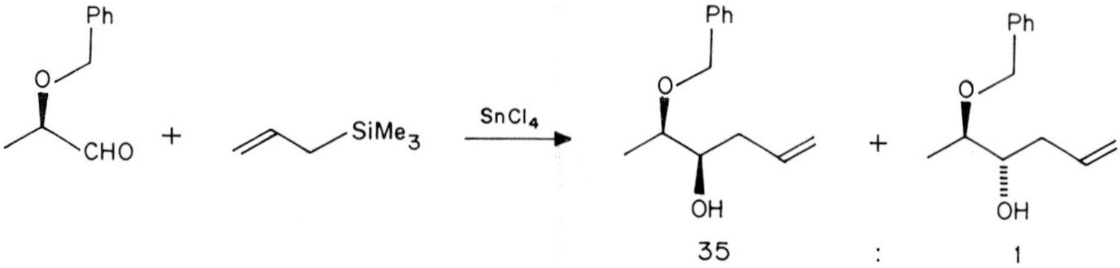

SCHEME 38

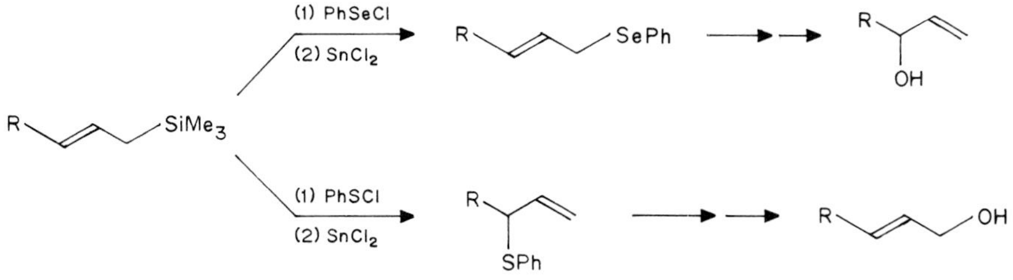

(1) PhSeCl
(2) $SnCl_2$
SePh
OH
R
$SiMe_3$
(1) PhSCl
(2) $SnCl_2$
SPh
OH

SCHEME 39

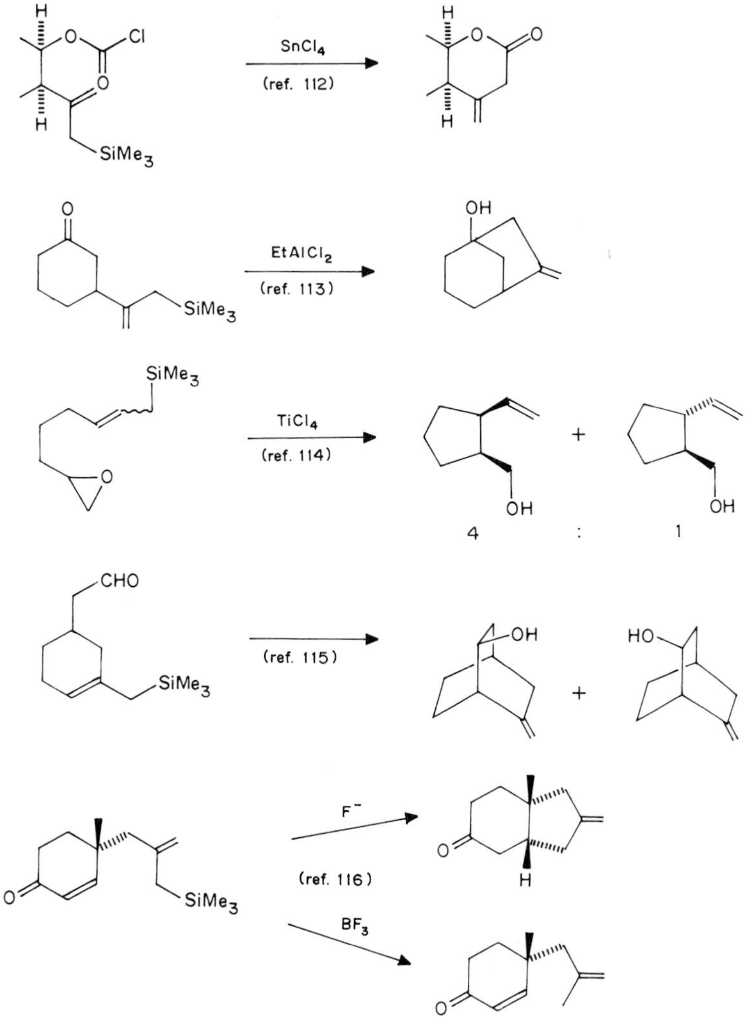

$SnCl_4$
(ref. 112)
$SiMe_3$

$EtAlCl_2$
(ref. 113)
$SiMe_3$
OH

$SiMe_3$
$TiCl_4$
(ref. 114)
OH OH
4 : 1

CHO
$SiMe_3$
(ref. 115)
OH HO

F^-
(ref. 116)
$SiMe_3$
H
BF_3

SCHEME 40

Strong evidence that the cation **20** is indeed an intermediate, stabilized by favourable overlap of the Si—C bonding orbital with the empty p-orbital of the cation, has been provided by an investigation[103] of the silanes **21** and **22**; protiodesilylation gave the same mixture of products, in the same ratios, in each case. Interestingly, the unsymmetrical allyldisilane **23** undergoes[104] chemospecific electrophilic attack.

Extensive studies[105] of such $S_{E'}$ reactions of allylsilanes have demonstrated high *anti* stereoselectivity with the majority of electrophiles employed (Scheme 37) in cases where steric effects do not play a dominant role. An investigation[106] of possible chirality transfer by electrophilic attack on a silicon-chiral allylsilane proved discouraging. However, Chow and Fleming[107] have demonstrated, in an exemplary synthesis, the full range of stereocontrol available using stereo-defined allylsilanes and related species.

Diastereoselective additions[108–110] of allylsilanes to chiral aldehydes and α-ketoamides have been reported; with aldehydes, the best results were obtained using α-alkoxyaldehydes and $SnCl_4$ as a Lewis acid catalyst (Scheme 38).

Certain allylsilanes react with PhSeCl to give products of apparent direct desilylation. Careful investigation[111] showed that the initial product was indeed the expected $S_{E'}$ internal selenide, but that this rearranged rapidly under the reaction conditions to give the terminal isomer. Coupled with the observation that the products from similar reactions with PhSCl were regiostable, this has provided the alternative allyl alcohol transformations shown (Scheme 39).

Several additional applications[112–116] of intramolecular attack, with resulting carbocyclization, are illustrated in Scheme 40; for such reactions, both Lewis acids and fluoride ion seem equally effective, although sometimes with complementary results[116].

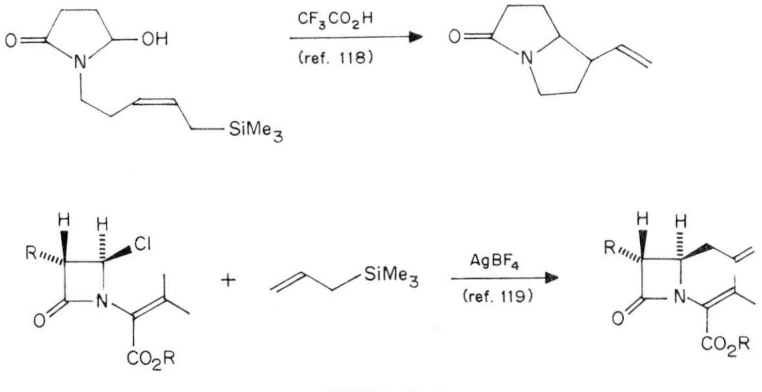

SCHEME 41

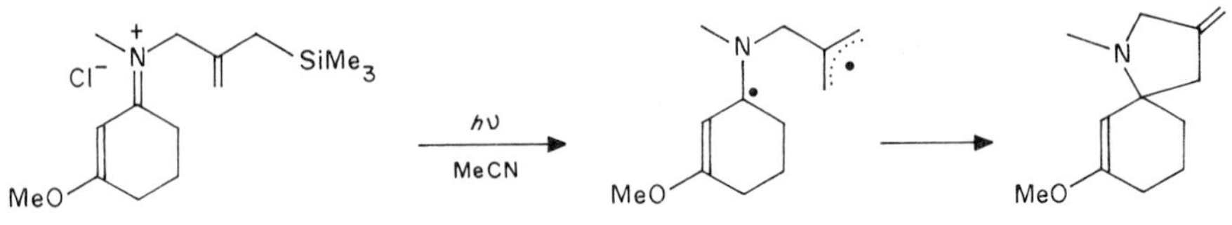

SCHEME 42

Chemoselectivity[117] in the palladium-mediated cycloaddition of substituted trimethylenemethanes, generated from substituted allylsilanes, has been investigated.

Iminium ions of various types, including those derived from β-lactams, react[118,119] with allylsilanes to give usefully functionalized products (Scheme 41). In showing this reactivity, allylsilanes once again demonstrate similar reactivity to silyl enol ethers, which, for example, also react[120] smoothly with β-lactam derived iminium ions (Section XIII.B.6). Additional electrophiles furthering this parallel behaviour include α,β-unsaturated nitro compounds[121], simple acetals[122], and the 1, 3-dithienium cation[123].

Iminium ions also take part in a photochemically induced single electron-transfer reaction[124] with allylsilanes to produce cyclic amines, providing, in intramolecular cases, spiro-fused products (Scheme 42).

Oxocarbenium ions and other stabilized carbenium ions efficiently attack allylsilanes. The anomeric *C*-allylation of carbohydrates[125] and related species[126], and of diaryl carbinols and ketones[127], attests to this. Sakurai[128] has reviewed his extensive contributions to allylsilane chemistry and he and coworkers have provided details[129] of further applications of 2-trimethylsilylmethylbuta-1,3-diene in synthesis, including its reactions

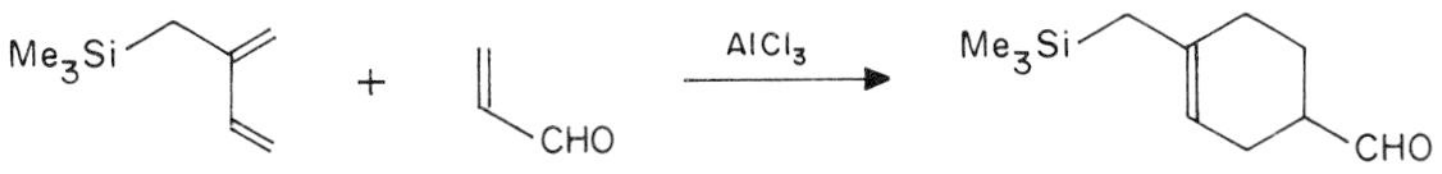

Me_3Si + CHO —$AlCl_3$→ Me_3Si … CHO

SCHEME 43

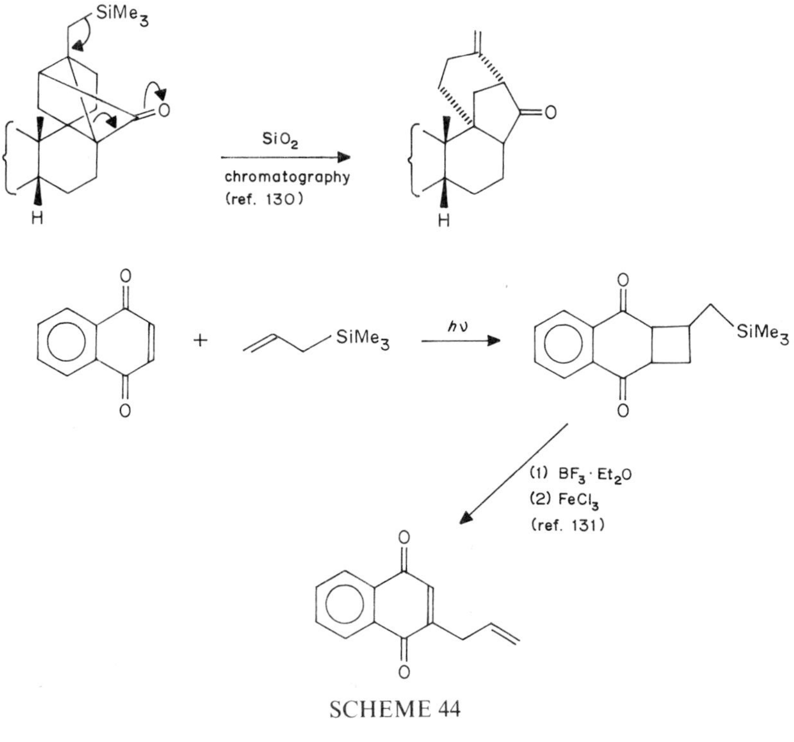

$SiMe_3$ —SiO_2, chromatography (ref. 130)→ O, H

O, O + $SiMe_3$ —$h\nu$→ O, O, $SiMe_3$ —(1) $BF_3 \cdot Et_2O$ (2) $FeCl_3$ (ref. 131)→ O, O

SCHEME 44

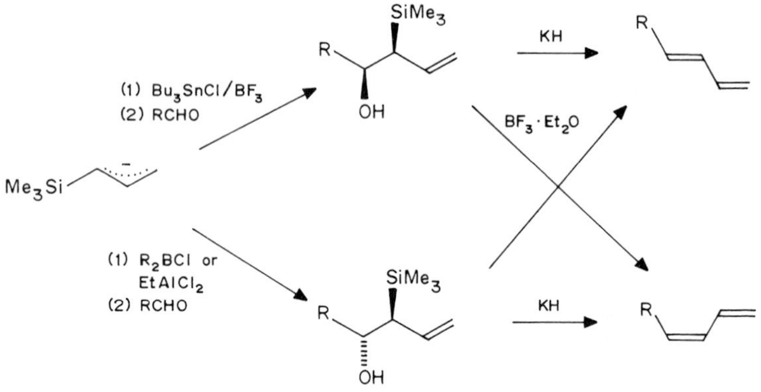

SCHEME 45

with electrophiles with Lewis acid activation, and its Diels–Alder cycloaddition reactions, in which it shows a high degree of *para*-regioselectivity (Scheme 43).

Allylsilanes can be converted by a variety of methods into trimethylsilylmethylcyclobutanones, which in turn undergo controlled fragmentation on exposure to mild acid. Applications of this methodology can be seen in an elegant synthesis[130] of (±)-aphidicolin, and in a photochemical method[131] for the allylation of naphthoquinones (Scheme 44).

Further evidence[132] has been presented for the general γ-regioselectivity in the reactions of simple deprotonated allylsilane anions with electrophiles, vinylsilanes being obtained as products. However, if the allylsilane anion is first complexed with certain metals, then α-regioselectivity predominates, and a high degree of diastereoselectivity[133] can be obtained with aldehyde substrates. For example, boron, aluminium, and titanium complexation all induce *threo* selectivity, whereas use of tin results in an *erythro* preference (Scheme 45), providing, *inter alia*, a stereocontrolled route to terminal dienes.

VII. ARYLSILANES

The preparation and many of the properties of aryl- and heteroaryl-silanes have been well reviewed[134,135] and only some recent advances will be discussed here.

A. Preparation

The most frequently employed route to arylsilanes involves quenching of a metallated, normally lithiated, arene with chlorotrimethylsilane. Site-selective *ortho*-lithiation can be achieved using tertiary benzamides[136] or η^6-anisole chromium tricarbonyl complexes[137] (Scheme 46). Metallation of related substrates carrying a trialkylsilyloxy substituent[138,139] induces an O → C migration, often in a preparatively useful manner. In the case illustrated, cross-over experiments have indicated an intermolecular mechanism.

Tertiary benzamides undergo double[140] *ortho*-lithiation: in a similar manner, the thiophene carboxamide[141] **24** undergoes 3, 5-dilithiation (Scheme 47). Phenylethyne[142] can also be doubly metallated.

η^6-Indole chromium tricarbonyl complexes undergo regiospecific lithiation at the 2-position or, if blocked, at the 7-position. However, when the N atom is functionalized by the bulky $Pr^i{}_3Si$ group, lithiation and thence silylation occurs regiospecifically[143] at

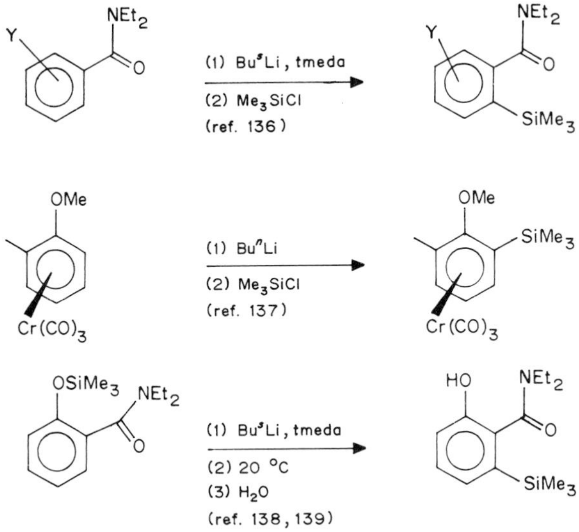

SCHEME 46

the 4-position (Scheme 48). Similar *N*-protection of pyrrole itself results in electrophilic attack[144] occurring regiospecifically at the 3-position under conditions of kinetic control.

Bromoarenes bearing a considerable range of additional functionality undergo a nickel-catalysed exchange reaction[46] with $(Me_3Si)_3Al$. The first synthesis of a 1, 2, 3-tristrimethylsilylarene was achieved by Halterman *et al.*[145]; other applications of the cobalt-mediated co-cyclization route to arylsilanes have been described[146]. Lithiation of oxazoles gives a mixture of C-2-lithiated products and the open-chain isocyano enolate

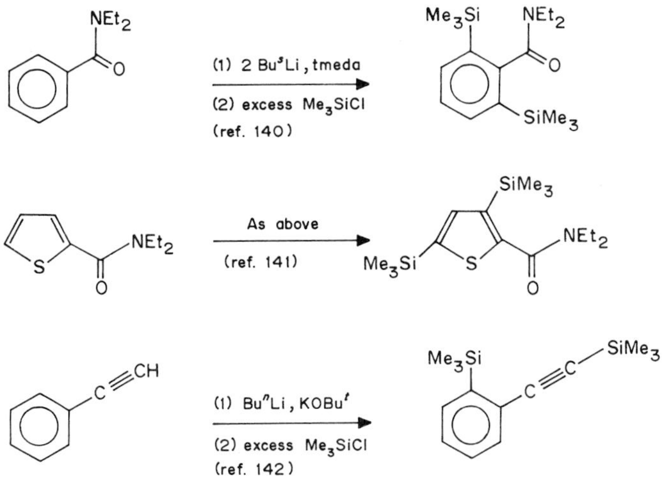

SCHEME 47

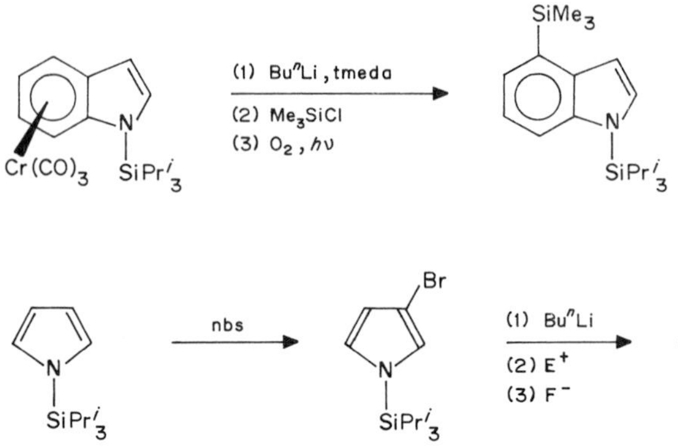

SCHEME 48

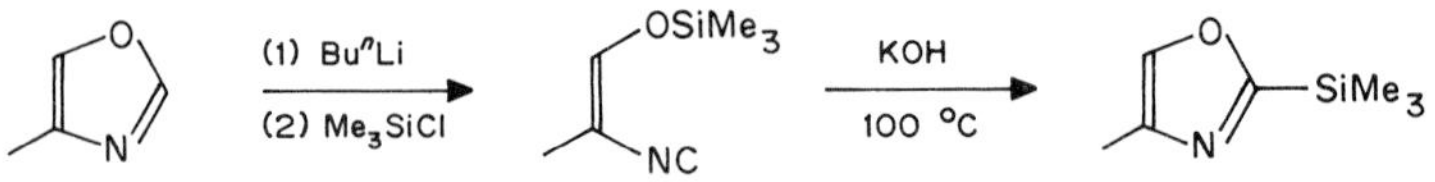

SCHEME 49

isomers, which can be trapped[147] by quenching with chlorotrimethylsilane. In the case of 4-methyloxazole, subsequent distillation in the presence of KOH resulted in intramolecular insertion of the isonitrile group[148] into the O—Si bond (Scheme 49).

B. Reactions

The advantages gained by replacing an aromatic/heteroaromatic proton with a trialkylsilyl group are manifold. The silyl group can mask a potentially acidic proton, it can act as a strong *para*-director in reactions of η^6-arene chromium tricarbonyl complexes, and it can be readily removed, either by electrophiles or nucleophiles, normally resulting in products of *ipso*-desilylation.

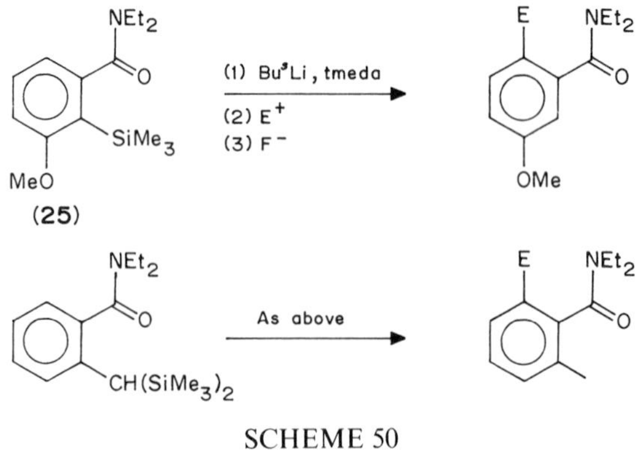

SCHEME 50

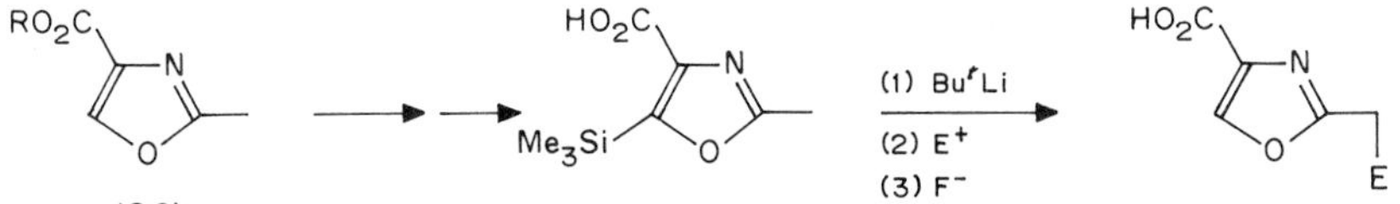

(26)

(1) Bu^tLi
(2) E^+
(3) F^-

SCHEME 51

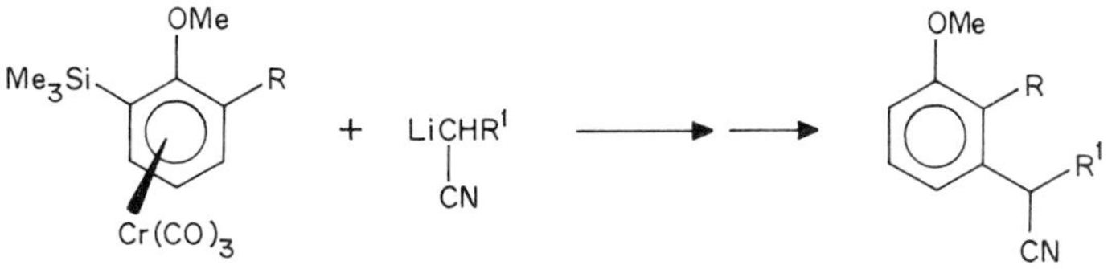

SCHEME 52

To exemplify each of these in turn, the substituted benzamide **25** undergoes directed lithiation and subsequent electrophilic attack[149] as shown in Scheme 50. This scheme also illustrates a case where potentially acidic benzylic protons can similarly be afforded temporary protection by silylation.

The hydrogen at the 5-position of 1,3-oxazoles such as **26** is much more acidic than a 2-alkyl substituent; masking by silylation has allowed[150] selective metallation and functionalization (Scheme 51). The benefit[151] of silyl substitution in the stereospecific transformation of thiophenes into enynes has been demonstrated.

A variety of nucleophiles add to η^6 arene chromium tricarbonyl complexes; trimethylsilyl substituents have been demonstrated[152] to act as strong *para*-directors in such reactions (Scheme 52).

All three isomeric aryl acetates **27** undergo clean *ipso*-desilylation allowing the preparation[153] of, *inter alia*, radiohalogen-labelled phenols; the free phenols and methyl ethers corresponding to **27** proved too reactive, giving mixtures of products of substitution and desilylation. Similarly, 2-trimethylsilylthiazole (**28**) reacts regiospecifically[154] at the 2-position with a range of electrophiles.

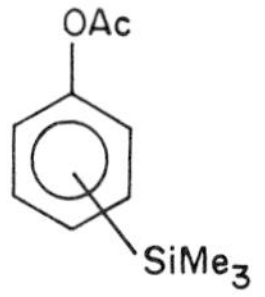

(27) 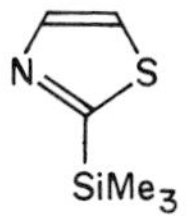 (28)

Aryltrimethylsilanes have been shown[155] to take part in a butoxide-catalysed condensation with electrophiles (Scheme 53). The order of reactivity correlates well with the σ^I substituent constants, but not with the expected order of aryl anion stability; this suggests that the electrophile participates significantly in the rate-limiting step, in contrast to normal nucleophilic desilylation[156].

$$ArSiMe_3 + E^+ \xrightarrow[\text{dmf or hmpa}]{\text{KOBu}^t} ArE$$

SCHEME 53

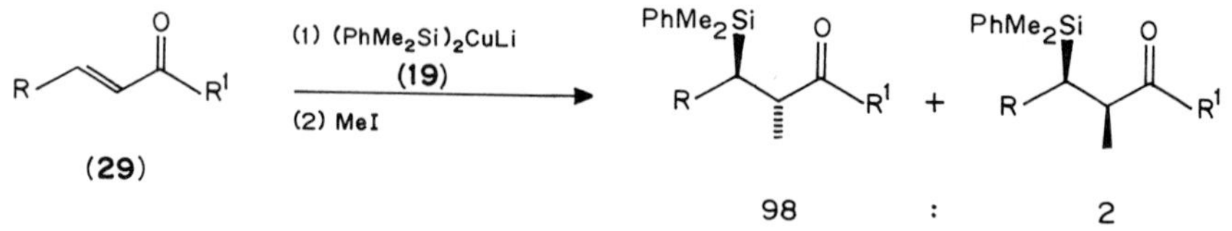

(1) $(PhMe_2Si)_2CuLi$ (**19**)
(2) MeI

(**29**)

98 : 2

SCHEME 54

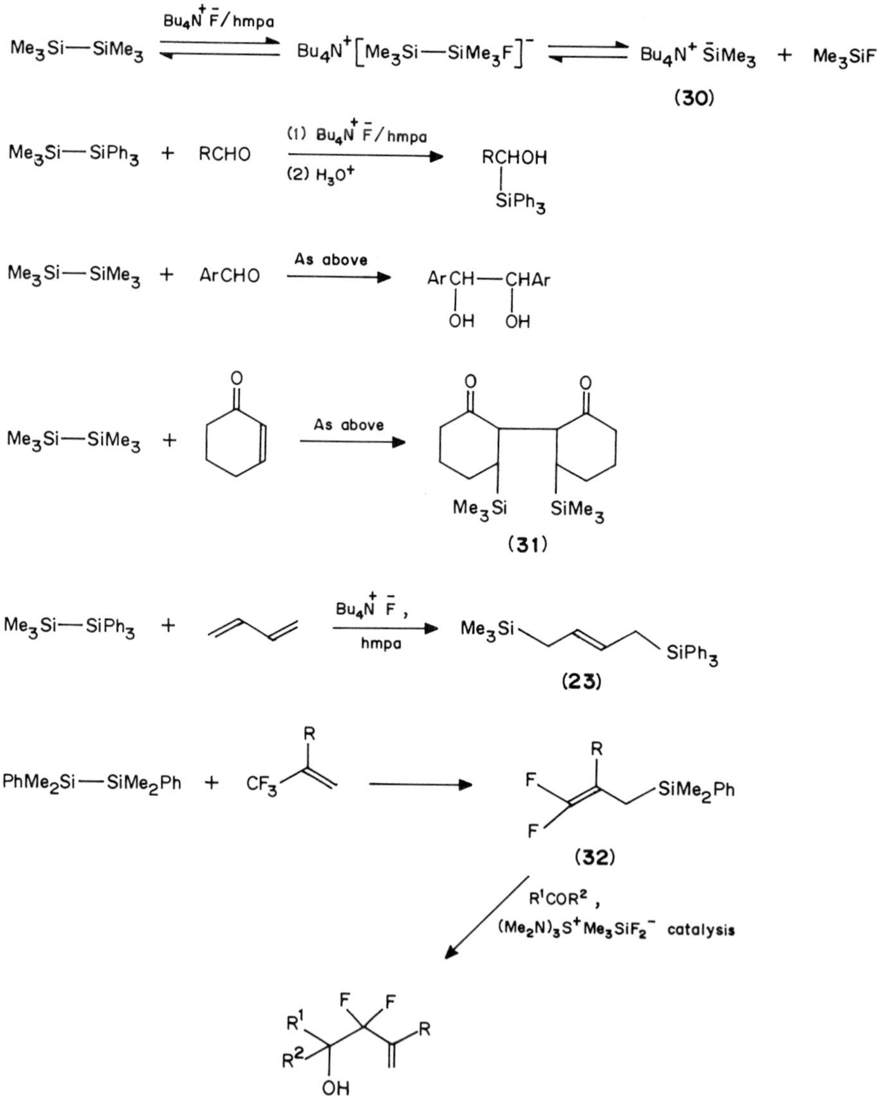

$Me_3Si—SiMe_3 \rightleftharpoons Bu_4N^+[Me_3Si—SiMe_3F]^- \rightleftharpoons Bu_4N^+\ \bar{S}iMe_3 + Me_3SiF$ ($Bu_4\overset{+}{N}\ \overset{-}{F}$/hmpa)

(**30**)

$Me_3Si—SiPh_3$ + RCHO → (1) $Bu_4\overset{+}{N}\ \overset{-}{F}$/hmpa (2) H_3O^+ → $RCH(OH)SiPh_3$

$Me_3Si—SiMe_3$ + ArCHO → As above → ArCH(OH)—CH(OH)Ar

$Me_3Si—SiMe_3$ + cyclohexenone → As above → (**31**)

$Me_3Si—SiPh_3$ + butadiene → $Bu_4\overset{+}{N}\ \overset{-}{F}$, hmpa → Me_3Si…$SiPh_3$ (**23**)

$PhMe_2Si—SiMe_2Ph$ + $CF_3C(R)=CH_2$ → $F_2C=C(R)CH_2SiMe_2Ph$ (**32**)

R^1COR^2, $(Me_2N)_3S^+Me_3SiF_2^-$ catalysis

SCHEME 55

VIII. ORGANOSILYL ANIONS/ANIONOIDS

Addition of the silyl cuprate **19** to an acyclic unsaturated system **29** produces a β-silyl enolate, which can be alkylated or, in suitable cases, protonated with high diastereoselectivity[157], an effect which seems to be mainly electronic in origin (Scheme 54).

$(Me_3Si)_3Al$ adds regioselectively[158] to enones, low temperatures favouring 1,4-addition; it has also found use[46,97] in routes to vinyl-, allyl-, and aryl-silanes. Tetrabutylammonium fluoride in hmpa cleaves hexaalkyl-/aryl-disilanes to produce 'naked' silyl anions[104] (**30**); in unsymmetrical cases, production of the electronically more favoured anions is observed. These anions react with aliphatic aldehydes to give the expected addition products (Scheme 55). Aromatic aldehydes give dimeric products, formed either by single electron-transfer processes, or by addition, Brook rearrangement, and further condensation. Although cyclohexenone gives the dimer **31**, conjugate addition to cyclopentenones[159] proceeds normally, in contrast to the attempted use of other silyl anion systems. Reaction with butadienes produces the doubly silylated allyl species **23**. Higher silane homologues behave analogously[160]. This same reagent system transforms[161] trifluoromethylallyl species into difluoroallylsilanes (**32**). These in turn react regioselectively with carbonyl compounds, the fluoride ion catalyst shown being the most effective.

IX. β-FUNCTIONAL ORGANOSILANES AND PETERSON OLEFINATION

The general process whereby organosilanes substituted at the β-position by an electronegative group can be induced to undergo 1,2-elimination has been subjected to continuing scrutiny. In those cases where the electronegative group is hydroxy, the reaction is known as the Peterson reaction, and this has been recently reviewed[162]. Elimination can be induced by either acid or base, with complementary stereochemical consequences (Scheme 56).

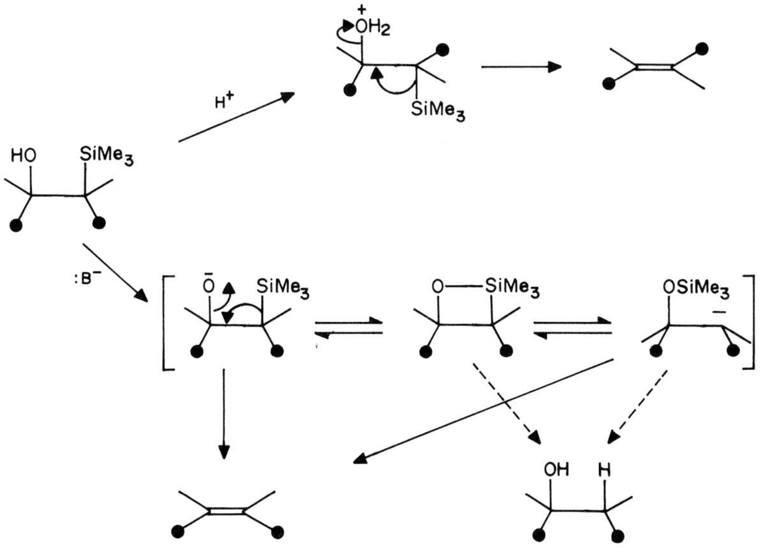

SCHEME 56

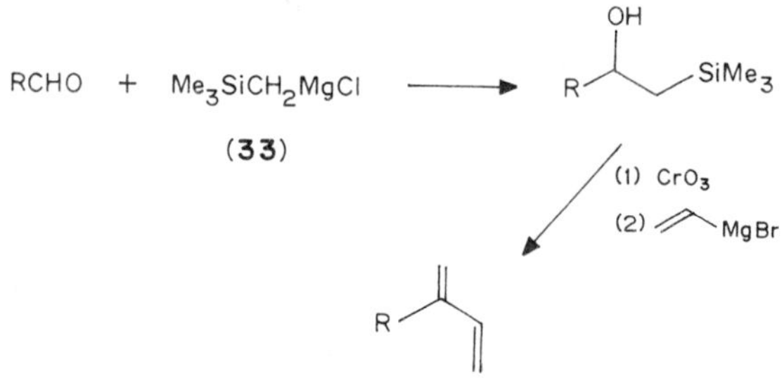

RCHO + Me_3SiCH_2MgCl (**33**) → R–CH(OH)–CH_2SiMe_3 → (1) CrO_3 (2) $CH_2{=}CHMgBr$ → R–C(=CH_2)–CH=CH_2

SCHEME 57

In base-induced rearrangements where X is an anion-stabilizing group, such as carbonyl, facile 1, 3-C → O migration of the silyl group is very likely to take part in the elimination pathway[163]. Although this pathway does not necessarily preclude a direct *syn*-elimination, stereoselective alkene formation can equally well be accounted for by *anti*-elimination of a tmso group. Hudrlik *et al.*[164] reported that simple unactivated β-hydroxysilanes can undergo protiodesilylation when treated with strong base in dmso; this homo-Brook rearrangement occurs, as expected, with retention of configuration at carbon.

The main route to β-hydroxysilanes involves reaction between carbonyl compounds and α-metallosilanes, normally under kinetic control. The simplest version of this uses the Grignard reagent **33** to introduce methylene units; a particularly nice example[165] creates substituted butadienes (Scheme 57).

In more complex cases, the metal of choice is lithium, but this imposes a considerable limitation in that, until recently, α-lithiosilanes have not been easy to prepare in the absence of another carbanion-stabilizing substituent. However, reductive lithiation of phenylthioacetals, using either lithium naphthalenide[166] or, more advantageously, lithium dimethylaminonaphthalenide[167], makes the general process outlined in Scheme 58 feasible.

In certain cases, simple deprotonation of a functionalized silane can suffice, as illustrated (Scheme 59) in a route[168] to asparenomycin C. Such examples normally carry a carbanion-stabilizing substituent, as mentioned earlier, and a selection of these is given in Table 3.

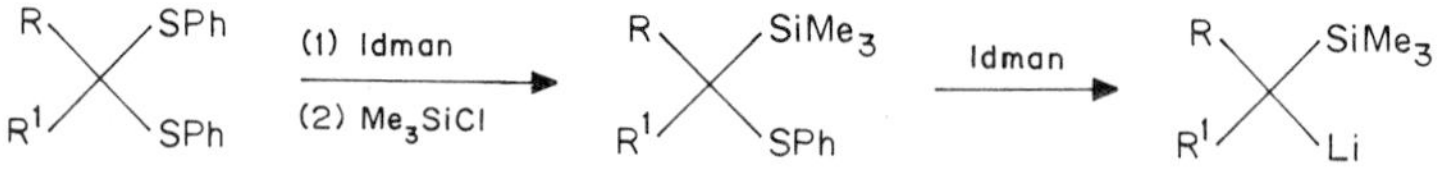

$RR^1C(SPh)_2$ → (1) ldman (2) Me_3SiCl → $RR^1C(SiMe_3)(SPh)$ → ldman → $RR^1C(SiMe_3)Li$

SCHEME 58

Me_3Si / $OSiMe_2Bu^t$ / $SiMe_2Bu^t$ → (1) lda (2) $CH_3COCH_2OCH_2SMe$ → O / SMe / $OSiMe_2Bu^t$ / $SiMe_2Bu^t$

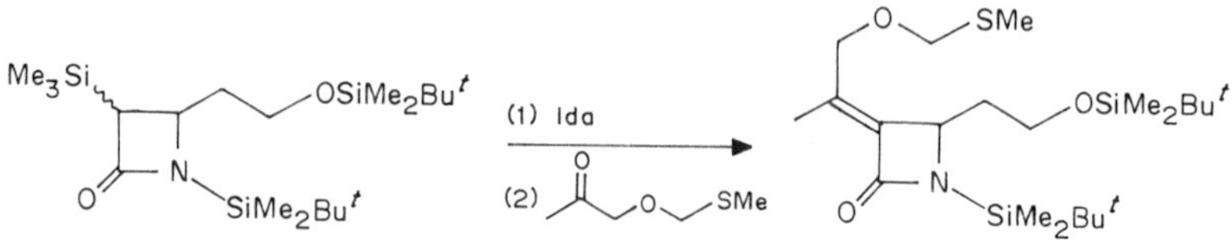

SCHEME 59

TABLE 3. Some additional examples of the Peterson reaction

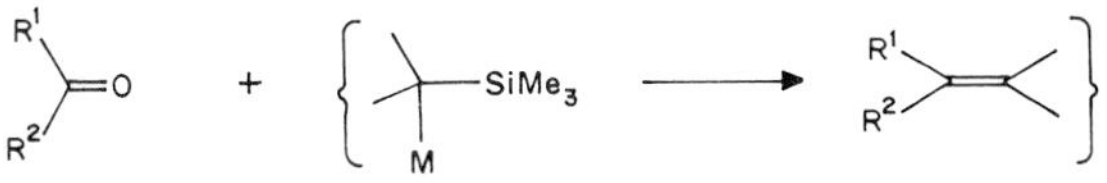

$R^1R^2C{=}O$ + { $SiMe_3$, M } → { R^1, R^2 }

Metallosilane	Products	Notes[a]	Ref.
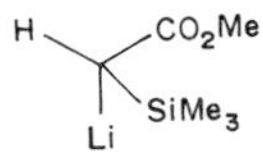 H, CO_2Me, $SiMe_3$, Li	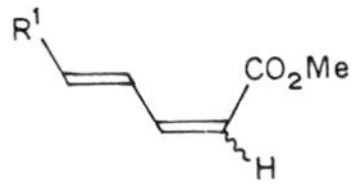 R^1, CO_2Me, H	1, 2	169
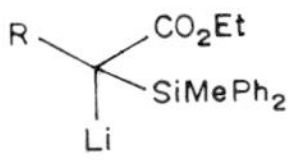 R, CO_2Et, $SiMePh_2$, Li	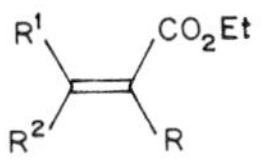 R^1, CO_2Et, R^2, R	3, 4, 5	170–172
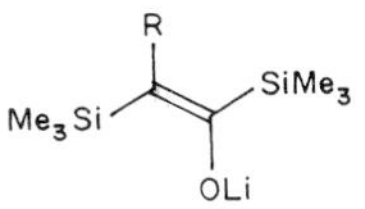 R, Me_3Si, $SiMe_3$, OLi	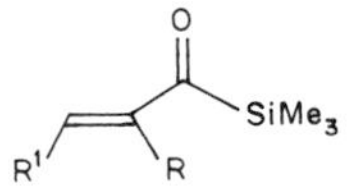 O, $SiMe_3$, R^1, R	6	173
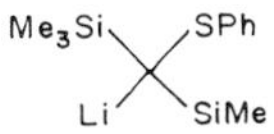 Me_3Si, SPh, Li, SiMe	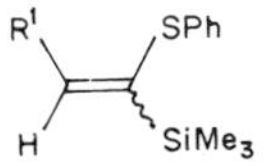 R^1, SPh, H, $SiMe_3$	6	174
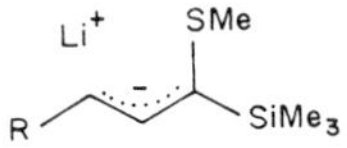 Li^+, SMe, R, $SiMe_3$	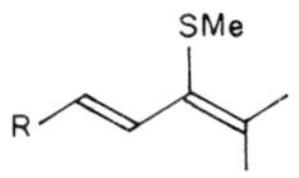 SMe, R	7, 8	175,176
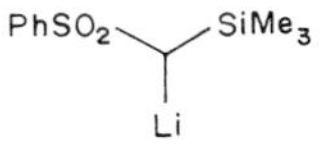 $PhSO_2$, $SiMe_3$, Li	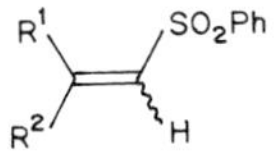 R^1, SO_2Ph, R^2, H	1, 9	177,178
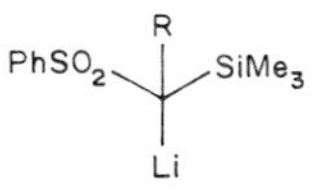 R, $PhSO_2$, $SiMe_3$, Li	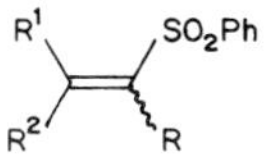 R^1, SO_2Ph, R^2, R	—	178
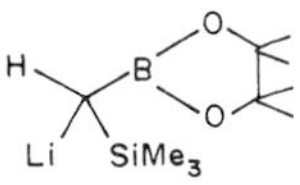 H, B, O, O, Li, $SiMe_3$	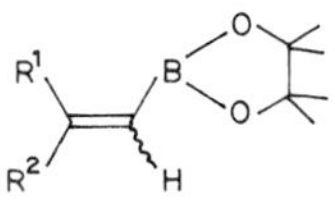 R^1, B, O, O, R^2, H	10	179
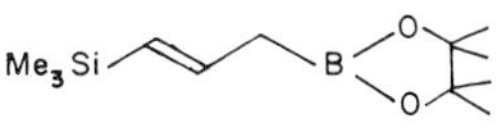 Me_3Si, B, O, O	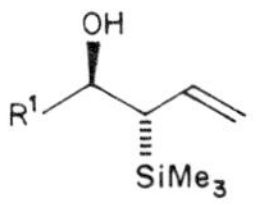 OH, R^1, $SiMe_3$	6, 11, 12, 13	180

TABLE 3. (*Contd.*)

Metallosilane	Products	Notes[a]	Ref.
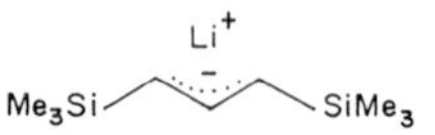 Me_3Si–allyl anion (Li^+)–$SiMe_3$	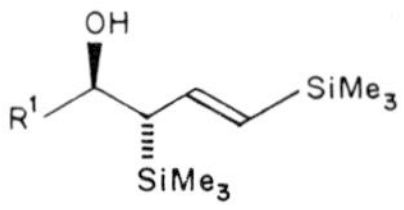 $R^1CH(OH)CH(SiMe_3)CH{=}CHSiMe_3$	6, 12, 13	181
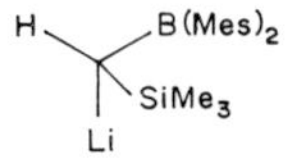 $H(Li)C(B(Mes)_2)SiMe_3$	$PhCH_2CHO$	14, 15	182
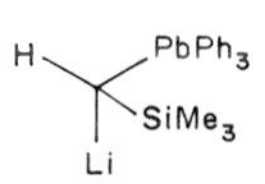 $H(Li)C(PbPh_3)SiMe_3$	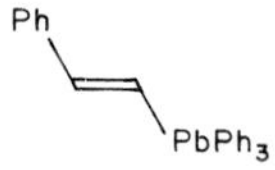 $PhCH{=}CHPbPh_3$	14	183
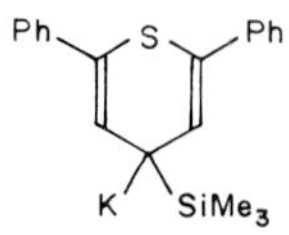 2,6-diphenyl-4-K-4-$SiMe_3$-4H-thiopyran	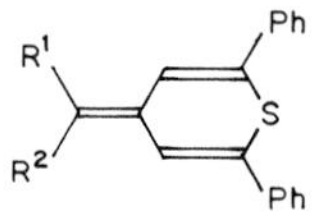 4-($R^1R^2C{=}$)-2,6-diphenyl-4H-thiopyran	—	184

[a] 1, Normally *ca.* 1:1 *E*:*Z*. 2, with (*E*)-$R^1CH{=}CHCHO$. 3, Mainly *Z*. 4, Reagent (R = H) adds 1,4 to cyclopentenones. 5, Reagent (non-metallated) also reacts with Grignard reagents to produce alkenes. 6, With R^1CHO. 7, With MeCOMe. 8, Simple allylic anion adds both 1,2 and 1,4 to enones, γ-regioselectively. 9, Also with R^1CHO. 10, With R^1CHO, mainly *Z*. 11, No base required. 12, Subsequent base or acid treatment yields (*Z*)- or (*E*)-dienes. 13, See also ref. 133. 14, With PhCHO. 15, After oxidative work-up.

An extension of this elimination process can be seen in the use of β-trimethylsilylethanol and its equivalents as protecting/masking groups for molecules with reactive hydrogen atoms. The parent alcohol can be obtained in a straightforward manner[185] from chloromethyltrimethylsilane (Scheme 60). The protection afforded is related to the functionality involved and, when desired, liberation is readily achieved using fluoride ion

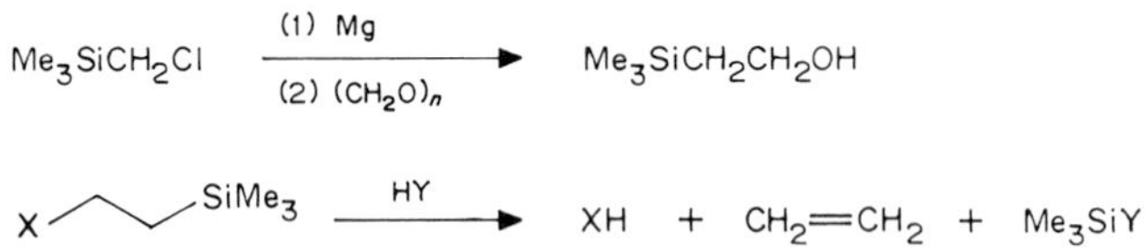

$$Me_3SiCH_2Cl \xrightarrow[(2)\ (CH_2O)_n]{(1)\ Mg} Me_3SiCH_2CH_2OH$$

$$XCH_2CH_2SiMe_3 \xrightarrow{HY} XH + CH_2{=}CH_2 + Me_3SiY$$

SCHEME 60

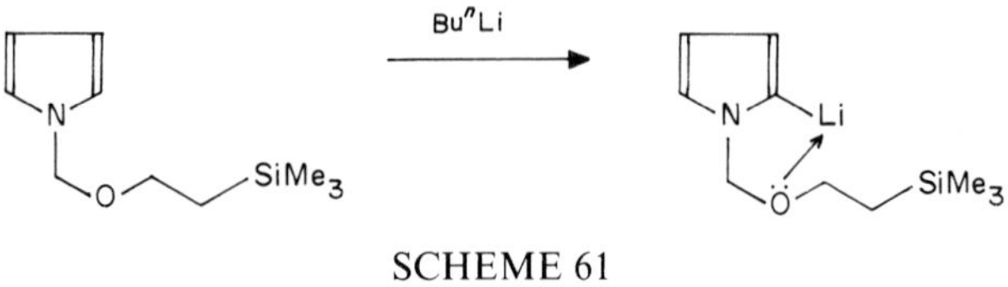

$$\text{1-}(Me_3SiCH_2CH_2OCH_2)\text{pyrrole} \xrightarrow{Bu^nLi} \text{2-lithio-1-}(Me_3SiCH_2CH_2OCH_2)\text{pyrrole (O}\rightarrow\text{Li)}$$

SCHEME 61

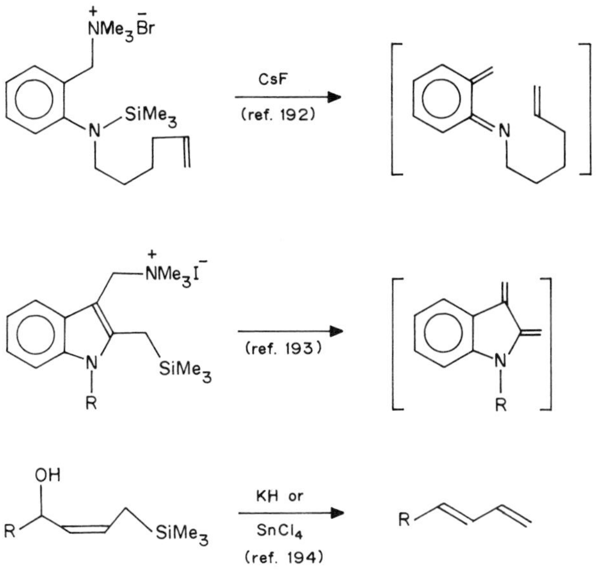

SCHEME 62

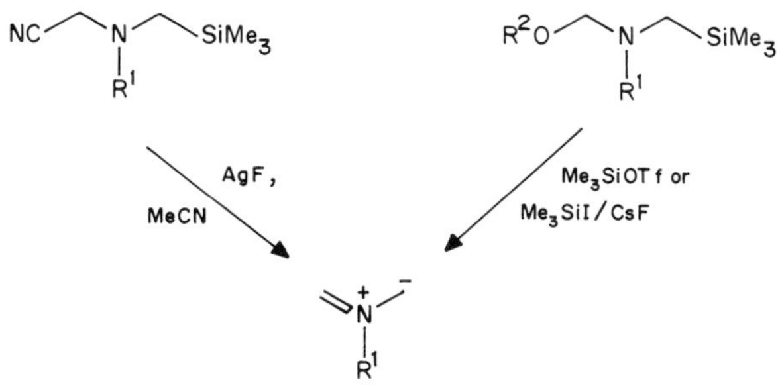

SCHEME 63

or a Lewis acid. Reactive groups to which such protection has been afforded include carboxylic[186] and phosphorus acids[187], alcohols[188], including sugar hemiacetals[189], and amines[190]. Pyrrole can be *N*-protected in this manner, encouraging α-lithiation and so providing a synthetic equivalent[191] of 2-lithiopyrrole (Scheme 61).

Vinylogous versions of this elimination have found use in the generation of *o*-quinodimethanes[192] and related species[193], and in the preparation[194] of terminally substituted butadienes (Scheme 62); other highly reactive species which have been generated include sulphenes[195], nitrile oxides[196], and azomethine ylids[197,198] (Scheme 63).

A remarkable rearrangement occurs when the tertiary seleno ether **34** is treated with

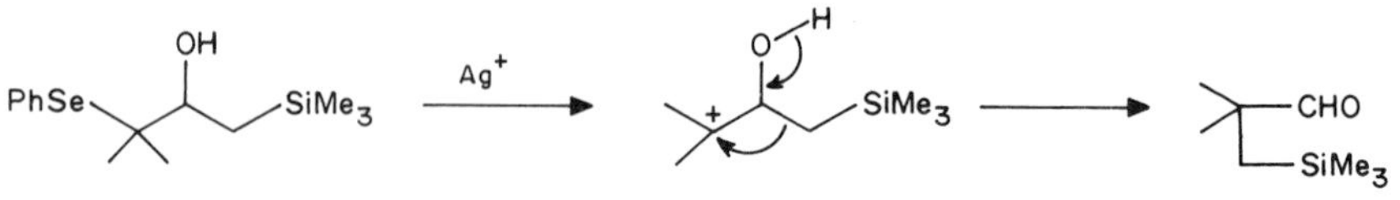

SCHEME 64

Ag^+ ions; the product **35** has been rationalized[199] in terms of a 1,2-migration of a Me_3SiCH_2 group from a secondary to a tertiary carbon atom (Scheme 64).

Fleming and co-workers have given full details[200] of the synthetic equivalence between β-silyl ketones and enones: the parent ketones arise from conjugate addition of various silyl cuprate reagents[99,201] to enones, or from conjugate addition of carbon nucleophiles to β-silyl enones[202]; demasking is effected by bromination/desilylbromination.

X. ALKYNYL- AND ALLENYL-SILANES

The physical and spectroscopic properties[203] of a wide range of ethynyl silanes have been tabulated.

Because of the tautomeric relationship between alkynyl- and allenyl-silanes and their corresponding metalloids, no strict separation of preparation and properties can be made. To exemplify the facile nature of this tautomerism, the allene **36**, on deprotonation with Bu^tLi, has been reported[204] to react with aldehydes with integrity; on treatment with acid, 2-silylfurans[205] are produced (Scheme 65). On the other hand, deprotonation using lda

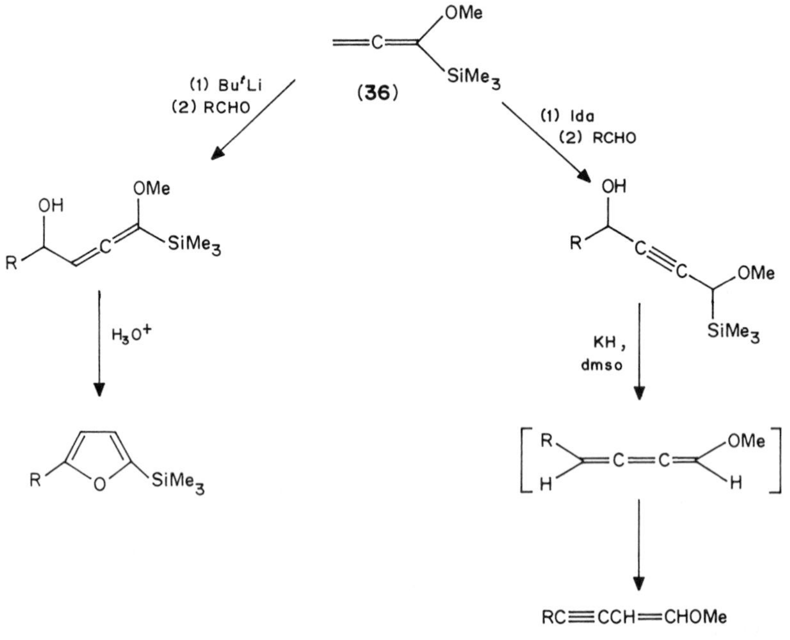

SCHEME 65

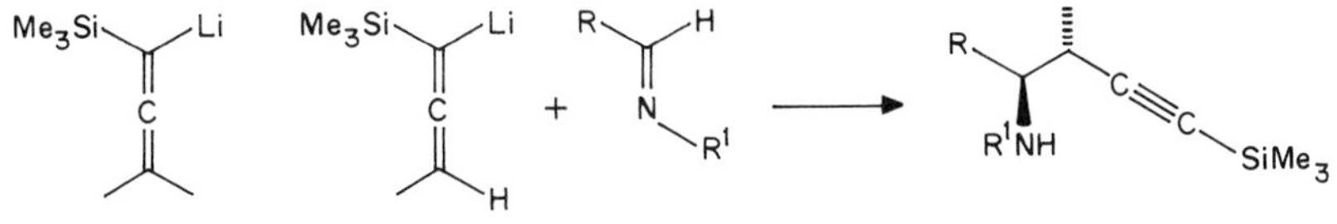

(37) (38)

SCHEME 66

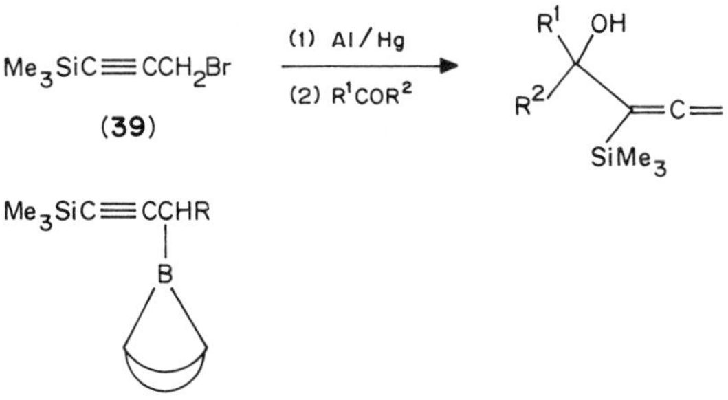

$Me_3SiC{\equiv}CCH_2Br$ (39) —(1) Al/Hg; (2) R^1COR^2→

(40)

SCHEME 67

has been reported[206] to lead to the propargyl tautomer; the products from aldehydes undergo a further base-induced Peterson elimination to give enynes[207], in spite of the remoteness of the groups being eliminated (see also Scheme 62).

The lithiated tms-allene **37** undergoes silylation with integrity, but reacts propargylically[208] with carbonyl compounds; the lower homologue (**38**) reacts similarly with imines[209], when it shows high *threo* diastereoselectivity (Scheme 66).

Propargyl bromide, as the corresponding aluminium organometallic, reacts with carbonyl compounds to give propargyl alcohols. However, use of the silyl derivative **39** leads[210] to allenyl alcohols (Scheme 67); similar results, especially with aldehydes[211], have been obtained with the 9-bbn derivative **40** and with related titanium species[212].

The ortho-ester **41** functions[213] as a synthetic equivalent of the 3-anion of ethyl propiolate. The preparation and some reactions of the dithiane derivative **42** have been described[214].

$Me_3SiC{\equiv}CC(OEt)_3$

(41) (42)

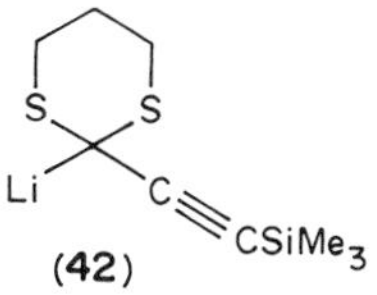

Alkynylsilanes are attacked by electrophiles at the α-position, normally regiospecifically (Scheme 68), in an extension of the β-effect. Recent examples of electrophiles being used for such a purpose include acid chlorides[215], β,γ-unsaturated acid chlorides[216], thioesters[217], and α,β-unsaturated acyl cyanides[218]; the employment of oxocarbenium ions can be seen in an elegant synthesis[219] of resistomycin.

$$RC{\equiv}CSiMe_3 + E^+ \longrightarrow RC{\equiv}CE$$

SCHEME 68

Ethynyltrimethylsilane reacts with dichloroketene in apparent contravention of this generalization (Scheme 69); this has been explained[220] in terms of FMO theory by invoking a concerted reaction pathway.

Alkynylsilanes have been used as radical traps in routes to pyrrolizidine alkaloids[221] and to β-agarofuran[222] (Scheme 70). The origin of this regiospecific directing effect is probably steric, since in the former case the corresponding *tert*-butyl derivative reacted analogously.

Allenylsilanes undergo a titanium-catalysed cycloaddition reaction[223] with electron-deficient alkenes (Scheme 71); the mechanistic explanation of the example shown is left to the ingenuity of the reader!

Terminal silylation of diynes permits selective reduction of the internal alkyne. Recent examples include the use of either activated Zn in EtOH[224] or, if there is a proximate

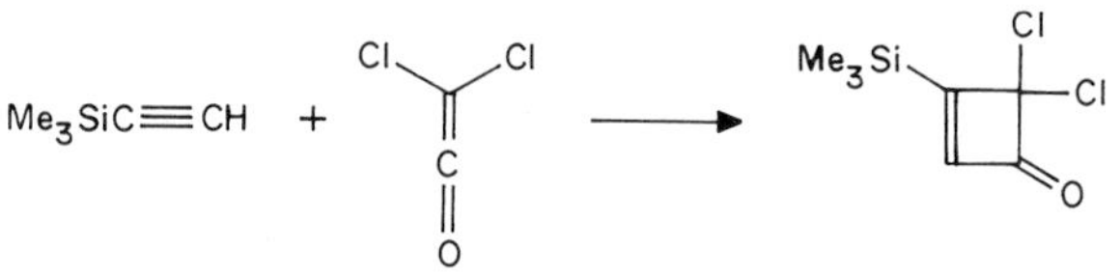

Me3SiC≡CH + Cl2C=C=O → (Me3Si-substituted dichlorocyclobutenone)

SCHEME 69

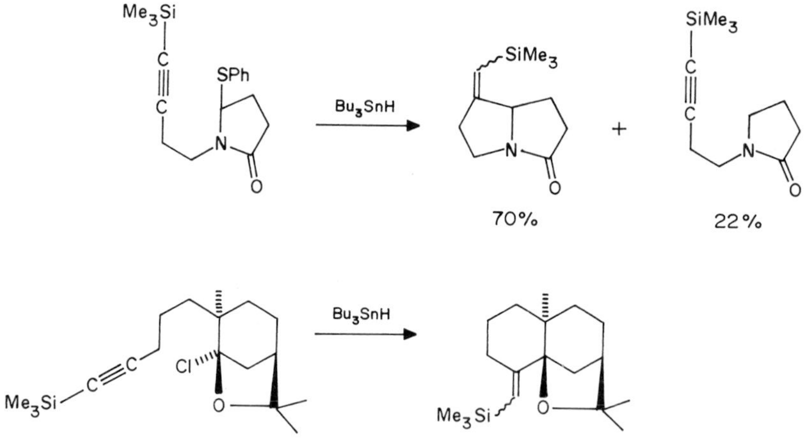

Bu3SnH

70%

22%

Bu3SnH

SCHEME 70

TiCl4

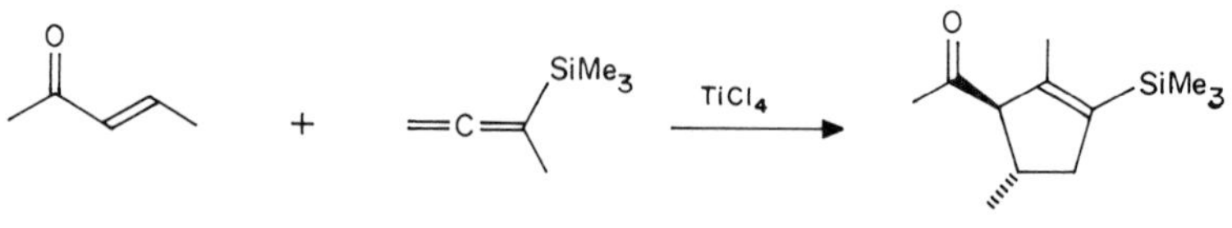

SCHEME 71

hydroxy group, of $LiAlH_4$[225] for reduction, both leading to (*E*)-enynes. Conjugate addition reactions of the ethyne anionoid **43** to α,β-unsaturated sulphones[226], and of the Grignard reagent **44** to enones[227], have been reported, as have several new routes[228] to propargylic silanes.

$$Me_3SiC{\equiv}C^-K^+ \quad \textbf{(43)} \qquad\qquad Me_3SiC{\equiv}CCH_2CH_2MgCl \quad \textbf{(44)}$$

Kuwajima *et al.*[229] have given full details of the use of fluoride ion to effect desilylation of alkynysilanes and so afford nucleophilic alkyne anion equivalents for reaction with carbonyl compounds.

XI. CYCLOPROPYL- AND CYCLOPROPYLCARBINYL-SILANES

These species, represented by **45** and **46**, can be obtained by carbene/carbenoid addition to suitable vinyl-[230] and allyl-silanes[231], respectively. Cyclopropylsilanes can also be introduced as intact units using the organolithium reagent **47**, generated by transmetallation of the corresponding bromo-[232], thio-[230], or seleno-cyclopropane[233]. Alternatively, more complex silanes[234] can be employed, such as the aldehyde **48**.

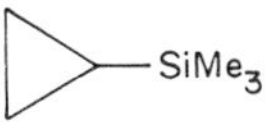

SiMe3

(**45**)

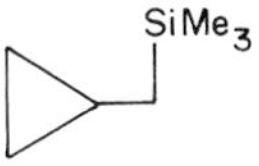

SiMe3

(**46**)

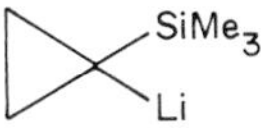

SiMe3
Li

(**47**)

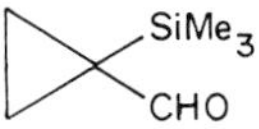

SiMe3
CHO

(**48**)

Paquette and co-workers[235] have summarized their extensive contributions in this area, particularly in the vinylcyclopropane/cyclopentene thermal rearrangement, which here produces vinylsilanes (Scheme 72). They have also demonstrated[236,237] that cyclopropylsilanes bearing anion-delocalizing substituents, such as carbonyl or vinyl, can provide synthetically useful sources of the corresponding cyclopropyl anions or their equivalents (Scheme 73).

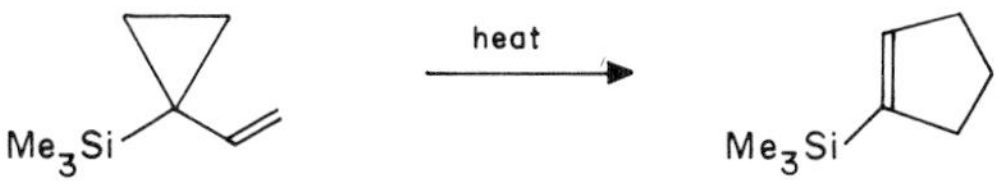

Me3Si — heat → Me3Si

SCHEME 72

SiMe3
X
— R^1COR^2 / F^- →
R^1 OH
R^2
X

X = CN, CO_2Me, COMe,
$CH{=}CR_2$

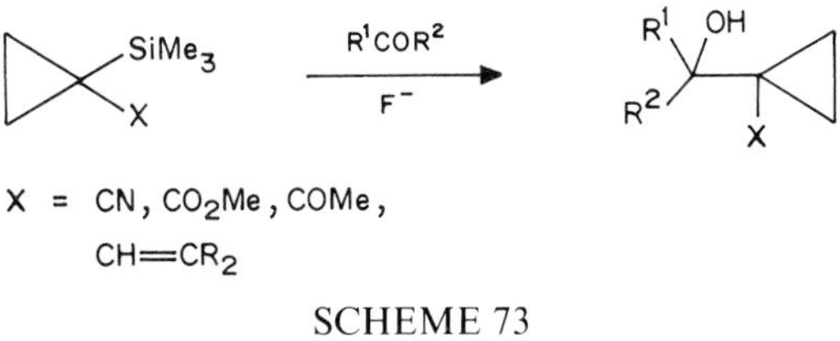

SCHEME 73

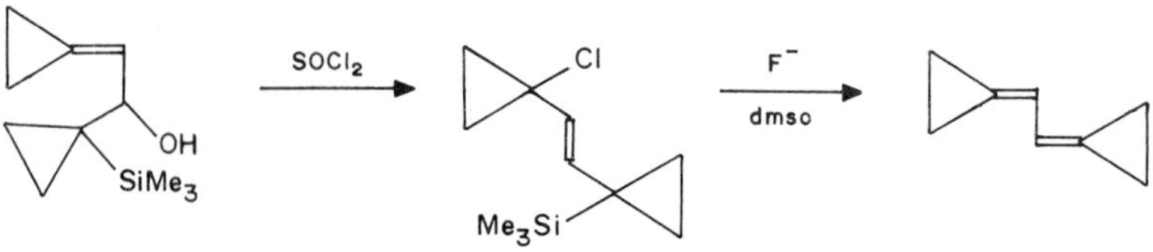

(49) (50) (51)

SCHEME 74

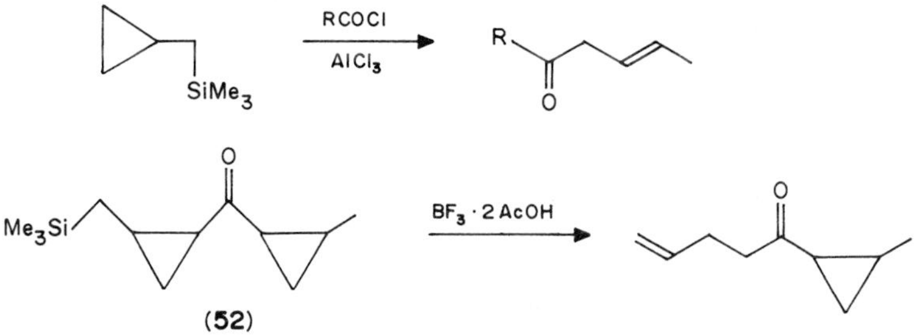

(52)

SCHEME 75

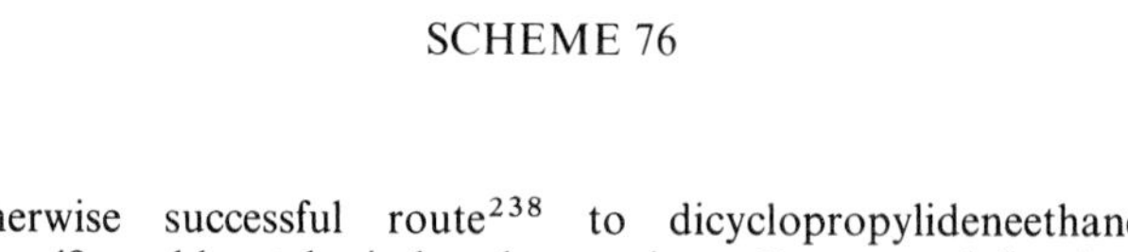

(53) X = O

(54) X = CH_2

SCHEME 76

In an otherwise successful route[238] to dicyclopropylideneethane (**51**), the β-hydroxysilane **49** could not be induced to undergo Peterson olefination. The allylic chloride **50**, on the other hand, reacted smoothly (Scheme 74). The influence of bridgehead silyl substitution in controlling displacement reactions on substituted bicyclo [4.1.0] heptane systems has been demonstrated[239].

Cyclopropylcarbinylsilanes behave as 'homoallyl' silanes, the outcome of electrophilic attack[240] being controlled by the β-effect (Scheme 75). This scheme also illustrates a particularly nice example of such regio-control, in the acid-catalysed fragmentation[241] of the biscyclopropyl ketone **52**.

As mentioned in Section VI.B, the reactions of allylsilanes and silyl enol ethers are comparable. This 'homoenol' ether **53** shows a higher degree of reactivity towards ring opening[242] using fluoride ion than does the 'homoallyl' silane **54** (Scheme 76). High-temperature fluoride ion-induced desilylation of cyclopropylcarbinylsilanes bearing

TABLE 4. Preparation of alkyl silyl ethers

$$ROH \longrightarrow ROSiR^1_3$$

(p = primary, s = secondary, t = tertiary alcohol)

	Reaction		Ref.
p,s,t	3-(trimethylsilyl)-2-oxazolidinone (O, N–SiMe$_3$, C=O), TfOH cat. →	$ROSiMe_3$	244
p,s,t	$Me_2C{=}CH{-}C(O)Me$ with OSiMe$_3$ →	$ROSiMe_3$	245
p,s,t	$Me_2C{=}C(OEt)(OSiEt_3)$, TfOH cat. for s,t →	$ROSiEt_3$	246
p,s	Pr^i_3SiOTf, 2,6-lutidine →	$ROSiPr^i_3$	247
p	$Me_2C{=}C(OSiMe_2Bu^t)(OMe)$, TsOH cat. →	$ROSiMe_2Bu^t$	248
p,s	Bu^tMe_2SiO–C(Me)=CH–C(O)Me, TsOH cat. →	$ROSiMe_2Bu^t$	249
p,s,t	Bu^tMe_2SiOTf, 2, 6-lutidine →	$ROSiMe_2Bu^t$	247
p,s	Bu^tMe_2SiCl, Pr^i_2NEt or dbu →	$ROSiMe_2Bu^t$	250, 251
diol (OH, OH)	$Pr^i_2Si(OTf)_2$, 2, 6-lutidine →	cyclic O–$SiPr^i_2$–O	252
	$Bu^t_2Si(OTf)_2$, 2, 6_lutidine →	cyclic O–$SiBu^t_2$–O	252
	or Bu^i_2SiCl, Et_3N, 1-hydroxy benzotriazole		253
	or $Bu^t_2SiCl_2$, imidazole, dmf		254
	$R'_2Si(OCR''\ \ CR''_2)_2$ →, R′ = Me, Et, Pr^i	cyclic O–SiR'_2–O	255

2-chloro-2-fluoro substituents has been reported[243] to lead to monofluorinated butadienes.

XII. ALKYL SILYL ETHERS

This section deals with the preparation, stability range, and cleavage of alkyl silyl ethers. The number of instances of the use of silyl ethers in synthesis is vast, and comprehensive coverage would be impossible; many interesting applications are contained in the preparative and cleavage references cited here. Additionally, many of the methods cited are equally applicable to other protic substrates, such as phenols, carboxylic acids, mercaptans, and amines (Section XV), although the individual stability ranges vary widely. As with other sections, only recent material is covered.

A. Preparation

A selection of recently reported preparative methods is given in Table 4. Emphasis has been placed on mild, often uncatalysed, routes of sufficient generality to be applicable to alcohols of all degrees of substitution, including diols.

B. Stability Range

Table 5 gives the relative rates of cleavage, by acid- or base-catalysed methanolysis, of a selection of aryl and alkyl silyl ethers. From even these limited data, the effect of substituent change on silicon is apparent. Additionally, it can be seen that alkyl silyl ethers are much more labile in acid than in base; the reverse is the case with aryl silyl ethers.

C. Cleavage

Trimethylsilyl ethers are readily and rapidly cleaved under a variety of mild conditions. The more stable silyl ethers require more vigorous, and sometimes more selective, conditions. A selection of these are given in Table 6. Aryl and alkyl *tert*-butyldimethylsilyl ethers can be cleaved[259] individually and selectively (Scheme 77); similar results were obtained with Bu^tPh_2Si ethers but, not surprisingly, no selectivity was observed with Me_3Si ethers. In an elegant synthesis[260] of the endiandric acids, selective cleavage of a Bu^tMe_2Si ether in the presence of a trimethylsilylalkyne proved possible (Scheme 78).

TABLE 5. Relative rates of cleavage

Compound	Base-induced	Acid-induced
	$ROSiR^1_3 \longrightarrow ROH$	
Me_3Si	1	10^5
Et_3Si	10^{-3}	2×10^4
Pr^i_3Si		14
$PhMe_2Si$	3	10^5
	$ArOSiR^1_3 \longrightarrow ArOH$	
Me_3Si	32	1
Et_3Si	0.2	0.02
Bu^tMe_2Si	1.6×10^{-3}	5.5×10^{-5}

TABLE 6. Cleavage of alkyl silyl ethers

$$ROSiR^1_3 \longrightarrow ROH$$

Reactant	Conditions	Ref.
$ROSiMe_2Bu^t$	HF, H_2O, MeCN	256
	KO_2, 18-crown-6	257
$ROSiBu^tPh_2$	py, HF	258
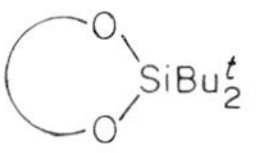 cyclic O,O-$SiBu^t_2$	KO_2, 18-crown-6	57
	HF, H_2O, MeCN	252
	py, HF	253

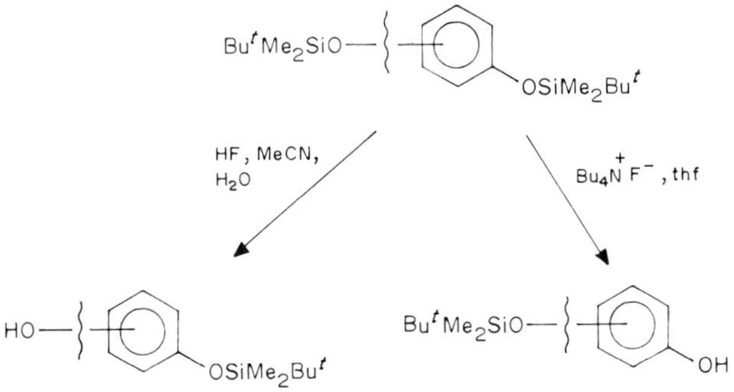

SCHEME 77

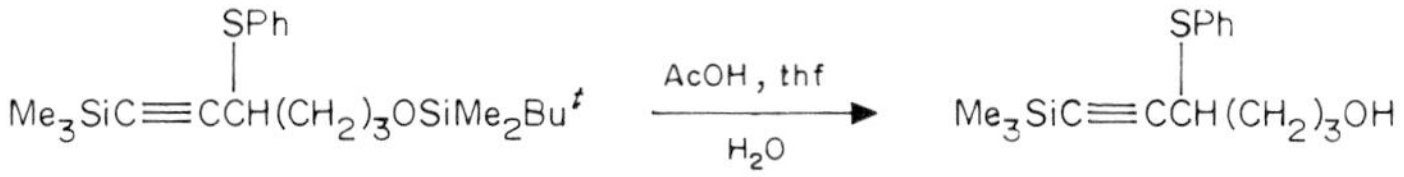

$$Me_3SiC{\equiv}C\overset{SPh}{\overset{|}{C}}H(CH_2)_3OSiMe_2Bu^t \xrightarrow[H_2O]{AcOH,\ thf} Me_3SiC{\equiv}C\overset{SPh}{\overset{|}{C}}H(CH_2)_3OH$$

SCHEME 78

XIII. SILYL ENOL ETHERS AND KETENE ACETALS

The preparation and properties of silyl enol ethers and ketene acetals have been reviewed[261] up to 1981, so only some of the more recent advances will be discussed here.

A. Preparation

It is comparatively simple to generate 'kinetic' silyl enol ethers in high regiochemical purity. Corey and Gross[262] have advocated the simultaneous presence of lda or a more hindered base and tms chloride at $-78\,^{\circ}C$ as a further improvement. 'Thermodynamic' enol ethers are more problematic. The use of boron enolates[263] or of magnesium amide bases[264] has been reported to be of value, giving virtually pure regioisomers; variation of the base in the boron enolate procedure can also yield 'kinetic' enolates of similar regiochemical purity (Scheme 79). A very simple method employing[265] triflic anhydride

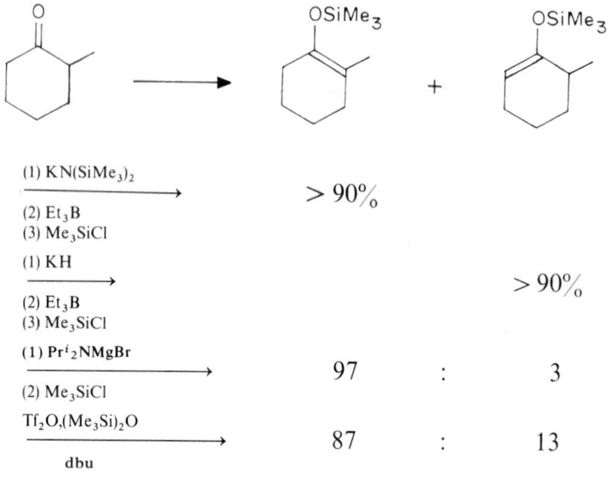

Conditions	Product A (OSiMe$_3$, $\Delta^{1,2}$ tetrasubstituted)		Product B (OSiMe$_3$, $\Delta^{1,6}$)
(1) $KN(SiMe_3)_2$; (2) Et_3B; (3) Me_3SiCl	> 90%		
(1) KH; (2) Et_3B; (3) Me_3SiCl			> 90%
(1) $Pr^i{}_2NMgBr$; (2) Me_3SiCl	97	:	3
$Tf_2O,(Me_3Si)_2O$; dbu	87	:	13

SCHEME 79

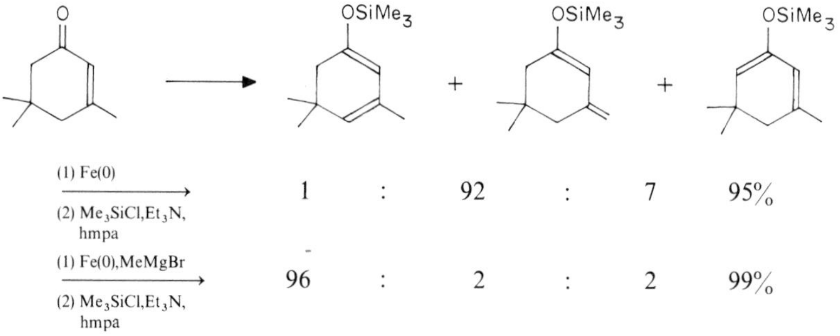

Conditions	OSiMe$_3$		OSiMe$_3$		OSiMe$_3$	
(1) Fe(0); (2) Me_3SiCl,Et_3N, hmpa	1	:	92	:	7	95%
(1) Fe(0),MeMgBr; (2) Me_3SiCl,Et_3N, hmpa	96	:	2	:	2	99%

SCHEME 80

and hexamethyldisiloxane, to generate tms OTf *in situ*, also gives a good thermodynamic bias.

Cyclic enones undergo kinetic deprotonation to give the cross-conjugated enolate ions[266], and hence enol ethers. Treatment of such enones with two different modifications[267] of the Kharasch reagent, Fe(0), gives individual access to the isomeric exo- and endo-cyclic 'thermodynamic' isomers, as illustrated for the case of isophorone (Scheme 80); the rapid process leading to the endocyclic isomer is very sensitive to the presence of O_2, suggesting a single electron-transfer pathway.

A careful study[262] of stereoselectivity in silyl enol ether generation, employing the 'internal quench' method described above, has shown that production of the (*E*)-enolate is kinetically favoured; the presence of hmpa allows equilibration to the thermodynamically more stable (*Z*)-enolate isomer. Additionally, the more hindered the base, the better was the kinetic stereoselectivity (Scheme 81), and hence the greater the preponderance of the (*E*)-isomer.

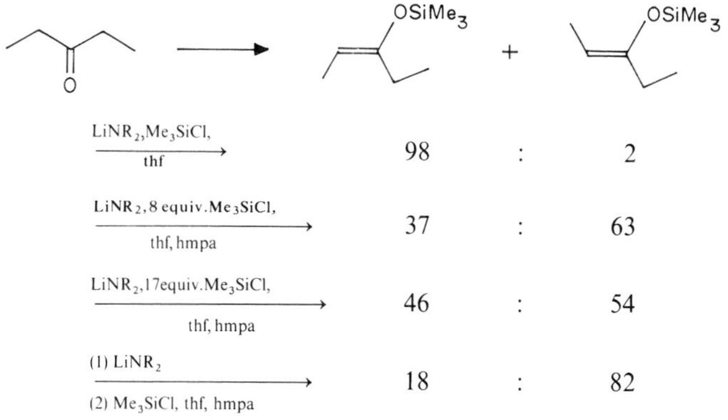

Conditions			
$LiNR_2$, Me_3SiCl, thf	98	:	2
$LiNR_2$, 8 equiv. Me_3SiCl, thf, hmpa	37	:	63
$LiNR_2$, 17equiv. Me_3SiCl, thf, hmpa	46	:	54
(1) $LiNR_2$ (2) Me_3SiCl, thf, hmpa	18	:	82

SCHEME 81

β-Ketosilanes undergo rearrangement to silyl enol ethers. As stated earlier (Section IIIB), this has rarely been viewed as a viable route to such species, largely owing to a paucity of general routes to the requisite β-ketosilanes; for example, dithiane-based routes[268] have been largely unsuccessful. This situation is now changing: in addition to the rearrangement routes discussed earlier, esters, especially of hindered acids[269], and acid chlorides[270] can both be converted into β-ketosilanes in good yield (Scheme 82). Alternatively, α-silylated esters[271] react with excess of Grignard reagent as shown. Simple β-ketosilanes undergo a tms triflate-catalysed rearrangement[270] of the above type to furnish 'kinetic' silyl enol ethers of methyl ketones.

Interestingly, β-ketosilanes undergo deprotonation[272] at the side remote from the silyl substituent, a selectivity ascribed to steric hindrance; Lewis acid-catalysed reactions proceed as expected (Scheme 83). Fluoride ion-induced reactions[273] of such species with electrophiles have also been described.

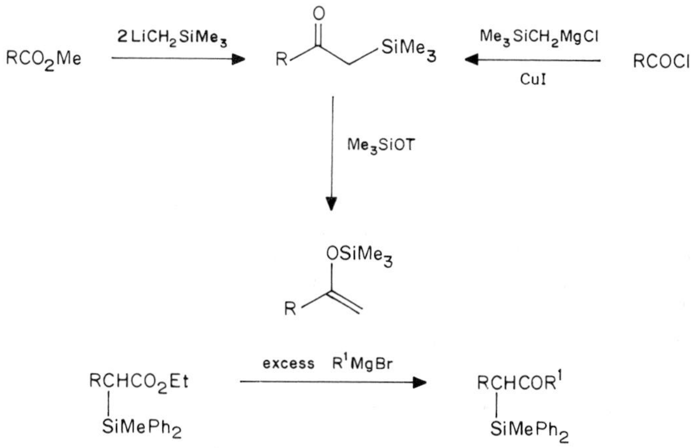

RCO_2Me —($2\,LiCH_2SiMe_3$)→ $RCOCH_2SiMe_3$ ←(Me_3SiCH_2MgCl, CuI)— $RCOCl$

$RCOCH_2SiMe_3$ —(Me_3SiOT)→ $RC(OSiMe_3)=CH_2$

$RCH(SiMePh_2)CO_2Et$ —(excess R^1MgBr)→ $RCH(SiMePh_2)COR^1$

SCHEME 82

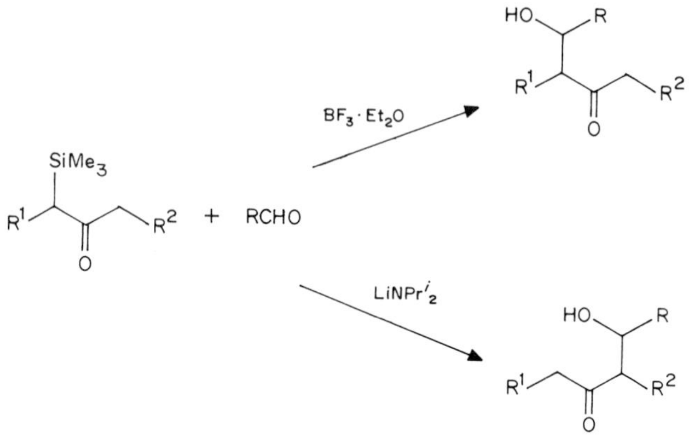

SCHEME 83

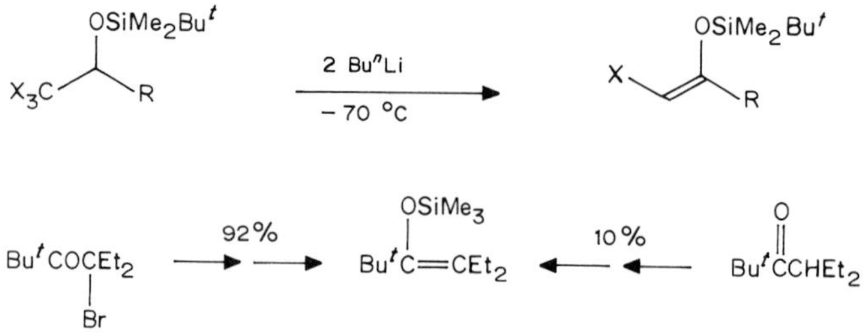

SCHEME 84

Trichloromethyl-substituted silyl ethers[274] can be converted into (Z)-haloenol ethers (Scheme 84). Blocked or highly substituted α-bromoketones give better yields of silyl enol ethers[275] than does simple deprotonation, lda being used in each case.

B. Reactions

The reactivity of silyl enol ethers and ketene acetals is based largely on their synthetic equivalence with enols/enolate ions, although recently some effort has been devoted to revealing a different spectrum of behaviour. Silyl enol ethers are, of course, highly nucleophilic in comparison with simple alkenes. In situations where such enhanced nucleophilicity might be disadvantageous, the use of enol carbonates[276] has been recommended; the nucleophilicity order is silyl enol ether > alkene > enol carbonate, and enol carbonates can still give rise to site-specific enolates. It would appear that, under Lewis acid conditions, silyl enol ethers are more nucleophilic[277] than allyl silanes (a frequently drawn reactivity parallel).

1. Direct reaction with strong electrophiles

A good example of this is reaction with the dithienium cation (Scheme 85); with dienolates, selective γ-attack is normally observed[278].

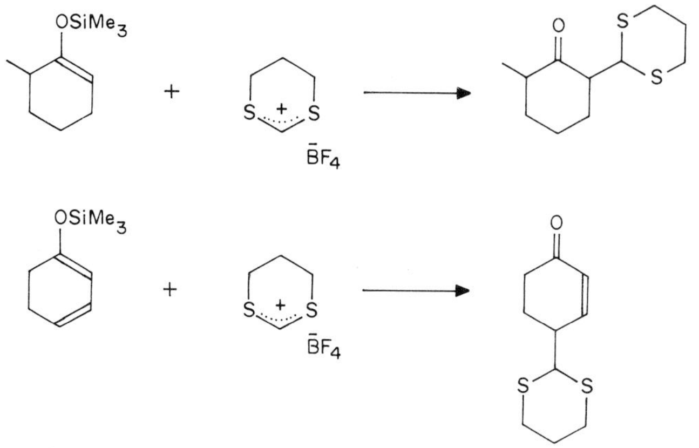

SCHEME 85

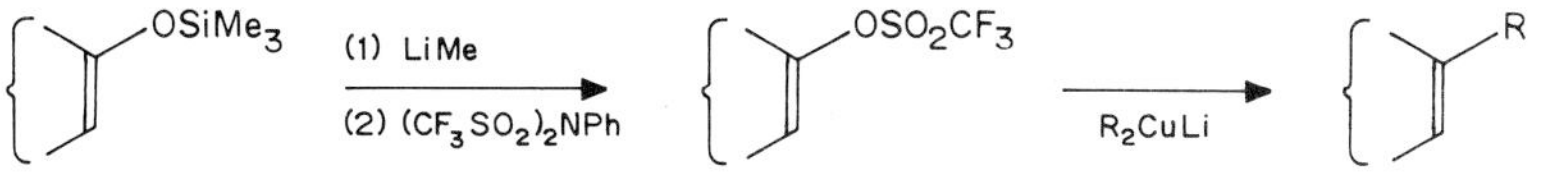

SCHEME 86

2. Enolate generation using MeLi

A useful example of this reaction allows the regioselective generation[279] of enol triflates from silyl enol ethers; the former species have been shown to undergo a regiospecific conversion into the corresponding alkenes (Scheme 86).

3. Formal enolate generation using F^-, increasing the nucleophilicity of the silyl enol ether

Mukaiyama[280] has provided an excellent overview of directed aldol reactions, and Kuwajima *et al.*[281] have given full details of the regiospecific monoalkylation of ketones. More information has been provided by Noyori *et al.*[282] on the use of the commercially available[283] (and expensive!) complex fluoride source **55**, which generates 'naked' enolate ions. These have been found to react with high *syn*-selectivity with aldehydes, regardless of

$$(Et_2N)_3S^+ \qquad Me_3SiF_2^-$$

(55)

the original enolate geometry; an extended acyclic transition state has been proposed. When applied to silyl ketene acetals, the same fluoride source induces[284] 1,4-addition to enones; interestingly, the same overall addition can be achieved in the absence of any catalyst, but only when MeCN or $MeNO_2$[285] is used as a solvent. Similar conjugate addition of ester enolates to acrylate esters has been explored[286] as a method of group-transfer polymerization. The reaction is rapid, gives a controllable narrow molecular weight distribution, and can additionally lead to block copolymers.

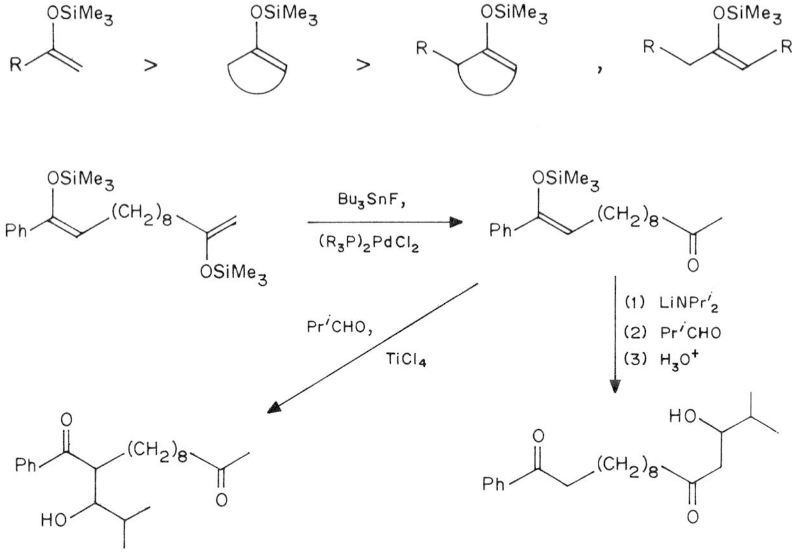

SCHEME 87

Tributyltin fluoride has been found[287] to cleave the silyl group from silyl enol ethers fairly selectively in the presence of a palladium catalyst. The reaction rate is highly dependent on the degree of steric hindrance around the enol double bond, with the relative order of cleavage being as shown in Scheme 87. Applications include selective desilylation in intramolecular cases, and in removal of the less substituted 'kinetic' enol ether from a 'thermodynamically' enriched mixture.

4. Lewis acid-induced reactions, increasing the electrophilicity of the electrophile

The Lewis acid most frequently employed in this context is $TiCl_4$, and its broad applicability in directed aldol reactions using silyl enol ethers has been discussed in Mukaiyama's review[280]. Lewis acid-induced alkylation, using S_N1 electrophiles, has been further advanced and reviewed by Reetz[288] and Fleming[2]. The factors controlling[289] the introduction of the *tert*-butyl group have been studied. Lewis acid-induced reactions of functionalized phenylthio systems such as **56**[290] and **57**[291] have been outlined; the species **56** reacts only with ketone-derived silyl enol ethers, and is γ-regioselective in its reactions with dienolates. An ingenious method for the formal alkylation[292] of esters with alkenes is illustrated in Scheme 88.

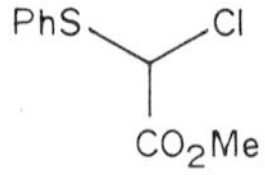

(**56**) 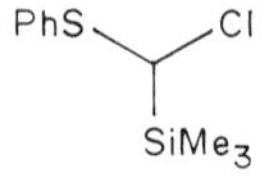 (**57**)

Oxocarbenium ions function efficiently as electrophiles towards silyl enol ethers. This has been employed to good effect by Kocienski and coworkers in their elegant approaches to pederin[293] and milbemycin[294], and can be seen to advantage in a synthesis[295] of

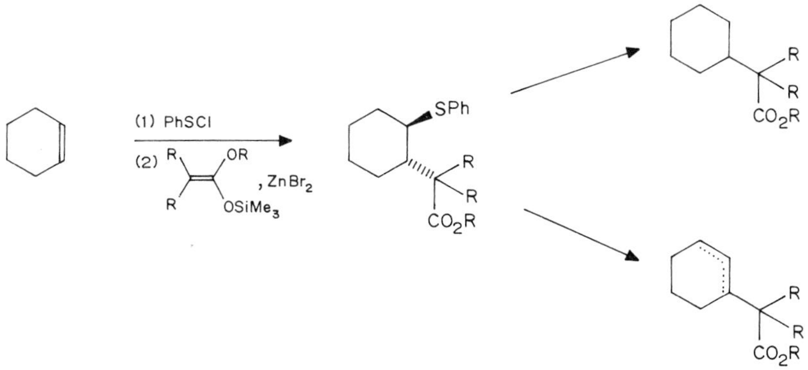

SCHEME 88

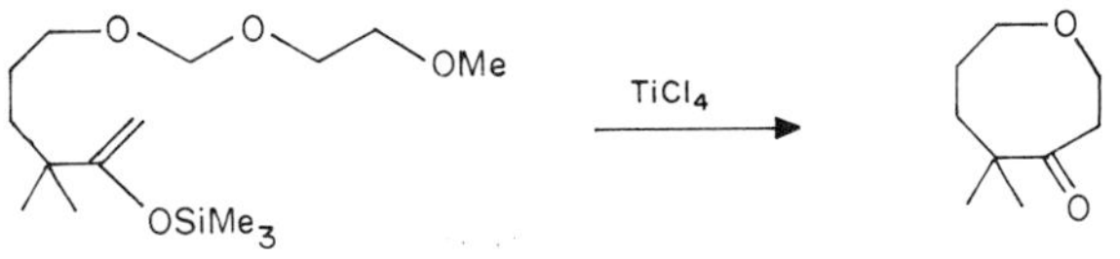

SCHEME 89

medium-sized rings (Scheme 89). Here, the formation of an eight-membered ring without the need for high dilution conditions has been ascribed to a Ti-template effect, with the silyl enol ether and acetal oxygens both coordinated to the metal catalyst in the transition state. When such reactions of silyl enol ethers are applied to sugar chemistry, axial, α-attack is observed to occur at the anomeric centre[296].

Miyashita *et al.*[297] have given full details of the Michael addition of silyl enol ethers and ketene acetals to conjugated nitroalkenes, when 1,4-diketones and γ-keto-esters, respectively, are produced.

5. Acylation

Contrary to general impression, silyl enol ethers do undergo smooth acylation[298] on reaction with simple acid chlorides in the presence of Lewis acids (Scheme 90). The application of chlorinated acid chlorides has been extended to ketene acetals, so providing a rapid route[299] to tetronic acids. Acyl cyanides react[300] with ketone-derived silyl enol ethers to give selectively protected β-diketones: reaction of such cyanides with allylsilanes similarly gives cyanohydrins of β,γ-unsaturated ketones.

6. Reactions with iminium ions: amido- and amino-methylation

The simplest version is the equivalent of the Mannich reaction, and can be performed using the alkoxymethylamine **58** in the catalytic presence of tms iodide or tms triflate as a substitute[301] for Eschenmoser's salt (**59**). Interestingly, this iminium ion reacts with *tert*-butyldimethylsilyl enol ethers to produce amines which retain[302] the silyl enol ether

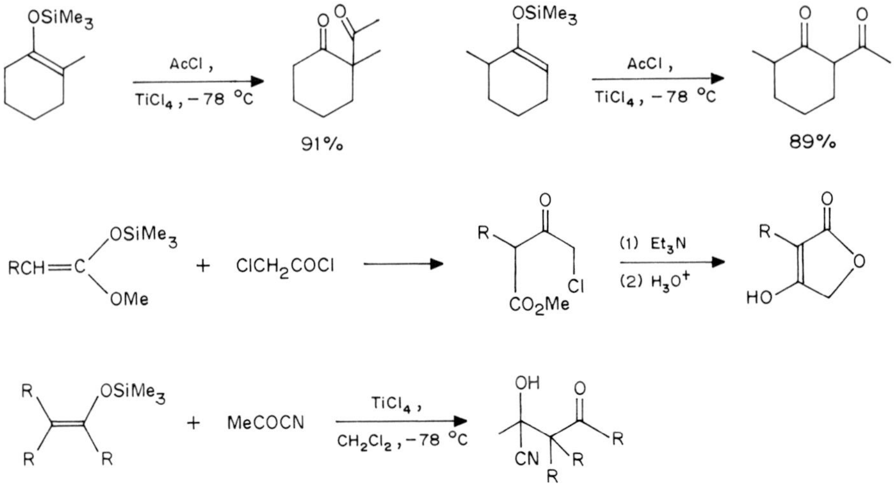

SCHEME 90

moiety (Scheme 91). This has been rationalized by proposing rapid inter- or intra-molecular deprotonation of the intermediate cation by the basic nitrogen atom; similar results[303] have been obtained on bromination in the presence of a base. Both of these observations highlight the different mechanisms[304] of hydrolysis of the *tert*-butyldimethylsilyl enol ethers, where the rate-determining step is protonation of the enol carbon atom, and their trimethylsilyl analogues, where the corresponding step is nucleophilic attack at silicon.

More complex examples of Lewis acid-catalysed reactions of alkoxymethyl-[305] and silyloxymethyl-amines[306] with silyl enol ethers and ketene acetals have been described. One of the major applications of such iminium ion chemistry is seen in the synthesis and synthetic manipulations of β-lactams. Further studies[307] on the Lewis acid-catalysed

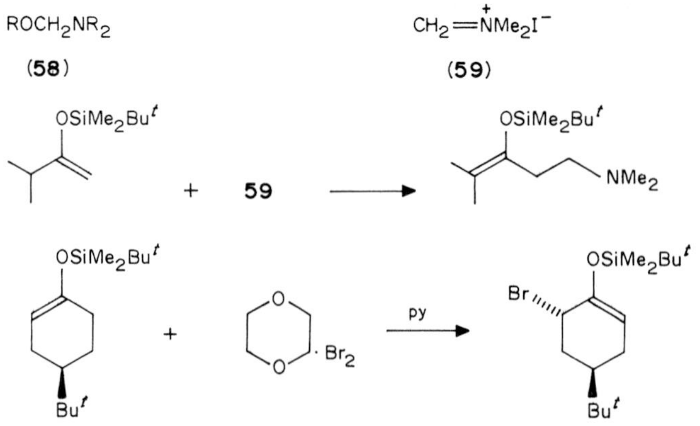

SCHEME 91

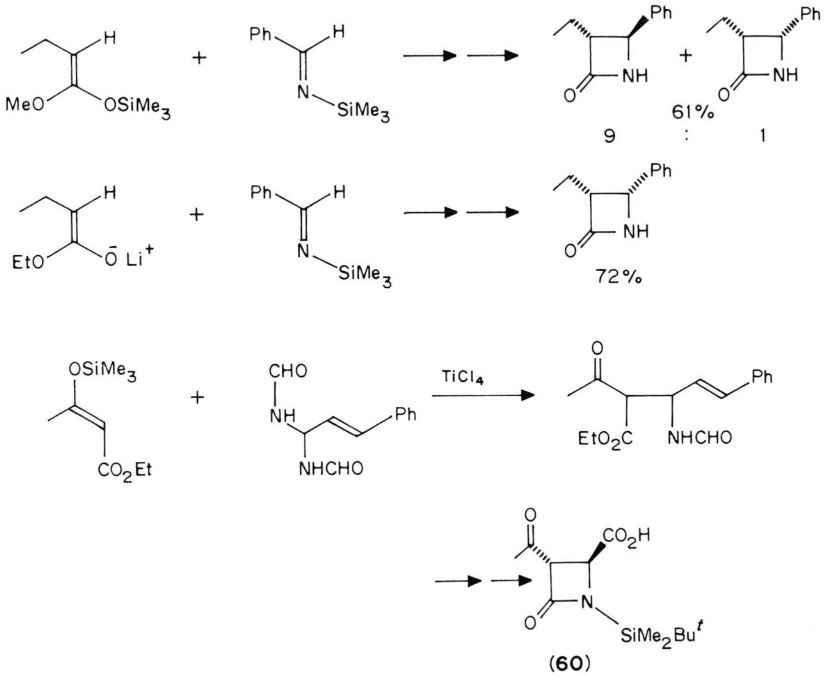

SCHEME 92

reaction of imines with ketene acetals have been described. In a significant advance, *N*-silylimines have been shown to react with silyl ketene acetals[308] in the presence of ZnI_2, or with the related lithium enolates[309], leading to *N*-unsubstituted β-lactams; the former method is *threo*-selective, preferentially producing *trans*-fused β-lactams, whereas the latter selectively yields the cis-fused isomers (Scheme 92). An elegant route[310] to the usefully functionalized azetidinone **60** has been described.

Attrill *et al.*[311] have given full details of the reaction of azetidinium ions with silyl enol ethers (Scheme 93). Several extensions[44,312] of this sequence have been reported, and various *N*-heterocycles[313] have been employed as precursors to iminium ions for similar trapping. The intermediates involved in the Beckman rearrangement of cyclic oxime sulphonates can be trapped[314] with silyl enol ethers, producing vinylogous amides (Scheme 94).

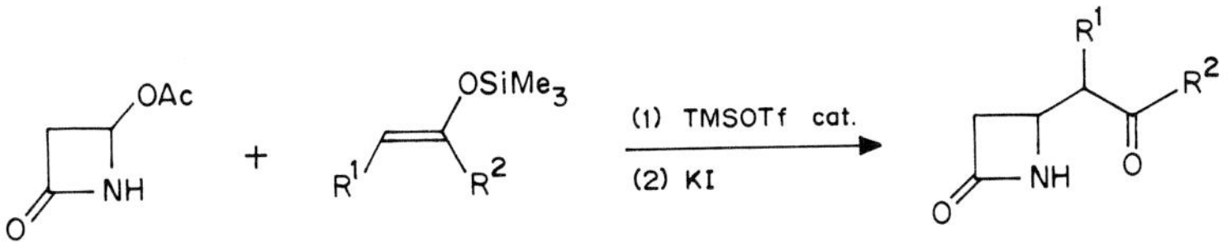

SCHEME 93

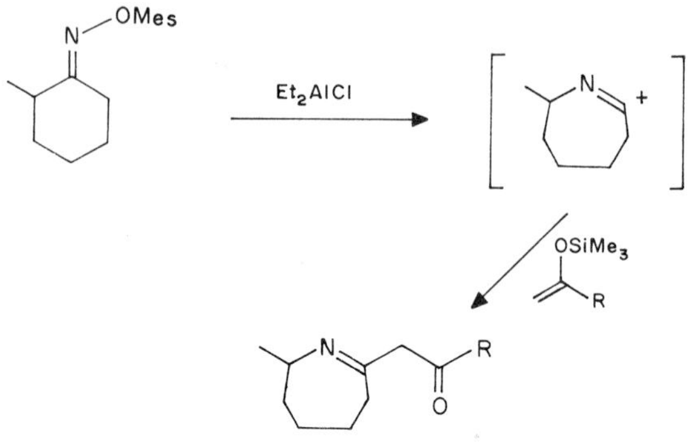

SCHEME 94

7. Diastereoselective aldol reactions

This area of silyl enol ether and ketene acetal chemistry has been discussed in an excellent review[315]. Silyl enol ethers are not sufficiently nucleophilic to react spontaneously with aldehydes; they do so, however, under the influence of either Lewis acids or fluoride ion, as mentioned earlier. The number of definitive studies in this area is limited, with few clear trends emerging, a situation not helped by certain ambiguities in diastereoisomeric assignment. The stereochemical outcome naturally depends on a number of factors, including the involvement of cyclic or extended transition state geometries, enol or ketene acetal structure and geometry, aldehyde (acetal) structure, especially if chiral, and kinetic *vs.* thermodynamic control. Cases involving the interplay of these factors have been reported[316], and it is to these primary sources that the reader is referred for further information.

8. Oxidation and reduction

Larson and Prieto[317] have extended their studies on the hydroboration of cyclic silyl enol ethers, and Rubottom *et al.*[318] have published full details on the oxidation of

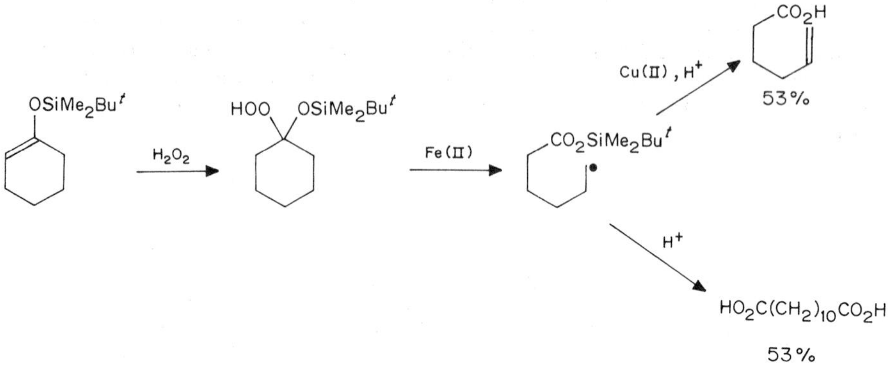

SCHEME 95

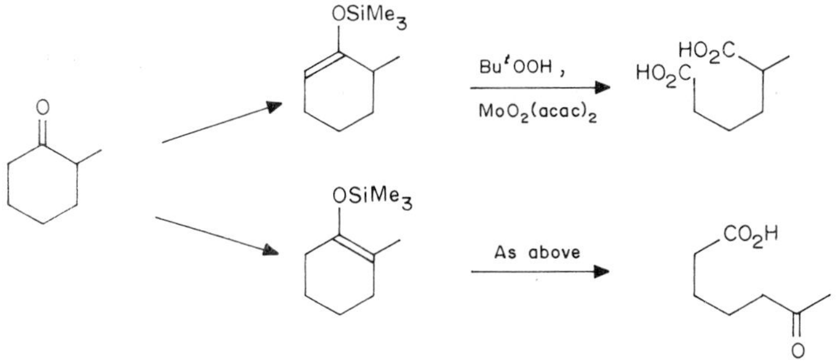

SCHEME 96

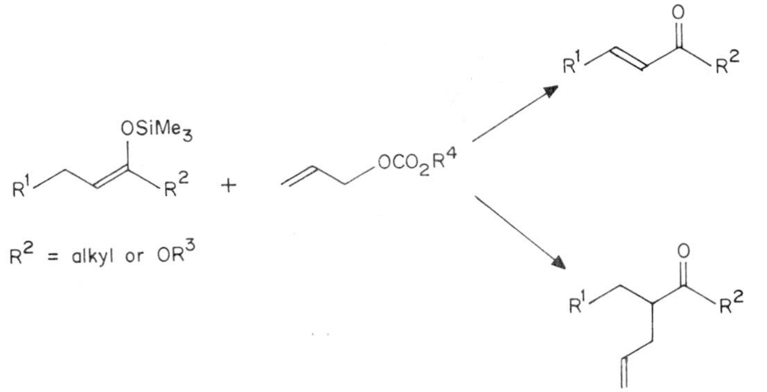

SCHEME 97

aldehyde-derived substrates with lead (IV) acetate, when α-acetoxyaldehydes are obtained as major products. The use of chromyl chloride, CrO_2Cl_2, has been recommended[319] for the regioselective production of α-hydroxyketones.

Cyclic silyl enol ethers react with H_2O_2 to produce hydroperoxy hemiacetals[320]. These in turn can be transformed[321], with ring scission, into ω-olefinic acids or dimeric diacids, generally in good yield; no unsymmetrical cases were described (Scheme 95). Similarly, cyclic silyl enol ethers undergo regioselective oxidative cleavage[322] on treatment with Bu^{t}OOH and a molybdenum catalyst (Scheme 96); simple double bonds are unaffected.

Silyl enol ethers and ketene acetals react with allyl alcohol carbonates under palladium catalysis[323] to give products of either allylation or dehydrogenation (Scheme 97), depending on the particular catalyst used.

9. Cycloaddition

The following reactions are classified either by the outcome of the cycloaddition process or by the substrate structure. No inference regarding concertedness or non-concertedness is intended.

a. [2 + 1]-Cycloaddition.

Cyclopropanone alkyl trimethylsilyl ketals can act as precursors to homoenolate ions, as exemplified by the isolation and characterization of the titanium[324] and zinc[325] homoenolates **61** and **62**. The zinc species **62** undergoes a copper(I) catalysed conjugate addition to unsaturated systems (Scheme 98) under carefully controlled conditions; in

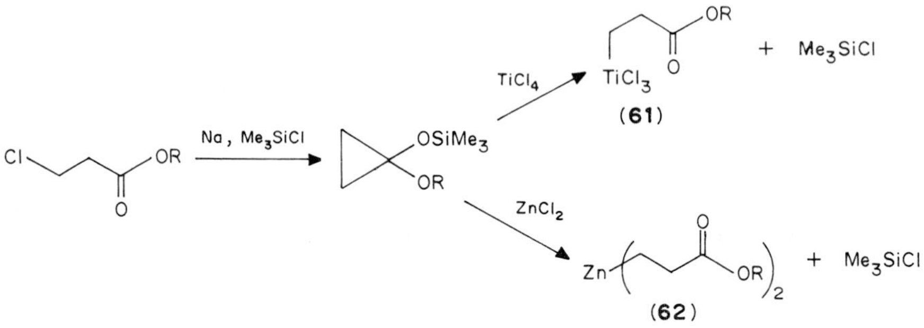

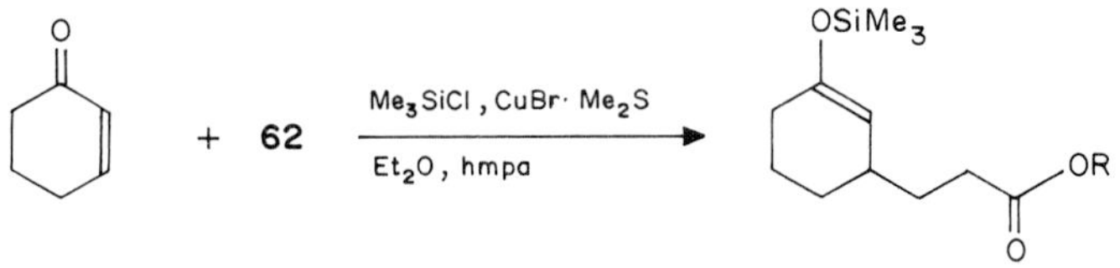

SCHEME 98

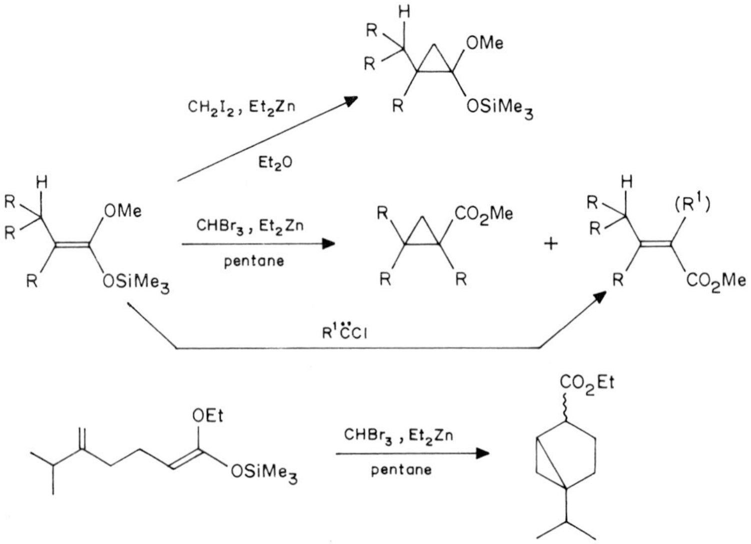

SCHEME 99

particular, the simultaneous presence of Me_3SiCl is mandatory, probably to activate the unsaturated system.

As shown, the precursor ketals were generated by reductive silylation. An alternative method[326] of wider potential application is cyclopropanation of ketene silyl acetals (Scheme 99). Using bromoform/Et_2Zn, mainly cyclopropropyl esters[327] were formed, accompanied by smaller amounts of α,β-unsaturated esters, which in turn are the sole products[328] when chlorocarbenes are used. The intermediacy of carbenoids in the formation of such cyclopropyl esters has been demonstrated by intramolecular trapping.

b. [2 + 2]-Cycloadditions.

In an interesting case of regioselectivity, silyl dienol ethers have been reported[329] to react with dichloroketene at the remote double bond (Scheme 100). Silyl ketene acetals, on the other hand, yield acyclic products[330]; other ketene acetals react differently. In an extended version of the de Mayo reaction, certain silyl enol ethers take part in a

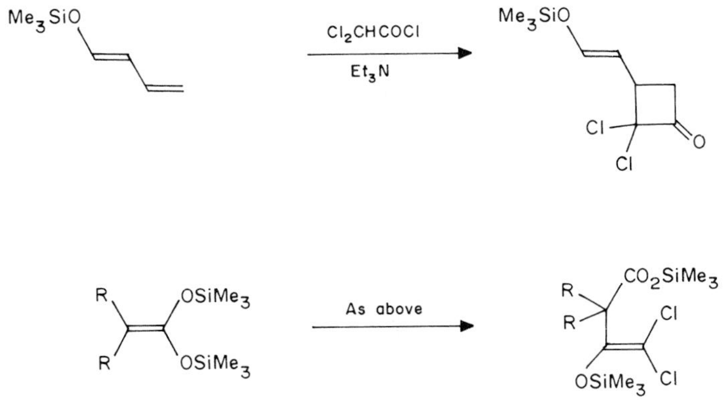

SCHEME 100

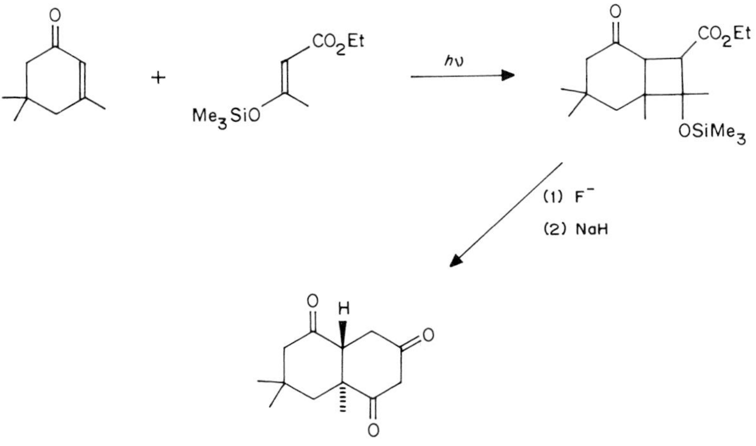

SCHEME 101

photochemical [2 + 2]-addition[331] with cyclohexenones, providing an efficient route to fused cyclohexane-1,3-diones (Scheme 101).

c. [3 + 2]-Cycloaddition.

Whereas nitrones undergo dipolar cycloaddition with dialkyl ketene acetals, with silyl ketene acetals[332] acyclic products are formed; a mechanism involving initial O ⟶ O silyl transfer has been proposed (Scheme 102). Silyl nitronates, hetero-analogues of ketene acetals, add to a range of alkenes[333], and have additionally been employed in the trapping of thioaldehydes[334] (Scheme 103).

d. [4 + 2]-Cycloadditions and silyloxybutadienes.

The most frequently encountered mode of reaction of silyloxybutadienes is Diels–Alder cycloaddition. Danishefsky[335] has reviewed his pioneering work in utilizing such reactions in natural product syntheses, and has additionally described preparations of the

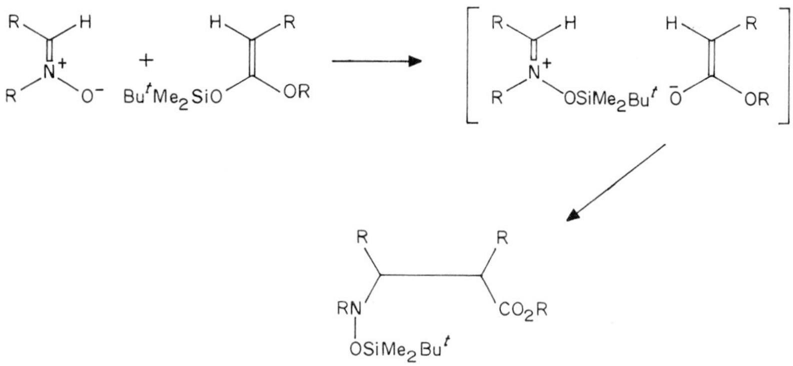

SCHEME 102

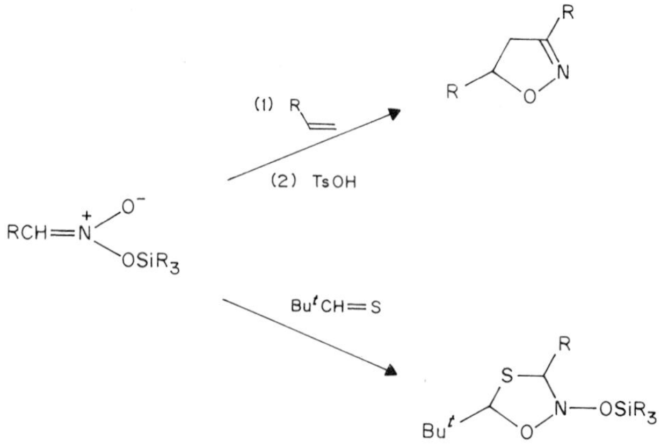

SCHEME 103

dienes **63**[336] and **64**[337]. Other new dienes whose preparation and use in this context have been described include **65**[338], **66**[339], **67**[45,340], and **68**[341]; a particularly simple preparation of the crotonate derivative **69** has been reported[342]. Diene **70** has been employed in routes to fomannosin[343] and illudol[344], **71** in an alternative route[345] to sodium prephenate, and **72** and **73** in anthracycline synthesis[346]. Intramolecular reactions of the dienes **74**[347] and **75**[348] have been reported, and nickel(0)-catalysed dimerization of the diene **76** leads[349] efficiently to the cyclooctadiene **77**; this last process is, of course, a formal [4 + 4]-cycloaddition.

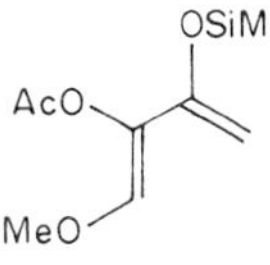

(63) 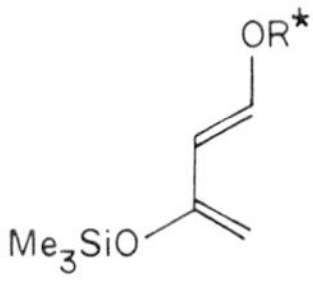 chiral R* **(64)** **(65)** 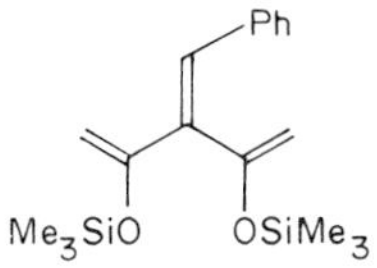 **(66)**

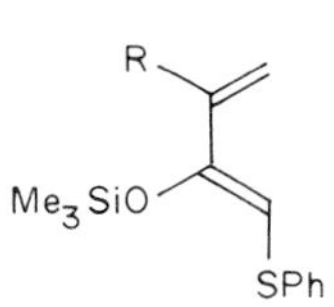

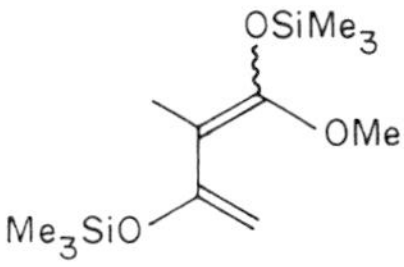

(67) 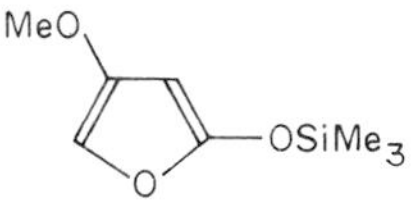 **(68)** 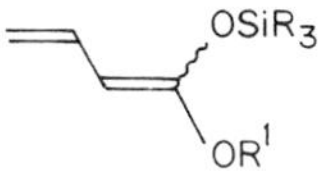 R_3 = Me_3, Bu^tMe_2 **(69)**

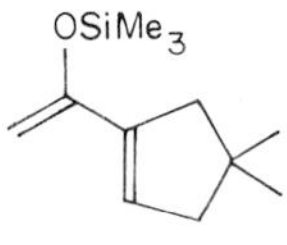

(70) 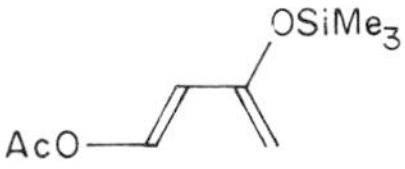 **(71)** 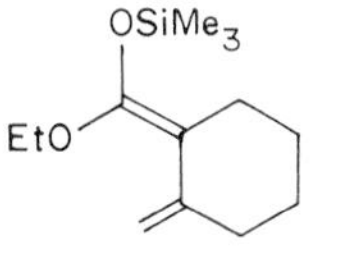 **(72)**

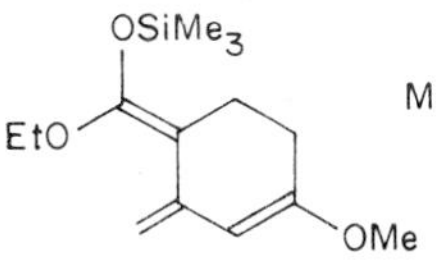

(73) 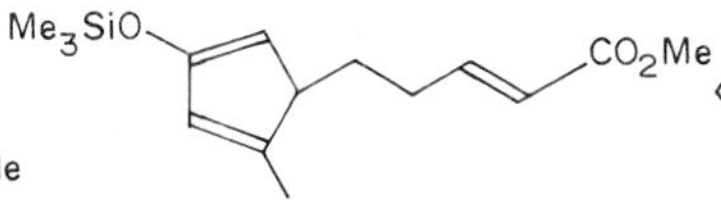 **(74)** 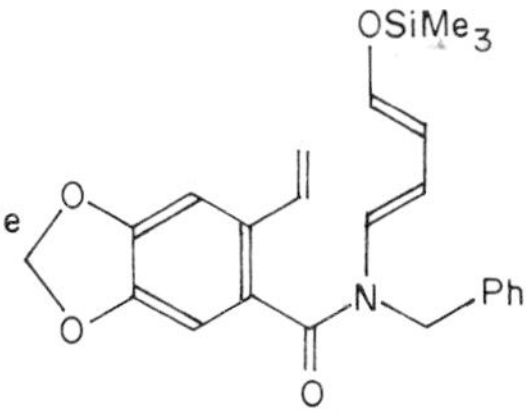 **(75)**

(76) 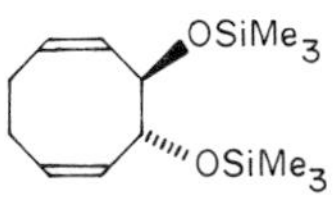 **(77)**

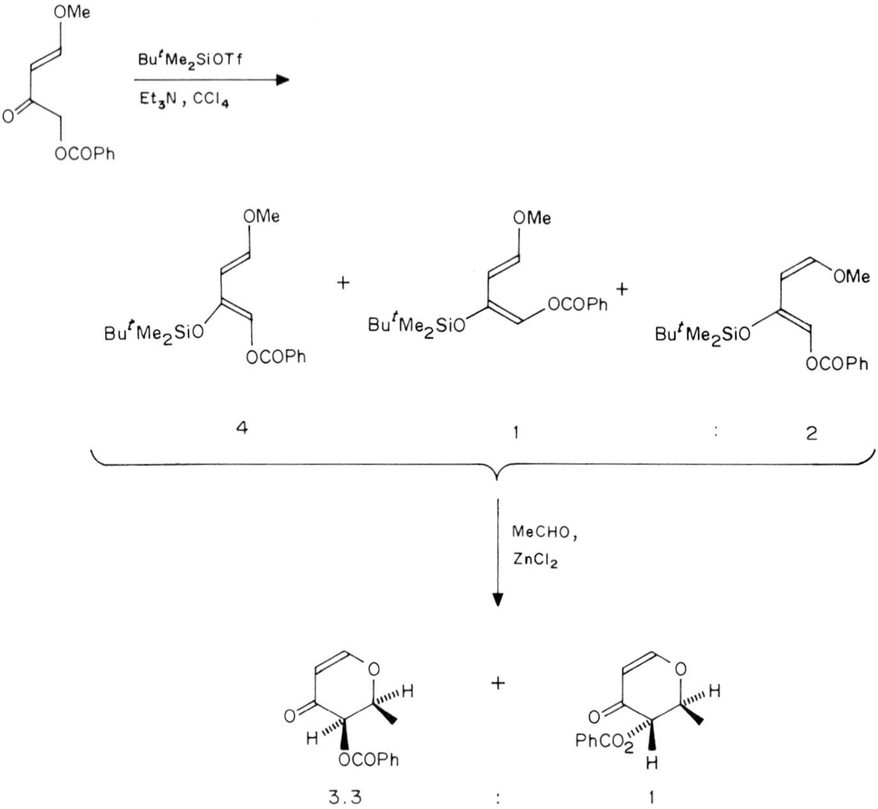

SCHEME 104

The nitro group functions as the stronger director[350] in cycloaddition reactions of cyclic β-nitroenones; with unsymmetrical dienes this gives the opposite substitution pattern in the product to that of enones themselves. Danishefsky and coworkers have also investigated extensively the reactions of oxygenated butadienes with heterodienophiles, particularly aldehydes. Here, Lewis acid catalysis is required, and substantial asymmetric induction can be achieved by using either an α-chiral aldehyde[351] or a chiral rare earth catalyst[352]. A typical example of such methodology can be seen in Scheme 104; further transformation[353] of the major adduct led to syntheses of ($\pm$)-fucose and ($\pm$)-daunosamine derivatives.

2,4-Bis(silyloxy)butadienes undergo a [4 + 2] non-concerted cycloaddition[355] reaction with ketenes, ultimately leading to 4-pyrones. Other reactions of bis(silyloxy)butadienes are discussed in Section XIII.B.11. Cycloaddition reactions of such oxygenated butadienes with singlet oxygen[356] have been investigated, as have further transformations of the products.

10. Sigmatropic rearrangements

Ireland's modification of the Claisen rearrangement of allyl alcohol esters has found further application, particularly in carbocycle synthesis, as illustrated by the two

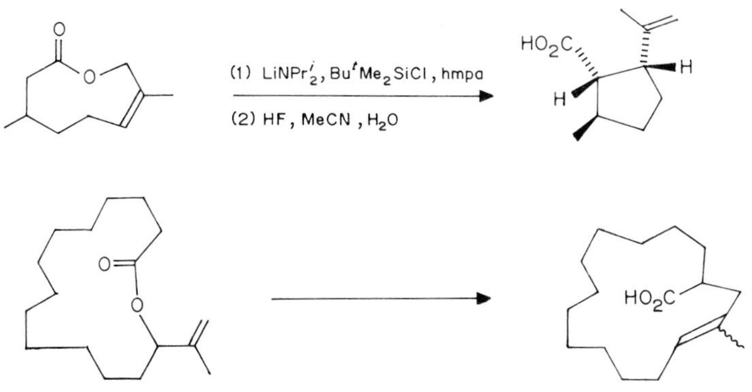
(1) LiNPr$^i_2$, Bu^{t}Me$_2$SiCl, hmpa
(2) HF, MeCN, H$_2$O
HO$_2$C
H
H
HO$_2$C

SCHEME 105

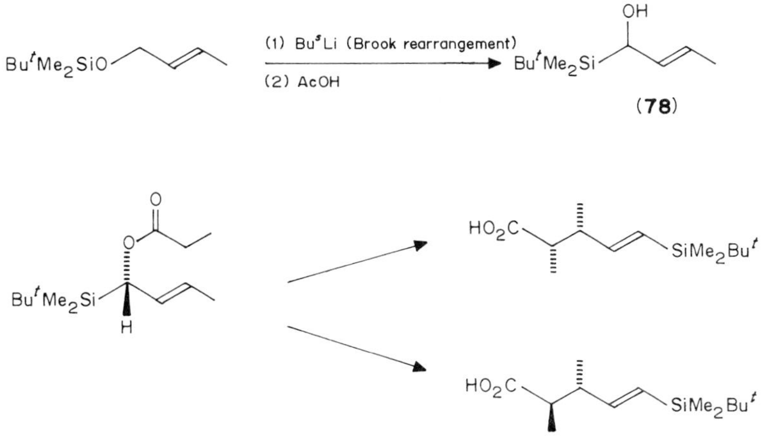
Bu^{t}Me$_2$SiO
(1) Bu^{s}Li (Brook rearrangement)
(2) AcOH
Bu^{t}Me$_2$Si
OH
(78)
Bu^{t}Me$_2$Si
H
HO$_2$C
SiMe$_2$Bu^t
HO$_2$C
SiMe$_2$Bu^t

SCHEME 106

examples[357,358] in Scheme 105. One of the many advantages of this reaction is the ability to control product stereochemistry by controlling the enolate/ketene acetal geometry[262]. Ireland and Varney[18] combined this with the (resolvable) chiral primary allyl alcohol equivalent **78** to achieve silyl-assisted asymmetric induction (Scheme 106).

11. Silyl dienol ethers and bis(silylenol)ethers

Chan and coworkers have provided full details of the preparation of the dienes **79, 80**, and **81** and their use in the synthesis[359] of substituted phenols, and of **82** in the synthesis[360] of cyclopropanes and cyclobutanes. The furan analogues **83**[361] and **68** both react with electrophiles showing the expected high γ-regioselectivity. The symmetrical pyrrole **84** and thiophene **85** analogues have been prepared and their properties explored[362].

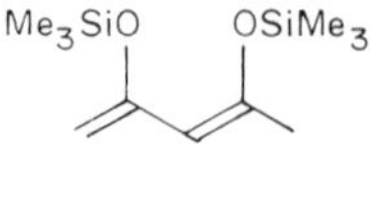

(79) 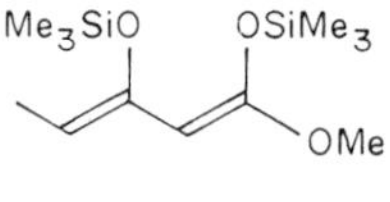 (80) 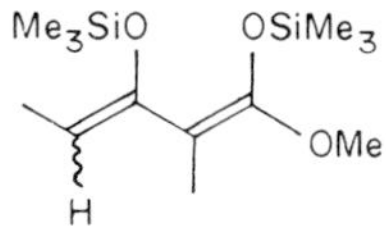 (81)

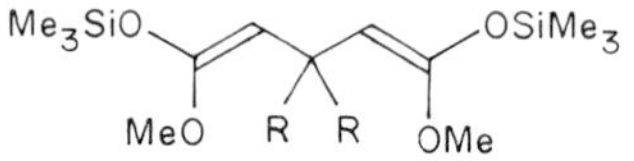

(82) 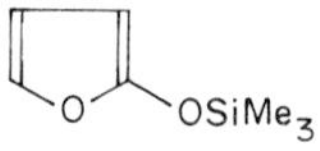 (83)

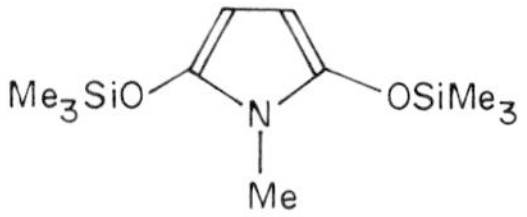

(84) 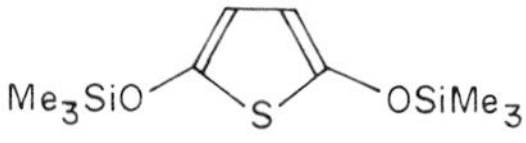 (85)

XIV. SILYL-BASED REAGENTS

Replacement of the proton of certain inorganic acids by a trialkylsilyl group gives a new family of reagents which behave as potent electrophiles, particularly when Si—O bond formation occurs. When the anion of the parent acid is also a good nucleophile, it can attack the cationic species generated in this manner or differently, resulting in overall processes such as ether and ester cleavage, addition to aldehydes and ketones, or replacement of heteroatom functional groups; if not a good nucleophile, then electrophile-induced rearrangements, etc., can occur. The preparation and properties of several of these reagents have been the subjects of recent reviews[363].

A. Preparation

Two routes for the *in situ* preparation of tms triflate, by reaction of triflic acid either with allyl trimethylsilane[364] or, very conveniently, with tetramethylsilane[365], have been described (Scheme 107). Other silyl triflates are discussed in Section XII. Methyltrichlorosilane has been suggested as an alternative to tms chloride for the *in situ* generation of the corresponding iodide[366] or cyanide[367].

B. Reactions

1. C—O Bond cleavage

Care must be exercized in the use of tms iodide for the cleavage of methyl ethers of polyfunctional substrates; adventitious HI or tms iodide itself can initiate further

$CH_2{=}CHCH_2SiMe_3 \xrightarrow{TfOH} Me_3SiOTf \xleftarrow{TfOH} Me_4Si$

SCHEME 107

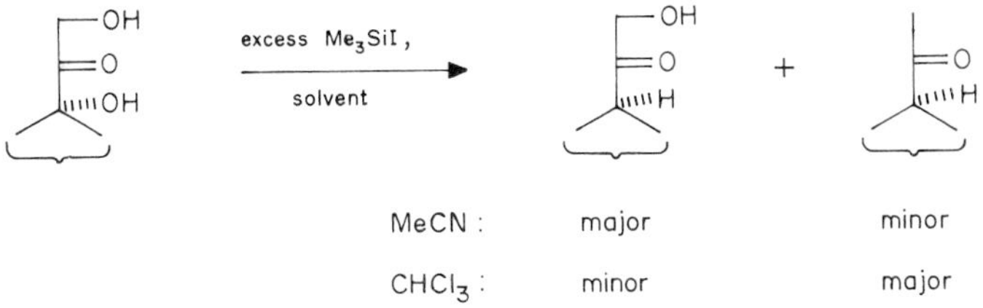
OH
O
OH
excess Me_3SiI,
solvent
OH
O
H
+
O
H
MeCN : major minor
$CHCl_3$: minor major

SCHEME 108

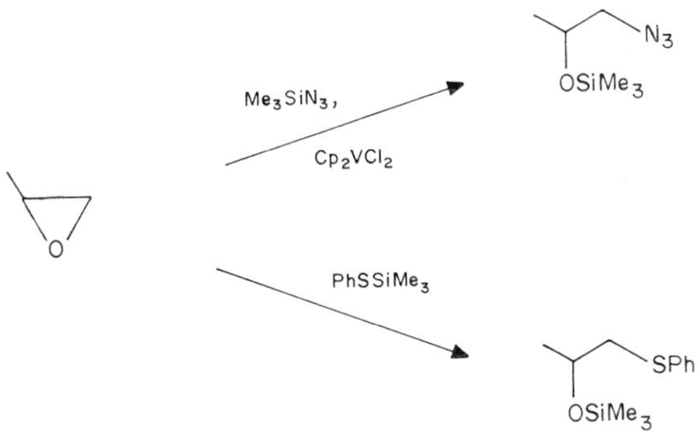
Me_3SiN_3,
Cp_2VCl_2
N_3
$OSiMe_3$
O
$PhSSiMe_3$
SPh
$OSiMe_3$

SCHEME 109

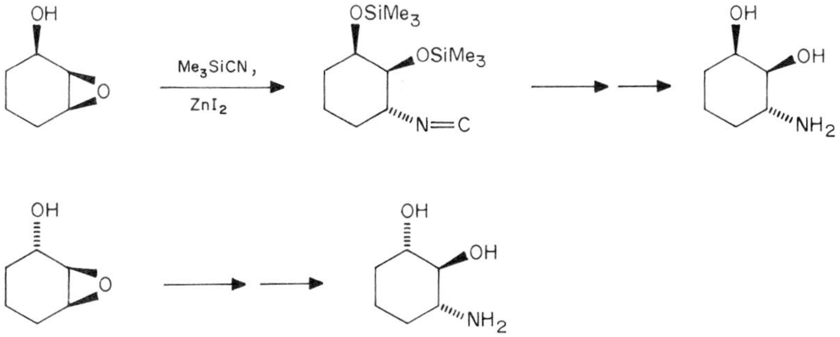
OH
O
Me_3SiCN,
ZnI_2
$OSiMe_3$
$OSiMe_3$
N=C
OH
OH
NH_2
OH
O
OH
OH
NH_2

SCHEME 110

cationic processes[368]. The combination of tms chloride and acetic anhydride has been recommended[369] for the cleavage of methyl and methylthiomethyl ethers: the corresponding acetates are produced with inversion of configuration. Tms iodide reductively deoxygenates[370] the dihydroxyacetone side-chain of corticosteroids in a process whose regioselectivity is solvent dependent (Scheme 108).

A similar critical choice of solvent system is involved in the conversion[371] of a wide range of alcohols into the corresponding nitriles, using *in situ* generated tms cyanide; in appropriate cases, inversion of stereochemistry is observed. The combination of tms iodide and hexamethyldisilazane transforms[372] dimethylketals into methyl enol ethers, by effective elimination of methanol.

Turning to ring-opening reactions of oxiranes, some controversy has arisen[373] over the regiochemistry of opening of terminal substrates using tms iodide. Both tms azide[374] and tms thiophenoxide[375] open oxiranes (Scheme 109) under Lewis acid-catalysed or thermal conditions, respectively. Tms triflate has been reported[376] to initiate a cation-induced cyclization of humulene 6, 7-oxide.

Interestingly, tms cyanide reacts with cyclic epoxy alcohols under Lewis acid catalysis to produce isonitriles, in a stereoselective route[377] to cyclohexane-1, 2, 3-aminodiols (Scheme 110).

2. Carbonyl addition processes

Full experimental details have been provided for the preparation of cyanohydrin silyl ethers[378] of aryl aldehydes (Scheme 111); with α-substituted ketones as substrates, KCN/18-crown-6 has been recommended[379] as a superior catalyst to ZnI_2.

Stabilized anions from aldehyde-derived cyanohydrin ethers undergo electrophilic

$$\text{ArCHO} \xrightarrow[\text{Zn(CN)}_2\text{, MeCN}]{\text{Me}_3\text{SiCl, KCN,}} \text{ArCH(OSiMe}_3\text{)CN}$$

SCHEME 111

$$\text{ArC}^-(\text{OSiMe}_3)(\text{CN}) \xrightarrow[\text{(2) H}_3\text{O}^+]{\text{(1) Ph}_2\text{P(O)ONMe}_2} \text{ArC(O)NMe}_2$$

SCHEME 112

$$\text{R}_2\text{C(OSiMe}_3\text{)CN} \xrightarrow[\text{(2) H}_3\text{O}^+]{\text{(1) R}'\text{MgX}} \text{R}_2\text{C(OH)C(O)R}'$$

$$\text{R}_2\text{C(OSiMe}_3\text{)CN} \xrightarrow[\text{(2) H}_3\text{O}^+]{\text{(1) 2 R}^1\text{Li}} \text{R}_2\text{C(OH)C(NH}_2\text{)R}^1_2$$

SCHEME 113

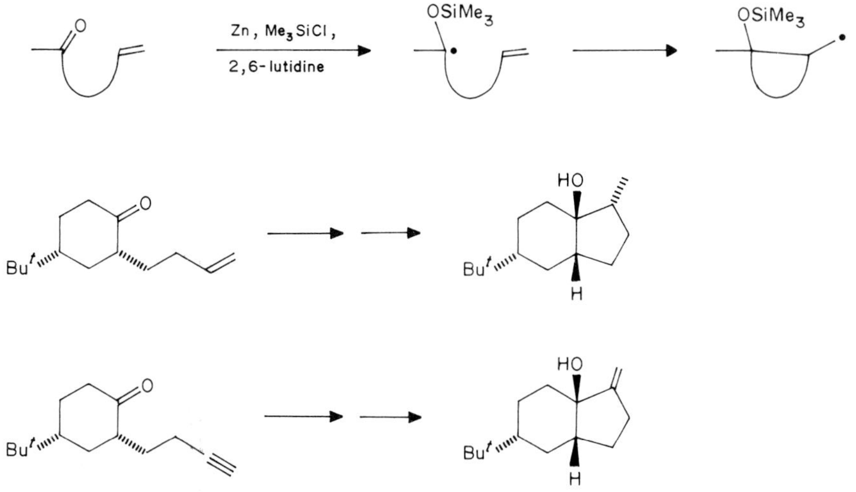

Zn, Me_3SiCl, 2,6-lutidine

$OSiMe_3$

SCHEME 114

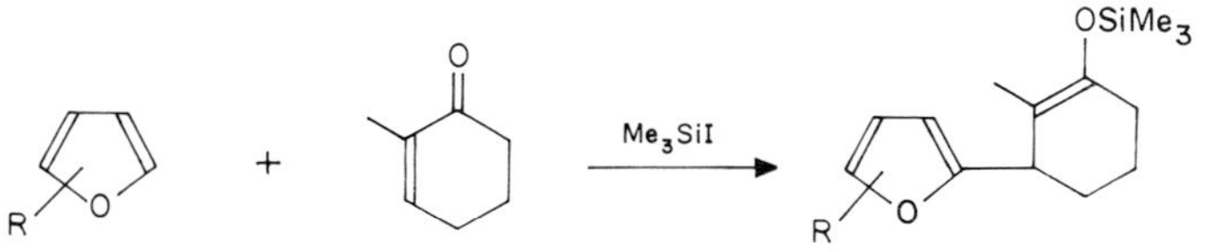

Me_3SiI

$OSiMe_3$

SCHEME 115

(1) Me_3SiCN, 0.1 equiv. $BF \cdot Et_2O$ (2) H_3O^+

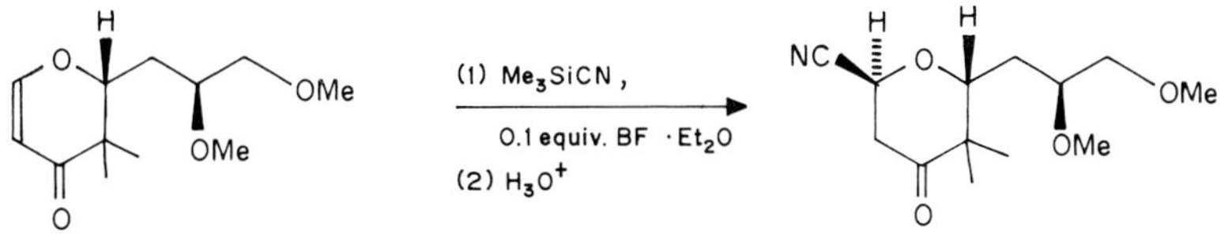

SCHEME 116

C-amination[380], with resulting overall oxidative conversion into the corresponding amides (Scheme 112).

Ketone-derived cyanohydrin ethers react with Grignard reagents[381] to produce acyloins in good yield; alternative use of alkyllithium reagents[382] leads to 1,2-amino alcohols (Scheme 113).

Extending the earlier work of Motherwell on the reductive activation of ketones, Corey and Pyne[383] have described a new method for the reductive cyclisation of ω-unsaturated ketones (Scheme 114); an intermediate silyloxy radical has been implicated.

Tms iodide induces the conjugate addition of furans to enones[384] whose substitution pattern resists normal acid catalysis (Scheme 115); similar conjugate addition[385] of silyl enol and dienol ethers has been reported. This reagent has also found use in a route[386] to

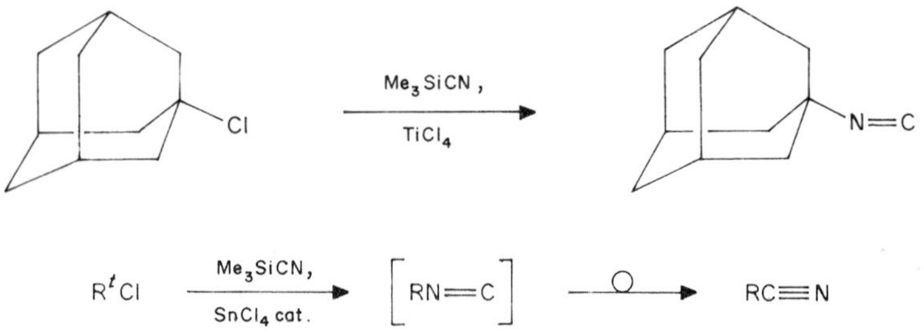

SCHEME 117

quassinoids. In all cases, the active intermediacy of a 3-iodosilyl enol ether (the product of conjugate addition of the reagent) has been involved.

Full details[387] of the Lewis acid-catalysed addition of tms cyanide to enones have been published; depending on the reaction conditions and the substrate structure, 1,2- or 1,4-addition can be favoured, as illustrated by the example[388] shown in Scheme 116.

3. Heteroatom exchange

S_N1-reactive halides have been found to give isonitriles[389] as initial products on treatment with tms cyanide in the presence of Lewis acids; in most cases, these rearrange spontaneously[390] to the more stable nitriles (Scheme 117). Tms azide reacts with such substrates to give the corresponding alkyl azides[391]; with aroyl chlorides as substrates,

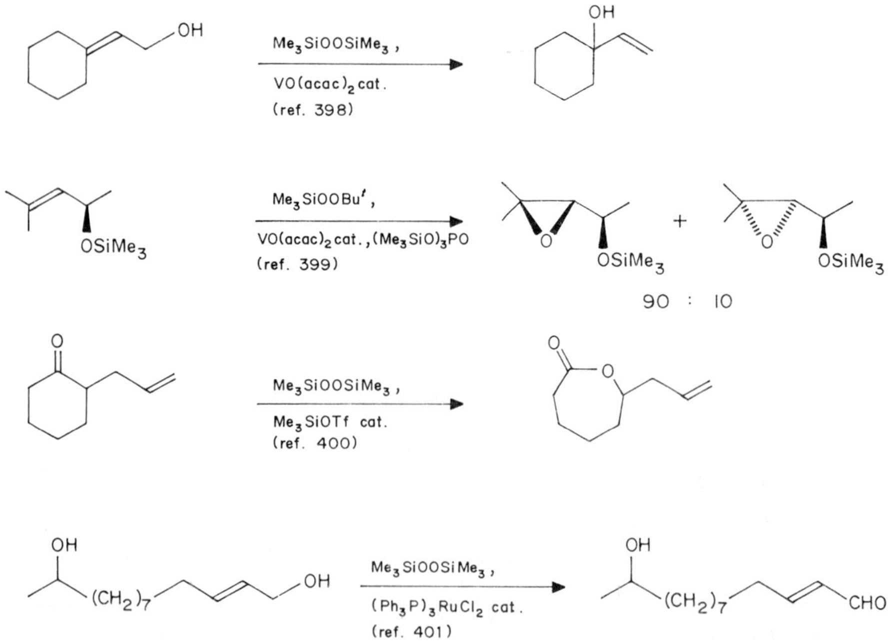

SCHEME 118

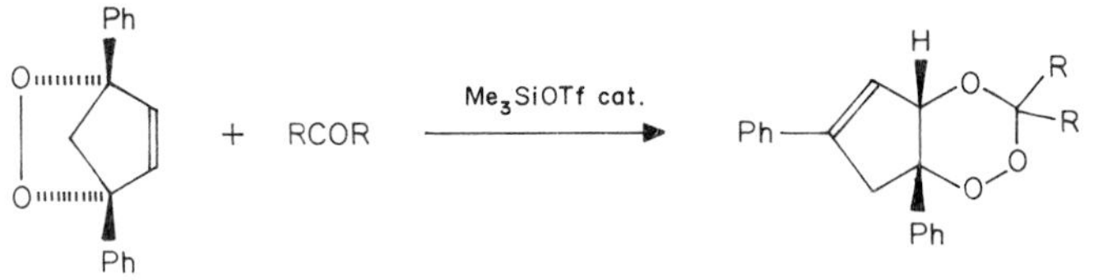

$$\text{(endo-peroxide, Ph-substituted)} + \text{RCOR} \xrightarrow{Me_3SiOTf\ cat.} \text{(1,2,4-trioxane)}$$

SCHEME 119

aroyl azides[392] are produced. A variety of oxocarbenium ions[387,393] react with TMS reagents in a similar manner. For example, tms bromide converts anomeric sugar acetates into the corresponding bromides[394] with inversion of configuration, whereas tms cyanide reacts analogously but with retention[395]. Tms iodide has been reported[396] to convert primary, secondary, and tertiary nitroalkanes into nitriles, oximes, and iodides, respectively.

4. Oxidizing agents

A short review[397] of the chemistry of silyl hydroperoxides has been published. Bis(tms) peroxide continues to find useful applications. With the help of vanadium catalysis, it mediates[398] the remarkable transformation of primary allylic alcohols into their tertiary isomers, without any accompanying double bond epoxidation; under similar conditions, tms *tert*-butylperoxide[399] epoxidizes allyl and homoallyl alcohol tms ethers, with mainly *syn*-stereoselectivity. Smooth Bayer–Villiger oxidation can be achieved[400] with tms triflate as catalyst, double bonds once again being left intact. Using a ruthenium catalyst, primary alcohols, especially allylic alcohols, are oxidized selectively[401] in the presence of secondary alcohols. Examples of these processes are shown in Scheme 118.

Tms triflate also catalyses a reaction between peroxides, including *endo*-peroxides[402], and carbonyl compounds, to produce 1, 2, 4-trioxanes (Scheme 119). This heterocyclic ring system is present in the antimalarial qinghaosu[84]. The preparation and oxidizing properties[403] of the dangerous chlorochromate **86** have been described.

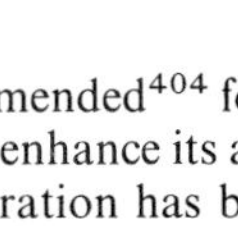

$$Me_3SiOCr(=O)_2Cl$$

(86)

5. Miscellaneous

Tms diazomethane has been recommended[404] for the esterification of carboxylic acids and the *O*-methylation of phenols. To enhance its advantages further over diazomethane, an improved method[405] for its preparation has been described (Scheme 120). Cycloaddition reactions with nitriles[406], and its involvement in a new route[407] to vinylsilanes, have also been reported. The preparation[408] and some reactions[409,410] of tms methyl azide have been described (Scheme 121).

$$Me_3SiCH_2MgCl + (PhO)_2P(O)N_3 \longrightarrow Me_3SiCHN_2$$

SCHEME 120

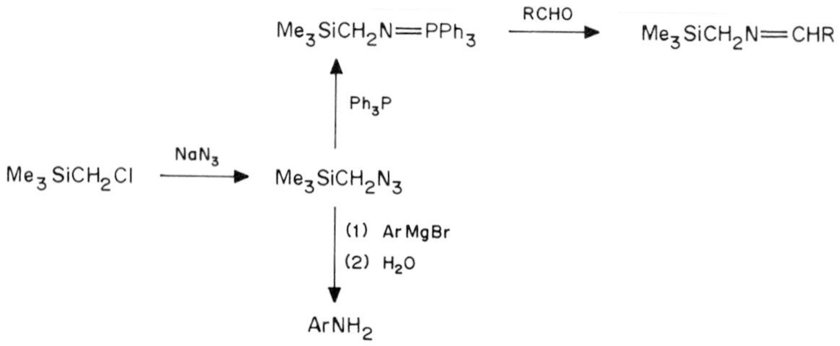

$$Me_3SiCH_2Cl \xrightarrow{NaN_3} Me_3SiCH_2N_3 \xrightarrow{Ph_3P} Me_3SiCH_2N{=}PPh_3 \xrightarrow{RCHO} Me_3SiCH_2N{=}CHR$$

$$Me_3SiCH_2N_3 \xrightarrow{(1)\ ArMgBr;\ (2)\ H_2O} ArNH_2$$

SCHEME 121

The combination of tms chloride and $AgBF_4$ has been suggested[411] for use in solvolysis reactions; the active species thought to be involved has been disputed[412]. An improved method[413] for the synthesis of methyl ketones from carboxylic acids has been described, using tms chloride to quench any excess of methyllithium prior to work-up. Certain silyl-based reagents show considerable regioselectivity[414] in the ring-opening reactions of cyclopropyl ketones and related species. Both tms iodide[415] and tms polyphosphate[416] have been recommended for use in Beckman rearrangement reactions.

XV. AMINOSILANES AND RELATED COMPOUNDS

The commercially available dichlorodisilane **87** converts primary amines into 'stabase' adducts (**88**) (Scheme 122). Such species are stable[417] to alkyllithiums, lithium dialkylamides, pyridinium dichromate, and, surprisingly, aqueous KF; regeneration is achieved using aqueous acid or alkali, or pyridinium chlorochromate. Reversing this concept, the siladiamine **89** has been reported[418] to react with diacid chlorides to give macrocyclic tetraamides; silicon apparently acts as a template, with high dilution conditions being unnecessary (Scheme 123).

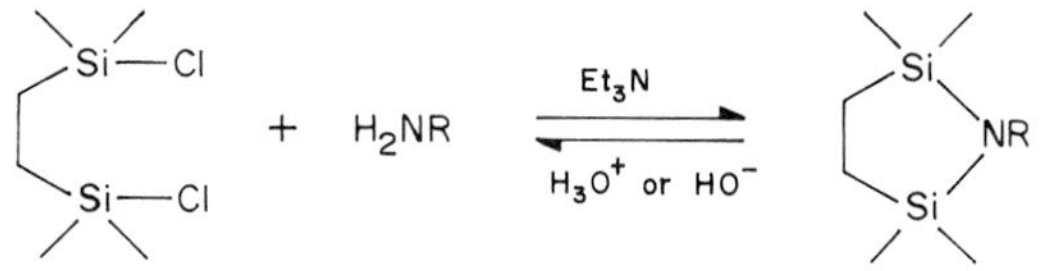

(87) + H_2NR ⇌ (88) [forward: Et_3N; reverse: H_3O^+ or HO^-]

SCHEME 122

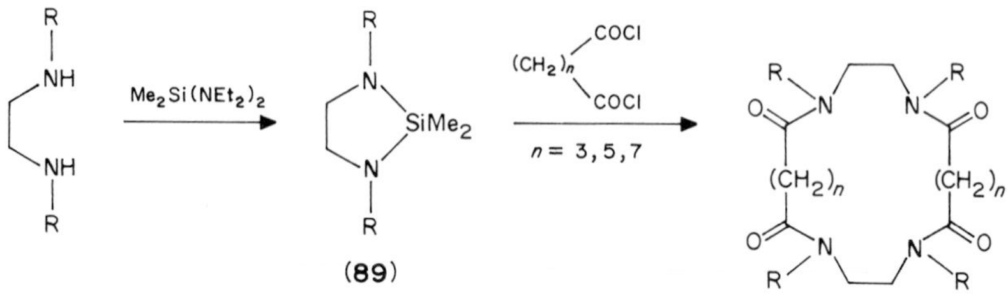

RNHCH₂CH₂NHR $\xrightarrow{Me_2Si(NEt_2)_2}$ (89) $\xrightarrow{(CH_2)_n(COCl)_2,\ n = 3,5,7}$ macrocyclic tetraamide

SCHEME 123

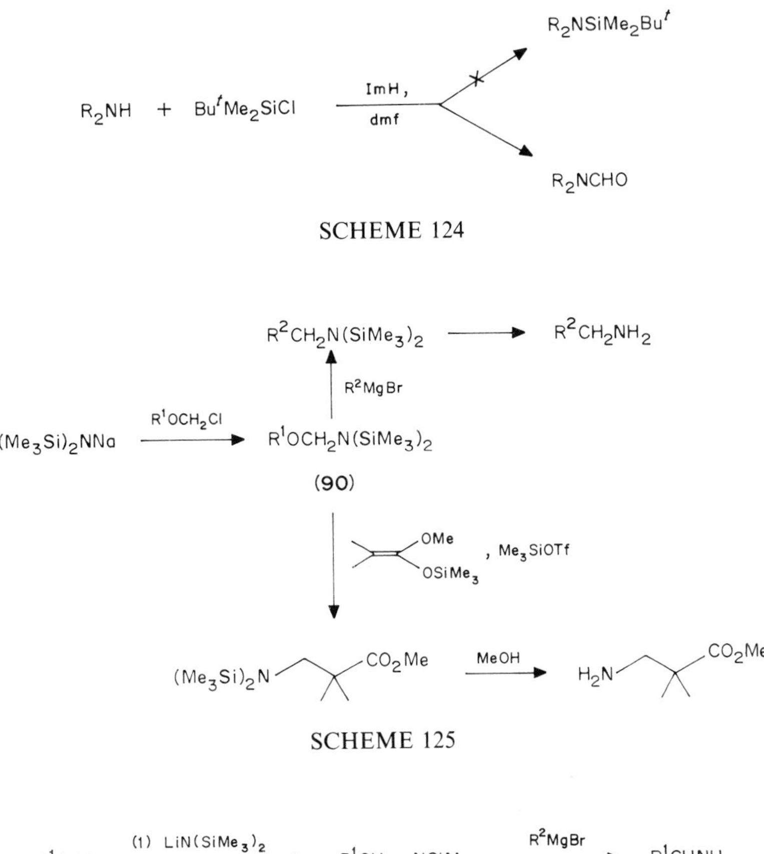

SCHEME 126

Secondary amines do not give the expected *N*-*tert*-butyldimethylsilyl derivatives under the conditions normal for such functionalization: instead, *N*-formamides are produced[419] in good yield through a dmf-derived Vilsmeier reagent (Scheme 124). The complete *N*-formyl unit can be introduced[420] at the 6-position of suitably activated penicillins using bis(tms)formamide. *N*-Tms carbodiimides undergo a cycloaddition reaction[421] with ketenes to give 4-iminoazetidinones.

Protected carbinolamines (**90**) react with a good range of Grignard reagents[422,423] and with ketene silyl acetals[424] in an overall process of electrophilic aminomethylation (Scheme 125). Improved Peterson-based (Section IX) methods have been presented[308,309] for the preparation of silyl imines from non-enolizable aldehydes (Scheme 126); such imines have found use in the synthesis of *N*-unsubstituted β-lactams (Section XIII.6) and in electrophilic aminomethylation[425].

Corriu and coworkers have given a detailed account[426] of their elegant studies on (*N*, *N*)-bissilylenamines, which can be prepared as shown[427] in Scheme 127 or from

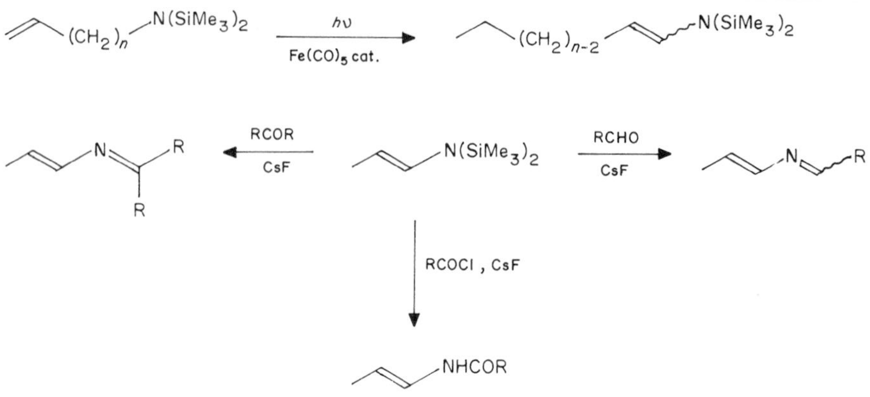

SCHEME 127

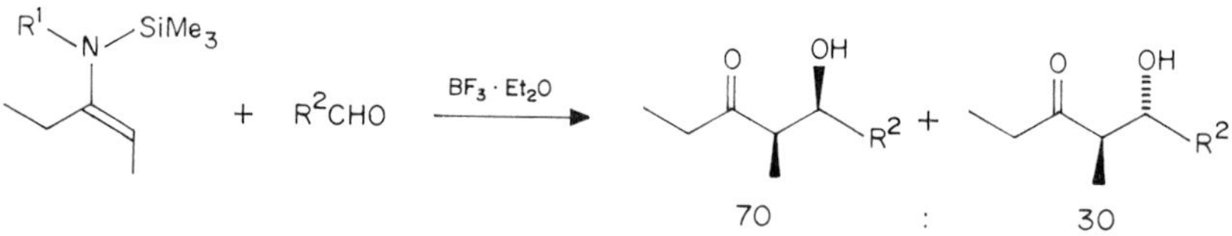

SCHEME 128

α-aminonitriles[428]. Under fluoride ion catalysis, they react with a variety of electrophiles[429] to produce, *inter alia*, azadienes by a modification of the Peterson reaction.

Ketone-derived silyl enamines have been reported[430] to undergo a Lewis acid-catalysed aldol condensation (Scheme 128) with a reasonable degree of *erythro* kinetic diastereoselectivity.

XVI. SILANES AS REDUCING AGENTS

A. Hydrosilylation

This term applies to the transition metal-catalysed addition of a hydrosilane to a multiply bonded system (see also Chapter 8). Further studies[431,432] of the rhodium(I)-catalysed hydrosilylation of alk-1-enes have been published; variable mixtures of vinyl and allylsilanes are produced, accompanied by saturated silanes. Selective reduction of carbonyl compounds with such catalytic systems has proved to be a fruitful area of investigation. Under suitable conditions[433], 4-*tert*-butylcyclohexanone can be reduced to the thermodynamically more stable diequatorial epimer with a high degree of selectivity (Scheme 129).

Full details have been published[434] on the selective 1,2- and 1,4-reduction of α,β-unsaturated ketones. The observation that dihydridosilanes give predominantly products of 1,2-addition, whereas 1,4-addition is favoured with bulky monohydridosilanes, has been rationalized in terms of competitive rates of hydrogen transfer from rhodium to carbon in the allyl complex **91** and of isomerization of **91** to **92**, with the latter process being accelerated by adverse steric interaction in **91** when bulky silanes are employed (Scheme 130).

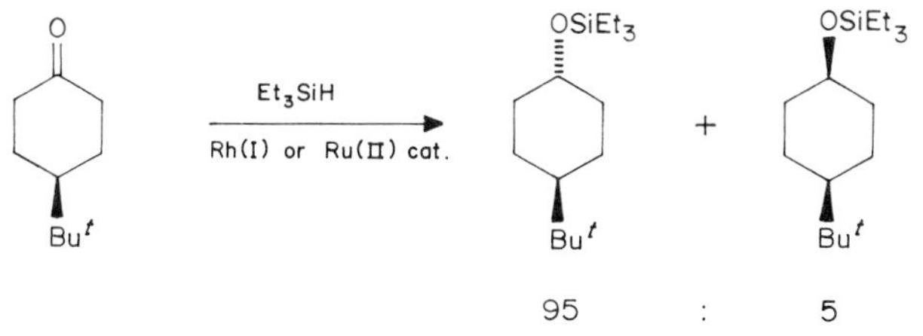

SCHEME 129

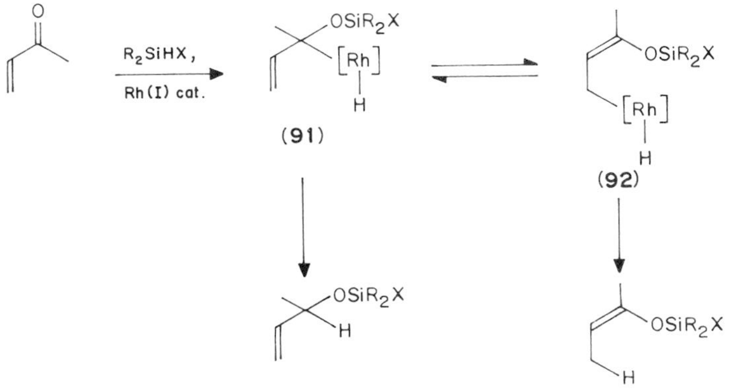

SCHEME 130

The enantioselective reduction of ketones[435] and oximes[436] using chiral catalyst systems has been studied further; use of the chiral silanes **93** derived from (−)-β-pinene resulted in modest chiral induction and enantioselectivity[437].

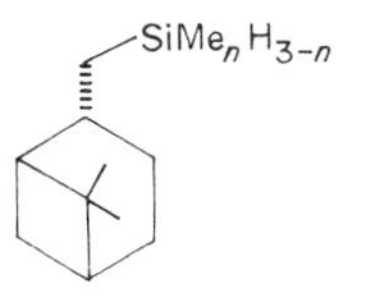

B. Fluoride Activation of Hydridosilanes

Under suitable conditions, fluoride ion coordinates with hydridosilanes to give a pentacoordinate silicon species in which the Si—H bond is weakened; KF, or better, CsF, are suitable fluoride sources in this context. Corriu and coworkers[426,438] have demonstrated the remarkable selectivity of this system using alkoxyhydridosilanes, especially for 1, 2-reduction of α, β-unsaturated ketones and aldehydes; most other functional groups are unaffected.

Using silicon-chiral silanes, a modest degree of enantioselectivity has been observed[439] with prochiral aralkyl ketones. Remarkably, α-chiral ketones can be reduced with extremely high diastereoselectivity[440] using a fluoride-activated system, when a Felkin

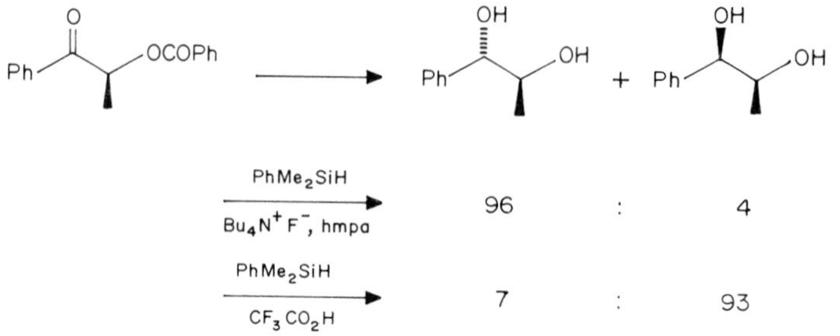

$PhMe_2SiH$, $Bu_4N^+F^-$, hmpa: 96 : 4

$PhMe_2SiH$, CF_3CO_2H: 7 : 93

SCHEME 131

transition-state model seems to be involved; the opposite diastereoselectivity is attained using trifluoroacetic acid as activator, i.e. ionic hydrogenation conditions, when a proton-bridged Cram cyclic model explains the observed results (Scheme 131).

C. Ionic Hydrogenation

This ability of hydridosilanes to transfer a hydride ion to 'onium ion systems has been used to trap an acyliminium ion intermediate[441] in an aza-Cope rearrangement. Anomeric sugars undergo exclusive axial (α) attack[442] at the intermediate oxonium ion (Scheme 132)

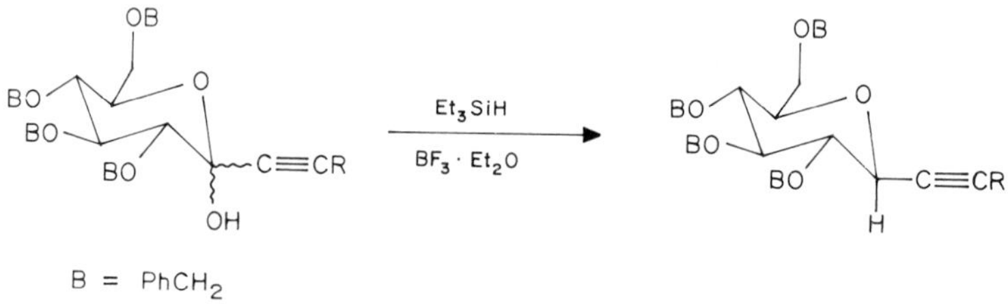

Et_3SiH, $BF_3 \cdot Et_2O$

B = $PhCH_2$

SCHEME 132

(94) + (95) → Et_3SiH, $BF_3 \cdot Et_2O$, −78 °C → + 95

Et_3SiH, $BF_3 \cdot Et_2O$, −78 °C

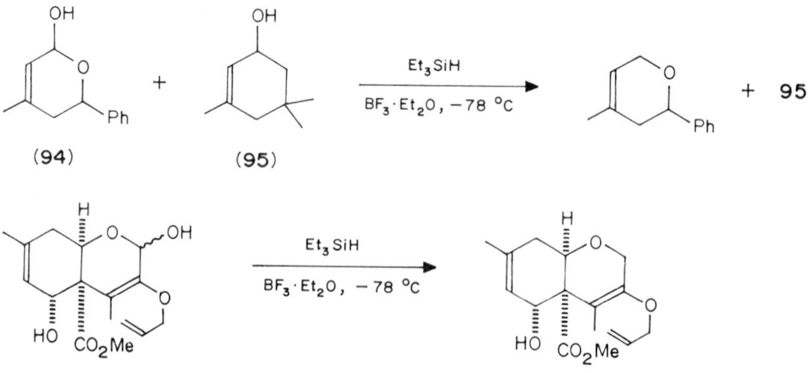

SCHEME 133

$$\text{ArCHO} \xrightarrow[\text{Me}_3\text{SiCl, (HSiMe}_2)_2\text{O}]{\text{NaI or LiBr}} \text{ArCH}_2\text{X}$$

SCHEME 134

on treatment with triethylsilane and boron trifluoride etherate. This reagent combination has proved to be extremely selective at low temperatures, reductively deoxygenating[443] the unsaturated lactol **94** without affecting the allylic alcohol **95**; the same n-selectivity has been observed[444] in intramolecular cases (Scheme 133).

Aryl aldehydes are converted reductively into the corresponding benzylic bromides or iodides by treatment[445] with tms halide, generated *in situ*, and a hydridosilane, with the plausible intermediacy of a silylated halohydrin (Scheme 134).

XVII. ACYLSILANES

The chemistry of acyl silanes has been reviewed[10].

A. Preparation

Reich *et al.*[446] have reviewed their extensive contributions to the preparation of α, β-unsaturated silanes of diverse types such as **96** and **97**. Two routes to saturated and

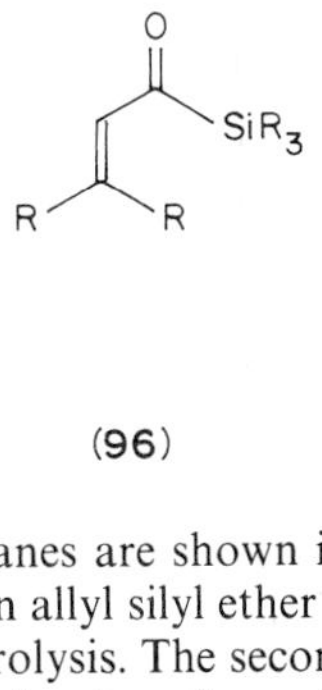

(96) 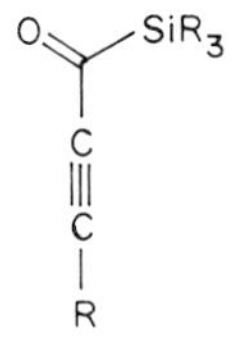 (97)

unsaturated acylsilanes are shown in Scheme 135. The first[447] of these involves Brook rearrangement of an allyl silyl ether anion, followed by catalysed isomerization to a silyl enol ether and hydrolysis. The second proceeds via hydroboration[448] of bis-tms-ethyne, with the product being transformed further[449] into (*E*)-α, β-unsaturated acylsilanes.

$$R^1_2C{=}CHCH_2OSiR^2_3 \xrightarrow[\text{(2) Me}_3\text{SiCl}]{\text{(1) Bu}^s\text{Li}} R^1_2C{=}CHCH(OSiMe_3)SiR^2_3 \xrightarrow[\text{MeOH or H}_2\text{O}]{10\%\ \text{Pd/C}} R^1_2CHCH_2C(O)SiR^2_3$$

$$Me_3SiC{\equiv}CSiMe_3 \xrightarrow[\text{(2) Me}_3\text{NO; (3) H}_3\text{O}^+]{\text{(1) Me}_2\text{S}\cdot\text{BH}_3} Me_3SiCH_2C(O)SiMe_3 \xrightarrow[\text{(2) R}^3\text{X; (3) LiNPr}^i_2\text{; (4) R}^4\text{CHO}]{\text{(1) LiNPr}^i_2} R^4CH{=}C(R^3)C(O)SiMe_3$$

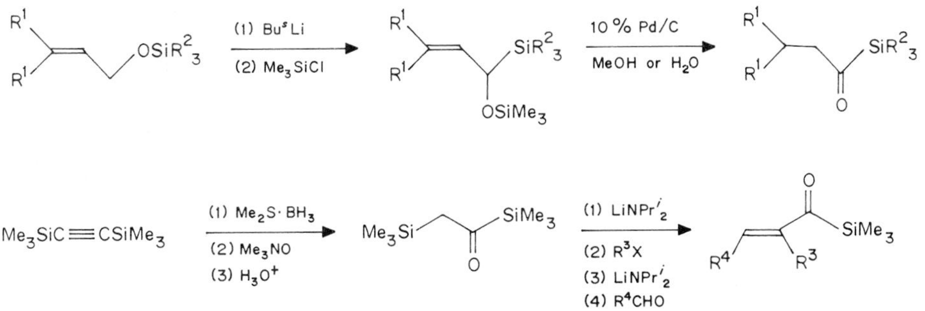

SCHEME 135

SCHEME 136

SCHEME 137

B. Reactions

In testament to the earlier work of Brook, several valuable reactions involving Si—C bond cleavage have been reported. Aromatic and, surprisingly, aliphatic acylsilanes can be converted into the corresponding aldehydes[450] by treatment with fluoride ion in the presence of a proton source, such as H_2O (Scheme 136). This transformation seems to proceed, at least in aromatic cases, via the acyl anion, which can be trapped[450,451] by other electrophiles. The (*E*)-α,β-unsaturated acylsilanes shown in Scheme 135 undergo clean oxidative cleavage to carboxylic acids.

Acylsilanes react, although sluggishly, with organolithium compounds to give α-silylalkoxides. In suitable cases (Scheme 137), these undergo Brook rearrangement to give simple alcohols[452] or, as Reich *et al.* have shown[446], to specific silyl enol ethers (see also Section III.A).

XVIII. α-SILYL RADICALS

α-Halosilanes undergo smooth reduction on treatment with organotin hydrides, in a reaction which does not affect Si—X bonds, but does, as expected, result in racemization

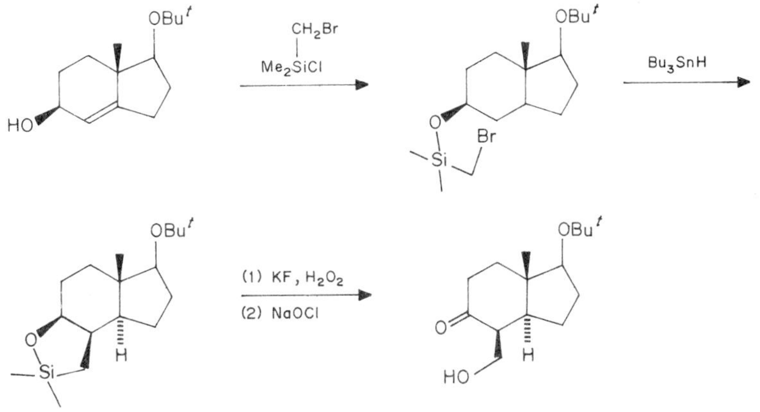

OBu^t HO → (CH_2Br / Me_2SiCl) → OBu^t O Si Br → (Bu_3SnH) →

OBu^t O Si H → (1) KF, H_2O_2 (2) NaOCl → OBu^t O HO H

SCHEME 138

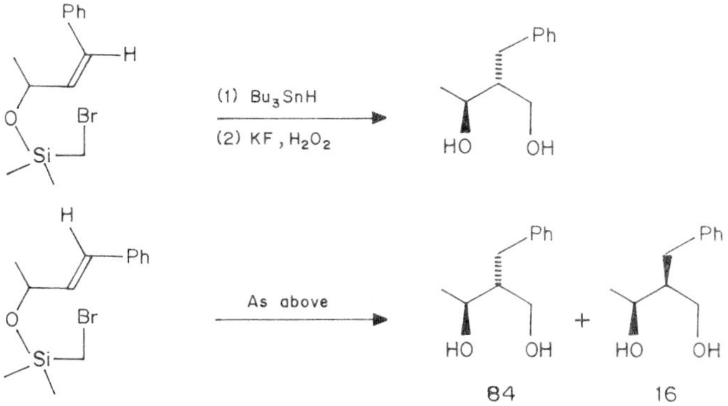

Ph H O Br Si → (1) Bu_3SnH (2) KF, H_2O_2 → Ph HO OH

H Ph O Br Si → As above → Ph HO OH + Ph HO OH

84 16

SCHEME 139

at carbon if appropriate. The observed rate of enhancement over all-carbon analogues has been ascribed[453] to a kinetic polar effect in the mechanism, and does not reflect *per se* any special thermodynamic stability of the intermediate α-silyl radicals.

Stork and Kahn[454] employed such α-silyl radicals to achieve the effective *trans*-addition of a functionalized alkane to the double bond of a cyclic alcohol (Scheme 138); critical steps are stereoselective quenching of the intermediate ring-fusion radical from the less hindered, α, face, and fluoride-assisted oxidative cleavage of the C—Si bonds (Section II.C). Such a protocol can also be seen in a new method[455] for the stereoselective synthesis of 1,3-diols, exemplified in Scheme 139.

XIX. CONCLUSIONS

This chapter has selectively reviewed some of the recent advances in organosilicon chemistry. It is hoped that it has demonstrated the wide applicability of such chemistry to organic transformations, capitalizing on the unique properties of silicon. As a measure of its respectability in the organic chemist's armoury, it has recently been subjected to a computer-assisted mechanistic evaluation[12] of its utility!

XX. REFERENCES

1. See, for example, G. Bertrand, G. Trinquier, and P. Mazerolles, *Organomet. Chem. Rev.*, **12**, 1 (1981); H. F. Schaefer, *Acc. Chem. Res.*, **15**, 283 (1982); P. A. Gaspar, in *Reactive Intermediates*, (Eds. M. Jones, Jr., and R. A. Moss), Vol. 2, Wiley–Interscience, New York, 1981, p. 335; M. S. Gordon and C. George, *J. Am. Chem. Soc.*, **106**, 609 (1984); C. Sosa and H. B. Schlegel, *J. Am. Chem. Soc.*, **106**, 5487 (1984).
2. I. Fleming, *Chem. Soc. Rev.*, **10**, 83 (1981).
3. E. Colvin, *Silicon in Organic Synthesis*, Butterworths, London, 1981; Krieger, Florida, 1985.
4. W. P. Weber, *Silicon Reagents for Organic Synthesis*, Springer-Verlag, Berlin, 1983.
5. L. A. Paquette, *Science*, **217**, 793 (1982); P. Magnus, *Aldrichim. Acta*, **13**, 43 (1980); P. D. Magnus, T. Sarkar, and S. Djuric, in *Comprehensive Organometallic Chemistry* (Eds. G. Wilkinson, F. G. A. Stone, and E. W. Abel) Vol. 7, Pergamon Press, Oxford, 1982, p. 515; D. A. Armitage, *ibid.*, Vol. 2, p. 1; T. J. Barton, *ibid.*, Vol. 2, p. 205; R. West, *ibid.*, Vol. 2, p. 365.
6. G. M. Rubottom, *Organomet. Chem. Rev.*, **11**, 267 (1981); **13**, 127 (1982); G. L. Larson, *Organomet. Chem. Rev.*, **14**, 267 (1984); G. L. Larson, *J. Organomet. Chem.*, **274**, 29 (1984); for a detailed listing of over 1800 organosilanes, see D. R. M. Walton, in *Dictionary of Organometallic Compounds*, Vol. 2, Chapman and Hall, London, 1984, p. 1907.
7. R. J. P. Corriu and C. Guerin, *Adv. Organomet. Chem.*, **20**, 265 (1982); R. J. P. Corriu, C. Guerin, and J. J. E. Moreau, *Top. Stereochem.*, **15**, 43 (1984).
8. V. Chvalovsky and J. M. Bellama (Eds.), *Carbon-Functional Organosilicon Compounds*, Plenum Press, New York, 1984.
9. T. A. Blinka, B. J. Helmer, and R. West, *Adv. Organomet. Chem.*, **23**, 193 (1984).
10. D. J. Ager, *Chem. Soc. Rev.*, **11**, 493 (1982).
11. Z. Parnes and G. I. Bolestova, *Synthesis*, 991 (1984).
12. C. E. Peishoff and W. L. Jorgensen, *J. Org. Chem.*, **48**, 1970 (1983).
13. R. Walsh, *Acc. Chem. Res.*, **14**, 246 (1981), and references cited therein.
14. R. K. Sharma and J. L. Fry, *J. Org. Chem.*, **48**, 2112 (1983).
15. R. Noyori, I. Nishida, and J. Sakata, *J. Am. Chem. Soc.*, **105**, 1598 (1983).
16. T. V. Rajan Babu, *J. Org. Chem.*, **49**, 2083 (1984).
17. R. F. Newton, D. P. Reynolds, M. A. W. Finch, D. R. Kelly, and S. M. Roberts, *Tetrahedron Lett.*, 3981 (1979).
18. R. E. Ireland and M. D. Varney, *J. Am. Chem. Soc.*, **106**, 3668 (1984).
19. K. Tamao, T. Kakui, M. Akita, T. Iwahara, R. Kanatani, J. Yoshida, and M. Kumada, *Tetrahedron*, **39**, 983 (1983).
20. (a) K. Tamao, N. Ishida, T. Tanaka, and M. Kumada, *Organometallics*, **2**, 1694 (1983). (b) I. Fleming, R. Henning, and H. Plaut, *J. Chem. Soc., Chem. Commun.*, 29 (1984).

21. E. Buncel and A. G. Davies, *J. Chem. Soc.*, 1550 (1958); G. A. Razuvaev, V. A. Yablokov, A. V. Ganyushkin, N. V. Yablokova, and G. S. Kalina, *J. Organomet. Chem.*, **165**, 281 (1979).
22. K. Tamao, N. Ishida, and M. Kumada, *J. Org. Chem.*, **48**, 2120 (1983).
23. K. Tamao and N. Ishida, *Tetrahedron Lett.*, **25**, 4245 (1984).
24. K. Tamao and N. Ishida, *Tetrahedron Lett.*, **25**, 4249 (1984); see also R. T. Taylor and J. G. Galloway, *J. Organomet. Chem.*, **220**, 295 (1981).
25. For an excellent discussion, see S. G. Wierschke, J. Chandrasekhar, and W. L. Jorgensen, *J. Am. Chem. Soc.*, **107**, 1496 (1985).
26. A. G. Brook and A. R. Bassindale, in *Rearrangements in Ground and Excited States* (Ed. P. de Mayo), Vol. 2, Academic Press, New York, 1980, pp. 149–227.
27. M. Kato, A. Mori, H. Oshino, J. Enda, K. Kobayashi, and I. Kuwajima, *J. Am. Chem. Soc.*, **106**, 1773 (1984); I. Kuwajima, *J. Organomet. Chem.*, **285**, 137 (1985).
28. G. Trimsis, S. Beers, J. Ridella, M. Carlon, D. Cullin, J. High, and D. Brutts, *J. Chem. Soc., Chem. Commun.*, 1088 (1984).
29. J. E. Baldwin, A. E. Derome, and P. D. Riordan, *Tetrahedron*, **39**, 2989 (1983).
30. J. E. Baldwin, J. C. Bottaro, P. D. Riordan, and A. E. Derome, *J. Chem. Soc., Chem. Commun.*, 942 (1982); for silicon-mediated routes to isocyanates, see A. E. Feoktistov, G. I. Orlov, V. P. Kozyukov, and V. F. Mironov, *J. Gen. Chem. USSR*, **54**, 780 (1984); V. F. Mironov, *J. Organomet. Chem.*, **271**, 207 (1984).
31. I. Matsuda, S. Sato, and U. Izumi, *Tetrahedron Lett.*, **24**, 2787 (1983); S. Sato, I. Matsuda, and Y. Izumi, *Tetrahedron Lett.*, **24**, 3855 (1983).
32. D. J. Ager, *Tetrahedron Lett.*, **22**, 587 (1981); for an interesting variant, see Y. Kita, Yasuda, O. Tamura, F. Itoh, and Y. Tamura, *Tetrahedron Lett.*, **25**, 4681 (1984).
33. J. D. White, M. Kang, and B. G. Sheldon, *Tetrahedron Lett.*, **24**, 4463 (1983).
34. I. Cutting and P. J. Parsons, *Tetrahedron Lett.*, **22**, 2021 (1981).
35. D. J. Hart and Y.-M. Tsai, *Tetrahedron Lett.*, **24**, 4387 (1983).
36. F. Sato, Y. Tanaka, and M. Sato, *J. Chem. Soc., Chem. Commun.*, 165 (1983).
37. I. Kuwajima and R. Takeda, *Tetrahedron Lett.*, **22**, 2381 (1981).
38. E. J. Corey and C. Rücker, *Tetrahedron Lett.*, **25**, 4345 (1984).
39. I. Fleming and C. D. Floyd, *J. Chem. Soc., Perkin Trans. 1*, 969 (1981).
40. T. Takeda, S. Naito, K. Ando, and T. Fujiwara, *Bull. Chem. Soc. Jpn.*, **56**, 967 (1983).
41. I. Urabe and I. Kuwajima, *Tetrahedron Lett.*, **24**, 4241 (1983).
42. F. Sato, M. Kusakabe, and Y. Kobayashi, *J. Chem. Soc., Chem. Commun.*, 1130 (1984): K. Suzuki, T. Ohkuma, and G. Tsuchihashi, *Tetrahedron Lett.*, **26**, 861 (1985).
43. C. Rücker, *Tetrahedron Lett.*, **25**, 4349 (1984).
44. S. Karady, J. S. Amato, R. A. Reamer, and L. M. Weinstock, *J. Am. Chem. Soc.*, **103**, 6765 (1981).
45. G. Anderson, D. W. Cameron, G. I. Feutrill, and R. W. Read, *Tetrahedron Lett.*, **22**, 4347 (1981).
46. B. M. Trost and J. Yoshida, *Tetrahedron Lett.*, **24**, 4895 (1983).
47. I. Fleming, T. W. Newton, and F. Roessler, *J. Chem. Soc., Perkin Trans. 1*, 2527 (1981).
48. H. Hayami, M. Sato, S. Kanemoto, Y. Morizawa, K. Oshima, and H. Nozaki, *J. Am. Chem. Soc.*, **105**, 4491 (1983).
49. E.-I. Negishi, and F.-T. Luo, *J. Org. Chem.*, **48**, 1560 (1983).
50. F. E. Ziegler and K. Mikami, *Tetrahedron Lett.*, **25**, 131 (1984).
51. F. Sato, H. Watanabe, Y. Tanaka, T. Yamaji, and M. Sato, *Tetrahedron Lett.*, **24**, 1041 (1983).
52. H. Hayami, K. Oshima, and H. Nozaki, *Tetrahedron Lett.*, **25**, 4433 (1984).
53. S. Fujikura, M. Inoue, K. Utimoto, and H. Nozaki, *Tetrahedron Lett.*, **25**, 1999 (1984); see also G. A. Molander, *J. Org. Chem.*, **48**, 5409 (1983).
54. F. Sato, H. Watanabe, Y. Tanaka, and M. Sato, *J. Chem. Soc., Chem. Commun.*, 1126 (1984).
55. J. Kang, W. Cho, and W. K. Lee, *J. Org. Chem.*, **49**, 1838 (1984).
56. M. Ochiai, T. Ukita, and E. Fujita, *Tetrahedron Lett.*, **24**, 4025 (1983).
57. S. E. Denmark and T. K. Jones, *J. Org. Chem.*, **47**, 4595 (1982); M. Mancini and J. F. Honek, *Tetrahedron Lett.*, **24**, 4295 (1983).
58. C.-N. Hsiao and H. Shechter, *Tetrahedron Lett.*, **24**, 2371 (1983).
59. T. Cuvigny, M. Julia, and C. Rolando, *J. Chem. Soc., Chem. Commun.*, 8 (1984).
60. K. Mikami, N. Kishi, and T. Nakai, *Chem. Lett.*, 1643 (1982).
61. H. Okumoto and J. Tsuji, *Synth. Commun.*, **12**, 1015 (1982); J. Enda and I. Kuwajima, *J. Chem. Soc., Chem. Commun.*, 1589 (1984); A. F. Sviridov, M. S. Ermolenko, D. V. Yashunsky, and N. K. Kochetkov, *Tetrahedron Lett.*, **24**, 4359 (1983).

62. R. K. Boeckman, *Tetrahedron*, **39**, 925 (1983); R. Urech, *J. Chem. Soc., Chem. Commun.*, 989 (1984).
63. P. Magnus, D. A. Quagliato, and J. C. Huffman, *Organometallics*, **1**, 1240 (1982); P. Magnus and D. Quagliato, *J. Org. Chem.*, **50**, 1621 (1985); for related Nazarov studies, see T. K. Jones and S. E. Denmark, *Helv. Chim. Acta*, **66**, 2377 (1983); G. Kjeldsen, J. S. Knudsen, L. S. Ravn-Petersen, and K. B. G. Torsell, *Tetrahedron*, **39**, 2237 (1983).
64. R. V. C. Carr, R. V. Williams, and L. A. Paquette, *J. Org. Chem.*, **48**, 4976 (1983).
65. D. Dhanak, C. B. Reese, and D. E. Williams, *J. Chem. Soc., Chem. Commun.*, 988 (1984).
66. F. E. Ziegler and K. Mikami, *Tetrahedron Lett.*, **25**, 127 (1984).
67. W. E. Fristad, T. R. Bailey, L. A. Paquette, R. Gleiter, and M. C. Böhm, *J. Am. Chem. Soc.*, **101**, 4420 (1979).
68. A. Padwa and J. G. MacDonald, *J. Org. Chem.*, **48**, 3189 (1983).
69. P. De Shong and J. M. Leginus, *J. Org. Chem.*, **49**, 3421 (1984).
70. R. F. Cunico, *J. Organomet. Chem.*, **212**, C51 (1981).
71. M. A. Tius, *Tetrahedron Lett.*, **22**, 3335 (1981).
72. H.-F. Chow and I. Fleming, *J. Chem. Soc., Perkin Trans. 1*, 1815 (1984).
73. B. M. Trost and E. Maruyama, *J. Am. Chem. Soc.*, **103**, 6529 (1981).
74. S. D. Burke, J. O. Saunders, J. A. Oplinger, and C. W. Murtiashaw, *Tetrahedron Lett.*, **26**, 1131 (1985).
75. L. E. Overman and T. C. Malone, *J. Org. Chem.*, **47**, 5297 (1982); L. E. Overman and R. M. Burke, *Tetrahedron Lett.*, **25**, 5739 (1984).
76. For example, J. E. Wrobel and B. Ganem, *J. Org. Chem.*, **48**, 3761 (1983).
77. R. G. Daniels and L. A. Paquette, *Organometallics*, **1**, 1449 (1982).
78. K. Tamao, R. Kanatani, and M. Kumada, *Tetrahedron Lett.*, **25**, 1905 (1984); K. Tamao, T. Iwahara, R. Kanatani, and M. Kumada, *Tetrahedron Lett.*, **25**, 1909 (1984); K. Tamao, R. Kanatani, and M. Kumada, *Tetrahedron Lett.*, **25**, 1913 (1984).
79. K. Tamao, M. Kumada, and M. Maeda, *Tetrahedron Lett.*, **25**, 321 (1984).
80. C. Burford, F. Cooke, G. Roy, and P. Magnus, *Tetrahedron*, **39**, 867 (1983).
81. F. Cooke, G. Roy, and P. Magnus, *Organometallics*, **1**, 893 (1982).
82. P. F. Hudrlik, A. M. Hudrlik, and A. K. Kulkarni, *Tetrahedron Lett.*, **26**, 139 (1985).
83. L. A. Paquette and Y.-K. Han, *J. Am. Chem. Soc.*, **103**, 1831 (1981).
84. G. Schmid and W. Hofheinz, *J. Am. Chem. Soc.*, **105**, 624 (1983).
85. G. L. Larson, S. Sandoval, F. Cartledge, and F. R. Fronczek, *Organometallics*, **2**, 810 (1983); see also J. W. Wilt, F. G. Belmonte, and P. A. Zieske, *J. Am. Chem. Soc.*, **105**, 5665 (1983).
86. G. Berti, S. Canedoli, P. Crotti, and F. Macchia, *J. Chem. Soc., Perkin Trans. 1*, 1183 (1984).
87. G. Nagendrappa, *Tetrahedron*, **38**, 2429 (1982).
88. I. Fleming and T. W. Newton, *J. Chem. Soc., Perkin Trans. 1*, 119 (1984).
89. P. Knochel and J. F. Normant, *Tetrahedron Lett.*, **25**, 4383 (1984).
90. I. Kuwajima and H. Urabe, *Tetrahedron Lett.*, **22**, 5191 (1981); S. P. Tanis and D. B. Head, *Tetrahedron Lett.*, **25**, 4451 (1984); and references cited therein.
91. F. Sato, Y. Tanaka, and H. Kanbara, *J. Chem. Soc., Chem. Commun.*, 1024 (1983).
92. E. Negishi, F.-T. Luo, and C. L. Rand, *Tetrahedron Lett.*, **23**, 27 (1982).
93. H. Nishiyama, M. Sasaki, and K. Itoh, *Chem. Lett.*, 905 (1981).
94. M. Bourgain-Commerçon, J. P. Foulin, and J. E. Normant, *Tetrahedron Lett.*, **24**, 5077 (1983).
95. J. G. Smith, S. E. Prozda, S. P. Petraglia, N. R. Quinn, E. M. Rice, B. S. Taylor, and M. Viswanathan, *J. Org. Chem.*, **49**, 4112 (1984); B. M. Trost and D. M. T. Chan, *J. Am. Chem. Soc.*, **104**, 3733 (1982); N. Shimuzu, F. Shibata, and Y. Tsuno, *Bull. Chem. Soc. Jpn.*, **57**, 3017 (1984).
96. Y. Tanigawa, Y. Fuse, and S.-I. Murahashi, *Tetrahedron Lett.*, **23**, 557 (1982).
97. B. M. Trost, J. Yoshida, and M. Lautens, *J. Am. Chem. Soc.*, **105**, 4494 (1983).
98. C. Biran, J. Dunoguès, R. Calas, J. Gerval, and T. Tskhovrebachvili, *Synthesis*, 220 (1981).
99. I. Fleming and N. K. Terrett, *Tetrahedron Lett.*, **24**, 4151 (1983); I. Fleming and D. Marchi, *Synthesis*, 560 (1981); I. Fleming and T. W. Newton, *J. Chem. Soc., Perkin Trans. 1*, 1805 (1984).
100. I. Fleming and D. Waterson, *J. Chem. Soc., Perkin Trans. 1*, 1809 (1984).
101. P. R. Jenkins, R. Gut, H. Wetter, and A. Eschenmoser, *Helv. Chim. Acta*, **62**, 1922 (1979); I. Kuwajima, T. Tanaka, and K. Atsumi, *Chem. Lett.*, 779 (1979); S. R. Wilson and M. F. Price, *J. Am. Chem. Soc.*, **104**, 1124 (1982); K. Mikami, N. Kishi, and T. Nakai, *Chem. Lett.*, 1643 (1982).
102. G. Deleris, J. Dunoguès, and A. Gadras, *Tetrahedron Lett.*, **25**, 2135 (1984).

103. I. Fleming and J. A. Langley, *J. Chem. Soc., Perkin Trans. 1*, 1421 (1981); see also I. Fleming, D. Marchi, and S. K. Patel, *J. Chem. Soc., Perkin Trans. 1*, 2519 (1981).
104. T. Hiyama, M. Obayashi, I. Mori, and H. Nozaki, *J. Org. Chem.*, **48**, 912 (1983).
105. T. Hayashi, K. Kabeta, T. Yamamoto, K. Tamao, and M. Kumada, *Tetrahedron Lett.*, **24**, 5661 (1983), and references cited therein; I. Fleming and N. K. Terrett, *Tetrahedron Lett.*, **24**, 4253 (1983); H. Wetter and P. Scherer, *Helv. Chim. Acta*, **66**, 118 (1983); G. Wickham and W. Kitching, *J. Org. Chem.*, **48**, 612 (1983); for some intramolecular studies, see S. R. Wilson and M. F. Price, *J. Am. Chem. Soc.*, **104**, 1124 (1982); S. R. Wilson and M. F. Price, *Tetrahedron Lett.*, **24**, 569 (1983).
106. S. J. Hathaway and L. A. Paquette, *J. Org. Chem.*, **48**, 3351 (1983); C. A. Maryanoff and B. E. Maryanoff, in *Asymmetric Synthesis* (Eds. J. D. Morrison and J. W. Scott), Vol. 4, Academic Press, New York and London, 1984, pp. 355–374; for an interesting use of the trimethylsilyl group itself as a detachable diastereoselectivity-inducing group in reactions of benzyl silanes (cleavage being achieved using $KOBu^t$/dmso), see C. G. M. Janssen and E. F. Godefroi, *J. Org. Chem.*, **49**, 3600 (1984).
107. H.-F. Chow and I. Fleming, *Tetrahedron Lett.*, **26**, 397 (1985); see also C. Santelli-Rouvier, *Tetrahedron Lett.*, **25**, 4371 (1984).
108. M. T. Reetz and K. Kesseler, *J. Chem. Soc., Chem. Commun.*, 1079 (1984); M. T. Reetz and A. Jung, *J. Am. Chem. Soc.*, **105**, 4833 (1983).
109. K. Soai and M. Ishizaki, *J. Chem. Soc., Chem. Commun.*, 1016 (1984).
110. S. Kiyooke and C. H. Heathcock, *Tetrahedron Lett.*, **24**, 4765 (1983); S. Danishefsky and M. de Ninno, *Tetrahedron Lett.*, **26**, 823 (1985).
111. H. Nishiyama, K. Itagaki, K. Sakuta, and K. Itoh, *Tetrahedron Lett.*, **22**, 5285 (1981); H. Nishiyama, S. Narimatsu, and K. Itoh, *Tetrahedron Lett.*, **22**, 5289 (1981); H. Nishiyama, H. Yokoyama, S. Narimatsu, and K. Itoh, *Tetrahedron Lett.*, **23**, 1267 (1982).
112. K. Isaac, P. Kocienski, and S. Campbell, *J. Chem. Soc., Chem. Commun.*, 249 (1983).
113. B. M. Trost and B. P. Coppola, *J. Am. Chem. Soc.*, **104**, 6879 (1982); see also B. M. Trost and M. Shimuzu, *J. Am. Chem. Soc.*, **104**, 4299 (1982); B. M. Trost and R. Remuson, *Tetrahedron Lett.*, **24**, 1129 (1983).
114. T. S. Tan, A. N. Mather, G. Proctor, and A. H. Davidson, *J. Chem. Soc., Chem. Commun.*, 585 (1984).
115. S. E. Denmark and E. J. Weber, *Helv. Chim. Acta*, **66**, 1655 (1983).
116. G. Majetich, R. Desmond, and A. M. Casares, *Tetrahedron Lett.*, **24**, 1913 (1983); G. Majetich, A. M. Casares, D. Chapman, and M. Behnke, *Tetrahedron Lett.*, **24**, 1909 (1983); see also D. Schinzer, S. Solyom, and M. Becker, *Tetrahedron Lett.*, **26**, 1831 (1985).
117. B. M. Trost, T. N. Nanninga, and T. Satoh, *J. Am. Chem. Soc.*, **107**, 721 (1985); see also B. M. Trost, D. M. T. Chan, and T. B. Nanninga, *Org. Synth.*, **62**, 58 (1984).
118. H. Hiemstra, W. J. Klaver, and W. N. Speckamp, *J. Org. Chem.*, **49**, 1149 (1984); J.-C. Gramain and R. Remuson, *Tetrahedron Lett.*, **26**, 327 (1985); A. P. Kozikowski and P. Park, *J. Org. Chem.*, **49**, 1674 (1984).
119. M. Aratani, K. Sawada, and M. Hashimoto, *Tetrahedron Lett.*, **23**, 3921 (1982); G. A. Kraus and K. Neuenschwander, *J. Chem. Soc., Chem. Commun.*, 134 (1982); see also K. Fujimoto, Y. Iwano, and K. Hirai, *Tetrahedron Lett.*, **26**, 89 (1982).
120. For example, S. Karady, J. S. Amato, R. A. Reamer, and L. M. Weinstock, *J. Am. Chem. Soc.*, **103**, 6765 (1981).
121. M. Ochiai, M. Arimoto, and E. Fujita, *Tetrahedron Lett.*, **22**, 1115 (1981).
122. H. Sakurai, K. Sasaki, and A. Hosomi, *Tetrahedron Lett.*, **22**, 745 (1981).
123. C. Westerlund, *Tetrahedron Lett.*, **23**, 4835 (1982).
124. P. S. Mariano, *Tetrahedron*, **39**, 3845 (1983); see also K. Misuno, M. Ikeda, and Y. Otsuji, *Tetrahedron Lett.*, **26**, 461 (1985).
125. M. D. Lewis, J. K. Cha and Y. Kishi, *J. Am. Chem. Soc.*, **104**, 4976 (1982); A. P. Kozikowski, K. L. Sorgi, B. C. Wang, and Z. Xu, *Tetrahedron Lett.*, **24**, 1563 (1983); S. Danishefsky and J. F Kerwin, *J. Org. Chem.*, **47**, 3803, 5428 (1982); A. P. Kozikowski and P. Park, *J. Am. Chem. Soc.*, **107**, 1763 (1985), and references cited therein.
126. H. Nishiyama, T. Naritomi, K. Sakuta, and K. Itoh, *J. Org. Chem.*, **48**, 1557 (1983); M. Wada, T. Shigehisa, and K. Akiba, *Tetrahedron Lett.*, **24**, 1711 (1983).
127. J. A. Cella, *J. Org. Chem.*, **47**, 2125 (1982).

128. H. Sakurai, *Pure Appl. Chem.*, **54**, 1 (1982).
129. H. Sakurai, A. Hosomi, M. Saito, K. Sasaki, H. Iguchi, J.-I. Sasaki, and Y. Araki, *Tetrahedron*, **39**, 883 (1983); see also D. Djahanbini, B. Cazes, J. Gore, and F. Gobert, *Tetrahedron*, **41**, 867 (1985).
130. R. E. Ireland, J. D. Godfrey, and S. Thaisrivongs, *J. Am. Chem. Soc.*, **103**, 2446 (1981).
131. M. Ochiai, M. Arimoto, and E. Fujita, *J. Chem. Soc., Chem. Commun.*, 460 (1981).
132. R. J. P. Corriu, C. Guerin, J. M'Boula, *Tetrahedron Lett.*, **22**, 2985 (1981); D. J. Kempf, K. D. Wilson, and P. Beak, *J. Org. Chem.*, **47**, 1610 (1982).
133. D. J. S. Tsai and D. S. Matteson, *Tetrahedron Lett.*, **22**, 2751 (1981); Y. Yamamoto, Y. Saito, and K. Maruyama, *Tetrahedron Lett.*, **23**, 4597 (1982); F. Sato, Y. Suzuki, and M. Sato, *Tetrahedron Lett.*, **23**, 4589 (1982); Y. Yamamoto, Y. Saito, and K. Maruyama, *J. Chem. Soc., Chem. Commun.*, 1326 (1982).
134. D. Häbich and F. Effenberger, *Synthesis*, 841 (1979); F. Effenberger and A. Krebs, *J. Org. Chem.*, **49**, 4687 (1984).
135. L. Birkofer and O. Stuhl, *Top. Curr. Chem.*, **88**, 33 (1980).
136. P. Beak and V. Snieckus, *Acc. Chem. Res.*, **15**, 306 (1982).
137. M. F. Semmelhack, G. R. Clark, J. L. Garcia, J. J. Harrison, Y. Thebtaranonth, W. Wulff, and A. Yamashita, *Tetrahedron*, **37**, 3957 (1981), and references cited therein.
138. R. J. Billedeau, M. P. Sibi, and V. Snieckus, *Tetrahedron Lett.*, **24**, 4515 (1983).
139. M. Fukui, T. Ikeda, and T. Oishi, *Tetrahedron Lett.*, **23**, 1605 (1982).
140. R. J. Mills, R. F. Horvath, M. P. Sibi, and V. Snieckus, *Tetrahedron Lett.*, **26**, 1145 (1985).
141. E. G. Doadt and V. Snieckus, *Tetrahedron Lett.*, **26**, 1149 (1985).
142. H. Hommes, H. D. Verkruijsse and L. Brandsma, *J. Chem. Soc., Chem. Commun.*, 366 (1981).
143. G. Nechvatal and D. A. Widdowson, *J. Chem. Soc., Chem. Commun.*, 467 (1982); P. J. Beswick, S. L. Leach, N. F. Masters, and D. A. Widdowson, *J. Chem. Soc., Chem. Commun.*, 46 (1984); see also A. G. M. Barrett, D. Dauzonne, I. A. O'Neil, and A. Renaud, *J. Org. Chem.*, **49**, 4409 (1984).
144. J. M. Muchowski and R. Naef, *Helv. Chim. Acta*, **67**, 1168 (1984); D. J. Chadwick and S. T. Hodgson, *J. Chem. Soc., Perkin Trans. 1*, 1833 (1982); A. Kozikowski and X.-M. Cheng, *J. Org. Chem.*, **49**, 3239 (1984).
145. R. L. Halterman, N. H. Nguyen and K. P. C. Vollhardt, *J. Am. Chem. Soc.*, **107**, 1379 (1985).
146. M. d'Alarcao and N. J. Leonard, *J. Am. Chem. Soc.*, **105**, 5958 (1983).
147. R. Schroeder, U. Schollkopf, E. Blume, and I. Hoppe, *Justus Liebigs Ann. Chem.*, 533 (1975).
148. A. Dondoni, T. Dall'Occo, G. Fantin, M. Fognagnolo, A. Medici, and P. Pedrini, *J. Chem. Soc., Chem. Commun.*, 258 (1984).
149. R. J. Mills and V. Snieckus, *J. Org. Chem.*, **48**, 1565 (1983).
150. R. D. Wood and B. Ganem, *Tetrahedron Lett.*, **24**, 4391 (1983); see also Y. Nagao, S. Yamada, and E. Fujita, *Tetrahedron Lett.*, **24**, 2291 (1983).
151. J. O. Karlsson, S. Gronowitz, and J. Frejd, *J. Org. Chem.*, **47**, 374 (1982).
152. M. F. Semmelhack and A. Zask, *J. Am. Chem. Soc.*, **105**, 2034 (1983), and references cited therein.
153. D. S. Wilbur, W. E. Stone, and K. W. Anderson, *J. Org. Chem.*, **48**, 1542 (1983).
154. A. Medici, P. Pedrini, and A. Dondoni, *J. Chem. Soc., Chem. Commun.*, 655 (1981); A. Medici, G. Fantin, M. Fogagnolo, and A. Dondoni, *Tetrahedron Lett.*, **24**, 2901 (1983).
155. E. Effenberger and W. Spiegler, *Angew. Chem., Int. Ed. Engl.*, **20**, 265 (1981).
156. C. Eaborn, J. G. Stamper, and G. Seconi, *J. Organomet. Chem.*, **150**, C23 (1978).
157. W. Bernhard, I. Fleming, and D. Waterson, *J. Chem. Soc., Chem. Commun.*, 28 (1984); W. Bernhard and I. Fleming, *J. Organomet. Chem.*, **271**, 281 (1984); I. Fleming, J. H. M. Hill, D. Parker, and D. Waterson, *J. Chem. Soc., Chem. Commun.*, 318 (1985).
158. G. Altnau and L. Rösch, *Tetrahedron Lett.*, **24**, 45 (1983).
159. M. Shibasaki, H. Fukasawa, and S. Ikegami, *Tetrahedron Lett.*, **24**, 3497 (1983).
160. T. Hiyama and M. Obayashi, *Tetrahedron Lett.*, **24**, 4109 (1983).
161. T. Hiyama, M. Obayashi, and M. Sawahata, *Tetrahedron Lett.*, **24**, 4113 (1983).
162. D. J. Ager, *Synthesis*, 384 (1984).
163. K. Yamamoto, Y. Tomo, and S. Suzuki, *Tetrahedron Lett.*, **21**, 2861 (1980); see also K. Yamamoto, T. Kimura, and Y. Tomo, *Tetrahedron Lett.*, **25**, 2155 (1984).
164. P. F. Hudrlik, A. M. Hudrlik, and K. Kulkarni, *J. Am. Chem. Soc.*, **104**, 6809 (1982).
165. P. A. Brown, P. R. Jenkins, J. Fawcett, and D. R. Russell, *J. Chem. Soc., Chem. Commun.*, 253 (1984).

166. D. J. Ager, *Tetrahedron Lett.*, **22**, 2923 (1981).
167. T. Cohen, J. P. Sherbine, J. R. Matz, R. R. Hutchins, B. M. McHenry, and P. R. Willey, *J. Am. Chem. Soc.*, **106**, 3245 (1984).
168. K. Okano, Y. Kyotani, H. Ishihama, S. Kobayashi, and M. Ohno, *J. Am. Chem. Soc.*, **105**, 7186 (1983).
169. D. B. Tulshian and B. Fraser-Reid, *J. Am. Chem. Soc.*, **103**, 474 (1981); for a discussion of kinetic *vs.* thermodynamic control, see also G. A. Larson, J. A. Prieto, and A. Hernandez, *Tetrahedron Lett.*, **22**, 1575 (1981).
170. G. L. Larson, C. F. de Kaifer, R. Seda, L. E. Torres, and J. R. Ramirez, *J. Org. Chem.*, **49**, 3385 (1984).
171. W. Oppolzer, M. Guo, and K. Haettig, *Helv. Chim. Acta.*, **66**, 2140 (1983).
172. D. Hernandez and G. L. Larson, *J. Org. Chem.*, **49**, 4285 (1984).
173. J. A. Miller and G. Zweifel, *J. Am. Chem. Soc.*, **103**, 6217 (1981).
174. M. Isobe, M. Kitamura, and T. Goto, *Tetrahedron Lett.*, **22**, 239 (1981); M. Isobe, Y. Funabashi, Y. Ichikawa, S. Mio, and T. Goto, *Tetrahedron Lett.*, **25**, 2021 (1984).
175. K. S. Kyler and D. S. Watt, *J. Am. Chem. Soc.*, **105**, 619 (1983).
176. K. S. Kyler, M. A. Netzel, S. Arseniyadis, and D. S. Watt, *J. Org. Chem.*, **48**, 383 (1983).
177. S. V. Ley, and N. S. Simpkins, *J. Chem. Soc., Chem. Commun.*, 1281 (1983).
178. D. J. Ager, *J. Chem. Soc., Chem. Commun.*, 487 (1984).
179. D. S. Matteson and D. Majumdar, *J. Chem. Soc., Chem. Commun.*, 39 (1980).
180. D. J. S. Tsai and D. S. Matteson, *Tetrahedron Lett.*, **22**, 2751 (1981); E. van Hulsen and D. Hoppe, *Tetrahedron Lett.*, **26**, 411 (1985).
181. T.-H. Chan and J.-S. Li, *J. Chem. Soc., Chem. Commun.*, 969 (1982).
182. M. V. Garad, A. Pelter, B. Singaram, and J. W. Wilson, *Tetrahedron Lett.*, **24**, 637 (1983).
183. A. Rensing, K.-J. Echsler, and T. Kauffmann, *Tetrahedron Lett.*, **21**, 2807 (1980).
184. C. H. Chen, J. J. Doney, and G. A. Reynolds, *J. Org. Chem.*, **47**, 680 (1982).
185. M. L. Mancini and J. F. Honek, *Tetrahedron Lett.*, **23**, 3249 (1982).
186. D. T. W. Chu, J. E. Hengeved, and D. Lester, *Tetrahedron Lett.*, **24**, 139 (1983); A. B. Smith and D. Boschelli, *J. Org. Chem.*, **48**, 1217 (1983); E. W. Logusch, *Tetrahedron Lett.*, **25**, 4195 (1984).
187. S. Honda and T. Hata, *Tetrahedron Lett.*, **22**, 2093 (1981); J. Binder and E. Zbiral, *Tetrahedron Lett.*, **25**, 4213 (1984); K. Schönauer and E. Zbiral, *Tetrahedron Lett.*, **24**, 573 (1983).
188. C. Gioeli, N. Balgovin, S. Josephson, and J. B. Chattopadhyaya, *Tetrahedron Lett.*, **22**, 969 (1981).
189. B. H. Lipshutz, J. J. Pegram, and M. C. Morey, *Tetrahedron Lett.*, **22**, 4603 (1981).
190. S. Björkman and J. Chattopadhyaya, *Chem. Scr.*, **20**, 201 (1982).
191. M. P. Edwards, S. V. Ley, S. G. Lister, and B. D. Palmer, *J. Chem. Soc., Chem. Commun.*, 630 (1983); J. M. Muchowski and D. R. Solar, *J. Org. Chem.*, **49**, 203 (1984).
192. Y. Ito, S. Miyata, M. Nakatsuka, and T. Saegusa, *J. Am. Chem. Soc.*, **103**, 5250 (1981), and references cited therein; J. P. Marino and S. L. Dax, *J. Org. Chem.*, **49**, 3671 (1984).
193. E. R. Marinelli, *Tetrahedron Lett.*, **23**, 2745 (1982).
194. A. G. Angoh and D. L. J. Clive, *J. Chem. Soc., Chem. Commun.*, 534 (1984).
195. E. Block and M. Aslam, *Tetrahedron Lett.*, **23**, 4203 (1982).
196. R. F. Cunico and L. Bedell, *J. Org. Chem.*, **48**, 2780 (1983).
197. A. Padwa, G. Haffmans, and M. Tomas, *J. Org. Chem.*, **49**, 3314 (1984); see also E. Vedejs and F. G. West, *J. Org. Chem.*, **48**, 4773 (1983); O. Tsuge, S. Kanemasa, S. Kuraoka, and S. Takenaka, *Chem. Lett.*, 279, 281 (1984).
198. A. Hosomi, Y. Sakata, and H. Sakurai, *Chem. Lett.*, 1117 (1984).
199. H. Nishiyama, T. Kitajima, A. Yamamoto, and K. Ito, *J. Chem. Soc., Chem. Commun.*, 1232 (1982); H. Nishiyama, K. Itagaki, N. Osaka, and K. Ito, *Tetrahedron Lett.*, **23**, 4103 (1982).
200. D. J. Ager, I. Fleming, and S. K. Patel, *J. Chem. Soc., Perkin Trans. 1*, 2520 (1981).
201. I. Fleming and T. W. Newton, *J. Chem. Soc., Perkin Trans. 1*, 1805 (1984).
202. I. Fleming and D. A. Perry, *Tetrahedron*, **37**, 4027 (1981).
203. Y. V. Kolodyazhnyi, N. I. Sizova, I. G. Lorents, L. I. Kuznetsova, A. P. Sadimenko, L. I. Ol'Khovskaya, and N. V. Komarov, *J. Gen. Chem. USSR*, **52**, 1644 (1982).
204. P. Pappalardo, E. Ehlinger, and P. Magnus, *Tetrahedron Lett.*, **23**, 309 (1982).
205. For a route to 3-silylfurans, see F. Sato and H. Katsuno, *Tetrahedron Lett.*, **24**, 1809 (1983).
206. I. Kuwajima, S. Sugahara, and J. Enda, *Tetrahedron Lett.*, **24**, 1061 (1983).

207. For another route to enynes, see Y. Yamakado, M. Ishiguro, N. Ikeda, and H. Yamamoto, *J. Am. Chem. Soc.*, **103**, 5568 (1981).
208. A. J. Bridges, V. Fedij, and E. C. Turowski, *J. Chem. Soc., Chem. Commun.*, 1093 (1983).
209. Y. Yamamoto, W. Ito, and K. Maruyama, *J. Chem. Soc., Chem. Commun.*, 1004 (1984).
210. R. G. Daniels and L. A. Paquette, *Tetrahedron Lett.*, **22**, 1579 (1981).
211. K. K. Wang, S. S. Nikam, and C. D. Ho, *J. Org. Chem.*, **48**, 5376 (1983).
212. M. Ishiguro, N. Ikeda, and H. Yamamoto, *J. Org. Chem.*, **47**, 2225 (1982).
213. G. Boche and J. Bigalke, *Tetrahedron Lett.*, **25**, 955 (1984).
214. N. Handese, *J. Org. Chem.*, **47**, 1145 (1982).
215. M. Ladika and P. J. Stang, *J. Chem. Soc., Chem. Commun.*, 459 (1981).
216. M. Karpf, *Helv. Chim. Acta*, **67**, 73 (1984), and references cited therein.
217. Y. Kawanami, T. Katsuki, and M. Yamaguchi, *Tetrahedron Lett.*, **24**, 5131 (1983).
218. A. Jellal, J.-P. Zahra, and M. Santelli, *Tetrahedron Lett.*, **24**, 1395 (1983).
219. B. A. Keay and R. Rodrigo, *J. Am. Chem. Soc.*, **104**, 4752 (1982).
220. R. L. Danheiser and H. Sard, *Tetrahedron Lett.*, **24**, 23 (1983).
221. J.-K. Choi, D. J. Hart, and Y.-M. Tsai, *Tetrahedron Lett.*, **23**, 4765 (1982).
222. G. Büchi and H. Wüest, *J. Org. Chem.*, **44**, 546 (1979).
223. R. L. Danheiser, D. J. Carini, D. M. Fink, and A. Basak, *Tetrahedron*, **39**, 935 (1983).
224. M. H. P. J. Aerssens and L. Brandsma, *J. Chem. Soc., Chem. Commun.*, 735 (1984).
225. A. B. Holmes, C. L. D. Jennings-White, and D. A. Kendrick, *J. Chem. Soc., Chem. Commun.*, 1594 (1984), and references cited therein.
226. P. R. Hamann and P. L. Fuchs, *J. Org. Chem.*, **48**, 914 (1983).
227. H. Schostarez and L. A. Paquette, *J. Am. Chem. Soc.*, **103**, 722 (1981); J. Drouin, F. Leyendecker, and J. M. Conia, *Tetrahedron*, **36**, 1203 (1980).
228. S. Rajagopalan and G. Zweifel, *Synthesis*, 111 (1984); J. Pornet, *Tetrahedron Lett.*, **22**, 453, 455 (1981); J. Pornet and B. Randianoelina, *Tetrahedron Lett.*, **22**, 1327 (1981).
229. I. Kuwajima, E. Nakamura, and K. Hashimoto, *Tetrahedron*, **39**, 975 (1983).
230. L. A. Paquette, G. J. Wells, K. A. Horn, and T.-H. Yan, *Tetrahedron*, **39**, 913 (1983), and references cited therein.
231. M. Grignon-Dubois, J. Dunogues, and R. Calas, *J. Chem. Res. (S)*, 6 (1979).
232. T. Hiyama, A. Kanakura, Y. Morizawa, and H. Nozaki, *Tetrahedron Lett.*, **23**, 1279 (1982).
233. S. Halazy, W. Dumont, and A. Krief, *Tetrahedron Lett.*, **22**, 4737 (1981).
234. G. J. Wells, T.-H. Yan, and L. A. Paquette, *J. Org. Chem.*, **49**, 3604 (1984).
235. L. A. Paquette, *Top. Curr. Chem.*, **119**, 1 (1984); *Isr. J. Chem.*, **21**, 128 (1981); L. A. Paquette, T.-H. Yan, and G. J. Wells, *J. Org. Chem.*, **49**, 3610 (1984).
236. T.-H. Yan and L. A. Paquette, *Tetrahedron Lett.*, **23**, 3227 (1982).
237. L. A. Paquette, C. Blankenship, and G. J. Wells, *J. Am. Chem. Soc.*, **106**, 6442 (1984).
238. L. A. Paquette, T.-H. Yan, and G. J. Wells, *J. Org. Chem.*, **49**, 3610 (1984).
239. T. Ishihara, T. Kudaka, and T. Ando, *Tetrahedron Lett.*, **25**, 4765 (1984).
240. M. Grignon-Dubois, J. Dunoguès, and R. Calas, *Can. J. Chem.*, **59**, 802 (1981).
241. M. Ochiai, K. Sumi, and E. Fujita, *Chem. Pharm. Bull.*, **31**, 3931 (1983).
242. I. Reichelt and H.-U. Reissig, *Justus Liebigs Ann. Chem.*, 828 (1984).
243. M. Schlosser, R. Dahan, and S. Cottens, *Helv. Chim. Acta*, **67**, 284 (1984).
244. J. M. Aizpurua, C. Palomo, and A. L. Palomo, *Can. J. Chem.*, **62**, 336 (1984).
245. T. Veysoglu and L. A. Mitscher, *Tetrahedron Lett.*, **22**, 1303 (1981).
246. E. Yoshii and K. Takeda, *Chem. Pharm. Bull.*, **31**, 4586 (1983).
247. E. J. Corey, H. Cho, C. Rucker, and D. H. Hua, *Tetrahedron Lett.*, **22**, 3455 (1981).
248. Y. Kita, J. Haruta, T. Fuji, J. Segawa, and Y. Tamura, *Synthesis*, 451 (1981).
249. T. Veysoglu and L. A. Mitscher, *Tetrahedron Lett.*, **22**, 1299 (1981).
250. L. Lombardo, *Tetrahedron Lett.*, **25**, 227 (1984).
251. J. M. Aizpurua and C. Palomo, *Tetrahedron Lett.*, **26**, 475 (1985).
252. E. J. Corey and P. B. Hopkins, *Tetrahedron Lett.*, **23**, 4871 (1982).
253. B. M. Trost and C. G. Caldwell, *J. Org. Chem.*, **48**, 3252 (1983); see, for example, C. J. W. Brooks, W. J. Cole, and G. M. Barrett, *J. Chromatogr.*, **315**, 119 (1984).
254. K. Furusawa and T. Katsura, *Tetrahedron Lett.*, **26**, 887 (1985).
255. Y. Kita, H. Yasuda, Y. Sugiyama, F. Fukata, J. Haruta, and Y. Tamura, *Tetrahedron Lett.*, **24**, 1273 (1983).

256. See ref. 17.
257. Y. Torisawa, M. Shibasaki, and S. Ikegami, *Chem. Pharm. Bull.*, **31**, 2607 (1983).
258. K. C. Nicolaou, S. P. Seitz, and M. R. Pavia, *J. Am. Chem. Soc.*, **103**, 1222 (1981).
259. E. W. Collington, H. Finch, and I. J. Smith, *Tetrahedron Lett.*, **26**, 681 (1985); for a similar advantageous use of *tert*-butylmethoxyphenylsilyl, $Bu^tPh(MeO)Si$, protection/deprotection, see Y. Gruindon, R. Fortin, C. Yoakim, and J. W. Gillard, *Tetrahedron Lett.*, **25**, 4717 (1984).
260. K. C. Nicolaou, R. E. Zipkin, and N. A. Petasis, *J. Am. Chem. Soc.*, **104**, 5558 (1982).
261. P. Brownbridge, *Synthesis*, 1, 85 (1983).
262. E. J. Corey and A. W. Gross, *Tetrahedron Lett.*, **25**, 495 (1984).
263. E. Negishi and S. Chatterjee, *Tetrahedron Lett.*, **24**, 1341 (1983).
264. M. E. Krafft and R. A. Holton, *Tetrahedron Lett.*, **24**, 1345 (1983).
265. J. M. Aizpurua and C. Palomo, *Synthesis*, 206 (1985).
266. C. Girard and J. M. Conia, *Tetrahedron Lett.*, 3327 (1974).
267. M. E. Krafft and R. A. Holton, *J. Am. Chem. Soc.*, **106**, 7619 (1984); *J. Org. Chem.*, **49**, 3669 (1984).
268. T. A. Hase and L. Lahtinen, *J. Organomet. Chem.*, **240**, 9 (1982).
269. M. Demuth, *Helv. Chim. Acta*, **61**, 3138 (1978).
270. Y. Yamamoto, K. Ohdoi, M. Nakatani, and K. Akiba, *Chem. Lett.*, 1967 (1984).
271. G. L. Larson, I. M. de Lopez-Cepero, and L. E. Torres, *Tetrahedron Lett.*, **25**, 1673 (1984).
272. T. Inoue, T. Sato, and I. Kuwajima, *J. Org. Chem.*, **49**, 4671 (1984).
273. M. Fiorenza, A. Mordini, S. Papaleo, S. Pastorelli, and A. Ricci, *Tetrahedron Lett.*, **26**, 787 (1985).
274. M. C. Pirrung and J. R. Hwu, *Tetrahedron Lett.*, **24**, 565 (1983).
275. J.-E. Dubois, C. Lion, and J.-Y. Dugast, *Tetrahedron Lett.*, **24**, 4207 (1983).
276. S. Danishefsky, M. Kahn, and M. Silvestri, *Tetrahedron Lett.*, **23**, 703 (1982).
277. B. M. Trost and M. J. Fray, *Tetrahedron Lett.*, **25**, 4605 (1984).
278. I. Peterson and L. G. Price, *Tetrahedron Lett.*, **22**, 2829, 2833 (1981).
279. J. E. McMurry and W. J. Scott, *Tetrahedron Lett.*, **24**, 979 (1983); **21**, 4313 (1980); see also E. Hirsch, S. Hünig, and H.-U. Reissig, *Chem. Ber.*, **115**, 3687 (1982).
280. T. Mukaiyama, *Org. React.*, **28**, 203 (1982).
281. I. Kuwajima, E. Nakamura, and M. Shimuzu, *J. Am. Chem. Soc.*, **104**, 1025 (1982).
282. See ref. 15.
283. Aldrich Catalogue, UK, 1985/86, compound 25,060-0.
284. See ref. 160.
285. Y. Kita, J. Segawa, H. Yasuda, and Y. Tamura, *J. Chem. Soc., Perkin Trans. 1*, 1099 (1982).
286. O. W. Webster, W. R. Kertler, D. Y. Sogah, W. B. Farnham, and T. V. Rajan Babu, *J. Am. Chem. Soc.*, **105**, 5706 (1983).
287. H. Urabe, Y. Takano, and I. Kuwajima, *J. Am. Chem. Soc.*, **105**, 5703 (1983); see also I. Kuwajima and H. Urabe, *J. Am. Chem. Soc.*, **104**, 6831 (1982).
288. M. T. Reetz, *Angew. Chem., Int. Ed. Engl.*, **21**, 96 (1982); *Org. Synth.*, **62**, 95 (1984).
289. C. Lion and J.-E. Dubois, *Bull. Soc. Chim. Fr.*, II, 375 (1982).
290. T. V. Lee and J. O. Okonkwo, *Tetrahedron Lett.*, **24**, 323 (1983); I. Fleming and J. Iqbal, *Tetrahedron Lett.*, **24**, 327 (1983).
291. I. Yamamoto, K. Okuda, S. Nagai, J. Motoyoshiya, H. Gotoh, and K. Matsuzaki, *J. Chem. Soc., Perkin Trans. 1*, 435 (1984).
292. S. K. Patel and I. Paterson, *Tetrahedron Lett.*, **24**, 1315 (1983).
293. K. Isaac and P. Kocienski, *J. Chem. Soc., Chem. Commun.*, 460 (1982).
294. P. Kocienski and S. D. A. Street, *J. Chem. Soc., Chem. Commun.*, 571 (1984).
295. G. S. Cockerill and P. Kocienski, *J. Chem. Soc., Chem. Commun.*, 705 (1983).
296. See, for example, R. M. Williams and A. O. Stewart, *Tetrahedron Lett.*, **24**, 2715 (1983).
297. M. Miyashita, T. Yanami, T. Kumazawa, and A. Yoshikoshi, *J. Am. Chem. Soc.*, **106**, 2149 (1984).
298. I. Fleming, J. Iqbal, and E.-P. Krebs, *Tetrahedron*, **38**, 841 (1983).
299. J. D. White, T. Nishiguchi, and R. W. Skeean, *J. Am. Chem. Soc.*, **104**, 3923 (1982).
300. G. A. Kraus and M. Shimagaki, *Tetrahedron Lett.*, **22**, 1171 (1981).
301. A. Hosomi, S. Iijima, and H. Sakurai, *Tetrahedron Lett.*, **23**, 547 (1982).
302. M. Wada, Y. Nishihara, and K. Akiba, *Tetrahedron Lett.*, **25**, 5405 (1984).
303. B. A. Brown and E. W. Colvin, unpublished work.
304. M. H. Novice, H. R. Seikaly, A. D. Seitz, and T. T. Tidwell, *J. Am. Chem. Soc.*, **102**, 5835 (1980).

305. T. Oida, S. Tanimoto, H. Ikehira, and M. Okano, *Bull Chem. Soc. Jpn.*, **56**, 645 (1983); K. Okano, T. Morimoto, and M. Sekiya, *J. Chem. Soc., Chem. Commun.*, 883 (1984); S. Miyazawa, K. Ikeda, K. Achiwa, and M. Sekiya, *Chem. Lett.*, 785 (1984).
306. N. K. Orlova, I. Y. Belavin, V. N. Sergeev, A. G. Shipov, and Y. I. Baukov, *J. Gen. Chem. USSR*, **54**, 635 (1984); N. A. Anisimova, N. K. Orlova, A. G. Shipov, I. Y. Belavin, and Y. I. Baukov, *J. Gen. Chem. USSR*, **54**, 1278 (1984).
307. J. E. Dubois and G. Axiotis, *Tetrahedron Lett.*, **25**, 2143 (1984); K. Ikeda, Y. Terao, and M. Sekiya, *Chem. Pharm. Bull.*, **29**, 1747 (1981).
308. E. W. Colvin and D. McGarry, *J. Chem. Soc., Chem. Commun.*, 539 (1985).
309. D.-C. Ha, D. J. Hart, and T. -K. Yang, *J. Am. Chem. Soc.*, **106**, 4819 (1984); see also G. Cainelli, M. Contento, D. Giacomini, and M. Panunzio, *Tetrahedron Lett.*, **26**, 937 (1985).
310. G. Gainelli, M. Contento, A. Drusiani, M. Panunzio, and L. Plessi, *J. Chem. Soc., Chem. Commun.*, 240 (1985).
311. R. P. Attrill, A. G. M. Barrett, P. Quayle, J. van der Westhuizen, and M. J. Betts, *J. Org. Chem.*, **49**, 1679 (1984).
312. P. J. Reider, R. Rayford, and E. J. J. Grabowski, *Tetrahedron Lett.*, **23**, 379 (1982); Y. Ueda, G. Roberge, and V. Vinet, *Can. J. Chem.*, **62**, 2936 (1984); Y. Tajima, A. Yoshida, N. Takeda, and S. Oida, *Tetrahedron Lett.*, **26**, 673 (1985).
313. For example, K. Akiba, M. Nakatani, M. Wada, and Y. Yamamoto, *J. Org. Chem.*, **50**, 63 (1985); A. Koskinen and M. Lounasmaa, *Tetrahedron Lett.*, **24**, 1951 (1983).
314. Y. Matsumura, J. Fujiwara, K. Maruoka, and H. Yamamoto, *J. Am. Chem. Soc.*, **105**, 6312 (1983).
315. C. H. Heathcock, in *Asymmetric Synthesis* (Ed. J. D. Morrison), Vol. 3, Academic Press, New York, 1984 pp. 111–212; see also D. A. Evans, J. V. Nelson, and T. R. Taber, *Top. Stereochem.*, **13**, 1 (1982).
316. For some recent studies, see J.-E. Dubois, G. Axiotis, and E. Bertounesque, *Tetrahedron Lett.*, **25**, 4655 (1984); B. D. Gray and J. D. White, *J. Chem. Soc., Chem. Commun.*, 20 (1985); C. Gennari, A. Bernardini, S. Cardani, and C. Scolastico, *Tetrahedron Lett.*, **26**, 797 (1985); C. Goasdoué and M. Gaudemar, *Tetrahedron Lett.*, **26**, 1015 (1985); Y. Tomo and K. Yamamoto, *Tetrahedron Lett.*, **26**, 1061 (1985); C. H. Heathcock and S. H. Montgomery, *Tetrahedron Lett.*, **26**, 1001 (1985); P. Kocienski and T. M. Willson, *J. Chem. Soc., Chem. Commun.*, 1011 (1984).
317. G. L. Larson and J. A. Prieto, *Tetrahedron*, **39**, 855 (1983).
318. G. M. Rubottom, R. Marrero, and J. M. Gruber, *Tetrahedron*, **39**, 861 (1983).
319. T. V. Lee and J. Toczek, *Tetrahedron Lett.*, **23**, 2917 (1982).
320. I. Saito, R. Nagata, K. Yuba, and T. Matsuura, *Tetrahedron Lett.*, **24**, 1737 (1983).
321. I. Saito, R. Nagata, K. Yuba, and T. Matsuura, *Tetrahedron Lett.*, **24**, 4439 (1983).
322. K. Kaneda, N. Kii, K. Jitsukawa, and S. Teranishi, *Tetrahedron Lett.*, **22**, 2595 (1981).
323. J. Tsuji, I. Minami, and I. Shimuzu, *Tetrahedron Lett.*, **24**, 5635 (1983); J. Tsuji, K. Takahashi, I. Minami, and I. Shimuzu, *Tetrahedron Lett.*, **25**, 4783 (1984); J. Tsuji, I. Minami, I. Shimuzu, and H. Kataoka, *Chemistry Lett.*, 1133 (1984).
324. E. Nakamura and I. Kuwajima, *J. Am. Chem. Soc.*, **105**, 651 (1983).
325. E. Nakamura and I. Kuwajima, *J. Am. Chem. Soc.*, **106**, 3368 (1984).
326. G. Rousseau and N. Slougui, *Tetrahedron Lett.*, **24**, 1251 (1983).
327. G. Rousseau and N. Slougui, *J. Am. Chem. Soc.*, **106**, 7283 (1984).
328. N. Slougui, G. Rousseau, and J. M. Conia, *Synthesis*, 58 (1982); N. Slougui and G. Rousseau, *Synth. Commun.*, **12**, 401 (1982).
329. W. T. Brady and R. M. Lloyd, *J. Org. Chem.*, **46**, 1322 (1981).
330. W. T. Brady and R. D. Watts, *J. Org. Chem.*, **46**, 4047 (1981).
331. H.-J. Liu and T. Dieck-Abularach, *Tetrahedron Lett.*, **23**, 295 (1982).
332. S. Tomoda, Y. Takeuchi, and Y. Nomura, *Chem. Lett.*, 1787 (1982).
333. N. B. Das and K. B. G. Torsell, *Tetrahedron*, **39**, 2227 (1983); S. K. Mukerji, K. K. Sharma, and K. B. G. Torsell, *Tetrahedron*, **39**, 2231 (1983); S. H. Andersen, K. K. Sharma, and K. B. G. Torsell, *Tetrahedron*, **39**, 2241 (1983); N. B. Das and K. B. G. Torsell, *Tetrahedron*, **39**, 2247 (1983).
334. E. Vedejs and D. A. Perry, *J. Am. Chem. Soc.*, **105**, 1683 (1983).
335. S. Danishefsky, *Acc. Chem. Res.*, **14**, 400 (1981).
336. S. Danishefsky and T. A. Craig, *Tetrahedron*, **37**, 4081 (1981).

337. S. Danishefsky, M. Bednarski, T. Izawa, and C. Maring, *J. Org. Chem.*, **49**, 2290 (1984).
338. A. P. Kozikowski and E. M. Huie, *J. Am. Chem. Soc.*, **104**, 2923 (1982).
339. O. Tsuge, E. Wada, and S. Kanemasa, *Chem. Lett.*, 239 (1983).
340. B. M. Trost and W. H. Pearson, *Tetrahedron Lett.*, **24**, 269 (1983).
341. A. Pelter, R. Al-Bayati, and W. Lewis, *Tetrahedron Lett.*, **23**, 353 (1982).
342. L. Lombardo, *Tetrahedron Lett.*, **26**, 381 (1985).
343. M. F. Semmelhack and S. Tomoda, *J. Am. Chem. Soc.*, **103**, 2427 (1981).
344. M. F. Semmelhack, S. Tomoda, and K. M. Hurst, *J. Am. Chem. Soc.*, **102**, 7567 (1980).
345. R. Ramage and A. M. McLeod, *J. Chem. Soc., Chem. Commun.*, 1008 (1984).
346. J.-P. Gesson, J.-C. Jacquesy, and M. Mondon, *Tetrahedron Lett.*, **22**, 1337 (1981); **21**, 3351 (1980).
347. R. L. Snowden, *Tetrahedron Lett.*, **22**, 97, 101 (1981).
348. G. E. Keck, E. Boden, and U. Sonnewald, *Tetrahedron Lett.*, **22**, 2615 (1981).
349. P. Brun, A. Tenaglia, and B. Waegell, *Tetrahedron Lett.*, **24**, 385 (1983).
350. E. J. Corey and H. Estreicher, *Tetrahedron Lett.*, **22**, 603 (1981).
351. S. J. Danishefsky, E. Larson, D. Askin, and N. Kato, *J. Am. Chem. Soc.*, **107**, 1246 (1985).
352. S. Danishefsky and J. F. Kerwin, *J. Org. Chem.*, **47**, 1597 (1982); S. Danishefsky, S. Kobayashi, and J. F. Kerwin, *J. Org. Chem.*, **47**, 1981 (1982).
353. S. J. Danishefsky, W. H. Pearson, D. F. Harvey, C. J. Maring, and J. P. Springer, *J. Am. Chem. Soc.*, **107**, 1256 (1985); M. Bednarski, C. Maring, and S. Danishefsky, *Tetrahedron Lett.*, **24**, 3451 (1983); M. Bednarski and S. Danishefsky, *J. Am. Chem. Soc.*, **105**, 6968 (1983).
354. S. J. Danishefsky and C. J. Maring, *J. Am. Chem. Soc.*, **107**, 1269 (1985); S. J. Danishefsky, E. Larson, and J. P. Springer, *J. Am. Chem. Soc.*, **107**, 1274 (1985).
355. W. T. Brady and M. O. Agho, *Synthesis*, 500 (1982).
356. E. L. Clennan and R. P. L'Esperance, *Tetrahedron Lett.*, **24**, 4291 (1983).
357. M. M. Abelman, R. L. Funk, and J. D. Munger, *J. Am. Chem. Soc.*, **104**, 4030 (1982).
358. R. K. Brunner and H.-J. Borschberg, *Helv. Chim. Acta*, **66**, 2608 (1983).
359. T. H. Chan and P. Brownbridge, *Tetrahedron*, **37**, 387 (1981).
360. I. H. M. Wallace and T. H. Chan, *Tetrahedron*, **39**, 847 (1983).
361. M. Asoaka, N. Yanagida, K. Ishibashi, and H. Takei, *Tetrahedron Lett.*, **22**, 4269 (1981).
362. L. Lozzi, A. Ricci, and M. Taddei, *J. Org. Chem.*, **44**, 3408 (1984).
363. Tms cyanide, iodide, azide, thiomethyl ether: W. C. Groutas and D. Felker, *Synthesis*, 861 (1980). Tms iodide, bromide: A. H. Schmidt, *Aldrichim. Acta*, **14**, 31 (1981). Tms iodide: G. A. Olah and S. C. Narang, *Tetrahedron*, **38**, 2225 (1982). Tms triflate: H. Emde, D. Domsch, H. Feger, U. Frick, A. Götz, H. H. Hergott, K. Hofmann, W. Kober, K. Krägeloh, T. Oesterle, W. Steppan, W. West, and G. Simchen, *Synthesis*, 1 (1982); R. Noyori, S. Murata, and M. Suzuki, *Tetrahedron*, **37**, 3899 (1981).
364. G. Olah, A. Husain, B. G. Balaram Gupta, G. F. Salem, and S. C. Narang, *J. Org. Chem.*, **46**, 5212 (1981).
365. M. Demuth and G. Mikhail, *Synthesis*, 827 (1982).
366. G. A. Olah, A. Husian, B. P. Singh, and A. K. Mehrotra, *J. Org. Chem.*, **48**, 3667 (1983).
367. F. Duboudin, P. Cazeau, O. Babot, and F. Moulines, *Tetrahedron Lett.*, **24**, 4335 (1983).
368. J. F. Ajao, C. W. Bird, and A. L. Brown, *Chem. Ind.* (*London*), 34 (1984).
369. N. C. Barua, R. P. Sharma, and J. N. Baruah, *Tetrahedron Lett.*, **24**, 1189 (1983).
370. M. Numazawa, M. Nagaoka, and Y. Kunitama, *J. Chem. Soc., Chem. Commun.*, 31 (1984).
371. R. Davis and K. G. Untch, *J. Org. Chem.*, **46**, 2985 (1981).
372. R. D. Miller and D. R. McKean, *Tetrahedron Lett.*, **23**, 323 (1982).
373. J. M. Denis and A. Krief, *Tetrahedron Lett.*, **22**, 1429 (1981); cf. H. Sakurai, K. Sasaki, and A. Hosomi, *Tetrahedron Lett.*, **21**, 2329 (1980); see also J. M. Aizpurua and C. Palomo, *Tetrahedron Lett.*, **25**, 3123 (1984).
374. C. Blandy, R. Choukroun, and D. Gervais, *Tetrahedron Lett.*, **24**, 4189, (1983).
375. M. A. Pudovik, Y. B. Mikhailov, and A. N. Pudovik, *J. Gen. Chem. USSR*, **53**, 413 (1983).
376. H. Shirahama, S. Murata, T. Fujita, B. R. Chhabra, R. Noyori, and T. Matsumoto, *Bull. Chem. Soc. Jpn.*, **55**, 2619 (1982).
377. P. G. Gassman and R. S. Gremban, *Tetrahedron Lett.*, **25**, 3259 (1984).
378. Org. Synth., **62**, 196 (1984).
379. W. J. Greenlee and D. G. Hangauer, *Tetrahedron Lett.*, **24**, 4559 (1983).
380. G. Boche, F. Bosolo, and M. Niessner, *Tetrahedron Lett.*, **23**, 3255 (1982).

381. L. R. Krepski, S. M. Heilmann, and J. K. Rasmussen, *Tetrahedron Lett.*, **24**, 4075 (1983).
382. R. Amoroux and G. P. Axiotis, *Synthesis*, 270 (1981).
383. E. J. Corey and S. G. Pyne, *Tetrahedron Lett.*, **24**, 2821 (1983).
384. G. A. Kraus and P. Gottschalk, *Tetrahedron Lett.*, **24**, 2727 (1983).
385. G. A. Kraus and P. Gottschalk, *J. Org. Chem.*, **49**, 1153 (1984).
386. M. Voyle, N. K. Dunlap, D. S. Watt, and O. P. Anderson, *J. Org. Chem.*, **48**, 3242 (1983).
387. K. Utimoto, Y. Wakabayashi, T. Horiie, M. Inour, Y. Shishiyama, M. Obayashi, and H. Nozaki, *Tetrahedron*, **39**, 967 (1983).
388. P. Kocienski and T. M. Willson, *J. Chem. Soc., Chem. Commun.*, 1011 (1984).
389. T. Sasaki, A. Nakanishi, M. Ohno, *J. Org. Chem.*, **46**, 5445 (1981).
390. M. T. Reetz, I. Chatziiosifidis, H. Kunzer, and H. Muller-Starke, *Tetrahedron*, **39**, 961 (1983).
391. K. Nishiyama and H. Karigomi, *Chem. Lett.*, 1477 (1982).
392. G. K. Surya Prakash, P. S. Iyer, M. Arvanaghi, and G. A. Olah, *J. Org. Chem.*, **48**, 3358 (1983).
393. T. H. Chan and S. D. Lee, *Tetrahedron Lett.*, **24**, 1225 (1983): for iminium ions, see N. Tokitoh and R. Okazaki, *Chem. Lett.*, 241 (1985).
394. J. W. Gillard and M. Israel, *Tetrahedron Lett.*, **22**, 513 (1981).
395. K. Utimoto and T. Horiie, *Tetrahedron Lett.*, **23**, 237 (1982).
396. G. A. Olah, S. C. Narang, L. D. Field, and A. P. Fung, *J. Org. Chem.*, **48**, 2766 (1983).
397. Y. A. Alexandrov, *J. Organomet. Chem.*, **238**, 1 (1982).
398. S. Matsubara, K. Tahai, and H. Nozaki, *Tetrahedron Lett.*, **24**, 3741 (1983).
399. T. Hiyama and M. Obayashi, *Tetrahedron Lett.*, **24**, 395 (1983).
400. M. Suzuki, H. Takada, and R. Noyori, *J. Org. Chem.*, **47**, 902 (1982).
401. S. Kanemoto, K. Oshima, S. Matsubara, K. Takai, and H. Nozaki, *Tetrahedron Lett.*, **24**, 2185 (1983).
402. C. W. Jefford, J. Boukovalas, and S. Kohmoto, *J. Chem. Soc., Chem. Commun.*, 523 (1984).
403. J. M. Aizpurua and C. Palomo, *Tetrahedron Lett.*, **24**, 4367 (1983).
404. N. Hashimoto, T. Aoyama, and T. Shioiri, *Chem. Pharm. Bull.*, **29**, 1475 (1981); T. Aoyama, S. Terasawa, K. Sudo, and T. Shioiri, *Chem. Pharm. Bull.*, **32**, 3759 (1984).
405. S. Mori, I. Sakai, T. Aoyama, and T. Shioiri, *Chem. Pharm. Bull.*, **30**, 3380 (1982).
406. T. Aoyama, K. Sudo, and T. Shioiri, *Chem. Pharm. Bull.*, **30**, 3849 (1982).
407. T. Aoyama, S. Toyama, N. Tamaki, and T. Shioiri, *Chem. Pharm. Bull.*, **31**, 2957 (1983).
408. O. Tsuge, S. Kanemasa, and K. Matsuda, *Chem. Lett.*, 1131 (1983).
409. O. Tsuge, S. Kanemasa, and K. Matsuda, *J. Org. Chem.*, **49**, 2688 (1984).
410. K. Nishiyama and N. Tanaka, *J. Chem. Soc., Chem. Commun.*, 1322 (1983).
411. R. Caputo, C. Ferreri, G. Palumbo, and E. Wenkert, *Tetrahedron Lett.*, **25**, 577 (1984).
412. A. R. Bassindale and T. Stout, *Tetrahedron Lett.*, **25**, 1631 (1984).
413. G. M. Rubottom and C. Kim, *J. Org. Chem.*, **48**, 1550 (1983).
414. M. Demuth and G. Mikhail, *Tetrahedron*, **39**, 991 (1983); S. P. Brown, B. S. Bal, and H. W. Pinnick, *Tetrahedron Lett.*, **22**, 4891 (1981).
415. M. E. Jung and Z. Long-Mei, *Tetrahedron Lett.*, **24**, 4533 (1983).
416. T. Imamoto, H. Yokoyama, and M. Yokoyama, *Tetrahedron Lett.*, **22**, 1803 (1981); for use in aldol reactions, see T. Imamoto, T. Matsumoto, H. Yokoyama, M. Yokoyama, and K. Yamaguchi, *J. Org. Chem.*, **49**, 1105 (1984).
417. S. Djuric, J. Venit, and P. Magnus, *Tetrahedron Lett.*, **22**, 1787 (1981).
418. E. Schwartz and A. Shanzer, *J. Chem. Soc., Chem. Commun.*, 634 (1981).
419. S. W. Djuric, *J. Org. Chem.*, **49**, 1311 (1984).
420. M. J. Pearson, *Tetrahedron Lett.*, **26**, 377 (1985).
421. L. Birkofer and W. Lückenhaus, *Justus Liebigs Ann. Chem.*, 1193 (1984).
422. H. J. Bestmann and G. Wölfel, *Angew. Chem., Int. Ed Engl.*, **23**, 53 (1984).
423. T. Morimoto, T. Takahashi, and M. Sekiya, *J. Chem. Soc., Chem. Commun.*, 794 (1984); see also T. Morimoto, M. Aono, and M. Sekiya, *J. Chem. Soc., Chem. Commun.*, 1055 (1984).
424. K. Okano, T. Morimoto, and M. Sekiya, *J. Chem. Soc., Chem. Commun.*, 883 (1984).
425. D. J. Hart, K. Kanai, D. G. Thomas, and T.-K. Yang, *J. Org. Chem.*, **48**, 289 (1983).
426. R. J. P. Corriu and J. J. E. Moreau, *Selectivity–a Goal for Synthetic Efficiency* (Eds. W. Bartmann and B. M. Trost), Verlag Chemie, Weinheim, 1984, pp. 21–42; R. J. P. Corriu, R. Perz, and C. Reye, *Tetrahedron*, **39**, 999 (1983).
427. R. J. P. Corriu, V. Huynh, J. J. E. Moreau, and M. Pataud-Sat, *J. Organomet. Chem.*, **255**, 359 (1983).

428. H. Albrecht and E.-O. Düber, *Synthesis*, 56 (1983).
429. R. J. P. Corriu, V. Huynh, J. J. E. Moreau, and M. Pataud-Sat, *Tetrahedron Lett.*, **23**, 3257 (1982).
430. W. Ando and H. Tsumaki, *Chem. Lett.*, 1409 (1983).
431. A. Millan, E. Towns, and P. M. Maitlis, *J. Chem. Soc., Chem. Commun.*, 673 (1981).
432. A. Onopchenko, E. T. Sabourin, and D. L. Beach, *J. Org. Chem.*, **48**, 5105 (1983); **49**, 3389 (1984).
433. M. Semmelhack and R. J. Misra, *J. Org. Chem.*, **47**, 2469 (1982).
434. I. Ojima and T. Kogure, *Organometallics*, **1**, 1390 (1982).
435. H. Brunner, R. Becker, and G. Riepl, *Organometallics*, **3**, 1354 (1984).
436. H. Brunner and R. Becker, *Angew. Chim., Int. Ed. Engl.*, **23**, 222 (1984).
437. D. Wang and T. H. Chan, *Tetrahedron Lett.*, **24**, 1573 (1983).
438. J. Boyer, R. J. P. Corriu, R. Perz, and C. Réyé, *J. Chem. Soc., Chem. Commun.*, 121 (1981).
439. J. L. Fry and M. A. McAdam, *Tetrahedron Lett.*, **25**, 5859 (1984).
440. M. Fujita and T. Hiyama, *J. Am. Chem. Soc.*, **106**, 4629 (1984).
441. D. J. Hart and Y.-M. Tsai, *Tetrahedron Lett.*, **22**, 1567 (1981).
442. J.-M. Lancelin, P. H. Amvam Zollo, and P. Sinay, *Tetrahedron Lett.*, **24**, 4833 (1983); see also ref. 296.
443. G. A. Kraus, K. A. Frazier, B. D. Roth, M. J. Taschner, and K. Neuenschwander, *J. Org. Chem.*, **46**, 2417 (1981); for an application, see G. A. Kraus and K. Frazier, *J. Org. Chem.*, **45**, 4820 (1980).
444. E. W. Colvin and I. G. Thom, unpublished results.
445. J. M. Aizpurua and C. Palomo, *Tetrahedron Lett.*, **25**, 1103 (1984).
446. H. J. Reich, M. J. Kelly, R. E. Olson, and R. C. Holtan, *Tetrahedron*, **39**, 949 (1983).
447. A. Hosomi, H. Hashimoto, and H. Sakurai, *J. Organomet. Chem.*, **175**, C1 (1979).
448. J. A. Miller and G. Zweifel, *Synthesis*, 288 (1981).
449. See ref. 173.
450. D. Schinzer and C. H. Heathcock, *Tetrahedron Lett.*, **22**, 1881 (1981).
451. A. Degl'Innocenti, S. Pike, D. R. M. Walton, G. Seconi, A. Ricci, and M. Fiorenza, *J. Chem. Soc., Chem. Commun.*, 1202 (1980); A. Ricci, A. Degl'Innocenti, M. Fiorenza, M. Taddei, M. A. Spartera, and D. R. M. Walton, *Tetrahedron Lett.*, **23**, 577 (1982).
452. S. R. Wilson, M. S. Haque, and R. N. Misra, *J. Org. Chem.*, **47**, 747 (1982).
453. J. W. Wilt, F. G. Belmonte, and P. A. Zieske, *J. Am. Chem. Soc.*, **105**, 5665 (1983).
454. G. Stork and M. Kahn, *J. Am. Chem. Soc.*, **107**, 500 (1985).
455. H. Nishiyama, T. Kitajima, M. Matsumoto, and K. Itoh, *J. Org. Chem.*, **49**, 2298 (1984).

Part 2

Use of Transition Metal Organometallics in Organic Synthesis

The Chemistry of the Metal—Carbon Bond, Vol. 4
Edited by F. R. Hartley

CHAPTER **7**

Use of organoiron compounds in organic synthesis

DIDIER ASTRUC

Laboratoire de Chimie Organique et Organométallique, U.A. CNRS No. 35, Université de Bordeaux I, 351 Cours de la Libération, 33405 Talence Cédex, France

I. INTRODUCTION

Iron plays an essential role in life and in modern industry and it also is the commonest and cheapest metal. However, the molecular chemistry of iron evolved slowly. The first carbonyl compound, $[Fe(CO)_5]$, was prepared by Berthelot and Mond[1] nearly a century ago (1891) and the first organoiron complex, [(butadiene)$Fe(CO)_3$] was found much later (1930) by Reihlen *et al.*[2]. On this time scale, the discovery of ferrocene, the first non-carbonyl organoiron compound, by Keally and Pauson[3] and by Miller *et al.*[4] appears relatively recent (1951). However, this latter date marks the beginning of the tremendous development of organoiron chemistry. This richness arises for two reasons. Firstly, iron has a strong tendency to form complexes with 18 electrons in the valence shell[5] and this rare gas configuration confers high thermodynamic stability. Secondly, in the organoiron complexes, the metal centre has no oxophilicity (in contrast to early transition metal complexes), which generally makes these complexes stable to air and water and thus easily handled.

The raw materials[6] for organoiron complexes are $FeCl_2$, ether-soluble $FeCl_3$, the toxic light yellow liquid [$Fe(CO)_5$], the orange solid [$Fe_2(CO)_9$] {available by photolysis of [$Fe(CO)_5$]}, cheap, commercially available ferrocene, and the more expensive dimer [$(\eta^5\text{-}C_5H_5)_2Fe_2(CO)_4$], often called in short form† Fp_2[13] still often made in the laboratory from [$Fe(CO)_5$] and dicyclopentadiene, as is the permethyl analogue[7]}.

The essential strategy in using iron complexes stoichiometrically in organic synthesis[8] involves (i) complexation of an organic molecule with one of these raw materials, (ii) reaction(s) of the iron complex, and (iii) decomplexation (Scheme 1).

However, two or even all three operations are sometimes carried out in a single reaction, which may be stoichiometric or, exceptionally, catalytic. For instance, many reactions of unsaturated substrates or of organic halides with [$Fe(CO)_5$] afford the transformation and/or the dehalogenation of the substrate, the resulting molecule being obtained free or as an iron complex. In the case of CO, homogeneous catalytic reduction, feasible using [$Fe(CO)_5$], may be compared with the heterogeneous, non-selective, Fischer–Tropsch process using iron[9]. Catalytic chemistry involving radical chain reactions of organoiron species will also be briefly mentioned.

Organoiron complexes have a three-dimensional structure affording regio- and stereo-control of reactions, a feature of considerable interest for the study of stereochemistry and asymmetric synthesis. The most common and useful class of compounds is the η^4-polyeneiron tricarbonyl series[10], synthesized by reaction of the polyene with an iron carbonyl (equation 2). [$Fe(CO)_5$] is commonly used under thermal or photolytic conditions but an improved method consists in using [(benzylideneacetone)$Fe(CO)_3$] instead of [$Fe(CO)_5$]. The latter is especially useful for the complexation of thermally and photolytically sensitive dienes, such as ergosteryl benzoate. In this case, *p*-methoxybenzylideneacetone is used as a catalytic transfer agent in the presence of [$Fe_2(CO)_9$] (see Section IV.B). In these robust, 18-electron complexes, the valence shell of iron is fulfilled by coordination with two conjugated double bonds in a *cis* configuration. Simple olefins give [$(\eta^2\text{-olefin})Fe(CO)_4$][11] (equation 1), whereas arenes do not give [$(\eta^6\text{-arene})Fe(CO)_2$][12] (equation 3).

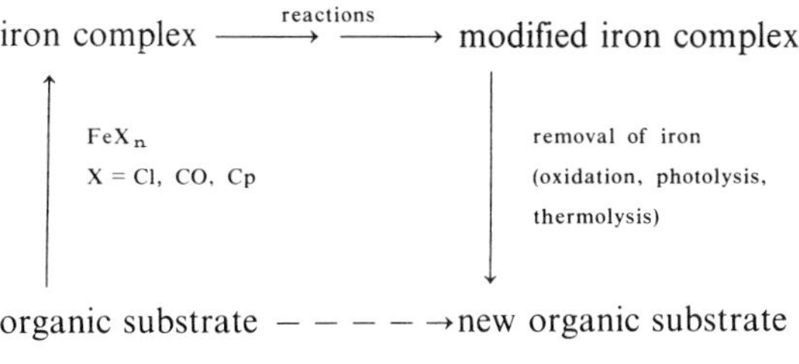

SCHEME 1

†$\eta^5\text{-}C_5H_5$ is abbreviated as Cp throughout this chapter and $CpFe(CO)_2$ is often written as Fp[13].

$[Fe(CO)_5]$, Δ (1)

$[Fe(CO)_5]$, Δ (2)

$[Fe_x(CO)_y]$ (3)

However, both olefinic[13] and arenic[14] ligands form monocationic complexes bearing the ancillary ligand Cp (equations 4 and 5). Useful bis(arene) complexes are also obtained from arenes, $AlCl_3$, and $FeCl_3$ according to a Fischer-type synthesis[15] (equation 6). In the

Ox, RT; Fe^+ X^- (4)

arene; $AlCl_3$, Al, 70–200 °C; Fe^+ X^- (5)

$FeCl_3$; $AlCl_3$, Al, 70–200 °C; Fe^{2+} $(X^-)_2$ (6)

two syntheses of sandwich complexes, the counter anion is $AlCl_4^-$ before hydrolysis, Cl^- after hydrolysis and PF_6^- after metathesis with aqueous $H^+PF_6^-$, which precipitates the convenient PF_6^- salts.

Cationic and neutral complexes are interconverted by hydride or proton addition or abstraction. This concept is essential since, in this way, the complexes can be reacted with electrophiles in their neutral forms and with nucleophiles as their cationic counterparts. Often, these operations can be combined and two-step sequences alternating the hapticity of the ligand have been used to obtain polyfunctionalization (Section III.A) (Scheme 2).

The reactivity of paramagnetic complexes (e.g. bearing 17 or 19 electrons in the valence shell, etc.) is considerably greater than that of diamagnetic complexes[16]. For instance, ligand-exchange reactions proceed at rates many orders of magnitude greater than with

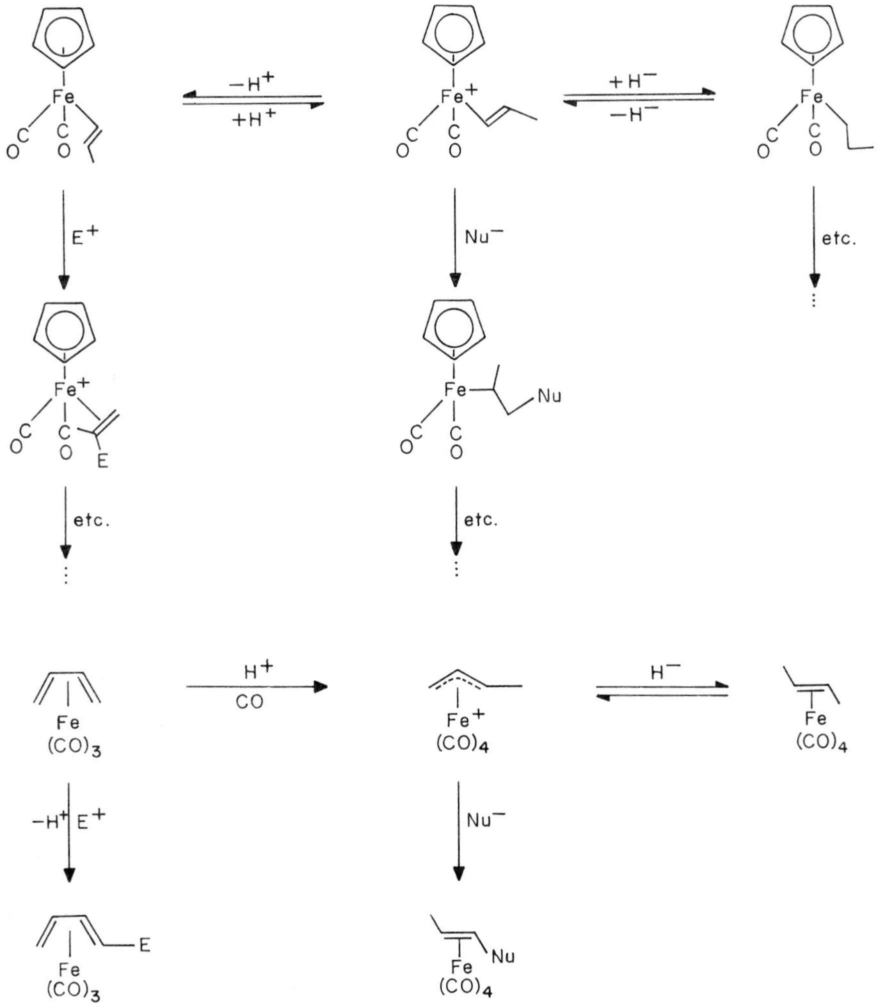

SCHEME 2a

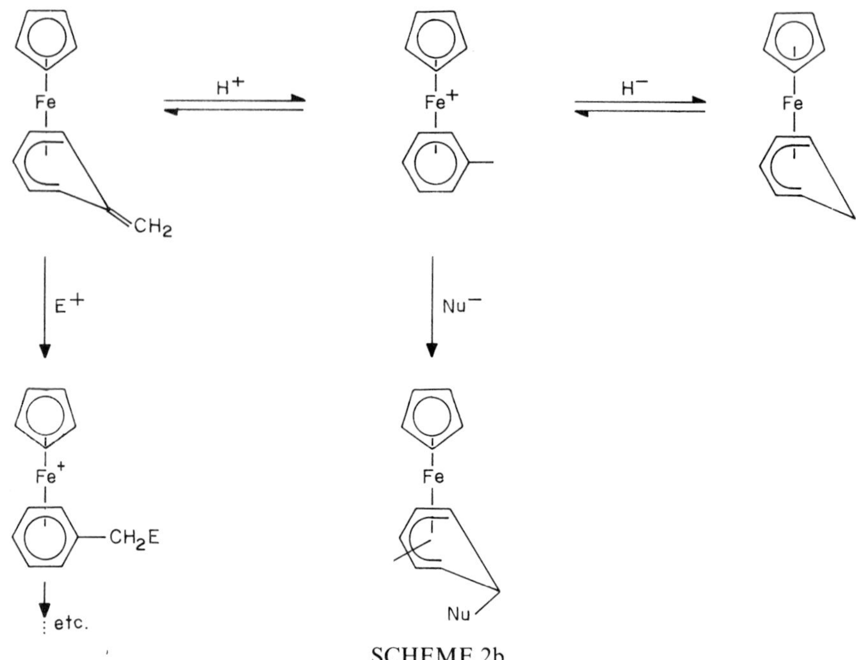

SCHEME 2b

their isostructural 18-electron counterparts[17]. This is the basis for electrocatalytic ligand-exchange[18] and decomplexation processes[19]. This concept is fundamental for the removal of organic substrates from complexes. Odd-electron complexes are also capable of effecting radical chain processes[20] and electron transfer leading to C—H bond activation and C—C bond formation[21] (cf. Section VIII).

II. REACTIONS OF IRON COMPLEXES WITH NUCLEOPHILES

The reaction of carbanions with iron-coordinated unsaturated ligands provides a powerful way of making C—C bonds. When the unsaturated hydrocarbon ligand is coordinated to a strongly electron-withdrawing group such as $Fe(CO)_3^+$, even neutral carbon nucleophiles can form C—C bonds. Depending on the nucleophile, the organometallics, and the reaction conditions, the reaction of a nucleophile with a transition metal complex can lead to: (i) nucleophilic addition; (ii) nucleophilic substitution; (iii) deprotonation of the side-chain of the ligand; (iv) electron transfer; or (v) ligand displacement by the nucleophile.

Carbon nucleophiles add exclusively to the *exo* face of the ligand (remote from iron) in the most favourable cases. Stabilized carbanions often do so whereas simple alkyl carbanions often transfer one electron as a side-reaction if they are used as alkali metal or Grignard reagents. In the latter case the best results for C—C bond formation are obtained using Cu, Zn, or Cd reagents. When the coordinated ligand bears β-hydrogen(s), O and N nucleophiles and also hydrides may also effect deprotonation. In any case, all these reagents can also effect electron transfer[22] (Scheme 3).

Sometimes what appears to be a simple hydride reduction is a masked process consisting of electron transfer followed by H atom transfer. In the case shown in

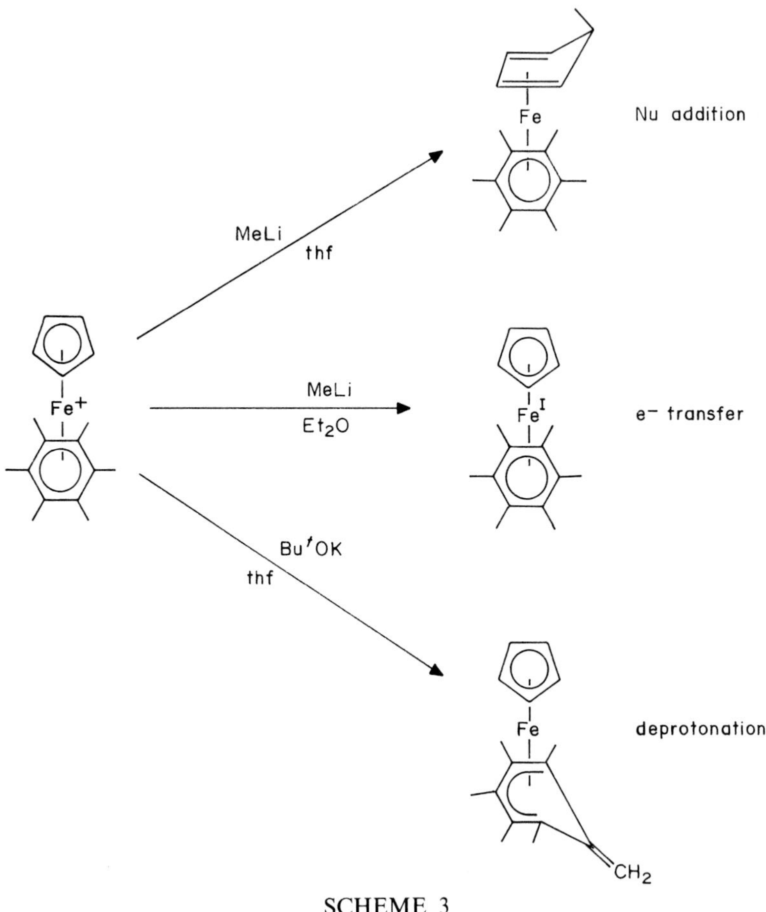

SCHEME 3

equation 7, the intermediate d^7Fe^I 19-electron species has been trapped and characterized quantitatively by Mössbauer spectroscopy.

$$\text{Fe}^+ \xrightarrow[\text{thf}, -60\ ^\circ\text{C}]{\text{LiAlH}_4} \text{Fe} \xrightarrow[\text{thf}, -40\ ^\circ\text{C}]{\text{LiAlH}_4} \text{Fe} \qquad (7)$$

Although it has been claimed that the reactions of nucleophiles with organometallic cations are charge-controlled[23], orbital control accounts equally well for the regioselectivities observed in nucleophilic additions, substitutions and deprotonations[24]. Moreover, in

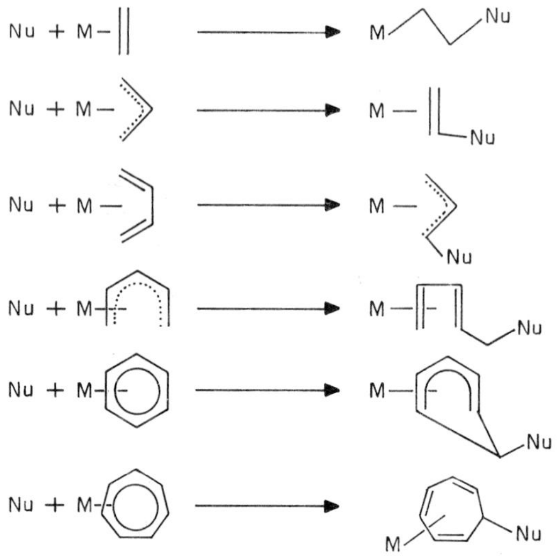

SCHEME 4

the series $[(\eta^6\text{-benzene})(\eta^5\text{-cyclohexadienyl})Fe]^+$, the orbital control theory predicts the correct regioselectivity of nucleophilic addition whereas the charge control theory does not[25]. It is probable that both types of control intervene more or less[26] depending on the series, but that orbital control predominates. This tendency is, of course, even more marked in neutral series.

On nucleophilic addition or deprotonation, the hapticity of the hydrocarbon ligand decreases by one unit and its parity changes (Scheme 4). Thus the temporary activation of an unsaturated substrate as an even ligand in an organoiron reaction necessitates two reactions of the coordinated ligand in the iron complex.

Nucleophilic addition may be followed or preceded by the reaction of an electrophile or, exceptionally, by that of another nucleophile. For instance, the well known substitution of cyclohexadiene, as in Scheme 5[27], requires altogether four reactions to make a single C—C bond. However, the iron complex is sometimes used twice to effect a double substitution as in bis(arene)iron complexes[28].

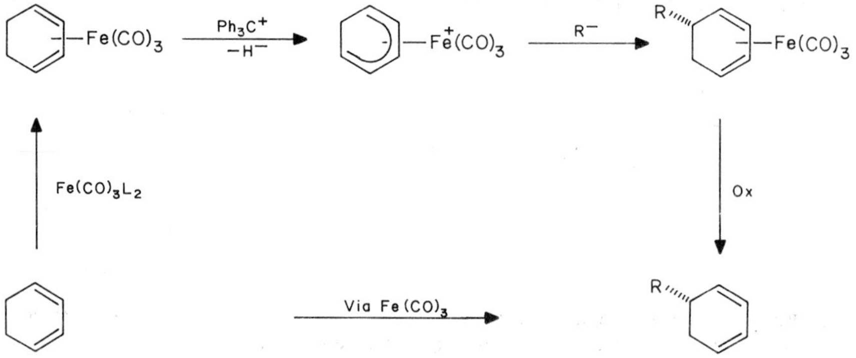

SCHEME 5

A. CO and Alkylidenes

The activation of CO by transition metals is relevant to the Fischer–Tropsch process transforming $CO + H_2$ into higher hydrocarbons and alcohols non-selectively

$$CO + H_2 \xrightarrow{Fe} \text{alcones} + \text{alcenes} + \text{alcohols} \tag{8}$$

(equation 8)[9]. Although many mechanisms for the Fischer–Tropsch process have been proposed, none is firmly established. However, the modelling of CO reduction on a single transition metal centre is fairly successful with Re[29] and even Fe[30] complexes. One may consider the activation of H_2 (by oxidative addition, giving a dihydride species) and of CO on the same or on two different metal centres. The first possibility involves the thermodynamically unfavourable insertion of CO into an M—H bond to provide a metal–formyl species (equation 9)[31]. The second possibility involves nucleophilic attack on coordinated CO by a metal hydride species[32]. Stable main group and also transition metal hydrides can serve as model reactants in the reduction of transtition metal carbonyl complexes. They react as nucleophiles at the carbonyl carbon to give Lewis acid-stabilized formyl complexes. Although the decomposition of these metal formyl species to metal hydrides is often a competitive process, further reduction to hydroxymethyl and methyl-metal species is also observed (Scheme 6)[30].

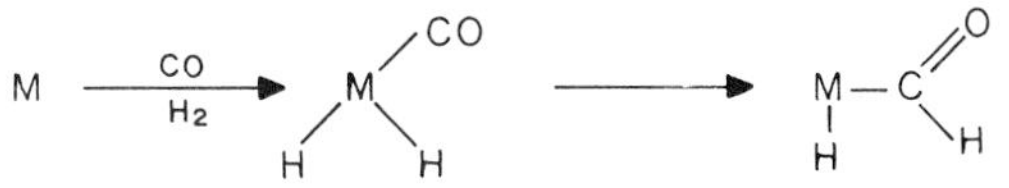

M —(CO, H_2)→ M(CO)(H)(H) → M(H)—C(=O)H (9)

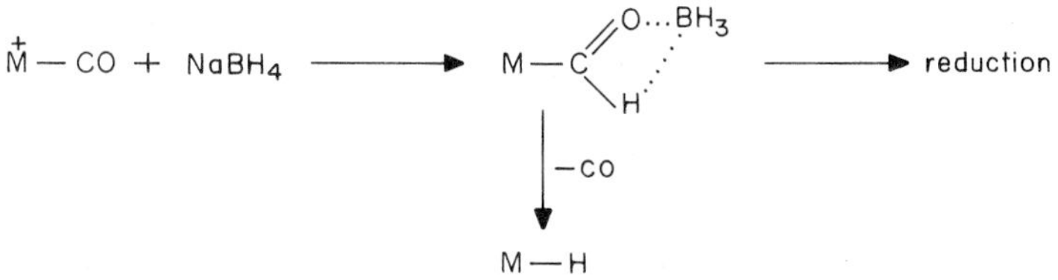

$\overset{+}{M}$—CO + $NaBH_4$ → M—C(=O···BH_3)(H) → reduction

(−CO) ↓ M—H

SCHEME 6

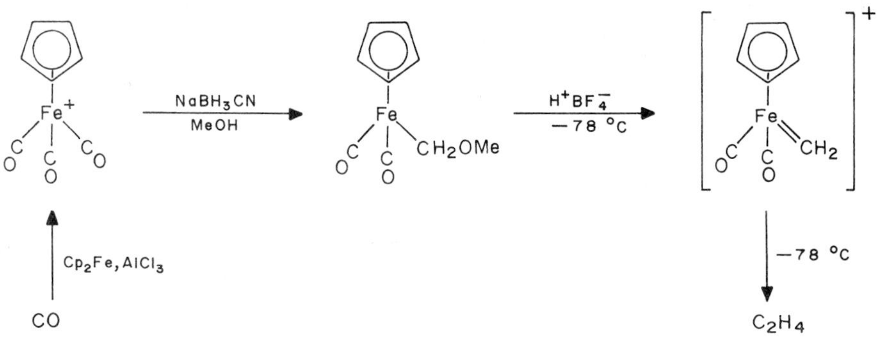

CO —($Cp_2Fe, AlCl_3$)→ $CpFe(CO)_3^+$ —($NaBH_3CN$, MeOH)→ $CpFe(CO)_2CH_2OMe$ —($H^+BF_4^-$, −78 °C)→ $[CpFe(CO)_2=CH_2]^+$ —(−78 °C)→ C_2H_4

SCHEME 7

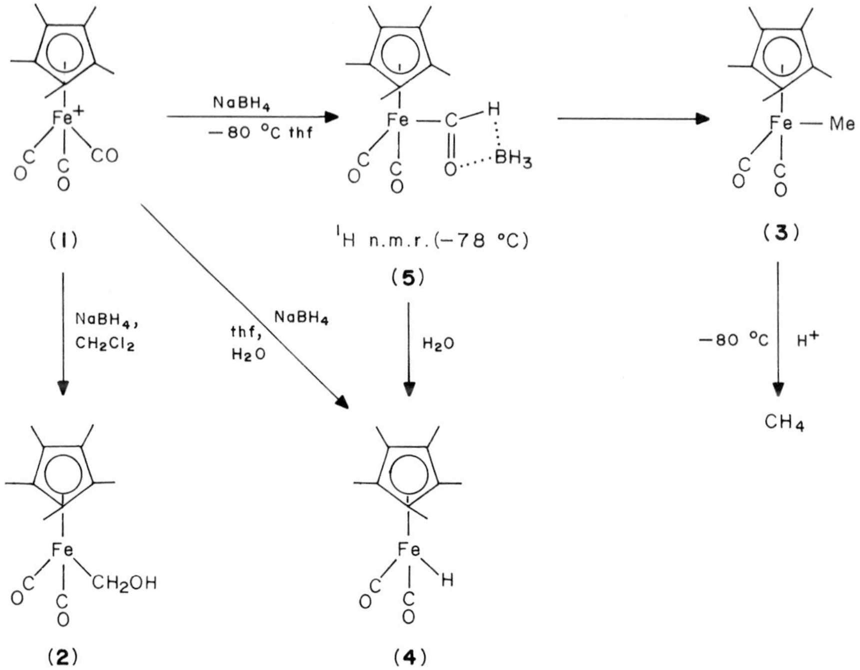

SCHEME 8

Bodnar *et al.*[300] found that the reduction of $[CpFe(CO)_3]^+$ by $NaBH_3CN$ in methanol gives the methoxymethyl complex $[CpFe(CO)_2CH_2OMe]$. It is known that protonation of the latter gives the transient methylene complex $[CpFe(CO)_2{=}CH_2]^+$, which decomposes to ethylene (Scheme 7)[31]. Using the pentamethylcyclopentadienyl analogue **1** and $NaBH_4$, all the steps of CO reduction can be observed[30b] as in the rhenium series (Scheme 8). The hydroxymethyl, methyl, and hydride complexes **2**, **3**, and **4** can be isolated under ambient conditions while the 'formyl' complex **5** is only observed free of BH_3 at $-80\,^{\circ}C$. When one CO is replaced with a phosphine, the formyl intermediates are also observed and the reaction product of the low-temperature reaction is again a methyl complex, $[C_5Me_5Fe(CO)(PR_3)Me]$ (R = Me, Bu^n, Ph), but the phosphine ligand can stabilize a carbonium attached to iron. This allows the observation of both the free formyl complex and its adduct $[C_5Me_5Fe(CO)(PR_3)(CHO{:}BH_3)]^+$ by ^1H and ^{13}C n.m.r. spectroscopy, but not the hydroxymethyliron complexes. The iron–acyl complex **6** is also reduced to the iron–ethyl complex **7** by reaction with BH_3 (equation 10) but not with

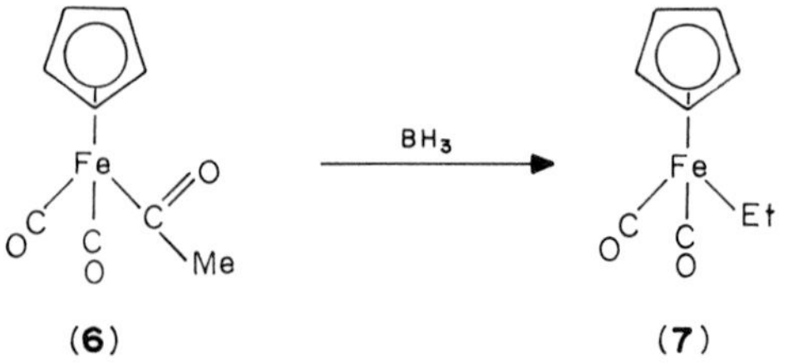

$$\text{CpFe(CO)}_2\text{C(=O)Me}\ (\mathbf{6}) \xrightarrow{BH_3} \text{CpFe(CO)}_2\text{Et}\ (\mathbf{7}) \qquad (10)$$

BH_4^-[35]. This is in contrast to the borohydride reduction of organic aldehydes, proceeding via alkoxyboranes, which are hydrolysed to alcohols without the formation of alkanes[29].

When two CO groups are replaced by phosphines, $NaBH_4$ does not react and $LiAlH_4$ reacts exclusively by electron transfer[34]. The metal hydrides, not the formyl complexes, are obtained in these cases. In the Cp series, the pioneering work by Green and Wilkinson indicated that $NaBH_4$ reduces $[CpFe(CO)_2(PPh_3)]^*$ to give $[(\eta^4\text{-}Cp)HFe(CO)_2(PPh_3)]$[35]. There is now n.m.r. evidence that the reduction of the cyclopentadienyl ring also proceeds via a formyl intermediate[30c] (equation 11). It is likely

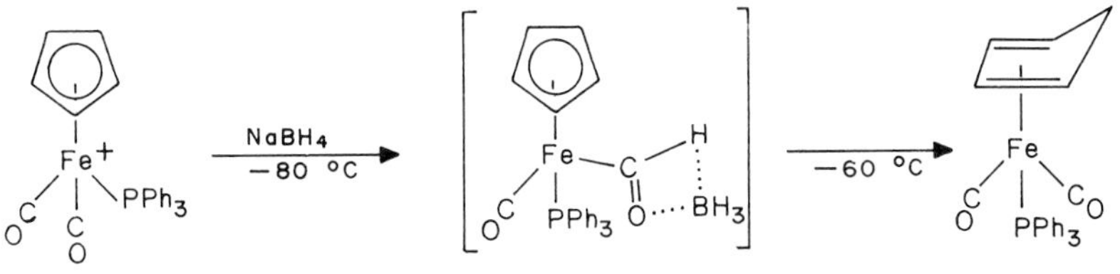

(11)

that such a carbonyl to hydrocarbon hydride shift is the true mechanism of reduction in many series containing both CO and a hydrocarbon ligand. These model studies show a high selectivity in transition metal-mediated CO reduction can be obtained.

Nucleophilic addition of OH^- at a carbonyl occurs in $[Fe(CO)_5]$ and $[C_5R_5Fe(CO)_3]^+$ (R = H, Me)[36,37], providing the hydrides $[HFe(CO)_4]^-$ and $[(C_5R_5)Fe(CO)_2H]$ via the metallocarboxylic acid intermediates; Pettit isolated such an acid starting from $[C_5H_5Fe(CO)_2(PPh_3)]^+$[36a]. Many other nucleophiles react in a similar fashion, in

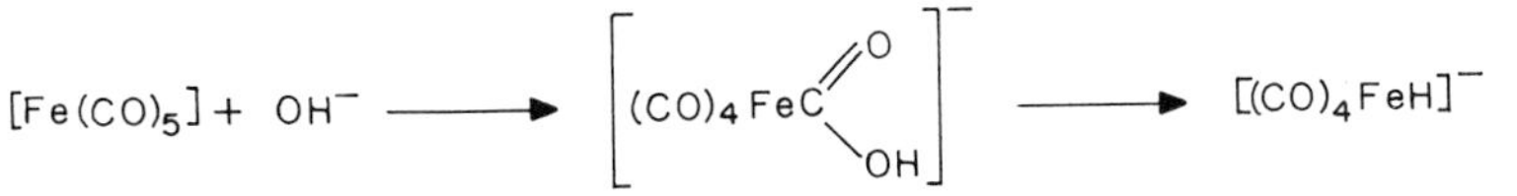

$$[Fe(CO)_5] + OH^- \longrightarrow \left[(CO)_4FeC{\begin{matrix}{}^{\nearrow O}\\ {}_{\searrow OH}\end{matrix}}\right]^- \longrightarrow [(CO)_4FeH]^- \qquad (12)$$

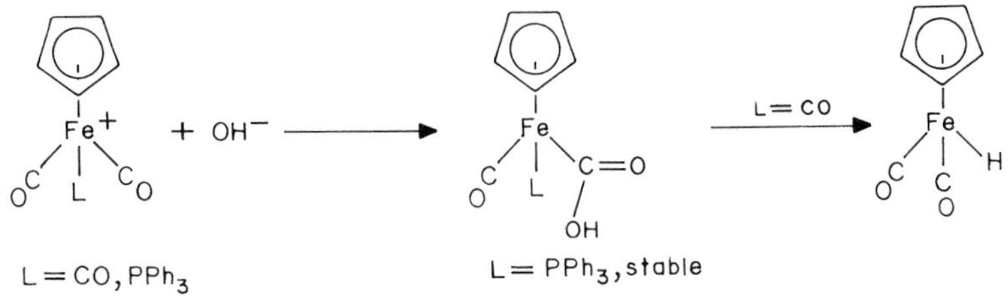

(13)

particular amines and carbanions[38]. The reaction of carbanions is especially useful (see Section V.B). All the species obtained by nucleophilic attack at the carbonyl have tautomeric zwitterionic alkylidene forms, which provide additional stabilization:

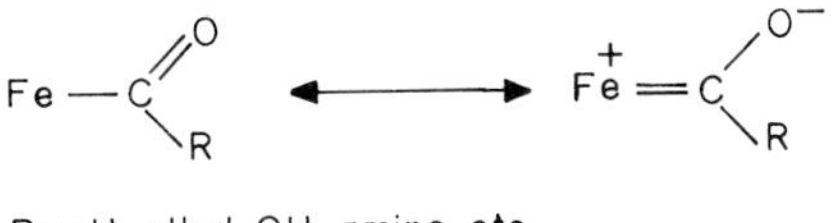

$$Fe{-}C(=O)R \longleftrightarrow \overset{+}{Fe}{=}C(O^-)R$$

R = H, alkyl, OH, amino, etc.

Isoelectronic to CO, the alkylidene ligand can similarly be attacked at the α-carbon by nucleophiles such as hydrides (equation 14)[39].

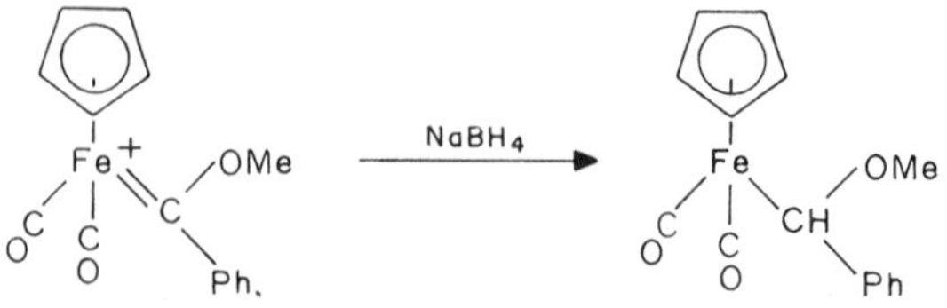

$$\text{CpFe}^{+}(\text{CO})_2{=}\text{C(OMe)Ph} \xrightarrow{\text{NaBH}_4} \text{CpFe}(\text{CO})_2\text{CH(OMe)Ph} \quad (14)$$

Of relevance to the Fischer Tropsch process is the finding that the hydroxymethyl complex **3** decomposes spontaneously to the methyl complex **4** in polar solvents. The latter favours the ionization of **3**:

$$\underset{(\mathbf{2})}{C_5Me_5Fe(CO)_2CH_2OH} \rightleftharpoons [C_5Me_5Fe(CO)_2\,({=}CH_2)]^+,\ OH^- \xrightarrow{(2)} (\mathbf{3}) \quad (15)$$

Hydride abstraction from (**2**) is easy, giving the Fischer-type carbene complex[15].

B. η^2-Olefin Complexes

Nucleophilic addition to olefin complexes followed by protonation of the metal–alkyl intermediate leads to reduction of the double bond in addition to C—C or C—element bond formation. The major activating organoiron group is $CpFe(CO)_2{}^+$, (Fp^+)[41], but some examples are also known with $Fe(CO)_4$.

$[Fp(olefin)]^+$ complexes are readily prepared by ligand exchange between free olefins and $[FpL]^+$ salts, L being a labile ligand such as isobutene[42], or more simple thf[43]. $[(Fpthf)]^+\ PF_6{}^-$ is best obtained by oxidation of Fp_2, e.g. with $[Cp_2Fe]^+PF_6{}^-$ in thf (Scheme 9)[44]. $[Fp(olefin)]^+$ complexes are also accessible by hydride abstraction from [Fp(alkyl)] complexes (as the isobutene complex above)[44], by protonation of [Fp(allyl)] complexes[45], or by deoxygenation of epoxides using Fp^-[46] followed by acidification (the stereochemistry of the epoxide is retained in the olefin complex (Scheme 10). The reaction of FpX with $AlCl_3$ in the presence of olefin, the first preparation of $[Fp(olefin)]^+$ complexes (equation 16)[47], is not applicable to functional olefins because of the presence of $AlCl_3$.

$$\tfrac{1}{2}Fp_2 \xrightarrow{X_2} FpX \xrightarrow[\text{olefin}]{AlX_3} [Fp(olefin)]^+X^- \quad (16)$$

$$(X = Cl,\ Br)$$

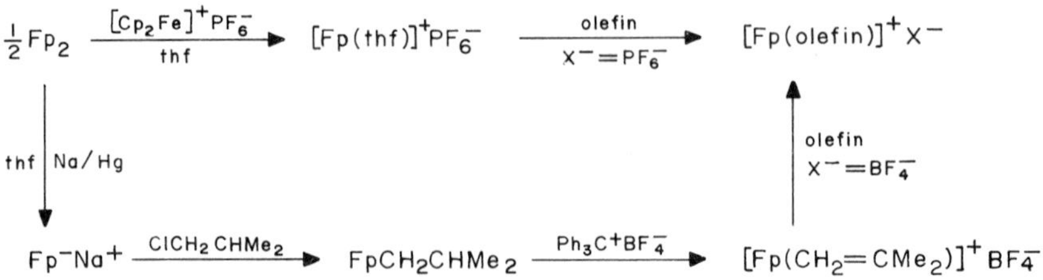

$$\tfrac{1}{2}Fp_2 \xrightarrow[\text{thf}]{[Cp_2Fe]^+PF_6^-} [Fp(thf)]^+PF_6^- \xrightarrow[X^-=PF_6^-]{\text{olefin}} [Fp(olefin)]^+X^-$$

$$\tfrac{1}{2}Fp_2 \xrightarrow{\text{thf, Na/Hg}} Fp^-Na^+ \xrightarrow{ClCH_2CHMe_2} FpCH_2CHMe_2 \xrightarrow{Ph_3C^+BF_4^-} [Fp(CH_2{=}CMe_2)]^+BF_4^- \xrightarrow[X^-=BF_4^-]{\text{olefin}} [Fp(olefin)]^+X^-$$

SCHEME 9

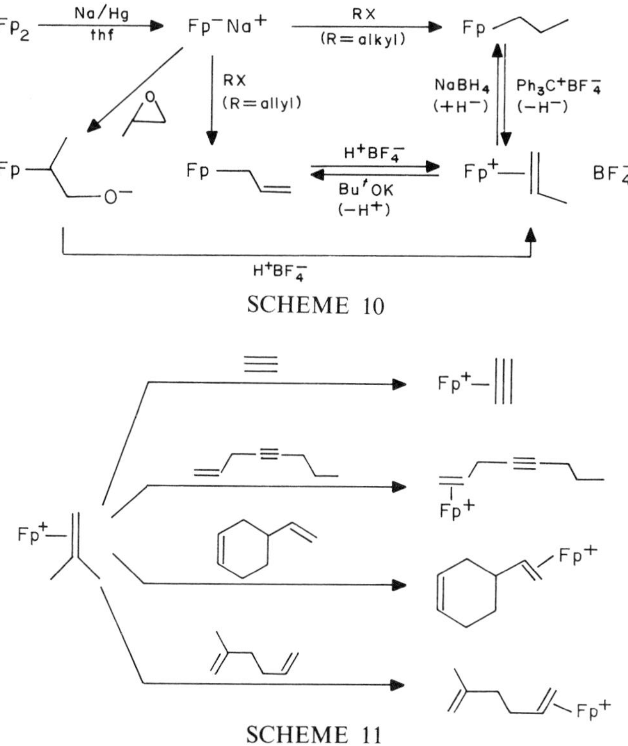

SCHEME 10

SCHEME 11

Although $[Fp(alkyne)]^+$ complexes are also prepared by the simple ligand-exchange reaction of alkynes and $[Fp(isobutene)]^+BF_4^-$, enynes are selectively complexed via the double bond because of the greater stability of olefin complexes (Scheme 11)[48]. Similarly, dienes are complexed at the less substituted double bond.

The preparation and synthetic applications of $[Fp(olefin)]^+$ complexes have mostly been developed by Cutler *et al.*[49]. These salts react with a wide variety of C and heteroatom nucleophiles. Reduction with $NaBH_4$, or better with $NaBH_3CN$, occurs regiospecifically at the less substituted carbon (equation 17), and stabilized carbanions

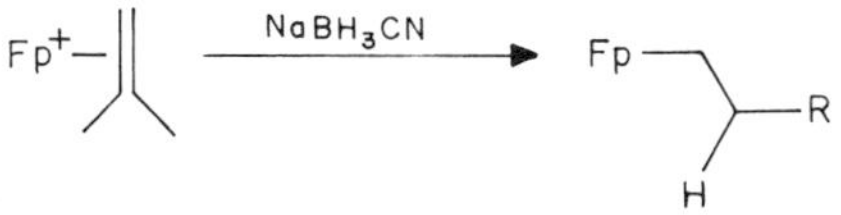

Fp⁺— $\xrightarrow{NaBH_3CN}$ Fp—, R, H (17)

R	Yield (%)
C_5H_{11}	81
$PhCH_2$	71
Me	96
Ph	56

add efficiently. However, the regioselectivity is not as good, unless the olefin substituent stabilizes a carbonium ion in the β-position or bears electron-withdrawing functionalities (equations 18 and 19).

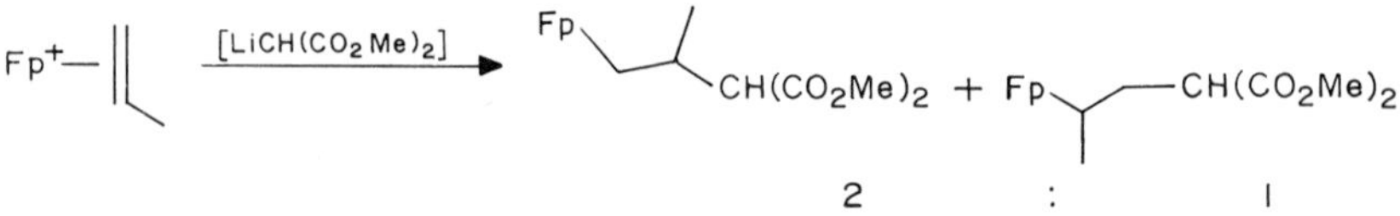

$$Fp^+\text{—}\|\text{(propene)} \xrightarrow{[LiCH(CO_2Me)_2]} FpCH_2CH(CH_3)CH(CO_2Me)_2 + FpCH(CH_3)CH_2CH(CO_2Me)_2 \quad 2 : 1 \tag{18}$$

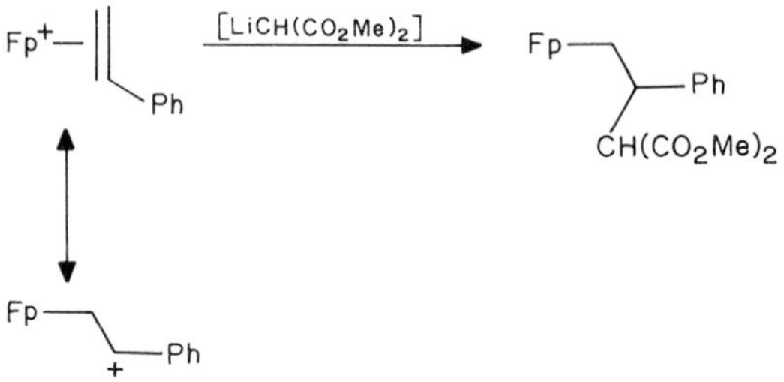

$$Fp^+\text{—}\|\text{(CH=CHPh)} \xrightarrow{[LiCH(CO_2Me)_2]} FpCH_2CH(Ph)CH(CO_2Me)_2 \tag{19}$$

$$\updownarrow$$

$$FpCH_2\overset{+}{C}HPh$$

Neutral organoiron complexes having nucleophilic methylenes also form C—C bonds by reaction with **8a** (Scheme 12). Unstabilized carbanions used as alkali metal or Grignard reagents give electron transfer leading to the formation of Fp_2. Milder alkylating agents such as $Li[CuR_2]$ must be used (Scheme 13).

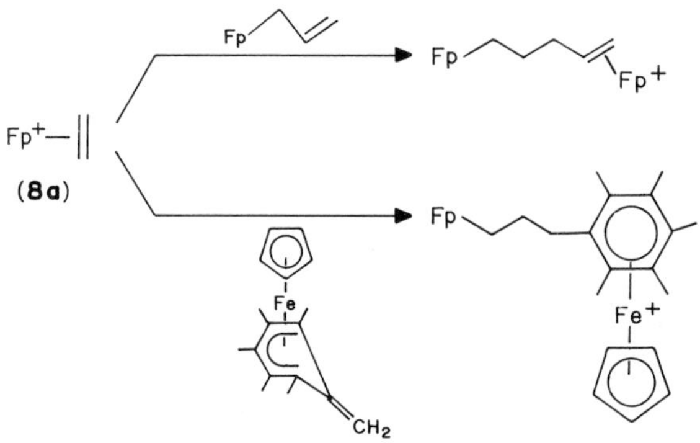

Fp⁺—‖ (**8a**) + Fp–allyl ⟶ Fp–(CH₂)₃–CH=CH₂–Fp⁺

Fp⁺—‖ (**8a**) + CpFe(η⁵-C₆Me₅=CH₂) ⟶ Fp–CH₂CH₂CH₂–C₆Me₅–Fe⁺Cp

SCHEME 12

$$Fp^+\text{—}\|\text{(propene)}\ (\mathbf{8b}) \xrightarrow{\text{MeLi or MeMgI}} [Fp^{\cdot}\text{—}\|,\ Me^{\cdot}]\ (19e) \longrightarrow Fp_2 + \text{propene} + MeH + Me_2$$

$$Fp^+\text{—}\|\text{(propene)}\ (\mathbf{8b}) \xrightarrow[70\%]{Li[CuMe_2]} FpCH_2CH(CH_3)_2 + FpCH(CH_3)CH_2CH_3 \quad 1 : 1$$

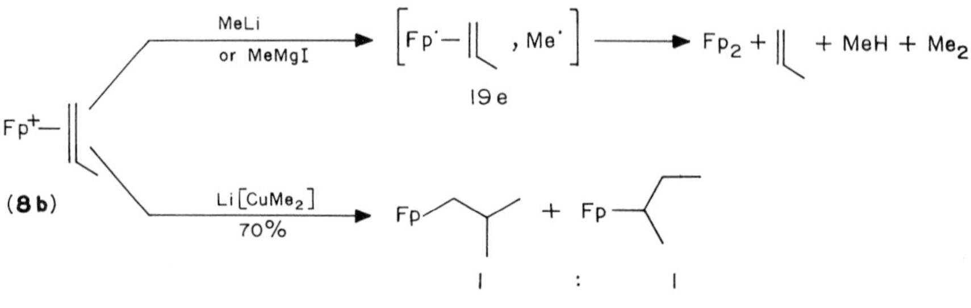

SCHEME 13

Addition of the nucleophile always proceeds *trans* with respect to iron, as shown below in the cyclopentene complex. The substituted olefin can be recovered by *trans* hydride abstraction using $Ph_3C^+BF_4^-$ followed by decomplexation with NaI. Alternatively, the new $[Fp(olefin)]^+$ complex can react with a nucleophile, bringing about a second functionalization (Scheme 14).

The methyl vinyl ketone complex[9], prepared from the corresponding epoxide and Fp^-, gives Michael addition regiospecifically at the unsubstituted olefinic carbon with lithium enolates and enamines (Scheme 15). Trimethylsilyl enol ethers can also be used as the

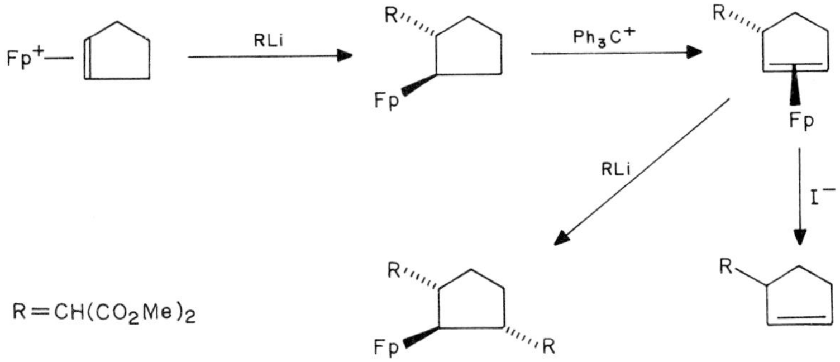

Fp+ RLi Ph3C+ RLi I−

R=CH(CO2Me)2

SCHEME 14

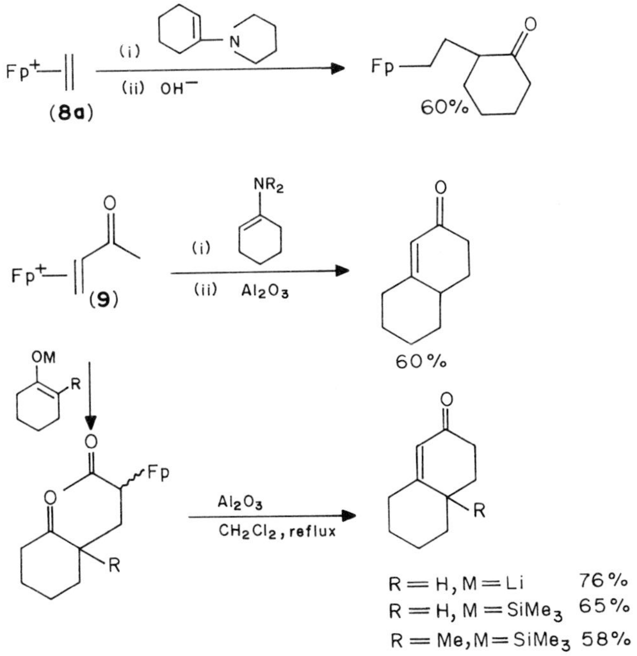

(8a) (i) (ii) OH− 60%

(9) (i) (ii) Al2O3 60%

Al2O3 CH2Cl2, reflux

R=H, M=Li 76%
R=H, M=SiMe3 65%
R=Me, M=SiMe3 58%

SCHEME 15

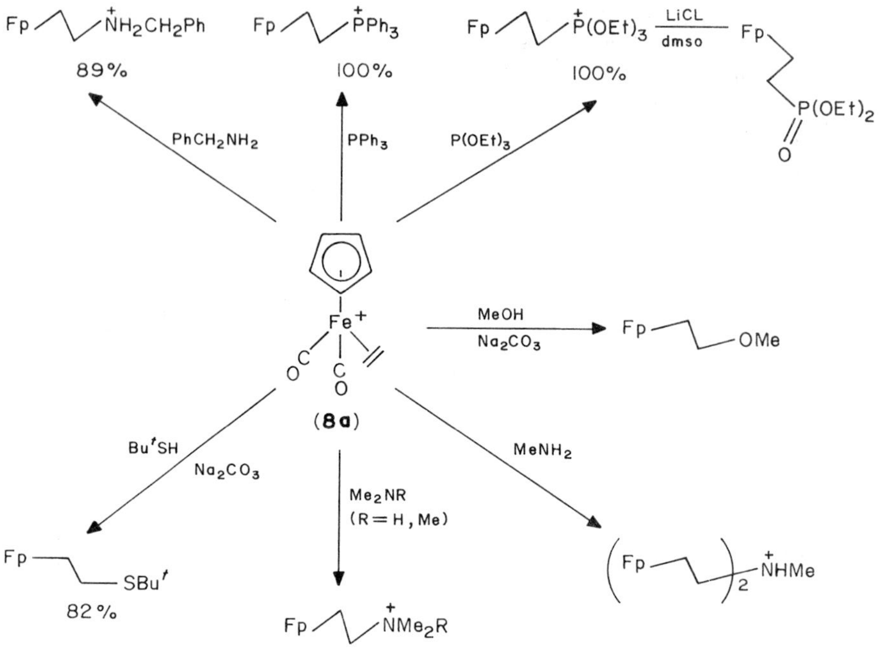

SCHEME 16

source of enolate anions. The $SiMe_3$ group is presumably removed by F^- derived from the counter anion BF_4^- or PF_6^- to generate the enolate. Further reaction with Al_2O_3 provides cyclization and decomplexation.

Several nucleophiles react with **8a** through their heteroatoms, including alkoxides, amines, thiols, phosphines, and phosphites (Scheme 16).

When the olefin bears an allyl proton, the reaction of basic O and N nucleophiles causes deprotonation of $[Fp(olefin)]^+$ to give $[Fp(\eta^1\text{-allyl})]$ complexes. The reactions of $X^- = CN^-$, NCO^-, N_3^-, and halides lead to removal of the olefin from the complex and formation of FpX (equation 20). This ligand-exchange reaction is useful for removing

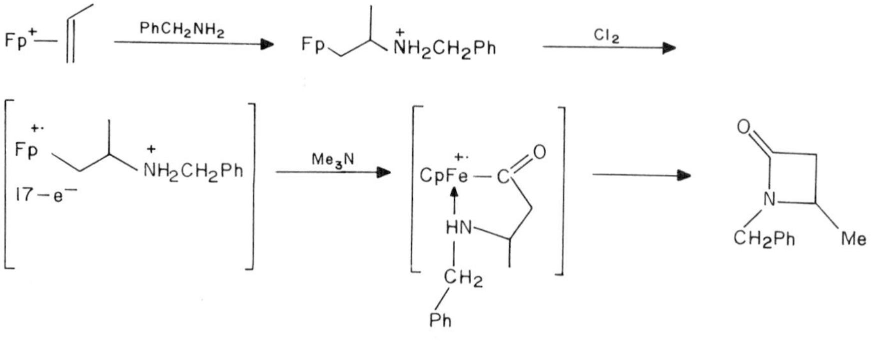

SCHEME 17

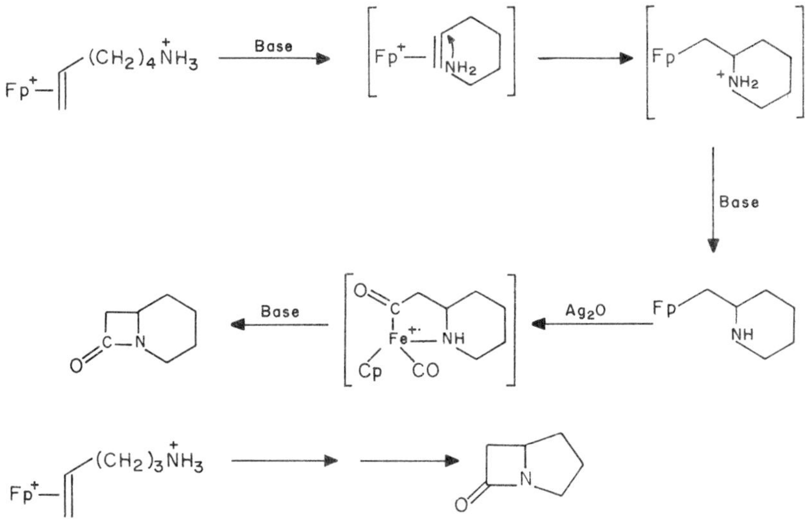

SCHEME 18

$$[Fp(olefin)]^+PF_6^- + Na^+I^- \longrightarrow FpI + olefin + Na^+PF_6^- \quad (20)$$

olefins from Fp complexes. When the addition of an amine to $[Fp(olefin)]^+$ is followed by oxidation, migratory insertion into the Fe—C bond ensues in the 17-electron intermediate. The free coordination site is then occupied by the nitrogen atom, leading to cyclization. Reductive elimination leads to the formation of a β-lactam ring (Scheme 17)[50].

It is also possible to synthesize bicyclic β-lactams from $[Fp(olefin)]^+$ complexes bearing an ammonium group. Deprotonation provides an amino group which effects cyclization by nucleophilic attack at the substituted olefinic carbon. Oxidative carbonylation using Ag_2O leads to formation of the second ring (Scheme 18). $[Fp(allene)]^+$ complexes also react with nucleophiles to give $[Fp(\eta^1\text{-allyl})]$ complexes (equation 21)[51].

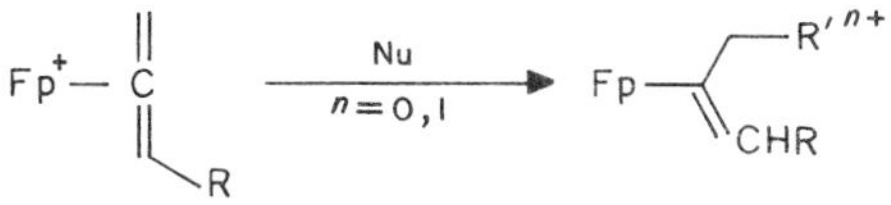

(21)

$NaBH_4$: $R' = H$, $n = 0$; R = Me : 10% ; Ph : 98%
Et_2NH : $R' = Et_2N$, $n = 0$; R = Me : 43% ; Ph : 42%
PPh_3 : $R' = PPh_3$, $n = 1$; R = Me : 88% ; Ph : 69%

Roberts and coworkers[52] showed that the reaction of carbanions with the neutral complexes $[Fe(CO)_4(olefin)]$ (even unactived ones) also generates C—C bonds (Scheme 19). The tetracarbonyl alkyl ferrate anions[50] obtained are of the same kind as those generated from $Na_2[Fe(CO)_4]$ and organic halides (cf. Section III.D.1). Protonation followed by oxidation leads to reductive elimination. Alternatively, the addition of an alkyl iodide under the appropriate conditions for CO insertion gives the expected asymmetric ketones.

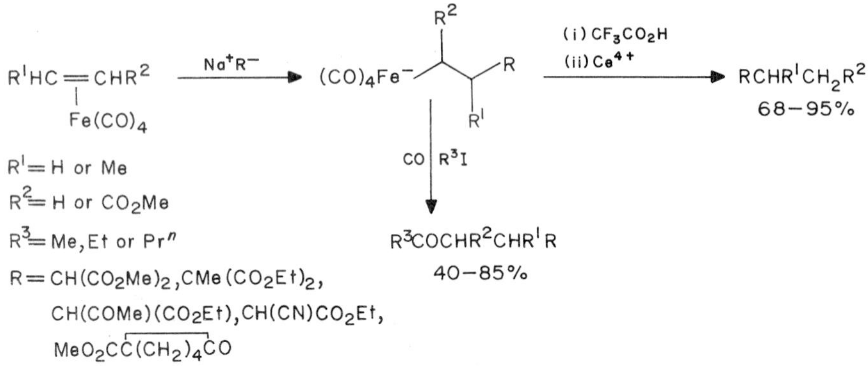

$$R^1HC{=}CHR^2(Fe(CO)_4) \xrightarrow{Na^+R^-} (CO)_4Fe^- \text{-alkyl}(R^2)(R)(R^1) \xrightarrow{(i)\,CF_3CO_2H;\ (ii)\,Ce^{4+}} RCHR^1CH_2R^2 \ (68–95\%)$$

$$\xrightarrow{CO \,|\, R^3I} R^3COCHR^2CHR^1R \ (40–85\%)$$

R^1 = H or Me
R^2 = H or CO_2Me
R^3 = Me, Et or Pr^n
R = $CH(CO_2Me)_2$, $CMe(CO_2Et)_2$, $CH(COMe)(CO_2Et)$, $CH(CN)CO_2Et$, $MeO_2C\overline{C(CH_2)_4C}O$

SCHEME 19

C. η^3-Allyl Complexes

The highly electrophilic cationic complexes $[Fe(CO)_4(\eta^3\text{-allyl})]^+$ react with nucleophiles to give either substituted olefin complexes or the free ligand directly[53]. These cations are prepared by protonation of $[(diene)Fe(CO)_3]$ under a CO atmosphere or by reaction of the neutral complexes $[Fe(CO)_3(\eta^3\text{-allyl})I]$ with $AgPF_6$, also under CO[54]. The latter crystalline complexes are obtained from $[Fe(CO)_5]$ or $[Fe_2(CO)_9]$ and the corresponding allyl iodide at elevated temperatures (Scheme 20).

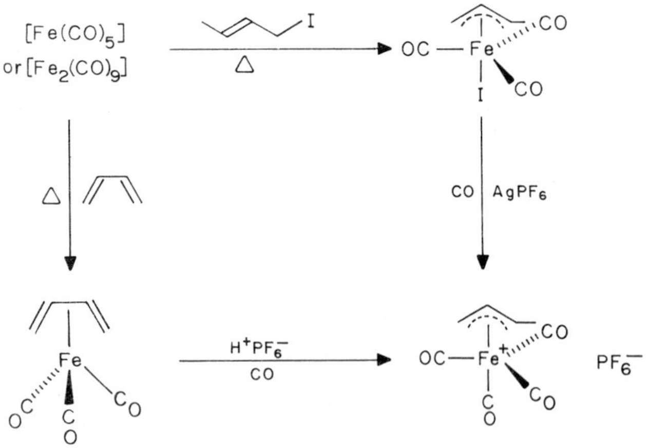

$[Fe(CO)_5]$ or $[Fe_2(CO)_9]$ —(allyl iodide, Δ)→ $[Fe(CO)_3(\eta^3\text{-allyl})I]$ —(CO, $AgPF_6$)→ $[Fe(CO)_4(\eta^3\text{-allyl})]^+ PF_6^-$

$[Fe(CO)_5]$ or $[Fe_2(CO)_9]$ —(Δ, butadiene)→ $[(butadiene)Fe(CO)_3]$ —($H^+PF_6^-$, CO)→ $[Fe(CO)_4(\eta^3\text{-allyl})]^+ PF_6^-$

SCHEME 20

Since $NaBH_4$ reduces $[Fe(CO)_4(\eta^3\text{-allyl})]^+$ complexes to alkenes, the temporary complexation of dienes by $Fe(CO)_3$ affords their regiospecific reduction on successive treatment of the complex with H^+ and H^- (Scheme 21). In the case of myrcene, cyclization of the allyl carbocation is also observed[55].

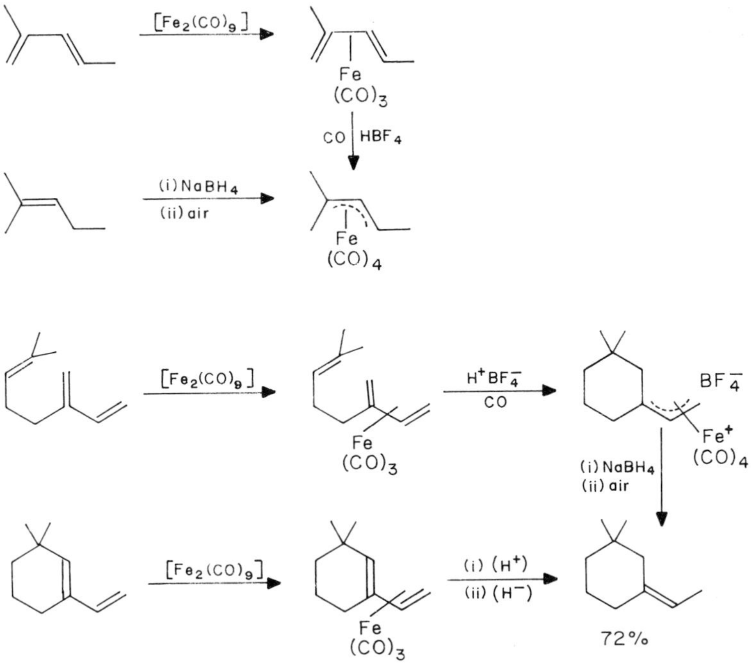

SCHEME 21

Alkylation of the complexes $[Fe(CO)_4(\eta^3\text{-allyl})]^+$ is not possible with lithium or Grignard reagents but proceeds well using organo-cadmium or -zinc reagents (equation 22)[53].

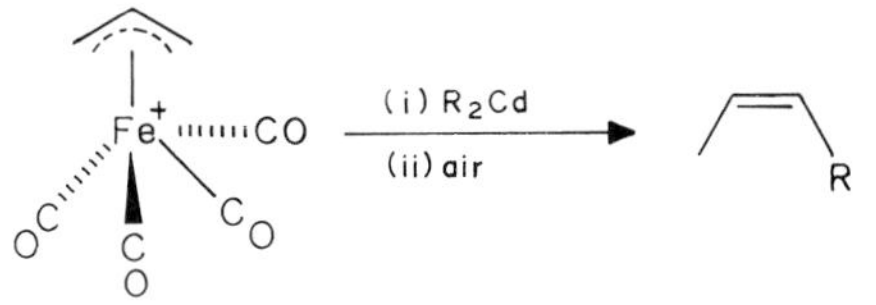

(22)

Neutral η^3-allyl complexes are also susceptible to nucleophilic attack by stabilized carbanions. Interestingly, the complexes $[Fe(CO)_3(NO)(\eta^3\text{-allyl})]$ generated *in situ* from $Na[Fe(CO)_3NO]$ were used in catalytic amounts by Roustan *et al.*[56] to form C—C bonds between diethyl malonate and allyl halides or acetates (equation 23). Alternatively, $[(\eta^3\text{-crotyl})Fe(CO)_2NO]$ could also be used as the catalyst.

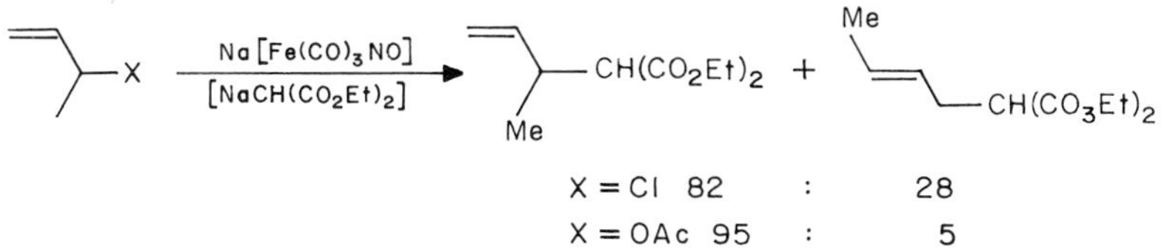

X = Cl	82	:	28
X = OAc	95	:	5

(23)

Nucleophilic addition also occurs regioselectively with stabilized carbanions and heteroatomic nucleophiles (Scheme 22).

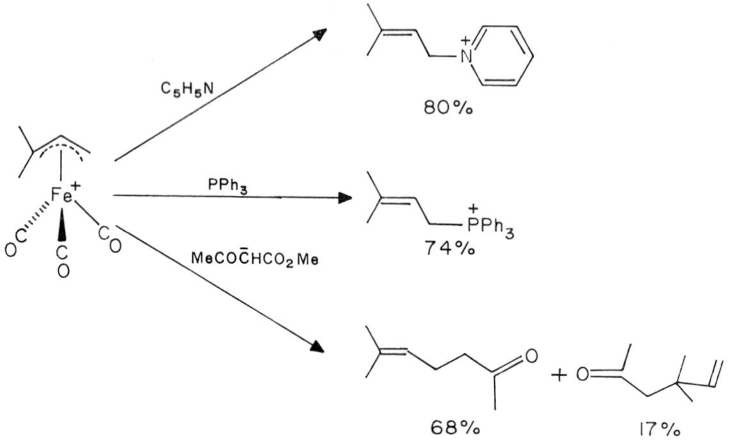

SCHEME 22

D. η^4-Diene Complexes

It is well known that $[(\eta^4\text{-diene})Fe(CO)_3]$ complexes react with electrophiles, but C—C bond formation is also possible on reaction with carbanions. Semmelhack *et al.*[57] showed that the kinetic product at $-78\,^\circ C$ results from attack at an internal carbon. However, rearrangement gives the thermodynamic product at $0\,^\circ C$, resulting from

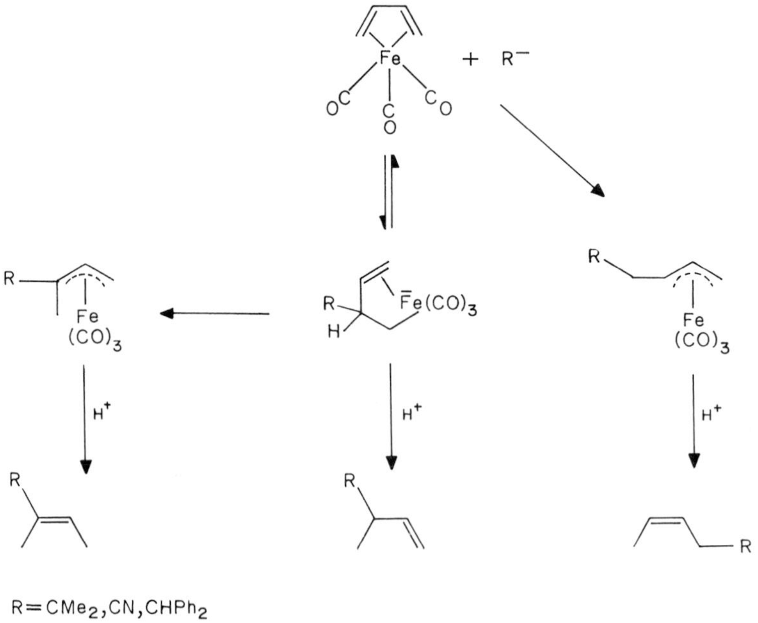

SCHEME 23

carbanion attack on the more stable allyl intermediate. Hence quenching with CF_3CO_2H at various temperatures provides different alkylated olefins (Scheme 23). However, if a lithium reagent and 12 are mixed at −78 °C in the presence of external CO, carbonylation also occurs via migration of CO and insertion into the Fe—C bond of the homoallylic intermediate **13** (Scheme 24). Further reaction with electrophiles gives *O*-protonation or alkylation, indicating the alkylidene nature of the carbonylated intermediate **13b**. Aldehydes, acids, and esters are isolated. It is assumed that alkylation of [(diene)Fe(CO)$_3$] by carbanions is kinetically favoured at C–2 over C–1. In the absence of CO, **13** rearranges to the thermodynamically more stable allylic complex **14** by a series of β-hydride eliminations and readditions. When the latter process is much favoured as in the reaction with [(η^4-cyclopentadiene)Fe(CO)$_3$], CO is not incorporated; substituted cyclopentenes are the only reaction products isolated.

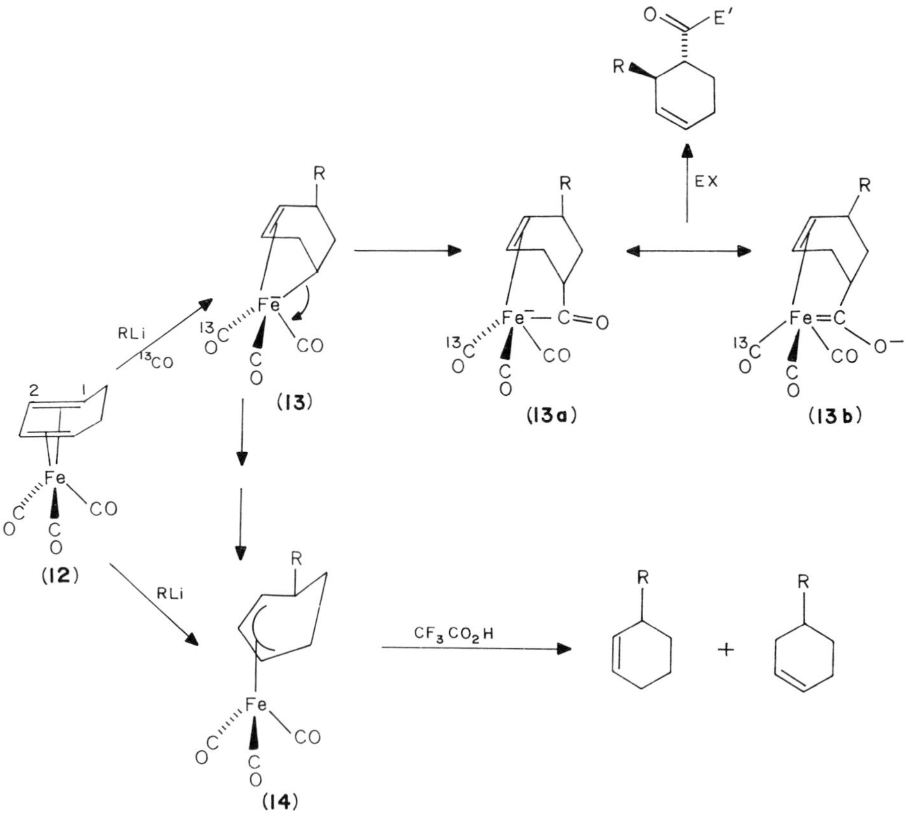

(12) (13) (13a) (13b) (14)

R = CMe_2CN, $CHMeCN$, CH_2CN, $CHMeCO_2Et$, $CHMeCO_2Bu^t$, $CH_2CO_2Bu^t$, CMe_2CO_2Li, 1, 3-dithiane

EX = CF_3CO_2H, CH_3I, R^1OSO_2F (R^1 = Me, Et), O_2 (E′ = OH)

SCHEME 24

This carbonylation reaction also occurs with complexes of open dienes (equations 24 and 25). Less reactive carbanions such as enolates do not attack $[(\eta^4\text{-diene})Fe(CO)_3]$;

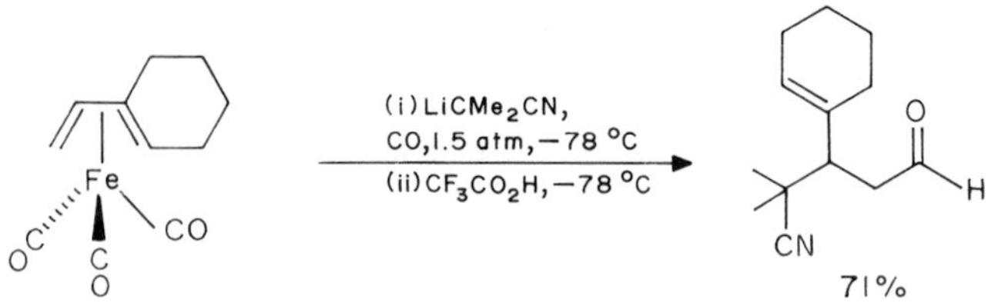

(i) $LiCMe_2CN$, CO, 1.5 atm, −78 °C
(ii) CF_3CO_2H, −78 °C
71% (24)

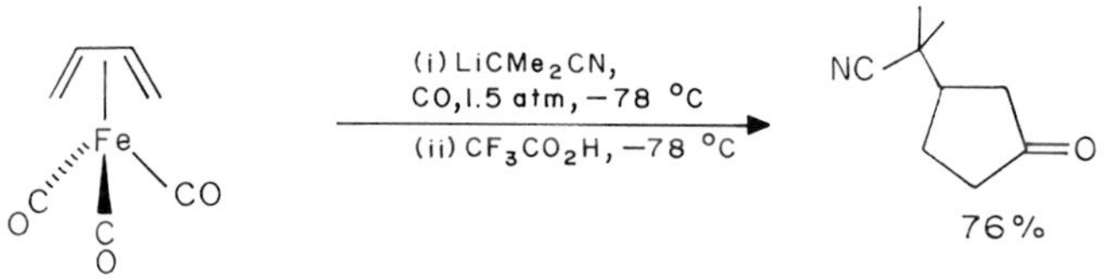

(i) $LiCMe_2CN$, CO, 1.5 atm, −78 °C
(ii) CF_3CO_2H, −78 °C
76% (25)

replacement of one carbonyl ligand by NO would give cationic complexes able to react with such mild nucleophiles.

E. η^5-Dienyl- and η^5-Polyenyl-Iron Tricarbonyl Cations

1. η^5-Pentadienyliron tricarbonyl cation

The complex $[(\eta^5\text{-pentadienyl})Fe(CO)_3]^+$ is prepared by acidification of the neutral pentadienol complex[58]. It reacts with various classical nucleophiles but the applications have not yet been developed. The action of zinc dust or lithium or Grignard reagents leads to dimerization (Scheme 25). Whereas nucleophilic addition always proceeds at the dienyl

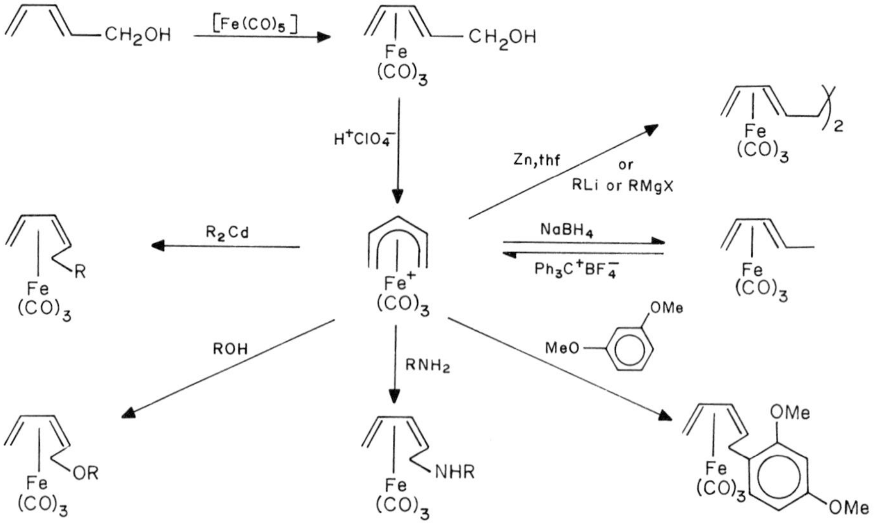

CH_2OH $[Fe(CO)_5]$ $H^+ClO_4^-$ Zn, thf or RLi or RMgX R_2Cd $NaBH_4$ $Ph_3C^+BF_4^-$ ROH RNH_2 MeO OMe Fe $(CO)_3$ R OR NHR

SCHEME 25

terminus (C-5 position) in the iron complex, the regioselectivity differs for the rhodium and iridium complexes [CpM(η^5-dienyl)]. In the latter complexes, attack occurs at C-3, giving [CpM(η^4-1, 4-dienes)][59].

2. η^5-Cyclohexadienyliron tricarbonyl cation

Nucleophilic additions to cyclohexadienyliron tricarbonyl cations have been developed by Birch and Jenkins[60] and Pearson[61] and have proved extremely useful for the syntheses

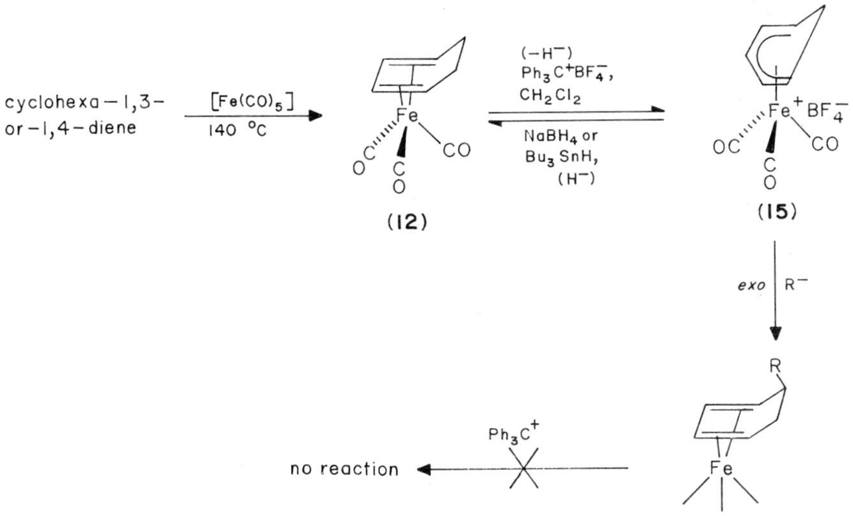

cyclohexa−1,3− or −1,4−diene → [Fe(CO)$_5$], 140 °C → (12) ⇌ [(−H$^-$) Ph$_3$C$^+$BF$_4^-$, CH$_2$Cl$_2$ / NaBH$_4$ or Bu$_3$SnH, (H$^-$)] (15) Fe$^+$ BF$_4^-$ → exo R$^-$ → R-substituted complex → Ph$_3$C$^+$ ✕ → no reaction

SCHEME 26

(16) MeO-substituted complex → Ph$_3$C$^+$ → (17) MeO, Fe$^+$ 94% + MeO, Fe$^+$ 6%

(17) → exo R$^-$ → MeO, R complex → (i) Me$_3$NO (ii) H$_3$O$^+$ → 4-R-cyclohex-2-enone

MeO, R complex → Ph$_3$C$^+$ ✕ → no reaction

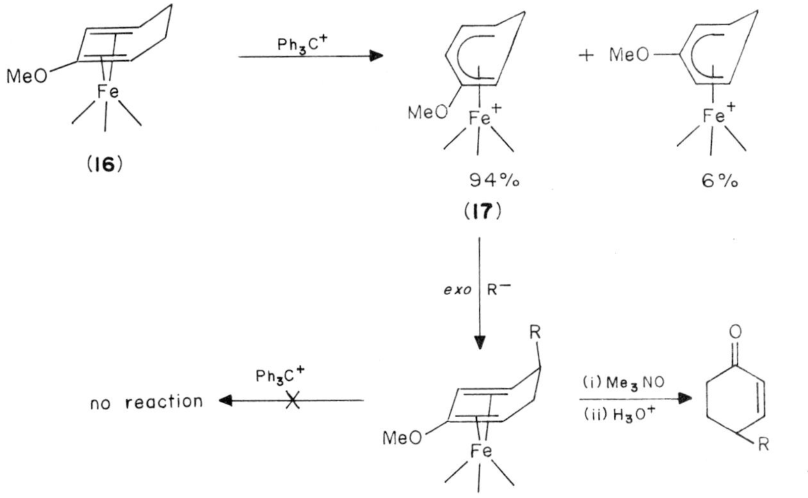

SCHEME 27

of natural products. The parent complex $[(\eta^5\text{-}C_6H_7)Fe(CO)_3]^+$ (**15**) is accessible by hydride abstraction from $[(\eta^4\text{-}C_6H_8)Fe(CO)_3$ (**12**) using $Ph_3C^+BF_4^-$ (Scheme 26)[62]. Hydride abstraction from [(substituted cyclohexadiene)$Fe(CO)_3$] complexes generally gives mixtures of isomers. However, the useful precursor [(2-methoxy-cyclohexadiene)$Fe(CO)_3$] (**16**) gives mainly hydride abstraction from C-5. The major cation **17** so obtained undergoes regio- and stereo-specific (*exo*) nucleophilic addition at the same C-5 position[63] (Scheme 27). For steric reasons, H^- cannot be abstracted from the substituted cyclohexadiene complexes obtained, which limits the possible synthetic strategies. Removal of $Fe(CO)_3$ is achieved using Me_3NO, a procedure discovered by Shvo and Hazum[64]. Acid hydrolysis of the enol ether gives the

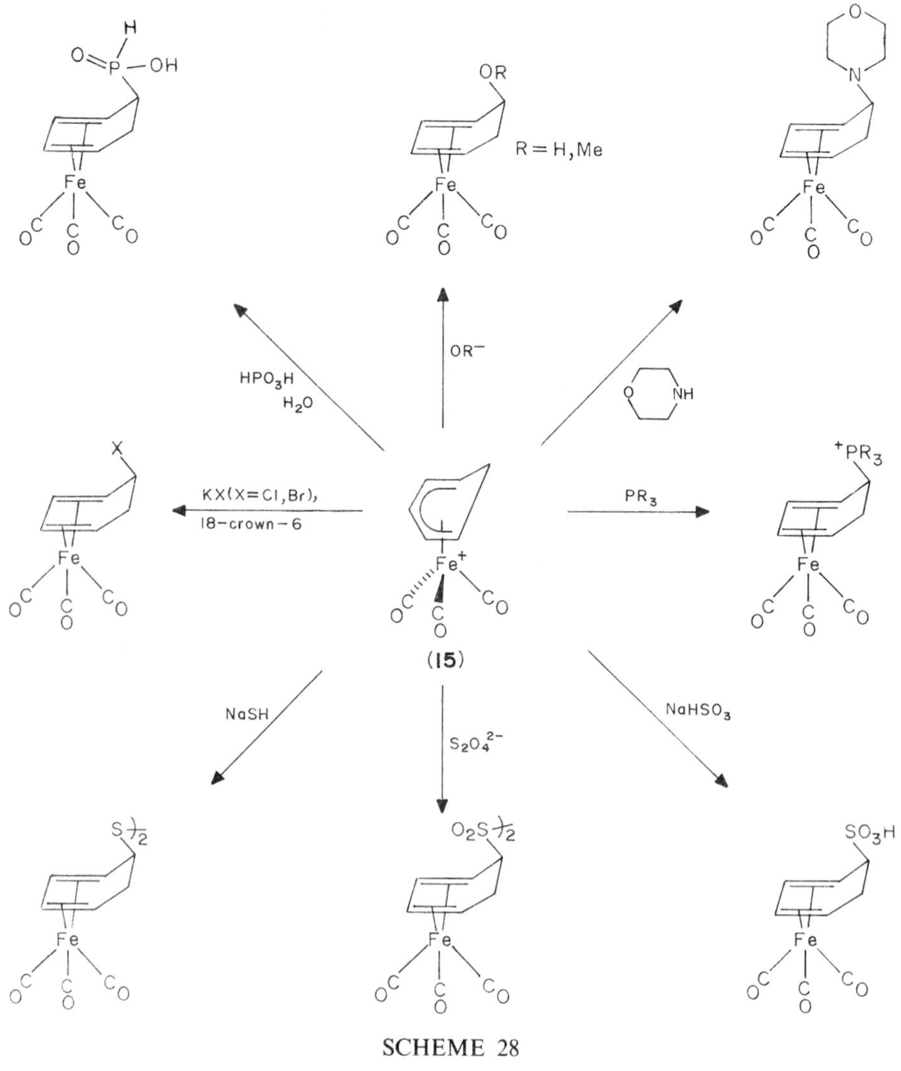

SCHEME 28

cyclohexenone[65]. Nucleophilic addition to **15** proceeds with charged or neutral heteroatomic nucleophiles[66] (Scheme 28).

Similarly, mild carbon nucleophiles react readily. Carbanions add in the form of zinc, cadmium, copper, and boron reagents, enolates, enamines, silylenol ethers, and trimethylvinyl or allyl silanes[67] (Scheme 29). *C*- or *N*-alkylation of **15** is observed with aromatic amines depending on the ring substituents. A 4-NO_2 group induces *N*-alkylation whereas alkyls, 3-OMe and 3-NR_2 groups favour *C*-alkylation. Reaction with indole leads to *C*-alkylation on the five-membered ring[68] (Scheme 30).

[(2-Methylcyclohexadienyl)Fe$(CO)_3$]$^+$ (18), prepared from 2-methylanisole, gives 80%

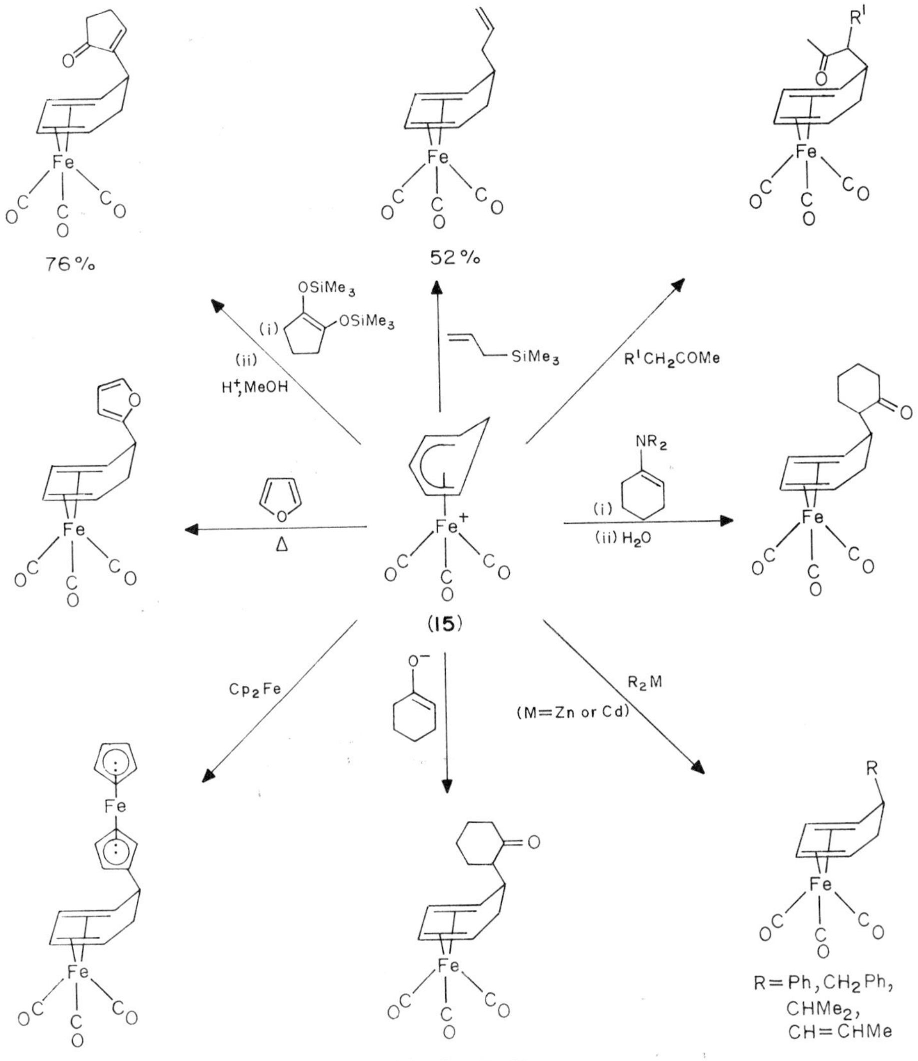

76%

52%

OSiMe3

(i)

OSiMe3

(ii)

H+, MeOH

SiMe3

R'CH2COMe

NR2

(i)

(ii) H2O

Δ

(15)

Cp2Fe

R2M

(M=Zn or Cd)

R=Ph, CH2Ph, CHMe2, CH=CHMe

SCHEME 29

SCHEME 30

SCHEME 31

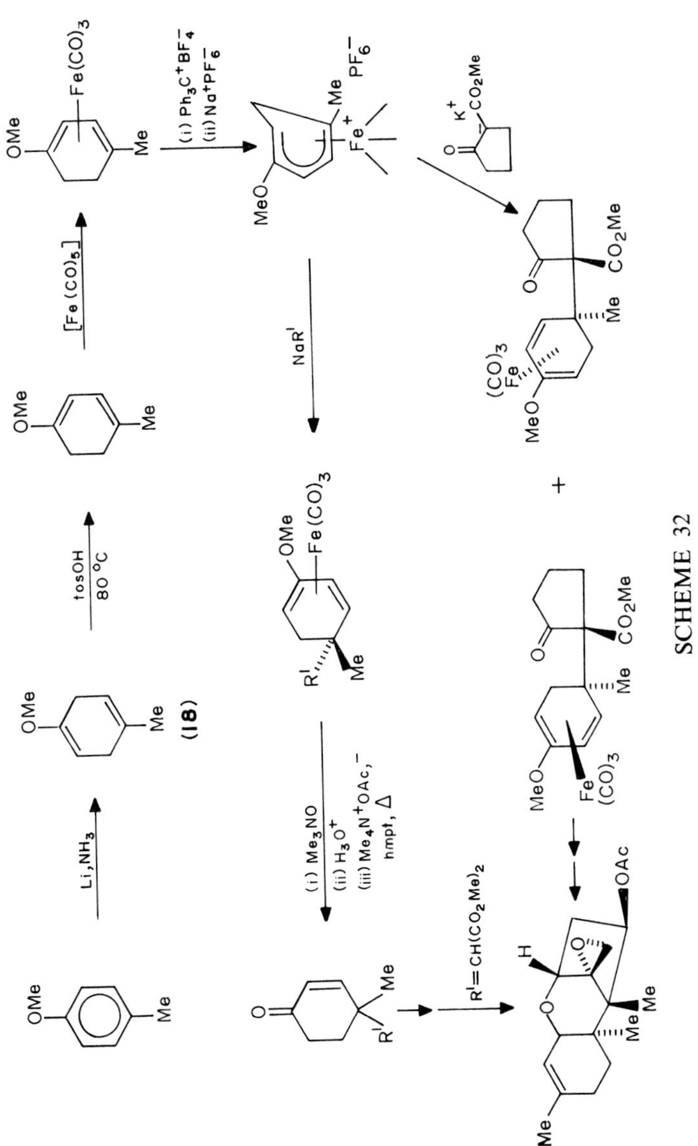
OMe
Me
Li, NH3
(18)
tosOH
80 °C
[Fe(CO)5]
Fe(CO)3
(i) Ph3C+BF4−
(ii) Na+PF6−
MeO
Fe+
PF6−
NaR'
K+
CO2Me
Fe(CO)3
R'
(i) Me3NO
(ii) H3O+
(iii) Me4N+OAc−,
hmpt, Δ
R'=CH(CO2Me)2
OAc
H
(CO)3
Fe
+

SCHEME 32

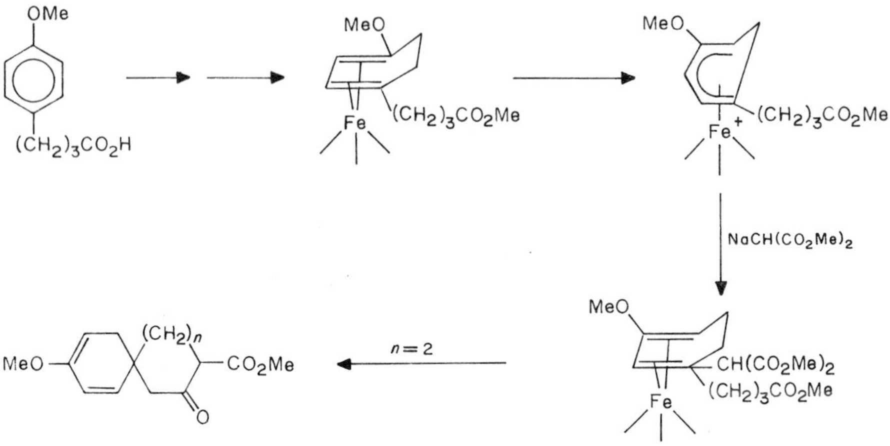

(**19a**) spiro[4.5]decane ($n = 1$)
(**19b**) spiro[5.5]undecane ($n = 2$)

SCHEME 33

selectivity for C-5 addition of bulky nucleophiles, e.g. organo-cadmium and-zinc reagents (the steric effect of the methyl group is much less marked in the borohydride reduction). The utility of the alkylation procedure is exemplified by the synthesis of zingiberene[69] (Scheme 31). Whereas the C-1/C-5 nucleophilic addition is *not* charge controlled in the

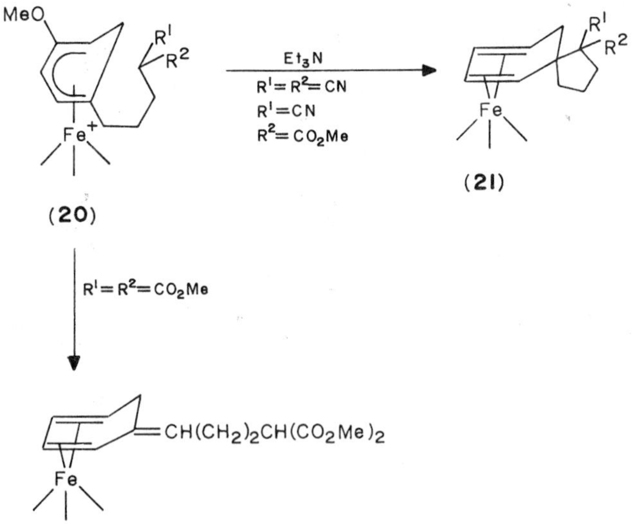

SCHEME 34

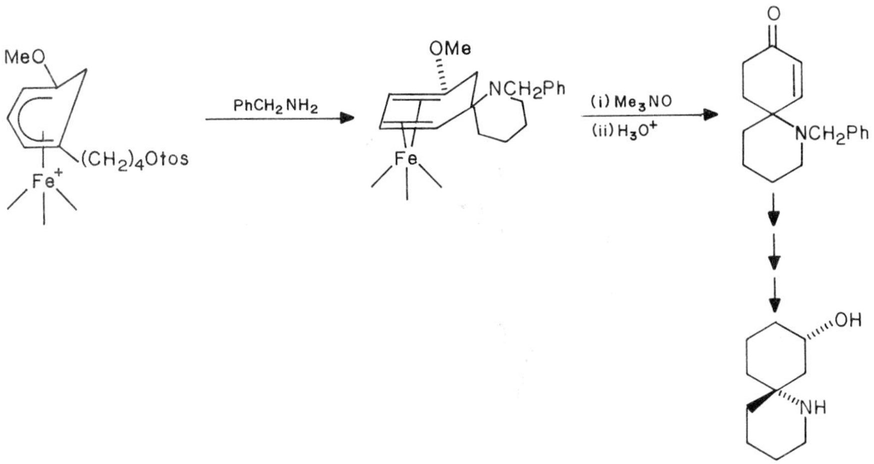

MeO (CH2)4Otos Fe+ PhCH2NH2 OMe NCH2Ph Fe (i) Me3NO (ii) H3O+ O NCH2Ph OH NH

(±)-Depentylperhydrohistrionicotoxin

SCHEME 35

parent complex, a methoxy substituent at C-2 induces a degree of charge control for nucleophilic addition at C-5. This allows the formation of quaternary carbon centres. The strategy used by Pearson[70] for the synthesis of cyclohexenones, of potential use in natural product synthesis, begins with the Birch reduction[71] of a 4-substituted anisole and reconjugation of the cyclohexadiene derivative **18**. Quaternary centres are formed at C-5 by addition of carbanions derived from *gem*-diesters, cyano esters, malononitrile, and β-keto esters[72] (Scheme 32).

Colvin *et al.*[73] previously used one of these cyclohexenones [$R_1 = CH(CO_2Me)_2$] for the synthesis of the trichothecane trichodermin. The potassium enolate of methyl cyclopentan-2-one carboxylate gives two interconvertible diastereoisomers also leading to the synthesis of trichothecane analogues. The synthesis of spirocyclic compounds, also related to natural products, has been achieved by Pearson[74] using two strategies. The first involves condensation of the two side-chains attached to the quaternary carbon centre. A spiro[5.5]undecane **19b** is accessible from p-MeOPh$(CH_2)_3CO_2H$, while a spiro[4.5]decane, **19a**, is synthesized starting from p-MeOPhCh$=$$CHCO_2H$ (Scheme 33). The other strategy used intramolecular nucleophilic attack for the construction of the quaternary carbon centre.

Because of the competition between deprotonation at the juxtacyclic position and at the derived chain γ-carbon in **20**, the γ-substituents must be very electron-withdrawing in order to achieve cyclization to **21** (Scheme 34). Intramolecular cyclization with a chain bearing an N-nucleophile leads to azaspirocyclic precursors of various alkaloids[75] (Scheme 35). Total syntheses of the alkaloid (±)-limaspermine (quaternary carbon centre at C-20) have been achieved[76] (Scheme 36).

The introduction of functionalized angular substituents into bicyclic systems is also possible[77] (equation 26). The presence of the uncomplexed ring does not perturb the regioselectivity of hydride abstraction and nucleophilic addition except when this ring bears an *exo*-substituent. On the other hand, the construction of the quaternary carbon centre at C-5 in the (2-methoxylcyclohexadienyl)iron tricarbonyl cation **17** is much

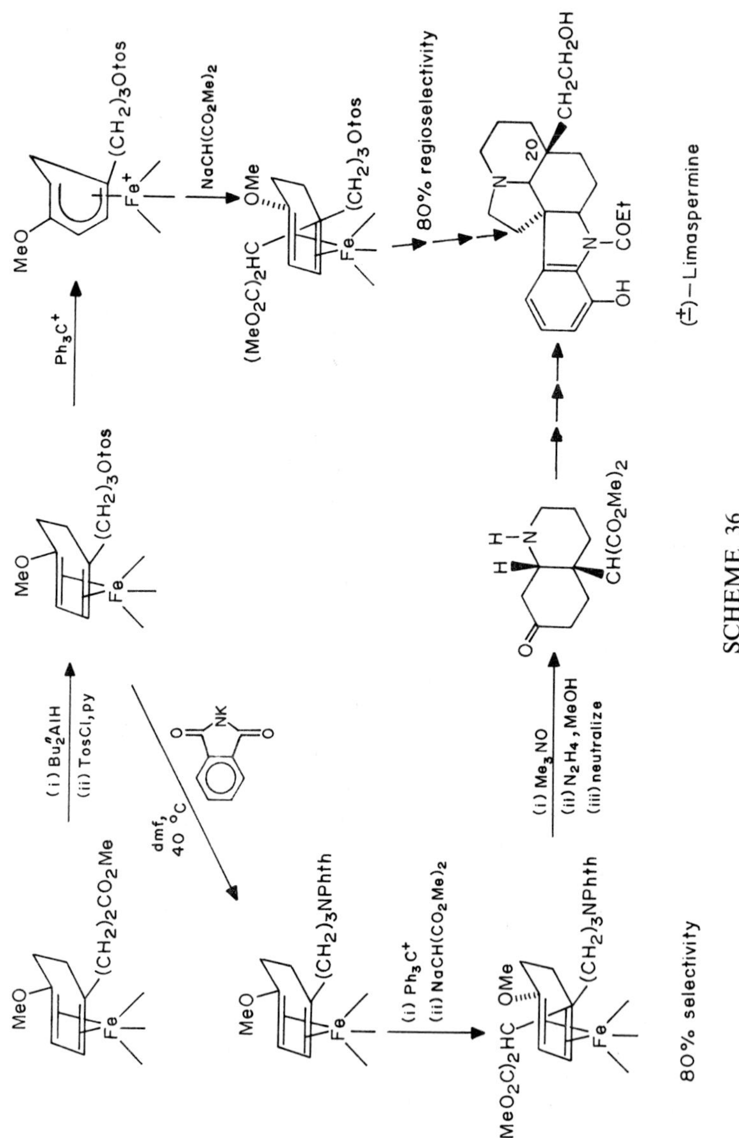
MeO
Fe
$(CH_2)_2CO_2Me$
(i) Bu^n_2AlH
(ii) TosCl, py
MeO
Fe
$(CH_2)_3Otos$
Ph_3C^+
MeO
Fe^+
$(CH_2)_3Otos$
dmf, 40 °C
NK
MeO
Fe
$(CH_2)_3NPhth$
(i) Ph_3C^+
(ii) $NaCH(CO_2Me)_2$
$(MeO_2C)_2HC$
OMe
Fe
$(CH_2)_3NPhth$
80% selectivity
$NaCH(CO_2Me)_2$
$(MeO_2C)_2HC$
OMe
Fe
$(CH_2)_3Otos$
80% regioselectivity
(i) Me_3NO
(ii) N_2H_4, MeOH
(iii) neutralize
O
H
H
N
$CH(CO_2Me)_2$
N
20
CH_2CH_2OH
N
COEt
OH
(±)−Limaspermine

SCHEME 36

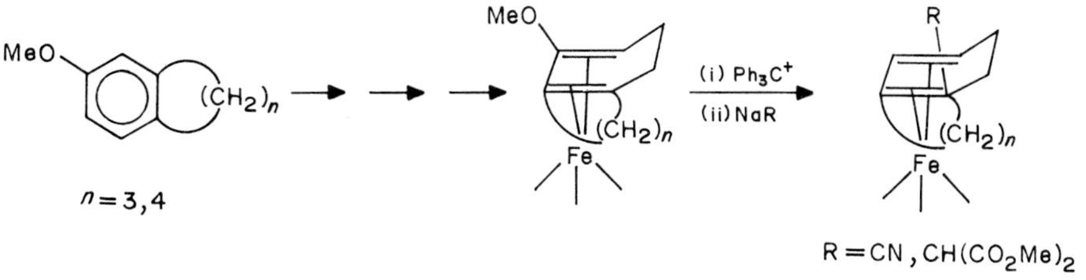

MeO (CH2)n → → → (i) Ph3C+ (ii) NaR

n = 3,4

R = CN, CH(CO2Me)2

(26)

inhibited or perturbed in the presence of C-5 chains bearing β-oxygenated substituents ($CH_2CO_2Me, CH_2CH_2OMe, CH_2CH_2OAc$). Deprotonation and poor selectivity were observed on reaction of nucleophiles.

3. η^5-Cycloheptatrienyliron tricarbonyl and dicarbonyl triphenylphosphine cations

Nucleophiles also give interesting reactions with $[(\eta^5\text{-cycloheptatrienyl})Fe(CO)_2L]^+$ ($L = CO, PPH_3$)[78]. Attack generally occurs regioselectively at C-1 on the tricarbonyl

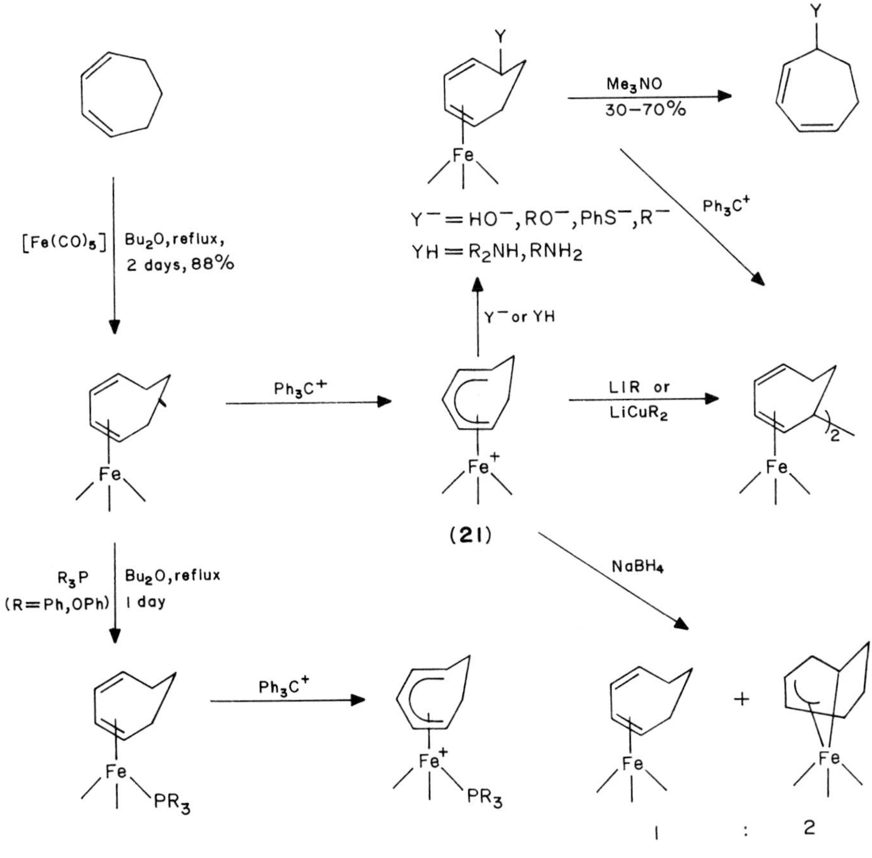

[Fe(CO)5] Bu2O, reflux, 2 days, 88%

Ph3C+

Me3NO 30–70%

Y⁻ = HO⁻, RO⁻, PhS⁻, R⁻
YH = R2NH, RNH2

Y⁻ or YH

Ph3C+

LiR or LiCuR2

(21)

NaBH4

R3P (R = Ph, OPh) Bu2O, reflux 1 day

Ph3C+

PR3 PR3

1 : 2

SCHEME 37

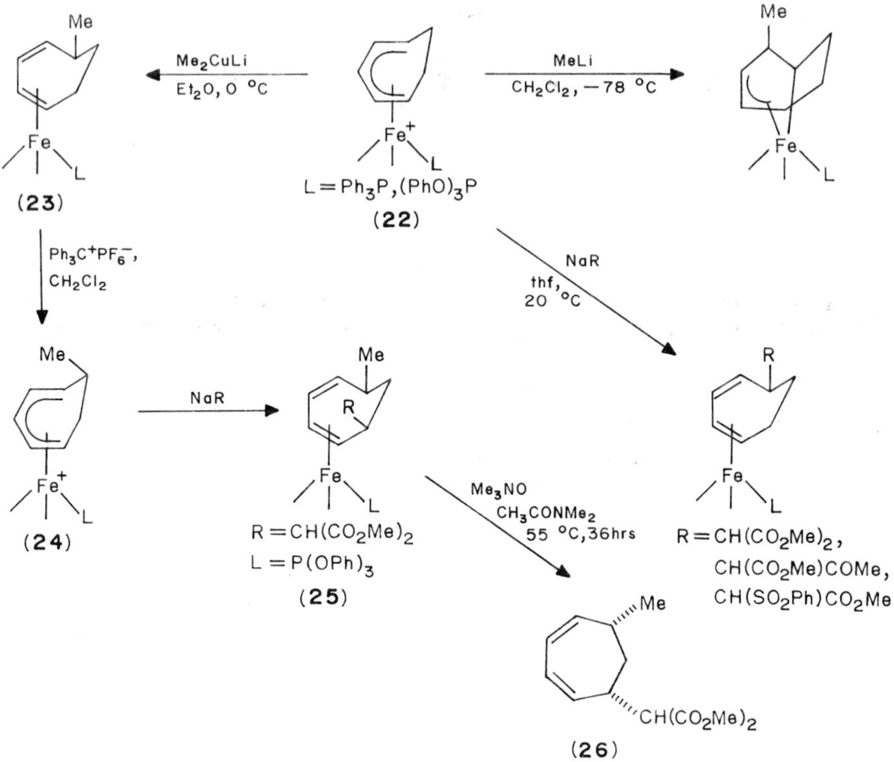

SCHEME 38

complex **21** with a variety of heteroatomic nucleophiles and dialkylcopper reagents. However, hydride reduction gives predominantly C-2 attack. Lithium reagents and lithium dialkylcuprates give both C-5 attack and dimerization resulting from electron transfer (Scheme 37). η^5-Cycloheptatrienyliron dicarbonyl phosphine cations, **22**, undergo clean C—C bond formation at C-5 on reaction of sodium reagents and lithium dimethylcuprate, but MeLi gives C-2 addition. The methyl-substituted cycloheptadiene iron dicarbonyl phosphine complexes **23** so obtained give regiospecific hydride abstraction and the cations **24** thus formed react again with carbanions to give **25**. Thus two successive hydride abstraction/nucleophilic additions provide stereochemically defined disubstituted cycloheptadienes, **26**, after decomplexation with Me_3NO[78c] (Scheme 38).

F. η^6-Arene Complexes

There are two main useful types of 18-electron η^6-areneiron complexes[12]. Several hundred complexes of the type $[Cp(\eta^6\text{-arene})Fe]^+$ have been synthesized according to equation 5. The other series, $[(\eta^6\text{-arene})_2Fe]^{2+}$, is limited to benzene and its methyl-substituted derivatives (equation 6). Both series are accessible via Fischer-type ligand-exchange reactions. from ferrocene and $FeCl_3$, respectively.

1. Nucleophilic additions

The reaction of carbanions (lithium and Grignard reagents) with $[Cp(\eta^6\text{-arene})Fe]^+$ complexes gives $[Cp(\eta^5\text{-cyclohexadienyl})Fe]$[79]. However, this process has not been used in aromatic synthesis because subsequent reaction with Ph_3C^+ gives both *exo* abstraction of the substituent and *endo* H abstraction[80]. Oxidants also favour the loss of the *exo* substituent. However, the directing effects of substituents are noteworthy. In the $NaBH_4$ reduction, electron-withdrawing groups favour *ortho* attack whereas electron- releasing ones favour *meta* attack[81] (equation 27). In this respect, complexation of arenes by $CpFe^+$

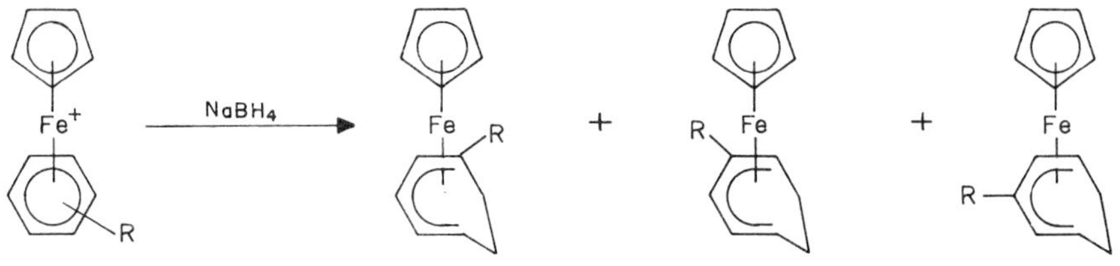

NaBH$_4$

R	*ortho*	*meta*	*para*
R = Me	1	1	0.5
Cl	4	1	0
OMe	0.2	1	0.6
CO_2Me	12.7	1	1

(27)

is much less useful than complexation with $Cr(CO)_3$. Semmelhack[82] demonstrated the formation of a series of substituted aromatics by reaction of $[(\eta^6\text{-arene})Cr(CO)_3]$ with carbanions followed by oxidation with I_2. The reaction of carbanions with $[(\eta^6\text{-arene})_2Fe]^{2+}$ fails to form carbon—carbon bonds except in the peculiar case of the mesitylene complex[83]. PhLi, Bu$^+$Li, and $CH_2{=}CHLi$ add to each ring of $[(\eta^6\text{-mesitylene})_2Fe]^{2+}$ to give $[(\text{cyclohexadienyl})_2Fe^{II}]$ complexes (**28**). Oxidative decomplexation gives substituted mesitylenes. Stabilized carbanions (LiCH, $LiCH_2NO_2$, $LiCHMeNO_2$) add to one ring only, leading to **27** (Scheme 39).

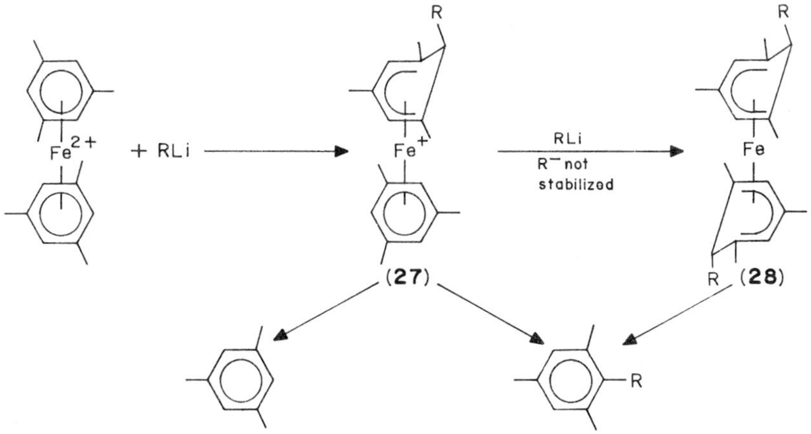

Fe^{2+} + RLi → (**27**) —RLi, R$^-$ not stabilized→ (**28**)

SCHEME 39

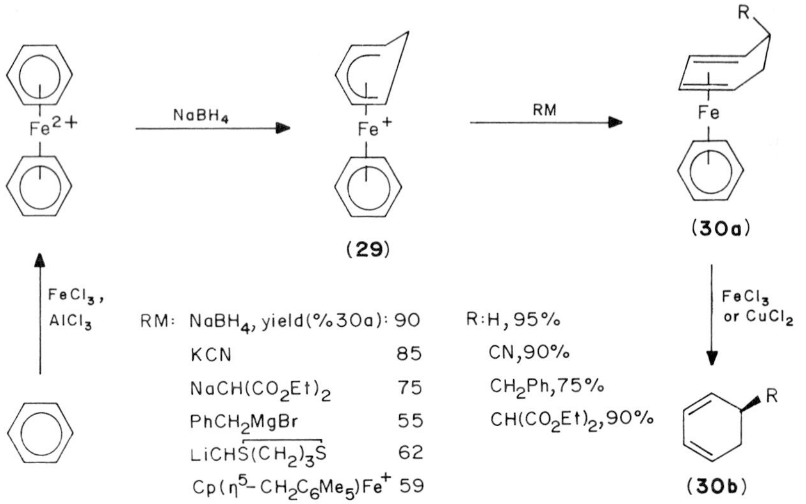

$[(\eta^6\text{-}C_6H_6)_2Fe]^{2+}$ $\xrightarrow{NaBH_4}$ (29) $\xrightarrow{RM}$ (30a) $\xrightarrow{FeCl_3 \text{ or } CuCl_2}$ (30b)

Benzene $\xrightarrow{FeCl_3, AlCl_3}$ $[(\eta^6\text{-}C_6H_6)_2Fe]^{2+}$

RM:	yield (% 30a)
$NaBH_4$	90
KCN	85
$NaCH(CO_2Et)_2$	75
$PhCH_2MgBr$	55
$LiCH\overline{S(CH_2)_3S}$	62
$Cp(\eta^5\text{-}CH_2C_6Me_5)Fe^+$	59

R: H, 95%; CN, 90%; CH_2Ph, 75%; $CH(CO_2Et)_2$, 90%

SCHEME 40

Although C, N, and O nucleophiles react with $[(\eta^6\text{-arene})_2Fe]^{2+}$ giving electron transfer, $NaBH_4$ gives the hydride-transfer product cleanly, whatever the arene ligand[84,85]. In contrast to the case of mesitylene, the benzene and hexamethylbenzene complexes give $[(\eta^6\text{-arene})(\eta^4\text{-cyclohexadiene})Fe^0]$ (**30a**) on addition of a second equivalent of hydride[28]. Alternatively, addition of a lithium, sodium, potassium, or Grignard reagent to the monohydrogenated complex **29** gives good yields of the functional cyclohexadiene complexes **30a** as the kinetic reaction products (Scheme 40)[25,28]. This regioselectivity may be accounted for by orbital control[25], whereas charge control predicts the wrong site of attack[23]. Decomplexation to free substituted cyclohexadienes, **30b**, proceeds with $FeCl_3$ or $CuCl_2$. This four-step synthesis starting from benzene is an alternative to the five-step route using the Birch reduction of benzene and complexation to the tricarbonyliron group. The latter also requires the use of zinc or cadmium reagents rather than lithium or Grignard reagents for reactive carbanions (cf. Section II.E).

Moreover, it is now possible to abstract a hydride from (**30a**) at $-40\,°C$ by an electron pathway using Ph_3C^+. Then, the addition of a second nucleophile (KCN) allows the formation of heterobi functional cyclohexadienes after decomplexation as in Scheme 40[85].

In the hexamethyl series, the second hydride reduction by $NaBH_4$ or $LiAlH_4$ proceeds by an outer-sphere electron transfer followed by H-atom transfer[22]. The intermediate 19-electron complex $[(\eta^6\text{-}C_6Me_6)(\eta^5\text{-}C_6Me_6H)Fe^I]$ is isolated in good yield in the course of this reduction. For steric reasons, hydride reduction of $[(\eta^6\text{-}C_6Me_6)(\eta^5\text{-}C_6Me_6H)Fe^{II}]^+$, **31**, does not proceed faster than electron transfer. Likewise, the reaction of carbanions with **31** does not occur at the cyclohexadienyl ligand but on the arene ring. Since hydride may be removed by Ph_3C^+ from the asymmetric cyclohexadienyl complex **32** thus formed, its role in this strategy is that of a protecting group[85]. Application to organic synthesis results from further deprotonation of a methyl substituent in **33** followed by acylation of **34**, another deprotonation and decomplexation on alumina to provide the trienone **35**[86] (Scheme 41). Thus, the addition of carbanions to the easily accessible complexes $[(\eta^5\text{-cyclohexadienyl})(\eta^6\text{-arene})Fe^{II}]^+$ occurs readily with a variety of reagents and is synthetically useful.

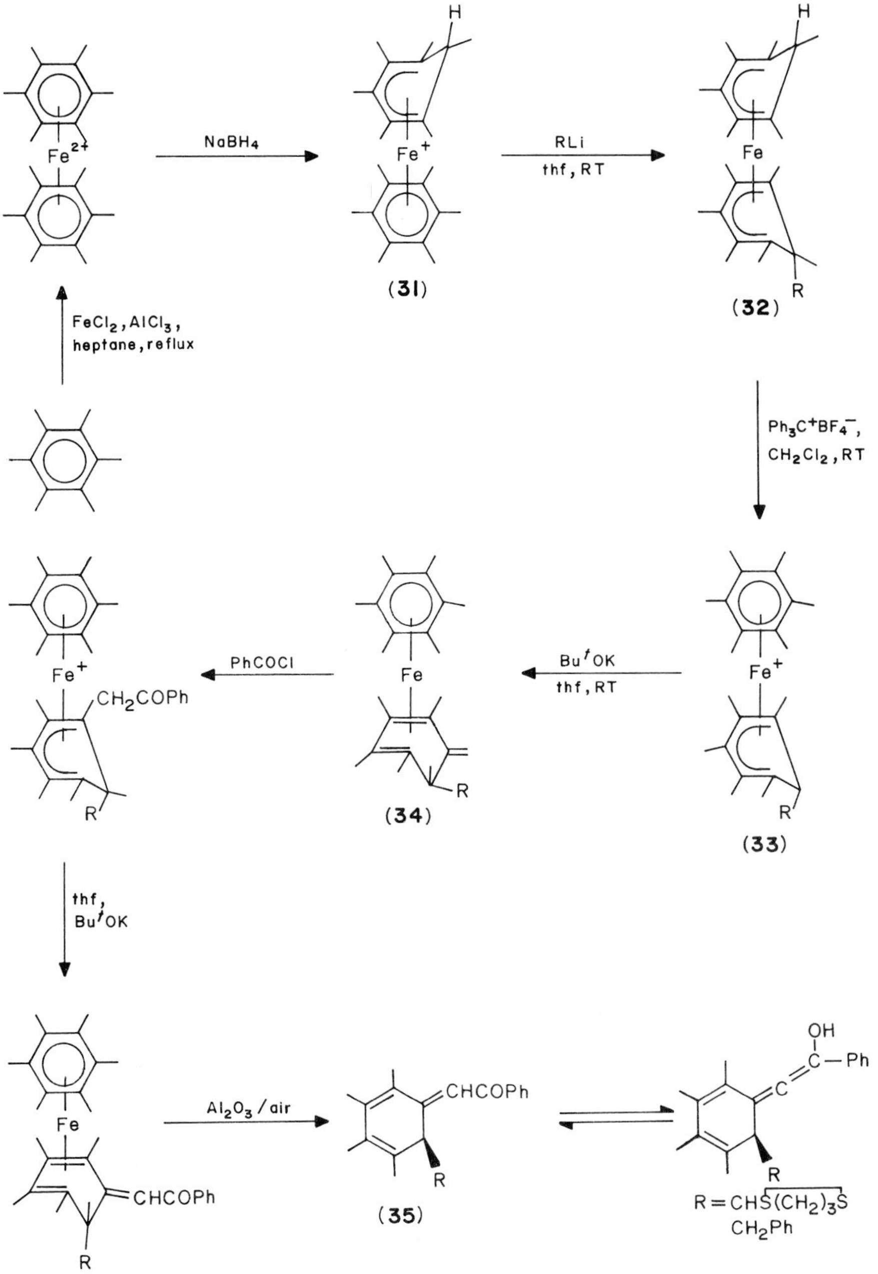

Fe^{2+} — $NaBH_4$ → Fe^+ (**31**) — RLi, thf, RT → Fe (**32**)

$FeCl_2$, $AlCl_3$, heptane, reflux

$Ph_3C^+BF_4^-$, CH_2Cl_2, RT

Fe^+ (**33**) — Bu^tOK, thf, RT → Fe (**34**) — PhCOCl → Fe^+ CH_2COPh

thf, Bu^tOK

Fe =CHCOPh — Al_2O_3/air → =CHCOPh (**35**) ⇌ OH C—Ph C

R = CHS$(CH_2)_3$S CH_2Ph

SCHEME 41

Electron-transfer side-reactions, often encountered in the $[Cp(\eta^6\text{-arene})Fe]^+$ series, depend on the solvent and reaction temperature[22]. Addition of carbanions to the latter cations in ether gives electron transfer whereas clean C—C bond formation is obtained in thf at low temperature. Interestingly, the solvent effect is opposite for the addition of hydrides. Thus, in monocationic iron–arene complexes, outer-sphere electron transfer can be circumvented by using thf for carbanions and diethyl ether for hydrides. On the other hand, no solvent allows C—C bond formation in the dicationic series $[(\eta^6\text{-}C_6R_6)_2Fe]^{2+}$ (R = H, Me)[85]. However, clean nucleophilic addition of hydride occurs readily in the latter series. Comparing the isolobal series $[(\eta^5\text{-cyclohexadienyl})(\eta^6\text{-arene})Fe^{II}]^+$ and $[(\eta^5\text{-cyclohexadienyl})Fe^{II}(CO)_3]^+$ the easily available lithium and Grignard reagents cleanly form C—C bonds with the former and give electron transfer with the latter. Hence decreased electron density favours electron transfer with carbanions. It is probable that the replacement of one or two carbonyls by phosphines in the tricarbonyl series would prevent electron transfer with these reagents, making the use of zinc and cadmium reagents unnecessary.

In several instances, it has been found that replacement of one CO by PPh_3 in the complexing group $Fe(CO)_3$ facilitates the desired process (cf. Sections II.E.3 and III.C.2). When nucleophilic attack and electron transfer are in competition, a low-temperature reaction always favours nucleophilic attack, and is thus advisable.

2. *Other reactions of nucleophiles, bases, and reducing agents with $[Cp(\eta^6\text{-arene})Fe]^+$ complexes.*

a. *Deprotonation of $[Cp(\eta^6\text{-arene})Fe]^+$ complexes.*

This deprotonation (arene = alkylbenzenes, anilines, phenol, and thiophenol) proceeds readily giving $[Cp(\eta^5\text{-cyclohexadienyl})Fe$ with exocyclic double bonds (C=C, C=N, C=O, C=S) which can be alkylated (cf. Section III.A)[87]. Deprotonation of $[(\eta^5\text{-cyclohexadienyl})FeL_3]^+$ $[L_3 = (CO)_3$ or arene] bearing exocyclic hydrogen(s) at the ligand terminus gives $(\eta^4\text{-triene})Fe^0$ complexes (cf. Section II.E.2.).

b. *Nucleophilic substitution at coordinated aryl halides and xanthone.*

Nesmeyanov's group found that O, S, and N nucleophiles can replace chlorine or fluorine in aryl halides coordinated to $CpFe^+$ (equation 28). Even NH_3 can displace

$$[CpFe^+(C_6H_5Cl)] + Nu^- \longrightarrow [CpFe^+(C_6H_5Nu)] + Cl^- \tag{28}$$

Nu = OH^-, OPh^-, CN^-, N-phthalimide (acetone), SH^- (acetonitrile),
OEt^-, $CH_3CO_2^-$, SPh^- (ethanol), N_3^- (dmf)

chloride at 50 °C in $[Cp(\eta^6\text{-}C_6H_5Cl)Fe]^+$ to give the aniline complex[88]. *N*-Substituted aniline complexes were later obtained using primary or secondary amines[89]. With methoxide ion, kinetic evidence has been obtained for the formation of a charge-transfer complex resulting in a reduction of the rate of nucleophilic substitution[90]. Starting from

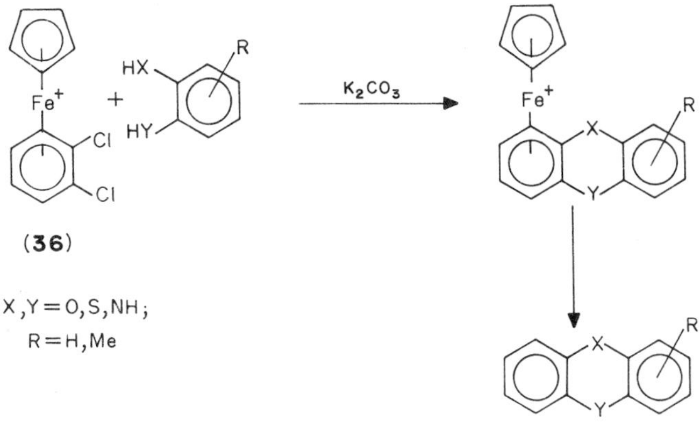

(36)

X,Y=O,S,NH;
R=H,Me

SCHEME 42

the *o*-dichlorobenzene complex **36**, double nucleophilic substitution gives various heterocycles[91] (Scheme 42).

With carbanions, nucleophilic addition is obtained *ortho* to chlorine[92], whereas nucleophilic substitution proceeds with the $Cr(CO)_3$ activating group[93]. However, N and C nucleophiles displace the oxygen bridge of xanthone in **37**, as is also known for free 2, 7-dinitroxanthone. This ring-opening reaction leads to *o*,*o*′-disubstituted benzophenones, **38**, after photolytic decomplexation[94] (equation 29).

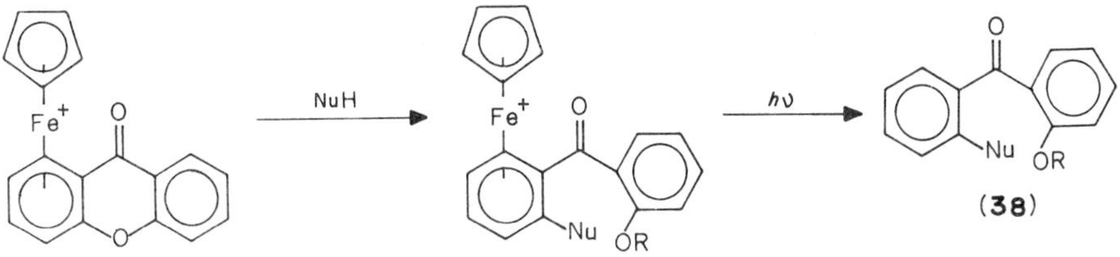

(37) NuH=pyriolidine (38)

(29)

c. Reduction of the side-chain.

The complexation of aromatics facilitates the reduction of unsaturated functions conjugated with the ring. For instance, $[Cp(\eta^6\text{-}C_6H_5CO_2H)Fe]^+$, **40**, easily synthesized by oxidation of $[Cp(\eta^6\text{-}C_6H_5CH_3)Fe]^+$, **39**[95], is reduced on a mercury cathode at −1.2 V *vs.* SCE in water (pH 0)[96] (equation 30). Since the carboxylic function in the iron complex

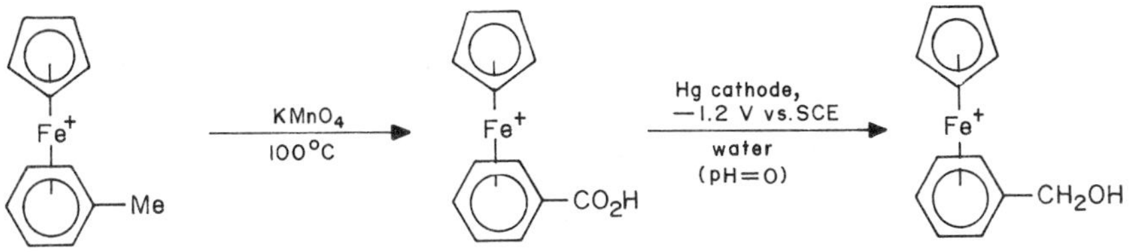

(39) (40)

(30)

can be transformed into ester, acid chloride, amide, and cyano functions[97], similar reductions could make available a number of reduced complexes. This principle has indeed been applied to the cathodic reduction of oximes to amines[98] (equation 31).

(31)

The cathodic reduction of carbonyl-containing side-chains in cyclic systems was shown to give *endo*-alcohols stereospecifically, the reduction occurring from the side remote from iron (*exo*)[99]. The hydride reduction of the aryl ketone complex **41** proceeds similarly[100] (equation 32). The *exo* addition of carbanions to the carbonyl carbon is limited to the case

(41)

(32)

of $[Cp(\eta^6\text{-fluorenone})Fe]^+$. Thus, in $[Cp(\eta^6\text{-arene})Fe]^+$ complexes bearing a conjugated unsaturated side-chain, the reduction generally proceeds on the side-chain rather than on the arene ring or at the metal centre. However, with a carboxymethyl substituent, hydride reduction gives a carboxymethylcyclohexadienyl ligand[101]. Electrochemical reduction is necessary to reduce the side-chain regiospecifically in this case[99], and probably in several others. In the absence of a reducible function in the side-chain, cathodic reduction and other one-electron reductions (including electron transfer from hydrides or carbanions) give the 19-electron complexes $[Cp(\eta^6\text{-arene})Fe^I]$[22]. The arene ligand is more labile in these species and is often removed thermally[102]. However, the interest of the unusual oxidation state of these complexes resides primarily in their electron-transfer chemistry, which is outlined in Section VIII.A.

III. REACTIONS OF IRON COMPLEXES WITH ELECTROPHILES

Neutral organometallic complexes react with neutral or cationic electrophiles at the metal centre or at a ligand. Although it is sometimes difficult to prove the site of attack, one may state that (i) attack generally occurs at an unsaturated ligand which is not coordinated, but conjugated with the coordinated part (equations 33 and 34)[103,104], (ii) if such a site does not exist {ferrocene, $[(diene)Fe(CO)_3]$ and $[Fe(CO)_4]^{2-}$}, electrophilic attack generally occurs at the metal centre (equation 35)[105]. When attack

PhCOCl

Fe^{0} → Fe^{II} + CH_2COPh Cl^- (33)

$FpC\equiv CR^1 \xrightarrow{R^2X} Fp{=}C{=}C(R^1)(R^2)$ X^- (34)

R^1=H, R^2=Me, X=SO_3F

R^1=Me, R^2=Me, X=SO_3F

R^3=Ph, R^2=H, X=BF_4

Fe $\xrightarrow{EX}$ Fe^{IV} + —E X^- (35)

EX=$H^+BF_4^-$, $Ac^+AlCl_4^-$

Stable for EX= $H^+BF_4^-$

occurs at a *ligand*, rearrangement ensues (CO insertion, cycloaddition, proton displacement, electroredistribution about the metal—ligand bond). Note, for instance, that formyl complexes are not usually obtained by protonating anionic metal carbonyls.

When attack occurs at a *metal* centre, reductive elimination proceeds more or less readily with H^+, alkyls, and non-carbon electrophiles (I_2, $HgCl_2$), while migratory CO insertion is observed with alkyls (not H^+ or acyl); ricochet-type migration from the metal to a ligand carbon followed by proton removal occurs with alkyl and acyl electrophiles[106].

A. Electrophilic Attack at an Uncoordinated Double Bond Conjugated with a Coordinated Hydrocarbon Fragment

This type of reaction occurs with closed or open polyenes coordinated by two double bonds only. The uncoordinated double bonds of cyclic ligands may be *endo*- or exo-cyclic. Examples are found in Section IV.B, devoted to the protection of a diene by $Fe(CO)_3$ (acylation or formylation of cyclooctatetraene, cycloheptatrienone, cycloheptatriene, azepine), and in Section V.C, concerning the stabilization of unstable hydrocarbons such as *o*-xylylene. In [(myrcene)$Fe(CO)_3$], the uncoordinated double bond is not conjugated with the coordinated ones but electrophilic attack on the former affords cyclization[107].

Deprotonation of arene ligands[87a–110] in 18-electron complexes $[Cp(\eta^6\text{-arene})Fe^{II}]^+$ or H-atom abstraction by O_2[110] in 19-electron Fe^I isostructural complexes gives $[Cp(\eta^5\text{-cyclohexadienyl})Fe^{II}]$ with an exocyclic double bond (complexes of η^5-benzyl). Reactions of the latter with many electrophiles[87c] lead to the formation of C—C bonds or C—element bonds (Scheme 43), giving back the cationic $[Cp(\eta^6\text{-arene})Fe^{II}]^+$ structure (Section I). Since photolysis of these cations rapidly liberates the free arene, temporary complexation by $CpFe^+$ provides a powerful means of modifying arenes according to this

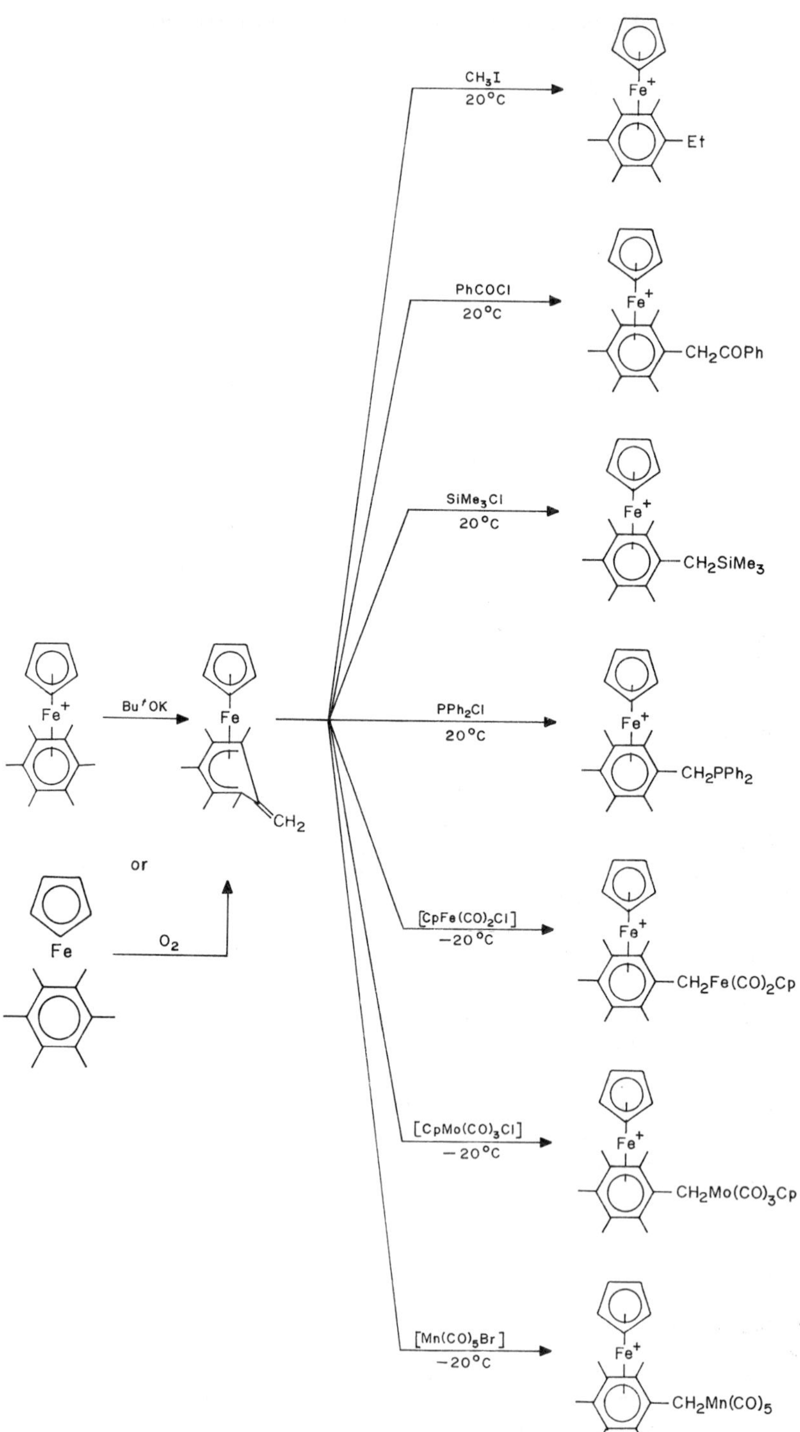

Fe+
Bu'OK
Fe
CH2
or
Fe
O2
CH3I
20°C
Et
PhCOCl
20°C
CH2COPh
SiMe3Cl
20°C
CH2SiMe3
PPh2Cl
20°C
CH2PPh2
[CpFe(CO)2Cl]
−20°C
CH2Fe(CO)2Cp
[CpMo(CO)3Cl]
−20°C
CH2Mo(CO)3Cp
[Mn(CO)5Br]
−20°C
CH2Mn(CO)5

SCHEME 43

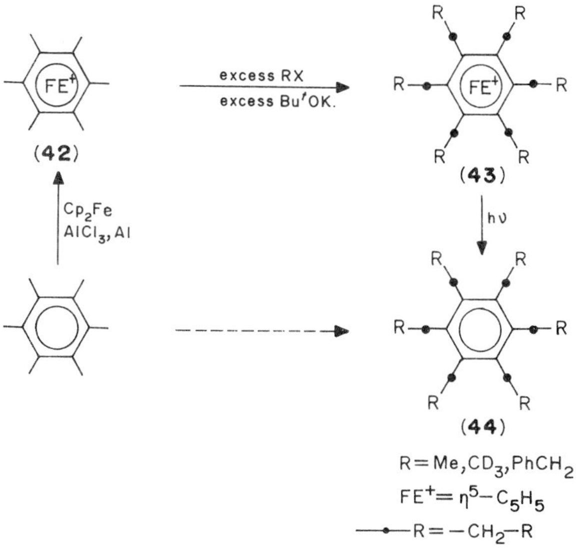

SCHEME 44

process (see below). It is possible to repeat these sequences several times in a single reaction. One-pot multiple C—C bond formation is possible if excess of base and alkylating reagent are reacted with $[Cp(\eta^6\text{-arene})Fe]^+$. Using the C_6Me_6 complex **42**, this peralkylation was performed with MeI, CD_3I, $PhCH_2Cl$, and $PhCH_2Br$ and peralkylated arenes **44** were obtained in good yields after photolysis of the iron complexes **43**[111] (Scheme 44). Indeed, $[Cp(C_6Et_6)Fe^{II}]^+$ (**45**) can be deprotonated by Bu^{t}OK {or more cleanly $[Cp(C_6Et_6)Fe^I]$ loses a H atom on contact with O_2} and the deprotonated complex **46** slowly reacts with MeI in refluxing thf to give one more alkylation leading to **47** (equation 36). This step is much slower than alkylation at an unsubstituted benzylic methyl group (Scheme 44) and is also slower than the competing reaction of Bu^{t}OK with the alkylating agent.

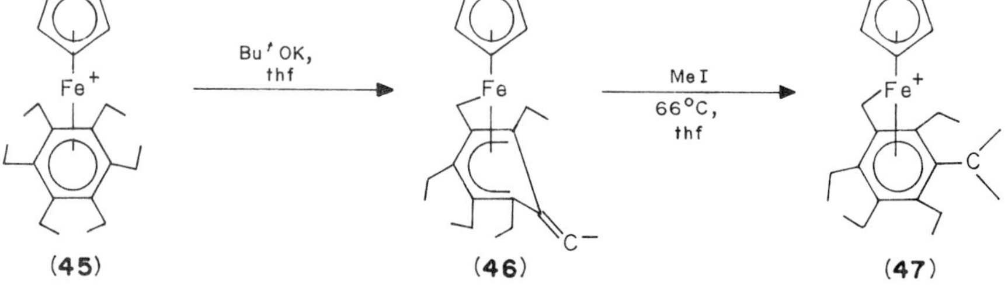

(36)

The number of alkylations by MeI in a polymethylbenzene complex on reaction of $[Cp(\eta^6\text{-arene})Fe^{II}]^+$ with excess of Bu^{t}OK + MeI depends on the steric bulk about an arene methyl group, e.g. on the number of neighbouring (*ortho*) methyls. Complete (tris)alkylation occurs on a methyl without neighbours (equation 37a), intermediate (double alkylation occurs on a methyl with one methyl neighbour (equations 37b and c), and only single alkylation occurs in methyls with two methyl neighbours (Scheme 44)[106]. Using NaOD in D_2O, all the 18 arene H atoms are replaced in **42** with

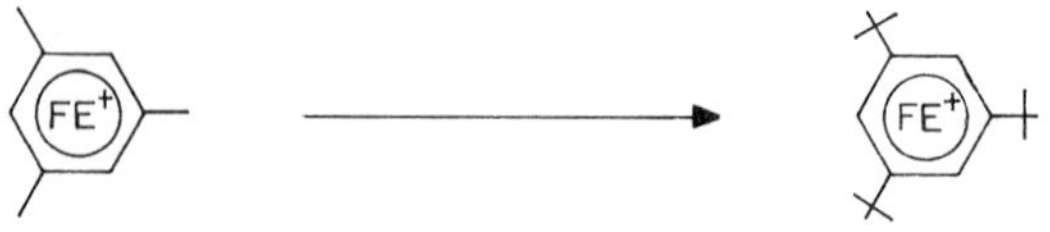

(37a)

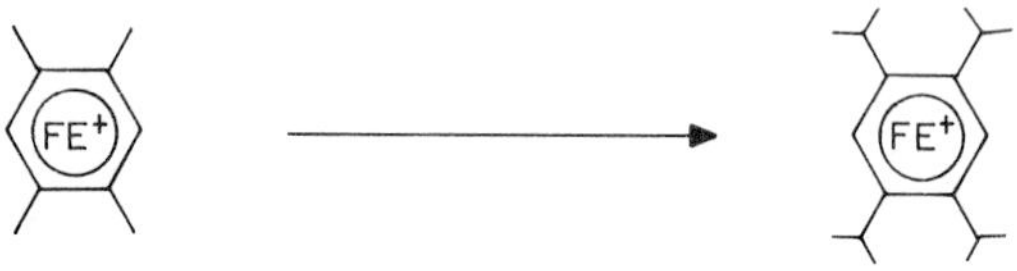

(37b)

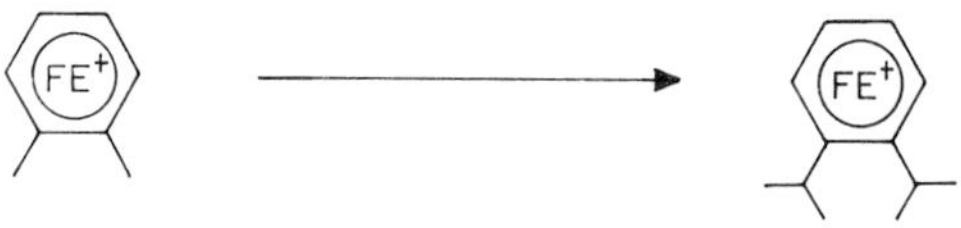

(37c)

18 D atoms after 12h at 80 °C. Permethylation of this deuteriated complex **48** as above with MeI followed by photolysis gives $C_6(CD_2Me)_6$ (**49**) (Scheme 45)[112].

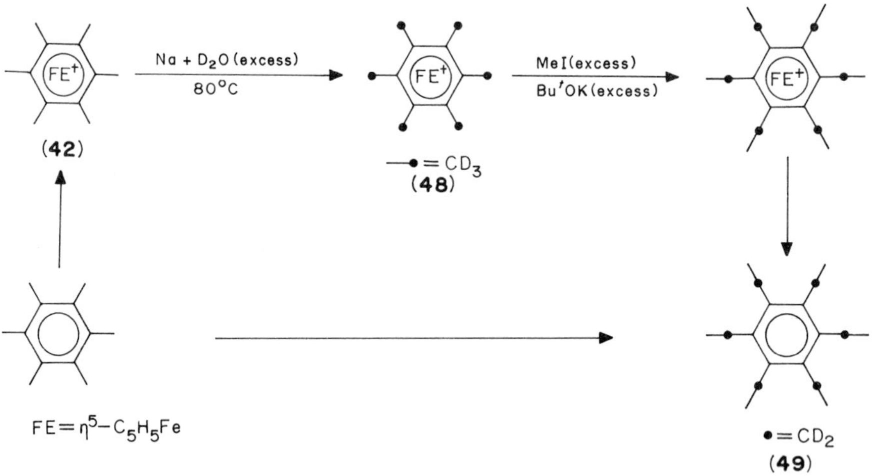

Na + D_2O (excess), 80°C

MeI(excess), Bu^{t}OK(excess)

(**42**)

—• = CD_3 (**48**)

FE = η^5–C_5H_5Fe

• = CD_2 (**49**)

SCHEME 45

B. Rosenblum's Electrophilic Attack at an Uncoordinated Multiple Bond, a Cyclopropyl Ring, or a Heteroatom Conjugated with Iron

Iron alkyl, cyclopropylmethyl, alkenyl, and alkynyl complexes react with neutral and cationic electrophiles. Neutral electrophiles give cycloadducts via dipolar intermediates, whereas cationic electrophiles give cationic η^2-olefin or carbene complexes. Virtually all of these studies were performed using [Fp(allyl)] complexes, a field developed in the early 1970's by Rosenblum and coworkers[113].

1. Cationic electrophiles

The iron η^1-allyl complex **50** and related complexes react readily with cationic electrophiles to give cationic olefin complexes[114] (Scheme 46). The rearrangement is the

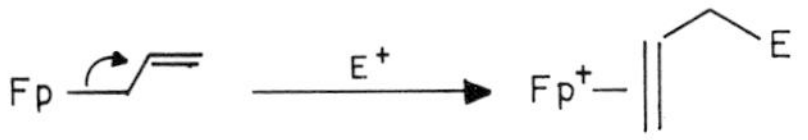

$$\text{Fp}\!-\!\text{CH}_2\text{CH}{=}\text{CH}_2 \xrightarrow{E^+} \text{Fp}^+\!-\!\|\!\!\diagup\!\!\diagdown E$$

driving force for C—C or C—element bond formation in these reactions. The olefinic ligand formed is easily liberated from the metal on displacement by NaI. Deprotonation of this olefin complex, which usually occurs with *trans* stereochemistry, gives the substituted η^1-allyl complex, which can react again with an electrophile.

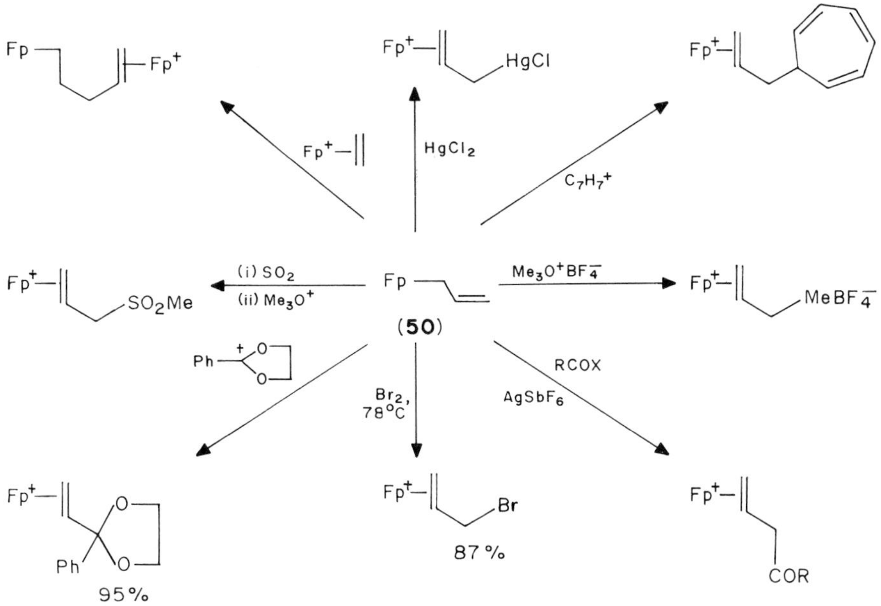

Fp(allyl) (**50**) with: Fp^+—olefin → Fp–CH₂CH₂CH₂–olefin–Fp^+; $HgCl_2$ → Fp^+–olefin–HgCl; $C_7H_7^+$ → Fp^+–olefin–CH₂–C₇H₇; (i) SO_2, (ii) Me_3O^+ → Fp^+–olefin–SO_2Me; $Me_3O^+BF_4^-$ → Fp^+–olefin–Me BF_4^-; Ph–C⁺(OCH₂CH₂O) → Fp^+–olefin–C(Ph)(OCH₂CH₂O) 95%; Br_2, 78°C → Fp^+–olefin–Br 87%; RCOX, $AgSbF_6$ → Fp^+–olefin–COR

SCHEME 46

The monodentate allenyl, butynyl, and cyclopentadienyl complexes react similarly[115] (equations 38a–d). Monohapto ligands bearing unsaturated groups such as CN[116],

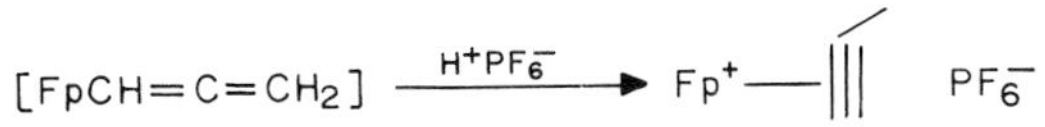

$$[\text{FpCH}{=}\text{C}{=}\text{CH}_2] \xrightarrow{H^+PF_6^-} \text{Fp}^+\text{—(allene)} \quad PF_6^- \tag{38a}$$

$$[\text{FpCH}_2\text{C}{\equiv}\text{CMe}] \xrightarrow{H^+PF_6^-} \text{Fp}^+\text{—(CH}_2{=}\text{C}{=}\text{CHMe)} \quad PF_6^- \tag{38b}$$

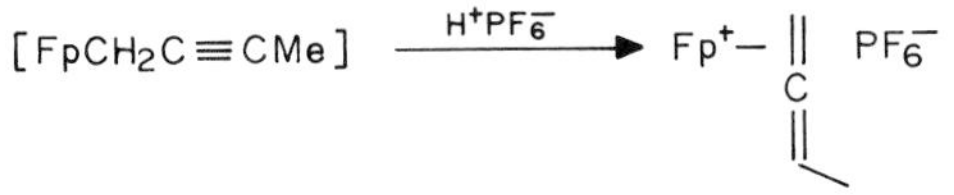

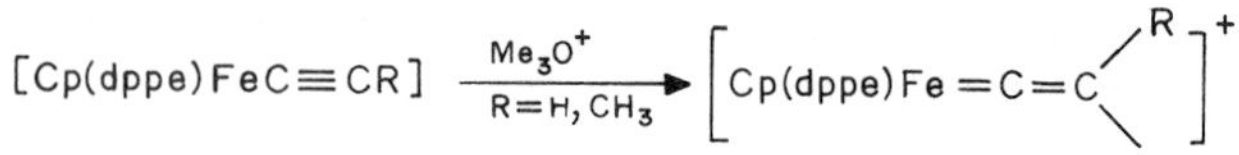
[Cp(dppe)FeC≡CR]
Me3O+
R=H, CH3
Cp(dppe)Fe=C=C
R
+

(38c)

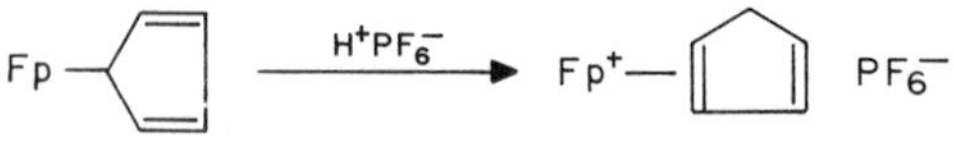
Fp
H+PF6−
Fp+
PF6−

(38d)

CO[117], or CS[118] behave analogously (equations 38e–g). The reactions of the Fp-η^1-alkyl, -η^1-butynyl, or -η^1-allenyl complexes with the tricarbonyl iron tropylium salts **51**, reported

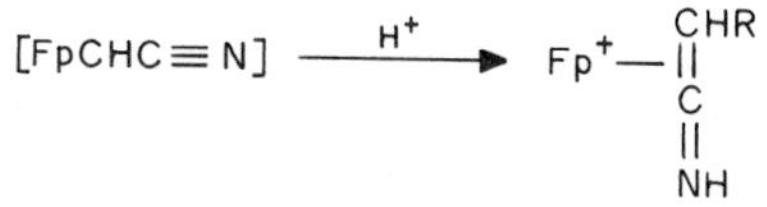
[FpCHC≡N]
H+
Fp+
CHR
C
NH

(38e)

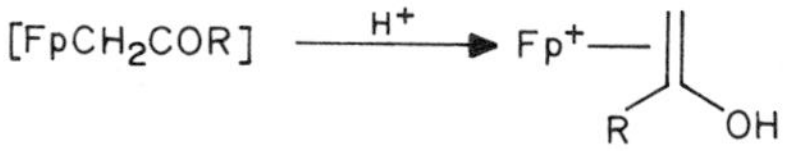
[FpCH2COR]
H+
Fp+
R
OH

(38f)

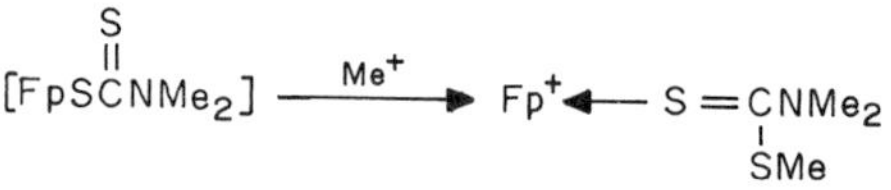
S
[FpSCNMe2]
Me+
Fp+
S=CNMe2
SMe

(38g)

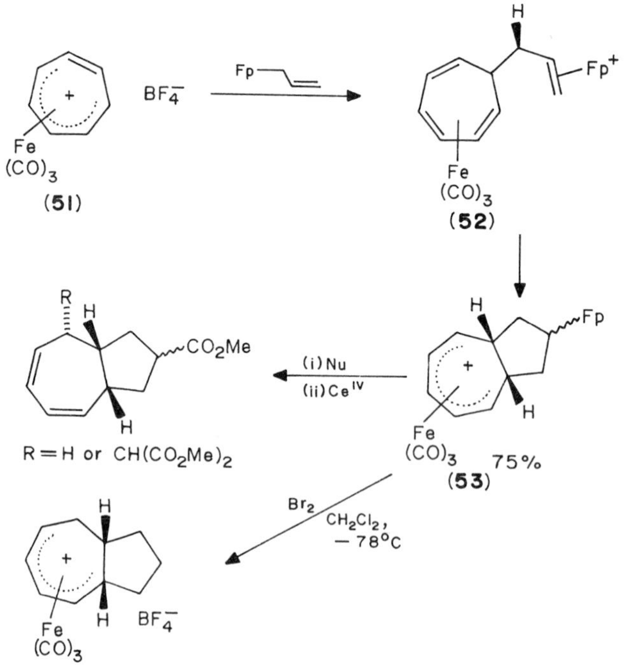
BF4−
Fp
Fe
(CO)3
(51)
H
Fp+
Fe
(CO)3
(52)
R
H
CO2Me
H
R=H or CH(CO2Me)2
(i) Nu
(ii) CeIV
H
Fp
H
Fe
(CO)3
75%
(53)
H
Br2
CH2Cl2,
− 78°C
H
BF4−
Fe
(CO)3

SCHEME 47

by Raghu and Rosenblum[119], lead to [Fp(η^2-olefin)]$^+$ intermediates **52**, which cyclize to cationic bimetallic hydroazulene complexes **53** (Scheme 47). The latter react with nucleophiles and the iron moieties can be removed by the usual decomplexing reagents.

2. Neutral electrophiles

Reaction of the [Fp(η^1-allyl)] complexes **54** with neutral electrophiles E gives an intermediate dipolar species **55**, which will cyclize by attack of the E^- unit on the metal (insertion of SO_2) or on the C-1 carbon bound to the metal (cycloaddition with tcne and other uncharged electrophiles)[13] (Scheme 48). The latter process affords the syntheses of heterocycles as indicated in Scheme 49.

Cyclic [Fp(allyl)] complexes **57** {obtained from cationic [Fp(cycloalkene)]$^+$ complexes **56** and NEt_3} react with tcne and isocyanate to give **58** and the bicyclic lactams **59** (Scheme 50). With the Fp-cyclopropylmethyl complex **60a** (obtained from Fp^-Na^+ and cyclopropylmethyl tosylate), the intermediate **60b** formed on reaction with neutral electrophiles is also dipolar, but contains one more carbon between the charged atoms (Scheme 51). The Fp-cyclopropyl complex **61** reacts with neutral electrophiles such as SO_2 giving a zwitterionic carbene intermediate **62**; the rearrangement of **62** gives both the insertion and cycloaddition products **63** and **64** (Scheme 52).

The Fp-allenyl complex **65** also gives interesting cycloadducts (**67**, **69**, **70**) via zwitterionic [Fp(η^2-alkyne)] intermediates **66** rather than [Fp(allene)], as confirmed by the isolation of the [Fp(η^2-propyne)]$^+$ complex **68** on protonation (Scheme 53). The [Fp(η^1-propargyl)] complex reacts with electrophiles in the same fashion as the [Fp(η^1-allyl)] complex. The non-stereoselectivity was pointed out (equation 39)[120].

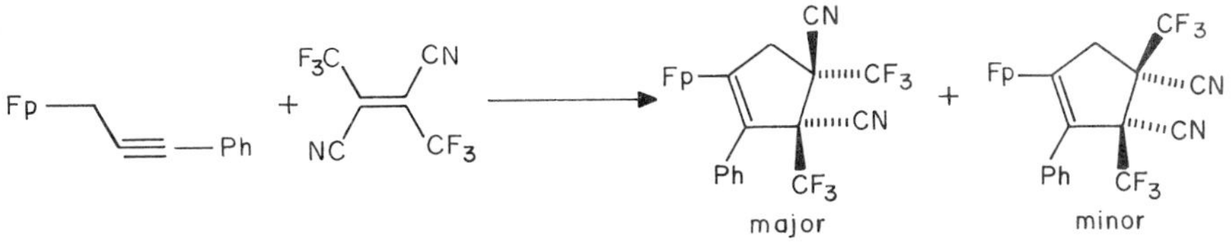

(39)

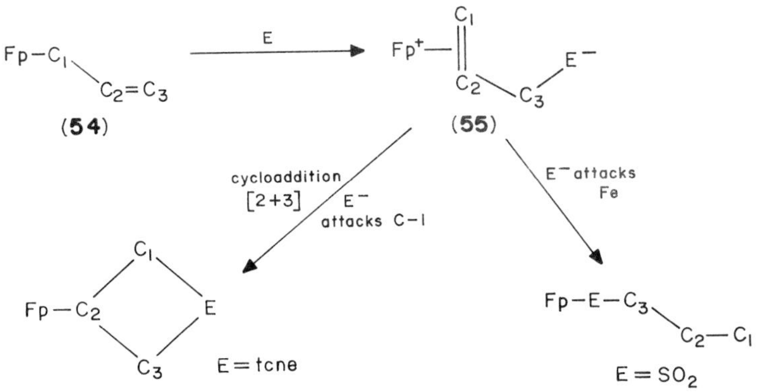

SCHEME 48

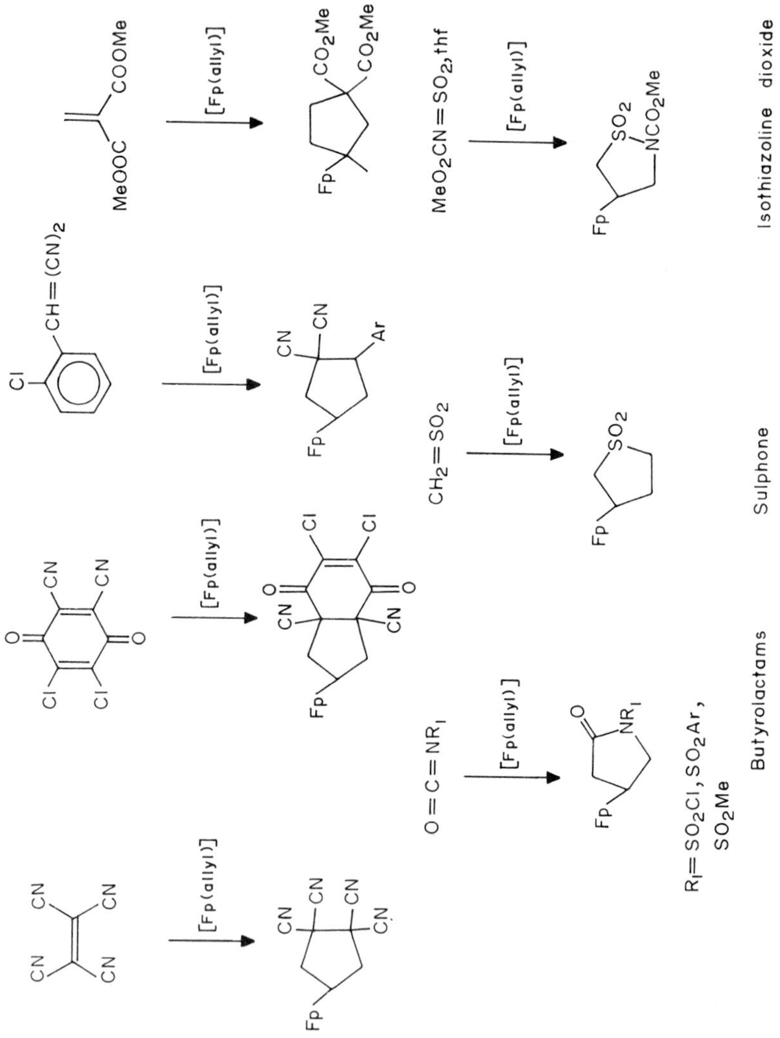

$(CN)_2C{=}C(CN)_2$ $\xrightarrow{[Fp(allyl)]}$ Fp-cyclopentane-$(CN)_4$

Cl, Cl, CN, CN quinone $\xrightarrow{[Fp(allyl)]}$ Fp, CN, CN, Cl, Cl, O, O

Cl-C_6H_4-$CH{=}(CN)_2$ $\xrightarrow{[Fp(allyl)]}$ Fp, CN, CN, Ar

MeOOC, COOMe $\xrightarrow{[Fp(allyl)]}$ Fp, CO_2Me, CO_2Me

$O{=}C{=}NR_1$ $\xrightarrow{[Fp(allyl)]}$ Fp, O, NR_1

$R_1{=}SO_2Cl, SO_2Ar,$ SO_2Me

Butyrolactams

$CH_2{=}SO_2$ $\xrightarrow{[Fp(allyl)]}$ Fp, SO_2

Sulphone

$MeO_2CN{=}SO_2$, thf $\xrightarrow{[Fp(allyl)]}$ Fp, SO_2, NCO_2Me

Isothiazoline dioxide

SCHEME 49

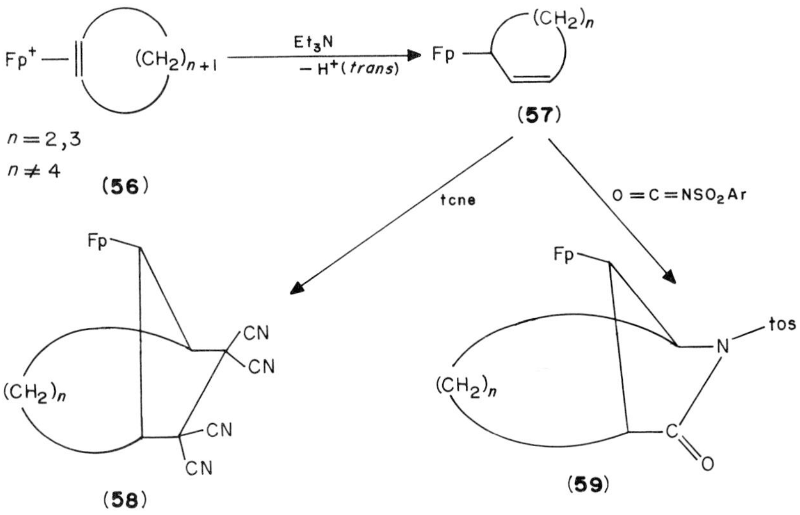

SCHEME 50

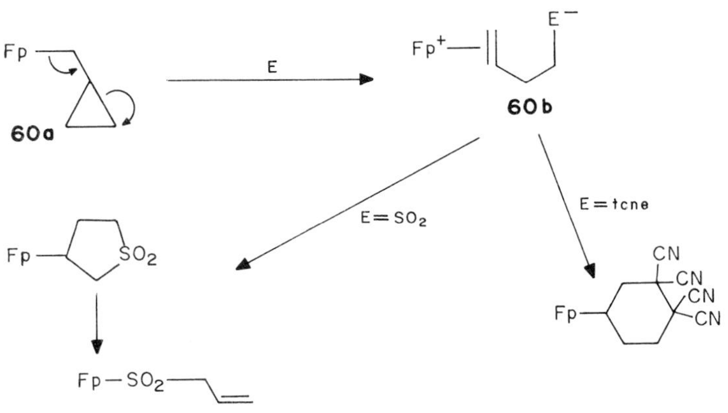

SCHEME 51

$Fp-CH{=}C{=}CH_2$ (65) $\xrightarrow{E}$ $[Fp^+$—(CH≡C—CH_2E^-)$]$ (66) ⟶ (67)

SCHEME 52a

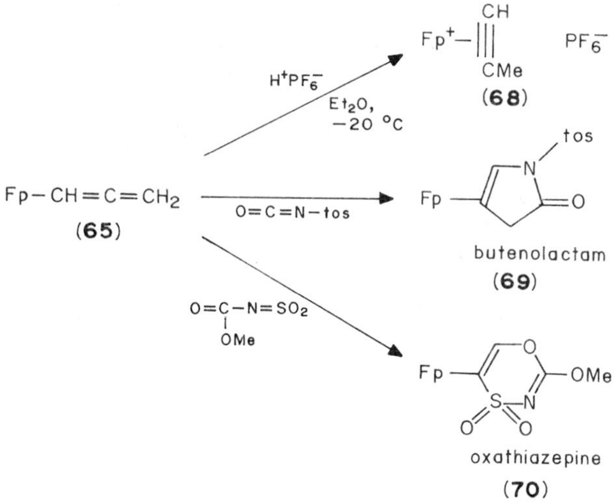

SCHEME 52b

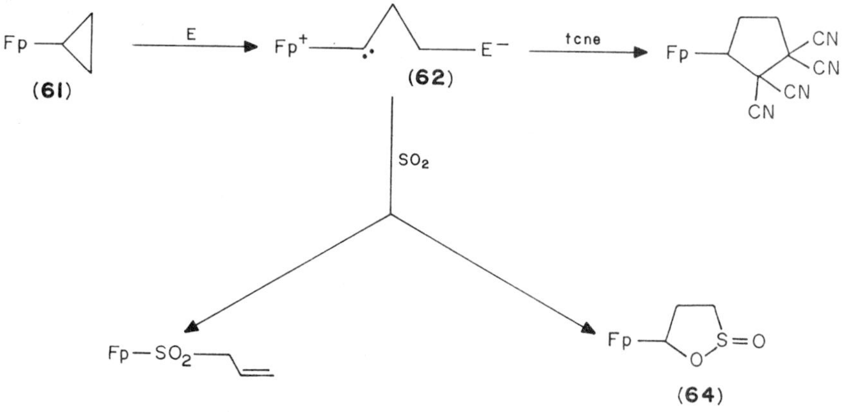

SCHEME 53

C. Reaction of Electrophiles with a π-coordinated Hydrocarbon Ligand Via Initial Attack at Iron

This type of reaction concerns organoiron compounds lacking a non-coordinated site susceptible to electrophilic attack, e.g. open or cyclic dieneiron tricarbonyl complexes and ferrocenes. In these series, the overall result is an electrophilic reaction on the unsaturated hydrocarbon ligand, and indeed the problem of ligand *vs.* metal attack by electrophiles has long been controversial.

1. Dieneiron tricarbonyl complexes. Syntheses of pheromones

Whereas acylation of dienes leads to polymerization, a smooth transformation was observed in dieneiron tricarbonyls; thus the reactivity of the diene is moderated in the complex. Nonetheless, butadieneiron tricarbonyl undergoes Friedel–Crafts acetylation 3800 times faster than benzene[121]. The product results from *endo*-acylation in which attack occurs on the same side as iron[122]. The intermediate, **71**, was isolated in 86% yield and subjected to X-ray analysis, which showed the allyl structure and coordination of the acyl oxygen to iron[123]. Careful work-up with dicyclohexylethylamine or cold aqueous ammonia effects deprotonation of **71** at the temporarily uncoordinated outer diene carbon and leads exclusively to the *cis*-acylated product **72** resulting from reaction at this carbon. Acidic or basic work-up gives the isomerized *trans*-acetyl complex **72a** (Scheme 54)[124]; thus, an easy route to dienones from dienes is provided. The norbornadiene complex leads to the methyl-substituted complex **73** (equation 40a). Graf and Lillya[125] noted that the

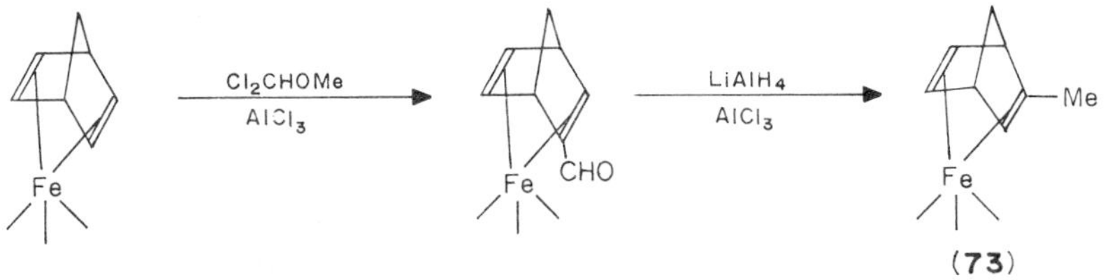

Cl_2CHOMe, $AlCl_3$ → (CHO); $LiAlH_4$, $AlCl_3$ → Me (**73**)

(40a)

regioselectivity obtained in the acetylation of (2-methoxybutadiene)iron tricarbonyl, **74**, is opposite to what would be expected from the polarization in the coordinated diene (equation 40b)[125].

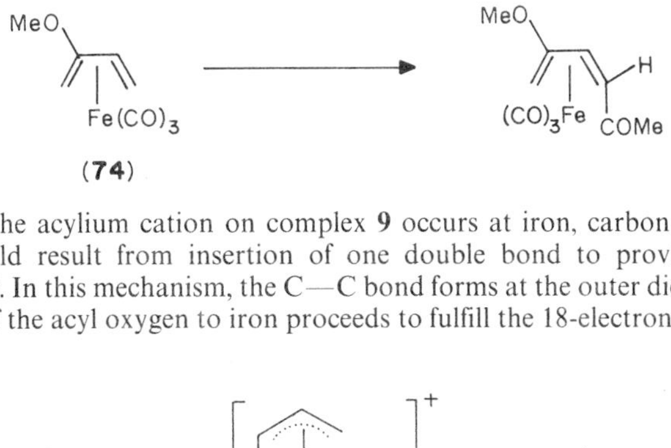

MeO, $Fe(CO)_3$ (**74**) → MeO, H, $(CO)_3Fe$, COMe

(40b)

If attack of the acylium cation on complex **9** occurs at iron, carbon—carbon bond formation would result from insertion of one double bond to provide the η^3-allyl intermediate **75**. In this mechanism, the C—C bond forms at the outer diene carbon. The coordination of the acyl oxygen to iron proceeds to fulfill the 18-electron valence shell of

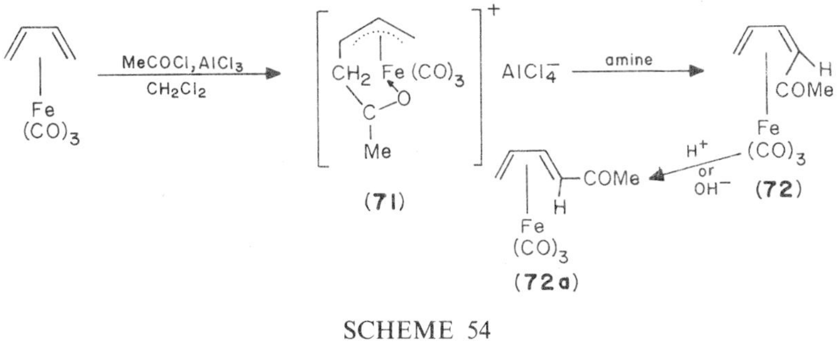

$Fe(CO)_3$ → $MeCOCl, AlCl_3$ / CH_2Cl_2 → $[CH_2\ Fe(CO)_3\ C{-}O\ Me]^+\ AlCl_4^-$ (**71**) → amine → H, COMe, Fe $(CO)_3$ (**72**) → H^+ or OH^- → COMe, H, Fe $(CO)_3$ (**72a**)

SCHEME 54

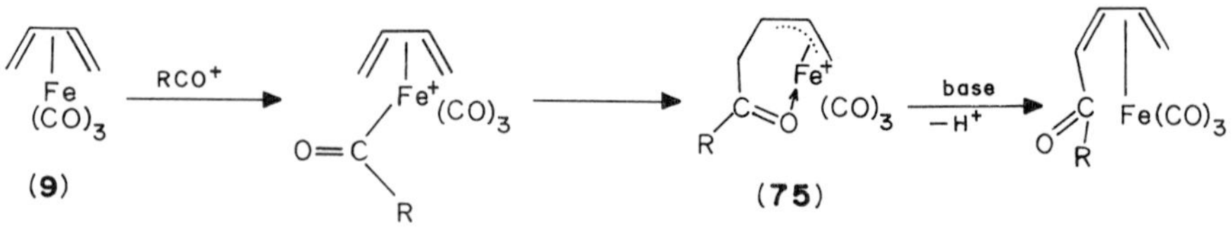

SCHEME 55

iron (Scheme 55). If protonation of [(diene)$Fe(CO)_3$] is effected in the presence of CO, the 16-electron η^3-allyl intermediate formed is stabilized by coordination of CO, which completes the valence shell of iron (equation 40c) [(η^3-allyl)$Fe(CO)_4$]$^+$ complexes **76** are obtained[126] and can be further subjected to nucleophilic attack leading to the formation of C—C bonds in olefinic structures (cf. Section II.C). The $Fe(CO)_3$ unit is easily removed from the [(diene)$Fe(CO)_3$] complexes to provide the free dienes by means of the classical oxidants Ce^{4+}, Cu^{2+}, and Fe^{3+}.

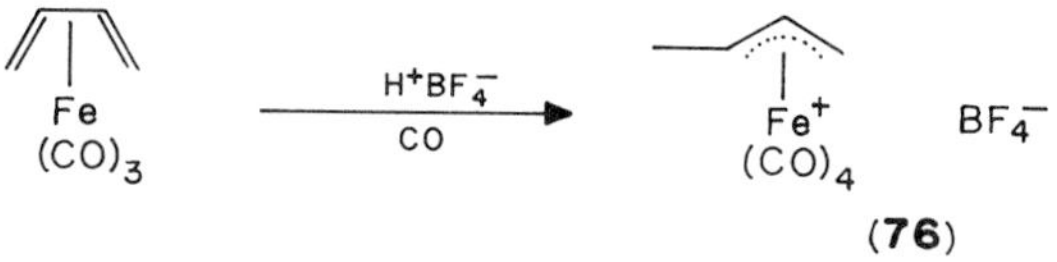

(40c)

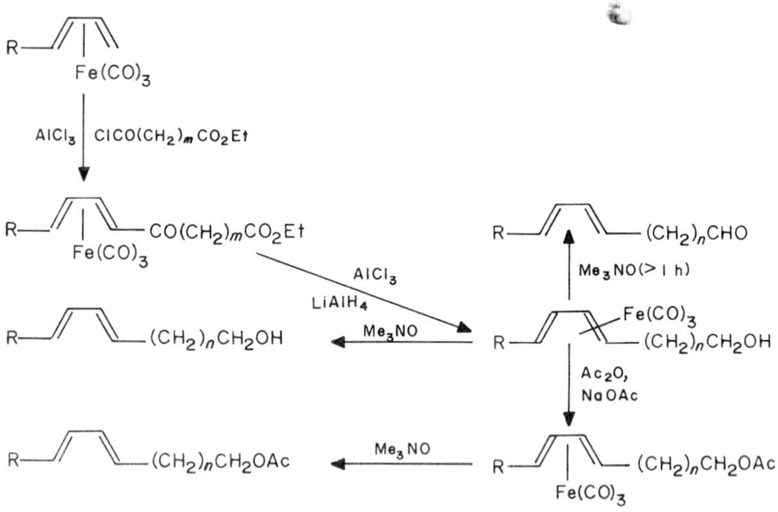

R	*m*	*n*	Moth species
H	6	7	Red bollworm moth
Et	—	—	Ligh-brown apple moth
Me	5	6	Pea moth, pitch pine tip moth, codling moth
Pr^n	7	8	Spiny bollworm moth

SCHEME 56

The stereospecificity of electrophilic addition to dieneiron tricarbonyl complexes was judiciously used by Knox and Thom[127] to synthesize insect pheromones with $>99\%$ stereochemical purity[127] (Scheme 56). This result is much superior to methods using Wittig-type syntheses or Grignard coupling[128] (80–96%). The $Fe(CO)_3$ complexation of dienes not only serves as a protecting and directing group for acylation, but can in some instances be used to elucidate the stereochemistry of a pheromone. Clelland and Knox[129] showed that the complexation of 3-methylnavenone-B, synthesized independently, gives mono- and bi-metallic complexes (Scheme 57). Since their decomplexation by Me_3NO gives back the pheromone, the all-*E*-configuration was established.

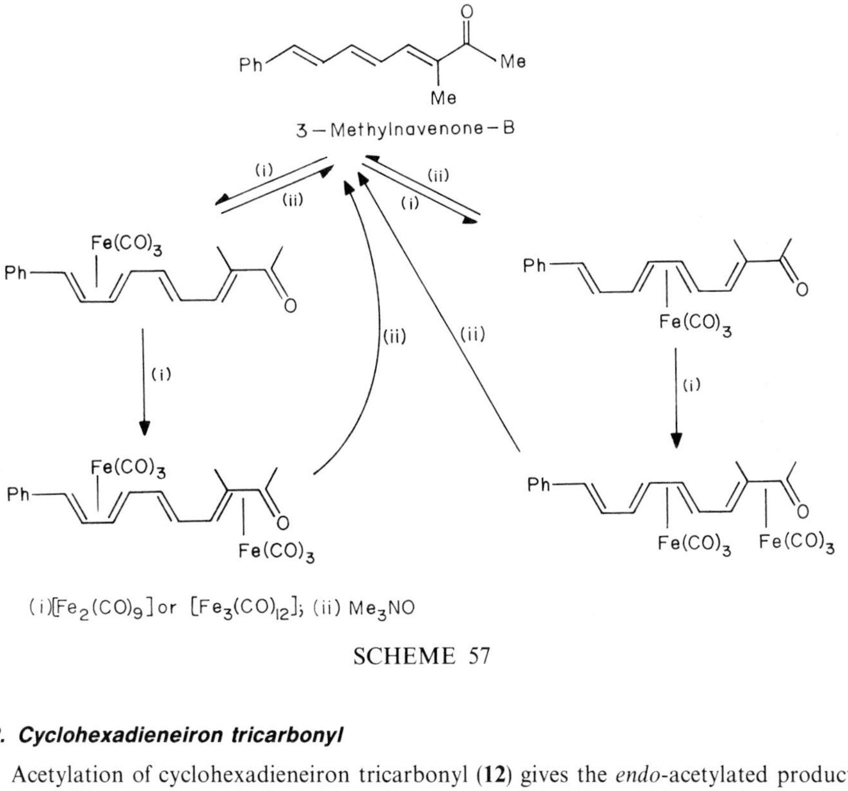

(i) $[Fe_2(CO)_9]$ or $[Fe_3(CO)_{12}]$; (ii) Me_3NO

SCHEME 57

2. Cyclohexadieneiron tricarbonyl

Acetylation of cyclohexadieneiron tricarbonyl (**12**) gives the *endo*-acetylated product **77a**, which rearranges during chromatography on alumina to the *exo*-isomer **77b** (equation 41)[130]. Acetylation of cyclohexadieneiron dicarbonyl triphenylphosphine also

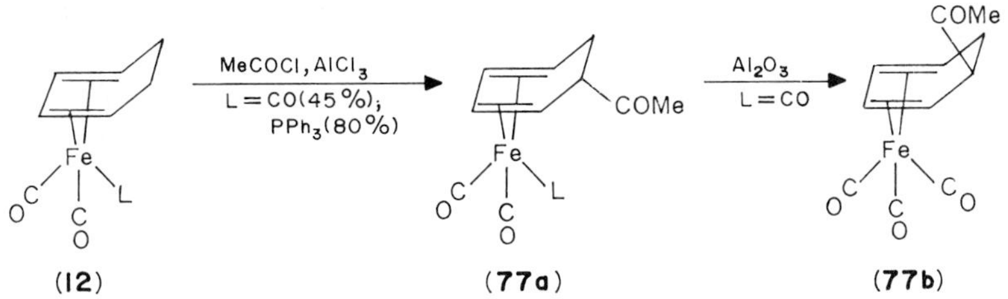

(41)

gives the *endo*-acetyl derivative **77a** in a cleaner reaction[131]. While this stereochemistry is consistent with attack of the electrophile at the metal centre, note that acetylation now occurs at a saturated carbon of cyclohexadiene, which was not the case in the acylation of $Fe(CO)_3$ complexes of open dienes. This distinction is probably the result of different rearrangements of the η^3-allyl intermediates. The reaction of electrophiles with cyclobutadieneiron tricarbonyl parallels those of other aromatics such as ferrocene.

3. Ferrocene chemistry

The best known metallocene has a special place in organoiron chemistry[132]. One should not regard it as a means to perform chemistry on C_5 rings via temporary complexation by iron, but rather as a three-dimensional organic molecule *per se*, for which a tremendous organic chemistry has been developed[133]. A large part of this chemistry is based on reactions of electrophiles. For instance, acylation reactions take place 3.3×10^6 times faster than with benzene[134]. Formylation[135] and aminomethylation[136] are also easy. The introduction of electron-withdrawing substituents deactivates the substituted ring towards further electrophilic reaction, but attack at the unsubstituted ring remains possible. Thus, depending on the RCO to ferrocene ratio and on the order of addition, mono- or 1, 1′-di-substituted ferrocenes are selectively obtained if the reactions are carried out carefully (equation 42).

Fe —[RCOCl, $AlCl_3$ / CH_2Cl_2, RT]→ Fe(COR) —[RCOCl, $AlCl_3$ / CH_2Cl_2, RT]→ Fe(COR)(COR) (42)

Phosphaferrocenes such as **78**, obtained by Mathey's group[138] from *p*-phenylphospholes and Fp_2, are acylated on the aromatic phospholyl ligand[138] (equation 43). α-Ferrocenylcarbinol[139], obtained from these ketones by $NaBH_4$ reduc-

P–Ph —[Fp_2 / 150 °C, xylene]→ P, Fe (**78**) —[MeCOCl, $AlCl_3$ / CH_2Cl_2, 25 °C]→ P, Fe, COMe (43)

tion, can be protonated to give stable and useful α-ferrocenylcarbonium cations[140] (equation 44).

$CHOHR^1$, Fe —[HBF_4]→ Fe, $\overset{+}{C}$(H)(R^1) BF_4^- —[NHR^2_2]→ Fe, C(H)(R^1)–NR^2_2

(44)

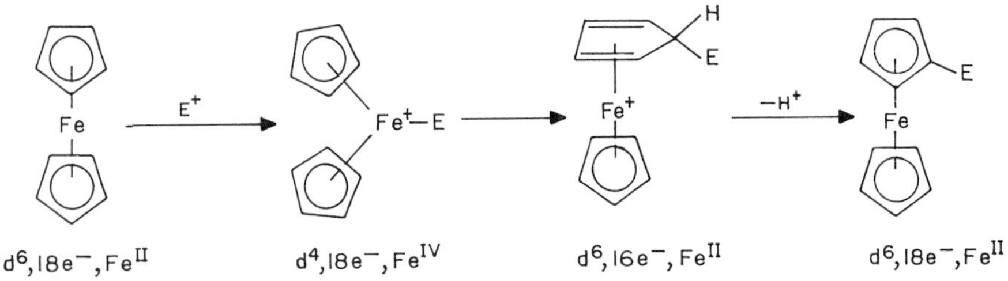

d^6, 18e$^-$, Fe^{II} → (E^+) → Fe$^+$—E, d^4, 18e$^-$, Fe^{IV} → Fe$^+$, d^6, 16e$^-$, Fe^{II} → ($-H^+$) → Fe, d^6, 18e$^-$, Fe^{II}

SCHEME 58

The 'ricochet' mechanism proposed for electrophilic substitution involves electrophilic attack at the metal centre to generate an Fe^{IV} intermediate followed by migration to the ring (Scheme 58). This intermediate is indeed isolated when the electrophile is $H^+BF_4^-$. However, the question of the site of electrophilic attack is controversial and may depend on the electrophile[140]. A useful application of ferrocene in aromatic synthesis found by Nesmeyanov and coworkers[141] is ligand exchange which allows complexation and activation of many arenes by the cationic group $(\eta^5\text{-}C_5H_5)Fe^+$ (equation 5). Chemistry related to that of ferrocene is encountered for cyclobutadieneiron tricarbonyl (Section V.C).

D. Alkylation and Acylation of $[Fe(CO)_4]^{2-}$, $[HFe(CO)_4]^-$ and $[CpFe(CO)_2]^-$ Electrophilic Cleavage of Fe—C Bonds

1. *$Na_2[Fe(CO)_4]$, Collman's reagent[142]: carbonylation of halides*

$Na_2[Fe(CO)_4]$ is obtained by reducing $[Fe(CO)_5]$ with Na/Hg or, better, with sodium benzophenone ketyl[143]. The use of dioxane results in the formation of a stable dioxane complex (Collman's reagent). It reacts with an alkyl or acyl halide or tosylate by an S_N2 substitution (second-order kinetics, inversion of configuration) and the alkyl or acyl tetracarbonylferrate is obtained[144] (equation 45a). The reaction of a second electrophile R′X gives the coupled organic product RR′ by reductive elimination from the intermediate $(R)(R')Fe(CO)_4$, **79**[145] (equation 45b). Alkanes are obtained from alkyl halides (R = alkyl, R′ = H) and aldehydes are obtained from acylhalides (R = acyl, R′ = H).

$$Na_2[Fe(CO)_4] + RX \longrightarrow Na[RFe(CO)_4] + NaX \qquad (45a)$$
$$R = \text{alkyl, acyl, etc.}$$

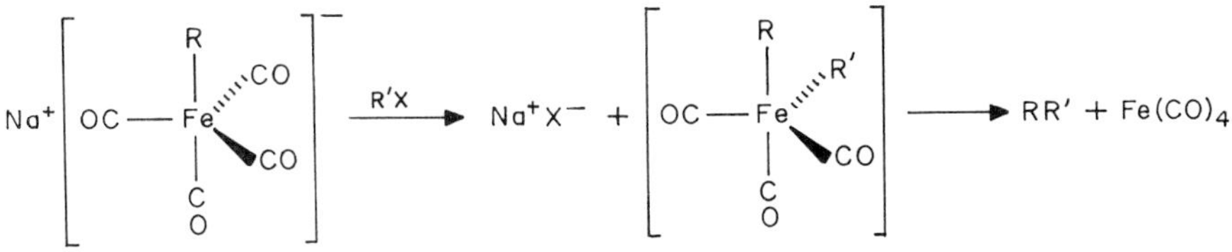

$$Na^+\left[OC{-}Fe(R)(CO)_3\right]^- \xrightarrow{R'X} Na^+X^- + \left[OC{-}Fe(R)(R')(CO)_3\right] \longrightarrow RR' + Fe(CO)_4 \qquad (45b)$$

(79)

The most remarkable and useful feature of this double electrophilic attack of $[Fe(CO)_4]^{2-}$ is the easy migratory insertion of CO into the Fe—C bond after the first

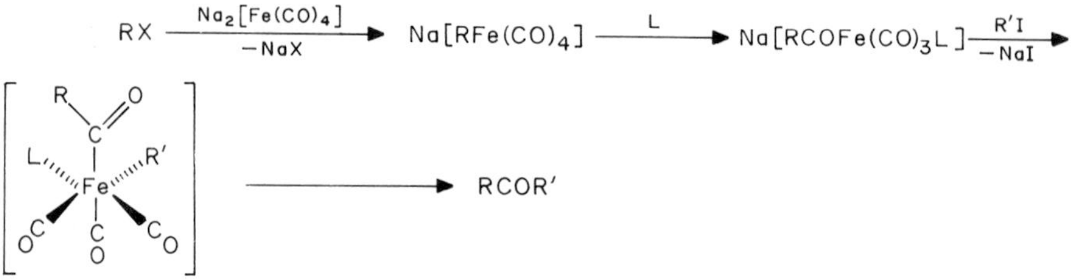

$$RX \xrightarrow[-NaX]{Na_2[Fe(CO)_4]} Na[RFe(CO)_4] \xrightarrow{L} Na[RCOFe(CO)_3L] \xrightarrow[-NaI]{R'I}$$

$$[(RCO)Fe(L)(R')(CO)_3] \longrightarrow RCOR'$$

SCHEME 59

step if a potential ligand L is added (L = CO or phosphine) (equation 46). Thus, in this case, starting from an alkyl halide also gives the acylferrate accessible from $[Fe(CO)_4]^{2-}$ and an acyl halide. This process is a high-yield conversion of alkyl halides to aldehydes. Instead of a proton, the second electrophile may be another alkyl iodide, which provides unsymmetrical ketones (Scheme 59). The nature of the counter cation and the reaction conditions direct Fe *vs.* C alkylation (see Section V.B). Another way to obtain acyltetracarbonylferrate is to react a carbanion with $(Fe(CO)_5]$. The three methods are summarized in Scheme 60.

$$Na^+[RFe(CO)_4]^- \xrightarrow{L} Na^+[RCOFe(CO)_3L]^- \xrightarrow{MeCO_2H} RCHO \qquad (46)$$

Oxidative cleavage of the alkyl ferrate with O_2 gives carboxylic acids while cleavage with halogens gives acyl halides. Depending on the work-up, the alkyl halides are converted into acids, esters, or amides (Scheme 61). These organic products result from the one-electron oxidation of the 18-electron alkyl or acyl ferrate to the 17-electron complex. Migratory CO insertion is fast in the latter; the acyl radical is cleaved and further oxidized or halogenated. All these reactions are synthetically useful[146] since they proceed in good yields and tolerate functionalities such as ester, ketone, nitriles, and olefin (equation 47a and 47b). The range of possible substrates RX is limited by the high basicity

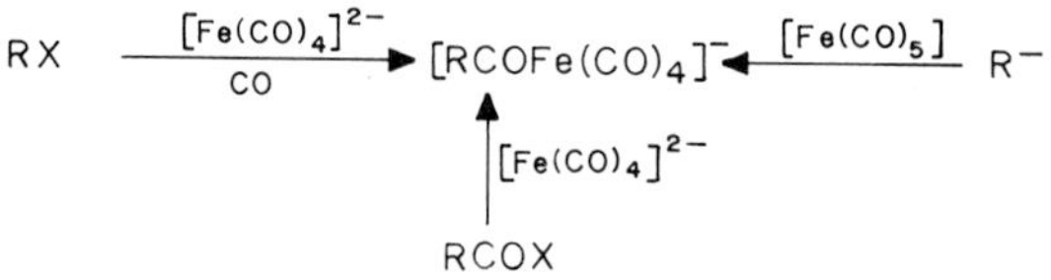

$$RX \xrightarrow[CO]{[Fe(CO)_4]^{2-}} [RCOFe(CO)_4]^- \xleftarrow{[Fe(CO)_5]} R^-$$

$$RCOX \xrightarrow{[Fe(CO)_4]^{2-}} [RCOFe(CO)_4]^-$$

SCHEME 60

$$Na^+[RFe(CO)_4]^- \text{ or } Na^+[RCOFe(CO)_3]^- \xrightarrow{O_2} RCOOH; \quad \xrightarrow{X_2} RCOX$$

$$RCOH(O) \xrightarrow[(R=H)]{H_2O} \ ; \ RCOX \xrightarrow{MeOH (R=Me)} RCOOR; \quad \xrightarrow{RR''NH} RCONR'R''$$

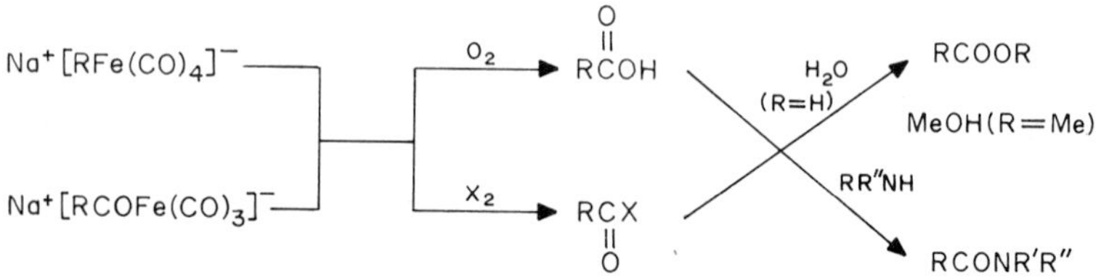

SCHEME 61

$$Na_2[Fe(CO)_4] + Br(CH_2)_3CH=CH_2 \longrightarrow [(CO)_4Fe(CH_2)_3CH=CH_2]^-$$

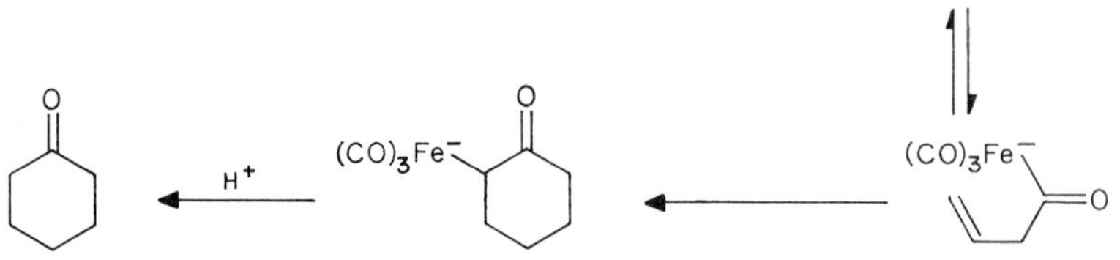

SCHEME 62

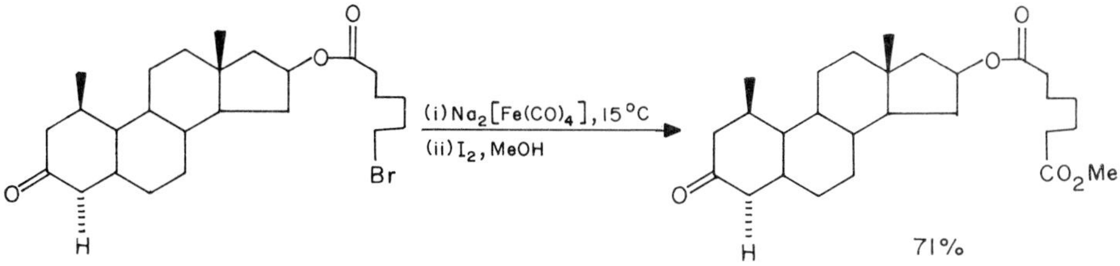

(47a)

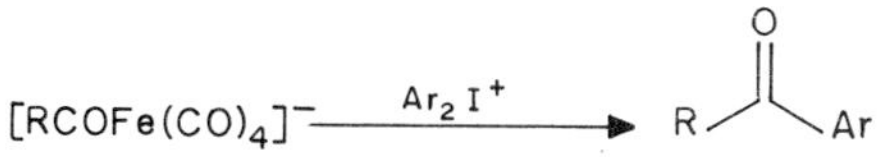

(47b)

of $Na_2[Fe(CO)_4]$ (competing E_2 elimination with secondary halides). Other features are noted with allylic halides bearing δ C—H bonds (which give stable diene complexes); the migratory–insertion step may fail when R bears an electronegative group.

Attractive ring formations are possible starting from halo- or tosyloxy-alkenes[147]. The unsaturated double bond plays the part of the added ligand, the coordination of which favours migratory insertion of a carbonyl; a second insertion, that of the coordinated double bond into the iron—acyl bond, leads to ring closure (equation 48 and Scheme 62).

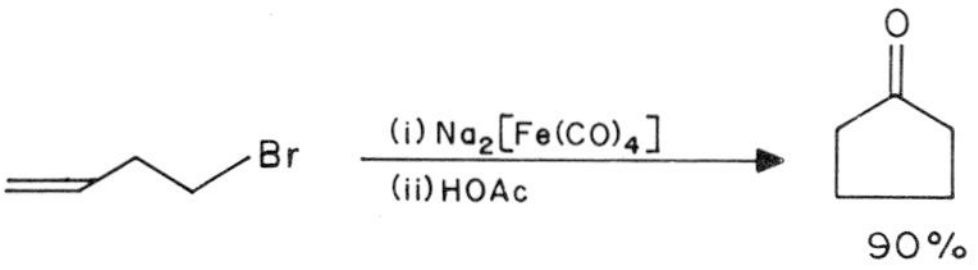

(48)

The acyl group migrates to the terminal carbon of the coordinated olefin rather than to the substituted one. A route to a key intermediate in the synthesis of the tetracyclic diterpene aphidicolin, showing antiviral activity, is an outstanding example of this ring formation[148] (Scheme 63).

Reaction of enones with $Na[HFe_2(CO)_8]$ {derived from $Na_2[Fe(CO)_4] + [Fe(CO)_5]$ followed by acidification} gives reduction of the carbon—carbon double bond[149] (equation 49). Reaction of allenes with $[RFe(CO)_4]$ followed by acidification and decomplexation by Me_3NO gives free enones[150] (equation 50).

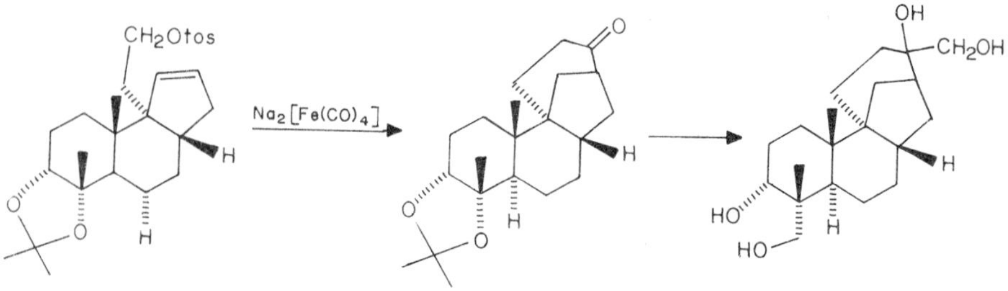

SCHEME 63

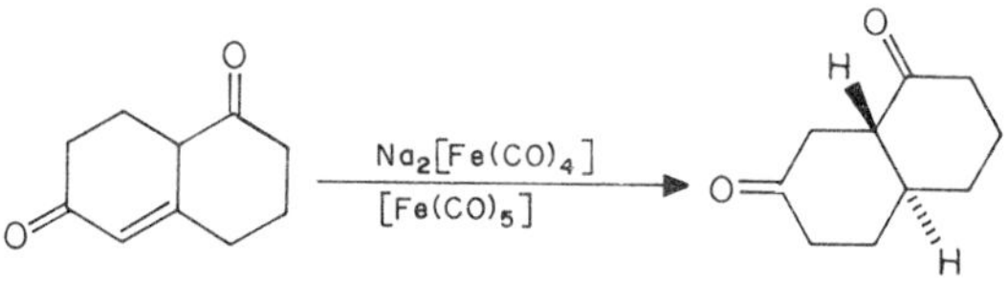

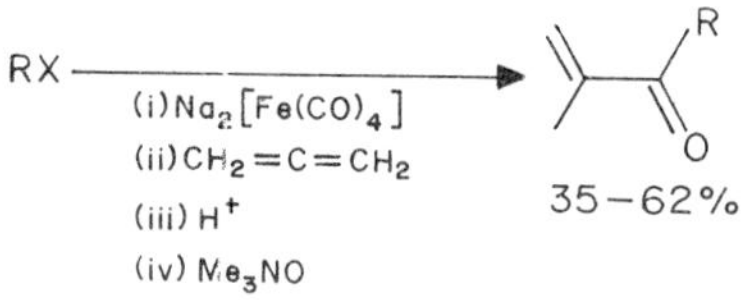

$$\text{RX} \xrightarrow[\text{(i) Na}_2[\text{Fe(CO)}_4]\text{; (ii) CH}_2\text{=C=CH}_2\text{; (iii) H}^+\text{; (iv) Me}_3\text{NO}]{} \quad 35\text{–}62\% \tag{50}$$

(49)

The reactions of $Na_2[Fe(CO)_4]$ and the migratory insertion of CO are dependent on ion pairing effects, the former being slowed and the latter accelerated by tight ion pairs $[Li^+ > Na^+ > (PPh_3)_2N^+]$[151]. The $(PPh_3)_2N^+$ salts of both $[RFe(CO)_4]^-$ and $[RCOFe(CO)_3L]^-$ have been isolated as air-stable crystals and thoroughly characterized. Des Abbayes[303] has elegantly shown that carbonylation of benzyl halides can proceed catalytically (in $Fe(CO)_5$) using phase transfer catalysis $\{CH_2Cl_2$, CO, aq, NaOHM, $(Bu_4N^+)_2SO_4{}^{=}\}$.

2. $[HFe(CO)_4]^-$: stoichiometric and catalytic reductions

$[HFe(CO)_4]^-$ is generated from $[Fe(CO)_5]$ and a base[152]. This species is an intermediate in catalytic processes affording $C_1 \rightarrow C_2$ or $C_2 \rightarrow C_3$ transformations. The oxo process[153], namely transformation of an olefin to a higher alcohol or aldehyde using $CO + H_2$, is catalysed by cobalt and rhodium carbonyls. In 1983, Reppe and Vetter[154] found an interesting variation using water as a hydrogen source in place of dihydrogen and $[Fe(CO)_5]$ as a catalyst in basic media and relatively mild conditions (equation 51a). Two

$$CH_2{=}CH_2 \xrightarrow[\text{[Fe(CO)}_5\text{], amine or K}_2\text{CO}_3\text{, 100 °C, 200 psi}]{2CO + H_2O} Pr^nOH \tag{51a}$$

species were found to be formed by reaction of $[Fe(CO)_5]$ with a base, $[HFe(CO)_4]^-$ and $[HFe_3(CO)_{11}]^-$. However, Pettit *et al.*[155] found that the reaction was extremely pH

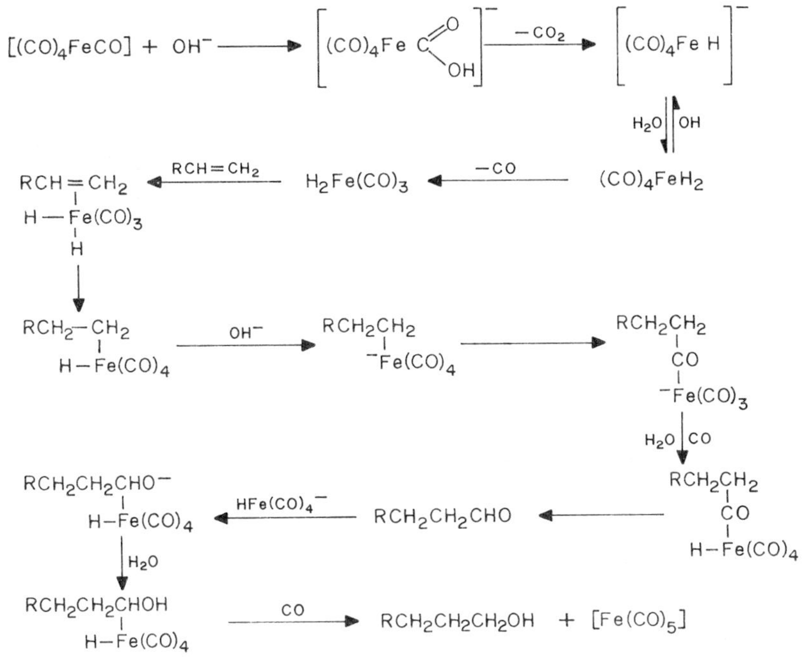

$$[(CO)_4FeCO] + OH^- \longrightarrow \left[(CO)_4Fe\,C\genfrac{}{}{0pt}{}{\nearrow O}{\searrow OH}\right]^- \xrightarrow{-CO_2} \left[(CO)_4Fe\,H\right]^- \underset{OH}{\overset{H_2O}{\rightleftharpoons}} (CO)_4FeH_2$$

$$(CO)_4FeH_2 \xrightarrow{-CO} H_2Fe(CO)_3 \xrightarrow{RCH=CH_2} RCH{=}CH_2 \cdot H{-}Fe(CO)_3{-}H \longrightarrow RCH_2{-}CH_2{-}Fe(CO)_4H$$

$$RCH_2CH_2{-}Fe(CO)_4H \xrightarrow{OH^-} RCH_2CH_2{-}Fe(CO)_4^- \longrightarrow RCH_2CH_2{-}CO{-}Fe(CO)_3^- \xrightarrow{H_2O,\ CO} RCH_2CH_2{-}CO{-}Fe(CO)_4H$$

$$RCH_2CH_2{-}CO{-}Fe(CO)_4H \longrightarrow RCH_2CH_2CHO \xrightarrow{HFe(CO)_4^-} RCH_2CH_2CHO^-{-}Fe(CO)_4H \xrightarrow{H_2O} RCH_2CH_2CHOH{-}Fe(CO)_4H \xrightarrow{CO} RCH_2CH_2CH_2OH + [Fe(CO)_5]$$

SCHEME 64

dependent, the catalyst being efficient in the pH range 8–10.7. $[H_2Fe(CO)_4]$ is the catalytically active species. The mechanism in Scheme 64 was proposed.

Pettit *et al.* also attempted to convert CO to methanol using water in an analogous system, a goal of considerable importance since it involves the preparation of a liquid fuel (methanol) according to equation 51b. Based on $[Fe(CO)_5]$, 10% MeOH, 150% H_2 and 700% HCO_2^- are produced in the reaction of excess of aqueous K_2CO_3 with $[Fe(CO)_5]$ under a pressure of 300 psi of CO at 100 °C. Pettit *et al.* proposed that the key step is not CO insertion into an Fe—H bond but rather nucleophilic attack of $[HFe(CO)_4]^-$ on $[Fe(CO)_5]$, giving a bimetallic intermediate[155] (Scheme 65). $[HFe(CO)_4]^-$ and $[HFe_3(CO)_{11}]^-$ are also used in stoichiometric and catalytic reductions of organic functions.

$$3CO + 2H_2O \longrightarrow CH_3OH + 2CO_2 \qquad (51b)$$

$NMe_4^+[HFe(CO)_4]^-$ reacts with acyl chlorides in CH_2Cl_2 at 25 °C to give aldehydes in excellent yields[156]. It is assumed that the mechanism again involves nucleophilic attack at

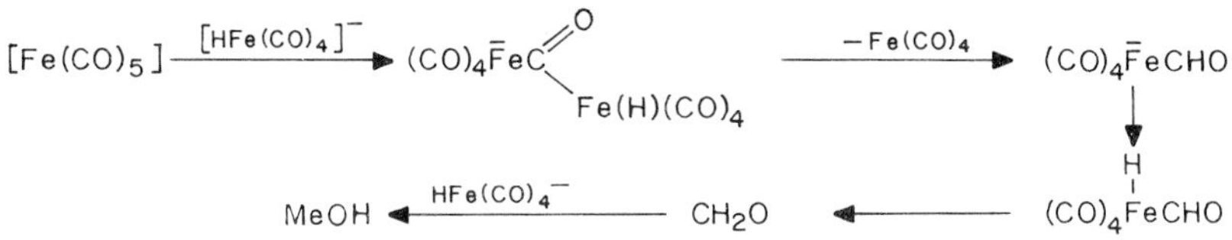

$$[Fe(CO)_5] \xrightarrow{[HFe(CO)_4]^-} (CO)_4\bar{F}eC(=O){-}Fe(H)(CO)_4 \xrightarrow{-Fe(CO)_4} (CO)_4\bar{F}eCHO \longrightarrow (CO)_4Fe(H)CHO \longrightarrow CH_2O \xrightarrow{HFe(CO)_4^-} MeOH$$

SCHEME 65

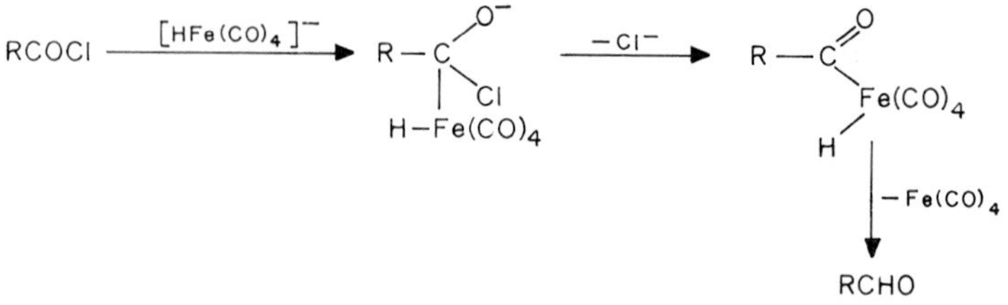

$$\text{RCOCl} \xrightarrow{[\text{HFe(CO)}_4]^-} \text{R–C(O}^-\text{)(Cl)(H–Fe(CO)}_4) \xrightarrow{-\text{Cl}^-} \text{R–C(=O)–Fe(CO)}_4\text{(H)} \xrightarrow{-\text{Fe(CO)}_4} \text{RCHO}$$

SCHEME 66

the carbonyl group (Scheme 66). This reaction also proceeds in excellent yields starting from $[Fe(CO)_4]^{2-}$ that has been protonated[157]: $[Fe(CO)_4]^{2-} + H^+ \rightarrow [HFe(CO)_4]^-$. CO and CN functionalities can be reduced catalytically by $[HFe(CO)_4]^-$ generated from $[Fe(CO)_5]$ and Et_3N using $CO + H_2O$ as the source of hydrogen[158]. This transformation was demonstrated for acetone and benzylideneaniline by Markó *et al.*[158] (equation 52).

$$\text{PhCH=NPh} + \text{CO} + \text{H}_2\text{O} \xrightarrow[100\,^\circ\text{C},\ 100\text{ bar CO}]{[\text{HFe(CO)}]^-\ \text{cat.}} \text{PhCH}_2\text{NHPh} + \text{CO}_2 \qquad (52)$$

K$[HFe(CO)_4]$, prepared by addition of $[Fe(CO)_5]$ to KOH in ethanol (ratio 1 : 3), is useful for the reductive alkylation and acylation of aldehydes, ketones, esters, nitriles, or amines with an active methylene group adjacent to the functional group[159]. It also reduces carbon—carbon double bonds in conjugated systems[160], as well as epoxides[161] and nitrogen compounds[162] (equations 53–58). Nitroaryls are reduced to anilines by

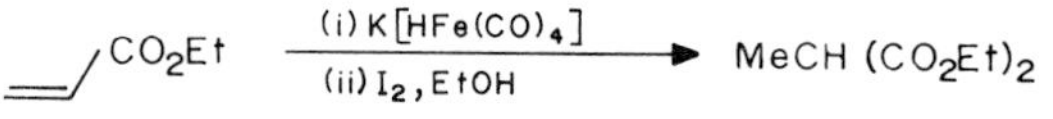

$$\text{CH}_2\text{=CHCO}_2\text{Et} \xrightarrow[\text{(ii) I}_2\text{, EtOH}]{\text{(i) K[HFe(CO)}_4]} \text{MeCH(CO}_2\text{Et)}_2 \qquad (53)$$

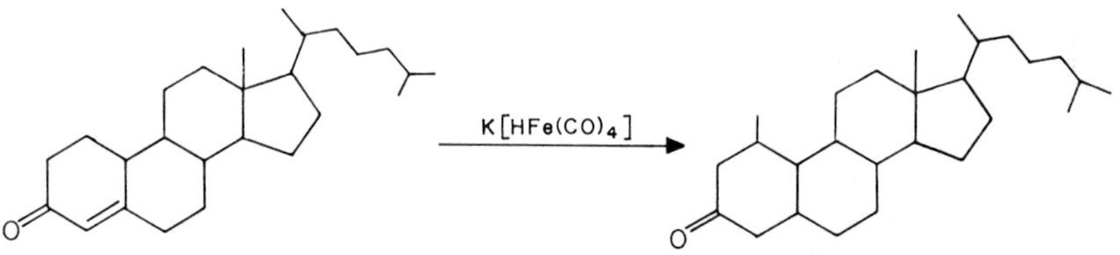

$$\text{cholest-4-en-3-one} \xrightarrow{\text{K[HFe(CO)}_4]} \text{coprostanone (32\%)} \qquad (54)$$

$$\text{ethylene oxide} \xrightarrow[\text{(ii) I}_2\ \text{(iii) HF/EtOH}]{\text{(i) K[HFe(CO)}_4]/\text{CO}} \text{HOCH}_2\text{CH}_2\text{CO}_2\text{Et} \qquad (55)$$

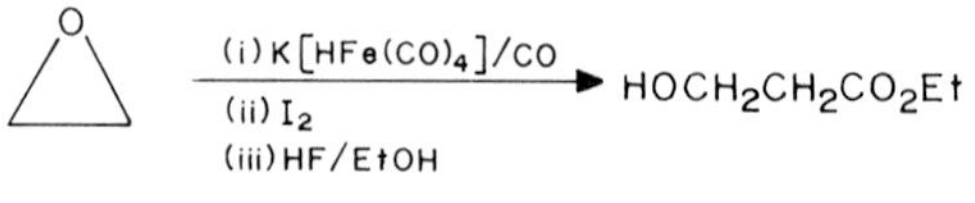

$$\text{RCHO} + \text{R}^1\text{COCH}_2\text{R}^2 \xrightarrow[\text{base catalysed}]{-\text{H}_2\text{O}} \text{R}^1\text{COC(=CHR)R}^2 \xrightarrow{\text{K[HFe(CO)}_4]} \text{R}^1\text{COCH(CH}_2\text{R)R}^2 \qquad (56)$$

$$O_2N\text{-}C_6H_{11} \xrightarrow{K[HFe(CO)_4]} H_2N\text{-}C_6H_{11} \quad (57)$$

$$o\text{-}O_2NC_6H_4CH_2CHO \xrightarrow{K[HFe(CO)_4]} \text{quinoline} \quad (58)$$

$[HFe_3(CO)_{11}]^-$ generated from $[Fe_3(CO)_{12}]$ and methanol[163]. Carbonyl, ester, amide, and olefin functionalities are unchanged (equation 59). $K[HFe(CO)_4]$ (as $[FeCO_4]^{2-}$) dehalogenates alkyl halides at 20 °C (S_N2 inversion)[164] (equations 60 and 61). It should be

$$O_2N\text{-}C_6H_4\text{-}COCH_3 \xrightarrow{[HFe_3(CO)_{11}]^-} H_2N\text{-}C_6H_4\text{-}COCH_3 \quad (59)$$

$$p\text{-}BrC_6H_4COCH_2Br \xrightarrow{K[HFe(CO)_4]} p\text{-}BrC_6H_4COCH_3 \quad 66\% \quad (60)$$

$$\text{3-bromocamphor (Br, H)} \xrightarrow{K[DFe(CO)_4]} \text{3-deuteriocamphor (D, H)} \quad 60\% \quad (61)$$

noted that reductions and dehalogenations of organic halides can be effected not only with iron complexes $\{[Fe(CO)_5]$[165], $[HFe(CO)_4]^-$, $[Fe(CO)_4]^{2-}$[166], $[C_5Me_5Fe(CO)_2H]$[167]$\}$ but also with many other main group or transition metal compounds (Sn, V, Mo, Nb, etc.)[168].

3. *Alkylation of $[CpFe(CO)_2]^-$ and electrophilic cleavage of the Fe—C bond in $[CpFe(CO)_2R]$*

There is a large overlap between the chemistry of iron carbonyl anions and that of $[CpFe(CO)_2]^-$ (Fp^-). Alkylation of this extremely nucleophilic anion proceeds at the metal centre with a variety of halides to give easily isolable FpR complexes[169]. In these complexes, migratory CO insertion can take place in two ways: addition of CO or phosphine[170] or one-electron oxidation[171] (Scheme 67). In the latter process, decomplexation is obtained owing to the instability of the 17-electron organoiron species. On decomplexation, the functionalization depends on the solvent[172]. The reactions are stereospecific with retention of configuration at carbon (equations 62 and 63). To

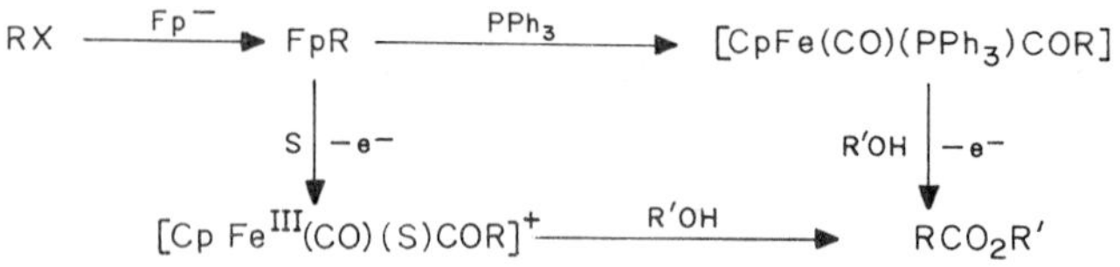

$$RX \xrightarrow{Fp^-} FpR \xrightarrow{PPh_3} [CpFe(CO)(PPh_3)COR]$$

$$FpR \xrightarrow[S]{-e^-} [Cp\,Fe^{III}(CO)(S)COR]^+ \xrightarrow{R'OH} RCO_2R'$$

$$[CpFe(CO)(PPh_3)COR] \xrightarrow[R'OH]{-e^-} RCO_2R'$$

SCHEME 67

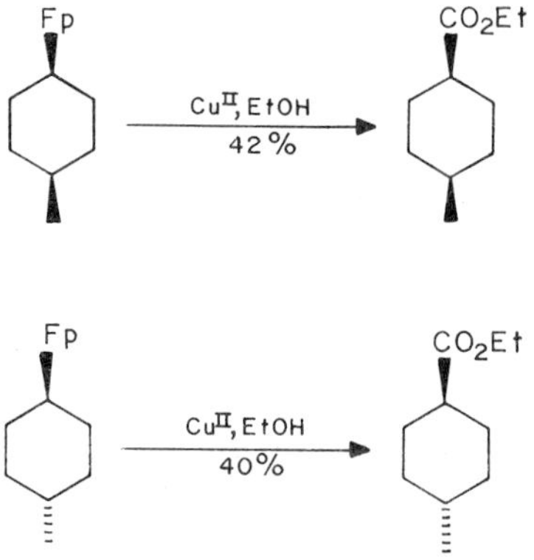

$$\text{Fp-cyclohexyl (cis)} \xrightarrow[42\%]{Cu^{II}, EtOH} \text{CO}_2\text{Et-cyclohexyl (cis)} \quad (62)$$

$$\text{Fp-cyclohexyl (trans)} \xrightarrow[40\%]{Cu^{II}, EtOH} \text{CO}_2\text{Et-cyclohexyl (trans)} \quad (63)$$

determine this stereochemistry at carbon. Whitesides and Boschetto[173] designed a simple and elegant technique based on the identification by ^1H n.m.r. of the *threo* and *erythro* derivatives $Me_3CCHDCHDX$. The two large groups Bu^t and X (X = Fp or carbonyl function) are in a *trans* conformation. Consequently (as easily seen in the Walden projections), the two H atoms are either *trans* (*erythro*) or *gauche* (*threo*). In the pure *trans* conformations of the two large groups, the vicinal H coupling constant is 14 Hz for *erythro* and 4 Hz for *threo*. Although the difference between the actual values of the vicinal coupling constants is not so large owing to some rotation, it is large enough to determine unambiguously the *threo* or *erythro* configuration. Thus, the dynamic stereochemistry can be determined in this way [for instance, oxidative cleavage of the Fp-*erythro* complexes above gives *erythro* organic derivatives ($J = 11 - 14$H), proving retention] (Scheme 68).

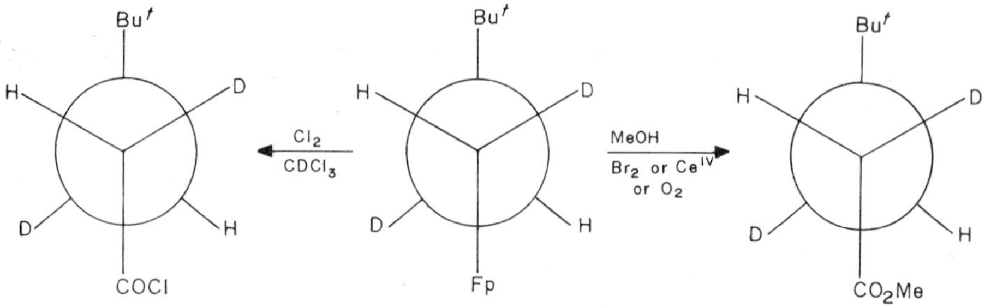

SCHEME 68

Oxidative cleavage of Fp—R compounds by electrophiles can proceed in two ways. Oxidants can induce outer-sphere electron transfer as above to give unstable 17-electron organoiron species. The other path is electrophilic attack at the 18-electron iron centre producing a square-pyramidal d^4 Fe^{IV} cation (equation 64). This intermediate can partially epimerize by pseudo-rotation and reductive elimination from the various stereoisomers gives the reaction products[174]. Knowledge of the dynamic stereochemistry at the metal centre is thus also necessary for mechanistic investigations[175]. Indeed, the stereochemistry, mechanism, and products in this case vary considerably with the nature of the electrophile.

d^0, Fe^{II} + EX ⟶ d^4, Fe^{IV} X^- ⟶ RE or RX (64)

EX = HI, I_2, ICl, HgI_2

IV. IRON PROTECTING GROUPS FOR MONO- AND DI-ENES

Complexation of olefins by an iron group inhibits the classical chemistry usually encountered in the absence of complexation, thus providing the possibility of reaction at other parts of the molecules such as other non-coordinated unsatured functional groups. The cationic 16-electron moiety $[\eta^5\text{-}C_5H_5Fe(CO)_2]^+$ protects one olefinic double bond in the complexes $[Fp(\eta^2\text{-olefin})]^+$, whereas the neutral 14-electron group $Fe(CO)_3$ protects two conjugated double bonds in the complexes $[Fe(CO)_3(\eta^4\text{-}cis\text{-}1,3\text{-diene})]$.

A. Protection of one Olefinic Double Bond in $[Fp(\eta^2\text{-olefin})]^+$

In diolefins, metallation by Fp^+ is governed by steric factors. This property has been used by Nicholas[176] for the selective reduction of the other double bond (equation 65).

Fp^+ ; H_2, Pd/C (65)

Hence this technique is complementary to the traditional halogenation–dehalogenation procedure in which halogens attack the more substituted double bond.

It is possible to prevent the reaction of electrophiles, such as in the norbornene complex **80**, which allows selective attack at the free double bond (Scheme 69). A double bond is complexed preferably to a triple bond, which affords the specific reduction of the latter in enynes (equation 66). By protecting the double bond, which is otherwise more reactive,

Fp^+ ; Pd / H_2 ; Fp^+ ; Fp^+

(66)

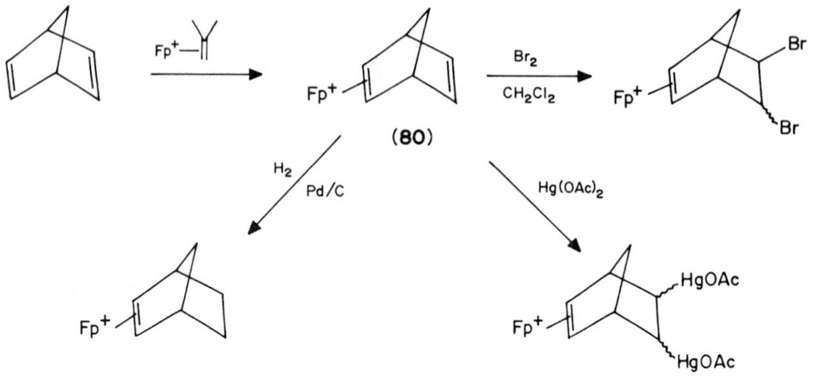

SCHEME 69

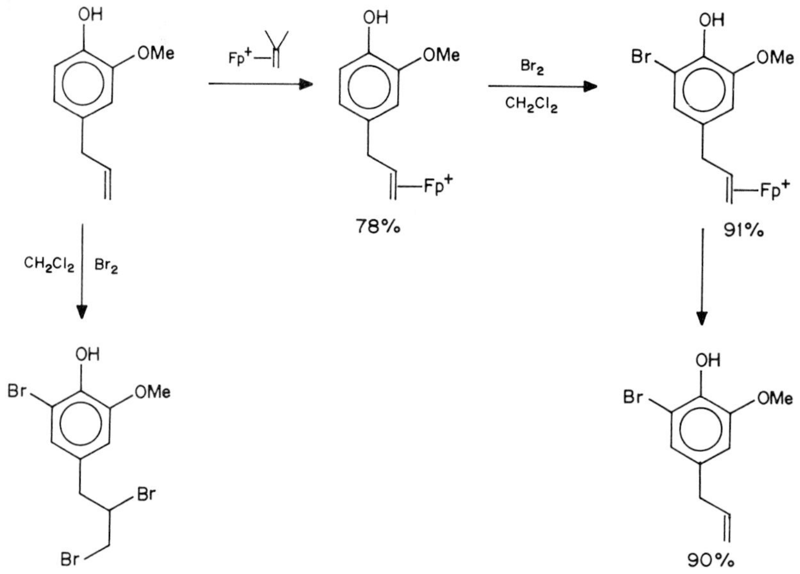

SCHEME 70

bromination of eugenol was effected on the aromatic ring (Scheme 70). All these examples, reported by Nicholas[176] in 1975, illustrate the use of Fp^+ as protecting group in organic synthesis. Indeed, the metal is easily removed from the organic substrate, after the desired modification, by reaction with NaI in acetone at room temperature (equation 20).

B. Protection of a Diene in [$Fe(CO)_3(\eta^4$-1,3-diene)]

Possibly the classic example of protection of double bonds concerns the reactivity of myrcene, reported by the group of Lewis[177] in 1973. A report on acylation by Birch and

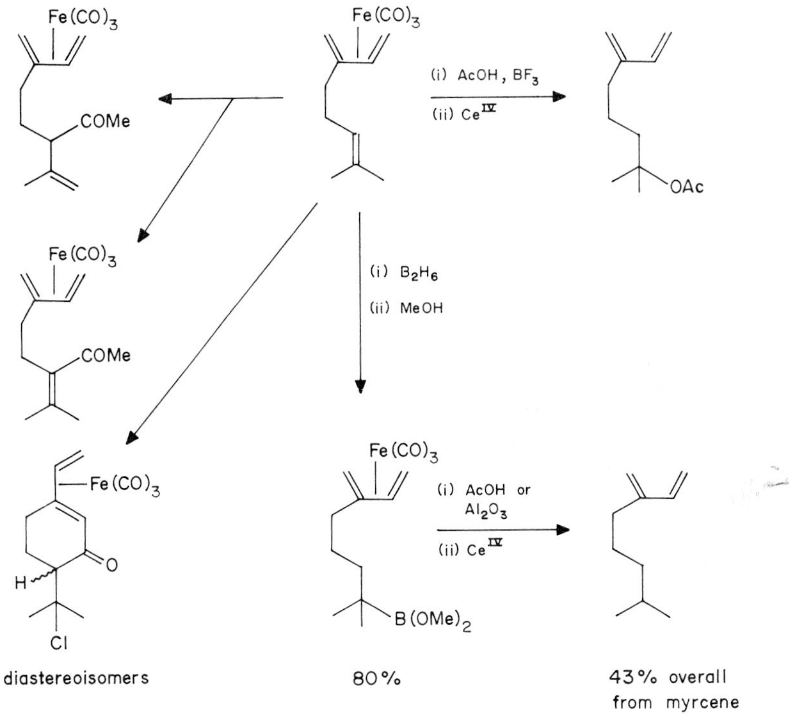

SCHEME 71

Pearson[107] also appeared in 1976 (Scheme 71). This group[178a] also showed that the $Fe(CO)_3$ complex of ergosteryl acetate (**81a**) can undergo hydration and oxidation of the free double bond; reduction of the latter was effected by Barton *et al.*[178b] (Scheme 72). Birch and Fitton[179] reported the *N*-cyanation of thebaïne via its $Fe(CO)_3$ complex **82** (Scheme 73); extensive rearrangement occurs if thebaïne is not coordinated to $Fe(CO)_3$[180].

Protection by $Fe(CO)_3$ is also achieved on partial coordination of conjugated polyenes and polyenones, such as cyclooctatriene, cyclooctatetraene, and cycloheptatrienone (tropone). In these cases, the role of $Fe(CO)_3$ is not only to protect two coordinated double bonds, but also to modulate the reactivity of electrophiles on the conjugated, non-coordinated double bonds. Thus, cycloheptatriene and cyclooctatetraene are usually polymerized by electrophiles. As early as 1972, Johson *et al.*[181] reported the mild acetylation and formylation of $[Fe(CO)_3(\eta^4\text{-cycloheptatriene})]$, **83a**, which proceeds via the relatively stable $[(\eta^5\text{-dienyl})Fe(CO)_3]^+$ intermediate, **83b** (from which an *exo* proton is removed upon hydrolysis (Scheme 74). The formation of this dienyl intermediate, characterized in the acetylation reaction, shows that the $Fe(CO)_3$ group plays more an active role in the reaction than that of a simple protecting group. Similarly, complexation of cyclooctatetraene by $Fe(CO)_3$ (in **84a**) affords protonation at a non-coordinated double bond, leading to homotropone, **84b**; formylation opens the route to a series of functional cyclooctatetraenes[182] (Scheme 75).

A homo-2,3-tropone, of the same family as that obtained by Pettit *et al.* from protonation of **84a**, was also synthesized by reaction of diazoethane with the tropone

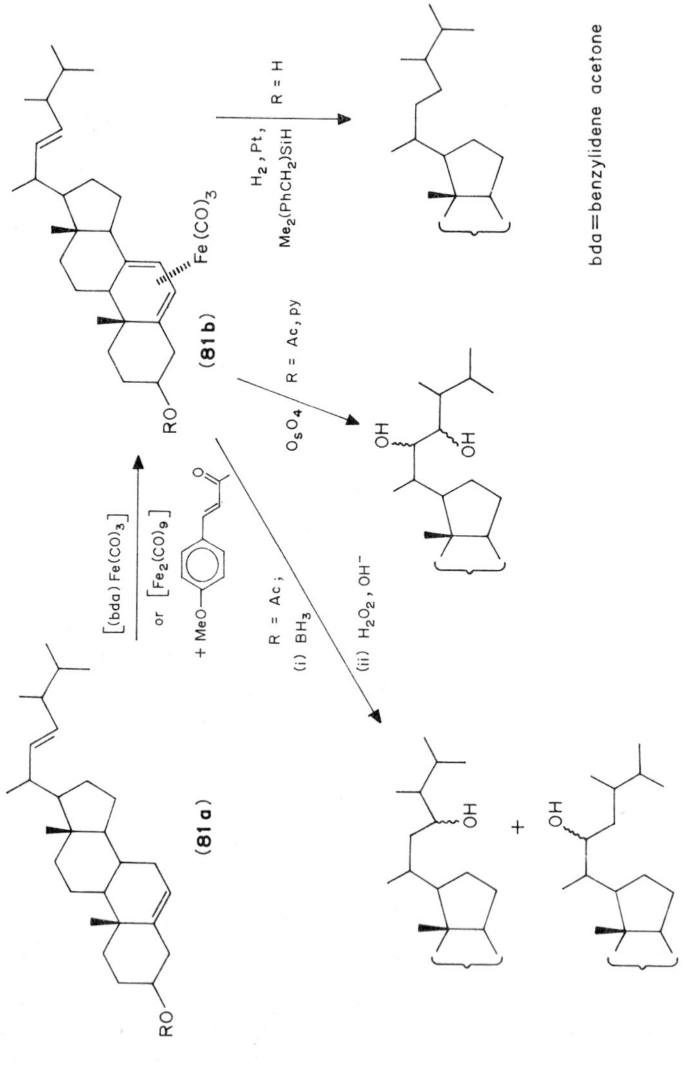

(81a)

$[(bda)Fe(CO)_3]$ or $[Fe_2(CO)_9]$ + MeO–C_6H_4–CH=CH–CO–Me

(81b)

H_2, Pt, $Me_2(PhCH_2)SiH$; R = H

OsO_4; R = Ac, py

R = Ac; (i) BH_3 (ii) H_2O_2, OH^-

bda=benzylidene acetone

SCHEME 72

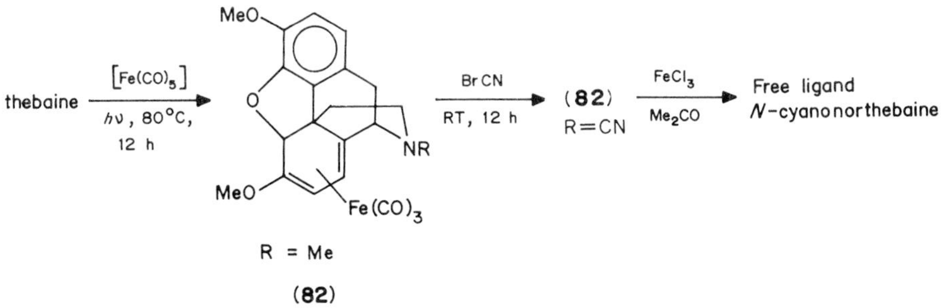

thebaine $\xrightarrow[h\nu,\ 80^{\circ}C,\ 12\ h]{[Fe(CO)_5]}$ (82) R = Me $\xrightarrow[RT,\ 12\ h]{BrCN}$ (82) R=CN $\xrightarrow[Me_2CO]{FeCl_3}$ Free ligand *N*-cyanonorthebaine

SCHEME 73

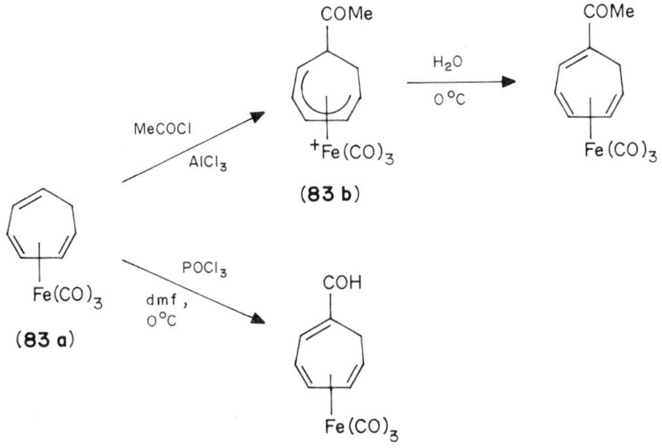

SCHEME 74

complex **85**. Frank-Neumann *et al.*[183] showed that tropone cannot be acylated unless it is coordinated to $Fe(CO)_3$, the driving force then being provided by the intermediacy of $[\eta^5\text{-dienyl})Fe(CO)_3]^+$. Reaction of diazopropane with the acetylated complex **86** gave a precursor of β-dolabrin and β-thujaplicin (note that a precursor of the latter can also be obtained from Noyori's condensation of 2-isopropylfuran with $\alpha,\alpha,\alpha',\alpha'$-tetrabromoacetone) (Scheme 76). The introduction of an acyl function into the azepine ring was not possible before partial complexation by $Fe(CO)_3$ was achieved[184] (equation 67).

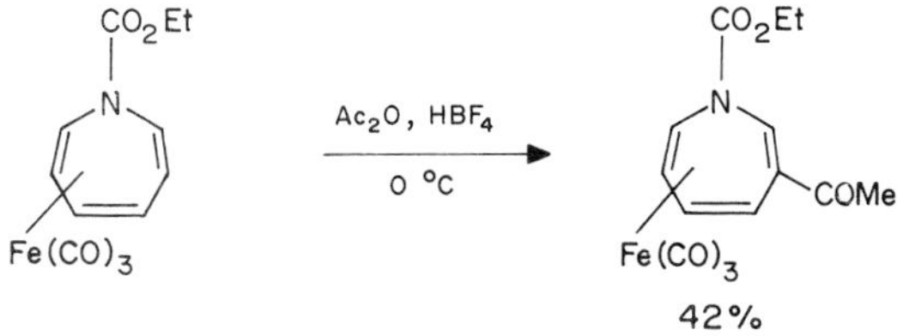

$$\xrightarrow[0\ ^{\circ}C]{Ac_2O,\ HBF_4}\ 42\% \tag{67}$$

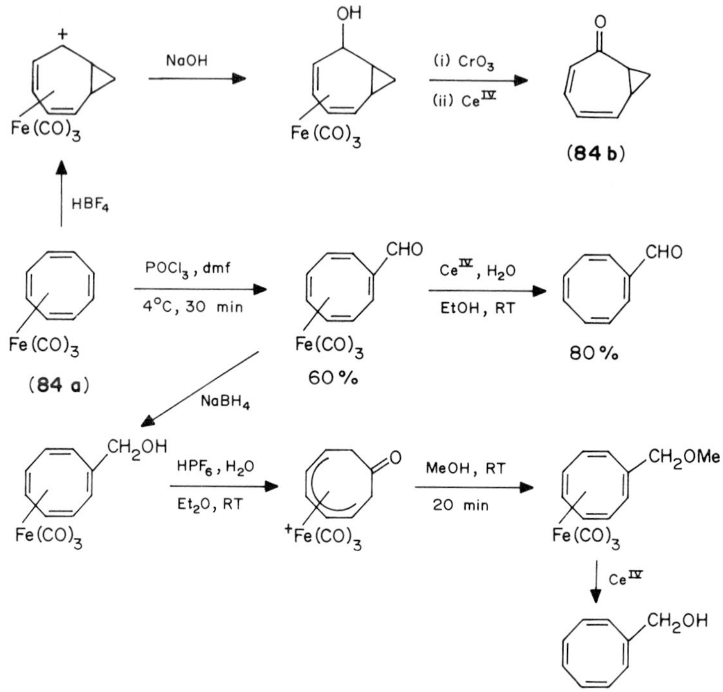

SCHEME 75

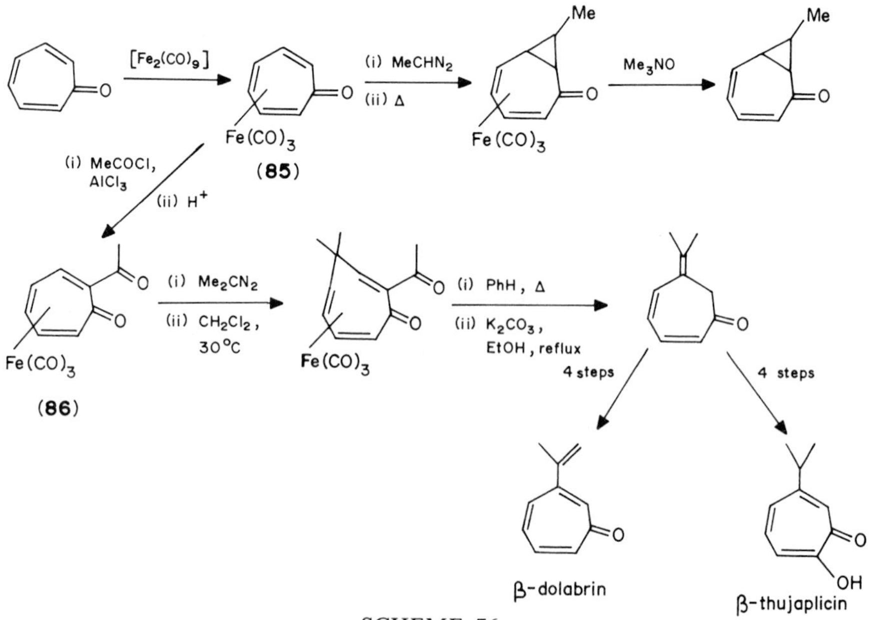

SCHEME 76

V. STABILIZATION OF UNSTABLE SPECIES AS EVEN LIGANDS IN IRON COMPLEXES (ALKYLIDENES, CYCLOBUTADIENE, o-XYLYLENE) AND THEIR TRANSFER TO ORGANIC SUBSTRATES

Although many unstable species can be stabilized by complexation, very few have provided synthetic applications. For instance, the iron carbonyl complexes of trimethylenemethane, penta- and hepta-fulvalene, cyclopentadiene, cyclopentadienone, norbornadienone, and benzyne (Scheme 77) gave either no synthetic application or disappointing results. In particular, the trimethylenemethane complex gave mixtures of many products on decomplexation[185], in contrast to the much better results obtained using palladium complexes[186].

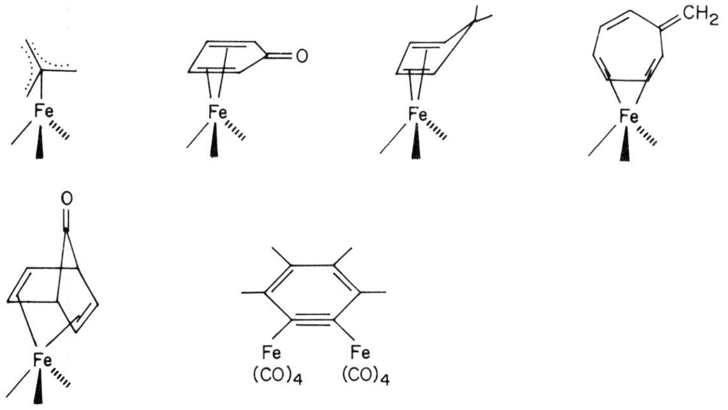

SCHEME 77

We shall therefore concentrate here on alkylidene and cyclobutadiene complexes, which give really useful applications; *o*-xylylene complexes will also be considered. Cyclohexadienone can be stabilized by $Fe(CO)_3$ and its complex **87** has been used for the arylation of amines[187].

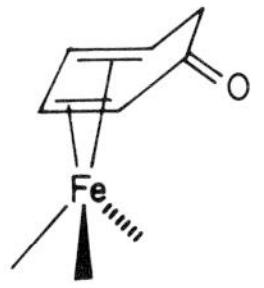

(87)

We can distinguish here *unstable* molecules as ligands in *unstable* complexes but for which complexation provides a suitable route to generate conveniently the desired species due to a certain degree of stabilization in the complex (Fp^+ complexes of alkylidenes and cyclobutadienes) and those forming *stable* complexes which require a decomplexation reagent for the generation of the free ligand [$Fe(CO)_3$ complexes of cyclobutadiene, *o*-xylylene, and cyclohexadienone].

A. Stabilization of Methylene and Alkylidenes by Fp^+ or $(CpFeL_1L_2)^+$ and the Cyclopropanation of Olefins

Singlet methylene CH_2, the simplest carbene, is isoelectronic with CO. Hence it can be regarded as a two-electron ligand, either donating its lone pair as with CO, **87a**, or else as half of ethylene, doubly bonded to the metal, **87b**. Although transition metal methylene complexes are scarce, mono- or di-alkyl substituted methylene, e.g. primary or secondary alkylidene complexes, respectively, are now more common[188]. Early transition metal alkylidene complexes with less than 18 valence electrons are catalysts for the metathesis[189] (generally through the popular Chauvin mechanism[190]) or the selective dimerization of olefins[191]. The polarity of the metal—alkylidene bond, essential for reaction with olefins, shifts predictably on moving along the first row transition elements, **87c** and **87d**. Thus, iron alkylidene complexes have the most electrophilic alkylidene fragments known, but the fraction of positive change on the alkylidene carbon can be diminished by replacing one or two carbonyls by phosphines, which also stabilizes the complexes. Jolly and Pettit[192] reported the first evidence for a metal–methylene intermediate, $(FpCH_2)^+$, but this species could not be characterized spectroscopically even at low temperature because of its fast disproportionation to Fp^+ and $[Fp(\eta^2\text{-}C_2H_4)]^+$ (equation 68).

$$\mathrm{M \leftarrow CH_2} \quad (\mathbf{87a}) \qquad \mathrm{M{=}CH_2} \quad (\mathbf{87b})$$

$$\mathrm{Cp_2(Me)\overset{\delta+}{Ta}{=}\overset{\delta-}{C}H_2} \quad (\mathbf{87c}) \qquad \mathrm{Cp(L)_2\overset{\delta-}{Fe}{=}\overset{\delta+}{C}H_2} \quad (\mathbf{87d})$$

$$[\mathrm{FpCH_2OCH_3}] \xrightarrow[-\mathrm{CH_3OH}]{\mathrm{H^+}} \text{‘}\mathrm{FpCH_2}\text{’} \longrightarrow [\mathrm{Fp}(\eta^2\text{-}\mathrm{C_2H_4})]^+ + \mathrm{Fp}^+ \qquad (68)$$

$$[\mathrm{CpFe(dppe)CH_2OEt}] \xrightarrow[-\mathrm{EtOH}]{\mathrm{H^+}} \mathrm{CpFe(dppe)(CH_2)}]^+ \qquad (69)$$

Brookhart *et al.*[193] obtained $[CpFe(dppe)CH_2]^+$ at 0 °C in CD_2Cl_2 solution from $CpFe(dppe)(CH_2OEt)$ and trifluoroacetic or triflic acid (equation 69), and carried out a successful n.m.r. study of this methylene complex. The ^{13}C spectrum displays a signal at 317 ppm, a low-field resonance characteristic of a terminal alkylidene carbon. The two methylene protons are not equivalent, as seen in the ^1H spectrum in CD_2Cl_2–SO_2 at − 90 °C, which shows two signals at 13.29 and 17.89 ppm; this confirms that the CH_2 plane lies perpendicular to the Cp plane as calculated by Hoffmann *et al.*[194] for $[CpFe(CO)_2CH_2]^+$. The free energy of activation for the rotation of the methylene was obtained from the variable-temperature ^1H spectra: $\Delta G^{\neq} = 43.5 \pm 0.4$ kJ mol^{-1}. This value is substantially higher than, albeit consistent with, the 25 kJ mol^{-1} calculated by Hoffmann *et al.* for $[FpCH_2]^+$, the back-bonding to the methylene provided by the diphos increasing the double bond character. As a result, the methylene carbon is less electrophilic and less reactive, providing stabilization which is further enhanced by steric protection.

Primary alkylidene complexes $[FpCHR]^+$, like the parent methylene complex, cannot be observed spectroscopically unless the atom of R attached to the carbene carbon is an heteroatom; in this case, its *p*-orbitals interact with those of the metal and the carbene carbon in a three-centre stabilizing conjugation. This occurs for R = phenyl, alkoxy, sulphoxy, amino, and other groups bearing an heteroatom. These stable complexes belong to the rich family of the Fischer-type alkylidene complexes[195]. In the Fp series, Cutler prepared these complexes as yellowish solids by hydride abstraction from alkoxymethyl complexes **88**[196] (equation 70a). Brookhart and coworkers[197] prepared $[FpCHPh]^+$

$$\underset{(\mathbf{88})}{[\mathrm{FpCH_2OR}]} \xrightarrow[-\mathrm{H^-}]{\mathrm{Ph_3C^+\,PF_6^-}} \left[\mathrm{Fp}\cdots\mathrm{C}\begin{smallmatrix}\mathrm{OR}\\ \mathrm{H}\end{smallmatrix}\right]^+ \mathrm{PF_6^-} + \mathrm{Ph_3CH} \qquad (70a)$$

(stable for 1 h at 25 °C) via the alkoxyalkylidene complex **89** according to Scheme 78. This route is of general interest and was used by Brookhart *et al.*[197a] and Bodnar and

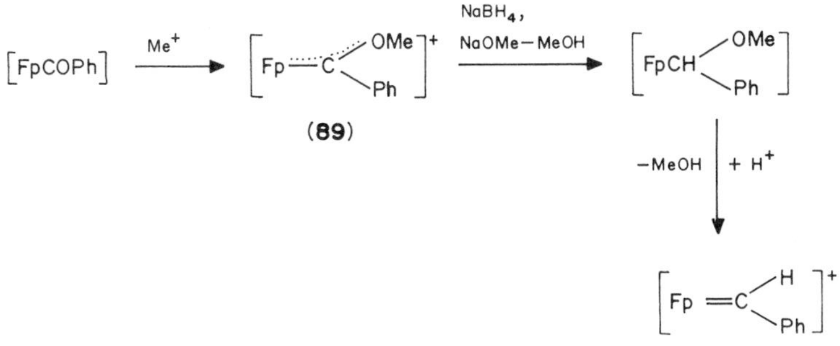

$$[FpCOPh] \xrightarrow{Me^+} \left[Fp \cdots C\begin{matrix} OMe \\ Ph \end{matrix}\right]^+ \xrightarrow{NaBH_4,\ NaOMe-MeOH} \left[FpCH\begin{matrix} OMe \\ Ph \end{matrix}\right] \xrightarrow{-MeOH,\ +H^+} \left[Fp{=}C\begin{matrix} H \\ Ph \end{matrix}\right]^+$$

(89)

SCHEME 78

Cutler[197b] to prepare the primary alkylidene complexes $[CpFe(CO)(PPh_3)(CHR)]^+$; they can be characterized spectroscopically at low temperature but, like $[FpCHR]^+$, isomerize below 25 °C via intramolecular hydride migration to give olefin complexes (equation 70b). However, the benzylidene analogue is fairly stable (50% decomposition in 60 h).

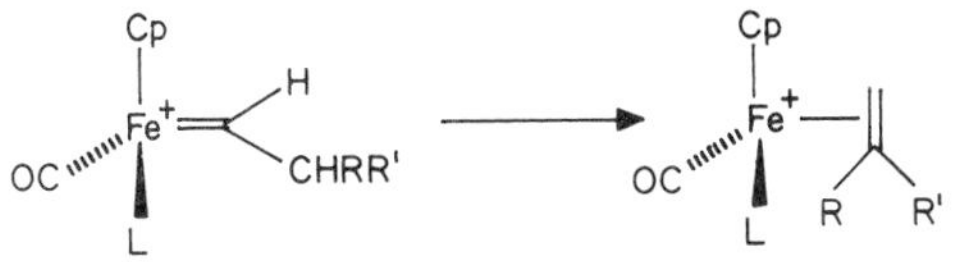

$$CpFe^+(CO)(L){=}C(H)(CHRR') \longrightarrow CpFe^+(CO)(L)\text{—}(CH_2{=}CRR') \qquad (70b)$$

The first observation of a methylene transfer to an olefin to give a cyclopropane mediated by an iron complex was reported by Jolly and Pettit[192]. Protonation of $[FpCH_2OMe]$ in the presence of cyclohexene gave norcarane (equation 71a). Davison

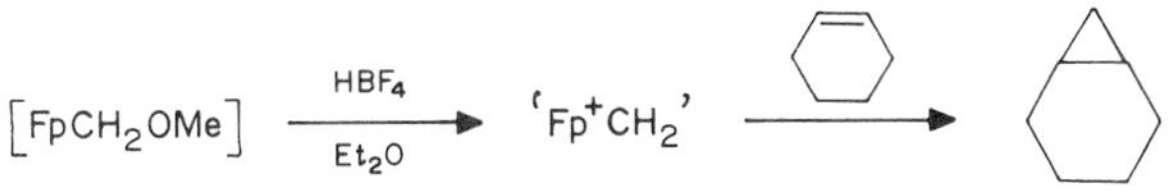

$$[FpCH_2OMe] \xrightarrow[Et_2O]{HBF_4} \text{'}Fp^+CH_2\text{'} \xrightarrow{\text{cyclohexene}} \text{norcarane} \qquad (71a)$$

et al.[198] used the optically active methylene precursor $[CpFe(CO)(PPh_3)(CH_2O$ menthyl)] to transfer the methylene unit to the prochiral olefin *trans*-1-phenylpropene, which yields optically active *trans*-1-methyl-2-phenylcyclopropane. The stable complex $[CpFe(dppe)(CH_2)]^+$ can also transfer its methylene unit to olefins, although the yield of cyclopropane formed depends strongly on stereoelectronic factors (ethylvinyl ether, 98%; hex-1-ene, 30%; cyclohexene, ~ 10%).

A useful methylene transfer reagent was reported in 1979 by Brand and Helquist[199]. Methylation of the thioether $[FpCH_2SMe]$ gives the sulphonium cation $[FpSMe_2]^+$, which can release CH_2 thermally (Scheme 79). Heating this cation in refluxing dioxane in the presence of various simple olefins gives good yields of cyclopropanes.

The transfer of simple primary alkylidene fragments such as ethylidene and benzylidene mediated by iron is also known $[FpCHMeSMe_2]^+BF_4^-$ and [FpCH(OMe)Me] were used as ethylidene precursors to form methylcyclopropanes on reaction with olefins (equation 71b). The yields of methylcyclopropane formed with various olefins compare favourably with those obtained with the procedures using $MeCHI_2$–Et_2Zn as the ethylidene transfer reagent. In view of the high electrophilicity of the alkylidene carbon, it is most readily transferred to the most electronically activated alkenes. Transfer of ethylidene from optically active ethylidene complexes $[Cp(CO)(Ph_2R^*P)Fe{=}CHCH_3]^+$

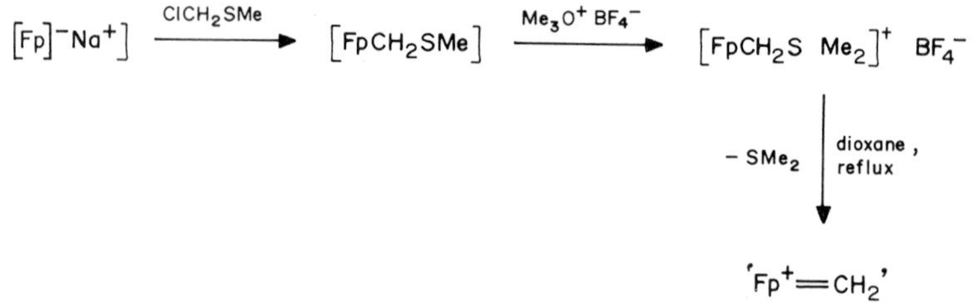

$[Fp]^-Na^+ \xrightarrow{ClCH_2SMe} [FpCH_2SMe] \xrightarrow{Me_3O^+BF_4^-} [FpCH_2SMe_2]^+ BF_4^-$

$\xrightarrow[\text{dioxane, reflux}]{-SMe_2}$ '$Fp^+{=}CH_2$'

SCHEME 79

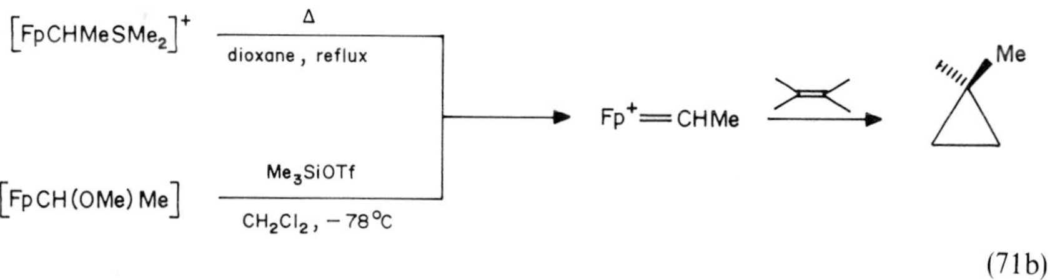

(71b)

to styrene, as achieved by Brookhart and coworkers[200], gives *cis*- and *trans*-1-methyl-2-phenylcyclopropanes in 70–75% yields and 84–90% optical yields. The two diastereoisomeric complexes give cyclopropanes of opposite configurations in almost identical optical purities, which indicates that the chirality at the iron centre, but not that at phosphorus, is responsible for asymmetric induction. The high optical yields were attributed by Brookhart and coworkers to the prochirality of the alkylidene carbon.

The mechanism proposed to account for the observed selectivity is attack of the more nucleophilic olefinic carbon at the ethylidene centre in the anticlinal conformation, followed by front-side collapse of the developing electrophilic centre at C-2 (Scheme 80). Brookhart and coworkers also obtained highly stereoselective benzylidene transfer from $[Fp{=}CHPh]^+$ to a wide variety of olefins ranging from ethylene to tetra-substituted olefins; syn- or *cis*-cyclopropanes are obtained.

In this peculiar case, Casey *et al.*'s model implying an interaction with the ipso phenyl carbon in the transition state is invoked, as for $[(CO)_5W = CHPh]$; the stereoselectivity is

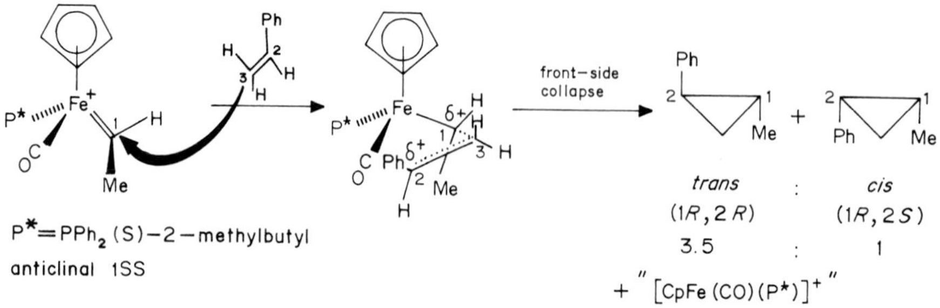

P* = PPh₂ (S)−2−methylbutyl

anticlinal 1SS

front−side collapse

trans (1*R*,2*R*) 3.5 : *cis* (1*R*,2*S*) 1

+ "$[CpFe(CO)(P^*)]^+$"

SCHEME 80

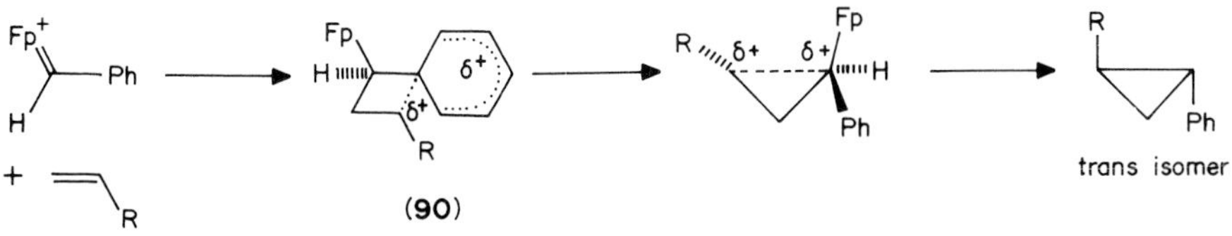

SCHEME 81

proposed to be induced by the *trans* configuration of R to Fp in this cyclobutane-like transition state 90 (Scheme 81). Acetylenes react to yield phenylcyclopropenes. These valuable alkylidene transfer studies show that iron is a promising agent for the preparation of cyclopropanes from olefins, especially in view of the rejection of the classical diazoalkane procedure for industrial applications.

B. Fischer-type Stabilized Secondary Alkylidene Complexes and their Coupling Reactions with Olefins

The reaction of carbanions with metal carbonyls gives acylate salts which can be *O*-alkylated to give heteroatom-stabilized alkylidene complexes[202]. This reaction, discovered by Fischer, is quite general and was first applied to chromium and tungsten carbonyls. However, alkylation of iron acylates obtained from [$Fe(CO)_5$] generally gives unsymmetrical ketones resulting from alkylation at iron (Collman's reaction[142]). Indeed, the iron acylate salts **91a** can also be written as acyl ferrates **91b** and the order of the metal—alkylidene bond is lower than in alkylidene complexes which do not bear heteroatoms (Scheme 82).

In some instances, however, *O*-alkylation can be obtained. A rationalization for the competition between *O*- and Fe-alkylation is provided by the hard–soft acid–base (HSAB) theory, a well known factor in organic chemistry for enolates (*O*- *vs.* *C*-alkylation). This comparison was emphasized recently by Semmelhack and Tamura[203], who obtained Fischer-type alkylidene complexes by taking into account the factors favouring the *O*-alkylation:

(i) addition of hmpa to the ether solution of the iron acylate;
(ii) use of NMe_4^+ instead of Li^+ as a counter cation;
(iii) Increasing the size of the alkylating agent;
(iv) use of a relatively unreactive leaving group (toluenesulphonate) instead of a soft one (iodide).

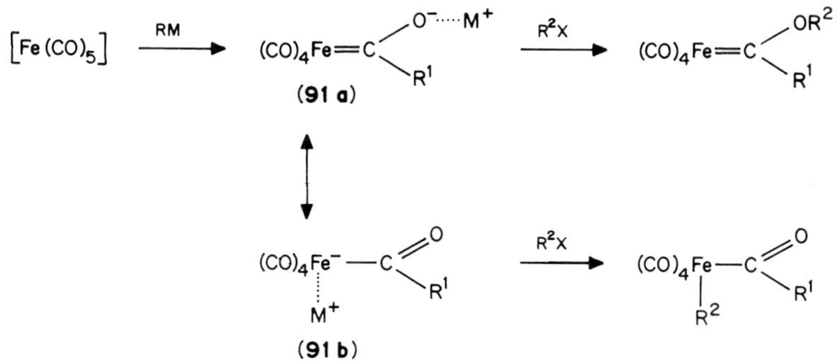

SCHEME 82

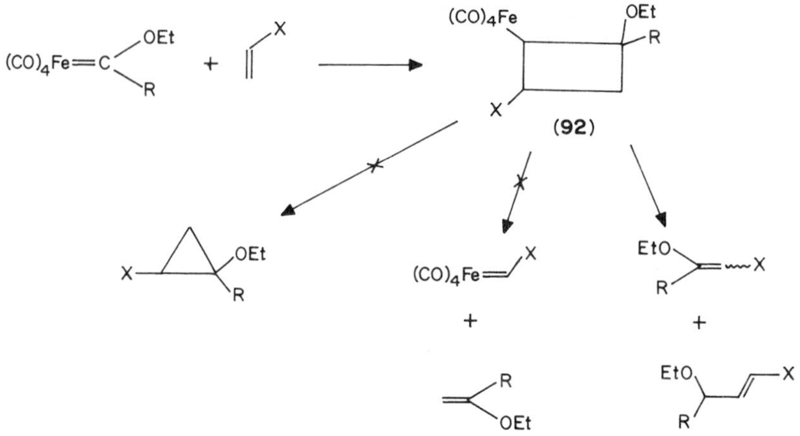

SCHEME 83

Reactions of these alkylidene complexes with olefins at 50 °C were shown to give regioselective formation of coupled products, as previously known with tantalum and niobium alkylidene complexes. In both the iron and the tantalum and niobium reactions, the intermediacy of metallacyclobutanes was postulated, although the metal—alkylidene bonds have opposite polarities. The two other decomposition modes known for metallacyclobutanes, reductive elimination to give a cyclopropane (chromium)[204] and metathesis according to Chauvin and Herrisson's mechanism (tungsten)[205], were not observed with iron. Reaction products result from coupling of the alkylidene unit with the unsubstituted end of the terminal olefin via decomposition of the metallacyclobutane **92** by β-elimination to an iron allyl hydride intermediate followed by regioselective elimination (Scheme 83).

The sulphur stabilized iron–carbene complexes **94a** synthesized from iron–CS_2 complexes, **93**, and acetylenes bearing at least one electron-withdrawing group, react with air giving dithiolene complexes **94b** and tetrathiafulvalenes **95**[206] (equation 72), but the mechanism is not known.

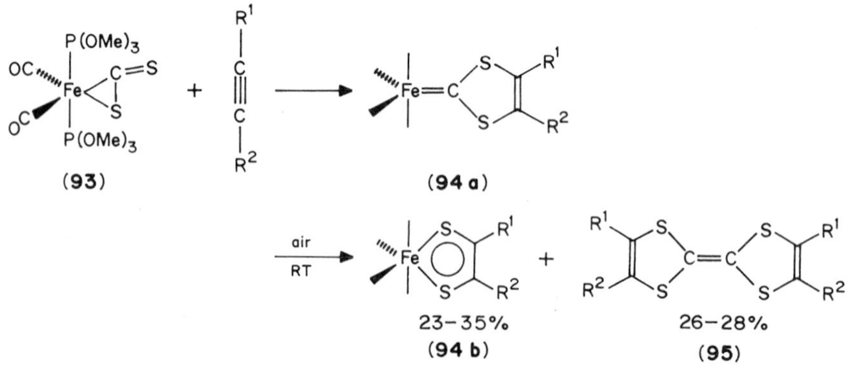

R^1, R^2 = H, CO_2Et; *p*-ClC_6H_4, CHO;
CO_2Me; CO_2Me; *p*-$O_2NC_6H_4$; COMe; (72)
Ph; CHO

C. Trapping of Cyclobutadiene by Organic Substrates on Decomplexation from Stable or Unstable Iron Complexes

Predicted by Longuet-Higgins and Orgel in 1956[207], the stabilization of the unstable cyclobutadiene molecule by complexation to a transition metal group was realized by Emerson *et al.* in 1956[208]. Reaction of $[Fe_2(CO)_9]$ with *cis*-dichlorocyclobutene or *cis*-dibromobenzocyclobutene gave the stable $Fe(CO)_3$ complexes **96** and **97** of cyclobutene and benzocyclobutene, respectively (Section VI.A, equation 74, page 699).

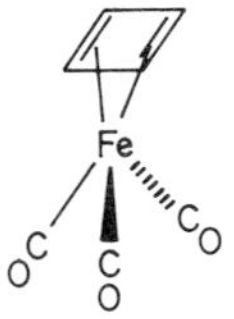

Fe CO CO CO

(96)

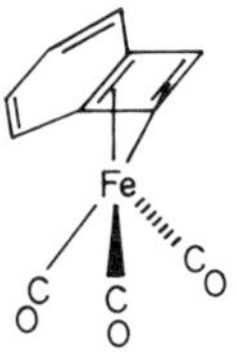

Fe CO CO CO

(97)

Antiaromatic as a free molecule, cyclobutadiene becomes aromatic in its stable $Fe(CO)_3$ complex, as indicated by its classical electrophilic reactivity acylation, Vilsmeir formylation, mercuration as for ferrocene). Free cyclobutadiene, obtained on decomplexation of **96** with Ce(IV) or Pb(IV), can be trapped with various dienophiles to provide original

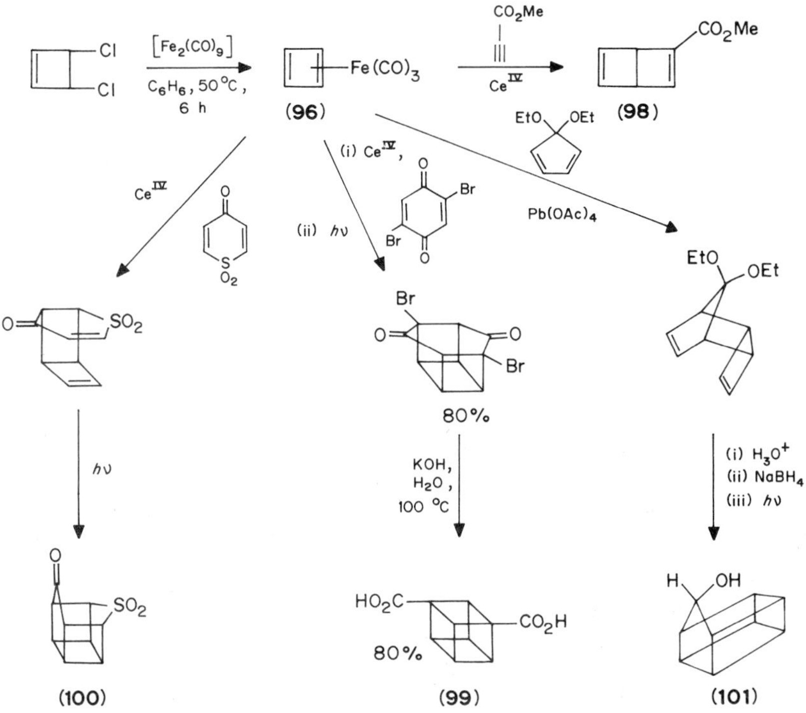

Cl Cl $[Fe_2(CO)_9]$ C_6H_6, 50 °C, 6 h → Fe(CO)$_3$ **(96)** → CO_2Me (alkyne), Ce^{IV} → CO_2Me **(98)**

(96) → Ce^{IV}, 4H-thiopyran-4-one S-dioxide → O, SO_2 → $h\nu$ → O, SO_2 **(100)**

(96) → (i) Ce^{IV}, dibromobenzoquinone; (ii) $h\nu$ → Br, O, O, Br 80% → KOH, H_2O, 100 °C → HO_2C, CO_2H 80% **(99)**

(96) → EtO OEt cyclopentadiene, $Pb(OAc)_4$ → EtO OEt → (i) H_3O^+ (ii) $NaBH_4$ (iii) $h\nu$ → H OH **(101)**

SCHEME 84

synthetic applications. Examples are the formation of Dewar benzene, **98**, and the very interesting synthesis of the cubane, **99**, cage keto sulphones, **100**, and 9-hydroxy-homocubanes, **101**[209] (Scheme 84).

Although cyclobutadiene generated from **96** most often behaves as a diene, there are cases where it acts as a dienophile, for instance in the synthesis of homopentaprismanone, **102**[210a] (equation 73). In this regard, an unstable Fp complex of cyclobutadiene, **103**, is

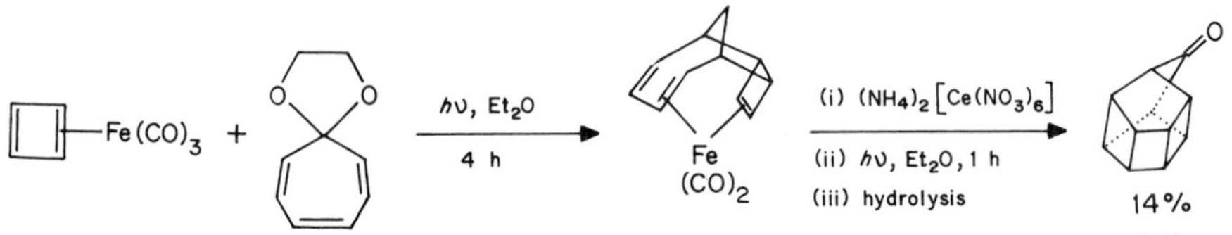

(73)

known which always provides the free ligand as a dienophile[210b], trapping with cyclopentadiene is feasible, but not with dimethyl fumarate. In **103**, the type of coordination, presumably dihapto, is necessarily different from that found in **96** (tetrahapto) (Scheme 85).

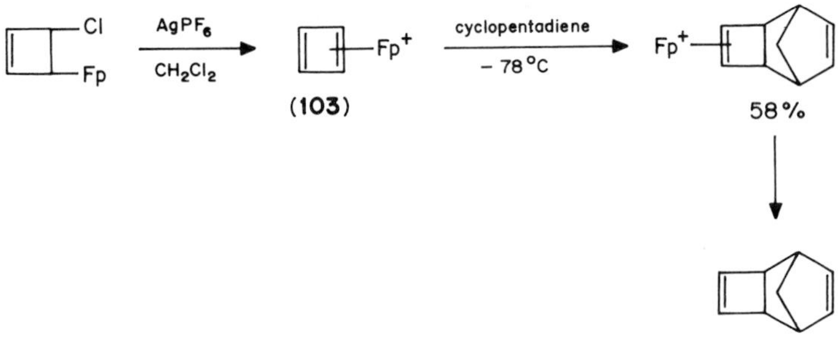

SCHEME 85

D. o-Xylylene

o-Xylylene (*o*-quinodimethane or 5, 6-dimethylenecyclohexa-1, 3-diene) is an unstable species and a key intermediate in steroid synthesis[211], which can be stabilized by coordination to iron moieties[212,213]. The reaction of α,α'-dibromo-*o*-xylene with $[Fe_2(CO)_9]$ in refluxing diethyl ether gives a 5% yield of a $Fe(CO)_3$ complex, **104**, in which the coordination is exocyclic. The yield can be increased to 35% by using $Na_2[Fe(CO)_4]$ instead of $[Fe_2(CO)_9]$. Several derivatives of this $Fe(CO)_3$ complex are known with an additional metal coordinated via the endocyclic double bonds. A free double bond in **104** can be acylated using MeCOCl + $AlCl_3$. The reaction of $AlCl_3$ alone leads to the indanone **105** (Scheme 86).

Recently, $[\{\eta^4\text{-}C_6Me_4(CH_2)_2\}\ (\eta^6\text{-}C_6Me_6)Fe^0]$, a permethylated iron(0) sandwich *o*-xylylene complex, stable up to $-20\,°C$, has been characterized by ^1H and ^{13}C n.m.r. in d_8-toluene[21]. This compound, synthesized by double hydrogen atom abstraction by O_2 from the 20-electron complex $[\eta^6\text{-}C_6Me_6)_2Fe^0]$ (cf. Section VIII.A), is the only iron complex known in which the *o*-xylylene ligand is coordinated exclusively in an endocyclic fashion. Its reaction with electrophiles is described in Section VIII.A.

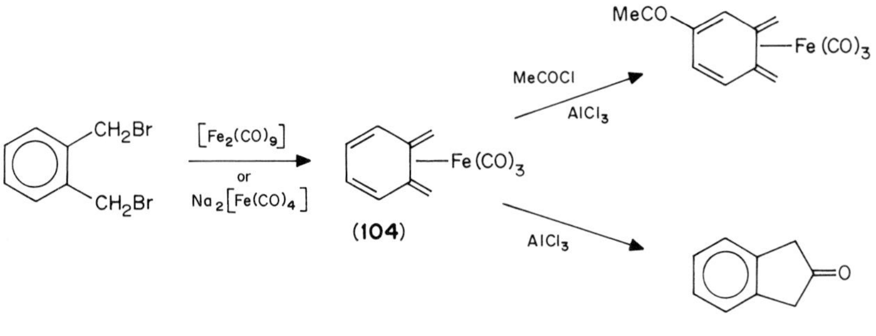

SCHEME 86

VI. REACTIONS OF ORGANIC COMPOUNDS WITH IRON CARBONYLS

A. Reactions of Organic Polyhalides with Iron Carbonyls and Noyori's Condensation of Oxoallyl Ferrate with Unsaturated Substrates

Dehalogenation of organic halides by reaction with iron carbonyls is useful for the preparation of diene iron tricarbonyl complexes which are not accessible by direct complexation of the dienes because of their instability in the free state. This is the case for cyclobutene, trimethylenemethane, and *o*-xylylene (equation 74–76). Decomplexation and trapping of the unstable ligand is often useful in organic synthesis (see Section V.C).

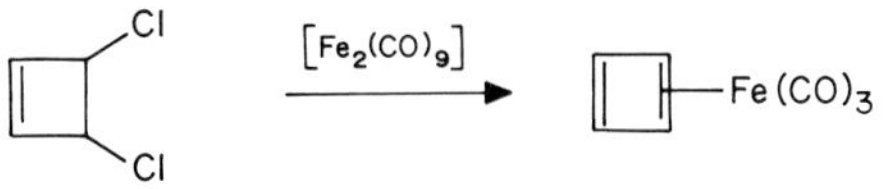

(74)

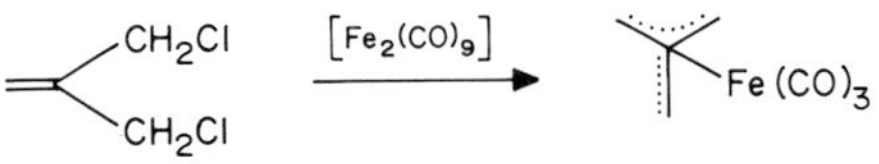

(75)

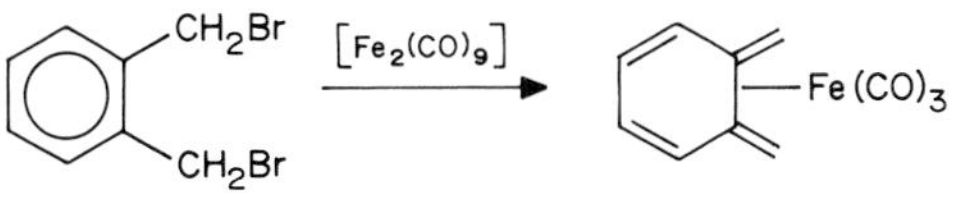

(76)

Useful ferrelactone complexes are formed on dehalogenation of chloroallyl alcohol with $[Fe(CO)_5]$ (equation 77; cf. Section VI.B).

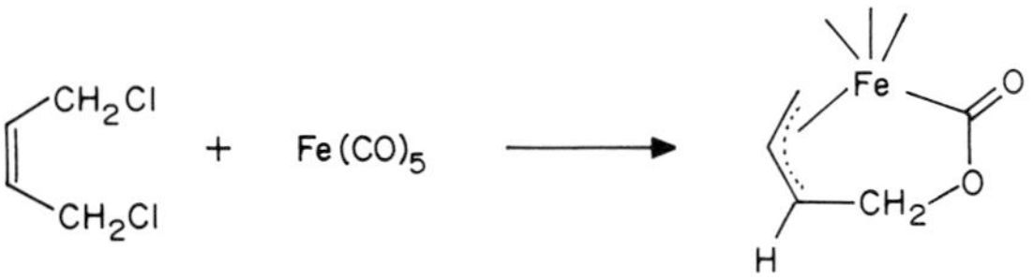

(77)

The first coupling reaction of certain *gem*-dihalides induced by $[Fe(CO)_5]$ was found by Coffey in 1961[214] (equation 78). Ordinary monohalides are much less reactive, but in 1967 Rhee *et al.*[215] reported the reaction of $[Fe_2(CO)_9]$ with $PhCH_2Cl$ (equation 79). In 1972, Alper and Keung[216] reported the reaction between α-haloketones and $[Fe(CO)_5]$ giving 1,4-diketones and β-epoxyketones or reduced monoketones (equation 80). In 1971, there appeared the first reports by Noyori and coworkers[217] of the reaction of $[Fe(CO)_5]$ with α,α′-dibromoketones and condensation with olefins (equation 81). A reactive oxoallyl ferrate intermediate, **106**, is formed, presumably through oxidative addition to the 16-electron moiety $Fe(CO)_4$ (generated from $[Fe_2(CO)_9]$, by elimination of bromide anion. The reaction is carried out in the presence of an unsaturated substrate which condenses with **106**, provided that certain conditions are met. This led Noyori and coworkers[218] to the construction of a variety of cyclic organic structures, many of which possess interesting pharmacological properties.

$$Ar_2CX_2 + [Fe(CO)_5] \xrightarrow{PhH,\ 80\ ^\circ C} Ar_2C{=}CAr_2\ (X = Cl, Br;\ Ar = aryl) \tag{78}$$

$$PhCH_2Cl + [Fe_2(CO)_9] \xrightarrow{PhH} \underset{56\%}{(PhCH_2)_2CO} \tag{79}$$

$$ArC(O)CR_2Br \xrightarrow[(ii)\ H_2O]{(i)\ [Fe(CO)_5],\ dmf,\ \Delta} ArC(O)CR_2CR_2C(O)Ar + ArCOCHR_2 \tag{80}$$

R, R, Br, Br, O; "Fe(CO)$_4$"; (CO)$_4$(Br)Fe—O; R, R, R, Br; − Br$^-$; (CO)$_4$(Br)Fe, O, +; **(106)**

(81)

Thus, **106** condenses with aromatic olefins giving 3-arylcyclopentanones (equation 82a). Stereospecific cycloaddition is obtained as indicated by the reaction with β-*cis*-deuteriostyrene.

O, Br, Br; + ; Ph, D, H, H; $[Fe_2(CO)_9]$; O, Ph, D

(82a)

The usefulness of this reaction was shown by the straightforward synthesis of (±)-α-cuparenone[218c] (equation 82b), a great improvement over the previously known

O, Br, Br; + ; $[Fe_2(CO)_9]$, PhH, 55 °C, 17 h; O

(82b)

seven-step synthesis[218d]. The cycloaddition of **106** with olefins is symmetry restricted and thus proceeds stepwise. In this $\pi^2 + \pi^2$ non-concerted process, the regiochemistry is directed by the stability of the intermediate cation[219]. Using morpholinoenamines, cyclopentenones (**107a**) are accessible in a one-pot synthesis, morpholine being simultaneously eliminated[220] (equation 83).

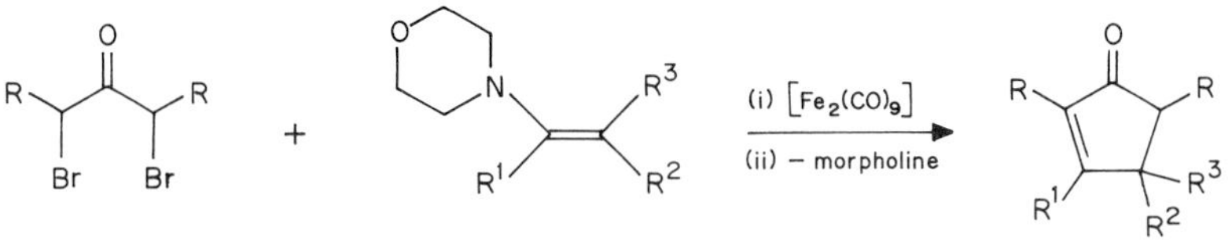

(i) $[Fe_2(CO)_9]$ (ii) − morpholine

(107a)

(83)

The cycloaddition of **106** with 1,3-dienes leads to 4-cycloheptenones, **107b**, precursors for troponoid chemistry (equation 84). This is a thermally allowed $\pi^2 + \pi^4$ concerted

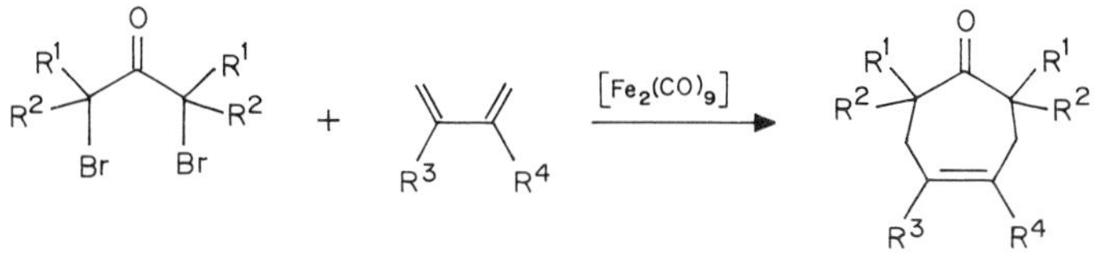

$[Fe_2(CO)_9]$

(107 b)

(84)

process and regioselectivity is governed by primary orbital overlap. The simplest α, α'-dibromoketones, those of acetone and methyl derivatives, fail to react, but α,α,α', α'-tetrabromoacetone and even α,α,α',-tribromoacetone work. Another problem in the synthesis of these cyclic compounds is that electrophilic substitution is a competitive reaction. For instance, thiophene and pyrrole or *N*-methylpyrrole give only **108**, resulting from electrophilic substitution (equation 85). Fortunately, this problem can be circumvented in the pyrrole series[221] using *N*-carbomethoxypyrrole, which gives only cycloaddition. After reductive dehalogenation using Zn–Cu, reduction of the cycloadduct by $Bu^i{}_2AlH$ gives a precursor **109** to all naturally occurring tropane alkaloids (Scheme 87).

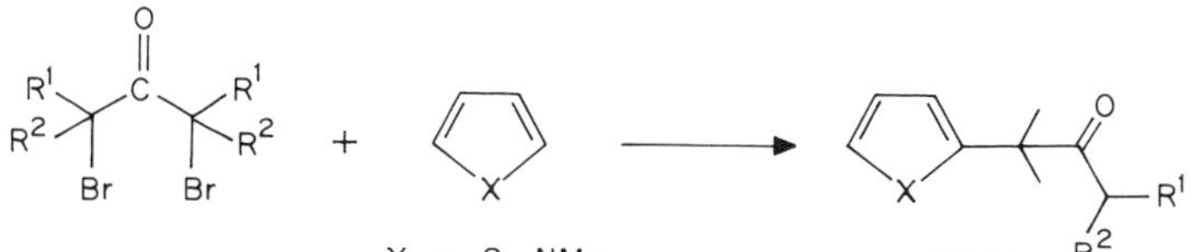

X = S, NMe

(108)

(85)

Another minor constraint is that simple 1,3-dienes must be held *cis* in the reaction in order to obtain good to high yields. Butadiene gives only a 33% yield of cycloadduct, whereas cyclopentadiene gives a 93% yield. Noyori and coworkers[222] solved this problem by using the $[Fe(CO)_3(diene)]$ complex in place of the free diene and $[Fe_2(CO)_9]$; the yield was increased to 90% in this way with butadiene.

After delineating and overcoming these few constraints, Noyori and coworkers developed useful syntheses of cycloadducts **110**, **111**, and **112** with furan and its isopropyl derivatives (Scheme 88 and 89); these cycloadducts are precursors of nezukone and α- and β-thujaplicin; moreover, stereocontrolled syntheses of *C*-nucleotides possessing important antibiotic and potent anticancer and antiviral activity were achieved[223].

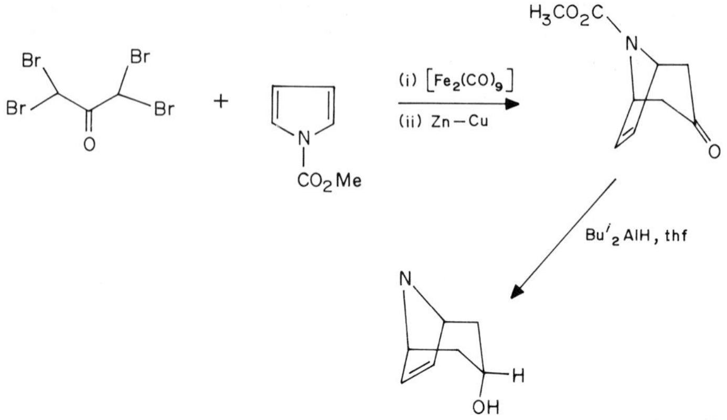

(109)

SCHEME 87

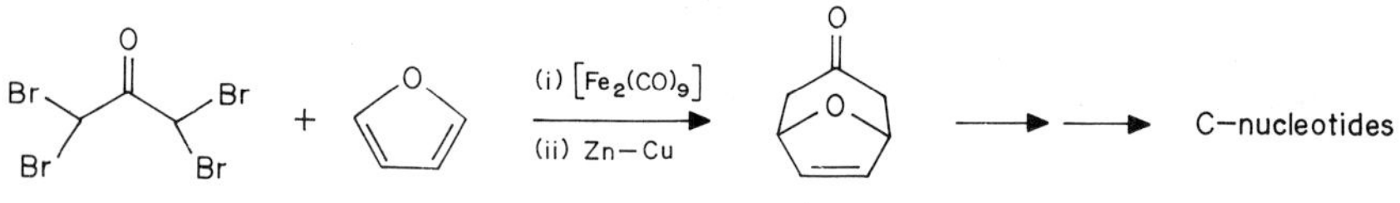

C–nucleotides

(110)

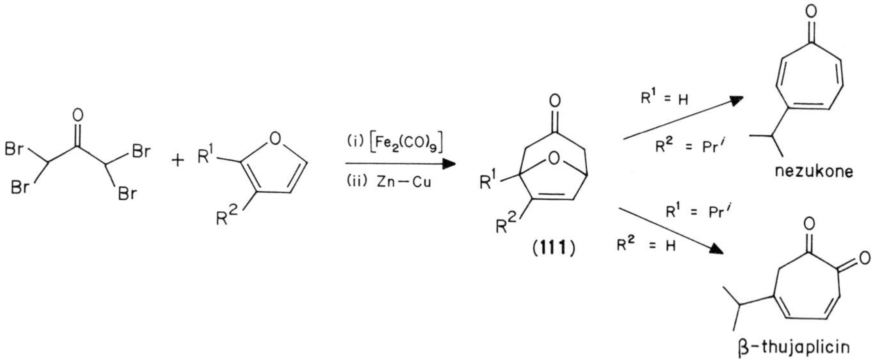

(111)

nezukone

β-thujaplicin

SCHEME 88

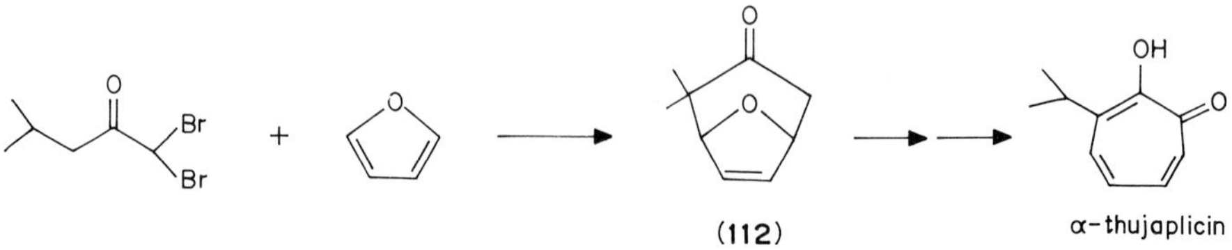

(112)

α-thujaplicin

SCHEME 89

Finally, other unsaturated substrates can be condensed with the oxoallylferrate, including carboxamides such as *N,N*-dimethylformamide, *N,N*-dimethylacetamide, *N,N*-dimethylbenzamide, and *N*-methylpyrrolidone. *N,N*-Dimethylformamide leads to a muscarine precursor **113**[224] (equation 86).

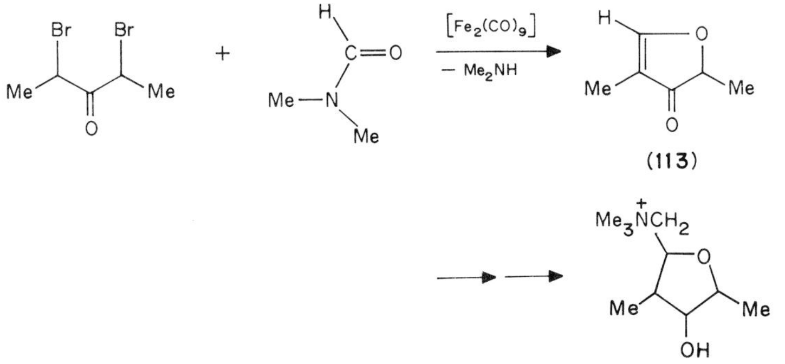

$[Fe_2(CO)_9]$, $-Me_2NH$

(113)

(86)

B. Ferrelactones and Ferralactams

Ferrelactones, **114**, can be obtained from an iron carbonyl and chloroallyl alcohols, vinyloxiranes, or oxazines. Their decomplexation by Ce(IV) generally yields β-lactones, **115**, whilst reaction with amines in the presence of alumina gives ferralactams, **116**[225] (Scheme 90). Similarly, the decomplexation of ferralactams by Ce(IV) gives β-lactams. For instance, **117**, related to the novel nocardicine antibiotics, was prepared from a ferrelactone in this way[226] (Scheme 91).

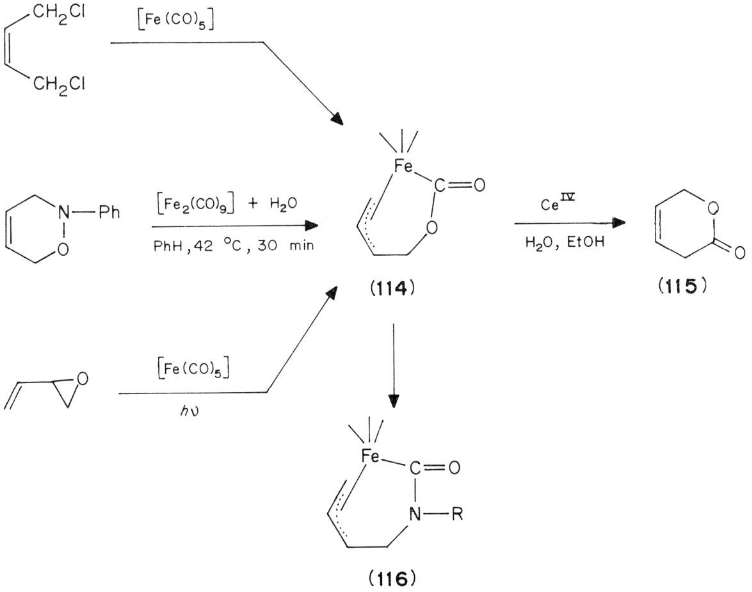

$[Fe(CO)_5]$

$[Fe_2(CO)_9] + H_2O$, PhH, 42 °C, 30 min

Ce^{IV}, H_2O, EtOH

$[Fe(CO)_5]$, $h\nu$

(114) **(115)** **(116)**

SCHEME 90

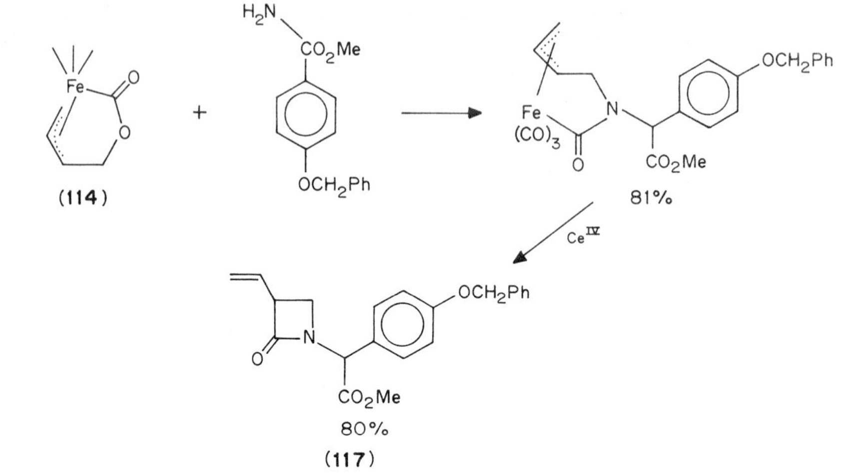

(114)

H_2N CO_2Me OCH_2Ph

Fe $(CO)_3$ N CO_2Me OCH_2Ph 81%

Ce^{IV}

N OCH_2Ph CO_2Me 80% (117)

SCHEME 91

C. Isomerizations of Olefins

The reaction of olefins with $[Fe(CO)_5]$ often results in rearrangement of the olefin. The mechanism consists in oxidative addition of a C—H bond following coordination of the olefin. The resulting allyliron hydride tetracarbonyl species, **118**, reductively eliminates, which gives the more stable isomer[227] (Scheme 92). Conjugated dienes may isomerize with or without double bond migration via dienyliron hydride, intermediates[228] (equations 87 and 88). Non-conjugated dienes are isomerized by $[Fe(CO)_5]$ to the thermodynamically more stable conjugated isomers[229] (equation 89).

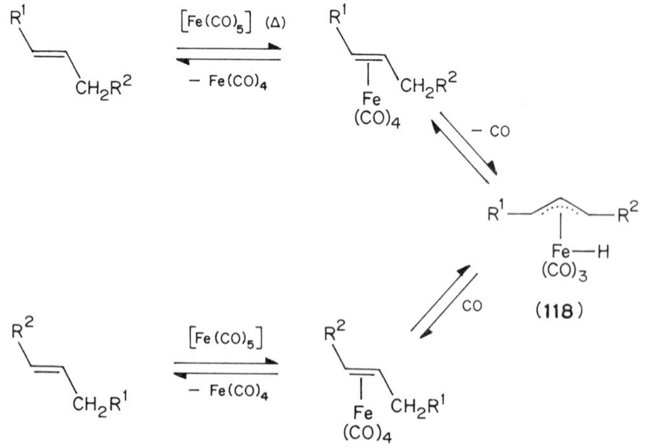

R^1 CH_2R^2 $[Fe(CO)_5]$ (Δ) $-Fe(CO)_4$ R^1 CH_2R^2 Fe $(CO)_4$ $-CO$

R^1 R^2 Fe—H $(CO)_3$ (118)

CO

R^2 CH_2R^1 $[Fe(CO)_5]$ $-Fe(CO)_4$ R^2 CH_2R^1 Fe $(CO)_4$

SCHEME 92

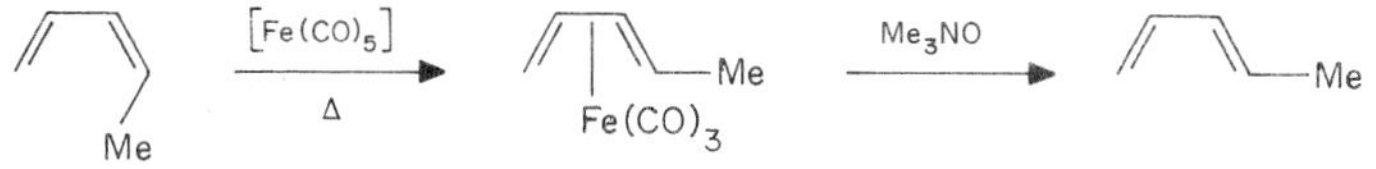

(87)

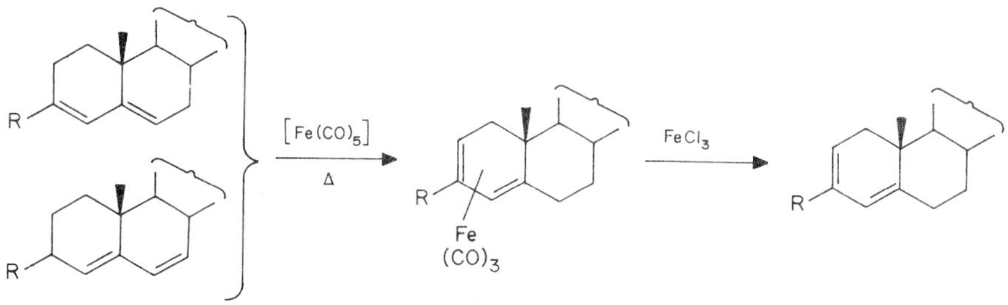

(88)

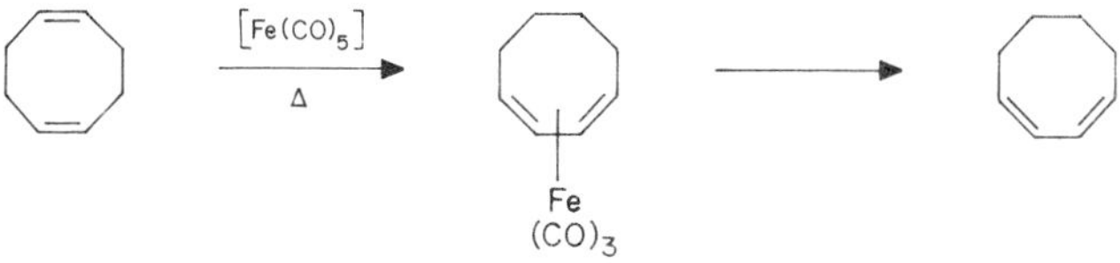

(89)

Cyclohexa-1,4-dienes, available by the Birch reduction of aromatics, are isomerized to cyclohexa-1,3-dienes on complexation to $Fe(CO_3)$[230]. However, 2,5-dihydrobenzoate is converted into the conjugated diene **119** in which the ester group remains unconjugated [in contrast with the base-catalysed treatments leading to a fully conjugated diene ester **120**[230] (equation 90)]. Decomplexation of the useful 1-methoxy derivative **121** proceeds with decomposition of the dienol ether[231] (equation 91, Schemes 27, 32, 35, 36). A useful example of the conjugation of dienes shown in Scheme 93 affords an entry to the prostaglandin C series[232].

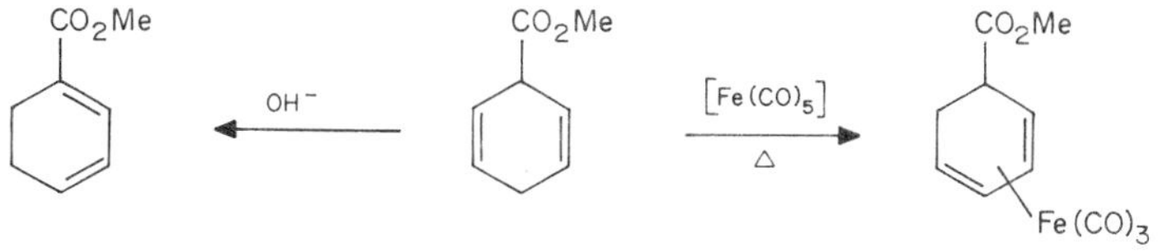

(120) (119) (90)

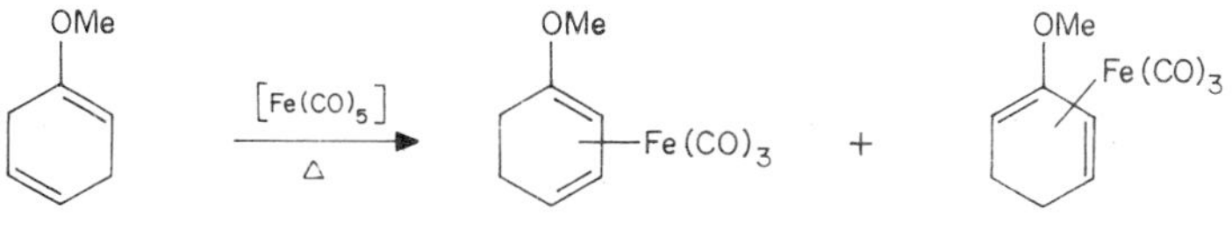

(121) (91)

Primary and secondary allylic alcohols are isomerized to aldehydes using catalytic amounts of $[Fe(CO)_5]$[233] (Scheme 94). Other functionalities such as esters, alcohols, and

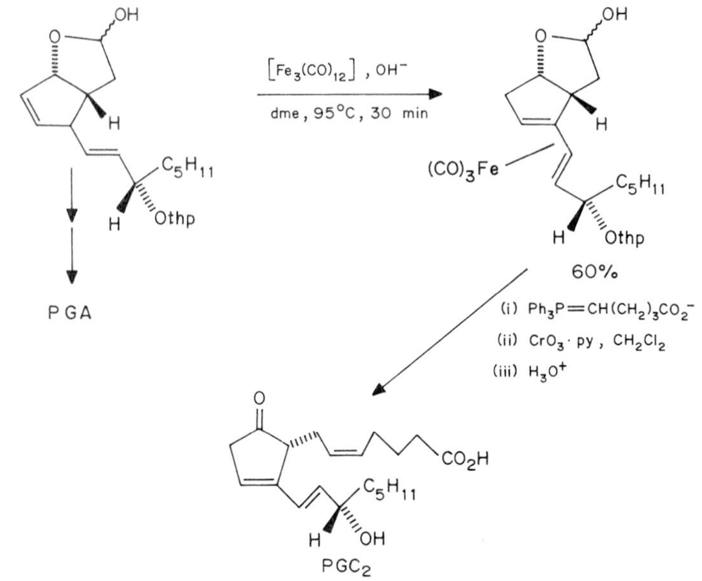

$[Fe_3(CO)_{12}]$, OH^-; dme, 95°C, 30 min; 60%; PGA; (i) $Ph_3P{=}CH(CH_2)_3CO_2^-$ (ii) $CrO_3 \cdot py$, CH_2Cl_2 (iii) H_3O^+; PGC_2

SCHEME 93

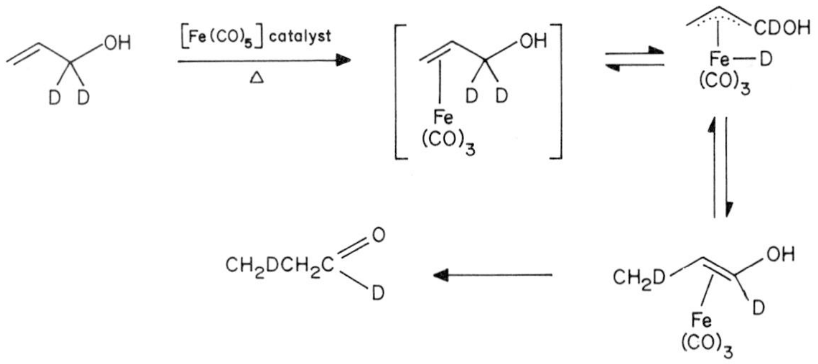

$[Fe(CO)_5]$ catalyst, Δ; CH_2DCH_2CDO

SCHEME 94

ketones withstand the reaction. Photolysis in the presence of $[Fe(CO)_5]$ induces the isomerization of *N*-allyl amides **122** to prop-2-enyl amides **123**[234] (equation 92).

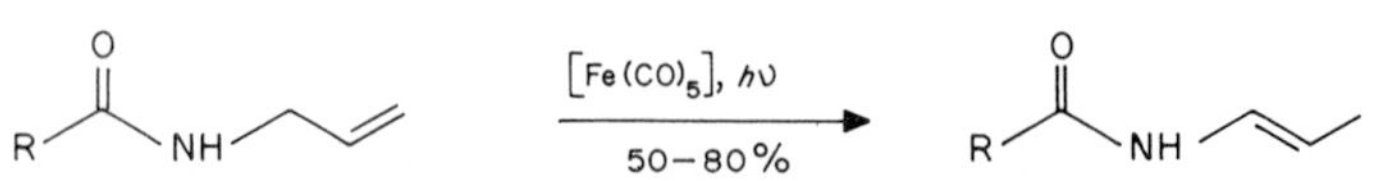

$$\mathrm{RC(O)NHCH_2CH{=}CH_2}\ (\mathbf{122}) \xrightarrow[50-80\%]{[Fe(CO)_5],\ h\nu} \mathrm{RC(O)NHCH{=}CHCH_3}\ (\mathbf{123}) \tag{92}$$

D. Transformations of Small Rings

Vinylcyclopropanes react with $[Fe(CO)_5]$ in thermal or photolytic reactions[234,235] (equation 93). The ring opening of epoxides by reaction with $[Fe(CO)_5]$ and the reaction

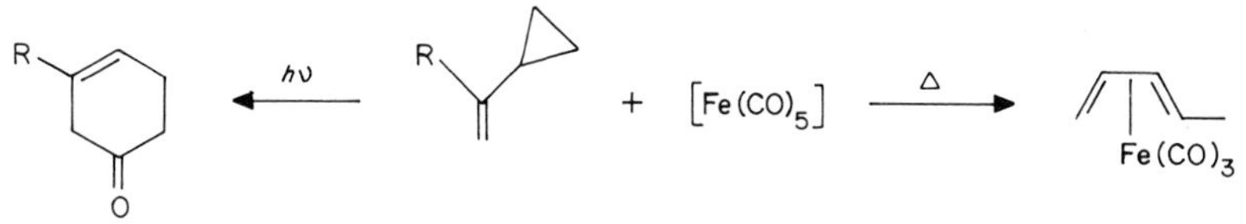

(93)

of oxazines with $[Fe_2(CO)_9]$ are routes to ferrelactones, precursors of lactones and of the synthetically important β-lactams (see Section VI.A). Epoxides are deoxygenated by $[CpFe(CO)_2]^-$, giving olefin complexes with either retention or inversion of the stereochemistry, depending on the reaction conditions[236] (see Section VI.E1).

The epoxide of cyclooctatetraene, **124**, is thermally isomerized by $(Fe_3(CO)_{12}]$[237]. Photolysis with $[Fe(CO)_5]$ followed by reaction with $[Fe(CO)_5]$ and decomplexation leads to 9-oxabicyclo[4.2.1]nona-2, 4, 7-triene, **125**[238] (Scheme 95).

$[Fe_2(CO)_9]$ converts epoxides to allylic alcohols under mild conditions[239], a reaction much more facile than treatment with a strong base (equation 94). 2-Arylazirines, **126**, are rearranged on reaction with Fp_2[240] (equation 95). The structure of the product obtained

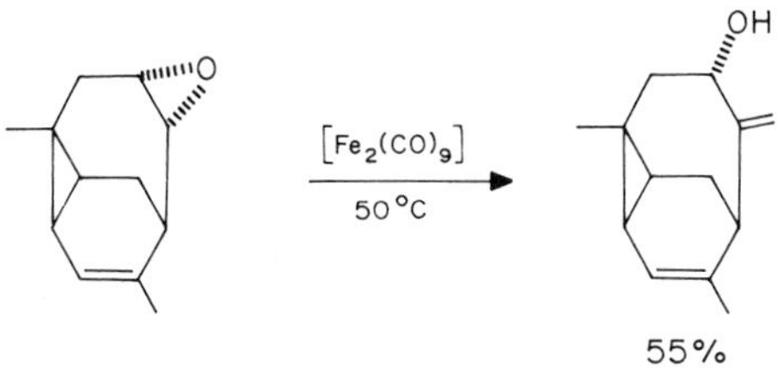

(94)

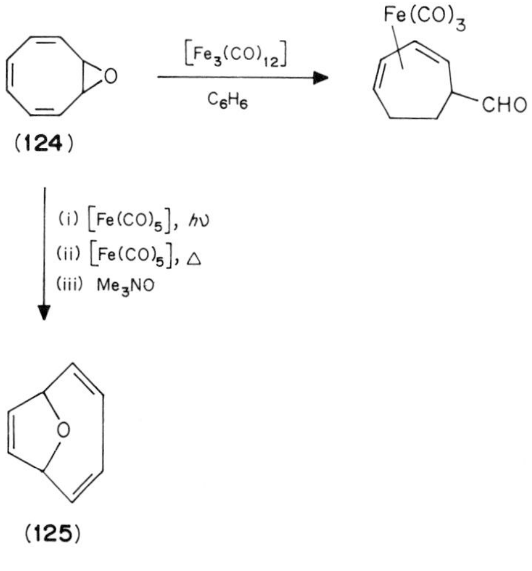

SCHEME 95

(126) → [Fp_2] → 39% (95)

varies markedly with the transition metal carbonyl used (Co, Mo, etc.). Diazatrienes (**127**) are rearranged to pyrroles (**128**) on reaction with $[Fe_2(CO)_9]$[241] (equation 96).

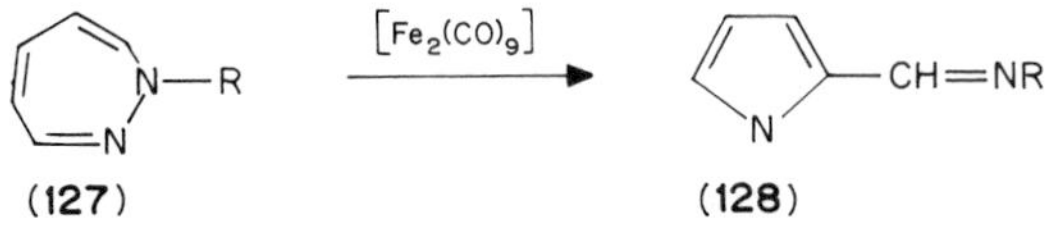

(127) → [$[Fe_2(CO)_9]$] → (128) CH=NR (96)

E. Deoxygenation, Desulphurization, and Dehalogenation

1. Deoxygenation

$[CpFe(CO)_2]^-$ reacts with epoxides to give alkoxide complexes **129** (Scheme 96) Thermolysis of **129** gives the free olefins with inversion of the stereochemistry, whereas protonation gives the olefin complexes with retention[236]. Since epoxidation of olefins proceeds with retention, the first reaction provides a method of inverting their stereochemistry, whereas the latter represents a protection technique together with a route to $[Fp(olefin)]^+$ complexes (Scheme 96). $[Fe(CO)_5]$ also deoxygenates epoxides but not stereospecifically (equation 97 and 98)[242].

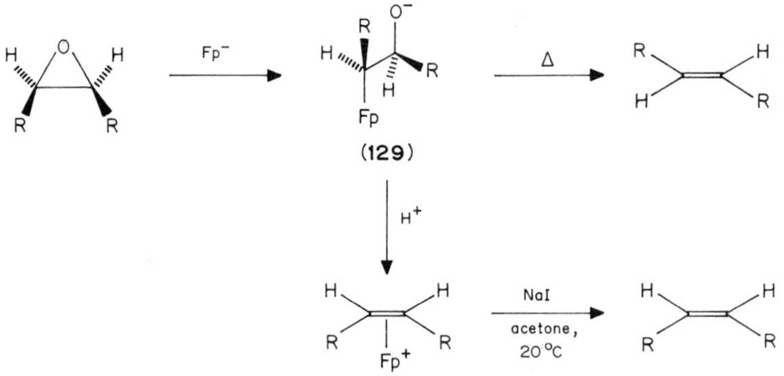

Fp⁻ → (129) → Δ ; H⁺ ; NaI, acetone, 20 °C

SCHEME 96

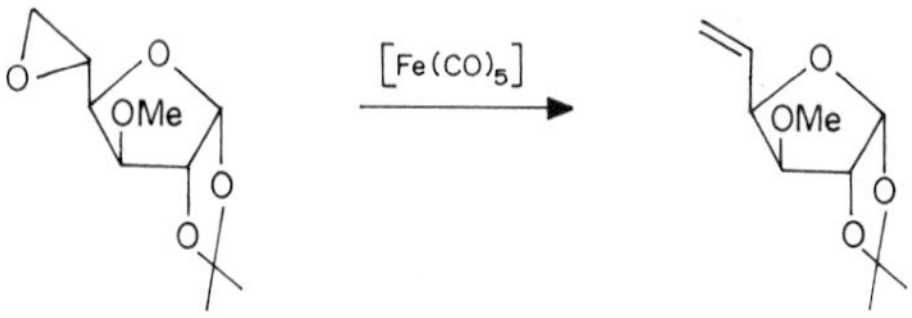

[$Fe(CO)_5$] (97)

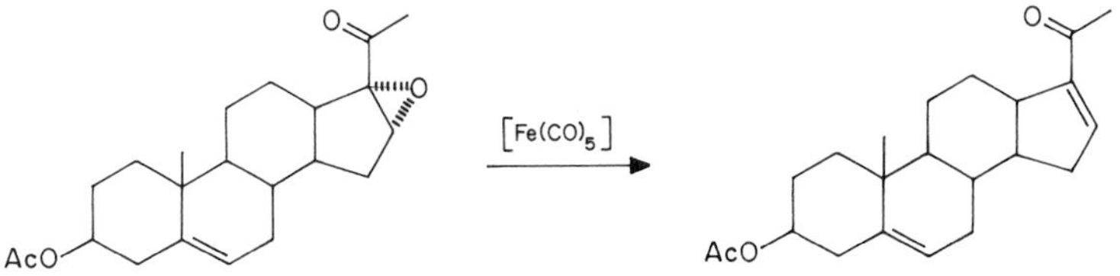

$$\xrightarrow{[Fe(CO)_5]} \quad 82\% \qquad (98)$$

Amine oxides and sulphoxides can be deoxygenated by $[Fe(CO)_5]$[243] (equation 99 and 100). Such nucleophilic oxides generally remove CO from metal carbonyls with

$$R_3NO \xrightarrow[Bu^n_2O,\ 140\,^{\circ}C]{[Fe(CO)_5]} R_3N \quad 45\text{–}80\% \qquad (99)$$

$$R_2SO \xrightarrow[130\,^{\circ}C]{[Fe(CO)_5]} R_2S \quad 57\text{–}91\% \qquad (100)$$

production of CO_2. Me_3NO is commonly used for synthetic purposes to disengage a metal carbonyl fragment from the desired organic ligand [diene from $Fe(CO)_3$][244]; it is also often used to replace one carbonyl by another ligand in organometallic chemistry[246]. The mechanism can be depicted as in equation 101. A detailed study of the deoxygenation of

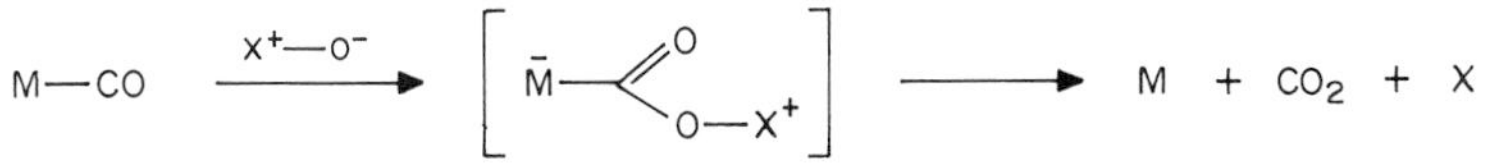

$$M{-}CO \xrightarrow{X^+{-}O^-} \left[\bar{M}{-}C(=O){-}O{-}X^+\right] \longrightarrow M + CO_2 + X \qquad (101)$$

N-oxides with various *N*-amine oxides was made by Alper and Edward[246], who obtained yields ranging from 50 to 80%. Similarly, nitrosoarenes give azoarenes (equation 102) whereas aromatic *N*-nitroso compounds give secondary amines (equation 103)[246]. The aliphatic derivatives *N*-nitrosopiperidine and *N*-nitrosomorpholine give the ureas resulting from CO insertion (equation 104) $[Fe_3(CO)_{12}]$, rather than $[Fe(CO)_5]$, cleanly

$$ArNO \longrightarrow ArN{=}NAr \quad 75\% \qquad (102)$$

$$Ar_2NNO \longrightarrow Ar_2NH \quad 85\text{–}92\% \qquad (103)$$

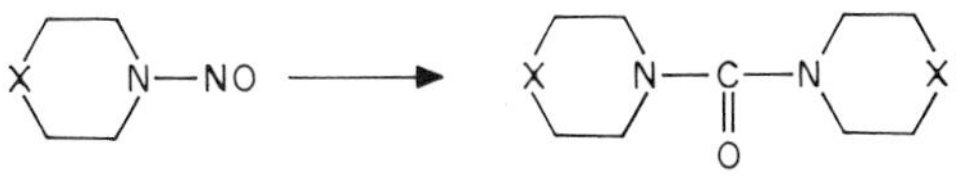

$$X(CH_2CH_2)_2N{-}NO \longrightarrow X(CH_2CH_2)_2N{-}C(=O){-}N(CH_2CH_2)_2X \qquad (104)$$

reduces a variety of *ortho*-, *meta*-, and *para*-substituted nitroaromatics to anilines in over 70% yields[247] (equation 105). Given the generality and the selectivity of the reaction (unaffected substituents include Me, Cl, NH_2, CO_2R, OMe, OH, Ac, NHAc), this method of reduction appears quite useful.

$$ArNO_2 \xrightarrow[PhH,\ MeOH]{[Fe_3(CO)_{12}]} ArNH_2 \qquad (105)$$

Amide oxides and oximes can be deoxygenated by $[Fe(CO)_5]$ to give amidines and imines, respectively[248]. (equations 106 and 107). Carbonyl derivatives can be regenerated from their oximes using this reaction[249] (equation 108). Some tertiary alcoholates can be deoxygenated by $[Fe(CO)_5]$ and subsequent treatment with HCl[250] (Scheme 97).

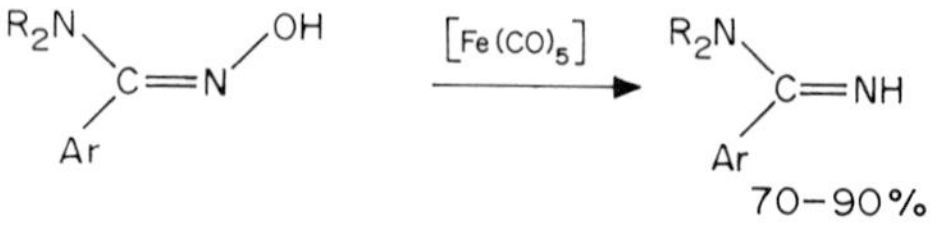

$$R_2N(Ar)C{=}N{-}OH \xrightarrow{[Fe(CO)_5]} R_2N(Ar)C{=}NH \quad (70\text{–}90\%) \tag{106}$$

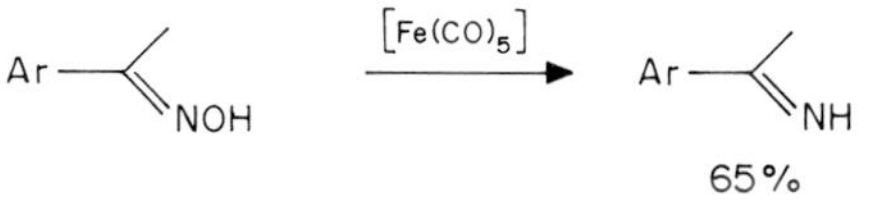

$$Ar{-}C(Me){=}NOH \xrightarrow{[Fe(CO)_5]} Ar{-}C(Me){=}NH \quad (65\%) \tag{107}$$

$$4\text{-}Cl_3C\text{-}4\text{-}Me\text{-cyclohexa-2,5-dien-1-one oxime } ({=}NOH) \xrightarrow{[Fe(CO)_5]} 4\text{-}Cl_3C\text{-}4\text{-}Me\text{-cyclohexa-2,5-dien-1-one } ({=}O) \quad (72\%) \tag{108}$$

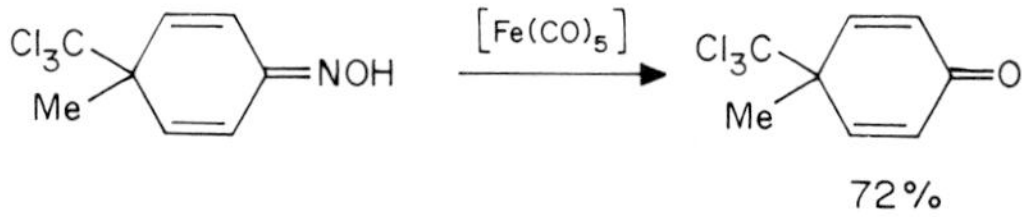

$$RR'_2CO^- \xrightarrow{[Fe(CO)_5]} \left[RR'_2C{-}O{-}C({=}O){-}\bar{F}e(CO)_4 \right] \longrightarrow RR'_2C^- + {}^{\bullet}Fe(CO)_4 + CO_2$$

$$RR'_2C^- \xrightarrow{HCl} RR'_2CH$$

R = Me ; R′ = Ar
R = Ar R′ = Me

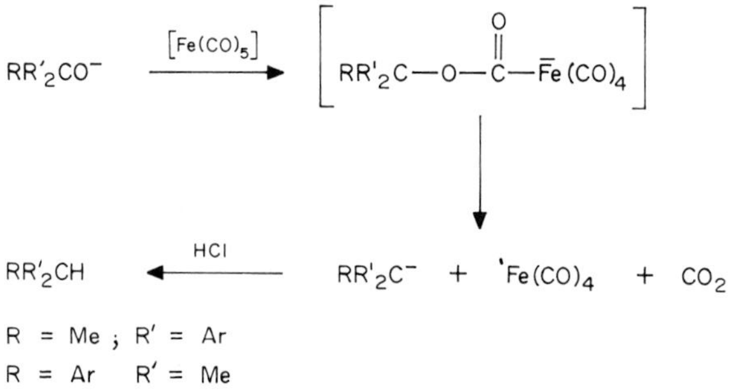

SCHEME 97

2. Desulphurization

Alkene episulphides are desulphurized efficiently by $[Fe_2(CO)_9]$[251], a reaction which parallels the deoxygenation of epoxides. The reaction proceeds with retention of stereochemistry (equation 109). Thioanhydrides may be similarly desulphurized using

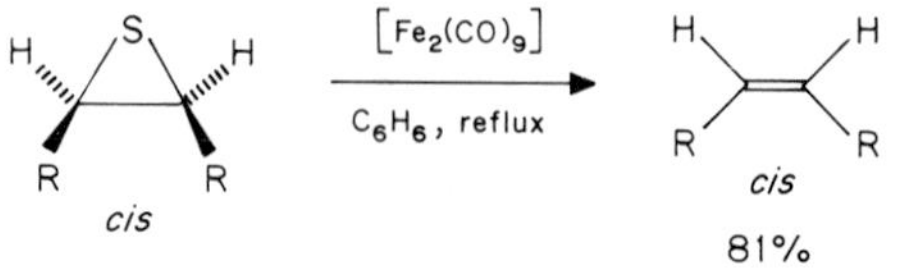

$$\textit{cis}\text{-}RHC{-}S{-}CHR \text{ (episulphide)} \xrightarrow[C_6H_6,\ \text{reflux}]{[Fe_2(CO)_9]} \textit{cis}\text{-}RHC{=}CHR \quad (81\%) \tag{109}$$

$[Fe(CO)_5]$ or $[Fe_2(CO)_9]$[252], which has potential use in the Corey–Winter procedure for olefin synthesis (equation 110). The reaction of thiocarbonates with $[Fe(CO)_5]$ also provides olefins[253] (equation 111). Thioamides are desulphurized and converted into amines by reaction with $[Fe(CO)_5]$–KOH[254] (equation 112a). Similarly, thioketones are

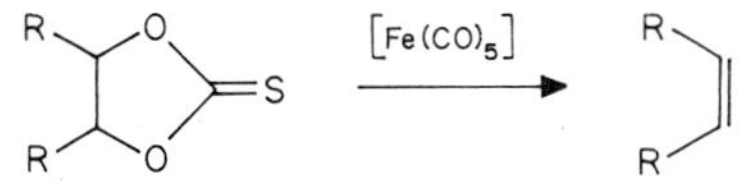

$$\text{cyclic thionocarbonate (R,R)} \xrightarrow{[Fe(CO)_5]} RCH{=}CHR \tag{110}$$

$$\text{cyclic thioanhydride} \xrightarrow{[Fe_2(CO)_9]} \text{norbornene} \tag{111}$$

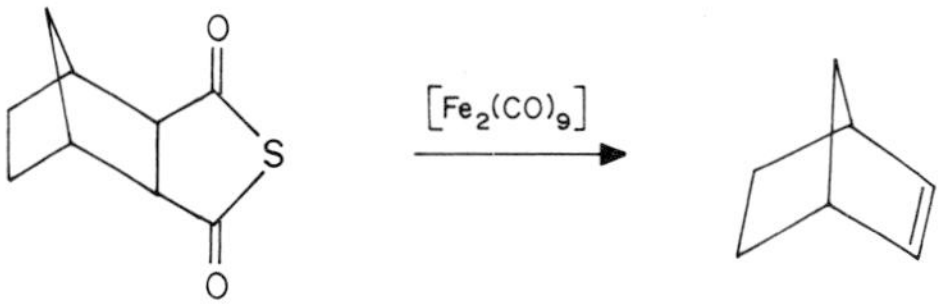

$$RC(S)NHPh \xrightarrow[KOH]{[Fe(CO)_5]} RCH_2NHPh \quad 38-51\% \tag{112a}$$

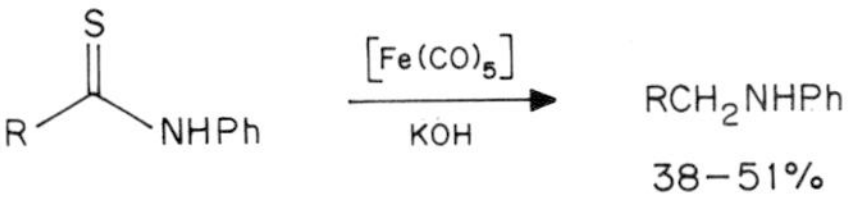

converted into the corresponding hydrocarbons by reaction with $Na[HFe(CO)_4]$[255]. Alper and coworkers[256] converted thiobenzophenone into a variety of desulphurized organic products via formation of an iron–sulphur cluster produced on reaction with $[Fe_2(CO)_9]$ (Scheme 98).

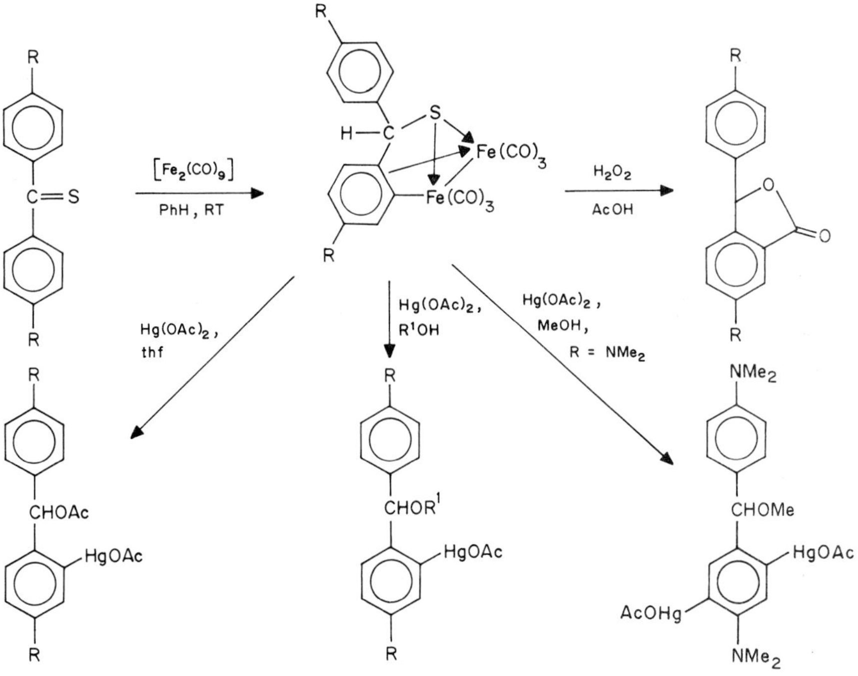

SCHEME 98

3. Dehalogenation

Halo derivatives are readily dehalogenated in a variety of ways by reaction with iron carbonyls[214–224], iron carbonyl anions[142–152], and iron hydrides[156–168]. Dehalogenation may proceed with functionalization using $Na_2[Fe(CO)_4]$ under CO or phosphine (Section III.D.1) or without functionalization using $[HFeCO_4]^-$ salts, $[(C_5Me_5)Fe(CO)_2H]$, or other hydrides (Section III.D.2. and 3). The coupling of *gem*-dihalides, haloketones, and Noyori's condensation of polyhalides with $[Fe(CO)_5]$ are described in Section VI.A. Dehalogenation of specific organic dihalides by $[Fe(CO)_5]$ affords the synthesis of cyclobutadiene-[208], trimethylenemethane-[257] and *o*-xylylene-iron tricarbonyl[212] complexes (Sections V.C and VI.A). The reaction of chloroallyl alcohols with $[Fe(CO)_5]$ is one of the routes to the useful precursors ferrelactones[225] (Section VI.A and B).

F. Miscellaneous Carbonylation Reactions

Well known rational routes for the carbonylation of organic compounds are the reactions of organic halides with Collman's reagent[142–168] (Section III.D.1) and Noyori's condensation of unsaturated substrates with oxyallyl ferrates generated from polyhalides and $[Fe(CO)_5]$[217–224] (Section VI.A). In addition, it was found that iron carbonyls react with a variety of unsaturated organic compounds to give miscellaneous carbonylation reactions. The synthesis of ferrelactones by dehalogenation–carbonylation of chloroallyl alcohol[225] (Section VI.B) and the carbonylation of *o*-xylylene by reaction of $AlCl_3$ with its tricarbonyl complex[212] (Section V.C) are examples already described in this Chapter.

$[Fe_2(CO)_9]$ reacts with acetylene to provide cycloheptatrienoneiron tricarbonyl, which, on decomplexation with $FeCl_3$, gives the free ligand[258] (equation 112b). The

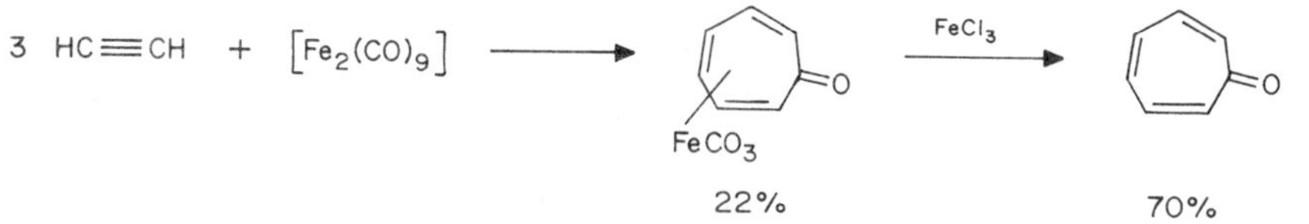

$3\ HC{\equiv}CH + [Fe_2(CO)_9] \longrightarrow$ (cycloheptatrienone)$FeCO_3$ 22% $\xrightarrow{FeCl_3}$ cycloheptatrienone 70% (112b)

(130) $\xrightarrow{\text{(i) } H^+ \text{ (ii) } H^-}$ (131) $\xrightarrow{CO}$ Fe–CO complex $\longrightarrow$ ketone 57% (overall)

(84a) $\xrightarrow{\text{(i) } H^+ \text{ (ii) } H^-}$ (132) $\xrightarrow{CO}$ (133) Fe—C=O $\longrightarrow$ ketone 60% (overall)

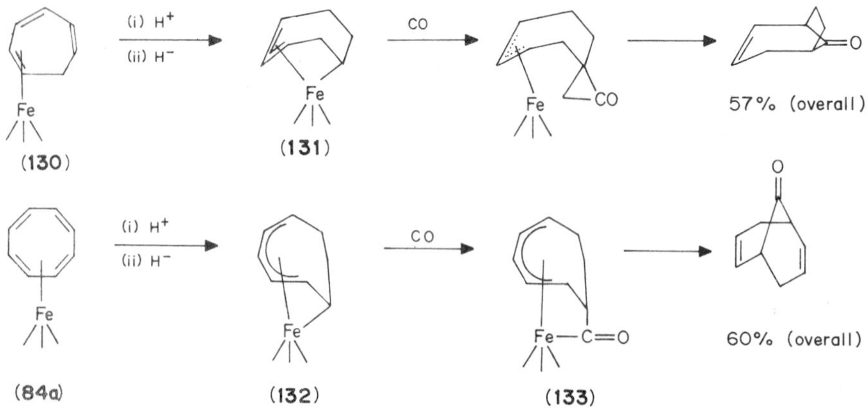

SCHEME 99

cycloheptatriene- and cyclooctatetraene-iron tricarbonyl complexes **130** and **84a**, when treated by H^+ followed by H^-, give the homo-allyl and -cyclopentadienyl complexes **131** and **132**, respectively[259] (Scheme 99). The latter are carbonylated under CO. These reactions are examples of CO insertion into iron—alkyl bonds. Another example is the reaction of vinylcyclopropane with iron carbonyls followed by carbonylation under CO[260]. This process is related to the conversion of vinyl epoxides to ferrelactones[225] (Section VIA, B) (Scheme 100). The cyclic polyeneiron tricarbonyl complexes **12** and **84a** react with $AlCl_3$ under CO to give CO insertion products[261]. Barbaralone is obtained from cyclooctatetraene in this way (Scheme 101).

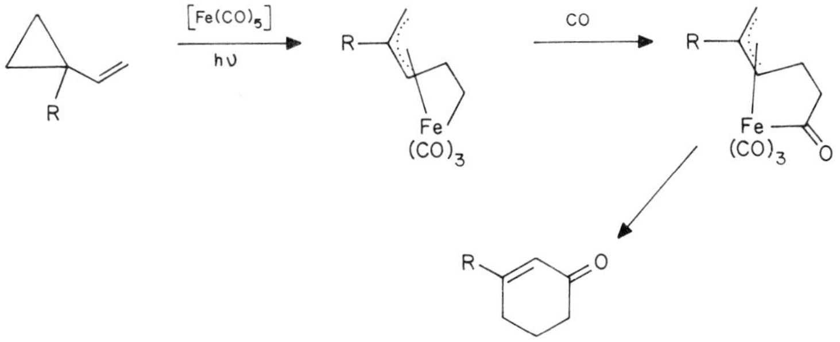

SCHEME 100

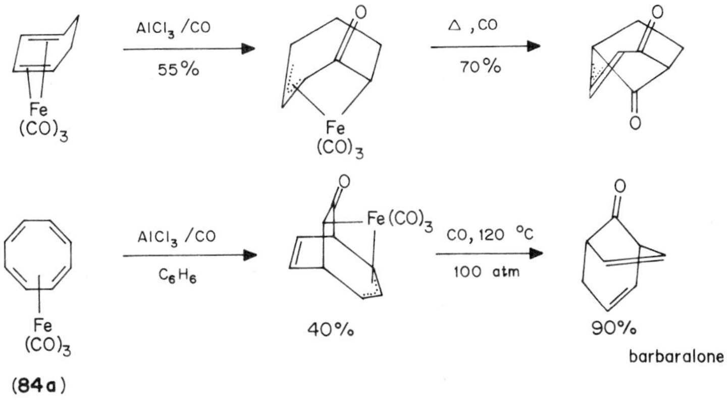

SCHEME 101

VII. ASYMMETRIC SYNTHESES

Complexation of potential ligands such as butadiene, cyclohexadiene, or arenes by an organometallic group renders the molecule asymmetric. Thus asymmetric syntheses are possible using complexes in their optically active forms. For instance, chiral butadiene $Fe(CO)_3$ complexes have been used for the asymmetric synthesis of formylcyclopropanes. In particular, optically active hemicaronaldehydes, key intermediates in the preparation of the very useful insecticides pyrethroides, were prepared by Frank-Neumann *et al.*[262] and

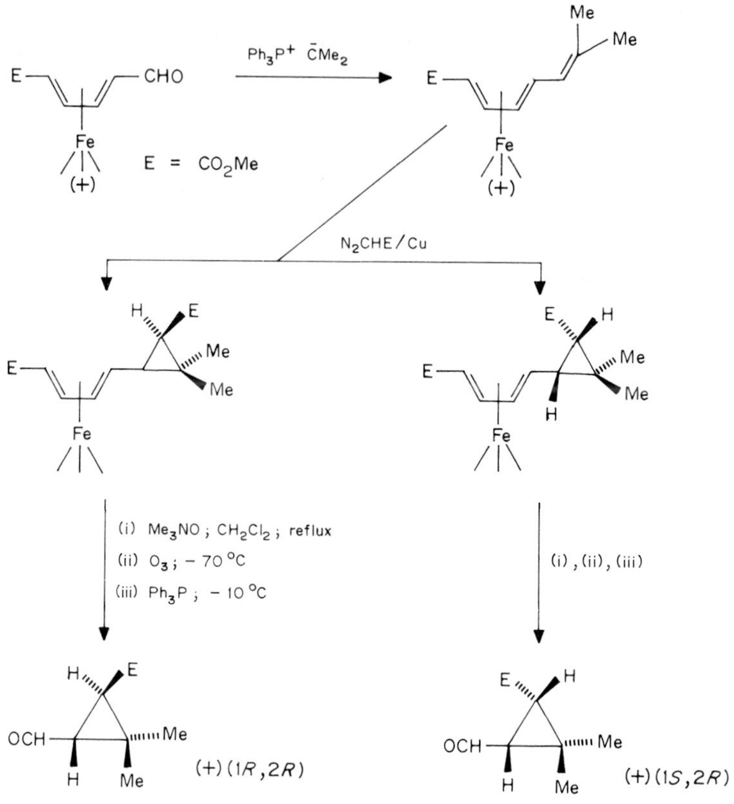

SCHEME 102

by Grée and Carrié's group[263]. Using the Scheme 102, the latter obtained these cyclopropanes with > 90% enantiomeric excess.

Recently, Howell and Thomas[264] performed enantioselective syntheses of cyclohexadieneiron complexes. The process involves the synthetically useful nucleophilic attack on cyclohexadienyliron complexes. Two methods were used: (i) reaction with chiral nucleophiles and (ii) replacement of a carbonyl ligand by a chiral phosphine. In the latter case, asymmetric induction is also possible by reaction with a chiral nucleophiles. Hence this field is promising. Deprotonation of acyl complexes give metalloenolates which undergo diastereoselective alkylation, with Fe[301] as with other metals[302].

VIII. NON-PAIRWISE CHEMISTRY OF IRON COMPLEXES

A. Electron-transfer Chemistry of 19- and 20-electron Arene Complexes: C—H Activation and C—C Bond Formation

1. C—H activation

Nineteen-electron [Cp(η^6-arene)Fe^I] complexes, **134**, synthesized by Na–Hg reduction of their cationic precursors[265,273], react rapidly with 0.5mol of O_2 at −80 °C to give benzylic H-atom abstraction and 0.5 mol of H_2O_2[266]. The organoiron reaction products **135** are the same as those obtained by deprotonation of the 18-electron cations[265b,267], but the H-atom abstraction is cleaner and proceeds in higher yields. The mechanism is an outer-sphere electron transfer from the Fe^I complex to O_2 providing the basic superoxide radical anion in the cage, as confirmed by e.p.r. spectroscopy[268]. In the absence of a benzylic hydrogen, nucleophilic addition of $O_2^{-\bullet}$ occurs to the arene ligand activated by $CpFe^+$ which gives **136**[269]. When salts such as $Na^+PF_6^-$ have not been removed from the reaction medium, the salt effect prevents the deprotonation or the nucleophilic attack and $O_2^{-\bullet}$ dismutates to O_2^{2-} (and O_2, which is consumed); the organoiron cations are recovered as PF_6^- salts **137** (Scheme 103).

The complexes obtained in the H-atom abstraction reactions are [Cp(η^5-cyclohexadienyl)Fe^{II}] with exocyclic double bonds which can be functionalized by a large variety of electrophiles[270] (Section III.A). Similarly, [Cp(η^6-$C_6Me_5NH_2$)Fe^I] reacts

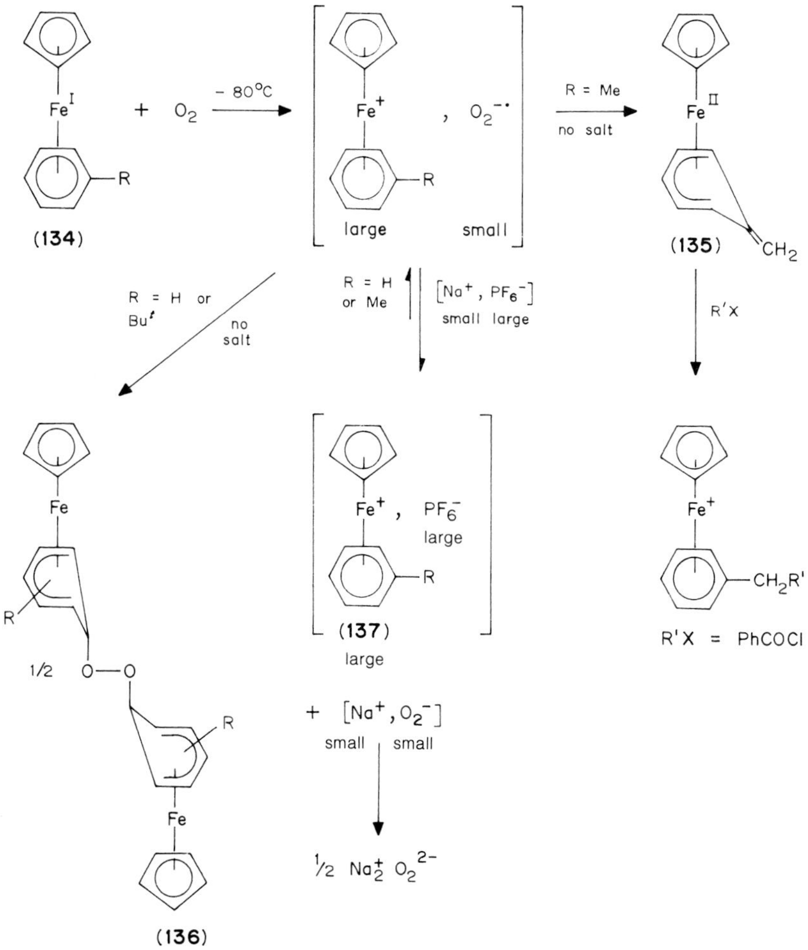

SCHEME 103

with O_2[271] to give [Cp(η^5-C_6Me_5NH)Fe^{II}], which, on reaction with CO_2, gives (Cp(η^6-$C_6Me_5NHCO_2^-$)Fe^+]. In the case of [Cp(η^6-C_6Me_6)Fe^I], the C—H activation product, obtained in 97% yield, was characterized by X-ray analysis and the crystal structure compared with that of the Fe^I complex. Since the $CpFe^+$ group is easily removed by photolysis in the 18-electron cations, the process provides a means of activating C—H bonds by O_2 under mild conditions and of functionalizing arenes.

The 20-electron complex [(η^6-C_6Me_6)$_2Fe^0$], **138**[272], obtained by Na–Hg reduction of the isostructural dication, reacts with O_2 at −40 °C to give the double C—H activation product, namely the 18-electron *o*-xylylene complex [(η^6-C_6Me_6){η^4-$C_6Me_4(CH_2)_2$}Fe^0], **139**[103], not accessible by deprotonation of [(η^6-C_6Me_6)$_2$Fe]$^{2+}$. Acylation with PhCOCl also proceeds at −40 °C, giving back the 18-electron dicationic structure in **140** [bis(arene)Fe^{2+} salts rapidly liberate the arenes upon photolysis] (equation 113a).

(138) —O_2, −40 °C; −2H•→ (139) —PhCOCl→ (140)

(113a)

The 19-electron complex [(η^6-C_6Me_6)(η^5-C_6Me_6H)Fe^I], **141**, is obtained by hydride reduction of [(η^6-C_6Me_6)$_2$Fe]$^{2+}$ followed by one-electron reduction using Na–Hg[22], or better from [Cp(η^6-C_6Me_6)Fe^I]. It reacts with 0.5mol of O_2 giving the single H-atom abstraction complex [(η^6-C_6Me_6){η^4-$C_6Me_5H(CH_2)$}Fe^0], **142** (equation 113b). Reaction of the latter with electrophiles yields functionalization of the exocyclic methylene group and further to the chemistry indicated in Section III.A.

(141) —O_2→ (142) —PhCOCl→ [Fe^+, CH_2COPh]

(113b)

2. C—C bond formation

[Cp(η^6-C_6H_6)Fe^I] reacts with some organic halides to give equal amounts of [Cp(η^5-cyclohexadienyl)Fe^{II}], resulting from C—C bond formation, and [Cp(η^6-C_6H_6)Fe]$^+$[274].

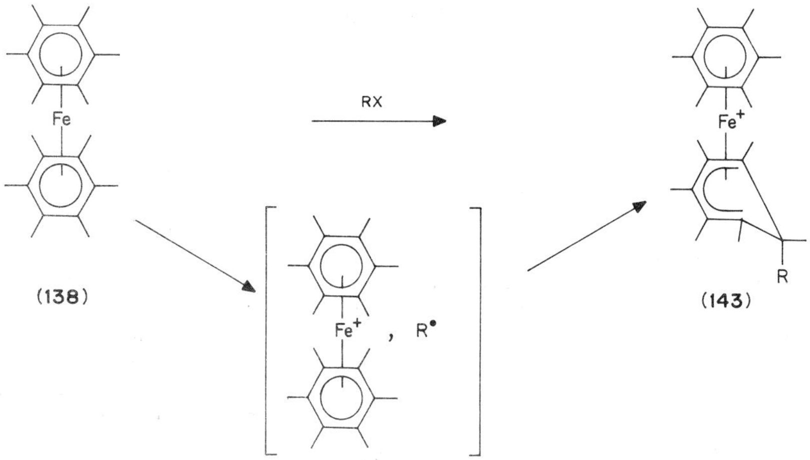

SCHEME 104

The 50% limit on the theoretical yield of C—C bond formation no longer holds if one starts from a 20-electron iron complex such as **138** since, in this case, the organic radical couples with an organometallic radical (19-electron Fe^{I}) in the cage subsequent to electron transfer (Scheme 104). Whereas carbanions do not form C—C bonds on reaction with $[(\eta^6\text{-}C_6Me_6)_2Fe]^{2+}$ because of electron transfer, the desired coupling products **143** may be simply obtained in the above process. Further implications in organic synthesis are detailed in Section III.A.

B. Iron-catalysed Alkyl Disproportionation in Kharash Reactions

Iron(I) species, generated *in situ* by reduction of iron(II) or iron(III) salts or complexes and Grignard reagents, catalyse the disproportionation of alkyl radicals in the reaction between Grignard reagent and organic halides[275] (equation 114). Side-reactions are alkyl exchange (equation 115) and hydrogen transfer from the olefin formed (equation 116). The reaction mechanism was carefully examined by Kochi[276]: (i) stoichiometric studies of the reduction of $FeCl_3$ by Grignard reagents and the e.p.r. spectrum of the iron product suggest that Fe(I) species is the active catalyst; (ii) the reactivities of alkyl bromides decrease in the order $Bu^t > Pr^i > Pr^n$; and (iii) disproportionation follows the kinetic law.

$$\frac{d[C_2H_4]}{dt} = k[FeCl_3][EtBr][EtMgBr]^0$$

$$EtMgBr + EtBr \xrightarrow{[Fe]} C_2H_6 + CH_2{=}CH_2 \tag{114}$$

$$EtMgBr + Pr^nBr \rightleftharpoons EtBr + Pr^nMgBr \tag{115}$$

$$Pr^nMgBr + CH_2{=}CH_2 \rightleftharpoons EtMgBr + CH_3CH{=}CH_2 \tag{116}$$

The probable mechanism, proposed by Kochi, is given in equation 117–120. In the presence of silver instead of iron as a catalyst, alkyl dimers are formed, but no disproportionation product is formed[277].

$$RMgBr + Fe^{II}Br \longrightarrow RFe^{II} + MgBr_2 \tag{117}$$

$$RFe^{II} + R'Fe^{II} \longrightarrow [RH, R'H] + [R(-H), R'(-H)] + 2Fe^{I} \quad (118)$$

$$Fe^{I} + R'Br \xrightarrow{\text{slow}} Fe^{II}Br + R''^{\bullet} \quad (119)$$

$$R''^{\bullet} + Fe^{I} \longrightarrow R'Fe^{II}, \text{ etc.} \quad (120)$$

C. Oxidations Using Fenton's Reagent

H_2O_2 reacts with Fe^{II} to produce $OH^{\bullet}$ radicals, which considerably enhances its reactivity[278]. The decomposition of H_2O_2 by Fe^{2+} is catalytic and proceeds via a free radical chain mechanism (equations 121–125). In the presence of organic substrates, organic radicals are formed by reaction with $OH^{\bullet}$ and dimerize or may be reduced or oxidized. For instance, *tert*-butyl alcohols reacts with Fenton's reagent to yield 84% of dimer[279] (Scheme 105). Analogous coupling also proceeds with aliphatic esters, ethers, nitriles, and carboxylic acids. In the presence of CO, carboxylation occurs[280] (equations 126 and 127). Maleic acid is hydrated via an α-carboxyalkyl radical and, in the presence of methanol, provides a γ-lactone[281] (equations 128–130).

$$[Fe^{II}] + H_2O_2 \longrightarrow [Fe^{II}OH] + OH^{\bullet} \quad (121)$$

$$RH + OH^{\bullet} \longrightarrow R^{\bullet} + H_2O \quad (122)$$

$$R^{\bullet} + R^{\bullet} \longrightarrow R_2 \quad (123)$$

$$R^{\bullet} + Fe^{III} \longrightarrow R^{+} + [Fe^{II}] \quad (124)$$

$$R^{\bullet} + Fe^{II} \longrightarrow R^{-} + Fe^{III} \quad (125)$$

$$RH + CO \xrightarrow[{[Fe^{II}]}]{H_2O_2} RCO_2H + H_2O \quad (126)$$

$$Me_3COH + CO \xrightarrow[{[Fe^{II}]}]{H_2O_2} Me_2C(OH)CH_2CO_2H + H_2O \quad (127)$$

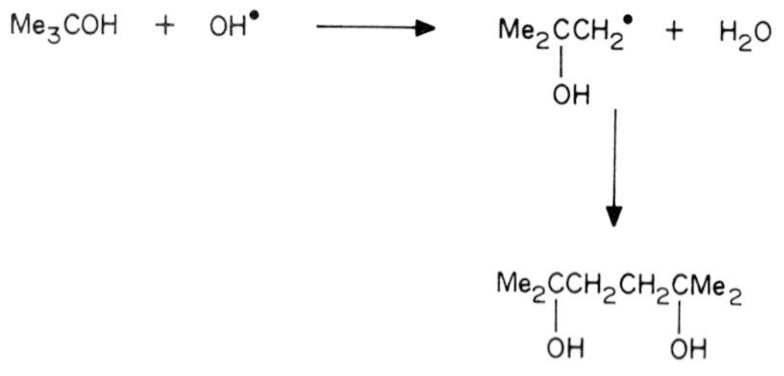

$$Me_3COH + OH^{\bullet} \longrightarrow Me_2C(OH)CH_2^{\bullet} + H_2O$$

$$Me_2C(OH)CH_2^{\bullet} \longrightarrow Me_2C(OH)CH_2CH_2C(OH)Me_2$$

SCHEME 105

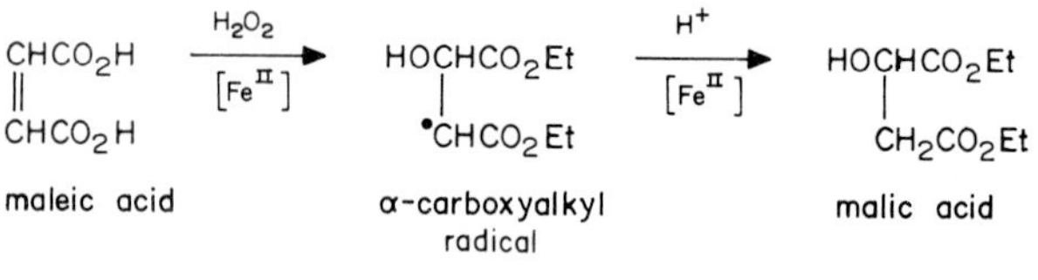

$$\underset{\text{maleic acid}}{CHCO_2H{=}CHCO_2H} \xrightarrow[{[Fe^{II}]}]{H_2O_2} \underset{\alpha\text{-carboxyalkyl radical}}{HOCHCO_2Et{-}{}^{\bullet}CHCO_2Et} \xrightarrow[{[Fe^{II}]}]{H^{+}} \underset{\text{malic acid}}{HOCHCO_2Et{-}CH_2CO_2Et} \quad (128)$$

$$OH^{\bullet} + CH_3OH \longrightarrow {}^{\bullet}CH_2OH + H_2O \quad (129)$$

$$CHCO_2H{=}CHCO_2H + {}^{\bullet}CH_2OH \xrightarrow[Fe^{II}]{H^+} \text{4-carboxy-}\gamma\text{-butyrolactone} \quad (130)$$

Aromatic substrates are hydroxylated via hydroxycyclohexadienyl radicals[282] (equation 131). This type of aromatic hydroxylation also proceeds with $[Fe^{II}]/S_2O_8{}^{2-}$[283]

$$RC_6H_5 + OH^{\bullet} \longrightarrow R\text{-}C_6H_5(OH)(H)^{\bullet} \xrightarrow{[Fe^{III}]} RC_6H_4OH + [Fe^{II}] + H^+ \quad (131)$$

or with $[Fe^{II}] + O_2$ + ascorbic acid[284], but the yields are poor to moderate (equations 132–134).

$$Fe^{II} + S_2O_8{}^{2-} \longrightarrow [Fe^{II}] + SO_4{}^{2-} + SO_4{}^{-\bullet} \quad (132)$$

$$ArH + SO_4{}^{-\bullet} \longrightarrow ArH^{+} + SO_4{}^{2-} \quad (133)$$

$$ArH^{+\bullet} + H_2O \longrightarrow R\text{-}C_6H_5(H)(OH)^{\bullet} \xrightarrow{[Fe^{III}]} RC_6H_4OH + [Fe^{II}] + H^+ \quad (134)$$

D. Reactions of Alkylperoxides Catalysed by Fe^{II}[285]

Oligomers are obtained by reaction of alkyl hydroperoxides with $[Fe^{II}]$ in the presence of butadiene[286] (equations 135 and 136). The radicals $RO^{\bullet}$ formed in the $[Fe^{II}]$-catalysed

$$RO_2H + [Fe^{II}] \longrightarrow RO^{\bullet} + Fe^{III}OH \quad (135)$$

$$RO^{\bullet} + CH_2{=}CHCH{=}CH_2 \longrightarrow ROCH_2\dot{C}HCH{=}CH_2 \longrightarrow ROCH_2CH{=}CHCH_2CH_2CH{=}C(OR)CH_2OR + ROCH_2CH{=}CHCH_2CH_2CH(OR)CH{=}CH_2 \quad (136)$$

decomposition often dimerize. However, in the presence of Cu^{II}, introduction of a double bond occurs by intramolecular hydrogen transfer[287] (equations 137–139).

$$(CH_3)_2CHCH_2CH_2C(CH_3)_2O_2H + [Fe^{II}] \longrightarrow (CH_3)_2CHCH_2CH_2C(CH_3)_2O^{\bullet} + [Fe^{III}OH] \quad (137)$$

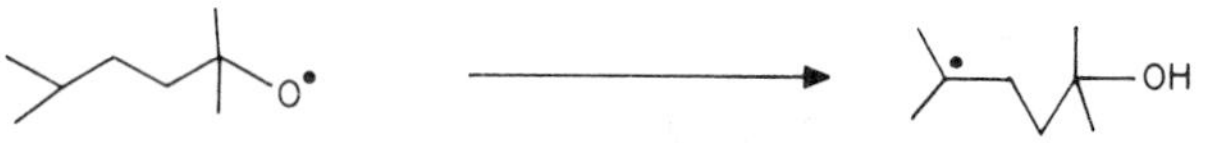

(138)

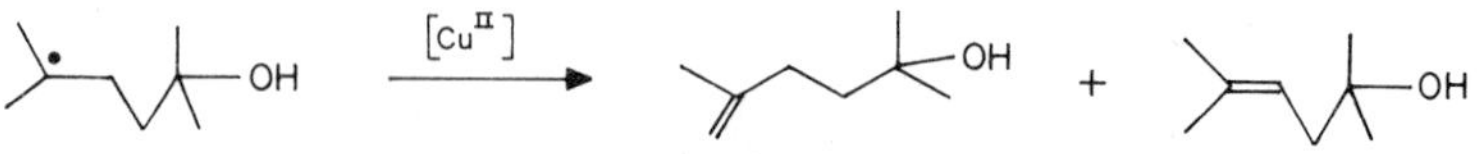

(139)

E. Oxidations by the Ferryl Group, $Fe^{V}{=}O$, and Biomimetic Process[278]

The common active oxidant of a number of hemoproteins (P–450 monooxygenase, catalase, peroxides, chloroperoxidase) having various functions is an oxoiron(V) (protoporphyrin IX) species (Scheme 106). Depending on the nature of the hemoprotein, the mechanism of formation of this $Fe^{V}{=}O$ species is totally different. Unactivated C—H bonds in alkanes can be hydroxylated by the $Fe^{V}{=}O$ oxidant, probably via a free-radical mechanism as proposed by Groves *et al.*[289,290c] (equation 140). C—H activation of aromatics and epoxidation of olefins are other typical reactions produced by ferryl species.

$$[O = Fe^{V}] + RH \longrightarrow [HOFe^{IV}R] \longrightarrow Fe^{III} + ROH \quad (140)$$

Biomimetic oxygenations using chemical models have been attempted with the aim of mimicking the performances of oxygenases. In particular, models of cytochrome P-450 monooxygenase have attracted a great deal of attention[290]. Simple chemical models do not necessarily involve haeme iron complexes. For instance, Fenton's reagent in aprotic solvents presumably generates ferryl species active in the hydroxylation reactions (equation 141). Similar results were also obtained with Udenfried's reagent, consisting of Fe(II), EDTA, ascorbic acid, and O_2[294], although the mechanism involving these two reagents, still unclear, differs significantly. In particular, a feature exhibited by enzymes *in vivo*, and also by Fenton's reagent, but not by Udenfried's reagent, is the NIH shift induced

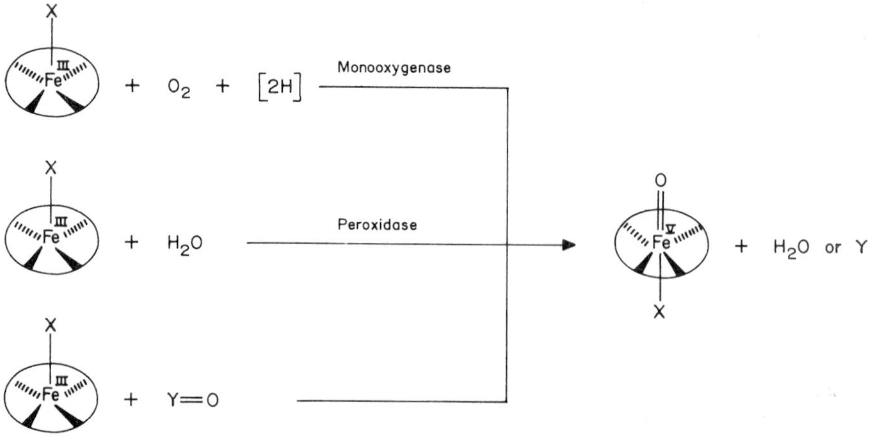

Y=O, single oxygen atom donor ClO^-, ClO_2^-, IO_4^-, PhIO

SCHEME 106

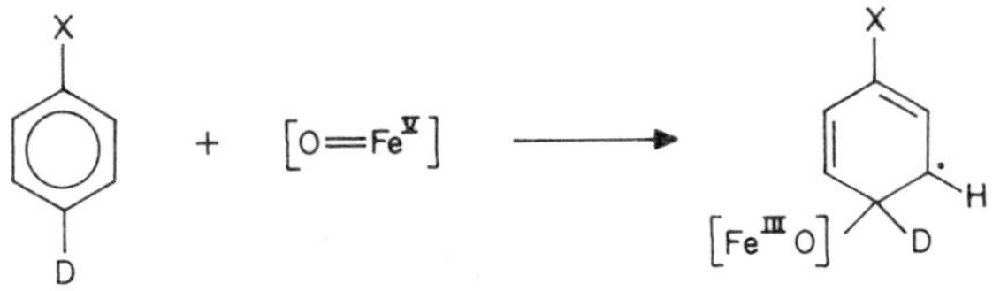

(141)

during aromatic hydroxylation[292] (Scheme 107). The NIH shift induced by Fenton's reagent increases with the amount of water present.

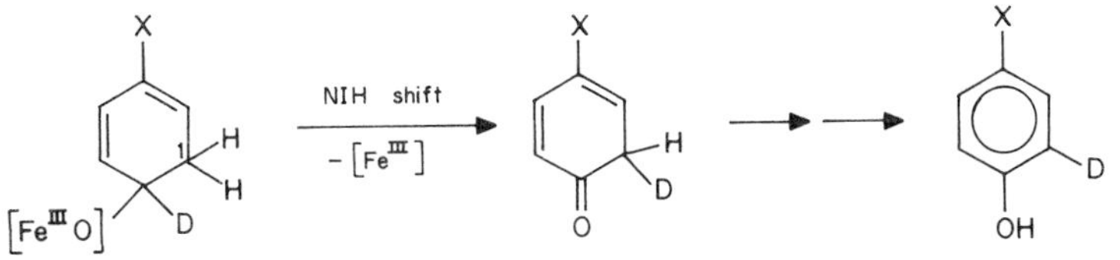

SCHEME 107

SalenFe^{III} complexes are also able to hydroxylate adamantane (the secondary position is favoured) in the presence of O_2 and an H donor[293] (equation 142). Peroxidase models

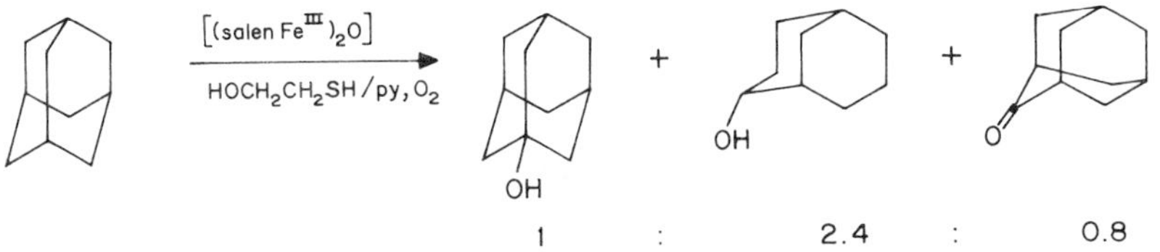

1 : 2.4 : 0.8

(142)

typically consist of porphinatoiron(III) chloride, H_2O_2, and a hydrogen donor or a single oxygen atom donor such as iodosoarenes[294,295]. Using the latter system, stereospecific epoxidation of olefins proceeds in good yield (equation 143); hydroxylation of cyclohex-

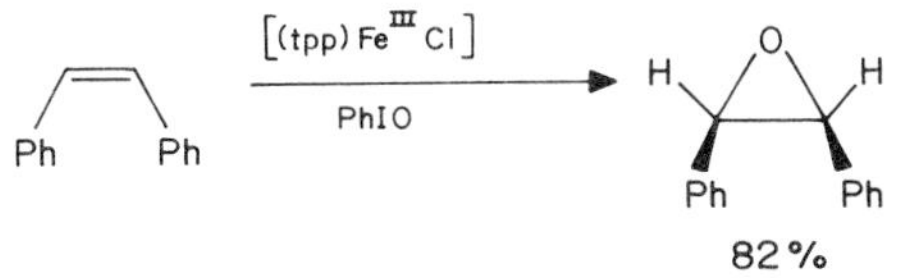

(143)

82%

ane gives only an 8% yield of cyclohexanol owing to a competing intramolecular hydroxylation of the porphirin[295]. Simple iron complexes such as bipyFe(II) can mimic dioxygenases, catalysing the oxidative cleavage of 3,5-di-*tert*-butylcatechol by O_2[296] (equation 144). In dioxygenase and probably also in simple models, the substrate reacts

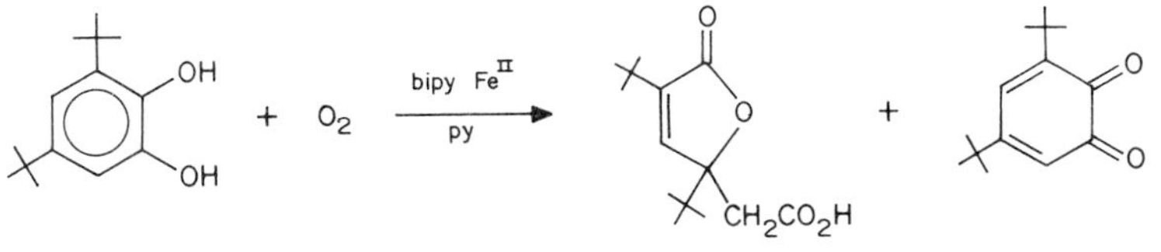

(144)

directly with a superoxo or μ-peroxoFe^{III} complex. Singlet oxygen may also be involved following extrusion from such a complex.

The convenient use of enzyme catalysis in biomimetic design is greatly improved by the immobilization of enzymes and coenzymes (e^- and H donors) on supports[297]. In this way, enzymes can be reused and handling is minimized. Such biocatalysis is extremely promising (i.e. synthesis of epoxides from H_2O_2 or NaOCl).

F. Catalytic Addition of Halides to Olefins

A number of redox reactions based on the generation and addition of free radicals can be made catalytic. For instance, the addition of chloroform across olefinic bonds is induced by iron complexes[298] (equation 145). The catalytic cycle is based on the interchange of Fe^{II} and Fe^{III} as a result of chlorine transfer (equations 146–148). Similarly, the addition of trichloroacetates gives lactones[299] (equation 149). However, the iron species actually responsible for catalysis is not known in the latter process. It is conceivable that the 17-electron radical $[CpFe^{I}(CO)_2]$ produced thermally from the dimer should work as Fe^{II} in the reaction of chloroform. If that is the case, the catalytic cycle would now involve Fe^{I}/Fe^{II} interchange with a similar mechanism. In the presence of trimethylamine oxide, $[Fe(CO)_5]$ gives CO_2 and $[Fe(CO)_4NMe_3]$. The latter catalyses the addition of CCl_4 to olefins, also according to a homolytic process[300]. Here again, the active catalytic species is unknown.

$$CHCl_3 + CH_2{=}CHR \xrightarrow{[Fe]} CHCl_2CH_2CHClR \tag{145}$$

$$Fe^{II} + HCCl_3 \longrightarrow Fe^{III}Cl + {}^{\bullet}CHCl_2 \tag{146}$$

$${}^{\bullet}CHCl_2 + CH_2{=}CHR \longrightarrow CHCl_2CH_2C^{\bullet}HR \tag{147}$$

$$CHCl_2CH_2C^{\bullet}HR + Fe^{III}Cl \longrightarrow CHCl_2CH_2CHClR + Fe^{II} \tag{148}$$

$$CCl_3CO_2Me + CH_2{=}CHR \xrightarrow{[Cp_2Fe_2(CO)_4]} \underset{\text{(five-membered lactone: } CH_2\text{—}CH(R)\text{—O—C(=O)—}CCl_2\text{)}}{} \tag{149}$$

IX. ACKNOWLEDGEMENTS

I am grateful to Drs. Alex M. Madonik (U. C. Berkeley) and Nicole Ardoin (U. Bordeaux I) for their helpful assistance in the preparation of the manuscript.

NOTE ADDED IN PROOF

Since this review has been achieved (December 1984), recent papers relevant to the subject deserve special attention; a few of these are quoted in refs. 85, 86, 133g, 141c, 268, 301–303 (until August 1986).

X. REFERENCES

1. (a) M. Berthelot, *C. R. Acad. Sci.*, **112**, 1343 (1891); (b) L. Mond and C. Langler, *J. Chem. Soc.*, **59**, 1090 (1891).
2. H. Reihlen, A. Gruhl, G. von Hessing, and O. Pfrengle, *Justus Leibigs. Ann. Chem.*, **482**, 161 (1930).

3. T. J. Keally and P. J. Pauson, *Nature (London)*, **168**, 1039 (1951), see also refs. 133–137.
4. S. A. Miller, J. A. Tebboth, and J. F. Tremaine, *J. Chem. Soc.* 632 (1952).
5. J. P. Collman and L. S. Hegedus, *Principles and Applications of Organotransition Metal Chemistry*, University Science Books, Mill Valley, CA, 1980.
6. (a) J. J. Ersch and R. B. King (Eds.), *Organometallic Synthesis*, Vol. 1, Academic Press, New York 1965; (b) R. B. King, *Transition Metal Organometallic Chemistry*, Academic Press, New York 1969; (c) E. I. Becker and M. Tsutsui, *Organometallic Reactions*, Vol. 4, Wiley–Interscience, New York, 1972.
7. D. Catheline and D. Astruc, *Organometallics*, **3**, 1094 (1984).
8. A. J. Pearson, in *Comprehensive Organometallic Chemistry* (Eds. G. Wilkinson, F. G. A. Stone, and W. E. Abel), Vol. 8, Pergamon Press, London, 1982, p. 939.
9. (a) C. K. Roper-de Poorter, *Chem. Rev.*, **81**, 447 (1981); (b) E. L. Muetterties and J. Stein, *Chem. Rev.*, **79**, 474 (1979); (c) C. Master, *Adv. Organomet. Chem.*, **17**, 61 (1979); (d) J. Gauthier-Lafaye. R. Perron, and Y. Coleuille, *Actual. Chim.*, Nov., 11 (1983); (e) V. Ponec, *Catal. Rev. Sci. Eng.*, **18**, 151 (1978).
10. R. B. King, in *The Organic Chemistry of Iron* (Eds. E. A. Koerner von Gusthorf, Friedrich-Wilhelm Grevels, and Ingrid Fischler), Vol. 1, Academic Press, New York, 1978, p. 525.
11. R. B. King, Ref. 10, p. 397.
12. D. Astruc, *Tetrahedron*, **39**, 4027 (1983).
13. M. Rosenblum, *Acc. Chem. Res.*, **7**, 122 (1974).
14. A. N. Nesmeyanov, *Pure Appl. Chem.*, **17**, 211 (1968).
15. (a) E. O. Fischer and J. Müller, *Z. Naturforsch., Teil B.* 776 (1962); *Chem. Ber.*, **96**, 3217 (1963); (b) D. M. Braitsch, *J. Chem. Soc., Chem. Commun.* 460 (1974).
16. J. K. Kochi, *Organometallic Mechanisms and Catalysis*, Academic Press, New York, 1978.
17. (a) M. Absi-Halabi, J. D. Atwood, N. P. Forbus, and T. L. Brown, *J. Am. Chem. Soc.*, **102**, 6248 (1980), and references cited therein; (b) G. J. Bezems, P. H. Rieger, and S. Visco, *J. Chem. Soc. Chem. Commun.*, 265 (1981); (c) D. P. Summers, J. C. Luong, and M. S. Wrighton, *J. Am. Chem. Soc.*, **103**, 5238 (1981).
18. J. W. Hershberger, R. J. Klinger, and J. K. Kochi, *J. Am. Chem. Soc.*, **105**, 61 (1983).
19. See Refs. 8, 13 and 16 (part III).
20. Ref. 16, Chapter 2 and 14.
21. A. M. Madonik and D. Astruc, *J. Am. Chem. Soc.*, **106**, 2437 (1984).
22. P. Michaud, D. Astruc, and J. H. Ammeter, *J. Am. Chem. Soc.*, **104**, 3755 (1982).
23. S. G. Davies, M. L. H. Green, and D. M. P. Mingos, *Tetrahedron*, **34**, 3047 (1978).
24. O. Eisenstein and R. Hoffmann, *J. Am. Chem. Soc.*, **103**, 4308 (1981).
25. D. Astruc, P. Michaud, A. M. Madonik, J.-Y. Saillard, and R. Hoffmann, *Nouv. J. Chim.*, **8**, 41 (1985).
26. M. F. Semmelhack, J. L. Garcia, D. Cortes, R. Farina, R. Hong, and B. K. Carpenter, *Organometallics*, **2**, 467 (1983).
27. Ref. 8, p. 984.
28. A. M. Madonik, D. Mandon, P. Michaud, C. Lapinte, and D. Astruc, *J. Am. Chem. Soc.*, **106**, 3381 (1984).
29. (a) C. P. Casey, M. A. Andrews, D. R. Mc Alister and J. E. Ruiz, *J. Am. Chem. Soc.*, **102**, 1927 (1982); (b) W. Tan, G. Ylin, W. K. Wong, W. A. Kiel, V. K. Wong, and J. A. Gladysz, *J. Am. Chem. Soc.*, **104**, 21 (1982); (c) J. R. Sweet and W. A. G. Graham, *J. Am. Chem. Soc.*, **104**, 2811 (1982).
30. (a) T. Bodnar, E. Coman, K. Menard, and A. Cutler, *Inorg. Chem.*, **21**, 127 (1982); (b) C. Lapinte and D. Astruc, *J. Chem. Soc., Chem. Commun.*, 430 (1983); (c) C. Lapinte, D. Catheline and D. Astruc, unpublished work; (d) P. W. Jolly and R. Pettit, *J. Am. Chem. Soc.*, **88**, 5044 (1966).
31. Ref. 5, Chapter 5.
32. R. Pettit, C. Mauldin, J. Cole, and H. Kang, *Ann. N. Y. Acad. Sci.*, **295**, 151 (1977).
33. J. A. Van Doorn, C. Masters, and H. C. Volger, *J. Organomet. Chem.*, **105**, 245 (1976).
34. C. Lapinte, C. Catheline, and D. Astruc, *Organometallics*, **3**, 817 (1984).
35. A. Davison, M. L. H. Green, and G. Wilkinson, *J. Chem. Soc.*, 3172 (1961).
36. (a) N. Grice, S. C. Kao, and R. Pettit, *J. Am. Chem. Soc.*, **101**, 1692 (1979); (b) Ref. 5, Chapter 8.
37. C. Lapinte and D. Astruc, unpublished work.
38. R. J. Angelici, *Acc. Chem. Res.*, **5**, 335 (1972).

39. A. Cutler, *J. Am. Chem. Soc.*, **101**, 604 (1979).
40. D. Catheline, C. Lapinte, and D. Astruc, *C. R. Acad. Sci.*, **301**, 479 (1985).
41. P. Lennon, A. M. Rosan, and M. Rosenblum, *J. Am. Chem. Soc.*, **99**, 8426 (1977).
42. W. P. Giering and M. Rosenblum, *J. Chem. Soc., Chem. Commun.*, 441 (1971).
43. D. Catheline and D. Astruc, *J. Organomet. Chem.*, **266**, C11 (1984).
44. D. Catheline, State Thesis, Rennes, 1985.
45. A. Cutler, D. Ehnholt, P. Lennon, K. Nicholas, D. F. Marten, M. Madhavarao, S. Raghu, A. Rosan, and M. Rosenblum, *J. Am. Chem. Soc.*, **97**, 3149 (1975).
46. (a) W. P. Giering, M. Rosenblum, and J. Tancrede, *J. Am. Chem. Soc.*, **94**, 7170 (1972); (b) M. Rosenblum, M. R. Saidi and M. Madhavarao, *Tetrahedron Lett.*, **46**, 400 (1975).
47. E. O. Fischer and K. Fichtel, Chem. Ber., 3172 (1961); see also Ref. 35.
48. K. M. Nicholas, *J. Am. Chem. Soc.*, **97**, 3254 (1975).
49. A. Cutler, D. Ehntholt, W. P. Giering, P. Lennon, S. Raghu, A. Rosan, M. Rosenblum, J. Trancrede, and D. Wells, *J. Am. Chem. Soc.*, **98**, 3495 (1976); see also Refs. 42 and 45.
50. P. Lennon, M. Madhavarao, A. Rosan, and M. Rosenblum, *J. Organomet. Chem.*, **112**, C17 (1976).
51. D. W. Lichtenberg and A. Wojcicki, *J. Organomet. Chem.*, **94**, 311 (1975).
52. (a) B. W. Roberts and J. Wong, *J. Chem. Soc., Chem. Commun.*, 20 (1977); (b) B. W. Roberts, M. Ross, and J. Wong, *J. Chem. Soc., Chem. Commun.*, 428 (1980).
53. T. H. Whitesides, R. W. Arhart, and R. W. Shaven, *J. Am. Chem. Soc.*, **95**, 5792 (1973).
54. (a) R. A. Plowman and F. G. A. Stone, *Z. Naturforsch., Teil B*, **17**, 575 (1962); (b) H. D. Murdoch and E. Weiss, *Helv. Chim. Acta*, **45**, 1927 (1962); (c) T. H. Whitesides, R. W. Arhart, and R. W. Slaven, *J. Am. Chem. Soc.*, **95**, 5792 (1973).
55. A. J. Pearson, *Aust. J. Chem.*, **29**, 1841 (1976).
56. J. L. Roustan, J. Y. Mérour, and F. Houlihan, *Tetrahedron Lett.*, 3721 (1979).
57. (a) M. F. Semmelhack, J. W. Herndon, and J. P. Springer, *J. Am. Chem. Soc.*, **105**, 2497 (1983); (b) M. F. Semmelhack, *Pure Appl. Chem.*, **53**, 2379 (1981).
58. (a) J. E. Mahler and R. Pettit, *J. Am. Chem. Soc.*, **85**, 3955 and 3959 (1963); (b) T. S. Sorensen and C. R. Jablonski, *J. Organomet. Chem.*, **25**, C62 (1970).
59. P. Powell, *J. Organomet. Chem.*, **165**, C43 (1979).
60. A. J. Birch and I. Jenkins, in *Transition Metal Organometallics in Organic Synthesis* (Ed. H. Alper), Vol. 1, Academic Press, New York, 1976, pp. 1–82.
61. A. J. Pearson, *Transition Met. Chem.*, **6**, 67 (1981).
62. E. O. Fischer and R. D. Fischer, *Angew. Chem.*, **72**, 919 (1960).
63. (a) A. J. Birch, P. W. Westerman, and A. J. Pearson, *Aust. J. Chem.*, **107**, C40 (1976); (b) D. W. Clack, M. Monshi, and L. A. P. Kane-Maguire, *J. Organomet. Chem.*, **107**, C40 (1976).
64. Y. Shvo and E. Hazum, *J. Chem. Soc., Chem. Commun.*, 829 (1975).
65. (a) A. J. Birch, A. S. Narula, P. Dahler, G. R. Stephenson, and L. F. Kelly, *Tetrahedron Lett.*, 979 (1980); (b) R. E. Ireland, G. G. Brown, R. H. Standford, and T. C. McKenzie, *J. Org. Chem.*, **39**, 51 (1974); see also Ref. 63a.
66. A. J. Birch, K. B. Chamberlain, M. A. Haas, and D. J. Thomson, *J. Chem. Soc., Perkin Trans. 1*, 1882 (1973); see also Ref. 65; (b) A. Pelter, K. J. Gould, L. A. P. Kane-Maguire, T. I. Odiaka, and A. Salzer, *J. Chem. Soc., Chem. Commun.*, 1029 (1974); (c) M. Gower, G. R. John, L. A. P. Kane-Maguire, T. I. Odiaka, and A. Salzer, *J. Chem. Soc., Dalton Trans.*, 2003 (1979).
67. (a) A. J. Birch, P. E. Cross, J. Lewis, D. A. White, and S. B. Wild, *J. Chem. Soc.*, 332 (1968); (b) L. F. Kelly, A. S. Narula and A. J. Birch, *Tetrahedron Lett.*, **21**, 871 and 2455 (1980); (c) L. A. P. Kane-Maguire and C. A. Mansfield, *J. Chem. Soc., Chem. Commun.*, 540 (1973); (d) C. A. Mansfield, K. M. Al-Kathumi, and L. A. P. Kane-Maguire, *J. Organomet. Chem.*, **71**, C11 (1974); (e) A. J. Birch and A. J. Pearson, *J. Chem. Soc., Perkin Trans. 1*, 954 (1976); *Tetrahedron Lett.*, 2379 (1975); (f) A. J. Pearson, *Aust. J. Chem.*, **29**, 1101 (1976); **30**, 345 (1977); (g) A. J. Birch, D. Jenkins, and A. J. Liepa, *Tetrahedron Lett.*, 1723 (1975).
68. (a) A. J. Birch, A. J. Liepa, and G. R. Stephenson, *Tetrahedron Lett.*, 3665 (1979); (b) G. R. John and L. A. P. Kane-Maguire, *J. Chem. Soc. Dalton Trans.*, 1196 (1979).
69. Ref. 8, p. 993.
70. A. J. Pearson, *J. Chem. Soc., Chem. Commun.*, 339 (1977); *Perkin Trans. 1*, 2069 (1977).
71. A. J. Birch, *Q. Rev. Chem. Soc.*, **4**, 69 (1950); see also Ref. 60.
72. A. J. Pearson, *J. Chem. Soc., Perkin Trans. 1*, 1255 (1979).
73. E. W. Colvin, S. Malchenko, R. A. Raphael, and J. S. Roberts, *J. Chem. Soc, Perkin Trans. 1*, 1989 (1973).

74. A. J. Pearson, *Acc. Chem. Res.*, **13**, 463 (1980); see also Refs. 8 and 61.
75. (a) A. J. Pearson, *J. Chem. Soc., Perkin Trans. 1*, 400 (1980); (b) A. J. Pearson, P. Ham and D. C. Rees, *Tetrahedron Lett.*, **21**, 4637 (1980); *J. Chem. Soc., Perkin Trans. 1*, 489 (1982).
76. A. J. Pearson and D. C. Rees, *Tetrahedron Lett.*, **21**, 3937 (1980); *J. Am. Chem. Soc.*, **104**, 1118 (1982).
77. A. J. Pearson, *J. Chem. Soc., Perkin Trans. 1*, 495 (1978); see also Ref. 70.
78. (a) R. Auman, *J. Organomet. Chem.*, **47**, C28 (1973); (b) R. Edwards, J. A. S. Howell, B. F. G. Johnson, and J. Lewis, *J. Chem. Soc., Dalton Trans.*, 2105 (1974); (c) A. J. Pearson, S. L. Kole, and B. Chen, *J. Am. Chem. Soc.*, **105**, 4483 (1983).
79. (a) I. U. Khand, P. L. Pauson and W. E. Watts, *J. Chem. Soc. C*, 2024 (1969); (b) A. N. Nesmeyanov and N. A. Vol'kenau, *Izv. Akad. Nauk SSSR, Ser. Khim.*, 1151 (1975).
80. A. N. Nesmeyanov, N. A. Vol'kenau, I. N. Bolesova, and L. S. Polovnikova, *Dokl. Akad. Nauk SSSR*, **236**, 1130 (1977).
81. (a) M. L. H. Green, L. Pratt and G. Wilkinson, *J. Chem. Soc.*, 989 (1960). (b) J. F. McGreer and W. E. Watts, *J. Organomet. Chem.*, **110**, 103 (1976). (c) M. Khan and W. E. Watts, *J. Organomet. Chem.*, **108**, C11 (1976); see also Ref. 96.
82. M. F. Semmelhack, *Ann. N. Y. Acad. Sci.*, **295**, 36 (1977).
83. J. F. Helling and D. M. Braitsch, *J. Am. Chem. Soc.*, **92**, 7207 and 720 (1970).
84. A. N. Nesmeyanov, N. A. Vol'kenau, I. N. Bolesova, and L. S. Polkovnikova, *Dokl. Akad. Nauk SSSR*, **236**, 1130 (1977).
85. D. Mandon, L. Toupet and D. Astruc, *J. Am. Chem. Soc.*, **108**, 1321 (1986).
86. D. Mandon and D. Astruc, *J. Organomet. Chem.*, 1986, 000.
87. (a) J. W. Johnson and P. M. Treichel, *J. Am. Chem. Soc.*, **99**, 1427 (1977); *J. Chem. Soc., Chem. Commun.*, 688 (1972); (b) J. F. Helling and W. A. Hendrickson, *J. Organomet. Chem.*, **99**, 1411 (1977); **87**, 168 (1979); (c) C. C. Lee, B. R. Steele, K. J. Demchuk, and R. G. Sutherland, *Can. J. Chem.*, **57**, 946 (1979); *J. Organomet. Chem.*, **57**, 946 (1979); (d) see also Refs. 92, 108, and 109.
88. A. N. Nesmeyanov, N. A. Vol'kenau and I. N. Bolesova, *Dokl, Akad. Nauk SSSR*, **175**, 606 (1967).
89. C. C. Lee, U. S. Gill, M. Iqbal, C. I. Azogu, and R. G. Sutherland, *J. Organomet. Chem.*, **231**, 151 (1982).
90. A. C. Knipe, S. J. McGuinness, and W. E. Watts, *J. Chem. Soc., Chem. Commun.*, 842 (1979).
91. R. G. Sutherland, A. Piorko, U. S. Gill and C. C. Lee, *J. Heterocycl. Chem.*, **19**, 801 (1982). See also C. C. Lee, U. S. Gill, R. G. Sutherland, *J. Organomet. Chem.*, **267**, 157 (1984).
92. J.-R. Hamon, D. Astruc, E. Román, P. Batail, and J. J. Mayerle, *J. Am. Chem. Soc.*, **103**, 2431 (1981).
93. B. Nicholls and M. C. Whiting, *J. Chem. Soc.*, 551 (1959).
94. C. C. Lee, A. Piórko, B. R. Steele, U. S. Gill, R. G. Sutherland, *J. Organomet. Chem.*, **256**, 303 (1983).
95. A. N. Nesmeyanov, N. A. Vol'kenau, E. I. Sirotkina, and V. V. Deryabin, *Dokl. Akad. Nauk. SSSR, Ser. Khim.*, 1605 (1968).
96. E. Román, D. Astruc and A. Darchen, *J. Chem. Soc., Chem. Commun.*, 512 (1976).
97. A. N. Nesmeyanov, *Pure Appl. Chem.*, **17**, 211 (1968); see also Ref. 88.
98. C. Moinet and E. Raoult, *J. Organomet. Chem.*, **229**, C13 (1982) and **310**, 225 (1986).
99. E. Román, D. Astruc, and A. Darchen, *J. Organomet. Chem.*, **219**, 221 (1981).
100. C. C. Lee, K. J. Demchuk, U. S. Gill, and R. G. Sutherland, *J. Organomet. Chem.*, **247**, 71 (1983).
101. J. F. McGreer and W. E. Watts, *J. Organomet. Chem.*, **110**, 103 (1976).
102. A. N. Nesmeyanov, N. A. Vol'kenau, L. S. Shilovtseva, and V. A. Petrakova, *J. Organomet. Chem.*, **62**, 329 (1973).
103. P. Michaud, State Thesis, Rennes, 1983; see also Ref. 21.
104. (a) A. Davison and J. P. Selegue, *J. Am. Chem. Soc.*, **102**, 2455 (1980); (b) S. B. Samuels, S. R. Berryhill and M. Rosenblum, *J. Organomet. Chem.*, **166**, C9 (1979); (c) D. F. Marten, *J. Chem. Soc., Chem. Commun.*, 341 (1980); (d) A. Davison and J. P. Selegue, *J. Am. Chem. Soc.*, **100**, 7763 (1978).
105. H. L. Lentzner and W. E. Watts, *J. Chem. Soc., Chem. Commun.*, 26 (1970).
106. G. Marr and B. W. Rockett, *J. Organomet. Chem.*, **189**, 163 (1980) and earlier annual reviews.
107. A. J. Birch and A. J. Pearson, *J. Chem. Soc., Chem. Commun.*, 601 (1976); A. J. Pearson, *Aust. J. Chem.*, **29**, 1841 (1976).
108. D. Astruc, E. Román, J.-R. Hamon and P. Batail, *J. Am. Chem. Soc.*, **101**, 2240 (1979).

109. D. Astruc, J.-R. Hamon, E. Román, and P. Michaud, *J. Am. Chem. Soc.*, **103**, 7502 (1981).
110. (a) A. N. Nesmeyanov, N. A. Vol'kenau, and L. S.. Shilovtseva, *Izv. Akad. Nauk SSSR, Ser. Khim.*, 726 (1969); (b) Ref. 270, footnote 57; (c) T. P. Gill and K. R. Mann, *Inorg.*, **19**, 3008 (1980); *J. Organomet. Chem.*, **216**, 65 (1981); (d) R. T. Swann and V. Boekelheide, *J. Organomet. Chem.*, **231**, 143 (1982); (e) E. D. Laganis, R. G. Finke, and V. Boekelheide, *Proc. Natl. Acad. Sci. U.S.A.*, **78**, 2657 (1981); (f) D. Catheline and D. Astruc, *J. Organomet. Chem.*, **248**, C9 (1983); **272**, 417 (1984).
111. J.-R. Hamon, J.-Y. Saillard, A. Le Beuze, M. McGlinchey, and D. Astruc, *J. Am. Chem. Soc.*, **104**, 7549 (1982).
112. J. R. Hamon, State Thesis, Rennes, 1982.
113. A. Rosan, M. Rosenblum, and J. Tancrede, *J. Am. Chem. Soc.*, **95**, 3062 (1973).
114. (a) M. L. H. Green and P. L. I. Nagy, *J. Chem. Soc.*, 189 (1963); (b) J. Y. Mérour and P. Cadiot, *C. R. Acad. Sci. Ser. C*, **271**, 83 (1970); (c) A. Cutler, D. Ehnthold, P. Lennon, K. Nicholas, D. F. Marten, M. Madhavarao, S. Raghu, A. Rosan, and M. Rosenblum, *J. Am. Chem. Soc.*, **97**, 3149 (1975); (d) L. J. Dizikes and A. Wojcicki, *J. Organomet. Chem.*, **137**, 79 (1977).
115. (a) N. Genco, D. Marten, S. Raghu, and M. Rosenblum, *J. Am. Chem. Soc.*, **98**, 848 (1976); (b) T. S. Abram, R. Baker, and C. M. Exon, *Tetrahedron Lett.*, 4103 (1979); (c) A. Cutler, S. Raghu, and M. Rosenblum, *J. Organomet. Chem.*, **77**, 381 (1974); (d) P. W. Jolly and R. Pettit, *J. Organomet. Chem.*, **12**, 491 (1968).
116. (a) J. K. P. Ariyatne and M. L. H. Green, *J. Chem. Soc.*, 2976 (1963); (b) R. Lazzaroni and B. E. Mann, *J. Organomet. Chem.*, **164**, 79 (1979).
117. J. K. P. Ariyatne and M. L. H. Green, *J. Chem. Soc.*, 1 (1964).
118. D. Catheline, E. Román and D. Astruc, *Inorg. Chem.*, **23**, 4508 (1984).
119. S. Raghu and M. Rosenblum, *J. Am. Chem. Soc.*, **93**, 3060 (1973).
120. J. P. Williams and A. Wojcicki, *Inorg. Chem.*, **16**, 2506 (1977).
121. (a) R. B. King, in *The Organic Chemistry of Iron* (Eds E. A. Koerner Von Gustorf, F. W. Grevels, and I. Fischler), Academic Press, New York, 1978, p. 525; (b) K. N. Anisinov, G. K. Magomedov, N. E. Kolobova, and A. G. Trufanov, *Izv. Akad. Nauk SSSR, Ser Khim.*, 2533 (1970).
122. (a) E. D. Greaves, G. R. Knox, and P. L. Pauson, *J. Chem. Soc., Chem. Commun.*, 1124 (1969); (b) A. D. U. Hardy and G. A. Sim, *J. Chem. Soc., Dalton Trans.*, 2305 (1972).
123. (a) E. O. Greanes, G. R. Knox, P. L. Pauson, S. Torna, G. A. Sim, and D. I. Woodhouse, *J. Chem. Soc., Chem. Commun.*, 257 (1974); (b) M. R. Churchill, J. Wormald, D. A. T. Young, and H. D. Kaesz, *J. Am. Chem. Soc.*, **91**, 7201 (1969); (c) M. R. Churchill and J. Wormald, *Inorg. Chem.*, **9**, 2430 (1970).
124. (a) R. Pettit, G. F. Emerson, and J. Mahler, *J. Chem. Educ.*, **40**, 175 (1963); (b) R. Pettit and G. F. Emerson, *Adv. Organomet. Chem.*, **1**, 1 (1964); (c) R. Pettit, *Ann. N.Y. Acad. Sci.*, **125**, 89 (1965).
125. R. E. Graf and C. P. Lillya, *J. Organomet. Chem.*, **122**, 377 (1976).
126. (a) F. J. Impastato and K. G. Ihrman, *J. Am. Chem. Soc.*, **83**, 3726 (1961); (b) T. H. Whitesides and R. W. Arhart, *J. Am. Chem. Soc.*, **93**, 5296 (1971); (c) G. F. Emerson and R. Pettit, *J. Am. Chem. Soc.*, **84**, 4591 (1962).
127. I. R. Knox and I. G. Thom, *J. Chem. Soc., Chem. Commun.*, 373 (1981).
128. K. Mori, *Tetrahedron*, **30**, 3807 (1974).
129. J. Clelland and G. R. Knox, *J. Chem. Soc., Chem. Commun.*, 1219 (1983).
130. (a) N. S. Nametkin, A. I. Nekhaev, V. D. Tyurin, and S. P. Gubin, *Izv. Akad. Nauk SSSR, Ser. Khim.*, 676 (1975); (b) B. F. G. Johnson, J. Lewis, and D. G. Parker, *J. Organomet. Chem.*, **141**, 319 (1977).
131. A. J. Birch, A. J. Pearson, and W. D. Raverty, unpublished work (Ref. 61, footnote 115).
132. G. Wilkinson, *J. Organomet. Chem.*, **100**, 273 (1975).
133. (a) E. I. Becker and M. Tsutsui, *Organometallic Reactions*, Vol. 4, Wiley–Interscience, New York, 1972, pp. 163–460; (b) D. Stc. Black, W. R. Jackson, and J. M. Swan, in *Comprehensive Organic Chemistry* (Eds. D. Barton and W. O. Ollis), Vol. 3, Pergamon Press, New York, 1979, Chapter 156, pp. 1127–1323; (c) G. R. Knox and W. E. Watts, in *International Review of Science, Inorganic Chemistry, Series 2, Vol. 6, Transition Metals* (Ed. M. J. Mays), Butterworth, London, 1976, p. 219; (d) W. L. Jolly, *Inorg. Synth.*, **11**, 120 (1968); (e) K. L. Rinehart, *Organic Reactions*, Vol. 17 Wiley, New York, 1969; (f) *Gmelin Handbuch der Anorganische Chemie*, Vol. 49, Part A; (g) Annual Reviews of J. Organomet. Chem., see ref. 106.

134. W. F. Little, in *Survey of Progress in Chemistry* (Ed. A. Scott), Vol. 1, Academic Press, London, 1963, p. 133.
135. W. Walozak, *J. Am. Chem. Soc.*, **100**, 6382 (1978).
136. E. G. Perevalova and T. V. Nikitina, *Organomet. React.*, **4**, 163 (1972).
137. M. Rosenblum, *Chemistry of the Iron Group Metallocenes, Part I*, Wiley, New York, 1965.
138. G. de Lanzon, F. Mathey, and M. Simalty, *J. Organomet. Chem.*, **156**, C33 (1978).
139. M. J. Nugent, R. Kummer, and J. H. Richards, *J. Am. Chem. Soc.*, **91**, 6141 (1969).
140. S. Allemark, *Tetrahedron Lett.*, 371 (1974).
141. (a) A. N. Nesmeyanov, N. A. Vol'kenau and I. N. Bolesova, *Dokl. Akad. Nauk SSSR*, **149**, 615 (1963); **166**, 607 (1966); *Tetrahedron Lett.*, 1725 (1963); (b) A. N. Nesmeyanov, *Adv. Organomet. Chem.*, **10**, 1 (1972); (c) R. G. Sutherland, *J. Organomet. Chem. Library*, **3**, 311 (1977); and R. G. Sutherland, M. Iqbal and A. Piorkó, *J. Organomet. Chem.*, **302**, 307 (1986); (d) D. Astruc and R. Dabard, *Bull. Soc. Chim. Fr.*, 2571 (1975); *J. Organomet. Chem.*, **111**, 339 (1976); *Tetrahedron*, **32**, 245 (1976).
142. J. P. Collman, *Acc. Chem. Res.*, **8**, 342 (1975).
143. M. P. Cooke, *J. Am. Chem. Soc.*, **92**, 6080 (1970).
144. J. P. Collman and L. S. Hegedus, *Principles and Applications of Organotransition Metal Chemistry*, University Science Books, Mill Valley, CA, 1980, Chapter 9.
145. H. O. House, *Modern Synthetic Reactions*, 2nd ed., Benjamin, New York, 1972, p. 174.
146. R. F. Heck, in *Organic Synthesis via Metal Carbonyls* (Eds. I. Wender and P. Pino, Vol. 1, Wiley-Interscience, New York, 1968.
147. J. Y. Mérour, J. L. Roustan, C. Charrier, J. Collins, and J. Benaim, *J. Organomet. Chem.*, **51**, C24 (1973).
148. J. E. McMurry, A. Andrews, G. M. Ksander, J. H. Musser, and M. A. Johnson, *J. Am. Chem. Soc.*, **101**, 1330 (1979).
149. Ref. 142, footnote 15 (pp. 343–344).
150. A. Guinot, P. Cadiot, and J. L. Roustan, *J. Organomet. Chem.*, **128**, C35 (1977).
151. J. P. Collman, R. G. Finke, J. N. Cawse, and J. I. Brauman, *J. Am. Chem. Soc.*, **100**, 4766 (1978).
152. J. P. Collman, R. G. Finke, J. N. Cawse, and J. I. Brauman, *J. Am. Chem. Soc.*, **99**, 2515 (1977).
153. (a) R. L. Privett, *Adv. Organomet. Chem.*, **17**, 1 (1979); (b) L. Markó, in *Aspects of Homogeneous Catalysis* (Ed. R. Ugo), Vol. II, Reidel, Dordrecht, 1974.
154. W. Reppe and H. Vetter, *Justus Liebigs Ann. Chem.*, **582**, 133 (1953).
155. R. Pettit, C. Mauldin, T. Cole, and H. Kang, *Ann. N. Y. Acad. Sci.*, **295**, 151 (1977).
156. Ref. 155, p. 157.
157. L. Markó, M. A. Radhi, and I. Otvös, *J. Organomet. Chem.*, **218**, 369 (1981).
158. M. A. Radhi, G. Palyi, and L. Markó, *J. Mol. Cat.*, **22**, 195 (1983).
159. H. Masada, M. Mizuno, S. Suga, Y. Watanabe, and Y. Takegami, *Bull. Chem. Soc. Jpn.*, **43**, 3824 (1970).
160. (a) R. Noyori, I. Umeda, and T. Ishigami, *J. Org. Chem.*, 37 (1977); (b) G. P. Boldrini and A. Umani-Ronchi, *Synthesis*. 596 (1976)
161. Y. Takegami, Y. Watanabe, H. Masada, and I. Kanaya, *Bull. Chem. Soc. Jpn.*, **40**, 1456 (1967).
162. Y. Watanabe, T. Mitsudo, M. Yamashita, and Y. Takegami, *Bull. Chem. Soc. Jpn.*, **48**, 1478 (1975).
163. (a) J. M. Landesberg, L. Katz, and C. Olsen, *J. Org. Chem.*, **37**, 930 (1972); (b) A. Hajos, *Complex Hydrides*, Elsevier, Amsterdam, New York, 1979, Chapter 9, pp. 249–281.
164. H. Alper, *Tetrahedron Lett.*, 2257 (1975).
165. C. E. Coffey, *J. Am. Chem. Soc.*, **83**, 1623 (1961).
166. Y. Watanabe, T. Mitsudo, M. Tanaka, K. Yamamato, T. Okajima, and Y. Takegami, *Bull. Chem. Soc. Jpn.*, **44**, 2569 (1971).
167. P. Michaud, C. Lapinte, and D. Astruc, *Ann. N. Y. Acad. Sci.*, **415**, 97 (1983).
168. S. G. Davies, *Organotransition Metal Chemistry: Application to Organic Synthesis*, Pergamon Press, Oxford, 1982, Chapters 5, 6, 8 and 9.
169. (a) M. Cousins and M. L. H. Green, *J. Chem. Soc.*, 889 (1963); (b) M. L. H. Green and A. N. Stear, *J. Organomet. Chem.*, **1**, 230 (1964); (c) W. P. Giering, M. Rosenblum, and J. Tancrede, *J. Am. Chem. Soc.*, **94**, 7170 (1972); (d) M. Rosenblum, M. R. Saidi, and M. Madhanarao, *Tetrahedron Lett.*, 4009 (1975).

170. (a) M. L. H. Green and C. R. Hurley, *J. Organomet. Chem.*, **10**, 188 (1967); (b) S. R. Su and A. Wojacki, *J. Organomet. Chem.*, **27**, 231 (1971).
171. K. M. Nicholas and M. Rosenblum, *J. Am. Chem. Soc.*, **95**, 4449 (1973).
172. (a) P. L. Bock, D. J. Boschetto, J. R. Rasmussen, J. P. Demers, and G. M. Whitesides, *J. Am. Chem. Soc.*, **96**, 2814 (1974); (b) P. L. Bock and G. M. Whitesides, *J. Am. Chem. Soc.*, **96**, 2826 (1974); (c) D. A. Slack and M. C. Baird, *J. Am. Chem. Soc.*, **98**, 5539 (1976).
173. G. M. Whitesides and D. J. Boschetto, *J. Am. Chem. Soc.*, **93**, 1529 (1971).
174. (a) T. C. Flood and D. L. Miles, *J. Organomet. Chem.*, **127**, 33 (1977); (b) T. G. Attig, R. G. Teller, S.-M. Wu, R. Baux, and A. Wojcik, *J. Am. Chem. Soc.*, **101**, 619 (1979); (c) W. N. Rogers and M. C. Baird, *J. Organomet. Chem.*, **182**, C65 (1979); (d) D. Dodd and M. D. Johnson, *J. Chem. Soc. B*, 662 (1971); (e) D. Dong, D. A. Slack, and M. C. Baird, *Inorg. Chem.*, **18**, 188 (1979).
175. M. D. Johnson, *Mononuclear Iron Compounds with* η^1-Hydrocarbon Ligands, in *Comprehensive Organometallic Chemistry*, (Eds. G. Wilkinson, W. E. Abel, and F. G. A. Stone), Pergamon Press, Oxford, 1982, Chapter 31.2.
176. K. M. Nicholas, *J. Am. Chem. Soc.*, **97**, 3254 (1975).
177. D. V. Banthorpe, H. Fitton, and J. Lewis, *J. Chem. Soc., Perkin Trans. 1*, 2051 (1973).
178. (a) G. Evans, B. F. G. Johnson, and J. Lewis, *J. Organomet. Chem.*, **102**, 507 (1975); (b) D. H. R. Barton, A. A. L. Guntilaka, T. Nakanishi, H. Patin, D. A. Widdowson, and B. R. Worth, *J. Chem. Soc., Perkin Trans. 1*, 821 (1976).
179. A. J. Birch and H. Fitton, *Aust. J. Chem.*, **22**, 971 (1979).
180. A. J. Birch and I. D. Jenkins, in *Transition Metal Organometallics in Organic Synthesis* (Ed. H. Alper), Vol. I, Academic Press, New York, 1976, p. 1.
181. B. F. G. Johnson, J. Lewis, P. McArdle and G. L. P. Randall, *J. Chem. Soc., Dalton Trans.*, 456 (1972).
182. (a) A. Davison, W. McFarlane, and G. Wilkinson, *J. Chem. Soc.*, 4821 (1962); (b) B. F. G. Johnson, J. Lewis, and G. L. P. Randall, *J. Chem. Soc. A*, 422 (1971).
183. M. Franck-Neumann, F. Brion, and D. Martina, *Tetrahedron Lett.*, 5033 (1978).
184. G. B. Gill, N. Gourlay, A. W. Johnson, and M. Mahendron, *J. Chem. Soc., Chem. Commun.*, 631 (1969).
185. J. S. Ward and R. Pettit, *J. Chem. Soc., Chem. Commun.*, 59 (1969).
186. B. M. Trost and D. M. T. Chan, *J. Am. Chem. Soc.*, **101**, 6429 (1979).
187. A. J. Birch and I. D. Jenkins, *Tetrahedron Lett.*, 119 (1975).
188. (a) R. R. Schrock, *Acc. Chem. Res.*, **12**, 98 (1979); (b) M. Brookhart, J. R. Tucker, and G. R. Husk, *J. Am. Chem. Soc.*, **105**, 258 (1983), and references cited therein.
189. R. R. Schrock, S. Rocklage, J. Wengrovius, G. Rupprecht, and J. Fellmann, *J. Mol. Catal.*, **8**, 73 (1980).
190. J. P. Soufflet, D. Commereuc, and Y. Chauvin, *C. R. Acad. Sci., Ser. C*, **276**, 169 (1973).
191. (a) G. N. Schrauzer, *Adv. Catal.*, **18**, 373 (1968); (b) P. Heinback, *Angew. Chem., Int. Ed. Engl.*, **22**, 975 (1973); (c) G. Lefebvre and Y. Chauvin, *Aspects Homogeneous Catal.*, **1**, 108 (1970).
192. P. W. Jolly and R. Pettit, *J. Am. Chem. Soc.*, **88**, 5044 (1966).
193. M. Brookhart, J. R. Tucker, T. C. Flood, and J. Jensen, *J. Am. Chem. Soc.*, **102**, 1203 (1980).
194. R. Hoffmann, G. N. Wilker and O. Eisenstein, *J. Am. Chem. Soc.*, **104**, 632 (1982).
195. (a) C. P. Casey in *Transition Metals Organometallics in Organic Synthesis* (Ed. H. Alper), Vol. I, Academic Press, New York, 1976, p. 190; (b) D. J. Cardin, B. Cetinkaya, and M. F. Lappert, *Chem. Ber.*, **72**, 575 (1972); (c) F. A. Cotton and C. M. Lukehart, *Prog. Inorg. Chem.*, **16**, 243 (1972); (d) D. J. Cardin, B. Cetinkaya, M. J. Doyle, and M. F. Lappert, *Chem. Soc. Rev.*, **2**, 99 (1973).
196. A. R. Cutler, *J. Am. Chem. Soc.*, **101**, 604 (1979).
197. (a) M. Brookhart, M. B. Humphrey, H. J. Kratzer, and G. O. Nelson, *J. Am. Chem. Soc.*, **102**, 7802 (1980); (b) T. Bodnar and A. R. Cutler, *J. Organomet. Chem.*, **213**, C31 (1981).
198. A. Davison, W. C. Krusell, and R. Michaelson, *J. Organomet. Chem.*, **72**, C7 (1974).
199. S. Brandt and P. Helquist, *J. Am. Chem. Soc.*, **101**, 6473 (1979).
200. (a) M. Brookhart, D. Timmer, J. R. Tucker, G. D. Williams, G. R. Husk, H. Brunner, and B. Hammer, *J. Am. Chem. Soc.*, **105**, 6721 (1983); (b) M. Brookhart, J. R. Tucker, and G. R. Husk, Ibid., **103**, 979 (1981).
201. C. P. Casey, S. W. Polichnowski, and A. J. Shusterman, *J. Am. Chem. Soc.*, **101**, 7282 (1979).
202. E. O. Fischer, *Adv. Organomet. Chem.*, **14**, 1 (1976); *Rev. Pure. Appl. Chem.*, **30**, 353 (1972); *Angew. Chem.*, **86**, 651 (1974).

203. M. F. Semmelhack and R. Tamura, *J. Am. Chem. Soc.*, **105**, 4099 (1983).
204. (a) R. H. Grubbs and A. Msiyashita, *J. Am. Chem. Soc.*, **100**, 7416 (1978); (b) S. J. McLain, J. Sancho and R. R. Schrock, *J. Am. Chem. Soc.*, **102**, 5610 (1980); (c) J. P. Collman and L. S. Hegedus, *Principles and Applications of Organotransition Metal Chemistry*, University Science Books, Mill Valley, CA, 1980, Chapter 10, pp. 506–535.
205. Y. Chauvin and J. L. Herrisson, *Makomol. Chem.*, **141**, 161 (1971).
206. H. Le Bozec, A. Gorgues, and P. H. Dixneuf, *J. Am. Chem. Soc.*, **100**, 3946 (1978); *Inorg. Chem.*, **20**, 2486 (1981).
207. H. C. Longuet-Higgins and L. E. Orgel, *J. Chem. Soc.*, 199 (1956).
208. G. F. Emerson, L. Watts, and R. Pettit, *J. Am. Chem. Soc.*, **92**, 131 (1965).
209. (a) L. Watts, J. D. Fitzpatrick, and R. Pettit, *J. Am. Chem. Soc.*, **87**, 3253 (1965); (b) R. Pettit, J. C. Barborak, and L. Watts, *J. Am. Chem. Soc.*, **88**, 1328 (1966).
210. (a) J. S. Ward and R. Pettit, *J. Chem. Soc., Chem. Commun.*, 419 (1970); (b) A. Sanders and W. P. Giering, *J. Am. Chem. Soc.*, **97**, 919 (1975).
211. (a) K. H. Grellman, J. Palmowski, and G. Quinkert, *Angew. Chem.*, **83**, 209 (1971); *Angew. Chem., Int. Ed. Engl.*, **10**, 196 (1971); (b) K. Alder and M. Fremery, *Tetrahedron*, **14**, 190 (1961).
212. (a) J. M. Landesberg, in *The Organic Chemistry of Iron* (Eds. E. A. Koerner Von Gustorf, F.-W. Grevels and I. Fischler), Vol. I. Academic Press, New York, 1978, p. 627; (b) R. L. Funk and K. P. C. Vollhardt, *Chem. Soc. Rev.*, **9**, 41 (1980); (c) W. R. Roth and J. D. Meter, *Tetrahedron Lett.*, 2053 (1967).
213. B. F. G. Johnson, J. Lewis, and D. J. Thompson, *Tetrahedron Lett.*, 3789 (1974).
214. C. F. Coffey, *J. Am. Chem. Soc.*, **83**, 1623 (1961).
215. I. Rhee, M. Ryang, and S. Tsutsumi, *J. Organomet. Chem.*, **9**, 361 (1967).
216. H. Alper and E. C. H. Keung, *J. Org. Chem.*, **37**, 2566 (1972).
217. (a) R. Noyori, S. Makino, and H. Takaya, *J. Am. Chem. Soc.*, **93**, 1272 (1971); (b) R. Noyori, Y. Baba, S. Makino, and H. Takaya, *Tetrahedron Lett.*, 1741 (1973); (c) R. Noyori, *Acc. Chem. Res.*, **12**, 61 (1979); in *Transition Metal Organometallics in Organic Synthesis* (Ed. H. Alper, Vol. I, Academic Press, New York, 1976, Chapter 2, p. 83.
218. (a) R. Noyori, T. Sato, and Y. Hayakawa, *J. Org. Chem.*, **40**, 2681 (1975); (b) T. Sato, R. Ito, Y. Hayakawa, and R. Noyori, *Tetrahedron Lett.*, 993 (1978); (c) Y. Hayakawa, F. Shimiza, and R. Noyori, *Tetrahedron Lett.*, 993 (1978); (d) W. Parker, R. Ramage, and R. A. Raphael, *J. Chem. Soc.*, 1558 (1962).
219. R. Noyori, F. Shimizu, K. Fukute, H. Tayaka, and Y. Hayakawa, *J. Am. Chem. Soc.*, **99**, 5196 (1977).
220. Y. Hayakawa, K. Yokoyama, and R. Noyori, *J. Am. Chem. Soc.*, **100**, 1799 (1978).
221. (a) Y. Hayakawa, Y. Baba, S. Makino, and R. Noyori, *J. Am. Chem. Soc.*, **100**, 1786 (1978); (b) R. Noyori, Y. Baba, and Y. Hayakawa, *J. Am. Chem. Soc.*, **96**, 3336 (1974); (c) R. Noyori, S. Makino, and Y. Hayakawa, *Tetrahedron Lett.*, 1049 (1974).
222. R. Noyori, S. Makino, and H. Takaya, *J. Am. Chem. Soc.*, **93**, 1272 (1971); (b) H. Takaya, S. Makino, Y. Hayakawa, and R. Noyori, *J. Am. Chem. Soc.*, **100**, 1765 (1978).
223. (a) R. Noyori, S. Makino, and H. Takaya, *Tetrahedron Lett.*, 1745 (1973); (b) R. Noyori, S. Makino, T. Okita, and Y. Hayakawa, *J. Org. Chem.*, **40**, 806 (1975); (c) R. Noyori, T. Sato, and Y. Hayakawa, *J. Am. Chem. Soc.*, **100**, 2561 (1978).
224. R. Noyori, Y. Hayakawa, S. Makino, N. Hayakawa, and H. Takaya, *J. Am. Chem. Soc.*, **95**, 4103 (1973).
225. (a) R. F. Heck and C. R. Boss, *J. Am. Chem. Soc.*, **86**, 2580 (1964); (b) H. D. Murdoch, *Helv. Chim. Acta.*, **47**, 936 (1964); (c) R. Aumann, K. Fröhlich, and H. Ring, *Angew. Chem., Int. Ed. Engl.*, **13**, 275 (1974); (d) K.-N. Chen, R. M. Moriarty, B. G. de Boer, M. R. Churchill, and H. J. C. Yeh, *J. Am. Chem. Soc.*, **97**, 5602 (1975).
226. G. D. Annis, E. M. Hebblethwaite, and S. V. Ley, *J. Chem. Soc., Chem. Commun.*, 297 (1980).
227. G. F. Emerson, J. E. Mahler, R. Kochbar, and R. Pettit, *J. Org. Chem.*, **29**, 3620 (1964).
228. R. Cramer, *Acc. Chem. Res.*, **1**, 186 (1968).
229. (a) J. E. Arnet and R. Pettit, *J. Am. Chem. Soc.*, **83**, 2954 (1961); (b) R. Pettit, G. F. Emerson, and J. Mahler, *J. Chem. Educ.*, **40**, 175 (1963).
230. (a) A. J. Birch, P. E. Cross, J. Lewis, D. A. White, and S. B. Wild, *J. Chem. Soc. A*, 332 (1968); (b) A. J. Birch and A. J. Pearson, *J. Chem. Soc., Perkin Trans. 1*, 638 (1978).
231. A. J. Birch and D. H. Williamson, *J. Chem. Soc., Perkin Trans. 1*, 1892 (1973).
232. E. J. Corey and G. Moinet, *J. Am. Chem. Soc.*, **95**, 7185 (1973).

233. G. F. Emerson and R. Pettit, *J. Am. Chem. Soc.*, **84**, 4591 (1962).
234. A. J. Hubert, P. Moniotte, G. Goebbells, R. Warin, and P. Teyssié, *J. Chem. Soc., Perkin Trans. 2*, 1954 (1973).
235. (a) S. Sarel, R. Ben-Shoshan and B. Kirson, *J. Am. Chem. Soc.*, **87**, 2517 (1965); (b) R. Victor, R. Ben-Shoshan, and S. Sarel, *Tetrahedron Lett.*, 4211 (1973).
236. (a) W. P. Giering, M. Rosenblum, and J. Tancrede, *J. Am. Chem. Soc.*, **94**, 7170 (1972); (b) M. Rosenblum, M. R. Saidi, and M. Madhavarao, *Tetrahedron Lett.*, 4009, (1975).
237. H. Maltz and G. Deganello, *J. Organomet. Chem.*, **27**, 383 (1971).
238. R. Aumann and H. Averbeck, *J. Organomet. Chem.*, **85**, C4 (1975).
239. K. Hayakawa and H. Schmidt, *Helv. Chim. Acta*, **60**, 1942 (1977).
240. F. Bellamy, J.-L. Schuppiser, and J. Streith, *Heterocycles*, **11**, 461 (1978).
241. H. Alper and T. Sakakibara, *Can. J. Chem.*, **57**, 1541 (1979).
242. H. Alper and D. Des Roches, *Tetrahedron Lett.*, 4155 (1977).
243. H. Alper and E. C. H. Keung, *Tetrahedron Lett.*, 53 (1970).
244. Y. Shvo and E. Hazum, *J. Chem. Soc., Dalton Trans.*, 336 (1974).
245. H. Alper, *J. Organomet. Chem.*, **96**, 95 (1975).
246. H. Alper and J. T. Edward, *Can. J. Chem.*, **48**, 1543 (1970).
247. J. M. Landesberg, L. Katz, and C. Olsen, *J. Org. Chem.*, **37**, 930 (1972).
248. A. Dondoni and G. Barbaro, *J. Chem. Soc., Chem. Commun.*, 761 (1975).
249. H. Alper and J. T. Edward, *J. Organomet. Chem.*, **16**, 342 (1969).
250. H. Alper and M. Salisova, *Tetrahedron Lett.*, 801 (1980).
251. (a) R. B. King, *Inorg. Chem.*, **2**, 326 (1963); (b) Y. Watanabe, M. Yamashita, T. Mitsudo, M. Tanaka, and Y. Takegami, *Inorg. Chem.*, **12**, 3535 (1973); (c) B. M. Trost and S. D. Ziman, *J. Org. Chem.*, **38**, 932 (1973).
252. B. M. Trost and F. Chen, *J. Org. Chem.*, **36**, 2603 (1971).
253. (a) J. Daub, Y. Trautz, and U. Erhardt, *Tetrahedron Lett.*, 4435 (1972); (b) J. Daub, U. Ehrardt, J. Kappler, and V. Trautz, *J. Organomet. Chem.*, **69**, 423 (1974).
254. H. Alper and Hang-Nam Paik, *J. Org. Chem.*, **42**, 3522 (1977).
255. H. Alper, *J. Org. Chem.*, **40**, 2694 (1975).
256. (a) H. Alper and A. S. K. Chan, *J. Am. Chem. Soc.*, **95**, 4905 (1973); (b) H. Alper and W. G. Root, *J. Chem. Soc., Chem. Commun.*, 956 (1974); *Tetrahedron Lett.*, 1611 (1974); (c) H. Alper and C. K. Foo, *Inorg. Chem.*, **14**, 2928 (1975).
257. (a) G. F. Emerson, K. Ehrlich, W. P. Giering, and P. C. Lauterbur, *J. Am. Chem. Soc.*, **88**, 3172 (1966); (b) K. Ehrlich and G. F. Emerson, *J. Am. Chem. Soc.*, **94**, 2464 (1972); (c) J. S. Ward and R. Pettit, *J. Chem. Soc., Chem. Commun.*, 1419 (1970).
258. E. Weiss and W. Hubel, *Chem. Ber.*, **95**, 1179 (1962).
259. R. Aumann and J. Knecht, *Chem. Ber.*, **109**, 174 (1976).
260. (a) R. R. Victor, R. Ben-Shoshan and S. Sarel, *Tetrahedron Lett.*, 4253 (1970); (b) R. Aumann, *J. Am. Chem. Soc.*, **96**, 2631 (1974).
261. (a) V. Heil, B. F. G. Johnson, J. Lewis, and D. J. Thompson, *J. Chem. Soc., Chem. Commun.*, 270 (1974); (b) B. F. G. Johnson, K. D. Karlin, J. Lewis, *J. Organomet. Chem.*, **145**, C23 (1978); (c) Review: A. J. Deeming, in "Comprehensive Organometallic Chemistry," Eds. G. Wilkinson, F. G. A. Stone, and W. E. Abel, Vol. 4, Pergamon Press, London, 1982, p. 377.
262. M. Franck-Neumann, D. Martina, and M. P. Heitz, *Tetrahedron Lett.*, **23**, 3493 (1982).
263. A. Monpert, J. Martelli, R. Grée, and R. Carrié, *Nouv. J. Chim.*, **7**, 345 (1983).
264. J. A. S. Howell and M. J. Thomas, *J. Chem. Soc., Dalton Trans.*, 1401 (1983).
265. D. Astruc, J. R. Hamon, G. Althoff, E. Román, P. Batail, P. Michaud, J.-P. Mariot, F. Varret and D. Cozak, *J. Am. Chem. Soc.*, **101**, 5445 (1979).
266. A. N. Nesmeyanov, N. A. Vol'kenau, L. S. Shilovtseva and V. A. Petrakova, *J. Organomet. Chem.*, **61**, 329 (1973).
267. D. Astruc, *Image de la Chimie*, CNRS, Paris, 1982, p. 10.
268. D. Astruc, *Acc. Chem. Res.*, in press and Comments Inorg. Chem., in press.
269. J.-R. Hamon and D. Astruc, *J. Am. Chem. Soc.*, **105**, 5951 (1983).
270. J.-R. Hamon, D. Astruc and P. Michaud, *J. Am. Chem. Soc.*, **103**, 758 (1981).
271. P. Michaud and D. Astruc, *J. Chem. Soc., Chem. Commun.*, 416 (1982).
272. P. Michaud, J.-P. Mariot, F. Varret, and D. Astruc, *J. Chem. Soc., Chem. Commun.*, 1383 (1982).
273. M. H. Desbois, D. Astruc, J. Guillin, J. P. Mariot, and F. Varret, *J. Am. Chem. Soc.*, **107**, 52 (1985) and *J. Chem. Soc., Chem. Commun.*, 447 (1985).

274. A. N. Nesmeyanov, N. A. Volkenau, and V. A. Petrakova, *J. Organomet. Chem.*, **136**, 363 (1977).
275. M. S. Karasch and G. Reinmuth, *Grignard Reactions of Non-Metallic Substances*, Prentice-Hall, Englewood Cliffs, NJ, 1954, p. 22.
276. J. K. Kochi, *Am. Chem. Soc. Symp. Ser.*, No. 55, 167 (1977).
277. M. Tamura and J. K. Kochi, *J. Am. Chem. Soc.*, **93**, 1483 (1971); *Synthesis*, 303 (1971).
278. (a) D. F. Sangster in *The Chemistry of the Hydroxyl Group, Part 1* (Ed. S. Patai), Wiley–Interscience, New York, 1971, p. 51; (b) N. Uri, *Chem. Rev.*, **50**, 375 (1952); (c) R. O. C. Norman and J. R. Lindsay-Smith, in *Oxidase and Related Redox Systems* (Eds. T. E. King, H. S. Mason, and M. Morrison), Vol. I, Wiley New York, 1965, p. 131; (d) G. Sosnovsky and D. J. Rawlinson, in *Organic Peroxides* (Ed. D. Swern), Wiley–Interscience, New York, 1971, Vol. II, p. 269 and Vol. III, p. 141.
279. (a) M. S. Karasch, F. Kawahara, and W. Nudenberg, *J. Org. Chem.*, **19**, 1977 (1954); (b) D. D. Coffman, E. L. Jenner, and R. D. lipsomb, *J. Am. Chem. Soc.*, **80**, 2864 (1958).
280. D. D. Coffman, R. Cramer and W. E. Mochel, *J. Am. Chem. Soc.*, **80**, 2882 (1958).
281. (a) C. Walling, *Acc. Chem. Res.*, **8**, 125 (1975); (b) C. Walling and G. M. El-Taliawi, *J. Am. Chem. Soc.*, **95**, 844 and 848 (1973).
282. (a) C. Walling, G. M. El-Taliawi, and R. A. Johnson, *J. Am. Chem. Soc.*, **96**, 133 (1974); (b) N. A. Milas, *J. Am. Chem. Soc.*, **59**, 2342 (1937); (c) N. A. Milas, P. F. Kurz, and W. P. Anslow, *J. Am. Chem. Soc.*, **59**, 543 (1937).
283. C. Walling, D. M. Camaioni and S. S. Kim, *J. Am. Chem. Soc.*, **100**, 4814 (1978).
284. G. A. Hamilton, *Adv. Enzymol.*, **32**, 55 (1969); *J. Am. Chem. Soc.*, **86**, 3391 (1964).
285. R. A. Sheldon and J. K. Kochi, *Metal-Catalysed Oxidations of Organic Compounds*, Academic Press, New York, 1981, Chapter 3.
286. (a) M. S. Kharash, F. S. Arimoto, and W. Nudenberg, *J. Org. Chem.*, **16**, 1556 (1951); (b) M. S. Kharash and A. Fono, *J. Org. Chem.*, **24**, 72 (1959).
287. Ž. Čeković and M. M. Green, *J. Am. Chem. Soc.*, **96**, 3000 (1974).
288. R. A. Sheldon and J. K. Kochi, *Metal-Catalyzed Oxidations of Organic Compounds*, Academic Press, New York, 1981, Chapter 6 and 8.
289. J. T. Groves, G. A. McClusky, R. E. White and M. J. Coon, *Biochem. Biophys. Res. Commun.*, **81**, 154 (1978).
290. (a) A. A. Akhrem, D. I. Metelisa, and M. E. Surko, *Russ. Chem. Rev.*, **44**, 398 (1975); (b) V. Ullrich, *Angew. Chem., Int. Ed. Engl.*, **84**, 701 (1972); (c) G. A. Hamilton in *Molecular Mechanisms of Oxygen Activation* (Ed. O. Hayaishi), Academic Press, New York, 1974, p. 401; (d) T. Matsuura, *Tetrahedron*, **33**, 2869 (1973).
291. S. Udenfried, C. T. Clark, J. Axelrod, and B. B. Brodie, *J. Biol. Chem.*, **208**, 731 (1954).
292. (a) L. Castle, J. R. Lindsay-Smith, and G. V. Buxton, *J. Mol. Catal.*, 7, 235 (1980); (b) D. M. Jerina, J. W. Daly, W. Landies, B. Witkop, and S. Udenfried, *J. Am. Chem. Soc.*, **89**, 3347 (1967).
293. I. Tabushi, T. Nakajima, and K. Seto, *Tetrahedron Lett.*, **21**, 2565 (1980).
294. J. T. Groves, T. E. Nemo, and R. S. Myers, *J. Am. Chem. Soc.*, **101**, 1032 (1979).
295. C. K. Chang and M. S. Kuo, *J. Am. Chem. Soc.*, **101**, 3413 (1979).
296. T. Funabiki, H. Sakamoto, S. Yoshida, and K. Tamara, *J. Chem. Soc., Chem. Commun.*, 754 (1979).
297. (a) M. L. Sinnott, *Chem. Ber.*, **15**, 293 (1979); (b) P. Brodelins, *Adv. Biochem. Eng.*, **10**, 99 (1978); (c) W. R. Vieth and K. Venkatasubramanian, *Chemtech.*, **3**, 667 (1973); (d) S S. Wang and C. K. King, *Adv. Biochem. Eng.*, **12**, 119 (1979).
298. F. Minisci, *Acc. Chem. Res.*, **8**, 165 (1975), and references cited therein.
299. Y. Mari and J. Tsuji, *Tetrahedron*, **28**, 29 (1972).
300. J. Elzinga and H. Horgeveen, *J. Chem. Soc., Chem. Commun.*, 705 (1977).
301. (a) L. S. Liebeskind, M. E. Welker, V. Goedken, *J. Am. Chem. Soc.*, **106**, 441 (1984); (b) S. G. Davies and J. C. Walker, *J. Chem. Soc., Chem. Commun.*, 209 (1985).
302. Review: C. A. Rusik, T. L. Tonker, and J. L. Templeton, *J. Am. Chem. Soc.*, **108**, 4652 (1986).
303. (a) G. Tanguy, B. Weinberger, and H. des Abbayes, *Tetrahedron Lett.*, **24**, 4005 (1983); (b) for an excellent review of the use of phase transfer catalytic with metal carbonyls, see H. des Abbayes, *Isr. J. Chem.*, **26**, 249 (1985).

The Chemistry of the Metal—Carbon Bond, Vol. 4
Edited by F. R. Hartley

CHAPTER **8**

Use of organorhodium compounds in organic synthesis

F. H. JARDINE

Department of Chemistry, North East London Polytechnic, Romford Road, London E15 4LZ, UK

I. INTRODUCTION

Rhodium complexes are pre-eminent in homogeneous catalysis. Many complexes of this element are very effective catalysts in hydroformylation, hydrogenation, and hydrosilylation reactions. In transfer hydrogenation and isomerization reactions their activity is slightly overshadowed by the complexes of other platinum group metals.

The rapid burgeoning of homogeneous catalysis has come about since the discovery of $[RhCl(PPh_3)_3]$, which was the first practically useful homogeneous hydrogenation catalyst that could operate under ambient conditions[1]. Almost simultaneously it was found to be an excellent hydrosilylation catalyst[2]. Derivatives of this complex have been found to be useful and commercially important hydroformylation catalysts. Other rhodium complexes are used in the large-scale catalytic carbonylation of methanol.

The practical applications of the above complexes focussed attention on the organometallic chemistry of the element with a view to the development of catalysts for a variety of reactions involving organic substrates. A vital step in the development of rhodium complexes as homogeneous catalysts was the stabilization of rhodium(I) complexes by ligands that also conferred solubility in non-polar solvents.

Many such complexes are only four coordinate. Hence they are coordinatively unsaturated or, in modern parlance, 16-electron species. The ready availability of an extra coordination site permits the activation of reactants in catalytic cycles.

The ease of oxidation to rhodium(III) complexes permits the oxidative addition of reactants. Further, the rhodium(III) complexes so formed are often themselves coordinatively unsaturated, thus permitting the activation of reactants at this stage in the catalytic cycle also. The stability of the univalent state conferred by the ligands present in the original complex also aids the reductive elimination of products and permits the resumption of the catalytic cycle.

The inferior catalytic activity of complexes of other transition metals arises from a poor balance between the stabilities of the different oxidation states involved in the catalytic process and the absence of a vacant coordination site where the reactants can be coordinated and activated.

Nevertheless, apart from the very important catalysts used in the hydroformylation of propene and in the carbonylation of methanol, complex rhodium catalysts are seldom used in industrial processes. Partly this is due to the high cost of the metal, but more usually it is due to the ease with which catalytically active complexes are oxidized to inactive species by trace impurities such as oxygen or hydroperoxides in the feedstocks. Accordingly, rhodium complexes find their greatest application in small-scale preparations in academic, pharmaceutical, or fine organic chemical laboratories.

The incorporation of chiral ligands into rhodium complexes has been the principal development of practical homogeneous catalysis in recent years. The resulting chiral complex catalysts can achieve high optical yields in the products derived from suitable prochiral substrates. Given the sensitivity of most rhodium catalysts to oxidants, this area still represents the most promising area for the future development of homogeneous rhodium complex catalysts.

All homogeneous catalytic systems suffer from one major practical disadvantage—normally it is impossible to separate the products from the catalyst. Unless the separation can be achieved by simple physical means, the catalytic systems are inapplicable to continuous industrial processes. Additionally, in small-scale work, separation of the catalyst from reactive products can result in a serious depletion of the yield.

To avoid such problems, supported and heterogenized catalysts have been devised. The former type has the catalyst physically adsorbed in the pores of an inert support. Many homogeneous catalysts retain at least some of their activity under these conditions, but elution of the catalyst complex from the support is a major disadvantage. In the latter type of catalyst, one or more ligands of the parent catalyst complex are replaced by similar ligand groups bound to an insoluble polymer, which anchors the complex to the surface of the polymer. This type of catalyst usually retains many of the catalytic attributes of the original complex, but it is less effective towards large substrate molecules. In asymmetric syntheses the optical yields obtained from polymer-bound chiral catalysts are usually markedly inferior to those obtained from the parent chiral catalysts. Since essentially the same mechanism is involved in both homogeneous and polymer-bound systems, only passing mention will be made of derived heterogenized catalysts in this chapter. They are discussed in detail in Chapter 14.

Nor will great emphasis be placed on the stoicheiometric reactions of the rhodium—carbon bond. The high cost of the metal and its salts makes it unsuitable for large-scale stoicheiometric preparations. However, it must be conceded that, in certain instances, some stoicheiometric reactions have provided pilot routes to compounds of pharmaceutical interest.

II. ISOMERIZATION

Many rhodium compounds are capable of bringing about the isomerization of hydrocarbons and their derivatives, often under very mild conditions. The simplest form of isomerization is double bond migration in alkenes, in which the carbon skeleton remains intact. The second type of isomerization is usually confined to strained cyclic or polycyclic species, and in this type of reaction the products have a different carbon skeleton to the reactants.

Often many other catalytic processes are accompanied by isomerization. The lower activity of rhodium compounds in isomerization reactions is frequently the reason for selecting rhodium complexes as catalysts in preference to those of other transition metals.

A. Double Bond Migration

1. Linear alkenes

Although allylic mechanisms have been proposed for several isomerization reactions of this type, most of the isomerizations can be satisfactorily explained by the successive β-hydride abstraction reactions of the hydroisomerization process (Scheme 1). The key feature of Scheme 1 is the addition of rhodium to the alkene to form a secondary alkyl complex. It is impossible for isomerization to occur within the confines of Scheme 1 if a primary alkyl complex is formed.

When acyclic alkenes are the substrate, the alkene bond migrates into the chain, since internal alkenes possess greater thermodynamic stability than terminal alkenes. Exceptionally, in the case of substituted alkenes, migration of the double bond outwards to the end of the chain may be observed if a stable, intermediate, chelated alkene complex can be formed.

Many non-hydridic rhodium complexes bring about isomerization provided that they can form hydrido complexes during the reaction. For example, solutions of the hydrido complex $[RhH(CO)(PPh_3)_3]$ isomerize hept-1-ene when stored under either hydrogen or nitrogen[3]. *trans*-Carbonylchlorobis(triphenylphosphine)rhodium(I) and its analogues can only isomerize the alkene in the presence of hydrogen[3–5]. As implied in Scheme 1, the hydrido complexes are the true catalysts.

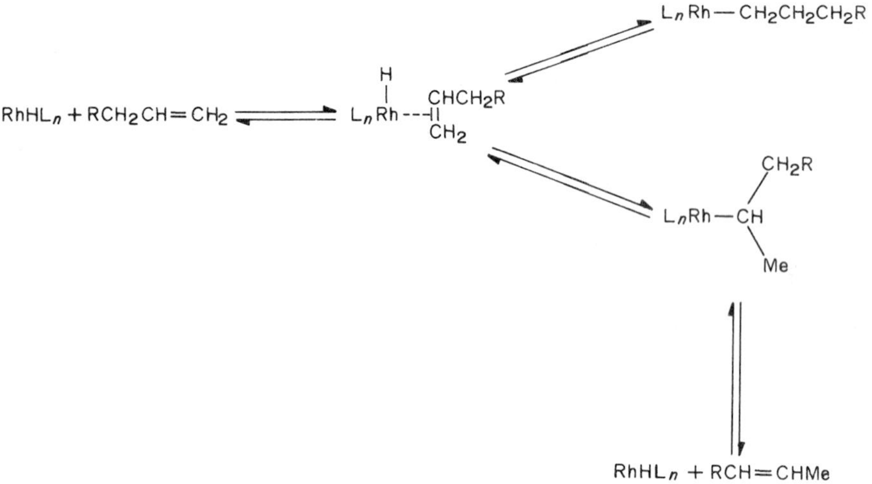

$$RhHL_n + RCH_2CH{=}CH_2 \rightleftharpoons L_nRh(H)\text{---}(CH_2{=}CHCH_2R)$$

$$L_nRh(H)\text{---}(CH_2{=}CHCH_2R) \rightleftharpoons L_nRh{-}CH_2CH_2CH_2R$$

$$L_nRh(H)\text{---}(CH_2{=}CHCH_2R) \rightleftharpoons L_nRh{-}CH(CH_2R)(Me)$$

$$L_nRh{-}CH(CH_2R)(Me) \rightleftharpoons RhHL_n + RCH{=}CHMe$$

SCHEME 1. The β-hydride mechanism of alkene isomerization.

Usually rhodium(I) species are better catalysts than rhodium(III) complexes[6], although the latter include such powerful catalysts as the five-coordinate hydridosilyl complexes $[RhH(SiR_3)Cl(PPh_3)_2]$. Alkenes do not compete effectively for coordination sites on rhodium. This is particularly true for internal alkenes. Accordingly, 16-electron species, such as $[RhH(SiR_3)Cl(PPh_3)_2]$, are usually more efficient catalysts for the reaction. The importance of the vacant site can be seen from the observation that aerated or peroxidized alkenes are more readily isomerized than the pure compounds by tertiary phosphine complexes[7]. Some of the tertiary phosphine is oxidized under these conditions and thereby releases a coordination site for alkene complexation. Activation of hydroperoxides by irradiation with ultraviolet light also brings about an increase in the rate of hydroisomerization[8].

Blocking the vacant coordination site by addition of ligands more readily coordinated than the alkene substrate reduces the rate of alkene isomerization[9]. The availability of a coordination site is probably the reason why $[RhH(CO)(PPh_3)_3]$, which dissociates to $[RhH(CO)(PPh_3)_2]$, acts as an isomerization catalyst for but-1-ene, whilst the chelated complex $[RhH(diphos)_2]$ is ineffective[10].

Electronic factors are also important. If the alkyl complex produced in the first stage of catalytic hydrogenation via the alkyl route (see Chapter 12) does not rapidly activate molecular hydrogen then the alkyl complex will probably decompose by β-hydride abstraction[11,12]. Thus, whereas $[RhH(PF_3)(PPh_3)_3]$ brings about rapid hydrogenation and hydroisomerization of terminal alkenes, $[RhH(PF_3)_2(PPh_3)_2]$ functions solely as an isomerization catalyst. It is likely that the latter complex, which has a lower electron density on rhodium, does not activate molecular hydrogen. Further, the lower steric requirements of the two PF_3 ligands enhance the probability of forming 2-alkyl complexes from alk-1-enes[13].

The low coordinating power of internal alkenes usually permits the isolation of intermediate products. The normal course of isomerization of alk-1-enes is

$$\text{alk-1-ene} \rightarrow \textit{cis}\text{-alk-2-ene} \rightarrow \textit{trans}\text{-alk-2-ene} \qquad (1)$$

Thus, but-1-ene[10], pent-1-ene[14,15], and hept-1-ene[3] are eventually isomerized by $[RhH(CO)(PPh_3)_3]$ to the *trans*-alk-2-enes. The isomerization of pent-1-ene is also catalysed by heterogenized derivatives of $[RhH(CO)(PPh_3)_3]$[16]. Both pent-1-ene[14,15] and hept-1-ene[17] are more rapidly isomerized to the *cis*-alk-2-ene than this product is to its *trans*-isomer. Because of the decreasing complexity constants of internal alkenes, it is unusual for further isomerization of *trans*-alk-2-enes to occur. However, both $[RhH(1,7\text{-}C_2B_9H_{11})(PPh_3)_2]$ and $[RhH(1,2\text{-}C_2B_9H_{11})(PPh_3)_2]$ isomerize hex-1-ene to *trans*-hex-3-ene at 20 °C[18].

The low coordinating power of internal alkenes is further illustrated by the regioselective isomerization of 3,7-dimethylocta-1,6-diene, which forms 3,7-dimethylocta-2,6-diene when allowed to isomerize in the presence of $[RhH(CO)(PPh_3)_3]$[19]. In the regioselective isomerizations of *N*,*N*-diethylnerylamine and *N*,*N*-diethylgeranylamine by a complex of (+)-bis-2,2'-(diphenylphosphino)-1,1'-dinaphthyl, virtually 100% optical yields were obtained (reactions 2 and 3). The chirality of the products could be reversed by employing the enantiomeric ligand[20].

$$[Rh(C_8H_{12})(+)\text{-binap}]^+,\ 40\,°C \qquad (2)$$

NEt2 → Me, H, NEt2

(3)

In contrast to the widely accepted β-hydride process, the η^3-allylic mechanism is invoked much less frequently. One reaction in which the latter mechanism undoubtedly occurs is the protonation of η^3-allylic rhodium(I) complexes (reaction 4)[21]. In one such reaction the methallyl ligand is converted into but-1-ene in the reaction.

$$2[Rh(\eta^3\text{-}C_3H_4R)(PF_3)_3] + 2HCl \longrightarrow 2RC_3H_5 + [RhCl(PF_3)_2]_2 + 2PF_3 \qquad (4)$$

It has also been proposed that the allylic mechanism is involved in the isomerization of but-1-ene by $[RhCl(C_2H_4)_2]_2$. The study of the reaction mechanism is complicated by the equilibrium between the alkene complexes (equation 5)[22]. When the reaction is carried out in the presence of PPh_3, the true catalyst is probably $[RhCl(PPh_3)_3]$[23]. N.m.r. spectrometry shows *trans*-$[RhCl(PPh_3)_2(C_4H_8)]$ to be an important species in the catalytic cycle[24]. Again, the kinetics of the system are complicated and defy analogue computer-fitting techniques[25]. Nevertheless, the authors consider that the high initial ratio of *cis*- to *trans*-but-2-ene among the products to be indicative of an η^3-allylic mechanism[26].

$$[RhCl(C_2H_4)_2]_2 + CH_2{=}CHEt \rightleftharpoons [Rh_2Cl_2(C_2H_4)_3(CH_2{=}CHEt)] + C_2H_4 \qquad (5)$$

The isomerization of allylbenzene to prop-1-enylbenzene proceeds much more slowly when catalysed by $[RhH(CO)(PPh_3)_3]$ than by $[RhCl(PPh_3)_3]$. Similarly, 4-phenylbut-1-ene is only slowly isomerized to 1-phenylbut-2-enes[27], and poor yields are obtained in the isomerization of the allyl ether when $[RhH(CO)(PPh_3)_3]$ is the catalyst (equation 6)[28].

(6)

One useful reaction is the conversion of allyl ethers into prop-1-enyl ethers (reaction 7)[29]. The prop-1-enyl group can be easily hydrolysed and the two reactions represent

$$ROCH_2CH{=}CH_2 \xrightarrow{[RhCl(PPh_3)_3]} ROCH{=}CHMe \xrightarrow[H_2O]{pH\,2}$$

$$ROH + CH_3CH_2CHO \qquad (7)$$

a simple method of removing allyl protecting groups[29–31]. However, the homologous 3-methylbut-2-enyl group is not readily isomerized by $[RhCl(PPh_3)_3]$ and this has been used in the selective isomerization of allyl and 3-methylbut-2-enyl groups (reaction 8).

$$\text{(sugar)-OCH}_2\text{CH=CH}_2 \xrightarrow{[RhCl(PPh_3)_3]} \text{(sugar)-OCH=CHMe}^* \qquad (8)$$

(* *E* and *Z* isomers obtained)

Allyl 3, 4, 6-tri-*O*-benzyl-2-*O*-(3-methylbut-2-enyl)-α-D-galactopyranoside reacted similarly[32].

Ketones are produced from secondary allylic alcohols on treatment with $[RhCl(CO)_2]_2$ and NaOH in the presence of benzyltriethylammonium chloride as a phase transfer catalyst. Under these conditions the reactions occur at room temperature, whereas $[RhH(CO)(PPh_3)_3]$, in the absence of a phase transfer catalyst, gives poorer yields at 70 °C[33]. The latter catalyst isomerizes primary allylic alcohols to aldehydes (reaction 9)[34]. In the presence of (−)-diop, which presumably forms the complex $[RhH(CO)(diop)(PPh_3)]$, chiral aldehydes result (equations 10 and 11)[35].

$$CH_2{=}CMeCH_2OH \xrightarrow[CF_3CH_2OH]{[RhH(CO)(PPh_3)_3]} Me_2C{=}CHOH \longrightarrow Me_2CHCHO \qquad (9)$$

$$\text{MeC(H)=C(Me)CH}_2\text{OH} \xrightarrow[(-)\text{-diop}]{[RhH(CO)(PPh_3)_3]} \text{MeC(H)(Et)CHO} \qquad (10)$$

$$\text{HC(H)=C(Et)CH}_2\text{OH} \xrightarrow[(-)\text{-diop}]{[RhH(CO)(PPh_3)_3]} \text{EtC(H)(Me)CHO} \qquad (11)$$

Allylic alcohols are also isomerized to aldehydes by $[RhCl(PPh_3)_3]$. The aldehyde produced is decarbonylated by the rhodium complex. This decarbonylation reaction provides some insight into the reaction mechanism. The formation of both *erythro*- and *threo*-alkanes from alcohols containing CD_2OH groups is strong evidence that η^3-allyl complexes are involved[36].

Chlorotris(triphenylphosphine)rhodium(I) also catalyses the isomerization of allyl tertiary amines to prop-1-enyl tertiary amines. The prop-1-enyl group is less readily solvolysed from nitrogen[37].

2. Cycloalkenes

Cycloalkenes are also isomerized by rhodium complexes. For example, bicyclo[4.2.0]oct-7-ene is isomerized by $[RhCl(PPh_3)_3]$ (equation 12)[38]. No mechanism has been proposed for this reaction. The apparent absence of any catalytic activity towards

$$\text{bicyclo[4.2.0]oct-7-ene} \xrightarrow{[RhCl(PPh_3)_3]} \text{bicyclo[4.2.0]oct-1(8)-ene} \qquad (12)$$

cycloalkenes by monohydrido species suggests that secondary alkyl complexes are involved.

The difference in thermodynamic stability of isomeric cycloalkadienes makes their isomerization more feasible, and there are several examples of these substrates undergoing isomerization reactions. Cycloocta-1, 5-diene forms both the 1, 3- and 1, 4-isomers when heated with $[RhCl(PPh_3)_3]$[39]. This complex also brings about the isomerization of substituted hexa-1, 4-dienes (equation 13). The major product has a double bond in each

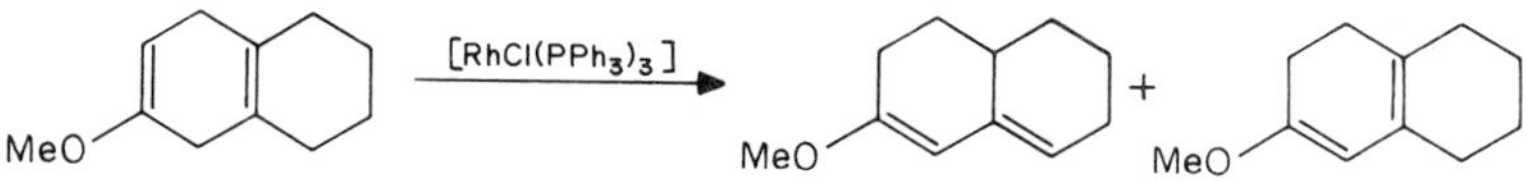

[RhCl(PPh$_3$)$_3$] (13)

ring[40]. The isomerization of 4-vinylcyclohexene to 3-ethylidenecyclohexene is also catalysed by the above complex[41,42].

Despite its inactivity towards cycloalkenes and cycloalkadienes, double bond migration in nitrogen heterocycles is catalysed by $[RhH(CO)(PPh_3)_3]$ (equations 14 and 15)[43].

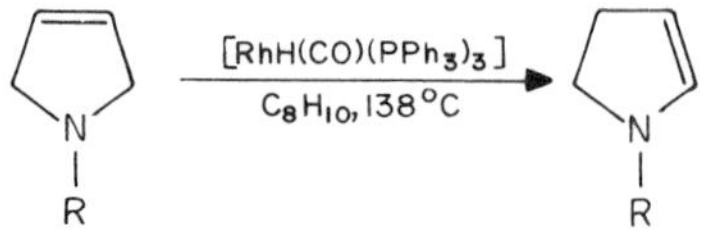

[RhH(CO)(PPh$_3$)$_3$], C_8H_{10}, 138 °C (14)

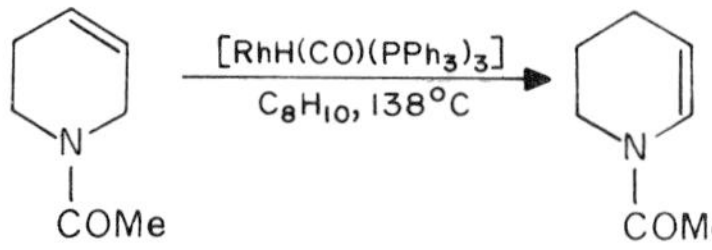

[RhH(CO)(PPh$_3$)$_3$], C_8H_{10}, 138 °C (15)

A dihydromethylpyridinium derivative has been isomerized to a compound containing conjugated double bonds by $[RhCl(PPh_3)_3]$ (equation 16)[44].

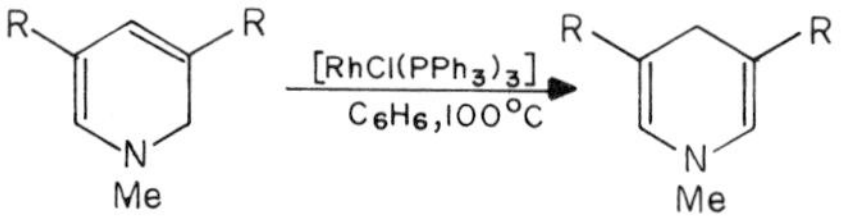

[RhCl(PPh$_3$)$_3$], C_6H_6, 100 °C (16)

B. Skeletal Rearrangements

1. *Cyclopropyl rings*

The strain energy of the cyclopropyl ring is high, and is an important factor in the ring-opening reactions catalysed by a variety of rhodium complexes. Both *cis*-1-methyl-2-phenylcyclopropane and the *trans*-isomer decompose when allowed to react with $[RhCl(CO)_2]_2$ (equation 17). Although the same proportions of products are formed in

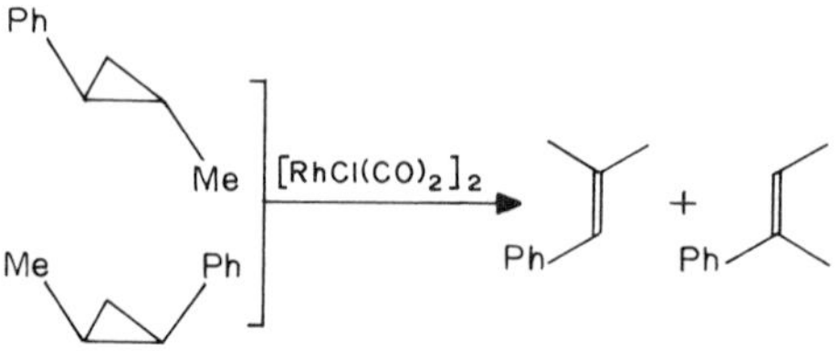

[RhCl(CO)$_2$]$_2$ (17)

both cases, the reactions occur at different rates. The proposed mechanism involves ring opening to give rhodium coordination to C-1 followed by hydride migration[45].

Ring opening of monosubstituted cyclopropanes is also catalysed by $[RhCl(CO)_2]_2$ (equations 18 and 19)[46].

$$\xrightarrow{[RhCl(CO)_2]_2} \quad + \tag{18}$$

$$\xrightarrow{[RhCl(CO)_2]_2} \quad + \tag{19}$$

Trisubstituted cyclopropanes also undergo ring opening in the presence of rhodium compounds. Rhodium(II) acetate or $[RhCl(CO)_2]_2$ both catalyse reaction 20. Whichever

$$\xrightarrow{RhL_n} \text{Ph(MeO)C=CH}_2\text{CH}_2\text{CO}_2\text{Et} \tag{20}$$

catalyst was used, twice as much *E* as *Z* product was formed. Replacement of the phenyl substituent by methyl gives rise to two products, since the latter group contains an α-hydrogen atom (equation 21). The two products are not interconverted during the

$$\longrightarrow \text{H}_2\text{C=C(MeO)CH}_2\text{CH}_2\text{CO}_2\text{Et} + \text{Me(MeO)C=CHCH}_2\text{CO}_2\text{Et} \tag{21}$$

reaction. More complex substituents, however, give products that themselves rearrange during the course of the reaction (equations 22 and 23)[47]. Under more severe conditions

$$\longrightarrow \text{(MeO)C=CHCH}_2\text{CO}_2\text{Et} \longrightarrow \text{MeCH=C(MeO)—CH=CHCO}_2\text{Et} \tag{22}$$

$$\longrightarrow \text{H}_2\text{C=C(OMe)—C(MeO)=CHCH}_2\text{CO}_2\text{Et} \longrightarrow \text{MeC(OMe)=C(MeO)—CH=CHCO}_2\text{Et} \tag{23}$$

rhodium(II) acetate opens the cyclopropyltosyl derivative (equation 24). The reaction is also catalysed by $[RhCl(CO)_2]_2$, but much lower yields are obtained. At 160 °C a cyanocyclopropane is converted to acyclic products including a ketal (equation 25).

$$\text{Ph(MeO)C(cyclopropyl)}\!\sim\!SO_2C_6H_4Me \xrightarrow{[Rh(O_2CMe)_2]_2} \text{Ph(MeO)C}{=}CHCH_2SO_2C_6H_4Me \qquad (24)$$

$$\text{Ph(MeO)C(cyclopropyl)}\!\sim\!CN \xrightarrow[\text{MeOH}]{\text{Rh(II)}} \text{Ph(MeO)C}{=}CHCH_2CN + Ph{-}C(OMe)_2{-}CH_2CH_2CN \qquad (25)$$

Similar isomerizations have been observed in other cyclopropyl compounds. Tetrasubstituted cyclopropanes undergo similar reactions to those shown in equation 21[47].

Trimethylbicylo[1.1.0]butane is isomerized in chloroform solution by either $[RhCl(CO)_2]_2$ or by the mononuclear complex $[Rh(acac)(CO)_2]$. The major product is acyclic and deuterium labelling shows rupture to have occurred at the bridgehead carbon (equation 26)[48,49]. Partial kinetic resolution of racemic trisubstituted bicyclobutanes has

$$\xrightarrow[\text{CHCl}_3]{[RhCl(CO)_2]_2} \qquad (26)$$

been achieved by using [RhCl(diop)], which preferentially catalyses the decomposition of the *R*-isomer[50]. Methanol is added to the vinylcyclopropane if the reaction is carried out in this solvent (equation 27)[48].

$$\xrightarrow[\text{MeOH}]{[RhCl(CO)_2]_2} \text{Me, OMe, Me, H} \qquad (27)$$

A similar dependence on solvent is found in the $[RhCl(CO)_2]_2$ catalysed isomerization of trimethylbicyclo[1.1.0]butane. In chloroform the product is 3,4-dimethylhexa-1,3-diene. In methanol the solvent adds across the terminal double bond of the product. However, under these conditions the acyclic ester is the minor product (equation 28)[51].

$$\xrightarrow[\text{MeOH}]{[RhCl(CO)_2]_2} \text{OMe} + \text{OMe H} \qquad (28)$$

In contrast, bicyclo[2.1.0]pentane is isomerized to cyclopentene[52]. Ring expansion to phenalene occurs when cycloprop[*a*]acenaphthylene is allowed to react with catalytic quantities of $[RhCl(CO)_2]_2$ (equation 29). Deuterium labelling at *exo*-C-7 shows that this atom can migrate to six different positions in the product[53].

$$\xrightarrow{[RhCl(CO)_2]_2} \qquad (29)$$

Although ring expansion does not occur in the isomerization of bicyclo[4.1.0] hept-2-ene (equation 30)[46], this process is involved in the formation of all three products

$[RhCl(CO)_2]_2$, $CHCl_3$, 90 °C (30)

from the $[RhCl(PPh_3)_3]$ catalysed isomerization of tricyclo[3.2.2.0^{2,4}]nona-6, 8-diene (equation 31)[54].

Rh(I) (31)

If the cyclopropyl ring is fused to a five-membered ring, then ring expansion takes place on reaction with catalytic quantities of rhodium complexes, particularly in the presence of oxygen (equation 32)[55,56]. The *endo*-isomer is opened more rapidly than the *exo*-isomer[54,56].

MeO_2C — $[RhCl(PPh_3)_3]$, fast → CO_2Me; slow → CO_2Me (32)

2. Quadricyclanes

The strained hydrocarbon quadricyclane is easily isomerized, under mild conditions, by several rhodium complexes, even though the isomerization to bicyclo[2.2.1]hepta-2, 5-diene contravenes the Woodward–Hoffmann rules. Nevertheless, the uncatalysed reaction occurs slowly at temperatures above 140 °C. Addition of catalytic quantities of $[RhCl(CO)_2]_2$ permits the reaction to take place at -26 °C[58]. The bicyclo[2.2.1]hepta-2, 5-diene complex $[RhCl(nbd)]_2$ also isomerizes quadricyclane to this bicyclic hydrocarbon (equation 33). The exothermic reaction takes place in chloroform solution at room

$[RhCl(nbd)]_2$ (33)

temperature[59]. Cassar and Halpern[60] suggested that one of the cyclopropane rings in quadricyclane adds oxidatively to rhodium when the reaction is catalysed by $[RhCl(CO)_2]_2$. Addition of PPh_3 resulted in the formation of a diacyl complex (equation 34)[60].

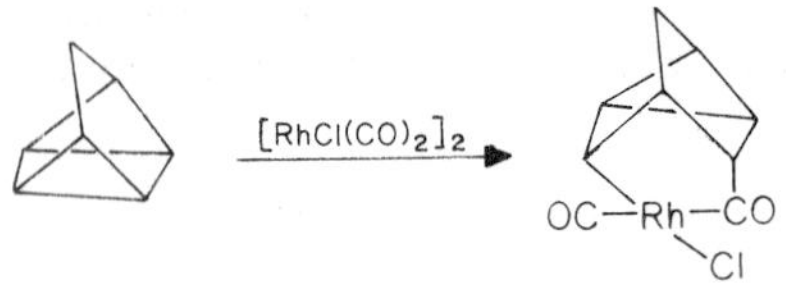

$$\xrightarrow{[RhCl(CO)_2]_2} \quad \text{(OC—Rh—CO, Cl)} \qquad (34)$$

Rapid isomerization of quadricyclane is also catalysed by $[Rh(O_2CMe)(nbd)]_2$. In these reactions, carried out at $-50\,°C$, it is believed that the cyclobutane complex **1** is an

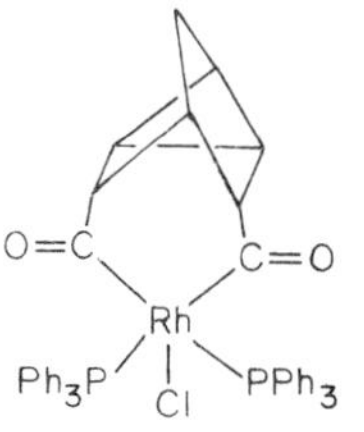

O=C C=O Rh Ph_3P Cl PPh_3

(1)

intermediate. It should also be noted that bicyclo[2.2.1]hepta-2, 5-diene is not the sole product of the reaction, as dimerization of the alkadiene also takes place (see Section XI.A)[61]. On the other hand, heteroquadricyclanes are isomerized to fulvenes. In the absence of substitution at a bridgehead, the reaction occurs much more slowly. In all cases the reaction is believed to occur via an alkadiene complex[62]. Paradoxically, a quadricyclane homologue can be obtained from the $[RhCl(CO)_2]_2$ catalysed isomerization of *exo*-tricyclo[3.2.1.0^{2,4}]oct-6-ene. Different products are obtained from the thermal and photochemical isomerization of the tricyclooctene. The *endo*-isomer does not react in the presence of $[RhCl(CO)_2]_2$[63]. The same catalyst converts tricycloheptane to 3-methylenecyclohexene in an exothermic reaction (equation 35)[59].

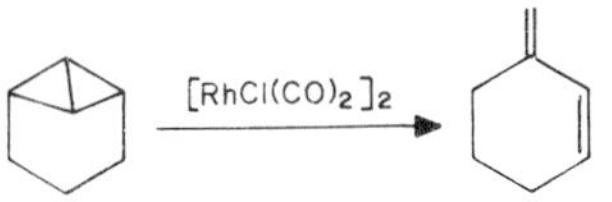

$$\xrightarrow{[RhCl(CO)_2]_2} \qquad (35)$$

3. Cyclobutane rings

Hexamethyl(Dewar)benzene reverts to hexamethylbenzene on heating to 100–140 °C. The reaction is brought about by catalytic quantities of $[RhCl(hmdb)]_2$ at 65–70 °C. Since the rhodium complex itself is stable at 100 °C, it has been proposed that the dimeric complex is cleaved by excess of hexamethyl(Dewar)benzene and that the rate-determining step is the elimination of hexamethylbenzene from $[RhCl(hmdb)_2]$ (reaction 36)[64,65]. The

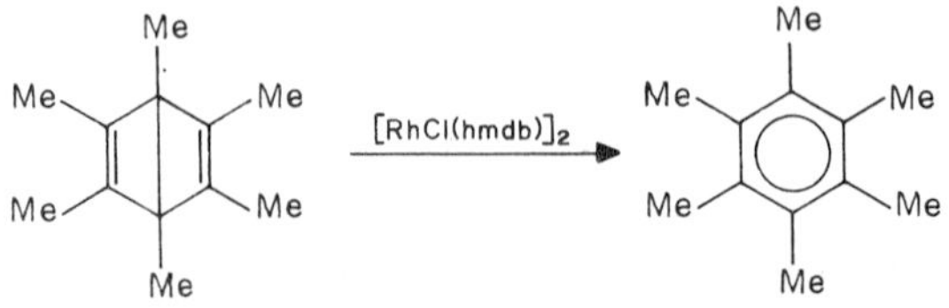

$$\xrightarrow{[RhCl(hmdb)]_2} \qquad (36)$$

Me Me Me Me Me Me — Me Me Me Me Me Me

isomerization of bicyclo[2.2.0]hexane and tetradeuteriobicyclo[2.2.1]hexane to give cyclohexene and its isotopomer, respectively (equations 37 and 38), are believed to involve

$[RhCl(CO)_2]_2$ (37)

$[RhCl(CO)_2]_2$ (38)

rhodium insertion across the bridge[66]. Insertion is also invoked when the reaction is catalysed by $[RhCl(CO)_2]_2$. In this case the acyl intermediate **2** was isolated. *syn*-

(2) (3)

Tricyclo[4.2.0.0^{2,5}]octane isomerizes similarly (reaction 39). Bis[chloro(norbornadiene)-rhodium(I)] is a superior catalyst to $[RhCl(cod)]_2$ for the reaction. It was not possible to

$[RhCl(nbd)]_2$ 70% 30% (39)

isolate an intermediate from the $[RhCl(CO)_2]_2$-catalysed reaction. The complex **3** could, however, be isolated from the stoichiometric reaction if this was carried out at 5 °C[66,67].

Although *endo*-tricyclo[5.2.1.0^{2,6}]deca-4, 8-diene-3-one forms a cage isomer when irradiated with ultraviolet light, the reaction can be reversed when the latter compound is heated in diphenyl ether in the presence of catalytic quantities of $[RhCl(PPh_3)_3]$ (equation 40)[69]. The cyclobutane ring is invariably disrupted and additionally the ether

$[RhCl(CO)_2]_2$ (40)

$[RhCl(CO)_2]_2$ (41)

bonds are broken in forming the minor products of the reaction (equation 41). However, if the reactant contains an octamethylene bridge, the different strain energies involved bring about a different product distribution (equation 42)[70].

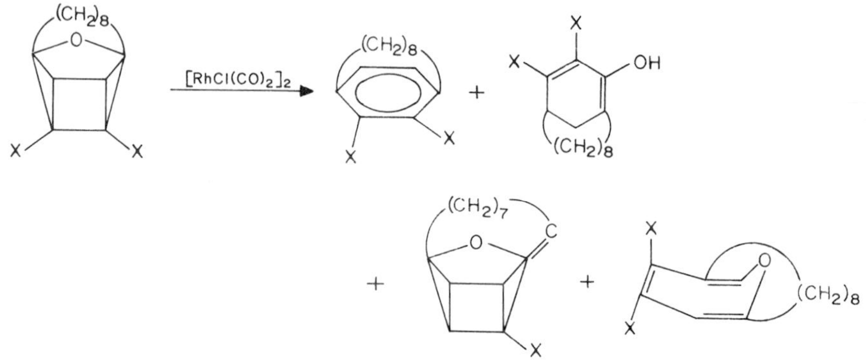

(42)

4. *Norbornane derivatives*

Norbornylenedioxin and oxanorbornylene undergo oxygen—oxygen bond cleavage in the presence of $[RhCl(CO)_2]_2$ (equation 43)[71,72]. Other related dioxins react similarly (equations 44–45)[66,73]. Oxanorbornadiene derivatives form enolic cyclopentadienyl compounds (equations 46–48). The first step in the reaction is believed to be the displacement of a carbonyl ligand by coordination of an alkene bond[73].

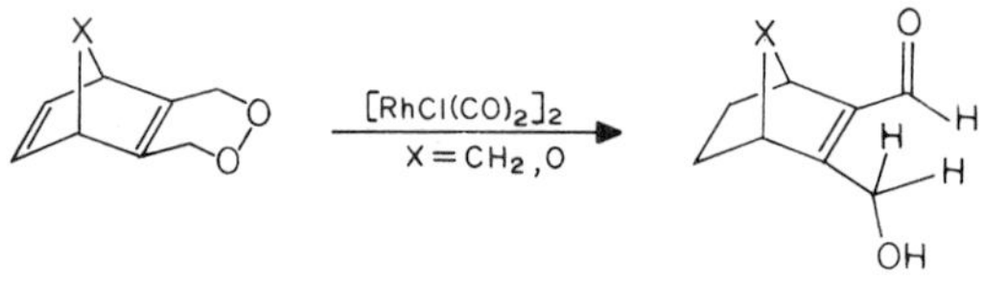

$[RhCl(CO)_2]_2$, X = CH$_2$, O (43)

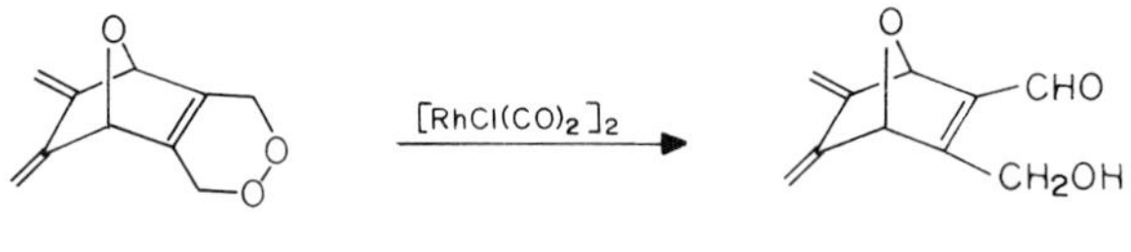

$[RhCl(CO)_2]_2$ (44)

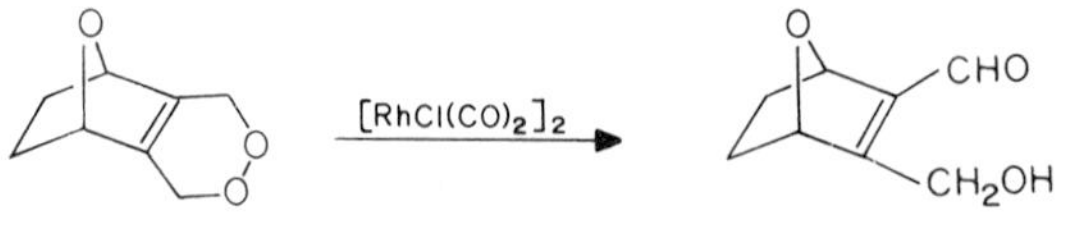

$[RhCl(CO)_2]_2$ (45)

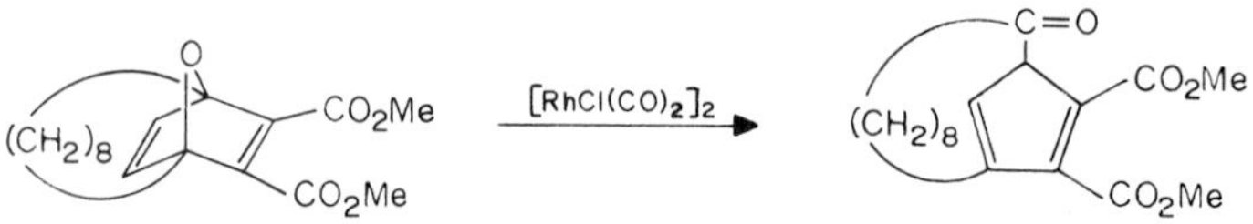

$[RhCl(CO)_2]_2$ (46)

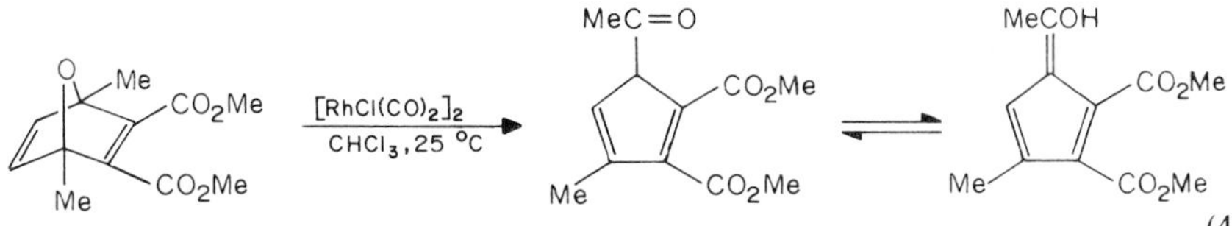

(47)

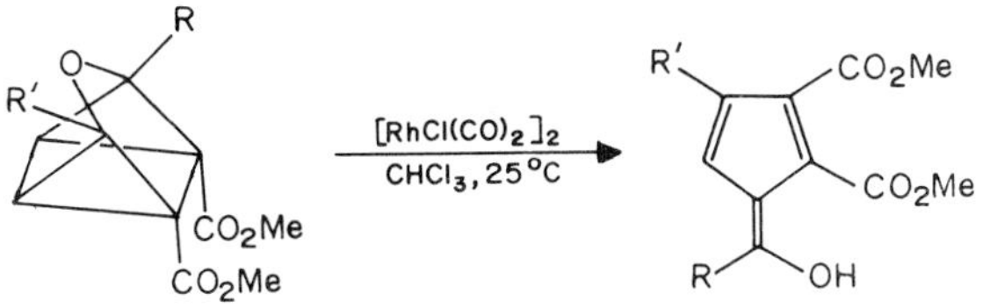

(48)

5. Cubanes

The isomerization of cubane to tricyclo[2.2.0.0]octadiene is catalysed by either $[RhCl(nbd)]_2$ or $[RhCl(tcod)]_2$ (equation 49). Dicarbomethoxycubane forms *syn*-dicarbomethoxytricyclooctadiene. Monosubstituted cubanes form two products (equation 50). The reactions are first order in both cubane and catalyst. The cubane skeleton is also disrupted by $[RhCl(CO)_2]_2$, but ketones can be formed by CO insertion. These reactions occur via a metallocycle[66].

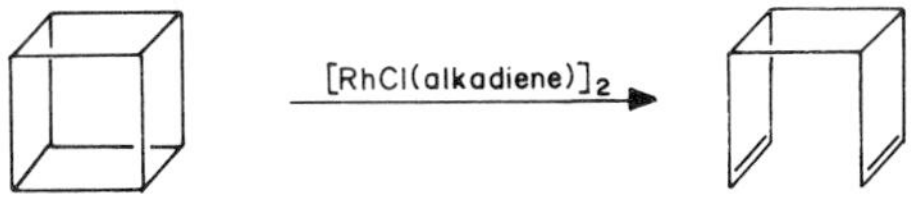

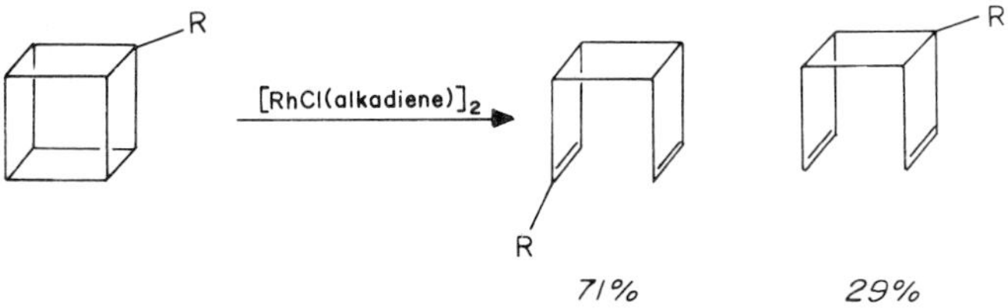

(49)

(50)

The valence isomerization of homocubane is also catalysed by $[RhCl(nbd)]_2$, which forms a tricyclononadiene (equation 51). It was found that the strain energy of the hydrocarbon dictates the course of the reaction[74].

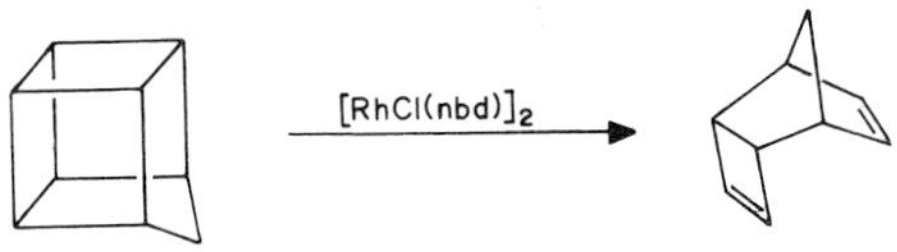

(51)

Methylphosphahomocubane forms two isomeric products when heated in chloroform with $[RhCl(nbd)]_2$ (equation 52)[75]. The behaviour of bishomocubane derivatives is similar[76], the major products being two isomeric tricyclodecadienes (equation 53). The

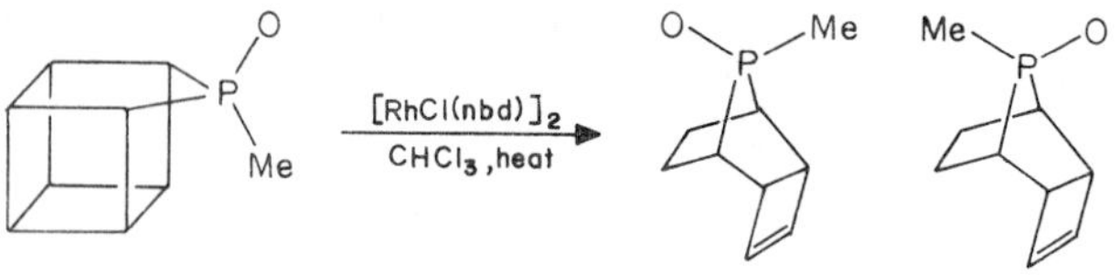

(52)

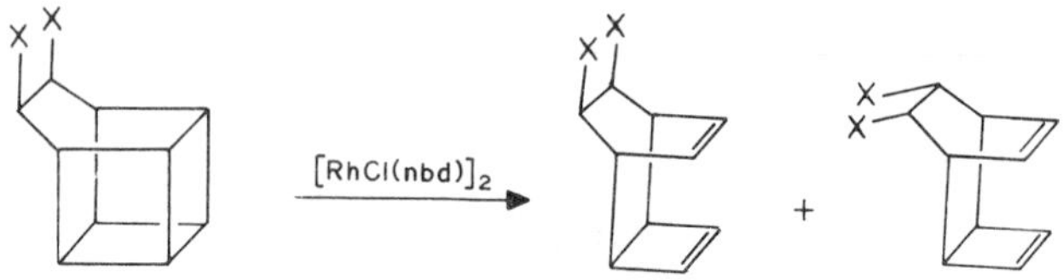

(53)

catalytic isomerization of 9, 10-dicarbomethoxybishomocubane and its derivatives also give two isomeric products in each case. However, in all rhodium-catalysed valence isomerizations of bishomocubanes, small quantities of snoutanes are obtained[77].

6. Other compounds

Homopentaprismane is isomerized by either $[RhCl(nbd)]_2$ or *trans*-$[RhCl(CO)(PPh_3)_2]$ to homohypostrophene (equation 54)[78]. The photoisomerization of

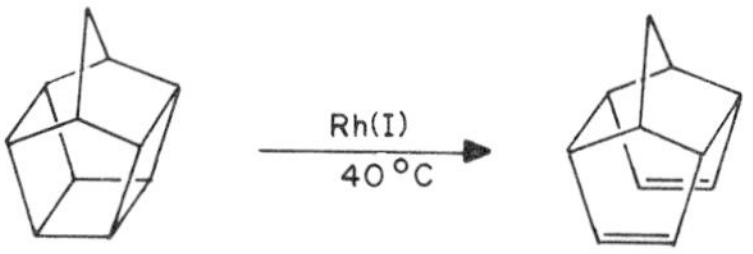

(54)

exo-dicyclopentadienone is reversed by treatment with the above two complexes (equation 55)[79].

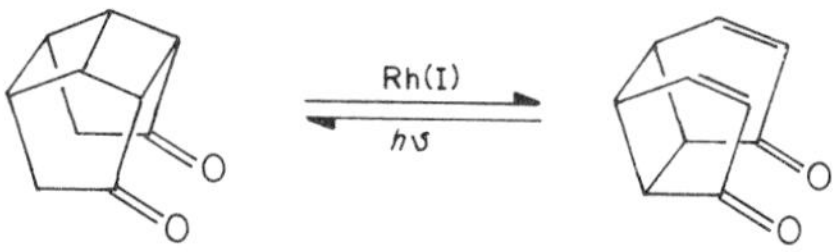

(55)

Although the major product from the $[RhCl(cod)]_2$-catalysed photo-rearrangement of cyclooct-1, 5-diene is cycloocta-1, 4-diene, a small amount of cyclooctene is formed, together with bicyclooctenes. A kinetic isotope effect is observed in the reaction and a greater proportion of bicyclooctene is formed from the tetradeuteriated reactant[80].

Several epoxides are opened catalytically by rhodium complexes. Butadiene monoepoxide forms both *cis*- and *trans*-crotonaldehyde[81]. Cyclooctatetrene monoepoxide undergoes ring contraction that progresses as the temperature is increased when allowed to react with catalytic quantities of $[RhCl(CO)_2]_2$ (equation 56)[82].

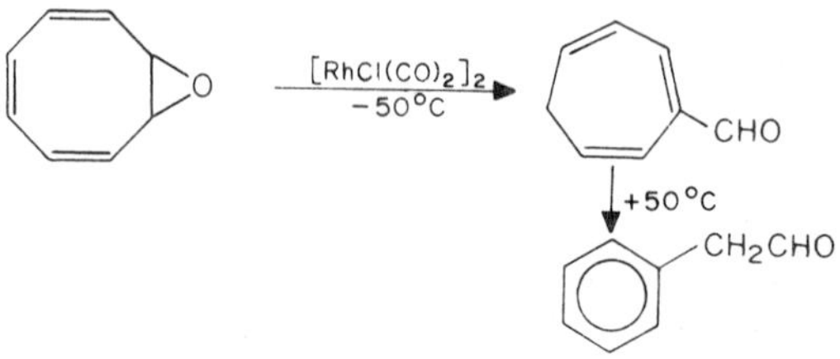

(56)

Ring expansion occurs if tricyclononadiene is heated with $[RhCl(CO)_2]_2$ (equation 57). However, this complex forms only bicyclo[4.2.1]nonadiene (cf. equation 31)[54].

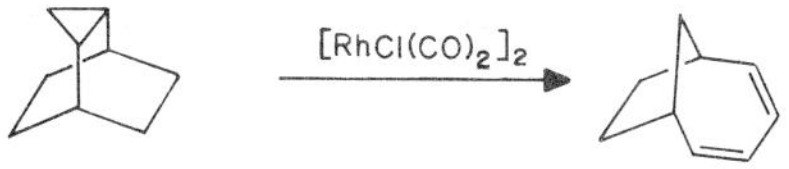

$[RhCl(CO)_2]_2$ (57)

Dimethylbicyclo[6.1.0]cyclonona-2, 4, 6-triene also undergoes ring equalization on gentle warming with $[RhCl(CO)_2]_2$ (equation 58)[83]. Salomon *et al.*[83] consider that the

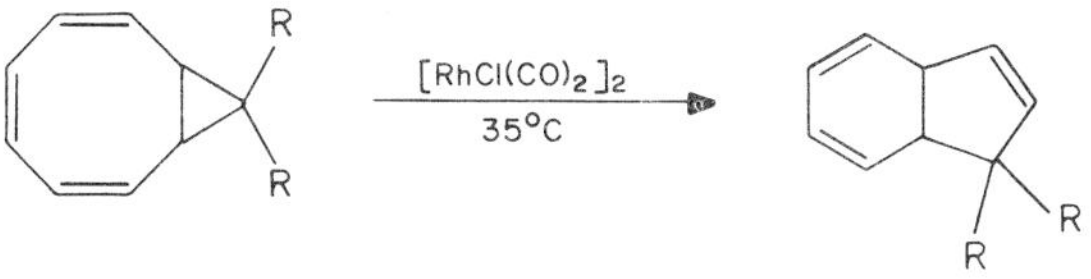

$[RhCl(CO)_2]_2$, 35°C (58)

catalysed reaction first produces cyclonona-1, 3, 5, 7-tetraene, which then undergoes an uncatalysed electrocyclic rearrangement to form *cis*-8, 9-dihydroindene. In support of this proposal they demonstrated that the cyclopropyl derivative **4** forms **5** and **6** directly and that **7** is formed directly from **8** or **9** and not as a result of the rhodium-catalysed isomerization of the intermediate **10**. The failure to produce the normally expected 1, 3-

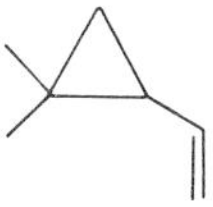

(4) 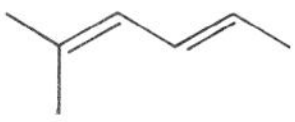 (5)

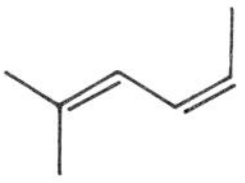

(6) 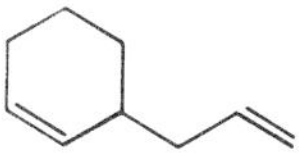 (7)

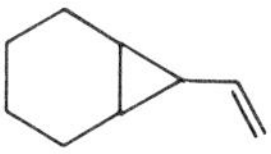

(8) 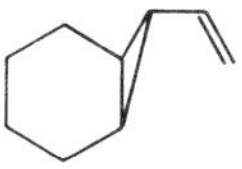 (9)

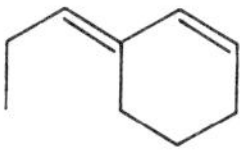

(10)

dienes from the ring cleavage reactions was ascribed to the participation of η^3-allylic intermediates which did not undergo the usual β-hydride abstraction reactions[83].

7. Cyclization reactions

In addition to effecting the ring opening of alicyclic compounds, rhodium complexes also catalyse cyclization reactions. Two brief reports have shown that acyclic diallylic compounds form compounds that contain five-membered rings. For example, hydrated rhodium trichloride catalyses the formation of 3-methyl-4-methylenetetrahydrofuran when allowed to react with diallyl ether. If the reaction is carried out in methanol solution the intermediate complex $[RhCl_2(C_6H_{11}O)]_n \cdot \frac{n}{2}MeOH$ can be isolated. This complex also catalyses the cyclization reaction[84].

Chlorotris(triphenylphosphine)rhodium(I) catalyses similar cyclizations of other diallylic compounds (equation 59)[85].

$$\xrightarrow{[RhCl(PPh_3)_3]} \tag{59}$$

$X = CO_2Me$, CO_2Et, COPh, $\frac{1}{2}$(5, 5-cyclohexa-1, 3-dienyl)

C. Isotope Exchange

Rhodium compounds are not good catalysts for isotope exchange. For example, chlorotris(triphenylphosphine)rhodium(I) does not catalyse specific deuterium incorporation into the *ortho*-positions of its triphenylphosphine ligands in the way that $[RuCl_2(PPh_3)_3]$ does[86]. There are, however, a few instances where rhodium compounds catalyse isotope exchange between alkenes and deuteriated species. Thus, rhodium trichloride brings about the incorporation of deuterium into allylbenzene when the alkene is allowed to react with *O*-deuteriomethanol (equation 60). The incorporation of deuterium at the terminal carbon atom and the parallel isomerization to prop-1-enylbenzene suggest that a 2-alkyl complex is involved in the reaction. The reaction is also catalysed by $[RhCl(C_2H_4)_2]_2$. The addition of chloride ion to the latter catalytic system, which may form mononuclear complexes such as $[RhCl_2(C_2H_4)_2]^-$, increases the rate of isomerization[87].

$$PhCH_2CH{=}CH_2 \xrightarrow{RhCl_3/MeOD} PhCH_2CH{=}CHD + (E)\text{-and } (Z)\text{-}PhCH{=}CH_2D \tag{60}$$

Isotope exchange frequently occurs during the homogeneous catalytic deuteration of alkenes[88-90]. The source of the exchange is the decomposition of an intermediate deuterioalkyl complex, the stability of which depends on the nature of the catalyst and on the lifetime of the alkyl complex in the reaction (equations 61 and 62).

$$RCH{=}CH_2 \underset{}{\overset{RhDL_n}{\rightleftharpoons}} L_nRhCH_2CHDR \underset{}{\overset{-RhHL_n}{\rightleftharpoons}} RCD{=}CH_2 \tag{61}$$

$$RCH{=}CH_2 \overset{RhDL_n}{\rightleftharpoons} L_nRh{-}CH(CH_2D)(R) \overset{-RhHL_n}{\rightleftharpoons} RCH{=}CHD \tag{62}$$

Isotope exchange is an inevitable consequence of deuterioisomerization reactions (equation 63)[91,92]. Scrambling of the deuterium atoms in *trans*-$C_2H_2D_2$ is catalysed by $[RhH(CO)(PPh_3)_3]$. No deuterium was incorporated in the catalyst[93].

$$RCH_2CH{=}CH_2 + D_2 \xrightarrow{RhL_n} RCH{=}CHCH_2D + HD \tag{63}$$

III. HYDROGENATION

The most widely studied homogeneous hydrogenation catalysts are all rhodium complexes. Different rhodium complexes provide examples of all three mechanistic pathways observed in the homogeneous catalytic hydrogenation of alkenes. These mechanistic pathways are discussed in Chapter 12. Pre-eminent amongst these complexes is $[RhCl(PPh_3)_3]$, which was the first effective homogeneous hydrogenation catalyst for a wide range of alkene substrates[1]. The catalyses of this complex proceed by the dihydride route[94].

Carbonylhydridotris(triphenylphosphine)rhodium(I) reacts with terminal alkenes to form alkyl complexes, and these in turn activate molecular hydrogen to provide an example of the alkyl route in the catalytic hydrogenation of alkenes[95]. Cationic complexes of rhodium(I), e.g. $[Rh(PPh_3)_2(nbd)][ClO_4]$, permit the hydrogenation of chelating alkenes to take place via the alkene route[96–98]. This route is of extreme importance in the asymmetric hydrogenation of prochiral alkenes (see also Section IV). In the latter instance the cationic complexes give the best optical yields in the reductions when the two tertiary phosphine ligands are replaced by a chiral bidentate ligand.

The excellent performance of the above catalysts has tended to discourage investigations into the catalytic properties of other rhodium complexes. Work on cognate systems to the well established complexes has always proved popular, but comparatively little progress has been made towards developing novel catalysts. All too often these investigations seem to be abandoned once the system in question appears to offer no advantages over the well established complex catalysts.

One of the few system investigated that clearly catalyses alkane production by the dihydride route is that of $[RhCl(C_8H_{14})_2]_2$ in the presence of 2-aminopyridine. It has been stated that this system catalyses the hydrogenation of alkenes much more rapidly than either $[RhCl(PPh_3)_3]$ or even $[RuHCl(PPh_3)_3]$[99].

The rhodium(III) complex $[RhCl_3L_3]$, where L is the very large tertiary phosphine **11**,

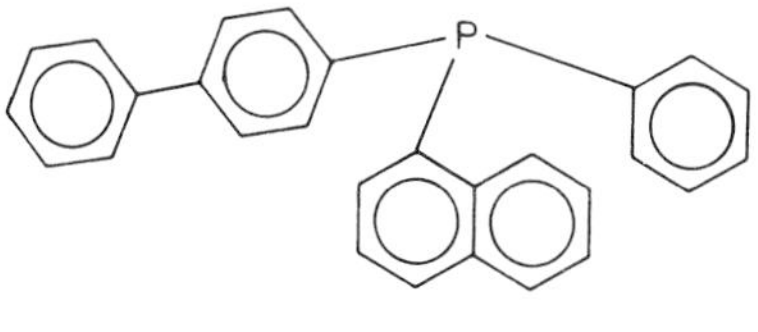

P

(11)

catalyses the reduction of unsaturated carboxylic acids at 50 °C and 5 atm hydrogen pressure. It seems very likely that the complex is reduced under these conditions to a rhodium(I) complex and that the hydrogenation is then brought about by the same mechanism as Wilkinson's catalyst. Usually lower yields were obtained than with the latter catalyst[100].

Another rhodium(III) complex that catalyses the hydrogenation of terminal, internal, and cyclic alkenes is $[Rh(\eta^5\text{-}C_5H_5)Cl_2]_2$. Hydrogen, in the presence of base, converts this complex into $[Rh_2H(C_5H_5)_2Cl_3$[101]. However, it is difficult to see how this species could be converted into a rhodium(V) complex by the oxidative addition of hydrogen if the reaction proceeds via the alkyl route, or how coordinative unsaturation occurs if the dimeric structure is retained throughout the catalytic cycle.

Dimeric rhodium(II) acetate does, however, retain its dimeric structure throughout the catalytic cycle. The reaction shows many similarities to the alkyl route, but in this system the alkyl complex is decomposed by the acetic acid liberated earlier in the catalytic cycle (Scheme 2)[102].

Although alkenes are known to coordinate to a vacant site on $[Rh_2(O_2CMe)_4]$, the

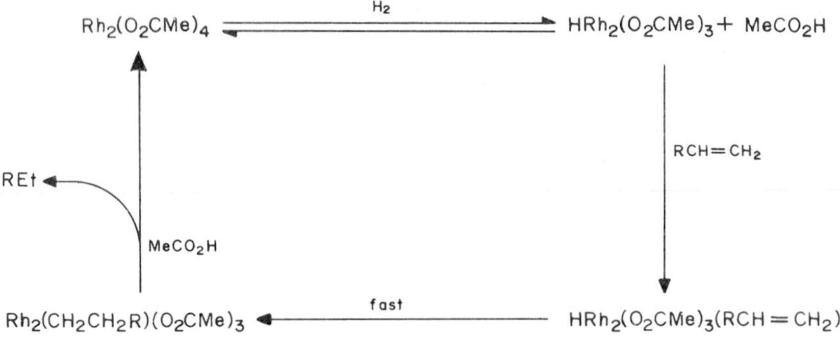

SCHEME 2. Alkene hydrogenation catalysed by rhodium(II) acetate.

cycle in Scheme 2 best fits the behaviour of the system[103]. Oxygen is reduced to water in the presence of the catalyst[102], so this system should be one of the least sensitive to this impurity.

There have been two reports of the catalytic activity of $[Rh(NO)(PPh_3)_3]$[104,105]. The catalyst has been used in the reduction of terminal and cyclic alkenes, alkadienes and even alkynes. However, although the reactions are carried out under 5 atm of hydrogen, the nitrosyl complex does not unequivocally form a hydrido complex in the system. Since both reduction and a small degree of isomerization are brought about by the catalyst, a hydrido species must participate at some point in the catalytic cycle. It is possible that this catalyst brings about the hydrogenation via the alkene route, which would not require a standing concentration of hydrido complex.

The cationic complex $[Rh(tfbb)(phen)][ClO_4]$ might be expected to catalyse the hydrogenation of alkene or alkyne substrates via the alkene route by analogy with the prototype complex $[Rh(PPh_3)_2(nbd)][ClO_4]$[96–98]. This would appear to be the case since alkynes and alkadienes are preferentially reduced to alkanes in the system[106].

The formation of cyclooctane on treating the diphenylphosphido-bridged complex $[Rh(\mu\text{-}PPh_2)(cod)]_2$ shows that the alkene route is followed at least initially in the hydrogenation of oct-1-ene[107]. However, it will be recalled that the cationic catalyst $[Rh(PPh_3)_2(nbd)][ClO_4]$ similarly reduced the coordinated norbornadiene before catalysing the reduction of non-chelating alkenes by the dihydride route[108]. To see if both rhodium atoms of the dimeric complex participated in the catalysis, the activity of $[(diphos)Rh(\mu\text{-}PPh_2)_2Rh(cod)]$ was investigated. The latter complex was more active in the hydrogenation reaction than the bis(cycloocta-1, 5-diene) complex, and it was concluded that electronic effects were more important than coordination site availability[107].

The catalytic properties of the rhodacycle $[(cod)Rh(CH_2)_2PMe_2]$ **12** have also been

Rh, H_2C, CH_2, PMe_2

(12)

investigated. The complex catalyses the hydrogenation of hex-1-ene, hex-2-ene, and cyclohexene. The ability to catalyse the reduction of internal alkenes implies that the induction period observed is not due to the formation of a monohydrido complex, but rather to the reduction of the cycloocta-1, 5-diene to provide a vacant coordination site. Since, in the absence of alkene, autoreduction to rhodium metal occurs, it seems likely that the alkene route is followed[109].

Other dinuclear complexes that act as homogeneous hydrogenation catalysts include $[RhCl(C_8H_{14})_2]_2$. When this complex is dissolved in dimethylacetamide in the presence of chloride ion or dimethyl sulphide, it catalyses the reduction of unsaturated carboxylic acids at 80 °C[110,111].

Bridged thiolato complexes of the type $[Rh_2(\mu\text{-SR})_2L_2(CO)][L = P(OMe)_3, P(OPh)_3]$ catalyse the hydrogenation of hex-1-ene and cyclohexene. The best yields of saturated product are obtained when the thiolato ligand's alkyl group is large (e.g. *tert* -butyl)[112].

The main advantage of homogeneous hydrogenation catalysts over conventional heterogeneous hydrogenation catalysts is their different regioselectivity. The stereochemistry about the alkene bond usually determines the rate of homogeneous catalytic hydrogenation. In heterogeneous hydrogenation the rate of reduction is usually determined by electronic factors, which permits different products to be obtained from the two processes. A classic example is provided by the reduction of eremophilone, which more readily forms 8, 9-dihydroeremophilone when the reaction is catalysed by palladium on charcoal. Chlorotris(triphenylphosphine)rhodium(I)-catalysed reductions form 13, 14-dihydroeremophilone[113]. The best regioselectivity in such cases is obtained from catalytic reductions which proceed by the alkyl route.

A further advantage of homogeneous catalytic hydrogenation is that it is usually specific for alkene or alkyne bonds. Other functional groups such as azo, carboxylic, ether, keto, nitro[1], cyano, ester, hydroxy[1,114], amide, or fluoro[114] are not reduced by $RhH_2Cl(PPh_3)_2$ under mild conditions. This selectivity has been used, for example, in the selective hydrogenation of alkenals to alkanals[115]. The hydrogenation of ketones to secondary alcohols, catalysed by rhodium complexes, is discussed in Chapter 12. The ability to reduce substrates containing sulphur is additionally an important advantage over heterogeneous catalysts[116]. It may also be noted that $RhCl(PPh_3)_3$ can be used to catalyse the reduction of compounds that are destroyed by heterogeneous catalysts[117,118] or that destroy the catalyst surface[119].

Polynuclear complexes are now being investigated as homogeneous catalysts in several reactions with a view to combining the desirable features of both heterogeneous and homogeneous catalysis. So far only $Rh_4(CO)_{12}$ and $Rh_6(CO)_{16}$ have been investigated as hydrogenation catalysts. Both the former and $[RhCl(CO)_2]_2$ show some selectivity towards pent-2-ene production in the hydrogenation of pent-2-yne[120].

The dinuclear A-frame complex $[RhCl(CO)_2(Ph_2PCH_2PPh_2)]^+$ and its acido or arsenic analogues can catalyse the hydrogenation of hex-1-ene or the reduction of phenylacetylene to styrene. The rate of hydrogenation of styrene in this system is much less than the rate of phenylacetylene hydrogenation[121]. This selectivity would appear to be one of the major advantages of dinuclear complexes, since the di-μ-hydrido complex $[RhH\{P(OPr^i)_3\}_2]_2$ is a useful catalyst for the production of *trans*-alkenes from dialkyl- or diaryl-alkynes. These catalyses proceed via vinyl intermediates whose stereochemistry dictates the nature of the final product[122,123].

Future developments in homogeneous catalytic hydrogenation by rhodium complexes would seem to lie in further study of polynuclear complexes. It seems most probable that rhodium will be the metal selected to catalyse hydrogenations if bimetallic catalysts are employed to bring about sequential multi-step reactions.

IV. ASYMMETRIC HYDROGENATION

A. Alkene Substrates

Asymmetric hydrogenation is currently the most active research area in homogeneous catalysis. The possible production of chiral compounds of pharmaceutical importance provides a powerful commercial incentive. This aim has already been realized in the production of L-DOPA (equation 64), which is used in the treatment of Parkinson's disease[124].

$$3\text{-MeO-}4\text{-(CO}_2\text{Me)C}_6\text{H}_3\text{-CH}{=}\text{C(NHCOMe)CO}_2\text{H} \xrightarrow{H_2} 3\text{-MeO-}4\text{-(CO}_2\text{Me)C}_6\text{H}_3\text{-CH}_2\text{-CH(NHCOMe)CO}_2\text{H} \qquad (64)$$

Asymmetric hydrogenation requires a chiral catalyst to act on a prochiral substrate. Chiral substrates or solvents do not bring about asymmetric induction[125]. Chirality may be incorporated into the catalyst in several ways. Either the anionic or neutral ligands can contain a chiral centre. Only one chiral anionic ligand has been employed in a rhodium complex catalyst. The optical yield obtained in the reduction of α-acetamidocinnamic acid was low[126], and did not encourage further investigations in this direction.

More usually, chiral neutral ligands are incorporated in the catalyst. Catalysts that use the three possible hydrogenation pathways (see Chapter 12) have been employed. Of these three routes, the least successful has been the alkyl route. Currently the alkene route seems to offer the most promise. Unlike the homogeneous catalytic hydrogenation of alkene hydrocarbons, neutral complex catalysts do not necessarily follow the dihydride route, particularly when potentially bidentate substituted monoalkenes are the substrates[127]. Indeed, both the alkene and dihydride pathways may participate simultaneously in the catalysis. The wide variations in optical yields on making minor changes in the reaction conditions or solvent composition may arise from the different contributions the two pathways may make to the overall reduction. In general, however, the best optical yields are obtained under mild conditions. The high hydrogen pressures or high temperatures required to achieve reasonable chemical yields with some catalysts invariably result in low optical yields.

For comparative purposes the efficiencies of the catalysts described in this section will be judged by their performance in the asymmetric hydrogenation of α-acetamidocinnamic acid, the most widely used substrate.

1. Neutral complexes

These are commonly generated *in situ* by allowing the ligands to react with dimeric rhodium(I) alkene complexes such as $[RhCl(C_8H_{14})_2]_2$, $[RhCl(nbd)]_2$ or $[RhCl(hexa\text{-}1,5\text{-diene})]_2$. Both monodentate and bidentate ligands may be used. They are usually added to the system in such quantities that they occupy only two coordination sites on each rhodium atom in order to permit ready coordination of substrate.

a. Monodentate tertiary phosphines

The chiral centre in these ligands can be either the phosphorus atom itself[128–130] or a chiral group (e.g. neomenthyl[131]) bound to phosphorus. In the former case three different organo groups are linked to phosphorus to give a ligand that does not racemize on coordination to rhodium. If the ligand contains two chiral centres then the chirality at phosphorus is dominant[132]. When the chiral centre is too far removed from phosphorus, asymmetric induction does not occur[133].

Usually only moderate optical yields can be obtained with monodentate tertiary phosphine ligands (Table 1), because the catalysis follows the stereochemically undemand-

TABLE 1. Optical yields obtained from α-acetamidocinnamic acid reductions catalysed by neutral rhodium complexes containing monodentate tertiary phosphines

Ligand	Solvent	pH_2 (atm)	T (°C)	Optical yield (%)	Product chirality	Ref.
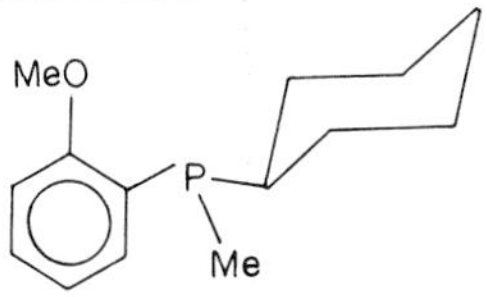 (13) MeO, P, Me (o-methoxyphenyl)(cyclohexyl)methylphosphine	95% EtOH	1.6	25	85		128
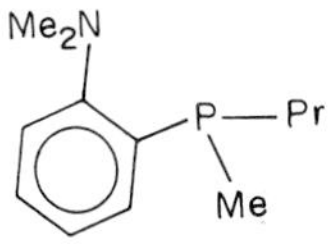 (14) Me_2N, P—Pr, Me				80		129
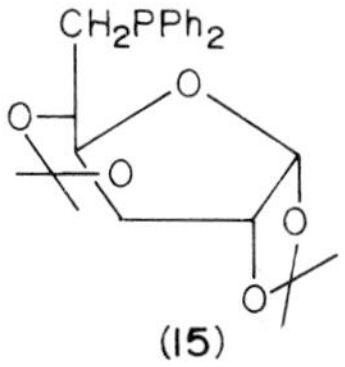 (15) CH_2PPh_2	C_6H_6–EtOH	1	25	9	*R*	134, 135
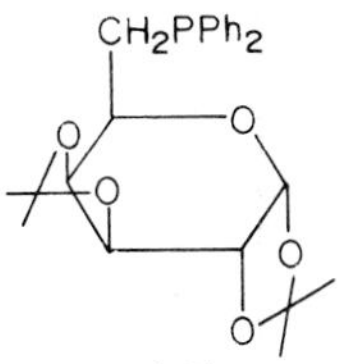 (16) CH_2PPh_2	C_6H_6–EtOH	1	25	67	*R*	134, 135
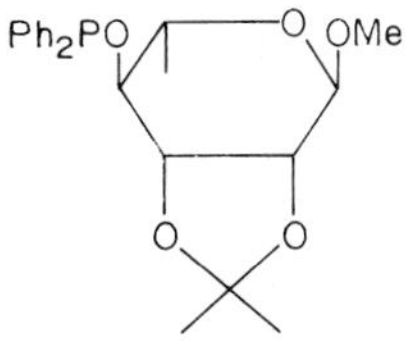 (17) Ph_2PO, OMe	C_6H_6–EtOH	1	25	13	*R*	135

ing dihydride route. There is little chiral discrimination since the rate-determining step is the coordination of alkene to rhodium. Under these circumstances there is little opportunity for the less stable alkene complex to dissociate before hydrogenation takes place. However, optical yields of up to 67% have been obtained with other substrates[135,137].

Neutral rhodium(0) catalysts derived from $[Rh_6(CO)_{16}]$ and (−)-diop gave poor chemical and optical yields[138,139].

b. *Bidentate ligands*

Better optical yields can be obtained when bidentate ligands are employed. Ligands containing PAr_2 groups attached to a chiral framework are the most commonly employed bidentate ligand type.

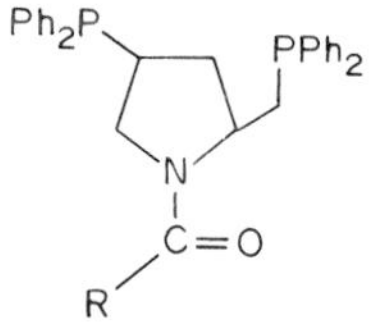

(18)

Similar strictures apply to the choice of the hydrogenation pathway to be followed when this type of catalyst is employed. This is well illustrated by the reversal of product chirality when complexes of either (−)-diop or the pyrrolidone ligand **18** (R = OBu^t) are used to catalyse the asymmetric hydrogenation of (Z)-PhCH=CCO_2H, ($NHCOCH_2Ph$). At 1 atm pressure (−)-diop complexes yield 55.2% of the *R*-product. This optical yield declines to 8.4% *R* at 5 atm hydrogen pressure, whilst the optical yield is 4.9% *S* at 50 atm hydrogen pressure. Complexes of **18** (R = OBu^t) give optical yields of 84% *R* and 14.4% *S* at 1 and 100 atm hydrogen pressure, respectively[139]. Increasing the hydrogen pressure increases the participation of the dihydride route[140].

The highest optical yields have been obtained when rigid, ditertiary phosphines were incorporated in the catalyst. The rigidity of the catalytic complex discriminates more effectively between the two modes of alkene coordination. This is well illustrated by the decrease in optical yields on going from a cyclobutane to a cyclohexane framework[141].

Phosphonite ligands give disappointing optical yields. Noteworthy in this respect is the phosphonite analogue of diop, which gives an optical yield approximately 20% of that of diop itself[142]. In contrast to both the phosphonite ligand above and their monodentate analogues[133], bidentate aminophosphine ligands generally give excellent results. The yields obtained in the asymmetric reduction of α-acetamidocinnamic acid by this class of catalyst are summarized in Table 2.

The neutral complexes containing bidentate ligands have also been used to catalyse the reductions of other prochiral substrates. The most frequently investigated have been esters of α-acetamidocinnamic acid. The size of the ester's alkyl group was reported to have very little effect on the optical yield[164,165]. On the other hand, increasing the bulk of the amide's alkyl or aryl group reduces the optical yield obtained[166]. The latter observation shows the importance of amide coordination to rhodium during the catalytic cycle. This feature of the catalytic cycle has been further investigated by replacing one of the phenyl groups attached to phosphorus in the bidentate ligand by an *o*-anisyl group[167]. The major effect of such a change is a decrease in optical yield when an *o*-anisyl-diop derivative is employed[168].

TABLE 2. Optical yields obtained from α-acetamidocinnamic acid reductions catalysed by neutral rhodium complexes containing bidentate ligands

Ligand	Solvent	pH_2 (atm)	T (°C)	Optical yield (%)	Product chirality	Ref.
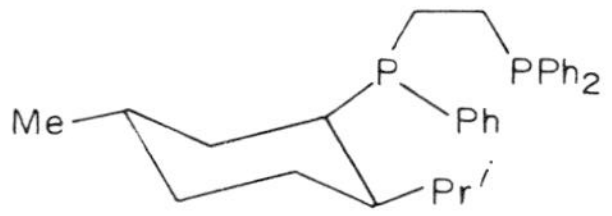 (19)		70	25	46	*R*	143
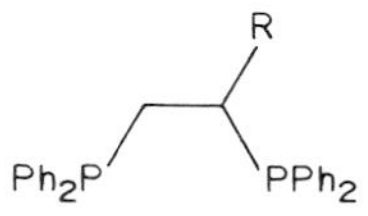 (20) R = Ph		70	30	82	*R*	144
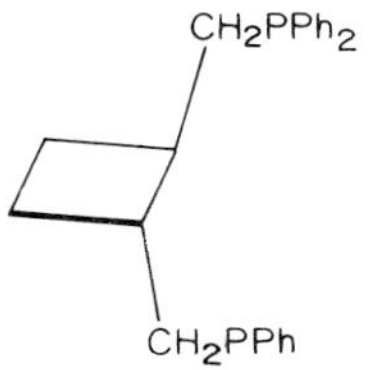 (1*R*,2*R*) (21)	C_6H_6–EtOH	1	25	86	*R*	140

TABLE 2. (*Contd.*)

Ligand	Solvent	pH_2 (atm)	T (°C)	Optical yield (%)	Product chirality	Ref.
CH_2PPh_2, CH_2PPh_2 (1*R*,2*R*) (**22**)				63	*R*	140
Ph_2PCH_2, CH_2PPh_2 (1*S*,2*S*) (**23**)				35	*S*	140, 145
PPh_2, PPh_2 (−) (**24**)	EtOH	1	25	95	*S*	146
Me, PPh_2, PPh_2 (+) (**25**)	MeOH	1	25	60	*R*	147

Ligand		Solvent					Ref.
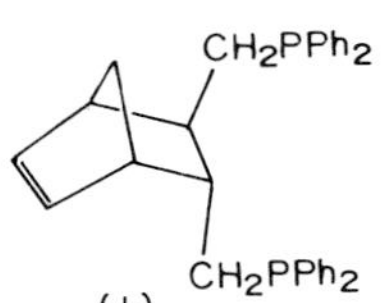 (+) (26) (CH_2PPh_2, CH_2PPh_2)		EtOH	1	25	50	*R*	148
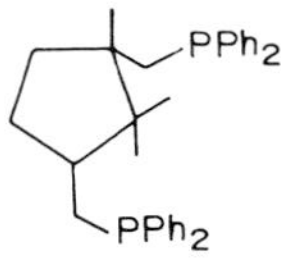 (27) (PPh_2, PPh_2)		EtOH	1	25	7	*S*	149
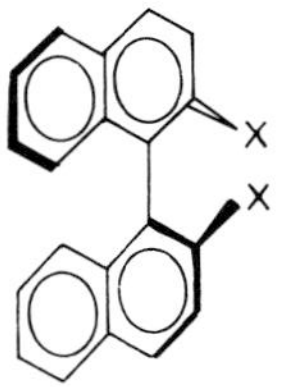 (28)	$X = CH_2PPh_2$		50	25	54	*S*	150
	$X = NHPPh_2$	C_6H_6–EtOH	30		85	*R*	142
		thf	30		8.7	*S*	142
	$X = OPPh_2$		102		9		151
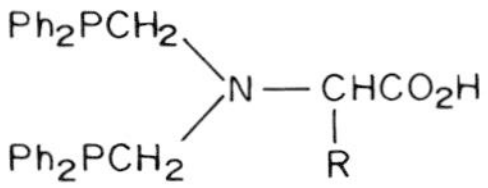 $(Ph_2PCH_2)_2N—CH(R)CO_2H$; (−); R = Me (29)			1	30	15		152

TABLE 2. (*Contd.*)

Ligand		Solvent	pH_2 (atm)	T (°C)	Optical yield (%)	Product chirality	Ref.
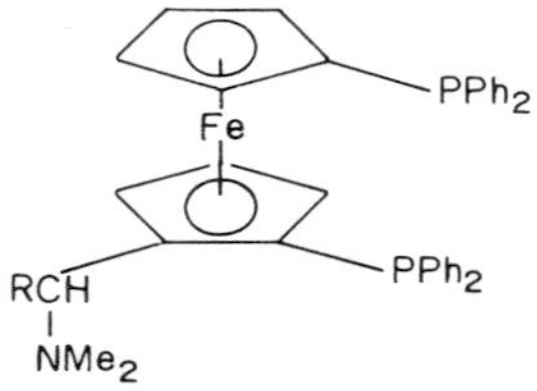 (30)	R = Me	C_6H_6–MeOH	20	25	86	*S*	153
		MeOH	50	25	93	*S*	154
		EtOH–H_2O	50	25	92	*S*	154
		MeOH–H_2O	50	25	89	*S*	154
	R = Pr^i	C_6H_6–MeOH	20	25	52	*R*	153
	R = Ph	C_6H_6–MeOH	20	25	86	*S*	153
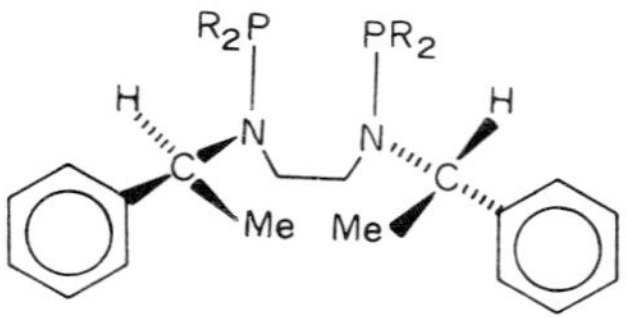 (31)	R = Ph	C_6H_6			49		155
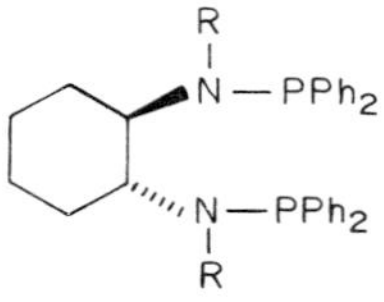 (32)		EtOH			73	*S*	156

Ligand		Solvent					Ref.
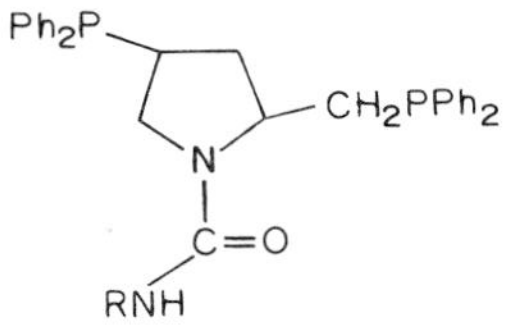 Ph_2P–pyrrolidine–CH_2PPh_2, N–C(=O)–RNH (**33**)	R = Me	EtOH	1	25	94.5	*S*	157
	R = $CH_2{=}CHCH_2$	EtOH	1	25	90.6	*S*	157
	R = Ph	EtOH	1	25	95.4	*S*	157
	R = C_6H_{11}	EtOH		25	92.8	*S*	157
	R = *p*-C_6H_4Cl	EtOH	1	25	96.1	*S*	157
	R = *p*-C_6H_4Br	EtOH	1	25	97.4	*S*	157
	R = 3,4-$C_6H_3Cl_2$	EtOH	1	25	97.9	*S*	157
	R = *p*-$C_6H_4NO_2$	EtOH	1	25	94.3	*S*	157
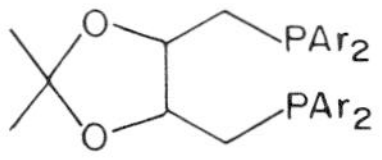 dioxolane–PAr_2, PAr_2; Ar = Ph (**34**)	(4*R*, 5*R*)	C_6H_6–EtOH	1	25	72[a]	*R*	158, 159
			1		60.5	*R*	139
			5		38.8	*R*	139
			20		16.5	*R*	139
			50		9.1	*R*	139
	(4*S*, 5*S*)	C_6H_6–EtOH	10	50	84.4[b]	*S*	160
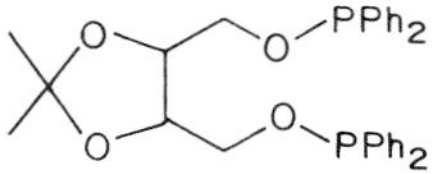 dioxolane–O–PPh_2, O–PPh_2 (**35**)					17	*R*	141
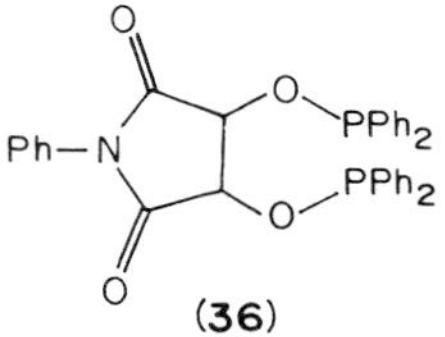 Ph–N succinimide–O–PPh_2, O–PPh_2 (**36**)					44	*S*	141

TABLE 2. (*Contd.*)

Ligand		Solvent	pH_2 (atm)	T (°C)	Optical yield (%)	Product chirality	Ref.
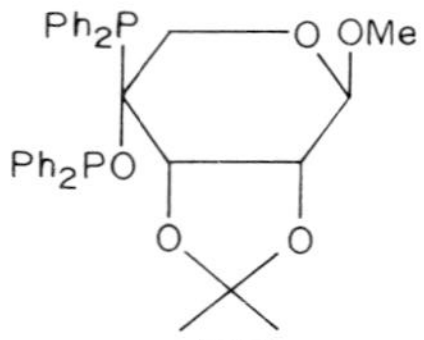 (37)		C_6H_6–EtOH			4.7	*R*	135
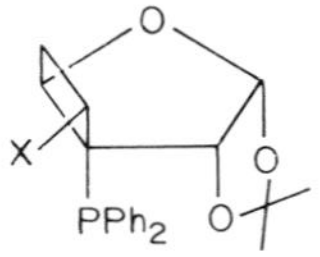 (38)	X = CN	MeOH	1	25	91.6	*S*	161
		MeOH	15	25	85.6	*S*	161
		MeOH	30	25	60.1	*S*	161
	X = CH_2NH_2	MeOH	1	25	39	*R*	161
		MeOH	15	25	14.8	*R*	161
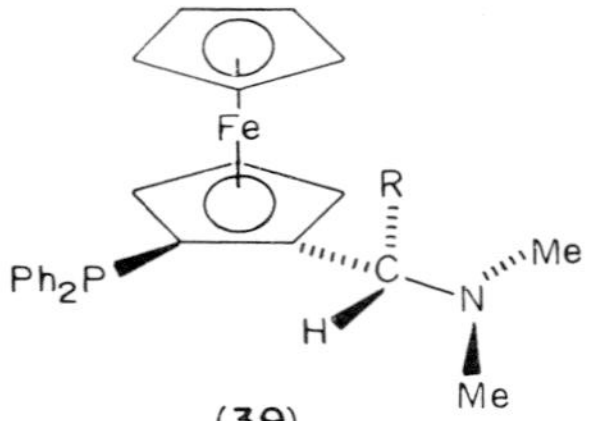 (39)	R = Me	C_6H_6–MeOH	20	25	67	*R*	153
	R = Ph	C_6H_6–MeOH	20	25	34	*R*	153

[a] 83% *R* when catalysed by $[RhCl(cod)]_2$ + 2(–)-diop[162].
[b] 81.1% *S* when catalysed by $[RhCl(C_2H_4)_2]_2$[163] or $[RhCl(cod)]_2$[162] + 2(+)-diop.

In view of the biochemical implications of asymmetric hydrogenation it is interesting to note the development of a water-soluble chiral catalyst[169].

2. *Cationic complexes*

Cationic complexes containing either two monodentate or one bidentate chiral ligand and an alkadiene ligand coordinated to a central rhodium(I) atom have proved to be the most generally useful catalysts for asymmetric hydrogenation reactions. The anion in these complexes is usually perchlorate or tetrafluoroborate, but it plays no part in the catalytic process and in general has little if any influence on the optical yields obtained[170]. The same cannot be said of the tetraphenylborate anion, a phenyl group of which may form an η^6-complex with rhodium and distort the optical yields obtained by its influence on substrate coordination[171].

a. Monodentate ligands

Comparatively few investigations have been made of the catalytic properties of this sub-group of complexes. It may be more probable that their catalyses follow the dihydride route, and thereby give lower optical yields than those obtainable from catalysts containing bidentate ligands. The optical yields obtained in their catalysed reductions of α-acetamidocinnamic acid are given in Table 3.

In the asymmetric hydrogenation of α-benzoylaminocinnamic acid, catalysed by perchlorate complexes of the ligands **40** and **41**, the optical yields were 5.8% *S* and 6.7% *S*, respectively. This reversal of product chirality shows the inherently low chiral discrimination of this type of catalyst[172].

b. Bidentate ligands

The catalytic properties of this sub-group of complexes has been exhaustively investigated. The optical yields obtained in their asymmetric reductions of α-acetamidocinnamic acid are given in Table 4. In general, there is little difference in the yields from cationic and neutral complexes containing the same chiral ligand[201], which makes the current fixation on the former difficult to understand since the latter are usually more accessible.

There is still some doubt about the route of asymmetric hydrogenation when this class of complex is employed. The variation of optical yield with increasing hydrogen pressure, or on lowering the temperature, when cationic complexes containing ligand **50** were employed was ascribed to increasing participation of the dihydride route[202]. However, it is generally agreed that the route showing the greatest asymmetric induction is the alkene route. Many investigations of the mechanistic details of this route have been made. Despite this, some dispute still exists as to which step is responsible for the asymmetric induction. One school of thought believes that product chirality arises from preferential coordination of one face of the alkene to the catalyst complex. This preferential coordination is assisted by simultaneous coordination via the alkene bond of α-acetamidocinnamic acid and its amide oxygen atom. However, it has been proposed by Chan *et al.*[203] that the least favoured isomer is the more rapidly hydrogenated. There is some support for this proposal in the observation that the *E*-isomer of α-acetamidocinnamic acid is isomerized to (*Z*)-α-acetamidocinnamic acid by diop complexes, which implies that hydrogen transfer is rate determining[204]. Isotope exchange has also been observed in the deuteriation reaction[205]. On the other hand, no exchange was reported in the deuteriation of α-benzoylaminocinnamic acid[206].

Again, the most popular alternative substrate to α-acetamidocinnamic acid has been its esters. Many investigations of the chiral hydrogenation of itaconic acid[207–209] or

TABLE 3. Optical yields obtained from α-acetamidocinnamic acid reductions catalysed by cationic rhodium complexes containing monodentate ligands

Ligand	Solvent	pH_2 (atm)	T (°C)	Optical yield (%)	Product chirality	Ref.
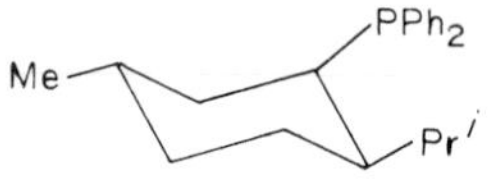 Me, PPh_2, Pr^i (40)		3	30	20.8	*R*	172
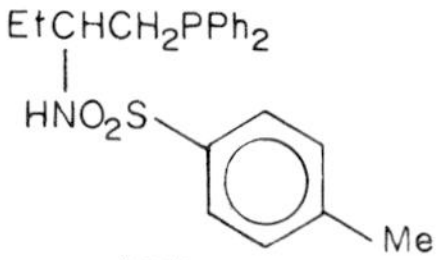 $EtCHCH_2PPh_2$, HNO_2S, Me (41)		3	50	6.2	*R*	172
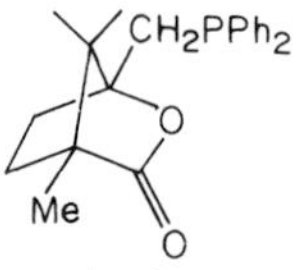 CH_2PPh_2, O, Me, O (42)		80	20	3.2	*S*	173
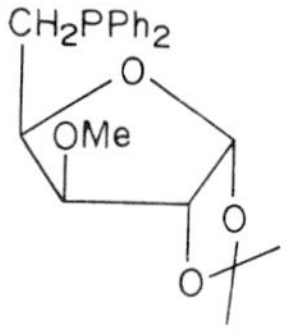 CH_2PPh_2, O, OMe, O, O (43)				20	*S*	174
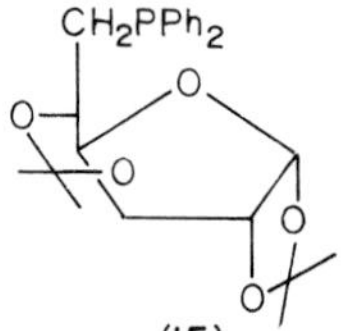 CH_2PPh_2, O, O, O, O, O (15)				21	*S*	174
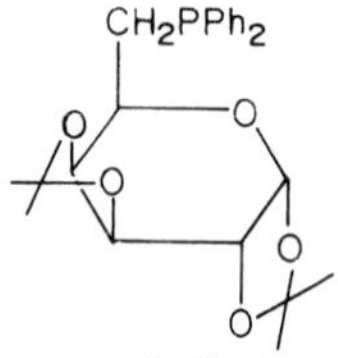 CH_2PPh_2, O, O, O, O, O (16)				15	*S*	174

TABLE 4. Optical yields obtained from α-acetamidocinnamic acid reductions catalysed by cationic rhodium complexes containing bidentate ligands

Ligand		Solvent	pH_2 (atm)	T (°C)	Optical yield (%)	Product chirality	Ref.
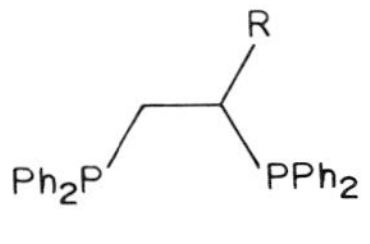 (20)	R = Me	MeOH	1		83	*S*	175
		thf			90	*S*	176
	R = Ph (*R*)	MeOH	1		79	*S*	175
	R = Ph (*S*)	MeOH	1	30	78	*R*	144
	R = C_6H_{11} (*R*)	MeOH	1		84	*S*	175
		thf	50		83	*S*	177
		MeOH	50		84	*S*	177
		EtOAc	50		91	*S*	177
	R = CH_2Ph (*R*)	C_6H_6–EtOH	20		99	*S*	178
	R = $PhCH_2OCH_2$	95% EtOH	1	25	87	*S*	179
	R = Bu^tOCH_2	EtOH	1	25	86	*S*	179
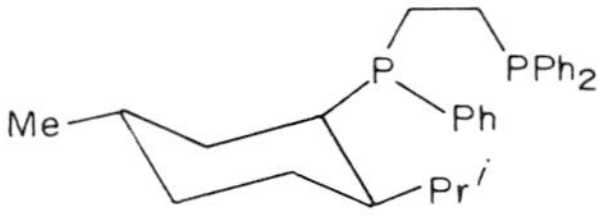 (19)	(+)		70	25	62	*R*	143
	(−)		70	25	60	*S*	143
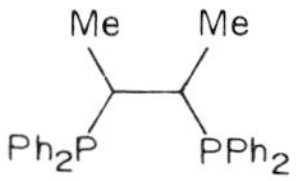 (2*R*,3*R*) (44)		EtOH	1–4		88		180

TABLE 4. (*Contd.*)

Ligand	Solvent	pH_2 (atm)	T (°C)	Optical yield (%)	Product chirality	Ref.
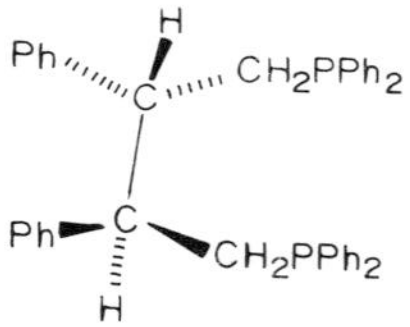 Ph, H, CH_2PPh_2 / Ph, CH_2PPh_2, H (*S*,*S*) (**45**)				54	*R*	181
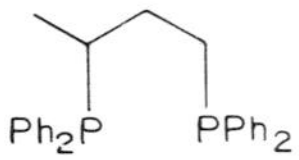 Ph_2P PPh_2 (*S*) (**46**)	EtOH	1	25	20	*S*	182
	thf	1	25	1	*S*	182
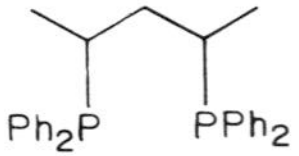 Ph_2P PPh_2 (*S*,*S*) (**47**)	thf			93	*R*	182
	EtOH			92	*R*	182
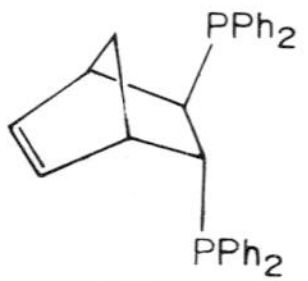 PPh_2 PPh_2 (+) (**24**)	MeOH	1	25	97	*R*	146

Ligand		Solvent					Ref.
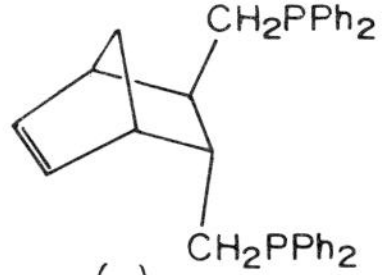 (−) (26)		EtOH	1	25	51	*S*	148
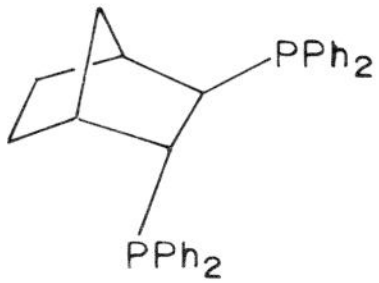 (48)	(+)	MeOH	1.6	24	95	*R*	183
		MeOH	65–70	24	30	*R*	183
	(−)	MeOH	1.6	24	95	*S*	183
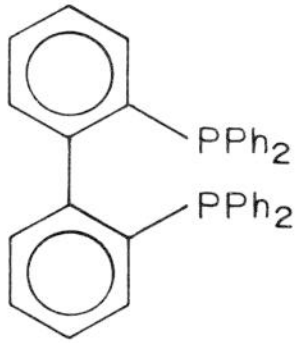 (49)			1		6.5	*R*	171
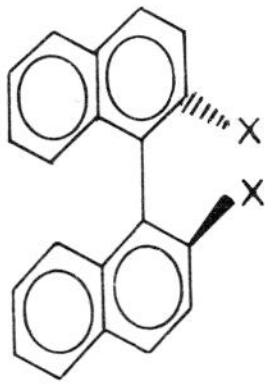 (28)	$X = PPh_2$	EtOH	3–4		84	*R*	184

TABLE 4. (*Contd.*)

Ligand		Solvent	pH_2 (atm)	T (°C)	Optical yield (%)	Product chirality	Ref.
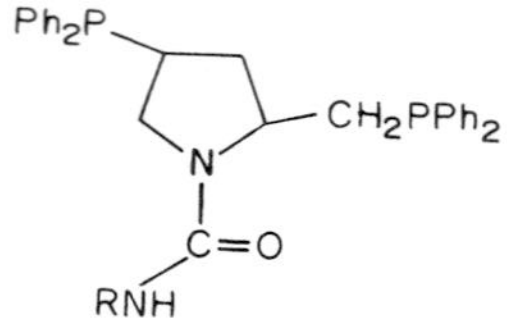 (33)	R = Me	EtOH	1	25	93.2	*S*	157
	R = $CH_2{=}CHCH_2$	EtOH	1	25	92.5	*S*	157
	R = Ph	EtOH	1	25	95.4	*S*	157
	R = C_6H_{11}	EtOH	1	25	93.2	*S*	157
	R = *p*-C_6H_4Cl	EtOH	1	25	95.4	*S*	157
	R = *p*-C_6H_4Br	EtOH	1	25	95.3	*S*	157
	R = 3, 4-$C_6H_4Cl_2$	EtOH	1	25	95.3	*S*	157
	R = *p*-$C_6H_4NO_2$	EtOH	1	25	95.7	*S*	157
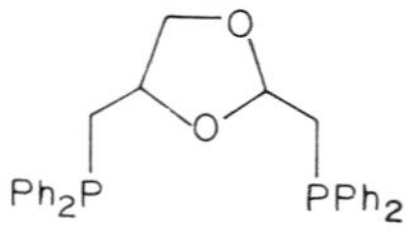 (2*R*, 4*R*) (50)			1	25	78	*S*	185, 186
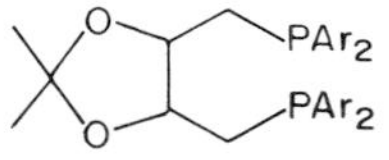 Ar = Ph (34)	Ar = Ph	C_6H_6–EtOH	1		81	*R*	187
		MeOH	1	30	78	*S*	188
		EtOH	1	30	84	*S*	188
		Pr^iOH	1	30	85	*S*	188
	Ar = *o*-C_6H_4Me	C_6H_6–EtOH	1		27	*R*	187
	Ar = *m*-C_6H_4Me	C_6H_6–EtOH	1		87.5	*R*	187
	Ar = 2, 5-$C_6H_3Me_2$	C_6H_6–EtOH	1		44	*R*	187
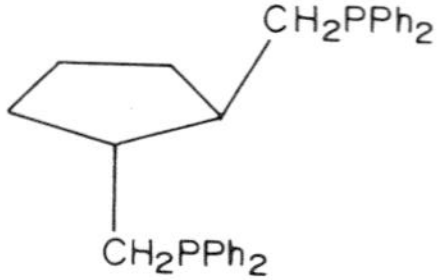 (1*R*,2*R*) (22)		C_6H_6–EtOH	1		63	*R*	187

Ligand		Solvent					Ref.
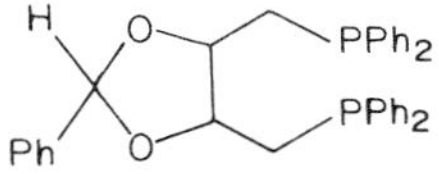 (51)		C_6H_6–EtOH	1		79.5	*R*	187
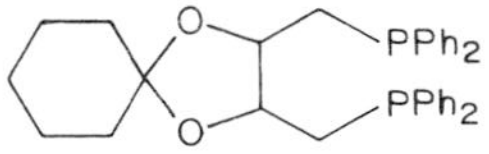 (52)			3	50	61	*R*	172
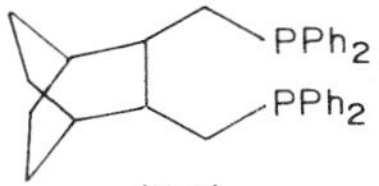 (53)		C_6H_6–EtOH	1		79.5	*R*	187
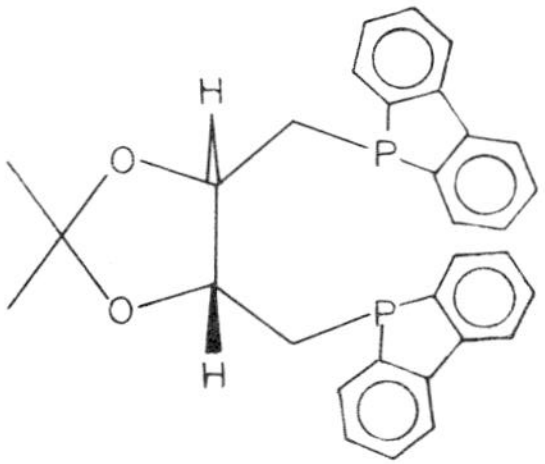 (54)		C_6H_6–EtOH	1		28	*S*	187
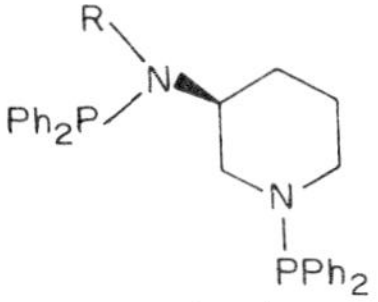 (55)	R = H	C_6H_6–EtOH			73	*S*	189
	R = Me	C_6H_6–EtOH	1		79	*S*	189

TABLE 4. (*Contd.*)

Ligand	Solvent	pH_2 (atm)	T (°C)	Optical yield (%)	Product chirality	Ref.
(56)		5		80.6	*S*	190
(57)		5		74.8	*S*	190
(31) R = Et		5		40.8	*S*	190
R = Ph		5		77.3	*R*	190
	MeOH	1	25	81.7	*R*	191

Ligand		Solvent					Ref.
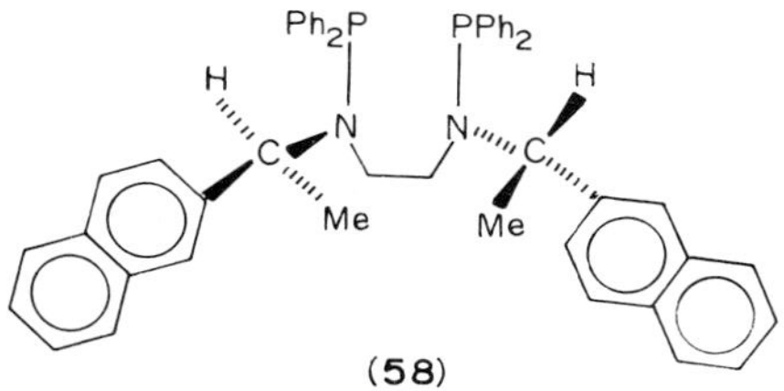 (58)			5		91.9	*R*	190
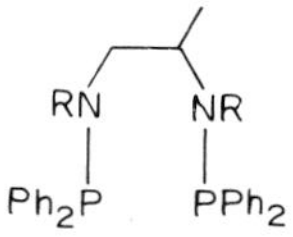 (59)	R = H	EtOH	1		28	*R*	192
	R = Me	EtOH	1		70	*S*	192
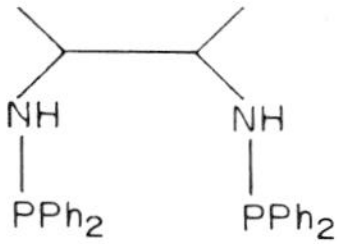 (60)		EtOH	1		45	*S*	192
		Pr^iOH	8	25	80	*S*	193
		$Pr^iOH–C_6H_6$	8	25	94	*S*	193
		C_6H_6–EtOH			62	*S*	194
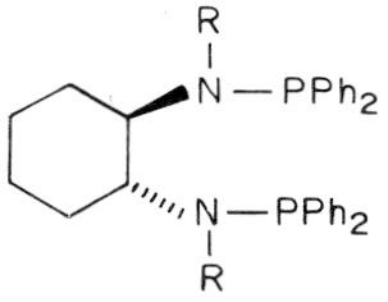 (32)	R = H		5		47	*R*	190
		EtOH	1		49	*R*	192
		EtOH	8	25	41	*R*	193
	(*S*, *S*)	EtOH	8	25	41	*S*	193
		Pr^iOH	8	25	71	*S*	193
		C_6H_6–EtOH			72	*S*	194
	R = Me		5		91.2	*S*	190
		EtOH	1	25	89	*S*	192

TABLE 4. (*Contd.*)

Ligand		Solvent	pH_2 (atm)	T (°C)	Optical yield (%)	Product chirality	Ref.
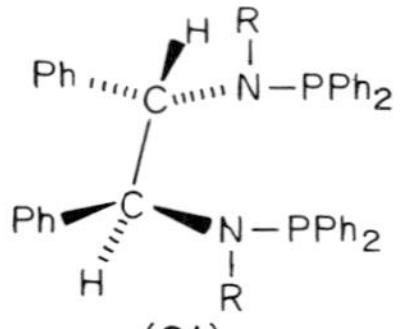 (61)	R = H	EtOH	5		93.8	*S*	190
			1		93	*S*	192
	R = Me	EtOH	5		68.4	*R*	190
			1		68	*R*	192
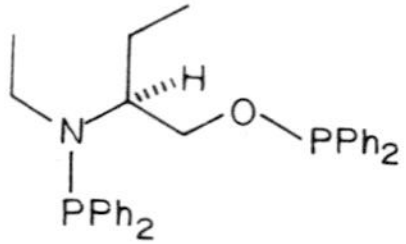 (62)		EtOH	1	20	23	*S*	195
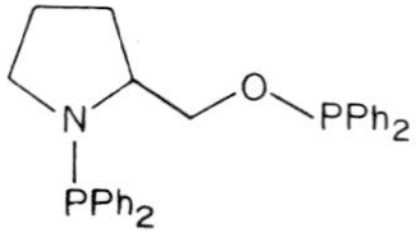 (63)		EtOH	1	20	50	*S*	195
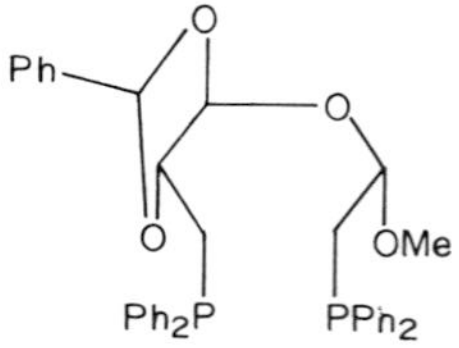 (64)		EtOH	1	20	78	*S*	174

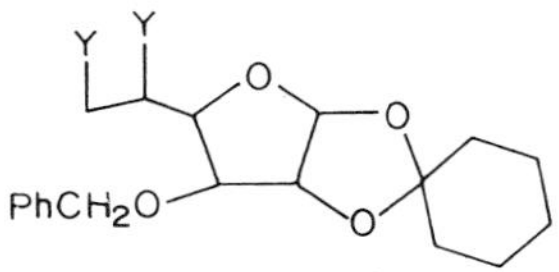 (65)				0	65	*S*	196
				30	60	*S*	196
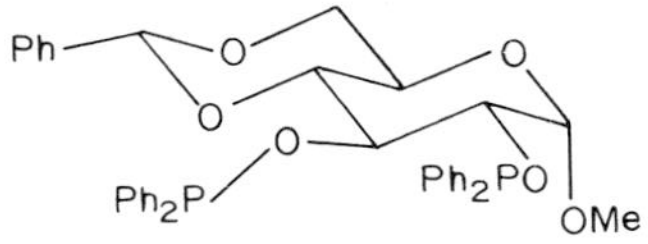 (66)		EtOH	10	25	60	*S*	197
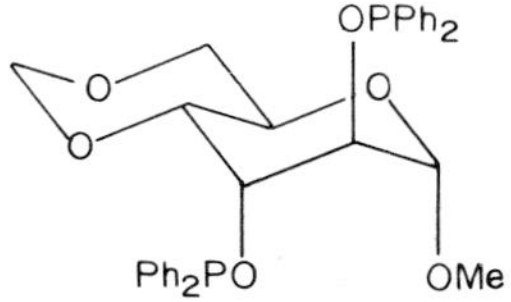 (67)		EtOH	10	25	52	*R*	197
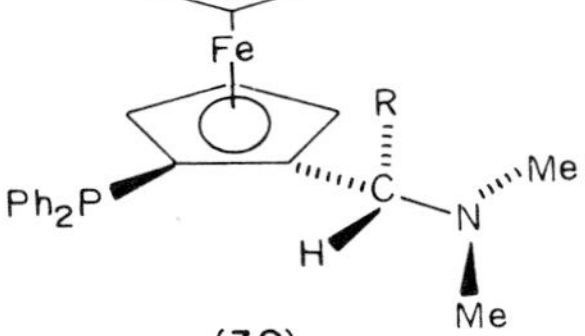 (39)	(*R*, *S*); *R* = Bu^t		1	30	76	*R*	198
	(*R*, *S*); *R* = Ph		1	30	84	*S*	198
	(+)	MeOH	1	32	73	*R*	199, 200
		EtOH	1	32	83	*R*	199, 200
		Pr^iOH	1	32	80	*R*	199 200
	(−)	MeOH	1	32	80	*S*	199, 200
		EtOH	1	32	75	*S*	199 200

amidoacrylic acid[210–213] have also been made. More recently, attempts have been made to bring about the chiral hydrogenation of dehydropeptides[214,215]. It has been stated that the chiral centre in the dipeptide precursor does not influence the direction of asymmetric induction at the alkene bond[216].

In all these cognate reactions there has been little or no attempt to tailor the chiral ligand to the needs of the new substrate. With the 'lock and key' nature of much chiral catalysis it is very probable that improved optical yields could be obtained by suitably modifying the chiral catalyst. One small step in this direction has been the incorporation of a long-chain aliphatic residue in the chiral ligand to enhance the solubility of the catalyst in hydrocarbon solvents[217].

3. *Heterogenized catalysts*

Although incorporation of a large alkyl group in the chiral ligand was claimed to have little influence on the optical yields obtained, other workers have found that comparatively minor changes in the support bring about large changes in the optical yields obtained from heterogenized systems[218]. In general, the optical yields obtained from heterogenized systems are much lower than those from homogeneous catalysts of similar composition[219–221]. Further, the optical yields declined when the heterogenized catalyst was recycled[222,223]. There is one claim that copolymerization of pyrrolidine ligands and methacrylate monomers form the basis of a catalyst capable of giving high optical yields[224]. There is also one report that binding tertiary phosphine residues to a chiral support, in this case cellulose, can form heterogenized species whose rhodium complexes can bring about chiral hydrogenation[225].

B. Ketone Substrates

Cationic rhodium(I) complexes of the type $[RhL_2(alkadiene)]^+$ can function as catalysts in the hydrogenation of ketones under mild conditions (see Chapter 12). If the ligand L is a chiral tertiary phosphine, then suitable ketonic substrates can give rise to chiral secondary alcohols. The chiral tertiary phosphines used have been chiral at phosphorus and the optical yields obtained have been far below those obtained in the asymmetric hydrogenation of α-acetamidocinnamic acid[226–229].

The general observation can be made that the optical yields apparently improve with increasing bulk of the ketone's substituents. One reason for the low optical yields may be the elevated temperatures and hydrogen pressures that have occasionally been used to effect the reduction[230]. It has been claimed that increasing the tertiary phosphine to rhodium ratio improves the optical yields[229].

In the few instances that catalysts containing one chiral ditertiary phosphine ligand have been employed, higher optical yields have been obtained.[227,229,231,232] Higher optical yields are obtained from substituted ketones, particularly if the substituent can complex to rhodium in conjunction with the keto group. Thus optical yields in the region of 90% can be obtained from ketones containing the 2-*N*, *N*-diethylaniline group[233]. Pyruvic esters give moderate to good optical yields even when reduced by hydrogen at 20 atm pressure. However, there are marked solvent effects on both the chemical and optical yields[234,235]. Moderate yields of (*R*)-pantolactone can be achieved in reaction 65

$$\text{(ketopantolactone)} \xrightarrow[\text{RhL}_n]{\text{H}_2} \text{(pantolactone, HO)} \qquad (65)$$

when this is catalysed by cationic rhodium complexes containing pyrrolidone ligands[236,237].

V. DEHYDROGENATION AND TRANSFER HYDROGENATION

A. Dehydrogenation

Rhodium complexes are not effective dehydrogenation catalysts and find little application as transfer hydrogenation catalysts. In both these respects they are markedly inferior to the ruthenium complex $RuCl_2(PPh_3)_3$[86]. The poor performance of rhodium complexes in these two areas enhances their applicability to other catalytic processes since the reactions described in this section often give rise to unwanted byproducts.

There would be considerable commercial benefit if rhodium complexes were able to dehydrogenate alkenes catalytically to terminal alkenes with the ease and efficiency so characteristic of their catalyses of the reverse reaction. However, very few substrates are dehydrogenated by complexes more properly regarded as hydrogenation catalysts, since alkanes do not coordinate to the metal and provide a pathway for the reaction. The very few species that have been dehydrogenated, often under severe conditions, are sundry hydroaromatic compounds. Above 200 °C low yields of aromatic compounds can be obtained from the $RhCl(PPh_3)_3$ catalysed reactions shown equations 66 and 67.

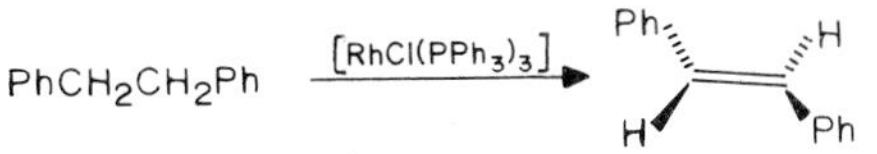

$$PhCH_2CH_2Ph \xrightarrow{[RhCl(PPh_3)_3]} PhCH{=}CHPh \quad (66)$$

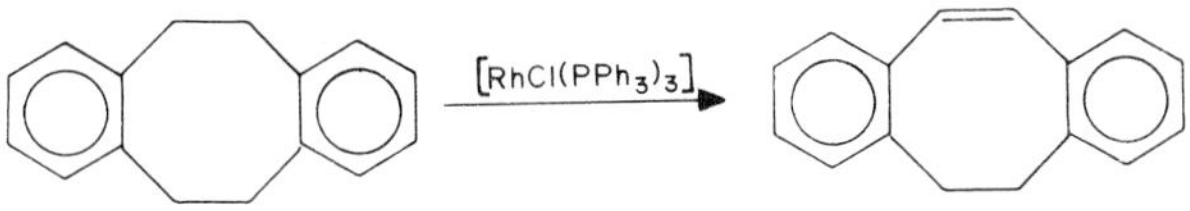

[RhCl(PPh₃)₃] (67)

Nevertheless, a 97% yield of anthracene can be obtained from 9, 10-dihydroanthracene at 225 °C after reaction for 15 h. Poorer yields result when the rhodium(III) complex $RhCl_3(AsPh_3)_3$ is employed[238].

Substrates containing a heteroaton that can coordinate to rhodium give better yields of dehydrogenated products (equation 68). The yields depend on the R substituent, being 22.7% for the parent compound but 74.5% if R = Ph[239].

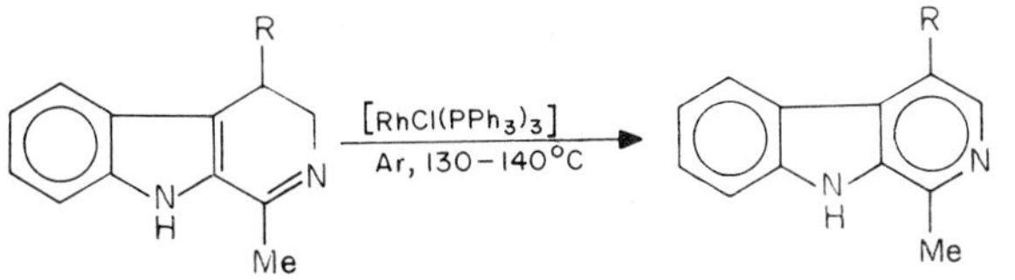

[RhCl(PPh₃)₃], Ar, 130–140 °C (68)

R = H, Me, Pr, Bu^i, Ph, Bz, 2-C_5H_4N, 3-C_5H_4N, 4-C_5H_4N

Several rhodium complexes can catalyse the dehydrogenation of secondary alcohols to ketones. A mixture of rhodium(II) acetate and PPh_3 catalyses the slow formation of acetone from propan-2-ol[240]. The removal of the acetone produced increases the rate of propan-2-ol dehydrogenation in the $[RhCl(SnCl_3)_2]_2$-catalysed system[241]. This may prove beneficial in other instances.

The photocatalytic dehydrogenation of this alcohol can take place in the presence of small quantities of $RhCl(PPh_3)_3$[242] or chloro(tetraphenylporphinato)rhodium[243].

A mere 1% optical yield has been achieved in the attempted chiral dehydrogenation of benzalacetone in the diphenyl(neomenthyl)phosphine/$[RhCl(C_2H_4)_2]_2$ system[244]. This catalyst has also been used to dehydrogenate (*R*)-1-phenylethanol preferentially in the presence of an alkene hydrogen acceptor. The *S*-isomer reacts more slowly and the initially racemic alcohol becomes enriched in this component as the reaction proceeds[245].

One of the more successful rhodium-based dehydrogenation catalysts is the *ortho*-metallated complex $[Rh\{(o\text{-}C_6H_4)PPh_2\}(PPh_3)_2]$. This decomposes formic acid by the sequence of reactions shown in equation 69–71[246].

$$Rh(o\text{-}C_6H_4PPh_2)(PPh_3)_2 + HCO_2H \longrightarrow Rh(O_2CH)(PPh_3)_3 \tag{69}$$

$$Rh(O_2CH)(PPh_3)_3 \longrightarrow RhH(PPh_3)_3 + CO_2 \tag{70}$$

$$RhH(PPh_3)_3 + HCO_2H \longrightarrow Rh(O_2CH)(PPh_3)_3 + H_2 \tag{71}$$

B. Transfer Hydrogenation

Despite the low efficiencies of rhodium complexes in dehydrogenation reactions, several complexes are fairly active in transfer hydrogenation reactions. It is possible that this dichotomy arises from the stability of the rhodium hydrido complexes produced in the dehydrogenation reactions and the failure of these complexes to release hydrogen except to a hydrogen acceptor. Thus, in transfer hydrogenation several criteria must be met simultaneously: the hydrogen donor must coordinate to the catalytic complex and be dehydrogenated; the dehydrogenated compound must not coordinate to the catalytic complex and thereby poison it; the hydrogen acceptor must next coordinate to the catalyst and the hydrogen atoms be transferred to it; and further, the complexity constants of the substrate and hydrogen donor with the catalyst should be approximately equal to avoid either of them inhibiting the reaction by preferentially occupying a catalyst site required by the other.

1. Alkenes

It is for the above reason that the equivalent of the alkene route is seldom proposed as the mechanism for transfer hydrogenation reactions. Prior coordination of the substrate to the catalyst would form a coordinatively saturated species unable to activate the hydrogen donor[247].

The presence of hydrido species is indicated by the hydroisomerization reactions that occur during certain transfer hydrogenation reactions. Both $[RhCl(PPh_3)_3]$ and $[RhH(PPh_3)_3]$ have been shown to catalyse the transfer hydrogenation of cycloalkenes. Imai and his coworkers[248] have shown $[RhH(PPh_3)_3]$ to be an effective catalyst for the reaction and superior even to the more commonly employed ruthenium catalysts. The rates of hydrogenation achieved with propan-2-ol as the hydrogen donor were an order of magnitude greater than when other primary or secondary alcohols were used as the hydrogen source. Linear alkenes were isomerized by $[RhH(PPh_3)_3]$ during the catalysis. Although addition of acetone suppressed the rate of transfer hydrogenation, it had no effect on the rate of isomerization[248]. This result implies that there is a standing concentration of a monohydrido complex during the reaction. Other monohydrido catalysts of the type $[RhH(CO)(PR_3)_3]$ ($R_3 = Et_3$, Bu_3, Et_2Ph, $MePh_2$) have been used to transfer hydrogen from benzyl alcohol to oct-1-ene, but no mention was made of isomerization occurring during the reactions[249].

Primary alcohols are seldom employed as hydrogen donors as aldehydes are formed from them in this reaction. The aldehydes produced can then bring about catalyst deactivation via carbonyl abstraction[250]. Indeed, the formation of $[RhCl(CO)(PPh_3)_2]$ from $[RhCl(PPh_3)_3]$ has been noted in these reactions[251].

The use of perdeuteriodioxane has confirmed that this compound is the hydrogen donor[252] in the $[RhCl(PPh_3)_3]$-catalysed transfer hydrogenation of cyclopentene[253]. The large kinetic isotope effect shows that dehydrogenation is the rate-determining step in the reaction. When triphenylphosphine ligands were replaced by other triarylphosphine ligands, it was found that electron-withdrawing substituents in the aryl groups brought about a large decrease in the rate of transfer hydrogenation[254].

Tetrahydroquinoline or indoline can be used as hydrogen donors in the $[RhCl(PPh_3)_3]$-catalysed hydrogenation of cycloheptene[255]. Acetatotris(triphenylphosphine)rhodium(I) has been claimed to catalyse the selective transfer hydrogenation of dienes to monoenes. The hydrogen donor was *p*-toluenesulphonic acid. The rate of hex-1-ene hydrogenation increased with increasing acid concentration and internal alkenes did not react except at high acid concentrations[256]. Generally, however, little effort has been expended in investigating the transfer hydrogenation of alkenes.

2. Ketones

The transfer hydrogenation of ketones has been more frequently studied (equation 72). This is not because of the difficulty of transferring hydrogen to alkenes since unsaturated ketones are usually transfer hydrogenated to saturated ketones[257–260]. In these reactions the rate-determining step has been found to be the scission of the O—H bond in the alcohol. This hydrogen atom is transferred selectively to the α-carbon atom of the ketone[259].

$$\mathrm{PhCH(OH)Me + PhCH{=}CHCOMe \xrightarrow{[RhH(PPh_3)_4]} PhCOMe + PhCH_2CH_2COMe} \qquad (72)$$

Chlorotris(triphenylphosphine)rhodium(I) differs from $[RhH(PPh_3)_4]$ in that it first converts cyclohex-2-enone into cyclohex-2-enol[261]. The $[RhCl(PR_3)_3]$ complexes prepared *in situ* from $[RhCl(C_2H_4)_2]_2$ or $[RhCl(cod)]_2$ have been used to catalyse the transfer hydrogenation of methylcyclohexanones. Two isomeric products were formed, the thermodynamically most stable being formed in excess (equations 73 and 74)[262].

3-methylcyclohexanone + Pr^iOH $\xrightarrow[\text{Ar, 80°C}]{RhClL_n}$ (OH, Me-substituted cyclohexanol) + (OH, Me-substituted cyclohexanol) (73)

4-methylcyclohexanone + Pr^iOH $\xrightarrow[\text{Ar, 80°C}]{RhClL_n}$ (Me, OH cyclohexanol) + (HO, Me cyclohexanol) (74)

When 4-*tert*-butylcyclohexanone is the substrate and $[RhCl(PCy_3)_2]$ the catalyst, the *cis*-product is produced in greater yield[263]. Similar product distributions have been observed when $[RhCl(PPh_3)_3]$ was the catalyst. This complex also catalyses the formation of *cis*-2-methylcyclohexanol from 2-methylcyclohexanone[264]. The cationic complexes $[Rh(cod)L_2]^+$[265,266], $[RhLL(1,5\text{-}C_6H_{10})][PF_6]$ (LL = phen, bipy)[267], $[Rh(nbd)(Pr^i_2P(CH_2)_3PPr^i_2)][ClO_4]$[263], and $[Rh(nbd)(PR_3)_2][ClO_4]$[268] all catalyse the transfer hydrogenation of ketones. The last complex also catalyses the transfer hydrogenation of hex-1-ene, cyclohexene, or styrene[268].

The mechanism of hydrogen transfer from secondary alcohols to ketones catalysed by [$RhCl(PPh_3)_3$] and KOH has been investigated. However, as these reagents form [$RhH(PPh_3)_3$], the true catalyst in the reactions is uncertain[269]. The nature of the secondary alcohol has little effect on the rate, but the rate of cycloalkanol formation could be correlated with the strain energies of the cycloalkanol rings[270]. Further evidence for the rate-determining step being transfer of hydrogen to the ketone comes from the lower reduction rates of substituted cyclic ketones[271,272].

There have been no reports of the transfer hydrogenation of aldehydes, but acetaldehyde disproportionates in the presence of [$Rh(C_5Me_5)]_2Cl_4$[273].

3. Other multiple interatomic bonds

Chlorotris(triphenylphosphine)rhodium(I) and KOH catalyse the transfer hydrogenation of several nitrogenous substrates, again using secondary alcohols as the hydrogen donors. Nitrobenzene can be reduced via azoxybenzene to aniline[274,275], and 4-alkylpiperindones are also reduced in the system[276]. Sundry Schiff bases can also be reduced to secondary amines[277].

Mechanistic studies have been confined to confirming that the source of transferred hydrogen is the hydroxyl proton by isotopic labelling[278]. The catalyst is poisoned by the addition of styrene or phenylacetylene[279]. This study again indicates the importance of a vacant site on the catalyst for substrate and donor activation. However, thiophene does not poison the catalyst since it is hydrogenolysed to but-2-ene and H_2S by transfer hydrogenation from decalin[280].

VI. HYDROFORMYLATION

Hydroformylation (see also Chapter 8 in Volume 3) is the addition of the elements of formaldehyde across the double bond of an alkene. The process has been catalysed by a variety of transition metal complexes.

Carbonylhydridotris(triphenylphosphine)rhodium(I) has now supplanted [$CoH(CO)_4$] as a commercial hydroformylation catalyst[95]. However, [$RhH(CO)(PPh_3)_3$] was not recognized for some time as the true catalyst in triphenylphosphine-stabilized rhodium systems. The catalysts were successively believed to be $RhCl(PPh_3)_3$[281] or $RhCl(CO)(PPh_3)_2$[282], until it was shown that these complexes were merely precursors of the true catalyst under hydroformylation conditions.

Many other rhodium(I) complexes are precursors of the catalytic species in the presence of PPh_3, CO, and H_2. These include [$RhCl(CO)_2]_2$[283,284], [$RhCl(nbd)]_2$[285], [$Rh(diene)_2][ClO_4]$[286], and [$Rh(cod)(PPh_3)_2][PF_6]$[287]. The hydrido complex [$RhH(PPh_3)_4$] is even more easily converted into [$RhH(CO)(PPh_3)_3$] than the above chloro or ionic complexes[288]. Ease of conversion and the avoidance of plant corrosion by HCl probably explains why [$Rh(acac)(CO)_2$] is the commercial catalyst precursor[289]. The rhodium(III) species [$RhCl(\eta^5\text{-}C_5H_5)_2]_2$ also forms the catalyst under severe operating conditions[283]. Carbonylhydridotris(triphenylphosphine)rhodium(I) is also believed to be formed from rhodium metal, again under severe operating conditions[290,291]. However, neither triphenylphosphine nor tertiary phosphines are essential to the catalytic activity of rhodium compounds in hydroformylation reactions. The tertiary phosphine complexes [$RhH(CO)\{PPh_2(m\text{-}C_6H_4SO_3Na)\}_3$][292], the tris(2-pyridyl)phosphine complex [$RhH(CO)(Ppy_3)(PPh_3)_2$][293], and the dibenzophosphole complex [$RhH(dbp)_4$][288] all function as hydroformylation catalysts under relatively mild conditions. The first of these complexes is of interest since it can function as a water-soluble catalyst[292]. The secondary

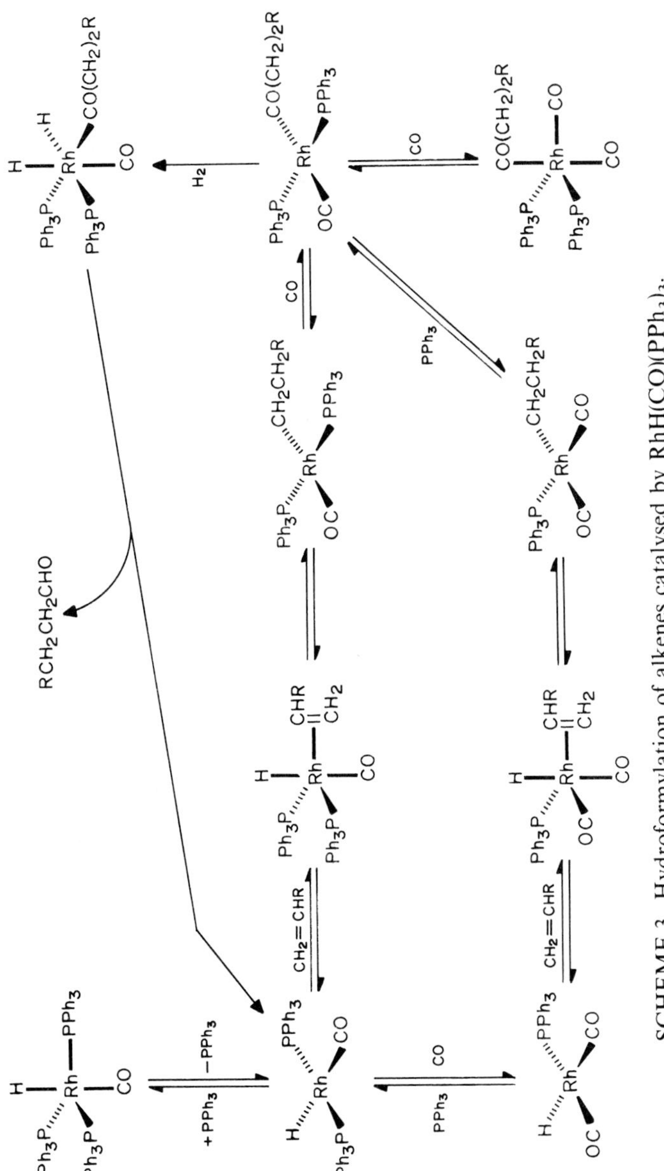

SCHEME 3. Hydroformylation of alkenes catalysed by $RhH(CO)(PPh_3)_3$.

phosphine complex $[RhCl(PHPh_2)_3]$[294] and the phosphido complex $[Rh(\mu\text{-}Bu^t{}_2P)(CO)_2]_2$[295] are also catalytically active, although both require pressures of several atmospheres to function effectively. Nitrogen[296,297] or arsenic[298] ligands also give rise to catalytically active species. Even hydrated rhodium trichloride in the presence of norbornadiene brings about the reaction. Unfortunately, this simple catalyst also permits the aldehyde produced to be reduced and the final products from the reaction are acetals[299]. Catalyst precursors and the product distribution from their catalyses have been reviewed[300].

It is important that the three reactants in hydroformylation are activated by the catalyst in the correct order. The monohydrido catalyst first coordinates the alkene to a vacant coordination site and forms an alkyl complex. The alkyl complex should react preferentially with carbon monoxide to form an acyl complex rather than undergo competing isomerization[301,302] or hydrogenation reactions which reduce the yield of aldehyde. Finally, the acyl complex undergoes oxidative addition of hydrogen followed by reductive elimination of aldehyde (Scheme 3).

Commercially, terminal aldehydes are of far greater value than internal aldehydes, and many modifications of the catalytic system have been made to improve the yield of terminal aldehyde at the expense of the latter product. Internal aldehydes arise from the formation of a 2-alkyl complex during the early part of the catalytic cycle. These then undergo a series of parallel reactions to the more stable 1-alkyl complexes. Improved yields of terminal aldehydes are obtained by arranging the reaction conditions such that a crowded alkyl complex is formed. The formation of 2-alkyl complexes is sterically inhibited and, if formed, the 2-alkyl complex is less stable than the 1-alkyl complex and may decompose before forming an acyl complex. The 2-alkyl complex can also arise from reaction of the catalyst with an alk-2-ene formed from the terminal alkene by catalytic isomerization[302].

Formation of branched-chain aldehydes is also favoured by impure feedstocks. Traces of oxygen or hydroperoxides can oxidize the tertiary phosphine ligands of the complex catalyst and form a species of low coordination number which favours 2-alkyl production. Normally these oxidants react with the excess of ligand or even the carbon monoxide present[303]. Eventually, however, they destroy the catalyst and are best removed[304].

Attempts have been made to improve the selectivity of the system by using ditertiary phosphine ligands, as these should remain coordinated to rhodium during the catalytic cycle[305–308]. However, their selectivity towards terminal aldehydes is not high. This could have been anticipated had due regard been paid to the importance of both triphenylphosphine ligands being mutually *trans* but *cis* to the alkyl ligand (Scheme 3). To this end, attention could have been better directed towards the beneficial effects of bulky tertiary phosphines[309].

A. Hydrocarbon Substrates

Whilst most interest has been concentrated on the commercially important hydroformylation of linear alk-1-enes, other substrates have attracted their share of attention. The regioselectivity of catalytic hydroformylation is well illustrated by the reactions of alkadienes. Octa-1, 6-diene only undergoes substitution at the terminal bond, both 1- and 2-aldehydes being formed. The latter is the minor product[310]. Only the vinyl group is hydroformylated in vinyldimethylcyclohexene[311]. Buta-1, 3-diene undergoes some 1,4-addition to form $MeCH{=}CHCH_2CHO$ which, unlike the other primary product $CH_2{=}CHCH_2CH_2CHO$, cannot react further to form a dialdehyde[312]. It has been claimed that secondary phosphines and secondary diphosphines favour 1,4-addition[313]. Analogous behaviour is observed with bis(alkenyl)amines (equation 75)[314].

(75)

Although internal alkenes are not readily hydroformylated, regioselectivity is observed in their reactions also (equation 76)[315]. Cycloalkadienes also form dialdehydes under

(76)

harsh conditions. An interesting byproduct in the hydroformylation of cyclohexa-1,4-diene arises from its isomerization to cyclohexa-1,3-diene, which then undergoes 1,4-addition and finally hydrogenation to form a monoaldehyde. Cycloocta-1,5-diene behaves similarly, forming the saturated monoaldehyde exclusively, as does cycloocta-1,3-diene[316]. Cycloheptatriene also forms large quantities of the monoaldehyde[317]. In the hydroformylation of cycloalkadienes that cannot isomerize in this way, the least hindered double bond reacts (equation 77)[318].

(77)

Few attempts have been made to hydroformylate alkynes[281]. Their formation of relatively stable vinyl complexes[319–321] with $[RhH(CO)(PPh_3)_3]$ inhibits the hydroformylation.

B. Substituted Alkenes

Unless the substrate reacts stoicheiometrically with the catalyst to form an inactive complex, substituted alkenes can be hydroformylated by $[RhH(CO)(PPh_3)_3]$[95]. Both ethers[322] and dinitriles[323] can be hydroformylated at elevated temperatures and pressures. Whilst the former yield mainly terminal aldehydes, the latter forms a 2-aldehyde. Allyl *tert*-butyl ether gives a similar product distribution[324–326]. Only a terminal aldehyde is formed in the hydroformylation of 2-methoxy-2,6-dimethyloct-7-ene[327].

The hydroformylation of the esters of unsaturated carboxylic acids can give rise to terminal aldehydes. However, the harsh reaction conditions often required favour the production of 2-aldehyde[328–330]. Harsh conditions are required because coordination of an ester's carbonyl group to rhodium's normally vacant sixth coordination site inhibits the reaction[331]. Indeed, this chelation can direct the course of substitution in the hydroformylation of vinyl esters. The more stable five-membered ring gives rise to the major product[332]. In view of the importance of chelation, it is interesting to note that the thioether $MeSCH_2CH{=}CH(CO_2Me)$ can be hydroformylated, albeit under severe conditions[333].

The hydroformylation of unsaturated alcohols can give rise to mixtures of products as allylic alcohols in particular tend to form substituted tetrahydrofurans[95,334]. However, a pyran is formed from isopulegol (equation 78)[335]. Non-allylic alcohols usually undergo straightforward hydroformylation and yield terminal aldehydes[95].

$$\text{isopulegol} \xrightarrow[\text{H}_2,\text{CO}]{[\text{RhCl(PPh}_3)_3]} \text{pyran product} \quad (78)$$

C. Heterogenized Catalysts

The industrial importance of the hydroformylation reaction has led to the development of heterogenized catalysts so that the catalyst may be separated more easily from involatile products. Both supported and polymer bound catalysts have been used.

In the former case the catalyst is absorbed on a porous inert support. The main disadvantage of this type is bleeding of the catalyst from the support during the reaction. The results are very dependent on the type of support and the degree of pore filling, since only those catalyst molecules which lie at a phase boundary are active. Generally, the selectivity towards terminal aldehydes is low and competition from catalytic hydrogenation more severe than in homogeneous systems. Dissolving $[RhH(CO)(PPh_3)_3]$ in triphenylphosphine before applying it to the support was claimed to minimize these disadvantages[336–340a].

Similar disadvantages attend the use of polymer-bound catalysts. Usually the selectivity towards linear aldehyde production is lower than in corresponding homogeneous systems, although there are important exceptions, as mentioned in Chapter 14 (section V.D.2)[340b,c]. Again, using a high tertiary phosphine to rhodium ratio improves the selectivity[341], and the support can contribute to this[340b].

D. Chiral Hydroformylation

In chiral hydroformylation reactions, the reverse regioselectivity to normal hydroformylation is usually sought, since only the branched-chain aldehyde produced from an achiral terminal alkene can be chiral (equation 79). A successful chiral hydroformylation catalyst should give high chemical and optical yields of the 2-aldehyde.

$$\text{RCH}{=}\text{CH}_2 \longrightarrow \text{R}\overset{*}{\text{C}}\text{HMe}(\text{CHO}) + \text{RCH}_2\text{CH}_2\text{CHO} \quad (79)$$

As is the case in chiral hydrogenation, these regio- and stereo-selectivities of the catalyst are more readily achieved by the incorporation of chiral ditertiary phosphine ligands. However, there have been two investigations of catalysts containing the chiral tertiary phosphine (*R*)-benzylmethylphenylphosphine. Since styrene normally gives high yields of hydrotropaldehyde, PhCH(CHO)Me, this alkene is usually the substrate in asymmetric syntheses[342]. Stereo- and regio-selectivities were found to be incompatible. As is the case in achiral hydroformylation, a low concentration of tertiary phosphine is conducive to the production of hydrotropaldehyde[343]. However, this low concentration of tertiary phosphine does not favour high optical yields. Harsh conditions, as in most asymmetric syntheses, reduce the optical yields[344]. The effect of temperature is particularly noteworthy when neomenthyldiphenylphosphine is the chiral ligand. Excess of (*S*)-hydrotropaldehyde is produced at 75 °C, but the (*R*)-isomer predominates at 20 °C[344]. A similar low optical yield of hydrotropaldehyde is obtained at 75 °C and 100 atm pressure when *trans*-[RhCl(CO){PPh_2(neomenthyl)}]$_2$ is the catalyst[345].

Better optical, if not chemical, yields have been obtained when chiral bidentate ligands have been employed. Thus, *cis*-[RhCl(CO)(diop)] in the presence of excess of diop gives hydrotropaldehyde in 57% optical yield. An even higher optical yield of 71% can be achieved using a larger excess of diop in the $[RhCl(C_2H_4)_2]_2$/diop catalytic system. As is the case with monodentate chiral ligands, harsher reaction conditions increase the conversion to hydrotropaldehyde but decrease its optical purity[346]. It has been claimed that diphol, **54**, used in conjunction with $[RhCl(CO)_2]_2$, gives even better optical yields than diop. Strangely, (−)-diphol, although closely related to (−)-diop, preferentially forms (*S*)-hydrotropaldehyde whilst the latter forms more (*R*)-hydrotropaldehyde[347].

The chiral hydroformylation of suitable internal alkenes gives rise to two isomeric products whose chirality differs (equation 80)[348]. The optical yields obtained from (*Z*)-alkene substrates are superior to those obtained from the (*E*)-alkene isomers. However, the same optical isomer is formed in excess whichever substrate is selected[349]. The regioselectivity of the reaction can give rise to different isomers. But-1-ene gives a different chiral product to *cis*-but-2-ene or *trans*-but-2-ene (equation 81)[347,349]. It has also been noted that the two products obtained from hex-2-enes have different chirality[349].

$$\mathrm{RCH{=}CHMe + H_2 + CO \xrightarrow[(-)\text{diop}]{[RhH(CO)(PPh_3)_3]} (\mathit{S})\text{-}RCH_2\underset{\displaystyle Me}{\underset{|}{C}}HCHO + (\mathit{R})\text{-}R\underset{\displaystyle CHO}{\underset{|}{C}}HCH_2Me} \qquad (80)$$

$$\mathrm{EtCH{=}CH_2 + H_2 + CO \xrightarrow[\text{diphol}]{[RhCl(CO)_2]_2} (\mathit{S})\text{-}EtC\underset{\displaystyle CHO}{\underset{|}{H}}Me} \qquad (81)$$

The only way to reverse the prevailing chirality is to employ a different catalyst. For example, (−)-diphol and (−)-diop again exhibit different stereoselectivities in the chiral hydroformylation of both but-1-ene and *cis*-but-2-ene[347]. These two chiral, bidentate ligands have also been employed in the chiral hydroformylation of vinyl acetate[350,351]. Two products can be formed from this unsaturated ester (equation 82). The terminal aldehyde decomposes to acrolein, which is reduced to propanal in the system[351]. When (−)-diop is the chiral ligand, optical yields of up to 23% have been obtained[350]; however, attempts to obtain higher stereoselectivity have not always been successful. Although diphol gave the best optical yields (*ca.* 56%), increasing the bulk of the substituents on phosphorus by replacing the four phenyl groups of diop by four 1-naphthyl substituents gave very poor yields[351].

$$\mathrm{H_2C{=}CH(OAc) + H_2 + CO \xrightarrow{RhL^*_n} Me\underset{\displaystyle CHO}{\underset{|}{C}}H(OAc) + \underset{\displaystyle CHO}{\underset{|}{C}}H_2{=}CH_2(OAc)} \qquad (82)$$

Rhodium/diop complexes also catalyse the hydroformylation of 2-methylbuta-1,3-diene. Only the unsubstituted alkene bond is hydroformylated, the other being hydrogenated. The monoaldehyde is obtained in 32.3% optical yield[352].

Alkynes can also be hydroformylated using this catalyst. Phenylacetylene, like styrene, forms both the terminal aldehyde and hydrotropaldehyde. But-2-yne eventually yields (*S*)-

2-methylbutanal when the reaction is catalysed by (−)-diop complexes. Hence both these alkynes give products of the same chirality as the corresponding alkene. However, oct-1-ene and oct-1-yne form different enantiomers. The intermediate (*E*)-alkenal was isolated from the but-2-yne reaction. Besides indicating the mode of addition, this reaction also shows that alkenals are reduced in the reaction rather than alkenes being first formed and then hydroformylated[353].

The hydroformylation of *N*-vinylsuccinimide is relatively facile and can be brought about by catalysts containing either diop or diphol. Interestingly, a change in the catalytic system is necessary in order to obtain both enantiomers of the product since catalysts containing either diop enantiomer preferentially form the *R*-product. Decreasing reactivity is shown by the substrates *N*-vinylphthalimide and *N*-prop-2-enylacetamide, whilst *N*-2-methylpropenylacetamide did not react[354]. Although *N*-acyl-2, 3-dihydropyrroles are readily hydroformylated[354], 2, 5-dihydrofuran requires severe conditions which impair the optical yields[355].

Heterogenized chiral hydroformylation catalysts have been prepared, but the optical yields are lower than those obtained from homogeneous systems[356].

VII. HYDROSILYLATION

Rhodium(I) complexes add hydrosilanes oxidatively. In many instances, particularly with $[RhCl(PPh_3)_3]$, the intermediate rhodium(III) silyl complexes have been isolated. The silyl complexes $[RhHCl(SiX_3)(PPh_3)_2]$ have been shown to be trigonal bipyramidal[357]. They are thus coordinatively unsaturated and are able to coordinate a substrate molecule. As in all homogeneous catalytic reactions, the availability of this vacant site is very important. It has been demonstrated that if all the reagents have been carefully purified to exclude oxygen and hydroperoxides, then $[RhCl(PPh_3)_3]$ is incapable of catalysing the addition of hydrosilanes to alkenes[358]. The oxidizing agents attack the PPh_3 ligands of the catalyst and promote the formation of coordinatively unsaturated bis(triphenylphosphine) complexes. Generation of hydroperoxides by irradiation of aerated solutions increases the rate of the hydrosilylation reaction[359].

In addition to $[RhCl(PPh_3)_3]$, many other rhodium complexes can also catalyse hydrosilylation reactions; the ease of H—SiX_3 bond fission by transition metal complexes has been noted previously[360]. Analogues of $[RhCl(PPh_3)_3]$ have been found to catalyse the reactions and the contributions of both their halo[361,362] and tertiary phosphine ligands[361] have been evaluated. Precursors of $[RhCl(PPh_3)_3]$ such as the alkene complexes $[RhCl(C_8H_{14})_2]_2$[363] and $[RhCl(C_2H_4)_2]_2$[364] have been employed, but the latter complex is not particularly effective in catalysing the reactions of chlorosilanes. The carbonyl complexes $[RhCl(CO)_2]_2$[365], $[RhH(CO)(PPh_3)_3]$[362,365], and *trans*-$[RhCl(CO)(PPh_3)_2]$[362,366] have been employed as catalysts, as has the hydrido complex $[RhH(PPh_3)_4]$[367]. Additionally, carbene complexes of the type **68**, **69**, and **70** have been used to catalyse the hydrosilylation of oct-1-ene[366]. Heterogenized catalysts have also been used[340c,368]. Rhodium(II) complexes such as $[RhCl_2\{P(o\text{-}C_6H_4Me)_3\}_2]$ or $[RhCl_2(PCy_3)_2]$ have been used to catalyse the addition of monohydrosilanes to alk-1-enes[369]. The hydrosilylation of alkenes has also been catalysed by the rhodium(III) complexes $[Rh(acac)_3]$[370] or $[Rh(\eta^5\text{-}C_5Me_5)]_2Cl_4$[371].

Monohydridorhodium(III) complexes have been employed as catalysts. Carbaborane complexes of general formula $[RR'C_2B_9H_9RhH(PPh_3)_2]$ are claimed to catalyse the hydrosilylation of alkenes, alkynes, and alkanols[372]. The dimeric rhodium(II) complex $[Rh(dmg)_2(PPh_3)]_2$ forms the hydridorhodium(III) complex $[RhH(dmg)_2(PPh_3)]$ when allowed to react with hydrosilanes. The latter then adds alkene to yield an alkyl complex which undergoes hydrosilylation[373]. If this interpretation of the reaction mechanism is correct then this is equivalent to the alkyl route in catalytic hydrogenation. Generally, it

(68) **(69)**

(70)

has been found that the order of hydrosilane reactivity is monohydro < dihydro < trihydro. In practice, this means that the reactions of $SiHX_3$ require heating, those of SiH_2X_2 occur at room temperature, and the few trihydrosilane reactions so far investigated have been carried out at 0 °C or below in order to moderate them.

As can be seen from Figure 1, many compounds containing either multiple bonds or active hydrogen atoms react cleanly with hydrosilanes in the presence of $[RhCl(PPh_3)_3]$, in contrast to similar reactions catalysed by complexes of other transition metals. The reactions of each type of substrate will be discussed in turn.

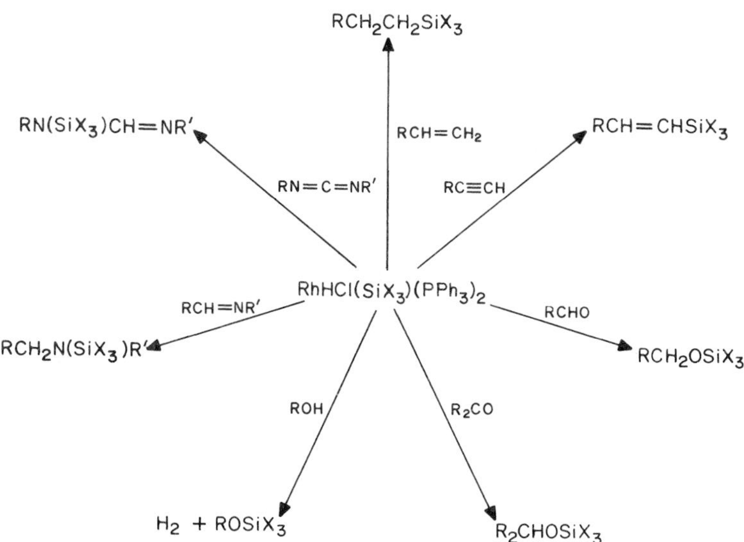

FIGURE 1. Hydrosilylation reactions catalysed by $RhHCl(SiX_3)(PPh_3)_2$.

A. Alkenes

Generally, good yields of organosilanes are obtained when hydrosilanes are allowed to react with terminal alkenes in the presence of $[RhCl(PPh_3)_3]$. However, total conversion is seldom achieved in the reaction, since a greater or lesser degree of isomerization occurs, via side reactions, in the system. These involve the intermediate formation of a 2-alkyl complex from the hydridorhodium(III) complex (Scheme 4).

Internal alkenes (e.g. cyclohexene[370]) are not hydrosilylated by the catalyst, so any isomerization product formed reduces the overall yield of organosilane. The degree of isomerization has been found to be greatly influenced by the hydrosilane employed. Thus the yields of organosilanes from hex-1-ene and $SiHPh_3$, $SiHEt_3$, or $SiHCl_3$ have been found to be 100%, 60% and 8%, respectively[374]. In the last two cases extensive isomerization of hex-1-ene was found to occur.

In catalysing hydrosilylation reactions, $[RhCl(PPh_3)_3]$ and its congeners have been found to be more effective than either of the carbonyl complexes *trans*-$[RhCl(CO)(PPh_3)_2]$ or $[RhH(CO)(PPh_3)_3]$[361]. This order may reflect the relative ease of formation of five-coordinate rhodium(III) species which are capable of activating the alkene after oxidative addition of hydrosilane.

The hydrosilylation of styrene using any rhodium complex catalyst gives rise to three products (equation 83)[361,375,376]. Unlike other terminal alkenes, addition of silyl groups to

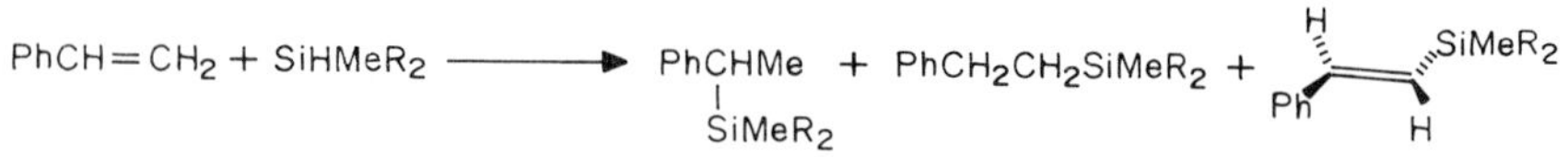
PhCH=CH2 + SiHMeR2
PhCHMe
SiMeR2
+ PhCH2CH2SiMeR2 +
H
SiMeR2
Ph
H

(83)

both the α- and β-carbon atoms of the substrate occurs. Additionally, some substitution product is formed. The formation of a silyl-substituted styrene is favoured by large

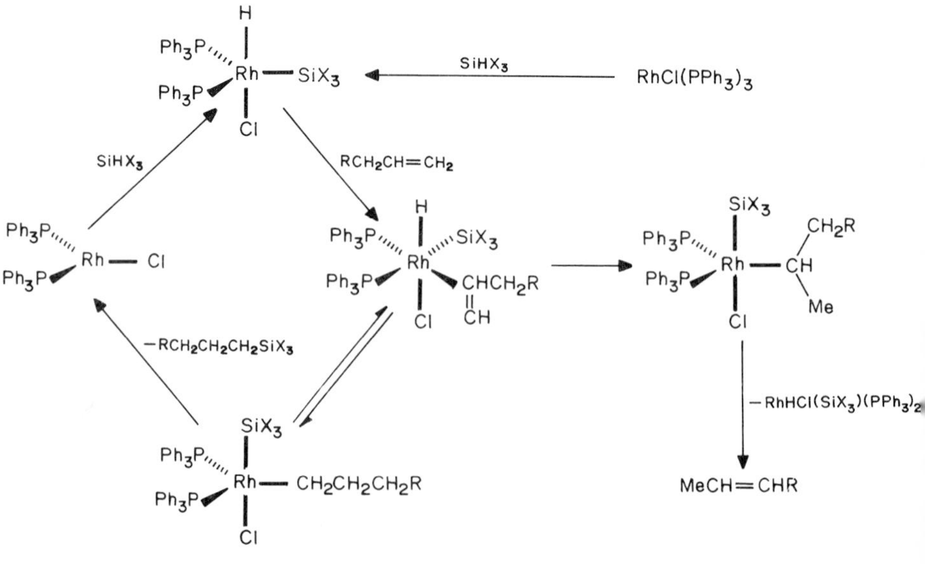
SiHX3
RhCl(PPh3)3
Ph3P
Rh
SiX3
H
Cl
RCH2CH=CH2
Rh — Cl
CHCH2R
CH
CH2R
CH
Me
−RCH2CH2CH2SiX3
−RhHCl(SiX3)(PPh3)2
CH2CH2CH2R
MeCH=CHR

SCHEME 4. Alkene isomerization during hydrosilylation.

substituents on the silyl group[375]. The proportions of α- and β-products formed in the reaction are influenced by both the solvent[376] and the substituents on the silyl group. The proportion of the β-isomer increases with increasing electronegativity of these substituents[361]. Further fundamental differences between the hydrosilylation of styrene and other alkenes are indicated by a reversal of the order of catalyst efficiencies in the former case. *trans*-Carbonylchlorobis(triphenylphosphine)rhodium(I) is superior to both $[RhH(CO)(PPh_3)_3]$ and $[RhCl(PPh_3)_3]$ when styrene is the substrate.

In the various hydrosilylations of styrene, ethylbenzene has not been reported as a product. When $[RhCl_2(\eta^5\text{-}C_5Me_5)]_2$ was used to catalyse the hydrosilylation of hex-1-ene, the small quantity of (*E*)-hex-1-enyl(triethyl)silane formed could be correlated with the yield of the disproportionation product, hexane (equation 84)[371].

$$BuCH{=}CH_2 + SiHEt_3 \xrightarrow{[Rh(\eta^5\text{-}C_5Me_5)]_2Cl_4} (E)\text{-}BuCH{=}CH(SiEt_3) + C_6H_{14} \quad (84)$$

In rare instances, some regioselectivity is observed in the hydrosilylation of alkadienes. The terminal double bond is preferentially hydrosilylated. Conjugated alkenes normally give rise to 1,4-addition. Carbonylhydridotris(triphenylphosphine)rhodium(I), *trans*-$[RhCl(CO)(PPh_3)_2]$, and $[RhX(PPh_3)_3]$ (X = Me, Cl, Br, I) all catalyse the 1,4-addition of hydrosilanes to buta-1,3-diene (equation 85). Isoprene reacts similarly, but both possible 1,4-addition products are obtained (equation 86)[377,378]. Regioselectivity is achieved in

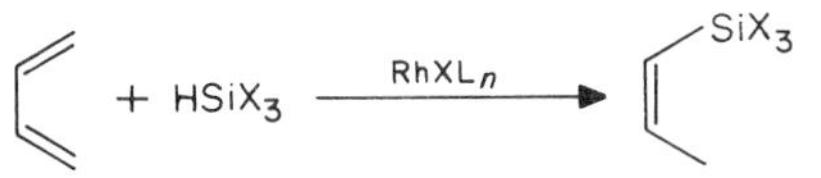

+ HSiX₃ —RhXLn→ SiX₃ (85)

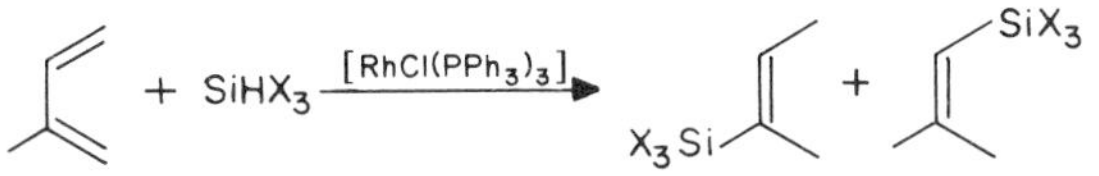

+ SiHX₃ —[RhCl(PPh₃)₃]→ X₃Si + SiX₃ (86)

the hydrosilylations of myrcene and ocimene[378], but only the latter forms any 1,2-addition product (equations 87 and 88).

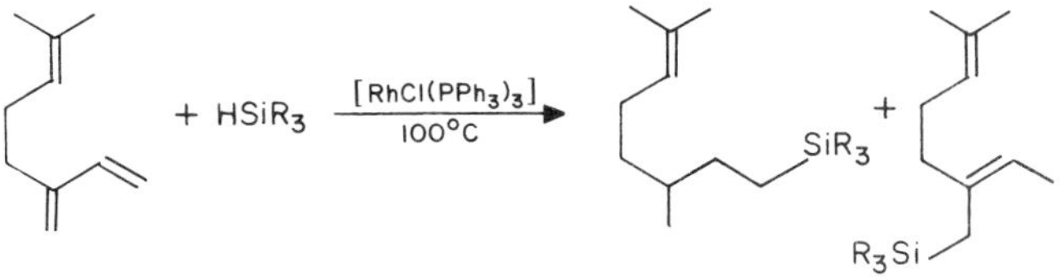

+ HSiR₃ —[RhCl(PPh₃)₃], 100°C→ SiR₃ + R₃Si (87)

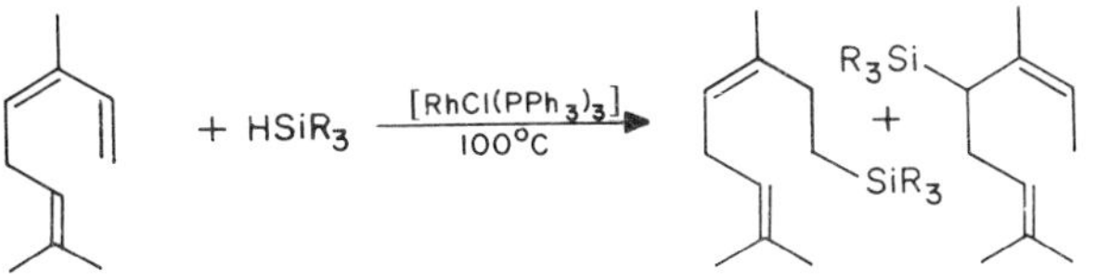

+ HSiR₃ —[RhCl(PPh₃)₃], 100°C→ R₃Si + SiR₃ (88)

Owing to the reactivity of other multiple interatomic linkages or active hydrogen atoms, very few hydrosilylations of substituted alkenes have been attempted. An illustration of the problems involved in such hydrosilylations is given by the reaction between tri(ethoxy)silane and $CH_2{=}CHSEt$. The major product (44%) is $EtSCH_2CH_2Si(OEt)_3$, but alkenyl sulphides and $(EtO)_3SiSEt$ also feature amongst the products[379]. Similarly, the attempted hydrosilylation of vinyl sulphide gives a mixture of products arising from simultaneous hydrogenation, hydrosilylation, and carbon—sulphur bond scission reactions[380,381].

B. Alkynes

The addition of hydrosilanes to terminal alkynes is catalysed by several rhodium complexes. Even rhodium(II) and rhodium(III) complexes have been used as the catalysts. The reaction usually ceases after one molecule of hydrosilane has been added across the triple bond. Exceptionally, ethyne forms between 10 and 30% of the diaddition product[382].

Rhodium(I) and rhodium(II) catalysts both bring about the addition of the silyl group to the terminal carbon atom but both E and Z products are obtained (equation 89)[369,363].

$$RC{\equiv}CH + SiHX_3 \xrightarrow{RhL_n} (Z)\text{-}RCH{=}CHSiX_3 + (E)\text{-}RCH{=}CHSiX_3 \qquad (89)$$

The Z-isomer usually predominates, but the proportion of E-isomer increases on addition of PPh_3 to the $[RhCl(C_8H_{14})_2]_2$ catalyst. The relative yields of these products also depend on the nature of the substituents on the silicon[383,384]. However, since $[RhCl(PPh_3)_3]$ has been shown to isomerize (Z)-$PhCH{=}CH(SiPhMe_2)$ to the (E)-silylalkene[385], discussion of product yields is not particularly meaningful.

Tris(pentan-2,4-dionato)rhodium(III) catalyses the production of both 1- and 2-silyl derivatives (equation 90). The 2-silyl compound is always the minor product[370]. It has been found that hex-1-yne can poison its own hydrosilylation when $[RhCl(PPh_3)_3]$ is used as the catalyst. To avoid this, the hex-1-yne should be added to the reaction mixture after the hydrosilane[358]. One of the more effective catalysts for alkyne hydrosilylation is $[RhH(CO)(PPh_3)]$[95]. Since this complex normally reacts stoicheiometrically with alkenes to form vinyl complexes, it does not simultaneously catalyse the hydrogenation or polymerization reactions brought about by other rhodium complexes.

$$RCH{\equiv}CH + SiHEt_3 \xrightarrow{[Rh(acac)_3]} RCH{=}CHSiEt_3 + RC(SiEt_3){=}CH_2 \qquad (90)$$

Chlorotris(triphenylphosphine)rhodium(I) permits the addition of monohydrogermanes to phenylacetylene. The bulk of the product is the E-isomer, but a very small yield of the α-isomer is also obtained (equation 91)[386].

$$PhC{\equiv}CH + HGeR_3 \xrightarrow[20\,^{\circ}C]{[RhCl(PPh_3)_3]} R_3GeCH{=}CHPh + R_3GeCH{=}CHPh + R_3GeC(Ph){=}CH_2 \qquad (91)$$

C. Aldehydes and Ketones

In the presence of catalytic quantities of $[RhCl(PPh_3)_3]$, monohydrosilanes readily add across the C=O group of ketones to give trialkylsilyl ethers in high yields (equation 92).

Unlike the addition of hydrosilanes to alkenes, this exothermic reaction takes place at room temperature with aliphatic ketones; however, higher temperatures are required for aromatic ketones[387]. The intermediate $[RhHCl(SiEt_3)(PPh_3)_2]$ has been isolated from some of its slower catalytic reactions[388].

$$R_2CO + SiHX_3 \xrightarrow{[RhCl(PPh_3)_3]} R_2CHOSiX_3 \qquad (92)$$

Reactions of ketones with dihydrosilanes take place below room temperature[389]. The addition of trihydrosilanes to ketones is also brought about by this catalyst[390]. The silyl ethers obtained in these reactions can be quantitatively solvolysed to secondary alcohols. Hence hydrosilylation followed by solvolysis is equivalent to reduction[387]. This sequence of reactions has been particularly exploited in chiral hydrosilylation (see Section VIII).

The catalytic hydrosilylation of aldehydes has been attempted only infrequently. Many transition metal complexes fail to catalyse the reaction cleanly. Surprisingly, in view of its ready decarbonylation of aldehydes, $[RhCl(PPh_3)_3]$ is an effective catalyst for the reaction[387,391].

The type of hydrosilane determines the course of the reaction with unsaturated ketones. Monohydrosilanes react with α,β-unsaturated ketones to give the 1,4-addition product, which on hydrolysis forms a ketoalkane[392–394]. Dihydrosilanes attack the keto group[393]. The reactions between monohydrosilanes and unsaturated esters are more complex and, depending on the silane employed, a 1- or 2-silyl ester or a ketene can be formed[395,396]. The last product is formed by methyl esters in which the alkene bond is sterically hindered, but some 1-silyl product is always formed[397].

The catalysed addition of hydrosilanes to α,β-unsaturated aldehydes gives 1,4-products[387,398]. The reaction is slow and both E- and *Z*-products can be formed[398]. Despite these disadvantages, hydrolysis of the product gives a saturated aldehyde and the overall reaction represents an effective, if circuitous, method of hydrogenating unsaturated aldehydes.

D. Alcohols

Alcohols also form silyl ethers when allowed to react with hydrosilanes. As with alkenes, dihydrosilanes are more reactive than monohydrosilanes. Phenols are also attacked by dihydrosilanes[399]. Trihydrosilanes are so reactive that di(alkoxy)silanes are formed[399,400] unless either the trihydrosilane or the alcohol contains a large alkyl or aryl group[400]. Although hydrodisilanes also form di(alkoxy)silanes when allowed to react with an excess of alcohol, the reaction involves a silene intermediate[401]. Thiophenols react with monohydrosilanes at 50 °C when the reaction is catalysed by $[RhCl(PPh_3)_3]$[402].

E. Nitrogenous Substrates

Primary[403,404] and secondary[404–406] and *N*-alkylamides[405] undergo dehydrogenative condensation with hydrosilanes in the presence of $[RhCl(PPh_3)_3]$. Imines, however, add hydrosilane across the C=N bond. The *N*-silyl products can be hydrolysed to secondary amines[407–409]. Despite employing severe conditions, formamidienes were only semi-hydrosilylated[410]. An interesting reaction occurs between aliphatic nitriles and bis-1,2-silylbenzene (equation 93). The alkene is formed by aminolysis of the intermediate while the saturated product arises from a second hydrosilylation reaction involving the C—N bond[411].

The hydrosilylation of vinyl cyanide, catalysed by $[RhH(CO)(PPh_3)_3]$, does not involve reduction of the C≡N bond. The alkene bond alone is attacked and the β-product is formed[412].

$$\mathrm{RR'CHCN} + \mathrm{C_6H_4(SiHMe_2)_2} \xrightarrow[\mathrm{C_7H_8}]{[\mathrm{RhCl(PPh_3)_3}]} \mathrm{C_6H_4(SiMe_2)_2NCH{=}CHRR'} + \mathrm{C_6H_4(SiMe_2)_2NCH_2CH_2RR'} \quad (93)$$

(RR′ = PhH, H_2, MeH, Me_2, Ph_2)

Only the alkyne bond undergoes hydrosilylation when tertiary alkynylamines are allowed to react with triethylsilane in the presence of catalytic quantities of $[RhI(PPh_3)_3]$, $[RhCl(SbPh_3)_3]$, or $[RhCl(CO)(AsPh_3)_2]$[413].

F. Other substrates

It was originally proposed that aroyl chlorides gave rise to ketones when allowed to react with monohydrosilanes in the presence of catalytic quantities of either *mer*-$[RhCl_3(PBu_2Ph)_3]$ or *trans*-$[RhCl(CO)(PEtPh_2)_2]$. However, $C_5H_{11}COCl$ forms the corresponding aldehyde in 68% yield[414]. Later it was stated that acyl chlorides preferentially formed aldehydes, and that the reaction was better carried out at 120 °C in the presence of the carbonyl complex[415].

VIII. ASYMMETRIC HYDROSILYLATION

Many chiral rhodium catalysts can be used to bring about chiral hydrosilylation of unsaturated substrates. However, fairly large optical yields can be obtained when the addition of bulky dihydrosilanes to highly unsymmetric ketones is catalysed by achiral $[RhCl(PPh_3)_3]$. Asymmetric hydrosilylation is virtually confined to that of unsymmetric ketones, since this is the first stage in the production of chiral alcohols. Other substrates include imines and keto esters.

There appears to be only one report of the hydrosilylation of a prochiral alkene being brought about by rhodium catalysts. Trimethyl- or dimethylphenyl-silane can be added to α-methylstyrene if these reagents are allowed to react at 120 °C in the presence of chiral rhodium complexes (equation 94). Disappointing chemical and optical yields were obtained with both catalysts[416].

$$\mathrm{PhMeC{=}CH_2 + SiHMe_2R \xrightarrow{RhL^*_n} PhMeCHCH_2SiMe_2R} \quad (94)$$

$$\mathrm{RhL_n^* = [RhH_2(PBzMePh)_2(solv.)_2][ClO_4], RhCl[(-)\text{-}diop]}$$

A. Ketones

1. Achiral catalysts

When dihydrosilanes containing two different alkyl or aryl groups are allowed to react with $[RhCl(PPh_3)_3]$, the resulting rhodium(III) complex contains a chiral centre at silicon. Since it is this species that activates the ketone in the catalytic reaction, prochiral ketones can give rise to an enantiomeric excess of the diasteriomeric silyl ether. The products are diasteriomeric since the silyl group retains its chirality during the transfer. Before the full development of chiral ligands had been achieved, this method had been used to prepare optically active alcohols from suitable ketones[94].

2. Chiral catalysts

Although it has been claimed that when $RhCl(PBzMePh)_2$(solv.) is the catalyst better optical yields are achieved in hydrosilylations using monohydrosilanes[417], most investigators have used dihydrosilanes. Possibly this choice was made in order to achieve the superior optical yields associated with the milder reaction conditions required for this class of silane.

Chiral bidentate ligands give lower optical yields with monohydrosilanes than chiral monodentate ligands[418,419]. Accordingly, benzylmethylphenylphosphine has been used to prepare both neutral and cationic complex catalysts. The best optical yields have been obtained when bulky silanes and bulky ketones have been allowed to react[418,420]. For example, a 61.8% optical yield was obtained in reaction 95. When the enantiomeric

$$RR'CO + PhMe_2SiH \xrightarrow{RhCl[(S)\text{-}BzMePhP]_2} RR'CHOSiMe_2Ph \xrightarrow{MeOH \mid MeONa} (R)\text{-}RR'CHOH \quad (95)$$

phosphine is incorporated in $[RhCl(PBzMePh)_2]$, (*R*)-alcohols are formed preferentially[421]. However, the latter complex forms (*S*)-alcohols in excess when dihydrosilanes are employed[422,423]. This reversal of predominant chirality is also seen in rhodium/diop systems.

It should also be noted that symmetric ketones still form chiral products when allowed to react with 1-naphthyldihydrosilanes in the presence of rhodium(III) complexes of PBzMePh. However, in these instances the chirality is at the silicon atom of the silyl ether[424]. In fact, dihydrosilanes containing two different organo groups invariably give rise to chirality at silicon in the resulting silyl ether. Optical yields of monohydrosilanes derived from these species are in the range 30–40%. This is lower than the optical yields of the alcohols obtained from the reaction, which are about 10% higher (equation 96)[425].

$$R^1R^2CO + SiH_2R^3R^4 \xrightarrow{RhL_n^*} R^3R^4SiH\text{–O–}CHR^1R^2 \xrightarrow{(i)\,R^5MgX \mid (ii)\,H_2O} R^3R^4R^5SiH + HOCHR^1R^2 \quad (96)$$

Despite these shortcomings, the ready availability of chiral ditertiary phosphines has brought about their widespread use in the preparation of catalysts for chiral hydrosilylation reactions.

By employing an excess of the chiral ligand **34** in conjunction with $[RhCl(C_2H_4)_2]_2$, improved chemical and optical yields were obtained in hydrosilylation reactions. This mirrors the slight superiority of $[Rh(diop)_2][ClO_4]$ over $[Rh(cod)(diop)][ClO_4]$ as a chiral catalyst for the production of silyl ethers[426]. However, it has been claimed that incorporation of the ligand (*S*)-1,2-$(Ph_2As)C_6H_4CHMe(NMe_2)$ into a catalyst of the type $[Rh(nbd)(LL)][ClO_4]$ results in higher optical yields than when (+)-diop is the bidentate ligand. The prevailing chirality of the product was reversed when the corresponding aminoarsine ligand was used, but the optical yields were either very low or zero in these cases[427]. The predominant product chirality can be reversed on making fairly minor changes in the chiral ligand[428].

Glucophinite and camphinite, which can be prepared from natural products, can be used in the form of their $[Rh(nbd)(LL)][BF_4]$ complexes to catalyse the hydrosilylation of a variety of aryl ketones. Generally, the former ligand, which is larger, gave superior optical yields to the latter. Likewise, the larger dihydrosilane SiH_2NpPh usually gave superior optical yields to SiH_2Ph_2. In contrast, the optical yields declined with increasing size of the alkyl group of the ketone[429].

Complexes of the bidentate ligands **71**, **72**[430], and **73**[431], which coordinate to rhodium

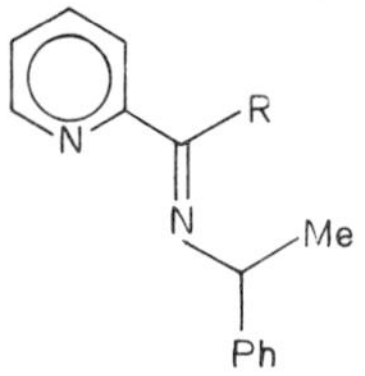

(71) 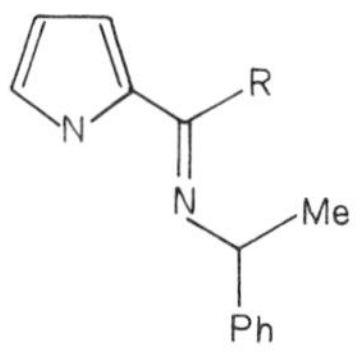 (72) 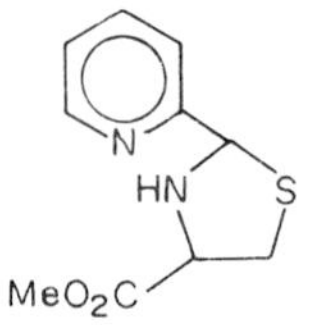 (73)

through their nitrogen atoms, also gave good optical yields when used to catalyse the hydrosilylation of benzophenone by dihydrosilanes. Better yields were obtained from neutral catalysts derived from $[RhCl(cod)]_2$ than from the ionic complexes $[Rh(cod)(LL)][PF_6]$[431].

Although the silyl tertiary phosphine ligand $(EtO)_3Si(CH_2)_3PPh(menthyl)$ gave an optical yield of 23% in the hydrosilylation of benzophenone, the prevailing chirality was reversed and the optical yield reduced to 4% when this ligand was used to incorporate the catalyst in a heterogenized system[432].

B. Substituted Ketones

Pyruvic acid has been converted into hydroxy esters via an intermediate silyl ether using 1-naphthylsilane. As is the case with unsubstituted ketones, PBzMePh and diop ligands gave complexes of opposite chirality. Attempts to reap the benefits of double asymmetric reduction failed, as lower optical yields were obtained than when achiral esters were employed[433]. This dihydrosilane has also been used in the production of α-(hydroxyacyl)amido esters (equation 97). Since the original ester itself contains one chiral centre, achiral catalysts such as $[RhCl(PPh_3)_3]$ can be used with some success. Almost invariably, however, superior optical yields are obtained when catalysts that are themselves chiral are employed[434].

$$\underset{\displaystyle\underset{R^2}{|}}{R^1COCONHCHCO_2Me} \xrightarrow[\text{(ii) MeOH/H}^{+*}]{\text{(i) RhL}_n\text{, SiH}_2\text{NhPh}} R^1\underset{\displaystyle\underset{OH}{|}}{C^*H}CONH\underset{\displaystyle\underset{R^2}{|}}{C^*H}CO_2Me$$

$$RhL_n = RhCl(PPh_3)_3,\ [RhCl(C_2H_4)_2]_2/\text{diop} \qquad (97)$$

$$Ph\underset{}{CO}\underset{\displaystyle\underset{Me}{|}}{CH}NMe_2 \xrightarrow[C_6H_6,\ 50\,^\circ C]{RhL_n^*,\ SiH_2Ph_2} Ph\underset{\displaystyle\underset{OSiHPh_2}{|}}{CH}CHMe(NMe_2) \qquad (98)$$

The hydrosilylation of racemic $PhCOCHMe(NMe_2)$ in the presence of rhodium–diop complexes brings about partial kinetic resolution of the substrate (equation 98). The (*S*)-enantiomer is preferentially hydrosilylated. If the reaction is stopped before completion, the (*R*)-isomer of the ketone can be isolated in 23% optical yield. On treatment of the hydrosilylated product with KOH in aqueous methanol, two diastereoisomers were formed, (1*S*,2*S*)-(+)-pseudomethylephedrine (27% optical yield) and (1*R*,2*S*)-(−)-methylephedrine (20% optical yield)[435].

1. *Unsaturated ketones*

There are two possible pathways for the hydrosilylation of unsaturated ketones, determined by the class of hydrosilane selected. It was noted in Section VII above that monhydrosilanes gave 1,4-addition, whereas dihydrosilanes merely reacted with the keto group. The chiral consequences of these two pathways are shown in Scheme 5[436]. Good chemical yields of the 1,4-addition product have been obtained when (*Z*)-PhMeC=CH(COR′) has been allowed to react with either trimethyl- or dimethylphenylsilane. However, despite the use of two different catalyst systems, $[RhH_2(PBzMePh)_2(solv.)_2][ClO_4]$ and $[RhCl(C_6H_{10})]_2/(-)$-diop, the optical yields were disappointingly low[437]. The 1,4-addition product is not formed if the alkene bond is substituted by a methyl group (equation 99). Both catalysts gave lower yields when mesityl oxide or β-ionone were the substrates[438].

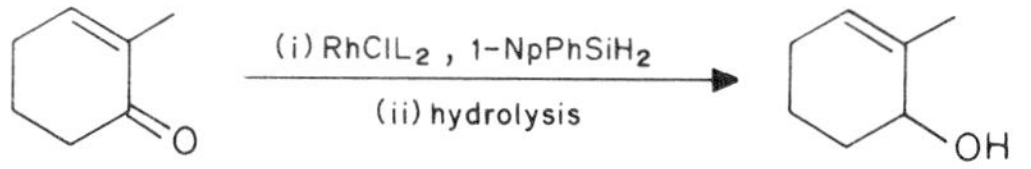

$$\text{2-methylcyclohex-2-enone} \xrightarrow[\text{(ii) hydrolysis}]{\text{(i) RhClL}_2\text{, 1-NpPhSiH}_2} \text{2-methylcyclohex-2-enol (OH)} \qquad (99)$$

$$R^1R^2C{=}CHC({=}O)R^3 \xrightarrow{RhL_n^*} \begin{cases} \xrightarrow{SiHR_3} R^1R^2CHCH{=}CR^3(OSiX_3) \xrightarrow{\text{hydrolysis}} R^1R^2\overset{*}{C}HCH_2C({=}O)R^3 \\ \xrightarrow{SiH_2X_2} R^1R^2C{=}CHCHR^3(OSiHX_2) \xrightarrow{\text{hydrolysis}} R^1R^2C{=}CH\overset{*}{C}HR^3(OH) \end{cases}$$

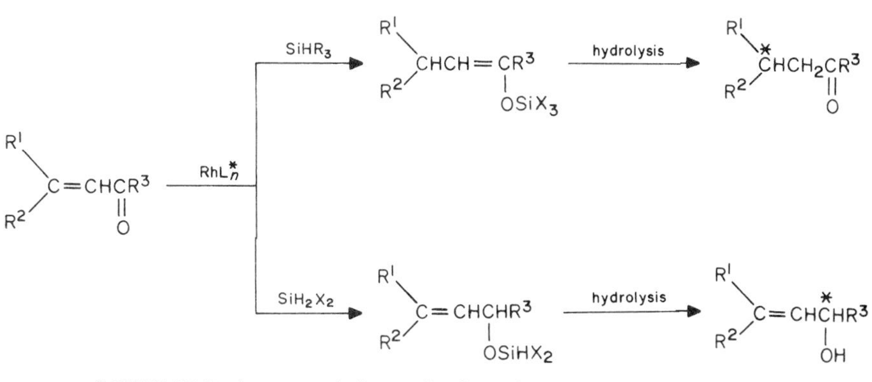

SCHEME 5. Asymmetric isomerization of α,β-unsaturated ketones.

C. Other Substrates

Very little work has been carried out on the chiral hydrosilylation of other substrates. There is only one report of the chiral hydrosilylation of imines. The *N*-silyl product (see Section VII) could be converted into a chiral secondary amine if the original imine contained two different carbon substituents[439].

IX. CARBONYLATION AND DECARBONYLATION

Both these reactions are closely related to hydroformylation since they involve acyl complexes at some stage of the catalytic cycle. Of the two reactions, only carbonylation is of any commercial importance, being used in the large-scale production of acetic acid from methanol. The latter remains a laboratory-scale reaction but represents an increasingly useful synthetic method.

A. Carbonylation

The most important carbonylation reaction is that in which methanol is converted into acetic acid (see also Chapter 7, Volume 3). The definitive paper on the subject is that by Forster[440], who showed that the key step in the catalytic cycle is the oxidative addition of methyl iodide to the dicarbonyldiiodorhodate(I) complex (Scheme 6). The methyl iodide arises from the reaction between hydrogen iodide and methanol. The most troublesome side reaction is the methanolysis of the six-coordinate acetyl complex, which yields methyl acetate rather than acetic acid produced by hydrolysis.

There have been many rhodium complexes proposed as catalysts for the carbonylation of methanol[441–443]. However, it has been demonstrated that many labile rhodium complexes give virtually identical yields in the presence of iodine at 140 °C[444]. In these

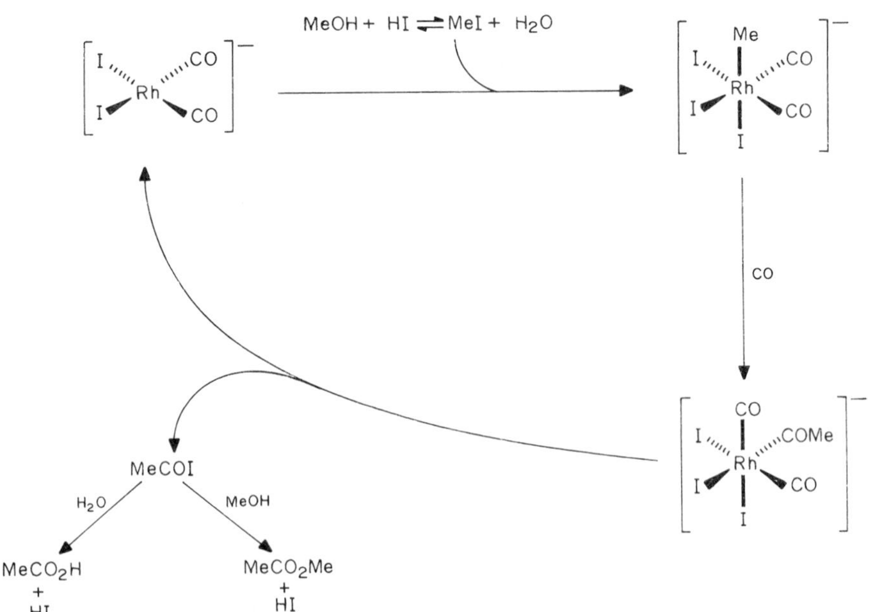

SCHEME 6. Catalytic carbonylation of methanol.

instances the true catalyst is again $[RhI_2(CO)_2]^-$. The exceptions to this behaviour are those complexes which contain bidentate ligands, which are not readily converted into the above anionic complex[445]. At higher temperatures even these bidentate ligands are displaced, since $[RhCl(diphos)_2]$ has been claimed to act as catalyst at 190 °C[443].

Despite the megatonne annual production of acetic acid by this process, it does not seem to have been widely applied to the production of its homologues. The carbonylation of propan-2-ol forms both butyric and isobutyric acids together with their *iso*- and *n*-propyl esters[446]. Levulinic acid has been prepared by the carbonylation of either 4-methoxy- or 4-ethoxy-butan-2-one; again the corresponding ester is also formed (equation 100)[447].

$$MeCOCH_2CH_2OR + CO \xrightarrow[MeI]{[RhCl(PPh_3)_3]} MeCOCH_2CH_2CO_2R \qquad (100)$$

The catalytic carbonylation of styrene epoxide forms a lactone (equation 101). This product is formed by the epoxide ring first forming a metallocyle, which then undergoes

$$PhCH\text{—}CH_2 \text{ (epoxide, bridging O)} \xrightarrow{[RhCl(CO)(PPh_3)_2]} \text{PhHC—CH}_2\text{, C(=O)—O (}\beta\text{-lactone)} \qquad (101)$$

carbonyl insertion and finally elimination[448]. Amines, such as cyclopropylamine, can be similarly carbonylated to lactams[449]. Butylamine undergoes carbonylation as in equation 102. If a tertiary phosphine is added to the system, the yield of amide is increased[450]. A symmetrical diarylurea is also formed in the carbonylation of aniline or *p*-nitroaniline. In the presence of ethanol, an ethyl ester is formed (equation 103)[451]. Allylamine, on the other hand, forms 2-pyrrolidinone on carbonylation in the presence of $[RhCl(PPh_3)_3]$ coated on alumina or silica microspheres[452].

$$Bu^nNH_2 + CO \longrightarrow Bu^nNHCHO + (Bu^nNH)_2CO \qquad (102)$$

$$p\text{-}XC_6H_4NH_2 + CO \xrightarrow[EtOH]{RhClL_n} p\text{-}XC_6H_4NHCO_2Et \qquad (103)$$

$$X = H,\ NO_2;\ RhClL_n = [Rh(diphos)_2]Cl,$$
$$[Rh(dppp)_2]Cl,\ \{RhCl(CO)(dppp)\}_2,\ RhCl(CO)(PPh_3)_2$$

The decarbonylation of acyl chlorides is reversed under high pressures of carbon monoxide. There is a patent claim that the carbonylation of benzal chloride takes place at 150 atm and 150 °C in the presence of $[RhCl(PPh_3)_3]$[453]. Obviously *trans*-$[RhCl(CO)(PPh_3)_2]$ is the true catalyst for the reaction.

There have been several attempts to prepare aryl cyanates by carbonylation of nitrogen-substituted aromatic species. Aryl azides undergo carbonylation with loss of nitrogen (equation 104)[451,454,455]. In the presence of aniline diarylureas are formed and, if the reaction is carried out in ethanol, esters are formed (cf. equation 103). The rhodium complexes that catalyse the reaction include $[RhCl(CO)(PPh_3)_2]$ and both cationic and neutral rhodium(I) complexes of $Ph_2P(CH_2)_nPPh_2$ $(n = 1, 2)$[454].

$$p\text{-}XC_6H_4N_3 + CO \xrightarrow[25-80\ ^\circ C]{RhL_n} p\text{-}XC_6H_4NCO + N_2 \qquad (104)$$

$$X = H,\ NO_2;\ RhL_n = [Rh(diphos)_2]Cl,$$
$$[Rh(dppp)_2]Cl,\ [RhCl(CO)(dppp)]_2,\ RhCl(CO)(PPh_3)_2{}^{451},$$
$$RhBr(CO)(PPh_3)_2,\ RhCl(CO)(PCy_3)_2,\ RhCl(CO)\{P(OPh)_3\}_2{}^{455}$$

Nitrosobenzenes can be carbonylated in the presence of $[RhCl(CO)_2]_2$ (equation 105)[456]. A similar reduction of nitrobenzene takes place in the presence of molybdenum(V) chloride as a promoter (equation 106)[457]. The role of the cocatalyst has been further investigated[458,459].

$$ArNO + 2CO \xrightarrow[1,2\text{-}C_6H_4Cl_2]{[RhCl(CO)_2]_2} ArNCO + CO_2 \tag{105}$$

$$ArNO_2 + 3CO \xrightarrow[MoCl_5]{[RhCl(CO)_2]_2} ArNCO + 2CO_2 \tag{106}$$

B. Decarbonylation

Chlorotris(triphenylphosphine)rhodium(I) has proved to be a useful stoicheiometric reagent in the decarbonylation of aldehydes[460]. It is believed that acetone can be slowly decarbonylated by $[RhCl(PPh_3)_3]$ at high temperatures[461], and α-alkynylketones are decarbonylated by this complex (equation 107). The best yields are obtained if both R and R′ are aryl groups, since neither PhC≡CCOMe nor MeC≡CCOPh is decarbonylated and only a 1% yield is obtained from PhC≡CCOC≡CMe[462].

$$RhCl(PPh_3)_3 + RC{\equiv}CCOR' \longrightarrow RC{\equiv}CR' + trans\text{-}RhCl(CO)(PPh_3)_2 \tag{107}$$
$$R = Ph,\ R' = C{\equiv}CPh,\ CH{=}CHPh,\ Ph$$

Acyldiphenylphosphines are decarbonylated at high temperatures (equation 108). The reaction is believed to proceed by the initial oxidative addition of acyl and diphenylphosphino fragments to rhodium, followed by alkyl migration to the phosphorus of the diphenylphosphido group[463]. Carbon monoxide is also reported to be abstracted from triallyl phosphite[464].

$$RhCl(PPh_3)_3 + RCOPPh_2 \longrightarrow RPPh_2 + trans\text{-}RhCl(CO)(PPh_3)_2 \tag{108}$$
$$R = Me,\ CF_3$$

The decarbonylation of *aci*-nitromethane has been achieved, but no organic products were identified[465].

Pentan-2,4-dione and certain other diketones are semidecarbonylated by $[RhCl(PPh_3)_3]$ in refluxing toluene, but some oxidative addition of pentan-2,4-dione also occurs. Diacetyl forms acetone and MeCOCOPh gives acetophenone. The semidecarbonylation of benzil occurs at temperatures above 110 °C, but pyruvic acid can be decarbonylated at room temperature. Selective removal of the acyl group occurs when acetoacetic esters are allowed to react with $[RhCl(PPh_3)_3]$ (equation 109)[466]. Diphenyl ketene is decarbonylated to the carbene $Ph_2C{=}$, which immediately reacts with other components of the $[RhCl(PPh_3)_3]$-catalysed system[467].

$$RhCl(PPh_3)_3 + MeCOCH_2CO_2R \longrightarrow EtCO_2R + trans\text{-}RhCl(CO)(PPh_3)_2 \tag{109}$$
$$R = Me,\ Et$$

Acetic anhydride is decarbonylated to methyl acetate[466]. The decarbonylation of benzoic anhydrides has been used as a route to fluorenones[468,469], as has the decarbonylation of naphthoic anhydrides to benzofluorenones[470]. The yields in all cases were only moderate. Acyl or aroyl chlorides, which are more thermally stable, can be catalytically decarbonylated at high temperatures[471,472]. The catalyst is $[RhCl(CO)(PPh_3)_2]$ in these reactions.

1. Acid halides

The mechanism of the catalytic decarbonylation of acid halides is better understood than that of the corresponding reaction with aldehydes. The isolation of several intermediates from the reactions of acid chlorides has helped to elucidate the reaction. The key intermediates are the fluxional five-coordinate acylrhodium(III) complexes[473]. Both square pyramidal[474] and trigonal bipyramidal[475] acyl complexes have been isolated. These complexes decompose by the migration of the alkyl or aryl group to the vacant sixth coordination site. The reaction is retarded by other ligands that can occupy this site. The migration of the alkyl group occurs with retention of configuration[476–480]. Reductive elimination of alkyl or aryl chloride, again with retention of configuration, completes the catalytic cycle.

Substituted acid chlorides are also decarbonylated (equation 110)[481]. However, if the

$$\text{R-C}_6\text{H}_4\text{-OCH}_2\text{COCl} \xrightarrow[160-180\,^\circ\text{C}]{[\text{RhCl(PPh}_3)_3]} \text{R-C}_6\text{H}_4\text{-OCH}_2\text{Cl} \tag{110}$$

R = H, 2-Me, 2-MeO, 4-Cl, 2, 4-Cl_2, 4-MeCO

alkyl group contains a hydrogen atom on its β-carbon atom, then an alkene and hydrogen chloride are eliminated (equation 111)[480,482,483]. The formation of the eliminated alkene has also evoked stereochemical attention[475,476,478,479,484]. It has been shown that Saytzeff elimination occurs. Since *threo*-and *erythro*-2, 3-diphenylbutanoyl chloride form (*Z*)- and (*E*)-1, 2-diphenylpropene, respectively, alkene elimination from rhodium is *cis*[478,479]. Unsaturated acyl halides can themselves be decarbonylated[485]. The high kinetic isotope effect observed in the decarbonylations of 3-phenylpropanoyl chloride suggests that the rate-determining step is the scission of a carbon—hydrogen bond[479].

$$\text{RhCl(PPh}_3)_3 + \text{PhCH}_2\text{CH}_2\text{COCl} \longrightarrow \text{PhCH}{=}\text{CH}_2 + \text{HCl} + trans\text{-RhCl(CO)(PPh}_3)_2 \tag{111}$$

The desulphonylation of aryl sulphonyl chlorides by [$RhCl(PPh_3)_3$] is very similar to the decarbonylation of aroyl chlorides. However, the reaction is of lower synthetic importance, since the temperatures required for the catalytic production of aryl chlorides are much higher than when the more accessible aroyl chlorides are the substrates[486].

2. Aldehydes

Aldehydes are catalytically decarbonylated when allowed to react with chlorotris(triphenylphosphine)rhodium(I) at 160 °C or above. Other catalysts that may be employed include [$RhCl(PF_2NMe_2)_3$], [$RhCl(PF_2NMe_2)_2]_2$, and [$RhCl(CO)_2]_2$[487]. Many aldehydes decompose at the elevated temperatures required and this limits the utility of the catalytic reaction. The catalyst [$RhCl(CO)\{Ph_2P(CH_2)_3PPh_2\}$] is more suitable, if less accessible, for the decarbonylation of thermally sensitive aldehydes, as it decomposes at a lower temperature[488]. This greater ease of decomposition arises from CO being *trans* to phosphorus rather than chloride as is the case with [$RhCl(CO)(PPh_3)_2$].

It would appear that the stoicheiometric decarbonylation of aldehydes follows a very similar mechanism to the catalytic decarbonylation of acid halides. The decarbonylation occurs with retention of configuration at carbon. The formation of a deuteriocyclopropane (equation 112) shows the reaction to be intramolecular[489]. Normally an

$$\text{Ph}_2\text{C}_3\text{H}_2(\text{Me})(\text{CDO}) \xrightarrow[\text{C}_8\text{H}_{10},\,140\,^\circ\text{C}]{[\text{RhCl(PPh}_3)_3]} \text{Ph}_2\text{C}_3\text{H}_2(\text{Me})\text{D} \tag{112}$$

alkane is formed in the catalytic decarbonylation of aldehydes. However, when the β-carbon atom is bound to a hydrogen atom, a small percentage of alkene is formed with elimination of dihydrogen (equation 113)[483]. Some long-chain aldehydes undergo cyclization in the reaction rather than decarbonylation (see Section I). The yield of cyclized product is increased when a potential ligand is added to the reaction mixture. This ligand blocks the sixth coordination site required for alkyl group migration[490].

$$RhCl(PPh_3)_3 + RCH_2CH_2CHO \longrightarrow RCH{=}CH_2 + H_2 + trans\text{-}RhCl(CO)(PPh_3)_2 \qquad (113)$$

Citronelal is also cyclized when its decarbonylation is attempted using $[RhCl(PPh_3)_3]$, but no cyclohexanones are formed (equation 114). The migration of the double bond

4 CHO $\xrightarrow{[RhCl(PPh_3)_3]}$ 3 OH OH (114)

implies that a η^3-allyl mechanism is operative[491]. This process is also involved in the production of an aldehyde tautomer from allylic alcohols. However, in these cases the intermediate is decarbonylated. The formation of both *erythro*- and *threo*-alkanes from alcohols containing CD_2OH groups (equation 115) also implies the participation of η^3-allylic intermediates[492].

R H R′ CD_2OH $\xrightarrow{[RhCl(PPh_3)_3]}$ R D H R′ H D + R D D R′ H H (115)

X. OXIDATION

The principal problem in the catalytic oxidation of organic substrates by rhodium complexes is the determination of the mechanism of the reactions. Two totally different general mechanisms have been proposed. The first requires that the rhodium complexes serve only to decompose the traces of hydroperoxides present in the substrate. This is the well established Haber–Weiss mechanism and involves free radical intermediates. It is outside the scope of the review since rhodium—carbon bonds are not involved. The second mechanism involves the participation of rhodium–dioxygen complexes. These complex with a substrate molecule and transfer the oxygen intramolecularly to the coordinated substrate, thus producing a low oxidation state, coordinatively unsaturated species to continue the catalytic cycle.

The difficulty of deciding between these two mechanisms is compounded by the poor quality of the experimental work in many instances. Two errors are commonly committed. The first error made is to add free radical scavengers to the system and to regard cessation of catalytic activity as evidence for a free radical mechanism. This assumption totally ignores the strong possibility that the scavenger can function as a ligand and block the vacant coordination site required for catalytic activity. The second fundamental error is the failure to consider possible reactions between the products and the catalyst. In many of the oxidation reactions aldehydes are formed. The decarbonylation of aldehydes by rhodium complexes was discussed above in Section IX. Thus, in these reactions, the true catalysts may be carbonyl complexes, and their formation may account for the induction periods observed. Only rarely is the decarbonylation reaction considered to be an important facet of the oxidation process[493]. However, it must be conceded that investigation of the reactions is difficult. There are several instances where it seems likely that both of the principal mechanistic pathways participate simultaneously.

Possibly the best understood reactions are those in which terminal, acyclic alkenes are oxidized to methyl ketones. It has been demonstrated that oxidation of these alkenes with $^{18}O_2$ results in the incorporation of ^{18}O in the ketone. However, it was not possible to distinguish between the attack of dioxygen on a rhodium alkene complex or the attack of alkene on a rhodium–dioxygen complex. This parallels the dichotomy of the dihydride and alkene routes in catalytic hydrogenation reactions (see Section III). The latter reaction initiates the catalytic cycle since $[RhO_2L_4]X$ complexes ($L = AsPh_3$, $AsPhMe_2$; $X = ClO_4$, PF_6) catalyse the reaction. However, since $[Rh(1,7\text{-octadiene})_2]X$ reacts with dioxygen to form oct-1-en-7-one, there are equally good grounds for believing the former route to comprise the catalytic cycle[494].

Chlorotris(triphenylphosphine)rhodium(I) and *trans*-$[RhCl(CO)(PPh_3)_2]$ similarly catalyse the oxidation of alkenes to ketones. In both cases oxidation of a triphenylphosphine ligand occurred. Aldehydes were minor products when either catalyst was employed[495]. It has been demonstrated that $[RhCl(PPh_3)_3]$ is a more active catalyst than its cyano[496], cyanato, or thiocyanato[496,497] analogues. Internal alkenes do not undergo catalytic oxidation[498].

Cyclooct-2-enone has been formed by oxidation of $[RhCl(cyclooctene)_2]_2$ (equation 116). The reaction was believed to proceed by intramolecular oxidation in an

$$\text{cyclooctene} + O_2 \xrightarrow{[RhCl(C_8H_{14})_2]_2} \text{cyclooct-2-enone} + \text{cyclooctanone} \qquad (116)$$

intermediate monomeric complex[499]. Other workers have detected cyclooctanone amongst the products. The formation of cycloocten-3-ol gives rise to both products. Isomerization of this intermediate forms the cyclooctanone whilst its further oxidation forms the cycloalkenone[500].

Cycloocta-1,5-diene is not oxidized in the presence of rhodium complexes, even under conditions where it is oxidized in their absence. The failure to oxidize the substrate was ascribed to it coordinating preferentially to rhodium and poisoning its own oxidation[501]. The importance of simultaneously coordinating both oxygen and substrate is further demonstrated by resistance of α- or β-substituted styrenes to oxidation[502].

The formation of epoxides in the oxidation of cycloalkenes has caused much controversy. On the one hand, cyclohexene epoxide is believed to result from the catalysed oxidation of cyclohexene since $[RhCl(PPh_3)_3]$ does not decompose cyclohexene hydroperoxide to the epoxide[503]. Again, the last experiment is ambiguous since in the catalytic system the hydroperoxide is formed in the presence of the alkene. It has been demonstrated elsewhere that hydroperoxides may catalytically react with alkenes to form epoxides (equation 117)[504,505]. On the other hand, it has been shown that the addition of a radical

$$\text{cyclohex-2-enyl-OOH} + \text{cyclohexene} \xrightarrow{[RhCl(PPh_3)_3]} \text{cyclohexene oxide} + \text{cyclohex-2-en-1-ol} \qquad (117)$$

inhibitor stopped the production of cyclooctene epoxide but not the formation of those products shown in equation 116[500]. Similarly, in the catalytic oxidation of styrene, styrene epoxide is formed in the absence of radical inhibitors[506] but not in their presence[502]. However, it has been claimed that the inhibitor and scavenger **74** totally stopped the reaction without poisoning the catalyst[507].

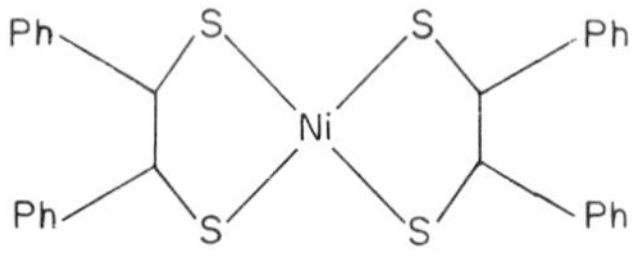

(**74**)

The solvent also plays an important part in determining the product distribution. The catalytic oxidation of styrene in ethanol gives PhCOMe as virtually the sole product, whereas in dioxane benzaldehyde is the major product[508]. The catalytic oxidation of methoxytetralin gives a ketone, but if the oxidation is carried out in benzene solution no ketone is produced and the corresponding alcohol is the product[509].

Generally, rhodium-catalysed oxidations are of little use as preparative methods owing to the variety of products formed.

XI. OLIGOMERIZATION AND POLYMERIZATION

Rhodium complexes are moderately active as catalysts in the dimerization and polymerization reactions of alkene and alkyne substrates. The main disadvantage to their use comes from the stability of the rhodium complexes formed. These complexes do not release the product readily[510] and in certain instances can only be decomposed by the addition of further reagents, thus making the reactions stoicheiometric. However, this latter feature has been turned to advantage by Müller and his school, who have prepared many important heterocyclic compounds by decomposing the rhodacycles formed in these reactions[94].

A. Alkene Substrates

The simplest oligomerization reaction is that discovered by Cramer about 20 years ago in which ethene is dimerized to but-1-ene. The catalytic cycle shown in Scheme 7 embodies the main features of this work[511–513]. The product, but-1-ene, is not further dimerized under the mild conditions of the Cramer process. However, at 50 °C the product is principally but-2-ene. Under these conditions the dimerization of propene can also be achieved. Additionally at this temperature the dimerization of buta-1, 3-diene is catalysed by $RhCl_3{\cdot}3H_2O$, particularly in combination with potassium acetate. It is believed that the initial product is octa-1, 3, 6-triene, which isomerizes in the course of the reaction to the more stable octa-2, 4, 6-triene[514]. At still higher temperatures substituted alkenes can be dimerized (equation 118).

$$2CH_2{=}CHCO_2Me \xrightarrow[\text{MeOH}]{[RhCl_3\cdot 3H_2O]} MeO_2CCH{=}CHCH_2CH_2CO_2Me \tag{118}$$

Copolymerization of ethene and butadiene can also take place in the system[515]. High conversions to hexa-1, 4-diene and hexa-2, 4-diene are achieved at 50 °C. At 100 °C ethene

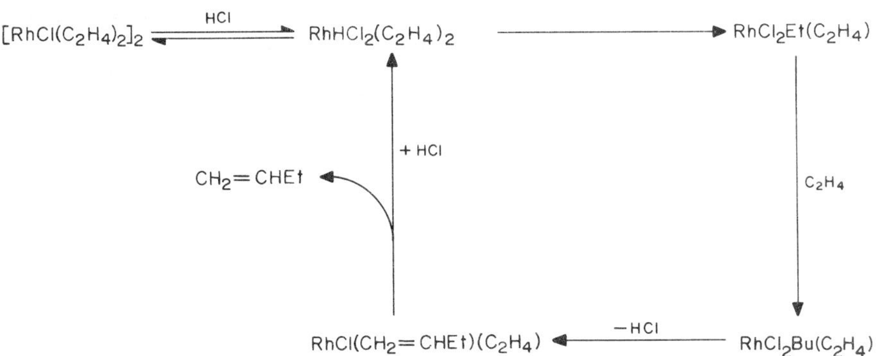

SCHEME 7. Ethene dimerization.

also adds to the hexadienes to yield C_8-alkadienes. Lower yields are obtained in the reaction between ethene and styrene, and even lower yields from the reaction of ethene and 2-methylbuta-1, 3-diene. However, propene adds smoothly to the latter alkadiene to form a heptadiene believed to be 2-methylhexa-1, 4-diene[514].

A more complex reaction of this type involves the oligomerization of 2-methylpropene and the subsequent isomerization of the oligomers (Scheme 8). The oligomerization is catalysed by the nitrosyl complex $[Rh(NO)(MeCN)_4][BF_4]_2$. The acetonitrile ligands are easily lost and the reactions involve coordination of an alkene to rhodium to effect both oligomerization and isomerization[516].

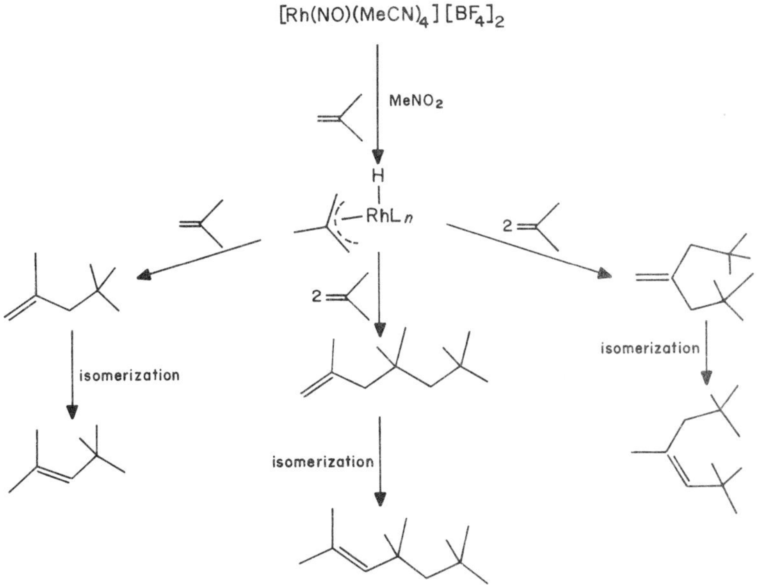

SCHEME 8. Oligomerization and isomerization of 2-methylpropene catalysed by $[Rh(NO)(MeCN)_4][BF_4]_2$.

In these and similar oligomerizations both Ziegler–Natta and η^3-allylic mechanisms have been invoked. Although the oligomerization of buta-1, 3-diene to 4-vinylcyclohexene and cycloocta-1, 5-diene and its polymerization to 1, 4-*trans*-polybutadiene were thought to take place by an η^3-allylic mechanism[517,518], the polymerization of penta-1, 3-diene[519] and propadiene[520,521] apparently occurs by Ziegler–Natta-type insertion reactions.

The carbonyl complex $[RhCl(CO)_2]_2$ polymerizes propadiene to 1, 2-polyallene[520] but Cramer's compound in the presence of PPh_3 brings about its oligomerization to tetramers (equation 119)[522]. Using Cramer's compound and triphenylphosphine is equivalent to

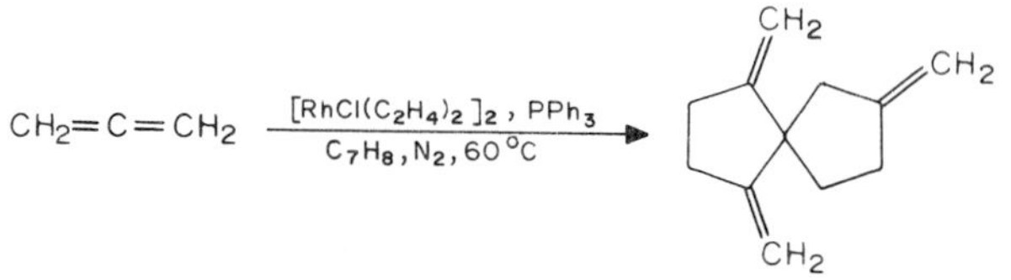

$$CH_2{=}C{=}CH_2 \xrightarrow[C_7H_8,\,N_2,\,60\,^\circ C]{[RhCl(C_2H_4)_2]_2,\ PPh_3} \quad (119)$$

using $[RhCl(PPh_3)_3]$, so it is not surprising that this complex and its bromo and iodo analogues also catalyse the reaction[523]. In the formation of the allene tetramer the presence of PPh_3 was considered essential, and this compound also catalyses the dimerization of a diquinone. The dimerization is also catalysed by $[RhCl(PPh_3)_3]$ (equation 120). The alkene complex **75** has been isolated from the stoicheiometric reaction, but it proved to be a poorer catalyst than $[RhCl(PPh_3)_3]$ for the reaction[524].

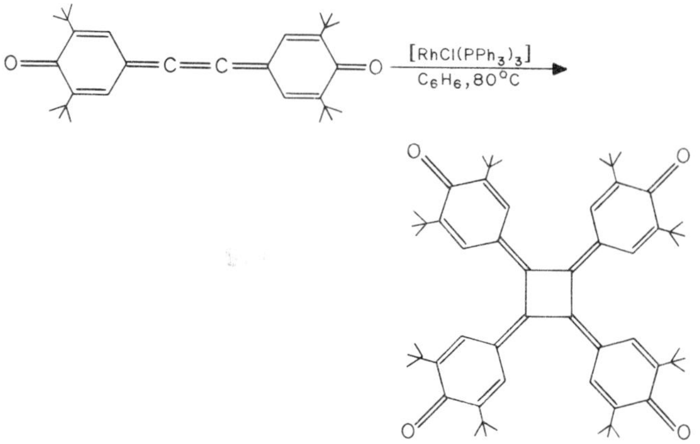

$$\xrightarrow[C_6H_6,\,80\,^\circ C]{[RhCl(PPh_3)_3]} \quad (120)$$

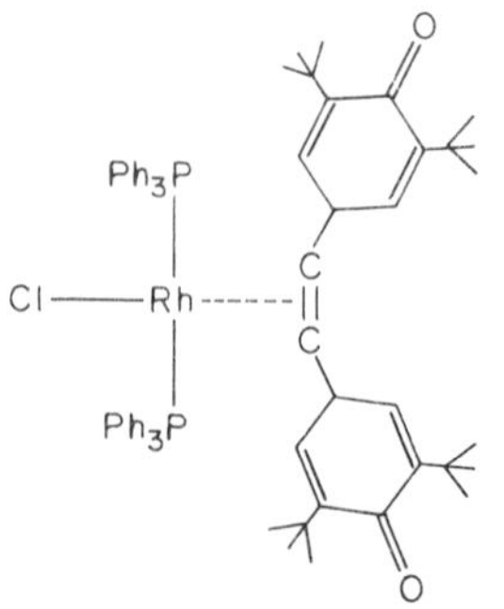

(75)

Rhodium complexes catalyse the polymerization of isoprene[525], chloroprene[526], and 2, 3-dimethylbuta-1, 3-diene[525]. These polymerizations are believed to occur via insertion into rhodium η^3-allyl complexes[526].

Copolymerizations can also be achieved by means of rhodium catalysts. *cis*-Hept-4-enal condenses with ethene in the presence of $[Rh(acac)(C_2H_4)]$ to form non-6-en-3-one[527]. A more complex reaction takes place between pent-3-enal and ethene in the presence of this catalyst, owing to isomerization reactions affecting both the reactant and products (equation 121)[528].

$$\text{pent-4-enal} + C_2H_4 \xrightarrow[CHCl_3]{[Rh(acac)(C_2H_4)_2]} \text{(mixture of unsaturated ketones and aldehydes)} \qquad (121)$$

Chlorotris(triphenylphosphine)rhodium(I) catalyses the addition of ω-bromostyrene to potassium butenoate (equation 122)[529]. The cationic complex $[Rh(cod)(PPh_3)_2][PF_6]$ catalyses a similar copolymerization between buta-1, 3-diene and but-3-enoic acid. The products are mainly octadienoic acids but some dodecatrienoic acids are also formed[530,531].

$$CH_2{=}CHCH_2CO_2K + (E)\text{-}PhCH{=}CHBr \xrightarrow[EtOH]{[RhCl(PPh_3)_3]} \qquad (122)$$

$$(E,E)\text{-} + (E,Z)\text{-}Ph(CH{=}CH)_2CH_2CO_2H$$

Most attention has been focussed on the polymerization of norbornadiene, probably because the polycyclic products can be used as high-energy rocket fuels. The products produced are complex, but some selectivity has been achieved by using $[RhCl(PPh_3)_3]$ in the presence of additives[532]. The first step in the reaction is complexation of norbornadiene to rhodium[533], whereupon an exothermic reaction takes place. Whilst the additives restrict the reaction to dimerization, the simple catalytic system gives polymeric products[534–536]. Dimerization of norbornadiene is also brought about by $[RhCl(cyclooctene)_2]_2$, but the product[537] differs from the dimer formed in the $[RhCl(PPh_3)_3]/BF_3$ system.

The chlororhodium(I) complex of spiro(bicyclo[2.2.1]hepta-2, 5-diene-7, 1′-cyclopropane catalyses the dimerization of this alkadiene (equation 123). The complex

$$\xrightarrow{[RhCl(cpnbd)_2]_2} \qquad (123)$$

also brings about its codimerization with norbornadiene, but all three possible dimers are formed[538].

B. Alkyne Substrates

Alkynes first form alkyne complexes with $[RhCl(PPh_3)_3]$, but these rearrange to form alkynyl rhodium(III) complexes (equation 124). These alkynyl complexes then react with a further molecule of alkyne, whereupon the rhodium(III) complexes reductively eliminate an alkenyne (equation 125). Sterically crowded alkynes form dimers, but those having small substituents show a greater tendency to polymerize in the presence of the rhodium catalysts, particularly at higher temperatures. However, the degree of substitution also influences the efficiency of the dimerization reaction. Thus, whereas phenylacetylene readily dimerizes in the presence of $[RhCl(PPh_3)_3]$, ethyne itself dimerizes in only 1% yield[539]. Hex-1-yne, hept-1-yne, and oct-1-yne similarly undergo dimerization in the presence of this catalyst[540].

$$\begin{aligned} &RhCl(PPh_3)_3 + RC{\equiv}CH \longrightarrow \\ &RhCl(RC{\equiv}CH)(PPh_3)_2 \longrightarrow \\ &RhHCl(C{\equiv}CR)(PPh_3)_2 \end{aligned} \tag{124}$$

$$\begin{aligned} &RhClH(C{\equiv}CR)(PPh_3)_2 + RC{\equiv}CH \longrightarrow \\ &RhHCl(C{\equiv}CR)(RC{\equiv}CH)(PPh_3)_2 \longrightarrow \\ &RhCl(PPh_3)_2 + RC{\equiv}CCH{=}CHR \end{aligned} \tag{125}$$

Hydroxyacetylenes can also be catalytically dimerized by this rhodium(I) complex. Their rate of reaction is slower than that of the acetylenic hydrocarbons. This has been demonstrated by their codimerization with the hydrocarbons where the bulk of the product contains no hydroxy groups[541]. Nevertheless, large alkynols can be dimerized by $[RhCl(PPh_3)_3]$[542].

Carbonylrhodium(I) complexes bring about the trimerization of alkynes. Thus phenylactylene forms triphenylbenzenes[543] and dimethyl acetylenedicarboxylate forms the hexasubstituted product (equations 126 and 127)[544,545]. Both of these cyclotrimeriz-

$$3PhC{\equiv}CH \xrightarrow[C_6H_6,\ 60\,^{\circ}C]{Rh_4(CO)_{12}} 1,2,4\text{-}C_6H_3Ph_3 + 1,3,5\text{-}C_6H_3Ph_3 \tag{126}$$

$$3RC{\equiv}CR \xrightarrow{[RhCl(CO)(AsPh_3)_2]} C_6R_6 \tag{127}$$

ations proceed via rhodacyclopentadienyl complexes. In support of this mechanism, the complex **76** has been shown to form the cyclotrimer when allowed to react with dimethyl acetylenedicarboxylate[544].

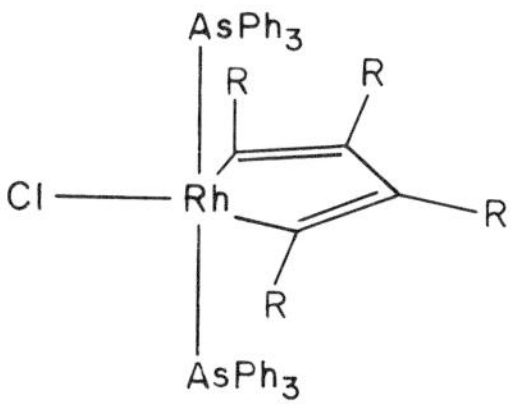
AsPh3
R
R
Cl
Rh
R
R
AsPh3

(76)

In all the dimerization reactions, two possible isomeric products, **77a** and **77b**, can be obtained. The branched dimer **77a** is formed by alkyl- and methoxy-acetylenes and also

R
C
C
H
C=C
R
H
R
C
C
H
C=C
H
R

(77a) (77b)

from phenylacetylene when the catalyst contains bulky PR_3 ligands[539]. Normally phenyl- or *tert*-butyl-acetylene form the linear dimer **77b**. The regioselectivity is controlled by the direction of alkynyl transfer (Scheme 9). As noted above, at temperatures above 80 °C the alkynes are polymerized owing to the dimers reacting further. The stereochemistry of the dimer is retained throughout the polymer chain[546,547].

The polymerization of alkynols is catalysed by [RhClL$_3$] complexes [L = PPh$_3$,[546,548] P(p-C$_6$H$_4$NMe$_2$)$_3$[549]].

Rhodium(I) complexes also catalyse the codimerization of alkenes and alkynes, although some alkyne cyclotrimerization is also observed in these reactions (equations 128 and 129)[550]. Codimerization of alkadiynes and alkynes has also been

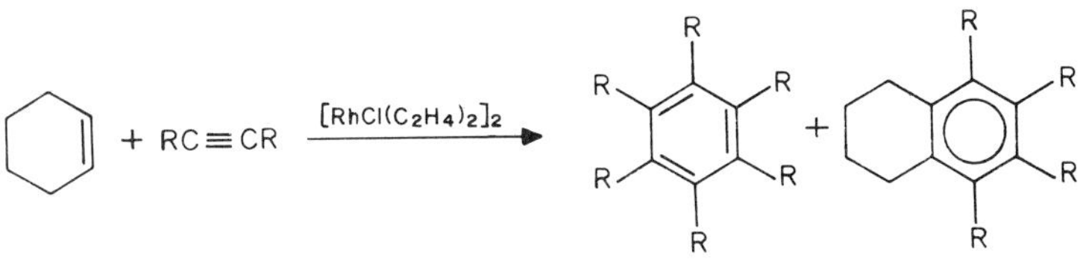
+ RC≡CR
[RhCl(C2H4)2]2
R
R
R
R
R
R
+
R
R
R
R

(128)

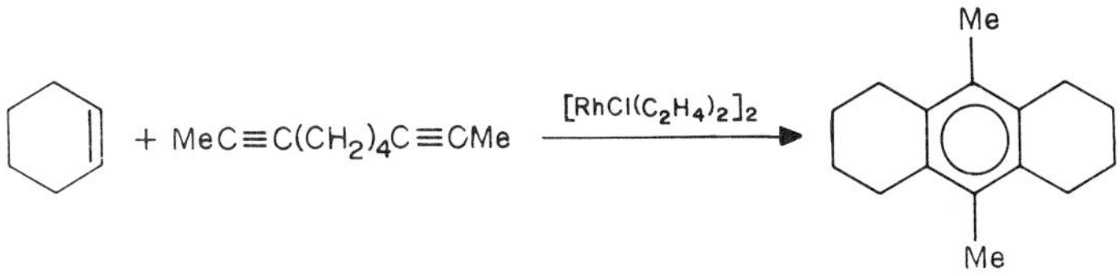
+ MeC≡C(CH2)4C≡CMe
[RhCl(C2H4)2]2
Me
Me

(129)

achieved (equations 130 and 131)[551]. A trialkynyl diether undergoes a very similar reaction that is strictly an isomerization (equation 132)[551].

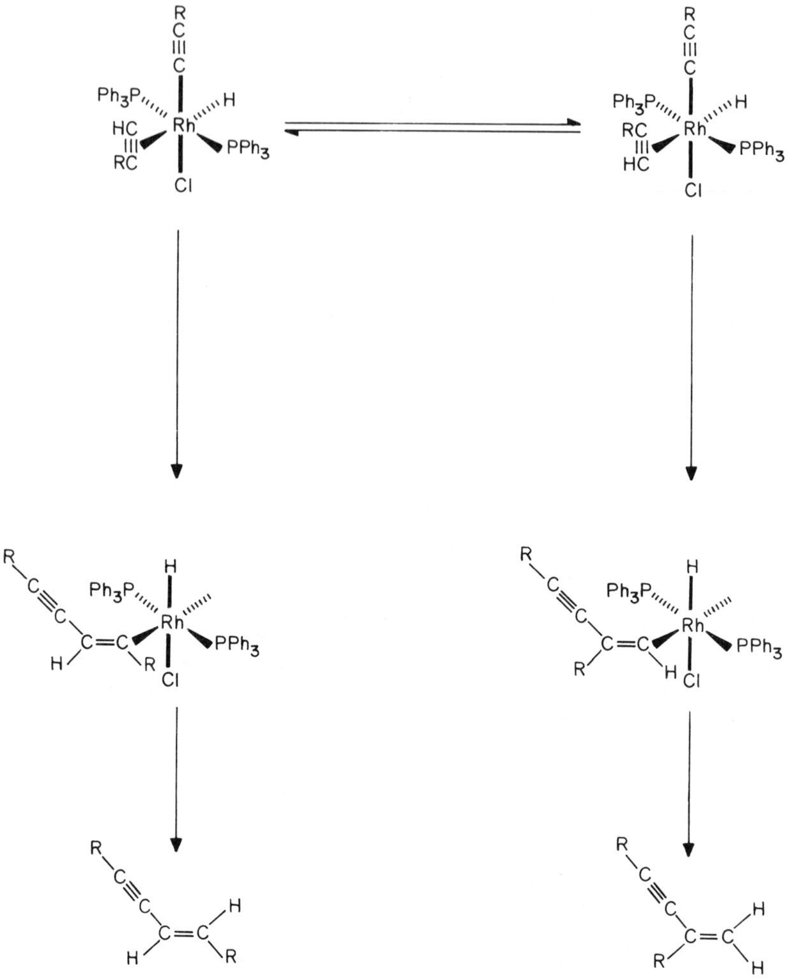

SCHEME 9. Alkyne polymerization catalysed by $RhCl(PPh_3)_3$.

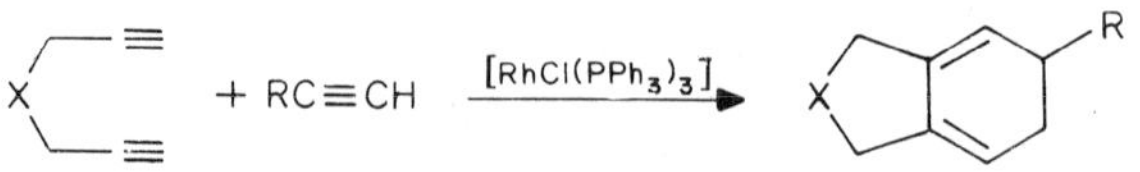

$$+ \ RC{\equiv}CH \xrightarrow{[RhCl(PPh_3)_3]} \tag{130}$$

$$+ \ RC{\equiv}CH \xrightarrow{[RhCl(PPh_3)_3]} \tag{131}$$

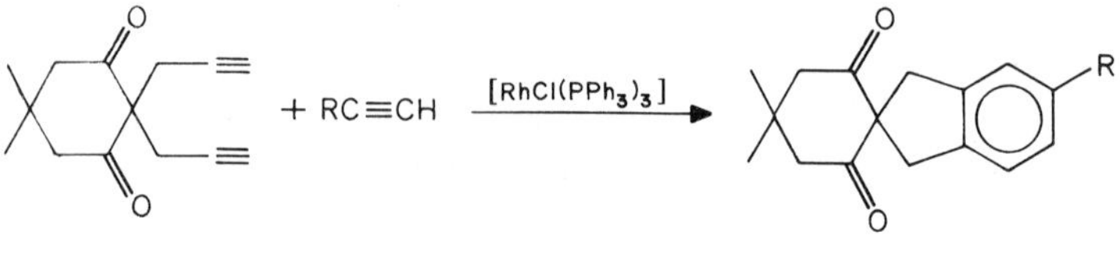

$$\xrightarrow[\text{EtOH, 25 °C}]{[RhCl(PPh_3)_3]} \qquad (132)$$

XII. REFERENCES

1. J. A. Osborn, F. H. Jardine, G. Wilkinson, and J. F. Young, *J. Chem. Soc. A*, 1711 (1966).
2. Imperial Chemical Industries, *Neth. Pat.*, 6602062 (1966); *Chem. Abstr.*, **67**, 64522 (1967).
3. W. Strohmeier, R. Fleischman, and W. Rehder-Stirnweiss, *J. Organomet. Chem.*, **47**, C37 (1973).
4. W. Strohmeier and W. Rehder-Stirnweiss, *J. Organomet. Chem.*, **19**, 417 (1969).
5. W. Strohmeier and W. Rehder-Stirnweiss, *J. Organomet. Chem.*, **18**, P28 (1969).
6. H. Hirai, H. Sawai, E. Ochai, and S. Makishima, *J. Catal.*, **17**, 119 (1970).
7. H. B. Tinker and D. E. Morris, *J. Organomet. Chem.*, **52**, C55 (1973).
8. W. Strohmeier, *J. Organomet. Chem.*, **60**, C60 (1973).
9. L. Horner, H. Buthe, and H. Siegel, *Tetrahedron Lett.*, 4023 (1968).
10. R. A. Schunn, *Inorg. Chem.*, **9**, 2567 (1970).
11. J. F. Biellmann and M. J. Jung, *J. Am. Chem. Soc.*, **90**, 1673 (1968).
12. H. Ruesch and T. J. Mabrey, *Tetrahedron*, **25**, 805 (1969).
13. J. F. Nixon and J. R. Swain, *J. Organomet. Chem.*, **72**, C15 (1974).
14. M. Yagupsky and G. Wilkinson, *J. Chem. Soc. A*, 941 (1970).
15. P. S. Hallman, D. Evans, J. A. Osborn, and G. Wilkinson, *Chem. Commun.*, **530**, 15 (1967).
16. W. Strohmeier and W. Rehder-Stirnweiss, *J. Organomet. Chem.*, **22**, C27 (1970).
17. G. Valentini, G. Sbrana, and G. Braca, *J. Mol. Catal.*, **11**, 383 (1981).
18. T. E. Paxon and M. F. Hawthorne, *J. Am. Chem. Soc.*, **96**, 4676 (1974).
19. F. J. McQuillin and D. G. Parker, *J. Chem. Soc., Perkin Trans. 1*, 2092 (1975).
20. K. Tani, T. Yamagata, S. Otsuka, S. Akutagawa, H. Kumobayashi, T. Taketomi, H. Tayaka, A. Miyashita, and R. Noyori, *Chem. Commun.*, 600 (1982).
21. J. F. Nixon and B. Wilkins, *J. Organomet. Chem.*, **44**, C25 (1972).
22. I. Okura, N. Takahashi, and T. Keii, *J. Mol. Catal.*, **4**, 237 (1978).
23. M. Turner, J. von Jouanne, H. D. Brauer, and H. Kelm., *J. Mol. Catal.*, **5**, 425 (1979).
24. F. D'Amico, J. von Jouanne, and H. Kelm, *J. Mol. Catal.*, **6**, 327 (1979).
25. M. Turner, J. von Jouanne, and H. Kelm, *J. Mol. Catal.*, **5**, 447 (1979).
26. M. Turner, J. von Jouanne, H. D. Brauer, and H. Kelm, *J. Mol. Catal.*, **5**, 443 (1979).
27. J. Blum and Y. Pickholz, *Isr. J. Chem.*, **7**, 723 (1979).
28. P. Goldbourn and F. Scheinman, *J. Chem. Soc., Perkin Trans. 1*, 2870 (1973).
29. E. J. Corey and J. W. Suggs, *J. Org. Chem.*, **38**, 3224 (1973).
30. P. A. Gent and R. Gigg, *Chem. Commun.*, 277 (1974).
31. P. A. Gent and R. Gigg, *J. Chem. Soc., Perkin Trans. 1*, 1835 (1974).
32. R. Gigg, *J. Chem. Soc., Perkin Trans. 1*, 738 (1980).
33. H. Alper and K. Hachem, *J. Org. Chem.*, **45**, 2269 (1980).
34. W. Strohmeier and L. Weigelt, *J. Organomet. Chem.*, **86**, C17 (1975).
35. C. Botteghi and G. Giacomeli, *Gazz. Chim. Ital.*, **106**, 1131 (1976).
36. A. Emery, A. C. Oehlschlager, and A. M. Unrau, *Tetrahedron Lett.*, 4401 (1970).
37. B. Moreau, S. Lavielle, and A. Marquet, *Tetrahedron Lett.*, 2591 (1977).
38. J. C. Duggan, *US Pat.*, 3660642 (1972); *Chem. Abstr.*, **77**, 19256 (1972).
39. T. Nishiguichi, H. Imai, and K. Fukuzumi, *J. Catal.*, **39**, 375 (1975).
40. A. J. Birch and G. S. R. Subba Rao, *Tetrahedron Lett.*, 3797 (1968).
41. W. B. Hughes, *US Pat.*, 3514497 (1970); *Chem. Abstr.*, **73**, 25016 (1970).
42. K. Madeja, W. Jabs, and U. Hahn, *Z. Chem.*, **17**, 235 (1977).
43. J. K. Stille and Y. Becker, *J. Org. Chem.*, **45**, 2139 (1980).
44. U. Eisner and M. M. Sadeghi, *Tetrahedron Lett.*, 299 (1978).
45. P.-W. S. Chum and J. A. Roth, *J. Catal.*, **39**, 198 (1975).
46. H. W. Voight and J. A. Roth, *J. Catal.*, **33**, 91 (1974).

47. M. P. Doyle and D. van Leusen, *J. Org. Chem.*, **47**, 5326 (1982).
48. P. G. Gassman and F. J. Williams, *Chem. Commun.*, 80 (1972).
49. P. G. Gassman and F. J. Williams, *J. Am. Chem. Soc.*, **92**, 7631 (1970).
50. P. G. Gassman, T. Sugawara, and L. G. Tillotson, *J. Org. Chem.*, **42**, 3785 (1977).
51. P. G. Gassman and T. Nakai, *J. Am. Chem. Soc.*, **94**, 5497 (1972).
52. P. G. Gassman, T. J. Atkins, and J. T. Lumb, *Tetrahedron Lett.*, 1643 (1971).
53. L. A. Paquette and R. Gree, *J. Organomet. Chem.*, **146**, 319 (1978).
54. T. J. Katz and S. A. Cerefice, *Tetrahedron Lett.*, 2561 (1969).
55. D. L. Beach and K. W. Barnett, *J. Organomet. Chem.*, **142**, 225 (1977).
56. D. L. Beach, D. L. Garin, L. A. Kaempfe, and K. W. Barnett, *J. Organomet. Chem.*, **142**, 211 (1977).
57. K. W. Barnett, D. L. Beach, D. L. Garin, and L. A. Kaempfe, *J. Am. Chem. Soc.*, **96**, 7127 (1974).
58. H. Hogeveen and H. C. Volger, *J. Am. Chem. Soc.*, **89**, 2486 (1967).
59. K. B. Wiberg and H. A. Connon, *J. Am. Chem. Soc.*, **98**, 5411 (1976).
60. L. Cassar and J. Halpern, *Chem. Commun.*, 1082 (1970).
61. M. J. Chen and H. M. Feder, *Inorg. Chem.*, **18**, 1864 (1979).
62. A. Bruggink and H. Hogeveen, *Tetrahedron Lett.*, 4961 (1972).
63. H. C. Volger, H. Hogeveen, and M. M. P. Gaasbeek, *J. Am. Chem. Soc.*, **91**, 218 (1969).
64. H. C. Volger and H. Hogeveen, *Recl. Trav. Chim. Pays-Bas*, **86**, 830 (1967).
65. H. C. Volger and M. M. P. Gaasbeek, *Recl. Trav. Chim. Pays-Bas*, **87**, 1290 (1968).
66. L. Cassar, P. Eaton, and J. Halpern, *J. Am. Chem. Soc.*, **92**, 3515 (1970).
67. J. Wristers, L. Brener, and R. Petit, *J. Am. Chem. Soc.*, **92**, 7499 (1970).
68. M. Sohn, J. Blum, and J. Halpern, *J. Am. Chem. Soc.*, **101**, 2694 (1979).
69. G. Jones and B. R. Ramachandran, *J. Org. Chem.*, **41**, 798 (1976).
70. H. Hogeveen and B. J. Nusse, *J. Am. Chem. Soc.*, **100**, 3110 (1978).
71. J. P. Hagenbuch and P. Vogel, *Tetrahedron Lett.*, 561 (1979).
72. J. P. Hagenbuch, J. L. Birbaum, J. L. Metral, and P. Vogel, *Helv. Chim. Acta*, **65**, 887 (1982).
73. H. Hogeveen and T. B. Middelkoop, *Tetrahedron Lett.*, 4325 (1973).
74. L. A. Paquette, R. A. Boggs, and J. S. Ward, *J. Org. Chem.*, **42**, 2299 (1977).
75. E. W. Turnblom and T. J. Katz, *J. Am. Chem. Soc.*, **95**, 4292 (1973).
76. W. G. Dauben and A. J. Kielbania, *J. Am. Chem. Soc.*, **93**, 7345 (1971).
77. L. A. Paquette, R. A. Boggs, W. B. Farnham, and R. S. Beckley, *J. Am. Chem. Soc.*, **97**, 1112 (1975).
78. A. P. Marchand, T.-C. Chou, J. D. Ekstrand, and D. van der Helm, *J. Org. Chem.*, **41**, 1438 (1976).
79. P. E. Eaton and S. A. Cerefice, *Chem. Commun.*, 1494 (1970).
80. R. G. Salomon and N. El Sanadi, *J. Am. Chem. Soc.*, **97**, 6214 (1975).
81. G. Adams, C. Bibby, and R. Grigg, *Chem. Commun.*, 491 (1972).
82. R. Grigg, R. Haynes, and A. Sweeney, *Chem. Commun.*, 1248 (1971).
83. R. G. Salomon, M. F. Salomon, and J. L. C. Katchinski, *J. Am. Chem. Soc.*, **99**, 1043 (1977).
84. J. F. Malone, A. Bright, J. K. Nicholson, and B. L. Shaw, *Chem. Commun.*, 712 (1971).
85. R. Grigg, T. R. B. Mitchell, and A. Ramasubbu, *Chem. Commun.*, 27 (1980).
86. F. H. Jardine, *Prog. Inorg. Chem.*, **31**, 265 (1984).
87. B. I. Cruikshank and N. R. Davies, *Aust. J. Chem.*, **26**, 1935 (1973).
88. J. G. Atkinson and M. O. Luke, *Can. J. Chem.*, **48**, 3580 (1970).
89. B. Bayerl, M. Wahren, and J. Grafe, *Tetrahedron*, **29**, 1837 (1973).
90. M. Güngör, F. H. Jardine, and J. D. Wheatley, unpublished results.
91. A. Stefani, G. Consiglio, C. Botteghi, and P. Pino, *J. Am. Chem. Soc.*, **99**, 1058 (1977).
92. T. Ueda, *Proc. 5th Int. Cong. Catal.*, **1**, 431 (1973).
93. B. Hudson, D. E. Webster, and P. B. Wells, *J. Chem. Soc., Dalton Trans.*, 1204 (1972).
94. F. H. Jardine, *Prog. Inorg. Chem.*, **28**, 63 (1981).
95. F. H. Jardine, *Polyhedron*, **1**, 569 (1982).
96. R. R. Schrock and J. A. Osborn, *J. Am. Chem. Soc.*, **98**, 2134 (1976).
97. R. R. Schrock and J. A. Osborn, *J. Am. Chem. Soc.*, **98**, 2143 (1976).
98. R. R. Schrock and J. A. Osborn, *J. Am. Chem. Soc.*, **98**, 4450 (1976).
99. M. Zuber, B. Banas, and F. Pruchnik, *J. Mol. Catal.*, **10**, 143 (1981).
100. R. E. Harmon, J. L. Parsons, and S. K. Gupta, *Chem. Commun.*, 1365 (1969).

101. P. M. Maitlis, C. White, D. S. Gill, J. W. Kang, and H. B. Lee, *Chem. Commun.*, 734 (1971).
102. B. C. Hui and G. L. Rempel, *Chem. Commun.*, 1195 (1970).
103. B. C. Y. Hui, W. K. Teo, and G. L. Rempel, *Inorg. Chem.*, **12**, 757 (1973).
104. J. P. Collman, N. W. Hoffman, and D. E. Morris, *J. Am. Chem. Soc.*, **91**, 5659 (1969).
105. G. Dolcetti, *Inorg. Nucl. Chem. Lett.*, **9**, 705 (1973).
106. R. Uson, L. A. Oro, R. Sariego, M. Valderrama, and C. Rebullida, *J. Organomet. Chem.*, **197**, 87 (1980).
107. P. E. Kreter and D. W. Meek, *Inorg. Chem.*, **22**, 319 (1983).
108. J. M. Brown, P. A. Chaloner, and P. N. Nicholson, *Chem. Commun.*, 646 (1978).
109. R. A. Grey and L. R. Anderson, *Inorg. Chem.*, **16**, 3187 (1977).
110. B. R. James and F. T. T. Ng, *Chem. Commun.*, 908 (1970).
111. B. R. James and F. T. T. Ng, *Can. J. Chem.*, **53**, 797 (1975).
112. P. Kalck, R. Poilblanc, R. P. Martin, A. Rovera, and A. Gaset, *J. Organomet. Chem.*, **195**, C9 (1980).
113. M. Brown and L. W. Piszkiewicz, *J. Org. Chem.*, **32**, 2013 (1967).
114. F. H. Jardine, J. A. Osborn, and G. Wilkinson, *J. Chem. Soc. A*, 1574 (1967).
115. F. H. Jardine and G. Wilkinson, *J. Chem. Soc. C*, 270 (1967).
116. A.-B. Hornfeldt, J. S. Gronowitz, and S. Gronowitz, *Acta Chem. Scand.*, **22**, 2725 (1968).
117. R. L. Krause, *US Pat.*, 3898208, (1975); *Chem. Abstr.*, **83**, 180774 (1975).
118. N. V. Shulyakovskaya, L. V. Vlasova, M. L. Khidekel', and A. I. Markushina, *Bull. Acad. Sci. USSR, Div. Chem. Sci.*, 1689 (1971).
119. W. C. Baird and J. H. Surridge, *J. Org. Chem.*, **40**, 1364 (1975).
120. P. M. Lausarot, G. A. Vaglio, and M. Vale, *J. Organomet. Chem.*, **204**, 249 (1981).
121. A. R. Sanger, *Can. J. Chem.*, **60**, 1363 (1982).
122. R. R. Burch, E. L. Muetterties, R. G. Teller, and J. M. Williams, *J. Am. Chem. Soc.*, **104**, 4257 (1982).
123. R. R. Burch, A. J. Shusterman, E. L. Muetterties, R. G. Teller, and J. M. Williams, *J. Am. Chem. Soc.*, **105**, 3546 (1983).
124. W. S. Knowles, M. J. Sabacky, and D. D. Vineyard, *US Pat.*, 4005127 (1977); *Chem. Abstr.*, **86**, 190463 (1977).
125. R. Glaser and J. Blumenfeld, *Tetrahedron Lett.*, 2525 (1977).
126. Z. Nagy-Magos, S. Vastag, B. Heil, and L. Marko, *J. Organomet. Chem.*, **171**, 97 (1979).
127. J. C. Poulin, Than Phat Dang, and H. B. Kagan, *J. Organomet. Chem.*, **84**, 87 (1975).
128. W. Knowles, M. J. Sabacky, and B. D. Vineyard, *Chem. Commun.*, 10 (1972).
129. L. Horner and B. Schlotthauer, *Phosphorus Sulfur*, **4**, 155 (1978).
130. W. S. Knowles and M. J. Sabacky, *Chem. Commun.*, 1445 (1968).
131. A. M. Aguiar, C. J. Morrow, J. D. Morrison, R. E. Burnett, W. F. Masler, and N. S. Bhacca, *J. Org. Chem.*, **41**, 1545 (1976).
132. C. Fisher and H. S. Mosher, *Tetrahedron Lett.*, 2487 (1977).
133. M. Fiorini, F. Marcati, and G. M. Giongo, *J. Mol. Catal.*, **4**, 125 (1978).
134. M. Yamada, M. Yamashita, and S. Inokawa, *Carbohydr. Res.*, **95**, C9 (1981).
135. M. Yamashita, K. Hiramatsu, M. Yamada, N. Suzuki, and S. Inokawa, *Bull. Chem. Soc. Jpn.*, **55**, 2917 (1982).
136. J. D. Morrison, R. E. Burnett, A. M. Aguiar, C. J. Morrow, and C. Phillips, *J. Am. Chem. Soc.*, **93**, 1301 (1971).
137. G. Balavoine, T. Dang, C. Eskenazi, and H. B. Kagan, *J. Mol. Catal.*, **7**, 531 (1980).
138. R. Mutin, W. Abboud, J. M. Basset, and D. Sinou, *Polyhedron*, **2**, 539 (1983).
139. I. Ojima, T. Kogure, and N. Yoda, *Chem. Lett.*, 495 (1979).
140. R. Glaser, M. Twaik, S. Geresh, and J. Blumenfeld, *Tetrahedron Lett.*, 4635 (1977).
141. J. Bourson, and L. Oliveros, *J. Organomet. Chem.*, **229**, 77 (1982).
142. S. Miyano, M. Nawa, and H. Hashimoto, *Chem. Lett.*, 729 (1980).
143. R. B. King, J. Bakos, C. D. Hoff, and L. Marko, *J. Org. Chem.*, **44**, 3095 (1979).
144. R. B. King, J. Bakos, C. D. Hoff, and L. Marko, *J. Org. Chem.*, **44**, 1729 (1979).
145. R. Glaser, S. Geresh, J. Blumenfeld, and M. Twaik, *Tetrahedron*, **34**, 2405 (1978).
146. H. Brunner, W. Pieronczyk, B. Schonhammer, K. Steng, I. Bernal, and J. Korp, *Chem. Ber.*, **114**, 1137 (1981).
147. H. Brunner and M. Probster, *Inorg. Chim. Acta*, **61**, 129 (1982).

148. H. J. Kreutzfeld and C. Dobler, *React. Kinet. Catal. Lett.*, **16**, 229 (1981).
149. W. Beck and H. Menzel, *J. Organomet. Chem.*, **133**, 307 (1977).
150. K. Tamao, H. Yamamoto, H. Matsumoto, N. Miyake, T. Hayashi, and M. Kumada, *Tetrahedron Lett.*, 1389 (1977).
151. R. H. Grubbs and R. A. De Vries, *Tetrahedron Lett.*, 1879 (1977).
152. K. Kellner, A. Tzschach, Z. Nagy-Magos, and L. Marko, *J. Organomet. Chem.*, **93**, 307 (1980).
153. K. Yamamoto, J. Wakatsuki, and R. Sugimoto, *Bull. Chem. Soc. Jpn.*, **53**, 1132 (1980).
154. T. Hayashi, T. Mise, S. Mitachi, K. Yamamoto, and M. Kamuda, *Tetrahedron Lett.*, 1133 (1976).
155. M. Fiorini, G. M. Giongo, F. Marcati, and W. Marconi, *J. Mol. Catal.*, **1**, 451 (1976).
156. K. Hanaki, K. Kashiwaba, and J. Fujita, *Chem. Lett.*, 489 (1978).
157. I. Ojima and N. Yoda, *Tetrahedron Lett.*, **21**, 1051 (1980).
158. T. P. Dang and H. B. Kagan, *Chem. Commun.*, 481 (1971).
159. H. B. Kagan and Dang Tuan Phat, *J. Am. Chem. Soc.*, **94**, 6429 (1972).
160. Y. Chauvin, D. Commereuc, and R. Stern, *J. Organomet. Chem.*, **146**, 311 (1978).
161. Y. Nakamura, S. Saito, and Y. Morita, *Chem. Lett.*, 7 (1980).
162. R. Glaser and S. Geresh, *Tetrahedron*, **35**, 2381 (1980).
163. G. Gelbard, H. B. Kagan, and R. Stern, *Tetrahedron*, **32**, 233 (1976).
164. R. Glaser and B. Vainas, *J. Organomet. Chem.*, **121**, 249 (1976).
165. R. Glaser, J. Blumenfeld, and M. Twaik, *Tetrahedron Lett.*, 4639 (1977).
166. R. Glaser and S. Geresh, *Tetrahedron Lett.*, 2527 (1977).
167. B. D. Vineyard, W. S. Knowles, M. J. Sabacky, G. L. Bachman, and D. J. Weinkauff, *J. Am. Chem. Soc.*, **99**, 5946 (1977).
168. J. M. Brown and B. A. Murrer, *Tetrahedron Lett.*, **21**, 581 (1980).
169. R. G. Nuzzo, D. Feitler, and G. M. Whitesides, *J. Am. Chem. Soc.*, **101**, 3683 (1979).
170. H. B. Kagan, in "Comprehensive Organometallic Chemistry", Ed. G. Wilkinson, Vol. 8, P463, Pergamon, Oxford, 1982.
171. A. Uehara and J. C. Bailar, *J. Organomet. Chem.*, **239**, 1 (1982).
172. J. Irrure, A. Bosch, and J. Capdevila, *Afinidad*, **38**, 201 (1981).
173. G. Comisso, A. Sega, and V. Sunjic, *Croat. Chem. Acta*, **53**, 445 (1980).
174. D. Lafont, D. Sinou, and G. Descotes, *J. Organomet. Chem.*, **169**, 87 (1979).
175. J. M. Brown and B. A. Murrer, *J. Chem. Soc., Perkin Trans. 2*, 489 (1982).
176. M. D. Fryzuk and B. Bosnich, *J. Am. Chem. Soc.*, **100**, 5491 (1978).
177. D. P. Riley and R. E. Shumate, *J. Org. Chem.*, **45**, 5187 (1980).
178. W. Bergstein, A. Kleeman, and J. Matens, *Synthesis*, 76 (1981).
179. J. P. Amma and J. K. Stille, *J. Org. Chem.*, **47**, 468 (1982).
180. J. Kottner and G. Greber, *Chem. Ber.*, **113**, 2323 (1980).
181. M. Fiorini and G. M. Giongo, *J. Mol. Catal.*, **7**, 411 (1980).
182. P. A. McNeil, N. K. Roberts, and B. Bosnich, *J. Am. Chem. Soc.*, **103**, 2273 (1981).
183. E. P. Kyba, R. E. Davis, P. N. Juri, and K. R. Shirley, *Inorg. Chem.*, **20**, 3616 (1981).
184. A. Miyashita, A. Yasuda, H. Takaya, K. Toriumi, T. Ito, T. Souchi, and R. Noyori, *J. Am. Chem. Soc.*, **102**, 7932 (1980).
185. G. Descotes, D. Lafont, and D. Sinou, *J. Organomet. Chem.*, **150**, C14 (1978).
186. D. Lafont, D. Sinou, G. Descotes, R. Glaser, and S. Geresh, *J. Mol. Catal.*, **10**, 305 (1981).
187. Tuan Phat Dang, J. C. Poulin, and H. B. Kagan, *J. Organomet. Chem.*, **91**, 105 (1975).
188. A. Levi, G. Modena, and G. Scorrano, *Chem. Commun.*, 6 (1975).
189. K. Osakada, T. Ikariya, M. Saburi, and S. Yoshikawa, *Chem. Lett.*, 1691 (1981).
190. M. Fiorini and G. M. Giongo, *J. Mol. Catal.*, **5**, 303 (1979).
191. G. Pracejus and H. Pracejus, *Tetrahedron Lett.*, 3497 (1977).
192. K. Kashiwabra, K. Hanaki, and J. Fujita, *Bull. Chem. Soc. Jpn.*, **53**, 2275 (1980).
193. K. Onuma, T. Ito, and A. Nakamura, *Chem. Lett.*, 905 (1979).
194. K. Onuma, T. Ito, and A. Nakamura, *Bull. Chem. Soc. Jpn.*, **53**, 2012 (1980).
195. E. Cesarotti, A. Ciesa, and G. D'Alfonso, *Tetrahedron Lett.*, **23**, 2995 (1982).
196. T. H. Johnson and G. Rangarajan, *J. Org. Chem.*, **45**, 62 (1980).
197. D. Sinou and G. Descotes, *React. Kinet. Catal. Lett.*, **14**, 463 (1980).
198. W. R. Cullen and J. D. Woolins, *Can. J. Chem.*, **60**, 1793 (1982).
199. W. R. Cullen and E.-S. Yeh, *J. Organomet. Chem.*, **139**, C13 (1977).
200. W. R. Cullen, F. W. B. Einstein, C.-H. Huang, A. C. Willis, and E.-S. Yeh, *J. Am. Chem. Soc.*, **102**, 988 (1980).

201. R. Glaser, S. Geresh, and J. Blumenfeld, *J. Organomet. Chem.*, **112**, 355 (1976).
202. D. Sinou, *Tetrahedron Lett.*, **22**, 2987 (1981).
203. A. S. C. Chan, J. J. Pluth, and J. Halpern, *J. Am. Chem. Soc.*, **102**, 5952 (1980).
204. C. Detellier, G. Gelbard, and H. B. Kagan, *J. Am. Chem. Soc.*, **100**, 7556 (1978).
205. J. M. Brown and D. Parker, *J. Org. Chem.*, **47**, 2772 (1982).
206. M. D. Fryzuk and B. Bosnich, *J. Am. Chem. Soc.*, **99**, 6262 (1977).
207. I. Ojima, T. Kogure, and K. Achiwa, *Chem. Lett.*, 567 (1978).
208. K. Achiwa, *Chem. Lett.*, 561 (1978).
209. K. Achiwa, Y. Ohga, Y. Iitakaka, and H. Saito, *Tetrahedron Lett.*, 4683 (1978).
210. W. Knowles, M. J. Sabacky, B. D. Vineyard, and D. J. Weinkauff, *J. Am. Chem. Soc.*, **97**, 2567 (1975).
211. K. Achiwa, *Chem. Lett.*, 777 (1977).
212. O. Samuel, R. Couffignal, M. Lauer, S. Y. Zhang, and H. B. Kagan, *Nouv. J. Chim.*, **5**, 15 (1981).
213. A. Uehara, T. Kubota, and R. Tsuchiya, *Chem. Lett.*, 441 (1983).
214. J. C. Poulin and H. B. Kagan, *Chem. Commun.*, 1261 (1982).
215. K. Onuma, T. Ito, and A. Nakamura, *Chem. Lett.*, 481 (1980).
216. I. Ojima and T. Suzuki, *Tetrahedron Lett.*, **21**, 1239 (1980).
217. K. Achiwa, *Tetrahedron Lett.*, 3735 (1977).
218. G. L. Baker, S. J. Fritschel, and J. K. Stille, *J. Org. Chem.*, **46**, 2960 (1981).
219. K. Ohkubo, M. Haga, K. Yoshinaga, and Y. Motozato, *Inorg. Nucl. Chem. Lett.*, **16**, 155 (1980).
220. K. Ohkubo, K. Fujimori, and K. Yoshinaga, *Inorg. Nucl. Chem. Lett.*, **15**, 231 (1979).
221. U. Nagel, H. Menzel, P. W. Lednor, W. Beck, A. Guyot, and M. Bartholin, *Z. Naturforsch., Teil B*, **36**, 578 (1981).
222. K. Achiwa, *Chem. Lett.*, 905 (1978).
223. K. Ohkubo, M. Haga, K. Yoshinaga, and Y. Motozato, *Inorg. Nucl. Chem. Lett.*, **17**, 215 (1981).
224. G. L. Baker, S. J. Fritschel, J. R. Stille, and J. K. Stille, *J. Org. Chem.*, **46**, 2954 (1981).
225. R. Selke, *React. Kinet. Catal. Lett.*, **10**, 135 (1979).
226. M. Tanaka, Y. Watanabe, T. Mitsudo, H. Iwane, and Y. Takegami, *Chem. Lett.*, 239 (1973).
227. B. Heil, S. Toros, S. Vastag, and L. Marko, *J. Organomet. Chem.*, **94**, C47 (1975).
228. P. Bonvincini, A. Levi, G. Modena, and G. Scorrano, *Chem. Commun.*, 1188 (1972).
229. K. Ohkubo, I. Setoguchi, and K. Yoshinaga, *Inorg. Nucl. Chem. Lett.*, **15**, 235 (1979).
230. J. Solodar, *Chem. Technol.*, **5**, 421 (1975).
231. M. Fiorini, F. Marcati, and G. M. Giongo, *J. Mol. Catal.*, **3**, 385 (1978).
232. S. Toros, B. Heil, L. Kollar, and L. Marko, *J. Organomet. Chem.*, **197**, 85 (1980).
233. S. Toros, B. Heil, L. Kollar, and L. Marko, *J. Organomet. Chem.*, **232**, C17 (1982).
234. I. Ojima, T. Kogure, and K. Achiwa, *Chem. Commun.*, 428 (1977).
235. I. Ojima and T. Kogure, *J. Organomet. Chem.*, **195**, 239 (1980).
236. K. Achiwa, T. Kogure, and I. Ojima, *Tetrahedron Lett.*, 4431 (1977).
237. K. Achiwa, T. Kogure, and I. Ojima, *Chem. Lett.*, 297 (1978).
238. J. Blum and S. Biger, *Tetrahedron Lett.*, 1825 (1970).
239. M. O. Broitman, A. V. Ablov, and K. I. Kuchkova, *Kinet. Catal.*, **18**, 537 (1977).
240. S. Shinoda, T. Kojima, and Y. Saito, *J. Mol. Catal.*, **18**, 99 (1983).
241. H. B. Charman, *J. Chem. Soc. B*, 584 (1970).
242. H. Arakawa and Y. Sugi, *Chem. Lett.*, 1323 (1981).
243. R. Irie, X. S. Li, and Y. Saito, *J. Mol. Catal.*, **18**, 263 (1983).
244. K. Ohkubo, T. Ohgushi, and K. Yoshinaga, *Chem. Lett.*, 775 (1976).
245. K. Ohkubo, K. Hirata, T. Ohgushi, and K. Yoshinaga, *J. Coord. Chem.*, **6**, 185 (1977).
246. S. H. Strauss, K. H. Whitmire, and D. F. Shriver, *J. Organomet. Chem.*, **174**, C59 (1979).
247. D. Beaupere, L. Nadjo, R. Uzan, and P. Bauer, *J. Mol. Catal.*, **20**, 195 (1983).
248. H. Imai, T. Nishiguchi, and K. Fukuzumi, *J. Org. Chem.*, **39**, 1622 (1974).
249. G. Gregorio, G. Pregaglia, and R. Ugo, *Inorg. Chim. Acta*, **3**, 89 (1969).
250. L. Kollar, S. Toros, B. Heil, and L. Marko, *J. Organomet. Chem.*, **192**, 253 (1980).
251. A. Yamuda, T. Fukuda, and M. Yamagita, *Bull. Chem. Soc. Jpn.*, **48**, 353 (1975).
252. T. Nishiguchi, K. Tachi, and K. Fukuzumi, *J. Am. Chem. Soc.*, **94**, 8916 (1972).
253. T. Nishiguchi and K. Fukuzumi, *J. Am. Chem. Soc.*, **96**, 1893 (1974).
254. C. Masters, A. A. Kiffen, and J. P. Visser, *J. Am. Chem. Soc.*, **98**, 1357 (1976).
255. T. Nishiguchi, K. Tachi, and K. Fukuzumi, *J. Org. Chem.*, **40**, 237 (1975).

256. A. Spencer, *J. Organomet. Chem.*, **93**, 389 (1975).
257. D. Beaupere, P. Bauer, and R. Uzan, *Can. J. Chem.*, **57**, 218 (1979).
258. D. Beaupere, L. Nadjo, R. Uzan, and P. Bauer, *J. Mol. Catal.*, **14**, 129 (1982).
259. D. Beaupere, P. Bauer, L. Nadjo, and R. Uzan, *J. Organomet. Chem.*, **238**, C12 (1982).
260. D. Beaupere, L. Nadjo, R. Uzan, and P. Bauer, *J. Mol. Catal.*, **18**, 73 (1983).
261. V. Z. Sharf, L. K. Freidlin, I. S. Shekoyan, and V. N. Krutii, *Bull. Acad. Sci. USSR, Div. Chem. Sci.*, 758 (1977).
262. P. Svoboda and J. Hetflejš, *Collect. Czech. Chem. Commun.*, **42**, 2177 (1977).
263. K. Tani, K. Suwa, E. Tanigawa, T. Yoshida, T. Okano, and S. Otsuka, *Chem. Lett.*, 261 (1982).
264. V. Z. Sharf, L. K. Freidlin, V. N. Krutii, and I. S. Shekoyan, *Bull. Acad. Sci. USSR, Div. Chem. Sci.*, 1251 (1974).
265. R. Spogliarich, G. Zassinovich, G. Mestroni, and M. Graziani, *J. Organomet. Chem.*, **179**, C45 (1979).
266. R. Spogliarich, G. Zassinovich, G. Mestroni, and M. Graziani, *J. Organomet. Chem.*, **198**, 81 (1980).
267. G. Zassinovitch, G. Mestroni, and A. Camus, *J. Organomet. Chem.*, **168**, C37 (1979).
268. R. Uson, L. A. Oro, R. Sariego, and M. A. Esteruelas, *J. Organomet. Chem.*, **214**, 399 (1981).
269. B. Ilmaier and R. S. Nyholm, *Naturwissenschaften*, **56**, 415 (1969).
270. V. Z. Sharf, L. K. Freidlin, and V. N. Krutii, *Bull. Acad. Sci. USSR, Div. Chem. Sci.*, 2207 (1973).
271. V. Z. Sharf, L. K. Freidlin, I. S. Shekoyan, and V. N. Krutii, *Bull. Acad. Sci. USSR, Div. Chem. Sci.*, 919 (1978).
272. V. N. Krutii, S. A. Chelmakova, A. S. Gurovets, T. V. Vasina, D. B. Furman, V. Z. Sharf, and A. L. Liberman, *Bull. Acad. Sci. USSR, Div. Chem. Sci.*, 1019 (1975).
273. J. Cook, J. E. Hamlin, A. Nutton, and P. M. Maitlis, *J. Chem. Soc., Dalton Trans.*, 2342 (1981).
274. V. Z. Sharf, L. K. Freidlin, B. M. Savchenko, and V. N. Krutii, *Izv. Akad. Nauk SSSR, Ser. Khim.*, 1134 (1979); *Chem. Abstr.*, **91**, 91268 (1979).
275. V. Z. Sharf, L. K. Freidlin, B. M. Savchenko, and V. N. Krutii, *Izv. Akad. Nauk. SSSR, Ser. Khim.*, 1393 (1979); *Chem. Abstr.*, **91**, 107737 (1979).
276. V. Z. Sharf, E. A. Mistryukov, L. K. Freidlin, I. S. Portyakova, and V. N. Krutii, *Izv. Akad. Nauk SSSR, Ser. Khim.*, 1411 (1979); *Chem. Abstr.*, **91**, 123152 (1979).
277. I. S. Shekoyan, G. V. Varnakova, V. N. Krutii, E. I. Karpeiskaya, and V. Z. Sharf, *Bull. Acad. Sci. USSR Div. Chem. Sci.*, 2700 (1975).
278. V. N. Krutii, V. Z. Sharf, I. P. Yakolev, T. N. Sedletskaya, A. S. Gurovets, and L. K. Freidlin, *J. Org. Chem. USSR*, **12**, 745 (1976).
279. V. Z. Sharf, L. K. Freidlin, and V. N. Krutii, *Bull. Acad. Sci. USSR, Div. Chem. Sci.*, 666 (1977).
280. G. Doyle, *Am. Chem. Soc. Div. Pet. Chem. Prepr.*, **21**, 165 (1976).
281. F. H. Jardine, J. A. Osborn, G. Wilkinson and J. F. Young, *Chem. Ind. (London)*, 560 (1965).
282. K. L. Oliver and F. B. Booth, *Am. Chem. Soc. Div. Pet. Chem. Prepr.*, **14**, A7 (1969).
283. J. H. Craddock, A. Hershman, F. E. Paulik, and J. F. Roth, *Ind. Eng. Chem., Prod. Res. Dev.*, **8**, 291 (1969).
284. T. Hatashi, M. Tanaka, and I. Ogata, *J. Mol. Catal.*, **13**, 323 (1981).
285. M. Royo, F. Melo, A. Manrique, and L. Oro, *Transition Met. Chem.*, **7**, 44 (1982).
286. L. Oro, A. Manrique, and M. Royo, *Transition Met. Chem.*, **3**, 383 (1978).
287. R. H. Crabtree and H. Felkin, *J. Mol. Catal.*, **5**, 75 (1979).
288. T. Hayashi, M. Tanaka, and I. Ogata, *J. Mol. Catal.*, **6**, 1 (1979).
289. G. Wilkinson, personal communication.
290. E. N. Frankel and F. L. Thomas, *J. Am. Oil Chem. Soc.*, **49**, 10 (1972).
291. E. N. Frankel, F. L. Thomas, and W. K. Rohweder, *Ind. Eng. Chem., Prod. Res. Dev.*, **12**, 47 (1973).
292. A. F. Borowski, D. J. Cole-Hamilton, and G. Wilkinson, *Nouv. J. Chem.*, **2**, 137 (1978).
293. K. Kurtev, D. Ribola, R. A. Jones, D. J. Cole-Hamilton and G. Wilkinson, *J. Chem. Soc., Dalton Trans.*, 55 (1980).
294. P. Svoboda, M. Čapka, and J. Hetflejš, *Collect. Czech. Chem. Commun.*, **38**, 1235 (1973).
295. R. A. Jones, T. C. Wright, J. L. Atwood, and W. E. Hunter, *Organometallics*, **2**, 47 (1983).
296. K. Kaneda, M. Yasumara, M. Hiraki, T. Imanaka, and S. Teranishi, *Chem. Lett.*, 1763 (1981).
297. B. Fell and E. Muller, *Monatsh. Chem.*, **103**, 1222 (1972).
298. G. Wilkinson, *Br. Pat.*, 1357735 (1974); *Chem. Abstr.*, **81**, 119889 (1974).

299. A. Saus, Tuyet Nhu Phu, M. J. Mirbach, and M. F. Mirbach, *J. Mol. Catal.*, **18**, 117 (1983).
300. C. Botteghi, S. Gladiali, V. Bellagamba, R. Ercoli, and A. Gamba, *Chim. Ind. (Milan)*, **62**, 604 (1980).
301. C. K. Brown and G. Wilkinson, *J. Chem. Soc. A*, 2753 (1970).
302. S. Franks and F. R. Hartley, *J. Mol. Catal.*, **12**, 121 (1981).
303. W. R. Cullen, B. R. James, and G. Strukul, *Can. J. Chem.*, **56**, 1965 (1978).
304. A. G. Abatioglu and D. R. Bryant, *Eur. Pat. Appl.* 60523 (1982); *Chem. Abstr.*, **98**, 4298 (1983).
305. J. D. Unruh and W. J. Wells, *Ger. Pat.*, 26617306 (1976); *Chem. Abstr.*, **86**, 22369 (1977).
306. O. R. Hughes and J. D. Unruh, *J. Mol. Catal.*, **12**, 71 (1981).
307. J. D. Unruh and J. R. Christenson, *J. Mol. Catal.*, **14**, 19 (1982).
308. M. Matsumoto and M. Tamura, *J. Mol. Catal.*, **16**, 209 (1982).
309. R. L. Pruett and J. A. Smith, *J. Org. Chem.*, **34**, 327 (1969).
310. R. Grigg, G. J. Reimer, and A. R. Wade, *J. Chem. Soc., Perkin Trans. 1*, 1929 (1983).
311. M. Takeda, H. Iwane, and T. Hashimoto, *Jpn. Pat.*, 80 28969 (1980); *Chem. Abstr.*, **93**, 114041 (1980).
312. B. Fell and H. Bahrmann, *J. Mol. Catal.*, **2**, 211 (1977).
313. H. Bahrmann and B. Fell. *J. Mol. Catal.*, **8**, 329 (1980).
314. M. Garst and D. Lukton, *J. Org. Chem.*, **46**, 4433 (1981).
315. A. de Jong, *Ger. Pat.*, 2719735 (1977); *Chem. Abstr.*, **88**, 89113 (1978).
316. A. Spencer, *J. Organomet. Chem.*, **124**, 85 (1977).
317. B. Falk and B. Fell, *J. Mol. Catal.*, **18**, 127 (1983).
318. Y. Inamoto, Y. Fujikura, and H. Ikeda, *Ger. Pat.*, 2654799 (1977); *Chem. Abstr.*, **88**, 6432 (1978)
319. B. L. Booth and A. D. Lloyd, *J. Organomet. Chem.*, **35**, 195 (1972).
320. J. Schwartz, D. W. Hart, and J. L. Holden, *J. Am. Chem. Soc.*, **94**, 9269 (1972).
321. R. A. Sanchez-Delgado and G. Wilkinson, *J. Chem. Soc., Dalton Trans.*, 804 (1977).
322. G. Caccia, G. Chelucci, and C. Botteghi, *Synth. Commun.*, **11**, 71 (1981).
323. Mitsubishi Petrochemical Co., *Jpn Pat.*, 81 87549 (1981); *Chem. Abstr.*, **95**, 203367 (1981).
324. A. J. Dennis, *Eur. Pat. Appl.*, **18**, 162 (1980); *Chem. Abstr.*, **94**, 156338 (1981).
325. N. Harris, A. J. Dennis, and G. E. Harrison, *Eur. Pat. Appl.*, **18**, 163 (1980); *Chem. Abstr.*, **94**, 156295 (1981).
326. N. Harris, A. J. Dennis, and G. E. Harrison, *Eur. Pat. Appl.*, **18**, 161 (1980); *Chem. Abstr.*, **94**, 156315 (1981).
327. F. J. McQuillin, *Br. Pat.*, 1555331 (1979); *Chem. Abstr.*, **93**, 12948 (1980).
328. C. U. Pittman, W. D. Honnick, and J. J. Yang, *J. Org. Chem.*, **45**, 684 (1980).
329. H. B. Tinker, *Ger. Pat.* 2623673 (1976); *Chem. Abstr.*, **86**, 105957 (1977).
330. K. Schwirten, H. W. Schneider, and R. Kummer, *Ger. Pat.*, 2643205 (1978); *Chem. Abstr.*, **89**, 5946 (1978).
331. M. Matsumoto and M. Tamura, *J. Mol. Catal.*, **16**, 195 (1982).
332. A. G. Abatjoglou, D. R. Bryant, and L. C. D'Esposito, *J. Mol. Catal.*, **18**, 381 (1983).
333. A. Kleeman and R. Fahnenstich, *Fr. Pat.*, 2500488 (1982); *Chem. Abstr.*, **98**, 53176 (1983).
334. C. Botteghi, G. Ceccacarelli, and G. Consiglio, *J. Prakt. Chem.*, **314**, 840 (1972).
335. M. Takeda, H. Iwane, and T. Hashimoto, *Jpn. Pat.*, 80 29920 (1980); *Chem. Abstr.*, **93**, 169128 (1980).
336. L. A. Gerritsen, J. M. Herman, W. Klut, and J. J. F. Scholten, *J. Mol. Catal.*, **9**, 157 (1980).
337. L. A. Gerritsen, J. M. Herman, and J. J. F. Scholten, *J. Mol. Catal.*, **9**, 241 (1980).
338. L. A. Gerritsen, W. Klut, M. H. Vreugdenhill, and J. J. F. Scholten, *J. Mol. Catal.*, **9**, 257 (1980).
339. L. A. Gerritsen, W. Klut, M. H. Vreugdenhill, and J. J. F. Scholten, *J. Mol. Catal.*, **9**, 265 (1980).
340. (a) J. Hjorkjaer, M. S. Scurrell, and P. Simonsen, *J. Mol. Catal.*, **6**, 405 (1979); (b) F. R. Hartley, S. G. Murray and P. N. Nicholson, *J. Mol. Catal.*, **16**, 363 (1982); (c) F. R. Hartley, *Supported Metal Complexes: A New Generation of Catalysts*, Reidel, Dordrecht, 1985, Chapter 8.
341. C. U. Pittman and R. M. Hanes, *J. Am. Chem. Soc.*, **98**, 5402 (1976).
342. A. Stefani and D. Tatone, *Helv. Chim. Acta*, **60**, 518 (1977).
343. I. Ogata and Y. Ikeda, *Chem. Lett.*, 487 (1972).
344. M. Tanaka, Y. Watanabe, T. Mitsudo, and Y. Takegami, *Bull. Chem. Soc. Jpn.*, **47**, 1698 (1974).
345. M. Tanaka, Y. Watanabe, T. Mitsudo, K. Yamamoto, and Y. Takegami, *Chem. Lett.*, 483 (1972).
346. R. Stern, A. Hirsauer, and L. Sajus, *Tetrahedron Lett.*, 3247 (1973).
347. M. Tanaka, Y. Ikeda, and I. Ogata, *Chem. Lett.*, 1115 (1975).

348. P. Pino, G. Consiglio, C. Botteghi, and C. Salomon, *Adv. Chem. Ser.*, No. 132, 295 (1974).
349. G. Consiglio, C. Botteghi, C. Salomon, and P. Pino, *Angew. Chem.*, **85**, 663 (1973).
350. Y. Watanabe, T. Mitsudo, Y. Yasunori, J. Kikuchi, and Y. Takegami, *Bull. Chem. Soc. Jpn.*, **52**, 2735 (1979).
351. C. F. Hobbs and W. S. Knowles, *J. Org. Chem.*, **46**, 4422 (1981).
352. C. Botteghi, M. Branca, and A. Saba, *J. Organomet. Chem.*, **184**, C17 (1980).
353. C. Botteghi and C. Salomon, *Tetrahedron Lett.*, 4285 (1974).
354. Y. Becker, A. Eisenstadt, and J. K. Stille, *J. Org. Chem.*, **45**, 2145 (1980).
355. C. Botteghi, M. Branca, G. Micera, F. Piacenti, and G. Menchi, *Chim. Ind. (Milan)*, **60**, 809 (1978).
356. S. J. Fritschel, J. J. H. Ackerman, and J. K. Stille, *J. Org. Chem.*, **44**, 3152 (1979).
357. K. W. Muir and J. A. Ibers, *Inorg. Chem.*, **9**, 440 (1970).
358. H. M. Dickers, R. N. Hazeldine, L. S. Malkin, A. P. Mather, and R. V. Parrish, *J. Chem. Soc., Dalton Trans.*, 308 (1980).
359. R. A. Faltynek, *Inorg. Chem.*, **20**, 1357 (1981).
360. A. J. Chalk and J. F. Harrod, *J. Am. Chem. Soc.*, **87**, 16 (1965).
361. J. Rehjon and J. Hetflejš, *Collect. Czech. Chem. Commun.*, **40**, 3680 (1975).
362. J. Rehjon and J. Hetflejš, *Collect. Czech. Chem. Commun.*, **40**, 3190 (1975).
363. K. A. Brady and T. A. Nile, *J. Organomet. Chem.*, **206**, 209 (1981).
364. P. Svoboda, M. Čapka, and J. Hetflejš, *Collect. Czech. Chem. Commun.*, **37**, 3059 (1972).
365. A. J. Chalk, *J. Organomet. Chem.*, **21**, 207 (1970).
366. J. E. Hill and T. A. Nile, *J. Organomet. Chem.*, **137**, 293 (1977).
367. H. Kono, M. Wakao, I. Ojima, and Y. Nagai, *Chem. Lett.*, 189 (1975).
368. B. Marciniec, W. Urbaniak, and P. Pawlak, *J. Mol. Catal.*, **14**, 323 (1982).
369. J. P. Howe, K. Lung, and T. Nile, *J. Organomet. Chem.*, **208**, 401 (1981).
370. A. J. Cornish, M. F. Lappert, G. L. Filatovs, and T. A. Nile, *J. Organomet. Chem.*, **172**, 153 (1979).
371. A. Millan, E. Towns, and P. M. Maitlis, *Chem. Commun.*, 673 (1981).
372. L. I. Zakharin, I. V. Pisareva, and T. B. Agakhanova, *Bull. Acad. Sci. USSR, Div. Chem. Sci.*, 2222 (1977).
373. V. B. Panov, M. L. Khidekel, and S. A. Shchepinov, *Bull. Acad. Sci. USSR, Div. Chem. Sci.*, 2272 (1968).
374. R. N. Hazeldine, R. V. Parrish, and D. J. Parry, *J. Chem. Soc. A*, 683 (1969).
375. G. Kuncova and V. Chvalovsky, *Collect. Czech. Chem. Commun.*, **45**, 2085 (1980).
376. G. Kuncova and V. Chvalovsky, *Collect. Czech. Chem. Commun.*, **45**, 2240 (1980).
377. I. Ojima and M. Kumagi, *J. Organomet. Chem.*, **134**, C6 (1977).
378. I. Ojima and M. Kumagi, *J. Organomet. Chem.*, **157**, 359 (1978).
379. M. G. Voronkov, N. N. Vlasova, S. A. Bolshakova, and S. V. Kirpichenko, *J. Organomet. Chem.*, **190**, 335 (1980).
380. M. G. Voronkov, V. Chvalovsky, S. V. Kirpichenko, N. N. Vlasova, S. A. Bolshakova, G. Kuncova, V. V. Keiko, and E. O. Tsetlina, *Collect. Czech. Chem. Commun.*, **44**, 742 (1979).
381. M. G. Voronkov, N. N. Vlasova, S. V. Kirpichenko, S. A. Bolshakova, V. V. Keiko, E. O. Tsetlina, S. V. Amosova, B. A. Trofimov, and V. Chvalovsky, *Bull. Acad. Sci. USSR, Div. Chem. Sci.*, 422 (1979).
382. H. Watanabe, M. Asami, and Y. Nagai, *J. Organomet. Chem.*, **195**, 363 (1980).
383. V. B. Pukhnarevitch, L. I. Kopylova, E. O. Tsetlina, V. A. Pestunovich, V. Chvalovsky, J. Hetflejš, and M. G. Voronkov, *Proc. Acad. Sci. USSR, Sect. Chem.*, **231**, 764 (1976).
384. V. B. Pukhnarevitch, L. I. Kopylova, M. Capka, J. Hetflejš, E. N. Satsuk, M. V. Sigalov, V. Chvalovsky, and M. G. Voronkov, *J. Gen. Chem. USSR*, **50**, 1259 (1980).
385. H. Watanabe, T. Kitahara, T. Motegi, and Y. Nagai, *J. Organomet. Chem.*, **139**, 215 (1977).
386. R. J. P. Corriu and J. J. E. Moreau, *Chem. Commun.*, 812 (1971).
387. I. Ojima, T. Kogure, and Y. Nagai, *Tetrahedron Lett.*, 5035 (1972).
388. I. Ojima, M. Nihonyanagi, and Y. Nagai, *Chem. Commun.*, 938 (1972).
389. I. Ojima, M. Nihonyanagi, and Y. Nagai, *Bull. Chem. Soc. Jpn.*, **45**, 3506 (1972).
390. I. Ojima, M. Nihonyanagi, and Y. Nagai, *Bull. Chem. Soc. Jpn.*, **45**, 3722 (1972).
391. C. Eaborn, K. Odell, and A. Pidcock, *J. Organomet. Chem.*, **63**, 93 (1973).
392. V. Z. Sharf, L. K. Freidlin, I. S. Shekoyan, and V. N. Krutii, *Bull. Acad. Sci. USSR, Div. Chem. Sci.*, 995 (1977).

393. I. Ojima, in *Organotransition Metal Chemistry* (Eds. Y. Ishii and M. Tsutsui), Plenum Press, New York, 1975, pp. 255–264.
394. H. Matsumoto, Y. Hoshino, and Y. Nagai, *Bull. Chem. Soc. Jpn.*, **54**, 1279 (1981).
395. Y. Nagai, I. Ojima, and M. Tajima, *Jpn. Pat.*, 76 98225 (1976); *Chem. Abstr.*, **86**, 106768 (1977).
396. I. Ojima, M. Kumagai, and Y. Nagai, *J. Organomet. Chem.*, **111**, 43 (1976).
397. E. Yoshii, Y. Kobayashi, T. Koizumi, and T. Oribe, *Chem. Pharm. Bull.*, **22**, 2767 (1974).
398. M. F. Lappert and T. A. Nile, *J. Organomet. Chem.*, **102**, 543 (1975).
399. I. Ojima, T. Kogure, M. Nihonyanagi, H. Kono, S. Inaba, and Y. Nagai, *Chem. Lett.*, 501 (1973).
400. R. J. P. Corriu and J. J. E. Moreau, *J. Organomet. Chem.*, **127**, 7 (1977).
401. I. Ojima, S. Inaba, T. Kogure, S. Matsumoto, H. Matsumoto, H. Watanabe, and Y. Nagai, *J. Organomet. Chem.*, **55**, C4 (1973).
402. T. Fujisawa, Y. Nagai, and I. Ojima, *Jpn. Pat.*, 74 75560 (1974); *Chem. Abstr.*, **82**, 57927 (1975).
403. K. A. Andrianov, V. I. Sidorov, and M. I. Filimonova, *J. Gen. Chem. USSR*, **47**, 446 (1977).
404. H. Kono, I. Ojima, and M. Matsumoto, *Org. Prep. Proced. Int.*, **5**, 135 (1973).
405. K. A. Andrianov, N. V. Vakhrameeva, K. S. Pushchevaya, and V. I. Sidorov, *J. Gen. Chem. USSR*, **46**, 1973 (1976).
406. Y. Nagai, I. Ojima, and H. Kono, *Jpn. Pat.*, 74 108024 (1974); *Chem. Abstr.*, **82**, 73160 (1975).
407. K. A. Andrianov, M. I. Filimonova, and V. I. Sidorov, *J. Gen. Chem. USSR*, **47**, 446 (1977).
408. Y. Nagai, I. Ojima, and T. Kogura, *Jpn. Pat.*, 74 110632 (1974); *Chem. Abstr.*, **82**, 156487 (1975).
409. I. Ojima, T. Kogure, and Y. Nagai, *Tetrahedron Lett.*, 2475 (1973).
410. I. Ojima, S. Inaba, and Y. Nagai, *J. Organomet. Chem.*, **72**, C11 (1974).
411. R. J. P. Corriu, J. J. E. Moreau, and M. Pataud-Sat, *J. Organomet. Chem.*, **228**, 301 (1982).
412. A. J. Chalk, *J. Organomet. Chem.*, **21**, 207 (1970).
413. L. I. Kopylova, M. V. Sigalov, E. N. Satsuk, M. Capka, V. Chvalovsky, V. B. Pukhnarevich, E. Lukevics, and M. G. Voronkov, *J. Gen. Chem. USSR*, **51**, 311 (1981).
414. S. P. Dent, C. Eaborn, and A. Pidcock, *Chem. Commun.*, 1703 (1970).
415. B. Courtis, S. P. Dent, C. Eaborn, and A. Pidcock, *J. Chem. Soc., Dalton Trans.*, 2460 (1975).
416. K. Yamamoto, T. Hayashi, Y. Uramoto, R. Ito, and M. Kamuda, *J. Organomet. Chem.*, **118**, 331 (1976).
417. I. Ojima, T. Kogura, M. Kumagi, S. Horiuchi, and T. Sato, *J. Organomet. Chem.*, **122**, 83 (1976).
418. T. Hayashi, K. Yamamoto, K. Kasuga, H. Omiju, and M. Kamuda, *J. Organomet. Chem.*, **113**, 127 (1976).
419. T. Hayashi, K. Yamamoto, and M. Kamuda, *Tetrahedron Lett.*, 4405 (1974).
420. K. Yamamoto, T. Hayashi, and M. Kamuda, *J. Organomet. Chem.*, **54**, C45 (1973).
421. I. Ojima, T. Kogura, and Y. Nagai, *Chem. Lett.*, 541 (1973).
422. I. Ojima and Y. Nagai, *Chem. Lett.*, 223 (1974).
423. I. Ojima, T. Kogure, and M. Kumagai, *J. Org. Chem.*, **42**, 1671 (1977).
424. T. Hayashi, K. Yamamoto, and M. Kamuda, *Tetrahedron Lett.*, 331 (1974).
425. R. J. P. Corriu and J. J. E. Moreau, *J. Organomet. Chem.*, **64**, C51 (1974).
426. I. Kolb and J. Hetflejš, *Collect. Czech. Chem. Commun.*, **45**, 2808 (1980).
427. N. C. Payne and D. W. Stephan, *Inorg. Chem.*, **21**, 182 (1982).
428. J. Benes and J. Hetflejš, *Collect. Czech. Chem. Commun.*, **41**, 2264 (1976).
429. T. H. Johnson, K. C. Klein, and S. Thomen, *J. Mol. Catal.*, **12**, 37 (1981).
430. H. Brunner and G. Reipl, *Angew. Chem., Ed. Engl. Int.*, **21**, 377 (1982).
431. H. Brunner, G. Reipl, and H. Weitzer, *Angew. Chem., Int. Ed. Engl.*, **22**, 331 (1983).
432. M. Čapka, *Collect. Czech. Chem. Commun.*, **42**, 3410 (1977).
433. G. Kuncova and V. Chvalovsky, *Collect. Czech. Chem. Commun.*, **45**, 2085 (1980).
434. I. Ojima, T. Tanaka, and T. Kogura, *Chem. Lett.*, 823 (1981).
435. K. Yamamoto, K. Tsuroka, and J. Tsuji, *Chem. Lett.*, 1115 (1977).
436. T. Kogure and I. Ojima, *J. Organomet. Chem.*, **234**, 249 (1982).
437. T. Hayashi, K. Yamamoto, and M. Kumada, *Tetrahedron Lett.*, 3 (1975).
438. I. Ojima, T. Kogure, and Y. Nagai, *Chem. Lett.*, 985 (1975).
439. N. Langlois, Tuan Phat Dang, and H. B. Kagan, *Tetrahedron Lett.*, 4865 (1973).
440. D. Forster, *J. Am. Chem. Soc.*, **98**, 846 (1976).
441. P. Uguagliati, A. Palazzi, G. Deganello, and U. Belluco, *Inorg. Chem.*, **9**, 724 (1970).
442. A. Hershman, K. K. Robinson, J. H. Craddock, and J. F. Roth, *Am. Chem. Soc. Div. Pet. Chem. Prepr.*, **17**, E7 (1972).
443. C. M. Bartish, *US Pat.*, 4102920 (1978); *Chem. Abstr.*, **90**, 54480 (1979).

444. J. Hjortkjaer and V. W. Jensen, *Ind. Eng. Chem., Prod. Res. Dev.*, **15**, 46 (1976).
445. D. Brodzki, C. Leclere, B. Denise, and G. Parmetier, *Bull. Soc. Chim. Fr.*, 61 (1976).
446. J. Hjortkjaer and J. C. E. Jorgensen, *J. Chem. Soc., Perkin Trans. 2*, 763 (1978).
447. Y. Matsui, *Tetrahedron Lett.*, 1107 (1976).
448. Y. Kamiya, K. Kawato, and H. Ota, *Chem. Lett.*, 1549 (1980).
449. A. F. M. Iqbal, *Tetrahedron*, **37**, 3381 (1971).
450. D. Durand and C. Lassau, *Tetrahedron Lett.*, 2329 (1969).
451. G. LaMonica, C. Monti, and S. Cenini, *J. Mol. Catal.*, **18**, 93 (1983).
452. J. J. McCoy and J. S. Yoo, *US Pat.*, 3714185 (1973); *Chem. Abstr.*, **78**, 147785 (1973).
453. J. Tsuji and K. Ono, *Jpn. Pat.*, 69 17128 (1969); *Chem. Abstr.*, **71**, 101570 (1969).
454. G. LaMonica and S. Cenini, *J. Organomet. Chem.*, **216**, C35 (1981).
455. F. Flemming, K. Unverferth, and K. Schwetlick, *J. Prakt. Chem.*, **317**, 455 (1975).
456. K. Unverferth, C. Ruger, and K. Schwetlick, *J. Prakt. Chem.*, **319**, 841 (1977).
457. K. Unverferth and K. Schwetlick, *React. Kinet. Catal. Lett.*, **6**, 231 (1977).
458. K. Unverferth, R. Hontsch, and K. Schwetlick, *J. Prakt. Chem.*, **321**, 921 (1979).
459. K. Unverferth, R. Hontsch, and K. Schwetlick, *J. Prakt. Chem.*, **321**, 86 (1979).
460. J. Tsuji and K. Ohno, *Tetrahedron Lett.*, 4713 (1966).
461. I. S. Akhrem, R. S. Vartanyan, L. E. Kotlyar, and M. E. Vol'pin, *Bull. Acad. Sci. USSR, Div. Chem. Sci.*, 227 (1977).
462. E. Muller and A. Segnitz, *Justus Liebigs Ann. Chem.*, 1583 (1973).
463. E. Lindner and A. Thasitis, *Chem. Ber.*, **107**, 2418 (1974).
464. J. M. Suarez Cardeso and R. S. Nyholm, *Ars Pharm.*, **15**, 111 (1974); *Chem. Abstr.*, **82**, 148863 (1975).
465. K. Schorpp and W. Beck, *Chem. Ber.*, **107**, 1371 (1974).
466. K. Kaneda, H. Azuma, M. Wayaku, and S. Teranishi, *Chem. Lett.*, 215 (1974).
467. P. Hong, K. Sonogashira, and N. Hagihara, *Nippon Kagaku Zasshi*, **89**, 74 (1968); *Chem. Abstr.*, **69**, 2599 (1968).
468. J. Blum and Z. Lipshes, *J. Org. Chem.*, **34**, 3076 (1969).
469. J. Blum, D. Milstein, and Y. Sasson, *J. Org. Chem.*, **35**, 3233 (1970).
470. J. Blum, M. Ashkenasy, and Y. Pickholz, *Synthesis*, 352 (1974).
471. J. Blum, *Tetrahedron Lett.*, 1605 (1966).
472. J. Blum, S. Krauss, and Y. Pickholz, *J. Organomet. Chem.*, **33**, 227 (1971).
473. D. Egglestone and M. C. Baird, *J. Organomet. Chem.*, **113**, C25 (1976).
474. D. Eggleston, M. C. Baird, C. J. L. Lock, and G. Turner, *J. Chem. Soc., Dalton Trans.*, 1576 (1977).
475. K. S. Y. Lau, Y. Becker, F. Huang, N. Baenziger, and J. K. Stille, *J. Am. Chem. Soc.*, **99**, 5664 (1977).
476. N. A. Dunham and M. C. Baird, *J. Chem. Soc., Dalton Trans.*, 774 (1975).
477. N. A. Dunham, D. Egglestone, D. Slack, and M. C. Baird, *Proc. XVI Int. Conf. Coord. Chem., Dublin*, R9 (1974).
478. J. K. Stille, *Ann. N. Y. Acad. Sci.*, **52**, 295 (1977).
479. J. K. Stille, F. Huang, and M. T. Regan, *J. Am. Chem. Soc.*, **96**, 1518 (1974).
480. J. Tsuji and K. Ohno, *Synthesis*, **1**, 157 (1969).
481. T. Benneche and K. Undheim, *Acta Chem. Scand.*, **B36**, 409 (1982).
482. J. K. Stille, M. T. Regan, R. W. Fries, F. Huang, and T. McCarley, *Adv. Chem. Ser.*, No. 132, 181 (1974).
483. K. Ohno and J. Tsuji, *J. Am. Chem. Soc.*, **90**, 99 (1968).
484. J. K. Stille and K. S. Y. Lau, *Acc. Chem. Res.*, **10**, 434 (1977).
485. J. A. Kampmeier, S. H. Harris, and R. M. Rodehorst, *J. Am. Chem. Soc.*, **103**, 1478 (1981).
486. J. Blum, *Tetrahedron Lett.*, 3401 (1966).
487. D. A. Clements and J. F. Nixon, *J. Chem. Soc., Dalton Trans.*, 195 (1973).
488. D. H. Doughty, M. F. McGuigan, H. Wang, and L. H. Pignolet, *Fundam. Res. Homogeneous Catal.*, **3**, 909 (1979).
489. A. J. Birch and K. A. M. Walker, *Aust. J. Chem.*, **24**, 513 (1971).
490. C. F. Lochow and R. G. Miller, *J. Am. Chem. Soc.*, **98**, 1281 (1976).
491. K. Sakai and I. Oda, *Tetrahedron Lett.*, 4375 (1972).
492. A. Emery, A. C. Oehlschlager, and A. M. Unrau, *Tetrahedron Lett.*, 4401 (1970).

493. K. Takao, M. Wayaku, Y. Fujiwara, T. Imanaka, and S. Teranishi, *Bull. Chem. Soc. Jpn.*, **45**, 1505 (1972).
494. F. Ingersheim and H. Mimoun, *Nouv. J. Chim.*, **4**, 161 (1980).
495. C. W. Dudley, G. Read, and P. J. C. Walker, *J. Chem. Soc., Dalton Trans.*, 1926 (1974).
496. L. Carlton and G. Read, *J. Mol. Catal.*, **10**, 133 (1981).
497. M. T. Attay, L. Carlton, and G. Read, *J. Mol. Catal.*, **19**, 57 (1983).
498. M. Bressan, F. Morandini, and P. Rigo, *J. Organomet. Chem.*, **247**, C8 (1983).
499. B. R. James and E. Ochai, *Can. J. Chem.*, **49**, 975 (1971).
500. D. Holland and D. J. Milner, *J. Chem. Soc., Dalton Trans.*, 2440 (1975).
501. J. Farrar, D. Holland, and D. J. Milner, *J. Organomet. Chem.*, **153**, 239 (1978).
502. J. Farrar, D. Holland, and D. J. Milner, *J. Chem. Soc., Dalton Trans.*, 815 (1975).
503. K. Kaneda, T. Itoh, Y. Fujiwara, and S. Teranishi, *Bull. Chem. Soc. Jpn.*, **46**, 3810 (1973).
504. H. Arzoumanian, A. Blanc, U. Hartig, and J. Metzger, *Tetrahedron Lett.*, 1011 (1974).
505. A. Fusi, R. Ugo, and G. M. Zanderighi, *J. Catal.*, **34**, 175 (1974).
506. K. Takao, M. Wayaku, Y. Fujiwara, T. Imanaka, and S. Teranishi, *Bull. Chem. Soc. Jpn.*, **43**, 3898 (1970).
507. L. W. Fine, M. Grayson, and V. H. Suggs, *J. Organomet. Chem.*, **22**, 219 (1970).
508. Y. Seki, S. Murai, A. Hidaka, and N. Sonoda, *Angew. Chem., Int. Ed. Engl.*, **16**, 881 (1976).
509. A. J. Birch and G. S. R. Subba Rao, *Tetrahedron Lett.*, 2917 (1968).
510. H. Eshtiagh-Hosseni, J. F. Nixon, and J. S. Poland, *J. Organomet. Chem.*, **164**, 107 (1979).
511. R. Cramer, *J. Am. Chem. Soc.*, **87**, 4717 (1965).
512. R. Cramer, *J. Am. Chem. Soc.*, **89**, 1633 (1967).
513. R. Cramer, *J. Am. Chem. Soc.*, **89**, 4621 (1967).
514. T. Alderson, E. L. Jenner, and R. V. Lindsay, *J. Am. Chem. Soc.*, **87**, 5638 (1965).
515. A. C. L. Su and J. W. Collette, *J. Organomet. Chem.*, **46**, 369 (1972).
516. N. G. Connelly, P. T. Draggett, and M. Green, *J. Organomet. Chem.*, **140**, C10 (1977).
517. P. S. Chekrii, M. L. Khidekel, I. V. Kalechitz, O. N. Eremenko, G. I. Karyakina, and A. S. Todozhkova, *Bull. Acad. Sci. USSR, Div. Chem. Sci.*, 1521 (1972).
518. V. A. Kormer, B. D. Babitsky, I. Y. Poddubnyi, and V. N. Sokolov, *Am. Chem. Soc. Div. Polym. Chem. Prepr.*, **7**, 548 (1966).
519. J. Zachoval, J. Kepelka, and M. Klimova, *Collect. Czech. Chem. Commun.*, **37**, 3271 (1972).
520. J. P. Scholten and H. J. Van der Ploeg, *J. Polymer Sci. Part A-1*, **10**, 3067 (1972).
521. J. P. Scholten and H. J. Van der Ploeg, *J. Polym. Sci. Polym. Chem. Ed.*, **11**, 3205 (1973).
522. S. Otsuka and A. Nakamura, *Jpn. Pat.*, 72 45330 (1972); *Chem. Abstr.*, **78**, 57872 (1973).
523. S. Otsuka, A. Nakamura, and H. Minamida, *Chem. Commun.*, 191 (1969).
524. L. Hagelee, R. West, J. Cabrese, and J. Norman, *J. Am. Chem. Soc.*, **101**, 4888 (1979).
525. A. Entezami, A. Geraudelle, F. Kaufmann, F. Schue, A. Deluzarche, and A. Maillard, *Eur. Polym. J.*, **7**, 889 (1971).
526. K. Bouchal, J. Skramova, J. Coupek, S. Pokorny, and F. Hrabak, *Makromol. Chem.*, **156**, 225 (1972).
527. K. P. Vora, *Synth. Commun.*, **13**, 99 (1983).
529. G. P. Chiusoli, W. Giroldini, and G. Salerno, *Eur. Pat. Appl.*, 15537 (1980); *Chem. Abstr.*, **94**, 120875 (1981).
530. G. P. Chiusoli, L. Pallini, and G. Salerno, *J. Organomet. Chem.*, **238**, C85 (1982).
531. G. P. Chiusoli, L. Pallini, and G. Salerno, *Eur. Pat. Appl.*, 54 954 (1982); *Chem. Abstr.*, **97**, 197850 (1982).
532. G. N. Schrauzer, R. K. Y. Ho, and O. Schlesinger, *Tetrahedron Lett.*, 543 (1970).
533. M. A. Bennett and P. A. Longstaff, *Chem. Ind.* (*London*), 846 (1965).
534. T. J. Katz, N. Acton, and I. C. Paul, *J. Am. Chem. Soc.*, **91**, 206 (1969).
535. T. J. Katz and S. A. Cerefice, *J. Am. Chem. Soc.*, **91**, 2405 (1969).
536. N. Acton, R. Roth, T. J. Katz, J. K. Franck, C. A. Maier, and I. C. Paul, *J. Am. Chem. Soc.*, **94**, 5446 (1972).
537. J. Kiji, S. Nishimura, S. Yoshikawa, E. Sasakawa, and J. Furukawa, *Bull. Chem. Soc. Jpn.*, **47**, 2523 (1974).
538. U. M. Zhemilov, R. I. Khusnutdinov, Z. S. Muslimov, L. V. Spirikhin, G. A. Tolstikov, and O. M. Nefedov, *Bull. Acad. Sci. USSR, Div. Chem. Sci.*, 1889 (1981).
539. S. Yoshikawa, J. Kiji, and J. Furukawa, *Makromol. Chem.*, **178**, 1077 (1977).

540. L. Carlton and G. Read, *J. Chem. Soc., Perkin Trans. 1*, 1631 (1978).
541. H. Singer and G. Wilkinson, *J. Chem. Soc. A*, 849 (1968).
542. Y. Fujita, T. Nishida, and K. Itoi, *Ger. Pat.*, 2520735 (1975); *Chem. Abstr.*, **84**, 104794 (1976).
543. Y. Iwashita and F. Tamura, *Bull. Chem. Soc. Jpn.*, **43**, 1517 (1970).
544. J. P. Collman, J. W. Kang, W. F. Little, and M. F. Sullivan, *Inorg. Chem.*, **7**, 1298 (1968).
545. B. L. Booth, R. N. Hazeldine, and I. Perkins, *J. Chem. Soc., Dalton Trans.*, 2593 (1981).
546. R. J. Kern, *Chem. Commun.*, 706 (1968).
547. R. J. Kern, *J. Polym. Sci. Part. A-1*, **7**, 621 (1969).
548. H. A. Schaefer, R. Marcy, T. Rueping, and H. Singer, *J. Organomet. Chem.*, **240**, 17 (1982).
549. J. W. Scott and D. Valentine, *Org. Prep. Proced. Int.*, **12**, 7 (1980).
550. D. M. Singleton, *Tetrahedron Lett.*, 1245 (1973).
551. R. Grigg, R. Scott, and P. Stevenson, *Tetrahedron Lett.*, **23**, 2691 (1982).

The Chemistry of the Metal—Carbon Bond, Vol. 4
Edited by F. R. Hartley

CHAPTER **9**

Use of organonickel compounds in organic synthesis

K. TAMAO AND M. KUMADA

Department of Synthetic Chemistry, Kyoto University, Kyoto 606, Japan

---

I. INTRODUCTION

Nickel complexes are among the leading members of synthetically useful transition metal complexes. Their potential utility has been demonstrated by a variety of carbon–carbon bond-forming reactions which have been developed mainly during the last two decades. This field has been comprehensively covered twice in the past in books published by Jolly and Wilke in 1974 and 1975[174] and reviews by Jolly in 1982[173].

Nickel-induced carbon—element bond-forming reactions have been classified into three categories: (1) oligomerization of alkenes, dienes, and alkynes; (2) carbonylation; and (3) coupling reactions. In Figure 1, the numbers of primary papers published on each aspect are plotted against intervals of two years from 1940 to 1980. There is an interesting trend, viz. that whereas studies on oligomerization and carbonylation slowed down from the late 1970s, only studies on the coupling reactions have steadily and rapidly increased every year from early 1970s. This chapter will therefore concentrate on the coupling reactions and related reactions. An attempt will be made to classify these reactions into five categories.

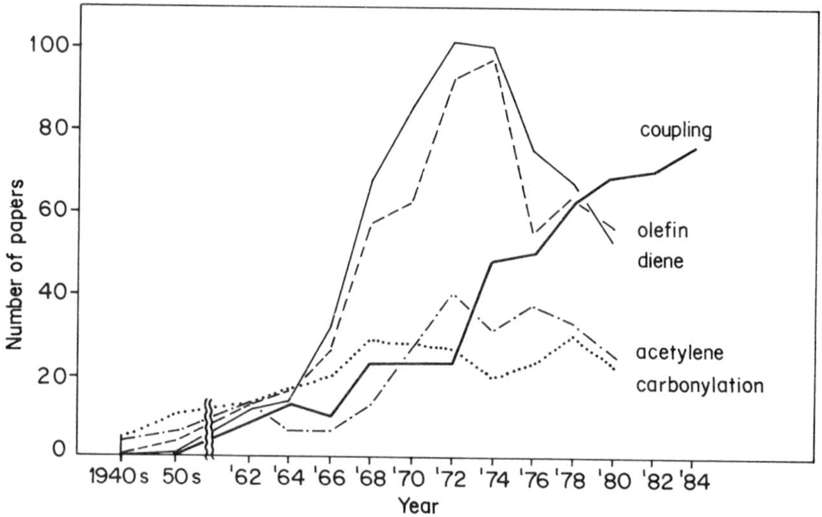

FIGURE 1. Number of publication on nickel-mediated carbon—element bond-forming reactions. Plotted are 2 years' total numbers from 1960 to 1980 against intervals of 2 years and 2 years' average 1940s and 1950s. (Sources: until 1980, refs. 173 and 174; 1981–1984, this work).

II. IMPORTANT ELEMENTARY PROCESSES

Although the formation and reactions of organonickel complexes have been well reviewed by Jolly and Wilke[172,174], Kochi[187], and Yamamoto and coworkers[353,356], a brief mention will be made here of several elementary processes pertinent to the nickel-promoted and catalysed reactions.

Organonickel complexes may be formed by (1) oxidative addition of various organic substrates to Ni(0) d^{10} complexes, (2) transmetallation between Ni(II) d^8 complexes and active organometallics, or (3) insertion of unsaturated species into nickel—element bonds. The organic products may be released via (4) β-elimination or (5) reductive elimination.

The most commonly used Ni(0) species are $[Ni(CO)_4]$, $[Ni(PPh_3)_4]$[257], and $[Ni(cod)_2]$[261]. Recent alternative approaches use Ni(0) species generated *in situ* from stable Ni(II) compounds[31,45,178,179,283,363]. In catalytic reactions, stable Ni(II) species are mostly used as catalyst precursors, but the real catalytically active species are believed to be low-valent Ni(0) or Ni(I) species.

(1) Oxidative addition to Ni(0) species has been observed with a variety of compounds. Representative examples are shown for the strained C—C bonds (equation 1)[218], olefins (equation 2)[27], dienes (equation 3)[341], acetylenes (equation 4)[30], C—O bonds (equations 5 and 6)[163,256,349], C—halogen bonds of alkyl (equation 7)[285], allyl (equation 8)[94,339], acyl (equation 9)[228], and aryl (equation 10)[80,135,315] halides, C—CN bonds (equation 11)[87,88], C—S bonds (equation 12)[229], and C—Hg bonds (equation 13)[162]. Reversible oxidative addition of PPh_3 to Ni(0) has also been observed[81].

The most extensively studied are the oxidative additions of organic halides, from both synthetic and mechanistic viewpoints. Mechanistic studies with aryl halides, for example, have shown that in addition to the normal arylnickel(II) halides (equation 10), paramagnetic nickel(I) halides are usually formed as side products, the ratio being dependent on the halides ($I < Br < Cl$)[315], the substituents on the aromatic ring[315], the ligands[276,322], and the solvent polarity[315]. Oxidative addition is accelerated by electron-withdrawing substituents on aromatic rings[82,96]. The proposed mechanism, shown in Scheme 1[315], involves the rate-limiting electron transfer from Ni(0) to aryl halides.

Nickel(I) complexes also undergo oxidative addition with aryl halides, although lower in reactivity compared with Ni(0) complexes; oxidative addition products are not Ni(III) complexes, but disproportionated Ni(II) species (equation 14)[9]. A rare, stable organonickel(III) complex, $[NiBr_2(CCl=CCl_2)(PPhMe_2)_2]$, has been reported[219].

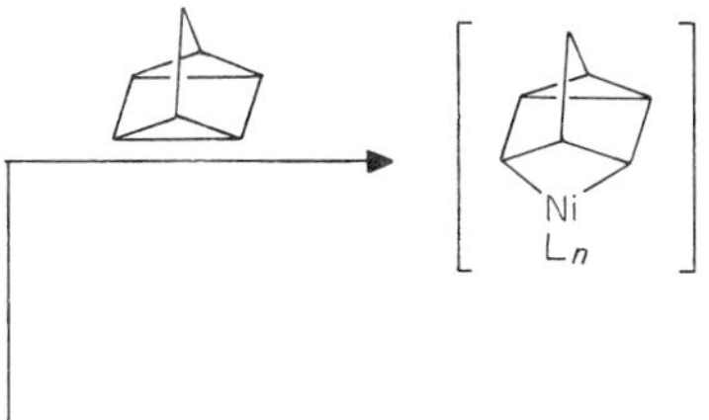

Ni
Ln
(1)

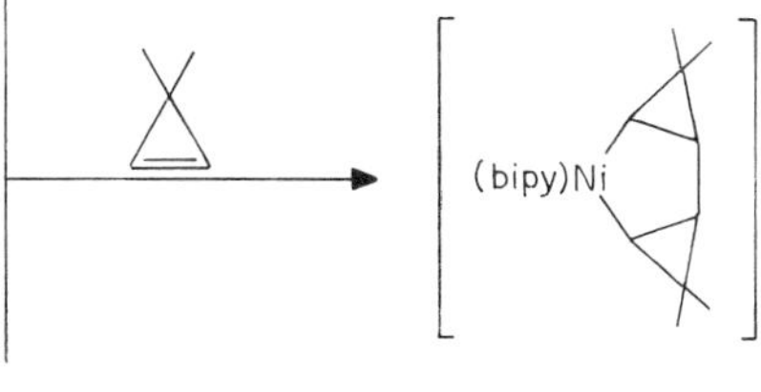

(bipy)Ni
(2)

$[L_nNi^0]$ →

- (butadiene) → $[Ph_3P-Ni(C_{12}H_{18})]$ (3)
- $CF_3C{\equiv}CCF_3$ → $[L_2Ni(C(CF_3)=C(CF_3))_3]$ (nickelacycloheptatriene with six CF_3 groups) (4)
- $CH_2{=}CHCH_2OPh$ → $[(\eta^3\text{-allyl})Ni(OPh)(PPh_3)]$ (5)
- succinic anhydride, $-CO$ → $[(bipy)Ni(CH_2CH_2C(=O)O)]$ (6)
- $Br(CH_2)_4Br$ → $[(bipy)Ni(CH_2)_4]$ (7)
- $CH_2{=}CHCH_2X$ (X=halogen) → $[(\eta^3\text{-allyl})NiX]_2$ (8)
- $ClCO_2C_2H_5$ → *trans*–$[NiCl(COOC_2H_5)(PPh_3)_2]$ (9)
- ArX (X=halogen) → *trans*–$[NiX(Ar)(PPh_3)_2]$ (10)
- $PhCH_2CN$ → $[Ni(CN)(CH_2Ph)(PCy_3)]$ (11)
- CS_2 → $[(triphos)Ni(\eta^2\text{-}CS_2)]$ (12)
- $(C_6F_5)_2Hg$ → $[Ni(HgC_6F_5)(C_6F_5)(PPh_3)_2]$ (13)

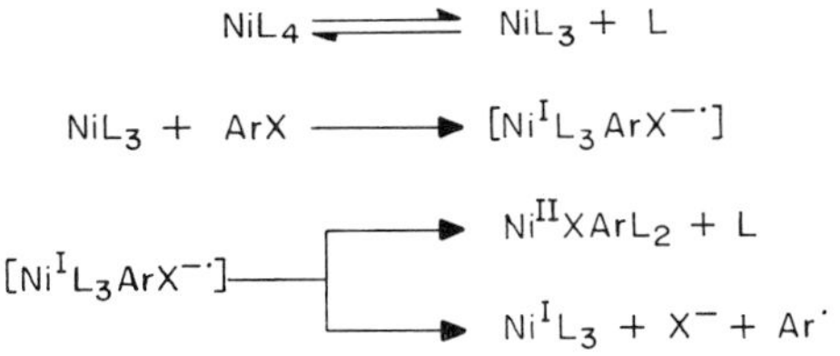

$$NiL_4 \rightleftharpoons NiL_3 + L$$

$$NiL_3 + ArX \longrightarrow [Ni^{I}L_3ArX^{-\cdot}]$$

$$[Ni^{I}L_3ArX^{-\cdot}] \longrightarrow Ni^{II}XArL_2 + L$$

$$[Ni^{I}L_3ArX^{-\cdot}] \longrightarrow Ni^{I}L_3 + X^- + Ar^{\cdot}$$

SCHEME 1. Mechanism of oxidative addition of aryl halides to Ni(0) complexes.

$$2[NiBr(PPh_3)_3] + ArX \longrightarrow [NiBr(Ar)(PPh_3)_2] + [NiBrX(PPh_3)_2] \quad (14)$$

Interaction of alkyl halides with Ni(0) complexes may generally result in the formation of Ni(I) species and alkyl radicals rather than the oxidative addition products, alkylnickel(II) species[209,210], with a few exceptions[228,285].

While oxidative addition of alkenyl halides proceeds with retention of configuration[40,82], an optically active benzyl chloride gives racemic products during the Ni(0)-induced reactions (equation 15)[275].

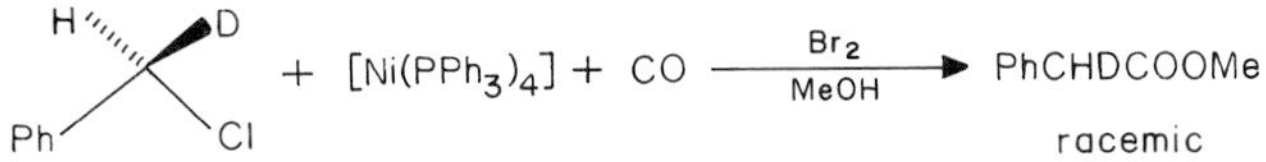

$$\text{PhC(H)(D)Cl} + [Ni(PPh_3)_4] + CO \xrightarrow[\text{MeOH}]{Br_2} \underset{\text{racemic}}{\text{PhCHDCOOMe}} \quad (15)$$

(2) Transmetallation is another important route to organonickel species (equation 16)[172]. It should be noted here that this route can introduce even alkyl groups on to the nickel centre, in contrast to the oxidative addition routes mentioned above.

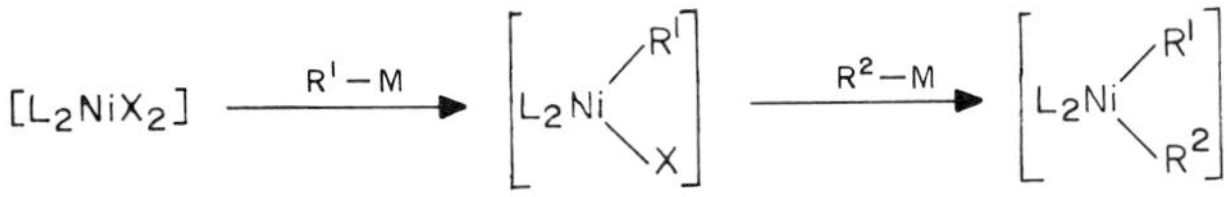

$$[L_2NiX_2] \xrightarrow{R^1-M} [L_2Ni(R^1)(X)] \xrightarrow{R^2-M} [L_2Ni(R^1)(R^2)] \quad (16)$$

M= Li, MgX, AlX_2, etc.

Disproportionation of mono- to di-organonickel complexes (symmetrization) is a sort of transmetallation and one of the key steps in the nickel-induced coupling reactions (equations 17 and 18)[40,350].

$$2[NiClEt(bipy)] \longrightarrow [NiEt_2(bipy)] + [NiCl_2(bipy)] \quad (17)$$

$$2[NiCl(CCl{=}CCl_2)(PMe_2Ph)_2] \xrightarrow{\text{bipy}} [Ni(CCl{=}CCl_2)_2(bipy)] + [NiCl_2(bipy)] \quad (18)$$

(3) Insertion into the nickel—carbon σ-bond has been observed with a variety of unsaturated molecules such as carbon monoxide[42,185,351], isonitriles[228], carbon dioxide[346], olefins[102], dienes[137,340], and acetylenes[38,137,150]. Insertion of carbon monoxide is reversible (equation 19)[185]. Acetylene insertion occurs stereoselectively in a *cis* fashion,

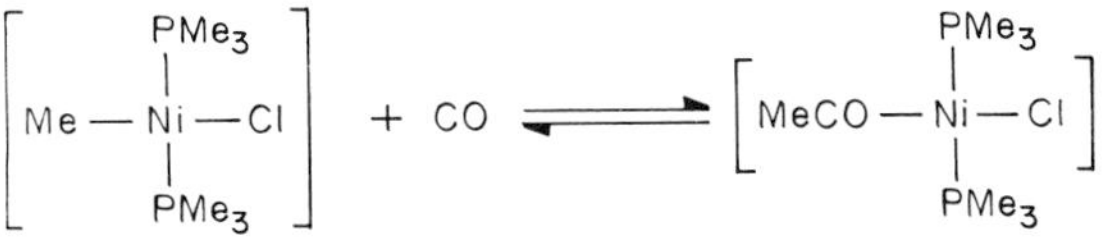

$$[Me{-}Ni(PMe_3)_2{-}Cl] + CO \rightleftharpoons [MeCO{-}Ni(PMe_3)_2{-}Cl] \quad (19)$$

but the predominant thermodynamic product has the larger group on the β-carbon atom *cis* to nickel (equations 20 and 21)[150]. Insertion of unsymmetrical acetylenes is highly regioselective to form the vinylnickel complexes with the larger group being nearest the nickel atom (equation 20).

$$[(acac)Ni(Me)(PPh_3)] \xrightarrow{RC\equiv CH} [(acac)Ni(PPh_3)\text{—}C(R){=}C(H)Me] \quad (20)$$

$$[(acac)Ni(Me)(PPh_3)] \xrightarrow{PhC\equiv CPh} [(acac)Ni(PPh_3)\text{—}C(Ph){=}C(Me)Ph] \quad (21)$$

(4) β-Elimination is a reverse course of the insertion reaction and one of the very common degradation processes of organonickel complexes (equation 22). Therefore, β-hydrogen-bearing alkylnickel complexes are hardly isolated, the isolable ethylnickel complexes being $[NiEt(Cp)(PPh_3)]$[251], $[NiEt_2(bipy)]$[359], and $[Ni(acac)(Et)(PPh_3)]$[63,344].

$$L_nNi\text{—}CH_2\text{—}CHR(H) \rightleftharpoons L_mNiH(CH_2{=}CHR) \quad (22)$$

This σ–π isomerization step sometimes causes the formation of by-products in coupling reactions. It should also be noted that the coordination number is changed during this σ–π conversion.

(5) Reductive elimination of diorganonickel complexes corresponds to the reverse of the oxidative addition and is of fundamental importance as a final product-releasing step in various nickel-promoted reactions. There is a general tendency that the weaker the σ-donor ligands *trans* to the leaving organic groups and the higher the electron-donating ability of the organic groups, then the more easily the reductive elimination takes place[304].

Reductive elimination of *cis*-diorganonickel complexes occurs thermally from tetracoordinate species themselves (equations 23 and 24)[107,189,191]. The ease of reductive

$$[(CH_2)_n(PPh_2)_2NiMe_2] \longrightarrow MeMe + (CH_2)_n(PPh_2)_2Ni \quad (23)$$

$$[(Ph_3P)_2Ni(CH_2)_4] \longrightarrow \text{cyclobutane} + (Ph_3P)_2Ni \quad (24)$$

elimination in reaction 23 depends on the chain length of the ligands; for example, dppp ($n = 3$) is much more effective than dppe ($n = 2$)[189]. Reductive elimination is accelerated by the addition of weak σ-electron donors, such as phosphites or phosphines (but not amines) (equation 25)[191], and olefins with electronegative substituents (equation 26)[342], by an associative mechanism involving pentacoordinate intermediates[305]. In contrast, reductive elimination of *trans*-diorganonickel complexes is retarded by the added neutral ligands in many cases[191,273,305]. However, acceleration has also been observed in some cases

$$L_2Ni(Ar)(R) + ArX \xrightarrow{\text{slow}} [L_2Ni(Ar)(R)^+ ArX^{-\bullet}]$$

$$L_2Ni(Ar)(R)^+ \xrightarrow{\text{fast}} L_2Ni^+ + ArR$$

$$L_2Ni^+ + ArX^{-\bullet} \xrightarrow{\text{fast}} L_2Ni(Ar)(X)$$

SCHEME 2

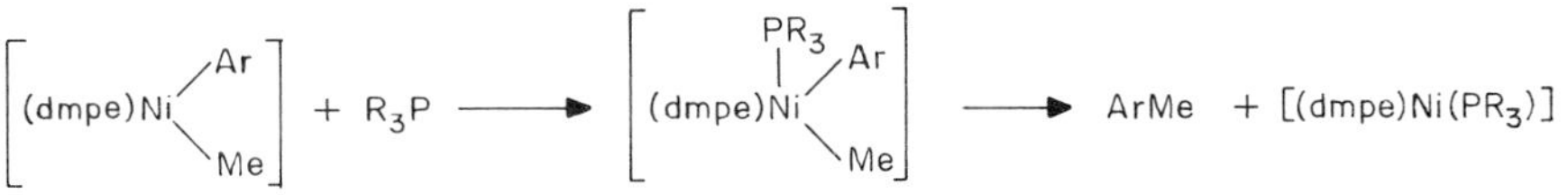

$$[(dmpe)Ni(Ar)(Me)] + R_3P \longrightarrow [(dmpe)Ni(PR_3)(Ar)(Me)] \longrightarrow ArMe + [(dmpe)Ni(PR_3)] \quad (25)$$

$$[(bipy)Ni(Et)_2] + CH_2{=}CHZ \longrightarrow [(bipy)Ni(CH_2{=}CHZ)(Et)_2] \longrightarrow EtEt + [(bipy)Ni(CH_2{=}CHZ)_n] \quad (26)$$

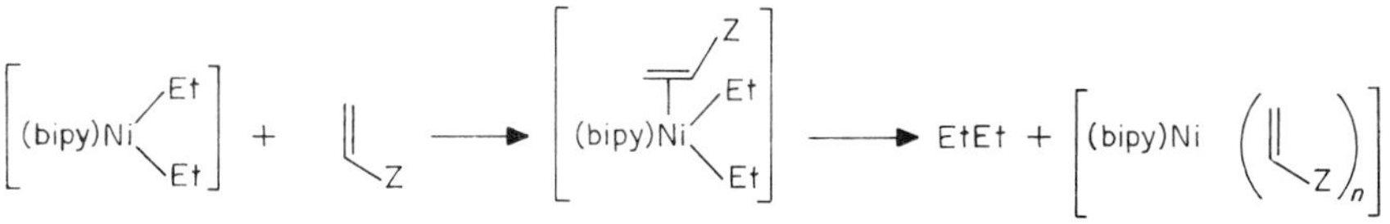

(equation 27)[86]. Reductive elimination is also accelerated by one-electron oxidants such as $[IrCl_6]^{2-}$, $CuBr_2$, Br_2, I_2, Ce(IV), and even O_2[208,314]. Certain organic halides not only accelerate the reduction elemination, but also undergo an almost instantaneous oxidative addition to the resulting Ni(0) species (equations 28 and 29)[208,319,326].

$$[Ni(CN)(Ph)(PPh_3)_2] \xrightarrow{PPh_3} PhCN + [Ni(PPh_3)_3] \quad (27)$$

$$[NiEt_2(bipy)] + PhCl \longrightarrow [NiCl(Ph)(bipy)] + EtEt \quad (28)$$

$$[NiR(C_6Cl_5)(PMe_2Ph)_2] + CCl_2{=}CCl_2 \longrightarrow [NiCl(CCl{=}CCl_2)(PMe_2Ph)_2] + RC_6Cl_5 \quad (29)$$

These halide-induced reductive eliminations may proceed through one-electron transfer from nickel complex to halide as an electron acceptor (Scheme 2). However, this reductive elimination–oxidative addition sequence can also occur intramolecularly (equation 30)[327]. Some theoretical studies have recently been reported[4,5,19,29,304,305].

$$[Ar{-}Ni(PMe_3)_2{-}CCl{=}CCl_2] \longrightarrow Cl{-}Ni(PMe_3)_2{-}C_2Cl_2Ar \quad (30)$$

III. CARBON—ELEMENT BOND-FORMING REACTIONS

Coupling reactions and related reactions may be conveniently classified into five general patterns, as shown in Table 1. For the sake of clarity, the general equations are shown by the typical transformations of an organic halide (RX) as a representative electrophile and an organometallic reagent (R—M) as a representative nucleophile.

A. Type 1: Nickel(0)-Promoted Homocoupling of Electrophiles

This type of coupling reaction proceeds through typical oxidative addition of substrates to Ni(0) species. Originally developed were stoichiometric reactions using air-sensitive

TABLE 1. Five Basic Patterns of Coupling and Related Reactions

Type	Reactions
1	Nickel(0)-promoted homocoupling of electrophiles
	$2RX + [NiL_4] \longrightarrow RR + [NiX_2L_2]$
2	Nickel(II)-promoted homocoupling of nucleophiles
	$2R{-}M + [NiX_2L_2] \longrightarrow RR + NiL_2 + 2M{-}X$
3	Reaction of organonickel compounds with electrophiles
	$RX + [R'NiXL_2] \longrightarrow RR' + [NiX_2L_2]$
4	Reaction of organonickel compounds with nucleophiles
	$R{-}M + [R'NiXL_2] \longrightarrow RR' + NiL_2 + M{-}X$
5	Nickel-catalysed cross-coupling between nucleophiles and electrophiles
	$R{-}M + R'X \xrightarrow{\text{Ni(II) or Ni(0)}} RR' + M{-}X$

Ni(0) species, but many of the processes can now be performed catalytically with nickel in the presence of appropriate reducing agents. The following are the *in situ* prepared Ni(0) or Ni(I) reagents so far developed: $[NiCl_2(PPh_3)_2]/Zn/I^-/dmf$[179,363] (or benzene)[168], $[NiX_2\{P(alkyl)_3\}_2]/Zn/I^-/hmpa$ (or nmp)[280,382], $NiX_2/Zn/I^-$ (or tu)/hmpa[281,283], NiX_2/Li/naphthalene/dme[178], $Ni(OAc)_2/NaH/t\text{-}AmONa/PPh_3$[31,45] (or bipy)[325], and electrochemically reduced species of $[NiCl_2(PPh_3)_2]/PPh_3$[207,312], $[Ni(acac)_2]/PPh_3$[171], $[Ni(teta)]^{2+}$ or [Ni(salen)][104,130].

1. Organic halides

Coupling reactions of alkyl, benzyl, allyl, alkenyl, aryl, heteroaryl, and acyl halides will be described in this order in two categories, stoichiometric and catalytic reactions. The mechanism is not simple, as will be described later (Section III.F).

a. Stoichiometric reaction

Unlike simple alkyl halides, α,ω-dibromoalkanes give the coupling products, cycloakanes, when treated with $[Ni(cod)_2]$ and bipy, which is an essential ligand (equation 31)[284]. When $n = 2$, the product is ethylene. When $n = 4$, since the nickelacyclopentane intermediate is fairly stable (see also Section III.C.1), the coupling product should be released by the action of oxygen or *p*-benzoquinone. Dibromomethane dimerizes to ethylene under the same conditions[284].

$$Br(CH_2)_nBr + [Ni(cod)_2] + 3\,bipy \xrightarrow{\text{thf}} \text{cyclo-}(CH_2)_n + [NiBr_2(bipy)_3] + 2\,cod$$

$n=2$, 100%; $n=3$, 91%; $n=4$, 52%; $n=5$, 80% (31)

α-Bromoketones and benzyl halides are dimerized in the presence of a nickelate salt containing Ni(I) species, $K_4[Ni_2(CN)_6]$, in aqueous acetone (equation 32)[108] or the *in situ* prepared metallic nickel (equation 33)[158]. In the latter case, certain functional groups including a nitro group can tolerate this coupling. The metallic nickel induces dimerization of benzylic dihalides and trihalides to symmetrical olefins (equation 34)[108]. α,α-Dibromoketones trimerize on treatment with $[Ni(cod)_2]$ (equation 35), whereas they dimerize with $[Ni\{P(OEt)_3\}_4]$[100]. Coupling of poly(bromomethyl)benzenes provides an efficient route to $[2_n]$cyclophanes (equation 36)[136,323].

$$2\ PhCOCH_2Br \xrightarrow[H_2O-acetone]{K_4[Ni_2(CN)_6]} PhCOCH_2CH_2COPh \quad 47\% \tag{32}$$

$$2\ p\text{-}O_2NC_6H_4CH_2Cl \xrightarrow[dme]{NiCl_2/Li/NpH} (p\text{-}O_2NC_6H_4CH_2\text{—})_2 \quad 78\% \tag{33}$$

$$2\ PhCHBr_2 \xrightarrow[dme]{NiCl_2/Li/NpH} PhCH{=}CHPh \tag{34}$$

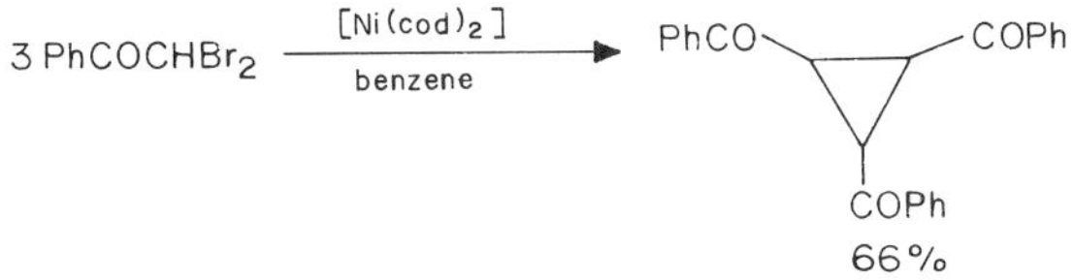

$$3\ PhCOCHBr_2 \xrightarrow[benzene]{[Ni(cod)_2]} \text{1,2,3-tribenzoylcyclopropane (PhCO, COPh, COPh)} \quad 66\% \tag{35}$$

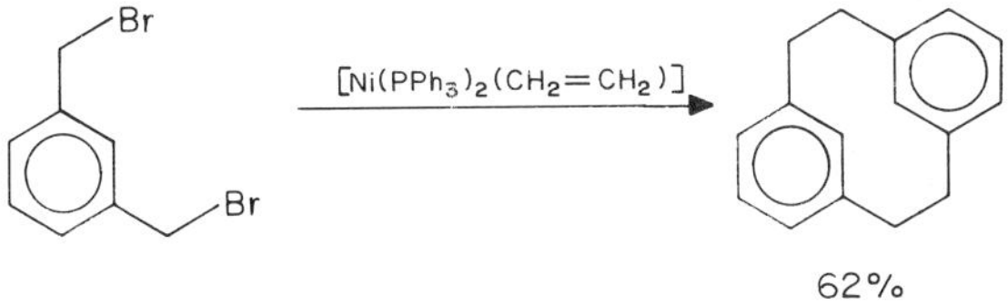

$$\text{1,3-bis(bromomethyl)benzene} \xrightarrow{[Ni(PPh_3)_2(CH_2{=}CH_2)]} [2.2]\text{metacyclophane} \quad 62\% \tag{36}$$

Homocoupling of allylic halides to 1,5-diene skeletons has been appreciated as pioneering work on Ni(0)-promoted coupling reactions from a synthetic point of view[57] and has been well reviewed[16,173,174,261]. Intramolecular allylic homocoupling has been widely applied to the synthesis of macrocyclic terpenoids and macrolides (equations 37 and 38)[60,70]. The *in situ* generated $[Ni(PPh_3)_4]$ is also effective for the allylic coupling[179].

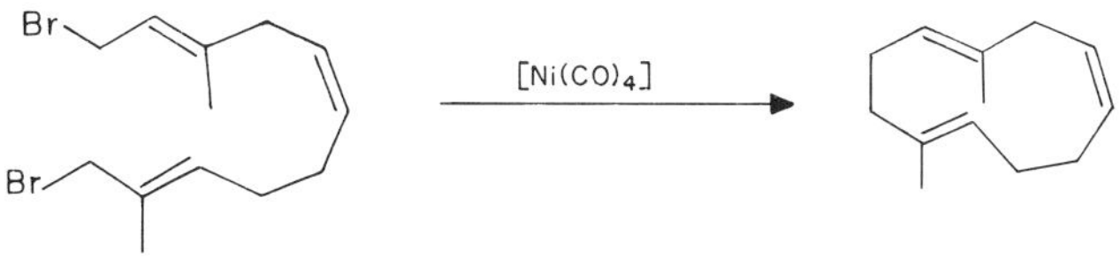

$$\text{Br—(bis-allylic dibromide)—Br} \xrightarrow{[Ni(CO)_4]} \text{macrocyclic triene} \tag{37}$$

$$\text{Br—(dienyl ester dibromide)—Br} \xrightarrow{[Ni(CO)_4]} \text{macrolide} \tag{38}$$

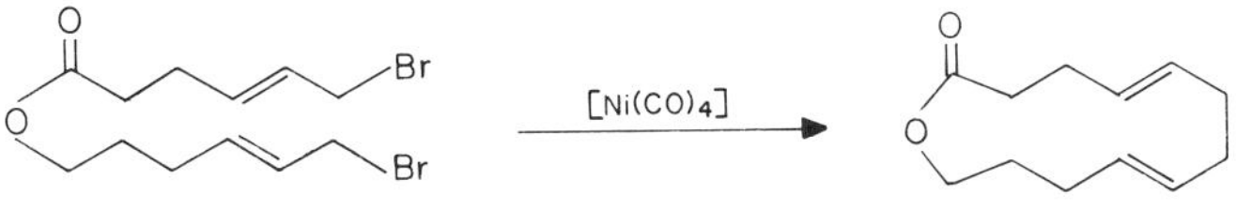

Homocoupling of alkenyl halides is induced by $[Ni(cod)_2]$ alone in dmf or together with R_3P in diethyl ether to give 1,3-dienes[263,270]. Electronegative substituents not only facilitate the coupling, but also bring about a high degree of stereospecificity with retention of configuration (equations 39 and 40)[470]. This mild procedure is suitable for the synthesis of highly reactive 1, 3-dienes (equation 41).

2 MeCH=CHBr —[Ni(cod)$_2$]/Ph_3P→ Me(CH=CH)$_2$Me (19) + Me(CH=CH)$_2$Me (81); 66% (39)

2 MeO_2CCH=CHBr —[Ni(cod)$_2$]/Bu^n_3P→ MeO_2C(CH=CH)$_2$$CO_2Me$; 91% (40)

2 CH_2=C(Br)CO_2Me —[Ni(cod)$_2$]/PPh_3→ CH_2=C(CO_2Me)C(CO_2Me)=CH_2; 99% (41)

A combination of a nickel complex reducing agent (abbreviated to NiCRA)[45] with bipy, consisting of NaH/*t*-AmONa/Ni(OAc)$_2$/bipy (2:2:1:2), is an efficient coupling agent for alkenyl halides[325]. The use of bipy as a ligand is essential for coupling; with PPh_3 or without a ligand only reduction of halides is observed. Remarkably, this is the only method that induces the coupling of alkenyl chlorides (equation 42). The stereospecificity is also

2 1-chlorocyclohexene —NiCRA/bipy→ 1,1'-bi(cyclohexenyl); 80% (42)

very high. Homocoupling of styryl bromide is also achieved with *in situ* generated [Ni(PPh$_3$)$_4$][179] or $K_4[Ni_2(CN)_6]$[108].

1, 1-Dibromoalkenes are dimerized with the *in situ* generated Ni(0) in benzene to butatrienes (equation 43)[168], while radialenes are formed from 2, 3-diiodo-1, 3-dienes under similar conditions (equation 44)[167]. In both cases, the nature of the solvent is critical, benzene being best and dmf the worst.

$Ph_2C{=}CBr_2$ —[NiBr$_2$(PPh$_3$)$_2$]/Zn, Et_4NI/C_6H_6→ $Ph_2C{=}C{=}C{=}CPh_2$; 60% (43)

2,3-diiodo-1,3-diene —[NiBr$_2$(PPh$_3$)$_2$]/Zn, C_6H_6→ [4]radialene; 48% (44)

Homocoupling of aryl halides is also promoted by $[Ni(cod)_2]$ in dmf[262,270], $[Ni(PPh_3)_4]$[262,270], and the *in situ* generated low-valent nickel species from $[NiX_2(PPh_3)_2]/PPh_3/Zn/dmf$[179,204], $NiCl_2/Zn/KI/hmpa$[280], NiCRA/bipy[199,325], $NiCl_2/Li/NpH/dme$[220], and electrochemically reduced $[Ni(PPh_3)_4]$[207,312]. The coupling reactions occur under mild conditions, usually from around room temperature up to 80 °C. The most significant feature is the compatibility of a wide range of functional groups such as ketone, aldehyde, ester, nitrile, and amino groups (equations 45–47)[179,270].

2 MeC(O)–C6H4–Br —[Ni(cod)2] / dmf→ MeC(O)–C6H4–C6H4–C(O)Me 93% (45)

2 H2N–C6H4–Br —[Ni(cod)2] / dmf→ H2N–C6H4–C6H4–NH2 54% (46)

2 MeO–C6H2(CH2OMe)(OCH2Ph)–I —[NiCl2(PPh3)2]/PPh3/Zn / dmf→ MeO(PhCH2O)(MeOCH2)C6H2–C6H2(OCH2Ph)(OMe)(CH2OMe) 68% (47)

Powerful electron-withdrawing substituents such as the nitro group, however, almost inhibit the coupling reaction[200,270]. The order of reactivity of the halides is I > Br > Cl. Only NiCRA/bipy, consisting of $NaH/t\text{-}AmONa/Ni(OAc)_2/bipy$ (4:2:1:2); seems to be reactive enough for coupling of aryl chlorides (equation 48)[325]. When triphenylphosphine

2 (2-Me-C6H4)–Cl —NiCRA/bipy→ 2,2'-Me2-biphenyl 84% (48)

is used as a ligand, undesirable phenylated products are formed in some cases as by-products through cleavage of the phosphorus—phenyl bond (equation 49)[204,324].

Br-substituted substrate —[NiCl2(PPh3)2]/2PPh3/Zn / dmf→ [ ]2 41% + Ph-substituted product 16% (49)

Two types of phosphine-free coupling agents, in addition to NiCRA/bipy, have also been developed (equations 50 and 51)[200,270]; in both cases iodide ions seem to play an

$$2\ \text{2-MeCOC}_6\text{H}_4\text{I} \xrightarrow[\text{hmpa}]{\text{NiCl}_2/\text{KI}/\text{Zn}} \text{2,2'-(MeCO)}_2\text{C}_6\text{H}_4\text{C}_6\text{H}_4 \quad 96\% \tag{50}$$

$$2\,C_6F_5I \xrightarrow[\text{dme}]{\text{NiCl}_2/\text{Li}/\text{NpH}} C_6F_5C_6F_5 \quad 100\% \tag{51}$$

important role for coupling (see also Section III.A.b). Intramolecular coupling provides an efficient procedure for cyclic biaryls[98,270,338] (equations 52 and 53)[270]. Some natural

$$\text{MeO-C}_6\text{H}_3(\text{I})\text{-(CH}_2)_n\text{-C}_6\text{H}_3(\text{I})\text{-OMe} \xrightarrow[\text{dme}]{[\text{Ni(PPh}_3)_4]} \text{MeO-C}_6\text{H}_3\text{-C}_6\text{H}_3\text{-OMe (bridged by (CH}_2)_n) \quad n=2,\ 81\%;\ n=5,\ 85\% \tag{52}$$

$$\xrightarrow[\text{dmf}]{[\text{Ni(PPh}_3)_4]} \quad 52\% \tag{53}$$

products have been prepared by this method[270,338]. Halopyridines and quinolines are coupled by the action of the *in situ* generated Ni(0) species (equations 54 and 55)[309]. Notably, all the 2-, 3-, and 4-chlorine and -bromine derivatives can be employed.

$$2\ \text{2-bromo-3-methoxypyridine} \xrightarrow[\text{dmf}]{\text{NiCl}_2/\text{PPh}_3/\text{Zn}} \text{3,3'-dimethoxy-2,2'-bipyridine} \quad 75\% \tag{54}$$

$$2\ \text{2-chloroquinoline} \xrightarrow[\text{dmf}]{\text{NiCl}_2/\text{PPh}_3/\text{Zn}} \text{2,2'-biquinoline} \quad 83\% \tag{55}$$

Nickel(0)-promoted reactions of acyl halides produce generally a mixture of products which involve diketones, ketones, and decarbonylated coupling products. Rarely are symmetrical ketones formed with a satisfactory selectivity (equations 56 and 57)[47,95].

$$2\,n\text{-}C_7H_{15}COCl \xrightarrow[Et_2O]{[Ni(CO)_2(PPh_3)_2]} n\text{-}C_7H_{15}C(=O)C_7H_{15}\text{-}n \quad 40\% \tag{56}$$

$$2\ \text{2-AcOC}_6H_4COCl \xrightarrow[thf]{[Ni(cod)_2]} (\text{2-AcOC}_6H_4)_2C{=}O \quad 65\% \tag{57}$$

Carbonylative coupling of organic halides is mentioned here for comparison. Unlike allylic halides (cf. equations 37 and 38), benzylic halides undergo carbonylative coupling with $[Ni(CO)_4]$ in dmf or acetonitrile to give symmetrical ketones (equations 58 and 59)[180,266,360]. In the presence of carbon monoxide, $K_4[Ni_2(CN)_6]$ also acts as a carbonylative coupling agent for benzyl halides (equation 60)[72,108].

Imidoyl chlorides are dimerized on treatment with $Ni(CO)_4/PhLi$ (equation 61)[8].

$$2\,PhCH_2Br + [Ni(CO)_4] \xrightarrow{dmf} PhCH_2C(=O)CH_2Ph + NiBr_2 \tag{58}$$

$$\text{1,8-bis(bromomethyl)naphthalene} + [Ni(CO)_4] \xrightarrow{MeCN} \text{1H-phenalen-2(3H)-one} \tag{59}$$

$$2\,PhCH_2Br + CO \xrightarrow{K_4[Ni_2(CN)_6]} PhCH_2C(=O)CH_2Ph \tag{60}$$

$$2\,\text{4-MeC}_6H_4C(Cl){=}NPh \xrightarrow{[Ni(CO)_4]/PhLi} \text{4-MeC}_6H_4C(=NPh){-}C(=NPh)C_6H_4\text{-4-Me} \tag{61}$$

b. Catalytic reaction

Homocoupling reactions of organic halides can also be performed catalytically with respect to the nickel species by regeneration of low-valent nickel species from the resultant nickel(II) species by appropriate reducing agents. Although zinc powder has chiefly been used[279–283,363], electrochemical reduction is also useful[104,171,207,313]. The overall catalytic cycle is presented in Scheme 3.

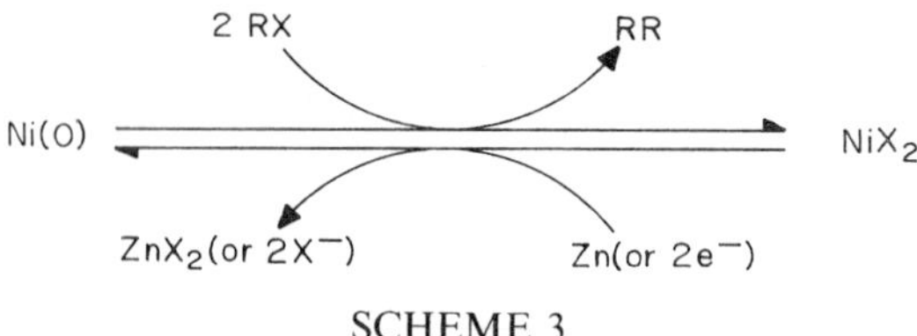

SCHEME 3

Homocoupling of alkyl bromides bearing β-hydrogens is satisfactorily achieved by the indirect cathodic reduction in the presence of [Ni(salen)] as a catalyst (equation 62)[104]. The proposed active species is Ni(I) rather than Ni(0) species. Catalytic coupling of benzyl halides is also attained electrochemically (equation 63)[171,312]. Propargyl chlorides undergo a novel cyclodimerization on treatment with a $[NiBr_2(PPh_3)_2]/Zn$ system (equation 64)[232].

$$2\,EtO_2CCH_2CH_2Br \xrightarrow[e^-]{[Ni(salen)](10\text{ mol-}\%)} (EtO_2CCH_2CH_2\text{—})_2 \quad 63\% \qquad (62)$$

$$2\,PhCH_2Cl \xrightarrow[PPh_3/e^-/dmf]{[Ni(acac)_2](20\text{ mol-}\%)} PhCH_2CH_2Ph \quad 85\% \qquad (63)$$

$$2\,Me_2C(Cl)C{\equiv}CR \xrightarrow[PPh_3/Zn/thf]{[NiBr_2(PPh_3)_2](20\text{ mol-}\%)} \qquad (64)$$

R=H, Me 65–75%

Catalytic homocoupling of alkenyl halides is achieved either with $[NiCl_2(PEt_3)_2]$ 4–10 mol-%)/Zn/KI[282] or with the phosphine-free counterpart $NiX_2/Zn/KI$(and/or tu)[278,281,283], the solvent being hmpa or nmp in both cases (equations 65 and 66). Potassium iodide is an essential additive for the catalytic coupling of alkenyl bromides, since it not only facilitates the reduction of Ni(II) species, but also acts as a reagent for the nickel-catalysed transformation of alkenyl bromides into more reactive alkenyl iodides (see also Section III.E). For coupling of alkenyl iodides, potassium iodide is not necessarily required, but instead tu or tmtu is used as an accelerator, which assists the reaction of Ni(II) species with zinc[283]. Alkenyl chlorides are mostly recovered. The stereoselectivity (equation 67) seems lower than that observed in the stoichiometric reactions (equation 40).

$$2\,Ph_2C{=}CHBr \xrightarrow[Zn/KI/nmp]{[NiCl_2(PEt_3)_2](4\text{ mol-}\%)} Ph_2C{=}CH{-}CH{=}CPh_2 \quad 97\% \qquad (65)$$

$$2\ \mathrm{Ph_2C{=}CHI} \xrightarrow[\mathrm{Zn/tu/hmpa}]{\mathrm{NiCl_2\ (40\ mol\text{-}\%)}} \mathrm{Ph_2C{=}CH{-}CH{=}CPh_2}\quad 77\% \tag{66}$$

$$2\ \mathrm{MeO_2C{-}CH{=}CH{-}Br} \xrightarrow[\mathrm{Zn/KI/hmpa}]{\mathrm{NiCl_2(40\ mol\text{-}\%)}} \mathrm{MeO_2C{-}CH{=}CH{-}CH{=}CH{-}CO_2Me}\quad 90\% \tag{67}$$

E,E:*E,Z*:*Z,Z* = 12:19:69

Aryl halides undergo homocoupling under similar nickel-catalysed conditions: $[NiCl_2(PPh_3)_2]$(5 mol − %)/PPh_3/Zn/KI/dmf[383] (equation 68), $[NiCl_2(PEt_3)_2]$(4 mol − %)/Zn/KI/hmpa (or nmp)[282] (equation 69), $NiBr_2$ (2.5 mol − %)/Zn/KI/hmpa[280], or electrochemically reduced nickel species[207,312] (equation 70). Acceleration effects by potassium iodide are also notable, as in the alkenyl cases mentioned above. While in the first system excess of triphenylphosphine must be added, in the second trialkylphosphine system extra phosphine retards the coupling. Coupling reactions usually proceed at 25–50 °C. Only the last electrochemical synthesis is effective for coupling of the least reactive aryl chlorides also.

$$2\ \mathrm{MeO_2C{-}C_6H_4{-}Br} \xrightarrow[\mathrm{PPh_3/Zn/dmf}]{\mathrm{[NiCl_2(PPh_3)_2](5\ mol\text{-}\%)}} \mathrm{MeO_2C{-}C_6H_4{-}C_6H_4{-}CO_2Me}\quad 83\% \tag{68}$$

$$2\ \mathrm{MeO_2C{-}C_6H_4{-}Br} \xrightarrow[\mathrm{Zn/KI/nmp}]{\mathrm{[NiCl_2(PEt_3)_2](4\ mol\text{-}\%)}} \mathrm{MeO_2C{-}C_6H_4{-}C_6H_4{-}CO_2Me}\quad 96\% \tag{69}$$

$$2\ \mathrm{C_6H_5{-}C_6H_4{-}Cl} \xrightarrow[\mathrm{PPh_3/e^-/thf/hmpa}]{\mathrm{[NiCl_2(PPh_3)_2](10\ mol\text{-}\%)}} \mathrm{C_6H_5{-}C_6H_4{-}C_6H_4{-}C_6H_5}\quad 71\% \tag{70}$$

Certain furyl and thienyl halides are coupled under the first two conditions just mentioned (equations 71 and 72)[282,363], but iodopyridines give no coupling products;

$$2\ \mathrm{MeO_2C{-}C_4H_2O{-}Br} \xrightarrow[\mathrm{Zn/KI/hmpa}]{\mathrm{[NiCl_2(PEt_3)_2](4\ mol\text{-}\%)}} \mathrm{MeO_2C{-}C_4H_2O{-}C_4H_2O{-}CO_2Me}\quad 90\% \tag{71}$$

$$2\ \text{3-iodothiophene} \xrightarrow[\text{Zn/KI/hmpa}]{[NiCl_2(PEt_3)_2]\ (4\ \text{mol-\%})} \text{3,3'-bithiophene}\ (83\%) \qquad (72)$$

compare with the stoichiometric reactions (equations 54 and 55). The nickel-promoted or -catalysed homocoupling reactions of aryl and alkenyl halides are thus being recognized as useful alternatives to the classical copper-induced Ullmann reaction.

2. Organosulphur compounds

Whereas heterogeneous Raney nickel has been widely used for desulphurization of organosulphur compounds, homogeneous Ni(0) complexes have rarely been applied. Only a few stoichiometric reactions have been reported.

Cyclic diphenyl sulphide derivatives are desulphurized by $[Ni(cod)_2]$/bipy to give coupling, ring-contraction products (equation 73)[77,78]. Whereas *S*-(2-pyridyl) aliphatic

$$\xrightarrow[\text{thf}]{[Ni(cod)_2]/\text{bipy}} \quad E = O, NH, S;\ 50\text{–}70\% \qquad (73)$$

thioates are reductively dimerized to a mixture of α-diketones and α-hydroxyketones on treatment with $[Ni(cod)_2]$[225], aromatic counterparts form symmetrical ketones selectively (equations 74 and 75)[105].

$$2\ \text{MeO-C}_6\text{H}_4\text{-CO-S-(2-pyridyl)} + [Ni(cod)_2] \xrightarrow{\text{dmf}} \text{(2-MeOC}_6\text{H}_4)_2\text{CO}\ (80\%) + Ni(\text{S-2-pyridyl})_2 + CO + 2\,\text{cod} \qquad (74)$$

$$2\ \text{2-thienyl-CO-S-(2-pyridyl)} \xrightarrow[\text{dmf}]{[Ni(cod)_2]} \text{(2-thienyl)}_2\text{CO}\ (71\%) \qquad (75)$$

B. Type 2: Ni(II)-Promoted Homocoupling of Nucleophiles

Nickel(II) salts or complexes have long been known to be useful stoichiometric coupling agents for organolithium or magnesium reagents. Synthetically, however, the procedure is applicable only to aryl and alkenyl organometallics (equations 76 and 77)[302,337]. The stereochemistry of the alkenyl group is retained[337]. The now classical method is not so frequently used as the copper-induced coupling reactions.

$$2\ PhMgBr + NiBr_2 \longrightarrow PhPh \quad (76)$$

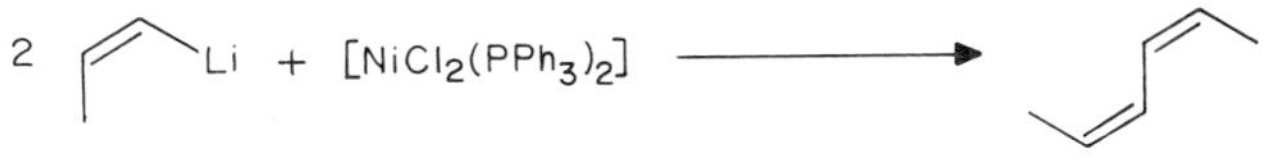

$$2\ \text{CH}_3\text{CH=CHLi} + [NiCl_2(PPh_3)_2] \longrightarrow \text{CH}_3\text{CH=CH–CH=CHCH}_3 \quad (77)$$

Organomercury(II) bromides are carbonylated to symmetrical ketones with $[Ni(CO)_4]$ in the presence of potassium iodide, which converts the bromides into more reactive organomercuric iodides[249]. Since this transformation proceeds through oxidative addition of the mercury—halogen bond to $[Ni(CO)_4]$ followed by the expulsion of metallic mercury to form an organonickel species, even β-hydrogen-bearing alkyl derivatives can be employed (equation 78).

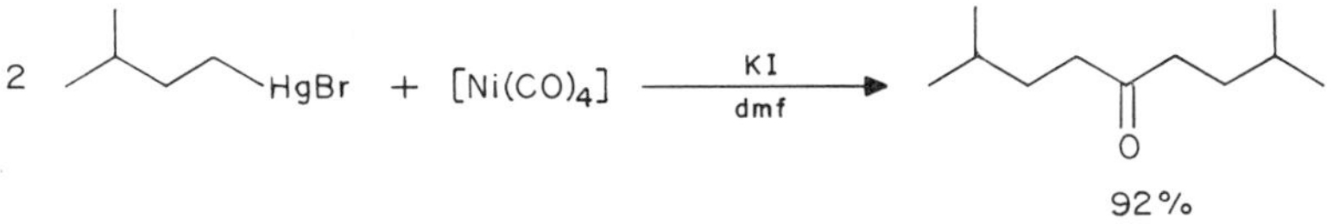

$$2\ (CH_3)_2CHCH_2CH_2HgBr + [Ni(CO)_4] \xrightarrow[\text{dmf}]{\text{KI}} (CH_3)_2CHCH_2CH_2COCH_2CH_2CH(CH_3)_2 \quad (92\%) \quad (78)$$

Two catalytic processes have recently been developed[175,236]. One is homocoupling of a phenyl Grignard reagent/bipy complex in the presence of a catalytic amount of $[NiX_2(bipy)]$ (equation 79)[236]. Notably in this catalytic reaction, bipy acts as a reoxidant

$$2\ MgBrPh(bipy) \xrightarrow[\text{thf}]{[NiCl_2(bipy)]} PhPh + 2\ MgBr^+(bipy^{-\cdot}) \quad (79)$$

of low-valent nickel species, a radical anion ($bipy^{-\cdot}$) being formed. The proposed catalytic cycle is shown in Scheme 4 (see also Section III.F). The other is dimerization and/or trimerization of α-sulphonyl anions in the presence of $[Ni(acac)_2]$ as a catalyst[175]. Whereas allylic derivatives give dimeric symmetrical trienes selectively (equation 80), alkyl

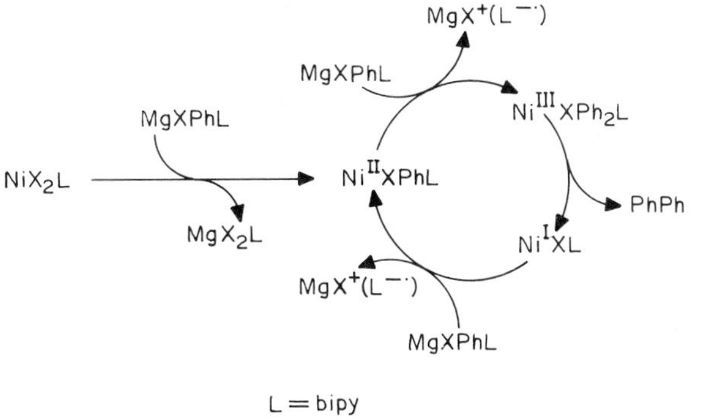

$NiX_2L \xrightarrow{MgXPhL,\ -MgX_2L} Ni^{II}XPhL$

$Ni^{II}XPhL + MgXPhL \longrightarrow Ni^{III}XPh_2L + MgX^+(L^{-\cdot})$

$Ni^{III}XPh_2L \longrightarrow Ni^{I}XL + PhPh$

$Ni^{I}XL + MgXPhL \longrightarrow Ni^{II}XPhL + MgX^+(L^{-\cdot})$

L = bipy

SCHEME 4

and benzyl counterparts form a mixture of olefins and cyclopropanes. Phytoene has been synthesized from geranylgeraryl sulphone. The reaction may proceed in two steps: the

$$2\ \text{(SO}_2\text{Ph, MgBr allyl reagent)} \xrightarrow[\text{thf}]{[\text{Ni(acac)}_2](2\ \text{mol-\%})} \text{diene product}\ (E{:}Z = 1{:}1)\ 80\% \tag{80}$$

Ni(II)-promoted coupling of anions produces a symmetrical vicinal disulphone and Ni(0) species, and the subsequent desulphurization of the disulphone by the resulting Ni(0) species gives the product olefin to regenerate the catalytically active Ni(II) species. This scheme is reminiscent of the mechanism of the nickel-catalysed Grignard cross-coupling of organosulphur compounds (see Section III.E).

C. Type 3: Reactions of Organonickel Compounds with Electrophiles

This section is concerned with coupling reactions of organonickel complexes, such as η^3-allyl-, alkyl-, aryl-, and acylnickel complexes, with electrophiles which include organic halides, carbonyl compounds, electrophilic olefins and acetylenes, and epoxides.

1. η^3-Allylnickel complexes

Reactions of η^3-allylnickel complexes have been extensively studied and covered by several reviews[16,125,173]. We therefore describe here only a few typical reactions and some applications to natural product synthesis.

η^3-Allylnickel halide dimers, obtainable by the reaction of allylic halides with $[Ni(CO)_4]$ or $[Ni(cod)_2]$ in non-polar solvents (cf. equation 8), react with alkyl, allyl, alkenyl, and aryl halides in polar solvents such as dmf or hmpa from ambient temperature to 50 °C to give the corresponding cross-coupling products (equation 81). Significant

$$RX + \tfrac{1}{2}\left[(\eta^3\text{-C}_3\text{H}_5)\text{NiBr}\right]_2 \longrightarrow R\text{CH}_2\text{CH}{=}\text{CH}_2 + NiX_2 \tag{81}$$

features are as follows. The carbon—carbon bond formation occurs at the less substituted site of the allyl group. Ester, nitrile, hydroxy, amino, and amide groups are compatible, but aldehydes and ketones may be allylated under more drastic conditions (see below). Secondary alkyl iodides are also allylated with no side reactions such as β-elimination, but with complete recemization of a chiral centre. Allylation of alkenyl bromides proceeds with retention of configuration (equations 82–85)[126,127,129,131]. For the related Ni(0) promoted homocoupling, see Section III.A.

$$\text{2-BrC}_6\text{H}_4\text{NH}_2 + \left[(\eta^3\text{-C}_3\text{H}_5)\text{NiBr}\right]_2 \xrightarrow{\text{dmf}} \text{2-(CH}_2{=}\text{CHCH}_2)\text{C}_6\text{H}_4\text{NH}_2 \tag{82}$$

$$\text{2-Br-3-CO}_2\text{Me-pyridine} + \left[(\eta^3\text{-MeO-C}_3\text{H}_4)\text{NiBr}\right]_2 \xrightarrow{\text{dmf}} \text{2-(CH}_2\text{COCH}_3)\text{-3-CO}_2\text{Me-pyridine}\ \ 78\% \tag{83}$$

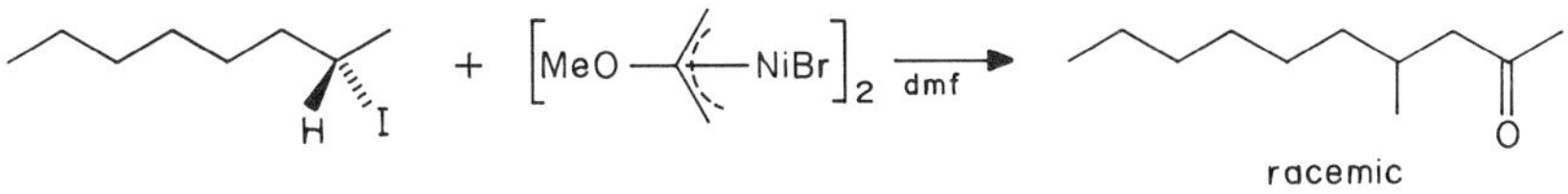

(84)

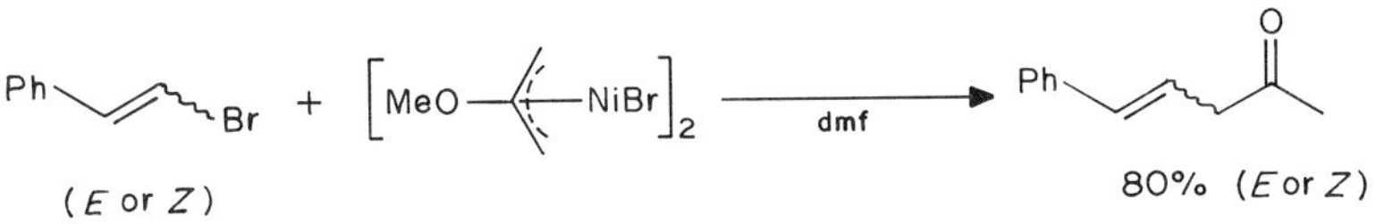

(85)

The synthetic utility of η^3-allylnickel halides has been shown by the synthesis of many natural products which include α-santalene (equation 86)[57], coenzyme Q (equation 87)[160], vitamin K[252], and oxygen-[159,192] and nitrogen-containing heterocycles[129]. The reaction

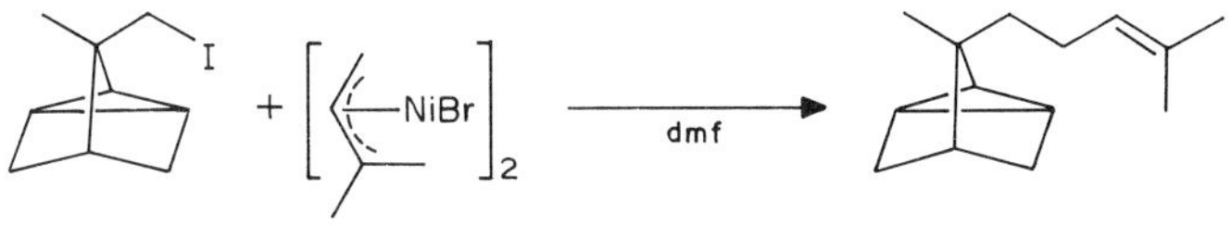

(86)

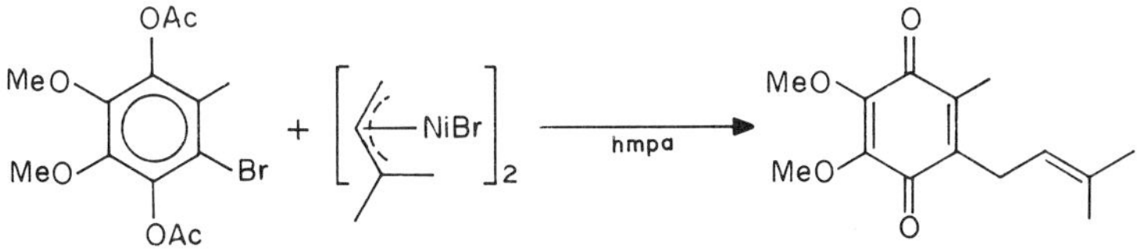

(87)

has been proposed to proceed through the initial one-electron transfer from an η^3-allylnickel complex to an organic halide[127]. The detailed mechanism will be discussed later (Section III.F).

η^3-Allylnickel halides also react with aldehydes and ketones under more vigorous conditions to form homoallyl alcohols[57,128]. In addition to simple allyl groups, 2-carbethoxyallyl[128], 2-methoxyallyl[126] and 2-vinylallyl groups[131] can be introduced to form α-methylene-γ-lactone skeletons, β-hydroxyketones and isoprenyl derivatives, respectively. 1,2-Diketones are the most reactive carbonyl substrates (equations 88–91).

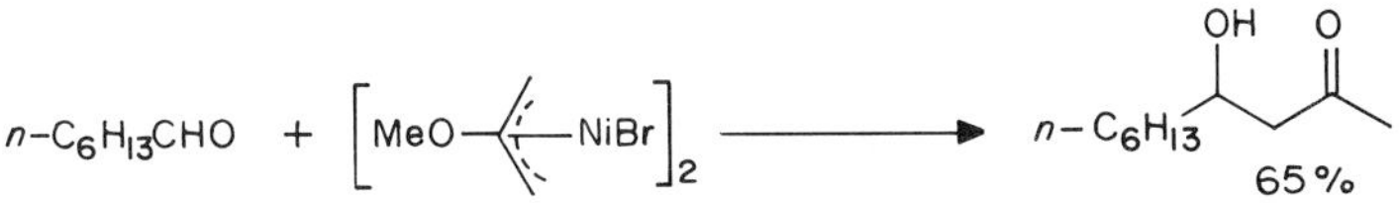

(88)

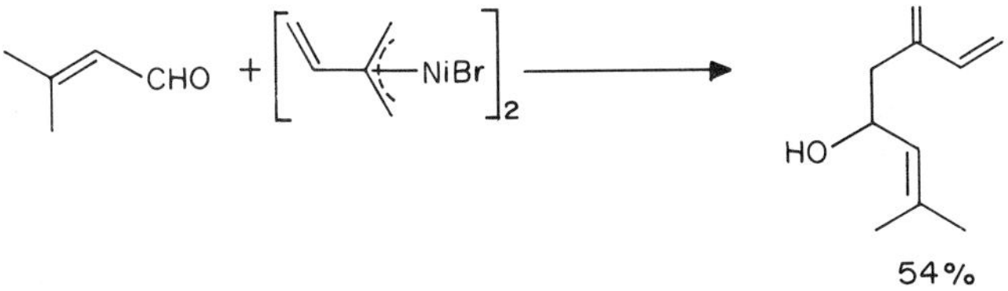

(89)

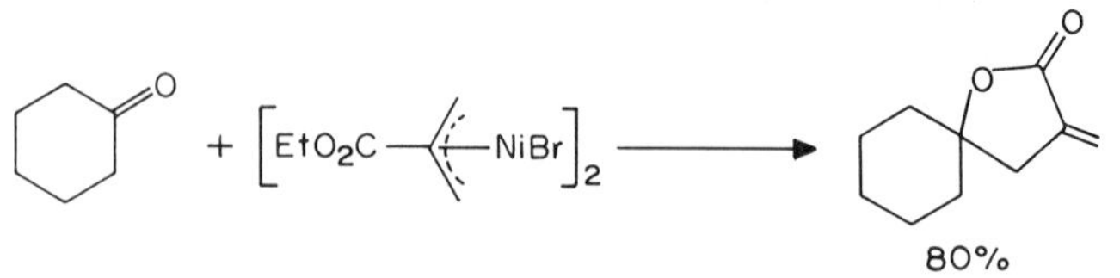

(90)

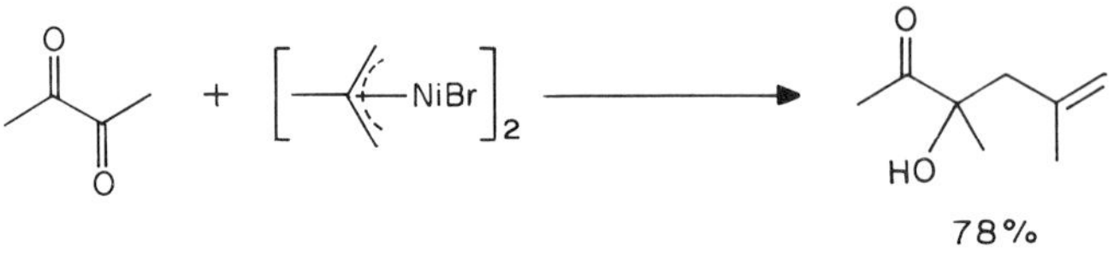

80%

(91)

78%

This type of reaction may be performed by the *in situ* generated η^3-allylnickel species from allylic esters[48,269] or allylic sulphonium salts[268] as well as allylic halides[267] (equation 92).

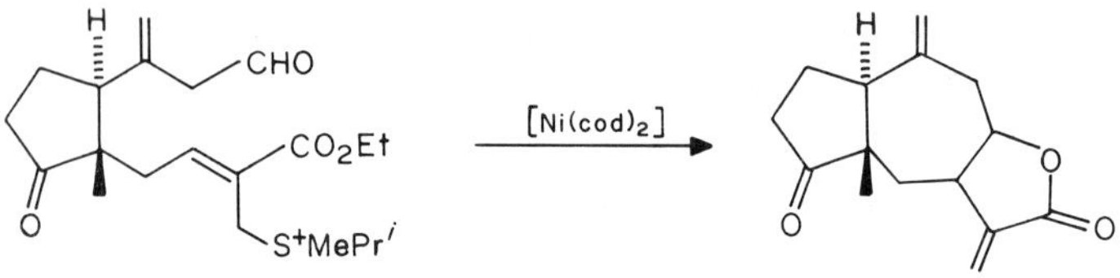

(92)

η^3-Allylnickel halides readily transfer the allyl group also to quinones in dmf even at $-50\,^{\circ}C$ to give allylhydroquinone derivatives directly (equation 93)[130]. According to the

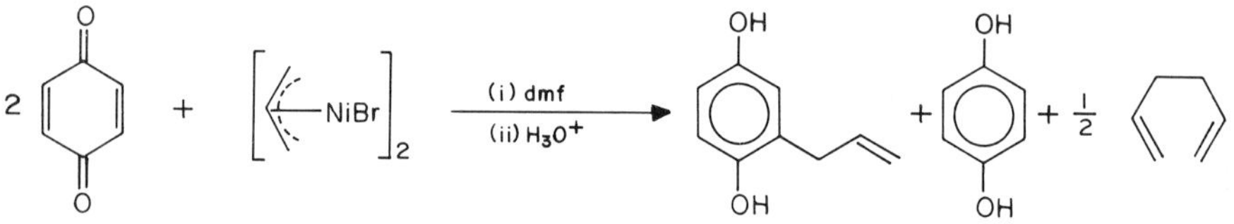

(93)

following stoichiometry, a half equivalent of each of the quinone and the allyl group is consumed, with the former being simply reduced and the latter self-coupled. A direct synthesis of coenzyme Q_1 has been achieved (equation 94)[130]; compare with the halide-coupling route (equation 87). A one-electron transfer from η^3-allylnickel complexes to

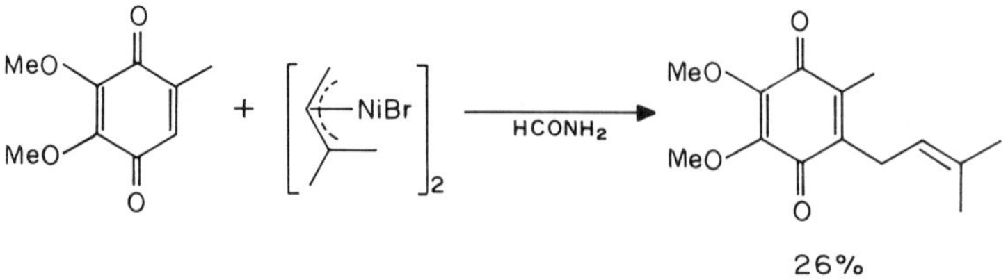

(94)

26%

quinones is considered to be involved, based mainly on the site selective allylation at the non-carbonyl carbon atom of highest spin density in the corresponding quinone radical anions[130].

Although η^3-allylnickel halides are inert to acyl halides and ordinary esters, they do react with 2-pyridyl carboxylates to form mainly β,γ-unsaturated ketones (equation 95)[224]. A reaction with an epoxide is also known (equation 96)[57].

$MeO_2C(CH_2)_3C(=O)-O-(2\text{-pyridyl}) + [(\eta^3\text{-allyl})NiBr]_2 \xrightarrow{\text{dmf}} MeO_2C(CH_2)_3C(=O)CH_2CH=CH_2$ (87) $+ MeO_2C(CH_2)_3C(=O)CH=CHCH_3$ (13) ; 71% (95)

Ph-epoxide $+ [(\eta^3\text{-2-methylallyl})NiBr]_2 \xrightarrow{\text{dmf}}$ $CH_2=C(CH_3)CH_2CH(Ph)CH_2OH$ 60% (96)

The reactivities of bis(η^3-allyl)nickel complexes are mentioned here for comparison. (η^3,η^3-Dodecatrienediyl)nickel, known as an intermediate in the nickel-catalysed trimerization of butadiene, reacts with aldehydes and activated organic halides such as allyl and benzyl bromides[17]. Noteworthily, unlike η^3-allylnickel halides mentioned above, it reacts also with acetyl chloride to form ketones[17]. The stepwise nature of the reaction with aldehydes and acyl halides makes possible unsymmetrical derivatization of the symmetrical bis(η^3-allyl)nickel complex (equation 97)[17].

[Ni complex] $\xrightarrow{\text{MeCHO}}$ [Ni–O complex] $\xrightarrow{CH_2=CHCH_2Br}$ product with OH (97)

Although bis(η^3-allyl)nickel complexes do not react with quinones[130], they do react with activated acetylenes in a stepwise manner to form finally diallylated products (equation 98)[18,34].

$[Ph_3P-Ni]$ complex $+ Z-C\equiv C-Z \longrightarrow$ [Ni complex] $\xrightarrow{\text{CO}}$ product

$Z = CO_2Et$ (98)

2. Alkyl- and aryl-nickel complexes

Fundamental, mechanistic work on the reactions of diorganonickel complexes with organic halides revealed that the reaction courses are fairly sensitive to the nature of the organic groups on nickel, the ligands, and the organic halides. At least five types of reactions have been observed[355], viz. reductive elimination with aryl and alkyl halides (equations 27 and 99), β-hydrogen transfer with alkyl halides (equation 100), cross-coupling with aryl and acyl halides (equations 101 and 102), and carbonylative reductive elimination with acyl halides (equation 103). These results promise a variety of synthetically useful applications.

$$NiEt_2(bipy) + [\text{polymer unit with Cl}]_n \longrightarrow EtEt + [\text{polymer unit with NiCl(bipy)}]_n \qquad (99)$$

$$[NiMe_2(PEt_3)_2] + Pr^nBr \longrightarrow MeH + MeCH{=}CH_2 + [NiBr_2(PEt_3)_2] \qquad (100)$$

$$[NiMe_2(dppe)] + PhCl \longrightarrow PhMe + [NiCl(Me)(dppe)] \qquad (101)$$

$$[NiEt_2(bipy)] + MeCOCl \longrightarrow MeCOEt + [NiCl(Et)(bipy)] \qquad (102)$$

$$[NiMe_2(dppe)] + PhCOCl \longrightarrow MeCOMe + [NiCl(Ph)(dppe)] \qquad (103)$$

Alkylnickel species generated *in situ* from low-valent nickel species and organic halides are useful reagents for cross-coupling with other organic halides. A nickelacyclopentane, readily available from $[Ni(cod)_2]$, bipy, and 1,4-dibromobutane (equation 104), reacts with alkyl halides, *gem*-dihalides, and acyl halides to form cross-coupling products[285]. Of particular interest are the cyclocoupling with *gem*-dihalides (equation 105) and the formation of octamethylene derivatives with acyl and sulphonyl halides (equation 106).

$$Br(CH_2)_4Br + 2[Ni(cod)_2] + 4\,bipy \longrightarrow [(bipy)Ni(CH_2)_4] + [Ni(bipy)_3]Br_2 + 4\,cod \qquad (104)$$

$$[(bipy)Ni(CH_2)_4] + \text{7,7-dibromobicyclo[4.1.0]heptane} \longrightarrow \text{spiro[bicyclo[4.1.0]heptane-7,1'-cyclopentane]}\ \ 69\% \qquad (105)$$

$$[(bipy)Ni(CH_2)_4] \xrightarrow[(Y=CO,\,SO_2)]{PhYX} PhY(CH_2)_8YPh\ \ \textit{ca.}\ 50\% \qquad (106)$$

An active metallic nickel induces the cross-coupling reaction of benzyl halides with α-haloacetonitrile (equation 107)[156] and with acyl halides (equation 108)[155]. Cross-

$$3\text{-}CF_3C_6H_4CH_2Cl + ClCH_2CN \xrightarrow[dme]{NiCl_2/Li/NpH} 3\text{-}CF_3C_6H_4CH_2CH_2CN \quad 58\% \qquad (107)$$

$$2\text{-}NpCH_2Br + MeCOCl \xrightarrow[dme]{NiCl_2/Li/NpH} 2\text{-}NpCH_2COMe \quad 62\% \qquad (108)$$

coupling of primary alkyl iodides with 2-pyridyl carboxylates is catalysed by $NiCl_2$ in the presence of excess of zinc to give unsymmetrical ketones (equation 109)[226].

$$MeO_2C(CH_2)_3I + Ph(CH_2)_3C(=O){-}O{-}C_5H_3N{-}(CH_2)_2OMe \xrightarrow[Zn/dmf]{NiCl_2(10\ mol-\%)} MeO_2C(CH_2)_3C(=O)(CH_2)_3Ph \quad (109)$$

81%

Reactions of organonickel intermediates with carbonyl compounds are also of synthetic value (equations 110–112)[73,157,227]. The last nickel-catalysed reaction may be one of the most efficient methods for the introduction of perfluoroakyl groups[227].

$$NCCH_2Br + PhCHO \xrightarrow{[Ni(CO)_4]} NCCH_2CH(OH)Ph \quad (110)$$

$$Cl{-}C_6H_4{-}CH_2Cl + PhC(=O){-}C(=O)Ph \xrightarrow[dme]{NiI_2/Li/NpH} Cl{-}C_6H_4{-}CH_2C(Ph)(OH){-}C(=O)Ph \quad (111)$$

83%

$$i\text{-}C_3F_7I + PhCHO \xrightarrow[Zn/dmf]{NiCl^2(PPh_3)_2(0.1\,mol-\%)} i\text{-}C_3F_7CH(OH)Ph \quad (112)$$

78%

Low-valent nickel species induce the addition of organic halides to activated olefins. Dibromomethane and several *gem*-dihalides react with electron-deficient olefins to form cyclopropane derivatives with complete retention of *E*-stereochemistry, but with substantial loss of *Z*-stereochemistry of olefins (equation 113)[177]. The reaction can be achieved catalytically in nickel in the presence of zinc powder[177]. Electrochemically reduced nickel

$$CH_2Br_2 + MeO_2C{-}CH{=}CH{-}CO_2Me \xrightarrow{NiBr_2/Zn/NaI/MeCN} \textit{trans-}1,2\text{-}(MeO_2C)_2\text{cyclopropane} \quad (113)$$

78%

species stoichiometrically promote conjugate addition of alkyl bromides, even secondary alkyl bromides, to activated olefins (equation 114)[124]. Similarly, arylnickel species also

$$(CH_3)_2CHBr + CH_2{=}CHC(=O)CH_2CH_3 \xrightarrow[e^-]{[Ni(teta)]^{2+}} (CH_3)_2CHCH_2CH_2C(=O)CH_2CH_3 \quad (114)$$

83%

add to olefins, which need not be electron deficient[205–207,247]. In particular, intramolecular reactions provide efficient methods for the synthesis of heterocyclic compounds[206] (equations 115 and 116).

$$PhI + CH_2{=}CH_2 \xrightarrow[e^-]{[NiCl_2(PPh_3)_2]/PPh_3} PhCH{=}CH_2 \quad (115)$$

80%

(116)

Nickel(0)-catalysed addition of a strained molecule, bicyclo[2.1.0]pentane, to elctron-deficient olefins should also be mentioned here; the reaction may proceed through a diorganonickel intermediate through the cleavage of the central σ-bond (equation 117)[286].

(117)

Nickel-catalysed electrosynthesis of aromatic carboxylic acids from aryl halides and carbon dioxide is achieved efficiently (equation 118)[85]. An acetylacetonato group in $[Ni(acac)_2]$ couples with reactive organic halides in dmf to give *C*-alkylation products exclusively (equation 119)[28].

$$p\text{-}FC_6H_4Br + CO_2 \xrightarrow[e^-]{[NiCl_2(dppe)](10\ mol\text{-}\%)} p\text{-}FC_6H_4COOH \quad (118)$$

ca. 60%

(119)

3. Acylnickel complexes

This section describes the reactions of two types of acylnickel complexes, acylnickel carbonylate anions and nickelacyclopentenedione species.

Acylnickel carbonylate anions, obtainable from nickel carbonyl and organolithium reagents (equation 120)[322], are now classical acyl anion equivalents. They undergo conjugate addition to α,β-enones in diethyl ether under very mild conditions to form 1,4-dicarbonyl compounds (equation 121)[59]. The conjugate addition and the

$$RLi + [Ni(CO)_4] \longrightarrow Li[RCONi(CO)_n] \quad (120)$$

(121)

subsequent alkylation of the resulting enolates provide an efficient methodology for the consecutive dialkylation at the 2,3-vicinal positions of enones, as demonstrated by the synthesis of skeletons of natural products such as deoxyfrenolicin (equation 122)[271]. Acylnickel anions also react with η^3-allylpalladium cationic complexes to form β, γ-unsaturated ketones (equation 123)[132].

(122)

(123)

Interactions of nickel carbonyl with *tert*-butoxide and dialkylamide anions have been considered to lead to the formation of carbalkoxynickel and carbamoylnickel carbonylate anions, respectively (equations 124 and 125)[58,99]. These species react with alkyl, alklenyl, aryl, and/or acyl halides to form the corresponding esters or amides (equations 126–128)[58,99]. The carbamoyl group is transferred also to a ketone (equation 129)[99].

$$[Ni(CO)_4] + Bu^tOK \longrightarrow K[Bu^tOCONi(CO)_3] \qquad (124)$$

$$[Ni(CO)_4] + Me_2NLi \longrightarrow Li[Me_2NCONi(CO)_3] \qquad (125)$$

$$n\text{-}C_7H_{15}I \xrightarrow{[Ni(CO)_4]/Bu^tOK} n\text{-}C_7H_{15}CO_2Bu^t \quad 66\% \qquad (126)$$

(127)

$$Bu^nCOCl \xrightarrow{[Ni(CO)_4]/Me_2NLi} Bu^nCOCONMe_2 \quad 97\% \qquad (128)$$

$$Ph_2C{=}O \xrightarrow{[Ni(CO)_4]/Me_2NLi} Ph_2C(OH)CONMe_2 \quad 30\% \qquad (129)$$

Nickelacyclopentanediones are readily formed by the reaction of $[Ni(CO)_2(bipy)]$ with acetylenes (equation 130)[133] and constitute a new class of synthetically useful acynickel complexes. They react with *gem*-dihalides and activated olefins to form cyclic products (equations 131 and 132)[134].

$$[Ni(CO)_2(bipy)] + RC{\equiv}CR' \xrightarrow[\text{thf}]{\text{CO(1 atm)}} \left[\text{nickelacycle: } R, R' \text{ cyclic dione } Ni(bipy)\right] \quad (130)$$

$$\text{[Me}_2\text{-nickelacycle Ni(bipy)]} \xrightarrow{Br_2CHMe} \text{trimethylcyclopentenedione, 66\%} \quad (131)$$

$$\text{[Me}_2\text{-nickelacycle Ni(bipy)]} \xrightarrow{\text{maleic anhydride}} \text{dimethyl-dihydroxyphthalic anhydride, 85\%} \quad (132)$$

4. Nickel-promoted reactions of unsaturated compounds with heterocumulenes

This section is concerned with fairly new types of reaction between π-complex intermediates generated from Ni(0) complexes with unsaturated compounds such as olefins and acetylenes, and electrophilic heterocumulenes such as carbon dioxide and isocyanates.

Olefins[144], allenes[146], 1, 3-dienes[145], and acetylenes[138,144] all react with carbon dioxide (1 atm) in the presence of a stoichiometric amount of $[Ni(cod)_2]$ and appropriate ligands to form nickel-containing lactone skeletons (equations 133–136). The nature of the ligand

$$[Ni(cod)_2] + CO_2 \xrightarrow[\text{dcpe, 60 °C}]{Bu^n\text{CH=CH}_2} [(dcpe)Ni\text{-lactone (}Bu^n\text{)}] + \text{regioisomer} \quad 25 : 1 \quad (133)$$

$$[Ni(cod)_2] + CO_2 \xrightarrow[\text{dcpe}]{\text{allene}} [(dcpe)Ni\text{-lactone (alkylidene)}] \quad (134)$$

$$[Ni(cod)_2] + CO_2 \xrightarrow[\text{dcpe}]{\text{1,3-diene}} [(dcpe)Ni\text{-lactone (}\pi\text{-allyl)}] \quad (135)$$

$$[Ni(cod)_2] + CO_2 \xrightarrow[\text{tmeda}]{MeC{\equiv}CMe} [(tmeda)Ni\text{-lactone (dimethyl)}] \quad (136)$$

is critical; bipy or dcpe is suitable for the first three substrates, whereas tmeda is best for the last acetylenes. Since the reactions of olefins and 1,3-dienes are reversible, these components are replaced by more reactive unsaturated compounds and the regioselectivity in the reaction of unsymmetrical olefins is dependent on the temperature. The carbon—nickel bonds therein are cleaved by acidolysis to form linear carboxylic acids or esters. With 1,3-dienes, dienoic acids are formed on treatment with maleic anhydride, the overall reaction being equivalent to the insertion of carbon dioxide into the olefinic carbon—hydrogen bond (equation 137)[145].

(dcpe)Ni → CO_2H 40% (137)

The nickelacycles from acetylenes react with alkyl halides to form β-alkylated, α,β-unsaturated carboxylic acids (equation 138), with *gem*-dihalides to form eight-membered cyclic acid anhydrides (equation 139), with electrophilic acetylenes to give pyrones (equation 140), and with triazolinedione to form pyrazolone derivatives with loss of one oxygen atom (equation 141)[138,140,144]. The formation of pyrone derivatives from

(tmeda)Ni

EtBr → Et CO_2H 78% (138)

CH_2Cl_2 → 80% (139)

ZC≡CZ ($Z = CO_2Me$) → 20% (140)

PhN → PhN 9% (141)

acetylenes and carbon dioxide is also achieved catalytically with respect to nickel (equation 142)[161], and may proceed through nickelalactone intermediates[140] rather than nickelacyclopentadienes proposed originally[161].

$$2\ EtC{\equiv}CEt + CO_2\ (50\ atm) \xrightarrow[\text{dppb}]{[Ni(cod)_2]\ (5\ mol\text{-}\%)} \text{tetraethyl-2-pyranone}\ 57\% \quad (142)$$

Isocyanates behave like carbon dioxide in nickel-promoted reactions with allenes[147] and acetylenes[137,141] to form nickelalactam derivatives, the reactions of which with alkyl halides and activated acetylenes also proceed similarly to produce α,β-unsaturated carboxamides and/or pyridones (equations 143 and 144). Isocyanates form similar

$$[Ni(cdt)] + \text{allene} + PhNCO \xrightarrow{\text{tmeda}} [\text{(tmeda)Ni nickelalactam}] \xrightarrow{MeI} \text{CONHPh}\ 29\% \quad (143)$$

$$[Ni(cod)_2] + PhC{\equiv}CPh + PhNCO \xrightarrow{\text{tmeda}} [\text{(tmeda)Ni nickelalactam}] \xrightarrow{PhC{\equiv}CPh} \text{pentaphenyl-2-pyridone} \quad (144)$$

nickela-heterocycles also with aldehydes[143] or imines[142] and the subsequent reactions with alkyl halides give carbamates or urea derivatives, respectively (equation 145). Reactions of imines are considered to proceed through nickel(0)-imine π-complexes as initial intermediates.

$$[Ni(cod)_2] + PhCH{=}X + PhNCO \xrightarrow{\text{tmeda}} [\text{(tmeda)Ni metallacycle}] \xrightarrow{MeI} PhCH(Me)XC(O)NHPh \quad (145)$$

X=O 55%

X=NPh 42%

Finally in this section we consider a nickel-promoted reaction of carbon disulphide. A nickel complex of carbon disulphide readily reacts with an electron-deficient acetylene to form a nickel–carbene complex. The carbene ligand therein is released by carbon monoxide to give a dimeric tetrathiafulvalene derivative (equation 146)[26]. The nickel complex may be used catalytically in principle, since the starting material is obtained from the resulting nickel(0)-carbonyl complex with carbon disulphide (cf. equation 12).

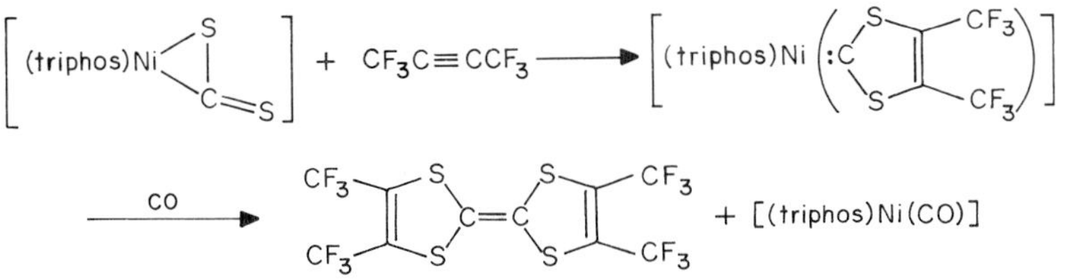

(146)

D. Type 4: Reactions of Organonickel Compounds with Nucleophiles

Although stoichiometric carbon—carbon bond-forming reactions of this type have been the subject of extensive studies from mechanistic viewpoints, there have been only a few synthetically useful reactions. Arylnickel halides react with metal enolates of esters to form α-arylated products (equation 147)[83,264]. The total synthesis of a cephalotaxinone derivatives represents a useful application of this type of reaction (equation 148)[265]. A variety of reactions which involve this type of reaction as a key step can now be achieved catalytically with nickel (see Section III.E).

$$[NiX(Ph)(PPh_3)_2] + MCH_2CO_2Et \longrightarrow PhCH_2CO_2Et \qquad (147)$$
$$M = Li \text{ or } ZnBr$$

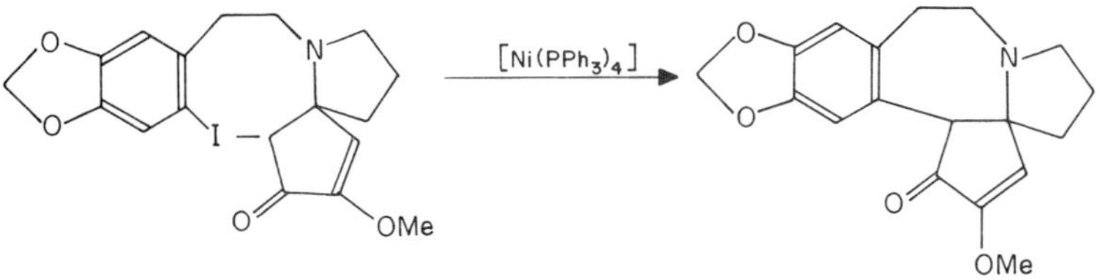

(148)

Carbonylative cross-coupling of aryl halides with arylmercurials[245] and enamines[260] are promoted by nickel carbonyl to form unsymmetrical ketones and 1,3-diketones, respectively (equations 149 and 150). These reactions are considered to proceed through acylnickel halide intermediates.

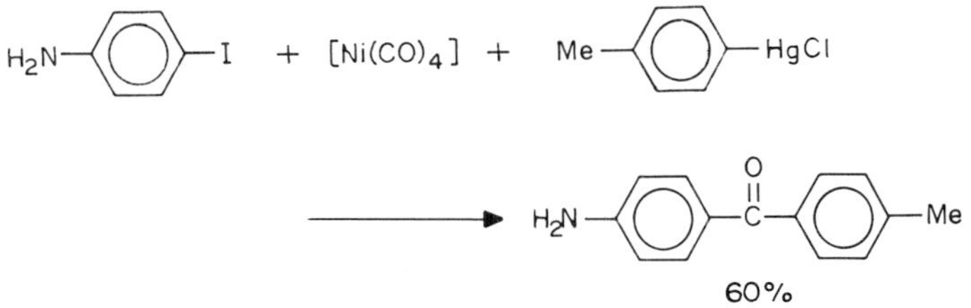

(149)

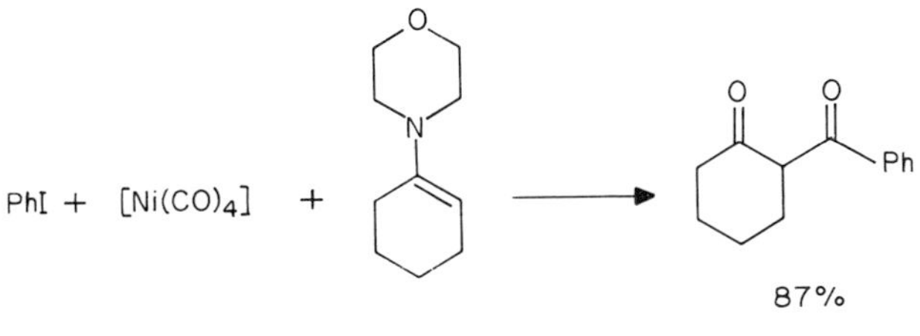

(150)

Nickel carbonyl-promoted reactions of organic halides in the presence of heteroatom nucleophiles such as water, alcohols, and amines have long been studied in order to obtain carboxylic acids, esters, and amides, respectively. Since this field has been well reviewed[173,174], only one mechanistic comment is given here. The reactions may involve the initial oxidative addition of organic halides to nickel carbonyl and the subsequent solvolytic decomposition (equation 151), whereas in the presence of strongly basic alkoxide or amide ions carbalkoxy- or carbamoyl-nickel anion species may be formed initially, as mentioned in the previous section (see Section III.C.3., equations 124–129). A similar reaction is observed also with epoxides (equation 152)[246].

$$\mathrm{RX} + [\mathrm{Ni(CO)_4}] \xrightarrow{-\mathrm{CO}} [\mathrm{RNi(CO)_2X}] \xrightarrow{\mathrm{CO}} [\mathrm{RCONi(CO)_2X}] \xrightarrow{\mathrm{R'OH\ or\ R'_2NH}} \mathrm{RCOOR'}\ \text{or}\ \mathrm{RCONR'_2} \quad (151)$$

PhCH=CHBr + $[Ni(CO)_4]$ + cyclohexene oxide ⟶ trans-2-bromocyclohexyl cinnamate (PhCH=CHC(=O)O–, Br) 78% (152)

Finally, a reaction which may be mentioned in this section is an unusual proton transfer/oxidative addition of electrophilic α, β-unsaturated carboxylic acids and amides to Ni(0) species in the presence of bulky ligands (equation 153)[347].

$[Ni(cod)_2]$ + $CH_2=C(CH_3)CONHMe$ + PCy_3 ⟶ $[(Cy_3P)Ni$ metallacycle with N–Me, C=O$]$ (ca. 65%) $\xrightarrow{\mathrm{CO}}$ N-methyl-3-methylsuccinimide (93%) (153)

E. Type 5: Nickel-Catalysed Cross-Coupling between Electrophiles and Nucleophiles

Nickel complex-catalysed cross-coupling reactions of organometallic reagents with organic halides and related compounds have now attained a history of nearly 15 years[61,289], and have been recognized as a useful practical method for the formation of carbon—carbon bonds. Several reviews have been published[89,173,174,193,194,211,212]. In addition to these reactions, nickel-catalysed conjugate addition of organometallics to α, β-enones and nickel-catalysed substitution of organic halides with heteroatom nucleophiles are also described in this section.

1. *Grignard cross-coupling reactions of organic halides and related reactions*

Known reaction patterns are listed in Table 2, where the cited references are restricted to those which deal with coupling reactions as their main subject; all the equations are numbered consecutively from those in the text. The most outstanding feature is the selective cross-coupling of aryl and alkenyl halides [C(sp^2)-halides] with almost any kind

TABLE 2. Nickel-catalysed Grignard cross-coupling reactions and related reactions

Reaction		References	Equation No.
Aryl-X + RMgX →	Aryl-R	61, 76, 89, 151–153, 194, 202, 241, 263, 289, 290, 294, 295, 308, 345	154
Aryl-X + $LiCH_2CO_2R'$ →	Aryl-CH_2CO_2R'	201	155
Aryl-X + RZnX →	Aryl-R	84, 214	156
Aryl-X + Alkenyl-$AlBu^i_2$ →	Aryl-Alkenyl	213	157
Aryl-X + Alkenyl-$ZrCp_2Cl$ →	Aryl-Alkenyl	71, 215	158
Aryl-X + CO + Aryl-MgX →	$(Aryl)_2CO$ + $(Aryl)_3COH$	343	159
Aryl-X + CO + Me_4Sn →	Aryl-COMe	303	160
Heteroaryl-X + RMgX →	Heteroaryl-R	24, 164, 203, 220, 234, 237, 238, 240, 248, 293, 300, 301, 307, 354, 357	161
Alkenyl-X + RMgX →	Alkenyl-R	61, 72, 92, 111, 242, 289, 291, 296, 297, 328	162
Alkenyl-X + $LiCH_2CO_2R'$ →	Alkenyl-CH_2CO_2R'	201	163
Alkenyl-X + Alkenyl-$AlBu^i_2$ →	Alkenyl-Alkenyl	11, 244	164
Alkenyl-X + RMgX →	Alkenyl-R	292	165
Alkenyl-X + R_3Al →	Alkenyl-R	103	166
(allyl*)-X + RMgX →	(*allyl*)-R	176	167
(butenyl)-Br + RMgX →	(1-methylallyl)-R (+ isomer)	362	168

TABLE 2. (*Contd.*)

Reaction		References	Equation No.
$R'O_2CCH_2Br + RZnX$ →	$R\text{-}CH_2CO_2R'$	186	169
$R'C{\equiv}CCR''_2X + RMgX$ →	$RR'C{=}C{=}CR''_2$	231	170
Aryl-OR′ + RMgX → [R′ = alkyl, $P(O)(OR'')_2$]	Aryl-R	115, 331, 336	171
Aryl-OP(O)$(OEt)_2$ + Alkenyl-Al$(Bu^i)_2$ →	Aryl-Alkenyl	115	172
Alkenyl-OR′ + RMgX → [R′ = alkyl, $SiMe_3$, P(O)$(OEt)_2$]	Alkenyl-R	111, 113, 250, 331, 335, 336	173
=O, O + RMgX →	R, CO_2H	1, 2, 166	174
=O, O + RZnX →	R, CO_2H	1, 2	175
OH, * + RMgX →	R, * *	33, 49, 53, 89–91	176
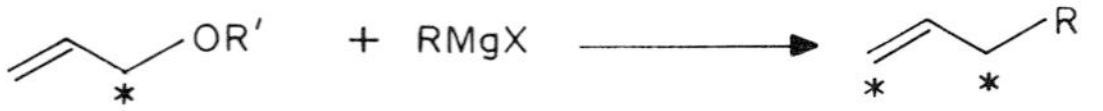 OR′, * + RMgX → [R′= alkyl, Ph, $SiMe_3$, thp]	R, * *	116, 227, 333, 334	177
OAc, * + (–) CHXY →	CHXY, * *	69, 318	178

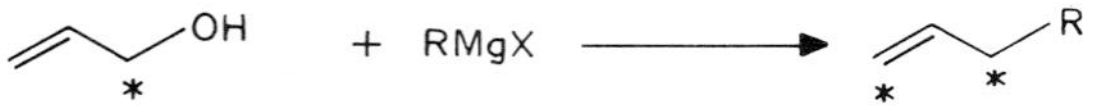

Reaction		Product	References	
ArCOOH + RMgX	→	ArCOR	93	179
Aryl-SR′ + RMgX	→	Aryl-R	239, 287, 307, 330, 332	180
Alkenyl-SR′ + RMgX	→	Alkenyl-R	92, 221, 288, 330, 332, 334, 335	181
$CH_2{=}CH\text{–}CH_2^{*}\text{–}SR'$ + RMgX	→	$CH_2^{*}{=}CH\text{–}CH_2^{*}\text{–}R$	222, 334	182
Aryl-SOR′ + RMgX	→	Aryl-R	330	183
Aryl-SO_2R' + RMgX	→	Aryl-R	330	184
Alkenyl-SO_2R' + RMgX	→	Alkenyl-R	79	185
Aryl-SeR′ + RMgX	→	Aryl-R	223	186
Alkenyl-SeR′ + RMgX	→	Alkenyl-R	223, 335	187
$CH_2{=}CH\text{–}CH_2^{*}\text{–}SeR'$ + RMgX	→	$CH_2^{*}{=}CH\text{–}CH_2^{*}\text{–}R$	223	188
Aryl-TeR′ + RMgX	→	Aryl-R	320, 321	189
Alkenyl-TeR′ + RMgX	→	Alkenyl-R	320, 321, 335	190

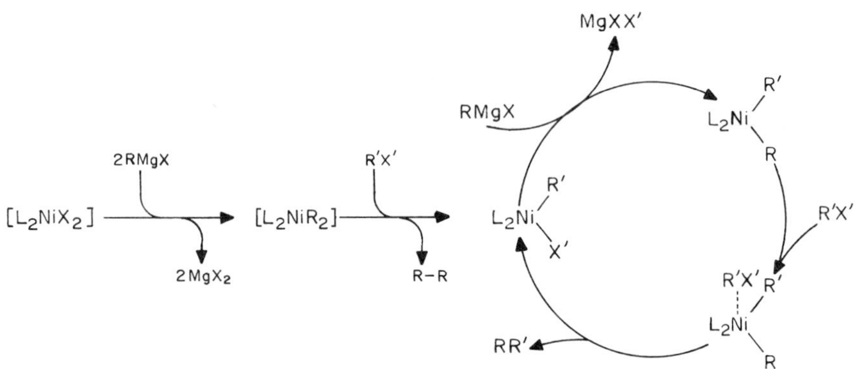

SCHEME 5

of Grignard reagents involving even alkyl Grignard reagents with β-hydrogens catalysed by nickel–phosphine complexes[289,295]. Another characteristic feature is the extension to a variety of substrates such as aryl—, alkenyl— and/or allyl—oxygen, —sulphur, —selenium and —tellurium compounds.

A catalytic cycle depicted in Scheme 5, which was originally proposed for the Grignard coupling with organic halides[289], may be generally applicable to all such coupling reactions. In Scheme 5, a dihalodiphosphinenickel as a catalyst precursor reacts with a Grignard reagent to form a diorganonickel intermediate, which is subsequently converted into a halo(organo)nickel complex by an organic halide. Reaction of this key intermediate with the Grignard reagent then forms a new diorganonickel complex from which the cross-coupling product is released by the attack of the organic halide, possibly via a pentacoordinate intermediate, and thereby the original key intermediate is regenerated to complete the catalytic cycle[230,295,316]. More detailed discussion is given in Section III.F.

a. Cross-coupling of organic halides with Grignard reagents (equations 154, 161, 162, 165, 167, 168, and 170)

Air-stable nickel(II)–phosphine complexes, $[NiX_2L_2]$, are usually used as a catalyst precursor in amounts of 1 mol-% or less. The catalytic activity of the nickel complexes depends strongly not only on the nature of the ligands, but also on the combination of Grignard reagent and organic halide[295]. Generally, bidentate are better than unidentate phosphine ligands. The catalytic activity is dependent on the chain length of the bidentate ligands $Ph_2P(CH_2)_nPPh_2$ in the order $n = 3$(dppp) > 2(dppe) > 4(dppb)[295]. This order may reflect a subtle balance between the stability and lability of the corresponding diorganonickel intermediates[189]; maybe those with dppe are too stable, whereas those with dppb are too labile. A recent paper has described a correlation between the inter-ligand angles and the catalytic activity in the similar palladium-catalysed cross-coupling reactions[123].

The following general tendencies have also been observed[295]. Whereas $[NiCl_2(dppp)]$ is most effective for primary and secondary alkyl and aryl Grignard reagents, $[NiCl_2(dmpe)]$, which contains an electron-donating bidentate ligand, is the most suitable catalyst for allylic and vinylic Grignard reagents. For the coupling of sterically hindered aryl Grignard reagents, $[NiCl_2(PPh_3)_2]$ is better than catalysts with bidentate ligands. Neutral phosphine ligand-free nickel salts, represented by $[Ni(acac)_2]$, seem to be among the most effective catalysts for aryl Grignard reagents[61,151]. The halides may be chlorides, bromides, or iodides, although chlorides usually give the most satisfactory results, since they exhibit a reasonable reactivity and give little side reaction such as simple reduction. Even aryl fluorides react with comparable facility in some cases. The reactivity of halobenzenes decreases in the order PhI > PhBr > PhCl > PhF. Diethyl ether is usually better than thf as a solvent.

Representative reactions and their applications are shown below. Aromatic polyhalides are easily polyalkylated without any positional scrambling (equation 191)[76,194,295]. *o*-Dibutylbenzene thus obtained has been used as a starting material for soluble

$$C_6H_4Cl_2 + Bu^nMgBr \xrightarrow{[NiCl_2(dppp)]} C_6H_4(Bu^n)_2 \quad (191)$$

ortho 79–83%
meta 94%
para 95%

phthalocyanines[68]. Precursors for *o*-quinodimethane have been prepared by coupling with the trimethylsilylmethyl Grignard reagent (equation 192)[165].

$$\text{MeO-C}_6\text{H}_3(\text{Cl})(\text{CH}_2\text{NMe}_2) + Me_3SiCH_2MgCl \xrightarrow{[NiCl_2(dppp)]} \text{MeO-C}_6\text{H}_3(\text{CH}_2SiMe_3)(\text{CH}_2\text{NMe}_2) \quad (192)$$

Aryl–aryl cross-couplings have potential utility for the synthesis of a variety of structurally interesting polyphenylenes and sterically hindered polyphenylenes (equations 193–196)[50,152,153,190,272,295], some of which have been used as precursors for the synthesis of spherands in host–guest chemistry (equation 197)[64–66].

$$\text{I}_2\text{C}_6\text{H}_3\text{-C}_6\text{H}_5 + \text{C}_6\text{H}_5\text{-C}_6\text{H}_4\text{MgBr} \xrightarrow{[Ni(acac)_2]} \text{product} \quad (193)$$

92%

$$o\text{-MeOC}_6\text{H}_4\text{Br} + \text{BrMg-C}_6\text{H}_2\text{Me}_3 \xrightarrow{[NiCl_2(dppp)]} o\text{-MeOC}_6\text{H}_4\text{-C}_6\text{H}_2\text{Me}_3 \quad (194)$$

74%

Br + Me MgBr $[NiCl_2(dppp)]$ Fe Fe Me 57% (195)

Br + PhMgBr $Ni(acac)_2$ (196)

I MeO OMe I + MgBr $[NiCl_2(PPh_3)_2]$

MeO OMe 75% (197)

Alkenyl halides are among the most reactive halides[295]. In particular, the high reactivity of vinyl chloride should be noted; isomerically pure 4-chlorostyrene has now been produced on an industrial scale in Japan (equation 198). The cross-coupling of alkenyl monohalides proceeds with retention of configuration[291] and should be useful for stereoselective olefin synthesis (equations 199 and 200)[75,148,243]. A variety of synthetically useful allylsilanes are readily obtained by this method (equations 200–202)[120,149,216,301]. Axially chiral,

Cl + MgX Cl $[NiCl_2(dppp)]$ Cl 74% (198)

Bu^n Cl + $Bu^tO(CH_2)_6MgCl$ $[Ni(PPh_3)_4]$ Bu^n $(CH_2)_6OBu^t$ 78% (199)

$$\text{MeCH=CHBr} + Me_3SiCH_2MgCl \xrightarrow{[NiCl_2(dppp)]} \text{MeCH=CHCH}_2SiMe_3 \quad 80\% \tag{200}$$

$$CH_2{=}C(Cl)CH{=}CH_2 + Me_3SiCH_2MgCl \xrightarrow{[NiCl_2(dppp)]} CH_2{=}C(CH_2SiMe_3)CH{=}CH_2 \quad 91\% \tag{201}$$

$$\text{1-bromocyclooctene} + (Pr^iO)_2MeSiCH_2MgCl \xrightarrow{[NiCl_2(dppp)]} \text{1-}(CH_2SiMe(OPr^i)_2)\text{cyclooctene} \quad 88\% \tag{202}$$

optically active alkenyl halides couple with Grignard reagents also with retention of configuration[274,328]; the reaction may be useful for the synthesis of optically active liquid crystals (equation 203)[274].

MeOCH2–cyclohexylidene–CHBr (R)-(–) + XMg–CH2–cyclohexyl–propyl $\xrightarrow{[NiCl_2(dppe)]}$ MeOCH2–cyclohexylidene–CH(H)–CH2–cyclohexyl–propyl (R)-(–) 90% (203)

Coupling of 1, 2-dihaloethylenes, however, proceeds non-stereospecifically; whether the (*Z*)- or (*E*)-dihalide is used, almost the same *E*/*Z* ratio is obtained, depending on the nature of the phosphine ligand (equations 204 and 205)[291,329].

$$ClCH{=}CHCl + PhMgBr \xrightarrow{[cat.]} PhCH{=}CHPh \tag{204}$$

[NiCl2(dppp)] (*E*:*Z* 70:30)
[NiCl2(dmpe)] (*E*:*Z* 1:99)
90%

$$ClCH{=}CHCl + Me_3SiC{\equiv}CMgBr \xrightarrow{[NiCl_2(dppp)]} Me_3SiC{\equiv}CCH{=}CHC{\equiv}CSiMe_3 \quad 70\% \tag{205}$$

Alkenyl Grignard reagents are virtually unreactive with $[NiCl_2(dppp)]$ as the catalyst, but smoothly undergo coupling in the presence of $[NiCl_2(dmpe)]$[295]. The stereoselectivity of the coupling of 1-alkenyl Grignard reagents is dependent on the nature of the organic halides and the halide to Grignard molar ratio, since a nickel-catalysed *E/Z* isomerization of the Grignard reagent competes with the coupling (equation 206)[361].

Br Me MgBr [NiCl₂(dmpe)] Me (206)

(*Z* 96%) (*Z* 95%)

87%

Haloacetylenes also undergo coupling readily to give disubstituted acetylenes (equation 207)[292].

$$Et_3SiC{\equiv}CBr + PhMgBr \xrightarrow{[NiCl_2(dppp)]} Et_3SiC{\equiv}CPh \qquad (207)$$

76%

Grignard cross-coupling reactions can be applied to heterocyclic halides, such as furan (equation 208)[240], thiophene (equation 209)[203,248,300], pyridine (equation 210)[164,234,237,293,300,301], quinoline (equation 211)[300,307], isoquinoline (equation 212)[238,300], pyrimidine (equation 213)[357], pyridazine (equation 214)[220], and purine derivatives of biological interest (equation 215)[23,24].

Br Me MgX [NiCl₂(dppe)] Me (208)

93%

Br S Ph + MeO MgBr [NiCl₂(dppp)] MeO S Ph

87%

(209)

Cl N Cl + BrMg(CH₂)₈CH(Me)CH₂MgBr [NiCl₂(dppp)] N Me

muscopyridine

20%

(210)

2-chloroquinoline + cyclohexyl-MgBr $\xrightarrow{[NiCl_2(dppe)]}$ 2-cyclohexylquinoline 90% (211)

MeO, MeO, N, Cl + MgBr, Cl $\xrightarrow{[NiCl_2(dppp)]}$ MeO, MeO, N, Cl 92% (212)

Cl, N, Me, N, Cl + EtMgBr $\xrightarrow{[NiCl_2(dppp)]}$ Et, N, Me, N, Et 85% (213)

N, N=N, Cl + S, MgBr $\xrightarrow{[NiCl_2(dppe)]}$ N, N=N, S 65% (214)

Cl, N, N, N, N, R'O, O, R'O R'O + $PhCH_2CH_2MgBr$ $\xrightarrow{[NiCl_2(dppp)]}$ CH_2CH_2Ph, N, N, N, N, R'O, O, R'O R'O 50% (215)

The coupling reaction can be extended to a new type of polycondensation of aromatic and heterocyclic polyhalides to form polyphenylenes[345], poly(2, 5-thienylene) (equation 216)[197,348,354], poly(2, 4-thienylene)[254,352,354], poly(2, 5-pyrrolylene)[181], and poly(2, 5-selenienylene)[25]. The most interesting polymer is poly(2, 5-thienylene), which exhibits high electrical conductivity when doped with iodine[348].

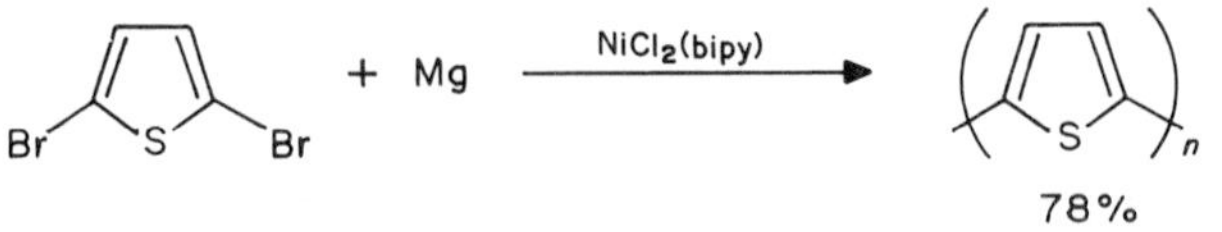

(216)

Alkyl group isomerization from secondary to primary is observed in the coupling of secondary Grignard reagents with organic halides (equation 217). The extent of the isomerization is strongly dependent on the electronic nature of both the phosphine ligands in the catalyst[290] and the halides employed[183]. Thus, the electron-releasing nature of phosphine ligands (dmpe *vs.* dppe or dppp) and the substituents on the aromatic ring favours the isomerization. It has also been pointed out that the isomerization may parallel the tendency of bidentate ligands bonded in a unidentated fashion[56]. With [$NiCl_2$(dmpe)]

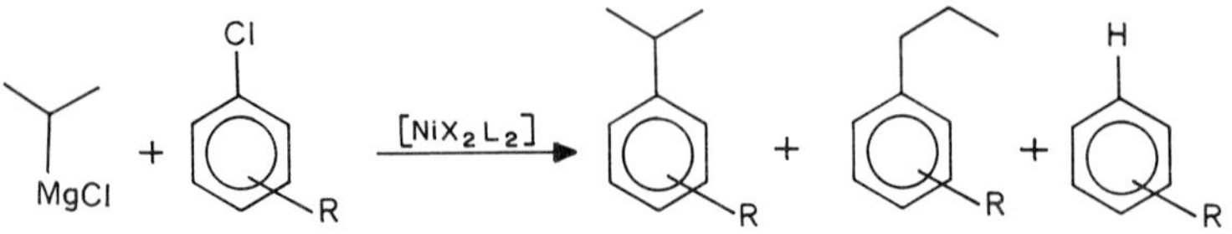

(217)

as a catalyst the isomerization easily occurs even along four carbon—carbon single bonds (equation 218)[182]. Tertiary to primary isomerization is also induced by this catalyst[182], whereas no isomerization occurs with [$NiCl_2$(dppf)] (equation 219)[110]. A mechanism for

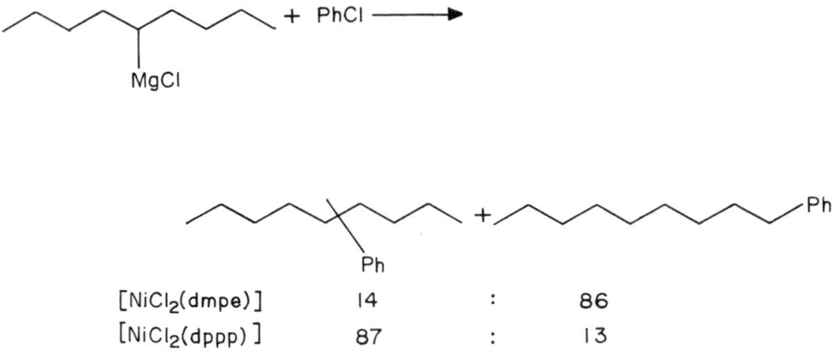

[$NiCl_2$(dmpe)]	14	:	86
[$NiCl_2$(dppp)]	87	:	13

(218)

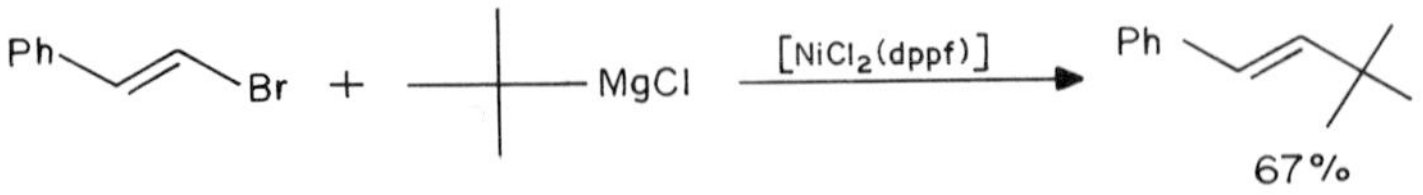

(219)

this alkyl group isomerization involves the formation of a (hydrido)(olefin)nickel intermediate arising from β-elimination of a secondary alkylnickel intermediate[183,290].

Whereas the primary to secondary isomerization is not observed for coupling of simple primary alkyl Grignard reagents, coupling of but-3-enyl bromide with the phenyl Grignard reagent is accompanied by such an alkyl group isomerization (equation 220)[362].

$$\text{CH}_2{=}\text{CHCH}_2\text{CH}_2\text{Br} + \text{PhMgBr} \xrightarrow{[\text{NiCl}_2(\text{dppp})]} \text{CH}_2{=}\text{CHCH(Ph)CH}_3 \; (48\%) + \text{CH}_3\text{CH}{=}\text{CHCH}_2\text{CH}_2\text{Ph} \; (15\%) \tag{220}$$

Nickel-catalysed coupling of isoprenylmagnesium with allylic chlorides may be useful for isoprenoid synthesis (equation 221)[176]. Propargyl chloride is alkylated at the γ-position to give an allene derivative (equation 222)[231].

$$(\text{-Mg-CH}_2\text{CH}{=}\text{C(Me)CH}_2\text{-})_n + \text{ClCH}_2\text{C}({=}\text{CH}_2)\text{CH}{=}\text{CH}_2 \xrightarrow[\text{(ii) } \text{H}_2\text{O}]{\text{(i) } [\text{NiCl}_2(\text{dppe})]} \text{product} \; (70\%) \tag{221}$$

$$\text{HC}{\equiv}\text{CCMe}_2\text{Cl} + \text{Bu}^n\text{MgBr} \xrightarrow{\text{NiBr}_2} \text{Bu}^n\text{CH}{=}\text{C}{=}\text{CMe}_2 \; (90\%) \tag{222}$$

b. Cross-coupling of organic halides with other organometallics (equations 155–160, 163, 164, 166, and 169)

Certain organolithium, organozinc, organoaluminium and organozirconium reagents also undergo coupling with C(sp^2)—halides. Lithium enolates couple with aryl and alkenyl halides in the presence of $NiBr_2$ without any phosphine ligands. Optimal yields are obtained with a stoichiometric amount of $NiBr_2$, but the reaction proceeds catalytically in nickel, forming the coupling product up to a 350% yield based on nickel bromide (equation 223)[201]. The coupling is stereospecific with respect to alkenyl halides (equation 224). Selective γ-arylation or vinylation is also a significant feature of this coupling (equation 225). The reaction is useful for the preparation of allylsilanes with an ester group at the α-position (equation 226)[6].

$$\underset{1}{\text{MeCH}{=}\text{CHBr}} + \underset{1}{\text{LiCH}_2\text{CO}_2\text{Bu}^t} + \underset{0.2}{\text{NiBr}_2} + \underset{0.2}{\text{BuLi}} \longrightarrow \underset{0.7}{\text{MeCH}{=}\text{CHCH}_2\text{CO}_2\text{Bu}^t} \tag{223}$$

$$(Z)\text{-PhCH}{=}\text{CHBr} + \text{LiCMe}_2\text{CO}_2\text{Et} \xrightarrow[\text{BuLi}]{\text{NiBr}_2} (Z)\text{-PhCH}{=}\text{CHCMe}_2\text{CO}_2\text{Et} \tag{224}$$

$$\text{PhI} + \text{LiCH}_2\text{CH}{=}\text{CHCO}_2\text{Et} \xrightarrow[\text{BuLi}]{\text{NiBr}_2} \text{PhCH}_2\text{CH}{=}\text{CHCO}_2\text{Et} \tag{225}$$

$$\text{CH}_2{=}\text{C(Me)Br} + \text{Li(SiMe}_3\text{)CHCO}_2\text{Et} \xrightarrow[\text{Bu}^n\text{Li}]{\text{NiBr}_2} \text{CH}_2{=}\text{C(Me)CH(SiMe}_3\text{)CO}_2\text{Et} \; (74\%) \tag{226}$$

Couplings with aryl- and benzyl-zinc[195,214] and Reformatsky reagents[84] are catalysed by Ni(0) complexes. The advantages of organozinc reagents over organomagnesium reagents reside in their compatibility with certain functional groups such as cyano, ester, and imino groups (equations 227 and 228)[195,214]. Hmpa is an essential co-solvent for the coupling of

$$NC-C_6H_4-Br + PhCH_2ZnBr \xrightarrow{[Ni(PPh_3)_4]} NC-C_6H_4-CH_2Ph \quad 92\% \tag{227}$$

$$\text{2-I-3,4,5-(MeO)}_3C_6H\text{-CH=NC}_6H_{11} + \text{3,4-methylenedioxyphenyl-ZnCl} \xrightarrow[\text{(ii) } H_3O^+]{\text{(i) } [Ni(PPh_3)_4]} \text{biaryl-CHO (MeO, MeO, OMe)} \quad (\rightarrow \text{steganone}) \tag{228}$$

the Reformatsky reagent (equation 229)[84]. Alternatively, arylzinc reagents couple with an α-bromoacetate to form the similar coupling products (equation 230)[186].

$$\text{1-bromonaphthalene} + BrZnCH_2CO_2Et \xrightarrow[(MeO)_2CH_2/\text{hmpa}]{[Ni(PPh_3)_4]} \text{1-}(CH_2CO_2Et)\text{naphthalene} \quad 69\% \tag{229}$$

$$\text{2-thienyl-ZnCl} + BrCH_2CO_2Et \xrightarrow{[Ni(acac)_2]/PPh_3} \text{2-thienyl-}CH_2CO_2Et \quad 60\% \tag{230}$$

(*E*)-Alkenyl-aluminium[11,213] and -zirconium[215] reagents, being obtainable by hydro-alumination and -zirconation of acetylenes (equation 231), couple also with aryl and alkenyl halides containing certain functional groups, with retention of configuration (equations 232–234). In the coupling of the alkenylzirconium reagents, a high catalytic ability of ligand-free Ni(I) species generated *in situ* from Ni(acac)$_2$/dibah (1:1) has been claimed[71]. Organoaluminium reagents also couple with alkynyl bromides in the presence of [Ni(mesal)$_2$] (equation 166)[103].

$$RC\equiv CH + M-H \longrightarrow \underset{H}{\overset{R}{>}}C=C\underset{m}{\overset{H}{<}} \qquad M = AlBu^i_2, ZrCp_2Cl \tag{231}$$

$$Bu^n\text{CH=CH}AlBu^i_2 + NC\text{-}C_6H_4\text{-}Br \xrightarrow{[Ni(PPh_3)_4]} Bu^n\text{CH=CH}C_6H_4CN\text{-}p \quad 64\% \qquad (232)$$

$$n\text{-}C_5H_{11}\text{CH=CH}AlBu^i_2 + Bu^n\text{CH=CH}I \xrightarrow{[Ni(PPh_3)_4]} Bu^n\text{CH=CH-CH=CH}C_5H_{11}\text{-}n \quad 70\% \qquad (233)$$

$$thpO\text{CH}_2\text{CH}_2\text{CH=CH}ZrCp_2Cl + PhI \xrightarrow{[Ni(PPh_3)_4]} thpO\text{CH}_2\text{CH}_2\text{CH=CH}Ph \quad 84\% \qquad (234)$$

Tetramethyltin reacts with aryl iodides under carbon monoxide pressure in the presence of nickel catalysts to give aryl methyl ketones selectively (equation 235)[303]. It may be mentioned that Grignard coupling under the carbon monoxide atmosphere gives a mixture of ketones and tertiary alcohols[343].

$$MeO_2C\text{-}C_6H_4\text{-}I + CO + Me_4Sn \xrightarrow{[Ni(CO)_2(PPh_3)_2]} MeO_2C\text{-}C_6H_4\text{-}C(O)Me \quad 95\% \qquad (235)$$

c. Grignard cross-coupling of C(sp²)—O, —S, —Se, and —Te compounds (equations 171, 175, 179–181, 183–187, 189, and 190)

Aryl ethers (equation 236)[331,336], aryl phosphates (equations 237 and 238)[115], alkenyl ethers (equation 239)[331,336], enol silyl ethers (equation 240)[111], and enol phosphates (equation 241)[113,250] react with Grignard reagents in the presence of nickel catalysts to form cross-coupling products through cleavage of $C(sp^2)$—oxygen bonds. Benzene is used as a superior solvent in many cases. While aryl ethers are inert to alkyl Grignard reagents, aryl phosphates couple with any kind of Grignard reagents and even with alkenylaluminium compounds (equation 238)[115]. A ketene dimer also undergoes similar

$$1\text{-}MeO\text{-}C_{10}H_7 + PhMgBr \xrightarrow[\text{benzene}]{[NiCl_2(PPh_3)_2]} 1\text{-}Ph\text{-}C_{10}H_7 \quad 70\% \qquad (236)$$

$$1\text{-}(EtO)_2P(O)O\text{-}C_{10}H_7 + Bu^nMgBr \xrightarrow{[NiCl_2(dppp)]} 1\text{-}Bu^n\text{-}C_{10}H_7 \quad 82\% \qquad (237)$$

$OP(O)(OEt)_2$ (1-naphthyl) + Bu^nCH=CH$AlBu^i_2$ $\xrightarrow{[Ni(acac)_2]}$ 1-(Bu^nCH=CH)naphthalene 99% (238)

(2,3-dihydrofuran) + PhMgBr $\xrightarrow{[NiCl_2(PPh_3)_2]}$ HO…Ph (239)

termite trail pheromone 71%

OSiMe (cyclohexenyl) + Me_3SiCH_2MgCl $\xrightarrow{[Ni(acac)_2]}$ ($SiMe_3$CH$_2$)cyclohexene 71% (240)

$OP(O)(OEt)_2$ Ph… + Me_3SiCH_2MgCl $\xrightarrow{NiBr_2}$ Ph…$SiMe_3$ 70% (241)

coupling reactions with Grignard, organozinc, and alkenylaluminium reagents (equations 242 and 243)[1,2,166].

(diketene) =O + Me_3SiCH_2MgCl $\xrightarrow{NiCl_2}$ Me_3Si…CO_2H 95% (242)

(diketene) =O + $Pr^nC{\equiv}CZnCl$ $\xrightarrow{[Ni(PPh_3)_4]}$ $Pr^nC{\equiv}C$…CO_2H 60% (243)

Similar Grignard coupling reactions are observed with a variety of C(sp^2)—sulphur compounds, such as aryl thiols[330], sulphides[221,239,330,332], sulphoxides[330], and sulphones[330] (equation 244), heteroaryl sulphides (equations 245 and 246)[239,287], and

$$PhX + MeMgBr \xrightarrow{[NiCl_2(PPh_3)_2]} PhMe \quad (244)$$

X = SH 64%; SMe 97%; SOPh 77%; SO_2Me 97%

$$\text{2-(MeS)benzothiazole} + Bu^nMgBr \xrightarrow{[NiCl_2(dppp)]} \text{2-}Bu^n\text{-benzothiazole} \quad (245)$$

93%

$$\text{2-mercaptopyridine} + Ph(CH_2)_3MgBr \xrightarrow{[NiCl_2(dppp)]} \text{2-}(CH_2)_3Ph\text{-pyridine} \quad (246)$$

94%

alkenyl sulphides[221,288,330,334] and sulphones[79]. Coupling reactions of both alkenyl sulphides and sulphones proceed with retention of configuration[221,330] and are applied to the stereocontrolled synthesis of olefins (equation 247)[277,288,311,332,334] and 1, 3-dienes (equation 248)[101,154,324] (see also the next section). The stereoselectivity in the sulphone cases depends on the halides associated with the Grignard reagents in the order Cl > Br > I (equation 249)[79]. Aryl and alkenyl selenides[223] and tellurides[320,321] also

$$\text{2-PyS–C(Me)=CH}(CH_2)_2Ph + Bu^nMgBr \xrightarrow{[NiCl_2(dppp)]} Bu^n\text{C(Me)=CH}(CH_2)_2Ph \quad (247)$$

79%

$$\text{C}_6\text{H}_{11}\text{CH=C(SPh)CH=CH}_2 + MeMgBr \xrightarrow{[NiCl_2(PPh_3)_2]} \text{C}_6\text{H}_{11}\text{CH=C(Me)CH=CH}_2 \quad (248)$$

70%

$$Bu^tSO_2\text{-diene} + MeMgX \xrightarrow{[Ni(acac)_2]} \text{product} \quad (249)$$

X=Cl 71% (Z 98%)
X=I 80% (Z 75%)

undergo coupling reactions (equation 250). The reactivity order is PhSeMe > > PhCl > PhSMe[223].

$$PhX\text{CH=CHPh} + PhMgBr \xrightarrow{[NiCl_2(PPh_3)_2]} Ph\text{CH=CHPh} + Ph_2 \quad (250)$$

X = Se	89%	100%
X = Te	60%	92%

Furan, thiophene, selenophene, and tellurophene and their derivatives undergo ring-opening coupling reactions with Grignard reagents in benzene to form (Z,Z)-1, 3-dienes with retention of configuration, with a few exceptions (equations 251 and 252)[330,335]. The reactivity increases in the above order from furan to tellurophene.

$$\text{C}_4\text{H}_4\text{E} + \text{PhMgBr} \xrightarrow{[\text{NiCl}_2(\text{PPh}_3)_2]} \text{PhCH=CHCH=CHPh} \tag{251}$$

E = O 49% (E,E 100%)
E = Te 88% (Z,Z 100%)

$$\text{C}_4\text{H}_4\text{S} + \text{Bu}^n\text{MgBr} \xrightarrow{[\text{NiCl}_2(\text{dppp})]} \text{Bu}^n\text{CH=CHCH=CHBu}^n \tag{252}$$

58%

It may be also mentioned here that carboxylic acids react with a large excess amount of Grignard reagent in the presence of a nickel catalyst to form unsymmetrical ketones (equation 253)[93].

$$\text{PhCOOH} + n\text{-C}_5\text{H}_{11}\text{MgX} \xrightarrow{[\text{NiCl}_2(\text{dppe})]} \text{PhCC}_5\text{H}_{11}\text{-}n \ (\text{C=O}) \tag{253}$$

75%

d. Grignard coupling of allylic O, S, and Se compounds and related reactions (equations 176–178, 182, and 188)

Allyl alcohols[33,49,53,54,89–91] and ethers[116,333,334] such as $OSiR_3$, Othp, or OPh, undergo nickel-catalysed coupling reactions with Grignard reagents. While $[NiCl_2(PPh_3)_2]$ is an effective catalyst for Grignard reagents having no β-hydrogens, $[NiCl_2(dppp)]$ or $[NiCl_2(dppe)]$ should be used for β-hydrogen-containing alkyl Grignard reagents to avoid the reduction of allylic alcohols[49,90]. A mechanism proposed for the Grignard coupling of allylic alcohols involves an organonickel–magnesium species via the oxidative-addition of a Grignard reagent to nickel(0) species and a η^3-allylnickel formation (Scheme 6)[33,49], being different from that proposed for the Grignard coupling of C(sp^2)-halides mentioned above (Scheme 5).

Although the coupling reaction occurs at both of the α- and γ-positions to form all possible regio and olefin-geometrical isomers, the regioselectivity is independent of the nature of the leaving group[116], but dependent on the structure of the allyl moiety[33,49] and the ligand on nickel[116]. There seems to be a tendency for the coupling to occur preferentially at the more highly substituted allylic position, especially when $[NiCl_2(dppf)]$ is used as a catalyst (equation 254)[116]. Allylic acetals form double alkylation products (equation 255)[333].

$$\text{MeCH=CHCH}_2\text{OSiMe}_3 + \text{PhMgBr} \xrightarrow{[\text{NiCl}_2(\text{dppf})]} \text{MeCH=CHCH}_2\text{Ph} + \text{CH}_2\text{=CHCH(Me)Ph} \tag{254}$$

ca. 20 : 80

100%

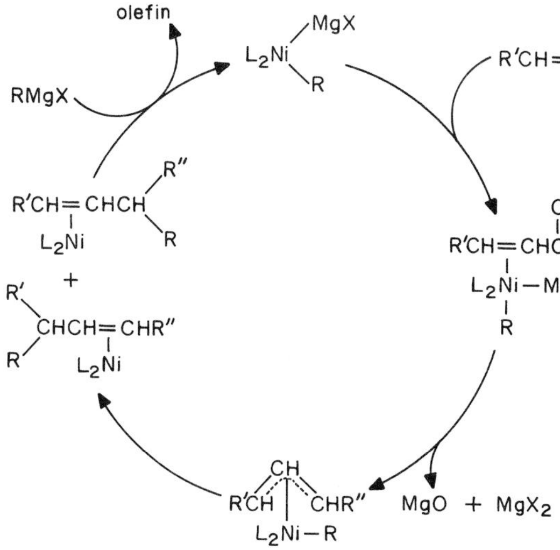

SCHEME 6

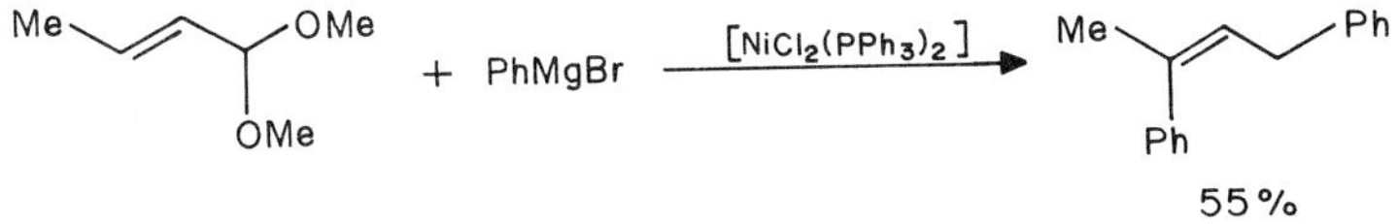

(255)

Couplings of allyl alcohols proceed with overall inversion of stereochemistry at a chiral carbon atom regardless of acyclic or cyclic systems (equations 256 and 257)[54,91]. The

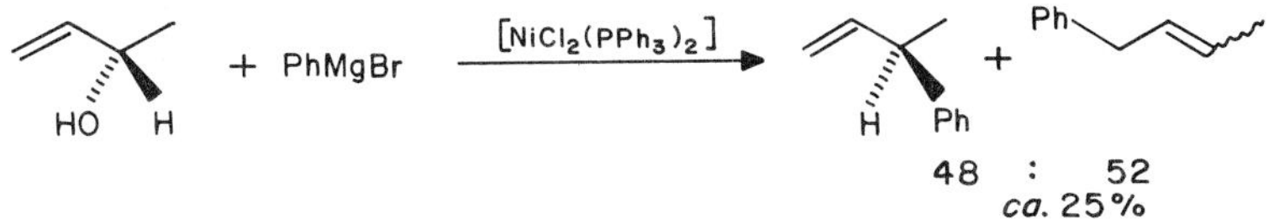

(256)

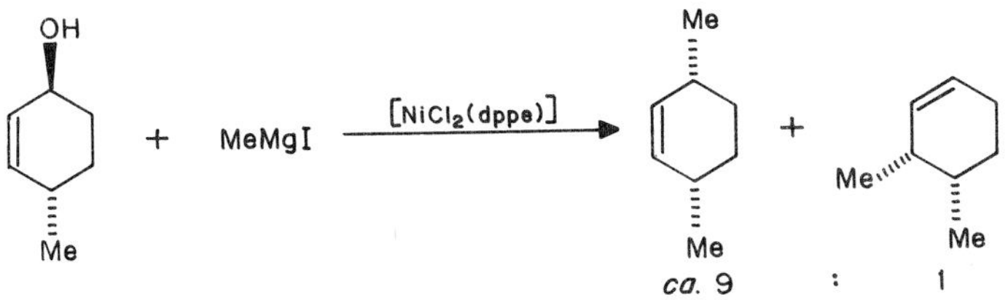

(257)

allylic coupling has been applied to the synthesis of terpenoids such as Δ^7-pimaradiene and hibaene (equation 258)[33].

+ MeMgBr $\xrightarrow{[NiCl_2(PPh_3)_2]}$ Δ^7-pimaradiene (258)

Cross-coupling of allylic acetates with stable enolates is also catalysed by low-valent nickel–phosphine complexes (equation 259)[69]. A similar reaction can also be achieved intramolecularly (equation 260)[318]. Allylic thiols[334], sulphides[222,334], selenides[223], and

AcO + NaCHXY $\xrightarrow{[NiCl_2(dppe)]}$ 97 : 3, 91%

X = CO_2Et
X = SO_2Ph

(259)

$\xrightarrow{[Ni\{P(OEt)_3\}_4]}$ 98% (260)

amines[222] as well as chlorides undergo similar coupling reactions (equation 261). Allyl—S and —Se bonds are more reactive than the C(sp^2)—S and —Se bonds (equation 262)[334].

Ph–CH=CH–CH$_2$SMe + $Ph(CH_2)_3MgBr$ $\xrightarrow{[NiCl_2(dppp)]}$ Ph–CH=CH–$(CH_2)_4Ph$ 83% (261)

CH$_2$=CH–CH$_2$SPh + PhMgBr (1 equiv.) $\xrightarrow{[NiCl_2(dppp)]}$ CH$_2$=CH–CH$_2$Ph (no Ph_2) 90% (262)

e. Polyfunctional compounds

Chemoselective coupling of polyfunctional compounds is of current synthetic interest. Nickel-catalysed Grignard coupling reactions of aromatic polyhalides, in contrast to similar palladium-catalysed coupling reactions[202], tend to result in the complete alkylation of all the halogen atoms present[202,263,295]. In the presence of [Ni(acac)$_2$] at low temperatures, however, a highly selective monoalkylation of 1, 3, 5-trichlorobenzene can be achieved; the subsequent Grignard coupling reaction under the usual conditions leads to the formation of trialkylbenzenes with different alkyl groups (equation 263)[76]. Monoalkylation of dichloroarenes is catalyzed by $[Ni(triphos)]PF_6$[244a].

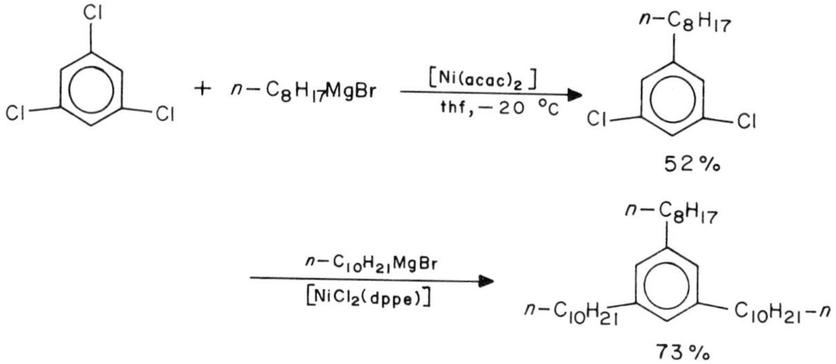

(263)

Monoalkylation of (*E*)- and (*Z*)-1, 2-dichloroethylene by Grignard reagents[74,242] or alkenylaluminium reagents[244] forms alk-1-enyl chloride selectively (equation 264).

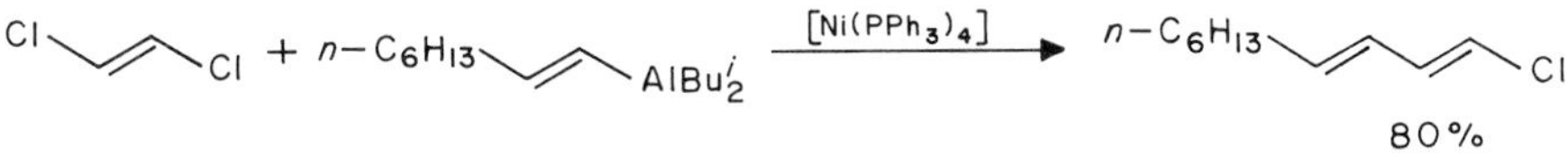

(264)

Certain substituents can either activate or deactivate specifically one adjacent functional group in polyfunctional compounds. Thus, an oxazolinyl group activates the *ortho*-chlorine atom in aromatic dichlorides, a stepwise Grignard coupling being possible

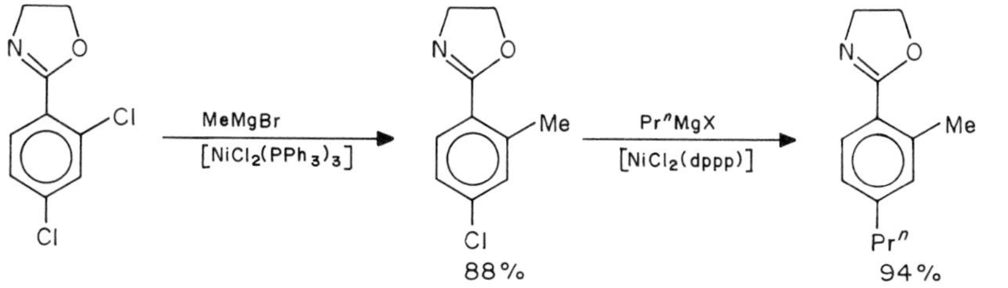

(265)

(equation 265)[241]. In the Grignard coupling reaction of ketene dithioacetals, one sulphide group is deactivated by the *cis* substituent (equation 266)[332]. A chemoselective

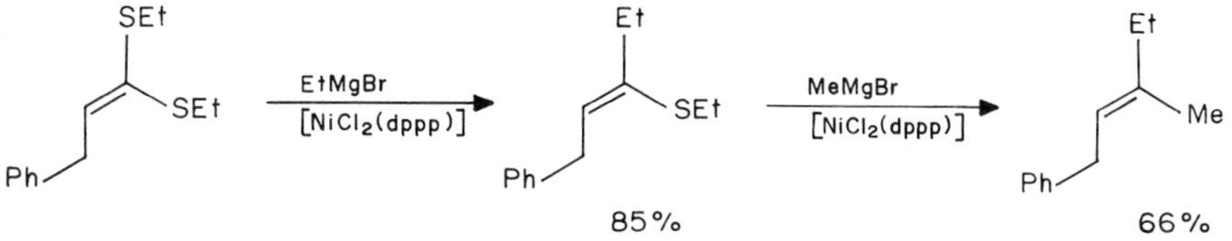

(266)

monoalkylation of two unlike leaving groups, e.g. Br > Cl > SR, is also useful for aromatic[308] and olefinic[92] systems (equations 267 and 268).

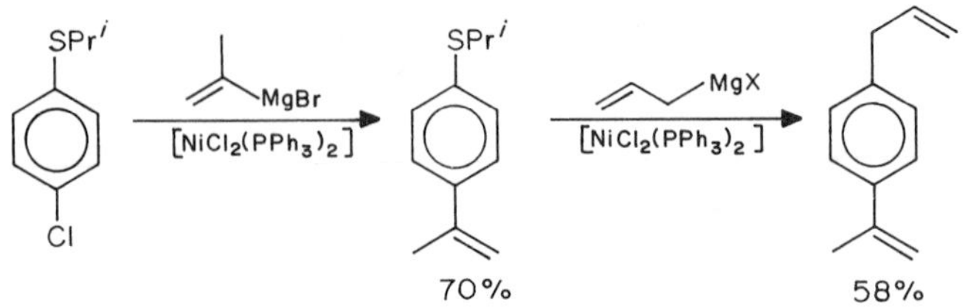

(267)

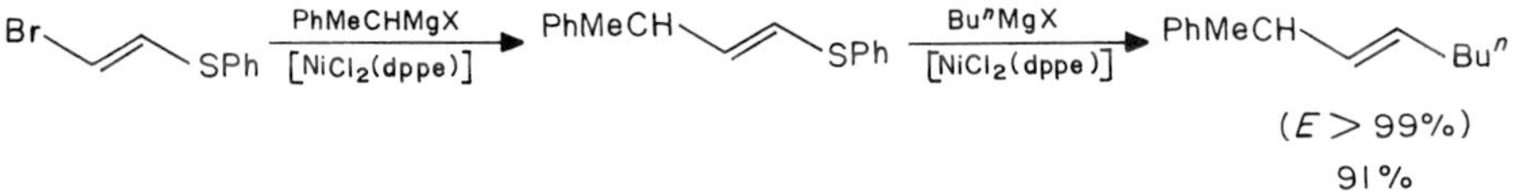

(268)

The Grignard coupling reactions of α-bromoenol ethers generally stop cleanly at the monocoupling stage to provide new routes to α-substituted carbonyl derivatives

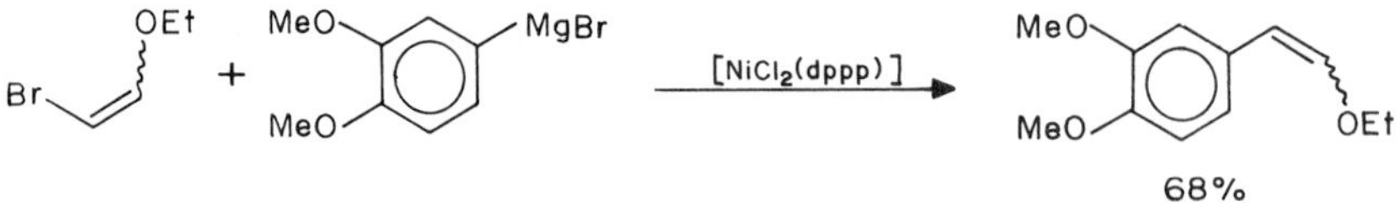

(269)

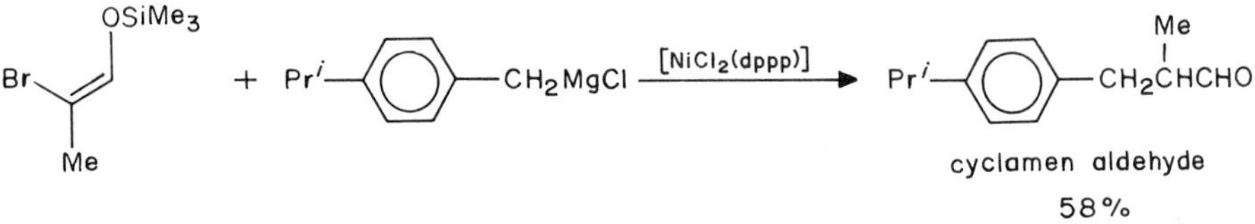

(270)

(equations 269 and 270)[296,297]. An allylic ether is more reactive than alkenyl sulphides (equation 271)[277].

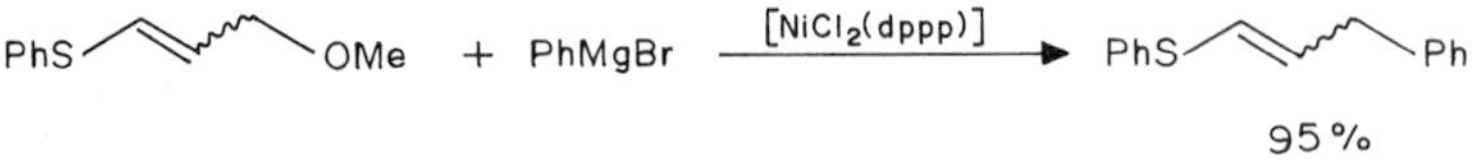

(271)

2. *Asymmetric Grignard cross-coupling reactions*

Catalytic asymmetric carbon—carbon bond formation can be achieved using optically active phosphine ligands in the nickel catalysts. Three types of asymmetric coupling reactions have been reported. The first is an asymmetric coupling of secondary alkyl Grignard reagents with aryl and alkenyl halides through a kinetic resolution of racemic *sec*-alkyl Grignard reagents, a chiral centre being present on the organometallic side. The second is an asymmetric coupling of allylic ethers and homoallyl halides with achiral Grignard reagents through the formation of η^3-allyl-nickel intermediates, a chiral centre being induced on the electrophilic substrates. The third is an asymmetric synthesis of biaryl atropisomers. Several reviews have already appeared[109,118,235].

a. Asymmetric coupling of sec-alkyl Grignard reagents

The type of asymmetric coupling reaction shown in Scheme 7 has been most extensively studied. Various types of chiral ligands have been examined for the coupling of (1-phenylethyl)magnesium chloride with vinyl bromide (equation 272)[112,117,119,121]. Representative results are listed in Table 3, where the enantiomeric excess, the configuration of the coupling product, and references are shown beside the chiral ligand employed.

$$\mathrm{PhMeCHMgCl} + \mathrm{CH_2{=}CHBr} \xrightarrow{[\mathrm{NiL*}]} \mathrm{PhMe\overset{*}{C}HCH{=}CH_2} \qquad (272)$$

Results obtained with a series of ferrocenylphosphines and aminoalkylphosphines indicate that the amino group on the phosphine ligand is the first requisite for high stereoselectivity and that the surroundings around the nitrogen atom, rather than the phosphorus atom, exert a strong effect on the stereoselectivity, even reversal of the preferred configuration being observed. The methoxy group also has nearly the same efficiency as the dimethylamino group. The important role of the amino or alkoxy group may be visualized by its strong ability to coordinate with the magnesium atom in the Grignard reagent as shown in Scheme 8.

Whereas (1-phenylethyl)magnesium chloride also couples with other alkenyl bromides to form optically active products of higher than 50% ee (equation 273)[119], asymmetric coupling of 2-butylmagnesium chloride with vinyl bromide in the presence of the same chiral ligand gives only moderate optical purity (equation 274)[119]. Asymmetric coupling

PhMeCHMgCl + PhCH=CHBr —(*S*)-(*R*)-ppfa/NiCl$_2$→ PhCH=CH–C(Me)(H)Ph (273)

52% ee (*R*)

2-butyl-MgCl + CH$_2$=CHBr —(*S*)-(*R*)-ppfa/NiCl$_2$→ EtC(Me)(H)CH=CH$_2$ (274)

30% ee (*R*)

of 2-butylmagnesium halides with aromatic halides has also been studied in detail by use of several chiral homologues of dppe, such as prophos and norphos (see Table 3), as ligands[56]. Optical yields up to 50% ee have been obtained with little influence of the substituents on the chiral ligand, but with great influence of the nature of the halides in both the Grignard reagent and the organic halide to such an extent that the configuration of the product can be reversed (equation 275)[56].

2-butyl-MgX + PhX′ —(−)-norphos/NiCl$_2$→ EtC(Me)(H)Ph (275)

X = X′ = Cl 26.7% ee (*R*)
X = X′ = Br 50.7% ee (*R*)

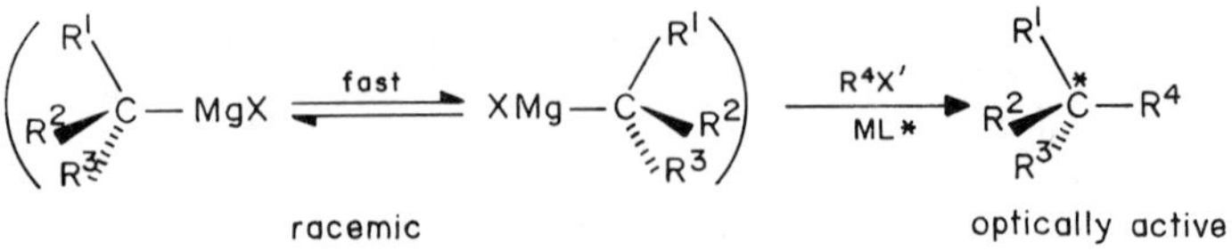

racemic → optically active

SCHEME 7

TABLE 3. Chiral ligands for asymmetric coupling reaction (equation 272)

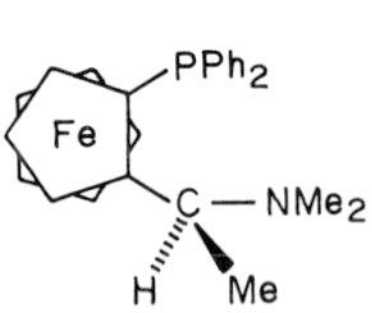

(S)-(R)-ppfa
59% ee (R)[119]

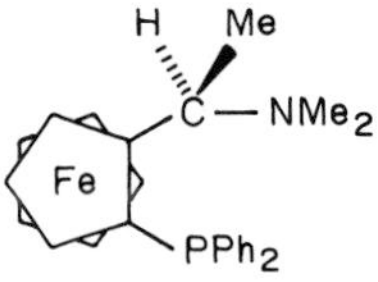

(R)-(S)-ppfa
63% ee (S)[119]

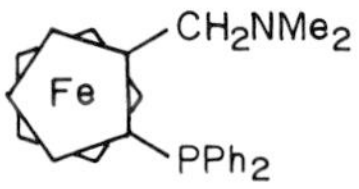

(S)-FcPN
60% ee (S)[119]

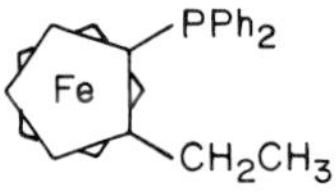

(R)-ppef
4% ee (S)[119]

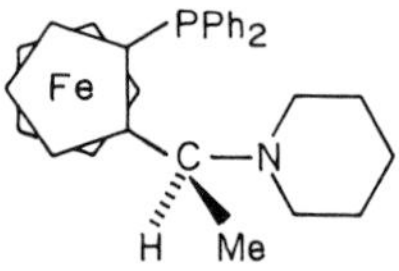

(S)-(R)
42% ee (S)[119]

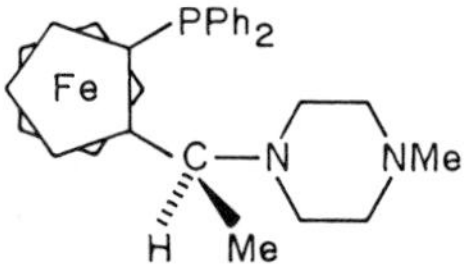

(S)-(R)
65% ee (R)[119]

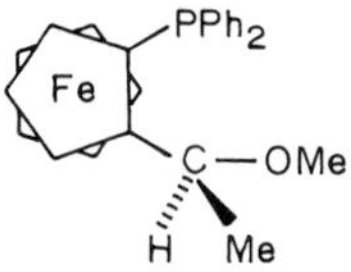

(S)-(R)
57% ee (R)[119]

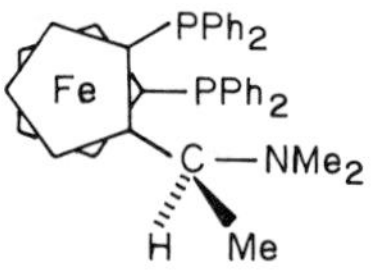

(S)-(R)-bppfa
65% ee (R)[119]

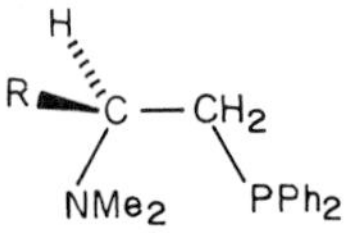

R = Me : (S)-alaphos[121]
37% ee (S)
R = Pr^i : (S)-valphos
81% ee (S)
R = $PhCH_2$: (S)-phephos
71% ee (S)

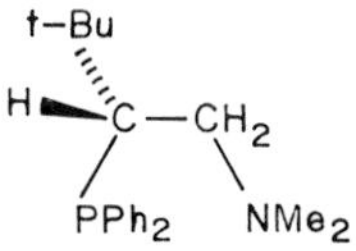

tert-leuphos
94% ee (R)[121]

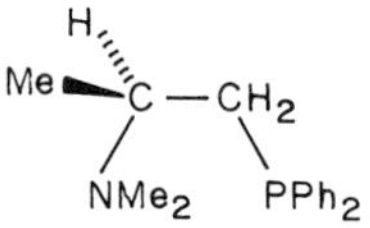

25% ee (S)[121]

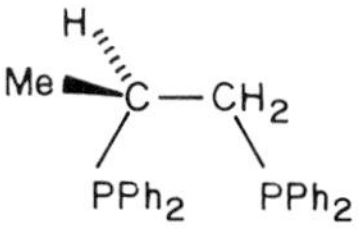

(S)-prophos
0% ee[121]

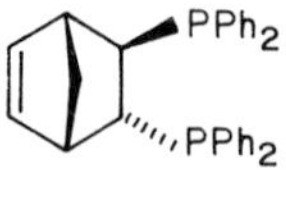

(−)-(R,R)-norphos
67% ee (S)[32]

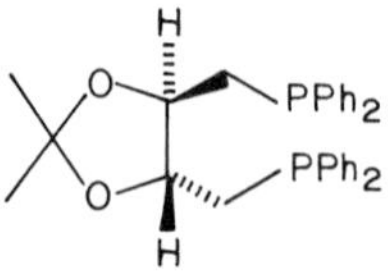

(−)-diop
<16% ee (R)[52,184]

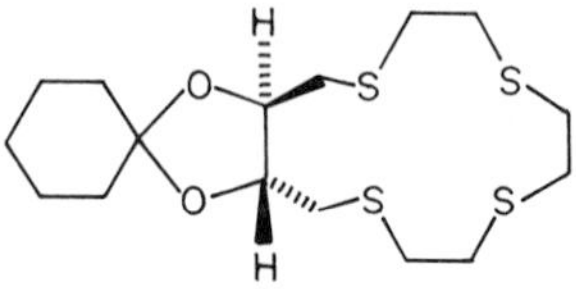

<17% ee (S)[196]

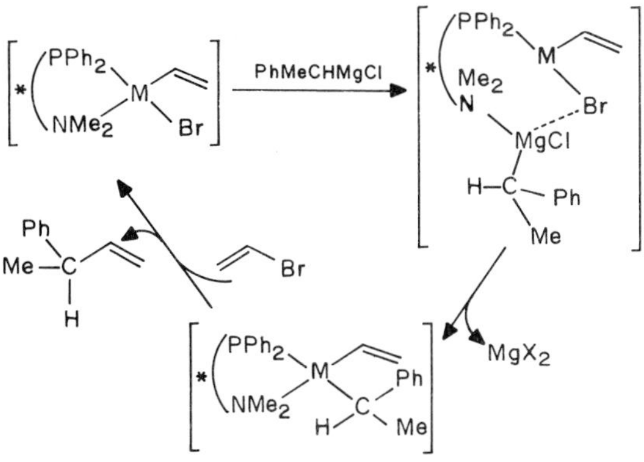

SCHEME 8

Grignard reagents which do not undergo racemization are kinetically resolved by asymmetric coupling in the presence of a chiral catalyst, the absolute configuration being opposite to that of the coupling product (equation 276)[114]. The nickel-catalysed

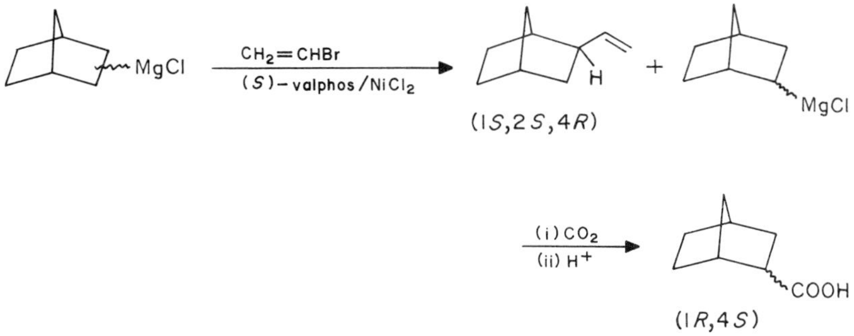

$CH_2{=}CHBr$, (S)-valphos/$NiCl_2$ → (1S,2S,4R) + MgCl; (i) CO_2 (ii) H^+ → COOH (1R,4S)

(276)

asymmetric cross-coupling between *sec*-alkyl Grignard reagents with vinyl bromide finds many applications in the synthesis of optically and biologically active substances, e.g. α-curcumene (equation 227)[298] and 2-arylpropionic acids (anti-inflammatory drugs) (equation 278)[118].

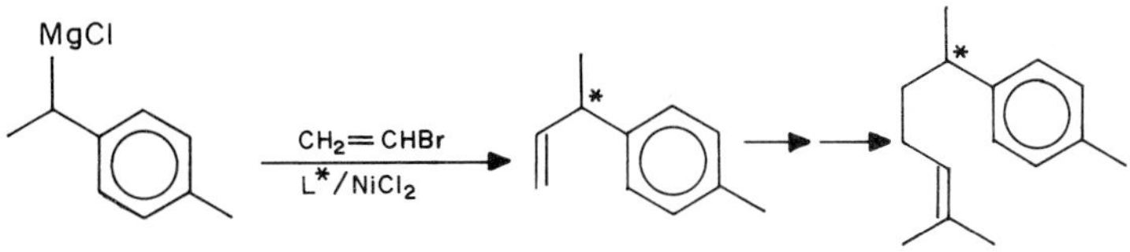

MgCl; $CH_2{=}CHBr$, $L^*/NiCl_2$

(277)

L^* = (S)-(R)-ppfa: 66% ee (R)

L^* = (S)-valphos: 83% ee (S)

Ar = p-Bu^{i}C$_6$H$_4$: 81% ee (R)
Ar = p-PhC$_6$H$_4$: 82% ee (R) (278)

b. Asymmetric Grignard coupling with allylic and homoallylic substrates

Allylic alcohols (equation 279)[46] and ethers (equations 280 and 281)[55] undergo asymmetric Grignard cross-coupling reactions in the presence of chiral phosphinenickel catalysts. Two regioisomeric allylic ethers give rise to the same chiral product through the formation of common η^3-allylnickel intermediates (equation 280). Asymmetric coupling of a cyclic allylic ether using a ligand having a C_2 axis, (−)-(S)-(S)-chiraphos, has afforded the highest optical yield ever reported for the catalysed carbon—carbon bond-forming reactions (equation 281).

+ MeMgBr → [NiCl$_2${(−)-diop}] (279)

14.9% ee (R)

+ PhMgBr → [NiCl$_2${(S)-(S)-chiraphos}]

58% ee (R)

(280)

+ EtMgBr → [NiCl$_2${(S)-(S)-chiraphos}]

97.7% ee (R)

(−)-(S)-(S)-chiraphos (281)

Another type of asymmetric coupling reaction has been observed in the reaction of 4-bromobut-1-ene with the phenyl Grignard reagent via alkyl group isomerization (equation 282)[362] (see also Section III.E, equation 220). The product is the same as that obtained via the alternative route shown in equation 272, but has the opposite configuration when the same chiral ligand is used (cf. Table 3).

+ PhMgBr → (R)-(S)-bppfa/NiCl$_2$ (282)

33% ee (R)

c. Asymmetric synthesis of biaryl atropisomers

Cross-coupling reaction of (2-methyl-1-naphthyl)magnesium bromide with 2-methyl-1-naphthyl bromide proceeds in the presence of chiral phosphinenickel catalysts to give an optically active binaphthyl atropisomer of up to 12.5% ee[294,299].

d. Aspects of chiral secondary alkylnickel complexes

A primary alkyl-nickelacycle undergoes ring contraction in the presence of a chiral phosphine ligand, (*S*)-(*S*)-chiraphos, to form a kinetically controlled diastereomeric mixture of chiral secondary alkyl-nickelacycle enriched in the *R*-isomer, which undergoes thermal epimerization to give a thermodynamically equilibrated mixture enriched in the *S*-isomer of 54% diastereomeric excess (equation 283)[255]. This is the first observation of dynamic behaviour of chiral alkylnickel species by n.m.r. and should be helpful for an insight into the machanism of nickel-catalysed asymmetric synthesis.

(bipy)Ni + (*S*)-(*S*)-chiraphos → *R*-isomer ⇌ *S*-isomer (283)

3. Nickel-catalysed conjugate addition of organometallics

Conjugate addition of organo-aluminium[10,13,14,169,258], -zinc[106,233] and -zirconium[71,198,259] reagents to α,β-enones is catalysed by [Ni(acac)$_2$] alone or by its 1:1 mixture with dibah.

Conjugate addition of alkyl groups has been achieved with trimethylaluminium (equation 284)[10] or organozinc reagents prepared *in situ* from alkyl halides, zinc bromide, and lithium under ultrasonic irradiation[233]. The latter allows the introduction of even tertiary alkyl groups (equation 285)[233], whereas tri(isobutyl)aluminium leads to the reduction of enones[37]. Arylzinc reagents also undergo conjugate addition, as exemplified by the synthesis of β-cuparenone (equation 286)[106]. Prostaglandin skeletons have been

+ Me_3Al —[Ni(acac)$_2$]→ 86% (284)

+ $(Bu^t)_2Zn$ —[Ni(acac)$_2$]→ 78% (285)

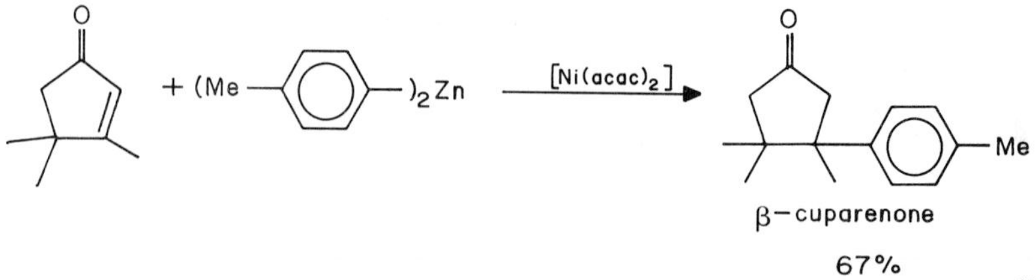

(286)

synthesized by conjugate addition of alkenylzirconium reagents (equation 287)[259]. The nickel-catalysed conjugate addition of alkynylalanes provides the first procedure which

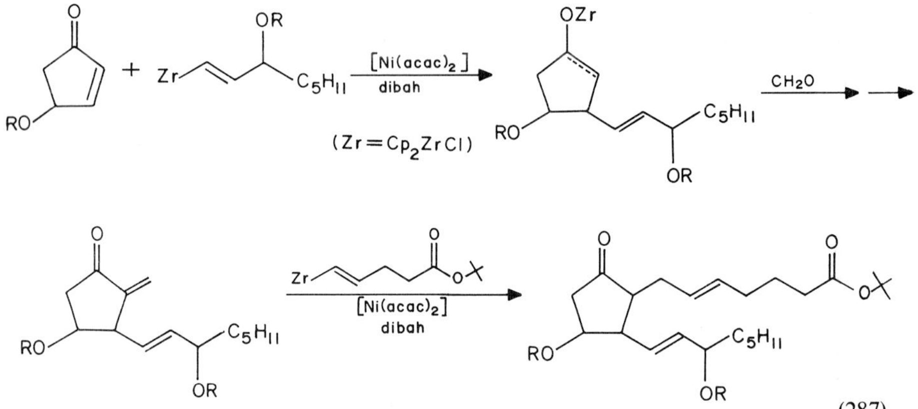

(287)

permits the introduction of terminal alkynyl group to ordinary *s-trans*-enones (equation 288)[258]. These nickel-catalysed procedures have been claimed to be much superior in many cases[106,258] to the more widely used, traditional copper-induced

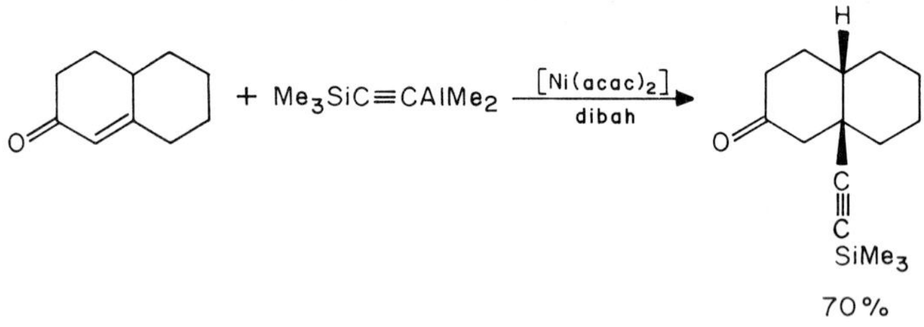

(288)

conjugate addition. Mechanistic studies show that the real active species may be a Ni(I) species which transfers an electron to enones as the first step of the addition[71], [$Ni(PPh_3)_4$] being inactive. More detailed discussion will be made in Section III.F.

A similar nickel-catalysed reaction of trimethylaluminium with cyclopropyl ketones causes the ring opening (equation 289)[15]. In connection with these reactions, it may be mentioned here that trimethylaluminium undergoes nickel-catalysed addition to ketones and nitriles[3,12,170]. Finally, [$Ni(acac)_2$] also catalyses conjugate addition of neutral acetylacetone to enones (equation 290)[217].

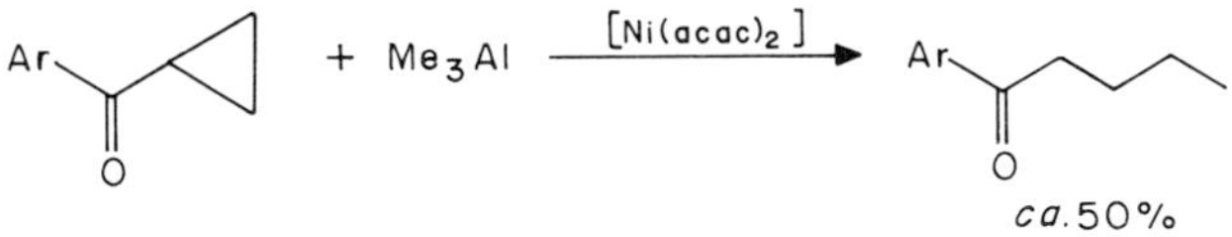

(289)

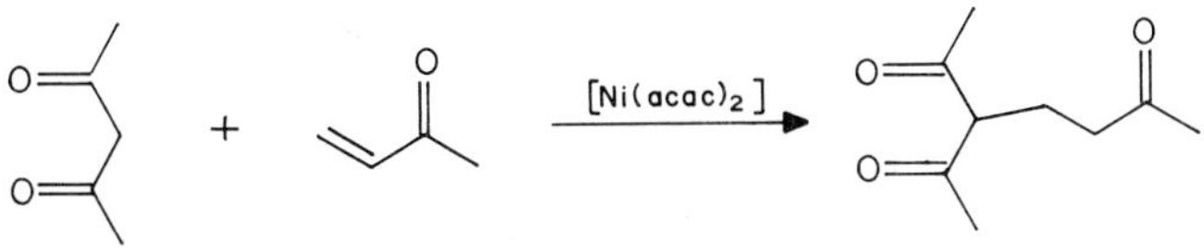

(290)

4. Carbon—heteroatom bond-forming reactions

The following nickel-catalysed reactions will be described in this section: (a) Grignard cross-coupling reactions with hydro-silanes and germanes and (b) cross-coupling of organic halides with heteronucleophiles of Groups IV–VII, which include silyl anions, amines, phosphines, alcohols, thiols, halides, and cyanides as pseudohalides.

a. Grignard cross-coupling with hydro-silanes and -germanes

Hydro-silanes and -germanes couple with Grignard reagents having no β-hydrogens in the presence of $[NiCl_2(PPh_3)_2]$ or $[Ni(acac)_2]$[39,51,62]. The reaction has been considered to proceed via the oxidative addition of the Si—H or Ge—H bond to low-valent nickel species[51]. The stereochemistry at silicon is retained (equation 291)[62]. In the coupling with

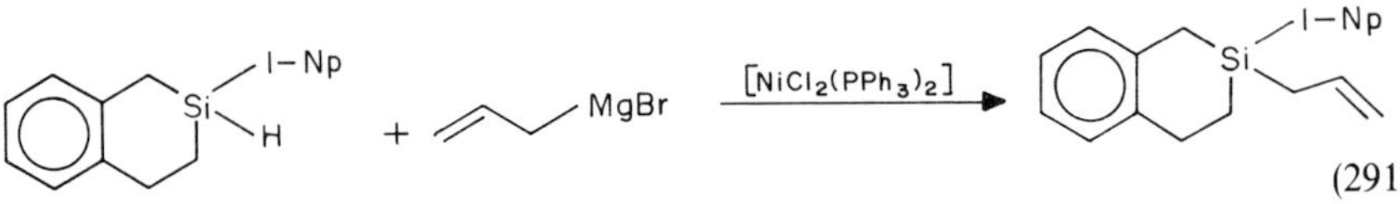

(291)

the crotyl Grignard reagent, a highly stereo- and regio-selective formation of (*E*)-crotylsilanes is attained with $[NiCl_2(dppf)]$ as catalyst (equation 292)[122].

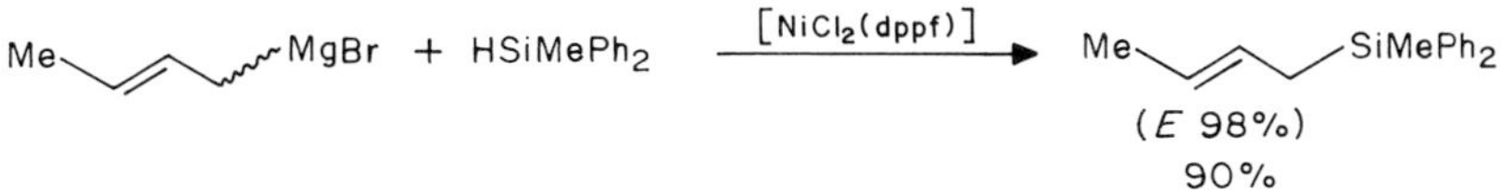

(292)

b. Cross-coupling of organic halides with heteronucleophiles

Aryl and alkenyl halides couple with tris(trimethylsilyl)aluminium in the presence of $[NiCl_2(PPh_3)_2]$ in dioxane to give the corresponding aryl- and alkenyl-trimethylsilanes

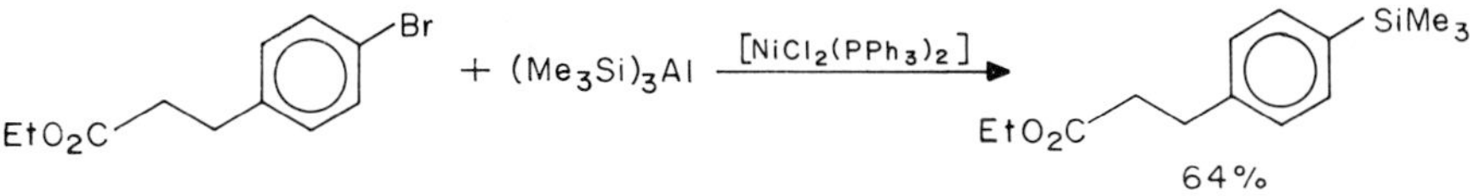

(293)

(equation 293)[310]. Substitution of aryl halides by amines occurs in the presence of nickel salts or Ni(0) complexes, but under forcing conditions (equation 294)[67].

$$\text{PhBr} + \text{pyrrolidine (N–H)} \xrightarrow[160\,^{\circ}\text{C}]{[\text{Ni(dppe)(CO)}_2]} \text{N-phenylpyrrolidine (N–Ph)}\quad 85\% \tag{294}$$

Quaternary arylphosphonium salts are formed by a nickel-catalysed reaction between tertiary phosphines and aryl halides (equation 295)[7,43,316]. While *ortho*-substituents usually inhibit the reaction, certain coordinating groups can activate the *ortho*-halide specifically (equation 296)[7]. A similar nickel-catalysed Arbuzov reaction is observed when dialkyl phosphonites or trialkyl phosphites are used (equation 297)[20–22,306]. In both cases, electron-withdrawing substituents on aryl halides retard the reaction[316]. It has been

$$\text{Me}_2\text{N–C}_6\text{H}_4\text{–Br} + \text{PPh}_3 \xrightarrow[\text{EtOH}]{[\text{Ni(PPh}_3)_4]} \text{Me}_2\text{N–C}_6\text{H}_4\text{–PPh}_3{}^{+}\text{Br}^{-}\quad 86\% \tag{295}$$

$$\text{2,5-dibromo-4'-(NMe}_2\text{)azobenzene (Br, Br; N=N; NMe}_2\text{)} + \text{PBu}_3 \xrightarrow[\text{EtOH}]{\text{NiBr}_2} \text{(Br; PBu}_3{}^{+}\text{Br}^{-}\text{; N=N; NMe}_2\text{)} \tag{296}$$

$$\text{2-bromonaphthalene (Br)} + \text{P(OEt)}_3 \xrightarrow[160\,^{\circ}\text{C}]{\text{NiCl}_2} \text{2-naphthyl–P(O)(OEt)}_2\quad 84\% \tag{297}$$

shown that Ni(II) salts are reduced by $P(OEt)_3$ to a Ni(0) complex, $[Ni\{P(OEt)_3\}_4]$, under the reaction conditions[22]. The catalytic mechanism is not clear yet, but an important role of Ni(I) species has been discussed[316]. It should be mentioned here that the phosphonium salt formation is the accompanying process in the biaryl synthesis (equation 298)[316].

$$[\text{ArNiBr(PEt}_3)_2] + 3\text{ArBr} \longrightarrow \text{ArAr} + (\text{ArPEt}_3{}^{+})_2[\text{NiBr}_4]^{2-} \tag{298}$$

Substitution reactions of aryl and alkenyl halides by thiolate anions[97] are also catalysed by nickel salts or nickel–phosphine complexes (equation 299)[97]. It may be mentioned here that conversion of alkyl, especially tertiary alkyl, chlorides into the corresponding ethers is also catalysed by nickel salts (equation 300)[35,36,358].

$$\text{PhCH=CHBr} + \text{PhSH} \xrightarrow[\text{NaOH/H}_2\text{O}]{[\text{NiCl(1-Np)(PPh}_3)_2]} \text{PhCH=CHSPh}\quad 75\% \tag{299}$$

$$\text{3,4-(MeO)}_2C_6H_3OH + ClCH_2CH_2CMe_2Cl \xrightarrow[125\ ^\circ C]{[Ni(acac)_2]} \text{6,7-dimethoxy-2,2-dimethylchroman}\ (67\%) \tag{300}$$

Transformation of aromatic halides to nitriles by treatment with potassium or sodium cyanide is catalysed by nickel–phosphine complexes, most conveniently in the presence of phase-transfer catalysts (equation 301)[41,44]. A similar cyanation of alkenyl halides cannot be induced by this method, but is catalysed by low-valent nickel species generated *in situ* from $[NiBr_2(PPh_3)_2]/Zn/PPh_3$/dmf (equation 302)[253]. Although the reaction proceeds with retention of configuration with *E*-isomers, considerable loss of stereochemistry is encountered with *Z*-isomers. Halogen exchange with aryl and alkenyl halides is catalysed

$$p\text{-MeCOC}_6H_4Cl + NaCN \xrightarrow[(C_8H_{17})_3NMeCl/H_2O]{[NiCl(1\text{-}Np)(PPh_3)_2]} p\text{-MeCOC}_6H_4CN\ (96\%) \tag{301}$$

$$\text{1-chlorocyclohexene} + KCN \xrightarrow[\text{dmf}]{[NiBr_2(PPh_3)_2]/Zn} \text{1-cyanocyclohexene}\ (92\%) \tag{302}$$

by various types of nickel salts or complexes[278,280,281,283,317]. The exchange can attain equilibrium (equation 303)[317].

$$PhI + Bu_4NBr \underset{K \gtrsim 3}{\overset{\text{Ni cat.}}{\rightleftharpoons}} PhBr + Bu_4NI \tag{303}$$

Conversion of aryl and alkenyl bromides into the iodides can be achieved by *in situ* generated Ni(0) species from $NiBr_2$ and zinc in the presence of potassium iodide (equation 304)[278,280,281], but this process is accompanied by the aryl–aryl homocoupling, as mentioned in Section III.A; the homocoupling contamination can be eliminated by a Ni(II)-catalysed reaction at elevated temperatures in the absence of zinc (equation 305)[280].

$$p\text{-ClC}_6H_4Br + KI \xrightarrow[\text{hmpa}/50\ ^\circ C]{NiBr_2/Zn} p\text{-ClC}_6H_4I\ (81\%) + (p\text{-ClC}_6H_4)_2\ (9\%) \tag{304}$$

$$PhBr + KI \xrightarrow[\text{dmf}/150\ ^\circ C]{[NiBr_2(PBu_3)_2]} PhI\ (88\%) \tag{305}$$

$$p\text{-MeOC}_6H_4Br + PhI \overset{\text{Ni cat.}}{\rightleftharpoons} p\text{-MeOC}_6H_4I + PhBr \tag{306}$$

$$[(ArX')NiXL_2] \rightleftharpoons [Ar\langle^{X'}_{X}\rangle NiL_2] \rightleftharpoons [(ArX)NiX'L_2] \tag{307}$$

The nickel catalysis is also effective for redistribution of halogens between two different aryl groups (equation 306)[317]. A suggested mechanism for these halogen exchange reactions involves a Ni(I)-assisted four-centre process (equation 307)[317].

F. Mechanistic Considerations

Of the variety of coupling reactions in the foregoing sections, those which are represented by equations 308–311 are most typical nickel-mediated reactions. Here we consider them altogether from mechanistic standpoints. Detailed mechanistic studies

Type 1:

$$2\ RX + Ni(0) \longrightarrow RR + NiX_2 \tag{308}$$

Type 3:

$$[(\pi\text{-allyl})NiX]_2 + RX \longrightarrow CH_2{=}CHCH_2R + NiX_2 \tag{309}$$

Type 5:

$$R{-}M + R'X \xrightarrow{[NiX_2L_2]} RR' + M{-}R \tag{310}$$

$$R{-}M + CH_2{=}CHCOCH_3 \xrightarrow[\text{dibah}]{[Ni(acac)_2]} RCH_2CH_2COCH_3 \tag{311}$$

mainly by Kochi and coworkers[187,188,208,315,316] Hegedus and coworkers[127,130], and Dayrit and Schwartz[71] have suggested that these seemingly completely different reactions may be explained by a common mechanism which involves Ni(I)⇌Ni(III) radical chain processes. The generalized catalytic cycles are shown in Schemes 9 and 10. Each of these reactions will be discussed briefly.

The Type 1 reaction (equation 308) may proceed in two steps: the oxidative addition of an organic halide to a Ni(0) species (equation 312) and the cross-coupling reaction of the resulting organonickel halide with an organic halide (equation 313). Thus, the crucial carbon—carbon bond-forming step in the Type 1 reaction amounts to the Type 3 coupling reaction.

$$RX + Ni^{0} \longrightarrow RNi^{II}X \tag{312}$$

$$RNi^{II}X + RX \longrightarrow RR + Ni^{II}X_2 \tag{313}$$

The Type 3 reaction may in turn be analysed as follows. In a typical case, an arylnickel(II) halide, e.g. $[ArNiBr(PEt_3)_2]$, gives no biaryl by itself, but undergoes an induced coupling on treatment with an aryl bromide to form the coupled biaryl[316]. This coupling reaction

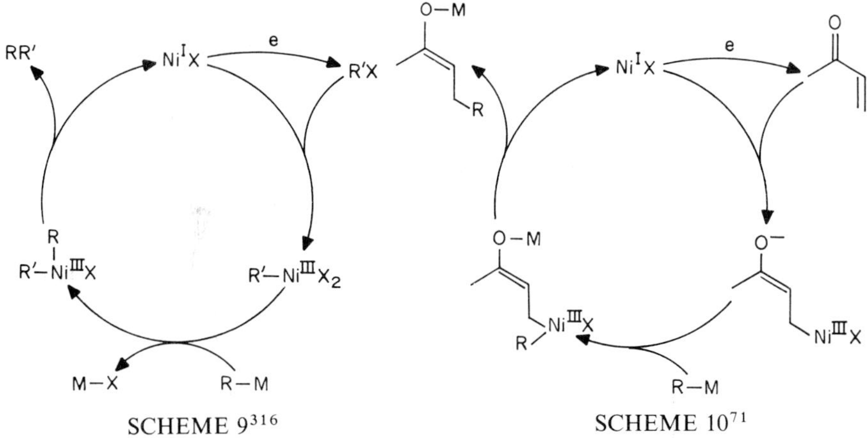

SCHEME 9[316] SCHEME 10[71]

has an induction period and is inhibited by nitroaromatics or other effective suppressors of chain reactions involving radical-anion intermediates[316]. A similar inhibition has also been observed in the cross-coupling of η^3-allylnickel halides with organic halides (equation 309)[127]. The observed inhibition may be associated with the destruction of paramagnetic Ni(I) and Ni(III) species via one-electron oxidation or reduction, respectively, to form inactive Ni(II) species. The radical chain process may be initiated by an electron transfer from an organonickel(II) halide to an organic halide. The catalytic cycle shown in Scheme 9 represents the propagation steps. The cycle involves the oxidative addition of an organic halide to Ni(I), possibly via one-electron transfer, to produce an organonickel(III) halide which undergoes transmetallation with an organonickel(II) halide to form a diorganonickel(III) species, from which the coupling product is released by reductive elimination to regenerate the Ni(I) species, where R-M represents $[ArNiXL_2]$ or $[(\eta^3\text{-allyl})NiX]_2$.

The Type 5 reactions, apparently nickel-catalysed reactions, may also proceed through the same catalytic cycle[187], where R-M represents RMgX, RZnX, etc. Also in conjugate addition, Ni(I) species generated from $[Ni(acac)_2]$/dibah have been shown to be the real catalytically active species, a similar radical chain processes being shown in Scheme 10[71].

It should be emphasized here that the radical chain process in Scheme 9 is not necessarily applicable to any kinds of substrate. For example, in the η^3-allylnickel coupling reactions, alkyl halides afford racemic coupling products (cf. equation 84), indicative of the intervention of free alkyl radicals. In those cases alternative radical chain processes should be considered, as exemplified by steps shown in Scheme 11[127].

Initiation:

$$[(C_3H_7)NiX] + RX \longrightarrow [(C_3H_7)NiX]^{+\bullet} + RX^{-\bullet}$$

Propagation:

$$RX^{-\bullet} \longrightarrow R^{\bullet} + X^{-}$$

$$R^{\bullet} + [(C_3H_7)NiX] \longrightarrow RC_3H_7 + NiX^{\bullet}$$

$$NiX^{\bullet} + RX \longrightarrow RX^{-\bullet} + NiX^{+}$$

SCHEME 11

For the nickel-catalysed Grignard cross-coupling reaction of Type 5, an alternative mechanism has already been mentioned (Scheme 5). It may be noted that whereas in Scheme 9 an organic halide undergoes oxidative addition to a Ni(I) species *after* reductive elimination of a coupling product from a diorganonickel(III) species, in Scheme 5 an organic halide interacts with a diorganonickel(II) intermediate *to promote* reductive elimination, which may involve an electron transfer process, as shown already in Section II, Scheme 2. The difference, however, cannot be distinguished experimentally.

As shown by these examples, the mechanisms have not been fully delineated, but Schemes 9 and 10 are still helpful in understanding a rapidly growing list of nickel-induced coupling reactions.

IV. REFERENCES

1. Y. Abe, H. Goto, M. Sato, and T. Kato, *Chem. Pharm. Bull.*, **31**, 1108 (1983).
2. Y. Abe, M. Sato, H. Goto, R. Sugawara, E. Takahashi, and T. Kato, *Chem. Pharm. Bull.*, **31**, 4346 (1983).
3. D. Abenhaim, G. Boireau, C. Bernardon, A. Deberly, and C. Germain, *Tetrahedron Lett.*, 993 (1976).
4. B. Åkermark, H. Johansen, B. Roos, and U. Wahlgren, *J. Am. Chem. Soc.*, **101**, 5876 (1979).
5. B. Åkermark and A. Ljungqvist, *J. Organomet. Chem.*, **182**, 59 (1979).
6. P. Albaugh-Robertson and J. A. Katzenellenbogen, *J. Org. Chem.*, **48**, 5288 (1983).
7. D. W. Allen, I. W. Nowell, L. A. March, and B. F. Taylor, *Tetrahedron Lett.*, **23**, 5479 (1982).
8. H. Alper, M. Tanaka, and K. Hachem, *J. Organomet. Chem.*, **190**, 95 (1980).
9. N. Anton, N. Clos, and G. Muller, *J. Organomet. Chem.*, **267**, 213 (1984).
10. E. C. Ashby and G. Heinsohn, *J. Org. Chem.*, **39**, 3297 (1974).
11. S. Baba and E. Negishi, *J. Am. Chem. Soc.*, **98**, 6729 (1976).
12. L. Bagnell, E. A. Jeffery, A. Meisters, and T. Mole, *Aust. J. Chem.*, **27**, 2577 (1974).
13. L. Bagnell, E. A. Jeffery, A. Meisters, and T. Mole, *Aust. J. Chem.*, **28**, 801 (1975).
14. L. Bagnell, A. Meisters, and T. Mole, *Aust. J. Chem.*, **28**, 817 (1975).
15. L. Bagnell, A. Meisters, and T. Mole, *Aust. J. Chem.*, **28**, 821 (1975).
16. R. Baker, *Chem. Rev.*, **73**, 487 (1973).
17. R. Baker, B. N. Blackett, R. C. Cookson, R. C. Cross, and D. P. Madden *J. Chem. Soc., Chem. Commun.*, 343 (1972).
18. R. Baker, P. Bevan, and R. C. Cookson, *J. Chem. Soc. Chem. Commun.*, 752 (1975).
19. A. C. Balazs, K. H. Johnson, and G. M. Whitesides, *Inorg. Chem.*, **21**, 2162 (1982).
20. T. M. Balthazor, J. A. Miles, and B. R. Stults, *J. Org. Chem.*, **43**, 4538 (1978).
21. T. M. Balthazor, *J. Org. Chem.*, **45**, 2519 (1980).
22. T. M. Balthazor and R. C. Grabiak, *J. Org. Chem.*, **45**, 5425 (1980).
23. D. E. Bergstrom, *Nucleosides Nucleotides*, **1**, 1 (1982).
24. D. E. Bergstrom and P. Anantha Reday, *Tetrahedron Lett.*, **23**, 4191 (1982).
25. M. Beroari, P. Kovacic, S. Gronowitz, and A.-B. Hornfieldt, *J. Polym. Sci., Polym. Lett. Ed.*, **19**, 347 (1981).
26. C. Bianchini and A. Meli, *J. Chem. Soc., Chem. Commun.*, 1309 (1983).
27. P. Binger and M. J. Doyle, *J. Organomet. Chem.*, **162**, 195 (1978).
28. M. Boya, M. Moreno-Manas and M. Prior, *Tetrahedron Lett.*, 1727 (1975).
29. P. S. Braterman and R. J. Cross, *Chem. Soc. Rev.*, **2**, 271 (1973).
30. J. Browning, M. Green, J. L. Spencer, and F. G. A. Stone, *J. Chem. Soc., Dalton Trans.*, 97 (1974).
31. J.-J. Brunet, D. Besazzi, A. Courtois, and P. Caubere, *J. Am. Chem. Soc.*, **104**, 7030 (1982).
32. H. Brunner and M. Proebster, *J. Organomet. Chem.*, **209**, C1 (1981).
33. B. L. Buckwalter, I. R. Burfitt, H. Felkin, M. Joly-Goudket, K. Maemura, M. F. Salomon, E. Wenkert, and P. M. Wovkulick, *J. Am. Chem. Soc.*, **100**, 6445 (1978).
34. B. Büssemeier, P. W. Jolly, and G. Wilke, *J. Am. Chem. Soc.*, **96**, 4726 (1974).
35. F. Camps, J. Coll, A. Messeguer, M. A. Pericas, and S. Ricart, *Synthesis*, 126 (1979).
36. F. Camps, J. Coll, and J. M. Moretó, *Synthesis*, 187 (1982).
37. A. M. Caporusso, G. Giacomelli, and L. Lardicci, *J. Org. Chem.*, **47**, 4640 (1982).

38. E. Carmona, E. Gutierrez-Puebla, A. Monge, J. M. Marin, M. Paneque, and M. L. Poveda, *Organometallics*, **3**, 1438 (1984).
39. F. H. Carre and R. J. P. Corriu, *J. Organomet. Chem.*, **74**, 49 (1974).
40. J. Carvajal, G. Muller, J. Sales, X. Solans, and C. Miravitlles., *Organometallics*, **3**, 996 (1984).
41. L. Cassar, *J. Organomet. Chem.*, **54**, C57 (1973).
42. L. Cassar, G. P. Chiusoli, and F. Guerrier, *Synthesis*, 509 (1973).
43. L. Cassar and M. Foa, *J. Organomet. Chem.*, **74**, 75 (1974).
44. L. Cassar, M. Foa, F. Montanari, and G. P. Marinelli, *J. Organomet. Chem.*, **173**, 335 (1979).
45. P. Caubere, *Angew. Chem., Int. Ed. Engl.*, **22**, 599 (1983).
46. M. Cherest, H. Felkin, J. D. Umpleby, and S. G. Davies, *J. Chem. Soc., Chem. Commun.*, 681 (1981).
47. G. P. Chiusoli, M. Costa, G. Pecchini, and G. Cometti, *Transition Met. Chem.*, **2**, 270 (1977).
48. G. P. Chiusoli, G. Salerno, U. Bersellini, F. Dallatomasina, and S. Preseglio, *Transition Met. Chem.*, **3**, 174 (1978).
49. C. Chuit, H. Felkin, C. Frajerman, G. Roussi, and G. Swierczewski, *J. Organomet. Chem.*, **127**, 371 (1977).
50. R. L. Clough, R. Mison, and J. D. Roberts, *J. Org. Chem.*, **41**, 2252 (1976).
51. E. Colomer, R. J. P. Corriu, and B. Meunier, *J. Organomet. Chem.*, **71**, 197 (1974).
52. G. Consiglio and C. Bottegi, *Helv. Chim. Acta*, **56**, 460 (1973).
53. G. Consiglio, F. Morandini, and O. Piccolo, *Helv. Chim. Acta*, **63**, 987 (1980).
54. G. Consiglio, F. Morandini, and O. Piccolo, *J. Am. Chem. Soc.*, **103**, 1846 (1981).
55. G. Consiglio, F. Morandini, and O. Piccolo, *J. Chem. Soc., Chem. Commun.*, 112 (1983).
56. G. Consiglio, F. Morandini, and O. Piccolo, *Tetrahedron*, **39**, 2699 (1983).
57. E. J. Corey and M. F. Semmelhack, *J. Am. Chem. Soc.*, **89**, 2735 (1967).
58. E. J. Corey and L. S. Hegedus, *J. Am. Chem. Soc.*, **91**, 1233 (1969).
59. E. J. Corey and L. S. Hegedus, *J. Am. Chem. Soc.*, **91**, 4926 (1969).
60. E. J. Corey and H. A. Kirst, *J. Am. Chem. Soc.*, **94**, 667 (1972).
61. R. J. P. Corriu and J. P. Masse, *J. Chem. Soc., Chem. Commun.*, 144 (1972).
62. R. J. P. Corriu, J. P. Masse, and B. Meunier, *J. Organomet. Chem.*, **55**, 73 (1973).
63. F. A. Cotton, B. A. Frenz, and D. L. Hunter, *J. Am. Chem. Soc.*, **96**, 4820 (1974).
64. D. J. Cram, G. M. Lein, T. Kaneda, R. C. Helgeson, C. B. Knobler, E. Maverick, and K. N. Trueblood, *J. Am. Chem. Soc.*, **103**, 6228 (1981).
65. D. J. Cram, M. deGrandpre, C. P. Knobler, and K. N. Trueblood, *J. Am. Chem. Soc.*, **106**, 3286 (1984).
66. D. J. Cram, I. B. Dicker, M. Lauer, C. B. Knobler, and K. N. Trueblood *J. Am. Chem. Soc.*, **106**, 7150 (1984).
67. R. Cramer and D. R. Coulson, *J. Org. Chem.*, **40**, 2267 (1975).
68. E. A. Cuellar and T. Marks, *Inorg. Chem.*, **20**, 3766 (1981).
69. T. Cuvigny and M. Julia, *J. Organomet. Chem.*, **250**, C21 (1983).
70. W. G. Dauben, G. H. Beasley, M. D. Broadhurst, B. Muller, D. J. Peppard, P. Pesnele, and C. Suter, *J. Am. Chem. Soc.*, **97**, 4993 (1975).
71. R. M. Dayrit and J. Schwartz, *J. Am. Chem. Soc.*, **103**, 4466 (1981).
72. R. Del Rosario and L. S. Stuhl, *Tetrahedron Lett.*, **23**, 3999 (1982).
73. R. J. dePasquale, *J. Chem. Soc., Chem. Commun.*, 157 (1973).
74. C. Descoins, M. Lettere, G. Linstrumelle, D. Michelot, and V. Ratovelomanana, *Synth. Commun.*, **14**, 761 (1984).
75. J. G. Duboudin, B. Jousseaume, and A. Bonakbar, *J. Organomet. Chem.*, **168**, 227 (1979).
76. K. C. Eapen, S. S. Dua, and C. Tamborski, *J. Org. Chem.*, **49**, 478 (1984).
77. J. J. Eisch and K. R. Im, *J. Organomet. Chem.*, **139**, C51 (1977).
78. J. J. Eisch, *Pure Appl. Chem.*, **56**, 35 (1984).
79. J.-L. Fabre, M. Julia, and J.-N. Verpeaux, *Tetrahedron Lett.*, **23**, 2469 (1982).
80. D. R. Fahey, *J. Am. Chem. Soc.*, **92**, 402 (1970).
81. D. R. Fahey and J. E. Mahan, *J. Am. Chem. Soc.*, **98**, 4499 (1976).
82. D. R. Fahey and J. E. Mahan, *J. Am. Chem. Soc.*, **99**, 2501 (1977).
83. J. F. Fauvarque and A. Jutand, *J. Organomet. Chem.*, **132**, C17 (1977).
84. J. F. Fauvarque and A. Jutand, *J. Organomet. Chem.*, **209**, 109 (1981).
85. J. F. Fauvarque, C. Chevret, A. Jutand, M. Francois, and J. Perichon *J. Organomet. Chem.*, **264**, 273 (1984).

86. G. Favero, A. Morvillo, and T. Turco, *J. Organomet. Chem.*, **162**, 99 (1978).
87. G. Favero, A. Morvillo, and A. Turco, *J. Organomet. Chem.*, **241**, 251 (1983).
88. G. Favero and A. Morvillo, *J. Organomet. Chem.*, **260**, 363 (1984).
89. H. Felkin and G. Swierczewski, *Tetrahedron*, **31**, 2735 (1975).
90. H. Felkin, E. Jampel-Costa, and G. Swierczewski, *J. Organomet. Chem.*, **134**, 265 (1977).
91. H. Felkin, M. Joly-Goudket, and S. G. Davies, *Tetrahedron Lett.*, **22**, 1157 (1981).
92. V. Fiandanese, G. Marchese, F. Naso, and L. Ronzini, *J. Chem. Soc., Chem. Commun.*, 647 (1982).
93. V. Fiandanese, G. Marchese, and Ronzini, *Tetrahedron Lett.*, **24**, 3677 (1983).
94. E. O. Fischer and G. Burger, *Chem. Ber.*, **94**, 2409 (1961).
95. T. C. Flood and A. Sarhangi, *Tetrahedron Lett.*, 3861 (1977).
96. M. Foa and L. Cassar, *J. Chem. Soc., Dalton Trans.*, 2572 (1975).
97. M. Foa, R. Santi and F. Garavaglia, *J. Organomet. Chem.*, **206**, C29 (1981).
98. Y. Fujioka, *Bull. Chem. Soc. Jpn.*, **57**, 3494 (1984).
99. S. Fukuoka, M. Ryan, and S. Tsutsumi, *J. Org. Chem.*, **36**, 2721 (1971).
100. J. Furukawa, A. Matsumura, Y. Matsuda, and J. Kiji, *Bull. Chem. Soc. Jpn.*, **49**, 829 (1976).
101. K. Furuta, Y. Ikeda, N. Meguriya, N. Ikeda, and H. Yamamoto, *Bull. Chem. Soc. Jpn.*, **57**, 2781 (1984).
102. M. C. Gallazzi, L. Porri, and G. Vitulli, *J. Organomet. Chem.*, **97**, 131 (1971).
103. G. Giacomelli and L. Lardicci, *Tetrahedron Lett.*, 2831 (1978).
104. C. Gosden and D. Pletcher, *J. Organomet. Chem.*, **186**, 401 (1980).
105. T. Goto, M. Onaka, and T. Mukaiyama, *Chem. Lett.*, 51 (1980).
106. A. E. Greene, J.-P. Lansard, J.-L. Luche, and C. Petrier, *J. Org. Chem.*, **49**, 931 (1984).
107. R. H. Grubbs, A. Miyashita, M. Liu, and P. Burk, *J. Am. Chem. Soc.*, **100**, 2418 (1978).
108. I. Hashimoto, N. Tsuruta, M. Ryang, and S. Tsutsumi, *J. Org. Chem.*, **35**, 3748 (1970).
109. T. Hayashi, in *Asymmetric Reactions and Processes in Chemistry* (Eds. E. L. Eliel and S. Otsuka), ACS Symp. Ser., No. 185, American Chemical Society, Washington, DC, 1982, Chapter 12.
110. T. Hayashi, M. Konishi, K. Yokota, and M. Kumada, *Chem. Lett.*, 767 (1980).
111. T. Hayashi, Y. Katsuro, and M. Kumada, *Tetrahedron Lett.*, **21**, 3915 (1980).
112. T. Hayashi, N. Nagashima, and M. Kumada, *Tetrahedron Lett.*, **21**, 4623 (1980).
113. T. Hayashi, T. Fujiwa, Y. Katsuro, and M. Kumada, *Synthesis*, 1001 (1981).
114. T. Hayashi, K. Kanehira, T. Hioki, and M. Kumada, *Tetrahedron Lett.*, **22**, 137 (1981).
115. T. Hayashi, Y. Katsuro, Y. Okamoto, and M. Kumada, *Tetrahedron Lett.*, **22**, 4449 (1981).
116. T. Hayashi, M. Konishi, K. Yokota, and M. Kumada, *J. Chem. Soc., Chem. Commun.*, 313 (1981).
117. T. Hayashi, M. Konishi, T. Hioki, M. Kumada, A. Ratajczak, and H. Niedbala, *Bull. Chem. Soc. Jpn.*, **54**, 3615 (1981).
118. T. Hayashi and M. Kumada, *Acc. Chem. Res.*, **15**, 395 (1982).
119. T. Hayashi, M. Konishi, M. Fukushima, T. Mise, M. Kagotani, M. Tajika, and M. Kumada, *J. Am. Chem. Soc.*, **104**, 180 (1982).
120. T. Hayashi, K. Kabeta, I. Hamachi, and K. Kumada, *Tetrahedron Lett.*, **24**, 2865 (1983).
121. T. Hayashi, M. Konishi, M. Fukushima, K. Kanehira, T. Hioki, and M. Kumada, *J. Org. Chem.*, **48**, 2195 (1983).
122. T. Hayashi, K. Kabeta, and M. Kumada, *Tetrahedron Lett.*, **25**, 1499 (1984).
123. T. Hayashi, M. Konishi, Y. Kobori, M. Kumada, T. Higuchi, and K. Hirotsu, *J. Am. Chem. Soc.*, **106**, 158 (1984).
124. K. P. Healy and D. Pletcher, *J. Organomet. Chem.*, **161**, 109 (1978).
125. L. S. Hegedus, *J. Organomet. Chem. Library*, **1**, 329 (1976).
126. L. S. Hegedus and R. K. Stiverson, *J. Am. Chem. Soc.*, **96**, 3250 (1974).
127. L. S. Hegedus and L. L. Miller, *J. Am. Chem. Soc.*, **97**, 459 (1975).
128. L. S. Hegedus, S. D. Wagner, E. L. Waterman, and K. Shiirala-Hansen, *J. Org. Chem.*, **40**, 593 (1975).
129. L. S. Hegedus, G. F. Allen, and E. L. Waterman, *J. Am. Chem. Soc.*, **98**, 2674 (1976).
130. L. S. Hegedus, B. R. Evans, D. E. Korte, E. L. Waterman, and K. Sjoberg, *J. Am. Chem. Soc.*, **98**, 3901 (1976).
131. L. S. Hegedus and S. Varaprath, *Organometallics*, **1**, 259 (1982).
132. L. S. Hegedus and R. Tamura, *Organometallics*, **1**, 1188 (1982).
133. A. Herrera, H. Hoberg, and R. Mynott, *J. Organomet. Chem.*, **222**, 331 (1981).
134. A. Herrera and H. Hoberg, *Synthesis*, 831 (1981).
135. M. Hidai, T. Kashiwagi, T. Ikeda, and Y. Uchida, *J. Organomet. Chem.*, **30**, 279 (1971).

136. B. Hipler and E. Uhlig, *J. Organomet. Chem.*, **199**, C27 (1980).
137. B. Hipler and E. Uhlig, *Z. Anorg. Allg. Chem.*, **510**, 11 (1984).
138. H. Hoberg, D. Schaefer, and G. Burkhart, *J. Organomet. Chem.*, **228**, C21 (1982).
139. H. Hoberg and B. W. Oster, *J. Organomet. Chem.*, **234**, C35 (1982).
140. H. Hoberg and D. Schaefer, *J. Organomet. Chem.*, **238**, 388 (1983).
141. H. Hoberg and B. W. Oster, *J. Organomet. Chem.*, **252**, 359 (1983).
142. H. Hoberg and K. Sümmermann, *J. Organomet. Chem.*, **253**, 383 (1983).
143. H. Hoberg and K. Sümmermann, *J. Organomet. Chem.*, **264**, 379 (1984).
144. H. Hoberg, D. Schaefer, G. Burkhart, C. Kruger, and M. J. Romao, *J. Organomet. Chem.*, **266**, 203 (1984).
145. H. Hoberg, D. Schaefer, and B. W. Oster, *J. Organomet. Chem.*, **266**, 313 (1984).
146. H. Hoberg and B. W. Oster, *J. Organomet. Chem.*, **266**, 321 (1984).
147. H. Hoberg and K. Sümmermann, *J. Organomet. Chem.*, **275**, 239 (1984).
148. R. W. Hoffmann and B. Landmann, *Tetrahedron Lett.*, **24**, 3209 (1983).
149. A. Hosomi, M. Saito, and H. Sakurai, *Tetrahedron Lett.*, 429 (1979).
150. J. M. Huggins and R. G. Bergman, *J. Am. Chem. Soc.*, **103**, 3002 (1981).
151. E. Ibuki, S. Ozasa, Y. Fujioka, M. Okada, and K. Terada, *Bull. Chem. Soc. Jpn.*, **53**, 821 (1980).
152. E. Ibuki, S. Ozasa, Y. Fujioka, and Y. Yanagihara, *Chem. Pharm. Bull.*, **30**, 802 (1982).
153. E. Ibuki, S. Ozasa, Y. Fujioka, M. Okada, and Y. Yanagihara, *Chem. Pharm. Bull.*, **30**, 2369 (1982).
154. Y. Ikeda, J. Ukai, N. Ikeda, and H. Yamamoto, *Tetrahedron Lett.*, **25**, 5177 (1984).
155. S. Inaba and R. D. Rieke, *Tetrahedron Lett.*, **24**, 2451 (1983).
156. S. Inaba and R. D. Rieke, *Synthesis*, 842 (1984).
157. S. Inaba and R. D. Rieke, *Synthesis*, 844 (1984).
158. S. Inaba, H. Matsumoto, and R. D. Rieke, *J. Org. Chem.*, **49**, 2093 (1984).
159. S. Inoue, K. Saito, K. Kato, S. Nozaki, and K. Sato, *J. Chem. Soc., Perkin Trans. 1*, 2097 (1974).
160. S. Inoue, R. Yamaguchi, K. Saito, and K. Sato, *Bull. Chem. Soc. Jpn.*, **47**, 3098 (1974).
161. Y. Inoue, Y. Itoh, H. Kazama, and H. Hashimoto, *Bull. Chem. Soc. Jpn.*, **53**, 3329 (1980).
162. L. S. Isaeva, L. N. Morozova, V. V. Bashilov, P. V. Petrovskii, V. I. Sokolov and O. A. Reutov, *J. Organomet. Chem.*, **243**, 253 (1984).
163. J. Ishizu, T. Yamamoto, and A. Yamamoto, *Chem. Lett.*, 1091 (1976).
164. K. Isobe, Y. Nakamura, and S. Kawaguchi, *Chem. Lett.*, 1383 (1977).
165. Y. Ito, M. Nakatsuka, and T. Saegusa, *J. Am. Chem. Soc.*, **103**, 476 (1981).
166. K. Itoh, T. Yoga, and Y. Ishii, *Chem. Lett.*, 103 (1977).
167. M. Iyoda, S. Tanaka, M. Nose, and M. Oda, *J. Chem. Soc., Chem. Commun.*, 1058 (1983).
168. M. Iyoda, M. Sakaitani, T. Miyazaki, and M. Oda, *Chem. Lett.*, 2005 (1984).
169. E. A. Jeffery, A. Meisters, and T. Mole, *J. Organomet. Chem.*, **74**, 365 (1974).
170. E. A. Jeffery, A. Meisters, and T. Mole, *Aust. J. Chem.*, **27**, 2569 (1974).
171. P. W. Jennings, D. G. Pillsbry, J. L. Hall, and V. T. Brice, *J. Org. Chem.*, **41**, 719 (1976).
172. P. W. Jolly, in *Comprehesive Organometallic Chemistry* (Eds. G. Wilkinson, F. G. A. Stone and E. W. Abel), Vol. 6, Pergamon Press, Oxford, 1982, Chapter 37.4, p. 37.
173. P. W. Jolly, in *Comprehensive Organometallic Chemistry* (Eds. G. Wilkinson, F. G. A. Stone and E. W. Abel), Vol. 8, Pergamon Press, Oxford 1982, Chapters 56.1–56.6, pp. 613–797.
174. P. W. Jolly and G. Wilke, *The Organic Chemistry of Nickel*, Vols. I and II, Academic Press, New York, 1974, 1975.
175. M. Julia and J.-N. Verpeaux, *Tetrahedron Lett.*, **23**, 2457 (1982).
176. Y. Kajihara, K. Ishikawa, H. Yasuda, and A. Nakamura, *Bull. Chem. Soc. Jpn.*, **53**, 3035 (1980).
177. H. Kanai, Y. Nishiguchi, and H. Matsuda, *Bull. Chem. Soc. Jpn.*, **56**, 1592 (1983).
178. A. V. Kavaliunas, A. Taylor, and R. D. Rieke, *Organometallics*, **2**, 377 (1983).
179. A. S. Kende, L. S. Liebeskind, and D. M. Braitsch, *Tetrahedron Lett.*, 3375 (1975).
180. A. S. Kende, R. Greenhouse, and J. A. Hill, *Tetrahedron Lett.*, 2867 (1979).
181. I. Khoury, P. Kovacic, and H. M. Gilov, *J. Polym. Sci., Polym. Lett. Ed.*, **19**, 395 (1981).
182. Y. Kiso, Dissertation, Kyoto University (1974).
183. Y. Kiso, K. Tamao, and M. Kumada, *J. Organomet. Chem.*, **50**, C12 (1973).
184. Y. Kiso, K. Tamao, N. Miyake, K. Yamamoto, and M. Kumada, *Tetrahedron Lett.*, 3 (1974).
185. H.-F. Klein and H. H. Karsch, *Chem. Ber.*, **109**, 2524 (1976).
186. T. Klingstedt and T. Frejd, *Organometallics*, **2**, 598 (1983).
187. J. K. Kochi, *Organometallic Mechanisms and Catalysis*, Academic Press, New York, (1978).

188. J. K. Kochi, *Pure Appl. Chem.*, **52**, 593 (1980).
189. T. Kohara, T. Yamamoto, and A. Yamamoto, *J. Organomet. Chem.*, **192**, 265 (1980).
190. K. Komatsu, N. Abe, K. Takahashi, and K. Okamoto, *J. Org. Chem.*, **44**, 2712 (1979).
191. S. Komiya, Y. Abe, A. Yamamoto, and T. Yamamoto, *Organometallics*, **2**, 1466 (1983).
192. D. E. Korte, L. S. Hegedus, and R. K. Wirth, *J. Org. Chem.*, **42**, 1329 (1977).
193. M. Kumada, *Pure Appl. Chem.*, **52**, 669 (1980).
194. M. Kumada, K. Tamao, and K. Sumitani, *Org. Synth.*, **58**, 127 (1978).
195. E. R. Larson and R. A. Raphael, *Tetrahedron Lett.*, 5041 (1979).
196. M. Lemaire, J. Buter, B. K. Vriesema, and R. M. Kellog, *J. Chem. Soc., Chem. Commun.*, 309 (1984).
197. J. W.-P. Lin and L. P. Dudek, *J. Polym. Sci., Polym. Chem. Ed.*, **18**, 2869 (1980).
198. M. J. Loots and J. Schwartz, *J. Am. Chem. Soc.*, **99**, 8045 (1977).
199. B. Loubinoux, R. Vanderesse, and P. Caubere, *Tetrahedron Lett.*, 3951 (1977).
200. H. Matsumoto, S.-I. Inaba, and R. D. Rieke, *J. Org. Chem.*, **48**, 840 (1983).
201. A. A. Millard and M. W. Rathke, *J. Am. Chem. Soc.*, **99**, 4833 (1977).
202. A. Minato, K. Tamao, T. Hayashi, K. Suzuki, and M. Kumada, *Tetrahedron Lett.*, **21**, 845 (1980).
203. A. Minato, K. Tamao, K. Suzuki, and M. Kumada, *Tetrahedron Lett.*, **21**, 4017 (1980).
204. R. H. Mitchell, M. Chaudhary, T. W. Dingle, and R. V. Williams, *J. Am. Chem. Soc.*, **106**, 7776 (1984).
205. M. Mori and Y. Ban, *Tetrahedron Lett.*, 1807 (1976).
206. M. Mori and Y. Ban, *Heterocycles*, **9**, 391 (1978).
207. M. Mori, Y. Hashimoto, and Y. Ban, *Tetrahedron Lett.*, **21**, 631 (1980).
208. D. G. Morrell and J. K. Kochi, *J. Am. Chem. Soc.*, **97**, 7262 (1975).
209. A. Morvillo and A. Turco, *J. Organomet. Chem.*, **208**, 103 (1981).
210. A. Morvillo and A. Turco, *J. Organomet. Chem.*, **244**, 387 (1982).
211. E. Negishi, *Pure Appl. Chem.*, **53**, 2333 (1981).
212. E. Negishi, *Acc. Chem. Res.*, **15**, 340 (1982).
213. E. Negishi and S. Baba, *J. Chem. Soc., Chem. Commun.*, 596 (1976).
214. E. Negishi, A. O. King, and N. Okukado, *J. Org. Chem.*, **42**, 1821 (1977).
215. E. Negishi and D. E. Van Horn, *J. Am. Chem. Soc.*, **99**, 3168 (1977).
216. E. Negishi, F.-T. Luo, and C. L. Rand, *Tetrahedron Lett.*, **23**, 27 (1982).
217. J. H. Nelson, P. N. Howells, G. C. DeLullo, G. L. Landen, and R. A. Henry, *J. Org. Chem.*, **45**, 1246 (1980).
218. R. Noyori, I. Umeda, H. Kawaguchi, and H. Takaya, *J. Am. Chem. Soc.*, **97**, 812 (1975).
219. K. Ogura, M. Wada, and N. Sonoda, *J. Organomet. Chem.*, **165**, C10 (1979).
220. A. Ohsawa, Y. Abe, and H. Igeta, *Chem. Pharm. Bull.*, **26**, 2550 (1978).
221. H. Okamura, M. Miura, and H. Takei, *Tetrahedron Lett.*, 43 (1979).
222. H. Okamura and H. Takei, *Tetrahedron Lett.*, 3425 (1979).
223. H. Okamura, M. Miura, K. Kosugi, and H. Takei, *Tetrahedron Lett.*, **21**, 87 (1980).
224. M. Onaka, T. Goto, and T. Mukaiyama, *Chem. Lett.*, 1483 (1979).
225. M. Onaka, Y. Matsuoka, and T. Mukaiyama, *Chem. Lett.*, 905 (1980).
226. M. Onaka, Y. Matsuoka, and T. Mukaiyama, *Chem. Lett.*, 531 (1981).
227. N. J. O'reilly, M. Maruta, and N. Ishikawa, *Chem. Lett.*, 517 (1984).
228. S. Otsuka, A. Nakamura, T. Yoshida, M. Naruto, and K. Ataka, *J. Am. Chem. Soc.*, **95**, 3180 (1973).
229. P. Papporto, S. Midollini, A. Orlandini, and L. Sacconi, *Inorg. Chem.*, **15**, 2768 (1976).
230. G. W. Parshall, *J. Am. Chem. Soc.*, **96**, 2360 (1974).
231. D. J. Pasto, S. K. Chou, A. Waterhouse, R. H. Shults, and G. F. Hennion, *J. Org. Chem.*, **43**, 1385 (1978).
232. D. J. Pasto and D. K. Mitra, *J. Org. Chem.*, **47**, 1381 (1982).
233. C. Petrier, J.-L. Luche, and C. Dupuy, *Tetrahedron Lett.*, **25**, 3463 (1984).
234. O. Piccolo and T. Martinengo, *Synth. Commun.*, **11**, 497 (1981).
235. P. Pino and G. Consiglio, *Pure Appl. Chem.*, **55**, 1781 (1983).
236. W. Poppitz and E. Uhlig, *J. Organomet. Chem.*, **244**, C1 (1983).
237. L. N. Pridgen, *J. Heterocycl. Chem.*, **12**, 443 (1975).
238. L. N. Pridgen, *J. Heterocycl. Chem.*, **17**, 1289 (1980).
239. L. N. Pridgen and L. B. Killmer, *J. Org. Chem.*, **46**, 5402 (1981).
240. L. N. Pridgen and S. S. Jones, *J. Org. Chem.*, **47**, 1590 (1982).

241. L. N. Pridgen, *J. Org. Chem.*, **47**, 4319 (1982).
242. V. Ratovelomanana and G. Linstrumelle, *Tetrahedron Lett.*, **22**, 315 (1981).
243. V. Ratovelomanana and G. Linstrumelle, *Synth. Commun.*, **14**, 179 (1984).
244. V. Ratovelomanana and G. Linstrumelle, *Tetrahedron Lett.*, **25**, 6001 (1984).
244a. G. S. Reddy and W. Tam, *Organometallics*, **3**, 630 (1984).
245. I. Rhee, M. Ryang, T. Watanabe, H. Omura, S. Murai, and N. Sonoda, *Synthesis*, 776 (1977).
246. I. Rhee, M. Ryang, H. Hasegawa, S. Murai, and N. Sonoda, *Chem. Lett.*, 15 (1978).
247. Y. Rollin, G. Meyer, M. Troupel, J. F. Fauvarque, and J. Perichon, *J. Chem. Soc., Chem. Commun.*, 793 (1983).
248. R. Rossi, A. Carpita, and A. Lezzi, *Tetrahedron*, **40**, 2773 (1984).
249. I. Ryu, M. Ryang, I. Rhee, H. Omura, S. Murai, and N. Sonoda, *Synth. Commun.*, **14**, 1175 (1984).
250. C. Sahlberg, A. Quader, and A. Claesson, *Tetrahedron Lett.*, **24**, 5137 (1983).
251. T. Saito, Y. Uchida, A. Misono, A. Yamamoto, K. Morifuji, and S. Ikeda, *J. Am. Chem. Soc.*, **88**, 5198 (1966).
252. K. Sato, S. Inoue, and K. Saito, *J. Chem. Soc., Perkin Trans. 1*, 2289 (1973).
253. Y. Sakakibara, N. Yadani, I. Ibuki, M. Sakai, and N. Uchino, *Chem. Lett.*, 1565 (1982).
254. K. Sanechika, T. Yamamoto, and A. Yamamoto, *J. Polym. Sci., Polym. Lett. Ed.*, **20**, 365 (1982).
255. K. Sano, T. Yamamoto, and A. Yamamoto, *Chem. Lett.*, 941 (1984).
256. K. Sano, T. Yamamoto, and A. Yamamoto, *Chem. Lett.*, 115 (1983).
257. R. A. Schunn, *Inorg. Synth.*, **13**, 124 (1972).
258. J. Schwartz, D. B. Carr, R. T. Hansen, and F. M. Dayrit, *J. Org. Chem.*, **45**, 3053 (1980).
259. J. Schwartz, M. J. Loots, and H. Kosugi, *J. Am. Chem. Soc.*, **102**, 1333 (1980).
260. Y. Seki, S. Murai, M. Ryang, and N. Sonoda, *J. Chem. Soc., Chem. Commun.*, 528 (1975).
261. M. F. Semmelhack, *Org. React.*, **19**, 117 (1972).
262. M. F. Semmelhack, P. M. Helquist, and L. D. Jones, *J. Am. Chem. Soc.*, **93**, 5903 (1971).
263. M. F. Semmelhack, P. M. Helquist, and J. D. Gorzynski, *J. Am. Chem. Soc.*, **94**, 9234 (1972).
264. M. F. Semmelhack, R. D. Stauffer, and T. D. Rogerson, *Tetrahedron Lett.*, 4519 (1973).
265. M. F. Semmelhack, B. P. Chong, R. D. Stauffer, T. D. Rogerson, A. Chong, and L. D. Jones, *J. Am. Chem. Soc.*, **97**, 2507 (1975).
266. M. F. Semmelhack and L. S. Ryono, *J. Am. Chem. Soc.*, **97**, 3873 (1975).
267. M. F. Semmelhack and E. S. C. Wu, *J. Am. Chem. Soc.*, **98**, 3384 (1976).
268. M. D. Semmelhack, A. Yamashita, J. C. Tamesch, and K. Hirotsu, *J. Am. Chem. Soc.*, **100**, 5565 (1978).
269. M. F. Semmelhack and S. J. Brickner, *J. Am. Chem. Soc.*, **103**, 3948 (1981).
270. M. F. Semmelhack, P. Helquist, L. D. Jones, L. Keller, L. Mendelson, L. S. Ryono, J. G. Smith, and R. D. Stauffer, *J. Am. Chem. Soc.*, **103**, 6460 (1981).
271. M. F. Semmelhack, L. Keller, T. Sato, and E. Spiess, *J. Org. Chem.*, **47**, 4382 (1982).
272. I. Shimizu, Y. Kamei, T. Tazuka, T. Izumi, and A. Kasahara, *Bull. Chem. Soc. Jpn.*, **56**, 192 (1983).
273. G. Smith and J. K. Kochi, *J. Organomet. Chem.*, **198**, 199 (1980).
274. G. Solladie and R. G. Zimmermann, *Tetrahedron Lett.*, **25**, 5769 (1984).
275. J. K. Stille and A. B. Cowell, *J. Organomet. Chem.*, **124**, 253 (1977).
276. P. Stoppioni, A. Biloitti, and R. Morassi, *J. Organomet. Chem.*, **236**, 119 (1982).
277. H. Sugimura and H. Takei, *Chem. Lett.*, 1505 (1984).
278. K. Takagi, N. Hayama, and S. Inokawa, *Chem. Lett.*, 1435 (1978).
279. K. Takagi, H. Hayama, and S. Inokawa, *Chem. Lett.*, 917 (1979).
280. K. Takagi, N. Hayama, and S. Inokawa, *Bull. Chem. Soc. Jpn.*, **53**, 3691 (1980).
281. K. Takagi and N. Hayama, *Chem. Lett.*, 637 (1983).
282. K. Takagi, N. Hayama, and K. Sasaki, *Bull. Chem. Soc. Jpn.*, **57**, 1887 (1984).
283. K. Takagi, H. Minura, and S. Inokawa, *Bull. Chem. Soc. Jpn.*, **57**, 3517 (1984).
284. S. Takahashi, Y. Suzuki, and N. Hagihara, *Chem. Lett.*, 1363 (1974).
285. S. Takahashi, Y. Suzuki, K. Sonogashira, and N. Hagihara, *J. Chem. Soc., Chem. Commun.*, 839 (1976).
286. H. Takaya, T. Suzuki, Y. Kumagai, M. Yamakawa, and R. Noyori, *J. Org. Chem.*, **46**, 2846 (1981).
287. H. Takei, M. Miura, H. Sugimura, and H. Okamura, *Chem. Lett.*, 1447 (1979).
288. H. Takei, H. Sugimura, M. Miura, and H. Okamura, *Chem. Lett.*, 1209 (1980).
289. K. Tamao, K. Sumitani, and M. Kumada, *J. Am. Chem. Soc.*, **94**, 4374 (1972).
290. K. Tamao, Y. Kiso, K. Sumitani, and M. Kumada, *J. Am. Chem. Soc.*, **94**, 9268 (1972).
291. K. Tamao, M. Zembayashi, Y. Kiso, and M. Kumada, *J. Organomet. Chem.*, **55**, C91 (1973).

292. K. Tamao, J. Toei, M. Zembayashi, and M. Kumada, unpublished results (1974).
293. K. Tamao, S. Kodama, T. Nakatsuka, Y. Kiso, and M. Kumada, *J. Am. Chem. Soc.*, **97**, 4405 (1975).
294. K. Tamao, A. Minato, N. Miyake, T. Matsuda, Y. Kiso, and M. Kumada, *Chem. Lett.*, 133 (1975).
295. K. Tamao, K. Sumitani, Y. Kiso, M. Zembayashi, A. Fujioda, S. Kodama, I. Nakajima, A. Minato, and M. Kumada, *Bull. Chem. Soc. Jpn.*, **49**, 1958 (1976).
296. K. Tamao, M. Zembayashi, and M. Kumada, *Chem. Lett.*, 1237 (1976).
297. K. Tamao, M. Zembayashi, and M. Kumada, *Chem. Lett.*, 1239 (1976).
298. K. Tamao, H. Yamamoto, H. Matsumoto, N. Miyake, T. Hayashi, and M. Kumada, *Tetrahedron Lett.*, 1389 (1977).
299. K. Tamao, T. Hayashi, M. Matsumoto, H. Yamamoto, and M. Kumada, *Tetrahedron Lett.*, 2155 (1979).
300. K. Tamao, S. Kodama, I. Nakajima, M. Kumada, A. Minato, and K. Suzuki, *Tetrahedron*, **38**, 3347 (1982).
301. K. Tamao, J. Ishida, and M. Kumada, *J. Org. Chem.*, **48**, 2120 (1983).
302. M. Tamura and J. K. Kochi, *Bull. Chem. Soc. Jpn.*, **44**, 3063 (1971).
303. M. Tanaka, *Synthesis*, 47 (1981).
304. K. Tatsumi, R. Hoffmann, A. Yamamoto, and J. K. Stille, *Bull. Chem., Soc. Jpn.*, **54**, 1857 (1981).
305. K. Tatsumi, A. Nakamura, S. Komiya, A. Yamamoto, and T. Yamamoto, *J. Am. Chem. Soc.*, **106**, 8181 (1984).
306. P. Tavs, *Chem. Ber.*, **103**, 2428 (1970).
307. E. Thorsett and F. R. Stermitz, *J. Heterocycl. Chem.*, **10**, 243 (1973).
308. M. Tiecco, L. Testaferri, M. Tingoli, D. Chianelli, and E. Wenkert, *Tetrahedron Lett.*, **23**, 4629 (1982).
309. M. Tiecco, L. Testaferri, M. Tingoli, D. Chianelli, and M. Montanucci, *Synthesis*, 736 (1984).
310. B. M. Trost and J. Yoshida, *Tetrahedron Lett.*, **24**, 4895 (1983).
311. B. M. Trost and A. C. Lavoie, *J. Am. Chem. Soc.*, **105**, 5075 (1983).
312. M. Troupel, Y. Rollin, S. Sibille, J. F. Fauvarque, and J. Perichon, *J. Chem. Res. (S)*, 26 (1980).
313. M. Troupel, Y. Rollin, S. Sibille, J. Perichon, and J. F. Fauvarque, *J. Organomet. Chem.*, **202**, 435 (1980).
314. T. T. Tsou and J. K. Kochi, *J. Am. Chem. Soc.*, **100**, 1634 (1978).
315. T. T. Tsou and J. K. Kochi, *J. Am. Chem. Soc.*, **101**, 6319 (1979).
316. T. T. Tsou and J. K. Kochi, *J. Am. Chem. Soc.*, **101**, 7547 (1979).
317. T. T. Tsou and J. K. Kochi, *J. Org. Chem.*, **45**, 1930 (1980).
318. J. Tsuji, I. Minami, and I. Shimizu, *Chem. Lett.*, 1721 (1984).
319. M. Uchino, K. Asagi, A. Yamamoto, and S. Ikeda, *J. Organomet. Chem.*, **84**, 93 (1975).
320. S. Uemura and S. Fukuzawa, *Tetrahedron Lett.*, **23**, 1181 (1982).
321. S. Uemura, S. Fukuzawa, and S. R. Patil, *J. Organomet. Chem.*, **243**, 9 (1983).
322. E. Uhlig and W. Poppitz, *Z. Anorg. Allg. Chem.*, **477**, 167 (1981).
323. E. Uhlig and B. Hipler, *Tetrahedron Lett.*, **25**, 5871 (1984).
324. J. Ukai, Y. Ikeda, N. Ikeda, and H. Yamamoto, *Tetrahedron Lett.*, **25**, 5173 (1984).
325. R. Vanderesse, J.-J. Brunet, and P. Caubere, *J. Organomet. Chem.*, **264**, 263 (1984).
326. M. Wada, K. Kusabe, and K. Oguro, *Inorg. Chem.*, **16**, 446 (1977).
327. M. Wada, K. Nishizaki, and M. Kumazoe, *J. Chem. Soc., Chem. Commun.*, 980 (1984).
328. H. M. Walborsky and R. B. Banks, *J. Org. Chem.*, **46**, 5074 (1981).
329. J. A. Walker, S. P. Bitler, and F. Wudl, *J. Org. Chem.*, **49**, 4733 (1984).
330. E. Wenkert, T. W. Ferreira, and E. L. Michelotti, *J. Chem. Soc., Chem. Commun.*, 637 (1979).
331. E. Wenkert, E. L. Michelotti, and C. S. Swindell, *J. Am. Chem. Soc.*, **101**, 2246 (1979).
332. E. Wenkert and T. W. Ferreira, *J. Chem. Soc., Chem. Commun.*, 840 (1982).
333. E. Wenkert and T. W. Ferreira, *Organometallics*, **1**, 1670 (1982).
334. E. Wenkert, J. B. Fernandes, E. L. Michelotti, and C. S. Swindell, *Synthesis*, 701 (1983).
335. E. Wenkert, M. H. Leftin, and E. L. Michelotti, *J. Chem. Soc., Chem. Chem.*, 617 (1984).
336. E. Wenkert, E. L. Michelotti, C. S. Swindell, and M. Tingoli, *J. Org. Chem.*, **49**, 4894 (1984).
337. G. M. Whitesides, C. P. Casey, and J. K. Krieger, *J. Am. Chem. Soc.*, **93**, 1379 (1971).
338. D. A. Whiting and A. F. Wood, *Tetrahedron Lett.*, 2335 (1978).
339. G. Wilke, B. Bogdanovic, P. Hardt, P. Heimbach, W. Kein, M. Kröner, W. Oberkirch, K. Tanaka, E. Steinvilcke, D. Walter, and H. Zimmermann, *Angew. Chem.*, **78**, 157 (1966).

340. G. Wilke, *J. Organomet. Chem.*, **200**, 349 (1980).
341. G. Wilke, *Pure Appl. Chem.*, **56**, 1635 (1984).
342. T. Yamamoto, A. Yamamoto, and S. Ikeda, *J. Am. Chem. Soc.*, **93**, 3350, 3360 (1971).
343. T. Yamamoto, T. Kohara, and A. Yamamoto, *Chem. Lett.*, 1217 (1976).
344. T. Yamamoto, T. Saruyama, Y. Nakamura, and A. Yamamoto, *Bull. Chem. Soc. Jpn.*, **49**, 589 (1976).
345. T. Yamamoto, Y. Hayashi, and A. Yamamoto, *Bull. Chem. Soc. Jpn.*, **51**, 2091 (1978).
346. T. Yamamoto and A. Yamamoto, Chem. Lett., 615 (1978).
347. T. Yamamoto, K. Igarashi, S. Komiya, and A. Yamamoto, *J. Am. Chem. Soc.*, **102**, 7448 (1980).
348. T. Yamamoto, K. Sanechika, and A. Yamamoto, *J. Polym. Sci., Polym. Lett. Ed.*, **18**, 9 (1980).
349. T. Yamamoto, J. Ishizu, and A. Yamamoto, *J. Am. Chem. Soc.*, **103**, 6863 (1981).
350. T. Yamamoto, T. Kohara, and A. Yamamoto, *Bull. Chem. Soc. Jpn.*, **54**, 2010 (1981).
351. T. Yamamoto, T. Kohara, and A. Yamamoto, *Bull. Chem. Soc. Jpn.*, **54**, 2161 (1981).
352. T. Yamamoto, K. Sanechika, and A. Yamamoto, *Chem. Lett.*, 1079 (1981).
353. A. Yamamoto, T. Yamamoto, and F. Ozawa, *J. Synth. Org. Chem. Jpn.*, **41**, 827 (1983).
354. T. Yamamoto, K. Sanechida, and A. Yamamoto, *Bull. Chem. Soc. Jpn.*, **56**, 1497 (1983).
355. T. Yamamoto, T. Kohara, K. Osakada, and A. Yamamoto, *Bull. Chem. Soc. Jpn.*, **56**, 2147 (1983).
356. A. Yamamoto, T. Yamamoto, S. Komiya, and F. Ozawa, *Pure Appl. Chem.*, **56**, 1621 (1984).
357 H. Yamanaka, K. Edo, F. Shoji, S. Konno, T. Sakamoto, and M. Mizugami, *Chem. Pharm. Bull.*, **26**, 2160 (1978).
358. M. Yamashita and Y. Takegami, *Synthesis*, 803 (1977).
359. H. Yamazaki and N. Hahihara, *Bull. Chem. Soc. Jpn.*, **37**, 907 (1964).
360. E. Yoshisato and S. Tsutsumi, *J. Org. Chem.*, **33**, 869 (1968).
361. M. Zembayashi, K. Tamao, and M. Kumada, *Tetrahedron Lett.*, 1719 (1975).
362. M. Zembayashi, K. Tamao, T. Hayashi, T. Mise, and M. Kumada, *Tetrahedron Lett.*, 1799 (1977).
363. M. Zembayashi, K. Tamao, J. Yoshida, and M. Kumada, *Tetrahedron Lett.*, 4089 (1977).

The Chemistry of the Metal—Carbon Bond, Vol. 4
Edited by F. R. Hartley

CHAPTER **10**

Transition metal-stabilized carbocations in organic synthesis

A. J. PEARSON

Department of Chemistry, Case Western Reserve University, Cleveland, Ohio 44106, USA

I. INTRODUCTION

Other chapters in this book deal with applications of specific transition metals in organic synthesis, via their organometallic complexes. This chapter will unavoidably cross their paths, since we shall meet a range of cationic organometallic systems and see how they have been, or might be, applied to complex organic synthesis. We shall tend to concentrate on the more strategic aspects, emphasizing the relationship between the organometallic species and potential 'synthons'. Consideration will be given to complexes which might ultimately be used in organic synthesis but which are currently somewhat problematic owing to certain difficulties associated with products obtained during carbon—carbon bond-forming reactions. We shall pinpoint these problems and show how they have been partly or fully solved.

This chapter is organized according to ligand type, starting with η^2-alkene complexes and running through to η^6-arene complexes. Recent developments indicate excellent possibilities for using a metal moiety attached to an olefinic ligand as a template for controlling stereochemistry during diverse synthetic operations. With cationic complexes, stereocontrol during the formation of C—C bonds can be accomplished, and these developments will be highlighted at appropriate points. Such reactions usually require stoichiometric, rather than catalytic, use of transition metal complexes. This is a fairly recent development, both conceptually and technically, and, whilst there are few actual synthetic applications at present, it seems likely that this is an area of great potential for future growth.

II. ROLE OF THE TRANSITION METAL IN CATIONIC COMPLEXES

Sometimes it may appear that there are disparities in the way cationic organometallic complexes are represented here, since in some complexes the positive charge is assigned to the organic ligand, whilst in others it is drawn on the metal. Compare, for example, the diene–molybdenum complex **1** with the dienyliron complex **2**.

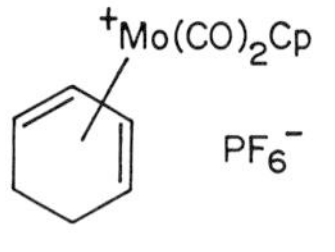

$^{+}Mo(CO)_2Cp$ PF_6^-

(1)

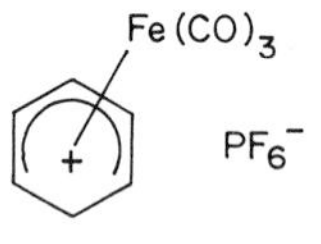

$Fe(CO)_3$ PF_6^-

(2)

Whilst this might appear confusing at first sight, the differentiation is logical, since it relates directly to the type of ligand. Both **1** and **2** undergo addition of nucleophile to the six-membered ring ligand, but if we relate **1** to the uncomplexed diene, it becomes very difficult to picture a corresponding positively charge organic species. On the other hand, **2** can be related easily to the uncomplexed cyclohexadienyl system, since it reacts as though it were such a cation, and the organic chemist is familiar with these. On this basis, we can draw a series of complexes, **3**–**7** as shown below, having even and odd numbers of atoms in the ligand. The 'even' compounds can be thought of as metal-activated olefinic systems and therefore we place the charge on the metal, whilst the 'odd' compounds readily relate to familiar organic species, and we place the charge on the ligand; we can think of these as metal-stabilized carbocations.

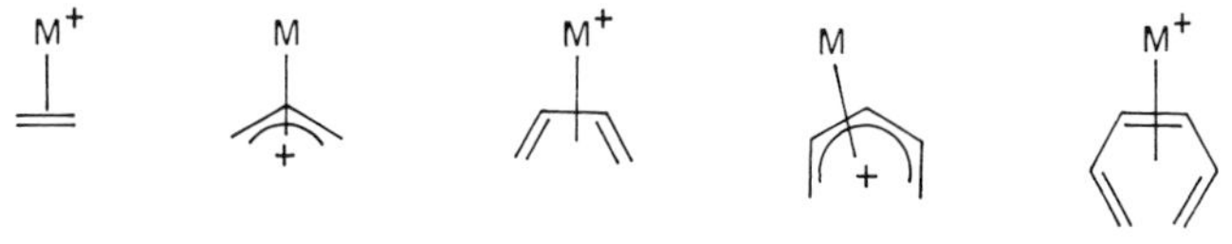

The placement of charge on these complexes leads to their considerable reactivity toward nucleophiles, and in most cases attack occurs at a terminal carbon atom[1]. We shall see specific examples later, but for the sake of clarity we can make a few generalizations at this point. Figure 1 shows a schematic representation of these complexes, re-drawing them in terms of their 'reaction equivalents'. In some cases reaction with a nucleophile leads to a complex which can be directly transformed, by demetallation, to the corresponding organic ligand without further alteration of that ligand. In other cases, nucleophile addition results in a complex which has no organic ligand equivalent, and any demetallation will have to be accompanied by changes in the ligand.

We can see that those complexes which give products having simple ligand equivalents also have a corresponding organic electrophile, at least in principle. For example, the

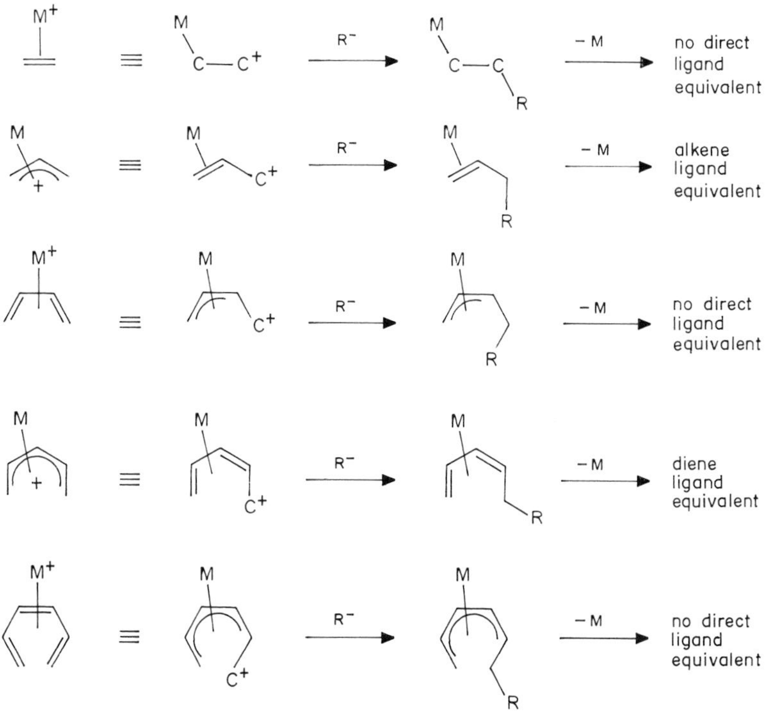

FIGURE 1. Schematic reactivity patterns of olefin and polyolefin metal complexes

reactivity patterns associated with η^3-allyl complexes are equivalent to nucleophilic displacement of halide from allylic halides, whereas dienyl complexes are equivalent to dienyl halides. While allylic halides have been used extensively in organic synthesis, dienyl halides have not. This poses a slight problem in terms of the organic chemist's familiarity with such electrophiles and their application in synthetic strategy. The use of transition metals in stabilizing such cations, having functional groups attached to the ligand, is of considerable potential, since this allows us to draw up a list of synthetic equivalents. Once compiled, this list would allow the organic chemist to choose a complex appropriate for a specific bond formation, bearing in mind any limitations which might exist. Complexes which do not have an organic equivalent are more difficult to handle, both conceptually and in terms of their conversion to organic molecules. However, it is these complexes which will ultimately allow further ligand functionalization, e.g. during demetallation, and, provided that this can be accomplished cleanly, such complexes will ultimately become extremely useful to the synthetic chemist, since they will provide a means of polyfunctionalization of readily accessible olefinic compounds in a manner which is directly controlled by the metal.

Throughout this chapter we shall try to draw attention to all of these aspects of organometallic carbocation chemistry.

III. η^2-ALKENE COMPLEXES

A. η^2-Alkene Complexes of Iron

The most studied complexes of this type are the (alkene)η^5-cyclopentadienyl)dicarbonyliron salts. We shall therefore devote more time to discussing these and related palladium complexes than those of other metals. The ruthenium and osmium analogues are known[2], but their high cost is prohibitive in terms of using them stoichiometrically. A range of methods are available for the preparation of $CpFe(CO)_2$–alkene complexes, largely owing to the efforts of Rosenblum[3] and coworkers, and these are summarized in equations 1–5.

$$[(\eta^5\text{-}C_5H_5)Fe(CO)_2Br] + CH_2{=}CHR \xrightarrow{AlBr_3} [(\eta^5\text{-}C_5H_5)Fe(CO)_2(CH_2{=}CHR)]^+ \; AlBr_4^- \qquad (1)$$

$$[CpFe(CO)_2]^- \; Na^+ \xrightarrow{RCH_2CH_2I} [Cp(CO)_2FeCH_2CH_2R] \xrightarrow{Ph_3C^+\,BF_4^-} [Cp(CO)_2Fe\text{—}(CH_2{=}CHR)]^+ \; BF_4^- \qquad (2)$$

$$[CpFe(CO)_2]^- \ Na^+ \xrightarrow{CH_2=CHCH_2Cl} [Cp(CO)_2FeCH_2CH{=}CH_2] \xrightarrow{HBF_4} [Cp(CO)_2Fe{-}(\eta^2\text{-propene})]^+ \ BF_4^- \quad (3)$$

$$[CpFe(CO)_2]^- \ Na^+ \xrightarrow[\text{(ii) } H_2O]{\text{(i) ethylene oxide}} [Cp(CO)_2FeCH_2CH_2OH] \xrightarrow{HBF_4} [Cp(CO)_2Fe{-}(\eta^2\text{-ethylene})]^+ \ BF_4^- \quad (4)$$

$$[Cp(CO)_2Fe{-}(\eta^2\text{-isobutene})]^+ \ BF_4^- + RCH_2CH{=}CH_2 \longrightarrow [Cp(CO)_2Fe{-}(\eta^2\text{-}CH_2{=}CHCH_2R)]^+ \ BF_4^- + \text{isobutene} \quad (5)$$

The above variety of methods of preparation indicates some of the potential of these complexes, since a wide range of organic substrates can be converted in to the alkene complexes. The last method is particularly interesting. Alkene exchange occurs between the readily prepared isobutene–Fp complex and an alkene to give an equilibrium which is displaced towards products by loss of (gaseous) isobutene. The utility of this method lies in the ability to complex selectively less highly substituted double bonds, and to complex olefinic double bonds in the presence of acetylenic groups[4], as shown by the simple examples in equations 6 and 7.

$$\text{CH}_2{=}\text{CHCH}_2\text{C}{\equiv}\text{CCH}_2\text{CH}_2\text{CH}_3 + (\text{isobutene}){-}\text{Fp}^+ \longrightarrow {}^+\text{Fp}{-}(\eta^2\text{-alkene}) \quad 65\% \quad (6)$$

$$\text{4-vinylcyclohexene} \xrightarrow{(\text{isobutene}){-}\text{Fp}^+} \text{Fp}^+{-}(\eta^2\text{-vinyl})\text{cyclohexene} \quad 76\% \quad (7)$$

These η^2-alkene–Fp complexes are very reactive electrophiles. Of particular note is the selectivity of nucleophile addition, which occurs entirely at the alkene ligand, despite the availability of CO and cyclopentadienyl ligands. Some discussion of this selectivity has been presented by Davies *et al.*[1]. A wide range of nucleophiles have been examined, and these are summarized in the equations 8–18[5].

$$\text{Fp}^+{-}\|\ + \ \text{Bu}^t\text{SH} \longrightarrow \text{FpCH}_2\text{CH}_2\text{SBu}^t \quad (8)$$

$$Fp^{+}\text{—}\| + R_3P \longrightarrow Fp\text{CH}_2\text{CH}_2\overset{+}{P}R_3 \tag{9}$$

(10) 48%

(11) 60%

(12) $LiCH(CO_2Et)_2$; $CH(CO_2Et)_2$; 82%

(13) $^{-}CH(CO_2Me)_2$; $CH(CO_2Me)_2$

(14) Ph_3C^+ ; I^- ; $\bar{C}H(CO_2R)_2$; $CH(CO_2R)_2$

(15) $^{-}CH(CO_2Me)_2$; $CH(CO_2Me)_2$; single isomer

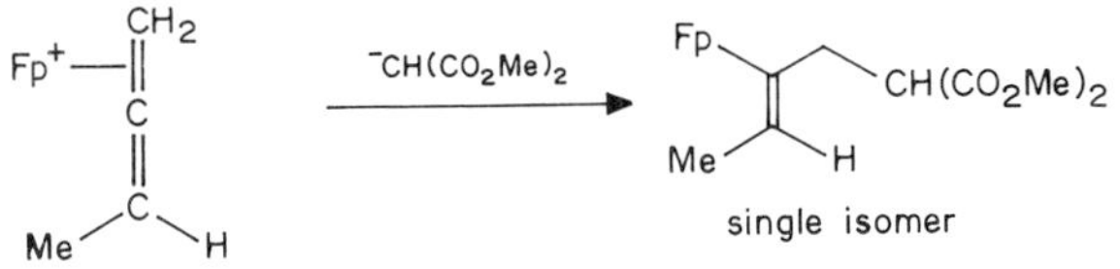

(16)

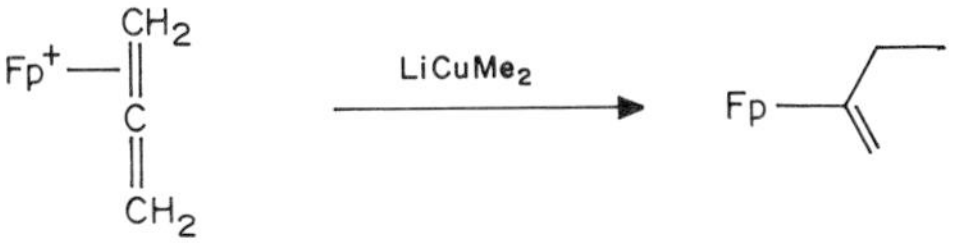

(17)

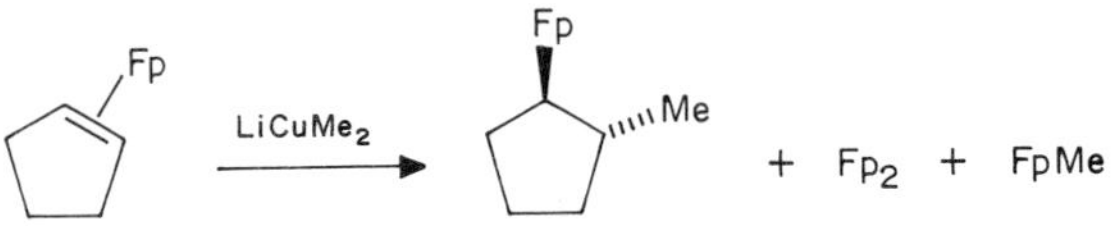

(18)

Sometimes, depending on the nature of the complex and the nucleophile, addition becomes difficult and deprotonation of the alkene–Fp complex occurs to give σ-allyl–Fp derivatives, equation 19[5].

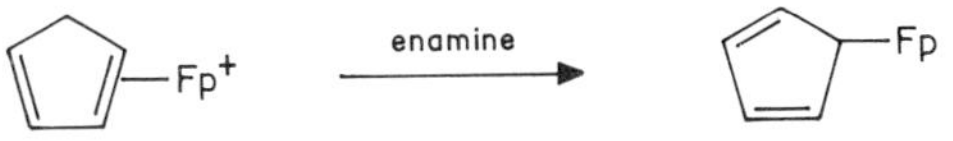

(19)

Complexes of enol ethers have been prepared and found to react with nucleophiles with a high degree of regioselectivity[6]; reaction of alkene–Fp complexes with Br^- or I^- leads to decomplexation, as shown by equation 20 and later examples. Thus, the alkoxy-

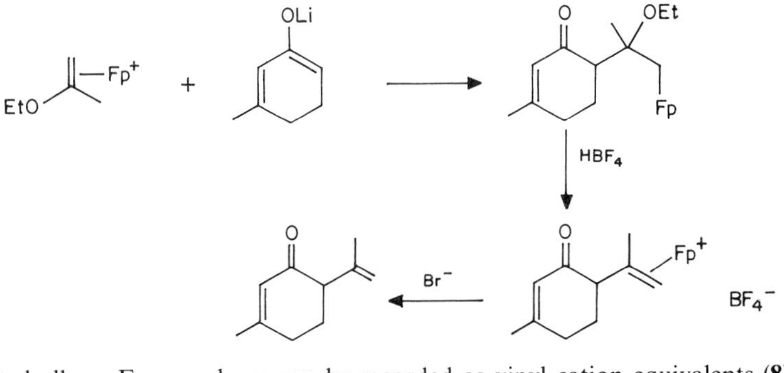

(20)

substituted alkene-Fp complexes can be regarded as vinyl cation equivalents (**8**). This might allow synthetic planning to utilize such electrophiles.

(**8**)

This regioselectivity is even observed in the presence of, for example, electron-withdrawing substituents which might otherwise cause opposite selectivity. A particularly interesting example is the complex of α-ethoxyacrylic ester[7] shown in reaction 21. This reacts with enolates to give a mixture of diastereomers **9** and **10**, despite the tendency of unsaturated ester complexes to undergo the equivalent of Michael addition.

(21)

The products of this reaction have been converted into the diastereomeric α-methylene-γ-lactones **11**, an example of a functional group which occurs in a range of important terpenes.

(22)

Of particular interest are the intramolecular variations on this nucleophile addition[8–10]. The transition metal in these reactions provides an opportunity for further carbon—carbon bond formation through the well established carbonyl insertion reaction, a nice illustration being provided by Rosenblum's synthesis of β-lactam derivatives (reactions 23–25)[8].

(23)

$$HC{\equiv}CCH_2CH_2OH \xrightarrow{Fp^+\text{-isobutylene}} \left[HC{\equiv}CCH_2CH_2OH \cdot Fp^+ \right] \longrightarrow \text{(dihydrofuranyl)}{-}Fp^+ + \text{(oxacyclopentylidene)}{=}Fp^+ \qquad (24)$$

$$\text{hex-3-yn-1-ol} + Fp^+\text{-isobutylene} \longrightarrow \text{(2-ethyl-4,5-dihydrofuran-3-yl)}{-}Fp^+ \qquad (25)$$

Although alkenes can be complexed selectively in the presence of carbon—carbon triple bonds, the latter can be converted in to Fp complexes in the usual way[10,11]. The resulting acetylene–Fp derivatives undergo stereospecific addition of nucleophiles to give σ-bonded vinyl–Fp complexes, but extensive studies towards further manipulation of these compounds have not been forthcoming. Cleavage of the C—Fp bond with iodine or bromine results in stereospecific formation of the vinyl halide, a very useful property, as shown in Fig. 2.

Attachment of the Fp^+ group of the olefinic bond of α,β-unsaturated ketones results in their effective activation towards Michael addition. This contrasts with the aforementioned additions to α-alkoxy–α,β-unsaturated ester complexes. The enone–Fp complexes are prepared via reaction of the Fp anion with corresponding epoxy ketone, as illustrated

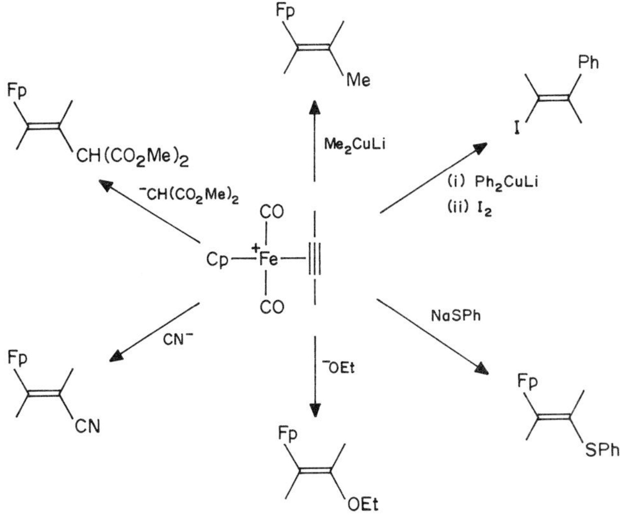

FIGURE 2. Reactions of acetylene–Fp cation with nucleophiles

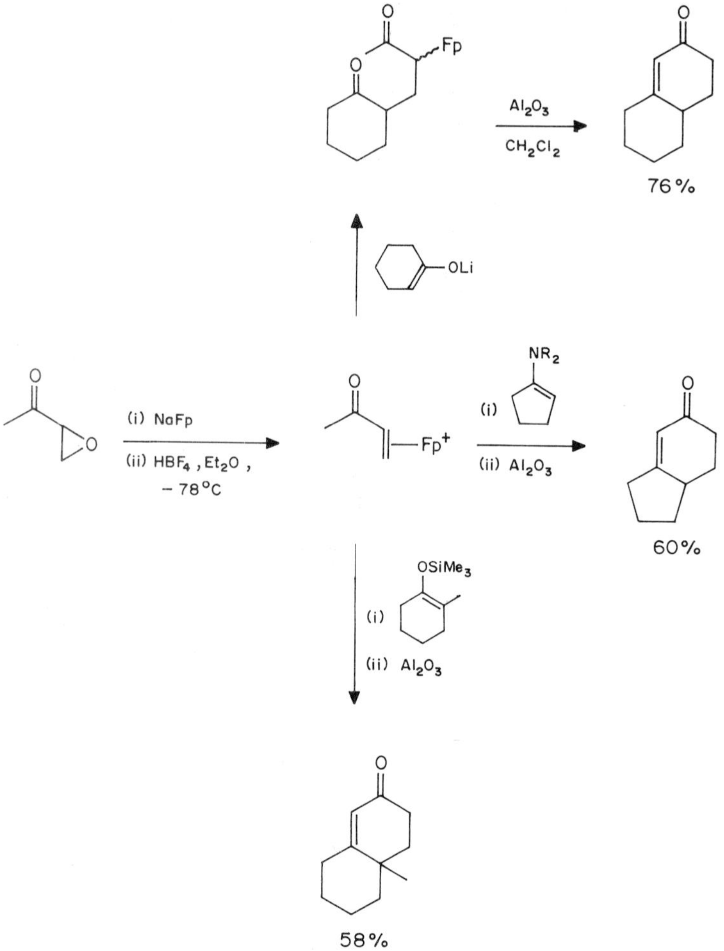

FIGURE 3. Fp-promoted Michael additions

in Fig. 3. It may be noted that silyl enol ethers undergo clean reaction with these activated enones, in contrast to the regiochemical scrambling of the enol ether which is often associated with their Lewis acid-catalysed Michael addition reactions[12]. These reactions provide a novel variation on the well known Robinson annulation.

The electrophilic reactivity of the alkene–Fp group is sufficient to allow its reaction with $[(\eta^4\text{-triene})Fe(CO)_3]$ complexes. The driving force here is the formation of the stable dienyl–$Fe(CO)_3$ cation (see later), and a nice illustration is provided by the hydroazulene synthesis outlined in reaction 26[13].

This set of reactions also illustrates the nucleophilic properties of σ-bonded allyl–Fp complexes and their conversion to cationic alkene–Fp systems. In this context, and

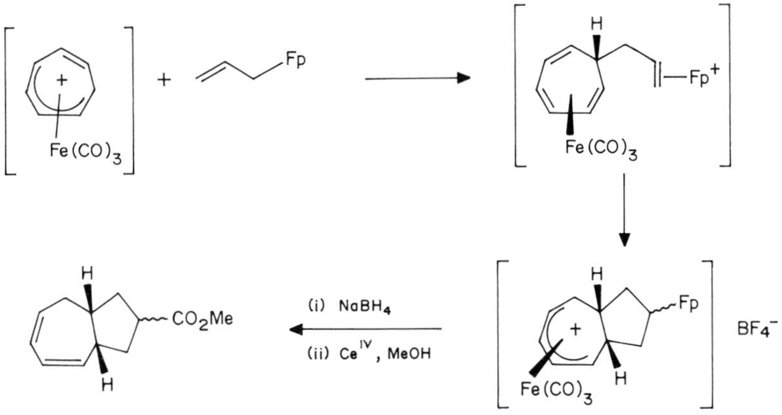

(26)

particularly in the above sequence, the allyl–Fp complex can be regarded as the synthetic equivalent of a three-carbon zwitterion, **12**.

(12)

All of the preceding discussion has centred around carbon—carbon bond-forming reactions using cationic alkene–Fp complexes. The same organometallic unit can also be used to protect carbon—carbon double bonds. Thus, whereas the alkene–Fp complexes are reactive towards nucleophiles, they are markedly unreactive towards electrophiles and reagents which normally attack uncomplexed alkenes. A number of reactions, such as bromination, hydrogenation, and mercuration, can be performed in the presence of the alkene–Fp group, and the metal can be removed easily at a later stage, by treatment with iodide anion in, e.g., acetone. Such neutral reaction conditions are expected to be very useful when sensitive functional groups are present in the molecule. Some examples of the reactions which can be performed are summarized in equations 27–29[14].

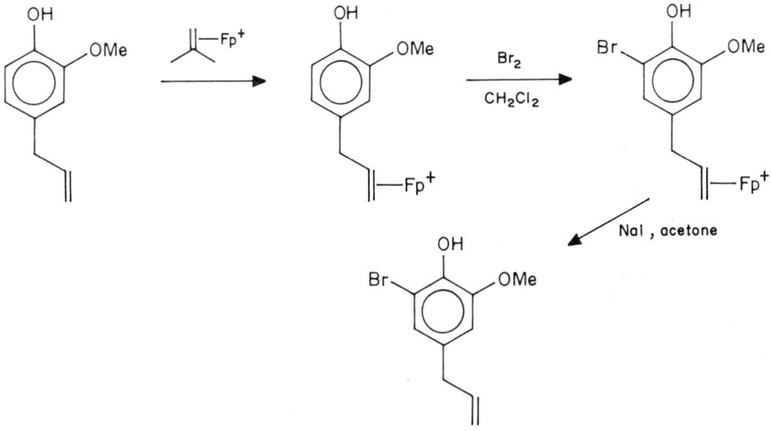

(27)

It may be noted that reactions of the non-complexed unsaturated portion of these molecules still leaves the alkene–Fp cation intact, so that further reactions, such as nucleophile addition, could be performed, leading to highly functionalized molecules as valuable synthetic intermediates. These aspects do not appear to have been fully explored.

B. η^2-Alkene Complexes of Palladium

Palladium compounds have been used for some time as catalysts for conversion of olefins into other useful compounds, perhaps the best known application being alkene hydrogenation to saturated carbon compounds using a palladium metal catalyst. Higher oxidation states of palladium catalyse a number of other transformations of olefins, invariably proceeding via intermediate cationic η^2-alkene complexes. This behaviour may be translated to the use of alkene–Pd complexes stoichiometrically and this offers certain advantages, such as double functionalization, which we shall discuss in this section. The main drawback in the use of palladium stoichiometrically is its high cost. Even though it is fully recoverable, any large-scale synthesis based on the use of such procedures involves high capital outlay.

The Wacker process[15] was originally developed as a method for converting ethylene to acetaldehyde. In this process the alkene is usually treated with oxygen in the presence of water (aqueous or modified aqueous solution), a catalytic amount of palladium(II) chloride, and a catalytic amount of copper(II) chloride. The oxygen atom which is introduced into the acetaldehyde product arises from the water, whilst the function of the oxygen (air), is to oxidize copper(I) chloride to copper(II) chloride, which in turn oxidizes palladium(0) to palladium(II). The catalytic scheme is shown in Figure 4, which illustrates each step.

The scheme shown in Figure 4a illustrates how the various substrates interact and also how the process depends on the intermediacy of a reactive ethylene–Pd complex. The finer details of the mechanism are not known, and the scheme is meant to provide a guideline for seeing how things fit together.

Laboratory syntheses of more complex molecules can also benefit from Wacker oxidation technology[16]. Terminal olefins can be efficiently oxidized to methyl ketones, whilst internal olefins invariably lead to mixtures of ketones. Usually, these reactions require prolonged times at elevated temperature, and often a co-solvent such as *N*, *N*-dimethylformamide is used. The reaction may not proceed in the desired direction if the

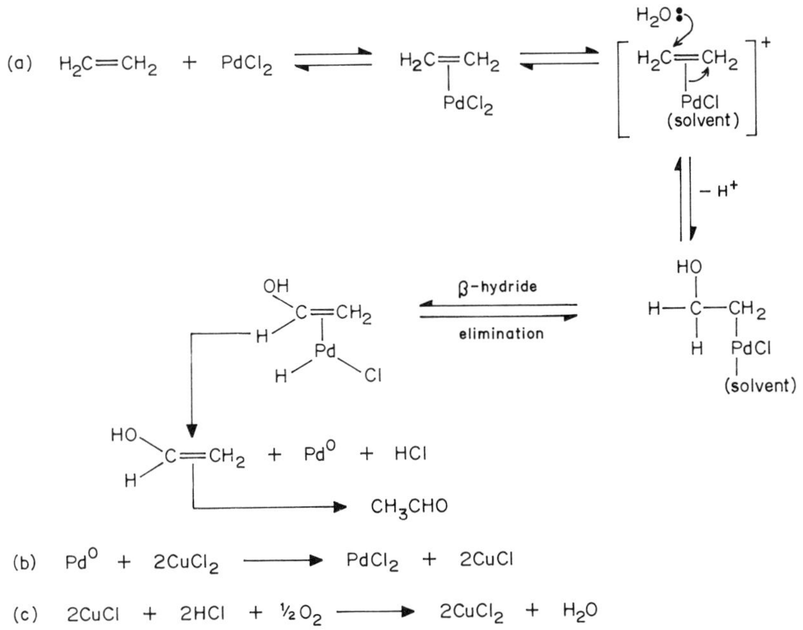

(a) $H_2C{=}CH_2 + PdCl_2 \rightleftharpoons H_2C{=}CH_2(PdCl_2) \rightleftharpoons [H_2C{=}CH_2 \cdot PdCl(\text{solvent})]^+$ (H_2O attack); $-H^+$ gives $HO{-}CH_2{-}CH_2{-}PdCl(\text{solvent})$; β-hydride elimination gives $HO(H)C{=}CH_2 \cdot Pd(H)Cl$ → $HO(H)C{=}CH_2 + Pd^0 + HCl$ → CH_3CHO

(b) $Pd^0 + 2CuCl_2 \longrightarrow PdCl_2 + 2CuCl$

(c) $2CuCl + 2HCl + \frac{1}{2}O_2 \longrightarrow 2CuCl_2 + H_2O$

FIGURE 4. Wacker process for the oxidative hydrolysis of ethylene to acetaldehyde

correct solvent combination is not used, as shown in equations 30 and 31 for the oxidation of 3, 7-dimethylocta-1, 6-diene.

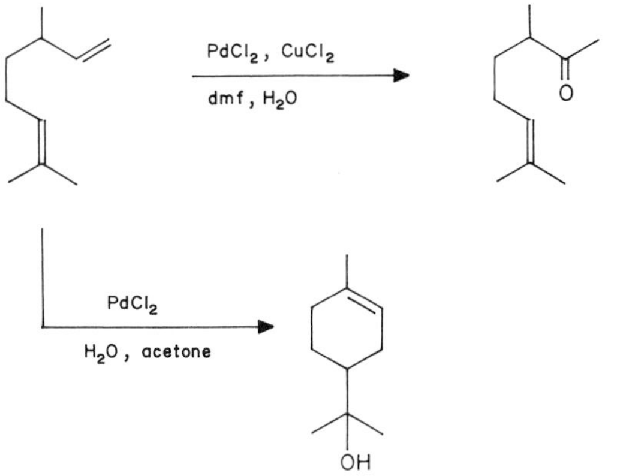

$PdCl_2$, $CuCl_2$; dmf, H_2O (30)

$PdCl_2$; H_2O, acetone (31)

However, high degrees of selectivity can be obtained under appropriate conditions, and terminal alkenes are oxidized more readily than internal double bonds in aqueous dmf solvent, as illustrated in equations 32–34[17].

$PdCl_2$, $CuCl_2$
O_2, H_2O, dmf
(32)

OAc
$PdCl_2$, $CuCl_2$
O_2, H_2O, dmf
OAc
(33)

CO_2R
$PdCl_2$, $CuCl_2$
O_2, H_2O, dmf
CO_2R
(34)

Often a terminal alkene group is very easily introduced into an organic molecule by means of, e.g., a Grignard reagent, and so the Wacker oxidation in conjunction with such strategies offers a superb means of constructing intermediates in which the (latent) ketone functionality can be utilized for further reaction, most commonly annulation. A very large number of natural products syntheses have been developed around this method, some examples of which are given in equations 35–40.

(ref. 18)
jasmone
(35)

CO_2Et
CO_2Et
(ref. 19)
muscone
(36)

OAc
OAc
(ref. 20)
HO
diplodialide
(37)

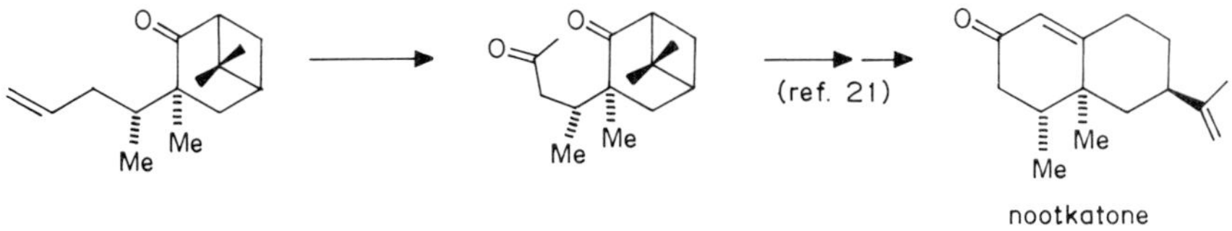

(ref. 21)

nootkatone

(38)

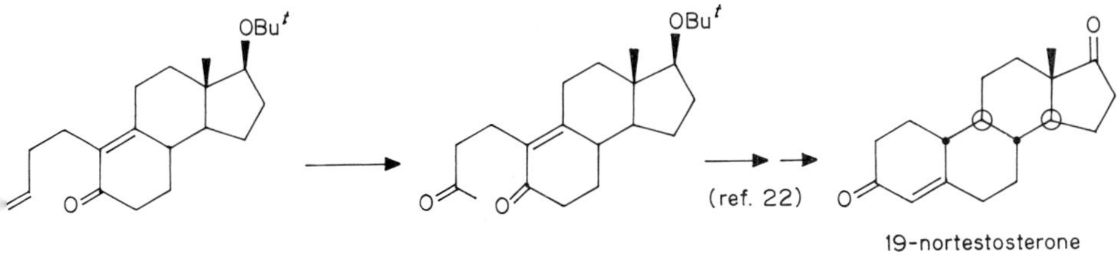

(ref. 22)

19-nortestosterone

(39)

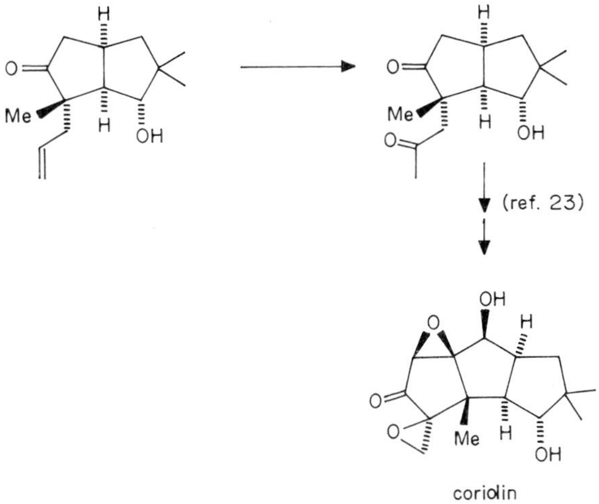

(ref. 23)

coriolin

(40)

The conversion of a terminal olefin to a methyl ketone can be performed in the presence of sensitive or oxidizable functional groups, so it displays high chemoselectivity. In the presence of a 1, 2-diol an acetal may be formed directly, instead of the usual ketone. Some examples of these aspects of the reaction are given in equations 41–45.[24]

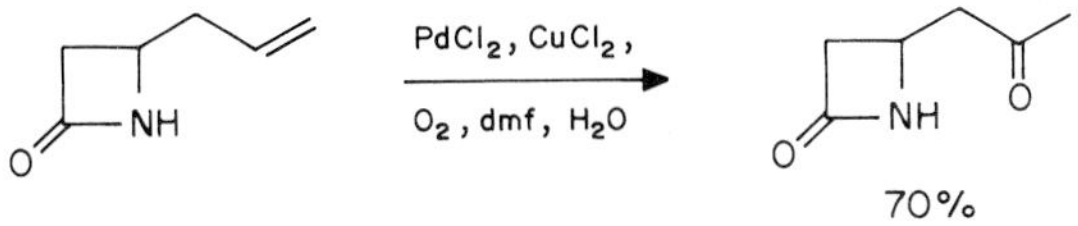

$PdCl_2$, $CuCl_2$, O_2, dmf, H_2O

70%

(41)

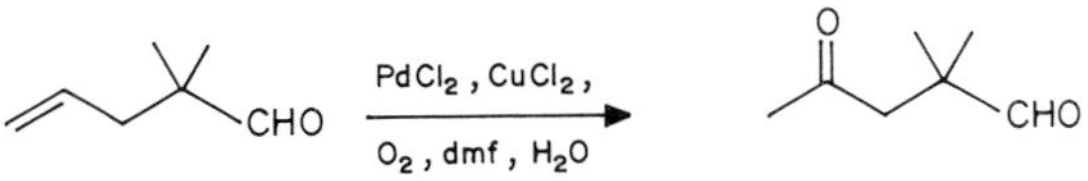

$PdCl_2$, $CuCl_2$, O_2, dmf, H_2O

(42)

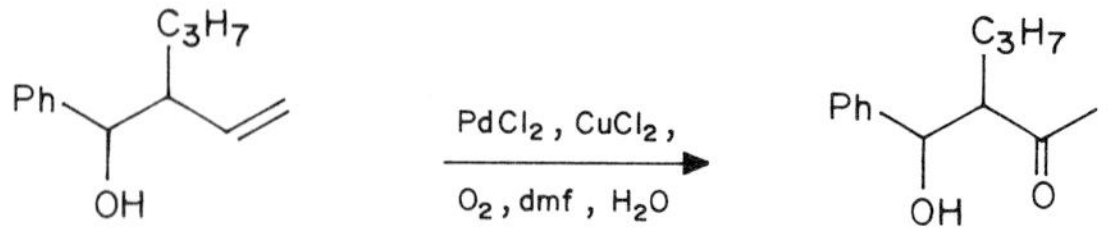

$$\text{(43)}$$

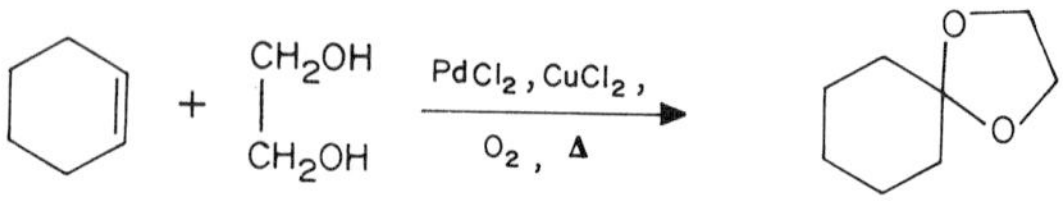

$$\text{(44)}$$

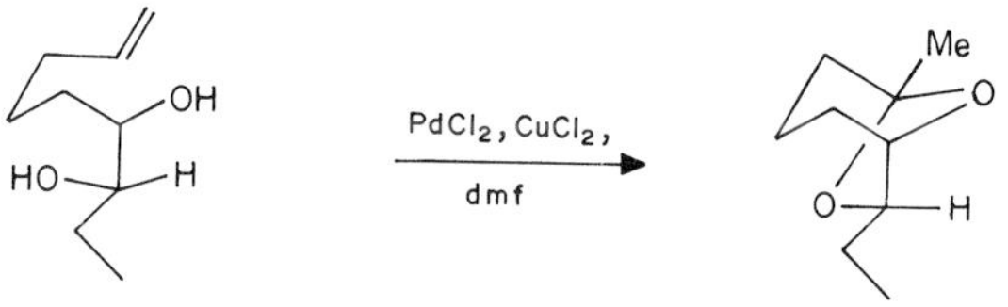

$$\text{(45)}$$

Careful attention to reaction conditions for any organic reaction often results in improvements in yield and or selectivity, and this is true for the Wacker oxidation. For example, by using *tert*-butyl hydroperoxide in aqueous acetic acid it is possible to effect high yield conversion of α,β-unsaturated esters to β-keto esters (equations 46 and 47)[25]. This contrasts with unconjugated internal olefins which only lead to π-allyl palladium complexes under similar conditions.

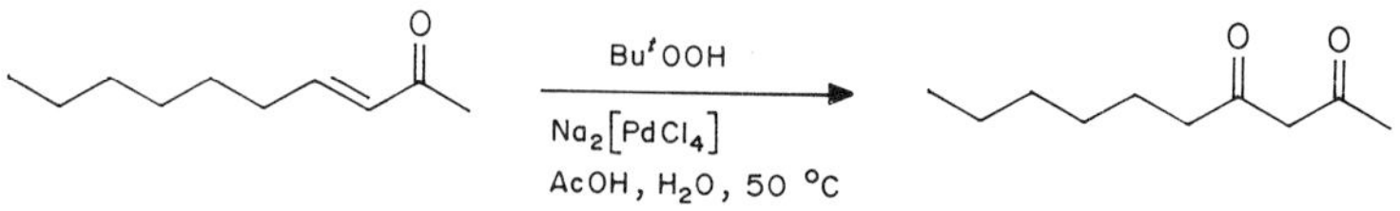

$$\text{(46)}$$

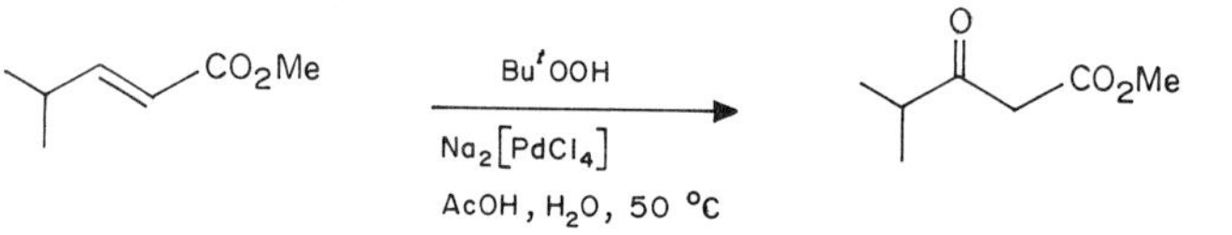

$$\text{(47)}$$

When the water is replaced with acetic acid and sodium acetate, ethylene is converted cleanly into vinyl acetate[26], but problems of regioselectivity arise when higher alkenes are employed[27], often leading to the allylic rather than the vinyl acetate. When a 1, 5-diene is treated in this manner, acetoxylation and cyclization occur to give substituted five-membered ring compounds[28]. Some examples of these type of reaction are shown in equations 48–53.

$$CH_2{=}CH_2 \xrightarrow[\text{or Pd(OAc)}_2]{\text{PdCl}_2\text{, AcONa, AcOH}} CH_2{=}CHOAc \tag{48}$$

$$RCH_2CH{=}CH_2 \xrightarrow{\text{Pd(OAc)}_2} RCH_2\underset{}{\overset{\text{OAc}}{\overset{|}{C}}}{=}CH_2 + RCH_2CHOAcCH_2OH + RCH_2CH{=}CHOAc + RCH{=}CHCH_2OAc + RCH_2CHOHCH_2OAc \tag{49}$$

$$\text{cyclohexene} \xrightarrow{Pd^{2+},\ AcOH} \text{3-acetoxycyclohexene} + \text{4-acetoxycyclohexene} \quad (50)$$

$$CH_2{=}CHMe \xrightarrow[AcOH,O_2]{Pd^{2+},Cu^{2+}} CH_2{=}CHCH_2OAc \quad (51)$$

$$\text{1,5-hexadiene} + Pd(OAc)_2 \longrightarrow \text{3-methylenecyclopentyl acetate} \ (64\%) \quad (52)$$

$$\text{1,5-cyclooctadiene} + Pd(OAc)_2 \longrightarrow \text{acetoxybicyclo[3.3.0]octene} \quad (53)$$

Further modification of the Pd-catalysed reaction of olefins with hydroxylic solvents can be accomplished by including carbon monoxide in the reaction medium, when carbonyl insertion occurs, usually resulting in carboxyalkylation[29].

$$\text{cyclopentene} + Pd^{2+} + CO \longrightarrow (\text{cyclopentene})Pd^{2+}(CO) \xrightarrow{MeOH} (\text{cyclopentene})Pd^{+}(CO_2Me) \longrightarrow \text{2-(}Pd^{+}\text{)cyclopentyl-}CO_2Me \longrightarrow (\text{cyclopentenyl-}CO_2Me)Pd^{2+} \longrightarrow\longrightarrow \textit{cis}\text{-1,2-(}CO_2Me)_2\text{-cyclopentane} + \textit{cis}\text{-1,3-(}CO_2Me)_2\text{-cyclopentane} + \text{minor products} \quad (54)$$

$$H_2C{=}CH_2 \xrightarrow[CO,\ MeCN]{PdCl_2,\ H_2O,} \beta\text{-propiolactone} \quad (55)$$

$$(\text{1,5-cyclooctadiene})PdCl_2 \xrightarrow[NaOAc]{(i)\ OH^-\ (ii)\ CO,\ H_2O,} \text{bicyclic }\beta\text{-lactone} \quad (56)$$

Semmelhack and Bodurow[30] combined intramolecular oxypalladation with the carbomethoxylation reaction to give high-yield conversions of hydroxyalkenes to tetrahydropyran derivatives with functionalized side-chains. These products were formed in a highly stereoselective manner, via η^2-alkene–Pd complexes, and are recognized as potential subunits for ionophore antibiotic synthesis[30].

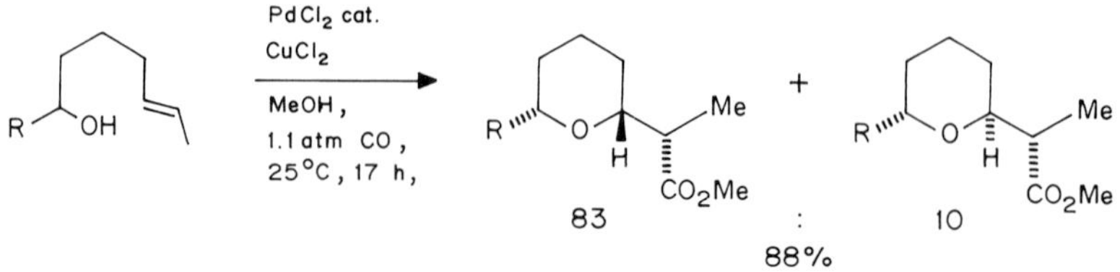

PdCl$_2$ cat.
CuCl$_2$
MeOH,
1.1 atm CO,
25 °C, 17 h,

83 : 10
88%

(57)

PdCl$_2$ cat.
CuCl$_2$, MeOH,
1.1 atm CO,
25 °C, 17 h

+ traces of isomers

82%

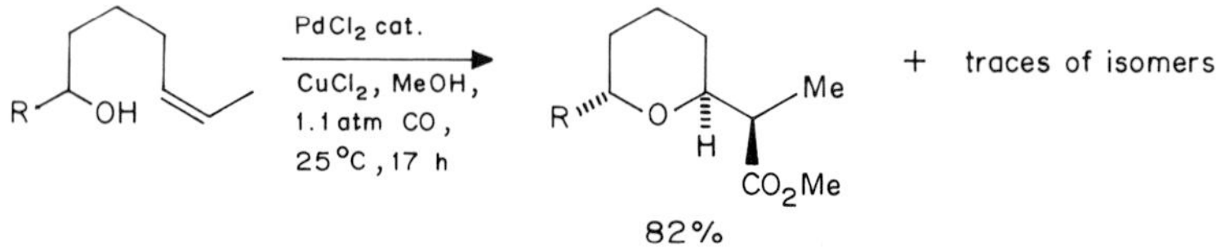

(58)

The above Wacker-type reactions involve addition of an oxygen nucleophile to a complexed alkene ligand. Nitrogen nucleophiles may also be used in similar Pd-catalysed reactions to give a variety of products, depending on the substrate and the reaction conditions. Although it has proved possible to isolate enamines and related products by appropriate choice of the reaction conditions, this is by no means general. A more useful procedure is to reduce the intermediate amino-substituted σ-complex to the amine using sodium borohydride or hydrogen, or to oxidize the Pd—C bond using, e.g., lead tetraacetate. Some examples of these processes are given in equations 59–61 (ref. 16, p. 32).

$CH_2=CH_2 \cdot PdCl_2$ + pyrrolidinone → N-vinylpyrrolidinone + Pd + HCl (59)

NH$_3$
benzene

NaOEt

NaBH$_4$

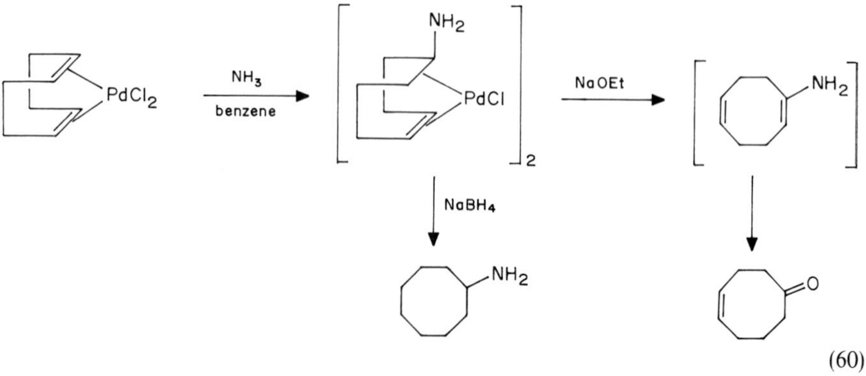

(60)

$$C_8H_{17}CH{=}CH_2 \xrightarrow[\text{[(PhCN)}_2\text{PdCl}_2]]{Me_2NH} C_8H_{17}CH(NMe_2)CH_2PdCl + C_8H_{17}CH(PdCl)CH_2NMe_2$$

$\xrightarrow{Pb(OAc)_4}$ $C_8H_{17}CH(NMe_2)CH_2OAc$ (major) + $C_8H_{17}CH(OAc)CH_2NMe_2$ (minor)

$\xrightarrow{NaBH_4 \text{ or } H_2}$ $C_8H_{17}CH(NMe_2)Me$ (major) + $C_8H_{17}CH_2CH_2NMe_2$ (minor)

(61)

If a primary amine is used in this reaction, the intermediate aminoalkylpalladium derivative can be induced to cyclize to give an aziridine by treatment with bromine[31].

R-CH=CH$_2$ $\xrightarrow[\text{Li[PdCl}_4], \text{ thf, } -50\,^\circ\text{C}]{MeNH_2}$ [RCH(NHMe)CH$_2$PdCl] $\xrightarrow{Br_2}$ 1-methyl-2-R-aziridine (62)

Intramolecular aminopalladation of alkenes is particularly interesting, since it allows access to heterocyclic systems present in a number of naturally occurring alkaloids. Some examples of these reactions, of varying complexity, are given in equation 63–66.

2-allylaniline (NH$_2$) $\xrightarrow[\text{(ii) Et}_3\text{N}]{\text{(i) [PdCl}_2\text{(MeCN)}_2\text{], thf}}$ 2-methylindole (N–H, Me) (63)

(ref.32)

acryloylurea (O, NH, O, NH$_2$) $\xrightarrow{PdCl_2}$ uracil (64)

(ref.33)

N-methyl-2-allylbenzamide (NHMe, O) $\xrightarrow[\text{NaH}]{\text{[PdCl}_2\text{(MeCN)}_2\text{]}}$ N-methyl-3-methylisoquinolinone (Me, NMe, O) (65)

(ref.34)

methylenedioxy aryl amine (NH, OMe, OMe) $\xrightarrow{\text{[PdCl}_2\text{(PhCN)}_2\text{]}}$ tetrahydroprotoberberine (N, OMe, OMe) (66)

(ref.35)

Alkene–palladium complexes are also reactive toward carbon nucleophiles, as might be expected. There are examples of the use of palladium both catalytically[36] and stoichiometrically[37], although obviously in most cases the former is more desirable. The use of pre-formed η^2-alkene complexes does give mechanistic information which is valuable for interpretation of the catalytic reactions. There are far too many important examples of carbon—carbon bond formation involving intermediate alkene–Pd cations in one form or another to give a complete coverage here, and the reader is referred to the more specialist chapters for further information.

The so-called Heck reaction involves coupling of an aryl or vinyl halide with a suitable olefinic substrate, usually an α,β-unsaturated ester. The reaction proceeds by oxidative addition of the halide to a palladium(0) catalyst to give a σ-aryl (or vinyl)–palladium(II) species. Coordination of the substrate olefin to this produces a transient η^2-alkene–palladium species, which undergoes intramolecular coupling with eventual expulsion of the product (reaction 67)[38].

ArBr + [Pd(PR$_3$)$_n$] ⟶ [(R$_3$P)(Br)Pd(Ar)(PR$_3$)] —CH$_2$=CHCO$_2$Me⟶ [R$_3$P—Pd(Ar)(Br)—(η^2-CH$_2$=CHCO$_2$Me)] ⟶ ArCH=CHCO$_2$Me (67)

In fact, it is not necessary to use an aryl halide in this reaction; using appropriate conditions the σ-aryl (or vinyl)–palladium intermediate can be generated from the hydrocarbon and coupled with the alkene. Equations 68 and 69 show examples of the

Benzene —Pd(OAc)$_2$, HOAc, 110 °C⟶ [PhPdOAc] —Ph-CH=CH$_2$⟶ PhCH=CHPh 90% (68)

Indole substrate —PdCl$_2$, AgBF$_4$, MeCN⟶ [2-PdX indole intermediate] ⟶ [PdX cyclized intermediate] —NaBH$_4$⟶ Iboga alkaloids (69)

simple basic reaction[39] and an interesting intramolecular variation which was used as a key step for construction of the complex alkaloids ibogamine and catharanthine[49].

There are a number of well characterized reactions involving the combination of stable enolate anions with isolated alkene–Pd complexes. The reaction of cycloocta-1, 5-dienepalladium dichloride with malonate anion is a particularly well known example[41].

PdCl2 → [Pd—Cl, CH(CO2Et)2]2; NaCH2SOCH3 → 51%; Na2CO3 → 77%

(70)

In these reactions, which involve 'soft' carbanion nucleophiles, the nucleophilic entity attacks the olefin–Pd complex *trans* to the metal groupling. When this fact is coupled with the ability of hetereoatoms, attached to olefinic ligand, to direct the stereochemistry of complex *formation*, some elegant stereocontrolled bond formations can be obtained, as illustrated in the approach to prostaglandin $PGF_{2\alpha}$ described by Holton[42], shown in Figure 5.

The reaction of enolates with alkenes can also be accomplished using a catalytic procedure. Usually these reactions have been performed using highly stabilized enolates, which require the presence of amine bases for best results, but incorporation of hexamethylphosphorotriamide into the reaction mixture allows successful vinylation of less stable enolates.

ONa, CO2Et + $CH_2{=}CH_2$ —[PdCl2(PhCN)2], Et3N, thf, −60 °C (ref. 43)→ CO2Et, 78% (71)

$$RCH{=}CH_2 + NaCMe(CO_2Et)_2 \xrightarrow[\text{Et}_3\text{N, thf, } -50 \text{ to } -25\ ^\circ\text{C (ref. 44)}]{[PdCl_2(MeCN)_2]} CH_2{=}C(R)CMe(CO_2Et)_2 \qquad (72)$$

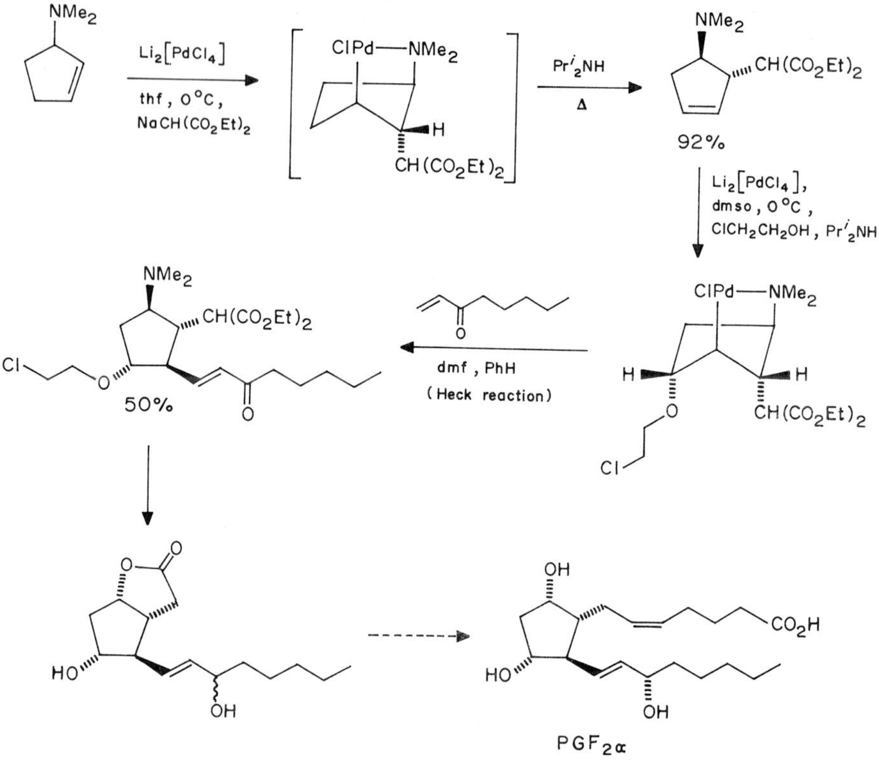

FIGURE 5. Synthesis of prostaglandin $PGF_{2\alpha}$ (from ref. 42)

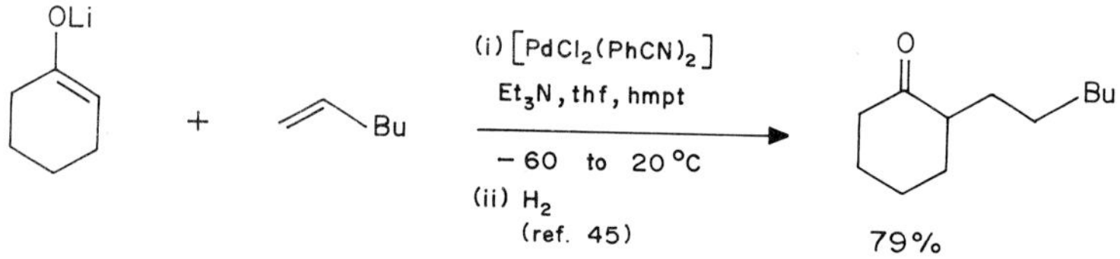

(73)

Although the above survey of this very important area of organic synthesis is not comprehensive, it does serve as an illustration of the types of reaction which may be accomplished using the technique of alkene activation by a palladium catalyst. A range of olefin complexes can be formed using other transition metals, but the synthetic potential of these compounds has not been explored to the same degree as the iron and palladium derivatives discussed above, so we shall not discuss them here.

IV. η^3-ALLYL COMPLEXES

A. η^3-Allylpalladium Complexes

1. Stoichiometric reactions

Again, palladium figures very prominently as a metal which can be used to stabilize allyl cations for synthetic purposes, and so we shall deal with this metal in some detail, presenting other potentially useful complexes later. As with the η^2-alkene–Pd complexes discussed in the previous section, η^3-allyl–Pd complexes can be formed either catalytically or as stoichiometric reagents. A discussion of stoichiometric reactions gives a good understanding of the chemistry involved, so this will be presented first. A number of methods exist for the preparation of the η^3-allyl–palladium complexes, usually isolated as stable chloride-bridged dimers. Perhaps the most useful in terms of synthetic application is the direct reaction of palladium chloride with an alkene. As noted in the previous section, this will initially result in a η^2-alkene–Pd complex. Provided that the correct reaction conditions are chosen, this complex will undergo intramolecular oxidative addition of an allylic C—H bond to the palladium moiety, resulting in an η^3-allylpalladium hydride which loses HCl to generate the η^3-allylpalladium chloride dimer. Naturally, there will be some stereochemical constraints on the formation of η^2-alkene complexes in certain cases, and this often results in a preferred regiochemistry during C—H bond cleavage to give the η^3-allyl complex. The overall process may be represented as in equation 74.

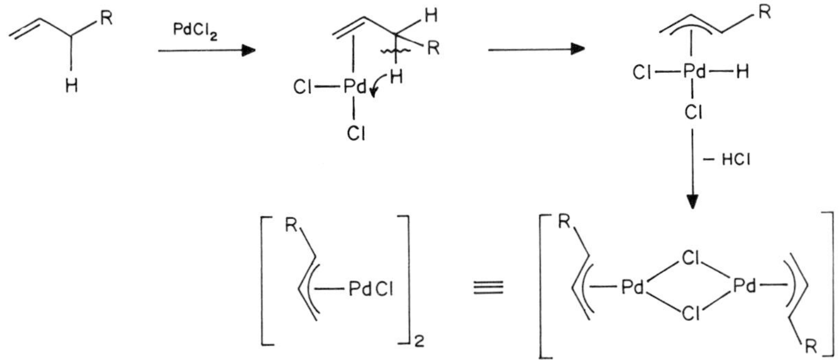

PdCl2; Cl—Pd—H; Cl; − HCl; PdCl; Pd; Cl; Cl; Pd; R

(74)

A number of cocktails have been developed for this method of preparation, since simple direct treatment of the alkene with $PdCl_2$–AcOH–H_2O usually results in low yield of η^3-allyl–Pd complex. The most successful mixture of reagents is probably $PdCl_2$–NaCl–$CuCl_2$–NaOAc–AcOH. Selected examples of this method of preparation (reactions 75–81) give some idea of the regiochemistry which can be expected and also some of the factors to keep in mind in designing a synthesis.

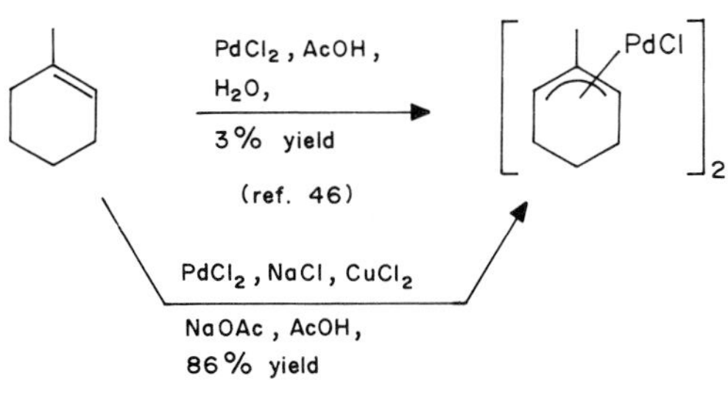

PdCl2, AcOH, H2O, 3% yield (ref. 46); PdCl2, NaCl, CuCl2, NaOAc, AcOH, 86% yield; PdCl

(75)

PdCl2, NaCl; CuCl2, NaOAc, AcOH (ref. 47); PdCl; 68%

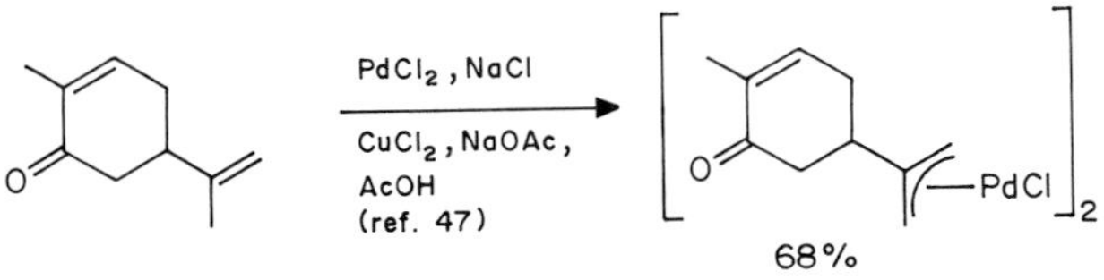

(76)

In certain cases, it may be found that best yields are obtained using, for example, palladium(II)trifluoroacetate, but the η^3-allyl complex product is not particularly stable. This is readily overcome by converting the complex into the more stable chloride-bridged dimer by treatment with, for example, tetrabutylammonium chloride.

Various η^3-allyl–palladium complexes can be prepared by transmetallation, reacting an allyl Grignard reagent[51] (or allylsilane[52]) with palladium(II) chloride, or directly from the

$PdCl_2$, NaCl, $CuCl_2$, NaOAc, AcOH (ref. 48)

AcO … PdCl … 2

92%

(77)

C_8H_{17}

$Na_2[PdCl_4]$, thf (ref. 49)

O … PdCl … 2

(78)

$PdCl_2$, NaCl, $CuCl_2$, NaOAc, AcOH (ref. 46)

N Ac … PdCl … N Ac … 2

71%

(79)

(i) $[Pd(O_2CCF_3)_2]$ (ii) Bu_4NCl

O … PdCl … O … 2

$PdCl_2$, $CuCl_2$, NaOAc

PdCl … O … 2 + PdCl … O … 2

(80)

C_8H_{17} PdCl$_2$, KOAc / AcOH, 55 °C (ref. 50) ClPd C_8H_{17}]$_2$

Only 1α-H (axial) abstracted by α-Pd group.

(81)

allylic chloride by treatment with sodium tetrachloropalladate (equations 82 and 83)[53]. There are also a number of variations on these procedures which are more appropriately discussed in specialist reviews[54].

MgCl + $PdCl_2$ → [Pd] (82)

Cl + $Na_2[PdCl_4]$ → [PdCl]$_2$ (83)

It has also been found that nucleophilic addition to dienes, and similar compounds such as vinyl- and methylene-cyclopropanes, in the presence of palladium salts results in the formation of η^3-allyl–palladium complexes, although these methods are probably less useful in terms of application to organic synthesis (equations 84–87)[55].

$[PdCl_2(PhCN)_2]$ → Cl^- (PhCN) Pd^+Cl → Cl (PhCN)Pd—Cl → [PdCl Cl]$_2$

(84)

PhHgCl / $Li_2[Pd_2Cl_6]$ MeCN → [Ph PdCl]$_2$ (85)

$Na_2[PdCl_4]$ / MeOH → [OMe PdCl]$_2$ (86)

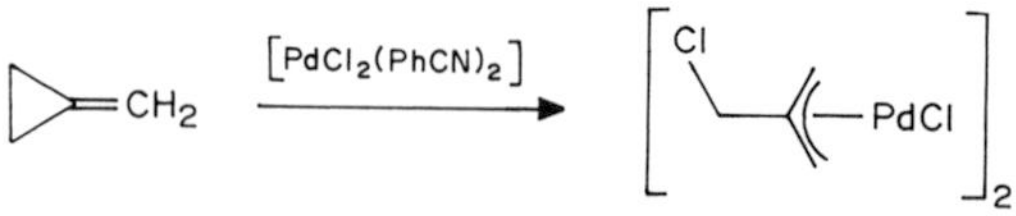

(87)

One aspect of the chemistry of η^3-allyl–palladium complexes which is important to understand if effective synthetic planning is required is their ability to undergo isomerization. For example, *anti*-isomers will usually be converted to *syn*-isomers during their preparation. Consequently, if an *anti*-complex is formed kinetically and is required as an intermediate, reaction conditions must be found where the reaction of the complex (e.g. with nucleophiles) is more rapid than *syn*/*anti* interconversion. This is especially true in catalytic reactions. The presence of other substituents may reverse this stability trend, due to steric effects, as indicated in equations 88 and 89.

(88)

(89)

The stoichiometrically-formed η^3-allyl complexes undergo a number of interesting and useful reactions, the most important being reaction with nucleophiles, giving access to new methods of carbon—carbon and carbon—heteroatom bond formation. During these reactions advantage may be taken of the stereo- and regio-chemical directing ability of the palladium group. First, let us examine some methods for C—C bond formation and see how they might be used synthetically. It is readily recognized that the chloride-bridged dimers described above do not carry a formal positive charge and, as expected, these complexes are not very reactive towards nucleophiles as they stand. However, activation is readily achieved on splitting the chloride bridge by reaction with an appropriate ligand. This might be a phosphine ligand added to the reaction mixture or it might be a molecule of solvent, such as dimethyl sulphoxide. The net result is to produce a positively charged η^3-allyl complex which is now much more reactive towards nucleophiles:

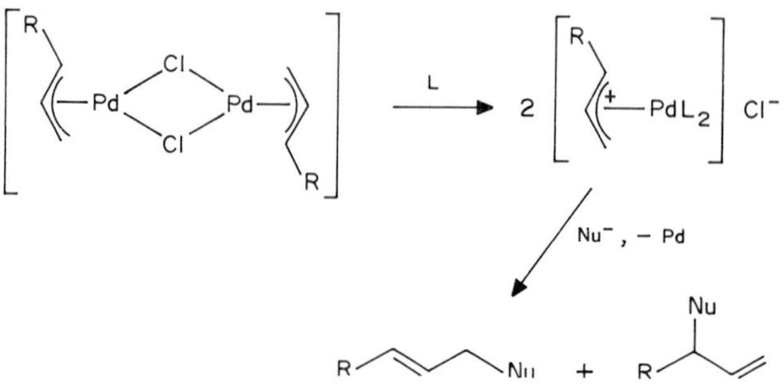

(90)

With unsymmetrical η^3-allyl complexes there are, of course, two possible sites of attack and the preferred regiochemistry is often determined by complex and fairly diverse factors: the electronic or steric requirements of the ligand *vs.* the steric requirements of the palladium moiety. The latter is often dependent on the nature of the activating ligand, making it possible to tailor the reaction conditions to give a desired product. A particularly good example of this effect is shown in equation 91[56].

NaCH(SO$_2$Ph)CO$_2$Me, L

CH(SO$_2$Ph)CO$_2$Me + CH(SO$_2$Ph)CO$_2$Me

L	Ratio	Total yield
(o-MeC$_6$H$_4$)$_3$P	15:85	79–90%
Ph$_3$P	62:38	75%
hmpt, dmso	100:0	90%

(91)

Thus, the sterically more demanding tri-*o*-tolyphosphine induces the nucleophile to attack at the more crowded allyl position so as to give a less sterically encumbered η^2-alkene–Pd initial product. A range of other examples of C—C bond formation are given in equations 92–95, from which it is readily observed that most reactions have been performed using softer carbanions.

OH, PdCl, NaCH(CO$_2$Et)$_2$, dmso, RT (ref. 57), OH, 90%, CH(CO$_2$Et)$_2$ (92)

MeO, PdCl, NaCH(SO$_2$Ph)CO$_2$Me, Ph$_2$PCH$_2$CH$_2$PPh$_2$, thf, 25 °C (ref. 58), H, SO$_2$Ph, CO$_2$Me, MeO, 82% (93)

CO$_2$Et, PdCl, NaCH(CN)CO$_2$Et, dmso, EtOH (ref. 59), CO$_2$Et, NC, CO$_2$Et, 79% (94)

(95)

A number of other, non-enolate, nucleophiles have been reacted successfully with η^3-allyl–palladium complexes, although these appear to be less general. Thus, addition of certain aryl–thallium species and dimethylcadmium proceed satisfactorily. With these hard nucleophiles the stereochemical course of the reaction is often reversed, possibly owing to initial addition of nucleophile to the coordinatively unsaturated metal followed by a coupling which is effectively a reductive elimination from the organopalladium intermediate. This is particularly evident in the reaction of stereochemically defined steroidal complexes with vinylzirconium reagents which give products having opposite stereochemistry at C-20 than is obtained using soft nucleophiles (equation 96)[61].

(96)

There are, of course, many other ways to manipulate η^3-allyl–palladium complexes to give useful products, such as oxidation, deprotonation, and reaction with amine nucleophiles, some examples of which are shown in equations 97–100.

(97)

(98)

(99)

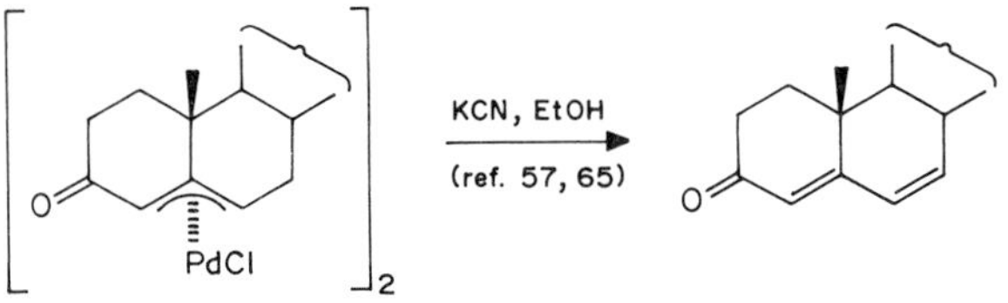
KCN, EtOH
(ref. 57, 65)
PdCl
O

(100)

2. Palladium-catalysed allylic substitution

Zerovalent palladium complexes react very readily with allylic substrates, such as allylic halides and acetates, to form η^3-allyl–palladium complexes. These behave as palladium-stabilized allyl cations in an analogous manner to the complexes dealt with above, and the process can be carried out using catalytic amounts of palladium complex. The general process of palladium-catalysed displacement of, e.g., allylic acetate group by a nucleophile is depicted in Scheme 1.

In this cycle, the palladium(0) catalyst is regenerated subsequent to nucleophile addition. The process is essentially equivalent to displacement of allylic halide by a nucleophile, but the intermediacy of the η^3-allyl–Pd complex allows many benefits, such as selectivity and stereocontrol. There are useful and interesting stereochemical consequences which will become apparent as our discussion proceeds.

We shall now consider a range of examples of application of this technology to natural products synthesis, illustrating various aspects of selectivity.

Reaction with carbon nucleophiles, mostly 'soft' enolate derivatives, allows selective carbon—carbon bond formation in the presence of other reactive groups which would themselves, under normal circumstances, react with the nucleophile. A good example is provided by the reactivity of 8-bromo-3-acetoxyoct-1-ene (reaction 101)[66]. The syntheses

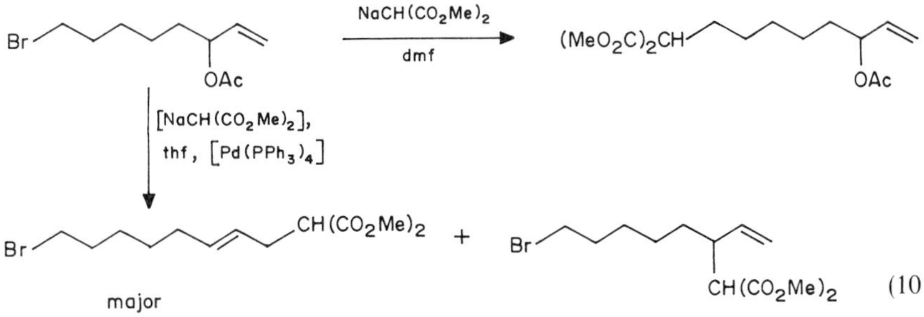
Br
OAc
NaCH(CO2Me)2
dmf
(MeO2C)2CH
OAc
[NaCH(CO2Me)2],
thf, [Pd(PPh3)4]
Br
CH(CO2Me)2
+
Br
major
CH(CO2Me)2

(101)

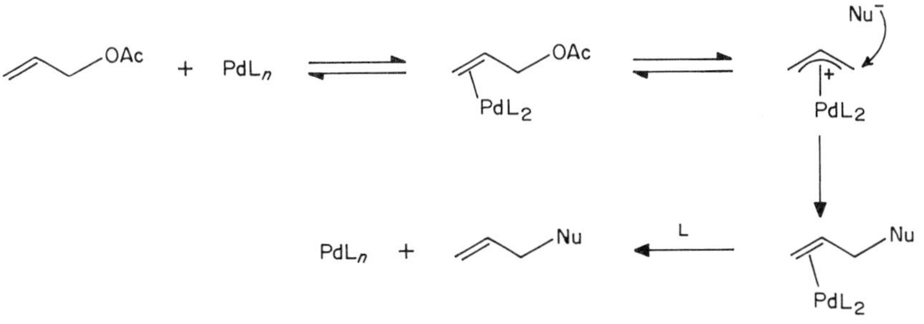
OAc
+ PdLn
OAc
PdL2
Nu
+
PdL2
PdLn +
Nu
L
Nu
PdL2

SCHEME 1

in reactions 102–112 give an indication of the types of carbon nucleophile, allylic substrate, and reaction conditions which can be used, and show the kinds of product which can be expected.

NaCH(SO$_2$Ph)CO$_2$Me, [Pd(PPh$_3$)$_4$], thf, Δ (ref. 66); product CH(SO$_2$Ph)CO$_2$Me, 79% (102)

NaCH(CO$_2$Me)$_2$, [Pd(PPh$_3$)$_4$], dmso, 120 °C (ref. 66); products CH(CO$_2$Me)$_2$ 43% + CH(CO$_2$Me)$_2$ 8% (103)

(i) NaCH(CO$_2$Me)$_2$, [Pd(dppe)$_2$], thf (ii) LiI, NaCN, dmf, Δ (ref. 67); starting material SO$_2$Ph, CO$_2$Me; product MeO$_2$C, CO$_2$Me, monarch butterfly pheromone diester (104)

CH$_2$(SO$_2$Ph)CO$_2$Me, dbu, toluene, [Pd(PPh$_3$)$_4$], 80 °C (ref. 68); starting material tbdmsO, OEt, OAc; product OEt, SO$_2$Ph, CO$_2$Me, Otbdms, 62%; → → pyrenophorin (105)

[Pd(PPh$_3$)$_4$], dbu, toluene (ref. 69); OEt, OAc; product OEt → → (106)

(i) NaCH(CO$_2$Me)$_2$, [Pd(PPh$_3$)$_4$] (ii) CH$_2$N$_2$ (ref. 66); product CO$_2$Me, CO$_2$Me, CO$_2$Me (107)

(i) $NaCH(SO_2Ph)_2$, $[Pd(PPh_3)_4]$, thf; (ii) CH_2N_2 (ref. 70)

CO_2Me, $CH(SO_2Ph)_2$

(108)

PhCOO, Me, from D-glucose

$NaCH(CO_2Me)_2$, $[Pd(PPh_3)_4]$, thf (ref. 71)

$CH(CO_2Me)_2$, OCOPh, OH, H, Me, O

95%

MeO_2C, Me, Me

vitamin E side-chain

(109)

OAc, CO_2Me

$NaCH(CO_2Me)_2$, $[Pd(PPh_3)_4]$, PPh_3, thf, Δ (ref. 72)

$CH(CO_2Me)_2$, CO_2Me

(110)

OAc, CO_2Me

$NaCH(CO_2Me)_2$, $[Pd(PPh_3)_4]$, PPh_3, thf, Δ (ref. 72)

$CH(CO_2Me)_2$, CO_2Me

(111)

H, Me, OAc, O, O, H

$NaCH(SO_2Ph)CO_2Me$, $[Pd(PPh_3)_4]$, PPh_3, thf

H, Me, $CH(SO_2Ph)CO_2Me$, O, O, H

H, Me, O, H

(112)

All of the above examples use highly stabilized, 'soft' enolate nucleophiles to effect carbon—carbon bond formation. Silyl enol ethers and lithium enolates of ketones give poor alkylation results[73]. In contrast, stannyl enol ethers and allyltin reagents have been found to react satisfactorily with the η^3-allyl–palladium complexes formed in these catalytic cycles (reactions 113 and 114)[74,75]. A range of heteroatom nucleophiles can also

OSnBu3 + OAc —[Pd(PPh3)4], thf, RT→ 91%

(113)

R⌒⌒OAc + SnBu3 —[Pd(PPh3)4], thf, Δ→ 69% (R = Ph)

(114)

be used in conjunction with the allylating system. Nitrogen nucleophiles allow access to a number of alkaloids and related compounds, some key examples being given in equations 115 and 116.

AcO, OAc, OEt + PhCH2NHMe —[Pd(PPh3)4], Ph3P, thf, Δ (ref. 76)→ Me, CH2Ph, N, OAc, OEt

(115)

CO2Me, OAc + H2N, OMe, OMe —[Pd(PPh3)4], thf, RT (ref. 77)→ CO2Me, NHCHAr2 → → CO2H, NH2 (±)-gabaculine

(116)

The addition of amine nucleophiles can also be carried out intramolecularly, leading to complex polycyclic systems. This has provided useful methodology for the preparation of intermediates for Iboga alkaloids synthesis, and for the synthesis of the neurotoxin perhydrohistrionicotoxin (reactions 117–120).

OAc, NHCH2Ph —[Pd(PPh3)4], Et3N, MeCN, Δ→ PhCH2N

(117)

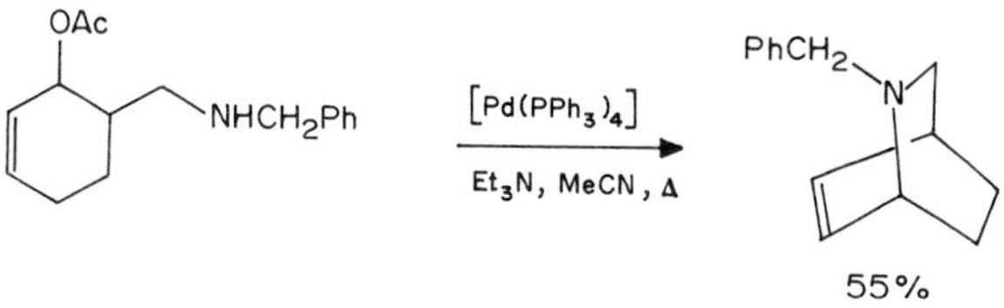

(118)

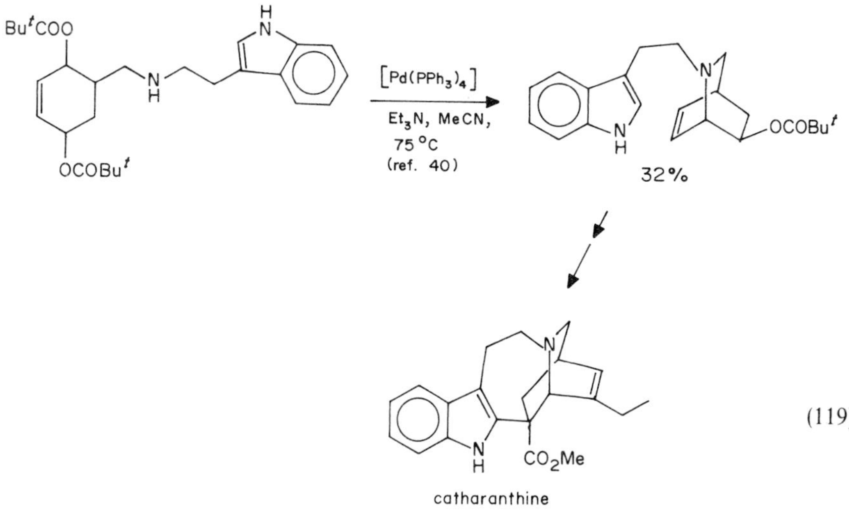

(119)

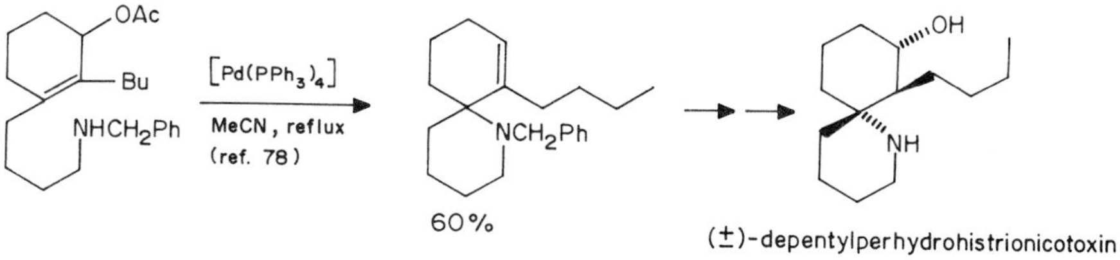

(120)

Acetate and other carboxylate anions can also be used as the nucleophilic portion in these reactions. This can lead to methods for the rearrangement of allylic acetates, alteration of stereochemistry, deprotection of allylic esters, and exchange of allylic ethers (reactions 121–123).

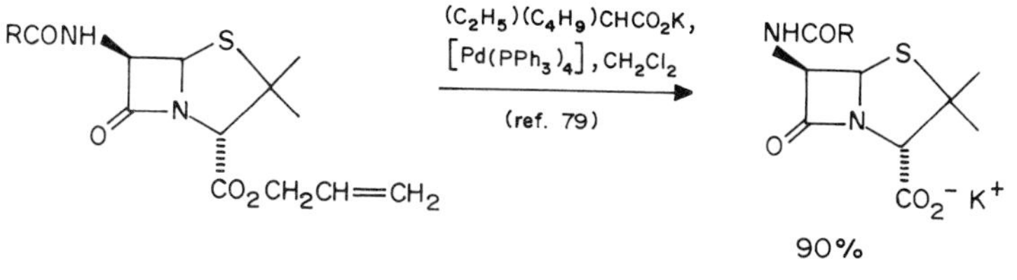

(121)

$$\mathrm{RCH{=}CHCH_2OPh} \xrightarrow[\mathrm{PPh_3,\ Ac_2O}]{[\mathrm{Pd(OAc)_2}]} \mathrm{RCH{=}CHCH_2OAc} \qquad (122)$$

(ref 80)

[PdCl₂(PhCN)₂] (ref. 81)

Prostaglandins

(123)

A considerable amount of work has recently been reported by Trost and coworkers in which stable enolate nucleophiles are reacted *intramolecularly* with allylic acetates in the presence of a palladium catalyst. This has led to new methods for ring formation which are complementary to the more usual 'organic' approaches and which, in many cases, lead to selectivity far different from that expected for organic reactions. For example, in many cases studied there is a preference for the formation of the larger of two possible ring sizes, e.g. eight- *vs.* six-membered rings (see below). In cases where carbocyclic ring formation is disfavoured over alternative pathways, palladium catalysis results in preference for carbon annulation, as shown in the formation of cyclopentane derivatives in reactions 124–126. These methods have allowed the development of novel and valuable approaches for the synthesis of macrocyclic lactones which have received considerable synthetic attention over the past few years, owing to their occurrence in many important (macrolide) antibiotics (reactions 124–127). Reaction 127 indicates a preference for formation of a six-

base

[Pd(PPh₃)₄]

(124)

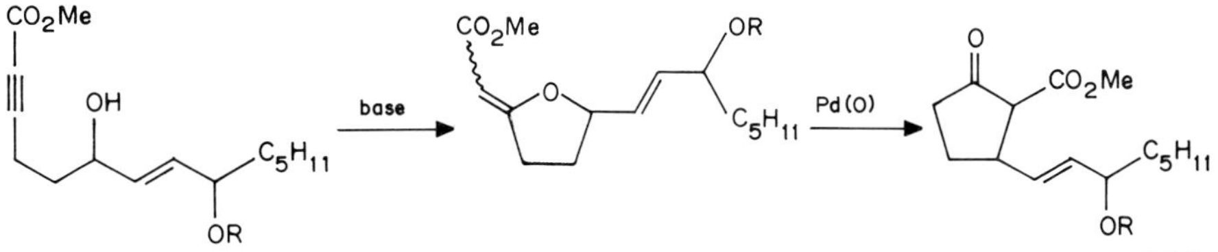

(125)

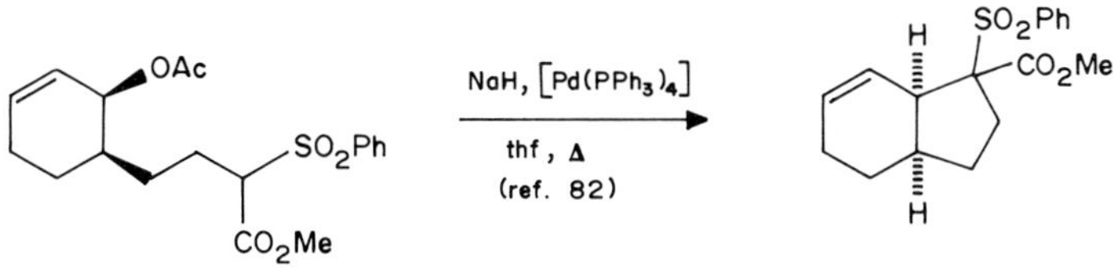

(126)

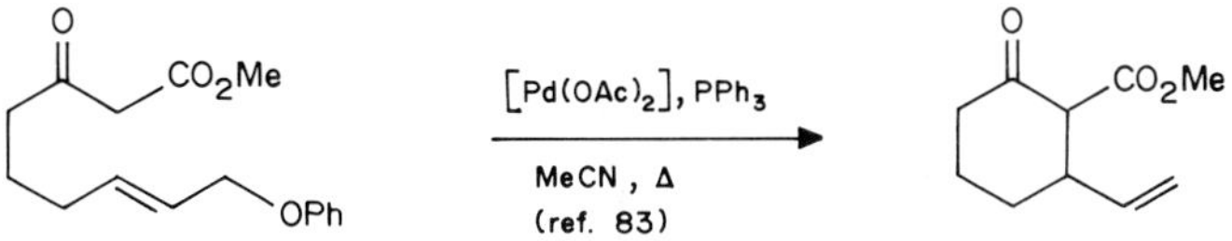

(127)

membered carbocyclic ring *vs.* and eight-membered ring, also a possible product. However, formation of oxygen-substituted rings leads to a preference for the larger ring as in reactions 128–131.

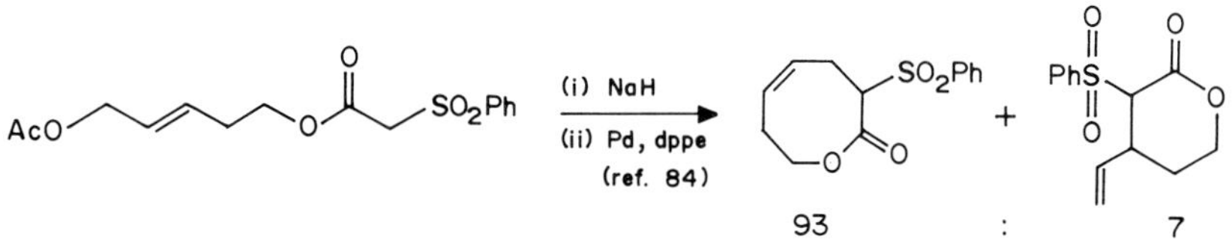

(128)

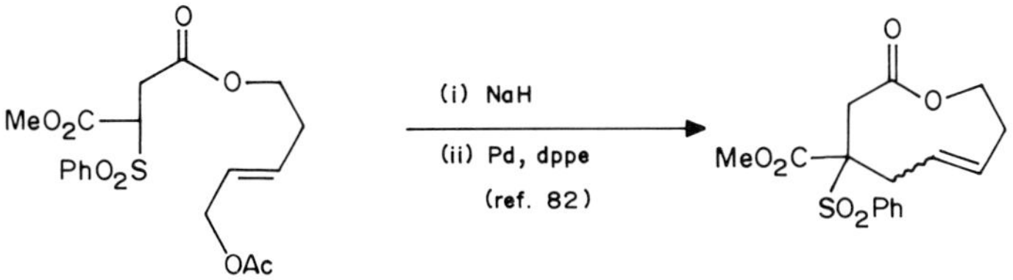

(129)

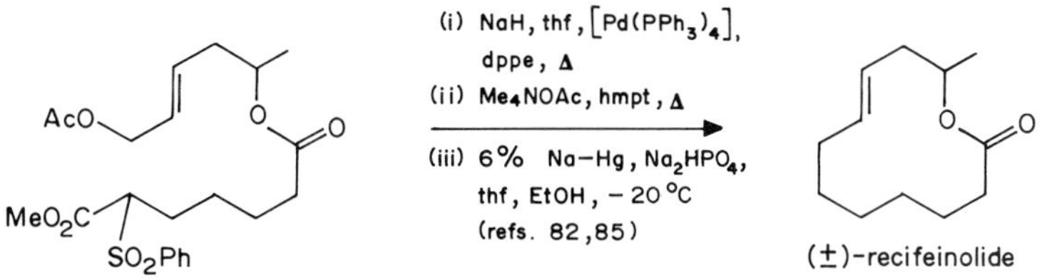

(130)

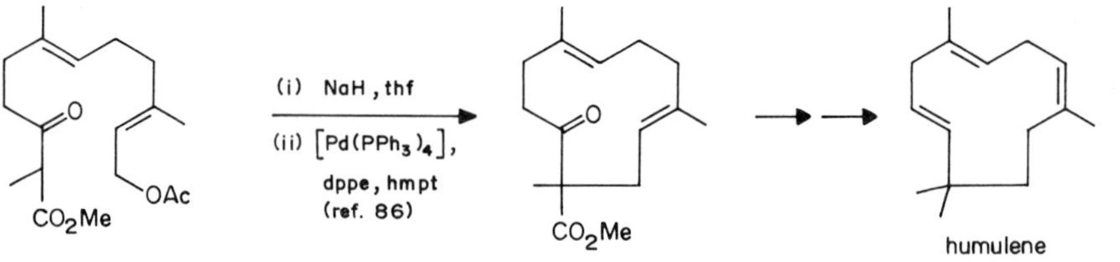

(i) NaH, thf
(ii) $[Pd(PPh_3)_4]$, dppe, hmpt
(ref. 86)

humulene

(131)

It has been found that the course of these cyclizations can be profoundly affected by the nature of the palladium catalyst. In particular, the steric bulk of the phosphine ligands present can affect the size of ring which is formed. For example, some of the reactions encountered could lead to either a five- or a seven-membered carbocycle, depending on the geometry of the η-allyl complex intermediate[87], as illustrated in Scheme 2.

In this example, the presence of a sterically demanding polymeric ligand on the catalyst disfavours the *syn/anti* interconversion of the η^3-allyl ligand, with the result that a five-membered ring is formed (the alternative cyclization would give a *trans*-cycloheptene derivative). When a sterically less demanding phosphine ligand is used, the *syn/anti* interconversion is rapid, and a seven-membered ring is formed in preference. Steric bulk of the ligands attached to the palladium catalyst also affects selectivity for formation of seven- *vs.* nine-membered carbocyclic rings. Thus, either ring size may be obtained depending on the reaction conditions (reaction 132).

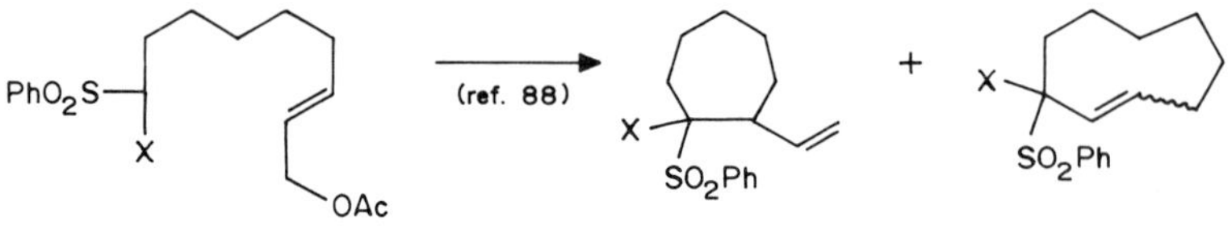

(ref. 88)

X = SO_2Ph; catalyst = $[Pd(dppe)_2]$ 0 : 100

X = CO_2Me; catalyst = $[Pd(PPh_3)_4]$ 93 : 7

(132)

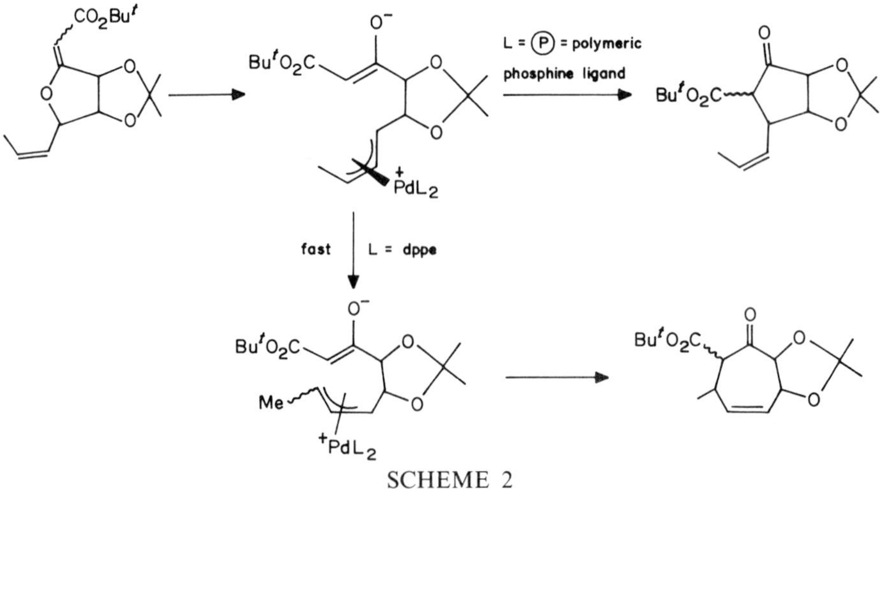

L = Ⓟ = polymeric phosphine ligand

fast L = dppe

SCHEME 2

η^3-Allyl–palladium complexes can undergo a number of other reactions apart from nucleophile addition. Thus, deprotonation will lead to a diene, whilst decarboxylation will occur with appropriate carboxylic acids, also giving a diene. We have already mentioned deprotonation of stoichiometrically formed complexes, and in reactions 133–136 we show these transformations using the metal catalytically. It is noteworthy that under these conditions conjugation with carbonyl groups is avoided.

CO_2Me OAc → [$Pd(PPh_3)_4$] Et_3N, thf, Δ (ref. 70) → CO_2Me (133)

OAc O O OAc O → [$Pd(PPh_3)_4$] Et_3N (ref. 70) → O O O 99% (134)

OAc CO_2H → [$Pd(PPh_3)_4$] Et_3N (ref. 89) → 71% (135)

thpO AcO CO_2H → (i) $Pd(PPh_3)_4$, Et_3N, thf (ii) H^+, H_2O → HO bombykol (136)

A recently developed method for accomplishing α-allylation of ketones utilizes the ability of allyl β-keto esters to give η^3-allyl–palladium complexes, coupled with decarboxylation of the palladium β-keto carboxylate. This generates an η^3-allyl cation and a ketone enolate, both bound to palladium, and the consequence is a coupling reaction (reaction 137).

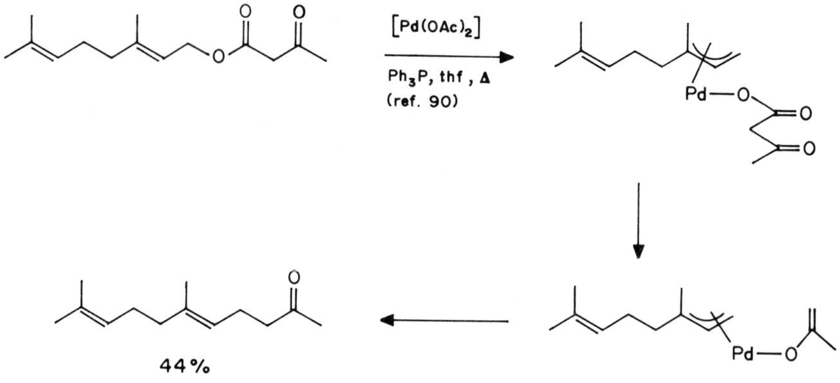

$[Pd(OAc)_2]$

Ph_3P, thf, Δ (ref. 90)

Pd—O

Pd—O

44%

(137)

B. η^3-Allylmolybdenum Complexes

These complexes are also well characterized. A range of derivatives have been prepared and these show interesting and potentially useful reactivity, although this has not been exploited to the same extent as the above palladium complexes. We shall focus our attention on the cationic, isolable, $[(\eta^3\text{-allyl})Mo(CO)(NO)Cp]$ complexes, which show some interesting reactivity toward nucleophiles, and some recent developments in which molybdenum carbonyl (and tungsten carbonyl) catalysts have been used to effect nucleophilic displacement of allylic acetate in much the same way as described for palladium. The most useful method for preparing $[(\eta^3\text{-allyl})Mo(CO)(NO)Cp]$ complexes is via the corresponding $[(allyl)Mo(CO)_2Cp]$ species, which are readily obtained from the allyl halide as shown in Scheme 3[91].

These complexes react with a variety of nucleophiles in the expected manner to give η^2-olefin complexes, which can be demetallated by mild oxidation (reactions 138 and 139)[92,93].

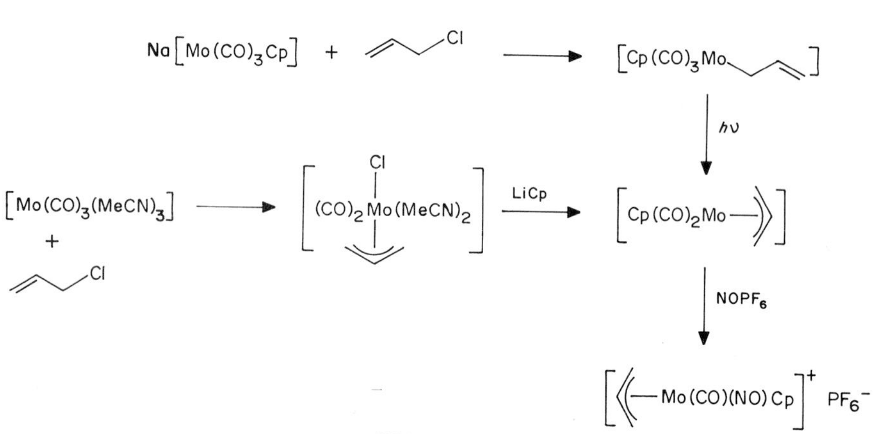

$Na[Mo(CO)_3Cp]$ + allyl chloride → $[Cp(CO)_3Mo$–allyl$]$

$h\nu$

$[Mo(CO)_3(MeCN)_3]$ + allyl chloride → $[(CO)_2Mo(MeCN)_2Cl(\eta^3\text{-allyl})]$ —LiCp→ $[Cp(CO)_2Mo(\eta^3\text{-allyl})]$

$NOPF_6$

$[(\eta^3\text{-allyl})Mo(CO)(NO)Cp]^+ PF_6^-$

SCHEME 3

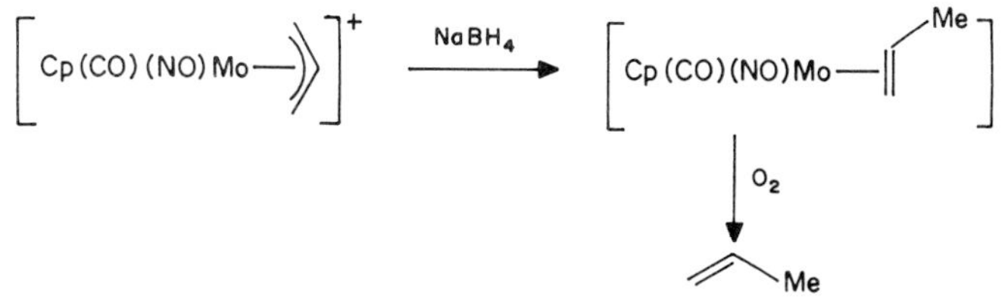

(138)

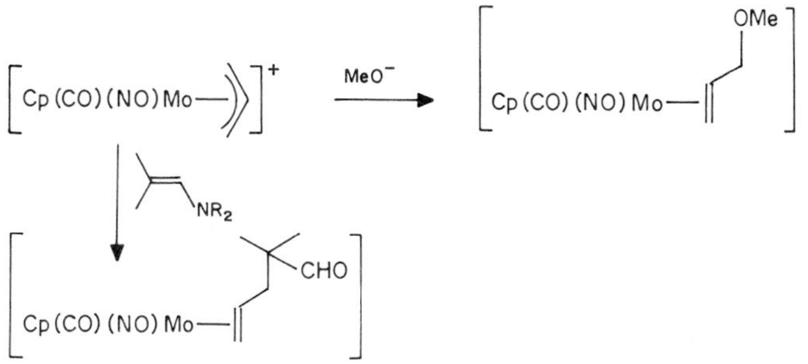

(139)

These η^3-allyl complexes show interesting conformation effects, leading to some unusual reactivity patterns. Since the metal centre has asymmetry and the olefin–Mo unit is also asymmetric, it is possible to obtain two diastereomeric complexes from nucleophilic addition. In fact, only one diastereomer is formed, and this is rationalized in terms of preferred addition modes on both conformations designated as *endo* or *exo*. The favoured products in each case turn out to be the same diastereomer, Scheme 4[94]. A theoretical interpretation of this behaviour, based on MO calculations, has appeared in the literature[95].

Although the above reactivity has been studied fairly extensively, no real synthetic applications have emerged. However, more recently efforts have been directed at using the ability of molybdenum and tungsten to form reactive η^3-allyl complexes to develop catalytic reactions which complement those discussed above for palladium. It was noted first that isolable η^3-allylmolybdenum chloride complexes reacted with stabilized enolate nucleophiles to give products of allylation of the enolate[96]. The regiochemistry of attack on the allyl ligand was dependent on the nature of the counter ligands attached to the metal (reactions 140 and 141). This type of reaction could be conducted using an allylic

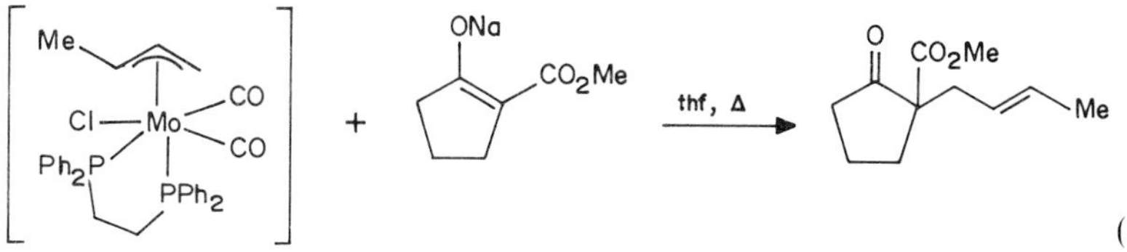

(140)

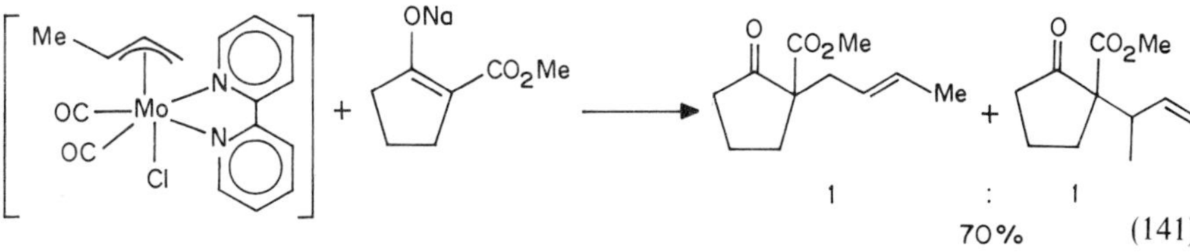

(141)

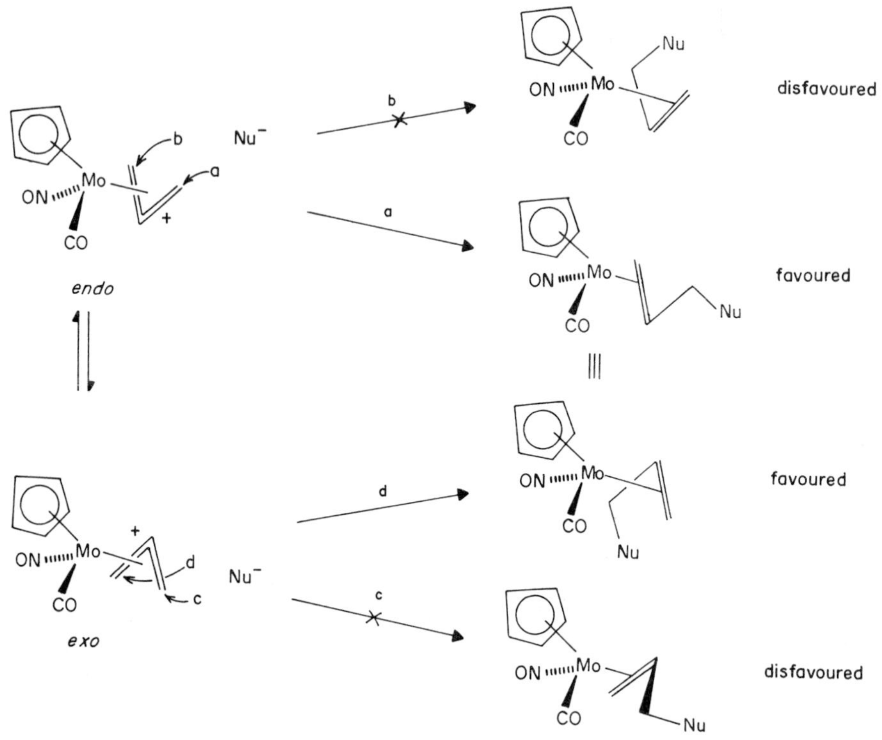

SCHEME 4

acetate and molybdenum catalyst. A number of catalysts were examined, and the best results were obtained with [$Mo(CO)_6$] or [$Mo(bipy)(CO)_4$]. In fact, some degree of complementarity exists between these two catalysts, as shown in equations 142–144.

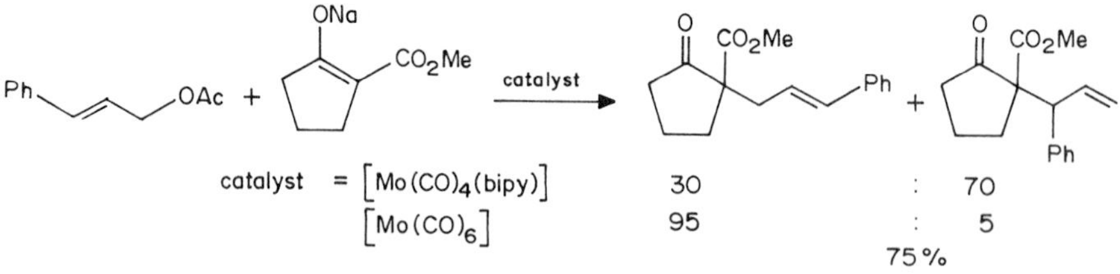

catalyst = [$Mo(CO)_4(bipy)$] 30 : 70
[$Mo(CO)_6$] 95 : 5
75%

(142)

[$Mo(CO)_6$], toluene, reflux, 6 h

82%

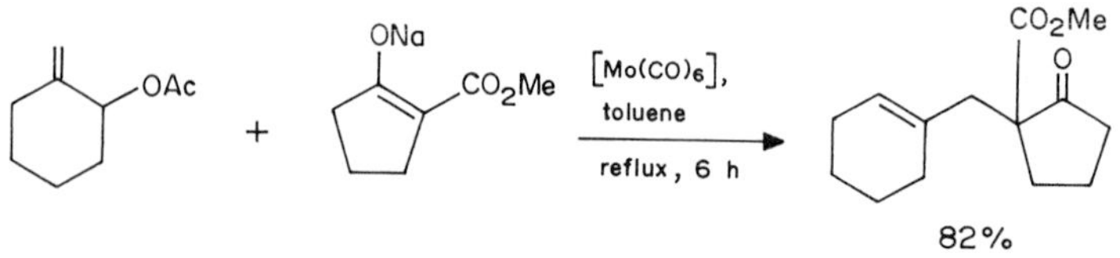

(143)

(A) OAc $NaCH(CO_2Me)_2$ [$Mo(CO)_6$] toluene, Δ

or

(B) OAc

$CH_2(CO_2Me)_2$ / OSiMe$_3$ NSiMe$_3$

[$Mo(CO)_6$]

CO_2Me CO_2Me + $CH(CO_2Me)_2$

(A)	15	:	85
		65–84%	
(B)	4	:	96
		84%	

(144)

Reaction 144 also indicates the importance of the exact nature of the enolate nucleophile, which will be seen again later. The solvent can also dramatically affect the selectivity observed in these reactions. Compare a complexing solvent (dimethoxyethane) with a non-complexing solvent (toluene) (reaction 145). Good stereocontrol can be achieved in

OAc + ONa CO_2Me [$Mo(CO)_6$] → O CO_2Me + O CO_2Me

Solvent			
= dme	1	:	6
PhMe	19	:	1

(145)

these reactions, as for the palladium analogues, but it is necessary to determine the best enolate to use for good results, for example reaction 146.

CO_2Me OAc + $CH_2(CO_2Me)_2$ + OSiMe$_3$ NSiMe$_3$

[$Mo(CO)_6$] → CO_2Me $CH(CO_2Me)_2$ + CO_2Me $CH(CO_2Me)_2$

95 : 5

(146)

A more detailed study of the dependence of these reactions on the nature of the nucleophile has appeared in the literature[97], and some key examples are given in equations 147–150.

$$Br(CH_2)_7CH(OAc)CH{=}CH_2 + \text{(sodium enolate of 2-methoxycarbonylcyclopentanone)} \xrightarrow[PhMe,\ \Delta]{[Mo(CO)_6]} \text{2-CO}_2\text{Me-2-[CH}_2\text{CH=CH(CH}_2)_7\text{Br]cyclopentanone} \quad 70\% \tag{147}$$

$$Br(CH_2)_7CH(OAc)CH{=}CH_2 \xrightarrow[\text{MeC(OSiMe}_3)\text{=NSiMe}_3,\ [Mo(CO)_6]]{CH_2(CO_2Me)_2} Br(CH_2)_7CH(CH(CO_2Me)_2)CH{=}CH_2 + Br(CH_2)_7CH{=}CHCH_2CH(CO_2Me)_2 \quad 5:1,\ 70\% \tag{148}$$

$$\text{4-}t\text{-Bu-1-vinylcyclohexyl acetate (either isomer)} \xrightarrow[[Mo(CO)_6]]{CH_2(CO_2Me)_2} \text{4-}t\text{-Bu-1-vinyl-1-CH(CO}_2\text{Me)}_2\text{cyclohexane} \quad 89\%;\quad \xrightarrow[[Mo(CO)_6]]{MeCH(CO_2Me)_2} \text{4-}t\text{-Bu-cyclohexylidene-CH}_2\text{CMe(CO}_2\text{Me)}_2 \tag{149}$$

$$\text{3-vinyl-3-OAc-5}\alpha\text{-steroid (C}_8\text{H}_{17}\text{)} \xrightarrow[[Mo(CO)_6],\ PhCH_3]{CH_2(CO_2Me)_2,\ \text{MeC(OSiMe}_3)\text{=NSiMe}_3} \text{3-vinyl-3-CH(CO}_2\text{Me)}_2\text{ steroid} \quad 69\% \tag{150}$$

Apparently, the differences in selectivity observed for dimethyl malonate and, for example, dimethyl methylmalonate are due to a delicate balance between a number of factors, such as reactivity and steric demand of the nucleophile, stability of the products (complexed olefin), and charge distribution in the intermediate η^3-allyl system. This methodology appears to have potential value for the construction of quaternary carbon centres[98].

Recognizing that regiocontrol might be affected by the charge distribution on the η^3-allyl intermediate and the steric and electronic demands of the metal, Trost and Hung[99] set out to examine allylations promoted by tungsten catalysts. Thus, the greater steric demand of a tungsten template can be expected to favour attack at the more substituted end of the allyl ligand. Again, an isolable η^3-allyltungsten halide was found to undergo nucleophilic attack, provided that dppe was present in the reaction mixture, presumably to promote ionization of the complex (reaction 151).

$$[(\eta^3\text{-allyl})W(CO)_2Br(dppe)] + NaCH(CO_2Me)_2 \xrightarrow[\Delta]{thf,\ dppe} CH_2{=}CHCH_2CH(CO_2Me)_2 \qquad (151)$$

For a catalytic reaction, a number of tungsten complexes were investigated, and the best results were obtained using $[W(CO)_3(MeCN)_3]$ in the presence of bipyridyl, a strong σ-donor, and when allyl carbonates were used as substrates. Compared with the molybdenum-catalysed reactions, a pronounced selectivity for the more substituted allyl terminus was found, some key examples being shown in reactions 152–156.

Ph–CH=CH–CH$_2$OCO$_2$Me + sodium enolate of 2-(methoxycarbonyl)cyclopentanone (ONa, CO$_2$Me) $\xrightarrow[\text{reflux, 15 h}]{[W(CO)_3(MeCN)_3],\ \text{bipy, thf}}$ 2-(1-phenylallyl)-2-(methoxycarbonyl)cyclopentanone (O, CO$_2$Me, Ph), 87% (152)

2-furyl–CH(OAc)–CH=CH$_2$ + NaCMe(CO$_2$Me)$_2$ $\xrightarrow[\text{reflux, 2 h}]{[W(CO)_3(MeCN)_3],\ \text{bipy, thf}}$ 2-furyl–CH(CMe(CO$_2$Me)$_2$)–CH=CH$_2$, 86% (153)

2-thienyl–CH(OCOOMe)–CH=CH$_2$ + sodium enolate of 2-(methoxycarbonyl)cyclopentanone (ONa, CO$_2$Me) $\xrightarrow[\text{reflux, 16.5 h}]{[W(CO)_3(MeCN)_3],\ \text{bipy, thf}}$ 2-thienyl–CH(CH=CH$_2$)–C(CO$_2$Me)(cyclopentanone, =O) + 2-thienyl–CH=CH–CH$_2$–C(CO$_2$Me)(cyclopentanone, O) (154)

84 : 16

76%

+ $LiCH(CO_2Me)_2$ → $[W(CO)_3(MeCN)_3]$, bipy, thf, reflux → 90 : 10, 70% (155)

+ $NaCH(CO_2Me)_2$ → $[W(CO)_3(MeCN)_3]$, bipy, thf, reflux → 71% (156)

It may be of some help to the organic chemist wishing to use some of these methods in synthesis to note that the order of reactivity of the η^3-allyl complexes discussed in the above sections is palladium > molybdenum > tungsten. There definitely exist different patterns for selectivity and so ultimately the choice of catalyst will be governed by the structure of the desired product.

C. η^3-Allyliron Complexes

Bearing in mind the cost of starting materials, the use of η^3-allyliron complexes for synthesis offers some advantages but, although the chemistry of some of these stable cationic species has been thoroughly investigated, their application to real synthetic problems has been lacking. Mostly, this stems from non-directed research—only fairly simple systems have been studied, and little or no attention has been given to the preparation and reactions of complexes carrying an array of functional groups. However, as we shall see later in this section, it is possible to use iron(0) catalysts in much the same way as the palladium, molybdenum, and tungsten systems, so more activity in this area can be expected.

A number of stable, diethyl ether-insoluble η^3-allyliron tetracarbonyl cations are available using the procedures outlined in equations 157–159. It is noteworthy that the parent $[(\eta^3\text{-allyl})Fe(CO)_3I]$ is available from allyl iodide and $[Fe_2(CO)_9]$, but little has been done to explore the reactivity of this complex or to prepare a range of related complexes with more diverse functionality present. These complexes are far more reactive

$[Fe_2(CO)_9]$ or $[Fe(CO)_5]$ → $[Fe(CO)_3I]$ → $AgBF_4$, CO, CH_2Cl_2 (refs. 100,101,102) → $[Fe(CO)_4]^+$ BF_4^- (157)

$$\xrightarrow[\text{CO (ref. 103)}]{\text{HBF}_4}\quad BF_4^-$$

R = R' = H
R = Me, R' = H
R = H, R' = Me

(158)

$$\xrightarrow[\text{(ref. 104)}]{\text{HBF}_4\text{, CO}}\quad BF_4^- \qquad (159)$$

towards nucleophiles than, for example, η^3-allyl-palladium complexes, and they usually react instantaneously with stable enolate anions at sub-zero temperatures to give a η^2-olefin complex which is readily oxidized to give the free ligand (reactions 160–162).

$$BF_4^- \xrightarrow[\text{(ii) air}]{\text{(i) NaCH(CO}_2\text{Me)COMe}}$$

(i) OH^-
(ii) Δ

68% 17%

(160)

$$BF_4^- \xrightarrow{Ph_3P} \quad BF_4^- \qquad (161)$$

$$BF_4^- \xrightarrow[\text{(ii) air}]{\text{(i) R}_2\text{Cd}} \qquad (162)$$

Perhaps the most significant advance in this area from the synthetic viewpoint is the development of catalytic reactions by Roustan *et al.*[106], summarized in reactions 163–165. Only a few examples have been reported so far, but this approach may prove complementary to the palladium-, molybdenum-, and tungsten-catalysed reactions discussed above.

$$\text{CH}_2{=}\text{CH}{-}\text{CH(Me)Cl} \xrightarrow[\text{Na[Fe(CO)}_3\text{NO]},\ 48\text{ h, RT}]{\text{NaCH(CO}_2\text{Et)}_2} \text{CH}_2{=}\text{CH}{-}\text{CH(Me)CH(CO}_2\text{Et)}_2 + \text{MeCH}{=}\text{CHCH}_2\text{CH(CO}_2\text{Et)}_2 \quad 82:28,\ 90\% \tag{163}$$

$$\text{CH}_2{=}\text{CH}{-}\text{CH(Me)OAc} \xrightarrow[\text{Na[Fe(CO)}_3\text{NO]}]{\text{NaCH(CO}_2\text{Et)}_2} \text{CH}_2{=}\text{CH}{-}\text{CH(Me)CH(CO}_2\text{Et)}_2 + \text{MeCH}{=}\text{CHCH}_2\text{CH(CO}_2\text{Et)}_2 \quad 95:5,\ 79\% \tag{164}$$

$$\text{PhCH}{=}\text{CHCH}_2\text{OAc} \xrightarrow[\text{Na[Fe(CO)}_3\text{NO]}]{\text{NaCH(CO}_2\text{Et)}_2} \text{PhCH}{=}\text{CHCH}_2\text{CH(CO}_2\text{Et)}_2 \quad 91\% \tag{165}$$

Of course, many other metals form η^3-allyl complexes. Often these are not electrophilic species and most have not been studied for their utility as allylating agents, so they cannot be discussed here.

V. CATIONIC η^4-DIENE COMPLEXES

Development of the reactivity of cationic diene–metal η^4-complexes towards organic synthesis has not progressed as far as that of η^3-allyl or -dienyl complexes. However, some interesting reactivity patterns are emerging which point the way for future applications.

There are basically two types of diene–metal complex: those having unconjugated, usually 1,4- or 1,5-diene ligands and those having conjugated, 1,3-diene, ligands. The former are probably better known for cationic complexes, exemplified by cycloocta-1, 5-diene metal complexes, whereas conjugated diene complexes such as $[(\text{diene})\text{Fe(CO)}_3]$ systems are better known in the uncharged series. This discussion does not include the latter complexes. We shall briefly present the chemistry of selected 1,4- and 1,5-diene complexes, followed by a discussion of cationic 1,3-diene complexes.

A. Complexes of Unconjugated Dienes

A variety of metals will form stable complexes with 1,4- or 1,5-dienes which, although they do not carry a formal positive charge, nevertheless show reactivity consistent with a cationic system, such as pronounced reactivity toward nucleophiles.

Platinum and palladium complexes of cycloocta-1,5-diene and norbornadiene can be obtained by direct reaction of the diene with an appropriate metal salt (reactions 166 and 167). Reaction of nickelocene with these dienes in the presence of tetrafluoroboric acid gives the corresponding diene–nickel cations (reaction 168).

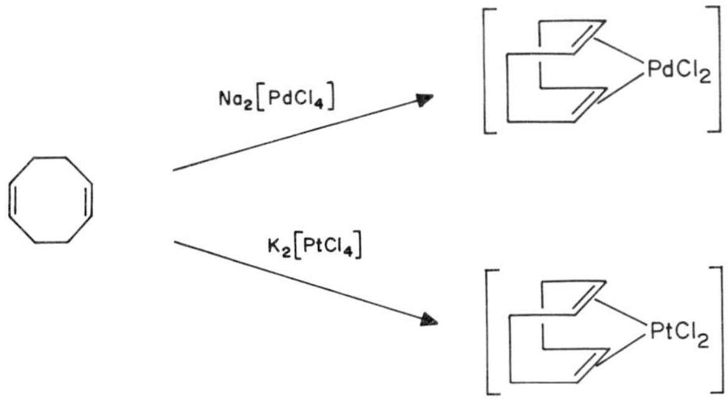

$Na_2[PdCl_4]$
$PdCl_2$
$K_2[PtCl_4]$
$PtCl_2$

(166)

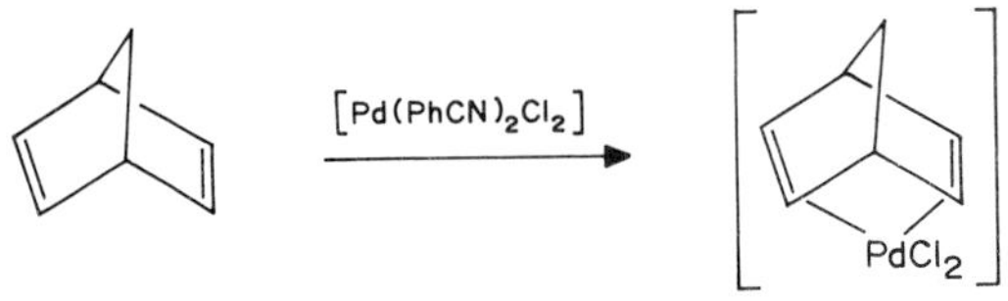

$[Pd(PhCN)_2Cl_2]$
$PdCl_2$

(167)

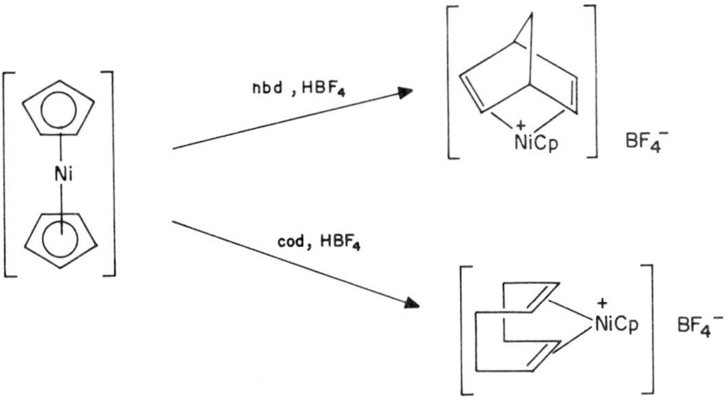

Ni
nbd , HBF_4
NiCp
BF_4^-
cod, HBF_4
NiCp
BF_4^-

(168)

In all of these complexes, nucleophilic attack occurs at the diene to give a substituted σ,η^2-complex. In the case of, for example, the palladium complexes, expulsion of chloride occurs with the formation of a chloride-bridged dimer. A variety of soft carbon and heteroatom nucleophiles have been found to react successfully by attack *trans* to the metal, and equations 169–171 and Scheme 5 summarize this reactivity and indicate how further manipulation of the product to give non-metallic species can be accomplished[107]. Some of these reactions have been met briefly in the section on η^2-alkene complexes.

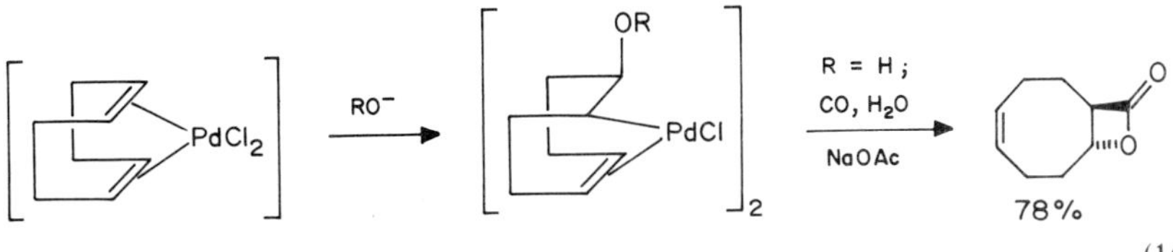

$PdCl_2$
RO^-
OR
PdCl
2
R = H;
CO, H_2O
NaOAc
O
O
78%

(169)

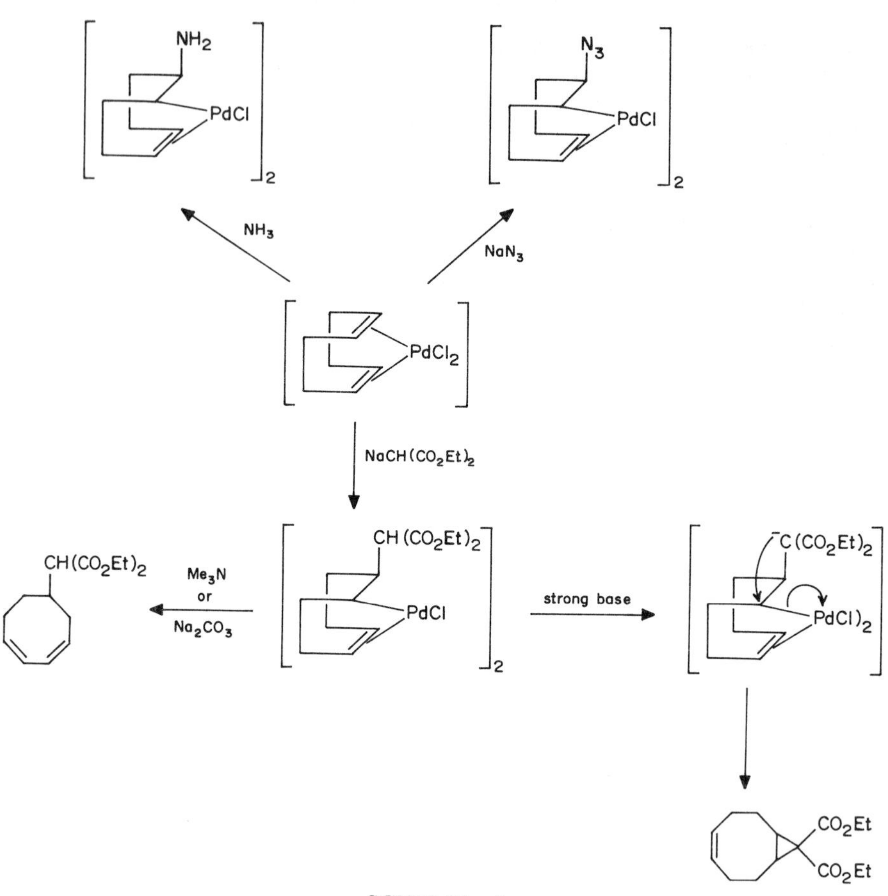

SCHEME 5

(170)

(171)

For the most part, these reactions have not been developed for organic synthesis purposes. However, with some recent interest being focussed on use of medium and large rings in organic synthesis[108], there may be some potential for the preparation of starting materials that are otherwise difficult to obtain. Clearly, this would have to involve the development of procedures which are catalytic with the more expensive metals.

B. Complexes of 1,3-Dienes

Complexes in which a 1,3-diene is attached to a cationic molybdenum moiety have been prepared and shown to be reactive towards simple nucleophiles[109]. The most usual method for the preparation of these complexes is hydride abstraction from an appropriate neutral η^3-allyl–molybdenum complex (reaction 172).

Mo(CO)$_2$Cp, R^1, R^4, R^2, R^3 —Ph$_3$C$^+$ BF$_4^-$→ $\overset{+}{\text{Mo}}$(CO)$_2$Cp, R^1, R^4, R^3 BF$_4^-$

$R^2 = Me,\quad R^1 = R^3 = R^4 = H$

$R^1 = R^2 = Me,\quad R^3 = R^4 = H$ (172)

$R^2 = R^4 = Me,\quad R^1 = R^3 = H$

These cations react with nucleophiles such as $NaBH_3CN$ and enamines to yield [(η^3-allyl)Mo(CO)$_2$Cp] complexes (reactions 173 and 174).

$^+$Mo(CO)$_2$Cp, Me BF$_4^-$ —NaBH$_3$CN, −78 °C→ Mo(CO)$_2$Cp, Me (173)

$^+$Mo(CO)$_2$Cp BF$_4^-$ + N —H$_2$O, work-up→ Mo(CO)$_2$Cp, CHO

(174)

Recently, studies have been directed at the preparation and reactivity of the cyclohexadiene–Mo(CO)$_2$Cp cation. This is an especially interesting example, since double nucleophile additions can be accomplished in a completely stereocontrolled manner. Some of these aspects, together with the preparation of the cyclohexadiene complex, are shown in equation 175 and Scheme 6.[110]

The above examples illustrate the possibility of carbon—carbon bond formation of potential synthetic utility. Of particular note is the fact that in these 18-electron complexes the nucleophile addition occurs stereospecifically *trans* to the metal. When the nucleophile is MeMgBr the methylated η^3-allyl complex product can be subjected to a second hydride abstraction and, since this reaction occurs by loss of hydride which is *trans* to the metal group, probably as a result of stereoelectronic factors, only one complex can be produced, as shown. Reaction of the methyl-substituted diene complex with MeMgBr occurs stereo-

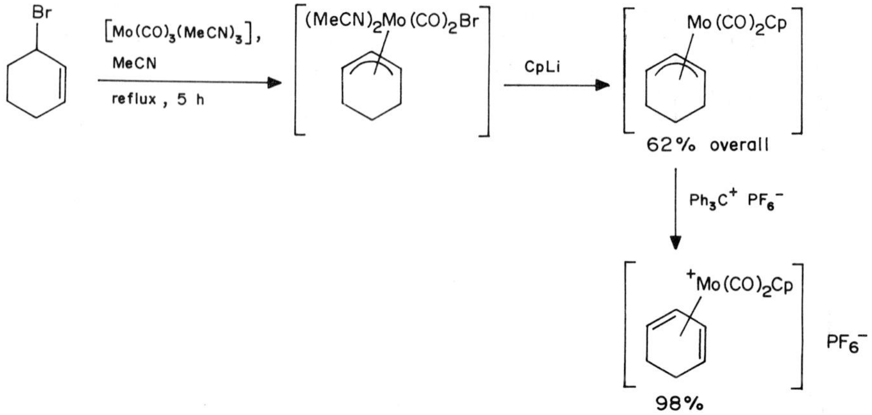

(175)

and regio-specifically but in low yield. Although these results are chemically interesting, they present a challenge in terms of synthetic application. How are we to manipulate the η^3-allyl complexes, resulting from nucleophile addition, in a controlled manner to give useful organic molecules? One answer to this question is to activate the $[(\eta^3\text{-allyl})Mo(CO)_2Cp]$ by conversion to the $[(\eta^3\text{-allyl})Mo(CO)(NO)Cp]$ cation and subject it to further nucleophile additions (reaction 176). This works for the dimethyl-substituted

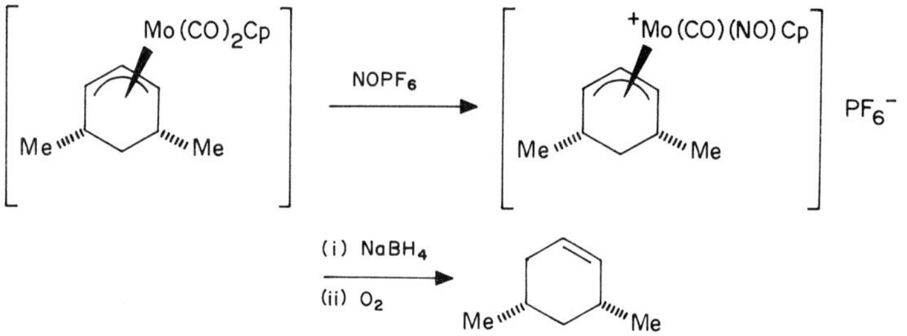

(176)

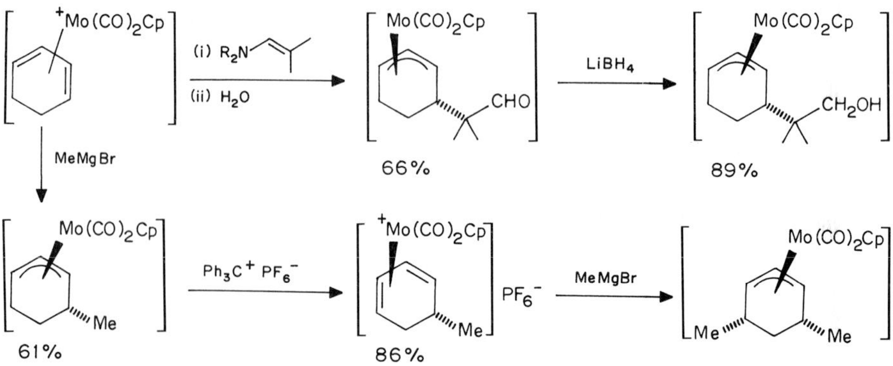

SCHEME 6

complex, but the method does not appear to be applicable when there are sensitive functional groups in the molecule. Further, poor regiochemical results can be anticipated when different substituents are present on the cyclohexenyl ligand.

A solution to these problems, and a demonstration of real synthetic utility, has recently been reported[111]. It was found that [(η^3-allyl)Mo(CO)$_2$Cp] complexes bearing a pendant nucleophilic group undergo cyclofunctionalization and demetallation on treatment with excess of iodine. The manipulations involved in the reaction sequence leading to these transformations, and the earlier work by Faller *et al.*[110], also show that the Mo(CO)$_2$Cp group is stable, allowing a range of functional group interconversions in its presence (Scheme 7).

The sequence in Scheme 7 results in a completely stereocontrolled approach to lactones. Of particular interest is the relationship of this type of lactone (R = Me), by ring cleavage, to the C-4—C-9 portion of the important macrolide antibiotic magnamycin B. If a

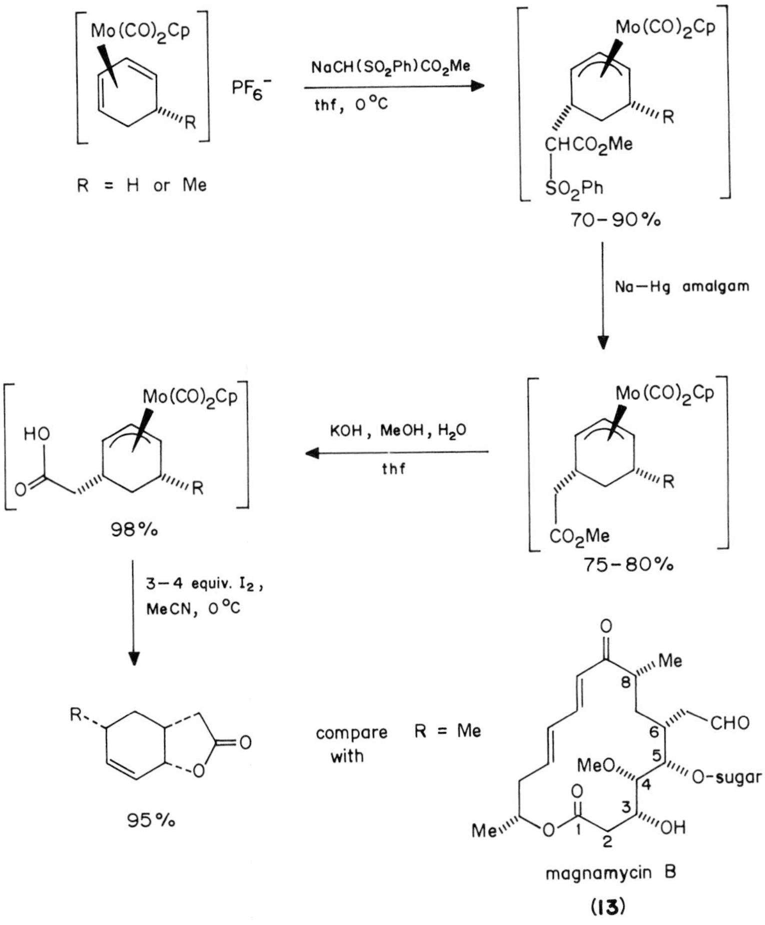

Mo(CO)$_2$Cp PF$_6^-$ R = H or Me

NaCH(SO$_2$Ph)CO$_2$Me, thf, 0 °C →

Mo(CO)$_2$Cp CHCO$_2$Me SO$_2$Ph 70–90%

Na—Hg amalgam →

Mo(CO)$_2$Cp CO$_2$Me 75–80%

KOH, MeOH, H$_2$O, thf →

HO O Mo(CO)$_2$Cp R 98%

3–4 equiv. I$_2$, MeCN, 0 °C →

R O O 95%

compare with R = Me

magnamycin B

(13)

SCHEME 7

nucleophilic group is not present in the η^3-allyl complex then iodide, generated in the reaction, acts as a nucleophile, presumably attacking an intermediate cationic $[(\eta^3\text{-allyl})MoI(CO)_2Cp]^+$ complex to give allyl iodides, again stereospecifically. This provides a means of selectively manipulating the $[(\eta^3\text{-allyl})Mo(CO)_2Cp]$ complexes in the presence of a wide range of sensitive functional groups. The allylic iodides can be further transformed to give a variety of allylic alcohols and cyclohexenones, some examples of which are given in equations 177 and 178.

$^+Mo(CO)_2Cp$ —$MeOC_6H_4MgBr$→ $Mo(CO)_2Cp$ ... OMe —I_2, MeCN→ I ... OMe (177)

$Mo(CO)_2Cp$ ... $CMe(CO_2Me)_2$ —I_2→ I ... $CMe(CO_2Me)_2$ —m-cpba→ OH ... $CMe(CO_2Me)_2$

AgOAc → AcO ... $CMe(CO_2Me)_2$ —(i) OH^- (ii) CrO_3→ O ... $CMe(CO_2Me)_2$

→ O, =O, CO_2Me, Me (178)

These results are only the beginnings of what appears to be a very fruitful area for future research.

VI. CATIONIC η^5-DIENYL COMPLEXES

This area is dominated by complexes of iron, especially in terms of their synthetic application, which has expanded considerably in the past decade. Since these complexes are also discussed in Chapter 7, we shall focus our attention on actual synthetic applications[112]. Mostly, these have been concerned with cyclohexadienyl complexes, with some recent promising work on cycloheptadienyl systems. There has been a reluctance on the part of many investigators to go beyond the reactions of these complexes with nucleophiles and removal of metal. In order to demonstrate some applicability of the systems to the organic chemist, it is necessary to explore the further transformations of dienes resulting from these simple operations, in anticipation that recognizable intermediates for, for example, natural products synthesis, can be generated.

We shall begin with the parent tricarbonylcyclohexadienyliron tetrafluoroborate, first reported by Fischer and Fischer in 1960[113], and prepared by reaction of tricarbonylcyclohexadieneiron with triphenylmethyl tetrafluoroborate. The reaction is carried out in dichloromethane at room temperature (40 min) and the product is precipitated with diethyl ether (reaction 179). The resultant cyclohexadienyliron complex

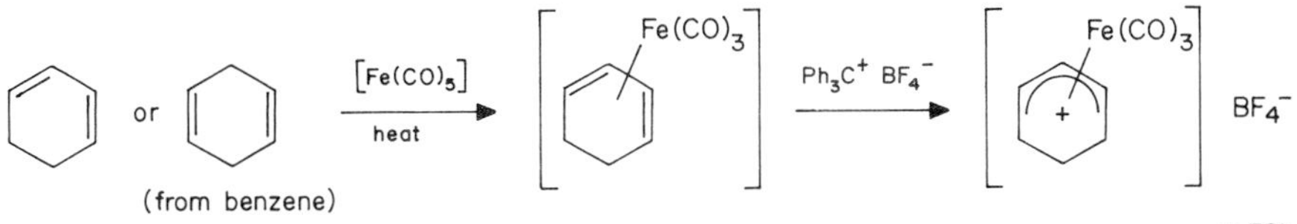

(179)

reacts stereo- and regio-specifically with a very wide range of nucleophiles, such as amines, alkoxides, stabilized enolate anions, enamines, silyl enol ethers, and allylsilanes. We shall see examples of these later. For the moment let us examine the consequences of reaction with a particular nucleophile, dimethyl sodiomalonate. Scheme 8 shows how this may be used to prepare cyclohexadienylacetic acid. This deceptively simple acid derivative is very easily prepared on a fairly large scale using this technique, but would be difficult to obtain using standard organic chemical procedures. This is an advantage to using the organometallic approach, provided that the diene can be further manipulated to give recognizable synthetic intermediates. It was found that a number of lactonization procedures using this and related dienes occur in a conjugate manner, and stereospecifically. The phenylselenolactonization[114] of such molecules is particularly useful, since advantage can be taken of the known[115] [2, 3]-sigmatropic rearrangement of allylic selenoxides to give allylic alcohols. These reactions, and similar reactions with related dienes obtained using organoiron technology, are summarized in equations 180–182[116].

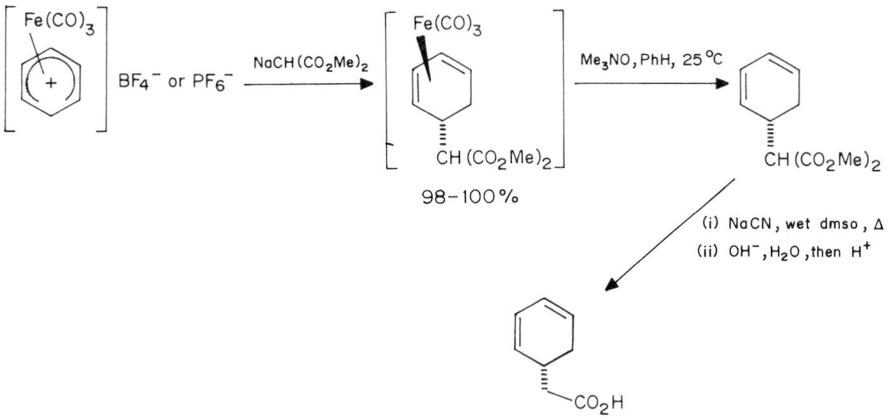

SCHEME 8

I_2, MeCN

CO_2H

$Hg(OAc)_2$

AcOHg

PhSeCl, Et_3N, CH_2Cl_2

PhSe

H_2O_2

OH

(180)

OMe

Fe (CO)$_3$

Me

(i) conc. H_2SO_4

(ii) NH_4PF_6 (ref. 117)

Me

Fe (CO)$_3$

PF_6^-

4 steps

CO_2H

PhSeCl, Et_3N, CH_2Cl_2

PhSe

(181)

(i) [Fe(CO)$_5$], Δ

(ii) Ph_3C^+ PF_6^-

Fe (CO)$_3$

PF_6^-

4 steps

CO_2H

PhSeCl, Et_3N

OH

H_2O_2

PhSe

Me

4 : 1 (*E* : *Z*)

(182)

Thus, the combination of simple dienyliron chemistry with other methods of alkene functionalization leads to a powerful methodology for converting readily available benzene derivatives, via their Birch reduction products (cyclohexadienes), into fairly complex substituted cyclohexenes in a stereocontrolled manner. The cyclohexene hydroxylactone produced in the above equations is useful since, on ozonolysis, it leads to an open-chain dialdehyde which corresponds stereochemically to a C-3—C-8 fragment of the macrolide antibiotic magnamycin B (**13**) (reaction 183).

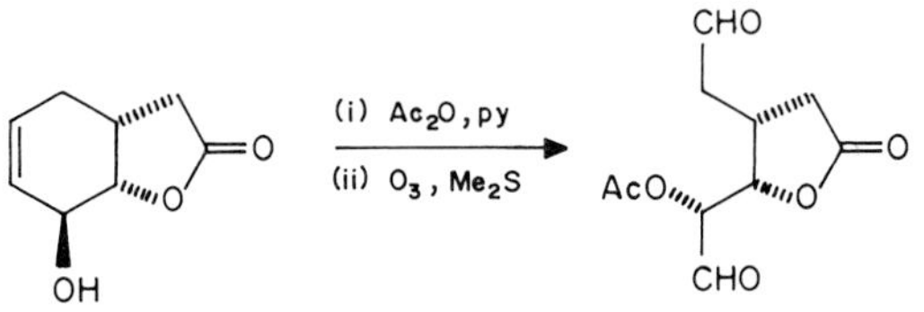

(i) Ac_2O, py
(ii) O_3, Me_2S

(183)

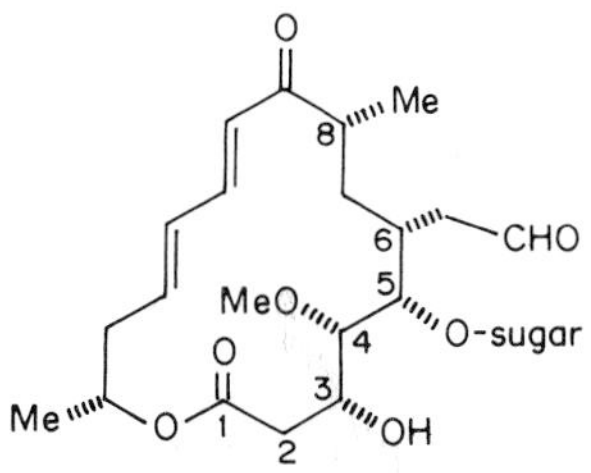

(**13**)

The ability to obtain cyclohexa-1, 4-dienes easily from benzene derivatives using metal–ammonia reduction led to fairly intensive investigations into the conversion of these dienes in to related [(1, 3-diene)$Fe(CO)_3$] complexes. These in turn could be transformed into a range of [(dienyl)$Fe(CO)_3$] cations whose synthetic utility could be probed. These complexes fall broadly into a number of categories according to the way we can relate them to various 'synthons'[119]. From the synthetic standpoint, the two most investigated reactivity patterns are those which allow us to think of the complexes in terms of (a) cyclohexenone γ-cation equivalents or (b) aryl cation equivalents, and these will now be discussed.

A. Cyclohexenone γ-cation Equivalents

Tricarbonyl(2-methoxycyclohexadienyl)iron tetrafluoroborate (or hexafluorophosphate) is readily prepared from anisole as shown in reaction 184. In this complex, the 2-methoxy group exerts a powerful directing effect on nucleophile addition, such that bond formation occurs entirely at the position *para* to the methoxy group. A number of nucleophiles have been investigated in this reaction and a list of typical examples is provided by equations 185–194[120]. Decomplexation of the product diene complexes is readily accomplished using Shvo and Hazum's amine oxide method[118], and the resulting dienol ethers can be hydrolysed to cyclohexenone derivatives. Alternatively, direct oxidation with Jones reagent affords the cyclohexenones in one step.

OMe

(i) Li, NH_3

(ii) $[Fe(CO)_5]$, Δ

OMe Fe(CO)$_3$ + OMe Fe(CO)$_3$

$Ph_3C^+ BF_4^-$

$Ph_3C^+ BF_4^-$

OMe + Fe(CO)$_3$ BF_4^-

tricarbonyl (1-methoxy-cyclohexadienyl) iron tetrafluoroborate

OMe + Fe(CO)$_3$ BF_4^-

tricarbonyl (2-methoxy-cyclohexadienyl) iron tetrafluoroborate (184)

OMe + Fe(CO)$_3$ $\xrightarrow[\Delta]{RCH_2COMe}$ OMe Fe(CO)$_3$ R O (185)

OMe + Fe(CO)$_3$ $\xrightarrow[\Delta]{MeCOCH_2CO_2Et}$ OMe Fe(CO)$_3$ O CO_2Et (186)

OMe + Fe(CO)$_3$ O O $\xrightarrow{\Delta}$ OMe Fe(CO)$_3$ O O (187)

(i) NR$_2$ (ii) H_2O; Jones oxidation

(188)

(i) OSiMe$_3$ OSiMe$_3$ (ii) H^+, MeOH

(189)

R–C$_6$H$_4$–NH$_2$, 25 °C → NHAr

(190)

R^1, R^2, NH$_2$; MeCN, reflux

(191)

SiMe$_3$, Δ; Ce^{IV}

(192)

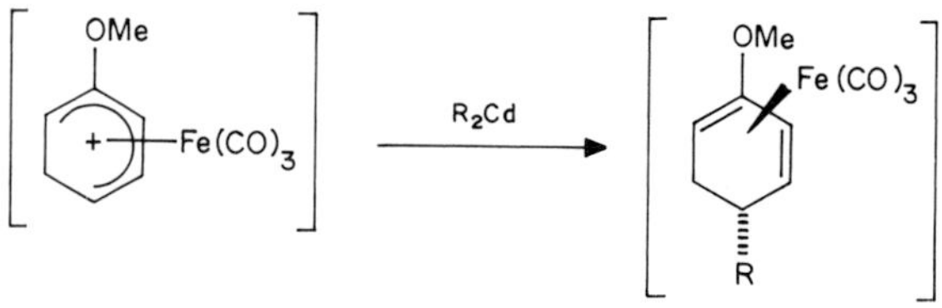

(193)

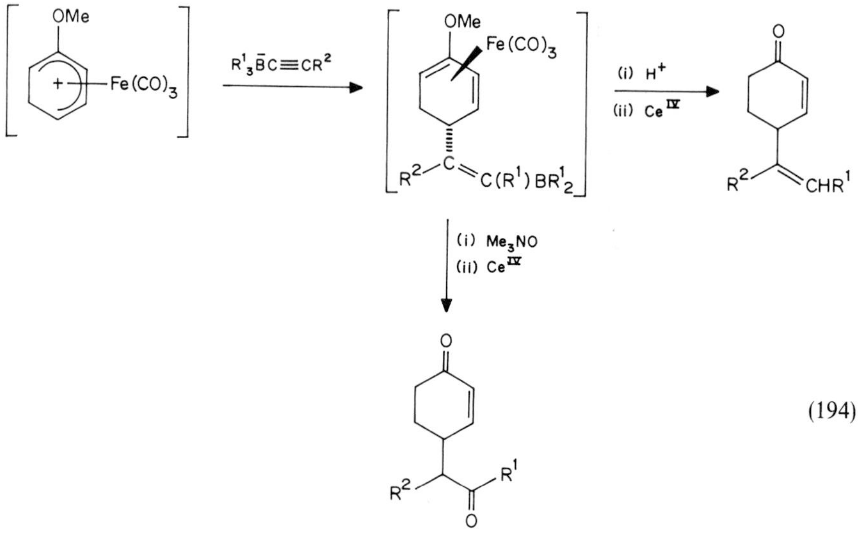

(194)

Effectively, we can view the overall sequence dienyl complex → diene complex → cyclohexenone as a nucleophilic γ-alkylation of cyclohexenone. It is convenient to regard the 2-methoxycyclohexadienyl complex as the synthetic equivalent of the cyclohexenone γ-cation, **14**. Realizing this, the question now arises as to how general this

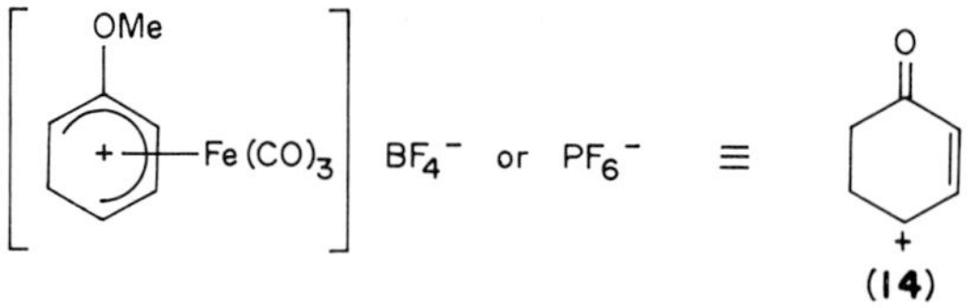

(**14**)

reactivity pattern is, i.e. is it successful when the dienyl ligand carries other substituents? In answer to this, it has been found that a range of di- and tri-substituted dienyl complexes will react at the enone γ-position with nucleophiles, but the nucleophiles which are well behaved are few in number. The reactivity pattern appears to be limited to soft enolate nucleophiles. We shall illustrate these results, and also their application in synthesis, by a series of schemes under appropriate headings. Of some interest is the range of functional group manipulations possible on the [(diene)Fe(CO)$_3$] complex without decomposition. This is important because in many instances it allows us to use the Fe(CO)$_3$ group as a diene protection or as a masking group for the latent enone. These aspects are emphasized in Scheme 9, but in most cases we have omitted the manipulation of compounds after decomplexation when this follows a fairly routine sequence of organic reactions.

SCHEME 9

SCHEME 10

1. *Synthesis of trichothecene analogues*[121]

Note that the hydride abstraction in this sequence occurs regioselectively: the hydride α- to the methoxy group is removed to the extent of at least 95%. This has been interpreted in terms of formation of the dienyl complex in which the preferred dienyl ligand has the higher energy HOMO and lower energy LUMO[112,122]. The reactions in Scheme 10 show how the product of nucleophile addition to the dienyl complex can be manipulated to give compounds having the skeleton of the cytotoxic sesquiterpenes known as trichothecenes[123]. The $Fe(CO)_3$ unit serves to prevent unwanted side reactions due to the presence of neighbouring enone functionality.

2. *Synthesis of steroids and C-nor-D-homosteroids*

The dienyl complex **15** (see Scheme 9) has been found to react regiospecifically with a wide range of fairly bulky keto ester enolates. Reaction with appropriate tetralone[124] and indanone[125] derivatives gives access to steroid precursors[126] and C-nor-D-homosteroid compounds which can serve as intermediates for the synthesis of Veratrum alkaloids such as veratramine (**16**) and verticine (**17**)[127]. The progress which has been made in these areas is summarized in Schemes 11 and 12.

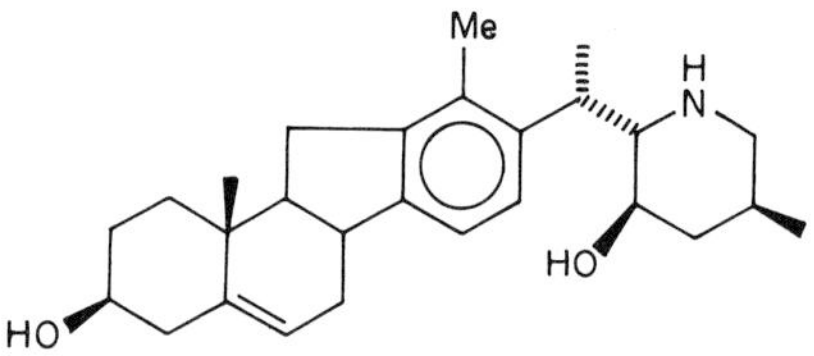

(16)

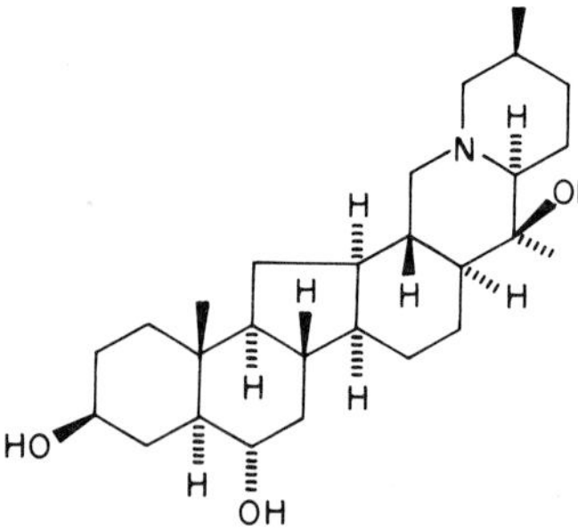

(17)

3. *Synthesis of Aspidosperma alkaloids*[123]

During the development of approaches to some Aspidosperma alkaloids, exemplified by aspidospermine (**18**) and limaspermine (**19**), a number of important technical improvements emerged in the control of regioselectivity of nucleophile addition to dienyl complexes bearing relatively bulky substituents. These will be presented during this discussion. As with the trichothecene synthesis described above, it was found that a number of useful transformations of functionality could be accomplished in the presence of the diene–$Fe(CO)_3$ group, again illustrating a potential for carrying a masked enone through a synthetic sequence. In some ways the importance of these discoveries outweighs the achievement of the final objective, i.e. the synthetic target.

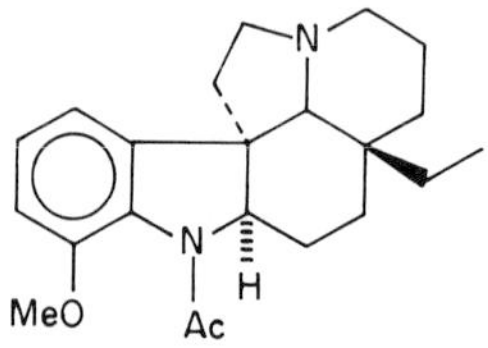

(18)

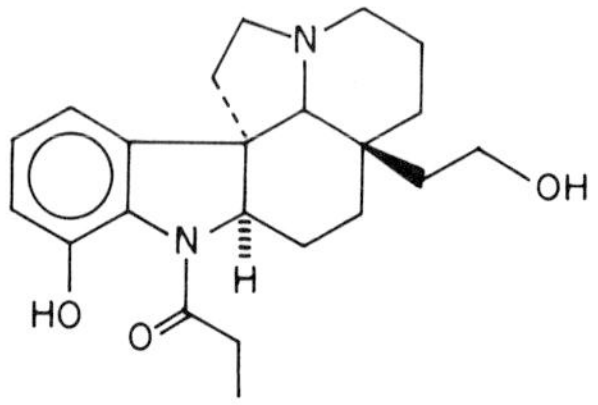

(19)

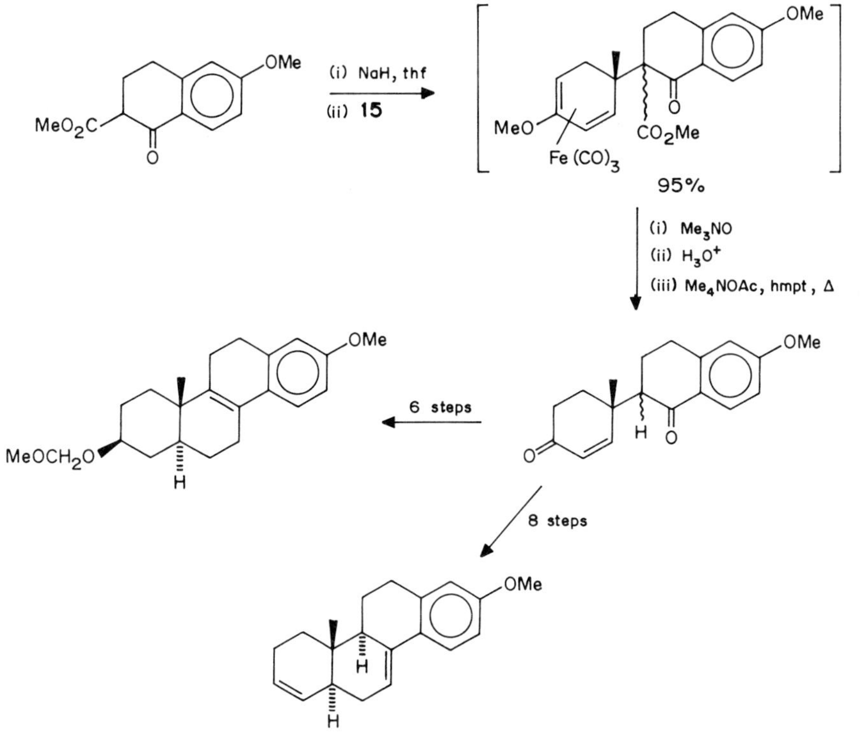

(i) NaH, thf
(ii) 15

95%

(i) Me_3NO
(ii) H_3O^+
(iii) Me_4NOAc, hmpt, Δ

6 steps

8 steps

SCHEME 11

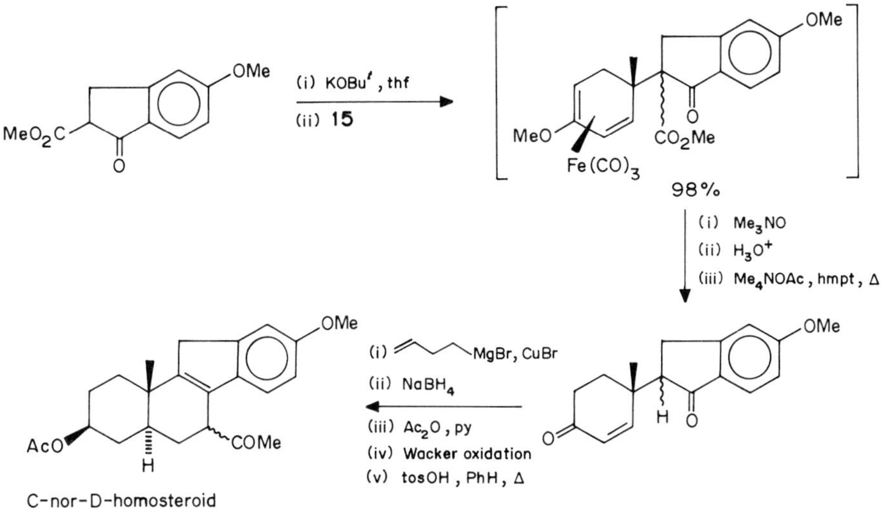

(i) $KOBu^t$, thf
(ii) 15

98%

(i) Me_3NO
(ii) H_3O^+
(iii) Me_4NOAc, hmpt, Δ

(i) $CH_2{=}CHCH_2CH_2MgBr$, CuBr
(ii) $NaBH_4$
(iii) Ac_2O, py
(iv) Wacker oxidation
(v) tosOH, PhH, Δ

C-nor-D-homosteroid

SCHEME 12

To the organometallic chemist, these targets (and also those in the preceding sections) look as formidable as some organometallic complexes look to the organic chemist. Bridging this gap between the two disciplines is therefore extremely important. Scheme 13 shows the application of cyclohexadienyliron complexes to the synthesis of a known racemic aspidospermine decahydroquinoline precursor. The fact that some of these dienyl complexes can now be prepared in optically active form[129] makes them very attractive precursors for asymmetric synthesis. During this study it was found that better regioselectivity for malonate carbanion addition at the substituted dienyl terminus, in the cationic iron complex, is obtained using the potassium enolate rather than the sodium or lithium enolate. In the latter stages of the study it was also found that replacement of the 2-methoxy substituent with 2-isopropoxy group also enhances the selectivity of nucleophile attack, owing to its greater steric requirement, but this does not outweigh the electronic effects which play a dominant part in controlling hydride abstraction from the precursor diene complexes. Thus, high yields of single product are obtained in the sequence of reactions in equation 195.

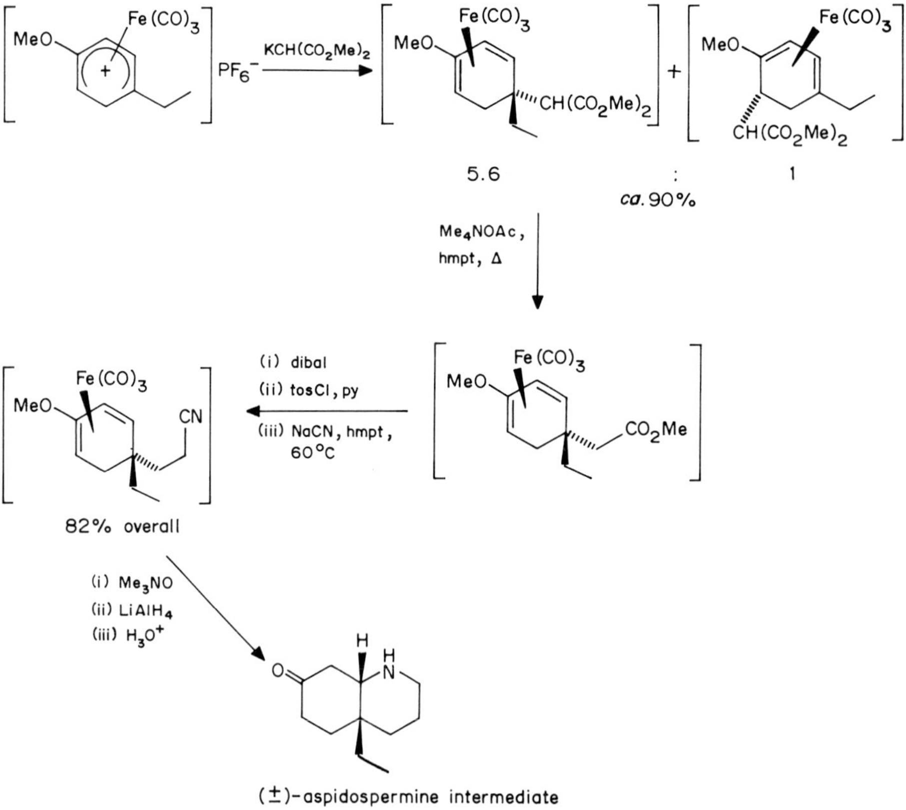

MeO Fe(CO)$_3$ + PF_6^- —KCH(CO$_2$Me)$_2$→ [MeO, Fe(CO)$_3$, CH(CO$_2$Me)$_2$] + [MeO, Fe(CO)$_3$, CH(CO$_2$Me)$_2$]

5.6 : 1

ca. 90%

Me$_4$NOAc, hmpt, Δ

[MeO, Fe(CO)$_3$, CO$_2$Me]

(i) dibal
(ii) tosCl, py
(iii) NaCN, hmpt, 60 °C

[MeO, Fe(CO)$_3$, CN]

82% overall

(i) Me$_3$NO
(ii) LiAlH$_4$
(iii) H$_3$O$^+$

O, H, H, N

(±)-aspidospermine intermediate

SCHEME 13

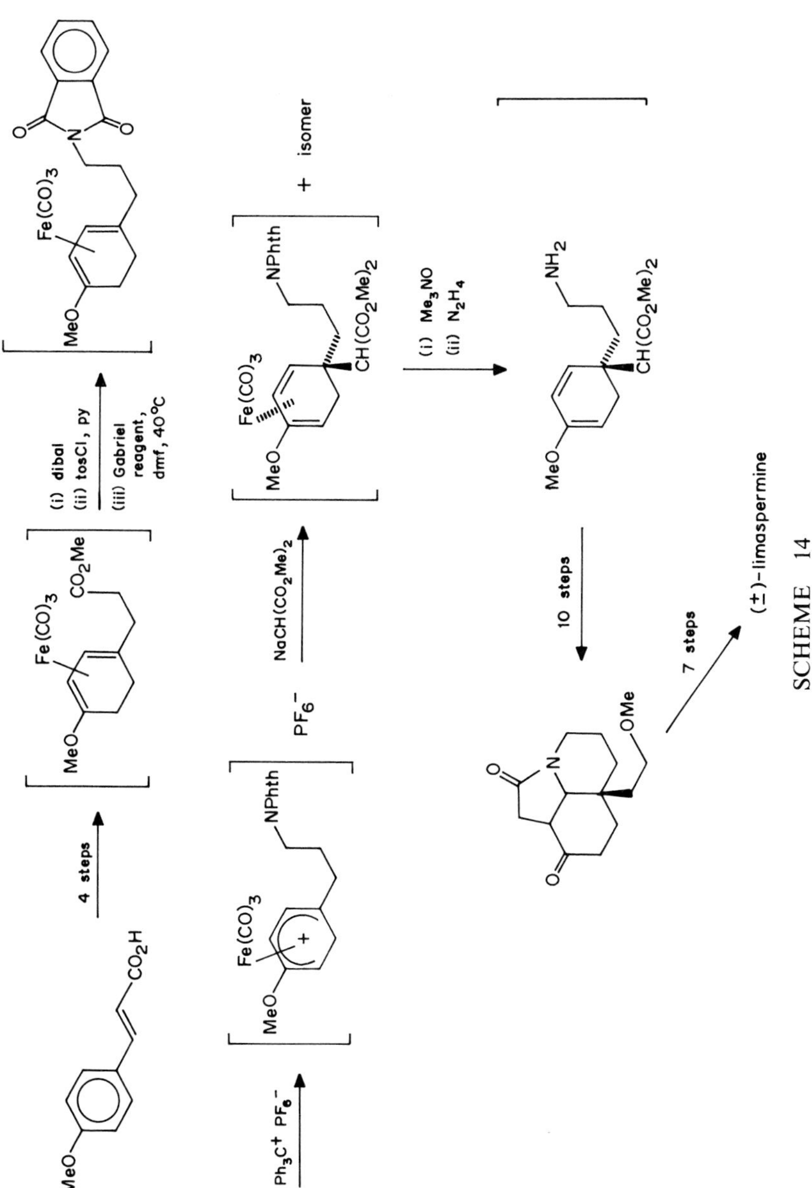
MeO
CO2H
4 steps
Fe(CO)3
CO2Me
(i) dibal
(ii) tosCl, py
(iii) Gabriel reagent, dmf, 40°C
Ph3C+ PF6−
NPhth
PF6−
NaCH(CO2Me)2
CH(CO2Me)2
+ isomer
(i) Me3NO
(ii) N2H4
NH2
10 steps
OMe
7 steps
(±)-limaspermine

SCHEME 14

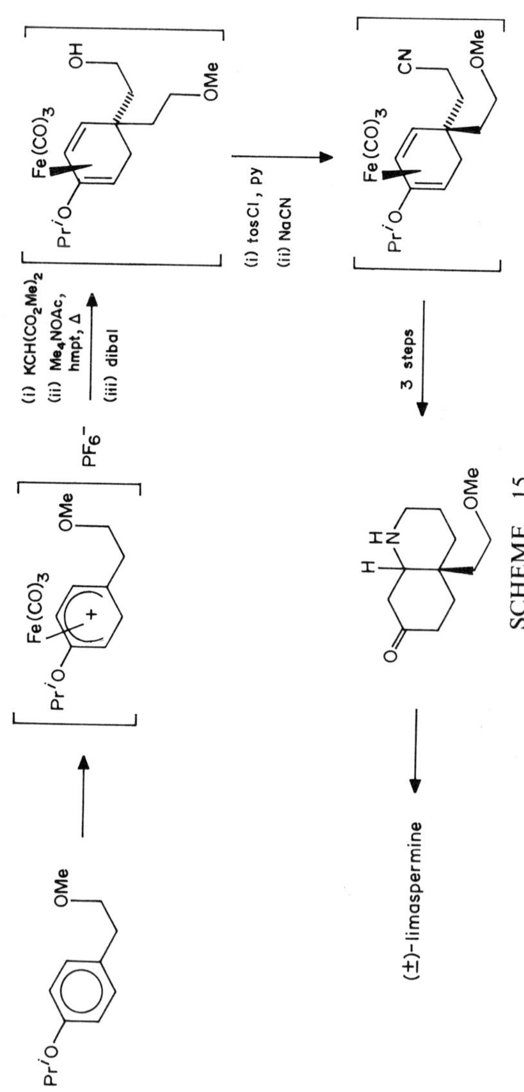

Pr^{i}O
OMe
Fe(CO)$_3$
PF$_6^-$
(i) KCH(CO$_2$Me)$_2$
(ii) Me$_4$NOAc, hmpt, Δ
(iii) dibal
OH
(i) tosCl, py
(ii) NaCN
CN
3 steps
H
N
(±)-limaspermine

SCHEME 15

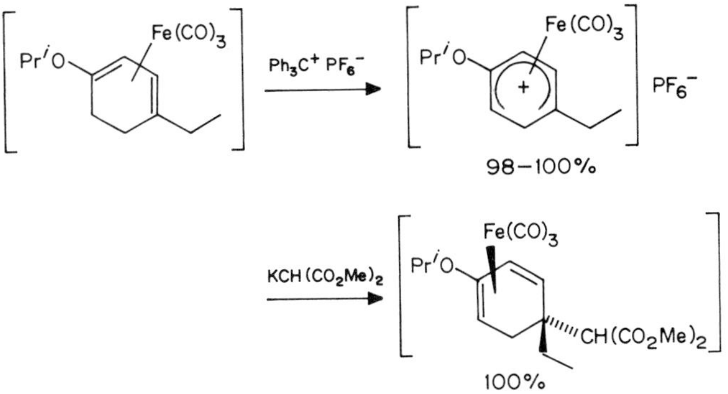
Fe(CO)3
Pr^{i}O
Ph3C+ PF6−
Fe(CO)3
Pr^{i}O
+
PF6−
98–100%
KCH(CO2Me)2
Fe(CO)3
Pr^{i}O
CH(CO2Me)2
100%

(195)

Two routes to (±)-limaspermine have been reported. The first, illustrated in Scheme 14, was lengthy, but again illustrates the potential for manipulating the iron complex towards natural product synthesis.

The second reported total synthesis of (±)-limaspermine took advantage of the use of isopropoxy group as a directing influence during nucleophile addition, and proceeded along similar lines to the aforementioned approach to (±)-aspidospermine (Scheme 15).

4. *Synthesis of spirocyclic compounds*[130]

Spiro[4.5]decane and spiro[5.5]undecane ring systems occur in a number of naturally occurring sesquiterpenes[131]. Some typical examples are given by the structures of β-vetivone (**20**), acorenone (**21**) and chamigrene (**22**). Whilst no total syntheses of these

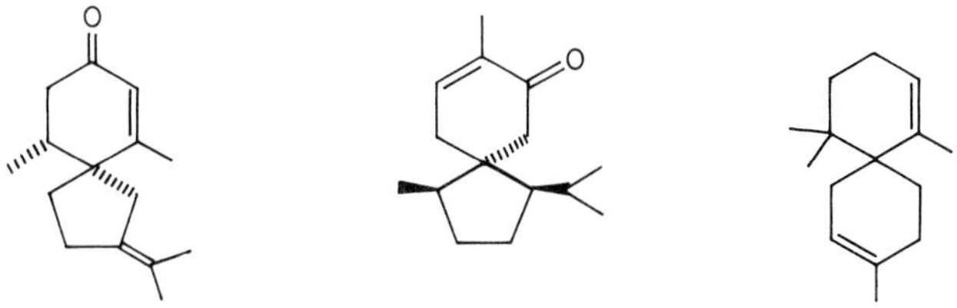

(**20**) (**21**) (**22**)

molecules have been reported using organoiron methodology, the possibility of using the dienyliron unit as a means of constructing a quaternary carbon centre destined to become a spiro centre has been developed, one involving addition of external nucleophile to the dienyl cation and the other using intramolecular nucleophilic attack, and resulting in the emergence of new methods for spiroannelation and additions to the chemistry of the organoiron system which promise to be more generally applicable to the formation of carbocyclic systems.

a. *External nucleophilic additions*

Reactions 196–198 illustrate how the above methodology can be adapted for synthesis of a spirocyclic nucleus. The procedure noted above using a *p*-toluenesulphonyloxyalkyl-

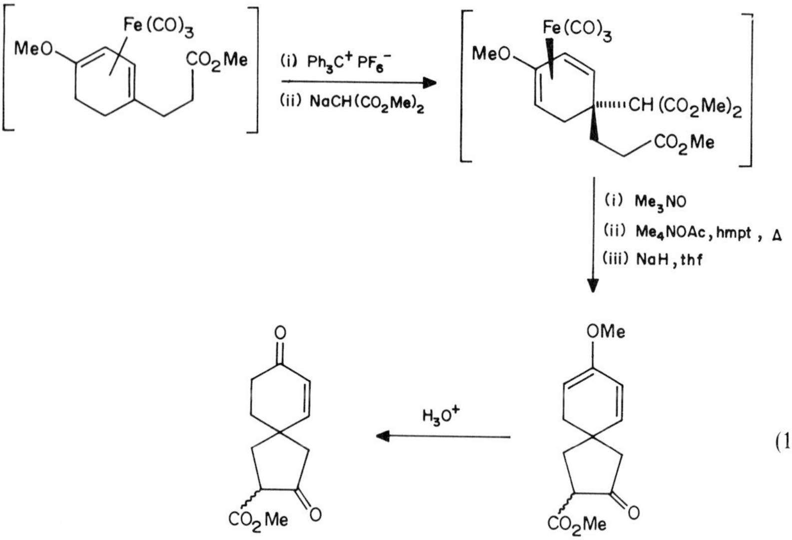

(196)

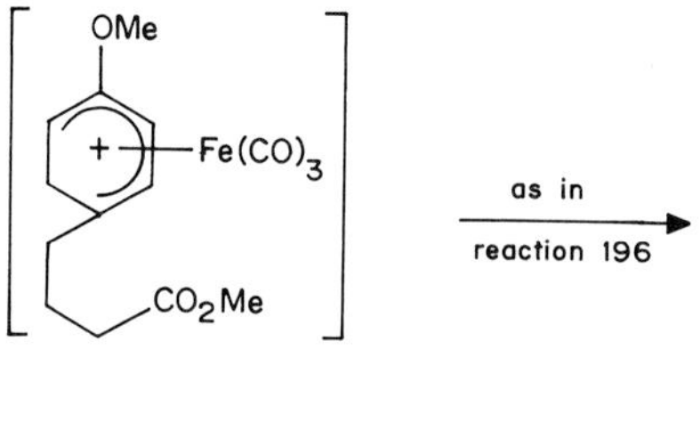

(197)

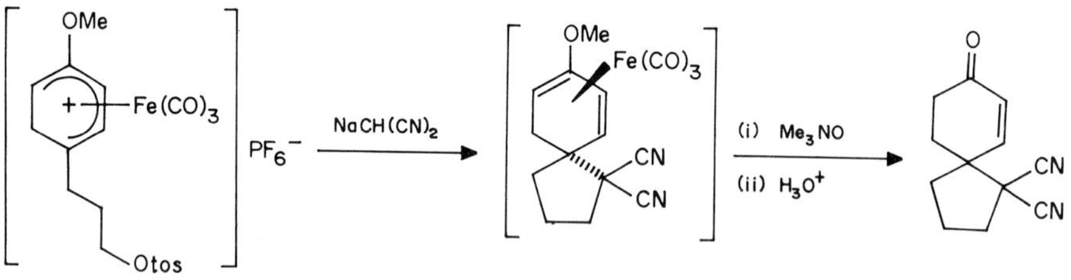

(198)

substituted dienyl complex as a double electrophile can be adapted to the synthesis of azaspirocyclic compounds. Reaction of these complexes with benzylamine occurs in a reversible fashion and is followed by intramolecular displacement of tosylate by the newly installed amino group. The combination of reversibility and cyclization overcomes any problems of regioselectivity during the nucleophilic addition reaction and results in very high yields of azaspirocycle. Decomplexation and further manipulation of these products result in interesting potential for synthetic applications, illustrated by a synthesis of (±)-depentylperhydrohistrionicotoxin (**23**) in reaction 199 and Scheme 16[132].

OMe Fe(CO)$_3$ Otos PhCH$_2$NH$_2$ OMe Fe(CO)$_3$ NCH$_2$Ph 95% (i) Me$_3$NO (ii) H$_3$O$^+$ O NCH$_2$Ph

(199)

OMe Fe(CO)$_3$ Otos (i) PhCH$_2$NH$_2$ (ii) Me$_3$NO (iii) H$_3$O$^+$ O NCH$_2$Ph

(i) Bu$_2$CuLi, Me$_3$SiCl (ii) PhSeCl, (iii) [O]

O NCH$_2$Ph (i) NaBH$_4$ (ii) LiAlH$_4$, AlCl$_3$ (iii) H$_2$, Pd–C (iv) BH$_3$·thf, H$_2$O$_2$, OH$^-$ OH NH

(23)

SCHEME 16

b. Intramolecular nucleophilic additions

A number of intramolecular additions of carbon nucleophiles to the cyclohexadienyl complex, resulting in spirocycle formation, have reported. These are the first examples of intramolecular nucleophilic addition to dienyliron cations and they serve as model studies for other annulation processes which might be conceived. Again, an interesting feature is the manipulation of functional groups which is possible without destruction of the organometallic moiety. Thus, from the examples shown below and the preceding examples, it can be seen that the diene–Fe(CO)$_3$ group can be carried unchanged through a very wide range of synthetic manipulations and removed selectively at the appropriate point in the synthesis (Schemes 17 and 18).

B. Aryl Cation Equivalents[133]

We have shown in the preceding sections that the Fe(CO)$_3$ group can be removed from diene complexes to generate either cyclohexadienes or cyclohexenones. If this demetall-

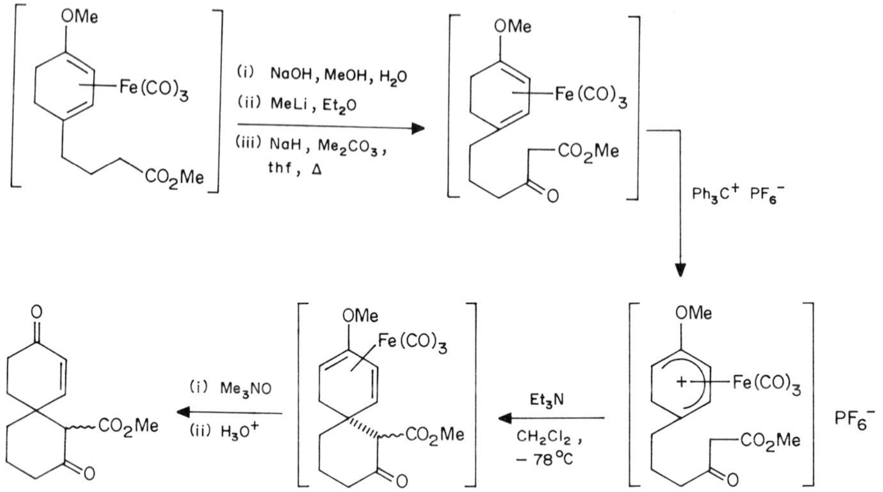

SCHEME 17

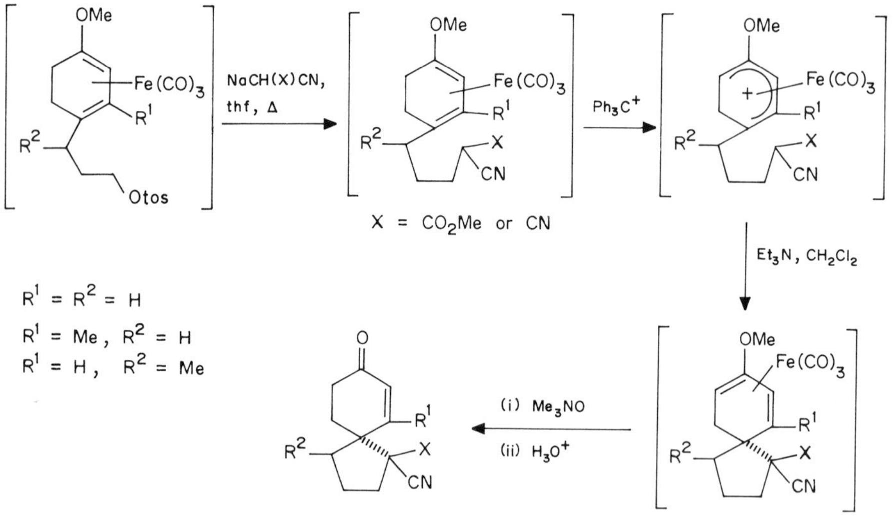

SCHEME 18

ation is carried out under strongly oxidizing (or dehydrogenating) conditions, or if the cyclohexadiene product is subjected to further oxidation, then a substituted aromatic compound is obtained. The overall transformation can be accomplished using either palladium on charcoal or 2, 3-dichloro-5, 6-dicyano-1, 4-benzoquinone (ddq) as oxidizing agent, either directly on the [(diene)$Fe(CO)_3$] complex or on the diene resulting from decomplexation. In this way the cyclohexadienyliron complexes may be regarded as aryl cation equivalents, as shown in the generalized sequence in equation 200. The overall

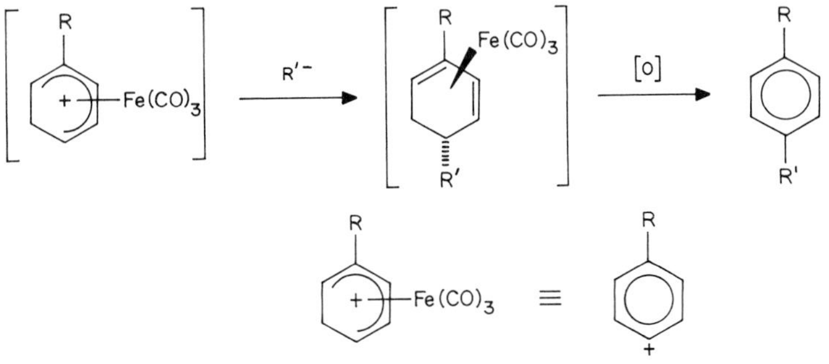

(200)

transformation corresponds to nucleophilic aromatic substitution, where hydride is the effective leaving group! Now, a very large number of substituted cyclohexadienyliron complexes are readily prepared, so that there is potential access to a wide range of aryl cation equivalents. Equations 201–208 summarize most of the available complexes, their corresponding aryl cations, and examples of conversion to substituted aromatics where this has been reported.

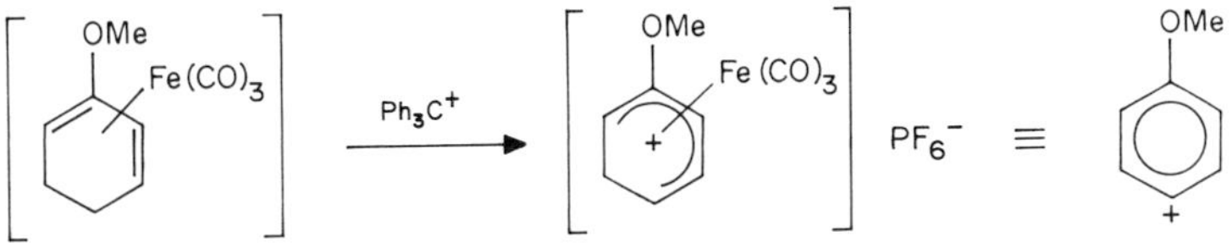

(201)

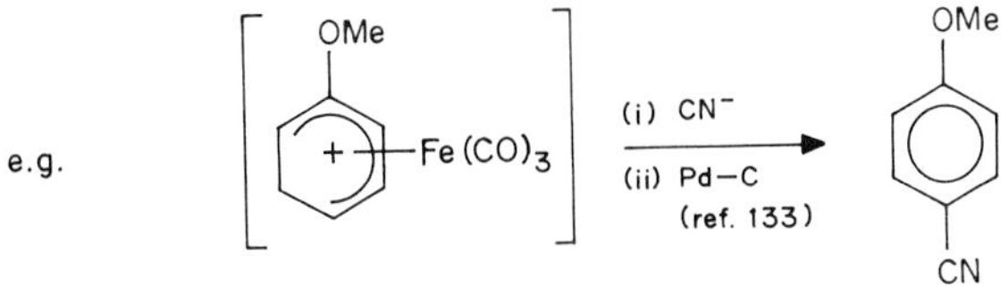

e.g. (i) CN^- (ii) Pd—C (ref. 133)

(202)

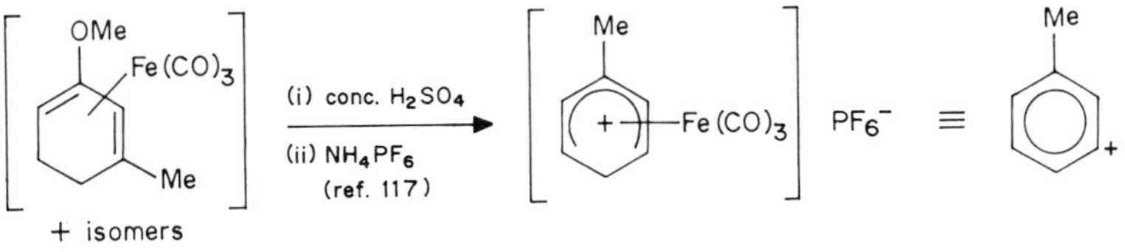

+ isomers (i) conc. H_2SO_4 (ii) NH_4PF_6 (ref. 117)

(203)

(i) conc. H_2SO_4 (ii) NH_4PF_6 (ref. 117)

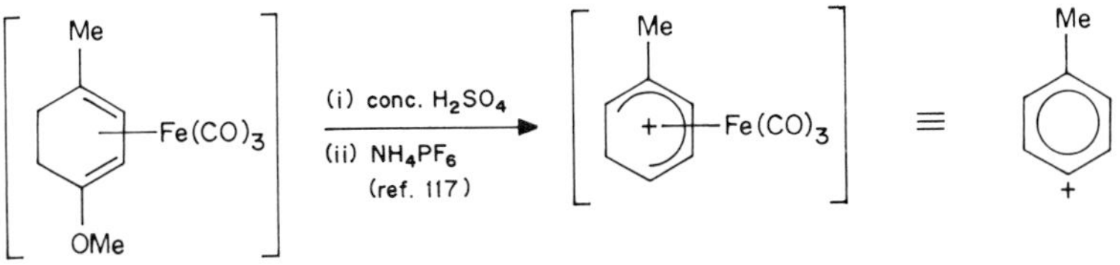

(204)

(i) conc. H_2SO_4 (ii) NH_4PF_6 (ref. 133) (205)

(i) conc. H_2SO_4 (ii) NH_4PF_6 (206)

e.g. PF_6^- + Me_3SiO $OSiMe_3$ CO_2Me

(i) MeCN, Δ (ii) H^+, MeOH (iii) Pd—C (ref. 134) (207)

$Ph_3C^+ BF_4^-$ BF_4^- (208)

Hence there are good possibilities for using this concept in synthesis. The examples in equations 201–208 could be easily extended to include di- and tri-substituted aryl cation equivalents since the appropriate diene complexes are readily prepared. A use of dienyl–iron complexes as aryl cation equivalents has recently been realized for a formal total synthesis of (±)-*O*-methyljoubertiamine (**24**), a member of the Sceletium alkaloids group. This synthesis is summarized in Scheme 19[135].

C. Control of Relative Stereochemistry Using Dienyliron Complexes

All irreversible nucleophile additions to the dienyl ligand in these complexes occur *trans* to the $Fe(CO)_3$ group. Therefore, provided that the diene complex resulting from

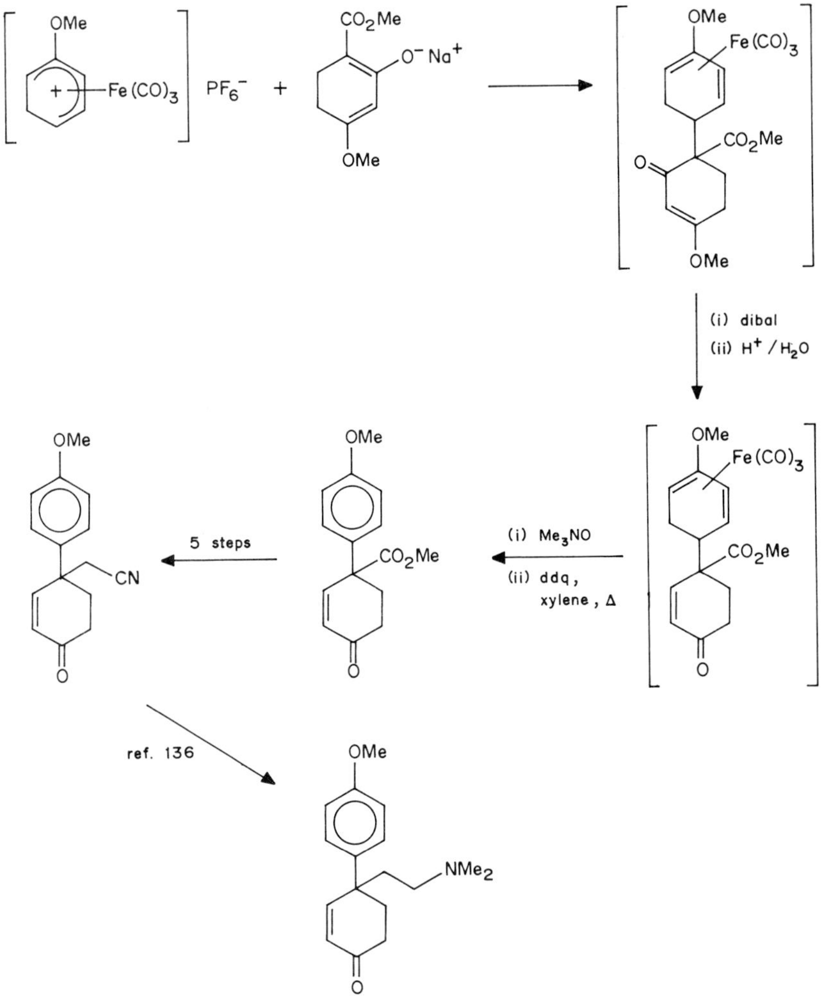

SCHEME 19

nucleophilic attack can be reconverted into a new reactive dienyl complex, it should be possible to add a second nucleophile *cis* to the first nucleophile, as in reaction 209.

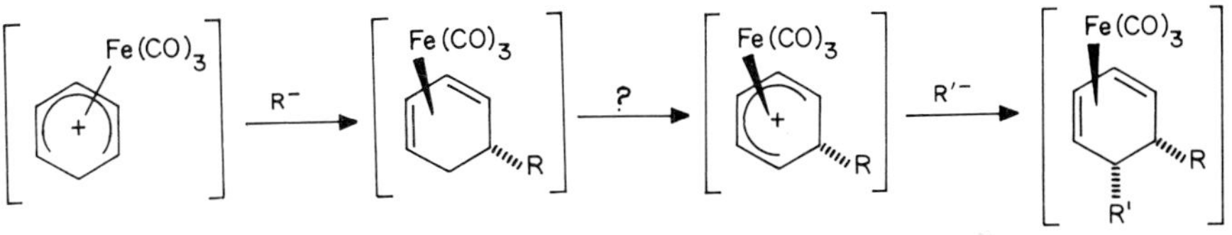

(209)

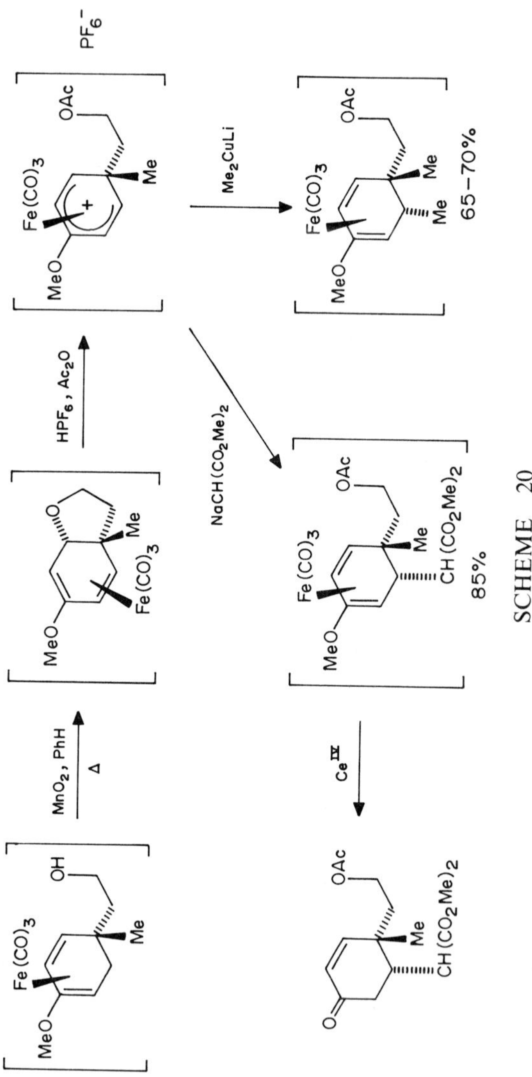
MeO
Fe(CO)3
OH
Me
MnO2, PhH
Δ
MeO
Fe(CO)3
Me
HPF6, Ac2O
MeO
Fe(CO)3
OAc
Me
PF6−
Me2CuLi
MeO
Fe(CO)3
OAc
Me
Me
65–70%
NaCH(CO2Me)2
MeO
Fe(CO)3
OAc
Me
CH(CO2Me)2
85%
CeIV
O
OAc
Me
CH(CO2Me)2

SCHEME 20

However, the steric requirements of the triphenylmethyl cation, the usual hydride abstracting reagent, are so great that for most cyclohexadiene–$Fe(CO)_3$ complexes, even when R = Me, no hydride abstraction occurs. Consequently, special methods have to be developed if we are to explore vicinal stereocontrol using the above sequence. It turns out that [(diene)$Fe(CO)_3$] complexes bearing OH groups in substituent R can undergo oxidative cyclization using a variety of oxidizing agents, and we can use this to advantage. Thus, treatment of a hydroxyethyl-substituted complex with manganese dioxide in refluxing benzene leads to a cyclic ether which undergoes oxygen ring opening on treatment with hexafluorophosphoric or tetrafluoroboric acid in acetic anhydride to give the desired substituted salt. This undergoes stereospecific addition of nucleophiles such as dimethyl sodiomalonate or lithium dimethyl cuprate, as illustrated in Scheme 20[137]. This sequence can be performed with complexes having moderately larger substituents, although there are obvious limitations. When the substituent is sterically too demanding, attempted nucleophile addition results in other reaction pathways being followed. Some examples are given in equations 210 and 211[138].

Fe(CO)3, OH; [Tl(CF3CO2)3], Na2CO3, EtOH; Fe(CO)3; HBF4, Ac2O; Fe(CO)3, OAc, BF4⁻; NaCH(CO2Me)2; PhS⁻; Fe(CO)3, OAc, CH(CO2Me)2, 35–40%; Fe(CO)3, OAc, SPh, 95%

(210)

MeO, Fe(CO)3, OH, Me, CO2Me; (i) Tl^III, (ii) HBF4, Ac2O; MeO, Fe(CO)3, OAc, Me, CO2Me, BF4⁻; nucleophiles; O, OAc, Me, CO2Me

(211)

Surprisingly, trimethylsilyl-substituted diene complexes bearing substituents *exo* to the $Fe(CO)_3$ group undergo hydride abstraction to give substituted dienyl complexes, although this requires prolonged reaction in refluxing dichloromethane. The product complexes undergo stereospecific nucleophile addition to give diene complexes which are converted into the free dienes in the usual way, as shown in equations 212 and 213[139].

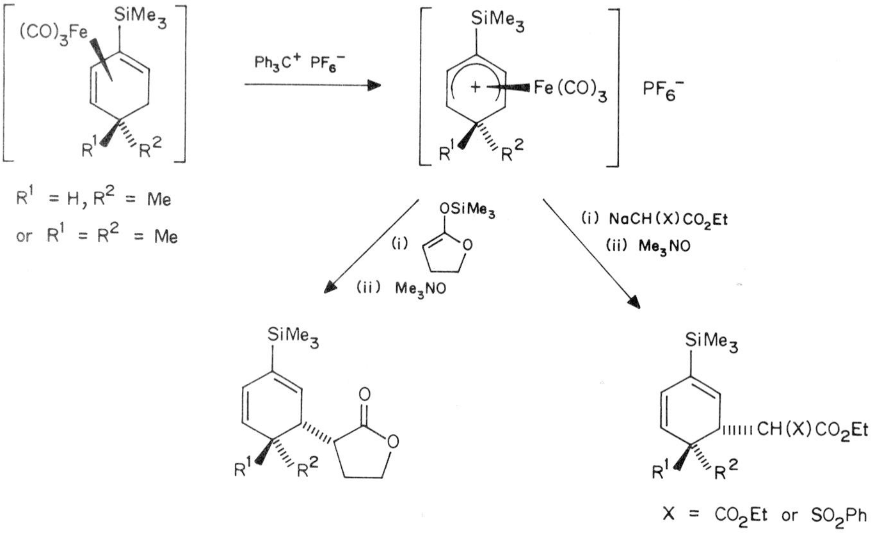

SiMe3
(CO)3Fe
R1
R2
R1 = H, R2 = Me
or R1 = R2 = Me
Ph3C+ PF6−
Fe(CO)3
PF6−
OSiMe3
(i)
(ii) Me3NO
(i) NaCH(X)CO2Et
(ii) Me3NO
CH(X)CO2Et
X = CO2Et or SO2Ph

(212)

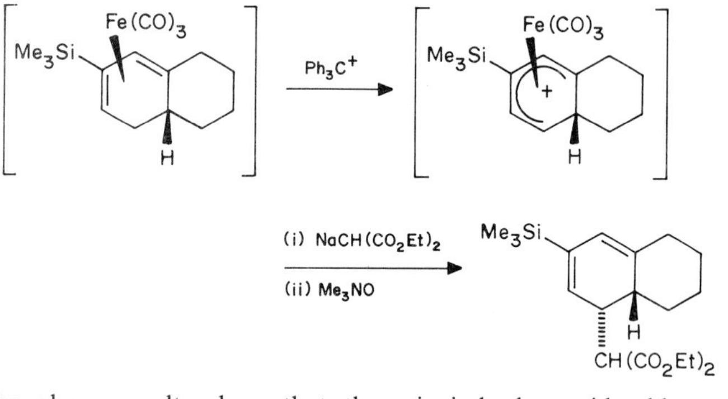

Fe(CO)3
Me3Si
H
Ph3C+
(i) NaCH(CO2Et)2
(ii) Me3NO
CH(CO2Et)2

(213)

All of the above results show that there is indeed considerable potential for accomplishing complete stereocontrol during the attachment of substituents at vicinal positions in the six-membered ring. When we move to larger ring sizes there emerges the possibility that stereocontrol can be achieved at positions which are separated by one or more methylene groups. The metal moiety might also exert a controlling effect in other ways, such as controlling the conformation of awkward ring sizes. However, there are problems which attend the departure from cyclohexadienyliron complexes, such as loss of regiocontrol during nucleophile addition and low yields. For example, tricarbonylcycloheptadienyliron tetrafluoroborate is very badly behaved towards nucleophiles, usually giving products resulting from addition to C-1 or C-2 of the dienyl ligand, or from attack at the metal with concomitant loss of carbonyl ligand, as in reaction 214[140]. The yield of the

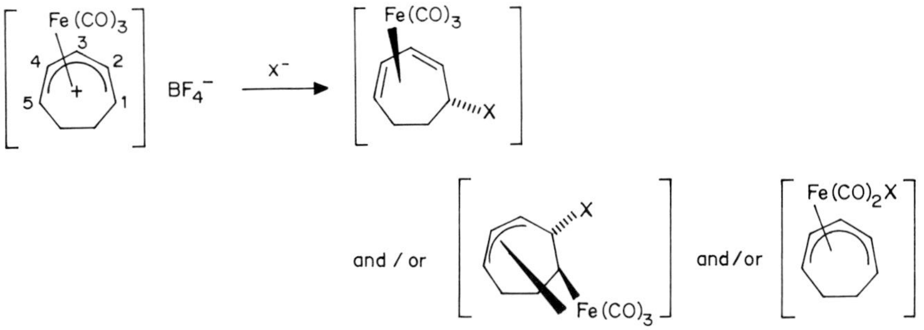

(214)

desired substituted cycloheptadiene complex is seldom greater than 30%. However, these problems can be overcome by altering the ligand environment of the metal, increasing the metal electron density by replacing one of the CO ligands with a poorer π-acceptor. Thus, dicarbonylcycloheptadienyltriphenylphosphiteiron hexafluorophosphate undergoes clean reaction with soft carbanion nucleophiles to give substituted diene complexes, while 'harder' nucleophiles add to the dienyl C-2 position (Scheme 21). Both of these modes of addition give very high yields[141].

The alkyl-substituted diene complexes which are available using this methodology undergo hydride abstraction followed by a second nucleophile addition with very high yields. The reaction with nucleophile occurs regio- and stereo-specifically to give disubstituted cycloheptadiene complexes which are readily demetallated to give the free diene as in Scheme 22. This provides very easy access to a range of stereodefined disubstituted cycloheptadienes which can be further manipulated in a number of ways. For example, diester derivatives can be decarboxylated to give the monocarboxylic acids, and these can be subjected to, for example, phenylselenolactonization, which can be performed in a stereo- and regio-controlled manner. The product allylic selenolactones

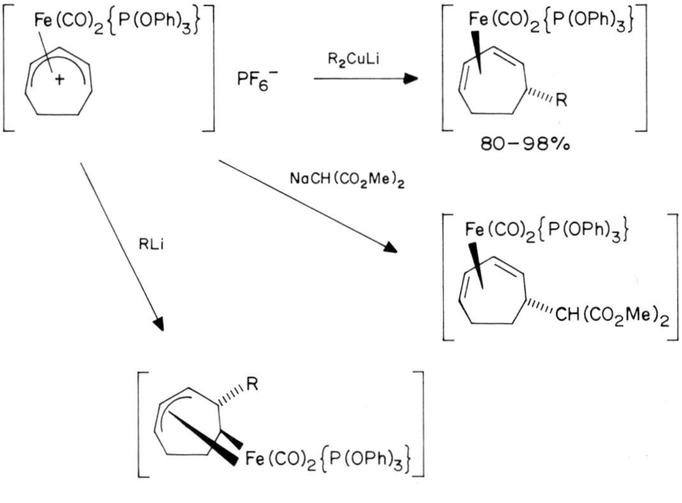

SCHEME 21

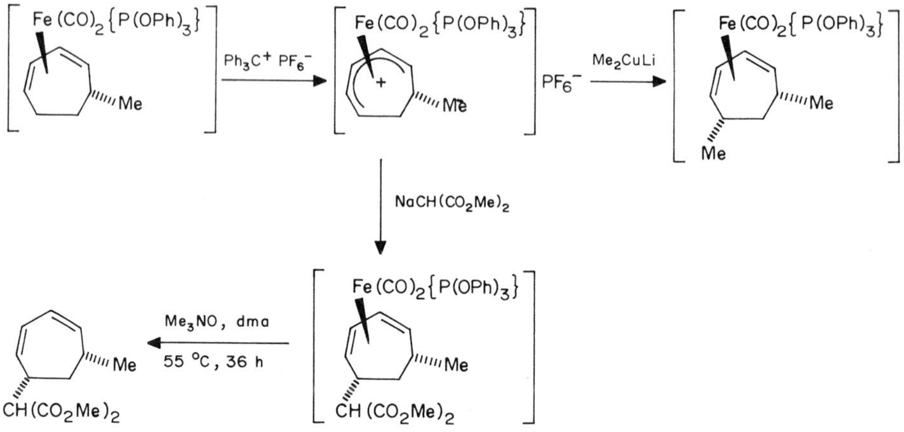

SCHEME 22

can be converted into hydroxylactones by oxidation and these can be further oxidized to cycloheptenone derivatives. This novel combination of organoiron chemistry with diene cyclofunctionalization promises to be an extremely useful method of functionalizing the seven-membered carbocyclic ring (reaction 215)[142].

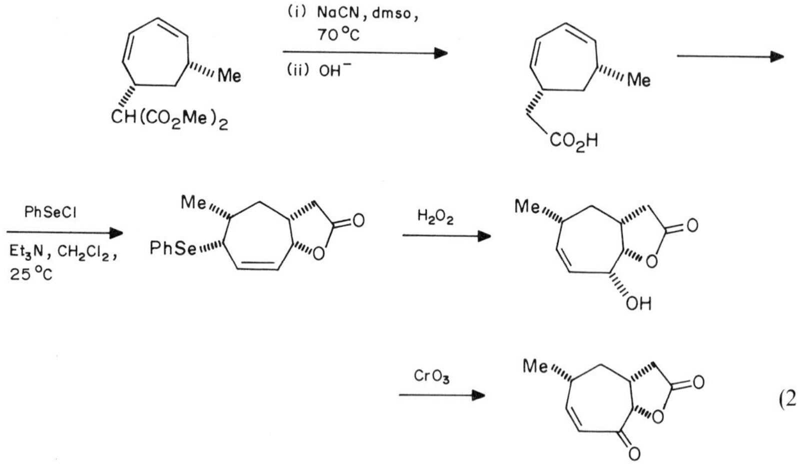

(215)

It may have been noted that the modes of stereocontrol which can be obtained using dienyliron complexes are complementary to those using dienemolybdenum systems. This complementarity will be particularly interesting with awkward ring sizes, where the metal can be used to create a bias. We can anticipate that a range of intermediates for complex organic synthesis with defined relative stereochemistry will eventually become available using this methodology.

VII. PROPARGYL CATIONS STABILIZED BY COBALT

Whilst these compounds may be regarded as similar to η^3-allyl–metal complexes, the fact that four π-electrons from the acetylene ligand are involved in bonding to (two) cobalt metal centres gives a sufficiently different bonding pattern to warrant their separate consideration. Reaction of an acetylene with dicobalt octacarbonyl gives an [(acetylene)$Co_2(CO)_6$] complex, in which the reactivity of the carbon—carbon triple bond is so moderated that this can be used as an effective means of protecting acetylenic groups[143]. The dicobalt hexacarbonyl moiety is removed by treatment with cerium(IV) ammonium nitrate or iron(III) nitrate (reactions 216–219).

RC≡CR' —[$Co_2(CO)_8$], 25 °C→ [RC≡CR' / $(CO)_3Co{-}Co(CO)_3$] ≡ [R≡R' / $Co_2(CO)_6$] —Ce^{IV} or $Fe(NO)_3$, EtOH, H_2O→ RC≡CR' (216)

Ph–C≡C–Ph —[$Co_2(CO)_8$]→ Ph–C≡C–Ph ($Co_2(CO)_6$) (217)

—[$Co_2(CO)_8$]→ ($Co_2(CO)_6$) —N_2H_2→ ($Co_2(CO)_6$) (218)

[OH, $Co_2(CO)_6$] —(i) $BH_3 \cdot$thf (ii) H_2O_2, OH^-→ [OH, OH, $Co_2(CO)_6$] (219)

Propargylic carbocations are found to be stabilized by the presence of the cobalt carbonyl moiety, presumably owing to overlap of the carbocation centre with cobalt d-orbitals. This effect is seen during electrophilic aromatic substitution on a neighbouring phenyl group, when the [(acetylene)$Co_2(CO)_6$] group acts as an *ortho/para* director (reaction 220).

A number of carbocation complexes have been prepared by reaction of a propargyl alcohol complex with tetrafluoroboric acid under anhydrous conditions. The salts are fairly stable and can be isolated by precipitation with diethyl ether. They are powerful electrophiles, undergoing facile alkylation with, for example, silyl enol ethers and allylsilanes, and the reaction occurs without the allenic rearrangements commonly

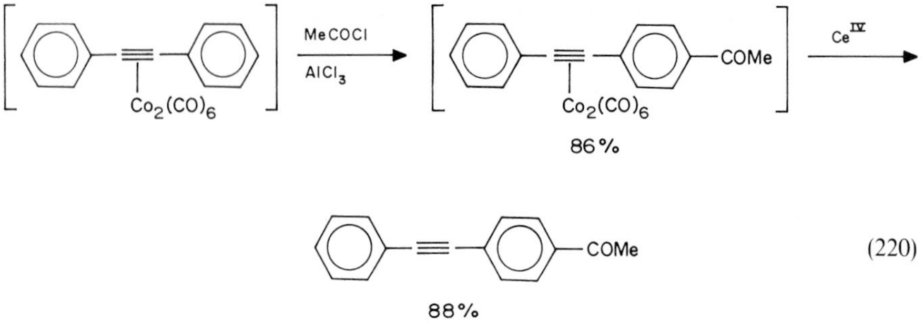

(220)

88%

encountered during reactions of propargylic halides. This offers a readily controlled means of introducing the propargyl group into various molecules, but so far it has not been exploited for natural products synthesis. Some examples of this reactivity are given in equations 221–226[114].

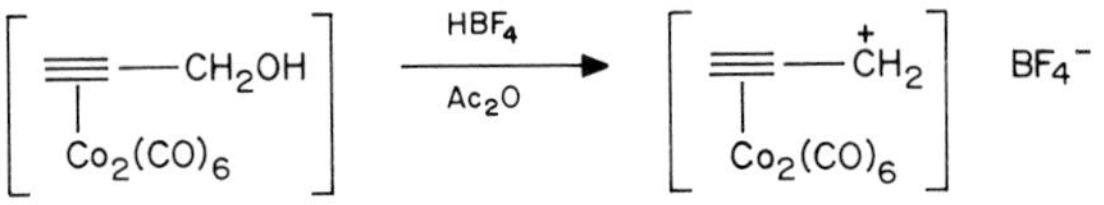

(221)

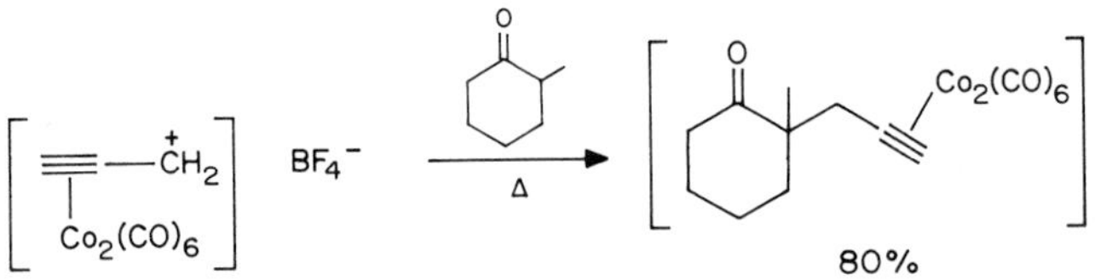

(222)

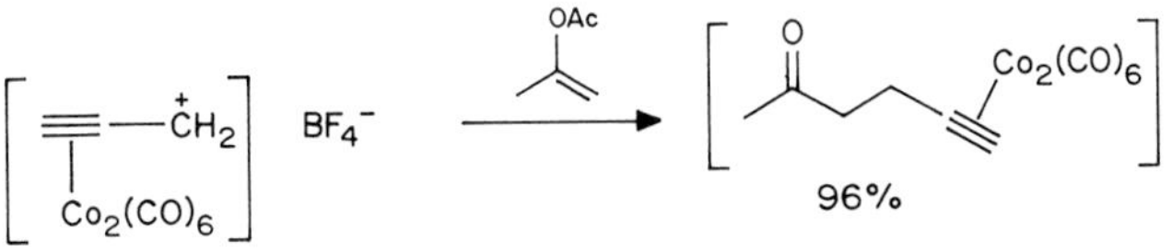

80%

(223)

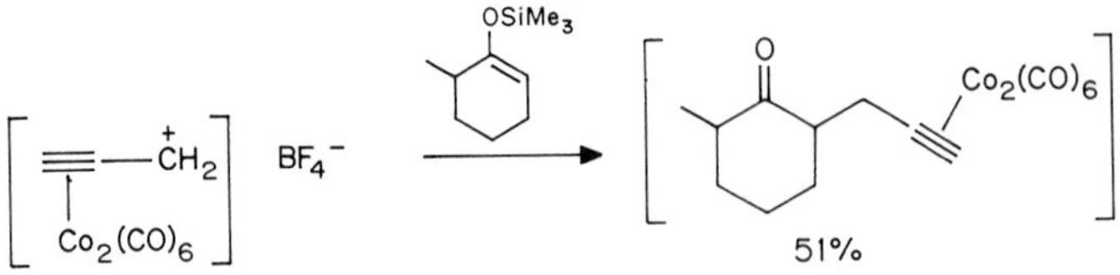

96%

(224)

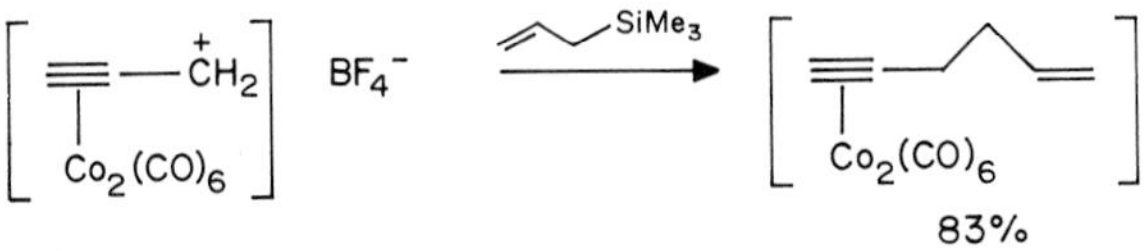

51%

(225)

83%

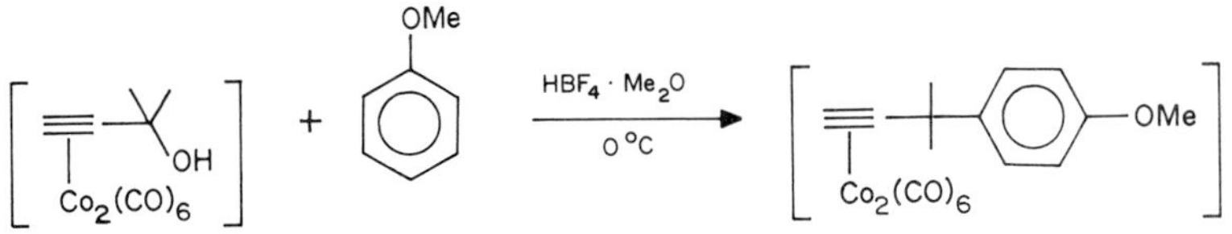

(226)

Although these complexes have not yet found application in organic synthesis problems, it should be borne in mind that, since the terminal acetylene is readily converted into methyl ketone, we can regard the complexes as equivalent to specific ketone α-cation equivalents, so that access is gained to unusual modes of ketone alkylation.

VIII. ARENE–METAL COMPLEXES

Although the uncharged arenechromium tricarbonyl complexes have been used for accomplishing the introduction of carbon substituents on to an aromatic ring with attention being paid to synthetic application[145], related cationic arene–metal complexes have not been explored as real synthetic intermediates. The chemistry of some of these complexes is fairly well developed, and they offer a useful series of electrophilic arylating reagents which are more reactive than the arenechromium tricarbonyl systems. The two metals of most interest for the present discussion are manganese and iron, since stable, reactive arene cationic complexes are readily prepared and isolated.

A. Arenemanganese Tricarbonyl Salts

Arenemanganese tricarbonyl cations are readily prepared from a number of simple aromatic compounds by their reaction with bromomanganese pentacarbonyl in the presence of aluminium trichloride at elevated temperature. The arene complex is initially formed as a tetrahaloaluminate, which is usually extracted into water and converted into the more easily manageable hexafluorophosphate by anion exchange (reaction 227)[146].

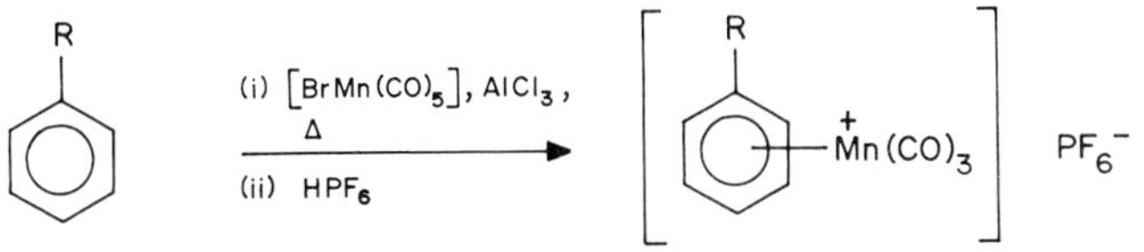

(227)

More recently, good yields of [(arene)$Mn(CO)_3$] complexes have been obtained by treatment of the arene with the inexpensive tricarbonyl(methylcyclopentadienyl)manganese in the presence of aluminium tribromide (reaction 228)[147]. Both of these methods use the aromatic compound as a solvent and

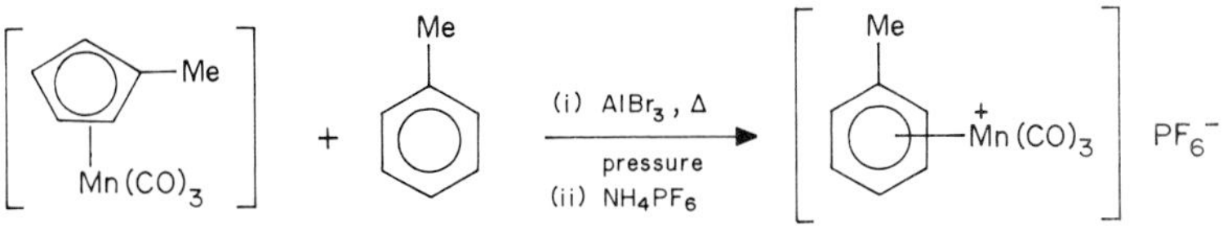

(228)

therefore are applicable only to readily available materials. Two similar methods have recently been developed for the preparation of [(arene)$Mn(CO)_3$] complexes under milder conditions and using dichloromethane as a solvent, which allows the use of stoichiometric amounts of arene and manganese complex starting material. These are illustrated in equations 229 and 230[148].

$$\text{RC}_6\text{H}_5 + [\text{Mn(CO)}_5]^+ \text{ClO}_4^- \xrightarrow{\text{CH}_2\text{Cl}_2} [(\text{RC}_6\text{H}_5)\overset{+}{\text{Mn}}(\text{CO})_3]\ \text{ClO}_4^- \qquad (229)$$

$$1{,}2\text{-(MeO)}_2\text{C}_6\text{H}_4 + [\text{Mn(CO)}_5]^+ \text{BF}_4^- \xrightarrow{\text{CH}_2\text{Cl}_2} [(1{,}2\text{-(MeO)}_2\text{C}_6\text{H}_4)\overset{+}{\text{Mn}}(\text{CO})_3]\ \text{BF}_4^- \qquad (230)$$

These cationic complexes are very reactive towards nucleophiles and lead to a method for nucleophilic aromatic substitution, provided that the initially formed [(cyclohexadienyl)Mn(CO)$_3$] complex can be oxidized to aromatic compound. It is noteworthy that this sequence of reactions is equivalent to nucleophilic aromatic substitution where *hydride* is the leaving group. Directing effects are observed when substituents are present on the aromatic ring. For example, methoxy and dimethylamino groups cause nucleophilic addition to occur at positions *meta* to themselves. The position *para* to a methoxy substituent is so deactivated that addition to the [(1,2-dimethoxybenzene)Mn(CO)$_3$]$^+$ cation occurs at the 3-position. When a leaving group for example, chloride, is present on the aromatic ring, reaction with *reversible* nucleophiles leads to its displacement, whilst *irreversible* nucleophiles add predominantly *meta* to the halogen. A range of nucleophiles can be combined with [(arene)Mn(CO)$_3$] complexes to give an interesting array of bond-formation methods. Examples of these aspects of chemistry, together with decomplexation of [(dienyl)Mn(CO)$_3$] complexes to give substituted aromatics, are given in equations 231–239[149].

$$[(\text{C}_6\text{H}_6)\overset{+}{\text{Mn}}(\text{CO})_3]\ \text{PF}_6^- \xrightarrow{\text{NaCH(CO}_2\text{Me)}_2} [(\text{C}_6\text{H}_6\text{CH(CO}_2\text{Me)}_2)\text{Mn(CO)}_3] \qquad (231)$$

$$[(\text{C}_6\text{H}_6)\overset{+}{\text{Mn}}(\text{CO})_3]\ \text{PF}_6^- \xrightarrow{\text{PhLi}} [(\text{C}_6\text{H}_6\text{Ph})\text{Mn(CO)}_3] \qquad (232)$$

$$[(C_6H_6)Mn(CO)_3]^+ PF_6^- \xrightarrow{N_3^-} [(exo\text{-}6\text{-}N_3C_6H_6)Mn(CO)_3] \tag{233}$$

$$[(1,3,5\text{-}Me_3C_6H_3)Mn(CO)_3]^+ PF_6^- \xrightarrow[20\,^{\circ}C,\ 4\ days]{(i)\ NaCN\ (ii)\ Ce^{IV},\ 2\,M\ H_2SO_4} 2,4,6\text{-}Me_3C_6H_2CN \tag{234}$$

$$[(MeOC_6H_5)Mn(CO)_3]^+ PF_6^- \xrightarrow{LiAlH_4} \underset{63}{[\ldots Mn(CO)_3]} + \underset{37}{[\ldots Mn(CO)_3]} \tag{235}$$

$$[(4\text{-}Me_2NC_6H_4Me)Mn(CO)_3]^+ PF_6^- \xrightarrow{LiAlH_4} \underset{97}{[\ldots Mn(CO)_3]} + \underset{3}{[\ldots Mn(CO)_3]} \tag{236}$$

$$[(4\text{-}ClC_6H_4Me)Mn(CO)_3]^+ PF_6^- \xrightarrow{MeLi} \underset{74}{[\ldots Mn(CO)_3]} + \underset{26}{[\ldots Mn(CO)_3]} \tag{237}$$

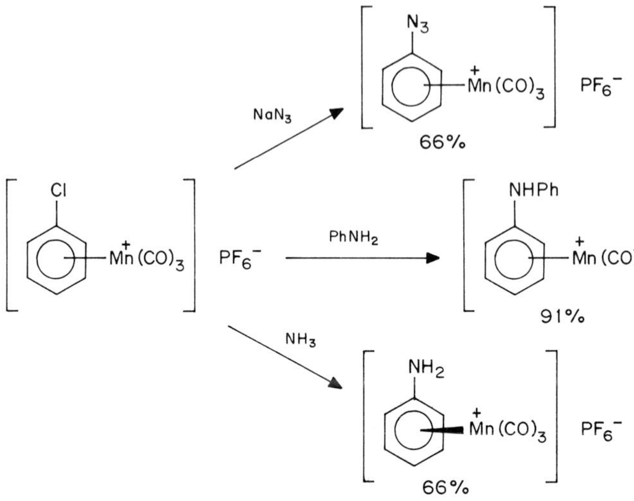

(238)

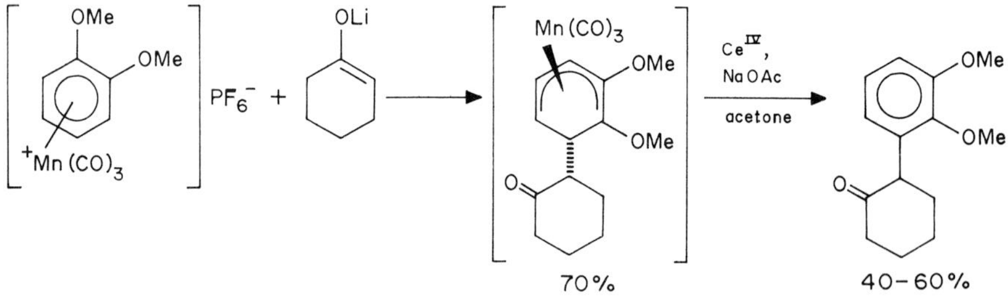

(239)

The neutral [(dienyl)$Mn(CO)_3$] complexes obtained from nucleophile addition can be further activated by replacing one CO ligand by NO^+, using nitrosonium hexafluorophosphate or tetrafluoroborate. The resulting cations are susceptible to nucleophile attack (reaction 240), but the application of this sequence in synthesis has not been

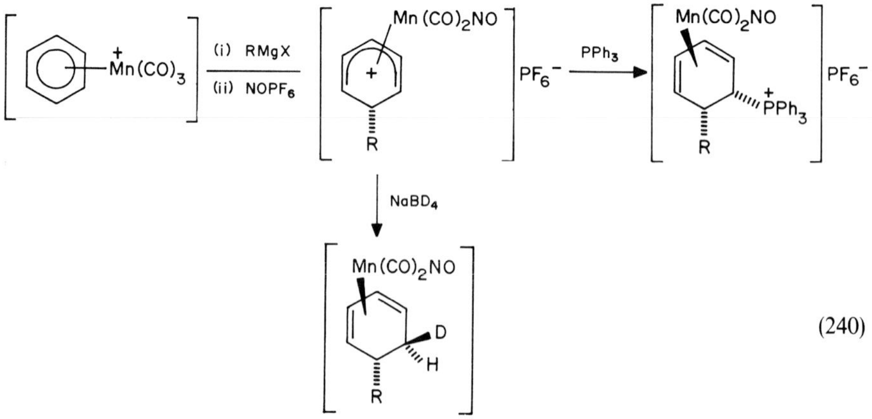

(240)

established[150]. All of the above work has been carried out on fairly simple arene derivatives. There is clearly a need to examine carefully the synthesis and reactions of a wider range of polysubstituted arene complexes, particularly those bearing functional groups which can be further utilized in synthesis.

B. Cyclopentadienyl(arene)iron Complexes

The preparation and reactivity of these complexes has been thoroughly reviewed[151]. A number of bis(arene)iron dications are known, but these have not been explored in any depth for synthesis and so they will not be discussed in this section.

The cyclopentadienyl(arene)iron complexes which we shall discuss are most readily prepared from ferrocene by ligand-exchange reactions. Typically, the arene to be complexed is used as the solvent, if it is a liquid. Aluminium trichloride is used as a Lewis acid catalyst, and powdered aluminium is incorporated in order to inhibit oxidation of ferrocene. For solid arene components, a hydrocarbon solvent such as cyclohexane or decalin can be used and the reaction is carried out at elevated temperature (70–190 °C) for a length of time depending on the particular reaction (equation 241). A large number of complexes have been prepared, in which the arene ranges from benzene, to anisole to halobenzenes, etc. Halobenzenes invariably give lower yields.

$$[\text{Cp}_2\text{Fe}] + \text{C}_6\text{H}_5\text{R} \xrightarrow[\text{Al powder}]{2\ \text{AlCl}_3} [(\text{RC}_6\text{H}_5)\overset{+}{\text{Fe}}\text{Cp}]\ \text{AlCl}_4^-$$

$$\xrightarrow[\text{(ii) HPF}_6]{\text{(i) H}_2\text{O}} [(\text{RC}_6\text{H}_5)\overset{+}{\text{Fe}}\text{Cp}]\ \text{PF}_6^- \qquad (241)$$

As with the $[(\text{arene})\text{Mn(CO)}_3]^+$ cations, these complexes are usually converted into the hexafluorophosphate salts for ease of handling. The arene complexes can also be prepared using $[\text{CpFe(CO)}_2\text{Cl}]$ in the presence of aluminium trichloride, although this is less convenient since ferrocene is a more readily available starting material (reaction 242).

$$[\text{CpFe(CO)}_2\text{Cl}] + \text{C}_6\text{H}_5\text{R} \xrightarrow[\text{(ii) HPF}_6]{\text{(i) AlCl}_3} [(\text{RC}_6\text{H}_5)\overset{+}{\text{Fe}}\text{Cp}]\ \text{PF}_6^- \qquad (242)$$

No mild methods suitable for more sensitive arenes appear to have been developed, and indeed this is an area for future research, together with optimization of procedures to obtain consistently high yields.

With condensed aromatics it is possible to accomplish selective complexation of one or more aromatic rings, depending on the reaction conditions employed as shown in reaction 243.

As might be expected from their cationic nature, these complexes undergo reactions with nucleophiles, although they are less reactive than the arene–manganese complexes

$$[Cp_2Fe] + \text{terphenyl} \xrightarrow[\text{1:1 stoichiometry}]{AlCl_3\text{, etc.}} [\text{(FeCp}^+\text{)terphenyl}]\,PF_6^-$$

2 : 1 stoichiometry

$$[\text{(FeCp}^+)_2\text{terphenyl}]\,(PF_6^-)_2 \qquad (243)$$

discussed in the preceding section. Some of these reactions are shown in equations 244–247. When the arene ligand is too highly substituted, nucleophilic addition occurs at the

$$[C_6H_6\text{Fe}^+\text{Cp}] \xrightarrow{NaBH_4} [\text{FeCp}(C_6H_7)] \qquad (244)$$

RLi → [FeCp(cyclohexadienyl-R)]

MeO⁻ → [(MeOC₆H₅)Fe⁺Cp]

[(ClC₆H₅)Fe⁺Cp]

MeLi → [Me, Cl cyclohexadienyl FeCp] + [Me, Cl cyclohexadienyl FeCp]

4 : 1

(245)

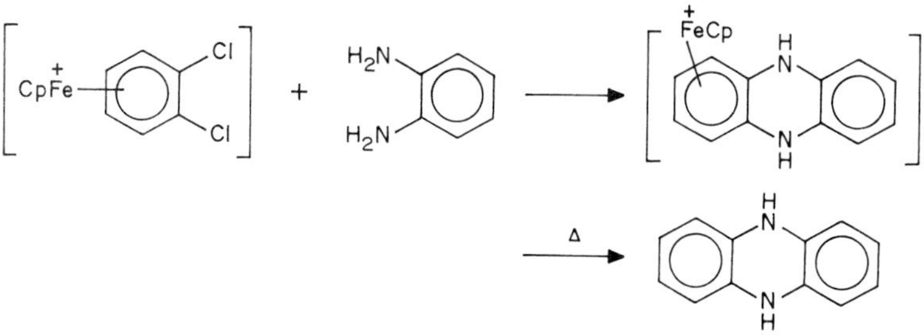

(246)

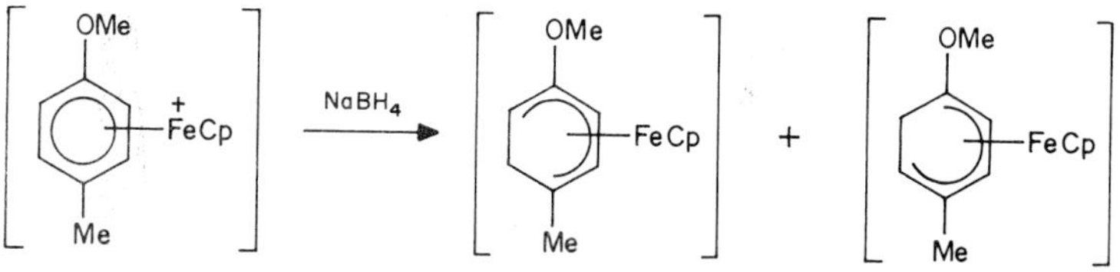

(247)

ca. 4 : 1

cyclopentadienyl ligand, although this has not been explored as a possible means of functionalizing the five-membered ring (reaction 248).

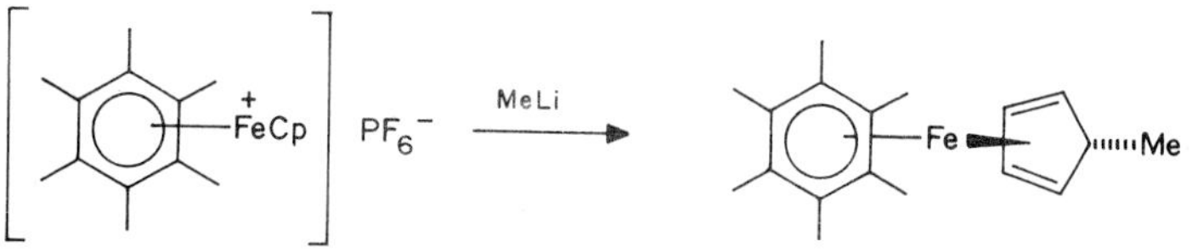

(248)

Reliable methods of decomplexation of arene complexes, suitable for use with sensitive groups, have not been developed. Usually the complexes are subjected to photolysis or vacuum thermolysis to liberate the aromatic compound. Careful attention does not appear to have been paid to the decomplexation of dienyl complexes resulting from carbon nucleophile addition.

Treatment of alkyl-substituted arene complexes with a strong base results in deprotonation of the side-chain α to the aromatic ring to give alkylidene-substituted cyclohexadienyl complexes. These in turn react readily with suitable electrophiles to produce appropriately substituted arene complexes. This might ultimately provide a reliable method for benzylic functionalization as in reactions 249 and 250.

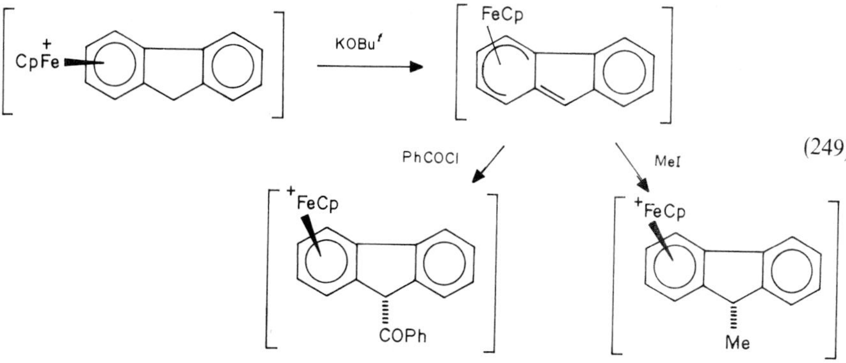

(249)

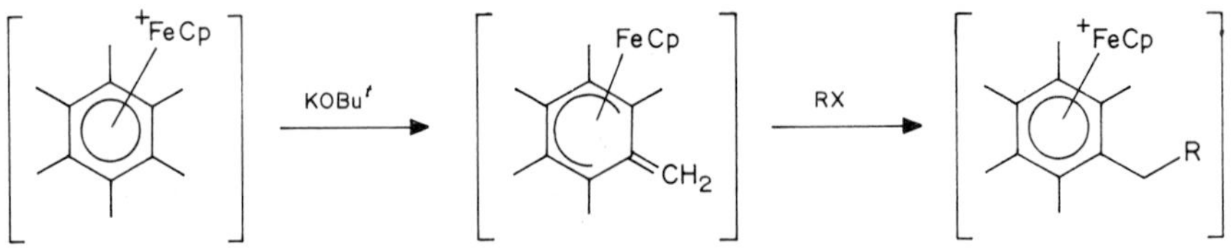

(250)

As can be seen from this brief discussion, with proper focus arene–metal cation complexes might well lead to reliable and efficient methods for introducing substituents into an aromatic ring. However, any future effort in this area must be made bearing in mind that, in order to find use in organic synthesis, methods must be found for producing real synthetic intermediates which are not easily available using standard procedures.

IX. REFERENCES

1. S. G. Davies, M. L. H. Green, and D. M. P. Mingos, *Tetrahedron*, **175**, 113 (1978).
2. A. Jungbaur and H. Behrens, *Z. Naturforsch., Teil B*, **33**, 1083 (1978); E. O. Fischer and A. Vogler, *Z. Naturforsch., Teil B*, **17**, 421 (1962); J. W. Faller, and B. V. Johnson, *J. Organomet. Chem.*, **88**, 101 (1975).
3. M. Rosenblum, *Acc. Chem. Res.*, **7**, 122 (1974).
4. P. F. Boyle and K. M. Nicholas, *J. Org. Chem.*, **40**, 2682 (1975).
5. P. Lennon, M. Madhavarao, A. Rosan, and M. Rosenblum, *J. Organomet. Chem.*, **108**, 93 (1976); A. Rosan, M. Rosenblum, and J. Tancrede, *J. Am. Chem. Soc.*, **95**, 3062 (1973). P. Lennon, A. M. Rosan, and M. Rosenblum, *J. Am. Chem. Soc.*, **99**, 8426 (1977); K. M. Nicholas, A. M. Rosan, and M. Rosenblum, *J. Organomet. Chem.*, **84**, 351 (1975).
6. T. C. T. Chang and M. Rosenblum, *J. Org. Chem.*, **46**, 4103 (1981).
7. T. C. T. Chang and M. Rosenblum, *J. Org. Chem.*, **46**, 4627 (1981).
8. S. R. Berryhill and M. Rosenblum, *J. Org. Chem.*, **45**, 1984 (1980).
9. D. F. Martin, *J. Chem. Soc., Chem. Commun.*, 341 (1980).
10. D. J. Bates, M. Rosenblum, and S. B. Samuels, *J. Organomet. Chem.*, **209**, C55 (1981).
11. D. L. Reger and P. J. McElligott, *J. Am. Chem. Soc.*, **102**, 5923 (1980).
12. A. M. Rosan and M. Rosenblum, *J. Org. Chem.*, **40**, 3621 (1975).
13. N. Genco, D. Marten, S. Raghu, and M. Rosenblum, *J. Am. Chem. Soc.*, **98**, 848 (1976).
14. K. M. Nicholas, *J. Am. Chem. Soc.*, **97**, 3254 (1975).
15. G. W. Parshall, *Homogeneous Catalysis*, Wiley, New York, 1980, Chapter 6.
16. J. Tsuji, *Organic Synthesis with Palladium Compounds*, Springer-Verlag, Berlin, 1980, Chapter 3.
17. J. Tsuji, M. Kaito, and T. Takahashi, *Bull. Chem. Soc. Jpn.*, **51**, 547 (1978).
18. C. S. Subraminiam, P. J. Thomas, V. R. Marndapur, and M. S. Chadha, *Synthesis*, 468 (1978).
19. J. Tsuji, M. Kaito, T. Yamada, and T. Mandai, *Bull. Chem. Soc. Jpn.*, **51**, 1915 (1978).
20. J. Tsuji and T. Mandai, *Tetrahedron Lett.*, 1817 (1978).
21. T. Yanami, M. Miyashita, and A. Yoshikoshi, *J. Org. Chem.*, **45**, 607 (1980).
22. J. Tsuji, I. Shimizu, H. Suzuki, and Y. Naito, *J. Am. Chem. Soc.*, **101**, 5070 (1979).
23. M. Shibasaki, K. Iseki, and S. Ikegami, *Tetrahedron Lett.*, **21**, 3587 (1980).
24. P. D. Magnus and M. S. Nobbs, *Synth. Commun.*, **10**, 273 (1980); H. Yatagai, Y. Yamamoto, and K. Maruyama, *J. Am. Chem. Soc.*, **102**, 4548 (1980); A. J. G. Baxter, K. H. Dickinson, P. M. Roberts, T. C. Smale, and R. Southgate, *J. Chem. Soc. Chem. Commun.*, 236 (1979); W. G. Lloyd and B. J. Luberoff, *J. Org. Chem.*, **34**, 3949 (1969); J. Tsuji, Y. Kobayashi, and T. Takahashi, *Tetrahedron Lett.*, **21**, 483 (1980).
25. J. Tsuji, M. Takahashi, and T. Takahashi, *Tetrahedron Lett.*, **21**, 849 (1980).
26. I. I. Moiseev, M. N. Vargaftik, and Y. K. Syrkin, *Dokl. Akad. Nauk. SSSR*, **133**, 377 (1960).
27. W. Kitching, Z. Rappoport, S. Winstein, and W. C. Young, *J. Am. Chem. Soc.*, **88**, 2054 (1966).
28. N. Adachi, K. Kikukawa, M. Takagi, and T. Matsuda, *Bull. Chem. Soc. Jpn.*, **48**, 521 (1975); C. B. Anderson and B. J. Burreson, *Chem. Ind. (London)*, 620 (1967).

29. D. E. James and J. K. Stille, *J. Am. Chem. Soc.*, **98**, 1810 (1976); J. K. Stille and D. E. James, *J. Am. Chem. Soc.*, **97**, 674 (1975); J. K. Stille and D. E. James, *J. Organomet. Chem.*, **108**, 401 (1976); J. K. Stille and R. Bivakaruni, *J. Am. Chem. Soc.*, **100**, 1303 (1978).
30. M. F. Semmelhack and C. Bodurow, *J. Am. Chem. Soc.*, **106**, 1496 (1984).
31. J. E. Backvall, *Tetrahedron Lett.*, 467 (1977).
32. L. S. Hegedus, G. F. Allen, and E. L. Waterman, *J. Am. Chem. Soc.*, **98**, 2674 (1976); L. S. Hegedus, G. F. Allen, J. J. Bozell, and E. L. Waterman, *J. Am. Chem. Soc.*, **100**, 5800 (1978).
33. A. Kasahara and N. Fukuda, *Chem. Ind.* (*London*), 485 (1976).
34. D. E. Korte, L. S. Hegedus, and R. K. Wirth, *J. Org. Chem.*, **42**, 1329 (1977).
35. S. I. Murahashi and T. Yano, *J. Am. Chem. Soc.*, **102**, 2456 (1980).
36. Y. Odaira, T. Oishi, T. Yukawa, and S. Tsutsumi, *J. Am. Chem. Soc.*, **88**, 4105 (1966); E. S. Brown and E. A. Rick, *Chem. Commun.*, 112 (1969); E. S. Brown, E. A. Rick, and F. D. Mendicino, *J. Organomet. Chem.*, **38**, 37 (1972).
37. B. M. Trost and T. R. Verhoeven, in *Comprehensive Organometallic Chemistry* (Eds. G. Wilkinson, F. G. A. Stone, and E. W. Abel), Vol. 8, Pergamon Press, Oxford, 1982, p. 881.
38. R. F. Heck, *Acc. Chem. Res.*, **12**, 146 (1979).
39. Y. Fujiwara, I. Moritani, S. Danno, R. Asano, and S. Teranishi, *J. Am. Chem. Soc.*, **91**, 7166 (1969); Y. Fujiwara, I. Moritani, R. Asano, H. Tanaka, and S. Teranishi, *Tetrahedron*, **25**, 4815 (1969); Y. Fujiwara, R. Asano, I. Moritani, and S. Teranishi, *J. Org. Chem.*, **41**, 1681 (1976).
40. B. M. Trost, S. A. Godleski, and J. P. Genet, *J. Am. Chem. Soc.*, **100**, 3930 (1978); B. M. Trost, S. A. Godleski, and J. Belletire, *J. Org. Chem.*, **44**, 2052 (1979).
41. H. Takahashi and J. Tsuji, *J. Am. Chem. Soc.*, **90**, 2387 (1968).
42. R. A. Holton, *J. Am. Chem. Soc.*, **99**, 8083 (1977).
43. L. S. Hegedus, T. Hayashi and W. H. Darlington, *J. Am. Chem. Soc.*, **100**, 7747 (1978).
44. T. Hayashi and L. S. Hegedus, *J. Am. Chem. Soc.*, **99**, 7093 (1977).
45. L. S. Hegedus, R. E. Williams, M. A. McGuire, and T. Hayashi, *J. Am. Chem. Soc.*, **102**, 4973 (1980).
46. R. Huttel and M. McNiff, *Chem. Ber.*, **106**, 1789 (1973); B. M. Trost, P. E. Strege, L. Weber, T. J. Fullerton, and T. J. Dietsche, *J. Am. Chem. Soc.*, **100**, 3407 (1978); P. S. Manchand, H. S. Wong, and J. F. Blount, *J. Org. Chem.*, **43**, 4769 (1978); K. Henderson and F. J. McQuillin, *J. Chem. Soc., Chem. Commun.*, 15 (1978); P. Boontanonda and R. Griff, *J. Chem. Soc., Chem. Commun.*, 583 (1977).
47. B. M. Trost and L. Weber, *J. Org. Chem.*, **40**, 3617 (1975).
48. B. M. Trost and P. J. Metzner, *J. Am. Chem. Soc.*, **102**, 3572 (1980).
49. D. J. Collins, W. R. Jackson, and R. N. Timms, *Aust. J. Chem.*, **30**, 2167 (1973).
50. J. Y. Sato and C. A. Horiuchi, *Bull. Chem. Soc. Jpn.*, **52**, 2653 (1979).
51. E. Samuel and M. Bigorgne, *J. Organomet. Chem.*, **19**, 9 (1969).
52. J. M. Kliegman, *J. Organomet. Chem.*, **29**, 73 (1971); K. Yamamoto, K. Shinohara, T. Ohuchi, and M. Kumada, *Tetrahedron Lett.*, 1153 (1974); S. Lin and P. Boudjouk, *J. Organomet. Chem.*, **187**, C11 (1980); K. Itoh, M. Fukui, and Y. Kurachi, *J. Chem. Soc., Chem. Commun.*, 500 (1977).
53. E. O. Fischer and G. Burger, *Z. Naturforsch.*, **616**, 702 (1961); W. T. Dent, R. Long, and G. Wilkinson, *J. Chem. Soc.*, 1585 (1964).
54. P. M. Maitlis, *The Organic Chemistry of Palladium*, Vols. 1 and 2, Academic Press, New York, 1971; see also ref. 16 and 37.
55. P. E. Slade, Jr., and H. B. Jonassen, *J. Am. Chem. Soc.*, **72**, 1277 (1957); E. O. Fischer and H. Werner, *Chem. Ber.*, **93**, 2075 (1962); S. D. Robinson and B. L. Shaw, *J. Chem. Soc.*, 4806 (1963); 5002 (1964); B. L. Shaw, *Chem. Ind.* (*London*), 1190 (1962); R. P. Hughes, D. E. Hunton and K. Schumann, *J. Organomet. Chem.*, **169**, C37 (1969); R. Noyori and H. Takaya, *Chem. Commun.*, 525 (1969).
56. B. M. Trost and P. E. Strege, *J. Am. Chem. Soc.*, **97**, 2534 (1975); B. M. Trost, L. Weber, P. E. Strege, T. J. Fullerton, and T. J. Dietsche, *J. Am. Chem. Soc.*, **100**, 3416 (1978).
57. D. J. Collins, W. R. Jackson, and R. N. Timms, *Tetrahedron Lett.*, 495 (1976).
58. B. M. Trost and T. R. Verhoeven, *J. Am. Chem. Soc.*, **98**, 630 (1976); **100**, 3435 (1978).
59. W. R. Jackson and J. V. Strauss, *Tetrahedron Lett.*, 2591 (1975); *Aust. J. Chem.*, **30**, 553 (1977).
60. See B. M. Trost *et al.*, ref. 46 and 47.
61. J. S. Temple and J. Schwartz, *J. Am. Chem. Soc.*, **102**, 7381 (1980).
62. R. A. Clement, V. Klabunde, and G. W. Parshall, *J. Mol. Catal.*, **4**, 87 (1978).
63. D. N. Jones and S. D. Knox, *J. Chem. Soc., Chem. Commun.*, 165 (1975).

64. F. G. Stankem and R. F. Heck, *J. Org. Chem.*, **45**, 3584 (1980); B. Akermark and K. Zetterberg, *Tetrahedron Lett.*, 3733 (1975); B. Akermark, J. E. Backvall, A. Lowenborg, and K. Zetterberg, *J. Organomet. Chem.*, **166**, C33 (1979).
65. I. T. Harrison, E. Kimura, E. Bohme, and J. H. Fried, *Tetrahedron Lett.*, 1589 (1969); W. R. Jackson and J. V. Strauss, *Aust. J. Chem.*, **31**, 1073 (1978).
66. B. M. Trost and T. R. Verhoeven, *J. Am. Chem. Soc.*, **102**, 4730 (1980).
67. See ref. 37, p. 827.
68. B. M. Trost and F. W. Gowland, *J. Org. Chem.*, **44**, 3448 (1979).
69. B. M. Trost and D. P. Curran, *J. Am. Chem. Soc.*, **102**, 5699 (1980).
70. B. M. Trost, T. R. Verhoeven, and J. M. Fortunak, *Tetrahedron Lett.*, 2301 (1979).
71. B. M. Trost and T. P. Klun, *J. Am. Chem. Soc.*, **103**, 1864 (1981).
72. B. M. Trost and T. R. Verhoeven, *J. Org. Chem.*, **41**, 3215 (1976).
73. See ref. 37, p. 826.
74. B. M. Trost and E. Keinan, *Tetrahedron Lett.*, 2591 (1980).
75. B. M. Trost and E. Keinan, *Tetrahedron Lett.*, 2595 (1980).
76. B. M. Trost and J. P. Genet, *J. Am. Chem. Soc.*, **98**, 8516 (1976); G. Hata, K. Takahashi and A. Miyake, *Chem. Commun.*, 1392 (1970); K. E. Atkins, W. E. Walker, and R. M. Manyik, *Tetrahedron Lett.*, 3821 (1970); K. Takahashi, A. Miyake, and G. Hata, *Bull. Chem. Soc. Jpn.*, **45**, 230 (1972); S. I. Murahashi, T. Shimamura and I. Moritani, *J. Chem. Soc., Chem. Commun.*, 931 (1974); H. H. Baer and Z. S. Hanna, *Carbohydr. Res.*, **78**, C11 (1980).
77. B. M. Trost and E. Keinan, *J. Org. Chem.*, **44**, 3451 (1979); **45**, 2741 (1980).
78. W. Carruthers and S. A. Cumming, *J. Chem. Soc., Chem. Commun.*, 360 (1983); S. A. Godleski, D. J. Heathcock, J. D. Meinhart, and S. Van Wallendael, *J. Org. Chem.*, **48**, 2101 (1983).
79. S. W. McCombie, *Eur. Pat.*, EPO-013-663-Al (1980).
80. K. Takahashi, G. Hata, and A. Miyake, *Bull. Chem. Soc. Jpn.*, **46**, 1012 (1973); D. G. Brady, *Chem. Commun.*, 434 (1970).
81. P. A. Grieco, T. Takigawa, S. L. Bongers, and H. Tanaka, *J. Am. Chem. Soc.*, **102**, 7588 (1980).
82. B. M. Trost and T. R. Verhoeven, *J. Am. Chem. Soc.*, **102**, 4743 (1980).
83. J. Tsuji, Y. Kobayashi, H. Kataoka, and T. Takahashi, *Tetrahedron Lett.*, 1475 (1980).
84. B. M. Trost and T. R. Verhoeven, *J. Am. Chem. Soc.*, **101**, 1595 (1979).
85. J. Tsuji, Y. Kobayashi, H. Kataoka, and T. Takahashi, *Tetrahedron Lett.*, 3393 (1980).
86. V. Kitagawa, A. Itoh, S. Hashimoto, H. Yamamoto, and H. Nozaki, *J. Am. Chem. Soc.*, **99**, 3864 (1977).
87. B. M. Trost, T. A. Runge, and L. N. Jungheim, *J. Am. Chem. Soc.*, **102**, 2840 (1980).
88. See ref. 37, p. 834.
89. B. M. Trost and J. Fortunak, *J. Am. Chem. Soc.*, **102**, 2841 (1980).
90. I. Shimizu, T. Yamada, and J. Tsuji, *Tetrahedron Lett.*, 3199 (1980).
91. R. G. Hayter, *J. Organomet. Chem.*, **13**, P1 (1963).
92. N. A. Bailey, W. G. Kita, J. A. McCleverty, A. J. Murray, B. E. Mann, and N. W. J. Walker, *J. Chem. Soc., Chem. Commun.*, 592 (1974).
93. J. W. Faller and A. M. Rosan, *J. Am. Chem. Soc.*, **98**, 3388 (1976).
94. R. D. Adams, D. F. Chodosh, J. W. Faller, and A. M. Rosan, *J. Am. Chem. Soc.*, **101**, 2590 (1979).
95. B. E. R. Schilling, R. Hoffman, and J. W. Faller, *J. Am. Chem. Soc.*, **101**, 5921 (1979).
96. B. M. Trost and M. Lautens, *J. Am. Chem. Soc.*, **104**, 5543 (1982).
97. B. M. Trost and M. Lautens, *J. Am. Chem. Soc.*, **105**, 3343 (1983).
98. S. F. Martin, *Tetrahedron*, **36**, 419 (1980).
99. B. M. Trost and M. H. Hung, *J. Am. Chem. Soc.*, **105**, 7757 (1983).
100. R. A. Plowman and F. G. A. Stone, *Z. Naturforsch., Teil B*, **17**, 575 (1962).
101. H. D. Murdoch and E. Weiss, *Helv. Chim. Acta*, **45**, 1927 (1962).
102. T. H. Whitesides, R. W. Arhart, and R. W. Slaven, *J. Am. Chem. Soc.*, **95**, 5792 (1973).
103. G. F. Emerson and R. Pettit, *J. Am. Chem. Soc.*, **84**, 4591 (1962); D. H. Gibson and R. L. Vonnahme, *J. Am. Chem. Soc.*, **94**, 5090 (1972); *J. Organomet. Chem.*, **70**, C33 (1974).
104. A. J. Pearson, *Aust. J. Chem.*, **29**, 1841 (1976).
105. A. J. Pearson, *Tetrahedron Lett.*, 3617 (1975).
106. J. L. Roustan, J. Y. Mérour and F. Houlihan, *Tetrahedron Lett.*, 3721 (1979).
107. R. Palumbo, A. De Renzi, A. Panunji, and G. Paiaro, *J. Am. Chem. Soc.*, **91**, 3874; 3879 (1969); J. Tsuji and H. Takahashi, *J. Am. Chem. Soc.*, **87**, 3275 (1965); M. Tada, Y. Kuroda, and T. Sato, *Tetrahedron Lett.*, 2871 (1969).

108. W. C. Still and I. Galynker, *Tetrahedron*, **37**, 3981 (1981); *J. Am. Chem. Soc.*, **104**, 1774 (1982).
109. J. W. Faller and A. M. Rosan, *Ann. N. Y. Acad. Sci.*, **295**, 186 (1977); *J. Am. Chem. Soc.*, **99**, 4858 (1977); M. L. H. Green, J. Knight, and J. A. Segal, *J. Chem. Soc., Dalton Trans.*, 2189 (1977).
110. J. W. Faller, H. H. Murray, D. L. White, and K. H. Chao, *Organometallics*, **2**, 400 (1983).
111. A. J. Pearson and Md. N. I. Khan, *J. Am. Chem. Soc.*, **106**, 1872 (1984); *Tetrahedron Lett.*, **25**, 3507 (1984).
112. A. J. Pearson, *Acc. Chem. Res.*, **13**, 463 (1980); A. J. Pearson in *Comprehensive Organometallic Chemistry* (Eds. G. Wilkinson, F. G. A. Stone, and E. W. Abel), Vol. 8, Pergamon Press, Oxford, 1982, Chapter 58; *Transition Met. Chem.*, **6**, 67 (1981); *Chem. Ind. (London)*, 741 (1982); *Science*, **223**, 895 (1984); *Pure Appl. Chem.*, **55**, 1767 (1983).
113. E. O. Fischer and R. D. Fischer, *Angew. Chem.*, **72**, 919 (1960).
114. M. de M. Campos and N. Petragnani, *Tetrahedron Lett.*, **6**, 11 (1959); *Chem. Ber.*, **93**, 317 (1960); K. C. Nicolaou, S. P. Seitz W. J. Sipio, and J. F. Blount, *J. Am. Chem. Soc.*, **101**, 3884 (1979); D. L. J. Clive, V. Farina, A. Singh, C. K. Wong, W. A. Kiel, and S. M. Menchen, *J. Org. Chem.*, **45**, 2120 (1980); K. C. Nicolaou, *Tetrahedron*, **37**, 4097 (1981).
115. K. B. Sharpless and R. F. Lauer, *J. Am. Chem. Soc.*, **94**, 7154 (1972); **95**, 2697 (1973); H. J. Reich, *J. Org. Chem.*, **40**, 2570 (1975).
116. A. J. Pearson, T. Ray, I. C. Richards, J. Clardy, and L. Silveira, *Tetrahedron Lett.*, **24**, 5827 (1983); A. J. Pearson and T. Ray, *Tetrahedron*, **41**, 5765 (1985).
117. A. J. Birch and M. A. Haas, *J. Chem. Soc. C*, 2465 (1971).
118. Y. Shvo and E. Hazum, *J. Chem. Soc., Chem. Commun.*, 336 (1974).
119. S. Warren, *Organic Synthesis: The Disconnection Approach*, Wiley, Chichester, 1982.
120. A. J. Birch, K. B. Chamberlain, M. A. Haas, and D. J. Thompson, *J. Chem. Soc., Perkin Trans. 1*, 1882 (1973); A. J. Birch and A. J. Pearson, *J. Chem. Soc., Perkin Trans. 1*, 954 (1976); R. E. Ireland, G. G. Brown, R. H. Stanford, Jr., and T. C. McKenzie, *J. Org. Chem.*, **39**, 51 (1974); A. J. Birch, A. S. Narula, P. Dahler, G. R. Stephenson, and L. F. Kelly, *Tetrahedron Lett.*, **21**, 979 (1980); A. J. Birch, A. J. Liepa, and G. R. Stephenson, *Tetrahedron Lett.*, 3565 (1979); G. R. John and L. A. P. Kane-Maguire, *J. Chem. Soc., Dalton Trans.*, 1196 (1979); L. A. P. Kane-Maguire and C. A. Mansfield, *J. Chem. Soc., Chem. Commun.*, 540 (1973); C. A. Mansfield, K. M. Al-Kathumi, and L. A. P. Kane-Maguire, *J. Organomet. Chem.*, **71**, C11 (1974); A. J. Birch and A. J. Pearson, *Tetrahedron Lett.*, 2379 (1975); A. J. Pearson, *Aust. J. Chem.*, **30**, 345 (1977); A. Pelter, K. J. Gould, and L. A. P. Kane-Maguire, *J. Chem. Soc., Chem. Commun.*, 1029 (1974); M. Gower, G. R. John, L. A. P. Kane-Maguire, T. I. Odiaka, and A. Salzer, *J. Chem. Soc., Dalton Trans.*, 2003 (1979); A. J. Birch, I. D. Jenkins, and A. J. Liepa, *Tetrahedron Lett.*, 1723 (1975); see also ref. 112.
121. A. J. Pearson and C. W. Ong, *J. Am. Chem. Soc.*, **103**, 6686 (1981).
122. O. Eisenstein, W. Butler, and A. J. Pearson, *Organometallics*, **3**, 1150 (1984).
123. Ch. Tamm, *Fortschr. Chem. Org. Naturst.*, **31**, 63 (1974); J. R. Bamburg and F. M. Strong, in *Microbial Toxins* (Eds. S. Kadis, A. Ciegler, and S. J. Ajl), Vol. 3, Academic Press, New York, 1971, p. 207.
124. A. J. Pearson, E. Mincione, M. Chandler, and P. R. Raithby, *J. Chem. Soc., Perkin Trans. 1*, 2774 (1980); A. J. Pearson and G. C. Heywood, *Tetrahedron Lett.*, **22**, 1645 (1981); E. Mincione, A. J. Pearson, P. Bovicelli, M. Chandler, and G. C. Heywood, *Tetrahedron Lett.*, **22**, 2929 (1981); A. J. Pearson, G. C. Heywood, and M. Chandler, *J. Chem. Soc., Perkin Trans. 1*, 2631 (1982).
125. G. R. Han, Ph.D. Dissertation, Case Western Reserve University, 1986.
126. W. S. Johnson, J. C. Collins, R. Pappo, and M. B. Rubin, *J. Am. Chem. Soc.*, **80**, 2585 (1958), and references cited therein; W. S. Johnson, W. A. Vredenburgh, and J. E. Pike, *J. Am. Chem. Soc.*, **82**, 3409 (1960).
127. Review: E. Brown and M. Ragault, *Tetrahedron*, **35**, 911 (1979).
128. A. J. Pearson and D. C. Rees, *Tetrahedron Lett.*, **21**, 3937 (1980); *J. Am. Chem. Soc.*, **104**, 1118 (1982); *J. Chem. Soc, Perkin Trans. 1*, 2467 (1982); A. J. Pearson, *Tetrahedron Lett.*, **22**, 4033 (1981); A. J. Pearson, D. C. Rees, and C. W. Thornber, *J. Chem. Soc., Perkin Trans. 1*, 625 (1983).
129. A. J. Birch, W. D. Raverty, and G. R. Stephenson, *J. Chem. Soc., Chem. Commun.*, 857 (1980); A. J. Birch and G. R. Stephenson, *Tetrahedron Lett.*, **22**, 779 (1981); A. J. Birch, W. D. Raverty, and G. R. Stephenson, *Tetrahedron Lett.*, **21**, 197 (1980); A. J. Birch and B. M. R. Bandara, *Tetrahedron Lett.*, **21**, 2981 (1980); B. M. R. Bandara, A. J. Birch, L. F. Kelly, and T. C. Khor, *Tetrahedron Lett.*, **24**, 2491 (1983).
130. A. J. Pearson, *J. Chem. Soc., Perkin Trans. 1*, 1255 (1979); 400 (1980); A. J. Pearson, P. Ham and

D. C. Rees, *J. Chem. Soc., Perkin Trans. 1*, 489 (1982); A. J. Pearson, T. R. Perrior and D. A. Griffin, *J. Chem. Soc., Perkin Trans. 1*, 625 (1983); A. J. Pearson and T. R. Perrior, *J. Organomet. Chem.*, **285**, 253 (1985).

131. Review: J. A. Marshall, S. F. Brady, and N. H. Andersen, *Fortschr. Chem. Org. Naturst.*, **31**, 283 (1974).
132. A. J. Pearson, P. Ham, and D. C. Rees, *Tetrahedron Lett.*, **21**, 4637 (1980); A. J. Pearson and P. Ham, *J. Chem. Soc. Perkin Trans. 1*, 1421 (1983).
133. A. J. Birch, P. E. Cross, J. Lewis, D. A. White, and S. B. Wild, *J. Chem. Soc. A*, 332 (1968); L. F. Kelly, A. S. Narula, and A. J. Birch, *Tetrahedron Lett.*, **21**, 2455 (1980); A. J. Birch, L. F. Kelly, and D. J. Thompson, *J. Chem. Soc., Perkin Trans. 1*, 1006 (1981); L. F. Kelly, P. Dahler, A. S. Narula, and A. J. Birch, *Tetrahedron Lett.*, **22**, 1433 (1981).
134. A. J. Birch, P. Dahler, A. S. Narula, and G. R. Stephenson, *Tetrahedron Lett.*, **21**, 3817 (1980).
135. A. J. Pearson, D. V. Gardner, and I. C. Richards, *J. Chem. Soc., Chem. Commun.*, 807 (1982); *J. Org. Chem.*, **49**, 3887 (1984).
136. I. H. Sanchez and F. R. Tallabs, *Chem. Lett.*, 891 (1981).
137. A. J. Pearson and C. W. Ong, *J. Org. Chem.*, **47**, 3780 (1982).
138. A. J. Pearson and C. W. Ong, *J. Chem. Soc., Perkin Trans. 1*, 1614 (1981); A. J. Pearson and M. Chandler, *Tetrahedron Lett.*, **21**, 3933 (1980); *J. Chem. Soc., Perkin Trans. 1*, 2467 (1982); A. J. Pearson, *J. Chem. Soc., Chem. Commun.*, 488 (1980).
139. L. A. Paquette, R. G. Daniels, and R. Gleiter, *Organometallics*, **3**, 560 (1984).
140. R. Aumann, *J. Organomet. Chem.*, **47**, C28 (1973); R. Edwards, J. A. S. Howell, B. F. G. Johnson, and J. Lewis, *J. Chem. Soc., Dalton Trans.*, 2105 (1974); R. Burton, M. L. H. Green, E. W. Abel, and G. Wilkinson, *Chem. Ind.* (*London*), 1952 (1958); R. Burton, L. Pratt, and G. Wilkinson, *J. Chem. Soc.*, 594 (1961); H. J. Dauben and D. Bertelli, *J. Am. Chem. Soc.*, **83**, 497 (1961); M. A. Hashmi, J. D. Munro, P. L. Pauson, and J. M. Williamson, *J. Chem. Soc. A*, 240 (1967); J. Evans, D. V. Howe, B. F. G. Johnson, and J. Lewis, *J. Organomet. Chem.*, **61**, C48 (1973); D. A. Brown, S. K. Chawla, and W. K. Glass, *Inorg. Chim. Acta*, **19**, L31 (1976); L. A. P. Kane-Maguire, T. I. Odiaka, and P. A. Williams, *J. Chem. Soc., Dalton Trans.*, 200 (1981); P. Hackett, B. F. G. Johnson, J. Lewis, and G. Jaouen, *J. Chem. Soc., Dalton Trans.*, 1247 (1982).
141. A. J. Pearson, S. L. Kole, and B. Chen, *J. Am. Chem. Soc.*, **105**, 4483 (1983).
142. A. J. Pearson, S. L. Kole, and T. Ray, *J. Am. Chem. Soc.*, **106**, 6060 (1984).
143. Review: K. M. Nicholas, M. O. Nestle, and D. Seyferth, in *Transition Metal Organometallics in Organic Synthesis* (Ed. H. Alper), Vol. 2, Academic Press, New York, 1978, Chapter 1.
144. K. M. Nicholas, M. Mulvaney and M. Bayer, *J. Am. Chem. Soc.*, **102**, 2508 (1980); J. O'Boyle and K. M. Nicholas, *Tetrahedron Lett.*, **21**, 1595 (1980); R. Lockwood and K. M. Nicholas, *Tetrahedron Lett.*, 4163 (1977).
145. G. Jaouen, in *Transition Metal Organometallics in Organic Synthesis* (Ed. H. Alper), Vol. 2, Academic Press, New York, 1978, Chapter 2; M. F. Semmelhack, *Tetrahedron*, **37**, 3956 (1981); *Ann. N. Y. Acad. Sci.*, **295**, 36 (1977); G. Jaouen, Ann. N. Y. Acad. Sci., **295**, 59 (1977).
146. T. H. Coffield, V. Sandel, and R. D. Closson, *J. Am. Chem. Soc.*, **79**, 5826 (1957); G. Winkhaus, L. Pratt, and G. Wilkinson, *J. Chem. Soc.*, 3807 (1961); P. L. Pauson and J. A. Segal, *J. Chem. Soc., Dalton Trans.*, 1677 (1975).
147. P. Bachmann, K.-R. Repp, and H. Singer, *Z. Naturforsch., Teil B*, **32**, 471 (1977).
148. K. K. Bhasin, W. G. Balkeen, and P. L. Pauson, *J. Organomet. Chem.*, **204**, C25 (1981); A. J. Pearson and I. C. Richards, *J. Organomet. Chem.*, **258**, C41 (1983).
149. A. Mawby, P. J. C. Walker, and R. J. Mawby, *J. Organomet. Chem.*, **55**, C39 (1973); P. J. C. Walker and R. J. Mawby, *Inorg. Chem.*, **10**, 404 (1971); *Inorg. Chim. Acta*, **7**, 621 (1973); *J. Chem. Soc., Dalton Trans.*, 622 (1973); P. L. Pauson and J. L. Segal, *J. Chem. Soc., Dalton Trans.*, 1683 (1975); G. A. M. Munro and P. L. Pauson, *Isr. J. Chem.*, **15**, 258 (1976/77).
150. Y. K. Chung, H. S. Choi, and D. A. Sweigart, *J. Am. Chem. Soc.*, **104**, 4245 (1982); Y. K. Chung, P. G. Willard, and D. A. Sweigart, *Organometallics*, **1**, 1053 (1982).
151. D. Astruc, *Tetrahedron*, **39**, 4027 (1983).

The Chemistry of the Metal—Carbon Bond, Vol. 4
Edited by F. R. Hartley

CHAPTER **11**

Hydrogenation

D. PARKER

Department of Chemistry, University Science Laboratories, South Road, Durham, DH1 3LE, UK

I. INTRODUCTION

A. Introductory Remarks

The hydrogenation of organic substrates using metal catalysts is easily the most studied reaction in both homogeneous and heterogeneous catalysis. To the synthetic organic chemist, catalytic hydrogenation is unique in that the controlled reduction of almost any unsaturated compound may be carried out selectively under a wide range of conditions and with very high yields[1-4]. Whereas the development of homogeneous hydrogenation has been phenomenal over the last 20 years[5], heterogeneous catalysis remains the more useful and versatile technique for the practising organic chemist. Heterogeneous catalysts generally exhibit good thermal stability, may be used in a wide range of solvents, and, most importantly, are easily separated and recovered from reaction products. Such advantages have maintained their pre-eminence in industrial chemical processes where cost and efficiency are of paramount importance. Despite early predictions to the contrary[6], the usage of many homogeneous systems remains limited owing to their sensitivity to molecular oxygen, their tendency to cause rearrangement of alkenes, and particularly the difficulties encountered in recovering and reusing the soluble and often expensive catalysts. On the other hand, homogeneous hydrogenation is much more amenable to mechanistic investigation and the intimate reaction pathway of many systems has been determined[7-11]. Since the catalysts often operate under mild conditions of temperature and pressure, they may be studied, *in situ*, by conventional spectroscopic and kinetic techniques. Studies of this kind have undoubtedly aided in the rational design of catalyst systems in which variations of the coordinating ligand, solvent, and temperature have been effected in order to improve selectivity[12]. A discussion of the mechanism of homogeneous hydrogenation lies outside the scope of this chapter and is dealt with in detail in Chapter 12.

In heterogeneous catalytic systems, the active site of the catalyst remains ill-defined. Mechanistic studies have tended to concentrate on two distinct methods. Using standard surface chemistry techniques [such as LEED, ESCA, EXAFS, and electron microscopy] the metal surfaces are examined in the absence or presence of the reacting molecules under high vacuum conditions completely unlike those used in catalysis[13-17]. Alternatively, a series of metals have been tested using a standard reaction or a series of organic substrates have been studied using a single metal[18-23]. These qualitative experiments have led to a qualitative understanding of the steric and electronic effects between the substrate and the metal surface, but the hypothetical mechanism promulgated by Horiuti and Polyani over 50 years ago remains generally accepted[24].

There is an extensive literature concerning hydrogenation and many texts[1,2,4,25,28,29] and authoritative reviews[3,26,27,30] have appeared, from which many of the original references may be traced. This review will necessarily be selective rather than comprehensive in covering the vast available literature (up to mid-1984) and pays attention to the use of hydrogenation in organic chemistry, in particular emphasizing chemoselective and stereospecific reductions.

B. Brief Historical Perspective

Whereas the growth of homogeneous catalysis began only 30 years ago, the hydrogenation of organic substrates using heterogeneous catalysts may be traced back even to the last century. Following the pioneering work of Sabatier and Senderens[31] using nickel powders as hydrogenation catalysts, Raney developed highly dispersed nickel catalysts on mineral supports. His early patents describe a process for the removal of silicon from a NiSi alloy by alkaline solutions[32], although the NiAl intermetallic compounds were found to be much more convenient for this purpose[33]. The series of Raney nickel catalysts primarily developed by Adkins and coworkers[34–37] and Mozingo[38,39] are still widely used industrially, for example in the catalytic hydrogenation of adiponitrile to hexamethylenediamine[40]. Soon afterwards Adkins and Cornor developed the copper–chromium oxide ('copper chromite') catalyst[41,42] which is generally considered to be complementary to Raney nickel in activity as it reduces esters and amides without affecting aromatic rings. Thus ethyl benzoate, **1**, may be reduced to benzyl alcohol, **2**, by a copper–barium–chromium oxide catalyst[43], while Raney nickel selectively reduces the aromatic ring to give **3** (equation 1)[44].

$$\textbf{(3)}\ C_6H_{11}CO_2Et \xleftarrow[50\,^\circ C,\ 1500\ psi]{Ni,\ H_2} \textbf{(1)}\ C_6H_5CO_2Et \xrightarrow[155\,^\circ C,\ 4000\ psi]{H_2,\ Cu{-}Ba{-}CrO_x} \textbf{(2)}\ C_6H_5CH_2OH \qquad (1)$$

The most useful catalysts for low-pressure hydrogenations employ the platinum metals, particularly platinum, palladium, rhodium, and ruthenium. Adams and coworkers early in the 1920s developed the use of PtO_2[45], which remains popular as it may be easily recycled[46]. These noble metal catalysts have been developed as dispersions on inert metal supports such as graphite, alumina, barium sulphate, or calcium carbonate. Supported palladium catalysts are particularly useful and remain the catalysts of choice for the hydrogenation of alkenes and the semi-hydrogenation of alkynes. Lindlar first described the use of palladium on calcium carbonate deactivated by the addition of lead acetate and quinoline[47] for the selective hydrogenation of alkynes to (*Z*)-alkenes.

In the past 25 years, ruthenium and especially rhodium supported on alumina or charcoal have been developed as important catalysts for the hydrogenation of aromatic and heteroaromatic systems under relatively mild conditions[48,49], without hydrogenolytic cleavage of amino or hydroxy groups[50,51]. Further heterogeneous hydrogenation catalysts developed in the last 20 years include the series of Urushibara catalysts, the very active nickel boride catalysts, and various enantioselective Raney nickel systems modified by treatment with tartaric acid. The Urushibara catalysts[52–54] are formed by precipitation of the catalyst metal (nickel, cobalt, or iron) from an aqueous solution of its salt by zinc dust or granular aluminium. The precipitated metal is digested by acid or alkali to give catalysts similar in activity to Raney nickel catalysts[55]. Nickel boride has been prepared by the reduction of various nickel salts with borohydride[56–59] and is more reactive than the related Raney nickel systems. Isoda *et al.*[60] were the first to report enantioselective hydrogenation over Raney nickel modified with chiral amino acids, but in the last few years tartaric acid-modified Raney nickel catalysts have allowed the reduction of various β-diketones and α- and β-keto esters to give the corresponding hydroxy ketones or esters in over 90% enantiomeric excess[61–65].

The first authenticated example of homogeneous hydrogenation was reported in 1938 by Calvin, who observed the reduction of benzoquinone by dissolved hydrogen in quinoline solution at 100 °C using copper(I) acetate[66,67]. Although Iguchi remarked upon the rapid absorption of hydrogen by cobalt(II) cyanide solutions[68], it was not until much later that the synthetic utility of the pentacyanocobaltate catalyst was discovered[69].

During the 1950s, while many transition metal hydrides were being isolated for the first time, for example *trans*-[PtHCl(PEt_3)$_2$] and [ReHCp_2][70], there was little reported of the reduction of organic substrates. In 1961, however, Halpern and coworkers reported the reduction of a series of activated alkenes using solutions of chlororuthenate(II)[71–73]. Simultaneously, workers at Du Pont discovered that a platinum–tin(II) chloride system in methanol solution catalysed the hydrogenation of ethene at room temperature and under 1 atm of hydrogen pressure[74].

The major breakthrough came in 1964 with the simultaneous discovery of the efficient catalytic activity of [RhCl(PPh_3)$_3$] by Wilkinson and coworkers[75,76] and Coffey[77]. In 1965 the related ruthenium complex [RuHCl(PPh_3)$_3$] was reported to be an active hydrogenation catalyst under ambient conditions[78]. Since then, a veritable plethora of literature has amassed discussing the catalytic activity of various rhodium, ruthenium and, iridium complexes with tertiary phosphine ligands, much of which is discussed in detail in recent reviews[5,25,26]. The more important recently developed catalysts include a family of cationic rhodium complexes, [RhL_2(S)$_2$]$^+$, in which L_2 represents two tertiary phosphines or a chelating diphosphine and S is a polar solvent such as methanol, tetrahydrofuran, or acetone. These catalysts, discovered by Schrock and Osborn[79,80], are generated *in situ* by hydrogenation of the accessible rhodium diene complexes such as **4** and **5**[81,82]. By using a chiral chelating diphosphine complex **6**[83], an efficient chiral catalyst may be obtained

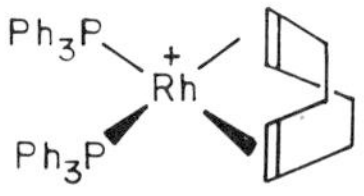

(**4**) 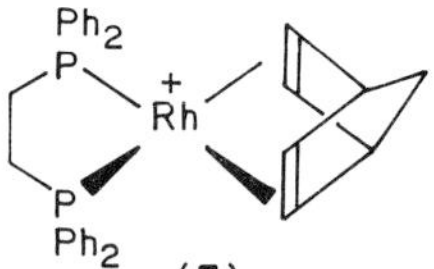 (**5**) 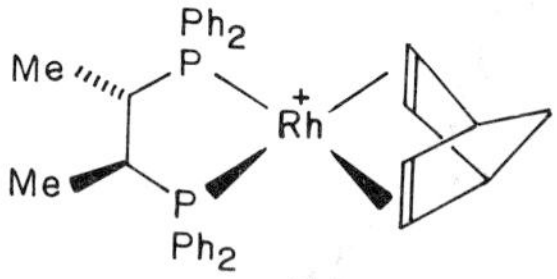 (**6**)

which catalyses the hydrogenation of *N*-acyldehydroamino acids to give *N*-acylamino acids in high enantiomeric purity. Monsanto have recently developed an industrial process which produces L-Dopa, used in the treatment of Parkinson's disease, by such a chiral catalytic process[84,85].

Another important recent advance has been made with the discovery of a series of active iridium complexes, for example **7**[86,87]. Under comparable conditions, **7** is nearly 100 times

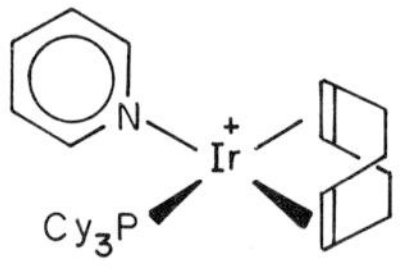

(**7**)

more active than Wilkinson's catalyst toward hex-1-ene reduction. Further, it reduces tetrasubstituted alkenes rapidly[88] and is reported to be insensitive toward oxygen or methyl iodide. Such properties have previously only been exhibited by heterogeneous catalysts[10,89].

During the past 15 years there have been many attempts to anchor homogeneous organometallic complexes to insoluble solid-phase supports[90–100] (see Chapter 14). In principle, such systems may combine the advantages of soluble homogeneous systems with certain valuable characteristics of heterogeneous systems, namely ease of product separation and catalyst recovery. The idea was reported initially by workers at Mobil in 1969[101], although the use of complexes adsorbed on to silica had been known for some time[102]. Unfortunately, useful catalysts have generally not been obtained and the typical problems encountered include the supervenient oxidation of phosphorus atoms, leaching of the metal into solution, and a reduced catalytic activity or selectivity[103–107]. A different approach to the problem of catalyst recovery involves the use of metal complexes of water-soluble ligands such as sulphonated triarylphosphines which can operate under phase-transfer catalysis[108–111].

The recent growth of interest in metal clusters has led to an intensification of comparative studies between heterogeneous and homogeneous systems[112,113]. Cluster and dinuclear catalysts[114–117] have been sought in attempts to emulate heterogeneous systems which must involve more than one metal atom at the surface. It is believed that cluster catalysts may afford selective reactions that are unavailable to mononuclear metal complexes by virtue of cooperativity between metal centres in a cluster. Recent results using dinuclear rhodium complexes, for example **8**, lend support to this premise, leading (for example) to a stereospecific *trans* reduction of alkynes[117,118].

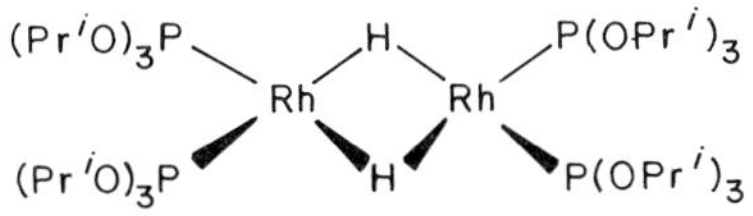

$(Pr^iO)_3P$ / $(Pr^iO)_3P$ — Rh(μ-H)₂Rh — $P(OPr^i)_3$ / $P(OPr^i)_3$

(**8**)

Finally hydrogenations not involving molecular hydrogen have attracted considerable recent interest. The catalytic transfer hydrogenation process[119] uses organic molecules of relatively low oxidation potential as sources of hydrogen, in the presence of homogeneous[120–123] or heterogeneous catalysts[124–126] to effect hydrogen transfer to an organic substrate (equation 2). Sources of hydrogen vary from the traditional cyclohexene[127] or propan-2-ol[126] to formic acid[128] or ammonium formate[129]. Such processes have proved to be synthetically useful in carbohydrate[128] and peptide chemistry[130] owing to the very mild conditions employed.

$$DH_n + A \xrightarrow[\text{solvent}]{\text{catalyst}} DH_{n-2} + AH_2 \qquad (2)$$

II. HETEROGENEOUS CATALYTIC HYDROGENATION

A. Introductory Remarks

Heterogeneous hydrogenation is a most versatile organic reaction as the reaction conditions employed may be 'tuned' to the particular reduction in question. Not only may the actual catalyst be varied, but also its state of dispersion on an inert support. Further, the solvent may be changed and the temperature and pressure of the hydrogenation varied according to the selectivity or reactivity required.

In early syntheses, the catalyst consisted of a noble metal in the form of a suspension or colloidal dispersion. Platinum and palladium have been widely used as suspension catalysts and Adams catalyst (finely divided platinum) is prepared *in situ* by reduction of PtO_2 with hydrogen. The series of Raney nickel catalysts are suspension catalysts and are most easily prepared by treatment of the inexpensive nickel–aluminium alloy with aqueous alkali. The different types of Raney nickel catalysts are designated W1 to W8 according to their method of preparation, which involves varying the amount of base used, the temperature and time of base treatment, and the method of catalyst washing. For example, the most active Raney nickel catalyst is the W6[36,37], which is prepared under a pressure of hydrogen with a continuous aqueous wash.

Supported metal catalysts are more frequently used as the metal is present in a more

TABLE 1. Common types of heterogeneous catalyst

Catalyst	Ref.
Raney Nickel W1	35
Raney Nickel W2	39
Raney nickel W3	34
Raney nickel W4	133
Raney nickel W5	36
Raney nickel W6	37
Raney nickel W7	36
Raney nickel W8	134
Nickel boride	58, 59
Urushibara nickel	53, 52
Palladium oxide	135
Pd–$SrCO_3$	136
Pd–C	39
Platinum oxide (Adams)	45
Platinum black	137
Pt–C	138
Rhodium on C	138
Rhodium on Al_2O_3	139

dispersed form and the soild support improves the thermal stability of the catalyst. Thus, palladium black typically has a surface area of 5–10m^2g^{-1} whereas for 10% palladium on carbon the surface area increases to 100–200m^2g^{-1}. Procedures for preparing supported catalysts have been thoroughly described[4,131,132] and most of the commercially available catalysts do not require pre-activation. The commonly used catalysts are listed in Table 1 together with references to their method of preparation.

The choice of solvent depends on a number of factors, including the substrate, the nature of the catalyst, and the temperature and pressure to be used. The most common solvents used are ethanol, acetic acid, water, tetrahydrofuran, 2-methoxyethanol, and cyclohexane and the nature of the solvent affects the rate and selectivity of reduction[4,140]. For example, reduction of the enone **9** gives the *cis* or *trans* fused 3-keto steroid **10** according to the solvent used (equation 3)[141]. Indeed, the acidity of the reaction medium has a particularly

(9) $\xrightarrow[\text{Pd}-\text{CaCO}_3,\ 25\,^\circ\text{C, solvent}]{\text{H}_2\text{(1 atm)}}$ *E* + *Z* **(10)**

Solvent	*Z/E* ratio
Pr^iOH	1:1
thf	3:2
C_6H_6	3:1
MeCN	19:1

(3)

marked effect on the outcome of hydrogenation reactions. For example, the rate of hydrogenation of the alkene bond in $\alpha\beta$-enones is generally increased in basic media, permitting chemo-selective hydrogenations, e.g. **11** → **12** (equation 4)[142]. The stereochemical outcome of hydrogenations is also very sensitive to solvent acidity. The hydrogenation of substituted cyclohexanones gives mainly the equatorial alcohol in basic media[143–145], but the axial epimer under acidic conditions. For example, 3-oxocholestanones (**13**) may be reduced to the axial 3-ols **14** using the acidic Urushibara–nickel(A) catalyst with cyclohexane co-solvent (equation 5)[146]. Increased temperature or pressure of reduction has the expected effect of generally increasing the reaction rate but often decreasing the selectivity. For example, Raney nickel reduction of **15** reduces only the alkene double bonds at 20 °C but all of the functional groups at 260 °C (equation 6)[147].

(11) $\xrightarrow{H_2, Pd-C,\ EtOH, 0{\cdot}3M\ KOH}$ (12) (4)

(13) $\xrightarrow{35\,°C,\ H_2,\ U-Ni(A)}$ (14) (5)

(15) $\xrightarrow{H_2, 20\,°C,\ Raney\ Ni}$ $C_6H_5(CH_2)_4COMe$

(15) $\xrightarrow{H_2, 260\,°C,\ Raney\ Ni}$ $C_6H_{11}(CH_2)_4CH(OH)Me$ (6)

B. Outline of Mechanistic Studies

The mechanism of heterogeneous hydrogenation remains poorly defined and there is much controversy over the kinetics of reduction, the features of the metals which account for their catalytic behaviour, and the specificity of the reduction of functional groups. There have been many studies of catalyst characterization, but even the basic surface chemistry of alkenes and alkynes on clean metals has not been rigorously established[148], despite extensive chemisorption studies of ethene and ethyne on single crystal surfaces of nickel, platinum, and iridium. The degree of metal dispersion on the catalyst support has

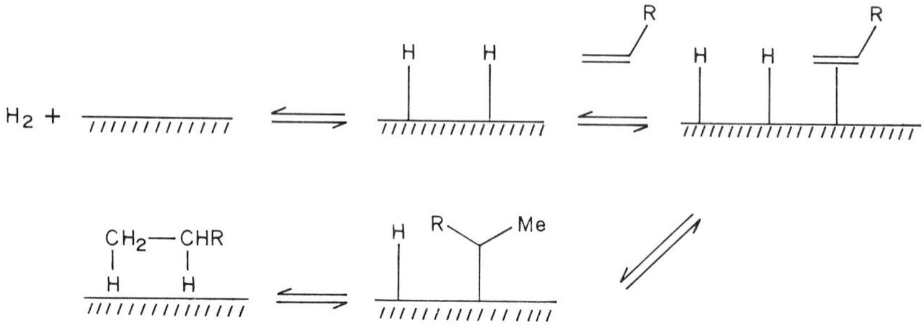

$$H_2 + \text{surface} \rightleftharpoons \text{H H (adsorbed)} \rightleftharpoons \text{H H + RCH=CH}_2\text{ (adsorbed)}$$

$$\text{CH}_2\text{—CHR (H H adsorbed)} \rightleftharpoons \text{H, R—CH—Me (adsorbed)}$$

SCHEME 1. Schematic mechanism of alkene hydrogenation

been well defined using classical chemisorption[149,150], X-ray diffraction[151], and electron microscopic[152] techniques. In addition, the surface concentration of catalyst active sites may be determined using temperature-programmed desorption[153], while the detailed surface structure may be analysed by techniques such as ion-scattering spectroscopy, Auger electron spectroscopy, secondary ion mass spectrometry, and ESCA. Despite the powerful armoury of surface techniques now available, the state of chemisorption of substrate alkenes or alkynes has not been clearly defined either structurally or stereochemically. In any event, to relate findings carried out at 10^{-8} Torr to working hydrogenation catalysts operating at pressures at least 10^9 times greater may well be misleading.

It is evident, however, that hydrogen is dissociatively chemisorbed on the catalyst and that the hydrogenation (or hydrogenolysis) reaction must involve a stepwise addition of two hydrogen atoms from the catalyst metal to the substrate[154–160]. This is consistent with the commonly accepted postulate of Horiuti and Polyani[24] for alkene hydrogenation which involves a series of equilibria implicating both π- and σ-bound intermediates (Scheme 1). However, it is clear that under certain circumstances and particularly when palladium catalysts are used, η^3-allyl intermediates are involved, their intervention explaining perhaps concomitant alkene isomerization[161,162]. An attractive alternative general mechanism has been proposed[163] in which hydrogenation is interpreted in terms of hydrogen transfer between an adsorbed hydrocarbon species $M—C_nH_x$ and the adsorbed alkene. This mechanism implies that the metal is of secondary importance and that alkene hydrogenation is intimately related to self-hydrogenation[164]. Indeed, in the absence of hydrogen, ethene is converted to a mixture of ethyne and ethane over transition metal surfaces[26].

Although none of the molecular features of the catalytic hydrogenation sequence have been validated experimentally, analysis of the nature of the products has established key points relating to the specificity and stereoselectivity of the reduction. Hydrogenation is *syn*-stereospecific with hydrogen adding to the least hindered face of the double bond. Hence hydrogenation of **16** affords the *cis*-hydrindanone **17** stereospecifically (equation 7)[165]. Arene hydrogenations also must involve a selective *cis* addition of

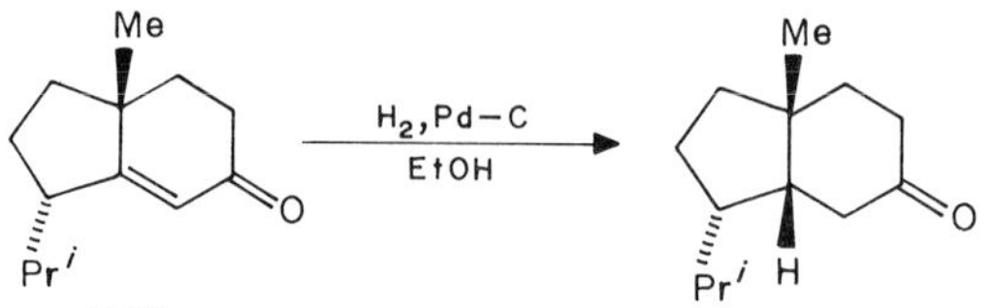

$$\textbf{(16)} \xrightarrow[\text{EtOH}]{H_2,\ Pd\text{–}C} \textbf{(17)} \qquad (7)$$

(16): Me, Pr^i; (17): Me, Pr^i, H

TABLE 2. Metal surface-catalysed hydrogenation of *p*-xylene

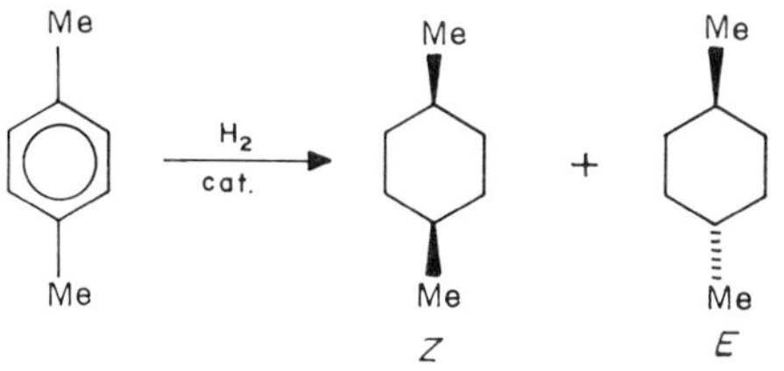

Metal surface	Temperature (°C)	Ratio of Z/E products
Pt black	20	4:1
PtO_2	20	7:3
Ru–C	100	7:3
PtO_2	85	1:2
Rh–C	100	1:2
Pt black	200	1:4
Nickel	180	1:4

hydrogen in forming the cyclohexane product, but the selectivities vary substantially for metal surfaces[166] (Table 2). The absence of all *cis* addition implies that there must be an intermediate unbound cyclohexene which is subsequently converted into a *trans*- or *cis*-cyclohexane[167,168]. The kinetics of arene reduction indicate that dienic structures are formed in the rate-controlling surface reaction[167], although the concentration of intermediate adsorbed dienes remains low[169]. Such intermediate dienes may desorb from the catalyst, but the reported isolation of a cyclohexadiene from catalytic hydrogenation of an arene[170] remains unconfirmed. In competitive hydrogenations between alkenes and dienes, the alkene is always selectively reduced by heterogeneous catalysts. It is apparent that the molecular basis for this selectivity is not related to preferential chemisorption of the alkene[148] as in competitive experiments there is a co-chemisorption of arene and alkene. Similarly, most metal surfaces will selectively hydrogenate alkynes in alkyne–alkene mixtures. The selective conversion of ethyne to ethene is illustrated in Table 3[175], although this does not seem to reflect preferential alkyne chemisorption[148].

There have been a number of relevant theoretical studies concerning the interaction of hydrogen with transition metal surfaces. This is the first step in any hydrogenation and has

TABLE 3. Selectivity in catalysed ethyne hydrogenation

Metal	Temperature (°C)	M Ethene in product (mol-%)
Pd	0	97
Pt	100	80
Rh	100	78
Ru	150	75
Os	150	60
Ir	150	55

been extensively modelled using films[171] or clusters[172–174]. Saillard and Hoffmann considered hydrogen activation on nickel and titanium surfaces and concluded that electron transfer from M to σ^* dominates the early stages of the reaction[176].

C. Synthetic Utility of Heterogeneous Hydrogenation

This section considers the reduction of the various functional groups, emphasizing some of the more recent developments in improving the selectivity and specificity of hydrogenation. The coverage is intended to be selective rather than comprehensive and further detailed discussions may be found in recent books[1,2,4,28] and reviews[26,177].

1. Alkenes and polyenes

Saturation of an alkene may be effected under ambient conditions using many different catalysts. Although palladium seems to have been the preferred catalyst historically, it also catalyses double bond isomerization, making it unsuitable in some cases. Catalytic activity generally decreases in the order $Pt > Pd > Rh > Ru \gg Ni$ for a given surface of metal. Platinum catalysts are particularly useful when double bond migration needs to be avoided and the nickel boride catalysts give considerably less double bond migration than Raney nickel.

The effect of alkene structure on the rate of hydrogenation is clearly defined. Reactivity decreases with increasing alkene substitution and an approximate order of reactivity is terminal > *cis*-internal > *trans*-internal > trisubstituted > tetrasubstituted. A typical example of this is shown by the relative rates of hydrogenation of a series of alkenes in ethanol using P-2 nickel (Table 4)[178]. Release of ring strain can also be an important factor determining reaction rate[178,181]. Exocyclic double bonds are generally reduced more rapidly than endocyclic bonds[182], permitting regioselective reductions. For example, **18** is selectively reduced to **19** over palladium on carbon (equation 8)[183]. However, there are exceptions to this generalization, so that with **20** selective endocyclic reduction occurs using a 5% rhodium on alumina catalyst to give **21** (equation 9)[184]. *Syn* stereospecific addition of hydrogen occurs almost exclusively in alkene hydrogenation, although there are some examples where *anti* addition contributes. Hydrogenation of **22** over rhodium on carbon gives exclusively **23**, but using platinum on carbon 7% of **27** is also formed (equation 10)[28]. In bridged bicyclic compounds, *exo* addition is favoured to give the thermodynamically less favoured *endo* product[185,186]. The hydrogenation of hexamethyl(Dewar)benzene (**25**), for example, over Raney nickel gives **26** selectively (equation 11)[187]. In steroid systems the axial angular methyl groups at the C-10 and C-13 positions inhibit binding of the β-side to the catalyst surface. α-Hydrogenation of alkenic steroids therefore predominates[140,188–190] so that reduction of 3-methoxyestra-1, 3, 5(10), 8(14)-tetraen-17β-ol (**27**) gives **28** selectively (equation 12)[191].

(8)

TABLE 4. Relative rates of hydrogenation of substituted alkenes[a]

Alkene	Relative rate
	1.00
	0.23
	7×10^{-2}
	3×10^{-2}
	4×10^{-3}
	3×10^{-3}
	$\ll 10^{-3}$
	0

[a] In ethanol at 298 K and 1 atm H_2 using P-2 nickel[178].

H_2, thf / Rh—Al_2O_3

(20) → (21) (9)

H_2 / M/C

(22) → (23) + (24) (10)

(11)

(12)

Outside these broad generalizations it is not easy to predict correctly the stereochemical outcome of alkene hydrogenations. In many instances the stereochemical outcome can only be anticipated if there is an extremely close analogy available. For example, it is not always straightforward to distinguish the least hindered face of an alkene and torsional strain and intramolecular non-bonding interactions may distort the geometry of the adsorbed species[192,193]. Moreover, binding of the alkene to the catalyst surface may not be product-determining, particularly if concomitant alkene isomerization is occurring. In certain cases the stereochemical outcome is reported to be a function of alkene purity, reaction conditions[194], and most notably the nature of the catalyst[195]. It is not surprising, therefore, that different workers have reported contradictory results with identical substrates[196–198].

a. Stereocontrolled hydrogenations

The use of a remote functional group to direct the stereochemical course of a reaction pathway is well known in reactions such as peroxy acid oxidations or Simmons–Smith cyclopropanations, but less well developed for hydrogenation. Such an effect (sometimes called the 'anchor effect' or haptophilicity) involves binding of a polar functional group, usually hydroxy, to the metal surface such that hydrogen addition is directed to that side of the molecule. The effect has been reported in isolation several times[199–213] but recently has been more systematically studied[215–217]. The stereospecificity observed is a sensitive function of solvent polarity as a result of competition between solvent and hydroxy groups for the catalyst site. The effect is demonstrated in Table 5 for the palladium-catalysed reduction of the substituted hexahydrophenanthrene derivative[216]. However, it seems clear that such stereocontrolled reductions may be effected more usefully using homogeneous catalysts (see Section III.B.1).

b. Regioselective hydrogenations

The selective partial hydrogenation of polyenes is important both commercially and synthetically. The selective saturation of the C-15 double bond in the triglyceride of

TABLE 5. Stereocontrolled reduction using hydroxy group[a]

H_2, solvent / Pd—C

trans

cis

Solvent	Dielectric constant	Product *cis/trans* ratio
Dimethylformamide	36.7	94:6
Ethanol	24.6	94:6
thf	7.6	82:18
Diglyme	7.2	81:19
$Bu^n{}_2O$	3.1	73:27
Dioxane	2.2	74:26
Hexane	1.9	39:61

[a] *Reprinted with permission from Thompson* et al., J. Org. Chem., **41**, *2903, Copyright* (1976) *American Chemical Society.*

(*Z, Z, Z*)-linolenic acid is of considerable interest to the food industry. The linolenate constituent [9, 12, 15-octadecatrienoates (**29**)] is found naturally in many unsaturated oils but has an unpleasant flavour and a catalyst for the selective reduction to linoleates [*Z, Z*-9, 12-octadecadienoates (**30**)] is still actively sought[218–224]. Industrially nickel and copper are used as catalysts and some selectivity is achieved through monopolization of the catalyst surface by linolenate[218].

$$\mathrm{MeCH_2\overset{15}{C}H{=}CHCH_2\overset{12}{C}H{=}CHCH_2\overset{9}{C}H{=}CH(CH_2)_7CO_2R}$$

(29)

Selectivity in the hydrogenation of non-conjugated polyenes depends on the degree of substitution of the alkenes, with the least substituted bond usually being selectively reduced. Undeca-1, 7-diene (**31**) is thus reduced to undeca-4-ene (**32**) with palladium on carbon[225] and the least substituted bond is also preferentially reduced in the reduction of β-ionone, **33**, over nickel P1[226]. Ring strain may determine the regioselectivity of reduction and hydrogenation of 5-methylenenorbornene, **34**, gives preferential hydrogenation of the more strained endocyclic double bond[178]. The results of hydrogenating conjugated dienes or polyenes are less clear, although 4-vinylcyclohexene (**35**) may be

$Me(CH_2)_4CH\overset{12}{=}CHCH_2CH\overset{9}{=}CH(CH_2)_7CO_2R$

(30)

$H_2C{=}CH(CH_2)_4CH{=}CH(CH_2)_2Me$

(31)

$Me(CH_2)_5CH{=}CH(CH_2)_2Me$

(32)

$H_2C{=}CMe(CH_2)_2CH{=}CH_2$

(33)

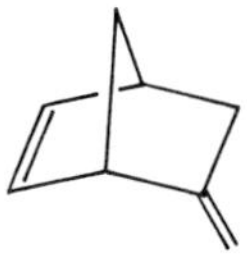

(34)

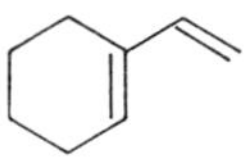

(35)

reduced to 1-ethylcyclohexene using a neutral Urushibara catalyst[55,227] and **36** may be selectively reduced to **37** using palladium on carbon (equation 13)[228].

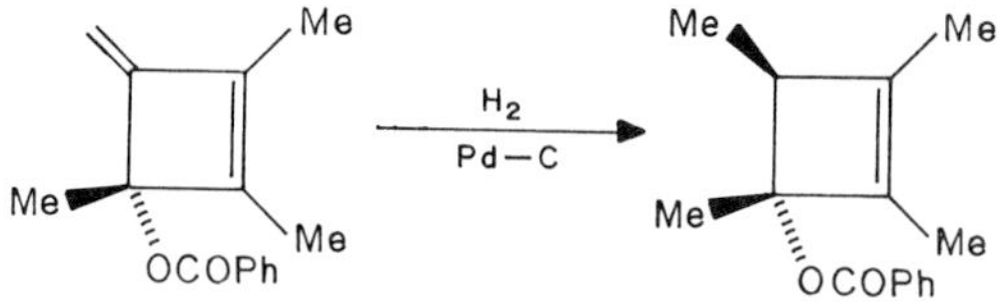

Me, Me, Me, OCOPh → (H_2, Pd—C) → Me, Me, Me, Me, OCOPh (13)

(36) **(37)**

2. Alkynes

Alkynes are important synthetic building blocks in organic chemistry, not only because terminal alkynes readily form carbon—carbon bonds but also because alkynes may be considered to be masked (*Z*)-alkenes[229]. In the nineteenth century Sabatier and Senderens discovered that palladium was the metal of choice for the semi-hydrogenation to (*Z*)-alkenes[31,230] and the Lindlar catalyst (palladium on calcium carbonate alloyed with lead and poisoned with quinoline)[47] continues to find wide application[231–233]. The potential anti-viral agent **38**, for example, has been prepared by Lindlar reduction of the corresponding alkyne[232]. Generally, reductions using this catalyst are best carried out at

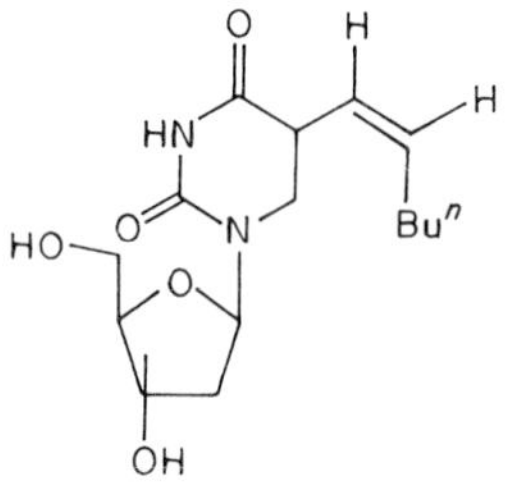

O, H, H, HN, HO, O, N, Bu^n, O, OH

(38)

low temperatures[229,234–236] and sometimes in the presence of added manganese(II) chloride[237]. Other catalysts are still sought[237–241] for the selective semi-hydrogenation and the nickel P2 catalysts seem particularly effective with ethylenediamine as promoter[178]. This catalyst has been used in the synthesis of the insect pheromone (*Z*)-6-heneicosen-11-one, **39**, via stereospecific hydrogenation of **40** followed by oxidation[238].

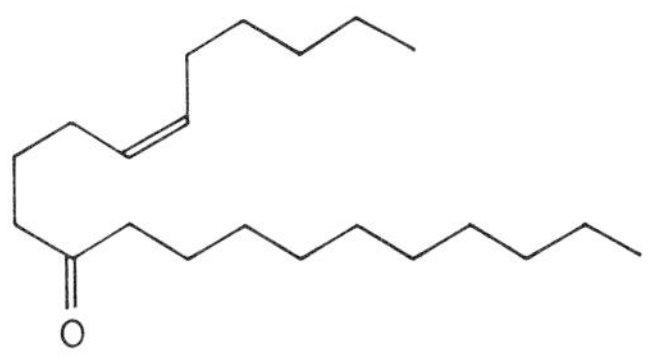

$C_5H_{11}C\equiv C(CH_2)_3CHOH(CH_2)_9Me$

(39) **(40)**

The reduction of MX_2 (M = Pd, Ni; X = Cl, Br) using the lamellar compound C_8K affords a highly dispersed metal(0) catalyst on the graphite surface. These catalysts have also been found to reduce alkynes to (*Z*)-alkenes effectively with high stereospecificity (> 96%)[239,240].

a. Polyynes and enynes

Half-hydrogenation of a polyyne gives the corresponding (*Z*)-polyene without selective formation of the enyne. The diyntriol **41** has been converted into the (*Z, Z*)-dientriol **42** using 5% palladium on barium sulphate[242] and hydrogenation of 5, 8, 11, 14-eicosatetraynoic acid (**43**) using a Lindlar catalyst affords arachidonic acid (**44**)[243]. Semi-

$Me_2(OH)CC\equiv CCH(OH)C\equiv CC(OH)Me_2$

(41) **(42)**

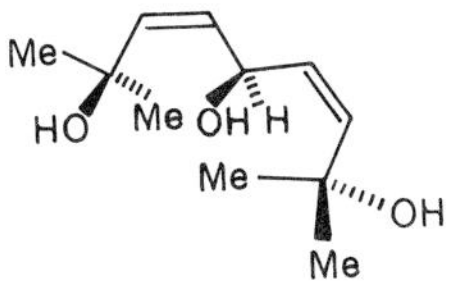

$Me(CH_2)_4C\equiv CC\equiv CCH_2C\equiv CC\equiv C(CH_2)_3CO_2H$

(43) **(44)**

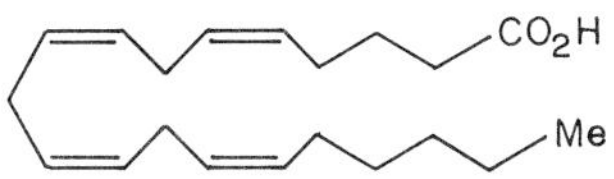

hydrogenation of 1-ethynylcyclohexene results in up to a 90% yield of vinylcyclohexene[24], but lower selectivities occur with a terminal double bond in the presence of an internal triple bond[245].

b. Functionalized alkynes

Reductions of alkynes bearing functional groups is chemoselective with the functional groups usually remaining intact. Methyl jasmonate, for example, may be prepared by selective hydrogenation of (*E*)-3-methoxycarbonylmethyl-2-(pent-2-ynyl)cyclopentanone using a Lindlar catalyst[246]. Reductions of alkynic alcohols and diols are notably less stereospecific than the corresponding unsubstituted alkynes. Although the (*Z*)-alkene remains the major product, considerable amounts of the (*E*)-alkene may also be formed, particularly in the presence of added base[225,247]. Hydrogenation of the propargylic alcohol **45** gave the corresponding allylic alcohol **46** in high yield, but with a 3:1 ratio of $Z{:}E$ diastereomers[248]. The ratio was insensitive to the catalyst and the reduction conditions, implying that the hydroxy group bound strongly to the metal surface, permitting isomerization of the intermediate *cis*-alkene. The hydroxysilane derivatives **47a** and **48** have been reduced to the corresponding (*Z*)-vinylsilanes[249,250] using palladium catalysts, although in the absence of the hydroxy group the isomerized product **49** was

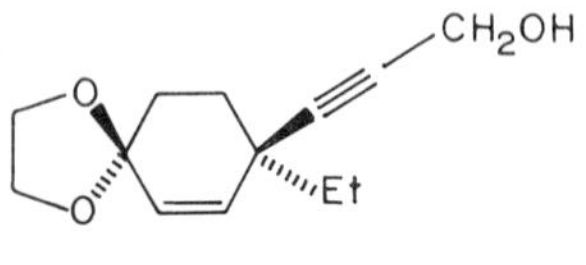

(**45**)

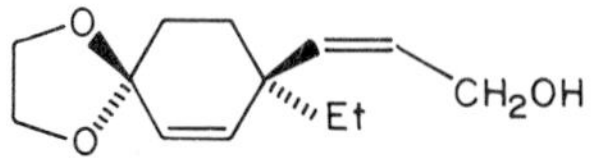

(**46**)

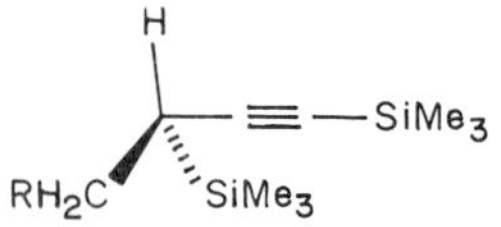

(**a**) R=OH
(**b**) R=Ac
(**47**)

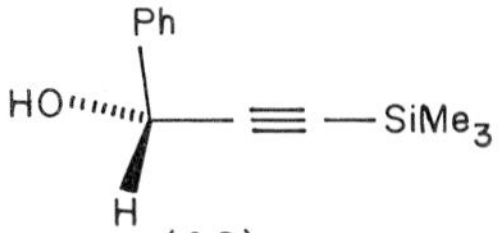

(**48**)

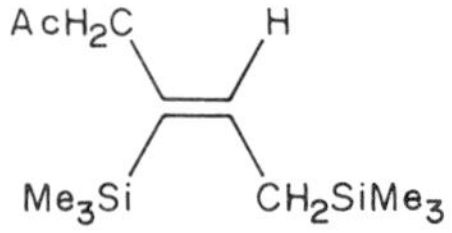

(**49**)

formed[250]. Hydrogenation of the hydroxy acid **50** using 5% palladium on calcium carbonate has permitted a synthesis of the pyrone (equation 14) **51**[251].

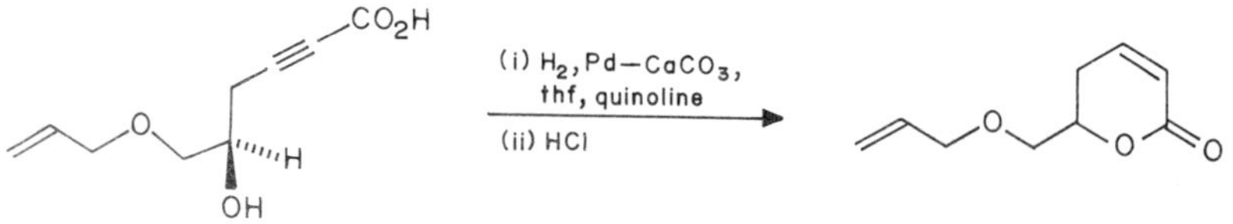

(i) H_2, Pd–$CaCO_3$, thf, quinoline
(ii) HCl

(**50**) (**51**) (14)

3. Saturated carbonyls

Under low-pressure conditions aldehydes are readily hydrogenated to alcohols, although the reduction of ketones is more difficult. Platinum, rhodium, or ruthenium catalysts are often used and Raney nickel has also found application. Sorbitol (**52**) is produced industrially by hydrogenation of glucose using a nickel catalyst[252–254], although ruthenium on carbon is a superior catalyst[255]. On a laboratory scale such

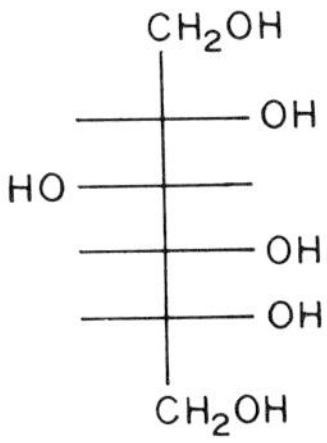

(52)

reductions are more conveniently effected using metal hydride reducing agents and few recent reports describe heterogeneous hydrogenation of simple aldehydes and ketones.

The effect of different catalysts on the hydrogenation of simple aliphatic ketones has been examined and different catalysts seem to affect the adsorption step rather than the subsequent reduction[250]. Rates of hydrogenation are highest in non-polar solvents[257] and both added acid and base may markedly affect the rate of reduction[258–260]. The steric course of ketone hydrogenation is not easy to predict, although various workers have tried in the past to lay down some useful guidelines[261–264]. The product stereochemistry is a sensitive function of the nature of the catalyst, solvent, temperature, and substrate structure. For the reduction of substituted cyclohexanones in ethanol, rhodium on carbon is the most effective for producing axial alcohols, and platinum oxide (Adams catalyst) is least effective[259]. With unhindered ketones, rhodium in tetrahydrofuran–hydrochloric acid is an excellent system for producing axial alcohols[265]. While reduction of 5α-cholestan-3-one, **53**, with platinum in *tert*-butyl alcohol gives the equatorial alcohol

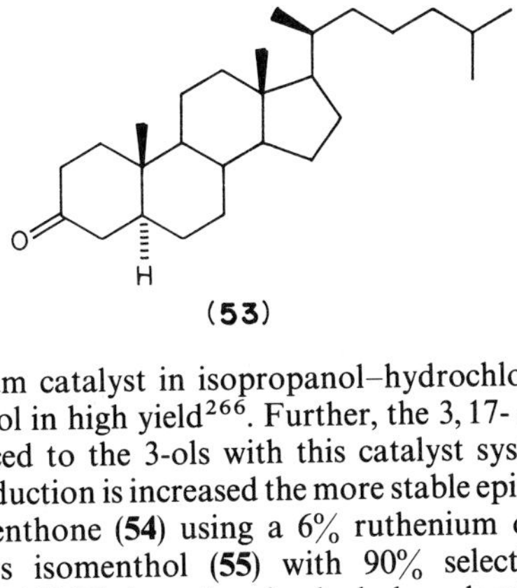

(53)

predominantly, a rhodium catalyst in isopropanol–hydrochloric acid affords the axial alcohol 5α-cholestan-3α-ol in high yield[266]. Further, the 3,17- and 3,20-steroidal diones may be selectively reduced to the 3-ols with this catalyst system[267]. Generally, as the temperature of ketone reduction is increased the more stable epimeric alcohol is produced. Hydrogenation of isomenthone (**54**) using a 6% ruthenium on carbon catalyst under neutral conditions gives isomenthol (**55**) with 90% selectivity, provided that the temperature is 150 °C; at 105 °C the epimeric alcohol predominated[268].

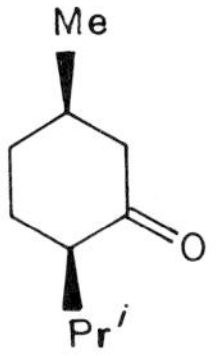

(54) 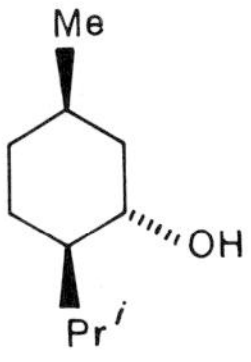 (55)

a. Keto esters and diketones

With substrates which enolize readily, such as β-keto esters and β-diketones, competitive hydrogenolysis of the C—O bond occurs[269–271]. Indeed, catalytic hydrogenolysis has been used as an alternative to the Wolff–Kishner reduction for the reduction of azabicyclic ketones such as **56**[272]. The degree of hydrogenolysis is dependent on the solvent, catalyst, and substrate structure.

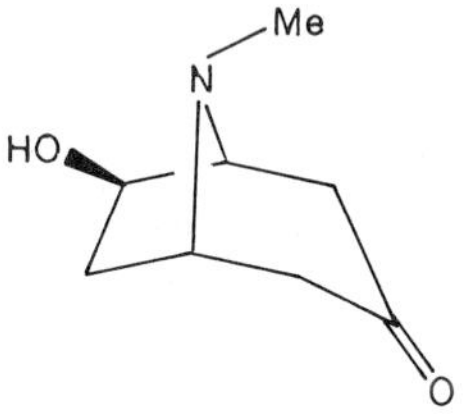

(**56**)

The asymmetric hydrogenation of α- and β-keto esters and β-diketones to give the corresponding hydroxy esters or hydroxy ketones has been thoroughly investigated[273,274]. The best catalyst system for such reductions is Raney nickel modified by

TABLE 6. Asymmetric ketone hydrogenations

Substrate	Conditions	Product Enantiomeric excess (%)	Ref.
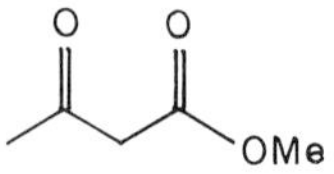	Ni–Pd/Kieselguhr (−)-tartaric acid/ HCO_2H	99	277
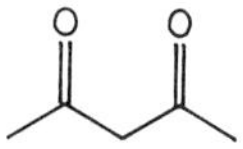	Ni–tartaric acid/ NaBr/thf	100%[a]	62
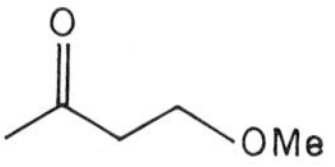	Ni–tartaric acid/NaBr	68	362
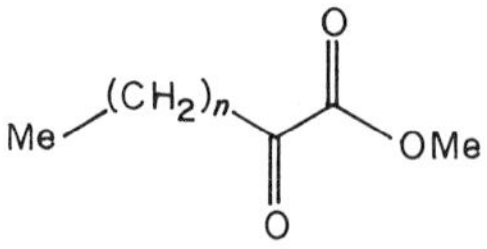	Ni–tartaric acid/NaBr	85	64
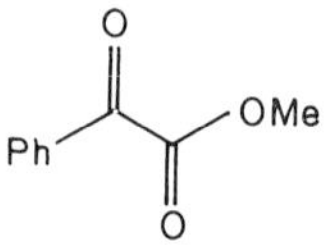	Pt–C, cinchonidine	82	278

[a] After one recrystallization.

treatment with tartaric acid. Modification is carried out by soaking the activated metal in a solution of the chiral modifying agent, filtering and washing the catalyst before use. Many modifying agents have been tested[63,275,276] but chiral α-hydroxy acids (particularly tartaric acid) and α-amino acids were found to give the highest optical yields. Both β-keto esters and β-diketones may be reduced with very high enantioselectivity under the appropriate conditions and some typical results are summarized in Table 6. Adding water to the reaction mixture appears to increase the rate of reduction but decreases the optical yield[279], although the effect of other additives is complex and poorly understood. The addition of acids with a pK_a higher than that of the modifying tartaric acid to a nickel–palladium catalyst on Kieselguhr has given enantiomer excesses of up to 89%[280].

Russian workers have considered a kinetic approach to the problem of selectivity[281–283]. Adsorption is proposed to be stereoselective as a result of interactions between the substrate, modifier, and catalyst and the subsequent rate-determining addition of hydrogen is non-stereospecific. The infrared spectra of nickel treated with various modified agents has been examined[284]. The results suggested that adsorption of modifier and reagent takes place on adjacent atoms with interaction between them occurring via hydrogen bonding. Notwithstanding these studies, a complete description of this enantioselective catalysis is lacking, although clearly both selective and non-selective adsorption sites exist on the catalyst surface. At the selective sites one diastereomeric complex is formed with high selectivity and the chiral differentiation may well occur on binding rather than during the rate-determining hydrogen addition to the bound substrate[285].

Despite the very high enantioselectivities obtained with these chiral catalysts, surprisingly few synthetic applications have been reported. In the light of the importance of the stereoselective formation of chiral 1, 3-diols for macrolide synthesis, perhaps there is scope here for future application[286].

4. Unsaturated carbonyls

With αβ-unsaturated carbonyls, reduction of either the carbonyl or the alkene double bond is possible, with the alkene double bond being the more easily reduced.

a. Reduction of carbonyl group

Allylic alcohols may be obtained in excellent yield using various modified platinum catalysts by reduction of the corresponding unsaturated aldehydes. Cobalt seems to be a particularly effective modifier and reduction of **57** gives the diunsaturated alcohol **58** in 95% yield[287]. Ruthenium on carbon modified with iron selectively reduced the aldehyde in citral (**59**) to give geraniol (**60**) selectively[288]. Using platinum oxide modified by nickel or iron, citral (**59**) may be further reduced to citronellol (**61**) with the conjugated alkene being selectively reduced[289–290]. Pre-reduced dirhenium heptoxide poisoned with pyridine is also an effective catalyst for the selective reduction to unsaturated alcohols. The yield of unsaturated alcohol falls marginally with increasing temperature up to 140 °C, above which the yield decreases markedly[291].

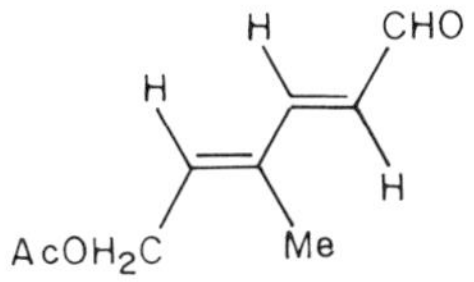

H CHO H H $AcOH_2C$ Me

(57)

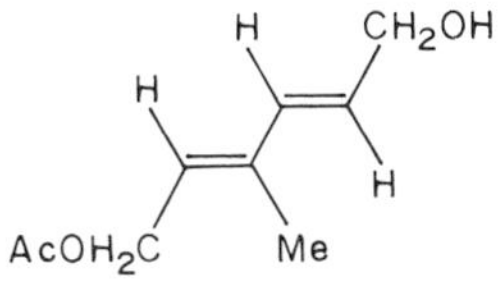

H CH_2OH H H $AcOH_2C$ Me

(58)

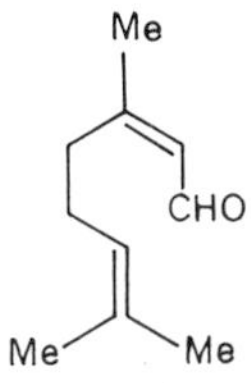

Me CHO Me Me

(59)

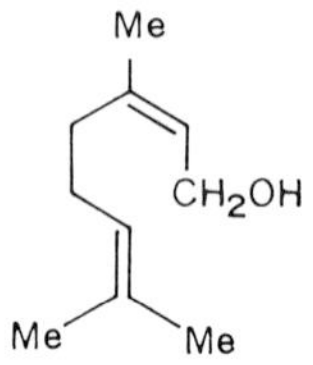

Me CH$_2$OH Me Me (60)

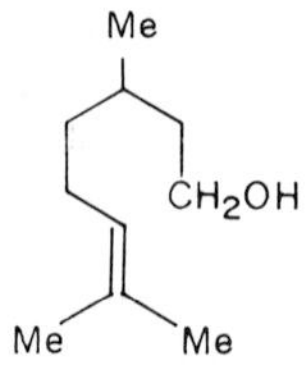

Me CH$_2$OH Me Me (61)

It is more difficult to reduce the carbonyl group in unsaturated ketones selectively. Usually the alkene is preferentially reduced, although the use of ruthenium[292] or platinum[293] catalysts has permitted certain selective ketone reductions.

b. Reduction of alkene double bond

With alphatic unsaturated carbonyl compounds, the alkene may be readily hydrogenated using palladium catalysts. Using palladium black, **62** may be selectively converted into the 5β-compound **63** under either acidic or basic conditions[294]. Unsaturated ketones have been formed and reduced in a one-pot procedure by use of a bifunctional catalyst[295,296]. For example, using an immobilized acid catalyst and palladium on carbon under hydrogen, propanone may be converted into 4-methylpentan-2-one, **64**, in high yield[297] and hexa-2, 5-dione may be transformed into **65** following reduction of the intermediate furan. (*E*)-$\alpha\beta$-Unsaturated-β-acyloxy crotonates such as **66** have been selectively reduced to the corresponding *threo* product **67** using a rhodium catalyst[298], and ascorbic acid, **68**, may be reduced in quantitative yield to *L*-gulono-1, 4-lactone (**69**) over 10% palladium on carbon[299].

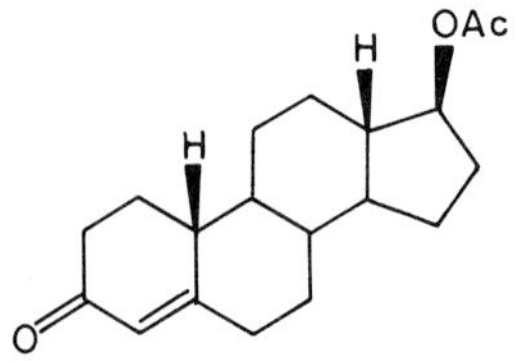

(62)

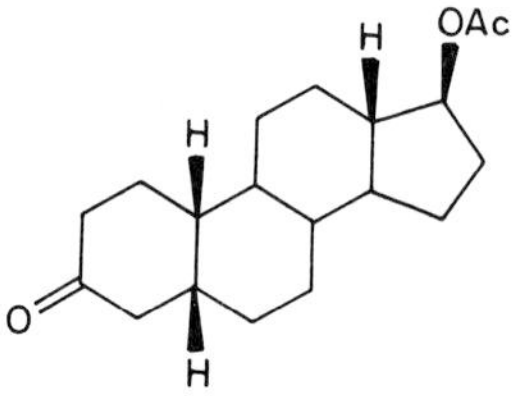

(63)

Pr^{i}CH$_2$COMe

(64)

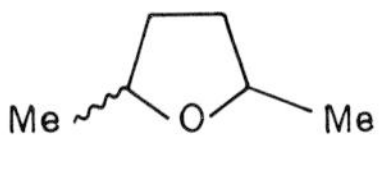

E and *Z*

(65)

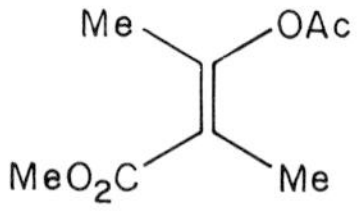

(66)

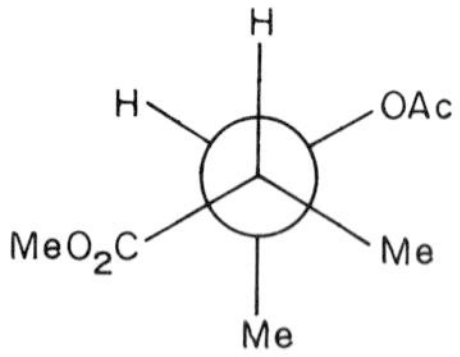

(67)

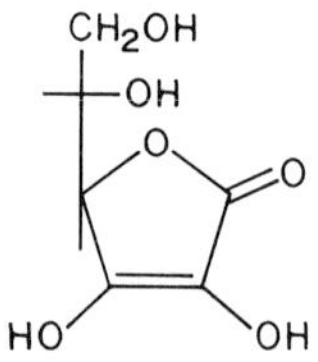

(68)

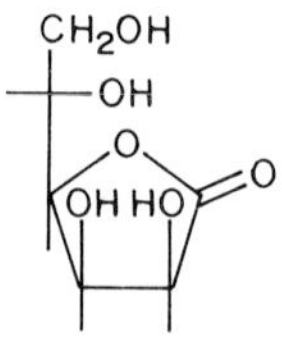

(69)

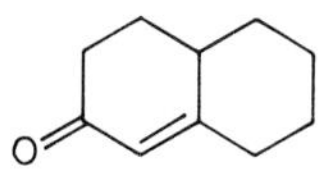

(70)

TABLE 7. Variation of product stereochemistry with solvent

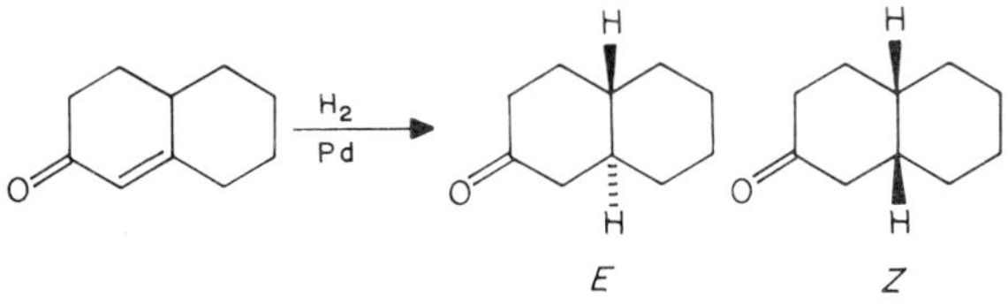

$\xrightarrow[Pd]{H_2}$ *E* *Z*

Solvent	Dielectric constant	Ratio E/Z
tert-Butanol	10.9	9:91
Methanol	33.6	59:41
Hexane	1.89	52:48
Dimethylformamide	38.0	21:79

The factors which determine the stereochemical outcome of the hydrogenation of $\alpha\beta$-unsaturated ketones in monocyclic and polycyclic systems are complex and interrelated. Attempts have been made to rationalize the observed effects of varying reaction parameters and comparing substrate structures[188]. In a model system such as β-octalone, **70**, the relative proportions of (*E*)- and (*Z*)-decalones produced may be related to the nature of the solvent and its dielectric constant. In aprotic solvents the E/Z ratio increases with decreasing dielectric constant, but is roughly proportional to the dielectric constant for protic solvents. Representative examples of this effect are given in Table 7. It has been proposed that 1,4-addition dominates in polar aprotic solvents, with 1,2-addition occurring mainly in non-polar aprotic solvents. In protic media, bulky solvents less effectively interact with the carbon—oxygen double bond and favour 1,4-addition.

The stereochemical outcome of enone hydrogenations is very sensitive to the substrate structure[300]. The hydrindanone **71a** gives mainly the *Z*-isomer, **25a**, but as the steric bulk of the R group increases in the 4-alkyl substituted derivatives, **72b**, more of the *E*-isomer is formed[301].

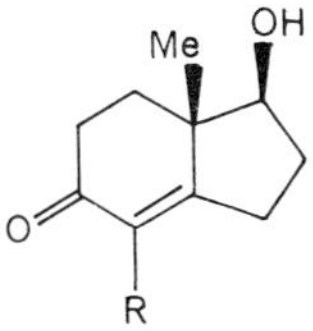

(**a**) R=H
(**b**) R=alkyl

(**71**)

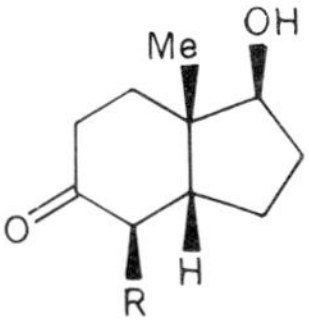

(**a**) R=H
(**b**) R=alkyl

(**72**)

c. Reduction of aromatic carbonyls

Aromatic aldehydes and ketones are reduced to the corresponding alcohols in high yield using palladium catalysts. Hydrogenolysis of the product alcohol occurs readily particularly in acidic or polar solvents and at higher temperatures[269,302–305]. In the hydrogenation of 3-methyl-2-carboxymethyl-1-tetralone, **73**, at 25 °C using palladium on carbon, only the carbonyl group is reduced, but as the temperature is increased to 80 °C hydrogenolysis dominates to form **74**[303]. Hydrogenolysis is suppressed by addition of

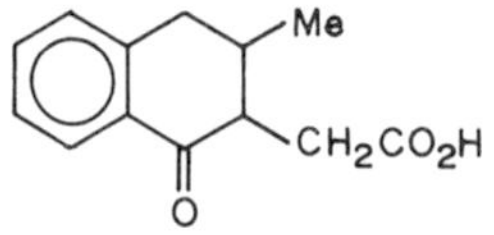

(73) 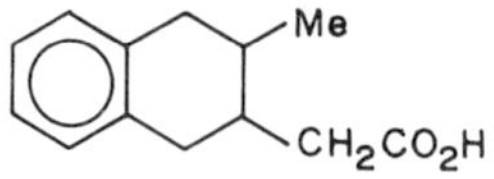 (74)

base[306–308]. The stereochemical course of aromatic ketone reduction using palladium catalysts is determined during hydrogen transfer from the catalyst surface to the bound substrate[309]. Using nickel catalysts, on the other hand, the first adsorption step appears to control the stereochemistry of reduction[310]. Aryl diketones bind more strongly to the catalyst surface than monoketones so that partial hydrogenations may be carried out successfully. In this manner **75** may be reduced to **76** selectively (equation 15)[311].

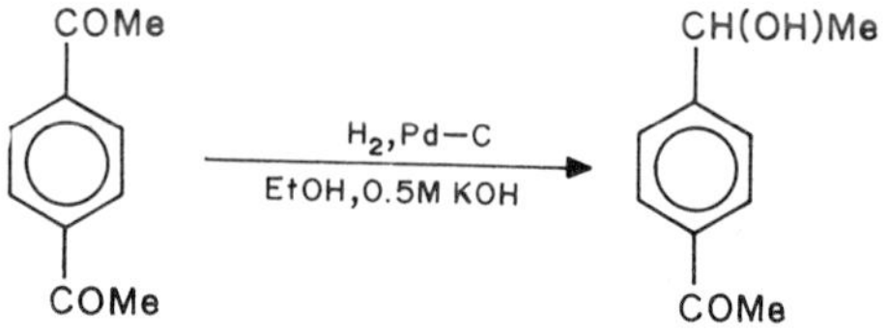

$$\text{(75)}\ p\text{-}C_6H_4(COMe)_2 \xrightarrow[\text{EtOH, 0.5M KOH}]{H_2,\ Pd\text{-}C} \text{(76)}\ p\text{-}MeCOC_6H_4CH(OH)Me \quad (15)$$

5. *Aromatics and heteroaromatics*

Although Sabatier and coworkers originally used nickel catalysts to reduce phenols, aminobenzenes, and benzene[312,313], the more common catalysts used nowadays are rhodium on a support (or PtO_2 for low-pressure reductions) and under more forcing conditions supported ruthenium catalysts. The main disadvantage of the rhodium and ruthenium catalysts is their tendency to saturate polycyclic systems fully.

a. *Selective arene reductions*

Aromatic ring reduction has been considered to involve consecutive transfer of six hydrogen atoms[314], so that various partially hydrogenated species are present as surface complexes. The desorption of partially hydrogenated products depends on the relative rates of hydrogen transfer to the bound species and the position of the chemisorption equilibrium. With a notable unconfirmed exception[170], the formation of cyclohexadienes from simple benzenes does not occur, although intermediate desorption of cyclohexenes is common[314–316]. The use of partial arene hydrogenation in synthesis is limited, however. The nature of the substituent on the aromatic ring determines the ease of ring hydrogenation, although this effect is not the same for all catalyst systems. For example, using rhodium in methanol the order of reactivity is $PhH > PhR > PhOH > PhNH_2$, whereas using platinum in acetic acid the order is $PhOH > PhNH_2 > PhH > PhR$[317,318]. Partial reduction of terephthalic acid with 5% ruthenium on carbon in water gave (*Z*)-1, 2, 3, 6-tetrahydroterephthalic acid (**77**) in 72% yield[319]. Unquaternized and quaternized pyridines may be partially reduced if there are electron-withdrawing groups in the 3-position. In this way 3-acetylpyridine was reduced in alcohol with 5% palladium on carbon to give **78** in 50% yield[320]. 2-Methylaminopyridine may be reduced to 2-methylamino-3, 4, 5, 6-tetrahydropyridine, **79**, in good yield using 5% rhodium on alumina in acetic acid[321].

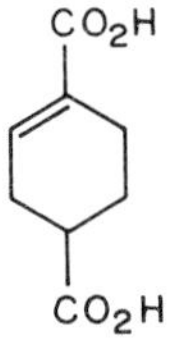

(77) 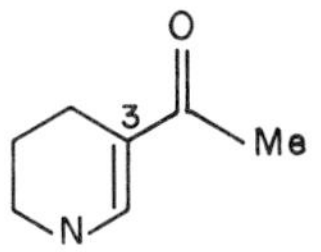 (78) 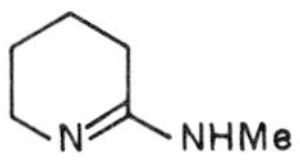 (79)

The hydrogenation of simple arenes does not proceed with all *cis* stereospecificity, consistent with the intermediate desorption of a cyclohexene[167,168]. However, using Raney nickel at 80 °C and 50 atm, some 2-alkylbiphenyls gave the corresponding (*Z*)-2-alkylcyclohexylcyclohexanes in up to 85% yield[322]. In a recent biotin synthesis, the surprisingly stereospecific hydrogenation of the thiophene diurethane, **80**, has been achieved using a palladium catalyst to give **81** in 95% yield[323].

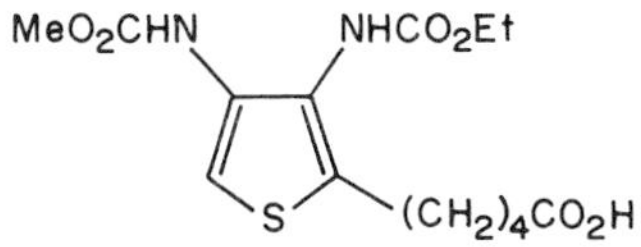

(80) 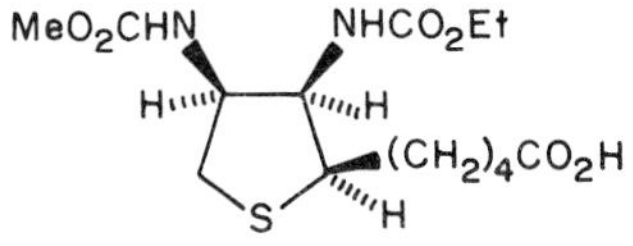 (81)

Generally, nitrogen and oxygen heterocyclic rings are reduced in preference to a benzene ring[323–326], although this selectivity may be reversed in very strong acid. Hydrogenation of the *N*-methylquinoline derivative **82** gives the benzopiperidine **83** in methanol, but the substituted pyridine **84** is produced in trifluoroacetic acid[326]. Under such strongly acidic conditions the benzene ring is activated to hydrogenation by formation of a protonated σ-intermediate. Pyridine rings are often reduced selectively in the presence of other heteroaromatics as in the reduction of azaindoles[327] and imidazole[1, 2-*a*]pyridines[328]. The reduction of **85**, however, using a 5% palladium on carbon catalyst in methanol gave the partially reduced triazine derivative **86** selectively[329]. A pyridine ring has also been reduced in the presence of an alkene; hydrogenation of **87** over platinum oxide gave the *N*-alkylpiperidine **88** with the double bond intact[330].

In general, aromatic rings are not reduced in preference to other unsaturated or hydrogenolysable groups, although of course there are exceptions. Rhodium catalysts are particularly effective at ring reduction without concomitant hydrogenolysis C—O

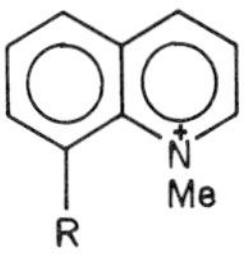

(82) 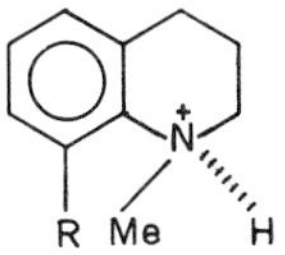 (83) 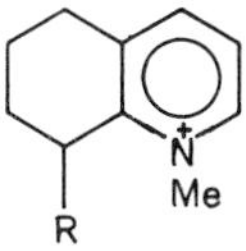 (84)

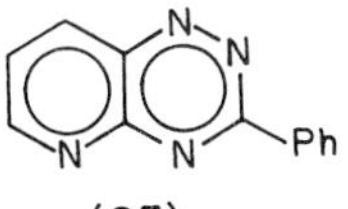

(85) 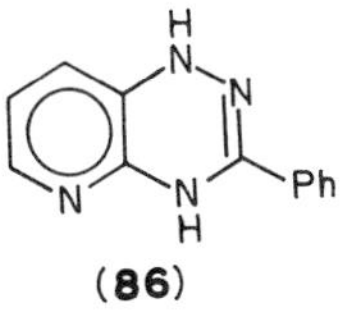 (86)

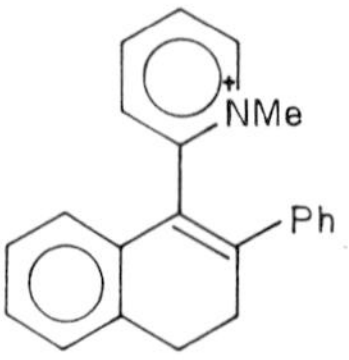

(87) 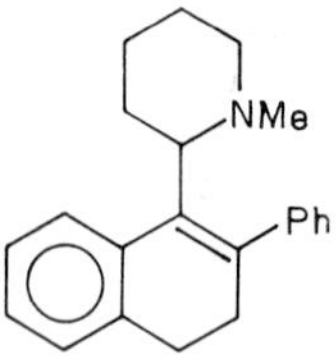 (88)

bonds[1,331]. Using a rhodium on carbon catalyst in ethyl acetate the quinoline **89** may be reduced to the substituted piperidine **90** without ester reduction or C—O hydrogenolysis[332]. Selective reduction of dibenzo-18-crown-6 (**91**) is only achieved with supported ruthenium catalysts to give the dicyclohexyl crown ether **92**[333]. Hydrogenolysis of benzylic C—O and C—N bonds occurs particularly easily with palladium and platinum catalysts but less readily with rhodium and ruthenium. Reduction of the dibenzyl **93** using ruthenium in dioxane under basic conditions gave[94] selectively[334].

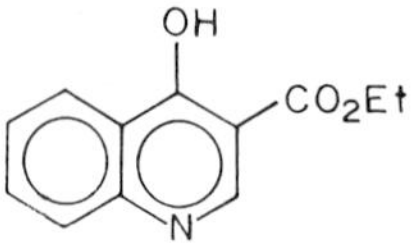

(89) 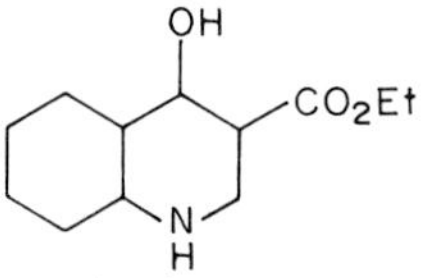 (90)

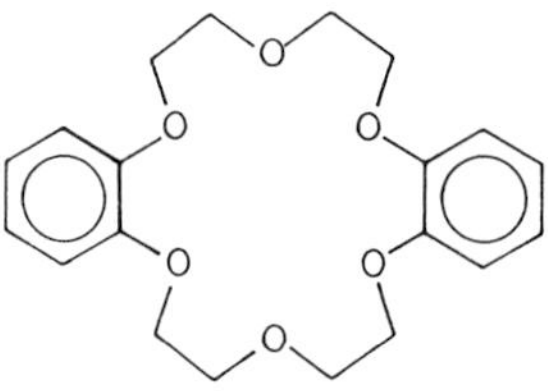

(91) 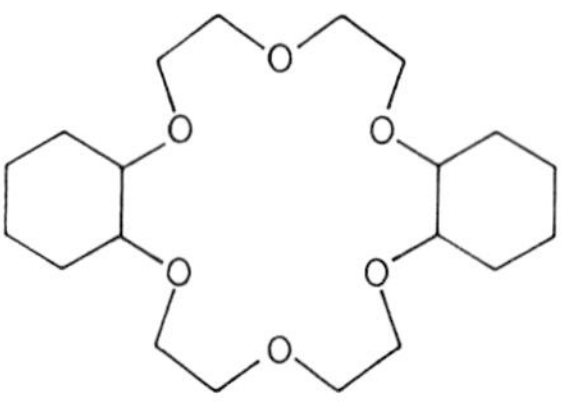 (92)

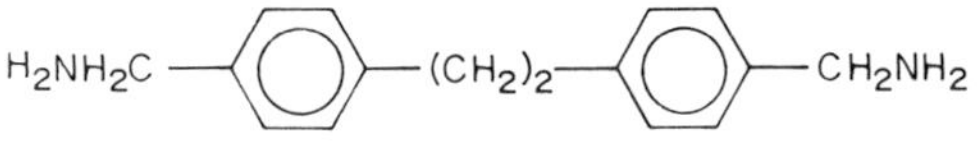

(93)

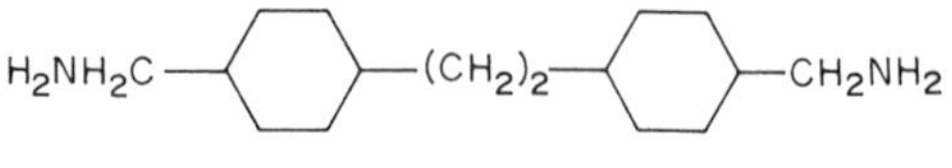

(94)

The reduction of polynuclear aromatics generally occurs in a stepwise manner. In the reduction of naphthalene, for example, a tetralin is formed initially, which is subsequently hydrogenated to the fully reduced decalin. The naphthalene is preferentially adsorbed, however, and is also reduced at a faster rate than the tetralin, so that little or no tetralin hydrogenation occurs until all of the naphthalene has been reduced[335]. The regioselectivity of the reduction of 1-alkylnaphthalenes varies with the bulk of the alkyl substituent. With large bulky groups the substituted ring is preferentially saturated, relieving unfavourable *peri* interactions (**95**→**96** + **97**) (equation 16)[335].

(95) → (96) + (97) (16)

R	% (96)	% (97)
Me	65	35
Et	55	45
Pr^i	32	68
Bu^t	3	97

A major development in the regioselective hydrogenation of polyarenes has been made following the observation that platinum catalysts under mild conditions selectively catalyse terminal arene rings, whereas palladium on carbon reduces in the so-called 'molecular K regions' to afford dihydroarenes[436–438]. In this way **98** may be reduced with platinum to give the tetrahydro derivative or with palladium on carbon to give the dihydro derivative (equation 17)[438].

← Pt, H_2 — (98) — Pd—C, H_2 →

(17)

b. Perhydrogenation

The favoured catalysts for complete hydrogenation of aromatic systems are rhodium and ruthenium on an inert support, although platinum oxide may be adequate with long reaction times in acidic solvents.

6. Nitriles and nitro groups

a. Reduction of nitriles

The catalytic hydrogenation of nitriles proceeds in a stepwise manner with intermediate formation of an unsubstituted imine (equation 18) to give a primary amine. The intermediate imine may add to the product amine to give a secondary amine after hydrogenolysis of the C—N bond (equation 19). Tertiary amines may also be formed similarly and in aqueous media the intermediate imine may be hydrolysed to give carbonyl compounds which may be directly reduced or reductively alkylated. Unless care is taken over the conditions of the reduction, therefore, complex mixtures of products may be obtained.

$$RC{\equiv}N \xrightarrow{H_2} RCH{=}HN \xrightarrow{H_2} RCH_2NH_2 \qquad (18)$$

$$RCH{=}NH + RCH_2NH_2 \longrightarrow RCH(NH_2)NHCH_2R \xrightarrow{H_2} RCH_2NHCH_2R \qquad (19)$$

The formation of secondary amines (equation 19) is suppressed in a variety of ways. Ammonia is effective at minimizing coupling reactions and reductions in alcohol–ammonia have been recommended for forming primary amines[336,337]. Reduction of the primary nitrile **99** gives the corresponding primary amine, tryptamine, under such conditions using a supported rhodium catalyst[338]. New catalysts have been sought which may form the primary amine selectively. Nickel boride and cobalt appear most useful for the conversion of low molecular weight aliphatic nitriles into primary amines[339], although the use of higher pressures certainly diminishes coupling reactions[340]. Strongly acidic conditions have also been used to inhibit the formation of coupling products and **100** has thus been selectively reduced to **101** using 10% palladium on carbon[341]. The related bromocyanamide **102** has been reduced to the bromoformamidine **103** under acidic conditions over 1% palladium on carbon without concomitant C—Br hydrogenolysis[342].

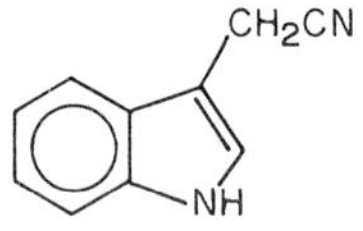

CH_2CN NH (**99**)

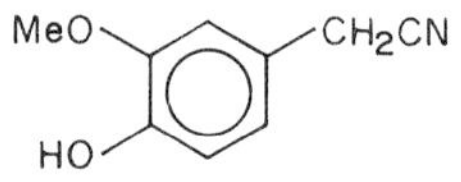

MeO HO CH_2CN (**100**)

MeO HO $CH_2CH_2NH_2$ (**101**)

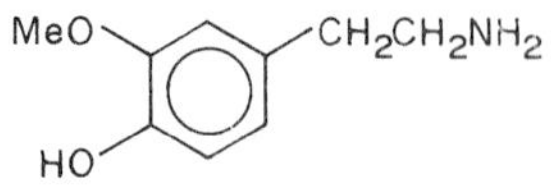

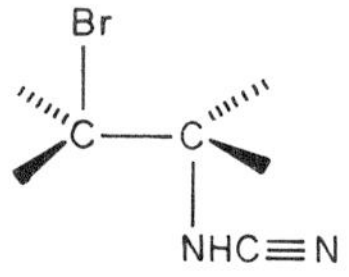

Br C—C $NHC{\equiv}N$ (**102**)

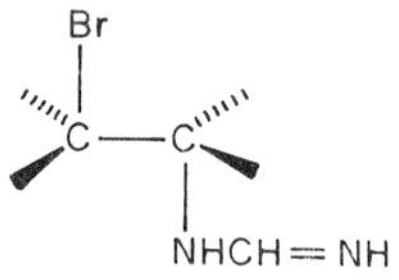

Br C—C $NHCH{=}NH$ (**103**)

Under certain conditions, the formation of the secondary amine (equation 19) may be favoured and two recent commercial syntheses illustrate the value of this reaction. Dibenzylamine may be manufactured by hydrogenation of benzonitrile over platinum on carbon in the presence of water[343]. Intramolecular coupling of the imine **104** leads to an industrial synthesis of piperidine derivatives, such as 2-methylpiperidine (**105**)[344,345]. Indeed, the hydrogenation of nitriles in molecules bearing an appropriately sited functional group has permitted many ring syntheses through cyclizations of the intermediate imine or amine[346–352]. Hydrogenation of the ketonitrile **106** over Raney nickel with short reaction times gives the imine **107**[353]. Reduction of **108** over palladium on carbon has led to a synthesis of a series of indole-3-carboxamides such as **109**[354]. In this reaction the aromatic nitro group is first reduced to an amine, which then cyclizes with the intermediate imine.

$N{\equiv}CCH_2CH_2CH_2C(Me){=}NR$

(104)

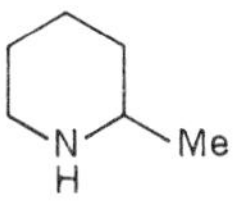

(105)

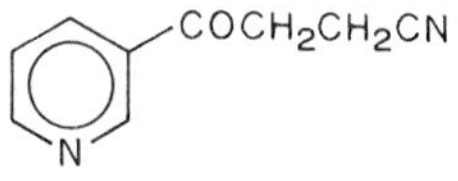

(106)

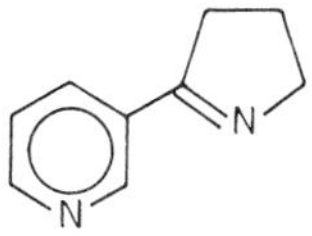

(107)

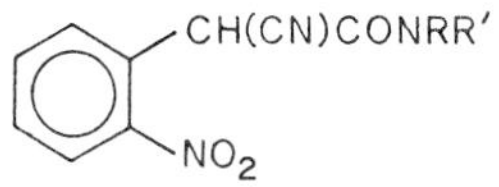

(108)

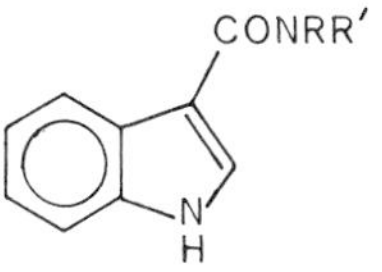

(109)

b. Reduction of nitro groups

Aromatic nitro groups are reduced readily with palladium, platinum, or nickel catalysts to give the corresponding amine in excellent yield. The intermediate aryl hydroxylamines may be the major reaction products under appropriate conditions. Nitroalkanes are reduced more slowly to give the amine, oxime, or hydroxylamine.

Aryl hydroxylamines are formed in good yield using platinum on carbon in methanol containing dimethyl sulphoxide as a catalyst promoter[355]. They have been trapped intramolecularly by remote carbonyl functions as in the preparation of 1-hydroxy-2-indolinones[356]. Reduction of **110** in dilute acid solution over a platinum on carbon catalyst gave an intermediate aryl hydroxylamine, which underwent a favourable 5-*exo*-trigonal ring closure[357] to give the *N*-hydroxy product **111** in 62% yield. Similarly, reduction of 2-nitrotrifluoraocetanilide (**112**) in ethanol over a palladium on carbon catalyst leads to formation of the 1-hydroxybenzimidazole (**113**)[358]. Indoles may be formed in good yield by hydrogenation of 2-nitrobenzyl ketones using a palladium on carbon catalyst[359].

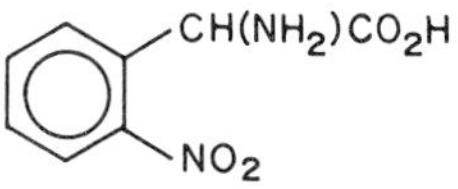

(110)

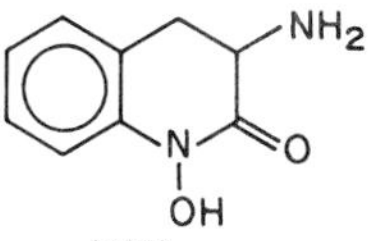

(111)

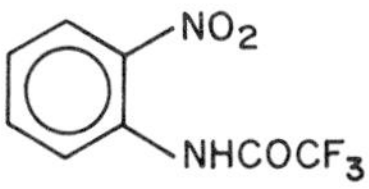

(112)

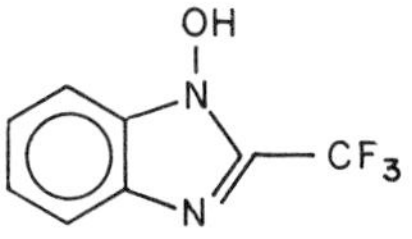

(113)

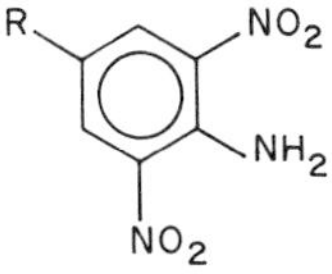

(114)

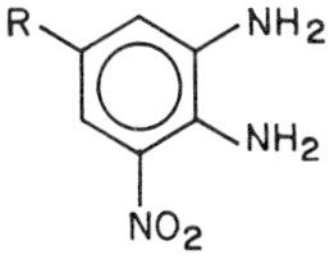

(115)

Dinitroaromatics may easily be reduced to the corresponding diamines but selective reduction of only one of the nitro groups is less easy. Nevertheless, certain 2,6-dinitroanilines may be selectively reduced over palladium on carbon in good yield (**114** → **115**[360], and the reduction of 2,4-dinitro-1-(*N*-piperidyl)benzene, **116**, over a Raney copper catalyst gave the 4-amino derivative **117** with over 99% selectivity[361].

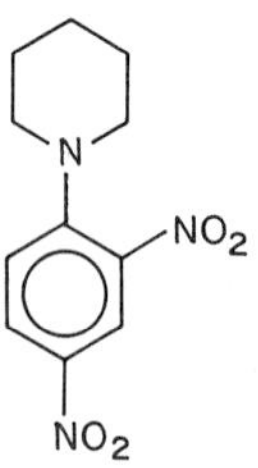

(116) 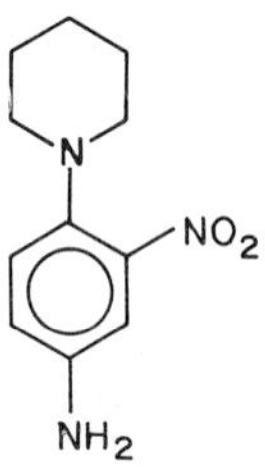 (117)

The reduction of aliphatic nitro compounds has often been used in the synthesis of amino sugars[363–366]. Hydrogenation of **118** over 10% palladium on carbon in acidic methanol gave high yields of 3-acetamido-3-deoxy-1, 2, 5, 6-di-*O*-isopropylidene-α-D-allofuranose (**119**) after treatment with acetic anhydride[366].

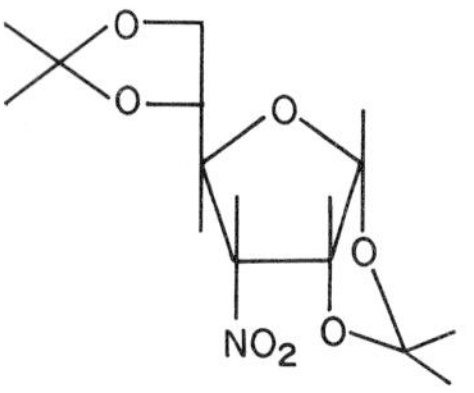

(118) 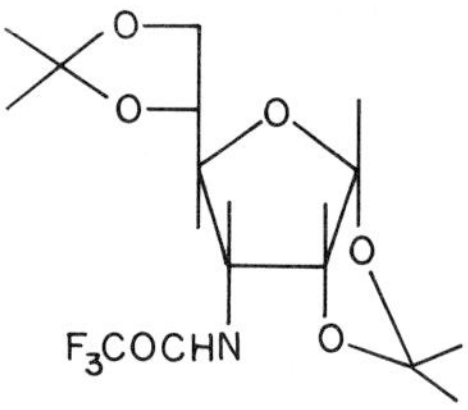 (119)

7. *Hydrogenolysis*

a. Carbon—oxygen bonds

The hydrogenolytic cleavage of C—O bonds occurs readily with palladium catalysts when the oxygen is bound to a phenyl or benzylic carbon[367], or when oxiranes are ring opened. The selective cleavage of C—O bonds is of some importance in organic synthesis, particularly in the removal of protecting groups in peptide synthesis, in the conversion of aryl ketones to aryl methylene groups, and in the preparation of polyhydric alcohols in sugar hydrogenolysis[368].

Aliphatic carbon—oxygen bonds are usually cleaved only under fairly forcing conditions, although trialkylsilyl ethers may be cleaved to the corresponding alkane under ambient conditions[369]. The *O*-alkylisourea **120** may also be cleaved over a palladium on carbon catalyst using mild conditions to give the alkane and dicyclohexylurea (**121**)[370].

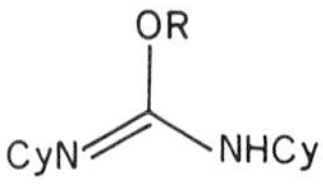

(120) 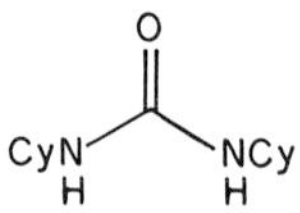 (121)

Oxiranes ring open readily and give the more substituted alcohol product in neutral or basic media[371], but the least substituted alcohol under acidic conditions[372]. However, the hydrogenolysis of **122** proceeds with preferential cleavage of the more substituted C—O bond[373]. The epoxytetrone **123** may be transformed into the dione derivative **124** using a palladium on carbon catalyst in dimethylformamide. This step is a key transformation in the synthesis of several deoxyanthracylinones[374].

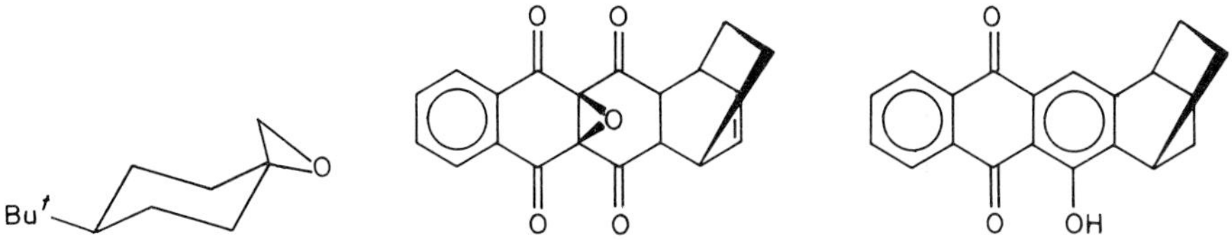

(122) (123) (124)

The hydrogenolytic cleavage of enol triflates to the corresponding saturated hydrocarbon has permitted the simple two-step conversion of aliphatic ketones to the corresponding methylene compounds[375] (equation 20). In this way androstane-3, 17-dione may be converted into androstane in 70% yield.

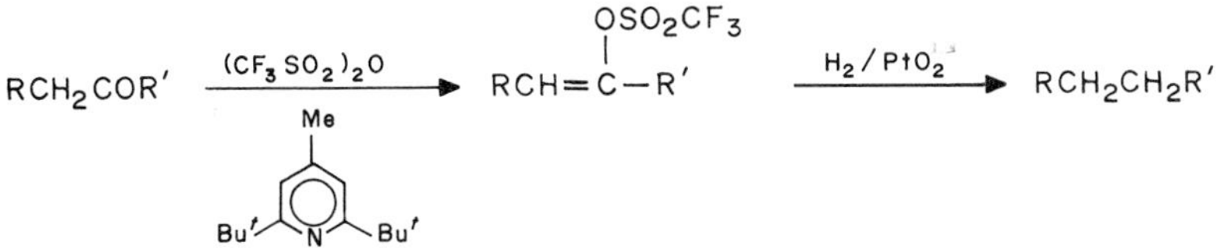

$$RCH_2COR' \xrightarrow[\text{2,6-di-}t\text{-butyl-4-methylpyridine}]{(CF_3SO_2)_2O} RCH{=}C(OSO_2CF_3){-}R' \xrightarrow{H_2/PtO_2} RCH_2CH_2R' \quad (20)$$

The cleavage of benzyl—oxygen bonds occurs smoothyl with palladium catalysts in polar solvents[376], particularly with added acid catalyst. Indeed, the benzyl group is a useful protecting group for alcohols[377] and acids[384] and the use of benzyloxycarbonyl groups in peptide chemistry is widespread[378,379,388] (equations 21 and 22).

$$PhCH_2OH + RCO_2H \xrightarrow{-H_2O} PhCH_2OCOR \xrightarrow{H_2/Pd} PhCH_3 + RCO_2H \quad (21)$$

$$PhCH_2OCOCl + RNH_2 \longrightarrow PhCH_2OCONHR \xrightarrow{H_2/Pd} PhCH_3 + CO_2 + RNH_2 \quad (22)$$

Primary alcohols have often been protected with trityl groups, which are also easily removed with hydrogen and palladium to regenerate the primary alcohol and liberate triphenylmethane[380]. Phosphoric acids may also be protected as their benzyl esters and the use of *meso*-hydrobenzoin has facilitated the synthesis of chiral $[^{16}O, ^{17}O, ^{18}O]$phosphate monoesters via hydrogenolysis of the 2-substituted-2-oxo-4, 5-diphenyl-1, 3, 2-dioxaphospholans, for example **125** → **126**[381–383].

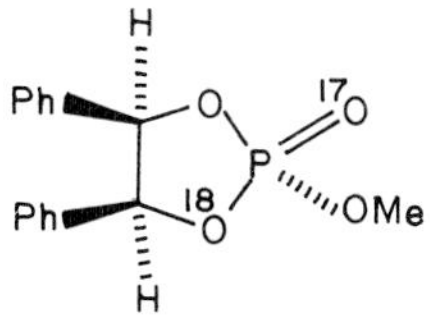

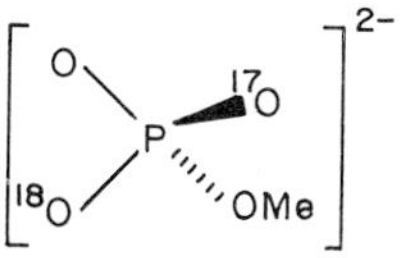

(125) (126)

The cleavage of aryl—oxygen bonds without affecting the aromatic ring also occurs readily particularly if there is a good leaving group produced. Cleavage of **127** occurs

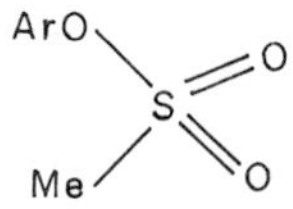

(127)

smoothly to give the arene and methanesulphonic acid in high yield[385] and **128** may be converted into **129** in benzene under mild conditions (equation 23)[386]. This latter procedure has been used to facilitate the specific tritium labelling of 1,2,3,4-tetrahydronaphthalene (equation 24)[387].

(128) (129) (23)

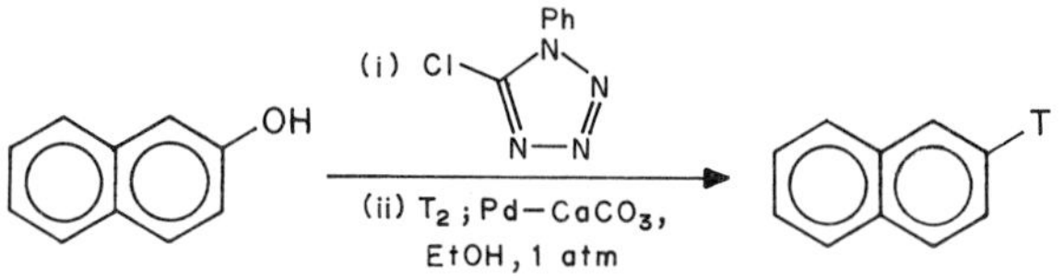

(i) Cl–(1-phenyltetrazol-5-yl)

(ii) T_2; Pd–$CaCO_3$, EtOH, 1 atm (24)

b. Carbon—nitrogen bonds

The hydrogenolytic cleavage of the C—N bond is similar to C—O cleavage but occurs less readily. Perhaps the most widely used C—N cleavage involves *N*-benzyl groups. These are removed with the following order of ease of hydrogenolysis: quaternary > tertiary > secondary > primary. The temporary protection and synthesis of tertiary amines may be effected via *N*-benzylation, with the benzyl group in **130** removed over palladium on carbon under ambient conditions to give **131**[389]. Tertiary benzylic groups may be cleaved in the presence of tertiary benzylic C—O bonds[390] and in the presence of carbon—halogen bonds[391], although this seems to depend on the basicity of the nitrogen[392]. With the *β*-lactam system in **132** the expected selective N—C-4 cleavage occurs to give **133** in quantitative yield[393].

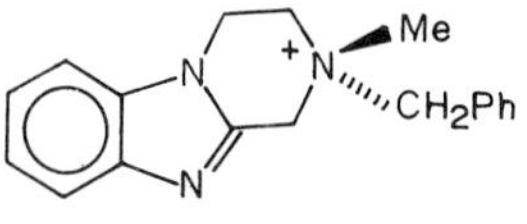

(130) 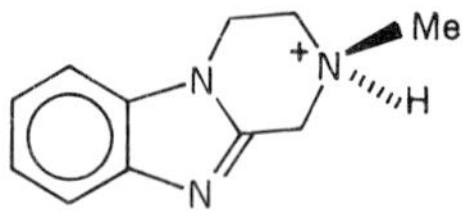 (131)

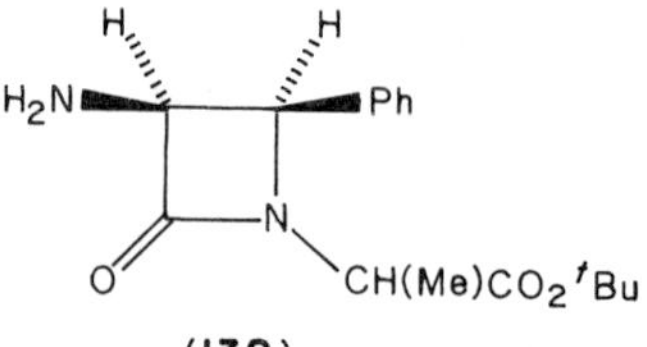

(132)

$PhCH_2CH(NH_2)CONHCH(Me)CO_2Bu^t$

(133)

c. Carbon—halogen bonds

The facile replacement of C—halogen by C—hydrogen in benzyl, allyl, and aryl positions permits the use of halogen as a protecting or directing group in synthesis and has also facilitated some specific deuteriations. The reduction of acid chlorides to aldehydes (Rosenmund reduction)[394] is an early example of this cleavage which has found wide application[395]. More recently this type of reduction has been found to be more conveniently effected with tertiary amines or acetate anion as acid receptors. For example, the transformation of **134** to **135** occurs smoothly under ambient conditions over palladium on carbon in the presence of diisopropylethylamine[396]. The ease of order of halogen removal follows the expected trend with the carbon—iodine bond most easily cleaved and the carbon—fluorine bond the most difficult to reduce. Benzylic and allylic halides are hydrogenolysed selectively in the presence of most reducible groups, but the reduction of vinyl, aryl, and alkyl halides seldom occurs selectively except in the presence of ketones, arenes, and cyclopropanes. Thus **136** may be reduced to **137** in high yield over a Raney nickel catalyst[397] and **138** is hydrogenolysed over palladium on carbon in basic media to give **139** selectively[398].

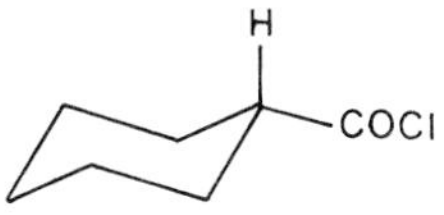

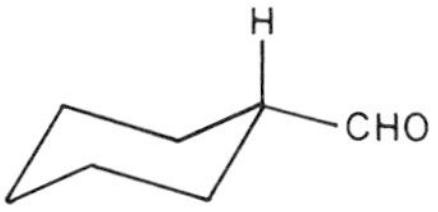

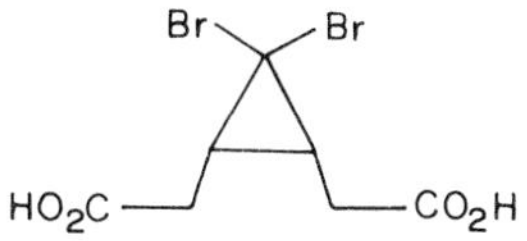

(134) (135)

(136) 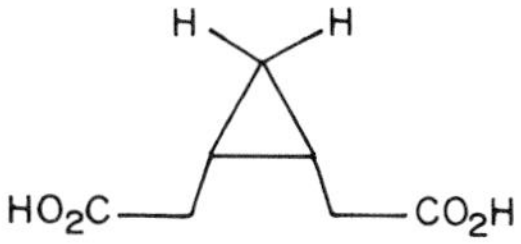 (137)

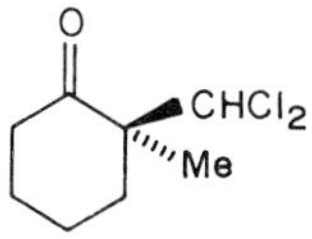

(138) 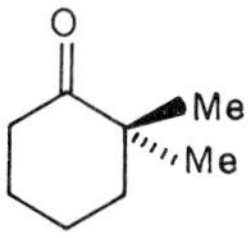 (139)

Deuteriolysis of the aryl carbon—halogen bond has facilitated several regiospecific labellings. Using a palladium catalyst generated *in situ* by reduction of $PdCl_2$ with $NaBD_4$, the labelled compound **140** has been prepared with $\geqslant 95\%$ isotopic purity[399].

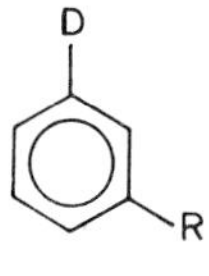

(140)

d. Carbon—carbon bonds

The cleavage of carbon—carbon σ-bonds requires very forcing conditions and is generally of little synthetic value to the organic chemist[400]. Notable exceptions to this generalization include the hydrogenolysis of activated or strained cyclobutanes[401–403] and particularly the ring opening of cyclopropyl derivatives.

The hydrogenolysis of cyclopropane rings may occur either via suprafacial hydrogen addition when double retention or inversion of configuration at each carbon occurs, or via antarafacial hydrogen addition where one carbon is inverted and one retains configuration (equation 25).

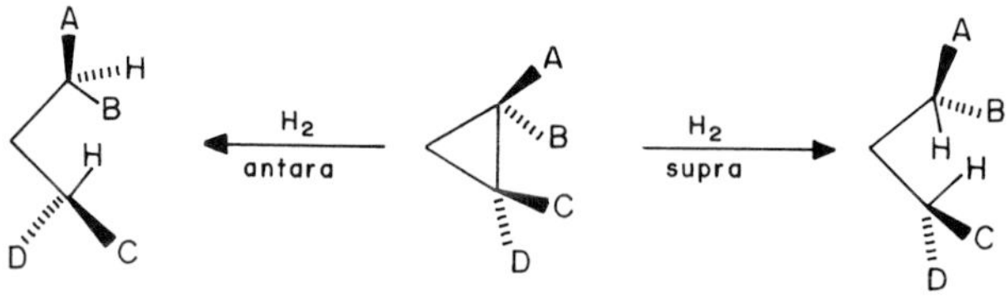

(25)

Suprafacial attack of hydrogen occurs for the deuteriolysis of **141** over platinum[404], whereas both senses of addition occur for 1,2-dimethyl-1,2-diphenylcyclopropane[405], generating a mixture of the racemic and *meso* diastereoisomers, **142** and **143**. Further examples of stereoselective ring cleavage include the hydrogenolytic opening of **144** catalysed by rhodium, platinum, or palladium which gives **146** specifically[466], and the ring expansion of **146** to generate **147** with a *cis* ring junction stereospecifically[407].

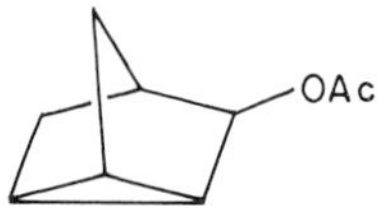

(**141**) 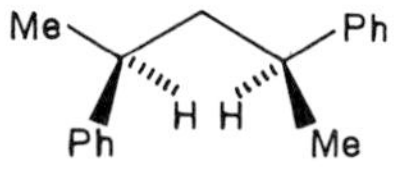 (**142**) 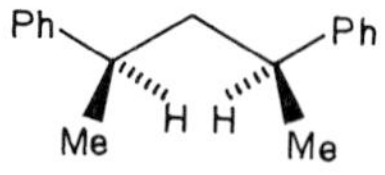 (**143**)

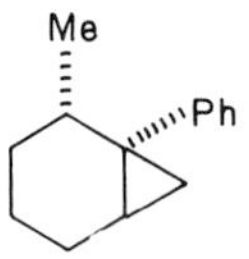

(**144**) 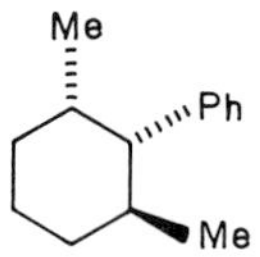 (**145**)

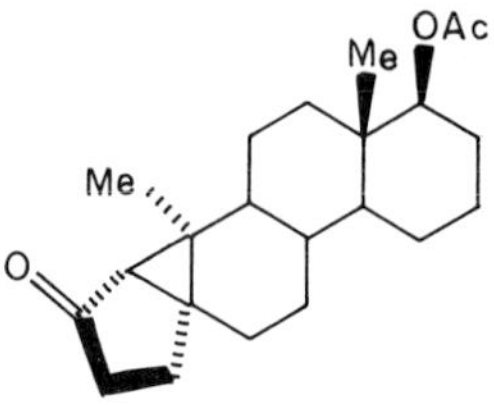

(**146**) 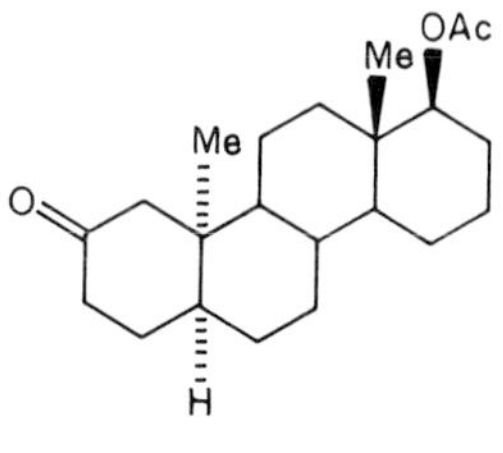 (**147**)

Alkylcyclopropanes tend to undergo hydrogenolytic cleavage at the least substituted bond[408–411], so that simple cyclopropanes may be usefully converted into *gem*-dimethyl groups, for example **148**–**149**[411,412]. However, preferential adsorption of the most strained carbon—carbon σ bond in **150** determines the rate and direction of hydrogen addition, (the 'anchor' effect), so that **151** is produced selectively[413].

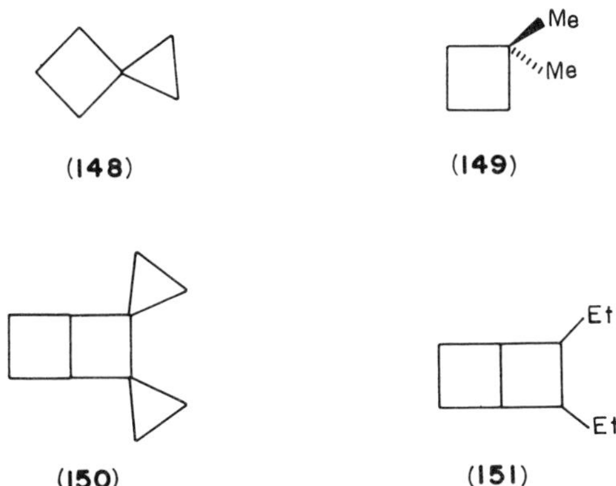

Vinyl- and phenyl-substituted cyclopropanes undergo preferential ring cleavage at the adjacent carbon—carbon bond[414–416], so that vinylcyclopropane (**152**) gives *n*-pentane specifically[417], and **153** may be hydrogenolysed over palladium on carbon to give **154**[418].

Ph CO_2H Ph Me

$(Ph)_2CHCH_2CH(Me)CO_2H$

(**152**) (**153**) (**154**)

III. HOMOGENEOUS CATALYTIC HYDROGENATION

A. Introductory Remarks

While there is a wealth of organometallic literature detailing the characterization and catalytic activity of transition metal complexes[5,25,26], there are perhaps only about a dozen practical homogeneous catalysts of interest to the synthetic organic chemist[1–4]. It is fair to say that most new organometallic complexes, particularly of rhodium, iridium, ruthenium, and cobalt, are tested for their ability to catalyse alkene hydrogenation[5,419–421] and with the development of the organometallic chemistry of f-block elements, these complexes are also being screened for potential catalytic activity[422–426]. The number of selective homogeneous catalysts is limited and Wilkinson's catalyst[75,76] is easily the most studied catalyst for alkene hydrogenation[3]. A summary of the useful practical procatalysts[a] is given in Table 8, all of which operate under ambient conditions of temperature and pressure unless indicated otherwise.

Although carbonyls, arenes, nitro groups, and imine double bonds may be reduced with homogeneous catalysts, the major application of such catalysts lies in the regioselective

[a]The active catalyst is derived from the given procatalyst under hydrogenation conditions.

TABLE 8. Practical homogeneous procatalysts

Catalyst	Substrates	Typical solvent	Comments	Ref.
$[RhCl(PPh_3)_3]$	Alkenes	Benzene, alcohols	Regioselective	3,76
$[RuHCl(Ph_3)_3]$	Alkenes, 1, 3-dienes	Toluene	Selective for terminal alkene	3, 78
$[RhL_2diene]^+$	Alkenes, alkynes, carbonyls	thf, MeOH	chiral catalysis with chiral phosphines; alkyne to alkene (*Z*)	79, 80
$[RhCl_2(BH_4)py_2]$/dmf	Alkenes, nitro groups	dmf	Asymmetric catalysis with chiral formamides	434
$[Ir(PCy_3)(cod)py]^+$	Alkenes	CH_2Cl_2	Hindered alkenes rapidly reduced; no O_2 poisoning	87, 435
$[Ir(cod)L_2]^{+a}$	Dienes	Acetone	Selective for diene	86
$[Cp_2MoH_2]$	Dienes	thf	140–180 °C, forms monoene	432
$[ArCr(CO)_3]$	Dienes	thf	Unconjugated dienes isomerized then reduced	430, 431
$[PtCl_2(PPh_3)_2]$/$SnCl_2$[b]	Alkenes	CH_2Cl_2	Elevated *T* and *P* required	429, 224
[PdPc]	Alkenes, carbonyls, nitro groups	H_2O	Substrate selectivity is pH sensitive	427
$[HCo(CN)_5]$	Conjugated alkenes	H_2O	C—X hydrogenolysis is easy	428

[a] L_2 = chelating diarylphosphine or two triarylmonophosphines.
[b] 4 atm. of H_2.

and stereoselective reduction of alkenes, polyenes, and alkynes. Such selectivity has been highlighted with the development of the catalytic asymmetric hydrogenation of prochiral enamides (see also Section III.d) to give substituted amino acids in $\geqslant 95\%$ enantiomeric purity[12,84,83].

The most common solvents for homogeneous hydrogenations are usually weakly coordinating alcohols, ethers, or arenes such as toluene, methanol, and tetrahydrofuran. The nature of the solvent may profoundly affect the course of reduction, for example the cationic iridium complex $[Ir(cod)(PCy_3)py]^+$ is only catalytically active in dichloromethane[87]. Usually chlorinated solvents are avoided; this is particularly true of chloroform and carbon tetrachloride, which function as efficient hydride abstractors[435]. Indeed, this reaction to form a metal—chlorine bond is commonly used to indicate the presence of a metal—hydrogen bond.

B. Synthetic Utility of Homogeneous Hydrogenation

This section will discuss the reduction of the common unsaturated functional groups and pays particular attention to more recent stereoselective and stereocontrolled hydrogenations. Coverage of alkene, polyene, and alkyne reduction in particular is selective rather than comprehensive and detailed summaries of such reductions may be found elsewhere[2,3,5].

1. Alkenes and polyenes

a. Stereoselectivity, chemoselectivity, and regioselectivity

Although there are a very large number of catalysts which are active towards alkene hydrogenation, most applications in synthesis have employed Wilkinson's catalyst, $[RhCl(PPh_3)_3]$[3], and increasingly the cationic iridium catalysts developed by Crabtree *et al.*[10]. Hydrogenation with these catalysts is *syn* stereospecific[439] and this was first demonstrated by the addition of deuterium to maleic acid (**155**) to give *meso*-1, 2-dideuteriosuccinic acid (**156**). There is no scrambling between deuterium and solvent or indeed between hydrogen and deuterium in reductions with 1:1 H_2–D_2 gas. This is in direct contrast to heterogeneous hydrogenations, where isotopic scrambling is the rule[442]. Stereospecific *syn* addition of deuterium has also been demonstrated with the cationic rhodium diphosphine catalyst **157**[440,441]. Stereospecific *syn* additions have been applied to label various compounds[444], including unsaturated 5α-spirotane steroid derivatives such as **158**[5] and in the synthesis of 'chiral methylvaline' by addition of tritium to labelled *N*-acetylisodehydrovaline (**159**)[443].

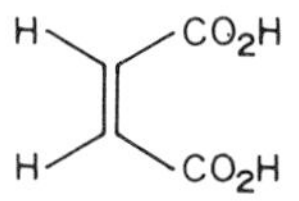

(**155**) 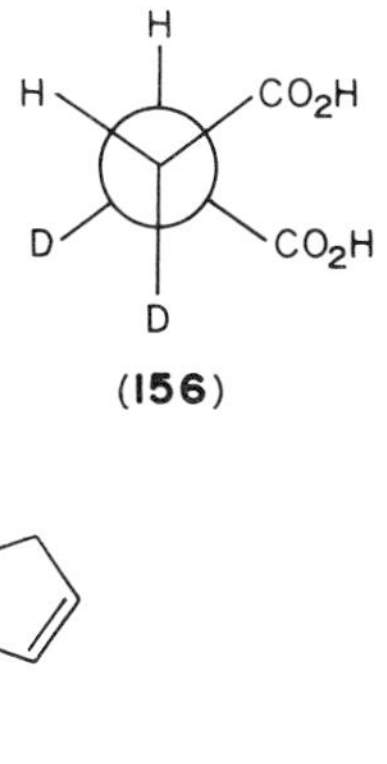 (**156**) 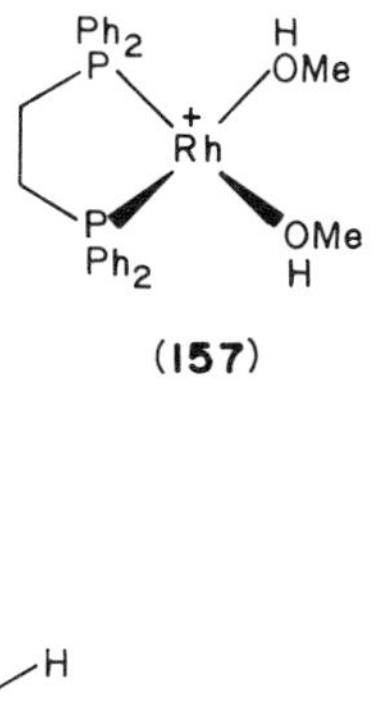 (**157**)

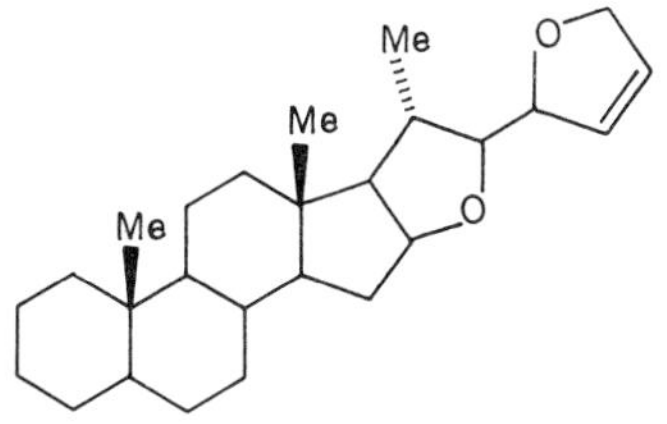

(**158**) 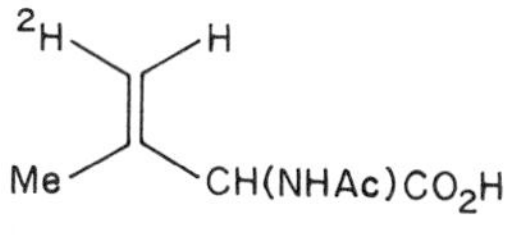 (**159**)

The relative rates of hydrogenation of non-conjugated alkenes follow their ability to bind to rhodium and so reflect the steric crowding afforded by the bulky phosphines. Terminal alkenes are readily hydrogenated and with cyclic alkenes the rate varies inversely with ring size. The order of reactivity in equation 26 is generally followed with Wilkinson's catalyst. However, $[RuClH(PPh_3)_3]$ rapidly reduces terminal alkenes in preference to

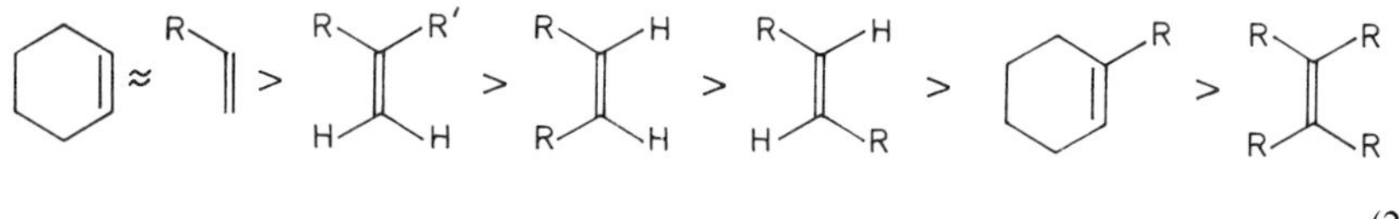

(26)

other substituted alkenes and the iridium complexes such as $[Ir(cod)(PCy_3)py]^+$ reduce tetrasubstituted alkenes very rapidly[10,445]. Indeed, tetramethylethylene is reduced 100 times faster than cyclohexene using $[RhCl(PPh_3)_3]$.

The differences in the rates of alkene hydrogenation have led to many regioselective hydrogenations[446–448]. The reduction of dihydroaromatic compounds occurs readily with selective reduction of the disubstituted alkene in **160** to give **161** in high yield[449,450]. The reduction proceeds without the disproportionation to aromatic and cyclohexane derivatives which usually occurs with palladium and platinum heterogeneous catalysts. The exocyclic double bonds are selectively reduced in **162**[451], **163**[452] and **164**[5], and with allenes the unsubstituted terminal alkene bond may be selectively reduced[453].

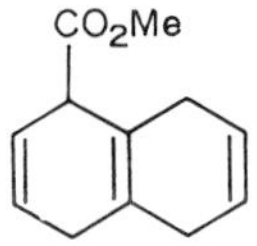

CO2Me CO2Me

(160) 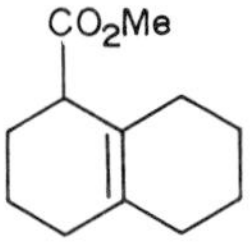 (161)

Me Me Me Me Me Me Me Me Me

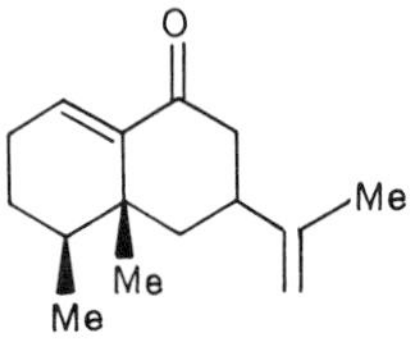

(162) 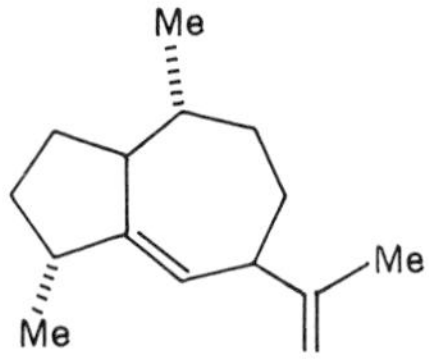 (163) 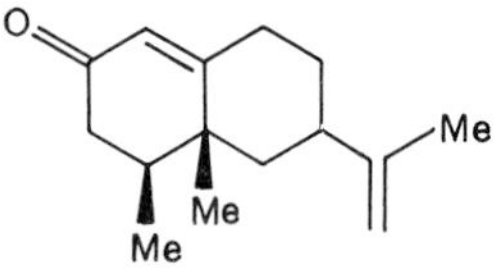 (164)

Homogeneous catalysts are particularly chemoselective and the following groups are not hydrogenated or hydrogenolysed under the usual conditions of alkene hydrogenation: carboxylic acids, ketones, arenes, esters, nitriles, amides, and ethers. Aza, hydroxy, chloro, and nitro compounds are also typically not reduced. The lack of cleavage of hydrogenolizable groups has been used in the reduction of dehydrogriseofulvin[450], **165** → **166**, with which facile C—Cl and ring-opening C—O hydrogenolysis occur with conventional heterogeneous catalysts[454,455]. The hydrogenolytic cleavage of carbon—carbon bonds is suppressed with homogeneous catalysts and **167** may be reduced to **168** in 97% yield. However, other simple cyclopropane systems show increased amounts of ring-opened products[456]. The selective saturation of the alkene double bond in nitrostyrenes occurs smoothly with $[RhCl(PPh_3)_3]$; for example, **169** may be reduced to **170** in high yield using 5 atm of hydrogen pressure[457]. Sulphur-containing compounds generally poison heterogeneous catalysts[458] but successful hydrogenations may be effected with homogeneous catalysts, for example in the reduction of alkenylthiophenes[459].

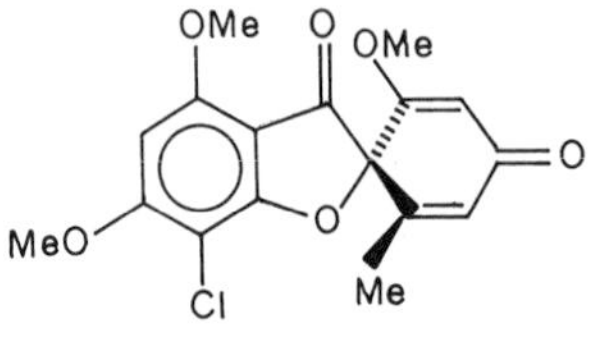

(165) 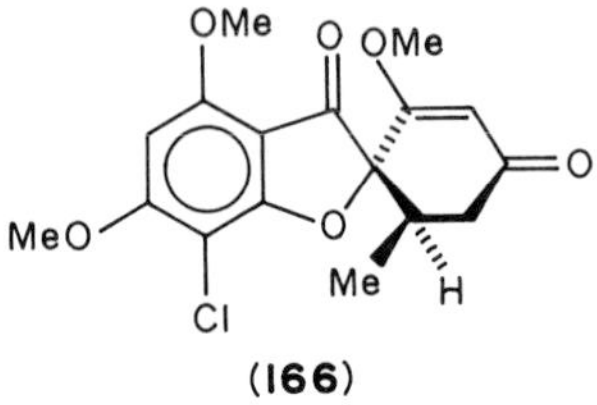 (166) 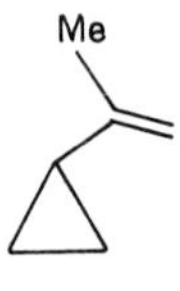 (167)

(168) 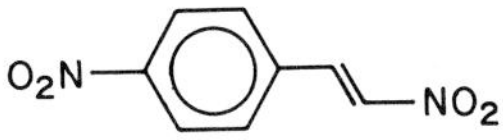 (169) 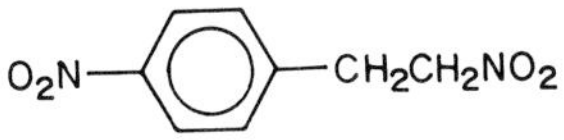 (170)

As with heterogeneous hydrogenation, the least hindered face of the alkene is bound to the metal and thereafter reduced. For example, in the deuteriation of ergosterol acetate (**171**) to 5α, 6α-[$^2H_2$]-ergost-7-en-3-β-ol acetate (**172**)[5], the least hindered alkene is reduced from the less crowded face of the steroid. Similarly, in the hydrogenation of **173** hydrogenation proceeds to give the *trans* fused decalone **174** stereospecifically[460].

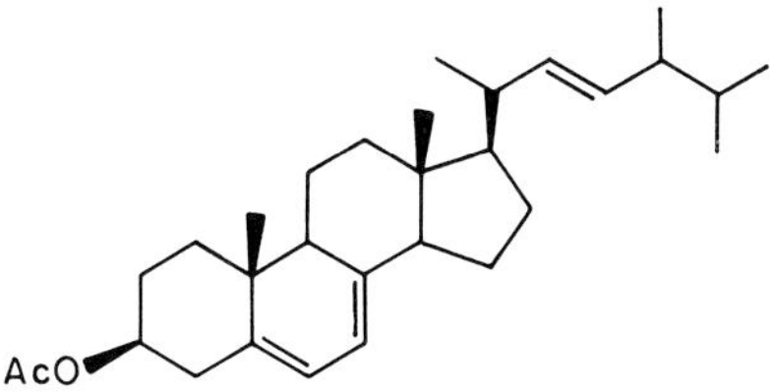

(171) 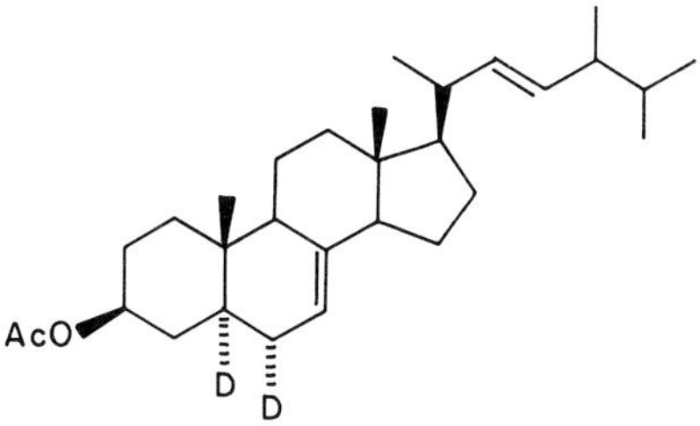 (172)

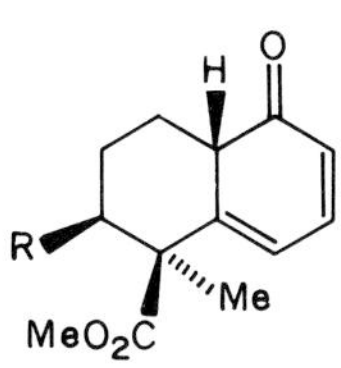

(173) 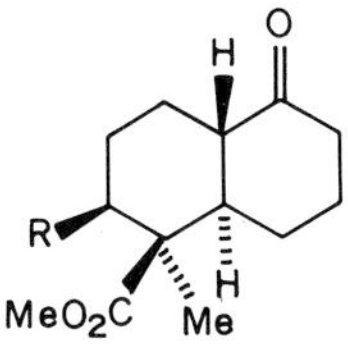 (174)

b. Stereocontrolled hydrogenation

Although there are several distinct examples in heterogeneous hydrogenation of the directing effect of a remote functional group (Section II.C.1.a), it is only recently that such effects have been discovered and applied in homogeneous systems. Hydrogenation of **175** with Wilkinson's catalyst gives the *cis* diastereoisomer **176** exclusively, presumably as a result of the preferential formation of the chelated intermediate **177**[461]. The hydrogenation of allylic and homoallylic alcohols with the cationic diphosphine rhodium

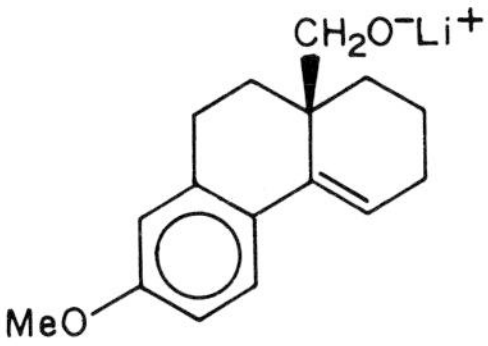

(175) 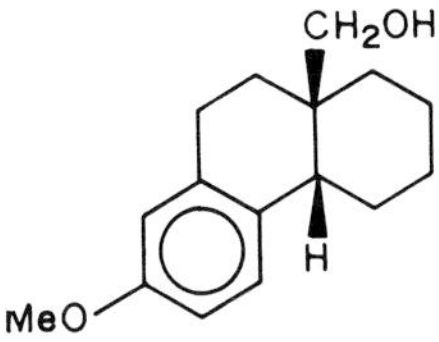 (176)

TABLE 9. Stereocontrolled hydrogenation of allylic and homoallylic alcohols[a] [463,464]

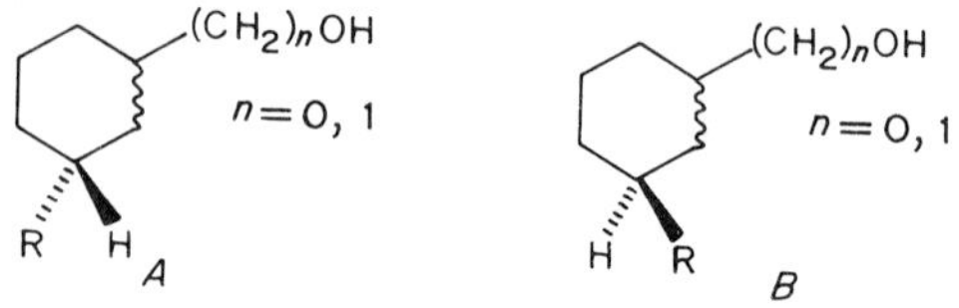

Entry	Substrate	Ratio $A:B^a$	Yield (%)
1	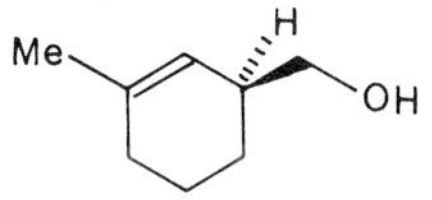	6:1 (1:3)	78
2	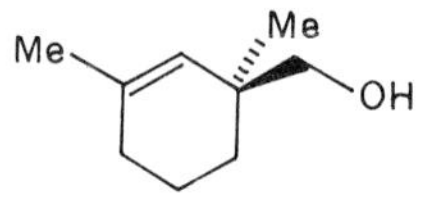	9:1 (1:1)	82
3	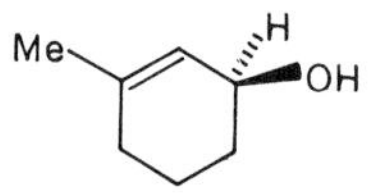	74:1 (1:3)	48
4	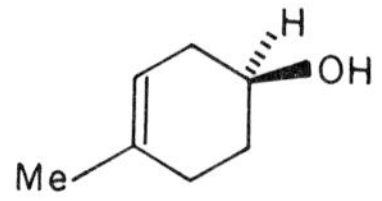	33:1 (1:3)	85
5	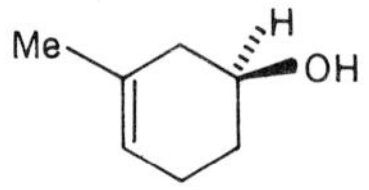	27:1 (1:5)	76
6	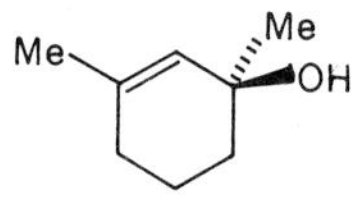	> 100:1 (1:1)	64
7	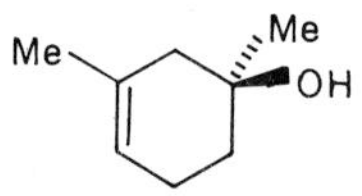	> 100:1 (1:1)	74
8	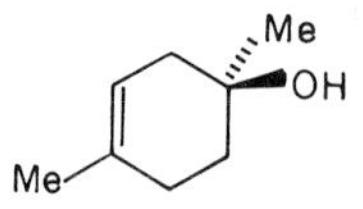	> 100:1 (1:1)	87
9[b]	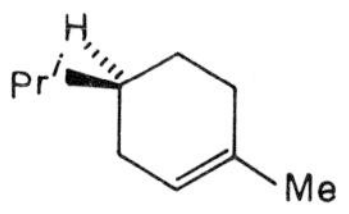	99.9:0.1 (1:1)	85

[a] Values in parentheses refer to reduction with 5% palladium–charcoal in methanol.
[b] Data from ref. 464; all other data from ref. 463.

procatalyst **178** has been reported[462]. Reduction of **179** proceeds smoothly in aprotic solvents such as dichloromethane or tetrahydrofuran to give (*R*, *S*)-*threo*-3-phenylbutan-2-ol (**180**) selectively over the *erythro* diastereoisomer (up to 30:1 selectivity). Such stereoselectivity is rationalized in terms of the model (**181**) in which the non-bonded interactions experienced by the methyl group are minimized in the transition state. The

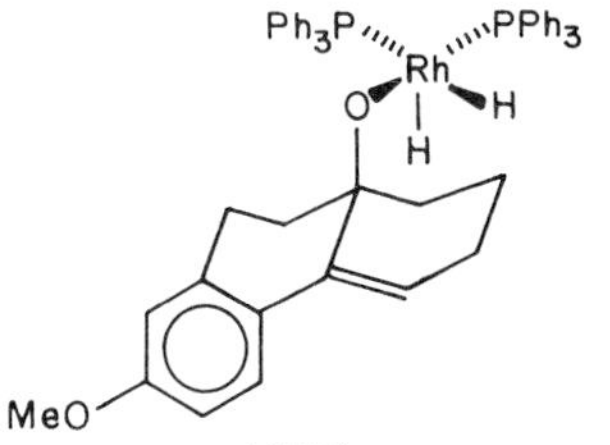

(**177**) 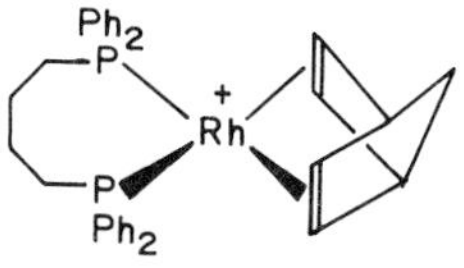 (**178**)

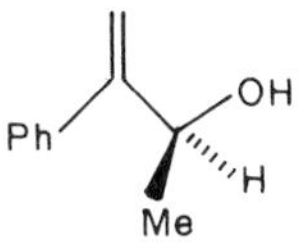

(**179**) 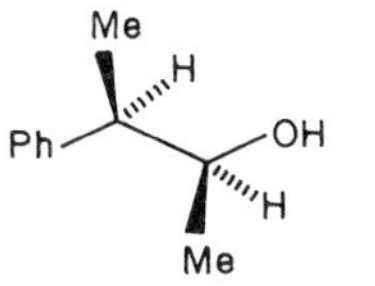 (**180**) 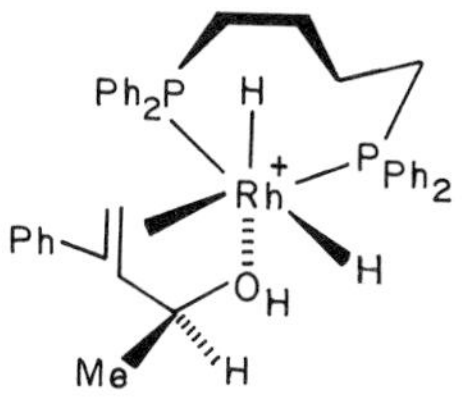 (**181**)

ability of an adjacent polar substituent to direct the stereochemical course of hydrogenation has also been observed with iridium catalysts. A series of allylic and homoallylic secondary and tertiary cyclohexenols have been reduced using $[Ir(cod)(PCy_3)py]^+$ as catalyst[463,464]. The results are presented in Table 9; high selectivities were observed (*ca.* 97:3) for the reduction of secondary alcohols, and even greater selectivity for the reduction of allylic tertiary alcohols. It is evident that the iridium binds to the hydroxy group and one face of the alkene and the substrate is reduced preferentially from that side. Table 9 also lists reductions of the same substrates with the heterogeneous palladium on carbon catalyst, for purposes of comparison. It is apparent that stereocontrolled reductions with cationic rhodium and iridium catalysts are reactions of considerable synthetic utility. It also seems that the diastereoselectivity observed with the cationic rhodium complexes is dependent on the hydrogen pressure and enhanced diastereoselectivities, with both cyclic and acylic allylic alcohols, have been observed on raising the hydrogen pressure[644]. The directing effect of adjacent polar substituents at a chiral centre in homogeneous hydrogenation has also been observed in the diastereoselective reduction of dehydrovalines[443] and in the hydrogenation of dehydrodipeptides [465,466].

2. Dienes

There are several homogeneous catalysts which selectively reduce dienes and conjugated alkenes[432,430,467–471]. A typical example is $[Cp_2MoH_2]$[432], which selectively reduces 1,3-and 1,4-dienes to monoalkenes. Cyclopentadiene may be reduced to cyclopentene at 180 °C and 160 atm in the absence of solvent. Arenechromium tricarbonyl complexes exhibit selectivity for the 1,4-hydrogenation of conjugated dienes which are able to take up a *cisoid* configuration[430]. For example, methyl sorbate (**182**) may be reduced to methyl hex-3-enoate (**183**) and addition of deuterium confirmed that 1,4-

addition was occurring exclusively[470]. These chromium–arene catalysts may find further application for the stereoselective formation of (*Z*)-alkenes.

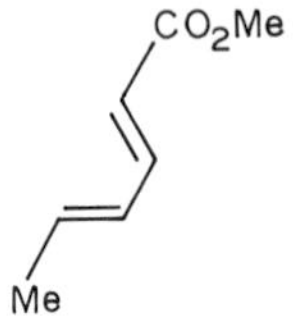

(182) 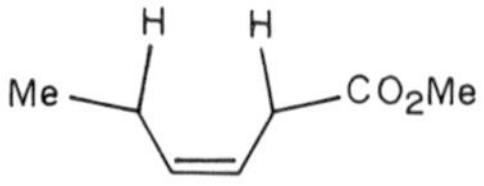 (183)

Ultraviolet irradiation has provided a simple method for the generation of active, coordinatively unsaturated catalysts for diene hydrogenation[472]. Irradiation of Group VI metal carbonyls (Cr, Mo, W) promotes the catalytic hydrogenation of 1, 3-dienes at room temperature[473]. (*Z*)-Alkenes are produced stereoselectively so that addition of deuterium to cyclohexa-1, 3-diene affords **184** specifically[474]. Again, an η^4-coordinated *cisoid* diene is an intermediate in these reductions[475,476].

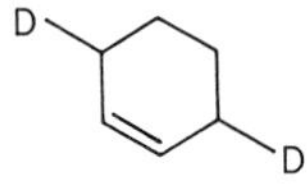

(184)

Pentacyabocobaltate(II) is a good catalyst for the selective hydrogenation of conjugated dienes to monoenes in aqueous or alcoholic solution[428]. The catalyst is relatively unreactive for non-conjugated dienes such as cycloocta-1, 5-diene. The catalyst is not very stereoselective and the product stereoisomer distribution is a function of solvent, catalyst ratios, and pH. Recently the catalyst has been used under phase-transfer conditions[469] and improved selectivities were observed. Using benzyltriethylammonium chloride as a phase-transfer catalyst, the reduction of **185** occurs stereospecifically to give the (*E*)-alkene **186**[469].

(185) 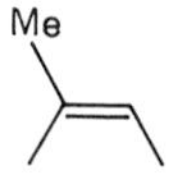 (186)

3. Alkynes

Despite the fact that many homogeneous catalysts will reduce alkynes to (*Z*)-alkenes, there have been very few reports of the application of such reductions in synthesis. Both $[RhCl(PPh_3)_3]$[439] and $[RuHCl(PPh_3)_3]$[477] will selectively reduce alkynes in alkene–alkyne mixtures but, as a result of the stronger binding of the alkyne to the metal, catalytic activities are reduced with respect to alkene hydrogenation. A related catalyst formed *in situ* from $RuCl_3$ and triphenylphosphine catalyses the reduction of 9-octadecynoic acid (**187**) to (*Z*)-oleic acid (**188**)[478], while $[RuHCl(PPh_3)_3]$ itself has been used to effect the conversion of **189** into the (*Z*)-alkene **190**[477].

$Me(CH_2)_7C{\equiv}C(CH_2)_7CO_2H$

(187)

$Me(CH_2)_7CH{=}CH(CH_2)_7CO_2H$

(188)

$AcOH_2CC{\equiv}CCH_2OAc$

(189)

AcO—CH₂CH=CHCH₂—OAc

(190)

The cationic rhodium complexes $[Rh(diene)L_2]^+$ and $[Rh(diene)Lpy]^+$ are particularly useful for the stereospecific *syn* reduction of internal and terminal alkynes in quantitative yield[79,80,479], although there have been no direct synthetic applications of this reduction. Indeed, despite the wealth of catalysts which reduce alkynes to (*Z*)-alkenes[480–491], heterogeneous catalysts are still commonly used preferentially in practice.

There have been isolated reports of the stereospecific *anti* reduction of alkynes to give an (*E*)-alkene[492,493], although in certain cases this may well be the consequence of an isomerization step involving the initially formed (*Z*)-alkene[148].

4. Carbonyls and unsaturated carbonyls

The carbon—oxygen double bond is a poor σ and π donor in organometallic chemistry and does not bind well to low-valent transition metal centres. Accordingly, there are few synthetically useful homogeneous catalysts for the reduction of these substrates, although recently several potentially very important asymmetric hydrogenations of the carbon—oxygen double bond have been reported (Section III.D). The most useful catalysts are cationic rhodium and iridium complexes typified by the series of procatalysts $[Rh(diene)L_2]^+$, although the use of the recently described palladium phthalocyanine catalyst merits further attention[427,494,495]. The iridium catalyst $[IrH_2(PPh_3)_2S_2]^+$ (S = solvent) in dioxane catalyses the reduction of butanone at 50 °C and under 1 atm of hydrogen[479]. The use of more basic tertiary phosphines such as PEt_3, PMe_2Ph, or $PBu^n{}_3$ has facilitated more general ketone reduction under mild conditions[496,497]. Such reductions proceed more rapidly in the presence of traces of water, suggesting that the active catalyst is a monohydride generated by proton loss. In the presence of water alkene hydrogenation is suppressed, although no selective reductions have been reported. For the reduction of aldehydes, competitive decarbonylation remains a serious problem, although this may be offset by working under higher partial pressures of hydrogen[497]. It seems that ketone reduction does not involve prior enolization and alkene coordination as deuterium addition gives the dideuteriated compound with no β-deuterium incorporation even in the presence of 1% water.

Other catalysts which have been reported to reduce carbon—oxygen double bonds include the cobalt catalyst $[Co(dmg)_2]$, which reduces 1,2-dicarbonyls and α-keto esters to the corresponding hydroxy carbonyls and α-hydroxy esters under ambient conditions[498]. Benzylic ketones are reduced to the corresponding alkanes using $[HCo(CO)_4]$ generated from $[Co_2(CO)_8]$ under hydrogen and carbon monoxide[499,500]. This reaction involves hydrogenolysis of the intermediate benzylic alcohol so that **191** may be successively reduced to **192**[499].

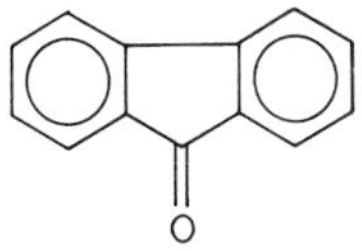

(191)

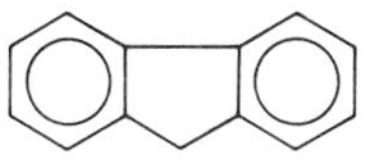

(192)

In $\alpha\beta$-unsaturated carbonyls the alkene double bond may be selectively reduced with impunity using most homogeneous catalysts. For example, reduction of **193** with $[Ir(cod)(PCy_3)py]^+$ in dichloromethane gives the *trans* ring fused product **194**, in which no carbonyl reduction has occurred[501]. However, whereas cinnamaldehyde (**195**) may be reduced to hydrocinnamaldehyde (**196**) using either $[RhCl(PPh_3)_3]$ or $[RhCl(CO)(PPh_3)_2]$ under oxo conditions[502], the use of dichlorotetracarbonyldirhodium(I) under similar conditions gives cinnamyl alcohol (**197**) in 94% yield[503].

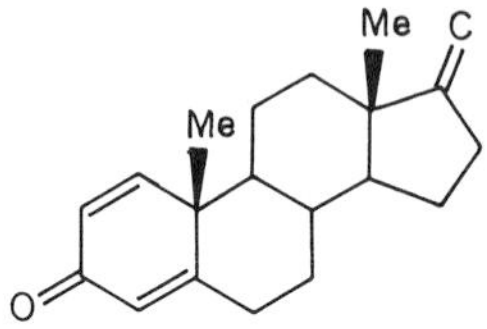

(**193**) 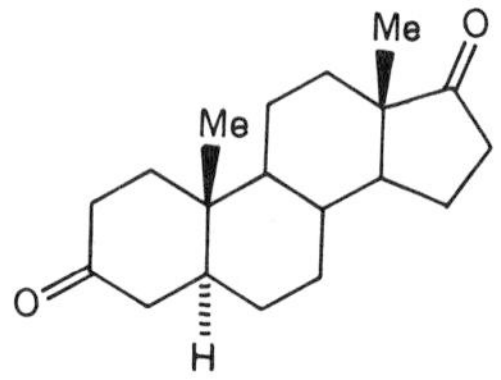 (**194**)

(*E*)–PhCH=CHCHO (**195**)

$PhCH_2CH_2CHO$ (**196**)

(*E*)–PhCH=$CHCH_2OH$ (**197**)

Selective reduction of the conjugated alkene in $\alpha\beta$-unsaturated carbonyl compounds that contain a remote carbon—carbon double bond has been achieved using triethylsilane and tris(triphenylphosphine)rhodium chloride, permitting the selective reduction of citral to citronellal in 97% yield[504]. The selective partial reduction of 3-oxo-1,4-diene steroids has been examined in detail[505–509]. The reduction of androsta-1, 4-diene-3, 17-dione (**193**) to androst-4-en-3, 17-dione (**198**) may be achieved using dichlorotris(triarylphosphine)ruthenium using low temperatures and high pressures in the presence of added triethylamine. Similarly, the reduction of α-santonin (**199**) using $[RhCl(PPh_3)_3]$ affords 1,2-dihydro-α-santonin (**200**) in high yield.

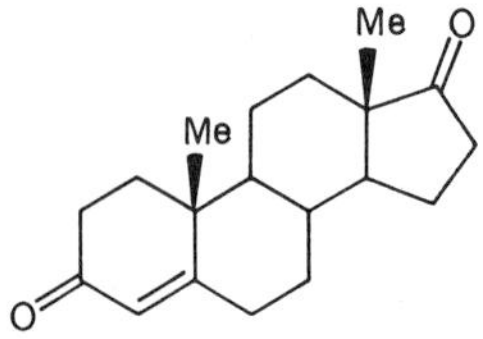

(**198**) 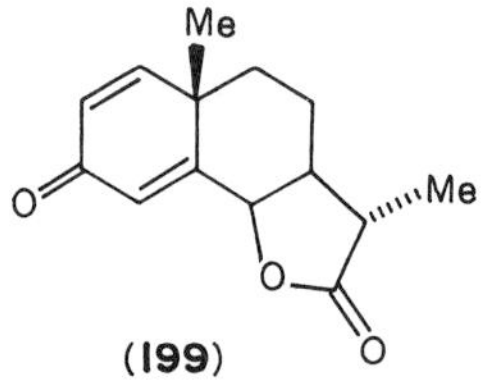 (**199**) 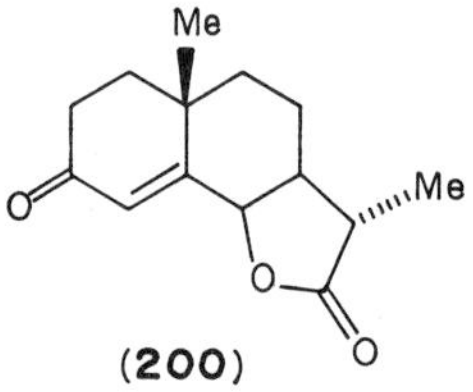 (**200**)

5. *Aromatics and heteroaromatics*

With the renewed interest in coal utilization, the hydrogenation of both polynuclear aromatic and heteroaromatics has become the subject of considerable interest, particularly as additional hydroprocessing is required to reduce the nitrogen and sulphur content[511–513]. Further, the traditional rhodium and ruthenium heterogeneous catalysts have limited use in synthetic organic chemistry because of their tendency to catalyse H/D exchange and their lack of stereospecificity. Despite these factors, homogeneous arene catalysts have so far not proved to be sufficiently reactive or stable to be used synthetically, despite some encouraging examples of stereoselectivity.

Among the earliest putative homogeneous arene catalysts were various Ziegler systems generated by the reduction of soluble metal complexes such as $[Ni(acac)_2]$ with metal hydrides or alkyls[514–519]. These systems require high pressure ($\geqslant$ 1000 psi) and elevated temperatures but they catalyse the reduction of dimethyl terephthalate (**201**) *syn* stereospecifically to give **202**. However, it has not been indisputably demonstrated that metastable metal particles are not involved in these systems. Another system which was investigated involved the use of $[HCo(CO)_4]$, which reduces anthracene derivatives via a two-step hydrogen atom transfer mechanism while leaving benzenoid and phenanthrenoid systems intact[520–522].

$MeO_2C-C_6H_4-CO_2Me$

(**201**)

H, CO_2Me, H, CO_2Me

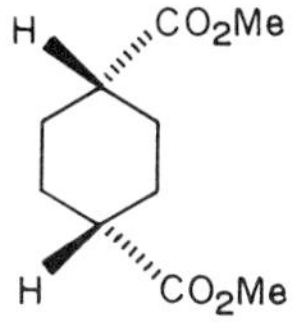

(**202**)

A highly active but poorly characterized rhodium(I) complex of *N*-phenyl anthranilate has been reported which catalyses the reduction of benzene to cyclohexane under ambient conditions and reduces anthracene to the 1, 2, 3, 4-tetrahydro product[523,524]. More recently, $[RuCl_2(CO)_2(PPh_3)_2]$ has been found to catalyse the reduction of both aromatic and heteroaromatics. In the presence of base, reduction of aromatics was enhanced although reduction of heteroaromatics was suppressed with base present[525]. The reduction of anthracene gives 1, 2, 3, 4-tetrahydroanthracene[526] in low yield, whereas reduction of benzoquinoline gave 1, 2, 3, 4-tetrahydro-5, 6-benzoquinoline **203** regiospecifically in high yield.

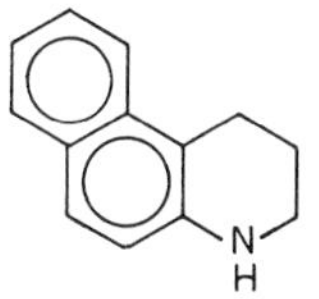

N
H

(**203**)

A series of allylcobalt phosphite complexes, for example $CoL_3(\eta^3\text{-}C_3H_5)$, have been intensively studied. Although these catalysts exhibit low activities toward arene reduction under ambient conditions, they are remarkably stereospecific[527–532]. Benzene may be deuteriated to give all-*cis*-$C_6D_6H_6$ with no competitive H/D exchange. Unfortunately, the low reactivity and limited catalyst lifetimes preclude the use of this catalyst in synthetic organic chemistry.

An air-stable procatalyst $[RhCl_2(C_5Me_5)]_2$ has been described which catalyses various arenes at 50 °C and 50 atm of hydrogen in the presence of base. Although the catalytic activity is fairly low, high stereospecificity is reported for the hydrogenation of benzene-d^6 to cyclohexane-d^6. Alkyl substituents retard the reduction rates but ether, ester, amide, and ketone groups are tolerated whereas hydroxy and carboxy groups suppress catalytic activity[533,534].

A series of η^6-arenemetal complexes have been described. $[(\eta^6\text{-}C_6Me_6)Ru(\eta^4\text{-}C_6Me_6)]$[535] is long-lived with moderate catalytic activity but the stereospecificity of reduction is low and extensive H/D exchange occurs[536]. It is significant, however, that substantial amounts of cyclohexenes may be isolated for alkylbenzene reductions[537]. The ruthenium hydride $[RuHCl(\eta^6\text{-}C_6Me_6)(PPh_3)]$ is reported to be a stable and long-lived arene hydrogenation catalyst, although slow decomposition of the catalyst occurs during hydrogenation[538]. Finally, $[(\eta^6\text{-}C_6Me_6)Ru(\mu\text{-}H)_2(\mu\text{-}Cl)Ru(\eta^6\text{-}C_6Me_6)]^+$ is claimed to be the most active arene catalyst. Hexadeuteriobenzene is hydrogenated to give more than

95% $C_6D_6H_6$ as an *E*/*Z* mixture of diastereoisomers. The catalyst is tolerant of carbomethoxy, hydroxy, and methoxy substituents and sulphur so that thiophene may be reduced[539,540].

6. Nitro groups, imines, and nitriles

Imines bind fairly well to transition metal centres, but such substrates have received little attention. Catalysts which have been reported to reduce imines include $[HCo(CN)_5]^{3-}$[542], $[RhCl_2(BH_4)dmf(py)_2]$[542] and the chiral chelating cationic complexes $[Rh(nbd)diop]^+$[543] and its ferrocenyldiphosphine analogue[544]. Dicobalt octacarbonyl under hydroformylation conditions reduces aryl- and alkyl-imines to the corresponding secondary amines in good yield[545–547], and no CO insertion to form the corresponding amides was observed.

Nitro groups are also reduced by several homogeneous catalysts but surprisingly this reaction has been little used synthetically. The hydrogenation of nitroalkanes and aromatic nitro compounds is catalysed by dichlorotris(triphenylphosphine)ruthenium(II)[548,549] to give the corresponding amines. At 135 °C and 80 atm of hydrogen this catalyst permits the selective reduction of only one of the nitro groups in *p*-dinitrobenzene to give *p*-nitroaniline.

The hydrogenation of aryl nitro compounds is also efficiently catalysed by $[RhCl_2(BH_4)dmf(py)_2]$[542] and pentacyanocobaltate(II) in aqueous solution[550]. The latter catalyst reductively dimerizes aryl nitro compounds to give azo and hydrazo derivatives.

The homogeneous hydrogenation of nitriles is virtually unknown, although dicobaltoctacarbonyl reduces aryl nitriles to the corresponding primary amines at elevated temperatures[551].

C. Heterogenized Homogeneous Hydrogenation Catalysis

A general drawback to the industrial utilization of homogeneous catalysts is the problem of separating the product from the catalyst and recovering the spent catalyst[552]. One commonly adopted solution to this problem is to anchor the homogeneous catalyst on a solid-phase support (see Chapter 14)[90–102]. In this way it was hoped that the advantages of soluble catalysts—high activity and stereoselectivity—could be combined with certain valuable characteristics of heterogeneous systems—ease of product separation and catalyst recovery and reuse. Indeed, it was hypothesized that such hybrid catalysts may even be superior to their homogeneous analogues in that catalyst aggregation and possible deactivation would be suppressed and even gas-phase processes made possible. In practice, such advantages have not been attained and many problems have been encountered with the preparation and characterization of the required catalysts, in addition to such drawbacks as leaching of metal ion into solution and reduced activities and selectivities[103–107]. Notwithstanding these problems, many important developments have been made and the technique may well become industrially significant in the not too distant future.

An alternative approach to the problem of product separation and catalyst recovery involves the use of metal complexes of polar, water-soluble ligands[108–111] such as sulphonated triarylphosphines. Aqueous solutions of these complexes catalyse reactions with water-insoluble substrates present in a second phase. Separation of the hydrogenated organic layer leaves the catalytic aqueous solution available for reuse[108]. Phase-transfer catalysis of such reactions looks particularly promising and has already been demonstrated with $[CoH(CN)_5]$ catalysts in water[469,553,554].

Within the same general domain some promising results have been obtained from hydrogenation within a bilayer as a method for controlling membrane fluidity. The degree of unsaturation of the fatty acid chains in the phospholipids of cell membranes determines the fluidity of their structure and hence modifies their physiological or biological action. Using sulphonated triarylphosphinerhodium complexes, selective hydrogenation of the polyunsaturated acyl chain has been effected and marked changes in membrane fluidity observed[555–559].

1. Polymer-supported catalysts

Many of the supported catalysts described have been attached to cross-linked polystyrene resins. Using microporous resins (prepared from *ca.* 2% divinylbenzene as a cross-linking agent) the interior is only accessible after swelling by a polar solvent, such as ethanol. Once swollen, the high internal mobility of the polymer means that the attached ligands which are widely dispersed along the polymer chain may coordinate with a single metal, but then tend to dimerize or aggregate reducing catalytic activity[560,561]. While macroporous resins are much less mobile internally they are equally much more difficult to functionalize and often the polymer may only be functionalized at the surface of the polymer bead or within the largest pores[27,562].

Some typical phosphine functionalized resins are **204**–**206**, with which metal complexes are formed by direct reaction with a metal halide or by ligand exchange reactions[90–98,563,564].

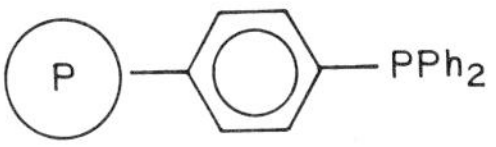

(**204**)

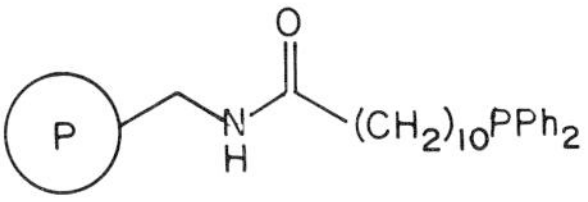

(**205**)

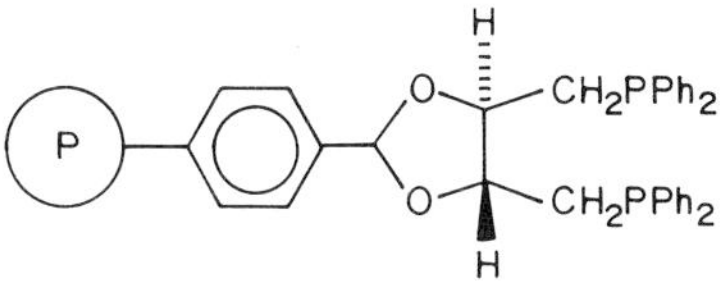

(**206**)

In most, but not all[563], of the examples described the catalytic activity of a polymer-bound hydrogenation catalyst is lower than its soluble counterpart[104,564,565]. With a ruthenium complex of a phosphinated polymer some selectivity for the reduction of short-chain over long-chain terminal alkenes has been reported[565]. There have been examples, however, where greater catalytic activity has been observed with highly cross-linked polystyrene resins. Using the titanium catalyst **207** the rate of alkene hydrogenation was claimed to be 100 times greater than that with the corresponding soluble catalyst[566]. In this case the enhanced activity occurs because the soluble catalyst tends to dimerize to form a catalytically inactive species and this reaction is inhibited with the supported metallocene system.

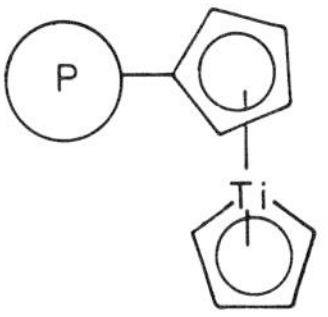

(**207**)

Other methods of producing functionalized polymers include copolymerization of a vinyl-substituted phosphine monomer or phosphine precursor, such as **208** or **209**, often with acrylate or acrylamide comonomers[567–570]. The phosphinated derivatives of **208** and **209** are both *cis*-chelating diphosphines, although there is some spectroscopic evidence with the **209**-derived system that, within the flexible polymer, the diphosphine is not *cis*-chelated and *trans*-phosphine coordination has been observed[571,572]. It seems clear that the supported chelating diphosphine catalysts may be superior to the monophosphine analogues with respect to stability and activity. The supported monophosphine catalysts are susceptible to reduction to metal(0) particles and also to catalyst leaching from the support[100,573,574,103–107]. Despite the enhanced stability of the diphosphine-derived catalysts some loss of metal may still occur[575].

PPh_2 CH_2PPh_2 N O

(208)

O O H H $TosOH_2C$ CH_2OTos

(209)

2. *Silica and zeolite functionalized catalysts*

Silica is less commonly used as a catalyst support owing to problems involving the functionalization and characterization of the heterogenized catalyst. However, it is intrinsically more attractive than polymeric supports as it has a high surface area, exhibits high thermal stability, and has a rigid structure which is not sensitive to solvent. The silica surface has free acidic hydroxy groups which may be esterified or silanized by a variety of methods, for example equations 27, 28, and 29[576,579–581]. The method illustrated in equation 27 generates two strong silicon—oxygen bonds and gives a more stable attachment to the surface. More recently a direct attachment of rhodium hydride complexes has been described involving a protolytic deposition of a soluble organo-rhodium complex, $[Rh(allyl)_3]$, on to the hydroxylated silica surface to form an oxidebound rhodium hydride complex (equation 30)[577,578].

$$\{Si\}\text{—}OH + (EtO)_3SiCH_2CH_2X \longrightarrow \{Si\}\text{—}O\text{—}\underset{\displaystyle OEt}{\overset{\displaystyle OEt}{\underset{|}{\overset{|}{Si}}}}\text{—}CH_2CH_2X \qquad (27)$$

$$X = SH,\ PPh_2,\ NR_2$$

$$\{Si\}\text{—}OH + [EtOSi(Me)_2CH_2CH_2PPh_2RhCl(cod)] \xrightarrow[\text{reflux}]{\text{PhMe}} [\{Si\}\text{—}O\text{—}SiMe_2CH_2CH_2PPh_2RhCl(cod)] \qquad (28)$$

$$\{Si\}—OH + HO(CH_2)_nX \longrightarrow \{Si\}—O(CH_2)_nX$$
$$X = SH, NR_2, PPh_2 \qquad (29)$$

$$\{Si\}—OH + [Rh(allyl)_3] \xrightarrow{\text{steps}} [\{Si\}—O—Rh(allyl)_2]$$
$$\xrightarrow{H_2} [\{Si\}—ORhH(allyl)] \qquad (30)$$

(210)

There have been few comparative studies between silica-bound complexes and their homogeneous counterparts. The anchored complex **211** was found to be as active for alkene hydrogenation as its soluble analogue. Being a chelated diphosphine complex, it showed a reduced tendency to form metal(0) particles compared with monophosphine analogues[583], as was observed with the polymer-supported systems[582]. The oxide-bound rhodium hydride complex **210** catalyses the reduction of alkenes and arenes. Benzene may be reduced to cyclohexane with a rate of seven turnovers per minute while naphthalene and aminobenzene were reduced more slowly[578].

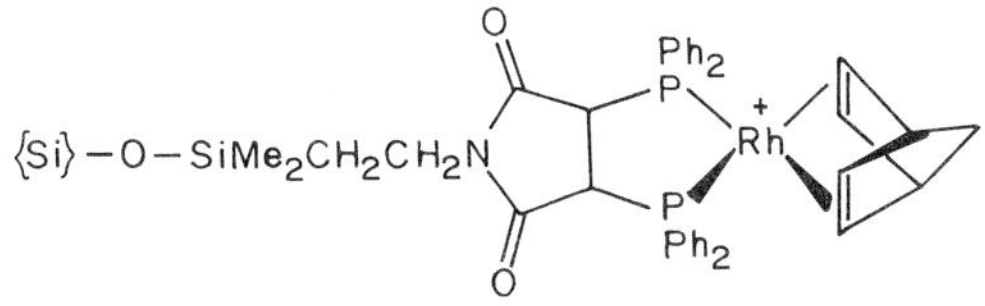

$\{Si\}-O-SiMe_2CH_2CH_2N$ (succinimide-bridged Ph_2P/PPh_2 chelate to Rh^+(norbornadiene))

(211)

Using the same deposition technique as for the formation of **210**, a partially proton-exchanged zeolite has been functionalized to give a zeolite-supported rhodium hydride, $[\{Z—X\}—ORhH(\eta^3\text{-allyl})]$[584,585]. Unlike the catalyst **210**, where the rate of alkene hydrogenation is a function of alkene substitution, the zeolite-supported catalyst exhibits high catalytic activity only for alkenes smaller in shape and size than cyclohexene. The 'molecular sieve' nature of the zeolite support precludes the hydrogenation of alkene substrates which are too bulky to pass through the crystalline channels to the catalyst sites within the zeolite cage.

Finally, it is worth noting that a cobalt carbonyl silica-supported catalyst has been described, **212**, which shows low activity for alkene hydrogenation under near-ultraviolet irradiation[586–588].

$$\{Si\}-O-Si(OEt)_2-Co(CO)_4$$

(212)

3. *Water-soluble catalyst systems*

Sulphonated triarylphosphine ligands are soluble in water and their metal complexes, such as $[\{(m\text{-}C_6H_4SO_3Na)PPh_2\}_3RhCl]$, may be used to effect two-phase reductions[108,589]. This complex is inactive in water alone, however, and fairly air-sensitive in two-phase systems. A more robust procatalyst for reductions in aqueous solution is **213**[111], which is equally as active for alkene hydrogenation as the triphenylphosphine analogue[80], may be used in aqueous solution for the hydrogenation of water-soluble alkenes such as maleic acid, and exhibits a reduced sensitivity to air in two-phase reductions. In such two-phase reductions no leaching of the metal complex from the aqueous solution was observed and the catalyst solution and product may be readily separated by decantation. The cationic complex **213** may also be adsorbed from solution on to a cation-exchange resin and the supported catalyst served to hydrogenate alkenes in acetone solution. The supported catalyst could be reused with little or no decrease in catalytic activity[111].

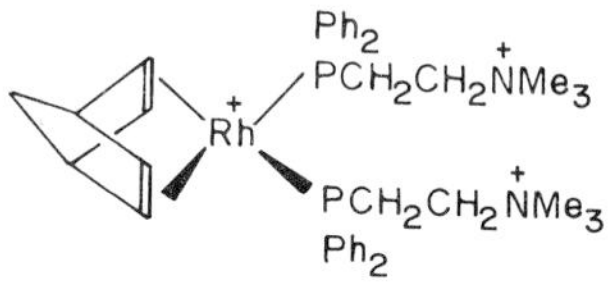

Ph_2 $PCH_2CH_2\overset{+}{N}Me_3$ $\overset{+}{Rh}$ $PCH_2CH_2\overset{+}{N}Me_3$ Ph_2

(213)

A series of chelating diphosphines which are soluble in water have been described. Cationic dienerhodium complexes of **214**, for example, rapidly hydrogenated enamide substrates under ambient conditions[590–593].

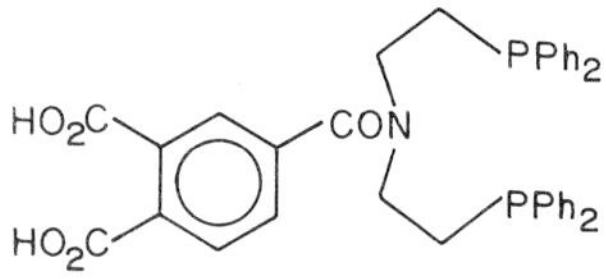

HO_2C HO_2C CON PPh_2 PPh_2

(214)

D. Dimetallic and Cluster Catalysis

Polynuclear complexes now constitute a new generation of organometallic complexes but their use as homogeneous catalysts is rare[594] and they have no synthetic utility at present. Moreover, there are only a handful of defined cluster-catalysed reactions[595–597,599–601], although molecular clusters often function as precursor complexes[5,598,602]. A more fruitful approach may be first to seek and define dinuclear catalysts which may permit new chemoselective and stereoselective reductions by virtue of cooperativity between the two metal centres. This line has been probed recently using the dinuclear complex $[(\mu\text{-H})Rh\{P(O\text{-}i\text{-}C_3H_7)_2\}_2]$, which is a procatalyst for the stereoselective hydrogenation of diarylalkynes and dialkylalkynes to the corresponding (*E*)-alkenes[117,118]. Unfortunately, the catalyst lifetimes were very short but the system described illustrates an important principle in dinuclear catalysis.

E. Catalytic Asymmetric Hydrogenation

The enantioselective catalytic hydrogenation of prochiral substrates using chiral catalysts has been the subject of intense research effort over the past 10 years[83–85,603–606]. Following the discovery of Wilkinson's catalyst, $[RhCl(PPh_3)_3]$, and the related cationic rhodium catalysts of Osborn, $[Rh(diene)(PR_3)_2]^+$, it was quickly realized that if chiral phosphines were used then the enantioselective hydrogenation of prochiral substrates was possible[607]. Initial results with simple resolved phosphines such as $PhPr^iMeP$ were disappointing and gave little selectivity and low rates[608]. It was soon apparent that several conditions needed to be fulfilled in order that high enantioselectivity could be attained. Firstly, the substrate (usually an alkene) needed to possess additional polar functional groups such as NHCOMe, OAc, or CO_2^- which could bind to the metal centre. Secondly, cationic rhodium procatalysts were found to be superior to neutral complexes, giving faster rates of reduction. Finally, the chiral phosphine should preferably be a chiral chelating diphosphine which binds tightly to the metal to give a rigid chelate. With these factors in mind, it is possible to achieve close to 100% enantioselectivity for the reduction of (*Z*)-dehydroamino acid derivatives such as **215**[12], up to 95% induction for the hydrogenation of carbonyl groups, and up to 90% selectivity for the reduction of various unsaturated monocarboxylic acids such as itaconic acid (**216**).

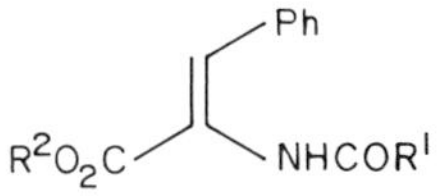

(**215**)

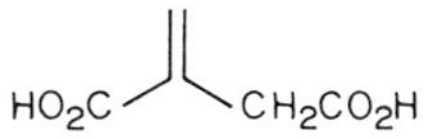

(**216**)

The most successful procatalysts are cationic dienerhodium complexes of chiral chelating diphosphines. Some of the more useful and versatile diphosphines are **217**–**222** their commonly used acronyms being given under the formulae. The number of useful chiral diphosphines is now over 100 and new phosphines continue to be reported[609–617].

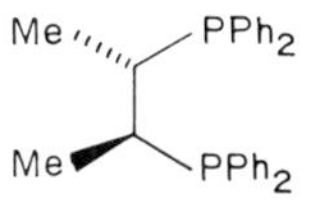

(*S*,*S*)– chiraphos

(**217**)

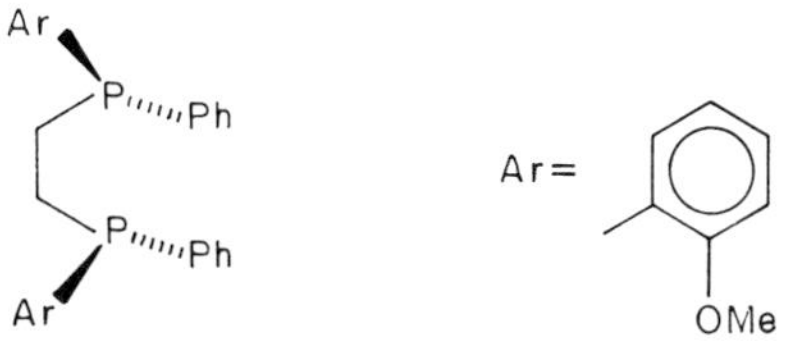

(*R*,*R*)– dipamp

(**218**)

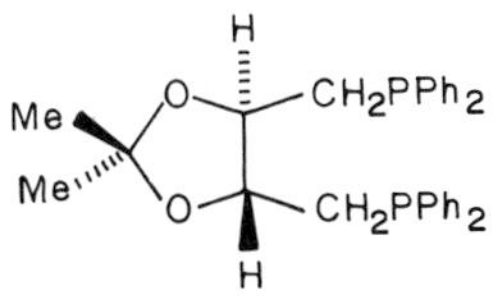

(*R*,*R*)– diop

(**219**)

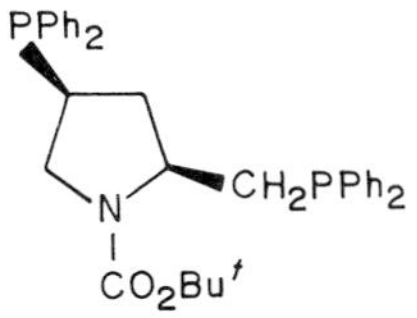

(*S*,*S*)– bppm

(**220**)

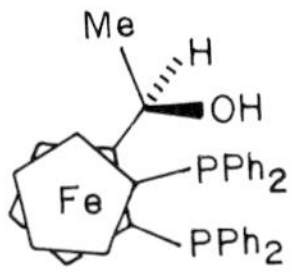

(*R*,*S*)–bppfoh

(**221**)

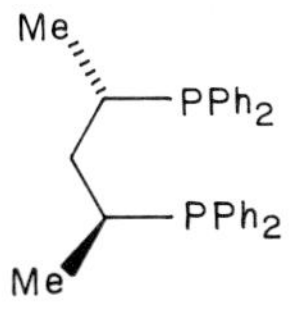

(*S*,*S*)– skewphos

(**222**)

TABLE 10. Asymmetric hydrogenation of typical enamide substrates

Substrate	Chiral phosphine	Product enantiomeric purity (%)	Ref.
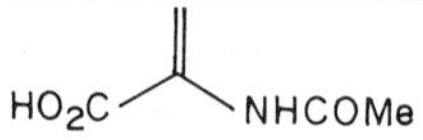 HO_2C NHCOMe	(*S*, *S*)-bppm	98 *R*	621
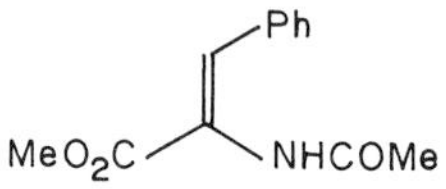 Ph MeO_2C NHCOMe	(*R*, *R*)-dipamp	96 *S*	618
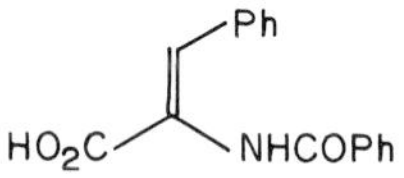 Ph HO_2C NHCOPh	(*R*, *R*)-chiraphos	95 *S*	619
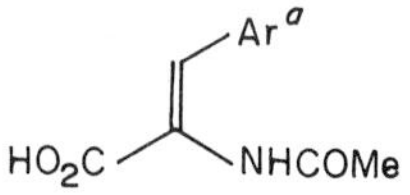 Ar[a] HO_2C NHCOMe	(*S*, *S*)-skewphos	92 *R*	615
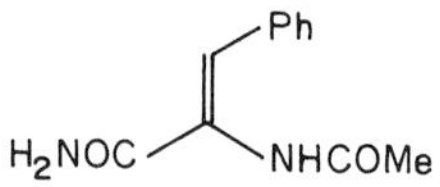 Ph H_2NOC NHCOMe	(*R*, *R*)-diop	71 *R*	620

[a] Ar = 3-methoxy-4-acetoxyphenyl.

The most useful substrates for these reductions are enamides and dehydroamino acids have been often used to test the utility of a new diphosphine. Hydrogenation yields α-amino acid derivatives and some representative examples are shown in Table 10. The hydrogenation of **223** has been investigated using rhodium complexes of **218** in some detail as the derived product, **224**, is L-dopa which is manufactured by Monsanto for the treatment of Parkinson's disease. A synthesis of the potential sweetening agent (*R*)-6-methyltryptophan (**225**) employs enantioselective hydrogenation of the (*Z*)-enamide precursor **226** in the key step[630].

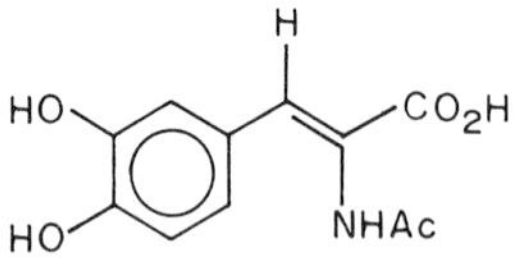

HO, HO, H, CO_2H, NHAc

(223)

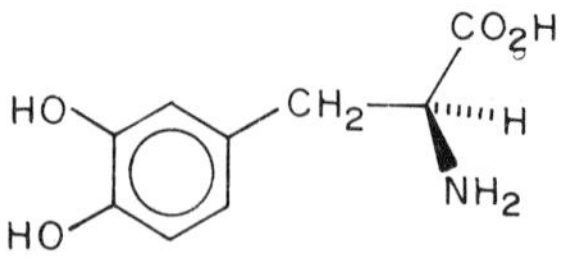

HO, HO, CH_2, CO_2H, H, NH_2

(224)

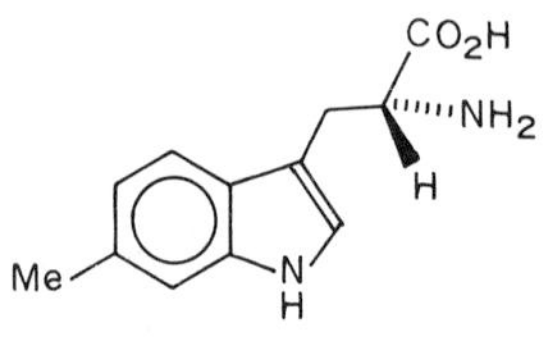

Me, N, H, CO_2H, NH_2, H

(225)

Me, N, H, H, CO_2H, NHAc

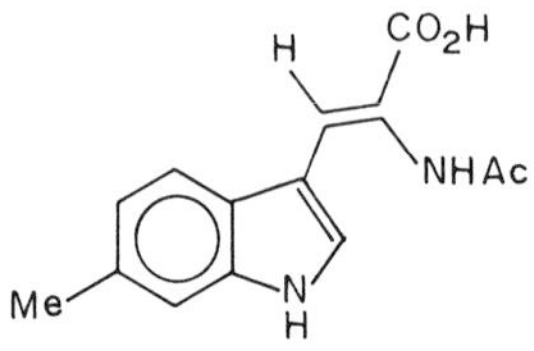

(226)

Five-membered chelate-ring complexes [such as those involved with dipamp (**218**) and chiraphos (**217**)] have not been shown to be successful over a broad range of substrates. Simple unsaturated acids, functioning as bidentate substrates, are reduced in fairly low optical yield although itaconic acid (**216**) and its derivatives may be reduced in fairly high optical yields[628]. Enol esters and enol phosphates possess the same relative disposition of alkene and carbonyl as enamides and these have been reduced with good enantioselectivity using dipamp complexes[629,631]. The hydrogenation of enol acetates has been used for the generation of chiral methyl chiral lactic acid. Using a rhodium complex of **227**, deuteriation of **228** was effected to give **229** in 81% enantiomeric purity[632]. Hydrolysis of **229** and product recrystallization gave optically pure (*S*, *S*)-chiral methyl chiral lactic acid (**230**). This is an important precursor for several labelled molecules and may be used to establish the stereospecificity of enzymic pathways[633].

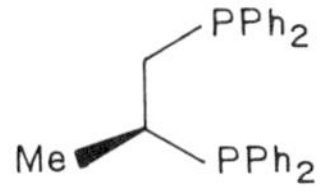

(**227**)

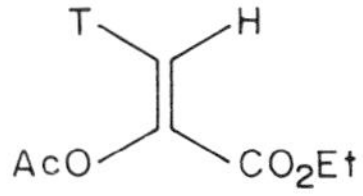

(**228**)

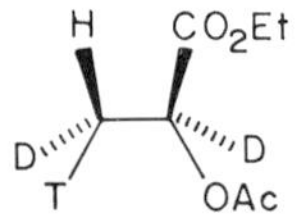

(**229**)

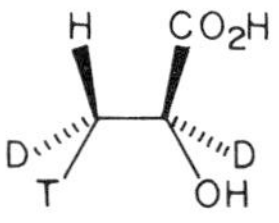

(**230**)

Larger ring chelate complexes, such as those of diop (**219**), bppm (**220**) and bppfoh (**221**), have proved to be more effective at hydrogenating non-enamide substrates. Some representative examples are given in Table 11. Using the ferrocenyl-derived diphosphine bppfoh (**221**), asymmetric hydrogenation of amino ketones proceeds with high enantioselectivity[626]. In this way β-amino alcohols, which function as adrenergic and cardiac stimulants, may be prepared in high optical purity (entry 2, Table 11). The chiral diphosphine bppm (**220**) has proved particularly versatile at hydrogenating non-enamide substrates with high enantioselectivity. The hydrogenation of α-keto esters gives α-hydroxy esters in over 70% enantiomeric purity and (*R*)-($-$)-pantolactone has been obtained in 87% optical yield by hydrogenation of the carbon—oxygen double bond (entries 6 and 5, Table 11.)

The asymmetric addition of deuterium to propenoic acid catalysed by cationic chiraphos–rhodium complexes provides a route to chiral α-^{2}H-propanoic acid (entry 7, Table 11). Indeed, the hydrogenation of $\alpha\beta$-unsaturated carboxylic acids has been studied in detail[627,628]. Using diop–rhodium complexes simple substrates may be reduced in up to 70% optical yield (entry 4, Table 11) and reduction of **231** gives the corresponding saturated acid in 88% enantiomeric purity[635].

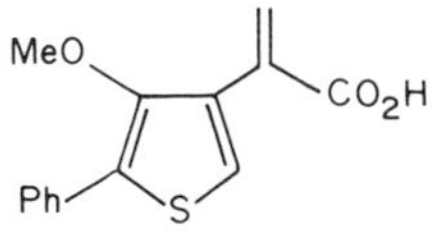

(**231**)

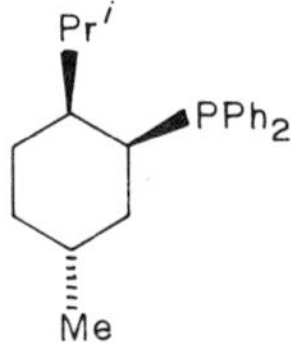

(**232**)

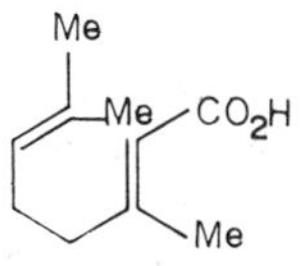

(**233**)

TABLE 11. Asymmetric hydrogenation of non-enamide substrates

Entry	Substrate	Chiral phosphine	Product enantiomeric purity (%)	Ref.
1	$HO_2C-C(=CH_2)-CH_2-CO_2H$	(*S*, *S*)-bppm	95 *R*	622
2	3,4-$(HO)_2C_6H_3-CO-CH_2\overset{+}{N}H_3$	(*R*, *S*)-bppfoh	95 *R*[a]	625, 626
3	$EtO_2C-C(=CH_2)-OCOMe$	(*R*, *R*)-dipamp	89 *R*	629
4	$Ph-C(=CH_2)-CO_2^-NMe_4^+$	(*R*, *R*)-diop	68 *S*	627
5	3,3-dimethyl-dihydrofuran-2,4-dione (Me, Me; O; O; O)	(*S*, *S*)-bppm	87 *R*	623
6	$Me-CO-CO_2Me$	(*S*, *S*)-bppm	76 *R*	624
7	$CH_2=CH-CO_2H$	(*S*, *S*)-chiraphos	58 *R*[b]	441, 440
8	$MeO_2C-C(=CH_2)-CH_2-CO_2Me$	(*R*, *R*)-dipamp	88 *R*	628

[a] In presence of triethylamine.
[b] Addition of deuterium.

Chiral monophosphines such as neomenthyldiphenylphosphine (**232**)[636] have also proved effective for the hydrogenation of $\alpha\beta$-unsaturated carboxylic acids[637]. Hydrogenation of 3, 7-dimethylocta-2,6-dienoic acid (**233**) with rhodium complexes of **232** selectively reduces the $\alpha\beta$-unsaturated double bond in 70% enantiomeric purity. Such reductions have been used in synthetic routes to chiral dihydrogeranic acid and other intermediates of importance in the synthesis of chiral vitamin E and citronellal[637].

In conclusion, asymmetric hydrogenation has proved to be one of the more successfully applied catalytic homogeneous reactions in synthetic organic chemistry. Although the reduction of dehydroamino acid substrates has been most intensively studied, other substrates such as $\alpha\beta$-unsaturated acids, α-keto esters, and β-amino ketones may also be reduced with high selectivity. Further, this subject has encouraged some very detailed mechanistic studies[7,11,638–643] of the intimate reaction pathway which have undoubtedly deepened our knowledge of some of the fundamental steps in homogeneous catalytic processes.

IV. CATALYTIC TRANSFER HYDROGENATION

A. Introduction

There is an important class of metal-catalysed hydrogenations which do not use molecular hydrogen[119]. The catalytic transfer hydrogenation process uses as a source of hydrogen organic molecules of relatively low oxidation potential in the presence of either homogeneous[120–123] or heterogeneous catalysts[124–126]. Hydrogen transfer to an organic substrate may thus be smoothly effected under ambient conditions (equation 31). Although water and alcohols (most often propan-2-ol) are the commonest sources of hydrogen[126], cyclohexene is also often used[127] and the ammonium and sodium salts of formic and phosphinic acid are becoming increasingly popular[128,129].

$$DH_n + A \xrightarrow[\text{solvent}]{\text{catalyst}} DH_{n-2} + AH_2 \qquad (31)$$

B. Heterogeneous Transfer Hydrogenation

Palladium on carbon or palladium black is the most commonly used catalyst for these reductions[124,125,654–647], although nickel catalysts are also often employed[648,649]. The reduction of alkenes and alkynes proceeds with *syn* stereospecificity. Compound **234** may be reduced selectively to the *endo* isomer **235** using 10% palladium on carbon and 1-methylcyclohexene as a hydrogen donor[647], and reduction of the alkyne **236** using sodium phosphinate with a mercury-modified palladium catalyst gave the (*Z*)-alkene **237**[650]. This reducing system was also found to reduce nitro groups faster than alkynes.

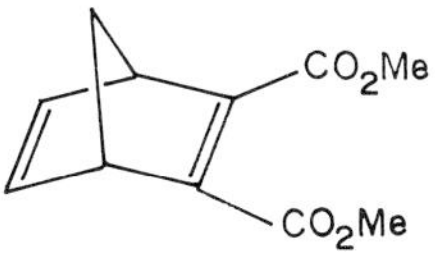

CO_2Me CO_2Me

(234)

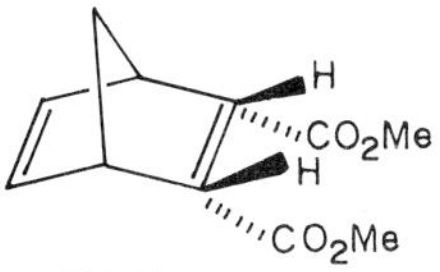

H CO_2Me H CO_2Me

(235)

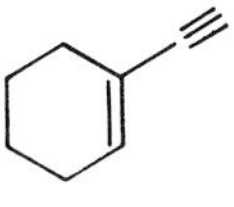

(236)

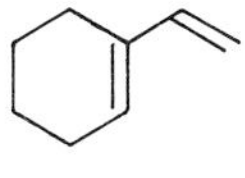

(237)

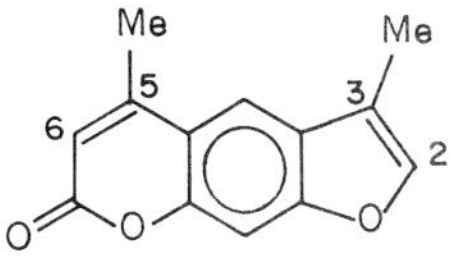

Me Me 5 3 6 2 O O O

(238)

In the furocoumarin derivative **238**, the 2, 3-double bond is selectively reduced using a palladium on carbon catalyst and cyclohexene[127,651], whereas a standard palladium-catalysed reduction of **238** using hydrogen also reduced the 5,6-double bond. The 2, 3-

double bond in various tryptophan derivatives may also be smoothly reduced using formic acid and palladium black[652].

The hydrogenolytic cleavage of benzylic carbon—oxygen bonds using transfer hydrogenation is commonly used. Using ammonium formate as the *in situ* source of hydrogen, the simultaneous deprotection and release of a pentadecapeptide analogue of ACTH from a Merrifield polystyrene resin has been effected[129]. Similarly, in the final deblocking step of another solid-phase peptide synthesis, hydrogenolytic cleavage of the benzylic carbon—oxygen bond in **239** was carried out using cyclohexadiene and a palladium on carbon catalyst[653,654].

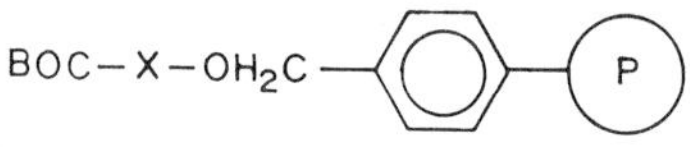

BOC—X—OH_2C—(C₆H₄)—(P)

(239)

In carbohydrate chemistry the selective cleavage of benzylic ethers in the presence of a benzylidene acetal has been reported using cyclohexene and 20% $Pd(OH)_2$ or carbon[655]. The benzylic ethers of several carbohydrates have also been cleaved at 25 °C with formic acid and palladium on carbon in methanol solvent[128,645].

C. Homogeneous Transfer Hydrogenation

A large number of homogeneous metal-catalysed transfer hydrogenations have been reported[656]. The most commonly used catalysts are complexes of ruthenium[120,121], rhodium[126,657–659], and iridium[660–663]. Of particular practical importance are the Henbest catalysts, in which an iridium salt is used with trimethyl phosphite under acidic conditions and with propan-2-ol as the source of hydrogen. This catalytic system has found some application for the stereospecific reduction of cyclic ketones to give axial alcohols. Simple cyclohexanones are reduced to the corresponding axial alcohols in good yield and with greater than 95% stereoselectivity. For example, reduction of **240** gives 97% of **241** and only 3% of the epimeric equatorial alcohol[662]. With steroidal substrates the 2-keto group may be selectively reduced in the presence of 11-, 17-, or 20-keto groups, and reduction of **242** gives the axial alcohol **243**[664].

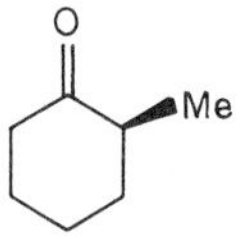

O Me

(240)

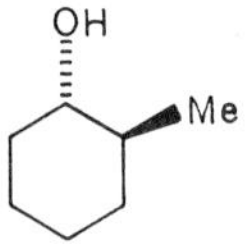

OH Me

(241)

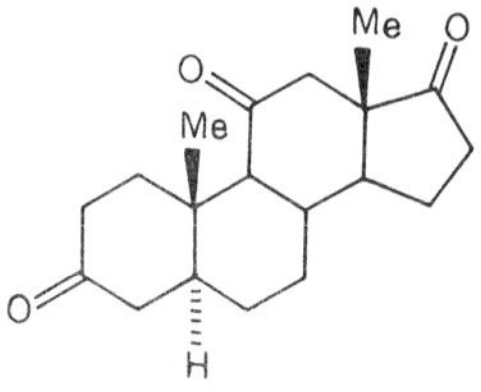

Me O O Me O H

(242)

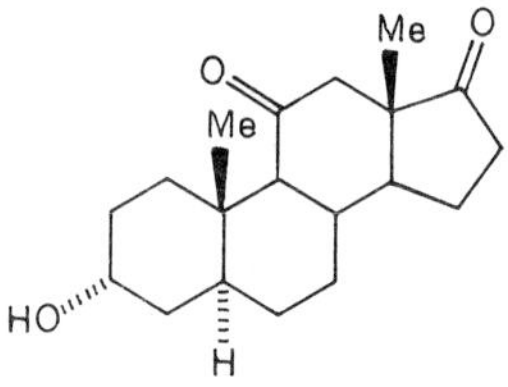

Me O O Me HO H

(243)

Hydrogen transfer with homogeneous catalysts is most common with substrates containing carbonyl groups, although transfer hydrogenation of alkenes does occur with, for example, $[IrClCO(PPh_3)_2]$[660,661]. The activity of this catalyst with formic acid as a hydrogen donor apparently increases when the catalyst is anchored to a diphenylphosphinated polystyrene resin. It is reported that the catalyst is leach-proof, air-stable and capable of an unlimited number of catalytic cycles[660]. Finally, the asymmetric transfer hydrogenation of prochiral alkenes[665] and ketones[126] has been reported using chiral phosphine complexes or chiral rhodium–imine complexes[126] with alcohol donors, but optical yields were low.

V. CONCLUSION AND OUTLOOK

There continues to be an intense research effort towards the development of more selective catalysts which may operate under mild conditions within both homogeneous and heterogeneous systems. The selective hydrogenation of unsaturated alkenes and arenes remains of considerable commercial importance with the petrochemical and fine chemical industries. Although many heterogeneous catalysts are well developed and their synthetic utility has been clearly defined, the mechanism of heterogeneous catalysis remains ill-defined at the molecular level. On the other hand, the intimate details of the reaction pathway of several homogeneous catalysts are now well understood, for example with $[RhCl(PPh_3)_3]$, $[Rh(diene)(PR_3)_2]^+$, and $[Irpy(cod)(PCy_3)]^+$, although their synthetic utility needs to be further developed, perhaps with the exception of $[RhCl(PPh_3)_3]$.

Some of the key problems which remain to be solved will involve the development and design of further chemoselective and stereospecific catalysts. The partial reduction of arenes to cyclic alkenes is one notable goal which is important to the polymer industry. Another key objective is the regioselective reduction of alkenic double bonds in unsaturated fats and oils, which continues to be the focus of much effort within the food and margarine industries. The development of practical catalysts for the selective reduction of carbonyls, nitro groups, and arenes in the presence of alkenes and alkynes is another target of interest to synthetic chemists.

It is worth noting, perhaps, some of the more important developments which have taken place recently. One of the major triumphs of homogeneous catalysis has been the application of the rhodium and iridium catalysts, notably $[RhCl(PPh_3)_3]$ and $[Rh(diene)(PR_3)_2]^+$, to synthetic organic chemistry. This has culminated in the development of asymmetric hydrogenation catalysts which can now operate in up to 100% enantioselectivity. Only the high cost of these complexes and problems in recovering and reusing the complexes have hindered their further usage in synthetic and industrial processes. With this in mind, the development of cheaper homogeneous catalysts based on cobalt and nickel complexes merits further study. Similarly, the use of water-soluble complexes for use in two-phase or phase-transfer catalytic systems looks promising and there have been some promising advances in anchoring many homogeneous catalysts to inert supports facilitating catalyst recovery and product isolation.

With the increasing interest in asymmetric synthesis that pervades most branches of organic chemistry, pharmacology, and microbiology, the search for better and more efficient enantioselective and diastereoselective catalysts is bound to continue. In addition to asymmetric homogeneous hydrogenation, the important development of chiral modified Raney nickel catalysts for the enantioselective heterogeneous hydrogenation of α-keto esters and β-diketones is still little appreciated and seldom applied, and merits wider attention. Similarly, it is only within the last few years that 'stereocontrolled' hydrogenations directed by a remote polar substituent have been explored. Such selective reductions may be expected to be more widely applied to the synthesis of important chiral natural products or pharmaceuticals in order to avoid more clumsy stoichiometric procedures or tedious stereoisomer separations or resolutions.

VI. REFERENCES

1. M. Freifelder, *Catalytic Hydrogenation in Organic Synthesis*, Wiley, New York, 1978.
2. P. N. Rylander, *Catalytic Hydrogenation in Organic Synthesis*, Academic Press, New York, 1979.
3. A. J. Birch and D. H. Williamson, *Org. React.*, **24**, 1 (1976).
4. R. L. Augustine, *Catalytic Hydrogenation*, Arnold, London, 1965.
5. B. R. James, in *Comprehensive Organometallic Chemistry* (Eds. F. G. A. Stone and E. W. Abel), Vol. 8, Pergamon Press, Oxford, 1982, pp. 285–353.
6. Science Research Council, *Report of the Organometallic Chemistry Panel*, H. M. Stationery Office, London, November 1968.
7. J. Halpern, *Inorg. Chim. Acta*, **50**, 11 (1981).
8. J. M. Brown, P. A. Chaloner, and D. Parker, *Adv. Chem. Ser.*, No. 196, 355 (1981).
9. R. R. Burch, A. J. Shusterman, E. L. Muetterties, R. G. Teller, and J. M. Williams, *J. Am. Chem. Soc.*, **105**, 3546 (1983).
10. R. H. Crabtree, P. C. Demou, D. Eden, J. M. Mihelcic, C. A. Parnell, J. M. Quirke, and C. E. Morris, *J. Am. Chem. Soc.*, **104**, 6994 (1982).
11. A. S. C. Chan, J. J. Pluth, and J. Halpern, *J. Am. Chem. Soc.*, **102**, 5952 (1980).
12. J. M. Brown, P. A. Chaloner, B. A. Murrer, and D. Parker, *Am. Chem. Soc. Symp. Ser.*, **119**, 169 (1980).
13. W. E. Spicer, I. Lindau, and C. R. Helms, *Res. Dev.*, **28**, (12), 20 (1977).
14. G. A. Somorjai, *Surf. Sci.*, **89**, 496 (1979).
15. A. T. Hubbard, *Acc. Chem. Res.*, **13**, 177 (1980).
16. R. J. Kokes, *Acc. Chem. Res.*, **6**, 226 (1973).
17. B. K. Teo, *Acc. Chem. Res.*, **13**, 412 (1980).
18. M. Boudart, *Chem. Eng. Prog.*, **57**, (8), 33 (1961).
19. A. P. G. Kieboom, *Bull. Chem. Soc. Jpn.*, **49**, 331 (1976).
20. I. Mochida and Y. Yoneda, *J. Catal.*, **11**, 183 (1968).
21. M. Kraus, *Adv. Catal.*, **17**, 75 (1967).
22. A. S. Hussey and G. P. Nowack, *J. Org. Chem.*, **34**, 439 (1969).
23. L. Cerveny, A. Prochazka, and V. Ruzicka, *Collect. Czech. Chem. Commun.*, **39**, 2463 (1974).
24. I. Horiuti and M. Polyani, *Trans. Faraday Soc.*, **30**, 1164 (1934).
25. B. R. James, *Homogeneous Hydrogenation*, Wiley, New York, 1973.
26. B. R. James, *Adv. Organomet. Chem.*, **17**, 319 (1979).
27. F. R. Hartley and P. N. Vezey, *Adv. Organomet. Chem.*, **15**, 189 (1977).
28. A. P. G. Kieboom and F. van Rantwijk, *Hydrogenation and Hydrogenolysis in Synthetic Organic Chemistry*, Delft University Press, Delft, 1977.
29. F. J. McQuillin, *Homogeneous Hydrogenation in Organic Chemistry*, Reidel, Dordrecht, 1976.
30. F. J. McQuillin, *Prog. Org. Chem.*, **8**, 314 (1973).
31. P. Sabatier and J. B. Senderens, *C. R. Acad. Sci.*, **124**, 1358 (1897).
32. M. Raney, *US Pat.*, 1563787 (1925).
33. M. Raney, *US Pat.*, 1 628 191 (1927).
34. H. Adkins and A. A. Pavlic, *J. Am. Chem. Soc.*, **69**, 3039 (1947).
35. L. W. Covert and H. A. Adkins, *J. Am. Chem. Soc.*, **54**, 4116 (1932).
36. H. Adkins and H. R. Billica, *J. Am. Chem. Soc.*, **70**, 695 (1948).
37. H. Adkins and H. R. Billica, *Org. Synth.*, Coll. Vol. 3, 176 (1955).
38. R. Mozingo, *Org. Synth.*, **21**, 15 (1941).
39. R. Mozingo, *Org. Synth.*, Coll. Vol. 3, 181 (1955).
40. F. A. Dufau, F. Eschard, A. L. Haddad, and C. H. Thonon, *Chem. Age*, **17**, 417 (1966).
41. H. Adkins and R. Connor, *J. Am. Chem. Soc.*, **53**, 1091 (1931).
42. G. Grundmann, in *Newer Methods of Preparative Organic Chemistry*, (ed. W. Foerst) Interscience, New York, 1948, p. 103.
43. K. Mitsui, *Mem. Coll. Sci. Kyoto Imp. Univ., Ser. A*, **18**, 329 (1935); *Chem. Abstr.*, **30**, 2930 (1936).
44. R. Mozingo and K. Folkers, *J. Am. Chem. Soc.*, **70**, 229 (1948).
45. R. Adams and J. R. Marshall, *J. Am. Chem. Soc.*, **50**, 1970 (1928).

46. R. Adams and F. L. Cohen, *Org. Synth.*, Coll. Vol. 1, 246 (1944).
47. H. Lindlar, *Helv. Chim. Acta*, **35**, 446 (1952).
48. G. Gilman and G. Cohn, *Adv. Catal.*, **9**, 733 (1957).
49. H. A. Smith and R. G. Thompson, *Adv. Catal.*, **9**, 727 (1957).
50. R. E. Ireland and P. W. Schiess, *J. Org. Chem.*, **28**, 6 (1963).
51. A. E. Barkdoll, D. C. England, H. W. Gray, W. Kirk, and G. M. Whitman, *J. Am. Chem. Soc.*, **75**, 1156 (1953).
52. I. Motayama, *Bull. Chem. Soc. Jpn.*, **33**, 232 (1960).
53. Y. Urushibara, *Ann. N. Y. Acad. Sci.*, **145**, 52 (1967).
54. K. Hata, I. Motoyama, and K. Sakai, *Org. Prep. Proced. Int.*, **4**, 180 (1972).
55. M. Shiota, *Stem Chem. III*, **1**, 13 (1975).
56. R. Paul, P. Buisson, and N. Joseph, *Ind. Eng. Chem.*, **44**, 1006 (1952).
57. H. C. Brown and C. A. Brown, *J. Am. Chem. Soc.*, **85**, 1003, 1005 (1963).
58. C. A. Brown, *J. Org. Chem.*, **35**, 1900 (1970).
59. V. K. Ahiya and C. A. Brown, *J. Chem. Soc., Chem. Commun.*, 553 (1973).
60. T. Isoda, A. Ichikawa, and T. Shimamoto, *Rikagaku Kenkyusho Hokoku*, **34**, 134 (1958); *Chem. Abstr.*, **54**, 287d (1960).
61. Y. Ohto, S. Imai, and S. Nuva, *Yuki Gosei Kagaku Kyokai Shi*, **35**, 753 (1977); *Chem. Abstr.*, **88**, 37219w (1978).
62. K. Ito, T. Harada, A. Tai, and Y. Izumi, *Chem. Lett.*, 1049 (1979).
63. A. Tai, T. Harada, Y. Hiraki, and S. Murakami, *Bull. Chem. Soc. Jpn.*, **56**, 1414 (1983).
64. M. Nakahata, M. Imaida, H. Ozaki, T. Harada, and A. Tai, *Bull. Chem. Soc. Jpn.*, **55**, 2186 (1982).
65. E. I. Klabunovskii, N. P. Sokolva, Y. I. Petrov, and V. V. Patrikeev, *Izv. Akad. Nauk SSSR, Ser, Khim.*, 838 (1983).
66. M. Calvin, *Trans. Faraday Soc.*, **34**, 1181 (1938).
67. M. Calvin, *J. Am. Chem. Soc.*, **61**, 2230 (1939).
68. M. Iguchi, *J. Chem. Soc. Jpn.*, **63**, 634, 1752 (1942).
69. J. Kwaitek and J. K. Seyler, *Adv. Chem. Ser.*, No. 70, 207 (1968).
70. H. D. Kaesz and R. B. Sailliant, *Chem. Rev.*, **72**, 231 (1972).
71. J. Halpern, J. F. Harrod, and B. R. James, *J. Am. Chem. Soc.*, **83**, 753 (1961).
72. J. Halpern, *Abstracts, 141st Meeting of the American Chemical Society, Washington*, 1962, p. 10Q.
73. J. Halpern and B. R. James, *Abstracts, 142nd Meeting of the American Chemical Society, Atlantic City, 1962*, p. 23N.
74. R. D. Cramer, E. L. Jenner, R. V. Lindsey, and U. G. Stolberg, *J. Am. Chem. Soc.*, **85**, 1691 (1963).
75. F. H. Jardine, J. A. Osborn, G. Wilkinson, and J. F. Young, *Chem. Ind.* (*London*), 560 (1965).
76. J. F. Young, J. A. Osborn, F. H. Jardine, and G. Wilkinson, *Chem. Commun.*, 131 (1965).
77. R. S. Coffey, to Imperial Chemical Industries, *Br. Pat.*, 1121642 (1965).
78. D. Evans, J. A. Osborn, F. H. Jardine, and G. Wilkinson, *Nature* (*London*), **208**, 1203 (1965).
79. R. R. Schrock and J. A. Osborn, *J. Am. Chem. Soc.*, **93**, 2397 (1971).
80. R. R. Schrock and J. A. Osborn, *J. Am. Chem. Soc.*, **98**, 2134 (1976).
81. J. M. Brown, P. A. Chaloner, and P. N. Nicholson, *J. Chem. Soc., Chem. Commun.*, 646 (1978).
82. J. M. Brown, P. A. Chaloner, A. G. Kent, B. A. Murrer, P. N. Nicholson, and D. Parker, *J. Organomet. Chem.*, **216**, 263 (1981).
83. M. D. Fryzuk and B. Bosnich, *J. Am. Chem. Soc.*, **99**, 6262 (1977).
84. W. S. Knowles, M. J. Sabacky, and B. D. Vineyard, *Adv. Chem. Ser.*, No. 132 274 (1974).
85. W. S. Knowles, M. J. Sabacky, and B. D. Vineyard, *US Pat.*, 4005127 (1977).
86. R. H. Crabtree, H. Felkin, T. Khan, and G. E. Morris, *J. Organomet. Chem.*, **144**, C15 (1978).
87. R. H. Crabtree, *Acc. Chem. Res.*, **12**, 331 (1979).
88. L. A. Oro, J. A. Cabeza, C. Cativiela, and E. Melendez, *J. Chem. Soc., Chem. Commun.*, 1383 (1983).
89. J. A. Ibers, R. DiCosimo, and G. Whitesides, *Organometallics*, **1**, 13 (1982).
90. R. H. Grubbs, *Chemtech*, 512 (1977).
91. J. C. Bailer, *Catal. Rev. Sci. Eng.*, **10**, 17 (1974).
92. P. Hodge, *Chem. Br.*, **14**, 237 (1978).
93. Z. M. Michalska and D. E. Webster, *Chemtech*, 117 (1975).
94. A. L. Robinson, *Science*, **194**, 1261 (1976).
95. J. Manassen, *Platinum Met. rev.*, **15**, 142 (1971).
96. E. M. Cernia and M. Graziani, *J. Appl. Polym. Sci.*, **18**, 2725 (1974).

97. L. L. Murrell, in *Advanced Materials in Catalysis* (Eds. J. J. Burton and R. L. Garton), Academic Press, New York, 1977, p. 235.
98. C. U. Pittman and G. O. Evans, *ChemTech*, 560 (1973).
99. D. C. Bailey and S. H. Langer, *Chem. Rev.*, **81**, 109 (1981).
100. C. Francesco, G. Braca, C. Carlini, G. Sbrana, and G. Valentini, *J. Mol. Catal.*, **14**, 1 (1982).
101. W. O. Haag and D. D. Whitehurst, *Ger. Pat.*, 1800371 (1969); *Chem. Abstr.*, **71**, 114951 (1969).
102. G. J. K. Acres, G. C. Bond, B. J. Cooper, and J. A. Dawson, *J. Catal.*, **6**, 139 (1966).
103. W. H. Lang, A. T. Jurewicz, W. O. Haag, D. D. Whitehouse, and L. D. Rollmann, *J. Organomet. Chem.*, **134**, 85 (1977).
104. R. H. Grubbs and E. M. Sweet, *J. Mol. Catal.*, **3**, 259 (1977).
105. S. C. Tang, T. E. Paxson, and L. Kim, *J. Mol. Catal.*, **9**, 313 (1980).
106. M. H. J. M. deCroon and J. W. E. Coenen, *J. Mol. Catal.*, **11**, 301 (1981).
107. L. Bemi, H. C. Clark, J. A. Davies, C. A. Fyfe, and R. E. Wasylishen, *J. Am. Chem. Soc.*, **104**, 438 (1982).
108. Y. Dror and J. Manassen, *J. Mol. Catal.*, **2**, 219 (1977).
109. F. Joo, Z. Toth, and M. T. Beck, *Inorg. Chim. Acta*, **25**, L61 (1977).
110. F. Joo and Z. Toth, *J. Mol. Catal.*, **8**, 369 (1980).
111. R. T. Smith, R. K. Ungar, L. J. Sanderson, and M. C. Baird, *Ann. N. Y. Acad. Sci.*, 259 (1983).
112. E. L. Muetterties, T. Rhodin, E. Band, C. Brucker, and W. R. Pretzer, *Chem. Rev.*, **79**, 91 (1979).
113. R. Ygo, R. Psaro, G. M. Zanderighi, J. M. Basset, A. Theolier, and A. K. Smith in *Fundamental Research in Homogeneous Catalysis* (Ed. M. Tsutsui), Vol. 3, Plenum, New York, 1979, p. 579.
114. E. L. Muetterties and J. Stein, *Chem. Rev.*, **79**, 479 (1979).
115. C. Masters, *Adv. Organomet. Chem.*, **17**, 61 (1979).
116. W. Keim, M. Berger, and J. Schlupp, *J. Catal.*, **61**, 359 (1980).
117. R. R. Burch, A. J. Shusterman, E. L. Muetterties, R. G. Teller, and J. M. Williams, *J. Am. Chem. Soc.*, **105**, 3546 (1983).
118. R. R. Burch, E. L. Muetterties, R. G. Teller, and J. M. Williams, *J. Am. Chem. Soc.*, **104**, 4257 (1982).
119. G. Brieger and T. J. Nestrick, *Chem. Rev.*, **74**, 567 (1974).
120. Y. Sasson and J. Blum, *J. Org. Chem.*, **40**, 1887 (1975).
121. G. Descotes, D. Sinou, and J. P. Praly, *Bull. Soc. Chim. Fr.*, 153 (1978).
122. H. Imai, T. Nishigucho, and K. Fukuzumi, *J. Org. Chem.*, **42**, 431 (1977).
123. K. Ohkubo, I. Terada, K. Sugahara, and H. Yoshinaga, *J. Mol. Catal.*, **7**, 421 (1980).
124. L. Szabo, L. Dobay, L. Radics, and C. Szantay, *Nouv. J. Chim.*, **4**, 199 (1980).
125. W. H. Chiu and M. E. Wolff, *Steroids*, **34**, 361 (1979).
126. G. Zassinovich and F. Grisani, *J. Organomet. Chem.*, **247**, C24 (1983).
127. N. D. Heindel, N. Foster, and M. Choudhuri, *J. Org. Chem.*, **48**, 3817 (1983).
128. V. S. Rao and S. A. Perlin, *J. Org. Chem.*, **47**, 367 (1982).
129. M. K. Anwer, A. F. Spatola, C. D. Bossinger, E. Flanigan, R. C. Liu, D. B. Olsen, and D. Stevenson, *J. Org. Chem.*, **48**, 3503 (1983).
130. S. A. Kahn, and K. M. Sivanandoriah, *Synthesis*, 750 (1978).
131. J. R. Anderson, *Structure of Metallic Catalysts*, Academic Press, London, 1975, p. 451.
132. R. J. Peterson, *Hydrogenation Catalysts*, Noyes Data Corp., Park Ridge, NY (1977).
133. A. A. Pavlic and H. Adkins, *J. Am. Chem. Soc.*, **68**, 1471 (1946).
134. N. A. Khan, *J. Am. Chem. Soc.*, **74**, 3018 (1952).
135. R. L. Shriner and R. Adams, *J. Am. Chem. Soc.*, **46**, 1683 (1924).
136. W. S. Johnson, E. R. Rogier, J. Szumuskovicz, H. I. Hadler, J. Ackerman, B. K. Bhattacharyya, B. M. Bloom, L. Stalmann, R. A. Clement, B. Bannister, and H. Wynberg, *J. Am. Chem. Soc.*, **78**, 6289 (1956).
137. R. Willstatter and E. Waldschmidt-Leitz, *Chem. Ber.*, **54**, 121 (1921).
138. R. Baltzly, *J. Org. Chem.*, **41**, 920 (1976).
139. G. E. Ham and W. P. Coker, *J. Org. Chem.*, **29**, 194 (1964).
140. R. L. Augustine, *Adv. Catal.*, **25**, 56 (1976).
141. M. G. Combe, H. B. Henbest, and W. R. Jackson, *J. Chem. Soc. C*, 2467 (1967).
142. R. Howe and F. J. McQuillin, *J. Chem. Soc.*, 1194 (1958).
143. A. Skita and W. Faust, *Chem. Ber.*, **64**, 2878 (1931).
144. R. Helg and H. Schinz, *Helv. Chim. Acta*, **35**, 2406 (1952).
145. N. Tsuji, J. Suzuki, M. Shiota and I. Takahashi, and S. Nishimura, *J. Org. Chem.*, **45**, 2729 (1980).

146. M. Ishige and M. Shiota, *Can. J. Chem.*, **58**, 1061 (1980).
147. Z. Zafiriadis, *C. R. Acad. Sci.*, **226**, 731 (1948).
148. E. L. Muetterties, *Inorg. Chim. Acta*, **50**, 1 (1981).
149. T. E. Whyte, *Catal. Rev.*, **8**, 117 (1973).
150. R. J. Farranto, *Chem. Eng. Prog.*, **71**, 37 (1975).
151. D. Pope, W. L. Smith, M. J. Eastlake, and R. L. Moss, *J. Catal.*, **22**, 72 (1971).
152. P. C. Flynn, S. E. Wanke, and P. S. Turner, *J. Catal.*, **33**, 233 (1974).
153. P. C. Aben, H. van der Eijk and J. M. Oelderik, *Proc. 5th Int. Congr. Catal.*, **1**, 717 (1973).
154. R. L. Burwell, *Acc. Chem. Res.*, **2**, 289 (1969).
155. H. A. Quinn, M. A. McKervey, W. R. Jackson, and J. J. Rooney, *J. Am. Chem. Soc.*, **92**, 2922 (1970).
156. S. Siegel, M. Bunkel, G. V. Smith, W. Halpern, and J. Cozort, *J. Org. Chem.*, **31**, 2802 (1966).
157. G. V. Smith and J. A. Roth, *J. Am. Chem. Soc.*, **88**, 3879 (1966).
158. G. V. Smith and R. L. Burwell, *J. Am. Chem. Soc.*, **84**, 925 (1962).
159. A. S. Hussey, T. A. Schenach, and R. H. Baker, *J. Org. Chem.*, **33**, 610 (1968).
160. I. Jardine and F. J. McQuillin, *J. Chem. Soc. C*, 458 (1966).
161. F. G. Gault, J. J. Rooney, and C. Kemball, *J. Catal.*, **1**, 255 (1962).
162. R. L. Augustine, J. F. Van Peppen, F. Yaghmaie, *J. Org. Chem.*, **49**, 1865 (1984).
163. S. J. Thomson and G. Webb, *J. Chem. Soc., Chem. Commun.*, 526 (1976).
164. B. A. Morrow and N. Sheppard, *Proc. R. Soc. London, Ser. A*, **311**, 391 (1969).
165. E. J. Corey and T. A. Engler, *Tetrahedron Lett.*, **25**, 149 (1984).
166. L. S. Stuhl, M. Rakowski-Dubois, F. J. Hirsekorn, J. R. Bleeke, A. E. Stevens, and E. L. Muetterties, *J. Am. Chem. Soc.*, **100**, 2405 (1978).
167. M. F. Madden and C. Kemball, *J. Chem. Soc.*, 302 (1961).
168. F. Hartog and P. Zwietering, *J. Catal.*, **2**, 79 (1962).
169. J. F. Outlaw, J. R. Cozort, N. Garti, and S. Siegel, *J. Org. Chem.*, **48**, 4186 (1983).
170. D. J. Cram and N. L. Allinger, *J. Am. Chem. Soc.*, **77**, 6289 (1955).
171. E. Shustorovich, R. Baetzold, and E. L. Muetterties, *J. Phys. Chem.*, **87**, 1100 (1983).
172. L. Salem and C. Leforestier, *Surf. Sci.*, **82**, 390 (1979).
173. B. I. Lundqvist, B. Hellsing, S. Holmstrom, P. Nordlander, M. Fersson, and J. K. Norskov, *Int. J. Quantum Chem.*, **23**, 1083 (1983).
174. A. I. Lundqvist, in *Vibrations at Surfaces* (Eds. R. Candano, J. M. Gilles, and A. A. Lucas), Plenum, New York, 1982, p. 541.
175. G. C. Bond, *C. R. Sem. Etud. Catal.* (*Extr. Mem. Soc. R. Soc. Liege*), **1** (6), 4 (1971).
176. J. Y. Saillard and R. Hoffmann, *J. Am. Chem. Soc.*, **106**, 2006 (1984).
177. P. N. Rylander and G. R. Patel, *Indian J. Technol.*, **21**, 404 (1983).
178. V. K. Ahujo and C. A. Brown, *J. Org. Chem.*, **38**, 2226 (1973).
179. O. L. Chapman, H. G. Smith, and P. A. Barks, *J. Am. Chem. Soc.*, **85**, 3171 (1963).
180. S. J. Cristol, T. W. Russell, and D. I. Davies, *J. Org. Chem.*, **30**, 207 (1965).
181. A. C. Cope and H. C. Campbell, *J. Am. Chem. Soc.*, **74**, 179 (1952).
182. H. A. Smith, J. F. Fuzek, and H. T. Meriwether, *J. Am. Chem. Soc.*, **71**, 3765 (1949).
183. T. Kametani, M. Tsubuki, and H. Nemoto, *J. Chem. Soc., Perkin Trans. 1*, 3077 (1981).
184. M. A. W. Finch, S. M. Roberts, G. T. Woolley, and R. F. Newton, *J. Chem. Soc., Perkin Trans. 1*, 1725 (1981).
185. H. C. Brown, I. Rothberg, and D. L. V. Jagt, *J. Org. Chem.*, **37**, 4098 (1972).
186. G. W. Eigenmann and R. T. Arnold, *J. Am. Chem. Soc.*, **81**, 3440 (1959).
187. H. van Bekkum, F. van Rantwijk, G. van Minnen-Pathuis, J. D. Remijnse, and A. van Veen, *Recl. Trav. Chim. Pays-Bas*, **88**, 911 (1969).
188. R. T. Blickenstaff, A. C. Gosh, and G. C. Wolf, *Total Synthesis of Steroids*, Academic Press, New York, 1974.
189. H. J. E. Loewenthal, *Tetrahedron*, **6**, 269 (1959).
190. R. L. Augustine, in *Organic Reactions in Steroid Chemistry* (Eds. J. Fried and J. A. Edwards), Vol. 1, Van Nostrand Reinhold, New York, 1979.
191. R. B. Woodward, F. Sondheimer, D. Taub, K. Heusler, and W. M. McLamore, *J. Am. Chem. Soc.*, **74**, 4223 (1952).
192. S. Siegel and J. R. Cozort, *J. Org. Chem.*, **40**, 3594 (1975).
193. J. A. Marshall, N. H. Andersen, and P. C. Johnson, *J. Org. Chem.*, **35**, 86 (1970).
194. S. Siegel, M. Foreman, and D. Johnson, *J. Org. Chem.*, **40**, 3589 (1970).

195. S. Mitsui, K. Gohke, H. Saito, A. Nanbu, and Y. Senda, *Tetrahedron*, **29**, 1523 (1973).
196. W. R. Nes, T. E. Varkey, and K. Krevitz, *J. Am. Chem. Soc.*, **99**, 260 (1977).
197. S. R. Schow and T. McMorris, *J. Org. Chem.*, **44**, 3760 (1979).
198. G. E. Dubois, *J. Org. Chem.*, **47**, 5035 (1982).
199. L. S. Minckler, A. S. Hussey, and R. H. Baker, *J. Am. Chem. Soc.*, **78**, 1009 (1956).
200. T. J. Howard, *Chem. Ind.* (*London*), 1899 (1963).
201. W. G. Dauben, J. W. McFarland, and J. B. Rogan, *J. Org. Chem.*, **26**, 297 (1961).
202. T. G. Halsall, W. J. Rodewald, and D. Willis, *J. Chem. Soc.*, 2798 (1959).
203. N. B. Haynes and C. J. Timmons, *Proc. Chem. Soc.*, 345 (1958).
204. M. C. Dart and H. B. Henbest, *J. Chem. Soc.*, 3563 (1960).
205. S. Mitsui, Y. Senda, and H. Saito, *Bull. Chem. Soc. Jpn.*, **39**, 694 (1966).
206. J. P. Minyard, A. C. Thompson, and P. A. Hedin, *J. Org. Chem.*, **33**, 909 (1968).
207. P. Beak and T. L. Chaffin, *J. Org. Chem.*, **35**, 2275 (1970).
208. Y. Watanabe, Y. Mizuhara, and M. Shito, *J. Org. Chem.*, **33**, 468 (1968).
209. J. E. McMurry, *Tetrahedron Lett.*, 3731 (1970).
210. K. Mori, K. Abe, M. Washida, S. Nishimura, and M. Schiota, *J. Org. Chem.*, **36**, 231 (1970).
211. R. D. Stolow and K. Sachdev, *J. Org. Chem.*, **36**, 960 (1971).
212. R. V. Madrigal, K. L. Mikolajczak, R. G. Powell, and C. R. Smith, *J. Org. Chem.*, **39**, 676 (1974).
213. J. R. Campbell and E. J. Warawa, *J. Org. Chem.*, **39**, 3511 (1974).
214. R. Fujimoto, Y. Kishi, and J. Blount, *J. Am. Chem. Soc.*, **102**, 7154 (1980).
215. H. W. Thompson and E. McPherson, *J. Am. Chem. Soc.*, **96**, 6232 (1974).
216. H. W. Thompson, E. McPherson, and B. L. Lences, *J. Org. Chem.*, **41**, 2903 (1976).
217. H. W. Thompson and R. E. Naipawar, *J. Am. Chem. Soc.*, **95**, 6379 (1973).
218. C. Okkerse, A. de Jonge, J. W. E. Coenen, and A. Rozendaal, *J. Am. Oil Chem. Soc.*, **44**, 152 (1967).
219. J. W. E. Coenen, *J. Am. Oil Chem. Soc.*, **53**, 382 (1976).
220. J. I. Gray and L. F. Russell, *J. Am. Oil Chem. Soc.*, **56**, 36 (1979).
221. H. J. Dutton, *Geom. Positional Fatty Acid Isomers*, 1 (1979); *Chem. Abstr.*, **93**, 184352 (1980).
222. Y. Murase, *Yakugaku Zasshi*, **28**, 689 (1979).
223. B. Drozdowski, *Zesz. Potespow. Nauk. Roln.*, **211**, 47 (1981).
224. J. C. Bailar, *Adv. Chem. Ser.*, No. 173, 1 (1979).
225. N. A. Dobson, G. Eglinton, M. Krishnamurti, R. A. Raphael, and R. G. Willis, *Tetrahedron*, **16**, 16 (1961).
226. P. Lombardi, *Gazz. Chim. Ital.*, **104**, 867 (1974).
227. M. Ishiga and M. Shiota, *Can. J. Chem.*, **53**, 1700 (1975).
228. C. F. Wilcox and D. L. Neely, *J. Org. Chem.*, **28**, 3446 (1968).
229. C. A. Henrick, *Tetrahedron*, **33**, 1845 (1972).
230. P. Sabatier and J. B. Senderens, *C. R. Acad. Sci., Ser. C*, **131**, 40 (1900).
231. E. N. Marvell and T. Li, *Synthesis*, 457 (1973).
232. M. J. Robins and P. J. Barr, *J. Org. Chem.*, **48**, 1854 (1983).
233. G. Han, J. Kong, and B. Jin, *Huaxue Shiji*, **1**, 49 (1982).
234. H. Guttmann and H. Lindlar, in *Chemistry of Acetylenes* (Ed. H. G. Viehe), Marcel Dekker, New York, 1969, p. 355.
235. W. L. Roelofs, J. P. Kochamsky, R. T. Carde, C. A. Henrick, J. N. Labovitz, and V. L. Corbin, *Life Sci.*, **17**, 699 (1975).
236. C. E. Descoins and D. Samain, *Tetrahedron Lett.*, 745 (1976).
237. J. Rajaram, A. P. S. Narula, H. P. S. Chauvla, and S. Dev, *Tetrahedron*, **39**, 2315 (1983).
238. R. G. Smith, G. D. Davies, and G. E. Daterman, *J. Org. Chem.*, **40**, 1593 (1975).
239. D. Savoia, E. Tagliavini, C. Trombini, and A. Umani-Ronchi, *J. Org. Chem.*, **46**, 5340 (1981).
240. D. Savoia, C. Trombini, A. Umani-Ronchi, and G. Verardo, *J. Chem. Soc., Chem. Commun.*, 540 (1981).
241. P. Mauret and P. Alphonse, *J. Org. Chem.*, **47**, 3322 (1982).
242. A. W. Burgstahlor and G. N. Widiger, *J. Org. Chem.*, **38**, 3652 (1973).
243. R. I. Fryer, N. W. Gilman, and B. C. Holland, *J. Org. Chem.*, **40**, 348 (1975).
244. J. J. Brunet, P. Gallois, and P. Caubere, *J. Org. Chem.*, **45**, 1937 (1980).
245. E. N. Marvell and J. Tashiro, *J. Org. Chem.*, **30**, 3991 (1965).
246. G. Buchi and B. Egger, *J. Org. Chem.*, **36**, 2021 (1971).
247. F. J. McQuillin and W. O. Ord, *J. Chem. Soc.*, 2902 (1959).

248. F. P. Darmory and S. S. Klioze, *J. Org. Chem.*, **40**, 1588 (1975).
249. G. Stork and E. Colvin, *J. Am. Chem. Soc.*, **93**, 2080 (1971).
250. K. Atsumin and I. Kuwajima, *Tetrahedron Lett.*, 2001 (1977).
251. R. M. Carlson, A. R. Oyler, and J. R. Peterson, *J. Org. Chem.*, **40**, 1610 (1975).
252. V. I. Sharkov, *Chem.-Ing.-Techn.*, **35**, 494 (1963).
253. W. S. Fedor, J. Millar, and A. J. Accola, *Ind. Eng. Chem.*, **52**, 282 (1960).
254. L. W. Wright, *ChemTech*, 42 (1974).
255. G. G. Boyers, *US Pat.*, 2808847 (1959).
256. K. Kochloefl, A. Ralkova and J. Simonikova, *J. Catal.*, **29**, 412 (1973).
257. S. Kishida and S. Teranishi, *J. Catal.*, **12**, 90 (1968).
258. Y. Takagi, *Sci. Pap. Inst. Phys. Chem. Res. Jpn.*, **57**, 105 (1963).
259. M. Kaminaga, S. Mitsui, H. Saito, Y. Senda, and Y. Yamashita, *Tetrahedron*, **29**, 1533 (1973).
260. T. Anderson, M. Freifelder, Y. H. Ng, and V. Papendick, *J. Pharm, Sci.*, **57**, 967 (1964).
261. D. H. R. Barton, *J. Chem. Soc.*, 1027 (1953).
262. Y. Takagi, K. Tanaka, and S. Tetratomi, in *Catalysis* (Ed. J. W. Hightower), Elsevier, New York, 1973, p. 757.
263. P. N. Rylander and D. R. Steele, *Engelhard Ind. Tech. Bull.*, **3**, 125 (1963).
264. S. P. Findlay, *J. Org. Chem.*, **24**, 1540 (1959).
265. M. Ishiga, S. Nishimura, and M. Shiota, *Chem. Lett.*, 535 (1977).
266. M. Katagiri, Y. Kunikata, and S. Nishimura, *Chem. Lett.*, 1235 (1975).
267. M. Ishiga, S. Nishimura, and M. Shiota, *Chem. Lett.*, 963 (1977).
268. J. Solodar, *J. Org. Chem.*, **41**, 3461 (1976).
269. G. N. Walker, *J. Org. Chem.*, **23**, 133 (1958).
270. J. S. Burton and R. Stevens, *J. Chem. Soc.*, 4382 (1963).
271. J. S. Burton, J. A. Elvidge, and R. Stevens, *J. Chem. Soc.*, 3816 (1964).
272. H. S. Aaron and L. P. Reiff, *Tetrahedron Lett.*, 2329 (1967).
273. E. I. Klabunovskii and A. A. Vedenyapin, *Russ. J. Phys. Chem.*, **51**, 1759 (1977).
274. Y. Izumi, *Adv. Catal.*, **32**, 215 (1983).
275. Y. Izumi, T. Harada, T. Tanabe, and K. Okuda, *Bull. Chem. Soc. Jpn.*, **44**, 1418 (1971).
276. Y. Izumi and K. Ohkubo, *Bull. Chem. Soc. Jpn.*, **44**, 1330 (1971).
277. Y. Ohto, S. Nuva, and S. Imai, *Yuki Gosei Kagaku Kyokai Shi*, **35**, 753 (1978).
278. Y. Ohto, S. Imai, S. Nuva, and N. Gia-Hung, *Yuki Gosei Kagaku Kyokai Shi*, **37**, 173 (1979).
279. F. Higashi, T. Ninomiya, and Y. Izumi, *Bull. Chem. Soc. Jpn.*, **44**, 1333 (1971).
280. Y. Ohto, S. Nuva and S. Imai, *Yuki Gosei Kagaku Kyokai Shi*, **35**, 672 (1977).
281. E. I. Klabunovskii and Y. I. Petrov, *Dokl. Akad. Nauk SSSR*, **173**, 1125 (1967).
282. E. I. Klabunovskii, N. P. Sokolova, Y. I. Petrov, and V. V. Patrikov, *Izv. Akad. Nauk SSSR, Ser. Khim.*, 838 (1983).
283. A. A. Vedenyapin, T. I. Kuznetsova and E. I. Klabunovskii, *Izv. Akad. Nauk SSSR, Ser. Khim.*, 1455 (1983).
284. J. A. Groenewegen and W. M. H. Sachtler, *J. Catal.*, **38**, 501 (1975).
285. H. Ozaki, A. Tai, S. Kobatake, H. Watanabe, and Y. Izumi, *Bull. Chem. Soc. Jpn.*, **51**, 3559 (1978).
286. T. G. Beck, *Tetrahedron*, **33**, 3041 (1977).
287. K. Steiner, *US Pat.*, 3953324 (1976).
288. M. Horner and M. Irgang, *Eur. Pat. Appl.*, EP71787 (1983); *Chem. Abstr.*, **99**, 54020n (1983).
289. Y. Ichikawa, T. Yamaji and T. Sawaki, *Jpn. Pat.*, **77** 46008 (1977); *Chem. Abstr.*, **87**, 136052x (1977).
290. Y. Ichikawa and Y. Teizo, *Jpn. Pat.* 77 12106 (1977); *Chem. Abstr.*, **87**, 52778a (1977).
291. W. E. Pascoe and J. F. Sternberg, in *Catalysis in Organic Synthesis* (*1978*), Vol. 7, (Ed. W. H. Jones), Academic Press, New York, 1980, pp. 1–9.
292. I. A. Kaye and R. S. Matthews, *J. Org. Chem.*, **29**, 1341 (1964).
293. D. J. France, J. J. Hand, and M. Los, *Tetrahedron*, **25**, 4011 (1969).
294. N. Tsuji, J. Suzuki, M. Shiota, I. Takahashi, and S. Nishimura, *J. Org. Chem.*, **45**, 2729 (1980).
295. S. Akiyama, Y. Izumi, Y. Mizutani, Y. Onoue, and Y. Watanabe, *ChemTech*, **1**, 36 (1977).
296. Y. Izumi, Y. Matsumura, Y. Mizutani, and Y. Watanabe, *Bull. Chem. Soc. Jpn.*, **47**, 2992 (1974).
297. C. U. Pittman and Y. F. Liang, *J. Org. Chem.*, **45**, 5048 (1980).
298. D. J. Rozzel, *Tetrahedron Lett.*, **23**, 1767 (1982).
299. G. C. Andrews, T. C. Crawford, and B. E. Bacon, *J. Org. Chem.*, **46**, 2976 (1981).
300. K. Mori, K. Abe, M. Washida, S. Nishimura, and M. Shiota, *J. Org. Chem.*, **36**, 231 (1971).

301. T. C. McKenzie, *J. Org. Chem.*, **39**, 629 (1974).
302. G. N. Walker, *J. Am. Chem. Soc.*, **78**, 3201 (1956).
303. J. G. Berger, S. R. Teller, and I. J. Pachter, *J. Org. Chem.*, **35**, 3122 (1970).
304. A. G. Anastassiou and G. W. Griffin, *J. Org. Chem.*, **33**, 3441 (1968).
305. J. W. Burnham and E. J. Eisenbraun, *J. Org. Chem.*, **36**, 737 (1971).
306. K. Kindler, H. G. Helling, and E. Sussner, *Justus Liebigs Ann. Chem.*, **605**, 200 (1957).
307. Y. Fujima, K. Sato, and A. Yamada, *Bull. Chem. Soc. Jpn.*, **41**, 442 (1968).
308. L. Hasbrouck and P. N. Rylander, *Engelhard Ind. Tech. Bull.*, **8**, 148 (1968).
309. K. Hanaya, *Bull. Chem. Soc. Jpn.*, **43**, 442 (1970).
310. S. Mitsui, H. Saito, and Y. Senda, *Bull. Chem. Soc. Jpn.*, **39**, 694 (1966).
311. H. van Bekkum, A. P. G. Kieboom, and K. J. G. van de Putte, *Recl. Trav. Chim. Pays-Bas*, **88**, 52 (1969).
312. P. Sabatier and J. B. Senderens, *Bull. Soc. Chim. Fr.*, 101 (1904).
313. L. Espil and P. Sabatier, *Bull. Soc. Chim. Fr.*, **15**, 228 (1914).
314. B. van de Graaf, H. van Bekkum, and B. M. Wepster, *Recl. Trav. Chim. Pays-Bas*, **87**, 777 (1968).
315. P. Dini, D. Dones, S. Montelatici, and N. Giordano, *J. Catal.*, **30**, 1 (1973).
316. H. van Bekkum, B. van de Graaf, J. A. Peters, and B. M. Wepster, *Recl. Trav. Chim. Pays-Bas*, **89**, 521 (1970).
317. J. H. Stocker, *J. Org. Chem.*, **27**, 2288 (1962).
318. J. H. Stocker, *J. Org. Chem.*, **29**, 3593 (1964).
319. P. N. Rylander and N. F. Rakconcza, *US Pat.*, 3162697 (1964).
320. M. Freifelder, *J. Org. Chem.*, **29**, 2895 (1964).
321. M. Freifelder, R. W. Matoon, and Y. H. Ng, *J. Org. Chem.*, **29**, 3730 (1964).
322. G. Descotes, P. Legrand-Berlebach, and J. Sabadic, *Bull. Soc. Chim. Fr.*, 1517 (1973).
323. P. N. Confalone, G. Pizzolato, and M. R. Uskokovic, *Helv. Chim. Acta*, **59**, 1005 (1976).
324. A. Bobst and M. Viscontini, *Helv. Chim. Acta*, **49**, 875 (1966).
325. H. I. X. Mager, R. Addink, and W. Berends, *Recl. Trav. Chim. Pays-Bas*, **86**, 833 (1967).
326. M. Honel and F. W. Vierhapper, *J. Chem. Soc., Perkin Trans. 1*, 2607 (1982).
327. B. Frydman, M. F. Despuy, and H. Rapoport, *J. Am. Chem. Soc.*, **87**, 3530 (1967).
328. L. M. Werbel and M. L. Zamora, *J. Heterocycl. Chem.*, **2**, 290 (1965).
329. J. Annand, K. Chekir, N. Ple, G. Queguiner, and M. P. Simmonin, *J. Org. Chem.*, **46**, 4754 (1981).
330. D. Lednicer, S. C. Lyster, and G. W. Duncan, *J. Med. Chem.*, **10**, 78 (1967).
331. J. H. Stocker, *J. Org. Chem.*, **27**, 2288 (1962).
332. K. Chebaane, M. Guyot, and D. Molko, *Bull. Soc. Chim. Fr.*, 244 (1975).
333. C. J. Pedersen, *J. Am. Chem. Soc.*, **89**, 7017 (1967).
334. H. van Bredorode, *Doctoral Thesis*, Delft University of Technology (1975).
335. H. van Bekkum, P. Klapwijk, and T. J. Nieuwstad, *J. Catal.*, **29**, 404 (1973).
336. M. Freifelder and Y. H. Ng, *J. Pharm. Sci.*, **54**, 1204 (1965).
337. M. Freifelder, *J. Am. Chem. Soc.*, **82**, 2386 (1960).
338. D. A. Baker and A. Rosenthal, *J. Org. Chem.*, **38**, 193, 198 (1973).
339. J. E. Cornelius, R. C. Hay, and T. W. Russell, *J. Org. Chem.*, **37**, 3552 (1972).
340. L. Hasbronck, I. Karpenko, and P. N. Rylander, *Ann. N. Y. Acad. Sci.*, **214**, 100 (1973).
341. I. Mami, M. A. Schwartz, B. Vishnuvajjala, and M. Zoda, *J. Org. Chem.*, **41**, 2502 (1976).
342. S. H. Jung and H. Kohn, *Tetrahedron Lett.*, **25**, 399 (1984).
343. H. Greenfield, *Ind. Eng. Chem., Prod. Res. Dev.*, **15**, 156 (1976).
344. J. J. M. Deumens and J. A. Thoma, *Ger. Offen.*, 2014837; *Chem. Abstr.*, **73**, 120505 (1970).
345. J. A. Thoma and J. J. M. Deumens, *US Pat.*, 3658824 (1971).
346. F. Bergel, A. L. Morrison, and H. Rinderknecht, *US Pat.*, 2446803 (1948).
347. H. O. House, H. C. Muller, and P. P. Wickham, *J. Am. Chem. Soc.*, **84**, 3139 (1962).
348. S. Trofimenko, *J. Org. Chem.*, **28**, 2755 (1963).
349. G. N. Walker, *J. Med. Chem.*, **8**, 583 (1965).
350. L. Mandell, J. U. Piper, and K. P. Singh, *J. Org. Chem.*, **28**, 3440 (1963).
351. M. Lamborg, W. J. Linn, V. Boekelheide, and P. O'Grady, *J. Am. Chem. Soc.*, **75**, 3243 (1953).
352. J. B. Hester, *J. Org. Chem.*, **29**, 1158 (1964).
353. G. B. Bodem, M. R. Chedekel, and E. Lette, *J. Org. Chem.*, **37**, 4465 (1972).
354. J. Bordais and C. Germain, *Tetrahedron Lett.*, 195 (1970).
355. I. M. Karpenko, G. R. Pond, and P. N. Rylander, *US Pat.*, 3694509 (1972).
356. A. L. Davis, T. J. McCord, and D. R. Smith, *J. Med. Chem.*, **16**, 1043 (1973).

357. J. E. Baldwin, *J. Chem. Soc., Chem. Commun.*, 734 (1976).
358. G. O. Doherty and K. H. Fuhr, *Ann. N. Y. Acad. Sci.*, **214**, 221 (1973).
359. R. L. Augustine, A. J. Gustavsen, K. S. Houghton, G. Koletar, I. C. Pattison, and S. F. Wanat, *J. Org. Chem.*, **38**, 3004 (1973).
360. R. E. Lyle and J. L. LaMattina, *Synthesis*, 726 (1974).
361. W. F. Benning, P. Davis, D. M. Mulvey, W. H. Jones, P. I. Pollak, J. C. Schaeffer, R. Tull, and L. M. Weinstock, *Ann. N. Y. Acad. Sci.*, **158**, 471 (1969).
362. A. Tai, T. Harada, H. Hiraki, and S. Murakami, *Bull. Chem. Soc. Jpn.*, **56**, 1414 (1983).
363. H. H. Baer and K. S. Ong, *J. Org. Chem.*, **34**, 560 (1969).
364. H. H. Baer and F. Kienzle, *J. Org. Chem.*, **34**, 3848 (1969).
365. H. H. Baer and T. Neilson, *J. Org. Chem.*, **32**, 1068 (1967).
366. T. Nakagawa, R. Sudoh, T. Takamoto, and Y. Yokota, *Bull. Chem. Soc. Jpn.*, **46**, 1532 (1973).
367. G. W. Brown, in *The Chemistry of the Hydroxyl Group, Part I*, (Ed. S. Patai), Interscience, London, 1971, p. 626.
368. S. R. Vidyarthi, P. C. Nigam, and J. K. Gehlawat, *J. Sci. Ind. Res.*, **42**, 268 (1983).
369. A. Holt, A. W. P. Jarvie, and J. J. Mallabar, *J. Organomet. Chem.*, **59**, 141 (1973).
370. I. Büthe and E. Vowinkel, *Chem. Ber.*, **107**, 1353 (1974).
371. G. V. Pigulevskii and Z. Y. Rubashko, *Zh. Obshch. Khim.*, **25**, 2227 (1955).
372. F. J. McQuillin and W. O. Ord, *J. Chem. Soc.*, 3169 (1959).
373. G. Accrombessi, P. Geneste, J. L. Olive, and A. A. Pavia, *Bull. Soc. Chim. Fr.*, 19 (1981).
374. J. O. Morley, N. P. Preston, and T. Winwick, *J. Chem. Soc., Chem. Commun.*, 307 (1984).
375. V. B. Jigajinni and R. H. Wightman, *Tetrahedron Lett.*, **23**, 117 (1982).
376. P. N. Rylander and D. R. Steele, *Engelhard Ind. Tech. Bull.*, **6**(2), 41 (1965).
377. J. W. Barton, in *Protective Groups in Organic Chemistry* (Ed. J. F. W. McOmie), Plenum, New York, 1973, p. 56.
378. R. Macrae and G. T. Young, *J. Chem. Soc., Chem. Commun.*, 446 (1974).
379. B. E. Watkins, J. S. Kiely, and H. Rapoport, *J. Am. Chem. Soc.*, **104**, 5702 (1982).
380. E. Cerutti, C. Morpain, M. Tisserand, and J. Y. Roualt, *C. R. Acad. Sci. Sci., Ser. C*, **277**, 787 (1973).
381. P. M. Cullis and G. Lowe, *J. Chem. Soc., Chem. Commun.*, 512 (1978).
382. R. L. Jarvest and G. Lowe, *J. Chem. Soc., Chem. Commun.*, 1145 (1980).
383. P. M. Cullis, R. L. Jarvest, G. Lowe, and B. V. L. Potter, *J. Chem. Soc., Chem. Commun.*, 245 (1981).
384. G. Lowe and S. Swain, *J. Chem. Soc., Chem. Commun.*, 1279 (1983).
385. K. Clauss and W. Jensen, *Angew. Chem.*, **85**, 981 (1973).
386. W. J. Musliner and J. W. Gates, *Org. Synth.*, **51**, 82 (1971).
387. P. J. van der Jagt, W. den Hollander, and B. van Zanten, *Tetrahedron*, **27**, 1049 (1971).
388. H. Kimura and C. H. Stammer, *J. Org. Chem.*, **48**, 2440 (1983).
389. F. Kunzle and J. Schmutz, *Helv. Chim. Acta*, **39**, 1144 (1956).
390. C. F. Chignell, N. J. Harper, and G. Kirk, *J. Med. Chem.*, **7**, 726 (1964).
391. W. Siedel, K. Sturm, and W. Scheurich, *Chem. Ber.*, **99**, 345 (1966).
392. M. Freifelder, *J. Org. Chem.*, **31**, 3875 (1966).
393. I. Ojima, N. Hatanaka, N. Yoda, R. Abe, M. Yatabe, and M. Yamashita, *Pept. Chem.*, **20**, 29 (1982).
394. K. W. Rosenmund and F. Zetzsche, *Chem. Ber.*, **51**, 578 (1915).
395. E. Mosettig and R. Mozingo, *Org. React.*, **4**, 362 (1948).
396. H. van Bekkum and J. A. Peters, *Recl. Trav. Chim. Pays-Bas.*, **90**, 1323 (1971).
397. K. Hofmann, G. A. Jeffrey, S. F. Orochena, and S. M. Sax, *J. Am. Chem. Soc.*, **81**, 992 (1959).
398. M. G. Reinecke, *J. Org. Chem.*, **29**, 299 (1964).
399. T. R. Bosin, A. R. Buckpitt, and M. G. Raymond, *Tetrahedron Lett.*, 4699 (1973).
400. J. H. Sinfelt, *Adv. Catal.*, **23**, 91 (1973).
401. H. Musso, *Chem. Ber.*, **108**, 337 (1975).
402. N. A. Sasuki, H. Musso, and R. Zunker, *Chem. Ber.*, **106**, 2992 (1973).
403. E. Osawa, P. von R. Schleyer, L. W. K. Chang, and V. V. Kane, *Tetrahedron Lett.*, 4189 (1974).
404. M. N. Ahktar and W. R. Jackson, *J. Chem. Soc., Chem. Commun.*, 813 (1972).
405. A. J. Breijer, H. van Bekkum, and A. P. G. Kieboom, *Recl. Trav. Chim. Pays-Bas*, **93**, 186 (1974).
406. S. Mitsui, Y. Sugi, M. Fujimoto, and K. Yokoo, *Tetrahedron*, **30**, 31 (1974).
407. R. L. Augustine and E. J. Reardon, *J. Org. Chem.*, **39**, 1627 (1974).

408. V. Buss, R. Gleiter, and P. von R. Schleyer, *J. Am. Chem. Soc.*, **93**, 3927 (1971).
409. I. I. Khochenko, E. M. Mil'vitskaya, and A. F. Plate, *J. Org. Chem. USSR*, **8**, 560 (1972).
410. V. Buss, C. Woodworth, and P. van R. Schleyer, *J. Chem. Soc., Chem. Commun.*, 569 (1968).
411. H. E. Simmons, T. L. Cairns, S. A. Vladuchick, and C. M. Hoiness, *Org. React.*, **20**, 1 (1973).
412. A. V. Egorov, E. S. Finkelstein, N. S. Nametkin, A. M. Popov, and V. M. Vdovin, *Izv. Akad. Nauk SSSR, Ser. Khim.*, 2806 (1973).
413. D. Kaufmann and A. De Meijere, *Chem. Ber.*, **116**, 833 (1983).
414. J. A. Roth, *J. Catal.*, **26**, 97 (1972).
415. C. Groger, H. Musso, and I. Rossnagel, *Chem. Ber.*, **113**, 3612 (1980).
416. C. H. Heathcock and C. R. Poulter, *Tetrahedron Lett.*, 2755 (1968).
417. D. Buchle-Kallfass, M. Gugel, C. Hanschild, H. Musso, I. Rossnagel, and C. Schreiber, *Isr. J. Chem.*, **21**, 190 (1981).
418. A. L. Schultz, *J. Org. Chem.*, **36**, 383 (1971).
419. D. W. Meek and J. Niewahner, *Inorg. Chim. Acta*, **64**, L123 (1982).
420. T. Yoshida, W. J. Youngs, T. Sakaeda, T. Ueda, S. Otsuka, and J. A. Ibers, *J. Am. Chem. Soc.*, **105**, 6273 (1983).
421. J. E. Hamlin, K. Hirai, V. C. Gibson, and P. M. Maitlis, *J. Mol. Catal.*, **15**, 337 (1982).
422. W. J. Evans, J. H. Meadows, W. E. Hunter, and J. L. Atwood, *J. Am. Chem. Soc.*, **106**, 1291 (1984).
423. W. J. Evans, I. Bloom, W. E. Hunter, and J. L. Atwood, *J. Am. Chem. Soc.*, **105**, 1401 (1983).
424. P. L. Watson and D. C. Roe, *J. Am. Chem. Soc.*, **104**, 6471 (1982).
425. W. J. Evans, S. C. Engerer, P. A. Piliero, and A. L. Wayda, *J. Chem. Soc., Chem. Commun.*, 1007 (1979).
426. W. J. Evans, S. C. Engerer, P. A. Piliero, and A. L. Wayda, in *Fundamental Research in Homogeneous Catalysis* (Ed. M. Tsutsui), Plenum, New York, 1979, p. 941.
427. H. Eckert, G. Fabry, Y. Kiesel, G. Raudaschl, and C. Seidel, *Angew. Chem., Int. Ed. Engl.*, **22**, 881 (1983).
428. G. S. R. Rao, J. Rajaram, S. Rathnamala, and S. Sivaramakrishnan, *Proc. Indian Acad. Sci., Sect. A*, **86**, 435 (1977).
429. H. A. Tayim and J. C. Bailar, *J. Am. Chem. Soc.*, **89**, 4330 (1967).
430. P. Le Maux, J. Y. Saillard, D. Grandjean, and G. Jaouen, *J. Org. Chem.*, **45**, 4524 (1980).
431. E. N. Fraenkel and R. O. Butterfield, *J. Org. Chem.*, **34**, 3930 (1969).
432. A. Nakamura and S. Otsuka, *Tetrahedron Lett.*, 4529 (1973).
433. R. Crabtree, J. H. Milhevic, and J. M. Quirke, *J. Am. Chem. Soc.*, **101**, 7738 (1979).
434. C. J. Love and F. J. McQuillin, *J. Chem. Soc., Perkin Trans. 1*, 2509 (1973).
435. H. D. Kaesz and R. B. Sailliant, *Chem. Rev.*, **72**, 231 (1972).
436. P. P. Fu and R. G. Harvey, *Tetrahedron Lett.*, 415 (1977).
437. P. P. Fu and R. G. Harvey, *Tetrahedron Lett.*, 551 (1978).
438. P. P. Fu and R. G. Harvey, *J. Org. Chem.*, **45**, 2797 (1980).
439. J. A. Osborn, G. H. Jardine, J. F. Young, and G. Wilkinson, *J. Chem. Soc., A*, 1711 (1966).
440. J. M. Brown and D. Parker, *Organometallics*, **1**, 950 (1982).
441. D. Parker, *D. Phil. Thesis*, University of Oxford (1980).
442. N. Dinh-Nguyen and R. Rhyage, *Acta Chem. Scand.*, **13**, 1032 (1959).
443. D. H. G. Crout, M. Lutstorf, P. J. Morgan, R. M. Adlington, J. E. Baldwin, and M. J. Crimmin, *J. Chem. Soc., Chem. Commun.*, 1175 (1981).
444. R. O. Adlof and E. A. Einken, *J. Labelled Compd. Radiopharm.*, **18**, 419 (1981).
445. L. A. Oro, J. A. Cabeza, C. Cativiela, and E. Melendez, *J. Chem. Soc., Chem. Commun.*, 1383 (1983).
446. A. J. Birch and K. A. M. Walker, *J. Chem. Soc. C*, 1894 (1966).
447. R. E. Ireland and P. Bey, *Org. Synth.*, **53**, 63 (1973).
448. F. H. Lincoln, W. P. Schneider, and J. E. Pike, *J. Org. Chem.*, **38**, 951 (1973).
449. J. J. Sims, V. K. Honward, and L. H. Selman, *Tetrahedron Lett.*, 87 (1969).
450. A. J. Birch and K. A. M. Walker, *Aust. J. Chem.*, **24**, 513 (1971).
451. M. Brown and L. W. Piszkiewicz, *J. Org. Chem.*, **32**, 2013 (1967).
452. J. F. Biellmann and H. Liesenfelt, *Bull. Soc. Chim. Fr.*, 4029 (1966).
453. M. M. Bhagwat and D. Devaprabhakara, *Tetrahedron Lett.*, 1391 (1972).
454. A. C. Day, J. Nabney, and A. I. Scott, *J. Chem. Soc.*, 4067 (1961).

455. J. F. Biellmann and H. Liesenfelt, *C. R. Acad. Sci., Ser. C*, **263**, 251 (1966).
456. C. H. Heathcock and S. R. Poulter, *Tetrahedron Lett.*, 2755 (1969).
457. R. E. Harman, J. L. Parsons, D. W. Cooke, S. K. Gupta, and J. Schoolenberg, *J. Org. Chem.*, **34**, 3684 (1969).
458. A. J. Birch and K. A. M. Walker, *Tetrahedron Lett.*, 1935 (1967).
459. A. B. Hornfeldt, J. S. Gronowitz, and S. Gronowitz, *Acta Chem. Scand.*, **22**, 2725 (1968).
460. J. T. A. Reuvers and A. De Groot, *J. Org. Chem.*, **49**, 1110 (1984).
461. H. W. Thompson and E. McPherson, *J. Am. Chem. Soc.*, **96**, 6232 (1974).
462. J. M. Brown and R. G. Naik, *J. Chem. Soc., Chem. Commun.*, 348 (1982).
463. G. Stork and D. E. Kahne, *J. Am. Chem. Soc.*, **105**, 1072 (1983).
464. R. H. Crabtree, D. R. Anton, and M. W. Davis, *Ann. N. Y. Acad. Sci.*, 415 (1983).
465. J. C. Poulin and H. B. Kagan, *J. Chem. Soc., Chem. Commun.*, 1261 (1982).
466. D. Sinou, D. Lafont, G. Descotes, and A. G. Kent, *J. Organomet. Chem.*, **217**, 119 (1981).
467. A. Andreeta, F. Conti, and G. F. Ferrari, *Aspects Homog. Catal.*, **1**, 204 (1970).
468. R. Bertani, G. Carturan, and A. Scrivanti, *Angew. Chem., Int. Ed. Engl.*, **22**, 246 (1983).
469. D. L. Roger, M. M. Habib, and D. J. Fauth, *J. Org. Chem.*, **45**, 3860 (1980).
470. E. N. Fraenkel, E. Selke, and C. A. Glass, *J. Am. Chem. Soc.*, **90**, 2446 (1968).
471. M. A. Schroeder and M. Wrighton, *J. Am. Chem. Soc.*, **95**, 5764 (1973).
472. R. G. Salomon, *Tetrahedron*, **4**, 485 (1983).
473. J. Nasielski, P. Kirsch, and L. Wilputte-Steinert, *J. Organomet. Chem.*, **27**, C13 (1971).
474. G. Platbrood and L. Wilputte-Steinert, *Bull. Soc. Chim. Belg.*, **32**, 733 (1973).
475. I. Fischler, M. Budzwait, and E. A. Koerner von Gustorff, *J. Organomet. Chem.*, **105**, 325 (1976).
476. M. A. Schroeder and M. Wrighton, *J. Organomet. Chem.*, **74**, C29 (1974).
477. E. F. Litvin, A. K. Freidlin, and K. K. Karimov, *Izv. Akad. Nauk. SSSR, Ser. Khim.*, 1853 (1972).
478. I. Jardine and F. J. McQuillin, *Tetrahedron Lett.*, 4871 (1966).
479. J. R. Shapley, R. R. Schrock, and J. A. Osborn, *J. Am. Chem. Soc.*, **91**, 2816 (1969).
480. S. Slater and E. L. Muetterties, *Inorg. Chem.*, **19**, 3337 (1980).
481. C. U. Pittman, R. C. Ryan, J. McGee, and J. P. O'Connor, *J. Organomet. Chem.*, **178**, C45 (1978).
482. J. L. Thomas and K. T. Brown, *J. Organomet. Chem.*, **111**, 297 (1976).
483. Y. A. Dorfman and L. V. Lyashenko, *Elektetrokhimiya*, **13**, 535 (1977).
484. E. B. Fleischer and M. Krishnamurthy, *J. Am. Chem. Soc.*, **94**, 1382 (1972).
485. C. Masters, W. S. McDonald, G. Raper, and B. L. Shaw, *J. Chem. Soc., Chem. Commun.*, 210 (1971).
486. T. M. Beloshyudova and C. A. Il'ina, *Kinet. Katal.*, **16**, 788 (1975).
487. W. A. Bulen, J. L. Corbin, W. E. Newton, and P. W. Schneider, *J. Am. Chem. Soc.*, **93**, 268 (1971).
488. L. S. Stuhl and E. L. Muetterties, *Inorg. Chem.*, **17**, 2148 (1978).
489. R. H. Crabtree, A. Gautier, G. Giordano, and T. Khan, *J. Organomet. Chem.*, **141**, 113 (1977).
490. T. Shimura, R. Iwata, I. Ogata, and Y. Arai, *Tokyo Kogyo Shikensho Hokoku*, **73**, 505 (1978); *Chem. Abstr.*, **92**, 128332 (1980).
491. A. J. Sivak and E. L. Muetterties, *J. Am. Chem. Soc.*, **101**, 4878 (1979).
492. P. Abley and F. J. McQuillin, *J. Chem. Soc., Chem. Commun.*, 1503 (1969).
493. T. Yoshida, W. J. Youngs, T. Sakaeda, T. Ueda, S. Otsuka, and J. A. Ibers, *J. Am. Chem. Soc.*, **105**, 6273 (1983).
494. H. Eckert, G. Fabry, Y. Kiesel, G. Raudaschl, and C. Seidel, *Angew. Chem.*, **95**, 894 (1983).
495. H. Eckert, G. Fabry, Y. Riesel, G. Raudaschl, and C. Seidel, *Angew. Chem., Suppl.*, 1291 (1983).
496. R. R. Schrock and J. A. Osborn, *J. Chem. Soc., Chem. Commun.*, 567 (1970).
497. H. Fujitsu, E. Matsumura, K. Takeshita, and I. Mochida, *J. Org. Chem.*, **46**, 5353 (1981).
498. Y. Ohgo, S. Takeuchi, and J. Yoshimura, *Bull. Chem. Soc. Jpn.*, **44**, 283 (1971).
499. I. Wender, H. Greenfield, and M. Orchin, *J. Am. Chem. Soc.*, **73**, 2656 (1951).
500. I. Ercoli and R. F. Torregross, *Chim. Ind. (Milan)*, **40**, 552 (1958).
501. W. J. Suggs, S. D. Cox, R. H. Crabtree, and J. M. Quirke, *Tetrahedron Lett.*, **22**, 303 (1981).
502. E. Ucciani, R. Lai, and L. Tanguy, *C. R. Acad. Sci., Ser. C*, **283**, 17 (1976).
503. T. Mizoroki, K. Seki, S. Meguro, and A. Ozaki, *Bull. Chem. Soc. Jpn.*, **50**, 2148 (1977).
504. I. Ojima, T. Kogure, and Y. Nagai, *Tetrahedron Lett.*, 5035 (1972).
505. C. Djerassi and J. Gutzwiller, *J. Am. Chem. Soc.*, **88**, 4537 (1966).
506. S. Nishimura and K. Tsuneda, *Bull. Chem. Soc. Jpn.*, **42**, 852 (1969).
507. S. Nishimura, T. Ichino, A. Akimoto, and K. Tsuneda, *Bull. Chem. Soc. Jpn.*, **46**, 279 (1973).

508. S. Nishimura, H. Mori, K. Tsuneda, and O. Yumoto, *Bull. Chem. Soc. Jpn.*, **48**, 2603 (1975).
509. A. Akimoto, T. Ichino, H. Mori, S. Nichimura, and K. Tsuneda, *Bull. Chem. Soc. Jpn.*, **48**, 2852 (1975).
510. A. E. Greene, J. C. Muller, and G. Ourisson, *J. Org. Chem.*, **39**, 186 (1974).
511. T. F. Yen and G. V. Chilingarian (Eds.), *Oil Shale*, Vol. 5, Elsevier, Amsterdam, (1976).
512. J. W. Larson (Ed.), *Organic Chemistry of Coal*, American Chemical Society, Washington, DC, *ACS Symp. Ser.*, No. 7 (1978).
513. D. D. Whitehurst (Ed.), *Coal Liquefaction Fundamentals*, American Chemical Society, Washington, DC, *ACS Symp. Ser.*, No. 139 (1980).
514. D. S. Breslow, A. S. Matlock, and M. F. Sloan, *J. Am. Chem. Soc.*, **95**, 4014 (1963).
515. W. R. Kroll, *J. Catal.*, **15**, 281 (1969).
516. S. J. Lapporte and W. R. Schuett, *J. Org. Chem.*, **28**, 1947 (1963).
517. S. J. Lapporte, *Ann. N. Y. Acad. Sci.*, **158**, 510 (1969).
518. V. C. Lipovich, F. K. Schmidt, and I. V. Kalechits, *Kinet. Katal.*, **8**, 939 (1968).
519. K. A. Klindist and M. Boudart, *J. Catal.*, **28**, 322 (1973).
520. H. M. Feder and J. Halpern, *J. Am. Chem. Soc.*, **97**, 7187 (1975).
521. J. Halpern and R. Sweany, *J. Am. Chem. Soc.*, **99**, 8335 (1977).
522. P. D. Taylor and M. Orchin, *J. Org. Chem.*, **37**, 3913 (1972).
523. N. L. Holy, *J. Org. Chem.*, **44**, 239 (1979).
524. I. Rajca, A. Borowski, and A. Rajca, *Symp. Rhodium Homog. Catal.* (*Proc.*), 58 (1978); *Chem. Abstr.*, **90**, 12835 (1979).
525. R. H. Fish, A. D. Thormodsen, and G. A. Cremer, *J. Am. Chem. Soc.*, **104**, 5234 (1982).
526. R. A. Grey, G. P. Pez, and H. A. Wall, *J. Am. Chem. Soc.*, **102**, 5948 (1980).
527. E. L. Muetterties and F. J. Hirsekorn, *J. Am. Chem. Soc.*, **96**, 4063 (1974).
528. E. L. Muetterties and F. J. Hirsekorn, *J. Am. Chem. Soc.*, **95**, 5419 (1973).
529. E. L. Muetterties and F. J. Hirsekorn, *J. Chem. Soc., Chem. Commun.*, 683 (1973).
530. M. C. Rakowski, F. J. Hirsekorn, L. S. Stuhl, and E. L. Muetterties, *Inorg. Chem.*, **15**, 2379 (1976).
531. L. S. Stuhl, M. PuBois, M. Rakowski, F. J. Hirsekorn, J. R. Bleeke, B. E. Stevens, and E. L. Muetterties, *J. Am. Chem. Soc.*, **100**, 2405 (1978).
532. E. L. Muetterties, M. C. Rakowski, F. J. Hirsekorn, B. W. Larson, V. J. Basus, and F. A. L. Anet, *J. Am. Chem. Soc.*, **97**, 1266 (1975).
533. M. J. Russel, C. White, and P. M. Maitlis, *J. Chem. Soc., Chem. Commun.*, 427 (1977).
534. P. M. Maitlis, *Adv. Chem. Ser.*, No. 173, 31 (1979).
535. E. O. Fischer and C. Eisenbroich, *Chem. Ber.*, **103**, 162 (1970).
536. M. Y. Darensbourg and E. L. Muetterties, *J. Am. Chem. Soc.*, **100**, 7425 (1978).
537. E. L. Muetterties and J. R. Bleeke, *Acc. Chem. Res.*, **12**, 324 (1979).
538. M. A. Bennett, T. N. Huang, A. K. Smith, and T. W. Turney, *J. Chem. Soc., Chem. Commun.*, 582 (1978).
539. M. A. Bennett, T. N. Huang, and T. W. Turney, *J. Chem. Soc., Chem. Commun.*, 312 (1979).
540. M. A. Bennett, *ChemTech*, 446 (1980).
541. M. Murakami and J. Kong, *Bull. Chem. Soc. Jpn.*, **36**, 763 (1963).
542. J. Jardine and F. J. McQuillin, *J. Chem. Soc., Chem. Commun.*, 626 (1970).
543. A. Levi, G. Modena, and G. Scorrano, *J. Chem. Soc., Chem. Commun.*, 6 (1975).
544. T. Hayashi, A. Katsamura, M. Konishi, and M. Kumada, *Tetrahedron Lett.*, 425 (1979).
545. S. Murahashi, S. Horiie, and T. Jo, *Nippon Kagaku Zasshi*, **79**, 68 (1958).
546. A. Nakamura and N. Hagihara, *Osaka Univ. Mem.*, **15**, 195 (1958).
547. S. Murahashi and S. Horiie, *Bull. Chem. Soc. Jpn.*, **33**, 78 (1960).
548. J. F. Knifton, *J. Org. Chem.*, **41**, 1200 (1976).
549. J. F. Knifton, in *Catalysis in Organic Syntheses* (Eds. P. N. Rylander and H. Greenfield), Academic Press, New York, 1976, p. 257.
550. M. Murakami, R. Kawai and K. Suzuki, *J. Chem. Soc. Jpn.*, **84**, 669 (1963).
551. M. Saito, Y. Onda, and Y. Murayama, *Eur. Pat. Appl.*, EP37716 (1980); *Chem. Abstr.*, **96**, 12101d (1980).
552. G. W. Parshall, *Homogeneous Catalysis*, Wiley–Interscience, New York (1980).
553. D. L. Reger and M. M. Habib, *J. Mol. Catal.*, **4**, 315 (1978).
554. D. L. Reger, M. M. Habib, and B. J. Fauth, *Tetrahedron Lett.*, 115 (1979).
555. D. Chapman, *Pure Appl. Chem.*, **50**, 627 (1978).

556. T. D. Madden and P. J. Quinn, *Biochem. Soc. Trans.*, **6**, 1345 (1978).
557. C. Restall, M. P. Percival, P. Williams, P. Q. Quina, and D. Chapman, *Biochim. Biophys. Acta*, **555**, 119 (1979).
558. W. E. Peel and A. E. R. Thomson, *Leuk. Res.*, **7**, 193 (1983).
559. C. Restall, D. Chapman, and P. J. Quinn, *Biochem. Soc. Trans.*, **7**, 366 (1979).
560. J. P. Collmann, L. S. Hegedus, M. P. Cooke, J. R. Norton, G. Dolcetti, and D. N. Marquardt, *J. Am. Chem. Soc.*, **94**, 1789 (1972).
561. P. Eisenberger, J. Reed, B. K. Teo, and B. M. Kincald, *J. Am. Chem. Soc.*, **100**, 2375 (1978).
562. D. J. A. McCaffrey, S. G. Murray, P. N. Nicholson, and F. R. Hartley, *J. Organomet. Chem.*, **206**, 347 (1981).
563. J. M. Brown and H. Molinari, *Tetrahedron Lett.*, 2933 (1979).
564. R. H. Grubbs, L. C. Kroll, and E. M. Sweet, *J. Macromol. Sci., Chem.*, **7**, 1047 (1973).
565. C. P. Nicolaides and N. J. Coville, *J. Organomet. Chem.*, **222**, 285 (1981).
566. W. D. Bonds, C. H. Brubaker, E. S. Chaidrasekaran, C. Gibbson, R. H. Grubbs, and L. C. Kroll, *J. Am. Chem. Soc.*, **97**, 2128 (1975).
567. G. L. Baker, S. J. Fritschel, J. R. Stille, and J. K. Stille, *J. Org. Chem.*, **46**, 2960 (1981).
568. G. L. Baker, S. J. Fritschel, J. R. Stille, and J. K. Stille, *J. Org. Chem.*, **46**, 2954 (1981).
569. T. Masuda and J. K. Stille, *J. Am. Chem. Soc.*, **100**, 268 (1978).
570. N. Takaishi, H. Imai, C. A. Bertelo, and J. K. Stille, *J. Am. Chem. Soc.*, **100**, 264 (1978).
571. A. R. Sanger and L. R. Schallig, *J. Mol. Catal.*, **3**, 101 (1978).
572. I. Tkatchenko, *C. R. Acad. Sci., Ser. C*, **282**, 229 (1976).
573. R. G. Allum, R. D. Hancock, I. V. Howell, R. C. Pitkelthy, and P. J. Robinson, *J. Catal.*, **43**, 322 (1976).
574. C. A. Pittmann, S. K. Wu, and S. E. Jacobson, *J. Catal.*, **44**, 87 (1976).
575. I. V. Howell, R. D. Hancock, R. C. Pitkelthy and P. J. Robinson, in *Catalysis, Homogeneous and Heterogeneous* (Eds. B. Delman and G. Jannes), Elsevier, Amsterdam, 1975, p. 349.
576. K. G. Allum, R. D. Hancock, N. Howell, S. McKenzie, R. C. Pitkelthy, and P. J. Robinson, *J. Organomet. Chem.*, **87**, 203 (1975).
577. M. D. Ward and J. Schwartz, *J. Mol. Catal.*, **11**, 397 (1981).
578. M. D. Ward and J. Schwartz, *J. Am. Chem. Soc.*, **103**, 5253 (1981).
579. K. G. Allum, R. D. Hancock, I. V. Howell, J. E. Lester, S. McKenzie, R. C. Pitkelthy, and P. J. Robinson, *J. Organomet. Chem.*, **107**, 393 (1976).
580. Z. M. Michalska, *J. Mol. Catal.*, **3**, 125 (1978).
581. A. K. Smith, J. M. Basset, and P. M. Maitlis, *J. Mol. Catal.*, **2**, 223 (1977).
582. K. D. Neuberg, *PhD Thesis*, University of Stanford (1978).
583. Z. C. Brzezinska, W. R. Cullen, and G. Strukul, *Can. J. Chem.*, **58**, 750 (1980).
584. T. N. Huang and J. Schwartz, *J. Am. Chem. Soc.*, **104**, 5244 (1982).
585. M. D. Ward and J. Schwartz, *Organometallics*, **1**, 1030 (1982).
586. C. L. Reichel and M. S. Wrighton, *J. Am. Chem. Soc.*, **103**, 7181 (1981).
587. C. L. Reichel and M. S. Wrighton, *Inorg. Chem.*, **19**, 3858 (1980).
588. J. B. Kinney, R. H. Staley, C. L. Reichel, and M. S. Wrighton, *J. Am. Chem. Soc.*, **103**, 4273 (1981).
589. A. F. Borowski, D. J. Cole-Hamilton, and G. Wilkinson, *Nouv. J. Chim.*, **2**, 137 (1978).
590. M. E. Wilson and G. M. Whitesides, *J. Am. Chem. Soc.*, **100**, 3061 (1978).
591. M. E. Wilson, R. G. Nozzo, and G. M. Whitesides, *J. Am. Chem. Soc.*, **100**, 2269 (1979).
592. R. G. Nozzo, D. Feitler, and G. M. Whitesides, *J. Am. Chem. Soc.*, **101**, 3683 (1979).
593. R. G. Nozzo, S. L. Haynic, M. E. Wilson, and G. M. Whitesides, *J. Org. Chem.*, **46**, 2861 (1981).
594. A. K. Smith and J. M. Basset, *J. Mol. Catal.*, **2**, 229 (1977).
595. A. J. Deeming and S. Hasso, *J. Organomet. Chem.*, **114**, 313 (1976).
596. B. Besson, A. Choplin, L. D'ornelas, and J. M. Basset, *J. Chem. Soc., Chem. Commun.*, 842 (1982).
597. R. Mutin, W. Abboud, J. M. Basset, and D. Sinou, *Polyhedron*, **2**, 539 (1983).
598. R. Laine, *J. Mol. Catal.*, **14**, 137 (1982).
599. W. Abboud, Y. Ben Taarit, R. Mutin, and J. M. Basset, *J. Organomet. Chem.*, **220**, C15 (1980).
600. W. Reimann, W. Abboud, J. M. Basset, R. Mutin, G. L. Rempel, and A. K. Smith, *J. Mol. Catal.*, **9**, 349 (1980).
601. J. B. Keister and J. R. Shapley, *J. Am. Chem. Soc.*, **98**, 1056 (1976).
602. A. J. Deeming, S. Hasso, and M. Underhill, *J. Chem. Soc., Dalton Trans.*, 1614 (1975).
603. J. Halpern, *Science*, **217**, 401 (1982).

604. L. Marko and J. Bakos, *Aspects. Homog. Catal.*, **4**, 145 (1981).
605. D. Valentine and J. W. Scott, *Synthesis*, 329 (1978).
606. P. A. Chaloner and D. Parker, in *Reactions of Coordinated Ligands* (Ed. P. S. Braterman), Plenum New York, 1986.
607. L. Horner, H. Buthe, and H. Siegel, *Tetrahedron Lett.*, 4023 (1968).
608. L. Horner, H. Siegel, and H. Buthe, *Angew. Chem., Int. Ed. Engl.*, **7**, 942 (1968).
609. W. Bergstein, A. Kleeman, and J. Martens, *Synthesis*, 76 (1981).
610. A. Miyashita, A. Yasuda, H. Tanaka, K. Toniimi, T. Ito, T. Souchi, and R. Noyori, *J. Am. Chem. Soc.*, 7932 (1980).
611. M. Fiorini and G. Giongo, *J. Mol. Catal.*, **7**, 41 (1980).
612. J. P. Amma and J. K. Stille, *J. Org. Chem.*, **47**, 468 (1982).
613. D. P. Riley and R. E. Shumate, *J. Org. Chem.*, **45**, 5187 (1980).
614. J. D. Oliver and D. P. Riley, *Organometallics*, **2**, 1032 (1983).
615. P. A. MacNeil, N. K. Roberts, and B. A. Bosnich, *J. Am. Chem. Soc.*, **103**, 2273 (1981).
616. K. Onuma, T. Ito, and N. Nakamura, *Bull. Chem. Soc. Jpn.*, **53**, 2012 (1980).
617. K. Osakada, T. Ikariya, M. Saburi, and S. Yoshikawa, *Chem. Lett.*, 1691 (1981).
618. B. D. Vineyard, W. S. Knowles, M. J. Sabacky, G. L. Bachman, and D. J. Weinkauff, *J. Am. Chem. Soc.*, **99**, 5946 (1977).
619. M. D. Fryzuk and B. Bosnich, *J. Am. Chem. Soc.*, **99**, 6262 (1977).
620. H. B. Kagan and T. P. Dang, *J. Am. Chem. Soc.*, **94**, 6429 (1972).
621. K. Achiwa, *Fundam. Res. Homog. Catal.*, **3**, 549 (1979).
622. K. Achiwa, *Chem. Lett.*, 561 (1978).
623. K. Achiwa, T. Kogure, and I. Ojima, *Chem. Lett.*, 297 (1978).
624. I. Ojima, T. Kogure, and K. Achiwa, *J. Chem. Soc., Chem. Commun.*, 428 (1977).
625. T. Hayashi, T. Mise, and M. Kumada, *Tetrahedron Lett.*, 4351 (1976).
626. T. Hayashi, A. Katsamura, M. Konishi, and M. Kumada, *Tetrahedron Lett.*, 425 (1979).
627. J. M. Brown and D. Parker, *J. Chem. Soc., Chem. Commun.*, 342 (1980).
628. W. C. Christopfel and B. D. Vineyard, *J. Am. Chem. Soc.*, **101**, 4406 (1979).
629. K. E. Koening, G. L. Bachman, and B. D. Vineyard, *J. Org. Chem.*, **45**, 2362 (1980).
630. U. Hengartner, D. Valentine, K. K. Johnson, M. E. Larscheid, F. Pigott, F. Schiedl, J. W. Scott, R. C. Sun, J. M. Townsend, and T. H. Williams, *J. Org. Chem.*, **44**, 3741 (1979).
631. T. Hayashi, K. Kanehira, and M. Kumada, *Tetrahedron Lett.*, **22**, 4417 (1981).
632. M. D. Fryzuk and B. Bosnich, *J. Am. Chem. Soc.*, **100**, 5491 (1978).
633. M. D. Fryzuk and B. Bosnich, *J. Am. Chem. Soc.*, **101**, 3043 (1979).
634. D. Parker, *J. Chem. Soc., Perkin Trans. 2*, 83 (1983).
635. A. P. Stoll and R. Suess, *Helv. Chim. Acta.*, **57**, 2487 (1974).
636. J. D. Morrison and W. F. Masler, *J. Org. Chem.*, **39**, 270 (1974).
637. D. H. Valentine, *US Pat.*, 4115417 (1978).
638. J. M. Brown and P. A. Chaloner, *J. Chem. Soc., Chem. Commun.*, 344 (1980).
639. J. M. Brown and D. Parker, *J. Org. Chem.*, **47**, 2722 (1982).
640. ASC Chan and J. Halpern, *J. Am. Chem. Soc.*, **102**, 838 (1980).
641. J. M. Brown and P. A. Chaloner, *Tetrahedron Lett.*, 1877 (1978).
642. K. Achiwa, P. A. Chaloner, and D. Parker, *J. Organomet. Chem.*, **218**, 249 (1981).
643. J. M. Brown, P. A. Chaloner, and G. E. Morris, *J. Chem. Soc., Chem. Commun.*, 644 (1983).
644. D. A. Evans and M. M. Morrissey, *J. Am. Chem. Soc.*, **106**, 3866 (1984).
645. V. S. Rao and A. S. Perlin, *Carbohydn. Res.*, **83**, 175 (1980).
646. G. Brieger, T. J. Nestrick, and T. H. Fu, *J. Org. Chem.*, **44**, 1876 (1979).
647. D. C. Tabor, F. H. White, L. W. Collier, and S. A. Evens, *J. Org. Chem.*, **48**, 1638 (1983).
648. L. A. Buralkina, V. P. Zhakin, N. U. Utesheva, and D. V. Sokolskii, *Khim. Tekhnol.* (*Alma-Ata*), **19**, 123 (1976); *Chem. Abstr.*, **91**, 39053 (1979).
649. M. J. Andrews and C. N. Pillai, *Indian J. Chem., Sect. B*, **16**, 465 (1978).
650. R. A. W. Johnstone and A. H. Wilby, *Tetrahedron*, **37**, 3667 (1981).
651. I. D. Entwistle, R. A. W. Johnstone, and T. J. Povall, *J. Chem. Soc., Perkin Trans. 1*, 1300 (1975).
652. Y. Kikugawa and M. Kashimura, *Chem. Pharm. Bull.*, **30**, 3386 (1982).
653. R. Colombo, *Chem. Lett.*, **9**, 1119 (1980).
654. R. Colombo, *J. Chem. Soc., Chem. Commun.*, 1012 (1981).
655. S. Hannessian, J. I. Liak, and B. Vanasse, *Synthesis*, **5**, 396 (1981).
656. H. Haertner, *Kontakte*, 3 (1980).

657. D. Beaupere, L. Nadia, R. Uzon, and P. Bauer, *J. Mol. Catal.*, **14**, 129 (1982).
658. T. Nishiguchi and K. Fukuzumi, *J. Am. Chem. Soc.*, **96**, 1893 (1974).
659. C. H. Cheng, D. E. Hendriksen, and R. Eisenberg, *J. Am. Chem. Soc.*, **99**, 2891 (1977).
660. J. Azrau, O. Buchanan, M. Orchin, and J. Blum, *J. Org. Chem.*, **49**, 1327 (1984).
661. J. Blum, Y. Sasson, and S. Iflah, *Tetrahedron Lett.*, 1015 (1972).
662. H. B. Henbest and T. R. B. Mitchell, *J. Chem. Soc. C*, 785 (1970).
663. Y. M. H. Haddad, H. B. Henbest, J. Husband S, T. R. B. Mitchell, and J. Trocha-Grimshaw, *J. Chem. Soc., Perkin Trans. 1*, 596 (1974).
664. J. C. Orr, M. Merserau, and A. Sanford, *J. Chem. Soc., Chem. Commun.*, 162 (1970).
665. K. Ohkubo, I. Terada, and K. Yoshinaga, *Inorg. Nucl. Chem. Lett.*, **15**, 421 (1979).

The Chemistry of the Metal—Carbon Bond, Vol. 4
Edited by F. R. Hartley

CHAPTER **12**

Mechanism of homogeneous hydrogenation

F. H. JARDINE

Department of Physical Sciences, North East London Polytechnic, Romford Road, London E15 4LZ, UK

I. INTRODUCTION

During the past two decades several effective homogeneous catalysts for alkyne and alkene hydrogenation have been discovered. These often exhibit many important advantages over the finely divided late transition metals that have customarily been used to catalyse these hydrogenations. The most immediately apparent advantage comes from the change to a homogeneous system, which eliminates all problems of reproducibility of catalyst particle size and surface properties.

Homogeneous catalysts are usually specific for alkene or alkyne hydrogenation, and other groups containing multiple interatomic bonds are rarely reduced during the hydrogenation. Thus, the specific hydrogenations shown in equation 1–4 can be achieved with homogeneous catalysts[1–3]. Exceptionally, under harsh conditions, nitro groups can be reduced (equations 5 and 6)[4,5]. However, the latter reaction may be heterogeneously catalysed by traces of rhodium since $[Rh_6(CO)_{16}]$ has been shown to decompose in the presence of hydrogen[6].

$$RCH{=}CHCHO \xrightarrow[\text{[RhCl(PPh}_3)_3]]{H_2} RCH_2CH_2CHO \qquad (1)$$

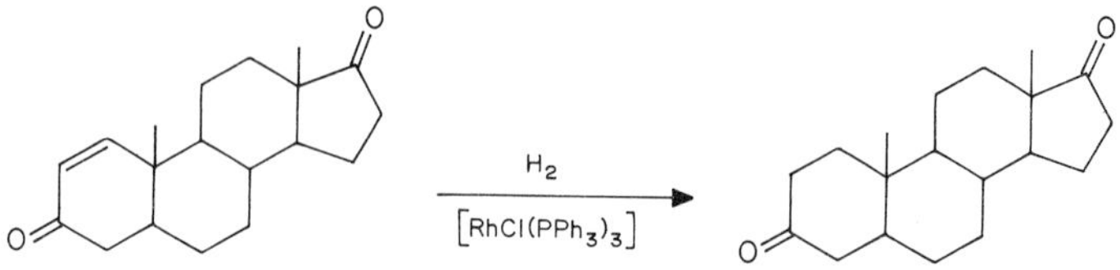

$$\xrightarrow[\text{[RhCl(PPh}_3)_3]]{H_2} \qquad (2)$$

$$CH_2{=}CHCH_2CN \xrightarrow[\text{[RhH(CO)(PPh}_3)_3]]{H_2} Pr^nCN \qquad (3)$$

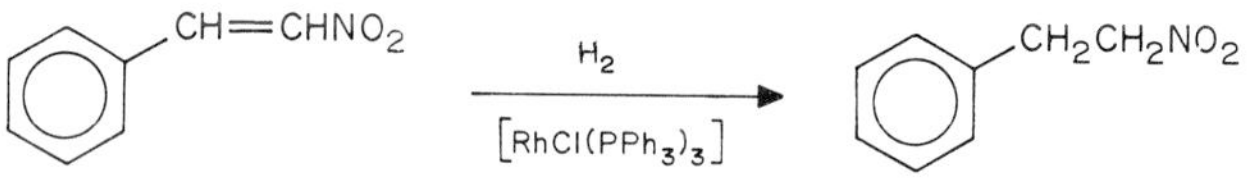

$$PhCH{=}CHNO_2 \xrightarrow[\text{[RhCl(PPh}_3)_3]]{H_2} PhCH_2CH_2NO_2 \qquad (4)$$

$$ArNO_2 + 3H_2 \xrightarrow[\text{80 atm, 130 °C}]{\text{[RuCl}_2\text{(PPh}_3)_3]} ArNH_2 + 2H_2O \qquad (5)$$

$$Ar = Ph, 4\text{-}C_6H_4NO_2, 1\text{-}C_{10}H_7$$

$$PhNO_2 + 3H_2 \xrightarrow[\text{PhCH}_2\text{NMe}_2]{\text{[Rh}_6\text{(CO)}_{16}\text{], 80 °C}} PhNH_2 + 2H_2O \qquad (6)$$

No intermediates have been detected in the reduction of aromatic nitro compounds catalysed by $[RuCl_2(PPh_3)_3]$, but most feasible intermediates have been shown to be reduced more rapidly than their parent nitro compounds[4]. Additionally, diazonium compounds are reduced by hydrogen in the presence of $[RhH(PPh_3)_3]$ or $[RhCl(PPh_3)_3]$[7].

The reduction of keto compounds by homogeneous catalysts is more widespread and is discussed in Section VII.A.

Aromatic nuclei are not reduced under most hydrogenation conditions. This permits aromatic hydrocarbons to be used as solvents for the reactions. Nevertheless, anthracene can be reduced at 60 °C (equation 7)[8].

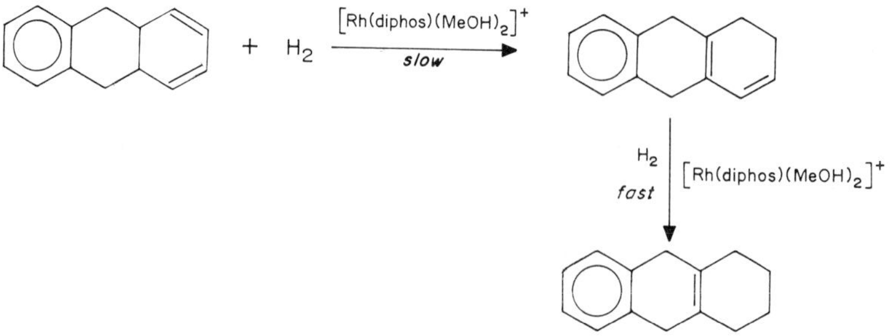

$$+ H_2 \xrightarrow[\text{slow}]{[Rh(diphos)(MeOH)_2]^+} \quad \xrightarrow[\text{fast}]{H_2,\ [Rh(diphos)(MeOH)_2]^+} \qquad (7)$$

Both $[RhH(PPr^i_3)_3]$ and $[Rh_2H_2(\mu\text{-}N_2)(PCy_3)_4]$ catalyse the hydrogenation of aliphatic nitriles under ambient conditions. However, when the reduction of unsaturated nitriles is attempted, the alkene bond is preferentially reduced[9].

The earliest homogeneous catalysts suffered from severe practical limitations. For example, aqueous solutions of $[Co(CN)_5]^{3-}$[10], $RhCl_3{\cdot}3H_2O$[11], or chlororuthenium(II) species[12] were inevitably restricted to the hydrogenation of water-soluble substituted alkenes such as maleic acid. Further, it was found that the rhodium and ruthenium salts underwent autoreduction to the metals after a few catalytic cycles. Obviously practical catalysts should be soluble in non-polar solvents, such as most alkene hydrocarbons and their derivatives, and be resistant to autoreduction.

Hydrido transition metal complexes were believed to participate in both the catalytic cycles and autoreductions. However, stable hydrido complexes of several platinum group metals had been isolated by employing π-acid ligands. It seemed likely, therefore, that stable catalysts that were soluble in organic solvents could be obtained in conjunction with ligands such as tertiary phosphines. However, although many such tertiary phosphine complexes were known at that time, few, if any, also retained the ability of heterogeneous transition metal catalysts to activate both hydrogen and alkenes.

Activation of these two reagents by a transition metal complex is facilitated by the central metal atom being in a low oxidation state and the complex as a whole being coordinatively unsaturated. The low oxidation state is important since it permits the oxidative addition of molecular hydrogen. Coordinative unsaturation is important since alkenes normally do not compete effectively for coordination sites. Alkene coordination is synonymous with alkene activation.

Many complexes are catalytically inactive because they fail to activate both reagents. Thus, *trans*-$[RhCl(CO)(PPh_3)_2]$ is a coordinatively unsaturated rhodium(I) complex which adds hydrogen oxidatively under mild conditions (equation 8)[13]. However, the resulting rhodium(III) dihydrido complex is a coordinatively saturated, 18-electron species which is unable to coordinate an alkene ligand. It is catalytically inactive at ambient temperatures. Above 40 °C the carbonyl complex loses one triphenylphosphine ligand to form a 16-electron species (equation 9). This five-coordinate complex can react with an alkene molecule and, since the hydrido ligands in the resulting alkene complex can be transferred to the alkene, *trans*-$[RhCl(CO)(PPh_3)_2]$ functions as a hydrogenation catayst at elevated temperatures.

$$[RhCl(CO)(PPh_3)_2] + H_2 \xrightarrow{C_6H_6} [RhCl(H)_2(CO)(PPh_3)_2] \tag{8}$$

$$[RhCl(H)_2(CO)(PPh_3)_2] \underset{+PPh_3}{\overset{-PPh_3}{\rightleftharpoons}} [RhCl(H)_2(CO)PPh_3] \tag{9}$$

There are three types of catalytic system which differ principally in the nature of the species that activates molecular hydrogen (Fig. 1). If the catalyst itself performs this task, then the reaction is said to follow the dihydride route. By analogy, if the dihydrogen is added oxidatively to the alkene complex, then this catalytic cycle is known as the alkene route. It can be seen in Fig. 1 that these two catalytic cycles have identical alkene and alkyl intermediates, and it is often difficult to determine by which route the reaction proceeds. Finally, if a monohydrido system gives rise to an alkyl complex which is capable of activating molecular hydrogen, then the third pathway, the alkyl route is followed. The three types of catalytic system will be considered in more detail in Sections III–V.

It has also proved possible to effect the hydrogenation of certain alkenes via a free radical mechanism using $[CoH(CO)_4]$ as the catalyst. This is outside the scope of this chapter since cobalt—carbon bonds are not directly involved in the final transfer of hydrogen to the alkyl radical[14].

The transition metal species involved in the catalytic cycles are present in very low concentrations (*ca.* 1 mM), and are, of necessity, highly reactive. Accordingly, the systems are very sensitive to traces of impurities. Hydroperoxides, which are easily and rapidly

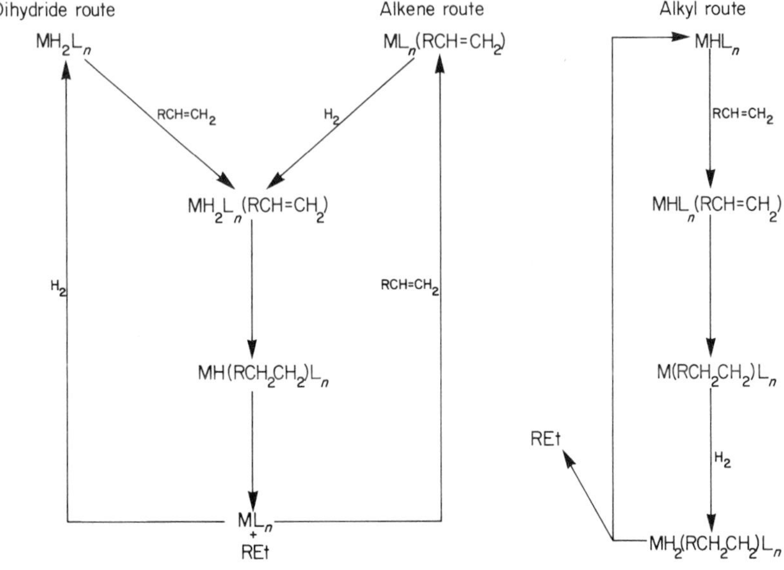

FIGURE 1. Three possible routes for homogeneous catalytic hydrogenation of alkenes

formed when alkenes are stored in contact with air, are the most troublesome impurity. When present in high concentrations these oxidize the central metal and its oxidizable ligands and thus destroy the catalytic properties of the complex. Paradoxically, low concentrations of hydroperoxides can sometimes induce an increase in the rate of hydrogenation. This is achieved by oxidation of one of the tertiary phosphine ligands (equation 10). The tertiary phosphine oxide does not recoordinate to the metal, which permits alkene to replace the original tertiary phosphine in the coordination sphere.

$$PR_3 + R'OOH \rightarrow OPR_3 + R'OH \qquad (10)$$

Particular attention should be paid to alkene purification in catalytic studies, especially those of a quantitative nature. Failure to remove hydroperoxides leads to spurious results[15-17].

The formation of reactive hydrido complexes during the catalytic cycle places some limitations on permissible solvents. Chlorinated solvents are often unsuitable since they are attacked by hydrido complexes (equation 11)[18]. Similarly, dimethyl sulphoxide is reduced by hydrogen to dimethyl sulphide in the presence of rhodium trichloride[19]. Conversely, successive hydrogen and carbonyl abstraction from primary alcohols can poison the catalyst by forming an inactive carbonyl complex[20].

$$[RhCl(H)_2(PPh_3)_2] + CHCl_3 \rightarrow [RhCl(PPh_3)_2] + CH_2Cl_2 + HCl \qquad (11)$$

The best solvents are probably aromatic hydrocarbons, although admixture of these with primary alcohols often increases the rate of hydrogenation.

II. ACTIVATION OF REACTANTS

A. Hydrogen

Although transition metal hydrido complexes are numerous and can be prepared from a wide variety of hydridic species[21], only those that can be prepared directly from molecular hydrogen can repeatedly participate in catalytic cycles.

The activation of hydrogen by transition metal complexes has been reviewed by Brothers[22]. She pointed out that there are three possible processes for hydrogen activation: oxidative addition, homolysis, or heterolysis, shown in general forms in equations 12, 13 and 14, respectively.

$$M^{n+}L_x + H_2 \rightarrow M^{(n+2)+}(H)_2L_y \tag{12}$$

$$2M^{n+}L_x + H_2 \rightarrow 2M^{(n+1)+}HL_y \tag{13}$$

$$M^{n+}L_x + H_2 \rightarrow M^{(n+1)+}HL_y + H^+ \tag{14}$$

In all these processes the hydrogen is nominally bound to the metal as the hydrido ligand, H^-. Of these processes, only the first is important in homogeneous catalytic cycles. Nevertheless, the second and third processes are often important in preparing, frequently *in situ*, the true hydrogenation catalyst from more accessible complexes. Examples of the last two reactions are provided by the catalyst preparations in equations 15 and 16[23,24].

$$[Co_2(CO)_8] + H_2 \rightarrow 2[CoH(CO)_4] \tag{15}$$

$$[RuCl_2(PPh_3)_3] + H_2 + Et_3N \xrightarrow{C_6H_6} [RuHCl(PPh_3)_3] + [Et_3NH]Cl \tag{16}$$

Oxidative addition of hydrogen, and more importantly the simultaneous ability of the higher oxidation state complex to undergo reductive elimination of alkane (see below), make it essential that the metal posesses two well defined oxidation states corresponding to d^n and d^{n-2} electronic configurations.

The stability of the hydrido complex can be increased if it also contains π-acid ligands. These ligands tend to inhibit the attack of electrophilic reagents at the metal—hydrogen bond. However, if too great a degree of stability is conferred upon the metal—hydrogen bond the efficiency of the catalyst will be impaired. Normally there is a fine balance of steric and electronic factors in a successful catalyst.

B. Alkenes

Alkene complexes, given the great number of potential alkene ligands, are even more numerous than hydrido complexes. Monoalkenes, nevertheless, do not readily displace other ligands from the coordination sphere. This can prevent catalytic activation of the alkene when strongly bound, involatile, π-acid ligands have been incroporated in the complex to preserve it from the ravages of autoreduction. Alkadienes coordinate more strongly to transition metals but their complexes commonly lack the ability to activate molecular hydrogen. Whilst this may arise from the same electronic source as the inability of many monoalkene complexes to add hydrogen oxidatively, many alkadienes chelate to the central metal through both C=C bonds and form coordinatively saturated complexes incapable of interacting with dihydrogen.

The most common solution to the problem of forming stable, coordinatively unsaturated complexes is to incorporate relatively bulky π-acid ligands. Such complexes

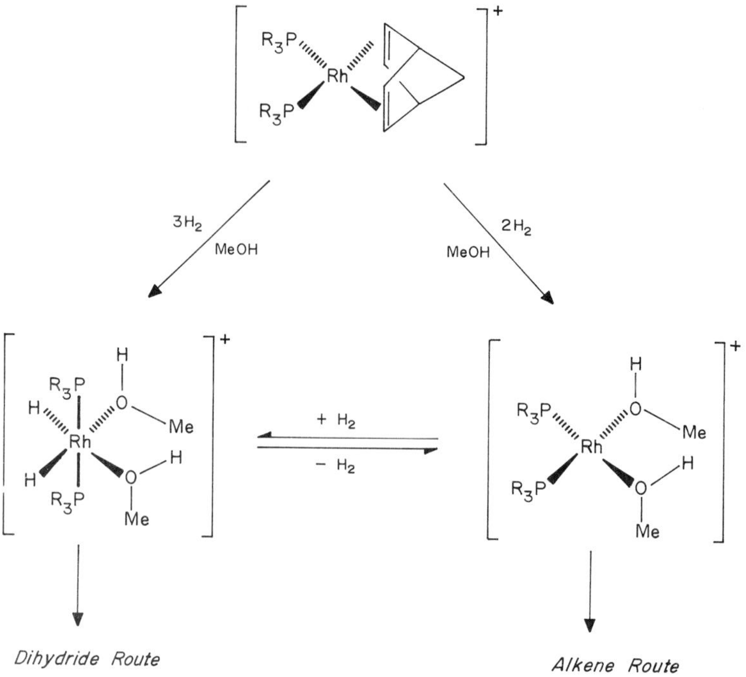

SCHEME 1. Generation of dihydride and alkene route catalysts from a common source.

give rise to coordinatively unsaturated or weakly solvated species on dissolution.

Another approach is to remove ligands from a coordinatively unsaturated complex in the presence of the alkene substrate. Cationic complexes of norbornadiene serve as catalytic precursors if the norbornadiene is removed by hydrogenation. Scheme 1 shows how the weakly solvated complexes produced can participate in either the dihydrido or the alkene route[25].

However, the formation of coordinatively unsaturated complexes capable of activating alkenes appears to be limited to late transition series elements, and is probably the reason why, to date, only these metals have been observed to exhibit widespread catalytic activity.

Alkene coordination to late transition metals is determined largely by the accessibility of the alkene bond. Classic studies on silver(I) complexes show that the complexity constants are directly related to the location, stereochemistry, and substitution of the C=C bonds[26].

The range of complexity constants encountered means that there is a wide variation of hydrogenation rates exhibited by a given catalyst. This variation in rates can be exploited in regioselective hydrogenations. Since this regioselectivity has steric origins, and the regioselectivity in heterogeneous catalysis often has electronic origins, different products can be obtained when the two types of catalyst are employed. Thus, the conjugated double bond in eremophilone is reduced in hydrogenations catalysed heterogeneously by palladium on charcoal (equation 17). The terminal double bond is hydrogenated when $[RhCl(PPh_3)_3]$ is the catalyst (equation 18)[27].

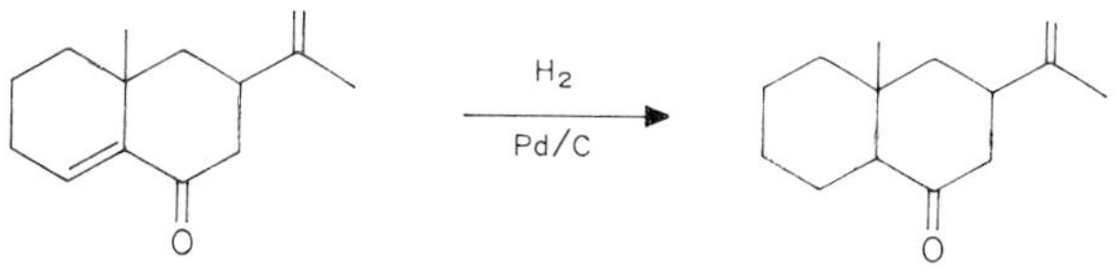

(17)

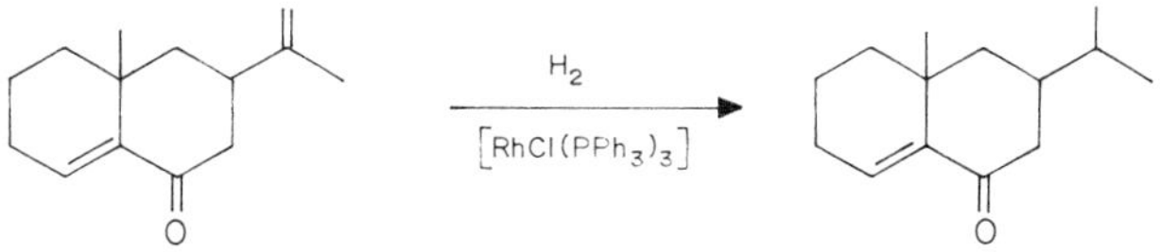

(18)

As noted above, alkene complexes are not particularly effective in activating molecular hydrogen. A consequence of this is that only a limited number of catalytic systems have, so far, been shown to follow the alkene route. Nevertheless, current research is concentrated on the alkene route since it has been found to give the best optical yields in the asymmetric hydrogenation of prochiral alkenes.

III. THE DIHYDRIDE ROUTE

The classic example of the dihydride route is afforded by Wilkinson's catalyst, $[RhCl(PPh_3)_3]$. This complex reacts with hydrogen to form a dihydrido species, $[RhCl(H)_2(PPh_3)_2]$. The rhodium(III) complex adds alkene and rapidly transfers the two hydrido ligands before eliminating alkane and recommencing the catalytic cycle[28]. The system is illustrated in Scheme 2.

The central catalytic cycle is very simple, but the parasitic side-reactions make studies of the system fairly difficult. The system has been frequently reinvestigated since its first discovery, but the only important change proposed in the mechanism has been the participation of an intermediate alkyl complex during the transfer of hydrogen from rhodium to alkene.

The major influence on the kinetic behaviour of the catalytic cycle is the complexation of alkene by the key intermediate $[RhCl(PPh_3)_2]$ in competition with its oxidative addition of hydrogen (equation 19). Only the ethene[28], trifluorochloroethene[29], and perfluoroethene[29] complexes have been isolated from the above reaction.

$$[RhCl(PPh_3)_2] + RCH{=}CHR' \rightleftharpoons [RhCl(PPh_3)_2(RCH{=}CHR')] \tag{19}$$

The undissociated ethene complex is incapable of activating molecular hydrogen. Indeed, ethene poisons its own reduction and that of other alkenes by the catalyst[28].

The rate-determining step in the hydrogenation is the attack of alkene on the dihydrido complex (equation 20). There is a wide variation in the rates of alkene hydrogenation in the system (Table 1). The rate of alkene hydrogenation can be approximately correlated with the steric hindrance created by substituents in the vicinity of the double bond. However, there is a slight compensatory effect arising from the complexity constants for alkenes varying in the same way for rhodium(I) species (equation 19) and rhodium(III) species (equation 20). Hence alkenes that are not readily activated by either complex result in an increase in the concentration of the dihydrido complex $[RhCl(H)_2(PPh_3)_2]$, since hydrogen competes more successfully for the common intermediate $[RhCl(PPh_3)_2]$.

$$[RhCl(H)_2(PPh_3)_2] + RCH{=}CHR' \rightarrow [RhCl(H)_2(PPh_3)_2(RCH{=}CHR')] \tag{20}$$

There are several objections to Halpern's proposal[30,31] that the rate-determining step is

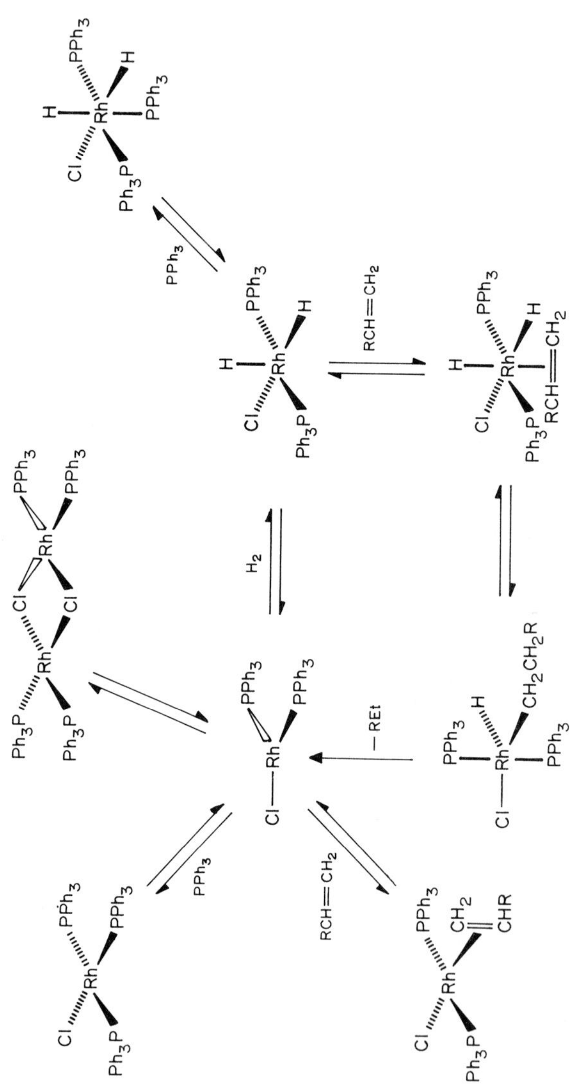

SCHEME 2. The dihydride route catalysed by $[RhCl(PPh_3)_3]$.

TABLE 1. Rates of hydrogen consumption by alkenes in $[RhCl(PPh_3)_3]$-catalysed reactions[a]

Alkene	Rate ($mmol\ H_2\ min^{-1}$)
Allyl alcohol	3.02
Styrene	2.56
Acenaphthylene	1.76
Cyclopentene	1.26
cis-pent-2-ene	1.01
Dodec-1-ene	1.01
Hex-1-ene	0.857
Cyclohexene	0.800
Octa-1, 7-diene	0.646
Methyl methacrylate	0.585
Cycloheptene	0.572
2-Methylpent-1-ene	0.500
cis-4-Methylpent-2-ene	0.458
Allyl cyanide	0.453
Acrylamide[b]	0.22
Hexa-1, 5-diene	0.209
Cycloocta-1, 3-diene	0.133
trans-4-Methylpent-2-ene	0.092
Penta-1, 3-diene	0.059
trans-Hex-3-ene	0.057
3-Chloro-2-methylpropene	0.034
1-Methylcyclohexene	0.026
3-Ethylpent-2-ene	0.017
2, 3-Dimethylbut-2-ene	0.002

[a]Under standard conditions: $[RhCl(PPh_3)_3]$, 1.25mM; alkene, 1.25 M; solvent, benzene; volume, 80 cm^3; temperature, 25 °C.
[b]Saturated solution.

the formation of an alkyl complex from the rhodium(III) alkene complex (equation 21). First, the dihydridoalkene complex has never been isolated whilst the existence of the complex $[RhCl(H)_2(PPh_3)_2]$ is well established, both in the catalytic system and as a compound in its own right. Further, if this is the rate-determining step then the wide variation in the rates of alkene hydrogenation is inexplicable. Finally, it may be noted that the activation energies determined for the hydrogenation of cycloalkenes do not reflect the strain energies that would be released if the formation of their cycloalkyl complexes were the rate-determining steps.

$$[RhCl(H)_2(PPh_3)_2(RCH{=}CHR')] \rightleftharpoons [RhCl(H)(RCH_2CHR')(PPh_3)_2] \quad (21)$$

The participation of an intermediate alkyl complex has some important consequences. It is the instability of this intermediate that gives rise to the hydroisomerization and isotope exchange reactions that are sometimes observed. Complexes of large alkyl ligands are inherently less stable than those containing small, unstrained alkyl ligands. The hydroisomerization of coronopilin occurs because of the instability of the alkyl complex[32].

$$\xrightarrow[\text{[RhCl(PPh}_3)_3]]{H_2} \qquad (22)$$

The secondary alkyl complex **1** formed during the hydrogenation of vinylcyclopropane is similarly unstable and decomposes to pent-2-ene[33]. However, the stability of the two alkyl complexes is sufficiently great to permit the formation of dihydrocoronopilin and ethylcyclopropane in the respective hydrogenations.

(1)

Similarly alkyl complexes are also of limited stability and can revert to dihydrido alkene complexes before transfer of the second hydrido ligand from rhodium can bring about reductive elimination of alkane. The reversion occurs via a β-hydride abstraction (Scheme 3). Such abstraction reactions can bring about the hydroisomerization if secondary alkyl complexes are formed. In deuteriations, isotope exchange can occur even if primary alkyl complexes are formed.

In general, $[RhCl(PPh_3)_3]$ is an excellent deuteriation catalyst and has been employed in the specific deuteriation of many alkenes that are not cleanly deuteriated by heterogeneous catalysts[34,35]. Nevertheless, it has been found that most alkenes undergo a small degree of scrambled addition of deuterium (Table 2)[36].

Despite the sequential addition of hydrogen or deuterium to alkene substrates, the overall addition is *cis*. This comes about by the synchronous addition of rhodium and hydrogen to the two carbon atoms that were previously linked by the alkene bond (equation 23). The second hydrogen atom is inserted into the rhodium—alkyl bond, thus

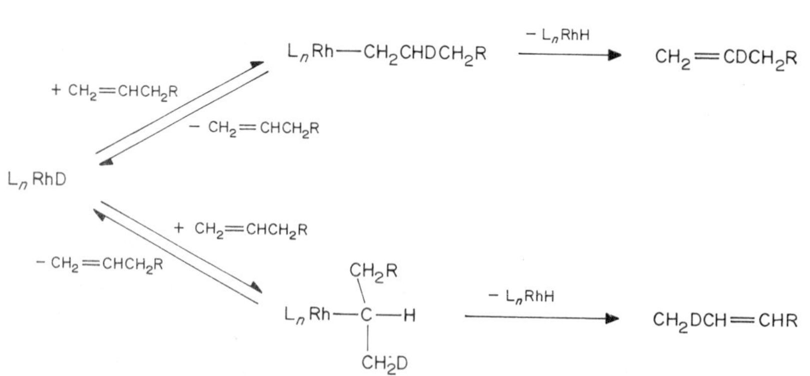

SCHEME 3. Alkene isomerization and isotope exchange arising from alkyl complex decomposition.

TABLE 2. Alkene deuteriation using $[RhCl(PPh_3)_3]$ catalyst

	Alkane-d_n (%)				
Alkene	d_0	d_1	d_2	d_3	d_4
Allylbenzene			99.4	0.6	
Allyl phenyl ether			100.0		
Cyclohexene			100.0		
Cyclohex-2-en-1-one			96.7	3.3	
Dec-1-ene	0.9	3.3	92.6	2.3	0.9
Dihydropyran			98.4	1.6	
3,4-Dihydro-2-methoxy-2*H*-pyran			100.0		
cis-Hept-2-ene	1.7	97.1	1.2		
trans-Hept-2-ene	2.9	95.4	1.7		
Hept-1-ene	3.1	2.1	91.1	1.6	2.1
Hex-1-ene			98.9	1.1	
Hex-5-en-2-one			100.0		
2-Methylbut-2-ene			97.4	2.6	
Methyl crotonate			85.7	9.5	4.8
6-Methylhept-5-en-2-one			80.0	15.8	4.2
2-Methylpent-1-ene			99.0	1.0	
Norbornylene			100.0		
Oct-1-ene	0.5	4.2	91.6	2.8	0.9
trans-Oct-4-ene	1.5	3.9	90.6	2.0	1.0
α-Pinene			93.5	3.7	2.8

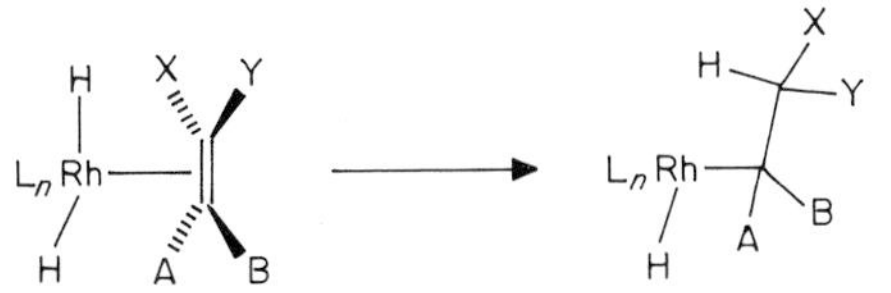

(23)

preserving the overall *cis* addition. The stereochemistry of the addition has been demonstrated by the deuteriation of substituted alkenes. Methyl cinnamate gives a *threo* product (equation 24)[37] whereas *cis*-alkenes give *erythro* products (equation 25)[38].

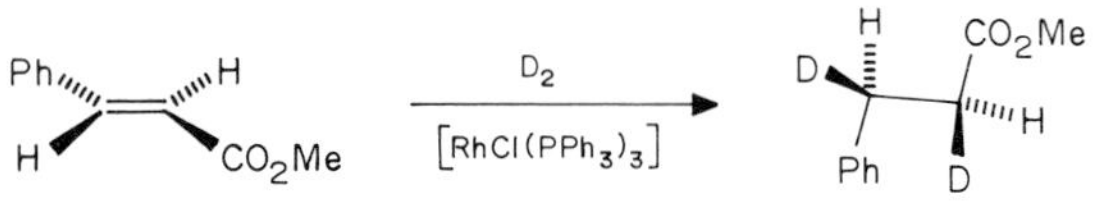

D_2, $[RhCl(PPh_3)_3]$ (24)

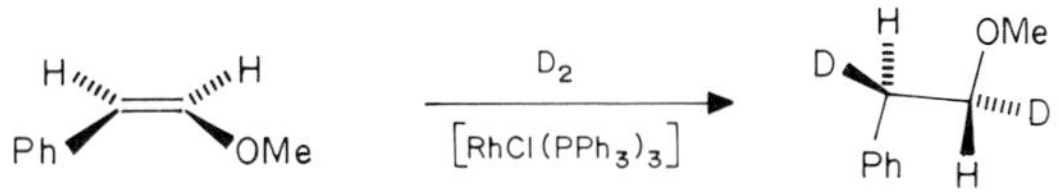

D_2, $[RhCl(PPh_3)_3]$ (25)

Although *cis* addition of hydrogen is always observed, both β-pinene[39] and 2-methylenebicyclo[2.2.1] heptane[40] yield two products from $[RhCl(PPh_3)_3]$-catalysed hydrogenations (equations 26 and 27).

$$[RhCl(PPh_3)_3] \quad (26)$$

90 : 10

$$[RhCl(PPh_3)_3] \quad (27)$$

73 : 27

Investigation of the product distributions from the hydrogenation of homologous exocyclic methylenehydrocarbons has demonstrated that hydrogen is preferentially added to the least hindered face of the alkene. However, the primary application of this concept has been in the asymmetric hydrogenation of prochiral alkene substrates.

The point at which the stereochemistry of unsymmetric alkene substrates becomes important in the reaction is not yet clear. In the above hydrocarbons the formation of the dihydrido(alkene)rhodium(III) complex was considered to determine the stereochemistry of the product. In the hydrogenation of alkylated cyclohex-2-enols the difference in stability between the two possible alkylrhodium(III) complexes was thought to exert the greatest effect on the product distribution[41].

The obvious advantages of $[RhCl(PPh_3)_3]$ as a homogeneous hydrogenation catalyst have not assisted the search for other catalysts that follow the dihydride route. Once a cursory investigation has failed to reveal any improvements that might be gained over the well established and accessible $[RhCl(PPh_3)_3]$, work on cognate systems has usually been abandoned. As a consequence, little mechanistic detail relating to other dihydride systems is known.

Many catalytically active dihydrido complexes have been considered to follow the dihydride route. This is not necessarily the case, particularly since most of these catalysts do not bring about rapid hydrogenation of the substrate. Although the lower oxidation state precursor of the hydrido complex may be capable of forming the dihydrido complex in uncompetitive reactions, in catalytic systems it may react preferentially with alkene and cause the catalysis to follow the alkene route. Other dihydride route catalysts include $[RuH_2(PPh_3)_4]$[42] and $[Co(N_2)(PPh_3)_3]$[43].

The bulk of the effort expended on investigation of dihydride route catalysts has not, therefore, been applied to the development of original systems but rather to seeking improvement of the $[RhCl(PPh_3)_3]$ system.

Usually improvement has been equated with faster hydrogenation rates. This can easily be achieved by reducing the concentration of free tertiary phosphine present. This free tertiary phosphine poisons the hydrogenation by occupying the sixth coordination site on $[RhCl(H)_2(PPh_3)_2]$ required for alkene coordination. Inevitably systems using $[RhX(PR_3)_3]$ catalysts have a PR_3 to Rh ratio of 3:1. Fortunately, lower ratios can be achieved by cleaving and displacing the alkene ligands from $[RhCl(alkene)_2]_2$ complexes using the minimum quantity of tertiary phosphine. In practice PR_3 to Rh ratios of about 2.2:1 have been found to give the most rapid rates of hydrogenation[44]. This clearly demonstrates that $[RhCl(H)_2(PR_3)_2]$ complexes are involved in alkene activation. Conversely, addition of tertiary phosphine to $[RhX(PR_3)_3]$ systems retards the rate of alkene hydrogenation.

Empirically it has been found that the replacement of the chloro ligand by heavier halides increases the rate of hydrogenation[28]. The corresponding fluoro complex is a less active catalyst[36].

Faster hydrogenation rates may also be achieved on replacing triphenylphosphine with other tri(aryl)phosphines. Electron-donating *para* substituents increase the rate. Tri-(*p*-anisyl)phosphine and tri(*p*-tolyl)phosphine have been found to be particularly effective[44]. Tri(*o*-anisyl)phosphine does not form an effective catalyst[36]. Presumably the oxygen atom of an anisyl group blocks the coordination site required for alkene coordination.

Replacement of triarylphosphines by trialkylphosphines does not result in an improvement in the rate of hydrogenation. Their dihydridorhodium(III) complexes are too stable to transfer hydrogen to alkenes. The much higher stability of the iridium analogue $[IrCl(H)_2(PPh_3)_3]$ also prevents the iridium system from bringing about rapid alkene hydrogenation under ambient conditions[45].

Dihydride catalytic systems offer several advantages over other homogeneous and heterogeneous catalytic systems. The short lifetime of the intermediate alkyl complex minimizes hydroisomerization reactions. This property also allows them to catalyse selective addition of deuterium[46] or tritium to alkene or alkyne substrates[47]. The limited regioselectivity exhibited by the $[RhCl(PPh_3)_3]$ system permits it to be employed in the catalytic hydrogenation of a wide range of substrates, particularly since other functional groups are unaffected.

IV. THE ALKENE ROUTE

The investigation of the catalytic properties of the cationic complexes $[Rh(PPh_3)_2(\text{alkadiene})]X$ (where X is a ligand of low coordinating power such as ClO_4^-) paved the way for the exploitation of the alkene route to alkanes[48]. These alkadiene complexes are unlike the similar neutral complexes $[RhCl(PPh_3)(\text{alkadiene})]$, which are unable to activate molecular hydrogen[28].

The alkadiene complexes are not the true catalysts but are converted into them in the presence of molecular hydrogen (Scheme 1). This stoicheiometric reaction must proceed by the alkene route. Depending on the tertiary phosphine present in the original complex, there is then a choice of routes for the hydrogenation of alkene substrates. Often both the alkene and dihydride routes are followed simultaneously in the hydrogenation of bulk alkene. It has been demonstrated that a slow equilibrium can exist between the two catalysts[25].

The solvated complexes that can be isolated after stoicheiometric hydrogenolysis of the original cationic complexes can also be obtained free of weakly bound solvent ligands. These latter species exhibit extreme coordinative unsaturation and can, for example, be obtained as dimers[49] or with coordinated BPh_4^- ligands[50]. In both these instances η^6-coordination of a phenyl group occurs. Accordingly, the nature of the alkene substrate also exerts a strong influence on the course of the catalytic hydrogenation. Monoalkene hydrocarbons are generally believed to be catalytically hydrogenated by the dihydride route. On the other hand, in the catalytic hydrogenation of norbornadiene all the alkadiene is reduced to norbornylene before hydrogenation of the latter compound commences. This provides strong evidence that hydrogenation of the alkadiene takes place by the alkene route[48]. It is also believed that the hydrogenation of functionally substituted alkenes which are capable of coordinating to rhodium through both the alkene bond and a heteroatom occurs via the alkene route.

Scheme 1 indicates that replacing the two tertiary monophosphine ligands by a single, chelating ditertiary phosphine should encourage participation of the alkene route, since the latter ligand cannot adopt the *trans*-bis(phosphine) structure of the dihydrido catalyst. However, since the loss of one weakly bound solvent ligand from the dihydrido complex would form a typically fluxional pentacoordinate complex, the dihydride route is still possible.

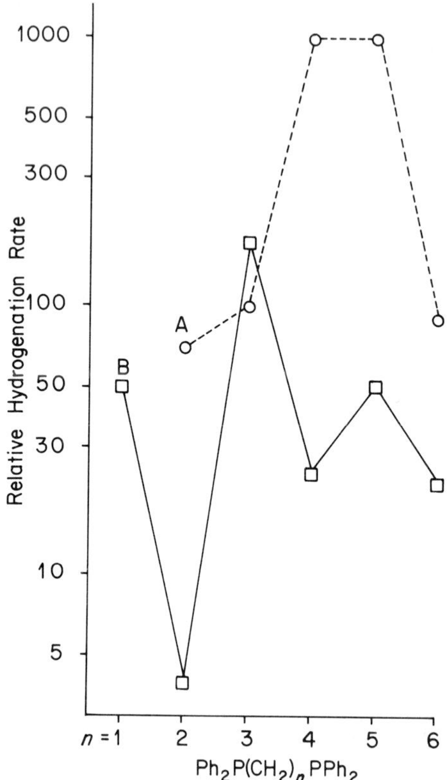

FIGURE 2. Variation of relative hydrogenation rates of α-acetamidocinnamic acid (A) and styrene (B) with ditertiary phosphine chelate ring size

Significant participation of the dihydride route when ditertiary phosphine complexes catalyse the addition of hydrogen to chelating alkadienes or substituted alkenes is unlikely, since both the ligand and the substrate favour the formation of an alkene complex.

The rate of hydrogenation of α-acetamidocinnamic acid (which coordinates additionally through the amide oxygen) varies steadily with diphosphine chelate ring size, and reaches a maximum for six- and seven-membered chelate rings (Fig. 2). However, the rate of hydrogenation of styrene (a monodentate alkene ligand) varies erratically with chelate ring size and does not have a common maximum rate with α-acetamidocinnamic acid. Moreover, the rate of hydrogenation of styrene seldom exceeds that of α-acetamidocinnamic acid, despite the greater substitution of the latter's alkene bond[51]. In dihydride systems the rate of reduction of styrene is usually rapid[52]. It seems probable that the hydrogenation of the two alkenes follows different pathways when cationic complexes catalyse their reductions.

Thus it can be seen that the alkene route enables chelating alkenes, which commonly poison their own reduction in dihydride systems, to be efficiently and rapidly hydrogenated.

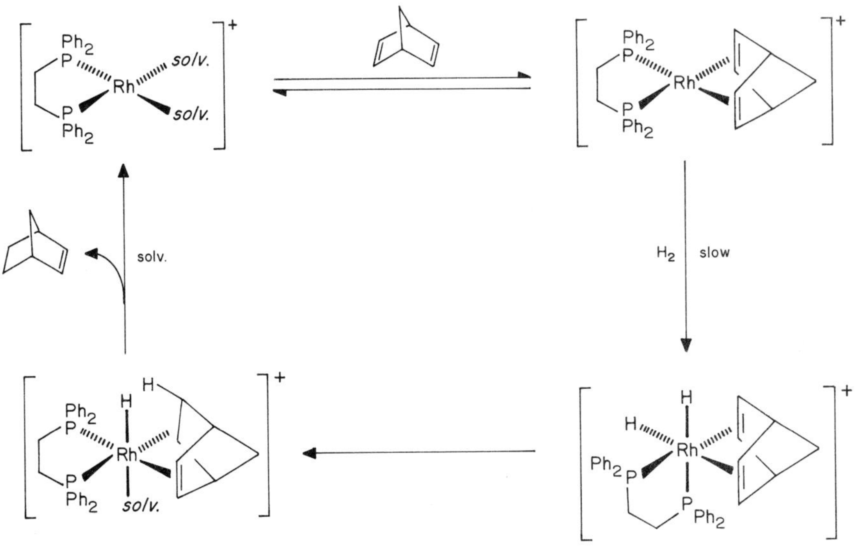

SCHEME 4. The alkene route catalysed by $[Rh(diphos)(nbd)]^+$ complexes.

The mechanism of the alkene route hydrogenation of norbornadiene to norbornylene is shown in Scheme 4. The rate-determining step in this catalytic cycle is the oxidative addition of hydrogen to the alkene complex. In systems where α-acetamidocinnamic acid is the substrate this step is also rate determining at temperatures above 0 °C. If the temperature is lowered to −40 °C then the reductive elimination of the saturated product becomes rate determining. At −78 °C the catalysis ceases and the intermediate alkyl complex, **2**, can be detected[53].

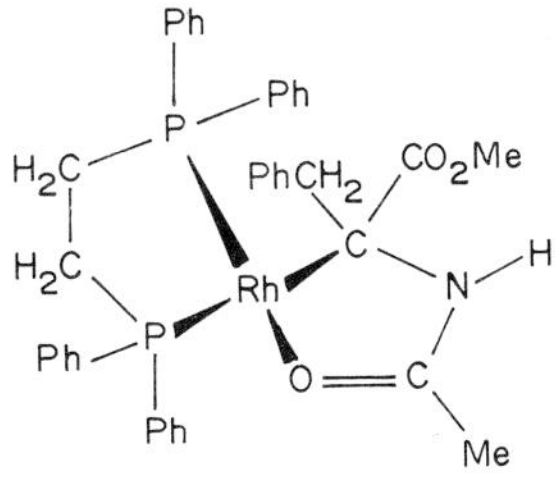

(2)

Except for chelating substrates, the cationic complexes offer few advantages over dihydride systems in the hydrogenation of non-prochiral alkenes. As noted above, the main application of the alkene route is in asymmetric hydrogenation. The preferred type of catalyst for this process contains a chiral, chelating, ditertiary phosphine ligand such as diop[54], **3**, or the pyrrolidine derivatives **4** ($R = CO_2Bu^t$[55], Bu^tCO, MeCO, PhCO[56]). The topic of asymmetric hydrogenation by rhodium complexes is discussed more fully in Chapter 8.

(3) (4)

V. THE ALKYL ROUTE

This route was first discovered during the investigation of the catalytic properties of $[RuCl_2(PPh_3)_3]$[57,58]. This complex was easily converted into the true catalyst $[RuHCl(PPh_3)_3]$ by dihydrogen in the presence of base (equation 16). The base commonly employed is triethylamine, but even ethanol can assist in the formation of the purple hydrido complex.

Chlorohydridotris(triphenylphosphine)ruthenium(II) is an exceptionally efficient catalyst for the homogeneous hydrogenation of terminal alkenes. The rates of hydrogen uptake at subatmospheric pressure and room temperature verge on the limit of diffusion control. For this reason, and because of the limited solubility of the catalyst, study of the mechanism of the hydrogenation reaction is very difficult[58].

The mechanism of this type of reaction is best illustrated by considering catalyses brought about by $[RhH(CO)(PPh_3)_3]$. This complex is more soluble in organic solvents and its rates of hydrogenation are less rapid than those of $[RuHCl(PPh_3)_3]$[58]. Many of the features of the ruthenium system are exhibited by $[RhH(CO)(PPh_3)_3]$, in particular a marked selectivity towards terminal alkenes. The proposed mechanism is illustrated in Scheme 5.

The rate-determining step is the same as in the alkene route, namely oxidative addition of hydrogen. The participation of an alkyl complex is implied by the small degree of isomerization and isotope exchange observed in the catalytic deuteriation of 2-methylpropene[59]. Both of these processes have their origin in a relatively long-lived alkyl species (cf. Scheme 3).

Unlike the dihydride system involving $[RhCl(PPh_3)_3]$ as catalyst, several important intermediates in the proposed catalytic cycle have been isolated from stoicheiometric reactions. Tetrafluoroethene reacts with the parent complex to form an alkyl complex (equation 28)[60,61]. The alkyl complex has been shown to react with hydrogen and triphenylphosphine to reform the original complex (equation 29)[61].

$$[RhH(CO)(PPh_3)_3] + C_2F_4 \xrightarrow[C_7H_8]{5\,\text{atm},\,25\,^\circ\text{C}} [Rh(CF_2CF_2H)(CO)(PPh_3)_2] + PPh_3 \quad (28)$$

$$[Rh(CF_2CF_2H)(CO)(PPh_3)_2] + H_2 + PPh_3 \xrightarrow{50\,\text{atm}} [RhH(CO)(PPh_3)_3] + HCF_2CF_2H \quad (29)$$

The marked regioselectivity of the catalyst has its origins in the stereochemistry of the alkyl complex. The alkyl complex illustrated in Scheme 5 shows the alkyl ligand to be *cis* to the two bulky triphenylphosphine ligands. Accordingly, secondary alkyl complexes are much less stable than primary alkyl complexes. Secondary alkyl complexes tend to

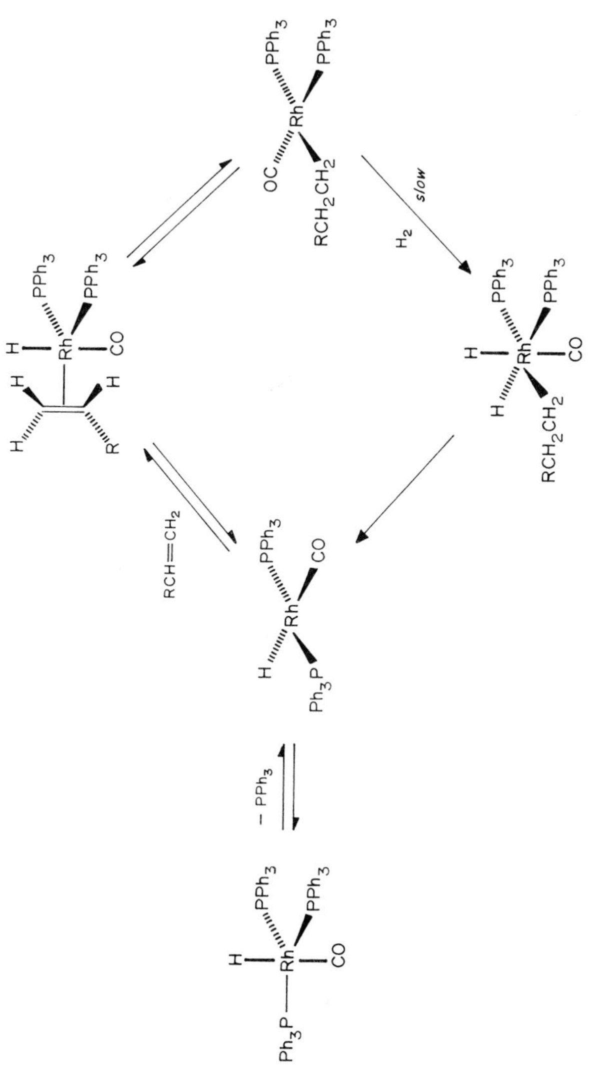

SCHEME 5. The alkyl route catalysed by $[RhH(CO)(PPh_3)_3]$.

decompose before the slow oxidative addition of hydrogen can occur. The alkyl route, therefore, offers the possibility of regioselective reductions such as that of 4-vinylcyclohexene (equation 30).

$$\text{4-vinylcyclohexene} + H_2 \xrightarrow{[RuCl_2(PPh_3)_3]} \text{4-ethylcyclohexene} \qquad (30)$$

However, at higher temperatures $[RhH(CO)(PPh_3)_3]$ can catalyse the hydrogenation of a wider range of substrates. Both internal and alicyclic alkenes can be hydrogenated. Whilst it is possible that internal alkenes undergo the thermodynamically unfavourable hydroisomerization to terminal alkenes, neither cyclohexene[62] nor cycloheptene[63] can be isomerized to such products.

Cyclohexene forms cyclohexane when allowed to react with hydrogen at 10 atm pressure and 50 °C in the presence of low concentrations of $[RhH(CO)(PPh_3)_3]$. However, the rate of hydrogenation at catalyst concentrations above 0.25 mM declines and is virtually zero when the concentration reaches 0.75 mM. It seems likely that further dissociation of the catalyst to a monotriphenylphosphine complex takes place (equation 31). At higher catalyst concentrations this monotriphenylphosphine complex is not formed since the concentration of free ligand released in the first equilibrium is too high[62].

$$[RhH(CO)(PPh_3)_3] \underset{+PPh_3}{\overset{-PPh_3}{\rightleftharpoons}} [RhH(CO)(PPh_3)_2] \underset{+PPh_3}{\overset{-PPh_3}{\rightleftharpoons}} [RhH(CO)PPh_3] \qquad (31)$$

Other rhodium complexes that follow the alkyl route include the related complexes $[RhH(PPh_3)_4]$[64] $[RhH(dbp)_4]$[65], and $[RhH(PPh_3)_3PF_3]$[66]. All of these complexes are reasonably effective catalysts for the hydrogenation of terminal alkenes. However, replacement of a second triphenylphosphine ligand in the last complex by a further PF_3 ligand results in a complex unable to catalyse alkene hydrogenation. Presumably the increased π-acidity of the PF_3 ligands so stabilizes the rhodium(I) complex that oxidative addition of hydrogen is impossible. This theory is supported by experiments which show the ability of the $[RhH(PPh_3)_2(PF_3)_2]$ complex to isomerize alkenes is retained[66].

The tetrakis (triphenylphosphine) complex $[RhH(PPh_3)_4]$ dissociates to coordinatively unsaturated $[RhH(PPh_3)_3]$ before taking part in the catalytic cycles. This is apparent since $[RhH(PPh_3)_3]$ is a more effective catalyst[64]. The dibenzophosphole complex $[RhH(dbp)_4]$ similarly dissociates to $[RhH(dbp)_2]$ before entering the catalytic cycle[67]. However, dissociation may not be essential to the catalytic activity of this type of complex since [RhH(triphos)] complexes, where triphos = $PhP(CH_2CH_2CH_2PPh_2)_2$ or $PhP(CH_2CH_2CH_2PCy_2)_2$, are very active catalysts for the hydrogenation of terminal alkenes. Even some internal, but not cyclic, alkenes can be hydrogenated using these catalysts. As might be expected from steric considerations, the reactions are more rapidly catalysed by the diphenylphosphido complex.

In addition to being able to revert to an alkene and hydrido complex, alkyl complexes can also undergo a second type of reaction that may destroy the catalyst. Two instances of decomposition to an *ortho*-metallated complex have been noted in stoicheiometric reactions (equations 32 and 33)[64,69]. Neither of these reactions is significant in catalytic cycles and both *ortho*-metallated products react with molecular hydrogen to reform hydrido complexes (equation 34). However, the original ruthenium complex is not reformed. Although normally of little consequence, these reactions may be of some importance if alkenes are added to catalyst solutions before hydrogen or if local hydrogen starvation occurs during the catalysis.

$$[RhH(PPh_3)_3] + C_2H_4 \rightarrow [RhEt(PPh_3)_3] \rightarrow C_2H_6 + [Rh\{(o\text{-}C_6H_4)PPh_2\}(PPh_3)_2] \quad (32)$$

$$[RuHCl(PPh_3)_3] + RCH{=}CH_2 \rightarrow RCH_2CH_3 + [RuCl\{(o\text{-}C_6H_4)PPh_2\}(PPh_3)] + PPh_3 \quad (33)$$

$$[RuCl\{(o\text{-}C_6H_4)PPh_2\}(PPh_3)] + H_2 \rightarrow [RuHCl(PPh_3)_2] \quad (34)$$

The main application of monohydrido catalysts which follow the alkyl route is undoubtedly regioselective reductions. Otherwise they offer little advantage over the dihydride route catalysts, apart from a more rapid reduction of terminal alkenes. As this class of substrate is usually hydrogenated fairly rapidly using the latter type of catalyst, the advantage is minimal.

VI. ALKYNE HYDROGENATION

Although basically similar to alkene hydrogenation, homogeneous catalytic hydrogenation of alkynes is much less readily achieved. Alkynes complex more strongly to transition metals than alkenes and can thereby poison the dihydride route. Monohydride catalysts often give rise to stable vinyl complexes when allowed to react with alkynes[70]. The vinyl complexes are frequently unable to activate molecular hydrogen under mild conditions. Potentially the most serious side-reaction in alkyne hydrogenation is polymerization of the alkyne substrate by the catalyst. Chlorotris(triphenylphosphine)rhodium(I), for example, is a fairly effective catalyst for alkyne polymerization[2].

Ideally, the hydrogenation of alkynes should stop at the alkene stage since alkenes are more useful as reagents than alkanes. In practice it has proved most difficult to stop the reaction at the alkene stage. In most catalytic systems the rate of alkyne hydrogenation is less than that of the alkene produced. However, since alkynes complex more strongly than alkenes to the central metal of the catalytic complex, some selectivity is observed in competitive reactions.

It has been claimed that the carbonyl complexes $[Rh_4(CO)_{16}]$ and $[RhCl(CO)_2]_2$ show some selectivity towards pent-2-ene formation when they are used to catalyse the hydrogenation of pent-2-yne[71]. Some selectivity towards alkenes has also been claimed for cationic complexes such as $[Rh(cod)(PPh_3)py]PF_6$, particularly in the presence of triethylamine[72,73]. The benzoato complex $[Rh(OCOPh)(cod)(PPh_3)]$ is also reputed to be a selective catalyst, and its behaviour may explain why addition of benzoic acid to the above cationic system increases the selectivity[72,73].

The most interesting selective reduction of alkynes is catalysed by the binuclear complexes **5**. When one hydrogen atom has been transferred the σ-alkenyl complex **6** is formed. The second atom of hydrogen is then transferred to yield an (*E*)-alkene[74]. Since

$$[(Pr^iO)_3P]_2Rh(\mu\text{-}H)_2Rh[P(OPr^i)_3]_2$$

(**5**)

$$[(Pr^iO)_3P]_2Rh(\mu\text{-}C(R){=}C(H)R)(\mu\text{-}H)Rh[P(OPr^i)_3]_2$$

(**6**)

these products are usually hydrogenated very slowly by all homogeneous hydrogenation catalysts, excellent selectivity is observed.

However, alkyne hydrogenation remains a neglected area of homogeneous catalytic hydrogenation. Demonstration of an active, selective catalyst is certainly required to restore interest to this area. In view of the experiments outlined above, it would seem that the most promising area of investigation is the alkyne route. In this analogue of the alkene route, maximum advantage would be made of the differences in the complexity constants of alkenes and alkynes. This should assist in stopping the reaction at the alkene stage.

VII. HYDROGENATION OF CARBON—OXYGEN BONDS

A. Ketones

Several cationic rhodium complexes, particularly those containing nitrogenous ligands, are able to catalyse the hydrogenation of ketones to secondary alcohols. Usually the reactions take place in alkaline solution under ambient conditions[75–77]. Even sterically hindered ketones such as 2-methylcyclohexanone can be reduced (equation 35)[77]. In the presence of excess of 2, 2′-bipyridyl, the keto group is selectively reduced in unsaturated ketones[76].

$$H_2 + \text{2-methylcyclohexanone} \xrightarrow{[Rh(bipy)(solv.)_2]^+} \text{2-methylcyclohexanol} \quad (35)$$

Addition of triethylamine to the reaction mixture allows the reduction to be catalysed by cationic tertiary phosphine complexes[78]. The addition of base implies that the reduction may be taking place via the enol form of the ketone. However, the reduction of moist acetone by deuterium in the presence of $[RhD_2(PPhMe_2)_2(Me_2CO)]^+$ complexes yields D_2-propan-2-ol (equation 36). The reduction of tetramethylcyclobutan-1, 3-dione (which cannot enolize) shows that enolization is not essential when this catalyst is used[79].

$$Me_2CO + D_2 \xrightarrow[1\%\ H_2O]{[RhD_2(PPhMe_2)_2(Me_2CO)_2]^+} Me_2CDOD \quad (36)$$

Neutral rhodium complexes containing tertiary phosphines also catalyse the hydrogenation of ketones. Hydridobis(tricyclohexylphosphine)rhodium(I) is an active catalyst in the reduction of cyclohexanone, acetone, and benzophenone[80]. Strongly alkaline conditions are required for the hydrogenation of acetone when $[RhCl(C_8H_{12})(PPh_3)]$ is the catalyst[81].

Ruthenium complexes require much more severe conditions of temperature and pressure if they are to catalyse the hydrogenation of ketones. Both $[RuCl_2(PPh_3)_3]$[82–84] and $[RuHCl(PPh_3)_3]$[85] have been used to catalyse the reductions. Probably neither of these complexes is the true catalyst since the formation of ruthenium carbonyl complexes has been noted in the reaction[83].

B. Aldehydes

The complexes of both rhodium and ruthenium catalyse the reduction of aldehydes, although in both cases the carbonyl abstraction impairs the activity of the catalysts. There is a claim that hydrated rhodium trichloride or $[RhCl(CO)_2]_2$ in the presence of triethylamine permits the hydrogenation of unsaturated aldehydes to unsaturated alcohols, but the selectivity is not high[86].

As in the case with ketones, ruthenium complexes require more severe conditions than rhodium complexes to catalyse the reduction of aldehydes. Hydrogen pressures between

10 and 30 atm are usually required[82–84]. One of the more interesting reductions that can be achieved is the reduction of glucose to sorbitol[87].

Epoxides are isomerized to aldehydes in the presence of the cationic rhodium(I) complex $[Rh(nbd)(PEt_3)_2]ClO_4$ (equation 37). The aldehydes produced are hydrogenated in the course of the reaction[88,89]. During the reaction acyl complexes are formed from the epoxides since only these intermediates undergo hydride transfer to form aldehydes[90].

$$\underset{\text{(epoxide: PhCH–CH}_2\text{ bridged by O)}}{PhCH{-}CH_2} + H_2 \xrightarrow{[Rh(nbd)(PEt_3)_2]ClO_4} PhCH_2CHO + PhCH_2CH_2OH \qquad (37)$$

VIII. CONCLUSION

It is difficult to avoid the belief that little remains to be achieved in homogeneous catalytic hydrogenation. This is particularly true in the area of alkene hydrogenation. Improvements to existing catalysts to permit more rapid hydrogenation of polysubstituted alkene bonds could still be made. The search for ever higher optical yields from a wider variety of prochiral alkene substrates will undoubtedly continue.

The selective reduction of alkynes to alkenes remains the greatest challenge in homogeneous catalysis. It would appear that the homogeneous reduction of organic carbonyl compounds offers no advantages over heterogeneous reduction unless complexes of much higher catalytic activity are synthesized.

On the industrial scale, homogeneous catalytic hydrogenation will never become important until catalytic systems less sensitive to traces of impurities come into use. Batch processes will also be the rule until the activity and selectivity problems associated with heterogenized catalysts have been overcome.

IX. REFERENCES

1. F. H. Jardine and G. Wilkinson, *J. Chem. Soc. C*, 270 (1967).
2. F. H. Jardine, *Prog. Inorg. Chem.*, **28**, 63 (1981).
3. C. O'Connor and G. Wilkinson, *J. Chem. Soc. A*, 2665 (1968).
4. J. F. Knifton, *J. Org. Chem.*, **41**, 1200 (1976).
5. R. C. Ryan, G. M. Wilemon, M. P. Delsanto, and C. U. Pittman, *J. Mol. Catal.*, **5**, 319 (1979).
6. W. Reimann, W. Abboud, J. M. Bassett, R. Mutin, G. L. Rempel, and A. K. Smith, *J. Mol. Catal.*, **9**, 349 (1981).
7. L. Toniolo, G. DeLuca, C. Panattoni, and G. Deganello, *Gazz. Chim. Ital.*, **104**, 961 (1974).
8. C. R. Landis and J. Halpern, *Organometallics*, **2**, 840 (1983).
9. T. Yoshida, T. Okano, and S. Otsuke, *Chem. Commun.*, 870 (1979).
10. J. Kwiatek and J. K. Seyler, *J. Organomet. Chem.*, **3**, 421 (1965).
11. M. Iguchi, *J. Chem. Soc. Jpn.*, **60**, 1287 (1939).
12. J. Halpern and B. R. James, *Can. J. Chem.*, **44**, 495 and 671 (1966); J. Halpern, J. F. Harrod, and B. R. James, *J. Am. Chem. Soc.*, **88**, 5150 (1966).
13. L. Vaska and R. E. Rhodes, *J. Am. Chem. Soc.*, **87**, 4970 (1965).
14. M. Orchin, *Acc. Chem. Res.*, **14**, 259 (1981).
15. W. Strohmeier and E. Hitzel, *J. Organomet. Chem.*, **110**, C72 (1976).
16. W. Strohmeier and G. Csontos, *J. Organomet. Chem.*, **67**, C27 (1974).
17. W. Strohmeier, and W. Rehder-Stirnweiss, *Z. Naturforsch., Teil B*, **25**, 1480 (1970).
18. F. H. Jardine, J. A. Osborn, and G. Wilkinson, *J. Chem. Soc. A*, 1574 (1967).
19. B. R. James, F. T. T. Ng, and G. L. Rempel, *Can. J. Chem.*, **47**, 4521 (1969).
20. L. Kollar, S. Toros, B. Heil, and L. Marko, *J. Organomet. Chem.*, **192**, 253 (1980).
21. D. S. Moore and S. D. Robinson, *Chem. Soc. Rev.*, **12**, 415 (1983).
22. P. J. Brothers, *Prog. Inorg. Chem.*, **28**, 1 (1981).

23. L. Kirch, I. J. Goldfarb, and M. Orchin, *J. Am. Chem. Soc.*, **78**, 5450 (1956).
24. R. A. Schunn and E. R. Wonchoba, *Inorg. Synth.*, **13**, 131 (1972).
25. J. M. Brown, P. A. Challoner, and P. N. Nicholson, *Chem. Commun.*, 646 (1978).
26. M. A. Muus and F. W. Weiss, *J. Am. Chem. Soc.*, **84**, 4697 (1962).
27. M. Brown and L. W. Piszkiewz, *J. Org. Chem.*, **32**, 2104 (1967).
28. J. A. Osborn, F. H. Jardine, G. Wilkinson, and J. F. Young, *J. Chem. Soc. A*, 1711 (1966).
29. M. J. Mays and G. Wilkinson, *J. Chem. Soc.*, 6629 (1965).
30. J. Halpern, in *Organotransition-Metal Chemistry* (Eds. Y. Ishii and M. Tsutsui), Plenum Press, New York, 1975, pp. 109–117.
31. J. Halpern, T. Okamoto, and A. Zakhariev, *J. Mol. Catal.*, **2**, 65 (1976).
32. H. Rüesch and T. J. Mabry, *Tetrahedron*, **25**, 805 (1969).
33. C. H. Heathcock and S. R. Poulter, *Tetrahedron Lett.*, 2755 (1969).
34. J. F. Biellmann, *Bull. Soc. Chim. Fr.*, 3055 (1968).
35. W. J. Dejarlais and E. A. Emken, *Lipids*, **11**, 594 (1976).
36. M. Gungor, F. H. Jardine, and J. D. Wheatley, unpublished results.
37. Y. Senda, S. Mitsui, H. Sugiyama, and S. Seto, *Bull. Chem. Soc. Jpn.*, **45**, 3498 (1972).
38. D. A. Slack and M. C. Baird, *J. Am. Chem. Soc.*, **98**, 5539 (1976).
39. M. Barthelemy, A. Gianfermi, and Y. Bessière, *Bull. Soc. Chim. Fr.*, 1821 (1976).
40. C. Rousseau, M. Evrard, and F. Petit, *J. Mol. Catal.*, **5**, 163 (1979).
41. Y. Senda, T. Iwasaki, and S. Mitsui, *Tetrahedron*, **28**, 4059 (1972).
42. S. Komiya, A. Yamamoto, and S. Ikeda, *J. Organomet. Chem.*, **42**, C65 (1972).
43. G. Speier and L. Marko, *Inorg. Chim. Acta*, **3**, 126 (1969).
44. S. Montelatici, A. van der Ent, J. A. Osborn, and G. Wilkinson, *J. Chem. Soc. A*, 3224 (1971).
45. W. Strohmeier and R. Endres, *Z. Naturforsch., Teil B*, **27**, 1415 (1972).
46. J. Morandi and H. B. Jensen, *J. Org. Chem.*, **34**, 1889 (1969).
47. G. K. Koch and J. W. Dalenberg, *J. Labelled Compd.*, **6**, 395 (1970).
48. R. R. Schrock and J. A. Osborn, *J. Am. Chem. Soc.*, **98**, 2134, 2143 and 4450 (1976).
49. J. Halpern, D. P. Riley, A. S. C. Chan, and J. J. Pluth, *J. Am. Chem. Soc.*, **99**, 8055 (1977).
50. P. Albano, M. Aresta, and M. Manassero, *Inorg. Chem.*, **19**, 1069 (1980).
51. J. C. Poulin, Than Phat Dang, and H. B. Kagan, *J. Organomet. Chem.*, **84**, 87 (1975).
52. F. H. Jardine, *PhD Thesis*, University of London 1967.
53. A. S. C. Chan, J. J. Pluth, and J. Halpern, *J. Am. Chem. Soc.*, **102**, 5952 (1980).
54. D. Sinou and H. B. Kagan, *J. Organomet. Chem.*, **114**, 325 (1976).
55. K. Achiwa, *Heterocycles*, **8**, 247 (1977).
56. K. Achiwa, T. Kogure, and I. Ojima, *Chem. Lett.*, 297 (1978).
57. D. Evans, J. A. Osborn, F. H. Jardine, and G. Wilkinson, *Nature (London)*, **208**, 1203 (1965).
58. P. S. Hallman, B. R. McGarvey, and G. Wilkinson, *J. Chem. Soc. A*, 3143 (1968).
59. E. Hiroto, M. Ito, and T. Ueda, *Proc. 6th. Int. Congr. Catal.*, **1**, 518 (1977).
60. G. Yagupsky, C. K. Brown, and G. Wilkinson, *J. Chem. Soc. A*, 1392 (1970).
61. G. Yagupsky, C. K. Brown, and G. Wilkinson, *Chem. Commun.*, 1244 (1969).
62. J. Hjortkjaer and Z. Kulicki, *J. Catal.*, **27**, 452 (1972).
63. W. Strohmeier, W. Rehder-Stirnweiss, and R. Fleischman, *Z. Naturforsch., Teil B*, **25**, 1481 (1970).
64. S. H. Strauss and D. F. Shriver, *Inorg. Chem.*, **17**, 3069 (1978).
65. D. E. Budd, D. G. Holah, A. N. Hughes, and B. C. Hui, *Can. J. Chem.*, **52**, 775 (1974).
66. J. F. Nixon and J. R. Swain, *J. Organomet. Chem.*, **72**, C15 (1974).
67. D. G. Holah, I. M. Hoodless, A. H. Hughes, B. C. Hui, and D. Martin, *Can. J. Chem.*, **52**, 3758 (1974).
68. J. Niewahner and D. W. Meek, *Inorg. Chim. Acta*, **64**, L123 (1982).
69. B. R. James, L. D. Markham, and D. K. W. Wang, *Chem. Commun.*, 439 (1974).
70. F. H. Jardine, *Prog. Inorg. Chem.*, **31**, 265 (1984).
71. P. M. Lausarot, G. A. Vaglio, and M. Vale, *J. Organomet. Chem.*, **204**, 249 (1981).
72. R. H. Crabtree, A. Gautier, G. Giordano, and T. Khan, *J. Organomet. Chem.*, **141**, 113 (1977).
73. R. H. Crabtree, *Chem. Commun.*, 647 (1975).
74. R. R. Burch, E. L. Muetterties, R. G. Teller, and J. M. Williams, *J. Am. Chem. Soc.*, **104**, 4257 (1982).
75. G. Zassinovich, A. Camus, and G. Mestroni, *Inorg. Nucl. Chem. Lett.*, **12**, 865 (1976).
76. G. Mestroni, G. Zassinovich, and A. Camus, *J. Organomet. Chem.*, **140**, 63 (1977).

77. G. Mestroni, R. Spogliarich, A. Camus, F. Martinelli, and G. Zassinovich, *J. Organomet. Chem.*, **157**, 345 (1978).
78. B. Heil, S. Toros, J. Bakos, and L. Marko, *J. Organomet. Chem.*, **175**, 229 (1979).
79. R. R. Schrock and J. A. Osborn, *Chem. Commun.*, 567 (1970).
80. K. Tani, K. Suwa, E. Tanigawa, T. Yoshida, T. Okano, and S. Otsuka, *Chem. Lett.*, 261 (1982).
81. M. Gargano, P. Gianoccaro, and M. Rosi, *J. Organomet. Chem.*, **129**, 239 (1977).
82. K. C. Dewhirst, *US Pat.*, 3454644 (1969); *Chem. Abstr.*, **71**, 91647 (1969).
83. R. A. Sanchez-Delgado and O. L. DeOchoa, *J. Mol. Catal.*, **6**, 303 (1979).
84. R. A. Sanchez-Delgado and O. L. DeOchoa, *J. Organomet. Chem.*, **202**, 247 (1980).
85. J. Tsuji and H. Suzuki, *Chem. Lett.*, 1085 (1977).
86. S. Meguro, T. Mizoroki, and A. Ozaki, *Chem. Lett.*, 943 (1975).
87. W. M. Kruse, *US Pat.*, 3935284, (1976); *Chem. Abstr.*, **84**, 136005 (1976).
88. I. Mochida, S. Shirahama, H. Fujitsu, and K. Takeshita, *Chem. Lett.*, 421 (1977).
89. H. Fujitsu, E. Matsumura, S. Shirahama, K. Takeshita, and I. Mochida, *J. Chem. Soc., Perkin Trans. 1*, 855 (1982).
90. D. Milstein, *J. Am. Chem. Soc.*, **104**, 5227 (1982).

The Chemistry of the Metal—Carbon Bond, Vol. 4
Edited by F. R. Hartley

CHAPTER **13**

Saturated carbon—hydrogen bond activation

J. R. CHIPPERFIELD and D. E. WEBSTER

Department of Chemistry, The University, Hull HU6 7RX, UK

I. INTRODUCTION

This chapter is concerned with the activation of sp^3 C—H bonds in alkanes and alkyl groups. sp^2 C—H bond activation in arenes is very well known but space does not permit anything more than passing mention to these reactions, except where the work on aromatic compounds illuminates the alkane studies.

The alkanes are the least reactive of organic compounds. This is reflected in the trivial names by which they have been called: 'paraffin' (Latin *Parum affinis*—having little affinity) and 'saturated hydrocarbon' (i.e. unable to undergo addition reactions like alkenes). Alkane C—H bonds are stronger (*ca.* $400\,kJ\,mol^{-1}$) than the C—C bonds in alkanes (and diamonds) (*ca.* $350\,kJ\,mol^{-1}$).

At high temperatures the alkanes react readily—their exothemic autoxidation in flames is at the heart of their great utility as fuels—but they are very unreactive at ambient temperature, and it is the current desire of many chemists to find ways of catalysing the reactions of alkanes at low temperatures in order to turn them into functional organic compounds. As we shall see, these researches have met with some success. An array of different types of reactions are now known that involve alkanes or alkyl groups. Here we concentrate on those reactions which involve a transition metal in a compound or as atoms or ions.

The first step in a transition metal-catalysed reaction of an alkane or alkyl group is an electronic interaction between the metal and the organic compound.

In a review in 1977[1], the transition metal complexes were divided into *hard* and *soft*[2,3]; this is a convenient division. A 'hard' metal ion will generally be in a high oxidation state and its properties will be directly related to its low polarizability and high polarizing power. These hard metals undergo one-electron oxidation–reductions and free radicals are formed as intermediates[4], and hence the transition metal is an initiator (catalyst) in these systems.

Reactions of this type are of considerable industrial interest[5,6], although often this type of reaction takes place on a solid (heterogeneous) catalyst. Catalytic cracking, dehydrogenation, oxidation, and other industrially important processes can all be of this type.

Whether or not catalysis will occur will depend on the ease of electron transfer from the substrate to the metal.

A 'soft' metal is one in a complex, usually in a low oxidation state. Here the metal can undergo two-electron oxidation–reductions; the substrate reacts while coordinated to the metal.

The 'hard' metals encourage *homolytic* fission ($C—H \rightarrow C^{\cdot} + H^{\cdot}$) and the 'soft' metals encourage *heterolytic* fission [$C—H \rightarrow C^{-} + H^{+}$ (or $C^{+} + H^{-}$)]. We discuss reactions of both types here. Section II is the major section of the chapter and reviews the activation of saturated C—H bonds by organometallic complexes in solution. Most of the work in this section was carried out in the past few years. Section III deals with the main features of activation of saturated C—H bonds by platinum(II) complexes, an area that has been extensively reviewed elsewhere[1,7–9].

The activation by metal atoms and ions is discussed in Section IV. In Section V the important features of alkane oxidation catalysed by transition metals are outlined. In the space available it is not possible to do full justice to this topic; an excellent review by Sheldon and Kochi[4] should be consulted by those who want to know more. We do not discuss here the oxidation of alkanes in superacids, a topic reviewed by Olah *et al.*[10]. Finally, in Section VI brief mention is made of oxidations by analogues of biological systems.

Carbon–hydrogen bond activation is a wide topic and we have tried to give a balanced picture without duplicating other reviews. Other reviews especially relevant to this subject are those of Parshall[11,12] and Shilov[9]. The latter's recently published book[9] is complementary to this review; some topics are duplicated, but our major section (Section II) occupies just 3 pages in Shilov's book.

II. ACTIVATION OF SATURATED C—H BONDS BY ORGANOMETALLIC COMPLEXES IN SOLUTION

A. Introduction

This section forms the major part of this chapter. In it we review the very rapidly growing body of work on the activation of saturated C—H bonds by organometallic complexes in solution. Examples going back to the early 1970s are discussed, but by far the bulk of the work relates to the last few years. Certainly the majority of the papers herein had not been written when the two most recent complete review of this topic were published in 1977[1,11].

This section has three parts. In the first (Section B) is a discussion of 'agostic' C—H bonds (see Section B for definition). This includes a comprehensive list of all the molecules in which a C—H bond is activated and interacts with a transition metal atom (an 'agostic' C—H bond).

If the activation becomes more complete then the C—H bond is broken and a chemical reaction occurs. If the C—H bond is part of an alkyl group in the same molecule as the metal atom (as it is in the 'agostic' bonded molecules) then reaction will be *intramolecular* and will result in the formation of a metallocycle. These cyclometallation reactions are discussed in Section C.

If the C—H bond that is activated by the metal is part of a different molecule, then the chemical rection will be an *intermolecular* reaction. Such reactions are discussed in Section D. Reactions of this latter type are particularly interesting—the ability of a transition metal complex to activate and react with a saturated hydrocarbon, even methane, is surprising to most chemists.

As will be seen, and as Halpern and coworkers have suggested[13], the problem of C—H bond activation is thermodynamic, and thermodynamic features are discussed where appropriate. The C—H bond has a standard free energy of formation of over

400 kJ mol^{-1}, and a typical M—H bond *ca.* 250 kJ mol^{-1}, and hence the free energy of the C—M bond would have to be at least 150 kJ mol^{-1}. The C—M bond energies of the third-row transition metals are the highest and hence these appear most likely to activate C—H bonds.

This is so: the metals that are most involved are the precious metals, platinum, iridium, palladium, and rhodium. The products are usually formed by oxidative addition to a coordinatively unsaturated complex to give the alkyl metal hydride. However, this is not always so and the lanthanide complexes that are discussed in Section D must activate the C—H bonds in a heterolytic, probably concerted, reaction. 'Electron-rich' metals were thought to be necessary to take part in these oxidative additions, although this criterion is now not so certain.

One striking feature of the complexes that activate C—H bonds is the large number of them that contain the pentamethylcyclopentadienyl ligand (η^5-C_5Me_5; Cp*). This has proved to be particularly suitable since it has substituents on the cyclopentadienyl ring (the methyl groups) that make self-metallation unlikely, and at the same time increase the electron density of the metal atom to which it is coordinated.

The features necessary for the C—H bond activations discussed here were also discussed in some detail, and compared with reactions at metal surfaces, in a review by Muetterties[14].

B. 'Agostic' Carbon—Hydrogen—Transition Metal Bonds

1. Introduction

The term 'agostic' has been proposed by Brookhart and Green[15] to describe situations in which a hydrogen atom is covalently bonded simultaneously to both a carbon atom and to a transition metal atom. 'Agostic' is derived from a Greek word which occurs in Homer and means 'to clasp or hold to oneself'[15].

The C—H group of saturated carbon centres is not normally thought of as a ligand. However, in recent years many examples have been reported in which a carbon—hydrogen group will interact with a transition metal centre to form a two-electron three-centre bond. In such compounds the interaction has a marked effect on the molecular and electronic structure of the molecule.

The agostic C—H⇀M bonds are similar to the well known B—H···B and M—H⇀M bonds, and it has been suggested[15] that they are probably much more common than hitherto suspected. We use here the 'half-arrow' convention suggested by Green and coworkers[16] to distinguish these two-electron from the four-electron bridging systems, as found in bridging chlorides for example, and to indicate that the two electrons of the C—H bond are donated to the metal (i.e. that the C—H group is a two-electron ligand).

2. Early reports

Most of the compounds reported during the 1960s and early 1970s, in which an interaction occurred between the metal and hydrogen attached to a carbon atom of an organic compound, were arylphosphine complexes[17–21], it being the *ortho*-hydrogens of the aryl ligands that interacted with the metals. The complexes concerned were of ruthenium(II), $[RuCl_2(PPh_3)_3]$ (**1**; X = Cl)[17] and $[RuClH(PPh_3)_3]$ (**1**; X = H)[18], of palladium(II), *trans*-$[Pd(PPhMe_2)_2I_2]$ (**2**)[19], of rhodium(III), $[Rh(SiCl_3)ClH(PPh_3)_2]$ (**3**)[20], and of the rhodium(I) cation, $[Rh(PPh_3)_3]^+$ (**4**)[21]. In all of these compounds the metal—hydrogen distance is in the range 2.5–2.8 Å and any interaction can only be very weak. La Placa and Ibers[17] at the time observed that there does not appear to be any basis for postulating any interaction between the metal and hydrogen.

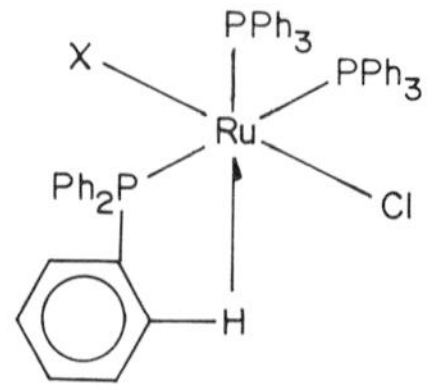

(1)

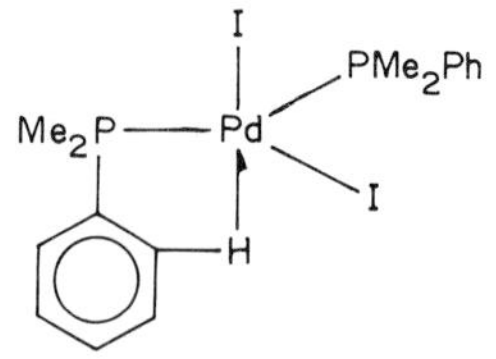

(2)

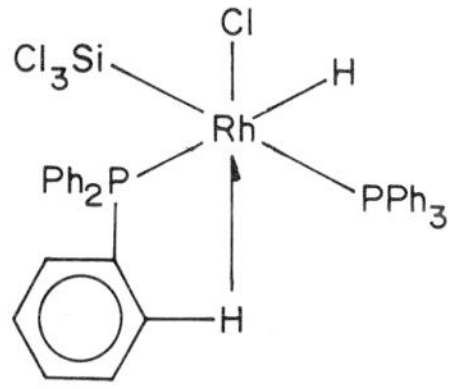

(3)

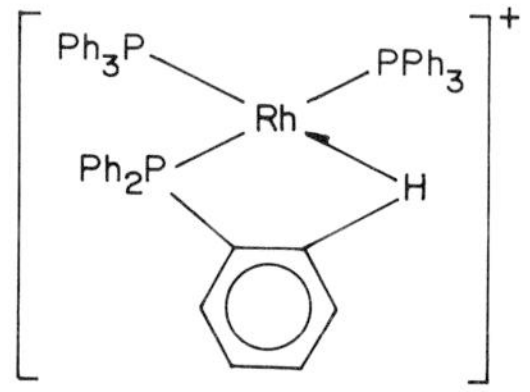

(4)

A clear indication of an agostic carbon—hydrogen bond was reported during a study of pyrazolylborato complexes of nickel and molybdenum by Trofimenko[22]. He observed that the methylene hydrogen atoms of an ethyl group had an abnormal lowfield shift in the ^{1}H n.m.r. spectrum and suggested that this was because the hydrogen was held close to the metal. Crystal structure determinations confirmed that this was so[23,24]. In (diethyldipyrazolylborato) (trihapto-2-phenylallyl) (dicarbonyl)molybdenum (**5**), the Mo—H distance is 2.15 Å, and the agostic hydrogen enables the molybdenum to achieve an 18-electron configuration. It is found[23] that this interaction is strong enough to compete with an extended interaction of the C_7H_7 ligand. The crystal structure of the compound where the C_7H_7 ring replaces the phenylallyl group shows a C—H→Mo interaction even stronger than in **5**, and the C_7H_7 ring is a three-electron donor, i.e. the compound prefers the structure with a C—H→Mo bond and an η^3-C_7H_7 to the structure without a C—H→Mo bond with a η^5-C_7H_7 ring.

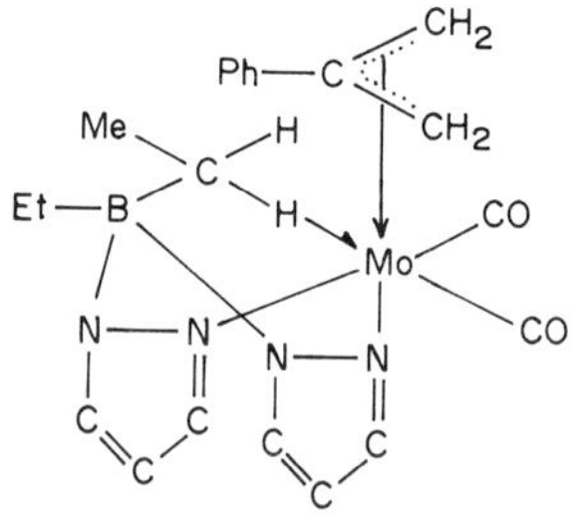

(5)

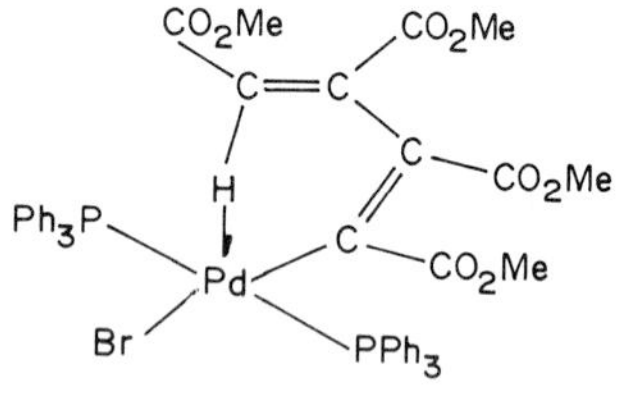

(6)

A palladium(II) diene complex, **6**, with an agostic C—H group was prepared by Maitlis and coworkers in 1972[25]. Here the Pd—H distance of 2.3 Å is much less than the sum of the van der Waals radii (3.1Å).

From the mid-1970s there was a rapid increase in reports of compounds with agostic C—H→M systems. They fall broadly into five groups, polyenyl complexes, alkylidene complexes, alkene complexes, alkyl complexes, and bi- and poly-nuclear complexes, and each of these will be discussed in turn. As will be evident (Table 1), the experimental evidence for such agostic bonds comes from three sources: (i) crystal structure determinations; (ii) n.m.r. data; and (iii) infrared data. This section will be completed with a summary of the essential points from these three methods of study and some general conclusions.

3. Polyenyl complexes

The first examples of polyenyl compounds in which the presence of an agostic C—H group was suggested were the 16-electron cyclohexenyl- and butenyl-tricarbonyliron anions (**7** and **8**:X = CO)[26,27]. Both of these compounds have small ^{13}C—H coupling

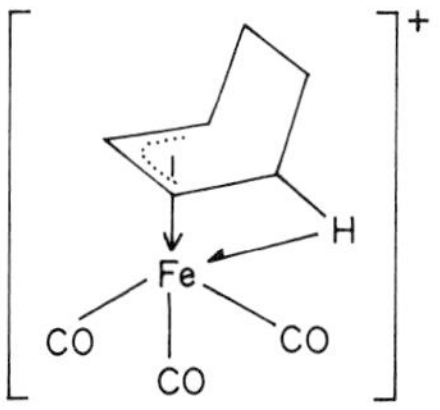

(**7**)

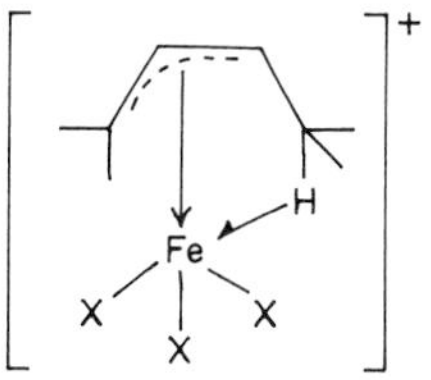

(**8**)

constants, indicative of this type of interaction. The butenyl complex with $P(OMe)_3$ ligands instead of CO [**8**: X = $P(OMe)_3$] and the related cycloheptenyl and octenyl complexes (**9** and **10**) have similar coupling constants[28,29]. The structure of the octenyl complex, determined by neutron diffraction, shows a short Fe—H distance (1.874 Å) and a long C—H bond (1.164 Å), as expected for an agostic interaction.

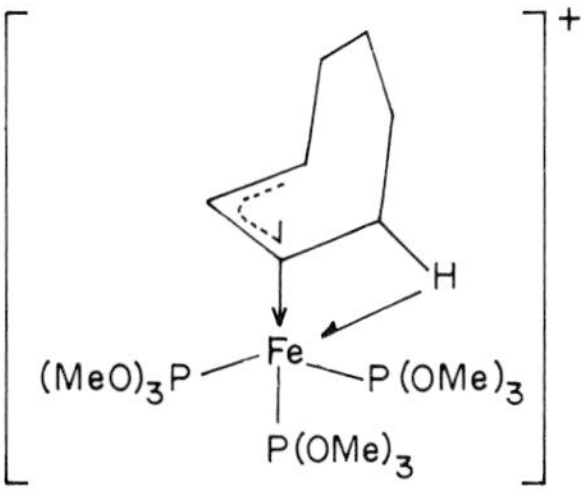

(**9**)

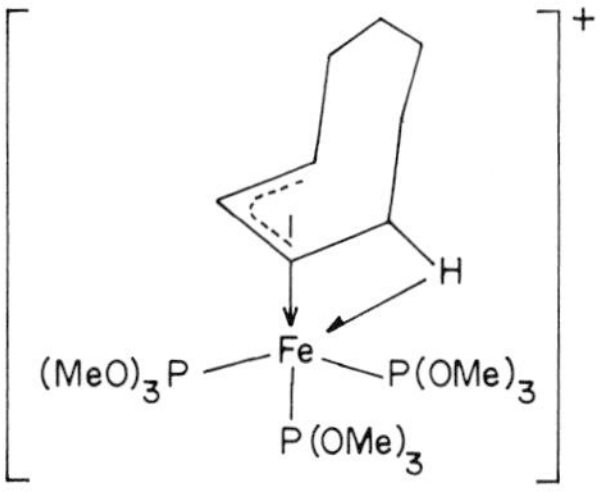

(**10**)

The 16-electron manganese compound analogous to the anion **7** has been studied in some detail by Brookhart and coworkers (**11**; R^1 = H, R^2 = CO)[30–33]. In both the cyclohexenyl complex (**11**; R^1 = H, R^2 = CO) and the methylcyclohexenyl complex (**11**; R^1 = Me, R^2 = CO), dynamic n.m.r. measurements show that an equilibrium exists between the structure with the agostic carbon—hydrogen bond (**B** in Figure 1) and the η^3-allylic (**A**) and diene hydride (**C**) isomers. The agostic hydrogen in **B** is acidic and can be readily deprotonated to the diene anion (**D**). Hence the agostic interaction results in a weakening of the C—H bond and as a consequence the hydrogen can be replaced by other groups. It is readily displaced (i.e. the agostic interaction destroyed) by donor ligands such as carbon monoxide or phosphines to form stable allylic compounds (E). Replacement of one of the CO ligands by $P(OMe)_3$ (**11**; R^1 = H, R^2 = $P(OMe)_3$) has almost no effect on the properties of the compound.

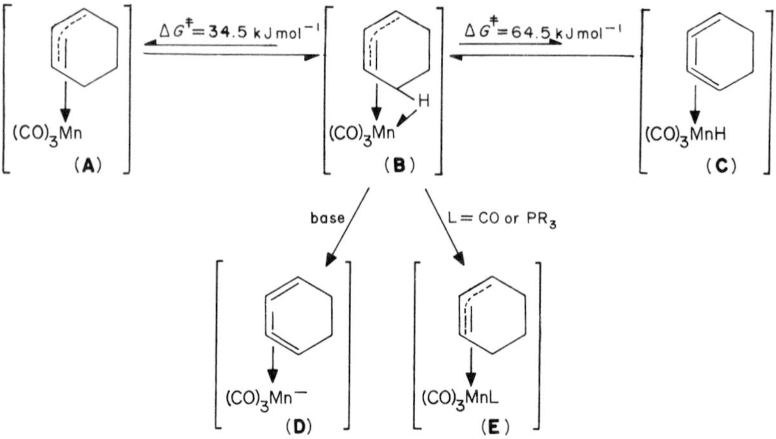

FIGURE 1. Chemistry of the agostic η^3-cyclohexenylmanganese tricarbonyl.

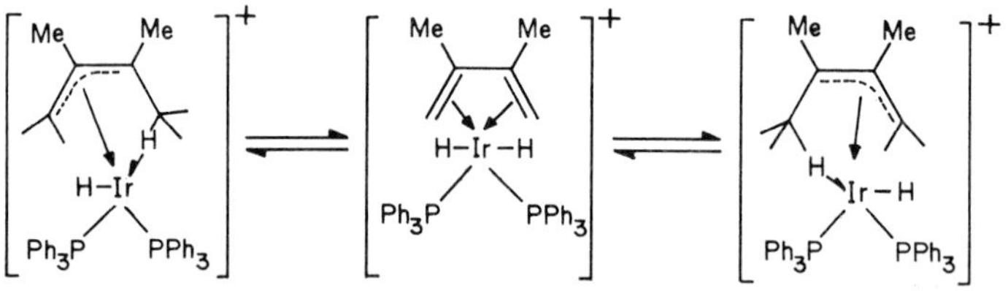

FIGURE 2. Fluxional behaviour of the η^3-2, 3-dimethylbutenyl complex of iridium(II) (**12**).

N.m.r. studies on deuteriated derivatives of the η^3-2, 3-dimethylbutenyl complex of iridium(II) (**12**) show both an agostic carbon—hydrogen group and the occurrence of fluxional behaviour involving the dienedihydride intermediate (Figure 2)[34]. A yellow air-

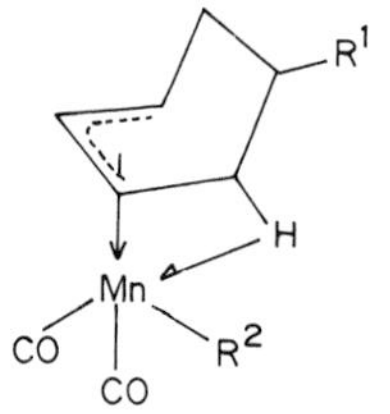

(11)

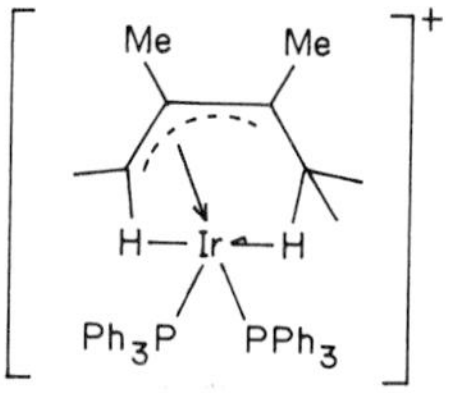

(12)

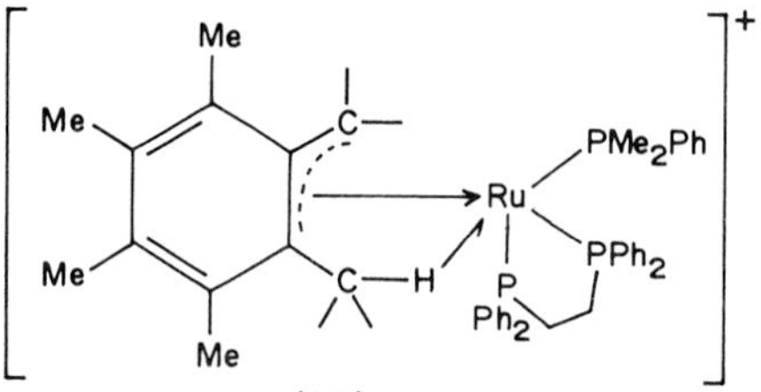

(13)

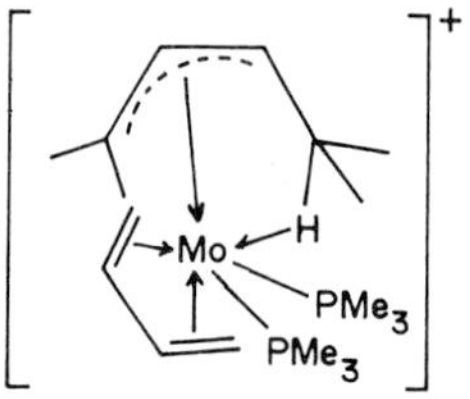

(14)

stable ruthenium(I) anion (**13**) and a molybdenum(I) anion (**14**) have both been reported, with large low-field chemically shifted agostic hydrogen atoms[13,35].

4. Alkylidene complexes

Schrock and coworkers have prepared and characterized an impressive number of neopentylidene, methylidene, and related complexes of tantalum, niobium, and tungsten and have reported on many aspects of their chemistry. For a number of their compounds (**15**–**24**), crystal structure data, from neutron diffraction and X-ray studies, and n.m.r.

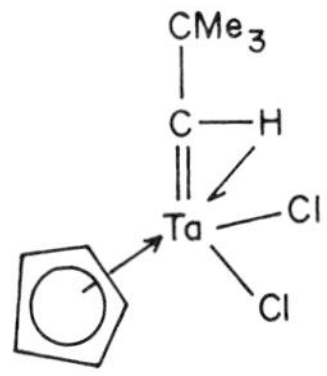

(15)

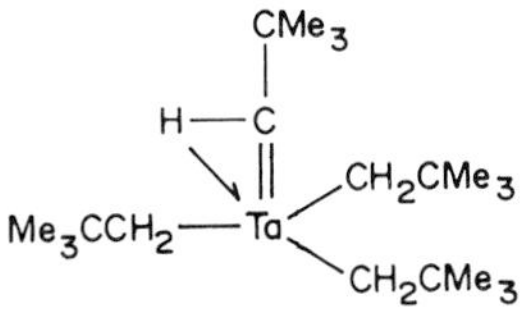

(16)

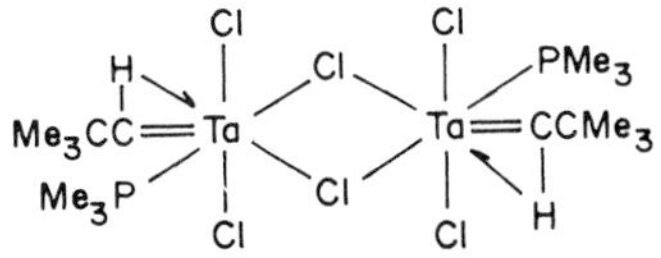

(17)

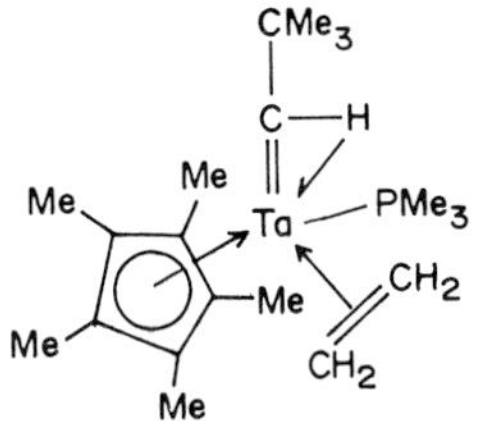

(18)

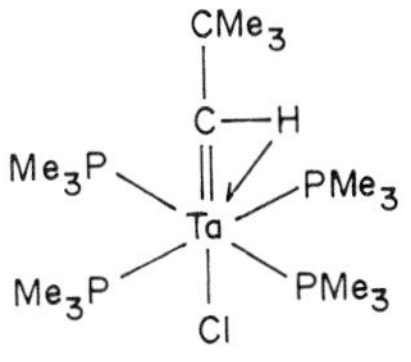

(19)

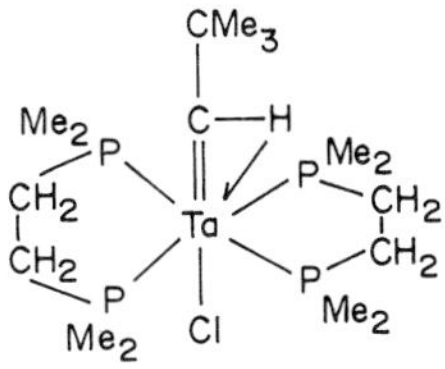

(20)

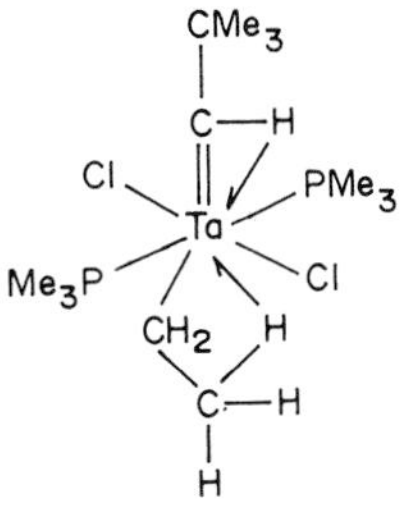

(21)

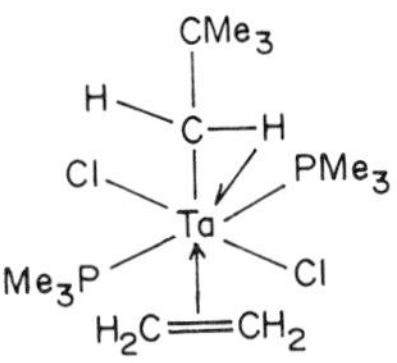

(22)

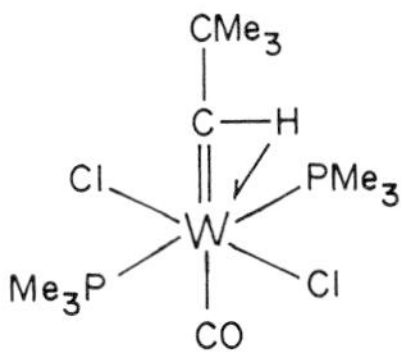

(23)

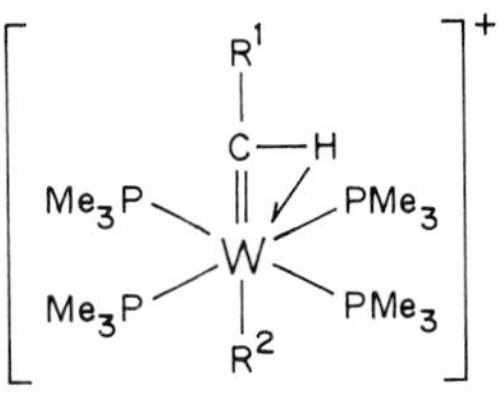

(24)

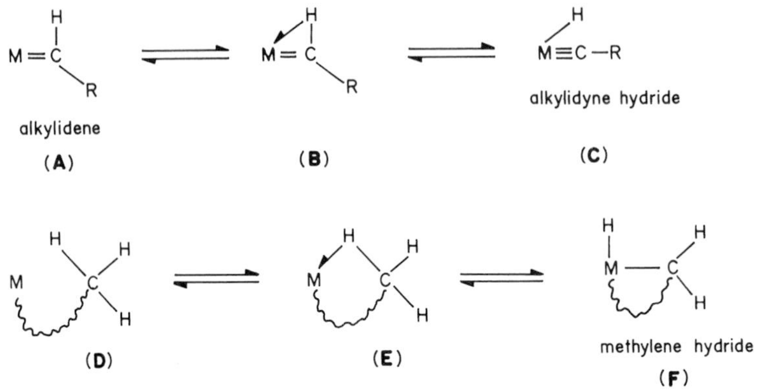

FIGURE 3. Equilibria between complexes with an agostic hydrogen for (a) an alkylidene complex and the isomeric alkylidynemetal hydride, and (b) a methyl complex and the isomeric methylenemetal hydride.

chemical shifts and coupling constants show that a structure with an agostic C—H group is the stable state[36–48].

The factors which determine whether the alkylidene complex (Figure 3, **A**), the alkylidyne metal hydride (**C**) or the complex containing the agostic hydrogen (**B**) will be the stable state depends on a number of factors (e.g. the oxidation state of the metal, and the other ligands), and is one of the features of these complexes being studied by Schrock and coworkers. For example, the complex **24** ($R^1 = H$, $R^2 = Cl$) exists as a methylidene complex with an agostic hydrogen, but the closely related complex where the four trimethylphosphine ligands have been replaced by two $Me_2PCH_2CH_2PMe_2$ ligands exists as the methylidenetungsten hydride. This is because the two phosphoethane ligands will allow a hydrogen to form a pentagonal bipyramid of ligands around the tungsten, a structure that is sterically impossible with the bulkier trimethylphosphine ligands[49,50].

The chemistry of these complexes is influenced by these agostic interactions. For example, the product of the reaction of trimethylphosphine with $[(Me_3CCH_2)_2Cl_2EtTa]$ is an approximately 1:1 mixture of complexes **21** and **22** which are in dynamic equilibrium. Compound **21** is a neopentylideneethyl complex, in which both the hydrogen on C_α of the neopentyl group and one of the C_β hydrogens of the ethyl group is thought to form an agostic link to the tantalum, and **22** is a neopentylethylene complex where one of the C_α hydrogens of the neopentyl group forms an agostic bond.

5. Alkene complexes

Two alkene complexes have been reported in which a metal—hydrogen distance is short enough for there to be an agostic interaction. In the norbornene-(diethylenetriamine)copper(I) cation (**25**), one of the hydrogen atoms on the bridge carbon of the norbornene is only 2 Å from the copper[51]; it should be noted that this is not to be expected as the complex has an 18-electron structure. In a ruthenium(II) complex of norbornadiene (**26**), n.m.r. evidence indicates that one of the hydrogen atoms in the condensed norbornadiene dimer forms an agostic bridge to the ruthenium[52,53].

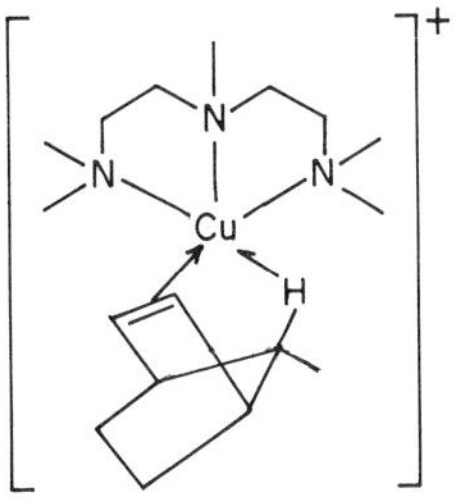

(25)

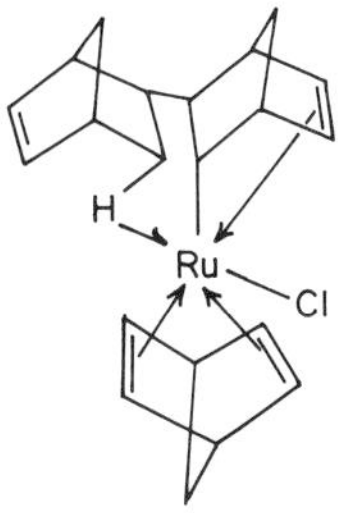

(26)

6. Alkyl complexes

A few compounds are now known in which a methyl, or other small alkyl group directly attached to a transition metal, also forms an agostic C—H→link.

Green and coworkers have reported the crystal structures of the compounds methyltrichloro(dimethylphosphoethane)titanium(IV) and its ethyl analogue (**27** and **28**)[54,55]. In both of these compounds an agostic interaction is clearly evident. These

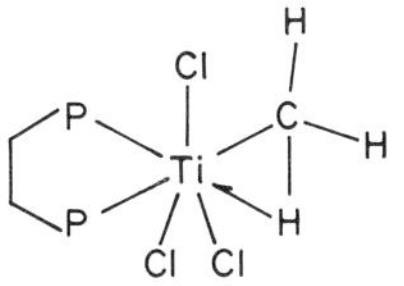

(27)

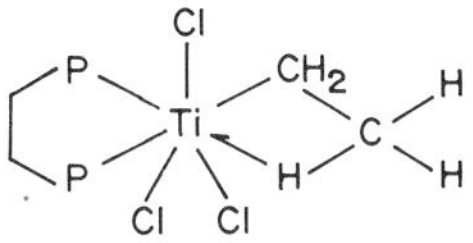

(28)

compounds were selected for study for several reasons. Titanium is a light metal with relatively few electrons and therefore hydrogen atoms in the molecule were more likely to be seen in the X-ray diffraction pictures than if a heavier metal were used. As an early transition element titanium is highly electron deficient, and in these molecules there was little likelihood of the titanium being sterically overcrowded. In the ethyl compound the Ti—C—C angle is only 86°. This indicates that it is the methyl hydrogens of the ethyl group that are forming the agostic link. In the methyl compound a three-membered C—H→Ti ring is formed. For this to occur one of the methyl hydrogens is severely distorted from its normal position, the Ti—C—H angle being reduced from 120° to 70°.

A recent theoretical study has looked at the structure of the six-coordinate [$TiEt(PH_3)_2X_2Y$] compounds (where X = H, F, Cl, or CF_3 and Y = H or Cl)[56], in an attempt to find theoretical evidence for the agostic interaction. *Ab initio* MO calculations have been carried out and it was found that the optimized geometry of [$TiEt(PH_3)_2Cl_2H$] (**29**) has a distorted ethyl group with a short ethyl H—Ti distance. The calculated

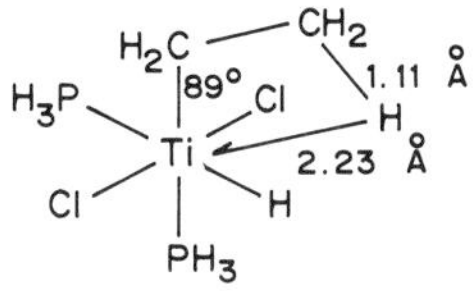

(29)

C—C—Ti bond angle of 89°, the H—Ti distance of 2.23 Å, and the length of the C—H_β bond of 1.11 Å are all in good agreement with the experimentally measured values of 86°,

2.20 Å, and 1.02 Å, respectively, found for the complex [$TiEt(dmpe)Cl_3$] (**28** and Table 1) which this study has chosen to simulate. This agreement indicates that the structural features found are caused by direct intramolecular interaction between the C—H_β bond and the titanium, and not by the effect of packing forces in the crystal. Analysis of the wave-function shows a low-lying unoccupied d_{xy} molecular orbital to which electron delocalization can take place from the C—H_β bonding orbital.

Further calculations show that the ligands make a large difference to the distortion of the ethyl group. The optimized geometry of [$TiEt(PH_3)H_3$], where the axial chlorines of **29** are replaced by hydrogen, has an undistorted alkyl group with a large M—H distance (3.01 Å), normal C—H bond distances and a normal C—C—Ti angle (114°). The optimized geometry of [$TiEt(PH_3)_2H_2Cl$], with the Cl in the equatorial position, also has a normal undistorted ethyl group, suggesting that the electron-withdrawing axial ligands enhance the C—H→M interaction.

Finally, calculations for [$TiEt(PH_3)_2X_2H$] (X = H, F, Cl, CF_3), with X ligands axial, of the energy difference between a structure with a distorted ethyl group and one with an undistorted ethyl group indicate a large axial ligand effect, the distorted structure being the most stable. For X = F the C—C—Ti bond angle is 88°, and the CF_3 complex might have an ethyl group even more distorted than found for the Cl complex, or it might possibly react by β-elimination.

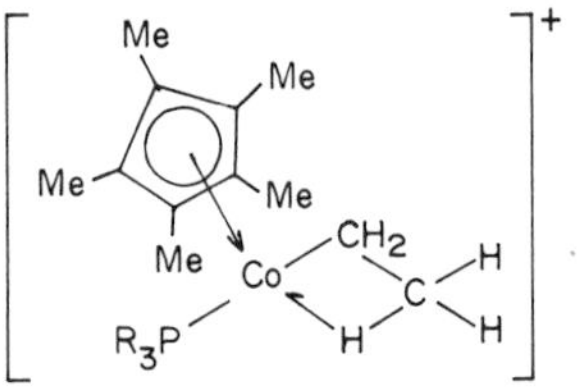

(30)

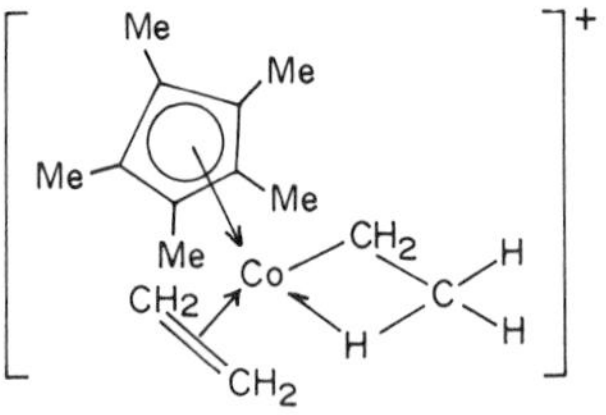

(31)

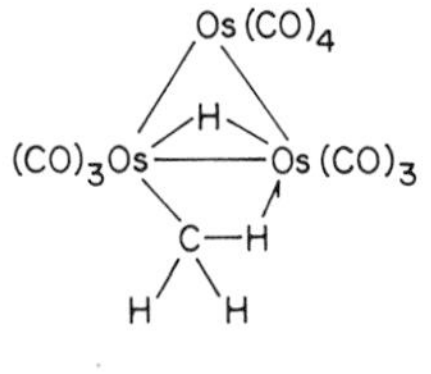

(32)

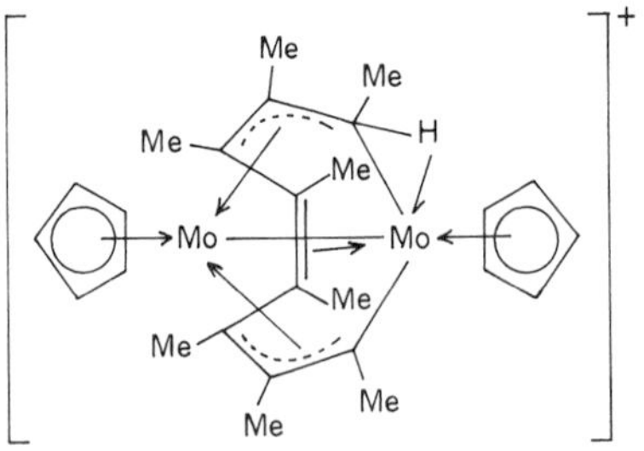

(33)

Two closely similar cationic cobalt(II) complexes containing an agostic ethyl group have also been reported (**30**[57] and **31**[46]). As in the titanium complex above, it is a hydrogen of the methyl groups that forms the agostic link.

7. Bi- and poly-nuclear complexes

There are a number of transition metal complexes containing two or more metal atoms in which agostic C—H bonds have been observed. The first such complex was the osmium carbonyl cluster compound **32** in which agostic C—H bonds were deduced from n.m.r.

studies of partially deuteriated methyl groups[58]. In this complex the C—H → M bonding is from an alkyl group attached to one osmium atom to a second osmium atom.

Agostic bonding occurs in the binuclear molybdenum complex **33**[59]. The eight-carbon ligand forms η^3-allylic bonds to one of the molybdenum atoms and is σ-bonded and forms an agostic C—H bond to the other molybdenum. Three iron carbonyl complexes have been reported. Neutron diffraction studies on the iron cluster compound [$HFe_4(\eta^2$-$CH)(CO)_{12}$] (**34**) indicate an agostic C—H → Fe bond to the iron atom not bonding the CH group[60,61]. Two closely related substituted iron carbonyl dimer anions both contain agostic methyl groups (**35** and **36**)[62,63].

8. Evidence for C—H → M bonds

It can be seen that experimental evidence for agostic C—H → M bonds comes from (i) crystal structure determinations, (ii) n.m.r. data, and (iii) infrared data.

a. Crystal structure determinations.

In the recent past, agostic hydrogens have been located in X-ray and neutron diffraction studies. With these H → M bond distances are measured directly. In earlier work the interaction between the C—H group and the metal was inferred from close M—C distances (see Table 1). Bridging C—H bond lengths are *ca.* 5–10% longer than for analogous non-bridging C—H bonds. The M—H bond distances are considerably less than the van der Waals radii, but are longer (*ca.* 20%) than normal M—H bonds.

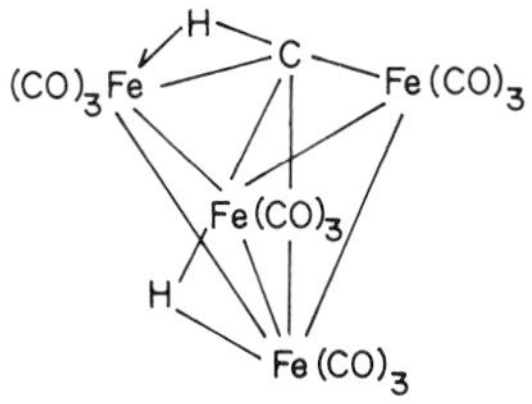

(CO)$_3$Fe, Fe(CO)$_3$, Fe(CO)$_3$, Fe(CO)$_3$, H, C, H

(**34**)

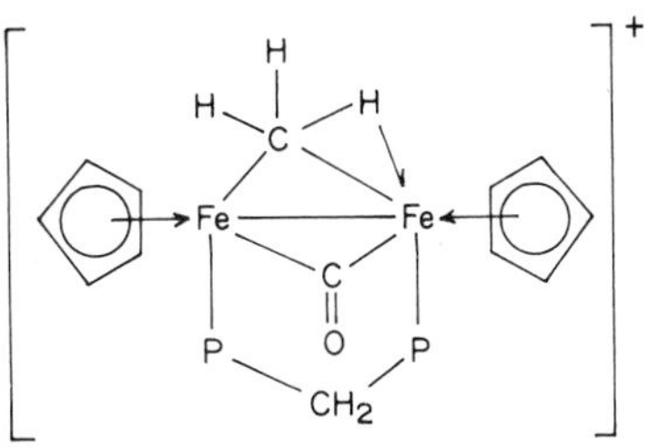

(**35**)

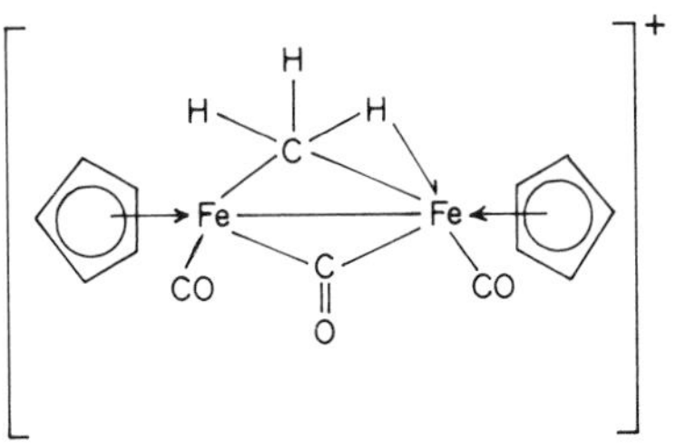

(**36**)

b. Nuclear magnetic resonance data

The presence of the agostic hydrogen in a metal complex has a number of effects on the n.m.r. spectrum. The agostic hydrogen and the ^{13}C atom to which it is attached usually show high-field shifts in their respective spectra. Also, the ^{13}C—H coupling constant (J) between these two atoms is also smaller than for a non-bridged C—H group. Many compounds of this type are fluxional and the average ^1H chemical shifts show a strong dependence on partial deuteriation at the bridging carbon.

i. Non-fluxional compounds. The high-field chemical shifts of agostic hydrogens, when attached to metals with d-electrons, are comparable to the ^1H shifts of metal hydrides and are not therefore diagnostic. However, in some of the alkylidene complexes (**15–19**) and in the d^0 titanium complexes (**27** and **28**) the high-field shift does not occur.

An agostic hydrogen reduces the C—H bond order with the result that the coupling constant $J(^{13}C—H)$ is found to be in the range 45–100 Hz. The equivalent value for a normal saturated (sp^3) C—H bond is in the range 120–130 Hz. This value of J also distinguishes the C—H→M from the C—M—H system as $J(^{13}C—H)$ for alkyl metal hydrides is normally much smaller (*ca.* 10 Hz).

ii. Fluxional compounds. In the n.m.r. spectrum of a fluxional compound only average values of chemical shifts and coupling constants can be obtained. The need is to distinguish between the compound containing the agostic hydrogen, that containing the normal C—H group, and the metal hydride (**E**, **D**, and **F**), respectively (see Figure 3).

For **D** the average ^1H chemical shift will be expected to be at lower field than for either **E** or **F**, except for d^0 compounds, when the shifts will be comparable. Hence the chemical shift might be able to distinguish **D** from **E** or **F** but will not be able to distinguish between **E** and **F**.

Coupling constants are not able to distinguish either. The average value of $J(^{13}C—H)$ expected for D is the normal alkyl value of 120–130 Hz. For **F** the average $J(^{13}C—H)$ will be in the range 80–90 Hz $\{J(H—M—^{13}C) = 0–10, J(^{13}C—H) = 120–130, J(^{13}C—H)_{av} = [(0–10) + 2(120–130)]/3 = 80–90\}$. For **E**$(J^{13}C—H)$ will be *ca.* 120 Hz (i.e. the same as for **D**) because, even though $J(^{13}C—H)_{agostic} = 80–100$ Hz, there is an increase in $J(^{13}C—H)_{non\text{-}bridging}$ up to *ca.* 140 Hz. This arises from the increase in s-character of the non-bridging C—H bonds. Hence $J(^{13}C—H)_{av} = [(80–100) + (2 \times 140)]/3 = ca.\ 120$ Hz.

Agostic C—H systems can be distinguished by n.m.r. if partial deuteriation of the alkyl groups is carried out. This method was first applied to the osmium complex **32** by Calvert and Shapley[58]. For the Me group and for the partially deuteriated analogues, CH_2D and CHD_2 in compounds with the agostic C—H→M link, both the ^1H chemical shifts and the $J(^{13}C—H)$ coupling constants fall in the order $Me > CH_2D > CHD_2$. Also, both the chemical shifts and coupling constants for the two partially deuteriated species are strongly temperature dependent. This is because there is a thermodynamic preference for hydrogen rather than deuterium in the agostic bond, as there is a smaller zero point energy difference between H and D in the C—H→M and C—D→M bonds relative to the difference in the non-bridging C—H and C—D bonds. Consequently, deuterium prefers the non-bridging positions. This temperature-dependent effect has now been observed for a number of compounds given in Table 1. It clearly distinguishes between a non-interacting methyl group (**D**) and agostic bridging (**E**) or a metal hydride (**F**), but it does not distinguish between systems **E** and **F** since the same argument can be applied to both.

In a fluxional molecule an agostic methyl is clearly indicated if both $J(^{13}C—H) \approx 110$ Hz and the temperature dependence of $J(^{13}C—H)$ and ^1H is observed in partially deuteriated compounds. Chemical shifts of $\delta < 0$ add further support.

c. Infrared data

The stretching frequencies have been reported for some agostic C—H bonds. As can be seen in Table 1, they are in the range 2200–2700 cm^{-1}, which is lower than for normal C—H stretching vibrations. Presumably this lowering is associated with the observed increase in length of the agostic C—H bonds.

TABLE 1. Compounds containing 'agostic' C—H → M bonds

No.	Compound	Distances M—H	Distances C—H	Distances M—C	N.m.r. (static) $J(^{13}C{-}H)$/Hz	N.m.r. (static) $\delta(H)$/ppm	N.m.r. (fluxional) $J(^{13}C{-}H)$/Hz	N.m.r. (fluxional) $\delta(H)$/ppm	I.r.: $\nu(C{-}H)/cm^{-1}$	Ref.
5	$[Mo(Et_2B(pz)_2)(2\text{-}PhC_3H_4)(CO)_2]$	2.27(8)	0.97(8)	3.06				− 2.4	2704, 2664	20–24
6	$[Pd\{(CCO_2Me)_4H\}(PPh_3)_2Br]$			2.3						25
7	$[Fe(\eta^3\text{-cyclohexenyl})(CO)_3]OSO_2F$				*ca.* 83					26
8	$[Fe(\eta^3\text{-butenyl})(CO)_3]OSO_2F$				74					26, 27
8	$[Fe(\eta^3\text{-butenyl})\{P(OMe)_3\}_3]BPh_4$				*ca.* 100	− 15.00	129	− 5.9		28
9	$[Fe(\eta^3\text{-cycloheptenyl})\{P(OMe)_3\}_3]BPh_4$				80	− 15.8		− 7.1		28
10	$[Fe(\eta^3\text{-octenyl})\{P(OMe)_3\}_3]BF_4$	1.874(3)	1.164(3)	2.384(4)				− 6.5		28
11	$[Mn(\eta^3\text{-cyclohexenyl})(CO)_3]$				85	− 12.8				31
11	$[Mn(\eta^3\text{-methylcyclohexenyl})(CO)_3]$	1.86(2)	1.07(2)	2.31(0)	85	− 13.6				31
11	$[Mn(\eta^3\text{-cyclohexenyl})(CO)_2\{P(OMe)_3\}]$				86	− 13.0				13
12	$[Ir(\eta^4\text{-}2,3\text{-dimethylbutenyl})(PPh_3)_2H]PF_6$							− 2.2		34
13	$[Ru(\eta^3\text{-}C_{12}H_{17})(dppe)(PMe_2Ph)]PF_6$							− 2.3		35
14	$[Mo(\eta^3\text{-butenyl})(\eta^4\text{-}C_4H_6)(PMe_3)_2]BF_4$					− 9.4		− 3.3		13
15	$[Ta(CHCMe_3)(\eta^5\text{-}C_5H_5)Cl_2]$				84	6.4			2510	36
16	$[Ta(CHCMe_3)(CH_2CMe_3)_3]$				90	1.9				37
17	$[Ta(CHCMe_3)Cl_3(PMe_3)]_2$	2.119	1.131	1.898	101	5.3			2605	28
18	$[Ta(CHCMe_3)(\eta^5\text{-}C_5Me_5)(\eta^2\text{-}C_2H_4)(PMe_3)]$	2.042	1.135	1.946	74				2520	39
19	$[Ta(CHCMe_3)(PMe_3)_4Cl]$				69	− 7.9			2200	47
20	$[Ta(CHCMe_3)(dmpe)_2Cl]$				57	− 8.49			2200	48
21	$[Ta(CHCMe_3)(PMe_3)_2Cl_2Et]$				80					64
22	$[Ta(CH_2CMe_3)(PMe_3)_2Cl_2(C_2H_4)]$				98					64
23	$[W(CHCMe_3)H(Cl)_2(CO)(PMe_3)_2]$	1.835(4)	1.053(4)	1.859(4)	84	1.4			2395	40
24	$[W(CH_2)(PMe_3)_4Cl]CF_3SO_3$					− 7.97				50
37	$[W(CHCMe_3)(PMe_3)_4Cl]CF_3SO_3$				45	− 8.3				50
38	$[W(CH_2)(PMe_3)_4I]CF_3SO_3$			1.83(2)						66
25	$[Cu(\eta^2\text{-}C_7H_{10})(dien)]BPh_4$	2.01	0.81	2.78(1)						51
26	$[Ru(C_{14}H_{17})(C_7H_8)Cl]$					− 3.7			2586	52
27	$[Ti(Me)Cl_3(dmpe)]$	2.03(4)	1.00(2)	2.149(5)				2.3		54
28	$[Ti(Et)Cl_3(dmpe)]$	2.29	1.02	2.516				2.7		55
30	$[Co(Et)(\eta^5\text{-}C_5Me_5)\{P(p\text{-tolyl})_3\}]BF_4$	2.128	1.31(4)	2.128	66	− 5.92				57
31	$[Co(Et)(\eta^5\text{-}C_5Me_5)(\eta^2\text{-}C_2H_4)]BF_4$				67	− 12.1				46
32	$[Os_3(CO)_{10}(Me)H]$						121	− 3.7		58
33	$[(\eta^5\text{-}C_5H_5)Mo(C_{16}H_{25})Mo(\eta^5\text{-}C_5H_5)](CF_3CO_2)_2H$	1.88(8)	0.89(7)	2.196(5)		− 9.4				59
34	$[HFe_4(\eta^2\text{-}CH)(CO)_{12}]$	1.80(4)	1.00(4)	1.926(5)	103	− 1.3				60
35	$[Fe_2(Me)(\mu\text{-}CO)(\mu\text{-dppm})(\eta^5\text{-}C_5H_5)_2]PF_6$[a]	1.64(4)	1.06(4)	2.108(3)			114	− 2.9		62
		1.78(3)	0.83(4)	2.118(3)						
36	$[Fe_2(Me)(\mu\text{-}CO)(CO)_2(\eta^5\text{-}C_5H_5)_2]BF_4$						121	− 2.0		63

[a] There are two crystallographically independent cations in the crystal.

9. Conclusions

a. General

A whole range of compounds containing agostic C—H→M bonding have now been characterized—it is likely that this type of interaction is more widespread than previously thought.

It has been recognized for some time that such agostic interactions may occur as transition-state intermediates in the formation of metal hydrides [$n = 18$ electrons (usually)] from the related $n-2$ electron complexes as shown in Figures 3 and 4. It is now evident that in certain systems the agostic structure can be the ground-state structure.

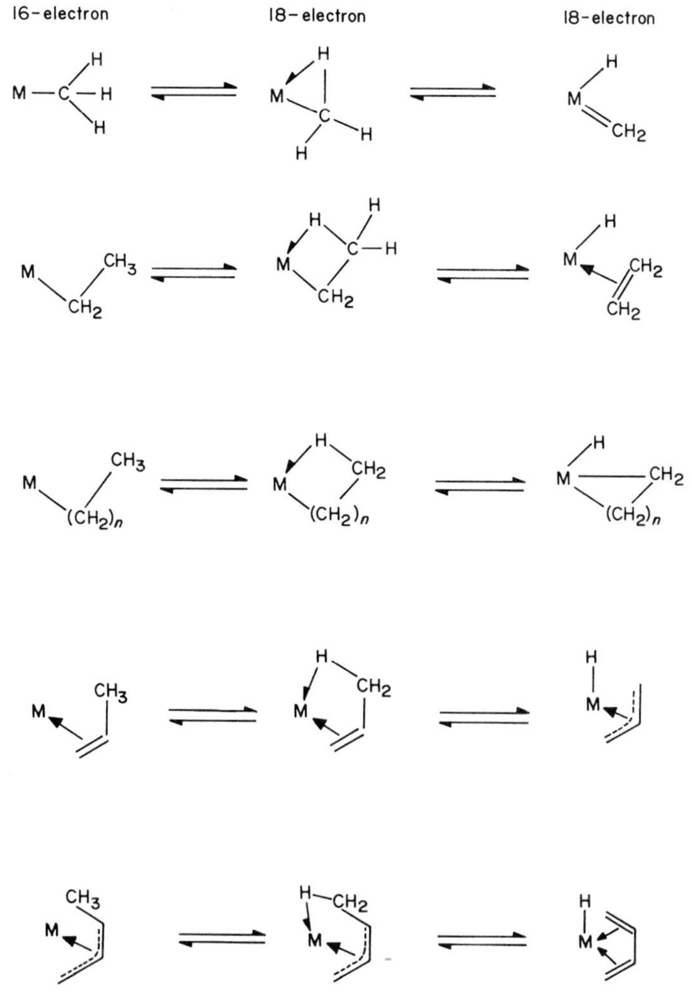

16-electron 18-electron 18-electron

FIGURE 4. Equilibria between 18-electron olefinmetal hydride, the related 16-electron complex, and the agostic intermediate.

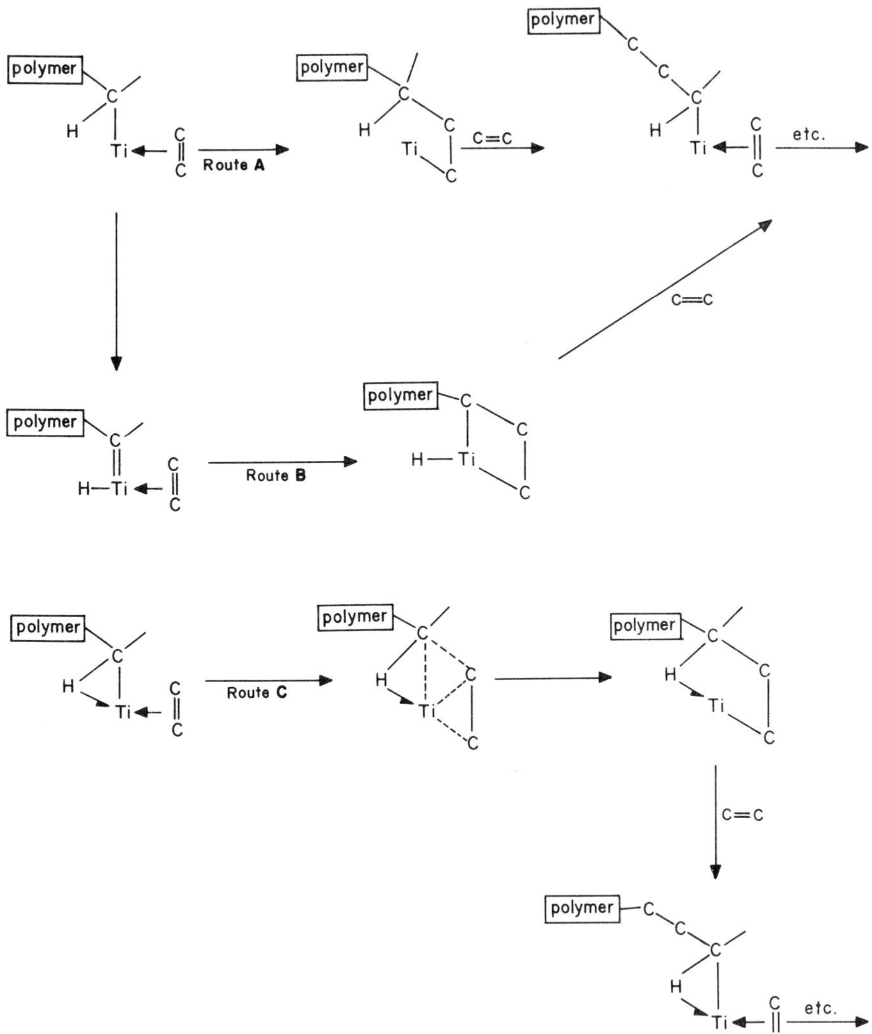

FIGURE 5. Ziegler–Natta polymerization by the Cossee mechanism (route **A**), a 1,2-hydrogen shift forming an alkylidene hydride (route **B**), and the formation of an agostic C—H→Ti bond (route **C**).

The minimum requirement for this is that the metal should have an empty orbital to receive the two electrons of the C—H bond, i.e. the metal should have 16 electrons or less. It also needs to be of the correct energy and orientated such that it can overlap with the C—H bond. This is illustrated by the 16-electron compound [Ti(η^7-C_7H_7)(dmpe)Et] (**37**)[42].

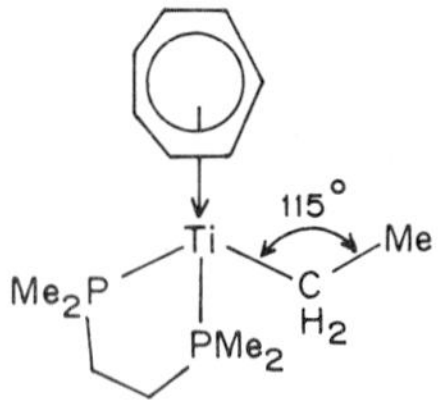

(37)

In contrast to $[TiCl_3(dmpe)Et]$ (**28**), **37** shows no interaction between the ethyl hydrogens and the titanium. This is because the lowest unoccupied molecular orbital that would be used in the formation of the C—H → Ti bridge will be the d_{z^2} and this is not of the correct orientation to interact with the ethyl group. Also, this orbital will be at a substantially higher energy than the ethyl electrons.

It is to be expected that agostic hydrogens can be displaced by a donor ligand. In many complexes the agostic hydrogen will have to compete with lone pairs on other ligands. In $[TiMe_3Cl_3(dmpe)]$, for example, the lone pairs on the chlorine ligands do not compete successfully with the agostic C—H electron pair.

b. The mechanism of olefin polymerization by Ziegler–Natta catalysts

Green and coworkers[13,54] have discussed the implications of the formation of agostic C—H → Ti links on the suggested mechanisms of Ziegler–Natta catalysts. The traditional mechanism for the olefin polymerization is that of simple alkyl migration to the coordinated olefin as proposed by Cossee[44] (Route **A** in Figure 5), in which there is no direct involvement of the C—H bonds of the alkyl chain. A suggested alternative mechanism (Route **B** in Figure 5)[45] involves the prior formation of a metallacarbene intermediate by a 1, 2-hydrogene shift from the alkyl polymer. This mechanism can explain the control of the stereochemistry of the polymerization since substituents could lie *cis* or *trans* to the metallacyclobutane ring, and implies a close connection between olefin polymerization and metathesis.

Schrock and coworkers have shown that the 1:1 mixture of complexes **21** and **22** will react with ethylene to give 4, 4-dimethylpent-1-ene and 4, 4-dimethyl-*trans*-pent-2-ene. These are products of the reaction of the neopentylidene ligand with ethylene[64]. Further, they found that the alkylidene hydride $[Ta(neopentylidene)(H)(PMe_3)_3I_2]$ with ethylene under mild pressure gives polyethylene[65]. These observations give support to an alkylidene hydride mechanism.

The possible agostic C—H → M interactions means that a mechanistic route (**C** in Figure 5) is attractive. This is intermediate between routes **A** and **B** and can control the stereochemistry of the polymerization, without the need for the formation of the alkylidene complex.

C. Intramolecular Activation of C—H Bonds in Alkyl Groups in Ligands—Cyclometallation

1. Introduction

In the previous section we surveyed a large number of transition metal complexes in which there is an electronic interaction between the metal and a carbon—hydrogen bond in a ligand—'agostic' bonding (**39** in equation 1).

$$\text{M}\cdots\text{H–CR}_2 \;(\mathbf{38}) \longrightarrow \text{M}\leftarrow\text{H–CR}_2 \;(\mathbf{39}) \longrightarrow \text{M–C(R}_2\text{)}\text{, M–H} \;(\mathbf{40}) \qquad (1)$$

Here we survey the molecules that are formed when this interaction, from a saturated, sp^3, carbon—hydrogen bond, results in the bond breaking and the carbon becoming bonded to the metal, i.e. with cyclometallation occurring (**40** in equation 1). If the reaction occurs by oxidative addition the hydrogen will be expected to become attached to the metal atom in the product. If the hydride so formed is unstable it may decompose and the hydrogen will not be in the recovered product. Also, if the mechanism is not oxidative addition then the hydrogen may never become attached to the metal.

This is a topic which has been extensively reviewed in the literature referred to in the Introduction to this chapter and in other reviews[9,67–77], although most of the work covered by other reviews concerns aromatic C—H bond (i.e. sp^2 C—H bond) activation. Here we concentrate on the less extensive, but by no means small, topic of cyclometallation by sp^3 C—H bond activation.

The most extensive range of cyclometallated compounds are complexes of the platinum metals—platinum, palladium, iridium, and rhodium. A number of distinctive classes of complexes can be identified and these will be examined in turn. Complexes with alkyl groups attached to the metal undergo a variety of cyclometallations. Complexes with bulky phosphine ligands constitute a most extensive class of compounds that undergo intramolecular C—H bond activation. Palladium–nitrogen complexes and an array of iridium–hydride complexes also cyclometallate. The metals from the middle of the transition series have not been extensively studied; the few examples that there are of cyclometallations are surveyed. The early transition elements also provide few examples, tantalum complexes being the most studied. A couple of reports of thorium and uranium complexes that cyclometallate complete this section.

We shall deal only briefly with work prior to 1977, when we[1] and Parshall[11] last reviewed this topic.

2. *Activation in alkyl groups bonded to the platinum metals*

a. *Platinum and palladium complexes*

The cyclometallations that occur in a series of platinum(II) alkyl complexes have been studied by Whitesides and coworkers[78–83].

If the complex dineopentylbis(triethylphosphine)platinum(II) is heated at 157 °C for 2 h in cyclohexane, activation of a γ-C—H bond occurs to form the four-membered ring compound bis(triethylphosphine)-3, 3-dimethylplatinacyclobutane[78,79] (equation 2). The

$$(Et_3P)_2Pt(CH_2CMe_3)_2 \xrightarrow[\text{cyclohexane}]{157\ ^\circ C/2\ h} \left[(Et_3P)_2Pt\begin{matrix}CH_2\\ \\ CH_2\end{matrix}CMe_2\right] + \text{neopentane} \qquad (2)$$

kinetics of this reaction show a dissociation of a phosphine ligand followed by C—H activation which occurs as oxidative addition to platinum(II), followed by reductive elimination of neopentane from the platinum(IV) intermediate. The rate-limiting step is either the phosphine dissociation or the reductive elimination. This implies that intramolecular oxidative addition is intrinsically rapid, in clear contrast to intermolecular C—H addition to platinum which is slow.

C—H bonds in δ- and ε-positions of the alkyl chain are also activated in appropriate complexes[81,82]. If the complex $[Pt(PEt)_2(CH_2CMe_2CH_2Me)]$ is heated to 126 °C there are two products: one, the minor product (2%), with the four membered ring platinacycle, where a γ-methyl C—H bond has been activated, and the major product (98%), with the five-membered ring platinacycle, where a δ-methyl C—H bond has been activated (equation 3).

$$(Et_3P)_2Pt(CH_2CMe_2CH_2Me)_2 \xrightarrow{126\,^{\circ}C} \left[(Et_3P)_2Pt\langle (CH_2)_2 \rangle C(Me)(CH_2Me)\right] \; (2\%) + \left[(Et_3P)_2Pt\langle CH_2CMe_2CH_2CH_2 \rangle\right] \; (98\%) \quad (3)$$

Extending the alkyl chain by another CH_2 group gives a complex which on heating to 146 °C produces three cyclometallated products, a four-membered ring complex (23%) (CH_3 activation), a five-membered ring complex (68%) (CH_2 activation), and also a six membered ring complex (9%) where ε-methyl C—H activation has occurred (equation 4). The bulkier alkyl chain with a *tert*-butyl end group cyclizes at the ε-methyl very readily.

$$(Et_3P)_2Pt(CH_2CMe_2CH_2CH_2Me)_2 \xrightarrow{146\,^{\circ}C} \left[(Et_3P)_2Pt\langle (CH_2)_2 \rangle C(Me)(CH_2CH_2Me)\right] \; (23\%) + \left[(Et_3P)_2Pt\langle CH_2CMe_2CH_2CH(Me) \rangle\right] \; (68\%) + \left[(Et_3P)_2Pt\langle CH_2CMe_2CH_2CH_2CH_2 \rangle\right] \; (9\%) \quad (4)$$

A 100% yield of the six-membered platinacycle ring product is obtained at 87 °C (equation 5).

$$(Et_3P)_2Pt(CH_2CMe_2CH_2CMe_3)_2 \xrightarrow{87\,^{\circ}C} \left[(Et_3P)_2Pt\langle CH_2CMe_2CH_2CMe_2CH_2 \rangle\right] \; (100\%) \quad (5)$$

If the neophylplatinum(II) complex (neophyl = CH_2CMe_2Ph) is heated it behaves completely differently[84], as there is now the opportunity for the more favoured reaction at the phenyl ring. At room temperature this complex cyclometallates at the *ortho*-CH of the phenyl ring (equation 6).

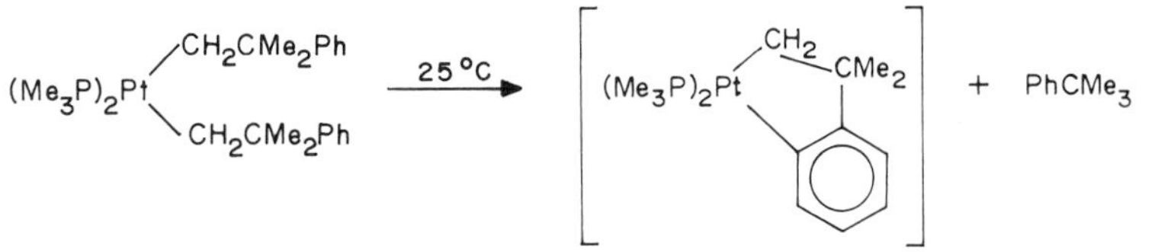

$$(Me_3P)_2Pt(CH_2CMe_2Ph)_2 \xrightarrow{25\,°C} [(Me_3P)_2Pt\text{—}CH_2CMe_2C_6H_4] + PhCMe_3 \qquad (6)$$

These results indicate that δ-C—H activation is the most favourable to give a platinacyclopentane, but they also show that platinacyclobutanes can be formed, suggesting that the ring strain is much smaller than in cyclobutane itself. There appear to be at least three factors which might make important contributions to these reactions: (i) the entropy increase that occurs on creating two molecules from one; (ii) the relief of non-bonding steric strain; and (iii) the favourable changes in local electronic energies that occur as the P—Pt—P bond angle widens during reaction[80]. In these complexes the intramolecular C—H activation always predominates over intermolecular C—H activation with solvent molecules, a feature that is discussed in Section II.E.

A platinacyclopentene is formed by δ-hydrogen abstraction from the o-$MeC_6H_4CH_2$ ligand. The $[Pt(PEt_3)_2(o\text{-}CH_2C_6H_4CH_3)_2]$ complex is prepared as shown in equation 7. This, after several hours in refluxing xylene, gives the cyclometallated complex **41**[85].

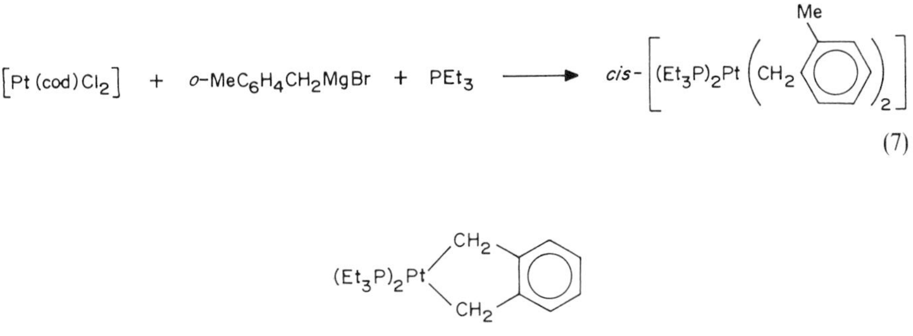

$$[Pt(cod)Cl_2] + o\text{-}MeC_6H_4CH_2MgBr + PEt_3 \longrightarrow cis\text{-}[(Et_3P)_2Pt(CH_2C_6H_4Me)_2] \qquad (7)$$

$$(Et_3P)_2Pt(CH_2)_2C_6H_4$$

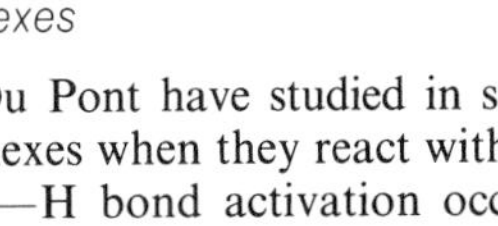

(**41**)

b. Iridium and rhodium complexes

Parshall and coworkers at Du Pont have studied in some detail the chemistry of trimethylphosphineiridium complexes when they react with compounds containing C—H bonds[86–88]. Intramolecular C—H bond activation occurs during the preparation of alkyliridium(I) complexes[84]. When $[(Me_3P)_4IrCl]$ reacts with MeLi the expected complex, $[(Me_3P)_4IrMe]$, is obtained. With EtLi, however, the ethyl complex that is presumably formed transfers a β-hydrogen to the iridium and gives ethylene as the product. With alkyllithium reagents that cannot undergo β-hydrogen elimination, γ-hydrogen transfer occurs to give the four-ring metallacycle (equation 8). Intramolecular metal activation of a remote (γ,δ,etc.) site in an hydrocarbon ligand has been called 'distal' C—H bond activation[84].

$$[(Me_3P)_4IrCl] + Me_3CCH_2Li \longrightarrow [Me_3CCH_2Ir(PMe_3)_n] \longrightarrow [(Me_3P)_3Ir(H)(CH_2CMe_2CH_2)] \qquad (8)$$

With neophyllithium (2-methyl-2-phenylpropyllithium)[84], the iridium complex gave an iridium(III) metallacycle formed by C—H activation of the *ortho*-bond in the phenyl ring of the ligand (equation 9) but the analogous rhodium complex gave the *tert*-butylphenylrhodium(I) complex (equation 9).

$$C_6H_5CMe_2CH_2Li + [(Me_3P)_3MCl] \longrightarrow [C_6H_5CMe_2CH_2M(PMe_3)_3]$$

M = Ir: $[C_6H_4CMe_2CH_2Ir(H)(PMe_3)_3]$ (metallacycle with H on Ir, ring CMe(Me))

M = Rh: $[2\text{-}Me_3CC_6H_4Rh(PMe_3)_3]$ (9)

For rhodium the initial product was presumably the analogue of the iridium product, but this was unstable and it transferred a hydrogen from rhodium to the CH_2 group to form the rhodium(I) product. This pair of reactions illustrates the general view that rhodium is often the more catalytically active of the two metals, but that iridium is more convenient for isolation of intermediates in a catalytic cycle[86].

It is evident that the products that are formed by these C—H activations depend critically on the metal and the ligands. When the iridium(I) complex is benzylated the products formed vary, depending on whether the ligands are trimethylphosphine or triethylphosphine, and in neither case is the benzyliridium complex, which is presumably formed initially, the final product[88]. With trimethylphosphine ligands a benzoirida-cyclobutene is formed analogous to the iridacycle formed with the neophyl group (equation 10), but with triethylphosphine as the ligand an *o*-tolyliridium complex is formed (equation 10), presumably because the iridacycle that would be formed transfers a hydrogen from the iridium to the CH_2 group, just as with the rhodium complex in equation 9.

$$C_6H_5CH_2MgBr \xrightarrow{L_3IrCl} [C_6H_5CH_2IrL_3]$$

L = PMe_3: $[C_6H_4CH_2Ir(H)(PMe_3)_3]$

L = PEt_3: $[2\text{-}MeC_6H_4Ir(PEt_3)_3]$ (10)

By using *trans*-$[RuCl_2(PMe_3)_4]$ instead of the iridium complex, the equivalent ruthenocyclobutane is formed[88], but attempts to prepare the rhodium analogue give the η^3-benzyl complex **42**. A similar transient complex in the other complexes could bring the *ortho*-C—H bond under the influence of the metal and initiate the cyclometallation.

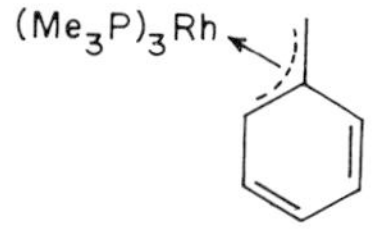

$(Me_3P)_3Rh$

(42)

If 2-phenylpropylmagnesium halide is used to alkylate the iridium complex, the products again differ depending on the phosphine ligand (equation 11). Here again, with triethylphosphine, hydrogen transfer occurs from the iridium to the methylene group to give the iridium(I) complex. In the intermediate here there is a β-CH bond, that would, by analogy with other systems, be expected to react to give a methylstyrene complex, but reaction of the δ-CH bond, on the phenyl ring, is preferred.

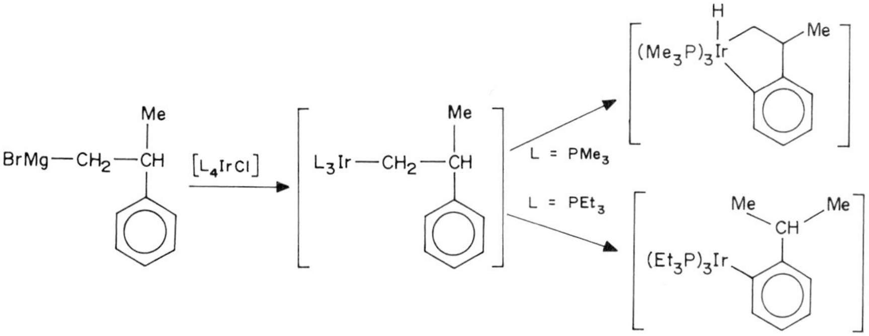

BrMg—CH_2—CH(Me)(Ph) $\xrightarrow{[L_4IrCl]}$ [L_3Ir—CH_2—CH(Me)(Ph)]

L = PMe_3: [$(Me_3P)_3Ir$(H) Me]

L = PEt_3: [$(Et_3P)_3Ir$ Me CH Me]

(11)

The preference for intramolecular aryl C—H activation probably results from η^2-arene precoordination. This would block the vacant coordination site required for β-elimination in complexes such as **43** as well as positioning an *ortho*-C—H bond in proximity to the metal. Aliphatic ligands cannot so precoordinate and this precludes β-elimination.

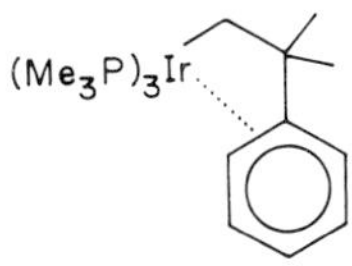

$(Me_3P)_3Ir$

(43)

3. *Activation in bulky phosphine ligands coordinated to the platinum metals*

a. *Monophosphines*

Shaw and coworkers have studied the chemistry of a large number of complexes of platinum and palladium containing bulky tertiary phosphine ligands[89]. In early work a number of phosphines containing aromatic groups, e.g. di-*tert*-butylphenylphosphine, were studied. On heating, intramolecular ring closure at the *ortho*-position on the phenyl ring occurred and HCl was eliminated[90,91] (equation 12).

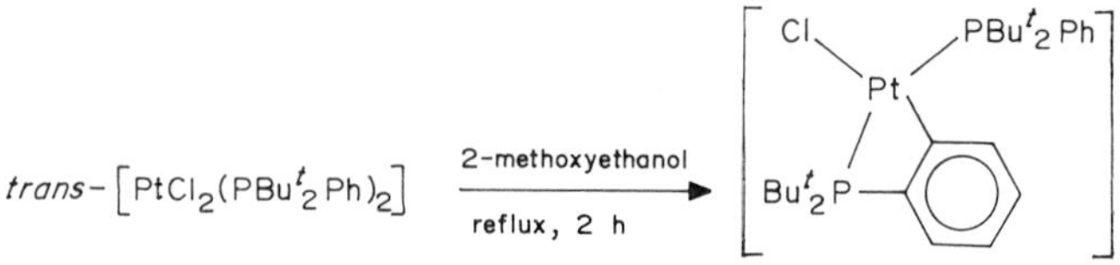

$$trans\text{-}[PtCl_2(PBu^t_2Ph)_2] \xrightarrow[\text{reflux, 2 h}]{\text{2-methoxyethanol}} [\text{cyclometallated Pt complex: Cl, } PBu^t_2Ph, Bu^t_2P\text{–}C_6H_4] \quad (12)$$

It soon became evident that bulky ligands were very important in causing such cyclometallations, since if the complex used had the di-*tert*-butylphenylphosphine ligand replaced by dimethylphenylphosphine then the ring closure (equation 12) did not occur.

The sp^3 C—H bonds in the methyl group of an *o*-tolyl group of a phosphine are activated and undergo cyclometallation when *trans*-$[PtCl_2\{P(o\text{-}C_6H_4Me)_2Ph\}_2]$ is heated in 2-methoxyethanol; complex **44** is formed.

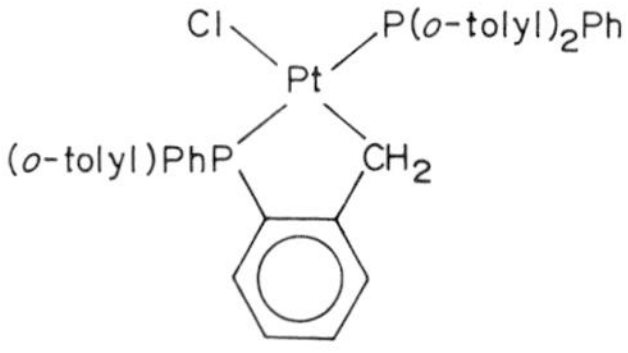

Cl, P(*o*-tolyl)$_2$Ph, Pt, (*o*-tolyl)PhP, CH_2

(**44**)

Intramolecular ring formation also occurs by activation of a C—H bond of an alkyl group attached directly to the phosphorus. When iridium trichloride reacts with the phosphine, PBu^t_2 (2-HO-Phenyl), the product is the purple five-coordinate iridium(III) hydride **45**. This in air is oxidized to the iridium(II) complex **46**, which in turn is slowly converted into the iridium(III) complex **47** in which one of the *tert*-butyl groups has been metallated (equation 13)[92]. This was the first example of the cyclometallation of a *tert*-butyl group; there have been many other examples since then.

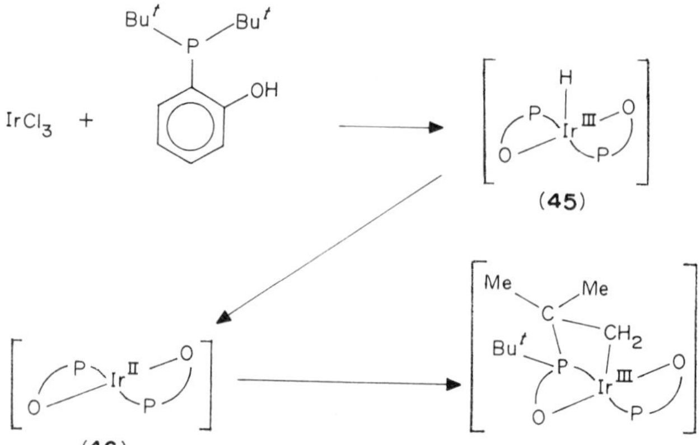

$IrCl_3$ + $Bu^t_2P(2\text{-}HOC_6H_4)$ → (**45**) → (**46**) → (**47**) (13)

(**45**) (**46**) (**47**)

Alkylphosphines, PR_3, have also been studied[93,94]. The reactions of platinum(II) complexes with the phosphines $PBu^t_2Pr^n$, $PBu^t_2Bu^i$, PBu^t_2 neopentyl, and PPh_2Bu^t show differences in reactivity that would be expected if steric bulk of the ligand were of crucial importance.

When $[PtCl_2(PhCN)_2]$ in methylene chloride reacts with $PBu^t{}_2Pr^n$ the product is *trans*-$[PtCl_2(PBu^t{}_2Pr^n)_2]$. Under forcing conditions, refluxing for 300 h in 2-methoxyethanol, cyclometallation occurs to give complex **48**.

If the neopentylphosphine $PBu^t{}_2(CH_2CMe_3)$ is used there is a remarkable increase in rate and after 20 min at 20 °C in methylene chloride solution cyclometallation has occurred to give the dimeric complex (**49**; R = Me)[93]. The isobutylphosphines exhibit intermediate behaviour; $PBu^t{}_2Bu^i$ with $[PtCl_2(Bu^tCN)_2]$ in refluxing 2-methoxyethanol forms the cyclometallated complex **49** (R = H). The less bulky PPh_2Bu^i does not cyclometallate, *trans*-$[PtCl_2(PPh_2Bu^i)_2]$ being formed.

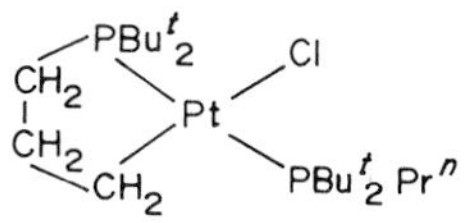

(48)

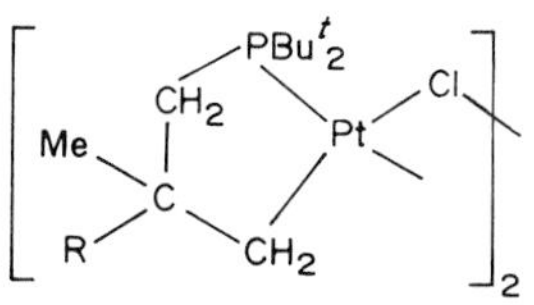

(49)

These two phosphines behave in an analogous manner with palladium, ($Na_2[PdCl_4]$). The differences in behaviour are a clear demonstration of the ability of an alkyl C—H bond in the phosphine to undergo cycloaddition to the metal when there is steric crowding in the phosphine and the cyclometallation can release this steric compression. The conformational and entropy factors are both favourable, particularly when there is a *gem*-di-*tert*-butyl and *gem*-dimethyl group in the molecule (see the conclusion to this section).

The phosphine $PBu^t{}_3$ also reacts with platinum(II) and palladium(II) compounds to give phosphine complexes that undergo intramolecular C—H activation[95–101]. The reaction products formed depend on the metal complex used and on the solvent. In benzene solution $PBu^t{}_3$ reacts with $PtCl_2$, $Na_2[PtCl_4]$, or $[PtCl_2(PhCN)_2]$ to give a mixture of the salt $[PBu^t{}_3H][PtCl_4]$ and the cyclometallated complex **50** (M = Pt). If the platinum(II) complex is $[PtCl_2(cod)]$ these products are not formed. Palladium complexes give the analogous cyclometallated products **50** (M = Pd). When methylene chloride is the solvent then the $PBu^t{}_3$ ligand in **50** is lost and the dimer **51** (M = Pt or Pd) is formed[96].

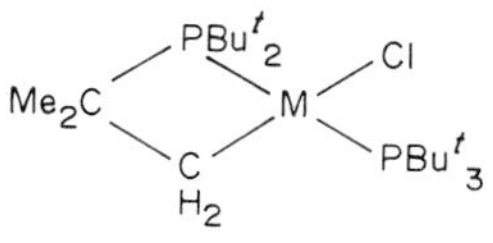

(50)

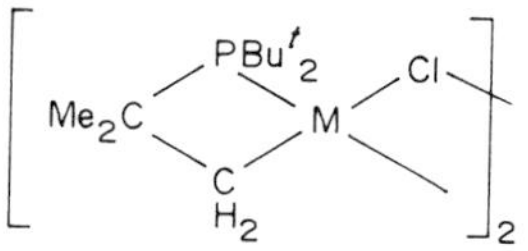

(51)

Both the platinum and the palladium complexes **51** can be converted to **50** by treatment with $PBu^t{}_3$[99]. When dimethylformamide is the solvent for palladium(II) compounds then it behaves as in methylene chloride and the complex **51** (M = Pd) is formed. With platinum(II) chloride, however, the reaction is different and the six-coordinate complexes (**52**; X = $PBu^t{}_3$ or H) are formed[100].

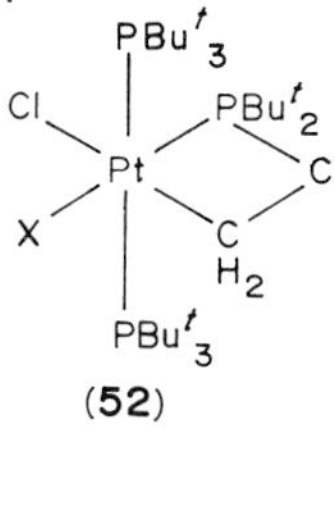

(52)

The palladium and platinum hydrides $[MH(X)(PBu^t_3)_2]$ ($M = Pd$ or Pt; X = halogen), which can be prepared by treating $[M(PBu^t_3)_2]$ with HX[101], undergo rapid intramolecular metallation in both benzene and methylene chloride solution. The products (**50**; M = Pd or Pt) and hydrogen are formed in a few hours at room temperature, and the reaction rate is enhanced if ethanol is added to the solvent[97,98]. The complex **50** is readily converted into the dimer **51**. If the PBu^t_3 ligand in **50** is replaced by a less bulky ligand (e.g. PEt_3, PPh_3, or $AsPh_3$) then this complex is stabilized. In contrast complex **50** more readily loses this phosphine to form the dimer **51** if it is bulky [e.g. $P(cyclohexyl)_3$, $P(o\text{-tolyl})_3$]. The facile elimination of hydrogen in this complex is thought to be the driving force for the cyclometallation, rather than oxidative addition of the C—H bond to palladium.

The effect on the course of the reaction of various alkylphosphines with iridium supports the work with platinum and palladium, and also illustrates the variety of the chemistry that these complexes display[102]. The reactions of $[IrCl(cyclooctene)_2]_2$ with $PBu^t_2Pr^n$, $PBu^t_2Bu^n$, PBu^n_3, PBu^t_3 and PPr^i_3 in the presence of γ-picolene or acetonitrile produce an array of cyclometallated products that contain both four- or five-membered iridacycle rings. No six-membered ring compounds are formed. The reactivity towards metallation is $Pr^i_3 \approx PBu^t_2Pr^n > PBu^t_2Bu^n > PBu^t_3$. PBu^n_3 did not give any metallated products.

Clearly the size of the phosphine (measurable by its 'cone-angle'[103]) determines the reactivity; when small no metallated products are obtained, and when very large the yields are low. The influence of methyl groups in the ligands on the course of the reaction is considerable. $PBu^t_2Pr^n$ gives as the major product the five-membered ring compound (equation 14) by metallation of the *n*-propyl group, and as a minor product the four-membered ring compound (equation 14) by metallation of the *tert*-butyl group. The four-

$$[(cyclooctene)_2IrCl]_2 + PBu^t_2Pr^n \xrightarrow[\gamma\text{-picoline}]{\text{hexane}} [IrHCl(\gamma\text{-picoline})(PBu^t_2Pr^n)\{PBu^t_2CH_2CH_2CH_2\}]\ (55\%) + [IrHCl(\gamma\text{-picoline})(PBu^t_2Pr^n)\{PBu^tPr^nCMe_2CH_2\}]\ (20\%) \quad (14)$$

membered ring complexes are formed almost exclusively by $PBu^t_2Bu^n$, because five-or six-membered rings would increase the crowding round the iridium. If acetonitrile is the sixth ligand, rather than γ-picoline, the crowding is slightly less and some five-membered ring product is formed.

The reactivity of these aliphatic C—H bonds towards iridium(I) does not differ much from the reactivity of allylic and aromatic C—H bonds. All three types of C—H bond have similar energies and it is suggested that when the steric conditions are well chosen the bond energies determine the metallation reactivity.

b. Diphosphines

Complexes of the diphosphine $Bu^t_2P(CH_2)_5PBu^t_2$ undergo intramolecular C—H activation[104–108]. When iridium trichloride reacts with the phosphine in boiling isopropanol for 3 days the product is a mixture of the orange–red cyclometallated five-coordinate iridium hydride **53** (M = Ir, R = H) and the red–purple binuclear hydride $[Ir(H)Cl_2Bu^t{}_2P(CH_2)_5PBu^t{}_2]_2$ containing a 16-atom ring. When the hydride **53** (M = Ir, R = H) is heated at *ca.* 170 °C and 15mm Hg pressure it partially decomposes to the dark brown complex **54**, which has a structure somewhere between the extremes of the iridium(I) carbene and the iridium(III) ylide **54**. The loss of dihydrogen is reversible; a benzene solution of **54** under dihydrogen at 20 °C gives **53** (M = Ir, R = H) after 3 h[104].

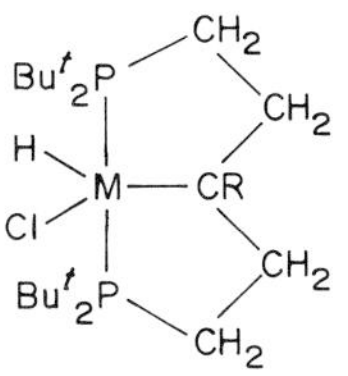

(53)

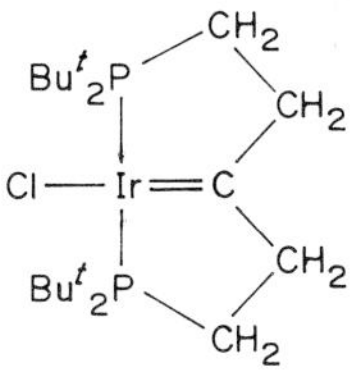

(54)

The analogous rhodium hydride product (**53**; M = Rh) is formed from the reaction of rhodium trichloride with the phosphine in tetrahydrofuran[105,106]. The rhodium cyclometallation does not occur as readily as that of iridium. In ethanol solvent only 40% of the product is the cyclometallated compound **53** (M = Rh), the other 60% is the dimer $[Rh_2H_2Cl_4(phosphine)_2]$. This gives the cyclometallated product when treated with the base 2-methylpyridine. With rhodium an olefinic complex (**55**) is formed if the phosphine with six CH_2 groups in the chain is used, i.e. two adjacent C—H bonds are activated[106]. A similar olefin complex is thought to be a minor impurity in the reaction with $Bu^t{}_2P(CH_2)_5PBu^t{}_2$.

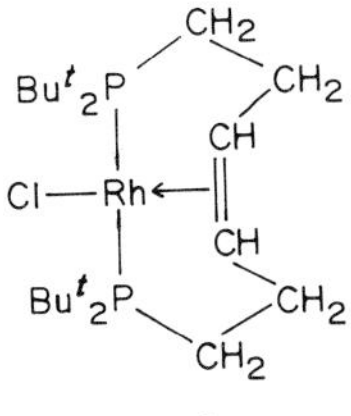

(55)

The phosphine $Ph_2P(CH_2)_6PPh_2$ also gives a cyclometallated olefin complex with iridium or rhodium[109]; refluxing the phosphine in mesitylene with $[M_2Cl_2(cod)_2]$ M = Ir or Rh) produces the complex of the ligand 1,6-bis(diphenylphosphino)-*trans*-hex-3-ene (**56**; M = Ir or Rh).

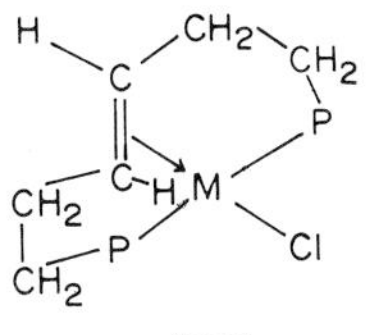

(56)

Palladium and platinum behave similarly[107]; the four-coordinate cyclometallated complexes **57** (M = Pd or Pt, R = H) are formed by reacting $[PdCl_2(PhCN)_2]$ or $[PtCl_2(PhCN)_2]$ with $Bu^t{}_2P(CH_2)_5PBu^t{}_2$. If the diphosphine with six CH_2 groups in the chain is used with platinum the reaction product is not the olefinic complex (analogue of

55) as formed by rhodium, but the complex **58** with a five-membered and a six-membered platinacycle ring. Palladium is less reactive than platinum and does not form a cyclometallated product with this phosphine. Neither platinum nor palladium form cyclometallated complexes with the diphosphine with seven CH_2 groups in the chain.

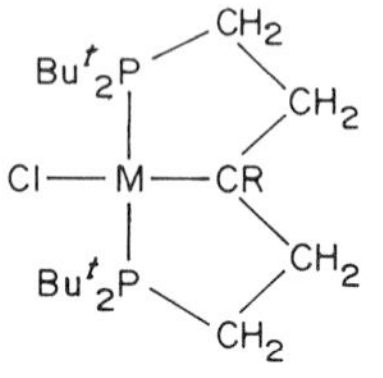

(57) 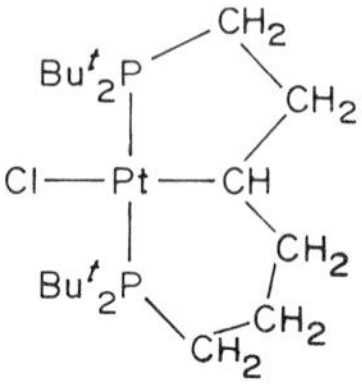 (58)

The diphosphine $Bu^t_2P(CH_2)_2CHMe(CH_2)_2PBu^t_2$ reacts with $RhCl_3$[105], $[PdCl_2(PhCN)_2]$, and $[PtCl_2(PhCN)_2]$[108] to give the cyclometallated complexes **53** (M = Rh, R = Me) and **57** (M = Pd or Pt, R = Me). The rhodium reaction again requires the use of 2 methylpyridine to aid the cyclometallation.

c. *Conclusions*

It is evident that in the cyclometallation of a tertiary phosphine ligand the steric requirements of the substituents on phosphorus play a dominant role. The ease with which intramolecular C—H bond activation occurs depends on the size of the groups R^1 and R^2 in equation 15. If they are small cyclometallation will not occur, if they are larger, e.g.

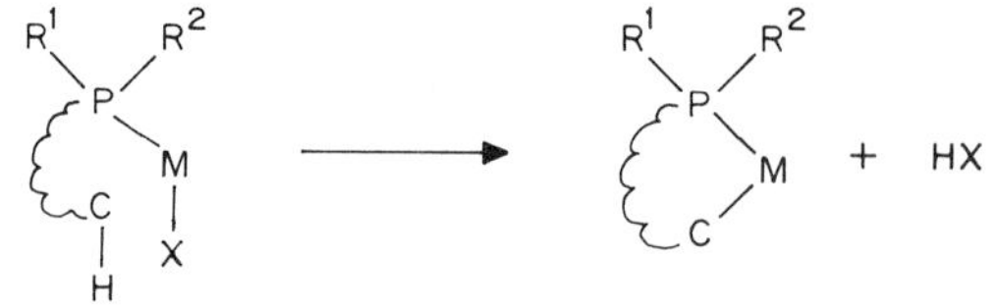

(15)

phenyls, cyclometallation will sometimes occur, and sometimes not, but if they are bulky, e.g. *tert*-butyl, then cyclometallation often occurs readily and rapidly. Clearly steric effects are dominant over electronic effects.

As Shaw[89] has pointed out, there is a close analogy between what happens here and the Thorpe–Ingold or *gem*-dialkyl effect, which has been known to organic chemists for many years. Thorpe and Ingold found that replacement of CH_2 by CMe_2 invariably increased the stability of small rings and the rate at which they were formed. As Ingold wrote in 1921[110], 'The reality of the phenomenon cannot be doubted'. Two factors are responsible for the effect: (a) an entropy component, first suggested by Hammond[111], who pointed out that a *gem*-dimethyl grouping would reduce the loss of internal rotational entropy which occurs on cyclization, and (b) an enthalpy effect, as suggested by Allinger and Zalkow[112], who showed that a *gem*-dialkyl substituent would reduce the number of extra *gauche* interactions which are introduced on cyclization. Methyl groups can have a large effect in carbon chemistry, but are not bulky enough to have a significant effect on the larger phosphorus atom, here the large *tert*-butyl group is required—a *gem*-*tert*-butyl effect[113].

These studies also show that it is more difficult for the second-row transition elements palladium and rhodium to activate the C—H bonds and form the metallocycle products than it is for the third-row elements platinum and iridium. The explanation for this is probably due, at least in part, to the greater difficulty in inducing the second-row elements to take part in oxidative addition when palladium(IV) and rhodium(V) must be formed.

4. Activation in alkylphosphine ligands coordinated to the platinum metals

a. Alkyl- and alkylaryl-phosphines

There are a few examples of cyclometallations where the tertiary phosphine ligand does not contain the bulky groups discussed in the previous section.

When the lithium derivatives of methyl or phenyl carboranes [$B_{10}C_2H_{10}R^-$ (R = Me or Ph)] react with *cis*- or *trans*-[$PtCl_2(PR_3)_2$] (R = Et or Pr^n) then cyclometallation occurs to give products that have been assigned the four-membered platinacycle structure **59** [R^1 = Me or Ph, R^2 = Et or Pr^n, R^3 = H or Me, carb = 1,2- or 1,7-dicarba-*closo*-dodecaborane(12)] on the basis of analytical and spectroscopic data[114,115].

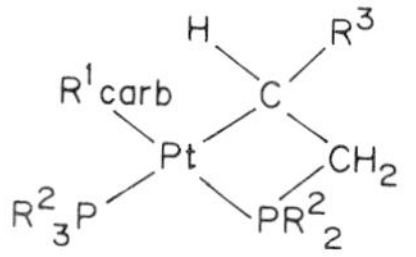

(59)

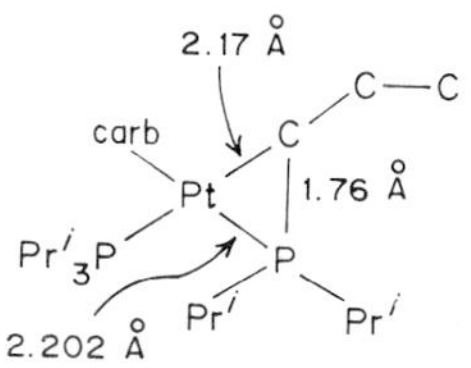

(60)

The palladium complex *trans*-[$PdCl_2(PEt_3)_2$] gave the analogous product. However, a crystal structure determination on the isopropyl phosphine complex shows that a three-membered ring platinacycle is formed (**60**), and the short phosphorus—carbon bond length in the crystal indicates that a coordinated P=C double bond is probably a better description of the bonding[116].

If the platinum(II) complex with the bulkier phosphine PPh_2Me is used in this reaction then it is the sp^3-methyl C—H bonds that are activated, the three-membered ring platinacycle **61** being formed[117]. It has been suggested that here the lithium alkyl directly

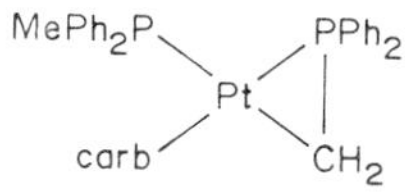

(61)

deprotonates the coordinated PPh_2Me to give the platinacycle without any C—H activation occurring[118], since in the less bulky phosphine PMe_2Ph, when it is coordinated to iridium it is an sp^2-phenyl C—H bond, not an sp^3-methyl bond, that is activated (equation 16).

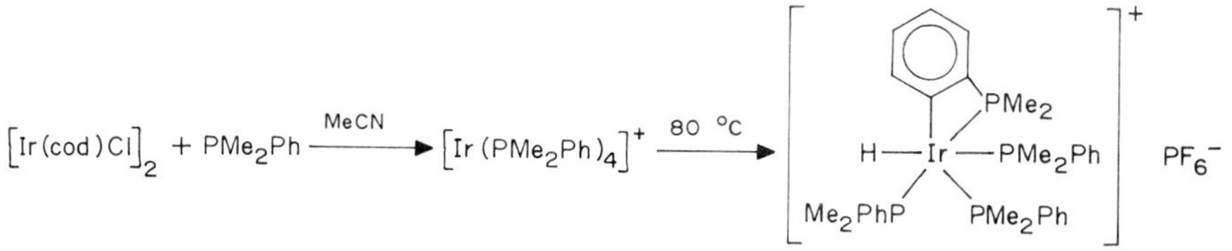

$$[Ir(cod)Cl]_2 + PMe_2Ph \xrightarrow{MeCN} [Ir(PMe_2Ph)_4]^+ \xrightarrow{80\ ^\circ C} [HIr(C_6H_4PMe_2)(PMe_2Ph)_3]^+ PF_6^- \qquad (16)$$

A three-membered cyclometallated ring is also formed if the ruthenium(II) complex [$Ru(PMe_3)_4Cl_2$] is treated with sodium amalgam in benzene solution; a C—H bond in a phosphinemethyl group is activated and undergoes oxidative addition to the ruthenium (equation 17)[119].

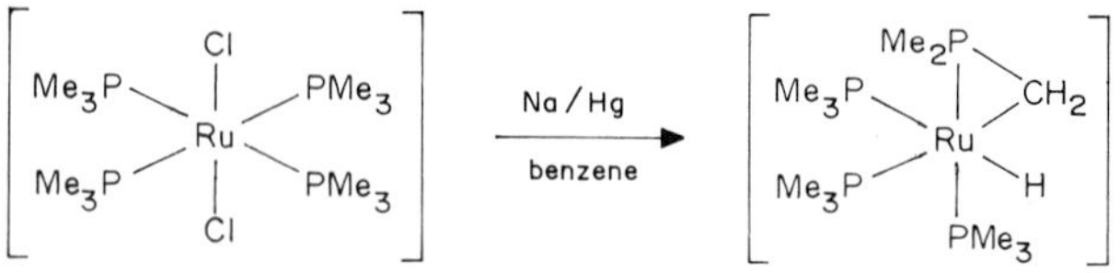

$$[RuCl_2(PMe_3)_4] \xrightarrow[\text{benzene}]{Na/Hg} [RuH(CH_2PMe_2)(PMe_3)_3] \quad (17)$$

A four-membered cyclometallated ring is also formed if the complex $[(\eta^6\text{-}C_6H_6)RuPPr^i_3H_2]$ is irradiated with u.v. light for 3h in cyclohexane solution (equation 18)[120]. Presumably a reactive intermediate is formed by loss of the hydride ligands, as in the work of Bergman and others discussed in Section II.D, which then undergoes oxidative addition of a C—H bond of the isopropyl group.

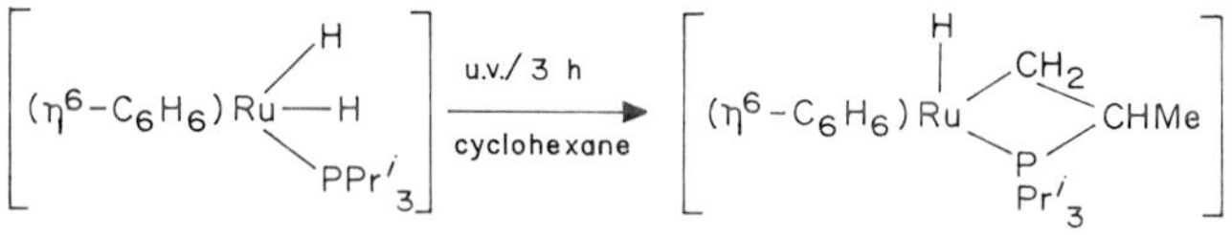

$$[(\eta^6\text{-}C_6H_6)RuH_2(PPr^i_3)] \xrightarrow[\text{cyclohexane}]{\text{u.v./3 h}} [(\eta^6\text{-}C_6H_6)RuH(CH_2CHMePPr^i_2)] \quad (18)$$

Bond activation of two adjacent C—H groups in the cyclohexyl group of cyclohexylphosphine occurs when the phosphine reacts with the iridium(I) complex $[IrCl(cyclooctene)_2]_2$. The reaction, in refluxing toluene, gives three products (equation 19)[121]. Two have cyclohexene groups bonded to the iridium by the double bond, i.e. two adjacent C—H's have been activated.

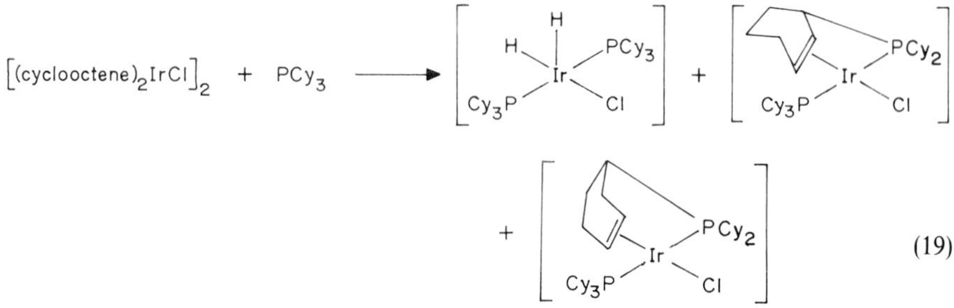

$$[(cyclooctene)_2IrCl]_2 + PCy_3 \longrightarrow [IrH_2Cl(PCy_3)_2] + [IrCl(\eta^2\text{-}C_6H_9PCy_2)(PCy_3)] + [IrCl(\eta^2\text{-}C_6H_9PCy_2)(PCy_3)] \quad (19)$$

b. Phosphines containing the cyclopropanyl group

The phosphine $PBu^t_2(CH_2$ cyclopropyl) reacts with platinum(II) chloride to form the complex *trans*-$[PtCl_2(phosphine)_2]$. This on heating in 2-methoxyethanol for 2 h forms a five-membered platinacycle by both C—H and C—C activation[122,123]. A similar palladium complex is also formed (**63**), but much less readily, 32 h reflux being required[124].

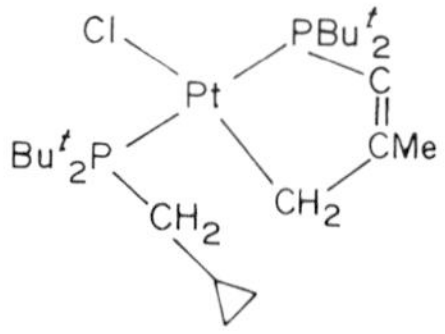

(**62**) $[PtCl(PBu^t_2CH_2C_3H_5)(PBu^t_2C{=}CMeCH_2)]$ (**63**) $[PdCl(PBu^t_2CH{=}CMeCH_2)]_2$

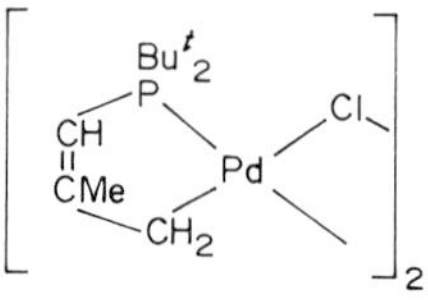

The mechanism by which this reaction takes place is not certain; it may be by C—C or C—H activation as shown in equation 20.

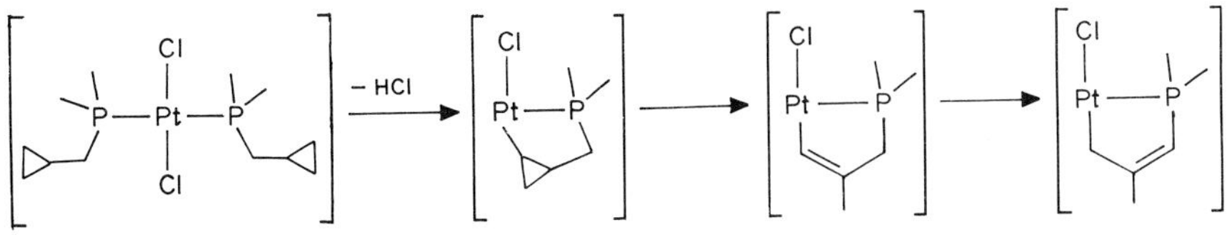

$$\left[\text{Cl}_2\text{Pt}(\text{PMe}_2\text{CH}_2\text{C}_3\text{H}_5)_2\right] \xrightarrow{-\text{HCl}} [\ldots] \longrightarrow [\ldots] \longrightarrow [\ldots] \qquad (20)$$

c. Hydrogen–deuterium exchange in phosphine ligands

The catalysed exchange of hydrogen in an organic molecule by deuterium from the solvent or deuterium gas has been much used to study alkane activation by transition metal complexes; it is discussed in detail in Section III.A. During such as exchange process activation of the C—H bond must occur. The metal hydride formed will exchange the hydrogen for deuterium by, for example, dissociation of a proton and association of a deuteron from an acidic solvent, or the complex may be a polyhydride and deuterium will be already present on the metal, giving the reaction as in equation 21.

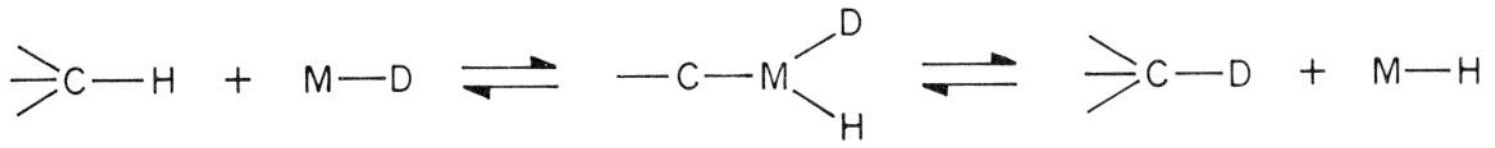

$$\geq\text{C—H} + \text{M—D} \rightleftharpoons \text{—C—M}(\text{D})(\text{H}) \rightleftharpoons \geq\text{C—D} + \text{M—H} \qquad (21)$$

If the organic molecule is already a ligand on the transition metal, then, although the metal will be a catalyst for hydrogen–deuterium exchange, the number of exchangeable C—H bonds is clearly limited to those in the ligands. Such exchange reactions are well known for aromatic carbon—hydrogen bonds in arylphosphine ligands[11,125], and this work is not discussed here. There are few examples of alkyl groups in phosphines undergoing exchange; deuterium from deuteriated ethanol solvent, or deuterium gas, enters the cyclohexyl groups of $[RuHCl(CO)(PCy_3)_2]$ or its osmium analogue,[126,127], and deuterium enters the methyl groups of $[IrH_5(PMe_3)_2]$ when the complex is heated in benzene solution under an atmosphere of deuterium gas[128]. It also deuteriates the benzene—clear evidence that the C—H bonds in the benzene are activated by the electron-rich iridium.

Using the $MeCO_2D$–D_2O system, extensively studied for catalyzed hydrogen–deuterium exchange (see Section III.A), Masters and coworkers obtained exchange in tertiary phosphine complexes of platinum(II)[129,130]. Dimeric complexes with a range of phosphine ligands (**64**; L = PEt_3, PPr_3, PBu_3, PBu^tPr_2, PBu^t_2Pr, $PPhEt_2$, PPh_2Et, PPh_2Pr, $PPhPr_2$, and PBu^tPh_2) have been studied. For the phosphines with *n*-alkyl groups [i.e. compounds **64** (L = PEt_3, PPr_3, PBu_3)], no exchange occurs in the ethyl group; the exchange is exclusively in the terminal methyl of the propyl group and predominantly at C-3 in the butyl group. After long reaction times, deuterium is also found at C-4 of the butyl group, but none is found at C-2 or C-1. This is clear evidence that the preferred activated complex has a five-membered ring (**65**; $n = 2$), that the six-membered ring (**65**; $n = 3$) is formed but much less readily, and that, for these compounds, the four-membered ring (**65**; $n = 1$) is not formed. The presence of phenyl groups in the phosphine has no effect: compounds **64** with L = $PPhEt_2$ or PPh_2Et, like that with L = PEt_3, do not undergo exchange, and compounds **64** with L = PPr_3, $PPhPr_2$, or PPh_2Pr all undergo H–D exchange exclusively at C-3 at approximately equal rates. The bulky *tert*-butyl group in

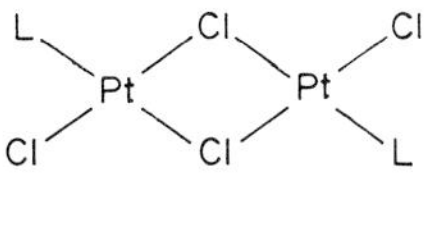

$$[\text{LClPt}(\mu\text{-Cl})_2\text{PtClL}] \quad \textbf{(64)} \qquad\qquad [\text{H}(\text{—})\text{Pt}(\text{CHR}(\text{CH}_2)_n\text{P})] \quad \textbf{(65)}$$

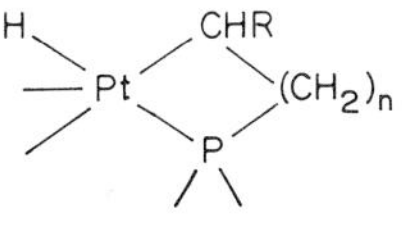

the ligand, however, has a marked accelerating effect, as it did in the compounds discussed in Section C.2. Compounds **64** with L = PBu^tPr_2 and PBu^t_2Pr undergo H–D exchange at the C-3 position of the propyl group, at rates 2.5 and 27 times faster, respectively, than when L = PPr_3. Also, when there are two *tert*-butyl groups in the ligand, then exchange at C-2 of the propyl group and of the hydrogen atoms of the methyl groups also occurs, i.e. complexes with four-membered rings (**65**; $n = 1$) are being formed under the influence of a pair of bulky *tert*-butyl groups in the phosphine. This work also indicated that a dimeric complex is necessary for H–D exchange to occur since exchange does not occur in the monomeric complexes *cis*- and *trans*-$[PtCl_2(PBu_3)_2]$. The mechanism suggested for this reaction is as shown in equation 22, involving oxidative addition to the platinum(II) to form a platinum(IV) hydride; this can then exchange hydrogen for deuterium with the solvent, leading to H–D exchange in the phosphine.

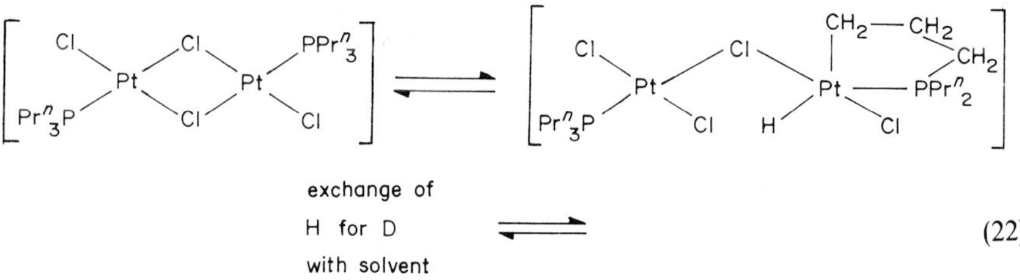

Cl Pt(PPr$^n_3$)(μ-Cl)$_2$Pt(PPr$^n_3$)Cl ⇌ [Pr^n_3P(Cl)Pt(μ-Cl)$_2$Pt(H)(Cl)($PPr^n_2CH_2CH_2CH_2$)]

exchange of H for D with solvent ⇌ (22)

5. Activation in nitrogen ligands coordinated to palladium

a. Quinoline derivatives

Cyclopalladation occurs by activation of the methyl C—H bonds in 8-methylquinoline when it reacts with lithium tetrachloropalladate in methanol solution. The compound formed is the dimer **66** (R = H)[131]. If 8-ethylquinoline is the reactant then cyclopalladation occurs at an α-CH_2 bond to form the chiral complex **66** (R = Me)[132]. If the reactant is

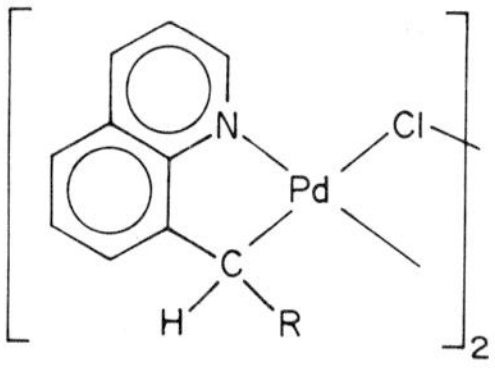

(**66**)

palladium(II) acetate then the acetate-bridged dimer is formed. Neither 8-isopropylquinoline nor any 2-methyl-substituted quinolines form a cyclometallated product, presumably owing to the inability of palladium and the reactant alkyl group being able to enter the coordination plane; the mechanism of reaction is thought to be as in equation 23[133].

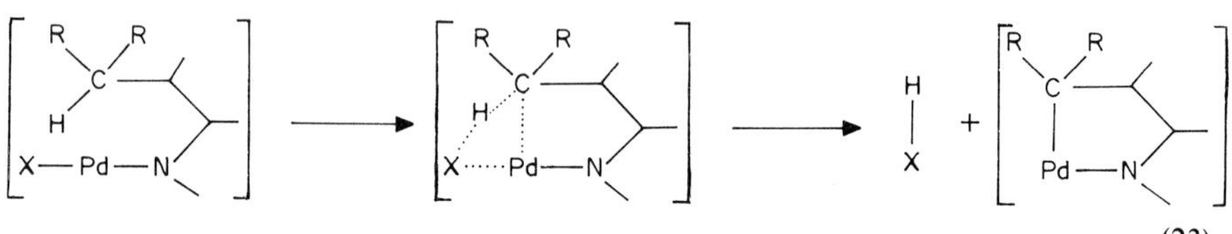

(23)

b. Other nitrogen ligands

A C—H bond in a methyl of the neopentyl group is activated in the complex formed between palladium and 2-neopentylpyridine, a six-membered palladacycle being formed[134] (equation 24).

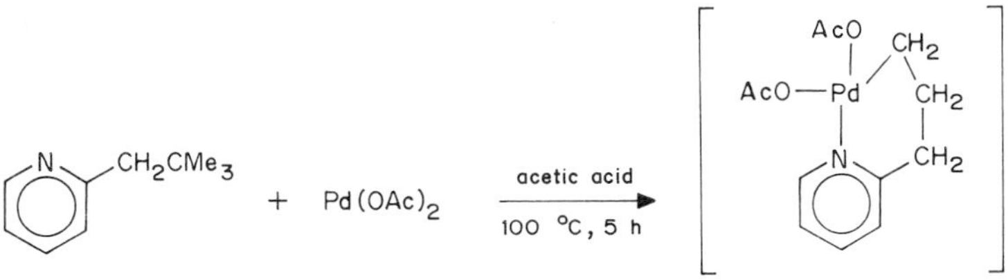

(24)

The compound *o*-*N*, *N*-dimethylaminotoluene reacts with palladium salts; the reaction products formed depend on the palladium salt and on the solvent[135]. With $Li_2[PdCl_4]$ in methanol demethylation of the base occurs to form the complex **67** (R = H). With $[PdCl_2(PhCN)_2]$ an unstable dimeric adduct (**67**; R = Me) is formed, and with $[Pd(OAc)_2]_3$ as the reactant cyclopalladation occurs to give the dimeric complex **68**.

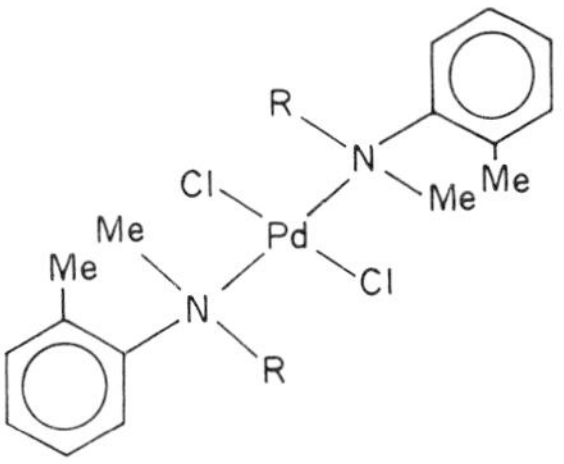

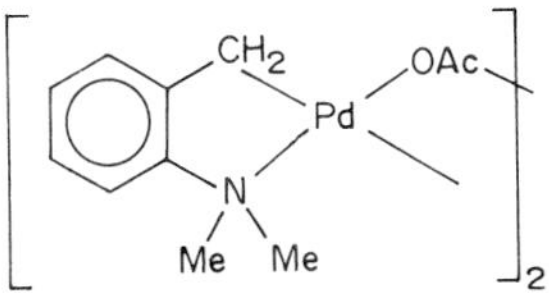

(67) (68)

(*E*)-Methyl-*tert*-butyl ketoxime with sodium acetate and sodium tetrachloropalladate in methanol solution forms the cyclopalladation product **69** (R = OH) in high yield after 3 days at 25 °C[136]. Under similar conditions methyl-*tert*-butyl-*N*, *N*-dimethylhydrazone forms complex **70**, and methyl-*tert*-butylketazine, which can cyclometallate at either the methyl or the *tert*-butyl group, does so exclusively at the *tert*-butyl to form the complex **69** [R = NC(Me)Bu^t].

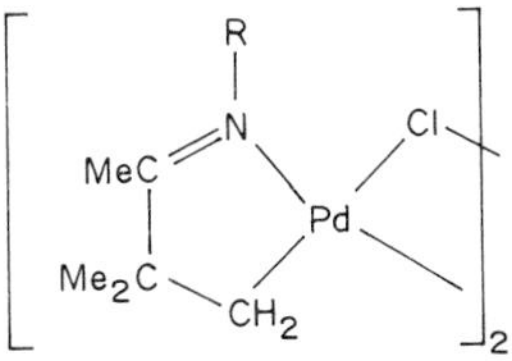

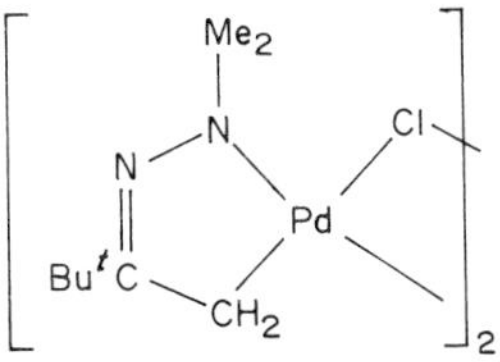

(69) (70)

6. Activation in miscellaneous complexes of the platinum metals

a. Trimethylsilylmethyl and neopentyl complexes

When $Mg(CH_2SiMe_3)_2$ and PMe_3 react with binuclear transition metal acetates in tetrahydrofuran solution, trimethylsilylmethylphosphine complexes are formed. If the metal is rhodium or ruthenium $\{[Rh_2(O_2CMe)_4]$ or $[Ru_2Cl(O_2CMe)_4]\}$ then cyclometallation occurs to form the four-membered ring complexes **71** (M = Rh, R = CH_2SiMe_3 or M = Ru, R = PMe_3)[137]. The ruthenium chloroacetate also reacts with bisneopentylmagnesium and PMe_3 in tetrahydrofuran solution to form the neopentyl-cyclometallated analogue (**72**).

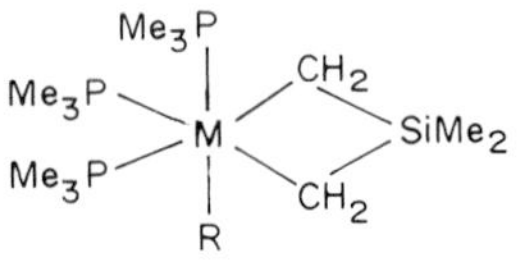

(**71**) 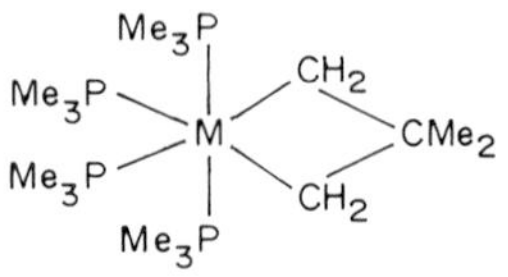 (**72**)

A γ-hydrogen abstraction also occurs when the rhodium complex $[RhCl_2Cp^*PPh_3]$, reacts with excess of neopentyllithium or neopentylmagnesium halide in diethyl ether solution[138]. Three products are formed, a rhodium–ethylene complex (which is not formed when pentane is the solvent and presumably therefore comes by reaction with the diethyl ether), a complex where aryl ring cyclometallation has occurred, and a complex where a C—H bond in the neopentyl group has been activated to give a four-membered rhodacycle ring (equation 25).

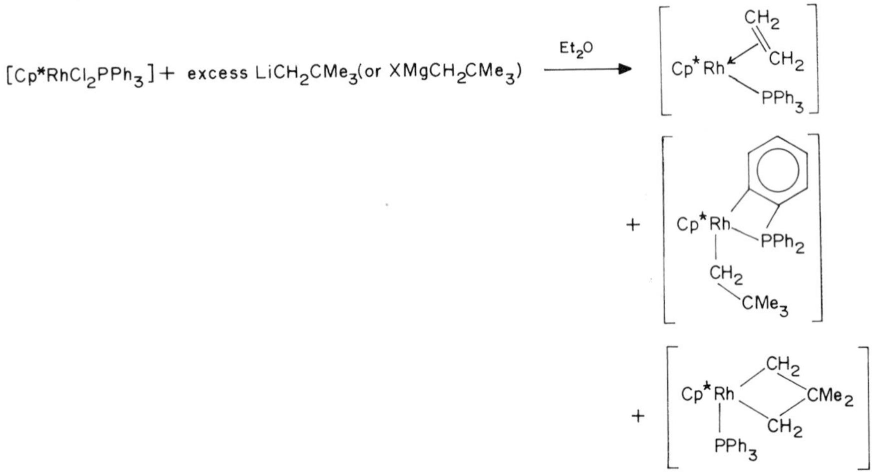

$$[Cp^*RhCl_2PPh_3] + \text{excess } LiCH_2CMe_3(\text{or } XMgCH_2CMe_3) \xrightarrow{Et_2O} [\ldots] + [\ldots] + [\ldots] \qquad (25)$$

b. *Arylmethyl complexes*

The iridium complexes *trans*-$[Ir(Me_nC_6H_{5-n})(CO)(PR^1{}_3)_2]$($PR^1{}_3 = PMe_3$, PEt_3, PMe_2Ph, or $PMePh_2$) react with phosphites $P(OR^2)_3$(R^2 = Me, Et, or Ph) to form products with three phosphite ligands in which cyclometallation at the ring *o*-methyl group has occurred to give complexes **73** (R = Me, Et, or Ph) where the aryl ring has one methyl substituent in the 3-, 4-, 5-, or 6-position or two in the 4- and 6-positions[139].

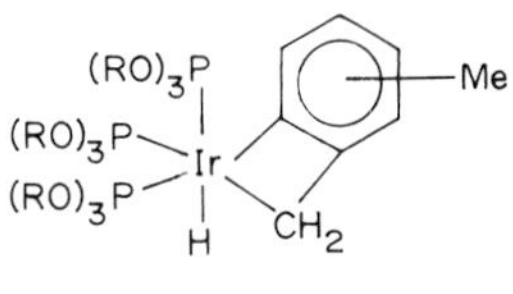

(**73**)

c. Osmium clusters

There is a facile interconversion of methyl groups attached to osmium in the carbonyl cluster [$HOs(CO)_{10}Me$] into CH_2 and CH groups[140]; the complexes (equation 26) are in thermal equilibrium. The interaction of the hydrogen atoms of the methyl group with the osmium have been studied by n.m.r. spectroscopy (see Section B.8.b)[58].

$$[(OC)_3Os(\mu\text{-}CH_3)(\mu\text{-}H)Os(CO)_3Os(CO)_4] \rightleftharpoons [(OC)_3Os(\mu\text{-}CH_2)(\mu\text{-}H)_2Os(CO)_3Os(CO)_4] \rightleftharpoons [(OC)_3Os(\mu\text{-}CH)(\mu\text{-}H)_3Os(CO)_3Os(CO)_4] \quad (26)$$

7. Activation in miscellaneous complexes of other late transition metals

a. Rhenium hydride complexes

Hydrogen–deuterium exchange in complexes similar to those used by Felkin (see Section D.5) with the bulky tricyclohexylphosphine ligand have been studied[141]. The septahydride complex [$(Cy_3P)_2ReH_7$] undergoes thermal dissociation, at 60 °C to give the reactive 16-electron compound [$(Cy_3P)_2ReH_5$]. The expected isotope exchange with hexadeuteriobenzene of the Re—H is found to occur. Deuterium is also found in the cyclohexyl groups of the ligands. This deuterium is only on C-2 and C-3 and on each carbon only one of the two hydrogen atoms undergoes exchange. The rate of exchange at C-2 is slower than at C-3, so there is no cyclohexenyl intermediate, as this would exchange the two hydrogens at the same rate. This is evidence for the exchange being an intramolecular process; the C-3 exchange would involve a five-membered ring (equation 27), the C-4 exchange a four-membered ring, and the five-membered ring would be expected to have a lower activation energy and hence a faster rate than the four-membered ring.

$$[Cy_2P\text{—}ReH_6PCy_3 \text{ (C-2 metallated)}] \underset{\text{C-2 exchange}}{\rightleftharpoons} [ReH_5(PCy_3)_2] \underset{\text{C-3 exchange}}{\rightleftharpoons} [Cy_2P\text{—}ReH_6PCy_3 \text{ (C-3 metallated)}] \quad (27)$$

These cyclometallated products are unstable and are not isolated, nor is a stable benzene complex formed by this system. However, if acenaphthalene (**74**) is used in place of benzene, then an [$(\eta^4\text{-}C_{12}H_n)ReH_3(PCy_3)_2$] complex is formed. Which four π electrons are involved in the bonding, and the exact structure of the organic ligand, are not yet known.

(**74**)

b. Nickel, iron, and molybdenum complexes

The diazadiene glyoxalbis(diisopropylmethylimine) (dad) reacts with anhydrous nickel bromide to form [$NiBr_2(dad)$]. When it reacts with the bulky *o*-tolylmagnesium bromide

γ-metallation of a diisopropylmethyl groups occurs to give the complex **75** (equation 28)[142]. When iron(II) chloride in tetrahydrofuran is reduced with sodium,

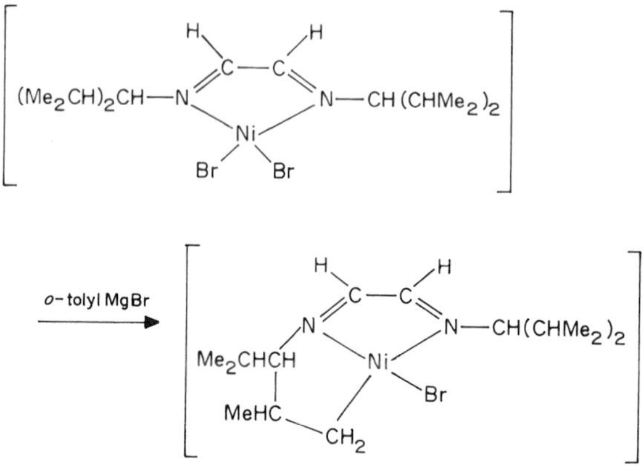

$$[\text{NiBr}_2\{(\text{Me}_2\text{CH})_2\text{CH—N=CH—CH=N—CH(CHMe}_2)_2\}] \xrightarrow{o\text{-tolyl MgBr}} \text{(75)} \quad (28)$$

(75)

potassium, magnesium, or n-butyllithium in the presence of trimethylphosphine, the iron(0) complex $[Fe(PMe_3)_4]$ is formed[143]. This complex in solution exists almost completely as the iron(II) hydride (**76**) owing to cyclometallation of a C—H group in a ligand (equation 29).

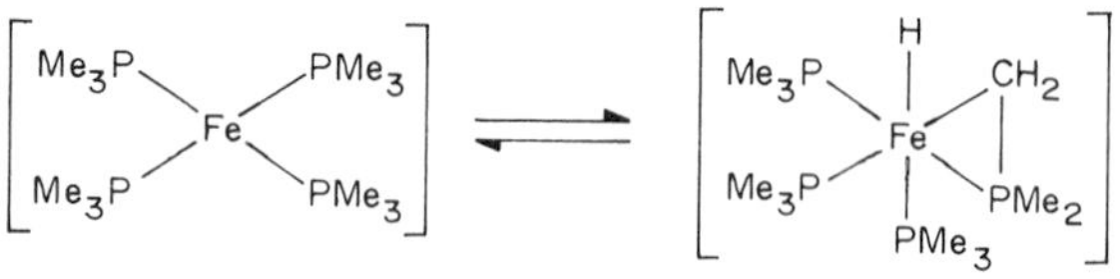

$$[Fe(PMe_3)_4] \rightleftharpoons [FeH(CH_2PMe_2)(PMe_3)_3] \quad (29)$$

(76)

The trimethylsilylmethyl group reacts with dimeric molybdenum acetates to give a product which differs from the rhodium and ruthenium products in Section II.C.6.a. The molybdenum—molybdenum bond is retained, the product being the dimeric complex **77**[137].

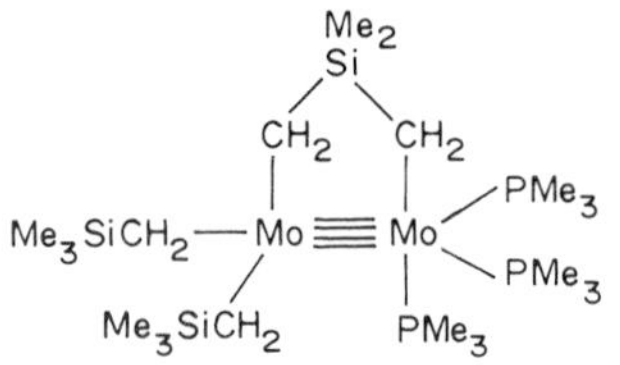

(77)

8. *Activation by complexes of the early transition metals*

a. *Tantalum and niobium complexes*

Niobium and tantalum mesityl complexes $\{[M(\text{mesityl})_nX_{5-n}];\ M = Nb \text{ or } Ta,\ X = Cl \text{ or } Br,\ n = 2 \text{ or } 3\}$ are considerably more stable than their phenyl analogues, presumably because they do not have β-hydrogen atoms. The tantalummesitylneopentyl complex $[Ta(mesityl)(CH_2CMe_3)X_3]$ reacts with PMe_3 to give the alkylidene complex **78**, but the mesitylmethyl complex $[Ta(mesityl)MeX_3]$ gives the benzylidene complex **79** and not the analogous methylene complex. This benzylidene complex is thought to be formed via the cyclometallated complex **80**[144].

$(Me_3P)_2X_3Ta{=}CHCMe_3$

(**78**)

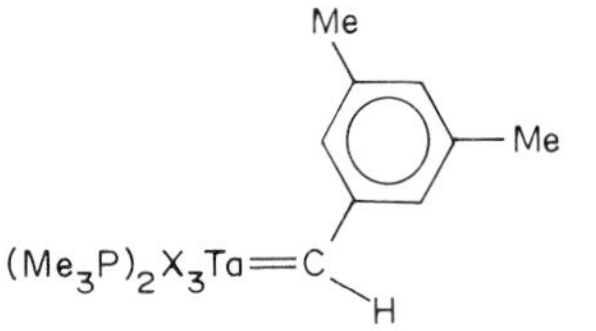

(**79**) 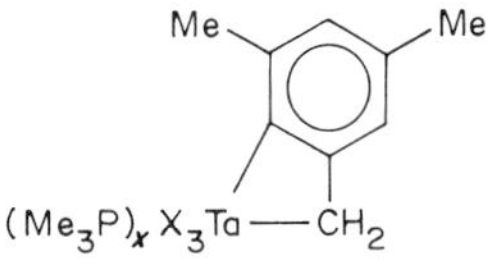 (**80**)

The tantalum hydridoethylene complex **81** is produced when $[Cp_2TaCl_2]$ reacts with 2 equiv. of EtMgBr in refluxing dimethoxyethane, i.e. β-C—H bond activation in the ethyl group has occurred[145]. N.m.r. studies show that the ethylene ligand does not rotate in solution up to 120 °C, and that the C—C axis lies in the plane of symmetry of the molecule. The formulation as a tantalacyclopropane as in **81** is thought to be more likely than as a ethylene complex.

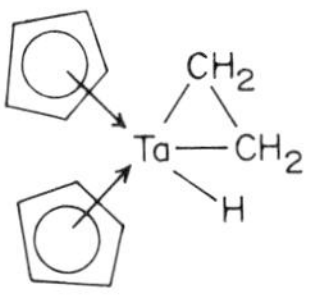

(**81**) 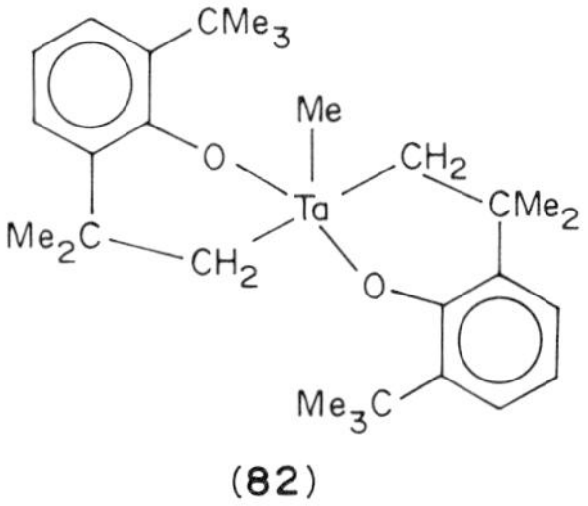 (**82**)

$TaCl_5$ reacts with excess of lithium 2,6-di-*tert*-butyl phenoxide (LiOR) in benzene solution to give the complex $[Ta(OR)_2Cl_3]$. This reacts with LiMe at 25 °C to give the complex $[Ta(OR)_2Me_3]$, which on heating in toluene to 120 °C cyclometallates in both ligands to form complex (**82**)[146]. Mild thermolysis (75 °C for 7 days) in toluene gives the analogous complex where cyclometallation has occurred in only one ligand (**83**)[147]. Photolysis of the complex $[Ta(OR)_2Me_3]$ forms the methylmethylidene complex **84**. This at 25 °C is smoothly converted into the monometallated complex **83**. Deuterium labelling experiments showed that the methylidene complex is not generated during the thermal reaction. Also, the fact that the methylidene complex **84** is converted into the cyclometallated complex **83** indicates that the methylidene group has a higher activity for aliphatic intramolecular C—H activation than simple alkyl groups.

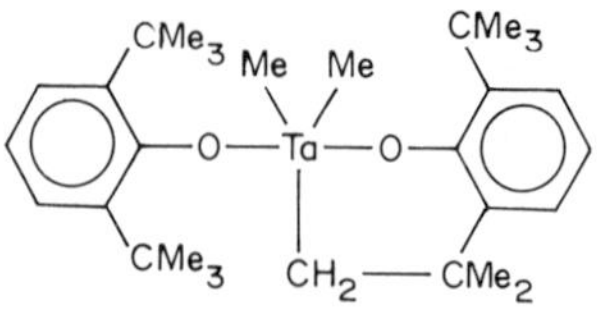

(83) 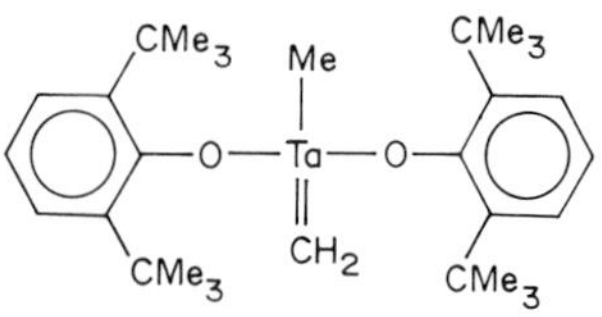 (84)

At elevated temperature, facile reversible cyclometallation of d^0 dialkylamido and alkoxo complexes occurs[148]. Treatment of *N*-deuteriodimethylamine with early transition metal dimethylamides at 140–180 °C produced rapid incorporation of deuterium into the methyl group (equation 30). The dimethylamides of niobium, tantalum, zirconium, and tungsten catalysed the exchange whereas those of titanium, hafnium and tin did not. The reaction is presumed to involve the reversible metallation of the dimethylamide ligand (equation 31). H–D exchange in ethanol-*d* was catalysed by metal ethoxides at 180–

$$Me_2ND \xrightleftharpoons{M(NMe_2)_n} Me(CH_2D)NH \qquad (30)$$

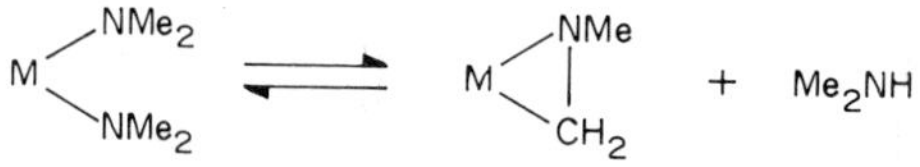

$$M(NMe_2)_2 \rightleftharpoons M(NMe)(CH_2) + Me_2NH \qquad (31)$$

220 °C, the deuterium being incorporated exclusively into the methyl group of ethanol. This can be understood if there is an oxametallocyclobutane intermediate (equation 32). Here the ethoxides of niobium, tantalum, and zirconium catalyse the exchange whereas those of titanium and tungsten do not.

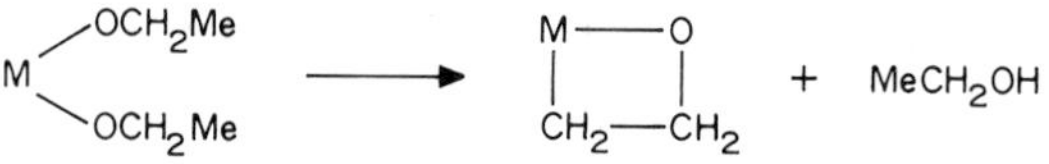

$$M(OCH_2Me)_2 \longrightarrow M(O)(CH_2CH_2) + MeCH_2OH \qquad (32)$$

b. Titanium and zirconium complexes

$[Cp_2TiCl_2]$ and $LiN(SiMe_3)_2$ react in pentane solution to give a deep red product in which activation of one of the methyl C—H groups has occurred to form the metallocycle **85**[149].

If $[(\eta^5\text{-}C_5Me_4CH_2CMe_3)_2ZrCl_2]$ is treated with sodium amalgam cyclometallation occurs with one of the methyl groups in the ligand (equation 33) to give complex **86**[150]. This can exist in equilibrium with the dihydride (**87**, equation 34), since with deuterium gas after 1 h at 70 °C the hydrogen atoms in both *tert*-butyl groups have almost completely exchanged for deuterium[151].

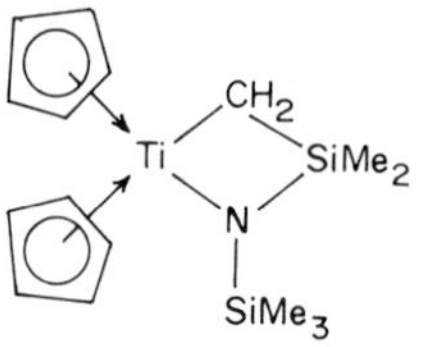

(85)

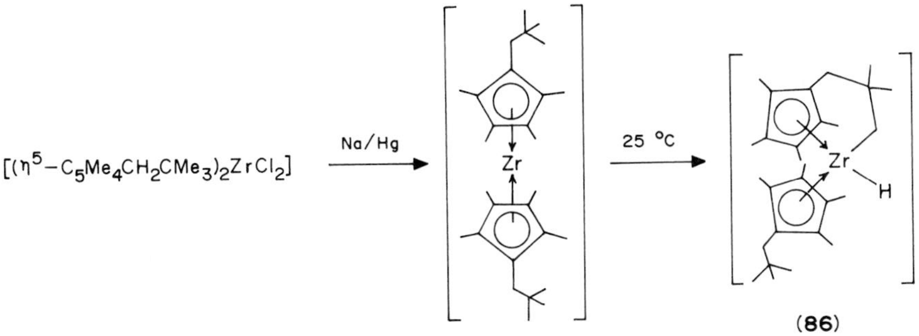

$$[(\eta^5\text{-}C_5Me_4CH_2CMe_3)_2ZrCl_2] \xrightarrow{Na/Hg} [\ldots] \xrightarrow{25\,°C} [\ldots]$$

(86)

(33)

$$H_2 + \mathbf{86} \rightleftharpoons [\ldots]$$

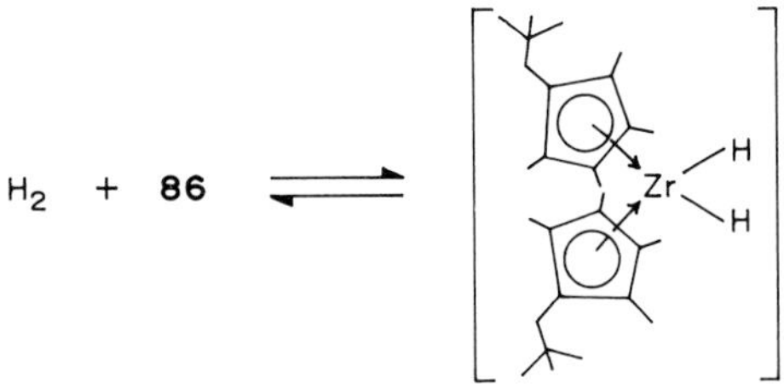

(87)

(34)

9. Activation by complexes of thorium and uranium

Thermolysis of the thorium complexes $[ThCp^*_2(CH_2CMe_2)_2]$ and $[ThCp^*_2(CH_2SiMe_2)_2]$ (**88**; M = C or Si) in hydrocarbon solvents at 50–75 °C yields the metallocyclobutanes (**89**; M = C or Si)[152,153].

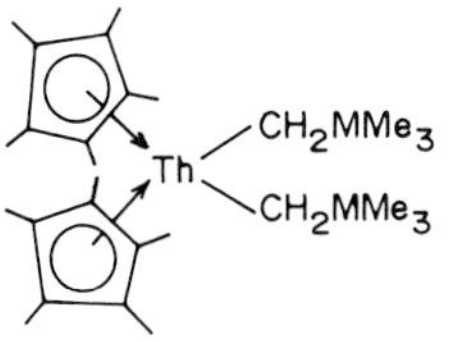

(88) 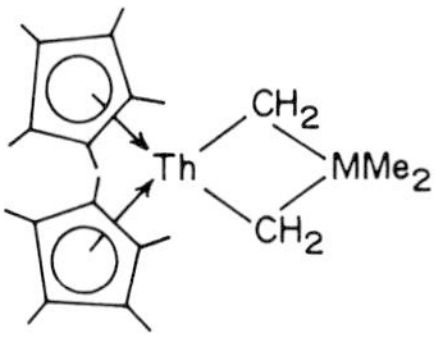 (89)

H–D exchanges occurs in the f^2 and f^0 complexes of uranium(IV) and thorium(IV), $[HU\{N(SiMe_3)_2\}_3]$ and $[HTh\{N(SiMe_3)_2\}_3]$[154,155]. A solution of $[HU\{N(SiMe_3)_2\}_3]$ in pentane stirred under deuterium gas at room temperature exchanges all of its hydrogen atoms for deuterium. Neither the methyl-, tetrahydroborato-, nor chlorotris-(hexamethyldisilylamido)uranium analogues, nor the uranium(III) complex $[U\{N(SiMe_3)_2\}_3]$, exchanges with deuterium under similar conditions. A mechanism for the exchange that involves oxidative addition and reductive elimination seems plausible since uranium(VI) is well known. However, the thorium analogue also undergoes complete exchange and thorium(VI) is unknown.

When the hydrides or their methyl analogues are heated to 150–190 °C a cyclometallated product (**90**; M = Th or U) is obtained. The reaction is reversible for the

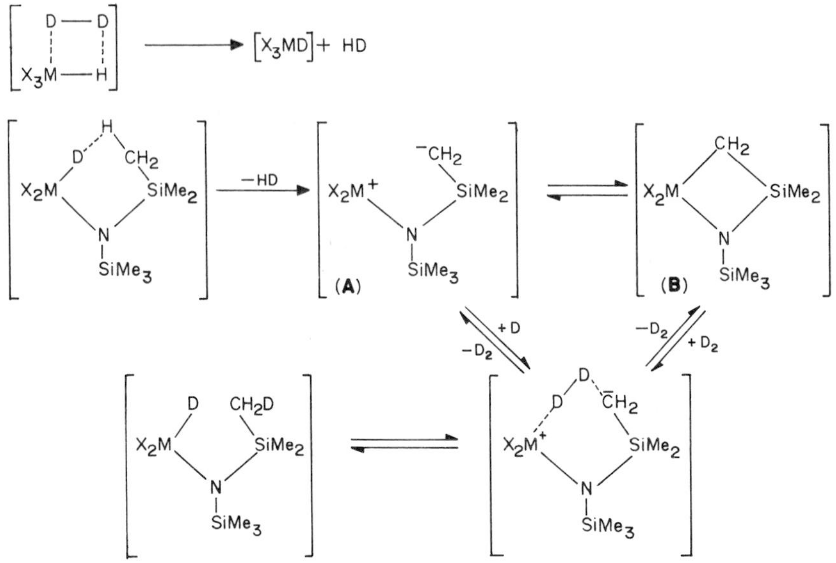

FIGURE 6. Mechanism of the H–D exchange in hydridotris(hexamethyldisilylamide)thorium(IV) and uranium(IV) [X = $(Me_3Si)_2N$].

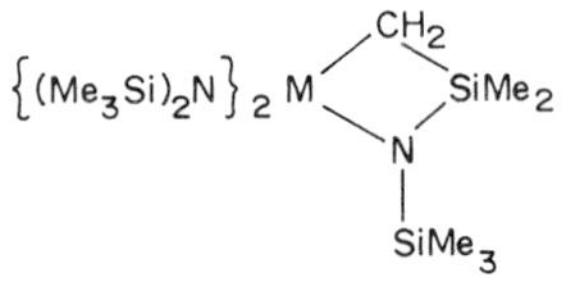

(90)

hydride; exposure of complex **90** to an atmosphere of dihydrogen gives the hydride back again and, further, exposure to an atmosphere of dideuterium give the perdeuteriodeuteride. The mechanism of the reaction is suggested to be as in Figure 6. The metal hydride–deuterium exchange occurs by a four-centred interaction. HD is eliminated to give the ylide (**A**) or its valence tautomer, the metallocyclobutane (**B**). This intermediate leads to incorporation of deuterium into the silylamido ligands.

D. Intermolecular Activation of C—H Bonds in Alkanes

1. *Introduction*

One of the most exciting advances in organometallic chemistry during the past few years has been the discovery that there are transition metal complexes which will react with the sp^3 C—H bonds in another molecule, an alkane, to give alkyl complexes that can, in some cases, react further to give functionalized products. These studies, as this section will show, involve highly reactive intermediates that are often obtained by photolysis under ultraviolet light or thermally by heating. Many of the reactions reported in this section have only been observed on a small scale, often only in a sample tube in an n.m.r.

spectrometer, and the products obtained have been difficult to extract. The reactions are stoichiometric, i.e. not catalytic, and there is a long way yet to go to produce chemical systems that are of any value in producing functionalized products from alkanes. Nevertheless, this is an area of intense current interest and the progress that will undoubtedly occur during the next few years is difficult to predict.

2. Activation by [Cp_2WH_2] and related complexes

The 16-electron molecule tungstenocene is able to undergo intermolecular insertions into both sp^2 and sp^3 C—H bonds[41,156,157]. In this molecule the two cyclopentadienyl ligands cause the tungsten d-orbitals to be higher in energy than in other complexes, i.e. the tungsten is electron rich. Also, cyclopentadienyl ligands are fairly inert to intramolecular reactions, and they are compact so there is room on the metal for up to three other ligands.

The reactive tungstenocene is produced *in situ* by photolysis or thermolysis of other tungsten complexes, u.v. irradiation of biscyclopentadienyltungsten dihydride, or thermolysis of the analogous methyl hydride being the most common.

The first example of C—H bond activation was of the sp^2 C—H bond in benzene[158,159] and fluorobenzene[160] (equation 35)·sp^3 C—H bonds in alkylbenzenes also react; for

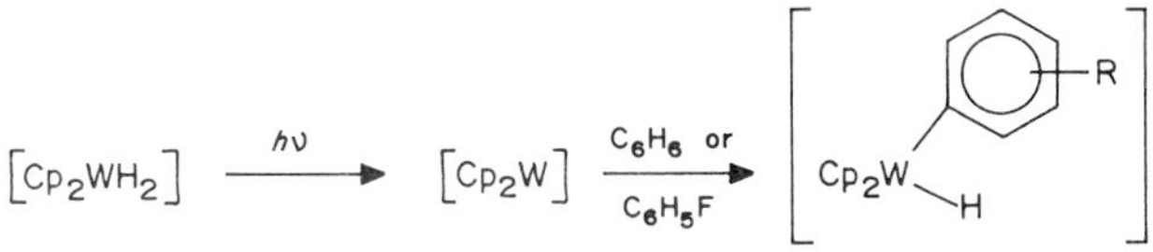

$$[Cp_2WH_2] \xrightarrow{h\nu} [Cp_2W] \xrightarrow[C_6H_5F]{C_6H_6 \text{ or}} \left[Cp_2W(H)(C_6H_4R) \right] \qquad (35)$$

R = H or *m*- or *p*-F

example, *p*-xylene, mesitylene, and anisole give the bisalkyl products (**91**; $R^1 = R^2 = CH_2C_6H_4Me$, $CH_2C_6H_3Me_2$, or $CH_2C_6H_4OMe$), and toluene gives a mixture of the tolyl hydride by metallation of the ring, and the tolylbenzyl complex **91** ($R^1 = C_6H_5Me$, $R^2 = CH_2Ph$)[157,161].

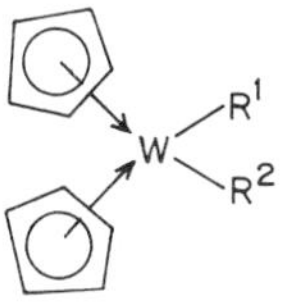

(**91**)

The reactant used for preparing tungstenocene thermally is the methyl hydride derivative **91** (R^1 = Me, R^2 = H)[162]. This on heating to 50–60 °C evolves methane. The tungstenocene formed reacts with *p*-xylene or mesitylene to give not the bisalkyl complexes but the alkyl hydrides **91** [$R^1 = CH_2(p\text{-}MeC_6H_4)$ or $CH_2(3,5\text{-}Me_2C_6H_3)$, R^2 = H]. The alkyl hydride from the mesitylene reaction when photolysed reacts with *p*-xylene to form the mixed dialkyl complex **91** [$R^1 = CH_2(p\text{-}MeC_6H_4)$, $R^2 = CH_2(3,5\text{-}Me_2C_6H_3)$] together with the bis-*p*-xylyl complex **91** [$R^1 = R^2 = CH_2(p\text{-}MeC_6H_4)$] and the dihydride **91** ($R^1 = R^2$ = H). The formation of the mixed alkyl complex strongly suggests that the alkyl hydrides are intermediates in the formation of the dialkyl complexes, and these differences in the photochemical and thermal products are explained by the two stage reaction scheme (equations 36 and 37).

$$[Cp_2WH_2] \longrightarrow [Cp_2W] + R^1H \xrightarrow[\text{or } h\nu]{\text{thermal}} \left[Cp_2W\begin{matrix}R^1\\H\end{matrix}\right] \quad (36)$$

$$\left[Cp_2W\begin{matrix}R^1\\H\end{matrix}\right] \xrightarrow{h\nu} \begin{matrix}\text{intermediate}\\\text{stages}\end{matrix} + R^2H \xrightarrow{-H_2} \left[Cp_2W\begin{matrix}R^1\\R^2\end{matrix}\right] \quad (37)$$

For an aryl ligand the aryl hydride is photochemically stable and these reactions stop at equation 36. However, when R^1 is CH_2Ph, p-$CH_2C_6H_4Me$, or $CH_2(3,5\text{-}Me_2C_6H_3)$ the monoalkyl complexes are thermally stable but photochemically unstable and on irradiation dialkyl compounds are formed.

If this is so, intermediates must be formed that are able to insert into sp^3 C—H bonds, and there are two possibilities. Either there is a reversible migration of the alkyl group from the tungsten to a ring carbon, as has been observed for $[Cp_2WEtCl]$[163], or a reversible η^5-$C_5H_5 \rightleftharpoons \eta^3$-$C_5H_5$ ring shift occurs.

Tungstenocene also inserts into a C—H bond of tetramethylsilane. Irradiation of the dihydride **91** ($R^1 = R^2 = H$) in tetramethylsilane gives **92** as both *cis-* and *trans-*isomers. This is thought to be formed by initial insertion of the tungstenocene to give the trimethylsilylmethyl hydride **91** ($R^1 = CH_2SiMe_3$, $R^2 = H$), which undergoes a further series of insertions to form the final product.

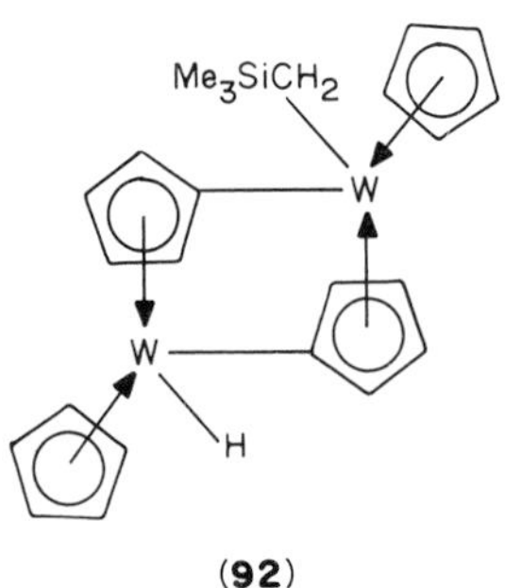

(**92**)

The C—H bonds of fully saturated hydrocarbons such as neopentane or cyclohexane are not activated by tungstenocene, probably because the alkyl hydride products that would be formed are photochemically and thermally unstable[41].

Matrix isolation methods have been used to examine the primary photochemical products of $[Cp_2ML_n]$ (M = Mo, W, or V). The common product from a range of Cp_2W complexes is tungstenocene $[Cp_2W]$[164]. This and $[Cp_2Mo]$ have been studied by magnetic circular dichroism spectroscopy and found to be paramagnetic in their ground state[165]. It is suggested that paramagnetic intermediates may be important in many C—H insertion reactions.

3. *Activation by [CpMo(Me$_2$PCH$_2$CH$_2$PMe$_2$)H$_3$] and related complexes*

In an attempt to find a molecule that would show such C—H activation, other high-energy 18-electron compounds that can undergo thermal or photoinduced ligand loss were studied. Complexes which are like the tungsten complex **91** ($R^1 = R^2 = H$) with two or more *cis*-orientated hydrogen ligands were studied. One such complex is that of molybdenum, $[CpMo(dmpe)H_3]$, **93**, which has been found to be an active catalyst for the photoinduced H–D exchange of C—H bonds in a wide variety of compounds with sp^2 C—H bonds, e.g. benzene and toluene, and sp^3 C—H bonds, e.g. methyl groups of toluene, mesitylene, and dimethyl ether and the ethyl group of ethylbenzene[166].

The related complexes of η^5-C_5H_4-Pr^i (**94**; $R^1{}_2 = 2PMe_3$ or $Pr^i{}_2PCH_2CH_2PPr^i{}_2$) were less effective at C—H bond activation, presumably owing to steric problems with the isopropyl group, and the ability of this group to undergo intramolecular C—H activation, as evidenced by the rapid exchange of all of the hydrogen atoms in the isopropyl group when either of these two complexes is pyrolysed in deuteriobenzene.

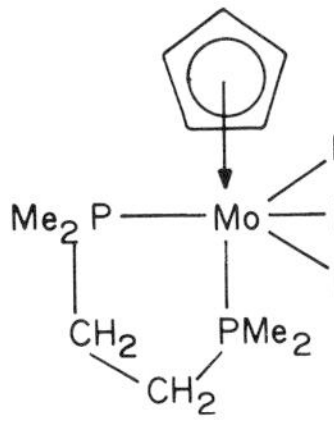

(**93**) 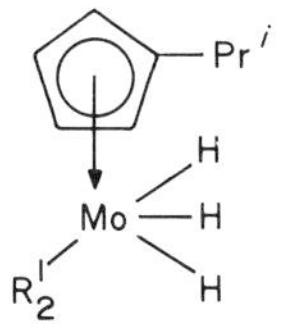 (**94**)

4. Activation by [Ir(acetone)$_2$(PPh$_3$)$_2$H$_2$]BF$_4$ and related complexes

A homogeneous dehydrogenation of alkanes was first reported by Crabtree *et al.* in 1979[167] and has been developed in a series of papers since then[168–172]. Alkane dehydrogenation is the reverse of alkene hydrogenation (equation 38) and it was suggested that any catalyst for the forward process should also speed up the reverse reaction. The activation enthalpy favours the alkane and a thermodynamically more favourable system for alkane dehydrogenation would be one with an alkene present as a hydrogen acceptor (equation 39).

$$H_2 + RHC{=}CHR \rightleftharpoons RH_2CCH_2R \qquad (38)$$

$$\text{(39)}$$

The initial reaction in such a process must be the activation of the alkane by oxidative addition to the metal (equation 40), and it might be expected that if any metal will take part in such a reaction it will be from the third-row of the Periodic Table as for these the M—C and M—H bonds are stronger than those to first- and second-row metals.

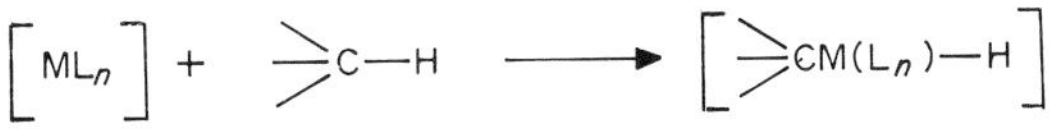

$$[ML_n] + \;>\!C{-}H \longrightarrow [\;>\!CM(L_n){-}H\;] \qquad (40)$$

Conventional homogeneous hydrogenation catalysts, such as $[RhCl(PPh_3)_3]$ or $[RuHCl(PPh_3)_3]$, do not dehydrogenate alkanes, presumably because they cannot compete with the PPh_3 for the metal; the C—H bonds in the ligands are preferentially activated. Catalysts that are best for hydrogenation of hindered alkenes were thought to be the best choice for alkane dehydrogenation. These are complexes of third-row transition metals with a low PPh_3 to metal ratio, and are used in a non-coordinating solvent. The complexes $[IrH_2(Me_2CO)_2(PPh_3)_2]BF_4$ and $[IrH_2(H_2O)_2(PPh_3)_2]BF_4$ were studied, since in refluxing CH_2Cl_2 at 40 °C they readily reacted with alkenes to dehydrogenate them and form chelate products (e.g. equation 41).

$$[IrH_2S_2L_2]^+ + 3\,\text{(bicyclooctene)} \longrightarrow [\text{(bicyclooctadiene)}IrL_2]^+ + 2\,\text{(bicyclooctane)} + 2S \quad (41)$$

Dehydrogenation of the corresponding alkanes did not occur, however, but cyclopentane gave a small yield (*ca.* 7%) of $[CpIrHL_2]^+$ at 80 °C. Many olefins were studied to find a hydrogen acceptor. Most olefins suppressed the little activity that had been observed, particularly those containing allylic hydrogens. Tetraphenylethylene and 1,2-di-*tert*-butylethylene were too bulky to react with the complex. *Tert*-butylethylene was suitable, however, 4 molar equivalents increasing the yield of $[CpIrHL_2]^+$ from 7% to 40%. Cyclooctane was unaffected in the absence of *tert*-butylethylene, but with 4 molar equivalents per iridium the cyclooctadiene complex $[Ir(cod)L_2]^+$ was obtained in 70% yield. However, addition of *tert*-butylethylene to [2.2.2]bicyclooctane did not initiate the reaction equivalent to equation 41.

A number of related complexes have also been studied. For example, the complexes $[Ir(cod)(PMePh_2)_2]BF_4$ and $[Ir(cod)(PPh_3)(amino)]BF_4$ are both more active hydrogenation catalysts than the bistriphenylphosphine compounds.

It was thought very necessary to show that these dehydrogenations were truly homogeneous and a number of tests were used to show that the reactions were not due to carbonium ions or free radicals or to precipitated or colloidal iridium. The possibility that traces of acid might protonate the *tert*-butylethylene to give a carbonium ion which would dehydrogenate the alkane are discounted since the carbonium ion so formed would rapidly rearrange, and such a rearrangement does not occur. Free radicals, if present, would be expected to abstract chlorine atoms from the chlorinated solvents, and this also does not occur. Colloidal iridium was shown to be absent from the reaction mixture by dynamic light-scattering experiments.

The proposed mechanism for the reaction is given in Figure 7[169]. It is the reverse of the proposed mechanism for hydrogenation using the same catalyst. The *tert*-butylethylene first hydrogenates and makes available coordination sites on the iridium to which the cyclopentane adds oxidatively. The hydrogen that is eliminated as the reaction proceeds is scavenged by the *tert*-butylethylene.

More recently, Crabtree and coworkers[171] have been able to extend the range of alkanes that react beyond cyclopentane, and cyclooctane by not using a halocarbon solvent and by using a slightly different complex. It has become apparent that halocarbon solvents react with the reactive iridium complex intermediates during the alkane dehydrogenation; approximately 40% of the two complexes $[(Ph_3P)_2(\mu\text{-Cl})_2(\mu\text{-X})Ir(PPh_3)H]BF_4$ (X = H or Cl) have been found in the reaction products. Using the alkane as the solvent for *tert*-butylethylene and the complex $[IrH_2(Me_2CO)_2L_2]SbF_6$ [L = PPh_3 or $P(p\text{-}FC_6H_4)_3$], in a molar ratio of 4:1, at 85–150 °C, yields of dehydrogenated products greatly improved and other alkanes reacted. Cyclopentane reacts at 90 °C to give $[IrCpH(p\text{-}FC_6H_4)_3P]_2SbF_6$ in 82% yield after 24 h and the previously unreactive methyl- and ethyl-cyclopentanes gave 78% and 36% yields of the corresponding cyclopentadienyliridium complexes after 14 h at 120 °C. Cyclohexane, inert under the former conditions, reacts at 80 °C for 20 h to give $[\eta^5\text{-(cyclohexadienyl)}IrH\{(p\text{-}FC_6H_4)_3P\}_2]^+$ (5%), $[(phenyl)Ir\{(p\text{-}FC_6H_4)_3P\}_2]^+$ (45%), and benzene (32%). At higher temperatures (150 °C) most of the product is free benzene as the iridium complexes pyrolyse at the higher temperature.

Decomposition of the iridium complexes probably prevent these reactions from being

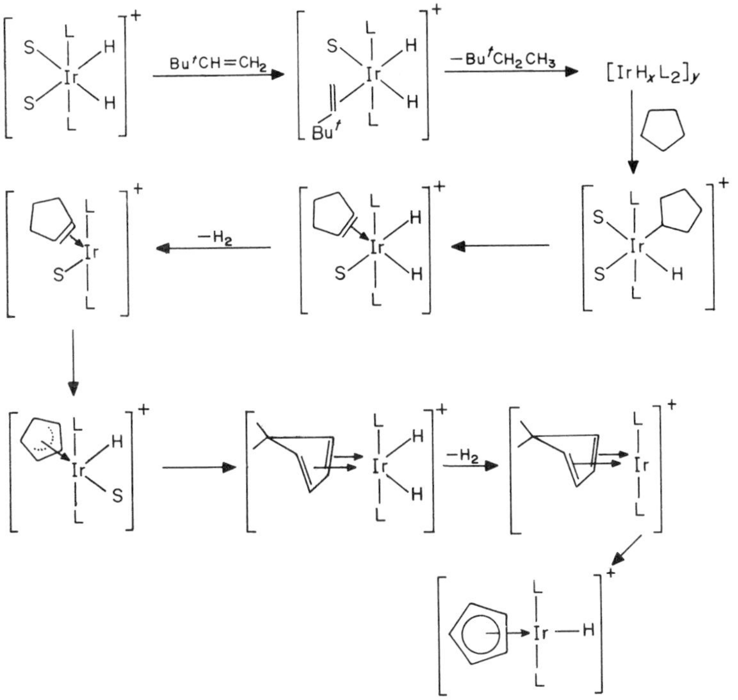

FIGURE 7. Proposed mechanistic scheme for alkane dehydrogenation [$L = PPh_3$ or $P(p\text{-}FC_6H_4)_3$].

catalytic. If cyclohexene is the reactant in the absence of *tert*-butylethylene then, at 80 °C, $[IrH_2Me_2(CO)_2(PPh_3)_2]BF_4$ catalyses the disproportionation to benzene and cyclohexane. If *tert*-butylethylene is added then benzene is selectively formed since the *tert*-butylethylene is hydrogenated preferentially to the cyclohexene.

It appears that the degree of alkane dehydrogenation that is achieved depends on the number of electrons on the metal. The 12-electron 'IrL_2^+' gives benzene, the neutral 12–14-electron fragments give olefins, and the 16-electron fragments give the alkyl hydride.

C—H bond cleavage by these iridium complexes has recently been extended to C—C bond cleavage[173]. Neat 1,1-dimethylcyclopentane reacted with $[IrH_2(Me_2CO)_2\{(p\text{-}FC_6H_4)_3P\}_2]SbF_6$ and *tert*-butylethylene at 150 °C for 8 h to give a 5,5-dimethylcyclopentadiene complex (**95**) in 50% yield, and the product resulting from C—C bond cleavage, the methylcyclopentadienyl complex (**96**), in 5% yield. On further heating complex **95** exchanged to **96** until after a further 12 h only **96** was present.

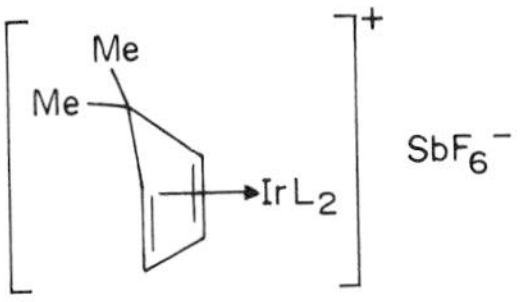

(**95**) $L = (p\text{-}FC_6H_4)_3P$ 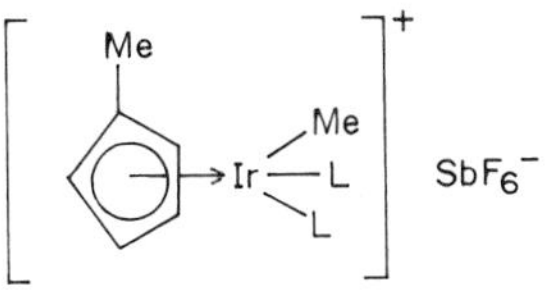 (**96**)

The analogous 1, 1-dimethylcyclohexene reacted with the same iridium complex and *tert*-butylethylene at 140 °C for 24 h to give the dimethylcyclohexadienyl complex **97**. This does not undergo C—C cleavage to give the analogous toluene complex, presumably because **97** is an 18-electron complex without a necessary coordination site where the C—C cleavage can occur.

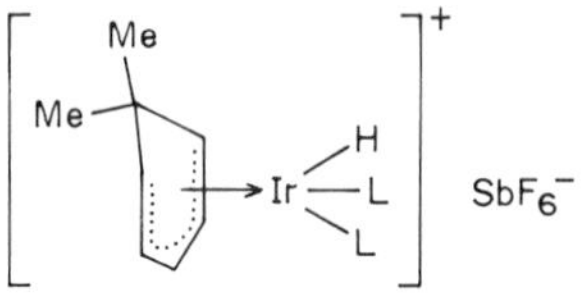

(**97**)

The driving force for the C—C bond activation, which one would expect to be less favourable than C—H bond activation, is undoubtedly the formation of the aromatic cyclopentadienyl ring. The need for the metal complex to have multiple sites for coordination is clear.

The effectiveness of a range of related complexes for the dehydrogenation of cyclopentane shows that the one most studied, the acetone complex, is the best. A variety of organic ligands all have lower activity, the order being $[IrH_2(Me_2CO)_2(PPh_3)_2]^+$ $> [Ir(\eta^5\text{-}C_6H_7)H(PPh_3)_2]^+ \approx [Ir(2,3\text{-dimethylbutadiene})H_2(PPh_3)_2]^+ > [Ir(\eta^6\text{-}C_6H_6)(PPh_3)_2]^+ \approx [Ir(\eta^5\text{-indenyl})H(PPh_3)_2]^+$ [172].

5. Activation by $[Re(PEt_2Ph)_2H_2]$ and related complexes

Reaction systems that are rather similar to those of Crabtree and coworkers are the rhenium hydride systems of Felkin and coworkers[174–177] and of Caulton and coworkers[141,178–180].

Felkin and coworkers used the complexes $[L_2ReH_7]$ (**98**; L = PPh_3 or PEt_2Ph) together with *tert*-butylethylene as a hydrogen acceptor to dehydrogenate cyclopentane at 50 °C to give an η^5-cyclopentadienyl complex (equation 42). The reaction (equation 43) is thought to take place via the formation of the 16-electron $[L_2ReH_5]$ or 14-electron $[L_2ReH_3]$, with the dehydrogenation occurring as a result of the imposed synperiplanar arrangement of the C—H and C—Re bonds in the initial cyclopentane complex[174]. This arrangement has been shown to be a requirement for facile β-elimination.

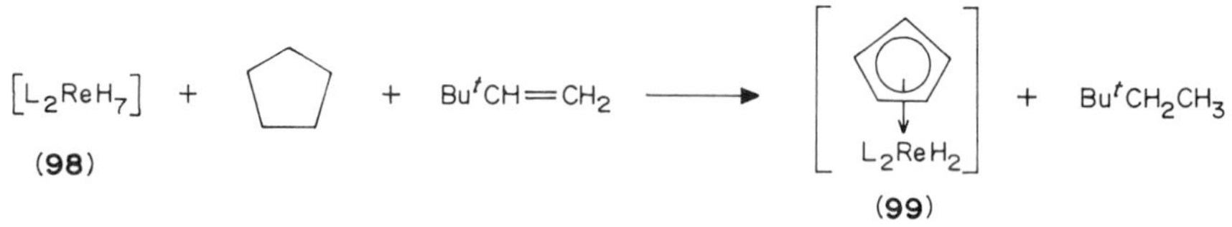

$[L_2ReH_7]$ (**98**) + cyclopentane + $Bu^tCH{=}CH_2$ ⟶ $[(\eta^5\text{-}C_5H_5)L_2ReH_2]$ (**99**) + $Bu^tCH_2CH_3$ (42)

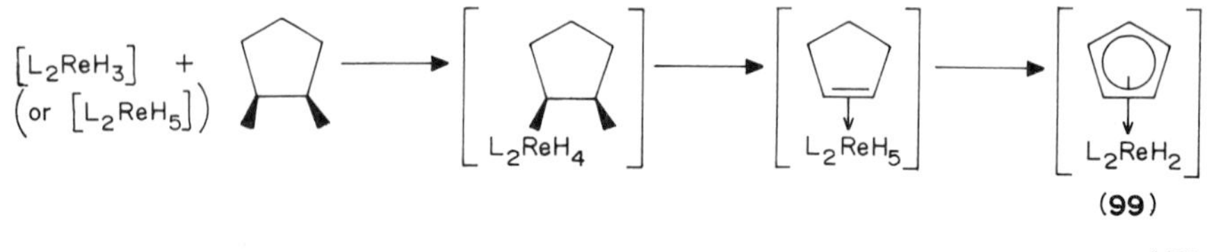

$[L_2ReH_3]$ (or $[L_2ReH_5]$) + cyclopentane ⟶ $[L_2ReH_4]$ ⟶ $[L_2ReH_5]$ ⟶ $[L_2ReH_2]$ (**99**) (43)

Other cycloalkanes do not dehydrogenate as fully as cyclopentane[176]. At 30–80 °C the appropriate cycloalkenes are formed from cyclohexane, cycloheptane, and cyclooctane in about 1 h. The yields of cycloalkenes, and of the cyclopentadienyl rhenium dihydride (**99**), depend on the phosphine (L) in the rhenium complex, $[L_2ReH_7]$ (**98**). It would appear that the more electron-releasing ligands give the more efficient dehydrogenation (but see later[177]), and for the ligands $(p\text{-}FC_6H_4)_3P$, Ph_3P, and $(p\text{-}MeC_6H_4)_3P$ the reported yields of cyclopentadienylrhenium dihydride are 10%, 25%, and 40%, respectively, and of cyclooctene 50%, 65%, and 80%, respectively. The larger the ring, the more efficient is the reaction—the yields of cyclohexene, cycloheptene, and cyclooctene are 25%, 30%, and 65%, based on the rhenium complex (**98**; $L = Ph_3P$).

These reactions are true homogeneous molecular processes, and not radical or metal atom catalysed. For the reaction of cyclooctane to cyclooctene the reaction continues unaffected if oxygen is bubbled through the solution. For the cyclohexane to cyclohexene reaction there is no bicyclohexyl formed. This would be expected if cyclohexyl radicals were involved. Also, cyclohexane would be expected to give benzene if metallic rhenium was the catalyst.

C—H bond activation also occurs in the linear alkane *n*-pentane, the product being pent-1-ene[175]. The reaction, given in equation 44, takes place at 80 °C. The pent-1-ene is formed by treating the penta-1,3-dienerhenium complex **100** with trimethyl phosphite. The yield of alkene is, as before, dependent on the phosphine, L, in the rhenium complex, $[L_2ReH_7]$. If L is PPh_3 the overall yield is 20%, but if L is the more electron releasing $P(p\text{-tolyl})_3$ the overall yield is 45%.

$$\text{n-pentane} + [L_2ReH_7] + Bu^tCH{=}CH_2 \xrightarrow[30\ \text{min}]{80\ ^\circ C} [(\eta^4\text{-penta-1,3-diene})L_2ReH_3]\ (\mathbf{100}) + Bu^tCH_2CH_3$$

$$\mathbf{100} \xrightarrow[90\ ^\circ C,\ 1\ h]{P(OMe)_3} \text{pent-1-ene} \qquad (44)$$

If these solutions are made very dilute then the system becomes catalytic[177]. If 3 mM solutions of the rhenium heptahydride are used at 80 °C then cyclooctane is dehydrogenated to cyclooctene with a turnover number of 9 (i.e. nine molecules of product are formed for each molecule of the rhenium hydride catalyst). The turnover numbers for cyclohexane to cyclohexene and for cycloheptane to cycloheptene are lower, 3.2 and 4, respectively. Interestingly in this study it is found that the most effective complex was that containing the phosphine $(p\text{-}FC_6H_4)_3P$, i.e. the most electron-withdrawing. The reason why this is so, in contrast to the earlier reports, is not clear.

If methylcyclohexane is the reactant then three different products are formed, 2-methylcyclohexene, 3-methylcyclohexene, and methylenecyclohexane. The relative yields of these three compounds depend on the phosphine ligands in the rhenium complex (Table 2). Several significant conclusions can be drawn from this observation. The alkenes formed do not interconvert since the most stable product that could be formed, 1-methylcyclohexene, is formed in only trace amounts. The reaction cannot involve radical intermediates since the major products, 3- and 4-methylcyclohexene cannot arise from the tertiary 1-methylcyclohexyl radical that is preferentially formed. Nor can the reaction take place on a rhenium colloid, as this would be the same for all three phosphine complexes.

The results in Table 2 suggest that the intrinsic reactivity is $CH_3 > CH_2 > CH$, and that this is counteracted by the strong steric effect of the bulky phosphines.

TABLE 2. Yields of products from the dehydrogenation of methylcyclohexane by $[L_2ReH_7]$ + $Bu^tCH{=}CH_2$[177]

L	Yield (%)		
	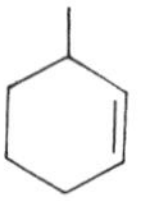 1-methylcyclohexene	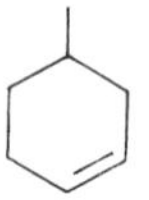 3-methylcyclohexene	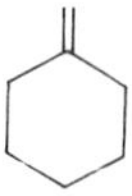 methylenecyclohexane
$(p\text{-}FC_6H_4)_3P$	29	65	6
Ph_3P	28	63	9
Et_3P	27	45	28

The related rhenium complex, $[(Me_2PhP)_3ReH_5]$, was used by Caulton and coworkers as the starting complex[178,179]. This is converted into an active 16-electron complex by u.v. irradiation when the complex loses a phosphine ligand to give $[(Me_2PhP)_2ReH_5]$[179]. This complex activates arene-sp^2 C—H bonds. If the irradiation is carried out on a solution of the starting material in deuteriobenzene then deuterium becomes attached to the rhenium in both $[(Me_2PhP)_3ReH_5]$ and in the dimeric rhenium complexes that are formed. For this to have happened, C—H bond activation must have occurred, although here the arylrhenium complex must be unstable and is not isolatable. A particularly interesting observation using this system was made when irradiating $[(Me_2PhP)_3ReD_5]$. Deuteriation of the aryl ring of the phosphine ligand occurred, but the deuterium was only in the *meta*- and *para*-positions on the ring, i.e. no *ortho*-hydrogen exchange occurred. This must mean that this H–D exchange is an intermolecular process, and that the intramolecular exchange does not occur. This is a surprising feature when one considers the ease of *ortho*-metallation even with this system[178].

If the irradiation is carried out with the complex dissolved in pure cyclopentane then the product is the dimer $[(Me_2PhP)_5Re_2H_6]$[178], i.e. the complex is not able to activate sp^3 C—H bonds.

Using *tert*-butylethylene as a hydrogen acceptor with this system alters the course of the reaction. If 8 vol.-% of *tert*-butylethylene is added then sp^3 C—H bond activation occurs, the cyclopentane is dehydrogenated, and the complex $[CpReH_2(PMe_2Ph)_2]$ is formed.

In benzene solution the *tert*-butylethylene coordinates to the rhenium hydride, which then hydrogenates it to give the rhenium alkyl complex $[(\eta^6\text{-}C_6H_6)Re(PMe_2Ph)_2CH_2CH_2CMe_3]$. This is not the only product from this reaction; complex **101** was isolated. This had incorporated a dinitrogen ligand and *ortho*-ring metallation had occurred.

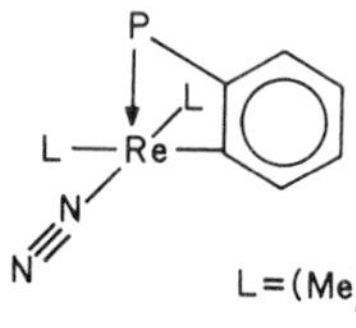

P
L—Re
N≡N
L=(Me₂Ph)₃P

(101)

Osmium hydrides also exchange deuterium with hexadeuteriobenzene[180]. If the complex $[(Me_2PhP)_3OsH_4]$ is irradiated with 254 nm u.v. light then the reactive intermediate $[(Me_2PhP)_3OsH_2]$ is formed. This exchanges hydrogen with C_6D_6, and also dimerizes to $[Os_2H_2(\mu\text{-}H)_2(PPhMe_2)_6]$. There are no reports, as yet, of osmium complexes activating saturated C—H bonds.

6. *Activation by [Cp*Ir(PMe$_3$)H$_2$] and related complexes*

In 1982 Janowicz and coworkers reported a homogeneous organotransition metal system in which oxidative addition of a C—H bond of saturated hydrocarbons occurs at room temperature[181–183]. The system is based on the dihydridoiridium(III) complex **102** (R^1 = Ph), which is prepared from the dimer, $[Cp^*IrCl_2]_2$, by treatment with triphenylphosphine and lithium triethylborohydride. Irradiation with u.v. light from a mercury lamp (λ_{max} 275 nm) of this complex in benzene solution gave an approximately equimolar mixture of the hydridophenyl complex **103** ($R^1 = R^2$ = Ph) and the *ortho*-metallated complex **104**.

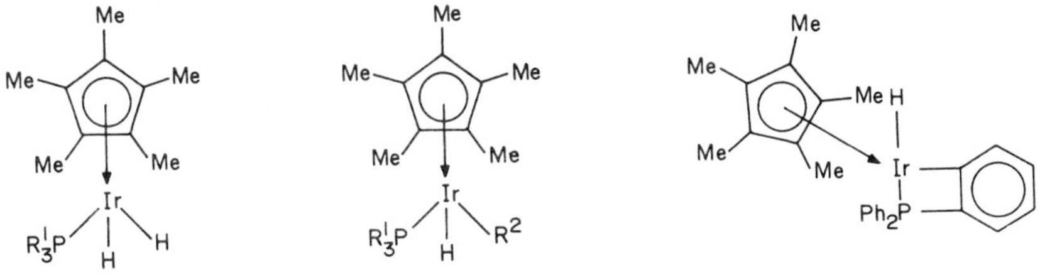

(102) (103) (104)

The analogous trimethylphosphine complex (**102**; R^1 = Me), having no phenyl groups, cannot readily *ortho*-metallate, and when this was used, not only did benzene give the analogous hydridophenyliridium complex (**103**; R^1 = Me, R^2 = Ph) but cyclohexane gave the hydridocyclohexyl complex (**103**; R^1 = Me, R^2 = cyclohexyl) in 90% yield. With neopentane as the solvent an 80% yield of the hydridoneopentyl complex **103** (R^1 = Me, R^2 = neopentyl) was obtained. The triphenylphosphineiridium complex in cyclohexane gave a mixture of the *ortho*-metallated product **104** and the triphenylphosphine complex **103** (R^1 = Ph, R^2 = cyclohexyl).

The proposed mechanism for this alkane activation is shown in Figure 8. The irradiation of the dihydride gives an electronically excited state that rapidly loses H_2 to form the coordinatively unsaturated 16-electron complex, which undergoes oxidative

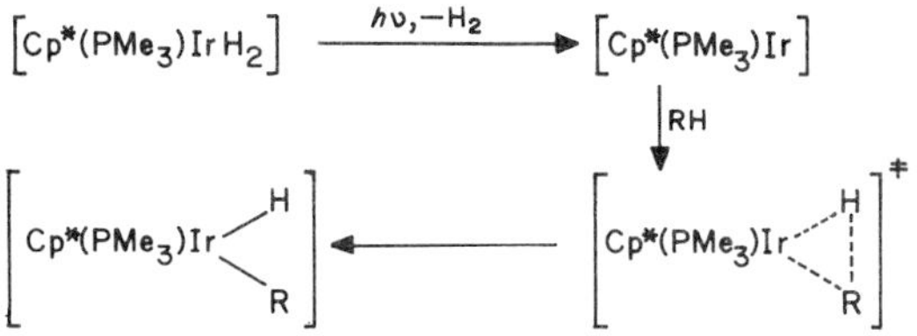

$$[Cp^*(PMe_3)IrH_2] \xrightarrow{h\nu,\,-H_2} [Cp^*(PMe_3)Ir] \xrightarrow{RH} [Cp^*(PMe_3)Ir\cdots(H)(R)]^{\ddagger} \longrightarrow [Cp^*(PMe_3)Ir(H)(R)]$$

FIGURE 8. Bergman's proposed mechanism for alkane activation (R = C_6H_{11} or CH_2Bu^t).

addition by a C—H bond of the alkane solvent, presumably by a three-centre transition state.

(**A**)

$$[Cp^*(L)IrH_2] \xrightarrow{h\nu} [Cp^*(L)IrH] + H^\bullet$$

$$H^\bullet + RH \rightarrow H_2 + R^\bullet$$

$$[Cp^*(L)IrH] + R^\bullet \rightarrow [Cp^*(L)Ir(R)(H)]$$

(**B**)

$$[Cp^*(L)IrH_2] \xrightarrow{h\nu} [Cp^*(L)Ir] + H_2$$

$$[Cp^*(L)Ir] + RH \rightarrow [Cp^*(L)IrH] + R^\bullet$$

$$[Cp^*(L)IrH] + R^\bullet \rightarrow [Cp^*(L)Ir(R)(H)]$$

FIGURE 9. Possible radical mechanisms for alkane activation.

The possibility of radical pathways needs to be considered. Two possible radical mechanisms are given in Figure 9. In the first (mechanism **A**) the irradiation has generated an excited state in which only one M—H bond is cleaved, giving a hydrogen atom. This might then be expected to abstract a hydrogen atom from an alkane and the alkyl radical produced would give the iridium hydridoalkyl product. The second mechanism (**B**) would occur if the unsaturated intermediate abstracted a hydrogen atom from an alkane molecule, generating an alkyl radical which would again give the iridium hydridoalkyl product.

Reaction by mechanism **A** (Figure 9), which predicts that a hydrogen atom of the reactant remains in the product, can be ruled out since the reaction of the dihydride **102** ($R^1 = Me$) in perdeuteriocyclohexane gives only $[Cp^*(PMe_3)Ir(C_6D_{11})D]$ and H_2 as the reaction products. Reaction by mechanism **B** (Figure 9), particularly if caged radical pairs are involved, is more difficult to eliminate. Evidence that strongly indicates that it does not occur is given in Figure 10. With *p*-xylene the reactive iridium intermediate can react with either the aromatic ring or at a benzylic C—H bond. If the mechanism involved radical formation then reaction at the benzylic position would be expected. In fact, reaction occurs preferentially (3.7 times more rapidly) at the aromatic ring.

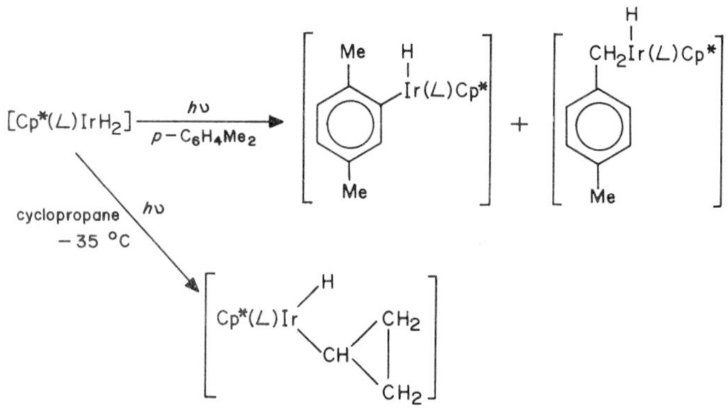

FIGURE 10. Reaction of *p*-xylene and cyclopropane with Bergman's iridium complex.

With cyclopropane, if radical intermediates were involved, insertion into the strained C—C bond would be expected. This does not occur. Also, the results of 'crossover' experiments, where a mixture of deuteriated and undeuteriated hydrocarbons are reacted together, should help decide whether radical routes are possible. The dihydride was irradiated with a 1:1 mixture of neopentane and cyclohexane-d_{12}, and at least 90% of the deuteriated reaction product was cyclohexyl-d_{11} deuteride, i.e. less than 10% was crossover product. This could arise from a reaction involving radicals and hence such a route cannot be conclusively eliminated, but the bulk of the evidence indicates it to be unlikely.

By using mixtures of different hydrocarbons as the solvent, selectivities of this reaction for different types of hydrogen have been established, and found to be as given in Table 3. For the cyclic alkanes the reactivity decreases as the ring size increases, presumably as the C—H bonds become sterically less accessible. Relative rates of reaction at primary, secondary, and tertiary C—H bonds in the same molecule were difficult to establish, but the results strongly indicate that primary is preferred over secondary C—H insertion, and tertiary C—H insertion is low, again presumably for steric reasons.

These alkyliridium hydride complexes are extremely hydrophobic, which makes them difficult to handle and to obtain pure. It also presents problems of identification, and throughout this work ^{1}H and ^{13}C n.m.r. spectroscopy has been the primary analytical tool.

To obtain functionalized organic products from the hydridoalkyl complexes, a complex was treated with bromoform and the bromoalkyl complex formed was further treated with mercury(II) chloride when, for the neopentyl complex, neopentylmercury(II) chloride was formed. This with bromine gave neopentyl bromide in high yield.

Heating of the hydridoalkyl complexes causes elimination of the alkane with reformation of the reactive intermediate, which can then react with more alkane solvent molecules[182]. This feature has enabled the constant for the equilibrium between a pair of alkanes and the hydridoalkyl complexes to be measured, and has been used to induce this system to activate methane[184]. Heating together the cyclohexyliridium hydride **103**, ($R^1 = Me$, R^2 = cyclohexyl) and *n*-pentane gives an equilibrium mixture containing these two

TABLE 3. Relative rates of reaction of [(Cp*M(PMe_3)] (M = Ir or Rh) with saturated hydrocarbons[82,183,186]

Hydrocarbon	Relative rate		
	M = Ir (0–10 °C)	M = Ir (−60 °C)	M = Rh (−60 °C)
Benzene	4.0	3.9	19.5
Cyclopropane	2.65	2.1	10.4
Cyclopentane	1.6	1.1	1.8
Cyclohexane	1.0	1.0	1.0
n-Hexane (primary C—H)	—	2.7	5.9
n-Hexane (sec. C—H)	—	0.2	0
Propane (primary C—H)	—	1.5	2.6
Propane (sec. C—H)	—	0.3	0
Ethane	—	—	2.0
Cycloheptane	—	—	0.14
Cyclodecane	0.23	—	—
Cyclooctane	0.09	0.09	0.06
Neopentane	1.14	—	—
Isobutane	—	—	3.6

compounds together with the n-pentyliridium hydride **103** (R^1 = Me, R^2 = n-pentyl) and cyclohexane (equation 45). The equilibrium constant (equation 46) was calculated, from the ratio of compounds in the equilibrium mixture, to be 10.8, and the primary metal—carbon bond energy in complex **B** (**103**; R^1 = Me, R^2 = n-pentyl) was calculated to be 23 kJ mol^{-1} higher than that of the secondary carbon—metal bond in complex **A** (**103**; R^1 = cyclohexyl).

$$\underset{\mathbf{A}}{[Cp^*(PMe_3)Ir(Cy)H]} + Me(CH_2)_3Me \overset{K_{eq}}{=} \underset{\mathbf{B}}{[Cp^*(PMe_3)Ir\{(CH_2)_4Me\}H]} + CyH \tag{45}$$

$$K_{eq} = \frac{[\mathbf{B}][\text{cyclohexane}]}{[\mathbf{A}][n\text{-pentane}]} \tag{46}$$

Methane activation did not occur with the dihydride **102** (R^1 = Me) in perfluoroalkane solvent under 4atm of methane gas, in contrast to the result with $[Cp^*Ir(CO)_2]$ discussed later, possibly because of the very low solubility of the dihydride (**102**; R^1 = Me) in perfluoroalkanes. Methane activation was achieved by taking advantage of the presumption that the hydridomethyl complex **103** ($R^1 = R^2$ = Me) would be thermodynamically more stable than other alkyl complexes. When the hydridocyclohexyl complex **103** (R^1 = Me, R^2 = cyclohexyl) was heated in cyclooctane solvent for 14h at 140–150 °C under 20 atm of methane, a 58% yield of the hydridomethyl complex **103** ($R^1 = R^2$ = Me) was obtained.

7. Activation by $[Cp^*Rh(PMe_3)H_2]$

a. Alkanes

The rhodium analogue of complex **102** (R^1 = Me) on irradiation also undergoes oxidative addition by alkane C—H bonds[185,186]. The rhodium complexes are less stable and reactions are carried out at lower temperatures (typically − 30 to − 60 °C compared with 0–10 °C). The rhodium system is also more selective than the iridium system. At the same temperature (− 60 °C) competitive studies give the relative rates reported in Table 3. The rhodium complex is much more discriminating, particularly between C—H bonds in the same molecule. With linear alkanes the rhodium complex inserts only into primary C—H bonds; the iridium complex favours primary over secondary, but inserts into both. No tertiary C—H bond insertion has been observed for either system.

Treatment of the product hydrides with bromoform at −60 °C converts them into the corresponding bromoalkylrhodium complexes. In contrast to the bromoiridium analogues, these react with bromine to give alkyl bromides in high yield. One of the bromoalkyl rhodium complexes, that with the cyclopropyl group, has had its X-ray crystal structure determined. This conclusively showed that the cyclopropyl ring is intact.

The hydridocyclopropylrhodium complex (Rh analogue of **103**; R^1 = Me, R^2 = cyclopropyl) behaves differently to the other alkyl complexes on warming to 20 °C. Others undergo reductive elimination of the alkane; the cyclopropyl complex rearranges to a rhodacyclobutane complex, i.e. C—C insertion occurs. This suggests that C—H insertion is favoured kinetically, but that the C—C insertion product is favoured thermodynamically.

b. Alkanes vs. arenes

A detailed comparison of C—H bond activation in alkanes and arenes by the dihydridorhodium complex (Rh analogue of **102**; R^1 = Me) has been reported by Jones and Feher[185,187–189]. They concluded (i) that the suggestion of Parshall [12] and Chatt and Davidson[190] that arenes precoordinate in an η^2 manner prior to activation occurs here and hence provides a lower energy pathway for oxidative addition of the C—H bond, a route unavailable to alkane, and (ii) that the activation of alkane and arene C—H bonds occurs at similar rates (with a slight preference for arenes), even though the C—H bond energies differ by 25–40 kJ mol^{-1}. They favour the alkane C—H reacting as shown in equation 47.

$$\mathrm{M + R{-}H \longrightarrow \left[M \cdots \begin{matrix} R \\ \vdots \\ H \end{matrix} \right]^{\ddagger} \longrightarrow M \begin{matrix} R \\ H \end{matrix}} \qquad (47)$$

8. Activation by [Cp*Ir(CO)$_2$] and related complexes

a. Alkanes

At about the same time as Janowicz and Bergaman's report[181], Hoyano and Graham[191] reported alkane C—H bond activation by a closely related, but significantly different, iridium complex [Cp*Ir(CO)$_2$], **105**, when it was irradiated with a mercury vapour lamp in an alkane solvent. In neopentane at room temperature the reaction given in equation 48 took place to give a high yield of the neopentyl hydride. This hydride

$$[(\eta^5\text{-}C_5Me_5)Ir(CO)_2] \xrightarrow[\text{RH}]{h\nu} [(\eta^5\text{-}C_5Me_5)Ir(CO)(H)(R)] \qquad (48)$$

(**105**) (**106**)

product (**106**; R = CH_2CMe_3) decomposed slowly and it was not possible to isolate it pure. The stable chloride (**106**; R = CH_2CMe_3 with Cl replacing H) was obtained by treating the hydride (**106**) with CCl_4. Using cyclohexane as the solvent the analogous cyclohexyl hydride (**106**; R = cyclohexyl) was formed. This complex is reactive enough to activate the C—H bonds in methane. A solution of complex **105** in the unreactive solvent perfluorohexane under 8 atm of methane after irradiation for 16 h at room temperature gave the methyl hydride (**106**; R = Me)[192].

When benzene is used as the solvent and the reaction mixture is treated with CCl_4, the known compound [Cp*Ir(CO)(Cl)(C_6H_5)] is formed, showing that the phenyl hydride complex **106** (R = Ph) had been formed. In 1977 Rausch *et al.*[193] had photolysed the cyclopentadienyl analogue (i.e., with Cp instead of Cp*) in benzene, and formed the analogous phenyliridium hydride complex. With this ligand, however, the product was very reactive and unstable and could not be isolated.

This cyclopentadienyl complex also activates methane and neopentane[192]. A solution of it in perfluorohexane under 10 atm of methane was irradiated for 6 h. The methyl hydride **106** (R = Me with Cp replacing Cp*) was detected in the n.m.r. spectrum, and after treatment with *N*-bromosuccinimide the bromo derivative was fully characterized.

b. Alkanes vs. arenes

Irradiation of **105** in an equimolar mixture of benzene and neopentane gave a product ratio which shows a 4-fold preference for the aromatic C—H over the aliphatic C—H bond. With benzene and cyclohexane this ratio is 2.5, very similar to the values obtained by Bergman and coworkers with their iridium complex[182,186]. Here as elsewhere a 16-electron coordinatively unsaturated iridium complex is thought to be the reactive intermediate, and the absence of ligands on the iridium that can metallate is obviously important.

A matrix isolation study has shown that C—H bond activation in methane by [Cp*Ir(CO)$_2$] (**105**) and related complexes occurs even at 12 K[194]. Infrared spectra at a 1:2000–1:5000 dilution of complex **105** in methane at 12 K were recorded with the sample under u.v. irradiation. New terminal C—O bands appeared at 1771.5 and 1971.4 cm^{-1}, together with a weak band at 2150.1 cm^{-1}. In tetradeuteriomethane the C—O bands were at 2136.8 and 1990.2 cm^{-1}. There was no weak band at 2150.1 cm^{-1}, but a new weak band appeared at 1548.8 cm^{-1}. This shift is appropriate for H–D exchange, and hence the 2150.1 cm^{-1} band can be assigned to Ir—H, i.e. C—H bond activation of methane has occurred. These values may be compared with the band positions in a solution of [Cp*IrCO(H)Me] (**106**; R = Me) where there is a very weak band at 2134 cm^{-1} and strong C—O bands around 1990 cm^{-1}.

If complex **105** or the cyclopentadienyl or rhodium analogue {[CpIr(CO)$_2$] or [Cp*Rh(CO)$_2$]} is irradiated in an argon or nitrogen matrix there are very few photodissociation products [LM(CO)]. This is in contrast to the related cobalt complex, [CpCo(CO)$_2$], which in nitrogen forms the complex [CpCo(CO)N$_2$]. In pure CO matrices photolysis produces new bands at those higher wave numbers which had been observed for [CpCo(CO)$_2$][195]. These can be assigned to [(η^3-C$_5$Me$_5$)Ir(CO)$_3$], which is formed by 'ring slippage'. In a 5% ^{13}CO in argon matrix the exchange of ^{13}CO is much slower for the iridium complex (**105**) than for the cobalt complex [CpCo(CO)$_2$], and in a 5% ^{13}CO in CH$_4$ matrix there is no ^{13}CO exchange with complex **105** or its rhodium analogue. It may be noted that in the thermal reaction at 25 °C in *n*-hexane, exchange of ^{13}CO with **105** occurs readily.

This failure to generate more than traces of [Cp*IrCO] in argon and nitrogen matrices and the slow exchange could indicate that the proposed dissociation to this reactive 16-electron species is incorrect. The alternative mechanism would be via a change in ring hapticity ($\eta^5 \rightleftharpoons \eta^3$), which has been demonstrated to occur for [CpCo(CO)$_2$][195].

This photoactivation at 12 K is the first example in matrix isolation studies of metal complexes activating methane.

9. Activation by [Cp$^*_2$LuH] and related complexes

Intermolecular C—H activations under mild conditions by complexes of the rare earth elements have been reported by Watson[196,197]. The hydride complex of the heaviest of the rare earth elements, lutetium, [Cp$^*_2$LuH] (**107**; R = H), undergoes H–D exchange when dissolved in deuteriated benzene or toluene. If hydrogen gas is purged from the solution, or if the Lu—Me complex (**107**; R = Me) is dissolved in benzene, then the product is the Lu—phenyl complex (**107**; R = Ph). This reacts, at a slower rate, with the hydride **107** (R = H) to

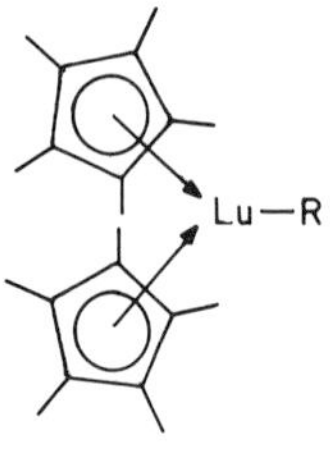

Lu—R

(**107**)

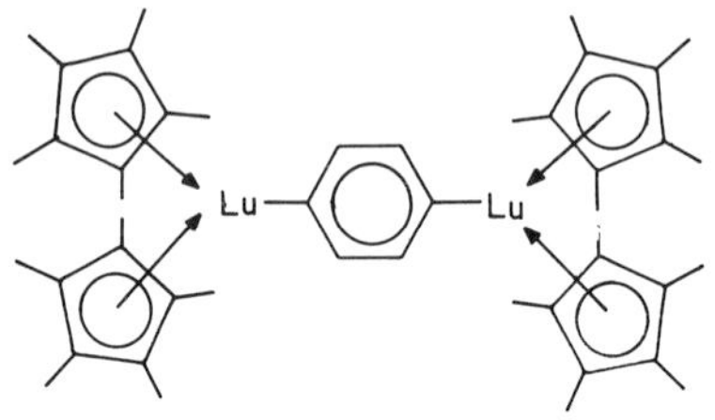

Lu—⟨C$_6$H$_4$⟩—Lu

(**108**)

Lu—(2-pyridyl)

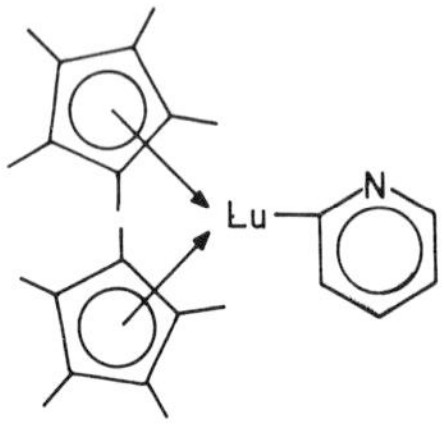

(**109**)

from the *p*-metallated complex **108**. Reaction of either the hydride **107** (R = H) or the methyl complex **107** (R = Me) with pyridine gives complex **109**, which is *o*-metallated in the pyridine ring.

With tetramethylsilane the hydride **107** (R = H) or methyl complex **107** (R = Me) gives the trimethylsilylmethyl complex **107** ($R = CH_2SiMe_3$). If the hydride is the reactant, the reaction is readily reversible, and therefore the hydride **107** (R = H) is a catalyst for H–D exchange between tetramethylsilane and D_2 or C_6D_6.

This complex is highly active and activates the C—H bonds in methane. Both the lutetium–methyl complex **107** (R = Me) and its yttrium analogue undergo exchange of methane with $^{13}CH_4$ (equation 49)[197]. Kinetic studies on the reaction in cyclohexane solution show the exchange reaction to be bimolecular. A symmetrical transition state (**110**) is the favoured one.

$$[Cp_2^*MMe] + {}^{13}CH_4 \rightleftharpoons [Cp_2^*M^{13}CH_3] + CH_4 \qquad (49)$$

$$M = Lu \text{ or } Y$$

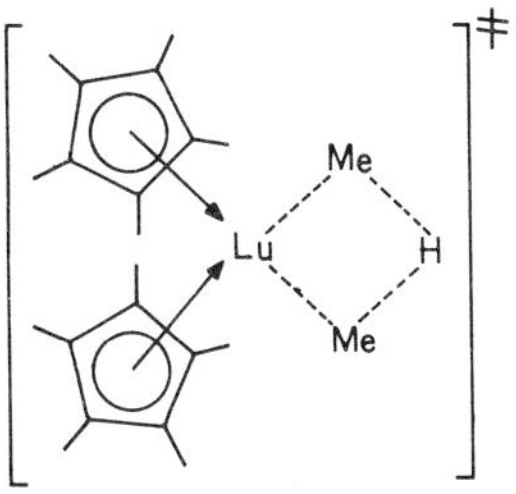

(**110**)

Both the lutetium (**107**; R = Me) and yttrium methyl complexes exist as asymmetric dimers in the solid state (equation 50). In hydrocarbon solution they dissociate rapidly and the coordinatively unsaturated monomers are the reactive species. It should be noted that the steric bulk of the Cp_2^*M group prevents the formation of the more stable symmetrical dimer with two bridging methyl groups, as is found for the yttrium and ytterbium complexes $[Cp_2MMe]_2$ (M = Y or Yb)[198].

$$2[Cp_2^*LuMe] \rightleftharpoons [Cp_2^*Lu{-}\underset{\underset{\textstyle Me}{|}}{Me}{-}LuCp_2^*] \qquad (50)$$

$$[Cp_2^*LuMe] \underset{-CH_4}{\overset{+CH_4}{\rightleftharpoons}} [H{-}\underset{\underset{\textstyle Me}{|}}{Me}{-}LuCp_2^*] \qquad (51)$$

The monomers are strong Lewis acids, and their electrophilicity can be satisfied by coordination of Lewis bases[199], or by three-centre interactions with σ-C—H bonds. The similarity of the initial coordination of a hydrocarbon (equation 51) and the bridging methyl group (equation 50) should be noted.

Other hydrocarbons, such as ethane and propane, also react with both the lutetium (**107**; R = Me) and yttrium complexes, but the products decompose by β-hydride elimination.

10. Activation by $[Cp_2^*Th(CH_2CMe_2CH_2)]$

The tetravalent Cp*Th–cycloalkyl complexes produced by Marks and coworkers[152] (Section II.C.9) are very reactive molecules and will react with the saturated C—H bonds in tetramethylsilane and methane[200] (equation 52). The compounds are very sensitive and the reactions are carried out under strictly anaerobic and anhydrous conditions in dark sealed n.m.r. tubes. Using cyclohexane-d_{12} as the solvent, tetramethylsilane reacts according to equation 52 at 30 °C. If the reaction mixture is heated to 60 °C, CMe_4 is expelled and the four-membered Th—C—Si—C cyclic complex **89** (M = Si) is formed. Methane (10 atm) reacts at 60 °C to give the product from reaction 52 (R = Me). In CD_4 the reaction is significantly slower; the kinetic isotope effect is *ca.* 6. This indicates that C—H bond breaking is the rate-limiting step.

$$\left[Cp^*_2Th\langle(CH_2)_2CMe_2\rangle\right] + RH \longrightarrow \left[Cp^*_2Th(R)(CH_2CMe_3)\right] \quad (52)$$

11. Activation by Ziegler–Natta catalysts and vanadocene

Russian workers have shown that methane can be activated in the presence of catalysts of the Ziegler–Natta type[201]. Thus, in the presence of a $TiCl_4$–$AlMe_2Cl$ catalyst deuteriomethane undergoes a hydrogen–deuterium exchange (equation 53). $[Cp_2TiCl_2]$, $[V(acac)_2Cl_2]$, VCl_3, or $[V(acac)_3]$ can be used in place of $TiCl_4$. These catalysts will also catalyse the methanation of ethylene (equation 54) and of acetylene (equation 55).

$$[MCH_3] + CD_4 \xrightarrow[0.4-1\ \text{atm}]{20-70\ ^\circ\text{C}} [MCD_3] + CH_3D \quad (53)$$

$$CH_4 + C_2H_4 \longrightarrow C_3H_8 \quad (54)$$

$$CH_4 + C_2H_2 \longrightarrow MeCH{=}CH_2 \quad (55)$$

Hydrogen–deuterium exchange between methane and ethylene catalysed by vanadocene (equation 56) was also reported by Grigoryan[201]. The benzene solvent and the cyclopentadienyl groups also undergo H–D exchange and the mechanism is thought to involve interactions of the type shown in equations 57–59.

$$CH_4 + C_2D_4 \xrightarrow[Cp_2V]{\text{benzene, } 70\ ^\circ\text{C}} CD_4 + C_2H_4 \quad (56)$$

$$>M + CH_2{=}CH_2 \rightleftharpoons \left[>M \leftarrow \begin{matrix} CH_2 \\ \| \\ CH_2 \end{matrix}\right] \underset{}{\overset{CH_4}{\rightleftharpoons}} \left[>M\begin{matrix} C_2H_5 \\ CH_3 \end{matrix}\right] \quad (57)$$

$$[(\eta^5\text{-}C_5H_5) \rightarrow M] \rightleftharpoons [C_5H_4{-}M(H)] \quad (58)$$

$$[(\eta^6\text{-}C_6H_6) \rightarrow M] \rightleftharpoons [C_6H_5{-}M(H)] \quad (59)$$

12. Activation by supported rhodium complexes

The *catalytic* activation of methane using rhodium(III) supported on silica as the catalyst has been reported[202]. This system is clearly different from most of those discussed in this section—they are 'electron-rich' and capable of entering into oxidative addition reactions; this system is 'electron-deficient' and is more akin to the organolutetium complexes (Section II.D.9). Silica-supported rhodium complexes have been shown to have electrophilic character[203,204] and to activate hydrogen. They also activate methane. The stoichiometric reactions (equations 60 and 61) have been reported. By using a mixture of methane and chlorine, chlorination of methane, catalysed by complex **111**, occurs since the dihydride product **112** reacts with chlorine to give the hydridochloride **113**. These reactions have only been observed on a small (millimolar) scale, and details of the mechanisms are uncertain. Further work is awaited with interest.

$$[\{Si\}-O-Rh(\eta^3\text{-}H_2C{\cdots}CH{\cdots}CH_2)_2] \; (\mathbf{111}) + CH_4 \longrightarrow [\{Si\}-O-Rh(\eta^3\text{-}H_2C{\cdots}CH{\cdots}CH_2)(Me)] + CH_2{=}CHMe \qquad (60)$$

$$[\{Si\}-O-Rh(H)(Cl)] \; (\mathbf{113}) + CH_4 \longrightarrow [\{Si\}-O-Rh(H)_2] \; (\mathbf{112}) + MeCl \qquad (61)$$

E. Conclusions

1. Intermolecular or intramolecular C—H activation?

An issue of some interest that has been the subject of a number of studies is that of the relative ease or difficulty of intra- or inter-molecular C—H bond activation by transition metal complexes. As pointed out in the previous section, there are numerous examples of both intra- and inter-molecular arene activation, but until the recent reports of intermolecular activation discussed in Section D, only the examples of intramolecular C—H bond activation in alkanes, as reported in Section C, were known. This may suggest that intermolecular alkane activation is kinetically, or thermodynamically, unfavourable.

To explain the many examples of intramolecular activation it has been suggested that the entropy plays a major role[78,82], the $-T\Delta S^{\neq}$ term being up to 40 kJ mol^{-1}. If it is assumed that the energies of the bonds broken and formed in inter- and intra-molecular activation (equations 62 and 63) are the same, and that there is negligible ring strain in the product of reaction 63, then there will be no difference in the enthalpy of activation contribution to the free energy of activation, which can also be up to 40 kJ mol^{-1}.

$$M + R{-}H \longrightarrow M(R)(H) \qquad (62)$$

$$M{-}R{-}H \longrightarrow M(R)(H) \;\text{(M–R ring)} \qquad (63)$$

If, however, the intermolecular reaction involves the solvent molecules, then it might be expected that their high concentration would help overcome this unfavourable entropy effect for intermolecular activation[83]. In the platinum(II) complexes studied by Whitesides and coworkers, activation of alkane solvent was not observed and, only intramolecular activation occurred[78,82]. However, the situation is not clear. Reamey and Whitesides[83] have shown that an intermolecular reaction involving the intermediate $[(Et_3P)Pt(CH_2CMe_3)_2]$ with H_2 can be fast, i.e. there appears to be no entropy barrier to the C—H bonds of the solvent reacting. The reason why intermolecular C—H activation is slower than expected may instead be due to the neglect of non-bonded steric interactions and to restrictions to vibrations in the sterically crowded transition state. It may also be due to the fact that intramolecular cyclometallations, reductive elimination of R_2, and migration of alkyl groups from phosphorus to platinum can occur rapidly in complexes such as $[(Et_3P)Pt(CH_2CMe_3)_2]$.

Certainly the reasons for a reaction being inter- or intra-molecular are no doubt subtle, as Marks and coworkers found[152,153]. The cyclometallated complex **89** (M = C in Section C.9 and equation 52 in Section D.10) underwent an intermolecular reaction with methane, but the related complex **88** (M = C in Section C.9), which could undergo an intramolecular reaction did so.

Jones and Feher[205] have carried out a detailed study to obtain the thermodynamic parameters of inter- and intra-molecular reactions at rhodium in a number of complexes. For the complex **113** (equation 64) in benzene solution the difference in the activation parameters for intramolecular [equation 64 (right)] compared with inter-molecular [equation 64 (left)] activation (intra- minus inter-) is $+8\,kJ\,mol^{-1}$ for the enthalpy of activation and $+20\,kJ^{-1}\,mol^{-1}$ for the entropy of activation. These figures suggest that for this arene C—H activation, the entropy term slightly favours intramolecular activation [equation 64 (right)] and intermolecular activation is favoured by the enthalpy term [equation 64 (left)].

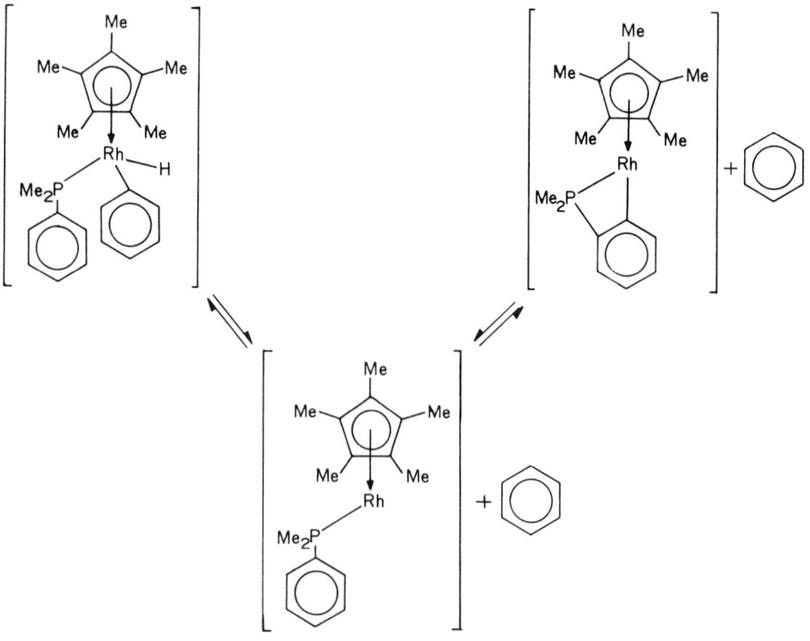

(64)

(113)

Similar experiments with alkanes were not possible owing to the instability of the alkylhydride complexes above $-20\,^{\circ}C$, but a comparison of the relative stabilities of the propylrhodium hydride **114** with the intramolecular hydride **115** was possible by studying their rates of decomposition. The intramolecular complex **115** is estimated to be at least $17\,kJ\,mol^{-1}$ more stable than the intermolecular complex **114** in neat propane solvent.

(114) $(C_5Me_5)Rh(H)(CH_2CH_2CH_3)(PMe_2Pr)$

(115) $(C_5Me_5)Rh(H)(CH_2CH_2CH_2PMe_2)$

Janowicz and Bergman[182] discussed the relationship between intra- and intermolecular activation in their system (Section II.D.6), and gave an estimate that for the zirconium complex (**87** in Section C.8) the intra-/inter-molecular rate constant ratio must be at least 10^6. This is much larger than that estimated for organic molecules $(10^3–10^5)$[206].

With intra-/inter-molecular rate ratios of this magnitude, systems that do undergo *intermolecular* activation must be rather special. In Janowicz and Bergman's iridium complex, intermolecular activation is comparable to intramolecular activation of the aromatic C—H bonds of triphenylphosphine, and much faster than that of the intramolecular activation of the C—H bonds in trimethylphosphine, or the C—H bonds in the pentamethylcyclopentadienyl group. Obviously some other factors, as yet not at all understood, in addition to entropy control the relative reactivities.

2. A theoretical study

A detailed study of C—H and H—H activation by transition metal complexes and on surfaces by Saillard and Hoffmann[207] addressed itself, in part, to the activation processes that are discussed here. They showed that during oxidative addition there is C—H σ-electron to metal transfer, metal to σ^*-electron transfer, a repulsive interaction between σ and filled metal orbitals (steric features), and a rearrangement of electron density on the metal.

Coordinative unsaturation is essential, and in the early stages of a reaction σ to metal electron transfer is dominant. Steric effects are also important. They also found that there are great similarities between the activation of C—H bonds at a metal complex and on a metal surface.

3. General conclusions

In this section we have seen that there are many compounds in which sp^3 C—H bonds can interact with a transition metal centre. We have also seen that it is possible to have a non-radical reaction of the sp^3 C—H bonds of an unstrained saturated hydrocarbon or alkyl group with a soluble transition metal complex. There are now many examples of such reactions, either intramolecular (Section C) when metallocyclic rings are formed, or intermolecular (Section D) where alkane molecules enter into reaction with the complex.

As in other areas of alkane C—H bond activation, the reactions parallel those of arenes. The alkanes are not as reactive, of course, as they lack the π-systems that can precoordinate an aromatic ring and bring the C—H bonds into the reaction zone.

There is an increasing understanding of the mechanisms of these reactions, although some of the details of the electronic interactions still remain obscure. An electron-rich metal looks desirable, but is clearly not essential. Coordinative unsaturation is obviously often essential. Oxidative addition appears to be the most common activation process, but not always so, as heterolytic fission of C—H bonds must sometimes occur, as examples are reported of active complexes that cannot increase their oxidation state as would be necessary during an oxidative addition process.

The metal complex will need to have a low-lying unfilled orbital to interact with the σ C—H bonding orbital of the alkane, and a high-lying filled orbital to interact with the σ C—H antibonding orbital of the alkane. It may be that photochemically or thermally excited states of the metal complex will be best for oxidative addition of saturated hydrocarbons by having such molecular energy levels.

The influence of other ligands, their electronic and steric properties, and their behaviour as a leaving groups are recognized and are being studied. As one might expect, the effects can often be subtle.

Most of the examples involve the late transition elements, particularly those of the lower series. Platinum and iridium, in particular, provide a rich collection of complexes where C—H bond activation has occurred.

This is a rapidly developing area of chemistry. There is little doubt that by the time this book is published the material herein will already need to be supplemented. It is a subject of intense interest, not only because the interactions (saturated C—H bonds with transition metals) are in themselves still fairly novel as they were thought to be impossible a decade or so ago, but because there are worthwhile rewards if *catalytic* as distinct from *stoichiometric* reactions (the bulk of those discussed here) can be developed. Such developments would open the way to alkane activations moving from the research laboratory to the industrial chemical plant.

III. ACTIVATION OF SATURATED CARBON—HYDROGEN BONDS BY PLATINUM(II) COMPLEXES

A. Catalysed Hydrogen–Deuterium Exchange

1. *Introduction*

This topic has been extensively reviewed[1,7–9], and only the main features are presented here. In 1967 Garnett and Hodges[208] reported that, in the presence of $[PtCl_4]^{2-}$ ions, aromatic compounds undergo catalysed exchange of hydrogen with deuterium in an acetic acid–deuterium oxide solvent. Temperatures of 80–100 °C were used, and a mineral acid was also added to stabilize the platinum(II) against disproportionation (equation 65). Garnett and Hodges found that in addition to aromatic hydrogen atoms, the hydrogen atoms in the side-chains of alkylbenzenes also underwent exchange. This included the normally 'inert' C—H bonds of the methyl group in ethylbenzene.

$$2[PtCl_4]^{2-} \longrightarrow Pt\downarrow + [PtCl_6]^{2-} + 2Cl^- \qquad (65)$$

It was soon shown by Shilov and coworkers[210] and in our laboratory[211] that alkanes also undergo hydrogen–deuterium exchange under similar conditions. Methane, ethane, and other alkanes when heated in sealed ampoules containing solutions of $K_2[PtCl_4]$ in a solvent mixture of deuterioacetic acid and deuterium oxide were found to exchange hydrogen atoms with the deuterium of the solvent.

These isotope exchanges take place at temperatures of 80–120 °C. It has been shown that under these conditions neither metallic platinum nor platinum(IV) compounds catalyse the hydrogen–deuterium exchange. These could be present in small amounts from disproportionation of the $[PtCl_4]^{2-}$ catalyst (equation 65). If a platinum(IV) compound is added as the sole potential catalyst, no exchange is observed until some platinum(II) is produced by reduction[212]. If a precipitate of metallic platinum is formed the exchange slows or even stops. Other additives such as mineral acids, molecular oxygen, and aromatic compounds such as pyrene are known to stabilize homogeneous platinum(II) solutions, and in the presence of such additives the kinetics of the isotope exchange become more reproducible. Most studies of the isotope exchange have been made in a 1:1 acetic acid–water solvent. Aqueous solutions are unsuitable because of the limited solubility of alkanes, whereas the catalyst has limited solubility in pure acetic acid. The rate of hydrogen–deuterium exchange of pentane has been shown to be a maximum using about 1:1 acetic acid–water, and to fall off at both lower and higher acetic acid concentrations[213]. In studies of hydrogen–deuterium exchange the solvent will usually consist of acetic acid-d_1 and deuterium oxide. However, at higher temperatures in the presence of platinum(II) the C—H bonds of acetic acid can also exchange, and it is necessary to use the fully deuteriated acid, CD_3CO_2D.

2. Kinetics

The rate of isotope exchange for ethane in acetic acid–water containing added perchloric acid is independent of both the acidity and the ionic strength for perchloric acid solutions above 0.2 M. The kinetics of the exchange reactions show a first-order dependence on the hydrocarbon concentration and a fractional (less than one) order with respect to the concentration of added $[PtCl_4]^{2-}$ catalyst. Addition of chloride ions markedly affects the rate (see Figure 11). This complex behaviour with additional chloride ions can be explained by considering solvation of the catalyst $[PtCl_4]^{2-}$ (equations 66 and 67; S = solvent). If the species $[PtCl_2S_2]$, $[PtCl_3S]^-$, and $[PtCl_4]^{2-}$ each has its own rate coefficient (k_1, k_2, and k_3) associated with the catalytic reaction, the overall measured rate coefficient, k^{overall}, can be derived as in equation 68:

$$[PtCl_4]^{2-} + S \overset{K'_1}{\rightleftharpoons} [PtCl_3S]^- + Cl^- \qquad (66)$$

$$[PtCl_3S]^- + S \overset{K'_2}{\rightleftharpoons} [PtCl_2S_2] + Cl^- \qquad (67)$$

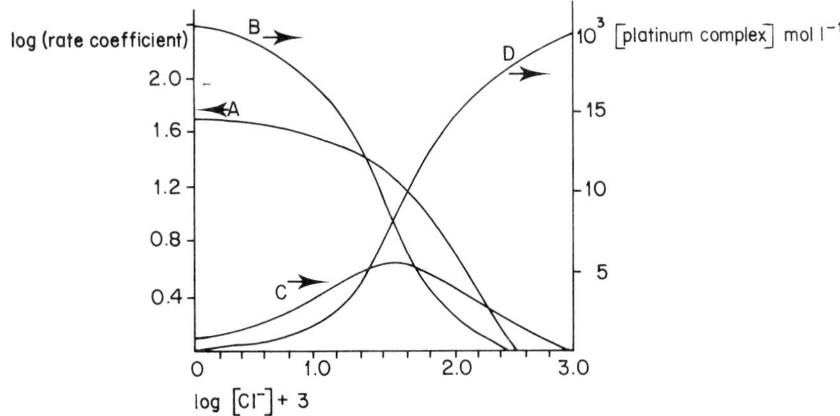

FIGURE 11. Dependence of the rate coefficient for platinum(II)-catalysed isotope exchange with concentration of chloride ions (curve A). Curves B, C and D show the equilibrium concentrations of $[PtCl_2S_2]$, $[PtCl_3S]^-$, and $[PtCl_4]^{2-}$, respectively. After ref. 212.

$$k^{\text{overall}} = \frac{(k_1 + k_2K_2[\text{Cl}^-] + k_3K_1K_2[\text{Cl}^-]^2}{1 + K_2[\text{Cl}^-] + K_1K_2[\text{Cl}^-]^2}[\text{Pt}^{\text{II}}{}_{\text{total}}] \qquad (68)$$

where $K_1 = 1/K'_2$ and $K_2 = 1/K'_1$). For cyclohexane reacting at 100 °C, $k_1:k_2:k_3 = 1:0.14:0.006$. This shows that the principal catalytic species is the uncharged $[PtCl_2S_2]$ species. The solvolysis sequence can be extended (equations 69 and 70). Both $[PtClS_3]^+$ and $[PtS_4]^{2+}$ (when $S = H_2O$) have been shown to react more slowly than $[PtCl_2S_2]$ with alkanes[214].

$$[\text{PtCl}_2\text{S}_2] + \text{S} \rightleftharpoons [\text{PtClS}_3]^+ + \text{Cl}^- \qquad (69)$$

$$[\text{PtClS}_3]^+ + \text{S} \rightleftharpoons [\text{PtS}_4]^{2+} + \text{Cl}^- \qquad (70)$$

The nature of the ligand attached to platinum(II) is of some importance. The effect of varying the ligand L on the catalytic properties of complexes $[PtLCl_3]^-$ and $[PtL_2Cl_2]$ has shown the general order of catalytic activity[215]: $PPh_3 < py < dmso < CN^- < NO_2^- < NH_3 < I^- < Br^- < Cl^- < F^- < H_2O$. This list shows two interesting features. The ligands are in an order from very 'soft' (left-hand side) to 'hard' (right-hand side), and it is 'hard' ligands which promote catalytic efficiency. This pattern may result from the well known ability of 'soft' ligands to stabilize platinum(II) complexes from oxidation to platinum(IV) complexes. As we shall see later, oxidative additions to give platinum(IV) complexes are an important part of the catalytic cycle, and ligands which discourage oxidation should impede catalytic ability. The second factor is the way in which the ligand L influences the initial complexing ability of the catalyst for the alkane. On the 'hard likes hard' principle a platinum(II) with 'hard' ligands would be more likely to react with an alkane (a 'hard', non-polarizable molecule) than a platinum(II) complex containing 'soft' ligands.

The results of studies of relative reactivity of various alkanes to hydrogen–deuterium exchange with the solvent show that isolated methyl groups have the greatest reactivity and the order primary > secondary > tertiary accounts well for the relative reactivities of various C—H groups[213]. This order is the reverse of that found for reactions of alkanes with free radicals (see Section V). The reactivity of methyl groups in compounds such as **116** and **117**, where they are attached to a quaternary carbon, is especially low. These low

Me—C(Me)(Me)—Me

(116)

Et—C(Me)(Me)—Me

(117)

reactivities could reflect steric interactions in the bonding of the alkane to a platinum(II) before hydrogen–deuterium exchange. Primary alkyl derivatives of metals are known to be more stable than secondary, and tertiary derivatives considerably less so.

Hodges *et al.*[213] noted that there was a linear correlation between log k (k = rate coefficient for hydrogen–deuterium exchange with solvent) and the ionization potential of the alkane. Arenes also obeyed this relationship. This suggests that the electron-donor properties of the hydrocarbons are important in determining the rate of hydrogen–deuterium exchange, and that donor–acceptor interaction is an important step in the mechanism. It is important not to read too much into this log k/ionization potential relationship. It is far from certain that the first electron removed from a gaseous alkane is the one which, together with a second electron, provides the donor electron pair for interaction with platinum(II) in a polar solvent. The application of the Taft correlation equation (equation 71) has produced further clues to the mechanism[216]. The σ^* value is the

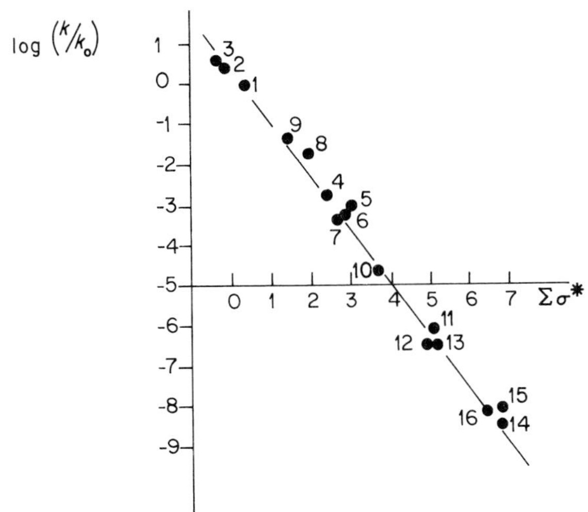

FIGURE 12. Dependence of the relative rate of H–D exchange on the Taft polar constant σ^*. 1 = Methane; 2 = ethane; 3 = propane; 4 = 1, 1, 1-trifluoroethane; 5 = fluoromethane; 6 = chloromethane; 7 = bromomethane; 8 = acetic acid; 9 = methanol; 10 = difluoromethane; 11 = dichloromethane; 12 = chlorofluoromethane; 13 = 1, 1-difluoroethane; 14 = trifluoromethane; 15 = trichloromethane; 16 = tribromomethane. After ref. 216.

polar parameter which characterizes the inductive effect of an alkyl substituent. Ψ is a 'resonance' term which characterizes the conjugation of the reacting centre with an α-substituent (n is the number of such α-substituents). The linear correlation obtained (Figure 12) with $\rho^* = 1.4$) indicates that the platinum(II) complex acts as a moderate electron acceptor with respect to the donor hydrocarbon.

$$\log(k/k^\circ) = \rho^*\sigma^* + n\psi \qquad (71)$$

Combination of the above observations leads to the proposed mechanism shown in Figure 13. At the heart of the exchange process is oxidative addition of the alkane, RH, to platinum(II) to give a platinum(IV) complex (steps 2 and 3). Hydrogen–deuterium exchange of the hydride ligand (steps 4 and 5) is followed by reductive elimination of the deuteriated alkane (steps 6 and 7). The exact nature of the preliminary coordination of RH to the platinum(II) (step 2) is uncertain. Compounds such as C in Figure 13 where RH is an arene are well established, but have not been characterized for alkanes. Interaction of the electron pair of a C—H bond with transition metals is discussed in Section II.B. It is not certain at present whether steps 2 and 3 are two separate steps as shown here or are combined into one concerted process.

Multiple exchange–Quantitative hydrogen–deuterium exchange experiments show that more than one hydrogen–deuterium atom can exchange while the alkane remains in the coordination sphere of the platinum(II) complex. The number of such exchanges taking place is expressed by the multiple exchange parameter, m, and for alkanes m values range from 1.3 to 2.0[213]. Multiple exchange will be noticed if complex G (Figure 13) reacts to

(a)

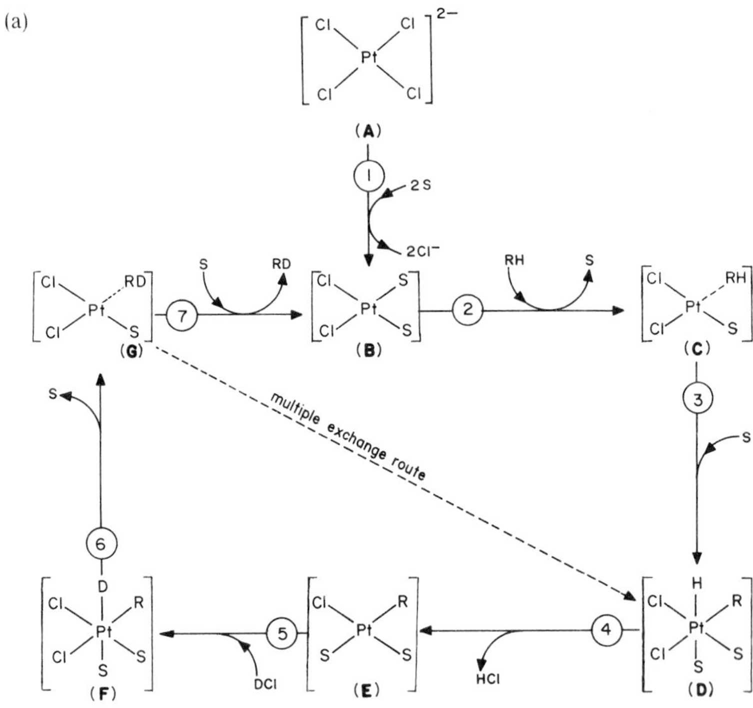

(b)

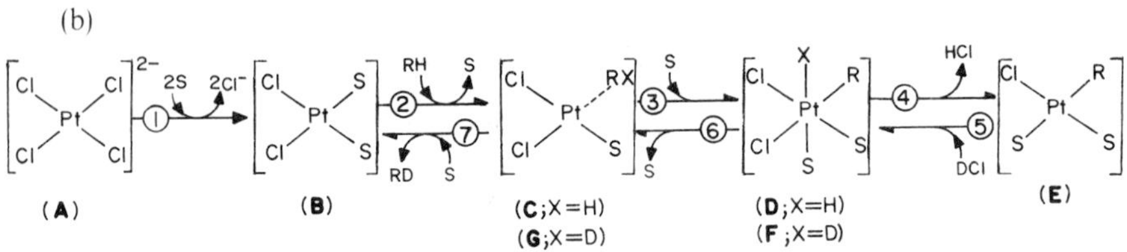

FIGURE 13. (A) Proposed mechanism of H–D exchange in alkanes and arenes catalysed by platinum(II), drawn as a cycle; (b) (a) drawn as an equilibrium.

form complex **D** by oxidative addition of a second C—H bond of the alkane (dotted line in Figure 13) rather than dissociation of the alkane to form complex **B**. This second hydrogen then can be replaced by deuterium before the alkane leaves the catalyst.

Another possibility that must be considered when explaining multiple exchange is the reversible formation of alkene complexes **119** from **118** (equivalent of **E** in Figure 13) with dissociation of a proton, H^+. This is thought to be unlikely as platinum(II)–alkene complexes are usually too stable to undergo reaction with a deuteron, D^+. Also, multiple exchange at methane cannot be explained by alkene complex formation, but could possibly result from formation of a carbene complex, **120**. A further suggestion for explanation of multiple exchange is the involvement of dimeric platinum complexes **121**, where two simultaneous Pt(II)–CH interactions can take place.

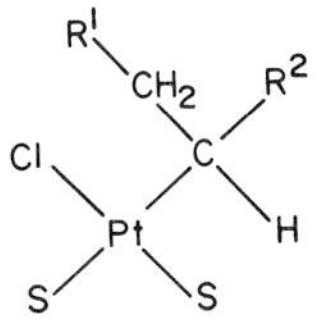

(118) 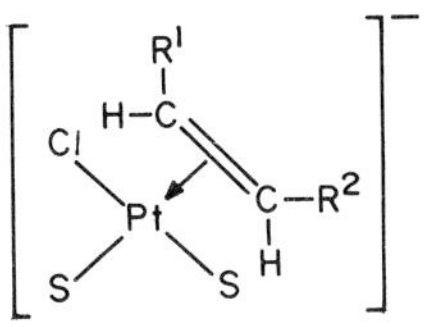 (119)

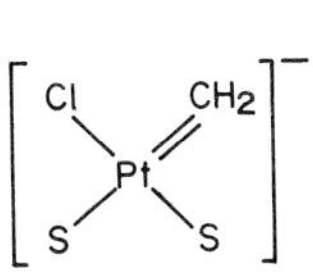

(120) 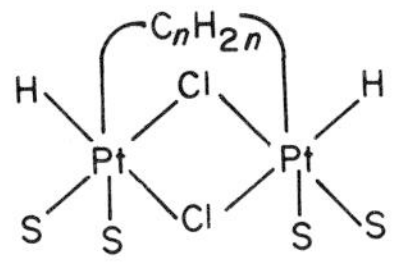 (121)

3. *Homogeneous or heterogeneous?*

Although there is a body of evidence to support the view that this reaction is a homogeneous process[9], there is clearly still doubt in the minds of a number of respected workers in this field who, when referring to this reaction, comment on its possible heterogeneity. The evidence for the process being homogeneous was discussed by Shilov[217], who concluded that 'experimental results leave no doubt that the H–D exchange is actually a homogeneous process'. Parshall[12] appears to be the first who 'has raised questions whether or not the exchange is catalysed by the metal', and others seem to have inherited his doubts. We are not aware of any experimental evidence that Shilov is wrong, but there is increasing interest in the possibility of catalysis by small clusters and colloids, and experimental methods are now being used to study them (see, for example, Crabtree *et al.*[168]) and further experimental light may be shed on this problem in the near future.

B. Alkane Oxidation

1. *Introduction*

The hydrogen–deuterium exchange of alkanes with the solvent discussed in Section A is of some interest, but the products of the reaction are not particularly useful. Activation of alkanes normally implies introduction of functional groups in place of a carbon—hydrogen bond. In 1968 Hodges and Garnett showed that benzene can be oxidized in a stoichiometric reaction to chlorobenzene by platinum(IV) (equation 72)[218]. They also showed that this reaction takes place only when platinum(II) is present.

$$[PtCl_6]^{2-} + PhH \longrightarrow PhCl + HCl + [PtCl_4]^{2-} \qquad (72)$$

2. *Alkane oxidation*

Alkanes also react with platinum(IV) in the presence of platinum(II) to form various products including mainly chloroalkanes, but also alcohols, esters, ketones, acids, and

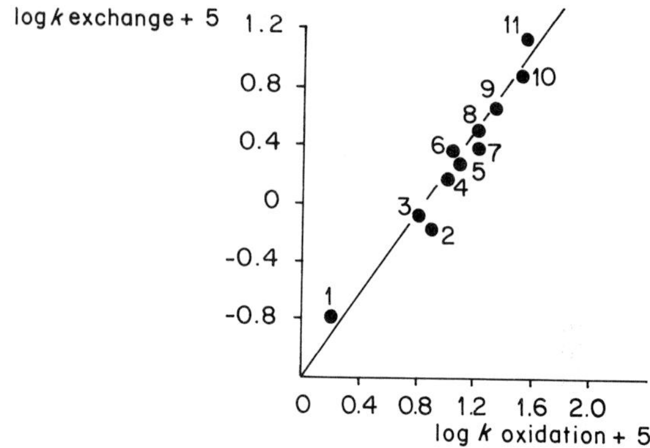

FIGURE 14. Correlation between the rates of H–D exchange and oxidation. 1 = Methane; 2 = ethane; 3 = isobutane; 4 = propane; 5 = 3-methylpentane; 6 = *n*-butane; 7 = *n*-pentane; 8 = *n*-hexane; 9 = cyclopentane; 10 = cyclohexane; 11 = benzene. After ref. 220.

products of oxidative coupling of two hydrocarbon radicals[219–223]. Cyclic alkanes such as cyclohexane and decalin give some aromatic products by a dehydrogenation reaction. As is common in oxidations the first product, here a chloroalkane, is more readily oxidized than the initial alkane reactant, and a variety of products are formed as a result of secondary oxidations. There is good evidence that chloroalkanes are the initial products. If chlorocyclohexane is oxidized by platinum(IV) in the presence of platinum(II), it is dehydrogenated and benzene is formed as a product. No chlorobenzene is formed. Subsequent reaction of the chlorocyclohexane is at the reactive carbon—chlorine bond, prior to reaction at the less reactive carbon—hydrogen bond. If a chloroalkane is formed which is particularly stable to oxidation, then it can be formed in significant amounts. An example is the formation of chloroacetic acid from the oxidation of acetic acid by platinum(IV). When hexane is oxidized with platinum(IV) no 1-chlorohexane is found in the products, as the chlorohexane oxidizes faster than the hexane. However, if a mixture of 1- , 2- , and 3-chlorohexane is oxidized with platinum(IV) the reaction mixture contains 1-chlorohexane, when the 2- and 3-chlorohexanes have been consumed, as non-terminal carbon—chlorine bonds are oxidized faster than terminal bonds[223].

The relative reactivity of a number of alkanes with respect to oxidation by platinum(IV) has been studied, and provides good evidence for the close relationship between the platinum(IV) oxidation of alkanes and the platinum(II)-catalysed hydrogen–deuterium exchanges discussed earlier. Figure 14 shows that there is an excellent correlation between the rates of hydrogen–deuterium exchange [catalysed by platinum(II)] and the rates of platinum(IV) oxidation [also catalysed by platinum(II)][220]. Such linear relationships often imply close similarity of the mechanisms of the reactions.

Carbon—hydrogen bonds which are adjacent to a quaternary carbon atom are inert to hydrogen–deuterium exchange, and are similarly resistant to platinum(IV) oxidation. Compounds **122** and **123** are thus not oxidized by platinum(IV). The other reactivity

$$\mathrm{Me{-}C(Me)_2{-}Me}\qquad\qquad \mathrm{Me{-}C(Me)_2{-}C(Me)_2{-}Me}$$

(122) **(123)**

TABLE 4. Rate coefficients for the oxidation of hydrocarbons in aqueous solution containing Pt^{II} and Pt^{IV}[9,224]. $[K_2PtCl_6] = 0.05$ mol dm^3; $[H_2PtCl_6] = 0.02$ mol dm^3; 98 °C

Hydrocarbon	$10^5 k_{ox}/s^{-1}$	n_0[a]	$10^5 k_{ox}/n_o$
Methane	1.6	1	1.6
Ethane	6.6	2	3.3
Propane	9.9	3	3.3
n-Butane	11.5	4	2.9
2-Methylpropane	8.2	3	2.7
n-Pentane	15.3	5	3.1
2, 2-Dimethylpropane	1.0	0	—
2-Methylbutane	11.2	4	2.8
n-Hexane	15.9	6	5.0
Cyclohexane	30	6	5.0
Benzene	35.8	6	6.0

[a]Number of 'attackable' C—H bonds = total number of C atoms – number of tertiary Cs – 5 (number of quaternary Cs).

trends (see Table 4) reflect the number of 'attackable' carbon—hydrogen bonds, taking into account possible hydrophilic 'rolling up' where in long-chain alkanes one section of the molecule screens another[224,225]. Overall, the range of rate coefficients relating to the oxidation of reactive carbon—hydrogen bonds in alkanes differs very little over a range of alkanes. The nature of the platinum(II) species necessary for oxidation by platinum(IV) to take place has been established. Of the possible platinum(II) species in solution, neither $[PtS_4]^{2+}$ nor $[PtCl_4]^{2-}$ is active whereas $[PtS_3Cl]^+$, $[PtS_2Cl_2]$, and $[PtSCl_3]^-$ are almost equally effective.

In a study of concurrent hydrogen–deuterium exchange and platinum(IV) oxidation it was shown[226] that (1) the sum of exchange and oxidation rates does not depend on the concentration of platinum(IV), but the oxidation–exchange ratio increases as the concentration of platinum(IV) is increased; (2) with increase in concentration of platinum(IV) [and constant concentrations of acid and platinum(II)] the rate of oxidation increases, but with increase in concentration of acid but constant concentrations of platinum(II) and platinum(IV) the rate of exchange reaction increases. These observations imply a common first step, followed by a choice of either exchange or oxidation (equation 73).

$$
\begin{array}{c}
Pt^{II} + RH \longrightarrow RPtCl + H^+ + Cl^- \\
\swarrow D^+ \qquad \searrow Pt^{IV} \\
Pt^{II} + RD \qquad \text{oxidation products} + Pt^{II}
\end{array} \tag{73}
$$

In the absence of added platinum(II) the reaction proceeds autocatalytically as platinum(II) is formed from the reduction of platinum(IV). As already suggested for the platinum(II)-catalysed hydrogen–deuterium exchange, it is likely that the platinum(II) species always has four ligands around it, and that solvent molecules (S) coordinate to platinum(II) where necessary to maintain this coordination. Similarly, it seems probable that platinum(IV) species are always 6-fold coordinated with S molecules occupying spare coordination sites. The suggested mechanism for the oxidation is shown in Figure 15. In

(a)

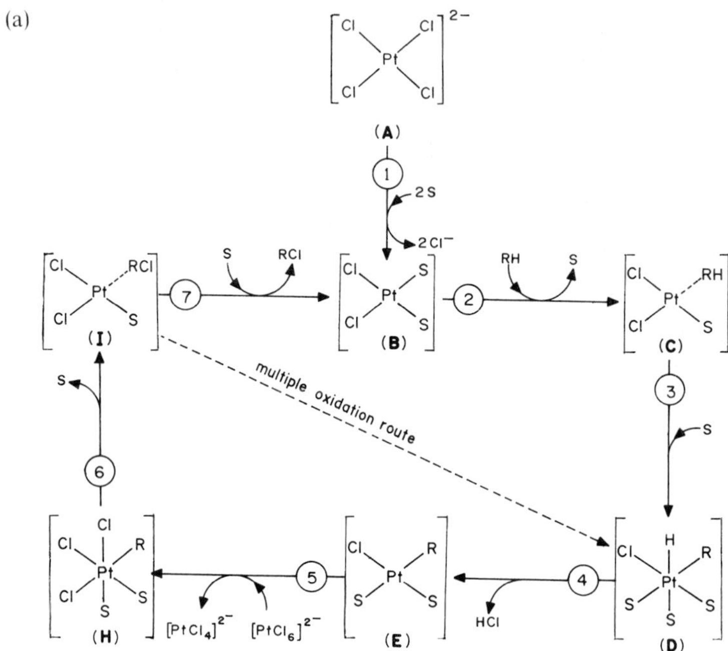

(b)

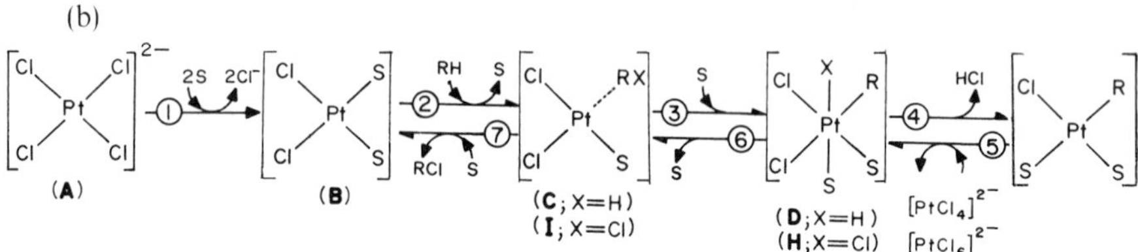

FIGURE 15. (A) Proposed mechanism of alkane and arene oxidations by platinum(IV), drawn as a cycle; (b) (a) drawn as an equilibrium. X = H or Cl.

this scheme reactions 1–4 are the same as for the hydrogen–deuterium exchange. In the oxidations the complex **E** reacts with $[PtCl_6]^{2-}$ to undergo an oxidative chlorination rather than DCl oxidative addition. The kinetics can be derived from the equations (74–77) associated with the reactions in Figure 15.

$$[PtCl_4]^{2-} + S \rightleftharpoons [PtCl_3S]^- + Cl^- \tag{74}$$

$$[PtCl_3S]^- + S \underset{\text{rapid}}{\overset{K}{\rightleftharpoons}} [PtS_2Cl_2] + Cl^- \tag{75}$$

$$[PtS_2Cl_2] + RH \underset{k_{-1}}{\overset{k_1}{\rightleftharpoons}} \underset{\textbf{(124)}}{[RPtClS_2]} + H^+ + Cl^- \tag{76}$$

$$[PtCl_6]^{2-} + \underset{\textbf{(124)}}{[RPtClS_2]} \xrightleftharpoons{k_2} RCl + 2[PtCl_3S]^- \tag{77}$$

Applying the usual 'steady-state' approximation for complex **124**, the rate of reaction is given by equation 78, where $[Pt^{II}]$ is the total added platinum(II) catalyst.

$$\text{rate} = \frac{k_1 k_2 K[RH][Pt^{II}][PtCl_6{}^{2-}]}{([Cl^-] + K)(k_{-1}[H^+][Cl^-] + k_2[PtCl_6{}^{2-}])} \tag{78}$$

As noted above for hydrogen–deuterium exchange, it is not clear whether steps 2, 3, and 4 and steps 5, 6, and 7 (Figure 15) are single concerted reactions (as used to calculate the rate expression above) or separate reactions. At high concentrations of platinum(IV) and low temperatures ($< 100\,°C$) reaction 76 is the rate-limiting step, whereas at higher temperatures and lower platinum(IV) concentrations reaction 77 is rate limiting.

It is possible that the reaction between $[PtCl_6]^{2-}$ and $[RPtS_2Cl]$ involves the formation of a binuclear intermediate.

3. *Isolation of intermediates*

Both the mechanism of the platinum(II)-catalysed hydrogen–deuterium exchange of alkanes and that of platinum(IV) oxidation of alkanes incorporate the formation of a platinum–alkyl complex. Normally, reaction intermediates in catalytic cycles are difficult to isolate. If they are stable enough to be isolated the catalytic cycle would be stopped. Nonetheless, it has been possible to use n.m.r. techniques to show that platinum alkyls are formed during these reactions[227]. Figure 16 shows the n.m.r. spectrum of a mixture of $[PtCl_6]^{2-}$ (0.32 mol l^{-1}) and $[PtCl_4]^{2-}$ (0.24 mol l^{-1}) in D_2O after reaction with methane for 30 min at 120 °C. Also shown is the spectrum of the reaction product of the reaction of methyl iodide with $[PtCl_4]^{2-}$ in water {a platinum(IV) derivative, either $[MePtCl_5]^{2-}$ or $[MePtCl_4(H_2O)]$, formed by oxidative addition to platinum(II)}. The similarity of these two spectra shows that oxidative addition of methane to platinum(II) does give a methylplatinum(IV) derivative. Further confirmation of this is the isolation of a complex $[MePt^{IV}(Ph_3P)_2Cl_3]$ when triphenylphosphine is added to a mixture of methane and platinum(IV) in aqueous acetic acid solution at 120 °C[228].

C. Arene Reactions

1. *Hydrogen–deuterium exchange*

In the study of activation of carbon—hydrogen bonds it is not surprizing that more work has been done on the activation of aromatic compounds than on the activation of saturated hydrocarbons. The work described above on platinum(II)-catalysed hydrogen–deuterium exchange and on platinum(II)-catalysed oxidation by platinum(IV) was preceded by similar studies on arenes. The hydrogen–deuterium exchange of arenes in acidic solution (equation 79) shows the same features as discussed above for alkanes. The

$$R\text{-}C_6H_4\text{-}H \xrightarrow[{[PtCl_4]^{2-}}]{D_2O,\ MeCO_2D} R\text{-}C_6H_4\text{-}D \tag{79}$$

rate of exchange is proportional to the concentration of benzene and to the concentration of $[PtCl_4]^{2-}$ and inversely proportional to the concentration of added chloride.

The mechanisms proposed for arene hydrogen–deuterium exchange have been discussed in depth[1]. The interaction of aryl derivatives with platinum(II) is well known and

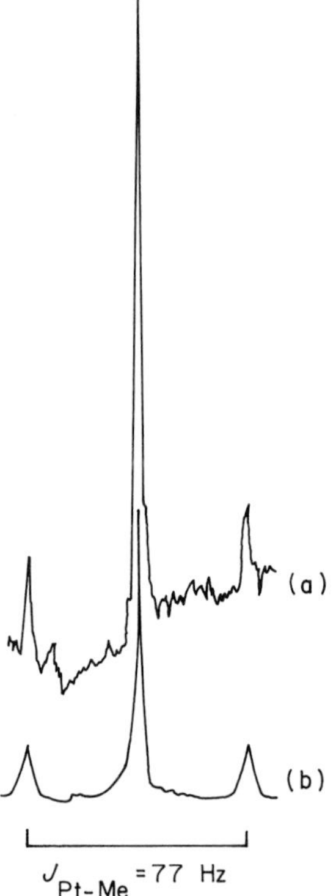

FIGURE 16. (a) The n.m.r. spectrum of a Me—Pt(IV) complex obtained by oxidation of methane; (b) the oxidative addition product of MeI and $[PtCl_4]^{2-}$. After ref. 227.

complexes such as **C** (Figure 13) can be written with more confidence for the initial interaction of platinum(II) with benzene than with an alkane. The hydrogen–deuterium exchange for benzene could well proceed via the mechanism shown in Figure 13. However, there are further possibilities for hydrogen–deuterium exchange and intermediates such as **125** and **126** have been proposed. The principal way in which hydrogen–deuterium

(**125**) (**126**)

exchange in benzene differs from that of alkanes is in the higher values (up to 3.5) found for the multiple exchange parameter[213]. This would naturally appear to fit in well with the greater stability of arylmetal derivatives compared with alkyl derivatives, so that once the arene has been joined to platinum (**C** in Figure 13) reaction to give complexes **D** and **E** (see Figure 13) is more likely than loss of hydrocarbon.

2. Oxidations

Oxidations also parallel those found for alkanes. Figure 17 shows time–concentration curves for the reaction of benzene with $H_2[PtCl_6]$ in aqueous trifluoroacetic acid at 120 °C[223]. Reaction is thought to proceed as shown in Figure 15. The profile of reaction components shows that **D** (benzene complex) and **H** (chlorobenzene complex) are stable enough to be present at fairly high concentrations. After long reaction times dichloro-

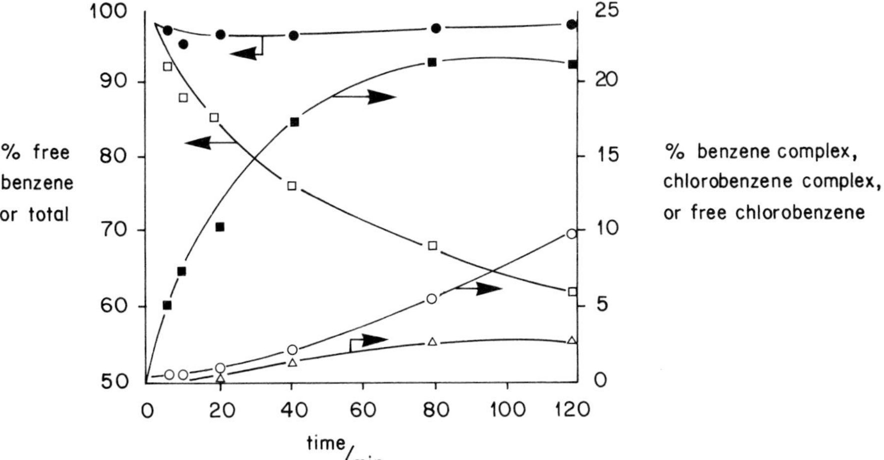

FIGURE 17. Molar percentages of aromatic compounds in a typical reaction mixture at 120 °C with reactants $H_2[PtCl_6]$ and benzene in water–trifluoroacetic acid solution. (□) Benzene; (■) benzene complex; (△) chlorobenzene complex; (○) chlorobenzene; (●) total. After ref. 223.

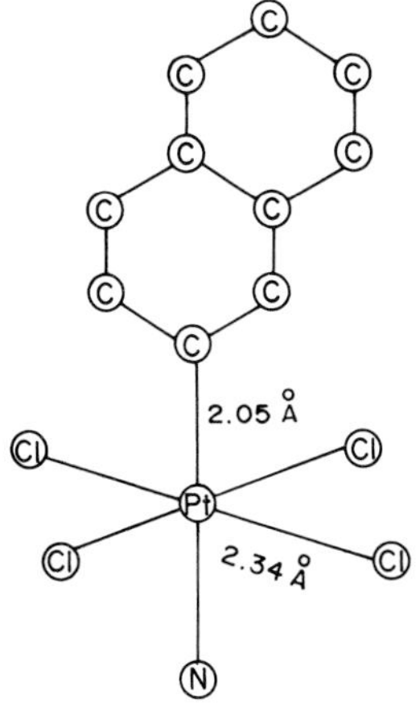

FIGURE 18. The platinum(IV) complex produced by the reaction of $H_2[PtCl_4]$ with naphthalene.

benzene can be detected. The formation of this is analogous to multiple exchange in the platinum(II)-catalysed hydrogen–deuterium exchange reaction of benzene. Instead of dissociation of RCl (step 7) from complex **I**, the reaction path follows the dotted line when a second carbon—hydrogen bond is replaced by a carbon—chlorine bond.

As, in general, aryl derivatives of platinum(II) are more stable than alkyl derivatives, it is not surprizing that there has been greater success in isolating them from arene oxidations[229–233]. Figure 18 shows the structure of the naphthylplatinum(IV) derivative formed in the reactions of $H_2[PtCl_6]$ with naphthalene (the nitrogen atom on the lower coordination position of platinum in Figure 18 results from an ammonia–water exchange during the chromatographic separation).

IV. ACTIVATION BY METAL ATOMS AND IONS

A. Activation by Metal Atoms

During the past decade, experimental techniques using metal atoms have been improved greatly. There have been spectroscopic studies of a range of metal atoms in a variety of matrices, and techniques have been developed to use metal atoms on a preparative scale to make organometallic compounds[234].

Metal atoms react readily with alkenes, polyenes, and arenes. For example, co-condensation of transition metals such as cobalt, nickel, or palladium with a large excess of ethylene at 50–77 K gives molecular complexes (equation 80), and direct co-condensation of the metal vapour with ethylene at $-196\,^{\circ}C$ gives good yields of $[M(C_2H_4)_3]$ in millimolar quantities. Alkanes, in general, do not react with metal atoms under these conditions. Indeed, alkanes ranging from methane to docosane, $C_{22}H_{46}$, have been used as inert media for co-condensing metal atoms and metal atom clusters from metal vapours[235].

$$M_{(g)} + C_2H_{4(g)} \xrightarrow[50-70\,K]{\text{co-condense}} [M(C_2H_4)_3]_{(s)} \qquad (80)$$

$$(M = Co, Ni, Pd)$$

By using higher concentrations of metal atoms it is possible to produce small clusters as well as the isolated atoms in a suitable matrix. Mössbauer spectroscopy has shown that isolated iron atoms in a methane matrix have the same isomer shifts as in an inert gas matrix. Higher concentrations of iron atoms lead to the formation of Fe_2 dimers as well as iron atoms, and the Mössbauer spectrum of these dimers in methane at 20 K shows lines characteristic of both covalent diamagnetic compounds of iron and ionic compounds of iron(II)[236]. Reactions 81 and 82 have been proposed. Infrared studies confirmed the presence of Fe—H bonds in the products, and suggested that at least two different products are formed at different Fe_2 to CH_4 ratios.

$$Fe_2 + CH_4 \xrightarrow{12\,K} HFeFeMe \text{ or } FeFe(H)Me \qquad (81)$$

$$FeMe \longrightarrow (H)Fe{=}CH_2 \qquad (82)$$

Small nickel clusters (<35 Å) prepared in alkane matrices at $-196\,^{\circ}C$ result in stable organonickel compounds of unknown nature[237,238]. Nickel atoms by themselves do not react with alkanes under these conditions. None of the metal atoms from magnesium, calcium, titanium, chromium, iron, cobalt, nickel, palladium, copper, silver, gallium, indium, or tin react with alkanes when the alkane and the metal are co-condensed at 10 K. However, co-deposition of methane with aluminium atoms at 10 K leads to HAlMe as a primary product[239]. The reason why aluminium atoms, and not atoms of other metals, will activate and cleave aliphatic C—H bonds in methane without photoactivation (see next paragraph) is thought to be because aluminium atoms are in a $^2P(3s^23p^1)$ ground state, which gives them radical-like properties. A second possible reason is that the Al—H and Al—Me bonds formed are comparatively strong (285 and 275 kJ mol^{-1}, respectively).

Rhenium atoms will react with alkyl-substituted benzenes to give products where two sp^3 C—H bonds have reacted[240]. For example, the product from the reaction of gaseous rhenium atoms with toluene is shown in structure **127** ($R^1 = R^2 = H$, $R^3 = Ph$). Similar products were observed with mesitylene or *p*-xylene. The products are all deep-red, air-sensitive compounds.

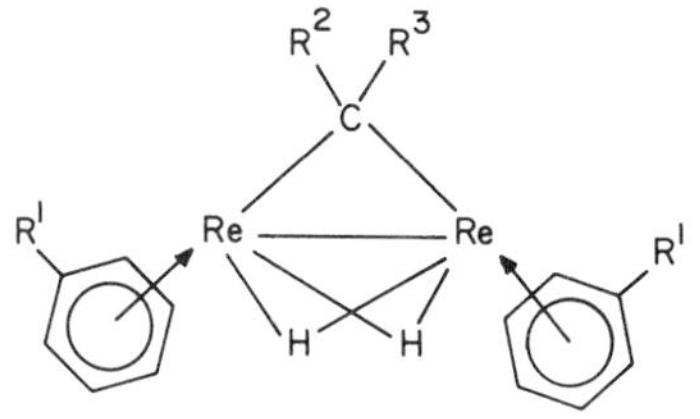

(**127**)

Although rhenium atoms do not react with alkanes when co-condensed, it has been shown that the alkane may be activated by mixing it with an arene. Thus, when rhenium atoms are allowed to react with a mixture of an alkane, such as ethane, butane, or cyclohexane, and benzene, complexes are formed with structures **127** where $R^1 = H$ and $CR^2R^3 = CHMe$, $CHPr^n$, and cyclohexylidine, respectively[241].

Although most metal atoms do not react when co-condensed with alkanes, reaction can be initiated by appropriate u.v. irradiation of the metal atom–alkane matrix. Thus, at 15 K iron, manganese, cobalt, copper, silver, gold, and zinc atoms react with methane when irradiated with u.v. radiation (wavelength <360 nm), and the products formed are those expected by an oxidative addition (equation 83)[242,243].

$$M + CH_4 \xrightarrow{15K, h\nu} HMMe \qquad (83)$$

In an ethane matrix, copper atoms under photoirradiation cleave the C—H bonds but not the C—C bonds (equation 84)[244]. Photoexcited copper atoms are in the same electronic state as aluminium atoms in the ground state and thus undergo a similar reaction.

$$Cu + C_2H_6 \xrightarrow[12K]{h\nu, 300-400\ nm} Cu(C_2H_6) \rightarrow HCuEt \qquad (84)$$

The photoinitiated reaction of alkanes with metal atoms is a reversible process. Thus, iron atoms will activate methane when irradiated at 300 nm, but if the product is irradiated at 420 nm the reverse reaction takes place (equation 85)[245].

$$Fe + CH_4 \underset{420\ nm}{\overset{300\ nm}{\rightleftharpoons}} HFeMe \qquad (85)$$

Photoactivation of metal atoms can be used at room temperature to activate alkanes. When a silver-loaded zeolite-Y containing both isolated silver atoms and low nuclearity silver clusters is irradiated at 220–300 nm in the presence of alkanes, the alkanes dimerize[246]. Methane gives ethane, ethane yields *n*-butane, and propane gives a mixture of hexane isomers.

B. Activation by Metal Ions

Owing to the great advances in instrumentation in recent years it is now possible to make very detailed studies of reactions in the gas phase. Among the large number of reactions which have been studied are the reactions of gaseous metal ions with various hydrocarbons including alkanes (equation 86).

$$M^+_{(g)} + C_nH_{2n+2(g)} \longrightarrow \text{products} \tag{86}$$

The techniques used for these studies include ion-beam methods, Fourier transform mass spectrometry, and ion cyclotron resonance spectroscopy. It is not appropriate to review these techniques here, but to comment that such experiments give useful information on the mode of interactions taking place. It must be noted that using these methods it is not possible to make useful quantities of compounds so that these are not preparative methods. The reactions of metal ions have been studied in the gas phase with linear alkanes[247–252], branched alkanes[253–255], and cyclic alkanes[256–258].

Cleavage of alkanes and dehydrogenation are the two most common processes observed in these reactions. This is in contrast with reactions in solution where, in the presence of transition metal compounds, C—C bond formation is more common than C—C cleavage. In the gas phase the ion–molecule complex is activated by excess internal energy. This energy together with the M—C or M—H bond energies may be sufficient to allow for oxidative addition across the C—C or C—H bonds of the alkane. Thus, when a gaseous metal ion interacts with an alkane, oxidative addition can take place involving a C—H bond (equation 87) or a C—C bond (equation 88).

$$M^+_{(g)} + C_2H_{6(g)} \longrightarrow [\text{HMEt}]_{(g)} \tag{87}$$

$$M^+_{(g)} + C_2H_{6(g)} \longrightarrow [\text{MeMMe}]_{(g)} \tag{88}$$

These initial reaction products then decompose with loss of hydrogen or alkane.

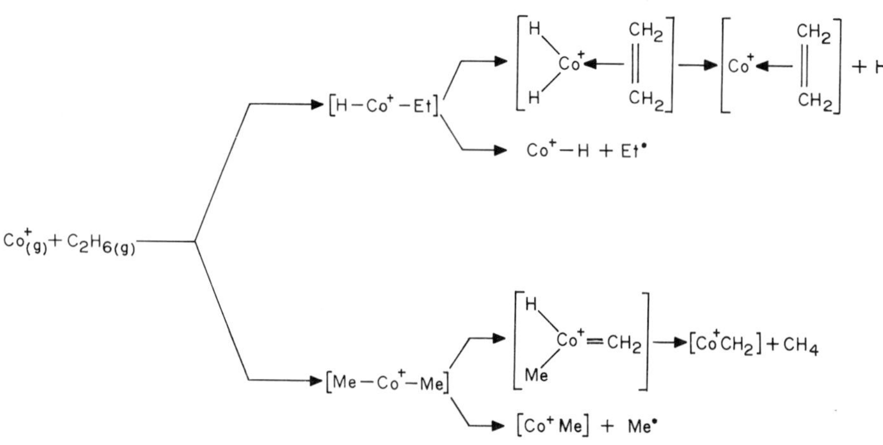

FIGURE 19. Reaction pathways for the interaction of Co^+ ions with ethane.

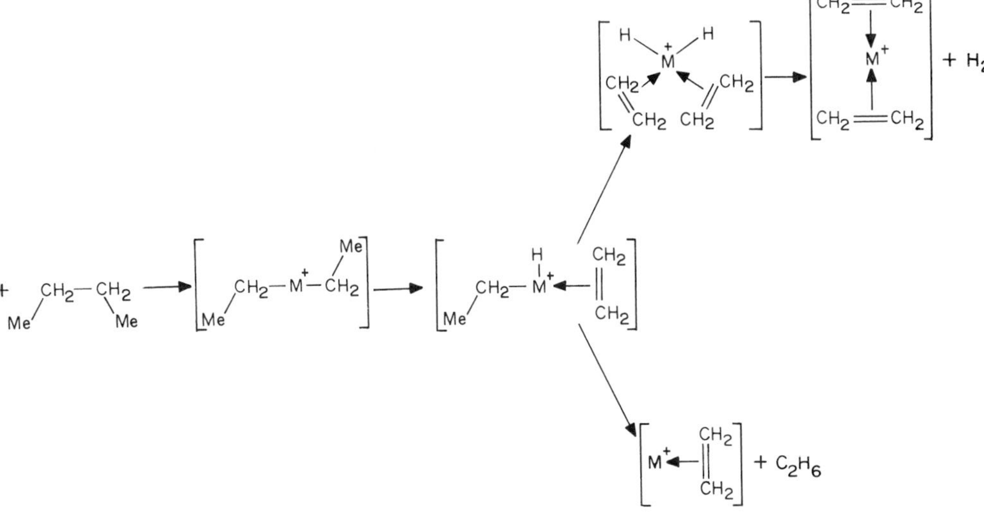

FIGURE 20. The reaction of $M^+_{(g)}$ (M = Fe or Ni) ions with *n*-butane.

Figure 19 shows the proposed reaction pathways for the interaction of the $Co^+_{(g)}$ ion with ethane[251]. The choice of route of either C—C or C—H cleavage is determined in part by the electronic configuration of the metal ion. The reactions of the $Ti^+_{(g)}$ ion are dominated by C—H insertions (equation 87)[249]. The s^1d^2 electronic configuration of the Ti^+ ion has more orbital vacancies than the $Fe^+_{(g)}$ ion, an s^1d^6 system, whose reactions with alkanes involve competition between the C—C and C—H cleavage routes. It is pertinent to recall that intermediates with a Ti—H bond are important species in the catalysis of the oligomerization of olefins. The $Co^+_{(g)}$ ion is found to undergo C—H insertion more readily than the $Fe^+_{(g)}$ ion and some hydrogen gas is produced in all cases when it reacts with an alkane.

The reaction of $M^+_{(g)}$ (M = Fe or Ni) ions with *n*-butane is shown in Figure 20. The initial reaction is C—C cleavage of butane into two C_2 units, with no C—H cleavage[259]. The electronic state of the metal ion is of importance. The Cr^+ ion usually reacts with methane to give CrH^+ as the only product in an endothermic reaction. When the Cr^+ ion is prepared in an electronically excited state by electron impact on chromium hexacarbonyl, it reacts with methane to give an abundance of the $CrCH_2^+$ ion in an exothermic reaction[260].

Although the dimeric $Co^+_{2(g)}$ ion is unreactive towards alkanes, addition of a carbon monoxide ligand to $Co^+_{2(g)}$ forms Co_2CO^+, which will attack alkane C—H bonds; its reaction with *n*-butane is shown in equation 89[261].

$$Co_2CO^+ + n\text{-}C_4H_{10} \xrightarrow{90\%} [Co_2COC_4H_6]^+ + 2H_2$$
$$Co_2CO^+ + n\text{-}C_4H_{10} \xrightarrow{10\%} [Co_2COC_4H_8]^+ + H_2 \qquad (89)$$

Both the ions $Co_{2(g)}^{+}$ and $Co_2CO_{(g)}^{+}$ are made by electron impact on dicobalt octacarbonyl. It is suggested that the role of the carbon monoxide in the oxidative addition of $Co_2(CO)^+$ with an alkane to give **128** is to help to concentrate positive charge on the

$$\mathrm{OCCo\overset{+}{Co}}\begin{matrix}\diagup \mathrm{H} \\ \diagdown \mathrm{R}\end{matrix}$$

(**128**)

cobalt atom undergoing the oxidative addition by electron withdrawal into the antibonding π^* orbitals of the carbon monoxide. The ion $FeCo_2^+$ is more reactive to alkanes than either $FeCo^+$ or Co_2^+. Although there is no reaction with methane, ethane, or neopentane, other alkanes (e.g., propane, butane, isobutane, or 2-methylbutane) react by insertion across a C—H bond followed by hydrogen elimination[262].

The ions MeM^+ (M = Fe, Co) have been studied in reactions with alkanes. The cobalt ion $MeCo^+$ reacts with all alkanes larger than ethane, whereas the $MeFe^+$ ion does not react with aliphatic alkanes. Both $MeCo^+$ and $MeFe^+$ ions react with cyclic alkanes to give ring cleavage products which involve the formation and subsequent breakdown of cyclometallated ring compounds[263,264].

V. OXIDATION OF ALKANES BY TRANSITION METAL COMPOUNDS

A. Introduction

The oxidation of alkanes can involve a transition metal compound in one of two ways. Either a compound such as chromic acid can directly oxidize an alkane, or a transition metal compound can act as a catalyst of the oxidation of an alkane by molecular oxygen in a catalysed autooxidation. These two processes often proceed by similar mechanisms, and will be discussed as appropriate below.

It is convenient to divide alkane oxidations into two groups: (a) those involving oxometal (M=O) reagents such as chromic acid and potassium permanganate, and (b) those involving other salts and complexes.

This topic has been extensively reviewed elsewhere[4], and the treatment given here is brief.

B. 'Hard' Metal Oxidations

1. Introduction

There are, in principle, two mechanisms for the oxidation of alkane C—H bonds by compounds such as chromic acid, chromyl chloride, and potassium permanganate. The first involves initial formation of an alkyl radical by hydrogen atom abstraction from the alkane (equation 90). This is followed by further reaction of the $R^\bullet$ radical with $[MOH]^{(n-1)+}$ (equation 91). The second possibility involves an electrophilic attack by $[M{=}O]^{n+}$ on the RH group (equation 92). At present the first mechanism is thought to take place in many reactions as the second involves the formation of an organometallic intermediate containing an alkyl group bonded to a transition metal in a high oxidation state.

$$[M{=}O]^{n+} + RH \rightarrow [MOH]^{(n-1)+} + R^\bullet \tag{90}$$

$$[MOH]^{(n-1)+} + R^\bullet \rightarrow M^{(n-2)+} + ROH \tag{91}$$

$$M{=}O \;(R{-}H) \longrightarrow M{-}OH \,(|\,R) \longrightarrow ROH + M^{(n-2)+} \tag{92}$$

2. Oxidation by chromium(VI) compounds

It has long been known that chromic acid and chromyl chloride are powerful oxidizing agents which are capable of oxidising $sp^3C{-}H$ groups. The oxidation of the methyl group of toluene to form benzaldehyde (equation 93) is the familiar Etard reaction. In these reactions the relative ease of oxidation of C—H bonds is primary < secondary < tertiary. It is likely that the first stage of the reaction is hydrogen atom abstraction from the alkane by the chromium(VI) species (equation 90)[265]. The alkyl radical formed must stay in the solvation sphere of the chromium, however, as experiments with chiral alkanes do not show the racemization of products expected if the alkyl radical is free in the solution. It is suggested that there is a 'solvent cage' associated with the chromium species in which the radical is constrained.

$$\mathrm{PhMe} \xrightarrow{\mathrm{CrO_2Cl_2}} \mathrm{PhCHO} \tag{93}$$

The oxidation of alkanes by oxymetal compounds is also discussed in Section VI, where alkane oxidation by analogues of cytochrome-P450 is shown to proceed via oxometal compounds.

3. Oxidation by cobalt(III) compounds

As indicated above, oxidation of alkanes by transition metal compounds can be a direct process (equation 94) or a catalysed autooxidation (equation 95). Thus cyclohexane is readily oxidized by cobalt(III) acetate in acetic acid at 80 °C to give cyclohexyl acetate (equation 96)[266], while autooxidation in the presence of a cobalt(II) catalyst yields adipic acid as the major product (equation 97)[267–270]. Methylcyclohexane is less reactive than cyclohexane, and the reactivity order is tertiary < secondary < primary.

$$\mathrm{RH} + \mathrm{M}^{n+} \longrightarrow \text{oxidation products} + \mathrm{M}^{(n-1)+} \tag{94}$$

$$\mathrm{RH} + \mathrm{O_2M}^{n+} \xrightarrow{\text{catalyst}} \text{oxidation products} \tag{95}$$

$$\text{cyclohexane} + \mathrm{Co^{III}(OAc)_3} \longrightarrow \text{cyclohexyl(OAc)(H)} \tag{96}$$

$$\text{cyclohexane} + \mathrm{O_2} \xrightarrow{\mathrm{Co^{II}(OAc)_2}} \mathrm{HO_2C(CH_2)_4CO_2H} \tag{97}$$

The proposed mechanism for such reactions is given by equations 98–101. The absence of any deuterium kinetic isotope effect indicates that the reaction rate is governed by equation 98 and not by equation 99.

$$\mathrm{RH} + \mathrm{Co^{III}} \underset{k_{-1}}{\overset{k_1}{\rightleftharpoons}} \mathrm{RH^{+\bullet}} + \mathrm{Co^{II}} \tag{98}$$

$$RH^{+\cdot} \xrightarrow{k_2} R^{\cdot} + H^+ \tag{99}$$

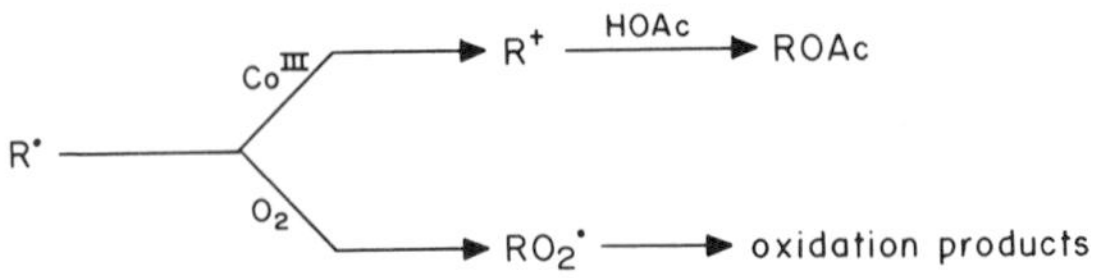

$$R^{\cdot} \xrightarrow{Co^{III}} R^+ \xrightarrow{HOAc} ROAc \tag{100}$$

$$R^{\cdot} \xrightarrow{O_2} RO_2^{\cdot} \longrightarrow \text{oxidation products} \tag{101}$$

For alkylarenes a deuterium isotope effect is observed, and probably equation 99 is rate limiting. Although benzene is usually significantly easier to oxidize than alkanes (the ionization potential of benzene is lower than that of cyclohexane) it cannot form a stable radical by proton loss (equation 99) and $k_{-1} > k_2$.

Studies of the relative reactivity of a number of cycloalkanes to oxidation by cobalt(III) were interpreted to imply that a complex is formed between cobalt(III) and the cycloalkane, and that the relative reactivities can be explained by steric hindrance in the formation of the complex (equation 102).

$$RH + CO^{III} \rightarrow RH \cdots Co(III) \rightarrow RH^+ + Co^{II} \tag{102}$$

This formation of a complex could be an example of a general class of electrophilic substitutions at a saturated carbon atom (equation 103). This type of reaction, with the

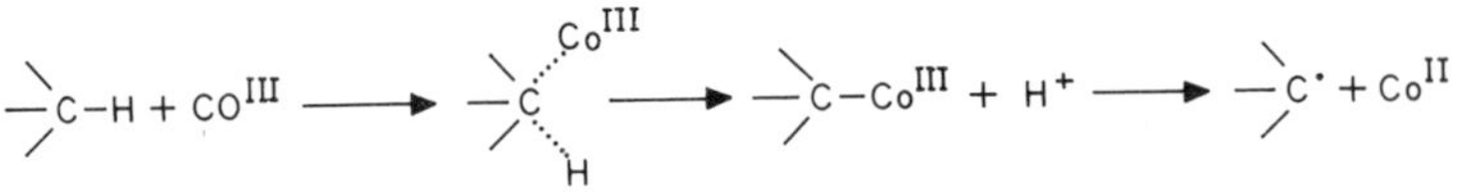

$$\rangle C{-}H + Co^{III} \longrightarrow \rangle C\begin{smallmatrix} \cdots Co^{III} \\ \cdots H \end{smallmatrix} \longrightarrow \rangle C{-}Co^{III} + H^+ \longrightarrow \rangle C^{\cdot} + Co^{II} \tag{103}$$

formation of an alkane–cobalt(III) complex, was also supported by a study of the oxidation of bridgehead hydrocarbons, the adamantanes (**129**; R = alkyl), by cobalt(III), manganese(III), and lead(IV) acetates in a trifluoroacetic acid solvent[271]. The product

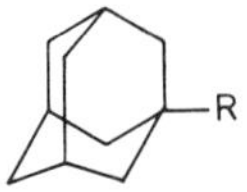

(**129**)

distribution from metal acetate oxidation was compared with that found in electrochemical oxidation, where a radical cation is formed which loses a proton to give an alkyl radical (equations 102 and 103). These studies indicate that the metal acetate oxidation proceeds by a different mechanism from that of the electrochemical oxidation (Figure 21). Adamantane derivatives have lower ionization potentials (adamantane, 9.20 eV) than cyclohexane (10.3 eV) and linear alkanes (hexane, 10.4 eV), and the balance between an electron transfer mechanism and complex formation mechanism could well change as the substrate undergoing oxidation is varied.

$$RH \longrightarrow RH^{+\cdot} + e^- \tag{104}$$

$$RH^{+\cdot} \longrightarrow R^{\cdot} + H^+ \tag{105}$$

Butane can be oxidized to acetic acid (equation 106) and the process is carried out commerically under two different sets of conditions[266,272], either (1) at 180 °C and 20 bar

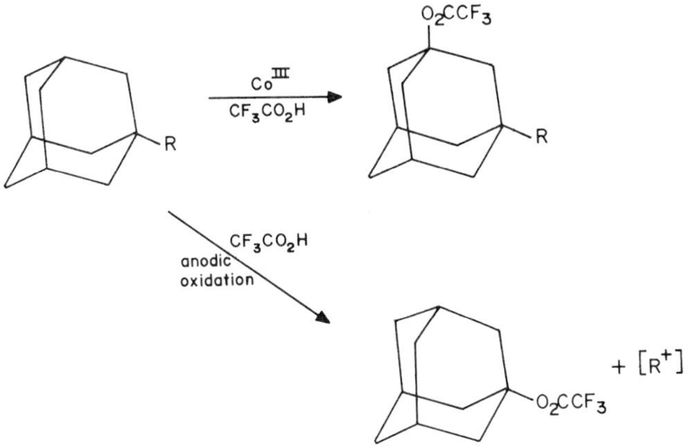

FIGURE 21. Electrochemical and chemical oxidation of adamantane.

pressure with a low concentration of cobalt(II), or (2) at 100–125 °C and 20 bar pressure with a high concentration of cobalt(II) (> 0.2 M).

$$n\text{-}C_4H_{10} + 5/2O_2 \longrightarrow 2MeCO_2H + H_2O \tag{106}$$

Under conditions (1), 57% of the product is acetic acid, but there are many other products including formic acid, acetaldehyde, methanol, dimethyl ether, butan-2-one, and various esters. Reaction under these conditions is mainly a free radical chain process with the cobalt assisting as a radical initiator. Under conditions (2), there is 87% selectivity for acetic acid as the product, with fewer by-products (propanoic acid, butanoic acid, and butan-2-one). A likely mechanism is direct oxidation of butane by cobalt(III) to produce a radical as the first step (equation 107), followed by formation of a peroxo radical and further reaction with cobalt(II) (equations 108 and 109). Here the maximum rate does not occur until the added cobalt(II) has been oxidized to cobalt(III).

$$\text{butane} + Co^{III} \longrightarrow \text{butan-2-yl radical} + Co^{II} + H^+ \tag{107}$$

$$\text{butan-2-yl radical} + O_2 \longrightarrow \text{CH}_3\text{CH}_2\text{CH(CH}_3\text{)O—O}^\bullet \tag{108}$$

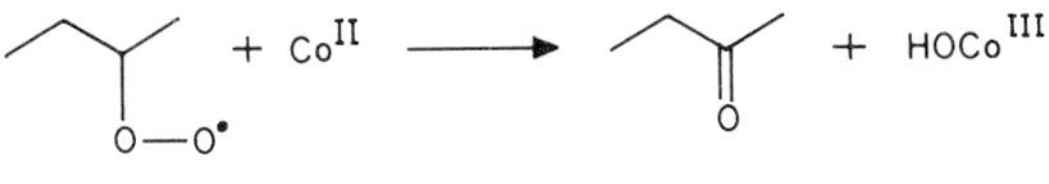

$$\text{CH}_3\text{CH}_2\text{CH(CH}_3\text{)O—O}^\bullet + Co^{II} \longrightarrow \text{butan-2-one} + HOCo^{III} \tag{109}$$

The rate of oxidation of alkanes by cobalt(III) acetate is markedly increased by the addition of bromide ions, or by the presence of strong acids such as trifluoroacetic acid. The evidence from studies of oxidation of arylalkanes in the presence of bromide ions is that there is a dramatic change of mechanism with formation of bromine atoms as intermediates (equations 110–112)[273–277], and it seems probable that the same process takes place with alkanes.

$$\text{Co}^{\text{III}}\text{ acetate} + \text{Br}^- \longrightarrow \text{AcOCo}^{\text{III}}\text{Br} \tag{110}$$

$$\text{AcOCo}^{\text{III}}\text{Br} \longrightarrow \text{AcOCo}^{\text{II}} + \text{Br}^{\bullet} \tag{111}$$

$$\text{Br}^{\bullet} + \text{RH} \longrightarrow \text{HBr} + \text{R}^{\bullet} \tag{112}$$

Studies of cobalt(III)-catalysed autoxidation reactions of the lower alkanes are limited as high-pressure equipment is required, and care is required to avoid explosions. In a kinetic study of the cobalt(III)-catalysed autoxidation of butane to give acetic acid (equation 106) in acetic acid solvent at 100–125 °C, Onopchenko and Schulz added cobalt in the form of cobalt(II) acetate[266,272]. It was also necessary to add some butan-2-one to assist in oxidizing cobalt(II) to cobalt(III). The maximum rate was not observed until all the cobalt(II) had been oxidized to cobalt(III). The reaction showed an induction period which could be shortened by adding the cobalt as cobalt(III) rather than as cobalt(II). The reaction proceeds via the formation of an alkyl radical as the result of an electron transfer between cobalt(III) and the alkane (equations 103 and 104). Under the conditions studied, about 85% selectivity and 80% conversion were achieved.

It is noteworthy that the autoxidation of butane at high temperatures can be catalysed by manganese(III) acetate in addition to cobalt(III) acetate. Manganese(III) is ineffective as a catalyst at temperature below 100 °C. Manganese(III) salts are known to break down to give radicals at high temperatures.

In 1970, Hanotier *et al.* observed that the activity of metal oxidants, especially cobalt(III) acetate, is considerably enhanced in the presence of strong acids[278]. They claimed that aliphatic hydrocarbons can be oxidized at temperatures as low as 20–40 °C, and that a high selectivity of product formation was found for both the stoichiometric reaction of cobalt(III) with alkanes and for cobalt(III)-catalysed autoxidations. At temperatures below 40 °C no hydrocarbon oxidation or catalyst reduction took place in the absence of strong acids, but addition of trichloroacetic acid or sulphuric acid increased both the decomposition rate of the cobalt(III) in the absence of alkane and the amount of hydrocarbon oxidation when alkanes were present. It is not clear whether the acid acts by (1) changing the proton donor power of the solvent or (2) altering the redox potential of the cobalt(III)–cobalt(II) system. In trifluoroacetic acid solvent cobalt(III)acetate is reduced to cobalt(II) seven times faster in the presence of an alkane. Experiments carried out in our laboratory[279] indicate that if such oxidations take place at low temperatures the yields are vanishingly small, and further work needs to be carried out on this complicated system to identify the important factors involved.

Results from photoelectron spectroscopy show that the electron most readily removed from most saturated hydrocarbons is an electron in a HOMO of mainly $\sigma_{C—C}$ character. In the usual electron transfer process (equation 94) for alkane oxidation it is a $\sigma_{C—H}$ electron which would need to be removed. This indicates that the reactions could be more complex than usually accepted[280].

4. *Oxidation by transition metal compounds in concentrated sulphuric acid*

In the Introduction it was noted that electrophiles in very stong acids will react with aliphatic C—H bonds. During the past decade Rudakov[281] has shown that alkanes can be oxidized by a number of transition metal compounds in solution in concentrated suphuric acid. These include platinum(III), manganese(III), palladium(II), and mercury(II). The metal compounds which normally are not sufficiently good oxidizing agents to oxidize alkanes are enhanced in their oxidation powers by interaction with the acidic medium.

In kinetic studies it has been established that the rate of oxidation of the alkane is first order in both alkane and the catalyst (equation 113). An alkane C—H bond is cleaved in the rate-limiting step as the kinetic isotope effect for the reaction (k_H/k_D) is 2. The relative rates of C—H cleavage are tertiary > secondary > primary, as expected for a radical reaction. The tertiary to secondary ratio is about 3000 when mercury(II) is the catalyst.

$$-\frac{d[RH]}{dt} = k_2[RH][M^{n+}] \qquad (113)$$

The mechanism suggested involves formation of an intermediate, **130**, where a ligand radical is formed but has not left the coordination shell of M^{n+} (equation 114). A hydrogen atom is then abstracted from the hydrocarbon (equation 115) leaving an alkyl radical. The nature of the products of these oxidations in concentrated sulphuric acid are not clear, but include carbocations, R^+, and alkenes.

$$[XM^{n+}L_m] \longrightarrow \underset{\textbf{130}}{[X^{\bullet}\cdots M^{(n-1)+}L_m]} \qquad (114)$$

$$RH + 130 \longrightarrow [R^{\bullet}\cdots HX\cdots M^{(n-1)+}L_m] \qquad (115)$$

VI. OXIDATION BY ANALOGUES OF BIOLOGICAL SYSTEMS

A. Introduction

Oxidations in biological systems involve electrophilic/radical oxidants. In the space available here we are only able to touch briefly on this topic. More extensive reviews have been published[4,9,282,283].

B. Cytochrome P450

In living systems, the oxidation of aliphatic or aromatic C—H bonds is hydroxylation (equation 116). This reaction is catalysed by monooxygenase enzymes. Dioxygen is the oxidizing agent and the source of hydrogen, (donor)H_2, is NADH. Many monooxygenases contain the same type of biocatalyst, the haemoprotein cytochrome P450. The structure of cytochrome P450 is that of a haem unit (**131**) in a single-chain protein, with the iron atom having its fifth coordination position linked to a cysteine thiolate sulphur atom. The sixth coordination position of the iron atom is involved in the oxidation.

$$RH + O_2 + (\text{donor})H_2 \xrightarrow{\text{enzyme}} ROH + (\text{donor}) + H_2O \qquad (116)$$

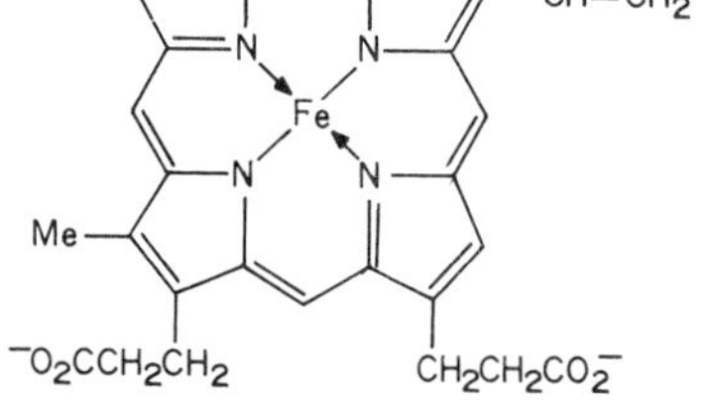

(**131**)

It is suggested that cytochrome P450 (**132**), after reduction to the iron(II) state (**133**), reacts with O_2 to give first an iron(II)$\cdots O_2$ species (**134**). By reduction and addition of H^+ the O—O bond splits, yielding an $Fe^V{=}O$ derivative (**135**) and H_2O (equation 117). The $Fe^V{=}O$ species **135** is the active species in reactions with hydrocarbons and can be written in a number of canonical forms, **135**–**138**.

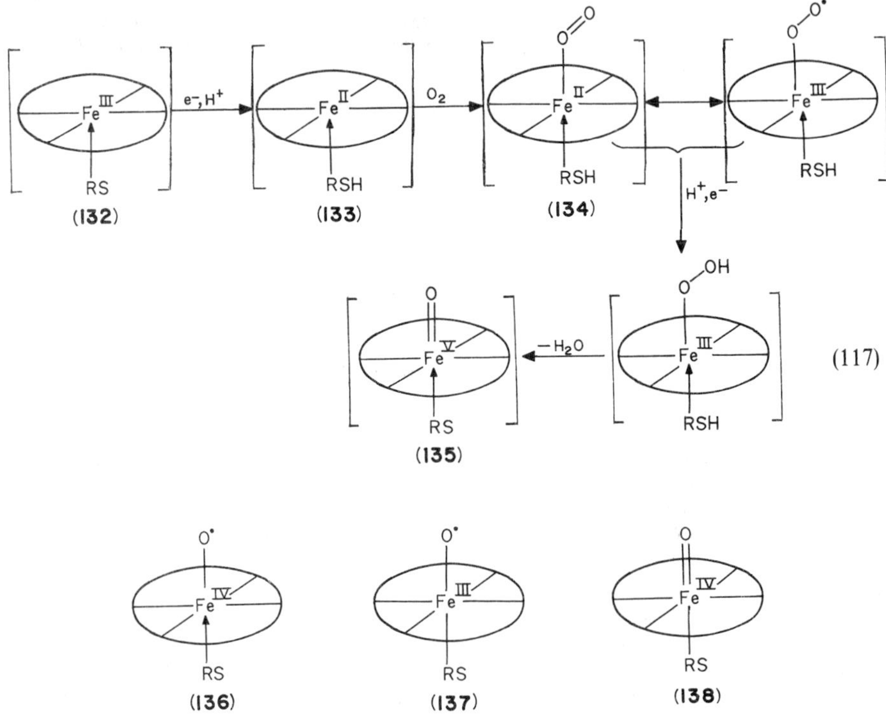

The interaction with the hydrocarbon is abstraction of an electron from the C—H bond by the active species **135** to give an OH bond and a radical. Next follows an extremely fast 'cage' reaction to produce the final hydroxylated carbon and P450 containing an Fe^{III} atom (**139**) (equation 118). The latter reaction must be fast as the configuration about the

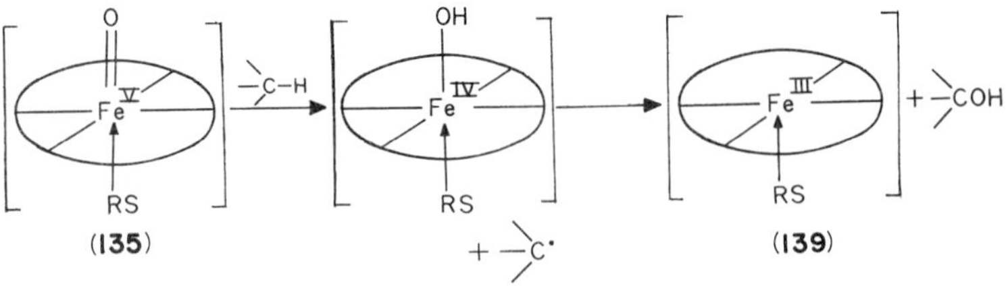

(135) (139)

(118)

carbon atom is retained. This indicates that the carbon radical is extremely short-lived. As is common with radical reactions, the activated oxygen complex attacks secondary in preference to primary C—H bonds, and tertiary in preference to secondary. As an example, *n*-heptane is oxidized by oxygen in the presence of rat microsomal cytochrome

P450 to give a mixture of heptan-1-, -2-, -3-, and -4-ols in relative percentage yields of 9, 74, 11, and 6%[284].

C. Synthetic Analogues

There have been a number of attempts to prepare synthetic electron-transfer catalysts using cytochrome P450 as a model[285]. These synthetic analogues can be made readily from the appropriate substituted porphin (a porphyrin) compound, **140**, and metal ions.

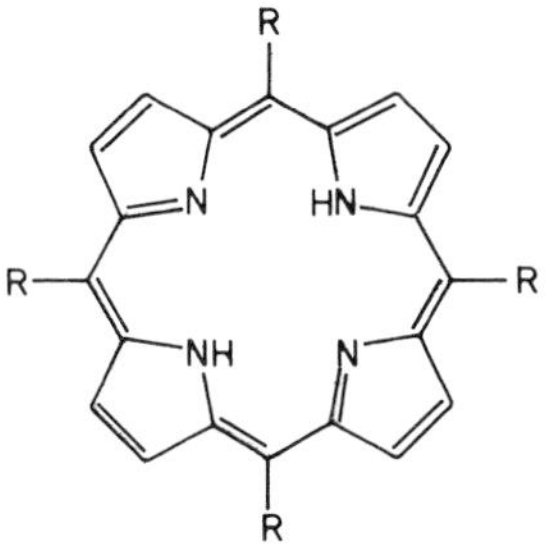

(**140**)

They are usually referred to by their abbreviated names. For example, the compound formed from tetraphenylporphin, a Mn^{III} ion, and a group X (such as Cl) also attached to the central metal is referred to as tpp–$Mn^{III}X$.

The catalytic activity of a number of synthetic porphyrins on the hydroxylation (by cumyl hydroperoxide) of cyclohexane (to give cyclohexanol) has been studied[286]. Tetraphenylporphyrin derivatives of Ni^{II}, Cu^{II}, Zn^{II}, Mg^{II}, V^{IV}, and Ti^{IV} show no catalytic activity. The compounds tpp–Co^{II} and [(tpp)Os(CO)(py)] catalysed the reaction but their structure was changed during reaction. Only tpp–$Fe^{III}Cl$ and tpp–$Mn^{III}Cl$ acted as true catalysts, and it is these that have been studied the most closely.

The P450 analogue tpp–$Fe^{III}Cl$ will act as a catalyst for the hydroxylation of cyclohexane to cyclohexanol[287,288]. In these non-biological systems oxygen is often replaced by a more convenient oxidizing agent such as hypochlorite ions or iodosylbenzene (PhIO) (formally an I^{III} derivative) (equation 119). The yield of cyclohexanol was 8%. Similar reactions give 12% adamantan-1-ol and 1% adamantan-2-ol from adamantane[287].

$$PhIO + C_6H_{12} \xrightarrow[\text{CH}_2\text{Cl}_2 \text{ under N}_2]{\text{room temp.}} C_6H_{11}OH \qquad (119)$$

In a series of studies, Hill and coworkers[289–291] used a manganese(III) porphyrin (tpp–$Mn^{III}X$) as an electron-transfer catalyst for alkane oxidation. A reaction similar to that in equation 119 takes place with C_6H_{12} and PhIO at room temperature[290]. The main product is $C_6H_{11}OH$ with a small amount of cyclohexanone. With *tert*-butane the same reactants yield **141** and **142**[291].

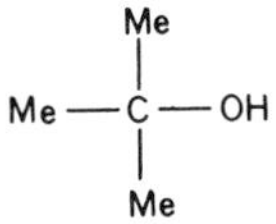

(**141**) 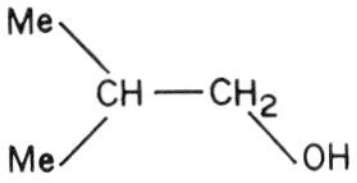 (**142**)

These reactions are thought to occur by a similar free radical mechanism to that described above. The oxidizing agent (PhIO) oxidizes the catalyst to a higher oxidation state (Fe^{IV} or Mn^{IV}), which then interacts with the hydrocarbon to abstract an electron, leaving a free radical (equation 120).

$$\text{PhCHMe}_2 \xrightarrow{\text{tpp-Mn}^{IV}} \text{PhC}^{\bullet}\text{Me}_2 \qquad (120)$$

Elegant syntheses can be carried out by using a two-phase system[292,293]. The alkane and the electron-transfer catalyst are dissolved in a suitable organic solvent such as dichloromethane and the oxidizing agent (PhIO or NaOCl) and the sodium salt of the anion to be incorporated are dissolved in water. A phase-transfer catalyst such as a trioctylmethylammonium salt is helpful. On vigorous agitation to equilibrate the mixture the alkane is oxidized. C—H bonds can be replaced by C—N, C—O, and C—halogen bonds. These reactions can take place with high yields of products. Under the above conditions, cyclohexane is converted into $C_6H_{11}Cl$ in 76% yield, with 4% of cyclohexanone also being produced.

The structure of the haem part of the catalyst plays a part in the efficiency of the oxidation. When tetraphenylporphyrin–Fe^{III} was replaced by tetra(pentafluorophenyl)porphyrin–Fe^{III} the yield of products for the PhIO oxidation of cyclohexane rose from 5 to 71%[294].

In an attempt to improve the yields of the oxidation reactions, tetraphenylhaemins have been prepared which (1) have groups present which by steric hindrance decrease the tendency of the porphyrins to form μ-oxo dimers and (2) contain electron-withdrawing groups which will reduce the rate of oxidative destruction of the catalyst[295]. Using PhIO as an oxidant, saturated hydrocarbons such as norbornane and cyclohexane can be oxidized at room temperature without extensive destruction of the catalyst.

VII. REFERENCES

1. D. E. Webster, *Adv. Organomet. Chem.*, **15**, 147 (1977).
2. R. G. Pearson, *Chem. Br.*, 103 (1967).
3. R. Ugo, *Chem. Ind. (Milan)*, **51**, 1319 (1969).
4. R. A. Sheldon and J. K. Kochi, *Metal Catalysed Oxidations of Organic Compounds*, Academic Press, New York, 1981.
5. H. Pines, *The Chemistry of Catalytic Hydrocarbon Conversions*, Academic Press, New York, 1981.
6. A. J. Hubert, in *Catalysis in C_1 Chemistry* (Ed. W. Keim), Vol. 4 of *Catalysis by Metal Complexes* (Eds. R. Ugo and B. R. James), Reidel, Dordrecht, 1983.
7. A. E. Shilov and A. A. Shteinman, *Coord. Chem. Rev.*, **24**, 97 (1977).
8. A. E. Shilov, *Sov. Sci. Rev., Sect. B*, **4**, 71 (1982).
9. A. E. Shilov, *Activation of Saturated Hydrocarbons by Transition Metal Complexes*, Reidel, Dordrecht, 1984.
10. G. A. Olah, D. G. Parker, and N. Yoneda, *Angew. Chem., Int. Ed. Engl.*, **17**, 909 (1978).
11. G. W. Parshall, in *Catalysis*, Specialist Periodical Report, Vol. 1, Chemical Society, London 1977, p. 335.
12. G. W. Parshall, *Homogeneous Catalysis*, Wiley, New York, 1980.
13. L. Abis, A. Sen, and J. Halpern, *J. Am. Chem. Soc.*, **100**, 2915 (1978).
14. E. L. Muetterties, *Chem. Soc. Rev.*, **11**, 283 (1982).
15. M. Brookhart and M. L. H. Green, *J. Organomet. Chem.*, **250**, 395 (1983).
16. M. Berry, N. J. Cooper, M. L. H. Green, and S. J. Simpson, *J. Chem. Soc., Dalton Trans.*, 29 (1980).
17. S. J. La Placa and J. A. Ibers, *Inorg. Chem.*, **4**, 778 (1965).
18. A. C. Skapski and P. G. H. Troughton, *Chem. Commun.*, 1230 (1968).

19. N. A. Bailey, J. M. Jenkins, R. Mason, and B. L. Shaw, *Chem. Commun.* 237 (1965).
20. K. W. Muir and J. A. Ibers, *Inorg. Chem.*, **9**, 440 (1970).
21. Y. W. Yared, S. L. Miles, R. Bau, and C. A. Reed, *J. Am. Chem. Soc.*, **99**, 7076 (1977).
22. S. Trofimenko, *J. Am. Chem. Soc.*, **89**, 6288 (1967); S. Trofimenko, *J. Am. Chem. Soc.*, **90**, 4754 (1968); S. Trofimenko, *Inorg. Chem.*, **9**, 2493 (1970).
23. F. A. Cotton and V. W. Day, *J. Chem. Soc., Chem. Commun.*, 415 (1974).
24. F. A. Cotton, T. LaCour, and A. G. Stanislowski, *J. Am. Chem. Soc.*, **96**, 754 (1974).
25. D. M. Roe, P. M. Bailey, K. Moseley, and P. M. Maitlis, *J. Chem. Soc., Chem. Commun.*, 1273 (1972).
26. M. Brookhart, T. H. Whitesides, and J. M. Crockett, *Inorg. Chem.*, **15**, 1550 (1976).
27. G. A. Olah, G. Liang, and S. H. Yu, *J. Org. Chem.*, **41**, 2227 (1976).
28. R. K. Brown, J. M. Williams, A. J. Schultz, G. D. Stucky, S. D. Ittel, and R. L. Harlow, *J. Am. Chem. Soc.*, **102**, 981 (1980).
29. J. M. Williams, R. K. Brown, A. J. Schultz, G. D. Stucky, and S. D. Ittel, *J. Am. Chem. Soc.*, **100**, 7407 (1978).
30. W. Lamanna and M. Brookhart, *J. Am. Chem. Soc.*, **103**, 989 (1981).
31. M. Brookhart, W. Lamanna, and M. B. Humphrey, *J. Am. Chem. Soc.*, **104**, 2117 (1982).
32. M. Brookhart, W. Lamanna, and A. R. Pinhas, *Organometallics*, **2**, 638 (1983).
33. M. Brookhart and A. Lukacs, *Organometallics*, **2**, 649 (1983).
34. O. W. Howarth, C. H. McAteer, P. Moore, and G. E. Morris, *J. Chem. Soc., Chem. Commun.*, 506 (1981).
35. M. A. Bennett, I. J. McMahon, and T. W. Turney, *Angew. Chem., Int. Ed. Engl.*, **21**, 379 (1982).
36. S. J. McLain, C. D. Wood, and R. R. Schrock, *J. Am. Chem. Soc.*, **99**, 3519 (1977).
37. R. R. Schrock and J. F. Fellmann, *J. Am. Chem. Soc.*, **100**, 3359 (1978).
38. A. J. Schultz, J. M. Williams, R. R. Schrock, G. A. Rupprecht, and J. D. Fellmann, *J. Am. Chem. Soc.*, **101**, 1593 (1979).
39. A. J. Schultz, R. K. Brown, J. M. Williams, and R. R. Schrock, *J. Am. Chem. Soc.*, **103**, 169 (1981).
40. J. H. Wengrovius, R. R. Schrock, M. R. Churchill, and H. J. Wasserman, *J. Am. Chem. Soc.*, **104**, 1739 (1982).
41. M. L. H. Green, *Pure Appl. Chem.*, **56**, 47 (1984).
42. M. L. H. Green, N. J. Hazel, P. D. Grebenik, V. S. B. Mtetwa, and K. Prout, *J. Chem. Soc., Chem. Commun.*, 356 (1983).
43. A. D. Caunt, in *Catalysis*, Specialist Periodical Report, Vol. 1 Chemical Society, London 1977, p. 234.
44. P. Cossee, *J. Catal.*, **3**, 80 (1964).
45. K. J. Ivin, J. J. Rooney, C. D. Stewart, M. L. H. Green, and R. Mahtab, *J. Chem. Soc., Chem. Commun.*, 604 (1978).
46. M. Brookhart, M. L. H. Green, and R. B. A. Pardy, *J. Chem. Soc., Chem. Commun.*, 691 (1983).
47. J. D. Fellmann, H. W. Turner, and R. R. Schrock, *J. Am. Chem. Soc.*, **102**, 6608 (1980).
48. M. R. Churchill, H. J. Wasserman, H. W. Turner, and R. R. Schrock, *J. Am. Chem. Soc.*, **104**, 1710 (1982).
49. S. J. Holmes and R. R. Schrock, *J. Am. Chem. Soc.*, **103**, 4599 (1981).
50. S. J. Holmes, D. N. Clark, H. W. Turner, and R. R. Schrock, *J. Am. Chem. Soc.*, **104**, 6322 (1982).
51. M. Pasquali, C. Floriani, A. Gaetani-Manfredotti, and A. Chiesi-Villa, *J. Am. Chem. Soc.*, **100**, 4918 (1978).
52. K. Itoh, N. Oshima, G. B. Jameson, H. C. Lewis, and J. A. Ibers, *J. Am. Chem. Soc.*, **103**, 3014 (1981).
53. K. Itoh and N. Oshima, *Chem. Lett.*, 1219 (1980).
54. Z. Dawoodi, M. L. H. Green, V. S. B. Mtetwa, and K. Prout, *J. Chem. Soc., Chem. Commun.*, 1410 (1982).
55. Z. Dawoodi, M. L. H. Green, V. S. B. Mtetwa, and K. Prout, *J. Chem. Soc., Chem. Commun.*, 802 (1982).
56. N. Koga, S. Obara, and K. Morokuma, *J. Am. Chem. Soc.*, **106**, 4625 (1984).
57. R. B. Cracknell, A. G. Orpen, and J. L. Spencer, *J. Chem. Soc., Chem. Commun.*, 326 (1984).
58. R. B. Calvert and J. R. Shapley, *J. Am. Chem. Soc.*, **100**, 7726 (1978).
59. M. Green, N. C. Norman, and A. G. Orpen, *J. Am. Chem. Soc.*, **103**, 1269 (1981).
60. M. A. Beno, J. M. Williams, M. Tachikawa, and E. L. Muetterties, *J. Am. Chem. Soc.*, **102**, 4542 (1980).

61. M. Tachikawa and E. L. Muetterties, *J. Am. Chem. Soc.*, **102**, 4541 (1980).
62. G. M. Dawkins, M. Green, A. G. Orpen, and F. G. A. Stone, *J. Chem. Soc., Chem. Commun.*, 41 (1982).
63. C. P. Casey, P. J. Fagan, and W. H. Miles, *J. Am. Chem. Soc.*, **104**, 1134 (1982).
64. J. D. Fellmann, R. R. Schrock, and D. D. Traficante, *Organometallics*, **1**, 481 (1982).
65. H. W. Turner and R. R. Schrock, *J. Am. Chem. Soc.*, **104**, 2331 (1982).
66. A. J. Schultz, J. M. Williams, R. R. Schrock, and S. J. Holmes, *Acta Crystallogr., Sect. C*, **40**, 590 (1984).
67. J. Dehand and M. Pfeffer, *Coord. Chem. Rev.*, **18**, 327 (1976).
68. M. I. Bruce, *Angew. Chem., Int. Ed. Engl.*, **16**, 73 (1977).
69. I. Omae, *Coord. Chem. Rev.*, **28**, 97 (1979).
70. I. Omae, *Chem. Rev.*, **79**, 287 (1979).
71. A. P. Humphries and H. D. Kaesz, *Prog. Inorg. Chem.*, **25**, 145 (1979).
72. R. J. Puddephatt, *Coord. Chem. Rev.*, **33**, 149 (1980).
73. I. Omae, *Coord. Chem. Rev.*, **32**, 235 (1980).
74. I. Omae, *Coord. Chem. Rev.*, **42**, 245 (1982).
75. S. D. Chappell and D. J. Cole-Hamilton, *Polyhedron*, **1**, 739 (1982).
76. E. C. Constable, *Polyhedron*, **3**, 1037 (1984).
77. R. Mason and D. W. Meek, *Angew. Chem., Int. Ed. Engl.*, **17**, 183 (1978).
78. P. Foley and G. M. Whitesides, *J. Am. Chem. Soc.*, **101**, 2732 (1979).
79. P. Foley, R. DiCosimo, and G. M. Whitesides, *J. Am. Chem. Soc.*, **102**, 6713 (1980).
80. G. M. Whitesides, *Pure Appl. Chem.*, **53**, 287 (1981).
81. S. S. Moore, R. DiCosimo, A. F. Sowinski, and G. M. Whitesides, *J. Am. Chem. Soc.*, **103**, 948 (1981).
82. R. DiCosimo, S. S. Moore, A. F. Sowinski, and G. M. Whitesides, *J. Am. Chem. Soc.*, **104**, 124 (1982).
83. R. H. Reamey and G. M. Whitesides, *J. Am. Chem. Soc.*, **106**, 81 (1984).
84. T. H. Tulip and D. L. Thorn, *J. Am. Chem. Soc.*, **103**, 2448 (1981).
85. S. D. Chappell and D. J. Cole-Hamilton, *J. Chem. Soc., Chem. Commun.*, 238 (1980).
86. G. W. Parshall, D. L. Thorn, and T. H. Tulip, *ChemTech*, 571 (1982).
87. G. W. Parshall, *ChemTech*, 628 (1984).
88. J. C. Calabrese, M. C. Colton, T. Herskovitz, U. Klabunde, G. W. Parshall, D. L. Thorn, and T. H. Tulip, *Ann. N. Y. Acad. Sci.*, **415**, 302 (1983).
89. B. L. Shaw, *J. Organomet. Chem.*, **200**, 307 (1980).
90. A. J. Cheney, B. E. Mann, B. L. Shaw, and R. M. Slade, *J. Chem. Soc., Chem. Commun.*, 1176 (1970).
91. A. J. Cheney, B. E. Mann, B. L. Shaw, and R. M. Slade, *J. Chem. Soc. A*, 3833 (1971).
92. H. D. Empsall, E. M. Hyde, and B. L. Shaw, *J. Chem. Soc., Dalton Trans.*, 1690 (1975).
93. R. Mason, M. Textor, N. Al-Salem, and B. L. Shaw, *J. Chem. Soc., Chem. Commun.*, 292 (1976).
94. A. R. H. Bottomley, C. Crocker, and B. L. Shaw, *J. Organomet. Chem.*, **250**, 617 (1983).
95. R. G. Goel and R. G. Montemayor, *Inorg. Chem.*, **16**, 2183 (1977).
96. H. C. Clark, A. B. Goel, R. G. Goel, S. Goel, and W. O. Ogini, *Inorg. Chim. Acta*, **31**, L441 (1978).
97. H. C. Clark, A. B. Goel, and S. Goel, *J. Organomet. Chem.*, **166**, C29 (1979).
98. H. C. Clark, A. B. Goel, and S. Goel, *Inorg. Chem.*, **18**, 2803 (1979).
99. H. C. Clark, A. B. Goel, R. G. Goel, and S. Goel, *Inorg. Chem.*, **19**, 3220 (1980).
100. R. G. Goel and W. O. Ogini, *Organometallics*, **1**, 654 (1982).
101. R. G. Geol, W. O. Ogini, and R. C. Srivastava, *Organometallics*, **1**, 819 (1982).
102. S. Hietkamp, D. J. Stufkens, and K. Vrieze, *J. Organomet. Chem.*, **139**, 189 (1977).
103. C. A. Tolman, *Chem. Rev.*, **77**, 313 (1977).
104. H. D. Empsall, E. M. Hyde, R. Markham, W. S. McDonald, M. C. Norton, B. L. Shaw, and B. Weeks, *J. Chem. Soc., Chem. Commun.*, 589 (1977).
105. C. Crocker, R. J. Errington, W. S. McDonald, K. J. Odell, and B. L. Shaw, *J. Chem. Soc., Chem. Commun.*, 498 (1979).
106. C. Crocker, R. J. Errington, R. Markham, C. J. Moulton, K. J. Odell, and B. L. Shaw, *J. Am. Chem. Soc.*, **102**, 4373 (1980).
107. N. A. Al-Salem, H. D. Empsall, R. Markham, B. L. Shaw, and B. Weeks, *J. Chem. Soc., Dalton Trans.*, 1972 (1979).

108. N. A. Al-Salem, W. S. McDonald, R. Markham, M. C. Norton, and B. L. Shaw, *J. Chem. Soc., Dalton Trans.*, 59 (1980).
109. P. W. Clark, *J. Organomet. Chem.*, **137**, 235 (1977).
110. C. K. Ingold, *J. Chem. Soc.*, **119**, 305 (1921); C. K. Ingold, *J. Chem. Soc.*, **119**, 951 (1921).
111. G. S. Hammond, in *Steric Effects in Organic Chemistry* (Ed. M. S. Newman), Wiley, New York, 1956, p. 468.
112. N. L. Allinger and V. Zalkow, *J. Org. Chem.*, **25**, 701 (1960).
113. B. L. Shaw, *J. Am. Chem. Soc.*, **97**, 3856 (1975).
114. S. Bresadola, P. Rigo, and A. Turco, *Chem. Commun.*, 1205 (1968).
115. S. Bresadola, A. Frigo, B. Longato, and G. Rigatti, *Inorg. Chem.*, **12**, 2788 (1973).
116. N. Bresciani, M. Calligaris, P. Delise, G. Nardin, and L. Randaccio, *J. Am. Chem. Soc.*, **96**, 5642 (1974).
117. S. Bresadola, B. Longato, and F. Morandini, *J. Organomet. Chem.*, **128**, C5 (1977).
118. R. H. Crabtree, J. M. Quirk, H. Felkin, T. Fillebeen-Khan, and C. Pascard, *J. Organomet. Chem.*, **187**, C32 (1980).
119. H. Werner and R. Werner, *J. Organomet. Chem.*, **209**, C60 (1981).
120. H. Kletzin and H. Werner, *Angew. Chem., Int. Ed. Engl.*, **22**, 873 (1983).
121. S. Heitkamp, D. J. Stufkens, and K. Vrieze, *J. Organomet. Chem.*, **152**, 347 (1978).
122. W. J. Youngs and J. A. Ibers, *J. Am. Chem. Soc.*, **105**, 639 (1983).
123. W. J. Youngs and J. A. Ibers, *Organometallics*, **2**, 979 (1983).
124. W. J. Youngs, J. Mahood, B. L. Simms, P. N. Swepston, J. A. Ibers, M. Shong, J. Huang, and J. Lire, *Organometallics*, **2**, 917 (1983).
125. G. W. Parshall, *Acc. Chem. Res.*, **8**, 113 (1975).
126. F. G. Moers, *J. Chem. Soc., Chem. Commun.*, 79 (1971).
127. F. G. Moers and J. P. Langhout, *Recl. Trav. Chim. Pays-Bas*, **91**, 591 (1972).
128. G. W. Parshall, *Acc. Chem. Res.*, **3**, 139 (1970).
129. C. Masters, *J. Chem. Soc., Chem. Commun.*, 191 (1973).
130. A. A. Kiffen, C. Masters, and L. Raynand, *J. Chem. Soc., Dalton Trans.*, 853 (1975).
131. G. E. Hartwell, R. V. Lawrence, and M. J. Smas, *J. Chem. Soc., Chem. Commun.*, 912 (1970).
132. V. I. Sokolov, J. A. Sorokina, L. L. Troitskaya, L. I. Solovieva, and O. A. Reutov, *J. Organomet. Chem.*, **36**, 389 (1972).
133. A. J. Deeming and I. P. Rothwell, *Pure Appl. Chem.*, **52**, 649 (1980).
134. Y. Fuchita, K. Hiraki, and T. Uchiyama, *J. Chem. Soc., Dalton Trans.*, 897 (1983).
135. C. Mutet and M. Pfeffer, *J. Organomet. Chem.*, **171**, C34 (1979).
136. A. G. Constable, W. S. McDonald, L. C. Sawkins, and B. L. Shaw, *J. Chem. Soc., Dalton Trans.*, 1992 (1980).
137. R. A. Andersen, R. A. Jones, G. Wilkinson, M. B. Hursthouse, and K. M. Abdul Malik, *J. Chem. Soc., Chem. Commun.*, 283 (1977); R. A. Andersen, R. A. Jones, and G. Wilkinson, *J. Chem. Soc., Dalton Trans.*, 446 (1978).
138. P. Diversi, G. Ingrosso, A. Lucherini, and D. Fasce, *J. Chem. Soc., Chem. Commun.*, 945 (1982).
139. L. Dahlenburg, V. Sinnwell, and D. Thoennes, *Chem. Ber.*, **111**, 3367 (1978).
140. R. B. Calvert and J. R. Shapley, *J. Am. Chem. Soc.*, **99**, 5225 (1977).
141. E. H. K. Zeiher, D. G. DeWit, and K. G. Caulton, *J. Am. Chem. Soc.*, **106**, 7006 (1984).
142. H. tom Dieck and M. Svoboda, *Chem. Ber.*, **109**, 1657 (1976).
143. H. H. Karsch, H.-F. Klein, and H. Schmidbauer, *Chem. Ber.*, **110**, 2200 (1977).
144. P. R. Sharp, D. Astruc, and R. R. Schrock, *J. Organomet. Chem.*, **182**, 477 (1979).
145. M. E. Eichner, H. G. Alt, and M. D. Rausch, *J. Organomet. Chem.*, **264**, 309 (1984).
146. L. Chamberlain, J. Keddington, I. P. Rothwell and J. C. Huffman, *Organometallics*, **1**, 1538 (1982).
147. L. R. Chamberlain, A. P. Rothwell, and I. P. Rothwell, *J. Am. Chem. Soc.*, **106**, 1847 (1984).
148. W. A. Nugent, D. W. Ovenall, and S. J. Holmes, *Organometallics*, **2**, 161 (1983).
149. C. R. Bennett and D. C. Bradley, *J. Chem. Soc., Chem. Commun.*, 29 (1974).
150. D. R. McAlister, D. K. Erwin, and J. E. Bercaw, *J. Am. Chem. Soc.*, **100**, 5966 (1978).
151. J. E. Bercaw and D. K. Erwin, personal communication, quoted in Ref. 182.
152. J. W. Bruno, T. J. Marks, and V. W. Day, *J. Am. Chem. Soc.*, **104**, 7357 (1982).
153. J. W. Bruno, T. J. Marks, and V. W. Day, *J. Organomet. Chem.*, **250**, 237 (1983).
154. S. J. Simpson, H. W. Turner, and R. A. Andersen, *J. Am. Chem. Soc.*, **101**, 7728 (1979).

155. S. J. Simpson, H. W. Turner, and R. A. Andersen, *Inorg. Chem.*, **20**, 2991 (1981).
156. M. L. H. Green, *Pure Appl. Chem.*, **50**, 27 (1978).
157. M. Berry, K. Elmitt, and M. L. H. Green, *J. Chem. Soc., Dalton Trans.*, 1950 (1979).
158. M. L. H. Green and P. J. Knowles, *J. Chem. Soc., Chem. Commun.*, 1677 (1970); M. L. H. Green and P. J. Knowles, *J. Chem. Soc., A*, 1508 (1971).
159. C. Giannotti and M. L. H. Green, *J. Chem. Soc., Chem. Commun.*, 1114 (1972).
160. M. L. H. Green, M. Berry, C. Couldwell, and K. Prout, *Nouv. J. Chim.*, **1**, 187 (1977).
161. K. Elmitt, M. L. H. Green, R. A. Forder, I. Jefferson, and K. Prout, *J. Chem. Soc. Chem. Commun.*, 747 (1974).
162. N. J. Cooper, M. L. H. Green, and R. Mahtab, *J. Chem. Soc., Dalton Trans.*, 1557 (1979).
163. F. W. S. Benfield and M. L. H. Green, *J. Chem. Soc., Dalton Trans.*, 1324 (1974).
164. J. Chetwynd-Talbot, P. Grebenik, and R. N. Perutz, *Inorg. Chem.*, **21**, 3647 (1982).
165. P. A. Cox, P. Grebenik, R. N. Perutz, M. D. Robinson, R. Grinter, and D. R. Stern, *Inorg. Chem.*, **22**, 3614 (1983).
166. P. D. Grebenik, M. L. H. Green, and A. Izquierdo, *J. Chem. Soc., Chem. Commun.*, 186 (1981).
167. R. H. Crabtree, J. M. Mihelcic, and J. M. Quirk, *J. Am. Chem. Soc.*, **101**, 7738 (1979).
168. R. H. Crabtree, M. F. Mellea, J. H. Mihelcic, and J. M. Quirk, *J. Am. Chem. Soc.*, **104**, 107 (1982).
169. R. H. Crabtree, P. C. Demou, D. Eden, J. M. Mihelcic, C. A. Parnell, J. M. Quirk, and G. E. Morris, *J. Am. Chem. Soc.*, **104**, 6994 (1982).
170. R.H. Crabtree, *ChemTech*, 506 (1982).
171. M. J. Burk, R. H. Crabtree, C. P. Parnell, and R. J. Uriarte, *Organometallics*, **3**, 816 (1984).
172. R. H. Crabtree and C. P. Parnell, *Organometallics*, **3**, 1727 (1984).
173. R. H. Crabtree and R. P. Dion, *J. Chem. Soc., Chem. Commun.*, 1260 (1984).
174. D. Baudry, M. Ephritikhine, and H. Felkin, *J. Chem. Soc., Chem. Commun.*, 1243 (1980).
175. D. Baudry, M. Ephritikhine, H. Felkin, and J. Zakrzewski, *J. Chem. Soc., Chem. Commun.*, 1235 (1982).
176. D. Baudry, M. Ephritikhine, and H. Felkin, *J. Chem. Soc., Chem. Commun.*, 606 (1982).
177. D. Baudry, M. Ephrititkhine, H. Felkin, and R. Holmes-Smith, *J. Chem. Soc., Chem. Commun.*, 788 (1983).
178. M. A. Green, J. C. Huffman, K. G. Caulton, W. K. Rybak, and J. J. Ziolkowski, *J. Organomet. Chem.*, **218**, C39 (1981).
179. M. A. Green, J. C. Huffman, and K. G. Caulton, *J. Am. Chem. Soc.*, **130**, 695 (1981).
180. M. A. Green, J. C. Huffman, and K. G. Caulton, *J. Organomet. Chem.*, **243**, C78 (1983).
181. A. H. Janowicz and R. G. Bergman, *J. Am. Chem. Soc.*, **104**, 352 (1982).
182. A. H. Janowicz and R. G. Bergman, *J. Am. Chem. Soc.*, **105**, 3929 (1983).
183. A. H. Janowicz, R. A. Periana, J. M. Buchanan, C. A. Kovac, J. M. Stryker, M. J. Wax, and R. G. Bergman, *Pure Appl. Chem.*, **56**, 13 (1984).
184. M. J. Wax, J. M. Stryker, J. M. Buchanan, C. A. Kovac, and R. G. Bergman, *J. Am. Chem. Soc.*, **106**, 1121 (1984).
185. W. D. Jones and F. J. Feher, *Organometallics*, **2**, 562 (1983).
186. R. A. Periana and R. G. Bergman, *Organometallics*, **3**, 508 (1984).
187. W. D. Jones and F. J. Feher, *J. Am. Chem. Soc.*, **106**, 1650 (1984).
188. W. D. Jones and F. J. Feher, *J. Am. Chem. Soc.*, **104**, 4240 (1982).
189. W. D. Jones and F. J. Feher, *Organometallics*, **2**, 686 (1983).
190. J. Chatt and J. M. Davidson, *J. Chem. Soc.*, 843 (1965).
191. J. K. Hoyano and W. A. G. Graham, *J. Am. Chem. Soc.*, **104**, 3723 (1982).
192. J. K. Hoyano, A. D. McMaster, and W. A. G. Graham, *J. Am. Chem. Soc.*, **105**, 7190 (1983).
193. M. D. Rausch, R. G. Gastinger, S. A. Gardner, R. K. Brown, and J. S. Wood, *J. Am. Chem. Soc.*, **99**, 7870 (1977).
194. A. J. Rest, I. Whitwell, W. A. G. Graham, J. K. Hoyano, and A. D. McMaster, *J. Chem. Soc., Chem. Commun.*, 624 (1984).
195. O. Crichton, A. J. Rest, and D. J. Taylor, *J. Chem. Soc., Dalton Trans.*, 167 (1980).
196. P. L. Watson, *J. Chem. Soc., Chem. Commun.*, 276 (1983).
197. P. L. Watson, *J. Am. Chem. Soc.*, **105**, 6491 (1983).
198. J. Holten, M. F. Lappert, D. G. H. Ballard, R. Pearce, J. L. Atwood, and W. E. Hunter, *J. Chem. Soc., Dalton Trans.*, 54 (1979).
199. G. A. Olah and G. K. S. Prakash, *Chem. Br.*, 916 (1983).
200. C. M. Fendrick and T. J. Marks, *J. Am. Chem. Soc.*, **106**, 2214 (1984).

201. E. A. Grigoryan, *Usp. Khim.*, **53**, 347 (1984); *Russ. Chem. Rev.*, **53**, 210 (1984).
202. N. Kitajima and J. Schwartz, *J. Am. Chem. Soc.*, **106**, 2220 (1984).
203. M. D. Ward and J. Schwartz, *Organometallics*, **1**, 1030 (1982).
204. N. Kitajima and J. Schwartz, *J. Mol. Catal.*, **26**, 165 (1984).
205. W. D. Jones and F. J. Feher, *J. Am. Chem. Soc.*, **107**, 620 (1985).
206. G. Illuminati and L. Mandolini, *Acc. Chem. Res.*, **14**, 95 (1981).
207. J-Y. Saillard and R. Hoffmann, *J. Am. Chem. Soc.*, **106**, 2006 (1984).
208. J. L. Garnett and R. J. Hodges, *J. Am. Chem. Soc.*, **89**, 4546 (1967).
209. O. Ginstrup and I. Leden, *Acta Chem. Scand.*, **21**, 2689 (1967).
210. N. F. Gol'dshleger, M. B. Tyabin, A. E. Shilov, and A. A. Shteinman, *Zh. Fiz. Khim.*, **43**, 2174 (1969); *Russ. J. Phys. Chem.*, **43**, 1222 (1969).
211. R. J. Hodges, D. E. Webster, and P. B. Wells, *J. Chem. Soc., Chem. Commun.*, 462 (1971).
212. M. B. Tyabin, A. E. Shilov, and A. A. Shteinman, *Dokl. Akad. Nauk SSSR*, **198**, 380 (1971).
213. R. J. Hodges, D. E. Webster, and P. B. Wells, *J. Chem. Soc. A*, 3230 (1971); R. J. Hodges, D. E. Webster, and P. B. Wells, *J. Chem. Soc., Dalton Trans.*, 2577 (1972).
214. A. A. Shteinman, *Thesis*, Chernogolovka (1978), quoted in Ref. 9; R. J. Hodges, D. E. Webster, and P. B. Wells, *J. Chem. Soc., Dalton Trans.*, 2571 (1972).
215. N. F. Gol'dshleger and A. A. Shteinman, *React. Kinet. Catal. Lett.*, **6**, 43 (1977).
216. L. F. Repka and A. A. Shteinman, *Kinet. Katal.*, **15**, 805 (1974).
217. Reference 9, p. 145.
218. R. J. Hodges and J. L. Garnett, *J. Phys. Chem.*, **72**, 1673 (1968).
219. J. L. Garnett and J. C. West, *Aust. J. Chem.*, **27**, 129 (1974).
220. N. F. Gol'dshleger, V. V. Es'kova, A. E. Shilov, and A. A. Shteinman, *Zh. Fiz. Khim.*, **46**, 1353 (1972); *Russ. J. Phys. Chem.*, **46**, 785 (1972).
221. V. V. Es'kova, A. E. Shilov, and A. A. Shteinman, *Kinet. Katal.*, **13**, 534 (1972).
222. J. L. Garnett and J. C. West, *Synth. React. Inorg. Metal-Org. Chem.*, **2**, 25 (1982).
223. J. R. Sanders, D. E. Webster, and P. B. Wells, *J. Chem. Soc., Dalton Trans.*, 1191 (1975).
224. V. P. Tretyakov, E. S. Rudakov, A. A. Galenin, and G. P. Zimtseva, *Dokl. Akad. Nauk SSSR*, **225**, 583 (1975).
225. E. S. Rudakov, V. P. Treyakov, A. A. Galenin, and R. I. Rudakova, *Dopov. Akad. Nauk Ukr. RSR Ser. B*, 146. (1977).
226. V. P. Tretyakov, E. S. Rudakov, A. V. Bogdanov, G. P. Zimtseva, and L. I. Kozhevnikova, *Dokl. Akad. Nauk SSSR*, **249**, 878 (1979).
227. L. A. Kushch, V. V. Lavrushko, Yu. S. Misharin, A. P. Moravskii, and A. E. Shilov, *Nouv. J. Chim.*, **7**, 729 (1983).
228. V. V. Lavrushko, S. A. Lermontov, and A. E. Shilov, *React. Kinet. Catal. Lett.*, **15**, 269 (1980).
229. G. B. Shul'pin, L. P. Rozenberg, R. P. Shibaeva, and A. E. Shilov, *Kinet. Katal.*, **20**, 1570 (1979).
230. G. B. Shul'pin, *Kinet. Katal.*, **22**, 520 (1981).
231. G. B. Shul'pin and G. V. Nizova, *Kinet. Katal.*, **22**, 1061 (1981).
232. G. B. Shul'pin, A. E. Shilov, A. N. Kitaigorodski, J. V. Zeile-Krevor, *J. Organomet. Chem.*, **201**, 319 (1980).
233. R. P. Shibaeva, L. P. Rozenberg, R. M. Lobkovskaya, A. E. Shilov, and G. B. Shul'pin, *J. Organomet. Chem.*, **220**, 271 (1981).
234. P. L. Timms and T. W. Turney, *Adv. Organomet. Chem.*, **15**, 53 (1977).
235. W. J. Power and G. A. Ozin, *Adv. Inorg. Chem. Radiochem.*, **23**, 79 (1980).
236. P. H. Barrett, M. Pasternak, and R. G. Pearson, *J. Am. Chem. Soc.*, **101**, 222 (1979).
237. S. C. Davies and K. J. Klabunde, *J. Am. Chem. Soc.*, **100**, 5973 (1978).
238. S. C. Davis, S. J. Severson, and K. J. Klabunde, *J. Am. Chem. Soc.*, **103**, 3024 (1981).
239. K. J. Klabunde, and Y. Tanaka, *J. Am. Chem. Soc.*, **105**, 3544 (1983).
240. F. G. N. Cloke, A. E. Derome, M. L. H. Green, and D. O.'Hare, *J. Chem. Soc., Chem. Commun.*, 1312 (1983).
241. J. A. Bandy, F. G. N. Cloke, M. L. H. Green, D. O'Hare, and K. Prout, *J. Chem. Soc., Chem. Commun.*, 240 (1984).
242. W. E. Billups, M. M. Konarski, R. H. Hauge, and J. L. Margrave, *J. Am. Chem. Soc.*, **102**, 7393 (1980).
243. G. A. Ozin, D. F. McIntosh, S. A. Mitchell, and J. Garcio-Prieto, *J. Am. Chem. Soc.*, **103**, 1574 (1981).
244. G. A. Ozin, S. A. Mitchell, and J. Garcia-Prieto, *Angew. Chem., Int. Ed. Engl.*, **21**, 211 (1982).

245. G. A. Ozin and J. G. McCaffrey, *J. Am. Chem. Soc.*, **104**, 7351 (1982).
246. G. A. Ozin, and F. Hugues, *J. Phys. Chem.*, **86**, 5174 (1982).
247. L. F. Halle, R. Houriet, M. M. Kappes, R. H. Staley, and J. L. Beauchamp, *J. Am. Chem. Soc.*, **104**, 6293 (1982).
248. D. B. Jacobson and B. S. Freiser, *J. Am. Chem. Soc.*, **105**, 5197 (1983).
249. G. D. Byrd, R. C. Burnier, and B. S. Freiser, *J. Am. Chem. Soc.*, **104**, 3565 (1982).
250. L. F. Halle, P. B. Armentrout, and J. L. Beauchamp, *Organometallics*, **1**, 963 (1982).
251. P. B. Armentrout and J. L. Beauchamp, *J. Am. Chem. Soc.*, **103**, 784 (1981).
252. R. Houriet, L. F. Halle, and J. L. Beauchamp, *Organometallics*, **2**, 1818 (1983).
253. J. Allison, R. B. Freas, and D. P. Ridge, *J. Am. Chem. Soc.*, **101**, 1332 (1979).
254. R. B. Freas and D. P. Ridge, *J. Am. Chem. Soc.*, **102**, 7129 (1980).
255. P. B. Armentrout and J. L. Beauchamp, *J. Am. Chem. Soc.*, **102**, 1736 (1980).
256. D. B. Jacobson and B. S. Freiser, *J. Am. Chem. Soc.*, **105**, 7492 (1983).
257. P. B. Armentrout and J. L. Beauchamp, *J. Am. Chem. Soc.*, **103**, 6628 (1981).
258. D. B. Jacobson and B. S. Freiser, *Organometallics*, **3**, 513 (1984).
259. D. B. Jacobson and B. S. Freiser, *J. Am. Chem. Soc.*, **105**, 736 (1983).
260. L. F. Halle, P. B. Armentrout, and J. L. Beauchamp, *J. Am. Chem. Soc.*, **103**, 962 (1981).
261. R. B. Freas and D. P. Ridge, *J. Am. Chem. Soc.*, **106**, 825 (1984).
262. D. B. Jacobson and B. S. Freiser, *J. Am. Chem. Soc.*, **106**, 5351 (1984).
263. D. B. Jacobson and B. S. Freiser, *J. Am. Chem. Soc.*, **106**, 3891 (1984).
264. D. B. Jacobson and B. S. Freiser, *J. Am. Chem. Soc.*, **106**, 3900 (1984).
265. K. B. Wiberg and G. Foster, *J. Am. Chem. Soc.*, **83**, 423 (1961).
266. A. Onopchenko and J. G. D. Schulz, *J. Org. Chem.*, **38**, 3729 (1973).
267. K. Tanaka, *ChemTech*, 555 (1974).
268. K. Tanaka, *Hydrocarbon Process.*, **53**, 114 (1974).
269. K. Tanaka, *Am. Chem. Soc. Div. Pet. Chem., Prepr.*, **19**, 103 (1974).
270. J. D. G. Schulz and A. Onopchenko, *J. Org. Chem.*, **45**, 3716 (1980).
271. S. R. Jones and J. M. Mellor, *J. Chem. Soc., Perkin Trans. 2*, 511 (1977).
272. A. Onopchenko and J. G. D. Schulz, *J. Org. Chem.*, **38**, 909 (1973).
273. Y. Kamiya, *J. Catal.*, **33**, 480 (1974).
274. D. A. S. Ravens, *Trans. Faraday Soc.*, **55**, 1768 (1959).
275. Y. Kamiya, *Tetrahedron*, **22**, 2029 (1966).
276. Y. Kamiya, *Adv. Chem. Ser.*, No. 76, 193 (1968).
277. C. E. H. Bawn and T. K. Wright, *Discuss. Faraday Soc.*, **46**, 164 (1968).
278. J. Hanotier, Ph. Camerman, M. Hanotier-Bridoux, and P. de Radzitzky, *J. Chem. Soc., Perkin Trans. 2*, 2247 (1972).
279. J. R. Chipperfield, C. R. Foulkes, S. Lau, and D. E. Webster, unpublished observations.
280. W. Schmidt, *Tetrahedron*, **29**, 2129 (1973).
281. E. S. Rudakov, *Izv. Sib. Otd. Akad. Nauk SSSR, Ser. Khim. Nauk*, 161 (1980).
282. J. T. Groves, *Adv. Inorg. Biochem.*, **1**, 119 (1979).
283. V. Ullrich, *Top. Curr. Chem.*, **83**, 67 (1979).
284. U. Frommer, V. Ullrich, Hj. Staudinger, and S. Orrenius, *Biochim. Biophys. Acta*, **280**, 487 (1972).
285. A. D. Adler, F. R. Longo, F. Kampas, and J. Kim, *J. Inorg. Nucl. Chem.*, **32**, 2443 (1970).
286. D. Mansuy, J. F. Bartoli, J. C. Chottard, and M. Lange, *Angew. Chem., Int. Ed. Engl.*, **19**, 909 (1980).
287. J. T. Groves, T. E. Nemo, and R. S. Myers, *J. Am. Chem. Soc.*, **101**, 1032 (1979).
288. J. T. Groves, W. J. Kruper, Jr., and R. C. Haushalter, *J. Am. Chem. Soc.*, **102**, 6375 (1980).
289. J. A. Smegal and C. L. Hill, *J. Am. Chem. Soc.*, **105**, 3515 (1983).
290. C. L. Hill and B. C. Schardt, *J. Am. Chem. Soc.*, **102**, 6374 (1980).
291. J. A. Smegal and C. L. Hill, *J. Am. Chem. Soc.*, **105**, 2920 (1983).
292. I. Tabushi and N. Koga, *Tetrahedron Lett.*, 3681 (1979).
293. C. L. Hill, J. A. Smegal, and T. J. Henly, *J. Org. Chem.*, **48**, 3277 (1983).
294. C. K. Chang and F. Ebina, *J. Chem. Soc., Chem. Commun.*, 778 (1981).
295. P. S. Traylor, D. Dolphin, and T. G. Traylor, *J. Chem. Soc., Chem. Commun.*, 279 (1984).

The Chemistry of the Metal—Carbon Bond, Vol. 4
Edited by F. R. Hartley

CHAPTER **14**

Supported metal complex catalysts

F. R. Hartley

Royal Military College of Science, Shrivenham, Swindon, Wiltshire SN6 8LA, UK

I. INTRODUCTION

Traditionally Haag and Whitehurst[1–4] are credited with describing the first supported metal complex catalysts. In 1969 they reported that $[Pt(NH_3)_4]^{2+}$ supported on sulphonated polystyrene was an effective olefin hydrogenation catalyst. After a slow start, work on supported metal complexes accelerated rapidly and by the mid-1970s virtually every homogeneous metal complex catalyst had been studied in supported form. The initial driving force for this work was the belief that by supporting a metal complex catalyst the high selectivity, specificity, activity, and ease of modification of such catalysts, which stem from their molecular nature, could all be retained and combined with the ease of separation of heterogeneous catalysts that industry had found to be so important in developing commercial processes. Subsequently, further advantages were discovered for supported metal complex catalysts, including in many cases their greater stability and resistance to deactivation compared with their homogeneous analogues. However, their activities were often lower than those of their homogeneous counterparts and in many cases the metal complexes did not bind as firmly to the supports as their originators would have wished; leaching was recognized to be a serious problem in many cases.

During the 1970s and into the 1980s it was increasingly realized that the full potential of supported metal complex catalysts would only be realized by carefully developing materials in which both the active metal centre and the support combined together during the catalytic reaction. Thus supports were no longer inert but played an important part in promoting the selectivity and specificity in addition to the activity of the catalyst. Although a tremendous amount of work has been done using polystyrene as the support, much of it based on Merrifield's resin originally developed for peptide synthesis and degradation[5,6], there are serious problems with this material when commercial processes are being considered. Polystyrene is not particularly strong in a mechanical sense so the polymer is broken down to 'fines' when agitated as in a stirred reactor. Many solvents swell polystyrene; although in a laboratory situation the swellability of polystyrene is an attractive phenomenon that can be used to enhance catalyst selectivity, it poses serious problems for the chemical engineer faced with designing a column.

The subject of supported metal complex catalysts has been reviewed on a number of occasions[7–15]. Recently the present author has completed a book on the subject[16]. This chapter will therefore not attempt to duplicate that book, but rather to summarize the present state of the art and to indicate the likely way forward for the future.

No commercial processes have yet been developed using supported metal complex catalysts, although the first is probably only a few years away because of the enhanced selectivity that can be achieved when both support and metal complex combine together. It has been reported[17] that Mobil did take a polymer-supported rhodium hydroformylation process as far as pilot-plant scale in the mid-1970s and that full commercial scale development was halted only because existing and projected markets were insufficient to justify the construction of a new plant. Had a new plant been built then a supported metal complex catalyst would have been the catalyst of choice.

A. Reasons for Supporting Metal Complexes

The best way to summarize the reasons for supporting metal complexes is to examine the advantages and disadvantages of supported (heterogeneous) and unsupported (homogeneous) metal complex catalysts.

1. Catalyst separation

The major disadvantage of homogeneous catalysts is the difficulty of separating the catalyst at the end of the reaction. By the very nature of the homogeneous system, separation must involve a very efficient distillation, ion exchange or solvent extraction. All of these are expensive relative to some kind of filtration which can be used with heterogeneous catalysts. Although distillation is inevitably endothermic and therefore expensive because of its significant energy requirements, it has nevertheless been the method of choice for most homogeneously catalysed commercial processes. Two of the commercially successful homogeneously catalysed processes, the Wacker process for the oxidation of ethylene to acetaldehyde[18] and the Monsanto process for the carbonylation of methanol to yield acetic acid[19], depend on the relatively low boiling points of the products (20.8 °C for acetaldehyde and 117.9 °C for acetic acid). In many cases, such as the hydrogenation of vegetable oils to yield components of margarine, distillation is not practicable because the products decompose below their boiling points. Distillation is also impossible for reactions which yield high-boiling side-products which would steadily build up in concentration if they were not removed.

2. Efficiency and activity

The efficiency and activity of a homogeneous catalyst in which all the metal complex centres are equally accessible to the reactants must necessarily be greater than those of most heterogeneous catalysts. However, by anchoring the metal complexes to the surface of the support through a long pendant chain, essentially the same effect can be achieved with heterogeneous systems[20]. Some homogeneous catalysts suffer deactivation owing to dimerization. The activity of such catalysts can be enhanced by supporting them on fairly rigid supports that isolate the individual metal centres, so preventing dimerization. However, care must be taken to ensure that the rigidity of the support is not compromised by the reaction conditions; thus a polymer which is fairly rigid in the presence of a solvent in which it is insoluble may become flexible in contact with a 'swelling solvent' (see Section I.B.5).

3. Reproducibility

The particular advantage of metal complex catalysts over supported metal catalysts such as Raney nickel is the total reproducibility of the former owing to the molecular nature of the complex, which ensures a unique stoichiometry and structure. By contrast, the structure of the surface of a supported metal catalyst is heavily dependent on both its method of preparation and its history subsequent to preparation. In theory, supported metal complex catalysts are as reproducible as their homogeneous analogues. However, great care is necessary in their preparation if this is to be achieved in practice. It will be necessary to ensure that the preparation and pretreatment of the support is reproducibly repeated every time. Where a multi-step series of reactions are used to link the metal complex to the support then great care will be necessary to ensure that each step has gone to completion at every site on the support. In practice this is very difficult, so that supported metal complex catalysts are generally less uniform than their formal representation indicates. As a result, their reproducibility tends to be lower than that of their homogeneous analogues.

4. Specificity

A given homogeneous metal complex catalyst will generally have only one type of active site and will often be more specific than a heterogeneous supported metal catalyst where several types of active site may be present in the form of different surface defects. These defects are extremely difficult to control and in many cases different defects promote different reactions. These may either be different reactions of the same substrate, or when more than one substrate is present one defect may preferentially promote reaction of one substrate whilst another defect may preferentially promote reaction of another substrate. Supported metal complexes can with care be made as specific as their homogeneous counterparts but, as just emphasized in Section I.A.3, this does need great care. The specificity of a homogeneous metal complex catalyst can often be selectively modified by altering the ligands present in such a way as to alter either the electronic or the steric requirements of the site. The specificity of supported metal complex catalysts can clearly be altered in exactly the same way.

5. Controllability

The specific structure of homogeneous metal complex catalysts enables them to be modified relatively easily in order to control a reaction. For example, altering $[Rh(acac)(CO)_2]$ to $[Rh(acac)(CO)(PPh_3)]$ results in an increase in the ratio of normal to branched aldehydes obtained when hex-1-ene is hydroformylated from 1.2:1 to 2.9:1[21]; replacement of PPh_3 by p-$CH_2{=}CHC_6H_4PPh_2$ further enhances this ratio to 3.9:1[22]. Exactly the same controllability applies to supported metal complex catalysts. By contrast, the ill-defined active sites of heterogeneous supported metal catalysts make systematic design and improvement very difficult.

6. Thermal stability

The thermal stability of heterogeneous supported metal catalysts is generally fairly high, whereas the thermal stability of metal complexes, either supported or unsupported, is usually lower. Since the rate of most reactions increases with increasing temperature, high operating temperatures can be disadvantageous. There are, however, at least two situations in which high operating temperatures are either of no advantage or are a positive disadvantage. The first is where high temperatures promote side-reactions either of the reactants or of the products. The second is in the case of reactions which involve a pre-equilibrium step which is disfavoured by increasing the temperature. This is exemplified by the reactions of olefins where the entropy change on metal–olefin complex formation is almost invariably negative so that the stability of metal–olefin complexes decreases with increasing temperature[23]. The lower thermal stability of homogeneous metal complex catalysts is often compensated for by their significantly higher activities than heterogeneous catalysts at lower temperatures and pressures.

7. Oxygen and moisture sensitivity

Homogeneous metal complex catalysts are often organometallic compounds with metals in low oxidation states. Accordingly, many of them are sensitive to oxygen and moisture. In many cases supporting such catalysts has been found to reduce this sensitivity. Heterogeneous supported metal catalysts are frequently subject to poisoning by 'soft' ligands such as mercaptans and thioethers to a much greater degree than homogeneous metal complex catalysts.

8. Solvent

The range of solvents suitable for a homogeneous catalyst is often limited by the solubility characteristics of the metal complex. Clearly this presents no problems for a heterogeneous supported metal catalyst which is insoluble in all solvents. In the case of supported metal complex catalysts, solvents have to be chosen with great care. If the support is a cross-linked polymer then the solvent may or may not enter the polymer and may alter its three-dimensional structure, swelling it or constricting it. Mixed solvents may give 'solutions' that have different compositions within the pores of the polymer compared with the bulk solvent owing to the preferential uptake of one component into the interstices of the polymer. Finally, solvents influence the reactivity of the actual catalytic site, although they do not of course permit aggregation of the metal complex sites to form precipitates as occurs with homogeneous catalysts when solvents in which they are insoluble are added.

9. Corrosion and plating out

The use of some homogeneous catalysts on a commercial scale has led to a number of practical problems, such as corrosion and plating out on the reactor walls, that are not immediately obvious when the reaction is carried out in all-glass apparatus on the laboratory scale. The oxidative acetylation of ethylene to vinyl acetate catalysed by palladium(II) (reaction 1) is an example of a process that suffers from severe corrosion problems under homogeneous conditions which can be eliminated by supporting the palladium(II). Similarly, replacing sulphuric acid by sulphonated Nafion, which is a tetrafluoroethylene–perfluorinated vinyl ether copolymer, gives a non-coorosive strong acid catalyst[23a].

$$C_2H_4 + NaOAc \xrightarrow[\text{HOAc}]{Pd^{II},\, Cu^{II},\, O_2} CH_2{=}CHOAc \tag{1}$$

B. Advantages of Supporting Metal Complexes

Although the original motivation for supporting homogeneous catalysts was to attempt to combine most of the advantages of homogeneous catalysts that arise from their molecular nature, especially their selectivity and controllability, with the ease of separation of the heterogeneous catalysts, experience has shown that the presence of both the support and the catalyst can have synergistically beneficial effects. Thus, attaching a metal complex to a support can have a number of effects, as follows.

1. *Preferential substrate orientation*

The support may not behave simply as an inert backbone. It may take a positive role in ensuring a particular orientation of the substrate at the catalytically active site, so promoting selectivity[24]. This, of course, is what the supporting backbones in many enzymes have been doing in nature for many millenia[25]. This effect is believed to be largely responsible for the 3.5-fold enhancement of the normal to branched aldehyde selectivity when hex-1-ene is hydroformylated over polypropylene-supported [Rh(acac)($Ph_2PC_6H_4CH{=}CH_2$-*p*)(CO)] compared with hydroformylation over the same complex unsupported[22].

2. *Change in chemical reactivity*

When organic functional groups are bound close to the surface of polymers or inorganic solids, they are subject to special constraints which can alter their chemical reactivities relative to the same groups in small mobile molecules[26]. In the same way, the properties of metal complexes can be altered when they are immobilized on supports.

3. *Multidentate ligands*

Functionalized supports are effectively multidentate ligands. Supporting metal complexes on such ligands can alter the stereochemical environment of the metal atom in a beneficial way. This is well illustrated by the selectivity of nylon-supported platinum benzene hydrogenation catalysts[27]. Platinum anchored on nylon 66, nylon 6, and nylon 610 catalyses the formation of cyclohexene, whereas platinum supported on nylon 3, although an active hydrogenation catalyst, results in cyclohexane being formed exclusively.

4. *Modification of metal–ligand equilibria*

Supporting a metal complex may alter the position of equilibria between a metal ion and its surrounding ligands. This effect is believed to be responsible for the fact that when rhodium(I) complexes are bound to phosphinated supports much lower phosphorus to rhodium ratios give greater selectivity enhancements in olefin hydroformylation when supported catalysts are used, compared with their homogeneous analogues[22,28].

5. *Stabilization of unstable structures: site isolation*

It is sometimes possible to stabilize metal complexes that are unstable in solution by supporting them on an inert matrix. Very often, although not always, this stabilization is achieved through site separation which prevents two extremely reactive monomeric complexes from combining together to form an unreactive dimer. This particular process is a major deactivation mechanism for many rhodium(I) complexes[29]. Activation through site isolation is the key to titanocene hydride hydrogenation catalysts, which are very active in supported form but virtually inactive when unsupported in homogeneous solution[30]. However, catalysts which rely on site isolation for their activity must be used with care if the supports are cross-linked polymers, since it has recently been shown that in swelling solvents there is considerable flexibility of the polymer. This results in sites, which in a non-swelling solvent are completely isolated, interacting with one another when a swelling solvent is added[31]. This has been demonstrated by a number of reactions, of which reaction 2 is an excellent example. Thus, when [Co(NO)(CO_3)] is reacted with either a low cross-linked or a 20% cross-linked phosphinated polystyrene in the absence of solvent or in the presence of *n*-hexadecene, which is a poor swelling solvent, only the 1:1 complex is formed, as shown by the infrared absorption at 1755 cm^{-1}. On adding *m*-xylene, which is a good swelling solvent, rapid site–site interaction occurs and a 1:2 complex is formed, demonstrating that cross-links are not in themselves sufficient to maintain site isolation in swelling solvents[31]. By contrast, inorganic supports such as silica are not usually susceptible to large structural changes on altering the solvent and so maintain site–site isolation[32].

$$\text{Ⓟ}\!-\!PPh_2 + [Co(NO)(CO)_3] \xrightarrow{n\text{-hexadecane}} [Co(\text{Ⓟ}\!-\!PPh_2)(NO)(CO)_2] \quad \nu_{NO} = 1755\ cm^{-1}$$

$$\xrightarrow{m\text{-xylene}} [Co(\text{Ⓟ}\!-\!PPh_2)_2(NO)(CO)] \quad \nu_{NO} = 1710\ cm^{-1} \qquad (2)$$

Ⓟ = polystyrene

6. Site cooperation

Enzymes frequently achieve their high selectivities by the simultaneous cooperative action of more than one type of catalytic site[33]. A similar effect can be achieved with supported catalysts. A good example of such site cooperation is found in the catalysis of the hydrolysis of unsaturated esters by acid ion-exchange resins in which some of the protons on the acidic sites have been replaced by silver(I) ions[34]. This substitution increases the rate of unsaturated ester hydrolysis owing to silver(I) binding the olefinic site, so tying down the ester whilst the hydrolysis is catalysed at nearby acidic sites.

7. Protection against poisons

Attaching a metal complex to a support can sometimes provide protection for the catalytically active species against poisons such as water or atmospheric oxygen. Thus, supporting aluminium(III) chloride, which is normally rapidly hydrolysed, on polystyrene results in a material that is almost completely insensitive to moisture during manipulation in air[35]. Similarly, rhodium(I)–phosphine complexes are generally unstable in the presence of air, although several supported rhodium(I) complexes have been found to be insensitive to oxygen and can be filtered and recycled in air without any need to take special precautions[36–38]. However, this is not true of all supported rhodium(I) catalysts, and most are usually best handled in the absence of air. Although many rhodium(I) homogeneous catalysts are sensitive to poisoning by thiols, *n*-butanethiol reacts with silica-supported Wilkinson's catalysts, $[(\{Si\}\text{-}CH_2CH_2PPh_2)_3RhCl]$, to reduce their activity but enhance their thermal stability[38a–c]. Similarly, rhodium(I)–anthranilic acid hydrogenation catalysts supported on chloromethylated polystyrene have long-term thermal stability and are fairly insensitive to poisoning[39,39a].

8. Cooperative effects of support and metal complex

Many supported metal complex catalysts are the exact analogues of their homogeneous counterparts, with one or more small molecule ligands being replaced by the same type of functional group bound to a support. This may well not be the best way to mimic the enzymes and achieve cooperation between the support and the metal complex. It may indeed be better to use totally different functional groups when these are bound to a support[39]. This is because a ligand that is ideal for a homogeneous metal complex catalyst is one that coordinates fairly strongly to the metal, so preventing metal–ligand bond cleavage and subsequent reduction of the metal ions to the free metal. However, on a support metal–ligand dissociation is more spacially restricted so that weaker bonding ligands may be used. These may have electronic and steric advantages that are not realizable in homogeneous situations for stability reasons.

One way of promoting high selectivity has been to support the metal complex within the interstices of a cross-linked polymer. Diffusion of the reactants into the polymer and up to the active site then provides for selectivity when the reactant is a mixture of components. However, such selectivity is necessarily achieved at the expense of activity.

C. Requirements of a Supported Catalyst

To be of commercial interest, a supported metal complex catalyst must possess a number of desirable features. First, it must be highly selective so that product separation is simple. It should have as high an activity per unit volume of reactor space as possible and the cost of the catalyst per unit of the product being produced should be low; these two imply a high turnover number.

From a commercial point of view, several of the highly desirable features are mutually exclusive. Thus, if the metal complex is supported solely on the surface of the catalyst the activity per unit volume of reactor space will be low; hence attention is often directed at porous supports. However, diffusion and mass transport within the support are usually much slower than in the bulk solution so that the activity may not be too greatly enhanced in this way.

Most commercial reactors involve considerable agitation of the catalyst. If the catalyst is not to be ground up to produce a lot of 'fines' then the support needs to be mechanically strong. Polystyrene, which is attractive from many points of view, particularly the ease with which it is functionalized, is unattractive from this point of view. Similarly, under pumped flow conditions polystyrene beads can pack down very tightly into a bed, so generating very high pressure drops[40]. Inorganic supports are less susceptible to this problem. Although polystyrene is by far the most widely studied polymer, more recently attention has been directed towards mechanically tougher polymers such as polypropylene[22,41] and poly(phenylene oxides)[42]. Industry has over many years developed the technology to handle catalysts based on inorganic supports, such as γ-alumina and silica, and would therefore be most receptive to metal complex catalysts supported on these materials. Inorganic supports can be combined with polymeric supports in two ways to enhance the strength of the latter. The first is to polymerize an organic material on to an inorganic base[40] and the second, which is applicable only in column operations, is to mix the polymer intimately with an inorganic matrix of similar bead dimensions, using the mechanical rigidity of the inorganic support to avoid clogging of the flow channels[43].

The possibility of swelling polymers is often an advantage in laboratory-scale reactions, but can be a disadvantage commercially where engineering for a material whose dimensions change during its lifetime presents major problems, particularly in flow columns. Clearly the support must be inert to the reactants and products of the catalytic reaction.

Since many reactions are exothermic, the supported catalyst must not only be stable to reasonable temperatures but must also have adequate heat transfer properties to disperse the heat generated at the active site. Catalytic centres located on the surface will aid this process owing to the free flow of solvent around them, whereas the removal of heat from a site within a support will necessarily be more difficult.

II. TYPES OF SUPPORT

There are two broad classes of supported metal complex catalysts, those in which the support is functionalized so that it contains donor groups which then bind to a metal and those in which the support forms a direct link to the metal complex. Phosphinated polymers or phosphinated silica both provide examples of the former, and the latter are exemplified by the reaction of organometallic complexes with silica or polystyrene (reactions 3 and 4).

$$\{Si\}—OH + MR_n \longrightarrow \{Si\}—OMR_{n-1} + RH \qquad (3)$$

$$-(CH_2-CH(C_6H_5))_n- + [Mo(CO)_6] \longrightarrow -(CH_2-CH(C_6H_5Mo(CO)_3))_n- + 3CO \qquad (4)$$

In both classes of catalyst two broad types of support are used, organic polymers and inorganic oxides. There are advantages and disadvantages to each of them, as considered below, but a significant difference between them is the degree to which they can be functionalized. Inorganic oxide matrices have an upper limit of monofunctional groups they can carry of 1–2 mequiv. per gram of matrix, whereas organic polymer matrices can carry up to 10 mequiv. per gram of matrix[14].

A. Organic Polymers

Organic polymers such as polystyrene, polypropylene, and to a lesser extent poly(vinyl chloride) have been widely used to support metal complex catalysts. They offer a number of advantages over inorganic oxide supports[44]:

1. They are easily functionalized, particularly polystyrene, which contains reactive phenyl side-chains.
2. They are chemically inert and, being hydrophobic, are not susceptible to reaction with water, so changing their nature and polarity. Their chemical inertness means that the support neither enters into undesirable chemical reactions nor interferes with the catalytic centre.
3. Polymers can be prepared with a wide range of physical properties enabling their porosity, surface area, and solution characteristics to be varied over a wide range, usually by varying the degree of cross-linking[24,45,46]. Polystyrene, for example, can be prepared as a material that is virtually soluble in solvents such as benzene (very little cross-linking) up to a completely insoluble material at 20% cross-linking.

The principal disadvantages of polymers are their poor heat transfer characteristics and, in the case of polystyrene, poor mechanical properties. In addition, unless the polymer is made under carefully controlled and recorded conditions it may not be well defined and may contain unknown impurities, both of which can lead to problems of reproducibility.

The physical properties of polymers vary widely, depending on their molecular weight, the chemical nature of the monomer or combinations of monomers, and the conditions of polymerization which affect the arrangement of the polymer molecules and their interactions with one another. There are essentially three major types of polymers, although in practice a continuous range of material is available in between these three extremes.

1. Gels

Gels, or microporous polymers, involve long strands of polymer molecules either lightly cross-linked or merely entangled together. Microporous polymers are produced when the polymerization is carried out in the absence of added inert polymers. As the polymerization proceeds the polymer chains are solvated by unreacted monomer which is used up as the polymerization occurs, causing the chains to aggregate, finally yielding a glassy product (Figure 1a). In the absence of solvent the pore size is approximately the distance between the polymer chains, hence the description microporous.

If a solvent with a high affinity for the polymer is present during the polymerization then considerable swelling will occur, giving rise to increased porosity and the formation of 'gels'. The degree of swelling depends on both the solvent used and the degree of cross-linking. In microporous polymers the molecules are in constant random motion. Hydrocarbon polymers such as polystyrene can accommodate high concentrations of chemically similar molecules such as other hydrocarbons, although they repel polar molecules such as water. Small molecules such as benzene can diffuse rapidly through the gel, encountering almost as little resistance as they would in solution. This facilitates the transport of reactants and products to and from the catalytically active sites as well as the removal of heat. Gellular polymers can be used in hydrocarbon solvents as essentially soluble supports that may be separated at the end of the reaction either by precipitation by changing the solvent, or by osmotic procedures such as membrane filtration[47–50]. In polar solvents, which are often used for such catalytic reactions as hydrogenation, gellular polymers are not swollen but instead tend to close up their pores[45].

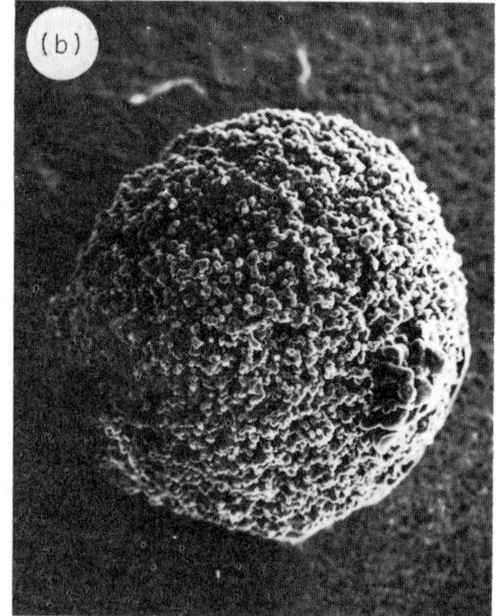

FIGURE 1. Scanning electron micrograph of (a) gel, bead diameter *ca.* 480 μm, and (b) macroreticular aminated polystyrene, bead diameter *ca.* 690 μm (photographs kindly provided by Dr. H. Widdecke)

2. Macroreticular polymers

Macroreticular or macroporous polymers have a carefully controlled regular cross-linking or reticulation which gives a high internal surface area to the polymer[14,24]. An inert solvent must be used during the polymerization process, together with carefully controlled amounts of difunctional and monofunctional monomers[24,51]. These polymers are easier to form on a large commercial scale than on a small laboratory scale because it is easier to maintain constant concentrations of reactants on a large scale. On a small scale there is a tendency for the difunctional monomers to be consumed preferentially at the start of the polymerization, giving rise to the initial formation of a highly cross-linked material which precipitates. These particles are then connected by polymer with a decreasing amount of cross-linking and within them are voids filled with monomer solution that steadily becomes depleted of monomer.

When polymerization is complete and the solvent has been removed, macroreticular resins may retain some porosity owing to their heterogeneous nature (see Figure 1b). They readily take up good solvents, but can also accommodate poor solvents owing to their macroporous nature. If the solvent used in the polymerization is one in which the polymer is insoluble then large permanent pores are formed. Such macroreticular polymers have high internal surface areas. Those that are highly cross-linked swell only slightly even in good solvents. Polystyrene cross-linked with divinylbenzene is a commonly used macroreticular polymer. Benzene swells all but the highest cross-linked polystyrenes. Donor ligands can be supported both on the surface of the polymer and within its pores. Clearly, reagents do not have the same access to the internal sites as to those on the surface and so such polymers can give high selectivity. For example, the rates of hydrogenation of a series of olefins in benzene solution in the presence of [Rh(Ⓟ-PPh_2)(PPh_3)$_2$Cl], where Ⓟ-PPh_2 is 100–200 mesh 2% cross-linked phosphinated polystyrene, decrease as the steric bulk of the olefin increases in the order hex-1-ene > > cyclohexene > > cyclooctene > cyclododecene > > Δ^2-cholestene[52]. The same effect can be used to promote the regioselective reduction of the side-chain double bond of the steroid **1** rather than reduction of the steroid nucleus (reaction 5), whereas the corresponding homogeneous catalyst promotes the reduction of both double bonds[53].

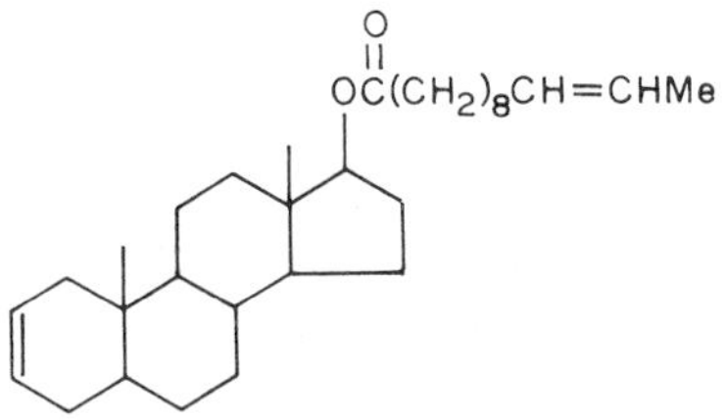

$\overset{O}{\|}$ OC(CH_2)$_8$CH=CHMe

(1)

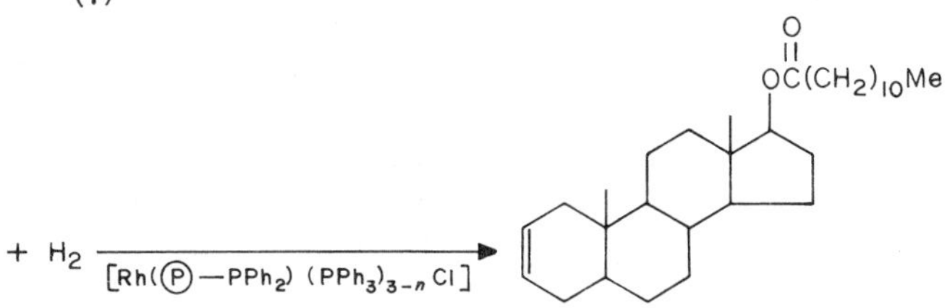

+ H_2 $\xrightarrow{[Rh(Ⓟ-PPh_2)(PPh_3)_{3-n}Cl]}$ [steroid with $\overset{O}{\|}$ OC(CH_2)$_{10}$Me] (5)

Ⓟ = polystyrene ; $n = 1,2$

A particularly selective catalyst for the hydrogenation of small olefins has been prepared by swelling a phosphinated polystyrene supporting rhodium(I) and then, after removing the solvent to contract the beads, poisoning the surface catalytic sites. On reswelling, the only catalytic sites that remain are those deep within the polymer and these are only accessible to small olefins[53a].

3. Proliferous polymers

Proliferous or 'popcorn' polymers[54,55] are formed spontaneously in the absence of initiator in butadiene and butadiene–styrene copolymerization plants. They are hard, opaque, microporous materials which swell in benzene and carbon tetrachloride but are insoluble in all common solvents.

One of the attractive features of organic polymer supports is the opportunity they offter to introduce extra selectivity into the catalyst as a result of the need for the reactants to diffuse through the polymer to reach the catalytically active sites. However, the diffusion necessarily reduces the activity, since diffusion rates within the polymer are typically an order of magnitude lower than in the bulk solution[14]. If the actual catalytic reaction is fast relative to diffusion then the full potential of the supported catalyst will not be reached and its activity will be less than that of its homogeneous analogue. This diffusion limitation of polymer-supported catalysts can be reduced by (i) reducing the particle size, (ii) increasing the overall surface area of the support, for example by using a gellular polymer, (iii) introducing a bimodal pore size distribution allowing rapid diffusion through a portion of the catalyst, and (iv) increasing the pore sizes. The pore sizes depend on the degree of cross-linking and also the degree of swelling which in turn depends on the nature of the solvent as well as of the reactants and products. Thus with polystyrene, solvents that are more polar than benzene (i) decrease the pore size as they give less swelling and (ii) create polar gradients between the bulk solvent and the local environment around the active site[52]. The first of these effects decreases the diffusion rate of large, bulky reactants whilst the second selectively enhances the diffusion rate of non-polar reactants within the polymer. These two conflicting effects give rise to the complex effect of increasing solvent polarity in the

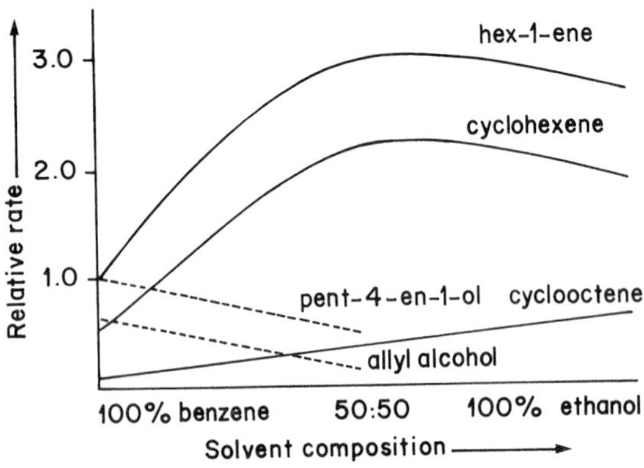

FIGURE 2. Influence of solvent polarity on the relative rates of hydrogenation of non-polar (solid lines) and polar (broken lines) olefins (reproduced with permission from ref. 52)

hydrogenation of olefins in the presence of rhodium(I) supported on phosphinated polystyrene (Figure 2). As the ethanol content of the benzene–ethanol solvent mixture is increased, the rate of reduction of non-polar olefins increases owing to enhanced diffusion rates until the ethanol content becomes sufficiently high that the pores begin to shrink significantly so that the molecular size begins to dominate. Polar olefins experience a steadily decreasing hydrogenation rate with increasing solvent polarity owing to suppression of the diffusion rate as the solvent polarity increases, combining with the reduction in pore size to make migration to the catalytic site increasingly difficult as the ethanol content is increased[52].

B. Inorganic Supports

The major advantages of inorganic supports over their organic counterparts are their better mechanical and thermal stabilities coupled with reasonable heat transfer properties. A further advantage to industry is that because of their use as supports for classical heterogeneous catalysts a great deal of technical 'know-how' and experience has been built up. Typical inorganic supports that have been used include silica, alumina, glasses, clays, and zeolities. Inorganic supports are not always as inert as organic polymers. For example, metal oxides such as silica and alumina pick up water reversibly, so giving rise to polarity changes which can influence reactions at catalytic metal complex centres. Supports such as silica lack the flexibility of organic polymers, whereas clays do have some flexibility. Although flexibility is often very desirable, this is not always so, particularly where site isolation is important in preventing catalyst deactivation through dimerization (see Section I.B.5). In such situations inorganic supports are particularly valuable. However, inflexibility can sometimes be a disadvantage, as in cases where the distance between the surface and the catalytic centre is critical to catalytic performance[56]. For example, rhodium(I) hydroformylation catalysts supported on phosphinated silica with short chains between the silica and the phosphine are less active than those with longer links. The same phenomenon has been observed with heterogenized enzymes[33].

There are many situations in which it is very advantageous to be able to control diffusion rates. This is often difficult with organic polymers owing to the degree of swelling being dependent on both the solvent and the precise temperature within the polymer, which may vary substantially within a polymer that is being used to support a catalyst for an exothermic reaction. Inorganic substrates, on the other hand, have fairly fixed diffusional properties under most reaction conditions[57]. Zeolites offer a wide range of well understood, well controlled pore sizes[58–63]. Smectites such as hectorite and montmorillonite are a group of naturally occurring silicates that swell in the presence of water, alcohols, and other organic solvents[64]. Smectites are made up of alternating layers of cations and negatively charged silicate sheets. The cations can be exchanged for cationic metal complexes such as rhodium complexes[64–69]. The degree of swelling depends on the nature of the cations, the solvent and the negative charge density on the silicate sheet.

Most inorganic oxides contain surface hydroxyl groups, which make them fairly polar. If this is undesirable it can be reduced by reaction with a chlorosilane such as trimethyl- or *tert*-butyldimethyl-chlorosilane to give a non-polar lipophilic surface[70–72].

III. FUNCTIONALIZATION OF SUPPORTS

There are sometimes several ways in which a particular functional group may be introduced on to a support. When considering which route to use an important consideration, apart from ease of reaction, is uniformity of functionalization. Any chemical reaction taking place on such a heterogeneous medium as an inorganic or organic support is likely to be incomplete unless elaborate precautions have been taken. Thus, if functionalization involves several steps, it is likely that some of the functional groups introduced in the early steps will still remain in the final functionalized support. If a particular functionalization reaction is rapid relative to the rate of diffusion of the reagents within the support, then a shell of progressive introduction of functionality from the exterior of the particle towards the centre will result. If diffusion is more rapid, then a more uniform functionalization will result.

Unless each step in the functionalization is taken to completion, unexpected side-reactions can occur. This is well illustrated by the side-reactions that can occur when phosphinated polystyrene is prepared by the chloromethylation route (see reaction 17). Chloromethyl side-chains can react with newly introduced phosphine groups to give quaternization (reaction 6)[73]. A useful technique for determining the distribution of

$$\text{Ⓟ}-CH_2Cl + \text{Ⓟ}-CH_2PPh_2 \longrightarrow (\text{Ⓟ}-CH_2)_2PPh_2^+Cl^- \qquad (6)$$

functional groups within a support is electron microprobe analysis of a microtome-sectioned support which allows the distribution of atoms such as phosphorus or metal to be determined[74].

The nature of the functional group introduced has often been assumed by analogy with reactions occurring on simple monomeric materials in homogeneous solution. This can be misleading and a complete characterization of the functionalized support should be undertaken. A number of techniques are available, including microanalysis, infrared and n.m.r. including magic-angle cross-polarization solid-state n.m.r. The interested reader is referred elsewhere for details of these other techniques[16], which for metal complexes themselves are described in Volume 1 of this series.

There are three broad routes for the functionalization of supports: functionalization of an organic polymer, polymerization of a functionalized monomer, and functionalization of an inorganic support. Each is considered in turn.

A. Functionalization of Organic Polymers

If a commercial polymer is to be functionalized it is often necessary first to remove the impurities that are left by the manufacturer[75]. In the case of polystyrene these may include alumina, Fuller's earth, carboxymethylcellulose, stearic acid, sodium lauryl sulphate, and sodium polyacrylamide[76]. If they were left they would inhibit the penetration of ionic reagents such as butyllithium or metal diphenylphosphides. A satisfactory procedure[75] for removing these surface impurities involves successive washings with 1 N NaOH (60 °C), 1 N HCl (60 °C), 1 N NaOH (60 °C), 1 N HCl (60 °C), H_2O (25 °C), dmf (40 °C), 1 N HCl (60 °C), H_2O (60 °C), MeOH (20 °C), 3:2 (v/v) MeOH–CH_2Cl_2, 1:3 (v/v) MeOH–CH_2Cl_2, 1:9 (v/v) MeOH–CH_2Cl_2, pure CH_2Cl_2, followed by drying to constant weight at 100 °C *in vacuo* (10 Torr overnight, then 0.1 Torr for several hours).

A wide range of functional groups may be introduced on to organic polymers. In many cases functionalization involves two stages, an initial functionalization of a hydrocarbon material followed by reaction of the initial group with a further reagent to yield the final material. There are three very important reactions that are widely used to achieve initial activation of organic polymers: lithiation, chloromethylation, and radiation grafting.

Lithiation is particularly valuable with polymers containing aromatic groups, such as polystyrene or poly(phenylene oxides). It can be achieved either by direct action of *n*-butyllithium, often complexed with *N,N,N′,N′*-tetramethylethylenediamine (reaction 7)[77–82], or by metal–halogen exchange on a ring-halogenated polymer (reactions 8 and 9)[42, 82–87]. Direct lithiation is now believed to yield a majority (*ca.* 66%) of *meta*-

$$\text{(P)}-\text{C}_6\text{H}_5 + \text{Bu}^n\text{Li} \xrightarrow{\text{tmeda}} \text{(P)}-\text{C}_6\text{H}_4\text{Li} + \text{Bu}^n\text{H} \quad (7)$$

$$\text{(P)}-\text{C}_6\text{H}_4-\text{X} + \text{RLi} \longrightarrow \text{(P)}-\text{C}_6\text{H}_4-\text{Li} + \text{RX} \quad (8)$$

$$(\text{C}_6\text{H(Br)(Ph)}_2-\text{O})_n + \text{Bu}^n\text{Li} \longrightarrow (\text{C}_6\text{H(Li)(Ph)}_2-\text{O})_n + \text{Bu}^n\text{Br} \quad (9)$$

product[79,80,88]. Lithium–halogen exchange involves an extra step, but its advantage lies in the fact that greater degrees of lithiation can be achieved by this route. Although at one time iodine was the preferred halogen, bromine is now more common, the bromination usually being carried out in the presence of a Lewis acid, thallic acetate being recommended[82, 89–92], although iron(III) chloride and aluminium chloride have also been widely used.

Chloromethylation has been extensively used to enable a given donor atom to be linked to an aromatic backbone through an aliphatic methylene group, so influencing the electronic properties of the ligand as experienced by a coordinated metal ion. Chloromethylation is usually achieved by a Friedel–Crafts reaction using chloromethyl methyl ether in the presence of tin(IV) chloride (reaction 10)[44, 93–95]. The product, commonly known as

$$\text{(P)}-\text{C}_6\text{H}_5 + \text{ClCH}_2\text{OMe} \xrightarrow{\text{SnCl}_4} \text{(P)}-\text{C}_6\text{H}_4\text{CH}_2\text{Cl} + \text{MeOH} \quad (10)$$

Merrifield's resin has a majority (*ca.* 85%) of the chloromethyl groups in the *para*-position[96,97]. Unless care is taken to keep solutions fairly dilute and reaction times short, further reaction (reaction 11) occurs to eliminate hydrogen chloride and effectively

$$\text{(P)}-\text{C}_6\text{H}_4-\text{CH}_2\text{Cl} + \text{(P)}-\text{C}_6\text{H}_5 \xrightarrow{\text{SnCl}_4} \text{(P)}-\text{C}_6\text{H}_4-\text{CH}_2-\text{C}_6\text{H}_4-\text{(P)} + \text{HCl} \quad (11)$$

introduce a cross-link into the polymer[98]. N.m.r. has recently revealed that some of the chloromethyl groups in 'chloromethylated polystyrene' are hydrolysed to hydroxymethyl groups in the course of normal storage[98a]. Polymers which already contain methyl side-chains can be brominated by using *N*-bromosuccinimide to yield bromomethylated aromatic rings (reaction 12)[42].

$$(\text{C}_6\text{H}_2(\text{Me})_2-\text{O})_n \xrightarrow[h\nu]{N\text{-bromosuccinimide}} (\text{C}_6\text{H}_2(\text{CH}_2\text{Br})(\text{Me})-\text{O})_n \quad (12)$$

Radiation grafting has proved a valuable route for functionalizing hydrocarbon polymers that are chemically rather inert such as polyethylene and polypropylene[41, 99–111]. Irradiation, usually using γ-radiation but also using u.v, high-energy (*ca.* 3–4 MeV) electrons, or plasmas, of hydrocarbon polymers results in the formation of radicals which in the presence of an unsaturated monomer can undergo

$$\sim\!CHR\!\sim \xrightarrow{\text{irradiation}} H^{\bullet} + \sim\!\dot{C}R\!\sim \xrightarrow[\text{absence of } O_2]{CH_2{=}CHX,} \sim\!\underset{\displaystyle CH_2\dot{C}HX}{\underset{|}{C}}R\!\sim \qquad (13)$$

carbon—carbon bound formation (reaction 13, where X = amine, phosphine, nitrile, OH or SH containing group). The resulting radical can then undergo one of three reactions:

(i) radical recombination with the original hydrogen radical displaced; this is relatively unlikely;

(ii) reaction with a solvent molecule to give chain termination accompanied by the formation of a new radical originating from the solvent;

(iii) reaction with a further molecule of the unsaturated monomer to yield a graft copolymer (reaction 14); this is the predominant reaction that occurs when 4-vinylpyridine[41] and *p*-styryldiphenylphosphine[109] are γ-radiation grafted onto polypropylene in benzene suspension.

$$\sim\!\underset{\displaystyle CH_2\dot{C}HX}{\underset{|}{C}}R\!\sim + CH_2{=}CHX \longrightarrow \sim\!\underset{\displaystyle CH_2CHXCH_2\dot{C}HX}{\underset{|}{C}}R\!\sim \longrightarrow \text{etc.} \qquad (14)$$

By far the commonest functional groups introduced on to supports are phosphines. Phosphination can be achieved in a number of ways, including those illustrated in reactions 15–21. Although formally the products of reactions 15 and 16 appear identical, they are

$$Ⓟ\!-\!C_6H_4\!-\!Br + MPPh_2 \xrightarrow[\text{(refs. 49, 112–121)}]{M = Li, Na, K} Ⓟ\!-\!C_6H_4\!-\!PPh_2 + MBr \qquad (15)$$

$$Ⓟ\!-\!C_6H_4\!-\!Li + Ph_2PCl \xrightarrow[R = \text{menthyl (refs. 125, 126)}]{R = Ph \text{ (refs. 81, 118, 122–124)}} Ⓟ\!-\!C_6H_4\!-\!PR_2 + LiCl \qquad (16)$$

$$Ⓟ\!-\!C_6H_4CH_2Cl + MPR_2 \xrightarrow[\substack{R = Ph \text{ (refs. 49, 113, 118, 127–132)} \\ R = \text{menthyl (refs. 125, 133, 134)}}]{M = Li, Na, K} Ⓟ\!-\!C_6H_4CH_2PR_2 + MCl \qquad (17)$$

$$Ⓟ\!-\!C_6H_4CH_2Cl + Li(CH_2)_nPRR' \xrightarrow{\text{(ref. 135)}} Ⓟ\!-\!C_6H_4(CH_2)_nPRR' + LiCl \qquad (18)$$

$$\text{Ⓟ-CH}_2\text{NH}_2 + \text{CH}_2\text{=CH(CH}_2)_8\text{COCl} \xrightarrow[\text{12 h}]{\text{py, 60 °C}} \text{Ⓟ-CH}_2\text{NHCO(CH}_2)_8\text{CH=CH}_2$$

(i) BH_3, SMe_2, 0 °C, 1 h
(ii) H_2O_2

$$\text{Ⓟ-CH}_2\text{NHCO(CH}_2)_{10}\text{OH}$$

tosCl

$$\text{Ⓟ-CH}_2\text{NHCO(CH}_2)_{10}\text{Otos}$$

$LiPPh_2$ in thf

$$\text{Ⓟ-CH}_2\text{NHCO(CH}_2)_{10}\text{PPh}_2$$

(19)

Ⓟ–C_6H_5 + PCl_3 —$BF_3 \cdot Et_2O$ (refs. 136–138)→ Ⓟ–C_6H_4–PCl_2 —RLi (ref. 139)→ Ⓟ–C_6H_4–PR_2

(20)

Ⓟ–C_6H_4–CH_2Cl —oxidize in dmso (ref. 140)→ Ⓟ–C_6H_4–CHO

HO, H, Otos, Otos, HO, H

Ⓟ–C_6H_4–CH(O)(O) ... H, Otos, Otos, H

$LiPPh_2$

Ⓟ–C_6H_4–CH(O)(O) ... H, PPh_2, PPh_2, H

(21)

not so in practice because the separate steps occur at different rates and so they result in different distributions of the phosphine groups within the support, as well as leaving different incompletely reacted intermediates on the supports. The reaction of $LiPPh_2$ with pvc is not as straightforward as might be expected[115,117,118]. Thus, although reaction 22

TABLE 1. Functional groups introduced into polystyrene

Functional group	Starting polymer[a]	Reagent	References
$—SO_3H$	Ⓟ	Conc. H_2SO_4	44, 95, 144, 145
$—CHO$	Ⓟ-CH_2Cl	dmso, $NaHCO_3$	93, 140, 146–150
$—COOH$	Ⓟ-Li	CO_2	82, 151
	Ⓟ-CH_2Cl	(i) dmso, $NaHCO_3$ (ii) $Na_2Cr_2O_7$, H_2SO_4, HOAc	148, 149
$—CH_2OH$	Ⓟ-CH_2Cl	NaOH	152
$—CH_2CH_2OH$	Ⓟ-Li	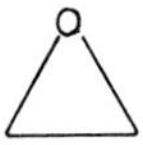 O	44, 82, 95
$—CH_2CH_2OP(OEt)_2$	Ⓟ-CH_2CH_2OH	$ClP(OEt)_2$	44, 95
$—CH_2$Ocrown ether	Ⓟ-CH_2Cl	Na salt of crown ether	153–156
$—CH_2NHCO(CH_2)_nBr$, used to support crown ethers	Ⓟ-CH_2NH_2	$ClCO(CH_2)_nBr$	157
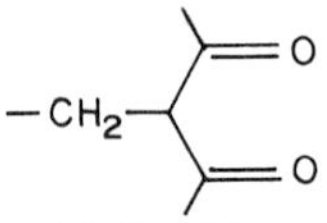 $—CH_2$ O O	Ⓟ-CH_2Cl	Either Hacac + trace of NaOEt in thf,	158
		or Naacac + NaI	159
$—CH(COOEt)_2$	Ⓟ-CH_2Cl	$CH_2(COOEt)_2 + NaHCO_3$	160
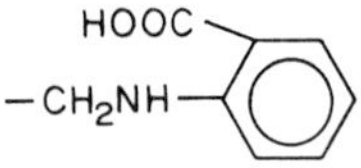 HOOC $—CH_2NH$	Ⓟ-CH_2Cl	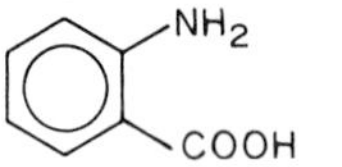 NH_2 COOH	163–163
$—NH_2$	Ⓟ	(i) HNO_3, Ac_2O, HOAc (ii) $SnCl_2$, HCl, HOAc (iii) KOH, MeOH	164
$—CH_2NH_2$	Ⓟ-CH_2Cl	(i) K phthalimide, dmf (ii) ethanolic hydrazine	157
$—CH_2NR_3^+Cl^-$	Ⓟ-CH_2Cl	NR_3	93, 165, 166
$—CH_2N\{(CH_2)_3NH_2\}_2$	Ⓟ-CH_2Cl	(i) NaI, Me_2CO (ii) $HN(CH_2CH_2CN)_2$, thf (iii) BH_3, thf	167, 168
$—CH_2NHCH_2CH_2NH_2$ $—CH_2NHCH_2CH_2NHCH_2CH_2NH_2$ $—CH_2NH(CH_2CH_2NH)_2CH_2CH_2NH_2$	Ⓟ-CH_2Cl	en dien trien	169
$—CH_2OCONH(CH_2)_nNH_2$	Ⓟ-CH_2OH	(i) $ClCOOC_6H_4NO_2$-*p* (ii) $H_2N(CH_2)_nNH_2$	170
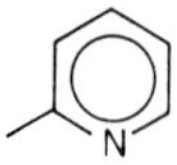 N	Ⓟ-Li	Pyridine	171
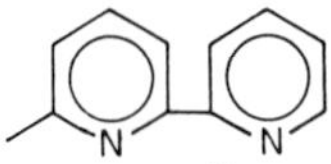 N N	Ⓟ-Li	Bipyridine	172–175
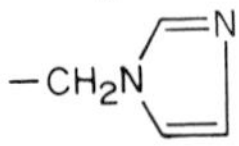 $—CH_2N$ N	Ⓟ-CH_2Cl	Li imidazolate in thf	176
		Na imidazolate in dmf	177
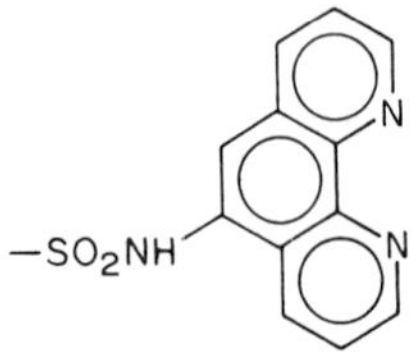 $—SO_2NH$ N N	Ⓟ-SO_2Cl	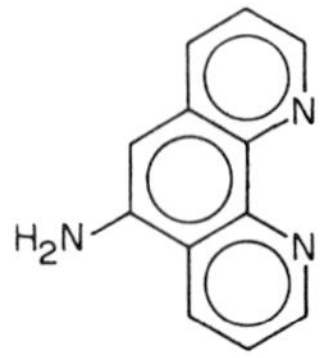 H_2N N N	178

TABLE 1. (*Contd.*)

Functional group	Starting polymer[a]	Reagent	References
$-SO_2Cl$	Ⓟ	$ClSO_3H$, CCl_4	178
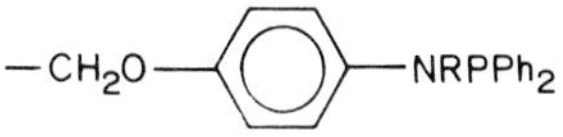 $-CH_2O-C_6H_4-NRPPh_2$	Ⓟ-CH_2Cl	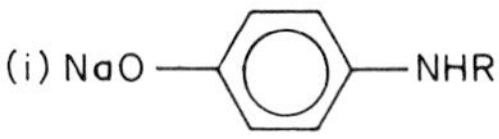 (i) $NaO-C_6H_4-NHR$ (ii) $ClPPh_2$	179
$-CH_2XTPP$ (X = NH, COO, CO; TPP = tetraphenylporphyrin)	Ⓟ-CH_2Cl	TPPXH (X = NH, COO) or TPPCOCl (X = CO)	180
—NHYTPP (Y = CO or SO_2)	Ⓟ-NH_2	$TPPH_2(YCl)_4$	164
$-CH_2CN$	Ⓟ-CH_2Cl	NaCN, dmso	181
$-CH_2CH_2NC$	Ⓟ-CH_2Cl	$LiCH_2NC$	182
$-AsPh_2$	Ⓟ-Li	$ClAsPh_2$	183
	Ⓟ-Br	$LiAsPh_2$	183
—SH	Ⓟ-Li	(i) Sulphur (ii) $LiAlH_4$	82, 184
—SMe	Ⓟ-Li	MeSSMe	89, 185
$-CH_2SH$	Ⓟ-CH_2Cl	$S{=}C(NH_2)_2$	186
$-CH_2SMe$	Ⓟ-CH_2Cl	KSMe	187
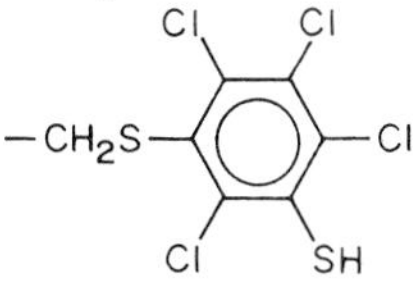 $-CH_2S-C_6Cl_4SH$ (Cl, Cl, Cl, Cl, SH)	Ⓟ-CH_2Cl	(i) C_6Cl_5SNa (ii) NaSH	188
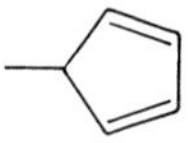 —C_5H_5 (cyclopentadienyl)	Ⓟ-Li	(i) 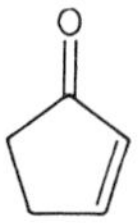 cyclopent-2-en-1-one (ii) H^+	189–191
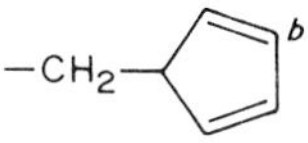 $-CH_2-C_5H_5$[b]	Ⓟ-CH_2Cl	NaCp	192, 193
$-CH_2C_2B_{10}H_{11}$	Ⓟ-CH_2Cl	$LiC_2B_{10}H_{11}$	194
$-CH_2C_2B_9H_{10}^-$	Ⓟ-CH_2Cl	$Na_2C_2B_9H_{10}$	195

[a] Ⓟ = Polystyrene; Ⓟ-Li and Ⓟ-CH_2Cl = functionalized polystyrenes prepared as described in the text; Ⓟ-NH_2, Ⓟ-CH_2NH_2, Ⓟ-CH_2OH, Ⓟ-SO_2Cl = functionalized polystyrenes prepared according to references in this Table.
[b] Material does not have the structural integrity once thought since it was believed[188,189] to be formed by

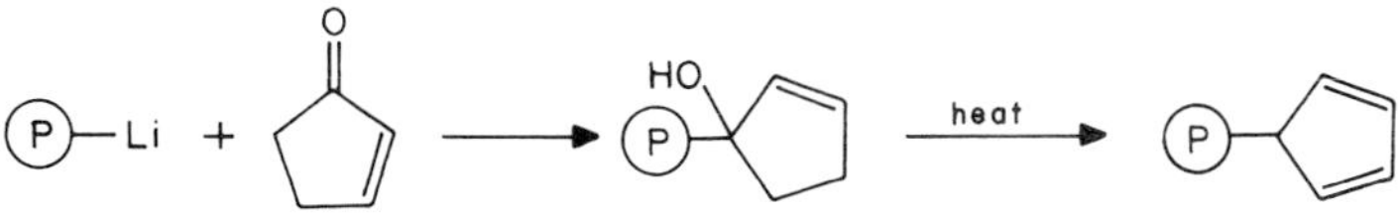

Ⓟ—Li + cyclopent-2-en-1-one ⟶ Ⓟ(HO)-cyclopentenyl —heat⟶ Ⓟ—cyclopentadienyl

whereas models show the initial reaction is more complex[191]:

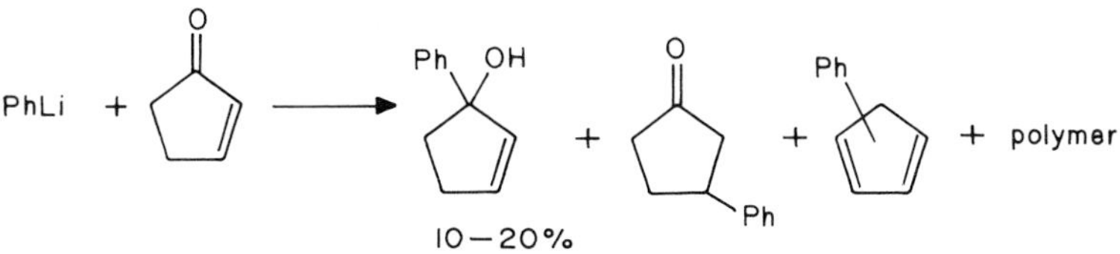

PhLi + cyclopent-2-en-1-one ⟶ 1-phenylcyclopent-2-en-1-ol (Ph, OH; 10–20%) + 3-phenylcyclopentanone (Ph) + phenylcyclopentadiene (Ph) + polymer does take place, it is accompanied by breakdown of the pvc[141,142]; complete replacement of all the chlorine atoms requires a 40 hour reflux, by which time the molecular weight has dropped to about 1500, corresponding to chains of 10–12 units[143].

$$-\!\!(\,CH_2-\underset{Cl}{\underset{|}{CH}}\,)_n- \;+\; LiPPh_2 \xrightarrow{thf} -\!\!(\,CH_2-\underset{Cl}{\underset{|}{CH}}\,)_{n-x}\!\!-\!\!(\,CH_2-\underset{PPh_2}{\underset{|}{CH}}\,)_x- \qquad (22)$$

A wide range of functional groups have been introduced on to polymers containing aromatic groups such as polystyrene. These are summarized in Table 1. A number of 2- and 4-vinylpyridine catalysts have been prepared either by polymerizing 2- or 4-vinylpyridine themselves[196–198], or by copolymerizing 4-vinylpyridine with styrene[199] or by radiation grafting 4-vinylpyridine on to polypropylene[41,99]. Poly(4-vinylpyridine) can be cross-linked with 1,2-dibromoethane (reaction 23), although the positive charges that this introduces can inhibit metal ion uptake[196].

$$-\!\!(\,CH_2-\underset{C_5H_4N}{\underset{|}{CH}}\,)_n- \;+\; BrCH_2CH_2Br \longrightarrow -\!\!(\,CH_2-\underset{C_5H_4N^+Br^-}{\underset{|}{CH}}\,)_n-\;\text{[}N^+-CH_2-CH_2-N^+\text{ linking two pyridinium rings, each }Br^-\text{]}\;-\!\!(\,CH_2-CH\,)_n- \qquad (23)$$

B. Polymerization of Functionalized Monomers

Functionalized polymers can be prepared either by homopolymerizing a functionalized monomer or by copolymerizing a functionalized monomer with another monomer that may be either inert (e.g. styrene) or designed to introduce specific properties (e.g. 2-hydroxyethyl methacrylate, which gives an asymmetric centre in addition to a hydrophilic matrix). In either case a difunctional monomer such as divinylbenzene can be used to introduce cross-linking. Copolymerization with styrene has been used successfully in the preparation of polystyrene-supported phosphines[123,135,200–206] and pyridines[175,176,200,207,208]. The optically active monomer **2** has been copolymerized with 2-hydroxyethyl methacrylate (reaction 24) to prepare as asymmetric hydrophilic support[209,210].

By copolymerizing styrene and divinylbenzene in the presence of silica either in aqueous emulsion or in methanolic solution, it is possible to coat the silica with cross-linked polystyrene. Such materials have a high active surface area and obviate the need for the reactants to diffuse through the polymer to the catalytically active sites[211–213].

As mentioned in footnote *b* in Table 1, cyclopentadienyl-substituted polystyrene prepared by reaction of lithiated polystyrene with cyclopent-2-enone may not be as simple as once thought[191]. The route outlined in Scheme 1 does lead to cyclopentadiene-substituted polystyrene[191].

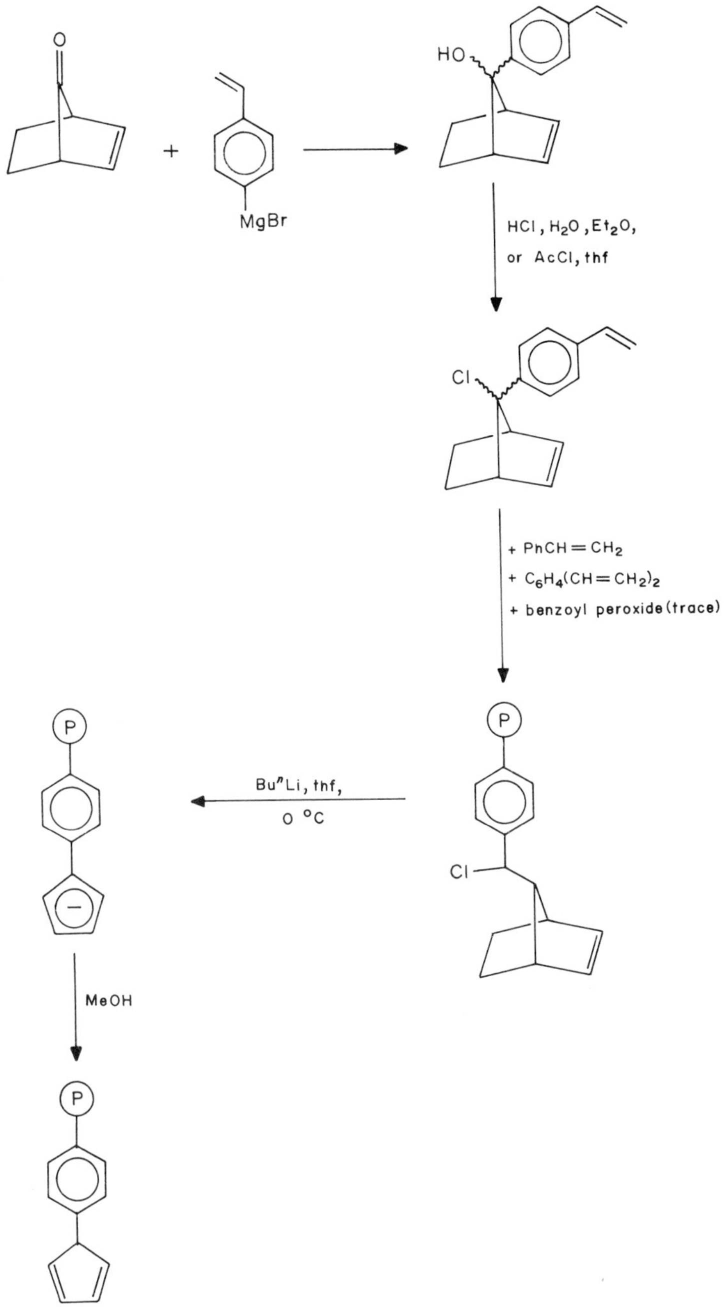

HO
MgBr
HCl, H_2O, Et_2O, or AcCl, thf
Cl
$+ PhCH{=}CH_2$
$+ C_6H_4(CH{=}CH_2)_2$
+ benzoyl peroxide (trace)
P
Cl
Bu^nLi, thf, 0 °C
P
MeOH
P

SCHEME 1

$$\text{(24)}$$

(2)

$m/n = 1/24$ + $CH_2{=}C(Me)COOCH_2CH_2OH$

C. Functionalization of Inorganic Supports

There are two major routes to the functionalization of metal oxide supports. The first involves the introduction of a functional group by a reaction such as 25, and the second involves the introduction of a 'functionalized metal complex' (reaction 26)[127,214–226]. There are several major advantages to the route shown in reaction 26 and one not insignificant disadvantage. The advantages are: (i) the range of bridging groups can readily be varied, for example to $—CH_2CH_2CH_2PPh_2$, p-$C_6H_4PPh_2$, $—(CH_2)_3C_6H_4(PPh_2)$-p, or 4-$(CH_2)_4C_5H_4N$; (ii) metal complexes that are unstable in solution, for example because they readily dimerize or are coordinatively unsaturated, may be prepared because the rigidity of the surface prevents molecular interaction; (iii) the surface of the silica after reaction with the bridging group will still be very polar owing to unreacted silanol groups, unless these have been removed by subsequent silylation of the surface. In this way the microenvironment of the catalytic centre can be carefully controlled as it is of course in a metalloenzyme, with consequent advantages in terms of the activity and selectivity of the catalyst. The problem with reaction 25 is the difficulty of determining the precise nature of the catalytic site since this is formed within the support. In principle, reaction 26 overcomes this problem because the metal complex is characterized as a molecular entity. However, many 'functionalized metal complexes' are not crystalline solids but oils which can only be purified chromatographically using non-hydroxylic phases.

$$\{M\}—OH + X_3SiY \xrightarrow[Y = \text{functional groups}]{X = \text{EtO, Cl}} \{M\}—O\text{-}Si\text{-}Y \qquad (25)$$

$$\{M\}—OH + [X_3Si(CH_2)_nPR_2ML_n] \longrightarrow [\{M\}—O\text{-}Si\text{-}(CH_2)_nPR_2ML_n] \qquad (26)$$

Some of the routes that have been described for the preparation of functionalized phosphines are illustrated in reactions 27–31[220,227–231].

$$Ph_2PCH_2PPh_2 \xrightarrow[\text{(ii) ClSiMe}_3]{\text{(i) Bu}^n\text{Li, tmeda}} Ph_2PCH(SiMe_3)PPh_2 \quad (27)$$

$$CH_2{=}CH(CH_2)_nCH{=}CH_2 \text{ (excess)} + 2HSiCl_3 \longrightarrow Cl_3Si(CH_2)_{n+2}CH{=}CH_2 \xrightarrow{HPR_2,\text{u.v.}} Cl_3Si(CH_2)_{n+4}PR_2 \quad (28)$$

$$(EtO)_3SiCH{=}CH_2 + Ph_2PH \xrightarrow{Bu^tOOBu^t} (EtO)_3SiCH_2CH_2PPh_2 \quad (29)$$

$$Br{-}C_6H_4{-}Br + \text{allylBr} \xrightarrow{\text{Mg, thf}} Br{-}C_6H_4{-}CH_2CH{=}CH_2 \xrightarrow[\text{(ii) Ph}_2\text{PCl}]{\text{(i) Mg, thf}} Ph_2P{-}C_6H_4{-}CH_2CH{=}CH_2 \xrightarrow{HSiCl_3,\ [Rh(PPh_3)_3Cl]} Ph_2P{-}C_6H_4{-}(CH_2)_3SiCl_3 \quad (30)$$

$$\text{EtOCO(H)C(OH)–C(OH)(H)OCOEt} + MeCO(CH_2)_2CH{=}CH_2 \longrightarrow \text{dioxolane: EtOCO, EtOCO; Me, }(CH_2)_2CH{=}CH_2 \xrightarrow{LiAlH_4} \text{HOCH}_2\text{, HOCH}_2\text{; Me, }(CH_2)_2CH{=}CH_2$$

$$\xrightarrow{\text{(i) tosCl; (ii) HSi(OEt)}_3 + [Rh(PPh_3)_3Cl]} \text{tosOCH}_2\text{, tosOCH}_2\text{; Me, }(CH_2)_4Si(OEt)_3 \xrightarrow{LiPPh_2} Ph_2PCH_2\text{, }Ph_2PCH_2\text{; Me, }(CH_2)_4Si(OEt)_3 \quad (31)$$

Other functional groups that have been introduced by reactions analogous to reaction 26 include amino, pyridyl, morpholino, piperidino, pyrrolidino, Schiff bases, cyano, acetylacetonato, SH and C_5H_5[220,221,232–249]. Silica has been functionalized by an initial chloromethylation (reaction 32)[250], enabling it subsequently to be subjected to most of the reactions applicable to chloromethylated polystyrene (see Table 1) and to functionalization with a bidentate phosphine (reaction 33)[250].

$$Me_2ClSiCH_2Cl + SiO_2 \xrightarrow{\text{toluene}} \{Si\}\text{—}Si(Me)_2CH_2Cl \qquad (32)$$

$$\{Si\}\text{—}Si(Me)_2CH_2Cl + NaCH(CH_2PPh_2)_2 \longrightarrow \{Si\}\text{—}Si(Me)_2CH_2CH(CH_2PPh_2)_2 \qquad (33)$$

A totally different approach to the use of silica is to form the silica *in situ* as in reaction 34. This reaction demonstrates the limitation of classifying supports as 'organic' or 'inorganic' since it involves an inorganic support prepared from an organic reagent.

$$(EtO)_3SiCH_2CH_2PPh_2 + Si(OEt)_4 \xrightarrow[\text{+ trace conc. HCl}]{\text{reflux in HOAc}} \{Si\}\text{—}CH_2CH_2PPh_2 \qquad (34)$$

Graphite has been used to support diop by first oxidizing the surface to yield aldehyde functional groups as in reaction 35 and then treating these in the same way as the aldehyde functional groups on polystyrene (reaction 21)[251].

$$Ⓖ \xrightarrow{KMnO_4} Ⓖ\text{—}COOH \xrightarrow{SOCl_2} Ⓖ\text{—}COCl \xrightarrow{HO\text{—}C_6H_4\text{—}CHO} Ⓖ\text{—}COO\text{—}C_6H_4\text{—}CHO$$

Ⓖ = graphite (35)

IV. INTRODUCTION OF METALS ON TO SUPPORTS

The first supported metal complex catalysts were prepared by reaction of cationic and anionic metal complexes with ion-exchange resins[1–4,252–257]. Although these are still used occasionally today, most supported systems involve a covalent link between support and metal. There are two broad types of covalent link. In the first a typical donor such as a phosphorus- or nitrogen-containing group is introduced on to the support and this is then used to bond to the metal. The second involves the formation of a direct covalent link between the support and the metal itself. Each is considered in turn.

A. Reaction of a Metal Complex with a Functionalized Support

Metal complexes can be introduced on to functionalized supports in four ways.

(i) Direct reaction between a metal salt and a functionalized support using essentially the same conditions as for the corresponding monomeric ligands (reactions 36 and 37).

$$MCl_2 + 2Ⓟ\text{—}PPh_2 \xrightarrow[\text{M = Co, Ni (ref. 118)}]{\text{reflux, thf}} [(Ⓟ\text{—}PPh_2)_2MCl_2] \qquad (36)$$

$$2Ⓟ\text{—}CH_2CN + PdCl_2 \xrightarrow[\text{(ref. 258)}]{} [(Ⓟ\text{—}CH_2CN)_2PdCl_2] \qquad (37)$$

(ii) The commonest route for preparing supported complex catalysts involves displacing a ligand already on the metal complex by one from the support (reactions 38–47). It should be appreciated that in many cases the products shown in reactions 38–47 are idealized; complete reaction of all donor groups within the support will depend on the reaction conditions. The ligand being displaced from the metal complex may have the same donor group as that on the support, as in reactions 40 and 42, or may be different. Whilst many research groups have used the equilibration of a phosphine complex with a phosphinated support, as in these two examples, there are serious drawbacks to the use of this technique. Phosphine ligands are not normally particularly volatile and therefore it is often difficult to remove all the liberated phosphine ligands from the support. If they are left within the support they will inevitably act as competitors to the anchored ligands and hence help to promote leaching of the metal complex. For this reason a number of groups, including my own, now prefer to displace volatile ligands such as carbon monoxide, which are readily removed from the support, as in reactions 38, 43 and 44[22,113,118,214,259,275]. Polynuclear complexes are often introduced in this way[276–285].

$$[Rh(acac)(CO)_2] + \text{Ⓟ-PPh}_2 \xrightarrow[\text{8 h (refs. 22,118,214,259)}]{\text{heptane, 20 °C}} [(\text{Ⓟ-PPh}_2)Rh(acac)(CO)] + CO \tag{38}$$

$$[Rh(acac)nbd] + 2\,\text{Ⓟ-PPR}' \xrightarrow[\text{70\% HClO}_4\text{ in thf (ref. 134)}]{\text{thf, 20 °C, then}} [(\text{Ⓟ-PPR}')_2Rh(nbd)] + acacH \tag{39}$$

$$[(PPh_3)_3RhCl] + \text{Ⓟ-PPh}_2 \xrightarrow[\text{2–4 weeks (refs. 52,114,130)}]{\text{benzene, 20 °C,}} [(\text{Ⓟ-PPh}_2)Rh(PPh_3)_2Cl] + PPh_3 \tag{40}$$

$$2\,\text{Ⓟ-PPh}_2 + [(PhCN)_2MCl_2] \xrightarrow[\text{60 h (M = Pd,Pt) (refs. 128,260,261)}]{\text{reflux in acetone}} trans\text{-}[(\text{Ⓟ-PPh}_2)_2MCl_2] + 2PhCN \tag{41}$$

$$\text{Ⓟ-PPh}_2 + [Pd(PPh_3)_4] \xrightarrow[\text{17 h (ref. 128)}]{\text{benzene, 25 °C,}} [(\text{Ⓟ-PPh}_2)Pd(PPh_3)_3] + PPh_3 \tag{42}$$

Ⓟ-PPh$_2$ $\xrightarrow[\text{3 h (ref. 113)}]{[Co_2(CO)_8]\text{ in toluene, 20 °C,}}$ [(support)(Ph$_2$P)$_2$Co(CO)$_3$]$^+$ [Co(CO)$_4$]$^-$ $\xrightarrow[\text{under }N_2\text{, 27 h}]{\text{heat at 69 °C}}$ [(support)Ph$_2$P—Co(CO)$_3$—Co(CO)$_3$—PPh$_2$]

Ⓟ-PPh$_2$ $\xrightarrow[\text{u.v. or heat (ref. 113)}]{+[Ni(CO)_2(PPh_3)_2]}$ [(Ⓟ—PPh$_2$)Ni(CO)$_2$(PPh$_3$)]

(43)

$$\text{(P)}\!-\!PPh_2 \xrightarrow[\text{(refs. 113, 262)}]{+[Fe(CO)_5],\ \text{heat or } h\nu} [(\text{(P)}\!-\!PPh_2)Fe(CO)_4] + [\text{(P)}\!-\!PPh_2\!-\!Fe(CO)_3\!-\!PPh_2\!-\!\text{(P)}]$$

$$\text{(P)}\!-\!PPh_2 \xrightarrow[\text{(refs. 263–267)}]{+[M(CO)_6],\ M = Cr, Mo, W} [(\text{(P)}\!-\!PPh_2)M(CO)_5] \qquad (44)$$

$$\text{(P)}\!-\!CH_2\!-\!C_5H_5 \xrightarrow[R = Me,\ Bu]{RLi} \uparrow RH + \text{(P)}\!-\!CH_2\!-\!C_5H_4^{-}\ Li^+$$

$$\text{(P)}\!-\!CH_2\!-\!C_5H_4^{-}\ Li^+ \xrightarrow[\text{(ref. 268)}]{[Rh(CO)_2Cl]_2} [\text{(P)}\!-\!CH_2\!-\!C_5H_4Rh(CO)_2]$$

$$\text{(P)}\!-\!CH_2\!-\!C_5H_4^{-}\ Li^+ \xrightarrow[\text{(refs. 269, 270)}]{[CpMCl_3],\ M = Ti, Zr} [\text{(P)}\!-\!CH_2\!-\!C_5H_4CpMCl_2] \qquad (45)$$

$$\text{(P)}\!-\!CH_2\!-\!C_5H_4^{-}\ Li^+ \xrightarrow[\text{(ref. 271)}]{CrCl_3\cdot 3thf} [\text{(P)}\!-\!CH_2\!-\!C_5H_4CrCl_2]$$

$$\text{(P)}\!-\!CH_2\!-\!C_5H_5 \xrightarrow[\text{(ref. 272)}]{[Fe_2(CO)_9],\ \text{thf, reflux}} [\text{(P)}\!-\!CH_2\!-\!C_5H_4Fe(CO)_2H]$$

$$\text{(P)}\!-\!CH_2\!-\!C_5H_5 \xrightarrow[\text{(refs. 272, 273)}]{[Co_2(CO)_8]} [\text{(P)}\!-\!CH_2\!-\!C_5H_4Co(CO)_2] \qquad (46)$$

$$\text{(P)}\!-\!CH_2\!-\!C_5H_5 \xrightarrow[RhCl_3\cdot 3H_2O]{\text{(ref. 268)}} [\text{(P)}\!-\!CH_2\!-\!C_5H_4RhCl_2] + HCl$$

$$\text{Ⓟ}-\text{C}_6\text{H}_4-\text{CH}_2\text{Cl} \xrightarrow[\text{(ref. 263)}]{\text{Na[Mn(CO)}_5]} [\text{Ⓟ}-\text{C}_6\text{H}_4-\text{CH}_2\text{Mn(CO)}_5] + \text{NaCl}$$

$$\text{Ⓟ}-\text{C}_6\text{H}_4-\text{CH}_2\text{Cl} \xrightarrow[\text{(refs. 263,274)}]{\text{Na[MoCp(CO)}_3]} [\text{Ⓟ}-\text{C}_6\text{H}_4-\text{CH}_2\text{MoCp(CO)}_3] + \text{NaCl} \quad (47)$$

The distribution of the metal complex can be altered by varying the reaction conditions. Thus, by using short reaction times and an excess of supported ligands a high concentration of metal complex in the outer regions of the support is achieved[74]. This promotes high activity but prevents diffusion within the support from being used to enhance specificity. Photochemical reactions also favour surface location of the metal complex. However, when such unevenly distributed metal complex catalysts are used, redistribution of the complex may occur during the course of the catalytic reaction.

(iii) Bridge splitting of dinuclear metal complexes can be used to introduce a mononuclear complex on to a functionalized support without the release of a displaced ligand. Most supports completely cleave rhodium(I) halide-bridged complexes (reactions 48 and 49)[28,122,286–289], although thiolated supports do not cleave the bridge, but rather support a bridged complex (reaction 50)[288].

$$[\text{Rh(CO)}_2\text{Cl}]_2 + 2\text{Ⓟ-Y} \xrightarrow[(\text{Y} = \text{NR}_2 \text{ or PPh}_2)]{\text{reflux, benzene, 6h}} 2\ cis\text{-}[(\text{Ⓟ-Y})\text{Rh(CO)}_2\text{Cl}] \quad (48)$$

$$\left(\text{CH}_2-\underset{\underset{\text{PPh}_2}{|}\text{O}}{\overset{\text{Me}}{\text{C}}}\right)_n \xrightarrow{[\text{Rh(CO)}_2\text{Cl}]_2} \left(\text{CH}_2-\overset{\text{Me}}{\text{C}}(\text{O})-\text{CH}_2-\overset{\text{Me}}{\text{C}}(\text{O})\right)_n \text{ with } \text{Ph}_2\text{P}-\text{Rh(Cl)(CO)}-\text{PPh}_2 \quad (49)$$

$$[\text{Rh(CO)}_2\text{Cl}]_2 + \text{Ⓟ}-\text{SH} \longrightarrow \left[(\text{OC})_2\text{Rh}(\mu\text{-Cl})(\mu\text{-S-Ⓟ})\text{Rh(CO)}_2\right] + \text{HCl} \quad (50)$$

(iv) Oxidative addition reactions provide a convenient route for preparing supported nickel(II), palladium(II), and platinum(II) complexes (reactions 51 and 52)[236,290–295]. In reaction 52 the active metal(0) species is formed by reductive elimination of two of the original ligands during the course of the reaction[295].

$$\text{Ⓟ}-\text{C}_6\text{H}_4-\text{X} + [\text{M(PPh}_3)_4] \xrightarrow[\text{M = Ni, Pd, Pt}]{\text{X = Br, I}} [\text{Ⓟ}-\text{C}_6\text{H}_4-\text{M(PPh}_3)_2\text{X}] + 2\text{PPh}_3 \quad (51)$$

$$\text{(P)}\!-\!C_6H_4\!-\!Cl + [(bipy)Ni(Et)_2] \longrightarrow [\text{(P)}\!-\!C_6H_4\!-\!Ni(bipy)Cl] + C_4H_{10} \quad (52)$$

B. Direct Reaction of a Metal Complex with a Support

Chromium, molybdenum, and tungsten hexacarbonyls react directly with the phenyl rings of polystyrene to form supported complexes (reaction 53)[296]. Similarly [Ru(cod)(cot)] reacts with polystyrene to eliminate both cycloocta-1, 5-diene and cycloocta-1, 3, 5-triene forming a supported diphenylruthenium(0) species[297].

$$\text{(P)}\!-\!C_6H_5 + [Mo(CO)_6] \xrightarrow{M=Cr,Mo,W} \left[\text{(P)}\!-\!C_6H_5M(CO)_3\right] + 3CO \quad (53)$$

Organometallic complexes, particularly allyl (Zr, Hf, Cr, Ni) benzyl (Ti, Zr, Hf), neopentyl (Ti, Zr), cyclopentadienyl (Cr), and arene (Cr) complexes, react with silica, alumina, and aluminosilicates to yield supported complexes (reaction 54)[298-321]. The reaction depends

$$\text{—OH} + MR_n \longrightarrow \text{—OMR}_{n-1} + RH \quad (54)$$

on the moisture sensitivity of the organometallic compounds and therefore depends very much on the nature of the support, in particular its past thermal and hydration/dehydration history. Thus, heating silica *in vacuo* for 3 h removes all the surface water. Further heating then successively removes more and more of the surface hydroxyl groups until at 1200 °C all these groups have been removed. Although not all hydroxyl groups in a metal oxide can react with metal alkyls, the number of groups that can react may readily be determined by reaction with methylmagnesium iodide[301] or methyllithium[322] followed by determination of the volume of methane evolved (reaction 55). The

$$\text{—OH} + MeM \xrightarrow[M=Li\text{ or }MgI]{\text{toluene}} \text{—OM} + CH_4\uparrow \quad (55)$$

importance of previous thermal history is well illustrated by the results in Table 2[320]. The nature of the support is critical: titanium and zirconium give more effective olefin polymerization catalysts when they are supported on alumina[298,301,323], whereas chromium catalysts are more active when supported on silica[324].

Although chromocene is inactive as an olefin polymerization catalyst, it is very active when deposited from hydrocarbon solution on to amorphous silica (reaction 56)[300,303,326-328]. The silica both anchors the chromium and stabilizes it in a coordinatively unsaturated state; anchoring prevents the mutual interaction and destruction of the coordinatively unsaturated species. The high silica dehydration temperature noted in Table 2 reflects the importance of having completely isolated chromium species.

$$\{Si\}\text{—}OH + [CrCp_2] \longrightarrow [\{Si\}\text{—}OCrCp] + C_5H_6 \quad (56)$$

In addition to organometallic complexes, metal hydrides can also react directly with silica and alumina (reaction 57)[329,330]. There is substantial interaction between the

TABLE 2. Dehydration temperatures necessary in the pretreatment of oxide supports to yield maximum activity in the subsequent supported organometallic ethylene polymerization catalyst (data from ref. 320)

Support	Metal complex	Optimum dehydration temperature (°C)
SiO_2	$[Cr(C_3H_5)_3]$	400
SiO_2	$[Cr(\eta^5\text{-}C_5H_5)_2]$	670
SiO_2	$[Zr(C_3H_5)_4]$	25
SiO_2	$[Zr(C_3H_5)_3X]$ (X = Cl, Br, I)	750
Al_2O_3	$[Zr(C_3H_5)_4]$	400
Al_2O_3	$[Ti(CH_2Ph)_4]$	600

triphenylphosphine ligands and the surface oxide ions in **3**, which promotes the dissociation of triphenylphosphine and hence enhances hydroformylation activity[330].

$$\text{—OH} + [RhH(CO)(PPh_3)_3] \longrightarrow [\text{—}ORh(CO)(PPh_3)_3] + H_2 \qquad (57)$$

(3)

Metal carbonyls can be immobilized by impregnation of metal oxides[12,331–334]. The critical variables are the nature of the pretreatment of the oxide, the contact time, the temperature and pH of the impregnating solution, and the post-impregnation treatment. After initial physical adsorption many metal carbonyls react with surface oxide sites as in reaction 58. This may be promoted by heating *in vacuo* or in the presence of either an inert

$$[M(CO)_n] + \text{—}O \longrightarrow [\text{—}OM(CO)_{n-1}] + CO \qquad (58)$$

gas or even in oxygen. The subcarbonyl species so formed may themselves react further with surface hydroxyl groups with evolution of carbon monoxide and hydrogen as in reaction 59[335] or even with formation of hydrocarbons[336–341]. A very valuable technique

$$[\text{—}OM(CO)_m] + 2\,\text{—}OH \longrightarrow [(\text{—}O)_2M^{II}] + H_2 + mCO \qquad (59)$$

for monitoring the progress of such reactions is temperature-programmed decomposition chromatography[336,342–349], in which the gaseous products are swept out in a stream of either an inert gas such as helium or an active gas such as hydrogen, and analysed using a gas chromatograph. This, combined with infrared spectrometry, provides a powerful technique for determining the nature of the supported metal complex species many of which are valuable Fischer–Tropsch catalysts.

V. THE USE OF SUPPORTED METAL COMPLEX CATALYSTS

A. Design of Supported Catalysts

The success of homogeneous transition metal complex catalysts, such as Wilkinson's catalyst[29], provided the early motivation for supporting them in order to combine the advantages inherent in their molecular nature with the ease of separation at the end of the reaction that an insoluble material should provide. Accordingly, many supported catalysts are essentially homogeneous catalysts that have been immobilized on a solid support. Whilst many of these catalysts have inferior properties to their homogeneous analogues, particularly low activities, sufficient of them have enhanced activity and selectivity and also greater stability and resistance to deactivation to justify further work in the area.

However, a number of workers have taken a different approach and decided that supported metal complex catalysts should attempt to combine the unique features offered by the combination of support and metal complex in creating a catalyst completely different from the homogeneous catalysts[350]. In creating such novel catalysts two features need consideration: the metal complex itself and particularly the ligands that surround the metal, and the support.

(i) Because of the uniquely different nature of supported metal complexes compared with their homogeneous analogues, the ideal ligands for a homogeneous catalyst may be far from ideal in a supported catalyst[39]. Hence the ideal ligand in a homogeneous catalyst is often one that does not dissociate too readily from the metal, because dissociation may lead to subsequent reduction of the metal ions to the free metal. In the interstices of a support, however, ligand dissociation is suppressed by special restrictions. Consequently, ligands that are not particularly effective in homogeneous catalysts can prove to be very effective in supported catalysts. Anthranilic acid ligands have rarely been used in homogeneous catalysts yet they have proved effective ligands in supported rhodium(I) and palladium(II) hydrogenation catalysts[39,39a,161,351–354]. Such rhodium(I) catalysts are very active, have long-term stability and are fairly inert to poisons, whilst the palladium(II)-supported anthranilic acid catalysts are very active benzene and nitrobenzene hydrogenation catalysts in contrast to their homogeneous analogues, which are inactive[353,355,356]. Polystyrene-supported cyclopentadienylcobalt(I) dicarbonyl, $[(P)\text{-}C_6H_4\text{-}C_5H_4\text{-}Co(CO)_2]$ is an active Fischer–Tropsch catalyst for the conversion of carbon monoxide–hydrogen mixtures to hydrocarbons, particularly methane[273,357,358], although its homogeneous analogue $[CoCp(CO)_2]$ is inactive owing to its instability. The activity of the supported catalyst is further enhanced by thermolysis to yield the coordinatively unsaturated monocarbonyl complex.

The ideal functionalized support will generally be one that promotes coordinative unsaturation on the metal complex, whilst binding the metal ion sufficiently tightly to prevent either leaching or decomposition to the free metal. The profound importance of the ligands surrounding the metal ion on the catalytic activity is well illustrated by the decreasing effectiveness in olefin hydrogenation of a series of rhodium(I) complexes supported on phosphinated polystyrene in the order $[Rh(PPh_3)_3(CO)H] > RhCl > RhCl_3 + PPh_3 > RhCl_3 + PHPh_2 > RhCl_3 + C_2H_4 > [Rh(PPh_3)_3Cl] > [Rh(PHPh_2)_3Cl]$[7,36]. Having the ligand donor atoms linked to a polymer can result in greatly enhanced selectivity in hydroformylation. Thus it is not necessary to swamp the metal complex with excess of phosphine in order to obtain a high ratio of normal to branched aldehyde product[22]. This is well illustrated in Table 3, where for hex-1-ene under identical conditions the homogeneous catalyst $[Rh(p\text{-styrylPPh}_2)(acac)(CO)]$ requires a phosphorus to rhodium ratio of 250:1 to yield a normal to branched aldehyde ratio of 5:1, whereas the corresponding supported catalyst in which all the phosphine ligands are bound to the polymer only requires a phosphorus to rhodium ratio of 2:1 to give the same selectivity. Since a large excess of phosphine inhibits olefin coordination and hence suppresses catalytic activity[360], it is possible to obtain the highest normal to branched ratios at higher activities using supported rather than homogeneous catalysts. A similar effect is observed with $[Ru(PPh_3)_2(CO)_3]$ supported on phosphinated polystyrene; the

TABLE 3. Effect of phosphorus to rhodium ratio on the normal to branched selectivity of the hydroformylation of hex-1-ene in benzene under 10 atm of $H_2 + CO$ (1:1) at 65 °C (data from ref. 22)

1. Supported catalyst [Rh(Ⓟ-PPh_2)(CO) (acac)], 300 min[a]:					
P:Rh ratio[b]	8.36	5.86	4.05	2.17	1.32
Selectivity[c]	16.0	14.1	13.2	5.2	2.5
2. Homogeneous catalyst [Rh(*p*-styryl PPh_2) (CO)(acac)], 100 min:					
P:Rh ratio[b]	250	80	8.0	2.0	1.0
Selectivity[c]	5.0	3.7	2.6	2.2	2.7

[a]Ⓟ-PPh_2 = *p*-styryl PPh_2, γ-radiation grafted on to polypropylene (prepared as in ref. 109).
[b]g-atom/g-atom
[c]Ratio of normal to branched aldehydes.

supported catalyst is more selective than its homogeneous counterpart[361]. The formation of large amounts of normal product depends on equilibrium 60 lying well to the left, which in turn depends on the phosphorus to ruthenium ratio and the degree of ligand mobility within the resin. Swelling drives equilibrium 60 to the right by pushing the phosphine groups further apart[361]. The enhancement of normal to branched selectivity in supported

$$[(\text{Ⓟ-PPh}_2)_2\text{RuH}_2(\text{olefin})(\text{CO})] \rightleftharpoons \text{Ⓟ-PPh}_2 + [(\text{Ⓟ-PPh}_2)\text{RuH}_2(\text{olefin})(\text{CO})] \quad (60)$$

catalysts in part arises from the different positions of the equilibria for the replacement of phosphine ligands by carbon monoxide in the metal coordination sphere in the two media. These different positions arise partly because the concentration of carbon monoxide within the polymer will probably be very different to that in the bulk solution. As would be expected from this, although supported catalysts normally have higher selectivities than their unsupported counterparts, this is not always so particularly at low phosphorus loadings[362].

The considerable influence of the metal complex itself is best illustrated with respect to Fischer–Tropsch synthesis of hydrocarbons from carbon monoxide–hydrogen mixtures. Thus ruthenium, osmium, and cobalt carbonyls supported on metal oxides promote alkane formation, iron carbonyl promotes alkane and olefin formation and rhodium, iridium, and platinum carbonyls supported on metal oxides promote alcohol formation[337,339,363–379].

(ii) Supports can have very major influences on the activity of the catalyst. Thus rhodium(I) hydrogenation catalysts supported on phosphinated poly(vinyl chloride) are fairly inactive[380], whereas phosphinated polystyrene catalysts are more active[53a] and phosphinated silica catalysts even more active, particularly when the substrates are cyclic olefins[360]. The greater activity of the silica-supported catalysts may be due to the rhodium(I) being largely located in accessible sites close to the surface. Certainly very active polystyrene supported catalysts can be obtained by supporting rhodium(I) on polystyrene with the aid of long alkyl chains that enable the supported rhodium(I) phosphine complexes to be freely available in solution[20]. [$Pd(PhCN)_2Cl_2$] is 100 times more active as an olefin isomerization catalyst when supported on silica than in the unsupported form, indicating a specific role for the support[381]. When palladium(0) complexes are supported on phosphinated silica gel or polystyrene, not only are they active in promoting the replacement of allylic acetate groups by other nucleophiles (reactions 61 and 62) but, because of 'steric steering' by the support, they may have greater selectivity than their homogeneous analogues[382]. Additionally, the supported catalysts can be stored for up to 2 months in air without loss of activity, in contrast to $[Pd(PPh_3)_4]$, which rapidly decomposes in air.

OAc + Na^+ $^-CH(SO_2Ph)COOMe$ —(i) Pd^0 cat. in thf; (ii) 6% Na–Hg→ (dienyl)COOMe + (branched diene)COOMe

$[Pd(PPh_3)_4]$:	69	:	31
Pd^0 on phosphinated polystyrene:	80	:	20

(61)

COOMe / OAc (cyclohexene) + Et_2NH —Pd^0 cat. in thf→ COOMe / H / NEt_2 + COOMe / H / NEt_2

$[Pd(PPh_3)_4]$	67	:	33
Pd^0 on phosphinated polystyrene:	100	:	0
Pd^0 on phosphinated silica gel:	100	:	0

(62)

In organic polymers it is possible to vary the degree of cross-linking, which can in turn have a profound influence on the reaction. For example, large beads (74–149 μm diameter) of 2% cross-linked phosphinated polystyrene exchanged with $[Rh(PPh_3)_3Cl]$ are only 0.06 times as active as the corresponding homogeneous hydrogenation catalyst[52,53a]. In contrast, smaller beads (37–74 μm diameter) of lower (1%) cross-linking are 0.8 times as active as the homogeneous catalyst, a 1300% improvement[383]. The promotion of very high activity is not always the best contribution that a support can make. For example, a polymeric support may be valuable in promoting high selectivity owing to the need for the substrate to diffuse within the support up to the active site. Within a polymer it is possible to increase the degree of cross-linking and so inhibit the migration of more bulky substrates. Thus the rates of hydrogenation of a series of olefins in benzene using 100–200 mesh 2% cross-linked phosphinated polystyrene carrying $[Rh(Ⓟ\text{-}PPh_2)_n(PPh_3)_{3-n}Cl]$ decrease in the order hex-1-ene ≫ cyclohexene ≫ cyclooctene > cyclododecene ≫ Δ^2-cholestene[52,384].

Polymeric supports sometimes undergo structural changes on heating or as a result of chemical reactions. These can lead to abrupt changes in catalytic reaction rate[385] or even complete deactivation of the catalyst[161,386]. Such changes are sometimes, but not always, reversible.

In methanol carbonylation, $[Rh(PPh_3)_3Cl]$, $[Rh(cod)Cl]_2$, and $[Rh(cod)(OMe)]_2$ supported on γ-alumina all had a similar activity and gave a selectivity of *ca.* 50% with respect to acetic acid formation, owing to dimethyl ether being formed as a by-product. $[Rh(PPh_3)_2(CO)Cl]$ on γ-alumina had a similar activity but a selectivity for acetic acid formation approaching 99%[387].

Many enzymes use the area around the active catalytic site to select and align the substrate and so achieve their remarkable powers of discrimination between substrates as well as their high regio- and stereo-selectivities[25]. Some attempts have been made to use supports in this way. For example, the introduction of an asymmetric alcoholic functional group adjacent to a rhodium(I)–diop catalytic site results in the optical yield in the asymmetric hydrogenation of α-acetamidoacrylic acid being dependent on the chirality of the alcoholic group (see Section V.D.3)[360]. Equilibration of $[Rh(PPh_3)_3Cl]$ with an optically active cellulose phosphinated with achiral Ph_2PCl yielded a catalyst that gave up to 28% enantiomeric excess in the reduction of *N*-α-phthalimidoacrylate to methyl *N*-phthaloyl-D-alaninate[388]. Thus the support can induce asymmetry in a catalyst where the catalytic metal centre is itself achiral. A similar result has been obtained in a system that is rather closer to nature. This involves an achiral rhodium(I) biotin complex bound within the chiral cavity of the enzyme avidin (reactions 63 and 64). The chiral cavity of the enzyme induces asymmetry in the catalysis with the result that an enantiomeric excess of about 40% can be obtained in the hydrogenation of α-acetamidoacrylic acid[389].

$$\text{enzyme with chiral cavity }(*) + ML_n \text{ (achiral homogeneous catalyst)} \longrightarrow \text{enzyme}(*)\text{—}ML_m + (n-m)L \qquad (63)$$

$$\text{biotin—C(=O)—O—N(succinimide)} \xrightarrow[\text{(ii) } [Rh(nbd)_2]^+X^-,\ thf]{\text{(i) } (Ph_2PCH_2CH_2)_2\overset{+}{N}H_2Cl^-,\ dmf,\ Et_3N} [\text{biotin—C(=O)—N}(CH_2CH_2PPh_2)_2Rh(nbd)]^+ X^- \qquad (64)$$

Copper and nickel complexes of polystyrene functionalized with $—CH_2$-L-Cys-$(CH_2CH_2NH_2)OH$ and $—CH_2$-L-Cys$(CH_2COOH)OH$ catalyse the hydrolysis of the methyl esters of L-phenylalanine and L-histidine more rapidly than the D-enantiomers[390]. This enables a facile enantiomeric enrichment to be carried out using the enantioselectivity of the ester hydrolysis reaction.

When metal oxide supports are used, residual hydroxyl groups markedly influence the activity of a supported catalyst. Thus the hydroformylation activity of *trans*-$[RhCl(CO)\{Ph_2PCH_2CH_2Si(OEt_3)_2\}_2]$ bound to γ-alumina is increased dramatically when triphenylphosphine is added[391]. This is believed to be due to triphenylphosphine blocking the deactivating Lewis acid sites on the γ-alumina. We have already noted (Section II.B) methods for removing or capping surface polar groups. It is sometimes possible when using mixtures of reactants to promote the activity of one relative to the others by using supports of different polarity, since a change in polarity can induce a change in the relative reactivities of the substrates. Thus in the Fischer–Tropsch reaction of hydrogen with carbon monoxide, $[Fe_3(CO)_{12}]$ supported on NaY faujasite promotes the formation of hydrocarbons up to C_{11}, whereas on silica gel or alumina only short-chain hydrocarbons up to C_4 are formed[363].

Supports which carry charged groups can have a profound influence on the catalytic ability of a supported metal complex. Thus copper(II) supported on poly-L-histidine promotes the oxidation of negatively charged and neutral organic substrates, but inhibits the oxidation of positively charged substrates[392,393].

B. Activity and Selectivity of Supported Catalysts

The activity and selectivity of a supported metal complex catalyst depend very much on the reaction conditions. As with any catalyst there are usually optimal reaction conditions which can only be determined empirically by varying such factors as the reaction temperature, the concentrations of the reactants, and the pressures of any gases. A particularly important variable is the solvent. Coordinating solvents, for example, often take part in the intimate mechanism of the reaction by being coordinated to and displaced from the active site during the catalytic cycle. A second way in which solvents can influence reactions is through their ability to swell the support. Non-polar polymeric supports such as polystyrene are swollen by non-polar solvents such as benzene, whereas polar supports such as silicates, including hectorite and montmorillonite, are swollen by polar solvents such as water or alcohols. The third way in which solvents can influence reactions is through their polarity, which can be used to control the polarity gradient between the bulk solvent and the local environment of the catalytic centre. Thus, by suitably adjusting the solvent polarity, it is possible to either enhance or inhibit substrate migration into the support up to the active site. These three solvent effects can often influence reactions in mutually conflicting ways, so that many catalytic reactions are very sensitive to the nature of the solvent. A good illustration of the complexity of solvent effects is the influence of solvent on the hydrogenation of a series of olefins by Wilkinson's catalyst supported on phosphinated polystyrene, Ⓟ–C_6H_4–CH_2PPh_2 discussed in Section II.A (Figure 1).

Although many supported catalysts are less active than their homogeneous analogues, there are an increasing number of cases where the reverse is true. Some of these depend for their high activity on the role of the support in depressing a deactivation mechanism and we have considered one case of this, deactivation through dimerization, in Section I.B.5. Others depend on the influence of the support in modifying the metal–ligand equilibria present. Thus, polystyrene-bound [(Ⓟ-$PPh_2)_2Ir(CO)Cl$] is more active as a hydrogenation catalyst for 4-vinylcyclohexene than its homogeneous counterpart because equilibrium 65 lies further to the right for the polymeric phosphine than for monomeric triphenylphosphine owing to steric constraints within the polymer that retard the reverse reaction[394]. Similarly, palladium(II)–phosphine complexes are more active hydrogenation catalysts when supported than in homogeneous solution because of having fewer than two phosphine ligands per palladium in the supported case[395]. Platinum(0)–phosphine complexes are 22 times as active as hydrosilylation catalysts when supported owing to equilibrium 66 lying further to the right as a consequence of the rigidity of the support[396].

$$[(\text{Ⓟ-PPh}_2)_2\text{Ir(CO)(olefin)Cl}] \rightleftharpoons [(\text{Ⓟ-PPh}_2)\text{Ir(CO)(olefin)Cl}] + \text{Ⓟ-PPh}_2 \qquad (65)$$

$$[(\text{Ⓟ-PPh}_2)_n\text{Pt(PPh}_3)_{4-n}] \rightleftharpoons [(\text{Ⓟ-PPh}_2)_{n-1}\text{Pt(PPh}_3)_{4-n}] + \text{Ⓟ-PPh}_2 \qquad (66)$$

The value of supports in promoting the formation of coordinatively unsaturated species in environments in which they are not immediately susceptible to dimerization and hence deactivation is illustrated by the fact that whereas [FeCp(CO)$_2$H] rapidly loses hydrogen in solution to form dinuclear [Cp(CO)$_2$FeFe(CO)$_2$Cp], [(Ⓟ-C$_6$H$_4$CH$_2$C$_5$H$_4$)Fe(CO)$_2$H] bound to polystyrene is stable for months at room temperature[272]. Similarly, when cobalt(II) tetraphenylporphyrin is exposed to 1-methylimidazole a six-coordinate complex is formed in which the solvent occupies the sixth position (reaction 67); such complexes do not bond to oxygen, whereas the corresponding five-coordinate complexes, which can be formed using polystyrene-bonded imidazole, readily form oxygen complexes on exposure to air (reaction 68)[397].

$$\text{Co(N}_4) + \text{MeN-imidazole} \xrightarrow{\text{solvent S}} \text{S–Co(N}_4)\text{–N-imidazole(NMe)} \qquad (67)$$

$$\text{Ⓟ–CH}_2\text{–C}_6\text{H}_4\text{–N-imidazole-N} + \text{Co(tpp)} \longrightarrow$$

$$[\text{Ⓟ–CH}_2\text{–C}_6\text{H}_4\text{–N-imidazole-N–Co(tpp)}] \underset{}{\overset{\text{O}_2}{\rightleftharpoons}} [\text{Ⓟ–CH}_2\text{–C}_6\text{H}_4\text{–N-imidazole-N–Co(tpp)O}_2] \qquad (68)$$

The selectivity[398] of supported catalysts can often be promoted by increasing the loading of catalytic centres on a support[53], because the polymer surrounding a catalytic centre imposes a diffusional barrier between the bulk solution and the active site. The more active the polymer bead the greater will be its demand on the bulk solution to supply substrate, which will result in an enhancement of the differences in the diffusion rates of the substrates. However, this effect will not apply in the case of reactions involving a free radical mechanism. Thus, when 2, 6-dimethylphenol is oxidized in the presence of supported Schiff base complexes (reaction 69), the ratio of 3, 3′, 5, 5′-tetramethyldiphenylquinone (dpq) to 2,6-dimethyl-1,4-benzoquinone (dmbq) is enhanced when the catalyst is supported[168]. Low concentrations of cobalt in the polymer increase the

$$\mathrm{HO{-}C_6H_3Me_2} \xrightarrow{\text{ox, cat.}} \mathrm{O{=}C_6H_2Me_2{=}C_6H_2Me_2{=}O} + \mathrm{O{=}C_6H_2Me_2{=}O} \qquad (69)$$

probability of two radicals combining to form dpq, whereas at higher cobalt concentrations on the support, and in solution, the probability of organic radical encounters with cobalt(II) is increased and more dmbq is formed. Thus, although free radical reactions normally have the disadvantage of yielding a variety of products, supporting the catalysts offers considerable potential for altering the selectivity.

C. Loss of Metal Complex

The loss of metal complexes from metal complex catalysts represents a major barrier to their application. Metal complex loss can arise from (i) decomposition to metal oxide, (ii) decomposition to the free metal, or (iii) leaching.

Decomposition to form the free metal is not always a disadvantage; indeed, a number of useful, very active catalysts have been prepared by supporting palladium and platinum complexes on a range of supports and then reducing the complexes to the free metals. The resulting very finely dispersed palladium and platinum systems are very active olefin hydrogenation and oxidative dehydrogenation catalysts[3,399,400].

Leaching of metal complexes off their supports is generally regarded as the greatest barrier to the widespread adoption of supported metal complex catalysts. This is particularly true where they are required for continuous, as opposed to batch, processes. Any supported metal complex catalyst is liable to suffer leaching in the presence of good donor solvents or reaction products, but it is obviously a particular problem where metal—ligand bond cleavage forms an intimate part of the reaction mechanism, as it does in rhodium(I)–phosphine complex-catalysed olefin hydrogenation and hydroformylation[401,402]. Leaching can be reduced by (i) using microporous resins which have been functionalized throughout, rather than macroreticular beads functionalized on the surface, (ii) increasing the number of donor groups on the functionalized support, (iii) increasing the number of functional groups per metal atom, (iv) avoiding the use of coordinating solvents, (v) using chelating functional groups, or (vi) ensuring the absence of oxygen in any reactions involving phosphine functionalized supports since these are readily oxidized to phosphine oxides which are generally poor ligands for metal ions that coordinate strongly to phosphines.

Not all of these possibilities are always practicable since their use may have important side effects which may be either advantageous or disadvantageous. For example, in the hydroformylation of olefins using supported rhodium(I)–phosphine complexes, increasing the number of phosphine groups per metal ion not only decreases the leaching but also enhances the selectivity as determined by the ratio of normal to branched aldehydes[15,22,403]. By contrast, the replacement of unidentate phosphines by bidentate $\text{Ⓟ}{-}\mathrm{C_6H_4}{-}\mathrm{PPhCH_2CH_2PPh_2}$ reduces the ratio of normal to branched aldehyde products and increases the amount of double-bond isomerization, neither of which is desirable[404].

The influence of reaction temperature on leaching can sometimes be unexpected. Thus, in olefin hydroformylation in the presence of supported rhodium(I)–phosphine complexes, leaching becomes increasingly important as the temperature decreases. This occurs because the relative equilibrium constants for rhodium–carbon monoxide and rhodium–phosphine coordination progressively favour rhodium–phosphine bonding as the temperature increases[405].

An interesting example of metal loss from a supported catalyst arises in the use of supported rhodium complexes in which tetrachlorobenzenedithiol promoters are also incorporated. When these catalysts are used to promote the carbonylation of methanol to acetic acid, oxidative addition of the tetrachlorobenzenedithiol to an adjacent rhodium complex results in the formation of a soluble rhodium mercaptide complex with consequent loss of rhodium (reaction 70)[188].

Cl Cl
CH_2–S– –Cl
Cl $SRhL_x$
S
$Rh(CO)_2$
S
oxidative addition
CO insertion
CH_2
C=O Cl Cl
S
Rh–S– –Cl
S
Cl $SRhL_x$
CH_2COOH
MeOH
S
$Rh(CO)_2$ + HS– –Cl
S
Cl Cl
Cl $SRhL_x$
(70)

D. Examples of the Application of Supported Metal Complex Catalysts

Supported metal complexes have been used to promote a very wide range of chemical reactions, including the following:

Hydrogenation and dehydrogenation
Reduction of inorganic molecules
Hydrosilylation
Hydroformylation
Methanol carbonylation
Fischer–Tropsch syntheses
Water gas shift reaction
Alkoxycarbonylation of olefins
Carbonylation of azides and nitro compounds
Dimerization, oligomerization, and polymerization of olefins, dienes, and acetylenes
Olefin disproportionation or metathesis
Olefin isomerization
Oxidation
Hydrolysis
Grignard cross-coupling
Michael reaction

In this section, no attempt is made to cover all of these reactions. Instead, a few well studied reactions that exemplify the strengths and weaknesses of supported catalysts are considered. For a recent comprehensive account of the application of supported catalysts, readers should consult ref. 16.

1. Cyclopentadienyltitanium-catalysed olefin hydrogenation

Although titanocene is an active olefin hydrogenation catalyst[406], it is readily deactivated through the formation of an inactive dimeric compound[407]. By supporting $[TiCp_2Cl_2]$ on a polymeric support and reducing this with butyllithium, a catalyst that is 20 times more active is formed[30,42,193,408,409]. The enhanced activity arises because attachment of the titanium to relatively rigid supports such as styrene–20% divinylbenzene prevents dimerization. In addition to their greater activity, the supported catalysts are more selective towards substrates of different degrees of bulkiness owing to the dispersion of the titanium complexes throughout the support[410]. The supported titanium catalyst can be prepared by reaction 71, although the route shown in reaction 72 is much preferred[189], for three reasons:

(i) it eliminates the possible presence of any potentially reactive benzyl groups;
(ii) it avoids the use of carcinogenic chloromethyl methyl ether;
(iii) the products of reaction 72 are between 1.25 and 6 times more reactive than those of reaction 71.

CH2Cl —NaCp→ CH2–C5H5 —LiMe→ CH2–C5H4⁻ Li⁺ —[TiCpCl3]→ CH2–Cp–TiCl2Cp

(71)

Li —(i) cyclopentanone, (ii) H⁺→ OH —gentle heating, −H2O→ —(i) LiMe, (ii) [TiCpCl3]→ Cp–TiCl2Cp

(72)

Supporting $[TiCp_2Cl_2]$ on silica using reaction 73 yields a product that has enhanced activity for terminal olefin hydrogenation relative to its homogeneous analogue, but has no enhanced activity for internal olefin hydrogenation unlike the polystyrene-supported complexes[411]. Clearly, the silica surface plays an important role in the internal olefin inhibition but it is not known exactly how it promotes this selectivity.

$$C_5H_5SiMe_2OEt \xrightarrow{+Na} Na^+[C_5H_4SiMe_2OEt]^- \xrightarrow{[TiCp_2Cl_2]} [\{EtOSi(Me)_2C_5H_4\}_nTiCp_{2-n}Cl_2] \; (n=1,2) \xrightarrow{+\,Silica} [\{\text{≹}-OSi(Me)_2C_5H_4\}_nTiCp_{2-n}Cl_2] \quad (73)$$

2. *Rhodium(I)-phosphine-catalysed olefin hydrogenation*

Since 1.75% of all the papers currently being recorded in *Chemical Abstracts* include the use of Wilkinson's catalyst $[Rh(PPh_3)_3Cl]$[411], it is not surprising that its application in supported form has been extensively investigated. In the course of any catalysis the complex loses a triphenylphosphine ligand[401,402,412], and in Section V.C we referred to the implications of this for rhodium loss. An interesting development of this has been the use of $[Rh(PPh_3)_3Cl]$ in conjunction with the silver salt of sulphonated polystyrene which enhances the homogeneous catalyst's activity by facilitating the removal of a triphenylphosphine ligand from the rhodium[413,414].

Wilkinson's catalyst has been supported on a wide range of materials, including both soluble and insoluble polystyrene[36,37,48], aromatic polyamides[415], phosphinated polydiacetylenes[416], phosphinated poly(vinyl chloride), which is fairly inactive[119,380], and silica, which tends to result in high activity owing to the predominant location of rhodium on the surface[212,360,417,418]. The importance of the support is well illustrated by the activity of materials formed by supporting $[Rh(nbd)Cl]_2$ on phosphinated polydiacetylene, phosphinated silica, and phosphinated polystyrene. The first two actively promote the reduction of arenes to cyclohexane derivatives, whereas that supported on polystyrene does not. This is believed to be due to the great ability of the more rigid polydiacetylene and silica to stabilize coordinatively unsaturated rhodium(I) as compared with the more flexible polystyrene[419]. The crucial importance of the detailed structure of the support on the activity of a supported catalyst is beautifully illustrated by the abrupt change in the ethylene hydrogenation activity of $[Rh(PPh_3)_3Cl]$ supported on a ternary copolymer of styrene, divinylbenzene, and *p*-styryldiphenylphosphine at 68 °C[385]. This arises from a change in the structure of the polymer at its glass transition temperature. In this particular instance the change is reversible, but not all such structural changes are reversible and many lead to catalyst deactivation[386].

The activity of supported rhodium(I) hydrogenation catalysts depends on the number and accessibility of the active sites. Fine beads of lightly cross-linked polystyrene give more active catalysts than larger beads of more highly cross-linked material[53a,383,420]; indeed, rhodium(I) complexes supported on completely uncross-linked polystyrene are more active and more stable than their homogeneous analogues[42]. This is almost certainly due to the role of the support in inhibiting dimerization as a deactivation mechanism[16,420], although this has been questioned[421]. Dimerization as the cause of deactivation probably accounts for the increasing rate of deactivation with increasing chain length, n, of rhodium(I) supported on silica functionalized with $—Si(CH_2)_nPPh_2$[422].

The ratio of phosphorus to rhodium in the catalyst has a major influence on activity. Ideally it should be less than 3:1 in order to promote coordinative unsaturation, but dimerization and even reduction to metallic rhodium may occur if insufficient phosphine groups are present[386,423]. The optimum P to Rh ratio depends on the exact nature of the solvent, more powerful coordinating solvents giving higher optimum P to Rh ratios as a consequence of solvent–phosphine competition[424]. We have already referred to the profound effect that solvents can exert on the catalytic ability of supported complexes as a consequence of their polarity in Section II.A (especially Figure 1).

The activity of phosphine-supported rhodium(I) catalysts is very sensitive to the presence of oxygen since phosphines bound through aliphatic carbon atoms with and without intermediate aryl groups, that is both $—CH_2PPh_2$ and $—CH_2C_6H_4(PPh_2)$-*p*, are very sensitive to oxidation to yield phosphorus(V) species which do not coordinate to rhodium(I)[22,425]. Rhodium(I) complexes, both supported and unsupported, are sensitive to poisoning by thiols, although *n*-butanethiol enhances the thermal stability of silica-supported $[(\{Si\}\text{-}OCH_2CH_2PPh_2)_3RhCl]$, albeit for reasons that are far from clear, although its activity is reduced[38a–c].

The selectivity of supported catalysts can be promoted by (i) the use of porous polymer supports, as discussed in Section II.A, (ii) altering the solvent, or (iii) altering the catalyst loading.

The influence of solvent can be very complex and can depend on the substrate concentration. Thus, in the hydrogenation of hex-1-ene and cyclohexene in the presence of Wilkinson's catalyst supported on 1 and 4% cross-linked polystyrene, a change of solvent from benzene to 1:1 benzene–ethanol leads to a significant enhancement of the rate of hydrogenation at low substrate concentrations, but little change in rate at high substrate concentrations[52,420]. At very high cyclohexene concentrations the rate decreases as benzene is replaced with ethanol. These apparently conflicting observations result from an increase in the substrate concentration within the resin when the solvent polarity is increased, coupled with the reduced swelling ability of ethanol compared with benzene. These effects, however, are significant only when the substrate concentration is low and hence rate limiting. Clearly the optimum solvent is one of good swelling ability and high polarity, such as tetrahydrofuran, which enhances the rate of hydrogenation of the sterically demanding cyclohexene when it is used instead of benzene[420,426].

The influence of catalyst loading on selectivity is well illustrated by the results in Table 4[53]. Higher loadings result in higher selectivities because the polymer surrounding the catalyst imposes a diffusion barrier between the bulk solution and the catalytic centre. The resulting concentration gradient across this diffusion barrier will be greatest for the most active catalyst. When comparing olefins of different bulk, the bulkiest should give rise to the greatest concentration gradient in the polymer. Consequently, the most active catalyst system will place the greatest demand on the bulk solution to supply substrate which will further enhance the differences in the diffusion rates of the two olefins.

Cationic rhodium(I) complexes have been anchored onto polystyrene by long alkyl chains in an attempt to suspend the rhodium(I) sites well away from the support in the bulk solution[20]. When the resulting catalysts, [Rh{ (P)—C_6H_4— NHCO$(CH_2)_{10}PPh_2\}_2(nbd)]^+BF_4^-$, were used to hydrogenate oct-1-ene in acetone it was found[20] that:

(i) the rate of hydrogenation was comparable or superior to that with the corresponding homogeneous catalyst based on $PMePh_2$ under comparable conditions;
(ii) the presence of excess of free phosphine groups on the support was detrimental,

TABLE 4. Relative rates of reduction of olefins in competition with cyclohexene in the presence of [(Ⓟ 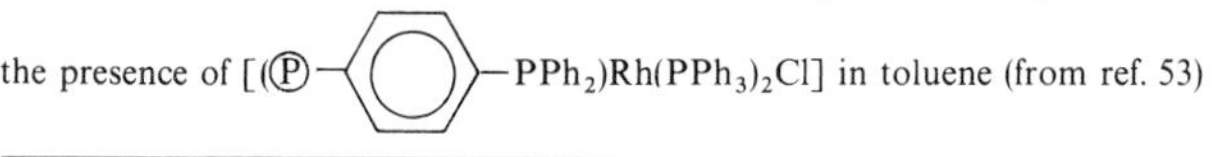 $-PPh_2)Rh(PPh_3)_2Cl]$ in toluene (from ref. 53)

Olefin	Relative hydrogenation rates on beads with	
	high Rh^I loading	low Rh^I loading
	1.75 ± 0.03	1.80 ± 0.1
	1.00 ± 0.05	1.00 ± 0.04
	0.805 ± 0.05	0.97 ± 0.06
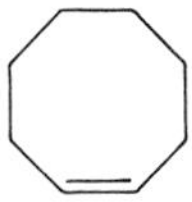	0.43 ± 0.08	0.64 ± 0.05
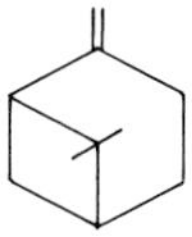	0.08 ± 0.003	0.35 ± 0.02

presumably because these were sufficiently free to coodinate to the rhodium(I), so preventing olefin access;

(iii) the supported catalyst normally remained light orange throughout the cycle but if excess of rhodium(I) was present or perchloric acid was added then rhodium metal was formed during the hydrogenation;

(iv) the decrease in hydrogenation rate due to isomerization of oct-1-ene to oct-2-ene was less for the supported than for the homogeneous catalyst containing $PMePh_2$.

3. Cobalt- and rhodium-catalysed hydroformylation

Hydroformylation of olefins (see Chapter 8 in Volume 3) results in the addition of H to one end and CHO to the other end of the double bond. In the case of terminal olefins either terminal (normal) aldehydes or internal (branched) aldehydes are formed, depending on the direction of addition. In the case of cobalt catalysts these aldehydes are frequently further reduced to alcohols, especially if tertiary phosphine ligands are present. The terminal (normal) products are of considerably greater commercial value than the branched products, so that interest in the use of supported hydroformylation catalysts has particularly centred on their ability to promote a high normal to branched ratio of products.

Although early work on the use of poly-2-vinylpyridine as a support for cobalt carbonyl was directed against the ease of separation of the catalyst at the end of the reaction[197,198], more recent work by the present author has shown that the use of poly-4-vinylpyridine grafted on to a polypropylene support can enhance the normal to branched selectivity of the cobalt carbonyl catalyst by a factor of 2.5[41]. The importance of the detailed nature of the support in emphasized by the fact that cobalt carbonyl supported on phosphinated polystyrene gives a comparable normal to branched selectivity to its homogeneous counterpart below 155 °C and a 2-fold lower selectivity above 155 °C[113].

Rhodium(I)–phosphine homogeneous hydroformylation catalysts are not only more active under milder conditions than their cobalt analogues, but also give much greater normal to branched selectivities. Several groups of workers have prepared supported rhodium(I) hydroformylation catalysts that are considerably more selective than their homogeneous analogues. We have already described our own system that shows this greatly enhanced selectivity in Section V.A (especially Table 3)[22]. That work and other workers' results show well the importance of the matrix[22,403,428], and also the importance of increasing the phosphorus to rhodium ratio in order to enhance the normal to branched selectivity[15,22,362,403,427–430]. As far as we are aware only one exception to the latter effect is known, viz. the hydroformylation of methyl methacrylate, where the reverse is observed[431].

The presence of bisphosphinerhodium(I) species promotes the formation of normal products more than monophosphinerhodium(I) species owing to steric crowding in the former. As a result, the higher the phosphine to rhodium concentration the greater is the proportion of normal product. However, polymer-bound phosphine groups are not freely mobile and thus the ability to promote bisphosphine complex formation depends on both the phosphorus to rhodium ratio (P:Rh) and the degree of phosphine loading (PL)[363,404,429], as can be seen from Figure 3. The normal to branched ratio was between 10 and 12.5 for the polymer with a phosphine loading of 40 (PL = percentage of the phenyl rings functionalized with —PPh_2) and P:Rh = 19. At low PL and low P:Rh the ratio fell to between 4.4 and 3.6, whereas intermediate PL and P:Rh gave intermediate normal to branched selectivities[362].

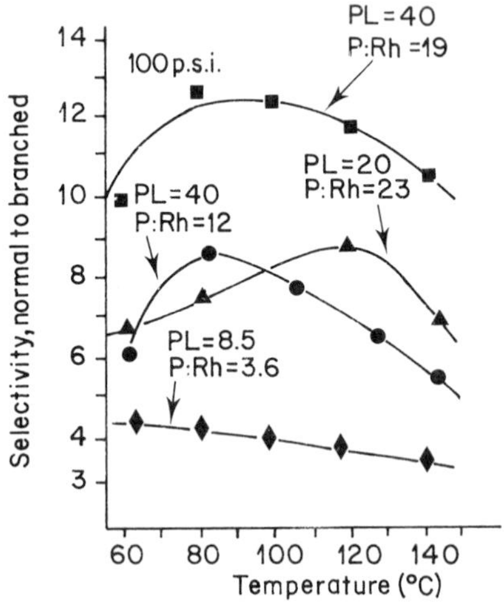

FIGURE 3. Effect of phosphorus to rhodium ratio (P:Rh) and degree of phosphorus loading (PL) on the ratio of normal to branched products in the hydroformylation of pent-1-ene using 1:1 H_2–CO at 6 atm pressure (reproduced with permission from ref. 362)

Many supported hydroformylation catalysts have comparable activities to those of their homogeneous analogues, provided that the comparisons are made at 60 °C or above. At lower temperature their activities are limited by the rates of diffusion of the reactants within the support[114,253,432,433]. Thus, at 40 °C and 17 atm [(Ⓟ-PPh_2)RhH(CO)(PPh_3)], where Ⓟ = polystyrene–1%-divinylbenzene, was 0.22 times as active as its homogeneous counterpart in pent-1-ene hydroformylation, whereas at 60 °C and 53 atm it was 1.08 times more active than its homogeneous analogue[114,433].

Supported catalysts sometimes show a decreasing normal to branched selectivity as the hydroformylation proceeds. There may be two causes for this. Firstly, many rhodium(I) catalysts promote olefin isomerization, leading to a build-up of internal olefins which are hydroformylated more slowly than terminal olefins, but which inevitably lead to branched products. Secondly, many commercial sources of hydrogen–carbon monoxide mixtures contain traces of oxygen which, if not removed, oxidize the phosphine groups to non-coordinating phosphorus(V) groups, so leading to a steady decrease in the phosphorus to rhodium ratio during the reaction.

We have already referred in Section V.C to the problem of leaching and noted in Section IV.A that there are significant advantages in ensuring that no free ligand remains in the support by displacing volatile ligands such as carbon monoxide. In homogeneous catalyses the presence of a bidentate α,ω-bis(diphenylphosphino)alkane markedly improves the stability of the homogeneous rhodium(I) catalyst and decreases the accompanying olefin isomerization[434–436]. Supported monodentate phosphines may well behave in a similar manner owing to the multidentate nature of the resulting material. Supported bidentate phosphine complexes such as polystyrene-bound [Rh(Ⓟ-$PPhCH_2CH_2PPh_2$)H(CO)(PPh_3)] give higher normal to branched selectivity than their homogeneous counterparts[437]; no rhodium leaching could be detected although isomerization is a problem[362].

4. Asymmetric hydrogenation and hydroformylation catalysts

The increasesing demand for optically pure chiral compounds in recent years has been boosted by the recognition that only one optical isomer of a chiral drug is usually active, and that the 'inactive' isomer can in some instances be positively harmful. Supported asymmetric catalysts can be subdivided into two groups: (i) catalysts supported on chiral supports such as cellulose[388], polypeptides[438,439], and chiral stereoregular polymers[44]; and (ii) catalysts in which a chiral complex is supported on an achiral support[38,125,126,137,138,209,210,231,251,360,441–448].

Most of the examples where chiral supports have been used involve supported metals rather than supported complexes. However, $RuCl_3$ supported on poly(L-methylenimine) has been used to hydrogenate methyl acetoacetate and mesityl oxide in low (< 6%) optical yield[438,439] and [Rh(PPh_3)$_3$Cl] supported on phosphinated cellulose catalysed the hydrogenation of the α-phthalimidoacrylic acid derivative of alanine in 28% optical yield[388]. Enzymes often have asymmetric sites. If an achiral catalyst could be immobilized in such a site the enzyme's chirality might effect a chiral catalytic reaction. This has now been achieved by binding an achiral biotin rhodium(I) complex into the chiral cavity of avidin to form a supported catalyst that hydrogenates α-acetamidoacrylic acid at 0 °C under 1.5 atmospheres of hydrogen in 34–41% enantiomeric excess[389] (see also Section V.A, especially reactions 63 and 64).

The first attempts to support rhodium(I)–diop asymmetric complexes were unsuccessful in promoting asymmetric hydrogenation of useful organic substrates because polystyrene was used as the support[38]. Polystyrene shrinks rather than swells in the polar solvents that have to be used for substrates such as acylamidoacrylic acids; even with substrates that are soluble in benzene these catalysts gave lower optical yields than their homogeneous counterparts[38]. Copolymerization of functionalized styrene with hydroxyethyl methacrylate (reaction 74) yielded a polymer that was swollen by alcohols and which catalysed the asymmetric hydrogenation of α-*N*-acylaminoacrylic acid in ethanol in the same optical yield and absolute configuration as its homogeneous counterpart, albeit at a slower rate[441,442]. When an optically active functional group was introduced on to the polymeric support adjacent to the rhodium(I)–diop centre (reaction 75), it was found that in a protic

HO—C(H)—CH_2Otos
HO—C(H)—CH_2Otos
+ $CH=CH_2$ / CHO (p-vinylbenzaldehyde) —C_6H_6→ acetal —$MPPh_2$→ polymer $(CH_2—CH)_n$ with CH_2PPh_2 CH_2PPh_2

$CH_2=C(CH_3)CO_2CH_2CH_2OH$, 1% aibn, C_6H_6

$(CH_2—CH)_{0.08}(CH_2—C(Me)(CO_2CH_2CH_2OH))_{0.92}$ with CH_2Otos CH_2Otos —(i) $NaPPh_2$ (ii) H_2O→ $(CH_2—CH)_{0.08}(CH_2—C(Me)(CO_2CH_2CH_2OH))_{0.92}$ with CH_2PPh_2 CH_2PPh_2

$[RhCl(C_2H_4)_2]_2$

$(CH_2—CH)_{0.08}(CH_2—C(Me)(CO_2CH_2CH_2OH))_{0.92}$ with CH_2PPh_2 CH_2PPh_2 Rh Cl (S)

(74)

$CH_2{=}CH{-}CO{-}Me$ + $CH_2{=}CH$–(C$_6$H$_4$)–dioxolane(CH$_2$Otos)$_2$ —aibn→ polymer

(i) [Rh(diop)Cl(S)] + PhArSiH$_2$
(ii) thf–H_2O, H^+

(i) $NaPPh_2$, thf
(ii) H_2O

$[(C_2H_4)_2RhCl]_2$

(75)

solvent such as benzene–ethanol the optical yield was independent of the absolute configuration of the copolymer alcohol group. However, in a non-protic solvent the absolute configuration of the alcohol did affect the optical yield; for example, in thf a 40% optical yield was obtained in the hydrogenation of α-*N*-acylaminoacrylic acid when the copolymer alcohol was in the *R*-configuration, whereas this was reduced to 25% when the *S*-configuration alcohol was present[360].

An investigation of the influence of the molecular weight of non-cross-linked polystyrenes on the ability of rhodium(I)–diop complexes to promote asymmetric hydrogenation showed that optical selectivity decreases as the molecular weight decreases, owing to the tendency of low molecular weight polymers to coil up and so depress the effective asymmetric interaction[443–445]. Thus a relatively rigid support such as silica gel is best, and indeed gives comparable asymmetric efficiency at an order of magnitude less activity than the homogeneous equivalent[231]. This explanation could be simplistic, however, since when graphite is used as the support the absolute configurations of the products of the graphite-supported catalyses are exactly the opposite of those obtained using the homogeneous counterpart[251], although the presence of surface phenolic groups on the graphite may be the cause of the reversal of configuration.

Polymer-supported chiral pyrrolidinephosphine ligands, **4a** and **4b**, have been used to support neutral **5**, or cationic, **6**, rhodium(I) complexes (reaction 76)[209,210]. Compound **5** in benzene catalysed the asymmetric hydrogenation of ketopantolactone to (*R*)-(−)-pantolactone (reaction 77) with optical yields of 73.4% (L = **4a**) and 75.7% (L = **4b**)[210].

Ph$_2$P, CH$_2$PPh$_2$, N–H + COCl → CO–N, PPh$_2$, CH$_2$PPh$_2$

MeC(=CH$_2$)COR →

CH, CH$_2$, Me, C, COOR, CH$_2$, CO–N, PPh$_2$, CH$_2$PPh, *m*, *n*

(**4a**) R = Me, *m*/*n* = 1/4

(**4b**) R = Bu^t, *m*/*n* = 1/4

[Rh(cod)(acac)] + HClO$_4$ → [RhL(cod)]$^+$ClO$_4^-$ (**6**) (L = **4a,4b**)

[Rh(cod)Cl]$_2$ → RhL(cod)Cl (**5**) (L = **4a,4b**)

(76)

$$\text{Me}_2\text{C(CO)(CO)O-CH}_2 \xrightarrow[+\text{catalyst}]{H_2,\text{benzene}} \text{Me}_2\text{C-CH(OH)-C(=O)O-CH}_2 \qquad (77)$$

The cationic supported complex, **6**, was more effective than the neutral supported complex, **5**, in the asymmetric hydrogenation of itaconic acid to methylsuccinic acid and (*Z*)-α-acetamidocinnamic acid; **6** was as effective as its homogeneous analogue under optimum conditions although those optimum conditions needed more careful control with the supported than with the unsupported catalyst[209].

When rhodium was supported on Ⓟ—C_6H_4—P(menthyl)$_2$ and Ⓟ—C_6H_4—CH_2P(menthyl)$_2$ (Ⓟ = polystyrene) optical yields as high as 58% were achieved in the hydrogenation of (*Z*)-α-acetamidocinnamic acid, although the activities of these supported catalysts were low. The optical activity of the product was heavily dependent on the solvent, varying from 58% enantiomeric excess in 1:1 benzene–ethanol, through 14% in dioxane to only 8% in thf. If oxygen was accidentally introduced during recycling, the activity and optical yield decreased remarkably[125,126]. Only one asymmetric catalysis has been reported using an asymmetric rhodium(I) complex, $[Rh(cod)\{(R)\text{-}(+)\text{-}PhCH_2CH(NHAc)COOH\}]^+ClO_4^-$, supported on the clay hectorite; this catalyses the asymmetric hydrogenation of (Z)-α-acetamidocinnamic acid[446].

Only limited success has so far been achieved in using supported rhodium(I)– or platinum(II)–diop catalysts in asymmetric hydroformylations. The best optical yields have been between 25 and 30%[49,126,428,448–452].

5. Methanol carbonylation

Attempts have been made to develop supported analogues of the homogeneous rhodium-catalysed carbonylation of methanol to acetic acid (Scheme 2) developed by Monsanto[453]. Rhodium catalysts have been supported on polystyrene[454,455], polystyrene–poly(4-vinylpyridine) copolymer[175], carbon[454,456–458], silica, alumina, magnesium oxide and type X zeolite molecular sieves[387,454,459–463]. However, all these catalysts, although many have very high selectivities, have activities typically two orders of magnitude less than the homogeneous system[454]. Since the homogeneous rhodium catalyst can be recovered and recycled fairly easily, this is likely to be preferred commercially for some time to come[464].

The rhodium methanol carbonylation catalyst requires a cocatalyst which in the homogeneous, and indeed several of the supported systems, is methyl iodide. However, this cannot readily be built on to the support and so an attempt has been made to develop a bifunctional catalyst be supporting the pseudo-alkyl halide **7** using reaction 78. On reaction with $RhCl_3$ and carbon monoxide, **7** forms **8**, which with methyl iodide yields **9**, which has methylthioether side groups that undergo oxidative addition to the rhodium(I) to yield **10**. This system is an active methanol carbonylation catalyst, although much less active than its homogeneous counterpart[466]. On continuous use the activity decreases steadily owing to loss of both rhodium and sulphur through reaction 70, in which it is the Ⓟ-CH_2—S– sulphur, rather than the Me—S– sulphur, that has coordinated to the rhodium(I). Indeed, since Ⓟ-CH_2—S– has a weaker C—S bond than Me—S–, if it were not for steric effects reaction 70 would totally dominate reaction 79.

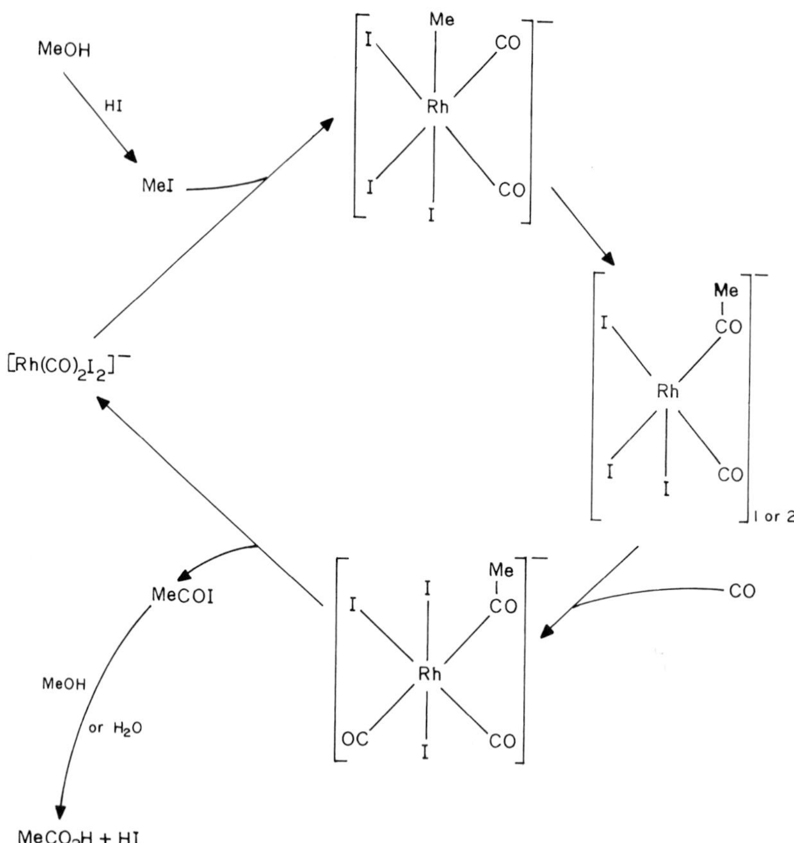

SCHEME 2

(P)—C_6H_4—CH_2Cl + Cl—C_6Cl_4—SNa → (P)—C_6H_4—CH_2—S—C_6Cl_4—Cl + NaCl

→ (P)—C_6H_4—CH_2—S—C_6Cl_3(Cl)—SH (78)

(**7**)

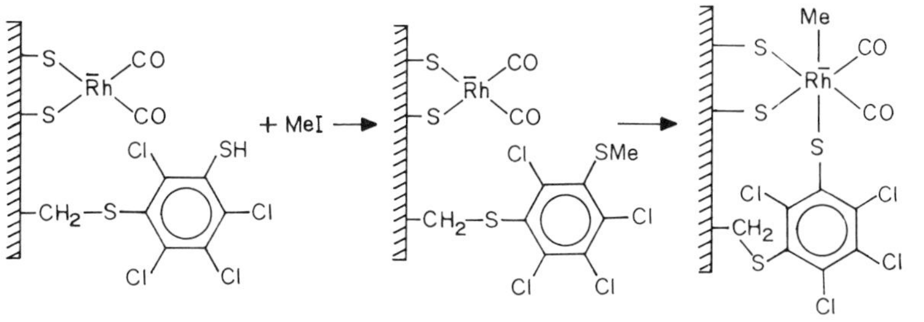

(79)

(8) (9) (10)

6. Sequential multi-step reactions

A number of chemically desirable reactions involve the sequential conversion of one compound into an intermediate that is further converted into the final product. There would be obvious advantages if this could be done in a single vessel with a single catalyst rather than separating off the catalyst and purifying between stages. This concept has been developed in reactions using immobilized enzymes[467] where, for example, glucose is converted first to glucose-1-phosphate and subsequently to glucose-6-phosphate using polystyrene supporting both hexokinase and glucose-6-phosphate isomerase.

Any attempt to introduce two catalytic centres on to a single support must necessarily introduce extra complications which will include the following:

(i) the catalysts may interfere with each other by the ligands from one complex coordinating to the second complex;

(ii) either catalyst may be destroyed by products and by-products formed due to the presence of the other catalyst;

(iii) extra side products may be formed by one catalyst promoting extra reactions of the products and by-products of the other;

(iv) reactions must be chosen with care if the conditions such as temperature, pressure, and solvent are to be identical for both catalysts.

In spite of these obvious difficulties, a number of sequential multi-step reactions have been examined to determine whether or not they are feasible[115,468–474]. Several of these involve an initial oligomerization or cyclooligomerization of butadiene followed by further reaction of the product:

(i) Cyclooligomerization–hydroformylation (reactions 80 and 81)[383,468–470].

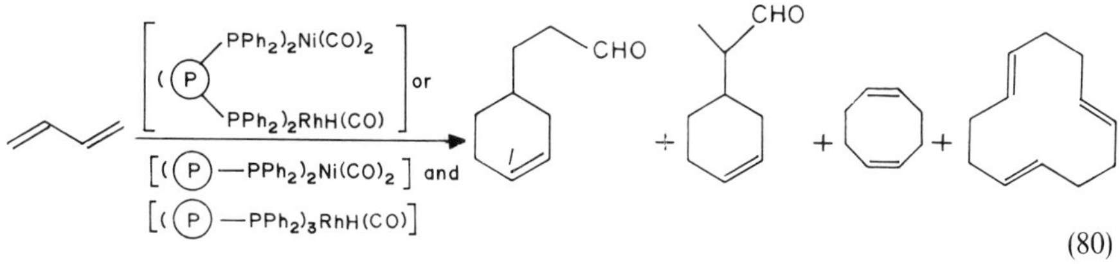

(80)

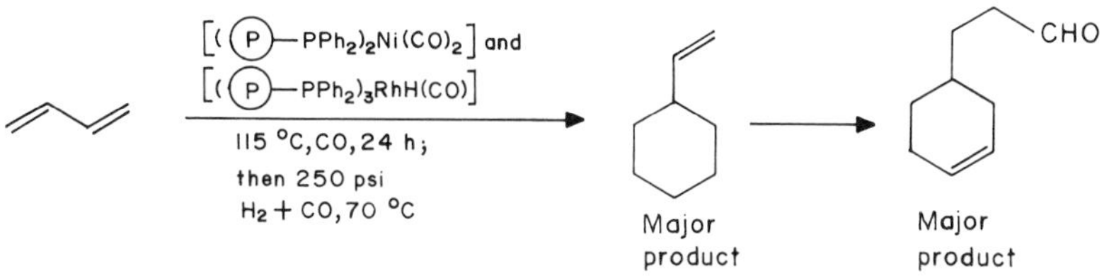

(81)

(ii) Cyclooligomerization–hydrogenation (reactions 82 and 83)[383,468–470].

[((P)–PPh$_2$)$_2$Ni(CO)$_2$ / PPh$_2$)$_3$RhCl] or [((P)–PPh$_2$)$_2$Ni(CO)$_2$] and [((P)–PPh$_2$)$_3$RhCl]

100 °C, 24 h; then 50 °C, 250 psi H_2, 4 h

25–29% (11) + 53–60% (12) + 1–12% (13)

(82)

[((P)–PPh$_2$)$_2$Ni(CO)$_2$ / PPh$_2$)$_3$Ru(CO)$_2$Cl$_2$] or [((P)–PPh$_2$)Ni(CO)$_2$] and [((P)–PPh$_2$)$_3$Ru(CO)$_2$Cl$_2$]

100 °C, 24 h; then 160 °C, 150 psi H_2

[11 + 12 + 13]

(83)

(iii) Oligomerization–hydroformylation (reaction 84)[472].

[((P)–PPh$_2$)$_2$NiBr$_2$] + NaBH$_4$ and [((P)–PPh$_2$)$_2$RhH(CO)]

100 °C, 24 h; then 65–70 °C 500 psi H_2 + CO

CHO + complex mixture of products

ca 50%

(84)

(iv) Oligomerization–acetoxylation–hydrogenation (reaction 85)[473].

[((P) $PPh_2)_2PdCl_2$ / $PPh_2)_3RhCl$] or [(P)— $PPh_2)_2PdCl_2$] and [(P)— $PPh_2)_3RhCl$] with HOAc, Et_3N, 100 °C, 24 h; then 350 psi H_2, 50 °C, 4 h

OAc + OAc

OAc + OAc

(85)

(v) Hydroformylation–aldol condensation–hydrogenation (reaction 86)[474].

+ H_2 + CO — [(P) $CH_2PPh_2)_2Rh(CO)Cl$ / CH_2NHEt] →

CHO — Aldol condensation, step 2 → CHO — H_2, step 3 → CHO

+

CHO — Aldol condensation, step 2 → CHO — H_2, step 3 → CHO

(86)

Multi-step reactions can be carried out either by attaching the two metal complexes to the same polymer bead or to two separate beads. In general the latter is more effective because it allows the ligand to metal ratios and the nature of the ligands to be varied independently. Reaction 86 is interesting in that it involves a rhodium-catalysed hydroformylation followed by an amine-catalysed aldol condensation, the products of which are hydrogenated in the presence of the rhodium catalyst. Competition between the phosphine and the amine for the rhodium might be expected. However, the absence of alcohol formation in the hydroformylation step indicates that such competition does not occur. Nevertheless, the presence of the amine groups within the same polymer bead does promote rhodium loss; it can be reduced by putting the rhodium on one set of beads and the amine on another, but when these two sets of beads are used the overall rate of the reaction is much reduced owing to two factors: firstly the extra diffusion barriers to be overcome when molecules have to diffuse into and out of three beads before reaction is complete, and secondly the amine groups increase the polarity within the polymer so enhancing the rate of the hydroformylation step[474].

VI. SUMMARY AND FUTURE DEVELOPMENTS

The work described in this chapter contains more than enough encouraging results to suggest that supported metal complex catalysts will have a valuable role to play in the future. It is also clear that the factors that lead to a successful active, specific, and selective supported catalyst are complex, and that merely supporting an effective homogeneous catalyst and trying to use it under the same conditions as when used homogeneously rarely lead to success. Great care needs to be taken in identifying the right ligands, which are often different to those in the equivalent homogeneous catalyst, and also the precise experimental conditions of solvent, concentrations, temperature, and pressure. The real future will lie in ensuring that the active metal centre and the support both take part in the catalytic reaction so that the supported catalyst is even more selective and specific than the equivalent homogeneous catalyst. In this way a new generation of man-made catalysts will arise that mimic the best of the metalloenzymes.

Clearly, the major development that everyone working in the field is aiming for is the commercial application of supported metal complex catalysts in large-scale industrial chemical reactions. They have already been used in small-scale fine chemical applications. There are potentially two reasons for choosing a supported catalyst: (i) either it has a high activity and long life, which means that it must suffer no deactivation by poisons or loss of metal complex due to leaching or degradation; or (ii) it gives rise to very high selectivity. The first property will require a thermally stable, mechanically durable support so that a great deal of further work is likely to concentrate on the use of inorganic supports of the type widely used in industry to support the classical heterogeneous catalysts. These inorganic supports will be functionalized with organic groups that give a flexible environment around the catalytic site. Inorganic supports that clearly merit further study include zeolites, clays, and glasses[475]. Their regular topological structure could be used for the entrapment of the catalyst and thus for the modification of its selectivity.

The importance of high selectivity will lead to far more studies of the detailed three-dimensional nature of the active sites. Thus the support will not be merely an insoluble support but rather an essential contributor to the total three-dimensional structure of the active site. It will be important for the synthetic work to be backed up by physical characterization. In addition to the traditional spectroscopic techniques, the development of solid-state n.m.r. spectroscopy[476–479] (described in Chapter 21 of Volume 1) will undoubtedly provide a major tool for studying the structures of active sites.

VII. REFERENCES

1. W. O. Haag and D. D. Whitehurst, *Ger. Pat.*, 1800371 (1969); *Chem. Abstr.*, **71**, 114951 (1969).
2. W. O. Haag and D. D. Whitehurst, *Ger. Pat.*, 1800379 (1969); *Chem. Abstr.*, **72**, 31192 (1970).
3. W. O. Haag and D. D. Whitehurst, *Ger. Pat.*, 1800380 (1969); *Chem. Abstr.*, **71**, 33823 (1969).
4. W. O. Haag and D. D. Whitehurst, *Belg. Pat.*, 721686 (1969).
5. R. B. Merrifield, Fed. Proc. Fed. Am. Soc. Exp. Biol., **21**, 412 (1962).
6. R. B. Merrifield, *Science*, **150**, 178 (1965).
7. Z. M. Michalska and D. E. Webster, *Platinum Met. Rev.*, **18**, 65 (1974).

8. J. C. Bailar, *Catal. Rev. Sci. Eng.*, **10**, 17 (1974).
9. F. R. Hartley and P. N. Vezey, *Adv. Organomet. Chem.*, **15**, 189 (1977).
10. Y. Chauvin, D. Commereuc, and F. Dawans, *Prog. Polym. Sci.*, **5**, 95 (1977).
11. R. H. Grubbs, *ChemTech.*, **7**, 512 (1977).
12. D. C. Bailey and S. H. Langer, *Chem. Rev.*, **81**, 109 (1981).
13. C. U. Pittman, in *Polymer-Supported Reactions in Organic Synthesis* (Eds P. Hodge and D. C. Sherrington), Wiley, Chichester, 1980, p. 249.
14. D. D. Whitehurst, *ChemTech.*, **10**, 44 (1980).
15. C. U. Pittman, in *Comprehensive Organometallic Chemistry* (Eds G. Wilkinson, F. G. A. Stone and E. W. Abel), Vol. 8, Pergamon Press, Oxford, 1982, Chapter 55.
16. F. R. Hartley, *Supported Metal Complexes: A New Generation of Catalysts*, Reidel, Dordrecht, 1985.
17. A. L. Robinson, *Science*, **194**, 1261 (1976).
18. J. Smidt, *Chem. Ind* (*London*), 54 (1962).
19. J. F. Roth, J. H. Craddock, A. Hershman and F. E. Paulik, *ChemTech.*, **1**, 600 (1971).
20. J. M. Brown and H. Molinari, *Tetrahedron Lett.*, 2933 (1979).
21. K. G. Allum, R. D. Hancock, S. Mackenzie, and R. C. Pitkethly, in *Catalysis* (Ed. J. W. Hightower), Vol. 1 North-Holland, Amsterdam, 1973, p. 477.
22. F. R. Hartley, S. G. Murray, and P. N. Nicholson, *J. Mol. Catal.*, **16**, 363 (1982).
23. F. R. Hartley, *Chem. Rev.*, **73**, 163 (1973).
23a. H. Widdecke, *Bri. Poly. J.*, **16**, 188 (1984).
24. W. Heitz, *Adv. Polym. Sci.*, **23**, 1 (1977).
25. W. N. Lipscomb, *Chem. Soc. Rev.*, **1**, 319 (1972).
26. S. Mazur, P. Jayalekshmy, J. T. Anderson, and T. Matusinovic, *Am. Chem. Soc. Symp. Ser.*, No. 192, 43 (1982).
27. D. P. Harrison and H. F. Rase, *Ind. Eng. Chem. Fundam.*, **6**, 161 (1967).
28. A. T. Jurewicz, L. D. Rollmann, and D. D. Whitehurst, *Adv. Chem. Ser.*, No. 132 240 (1974).
29. F. H. Jardine, *Prog. Inorg. Chem.*, **28**, 63 (1981).
30. R. H. Grubbs, C. Gibbons, L. C. Kroll, W. D. Bonds, and C. H. Brubaker, *J. Am. Chem. Soc.*, **95**, 2373 (1973).
31. S. L. Regen and D. P. Lee, *Macromolecules*, **10**, 1418 (1977).
32. D. K. Liu and M. S. Wrighton, AD-A110082 Report (1982); *Chem. Abstr.*, **97**, 64064 (1982).
33. J. Manassen, in *Catalysis, Progress in Research* (Eds. F. Basolo and R. L. Burwell), Plenum, New York, 1977, p. 177.
34. S. Affrossman and J. P. Murray, *J. Chem. Soc. B*, 1015 (1966).
35. D. C. Neckers, D. A. Kooistra, and G. W. Green, *J. Am. Chem. Soc.*, **94**, 9284 (1972).
36. G. Bernard, Y. Chauvin, and D. Commereuc, *Bull. Soc. Chim. Fr.*, 1163 (1976).
37. G. Bernard, Y. Chauvin, and D. Commereuc, *Bull. Soc. Chim. Fr.*, 1168 (1976).
38. W. Dumont, J.-C. Poulin, T.-P. Dang, and H. B. Kagan, *J. Am. Chem. Soc.*, **95**, 8295 (1973).
38a. I. V. Howell, R. D. Hancock, R. C. Pitkethly, and P. J. Robinson, in *Catalysis, Heterogeneous and Homogeneous* (Eds. B. Delmon and B. Jannes), Elsevier, Amsterdam, 1975, p. 349.
38b. K. G. Allum, R. D. Hancock, I. V. Howell, T. E. Lester, S. McKenzie, R. C. Pitkethly, and P. J. Robinson, *J. Organomet. Chem.*, 1976, **107**, 393.
38c. K. G. Allum, R. D. Hancock, I. V. Howell, T. E. Lester, S. McKenzie, P. C. Pitkethly, and P. J. Robinson, *J. Catal.*, 1976, **43**, 331.
39. N. L. Holy, in *Fundamental Research in Homogeneous Catalysis* (Ed. M. Tsutsui), Vol. 3, Plenum, New York, 1979, p. 691.
39a. N. L. Holy, *J. Org. Chem.*, 1979, **44**, 239.
40. E. Atherton, E. Brown, R. C. Sheppard, and A. Rosevear, *J. Chem. Soc., Chem. Commun.*, 1151 (1981).
41. F. R. Hartley, D. J. A. McCaffrey, S. G. Murray, and P. N. Nicholson, *J. Organomet. Chem.*, **206**, 347 (1981).
42. L. Verdet and J. K. Stille, *Organometallics*, **1**, 380 (1982).
43. J. H. Barnes, C. Bates, and F. R. Hartley, *Hydrometallurgy*, **10**, 205 (1983).
44. J. Lieto, D. Milstein, R. L. Albright, J. V. Minkiewitz, and B. C. Gates, *ChemTech.*, **13**, 46 (1983).
45. D. E. Bergbreiter, *Am. Chem. Soc. Symp. Ser.*, No. 192, 1 (1982).
46. A. Guyot and M. Bartholin, *Prog. Polym. Sci.*, **8**, 277 (1982).
47. B. C. Gates, *Nato Adv. Study Inst., Ser. E*, **39**, 437 (1980).

48. E. Bayer and V. Schurig, *Angew. Chem., Int. Ed. Engl.*, **14**, 493 (1975).
49. V. Schurig and E. Bayer, *ChemTech.*, **6**, 212 (1976).
50. V. A. Kabanov and V. I. Smetanyuk, *Macromol. Chem. Phys. Suppl.*, **5**, 121 (1981).
51. D. C. Sherrington, in *Polymer-Supported Reactions in Organic Synthesis* (Eds. P. Hodge and D. C. Sherrington), Wiley, London, 1980, and references cited therein.
52. R. H. Grubbs, L. C. Kroll, and E. M. Sweet, *J. Macromol. Sci., Chem.*, **7**, 1047 (1973).
53. R. H. Grubbs, E. M. Sweet, and S. Phisanbut, in *Catalysis in Organic Syntheses 1976* (Eds. P. N. Rylander and H. Greenfield), Academic Press, New York, 1976, p. 153.
53a. R. H. Grubbs and L. C. Kroll, *J. Am. Chem. Soc.*, **93**, 3062 (1971).
54. H. Standinger and E. Husemann, *Chem. Ber.*, **68**, 1618 (1935).
54a. M. S. Karasch, W. Nudenberg, E. V. Jensen, P. E. Fischer, and D. L. Mayfield, *Ind. Eng. Chem.*, **39**, 830 (1947).
55. R. L. Letsinger and M. J. Kornet, *J. Am. Chem. Soc.*, **85**, 3045 (1963).
56. L. J. Boucher, A. A. Oswald, and L. L. Murell, *Am. Chem. Soc. Div. Pet. Chem. Prepr.*, 162 (1974).
57. C. H. Brubaker, in *Encyclopaedia of Polymer Science and Technology*, Vol. Suppl.I, Wiley, New York, 1976, p. 116.
58. J. A. Rabo, *Zeolite Chemistry and Catalysis*, American Chemical Society Washington, DC, 1976 p. 171.
59. R. Le Van Mao, *Rev. Inst. Fr. Pét.*, **34**, 429 (1979).
60. L. V. C. Rees (Ed.), *Proceedings of 5th International Conference on Zeolites*, Heyden, London, 1980.
61. L. B. Sand and F. A. Mumpton (Eds.), *Natural Zeolites: Occurrence, Properties, Use*, Pergamon Press, Oxford, 1978.
62. D. W. Breck, *Zeolite Molecular Sieves: Structure, Chemistry and Use*, Wiley, London, 1974.
63. See also *Zeolites*, an international journal first published in 1981.
64. T. J. Pinnavaia and P. K. Welty, *J. Am. Chem. Soc.*, **97**, 3819 (1975).
65. T. J. Pinnavaia, P. Raythatha, J. G. S. Lee, L. J. Halloran, and J. F. Hoffman, *J. Am. Chem. Soc.*, **101**, 6891 (1979).
66. W. H. Quayle and T. J. Pinnavaia, *Inorg. Chem.*, **18**, 2840 (1979).
67. M. Mazzei, M. Riocci, and W. Marconi, *Ger. Pat.*, 2845216 (1979); *Chem. Abstr.*, **91**, 56329 (1979).
68. J. F. Hoffman, *Diss. Abstr. B*, **37**, 4454 (1977).
69. T. J. Pinnavaia, *Am. Chem. Soc. Symp. Ser.*, No. 192, 241 (1982).
70. E. J. Corey and A. J. Venkateswarlu, *J. Am. Chem. Soc.*, **94**, 6190 (1972).
71. W. R. Supina, in *The Packed Column in Gas Chromatrography*, Supelco, Bellefonte, PA, 1974, pp. 106–107.
72. P. H. Bach, *Lab. Pract.*, **24**, 817 (1975).
73. T. O. Mitchell and D. D. Whitehurst, *Proc. 3rd North Am. Conf. Catal. Soc.*, 1974, paper 27.
74. R. H. Grubbs and E. M. Sweet, *Macromolecules*, **8**, 241 (1975).
75. H. M. Relles and R. W. Schluenz, *J. Am. Chem. Soc.*, **96**, 6469 (1974).
76. G. A. Stark (Ed.), *Biochemical Aspects of Reactions on Solid Supports*, Academic Press, New York, 1971, pp. 197–198.
77. A. J. Chalk, *J. Polym. Sci., Part B*, **6**, 649 (1968).
78. D. C. Evans, M. H. George and J. A. Barrie, *J. Polym. Sci., Polym. Chem. Ed.*, **12**, 247 (1974).
79. N. Plate, M. A. Jampolskaya, S. L. Davydova, and V. A. Kargin, *J. Polym. Sci., Part C*, **22**, 547 (1969).
80. A. T. Bullock, G. G. Cameron, and P. M. Smith, *Polymer*, **14**, 525 (1974).
81. R. H. Grubbs and S. C. H. Su, *J. Organomet. Chem.*, **122**, 151 (1976).
82. M. J. Farrall and J. M. Frechet, *J. Org. Chem.*, **41**, 3877 (1976).
83. A. T. Bullock, G. G. Cameron, and J. M. Elsom, *Polymer*, **18**, 930 (1977).
84. D. Braun, *Makromol. Chem.*, **30**, 85 (1959).
85. D. Braun, *Makromol. Chem.*, **33**, 181 (1959).
86. D. Braun, *Makromol. Chem.*, **44**, 269 (1961).
87. F. C. Leavitt and L. H. Malternas, *J. Polym. Sci.*, **45**, 249 (1960).
88. D. C. Evans, L. Phillips, J. A. Barrie, and M. H. George, *J. Polym. Sci., Polym. Lett. Ed.*, **12**, 199 (1974).
89. G. A. Crosby, N. M. Weinshenker, and H. S. Uh, *J. Am. Chem. Soc.*, **97**, 2232 (1975).
90. N. M. Weinshenker, G. A. Crosby, and J. Y. Wong, *J. Org. Chem.*, **40**, 1966 (1975).

91. F. Camps, J. Castells, M. Ferrando, and J. Font, *Tetrahedron Lett.*, 1713 (1971).
92. L. T. Scott, J. Rebek, L. Ovsyanko, and C. L. Sims, *J. Am. Chem. Soc.*, **99**, 625 (1977).
93. K. W. Pepper, H. M. Paisley, and M. A. Young, *J. Chem. Soc.*, 4097 (1953).
94. G. D. Jones, *Ind. Eng. Chem.*, 44 (1952).
95. J. V. Minkiewicz, D. Milstein, J. Lieto, B. C. Gates, and R. L. Albright, *Am. Chem. Soc. Symp. Ser.*, No. 192, 9 (1982).
96. S. K. Freeman, *J. Org. Chem.*, **26**, 212 (1960).
97. T. Altares, D. P. Whyman, V. R. Allen, and K. Meyersen, *J. Polym. Sci., Part A*, **3**, 4131 (1965).
98. B. Green and L. R. Garson, *J. Chem. Soc. C*, 401 (1969).
98a. W. T. Ford, S. Mohanraj and M. Periyasany, *Bri. Polym. J.*, **16**, 179 (1984).
99. F. R. Hartley and D. J. A. McCaffrey, in *Fundamental Research in Homogeneous Catalysis* (Ed. M. Tsutsui), Vol. 3, Plenum, New York, 1979, 707.
100. D. A. Kritskaya, A. N. Ponomarev, A. D. Pomogailo, and F. S. Dyachovski, *J. Polym. Sci., Poly. Symp.*, **68**, 23 (1980).
101. F. R. Hartley, S. G. Murray, and P. N. Nicholson, *UK Pat. Appl.*, 8028823 (1980).
102. D. A. Kritskaya, A. D. Pomogailo, A. N. Ponomarev, and F. S. Dyaschkovskii, *J. Appl. Polym. Sci.*, **25**, 349 (1980).
103. D. A. Kritskaya, A. D. Pomogailo, A. N. Ponomarev, and F. S. Dyaschkovskii, *J. Pure Appl. Polym. Sci.*, 17 (1978).
104. Z. S. Kiyashkina, A. D. Pomogailo, A. I. Kuzaev, G. V. Lagodzinskaya, and F. S. Dyachkovskii, *J. Polym. Sci., Poly. Symp.*, **68**, 13 (1980).
105. H. Barker, J. L. Garnett, R. S. Kenyon, R. Levot, M. S. Liddy, and M. A. Long, *Proc. 6th Int. Congr. Catal.*, 551 (1976).
106. H. Barker, J. L. Garnett, R. Levot, and M. A. Long, *J. Macromol. Sci. Chem.*, **12**, 261 (1978).
107. Unisearch, *Br. Pat.*, 1519462 (1978).
108. J. L. Garnett, R. G. Levot, and M. A. Long, *Eur. Pat. Appl.*, 32455 (1981).
109. F. R. Hartley, S. G. Murray, and P. N. Nicholson, *J. Polym. Sci., Polym. Chem. Ed.*, **20**, 2395 (1982).
110. J. L. Garnett, R. S. Kenyon, and M. J. Liddy, *J. Chem. Soc., Chem. Commun.*, 735 (1974).
111. C. H. Ang, J. L. Garnett, R. Levot, M. A. Long, N. T. Yen, and K. J. Nicol, *Stud. Surf. Sci. Catal.*, **B7**, 953 (1981).
112. C. Tamborski, F. E. Ford, W. L. Lehn, G. J. Moore, and E. J. Soloski, *J. Org. Chem.*, **27**, 619 (1962).
113. G. O. Evans, C. U. Pittman, R. McMillan, R. T. Beach, and R. Jones, *J. Organomet. Chem.*, **67**, 295 (1974).
114. C. U. Pittman, L. R. Smith, and R. M. Hanes, *J. Am. Chem. Soc.*, **97**, 1742 (1975).
115. British Petroleum, *Neth. Pat.*, 7006740 (1970).
116. Y. Nonaka, S. Takahashi, and N. Hagihara, *Mem. Inst. Sci. Ind. Res. Osaka Univ.*, **31**, 23 (1974); *Chem. Abstr.*, **80**, 140678 (1974).
117. K. G. Allum, R. D. Hancock, and P. J. Robinson, *Br. Pat.*, 1277736 (1972).
118. K. G. Allum, R. D. Hancock, I. V. Howell, R. C. Pitkethly, and P. J. Robinson, *J. Organomet. Chem.*, **87**, 189 (1975).
119. K. G. Allum, R. D. Hancock, and R. C. Pitkethly, *Br. Pat.*, 1295675 (1972).
120. A. Guyot, C. Graillat, and M. Bartholin, *J. Mol. Catal.*, **3**, 39 (1977).
121. R. B. King and R. M. Hanes, *J. Org. Chem.*, **44**, 1092 (1979).
122. J. P. Collman, L. S. Hegedus, M. P. Cooke, J. R. Norton, G. Dolcetti, and D. N. Marquardt, *J. Am. Chem. Soc.*, **94**, 1789 (1972).
123. S. V. McKinley and J. W. Rakshays, *Chem. Commun.*, 134 (1972).
124. J. Moreto, J. Albaiges, and F. Camps, *An. Quim.*, 1974, **70**, 638; *Chem. Abstr.*, **82**, 72485 (1975).
125. H. W. Krause, *React. Kinet. Catal. Lett.*, **10**, 243 (1979).
126. H. W. Krause and H. Mix, *East Ger. Pat.*, 133199 (1978); *Chem. Abstr.*, **91**, 63336 (1979).
127. K. G. Allum, R. D. Hancock, S. McKenzie, and R. C. Pitkethly, *Proc. 5th Int. Congr. Catal.*, 477 (1972); published as *Catalysis* (Ed. J. W. Hightower), Vol. 1, North-Holland, Amsterdam, 1973.
128. M. Capka, P. Svoboda, M. Cerny, and J. Hetflejs, *Collect. Czech. Chem. Commun.*, **38**, 1242 (1973).
129. M. Capka, P. Svoboda, M. Kraus, and J. Hetflejs, *Chem. Ind. (London)*, 650 (1972).
130. R. H. Grubbs and L. C. Kroll, *J. Am. Chem. Soc.*, **93**, 3062 (1971).
131. R. H. Grubbs, L. C. Kroll, and E. M. Sweet, *J. Macromol. Sci.*, **7**, 1047 (1973).

132. D. Tatarsky, D. H. Kohn, and M. Cais, *J. Polym. Sci., Polym. Chem. Ed.*, **18**, 1387 (1980).
133. P. Svoboda, V. Vaisarova, M. Capka, J. Hetflejs, M. Kraus, and V. Bazant, *Ger. Pat.*, 2260260 (1973); *Chem. Abstr.*, **79**, 78953 (1973).
134. G. Strukul, M. Bonivento, M. Graziani, E. Cernia, and N. Palladino, *Inorg. Chim. Acta*, **12**, 15 (1975).
135. S. V. McKinley and J. W. Rakshays, *US Pat.*, 3708462; *Chem. Abstr.*, **78**, 85310 (1973).
136. W. Dumont, J. C. Poulin, T. P. Dang, and H. B. Kagan, *J. Am. Chem. Soc.*, **95**, 8295 (1973).
137. T. P. Dang and H. B. Kagan, *Fr. Pat*, 2199756 (1974); *Chem. Abstr.*, **82**, 22291 (1975).
138. J. C. Poulin, W. Dumont, T. P. Dang, and H. B. Kagan, *C. R. Acad. Sci., Sect. C*, **277**, 41 (1973).
139. N. A. De Munck, M. W. Verbruggen, and J. J. F. Scholten, *J. Mol. Catal.*, **10**, 313 (1981).
140. J. M. Frechet and C. Schuerch, *J. Am. Chem. Soc.*, **93**, 492 (1971).
141. N. P. Allen, R. P. Burns, J. Dwyer, and C. A. McAuliffe, *J. Mol. Catal.*, **3**, 325 (1978).
142. N. P. Allen, J. Dwyer, and C. A. McAuliffe, *17th Int. Coord. Chem. Conf., Hamburg*, **34** (1976).
143. K. A. Abdulla, N. P. Allen, A. N. Badrun, R. P. Burns, J. Dwyer, C. A. McAuliffe, and N. D. A. Toma, *Chem. Ind.* (*London*), 273 (1976).
144. J. R. Millar, D. G. Smith, W. E. Marr, and T. R. E. Kressman, *J. Chem. Soc.*, 218 (1963).
145. R. W. Wheaton and D. F. Harrington, *Ind. Eng. Chem.*, **44**, 1796 (1952).
146. R. L. Letsinger, M. J. Kornet, V. Mahadevan, and D. M. Jerina, *J. Am. Chem. Soc.*, **86**, 5163 (1964).
147. J. M. Frechet and K. E. Hague, *Macromolecules*, **8**, 131 (1975).
148. J. T. Ayres and C. K. Mann, *J. Polym. Sci., Poly. Lett.*, **3**, 505 (1965).
149. C. R. Harrison and P. Hodge, *J. Chem. Soc., Chem. Commun.*, 1009 (1974).
150. C. R. Harrison and P. Hodge, *J. Chem. Soc., Perkin Trans. 1*, 605 (1976).
151. T. M. Fyles and C. C. Leznoff, *Can. J. Chem.*, **54**, 935 (1976).
152. J. W. Stewart and J. D. Young, *Solid Phase Peptide Synthesis*, Freeman, San Francisco, 1979, pp. 27–28.
153. M. Tomoi, O. Abe, M. Ikeda, K. Kihara, and H. Kakiuchi, *Tetrahedron Lett.*, 3031 (1978).
154. A. Warshawsky, R. Kalir, A. Deshe, H. Berkovitz, and A. Patchornick, *J. Am. Chem. Soc.*, **101**, 4249 (1979).
155. F. Montanari and P. Tundo, *Tetrahedron Lett.*, 5055 (1979).
156. F. Montanari and P. Tundo, *J. Org. Chem.*, **46**, 2125 (1981).
157. H. Molinari, F. Montanari, and P. Tundo, *J. Chem. Soc. Chem. Commun.*, 639 (1977).
158. S. Bhaduri, A. Ghosh, and H. Khwaja, *J. Chem. Soc., Dalton Trans.*, 447 (1981).
159. G. L. Linden and M. F. Farona, *J. Catal.*, **48**, 284 (1977).
160. A. Ghosh and S. Bhaduri, *Indian Pat.*, 150033 (1982); *Chem. Abstr.*, **98**, 90459 (1983).
161. N. L. Holy, *J. Org. Chem.*, **43**, 4686 (1978).
162. N. L. Holy, in *Fundamental Research in Homogeneous Catalysis* (Ed. M. Tsutsui), Vol. 3, Plenum, New York, 1979, p. 691.
163. E. N. Frankel, J. P. Friedrich, T. R. Bessler, W. F. Kwolek, and N. L. Holy, *J. Am. Oil Chem. Soc.*, **57**, 349 (1980).
164. R. B. King and E. M. Sweet, *J. Org. Chem.*, **44**, 385 (1979).
165. H. Mueller, *East Ger. Pat.*, 24439 (1962).
166. M. Mejstrikova, R. Rericha, and M. Kraus, *Collect. Czech, Chem. Commun.*, **39**, 135 (1974).
167. R. S. Drago and J. H. Gaul, *J. Chem. Soc., Chem. Commun.*, 746 (1979).
168. R. S. Drago, J. H. Gaul, A. Zombeck, and D. K. Straub, *J. Am. Chem. Soc.*, **102**, 1033 (1980).
169. M. B. Shambhu, M. C. Theodorakis and G. A. Digenis, *J. Polym. Sci., Polym. Chem. Ed.*, **15**, 525 (1977).
170. D. M. Dixit and C. C. Leznoff, *Isr. J. Chem.*, **17**, 248 (1979).
171. H. C. Meinders, N. Prak, and G. Challa, *Makromol. Chem.*, **178**, 1019 (1977).
172. R. J. Card and D. C. Neckers, *Inorg. Chem.*, **17**, 2345 (1978).
173. R. J. Card and D. C. Neckers, *J. Am. Chem. Soc.*, **99**, 7733 (1977).
174. M. Kaneko, S. Nemoto, A. Yamada, and Y. Kurimura, *Inorg. Chim. Acta*, **44**, L289 (1980).
175. R. S. Drago, E. D. Nyberg and A. G. El A'mma, *Inorg. Chem.*, **20**, 2461 (1981).
176. R. S. Drago and J. H. Gaul, *Inorg. Chem.*, **18**, 2019 (1979).
177. F. B. Hulsbergen, J. Manassen, J. Reedijk, and J. A. Welleman, *J. Mol. Catal.*, **3**, 47 (1977).
178. J. Rebek and F. Gavina, *J. Am. Chem. Soc.*, **97**, 3453 (1975).
179. X. Cochet, A. Mortreux, and F. Petit, *C. R. Acad. Sci., Sect. C*, **288**, 105 (1979).
180. L. D. Rollmann, *J. Am. Chem. Soc.*, **97**, 2132 (1975).

181. M. Kraus, *Collect. Czech. Chem. Commun.*, **39**, 1318 (1974).
182. J. A. S. Howell and M. Berry, *J. Chem. Soc., Chem. Commun.*, 1039 (1980).
183. W. Tao and X. Hu, *Huaxue Shiji*, **4**, 197 (1982); *Chem. Abstr.*, **98**, 54618 (1983).
184. J. M. Frechet and M. J. Farrall, in *The Chemistry and Properties of Crosslinked Polymers* (Ed. S. S. Labana), Academic Press, New York, 1979, p. 59.
185. E. Seymour and J. M. J. Frechet, *Tetrahedron Lett.*, 1149 (1976).
186. J. M. Frechet, M. D. de Smet, and M. J. Farrall, *Polymer*, **20**, 675 (1979).
187. G. A. Crosby, *US Pat.*, 3928293 (1975); *Chem. Abstr.*, **84**, 106499 (1976).
188. K. M. Webber, B. C. Gates, and W. Drenth, *J. Mol. Catal.*, **3**, 1 (1977).
189. W. D. Bonds, C. H. Brubaker, E. S. Chandrasekaran, C. Gibbons, R. H. Grubbs, and L. C. Kroll, *J. Am. Chem. Soc.*, **97**, 2128 (1975).
190. J. E. Frommer and R. G. Bergman, *J. Am. Chem. Soc.*, **102**, 5227 (1980).
191. A. Sekiya and J. K. Stille, *J. Am. Chem. Soc.*, **103**, 5096 (1981).
192. E. S. Chandrasekaran, R. H. Grubbs, and C. H. Brubaker, *J. Organomet. Chem.*, **120**, 49 (1976).
193. E. S. Chandrasekaran, *Diss. Abstr. B*, **36**, 6143 (1976).
194. E. S. Chandrasekaran, D. A. Thompson, and R. W. Rudolph, *Inorg. Chem.*, **17**, 760 (1978).
195. B. A. Sosinsky, W. C. Kalb, R. A. Grey, V. A. Uski, and M. F. Hawthorne, *J. Am. Chem. Soc.*, **99**, 6768 (1977).
196. M. Chanda, K. F. O.'Driscoll, and G. L. Rempel, *J. Mol. Catal.*, **12**, 197 (1981).
197. A. J. Moffat, *J. Catal.*, **18**, 193 (1970).
198. A. J. Moffat, *J. Catal.*, **19**, 322 (1970).
199. R. S. Drago, E. D. Nyberg, and A. G. El A'mma, *Inorg. Chem.*, **20**, 2461 (1981).
200. D. C. Sherrington, D. J. Craig, J. Dalgleish, G. Domin, J. Taylor and G. U. Meehan, *Eur. Polym. J.*, **13**, 73 (1977).
201. J. Manassen, *Isr. Pat.*, 30505; quoted in J. Manassen, *Platinum Met. Rev.*, **15**, 142 (1971).
202. R. Rabinowitz, R. Marcus, and J. Pellon, *J. Polym. Sci., Part A*, **2**, 1241 (1964).
203. R. Rabinowitz and R. Marcus, *J. Org. Chem.*, **26**, 4157 (1961).
204. J. Lieto, J. J. Rafalko, and B. C. Gates, *J. Catal.*, **62**, 149 (1980).
205. W. M. McKenzie and D. C. Sherrington, *J. Polym. Sci., Polym. Chem. Ed.*, **20**, 431 (1982).
206. F. Camps, J. Castells, J. Font, and F. Velo, *Tetrahedron Lett.*, 1715 (1971).
207. J. A. Greig and D. C. Sherrington, *Polymer*, **19**, 163 (1978).
208. W. Heitz and K. L. Platt, *Makromol. Chem.*, **127**, 113 (1969).
209. K. Achiwa, *Chem. Lett.*, 905 (1978).
210. K. Achiwa, *Heterocycles*, **9**, 1539 (1978).
211. H. Arai, *J. Catal.*, **51**, 135 (1978).
212. J. Conan, M. Bartholin, and A. Guyot, *J. Mol. Catal.*, **1**, 375 (1976).
213. M. Bartholin, J. Conin and A. Guyot, *J. Mol. Catal.*, **2**, 307 (1977).
214. K. G. Allum, R. D. Hancock, I. V. Howell, S. McKenzie, R. C. Pitkethly, and P. J. Robinson, *J. Organomet. Chem.*, **87**, 203 (1975).
215. K. G. Allum, S. McKenzie, and R. C. Pitkethly, *US Pat.*, 3726809 (1973).
216. A. A. Oswald, L. L. Murrell, and L. J. Boucher, *Abstr: 167th Am. Chem. Soc. Meeting, 1974*, PETR 34 and PETR 35.
217. K. G. Allum, S. McKenzie, and R. C. Pitkethly, *Ger. Pat.*, 2062351 (1971); *Chem. Abstr.*, **75**, 122751 (1971).
218. S. Shinoda, K. Nakamura, and Y. Saito, *J. Mol. Catal.*, **17**, 77 (1982).
219. V. M. Vdovin, V. E. Fedorev, N. A. Pritula, and G. K. Fedorova, *Izv. Akad. Nauk SSSR, Ser. Khim.*, 2663 (1981).
220. L. L. Murrell, in *Advanced Materials in Catalysis* (Eds. J. J. Burton and R. L. Garten), Academic Press, New York, 1977, Chapter 8.
221. S. C. Brown and J. Evans, *J. Chem. Soc., Chem. Commun.*, 1063 (1978).
222. J. Pelz, K. Unverferth, and K. Schwetlick, *Z. Chem.*, **14**, 370 (1974).
223. M. Capka, *Collect. Czech. Chem. Commun.*, **42**, 3410 (1977).
224. M. H. J. M. De Croon and J. W. E. Coenen, *J. Mol. Catal.*, **11**, 301 (1981).
225. K. Schwetlick, J. Pelz, and K. Unverferth, *Proc. 16th Int. Coord. Chem. Conf., Dublin*, 1974, p. 4.4.
226. M. Capka and J. Hetflejs, *Collect. Czech. Chem. Commun.*, **39**, 154 (1974).
227. A. K. Smith, paper presented at 12th Sheffield–Leeds International Symposium on Organometalic, Inorganic and Catalytic Chemistry, 1983.
228. F. G. Young, *Ger. Pat.*, 2330308 (1974); *Chem. Abstr.*, **80**, 121737 (1974).

229. J. M. Moreto, J. Albaiges, and P. Camps, in *Catalysis: Homogeneous and Heterogeneous* (Eds. B. Delmon and G. Jannes), Elsevier, Amsterdam, 1975, p. 339.
230. M. Cerny, *Collect. Czech. Chem. Commun.*, **42**, 3069 (1977).
231. I. Kolb, M. Cerny, and J. Hetflejs, *React. Kinet. Catal. Lett.*, **7**, 199 (1977).
232. D. C. Locke, *J. Chromatogr. Sci.*, **11**, 120 (1973).
233. O.-E. Brust, I. Sebestian, and I. Halasz, *J. Chromatogr.*, **83**, 15 (1973).
234. C. R. Hastings, W. A. Aue, and F. N. Larsen, *J. Chromatogr.* **60**, 329 (1971).
235. J. J. Kirkland and J. J. Destefano, *J. Chromatogr. Sci.* **8**, 309 (1970).
236. D. C. Locke, J. T. Schmermund, and B. Banner, *Anal. Chem.*, **44**, 90 (1972).
237. T. G. Waddell, D. E. Leyden, and M. T. De Bello, *J. Am. Chem. Soc.*, **103**, 5303 (1981).
238. V. Z. Sharf, A. S. Gurovets, L. P. Finn, I. B. Slinyakova, V. N. Krutii, and L. Kh Freidlin, *Izv. Akad. Nauk SSSR*, 104 (1979).
239. K. Tanaka, S. Shinoda, and Y. Saito, *Chem. Lett.*, 179 (1979).
240. S. Shinoda and Y. Saito, *Inorg. Chim. Acta*, **63**, 23 (1982).
241. T. Catrillo, H. Knözinger, and M. Wolf, *Inorg. Chim. Acta*, **45**, L235 (1980).
242. F. R. W. P. Wild, G. Gubitosa, and H. H. Brintzinger, *J. Organomet. Chem.*, **148**, 73 (1978).
243. W. Parr and M. Novotny, *Bonded Stationary Phases in Chromatography*, (ed. by E. Grishka) Ann Arbor Science, Ann Arbor, Michigan, 1974, p. 173.
244. B. Marciniec, Z. W. Kornetka, and W. Urbaniak, *J. Mol. Catal.*, **12**, 221 (1981).
245. G. D. Shields and L. J. Boucher, *J. Inorg. Nucl. Chem.*, **40**, 1341 (1978).
246. E. D. Nyberg and R. S. Drago, *J. Am. Chem. Soc.*, **103**, 4966 (1981).
247. V. A. Semikolenov, V. A. Likholobov, and Yu I. Ermakov, *Kinet. Katal.*, **20**, 269 (1979).
248. V. A. Semikolenov, V. A. Likholobov, and Yu I. Ermakov, *Kinet. Katal.*, **18**, 1294 (1977).
249. V. V. Skopenko, T. P. Lishko, T. A. Sukhan, and A. K. Trofimchuk, *Ukr. Khim. Zh.*, **46**, 1028 (1980); *Chem. Abstr.*, **94**, 10560 (1981).
250. S. Lamalle, H. Mortreux, M. Evrard, F. Petit, J. Grimblot, and J. P. Bonnelle, *J. Mol. Catal.*, **6**, 11 (1979).
251. H. B. Kagan, T. Yamagishi, J. C. Motte, and R. Setton, *Isr. J. Chem.*, **17**, 274 (1979).
252. W. O. Haag and D. D. Whitehurst, *Proc. 2nd North Am. Meeting Catal. Soc.*, 16 (1971).
253. W. O. Haag and D. D. Whitehurst, *Proc. 5th Int. Congr. Catal.*, 465 (1972); published as *Catalysis* (Ed. J. H. Hightower), Vol. 1, North-Holland, Amsterdam, (1973).
254. D. L. Hanson, J. R. Kratzer, B. C. Gates, and G. C. A. Schuit, *J. Catal.*, **32**, 204 (1974).
255. R. Linarte-Lazcano, M. P. Pedrosa, J. Sabadie, and J. E. Germain, *Bull. Soc. Chim. Fr.*, 1129 (1974).
256. J. F. McQuillan, W. O. Ord, and P. L. Simpson, *J. Chem. Soc.*, 5996 (1963).
257. H. Dupin, J. Sabadie, D. Barthomeuf, and J. E. Germain, *Bull. Soc. Chim. Fr.*, Ser. I. 86 (1979).
258. M. Kraus and D. Tomanova, *J. Polym. Sci., Polym. Chem. Ed.*, **12**, 1781 (1974).
259. K. G. Allum and R. D. Hancock, *Br. Pat.*, 1295673 (1972).
260. H. S. Bruner and J. C. Bailar, *Inorg. Chem.*, **12**, 465 (1973).
261. A. R. Sanger, L. R. Schallig, and K. G. Tan, *Inorg. Chim. Acta*, **35**, L325 (1979).
262. R. D. Sanner, R. G. Austin, M. S. Wrighton, W. D. Honnick, and C. U. Pittman, *Inorg. Chem.*, **18**, 928 (1979).
263. W. Beck, R. Höfer, J. Erbe, H. Menzel, U. Nagel, and G. Platzen, *Z. Naturforsch., Teil B*, **29**, 567 (1974).
264. C. U. Pittman, B. T. Kim, and W. M. Douglas, *J. Org. Chem.*, **40**, 590 (1975).
265. S. Warwel and P. Buschmeyer, *Angew. Chim., Int. Ed. Engl.*, **17**, 131 (1978).
266. R. A. Awl, E. N. Frankel, J. P. Friedrich, and E. H. Pryde, *J. Am. Oil Chem. Soc.*, **55**, 577 (1978).
267. H. B. Gray and C. C. Frazier, *US Pat.*, 4228035 (1980); *Chem. Abstr.*, **94**, 37128 (1981).
268. B. H. Chang, R. H. Grubbs, and C. H. Brubaker, *J. Organomet. Chem.*, **172**, 81 (1979).
269. R. H. Grubbs, C. Gibbons, L. C. Kroll, W. D. Bonds, and C. H. Brubaker, *J. Am. Chem. Soc.*, **95**, 2373 (1973).
270. C. H. Brubaker, in *Catalysis in Organic Synthesis* (Ed. G. V. Smith), Academic Press, New York, 1977, p. 25.
271. G. Gubitosa and H. H. Brintzinger, *J. Organomet. Chem.*, **140**, 187 (1977).
272. G. Gubitosa, M. Boldt, and H. H. Brintzinger, *J. Am. Chem. Soc.*, **99**, 5174 (1977).
273. P. Perkins and K. P. C. Vollhardt, *J. Am. Chem. Soc.*, **101**, 3985 (1979).
274. C. U. Pittman and R. F. Felis, *J. Organomet. Chem.*, **72**, 389 (1974).
275. A. Luchetti, L. F. Wieserman, and D. M. Hercules, *J. Phys. Chem.*, **85**, 549 (1981).

276. B. C. Gates and J. Lieto, *ChemTech.*, 195 (1980).
277. B. C. Gates and J. Lieto, *ChemTech.*, 248 (1980).
278. D. F. Foster, J. Harrison, B. S. Nicholls, and A. K. Smith, *J. Organomet. Chem.*, **248**, C29 (1983).
279. J. Kohnle, *Ger. Pat.*, 1938613 (1971); *Chem. Abstr.*, **74**, 142612 (1971).
280. R. Pierantozzi, K. J. McQuade, B. C. Gates, M. Wolf, H. Knözinger, and W. Ruhmann, *J. Am. Chem. Soc.*, **101**, 5436 (1979).
281. M. S. Jarrell, B. C. Gates, and E. D. Nicholson, *J. Am. Chem. Soc.*, **100**, 5727 (1978).
282. E. W. Thornton, H. Knözinger, B. Tesche, J. J. Rafalko, and B. C. Gates, *J. Catal.*, **62**, 117 (1980).
283. K. Iwatate, S. R. Dasgupta, R. L. Schneider, G. C. Smith, and K. L. Watters, *Inorg. Chim. Acta*, **15**, 191 (1975).
284. J. J. Rafalko, J. Lieto, B. C. Gates, and G. L. Schrader, *J. Chem. Soc., Chem. Commun.*, 540 (1978).
285. T. Castrillo, H. Knözinger, J. Lieto, and M. Wolf, *Inorg. Chim. Acta*, **44**, L239 (1980).
286. G. Sbrana, G. Braca, G. Valentine, G. Pazienza, and A. Altomare, *J. Mol. Catal.*, **3**, 11 (1978).
287. I. Dietzmann, D. Tomanova, and J. Hetflejs, *Collect. Czech. Chem. Commun.*, **39**, 123 (1974).
288. L. D. Rollmann, *Inorg. Chim. Acta*, **6**, 137 (1972).
289. W. R. Cullen, D. J. Patmore, A. J. Chapman, and A. D. Jenkins, *J. Organomet. Chem.*, **102**, C12 (1975).
290. N. Kawata, T. Mizoroki, A. Ozaki, and M. Ohkawara, *Chem. Lett.*, 1165 (1973).
291. N. Kawata, T. Mizoroki, and A. Ozaki, *J. Mol. Catal.*, **1**, 275 (1976).
292. T. Mizoroki, N. Kawata, S. Hinata, K. Maruya, and A. Ozaki, in *Catalysis: Heterogeneous and Homogeneous* (Eds. B. Delmon and G. Jannes), Elsevier, Amsterdam, 1975, p. 319.
293. N. Kawata, T. Mizoroki, and A. Ozaki, *Bull. Chem. Soc. Jpn.*, **47**, 1807 (1974).
294. J. Manassen, *Isr. J. Chem.*, **8**, 5 (1970).
295. S. Ikeda and T. Harimoto, *J. Organomet. Chem.*, **60**, C67 (1973).
296. C. U. Pittman, P. L. Grube, O. E. Ayers, S. P. McManus, M. D. Rausch, and G. A. Moser, *J. Polym. Sci., Polym. Chem. Ed.*, **10**, 379 (1972).
297. P. Pertici, G. Vitulli, C. Carlini, and F. Ciardelli, *J. Mol. Catal.*, **11**, 353 (1981).
298. D. G. H. Ballard, *23rd Int. Congr. Pure Appl. Chem. Spec. Lect.*, **6**, 219 (1971).
299. Yu I. Yermakov, A. M. Lazutkin, E. A. Demin, V. A. Zakharov, and Yu P. Grabovskii, *Kinet. Katal.*, **13**, 1422 (1972).
300. F. J. Karol, G. L. Karapinka, C. Wu, A. W. Dow, R. N. Johnson, and W. L. Carrick, *J. Polym. Sci., Polym. Chem. Ed.*, **10**, 2621 (1972).
301. D. G. H. Ballard, *Adv. Catal.*, **23**, 263 (1973).
302. J. P. Candlin and H. Thomas, *Adv. Chem. Ser.*, No. 132, 212 (1974).
303. G. L. Karapinka, *Ger. Pat.*, 1808388 (1970); *Chem. Abstr.*, **72**, 80105 (1970).
304. D. G. H. Ballard, E. Jones, A. J. P. Piolo, P. A. Robinson, and R. J. Wyatt, *Ger. Pat.*, 2040353 (1971); *Chem. Abstr.*, **74**, 126364 (1971).
305. Union Carbide, *Br. Pat.*, 1264393 (1972); *Chem. Abstr.*, **76**, 154465 (1972).
306. Yu I. Ermakov, A. M. Lazutkin, E. A. Demin, V. A. Zakharov, E. G. Kushareva, and Yu P. Grabovskii, *USSR Pat.*, 334738 (1972); *Chem. Abstr.*, **78**, 44261 (1973).
307. R. N. Johnson and F. J. Karol, *Ger. Pat.*, 1963256 (1970); *Chem. Abstr.*, **73**, 77820 (1970).
308. Yu I. Yermakov and V. Zakharov, *Adv. Catal.*, **24**, 173 (1975).
309. Yu I. Yermakov, *Catal. Rev. Sci. Eng.*, **13**, 77 (1976).
310. F. J. Karol, in *Encyclopaedia of Polymer Science and Technology* Vol. Supp. 1, Wiley, New York, 1976, p. 120.
311. Yu I. Ermakov, Yu P. Grabovskii, A. M. Lazutkin, and V. A. Zakharov, *Kinet. Katal.*, **16**, 787 (1975).
312. Yu I. Ermakov, Yu P. Grabovskii, A. M. Lazutkin, and V. A. Zakharov, *Kinet. Katal.*, **16**, 911 (1975).
313. Yu I. Ermakov and B. N. Kuznetsov, *Kinet. Katal.*, **18**, 1167 (1977).
314. Yu I. Ermakov, B. N. Kuznetsov, Yu P. Grabovskii, A. N. Startsev, A. M. Lazutkin, V. A. Zakharov, and A. I. Lazutkina, *J. Mol. Catal.*, **1**, 93 (1976).
315. W. Skupinski and S. Malinowski, *J. Organomet. Chem.*, **117**, 183 (1976).
316. W. Skupinski and S. Malinowski, *J. Mol. Catal.*, **4**, 95 (1978).
317. Yu I. Ermakov and V. A. Likholobov, *Kinet. Katal.*, **21**, 1208 (1980).
318. A. P. Shelepin, A. P. Chernyshev, V. I. Koval'chuk, P. A. Zhdan, E. N. Yurchenko, B. N. Kuznetsov, and Yu I. Ermakov, *Kinet. Katal.*, **22**, 716 (1981).
319. V. A. Zakharov and Yu I. Yermakov, *Catal. Rev. Sci. Eng.*, **19**, 67 (1979).

320. Yu I. Yermakov, *Stud. Surf. Sci. Catal.*, **7A**, 57 (1981).
321. Yu I. Yerkamov, B. N. Kuznetsov, and V. A. Zakharov, *Catalysis by Supported Complexes*, Elsevier, Amsterdam, 1981, (a) Chapter 3; (b) Chapter 2.
322. J. Schwartz and M. D. Ward, *J. Mol. Catal.*, **8**, 465 (1980).
323. D. G. H. Ballard, *J. Polym. Sci., Polym. Chem. Ed.*, **13**, 2191 (1975).
324. V. A. Zakharov, V. K. Dudchenko, V. I. Babenko, and Yu I. Yermakov, *Kinet. Katal.*, **17**, 738 (1976).
325. G. L. Karapinka, *US Pat.*, 3709853 (1973); *Chem. Abstr.*, **78**, 85087 (1973).
326. F. J. Karol and R. N. Johnson, *J. Polym. Sci., Polym. Chem. Ed.*, **13**, 1607 (1975).
327. F. J. Karol, C. Wu, W. T. Reichle, and N. J. Maraschin, *J. Catal.*, **60**, 68 (1979).
328. F. J. Karol, W. L. Munn, G. L. Goeke, B. E. Wagner, and N. J. Maraschin, *J. Polym. Sci., Polym. Chem. Ed.*, **16**, 771 (1978).
329. S. A. Panichev, G. V. Kudryavtsev, and G. V. Lisichkin, *Koord. Khim.*, **5**, 1141 (1979).
330. J. Hjortkhaer, M. S. Scurrell, and P. Simonsen, *J. Mol. Catal.*, **10**, 127 (1980).
331. J. Evans, *Chem. Soc. Rev.*, **10**, 159 (1981).
332. F. G. Ciapetta and C. J. Plank, in *Catalysis* (Ed. P. H. Emmett), Vol. 1, Reinhold, New York, 1954, p. 315.
333. G. W. Higginson, *Chem. Eng.*, **81**(20), 98 (1974).
334. W. B. Innes, in *Catalysis* (Ed. P. H. Emmett), Vol. 1, Reinhold, New York, 1954, p. 245.
335. A. Brenner and R. L. Burwell, *J. Catal.*, **52**, 353 (1978).
336. D. A. Hucul and A. Brenner, *J. Am. Chem. Soc.*, **103**, 217 (1981).
337. R. Psaro, R. Ugo, G. M. Zanderighi, B. Besson, A. K. Smith, and J. M. Basset, *J. Organomet. Chem.*, **213**, 215 (1981).
338. M. Deeba and B. C. Gates, *J. Catal.*, **67**, 303 (1981).
339. M. Deeba, J. P. Scott, R. Barth, and B. C. Gates, *J. Catal.*, **71**, 373 (1981).
340. R. Ganzerla, M. Lenarda, F. Pinna, and M. Graziani, *J. Organomet. Chem.*, **208**, C43 (1981).
341. H. Knözinger and Y. Zhao, *J. Catal.*, **71**, 337 (1981).
342. D. A. Hucul, *Diss. Abstr. B*, **41**, 1362 (1980).
343. A. Brenner and D. A. Hucul, *J. Catal.*, **61**, 216 (1980).
344. A. Brenner and D. A. Hucul, *J. Am. Chem. Soc.*, **102**, 2484 (1980).
345. A. Brenner, D. A. Hucul, and S. J. Hardwick, *Inorg. Chem.*, **18**, 1478 (1979).
346. E. Guglielminotti, *J. Mol. Catal.*, **13**, 207 (1981).
347. A. Brenner and D. A. Hucul, *Inorg. Chem.*, **18**, 2836 (1979).
348. A. Brenner, *J. Mol. Catal.*, **5**, 157 (1979).
349. D. A. Hucul and A. Brenner, *J. Phys. Chem.*, **85**, 496 (1981).
350. F. R. Hartley, *Br. Polym. J.*, **16**, 199 (1984).
351. E. N. Frankel, J. P. Friedrich, T. R. Bessler, W. F. Kwolek, and N. L. Holy, *J. Am. Oil Chem. Soc.*, **57**, 349 (1980).
352. N. L. Holy, W. A. Logan, and K. D. Stein, *US Pat.*, 4313018 (1982); *Chem. Abstr.*, **96**, 141870 (1982).
353. N. L. Holy and S. R. Shelton, *Tetrahedron*, **37**, 25 (1981).
354. N. L. Holy, *ChemTech.*, **10**, 366 (1980).
355. N. L. Holy, *J. Chem. Soc., Chem. Commun.*, 1074 (1978).
356. M. Terasawa, K. Kaneda, T. Imanaka, and S. Teranishi, *J. Catal.*, **5**, 406 (1978).
357. K. P. C. Vollhardt and P. Perkins, *US Pat. Appl.*, 39986 (1979); *Chem. Abstr.*, **94**, 128169 (1981).
358. L. S. Benner, P. Perkins, and K. P. C. Vollhardt, *Am. Chem. Soc. Symp. Ser.*, No. 152, 165 (1981).
359. M. Capka, P. Svoboda, M. Cerny, and J. Hetflejs, *Tetrahedron Lett.*, 4787 (1971).
360. T. Matsuda and J. K. Stille, *J. Am. Chem. Soc.*, **100**, 268 (1978).
361. C. U. Pittman and G. Wilemon, *J. Org. Chem.*, **46**, 1901 (1981).
362. C. U. Pittman, A. Hirao, C. Jones, R. M. Hanes, and Q. Ng, *Ann. N. Y. Acad. Sci.*, **295**, 15 (1977).
363. D. Ballivet-Tkatchenko, N. D. Chau, H. Mozzanega, M. C. Roux and I. Tkatchenko, *Am. Chem. Soc. Symp. Ser.*, No. 152, 187 (1981).
364. A. L. Lapidus, A. Yu. Krylova, and L. T. Kondrat'ev, *Izv. Akad. Nauk SSSR, Ser. Khim.*, 1432 (1980).
365. A. L. Lapidus, A. Yu. Krylova, and L. T. Kondrat'ev, *Neftekhimiya*, **21**, 397 (1981).
366. K. Lazar, Z. Schay, and L. Guczi, *J. Mol. Catal.*, **17**, 205 (1982).
367. D. Vanhoue, P. Makambo, and M. Blanchard, *J. Chem. Soc., Chem. Commun.*, 605 (1979).
368. V. L. Kuznetsov, A. T. Bell, and Y. I. Yermakov, *J. Catal.*, **65**, 374 (1980).

369. J. Robertson and G. Webb, *Proc. R. Soc. London, Ser. A*, **341**, 383 (1974).
370. J. G. Goodwin and C. Nacoache, *J. Mol. Catal.*, **14**, 259 (1982).
371. A. K. Smith, A. Theolier, J. M. Basset, R. Ugo, D. Commereuc, and Y. Chauvin, *J. Am. Chem. Soc.*, **100**, 2590 (1978).
372. F. Hugues, A. K. Smith, Y. B. Taarit, J. M. Basset, D. Commereuc, and Y. Chauvin, *J. Chem. Soc., Chem. Commun.*, 68 (1980).
373. R. Hemmerich, W. Keim, and M. Roper, *J. Chem. Soc., Chem. Commun.*, 428 (1983).
374. F. Hugues, B. Besson, and J. M. Basset, *J. Chem. Soc., Chem. Commun.*, 719 (1980).
375. F. Hugues, P. Bussiere, J. M. Basset, D. Commereuc, Y. Chauvin, L. Bonneviot, and D. Oliver, *Stud. Surf. Sci. Catal.*, **7A**, 418 (1981).
376. F. Hugues, B. Besson, P. Bussiere, J.-A. Dalmon, M. Leconte, J. M. Basset, Y. Chauvin and D. Commereuc, *Am. Chem. Soc. Symp. Ser.*, No. 192 255 (1982).
377. M. Ichikawa and K. Shikakura, *Stud. Surf. Sci. Catal.*, **7B**, 925 (1981).
378. M. Ichikawa, *ChemTech.*, **12**, 674 (1982).
379. M. Ichikawa, K. Sekizawa, K. Shikakura, and M. Kawai, *J. Mol. Catal.*, **11**, 167 (1981).
380. British Petroleum, *Br. Pat.*, 1295675 (1972).
381. Yu M. Zhorov, A. V. Shelkov, and G. M. Panchenkov, *Kinet. Katal.*, **15**, 1091 (1974).
382. B. M. Trost and E. Keinan, *J. Am. Chem. Soc.*, **100**, 7779 (1978).
383. C. U. Pittman, L. R. Smith, and R. M. Hanes, *J. Am. Chem. Soc.*, **97**, 1742 (1973).
384. E. M. Sweet, *Diss. Abstr. B*, **38**, 4821 (1978).
385. T. Uematsu, F. Saito, M. Muira, and H. Hashimoto, *Chem. Lett.*, 113 (1977).
386. Z. C. Brzezinska, W. R. Cullen, and G. Strukul, *Can. J. Chem.*, **58**, 750 (1980).
387. A. Krzywicki and G. Pannetier, *Bull. Soc. Chim. Fr.*, 1093 (1975).
388. H. Pracejus and M. Bursian, *East Ger. Pat.*, 92031 (1972); *Chem. Abstr.*, **78**, 72591 (1973).
389. M. E. Wilson and G. M. Whitesides, *J. Am. Chem. Soc.*, **100**, 306 (1978).
390. S. J. Hardwick, *Diss. Abstr. B*, **42**, 624 (1981).
391. L. F. Wieserman, *Diss. Abstr. B*, **42**, 3253 (1982).
392. A. Levitzki, I. Pecht, and M. Anbar, *Nature (London)*, **207**, 1386 (1965).
393. I. Pecht, A. Levitzki, and M. Anbar, *J. Am. Chem. Soc.*, **89**, 1587 (1967).
394. S. Jacobson, W. Clements, H. Hiramoto, and C. U. Pittman, *J. Mol. Catal.*, **1**, 73 (1975).
395. C. Andersson and R. Larsson, *J. Am. Oil Chem. Soc.*, **58**, 675 (1981).
396. Z. M. Michalska, *J. Mol. Catal.*, **3**, 125 (1977).
397. J. P. Collman, R. R. Gagne, J. Kouba, and H. Ljusberg-Wohren, *J. Am. Chem. Soc.*, **96**, 6800 (1974).
398. B. M. Trost, *Chem. Brit.*, **20**, 315 (1984).
399. W. O. Haag and D. D. Whitehurst, *US Pat.*, 3578609 (1971).
400. I. Mochida, T. Jitsumatsu, A. Kato, and T. Seiyama, *Bull. Chem. Soc. Jpn.*, **44**, 2595 (1971).
401. G. W. Parshall, *Homogeneous Catalysis*, Wiley, New York, 1980.
402. C. Masters, *Homogeneous Transition-Metal Catalysis*, Chapman and Hall, London, 1981.
403. C. U. Pittman and R. M. Hanes, *J. Am. Chem. Soc.*, **98**, 5402 (1976).
404. C. U. Pittman and A. Hirao, *J. Org. Chem.*, **43**, 640 (1978).
405. W. H. Lang, A. T. Jurewicz, W. O. Haag, D. D. Whitehurst, and L. D. Rollman, *J. Organomet. Chem.*, **134**, 85 (1977).
406. Y. Tajima and E. Kunioka, *J. Org. Chem.*, **33**, 1689 (1968).
407. J. E. Bercaw, R. H. Marvich, L. G. Bell, and H. H. Brintzinger, *J. Am. Chem. Soc.*, **94**, 1219 (1972).
408. C. H. Brubaker, W. D. Bonds, and S. Chandrasekaran, *166th American Chemical Society Meeting, Industrial and Engineering Chemistry Division Chicago*, 1973, paper 31.
409. R. H. Grubbs, C. P. Lau, R. Cukier, and C. Brubaker, *J. Am. Chem. Soc.*, **99**, 4517 (1977).
410. E. S. Chandrasekaran, R. H. Grubbs, and C. H. Brubaker, *J. Organomet. Chem.*, **120**, 49 (1976).
411. R. Jackson, J. Ruddlesden, D. J. Thompson, and R. Whelan, *J. Organomet. Chem.*, **125**, 57 (1977).
412. F. H. Jardine, *Prog. Inorg. Chem.*, **28**, 63 (1981).
413. D. E. Bergbreiter and M. S. Bursten, in *Advances in Organometallics and Inorganic Polymer Science* (Eds. C. E. Carraher, J. E. Sheats, and C. U. Pittman), Marcel Dekker, New York, 1982, p. 369.
414. D. E. Bergbreiter, M. S. Bursten, K. Cook, and G. L. Parsons, *Am. Chem. Soc. Symp. Ser.*, No. 192, 31 (1982).
415. T. H. Kim, *Diss. Abstr. B*, **37**, 356 (1976).
416. J. Kiji, S. Kadoi, and J. Furukawa, *Angew. Makromol. Chem.*, **46**, 163 (1975).

417. K. Kochloefl, W. Liebelt, and H. Knözinger, *J. Chem. Soc., Chem. Commun.*, 510 (1977).
418. M. Bartholin, C. Graillat, A. Guyot, G. Coudurier, J. Bandiera, and C. Naccache, *J. Mol. Catal.*, **3**, 17 (1977).
419. T. Okano, K. Tsukiyama, H. Konishi, and J. Kiji, *Chem. Lett.*, 603 (1982).
420. G. Innorta, A. Modelli, F. Scagnolari, and A. Foffani, *J. Organomet. Chem.*, **185**, 403 (1980).
421. J. Halpern, *Colloq. Int. CNRS*, **281**, 27 (1977); *Chem. Abstr.*, **91**, 174549 (1979).
422. M. Czakova and M. Capka, *J. Mol. Catal.*, **11**, 313 (1981).
423. G. Strukul, P. D'Olimpio, M. Bonivento, F. Pina, and M. Graziani, *J. Mol. Catal.*, **2**, 179 (1977).
424. J. Manassen and Y. Dror, *J. Mol. Catal.*, **3**, 227 (1978).
425. K. G. Allum, R. D. Hancock, I. V. Howell, R. C. Pitkethly, and P. J. Robinson, *J. Catal.*, **43**, 322 (1976).
426. A. J. Naaktgeboren, R. J. M. Nolte, and W. Drenth, *J. Mol. Catal.*, **11**, 343 (1981).
427. C. U. Pittman and W. D. Honnick, *J. Org. Chem.*, **45**, 2132 (1980).
428. C. U. Pittman, Q. Ng, A. Hirao, W. D. Honnick, and R. Hanes, *Colloq. Int. CNRS*, **281**, 49 (1977); *Chem. Abstr.*, **92**, 136004 (1980).
429. C. Carlini, G. Sbrana, and L. Casucci, *Chim. Ind.* (*Milan*), **57**, 130 (1975).
430. T. Sasaki and F. Matsunaga, *Bull. Chem. Soc. Jpn.*, **41**, 2440 (1968).
431. C. U. Pittman, W. D. Honnick, and J. J. Yang, *J. Org. Chem.*, **45**, 684 (1980).
432. W. O. Haag and D. D. Whitehurst, *Proc. 2nd North Am. Catal. Soc.*, 16 (1971).
433. C. U. Pittman and R. M. Hanes, *Ann. N.Y. Acad. Sci.*, **239**, 76 (1974).
434. M. Matsumoto and M. Tamura, *J. Mol. Catal.*, **16**, 209 (1982).
435. M. Matsumoto and M. Tamura, *Br. Pat. Appl.*, 2056874A (1980).
436. M. Matsumoto and M. Tamura, *Br. Pat. Appl.*, 2014138A (1980).
437. C. U. Pittman and C.-C. Lin, *J. Org. Chem.*, **43**, 4928 (1978).
438. H. Hiari and T. Furuta, *J. Polym. Sci., Polym. Lett. Ed.*, **9**, 459 (1971).
439. H. Hiari and T. Furuta, *J. Polym. Sci., Polym. Lett. Ed.*, **9**, 729 (1971).
440. F. Ciardelli, E. Chiellini, C. Carlini, and R. Nocci, *Polym. Prepr., Am. Chem. Soc. Div. Polym. Chem.*, **17**, 188 (1976).
441. N. Takaishi, H. Imai, C. A. Bertelo, and J. K. Stille, *J. Am. Chem. Soc.*, **98**, 5400 (1976).
442. N. Takaishi, H. Imai, C. A. Bertelo, and J. K. Stille, *J. Am. Chem. Soc.*, **100**, 264 (1978).
443. K. Ohkubo, K. Fujimori, and K. Yoshinaga, *Inorg. Nucl. Chem. Lett.*, **15**, 231 (1979).
444. K. Ohkubo, M. Haga, K. Yoshinaga, and Y. Motozato, *Inorg. Nucl. Chem. Lett.*, **16**, 155 (1980).
445. K. Ohkubo, M. Haga, K. Yoshinaga, and Y. Motozato, *Inorg. Nucl. Chem. Lett.*, **17**, 215 (1981).
446. M. Mazzei, M. Riocci, and W. Marconi, *Ger. Pat.*, 2845216 (1979); *Chem. Abstr.*, **91**, 56329 (1979).
447. E. Bayer and V. Schurig, *ChemTech.*, **6**, 212 (1976).
448. C. U. Pittman, A. Hirao, J. J. Yang, Q. Ng, R. Hanes, and C. C. Lin, *Am. Chem. Soc. Div. Pet. Chem. Prepr.*, **22**, 1196 (1977).
449. A. Bortinger, *Diss. Abstr.*, **38**, 1714 (1977).
450. S. J. Fritschel, J. J. H. Ackerman, T. Keyser, and J. K. Stille, *J. Org. Chem.*, **44**, 3152 (1979).
451. J. K. Stille, *27th International Macromolecular Symposium*, (Eds. H. Benoit and P. Rempp), Pergamon Press, Oxford, 1982, p. 99.
452. C. U. Pittman, Y. Kawabata, and L. I. Flowers, *J. Chem. Soc., Chem. Commun.*, 473 (1982).
453. F. E. Paulik and J. F. Roth, *Chem. Commun.*, 1578 (1968).
454. M. S. Scurrell, *Platinum Met. Rev.*, **21**, 92 (1977).
455. M. S. Jarrell and B. C. Gates, *J. Catal.*, **40**, 255 (1975).
456. R. G. Schultz and P. D. Montgomery, *J. Catal.*, **13**, 105 (1969).
457. R. G. Schultz and P. D. Montgomery, *Am. Chem. Soc. Div. Pet. Chem. Prepr.*, **17**, 13 (1972).
458. K. K. Robinson, A. Hershman, J. H. Craddock, and J. F. Roth, *J. Catal.*, **27**, 389 (1972).
459. M. S. Scurrell, *J. Res. Inst. Catal.* (*Jpn.*), **25**, 189 (1977).
460. B. K. Nefedov, N. S. Sergeeva, T. V. Zueva, E. M. Shutkina, and Ya T. Eidus, *Izv. Akad. Nauk SSSR, Ser. Khim.*, 582 (1976).
461. M. S. Scurrell and R. F. Howe, *J. Mol. Catal.*, **7**, 535 (1980).
462. M. S. Scurrell, *J. Mol. Catal.*, **10**, 57 (1981).
463. S. L. T. Andersson and M. S. Scurrell, *J. Catal.*, **71**, 233 (1981).
464. D. T. Thompson, in *Catalysis and Chemical Processes* (Eds. by R. Pearce and W. R. Patterson), Blackie, Glasgow, 1981, Chapter 8.

465. B. C. Gates, in *Chemistry and Chemical Engineering of Catalytic Process* (Eds. R. Prins and G. C. A. Schuit), Sijthoff and Noordhoff, Alpher ann den Rijr, Netherlands, 1980, p. 437.
466. K. M. Webber, B. C. Gates, and W. Drenth, *J. Catal.*, **47**, 269 (1977).
467. K. Mosbach, *Sci. Am.*, No. 224 (March), 26 (1971).
468. C. U. Pittman and L. R. Smith, *J. Am. Chem. Soc.*, **97**, 1749 (1975).
469. C. U. Pittman, L. R. Smith, and S. E. Jacobson, in *Catalysis: Heterogeneous and Homogeneous* (Eds. B. Delmon and G. Jannes), Elsevier, Amsterdam, 1975, p. 939.
470. C. U. Pittman and L. R. Smith, in *Organotransition-Metal Chemistry* (Eds. Y. Ishii and M. Tsutsui), Plenum, New York, 1975, p. 143.
471. S. E. Jacobson and C. U. Pittman, *J. Chem. Soc., Chem. Commun.*, 187 (1975).
472. C. U. Pittman and L. R. Smith, *J. Am. Chem. Soc.*, **97**, 341 (1975).
473. C. U. Pittman, S. K. Wuu, and S. E. Jacobson, *J. Catal.*, **44**, 87 (1976).
474. R. F. Batchelder, B. C. Gates, and F. P. J. Kuijpers, *6th International Congress on Catalysis, London*, 1976, preprint A40.
475. J. M. Basset and J. Norton, in *Fundamental Research in Homogeneous Catalysis* (Eds. M. Tsutsui and R. Ugo), Plenum, New York, 1977, p. 215.
476. E. R. Andrew, *Phil. Trans. R. Soc. London, Ser. A*, **299**, 505 (1981).
477. H. C. Clark, J. A. Davies, C. A. Fyfe, P. J. Hayes, and R. E. Wasylichen, *Organometallics*, **2**, 197 (1983).
478. L. Bemi, H. C. Clark, J. A. Davies, D. Drexler, C. A. Fyfe, and R. E. Wasylichen, *J. Organomet. Chem.*, **224**, C5 (1982).
479. L. Bemi, H. C. Clark, J. A. Davies, C. A. Fyfe, and R. E. Wasylichen, *J. Am. Chem. Soc.*, **104**, 438 (1982).

Author Index

This author index is designed to enable the reader to locate an author's name and work with the aid of the reference numbers appearing in the text. The page numbers are printed in normal type in ascending numerical order, followed by the reference numbers in parentheses. The numbers in *italics* refer to the pages on which the references are actually listed.

Subject Index